The Prokaryotes

Second Edition

The Prokaryotes

Second Edition

A Handbook on the Biology of Bacteria:
Ecophysiology, Isolation, Identification,
Applications

Edited by

ALBERT BALOWS, Emory University
HANS G. TRÜPER, University of Bonn
MARTIN DWORKIN, University of Minnesota
WIM HARDER, TNO Institute of Environmental Sciences, Delft
KARL-HEINZ SCHLEIFER, Technical University, Munich

Volume II

Springer-Verlag
New York Berlin Heidelberg London Paris
Tokyo Hong Kong Barcelona Budapest

Albert Balows, 105 Bay Colt Road, Alpharetta, GA 30201, USA
Hans G. Trüper, Institute for Microbiology, University of Bonn, D-5300 Bonn 1, Germany
Martin Dworkin, Department of Microbiology, University of Minnesota, Minneapolis, MN 55455, USA
Wim Harder, TNO Institute of Environmental Sciences, 2600 AE Delft, The Netherlands
Karl-Heinz Schleifer, Department of Microbiology, Technical University, D-8000 Munich 2, Germany

Library of Congress Cataloging-in-Publication Data
The Prokaryotes: a handbook on the biology of bacteria:
    ecophysiology, isolation, identification, applications/editors, Albert Balows,
    2nd ed.
    Hans G. Trüper ... [et al.].
    p. cm.
    Includes bibliographical references and indexes.
    ISBN 0-387-97258-7 (set).—ISBN 3-540-97258-7 (set)
    1. Prokaryotes—Handbooks, manuals, etc. I. Balows, Albert.
QR72.5.P76   1991
589.9—dc20   91-17256

Printed on acid-free paper.

Production managed by Linda H. Hwang.
Editorial production management by Science Tech Publishers, Madison, WI.
Typeset by Impressions, Inc., Madison, WI.
Printed and bound by Arcata Graphics/Halliday, West Hanover, MA.
Printed in the United States of America.

9 8 7 6 5 4 3 2 1

ISBN 0-387-97258-7 Springer-Verlag New York Berlin Heidelberg
ISBN 3-540-97258-7 Springer-Verlag Berlin Heidelberg New York

Foreword

The purpose of this brief foreword is unchanged from the first edition; it is simply to make you, the reader, hungry for the scientific feast that follows. These four volumes on the prokaryotes offer an expanded scientific menu that displays the biochemical depth and remarkable physiological and morphological diversity of prokaryote life. The size of the volumes might initially discourage the unprepared mind from being attracted to the study of prokaryote life, for this landmark assemblage thoroughly documents the wealth of present knowledge. But in confronting the reader with the state of the art, the Handbook also defines where more work needs to be done on well-studied bacteria as well as on unusual or poorly studied organisms.

This edition of *The Prokaryotes* recognizes the almost unbelievable impact that the work of Carl Woese has had in defining a phylogenetic basis for the microbial world. The concept that the ribosome is a highly conserved structure in all cells and that its nucleic acid components may serve as a convenient reference point for relating all living things is now generally accepted. At last, the phylogeny of prokaryotes has a scientific basis, and this is the first serious attempt to present a comprehensive treatise on prokaryotes along recently defined phylogenetic lines. Although evidence is incomplete for many microbial groups, these volumes make a statement that clearly illuminates the path to follow.

There are basically two ways of doing research with microbes. A classical approach is first to define the phenomenon to be studied and then to select the organism accordingly. Another way is to choose a specific organism and go where it leads. The pursuit of an unusual microbe brings out the latent hunter in all of us. The intellectual challenges of the chase frequently test our ingenuity to the limit. Sometimes the quarry repeatedly escapes, but the final capture is indeed a wonderful experience.

For many of us, these simple rewards are sufficiently gratifying so that we have chosen to spend our scientific lives studying these unusual creatures. In these endeavors many of the strategies and tools as well as much of the philosophy may be traced to the Delft School, passed on to us by our teachers, Martinus Beijerinck, A. J. Kluyver, and C. B. van Niel, and in turn passed on by us to our students.

In this school, the principles of the selective, enrichment culture technique have been developed and diversified; they have been a major force in designing and applying new principles for the capture and isolation of microbes from nature. For me, the "organism approach" has provided rewarding adventures. The organism continually challenges and literally drags the investigator into new areas where unfamiliar tools may be needed. I believe that organism-oriented research is an important alternative to problem-oriented research, for new concepts of the future very likely lie in a study of the breadth of microbial life. The physiology, biochemistry, and ecology of the microbe remain the most powerful attractions. Studies based on classical methods as well as modern genetic techniques will result in new insights and concepts.

To some readers, this edition of the *The Prokaryotes* may indicate that the field is now mature, that from here on it is a matter of filling in details. I suspect that this is not the case. Perhaps we have assumed prematurely that we fully understand microbial life. Van Niel pointed out to his students that—after a lifetime of study—it was a very humbling experience to view in the microscope a sample of microbes from nature and recognize only a few. Recent evidence suggests that microbes have been evolving for nearly 4 billion years. Most certainly those microbes now domesticated and kept in captivity in culture collections represent only a minor portion of the species that have evolved in this time span. Sometimes we must remind ourselves that evolution is actively taking place at the present moment. That the eukaryote cell evolved as a chimera of certain prokaryote parts is a generally accepted concept today. Higher as well as lower eukaryotes evolved in contact with prokaryotes, and evidence surrounds us of the complex interactions between eukaryotes and prokaryotes as well as among prokaryotes. We have so far only scratched the surface of these biochemical interrelationships. Perhaps the legume nodule is a pertinent example of nature caught in the act of evolving the "nitrosome," a unique nitrogen-fixing organelle. Study of prokaryotes is proceeding at such a fast pace that major advances are occurring yearly. The increase of this edition to four volumes documents the exciting pace of discoveries.

To prepare a treatise such as *The Prokaryotes* requires dedicated editors and authors; the task has been enormous. I predict that the scientific community of microbiologists will again show its appreciation through use of these volumes—such that the pages will become "dog-eared" and worn as students seek basic information for the hunt. These volumes belong in the laboratory, not in the library. I believe that a most effective way to introduce students to microbiology is for them to isolate microbes from nature, i.e., from their habitats in soil, water, clinical specimens, or plants. *The Prokaryotes* enormously simplifies this process and should encourage the construction of courses that contain a wide spectrum of diverse topics. For the student as well as the advanced investigator these volumes should generate excitement.

Happy hunting!

Ralph S. Wolfe
Department of Microbiology
University of Illinois at Urbana-Champaign

Preface

In 1962 R. Y. Stanier and C. B. van Niel formulated the view that bacteria represented a definable, biologically coherent group of organisms called the prokaryotes. They pointed out that, up to that time, "the abiding intellectual scandal of bacteriology has been the absence of a clear concept of a bacterium" (Archiv für Mikrobiologie 42:17–35, 1962). In addition to formalizing the distinction between the eukaryotes and the prokaryotes, Stanier and van Niel also emphasized the diversity of the prokaryotes.

The early view of Ferdinand Cohn that bacteria were simply a group of unicellular microorganisms that divide by binary fission had gradually been supplanted, so that by 1962, Stanier and van Niel could point out that bacteria were "photosynthetic or non-photosynthetic; motile by any one of three different mechanisms or permanently immotile; unicellular, multicellular or coenocytic; multiplying by binary transverse fission, or by formation of gonidia or conidia." Although Stanier and van Niel never actually defined the bacteria as a group, their seminal article explicitly emphasized the wide variety of metabolic, physiological, and morphological types among the bacteria, a variety which was reflected in cellular organization, modes of cell division, mechanisms of locomotion, and patterns of energy-yielding metabolism. Nevertheless, the succeeding decades saw a narrowing of the scope of research on the bacteria, since the incredible power and successes of molecular biology required intense study of only a few suitable model organisms. The hypothesis that there was a small group of typical bacteria whose mechanisms and processes were accurately representative of the bacteria as a whole became tacitly accepted. Most areas of human activity have a dialectic quality, and the evolution of scientific ideas is no exception. Thus, the narrow focus on a small group of bacteria is now being broadened to recognize new mechanisms, new strategies for coping with the environment, newly expanded limits to the abilities of the microbe, and new experimental systems. The publication in 1981 of the first edition of *The Prokaryotes* played an important role in this broadening of the perspective of the microbial world. Ten years later, the second edition continues to emphasize the diversity of the prokaryotes while adding the entirely new perspective of prokaryotic phylogeny. As far as possible, in this edition the chapters dealing with individual groups and genera of bacteria are arranged in strictly phylogenetic order.

The pioneering work of Carl Woese in cataloging and sequencing the ribosomal ribonucleic acid of prokaryotes has, for the first time in the history of biology, provided a means of establishing a truly phylogenetic system for living organisms—a goal previously thought to be impossible. The use of the oligonucleotide sequences of 16S rRNA as a molecular/evolutionary chronometer has revealed unsuspected phylogenetic affiliations. Furthermore, it has shown that the uniform concept of the prokaryotes must give way to a dichotomy dividing the prokaryotes into two groups (the archaebacteria, or Archaea, and the eubacteria, or Bacteria) no more closely related to each other than either of them is to the eukaryotes. In essence, the molecular approach has added an entirely new dimension to prokaryotic diversity, providing a fascinating opportunity for gaining insight into the origin of life.

The work of Woese and his coworkers has uncovered so many inconsistencies in classical prokaryotic systematics that an entirely new system for the taxonomy of the prokaryotes will undoubtedly emerge. These phylogenetic conclusions, drawn mainly from 16S rRNA analyses, have been strongly supported by nucleic acid hybridization studies and comparative sequence analysis of 23S rRNA, of the beta sub-unit of adenosine triphosphatase, and of the elongation factor Tu. The phylogenetic relationships of bacteria are also reflected by studies of chemotaxonomic markers such as peptidoglycan and lipid.

A reader familiar with the first edition of *The Prokaryotes* will note that, in the new edition, the subtitle has been changed from "A Handbook on Habitats, Isolation, and Identification of Bacteria" to "A Handbook on the Biology of Bacteria: Ecophysiology, Isolation, Identification, Applications." This change reflects developments in bacteriology over the past 10 years as well as the growing awareness of the important role that prokaryotes play in determining and stabilizing the global environment. The term "Habitats" has been changed to "Ecophysiology" to emphasize the interactions of microorganisms both with their habitats and with other organisms that occupy those habitats. In this context, the number of chapters dealing with symbiotic and syntrophic associations of prokaryotes with other prokaryotes and with eukaryotes has been substantially increased in the new edition.

The terms "Isolation" and "Identification" have been maintained because they reflect essential aspects of the second edition of *The Prokaryotes*. However, the individual chapters have undergone considerable change so as to include new techniques for the handling of fastidious anaerobes, for chemotaxonomic analysis, for molecular genetic methods of identification, etc. In addition, the results of another 10 years of experience in preserving and maintaining cultures have been included in many of the chapters. An important addition has been the new key term "Applications." This reflects the conscious move by microbiology into biotechnology. The use of microbes to produce new and useful products and to recycle undesirable products is now of worldwide economic and ecological importance.

But the organization of the second edition of *The Prokaryotes* differs even more fundamentally than the subject matter of the first edition. The decision to organize the Handbook on the basis of phylogenetic relationships generated two problems. First, bacteria have traditionally been divided into groups that share one or more phenotypic properties. Thus, microbiologists are used to dealing with collective entities such as the "gliding bacteria," "the autotrophs," and

"the photosynthetic bacteria." These are groups that are similar in their overall physiology, behavior, or metabolism, or that inhabit similar ecological niches. However, the new phylogenetic analyses have made it clear that these groups are not necessarily related to each other in any evolutionary sense. Thus, an organization that reflected the true evolutionary relations among the prokaryotes would necessarily have sacrificed the familiar and intellectually useful associations of phenotypic properties, but retention of the traditional arrangements would have ignored the phylogenetic revolution that has taken place. A second problem is that there are still a number of genera or groups whose position in the phylogenetic scheme have not been determined.

The solution to this dilemma was to divide the present edition of *The Prokaryotes* into six parts. Part I contains introductory essays dealing with the broader aspects of microbiology that underlie the rest of the Handbook. Part II contains general chapters that cover life cycles, prokaryotic behavior, anaerobic growth, syntrophism, and a series of synoptic chapters that describe the familiar phenotypically organized groups. This has allowed us to retain discussions of those useful physiological, metabolic, and ecological generalizations that characterize the bacteria. Parts III and IV of the Handbook consist of phylogenetically arranged chapters devoted to genera, families, and sometimes higher taxa of related prokaryotes. For example, the archaebacteria (Archaea) have been strictly separated from the eubacteria, and within these major groups the individual genera are arranged phylogenetically.

These phylogenetic groupings have generated some phenotypically strange bedfellows. Nevertheless, the editors hope that the juxtaposition of organisms that are phylogenetically related, but appear phenotypically dissimilar, may stimulate microbiologists to seek taxonomically relevant similarities that otherwise would not have been evident. Following the major groups of generic chapters, Part V covers those microorganisms that have established firm symbiotic relationships. Finally, those genera that have not been phylogenetically allocated as yet are dealt with in Part VI. In order to help the reader find a specific genus, an alphabetical listing of genera and other higher taxa follows the Contents.

Microbiology is in the midst of the most significant conceptual revolution it has experienced in the past few decades. We have the privilege of a new way of thinking about the evolution of bacteria that may even help us to understand the origin of life. We also have a set of molecular tools that allow us to probe the innermost details of the workings of the cell and that may also allow us to ask ecological questions, the answers to which have thus far eluded us. Nevertheless, in the final analysis, it is only the ability to isolate and cultivate microbes individually and as consortia that will lead to a full understanding of the strategies, mechanisms, and processes of biodiversity. We dedicate this Handbook to the efforts of those who pursue these goals.

A scientific work of this magnitude requires support, advice, and assistance from many people with a wide range of skills and knowledge. We would be remiss if we failed to express our thanks and appreciation for the suggestions and hard work provided by many secretaries, typists, graphic artists, and photographers who worked with all of the authors of *The Prokaryotes,* second edition. We are indebted to the staff of Springer-Verlag who encouraged and indulged us and demonstrated remarkable patience with us. We are equally grateful to Thomas D. Brock and the staff of Science Tech Publishers, especially Carol Bracewell and Ruth Siegel, whose copyediting expertise and problem-solving capabilities lightened our load considerably. Finally, we are beholden to the more than 300 people throughout the world who gave of their time, knowledge, and experience to serve as authors in this international effort.

Albert Balows
Hans G. Trüper
Martin Dworkin
Wim Harder
Karl-Heinz Schleifer

Contents

Volume I

Part I: Introductory Essays

Part II: Ecophysiological and Biochemical Aspects of Phylogenetically Diverse Groups

Part III: Archaea

Part IV: Bacteria

Division A: Firmicutes (Gram-positive Bacteria)

Subdivision A1: Firmicutes with High GC Content of DNA

Volume II

Subdivision A2: Firmicutes with Low GC Content of DNA

Volume III

Subdivision C1: Alpha Subclass

Division F: Bacteroides and Cytophaga Group

Division G: Chlamydia

Division H: Planctomyces and Related Bacteria

Division I: Deinococcaceae and Thermus

Division J: Chloroflexaceae and Related Bacteria

Alphabetical List of Genera and Groups*

Genus or group	Chapter	Senior author	Page	Volume
Acetoanaerobium	87	Bernhard Schink	1925	II
Acetobacter	111	Jean Swings	2268	III
Acetobacterium	87	Bernhard Schink	1925	II
Acetogenium	87	Bernhard Schink	1925	II
Acetogens	21	Gabriele Diekert	517	I
Acholeplasma	88	Shmuel Razin	1937	II
Achromatium	218	Jan la Rivière	3934	IV
Achromobacter	128	Hans-Jürgen Busse	2544	III
Acinetobacter	164	Kevin J. Towner	3137	IV
Actinobacillus	182	James E. Phillips	3342	IV
Actinomaduraceae	46	R.M. Kroppenstedt	1085	II
Actinomyces	38	K.P. Schaal	850	I
Actinomycetales	35	Jerry Ensign	811	I
Actinoplanaceae	43	Gernot Vobis	1029	II
Aegyptianella	225	Julius P. Kreier	3994	IV
Aerobacter	149	Riichi Sakazaki	2816	III
Aerococcus	68	Norbert Weiss	1502	II
Aeromonas	158	J.J. Farmer III	3012	III
Agarbacterium	168	Karel Kersters	3189	IV
Agrobacterium	108	Allen Kerr	2214	III
Agromyces	62	Matthew D. Collins	1355	II
Alteromonas	159	Michel J. Gauthier	3046	III
Alysiella	139	Daisy Kuhn	2658	III
Ammonia oxidizers	137	Hans-Peter Koops	2625	III
Amoebobacter	170	Norbert Pfennig	3200	IV
Anaerobic Gram-positive cocci	83	Takayuki Ezaki	1879	II
Anaeroplasma	88	Shmuel Razin	1937	II
Anaerorhabdus	196	Haroun N. Shah	3593	IV
Anaplasmataceae	225	Julius P. Kreier	3994	IV
Ancalomicrobium	103	James Staley	2160	III
Aquaspirillum	131	Bruno Pot	2569	III
Arcanobacterium	38	K.P. Schaal	850	I
Archaeoglobus	31	Karl O. Stetter	707	I
Arthrobacter	55	Dorothy Jones	1283	II
Asteroleplasma	88	Shmuel Razin	1937	II
Asticcacaulis	106	Jeanne Poindexter	2176	III
Aureobacterium	62	Matthew D. Collins	1355	II
Azorhizobium	107	Gerald Elkan	2197	III
Azospirillum	109	Johanna Döbereiner	2236	III
Azotobacteraceae	165	Jan H. Becking	3144	IV

*Only the number of the primary chapter is given here; for further information, see the index.

Contributors

B. Ahlers
Institut für Allgemeine Botanik
Universität Hamburg
D-2000 Hamburg 52
Germany

Antoon D.L. Akkermans
Department of Microbiology
Agriculture University
6703 CT Wageningen
The Netherlands

Jan R. Andreesen
Institut für Mikrobiologie
Universität Göttingen
D-3400 Göttingen
Germany

Robert E. Andrews
Department of Microbiology
University of Iowa
Iowa City, IA 52242
USA

Michel Aragno
Institut de Botanique
Université de Neuchâtel
CH-2007 Neuchâtel
Switzerland

M.J. Arduino
Center for Infectious Diseases
Centers for Disease Control
Atlanta, GA 30333
USA

Judith P. Armitage
Department of Biochemistry
University of Oxford
Oxford OX1 3QU
England

Georg Auling
Institut für Mikrobiologie
Universität Hannover
D-3000 Hannover 1
Germany

Friedhelm Bak
Max-Planck-Institut für Terrestrische
  Mikrobiologie
D-3550 Marburg/Lahn
Germany

Dwight D. Baker
School of Forestry and
  Environmental Studies
Yale University
New Haven, CT 06511
USA

Albert Balows
105 Bay Colt Rd.
Alpharetta, GA 30201
USA

Robert C. Barnes
Center for Infectious Disease
Centers for Disease Control
Atlanta, GA 30333
USA

Jan Hendrik Becking
Stichting ITAL
Research Institute of the Ministry of
  Agriculture and Fisheries
6700 AA Wageningen
The Netherlands

Ulrich Berger
Georg-Friedrich-Händel-Strasse 18
D-6904 Eppelheim/Heidelberg
Germany

Karl A. Bettelheim
Department of Pathology
Fairfield Hospital
Fairfield, Victoria 3078
Australia

Bruno Biavati
Instituto di Microbiologia Agraria e
  Tecnica
Università degli Studi di Bologna
40126 Bologna
Italy

Eberhard Bock
Institut für Allgemeine Botanik
Universität Hamburg
D-2000 Hamburg 52
Germany

Martin Bomar
Lehrstuhl Mikrobiologie I
Eberhard-Karls-Universität
D-7400 Tübingen 1
Germany

David R. Boone
Department of Environmental
  Science and Engineering
Oregon Graduate Institute
Beaverton, OR 97006
USA

Edward J. Bottone
Clinical Microbiology Laboratories
Mt. Sinai Hospital
New York, NY 10029
USA

Timothy L. Bowen
Department of Microbiology
University of Georgia
Athens, GA 30602
USA

John F. Bradbury
Commonwealth Mycological Institute
Kew, Richmond TW93AF
England

Violette A. Breittmayer
INSERM Unité 303
1 Avenue Jean-Lorrain
06300 Nice
France

Don J. Brenner
Center for Infectious Diseases
Centers for Disease Control
Atlanta, GA 30333
USA

John A. Breznak
Department of Microbiology and
  Public Health
Michigan State University
East Lansing, MI 48824
USA

Stephanie Bringer-Meyer
Institut für Biotechnologie I
KFA Jülich GmbH
D-5170 Jülich
Germany

Thomas D. Brock
Department of Bacteriology
University of Wisconsin-Madison
Madison, WI 53706
USA

Clara R. Bunn

Department of Biology
Meredith College
Raleigh, NC 27607
USA

Tineke Burger-Wiersma

Laboratorium voor Microbiologie
University of Amsterdam
1018 WS Amsterdam
The Netherlands

Hans-Jürgen Busse

Institut für Mikrobiologie
Universität Hannover
D-3000 Hannover 1
Germany

Ercole Canale-Parola

Department of Microbiology
University of Massachusetts
Amherst, MA 01003
USA

Richard W. Castenholz

Biology Department
University of Oregon
Eugene, OR 97403
USA

Dieter Claus

Deutsche Sammlung von
  Mikroorganismen
D-3300 Braunschweig-Stockheim
Germany

Yehuda Cohen

Division of Microbial and Molecular
  Ecology
Life Sciences Institute
Hebrew University
Jerusalem
Israel

Matthew D. Collins

AFRC Institute of Food Research
Reading Laboratory
Schinfield, Reading RG2 9AT
England

Michael P. Coughlan

Department of Biochemistry
University College
Galway
Ireland

Cecil S. Cummins

Department of Anaerobic
  Microbiology
Virginia Polytechnic Institute and
  State University
Blacksburg, VA 24061
USA

Milton S. da Costa

Departmento de Zoologia
Universidade de Coimbra
3049 Coimbra Codex
Portugal

Gregory A. Dasch

Rickettsial Diseases Division
Naval Medical Research Institute
Bethesda, MD 20814
USA

Michael J. Davis

Institute of Food and Agricultural
  Sciences
University of Florida
Homestead, FL 33031
USA

Jan A.M. de Bont

Department of Food Science
Agricultural University
6700 EV Wageningen
The Netherlands

Jozef De Ley

Laboratorium voor Microbiologie en
  Microbiële Genetica
Rijksuniversiteit Ghent
B-9000 Ghent
Belgium

Paul De Vos

Laboratorium voor Microbiologie en
  Microbiële Genetica
Rijksuniversiteit Ghent
B-9000 Ghent
Belgium

Maria H. Deinema

Laboratory of Microbiology
Agricultural University
6703 CT Wageningen
The Netherlands

Luc A. Devriese

Faculty of Veterinary Medicine
University of Ghent
B-9000 Ghent
Belgium

Floyd E. Dewhirst

Forsyth Dental Center
140 Fenway
Boston, MA 02115
USA

Gabriele Diekert

Institut für Mikrobiologie
Universität Stuttgart
D-7000 Stuttgart 1
Germany

Lubbert Dijkhuizen

Department of Microbiology
University of Groningen
9751 NN Haren
The Netherlands

Daniel L. Distel

Scripps Institution of Oceanography
University of California-San Diego
La Jolla, CA 92093
USA

Johanna Döbereiner

EMPRABA
23851 Seropedica, Rio de Janeiro
Brazil

Gary V. Doern

Department of Clinical Microbiology
University of Massachusetts Medical
  Center
Worcester, MA 01605
USA

Patrick R. Dugan

Idaho National Engineering
  Laboratory
EG & G Idaho
Idaho Falls, ID 83415
USA

Martin Dworkin

Department of Microbiology
University of Minnesota
Minneapolis, MN 55455
USA

Robert R. Eady

AFRC Institute of Plant Science
  Research
The University of Sussex
Brighton BN1 9RQ
England

Jürgen Eberspächer

Institut für Mikrobiologie
Universität Hohenheim
D-7000 Stuttgart 70
Germany

Gerald H. Elkan

Department of Microbiology
North Carolina State University
Raleigh, NC 27695
USA

T. Martin Embley

Department of Paramedical Sciences
North East London Polytechnic
London E15 4LZ
England

Jerry Ensign

Department of Bacteriology
University of Wisconsin-Madison
Madison, WI 53706
USA

Takayuki Ezaki

Department of Microbiology
Gifu University School of Medicine
Gifu 500
Japan

Solly Faine
Department of Microbiology
Monash University
Clayton, Melbourne, Victoria 3168
Australia

J.J. Farmer III
Center for Infectious Diseases
Centers for Disease Control
Atlanta, GA 30333
USA

W. Edmund Farrar
Department of Medicine
Medical University of South
  Carolina
Charleston, SC 29425
USA

Horst Felbeck
Scripps Institution of Oceanography
University of California-San Diego
La Jolla, CA 92093
USA

Cynthia L. Fennell
Department of Medicine
Harborview Medical Center
Seattle, WA 98105
USA

Patricia I. Fields
Center for Infectious Diseases
Centers for Disease Control
Atlanta, GA 30333
USA

Wilhelm Frederiksen
Department of Diagnostic
  Bacteriology and Antibiotics
Statens Seruminstitut
DK-2300 Copenhagen S
Denmark

Dagmar Fritze
Deutsche Sammlung von
  Mikroorganism
D-3300 Braunschweig-Stockheim
Germany

John Fuerst
Department of Microbiology
University of Queensland
St. Lucia, Queensland 4067
Australia

Michel J. Gauthier
INSERM Unité 303
1 Avenue Jean-Lorrain
06300 Nice
France

Arnold Geis
Institut für Mikrobiologie
Bundesanstalt für Milchforschung
D-2300 Kiel
Germany

Monique Gillis
Laboratorium voor Microbiologie
  and Microbiële Genetica
Rijksuniversiteit Ghent
B-9000 Ghent
Belgium

Stephen Giovannoni
Department of Microbiology
Oregon State University
Corvallis, OR 97331
USA

Robert C. Good
Center for Infectious Diseases
Centers for Disease Control
Atlanta, GA 30333
USA

Michael Goodfellow
Department of Microbiology,
  The Medical School
The University of Newcastle upon
  Tyne
Newcastle upon Tyne NE2 4HH
England

Hans-Dieter Görtz
Zoologisches Institut
Westfälische Wilhelms Universität
  Münster
D-4400 Münster
Germany

Rainer Gothe
Institut für Vergleichende
  Tropenmedizin und Parasitologie
Universität München
8000 München 40
Germany

Jan C. Gottschal
Department of Microbiology
University of Groningen
9751 NN Haren
The Netherlands

Gerhard Gottschalk
Institut für Mikrobiologie
Universität Göttingen
D-3400 Göttingen
Germany

Friedrich Götz
Universität Tübingen
D-7400 Tübingen 1
Germany

Paul A. Granato
Department of Pathology
Crouse Irving Memorial Hospital
Syracuse, NY 13210
USA

Peter N. Green
NCIMB Ltd. Torry Research Station
P.O. Box 31
Aberdeen AB9 8DC
Scotland

James R. Greenwood
Bio-Diagnostics Laboratories
3555 Voyager Street
Torrance, CA 90503
USA

Francine Grimont
Unité 199 INSERM
Institut Pasteur
75724 Paris Cedex 15
France

Patrick A.D. Grimont
Unité 199 INSERM
Institut Pasteur
75724 Paris Cedex 15
France

Michael Gurevitz
Department of Botany
Life Sciences Institute
Tel Aviv University
Ramat Aviv 69978
Israel

David L. Gutnick
Department of Microbiology
Tel Aviv University
Ramat Aviv 69978
Israel

Dittmar Hahn
Department of Microbiology
Agriculture University
6703 CT Wageningen
The Netherlands

Auli Haikara
Valtion Teknillinen Tutkimuskeskus
Biotekniikan Laboratory
SF-02150 Espoo
Finland

Walter P. Hammes
Institut für Lebensmitteltechnologie
Universität Hohenheim
D-7000 Stuttgart 70
Germany

Hans Helmut Hanert
Institut für Mikrobiologie
Technische Universität Braunschweig
D-3300 Braunschweig
Germany

Theo A. Hansen
Laboratorium voor Microbiologie
Rijksuniversiteit Groningen
NL-9751 NN Haren
The Netherlands

Richard S. Hanson

Gray Freshwater Biology Research
  Institute
University of Minnesota
Navarre, MN 55392
USA

W. Harder

Hoofdgroep Maatschappelijke
  Technologie
MT-TNO Delft
2600 AE Delft
The Netherlands

Jeremy M. Hardie

The London Hospital Medical
  College Dental School
University of London
London E1 2AD
England

H. Harms

Institut für Allgemeine Botanik
Universität Hamburg
D-2000 Hamburg 52
Germany

Sybe Hartmans

Department of Food Science
Agricultural University Wageningen
6700 EV Wageningen
The Netherlands

Sydney M. Harvey

Nichols Institute Reference
  Laboratories
32961 Calle Perfecto
San Juan Capistrano, CA 92675
USA

J. Woodland Hastings

Department of Biology
Harvard University
Cambridge, MA 02138
USA

William J. Hausler, Jr.

Hygienic Laboratory
University of Iowa
Iowa City, IA 52242
USA

Klaus Heckmann

Zoologisches Institut
Westfälische Wilhelms-Universität
  Münster
D-4400 Münster
Germany

H. Ernest Hemphill

Department of Biology
Syracuse University
Syracuse, NY 13244
USA

Robert B. Hespell

Northern Regional Research Center,
  ARS
US Department of Agriculture
Peoria, IL 61604
USA

F.W. Hickman-Brenner

Center for Infectious Diseases
Centers for Disease Control
Atlanta, GA 30333
USA

Donald C. Hildebrand

Department of Plant Pathology
University of California-Berkeley
Berkeley, CA 94720
USA

Gregory Hinkle

Department of Botany
University of Massachusetts
Amherst, MA 01003
USA

Hans Hippe

Institut für Mikrobiologie
Universität Göttingen
D-3400 Göttingen
Germany

Peter Hirsch

Institut für Allgemeine Mikrobiologie
Universität Kiel
D-2300 Kiel
Germany

Tor Hofstad

Department of Microbiology and
  Immunology
University of Bergen
N-5021 Bergen
Norway

Susan K. Hoiseth

Department of Microbiology
Georgetown University School of
  Medicine
Washington, D.C. 20007
USA

Barry Holmes

Central Public Health Laboratory
National Collection of Type Cultures
London NW9 5HT
England

Wilhelm H. Holzapfel

Bundesforschungsanstalt für
  Ernährung
Institut für Hygiene und Toxologie
D-7500 Karlsruhe 1
Germany

Robert Huber

Institut für Mikrobiologie
Universität Regensburg
D-8400 Regensburg
Germany

Carrie A. Norton Hughes

Department of Microbiology
University of Minnesota Medical
  School
Minneapolis, MN 55455
USA

Garret M. Ihler

Department of Medical Biochemistry
  and Genetics
Texas A & M College of Medicine
College Station, TX 77843
USA

Johannes F. Imhoff

Institut für Mikrobiologie
Rheinische Friedrich-Wilhelms-
  Universität
D-5300 Bonn 1
Germany

Mahendra K. Jain

Michigan Biotechnology Institute
3900 Collins Road
Lansing, MI 48909
USA

Holger W. Jannasch

Biology Department
Woods Hole Oceanographic
  Institution
Woods Hole, MA 02543
USA

Kwang W. Jeon

Department of Zoology
University of Tennessee
Knoxville, TN 37996
USA

John L. Johnson

Department of Anaerobic
  Microbiology
Virginia Polytechnic Institute and
  State University
Blacksburg, VA 24601
USA

Russell C. Johnson

Department of Microbiology
University of Minnesota Medical
  School
Minneapolis, MN 55455
USA

Dorothy Jones

Department of Microbiology
University of Leicester, School of
  Medicine
Leicester LE1 9HN
England

Elliot Juni

Department of Microbiology and
    Immunology
University of Michigan
Ann Arbor, MI 48109
USA

Clarence Kado

Department of Plant Pathology
University of California-Davis
Davis, CA 95616
USA

Ronald M. Keddie

Craigdhu, 3 Ness Way
Fortrose
Ross-shire IV10 8SS
Scotland

Donovan P. Kelly

Natural Environment Research
    Council
Polaris House
Swindon SN2 1EU
England

Allen Kerr

Waite Agricultural Research Institute
The University of Adelaide
Glen Osmond 5064
South Australia

Karel Kersters

Laboratorium voor Microbiologie en
    Microbiële Genetica
Rijksuniversiteit Ghent
B-9000 Ghent
Belgium

Bruce C. Kirkpatrick

Department of Plant Pathology
University of California-Davis
Davis, CA 95616
USA

Wesley E. Kloos

Department of Genetics
North Carolina State University
Raleigh, NC 27695
USA

Joan S. Knapp

Center for Infectious Diseases
Centers for Disease Control
Atlanta, GA 30333
USA

Miroslav Kocur

Czechoslovak Collection of
    Microorganisms
J.E. Purkyně University
662 43 Brno
Czechoslovakia

Paul E. Kolenbrander

National Institute of Dental Research
National Institutes of Health
Bethesda, MD 20892
USA

Elena N. Kondratieva

Department of Microbiology
Moscow State University
Moscow B-234
U.S.S.R.

Hans-Peter Koops

Institut für Allgemeine Botanik
Universität Hamburg
D-2000 Hamburg 52
Germany

Felicitas Korn-Wendisch

Institut für Mikrobiologie
Technische Hochschule
D-6100 Darmstadt
Germany

Heinz Eberhard Krampitz

Universität München
D-8000 München 90
Germany

Thomas Krech

Risch Medical Laboratories
Landstrasse 97
FL-9494 Schaan
Liechtenstein

Julius P. Kreier

The Ohio State University
2047 Iuka Avenue
Columbus, OH 43201
USA

Noel R. Krieg

Department of Biology
Virginia Polytechnic Institute and
    State University
Blacksburg, VA 24061
USA

Reiner Michael Kroppenstedt

Deutsche Sammlung von
    Mikroorganism und Zellkulturen
D-3300 Braunschweig
Germany

J. Gijs Kuenen

Kluyver Laboratorium voor
    Microbiologie
Delft University of Technology
Julianalaan 67
2628 BC Delft
The Netherlands

Daisy Kuhn

Department of Biology
California State University
Northridge, CA 91330
USA

Hans Jürgen Kutzner

Institut für Mikrobiologie
Technische Hochschule
D-6100 Darmstadt
Germany

David P. Labeda

Northern Regional Research Center,
    ARS
US Department of Agriculture
Peoria, IL 61604
USA

Bart Lambert

Plant Genetic Systems N.V.
J. Plateaustraat 22
B-9000 Ghent
Belgium

Jan W.M. la Rivière

Institut für Mikrobiologie
Universität Göttingen
D-3400 Göttingen
Germany

John W. Lawson

Department of Microbiology
Clemson University
Clemson, SC 29632
USA

Léon Le Minor

WHO Collaborating Center for
    Reference and Research on
    Salmonella
Institut Pasteur
75724 Paris Cedex 15
France

Mary Lidstrom

Environmental Engineering Science
California Institute of Technology
Pasadena, CA 91125
USA

Wolfgang Liebl

Lehrstuhl für Mikrobiologie
Technische Universität München
D-8000 München
Germany

Franz Lingens

Institut für Mikrobiologie
Universität Hohenheim
D-7000 Stuttgart 70
Germany

Reggie Y.C. Lo

Department of Microbiology
University of Guelph
Guelph, Ontario N1G 2W1
Canada

Michael T. Madigan

Department of Microbiology
Southern Illinois University
Carbondale, IL 62901
USA

Sarabelle Madoff

Department of Bacteriology
Massachusetts General Hospital
Boston, MA 02114
USA

Lynn Margulis

Department of Botany
University of Massachusetts
Amherst, MA 01003
USA

Kevin C. Marshall

School of Microbiology
University of New South Wales
Kensington, New South Wales 2033
Australia

Frank Mayer

Institut für Mikrobiologie
Universität Göttingen
D-3400 Göttingen
Germany

Michael J. McInerney

Department of Botany and
  Microbiology
University of Oklahoma
Norman, OK 73109
USA

Glenda R. Mernaugh

Department of Medical Biochemistry
  and Genetics
Texas A & M College of Medicine
College Station, TX 77843
USA

James N. Miller

Department of Microbiology and
  Immunology
University of California-Los Angeles
Los Angeles, CA 90024
USA

Uwe Christian Möller

Institut für Allgemeine Botanik
Universität Hamburg
D-2000 Hamburg 52
Germany

Lillian V.H. Moore

Department of Anaerobic
  Microbiology
Virginia Polytechnic Institute and
  State University
Blacksburg, VA 24061
USA

Stephen A. Morse

Center for Infectious Diseases
Centers for Disease Control
Atlanta, GA 30333
USA

Nelson P. Moyer

Hygienic Laboratory
University of Iowa
Iowa City, IA 52242
USA

Eppe Gerke Mulder

Laboratory of Microbiology
Agricultural University
6703 CT Wageningen
The Netherlands

Luuc R. Mur

Laboratorium voor Microbiologie
University of Amsterdam
1018 WS Amsterdam
The Netherlands

Robert G.E. Murray

Department of Microbiology and
  Immunology
University of Western Ontario
London, Ontario N6A 5C1
Canada

Francis E. Nano

Department of Biochemistry and
  Microbiology
University of Victoria
Victoria, British Columbia V8W 2Y2
Canada

Kenneth H. Nealson

Center for Great Lakes Studies
University of Wisconsin-Milwaukee
Milwaukee, WI 53204
USA

Douglas Nelson

Department of Bacteriology
University of California-Davis
Davis, CA 95616
USA

A.I. Netrusov

Microbiology Department
Moscow University
Moscow 119899
U.S.S.R.

Horst Neve

Institut für Mikrobiologie
Bundesanstalt für Milchforschung
D-2300 Kiel
Germany

Steven J. Norris

Department of Pathology and
  Laboratory Medicine
University of Texas Medical School
  at Houston
Houston, TX 77225
USA

Noboru Okamura

Department of Microbiology
Institute of Public Health
Shirokanedai, Minato-ku
Tokyo
Japan

Aharon Oren

Institute of Life Sciences
The Hebrew University of Jerusalem
Jerusalem 91904
Israel

Paul Orndorff

School of Veterinary Medicine
North Carolina State University
Raleigh, NC 27606
USA

Hiroshi Oyaizu

Department of Microbiology
Gifu University School of Medicine
Gifu 500
Japan

Norberto J. Palleroni

Department of Microbiology
New York State University School of
  Medicine
New York, NY
USA

Guy H. Palmer

Department of Veterinary
  Microbiology and Pathology
Washington State University
Pullman, WA 99164-7040
USA

Nickolas Panopoulos

Department of Plant Pathology
University of California-Berkeley
Berkeley, CA 94720
USA

A. William Pasculle

Department of Pathology
University of Pittsburgh School of
  Medicine
Pittsburgh, PA 15261
USA

Bruce J. Paster

Forsyth Dental Center
140 Fenway
Boston, MA 02115
USA

Susanne Peinemann

Institut für Mikrobiologie
Universität Göttingen
D-3400 Göttingen
Germany

John L. Penner

Department of Medical Microbiology
University of Toronto
Toronto, Ontario M5G 1L5
Canada

Michel C.M. Pérombelon

Scottish Crop Research Institute
Invergowrie, Dundee DD2 5DA
Scotland

Jerome J. Perry

Department of Microbiology
North Carolina State University
Raleigh, NC 27695
USA

Norbert Pfennig

Fakultat für Biologie
Universität Konstanz
D-7750 Konstanz
Germany

James E. Phillips

Royal (Dick) School of Veterinary
    Studies
University of Edinburgh
Summerhall, Edinburgh EH9 1QH
Scotland

M.J. Pickett

Department of Microbiology
University of California-Los Angeles
Los Angeles, CA 90024
USA

Harvey M. Pickrum

Proctor and Gamble Co.
Miami Valley Laboratories
Cincinnati, OH 45239
USA

Beverly K. Pierson

Department of Biology
University of Puget Sound
Tacoma, WA 98416
USA

Jeanne S. Poindexter

Biology Department
Long Island University
Brooklyn, NY 11201
USA

Philip S. Poole

Department of Biochemistry
University of Oxford
Oxford OX1 3QU
England

Bruno Pot

Laboratorium voor Microbiologie en
    Microbiële Genetica
Rijksuniversiteit Ghent
B-9000 Ghent
Belgium

Helmut Prauser

Zentralinstitut für Mikrobiologie und
    Therapie
6900 Jena
Germany

Rudolf A. Prins

Department of Microbiology
University of Groningen
9751 NN Haren
The Netherlands

Thomas J. Quan

Center for Infectious Diseases
Centers for Disease Control
Ft. Collins, CO 80522
USA

Shmuel Razin

Department of Membrane and
    Ultrastructure Research
The Hebrew University,
    Haddassah Medical School
Jerusalem 91010
Israel

Annette C. Reboli

Department of Medicine
Hahneman University Hospital
Philadelphia, PA 19102
USA

Hans Reichenbach

Gesellschaft für Biotechnologische
    Forschung
Arbeitsgruppe Mikrobielle
    Sekundärstoffe
D-3300 Braunschweig
Germany

Claude Richard

Service des Enterobacteries
Institut Pasteur
75724 Paris Cedex 15
France

Lesley A. Robertson

Kluyver Laboratorium voor
    Microbiologie
Delft University of Technology
Julianalaan 67
2628 BC Delft
The Netherlands

Eugene Rosenberg

Microbiology Department
Tel Aviv University
Ramat Aviv 69978
Israel

Edward Ruby

Department of Biological Science
University of South California
Los Angeles, CA 90089
USA

Kathryn L. Ruoff

Francis Black Bacteriology
    Laboratory
Massachusetts General Hospital
Boston, MA 02114
USA

Hermann Sahm

Institut für Biotechnologie I
KFA Jülich GmbH
D-5170 Jülich
Germany

Riichi Sakazaki

Department of Microbiology
Tokai University
Bouseidai, Isehara, Kanagawa 259-11
Japan

Vittorio Scardovi

Instituto di Microbiologia Agraria e
    Tecnica
Università degli Studi di Bologna
40126 Bologna
Italy

Klaus P. Schaal

Institut für Medizinische
    Mikrobiologie
Universität Bonn
D-5300 Bonn 1
Germany

Hainfried E.A. Schenk

Institut für Biologie I
Universität Tübingen
D-7400 Tübingen 1
Germany

Ulrich Schillinger

Bundesforschungsanstalt für
    Ernährung
Institut für Hygiene und Toxologie
D-7500 Karlsruhe 1
Germany

Bernhard Schink

Lehrstuhl Mikrobiologie I
Eberhard-Karls-Universität
D-7400 Tübingen 1
Germany

Hans G. Schlegel

Institut für Mikrobiologie
Universität Göttingen
D-3400 Göttingen
Germany

Karl-Heinz Schleifer

Lehrstuhl für Mikrobiologie
Technische Universität München
D-8000 München 2
Germany

Heinz Schlesner

Institut für Allgemeine Mikrobiologie
Christian-Albrechts-Universität
D-2300 Kiel 1
Germany

Jean M. Schmidt

Department of Botany and
  Microbiology
Arizona State University
Tempe, AZ 85287
USA

Karin Schmidt

Institut für Mikrobiologie
Georg-August-Universität
D-3400 Göttingen
Germany

Milton N. Schroth

Department of Plant Pathology
University of California-Berkeley
Berkeley, CA 94720
USA

H.P.R. Seeliger

Institute für Hygiene und
  Mikrobiologie
D-8700 Wurzburg
Germany

Andreas H. Segerer

Institut für Mikrobiologie
Universität Regensburg
D-8400 Regensburg
Germany

Barbara Sgorbati

Instituto di Microbiologia Agraria e
  Tecnica
Università degli Studi di Bologna
40126 Bologna
Italy

Haroun N. Shah

Department of Oral Microbiology
The London Hospital Medical
  College
London E1 2AD
England

Patricia E. Shewen

Department of Veterinary
  Microbiology and Immunology
University of Guelph
Guelph, Ontario N1G 2W1
Canada

Tsuneo Shiba

Ocean Research Institute
University of Tokyo
Akahama, Otsuchi, Iwate 02811
Japan

Thomas M. Shinnick

Center for Infectious Diseases
Centers for Disease Control
Atlanta, GA 30333
USA

Ralph Slepecky

Department of Biology
Syracuse University
Syracuse, NY 13244
USA

Robert M. Smibert

Department of Anaerobic
  Microbiology
Virginia Polytechnic Institute and
  State University
Blacksburg, VA 24061
USA

Louis D.S. Smith

2449 N. Baker Avenue
East Wenatchee, WA 98802
USA

Elizabeth Sockett

Department of Biochemistry
University of Oxford
Oxford OX1 3QU
England

Per Søgaard

Department of Clinical Microbiology
Århus Kommunehospital
8000 Århus C
Denmark

Carol Spiegel

University of Wisconsin Hospitals
  and Clinics
A4/204 CSC
Madison, WI 53792
USA

Georg Sprenger

Institut für Biotechnologie I
KFA Jülich GmbH
D-5170 Jülich
Germany

Erko Stackebrandt

Department of Microbiology
University of Queensland
St. Lucia, Queensland 4067
Australia

Donald P. Stahly

Department of Microbiology
University of Iowa
Iowa City, Iowa 52242
USA

James T. Staley

Department of Microbiology
University of Washington
Seattle, WA 98105
USA

Karl O. Stetter

Institut für Mikrobiologie
Universität Regensburg
D-8400 Regensburg
Germany

Daphne L. Stoner

Idaho National Engineering
  Laboratory
EG & G Idaho
Idaho Falls, ID 83415
USA

Adriaan H. Stouthamer

Biologisch Laboratorium
Vrije Universiteit
1081 HV Amsterdam
The Netherlands

James R. Swafford

Department of Botany and
  Microbiology
Arizona State University
Tempe, AZ 85287
USA

Jean Swings

Laboratorium voor Microbiologie en
  Microbiële Genetica
Rijksuniversiteit Ghent
B-4000 Ghent
Belgium

Kazumichi Tamura

Enterobacteriology Laboratory
National Institute of Health
Kamiosaki, Shinagawa-ku, Tokyo
  141
Japan

Anne Tanner

Departments of Periodontology and
  Microbiology
Forsyth Dental Center
Boston, MA 02115
USA

Fred C. Tenover

Center for Infectious Diseases
Centers for Disease Control
Atlanta, GA 30333
USA

Michael Teuber

Laboratory of Food Microbiology,
  Institute of Food Science
Swiss Federal Institute of Technology
CH-8092 Zürich
Switzerland

Brian J. Tindall

Deutsche Sammlung von
  Mikroorganismen
D-3300 Braunschweig
Germany

Kevin J. Towner

University Hospital
Queen's Medical Centre
Nottingham NG7 2UH
England

Hans G. Trüper

Institut für Mikrobiologie
Rheinische Friedrich-Wilhelms-
    Universität
D-5300 Bonn 1
Germany

K. Tsuji

Gray Freshwater Biology Research
    Institute
University of Minnesota
Navarre, MN 55392
USA

Joseph G. Tully

National Institute for Allergy and
    Infectious Diseases
Frederick Cancer Research Center
Frederick, MD 21701
USA

Olli H. Tuovinen

Department of Microbiology
The Ohio State University
Columbus, OH 43210
USA

Henk W. Van Verseveld

Biologisch Laboratorium
Vrije Universiteit
1081 HV Amsterdam
The Netherlands

Gernot Vobis

Centro Regional Universitario
    Bariloche
Universidad Nacional del Comahue
Barioloche 8400, Rio Negro
Argentina

Alexander von Graevenitz

Department of Medical Microbiology
University of Zurich
CH 8028 Zurich
Switzerland

Russell H. Vreeland

Department of Biology
West Chester University
West Chester, PA 19383
USA

Haruo Watanabe

Department of Bacteriology
The National Institute of Health
Shinagawa-ku, Tokyo
Japan

John B. Waterbury

Woods Hole Oceanographic
    Institution
Woods Hole, MA 02543
USA

Alison Ann Weiss

Department of Microbiology and
    Immunology
Medical College of Virginia
Richmond, VA 23298
USA

Emilio Weiss

Infectious Disease Department
Naval Medical Research Institute
Bethesda, MD 20889
USA

Norbert Weiss

Deutsche Sammlung von
    Mikroorganism und Zellkulturen
D-3300 Braunschweig
Germany

Robert A. Whiley

The London Hospital Medical
    College Dental School
University of London
London E1 2AD
England

Robert F. Whitcomb

Insect Pathology Laboratory
Agricultural Research Service
USDA
Beltsville, MD 20705
USA

William B. Whitman

Department of Microbiology
University of Georgia
Athens, GA 30602
USA

Friedrich Widdel

Institut für Genetik und
    Mikrobiologie
Universität München
D-8000 München 19
Germany

Juergen Wiegel

Department of Microbiology
University of Georgia
Athens, GA 30602
USA

Anne Willems

Laboratorium voor Microbiologie en
    Microbiële Genetica
Rijksuniversiteit Ghent
B-9000 Ghent
Belgium

J.C. Williams

Department of Intracellular
    Pathogens, USAMRIID
Fort Detrick
Frederick, MD 21701
USA

R.A.D. Williams

Department of Biochemistry
Queen Mary and Westfield College
London E1 4NS
United Kingdom

Reinhard Wirth

Institut für Genetik und
    Mikrobiologie
Universität München
D-8000 München 19
Germany

Carl Woese

Department of Microbiology
University of Illinois
Urbana, IL 61801
USA

Ralph S. Wolfe

Department of Microbiology
University of Illinois
Urbana, IL 61801
USA

Eiko Yabuuchi

Department of Microbiology
Gifu University School of Medicine
Gifu 500
Japan

Allan A. Yousten

Biology Department
Virginia Polytechnic Institute and
    State University
Blacksburg, VA 24061
USA

Georgi A. Zavarzin

Institute of Microbiology
Academy of Sciences of the USSR
117312 Moscow
USSR

J. Gregory Zeikus

Michigan Biotechnology Institute
3900 Collins Road
Lansing, MI 48909
USA

Wolfram Zillig

Max-Planck-Institut für Biochemie
D-8033 Martinsried bei München
Germany

W. Zumft

Lehrstuhl für Mikrobiologie
Universität Karlsruhe
D-7500 Karlsruhe
Germany

The Genus *Actinoplanes* and Related Genera

GERNOT VOBIS

Sporeforming actinomycetes related to the genus *Actinoplanes* are combined at present under the suprageneric group named "actinoplanetes" (Goodfellow, 1989). Originally, the term actinoplanetes was used ecologically typical to describe the strains of *Actinoplanes* and a small number of their relatives (Nonomura and Takagi, 1977). The members of the genus *Actinoplanes* all produce sporangia that release actively swimming spores. Later, the term actinoplanetes was expanded to encompass also the five genera *Amorphosporangium, Ampullariella, Pilimelia, Dactylosporangium,* and *Micromonospora,* which all have a common cell wall chemotype containing *meso-* and/or 3-hydroxy diaminopimelic acid and glycine in combination with xylose and arabinose as the characteristic sugars in the hydrolysates of whole organisms (Goodfellow and Cross, 1984).

The GC content of their DNA is in general 71–73 mol% (Vobis, 1989a), but recent studies have expanded the range to 67–76 mol% (Kothe, 1987). All genera of the actinoplanetes studied so far belong to one RNA homology cluster (Stackebrandt and Woese, 1981), excluding the sporangiate genera *Streptosporangium, Spirillospora, Planomonospora,* and *Planobispora,* which were classified for long time within one family together with *Actinoplanes* (Bland and Couch, 1981). The genera excluded were placed temporarily into a suprageneric group called the "maduromycetes" (Goodfellow, 1989). According to Goodfellow et al. (1990) and Kothe (1987), the members of the actinoplanetes can now be harbored taxonomically in the newly defined family Micromonosporaceae, and those of the maduromycetes in the family Streptosporangiaceae (Chapter 47).

The genus *Actinoplanes* was first described by Couch (1950), who called attention to its close similarities in colonial characteristics to the genus *Micromonospora* (Ørskov, 1923), which produces single nonmotile spores. Our knowledge of the sporangiate members was extended by subsequent descriptions of further genera: *Ampullariella* and *Amorphosporangium* (Couch, 1963), *Pilimelia* (Kane, 1966), and *Dactylosporangium* (Thiemann et al., 1967). In recent years, two new genera have been added: *Glycomyces* (Labeda et al., 1985) and *Catellatospora* (Asano and Kawamoto, 1986). Both genera form nonmotile spores in chains. Although their chemotaxonomic features suggest convincing arguments for placing them into the actinoplanetes group, their phylogenetic relationships still need to be investigated (E. Stackebrandt, personal communication).

Ecophysiology

Without applying selective procedures, it is difficult to isolate the members of *Actinoplanes* and related genera from soil or other natural substrates. Although they are mesophilic and aerobic organisms, the growth rate of their colonies is often very slow, and on routine isolation plates, the fast-growing streptomycetes can overrun them before they have developed conspicuous mycelia. Furthermore, they exhibit quite varied types of reproductive structures, which are connected with distinct life cycles: 1) the genera *Actinoplanes, Ampullariella, Pilimelia,* and *Dactylosporangium* form sporangia with flagellated spores and are adapted to a periodic wetting and drying of the habitat; 2) the hydrophilic, nonmotile spores of *Micromonospora* are passively disseminated in water and in soil, again distributed by wind or rain; and 3) the genera *Catellatospora* and *Glycomyces* produce nonmotile, hydrophobic spores in chains, presumably associated with other survival strategies.

Sporangiate Actinoplanetes

The life cycles of the sporangiate genera are based on an alternation between terrestrial and "aquatic" habitats. The growth of vegetative mycelium on plant or animal residues culminates in the differentiation into sporangia, which are produced in general on the surface of the substrate, directly in contact with the air (Figs. 1F, 2C, 4A, and 7B) (Bland and Couch, 1981). The sporangia can easily lose their connection to the degenerating mycelium and are disseminated as diaspores by the wind or by soil fauna such as mites, collembola, or arthropods. The sporangia can withstand prolonged desiccation and survive for many years (Makkar and Cross, 1982). The sporangial envelope is usually water repellent. But if sporangia become rehydrated by sufficient moisture, e.g.,

during periods of fog or rain, the spores inside the sporangia begin to swell, the sporangial envelope bursts, and the flagellated spores are released. Under laboratory conditions, this process takes 10 to 60 min (Higgins, 1967; Vobis, 1984, 1987). Spores of *Actinoplanes brasiliensis* retain their motility for more than one day in liquid mineral medium with glucose (Palleroni, 1983). Even without an exogenous source of energy, the zoospores of *Dactylosporangium* species are motile for two to three days before they germinate (Vobis, 1987). Reserve substances are included in their cytoplasm (Fig. 4C), pos-

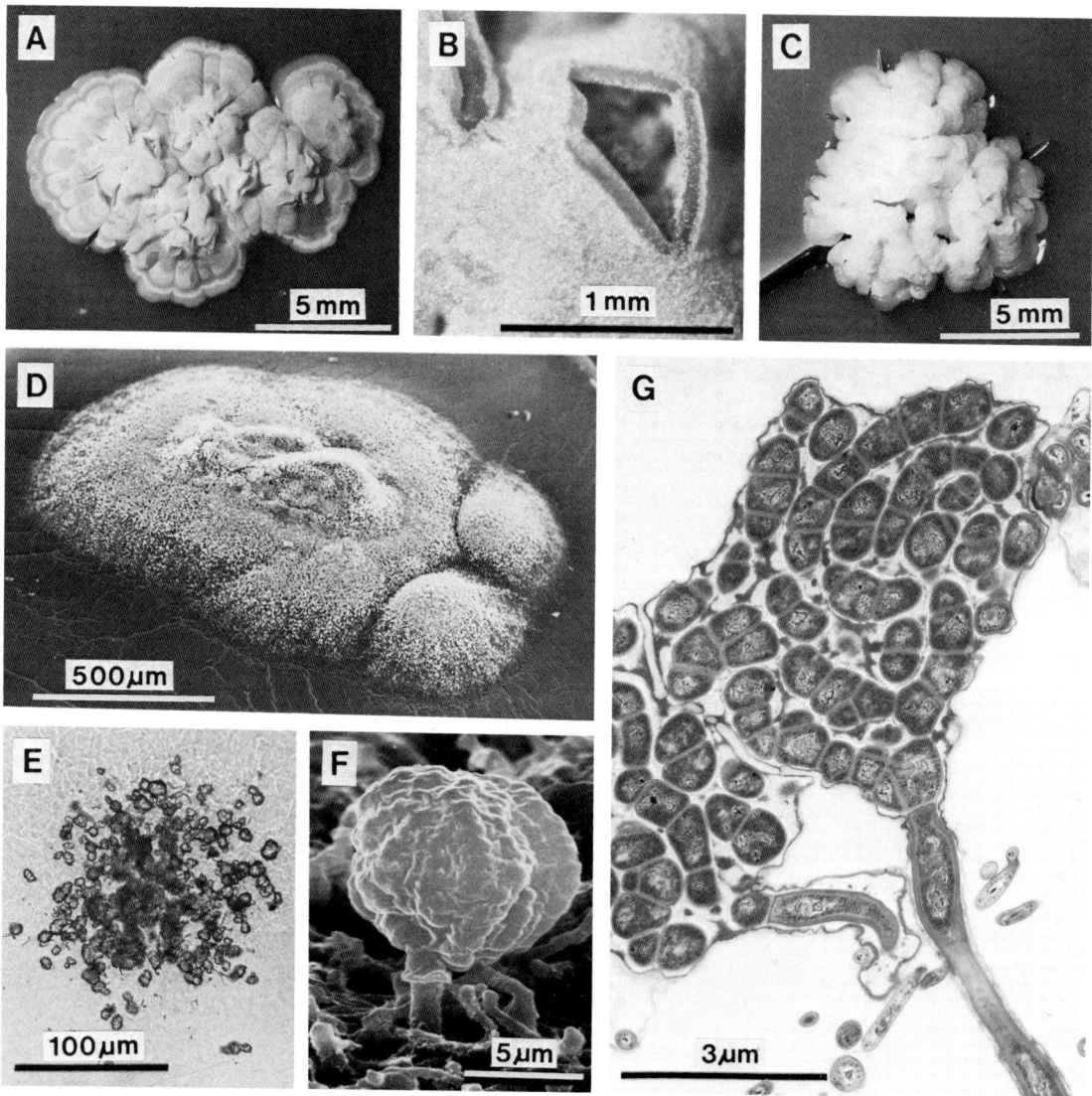

Fig. 1. Features of the genus *Actinoplanes*. (A) Colony with rough surface; marginal areas divided into radial and concentric sections, dissecting microscope (DM). (B) Burst substrate mycelium covered with a mass of sporangia (DM). (C) Elevated colony with squamules; smooth surface without sporangia (DM). (D) Flat colony with abundant sporangia visible on the substrate mycelium, scanning electron microscope (SEM). (E) Irregularly shaped sporangia on agar medium, light microscope (LM). (F) Globose sporangium at the tip of a palisade hypha (SEM). (G) Section of a sporangium with coiled chains of spores, transmission electron microscope (TEM). (G from Kothe, 1987; with permission.)

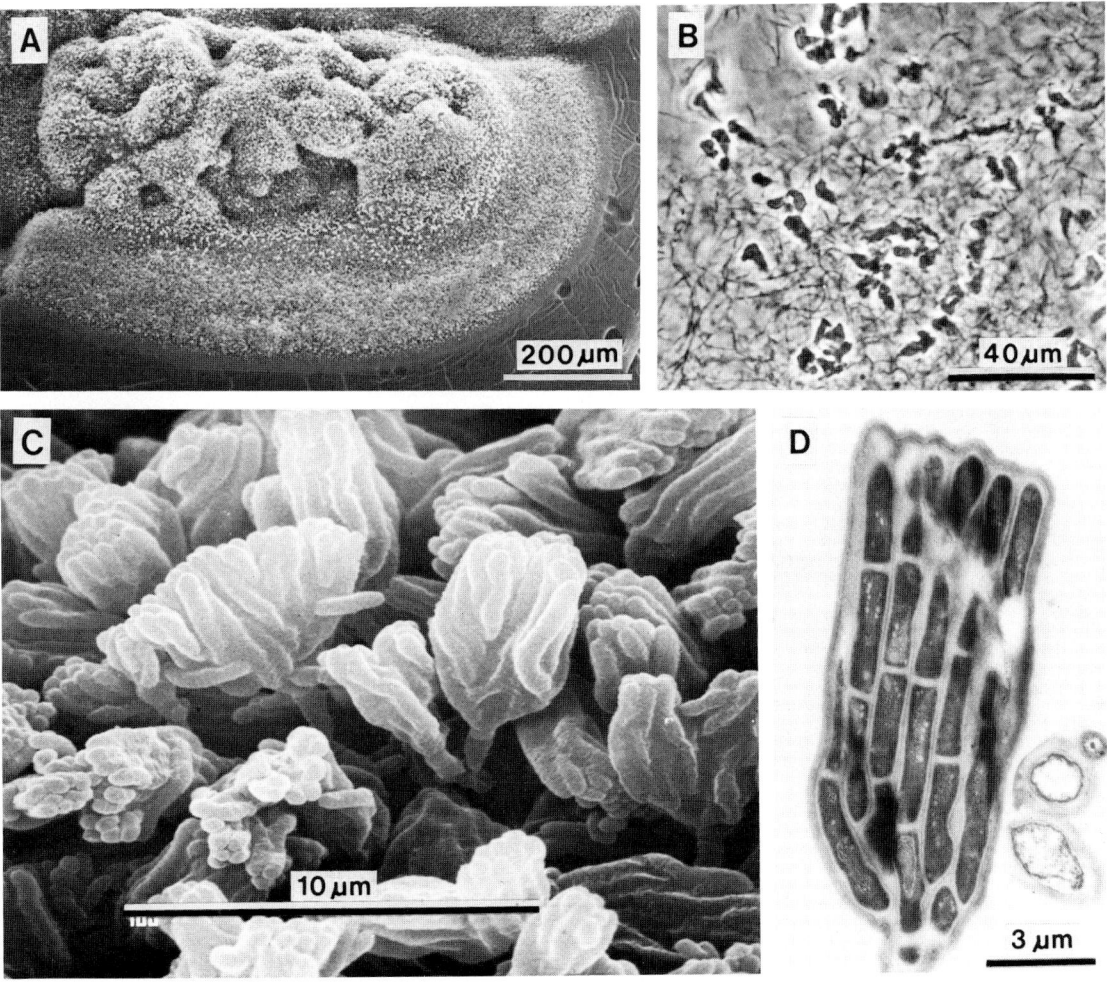

Fig. 2. Features of the genus *Ampullariella*. (A) Colony on agar medium, covered with a bloom-like layer of sporangia (SEM). (B) Sporangia on substrate mycelium, phase contrast (PHACO). (C) Campanulate sporangia; sporogenous hyphae are recognizable under the sporangial envelope (SEM). (D) Section of a cylindrical sporangium with parallel rows of bacilliform spores (TEM). (D from Vobis, 1987; with permission.)

sibly consisting of trehalose (J. C. Ensign, personal communication).

The zoospores of *Actinoplanes* exhibit chemotactic properties. In *A. brasiliensis,* Palleroni (1976) found bromide and chloride ions acting as attractants at a relatively high concentration (0.1 M). Addition of methionine stimulated this chemotactic effect, suggesting that protein methylation may be involved (Palleroni, 1983). Not all species of *Actinoplanes* are attracted by halides. Spores of *A. missouriensis* were attracted to fungal conidia, chlamydospores, and sclerotia, and to exudates of them (Arora, 1986). Several sugars had the same function. An extract of cattle horn meal can be more attractive to the spores than chloride ions (G. Vobis, unpublished observations). Phototactic effects could not be observed, but an apparent microaerophilic behavior was seen in *A. brasi-*

liensis (Palleroni, 1976). In baiting experiments simulating an aquatic microhabitat, pollen or hair are exposed to the surface of water. The zoospores, once released from the submerged sporangia, are able to swim to the surface, fasten to the natural substrates, germinate, and colonize them within several days (Couch, 1963; Vobis, 1984). This may be a result of aerotactic and chemotactic behavior of the spores (Cross, 1986). Although the chemotactic response is used effectively in the isolation method of Palleroni (1980), the exact physiological explanation is not yet known.

The members of the genus *Dactylosporangium* have an additional variant in their life cycle. Beside the typical few-spored sporangia (Fig. 4A–C), they can develop "globose bodies" (Thiemann et al., 1967). These are spores that are borne singly on substrate hyphae (Fig. 4D-

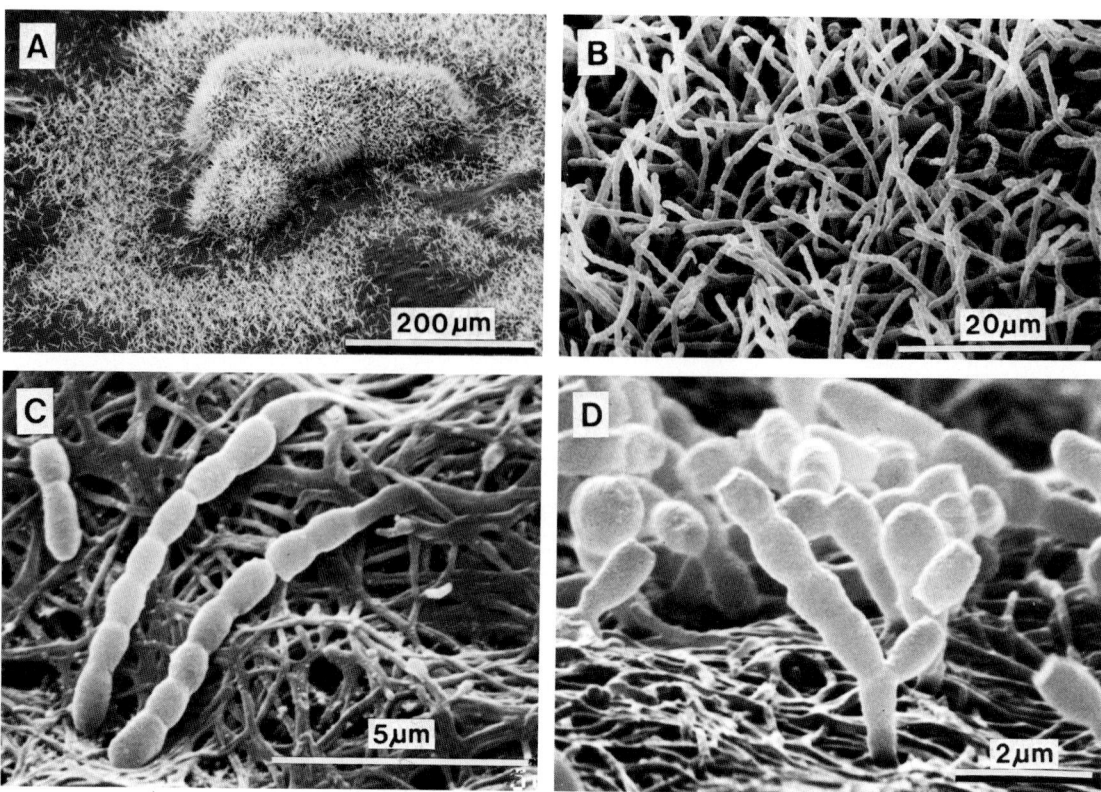

Fig. 3. Features of the genus *Catellatospora*. (All SEMs.) (A) A sporulating colony. (B) Sporeforming hyphae on the surface of agar medium. (C) Short spore chains emerging from substrate hyphae. (D) Branched spore chains on rudimentary sporophores containing cylindrical to ovoid spores.

F). Ensign (1978) demonstrated that these spores are able to germinate on 1.0% (w/v) yeast extract agar medium. The zoospores may function to insure a prompt dissemination in water; and the globose spores, which are constitutively dormant, may survive long periods of starvation or desiccation (Ensign, 1978). Likewise, treatment with dry heat (1 h at 120°C) seems to have no adverse effect on the dormant globose spores of *Dactylosporangium,* in contrast to the sporangiospores of other actinoplanetes (Shearer, 1987).

In general, the sporangiate actinoplanetes can be considered as normal inhabitants of the soil and leaf litter (Cross, 1981a), although they can also be isolated directly from lake or river water (Willoughby, 1969b, 1971). They colonize vegetable or animal debris preferably (Cross, 1981a). A frequent drying and wetting of the substrate increases their occurrence. Favored habitats are the edges of ponds, drainage ditches, and barnyards (Shearer, 1987). Sediments of rivers seem to be also a good source for the isolation of *Actinoplanes* species (Goodfellow et al., 1990). They can also be frequently isolated from twigs submerged in streams (Willoughby, 1971), muddy dead leaves that are

caught and dried on branches of overhanging trees (Cross, 1981a), and from allochthonous leaf litter cast up on the shores of lakes (Willoughby, 1969a). Rarely, strains of *Dactylosporangium* can be encountered in lake sediments. Possibly survival depends on the globose spores (Johnston and Cross, 1976). Also, marine sediments are not very productive habitats for sporangiate actinoplanetes (Goodfellow and Haynes, 1984).

The genus *Actinoplanes* has a world-wide distribution (Couch, 1963; Gaertner, 1955; Vobis, 1987; Schäfer, 1973; Nonomura and Takagi, 1977). Strains of *Actinoplanes* occur in all types of soil, arid desert areas (Makkar and Cross, 1982), sand dune systems close to seashores (Palleroni, 1976), and subtropical and tropical regions. The most productive samples seem to originate from the latter (Shearer, 1987). In a large-scale investigation of the distribution of actinoplanetes in soil in Japan, Nonomura and Takagi (1977) demonstrated a correlation between their abundance, the type of soil, its pH value, and the content of organic matter. Relatively few actinoplanetes occurred in soils with pH 4.0 to 5.0 and abundant organic matter content. Their number increased with lower humus

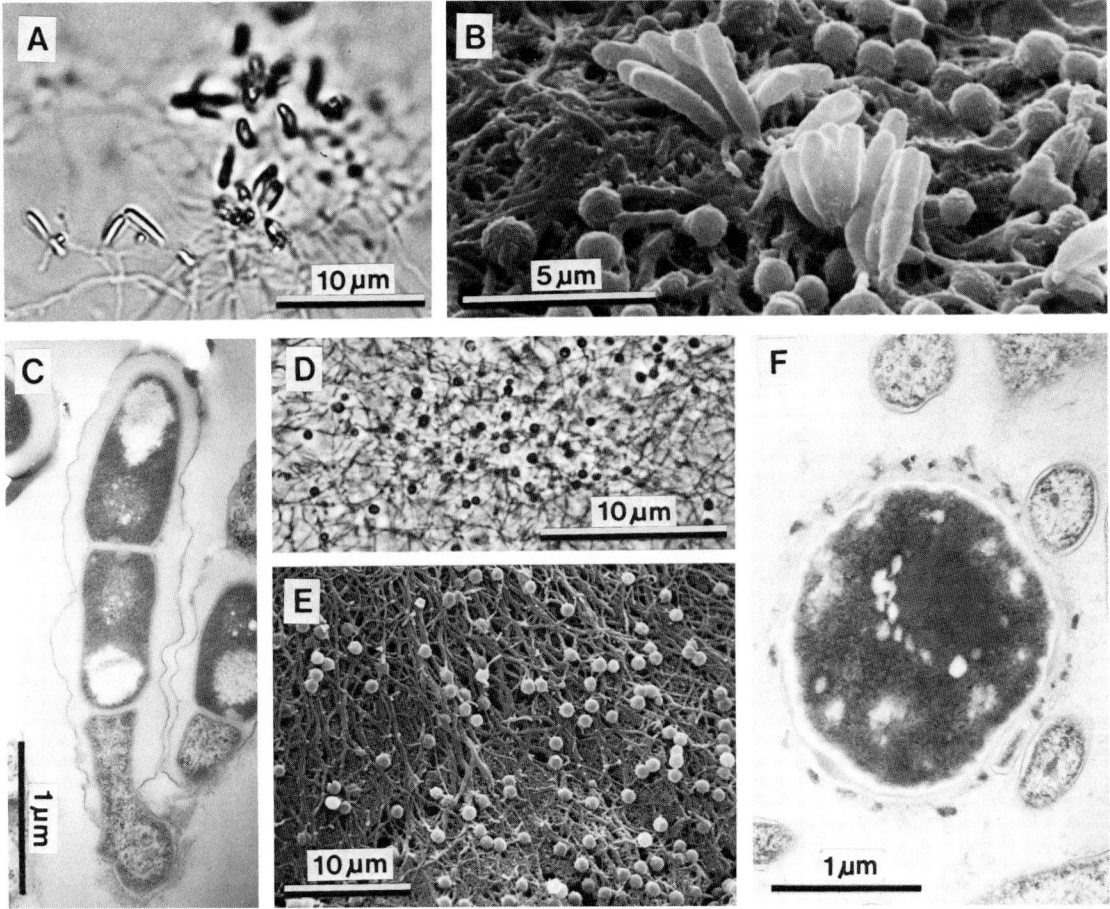

Fig. 4. Features of the genus *Dactylosporangium*. (A) Finger-like sporangia on agar medium (LM). (B) Bundles of sporangia and globose spores (SEM). (C) Section of a sporangium with two spores containing reserve material; sporangial envelope is thin and wavy (TEM). (D) Refractile globose spores dispersed in substrate mycelium (PHACO). (E) Globose spores on the surface of substrate mycelium (SEM). (F) Section of a globose spore with paracrystalline inclusion bodies and perispherical deposits (TEM). (C, D, and F from Vobis, 1987; with permission.)

content and a pH value between 6.4 to 7.2. Soil with a permanent high content of water (e.g., paddy rice fields) have no advantage compared with cultivated fields, which are dry for longer periods.

Strains of *Ampullariella* are widely distributed throughout the world (Couch, 1963). A little less than 10% of the isolated sporangiate actinoplanetes are represented by this genus (Schäfer, 1973; Vobis, 1987). *Ampullariella* could also be obtained from freshwater habitats (Willoughby, 1969b).

The function of sporangiate actinoplanetes in soil ecosystems is not really known. With the exception of a few strains, the *Actinoplanes* species cannot decompose cellulose (Parenti and Coronelli, 1979; Palleroni, 1989). Although they can be isolated on colloidal chitin agar medium, degradation tests with chitin from insects and fungi gave negative results (Schäfer, 1973). Since they exhibit good growth on xylose and

arabinose, it is conceivable that they play a role in decomposing pentosans of plant origin (Parenti and Coronelli, 1979). *Pilimelia* strains are able to colonize keratinic substrates like hair of mammalia or snake skin (Karling, 1954; Gaertner, 1955; Tribe and Abu El-Souod, 1979). They are distributed worldwide and occur statistically in about one of every five soil samples (Schäfer, 1973; Vobis et al., 1986). Although they can aggressively attack the scleroproteins of animals (Fig. 7C), they are not known as dermatophytes.

Micromonospora

Members of the genus *Micromonospora* can be commonly isolated from neutral and alkaline soils (Jensen, 1930; 1932) and according to Shearer (1987), two or three strains may be expected from most soil samples. However their predominant incidence seems to be in aquatic

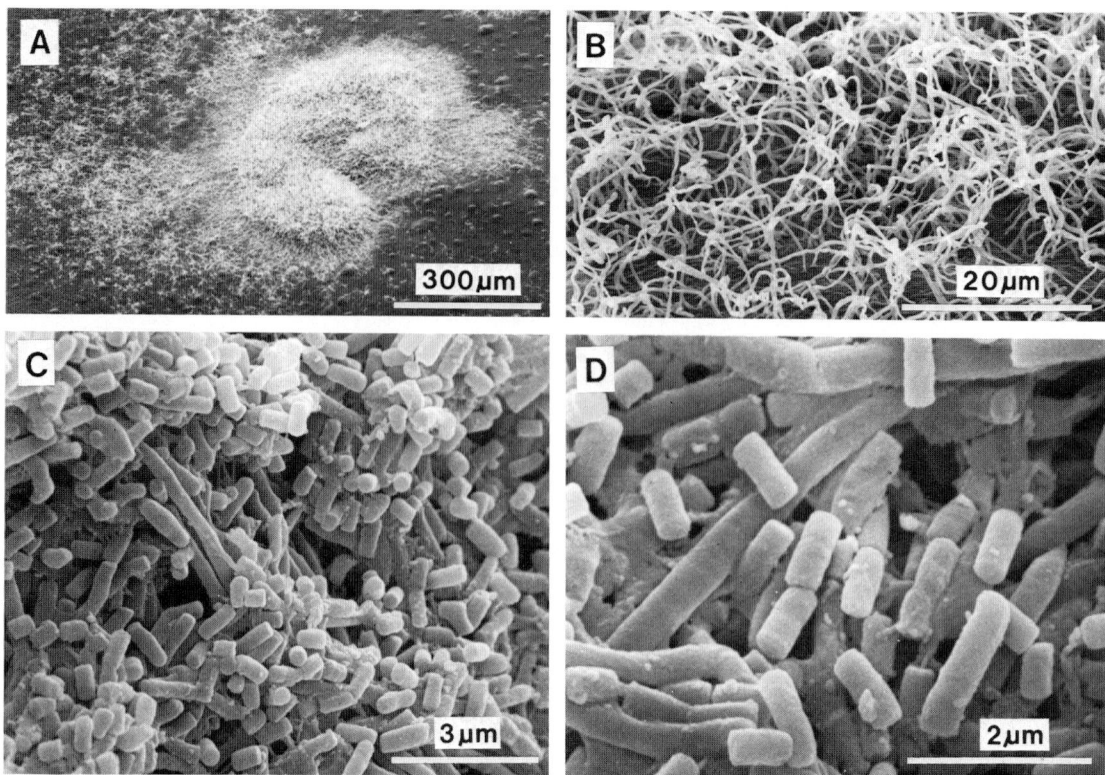

Fig. 5. Features of the genus *Glycomyces*. (All SEMs.) (A) A colony with aerial mycelium. (B) Irregularly oriented and branched aerial hyphae. (C) Mass of spores produced by fragmentation of aerial hyphae. (D) Square-ended spores of various lengths (0.5 × 0.7–1.8 μm).

ecosystems, including both freshwater and marine habitats (Cross, 1981a; Goodfellow and Haynes, 1984). Since they decompose chitin, cellulose, and lignin of lake sediments they might play an important role in lacustrine ecology (Erikson, 1941). The presence of *Micromonospora* in lake systems has been shown by investigations in many countries and was comprehensively reviewed by Cross (1981a, 1981b). Beside streptomycetes and nocardioforms, the micromonosporas were the predominant actinomycetes in the bottom sediments of Blelham Tarn UK, with numbers increasing from littoral to profundal mud samples (Willoughby, 1969b). This dominance was even more striking in deeper mud layers, as could be shown in studies of other lakes of the English Lake District (Johnston and Cross, 1976). Similar observations were made by Fernandez (1984) at a thermal lake, Lake Hévíz, in Hungary. Compared with the surface of the mud, the number of micromonosporas increases twofold in 20 cm of depth, whereas the number of streptomycetes decreases significantly in the same layer. Under those conditions, the spores of *Micromonospora* seem to be more resistant than the propagules of *Streptomyces* and nocardioform actinomy-

cetes. This could be confirmed in investigations on the longevity of actinomycete spores in deep mud cores. Viable spores of *Micromonospora* were recorded from sediments deposited at least 100 years before (Cross and Attwell, 1974). Populations of *Micromonospora* species, accompanied by other actinomycetes, have also been frequently found in streams and rivers (Rowbotham and Cross, 1977; Al-Diwany and Cross, 1978). Their spores are hydrophilic and wettable (Fig. 6C) and can easily be removed from soil by the passage of water (Ruddick and Williams, 1972). The spores withstand ultrasonication, moist heat treatment (20 min at 60°C), and dry heat up to 75°C, and they are resistant to various chemical solutions. However, they are somewhat sensitive to acidic pH (Kawamoto, 1989). Thus, the conclusion can be drawn that the spores of *Micromonospora* are washed into the streams, rivers, and lakes where they can survive as dormant propagules for many years (Cross, 1981a).

Micromonospora species have been isolated from many different marine habitats, ranging from coastal regions to deep-sea sediments. Hunter et al. (1981) found abundant micromonosporas in salt marsh ecosystems in New

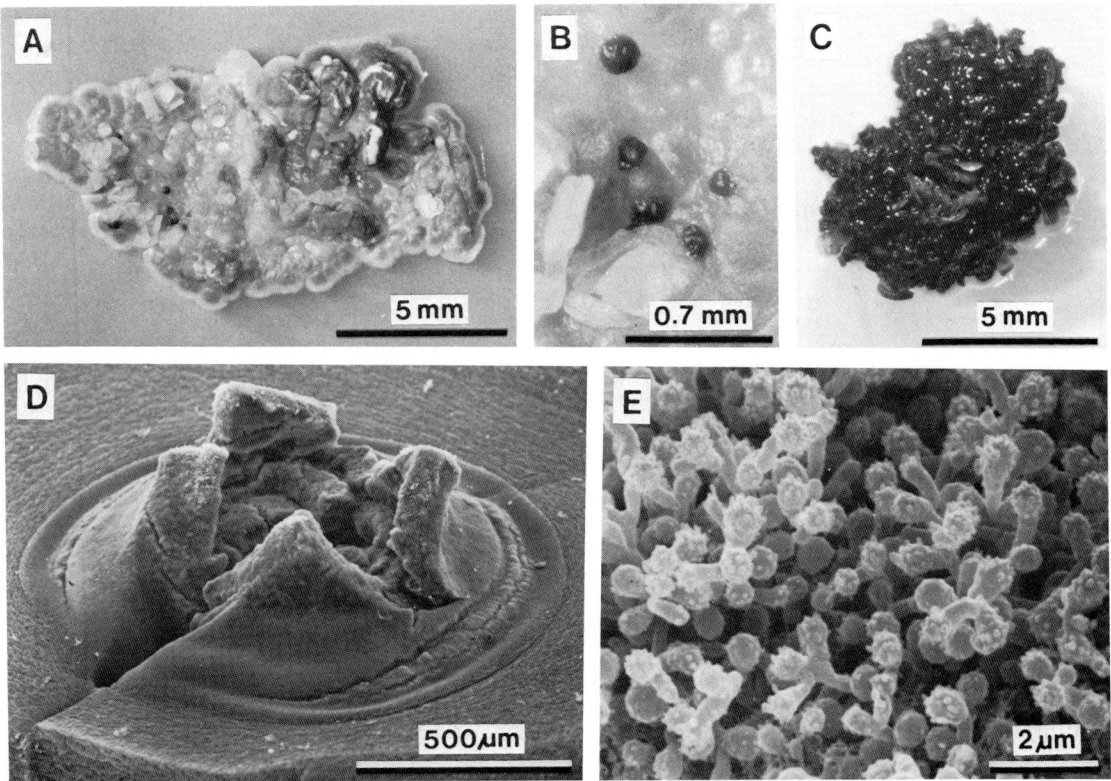

Fig. 6. Features of the genus *Micromonospora*. (A) Raised and folded colony with areas of different colors (DM). (B) Clusters of dark spore masses (DM). (C) Colony completely covered by a mucoid, black mass of spores (DM). (D) Crosswise-burst colony exposing the sporeforming substrate mycelium (SEM). (E) Cluster of spores formed on short side-branches of substrate hyphae; surface of the spores is covered with blunt spines (SEM).

Jersey (USA) with seasonal fluctuations in quantity. In a study at the San Francisco Bay (USA) National Wildlife Refuge, Hunter et al. (1984) showed that micromonosporas occur more frequently in rhizospheric soils of seashore plants than in mud samples obtained from plant-free areas. Watson and Williams (1974) studied the actinomycetes in a coastal sand belt near Formby, Lancashire (UK). In sea water and beach strand, the *Micromonospora* strains predominated. They grew well on freshwater media and most of them tolerated seawater salinity. Okazaki and Okami (1972) and Okami and Okazaki (1978) isolated micromonosporas from littoral muds and from samples collected in shallow sea areas of the Pacific Ocean, occurring more frequently at the bottom than in the sea water. Weyland (1969, 1981) found that the micromonosporas predominated in the deep-sea sediments, and his results were confirmed by Goodfellow and Haynes (1984). The ability of *Micromonospora* strains to tolerate reduced oxygen tensions (Watson and Williams, 1974) favored the view that after the spores sink into the sea, they settle into littoral

or marine sediments where they can survive for long periods of time (Goodfellow and Williams, 1983). Other authors suggest that the actinomycetes are a part of the indigenous marine microflora, able to grow in seawater and its sediments (Okami and Okazaki, 1978; Weyland, 1981).

The presence of *Micromonospora* species could also be established in terrestrial habitats like straw and hay, in grain stores, or as pollutants of the aerial environment in homes. They also can be found in structural timbers, chiefly those found below the watertable in the foundation piles of buildings. They have also been found in the trunks of trees of the genera *Picea* and *Pinus*, and in pulp from paper mills (Lacey, 1988).

Most *Micromonospora* species probably degrade biopolymers (Erikson, 1941), and they can even attack lignin complexes (McCarthy and Broda, 1984). Many of the salt marsh isolates of Hunter et al. (1981) were active in the decomposition of chitin and cellulose. In particular, cellulose is frequently utilized as substrate (Jensen, 1930; Sandrak, 1977; Kawamoto,

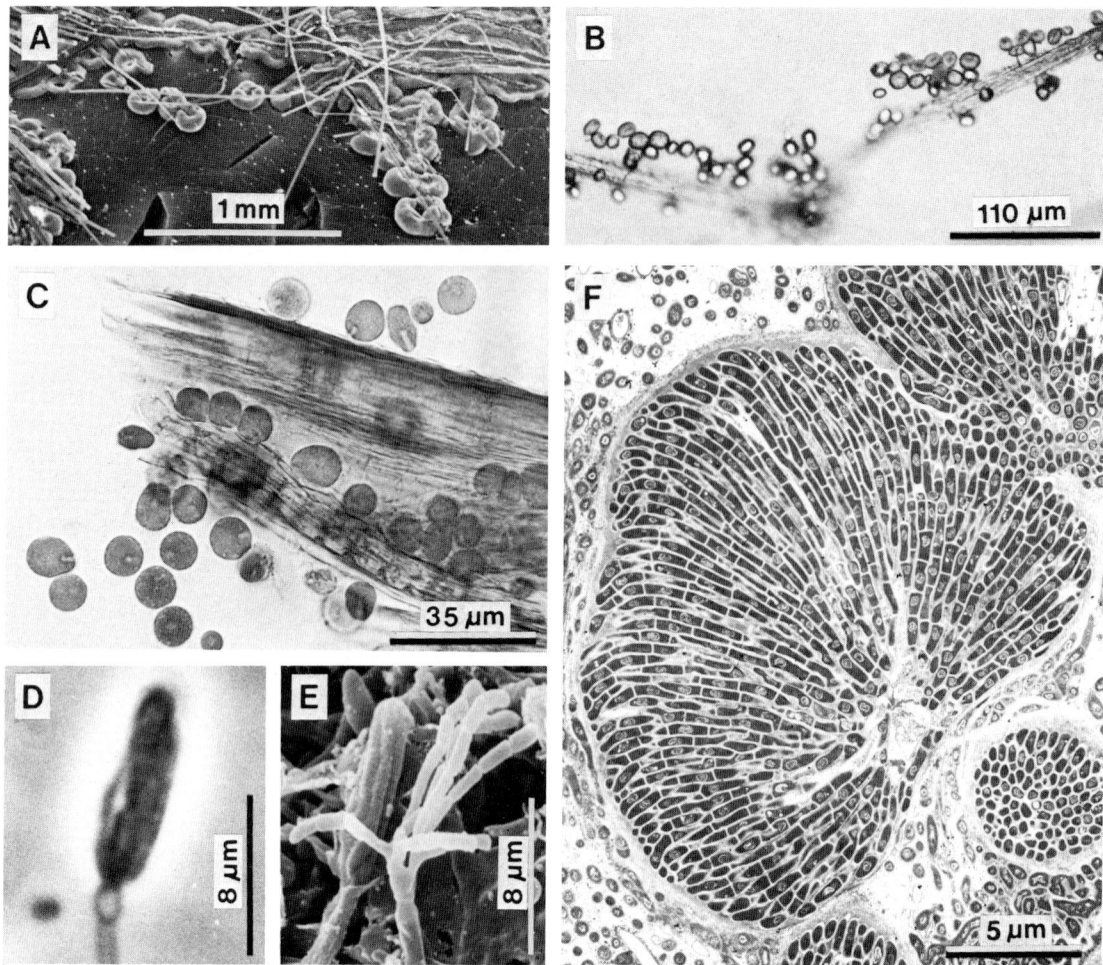

Fig. 7. Features of the genus *Pilimelia*. (A) Compact, small colonies on agar medium; hairs added as natural substrate (SEM). (B) Bundles of sporangia formed on hair (LM). (C) Globose to pyriform sporangia with internal columella; structure of the colonized part of the hair was destroyed (LM). (D) Cylindrical sporangium with an annulus at the base (PHACO). (E) Penicillate conidiophore with bacilliform conidia; the sporangium behind it has parallel-arranged sporogenous hyphae (SEM). (F) Section of a campanulate sporangium with branched spore chains (TEM). (C and D from Vobis et al., 1986; E from Vobis, 1987; F from Vobis, 1984; with permission.)

1989). The cellulase studied from *Micromonospora melanosporea* was found to be more heat stable than those of the imperfect fungus *Trichoderma,* but less stable than the enzymes of thermophilic actinomycetes. The principle sugar released by *Micromonospora* cellulase from ball-milled bagasse and filter paper was cellobiose (Van Zyl, 1985).

It is possible that micromonosporas are able to grow under microaerophilic conditions, as would be found in wet soils (Goodfellow and Williams, 1983). However, the strictly anaerobic *Micromonospora* strains isolated from the intestinal tract of termites (Hungate, 1946) and from the rumen of sheep (Maluszyńska and Janota-Bassalik, 1974) urgently require taxonomical studies to clarify their affiliation to the genus *Micromonospora.*

Catellatospora and *Glycomyces*

As Asano et al. (1989b) emphasized, all their isolates of *Catellatospora* species were recovered from woodland soils collected at various localities from the northern to the western part of Japan; no strains could be obtained from soils of agricultural fields or from sediments of lakes and rivers. The authors suggest that members of this genus are probably distributed widely in woodlands of temperate regions. *Catellatospora* has been isolated from a soil sample that originated from the pampas grassland south of Buenos Aires, Argentina (Vobis, 1987). Earlier descriptions of "nonmotile strains related to *"Dactylosporangium"* (Lechevalier and Lechevalier, 1970a; Thiemann, 1970) and the recent report of Meyertons et al. (1987) suggest a global distribution of these rare organisms.

The role of the singly produced globose spores, which can occur in addition to the nonmotile spores in chains (Asano et al., 1989b), has not been studied yet, but they may have interesting ecophysiological aspects similar to those of the *Dactylosporangium* spores.

It is premature to give ecological data for *Glycomyces* species. The two type strains were isolated from soil samples from Harbin, China, and from a greenhouse in Trenton, New Jersey, USA, respectively (Labeda et al., 1985).

Isolation

In agreement with the wide distribution of actinoplanetes in nature, samples from various habitats have proved to be favorable sources of inoculum: soil, sediment, mud, water, and plant material. Microbial populations vary greatly, so that only about one in four samples may prove successful (Shearer, 1987). Special pretreatments of the freshly collected samples enhance the numbers of actinoplanetes and reduce the nondesirable concomitant microorganisms. Selective isolation media (see "Media for Isolation and Cultivation") favor their development and limit the growth of other bacteria and common soil actinomycetes like *Streptomyces* species. The addition of humic acid, for example, activates the germination of spores (Hayakawa and Nonomura, 1987), and media with low nutrient concentration favors the growth of *Micromonospora* (Rowbotham and Cross, 1977).

To avoid the growth of fungi on the isolation plates, cycloheximide (50 μg/ml) and/or nystatin (100 units/ml) can be added to the isolation medium (Shearer, 1987). The selective effect can be enhanced by the addition of other antibiotics like novobiocin, streptomycin, gentamicin, or tunicamycin, either supplemented solely or in combination (Table 1). *Actinoplanes, Ampullariella,* and *Micromonospora* strains could be isolated very selectively using the agar plate method with additions of cycloheximide, nystatin, or novobiocin (Torikata et al., 1978, 1983). A combination of tunicamycin and nalidixic acid permitted a selective growth of *Micromonospora* species (Nonomura and Hayakawa, 1988). Instead of antibiotic supplements, a special membrane filter technique can be used, since the branching hyphae of the actinoplanetes can penetrate the small filter pores, whereas bacterial growth is restricted to one side of the membrane (Hirsch and Christensen, 1983).

The differing behavior of the spores permits the use of two isolation methods: 1) direct isolation on selective agar media for the actinoplanetes having nonmotile spores; and 2) techniques using spore motility to specifically enrich the sporangiate members. Obviously, strains belonging to one group can appear also on the isolation plates intended for the others.

The isolation plates and the enrichment cultures are usually incubated at 22 to 28°C. Because of the very slow growth rate of the actinoplanetes, the incubation time has to be extended up to several week in some cases. The use of dissecting microscope is recommended to select the colonies grown on the isolation plates. The mycelia can be picked up and transferred with toothpicks or with a thin metal needle.

Isolation from Water

The freshly collected water samples should be stored at 4°C until processed. If necessary, the spores can be concentrated from a relatively large volume of water either by the membrane filtration technique described by Burman et al. (1969) or by centrifugation (Okami and Okazaki, 1972). To reduce the numbers of the concomitant vegetative bacterial cells, a pretreatment either with 1) mild heating or 2) chemical substances is recommended.

1) For the heat treatment procedure, 2 ml of the water sample is placed in a glass tube which is sealed and heated in a water bath. Various periods of incubation and temperatures have been used: 6 min at 55°C (Rowbotham and Cross, 1977), 10 min at 70°C (Cross, 1981b); or 60 min at 44°C (Burman et al., 1969).

2) An alternative pretreatment with chlorine was suggested by Burman et al. (1969) and Willoughby (1969b): the samples are first treated with 4 mg/l ammonia, followed by 2 mg/l chlorine (added as 1 ml of a hypochlorite solution containing 200 mg/l of available chlorine. Samples are allowed to stand for 10 to 30 min; then the chlorine is neutralized with sodium thiosulfate. The correct amount has to be calculated from titration of a blank sample.

After brief mixing of the pretreated samples, (either heat or chlorine) spreading can be carried out immediately with 0.2 ml of the sample on each agar plate (Rowbotham and Cross, 1977). If necessary, dilutions can be made, either with sterile buffer (0.5 M KH_2PO_4 adjusted with NaOH to pH 7.2; Hsu and Lockwood, 1975), or with quarter-strength Ringer's solution containing gelatin (0.01% w/v; pH 7.0; Rowbotham and Cross, 1977). The inoculated plates are incubated at 28 or 30°C for 3 to 4 weeks.

As shown in Table 1, using selective media, *Micromonospora* strains are mainly obtained on the isolation plates from water sources. *Ac-*

Table 1. Examples of the direct isolation of actinoplanetes using selective media.

Source	Isolation method	Isolation medium	Agent added	Genus selected	Reference
Water	Dilution plate	Water agar or colloidal chitin agar	Potassium tellurite (0.1% w/v)	*Actinoplanes*	Willoughby (1971)
Water, soil, and sediment	Dilution plate	Colloidal chitin agar	—	*Micromonospora*	Hsu and Lockwood (1975)
	Dilution plate	M3 agar	Cycloheximide (50 mg/l) and thiamine (4.0 mg/l)	*Micromonospora*	Rowbotham and Cross (1977)
	Dilution plate	Kadota's cellulose benzoate agar	Sodium benzoate (20 g/l)	*Micromonospora*	Sandrak (1977)
Marine sediment	Dilution plate	Cellulose asparagine agar (+ artificial seawater)	Cycloheximide (50 mg/l) and novobiocin (50 mg/l)	*Micromonospora*	Goodfellow and Haynes (1984)
Soil	Dilution plate	Humic acid-vitamin agar	Cycloheximide (50 mg/l)	*Micromonospora* and *Dactylosporangium*	Hayakawa and Nonomura (1987)
Soil	Dilution plate	Humic acid-vitamin agar	Tunicamycin (20 mg/l) and nalidixic acid (30 mg/l)	*Micromonospora*	Nonomura and Hayakawa (1988)
Soil	Dilution plate	Bennet's agar (modified)	Tunicamycin (25 mg/l) and cycloheximide (30 mg/l)	*Micromonospora*	Wakisaka et al. (1982)
Soil	Dilution plate	Gause no. 1 agar or Czapek sucrose agar	Novobiocin (25 mg/l) and streptomycin (15 mg/l) (+ antifungal agent)	*Glycomyces*	Labeda (1989); Nolan and Cross (1988)
Soil	Dilution plate	ND	ND	*Catellatospora*	Asano and Kawamoto (1986)
Soil	Dry-heat treatment and dilution plate	Starch-casein-nitrate agar (+ B-vitamins)	— Gentamicin (5.0 mg/l)	*Micromonospora Dactylosporangium*	Shearer (1987)
Soil and mud of salt-marsh	Stamping	Arginine glycerol salts agar	Cycloheximide (75 mg/l) and nystatin (75 mg/l) (+ artificial seawater)	*Micromonospora*	Hunter et al. (1984); Hunter-Cevera et al. (1986)
Plant material	Washing	Colloidal chitin agar	Cycloheximide (50 mg/l)	*Actinoplanes, Ampullariella,* and *Micromonospora*	Willoughby (1968, 1969a)

ND, no data.

tinoplanes may also be isolated from water samples (Willoughby, 1971; Torikata et al., 1978).

Isolation from Soil and Sediments from Freshwater and Marine Habitats

SOIL-DILUTION-PLATE TECHNIQUES. Soil samples, marine sediments, or mud from lakes and rivers are air-dried at room temperature and then ground in a mortar (Shearer, 1987). About 1 g of the sample is added to 10 ml of saline in a 25-mm test tube. The suspension is mixed (vortex mixer) for 1 min and diluted in series with a sterile salt solution. The salt solution proposed by Wakisaka et al. (1982) contains 0.01% $MgSO_4 \cdot 7 H_2O$ and 0.002% Tween, from which air is eliminated by use of a vacuum desiccator for about 30 min. Instead of a salt solution, sterile water can also be used for suspension and serial dilutions (Hayakawa and Nonomura, 1987). Petridishes are prepared one day before plating and incubated at 37°C overnight to eliminate films of moisture on the agar surface (Shearer, 1987). An 0.1-ml inoculum of the proper dilution is placed on each plate and spread with a sterile glass rod. Plates are incubated at 28 to 30°C for 4 to 5 weeks.

Cellulose-decomposing micromonosporas can be isolated from soil adjacent to the roots of wheat and maize according to the method of Sandrak (1977). One ml of soil suspension is mixed with 0.67 g sterile cellulose powder (as used for thin-layer chromatography) and 2 ml of liquid Kadota's benzoate medium (Sandrak, 1977). The mixture is spread on plates with Kadota's benzoate agar. The cellulose layer is allowed to dry before the plates are incubated for 25–30 days at 28°C (Cross, 1981b).

For *Micromonospora* species from marine sediments, Goodfellow and Haynes (1984) incubated the isolation plates at 18°C for 10 weeks (duplicates at 4°C for 6 months).

Heat or chlorine treatment, as described for water samples, can also be used with soil and sediment dilutions (Cross, 1981b). An alkaline pretreatment method is suggested by Wakisaka et al. (1982): One ml of the diluent is mixed with 9 ml of 0.01 N NaOH. After standing for 5 to 10 min, the mixture is neutralized with 0.1 N HCl to pH 6 to 7 (with cooling) before serial dilution and plating. Nonomura and Hayakawa (1988) treated the soil-water suspension with 1.5% phenol at 30°C for 30 min.

The routine plating technique is preferable for the isolation of strains of the genera *Micromonospora*, *Dactylosporangium*, *Catellatospora*, and *Glycomyces*. Table 1 lists the results of the use of various selective agar media in combination with the additions of antibiotic agents.

DRY HEAT TECHNIQUE. A procedure which involved dry heating of soil samples at extreme temperatures was originally developed by Nonomura and Ohara (1969) for the isolation of *Microbispora* and *Streptosporangium* species. Shearer (1987) demonstrated, that this technique is also useful for the isolation of *Dactylosporangium* and *Micromonospora* strains (Table 1). The samples are first air-dried at room temperature and ground in a mortar. Then they are heated in a drying oven at 120°C for 60 min. One g of the heat-treated soil is added to 10 ml of saline solution and then processed as described for the routine dilution-plating technique. Inoculated plates are incubated at 28°C

for 4 to 5 weeks. Arginine-vitamin agar and starch-casein-nitrate agar with B vitamins are used as selective media (Shearer, 1987).

STAMPING TECHNIQUE. The stamp technique was used successfully in the study of actinomycete populations of salt marsh ecosystems (Hunter et al., 1984). Depending on the moisture content, the samples of soil or mud are air-dried in petri dishes for several days at room temperature. Two methods of further pretreatment were suggested by Hunter-Cevera et al. (1986): 1) the dried samples are ground with a pestle in a mortar and heated for 2 hours at 60 to 65°C; and 2) dried samples are mixed with powdered chitin (1:1) and incubated for 2 to 3 weeks at 26°C.

The pretreated and ground samples are stamped onto the isolation plates using following procedure: A small circular sponge (Dispo culture plug, 16 mm; Scientific Products) is pressed into the powdered sample and removed. The excess small crumbs are shaken off. A stack of a dozen plates with various different alternating selective media are then inoculated by successively "stamping" (lightly touching) the sponge to the agar surface; 10 times in a circle around the perimeter and 3 times in the middle of each plate (Hunter et al., 1984). Continuously stamping with the same plug yields the desired dilution effect. Plates are incubated at 26–28°C for 2 weeks. Hunter-Cevera et al. (1986) recommend arginine-glycerol salts agar, starch-casein-nitrate agar, and thin pablum agar as the selective media for the isolation of *Micromonospora* strains (Table 1).

Isolation from Plant Material

A special wash technique was employed by Willoughby (1968, 1969a) for the investigation of actinomycetes populations on decomposing leaf litter. The leaves are collected at a fairly early stage of the decomposition. Small pieces of approximately 3 cm² are cut out, and each piece transferred to a 100-ml conical flask containing 25 ml of sterile-filtered lake water. After 2 min of agitation on a rotary shaker, small aliquots of the leaf washing liquid are either incorporated into molten agar (0.5 ml/plate) or spread onto the surface of agar, 0.2 ml for each plate, using a right-angled glass rod. The plates are incubated at 25°C for 3 to 5 weeks (Willoughby, 1968). The most successful isolation medium for strains of *Actinoplanes*, *Ampullariella*, and *Micromonospora* was colloidal chitin agar with cycloheximide as the antifungal agent (Table 1).

Special Isolation Methods Using Motile Spores

The following very selective isolation methods is used when dormant sporangia are present in the substrate to be tested. The sporangia can release actively swimming spores when submerged in water. The individual spores must be motile for many hours and must show positive chemotaxis to specific chemical substances. Once fastened to a natural or cultural substrate, they must be able to germinate and form new mycelia and, for use of the baiting technique, produce a new generation of sporangia.

BAITING TECHNIQUE. The baiting technique is the classical isolation method for *Actinoplanes,* which made possible the first discovery of actinomycetes with motile spores (Couch, 1949). Although other powerful techniques are available, baiting is still the only way to isolate keratinophilic *Pilimelia* strains.

0.5 to 1.0 g of the sample is placed in a small, sterilized petri dish (3 or 4 cm in diameter) or in a chamber of a multi-well microtiter plate, which is then halfflooded with sterile demineralized water. After cautiously stirring, the particles settle to the bottom. Natural baits are exposed singly or in combination on the surface of the water: pollen of *Pinus, Liquidambar,* or *Sparganium,* boiled *Paspalum* grass leaves, hair of mammalia (human, dog, deer, cattle, white mice, etc.), or bits of snake skin (Couch, 1949, 1954; Karling, 1954; Gaertner, 1955; Kane, 1966; Schäfer, 1973; Tribe and Abu El-Souod, 1979; Makkar and Cross, 1982). The baits must be presterilized, depending on their consistency, either chemically with ethanol or propylene dioxide or by autoclaving (Gaertner, 1955; Schäfer, 1973; Makkar and Cross, 1982). The baiting enrichment cultures are closed and stored undisturbed at room temperature for several weeks. The water level can be regulated by additions of sterile distilled water. The examination for actinoplanetes can begin after one week with a dissecting microscope using 100-X magnification and horizontal lighting (Bland and Couch, 1981). Further examination after 3 to 4 weeks is recommended for keratinophilic organisms (Schäfer, 1973).

Sporangia of actinoplanetes are recognizable as glistening beads on the air-exposed sides of the baits. Such baits are then removed carefully from the water and transferred to a 3% agar plate (Bland and Couch, 1981). Individual sporangia are separated from the bait and rolled several centimeters over the surface of agar, using a thin-pointed tungsten needle, which has a tip curved like a hockey stick. In this way, contam-

inants are removed from the sporangial surface. Cleaned sporangia can be transferred either directly, or together with a small, cut-out agar block, onto a petri dish with suitable agar medium. Media for isolation from pollen and grass leaves include Czapek sucrose agar, peptone-Czapek agar (Bland and Couch, 1981), half-concentrated casamino acids-peptone Czapek agar (Schäfer, 1973), or Emerson's yeast extract-starch agar. Sporangia from keratinic baits should be transferred to highly diluted skim milk-cattle horn-meal agar (Vobis, 1984).

The colonies originating from the individual sporangia are visible with the naked eye after 1 to 4 weeks of incubation, and can partly be used as the inoculum for the new strain on slant cultures. The other part of the mycelium can be transferred onto sporulation agar for morphological identification.

DEHYDRATION – REHYDRATION TECHNIQUE. This technique utilizes the ability of the sporangia to withstand desiccation and to release motile spores when they are subsequently in contact with water. Beside soil samples, it is also applicable to leaf litter, decaying plant material from aquatic habitats, organic debris, etc. (Makkar and Cross, 1982).

The samples are dried at 28 to 30°C for 7 days. For rehydration, 0.5 g of soil or corresponding substrate is mixed with 50 ml of sterile tap water in a 150-ml beaker or Erlenmeyer flask, which is covered with sterile aluminium foil (Shearer, 1987; Vettermann and Prauser, 1979). The suspension is incubated at 20 to 30°C for about 1 hour. During the first 30 minutes, the vessel can be shaken at irregular intervals. After that, the particles should be permitted to settle. From the supernatant, 0.5 to 1.0 ml are removed with a sterile Pasteur pipette and spread onto agar plates (Shearer, 1987). If it is necessary, dilutions can be prepared from the inoculation fluid (Makkar and Cross, 1982). For cultivation, the following media can be used: soil extract agar; colloidal chitin agar containing cycloheximide and nystatin (Makkar and Cross, 1982); oatmeal-soil extract agar; and starch-casein-sulfate agar (Shearer, 1987). Plates are incubated at 28°C for 2 to 4 weeks. Colonies of actinoplanetes can be selected under a dissecting microscope at 60-X magnification and used as inoculum for new strains.

CHEMOTACTIC METHOD. The spores of *Actinoplanes* exhibit an apparently microaerophilic reaction and are attracted to chloride and bromide ions (Palleroni, 1976). Therefore, as chemotactic method can be used to isolate these strains. An essential part of this technique is a simple isolation chamber, a sterilizable plastic block (80 × 40 × 12 mm) with two circular holes (9 mm deep and 24 mm in diameter) whose centers are 32 mm apart. They are connected by a channel that is 2 mm wide and 3 mm deep (Palleroni, 1980). One gram of a soil sample is divided into two equal parts and then placed in each compartment. Sterile water is added nearly to the rim and stirred cautiously. After incubation for 1 hour at 30°C, the spores are released from the sporangia and move freely in the water. Using a sterilized tweezer, a sterile 1-μl glass capillary about 32-mm long is filled with 0.01 M phosphate buffer (pH 7.0) containing 0.01 M KCl, and placed in the channel. The capillary must be submerged, connecting the two suspensions. After incubation at 30°C for 1 hour more, the attracted spores are concentrated in the lumen of the capillary, which is then removed and washed from the outside with a jet of sterile water. The contents of the capillary are blown into 1-ml sterile water or buffer. Portions of the dilution are taken with a sterile pipette and spread onto carefully dried agar plates. The plates are then incubated at 28°C. Starch-casein-sulfate agar is recommended as the isolation medium (Palleroni, 1980). Although colonies can be selected after 4 days, slowly growing actinomycetes may only be detectable after 3 weeks.

MOIST INCUBATION TECHNIQUE. This method is suitable for the direct detection of actinoplanetes on natural substrate. Although the ability to produce motile spores obviously plays no role, *Actinoplanes* strains can be readily enriched (Willoughby, 1968). Portions of decaying leaves or other biological substrates, freshly collected from the field, are washed with sterile water to remove adhering detritus. They are placed in prepared petri dishes, the bottoms of which have been covered with very moist filter paper or layers of cellulose before autoclaving. The petri dishes, working as moist chambers, are sealed and incubated for about 4 weeks at 25°C. Examination with both dissecting and light microscopes is necessary to identify the sporangia of the actinoplanetes (Willoughby, 1969a).

Identification

As outlined by Labeda (1987) and Lechevalier (1989), a combined use of morphological and chemical criteria is still the best way to identify actinomycetes at the genus level. To start with simple and time-saving morphological studies can sometimes shorten large-scale chemotaxo-

nomical procedures. However, to verify the determination, biochemical analyses are still absolutely necessary.

Morphological Criteria

Because members of the actinoplanetes are highly differentiated morphologically, (Bland and Couch, 1981; Luedemann and Casmer, 1973; Vobis and Kothe, 1985), it is possible to use morphological criteria alone for identification at the genus level. Table 2 is a dichotomous key that can be used to determine the genus of the isolate being studied.

For morphological studies in general, the short guidelines compiled by Cross (1989) are useful. Tiny pieces of the mycelium from vigorously developing colonies are transferred to freshly prepared sporulation medium in small petri dishes. The temperature of incubation should be between 20 and 30°C, and observations are made after 2 days to 4 weeks of incubation (depending on the individual strain). Preliminary observations should be made with a dissecting microscope (Figs. 1A, B, and C; 6A, B, and C), and directly with a light microscope (Figs. 1E; 4A; 7B) before water mounts are prepared. The best magnification of undisturbed colonies is with 20 or 40× objectives so that appropriate structures can be accurately studied (Okami and Suzuki, 1958; Cross, 1981b). Spore chains and single spores can be detached by slight pressure on the colony with a cover slip.

Table 2. Determinative key for the actinoplanetes genera using morphological criteria.

1a Spores in chains	2
1b Spores not in chains	3
2a Spore chains long, on aerial hyphae	*Glycomyces*
2b Spore chains short, emerging from substrate hyphae	*Catellatospora*
3a Spores spherical, single on substrate hyphae, nonmotile	4
3b Spores enclosed in sporangia, motile by flagella	5
4a Diameter of spores: 0.7–1.5 μm, in clusters	*Micromonospora*
4b Diameter of spores: 1.7–2.8 μm, dispersed	*Dactylosporangium*
4c Diameter of spores: 0.4–0.6 μm, dispersed	*Catellatospora*
5a Sporangia oligosporous, with one row of spores	*Dactylosporangium*
5b Sporangia polysporous, with several rows of spores	6
6a Sporangiospores globose to subglobose	*Actinoplanes*
6b Sporangiospores bacilliform	7
7a Spores distinct rod-shaped, 2.0–4.0 μm long, tuft of flagella polar	*Ampullariella*
7b Spores rod-shaped to reniform, 0.7–1.5 μm long, tuft of flagella lateral	*Pilimelia*

Sporangia should be cut out and mounted with water. Although observation with a bright-field microscope is generally sufficient, phase-contrast is often also helpful (Figs. 2B; 4D; 7D).

For observing spore release from the sporangia, water-mounted preparations should be observed continuously for 1 to 2 h; unchlorinated tap water or distilled water should be used. To avoid evaporation, slides should be stored in moist chambers. Actively swimming spores are easy to distinguish from Brownian molecular movement since the action of the flagellar tuft causes the spores to move and rotate rapidly. Based on the type of flagellation, the spores of *Ampullariella* rotate around the longitudinal axis, while those of *Pilimelia* rotate around the lateral axis (Fig. 8). To observe the flagella with the light microscope, staining methods must be used (Couch, 1950), but the exact location of flagellar insertion can only be determined by transmission electron microscopy.

Although many strains sporulate readily on isolation media or on rich media used for subculturing, special agar media have to be used to obtain good sporulation of certain strains: Czapek sucrose agar for *Glycomyces harbinensis* (Labeda et al., 1985), *Dactylosporangium vinaceum* (Shomura et al., 1983), and *Ampullariella* spp. (Nonomura et al., 1979); calcium malate agar for *Dactylosporangium* spp. (Thiemann et al., 1967), *Actinoplanes* spp. (Palleroni, 1989), and *Catellatospora* spp. (Asano and Kawamoto, 1986); inorganic salts-starch agar for *Dactylosporangium* spp. (Vobis, 1989c) and *Ampullariella* spp. (Nonomura et al., 1979); humic acid-vitamin agar for *Dactylosporangium* spp. and *Micromonospora* spp. (Hayakawa and Nonomura, 1987); humic acid-ion agar

for *Actinoplanes* and *Ampullariella* spp. (Willoughby et al., 1968); soil extract agar for *Dactylosporangium* spp. (Thiemann et al., 1967) and *Actinoplanes* spp. (Parenti et al., 1975); diluted skim milk-hornmeal agar for *Pilimelia* spp. (Vobis, 1984); oatmeal agar for *Actinoplanes* spp. (Bland and Couch, 1981), and M3 agar, supplemented with 0.1% (w/v) fructose (Goodfellow et al., 1990). An excellent routine sporulation agar is artificial soil agar (Henssen and Schäfer, 1971). The compositions of two of the above, inorganic salts-starch agar and oatmeal agar are described by Shirling and Gottlieb (1966). The formulations of some of the other media are given below. For morphological studies, the same natural substrate can be used as that used for the isolation by baiting or moist incubation. (Fig. 7B and C).

Individual strains of *Actinoplanes, Ampullariella,* and *Pilimelia* may produce incomplete sporangiate structures. The sporogenous hyphae are still developed at the tip of the supporting hypha, but the sporangial envelope appears to be absent. Sometimes, the division into spores is not complete, so that the hyphae look like brushes or little trees. If the sporogenous hyphae are fragmented, the resulting spores are always arranged in chains (Fig. 7E). It is remarkable that these free, developed spores never have flagella. Therefore, they have been called "conidia" and the supporting hyphae "conidiophores" (Couch, 1954, 1963; Kane, 1966; Willoughby, 1966).

Isolates of the *Actinoplanes* group exhibiting marked morphological characteristics can be identified to the genus level with aid of the simplified determinative key given in Table 2.

Chemotaxonomical Criteria

The taxonomy of the actinoplanetes can be established; 1) by the composition of the amino acids of the cell walls; 2) by the sugar pattern of whole cell hydrolysates; and 3) by the absence of mycolic acids. Further chemotaxonomical markers, such as the composition of: 4) the phospholipids; 5) menaquinones; and 6) fatty acids, also have, in many cases, an important diagnostic value. Table 3 lists the most useful chemotaxonomical properties of the actinoplanetes. The chemical features are not only good indicators for identification, but they are also very important with regard to suprageneric groupings and phylogenetic relationships (Goodfellow and Cross, 1984; Stackebrandt and Schleifer, 1984).

CELL WALL CHEMOTYPE. According to the classification scheme of Lechevalier and Lechevalier (1970b), the members of the *Actinoplanes*

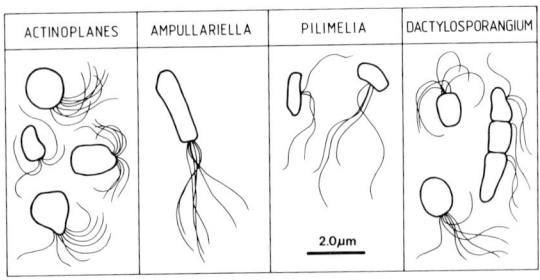

Fig. 8. Morphology of the motile sporangiospores of the genera of the actinoplanetes. *Actinoplanes:* globose, ellipsoidal, short bacilliform (0.8–2.0 μm), with a polar tuft of flagella. *Ampullariella:* bacilliform (0.5–1.0 × 2.0–4.0 μm), with a polar tuft of flagella. *Pilimelia:* bacilliform, reniform (0.3–0.7 × 0.7–1.5 μm), with a lateral tuft of flagella. *Dactylosporangium:* cylindrical, ovoid or pyriform (0.4–1.3 × 0.5–1.8 μm), with a polar or subpolar tuft of flagella. (From Vobis, 1987; with permission.)

Table 3. Chemotaxonomical characters of the genus *Actinoplanes* and related genera.[a]

Genus	Cell wall chemotype[b]	Whole cell sugar pattern[c]	Phospholipid pattern[d]	MK-9 H_2	MK-9 H_4	MK-9 H_6	MK-9 H_8	MK-10 H_2	MK-10 H_4	MK-10 H_6	MK-10 H_8	MK-12 H_4	MK-12 H_6	MK-12 H_8	S	U	I	A	T
Actinoplanes	II	D	P II	V	+	V									+	+	+	+	−
Ampullariella	II	D	P II	V	+	V			+						+	+	ND	+	−
Catellatospora	II	D	P II		+	+				+	+				V	V	+	+	−
Dactylosporangium	II	D	P II		+	+	+			+							ND		
Glycomyces	II	D	P I					+	+						+	+	+	+	+
Micromonospora	II	D	P II		+				+	+		+	+	+	+	+	+	V	−
Pilimelia	II	D	P II	+															

[a] Compiled from: Asano et al. (1989b); Collins et al. (1984); Goodfellow and Cross (1984); Kroppenstedt (1985); Labeda (1989); Labeda et al. (1985); Lechevalier et al. (1981); Stackebrandt and Kroppenstedt (1987); Vobis (1989b). For further details and exceptions, see text.
[b] Cell wall chemotype II, glycine and *meso*- and/or 3-hydroxy diaminopimelic acid (A_2pm).
[c] Whole cell sugar pattern D, xylose and arabinose (Lechevalier and Lechevalier, 1970b).
[d] Phospholipid pattern: P I, nitrogenous phospholipid absent; P II, phosphatidylethanolamine present (Lechevalier et al., 1977).
[e] S, saturated; U, unsaturated; I, iso-; A, anteiso-; T, tuberculostearic acid (10-methyl); V, variable; +, present; −, absent; ND, no data.

group are characterized by cell wall chemotype II. Glycine and *meso-* diaminopimelic acid (A$_2$pm) and/or 3-hydroxy-diaminopimelic acid are the amino acids characteristic of the peptidoglycan. The primary structure of the peptidoglycan was described by Kawamoto et al. (1981). Glycine, rather than L-alanine, is linked to muramic acid, and *meso*-diaminopimelic acid or its hydroxylated derivative are directly cross-linked to the D-alanine of an adjacent peptide subunit: Muramic acid is *N*-glycolated. The *N*-glycolyl muramic acid is a characteristic unique to the genera with cell wall type II and was found in many *Actinoplanes* and *Ampullariella* species (Stackebrandt and Kroppenstedt, 1987). It can be easily identified by a colorimetric method (Uchida and Aida, 1977).

Deviations from the typical wall chemotype II are found in some species of *Micromonospora,* which also contain LL-diaminopimelic acid (Kawamoto et al., 1981). A further exception concerns *Actinoplanes caeruleus,* which has lysine and serine as amino acids, instead of A$_2$pm (Horan and Brodsky, 1986a; Tille et al., 1982). The lack of A$_2$pm in a morphologically incontestable *Actinoplanes* species raises a lot of taxonomical problems (Stackebrandt and Kroppenstedt, 1987).

WHOLE CELL SUGAR PATTERN. The sugars used for classification are, in particular, the components of the cell wall polysaccharides. They can be detected after hydrolyzation of whole cells. Based on the pattern of diagnostic sugars, the actinomycetes containing *meso*-A$_2$pm can be divided into further groups (Lechevalier and Lechevalier, 1970a). In general, the actinoplanetes have wall chemotype II and the sugar pattern D in common (Table 3). The pentoses xylose and arabinose are the characteristic sugars.

Some species of *Micromonospora* also contain other diagnostic sugars, such as galactose and madurose (3-O-methyl-D-galactose) (Meyertons et al., 1987). In *Catellatospora* strains, an unusual sugar detected was 3-O-methylrhamnose (Asano et al., 1989a).

The separation of whole cell sugars is usually carried out by thin-layer chromatography (Hasegawa et al., 1983; Meyertons et al., 1987).

MYCOLIC ACIDS. The cell walls of the actinoplanetes have the *N*-glycolated muramic acid type in common with those of the *Mycobacterium-Nocardia-Rhodococcus-* group (Minnikin and O'Donnell, 1984). The mycelium of this group fragments into coccoid or bacillary elements. They can also be further characterized by the presence of mycolic acids, which are not present in the *Actinoplanes* group. For routine

analysis, thin-layer chromatography is usually used (Minnikin et al., 1975).

PHOSPHOLIPIDS. The composition of phospholipids is of high taxonomical value in actinomycetes in general, and five very distinct phospholipid types (P I to P V) have been observed (Lechevalier et al., 1977, 1981). Most of the genera of the actinoplanetes are of phospholipid type P II (phosphatidylethanolamine present, phosphatidylcholine and the unknown glucoseamine-containing phospholipid absent) (Table 3). However, *Glycomyces* is distinct since it has P I (phosphatidylcholine, phosphatidylethanol and the unknown glucoseamine-containing phospholipid absent) (Labeda et al., 1985). The phospholipid types are also useful taxonomic markers for separating other sporangia-forming genera with motile spores from the *Actinoplanes* group (Hasegawa et al., 1979). The unusual absence of phosphatidylethanolamine in *Ampullariella regularis* (Stackebrandt and Kroppenstedt, 1987) may indicate a peculiar variability in the phospholipid pattern in the genus *Ampullariella.*

Several methods for characterizing the phospholipids have been described (Lechevalier et al., 1977; Vitiello and Zanetta, 1978; O'Donnell et al., 1982).

MENAQUINONES. Determination of the composition of menaquinones is helpful for differentiation at the genus level (Table 3). The genera *Actinoplanes, Ampullariella,* and many of the *Micromonospora* species contain menaquinones with 9 to 10 tetrahydrogenated isoprene side chains, MK-9 (H$_4$) and MK-10 (H$_4$), thus fitting into type 3b of the classification scheme of Kroppenstedt (1985). In *Actinoplanes auranticolor* and in *Micromonospora carbonaceae* subsp. *aurantiaca* only MK-9 (H$_4$) occurs (Stackebrandt and Kroppenstedt, 1987). Type 4 occurs in *Micromonospora echinospora* subsp. *pallida,* which has tetra-, hexa-, and octahydrogenated quinones with 12 isoprene units. Many other *Micromonospora* species also have type 4 menaquinones, with MK-10 (H$_4$) and MK-10 (H$_6$) (Collins et al., 1984; Kawamoto, 1989).

The menaquinone type 4b, with MK-9 (H$_4$), (H$_6$) and (H$_8$), is characteristic for *Dactylosporangium* species and type 4a with MK-9 (H$_2$) and (H$_4$) for *Pilimelia* (Kroppenstedt, 1985). *Catellatospora* species have either type 4a with tetra- and hexahydrogenated menaquinones with nine isoprene units or type 4c with hexa- and octahydrogenated menaquinones with 10 units (Asano et al., 1989b). According to Labeda (1989), the genus *Glycomyces* has MK-10 (H$_2$) and MK-10 (H$_6$). In the species *Glycomyces rutgersensis,*

MK-9 (H_4) could also be detected (R. M. Kroppenstedt, unpublished observations). This indicates that the common type 3b for *Actinoplanes* and *Micromonospora* species is also present in *Glycomyces*.

FATTY ACIDS. The usefulness of fatty acids as chemotaxonomical markers was demonstrated by Kroppenstedt and Kutzner (1978) and Kroppenstedt (1979). In the classification scheme developed by Kroppenstedt (1985), three main types of fatty acid pattern were proposed. Among the actinoplanetes (Table 3), the genus *Micromonospora* has a separate position with type 3b, characterized by the occurrence of 10-methyl branched fatty acids (tuberculostearic acid). Recent studies indicate that some strains of *Dactylosporangium* belong to the same type (Stackebrandt and Kroppenstedt, 1987). In general, *Dactylosporangium* species have fatty acid type 2, as is also characteristic for *Ampullariella* (type 2d) and *Actinoplanes* (type 2c): iso/anteiso branched- and monosaturated, *cis*-9,10-octadecanonic acids are the predominant fatty acids (Kroppenstedt, 1985; Stackebrandt and Kroppenstedt, 1987). A similar iso/anteiso fatty acid pattern is found in *Glycomyces rutgersensis* (R. M. Kroppenstedt, unpublished observation). Members of *Pilimelia* have fatty acid type 2b with a high content of iso-C_{15}, iso-$C_{15:1}$, and iso-$C_{17:1}$ fatty acids, lacking substantial amounts of C_{18} fatty acid (Stackebrandt and Kroppenstedt, 1987).

Cultural Characteristics of the Differentiation of the Actinoplanetes Genera and Species

For subculturing new isolates, several complex agar media should be tested, including the isolation medium, because the individual strains might have quite different nutrient requirements. In our laboratory, we usually start with Emerson's yeast extract-starch agar or casamino acids-peptone-Czapek agar. Vegetative growth is also supported by Bennett agar, Hickey-Tresner agar, Czapek sucrose agar, peptone-Czapek agar, and glucose-asparagine agar. For keratinic isolates, half-concentrated skim milk agar is used (Schäfer, 1973). For the compositions of the above-mentioned media, see "Media for Isolation and Cultivation" below.

Actinoplanes (Couch) Palleroni 1989

Isolates of *Actinoplanes* form compact colonies on solid agar media (Fig. 1A and C). Aerial mycelium is usually absent or only rudimentarily developed. Colonies can be covered with a whitish bloom, if abundant sporangia are produced on the surface of the substrate mycelium (Fig. 1B and D). The basic color of the mycelium is orange, presumably due to a carotenoid pigment (Szaniszlo, 1968), but a great variety of colors exists, depending on the individual strains: cream to yellow, brown, rusty brown, red, blue, violet, green, or black (Palleroni, 1989; Vobis, 1987).

In general, sporangia develop directly on the surface of a colony (Fig. 1D, E, and F). In *A. minutisporangius,* they are also submerged (Ruan et al., 1986). Frequently, the sporangia arise terminally from "palisade" hyphae, which are thicker in diameter and vertically oriented (Bland and Couch, 1981). Inside the sporangia, the spores are arranged in coils (Fig. 1G); but in *A. rectilineatus,* they run in parallel rows (Lechevalier and Lechevalier, 1975). If the sporangial envelopes are very thin and transient, individual sporangia may be attached (Fig. 1E). The spores of *Actinoplanes* are globose or subglobose to short bacilliform, and possess a tuft of flagella (Fig. 8).

Species differentiation is based on a combination of morphological and physiological characters (Palleroni, 1989). The shape of the sporangia ranges from globose, subglobose, oval, umbelliform, cylindrical, or lobate to irregular. The average size of a sporangium is from 4 to 25 μm in diameter. Extreme dimensions of sporangia exist in *A. minutisporangius,* with 2 μm as a minimum (Ruan et al., 1986) and 47 μm as a maximum (G. Vobis, unpublished observations). The unusual presence of aerial hyphae is characteristic for *A. rectilineatus, A. ferrugineus, A. garbadinensis,* and *A. teichomyceticus* (Palleroni, 1989).

The color of the mycelium is another characteristic used in species differentiation: for instance, tan to blue in *A. caeruleus* (Horan and Brodsky, 1986a), violet in *A. ianthinogenes* (Coronelli et al., 1974), and rusty brown in *A. ferrugineus* (Palleroni, 1979). Beside yellowish or brownish soluble pigments in several species, a cherry red pigment is the distinctive mark of *A. italicus* (Beretta, 1973) and a soluble blue pigment in *A. cyaneus* (Terekhova et al., 1977). Physiological tests like degradation, hydrolysis, coagulation, peptonization, or liquefaction of various compounds or media can also be used (Palleroni, 1989). The assimilation tests with different carbon sources, routinely carried out with sugars, were extended to many more organic compounds (Palleroni, 1979). Palleroni (1989) listed a total of 15 species for the genus, including the two members of the genus formerly called *Amorphosporangium* (Couch, 1963). Using numerical and chemical classification methods, recently the numbers of *Actinoplanes* species could be enlarged by the description of five new species (Goodfellow et al., 1990).

Ampullariella Couch 1963

The colonies of *Ampullariella* strains grown on solid agar media are often soft and can be pulled apart easily with a needle. They are elevated, with protuberances in the center of the colony. Their marginal areas can be ridged or flat (Fig. 2A). The basic orange color of the substrate mycelium is frequently overlayed by darker pigments. Depending on the individual strain, red, brown, olive or black variations can be observed. Different-colored segments may even occur in a single colony. Rudimentary aerial mycelium can be produced by several strains. Their colonies are consequently covered by a whitish gray layer.

The production of sporangia may occur in such extensive masses that the whole surface of the colony is completely covered (Fig. 2A and C). Other strains only develop a very few sporangia, which can be detected only by direct observation with the light microscope. Palisade hyphae supporting the sporangia have never been observed. The species of the genus *Ampullariella* have to be identified by the morphology of their spores and sporangia. The spores are rod-shaped and are usually four times longer than they are wide (Table 2; Fig. 2D). They have a polarly inserted tuft of flagella (Fig. 8). The spores are arranged within the sporangia in parallel rows. *A. regularis* has cylindrical or bottle-shaped sporangia, measuring 5–14 × 8–30 μm (Fig. 2D). Pyriform or bell-shaped sporangia, sometimes with irregularly rounded protuberances at the top, can be split one or several times along the longitudinal axis (Fig. 2B and C). Such sporangia with lobes occur in *A. campanulata* (5–15 × 6–12 μm) and in *A. lobata* (4–20 × 12–32 μm). The sporangia of *A. digitata* are smaller (3–9 × 6–12 μm) and very deeply split (Couch, 1963; Vobis and Kothe, 1989). Beside the morphological features, a few physiological criteria can also be used for species differentiation (Schäfer, 1973; Vobis and Kothe, 1989). Melanoid pigments are produced in *A. digitata,* which has an olive- to-black substrate mycelium. Yellowish to greenish soluble pigments occur in several strains (Couch, 1963). *"A. violaceochromogenes"* can be identified by the presence of a violet-to-dark purple diffusible pigment (Nonomura et al., 1979).

Eight species and subspecies of *Ampullariella* are currently known (Vobis and Kothe, 1989)

although half of them are still "species incertae sedis." Whether *Ampullariella* should be a separate genus from *Actinoplanes* is unclear. The molecular systematic data favor a unification of the two genera (Stackebrandt and Kroppenstedt, 1987; Kothe, 1987), whereas following traditional morphological concepts (Vobis, 1989a), two genera can be distinguished clearly by the shape and size of spores (Table 2; Figs. 2D and 8).

Catellatospora Asano and Kawamoto 1986

Colonies of *Catellatospora* strains always produce substrate mycelium; true aerial mycelium is not developed. The color of the colonies varies from light wheat to yellow, mustard gold, orange, and reddish brown. Cultures grown on agar media that support poor growth can be crowded with short chains of spores (Fig. 3A), with 5 to 30 spores forming a chain which emerges directly from the substrate hypha (Fig. 3C). The spore chains are straight to flexuous, vertically oriented, and may be branched (Fig. 3B and D). The spores are nonmotile, cylindrical, barrel-shaped to ovoid, with smooth to slightly rough surfaces, measuring 0.6–0.8 × 0.9–1.2 μm (Asano and Kawamoto, 1986).

In addition to spore chains, a few strains of *C. tsunoense* and *C. ferruginea* produce single spores ("globose bodies") terminally on branches of substrate hyphae (Asano et al., 1989b). These spores are morphologically very similar to the globose spores of the genus *Dactylosporangium* (Fig. 4E). In *Catellatospora,* they are 0.4–0.6 μm in diameter, whereas in *Dactylosporangium,* they are more than double that size (Table 2).

The color of the substrate mycelium, the type of menaquinone, and the presence or absence of the specific cell wall sugar 3-O-methylrhamnose are used for species differentiation (Asano et al., 1989a, 1989b). *C. citrea* and *C. tsunoense* have major menaquinones with nine isoprene units (MK-9) and yellow-colored mycelium. The species *C. ferruginea* and *C. matsumotoense* have menaquinones with 10 isoprene units (MK-10) and are usually orange to reddish brown. Their cell walls also contain 3-O-methylrhamnose, and they are resistant to novobiocin (Asano et al., 1989b). Further species differentiation can be made by the requirement for thiamine in *C. tsunoense* and *C. ferrugineae,* and the requirement for methionine in *C. citrea* subsp. *methionotrophica* (Asano and Kawamoto, 1988). The utilization pattern of carbon sources also supports the differentiation into the four species and subspecies mentioned above.

Dactylosporangium Thiemann, Pagani and Beretta 1967

Dactylosporangium strains form compact colonies on agar media. The colonies are mostly flat with smooth to wrinkled surface, and look somewhat tough and leathery. Formation of coremia may occur but true aerial mycelium is not developed. The color of the substrate mycelium varies from pale orange to deep orange, amber, brownish with rose tinge, rose, and wine, to brown or yellowish brown (Shomura et al., 1986; Vobis, 1989c).

Two completely different types of spores can be formed in *Dactylosporangium* strains: motile spores inside the sporangia and nonmotile spores, which are borne singly on substrate hyphae. The sporangia sit directly on the substrate, either singly or in bundles (Fig. 4A and B). They are finger-shaped or claviform, containing only one row of no more than four spores (Fig. 4C). The sporangiospores are variable in shape and are sometimes still connected while swimming with the aid of flagella (Fig. 8). The nonmotile, substrate spores are spherical (1.7–2.8 μm in diameter) and exhibit a typical phase-brightness (Fig. 4D). They arise terminally on short side-branches of substrate hyphae (Fig. 4E). Amorphous material can be deposited outside the spore wall (Fig. 7F). The cytoplasm includes crystalline proteins and structured bodies (Sharples and Williams, 1974). Morphologically similar globose spores may also occur in *Catellatospora* strains (Asano et al., 1989b), but they are smaller in diameter (Table 2). In *D. fulvum,* the sporangia and the globose spores may also be visible on coremia (Shomura et al., 1986).

Because of the minimal differences in morphology, chemotaxonomy, and physiology between the various *Dactylosporangium* strains, speciation is mainly based on the color of the substrate mycelium and the diffusible pigments (Vobis, 1989c). *D. aurantiacum* has orange-colored substrate mycelium without soluble pigments. The mycelium of *D. thailandense* is orange to brown, producing an amber to brown soluble pigment (Thiemann et al., 1967). *D. matsuzakiense* also has orange colonies, but is recognizable by a light brownish-pink diffusible pigment (Shomura et al., 1980). The substrate mycelium of *D. vinaceum* is red to brown, and the soluble pigments are red to deep red (Shomura et al., 1983). *D. roseum* is characterized by a rose-colored mycelium, and *D. fulvum* by a yellowish-brown-colored one (Shomura et al., 1985, 1986).

At present, eight species of *Dactylosporangium* are described, but two of them are "species incertae sedis": *"D. variesporium"* and *"D. salmoneum"* (Vobis, 1989c).

Glycomyces Labeda, Testa, Lechevalier, and Lechevalier, 1985

Colonies of *Glycomyces* develop both substrate and aerial mycelium on agar media (Fig. 5A and B). The substrate mycelium is pale yellowish-white to tan. The aerial mycelium is white, and is produced abundantly to sparsely, depending on the medium used (Labeda, 1989). Slightly yellowish to brown colored soluble pigments can be produced (Labeda et al., 1985).

Long chains of spores are formed by fragmentation of aerial hyphae. The spores are nonmotile, cylindrical with square ends, measuring $0.5 \times 0.7–1.8$ μm (Fig. 5C and D).

Only two species have been described. They can be differentiated by their utilization pattern of organic compounds (Labeda, 1989). Furthermore, *G. rutgersensis* produces acid from adonitol and melibiose, whereas *G. harbinensis* does not (Labeda et al., 1985).

The glycolipid-containing genus *Glycomyces* is tentatively classified within the *Actinoplanes* group because of the chemotaxonomic similarities in the composition of the cell wall, the menaquinones, and the fatty acids (R.M. Kroppenstedt, unpublished observations), but because it has phospholipid type P I instead of type P II, its position remains exceptional among the actinoplanetes (Table 3). Phospholipid type P I is characteristic of the genus *Actinomadura* and of some species of *Microtetraspora* and *Nocardioides*. This correlates with the production of true aerial mycelium and of nonmotile spores in chains, which *Glycomyces* has in common with the above-mentioned genera, but not with the actinoplanetes.

Micromonospora Ørskov 1923

Colonies growing on solid media form only substrate mycelium, which is raised and folded (Fig. 6A). Some strains can develop short, sterile aerial hyphae, giving a pruinose surface. The color of young colonies is pale yellow to light orange, becoming deep orange, red, purple, brown, or blue-green with age, depending on the individual strain (Kawamoto, 1989). The upper mycelial layers may burst open (Fig. 6B and D) or may be completely covered by a mucous mass of spores, giving the colonies bright brown-black, green-black, or black-colored surfaces (Fig. 6C). Other strains produce the spores accumulated in distinct areas on the surface (Fig. 6B), or completely immersed in the substrate (Fig. 6, D).

The spores of *Micromonospora* are formed singly on substrate mycelium (Fig. 6E). They are nonmotile, and spherical to oval in shape, with a diameter of 0.7–1.5 μm. The sporophores, often together in clusters, produce spores terminally on short hyphal branches (Luedeman and Casmer, 1973). The spore surfaces of almost all species have blunt-spiny projections (Kawamoto, 1989).

The description of many species of this genus began with the taxonomical studies of the gentamicin-producing *Micromonospora* strains (Luedeman and Brodsky, 1964). More than 12 species and subspecies could be listed recently (Kawamoto, 1989). The species concept is based on chemotaxonomical markers, pigment production, and physiological characteristics. The presence or absence of the 3-hydroxy-A$_2$pm isomer in the peptidoglycans of the cell walls divides the genus into two species groups (Kawamoto et al., 1981). The predominant menaquinones of *M. carbonacea* and *M. halophytica* have nine isoprene units (MK-9) and those of *M. echinospora* subsp. *pallida* have 12 units (MK-12). All the other remaining species have MK-10 menaquinones (Collins et al., 1984; Kawamoto, 1989).

Mycelial pigments are also of diagnostic value: *M. coerulea* has a blue-green colored mycelium. *M. echinospora* subsp. *echinospora* and subsp. *ferruginea* a maroon-purple pigment in the substrate mycelium. The following species are characterized by soluble pigments of various colors: *M. halophytica* subsp. *halophytica* (red-brown), *M. chalcea* (yellow), *M. purpureochromogenes* and *M. olivasterospora* (olive-green), and *M. rosaria* (wine-red) (Horan and Brodsky, 1986b; Kawamoto, 1989). Further physiological parameters, such as growth on special media, carbon utilization profiles, glycosidase activity, nitrate reduction, and NaCl tolerance, support the identification of species and subspecies (Kawamoto, 1989).

Pilimelia Kane 1966

Strains of *Pilimelia* only form small, compact colonies, which are about 5 mm in diameter after 4 weeks of incubation. Growth is supported by keratinic substances like hair (Fig. 7A) and cattle-horn meal (Vobis, 1984). Aerial mycelium is not developed. The color of the substrate mycelium ranges from pale to lemon yellow, yellow-gray, and golden yellow to orange (Vobis, 1989b).

The sporangia are developed directly on the surface of agar medium or on natural substrates (Fig. 7B). The shape of the sporangia is globose, ovoid, pyriform, campanulate, or cylindrical and approximately 10–15 μm in size. In some strains, each sporangium contains up to a thousand spores (Fig. 7F). The spores are rod like to reniform, equipped with a laterally inserted tuft of flagella (Fig. 8). Nonmotile spores ("conidia") may be also produced (Fig. 7E).

Species differentiation for *Pilimelia* is based on morphological and colonial characters. *P. terevasa* has spherical to campanulate sporangia with parallel rows of abundantly branched spore chains (Fig. 7B and F). The colonies have a soft consistency and are yellow to yellow-gray. *P. anulata* has cylindrical sporangia (Fig. 7E). The top segment of the sporangiophore is expanded to form a small ring-like structure ("annulus") (Fig. 7D). The mycelium has a yellowish color and is soft and pasty. In contrast, the colonies of *P. collumellifera* are very solid and spherical to pyriform sporangia are produced, with the spore chains inside arranged in swirls (Vobis, 1984). The sporangiophores are unseptate and reach into the lumen of the sporangium, where they are visible as small columns (Fig. 7C). The substrate mycelium of *P. collumellifera* is either golden yellow to orange, or colorless to pale brownish in the subspecies *pallida* (Vobis et al., 1986).

Pilimelia strains were considered for a long time to be keratinophilic members of *Ampullariella*, but recent studies on the chemotaxonomy, ultrastructure, and physiology of the three species (Vobis, 1989b) have supported recognition of them as a separate genus (Kroppenstedt, 1979; Schäfer, 1973; Vobis et al., 1986).

Media for Isolation and Cultivation

The recipes for the media used for isolation, sporulation, and vegetative growth are listed below. However, the media developed by the "International Streptomyces Project" (Shirling and Gottlieb, 1966) are not included. Routine sterilization is carried out at 121°C for 15 to 20 min by autoclaving. The vitamins and the antibiotic agents are separately sterilized by filtration and added to cooled (50°C) agar media. If necessary, after autoclaving, the pH should be adjusted with sterile acid or base.

Additions to the Agar Media

ANTIBIOTICS. Many isolation media contain antifungal or antibacterial agents. The concentrations and/or combinations of these supplements are given either in Table 1, in the text, or in the corresponding recipes. To dissolve cycloheximide, warm water (45°C) is recommended (Hunter-Cevera et al., 1986). Nystatin can be dissolved initially in dimethylsulfoxide and then diluted in 95% alcohol (Makkar and Cross, 1982). As nystatin is not totally soluble in water at pH 7.0 and is unstable at high pH, the pH of the aquatic solution can be increased with 1 M NaOH to 11.0, filter sterilized, and immediately lowered to 7.0 with HCl (Hunter-Cevera et al., 1986).

VITAMINS. Some media require the addition of B vitamins (Nonomura and Ohara, 1969; Hayakawa and Nonomura, 1987; Shearer, 1987). The amounts indicated below are per liter of the corresponding medium.

B-Vitamins (Nonomura and Ohara, 1969)

Thiamine hydrochloride	0.5 mg
Riboflavin	0.5 mg
Niacin	0.5 mg
Pyridoxine hydrochloride	0.5 mg
Inositol	0.5 mg
Calcium pantothenate	0.5 mg
para-Aminobenzoic acid	0.5 mg
Biotin	0.25 mg

Recipes for Agar Media

The quantities below are all per liter of distilled water, unless otherwise stated.

Arginine Glycerol Salts Agar (El-Nakeep and Lechevalier, 1963)

Arginine monohydrochloride	1.0 g
Glycerol (specific gravity not less than 1.249 at 25°C)	12.5 g
K_2HPO_4	1.0 g
$MgSO_4 \cdot 7H_2O$	0.5 g
$Fe_2(SO_4)_3 \cdot 6H_2O$	0.01 g
$CuSO_4 \cdot 5H_2O$	0.001 g
$ZnSO_4 \cdot 7H_2O$	0.001 g
$MnSO_4 \cdot H_2O$	0.001 g
Agar	15.0 g
Adjust to pH 6.9 to 7.1.	

Arginine Vitamin Agar (Nonomura and Ohara, 1969)

L-Arginine	0.3 g
Glucose	1.0 g
Glycerol	1.0 g
K_2HPO_4	0.3 g
$MgSO_4 \cdot 7H_2O$	0.2 g
NaCl	0.3 g
$Fe_2(SO_4)_3$	10.0 mg
$MnSO_4 \cdot 7H_2O$	1.0 mg
$CuSO_4 \cdot 5H_2O$	1.0 mg
$ZnSO_4 \cdot 7H_2O$	1.0 mg
Agar	15.0 g
Adjust to pH 6.4, and then add:	
B-Vitamins	(see above)
Cycloheximide	50.0 mg

Nystatin	50.0 mg

Artificial Soil Agar (Henssen and Schäfer, 1971)

$CaSO_4 \cdot 2H_2O$	1.01 g
$Ca(NO_3)_2 \cdot 4H_2O$	0.49 g
$MgSO_4 \cdot 7H_2O$	0.70 g
K_2SO_4	0.025 g
K_2HPO_4	0.005 g
$NaHCO_3$	0.2 g
$FeCl_3$	trace
Yeast extract	0.1 g
Glucose	0.01 g
Soil extract	50 ml
Agar	15.0 g
Adjust to pH 6.6 to 6.8.	

To prepare soil extract, equal volumes of leafy soil and tap water are boiled for 2 hours and then cleared by centrifugation.

Bennett's Agar (Waksman, 1961)

Glucose	10.0 g
Yeast extract	1.0 g
Beef extract	1.0 g
N-Z-Amine A (casein digest; Sheffield Farms Co.)	2.0 g
Agar	15.0 g
Adjust to pH 7.3.	

Bennett's Agar (Modified) (Wakisaka et al., 1982)

Glucose	10.0 g
Casamino acids	2.0 g
Yeast extract	2.0 g
Beef extract	1.0 g
Agar	15.0 g
Adjust to pH 7.3, and then add:	
Cycloheximide	30.0 mg

Casamino Acids-Peptone-Czapek Agar (Henssen and Schäfer, 1971)

Casamino acids	1.0 g
Peptone	2.0 g
K_2HPO_4	1.0 g
KCl	0.5 g
$MgSO_4 \cdot 7H_2O$	0.5 g
$FeSO_4 \cdot 7H_2O$	0.01 g
Sucrose	30.0 g
Agar	15.0 g
Adjust to pH 7.0.	

Calcium Malate Agar (Waksman, 1961)

Glycerol	10.0 g
Calcium malate	10.0 g
NH_4Cl	0.5 g
K_2HPO_4	0.5 g
Agar	15.0 g
Adjust to pH 7.0.	

Colloidal Chitin Agar (Composition of Minerals According to Willoughby, 1968)

Colloidal chitin (dry weight)	2.0 g
$CaCO_3$	0.02 g
$FeSO_4 \cdot 7H_2O$	0.01 g
KCl	1.71 g

MgSO$_4$·7H$_2$O   0.05 g
Na$_2$HPO$_4$·12H$_2$O   4.11 g
Agar   18.0 g
Adjust to pH 7.0.

Collodial Chitin Agar (Composition of Minerals According to Hsu and Lockwood, 1975)

Collodial chitin (dry weight)   2.0 g
K$_2$HPO$_4$ 0.7 g
KH$_2$PO$_4$ 0.3 g
MgSO$_4$·5H$_2$O   0.5 g
FeSO$_4$·7H$_2$O   0.01 g
ZnSO$_4$ 0.001 g
MnCl$_2$ 0.001 g
Agar   20.0 g
Adjust to pH 7.0.

Preparation of Colloidal Chitin (Makkar and Cross, 1982)

Crude chitin is washed alternately in 1 N NaOH and 1 N HCl for 24-h periods each, five times. Then, it is washed four times with 95% (v/v) ethanol. 15 g of the purified white chitin is dissolved with 100 ml of concentrated HCl and stirred in an ice bath for 20 min. The mixture is filtered through glass wool, and the solution is poured into cold distilled water to precipitate the chitin. The insoluble chitin on the glass wool is treated again with HCl, and the process is repeated until no more precipitate is obtained when the filtrate is added to cold water. The colloidal chitin is allowed to settle overnight and the supernatant is decanted. The remaining suspension is neutralized to pH 7.0 with NaOH. The precipitated chitin is centrifuged, washed, and stored as a paste at 4°C.

Various other procedures for preparing colloidal chitin can be found in Willoughby (1968), Hsu and Lockwood (1975), Johnston and Cross (1976), and Hunter-Cevera et al. (1986).

Czapek Sucrose Agar (Bland and Couch, 1981)

Sucrose   30.0 g
NaNO$_3$ 3.0 g
K$_2$HPO$_4$ 1.0 g
MgSO$_4$·7H$_2$O   0.5 g
KCl   0.5 g
FeSO$_4$ 0.01 g
Agar   15.0 g
Adjust to pH 7.0 to 7.3.

Emerson's Yeast Extract-Starch Agar (Emerson, 1958)

Yeast extract   4.0 g
Soluble starch   15.0 g
K$_2$HPO$_4$ 1.0 g
MgSO$_4$·7H$_2$O   0.5 g
Agar   20.0 g

Gause Number 1 Agar (Gause, 1958)

KNO$_3$ 1.0 g
K$_2$HPO$_4$ 0.5 g
MgSO$_4$ 0.5 g
NaCl   0.5 g
FeSO$_4$ 0.01 g

Starch   20.0 g
Agar   30.0 g
Tap water   1 liter

Glucose Asparagine Agar (Gordon and Smith, 1955)

Glucose   10.0 g
Asparagine   0.5 g
K$_2$HPO$_4$ 0.5 g
Agar   15.0 g
Adjust to pH 6.8.

Hickey-Tresner Agar (Waksman, 1961)

Dextrin   10.0 g
Yeast extract   1.0 g
Beef extract   1.0 g
N-Z-Amine A (casein digest; Sheffield Farms   2.0 g
   Co.)
CoCl$_2$·7H$_2$O   0.02 g
Agar   20.0 g
Adjust to pH 7.3.

Humic Acid-Ion Agar (Willoughby et al., 1968)

5 g of humic acid in solid form (see below) is dissolved in 10 ml of 0.2 N NaOH, and the solution is made up to 1 liter with distilled water; pH 7.0 to 7.3. Then 20 g of agar (Oxoid Ionagar No. 2) is added.

Preparation of Humic Acid (Willoughby et al., 1968):

Air-dried (90°C for 24 hours) peat from a blanket bog or from compacted, decomposing tree leaves (oak and alder) are first extracted with acetone for 3 hours and then redried at 50°C overnight. The humic acids are extracted from the material (100 g) with 2 liters of 0.2 N NaOH by shaking occasionally, standing overnight at room temperature, and filtering off the insoluble residue on a no. 4 sintered glass funnel. The filtrate is acidified to pH 1.0 with concentrated HCl. The resulting precipitate is centrifuged off, washed twice with water, and left overnight at −20°C to freeze. After thawing, the now-granular particles are filtered off from the remaining liquors, washed, and air-dried at 50°C.

Humic Acid-Vitamin Agar (Hayakawa and Nonomura, 1987)

Humic acid (see below) (dissolved in 10 ml   1.0 g
   of 0.2 N NaCl)
Na$_2$HPO$_4$ 0.5 g
KCl   1.71 g
MgSO$_4$·7H$_2$O   0.05 g
FeSO$_4$·7H$_2$O   0.01 g
CaCO$_3$ 0.02 g
Agar   18.0 g
Adjust to pH 7.2, and then add:
B Vitamins (see above)
Cycloheximide   50.0 mg

Preparation of Humic Acid (Hayakawa and Nonomura, 1987)

500 g of soil sample (A-horizon of a forest) is suspended in 1 liter of 0.5% NaOH solution and left standing at room temperature for 24 hours, with occasionally stir-

ring. The precipitate of the suspension is removed by centrifugation (20 min at 7,000 rpm), and the supernatant is acidified to pH 1.0 with concentrated HCl. The resulting precipitate is centrifugated (20 min at 3,000 rpm), washed three times by centrifugation with 150 ml of water and suspended again in 150 ml of water. The suspension is frozen overnight at −20°C. After thawing, the granulated humic acid is filtered, washed, and air-dried.

Kadota's Cellulose Benzoate Agar (Sandrak, 1977)

1) Basal medium

NaNO$_3$	0.5 g
K$_2$HPO$_4$	1.0 g
MgSO$_4$·7H$_2$O	0.5 g
FeSO$_4$·7H$_2$O	0.01 g
Sodium benzoate	20.0 g
Agar	20.0 g
Adjust to pH 7.2	

2) Cellulose powder (0.67g) is mixed with 1 ml soil suspension and 2 ml liquid basal medium before inoculation of the plates.

M3 Agar Medium (Rowbotham and Cross, 1977)

KH$_2$PO$_4$	0.466 g
Na$_2$HPO$_4$	0.732 g
KNO$_3$	0.10 g
NaCl	0.29 g
MgSO$_4$·7H$_2$O	0.10 g
CaCO$_3$	0.02 g
Sodium propionate	0.20 g
FeSO$_4$·7H$_2$O	200 μg
ZnSO$_4$·7H$_2$O	180 μg
MnSO$_4$·4H$_2$O	20 μg
Agar	18.00 g
Ajust to pH 7.0, and then add:	
Cycloheximide	50.0 mg
Thiamine hydrochloride	4.0 mg

Oatmeal-Soil Extract Agar (Shearer, 1987)

Oatmeal agar (Difco)	5.5 g
Agar	16.0 g
Soil extract	500 ml
Deionized water	500 ml
Adjust to pH 7.2.	

Peptone-Czapek Agar (Bland and Couch, 1981)

Sucrose	30.0 g
Peptone	5.0 g
K$_2$HPO$_4$	1.0 g
MgSO$_4$·7H$_2$O	0.5 g
KCl	0.5 g
FeSO$_4$·7H$_2$O	0.01 g
Agar	15.0 g
Adjust to pH 7.3.	

Skim Milk Agar (Gordon and Smith, 1955)

Solution a:

Skim milk powder	50 g
Distilled water	500 ml

Solution b:

Agar	20 g
Distilled water	500 ml

Solutions a and b are autoclaved separately, cooled, mixed at 45°C, and the pH is adjusted to 7.2.

Skim Milk-Cattle Hornmeal Agar (Vobis, 1984)

Solution a:

Cattle hornmeal (powdered)	10.0 g
Ca(NO$_3$)$_2$·4H$_2$O	0.5 g
MgSO$_4$·7H$_2$O	0.7 g
K$_2$HPO$_4$	0.005 g
NaHCO$_3$	0.2 g
FeCl$_3$	trace
Agar	20.0 g
Distilled water	900 ml

Solution b:

Skim milk powder	2.5 g
Distilled water	100 ml

Solutions a and b are autoclaved separately, cooled, mixed at 48°C, and the pH is adjusted to 7.2.

Soil Extract Agar (Makkar and Cross, 1982)

First 150 g of garden soil is stirred in 600 ml of tap water, filtered immediately through Whatman No. 1 filter, and made up to 1 liter with tap water. To this is added:

Agar	18.0 g
Adjust to pH 7.2, and then add:	
Cycloheximide	50 mg
Nystatin	50 mg

Starch-Casein-Nitrate Agar (Küster and Williams, 1964)

Starch	10.0 g
Casein (Difco, vitamin-free)	0.3 g
KNO$_3$	2.0 g
NaCl	2.0 g
K$_2$HPO$_4$	2.0 g
MgSO$_4$·7H$_2$O	0.05 g
CaCO$_3$	0.02 g
FeSO$_4$·7H$_2$O	0.01 g
Agar	18.0 g
Adjust to pH 7.0 to 7.2.	

Hunter-Cevera et al. (1986) added:

Cycloheximide	75 mg
Nystatin	75 mg

Shearer (1987) added:
B-Vitamins (see above)

Starch-Casein Sulfate Agar (Palleroni, 1980)

K$_2$HPO$_4$	0.5 g
MgSO$_4$ (anhydrous)	5.0 g
Soluble starch	10.0 g
Casein (dissolved in diluted NaOH)	1.0 g
Agar	15.0 g
Adjust to pH 7.0 to 7.5.	

Thin Pablum Agar (Hunter-Cevera et al., 1986)

Pablum (boiled in cheesecloth for 20 to 30 min)	7.5g
Agar	15.0 g
Tap water	1 liter

Adjust to pH 6.8 to 7.0, and then add:
Cycloheximide   75 mg
Nystatin   75 mg

Water Agar (Hunter-Cevera et al., 1986)
Tap water   1 liter
Crude agar flakes   17.5 g
Adjust to pH 6.8, and then add:
Cycloheximide   75 mg
Nystatin   75 mg

Applications

Many members of the actinoplanete genera produce useful enzymes and secondary metabolites, so they may have important applications in industry, biotechnology, and agriculture. For example, vitamin B_{12} can be produced by a *Micromonospora* strain (Florent and Ninet, 1979), and strains of *Actinoplanes* and *Micromonospora* can be used for the bioconversion of complex compounds (Okami and Hotta, 1988; Boek et al., 1988). Applications which have been established already or might be used commercially in the future are described below.

Glucose Isomerase in the Food Industry

The enzyme glucose isomerase, which is primary a xylose isomerase, can be obtained from strains of *Actinoplanes, Ampullariella* and *Micromonospora,* in addition to strains of *Streptomyces, Bacillus,* and *Arthrobacter* (Crueger and Crueger, 1982; Peczyńsca-Czoch and Mordarski, 1988). The glucose isomerase converts D-glucose into D-fructose and is used commercially in the starch industry to obtain high-fructose corn syrup (Aunstrup et al., 1979). Starting from about 95% glucose syrup, a twice sweeter fructose syrup is produced which is usually composed of 53% of D-glucose, 42% of D-fructose, and 5% of oligosaccharides (Crueger and Crueger, 1982).

Actinoplanes missouriensis strain ATCC 14538 produces an intracellular, soluble glucose isomerase with a molecular weight of about 80,000 daltons. The optimal pH of the enzyme is 7.0 at temperatures between 60 and 65°C. A requirement for cobalt ions for optimal activity is eliminated if the proper amount of magnesium ions is used (Gong et al., 1980). The xylose isomerase of *Ampullariella* strain ATCC 31354 exhibits superior thermostability and activity over a wide range of conditions. However, the strain itself is difficult to use as a production organism, which makes it desirable to clone and express its enzyme in a more convenient microorganism (Saari et al., 1987).

Inhibitors of α-Glucosidase as Pharmaceutical Drugs

In the course of screening for inhibitors of amylases and other mammalian intestinal carbohydrate-splitting enzymes, strains of *Actinoplanes* exhibited higher amounts of activity than did those of *Streptomyces* and *Streptosporangium* (Frommer et al., 1979). Applied orally together with food carbohydrates like starch and other oligosaccharides, these glycoside hydrolase inhibitors slow down oligosaccharide decomposition and reduce or avoid postprandial hyperglycemia and hyperinsulinemia of type IV. Therefore, they may be useful to treat metabolic illnesses such as diabetes mellitus, adipositas, and hyperlipoproteinaemia (Truscheit et al., 1981; Creutzfeldt, 1988). Pseudo-oligosaccharides with an essential core consisting of an unsaturated cyclitol and 4-amino-4,6-dideoxyglucose are the most important group of α-glucosidase inhibitors. In culture filtrates of the *Actinoplanes* strain SE 50, a very effective pseudotetrasaccharide with the generic name "acarbose" could be found. This low-molecular-weight compound is stabile to acid, alkali, and heat treatment, and exhibits pronounced inhibition of sucrase, maltase, and amylase (Truscheit et al., 1981).

Antibiotics

The "rare" actinomycetes (i.e., nonstreptomycetes) have become increasingly interesting in the search for new active secondary metabolites (Nara et al., 1977; Okami and Hotta, 1988). In 1984, the number of known antibiotic compounds produced by all the actinomycetes together amounted to about 4,200 (Bérdy, 1984), and the proportion produced by actinoplanetes *(Micromonospora, Actinoplanes, Dactylosporangium,* and *Ampullariella)* increased from less than 1% in the year 1966 up to 10% in 1984 (Lechevalier and Lechevalier, 1967; Wagman and Weinstein, 1980; Bérdy, 1984). As can be seen in Table 4, each actinoplanetes genus covers only a part of all the chemical groups of antibiotics that are known from the genus *Streptomyces.* But they complement one another, so only the β-lactam antibiotics seem to be absent. Until now, members of *Catellatospora* and *Pilimelia* have not been included to any extent in antibiotic screening procedures.

ACTINOPLANES. More than 120 antibiotics are known from *Actinoplanes* species. The most common are peptides/depsipeptides, polyene-type macrolides, and aromatic compounds, whereas the aminoglycosides, streptothricins,

Table 4. Antibiotic groups produced by the genera of the actinoplanetes and by the streptomycetes.

Genus	Aminoglycoside	Macrolide	Ansamacrolide	β-Lactam	Peptide	Glycopeptide	Anthracycline	Tetracycline	Nucleoside	Polyene	Quinone
Actinoplanes					+	+			+	+	+
Ampullariella							+		+		
Catellatospora											
Dactylosporangium		+			+			+	+	+	
Glycomyces					+						
Micromonospora	+	+	+		+		+		+		+
Pilimelia	+	+	+	+	+	+	+	+	+	+	
Streptomyces	+	+	+	+	+	+	+	+	+	+	+

Data mainly based on Okami and Hotta (1988) and Nara et al. (1977); the additions made are discussed in the text.

macrolides, and ansamycine seem to be absent (Table 4). About 50% of the antibiotics found in *Actinoplanes* are amino acid derivatives. The proline antimetabolite L-acetidine-2-carboxylic acid could be isolated from *A. ferrugineus.* This amino acid has not been found in any other prokaryotes and has only been found in eukaryotes (Palleroni, 1979). Additional amino acid derivatives are 5-azacytidin, azaserine, and the chromopeptide actinomycin, which are all properly typical *Streptomyces* antibiotics (Torikata et al., 1978). They are active as antitumor agents. The polypeptides generally exhibit activity against Gram-positive bacteria. The acidic peptide 41.012 is also an agent against mycobacteria (Celmer et al., 1977). The antibiotics A-10947, A-7413, taitomycin, and gardimicin are sulfur-containing polypeptides; the latter two are also active against anaerobic bacteria (Yaginuma et al., 1979; Parenti and Coronelli, 1979). The cyclic polypeptides A/287 and mycoplanecin show growth-promoting and antituberculosis effects respectively (Hamill and Stark, 1974; Nakajima et al., 1983). The depsipeptide antibiotic plauracins A 17002 and A 2315 belong to the virginiamycin group, composed of a mixture of macrocyclic lactones and depsipeptides. They can be used for growth promotion in chicken, swine, and ruminants (Hamill and Stark, 1975; Parenti and Coronelli, 1979).

A well-studied example of the antifungal polyenic macrolides is the antibiotic 67–121 (Sch 16656). It is a complex of four polyene heptaenes produced by *A. caeruleus* (Horan and Brodsky, 1986a). It is also produced by *A. azureus,* a strain which also produces the plauracins (Parenti and Coronelli, 1979).

Some metabolites belonging to various other chemical groups are also found in strains of *Actinoplanes.* Purpuromycin is a naphthoquinone antibiotic of the rubromycin type, effective against bacteria and fungi (Coronelli et al., 1974). *A. teichomyceticus* produces the glycopetide teicoplanin (formerly called teichomycin A$_2$), which is composed of six factors. It belongs to the vancomycin family (Malabarba et al., 1984). The same strain also produces a phosphorus-containing glycolipid (teichomycin A$_1$). Both carbohydrate antibiotics are active against Gram-positive bacteria (Parenti et al., 1978). The recently described polycyclic xanthones actinoplanone A and B were found to be potent cytotoxins in in vitro assays with Hela cells (Kobayashi et al., 1988).

Some chemical novelties have been detected in *Actinoplanes* strains. For example, chuangxinmycin, clinically effective in cases of septicemia and urinary and biliary infections caused by *Escherichia coli,* is a completely new antibiotic composed of an unique bicyclic system formed of an indole nucleus fused to a thiopyran residue (Parenti and Coronelli, 1979). The antibiotic A/15104 Y is a chlorophenol derivative, active against bacteria and fungi. It represents the first example of a halogenated pyrrole from actinomycetes; the other biological sources have been sponges and pseudomonads (Cavalleri et al., 1978).

AMPULLARIELLA. Only a dozen antibiotics are known to be produced by *Ampullariella* strains. Nevertheless, they exhibit quite a range of different chemical structures and antibiotic effects. Viriplanin is an anthracyclic antibiotic isolated from *A. regularis,* which shows activities against herpes simplex viruses (Hütter et al., 1986). The neplanecins are nucleosides produced by another *A. regularis* strain. They are antitumor antibiotics with additional activities against phytopathogenic fungi (Yaginuma et al., 1981). Another antifungal antibiotic, candiplanecin, could be isolated from *A. regularis* subsp. *mannitophila* (Itoh et al., 1981).

DACTYLOSPORANGIUM. At present, about 30 antibiotics are known from *Dactylosporangium* strains, belonging to several chemical divisions (Table 4). As with *Micromonospora,* the major antibiotics seem to be aminoglycosides. Dactimicin is produced by *D. matsuzakiense* and *D. vinaceum* (Shomura et al., 1980, 1983). The closely related aminoglycosides gentamicin, sisomicin, fortimicin, and antibiotic G-367 could also be isolated. Another carbohydrate antibiotic is known from *D. roseum,* namely the orthosomycin complex SF-2107 (Shomura et al., 1985). All the above-mentioned antibiotics are generally active against Gram-positive and Gram-negative bacteria. Capreomycin, a polypeptide compound previously known from *Streptomyces capreolus,* could be obtained also from "*D. variesporium.*" It is of primary interest for its use as an antituberculosis agent (Tomita et al., 1977). "*D. salmoneum*" produces the polyether antibiotic compound 44,161, which is useful for the control of coccidiosis in poultry and improving feed efficiency in ruminants (Celmer et al., 1978).

The tetracycline antibiotic compound Sch 34164 (Patel et al., 1987), and the macrolide tiacumicin (Hochlowski et al., 1986) have been isolated from other *Dactylosporangium* species, suggesting that the capacity for producing antibiotic metabolites in this genus may be very large.

GLYCOMYCES. *Glycomyces harbinensis* produces the amino acid derivative azaserine and the antibiotic LL-D05139-beta (Labeda, 1989). No other antibiotics have been identified in any *Glycomyces* strains so far.

MICROMONOSPORA. Among the antibiotics produced by actinoplanete genera, those of *Micromonospora* occupy the most important commercial position (Crueger and Crueger, 1982). An intensive screening of *Micromonospora* species as sources for new antibiotics began in 1963 with the discovery of gentamicin. Over 300 different antibiotics have been described (Bérdy, 1984), and the range of chemical structures is quite large (Table 4). Examples include the aminoglycosides, represented by gentamicins, sisomicin, and verdamicin; the antibiotics G-52, G-418, and JI-20; mannosidostreptomycin, kanamycin, neomycin B (antibiotic 460), sagamicin (gentamicin C_{2b}), paromamine, fortimicins, and antibiotics 66–40 and SF 1854 (Nara et al., 1977; Wagman and Weinstein, 1980). The macrolides comprise megalomycins, rosaramicin, juvenimicins, the M-4365 complex, erythromycins, and antibiotic XK 41-B-2. Examples for ansa macrolides (ansamycins) are halomicins, rifamycins, and compound 32656. Bottromycin, microsporonin, the 70591 complex, and actinomycin should be cited as representatives of the peptide antibiotics. Other miscellaneous antibiotics isolated from *Micromonospora* species include the oligosaccharides everninomycin and antlermicin, the nucleosides PA-1322 and XK-101-2, and the quinone PA-2046 (Wagman and Weinstein, 1980).

Aminoglycosides of *Micromonospora* show antibiotic effects against both Gram-positive and Gram-negative bacteria and have been introduced into clinical practice. The gentamicins C_1, C_{1a}, and C_2 are produced by *M. purpurea* and *M. echinospora* and exhibit excellent activity against *Staphylococcus aureus* and species of *Pseudomonas* and *Proteus*. Because of their vestibular and nephrotoxicity, they are used in human therapy only for severe infections (Wagman and Weinstein, 1980; Crueger and Crueger, 1982; Lancini and Parenti, 1982). Sisomycin and fortimicins, derived from *M. inyonensis* and *M. olivoasterospora*, respectively, have a similar spectrum of effectivity as gentamicin and can be used against gentamicin-resistant organisms (Wagman and Weinstein, 1980).

Biological Control in Agriculture

Reports of hyperparasitism by actinoplanetes on parasitic *Peronosporales* and *Saprolegniales* demonstrate a possible biological control of serious diseases of economic plants (Lechevalier, 1988). The oospores of pink rot causing *Phytophthora megasperma* var. *sojae* or f. sp. *glycinea* can be parasitized by certain strains of *Actinoplanes, Ampullariella,* and *Micromonospora* (Sneh et al., 1977). The hyphae of *Actinoplanes missouriensis* penetrate the walls of the oogonia and the oospores without forming appressoria or haustoria or changing the morphological and internal structures of the oospores (Sutherland et al., 1984). In greenhouse experiments, the root rot of soybeans caused by *Phytophthora* could be reduced by *Actinoplanes missouriensis, A. utahensis,* and *Micromonospora* sp. (Filinow and Lockwood, 1985).

Acknowledgments

I wish to thank Dr. H.-W. Kothe (Burlington), Dr. R. M. Kroppenstedt (Braunschweig), Dr. H. Prauser (Jena), Dr. G. S. Saddler (Gerenzano), Prof. Dr. O. Salcher (Wuppertal), Dr. J.-J. Sanglier (Basel), and Dr. J. Wink (Frankfurt) for helpful cooperation.

Literature Cited

Al-Diwany, L. J. and T. Cross. 1978. Ecological studies on nocardioforms and other actinomycetes in aquatic habitats. Zentralbl. Bacteriol. Parasitenkd. Infektionskr. Hyg. Abt. 1, Suppl. 6:153–160.

Arora, D. K. 1986. Chemotaxis of *Actinoplanes missouriensis* zoospores to fungal conidia, chlamydospores and sclerotia. J. Gen. Microbiol. 132:1657–1663.

Asano, K. and I. Kawamoto. 1986. *Catellatospora,* a new genus of the Actinomycetales. Int. J. Syst. Bacteriol. 36:512–517.

Asano, K. and I. Kawamoto. 1988. *Catellatospora citrea* subsp. *methionotrophica* subsp. nov., a methionine-deficient auxotroph of the Actinomycetales. Int. J. Syst. Bacteriol. 38:326–327.

Asano, K., H. Sano, I. Masunaga, and I. Kawamoto. 1989a. 3-0-methylrhamnose: identification and distribution in *Catellatospora* species and related actinomycetes. Int. J. Syst. Bacteriol. 39:56–60.

Asano, K., I. Masunaga, and I. Kawamoto. 1989b. *Catellatospora matsumotoense* sp. nov. and *C. tsunoense* sp. nov., actinomycetes found in woodland soils. Int. J. Syst. Bacteriol. 39:309–313.

Aunstrup, K., O. Andresen, E. A. Falch, and T. K. Nielsen. 1979. Production of microbial enzymes, p. 281–309. In: H. J. Peppler and D. Perlman (ed.), Microbial technology, vol. 1. Academic Press, New York.

Bérdy, J. 1984. New ways to obtain new antibiotics. Chin. J. Antibiot. 7:348–360.

Beretta, G. 1973. *Actinoplanes italicus,* a new red-pigmented species. Int. J. Syst. Bacteriol. 23:37–42.

Bland, C. E., and J. N. Couch. 1981. The family Actinoplanaceae, p. 2004–2010. In: M. P. Starr, H. Stolp,

H. G. Trüper, A. Balows, and H. G. Schlegel (ed.), The prokaryotes: A handbook on habitats, isolation, and identification of bacteria. Springer-Verlag, Berlin.

Boeck, L. D., D. F. Fukuda, B. J. Abbott, and M. Debono. 1988. Deacylation of A21978C, an acidic lipopeptide antibiotic complex, by *Actinoplanes utahensis.* J. Antibiot. 41:1085–1092.

Burman, N. P., C. W. Oliver, and J. K. Stevens. 1969. Membrane filtration techniques for the isolation from water, of coli-aerogenes, *Escherichia coli,* faecal streptococci, *Clostridium perfringens,* actinomycetes and microfungi, p. 127–134. In: D. A. Shapton, and G. W. Gould (ed.), Isolation methods for microbiologists. Academic Press, London.

Cavalleri, B., G. Volpe, G. Tuan, M. Berti, and F. Parenti. 1978. A chlorinated phenylpyrrole antibiotic from *Actinoplanes.* Curr. Microbiol. 1:319–324.

Celmer, W. D., W. P. Cullen, C. E. Moppett, J. B. Routien, M. T. Jefferson, R. Shibakawa, and J. Tone. 1978. Polycyclic ether antibiotic produced by new species of *Dactylosporangium.* U.S. Patent 4,081,532, Mar. 28, 1978.

Celmer, W. D., C. E. Moppett, W. P. Cullen, J. B. Routien, M. T. Jefferson, R. Shibakawa, and J. Tone. 1977. Antibiotic compound 41,012. U.S. Patent 4,001,397, Jan. 4, 1977.

Collins, M. D., M. Falkner, and R. M. Keddie. 1984. Menaquinone composition of some sporeforming actinomycetes. System. Appl. Microbiol. 5:20–29.

Coronelli. C., H. Pagani, M. R. Bardone, and G. C. Lancini. 1974. Purpuromycin, a new antibiotic isolated from *Actinoplanes ianthinogenes* n. sp. J. Antibiot. 27:161–168.

Couch, J. N. 1949. A new group of organisms related to *Actinomyces.* J. Elisha Mitchell Sci. Soc. 65:315–318.

Couch, J. N. 1950. *Actinoplanes,* a new genus of the Actinomycetales. J. Elisha Mitchell Sci. Soc. 66:87–92.

Couch, J. N. 1954. The genus *Actinoplanes* and its relatives. Trans. N.Y. Acad. Sci. 16:315–318.

Couch, J. N. 1963. Some new genera and species of the Actinoplanaceae. J. Elisha Mitchell Sci. Soc. 79:53–70.

Creutzfeldt, W. 1988. Acarbose for the treatment of diabetes mellitus. Springer-Verlag, Berlin.

Cross, T. 1981a. Aquatic actinomycetes: a critical survey of the occurrence, growth and role of actinomycetes in aquatic habitats. J. Appl. Bacteriol. 50:397–423.

Cross, T. 1981b. The monosporic actinomycetes, p. 2091–2102. In: M. P. Starr, H. Stolp, H. G. Trüper, A. Balows, and H. G. Schlegel (ed.), The prokaryotes: A handbook on habitats, isolation, and identification of bacteria. Springer-Verlag, Berlin.

Cross, T. 1986. The occurrence and role of actinoplanetes and motile actinomycetes in natural ecosystems, p. 265–270. In: F. Megusar, and M. Gantar (ed.), Perspectives in microbial ecology. Slovene Soc. Microbiol., Ljubljana.

Cross, T. 1989. Growth and examination of actinomycetes—some guidelines, p. 2340–2343. In: S. T. Williams (ed.), Bergey's manual of systematic bacteriology, vol. 4. Williams & Wilkins, Baltimore.

Cross, T. and R. W. Attwell. 1974. Recovery of viable thermoactinomycete endospores from deep mud cores, p. 11–20. In: A. N. Barker, G. W. Gould, and J. Wolf (ed.), Spore research 1973. Academic Press, London.

Crueger, W., and A. Crueger. 1982. Lehrbuch der Angewandten Mikrobiologie. Akademische Verlagsgesellschaft, Wiesbaden.

El-Nakeeb, M., and H. A. Lechevalier. 1963. Selective isolation of aerobic actinomycetes. Appl. Microbiol. 11:75–77.

Emerson, R. 1958. Mycological organization. Mycologia 50:589–621.

Ensign, J. C. 1978. Formation, properties, and germination of actinomycete spores. Ann. Rev. Microbiol. 32:185–219.

Erikson, D. 1941. Studies on some lake-mud strains of *Micromonospora.* J. Bacteriol. 41:277–300.

Fernandez, C. 1984. Studies on the microflora of the curative bottom mud of the thermal lake Héviz (W. Hungary). Acta Bot. Hung. 30:257–268.

Filinow, A. B., and J. L. Lockwood. 1985. Evaluation of several actinomycetes and the fungus *Hyphochytrium catenoides* as biocontrol agents for phytophthora root rot of soybean. Plant disease 69:1033–1036.

Florent, J., and L. Ninet. 1979. Vitamin B_{12}, p. 497–519. In: H. J. Peppler, and D. Perlman (ed.), Microbial technology, vol. 2. Academic Press, New York.

Frommer, W., B. Junge, L. Müller, D. Schmidt, and E. Truscheit. 1979. Neue Enzyminhibitoren aus Mikroorganismen. Planta Medica 35:195–217.

Gaertner, A. 1955. Über zwei ungewöhnliche keratinophile Organismen aus Ackerböden. Arch. Mikrobiol. 23:28–37.

Gause, G. F. 1958. Zur Klassifizierung der Actinomyceten. VEB Gustav Fischer Verlag, Jena.

Goodfellow, M. 1989. Suprageneric classification of actinomycetes, p. 2333–2339. In S. T. Williams (ed.), Bergey's manual of systematic bacteriology, vol. 4. Williams & Wilkins, Baltimore.

Goodfellow, M., and T. Cross. 1984. Classification, p. 7–164. In: M. Goodfellow, M. Mordarski, and S. T. Williams (ed.), The biology of the actinomycetes. Academic Press, London.

Goodfellow, M., and J. A. Haynes. 1984. Actinomycetes in marine sediments, p. 453–472. In: L. Ortiz-Ortiz, L. F. Bojalil, and V. Yakoleff (ed.), Biological, biochemical, and biomedical aspects of actinomycetes. Academic Press, Orlando.

Goodfellow, M., and S. T. Williams. 1983. Ecology of actinomycetes. Ann. Rev. Microbiol. 37:189–216.

Goodfellow, M., L. J. Stanton, K. E. Simpson, and D. E. Minnikin. 1990. Numerical and chemical classification of *Actinoplanes* and some related actinomycetes. J. Gen. Microbiol. 136:19–36.

Gong, C.-S., L. F. Chen, and G. T. Tsao. 1980. Purification and properties of glucose isomerase of *Actinoplanes missouriensis.* Biotechnol. Bioeng. 22:833–845.

Gordon, R. E., and M. M. Smith. 1955. Proposed group of characters for the separation of *Streptomyces* and *Nocardia.* J. Bacteriol. 69:147–150.

Hamill, R. L., and W. M. Stark. 1974. Antibiotic A-287 and process for preparation thereof. U.S. Patent 3,824,305, July 16, 1974.

Hamill, R. L., and W. M. Stark. 1975. Antibiotic A-2315 and process for preparation thereof. U.S. Patent 3,923,980, Dec. 2, 1975.

Hasegawa, T., M. P. Lechevalier, and H. A. Lechevalier. 1979. Phospholipid composition of motile actinomycetes. J. Gen. Appl. Microbiol. 25:209–213.

Hasegawa, T., M. Takizawa, and S. Tanida. 1983. A rapid analysis for chemical grouping of aerobic actinomycetes. J. Gen. Appl. Microbiol. 29:319–322.

Hayakawa, M., and H. Nonomura. 1987. Humic acid-vitamin agar, a new medium for the selective isolation of soil actinomycetes. J. Ferment. Technol. 65:501–509.

Henssen, A., and D. Schäfer. 1971. Emended description of the genus *Pseudonocardia* Henssen and description of a new species *Pseudonocardia spinosa* Schäfer. Int. J. Syst. Bacteriol. 21:29–34.

Higgins, M. L. 1967. Release of sporangiospores by a strain of *Actinoplanes*. J. Bacteriol. 94:495–498.

Hirsch, C. F., and D. L. Christensen. 1983. Novel method for selective isolation of actinomycetes. Appl. Environ. Microbiol. 46:925–929.

Hochlowski, J. E., S. J. Swanson, D. N. Whittern, A. N. Buko, and J. B. McAlpine. 1986. Tiacumicins, a novel series of 18-membered macrolide antibiotics. II. Isolation and elucidation of structures. 26th Interscience Congress on Antimicrobial Agents and Chemotherapy, Abstract 937.

Horan, A. C., and B. Brodsky. 1986a. *Actinoplanes caeruleus* sp. nov., a bluepigmentd species of the genus *Actinoplanes*. Int. J. Syst. Bacteriol. 36:187–191.

Horan, A. C., and B. C. Brodsky. 1986b. *Micromonospora rosaria* sp. nov., nom. rev., the rosaramicin producer. Int. J. Syst. Bacteriol. 36:478–480.

Hsu, S. C., and J. L. Lockwood. 1975. Powdered chitin agar as a selective medium for enumeration of actinomycetes in water and soil. Appl. Microbiol. 29:422–426.

Hungate, R. E. 1946. Studies on cellulose fermentation. II. An anaerobic cellulose-decomposing actinomycete, *Micromonospora propionici* n. sp. J. Bacteriol. 51:51–56.

Hunter, J. C., D. E. Eveleigh, and G. Casella. 1981. Actinomycetes of a salt marsh. Zentralbl. Bakteriol. Mikrobiol. Hyg. Abt. 1, Suppl. 11:195–200.

Hunter, J. C., M. Fonda, L. Sotos, B. Toso, and A. Belt. 1984. Ecological approaches to isolation. Dev. Ind. Microbiol. 25:247–266.

Hunter-Cevera, J. C., M. E. Fonda, and A. Belt. 1986. Isolation of cultures, p. 3–23. In: A. L. Demain, and N. A. Solomon (ed.), Manual of industrial microbiology and Biotechnology. Am. Soc. Microbiol., Washington, D.C.

Hütter, K., E. Baader, K. Frobel, A. Zeek, K. Bauer, W. Gau, J. Kurz, T. Schröder, C. Wünsche, W. Karl, and D. Wendisch. 1986. Viriplanin, a new anthracycline antibiotic of the nogalamycin group. J. Antibiot. 39:1195–1204.

Itoh, Y., A. Torikata, C. Katayama, T. Haneishi, and M. Arai. 1981. Candiplanecin, a new antibiotic from *Ampullariella regularis* subsp. *mannitophila* subsp. nov. II. Isolation, physico-chemical characterization and biological activities. J. Antibiot. 34:934–937.

Jensen, H. L. 1930. The genus *Micromonospora* Ørskov, a little known group of soil microorganisms. Proc. Linn. Soc. N.S. Wales 55:231–248.

Jensen, H. L. 1932. Contributions to our knowledge of the actinomycetales. III. Further observations on the genus *Micromonospora*. Proc. Linn. Soc. N.S. Wales 57:173–180.

Johnston, D. W., and T. Cross. 1976. The occurrence and distribution of actinomycetes in lakes of the English Lake District. Freshw. Biol. 6:457–463.

Kane, W. D. 1966. A new genus of Actinoplanaceae, *Pilimelia*, with a description of two species, *Pilimelia terevasa* and *Pilimelia anulata*. J. Elisha Mitchel Sci. Soc. 82:220–230.

Karling, J. S. 1954. An unusual keratinophilic microorganism. Proc. Indiana Acad. Sci. 63:83–86.

Kawamoto, I. 1989. Genus *Micromonospora* Ørskov, p. 2442–2450. In: S. T. Williams (ed.), Bergey's manual of systematic bacteriology, vol. 4. Williams & Wilkins, Baltimore.

Kawamoto, I., T. Oka, and T. Nara. 1981. Cell wall composition of *Micromonospora olivoasterospora, Micromonospora sagamiensis,* and related organisms. J. Bacteriol. 146:527–534.

Kobayashi, K., C. Nishino, J. Ohya, S. Sato, T. Mikawa, Y. Shiobara, and M. Kodama. 1988. Actinoplanones A and B, new cytotoxic polycyclic xanthones from *Actinoplanes* sp. J. Antibiot. 41:502–511.

Kothe, H. -W. 1987. Die Gattung *Actinoplanes* und ihre Stellung innerhalb der Actinomycetales. Dissertation, Marburg.

Kroppenstedt, R. M. 1979. Chromatographische Identifizierung von Mikroorganismen, dargestellt am Beispiel der Actinomyceten. Kontakte (Merck) 2:12–21.

Kroppenstedt, R. M. 1985. Fatty acid and menaquinone analysis of actinomycetes and related organisms, p. 173–199. In: M. Goodfellow, and D. E. Minnikin (ed.), Chemical methods in bacterial systematics. Academic Press, New York.

Kroppenstedt, R. M., and H. J. Kutzner. 1978. Biochemical taxonomy of some problem actinomycetes. Zentralbl. Bakteriol. Parasitenkd. Infektionskr. Hyg. Abt. 1, Suppl. 6:125–133.

Küster, E., and S. T. Williams. 1964. Selection of media for isolation of streptomycetes. Nature 202:928–929.

Labeda, D. P. 1987. Actinomycete taxonomy: generic characterization. Dev. Ind. Microbiol. 28:115–121.

Labeda, D. P. 1989. Genus *Glycomyces* Labeda, Testa, Lechevalier and Lechevalier, p. 2586–2589. In: S. T. Williams (ed.), Bergey's manual of systematic bacteriology, vol. 4. Williams & Wilkins, Baltimore.

Labeda, D. P., R. T. Testa, M. P. Lechevalier, and H. A. Lechevalier. 1985. *Glycomyces,* a new genus of the Actinomycetales. Int. J. Syst. Bacteriol. 35:417–421.

Lacey, J. 1988. Actinomycetes as biodeteriogens and pollutants of the environment, p. 359–432. In: M. Goodfellow, S. T. Williams, and M. Mordarski (ed.), Actinomycetes in biotechnology. Academic Press, London.

Lancini, G., and F. Parenti. 1982. Antibiotics. An integrated view. Springer-Verlag, New York.

Lechevalier, H. A. 1989. A practical guide to generic identification of actinomycetes, p. 2344–2347. In: S. T. Williams (ed.), Bergey's manual of systematic bacteriology, vol. 4. Williams & Wilkins, Baltimore.

Lechevalier, H. A., and M. P. Lechevalier. 1967. Biology of actinomycetes. Ann. Rev. Microbiol. 21:71–100.

Lechevalier, M. P. 1988. Actinomycetes in agriculture and forestry, p. 327–358. In: M. Goodfellow, S. T. Williams, and M. Mordarski (ed.), Actinomycetes in biotechnology. Academic Press, London.

Lechevalier, M. P., and H. A. Lechevalier. 1970a. Composition of whole-cell hydrolysates as a criterion in the classification of aerobic actinomycetes, p. 311–316. In: H. Prauser (ed.), The Actinomycetales. VEB Gustav Fischer Verlag, Jena.

Lechevalier, M. P., and H. Lechevalier. 1970b. Chemical composition as a criterion in the classification of aerobic actinomycetes. Int. J. Syst. Bacteriol. 20:435–443.

Lechevalier, M. P., and H. A. Lechevalier. 1975. Actinoplanete with cylindrical sporangia, *Actinoplanes rectilineatus* sp. nov. Int. J. Syst. Bacteriol. 25:371–376.

Lechevalier, M. P., C. de Bievre, and H. Lechevalier. 1977. Chemotaxonomy of aerobic actinomycetes: phospholipid composition. Biochem. Syst. Ecol. 5:249–260.

Lechevalier, M. P., A. E. Stern, and H. A. Lechevalier. 1981. Phospholipids in the taxonomy of actinomycetes. Zentralbl. Bakteriol. Mikrobiol. Hyg. Abt. 1, Suppl. 11:111–116.

Luedemann, G. M., and B. C. Brodsky. 1964. Taxonomy of gentamicin-producing *Micromonospora*. Antimicrob. Agents Chemother. 1963:116–124.

Luedemann, G. M., and C. J. Casmer. 1973. Electron microscope study of whole mounts and thin sections of *Micromonospora chalcea* ATCC 12452. Int. J. Syst. Bacteriol. 23:243–255.

Makkar, N. S., and T. Cross. 1982. Actinoplanetes in soil and on plant litter from freshwater habitats. J. Appl. Bacteriol. 52:209–218.

Malabarba, A., P. Strazzolini, A. Depaoli, M. Landi, M. Berti, and B. Cavalleri. 1984. Teicoplanin, antibiotics from *Actinoplanes teichomyceticus* nov. sp. J. Antibiot. 37:988–999.

Maluszyńska, G. M., and L. Janota-Bassalik. 1974. A cellulolytic rumen bacterium, *Micromonospora ruminantium* sp. nov. J. Gen. Microbiol. 82:57–65.

McCarthy, A. J., and P. Broda. 1984. Screening for lignin-degrading actinomycetes and characterization of their activity against ¹⁴C-lignin labelled wheat lignocellulose. J. Gen. Microbiol. 130:2905–2913.

Meyertons, J. L., D. P. Labeda, C. L. Cote, and M. P. Lechevalier. 1987. A new thin-layer chromatographic method for whole-cell sugar analysis of *Micromonospora* species. The Actinomycetes 20:182–192.

Minnikin, D. E., L. Alshamaony, and M. Goodfellow. 1975. Differentiation of *Mycobacterium, Nocardia,* and related taxa by thin-layer chromatographic analysis of whole-organism methanolysates. J. Gen. Microbiol. 88:200–204.

Minnikin, D. E., and A. G. O'Donnell. 1984. Actinomycete envelope lipid and peptidoglycan composition, p. 337–388. In: M. Goodfellow, M. Mordarski, and S. T. Williams (ed.), The biology of the actinomycetes. Academic Press, London.

Nakajima, M., A. Torikata, Y. Ichikawa, T. Katayama, A. Shiraishi, T. Haneishi, and M. Arai. 1983. Mycoplanecins, novel antimycobacterial antibiotics from *Actinoplanes awajinensis* subsp. *mycoplanecinus* subsp. nov. J. Antibiot. 36:961–964.

Nara, T., I. Kawamoto, R. Okachi, and T. Oka. 1977. Source of antibiotics other than *Streptomyces.* Japanese J. Antibiot. Suppl. 30:174–189.

Nolan, R. D., and T. Cross. 1988. Isolation and screening of actinomycetes, p. 1–32. In: M. Goodfellow, S. T. Williams, M. Mordarski (ed.), Actinomycetes in biotechnology. Academic Press, London.

Nonomura, H., and M. Hayakawa. 1988. New methods for the selective isolation of soil actinomycetes, p. 288–293. In: Y. Okami, T. Beppu, and H. Ogawara (ed.), Biology of actinomycetes '88. Japan Sci. Soc. Press, Tokyo.

Nonomura, H., and Y. Ohara. 1969. Distribution of actinomycetes in soil (VI) A culture method effective for both preferential isolation and enumeration of *Micro-*

bispora and *Streptosporangium* strains in soil (part I). J. Ferment. Technol. 47:463–469.

Nonomura, H., and S. Takagi. 1977. Distribution of actinoplanetes in soils of Japan. J. Ferment. Technol. 55:423–428.

Nonomura, H., S. Iino, and M. Hayakawa. 1979. Classification of actinomycetes of genus *Ampullariella* from soils of Japan. Hakkokogaku 57:79–85.

O'Donnell, A. G., M. Goodfellow, and D. E. Minnikin. 1982. Lipids in the classification of *Nocardioides:* reclassification of *Arthrobacter simplex* (Jensen) Lochhead in the genus *Nocardioides* (Prauser) emend. O'Donnell et al. as *Nocardioides simplex* comb. nov. Arch. Microbiol. 133:323–329.

Okami, Y., and K. Hotta. 1988. Search and discovery of new antibiotics, p. 33–67. In: M. Goodfellow, S. T. Williams, and M. Mordarski (ed.), Actinomycetes in biotechnology. Academic Press, London.

Okami, Y., and T. Okazaki. 1972. Studies on marine microorganisms. I. Isolation from the Japan sea. J. Antibiot. 25:456–460.

Okami, Y., and T. Okazaki. 1978. Actinomycetes in marine environments. Zentralbl. Bakteriol. Parasitenkd. Infektionskr. Hyg. Abt. 1, Suppl. 6:145–151.

Okami, Y., and M. Suzuki. 1958. A simple method for microscopical observation of streptomycetes and critique of *Streptomyces* grouping with reference to aerial structure. J. Antibiot. 11:250–253.

Okazaki, T., and Y. Okami. 1972. Studies on marine microorganisms. II. Actinomycetes in Sagami Bay and their antibiotics substances. J. Antibiot. 25:461–466.

Ørskov, J. 1923. Investigations into the morphology of the ray fungi. Levin & Munksgaard Publishers, Copenhagen.

Palleroni, N. J. 1976. Chemotaxis in *Actinoplanes.* Arch. Microbiol. 110:13–18.

Palleroni, N. J. 1979. New species of the genus *Actinoplanes, Actinoplanes ferrugineus.* Int. J. Syst. Bacteriol. 29:51–55.

Palleroni, N. J. 1980. A chemotactic method for the isolation of Actinoplanaceae. Arch. Microbiol. 128:53–55.

Palleroni, N. J. 1983. Biology of *Actinoplanes.* The Actinomycetes 17:46–65.

Palleroni, N. J. 1989. Genus *Actinoplanes* Couch, p. 2419–2428. In: S. T. Williams (ed.), Bergey's manual of systematic bacteriology, vol. 4. Williams & Wilkins, Baltimore.

Parenti, F., G. Beretta, M. Berti, and V. Arioli. 1978. Teichomycins, new antibiotics from *Actinoplanes teichomyceticus* nov. sp. I. Description of the producer strain, fermentation studies and biological properties. J. Antibiot. 31:276–283.

Parenti, F., and C. Coronelli. 1979. Members of the genus *Actinoplanes* and their antibiotics. Ann. Rev. Microbiol. 33:389–411.

Parenti, F., H. Pagani, and G. Beretta. 1975. Lipiarmycin, a new antibiotic from *Actinoplanes* I. Description of the producer strain and fermentation studies. J. Antibiot. 28:247–252.

Patel, M., V. P. Gullo, V. R. Hedge, A. C. Horan, J. A. Marquez, R. Vaughan, M. S. Puar, and G. H. Miller. 1987. A new tetracyclone antibiotic from a *Dactylosporangium* species. J. Antibiot. 40:1414–1418.

Peczyńska-Czoch, W., and M. Mordarski. 1988. Actinomycete enzymes, p. 219–283. In: M. Goodfellow, S. T.

Williams, and M. Mordarski (ed.), Actinomycetes in biotechnology. Academic Press, London.

Rowbotham, T. J., and T. Cross. 1977. Ecology of *Rhodococcus coprophilus* and associated actinomycetes in fresh water and agricultural habitats. J. Gen. Microbiol. 100:231–240.

Ruan, J., M. P. Lechevalier, C. Jiang, and H. A. Lechevalier. 1986. A new species of the genus *Actinoplanes: Actinoplanes minutisporangius* n. sp. The Actinomycetes 19:163–175.

Ruddick, S. M., and S. T. Williams. 1972. Studies on the ecology of actinomycetes in soil V. Some factors influencing the dispersal and adsorption of spores in soil. Soil Biol. Biochem. 4:93–103.

Saari, G. C., A. A. Kumar, G. H. Kawasaki, M. Y. Insley, and P. J. O'Hara. 1987. Sequence of the *Ampullariella* sp. strain 3876 gene coding for xylose isomerase. J. Bacteriol. 169:612–618.

Sandrak, N. A. 1977. Degradation of cellulose by micromonosporas. Mikrobiologiya 46:478–481.

Schäfer, D. 1973. Beiträge zur Klassifizierung und Taxonomie der Actinoplanaceen. Dissertation, Marburg.

Sharples, G. P., and S. T. Williams. 1974. Fine structure of the globose bodies of *Dactylosporangium thailandense (Actinomycetales).* J. Gen. Microbiol. 84:219–222.

Shearer, M. C. 1987. Methods for the isolation of non-streptomycete actinomycetes. J. Indus. Microbiol. Suppl. 2:91–97.

Shirling, E. B, and D. Gottlieb. 1966. Methods for characterization of *Streptomyces* species. Int. J. Syst. Bacteriol. 16:313–340.

Shomura, T., S. Amano, H. Tohyama, J. Yoshida, T. Ito, and T. Niida. 1985. *Dactylosporangium roseum* sp. nov. Int. J. Syst. Bacteriol. 35:1–4.

Shomura, T., S. Amano, J. Yoshida, and M. Kojima. 1986. *Dactylosporangium fulvum* sp. nov. Int. J. Syst. Bacteriol. 36:166–169.

Shomura, T., M. Kojima, J. Yoshida, M. Itó, S. Amano, K. Totsugawa, T. Niwa, S. Inouye, T. Itó, and T. Niida. 1980. Studies on a new aminoglycoside antibiotic, dactimicin I. Producing organism and fermentation. J. Antibiot. 33:924–930.

Shomura, T., J. Yoshida, S. Miyadoh, T. Ito, and T. Niida. 1983. *Dactylosporangium vinaceum* sp. nov. Int. J. Syst. Bacteriol. 33:309–313.

Sneh, B., S. J. Humble, and J. L. Lockwood. 1977. Parasitism of oospores of *Phytophthora megasperma* var. *sojae, P. cactorum, Pythium* sp. and *Aphanomyces euteiches* in soil by oomycetes, chytridiomycetes, hyphomycetes, actinomycetes and bacteria. Phytopathology 67:622–628.

Stackebrandt, E., and R. M. Kroppenstedt. 1987. Union of the genera *Actinoplanes* Couch, *Ampullariella* Couch, and *Amorphosporangium* Couch in a redefined Genus *Actinoplanes.* Syst. Appl. Microbiol. 9:110–114.

Stackebrandt, E., and K-H. Schleifer. 1984. Molecular systematics of actinomycetes and related organisms, p. 485–504. In: L. Ortiz-Ortiz, L. F. Bojalil, and V. Yakoleff (ed.), Biological, biochemical, and biomedical aspects of actinomycetes. Academic Press, Orlando.

Stackebrandt, E., and C. R. Woese. 1981. The evolution of prokaryotes, p. 1–31. In: M. J. Carlile, J. F. Collins, and B. E. B. Moseley (ed.), Molecular and cellular aspects of microbial evolution. Cambridge University Press, Cambridge.

Sutherland, E. D., K. K. Baker, and J. L. Lockwood. 1984. Ultrastructure of *Phytophthora megasperma* f. sp. *glycinea* oospores parasitized by *Actinoplanes missouriensis* and *Humicola fuscoatra*. Trans. Br. Mycol. Soc. 82:726–729.

Szaniszlo, P. J. 1968. The nature of the intramycelial pigmentation of some Actinoplanaceae. J. Elisha Mitchell Sci. Soc. 84:24–26

Terekhova, L. P., O. A. Sadikova, and T. P. Preobrazhenskaya. 1977. *Actinoplanes cyaneus* sp. nov. and its antagonistic properties. Antibiotiki 22:1059–1063.

Thiemann, J. E. 1970. Study of some new genera and species of the Actinoplanaceae, p. 245–257. In: H. Prauser (ed.), The Actinomycetales. VEB Gustav Fischer Verlag, Jena.

Thiemann, J. E., H. Pagani, and G. Beretta. 1967. A new genus of the Actinoplanaceae: *Dactylosporangium,* gen. nov. Arch. Mikrobiol. 58:42–52.

Tille, D., R. Vettermann, and H. Prauser. 1982. DNA-DNA reassociation between DNAs of actinoplanetes and related genera, p. 182. Fifth Int. Sym. Actino. Biol., Abstracts. Oaxtepec, México.

Tomita, K., S. Kobaru, M. Hanada, and H. Tsukiura. 1977. Fermentation process. U.S. Patent 4,026,766, May 31, 1977.

Torikata, A., R. Enokita, H. Imai, Y. Itoh, M. Nakajima, T. Haneishi, and M. Arai. 1978. Studies on the antibiotics from genus *Actinoplanes*. I. Taxonomy of the producers of three antibiotics and their isolation and identification with azaserine, 5-azacytidine and actinomycins. Ann. Rep. Sankyo Res. Lab. 30:84–97.

Torikata, A., R. Enokita, T. Okazaki, M. Nakajima, S. Iwado, T. Haneishi, and M. Arai. 1983. Mycoplanecins, novel antimycobacterial antibiotics from *Actinoplanes awajinensis* subsp. *mycoplanecinus* subsp. nov. I. Taxonomy of producing organism and fermentation. J. Antibiot. 36:957–960.

Tribe, H. T., and S. M. Abu El-Souod. 1979. Colonization of hair in soil-water cultures, with especial reference to the genera *Pilimelia* and *Spirillospora* (Actinomycetales). Nova Hedwigia 31:789–805.

Truscheit, E., W. Frommer, B. Junge, L. Müller, D. D. Schmidt, and W. Wingender. 1981. Chemie und Biochemie mikrobieller α-Glucosidasen-Inhibitoren. Angew. Chem. 93:738–755.

Uchida, K., and K. Aida. 1977. Acyl type of bacterial cell wall: its simple identification by colorimetric method. J. Gen. Appl. Microbiol. 23:249–260.

Van Zyl, W. H. 1985. A study of the cellulases produced by three mesophilic actinomycetes grown on bagasse as substrate. Biotechnol. Bioeng. 27:1367–1373.

Vettermann, R., and H. Prauser. 1979. Comparative studies on the isolation of actinoplanetes. Fourth Int. Sym. Actino. Biol., Poster-presentation, Cologne.

Vitiello, F., and J. P. Zanetta. 1978. Thin-layer chromatography of phospholipids. J. Chromatogr. 166:637–640.

Vobis, G. 1984. Sporogenesis in the *Pilimelia* species, p. 423–439. In: L. Ortiz-Ortiz, L. F. Bojalil, and V. Yakoleff (ed.), Biological, biochemical, and biomedical aspects of actinomycetes. Academic Press, Orlando.

Vobis, G. 1987. Sporangiate Actinoplaneten, Actinomycetales mit aero-aquatischem Lebenszyklus. Forum Mikrobiol. 11:416–424.

Vobis, G. 1989a. Actinoplanetes, p. 2418–2419. In: S. T. Williams (ed.), Bergey's manual of systematic bacteriology, vol. 4. Williams & Wilkins, Baltimore.

Vobis, G. 1989b. Genus *Pilimelia* Kane, p. 2433–2437. In: S. T. Williams (ed.), Bergey's manual of systematic bacteriology, vol. 4. Williams & Wilkins, Baltimore.

Vobis, G. 1989c. Genus *Dactylosporangium* Thiemann, Pagani and Beretta, p. 2437–2442. In: S. T. Williams (ed.), Bergey's manual of systematic bacteriology, vol. 4. Williams & Wilkins, Baltimore.

Vobis, G., and H.-W. Kothe. 1985. Sporogenesis in sporangiate actinomycetes, p. 25–47. In: K. G. Mukerji, N. C. Pathak, and V. P. Singh (ed.), Frontiers in applied microbiology, vol. 1. Print House (India), Lucknow.

Vobis, G., and H.-W. Kothe. 1989. Genus *Ampullariella* Couch, p. 2429–2433. In: S. T. Williams (ed.), Bergey's manual of systematic bacteriology, vol. 4. Williams & Wilkins, Baltimore.

Vobis, G., D. Schäfer, H.-W. Kothe, and B. Renner. 1986. Descriptions of *Pilimelia columellifera* (ex Schäfer 1973) nom. rev. and *Pilimelia columellifera* subsp. *pallida* (ex Schäfer 1973) nom. rev. Syst. Appl. Microbiol. 8:67–74.

Wagman, G. H., and M. J. Weinstein. 1980. Antibiotics from *Micromonospora*. Ann. Rev. Microbiol. 34:537–557.

Wakisaka, Y., Y. Kawamura, Y. Yasuda, K. Koizumi, and Y. Nishimoto. 1982. A selective isolation procedure for *Micromonospora*. J. Antibiot. 35:822–836.

Waksman, S. A. 1961. The actinomycetes, vol. 2. Williams & Wilkins, Baltimore.

Watson, E. T., and S. T. Williams. 1974. Studies on the ecology of actinomycetes in soil—VII. Actinomycetes in a coastal sand belt. Soil Biol. Biochem. 6:43–52.

Weyland, H. 1969. Actinomycetes in North Sea and Atlantic Ocean sediments. Nature 223:858.

Weyland, H. 1981. Distribution of actinomycetes on the sea flor. Zentralbl. Bacteriol. Microbiol. Hyg. Abt. 1. Suppl. 11:185–193.

Willoughby, L. G. 1966. A conidial *Actinoplanes* isolate from Blelham Tarn. J. Gen. Microbiol. 44:69–72.

Willoughby, L. G. 1968. Aquatic Actinomycetales with particular reference to the Actinoplanaceae. Veroeff. Inst. Meeresforsch. Bremerh. Sonderband 3:19–26.

Willoughby, L. G. 1969a. A study on aquatic actinomycetes—the allochthonous leaf component. Nova Hedwigia 18:45–113.

Willoughby, L. G. 1969b. A study of the aquatic actinomycetes of Blelham Tarn. Hydrobiologia 34:465–483.

Willoughby, L. G. 1971. Observations on some aquatic actinomycetes of streams and rivers. Freshwater Biol. 1:23–27.

Willoughby, L. G., C. D. Baker, and S. E. Foster. 1968. Sporangium formation in the Actinoplanaceae induced by humic acids. Experientia 24:730–731.

Yaginuma, S., N. Muto, and M. Otani. 1979. A-10947, a new peptide antibiotic from *Actinoplanes*. J. Antibiot. 32:967–969.

Yaginuma, S., N. Muto, M. Tsujino, Y. Sudate, M. Hayashi, and M. Otani. 1981. Studies on Neplanocin A, new antitumor antibiotic. I. Producing organism, isolation and characterization. J. Antibiot. 34:359–366.

CHAPTER 44

The Genus *Saccharothrix*

DAVID P. LABEDA

The genus *Saccharothrix* contains nocardio-form actinomycetes that lack mycolic acids but have the *meso*-isomer of diaminopimelic acid in their cell walls and rhamnose and galactose as diagnostic whole-cell sugars. Aerial mycelia are produced by most strains, and both substrate and aerial hyphae fragment into ovoid nonmotile elements. The genus is taxonomi-cally very similar to the genus *Nocardiopsis,* which it resembles morphologically, but it can be distinguished by chemotaxonomic criteria since *Nocardiopsis* strains lack rhamnose as a diagnostic sugar in cell hydrolysates. When the genus was first described (Labeda et al., 1984), it consisted of only one species, *Saccharothrix australiensis*. The genus, as it is currently de-

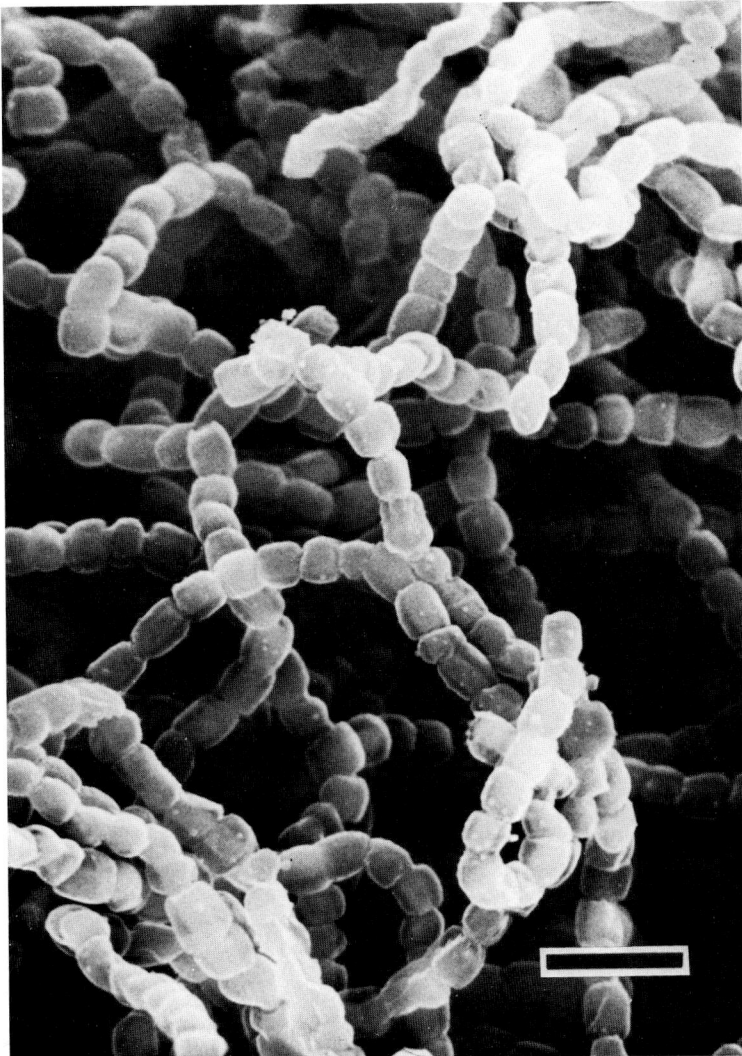

Fig. 1. Scanning electron micrograph of vegetative hyphae of a 28-day culture of *Saccharothrix australiensis* on ATCC Medium No. 172. Note fragmentation of mycelium into ovoid elements. Bar = 2 μm. (From Labeda et al., 1984.)

scribed (Labeda and Lechevalier, 1989; Grund and Kroppenstedt, 1989), consists of eleven described species and one subspecies, representing several new isolates as well as eight species transferred into the genus *Saccharothrix* from morphologically similar genera, i.e., *Saccharothrix aerocolonigenes* and *Saccharothrix mutabilis* subsp. *capreolus,* transferred from the genus *Nocardia;* and *Saccharothrix mutabilis, Saccharothrix cryophilis, Saccharothrix coeruleofusca, Saccharothrix flava, Saccharothrix longispora,* and *Saccharothrix syringae,* transferred from the genus *Nocardiopsis.* It has recently been proposed that the genus *Saccharothrix* be included in the family *Pseudonocardiaceae* based on a comparison of 16S rRNA sequence data from *Saccharothrix* with other members of this family (Bowen et al., 1989).

Habitats

The original strain of *Saccharothrix australiensis,* the type species of the genus, was isolated from a soil sample from Australia. The genus now appears to be ubiquitous in soil and has a worldwide distribution. Other described isolates of this genus have come from soil samples collected in the United States, Japan, Panama, Africa, and the Soviet Union.

Isolation

Selective Media

Many of the strains of *Saccharothrix* isolated from soils have been obtained on relatively routine selective media used for the general isolation of actinomycetes, such as 1.5% crude agar and 0.4% casein hydrolysate in tap water. The use of antibiotics to selectively isolate members of this genus has also been reported recently (Shearer, 1985; Takahashi et al., 1986).

Typical actinomycete isolation media, such as AV (arginine-vitamin) agar or starch-casein agar, amended with a combination of penicillin G (5–10 μg/ml) and nalidixic acid (15 μg/ml), have been used to selectively isolate *Saccharothrix* strains (Shearer, 1985).

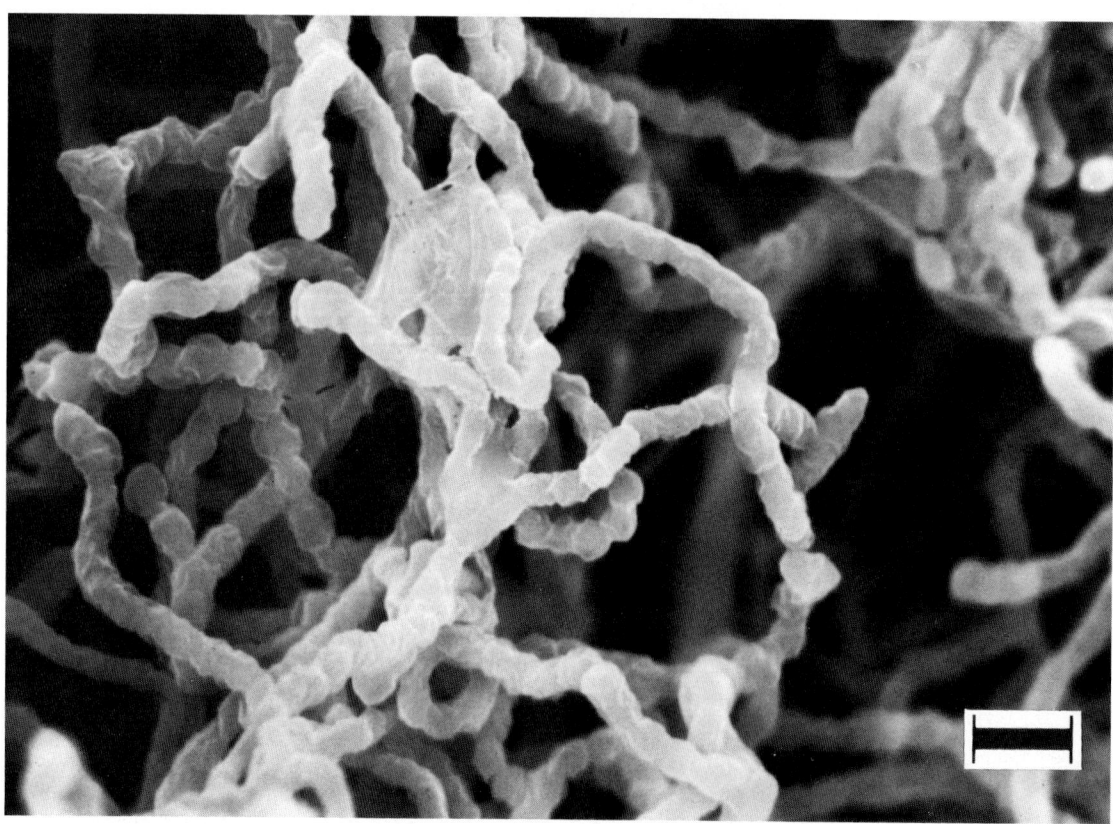

Fig. 2. Scanning electron micrograph of aerial hyphae of a 14-day culture of *Saccharothrix australiensis* on ATCC Medium No. 172. The aerial mycelium has fragmented into ovoid "spores" and demonstrates typical "zig-zag" morphology. Bar = 1 μm. (From Labeda et al., 1984.)

AV Agar (Nonomura and Ohara, 1969)

AV agar consists of the following in grams per liter of distilled water:

L-arginine	0.3
Glucose	1.0
Glycerol	1.0
K_2HPO_4	0.3
$MgSO_4 \cdot 7H_2O$	0.2
NaCl	0.3
Thiamine hydrochloride	0.5 mg
Riboflavin	0.5 mg
Niacin	0.5 mg
Pyridoxine hydrochloride	0.5 mg
Inositol	0.5 mg
Calcium pantothenate	0.5 mg
p-amino-benzoic acid	0.5 mg
Biotin	0.25 mg
$Fe_2(SO_4)_3$	10.0 mg
$CuSO_4 \cdot 5H_2O$	1.0 mg
$MnSO_4 \cdot H2O$	1.0 mg
$ZnSO_4 \cdot 7H_2O$	1.0 mg
Agar	15.0

The pH of this medium is adjusted to 6.4 prior to autoclaving. The vitamins should be filter sterilized and added to the sterilized, tempered medium just prior to dispensing.

Starch-Casein Agar (Kuster and Williams, 1964)

Starch-casein agar consists of the following in grams per liter of distilled water:

Soluble starch	10.0
KNO_3	2.0
Casein (vitamin free)	0.3
K_2HPO_4	2.0
$MgSO_4 \cdot 7H_2O$	0.05
NaCl	2.0
$FeSO_4 \cdot 7H_2O$	0.01
$CaCO_3$	0.02
Agar	18.0

The pH of this medium is adjusted to 7.0 to 7.2 prior to autoclaving.

The antibiotic dopsisamine-producing strain of *S. cryophilis* was isolated from soil using 4PC medium amended with the aminoglycoside antibiotics istamycin B and sisomycin at 20 μg/ml (Takahashi et al., 1986).

Medium 4 PC (Takahashi et al., 1986)

Medium 4PC consists of one volume of International Streptomyces Project medium ISP-4 (Shirling and Gottlieb, 1966), yeast extract-malt extract agar, and two volumes of potato-carrot extract. It consists of the following in grams per liter:

Yeast extract	1.3
Malt extract	3.3
Glucose	1.3
Agar	20.0
Potato-carrot extract	667 ml
Distilled water	to 1 lt

Potato-carrot extract is prepared by boiling 30 g of potatoes and 2.5 g of carrots in 1 liter of water for 30 minutes and then filtering.

Cultivation

Saccharothrix strains are rather nonfastidious in their growth requirements and can be cultivated on a range of standard media typically used for actinomycetes, such as the media described by Pridham et al. (1957) and by the International Streptomyces Project (Shirling and Gottlieb, 1966): inorganic salts-starch agar (ISP-2), yeast extract-malt extract agar (ISP-3), or glycerol-asparagine agar.

Two excellent media for the cultivation of *Saccharothrix* strains are ATCC Medium No. 172 (Cote et al., 1984) and ATCC Medium No. 174 (Cote et al., 1984).

Medium No. 172

The formulation of ATCC Medium No. 172 consists of the following in grams per liter:

NZamine Type A	5
Yeast extract	5
Glucose	10
Soluble starch	20
Calcium carbonate	1

Table 1. Chemotaxonomic properties of *Saccharothrix* and other nocardioform actinomycetes that lack mycolic acids.

Genus	Cell wall type	Whole-cell sugars	Phospholipid pattern	Principal menaquinones	GC content (mol%)
Actinomadura	III[a]	Madurose	PI[b]	MK-9(H_4,H_6)	66–70
Amycolata	IV	Arabinose, galactose	PIII	MK-8(H_4)	68–71
Amycolatopsis	IV	Arabinose, galactose	PII	MK-9(H_4)	66–69
Nocardioides	I	None	PI	MK-8(H_4)	66
Nocardiopsis	III	None	PIII	MK-10(H_4,H_6)	64–69
Saccharothrix	III	Rhamnose, galactose	PII or PIV	MK-9(H_4)	70–76

[a]Diagnostic constituents of cell wall types: Type I, LL-diaminopimelic acid (A_2pm); Type III, *meso*-A_2pm; Type IV, *meso*-A_2pm, arabinose, and galactose.
[b]Diagnostic constituents of phospholipid patterns: PI, no nitrogenous phospholipids; PII, phosphatidylethanolamine; PIII, phosphatidylcholine; PIV, phosphatidylcholine and/or glucoseamine-containing phospholipids.

Table 2. Physiological properties of *Saccharothrix* species.

	S. australiensis[d] (NRRL 11239)	*S. aerocolonigenes*[e] (NRRL B-3298)	*S. coeruleofuscus*[i] (DSM 43679)	*S. cryophilis*[g,i] (NRRL B-16238)	*S. espanaensis*[g] (NRRL 15764)	*S. flava*[j] (DSM 43885)	*S. longispora*[j] (DSM 43749)	*S. mutabilis*[g,h] (NRRL B-16077)	*S. mutabilis* sbsp. *capreolus*[i] (DSM 40225)	*S. syringae*[j] (DSM 43886)	*S. texasensis*[f] (NRRL B-16134)	*S. waywayandensis*[f] (NRRL B-16159)
Decomposition of												
Adenine	−	−	−	−	−	−	+	−	−	−	+	−
Allantoin	−	−	ND[a]	−	−	ND	ND	−	ND	ND	V[b]	−
Casein	+	+	+	−	+	−	+	+	ND	+	+	+
Hippurate	−	−	ND	−	+	ND	ND	+	ND	ND	+	+
Hypoxanthine	−	+	−	−	+	+	−	+	+	−	+	+
Potato starch	−	+	+	−	−	ND	+	+	ND	+	+	+
Tyrosine	+	+	−	+	−	+	+	+	+	+	+	+
Urea	−	V	ND	+	−	ND	ND	−	ND	ND	V	+
Acid from												
Adonitol	−	−	−	−	−	−	−	−	−	−	−	+
Arabinose	−	+	+	−	−	+	+	+	+	+	+	+
Cellobiose	+	+	+	−	+	+	+	+	+	+	+	+
Dextrin	+	+	+	−	−	+	+	+	+	+	+	+
Erythritol	+	−	ND	−	−	ND	ND	−	ND	ND	−	+
Glycerol	+	+	ND	W[c]	+	ND	ND	+	ND	ND	+	+
Inositol	−	+	−	+	−	+	−	+	+	−	+	+
Lactose	−	+	+	+	−	+	+	+	−	+	+	+
Maltose	+	+	ND	+	V	ND	ND	+	ND	ND	+	+
Melibiose	−	+	−	−	−	+	−	+	+	+	+	+
Raffinose	−	+	+	W	−	+	−	+	−	+	−	+
Rhamnose	−	+	+	−	−	+	+	−	−	+	+	+
Salicin	−	+	ND	−	−	ND	ND	+	ND	ND	+	+
Sorbitol	+	−	+	−	−	−	−	−	−	−	−	−
Sucrose	−	+	+	−	+	+	+	+	−	+	+	+
Xylose	−	+	+	−	V	−	+	+	+	+	+	+
α-Methyl-D-glucoside	−	V	ND	−	−	ND	ND	+	ND	ND	+	−
β-Methyl-D-xyloside	−	V	ND	−	−	ND	ND	ND	ND	ND	V	−

continued

Agar   15

The pH of this medium is adjusted to 7.3 prior to autoclaving. Sigma casein enzymatic hydrolysate type II is a good substitute for NZamine Type A.

ATCC Medium No. 174

The formulation for Bennett's Medium, ATCC Medium No. 174, consists of the following in grams per liter:

Yeast extract	
Beef extract	1
	1
NZamine Type A	2
Glucose	
Agar	10
	15

Adjust the pH of the medium to 7.3 prior to autoclaving. Sigma casein enzymatic hydrolysate type II is a good substitute for NZamine Type A.

All species are aerobic and have a growth temperature optimum around 28°C, except *S. cryophilis* which grows best around 25°C.

Identification

Morphology

All strains of *Saccharothrix* are quite similar in morphological appearance. Both substrate and aerial hyphae are approximately 0.5 to 0.7 μm

Table 2. *Continued*

	S. australiensis[d] (NRRL 11239)	S. aerocolonigenes[e] (NRRL B-3298)	S. coeruleofuscus[j] (DSM 43679)	S. cryophilis[g,i] (NRRL B-16238)	S. espanaensis[g] (NRRL 15764)	S. flava[j] (DSM 43885)	S. longispora[j] (DSM 43749)	S. mutabilis[g,h] (NRRL B-16077)	S. mutabilis sbsp. capreolus[j] (DSM 40225)	S. syringae[j] (DSM 43886)	S. texasensis[f] (NRRL B-16134)	S. waywayandensis[f] (NRRL B-16159)
Utilization of												
Citrate	−	+	−	−	V	−	+	+	−	−	−	+
Lactate	V	+	−	+	+	−	+	+	−	−	+	+
Malate	+	+	−	−	+	+	+	+	+	+	+	+
Oxalate	−	+	ND	+	−	ND	ND	−	ND	ND	−	+
Propionate	+	+	ND	+	V	ND	ND	+	ND	ND	V	+
Tartrate	−	+	ND	−	−	ND	ND	−	ND	ND	−	−
Production of												
Soluble pigments	+	−	−	−	−	−	−	−	−	+	−	−
Nitrate reductase	+	+	−	+	W	−	+	+	ND	−	+	−
Phosphatase	−	+	ND	+	+	ND	ND	+	ND	ND	+	+
Growth in presence of												
4% NaCl	+	+	ND	+	+	ND	ND	−	ND	ND	−	+
5% NaCl	−	W	+	−	−	+	+	−	+	+	−	+
Growth at												
37°C	+	+	ND	−	+	ND	ND	+	ND	ND	+	+
42°C	+	V	ND	−	+	ND	ND	+	ND	ND	+	−
45°C	+	−	ND	−	−	ND	ND	+	ND	ND	−	−

[a]ND = test not done.
[b]V = variable response.
[c]W = weak response.
[d]From Labeda et al., 1984.
[e]From Labeda, 1985.
[f]From Labeda and Lyons, 1989.
[g]From Labeda and Lechevalier, 1989.
[h]NRRL B-16077 = ATCC 31520 = SKF-AAA14475.
[i]NRRL B-16238 = IFO 14475.
[j]From Grund and Kroppenstedt, 1989.

in diameter and fragment into ovoid bacillary units typical of nocardioform actinomycetes (Fig. 1). The substrate mycelium of *Saccharothrix* species ranges from yellow to yellowish-brown in color, and the aerial mycelium tends to be white to yellowish-white or light gray in color, except for *S. coeruleofusca* and *S. longispora,* which produce blue aerial mycelia on glycerol-nitrate agar: *S. mutabilis* is photochromogenic and tends to produce orange-yellow aerial mycelia on many media when cultivated in the light, while the aerial mycelia are white when cultured in the dark. Soluble pigments are produced by very few species (e.g., *S. australiensis* and *S. syringae*). Aerial hyphae often display the "zig-zag" morphology, which is also typical of the genus *Nocardiopsis* (Fig. 2), thus making differentiation between these genera rather difficult based on morphological criteria. Aerial mycelium is best observed on minimal media such as 1.5% crude agar in tap water or Czapek's solution agar (Pridham and Lyons, 1980).

Chemotaxonomy

Determination of the chemotaxonomic profile of strains, particularly by use of cell wall type and whole-cell sugar pattern (*sensu* Lechevalier and Lechevalier, 1970), is absolutely essential for the identification of strains as members of the genus *Saccharothrix*. The chemotaxonomic characteristics of the typical nocardioform genera is shown in Table 1. Note that all *Saccharothrix* strains contain the *meso*-isomer of diaminopimelic acid (type III cell wall) and diagnostic quantities of rhamnose and galactose in whole-cell hydrolysates. The phospholipid profile of *Saccharothrix* strains is either a PII or PIV pattern *sensu* Lechevalier et al. (1977) and consists of phosphatidylethanolamine either with (for PIV) or without (for PII) glucosamine-containing phospholipids as the diagnostic constituents. Phosphatidylmethylethanolamine may be variably present as a diagnostic phospholipid, but phosphatidylcholine is not present in *Saccharothrix* strains. The described species *S. australiensis, S. aerocolonigenes, S. texasensis,* and *S. waywayandensis* have a PII phospholipid pattern, while *S. espanaensis, S. mutabilis,* and *S. cryophilis* have a PIV phospholipid pattern. The menaquinone profile of members of this genus consists predominantly of tetrahydrogenated menaquinones with nine isoprenoid repeating units called MK-9(H$_4$) (Labeda et al., 1984; Labeda and Lechevalier, 1989).

The chemotaxonomic properties of strains can be determined following the detailed procedures outlined by Lechevalier and Lechevalier (1980) and the thin-layer chromatography procedures suggested by Staneck and Roberts (1974) and Meyertons et al. (1988) for isomers of diaminopimelic acid and whole-cell sugars, respectively. Menaquinone analyses can be performed using the procedures outlined by Collins et al. (1977), Kroppenstedt (1982), and Tamaoka et al. (1983).

Physiological Properties

The physiological properties of *Saccharothrix* are best evaluated using the methods described by Gordon et al. (1974), Kurup and Schmitt (1973), and Goodfellow (1971). The described species of this genus can be clearly distinguished based on physiological properties as shown in Table 2. All species share some physiological properties, such as resistance to lysozyme, growth in the presence of 1.5% NaCl, and production of acid from both fructose and glucose.

Molecular Systematic Studies

Extraction of nucleic acids from cells of *Saccharothrix* is made more difficult by the lysozyme-resistance characteristic of the genus. Cells can be successfully disrupted, however, by passage through a French pressure cell at 10,000 p.s.i. or by freeze-drying whole cells and grinding them in a mortar and pestle. The DNA can be then extracted and purified by the method of Marmur (1961). The GC content of the DNA of *Saccharothrix* species ranges from 70 to 76 mol% as determined from the midpoint of thermal denaturation (T$_m$) by the method of Marmur and Doty (1962). The DNA relatedness of *Saccharothrix* strains has been determined (Labeda, 1988; Labeda and Lechevalier, 1989) from $C_0t_{0.5}$ values for renaturation in 5X standard saline citrate (SSC) (1X SSC is 0.15 M NaCl and 0.015 M Na$_3$citrate) and 20% dimethylsulfoxide at T$_m$ $-25°C$ (66°C) using the method of Seidler and Mandel (1971) and Seidler et al. (1975) as modified by Kurtzman et al (1980). Although this genus is well defined based on morphological and chemotaxonomic criteria, very low DNA relatedness is generally observed between species in the genus. *S. aerocolonigenes* NRRL B-3298 exhibits 34–44% DNA relatedness to *S. waywayandensis* NRRL B-16158 and NRRL B-16159, which may indicate that these are sibling species.

Culture Preservation

Saccharothrix strains can be maintained on slants of yeast extract-malt extract agar (Shirling and Gottlieb, 1966), ATCC Medium No. 174 (Bennett's agar) (Cote et al., 1984), or ATCC Medium No. 172 (Cote et al., 1984) with transfer every 2 months and storage at 4°C between transfers. Long-term storage of strains is best achieved by lyophilization of mycelial suspen-

Table 3. Antibiotics produced by *Saccharothrix* strains.

Species	Strain no.	Antibiotic	Reference
S. australiensis	NRRL 11239	BM782 complex	(Tresner et al., 1980)
S. aerocolonigenes	ATCC 39243	Rebeccamycin	(Bush et al., 1987)
S. cryophilis	IFO 14475	Dopsisamine	(Takahashi et al., 1986)
S. espanaensis	NRRL 15764	LL-C19004	(Kirby et al., 1987)
S. flava	INA 2171	Madumycin	(Gauze et al., 1974)
S. mutabilis	ATCC 31520	Polynitroxin	(Jain et al., 1982)
S. syringae	INA 2240	Nocamycin	(Gauze et al., 1977)

sions in sterile beef serum, sucrose and gelatin, or skim milk, with subsequent storage of ampules at −20°C to 4°C. Strains can also be stored frozen for shorter periods of time as stationary-phase broth cultures or mycelial suspensions in 20% glycerol at −70°C to −20°C.

Applications

New strains of the genus *Saccharothrix* that may be isolated from nature in the future should have considerable biotechnological potential as sources of new and useful secondary metabolities in natural products screening programs. The biosynthetic versatility of members of this genus is indicated by the list of antibiotics produced by *Saccharothrix* species (Table 3).

Literature Cited

Bowen, T., E. Stackebrandt, M. Dorsch, and T. M. Embley. 1989. The phylogeny of *Amycolata autotrophica, Kibdelosporangium aridum,* and *Saccharothrix australiensis.* J. Gen. Microbiol. 135:2529–2536.

Bush, J. A., B. H. Long, J. J. Catino, W. T. Bradner, and K. Tomita. 1987. Production and biological activity of rebeccamycin, a novel antitumor agent. J. Antibiot. 40:668–678.

Collins, M. D., T. Pirouz, M. Goodfellow, and D. E. Minnikin. 1977. Distribution of menaquinones in actinomycetes and corynebacteria. J. Gen. Microbiol. 100:221–230.

Cote, R., P. -M. Daggett, M. J. Gantt, R. Hay, S. -C. Hay, and P. Pienta. 1984. ATCC Media Handbook. 1st ed. American Type Culture Collection, Rockville, Maryland.

Gauze, G. F., T. S. Maksimova, O. L. Olkhovatova, M. A. Sveshnikova, G. V. Kochetkova, and G. B. Ilchenko. 1974. Production of madumycin, an antibacterial antibiotic, by *Actinomadura flava* sp. nov. Antibiotiki 9:771–775.

Gauze, G. F., M. A. Sveshnikova, R. S. Ukholina, G. N. Komorova, and V. S. Bashanov. 1977. Production of nocamycin, a new antibiotic, by *Nocardiopsis syringae* sp. nov. Antibiotiki 22:483–486.

Goodfellow, M. 1971. Numerical taxonomy of some nocardioform bacteria. J. Gen. Microbiol. 69:33–80.

Gordon, R. A., D. A., Barnett, J. E. Handerhan, and C. H. Pang. 1974. *Nocardia coeliaca, Nocardia autotrophica,* and the nocardin strain. Int. J. Syst. Bacteriol. 24:54–63.

Grund, E., and R. M. Kroppenstedt. 1989. Transfer of five *Nocardiopsis* species to the genus *Saccharothrix* Labeda et al. 1984. Syst. Appl. Microbiol. 10:267–274.

Jain, T. K. 1982. Polynitroxin antibiotics produced by *Nocardiopsis mutabilis* Shearer, sp. nov. U.S. Patent 4,317,812.

Kroppenstedt, R. M. 1982. Separation of bacterial menaquinones by HPLC using reverse phase (RP18) and a silver loaded ion exchanger as stationary phases. J. Liq. Chromatogr. 5:2357–2359.

Kurtzman, C. P., M. J. Smiley, C. J. Johnson, L. J. Wickerham, and G. B. Fuson. 1980. Two new and closely related heterothallic species, *Pichia amylophila* and *Pichia mississippiensis:* Characterization by hybridization and deoxyribonucleic acid reassociation. Int. J. Syst. Bacteriol. 30:208–216.

Kurup, P. V., and J. A. Schmitt. 1973. Numerical taxonomy of *Nocardia.* Can. J. Microbiol. 19:1035–1048.

Kuster, E., and S. T. Williams. 1964. Selective media for the isolation of streptomycetes. Nature 202:928–929.

Labeda, D. P. 1986. Transfer of *"Nocardia aerocolonigenes"* (Shinobu and Kawato 1960) Pridham 1970 into the genus *Saccharothrix* Labeda, Testa, Lechevalier, and Lechevalier 1984 as *Saccharothrix aerocolonigenes* sp. nov. Int. J. Syst. Bacteriol. 36:109–110.

Labeda, D. P. 1988. An evaluation of strains of the genus *Saccharothrix* by numerical taxonomic and molecular taxonomic methods, p. 227–232. In: Y. Okami, T. Beppu, and H. Ogawara (ed.), Biology of Actinomycetes '88. Japan Scientific Society Press, Tokyo.

Labeda, D. P., and M. P. Lechevalier. 1989. Amendment of the genus *Saccharothrix* Labeda et al., 1984 and description of new species of the genus, *Saccharothrix espanaensis,* sp. nov., *Saccharothrix cryophilis,* sp. nov., and *Saccharothrix mutabilis,* comb. nov. Int. J. Syst. Bacteriol. 39:420–423.

Labeda, D. P., and A. J. Lyons. 1989. *Saccharothrix texasensis,* sp. nov., and *Saccharothrix waywayandensis,* sp. nov. Int. J. Syst. Bacteriol. 39:355–358.

Labeda, D. P., R. T. Testa, M. P. Lechevalier, and H. A. Lechevalier. 1984. *Saccharothrix:* a new genus of the *Actinomycetales* related to *Nocardiopsis.* Int. J. Syst. Bacteriol. 34:426–431.

Lechevalier, M. P., C. DeBievre, and H. A. Lechevalier. 1977. Chemotaxonomy of aerobic actinomycetes: phospholipid composition. Biochem. Syst. Ecol. 5:249–260.

Lechevalier, M. P., and H. A. Lechevalier. 1970. Chemical composition as a criterion in the classification of aerobic actinomycetes. Int. J. Syst. Bacteriol. 20:435–444.

Lechevalier, M. P., and H. A. Lechevalier. 1980. The chemotaxonomy of actinomycetes. p. 227–291. In: A. Dietz, and D. W. Thayer (ed.), Actinomycete Taxonomy. Special Publication No. 6. Society for Industrial Microbiology, Arlington, VA.

Marmur, J. 1961. A procedure for the isolation of deoxyribonucleic acid from microorganisms. J. Mol. Biol. 3:208–218.

Marmur, J., and P. Doty. 1962. Determination of the base composition of deoxyribonucleic acid from its thermal denaturation temperature. J. Mol. Biol. 5:109–118.

Meyertons, J. L., D. P. Labeda, G. L. Cote, and M. P. Lechevalier. 1988. A new thin-layer chromatographic method for whole-cell sugar analysis of *Micromonospora* species. The Actinomycetes 20:182–192.

Nonomura, H., and Y. Ohara. 1969. The distribution of actinomycetes in soil. VI. A selective plate-culture isolation method for *Microbispora* and *Streptosporangium* strains. Part 1. J. Ferment. Technol. 47:463–469.

Pridham, T. G., P. Anderson, C. Foley, F. A. Lindenfelser, C. W. Hesseltine, and R. G. Benedict. 1957. A selection of media for maintenance and taxonomic study of *Streptomyces.* Antibiotic Ann. 1956/57:947–953.

Pridham, T. G., and A. J. Lyons. 1980. Methodologies for *Actinomycetales* with special reference to streptomycetes and streptoverticillia. p. 153–224. In: A. Dietz, and D. W. Thayer (ed.), Actinomycete Taxonomy. Special Publication No. 6. Society for Industrial Microbiology, Arlington, VA.

Seidler, R. J., M. D. Knittel, and C. Brown. 1975. Potential pathogens in the environment: cultural reactions and nucleic acid studies on *Klebsiella pneumonia* from chemical and environmental sources. Appl. Microbiol. 29:819–825.

Seidler, R. J., and M. Mandel. 1971. Quantitative aspects of deoxyribonucleic acid renaturation: base composition, state of chromosome replication, and polynucleotide homologies. J. Bacteriol. 106:608–614.

Shearer, M. C. 1987. Methods for the isolation of non-streptomycete antinomycetes. Dev. Ind. Microbiol. 28:91–97.

Shearer, M. C., P. M. Colman, and C. H. Nash III. 1983. *Nocardiopsis mutabilis,* a new species of nocardioform bacteria isolated from soil. Int. J. Syst. Bacteriol. 33:369–374.

Shirling, E. B., and D. Gottleib. 1966. Methods for characterization of *Streptomyces* species. Int. J. Syst. Bacteriol. 16:313–340.

Staneck, J. .L., and G. D. Roberts. 1974. Simplified approach to identification of aerobic actinamycetes by thin-layer chromatography. Appl. Microbiol. 28:226–231.

Takahashi, A., K. Hotta, N. Saito, M. Morioka, Y. Okami, and H. Umezawa. 1986. Production of novel antibiotic, dopsisamine, by a new species of *Nocardiopsis mutabilis* with multiple antibiotic resistance. J. Antibiot. 39:175–183.

Tamaoka, J., Y. Katayama-Fujimura, and H. Kuraishi. 1983. Analysis of bacterial menaquinone mixtures by high performance liquid chromatography. J. Appl. Bacteriol. 54:31–36.

Tresner, H. D., A. A. Fantini, D. B. Borders, and W. J. McGahren. 1980. Antibacterial antibiotic BM782. U.S. Patent 4,234,717.

CHAPTER 45

The Family Frankiaceae

ANTOON D. L. AKKERMANS, DITTMAR HAHN, and DWIGHT D. BAKER

Introduction to the Family

Various woody plants, e.g. *Alnus* spp. in temperate and *Casuarina* spp. in tropical regions, are able to form root nodules with a N_2-fixing actinomycete as endosymbiont. These nodules are called actinorhizal root nodules, to distinguish them from leguminous nodules induced by *Rhizobium*. The endophytes of actinorhizal root nodules are placed within the genus *Frankia* and are characterized by the ability to fix N_2, to induce actinorhizal root nodules, and to form unique cell differentiation of hyphae into sporangia and spherical cells, which are called vesicles.

The first attempts to classify these actinomycetes were made before pure cultures were available. Becking (1970) classified all N_2-fixing, nodule-forming actinomycetes in the genus *Frankia* within the new family Frankiaceae. The species concept in *Frankia* was based on host specificity and ultrastructure of the endophytes (Becking, 1970, 1974, 1981). A close relation to the members of the Dermatophilaceae was proposed based on the pattern of spore formation (van Dijk and Merkus, 1976), the presence of meso-diaminopimelic acid (A_2pm) in the nodule endophyte of *Alnus glutinosa* (van Dijk, 1978), and later on sugar and A_2pm analyses of pure cultures of *Frankia* (Lechevalier and Lechevalier, 1979). The family Dermatophilaceae consisted of two species classified on the basis of structural characteristics: both type species, *Dermatophilus* and *Geodermatophilus,* divide in both transverse and longitudinal planes to form clusters of coccoid cells (Luedemann, 1968).

Another organism, *"Blastococcus"* sp., an invalidly described isolate from the Baltic Sea (Ahrends and Moll, 1970), is reported to show *Geodermatophilus*-like reproduction. Studies on the ultrastructure of this organism also indicate a close relationship to *Geodermatophilus* (Ishiguro and Wolfe, 1970). However, data on chemotaxonomic properties that could indicate a possible membership of *"Blastococcus"* to frankiae, dermatophili or geodermatophili are not available.

The heterogeneity of the family Dermatophilaceae was demonstrated by significant differences in physiological characteristics (Goodfellow and Pirouz, 1982), the menaquinone composition (Collins et al., 1984), and the primary structures of the 16S rRNAs of both type species of *Dermatophilus* and *Geodermatophilus* (Stackebrandt et al., 1983) (Table 1, Fig. 1). Oligonucleotide cataloging of 16S ribosomal RNA of *G. obscurus* has shown a distinct relationship to *"Blastococcus"* and *Frankia*, giving sense to a phylogenetic classification of these three genera within the family Frankiaceae (Hahn et al., 1989b).

Geodermatophilus

Isolation

Geodermatophilus obscurus is a soil-borne microorganism that has been isolated from several desert soil samples, where it can form a significant part of the manganese-oxidizing bacterial population (Leudemann, 1968; Hungate et al., 1987). Isolation and growth media of the pink to black colored *Geodermatophilus* spp. may contain 0.5% yeast extract, 1.5% malt extract broth, 1% soluble starch, 1% sucrose and 0.2% $CaCO_3$ (Luedemann, 1968), or 2% tryptone, 0.2% yeast extract, 0.2% glucose, and 0.5% NaCl at pH 7.0 (Ishiguro and Wolfe, 1970). Organisms of this genus can be grown on agar plates or in liquid culture at about 30°C.

Identification

Geodermatophilus occurs in two major forms. One is a motile rod that multiplies exclusively by budding and the other is a nonmotile irregularly shaped aggregate of coccoid cells, representing a zoospore and a thallus stage (Luedemann, 1968; Ishiguro and Wolfe, 1970) (Fig. 2). Morphogenesis can be controlled by mono- or divalent cations and by organic amines to in-

Table 1. Phenotypic characteristics of members of the genera *Frankia*, *Geodermatophilus* and *Dermatophilus*.

Phenotypic characteristics	*Frankia*	*Geodermatophilus*	*Dermatophilus*
Hyphae, well-developed	+	−	−
Multilocular sporangia[a]	+	+	+
Vesicles	+	−	−
Outer membraneous spore layer	+	−	−
Capsule	−	−	+
Motile spores	−	+	+
Relation to O_2	A/MA[b]	A	MA
Cell wall type	III[c]	III[c]	III[c]
Whole cell sugars[d]	B, C, D, and fucose	C	B
Major menaquinone	MK 9 (H_4)[e]	MK 9 (H_4)[f]	MK 8 (H_4)[f]
GC mol%[g]	67–72	70–75	56–59
Habitat	Angiospermous nodules, soil	Soil, sea	Mammalian epidermis

Adapted from Hahn et al. (1989b).

[a]The entire thallus is a sporangium in the case of *Geodermatophilus* and *Dermatophilus*.

[b]A = aerobic; MA = microaerophilic.

[c]meso-diaminopimelic acid.

[d]Lechevalier and Lechevalier, 1970, B = madurose, no arabinose or xylose; C = none; D = xylose and arabinose.

[e]Lechevalier et al., 1987.

[f]Collins et al., 1984.

[g]Samsonoff et al., 1977; An et al., 1985, 1987.

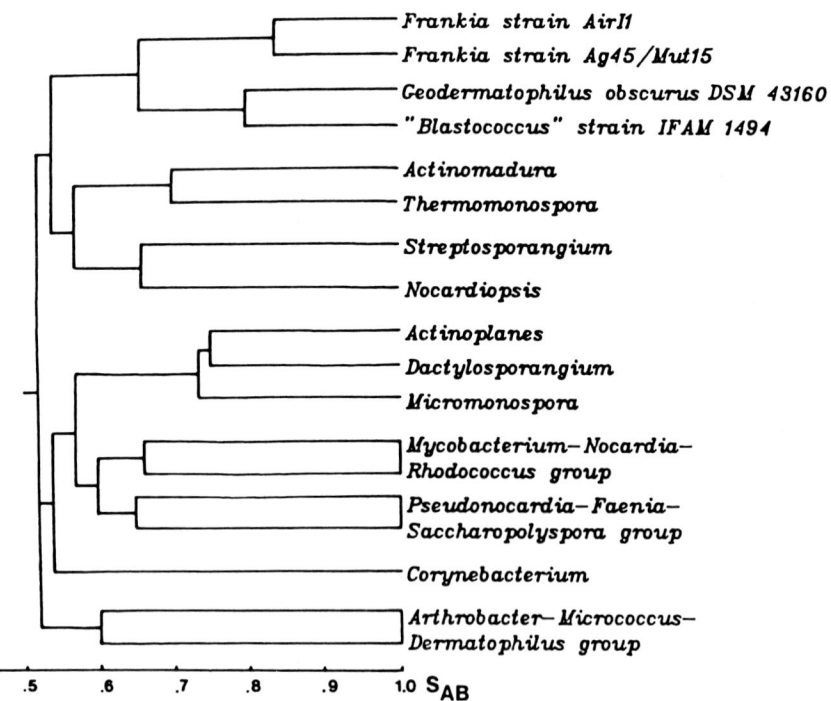

Fig. 1. Phylogenetic position of the genera *Frankia*, *Geodermatophilus*, and *"Blastococcus"* within the dendrogram of actinomycetes based on the analysis of 16S rRNA oligonucleotide catalogs. (From Hahn et al., 1989b.)

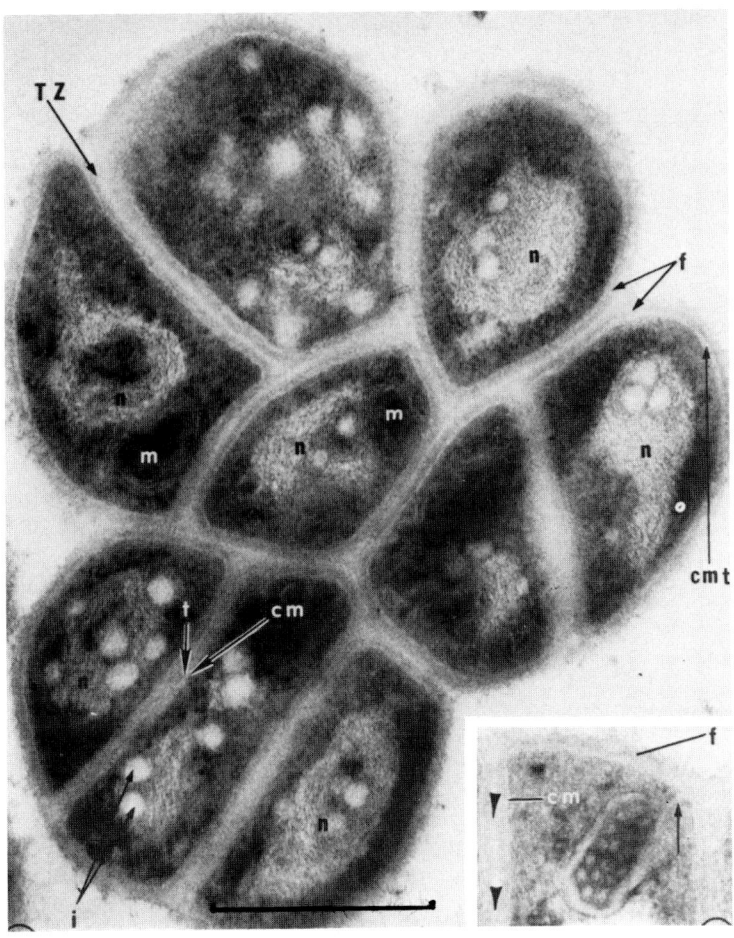

Fig. 2. Transmission electron micrograph of the coccoid form of *Geodermatophilus obscurus*. Irregularly arranged coccoid cells individually bound by cytoplasmic membrane (cm) and cell wall with at least two layers, a fibrous outer layer (f) and a transparent inner layer (t). The latter layer is not resolved from cm in some areas (cmt). Note transparent zone between cells (TZ). Dense cytoplasm contains fibrillar nucleoplasm (n), densely stained tubular mesosomes (m), and granular inclusions (i). The section has been overstained with lead to resolve surface layers. Bar = 1 μm. (From Ishiguro and Wolfe, 1970.)

duce stabilization of the coccoid form (Ishiguro and Wolfe, 1974). So far only a few strains have been obtained in pure culture.

"*Blastococcus*"

Isolation

"*Blastococcus*" is an invalidly described organism isolated from the Baltic Sea (Ahrends and Moll, 1970). Small red colonies form after incubation at 20°C with a generation time of 2 h on peptone-yeast extract agar.

Identification

"*Blastococcus*" is similar to *Geodermatophilus* in that two generation forms exist, represented by motile rods that multiply by budding and by thallic aggregates of coccoid cells. The manifestation of one of the stages is influenced by environmental circumstances. Low temperature, i.e. 5–10°C, low salt concentrations, and anaerobic conditions support the establishment of the motile rod stage. Increased temperatures

and salt concentrations induce the formation of nonmotile coccoid aggregates. Only one isolate has been described so far.

No data are available on the biochemistry and genetics of "*Blastococcus*." Further data on related isolates are needed to judge the taxonomic position of the taxon.

Frankia

Isolation

Attempts to isolate *Frankia* from root nodules have been made since the mid-1800s (Baker and Torrey, 1979). None of these early studies were successful, and the organism was generally viewed as an obligate root-nodule symbiont. The first isolation of an organism capable of reinfecting *Alnus glutinosa* was reported in 1959 (Pommer, 1959). This report included very detailed drawings (Fig. 3) and data from reinfection studies that we now interpret as the first correct report of a successful isolation of *Frankia* in pure culture. However, this report

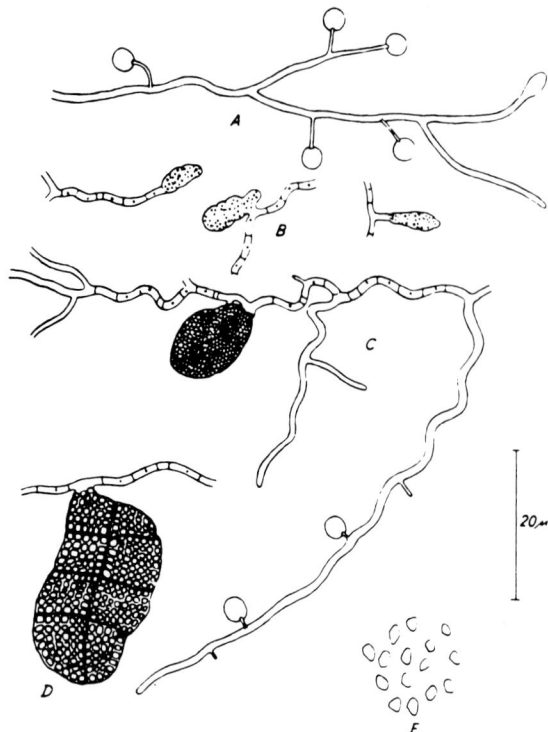

Fig. 3. Root-nodule endophyte of *Alnus glutinosa* grown on glucose-asparagine agar. (a) hyphae with vesicles; (b–d) different stages of "bacteroid" formation; (e) released "bacteroids." (From Pommer, 1959.) The "bacteroids" described by Pommer are spores (van Dijk and Merkus, 1976; Callaham et al., 1978).

tivated in media free of combined nitrogen (Tjepkema et al., 1980). Currently, isolations of this type of microbe are considered to be routine even though infective strains have not been isolated in pure culture from numerous host plants (Stowers, 1987; Baker, 1989). Isolations are best achieved from actinorhizal root nodules through selective enrichment, and thus far direct isolation from soil has been reported only once (Baker and O'Keefe, 1984). Widely used techniques for isolation include microdissection, serial filtration, or density fractionation (see Stowers, 1987 or Baker, 1989 for full descriptions). Pieces of surface sterilized nodule tissue or enriched fractions of *Frankia* from nodule tissue are inoculated on agar media. A nitrogen deficient medium containing simple inorganic salts and sodium propionate as the sole carbon source (Blom, 1981; Shipton and Burggraaf, 1982; Baker and O'Keefe, 1984; Murry et al., 1984) has been the most useful. Some strains will not grow well on simple media and the addition of lipids may be suitable (Lalonde and Calvert, 1979; Lechevalier et al., 1982). Quispel et al. (1988, 1989) have studied the lipid composition of root extracts to determine possible compounds for supplementing growth media to improve frankiae growth. They have identified a terpenoid compound that permits the cultivation of some fastidious strains.

CULTIVATION. The majority of *Frankia* strains existing in cultivation at present can be maintained on a simple minimal salts medium. A representative medium is the DPM (defined propionate minimal) medium of Baker and O'Keefe (1984) with the following composition (in amounts per liter for final solution):

DPM Medium
KH$_2$PO$_4$	1.00 g
MgSO$_4$·7H$_2$O	0.10 g
CaCl$_2$·2H$_2$O	0.01 g
Sodium propionate	1.20 g
Chelated iron solution (see below)	1.8 ml
Trace element solution (see below)	1.0 ml

Chelated iron solution:
FeSO$_4$·5H$_2$O	5.54 g
Disodium EDTA	7.56 g

Trace element solution (modified from Hoagland and Arnon, 1950):
CoCl$_2$	0.025 g
CuSO$_4$·5H$_2$O	0.08 g
H$_3$BO$_3$	2.86 g
MnCl$_2$·4H$_2$O	1.81 g
NaMoO$_4$·2H$_2$O	0.025 g
ZnSO$_4$·7H$_2$O	0.22 g

The pH of the final solution is adjusted to 6.8.

was generally dismissed and the cultures were lost.

Even without valid cultures, the root-nodule endophytes have been recognized microscopically as actinomycetes and placed in the family Frankiaceae with one genus, *Frankia* (Becking, 1970). In 1978 two research groups independently reconfirmed the facultatively symbiotic relationship of the endophyte of actinorhizal root nodules of *Alnus glutinosa* (Quispel and Tak, 1978) and *Comptonia peregrina* (Callaham et al., 1978) by obtaining pure cultured isolates. These bacteria have been placed within the genus *Frankia* with strain Cpl1 from *C. peregrina* as the type strain. Koch's postulates were fulfilled by reinfection of seedlings, and the isolation and infectivity of strain Cpl1 was independently verified (Lalonde, 1978). Subsequently, numerous isolates from other N$_2$-fixing root nodules were obtained in different laboratories (Baker et al., 1979; Berry and Torrey, 1979; Burggraaf et al., 1981; Lalonde et al., 1981; Diem et al., 1982; Diem and Dommergues, 1983; Lechevalier and Ruan, 1984). These isolates were morphologically quite similar and able to fix atmospheric N$_2$ when cul-

Propionate has been shown to be the best carbon source for one of the first *Frankia* isolates, Avcl1 (Blom, 1981), and this fatty acid was subsequently recognized as the major carbon source of most strains (Shipton and Burggraaf, 1982), although other organic acids (e.g., acetate, succinate) may also be used. For some strains, complex media may be required. Numerous formulations have been described (see Baker, 1989). One complex medium commonly used in several laboratories is Qmod (Lalonde and Calvert, 1979) with the following composition (in amounts per liter for final solution):

Qmod Medium

K_2HPO_4	0.3 g
NaH_2PO_4	0.2 g
$MgSO_4 \cdot 7H_2O$	0.2 g
KCl	0.2 g
Yeast extract	0.5 g
Bacto-peptone	5.0 g
Glucose	10.0 g
Ferric citrate (citric acid and ferric citrate, 1% solution)	1.0 ml
Minor salts (see below)	1.0 ml
Lecithin solution (see below)	1.0 ml

Minor salts solution:

$CoSO_4 \cdot 7H_2O$	0.01 g
$CuSO_4 \cdot 5H_2O$	0.10 g
H_3BO_3	1.50 g
$MnSO_4 \cdot 7H_2O$	0.80 g
$(NH_4)6Mo_7O_2 \cdot 4H_2O$	0.20 g
$ZnSO_4 \cdot 7H_2O$	0.60 g

The lecithin solution is made by dissolving 50 mg lecithin in 50 ml of ethanol and adding 50 ml of distilled water.
The pH of the solution is 6.8–7.0.

Frankia strains are routinely cultivated in standing liquid cultures having approximately 50 ml of medium in a 125 ml bottle or flask. Gyratory shaking at very low speeds (100 rpm) may be beneficial but is not required. *Frankia* will grow in submerged agar cultures but slower relative to liquid cultures. In all media, *Frankia* grows as a mycelial mat and over time this becomes quite dense. Subcultures are made by disrupting this mycelial mat with a Potter-Elvehjem or Ten-Broeck tissue grinder or by passage through a syringe needle numerous times. The disruption of the mycelium creates new growing tips and thus increases growth rates.

Identification

MORPHOLOGY. *Frankia* strains are similar to most other aerobic actinomycetes, producing a septate, filamentous mycelium that can differentiate into sporangia and vesicles (Fig. 4). Cells are routinely Gram positive, although older cells may be Gram variable. The hyphae range in diameter from 0.5 to 2.0 μm. Unlike other Gram-positive bacteria, the *Frankia* has a discontinuous membranous layer (Newcomb et al., 1979) which is particularly evident when cells are fixed in $KMnO_4$. The development or function of this membranous layer is unknown, but it occurs on hyphae, sporangia, spores, and vesicles both in culture and in the plant. *Frankia* produces sporangiophores within large terminal sprongia (Fig. 5) or smaller intercalary sporangia (Fig. 6). The morphology of sporangia is superficially similar to members of the family Actinoplanaceae. However, the sporangia are only produced on substrate mycelia, whereas the sporangia of the actinoplanetes are produced only on aerial mycella. Sporangia develop as swellings of hyphal tips or within hyphae. Sporangia are amorphous and range in size from 10 to 75 μm. Spores are produced by the division of swollen hyphae along numerous axes in a pattern resembling that of the family Dermatophilaceae (Lechevalier and Lechevalier, 1984). Spores can often be observed in pairs or tetrads, reflecting their development (Baker and Torrey, 1980). Within the sporangium, spores develop at different times and it is therefore common to observe both mature and developing spores within the same sporangium. They may be released by autolysis in some strains or by physical disruption of the sporangial wall in others.

Frankia also produces another specialized cell structure called a "vesicle" because of its globose shape. These structures are borne on short hyphae arising from the substrate mycelium (Fig. 7) or sometimes on the terminal ends of the mycelium. Since vesicles are the sites for N_2 fixation (Tjepkema et al., 1980; Murry et al., 1984; Meesters, 1987, 1988; Meesters et al., 1987), their morphology, development, and physiology have been studies extensively. Vesicles have multilayered, thick walls that prevent inactivation of the enzyme nitrogenase by oxygen. Vesicles produced in culture are spherical, 2–4 μm in diameter, and are attached to the substrate mycelium by a stalk cell. The cell walls of both the vesicle proper and the stalk cell are composed of multiple layers of lipid membranes, and the number of layers increases with the ambient oxygen concentration (Parsons et al., 1987).

PHYSIOLOGY AND GENETICS. All *Frankia* strains that have been investigated are slow-growing, filamentous microorganisms with generation times ranging from 20 h to several days. Most strains are able to grow on propionate. Some strains may utilize other short-chain fatty acids,

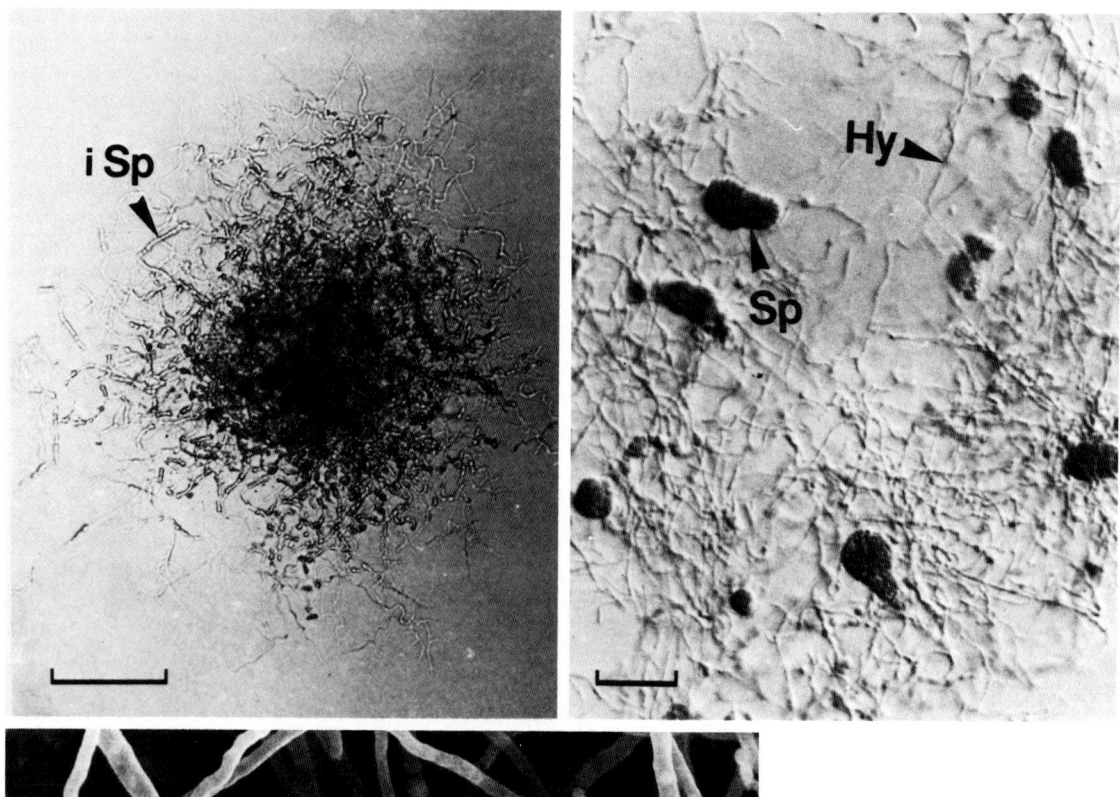

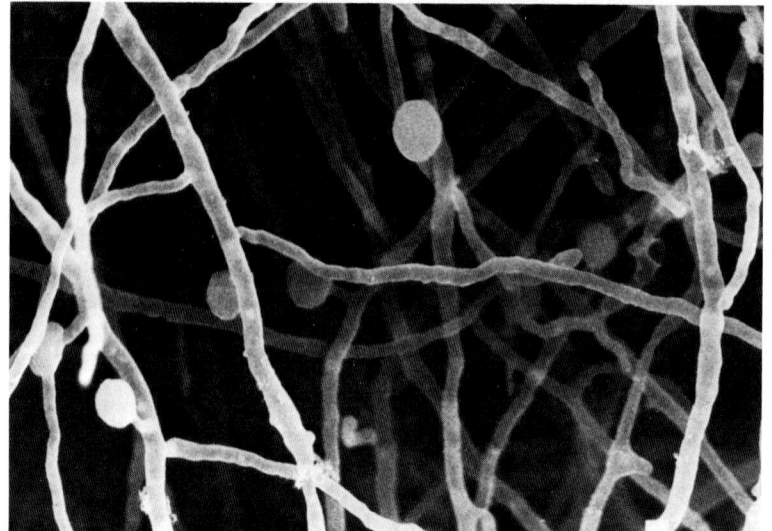

Fig. 4. (a) *Frankia* colony grown on agar supplemented with mineral salts and pyruvate as C source (Benson, 1982), showing hyphae with intercalar sporangia (iSp) (Courtesy of D. Hahn and S. Uhlenhaut.) Bar = 100 μm. (b) Light micrograph of hyphae (Hy) and terminal sporangia (Sp) of *Frankia* isolated from *Alnus glutinosa* (Courtesy of D. Hahn and S. Uhlenhaut.) Bar = 25 μm. (c) Scanning electron micrograph of vesicles and hyphae of *Frankia* (Courtesy of D. Hahn and S. Uhlenhaut.)

glucose, and other monosaccharides, as well as carboxylic acids such as succinate. A few additional reports mention the degradation of lipids, such as Tween-80 (Blom et al., 1980), pectin (Lalonde, 1977; Simonet et al., 1988a), and cellulose (Safo-Sampah and Torrey, 1988). Blom and Harkink (1981) reported the existence of β-oxidation of fatty acids and the presence of a glyoxylate cycle when cells of *Frankia* strain Avcl1 were grown on acetate or Tween-80. In cells grown on propionate, two key glyoxylate cycle enzymes, isocitrate lyase and malate synthase, were repressed. Enzymes involved in the degradation of monosaccharides have been re-

ported in strains grown on glucose (Lopez and Torrey, 1985).

Most strains are able to fix atmospheric N_2 in pure culture. The presence of nitrogenase in pure cultures has been extensively demonstrated by the C_2H_2 reduction method (Tjepkema et al., 1980). Nitrogenase genes are highly conserved and the *Frankia* nitrogenase enzymes strongly resemble those of other N_2-fixing organisms (Hennecke et al., 1985; Bomar et al., 1989). Immunogold labelling studies, as well as studies on purified vesicle suspensions indicate that the enzyme is localized in the vesicles of *Frankia* (Fig. 8) and no nitrogenase has been

Fig. 5. (a) Transmission electron micrographs of a terminal sporangium of *Frankia* strain Cpl.1 with compactly arranged immature spores at proximal end and loosely packed mature spores at the distal end. Outer layer (arrows) of sporangium appears thicker and denser at proximal end; much extracellular material, including numerous vesicular structures (VS) that are possibly droplets, is present at the distal end. Bar = 1 μm. (From Newcomb et al., 1979.) (b) Scanning electron micrograph of a terminal sporangium of *Frankia* strain Cpl.1. Bar = 1 μm. (From Baker et al., 1979.)

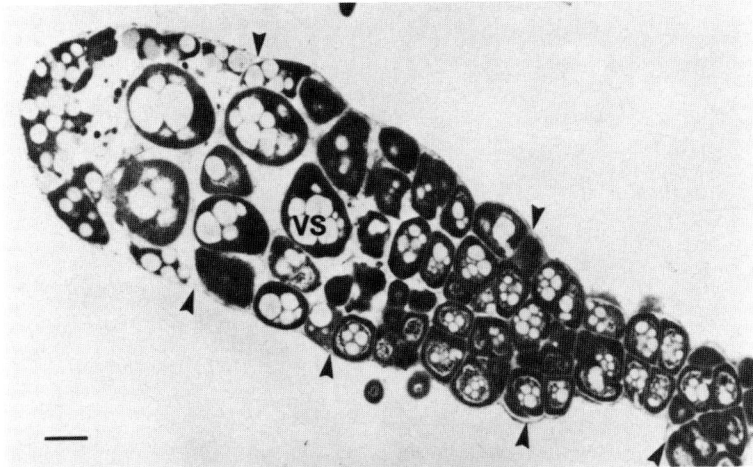

a

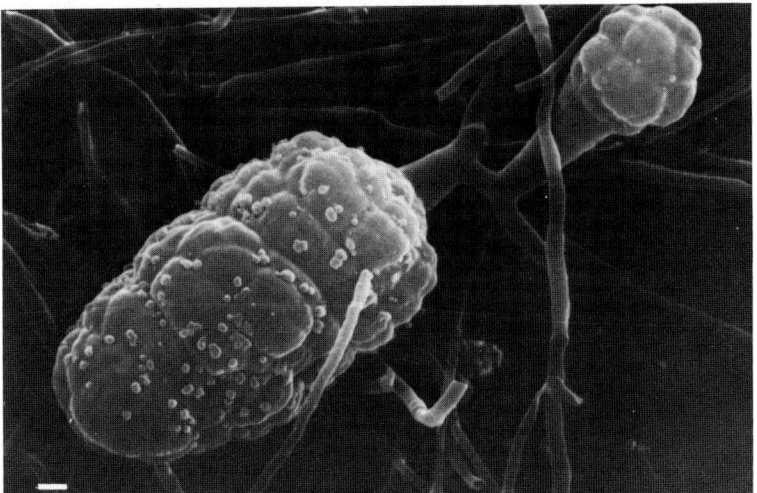

b

detected in the hyphae (Noridge and Benson, 1986; Meesters, 1987, 1988; Meesters et al., 1987). These experiments indicated a transport of combined nitrogen through the culture when grown under N_2-fixing conditions. The structural genes *nifK*, *nifD*, and *nifH*, encoding for nitrogenase, have been cloned and sequenced in two strains (Ligon and Nakas, 1987; Normand et al., 1988). The studies can be summarized as follows (Simonet et al., 1990):

1. Among *Frankia* genes, the *nifH* genes appear to be more conserved than the *nifD* gene, and the *nifD* gene is more conserved than the *nifK* gene. These observations agree with those described for other N_2-fixing systems (Hennecke et al., 1985).
2. The *nifK* gene exhibits some similarity to the *nifD* gene.
3. *Frankia* strain ARgP5Ag (ULQ0132105009) has a different organization of *nif* genes compared to the typical *Frankia* strains (Simonet et al., 1986). *Nif* probes were found to hybridize to both a 10-kb *EcoR1* fragment on total DNA blots and to a 5.6-kb *EcoR1* fragment from a large (190-kb) indigenous plasmid.

Other studies have shown that some *Frankia* strains, similar to some rhizobia, possess an active hydrogen uptake system, which may function in recycling electrons derived from nitrogenase and/or in maintaining low redox potential (Benson et al., 1980; Sellstedt et al., 1986).

Frankia convert the ammonia derived from nitrogenase into glutamine through glutamine synthetase (Blom et al., 1981; Akkermans et al., 1983). The ammonium assimilation of N_2-fixing cultures is localized in the vesicles, i.e., it is closely related to the nitrogenase system (Edmands et al., 1987). No glutamate dehydrogenase has been detected (Blom, 1981).

Frankia cultures contain significant levels of Fe- and Mn-containing superoxide dismutase (Steele and Stowers, 1986) in different isozymes (Puppo et al., 1985), which may function as additional protection against oxygen.

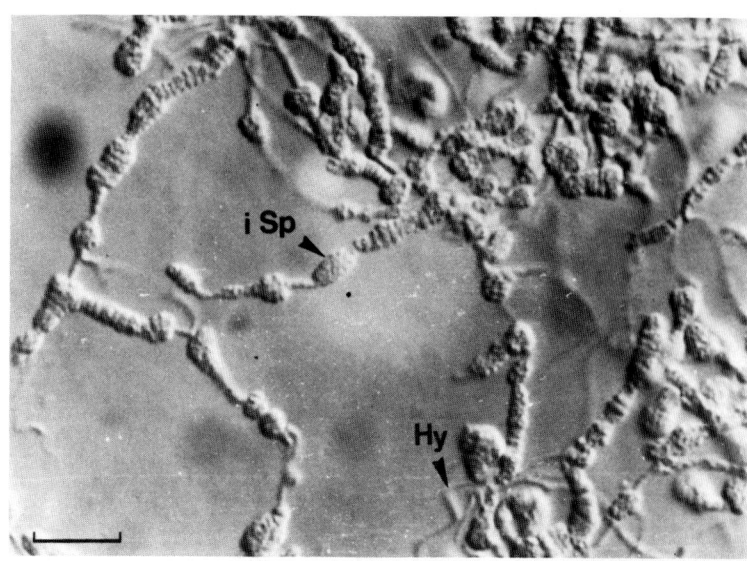

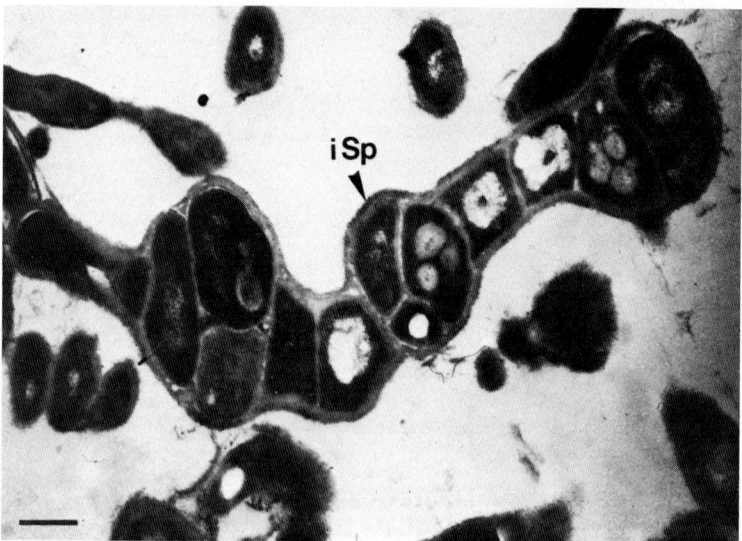

Fig. 6. (a) Light micrograph (Bar = 10 μm) and (b) transmission electron micrograph (Bar = 0.5 μm) of intercalar sporangia (iSp) and hyphae (Hy) of *Frankia*. (Courtesy of D. Hahn and S. Uhlenhaut.)

Recently, *Frankia*-like isolates have been obtained from actinorhizal nodules that are unable to fix N₂ sporadically sporulate and do not form vesicles. Some of these strains form ineffective nodules on certain hosts and remain noninfective on related plant clones, due to partial incompatibility of the host (Baker et al., 1980; Hahn et al., 1988; van Dijk and Sluimer, in press). This indicates the existence of a new group of free-living actinomycetes in soil.

MOLECULAR TAXONOMY. *Frankia* strains resemble *Geodermatophilus obscurus* and *Dermatophilus congolensis* in certain morphological features, e.g., hyphae divide in more than one plane, but differ from them in the possession of sporangia. Morphological features and cell wall type of both *Frankia* and *Geodermatohilus* were the basis for the previous classifi-

cation in a taxon "Multilocular sporangia" (Goodfellow, 1986). Even though *Frankia* show many more phenotypic similarities to dermatophili than to geodermatophili, their dissimilar whole cell sugars, host specificities, and serology indicated caution in making conclusions about their close relationship. Investigations on DNA base composition showed that the GC content of frankiae and dermatophili were very different. Furthermore, there was a lack of DNA homology between representatives of the two genera (Samsonoff et al., 1977; An et al., 1985, 1987). Detailed information on the phylogenetic position of *Frankia* has been obtained by sequencing and/or oligonucleotide cataloging of their 16S ribosomal RNA. Two *Frankia* strains analyzed are highly related, showing a distinct relationship to *Geodermatophilus obscurus* and to a strain of *"Blastococcus."* These organisms

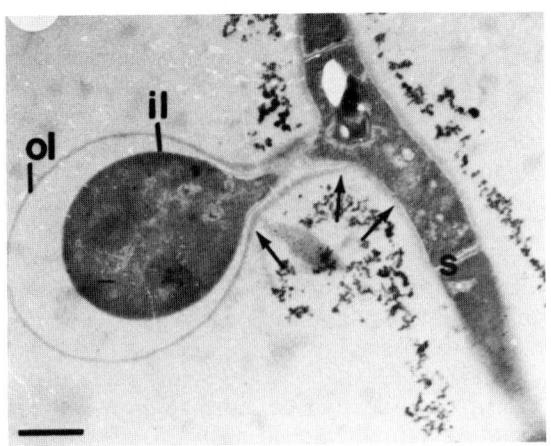

Fig. 7. Transmission electron micrograph of a terminal vesicle of *Frankia* strain Cc1.17 cultivated in the presence of NH₄Cl (0.2 g/l). Bar = 0.5 μm. il = cell wall inner layer, ol = cell wall outer layer. Note the constriction of the stalked hyphae adjacent to the vesicle. Similar structures of vesicles are found in medium without combined nitrogen (Meesters et al., 1985). (From Meesters, 1988.)

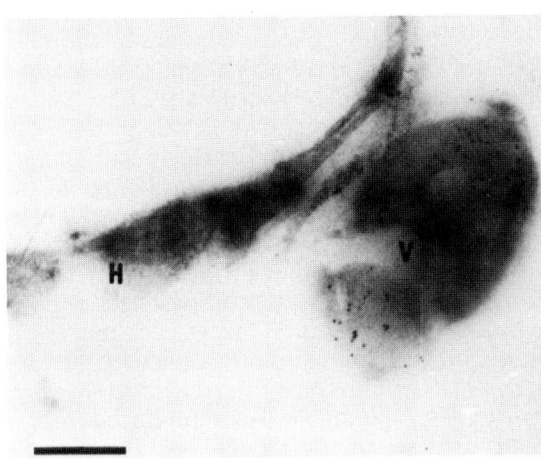

Fig. 8. Electron micrograph of immunogold-labelled *Frankia* Cc1.17 grown in medium free of combined nitrogen under 2% oxygen. Bar = 0.5 μm. Significant gold label is visible as black spots in vesicles (v) and not in hyphae (h). (From Meesters, 1987.)

constitute a main subline of descent within the phylogenetic radiation of the order Actinomycetales (Fig. 1).

The specific composition of 16S rRNA has been shown to be useful for unravelling large phylogenetic relationships. However, the phylogenetic investigations of *Frankia* by Hahn et al. (1989b) are restricted only to two typical strains belonging to one group, type B, according to the subdivision of Lechevalier and Lechevalier (1984).

The first classification within the genus *Frankia* was based on host plant relationships using crushed nodules as inocula for cross-infection studies sources (Becking, 1970). However, subsequent investigations with pure cultures indicated that the host specificity groups were quite different (Normand and Lalonde, 1986; Baker, 1987). Physiologically, *Frankia* strains have been classed into at least two groups (Lechevalier and Lechevalier, 1984): group A is a heterogenous group of morphologically and chemically diverse strains, whereas group B is a more homogenous group of colorless, microaerophilic strains.

Cytochemical criteria have proven to be helpful in studying close phylogenetic relationships. All *Frankia* strains studied so far contain 2-O-methyl D-mannose, a cell wall sugar specific for the genus *Frankia* (Mort et al., 1983). Quantitative analyses of sugars from whole-cell hydrolysates allowed the separation of the tested *Frankia* strains into their host-specificity groups, the *Alnus* and the *Elaeagnus* host specificity group, respectively, with 2-O-methyl D-mannose as diagnostic sugar (St-Laurent et al., 1987). Quantitative analyses of total fatty acid composition have also been used for taxonomical investigations and discrimination between type species *F. alni* and *F. elaeagni* and the subspecies *pommerii* and *vandijkii* of *F. alni* have been proposed (Lalonde et al., 1988). Fatty acid methyl ester analysis of 30 selected *Frankia* strains were used to discriminate between strains belonging to *F. elaeagni* and *F. alni* subsp. *pommerii* (Simon et al., 1989). The subdivision of *F. alni* into two subspecies has been disputed on the basis of DNA relatedness and DNA base composition of 43 isolates of different origins (Fernandez et al., 1989). At least nine genomic species were differentiated, including three genomic species among *Alnus*-compatible strains *("F. alni")*, five among Elaeagnaceae-compatible strains *("F. elaeagni")* and one among *Casuarina*-compatible strains. According to the recommendation that a bacterial species is defined as a DNA relatedness group (genomic species) that can be identified by phenotypic tests (Wayne et al., 1987), there is yet only one genomic species, *F. alni*, with strain Cp11 as the type strain. Available phenotypic data are not comprehensive enough to separate other genomic species at this moment.

Methods to measure small phylogenetic distances among closely related species include serological studies (Baker et al., 1981; Lechevalier et al., 1983) as well as electrophoretic patterns of proteins and isozymes (Benson and Hanna, 1983; Benson et al., 1984; Gardes and Lalonde, 1987; Gardes et al., 1987). Both serological and cytochemical methods show that the genetic diversity among *Frankia* strains is very large.

Based on immunodiffusion assays and one-dimensional gel electrophoresis (SDS-PAGE) two groups of *Frankia* strains could be distinguished, mainly comprising the *Alnus*-compatibility group and the *Elaeagnus*-compatibility group. However, immunofluorescence studies on the same strains failed to show reliable patterns. Furthermore, the electrophoretic characteristics of some strains did not correlate with their original compatibility group, indicating the insufficiency of nodulation tests for taxonomic purposes. Further subgrouping of the strains on the basis of SDS-PAGE within the compatibility groups is also possible. Related strains could also be distinguished on the basis of electrophoretic separation of isoenzymes and delineated in groups of strains.

Molecular methods such as determination of DNA base composition and DNA/DNA homology (An et al., 1985, 1987; Normand and Lalonde, 1986; Simonet et al., 1988a, 1988b; Normand et al., 1988; Meesters, 1988) also demonstrated the genetic diversity among *Frankia* strains because only low levels of homology were obtained. Ratios of homology in genomic DNA/DNA hybridizations between 67 and 94% within one compatibility group or levels lower than 50% between different compatibility groups were obtained. It has been demonstrated that *nif* genes of *Frankia* have a conserved character and could therefore be used to investigate large phylogenetic relationships. However, different restriction sites within these genes can supply specific hybridization patterns of digested total DNA with *nif* gene probes indicating a possible application of these patterns in taxonomical investigations. Another conserved molecule is 16S rRNA. Variable regions on this molecule were shown to be useful for distinguishing between closely related species of a phylogenetically conserved genus. Synthetic oligonucleotide probes against short sequences of these variable regions show specific hybridization patterns comprising compatibility groups, physiologically different groups (Nif$^+$ and Nif$^-$) and strains (Hahn et al., 1989a; Hahn et al., 1990).

Root Nodule Symbiosis

MORPHOLOGY. The majority of actinorhizal plants of those studies are infected by a root-hair-mediated mechanism (Callaham et al. 1979). An alternative mechanism involving direct penetration of the root via the intercellular spaces of the epidermis and cortex has been reported for the host genera of the families Elaeagnaceae (Miller and Baker 1985) and some Rhamnaceae (Liu and A. M. Berry, personal communication). In both cases, *Frankia* hyphae penetrate the host cells but remain enclosed by a polysaccharide capsule produced by the host (Lalonde et al., 1975). As a result of the root infection, a root meristem is induced from the pericicle near the site of infection. Newly formed cells of the root meristem are invaded by *Frankia*. As a result of this symbiosis, a deformed root with dichotomous branching is formed; the actinorhizal root nodule. *Frankia* grow as endosymbionts occupying the major part of the living host cell. Typical structures of the endophytes are the vesicle clusters, in which the tips of the hyphae develop into swollen structures with thickened cell walls, the vesicles. The morphology of the endophytes is host dependent; vesicles of *Frankia* Cpl1 are club-shaped in nodules of *Comptonia peregrina* and spherical when grown in root nodules of *Alnus glutinosa* (Lalonde, 1979). The vesicle morphologies observed in actinorhizal nodules can be classified into four major groups (Baker and Selling, 1984). Spherical vesicles are observed in *Alnus* spp. (Fig. 9a), and the Elaeagnaceae, lanceolate or fingerlike vesicles are observed in *Myrica*, *Coriari*, and *Datisca* spp., (Fig. 9b), and obovate or pearshaped vesicles are observed in *Ceanothus*, *Purshia* spp., and other species of the families Rhamnaceae or Rosaceae. The nature of this host-dependent morphology of the microsymbiont is unknown. No structure resembling a distinct vesicle has been observed in *Casuarina*, even though the microsymbiont fixes large amounts of N$_2$ and produces vesicles when grown in vitro. Whatever cellular entity is identified as the site of nitrogenase in *Casuarina*, it would constitute the fourth vesicle morphology group. Vesicles produced in the plant also bear multilaminar walls (Newcomb et al., 1987) and are the site for active nitrogenase (Sasakawa et al., 1988).

Frankia strains in ineffective nodules do not fix N$_2$ and may lack vesicles (Fig. 9c) or sporangia (Mian et al., 1976; Baker et al., 1980; Hahn et al., 1988). Ineffectivity was found either due to the lack of (at least) *nifH*, *nifD*, and *nifK* genes (*nif$^-$* strains, Hahn, unpublished data) or due to host-induced incompatibility of *nif$^+$* strains (Mian et al., 1976; VandenBosch et al., 1983; van Dijk et al., 1988).

In most actinorhizal nodules, the endophyte belongs to the spore negative (Sp$^-$) type and is unable to form spores. All isolates of this type, however, form spores in pure culture. Some host plants, e.g., *Alnus glutinosa* and *Myrica* spp., may contain *Frankia* strains of the spore positive (Sp$^+$) type, which produces masses of spores in the nodule. These strains have been incompletely described (see "Ecology" below).

Fig. 9. (a) Scanning electron micrograph of a naturally occurring (sp+ type), effective *Frankia* strain in root nodules of *Alnus glutinosa* (Akkermans, unpublished data). Bar = 10 μm. (b) Scanning electron micrograph of a naturally occurring effective *Frankia* in root nodules of *Datisca cannabina*. Bar = 10 μm. (From Akkermans et al., 1984.) (c) Scanning electron micrograph of the ineffective *Frankia* strain Eu11 in cells of *Elaeagnus umbellata*. Note the lack of vesicles. Bar = 10 μm. (From Baker et al., 1979.)

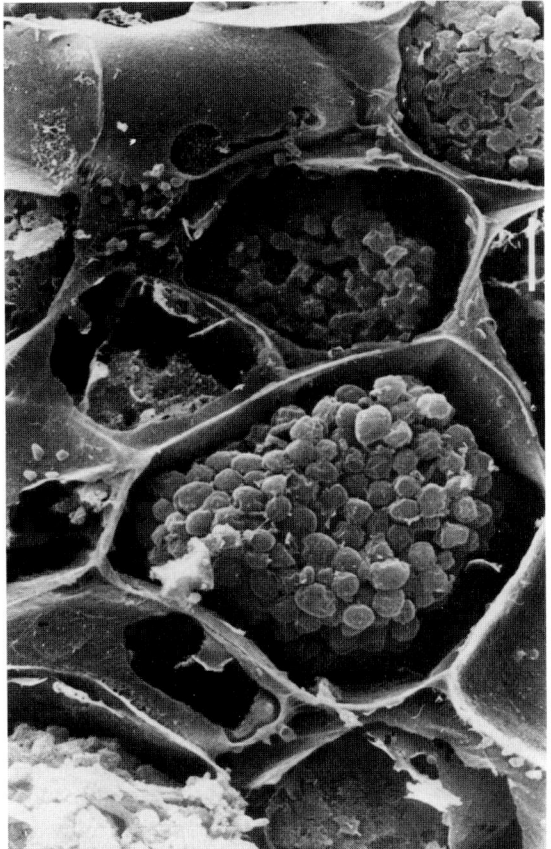

a

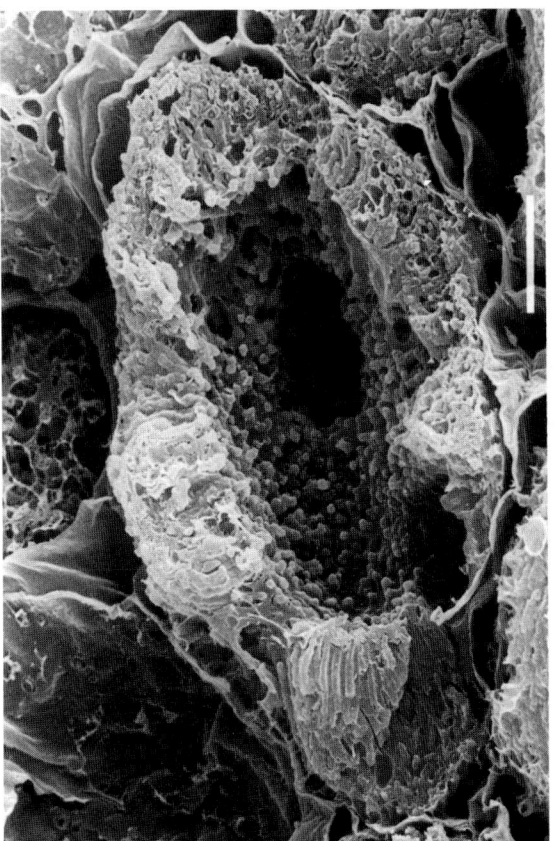

b

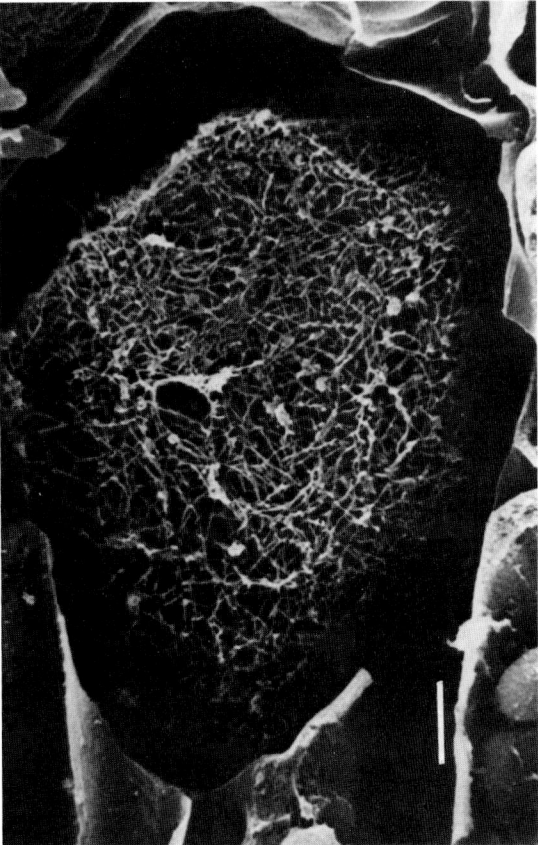

c

PHYSIOLOGY. Effective *Frankia* strains in actinorhizal root nodules fix atmospheric N_2 at the expense of organic carbon supplied by the host. Root nodule tissue usually comprises 1–5% of the total dry weight of the plant and contain a slightly (up to 2 times) higher nitrogen content than the rest of the plant (Akkermans and van Dijk, 1976). The major part of the ammonia produced by the microsymbiont is exuded into the cytoplasm of the infected plant cells and is converted into amino acids. The major amino acid transported through the plant is host-dependent, e.g., citrulline in the case of ureide plants, similar to *Alnus* spp. (Miettienen and Virtanen, 1953). During the growth in symbiosis, the synthesis of glutamine synthetase is largely suppressed (Blom et al., 1981), as observed in other types of N_2-fixing symbioses. The amount of N_2 fixed by the nodules is sufficient to sustain growth of the whole plant. During the reduction of N_2 a significant part of the electrons flow to protons and are released through H_2 gas. Many *Frankia* strains recycle H_2 by a hydrogenase (Benson et al., 1980).

ECOLOGY. One of the first ecological studies on *Frankia* was reported by van Dijk (1978) before routine in vitro grown cultures existed. At a small study site in The Netherlands, root nodules were sampled on naturally occurring *Alnus glutinosa* trees. By grouping endophytes on the basis of anatomical observations for the presence (Sp+) or absence (Sp−) of bacterial sporangiospores, the distribution of what were assumed to be genetically distinct *Frankia* strains was mapped. Subsequent studies on the ecological distribution of Sp+ and Sp− nodules of *A. incana* subsp. *rugosa* (Holman and Schwintzer, 1987) and *Myrica gale* (Kashanski and Schwintzer, 1987) were initiated in Maine, in an attempt to determine whether the presence or absence of spores could be correlated with soil pH, texture, moisture, composition, or site history. In these studies, as in that of van Dijk (1978), it was not possible to determine the relationships of the Sp+ and Sp− endophytes to one another. Van Dijk et al. (1988) directly compared the nodulation efficiencies of these two endophyte types by laboratory-based cross-inoculation studies, but were unable to demonstrate strict stability of the Sp+ endophyte type. It is not likely that the controversy surrounding Sp+/Sp− organisms will be resolved unless a consistently nodulating Sp+ strain is cultivated in vitro or a DNA or RNA study is undertaken directly on samples collected from field sites.

Frankia is often documented in soils where host plant species do not exist (Houwers and Akkermans, 1981; Huss-Danell and Frej, 1986; Smolander and Sundman, 1987). The *Frankia* on such sites may benefit from other plant genera in commensal or parasitic symbioses in which root nodules are not manifested. Smolander and Sundman (1987) found greater populations of *Frankia* in soils in Finland under the nonactinorhizal *Betula* than under the actinorhizal *Alnus*. They also observed a strong positive correlation between pH and *Frankia* population based on infectivity studies.

HOST SPECIFICITY. The infective abilities of each *Frankia* strain are limited to one or a few host plant genera. Studies of the *Frankia* strains isolated in pure culture suggest that strains can generally be divided into groups having similar host specificities. All of the *Frankia* strains belonging to any one of these host-specificity groups can infect the host plant species of that group, but not host plant species of other groups. Specificity among *Frankia* strains appears to be quite conservative. The host-specificity groups as currently defined are: 1) strains infective on *Alnus, Comptonia,* and *Myrica* species; 2) strains infective on species of the Casuarinaceae family; 3) flexible strains (sensu Miller and Baker 1986) infective on species of the Elaeagnaceae family and the promiscuous species of *Myrica* and *Gymnostoma;* and 4) strains infective solely on species of the family Elaeagnaceae.

A major difference accounting for one aspect of host specificity is the mode of infection of the host plant. *Frankia* strains belonging to specificity groups 1 and 2 infect the host plant via the root hair, whereas strains in specificity group 4 infect the host plant via intercellular penetration. It is interesting to note that some strains can be infected by either of the two modes of entry. Miller and Baker (1986) defined these strains as "flexible" and they are separated into their own specificity group 3.

Complicating the host-specificity relationships of *Frankia* strains is the recent description of promiscuity in several actinorhizal host plant genera (Baker, 1987). Several species of *Myrica* (Myricaceae) and *Gymnostoma* (Casuarinaceae) can be nodulated by *Frankia* strains belonging to several specificity groups. Some degree of ineffectivity is common among these promiscuous symbioses. The mechanism for such promiscuity has not been studied but cannot be explained by anatomical differences in the infection processes. Chemical recognition factors and its genetic background need to be studied to better understand the promiscuity phenomenon.

Literature Cited

Akkermans, A. D. L., F. Hafeez, W. Roelofsen, A. H. Chaudhary, and R. Baas. 1984. Ultrastucture and nitrogenase activity of *Frankia* grown in pure culture and in actinorrhizae of *Alnus, Colletia* and *Datisca* spp., p. 311–319. In: C. Veeger and W. E. Newton (ed.), Advances in nitrogen fixation research. Martinus Nijhoff/Dr. W. Junk Publishers, The Hague/Pudoc, Wageningen, The Netherlands.

Akkermans, A. D. L., W. Roelofsen, J. Blom, K. Huss-Danell, and R. Harkink. 1983. Utilization of carbon and nitrogen compounds by *Frankia* in synthetic media and in root nodules of *Alnus glutinosa, Hippophae rhamnoides* and *Datisca cannabina.* Can. J. Bot. 61:2793–2800.

Akkermans, A. D. L., and C. van Dijk. 1976. The formation and nitrogen-fixing activity of the root nodules of *Alnus glutinosa* under field conditions, p. 511–520. In: P. S. Nutman (ed.), Symbiotic nitrogen fixation in plants, vol. 7. Cambridge University Press, Cambridge, MA.

Ahrends, R., and G. Moll. 1970. Ein neues knospendes Bakterium aus der Ostsee. Arch. Mikrobiol. 70:243–265.

An, C. S., W. S. Riggsby, and B. C. Mullin. 1985. Relationship of *Frankia* isolates based on deoxyribonucleic acid homology studies. Int. J. System. Bacteriol. 35:140–146.

An, C. S., W. S. Riggsby, and B. C. Mullin. 1987. DNA relatedness of *Frankia* isolates Arl4 and Eul1 to other actinomycetes of cell wall type III. The Actinomycetes 20:50–54.

Baker, D., W. H. Pengelly, and J. G. Torrey. 1981. Immunochemical analysis of relationships among isolated Frankiae (Actinomycetales). Int. J. Syst. Bact. 31:148–151.

Baker, D. D. 1987. Relationships among pure cultured strains of *Frankia* based on host specificity. Physiol. Plant. 70:245–248.

Baker, D. D. 1989. Methods for the isolation, culture and characterization of the Frankiaceae: Soil actinomycetes and symbionts of actinorhizal plants, p. 213–236. In: D. Labeda (ed.), Isolation of microorganisms for biotechnological applications. MacGraw-Hill, New York.

Baker, D. D., W. Newcomb, and J. G. Torrey. 1980. Characterization of an ineffective actinorhizal microsymbiont, *Frankia* sp. Eul1 (Actinomycetales). Can. J. Microbiol. 26:1072–1089.

Baker, D. D., and D. O'Keefe. 1984. A modified sucrose fractionation procedure for the isolation of frankiae from actinorhizal root nodules and soil samples. Plant and Soil 78:23–28.

Baker, D. D., and E. Seling. 1984. *Frankia:* new light on an actinomycete symbiont, p. 563–574. In: L. Ortiz-Ortiz, L. G. Bojalil and V. Yakoleff (ed.), Biological, biochemical, and biomedical aspects of Actinomycetes. Academic Press, Orlando, FL.

Baker, D. D., and J. G. Torrey. 1979. The isolation and cultivation of actinomycetous root nodule endophytes, p. 38–56. In: J. G. Gordon, C. T. Wheeler and D. A. Perry (ed.), Symbiotic nitrogen fixation in the management of temperate forests. Forest Research Laboratory, Orgeon State University, Corvallis, OR.

Baker, D. D., and J. G. Torrey. 1980. Characterization of an effective actinorhizal microsymbiont, *Frankia* sp.

Avcl1 (Actinomycetales). Can. J. Microbiol. 26:1066–1071.

Baker, D. D., J. G. Torrey, and G. H. Kidd. 1979. Isolation by sucrose-density fractionation and cultivation in vitro of actinomycetes from nitrogen-fixing root nodules. Nature 281:76–78.

Becking, J. H. 1970. Frankiaceae fam. nov. *(Actinomycetales)* with one new combination and six new species of the genus *Frankia* Brunchorst 1896, 174. Int. J. Syst. Bact. 20:201–220.

Becking, J. H. 1974. The family Frankiaceae, p. 701–706. In: R. E. Buchanan and N. E. Gibbons (ed.). Bergey's manual of determinative bacteriology. Williams and Wilkins Co., Baltimore.

Becking, J. H. 1981. The genus *Frankia*, p. 1991–2003. In M. P. Starr, H. Stolp, H. G. Trueper, A. Balows, and H. G. Schlegel (ed.), The prokaryotes: A handbook on habitats, isolation and identification of bacteria, vol. 2. Springer-Verlag, Berlin.

Benson, D. R. 1982. Isolation of *Frankia* strains from alder actinorhizal root nodules. Appl. Environ. Microbiol. 44:461–465.

Benson, D. R., D. J. Arp, and R. H. Burris. 1980. Hydrogenase in actinorhizal root nodules and root nodule homogenates. J. Bacteriol. 142:138–144.

Benson, R. D., S. E. Buchhoiz, and D. G. Hanna. 1984. Identification of *Frankia* strains by two-dimensional polyacrylamide gel electrophoresis. Appl. Environ. Microbiol. 47:489–494.

Benson, D. R., and D. Hanna. 1983. *Frankia* diversity in an alder stand as estimated by sodium dodecyl sulfate-polyacryamide gel electrophoresis of whole-cell proteins. Can. J. Bot. 61:2919–2923.

Berry, A. M., and J. G. Torrey. 1979. Isolation and characterization in vivo and in vitro of an actinomycetous endophyte from *Alnus rubra* Bong, p. 69–83. In: J. C. Gordon, C. T. Wheeler and D. A. Perry (ed.), Nitrogen fixation in the management of temperate forests. Forestry Sciences Laboratory, Oregon State University, Corvallis, OR.

Blom, J. 1981. Utilization of fatty acids and NH_4^+ by *Frankia* Avcl1. FEMS Microbiol. Lett. 10:143–145.

Blom, J., and R. Harkink. 1981. Metabolic pathways for gluconeogenesis and energy generation in *Frankia* Avcl1. FEMS Microbiol. Lett. 11:221–224.

Blom, J., W. Roelofsen, and A. D. L. Akkermans. 1980. Growth of *Frankia* Avcl1 on media containing Tween-80 as C-source. FEMS Microbiol. Lett. 9:131–135.

Blom, J., W. Roelofsen, and A. D. L. Akkermans. 1981. Assimilation of nitrogen in root nodules of alder *(Alnus glutinosa).* New Phytol. 89:321–326.

Bomar, M., J. Hermans, T. M. Meesters, R. C. van den Bos, and A. J. B. Zehnder. 1989. Similarities among nitrogenase proteins and among *nif*-HDK genes of *Methanosarcina barkeri* and of other diazotrophic bacteria. Current Microbiology 19:35–38.

Burggraaf, A. J. P., A. Quispel, T. Tak, and I. Valstar. 1981. Methods of isolation and cultivation of *Frankia* species from actinorhizas. Plant and Soil 61:157–168.

Callaham, D., P. Del Tredici, and J. G. Torrey. 1978. Isolation and cultivation in vitro of the actinomycete causing root nodulation in *Comptonia.* Science 199:899–902.

Callaham, D., W. Newcomb, J. G. Torrey, and T. L. Peterson. 1979. Root hair infection in actinomycete induced root nodule initiation in *Casuarina, Myrica,* and *Comptonia.* Bot. Gaz. 140(S):1-9.

Collins, M. D., M. Paulkner, and R. M. Keddie. 1984. Menaquinone composition of some sporeforming actinomycetes. System. Appl. Microbiol. 5:20-29.

Diem, H. G., and Y. R. Dommergues. 1983. The isolation of *Frankia* from nodules of *Casuarina.* Can. J. Bot. 61:2822-2825.

Diem, H. G., D. Gauthier, and Y. R. Dommergues. 1982. Isolation of *Frankia* from nodules of *Casuarina equisetifolia.* Can. J. Microbiol. 28:526-530.

Edmands, J., N. A. Noridge, and D. R. Benson. 1987. Actinorhizal root-nodule symbiont *Frankia*-sp. strain Cpl1 has two glutamine synthetases. Proc. Natl. Acad. Sci. USA 84:6126-6130.

Fernandez. M. P., H. Meugnier, P. A. D. Grimont and R. Bardin. 1989. Deoxyribonucleic acid relatedness in the genus *Frankia.* Int. J. System. Bact. 39:424-429.

Gardes, E., J. Bousquet, and M. Lalonde. 1987. Isozyme variation among 40 *Frankia* strains. Appl. Environ. Microbiol. 53:1596-1603.

Gardes, M., and M. Lalonde. 1987. Identification and subgrouping of *Frankia* strains using sodium dodecyl sulfate-polyacrylamide gel electrophoresis. Physiol. Plant. 70:237-244.

Goodfellow, M. 1986. Actinomycete systematics: present state and future prospects, p. 487-496. In: G. Szabo, S. Biro, M. Goodfellow (ed.), Biological, biochemical and biomedical aspects of Actinomycetes. FEMS Symposium No. 34, Akademiai Kiado, Budapest.

Goodfellow, M., and Pirouz, T. 1982. Numerical classification of sporoactinomycetes containing meso-diaminopimelic acid in the cell wall. J. Gen. Microbiol. 128:503-527.

Hahn, D., M. Dorsch, E. Stackebrandt, and A. D. L. Akkermans. 1989a. Synthetic oligonucleotide probes in identification of *Frankia* strains. Plant and Soil 118:211-219.

Hahn, D., M. P. Lechevalier, A. Fischer, and E. Stackebrandt. 1989b. Evidence for a close phylogenetic relationship between members of the genera *Frankia, Geodermatophilus,* and *"Blastococcus"* and emendation of the family *Frankiaceae.* System. Appl. Microbiol. 11:236-242.

Hahn, D., M. J. C. Starrenburg, and A. D. L. Akkermans. 1988. Variable compatibility of cloned *Alnus glutinosa* ecotypes against ineffective *Frankia* strains. Plant and Soil 106:233-243.

Hahn, D., M. J. C. Starrenburg, and A. D. L. Akkermans. 1990. Oligonucleotide probes that hybridize with rRNA as a tool to study *Frankia* strains in root nodules. Appl. Environ. Microbiol. 56:1342-1346.

Hennecke, H., K. Kaluza, B. Thony, M. Fuhrmann, W. Ludwig, and E. Stackebrandt. 1985. Concurrent evolution of nitrogenase genes and 16S rRNA in *Rhizobium* species and other nitrogen fixing bacteria. Arch. Microbiol. 142:342-348.

Hoagland, D. R., and D. I. Arnon. 1950. The water culture method for growing plants without soil. Calif. Agric. Exp. Stn. Bulletin 347.

Holman, R. M., and C. R. Schwintzer. 1987. Distribution of spore-positive and spore-negative nodules of *Alnus incana* ssp. *rugosa* in Maine, USA. Plant and Soil 104:103-111.

Houwers A., and A. D. L. Akkermans. 1981. Influence of inoculation on yield of *Alnus glutinosa* in the Netherlands. Plant and Soil 61:189-202.

Hungate, B., A. Danin, N. B. Pellerin, J. Stemmler, P. Kjellander, J. B. Adams, and J. T. Staley. 1987. Characterization of manganese-oxidizing (MnII- MnIV) bacteria from Negev Desert rock varnish: implications in desert varnish formation. Can. J. Microbiol. 33:939-943.

Huss-Danell, K., and A.-K. Frej. 1986. Distribution of *Frankia* in soils from forest and afforestation sites in northern Sweden. Plant and Soil 90:407-418.

Ishiguro, E. E., and R. S. Wolfe. 1970. Control of morphogenesis in *Geodermatophilus:* ultrastructural studies. J. Bact. 104:566-580.

Ishiguro, E. E., and R. S. Wolfe. 1974. Induction of morphogenesis in *Geodermatophilus* by inorganic cations and by organic nitrogenous cations. J. Bact. 117:189-195.

Kashanski, C. R., and C. R. Schwintzer. 1987. Distribution of spore-positive and spore-negative nodules of *Myrica gale* in Maine, USA. Plant and Soil 104:113-120.

Lalonde, M. 1977. Infection process of the *Alnus* root nodule symbiosis, p. 586-589. In: W. E. Newton, J. R. Postgate, and C. Rodriguez-Barrueco (ed.), Recent developments in nitrogen fixation. Academic Press, London.

Lalonde, M. 1978. Confirmation of the infectivity of a free-living actinomycete isolated from *Comptonia peregrina* root nodules by immunological and ultrastructural studies. Can. J. Bot. 56:2621-2635.

Lalonde M. 1979. Immunological and ultrastructural demonstration of nodulation of the European *Alnus glutinosa* (L.) Gaertn. host plant by an actinomycetal isolate from the North American *Comptonia peregrina* (L.) Coult. root nodule. Bot. Gaz. 140:35-43.

Lalonde, M., and H. E. Calvert. 1979. Production of *Frankia* hyphae and spores as an infective inoculant for *Alnus* species, p. 95-110. In: J. C. Gordon, C. T. Wheeler, and D. A. Perry (ed.), Symbiotic nitrogen fixation in the management of temperate forests. Forestry Sciences Laboratory, Oregon State University, Corvallis, OR.

Lalonde, M., H. E. Calvert, and S. Pine. 1981. Isolation and use of *Frankia* strains in actinorhizae formation, p. 296-299. In: A. H. Gibson, and W. E. Newton (ed.), Current perspectives in nitrogen fixation. Australian Academy of Sciences, Canberra, Australia.

Lalonde, M., R. Knowles, and J.-A. Fortin. 1975. Demonstration of the isolation of non-infective *Alnus crispa* var. *mollis* Fern. nodule endophyte by morphological immunolabelling and whole cell composition studies. Can. J. Microbiol. 21:1901-1920.

Lalonde, M., L. Simon, J. Bousquet, and A. Seguin. 1988. Advances in the taxonomy of *Frankia:* recognition of species *alni* and *elaeagni* and novel subspecies *pommeri* and *vandijkii,* p. 671-680. In: H. Bothe, F. J. de Bruin, W. E. Newton (ed.), Nitrogen fixation: hundred years after. Gustav Fischer Verlag, Stuttgart, FRG.

Lechevalier. M. P., D. D. Baker, and R. Horrière. 1983. Physiology, chemistry, serology, and infectivity of two *Frankia* isolates from *Alnus incana* subsp. *rugosa.* Can. J. Bot. 61:2826-2833.

Lechevalier, M. P., F. Horrière, and H. A. Lechevalier. 1982. The biology of *Frankia* and related organisms. Devel. Indust. Microbiol. 23:51-60.

Lechevalier, M. P., D. P. Labeda, and J. -S. Ruan. 1987. Studies on *Frankia* sp. LLR 02022 from *Casuarina cunninghamiana* and its mutant LLR 02023. Physiol. Plantarum 70:249–254.

Lechevalier, M. P. and H. A. Lechevalier. 1970. Chemical composition as a criterion in the classification of aerobic actinomycetes. Int. J. System. Bact. 20:435–443.

Lechevalier, M. P., and H. A. Lechevalier. 1979. The taxonomic position of the actinomycetic endophytes, p. 111–122. In: J. C. Gordon, C. T. Wheeler, and D. A. Perry (ed.), Symbiotic nitrogen fixation in the management of temperate forests. Forestry Sciences Laboratory, Oregon State University, Corvallis, OR.

Lechevalier, M. P., and H. A. Lechevalier. 1984. Taxonomy of *Frankia* p. 575–582. In: L. Ortiz-Ortiz, L. F. Bojalil, and V. Yakoleff (ed.), Biological, biochemical and biomedical aspects of Actinomycetes. Academic Press, New York.

Lechevalier, M. P., and J. S. Ruan. 1984. Physiology and chemical diversity of *Frankia* spp. isolated from nodules of *Comptonia peregrina* (L.) Coult. and *Ceanothus americanus* L. Plant and Soil 78:15–22.

Ligon, J. M., and J. P. Nakas. 1987. Isolation and characterization of *Frankia* sp. strain FaC1 genes involved in nitrogen fixation. Appl. Environ. Microbiol. 53:2321–2327.

Lopez, M. F., and J. G. Torrey. 1985. Enzymes of glucose metabolism in *Frankia* sp. J. Bact. 162:110–116.

Luedemann, G. M. 1968. *Geodermatophilus*, a new genus of the Dermatophilaceae (Actinomycetales). J. Bact. 96:1848–1858.

Meesters, T. M. 1987. Localization of nitrogenase in vesicles of *Frankia* sp. Cc1.17 by immunogoldlabelling on ultrathin cryosections. Arch. Microbiol. 146:327–331.

Meesters, T. M. 1988. The function of vesicles in the actinomycete *Frankia*. PhD thesis, Univ. Wageningen, The Netherlands.

Meesters, T. M., S. Th. van Genesen, and A. D. L. Akkermans. 1985. Growth, acetylene reduction activity and localization of nitrogenase in relation to vesicle formation in *Frankia* strains Cc1.17 and Cp1.2. Arch. Microbiol. 145:137–142.

Meesters, T. M., W. M. van Vliet, and A. D. L. Akkermans. 1987. Nitrogenase is restricted to the vesicles in *Frankia* strain EAN1pec. Physiol. Plantarum 70:267–271.

Mian, S., G. Bond, and C. Rodriguez-Barrueco. 1976. Effective and ineffective root nodules in *Myrica faya*. Proc. R. Soc. London B. 194:285–293.

Miettienen J. K., and A. I. Virtanen. 1953. The free amino acids in the leaves, roots and root nodules of the alder *(Alnus)*. Physiol. Plantarum 5:540–557.

Miller, I. M., and D. D. Baker. 1985. Initiation, development and structure of root nodules in *Elaeagnus angustifolia* L. (Elaeagnaceae). Protoplasma 128:107–119.

Miller, I. M., and D. D. Baker. 1986. Nodulation of actinorhizal plants by *Frankia* strains capable of both root hair infection and intercellular penetration. Protoplasma 131:82–91.

Mort, A., P. Normand, and M. Lalonde. 1983. 2-*O*-Methyl-D-mannose, a key sugar in the taxonomy of *Frankia*. Can. J. Microbiol. 29:993–1002.

Murry, M. A., M. S. Fontaine, and J. G. Torrey. 1984. Growth kinetics and nitrogenase induction in *Frankia* sp. HFPArI3 grown in batch culture. Plant and Soil 78:61–78.

Newcomb, W., D. Baker, and J. G. Torrey. 1987. Ontogeny and fine structure of effective root nodules of the autumn olive *(Elaeagnus umbellata)*. Can. J. Bot. 65:80–94.

Newcomb, W., D. Callaham, J. G. Torrey, and R. L. Peterson. 1979. Morphogenesis and fine structure of the actinomycetous endophyte of nitrogen-fixing root nodules of *Comptonia peregrina*. Bot. Gaz. 140(S):22–34.

Noridge, N. H., and D. R. Benson. 1986. Isolation and nitrogen-fixing activity of *Frankia* sp. strain Cpl1 vesicles. J. Bact. 166:301–305.

Normand, P., and M. Lalonde. 1986. The genetics of actinorhizal *Frankia*: a review. Plant and Soil 90:429–453.

Normand, P., P. Simonet, and R. Bardin. 1988. Conservation of Nif sequences in *Frankia*. Mol. Gen. Genet. 213:238–246.

Parsons, R., W. B. Silvester, S. Harris, W. T. M. Gruijters, and S. Bullivant. 1987. *Frankia* vesicles provide inducible and absolute oxygen protection for nitrogenase. Plant Physiol. 83:728–731.

Pommer, E. H. 1959. Ueber die Isolierung des Endophyten aus den Wurzelknoellchen *Alnus glutinosa* Gaertn. und ueber erfolgreiche Re-infektionsversuche. Berichte der Deutschen Botanischen Gesellschaft 72:138–150.

Puppo, A., L. Dimitrijevik, H. G. Diem, and Y. R. Dommergues. 1985. Homogeneity of superoxyde dismutase patterns in *Frankia* strains from Casuarinaceae. FEMS Microbiol. Lett. 30:43–46.

Quispel, A., A. J. P. Burggraaf, A. Baerheim Svendsen, J. Schripsema, J. Lugtenburg, C. Erkelens, and W. J. Baas. 1988. Identification of an isolation factor for the endophytic phase of *Frankia* in root nodules of *Alnus glutinosa* as dipterocarpol, p. 697. In: H. Bothe, F. J. de Bruijn, and W. E. Newton (ed.), Nitrogen fixation: Hundred years after. Gustav Fischer, Stuttgart, FRG.

Quispel, A., A. B. Svendsen, J. Schripsema, W. J. Baas, C. Erkelens, and J. Lugtenburg. 1989. Identification of dipterocarpol as isolation factor for the induction of primary isolation of *Frankia* from root nodules of *Alnus glutinosa* (L.) Gaertner. Molecular Plant-Microbe Interaction 2:107–112.

Quispel A., and T. Tak. 1978. Studies on the growth of the endophyte of *Alnus glutinosa* (L.) Vill. in nutrient solutions. New Phytol. 81:587–600.

Safo-Sampah, S. and J. G. Torrey. 1988. Polysaccharide hydrolyzing enzymes of *Frankia* (Actinomycetales). Plant and Soil 112:89–97.

Samsonoff, W. A., M. A. Detlefsen, A. F. Fonseca, and M. R. Edwards. 1977. Deoxyribonucleic acid base composition of *Dermatophilus congolensis* and *Geodermatophilus obscurus*. Int. J. System. Bact. 27:22–25.

Sasakawa, H., T. Hiyoshi, and T. Sugiyama. 1988. Immunogold localization of nitrogenase in root nodules of *Elaeagnus pungens* Thunb. Plant Cell Physiol. 29:1147–1152.

Sellstedt, A., K. Huss-Danell, and A.-S. Ahlqvist. 1986. Nitrogen fixation and biomass production in symbioses between *Alnus incana* and *Frankia* strains with different hydrogen metabolism. Physiol. Plant. 66:99–107.

Shipton, W. A., A. J. P. Burggraaf. 1982. A comparison of the requirements for various carbon and nitrogen sources and vitamins in some *Frankia* isolates. Plant and Soil 69:149–161.

Simon, L., J. Jabaji-Hare, J. Bousquet, and M. Lalonde. 1989. Confirmation of *Frankia* species using cellular fatty acid analysis. System. Appl. Microbiol. 11:229–235.

Simonet, P., J. Haurat, P. Normand, R. Bardin, and A. Moiroud. 1986. Localisation of *nif* genes on a large plasmid in *Frankia* sp. strain ULQ0132105009. Mol. Gen. Genet. 204:492–495.

Simonet, P., P. Normand, and R. Bardin. 1988a. Heterologous hybridization of *Frankia* DNA to *Rhizobium meliloti* and *Klebsiella pneumoniae* nif genes. FEMS Microbiol. Lett. 55:141–146.

Simonet, P., P. Normand, A. M. Hirsch, and A. D. L. Akkermans. 1990. The genetics of the *Frankia*-actinorhizal symbiosis, p. 77–109. In: P. M. Gresshoff (ed.), The molecular biology of nitrogen fixation, C.R.C. Critical reviews, C.R.C. Press, Inc. Boca Raton, FL.

Simonet, P., N. Thi Le, E. Teissier du Cros, and R. Bardin. 1988b. Identification of *Frankia* strains by direct DNA hybridization of crushed nodules. Appl. Environ. Microbiol. 54:2500–2503.

Smolander, A., and V. Sundman. 1987. *Frankia* in acid soils of forests devoid of actinorhizal plants. Physiol. Plantarum 70:297–303.

Stackebrandt, E., R. M. Kroppenstedt, and V. J. Fowler. 1983. A phylogenetic analysis of the family Dermatophilaceae. J. Gen. Microbiol. 129:1831–1838.

Steele, D. B., and M. D. Stowers. 1986. Superoxide dismutase and catalase in *Frankia*. Can. J. Microbiol. 32:409–413.

St-Laurent, L., J. Bousquet, L. Simon, and M. Lalonde. 1987. Separation of various *Frankia* strains in the *Alnus* and *Elaeagnus* host specificity groups using sugar analysis. Can. J. Microbiol. 33:764–772.

Stowers, M. D. 1987. Collection, isolation, cultivation, and maintenance of *Frankia*, p. 29–53. In: G. H. Elkan (ed.), Symbiotic nitrogen fixation technology. Marcel Dekker, New York.

Tjepkema, J. D., W. Ormerod, and J. G. Torrey. 1980. Vesicle formation and acetylene reduction activity in *Frankia* sp. CpI1 cultured in defined nutrient media. Nature 287:633–635.

VandenBosch, K. A. and J. G. Torrey. 1983. Host-endophyte interactions in effective and ineffective nodules induced by the endophyte of *Myrica gale*. Can. J. Bot. 61:2898–2909.

van Dijk, C. 1978. Spore formation and endophyte diversity in root nodules of *Alnus glutinosa* (L.) Vill. New Phytol. 81:601–615.

van Dijk, C., and E. Merkus. 1976. A microscopical study of the development of a spore-like stage in the life cycle of the root-nodule endophyte of *Alnus glutinosa* (L.) Gaertn. New Phytol. 77:73–91.

van Dijk, C., A. Sluimer, and A. Weber. 1988. Host range differentiation of spore-positive and spore-negative strain types of *Frankia* in stands of *Alnus glutinosa* and *Alnus incana* in Finland. Physiol. Plantarum 72:349–358.

van Dijk, C., and A. Sluimer-Stolk. 1990. An ineffective strain type of *Frankia* in the soil of natural stands of *Alnus glutinosa* (L.) Gaertn. Plant and Soil 123: in press.

Wayne, L. G., D. J. Brenner, R. R. Colwell, P. A. D. Grimont, O. Kandler, M. I. Krichevsky, L. H. Moore, W. E. C. Moore, R. G. E. Murray, E. Stackebrandt, M. P. Starr, and H. G. Trueper. 1986. Report of the ad hoc committee on recognition of approaches to bacterial systematics. Int. J. Syst. Bacteriol. 37:463–464.

CHAPTER 46

The Family Thermomonosporaceae

REINER MICHAEL KROPPENSTEDT and MICHAEL GOODFELLOW

Introduction

The family Thermomonosporaceae was introduced for a morphologically diverse group of organisms classified in the genera *Actinomadura, Microbispora, Microtetraspora, Saccharomonospora*, and *Thermomonospora* (Cross and Goodfellow, 1973). This avowedly heterogeneous assemblage was formed in order to gather together genera that had more in common than the assortment of organisms previously classified in the family Nocardiaceae, an area long regarded as a dumping ground for aerobic actinomycetes (M. P. Lechevalier, 1976).

The new family Thermomonosporaceae encompassed mesophilic and thermophilic actinomycetes that formed heat-sensitive, nonmotile spores that were carried singly, in pairs, or in short chains on aerial hyphae or on both aerial and substrate mycelia, and which had a cell wall containing *meso*-diaminopimelic acid (*meso*-A$_2$pm) but no other characteristic amino acid or sugars (wall chemotype III; M. P. Lechevalier and H. A. Lechevalier, 1970a, 1970b). The genus *Saccharomonospora*, which also has *meso*-A$_2$pm as the diamino acid of the cell-wall peptidoglycan and a polysaccharide fraction rich in arabinose and galactose (wall chemotype IV), was included in the family, as it was considered to share some morphological similarities with Thermomonospora species. Other wall chemotype III sporoactinomycetes subsequently thought to be clearly related to *Thermomonospora* included the genera *Actinobifida, Actinosynnema, Nocardiopsis, Saccharothrix*, and *Streptoalloteichus* (Goodfellow, 1989; Goodfellow and Cross, 1984; Küster, 1974; McCarthy, 1989).

The application of chemical and molecular systematic methods has provided invaluable data for the classification of actinomycetes above the genus level (Goodfellow and Minnikin, 1985; Stackebrandt and Goodfellow, 1990; Stackebrandt and Woese, 1981) and shows that morphological features previously weighted for the circumscription of families and genera have little predictive value (Goodfellow, 1989; Stackebrandt, 1986; Stackebrandt and Schleifer, 1984). It soon became evident that the family Thermomonosporaceae (Cross and Goodfellow, 1973) contained markedly diverse taxa and that the genus *Actinomadura* (H. A. Lechevalier and M. P. Lechevalier, 1970) included two groups of organisms with little in common (Fischer et al,. 1983; Poschner et al., 1985). Ribosomal RNA partial-oligonucleotide sequencing (Fowler et al., 1985; Goodfellow et al., 1988) underlined this heterogeneity and showed that *A. madurae*, the type species of the genus, showed a closer affinity to *Thermomonospora curvata* than to the *A. pusilla* group; members of the latter were related to *Streptosporangium roseum*.

The division of the genus *Actinomadura* into two disparate groups was formally recognized when Kroppenstedt et al. (1990) proposed that the genus be retained for *A. madurae* and related species, and that the *A. pusilla* group be reclassified in the genus *Microtetraspora*. Similarly, several genera formerly associated with *Thermomonospora* have been transferred to other suprageneric groups. *Microbispora* and *Microtetraspora* have been assigned to the family Streptosporangiaceae and *Saccharomonospora* and *Saccharothrix* to the family Pseudonocardiaceae (see Chapter 42). The genus *Nocardiopsis* can be sharply separated from *Thermomonospora* (Fowler et al., 1985; Kroppenstedt et al., 1990), which can readily be distinguished from the poorly studied genera *Actinosynnema* and *Streptoalloteichus*. These developments leave *Actinomadura* and *Thermomonospora* as a relatively homogeneous group corresponding to other actinomycete families (Kroppenstedt et al., 1990; Stackebrandt et al., 1983a, 1983b).

It is becoming increasingly clear that the family Thermomonosporaceae should be retained for aerobic, Gram-positive, non-acid-alcohol-fast, chemoorganotrophic actinomycetes which produce a branched substrate mycelium bearing aerial hyphae that undergo differentiation into single or short chains of arthrospores. Constituent strains have meso-A₂pm acid in a wall peptidoglycan that lacks characteristic sugars (M. P. Lechevalier and H. A. Lechevalier, 1970a, 1970b), an Al-γ-peptidoglycan type (Schleifer and Kandler, 1972), N-acetylated muramic acid (Uchida and Aida, 1977), mixtures of straight and branched fatty acids, hydrogenated menaquinones with nine isoprenic units, and major amounts of phosphatidylglycerol, phosphatidylinositol, and phosphatidylinositol mannosides. Minor amounts of phosphatidylglycerol may be present in some strains (Kroppenstedt, 1985, 1987; Kroppenstedt et al., 1990). In addition, whole-organism hydrolysates of actinomadurae usually contain madurose (3-O-methyl-D-galactose; M. P. Lechevalier and Gerber, 1970). The GC content of the DNA lies within the range of 66 to 72 mol% (Fischer et al., 1983; Hasegava et al., 1986; Miyadoh et al., 1987; Poschner et al., 1985).

The revised genus Actinomadura (Kroppenstedt et al., 1990) currently accommodates 19 validly described species (Table 1) that characteristically form nonfragmenting, extensively branched substrate mycelia and aerial hyphae which carry up to fifteen arthrospores. Spore chains may be straight, hooked (open loops), or irregular spirals (1–4 turns) and spore surfaces may be folded, irregular, smooth, spiny, or warty. The growth temperature range is from 10 to 60°C. Actinomadurae contain major proportions of hexahydrogenated menaquinones with nine isoprenic units saturated at sites II, III, and VIII, and complex mixtures of fatty acids with hexadecanoic, 14-methylpentadecanoic, and 10-methyloctadecanoic acid predominating. Whole-organism hydrolysates contain the following sugars: galactose, glucose, mannose, ribose, and madurose (the latter sometimes only in trace amounts).

The genus Excellospora (Agre and Guzeva, 1975) was proposed for thermophilic actinomycetes that were distinguished from actinomadurae primarily by differences in fatty acid composition. The genus contained three species: the type species Excellospora viridilutea, which is cited in the Approved Lists of Bacterial Names (Skerman et al., 1980), and E. rubrobrunea and E. viridinigra, which had previously been classified in the genus Micropolyspora (Krassilnikov et al., 1968). Further studies (Greiner-Mai et al., 1987; Kroppenstedt et al., 1990; Meyer, 1989a) showed that E. rubrobrunea and E. viridinigra had many properties in common with one another and with Actinomadura madurae and related species. It is, however, true that excellosporae can be separated from actinomadurae as they contain relatively

high proportions of iso-branched fatty acids (high melting point) and low amounts of 10-methyl branched acids (low melting point) but these differences could be attributed to the thermophilic nature of Excellospora strains. Kroppenstedt et al. (1990) proposed that E. viridinigra be recognized as a synonym of E. rubrobrunea and that the latter be transferred to the amended genus Actinomadura as A. rubrobrunea. It is also clear that Excellospora viridilutea has properties (Table 2) consistent with its proposed inclusion in the genus Actinomadura. In addition, Table 2 shows the inhomogeneity of the thermomonosporas. Based on their chemical markers they form three well defined unrelated clusters which are very clearly separated from all other sporoactinomycetes. These data correspond very well with molecular genetic results (Fischer et al., 1983; Poschner et al., 1985; Stackebrandt, personal communication).

The generic name Thermomonospora was proposed by Henssen (1957) for thermophilic actinomycetes isolated from composted horse manure. The genus then contained three species which formed single spores on the aerial mycelium only. All produced colorless-to-pale yellow colonies and a white aerial mycelium, but were distinguished from one another by aerial mycelium morphology and the type of branching shown by the substrate hyphae. Thermomonospora curvata, the only species isolated and maintained in pure culture, was later named as the type species of the genus (Henssen and Schnepf, 1967). The description of the remaining two species, T. fusca and T. lineata, was based on morphological properties in contaminated preparations. Neither of these species were included in the Approved Lists of Bacterial Names (Skerman et al., 1980), even though T. fusca had in the meantime been isolated in pure culture and described in detail (Crawford, 1975; Crawford and Gonda, 1977). Mesophilic monosporic actinomycetes were subsequently assigned to the genus as T. mesophila (Nonomura and Ohara, 1971c) and T. mesouviformis (Nonomura and Ohara, 1974). A third mesophilic species, T. formosensis, was described by Hasegawa et al. (1986); this taxon is cited as a species incertae sedis in Bergey's Manual of Systematic Bacteriology (McCarthy, 1989).

Krassil'nikov and Agre (1964b) proposed the genus Actinobifida for actinomycetes forming single spores on dichotomously branched sporophores but failed to mention that dichotomous branching had previously been observed in the genus Thermomonospora (Henssen, 1957) and in species of Micromonospora (Jen-

Table 1. Taxonomic history of actinomycetes designated or once assigned to the genera *Actinomadura*, *Excellospora*, *Microtetraspora*, *Nocardiopsis*, *Saccharothrix*, and *Thermomonospora*.[a]

Original assignment	Type strain (DSM number[b])	Current designation in: Bergey's Manual of Systematic Bacteriology, vol. 4 (Williams et al., 1989)	Current designation in: Approved Lists of Bacterial Names (Skerman et al., 1980 plus subsequent listings)
Genus *Actinobifida*			
Actinobifida alba (1)	43795	*Thermomonospora alba* (2)	*Thermomonospora alba*
Actinobifida chromogena (3)	43794	*Thermomonospora chromogena* (4)	*Thermomonospora chromogena*
Genus *Actinomadura*			
A. africana (5, 6)	43748	*Nocardiopsis africana* (7)	*Microtetraspora africana* (8)
A. atramentaria (9)	43919	—[f]	*A. atramentaria*
A. aurantiaca (10)	43924	*A. aurantiaca*	*A. aurantiaca*
A. carminata (11)	INA 4281[c]	"*A. carminata*"[g]	*A. carminata*
A. citrea (12)	43461	*A. citrea*	*A. citrea*
A. coerulea (13)	43675	*A. coerulea*	*A. coerulea*
A. coeruleofusca (5)	43679	*Nocardiopsis coeruleofusca* (7)	*Saccharothrix coeruleofusca* (14)
A. coeruleoviolacea (15, 16)	43935	"*A. coeruleoviolacea*"	*Saccharothrix coeruleoviolacea* (8)
A. cremea (13)	43676	*A. cremea*	*A. cremea* subsp. *cremea* (17)
A. cremea subsp. *rifamicini* (17)	43936	—	*A. cremea* subsp. *rifamicini*
A. dassonvillei (18)	43111	*Nocardiopsis dassonvillei* subsp. *dassonvillei* (19, 20)	*Nocardiopsis dassonvillei* (21)
Nocardia dassonvillei (22)			
Streptothrix dassonvillei (23)			
A. fastidiosa (24)	43674	*A. fastidiosa*	*Microtetraspora fastidiosa* (8)
A. ferruginea (25, 26)	43563	*A. ferruginea*	*Microtetraspora ferruginea* (8)
A. fibrosa (27)	NRRL 18348[d]	—	*A. fibrosa*
A. flava (5)	43885	*Nocardiopsis flava* (7)	*Saccharothrix flava* (14)
A. fulvescens (28, 29)	43923	—	*A. fulvescens*
A. helvata (30)	43142	*A. helvata*	*Microtetraspora helvata* (8)
A. kijaniata (31)	43764	*A. kijaniata*	*A. kijaniata*
A. libanotica (25, 26)	43554	*A. libanotica*	*A. libanotica*
A. livida (10)	43677	*A. livida*	*A. livida*
A. longispora (5)	43749	*Nocardiopsis longispora* (7)	*Saccharothrix longispora* (14)
A. macra (56)	43862	*A. macra*	*A. macra*
A. madurae (18)	43067	*A. madurae* (18)	*A. madurae*
Nocardia madurae (32)			
Streptomyces madurae (33)			
Streptothrix madurae (34)			
A. malachitica (12)	43462	*A. malachitica*	*A. viridis* (8, 35)
A. oligospora (36)	43930	"*A. oligospora*"	*A. oligospora*
A. pelletieri (18)	43383	*A. pelletieri* (18)	*A. pelletieri*
Micrococcus pelletieri (37)			
Nocardia pelletieri (38)			
Streptomyces pelletieri (33)			
A. polychroma (39, 40)	43925	—	*Microtetraspora polychroma* (8)
A. pusilla (30)	43357	*A. pusilla*	*Microtetraspora pusilla* (8)
A. recticatena (16, 41)	43937	"*A. recticatena*"	*Microtetraspora recticatena* (8)
A. roseola (10)	43767	*A. roseola*	*Microtetraspora roseola* (8)
A. roseoviolacea (30)	43144	*A. roseoviolacea*	*Microtetraspora roseoviolacea* (8)
A. salmonea (13)	43678	*A. salmonea*	*Microtetraspora salmonea* (8)
A. spadix (30)	43459	*A. spadix*	*A. spadix*
A. spiralis (25, 26)	43555	*A. spiralis*	*Microtetraspora spiralis* (8)
A. turkmeniaca (28, 29)	43926	—	*Microtetraspora turkmeniaca* (8)

(*continued*)

Table 1. *Continued*

| | | Current designation in: | |
| | | | |
Original assignment	Type strain (DSM number[b])	*Bergey's Manual of Systematic Bacteriology, vol. 4* (Williams et al., 1989)	*Approved Lists of Bacterial Names* (Skerman et al., 1980 plus subsequent listings)
A. umbrina (39, 40)	43927	—	*A. umbrina* (8)
A. verrucosospora (30)	43358	*A. verrucosospora*	*A. verrucosospora*
A. vinacea (10)	43765	*A. vinacea*	*A. vinacea*
A. yumaensis (42)	43931	*A. yumaensis*	*A. yumaensis*
Genus *Microbispora*			
Microbispora echinospora (43)	43163	*M. echinospora*	*A. echinospora* (8, 35)
Genus *Micromonospora*			
Micromonospora rubra (44)	43768	*A. rubra* (45)	*Microtetraspora rubra* (8)
Genus *Micropolyspora*			
Micropolyspora rubrobrunea (46)	43750	"*E. rubrobrunea*"	*A. rubrobrunea* (6, 8)
Excellospora rubrobrunea (47)			*A. rubrobrunea* (6, 8)
Micropolyspora viridilutea (46)	INMI 187[c]	"*E. viridilutea*"	
Excellospora viridilutea (47)		—	*Excellospora viridilutea*
Micropolyspora viridinigra (46)	43751	"*E. viridinigra*"	*A. rubrobrunea* (6)
Excellospora viridinigra (47)			*A. rubrobrunea* (6)
Genus *Microtetraspora*			
Microtetraspora viridis (48)	43175	*Microtetraspora viridis*	*A. viridis* (8, 35)
Genus *Streptomyces*			
Streptomyces luteofluorescens (49)	40398	*A. luteofluorescens* (50)	*A. luteofluorescens*
Genus *Thermomonospora*			
T. curvata (51)	43183	*T. curvata*	*T. curvata*
T. formosensis (52)	43997	*T. formosensis*	*T. formosensis*
T. fusca (51)	43792	*T. fusca* (4)	*T. fusca*
T. mesophila (53)	43048	*T. mesophila*	*T. mesophila*
T. mesouviformis (54)	43185	*T. alba* (4)	*T. mesouviformis*
Genus *Thermopolyspora*			
Thermopolyspora flexuosa (55)	43186	*A. flexuosa* (2)	*Microtetraspora flexuosa* (8)

[a]The numbers in parentheses indicate the reference as follows: (1) Locci et al. (1967); (2) Cross and Goodfellow (1973); (3) Krassil'nikov and Agre (1965); (4) McCarthy and Cross (1984a); (5) Preobrazhenskaya and Sveshnikova (1974); (6) Greiner-Mai et al. (1987); (7) Preobrazhenskaya et al. (1982); (8) Kroppenstedt et al. (1990); (9) Miyadoh et al. (1987); (10) Lavrova and Preobrazhenskaya (1975); (11) Gauze et al. (1973); (12) Lavrova et al. (1972); (13) Preobrazhenskaya et al. (1975b); (14) Grund and Kroppenstedt (1989); (15) Preobrazhenskaya et al. (1976); (16) Preobrazhenskaya et al. (1987); (17) Gauze et al. (1987); (18) H. A. Lechevalier and M. P. Lechevalier (1970); (19) Miyashita et al. (1984); (20) Meyer (1976); (21) Grund and Kroppenstedt (1990); (22) Liegard and Landrieu (1911); (23) Brocq-Rousseu (1904); (24) Soina et al. (1975); (25) Meyer (1981); (26) Meyer (1989a); (27) Mertz and Yao (1990); (28) Terekhova et al. (1982); (29) Terekhova et al. (1987); (30) Nonomura and Ohara (1971d); (31) Horan and Brodsky (1982); (32) Blanchard (1896); (33) Waksman and Henrici (1948); (34) Vincent (1894); (35) Miyadoh et al. (1989); (36) Mertz and Yao (1986); (37) Laveran (1906); (38) Pinoy (1912); (39) Galatenko et al. (1981); (40) Galatenko et al. (1987); (41) Gauze et al. (1984); (42) Labeda et al. (1985); (43) Nonomura and Ohara (1971b); (44) Sveshnikova et al. (1969); (45) Meyer and Sveshnikova (1974); (46) Krassil'nikov et al. (1968); (47) Agre and Guzeva (1975); (48) Nonomura and Ohara (1971a); (49) Shinobu (1962); (50) Preobrazhenskaya et al. (1975a); (51) Henssen (1957); (52) Hasegawa et al. (1986); (53) Nonomura and Ohara (1971c); (54) Nonomura and Ohara (1974); (55) Krassil'nikov and Agre (1964a); and (56) Huang (1980).
[b]DSM, Deutsche Sammlung von Mikroorganismen und Zellkulturen, Braunschweig, Germany.
[c]INA, Institute for Research on New Antibiotics, Moscow, USSR.
[d]NRRL, U.S. Department of Agriculture, Northern Regional Research Laboratory Collection, Peoria, Illinois, USA.
[e]INMI, Institute for Microbiology, USSR Academy of Sciences, Moscow, USSR.
[f]Species not listed in *Bergey's Manual of Systematic Bacteriology* (Williams et al., 1989).
[g]Species in quotation marks are listed in *Bergey's Manual* as species incertae sedis.

sen, 1930, 1932; Krassilnikov, 1941). The following year these workers introduced a second species, *Actinobifida chromogena,* and in doing so suggested that all actinomycetes showing dichotomous branching be transferred to the genus *Actinobifida.* A third species, *A. alba,* was proposed by Locci et al. (1967). *A. dichotomica,* the type species of the genus, was later transferred to the genus *Thermoactinomyces* because of its ability to produce endospores, and *A. alba* to the genus *Thermomonospora* because it formed heat-sensitive spores on substrate and aerial hyphae (Cross and Goodfellow, 1973).

A comprehensive numerical taxonomic survey of *Thermomonospora* and related organisms (McCarthy and Cross, 1984a, 1984b) confirmed the status of *T. curvata* and provided strong evidence for the formal recognition of *T. fusca.* In contrast, *T. mesouviformis* was considered to be a synonym of *T. alba.* These taxa, termed the "white *Thermomonospora* group" because of their white aerial mycelium, were sharply distinguished from organisms which included the type strains of *Actinobifida chromogena* (Krassil'nikov and Agre, 1965), *"Thermomonospora falcata"* (Henssen, 1970) and similar actinomycetes from mushroom compost (McCarthy and Cross, 1981). The *"chromogena"* strains with reddish-brown colonies and a light-brown aerial mycelium had been provisionally included in the genus *Thermomonospora* (Cross, 1981) because of their wall composition and morphology. In the meantime *Thermomonospora viridis* (Küster and Locci, 1963) had been transferred to the genus *Saccharomonospora* (Nonomura and Ohara, 1971c).

The discontinuous distribution of chemical markers underlines the heterogeneity of the genus *Thermonospora* and suggests that the constituent species can be assigned to three distinct groups (Tables 2 and 5). The first group, which contains *T. curvata,* the type species, and *T. formosensis,* should also perhaps provide a home for *Actinomadura* species. Members of the genus *Actinomadura* have many chemical features in common with *T. curvata* (Greiner-Mai et al. 1987; Kroppenstedt, 1987) and have been shown to be clearly related to the latter in 16S rRNA oligonucleotide-sequencing studies (Fowler et al., 1985; Kroppenstedt et al., 1990). Similarly, chemical data support the transfer of *T. chromogena* and *T. mesophila* to the revised genus *Microtetraspora* (Kroppenstedt et al., 1990). These proposals, if supported by other lines of taxonomic evidence would leave *T. alba* (including *T. mesouviformis)* and *T. fusca,* as related species meriting generic status.

The family Thermomonosporaceae, as defined above, encompasses *Actinomadura* species (including *Excellospora viridilutea*), *Thermomonospora curvata,* and *T. formosensis. T. chromogena* and *T. mesophila* have properties consistent with their classification in the family Streptosporangiaceae (Goodfellow et al., 1990) though the description of the latter would need to be amended to include monosporic actinomycetes. Further comparative studies are needed to determine the exact suprageneric relationships of *T. alba* and *T. fusca.* The taxonomic history of actinomycetes designated or once assigned to the genera *Actinomadura, Excellospora, Thermomonospora,* and related taxa are shown in Table 1.

Habitats of *Actinomadura*

Actinomadura species are widely distributed in soil (Labeda et al., 1985; Meyer, 1979; Miyadoh et al., 1987; Nonomura and Ohara, 1971a, 1971d; Preobrazhenskaya et al., 1975b) where they probably have a role in the turnover of organic matter. *Actinomadura madurae* and *A. pelletieri* are agents of actinomycete mycetoma. *A. pelletieri* has only been isolated from clinical material but *A. madurae* seems to be widespread in soil. Isolates of the latter from environmental samples lack the red endopigment of clinical isolates and sporulate more readily. The red pigments are characterized by a tripyrrole skeleton and have been identified as cyclononylprodiginine, nonylprodiginine, and undecylprodiginine (Gerber, 1971, 1973; H. A. Lechevalier et al., 1971).

Information on the occurrence and frequency of *Actinomadura* species in different soils has mainly been provided by Preobrazhenskaya and her coworkers (Chormonova and Preobrazhenskaya, 1981; Galatenko et al., 1981; Preobrazhenskaya et al., 1978; Terekhova et al., 1982) who found that the total number of *Actinomadura* strains were higher in cultivated than in uncultivated soil. The highest number of actinomadurae were isolated from chernozem soil (Kazakhstan) and the lowest from sierozem soil (Turkmenistan). Dark chestnut soil (cultivated) contained half the number of actinomadurae found in the chernozem soil, but the species diversity in each of the soil types was similar. *Actinomadura citrea* was the most frequent species in the uncultivated and cultivated soils followed by *A. cremea* and *A. verrucosospora. A. (Excellospora) viridilutea* was isolated from an arid soil from the Soviet Union, and *A. rubrobrunea* from Egyptian soils under maize and rice (Agre and Guzeva, 1975).

Table 2. Guide to the separation of *Actinomadura*, *Excellospora*, and *Thermomonospora* species from morphologically related sporoactinomycetes using chemotaxonomic markers.[a]

	Wall chemotype[b]	Diagnostic sugar	Sugar type[c]	GC content (mol%)	Diagnostic phospholipids[d]	Phospholipid type[e]	Principal menaquinones[f]	Menaquinone type[g]	Fatty acid type[h]
Actinomadura	III	Madurose	B	66–70	PIM, PI, PG, DPG	I	$9/4, 9/6^1, 9/8$	4B2	3a
Excellospora viridilutea	III	Madurose	B	ND	PIM, PI, PG, DPG	I	$9/4, 9/6^1, 9/8$	4B2	3a
Thermomonospora curvata	III	None	C	ND	PIM, PI, PG, DPG	I	$9/4, 9/6^1, 9/8$	4B2	3a
T. formosensis	III	Madurose	B	72	PIM, PI, PG, DPG	I	$9/2, 9/4, 9/6$	4B	3a
Microtetraspora	III	Madurose	B	66–69	PE, OH-PE, NPG	IV	$9/2, 9/4^2, 9/6$	4A2	3c
Thermomonospora chromogena	III	None	C	ND	PME, OH-PE, NPG	IV	$9/2, 9/4^2, 9/6$	4A2	3c
T. mesophila	III	None	C	ND	PME, OH-PE, NPG	IV	$9/2, 9/4^2, 9/6$	4A2	3c
Thermomonospora alba	III	None	C	ND	ND		$10/6, 10/8, 11/6$	4D	3e
T. fusca	III	None	C	ND	PE, PME, GL	II	$10/6, 10/8, 11/6$	4D	3e
T. mesouviformis	III	None	C	ND	PE, PME, GL	II	$10/6, 10/8, 11/6$	4D	3e
Actinosynnema	III	None	C	71–73	PE, OH-PE, LPE	II	$9/4$	2D	3f
Nocardiopsis	III	None	C	64–69	PC, PME, GL	III	10/0 to 10/6	4C	3d
Streptoalloteichus	III	None	C	ND	PE, PI, DPG	II	$9/4, 10/4$	3B	ND
Saccharothrix	III	Rhamnose	E	70–76	PE, PI, DPG	II	$9/6, 10/6$	3C	3g
Faenia	IV	Arabinose	A	70–72	PC, PME, LPE	III	$9/2, 9/4, 10/4$	3/B	3e
Nocardia	IV	Arabinose	A	64–72	PE,	II	$8/4c^3$	2E	1b
Saccharomonospora	IV	Arabinose	A	69–74	PE, OH-PE	II	$8/4, 9/4$	3A	2a
Saccharopolyspora	IV	Arabinose	A	77	PC, PME, LPE	III	$9/2, 9/4^1, 10/4$	3B/4A1	3e
Streptomyces	I	Not characteristic		66–75	PE, OH-PE var.	II	$9/4, 9/6^2, 9/8$	4B1	2c

(continued)

[a]All organisms lack mycolic acids and contain N-acetylated muramic acid[o] except for the *Nocardia* strains, which contain mycolic acids[i] and have N-glycolated muramic acid[i] in the wall peptidoglycan. ND, not determined. Principal menaquinones are given in italics. Dashed lines separate the three chemotypes found among members of the family Thermomonosporaceae.

[b]Major constituents in cell walls of chemotype: I, LL = A_2pm; II, *meso*-diaminopimelic acid (*meso*-A_2pm) and glycine; III, *meso*-A_2pm, and IV, *meso*-A_2pm, arabinose, and galactose (H. A. Lechevalier and M. P. Lechevalier, 1970). The wall chemotypes correspond to the following peptidoglycan types (Schleifer and Kandler, 1972): I with A3γ, and III and IV with A1γ.

[c]Whole-organism sugar patterns of actinomycetes containing *meso*-A_2pm: A, arabinose and galactose; B, madurose (3-O-methyl-D-galactose; C, no diagnostic sugars; D, xylose; E, rhamnose (Labeda et al., 1984; H. A. Lechevalier and M. P. Lechevalier, 1970; M. P. Lechevalier and H. A. Lechevalier, 1970a, 1970b; H. A. Lechevalier et al., 1971).

[d]Abbreviations: PC, phosphatidylcholine; PE, phosphatidylethanolamine; OH-PE, hydroxy-phosphatidylethanolamine; PME, phosphatidylmethylethanolamine; NPG, phosphatidyl-N-acetylglucoseamin; GL, glycolipid. Other nondiagnostic phospholipids may also be found, i.e., DPG, diphosphatidylglycerol; PG, phosphatidylglycerol; PI, phosphatidylinositol, and PIM, phosphatidylinositol mannosides. The occurrence of OH-PE is species specific in *Streptomyces* (R. M. Kroppenstedt, unpublished observations).

[e]Phospholipid types according to M. P. Lechevalier et al. (1977, 1981).

[f]The numbers indicate as follows: for example, 9/6 = MK-9(H_6), menaquinones having three of the nine isoprene units saturated (Athalye et al., 1984; Collins and Jones, 1981; Collins et al., 1982, 1984, 1988; Kroppenstedt et al., 1981). 9/6[1] = MK-9(II, III, VIII H_6), saturated isoprene units in positions 2, 3, and 8 (Batrakov and Bergelson, 1978; Batrakov et al., 1976; Kroppenstedt, 1982c; Yamada et al., 1982a). 9/4[2] = MK-9(III, VIII H_4), saturated isoprene units in positions 3 and 8 (Collins et al., 1988). 8/4c[3] = MK-8(H_4-ω-cycl.), i.e., II, III-tetrahydro-ω-(2,6,6-trimethylcyclohex-2-enylmethyl) menaquinone-6 (Collins et al., 1988; Howard et al., 1986; Kroppenstedt 1982b, 1982c). 9/6[4] = MK-9(II, III, IX H_6) (Yamada et al., 1982b).

[g]Menaquinone types according to Kroppenstedt (1985) indicating the different types of biosynthesis pathways, exemplified by type 4: here menaquinones with the same isoprene chain length but different degree of saturation are synthesized i.e. MK-9($H_{2(1-4)}$) or MK-10($H_{2(1-4)}$). The positions of saturated units are taxon specific too and can be taken for further differentiation. 4A1 = MK-9(H_0), MK-9(H_2), MK-9(H_4); 4A2 = MK-9(H_0), MK-9(H_2), MK-9(III, VIII H_4); 4B1 = MK-9(H₄), MK-9(II, III, IX H_6), MK-9(H₈); 4B2 = MK-9(H_4), MK-9(II, III, VIII H_6), MK-9(H_8); 4C = MK-10(H_4), MK-9(H_6); 4D = MK-10(H_6), MK-10(H_8).

[h]Fatty acid types according to Kroppenstedt (1985) and Kroppenstedt and Kutzner (1978). See Table 4, Chapter 47 for explanation.

[i]Acyl group detected using the simple glycolate test (Uchida and Aida, 1977).

[j]α-Branched-β-hydroxylated long-chain fatty acids (Minnikin et al., 1975).

Data taken from Kroppenstedt (1987) and *Bergey's Manual of Systematic Bacteriology*, vol. 4 (Williams et al., 1989).

In humans, *A. madurae* and *A. pelletieri* cause mycetoma, a chronic suppurative granulomatous disease of subcutaneous tissue and bones that is characterized by swelling, abscesses, and discharge of pus. The pus contains granules which, in *A. madurae,* are soft, white- to-yellow, or reddish, usually 1–5 mm in diameter, and spherical or angular with lobes; in infections with *A. pelletieri,* they are hard, dark red, 0.3–0.5 mm in diameter, and usually spherical to oval. Infection usually occurs through skin lesions contaminated with dust, soil, or vegetable matter and remains localized with enlargement by direct extension of the organisms through the tissues. As infection becomes increasingly chronic, the sinus tracts extend more deeply into the body with involvement of both muscle and bone. Animal models of infection with *Actinomadura* have not been developed. However, Rippon (1968) reported that virulent strains of *A. madurae* produce a collagenase that has a significant role in the pathogenicity of the organisms.

Actinomadura pelletieri occurs mainly in Africa but *A. madurae* is more widespread, especially in tropical and subtropical areas (Develoux et al., 1988; Gumaa et al., 1986; Schaal and Beaman, 1984). Colonies resembling members of these species have been isolated from water (Lawson and Davey, 1972). There is evidence that climatic conditions may have a major role in the distribution of mycetomas. In Venezuela, most cases of actinomycetoma have been reported from semi-arid areas colonized by trees and bushes such as *Acacia tortuosa, Amarantus espinosus,* and *Prosepis fuliflora,* and cacti-like *Opuntia caribea* and *Opuntia wentiana* (Serrano et al., 1986). Since these trees and cacti have spines and thorns, it is very common for agricultural workers to puncture their skin.

Habitats of *Thermomonospora*

The ability of *Thermomonospora* species to secrete a variety of thermostable extracellular enzymes enables them to become established as dominant populations during high-temperature composting of plant residues and other wastes (Fergus, 1964; Stutzenberger, 1971; Bernier et al., 1988). Thermophilic thermomonosporas are common in overheated substrates such as bagasse, compost, fodders, and manures. They are especially abundant in mushroom compost (Fergus, 1964; Lacey, 1974, 1977; McCarthy and Cross, 1981, 1984a, 1984b). They are highly cellulolytic (McCarthy, 1987; Ball and McCarthy, 1988) and/or hemicellulolytic (Mc-

Carthy et al., 1985, 1988). *Thermomonospora curvata* and *T. fusca* produce endogluconase activity and attack cellulose. The production of endogluconase (e.g., cellobiohydrolase) by thermomonosporas is currently regarded as unsubstantiated. *Thermomonospora* species also grow on rapeseed that has been subject to heat damage during storage (Mills and Bollen, 1976).

Thermomonosporas are capable of extensively degrading the cellulose and lignocellulose residues that make up the bulk of agricultural and urban wastes (McCarthy, 1987). *Thermomonospora curvata* is active in the decomposition of municipal waste compost (Stutzenberger, 1971, 1972a, 1972b; Stutzenberger et al., 1970), where heavy metals have been shown to inhibit cellulase production (Stutzenberger and Sterpu, 1978). Similarly, *T. fusca* strains are very active against the bentose polysaccharide arabinoxylan and produce enzymes that have thermostability properties (McCarthy et al., 1985). Evidence that cellulose decomposition in mushroom compost involves a positive interaction between cellulolytic actinomycetes and certain bacteria (Stanek, 1972) underlines the importance of considering natural lignocellulose degradation as the product of an interactive heterogeneous microflora.

The growth of thermophilic actinomycetes in high-temperature environments leads to the release of spores that can cause allergic alveolitis. There are currently no grounds for implicating the spores of *Thermomonospora* species in such respiratory disorders (Lacey, 1988), though the cause of mushroom worker's disease, a form of allergic alveolitis, is still a matter for conjecture. Precipitins to *Faenia rectivirgula* and to *Thermoactinomyces* species have been found (Moller et al., 1976), and *Thermoactinomyces dichotomicus* has been implicated (Molina, 1982), but these species have seldom been found in compost (Lacey, 1973). Conversely, precipitins have not been demonstrated to any of the other actinomycetes tested, including the predominant *T. chromogena* and *T. fusca,* although the antigenic extracts used may have been unsatisfactory because of the poor growth of these organisms in culture (Lacey, 1988). Both mesophilic and thermophilic *Thermomonospora* strains have been isolated from soil (Krassil'nikov and Agre, 1965; Locci et al., 1967; Nonomura and Ohara, 1971c, 1974).

Isolation and Cultivation of *Actinomadura*

Actinomadura madurae and *A. pelletieri* can be inoculated from noncontaminated clinical samples, such as pus and biopsy material, using brain heart infusion (Oxoid CM261; Schaal, 1972), Sabouraud dextrose (Gordon, 1974) and yeast extract agars (Pridham et al., 1956–1957). (Recipes for media that are not commercially available are given in "Media and Techniques," this chapter.) Media should be transparent to facilitate direct microscopic examination of colonies. Selective isolation media need to be used for the examination of clinical material that contains large numbers of other microbes. Actinomadurae form small colonies on media selective for nocardiae (see Chapter 51). All cultures should be inoculated aerobically at 25 to 27°C and at 36°C for up to 3 weeks and examined both macroscopically and microscopically for growth every 2 days (Schaal, 1984). Actinomadurae can be recognized by their filamentous appearance, leathery colonies, and by the production of red prodiginine pigments. Actinomycetoma granules should be washed in sterile tap water before they are crushed to obtain material for inoculation of culture media.

Media formulations supplemented with antifungal antibiotics such as cycloheximide can be used to isolate actinomadurae from environmental samples. Media that have proved to be effective include egg albumin (Lawson and Davey, 1972), glucose-yeast extract (Athalye et al., 1981), glycerol-asparagine (International *Streptomyces* Project [ISP] medium 5; Shirling and Gottlieb, 1966), inorganic salts-starch (ISP medium 4; Shirling and Gottlieb, 1966), oatmeal (ISP medium 3; Shirling and Gottlieb, 1966) and yeast extract agars (ISP medium 2; Shirling and Gottlieb, 1966). Strains may be isolated from agar plates after incubation for 14 to 21 days.

Isolation of actinomadurae from enriched soil can be achieved using pretreatment regimes and selective agents. Nonomura and Ohara (1971d) reduced the number of unwanted microorganisms by air-drying soil and applying dry heat at 100°C for 1 h before plating onto various media, such as arginine-vitamins (AV) agar and mineral-glucose-asparagine (MGA) agar, and incubating for several weeks at 28° to 30°C. Lavrova et al. (1972) and Preobrazhenskaya et al. (1975b) increased the number of actinomadurae isolated by the addition of antibiotics to medium 2 of Gauze et al. (1957); the antibiotics were used to inhibit the growth of bacteria and the more frequently occurring streptomycetes, thereby providing more favorable conditions for the growth of the slow-growing actinomadurae. Bruneomycin (0.5, 1.0, or 2.0 μ/ml), rubromycin (5.0, 10.0, or 20 μg/ml), and streptomycin (0.5, 1.0, or 2.0 μg/ml) were the most successful antibiotics used. Athalye et al. (1981) combined drying and heat pretreatment regimes with the use of rifampicin (5 μg/ml) as the selective agent. *Actinomadura rubrobrunea* was isolated by scattering crushed soil over the medium of Kosmachev (1960) and incubating at 55°C.

Actinomadurae generally grow well on modified Bennett's agar (Jones, 1949), glucose-yeast extract agar (Athalye et al., 1981), and on formulations used for the cultivation of streptomycetes in the International *Streptomyces* Project (ISP; Shirling and Gottlieb, 1966). Most strains show abundant growth on oatmeal agar (ISP medium 3) at 30°C, but *A. kijaniata* and *A. macra* grow better on yeast extract-malt extract agar (ISP medium 2). *Actinomadura spadix* requires vitamin B_{12} for good growth (Nonomura and Ohara, 1971d), and *A. vinacea* grows well on glucose-peptone agar but poorly on the media mentioned above. *Actinomadura (Excellospora) rubrobrunea* and *Excellospora viridilutea* grow well on oatmeal- and peptone-maize agars at 50°C.

Isolation and Cultivation of *Thermomonospora*

Thermophilic *Thermomonospora* strains can be isolated from composts and heated vegetable material by diluting plating on nonselective media, but recovery is poor due to the rapid competing growth of *Bacillus* and *Thermoactinomyces* strains. The most effective isolation methods are those based on the use of a sedimentation chamber and Andersen air sampler (Andersen, 1958; Lacey and Dutkiewicz, 1976; McCarthy and Broda, 1984; McCarthy and Cross, 1981). Dried environmental samples are shaken within the sedimentation chamber to create an aerosol of propagules that, after 1 to 2 hours of sedimentation, still contains many actinomycete spores but relatively few other bacteria. Actinomycetes are isolated from this spore suspension using an Andersen sampler loaded with half-strength tryptone-soy agar plates supplemented with cycloheximide (50 μg/ml) to prevent growth of fungi. The recovery of white *Thermomonospora* strains can be enhanced by adjusting the isolation medium to pH 11.0 (Cross, 1982) and cellulolytic isolates can be detected by incorporating cellulose powder or ball-milled straw into the agar (McCarthy and Broda, 1984; Stutzenberger et al., 1970).

Thermomonospora colonies can usually be recognized after 3 to 5 days incubation at 50°C. Care must be taken to distinguish these target colonies from other actinomycetes with white aerial mycelium, for example, *Streptomyces* and

Thermoactinomyces strains. However, *Thermomonospora* colonies can be spotted by carefully searching isolation plates with a 40-×, long-working-distance objective and looking for the characteristic single spores on the aerial mycelium. The morphology of pure cultures can be confirmed by observing plate cultures or inclined cover slip preparations, as described by Williams and Cross (1971).

Thermomonospora chromogena is readily isolated on selective media containing kanamycin (25 μg/ml; McCarthy and Cross, 1981) or rifampicin (5 μg/ml; Athalye et al., 1981). Similarly, *T. formosensis* was obtained by plating soil dilutions onto starch-casein agar (Waksman, 1961) supplemented with kabicidin (6.25 μg/ml) and rifampicin (12.5 μg/ml; Hasegawa et al., 1986). Other mesophilic thermomonosporas have been isolated from soil samples using the selective procedure that involves heating environmental samples at 100°C prior to preparing suspensions for plating out (Nonomura and Ohara, 1971c, 1974). Apart from *T. formosensis* and *T. mesophila,* strains isolated at 30°C are usually able to grow at 50°C. Mixtures of organic matter and soil, inoculated in partially sealed polyethylene bags for a day, yield samples enriched in thermophilic actinomycetes, including thermomonosporas (McCarthy, 1989). Henssen (1957) used an anaerobic enrichment method to isolate her original *Thermomonospora* strains but the procedure and apparatus involved were unwieldy and may not have produced anything like anaerobic conditions.

Thermomonosporas grow well on any nutrient medium at pH 6.5 to 7.0 provided it contains some yeast extract. They are readily cultivated on Czapek peptone, glucose-yeast extract-malt extract (Greiner-Mai et al., 1987), Hickey and Tresner (1952), nutrient, or peptone-maize (Greiner-Mai et al., 1987) agars, and on ISP media (Shirling and Gottlieb, 1966) after 2 to 4 days incubation at their optimal temperature. *Thermomonospora alba* grows optimally at 37°C; *T. chromogena, T. curvata,* and *T. fusca* at 50°C; *T. formosensis* and *T. mesophila* at 28°C; and *T. mesouviformis* at 45°C (Greiner-Mai et al., 1987).

Media and Techniques for Isolation and Cultivation

The following are the formulae for the media mentioned in the preceding two sections that are not commercially available.

AV Agar + Vitamins (Nonomura and Ohara, 1969)

L-Arginine	0.3 g
Glucose	1.0 g
Glycerol	1.0 g
K₂HPO₄	0.3 g
MgSO₄·7H₂O	0.2 g
NaCl	0.3 g
Agar	15 g
Distilled water	1 liter

Adjust to pH 8.0 with NaOH.

Add to the sterilized basal medium aseptically, 1 ml of trace salt solution, 10 ml vitamin solution and 10 ml antibiotic solution.

Trace salts solutions: Dissolve in 10ml distilled water: $CuSO_4 \cdot 5H_2O$, 0.1 g; $Fe_2(SO_4)_3$, 1.0 g; $MgSO_4 \cdot 7H_2O$, 0.1 g; $ZnSO_4 \cdot 7H_2O$, 0.1 g.

Vitamin solution: Dissolve in 10 ml distilled water: 0.5 mg each of *p*-aminobenzoic acid, calcium pantothenate, inositol, niacin, pyridoxine HCl, riboflavin, thiamine HCl; 0.25 mg biotin; sterilize by filtration.

Antibiotic solution: Dissolve in 10 ml distilled water: Cycloheximide, 50 mg; nystatin, 50 mg; nalidixic acid, none or 20 mg; penicillin G, none or 0.8 mg; polymyxin, none or 4 mg; sterilize by filtration.

Bennett's Agar (Jones, 1949)

Beef extract	1 g
Glucose	10 g
N-Z amine A (enzymatic digest of casein)	2 g
Yeast extract	1 g
Agar	15 g
Distilled water	1 liter

Adjust to pH 7.3 with NaOH.

Modified Bennett's agar can also be used. In it, glucose is replaced by glycerol (10 g) and N-Z amine by Bacto-casein (2 g, Difco).

Czapek Peptone Agar (DSM Medium 83)

Peptone	5 g
Sucrose	30 g
Yeast extract	2 g
FeSO₄·7H₂O	0.01 g
KCl	0.5 g
K₂HPO₄	1 g
KNO₃	2 g
MgSO₄·7H₂O	0.5 g
Agar	15 g
Distilled water	1 liter

Adjust to pH 7.3.

Glucose-Yeast Extract-Malt Extract (GYM) Agar (Greiner-Mai et al., 1987)

Glucose	4 g
Yeast extract	10 g
CaCO₃	2 g
Distilled water	1 liter

Adjust to pH 7.2

Glucose-Peptone Agar (DSM Medium 85)

Bacto-peptone (Difco)	20 g
Glucose	10 g

Yeast extract	10 g
CaCO₃	10 g
Agar	15 g
Tap water	1 liter

Adjust to pH 7.3.

Glucose-Yeast Extract Agar (Waksman, 1950)

Glucose	10 g
Yeast extract (50% solution)	20 ml
Agar	15 g
Distilled water	980 ml

Adjust to pH 6.8.

Hickey and Tresner Agar (Hickey and Tresner, 1952)

Amidax (Corn Products Refining Co., Argo, IL, USA) or dextrin	10 g
Beef extract (Difco)	1 g
NZ amine A or tryptone (Difco)	2 g
Yeast extract (Difco)	1 g
CoCl₂·6H₂O	2 mg
Agar	20 g
Distilled water	1 liter

Adjust to pH 7.3.

ISP Medium 2: Yeast Extract-Malt Extract Agar (Pridham et al., 1956–1957)

Dextrose (Difco)	4 g
Malt extract (Difco)	10 g
Yeast extract (Difco)	4 g
Distilled water	1 liter

Adjust to pH 7.3, then add 20 g of Bacto-agar. Liquify agar by steaming at 100°C for 15 to 20 min.

ISP Medium 3: Oatmeal Agar (Küster, 1959a)

Cook or steam 20 g oatmeal in 1 liter distilled water for 20 min. Filter through cheesecloth. Add distilled water to restore volume of filtrate to 1 l. Add 1 ml trace salt solution (see below). Adjust to pH 7.5 with NaOH. Add 18 g agar; liquefy by steaming at 100°C for 15 to 20 min.

Trace salt solution: FeSO₄·H₂O, 0.1 g; MnCl₂·H₂O, 0.1 g; ZnSo₄·7H₂O, 0.1 g; distilled water, 100 ml.

ISP Medium 4: Inorganic Salt-Starch Agar (Küster, 1959b)

Solution I: Difco soluble starch, 10.0 g. Make a paste of the starch with a small amount of cold distilled water and bring to a volume of 500 ml.

Solution II:

CaCO₃	2 g
K₂HPO₄ (anhydrous)	1 g
MgSO₄·7H₂O	1 g
NaCl	1 g
(NH₄)₂SO₄	2.0 g
Distilled water	500 ml
Trace salt solution (see ISP Medium 3)	1.0 ml

The pH should be between 7.o and 7.4. Do not adjust if it is within this range. Mix solutions I and II together. Add 20 g agar. Liquify agar by steaming at 100°C for 10 to 20 min.

ISP Medium 5: Glycerol-Asparagine Agar (Pridham and Lyons, 1961)

L-Asparagine (anhydrous)	1.0 g
Glycerol	10.0 g
K₂HPO₄ (anhydrous)	1.0 g
Distilled water	1 liter
Trace salt solution (see ISP Medium 3)	1.0 ml

The pH of this solution is about 7.0 to 7.4. Do not adjust if it is within this range. Add 20 g agar. Liquify agar by steaming at 100°C for 15 to 20 min.

ISP Medium 7: Tyrosine Agar (Shinobu, 1958)

L-Asparagine	1.0 g
Glycerol	15.0 g
L-Tyrosine	0.5 g
FeSO₄·7H₂O	0.01 g
K₂HPO₄ (anhydrous)	0.5 g
MgSO₄·7H₂O	0.5 g
NaCl	0.5 g
Distilled water	1 liter
Trace salt solution (see ISP medium 3)	1.0 ml

Adjust to pH 7.2 to 7.4. Add 20 g agar. Liquefy agar by steaming at 100°C for 15 to 20 min.

Kosmachev's Medium (Kosmachev, 1960)

30% Yeast autolysate	15 ml
CaCO₃	4 g
FeSO₄·7H₂O	0.01 g
KNO₃	1 g
MgSO₄·7H₂O	0.5 g
Na₂HPO₄	1 g
(NH₄)₂SO₄	1 g
Agar	15 g
Distilled water	1 liter

MGA Agar (Nonomura and Ohara, 1971b)

L-Asparagine	1 g
Glucose	2 g
K₂HPO₄	0.5 g
MgSO₄·7H₂O	0.5 g
Trace salt solution (see Agar)	
Agar	20 g
Distilled water	1 liter

Add these antibiotics: cycloheximide, 50 mg; nystatin, 50 mg; benzylpenicillin, 0.8 mg; polymyxin B, 4 mg.

Peptone-Maize Agar (Greiner-Mai et al., 1987)

Corn meal agar	17 g
Peptone	5 g
Starch	10 g
CaCl₂·2H₂O	0.5 g
NaCl	5 g
Distilled water	1 liter

Adjust to pH 7.2 before sterilization.

Starch-Casein Agar (Waksman, 1961)

Casein	1 g
Soluble starch	10 g
K₂HPO₄	0.5 g
Agar	15 g

| Distilled water   1 liter
| Adjust to pH 7.0 to 7.5.

Preservation

Heavily sporulated cultures are needed to maintain high viability irrespective of the preservation method. The highest survival rates for strains that do not sporulate are achieved using cells from the exponential growth phase. The culture age of the organisms to be preserved is very important, especially for the preservation of thermophilic strains that lyse readily during the stationary phase of growth.

Sporulated *Actinomadura* strains can be kept on oatmeal agar or other suitable agar slants at 4°C and transferred every 4 months for short-term preservation. The tubes should be sealed with silicone stoppers to prevent drying. The same procedure can be used for strains of *Thermomonospora* but in this case cultures should be transferred every 1 to 2 months using Czapek Dox-yeast extract-casamino acids agar at pH 8.0 (McCarthy, 1989). Medium-term preservation for up to 4 years can be achieved by mixing spore suspensions or homogenized mycelia with glycerol (45%, v/v) and storing at 25°C (Zippel and Neigenfind, 1988). Lyophilization of spores and mycelia suspended in 10% skim milk is a convenient method for long-term storage. An alternative simple, reliable, and quick method involves nitrogen cryopreservation of living cells in small polyvinyl chloride tubes ("straws") at −196°C (Hoffmann, 1989a, 1989b). This procedure has been shown to be reliable for actinomycetes, including *Actinomadura* and *Thermomonospora*.

Identification

Actinomadura species, *T. curvata,* and *T. formosensis,* which may form the nucleus of a revised family Thermomonosporacea, can be distinguished from all other actinomycetes using a combination of chemical and morphological features. Primary diagnostic information can be gained by the detection of marker amino acids, fatty acids, menaquinones, sugars, and polar lipids (Table 2). Simple procedures have been devised for the detection of these chemical features (Kroppenstedt, 1985; M. P. Lechevalier and H. A. Lechevalier, 1980; Minnikin et al., 1984; Staneck and Roberts, 1974). One-dimensional thin-layer chromatography will determine if an organism contains diaminopimelic acid (A_2pm) and whether the latter is in the LL-

or *meso*-form. The presence of LL-A_2pm will lead to *Streptomyces* and related genera (see Chapter 41). The detection of various combinations of *meso*-A_2pm, mixtures of straight and branched-chain fatty acids, predominant amounts of hexahydrogenated menaquinones with nine isoprene units, and major amounts of diphosphatidylglycerol, phosphatidylglycerol, and phosphatidylinositol can be used to separate *T. curvata, T. formosensis,* and *Actinomadura* strains from other sporoactinomycetes, notably those classified in the families Actinoplanaceae, Nocardiaceae, Pseudonocardiaceae, including *Saccharothrix,* and in the genus *Thermoactinomyces.*

In morphology, cell wall chemotype, and natural habitat, thermomonosporas are in the broad sense similar, but not related, to thermoactinomycetes. The detection of endospores in the latter separates *Thermoactinomyces* from *Thermomonospora* and all other monosporic actinomycetes. The morphology of *Actinosynnema* and *Streptoalloteichus* strains is complex and differs fundamentally from *Actinomadura* and *Thermomonospora,* since the former produces motile spores. In *Actinosynnema,* peritrichously flagellated zoospores originate from aerial spore chains borne on synnemata whereas in *Streptoalloteichus,* spores from single polar flagella are formed in vesicles and sporangia are carried on substrate hyphae. *Streptoalloteichus* also produces chains of *Streptomyces*-like arthrospores on aerial hyphae. The genera *Actinosynnema* (Hasegawa et al., 1978) and *Streptoalloteichus* (Tomita et al., 1987) were proposed for actinomycetes considered to be morphologically unusual.

The genus *Actinomadura* (including *Excellospora*) can be distinguished from other sporoactinomycete taxa using a judicious selection for chemical and morphological features (Table 2; Kroppenstedt et al., 1990; Meyer, 1989a). Quantitative analyses of fatty acid data allow the separation of mesophilic actinomadurae from corresponding members of the genera *Microtetraspora, Nocardiopsis, Saccharothrix,* and *Thermomonospora* (Fig. 1). Actinomadurae and microtetrasporas can also be distinguished on the basis of their menaquinone and polar lipid profiles (Figs. 2 and 3).

Actinomadura species can be separated using color, and morphological, biochemical, and physiological properties (Tables 3 and 4). However, identification of most of these species is difficult because, in many instances, only one strain, usually the type strain, has been examined. Even when several representatives of species have been studied, the results of biochem-

Fig. 1. Dendrogram showing relationships among mesophilic representatives of *Actinomadura, Microtetraspora, Nocardiopsis, Saccharothrix,* and *Thermomonospora* species. The dendrogram, which is based on differences in the fatty acid patterns of the organisms, was generated by treating the Euclidian distances of the fatty acids with the unweighted pair group method with arithmetic average algorithm. Fatty acids were prepared after Kroppenstedt et al. (1990), and the numerical analyses were done using the standard MIS software (Microbial ID, Inc., Newark, NJ, USA). Statistical procedures are described by Eerolen and Lehtonen (1988) and O'Donnell (1985).

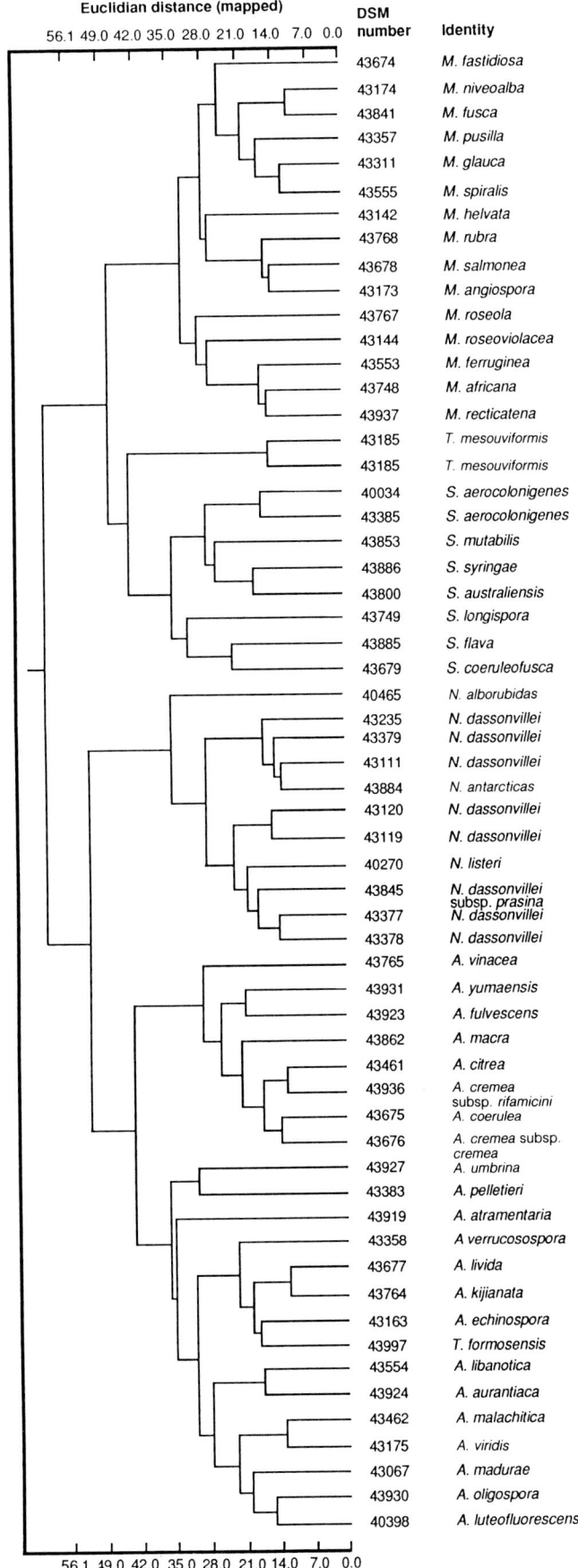

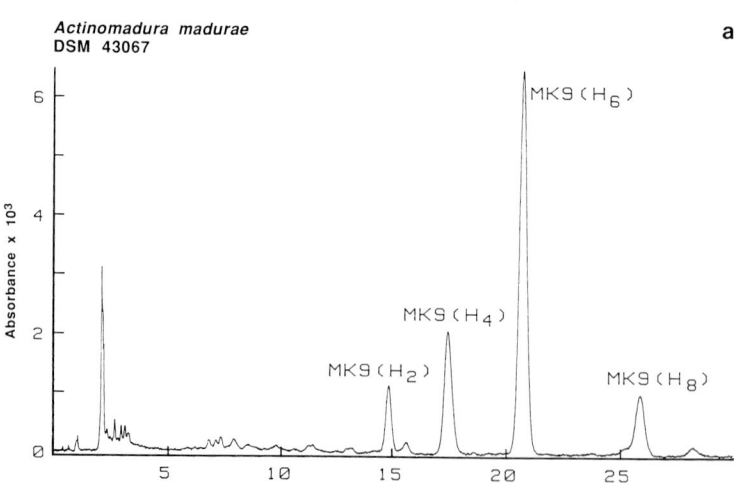

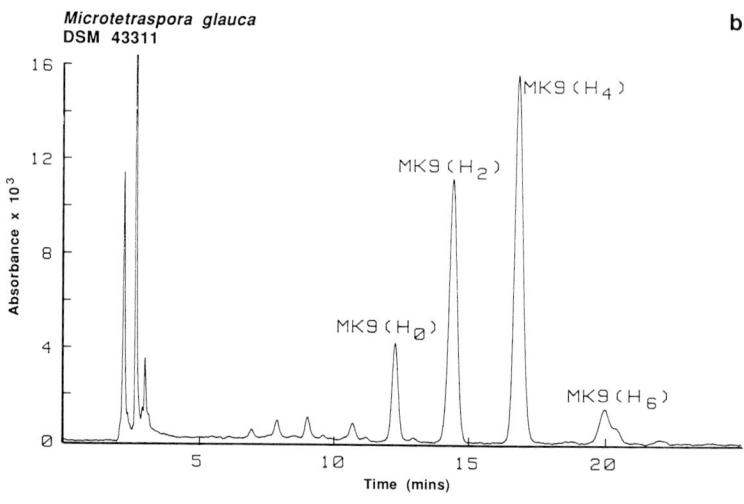

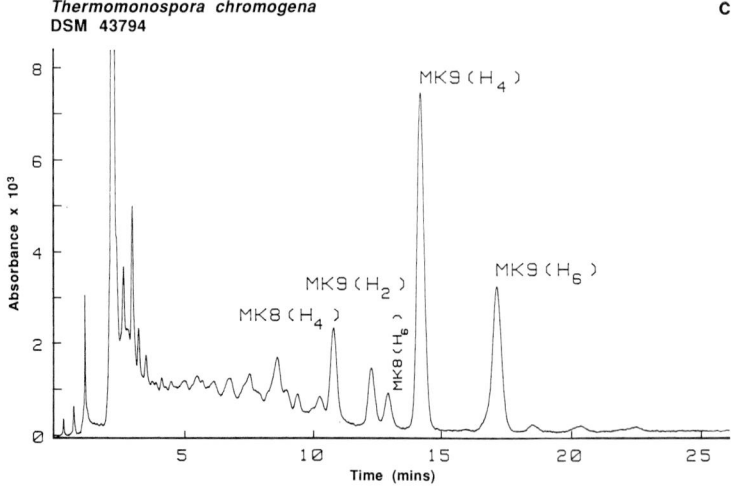

Fig. 2. Menaquinone profiles of: (a) *Actinomadura madurae* DSM 43067[T]; (b) *Microtetraspora glauca* DSM 43311[T]; (c) *Thermomonospora chromogena* DSM 43794[T]. (Continued on next page.)

ical and physiological tests have proved to be variable or inconsistent when data from the literature are compared. Even so, numerical taxonomic evidence shows that many of the validly described taxa do merit species status (Athalye et al., 1985; Goodfellow and Pirouz, 1982;

Goodfellow et al., 1979). It is also encouraging that some *Actinomadura* species have been shown to form distinct genomic species (Fischer et al., 1983; Poschner et al., 1985).

Thermomonospora species can be assigned to three groups solely on the basis of polar lipid

Fig. 2 (continued). (d) *T. curvata* DSM 43183ᵀ; and (e) *T. formosensis*ᵀ 43997. Chromatographic conditions: column, 250 mm × 4 mm; stationary phase, RP 18, size 5 μm; mobile phase, acetonitrile-isopropanol (65:35, v/v); flow rate, 1 ml/min; temperature, 40°C, injection volume, 5 μl of extract prepared after Minnikin et al. (1984) without further purification. Key to abbreviations: see Table 2.

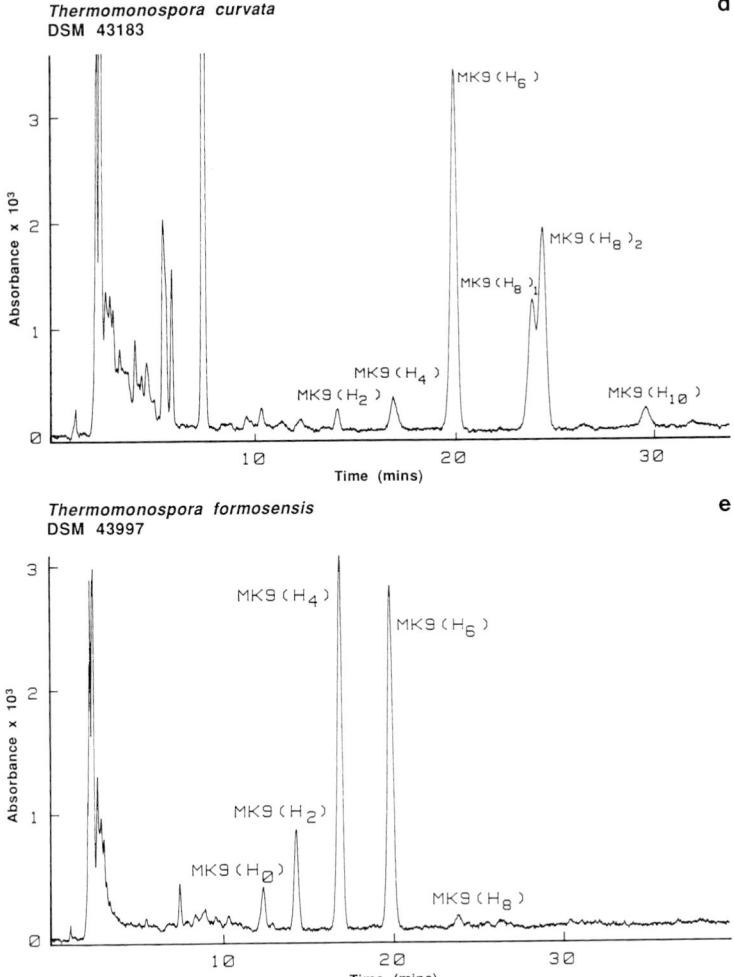

patterns (Table 5; Fig. 3). The taxonomic integrity of the group containing *T. curvata* and *T. formosensis* is also supported by fatty acid and menaquinone data. Similarly, organisms assigned to the group containing *T. alba* and *T. fusca* are characterized by menaquinones with long, highly saturated isoprenyl chains and major proportions of iso- and anteiso-branched pentadecanoic and heptadecanoic ($C_{15/17}$) acids. This combination of chemical markers separates these species from all other sporoactinomycete taxa. Minor variations in the fatty acid and menaquinone profiles of *T. chromogena* and *T. mesophila*, which comprise this group, may be attributed to differences in their temperature requirements for growth (Kroppenstedt et al., 1990).

The menaquinone data of most species support the findings of Collins et al. (1982), who also found complex mixtures of menaquinones with 9, 10, and 11 isoprene units in representative strains of *Thermomonospora*. Discrepancies have been found between the menaqui-

none analyses of *T. curvata* strains. The type strains of this species (DSM 43183ᵀ and JCM 3042ᵀ) showed MK-9(H_6) and MK-9(H_8) as principal menaquinones (Kroppenstedt, 1987) whereas Collins et al. (1982) reported menaquinones of the MK-10 series from a *T. curvata* isolate. The constituent markers of the three chemical groups can be distinguished using a judicious selection of biochemical, morphological, and physiological properties (Table 6, Figs. 4 and 5).

Metabolism and Genetics

Very little is known about the genetics or metabolism of *Actinomadura* strains. Current and earlier work on these organisms is difficult to interpret because of the uncertain taxonomic status of the organisms examined. Indeed, without further characterization it is not possible to know whether organisms simply labelled *Actinomadura* are correctly named or actually be-

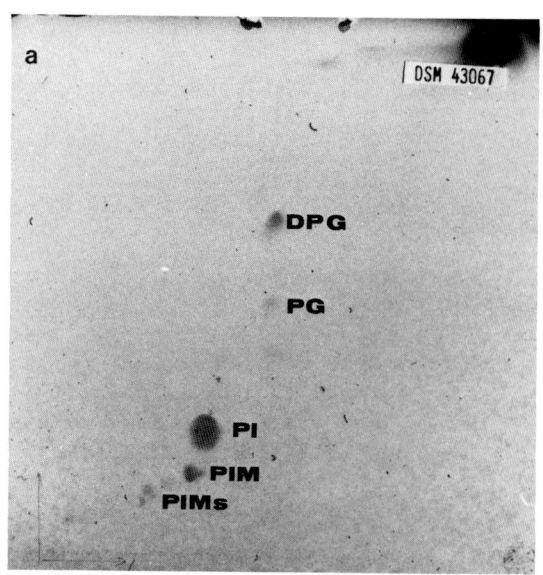

Fig. 3. Two-dimensional thin-layer chromatograms of polar lipids: (a) *Actinomadura madurae* DSM 43067[T]; (b) *Microtetraspora glauca* DSM 43311[T]; (c) *Thermomonospora mesouviformis* DSM 43185[T]; (d) *T. curvata* DSM 43183[T]; and (e) *T. mesophila* DSM 43048[T]. Chromatographic conditions: 10 cm × 10 cm silica-gel-60, thin-layer plates were spotted with 20 μl of the extracts from the test strains using the procedures of Minnikin et al. (1984); solvent I: chloroform-methanol-water (65:25:4, v/v); solvent II: chloroform-acetic acid-methanol-water (80:15:12:4, v/v). The developed plates were sprayed with 50% sulfuric acid or ammonium sulfate and charred at 180°C. Key to abbreviations: see Table 2.

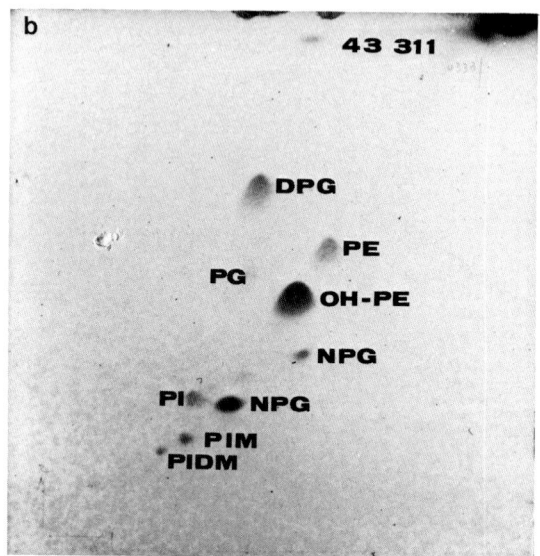

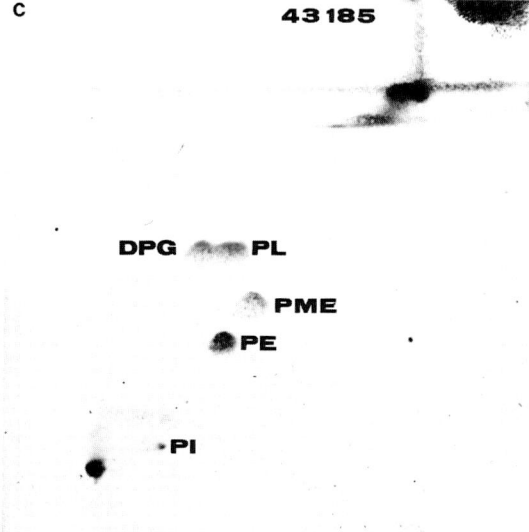

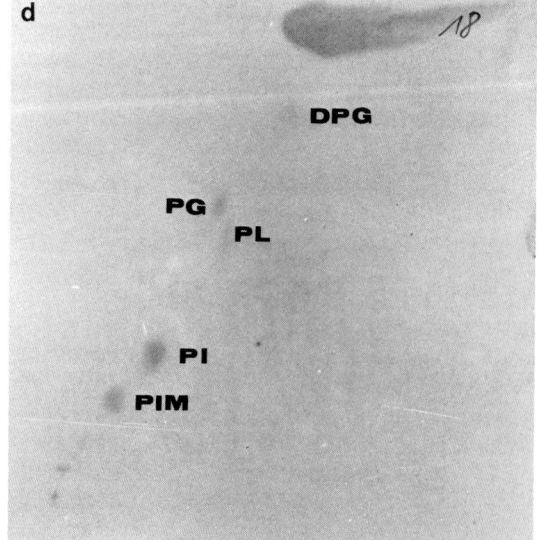

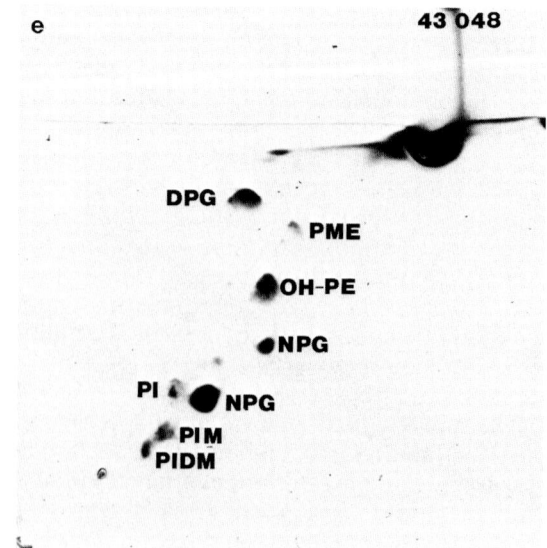

Table 3. Macroscopic and microscopic characters separating the species of *Actinomadura*.

Species	Color (on oatmeal agar)			Spore chain type	Spore surface type
	Aerial mycelium	Substrate mycelium	Excreted pigment		
A. madurae	Trace	Colorless	—	Hooks to spirals[a]	Warty
A. atramentaria	White	Colorless	Inky brown[b]	Cluster	Smooth
A. aurantiaca	Cream to pink	Yellow	—	Hooks to spirals	Warty
A. carminata	Pink	Red to violet	—	Pseudosporangia	Smooth
A. citrea	White to blue	Yellow	Yellow	Hooks or curled	Uneven
A. coerulea	Pink to blue	Colorless	—	Hooks to spirals	Warty
A. cremea subsp. cremea	White to yellow	Colorless	—	Hooks to spirals	Warty
A. cremea subsp. rifamycini	White to yellow	Colorless	—	Hooks to spirals	Warty
A. echinospora	Yellow to pink	Orange	Yellow	Cluster	Spiny
A. fibrosa	White	Brown	Brown	No spores formed	—
A. fulvescens	Absent	Red-brown	—	ND	ND
A. kijaniata	Trace	Gray	—	Spirals	Smooth
A. libanotica	White to pink	Yellow to brown	—	Hooks, curled	Folded
A. livida	Trace	Gray to brown	Violet	Hooks to spirals	Uneven
A. luteofluorescens	Yellow to blue	Yellow to green	Yellow/green	Hooks, curled	Warty
A. macra	Cream to pink	Cream to pink	—	Hooks, curled	Smooth
A. oligospora	Absent	Brown	Light brown	Straight	Uneven
A. pelletieri	Absent	Pink to brown	—	Hooks to spirals	Warty
A. rubrobrunea	Gray to blue	Orange	—	Spirals[c]	Spiny
A. spadix	Yellow/brown	Yellow/red/brown	—	Pseudosporangia	Smooth
A. umbrina	Absent	Colorless	—	ND	ND
A. verrucosospora	Pink to blue	Orange to pink	—	Hooks to spirals	Warty
A. vinacea	Absent	Pink to red	—	Straight	Uneven
A. viridis	Green	Yellow-brown to green	—	Straight	Smooth
A. yumaensis	Gray to yellow	Gray to yellow	—	Hooks	Smooth
E. viridilutea	Blue-green	Orange to yellow	—	Hooks to spirals	Spiny

Symbols: —, absent; ND, not determined.
[a]Spirals have two to four turns.
[b]Pigment is found on tyrosine agar (ISP medium 7) only.
[c]Coiled, 2 to 20 spores, some single.
Data from Athalye et al. (1985), Mertz and Yao (1990), Meyer (1989), Preobrazhenskaya et al. (1977).

long to the *A. pusilla* group that is now classified in the genus *Microtetraspora* (Kroppenstedt et al., 1990).

It is clear, nevertheless, that actinomadurae are chemoorganotrophic actinomycetes with an oxidative type of metabolism. Most species grow well between 25 and 40°C but several are thermophilic, with an optimum growth temperature between 50 and 65°C. Actinomadurae can metabolize a wide range of sugars as sole carbon sources for energy and growth; proteolytic activity is shown by the capacity of most strains to attack casein and gelatin (Athalye et al., 1985; Goodfellow and Pirouz, 1982; Goodfellow et al., 1979). In contrast, neither chitin, xanthine, nor xylan is metabolized, but an uncharacterized *Actinomadura* strain has been implicated in the degradation of wheat lignocellulose (McCarthy and Broda, 1984). Pathogenic strains of *A. madurae* produce collagenolytic

enzymes (Rippon, 1968), and *Actinomadura* strain R39 produces serine peptidases (Schindler et al., 1986).

All *Thermomonospora* strains can grow in the temperature range of 40 to 48°C and the pH range of 7.0 to 9.0. However, the most commonly isolated species are facultative thermophiles with a growth temperature range of 30 to 55°C, with an optimum at 50°C. *Thermomonospora mesophila* has a lower growth temperature range, from 25 to 48°C, with an optimum temperature for growth and spore formation between 35 and 40°C. The growth temperature range of *T. formosensis* is 23 to 41°C.

Thermomonospora strains have several enzymes in common, including catalase, carboxymethylcellulose, deaminase, β-glucosidase, β-galactosidase, and β-xylanase (McCarthy, 1989; McCarthy and Cross, 1984a, 1984b). Esculin, arbutin, and testosterone are degraded but chi-

Table 4. Characters differentiating the species of *Actinomadura*.

Species	Esculin hydrolysis	Arbutin hydrolysis	Casein hydrolysis	DNase	Elastin hydrolysis	Gelatin hydrolysis	Guanine degradation	Hypoxanthine degradation	Nitrate reduction	Starch hydrolysis	Testosterone degradation	Tyrosine degradation	Urease production	Xanthine degradation
A. madurae	+	+	+	d	+	+	d	+	+	d	+	+	−	−
A. atramentaria	ND	ND	ND	ND	ND	−	ND	ND	+	−	ND	+	ND	ND
A. aurantiaca	−	ND	−	V	ND	+	ND	−	ND	−	+	−	ND	−
A. citrea	+	+	+	−	+	+	d	+	+	+	+	+	−	−
A. coerulea	−	+	+	−	+	+	+	+	+	−	+	−	−	−
A. cremea subsp. *cremea*	+	+	+	V	+	+	−	−	+	−	+	−	+	−
A. fibrosa	+	ND	+	ND	ND	+	−	+	ND	+	+	+	+	−
A. kijaniata	+	ND	+	+	+	+	ND	+	ND	+	−	+	+	+
A. libanotica	+	ND	−	−	ND	−	ND	−	+	+	+	−	ND	−
A. livida	+	+	+	+	−	+	+	−	+	−	+	+	−	−
A. luteofluorescens	−	d	+	+	−	+	+	+	d	+	+	+	−	+
A. macra	−	ND	−	ND	ND	+	ND	+	+	−	ND	+	−	−
A. oligospora	+	ND	+	+	−	+	−	−	−	−	−	−	+	−
A. pelletieri	−	−	+	−	+	+	−	d	+	−	−	+	−	−
A. spadix	+	+	+	−	−	−	−	−	+	+	−	−	−	−
A. verrucosospora	+	+	+	+	+	+	d	+	d	+	+	+	−	−
A. vinacea	+	ND	+	−	ND	+	ND	−	ND	−	−	−	ND	−
A. viridis	ND	ND	ND	ND	ND	+	ND	ND	ND	+	ND	ND	ND	ND
A. yumaensis	+	ND	+	ND	ND	+	−	d	+	+	ND	+	ND	+

Symbols: +, 90% or more strains positive; −, 10% or less strains positive; d, 11 to 89% of strains positive; V, variable; ND, not determined.

Based on data from Athalye (1981), Athalye et al. (1985), Goodfellow et al. (1979), Horan and Brodsky (1982), Labeda et al. (1985), and Mertz and Yao (1986, 1990). Data were not available for *A. carminata*, *A. cremea* subsp. *rifamycini*, *A. echinospora*, *A. fulvescens*, *A. rubrobrunea*, *A. umbrina*, or *Excellospora viridilutea*.

Table 5. Chemical markers separating *Thermomonospora* species.

| | Diagnostic fatty acids | | | | |
	Iso/anteiso C$_{15/17}$	Hydroxy	10-Methyl	Menaquinones (MK-)[a]	Diagnostic phospholipids
T. curvata	−	−	++	9(H$_4$), *9(H$_6$)*, 9(H$_8$)	PIM, PI, DPG
T. formosensis	−	−	++	9(H$_2$), 9(H$_4$), *9(H$_6$)*	PIM, PI, DPG
T. chromogena	++	+	+	9(H$_2$), *9(H$_4$)*, 9(H$_6$)	PI, NPG, PME, OH-PE, LPE
T. mesophila	(+)	+	+	9(H$_0$), *9(H$_2$)*, 9(H$_4$)	PI, NPG, PME, OH-PE, LPE
T. alba	+++	−	(+)	10(H$_6$), *10(H$_8$)*, 10(H$_{10}$)	PI, PE, PME, DPG
T. fusca	+++	−	(+)	10(H$_6$), *10(H$_8$)*, 11(H$_6$), 11(H$_8$)	PI, PE, PME, DPG
T. mesouviformis[b]	+++	−	(+)	10(H$_6$), *10(H$_8$)*, 10(H$_{10}$)	PI, PE, PME, DPG

Symbols: (+), 1–5%; +, 5–15%; ++, 15–30%; +++, >30%; −, <1%.

[a]Abbreviations: See Table 2.

[b]This species is considered to be a synonym of *T. alba* (McCarthy et al., 1989).

R. M. Kroppenstedt, unpublished observations.

Table 6. Characters differentiating the species of *Thermomonospora*.

Characteristic	T. curvata	T. alba	T. chromogena	T. formosensis	T. fusca	T. mesophila
Color of substrate mycelium	Yellow/orange	Pale yellow	Brown	Light orange	Pale yellow	Brown
Spores on both aerial and substrate mycelia	−	d	−	+	d	−
Biochemical tests:						
Nitrate reduction	+	d	+	−	−	+
Oxidase	−	−	+	ND	−	+
Phosphatase	+	+	d	ND	+	−
Degradation of:						
Agar	+	+	−	ND	+	+
Cellulose powder (MN300)	+	+	−	ND	+	−
DNA	−	−	−	ND	−	+
Elastin	−	d	+	ND	+	+
Pectin	−	+	d	ND	+	+
Starch	+	+	−	−	+	+
Tween 20 and 80	+	+	+	ND	+	−
Tyrosine, hypoxanthine, xanthine	−	−	+	ND	−	+
Growth at:						
25°C	−	−	−	+	−	+
30°C	d	d	−	+	−	+
53°C	d	d	+	−	+	+
pH 11.0	+	+	−	−	+	−
Growth in presence of:						
Crystal violet (0.00002%)	+	−	+	ND	+	+
Tetrazolium chloride (0.002%)	d	−	d	ND	+	
Thallous acetate (0.001%)	d	−	d	ND	+	+
Kanamycin (25 µg/ml)	−	−	+	ND	−	−
Novobiocin (25 µg/ml)	−	−	−	ND	−	+
Growth on sole carbon source (1%):						
L-Arabinose	−	−	−	−	−	+
D-Galactose	−	d	+	+	+	+
Lactose	−	d	−	−	+	−
D-Mannitol	−	−	−	+	−	−
D-Ribose	+	−	d	ND	−	−
Sucrose	+	+	−	+	+	−

Symbols: +, 90% or more strains positive; −, 10% or less strains positive; d, 11 to 89% of strains positive; ND, not determined.
Based on data from Hasegawa et al. (1986), McCarthy (1989), and McCarthy and Cross (1984).

tinase activity is universally absent. All strains attack casein and gelatin, and usually agar, keratin, starch, and the Tweens and, apart from *T. curvata*, pectin. The capacity to degrade cellulosic substrates is an important property of many strains, and *Thermomonospora* cellulases (Crawford and McCoy, 1972; Hägerdal et al., 1978; Stutzenberger, 1972b) and xylanases (Bachmann and McCarthy, 1989; Ball and McCarthy, 1989; Ball and McCarthy et al., 1985) have been partially characterized. Amylases from *Thermomonospora* species, including *T. curvata,* are extremely active and stable at 60 to 70°C and at slightly acid to neutral pH values (Kuo and Hartman, 1967; Stutzenberger and Carnell, 1977). *Thermomonospora fusca* produces a heat-stable serine proteinase (Gusek and Kinsella, 1987) and an inducible thermoalkalophilic polygalacturonate lyase (Stutzenberger, 1987).

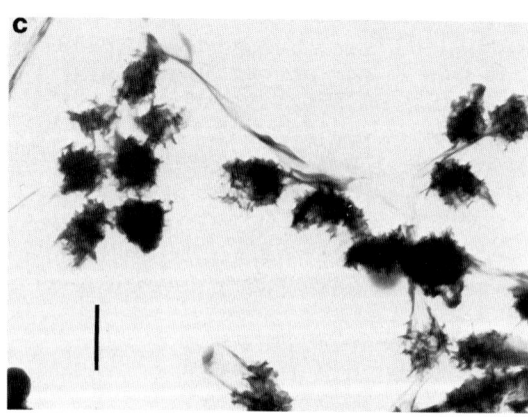

Fig. 4. Transmission electron photomicrographs showing spore formation in *Thermomonospora*. Bars = 1 μm. (A) *T. curvata*, single spores on branched and unbranched sporophores. (From E. Greiner-Mai et al., 1987.) (B) *T. alba*, single spores on relatively stiff sporophores. (Courtesy of F. Korn-Wendisch.) (C) *T. chromogena*, single spores in clusters. (Courtesy of F. Korn-Wendisch.)

Cellulases from *Thermomonospora* strains are composed of multiple extracellular endoglucanases and have temperature optima for activity between 55 and 70°C and pH optima between 6 and 7 (McCarthy, 1987). A typical example is the cellulase of *T. fusca* YX, which consists of five or more endoglucanases that are

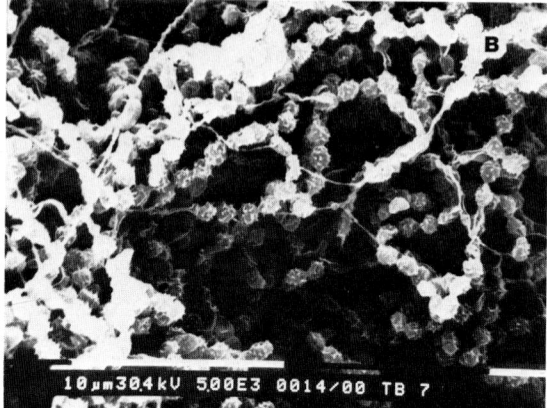

Fig. 5. Scanning electron photomicrographs showing spore formation. Bars = 1 μm. (Both courtesy of A. Kempf.) (A) Spore clusters of *Thermomonospora fusca*. (B) Long arthrospore chains of *Actinomadura rubrobrunea*.

immunologically distinct, differ in N-terminal amino acid sequences, have molecular weights ranging between 45 and 108 kDa, and vary greatly in specific activity towards different cellulase substrates (Calza et al., 1985; Wilson, 1988). *Thermomonospora fusca* (née *curvata*) produces at least five distinct endoglucanases that are age dependent (Lupo and Stutzenberger, 1988). One of these enzymes, Al, had an identical molecular weight to a highly purified endoglucanase from *T. fusca* that was characterized by Calza et al. (1985). Separation of endoglucanase and exoglucanase activity in culture filtrates of *T. curvata* has been described (Stutzenberger, 1972b; Stutzenberger and Lupo, 1986).

Actinomycete cellulase are inducible enzymes produced in response to growth on cellulose. Cellobiose is comparable to cellulose as a substrate for cellulase production in *T. fusca* (Moreira et al., 1981). In *T. curvata*, cellobiose has been reported as both a poor (Fennington et al., 1984) and good (Stutzenberger and Kahler, 1986) cellulase inducer. One possible explanation for this apparent discrepancy is that the role of cellobiose as an inducer or repressor of cellulase production is concentration-dependent and that thresholds vary in different strains. In *T. curvata* it has been shown that a sufficient concentration of cellobiose to support rapid growth leads to repression of cellulase biosynthesis (Wood et al., 1984). These cultures also provided evidence which implicated cyclic AMP as the mediator of catabolic repressor of cellulase biosynthesis in *T. curvata*.

Optimization of cellulase production is usually achieved by empirical manipulation of growth conditions and is often species- or strain-specific (McCarthy, 1987). Cellulose degradation or cellulase production can be improved by altering the cellulose source and nitrogen component employed (Fennington et al., 1984; Moreira et al., 1981). Comparative levels of cellulase activity have been reported between *Thermomonospora* strains and the cellulytic fungus *Trichoderma reesii* which is commonly used as the reference standard (Zerchak et al., 1980). However, different sources of cellulose can greatly affect the extent of hydrolysis realized (Zerchak et al., 1980; Stutzenberger, 1979). *Thermomonospora curvata* releases sixteen times more β-glucosidases when grown on protein-extracted lucerne fibre compound compared with growth on cellulose or purified cellulose (Bernier et al., 1988). Enhanced cellulase production has been achieved by mutation (Meyer and Humphrey, 1982) and in *T. curvata* (Fennington et al., 1984) the mutants have been identified as catabolite repressor-resistant.

Xylanase activity has received comparatively little attention even though it is more widespread amongst actinomycetes than cellulose activity. This imbalance is being redressed using *Thermomonospora* strains that have been extensively studied in relation to their cellulolytic activity (Ball and McCarthy, 1989; McCarthy et al., 1985, Ristroph and Humphrey, 1985a, 1985b). The production of xylanases is correlated with the growth stage and degradation is the result of endoxylanase activity. The pH and temperature relationship of *Thermomonospora* xylanases have much in common with those of the corresponding cellulases (McCarthy, 1987). Similarities include optimal activity within the pH range 5 to 8, increased thermostability of enzymes from thermophilic strains and lower temperature optima for all-bound dimer hydrolases. *Thermomonospora* xylanases are sufficiently active to produce same direct saccharification of lignocellulose as shown by the generation of xylose from straw (McCarthy et al., 1985). Xylose inhibition of β-xylosidase has been demonstrated in a *Thermomonospora* strain (Ristroph and Humphrey, 1985b).

T. fusca strains also attack arabinoxylan using thermostable enzymes (McCarthy et al., 1988). Two of the enzymes concerned, β-xylosidase and endoxylanase, were shown to be contrasting in their complexity. β-Xylosidase activity was predominantly intracellular and due to a single protein species of 168 kDa apparently comprising three inactive subunits. Separations of extracellular endoxylanase activity, by anion exchange and gel-permeation-fast-protein liquid chromatography, demonstrated the presence of two active proteins (approximately 24 and 32 kDa). However, with polyacrylamide gel electrophoresis, two additional proteins (48 and 66 kDa) were detected. Six active proteins were found on isoelectric focusing gels developed as zymograms; three of these probably represented variants of the 32-kDa protein which was separated into three fractions by chromatofocusing.

Some strains of the genus *Thermomonospora* also degrade lignin (McCarthy and Broda, 1984; McCarthy et al., 1986). McCarthy et al. (1986) found that *T. mesophila* degraded lignocellulose with a water-soluble polymeric comples as the main product; the latter is similar chemically to the water-soluble lignocarbohydrate fragments produced by ligninolytic *Streptomyces* strains. The capacity to solubilize lignin is common or unusual among lignocellulose-degrading actinomycetes (Ball et al., 1989).

Little information is available on the genetics of *Thermomonospora* species. Plasmids have been isolated from *Thermomonospora* isolates (Pidcock et al., 1985) and from *T. chromogena* (McCarthy, 1989). Protoplast formation, regeneration, and transformation of *T. fusca* with plasmid pIJ702 have also been reported (Pidcock et al., 1985). Plasmids such as pIJ702 (Katz et al., 1982) are now being used routinely for cloning antibiotic resistance and catabolic and biosynthetic genes of species of *Streptomyces,* with the most common gene recipient being *Streptomyces lividans* (Chater and Hopwood, 1984). A gene coding for one of the endoglucanases of *T. fusca* has been cloned into and expressed in *Bacillus subtilis, Escherichia coli,* and *Streptomyces lividans* (Collmer and Wilson, 1983; Ghangas and Wilson, 1987) as a first step towards construction of a vehicle for the production of high levels of thermostable cellulases. A number of bacteriophages that infect cellulolytic *Thermomonospora* strains (Lawrence et al., 1986) and members of *Faenia* and *Thermoactinomyces* species (Kurup and Heinzen, 1978) have been characterized.

Applications

Actinomadura and *Thermomonospora* strains are a source of new antibiotics, enzymes, and products with pharmacological activity: *Actinomadura carminata* produces carminomycin (Gauze et al., 1973); *A. citrea,* rifamycin (Gauze et al., 1975); *A. kijianata,* kijanimicin (Waitz et al., 1981); *A. macra,* polycyclic ether antibiotics (Huang, 1980); and *A. oligospora* and *A. yumaensis,* polyether antibiotics (Labeda et al., 1985; Mertz and Yao, 1986). In addition, novel antitumor agents (Matson et al., 1989; Takahashi et al., 1988) and glycolipid (Goldstein et al., 1987), nucleoside (Suhadolnik et al., 1989), tetracycline (Patel et al., 1987), and chlorine-containing polyether antibiotics (Cullen et al., 1987) have been isolated from poorly characterized strains labelled *Actinomadura.* Thermophilic thermomonosporas are the source of a new naphthoquinone antibiotic (Hedge et al., 1986). An *Excellospora viridilutea* strain produces a compound that inhibits reverse transcriptase of avian myeloblastosis virus (Sasaki et al., 1988), a mesophilic *Thermomonospora* strain of phospholipid type II produces a novel isochromane-quinone which inhibits thrombin-induced aggregation of human platelets (Patel et al., 1989), and an organism labelled *A. spiculosospora* produces a specific and reversible inhibitor of angiotensin I converting enzyme (Koguchi et al., 1986).

Some actinomycetes, including *Actinomadura* strain R39, excrete large amounts of DD-carboxypeptidases, enzymes that are generally membrane-bound. When isolated and characterized, DD-carboxypeptidases serve as model enzymes for elucidating the mechanism of β-lactam antibiotic action (Ghuysen et al., 1979). The inactivation of DD-carboxypeptidase from *Actinomadura* strain R39 by β-lactum antibiotics facilitated the development of a rapid and sensitive method for the qualitative determination of these antibiotics in biological fluids such as human serum and cows' milk (Frere et al., 1980, Schindler et al., 1986). *Actinomadura* strains are also a source of novel restriction endonucleases (Kessler et al., 1985; Roberts, 1984). Proteases from *T. fusca* have been used on an industrial scale (Desai and Dhala, 1969).

DNA fragments containing the gene that encodes the extracellular lactamases of *Actinomadura* strain R39 have been introduced into *Streptomyces lividans* TK24 via the high-copy-number, promotor-probe plasmid pIJ424 (Piron-Fraipont et al., 1989). Maximal level of β-lactamase secretion was observed with *S. lividans* CM3 harboring the recombinant plasmid pDML150. The DNA segment in pDML 150 that encodes the β-lactamase has been sequenced (Houba et al., 1989); the β-lactamase precursor is a 304-amino acid polypeptide, and its amino terminal region has the characteristic features of a signal peptide. The *Actinomadura* strain R39 β-lactamase contains a marker of the class A β-lactamases which shows a particularly high homology with the β-lactamases of *Bacillus licheniformis.*

Thermophilic thermomonosporas have been evaluated for cellulose bioconversion processes designed to produce single-cell protein (Bellamy, 1974, 1977; Crawford, 1974, 1988; Humphrey et al., 1977; Wood, 1985) or sugar syrups for fermentation to ethanol (Hägerdal et al., 1979; Lee and Humphrey, 1979; Ball and McCarthy, 1988). Several of these publications have referred to a cellulolytic *Thermoactinomyces* strain which has been labelled *T. cellulosae* (Vacca and Bellamy, 1976) but this organism was subsequently shown to be a member of the genus *Thermomonospora* (Hägerdal et al., 1980).

Thermomonospora fusca has been grown at 55°C on a pilot scale to convert high-cellulose, low-lignin pulp-mill waste into single-cell protein (Crawford et al., 1973). The proteinaceous product had a high nutritional value as a feed supplement in the diet of chicks (Harkin et al., 1974). Thermostable cellulases are also attractive as catalysts in biomass conversion, since their temperature optima can reduce contamination problems in economically feasible nonaseptic applications (Wood, 1985).

A cellulase-enhanced *Thermomonospora* mutant grown on chemically pretreated poplar wood was found to use 70% of the woody biomass after 62 h at 55°C, but failed to grow well on wood that had not been chemically pretreated. Bellamy (1977) used the parental strain of the same organism, then known as *Thermoactinomyces cellulosae,* for the bioconversion of chemically pretreated lignocellulosic feedlot waste into microbial biomass. The culture degraded up to 85% of the cellulosic and over 90% of the hemicellulosic fraction of the substrate. The process led to the production of a high-protein-content product, but it was not economical due to the high cost of the chemical pretreatment needed to obtain a suitable degradable substrate.

Some success has been achieved in the use of thermophilic actinomycetes in processes designed for the saccharification of purified celluloses. The cellulase complex of Bellamy's organism, *T. fusca* YX, readily saccharifies finely ground, acid-swollen Avicel, a purified chemically pretreated cellulose (Su and Paulavicius, 1975; Ferchak and Pye, 1980). In subsequent work (Ferchak and Pye, 1983), a process was developed for saccharification of cellulose using a cellulose/cellobiose mixture. In a 7-day, 50°C saccharification procedure, 15 to 20% glucose syrups were generated from acid-swollen Avicel. The availability of cellulase-overproducing strains of *Thermomonospora* species (Fennington et al., 1982; Meyer and Humphrey, 1982) should further improve the commercial potential of saccharification processes.

Ligninolytic *Thermomonospora* strains may have potential for the biological bleaching of pulps or in biological pulping processes, given their capacity to solubilize lignin and to separate it from the cellulose component of lignocellulose (McCarthy et al., 1986). It has also been suggested (Crawford, 1988) that the ability of some *Thermomonospora* strains to simultaneously attack all of the major components of lignocellulose, that is, cellulose, lignin, and hemicellulose, may be used to develop direct fermentations for the conversion of lignocellulosic wastes into valuable chemicals and for single-cell protein. The development of efficient, industrial-size bioreactors for solid substrate fermentations will be very important for commercially successful actinomycete-adapted bioconversion processes.

Acknowledgments

We are grateful to Dr. A. J. McCarthy (Liverpool University) for critically reviewing this chapter.

Literature Cited

Agre, N. S., and L. N. Guzeva. 1975. New genus of the actinomycetes: *Excellospora* gen. nov. Mikrobiologiya 44:518–522.

Andersen, A. A. 1958. A new sampler for the collection, sizing and enumeration of viable airborne particles. J. Bacteriol. 76:471–484.

Athalye, M. 1981. Classification and isolation of actinomadurae. Ph. D. Thesis, University of Newcastle upon Tyne, UK.

Athalye, M., M. Goodfellow and D. E. Minnikin. 1984. Menaquinone composition in the classification of *Actinomadura* and related taxa. J. Gen. Microbiol. 130:817–823.

Athalye, M., M. Goodfellow, J. Lacey, and R. P. White. 1985. Numerical classification of *Actinomadura* and *Nocardiopsis*. Int. J. Syst. Bacteriol. 35:86–98.

Athalye, M., J. Lacey, and M. Goodfellow. 1981. Selective isolation and enumeration of actinomycetes using rifampicin. J. Appl. Bacteriol. 51:289–297.

Bachmann, S. L., and A. J. McCarthy. 1989. Purification and characterization of a thermostable β-xylosidase from *Thermomonospora fusca*. J. Gen. Microbiol. 135:293–299.

Ball, A. S., and A. J. McCarthy. 1988. Saccharification of straw by actinomycete enzymes. J. Gen. Microbiol. 134:2139–2147.

Ball, A. S., W. B. Betto, and A. J. McCarthy. 1989. Degradation of lignin related compounds by actinomycetes. Appl. Environ. Microbiol. 55:1642–1644.

Ball, A. S. and A. J. McCarthy. 1989. Production and properties of xylanases from actinomycetes. J. Appl. Bacteriol. 66:439–444.

Batrakov, S. G. and L. D. Bergelson. 1978. Lipids of the streptomycetes. Structural investigation and biological interrelation. Chemistry and Physics 21:1–29.

Batrakov, S. G., A. G. Panosyan, B. V. Rozynov, I. V. Konova, and L. D. Bergelson. 1976. Menaquinones of *Actinomyces olivaceus:* the structure of MK-9(H$_6$), Mk-9(H$_8$), MK-8(H$_6$), MK-8(H$_8$). Bioorganicheskaya Khimiya 2:1538–1546. (English translation).

Bellamy, W. D. 1974. Single cell proteins from cellulosic wastes. Biotech. Bioeng. 16:869–880.

Bellamy, W. D. 1977. Cellulose and lignocellulose digestion by thermophilic actinomycetes for single cell protein production. Dev. Ind. Microb. 18:249–254.

Bernier, R., M. Kopp, B. Trakas, and F. Stutzenberger. 1988. Production of extracellular enzymes by *Thermomonospora curvata* during growth on protein-extracted lucerne fibers. J. Appl. Bacteriol. 65:411–418.

Bernier, R., and F. Stutzenberger. 1988. Extracellular and cell-associated forms of beta-glucosidase in *Thermomonospora curvata*. Lett. Appl. Microbiol. 7:103–107.

Blanchard, R. 1896. Parasites végétaux à l'exclusion des bactéries, p. 811–932. In: Bouchard (ed.), Traité de pathologie générale, vol. 2. G. Masson, Paris.

Brocq-Rousseu, D. 1904. Sur un *Streptothrix* cause de l'alteration des avoines moisies. Rev. Bot. 16:219–230.

Calza, R. E., D. C. Erwin, and D. B. Wilson. 1985. Purification and characterization of two β-1, 4-endoglucanases from *Thermomonospora fusca*. Biochemistry 24:7797–7804.

Chater, K. F. and D. A. Hopwood. 1984. *Streptomyces* genetics, p. 229–286. In: M. Goodfellow, M. Mordarski, and S. T. Williams (ed.), The biology of actinomycetes. Academic Press, London.

Chormonova, N. T., and T. P. Preobrazhenskaya. 1981. Occurrence of *Actinomadura* in Kazakhstan soils. Antibiotiki 26:341–345.

Collins, M. D., M. Faulkner, and R. M. Keddie. 1984. Menaquinone composition of some sporeforming actinomycetes. System. Appl. Microbiol. 5:20–29.

Collins, M. D. and D. Jones. 1981. Distribution of isoprenoid quinone structural types in bacteria and their taxonomic implications. Microbiol. Rev. 45:316–354.

Collins, M. D., A. J. McCarthy, and T. Cross. 1982. New highly saturated members of the vitamin K$_2$ series from *Thermomonospora*. Zbl. Bakt. Hyg. I. Abt. Orig. C3, 358–363.

Collins, M. D., R. M. Kroppenstedt, J. Tamaoka, K. Komagata, and T. Kinoshita. 1988. Structures of the tetrahydrogenated menaquinones from *Actinomadura angiospora, Faenia rectivirgula* and *Saccharothrix australiensis*. Curr. Microbiol. 17:275–279.

Collmer, A., and D. B. Wilson. 1983. Cloning and expression of a *Thermomonospora* YX endoglucanase in *Escherichia coli*. Biol. Technology 1:494–501.

Crawford, D. L. 1974. Growth of *Thermomonospora fusca* in lignocellulosic pulps of varying lignin content. Can. J. Microbiol. 20:1069–1072.

Crawford, D. L. 1975. Cultural, morphological and physiological characteristics of *Thermomonospora fusca* (strain 190Th). Can. J. Microbiol. 21:1842–1848.

Crawford, D. L. 1988. Biodegradation of agricultural and urban wastes, p. 433–439. In: M. Goodfellow, S. T. Williams, and M. Mordarski (ed.), Actinomycetes in biotechnology. Academic Press, London.

Crawford, D. L., and M. A. Gonda. 1977. The sporulation process in *Thermomonospora fusca* as revealed by scanning electron microscopy and transmission electron microscopy. Can. J. Microbiol. 23:1088–1095.

Crawford, D. L., and E. McCoy. 1972. Cellulases of *Thermomonospora fusca* and *Streptomyces thermodiastaticus*. Appl. Microbiol. 24:150–152.

Crawford, D. L., E. McCoy, J. M. Harkin, and P. Jones. 1973. Production of microbial protein from waste cellulose by *Thermomonospora fusca*, a thermophilic actinomycete. Biotech. Bioeng. 15:833–843.

Cross, T. 1981. The monosporic actinomycetes, p. 2091–2100. In M. P. Starr, H. Stolp, H. G. Trüper, A. Balows, and H. G. Schlegel (ed.), The prokaryotes: a handbook of habitats, isolation and identification of bacteria. Springer-Verlag, Berlin.

Cross, T. 1982. Actinomycetes: A continuing source of new metabolites. Dev. Ind. Microbiol. 23:1–18.

Cross, T. and M. Goodfellow. 1973. Taxonomy and classification of actinomycetes, p. 11–112. In: G. Sykes and F. A. Skinner (ed.), Actinomycetales: characteristics and practical importance. Academic Press, London.

Cullen, W. P., N. D. Celmer, L. R. Chappel, L. H. Huang, H. Maeda, S. Nishijama, R. Shibakawa, J. Sone, and P. C. Watts. 1987. CP-54–883 a novel chlorine-containing polyether antibiotic produced by a new species of *Actinomadura*: Taxonomy of the producing culture, fermentation, physiochemical and biological properties of the antibiotic. J. Antibiot. 40:1490–1495.

Desai, A. J. and S. A. Dhala. 1969. Purification and properties of proteolytic enzymes from thermophilic actinomycetes. J. Bacteriol. 100:149–155.

Deutsche Sammlung von Mikroorganismen und Zellkulturen GmbH, DSM Catalogue of Strains, 1989, 4th. ed., German Collection of Microorganisms and Cell Cultures, Braunschweig, Germany.

Develoux, M., J. Audvin, T. Freguer, T. M. Vetter, A. Warter, and A. Cenac. 1988. Mycetoma in the Republic of Niger: Clinical features and epidemiology. Clin. J. Trop. Med. Hyg. 38:386–390.

Eerolen, E., and O-P. Lehtonen. 1988. Optimal data processing procedure for automating bacterial identification by gas-liquid chromatography of cellular fatty acids. J. Clin. Microbiol. 26:1745–1753.

Fennington, G., D. Lupo, and F. J. Stutzenberger. 1982. Enhanced cellulase production in mutants of *Thermomonospora curvata*. Biotech. Bioeng. 24:2487–2497.

Fennington, G., D. Neubauer, and F. J. Stutzenberger. 1984. Cellulase biosynthesis in a catabolite repression-resistant mutant of *Thermomonospora curvata*. Appl. Environ. Microbiol. 47:201–204.

Ferchak, J. D., B. Hägerdal, and E. K. Pye. 1980. Saccharification of cellulose by the cellulytic enzyme system of *Thermomonospora* sp. II. Hydrolysis of cellulosic substrates. Biotech. Bioeng. 22:1527–1542.

Ferchak, J. D., and E. K. Pye. 1980. Saccharification of cellulose by the cellulolytic enzyme system of *Thermomonospora* species. I. Stability of cellulolytic activities with respect to time, temperature and pH. Biotech. Bioeng. 22:1515–1526.

Ferchak, J. D., and E. K. Pye. 1983. Effects of cellulose, glucose, ethanol, and metal ions on the cellulose complex of *Thermomonospora fusca*. Biotech. Bioeng. 25:2865–2872.

Fergus, C. L. 1964. Thermophilic and thermotolerant molds and actinomycetes in mushroom compost during peak heating. Mycologia 56:267–284.

Fischer A., R. M. Kroppenstedt, and E. Stackebrandt. 1983. Molecular-genetic and chemotaxonomic studies on *Actinomadura* and *Nocardiopsis*. J. Gen. Microbiol. 129:3433–3446.

Fowler, V. J., W. Ludwig, and E. Stackebrandt. 1985. Ribosomal ribonucleic acid cataloging in bacterial systematics: The phylogeny of *Actinomadura*, p. 17–40. In: M. Goodfellow and D. E. Minnikin (ed.), Chemical methods in bacterial systematics. Academic Press, London.

Frere, J. M., D. Klein, and J. M. Ghuysen. 1980. Enzymatic method for rapid and sensitive determination of β-lactam antibiotics. Antimicrob. Agents Chemother. 18:506–510.

Galatenko, O. A., L. P. Terekhova, and T. P. Preobrazhenskaya. 1981. New *Actinomadura* species isolated from Turkmen soil samples and their antagonistic properties. Antibiotiki 26:803–807.

Galatenko, O. A., L. P. Terekhova, and T. P. Preobrazhenskaya. 1987. Validation of the publications of new names and new combinations previously effectively published outside the IJSB. List no. 23. Int. J. Syst. Bacteriol. 37:179–180.

Gauze, G. F., T. P. Preobrazhenskaya, E. S. Kudrina, N. O. Blinov, I. D. Ryabova and M. A. Sveshnikova. 1957. Problems in the classification of antagonistic actinomycetes. Medzig, Moscow State Publishing House for Medical Literature.

Gauze, G. F., M. A. Sveshnikova, R. S. Ukholina, G. N. Gavrilina, V. A. Filicheva and E. G. Gladkikh. 1973. Production of antitumor antibiotic carminomycin by *Actinomadura carminata* sp. nov. Antibiotiki 18:675–678.

Gauze, G. F., T. P. Preobrazhenskaya, N. V. Lavrova, R. S. Ukholina, G. V. Kochetkova, N. P. Nechaeva, N. V. Konstantinova, and I. V. Tolstykh. 1975. *Actinomadura cremea* var. *rifamycini*, a rifamicin-producing organism. Antibiotiki 20:963–966.

Gauze, G. F., L. P. Terekhova, O. A. Galatenko, T. P. Preobrazhenskaya, V. N. Borisova, and G. B. Fedorova. 1984. *Actinomadura recticatena* sp. nov., a new species and its antibiotic properties. Antibiotiki 29:3–7.

Gauze, G. F., L. P. Terekhova, O. A. Galatenko, T. P. Preobrazhenskaya, V. N. Borisova, and G. B. Federova. 1987. Validation of the publications of new names and new combinations previously effectively published outside the IJSB. List no. 23. Int. J. Syst. Bacteriol. 37:179–180.

Gerber, N. N. 1971. Prodigiosin-like pigments from *Actinomadura (Nocardia) pelletieri*. J. Antibiot. 24:636.

Gerber, N. N. 1973. Minor prodiginine pigments from *Actinomadura madurae* and *Actinomadura pelletieri*. J. Heterocycl. Chem. 10:925.

Ghangas, G. S., and D. B. Wilson. 1987. Expression of a *Thermomonospora fusca* cellulase gene in *Streptomyces lividans* and *Bacillus subtilis*. Appl. Environ. Microbiol. 53:1470–1475.

Ghuysen, J. M., J. M. Frère, M. Levy-Bouille, J. Coyette, J. Dusart, and M. Nguyen-Disteche. 1979. Use of model enzymes in the determination of the mode action of penicillins and delta³-cephalosporins. Chem. Rev. Biochem. 48:73–101.

Goldstein, B. P., E. Selva, L. Gastaldo, M. Berti, R. Pallanza, F. Ripamonti, P. Ferrara, M. Denaro, V. Arioli, and G. Cassani. 1987. A40926, a new glycolipid antibiotic with anti-*Neisseria* activity. Antimicrob. Ag. Chemother. 31:1961–1965.

Goodfellow, M. 1984. Validation of the publication of new names and combinations previously effectively published outside the IJSB. List. no. 10 Int. J. Syst. Bacteriol. 34:503–504.

Goodfellow, M. 1989. The actinomycetes I, Supragenetic classification of actinomycetes, p. 2333–2339. In: S. T. Williams, M. E. Sharpe and J. G. Holt (ed.), Bergey's manual of systematic bacteriology, vol. 4. Williams and Wilkins, Baltimore.

Goodfellow, M., G. Alderson, and J. Lacey. 1979. Numerical taxonomy of *Actinomadura* and related actinomycetes. J. Gen. Microbiol. 112:95–111.

Goodfellow, M. and T. Cross. 1984. Classification, p. 7–164. In: M. Goodfellow, M. Mordarski and S. T. Williams (ed.), The biology of actinomycetes. Academic Press, London.

Goodfellow, M., and D. E. Minnikin, ed. 1985. Chemical methods in bacterial systematics. Academic Press, London.

Goodfellow, M., and T. Pirouz. 1982. Numerical classification of sporoactinomycetes containing *meso*-diaminopimelic acid in the cell wall. J. Gen. Microbiol. 128:503–507.

Goodfellow, M., E. Stackebrandt and R. M. Kroppenstedt. 1988. Chemotaxonomy and actinomycete systematics. In: Y. Okami, T. Beppu, and H. Ogawara (ed.), Biology of actinomycetes '88. Japan Scientific Societies Press, Tokyo.

Goodfellow, M., L. J. Stanton, K. E. Simpson, and D. E. Minnikin. 1990. Numerical and chemical classification of *Actinoplanes* and some related actinomycetes. J. Gen. Microbiol. 136:19–34.

Gordon, M. A. 1974. Aerobic pathogenic Actinomycetaceae, p. 175–188. In: E. H. Lennette, E. H. Spaulding, and T. P. Truant (ed.), Manual of clinical microbiology. American Society for Microbiology, Washington, D.C.

Greiner-Mai, E., R. M. Kroppenstedt, F. Korn-Wendisch, and H. J. Kutzner. 1987. Morphological and biochemical characterization and amended descriptions of thermophilic actinomycetes species. Syst. Appl. Microbiol. 9:97–109.

Grund, E. and R. M. Kroppenstedt. 1989. Transfer of five *Nocardiopsis* species to the genus *Saccharothrix* Labeda et al. 1984. J. Syst. Appl. Microbiol. 12:267–274.

Grund, E. and R. M. Kroppenstedt. 1990. Chemotaxonomy and numerical taxonomy of the genus *Nocardiopsis* Meyer 1976. Int. J. Syst. Bact. 40:5–11.

Gumaa, S. A., E. S. Mahgoub, and M. A. El Sid. 1986. Mycetoma of the head and neck. Am. J. Trop. Med. Hyg. 35:594–600.

Gusek, T. W., and J. E. Kinsella. 1987. Purification and characterization of a heat-stable serine proteinase from *Thermomonospora fusca* YX. Biochem. J. 246:511–517.

Hägerdahl, B. G. R., J. D. Ferchak, and E. K. Pye. 1978. Cellulolytic enzyme system of *Thermoactinomyces* sp. grown on microcristalline cellulose. Appl. Environ. Microbiol. 36:606–612.

Hägerdal, B. G. R., H. Harris, and E. K. Pye. 1979. Association of β-glucosidase with intact cells of *Thermoactinomyces*. Biotech. Bioeng. 21:345–356.

Hägerdal, B. G. R., J. D. Ferchak, and E. K. Pye. 1980. Saccharification of cellulose by the cellulolytic enzyme system of *Thermononospora* sp. I. Stability of cellulolytic activities with respect to time, temperature and pH. Biotech. Bioeng. 22:1515–1528.

Harkin, J. M., D. L. Crawford, and E. McCoy. 1974. Bacterial protein from pulps and papermill sludge. TAPPI 57:131–134.

Hasegawa, T., M. P. Lechevalier, and H. A. Lechevalier. 1978. A new genus of the Actinomycetales: *Actinosynnema* gen. nov. Int. J. Syst. Bacteriol. 28:304–310.

Hasegawa, T., S. Tanida, and H. Ono. 1986. *Thermomonospora formosensis* sp. nov. Int. J. Syst. Bacteriol. 36:20–23.

Hedge, V., T. Barrett, V. Gullo, A. Horan, A. T. McPhail, J. Marquez, M. Patel, and M. Puar. 1986. A novel naphthoquinone antibiotic Sch 3819 produced by a Thermomonospora. Abstract 908, 28th Interscience Congress on Antimicrobial Agents and Chemotherapy.

Henssen, A. 1957. Beiträge zur Morphologie und Systematik der thermophilen Actinomyceten. Arch. Microbiol. 26:373–414.

Henssen, A. 1970. Spore formation in thermophilic actinomycetes, p. 205–210 In: H. Prauser (ed.), The Actinomycetales. Gustav Fischer Verlag, Jena, Germany.

Henssen, A., and E. Schnepf. 1967. Zur Kenntnis thermophiler Actinomyceten. Archiv für Mikrobiologie 57:214–231.

Hickey, R. T., and H. D. Tresner. 1952. A cobalt containing medium for sporulation of *Streptomyces* species. J. Bacteriol. 64:891–892.

Hoffmann, P. 1989a. Cryopreservation of fungi. World Federation for Culture Collections. Technical information sheet No. 5. UNESCO/WFCC/Education Committee. Braunschweig, Germany.

Hoffmann, P. 1989b. Cryopreservation of basidiomycete cultures. Mushroom. Science XII. (Part I). Proceeding of the Twelfth International Congress on the Science and Cultivation of Edible Fungi (1987), Braunschweig, Germany.

Horan, A. C., and B. C. Brodsky. 1982. A novel antibiotic-producing *Actinomadura, Actinomadura kijaniata* sp. nov. Int. J. Syst. Bacteriol. 32:195–200.

Houba, S., S. Willem, C. Dueng, C. Molitor, J. Dusart, J.-M. Frère, and J. M. Ghuysen. 1989. Cloning and amplified expression in *Streptomyces lividans* of the gene encoding the extracellular β-lactamase of *Actinomadura* R39. FEMS Microbiol. Lett. 65:241–246.

Howard, O. W., E. Grund, R. M. Kroppenstedt, and M. D. Collins. 1986. Structural determination of a novel naturally occurring cyclic vitamin K. Biochem. Biophys. Res. Commun. 140:916–923.

Huang, L. H. 1980. *Actinomadura macra* sp. nov., the producer of antibiotics CP-47, 433 and CP-47434. Int. J. Syst. Bacteriol. 30:565–568.

Humphrey, A. E., A. Moreira, W. Armiger, and D. Zabriskie. 1977. Production of single cell protein from cellulose wastes. Biotech. Bioeng. Symp. 7:45–64.

Jensen, H. L. 1930. The genus *Micromonospora* Orskov, a little known group of soil microorganisms. Proc. Linnean Soc. N.S.W. 55:231–248.

Jensen, H. L. 1932. Contribution to our knowledge of Actinomycetales. III. Further observations on the genus *Micromonospora*. Proc. Linnean Soc. N.S.W. 57:173–180.

Jones, K. L. 1949. Fresh isolates of actinomycetes in which the presence of sporogenous aerial mycelium is a fluctuating characteristic. J. Bacteriol. 64:891–892.

Katz, E., C. J. Thompson, and D. A. Hopwood. 1983. Cloning and expression of the tyrosinase gene from *Streptomyces antibioticus* in *Streptomyces lividans*. J. Gen. Microbiol. 129:2703–2714.

Kessler, Ch., P. S. Neumaier, and W. Wolf. 1985. Recognition sequences of restriction endonucleases and methylases, a review. Gene 33:1–102.

Koguchi, T., K. Yamada, M. Yamato, R. Okachi, K. Nakayama, and H. Kase. 1986. K-4, a novel inhibitor of angiotensin I converting enzyme produced by *Actinomadura spiculosospora* J. Antibiot. 39:364–371.

Kosmachev, A. K. 1960. Preservation of viability of thermophilic actinomycetes. Microbiology 29:160

Kosmachev, A. K. 1964. A new thermophilic actinomycete *Micropolyspora thermovirida* n. sp. Microbiology (English translation of Microbiologiya) 33:235–237.

Krassil'nikov, N. A. 1941. Keys to Actinomycetales (In Russian) Izvest. Akad, Nauk. SSSR. Moscow (English translation: ed. E. Rabinovitz. Israel Program for Scientific Translations, Jerusalem, 1966).

Krassil'nikov, N. A., and N. S. Agre. 1964a. On two new species of *Thermopolyspora*. Hind. Antibiot. Bull. 6:97–107.

Krassil'nikov, N. A., and N. S. Agre. 1964b. A new genus of the actinomycetes—*Actinobifida*. The yellow group *Actinobifida dichotomica*. Mikrobiologiya 33:935–943.

Krassil'nikov, N. A., and N. S. Agre. 1965. The brown group of *Actinobifida chromogena* n. sp. Mikrobiologiya 34:284–291.

Krassil'nikov, N. A., N. S. Agre, and G. I. El-Registan. 1968. New thermophilic species of the genus *Micropolyspora*. Mikrobiologiya 37:1065–1072.

Kroppenstedt, R. M. 1982a. Anwendung von Dünnschicht- und Gas-Chromatographie in der Bakterientaxonomie. GIT-Chromatographie. Suppl. 1982:34–40

Kroppenstedt, R. M. 1982b. Separation of bacterial menaquinones by HPLC using reverse phase (RP-18) and a silver loaded ion exchanger. J. Liquid Chromat. 5:2359–2367.

Kroppenstedt, R. M. 1982c. Anwendung chromatographischer HP-Verfahren (HPTLC und HPLC) in der Bakterien-Taxonomie. GIT-Labor-Medizin. 5:266–275.

Kroppenstedt, R. M. 1985. Fatty acid and menaquinone analysis of actinomycetes and related organisms, p. 173–199. In: M. Goodfellow and D. E. Minnikin (ed.), Chemical methods in bacterial systematics. Academic Press, London.

Kroppenstedt, R. M. 1987. Chemische Untersuchungen an *Actinomycetales* und verwandten Taxa, Korrelation von Chemosystematik und Phylogenie. Habilitationsschrift. Technische Hochschule Darmstadt, Darmstadt, Germany.

Kroppenstedt, R. M., F. Korn-Wendisch, V. J. Fowler, and E. Stackebrandt. 1981. Biochemical and molecular genetic evidence for a transfer of *Actinoplanes armeniacus* into the family Streptomycetaceae. Zbl. Bakt. Hyg., I. Abt. Orig. C2, 254–262.

Kroppenstedt, R. M., and H. J. Kutzner. 1978. Biochemical taxonomy of some problem actinomycetes. Zentralbl. Bakteriol. Parasitenkr. Infektionskr. Hyg. Abt. 1 Orig. Reihe C Suppl. 6:125–133.

Kroppenstedt, R. M., E. Stackebrandt and M. Goodfellow. 1990. Taxonomic revision of the actinomycete genera *Actinomadura* and *Microtetraspora*. Syst. Appl. Microbiol. 13:148–160.

Kuo, M. J., and P. A. Hartman. 1967. Purification and partial characterization of *Thermomonospora vulgaris* amylases. Can. J. Microbiol. 13:1157–1163.

Kurup, V. P., and R. J. Heinzen. 1978. Isolation and characterization of actinophages of *Thermoactinomyces* and *Micropolyspora*. Can. J. Microbiol. 24:794–797.

Küster, E. 1959a. Outline of comparative study of criteria used in characterization of the actinomycetes. Int. Bull. Bact. Nomen. and Taxon. 9:98–104.

Küster, E. 1959b. Introductory remarks and welcome, "Round table conference on streptomycetes." 7th International Congress of Microbiology, Stockholm, August 4–5, 1958. Int. Bull. Bact. Nomen. and Taxon. 9:57–61.

Küster, E. 1974. Family VIII. *Micromonosporaceae* Krasil'nikov 1938, 272, p. 846. In: R. E. Buchanan, and N. E. Gibbon (ed.), Bergey's manual of determinative bacteriology, 8th ed. Williams and Wilkins, Baltimore.

Küster, E., and R. Locci. 1963. Transfer of *Thermoactinomyces viridis* Schuurmans et al. 1956 to the genus *Thermomonospora* as *Thermomonospora viridis* comb. nov. Int. Bull. Bacteriol. Nomencl. Taxon. 13:214–216.

Labeda, D. P., R. T. Testa, M. P. Lechevalier, and H. A. Lechevalier. 1984. *Saccharothrix:* A new genus of the Actinomycetales related to *Nocardiopsis*. Int. J. Syst. Bacteriol. 34:426–431.

Labeda, D. P., R. T. Testa, M. P. Lechevalier, and H. A. Lechevalier. 1985. *Actinomadura yumaensis* sp. nov. Int. J. Syst. Bacteriol. 35:333–336.

Lacey, J. 1973. Actinomycetes in soils, composts and fodders, p. 231–251. In: F. A. Skinner and G. Sykes (ed.), Actinomycetales: Characteristics and practical importance. Academic Press, London.

Lacey, J. 1974. Allergy in mushroom workers. Lancet i: 366.

Lacey, J. 1977. The ecology of actinomycetes in fodders and related substrates. Zentralbl. Bakteriol. Parasitenkd. Infektionskr. Hyg. Abt. 1, Suppl. 6:161–170.

Lacey, J. 1988. Actinomycetes as biodeteriogens and pollutants of the environment, p. 359–432. In: M. Goodfellow, S. T. Williams, and M. Mordarski (ed.), Actinomycetes in biotechnology. Academic Press, London.

Lacey, J., and J. Dutkiewicz. 1976. Isolation of actinomycetes and fungi using a sedimentation chamber. J. Appl. Bacteriol. 41:315–319.

Laveran, M. 1906. Tumeur provoquee par un microcoque rose en zooglées. C. R. Hebd. Soc. Biol. 2:340–341.

Lavrova, N. V. and T. P. Preobrazhenskaya. 1975. Isolation of new species of *Actinomadura* on selective media with rubromycin. Antibiotiki 20:483–488. (in Russian).

Lavrova, N. V., T. P. Preobrazhenskaya, and M. A. Svesh-nikova. 1972. Isolation of soil actinomycetes on selective media with rubromycin. Antibiotiki 11:965–970 (in Russian).

Lawrence, H. M., H. Merivuori, J. A. Sands, and K. A. Pidcock. 1986. Preliminary characterization of bacteriophages infecting the thermophilic actinomycete *Thermomonospora*. Appl. Environ. Microbiol. 52:631–636.

Lawson, E. N., and L. N. Davey. 1972. A waterborn actinomycete resembling strains causing mycetoma. J. Appl. Bacteriol. 35:389–394.

Lechevalier, H. A., and M. P. Lechevalier. 1970. A critical evaluation of the genera of aerobic actinomycetes, p. 393–405. In: H. Prauser (ed.), The Actinomycetales. Gustav Fischer Verlag, Jena, Germany.

Lechevalier, H. A., M. P. Lechevalier, and N. N. Gerber. 1971. Chemical composition as a criterion in the classification of actinomycetes. Adv. Appl. Microbiol. 14:47–72.

Lechevalier, M. P. 1976. The taxonomy of the genus *Nocardia:* Some light at the end of the tunnel? p. 1–38. In: M. Goodfellow, G. H. Brownell, and J. A. Serrano (ed.), The biology of the nocardiae. Academic Press, London.

Lechevalier, M. P., and N. N. Gerber. 1970. The identity of 3-O-methyl-D-galactose with madurose. Carb. Res. 13:451–454.

Lechevalier, M. P., and H. A. Lechevalier. 1970a. Composition of whole-cell hydrolysates as a criterion in the classification of aerobic actinomycetes, p. 311–316. In: H. Prauser (ed.), The Actinomycetales. Gustav Fischer Verlag, Jena, Germany.

Lechevalier, M. P. and H. A. Lechevalier. 1970b. Chemical composition as a criterion in the classification of aerobic actinomycetes. Int. J. Syst. Bacteriol. 20:435–443.

Lechevalier, M. P., and H. A. Lechevalier. 1980. The chemotaxonomy of actinomycetes. p. 227–291. In: A. Dietz and D. Thayer (ed.), Actinomycete taxonomy, special publication 6. Society for Industrial Microbiology, Arlington, VA.

Lechevalier, M. P., C. de Biévre, and H. A. Lechevalier. 1977. Chemotaxonomy of aerobic actinomycetes: Phospholipid composition. Biochem. Ecol. Syst. 5:249–260.

Lechevalier, M. P., A. E. Stern, and H. A. Lechevalier. 1981. Phospholipids in the taxonomy of actinomycetes. In: K. P. Schaal, and G. Pulverer (ed.), Actinomycetes. Gustav Fischer Verlag, Stuttgart.

Lee, S. E., and A. E. Humphrey. 1979. Use of continious culture techniques for determining the growth kinetics of celluloytic *Thermoactinomyces* sp. Biotech. Bioeng. 21:1277–1288.

Liegard, H., and M. Landrieu. 1911. Un cas de mycose conjonctivale. Ann. Ocul. 146:418–426.

Locci, R., E. Baldacci, and B. Petrolini. 1967. Contribution to the study of oligosporic actinomycetes. I. Description of new species of *Actinobifida: Actinobifida alba* sp. nov. and revision of the genus. G. Microbiol. 15:79–91.

Lupo, D., and F. Stutzenberger. 1988. Changes in endoglucanase patterns during growth of *Thermomonospora curvata* on cellulose. Appl. Environ. Microbiol. 54:588–589.

Matson, J. A., C. Claridge, J. A. Bush, J. Titus, W. T. Bradner, and T. W. Doyle. 1989. AT 2433-A1, AT 2433-A2, AT 2433-B1, and AT 2433-B2, novel antitumor anti-biotic compounds produced by *Actinomadura melliaura*. Taxonomy, fermentation, isolation and biological properties. J. Antibiot. 42:1547–1555.

McCarthy, A. J. 1987. Lignocellulose-degrading actinomycetes. FEMS Microbiol. Rev. 43:145–163.

McCarthy, A. J. 1989. Genus *Thermomonospora* Henssen 1957, p. 2553–2559. In: S. T. Williams, M. E. Sharpe, and J. G. Holt (ed.), Bergey's manual of systematic bacteriology, vol. 4. Williams and Wilkins, Baltimore.

McCarthy, A. J., A. S. Ball, and S. L. Bachmann. 1988. Ecological and biotechnological implication of lignocellulose degradation by actinomycetes, p. 283–287. In: Y. Okami, T. Beppu, and H. Ogawara (ed.), Biology of actinomycetes '88. Japan Scientific Society Press, Tokyo.

McCarthy, A. J., and P. Broda. 1984. Screening for lignin-degrading actinomycetes and characterisation of their activity against ^{14}C-lignin-labelled wheat lignocellulose. J. Gen. Microbiol. 130:2905–2913.

McCarthy, A. J., and T. Cross. 1981. A note on a selective isolation medium for the thermophilic actinomycete *Thermomonospora chromogena*. J. Appl. Bacteriol. 51:299–302.

McCarthy, A. J., and T. Cross. 1984a. A taxonomic study of *Thermomonospora* and other monosporic actinomycetes. J. Gen. Microbiol. 130:5–25.

McCarthy A. J.,and T. Cross. 194b. Taxonomy of *Thermomonospora* and related oligosporic actinomycetes, p. 521–536. In: L. Ortiz-Ortiz, L. F. Bojalil, and V. Yakoleff (ed.), Biological, biochemical and biomedical aspects of actinomycetes. Academic Press, San Diego.

McCarthy, A. J., A. Paterson, and P. Broda. 1986. Lignin solubilization by *Thermomonospora mesophila*. Appl. Microbiol. 24:347–352.

McCarthy, A. J., E. Peace, and P. Broda. 1985. Studies on extracellular activities of some thermophilic actinomycetes. Appl. Microbiol. 21:238–244.

Mertz, F. P., and R. C. Yao. 1986. *Actinomadura oligospora* sp. nov., the producer of a new polyether antibiotic. Int. J. Syst. Bacteriol. 36:179–182.

Mertz, F. P., and R. C. Yao. 1990. *Actinomadura fibrosa* sp. nov. isolated from soil. Int. J. Syst. Bacteriol. 40:28–33.

Meyer, H. P., and A. E. Humphrey. 1982. Cellulase production by a wild type and a new mutant strain of *Thermomonospora* sp. Biotech. Bioeng. 24:1901–1904.

Meyer, J. 1976. *Nocardiopsis*, a new genus of the order Actinomycetales. Int. J. Syst. Bacteriol. 26:487–493.

Meyer, J. 1979. New species of the genus *Actinomadura*. Z. Allg. Microbiol. 19:37–44.

Meyer, J. 1981. Validation of the publication of new names and new combinations previously effectively published outside the IJSB. List No. 6. Int. J. Syst. Bacteriol. 31:215–218.

Meyer, J. 1989a. Genus *Actinomadura* Lechevalier and Lechevalier 1970, p. 2511–2526. In: S. T. Williams, M. E. Sharpe, and J. G. Holt (ed.), Bergey's manual of systematic bacteriology, vol. 4. Williams and Wilkins, Baltimore.

Meyer, J. 1989b. Genus *Nocardiopsis* (Brocq-Rousseu) Meyer 1976, p. 2562–2569. In: S. T. Williams, M. E. Sharpe and J. G. Holt (ed.), Bergey's manual of systematic bacteriology, vol. 4. Williams and Wilkins, Baltimore.

Meyer, J., and M. Sveshnikova. 1974. *Micromonospora rubra* Sveshnikova et al. *Actinomadura rubra* comb. nov. Z. Allg. Microbiol. 14:167–170.

Mills, J. T. and G. J. Bollen. 1976. Microflora of heat damaged rapeseed. Can. J. Bot. 54:2893–2902.

Minnikin, D. E., L. Alshamaony, and M. Goodfellow. 1975. Differentiation of *Mycobacterium*, *Nocardia* and related taxa by thin-layer chromatographic analysis of whole-organism methanolysates. J. Gen. Microbiol. 88:200–204.

Minnikin, D. E, A. G. O'Donnell, M. Goodfellow, G. Alderson, M. Athalye, A. Schoal, and J. H. Parlett 1984. An integrated procedure for extraction of bacterial isoprenoid quinones and polar lipids. J. Microbiol. Methods. 2:233–241.

Miyadoh, S., S. Amano, H. Tohyama, and T. Shomura. 1987. *Actinomadura atramentaria*, a new species of the *Actinomycetales*. Int. J. Syst. Bacteriol. 37:342–346.

Miyadoh, S., H. Anzai, S. Amano, and T. Shomura. 1989. *Actinomadura malachitica* and *Microtetraspora viridis* are synonyms and should be transferred as *Actinomadura viridis* comb. nov. Int. J. Syst. Bacteriol. 39:152–158.

Miyashita, K., Y. Mikami, and T. Arai. 1984. Alkalophilic actinomycete, *Nocardiopsis dassonvillei* subsp. *prasina* subsp. nov., isolated from soil. Int. J. Syst. Bacteriol. 34:405–409.

Molina, C. 1982. Acquisition récentes sur les alvéolites allergiques professionelles. Schweiz. med. Wochen. 112:192–197.

Moller, B. B., P. Holberg, S. Gravesen, and B. Weeke. 1976. Precipitating antibodies against *Micropolyspora faeni* in sera from mushroom workers. Acta Allerg. 31:69–70.

Moreira, A. R., J. A. Phillips, and A. E. Humphrey. 1981. Production of cellulases by *Thermomonospora* sp. Biotechnol. Bioeng. 23:1339–1347.

Naumova, I. B. 1988. The teichoic acids of actinomycetes. Microbiol. Science 5:275–279.

Naumova, I. B., N. V. Potekhina, L. P. Terekhova, T. P. Preobrazhenskaya, and K. Digimbay. 1986. Cell wall polyol phosphate polymeres of bacteria belonging to the genus *Actinomadura*, p. 561–566. In: G. Szabó, S. Biró, and M. Goodfellow (ed.), Biological, biochemical and biomedical aspects of actinomycetes. Akadémiai Kiadó, Budapest.

Nonomura, H., and Y. Ohara. 1969. Distribution of actinomycetes in soil. VI. A culture method effective for both preferential isolation and enumeration of *Microbispora* and *Microtetraspora* strains in soil. J. Ferment. Technol. 47:463–469.

Nonomura, H., and Y. Ohara. 1971a. Distribution of actinomycetes in soil. VIII. Green-spore group of *Microtetraspora*, its preferential isolation and taxonomic characteristic. J. Ferment. Technol. 49:1–7.

Nonomura, H., and Y. Ohara. 1971b. Distribution of actinomycetes in soil. IX. New species of the genus *Microbispora* and *Microtetraspora* and their isolation methods. J. Ferment. Technol. 49:887–894.

Nonomura, H., and Y. Ohara. 1971c. Distribution of actinomycetes in soil. X. New genus and species of monosporic actinomycetes in soil. J. Ferment. Technol. 49:895–903.

Nonomura, H., and Y. Ohara. 1971d. Distribution of actinomycetes in soil. XI. Some new species of the genus

Actinomadura Lechevalier et al. J. Ferment. Technol. 49:904–912.

Nonomura, H., and Y. Ohara. 1974. A new species of actinomycetes, *Thermomonospora mesouviformis* sp. nov. J. Ferment. Techn. 53:10–13.

O'Donnell, A. G. 1985. Numerical analysis of chemotaxonomic data, p. 403–414. In: M. Goodfellow, D. Jones, and F. G. Priest (ed.), Computer-assisted bacterial systematics. Academic Press, London.

Patel, M., V. P. Gullo, V. R. Hedge, A. C. Horan, F. Gentile, J. A. Marquez, G. H. Miller, M. S. Puar, and J. A. Waitz. 1987. A novel tetracycline from *Actinomadura brunnea*. Fermentation, isolation and structure eluciation. J. Antibiot. 40:1408–1413.

Patel, M., V. Hedge, A. C. Horan, T. Barrett, R. Bishop, A. King, J. Marquez, R. Hare, and V. Gullo. 1989. Sch 38519, a novel platelet aggregation inhibitor produced by a *Thermomonospora* sp. Taxonomy, fermentation, isolation, physico-chemical properties, structure and biological properties. J. Antibiot. 42:1063–1069.

Pidcock, K. A., B. S. Montenecourt, and J. A. Sands. 1985. Genetic recombination and transformation in protoplasts of *Thermomonospora fusca*. Appl. Environ. Microbiol. 50:693–695.

Pinoy, E. 1912. Isolement et culture d'une nouvelle oospora pathogene. Mycetome a grains rouges de la paroi thoracique. Bull. Soc. Path. Exot. 5:585–589.

Piron-Fraipont, C., C. Duez, A. Matagne, C. Molitor, J. Dusart, J.-M. Frère, and J. M. Ghuysen. 1989. Cloning and amplified expression in *Streptomyces lividans* of the gene encoding the extracellular β-lactamase of *Actinomadura* R39. Biochem. J. 262:849–854.

Poschner, J., R. M. Kroppenstedt, A. Fischer, and E. Stackebrandt. 1985. DNA-DNA-reassociation and chemotaxonomic studies on *Actinomadura*, *Microbispora*, *Microtetraspora*, *Micropolyspora*, and *Nocardiopsis*. Syst. Appl. Microbiol. 6:264–270.

Preobrazhenskaya, T. P., and M. A. Sveshnikova. 1974. New species of the genus *Actinomadura*. Mikrobiologiya 43:864–868.

Preobrazhenskaya, T. P., N. V. Lavrova, and N. O. Blinov. 1975a. Taxonomy of *Streptomyces luteofluorescens*. Mikrobiologiya 44:524–527.

Preobrazhenskaya, T. P., N. V. Lavrova, R. S. Ukholina, and N. P. Nechaeva. 1975b. Isolation of new species of *Actinomadura* on selective media with streptomycin and bruneomycin. Antibiotiki 20:404–409.

Preobrazhenskaya, T. P., M. A. Sveshnikova, and L. P. Terekhova. 1977. Key for the identification of the species of the genus *Actinomadura*. Biol. Actin. Rel. Org. 12:30–38.

Preobrazhenskaya, T. P., M. A. Sveshnikova and G. F. Gauze. 1982. On the transfer of certain species of the genus *Actinomadura* Lechevalier et Lechevalier 1970 to the genus Nocardiopsis Meyer 1976. Mikrobiologiya 51:111–113.

Preobrazhenskaya, T. P., M. A. Sveshnikova, L. P. Terekhova, and N. T. Chormonova. Selective isolation of soil actinomycetes. 1978. p.119–123. In: M. Mordarski, W. Kurylowicz, and J. Jeljaszewicz (ed.), Nocardia and Streptomyces. Gustav Fischer Verlag, Stuttgart.

Preobrazhenskaya, T. P., L. P Terekhova, A. V. Laiko, T. I. Selezneva, V. A. Zenkova, and N. O. Blinov. 1976. *Actinomadura coeruleoviolacea* sp. nov. and its antagonistic properties. Antibiotiki 21:779–784.

Preobrazhenskaya, T. P., L. P Terekhova, A. V. Laiko, T. I. Selezneva, V. A. Zenkova, and N. O. Blinov. 1987. Validation of the publication of new names and new combinations previously effectively published outside the IJSB. List No. 23. Int. J. Syst. Bacteriol. 37:179–180.

Pridham, T. G. and A. T. Lyons, Jr. 1961. *Streptomyces albus* (Rossi Doria) Waksman and Henrici: Taxonomic study of strains labeled *Streptomyces albus*. J. Bacteriol. 81:431–441.

Pridham, T. G., P. Anderson, C. Foley, L. A. Lindenfelser, C. W. Hesseltine, and R. G. Benedict. 1956–1957. A selection of media for maintainance and taxonomic study of streptomycetes. Antibiot. Ann. 1956/57:947–953.

Rippon, T. W. 1968. Extracellular collagenase produced by *Streptomyces madurae*. Biochim. Biophys. Acta 159:147–152.

Ristroph, D. L. and A. E. Humphrey. 1985a. Kinetic characterization of the extracellular xylanases of *Thermomonospora* sp. Biotechnol. Bioeng. 27:832–836.

Ristroph, D. L., and A. E. Humphrey. 1985b. The β-xylosidase of *Thermomonospora*. Biotechnol. Bioeng. 27:909–913.

Roberts, R. J. 1984. Restriction and modification enzymes and their recognition sequences. Nucl. Acids Res. 12:167–204.

Sasaki, T., J. Yoshida, M. Itoh, S. Gomi, T. Shomura, and M. Senzaki. 1988. New antibiotic SF 2315A and B produced by an *Excellospora* sp. I. Taxonomy of the strain, isolation and characterization of antibiotic. J. Antibiot. 41:835–842.

Schaal, K. P. 1972. Zur mikrobiologischen Diagnostik der Nocardiose. Zbl. Bakt. Hyg. I. Abt. Orig. A 220:242–246.

Schaal, K. P. 1984. Laboratory diagnosis of actinomycete diseases, p. 425–456. In: M. Goodfellow, M. Mordarski, and S. T. Williams (ed.), The biology of the actinomycetes. Academic Press, London.

Schaal, K. P., and B. L. Beaman. 1984. Clinical significance of actinomycetes, p. 389–424. In: M. Goodfellow, M. Mordarski, and S. T. Williams (ed.), The biology of the actinomycetes. Academic Press, London.

Schindler, P. W., W. König, S. Chatterjee, and B. N. Ganguli. 1986. Improved screening for β-lactam antibiotics. A sensitive, high-throughput assay using DD-carbopeptidase and a novel chromophore labelled substrate. J. Antibiot. 39:53–57.

Schleifer, K. H. and O. Kandler. 1972. Peptidoglycan types of bacterial cell walls and their taxonomic implications. Bacteriol. Rev. 36:407–477.

Serrano, J. A., B. L. Beaman, T. E. Viloria, M. A. Mejia, and R. Zamora. 1986. Histological and ultrastrucural studies on human actinomycetomas, p. 647–662. In: G. Szabó, S. Biró, and M. Goodfellow (ed.), Biological, biochemical and biomedical aspects of actinomycetes. Akadémiai Kiadó, Budapest.

Shinobu, R. 1958. Physiological and cultural study for the identification of soil actinomycete species. Mem. Osaka Univ. Lib. Arts. Educ. Ser. B. Nat. Sci. 7:1–76.

Shinobu, R. 1962. A new *Streptomyces* species producing-fluorescent-yellow soluble pigment. Mem. Osaka Univ. Lib. Arts Educ. Ser. B. Nat. Sci. 11:115–122.

Shirling, E. B., and D. Gottlieb. 1966. Methods for the characterization of *Streptomyces* species. Int. J. Syst. Bacteriol. 16:313–340.

Skerman, V. B. D., V. McGowan, and P. H. A. Sneath. 1980. Approved lists of bacterial names. Int. J. Syst. Bacteriol. 30:225–420.

Soina, V. S., A. A. Sokolov, and N. S. Agre. 1975. Ultrastructure of mycelium and spores of *Actinomadura fastidiosa* sp. nov. Mikrobiologiya 44:883–887.

Stackebrandt, E. 1986. The significance of "Wall Types" in phylogenetically based taxonomic studies on actinomycetes, p. 497–506. In: G. Szabó, S. Biró, and M. Goodfellow (ed.), Biological, biochemical and biomedical aspects of actinomycetes. Akadémiai Kiadó, Budapest.

Stackebrandt, E., and M. Goodfellow. 1990. Nucleic acid techniques in bacterial systematics. John Wiley & Sons, Chichester, UK.

Stackebrandt, E., R. M. Kroppenstedt, and V. J. Fowler. 1983a. A phylogenetic analysis of the family Dermatophilaceae. J. Gen. Microbiol. 129:1831–1838.

Stackebrandt, E., W. Ludwig, E. Seewald, and K. H. Schleifer. 1983b. Phylogeny of sporeforming members of the order Actinomycetales. Int. J. Syst. Bacteriol. 33:173–180.

Stackebrandt, E., W. Ludwig, M. Weizenegger, M. Dom, S. Gill, T. J. McGill, G. E. Fox, C. R. Woese, H. Schubert, and K. H. Schleifer. 1987. Comperative 16S rRNA oligonucleotide analyses and murein types of round-sporeforming bacilli and nonsporeforming relatives. J. Gen. Microbiol. 133:2523–2529.

Stackebrandt, E. and K. H. Schleifer. 1984. Molecular systematics of actinomycetes and related organisms, p. 485–504. In: L. Ortiz-Ortiz, L. F. Bojalil, and V. Yakoleff (ed.), Biological, biochemical, and biomedical aspects of actinomycetes. Academic Press, Orlando, FL, USA.

Stackebrandt, E., and C. R. Woese. 1981. The evolution of prokaryotes, p. 1–31. In: M. J. Carlile, J. F. Collins, and B. E. B. Moseley (ed.), Molecular and cellular aspects of microbial evolution. University Press, Cambridge.

Stanek, M. 1972. Microorganisms inhabiting mushroom compost during fermentation. Mushroom Sc. 8:797–811.

Staneck, J. L., and G. D. Roberts. 1974. Simplified approach to identification of aerobic actinomycetes by thin-layer chromatography. Appl. Microbiol. 28:226–231.

Stutzenberger, F. J. 1971. Cellulase production by *Thermomonospora curvata* isolated from municipal solid waste compost. Appl. Microbiol. 22:147–152.

Stutzenberger, F. J. 1972a. Cellulolytic activity of *Thermomonospora curvata*. I. Nutritional requirements for cellulase production. Appl. Microbiol. 24:77–82.

Stutzenberger, F. J. 1972b. Cellulolytic activity of *Thermomonospora curvata*. 2. Optimal conditions, partial purification and product of the cellulase. Appl. Microbiol. 24:83–90.

Stutzenberger, F. J. 1979. Degradation of cellulosic substances by *Thermomonospora curvata*. Biotechnol. Bioeng. 21:909–913.

Stutzenberger, F. J. 1987. Inducible thermoalkalophilic polygalacturonate lyase from *Thermomonospora fusca*. J. Bacteriol. 169:2774–2780.

Stutzenberger, F. J. and R. Carnell. 1977. Amylase production by *Thermomonospora curvata*. Appl. Environ. Microbiol. 34:234–236.

Stutzenberger, F. J., and G. Kahler. 1986. Cellulase biosynthesis during degradation of cellulose derivatives by *Thermomonospora curvata*. J. Appl. Bact. 61:225–233.

Stutzenberger, F. J. and D. Lupo. 1986. pH-dependent thermal activation of endo-1,4-β-glucanase in *Thermomonospora curvata*. Enzyme Microb. Technol. 8:205–208.

Stutzenberger, F. J., and I. Sterpu. 1978. Effect of municipal refuse metals in cellulase production by *Thermomonospora curvata*. Appl. Environm. Microbiol. 36:201–204.

Stutzenberger, F. J., A. J. Kaufman, and R. D. Lossin. 1970. Cellulolytic activity in municipal solid waste compost. Can. J. Microbiol. 16:553–560.

Su, T. M., and D. Paulavicius. 1975. Enzymatic saccharification of cellulose by thermophilic actinomycetes. Appl. Polymer Symp. 28:221–236.

Suhadolnik, R. J., S. Pornbanlualap, D. C. Baker, K. N. Tiwari, and A. K. Hebbler. 1989. Stereospecific 2′-amination and 2′-chlorination of adenosine by *Actinomadura* in the biosynthesis of 2′-amino-2′-deoxyadenosine and 2′-chloro-2′-deoxycoformycin. Arch. Biochem. Biophys. 270:374–382.

Sveshnikova, M. A., T. S. Maksimova, and E. S. Kudrina. 1969. The species belonging to the genus *Micromonospora* Orskov, and their taxonomy. Mikrobiologiya 38:883–893.

Takahashi, I., K-I. Takahashi, K. Asano, I. Kawamoto, T. Yasuzawa, T. Ashizawa, F. Tomita, and H. Nakano. 1988. DC92-B, a new antitumor antibiotic from *Actinomadura*. J. Antibiot. 41:1151–1153.

Terekhova, L. P., O. A. Galatenko, and T. P. Preobrazhenskaya. 1982. *Actinomadura fulvescens* sp. nov. and *A. turkmeniaca* sp. nov. and their antagonistic properties. Antibiotiki 27:87–92.

Terekhova, L. P., O. A. Galatenko, and T. P. Preobrazhenskaya. 1987. In: Validation of publication of new names and new combinations previously effectively published outside the IJSB. List No. 23. Int. J. Syst. Bacteriol. 37:179–180.

Tomita, K., Y. Nakakita, Y. Hoshino, K. Numata, and H. Kawaguchi. 1987. New genus of the *Actinomycetales: Streptoalloteichus hindustanicus* gen. nov., nom. rev.; sp. nov., nom. rev. Int. J. Syst. Bacteriol. 37:211–213.

Uchida, K., and K. Aida. 1977. Acyl type of bacterial cell wall: its simple identification by colorimetric method. J. Gen. Appl. Microbiol. 23:249–260.

Vacca, J. G., and W. D. Bellamy. 1976. Classification and morphological properties of a high temperature, cellulolytic actinomycete. Abstracts of the Annual meeting of the American Society for Microbiology.

Vincent, H. 1894. Étude sur le parasite du pied le madura. Ann. Inst. Pasteur 8:129–151.

Waksman, S. A. 1950. The Actinomycetes: Their nature, occurrence, activities and importance. Ann. Cryp. Phytopath. 8:1–230.

Waksman, S. A. 1961. The Actinomycetes, vol 2. Classification, identification and descriptions of genera and species. Williams and Wilkins, Baltimore.

Waksman, S. A. and A. T. Henrici. 1948. Family Actinomycetaceae Buchanan and family Streptomycetaceae Waksman and Henrici, p. 961. In: R. S. Breed, E. G. D. Murray, and A. P. Hitchens (ed.), Bergey's manual of determinative bacteriology, 6th ed. Williams and Wilkins, Baltimore.

Waitz, J. A., A. C. Horan, M. Kalyanpur, B. K. Lee, D. Loebenberg, J. A. Marquez, G. Miller, and M. G. Patel. 1981. Kijanimicin (Sch 25663), a novel antibiotic produced by *Actinomadura kijaniata* SCC 1256. Fermentation, isolation, characterization and biological properties. J. Antibiot. 34:1101–1106.

Williams, S. T., and T. Cross. 1971. Actinomycetes. Methods Microbiol. 4:295–334.

Williams, S. T., E. M. Sharpe, and J. G. Holt. 1989. Bergey's manual of systematic bacteriology, vol. 4. Williams and Wilkins, Baltimore.

Wilson, D. B. 1986. Cellulase of *Thermomonospora fusca*. Meth. Enzymol. 160:314–323.

Wood, W. A. 1985. Useful biodegradation of cellulose. Ann. Phythochem. Soc. Eur. 26:295–309.

Wood, W. E., D. G. Neubauer, and F. J. Stutzenberger. 1984. Cyclic AMP levels during induction and repression of cellulase biosynthesis in *Thermomonospora curvata* J. Bacteriol. 160:1047–1054.

Yamada, Y., K. Aoki, and Y. Tahara. 1982a. The structure of the hexahydrogenated isoprenoid side-chain menaquinone with nine isoprene units isolated from *Actinomadura madurae*. J. Gen. Microbiol. 28:321–429.

Yamada, Y., C. F. Hou, J. Sasaki, Y. Tahara, and H. Yoshioka. 1982b. The structure of the octahydrogenated isoprenoid side-chain menaquinone with nine isoprene units isolated from *Streptomyces albus*. J. Gen. Appl. Microbiol. 28:519–529.

Zippel, M. and M. Neigenfind. 1988. Preservation of streptomycetes. J. Gen. Appl. Microbiol. 34:7–14.

The Family Streptosporangiaceae

MICHAEL GOODFELLOW

The family Streptosporangiaceae was proposed for organisms belonging to the genera *Microbispora, Microtetraspora, Planobispora, Planomonospora,* and *Streptosporangium* (Goodfellow et al., 1990). Members of the family are all aerobic, Gram-positive, non-acid-alcohol fast, chemoorganotrophic actinomycetes. They form a branched, stable (i.e., nonfragmenting) substrate mycelium that does not carry spores but bears aerial hyphae that can differentiate into either short chains of arthrospores or into spore vesicles (sporangia) containing one to many spores. These spores may be either motile or nonmotile. Constituent strains have *meso*-diaminopimelic acid (*meso*-A$_2$pm) in a wall peptidoglycan that lacks characteristic sugars (wall chemotype III of Lechevalier and Lechevalier, 1970a, 1970b), an A3γ peptidoglycan type (Schleifer and Kandler, 1972), *N*-acetylated muramic acid (Uchida and Aida, 1977), major amounts of glucosamine-containing polar lipids (phospholipid type 4 of Lechevalier et al., 1977), and partially hydrogenated menaquinones with nine isoprene units as the predominant isoprenolog (Kroppenstedt et al., 1990). Whole-organism hydrolysates yield madurose (3-O-methyl-D-galactose; Lechevalier and Gerber, 1970). The GC content of the DNA lies within the range of 66 to 74 mol%. The type genus is *Streptosporangium* Couch 1955.

Members of the family are chemically homogeneous and morphologically diverse (Table 1). However, strains that bear spore vesicles *(Planobispora, Planomonospora,* and *Streptosporangium)* are closely related to organisms forming paired or longer spore chains *(Microbispora, Microtetraspora,* and the *Actinomadura pusilla* group) but have little in common with sporangiate actinomycetes *(Actinoplanes, Dactylosporangium,* and *Pilimelia)* classified in the family Micromonosporaceae (Fig. 1; Goodfellow et al., 1988, 1990; Stackebrandt et al., 1981, 1983). The genus *Microtetraspora* has been redescribed, since it now encompasses the *Actinomadura pusilla* group (Kroppenstedt et al., 1990). The redefined genus *Actinomadura,* which is restricted to *Actinomadura madurae* and related species, belongs to the family Thermomonosporaceae (see Chapter 46).

Members of the genus *Spirillospora* may easily be confused with some species of the genus *Streptosporangium.* All strains belonging to these taxa have multi-spored, usually spherical, spore vesicles carried on the aerial mycelium. Nevertheless, it is evident from nucleic acid reassociation and chemical data (Goodfellow, 1989a; Stackebrandt et al., 1981) that the genus *Spirillospora* can only be assigned to the family Streptosporangiaceae as a matter of convenience. Thus, unlike other members of the family, the type strain of *Spirillospora albida* has a type I polar lipid pattern (Lechevalier et al., 1981) and a type 3 fatty acid profile (Kroppenstedt, 1985); *S. albida* ATCC 14541 has a type II phospholipid pattern (Hasegawa et al., 1979).

Couch (1955) proposed the genus *Streptosporangium* for sporangiate actinomycetes that formed nonmotile sporangiospores on an abundant aerial mycelium. Initially, only one species, *Streptosporangium roseum,* was recognized. The genus currently contains 13 validly described species (Nonomura, 1989a) all of which characteristically form aerial hyphae that carry, on either short or long sporangiophores, single or clustered spore vesicles that may be up to 40 μm in diameter (Figs. 2 and 3). Each spore vesicle contains a coiled chain of arthrospores formed by septation of an unbranched, spiral hypha within an expanded sporangiophore sheath (Vobis and Kothe, 1985). Since spore formation is not endogenous, it has been suggested that the term "spore vesicle" should replace "sporangium" (Cross, 1970; Sharples et al., 1974). Studies on spore maturation have shown that spores in both sporangia and chains are formed in essentially the same way. In both cases, spores are formed by fragmentation of a hypha within its sheath, which either expands to form the sporangial envelope or remains around the spore chain (Lechevalier et al., 1966;

Table 1. Characteristics differentiating the genera classified in the family Streptosporangiaceae.

	Microbispora	Microtetraspora	Planobispora	Planomonospora	Spirillospora	Streptosporangium
Morphological characters						
Substrate mycelium	Stable	Stable	Stable	Stable	Stable	Stable
Aerial mycelium and spores:						
Absent or in chains	+	+	−	−	−	−
Sporangiospores	−	−	+	+	+	+
Spores per chain/sporangium	Two	Two to many	Two	One	Many	Many
Spore motility	−	−	+	+	+	−
Temperature range	Mesophilic and thermophilic	Mesophilic and thermophilic	Mesophilic	Mesophilic	Mesophilic	Mesophilic
Chemical characters						
Wall chemotype[a]	III	III	III	III	III	III
Peptidoglycan type[b]	A1γ	A1γ	A1γ	A1γ	A1γ	A1γ
Characteristic sugar	Madurose	Madurose	Madurose	Madurose	Madurose	Madurose
Fatty acid profile[c]	3c	3c	3c	3c	3a	3c
Predominant menaquinone[d]	MK-9(H$_4$)	MK-9(H$_4$)	MK-9(H$_2$,H$_4$)	MK-9(H$_2$)	MK-9(H$_4$,H$_6$)	MK-9(H$_2$,H$_4$)
Phospholipid pattern[e]	IV	IV	IV	IV	I/II	IV
GC content (mol%)	67–74	64–69	70–71	72	71–73	69–71

Symbols: +, present; −, absent.

[a] Major constituents: alanine, glutamic acid, glucosamine, meso-diaminopimelic acid, and muramic acid (Lechevalier and Lechevalier, 1970b).

[b] A, cross-linkage between positions 3 and 4 of adjacent peptide subunits; 1, peptide bridge absent; γ, meso-diaminopimelic acid at position 3 of the tetrapeptide subunits (Schleifer and Kandler, 1972).

[c] Saturated, unsaturated, iso-, anteiso- (variable), and methyl-branched fatty acids (Kroppenstedt, 1985).

[d] Organisms contain major proportions of tetrahydrogenated menaquinones with nine isoprene units saturated at sites III and VIII (Kroppenstedt, 1982), apart from Planomonospora where MK-9(H$_2$) is the predominant isoprenolog (Kroppenstedt, 1985) and Spirillospora which contains major amounts of MK-9(H$_4$) and MK-9(H$_6$) (Collins et al., 1984).

[e] Phospholipid patterns: I, phosphatidylglycerol (variable); II, only phosphatidylethanolamine, and IV, phospholipids containing glucosamine (with phosphatidylethanolamine and phosphatidylmethylethanolamine variable) (Lechevalier et al., 1977).

Taken from Goodfellow (1989a, 1989b); Goodfellow et al. (1990); and Kroppenstedt et al. (1990).

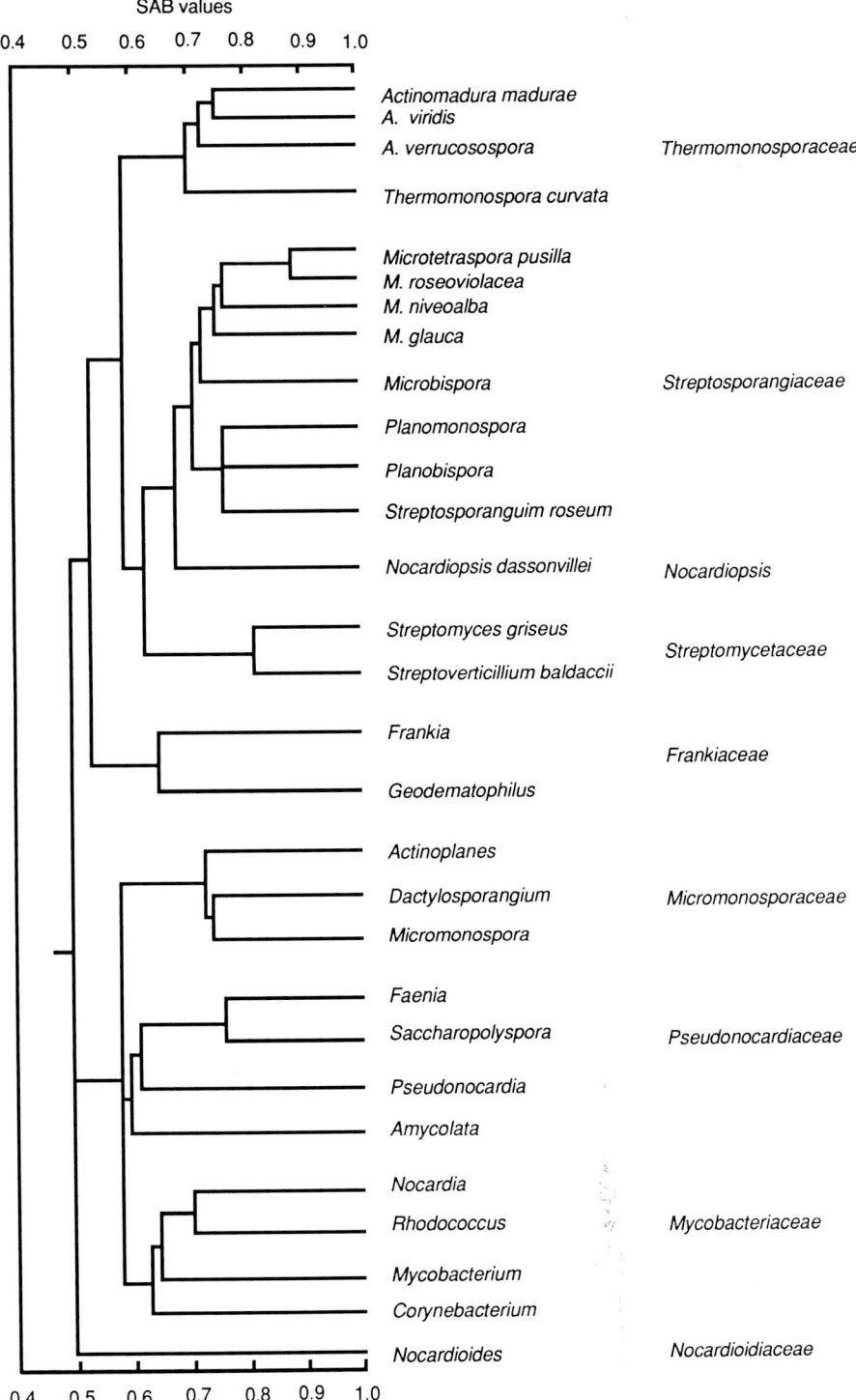

Fig. 1. Suprageneric relationships of actinomycetes based on 16S ribosomal RNA sequencing data.

Sharples et al., 1974; Vobis and Kothe, 1985). Spore vesicles burst when immersed in water, due to pressure exerted by the swelling intersporal matrix on the sporangial membrane, releasing nonmotile spores (Sharples et al., 1974).

The aplanospores are spherical, oval, or rod-shaped, with mainly smooth surfaces. Most species are mesophilic, but a few are thermotolerant. Some species require B vitamins for growth.

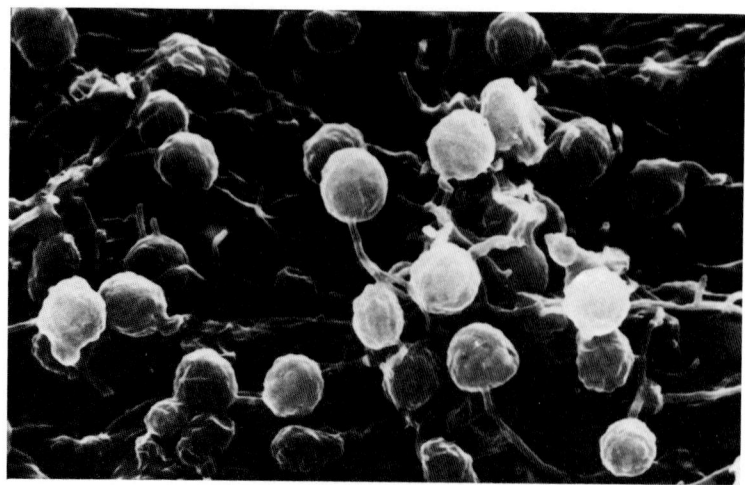

Fig. 2. *Streptosporangium album* CBS 426.61 on oatmeal agar. Scanning electron microscopy (SEM), gold splattered. Sporangiophores are short. (From Nonomura, 1989a, with permission.)

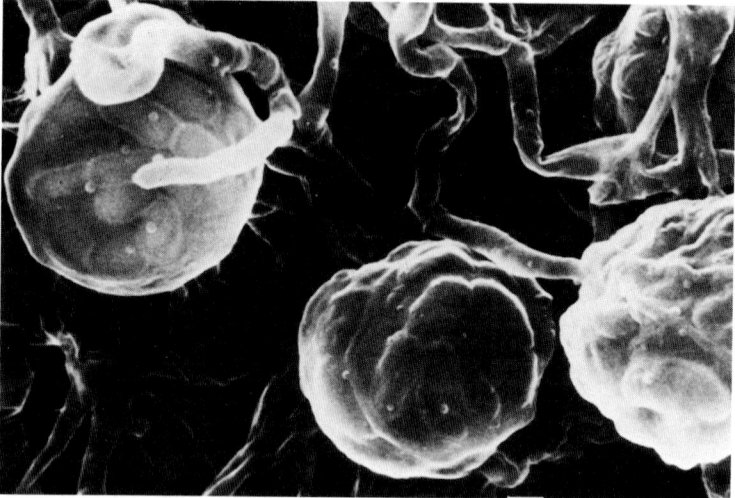

Fig. 3. *Streptosporangium album.* SEM, gold splattered. Sporangial walls (membranes) are thin. (From Nonomura, 1989a, with permission.)

The genus *Microbispora* was proposed by Nonomura and Ohara (1957) with *Microbispora rosea* as the type species. The genus contains 11 validly described species (Nonomura, 1989b; Rao et al., 1987), excluding *M. echinospora,* which has been reclassified as *Actinomadura echinospora* (Kroppenstedt et al., 1990), and *Microbispora viridis,* which has chemical properties consistent with an assignment to the genus *Microtetraspora* (Miyadoh et al., 1985). Microbisporae typically form a conspicuous aerial mycelium bearing longitudinal pairs of spores (Fig. 4) that may be closely arranged along the aerial hyphae, giving the hyphae the appearance of catkins. In some cases, the spores are borne at longer intervals (Fig. 5). They first appear as club-shaped initials that later become transformed into the paired spores visible under the light microscope. The spores are either sessile or on short sporophores, spherical to oval, and nonmotile. The spore surface

is smooth. Mature spores are easily detached from sporophores and from each other when placed in water. The B vitamins, particularly thiamine, are essential for the growth of *Microbispora* species on synthetic media. Both mesophilic and thermophilic species have been described.

Thiemann et al. (1968) proposed the genus *Microtetraspora* for those actinomycetes that formed short, sparsely branched, aerial hyphae carrying chains of four spores (Fig. 6). This morphological trait was considered typical of the genus but chains of two or three spores, and more rarely five spores, have also been reported. Until recently, only four species were recognized, including *Microtetraspora glauca,* the type species, and *M. viridis* (Nonomura and Ohara, 1971a), which has been reclassified as *Actinomadura viridis* (Miyadoh et al., 1989). Many of the 18 species that belong to the genus *Microtetraspora* were previously classified in

Fig. 4. *Microbispora rosea* ATCC 12950. SEM, gold splattered. Paired spores on hyphae. (From Nonomura, 1989b, with permission.)

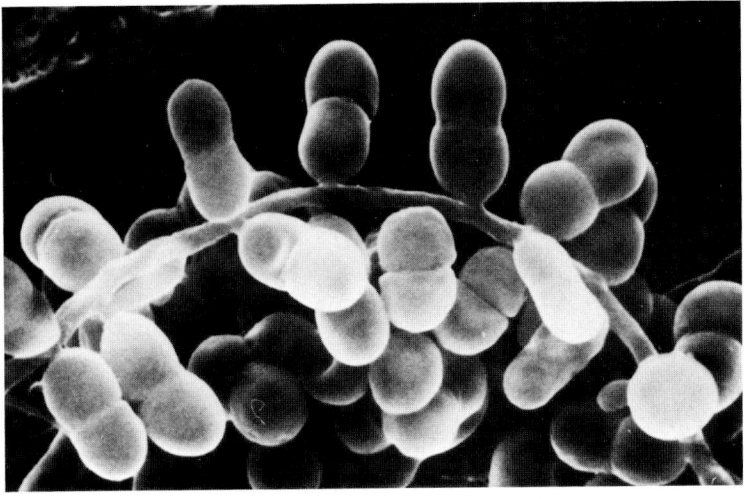

Fig. 5. Morphology of *Microbispora rosea* ATCC 12950 on oatmeal agar. SEM, gold splattered. Spores on entire mycelium. (From Nonomura, 1989b, with permission.)

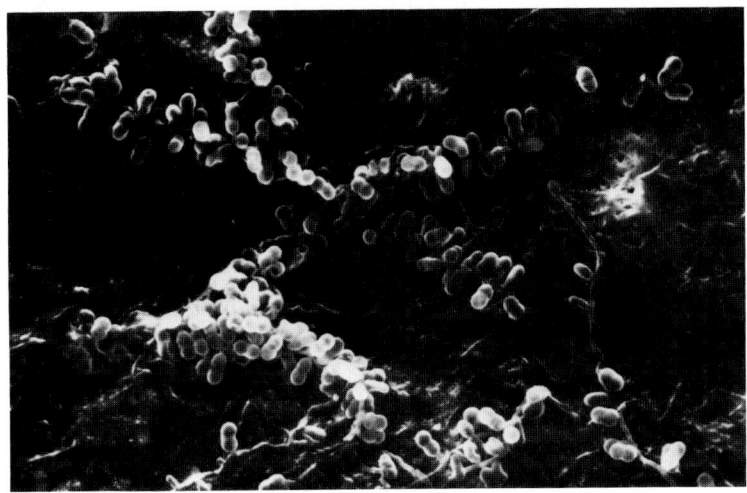

Fig. 6. Morphology of *Microtetraspora niveoalba* ATCC 27301 on inorganic salts-starch agar. SEM, gold splattered. (From Nonomura, 1989b, with permission.)

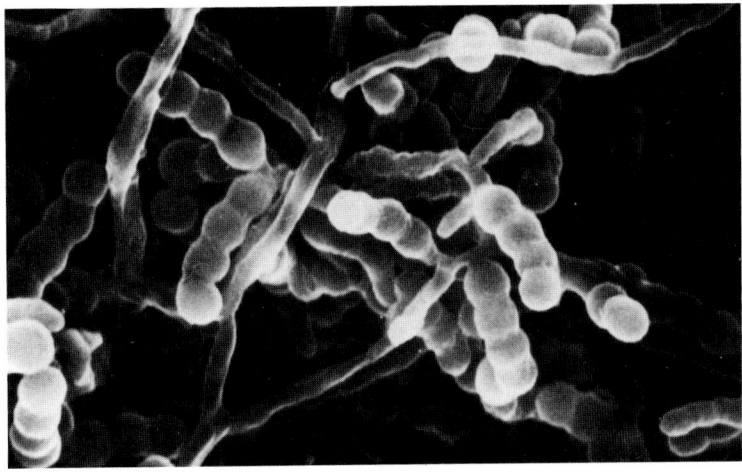

the *Actinomadura pusilla* group (Fischer et al., 1983; Kroppenstedt et al., 1990; Poschner et al., 1985). Most species are mesophilic, but some are thermophilic. Some species require B vitamins for growth.

The genus *Planobispora* (Thiemann and Beretta, 1968) encompasses the type species *Planobispora longispora,* and *P. rosea.* Planobisporae grow well between 28 and 40°C, but do not grow at 20 or 45°C (Goodfellow and Pirouz, 1982). Both substrate and aerial mycelia are formed. The substrate mycelium is either without distinctive color or is rose colored. The aerial mycelium, which develops only on certain agar media, is white or has a light rose tinge. Cylindrical to clavate sporangia, each containing a longitudinal pair of spores, are formed singly or in bundles on short ramifications of the aerial hyphae. The spores are straight or slightly curved with rounded ends and are motile by means of peritrichous flagella. They become motile only after being dispersed for some time and usually germinate with one or two polar germ tubes. The question remains as to whether the spores are formed endogenously (Williams and Wellington, 1980) or by single transformation of sporangeous hyphae (Bland and Couch, 1981). A transverse septum or diaphragm connected to the sporangial envelope divides the two spores (Fig. 7; Thiemann, 1970; Vobis and Kothe, 1985). The sporangial envelope is smooth and contains fibrillar elements (Vobis and Kothe, 1985) that resemble those present in *Planomonospora* (Sharples et al., 1974).

Members of the genus *Planomonospora* (Thiemann et al., 1967) develop substrate and aerial mycelia on solid media. Cylindrical or clavate sporangia, each containing a single spore, are formed only on the aerial hyphae. In *Planomonospora parontospora,* the type species, the sporangia are sessile and occur in double parallel rows on curved sporangiophores (Fig. 8). A single sporangiophore can bear up to sixty sporangia. In the second species, *P. venezuelensis,* the sporangia are developed singly or in groups on short lateral branches (Fig. 9). They form a characteristic palm leaf pattern (Thiemann, 1970). The spores may be formed endogenously (Sharples et al., 1974) but in *P. parontospora,* the sporangial development begins with the growth of a sporangeous hypha inside a thin expanding sheath (Vobis, 1985; Vobis and Kothe, 1985). Through thickening, the sheath becomes a massive sporangial envelope. The spores, which are released through apical pores, become motile by peritrichous flagella about 30 min after being expelled. They

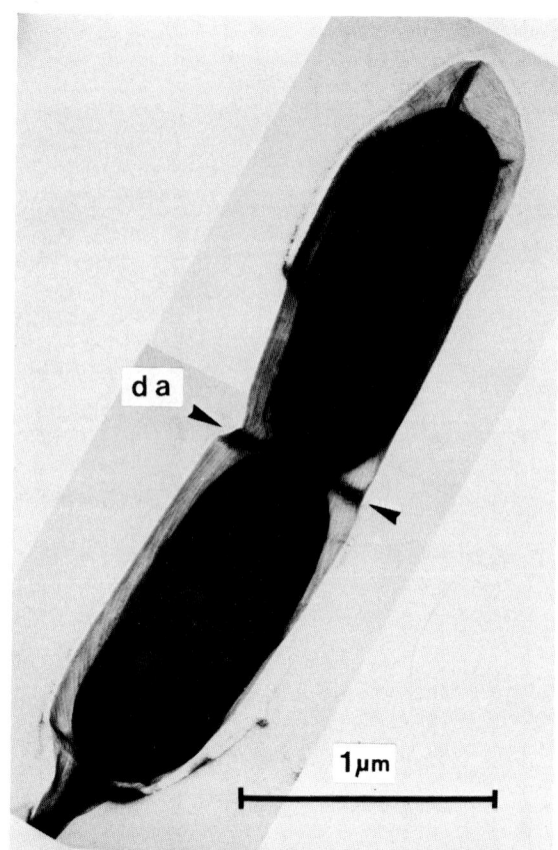

Fig. 7. Sporangium of *Planobispora rosea* strain MB-SE 893. Negative stained: transmission electron micrograph. A diaphragm (da) is visible between the two spores and the longitudinal fibrillar elements in the sporangial envelope. (From Vobis, 1989b, with permission.)

remain motile for up to a day, during which time spore germination may begin (Thiemann, 1970).

Spirillospora Couch 1963 contains two species, *Spirillospora albida,* the type species, and *S. rubra* (Vobis and Kothe, 1989). The substrate mycelium of these organisms is usually white, pale yellow, pale buffy pink, or red to reddish-brown; the aerial mycelium is white (Couch, 1963; Schäfer, 1973). Spherical to vermiform sporangia (5 to 24 μm in diameter) are formed on the aerial mycelium (Fig. 10). On initiation of sporangial development, the end of an aerial hypha coils first and then winds and branches within a common sheath (Lechevalier et al., 1966; Vobis, 1985). Small, two-layered cross-walls divide sporangeous hyphae into oblong segments that differentiate into spores. The latter, which are rod-shaped and curved, are motile by means of one to seven subpolarly inserted flagella. The sporangia are considered to be resistant to desiccation. When flooded with water,

Fig. 8. *Planomonospora parontospora* ATCC 23863. SEM. Numerous monosporous sporangia in double parallel rows, arranged directly on bent aerial hyphae. The strain was cultivated for 10 days on soil agar. (From Vobis, 1989b, with permission.)

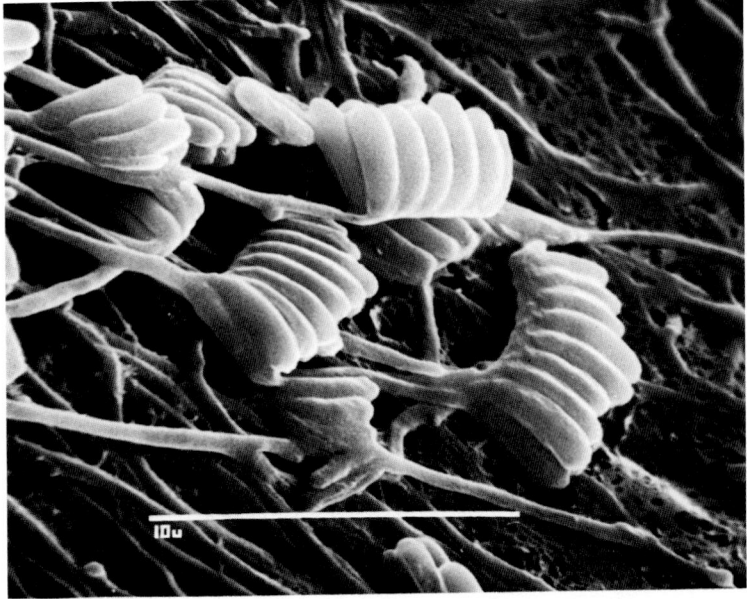

Fig. 9. *Planomonospora venezuelensis* ATCC 23865. SEM. Monosporous sporangia on aerial hyphae in young stages of formation of palm leaf pattern. Cultivation as in Fig. 8. (From Vobis, 1989a, with permission.)

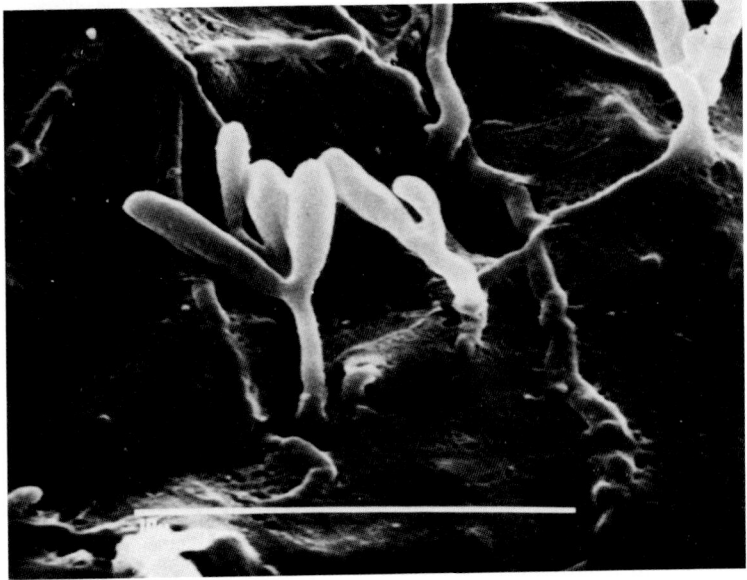

zoospores are released from the sporangium through a rupture in the envelope or through a large irregular pore (Couch, 1963).

Habitats

Members of the family Streptosporangiaceae are usually associated with soil but little is known about their role within this milieu. However, improvements in selective isolation procedures are beginning to cast light on the occurrence, distribution, numbers, and activity of actinomycete taxa in natural habitats (Goodfellow and O'Donnell, 1989). It seems likely that members of the Streptosporangiaceae will be involved in the primary decomposition of plant material in soils.

Streptosporangia were associated with leaf litter (Potekhina, 1865; Van Brummelen and Went, 1957) as well as soil and dung (Nonomura and Ohara, 1969a) until the introduction of a selective isolation procedure (Nonomura and Ohara, 1969a) showed that these organisms were actually an integral part of the actinomycete community in soils. The number of streptosporangia in various soils in Japan has been estimated at 10^4 to 10^6 colony-forming-units (CFU)/g dry weight of soil (Nonomura, 1984; Nonomura and Ohara, 1969a). Slightly

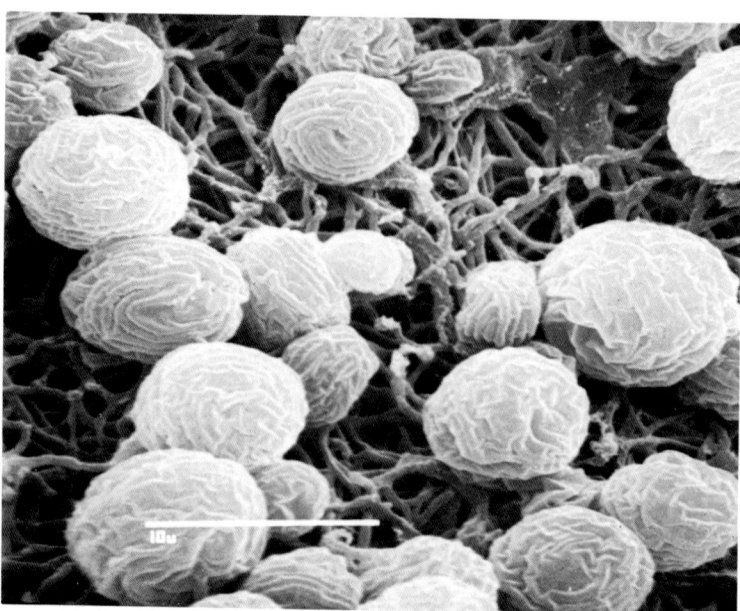

Fig. 10. *Spirillospora albida* ATCC 15331. SEM. Spherical sporangia formed on aerial mycelium after 15 days of incubation. (From Vobis and Kothe, 1989, with permission.)

acid, humus-rich garden soils are a favorite habitat. They have also been isolated from lake sediments (Johnstone and Cross, 1976; Willoughby, 1969a), beach sand (Williams and Sharples, 1976), and pasture and woodland soils (Whitham, 1988), but organisms labelled *Streptosporangium* type I from stream water (Willoughby, 1969b), given their morphological properties and capacity to form motile spores, probably belong to the genus *Actinoplanes*. "*Streptosporangium bovinum*" was isolated from infected bovine hooves (Chaves Batista et al., 1963).

Microbispora strains are common in soils. Using selective isolation procedures, counts of between 10^4 and 10^6 CFU/g dry weight of soil have been reported from various Japanese soils (Nonomura and Ohara, 1971b). Larger populations have been found in slightly acid (pH 5–6), humus-rich garden soils (Nonomura and Ohara, 1969b; Nonomura and Hayakawa 1988). Microbisporae have also been isolated from marine sediments (Weyland, 1969). One species, *Microbispora rosea,* has been implicated in a case of pericarditis and pleuritis in a human (Louria and Gordon, 1960). Microtetrasporae are also widely distributed in soil (Nonomura and Ohara, 1971a, 1971b; Thiemann et al., 1968) and seem to be common in mountain forest soils (Hayakawa et al., 1988). *Microtetraspora fusca* and *M. glauca* have been isolated from soils collected in Brazil, Italy, and Thailand, and *M. niveoalba* from stream mud and mulberry fields. Most of the species previously assigned to the *Actinomadura pusilla* group also originated in soil (Agre and Guzeva, 1975; Gal-atenko et al., 1981; Meyer, 1979; Nonomura and Ohara, 1971c).

Planomonospora strains have a worldwide distribution in the soils of arid, temperate, and tropical regions. Thiemann (1970) isolated 42 strains from 10 out of the 454 samples examined. These soil samples, which originated from South America, Italy, and India (Thiemann, 1974), had pH values ranging from 5.3 to 7.8. Additional strains have been isolated from soil samples collected in Africa, Europe, and Central and North America (Vobis, 1989a). In contrast, *Planobispora* appears to be a rare actinomycete, known from only two localities in the world. Several strains, including the type strains of the two validly described species, were isolated from soil taken from a river bank in Venezuela. The remaining organisms were isolated from a soil sample collected from near Windhoek, Namibia (Vobis, 1989b).

Members of the genus *Spirillospora* occur in soil, albeit infrequently. Using the baiting technique outlined below, Schäfer (1973) found that less than 1.0% of the isolates of sporangiate actinomycetes were represented by this genus.

Isolation and Cultivation

Selective Isolation

Dry heat treatment of air-dried soil samples and dilution plate culture with selective synthetic media are used for the preferential isolation and enumeration of members of the family Streptosporangiaceae. The procedures outlined below have been developed for the selective isolation of the genera *Microbispora* and *Streptosporangium* (Nonomura and Ohara, 1969a, 1969b) and with modifications for the isolation of other actinomycete genera, notably *Microtetraspora* (Nonomura and Ohara, 1971b, 1971c). There is also evidence (Nonomura and Hayakawa, 1988) that pretreatment of soil suspensions with yeast extract (6%, w/v) and sodium dodecyl sulfate (0.05%, w/v) at 40°C for 20 minutes, followed by dilution with water, activates actinomycete spores but kills vegetative cells of other soil bacteria in the suspensions. This practice leads to an increase in the counts of actinomycetes on isolation plates.

After soil samples are dried slowly at room temperature, passed through a 2-mm sieve, ground slightly in a mortar, spread on filter paper, and heated in a hot air oven at 120°C for 1 h, the number of bacteria and streptomycetes are dramatically reduced, and the isolation frequency of *Microbispora, Microtetraspora,* and *Streptosporangium* is enhanced. Heated soil is either incorporated directly onto isolation media or a suspension is used to make dilution plates. Initially, arginine-vitamins (AV) agar and minerals-glucose-asparagine plus soil extract (MGA-SE) agar were recommended for the selective isolation of microbisporae and streptosporangia, but two additional formulations, chitin-V and humic-vitamins (HV) agars (Hayakawa and Nonomura, 1987a, 1987b; Nonomura, 1989a) have also been developed. These media are supplemented with antifungal antibiotic(s); sometimes penicillin and polymixin B are also used. Inoculated plates are incubated for 4 to 6 weeks at 30°C or 2 to 3 weeks at 50°C, and examined using a light microscope with a long-working-distance objective. The highest counts and cleanest plates are usually obtained with HV agar.

Microbispora strains can be preferentially isolated by treating suspensions of dry heated soil samples with 1.5% phenol at 30°C for 30 min, diluting in water, and plating onto HV agar supplemented with nalidixic acid (20 mg/l). *Microbispora karnatakensis* (Rao et al., 1987) was isolated singly by plating a suspension of natural soil onto inorganic salts-starch agar (Küster, 1959).

Microtetraspora fusca and *M. glauca* were isolated from soil using a method that has not been disclosed (Thiemann et al., 1968). However, the pretreatment procedure described above has also been used to isolate several *Microtetraspora* species, including *M. helvata, M. pusilla, M. roseoviolacea,* and *M. spadix* (Nonomura and Ohara, 1971c). Dry *Microtetraspora*

spores appear to be particularly resistant to dry heat at 100 to 120°C, thereby allowing the slow-growing microtetrasporae to develop into recognizable colonies on dilution plates. Soil dilutions are plated onto various media, including AV and MGA-SE agars, and incubated for several weeks at 28 to 30°C (Nonomura and Ohara, 1971b). *Microtetraspora niveoalba* was isolated from dry-heated soil on MGA-SE agar incubated at 40°C for 1 month. Similarly, *M. glauca* strains have been isolated on plates of this medium incubated at 30°C.

Other *Microtetraspora* species, such as *M. roseola* and *M. salmonea,* have been isolated from soil by Soviet investigators who supplemented media with antibiotics to improve their selectivity. Lavrova et al. (1972) added rubomycin (5, 10, or 20 μg/ml) to medium no. 2 of Gauze et al. (1957). Preobrazhenskaya et al. (1975) added bruneomycin (0.5, 1, or 2 μg/ml) or streptomycin (0.5, 1, or 2 μg/ml). These antibiotics led to the growth of more *Microtetraspora* colonies on isolation plates while reducing the number of streptomycetes. In contrast, *M. ferruginea* and *M. spiralis* were isolated by plating soil suspensions onto oatmeal agar or Gauze's No. 1 medium without addition of selective antibiotics (Meyer, 1979).

Microbisporae, microtetrasporae, and streptosporangias grow well on rich media, including Bennett's (Jones, 1949), glucose-yeast extract (Waksman, 1950), oatmeal (ISP3; Difco 0771) and yeast extract-malt extract agars (Shirling and Gottlieb, 1966; ISP medium 2 [Difco 0770]). Oatmeal-yeast extract agar is recommended for the growth of mesophilic microbisporae and glycerol agar for the corresponding thermophilic strains (Nonomura, 1989b). Streptosporangiae also grow well and produce aerial spore mass on oatmeal-yeast extract agar (Nonomura and Ohara, 1960). Good growth of vegetative and sporing aerial mycelia was obtained for *Microtetraspora fusca* and *M. glauca* on Hickey and Tresner (1952) agar. *M. niveoalba* requires B vitamins for growth on synthetic media (Nonomura and Ohara, 1971b).

Baiting with Natural Substrates

The procedures used to isolate *Planobispora* and *Planomonospora* from soil were not described by Thiemann and his colleagues (Thiemann and Beretta, 1968; Thiemann et al., 1967). However, members of these taxa and *Spirillospora* strains have been recovered from soil by baiting with natural substrates (Bland and Couch, 1981; Couch, 1954), as follows: a small amount of soil, approximately one level teaspoonful, is placed in a sterile petri dish and flooded with sterile water (distilled water or filtered soil or charcoal water extracts may be used). Added pollen and hair float at the water surface; various types of pollen have been employed including that from members of the genera *Liquidamber, Pinus,* and *Sparganium* (Schäfer, 1973). After 1 to 4 weeks, examination of the water surface with a dissecting microscope ($\times$ 100) and strong horizontal lighting should reveal the white glistening sporangia formed in the air at the surface of the water by sporangiate members of the families Micromonosporaceae and Streptosporangiaceae present. The characteristic two-spored sporangia of *Planobispora* develop on long aerial hyphae growing between the baits. Similarly, the sporulating aerial hyphae of *Planomonospora* and *Spirillospora* grow on pollen grains. Single sporangia or bundles of sporangia can be picked up with a thin needle and placed on the surface of agar media in small petri dishes. After 2 to 4 weeks, the young colonies can be transferred to slant cultures.

Planobisporae and planomonosporae grow on standard media used for cultivating streptomycetes (Waksman, 1961); the first signs of visible growth appear after 3 to 4 days at 28 to 30°C. *Planobispora longispora* produces aerial hyphae and abundant sporangia on calcium malate, soil extract, and yeast extract-malt extract agars (Shirling and Gottlieb, 1966). Sporangial development in *P. rosea* is promoted by all media on which aerial mycelium is formed, such as soil extract and Hickey-Tresner agars. Similarly, sporangia development in *Planomonospora* strains is especially abundant on Bennett's, Hickey-Tresner, oatmeal, and soil extract agars (Thiemann et al., 1967; Vobis, 1989a). Various complex media support the growth of *Spirillospora* strains. These organisms grow well on Czapek, peptone-Czapek, and oatmeal agars.

Media for Isolation and Cultivation

The following are the recipes for the media mentioned in the preceding section. Quantities are for 1 liter of distilled water unless otherwise stated.

ISOLATION MEDIA.

Arginine-Vitamins (AV) Agar (Nonomura and Ohara, 1969a)

L-Arginine	0.3 g
Glucose	1.0 g
Glycerol	1.0 g
K$_2$HPO$_4$	0.3 g
MgSO$_4$·7H$_2$O	0.2 g
NaCl	0.3 g
Agar	15 g

Trace salts solution:

CuSO$_4$·5H$_2$O	1.0 mg/ml
Fe$_2$(SO$_4$)$_3$	10.0 mg/ml
MnSO$_4$·7H$_2$O	1.0 mg/ml
ZnSO$_4$·7H$_2$O	1.0 mg/ml

Adjust to pH 8.0

Vitamins (final weight in medium): 0.5 mg each of *p*-aminobenzoic acid, calcium pantothenate, inositol, niacin, pyridoxine HCl, riboflavin, thiamine HCl; 0.25 mg biotin.
Antibiotics: Cycloheximide, 50 mg; nystatin, 50 mg; nalidixic acid, none or 20 mg; penicillin G, none or 0.8 mg.; polymixin, none or 4 mg. Adjust pH to 6.4.

Chitin-V Agar (Hayakawa and Nonomura, 1984)

Colloidal chitin	2 g (dry weight)
CaCO$_3$	0.02 g
FeSO$_4$·7H$_2$O	10 mg
K$_2$HPO$_4$	0.35 g
KH$_2$PO$_4$	0.15 g
MgSO$_4$·7H$_2$O	0.2 g
MnCl$_2$	1 mg
NaCl	0.3 g
ZnSO$_4$·7H$_2$O	1 mg
Agar	18 g

B vitamins, as for AV agar; cycloheximide 50 mg; pH 7.2

Humic-Vitamins (HV) Agar (Hayakawa and Nonomura, 1984; Nonomura, 1984)

Humic acid (see below)	1 g
CaCO$_3$	0.02 g
FeSO$_4$·7H$_2$O	0.01 g
KCl	1.7 g
MgSO$_4$·7H$_2$O	0.05 g
Na$_2$HPO$_4$	0.5 g
Agar	18 g

Humic acid (lg) is used in the form of an alkaline solution. Artificial humic acid prepared from glycine and urea may be employed, as may natural humic acid from soil humus, but the pale brown humic acid designated as R$_p$ type gives the best result.

B vitamins, as for AV agar; cycloheximide, 50 mg; adjust to pH 7.2.

Inorganic Salts-Starch Agar (Küster, 1959)

Solution I: Difco soluble starch, 10 g. Make a paste of the starch with a small amount of cold distilled water and bring to a volume of 500 ml.

Solution II:

CaCO$_3$	2 g
K$_2$HPO$_4$ (anhydrous salt)	1 g
MgSO$_4$·7H$_2$O	1 g
NaCl	1 g
(NH$_4$)$_2$SO$_4$	2 g
Distilled water	500 ml
Trace salts solution*	1 ml

Adjust to pH 7–7.4. Mix solutions I and II and add 20 g agar. *Trace salts: FeSO$_4$·7H$_2$O, 0.1 g; MnCl$_2$·4H$_2$O, 0.1 g; ZnSO$_4$·7H$_2$O, 0.1 g; distilled water, 100 ml.

Medium No. 1 (Gauze et al., 1957)

Starch	20 g
FeSO$_4$·7H$_2$O	0.01 g
KNO$_3$	1 g
K$_2$HPO$_4$	0.5 g
MgSO$_4$	0.5 g
NaCl	0.5 g
Agar	30 g

Adjust to pH 7.2–7.4.

MGA-SE Agar (Nonomura and Ohara, 1971b)

L-Asparagine	1 g
Glucose	2 g
K$_2$HPO$_4$	0.5 g
Soil extract	200 ml
Distilled water	800 ml
Agar	20 g

Antibiotics: cycloheximide, 50 mg; nystatin, 50 mg; benzylpenicillin, 0.8 mg; polymixin B, 4 mg. Soil extract: 1,000 g soil in 1 liter water, autoclaved for 30 minutes, then decanted and filtered. Final pH 8.0.

CULTIVATION MEDIA.

Bennett's Agar (Jones, 1949)

Beef extract	1 g
Glucose	10 g
N-Z amine A (enzymatic digest of casein)	2 g
Yeast extract	1 g
Agar	15 g

Adjust to pH 7.3 with NaOH.

Modified Bennett's agar can be used where glucose is replaced by glycerol (10 g) and N-Z amine A by Bacto-casitone (2 g; Difco).

Calcium Malate Agar (Waksman, 1961)

Calcium malate	10 g
K$_2$HPO$_4$	0.5 g
NH$_4$Cl	0.5 g
Agar	15 g

Czapek Agar (Waksman, 1950)

Sucrose	30 g
FeSO$_4$·7H$_2$O	0.01 g
KCl	0.5 g
K$_2$HPO$_4$	1 g
MgSO$_4$·7H$_2$O	0.5 g
NaNO$_3$	3 g
Agar	15 g

After components are dissolved, the medium is dispensed into flasks or tubes then autoclaved. For peptone-Czapek agar, substitute 5 g peptone for 3 g NaNO$_3$ (Bland and Couch, 1981).

Glucose-Yeast Extract Agar (Waksman, 1950)

Glucose	10 g
Yeast extract (50% solution)	20 ml

Agar	15 g

Adjust to pH 6.8

Glycerol Agar (Nonomura and Ohara, 1969b)

Casamino acids	2 g
Glycerol	5 g
K$_2$HPO$_4$	0.3 g
MgSO$_4$·7H$_2$O	0.5 g
NaCl	0.3 g
Agar	20 g

Trace salts and B vitamins as for AV agar; final pH 7.2.

Hickey-Tresner Agar (Hickey and Tresner, 1952)

Amidax (Corn Products Refining, Argo, IL) or dextrin	10 g
Beef extract (Difco)	1 g
N-Z amine A or Tryptone (Difco)	2 g
Yeast extract (Difco)	1 g
CoCl$_2$·6H$_2$O	2 mg
Agar	20 g

Adjust to pH 7.3

Medium No. 2 (Gauze et al., 1957)

Glucose	10 g
Peptone	5 g
NaCl	10 g
Hottinger's broth	30 ml
Agar	15 g

Oatmeal Agar (Bland and Couch, 1981)

Baby oatmeal	65 g
Agar	15 g

Autoclave for 30 minutes, leave for 24 hours, then autoclave again before dispensing.

Soil Extract Agar (Thiemann et al., 1968)

Air-dried garden soil	30 g
Agar	20 g
Tap water	1 liter

pH 7.0

Autoclave for 20 minutes at 120°C.

Yeast Extract-Malt Extract Agar (Pridham et al., 1956–57)

Bacto-dextrose (Difco)	4 g
Bacto-malt extract (Difco)	10 g
Bacto-yeast extract (Difco)	4 g

Adjust to pH 7.3, then add:

Bacto-agar	20 g

Liquify agar by steaming at 100°C for 15 to 20 minutes.

Preservation of Cultures

The most convenient method for short-term storage is by serial transfer from agar slants of appropriate media (see above) every two months (Meyer, 1989). The tubes should be tightly closed with cotton plugs dipped in melted paraffin wax. Sporulated spore cultures can also be stored at 5°C and at room temperature. Lyophilization, storage in liquid nitrogen, and freezing in glycerol can be used for long-term preservation (Meyer, 1989; Wellington and Williams, 1978).

For lyophilization, the spore suspension or vegetative mycelium is suspended in a suitable fluid, such as serum plus 7.5% (w/v) glucose, or skimmed milk plus 7.5% (w/v) glucose. For storage in liquid nitrogen, the microorganisms are inoculated into small test tubes containing the appropriate medium and incubated until satisfactory growth is visible. The tubes are then closed with cotton plugs dipped in melted paraffin wax and placed in a liquid nitrogen container. Glycerol suspensions are prepared by scraping aerial growth and/or substrate mycelium from heavily inoculated plates and making heavy suspensions in 3 ml of aqueous glycerol in small (e.g. bijoux) bottles, which are stored at −20°C. The frozen glycerol suspensions serve both as a convenient means of long-term preservation and as convenient source of inoculum. Working inocula are obtained by thawing suspensions at room temperature prior to treating as for broth cultures. After use, glycerol suspensions are promptly frozen and stored again at −20°C.

Identification

Members of the family Streptosporangiaceae can be distinguished from all other actinomycetes using a combination of chemical and morphological features. Primary diagnostic information can be gained by examination of whole-organism hydrolysates (Lechevalier and Lechevalier, 1980). Unidimensional thin layer chromatography will determine whether an organism contains diaminopimelic acid (A$_2$pm) and whether the latter is in the L- or meso- form. The presence of L-A$_2$pm will be found in Streptomyces and related genera. The detection of meso-A$_2$pm and madurose with the absence of characteristic sugars serves to separate strains of Streptosporangiaceae from those of Actinoplanes and related genera, Nocardia and related genera, Pseudonocardia and related genera, Nocardiopsis, and Thermomonospora, but not from the genera Dermatophilus and Frankia. The latter can readily be distinguished from Streptosporangium and allied taxa on morphological grounds. To date, the presence of madurose is associated with wall chemotype III actinomycetes, although there is an unconfirmed report of this sugar from a wall chemotype I actinomycete with a streptomycete morphology

(Weyland et al., 1982) The discovery of 3-O-methylgalactosyl (madurosyl) units in the structure of teichoic acids of an Actinomadura carminata strain (Naumova et al., 1986) is also noteworthy as madurose is not perceived to be a cell wall constituent (Lechevalier and Lechevalier, 1981).

Wall chemotype III sporangiate actinomycetes can be identified to the genus level using morphological features (Table 1), though care must be taken to distinguish between Spirillospora and Streptosporangium even though only the former produces motile spores. It should also be noted, however, that the sporogenic hyphae in the sporangia of spirillosporae are branched, whereas those in streptosporangiae are unbranched. Additional comparative studies will be necessary to determine whether the genera Planobispora and Planomonospora actually merit generic status. Chemical analyses are required if the genera Microbispora and Microtetraspora are to be reliably separated from the genus Actinomadura (Kroppenstedt et al., 1990). Thus, actinomadurae contain mostly hexahydrogenated menaquinones with nine isoprene units (MK-9[H$_6$]) and diphosphatidylglycerol and phosphatidylinositol as predominant polar lipids (phospholipid type I sensu Lechevalier et al., 1977), whereas Microbispora and Microtetraspora strains have major amounts of MK-9(H$_4$) saturated at positions III and VIII, with major amounts of diphosphatidylglycerol, hydroxylated phosphatidylethanolamine, uncharacterized glycolipids, and a glucosamine-containing phospholipid (type 4 phospholipid pattern).

The continued separation of the genera Microbispora and Microtetraspora solely on morphological criteria is questionable. Microbispora is distinguished from related genera (Table 1) primarily by the formation of paired spores on aerial hyphae. These are either sessile or borne on short sporophores. The latter in M. rosea were found to attach to the base of the spore by a ball and socket arrangement (Williams, 1970). Microbisporae do not contain mycolic acids, but tuberculostearic acid and its homologs have been found in all species examined to date (Lechevalier, 1977). It seems probable that the genus is grossly overclassified, as species are currently separated by characters such as temperature requirements and a few biochemical and physiological features (Table 2).

Cultures of Microbispora aerata, M. amethystogenes, and M. parva deposit crystals with a metallic sheen in the medium, particularly when grown on Pablum extract agar (Lechevalier and Lechevalier, 1957) for about 10 days

Table 2. Characteristics differentiating the species of the genus *Microbispora*.

	M. aerata	*M. amethystogenes*	*M. bispora*	*M. chromogenes*	*M. diastatica*	*M. indica*	*M. karnatakensis*	*M. parva*	*M. rosea*	*M. thermodiastatica*	*M. thermorosea*
Color of aerial mycelium	Pink	Pink	White	Pink	Pink	Pinkish-white	White	Pink	Pale pink	Pink	Pink
Color of substrate mycelium	Yellowish-brown	Light brown	Yellowish-brown	Orange	Yellowish-brown	Violet-orange	Yellowish-pink	Light brown	Orange	Yellowish-brown	Yellowish-brown
Growth at:											
25°C	−	+	−	+	+	+	+	+	+	−	−
50°C	+	−	+	−	−	+	+	+	−	+	+
55°C	+	−	+	−	−	−	−	−	−	+	+
Iodinin production	+	+	−	−	−	−	−	(+)	−	−	−
Nitrate reduction	+	+	−	+	−	+	+	−	+	−	−
Starch hydrolysis	+	−,(+)	−	+	+	−	+	−	−	+	−
Soluble pigments[a]	−	−	−	Light yellow-pink	−	Deep orange-yellow	Deep orange-yellow	−	−	−	−
Utilization of:											
Arabinose	+	+	−	+	+	+	−	+	+	+	+
Glycerol	+	+	+	+	+	−	−	+	+	(+)	+
Inositol	−	(+)	+	+	−	−	+	−	−	−	−
Rhamnose	−	−	(+)	−	+	+	+	(+)	+	−	−

Symbols: +, positive reaction; (+), weak reaction; −, negative reaction.

[a]Other than yellow-brown.

Taken from Nonomura (1989b) and Rao et al. (1987).

(Gerber and Lechevalier, 1964). These crystals are composed of iodinin (1,6-phenazinediol-5,10-dioxide), a red, water-soluble pigment. In addition, *M. aerata* produces two brown-yellow pigments (2-aminophenoaxazine-3-one and 1,6-phenazinediol), a yellow pigment (2-acetamidophenoxazine-3-one) (Gerber and Lechevalier, 1964), and an orange pigment (1,6-phenazinediol-5-oxide) (Gerber and Lechevalier, 1965).

Microtetraspora is distinguished from related genera by its ability to form chains of spores on aerial hyphae (Figs. 11, 12, and 13). The constituent species may be distinguished by means of spore chain morphology, spore wall ornamentation, color of mature sporulated aerial mycelium, and substrate mycelium pigmentation (Table 3). Nevertheless, identification of many of these species is difficult because in most instances only one (the type) strain or a few strains have been examined. Even when several strains have been studied, the results of biochemical and physiological tests have proved to be variable or inconsistent when data from the literature are compared. Even so, numerical taxonomic evidence does show that most of the validly described taxa merit species status (Goodfellow et al., 1979; Goodfellow and Pirouz, 1982).

Streptosporangium species may be separated by spore vesicle size, sporangiophore length, spore shape, aerial spore mass, and substrate mycelium pigmentation (Table 4). The species can also be subdivided according to the nature of the vesicular wall. At one extreme, the spore vesicle membrane of *S. fragile* is so thin that it cannot be detected by light microscopy (Shearer et al., 1983). This may lead to difficulty in differentiating *Streptosporangium* from *Actino-* *madura* and *Microtetraspora,* as some species of the latter produce "pseudosporangia" covered by a slimy substance (Nonomura and Ohara, 1971c). In contrast, the spore vesicles of *S. albidum* and *S. viridogriseum* have thick and strong walls that enclose a "sheathed" chain of spores (Nonomura and Ohara, 1969b). The spore vesicles are not easily disrupted in wet mounts even by pressing the cover glass onto the slide. The genus *Kibdelosporangium* (Shearer et al., 1986) bears a close morphological resemblance to these organisms but has wall chemotype IV. The wall chemotype of *S. albidus* and *S. viridogriseus* needs to be determined to clarify their relationship with *Kibdelosporangium. Streptosporangium corrugatum* produces characteristically small, club-shaped spore vesicles, and the remaining species of *Streptosporangium* form thin sporangial membranes that are readily disrupted in water. "*Streptosporangium album* subsp. *thermophilus*" (Manachini et al., 1965) is a thermophilic organism, but has been shown to belong to the genus *Thermoactinomyces* (Goodfellow and Cross, 1984).

Differences in gelatin liquefaction, nitrate reduction, and starch hydrolysis have also been used for the identification of *Streptosporangium* species (Table 4). *Streptosporangium amethystogenes* produces violet crystals of iodinin after 1 month of incubation at 30°C on oatmeal-yeast extract agar. However, little credence can be placed in the predictiveness of such properties, given the small sample of strains examined. Nevertheless, the status of most species of *Streptosporangium* has been underpinned in a numerical taxonomic survey (Whitham, 1988); *S. viridogriseum,* however, clustered apart from all of the other streptosporangia studied.

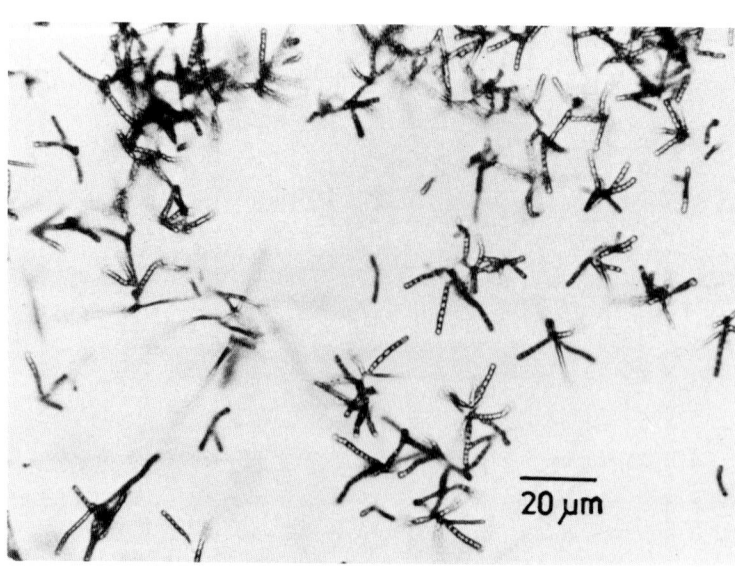

20 μm

Fig. 11. *Microtetraspora roseola,* type, strain INA 1671. Sporulating aerial mycelium. Cultivation on oatmeal-nitrate agar, 28°C, 12 days. (From Meyer, 1989, with permission.)

Fig. 12. *Microtetraspora spirilis,* type strain IMET 9621. Sporulating aerial mycelium. Oatmeal-nitrate agar, 28°C, 18 days. (From Meyer 1989, with permission.)

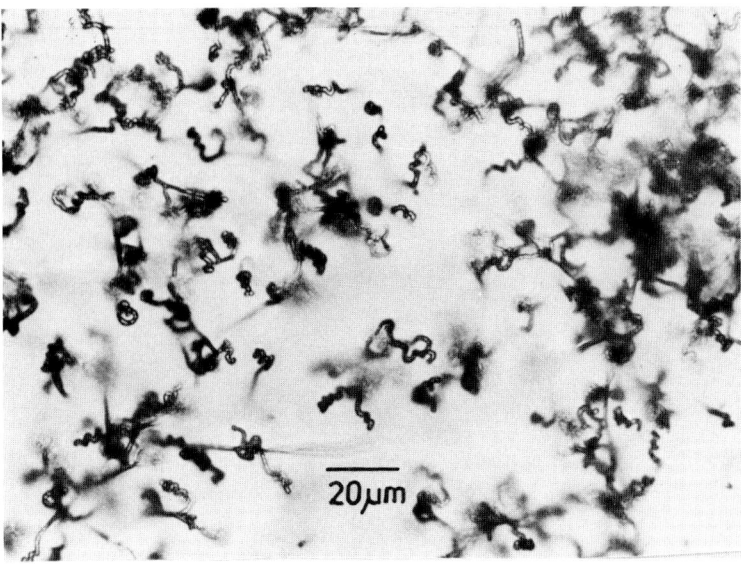

Fig. 13. *Microtetraspora pusilla,* type strain ATCC 27296. Sporulating aerial mycelium with "pseudosporangia." Cultivation on yeast extract-malt agar, 28°C, 9 days. (From Meyer 1989, with permission.)

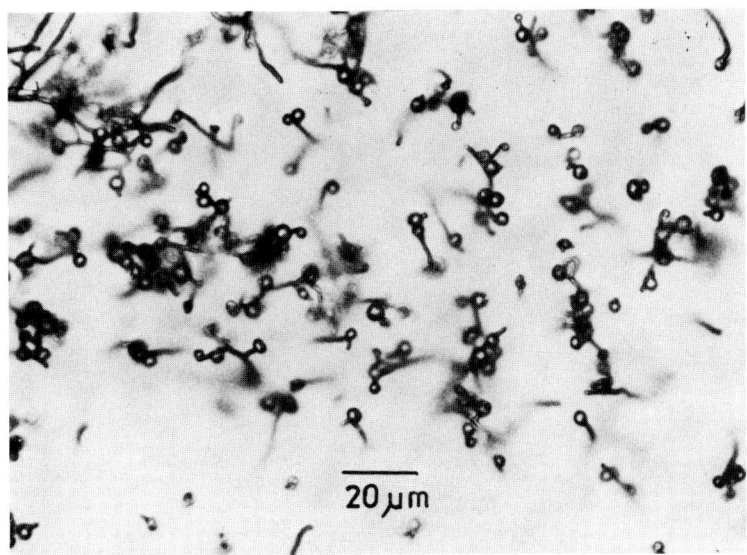

The genera *Planobispora, Planomonospora,* and *Spirillospora* may be distinguished by means of sporangial shape and the number of spores formed, and the constituent species can be separated using a few biochemical and morphological features (Table 5). *Planobispora longispora* and *P. rosea* have many properties in common (Goodfellow and Pirouz, 1982) but can be separated using cultural characteristics. Thus, *P. longispora* produces a hyaline- to creamish-colored substrate mycelium and a white aerial mycelium, whereas *P. rosea* has a rose-colored substrate mycelium and an aerial mycelium with a light rose tinge.

Planomonospora parontospora and *P. venezuelensis* can be distinguished by the morpho-

logical arrangement of sporangia (Figs. 8 and 9), different menaquinone profiles (Collins et al., 1984), and the characteristic color of the mycelium (Thiemann, 1974b). It is difficult to highlight biochemical and physiological tests for the separation of these species, given the inconsistent results obtained in different studies (Table 5). The two *Spirillospora* species can be differentiated by substrate mycelial color. The colonies of *S. albida* are white to pale yellow, whereas those of *S. rubra* are red to reddish brown. *Spirillospora albida* forms sporangia on Czapek, peptone-Czapek, and oatmeal agars, and *S. rubra* sporulates on cornmeal-soil agar (Vobis and Kothe, 1989) and artificial soil agar (Henssen and Schäfer, 1971).

Table 3. Characteristics differentiating the species of the genus *Microtetraspora*.

	M. afri-cana	M. an-giospora	M. fasti-diosa	M. ferru-ginea	M. flexu-osa	M. fusca	M. glauca	M. hel-vata	M. ni-veoalba
Spore chain morphology									
Hooks, curled	−	−	−	+	+	−	−	+	−
Pseudosporangia	−	−	−	−	−	−	−	+	−
Spirals, 1–2 turns	−	−	+	+	+	−	−	−	−
3–5 turns	−	+	+	−	−	−	−	−	−
Straight, two spores	−	−	−	−	−	−	−	−	−
many spores	+	−	−	−	−	−	−	−	−
Spore surface morphology									
Folded	−	−	−	+	−	−	−	−	−
Irregular, uneven	−	−	+	−	−	−	−	−	−
Ridged	−	+	−	−	−	−	−	−	−
Smooth	+	−	+	+	−	+	+	+	+
Warty	−	−	−	−	+	−	−	−	−
Wrinkled	−	−	−	−	−	−	−	−	−
Yeast extract-malt extract (ISP medium 2)									
Aerial mycelium	ND	ND	Trace	Orange-pink	−	ND	ND	White-yellow	ND
Substrate mycelium	ND	ND	Brown	Orange	−	ND	ND	Yellow-brown	ND
Oatmeal agar (ISP medium 3)									
Aerial mycelium	Grayish-blue	White	White-pink	White-pink	White-yellow	Trace	Blue-gray	Trace, white	ND
Substrate mycelium	Yellow	White-ochre	Colorless	Pink	Brown	Colorless	Blue-green	Yellow-brown	ND
Soluble pigment	Yellowish-brown	−	−	−	−	−	−	−	ND
Inorganic salts-starch agar (ISP medium 4)									
Aerial mycelium	Grayish-blue	ND	Colorless, pink	−	−	ND	ND	White-cream	ND
Substrate mycelium	Red, brownish-violet	ND	Colorless	Colorless, brown	−	ND	ND	Yellow-brown	ND
Degradation of:									
Aesculin	ND	+	+	−	+	+	+	+	+
Casein	ND	+	+	+	+	ND	+	−	+
DNA	ND	+	+	+	+	ND	+	−	+
Elastin	ND	+	ND	+	ND	ND	−	ND	ND
Gelatin	ND	+	+	−	+	−	+	−	+
Hypoxanthine	ND	+	+	+	−	−	+	−	+
Starch	ND	−	−	+	+	−	+	−	+
Testosterone	ND	+	−	+	+	−	+	+	+
Tyrosine	ND	+	−	+	+	+	+	−	+
Xanthine	ND	−	−	−	−	−	+	−	+
Reduction of nitrate	ND	−	+	+	+	−	+	+	+

Symbols: +, positive reaction; (+), weak reaction; −, negative reaction; ND, not determined.
[a]Currently classified as *Microbispora viridis*.

Metabolism and Genetics

Little is known about the metabolism and genetics of members of the family Streptosporangiaceae. Protoplasting and regeneration protocols have been developed for *Streptosporangium viridogriseum* (Oh et al., 1980) and plasmids have been isolated from *S. brasiliense* and *S. viridogriseum* (Fare et al., 1983); attempts to isolate phage have been unsuccessful (Prauser, 1984). Plasmid p Sg B-1 from *S. brasiliense* was characterized by electron microscopy and agarose gel electrophoresis and shown to be a closed circular DNA molecule of 9.4×10^6 Da. The restriction endonuclease map generated from this plasmid had unique sites for *Eco*R1, *Cla*I, *Xba*I, and *MSt*II. Plasmid p Sg V-1 from *S. viridogriseum* had an estimated M_r of 54×10^6. The p Sg B-1 plasmid is phenotypically cryptic, but an unusual expressed trait re-

M. poly-chroma	M. pusilla	M. rectica-tena	M. roseola	M. roseo-violacea	M. rubra	M. salmo-nea	M. spiralis	M. turk-meniaca	"M" viri-dis[a]
−	−	−	−	−	+	+	+	−	−
−	+	−	−	+	−	−	−	−	−
−	−	−	−	−	+	+	+	+	−
−	−	−	+	−	+	−	−	−	+
−	−	+	−	−	−	−	−	+	−
−	−	−	+	−	−	−	−	−	−
−	−	−	−	−	+	−	−	−	−
−	−	−	−	−	−	−	−	−	+
−	+	−	−	+	+	−	−	+	−
−	−	−	−	−	−	+	+	−	−
−	+	−	−	−	−	−	−	−	−
ND	Trace	ND	Pink	White-pink	Trace	Cream-pink	White-yellow	ND	ND
ND	Brown-red	ND	Red-brown	Purple-red	Red-brown	Brown	Yellow-brown	ND	ND
Trace	White-cream	White-cream	Pink	Pink-violet	Trace	Pink	White-yellow	Trace	Grayish-green
Colorless, brown	Gray-brown	Dark yellow-brown	Brown-red	Violet	Orange, red	Red	Yellow-brown	Violet-red	Yellow-brown
−	−	−	−	Violet	Red	−	−	Pink-violet	−
−	Trace	−	−	White	−	Trace	−	−	−
Colorless, yellow-brown	Colorless, brown	Yellow-pink	−	White, pink	−	Brown	Yellow-brown	Gray-violet	Yellow-brown
+	+	ND	+	+	−	+	+	+	ND
−	−	ND	−	−	−	+	−	+	ND
−	+	ND	−	+	−	+	+	−	ND
ND	−	ND	ND	−	+	+	ND	ND	ND
+	+	ND	+	+	+	+	−	ND	ND
+	+	ND	+	+	+	+	−	+	ND
−	−	ND	−	−	+	−	−	−	+
−	+	ND	+	−	+	+	+	−	ND
−	+	ND	+	−	+	+	+	−	ND
−	−	ND	−	−	−	−	−	−	ND
−	+	ND	+	+	+	+	+	−	−

Data taken from Athalye et al. (1985); Gauze et al. (1984); Greiner-Mai et al. (1987); Meyer (1989); Miyadoh et al. (1985); Nonomura (1989b, 1989c); Terekhova et al. (1982); Thiemann et al. (1968); and J. Meyer (personal communication).

sembling phage plaques may be associated with the *S. viridogriseum* plasmid p Sg V-1.

Applications

Members of the family Streptosporangiaceae can be expected to be an increasingly rich source of commercially significant products, notably antibiotics and enzymes. *Microbispora rosea* produces deoxycephalomycin B (Okazaki and Naito, 1985); *Microbispora* strain SCC 1438, a novel antifungal antibiotic (Patel et al., 1988); *Microtetraspora pusilla*, actinotiocin (Tamura et al., 1973); *M. roseoviolacea*, carminomycins (Nakagawa et al., 1983, 1989); and *M. rubra*, maduromycin (Fleck et al., 1978). Similarly, *Streptosporangium albidum* produces aculeximycin (Murata et al., 1989); *S. pseudovulgare* sporamycin (Komiyama et al., 1977); *S. violaceochromogenes*, platomycins A and B (Takasawa et al., 1975) and victomycin

Table 4. Characteristics differentiating the species of the genus *Streptosporangium*.

	S. album	*S. albidum*	*S. amethystogenes*	*S. corrugatum*	*S. fragile*	*S. longisporum*	*S. nondiastaticum*	*S. pseudovulgare*	*S. roseum*	*S. violaceochromogenes*	*S. viridialbum*	*S. viridogriseum* subsp. *viridogriseum*	*S. viridogriseum* subsp. *kofuense*	*S. vulgare*
Color of substrate mycelium														
Brown-black	−	−	−	−	+	−	−	−	−	−	−	−	−	−
Red or orange	−	−	−	−	−	+	+	+	+	−	−	−	−	+
Yellowish brown to brown	+	+	+	+	−	−	+	+	+	+	+	+	+	+
Color of spore mass														
Greenish gray	−	−	−	−	−	−	−	−	−	−	+	+	+	−
Pink	−	−	+	−	+	+	+	+	+	+	−	−	−	+
White	+	+	−	+	−	−	−	−	−	−	−	−	−	−
Spore vesicle size														
1–5 μm	−	−	−	+	−	−	−	−	−	−	−	−	−	−
6–10 μm	+	−	+	−	+	+	−	+	+	+	+	−	−	+
11–20 μm	−	(+)	−	−	+	+	+	−	(+)	−	−	−	+	−
21–30 μm	−	+	−	−	−	−	−	−	−	−	−	+	−	−
31–50 μm	−	−	−	−	−	−	−	−	−	−	−	+	−	−
Sporangiophore size														
Short (10 μm)	+	−	+	+	+	+	+	+	+	+	+	−	+	+
Long (50 μm)	−	+	−	−	−	−	−	−	−	−	−	+	+	−
Spore shape														
Spherical-oval	+	+	+	+	+	−	+	+	+	+	+	+	−	+
Rod	−	−	−	−	−	+	−	−	−	−	−	−	+	−
Soluble pigments[a]	−	−	−	−	+	−	−	−	+	+	−	−	−	−
B vitamins required	+	−	+	−	−	−	+	+	+	−	+	−	−	+
Growth at:														
42°C	−	−	−	−	+	−	+	+	−	−	−	+	+	−
50°C	−	−	−	−	−	−	−	−	−	−	−	+	(+)	−
Gelatin liquefaction	+	−	−	ND	−	−	+	+	+	(+)	d	+	+	d
Iodinin production	−	−	+	−	−	−	−	−	−	−	−	−	−	−
Nitrate reduction	−	+	+	−	+	(+)	+	+	+	+	d	+	−	−
Starch hydrolysis	−	−	+	−	+	+	−	+	+	+	+	+	+	+
Utilization of:														
Adonitol	+	ND	+	+	−	+	+	+	+	ND	−	−	−	+
Arabinose	+	ND	ND	+	+	+	+	+	+	ND	−	−	−	+
Galactose	+	ND	−	−	+	−	+	−	+	ND	−	+	+	+
Glycerol	−	ND	ND	−	−	−	−	+	+	ND	−	+	+	+
Inositol	−	ND	+	−	−	−	−	−	+	(+)	+	+	+	+
Mannitol	+	ND	ND	−	+	−	+	+	−	ND	+	+	+	−
Rhamnose	−	ND	+	−	+	−	−	−	+	(+)	+	+	(+)	+
Turanose	+	ND	ND	−	+	−	+	+	+	ND	+	−	+	+

Symbols: +, positive reaction; (+), weak reaction; −, negative reaction; ND, not determined.
[a]Other than pale yellow-brown.
Taken from Nonomura (1989a) and Whitham (1988).

Table 5. Characteristics differentiating the species of the genera *Planobispora*, *Planomonospora*, and *Spirillospora*.

| | Planobispora | | Planomonospora parontospora | | | Spirillospora | |
	longispora	rosea	subsp. parontospora	subsp. antibiotica	venezuelensis	albida	rubra
Shape of sporangium	Cylindrical to clavate	Cylindrical to clavate	Cylindrical to clavate	Cylindrical to clavate	Cylindrical to clavate	Spherical to vermiform	Spherical
Number of spores	2	2	1	1	1	Many	Many
Degradation of:							
Aesculin	−	+	+	ND	+	+	ND
Chitin	−	+	+	ND	+	−	ND
Hypoxanthine	+	−	+	ND	+	−	ND
Reduction of nitrate	+	+	$-^a, +^b$	+	$-^a, +^c$	−	−
Coagulation of litmus milk	+	−	−	−	−	ND	ND
Peptonization of litmus milk	+	−	+	−	−	ND	ND
Utilization of:							
Adonitol	−	−	+	ND	+	−	ND
Galactose	−	+	+	ND	+	−	ND
Inositol	+	+	−	−	−	−	ND
Lactose	−	−	+	ND	+	−	ND
Melezitose	−	+	+	ND	−	+	ND
Rhamnose	+	−	+	+	+	+	ND
Trehalose	−	−	+	ND	+	+	ND

Symbols: +, positive reaction; −, negative reaction; ND, not determined.
[a]Taken from Goodfellow and Pirouz (1982).
[b]Taken from Thiemann (1970).
[c]Taken from Thiemann et al. (1967).
Taken from Goodfellow and Pirouz (1982), Thiemann (1970), Thiemann et al. (1967), Vobis (1989a, 1989b), Vobis and Kothe (1989), and Whitham (1988).

(Kawamoto et al., 1975); and *S. viridogriseum*, chloramphenicol (Tamura et al., 1971) and sporaviridin (Okuda et al., 1966). A novel anthracycline antibiotic has been isolated from *S. fragile* (Shearer et al., 1983), an antitumor antibiotic from an organism resembling *S. pseudovulgare* (Umezawa et al., 1976), and unspecified antimicrobial agents from *S. indica* and *S. karnatakensis* (Rao et al., 1987).

Microbispora rosea is an excellent source of D-xylose (glucose) isomerase (Crueger and Crueger, 1982), which converts D-glucose into D-fructose. The enzyme is used to produce D-fructose on a commercial scale, and its biosynthesis, purification, and immobilization, as well as its application for the production of high fructose syrup, have been the subject of many reports and patents (Crueger and Crueger, 1984). Similarly, cystathionine γ-lyase has been detected in strains of *Streptosporangium* (Nagasawa et al., 1984). This enzyme has been shown to catalyse the α, γ-elimination reaction of L-cystathionine, and also the γ-replacement of L-homoserine in the presence of various thiol compounds (Kanzaki et al., 1986a). An efficient method based on the reaction of γ-replacement

has been developed (Kanzaki et al., 1986b) for the preparation of L-cystathionine, a product that may be useful because a deficiency of this has been observed in the brains of homocystinuric patients (Gerritsen and Waisman, 1964). The procedure allowed the total conversion of O-succinyl-L-homoserine and L-cysteine into L-cystathionine.

Thermophilic microbisporae produce a wide range of enzymes that are involved in the degradation and modification of heteropolysaccharides, notably celluloses, lignocelluloses, and hemicelluloses (Henssen and Schnepf, 1967; McCarthy, 1987; Crawford, 1988). These enzyme systems have the potential for novel applications in biotechnological processes, particularly for the enzymatic generation of fermentable sugars from agricultural residues (McCarthy et al., 1988; Zimmermann, 1989). The cellulolytic enzyme system of *Microbispora bispora* has been used to saccharify bleached hardwood Kraft pulp (Waldron et al., 1986). The maximum efficiency of saccharification achieved in a 24-hour hydrolysis at 58°C was just over 50% when the suspension contained 13.2 filter paper units of cellulase activity per ml. The enzyme system was less efficient on ground newsprint (11.5–13.1% hydrolysis).

The endoglucanase and exoglucanase components of the cellulolytic activity in *Microbispora bispora* have been separated (Bartley et al., 1984). The endoglucanase MbcellA of this organism has also been shown to be closely related to the endoglucanase CfconA and the exoglucanase Cfeex of *Cellulomonas fimi* (Yablonsky et al., 1989). The compounds α-L-arabinofuranosidases, endoxylanases (1,4-β-D-xylanase), and β-xylosidases have been detected in *Microbispora bispora* (Ball and McCarthy, 1988). The mesophilic *M. rosea* also produces exoxylanases (Kusakabe et al., 1969).

Literature Cited

Agre, N. S., and L. N. Guzeva. 1975. New genus of actinomycetes, *Excellospora* gen. nov. Microbiology [English translation of Mikrobiologiya] 44:459–463.

Athalye, M., M. Goodfellow, J. Lacey, and R. P. White. 1985. Numerical classification of *Actinomadura* and *Nocardiopsis.* Int. J. Syst. Bacteriol. 35:86–98.

Ball, A. S., and A. J. McCarthy. 1988. Saccharification of straw by actinomycete enzymes. J. Gen Microbiol. 134:2139–2147.

Bartley, T., C. Waldron, and D. Eveleigh. 1984. A cellobiohydrolase from a thermophilic actinomycete, *Microbispora bispora.* Appl. Biochem. Biotech. 9:337–338.

Bland, C. E., and J. N. Couch. 1981. The family *Actinoplanaceae,* p. 2004–2010. In: M. P. Starr, H. Stolp, H. G. Trüper, A. Balows, and H. G. Schlegel (ed.), The prokaryotes, vol. 2. Springer-Verlag, Berlin.

Chaves Batista, A., S. K. Shome, and J. Americo de Lima. 1963. *Streptosporangium bovinum* sp. nov. from cattle hoofs. Dermat. Trop. 2:49–54.

Collins, M. D., M. Faulkner, and R. M. Keddie. 1984. Menaquinone composition of some sporeforming actinomycetes. Syst. Appl. Microbiol. 5:20–29.

Couch, J. N. 1954. The genus *Actinoplanes* and its relatives. Trans. N.Y. Acad. Sci. 16:315–318.

Couch, J. N. 1955. A new genus and family of the *Actinomycetales* with a revision of the genus *Actinoplanes.* J. Elisha Mitchell Sci. Soc. 71:148–155.

Couch, J. N. 1963. Some new genera and species of the *Actinoplanaceae.* J. Elisha Mitchell Sci. Soc. 79:53–70.

Crawford, D. L. 1988. Biodegradation of agricultural and urban wastes, p. 433–459. In: M. Goodfellow, S. T. Williams, and M. Mordarski (ed.), Actinomycetes in biotechnology. Academic Press, London.

Cross, T. 1970. The diversity of bacterial spores. J. Appl. Bacteriol. 33:95–102.

Crueger, A., and W. Crueger. 1984. Carbohydrates, p. 421–457. In: H. J. Rehm and D. Reed (ed.), Biotechnology, vol. 6a. Verlag Chemie, Weinheim, Germany.

Crueger, W., and A. Crueger. 1982. Glukose isomerasen, p. 166–184. In: Lehrbuch der Angewandten Mikrobiologie. Akademische Verlagsgesellschaft, Weisbaden, Germany.

Fare, L. R., D. P. Taylor, M. J. Toth, and C. H. Nash. 1983. Physical characterization of plasmids isolated from *Streptosporangium.* Plasmid 9:240–246.

Fischer, A., R. M. Kroppenstedt, and E. Stackebrandt. 1983. Molecular-genetic and chemotaxonomic studies on *Actinomadura* and *Nocardiopsis.* J. Gen. Microbiol. 129:3433–3446.

Fleck, W. F., D. G. Strauss, J. Meyer, and G. Porstendorfer. 1978. Fermentation, isolation, and biological activity of maduramycin: a new antibiotic from *Actinomadura rubra.* Z. Allg. Mikrobiol. 18:389–398.

Galatenko, O. A., L. P. Terekhova, and T. P. Preobrazhenskaya. 1981. New *Actinomadura* species isolated from Turkmen soil samples and their antagonistic properties. (In Russian.) Antibiotiki 26:803–807.

Gauze, G. F., T. P. Preobrazhenskaya, E. S. Kudrina, N. O. Blinov, I. D. Ryabova, and M. A. Sveshnikova. 1957. Problems in the classification of antagonistic actinomycetes. Moscow State Publishing House for Medical Literature, Medgiz.

Gauze, G. F., L. P. Terekhova, O. A. Galatenko, T. P. Preobrazhenskaya, V. N. Borisova, and G. B. Federova. 1984. *Actinomadura recticatena* sp. nov., a new species and its antibiotic properties. (In Russian.) Antibiotiki 29:3–7.

Gerber, N. N., and M. P. Lechevalier. 1964. Phenazones and phenoxazinones from *Waksmania aerata* sp. nov. and *Pseudomonas iodina.* Biochemistry 3:598–602.

Gerber, N. N., and M. P. Lechevalier. 1965. 1-6-Phenazinediol-5-oxide from microorganisms. Biochemistry 4:176–180.

Gerritsen, T., and H. A. Waisman. 1964. Homocystonuria: Absence of cystathionine in the brain. Science 145:588.

Goodfellow, M. 1989a. Suprageneric classification of actinomycetes, p. 2333–2339. In: S. T. Williams, M. E. Sharpe, and J. G. Holt (ed.), Bergey's manual of systematic bacteriology, vol. 4. Williams and Wilkins, Baltimore.

Goodfellow, M. 1989b. Maduromycetes, p. 2509–2510. In: S. T. Williams, M. E. Sharpe, and J. G. Holt (ed.), Bergey's manual of systematic bacteriology, vol. 4. Williams and Wilkins, Baltimore.

Goodfellow, M., and T. Cross. 1984. Classification, p. 7–164. In: M. Goodfellow, M. Mordarski and S. T. Williams (ed.), The biology of the actinomycetes. Academic Press, London.

Goodfellow, M., and A. G. O'Donnell. 1989. Search and discovery of industrially significant actinomycetes, p. 343–383. In: S. Baumberg, M. Rhodes, and I. Hunter (ed.), Microbial products: new approaches. Cambridge University Press, Cambridge.

Goodfellow, M., and T. Pirouz. 1982. Numerical classification of sporoactinomycetes containing *meso*-diaminopimelic acid in the cell wall. J. Gen Microbiol. 128:503–527.

Goodfellow, M., G. Alderson, and J. Lacey. 1979. Numerical taxonomy of *Actinomadura* and related actinomycetes. J. Gen. Microbiol. 112:95–111.

Goodfellow, M., E. Stackebrandt, and R. M. Kroppenstedt. 1988. Chemotaxonomy and actinomycete systematics, p. 233–238. In: Y. Okami, T. Beppu, and H. Ogawara (ed.), Biology of actinomycetes. Japan Scientific Societies Press, Tokyo.

Goodfellow, M., L. J. Stanton, K. E. Simpson, and D. E. Minnikin. 1990. Numerical and chemical classification of *Actinoplanes* and some related actinomycetes. J. Gen Microbiol. 136:19–34.

Greiner-Mai, E., R. M. Kroppenstedt, F. Korn-Wendisch, and H. J. Kutzner. 1987. Morphological and biochemical characterization and emended descriptions of thermophilic actinomycetes species. System. Appl. Microbiol. 9:97–109.

Hasegawa, T., M. P. Lechevalier, and H. A. Lechevalier. 1979. Phospholipid composition of motile actinomycetes. J. Gen. Appl. Microbiol. 25:209–213.

Hayakawa, M., and H. Nonomura, 1984. HV agar, a new selective medium for isolation of soil actinomycetes, p. 6. In: Abstracts of papers presented at the annual meeting of the Actinomycetologists. Osaka, Japan.

Hayakawa, M., and H. Nonomura. 1987a. Humic acid-vitamin agar, a new medium for the selective isolation of soil actinomycetes. J. Fermen. Technol. 65:501–509.

Hayakawa, M., and H. Nonomura. 1987b. Efficacy of artificial humic acid as a selective nutrient in HV agar used for the isolation of soil actinomycetes. J. Fermen. Technol. 65:609–616.

Hayakawa, M., K. Ishizawa, and H. Nonomura. 1988. Distribution of rare actinomycetes in Japanese soils. J. Fermen. Technol. 66:367–373.

Henssen, A., and D. Schäfer. 1971. Emended description of the genus *Pseudonocardia* Henssen and description of a new species *Pseudonocardia spinosa* Schäfer. Int. J. Syst. Bacteriol. 21:29–34.

Henssen, A., and E. Schnepf. 1967. Zur Kenntnis thermophiler Actinomyceten. Arch. Mikrobiol. 57:214–231.

Hickey, R. J., and H. D. Tresner. 1952. A cobalt-containing medium for sporulation of *Streptomyces* species. J. Bacteriol. 64:891–892.

Johnstone, D. W., and T. Cross. 1976. The occurrence and distribution of actinomycetes in lakes of the English Lake District. Freshwater Biology 6:457–463.

Jones, K. L. 1949. Fresh isolates of actinomycetes in which the presence of sporangous aerial mycelia is a fluctuating characteristic. J. Bacteriol. 57:141–145.

Kanzaki, H., M. Kobayashi, T. Nagasawa, and H. Yamada. 1986a. Synthesis of S-substituted L-homocysteine derivatives by cystathionine γ-lyase of *Streptomyces phaeochromogenes*. Agric. Biol. Chem. 50:391–397.

Kanzaki, H., T. Nagasawa, and H. Yamada. 1986b. Highly efficient production of L-cystathionine from O-succinyl-L-homoserine and L-cysteine by *Streptomyces* cystathionine γ-lyase. Appl. Microbiol. Biotech. 25:97–100.

Kawamoto, I., S. Takasawa, R. Okachi, M. Kohakura, I. Takahashi, and T. Nara. 1975. A new antibiotic victomycin (XK 49-1-B-2). I. Taxonomy and production of the producing organisms. J. Antibiotics 28:358–365.

Komiyama, K., K. Sugimoto, H. Takeshima, and I. Umezawa. 1977. A new antitumour antibiotic, sporamycin. J. Antibiot. 30:202–208.

Kroppenstedt, R. M. 1982. Separation of bacterial menaquinones by HPLC using reverse phase (RP18) and a silver loaded ion exchanger as stationary phases. J. Liquid Chromat. 5:2359–2367.

Kroppenstedt, R. M. 1985. Fatty acid and menaquinone analysis of actinomycetes and related organisms, p. 173–199. In: M. Goodfellow, and D. E. Minnikin (ed.), Chemical methods in bacterial systematics. Academic Press, London.

Kroppenstedt, R. M., E. Stackebrandt, and M. Goodfellow. 1990. Taxonomic revision of the actinomycete genera *Actinomadura* and *Microtetraspora*. Syst. Appl. Microbiol. 13:148–160.

Kusakabe, I., T. Yasui, and T. Kobayashi. 1969. Some properties of extracellular xylanase from *Streptomyces*. J. Agric. Chem. Soc. Japan 43:145–153.

Küster, E. 1959. Outline of a comparative study of criteria used in characterization of the actinomycetes. Int. Bull. Bacteriol. Nomencl. Taxon 9:98–104.

Lavrova, N. V., T. P. Preobrazhenskaya, and M. A. Sveshnikova. 1972. Isolation of soil actinomycetes on selective media with rubomycin. (In Russian.) Antibiotiki 17:965–970.

Lechevalier, H. A., and M. P. Lechevalier. 1970a. A critical evaluation of the genera of aerobic actinomycetes, p. 393–405. In: H. Prauser (ed.), The *Actinomycetales*. Gustav Fischer Verlag, Jena, Germany.

Lechevalier, H. A., and M. P. Lechevalier. 1981. Introduction to the order *Actinomycetales*. p. 1915–1922. In: M. P. Starr, H. Stolp, H. G. Trüper, A. Balows, and H. G. Schlegel (ed.), The prokaryotes, vol. 2. Springer-Verlag, Berlin.

Lechevalier, H. A., M. P. Lechevalier, and P. E. Holbert. 1966. Electron microscopic observation of the sporangial structure of strains of *Actinoplanaceae*. J. Bacteriol. 92:1228–1235.

Lechevalier, M. P. 1977. Lipids in bacterial taxonomy—a taxonomist's view. CRC Crit. Rev. Microbiol. 5:109–210.

Lechevalier, M. P., and N. N. Gerber. 1970. The identity of 3-O-methyl-D-galactose with madurose. Carb. Res. 13:451–454.

Lechevalier, M. P., and H. A. Lechevalier. 1957. A new genus of the *Actinomycetales*: *Waksmania* gen. nov. J. Gen. Microbiol. 17:104–111.

Lechevalier, M. P., and H. A. Lechevalier. 1970b. Chemical composition as a criterion in the classification of aerobic actinomycetes. Int. J. Syst. Bacteriol. 20:435–443.

Lechevalier, M. P., and H. A. Lechevalier. 1980. The chemotaxonomy of actinomycetes, p. 227–291. In: A. Dietz and D. Thayer, Actinomycete taxonomy, special publication 6, Society for Industrial Microbiology, Arlington, VA.

Lechevalier, M. P., C. De Biévre, and H. A. Lechevalier. 1977. Chemotaxonomy of aerobic actinomycetes: phospholipid composition. Biochem. Syst. Ecol. 5:249–260.

Lechevalier, M. P., A. E. Stern, and H. A. Lechevalier. 1981. Phospholipids in the taxonomy of actinomycetes. Zbl. Bakt. Suppl. 11:111–116.

Louria, D. B., and R. E. Gordon. 1960. Pericarditis and pleuritis caused by a recently discovered microorganism, *Waksmania rosea*. Am. Rev. Resp. Dis. 81:83–88.

Manachini, P. L., A. Ferrari, and R. Craveri. 1965. Forme termofile de *Actinoplanaceae*. Isolamento et caracteristiche di *Streptosporangium album* var. *thermophilum*. Ann. Microbiol. Enzymol. 15:129–144.

McCarthy, A. . 1987. Lignocellulose-degrading actinomycetes. FEMS Microbiol. Rev. 46:145–163.

McCarthy, A. J., A. S. Ball, and S. L. Bachmann. 1988. Ecological and biotechnological implications of lignocellulose degradation by actinomycetes. p. 283–287 In: Y. Okami, T. Beppu and H. Ogawara (ed.), Japan Scientific Societies Press, Tokyo.

Meyer, J. 1979. New species of the genus *Actinomadura*. Z. Allg. Mikrobiol. 19:37–44.

Meyer, J. 1989. Genus *Actinomadura* Lechevalier and Lechevalier 1970, 400^AL, p. 2511–2526. In: S. T. Williams, M. E. Sharpe and J. G. Holt (ed.), Bergey's manual of systematic bacteriology, vol. 4. Williams and Wilkins, Baltimore.

Miyadoh, S., H. Anzai, S. Amano, and T. Shomura. 1989. *Actinomadura malachitica* and *Microtetraspora viridis* are synonyms and should be transferred as *Actinomadura viridis* comb. nov. Int. J. Syst. Bacteriol. 39:152–158.

Miyadoh, S. T., H. Tohyama, S. Amano, T. Shomura, and T. Niida. 1985. *Microbispora viridis*, a new species of *Actinomycetales*. Int. J. Syst. Bacteriol. 37:342–346.

Murata, H., N. Kojima, K-I. Harada, M. Suzuki, T. Ikemoto, T. Shibuya, T. Haneishi, and A. Torikata. 1989. Structural elucidation of aculescimycin. I. Further purification and glycosidic bond cleavage of aculescimycin. J. Antibiot. 42:691–700.

Nagasawa, T., H. Kanzaki, and N. Yamada. 1984. Cystathionine γ-lyase of *Streptomyces phaeochromogenes*—the occurrence of cystathionine γ-lyase in filamentous bacteria and its purification and characterisation. J. Biol. Chem. 259:10393–10403.

Nakagawa, M., Y. Hayakawa, H. Kawai, K. Imamura, H. Inoue, A. Shimazu, H. Seto, and N. Otake. 1983. A new anthracycline antibiotic N-formyl-13-dehydrocarminomycin. J. Antibiot. 36:457–458.

Nakagawa, M., Y. Hayakawa, K. Imamura, M. Seto, and N. Otake. 1989. Microbial conversion of anthracyclinones to carminomycins by a blocked mutant of *Actinomadura meovirlacea*. J. Antibiot. 42:1698–1703.

Naumova, I. B., N. V. Potekhina, L. P. Terekhova, T. P. Preobrazhenskaya, and K. Digimbay. 1986. Wall polyol phosphate polymers of bacteria belonging to the genus *Actinomadura*, p. 561–566. In: G. Szabó, S. Biró and M. Goodfellow (ed.), Biological, biochemical and biomedical aspects of actinomycetes. Akadémiai Kiadó, Budapest.

Nonomura, H. 1984. Design of a new medium for isolation of soil actinomycetes. The Actinomycetes 18:206–209.

Nonomura, H. 1989a. Genus *Streptosporangium* Couch 1955, 148^AL, p. 2545–2551. In: S. T. Williams, M. E. Sharpe, and J. G. Holt (ed.), Bergey's manual of systematic bacteriology, vol. 4. Williams and Wilkins, Baltimore.

Nonomura, H. 1989b. Genus *Microbispora* Nonomura and Ohara 1957, 307^AL, p. 2526–2531. In: S. T. Williams, M. E. Sharpe, and J. G. Holt (ed.), Bergey's manual of systematic bacteriology, vol. 4. Williams and Wilkins, Baltimore.

Nonomura, H. 1989c. Genus *Microtetraspora* Thiemann, Pagani and Beretta 1968, 296^AL, p. 2531–2536. In: S. T. Williams, M. E. Sharpe, and J. G. Holt (ed.), Bergey's manual of systematic bacteriology, vol. 4. Williams and Wilkins, Baltimore.

Nonomura, H., and M. Hayakawa. 1988. New methods for the selective isolation of soil actinomycetes, p. 288–293. In: Y. Okami, T. Beppu, and H. Ogawara (ed.), Biology of Actinomycetes. Japan Scientific Societies Press, Tokyo.

Nonomura, H., and Y. Ohara. 1957. Distribution of actinomycetes in the soil. II. *Microbispora*, a new genus of the *Streptomycetaceae*. J. Ferment. Technol. 35:307–311.

Nonomura, H., and Y. Ohara. 1960. Distribution of actinomycetes in soil. IV. The isolation and classification of the genus *Microbispora*. J. Ferment. Technol. 38:401–405.

Nonomura, H., and Y. Ohara. 1969a. Distribution of actinomycetes in soil. VI. A culture method effective for both preferential isolation and enumeration of *Microbispora* and *Streptosporangium* strains in soil (part 1). J. Ferment. Technol. 47:463–469.

Nonomura, H., and Y. Ohara. 1969b. Distribution of actinomycetes in soil. VII. A culture method effective for both preferential isolation and enumeration of *Microbispora* and *Streptosporangium* strains in soil (part 2). Classification of isolates. J. Ferment. Technol. 47:701–709.

Nonomura, H., and Y. Ohara. 1971a. Distribution of actinomycetes in soil. VIII. Green-spore group of *Microtetraspora*, its preferential isolation and taxonomic characteristics. J. Ferment. Technol. 49:1–7.

Nonomura, H., and Y. Ohara. 1971b. Distribution of actinomycetes in soil. IX. New species of the genera *Microbispora* and *Microtetraspora*, and their isolation method. J. Ferment. Technol. 49:887–894.

Nonomura, H., and Y. Ohara. 1971c. Distribution of actinomycetes in soil. XI. Some new species of the genus *Actinomadura* Lechevalier et al. J. Ferment. Technol. 49:904–912.

Oh, Y. K., J. L. Speth, and C. H. Nash. 1980. Protoplast fusion with *Streptosporangium viridogriseum*. Dev. Ind. Microbiol. 21:219–226.

Okazaki, T., and A. Naito. 1985. Studies in actinomycetes isolated from Australian soils, p. 739–741. In: G. Szabó, S. Biró, and M. Goodfellow (ed.), Biological, biochemical and biomedical aspects of actinomycetes. Akadémiai Kiadó, Budapest.

Okuda, T., Y. Ito, T. Yamaguichi, T. Furumai, M. Suzuki, and M. Tsuruoka. 1966. Sporaviridin, a new antibiotic produced by *Streptosporangium viridogriseum* nov. sp. J. Antibiotics Ser. A 19:85–87.

Patel, M., M. Conover, A. Horan, D. Loebenberg, J. Marquez, R. Mierzwa, M. S. Puar, R. Yarborough, and J. A. Waitz. 1988. Sch 31828, a novel antibiotic from a *Microbispora* sp.: taxonomy, fermentation, isolation and biological properties. J. Antibiot. 41:794–797.

Poschner, J., R. M. Kroppenstedt, A. Fischer, and E. Stackelbrandt. 1985. DNA:DNA reassociation and chemotaxonomic studies on *Actinomadura*, *Microbispora*, *Microtetraspora*, *Micropolyspora* and *Nocardiopsis*. System. Appl. Microbiol. 6:264–270.

Potekhina, L. L. 1965. *Streptosporangium rubrum* n. sp.—a new species of the *Streptosporangium* genus. Mikrobiologiya 34:292–299.

Prauser, H. 1984. Phage host ranges in the classification and identification of Gram-positive branched and related bacteria, p. 617–633. In: L. Ortiz-Ortiz, L. F. Bojalil, and V. Yakoleff (ed.), Biological, biochemical and biomedical aspects of actinomycetes. Academic Press, Orlando, Fl.

Preobrazhenskaya, T. P., N. V. Lavrova, R. S. Ukholina, and N. P. Nechaeva. 1975. Isolation of new species of *Actinomadura* on selective media with streptomycin and bruneomycin [in Russian]. Antibiotiki 20:404–409.

Pridham, T. G., P. Anderson, G. Foley, L. A. Lindenfelser, C. W. Hesseltine, and R. G. Benedict. 1956–57. A selection of media for maintenance and taxonomic study of streptomycetes. Antibiotics Ann. 1956/57:947–953.

Rao, V. A., K. K. Prabhu, B. P. Sridhar, A. Venkateswarlu, and P. Actor. 1987. Two new species of *Microbispora* from Indian soils: *Microbispora karnatakensis* sp. nov. and *Microbispora indica* sp. nov. Int. J. Syst. Bacteriol. 37:181–185.

Schäfer, D. 1973. Beitrage zur Klassifizerung and Taxonomie der Actinoplanaceen. Ph.D. Dissertation, University of Marburg/Lahn, Federal Republic of Germany

Schleifer, K. H., and O. Kandler. 1972. Peptidoglycan types of bacterial cell walls and their taxonomic implications. Bacteriol. Rev. 36:407–477.

Sharples, G. P., S. T. Williams, and R. M. Bradshaw. 1974. Spore formation in the *Actinoplanaceae (Actinomycetales)*. Arch. Microbiol. 101:9–20.

Shearer, M. C., P. M. Colman, and C. H. Nash. 1983. *Streptosporangium fragile* sp. nov. Int. J. Syst. Bacteriol. 33:364–368.

Shearer, M. C., P. M. Colman, R. M. Ferrari, L. J. Nisbet, and C. H. Nash, III. 1986. A new genus of the *Actinomycetales: Kibdelosporangium aridum* gen. nov., sp. nov. Int. J. Syst. Bacteriol. 36:47–54.

Shirling, E. B., and D. Gottlieb. 1966. Methods for characterization of *Streptomyces* species. Int. J. Syst. Bacteriol. 16:313–340.

Stackebrandt, E., B. Wunner-Füssl, V. J. Fowler, and K. H. Schleifer. 1981. Deoxyribonucleic acid homologies and ribosomal ribonucleic acid similarities among spore-forming members of the order *Actinomycetales*. Int. J. Syst. Bacteriol. 31:420–431.

Stackebrandt, E., W. Ludwik., E. Seewaldt, and K. H. Schleifer. 1983. Phylogeny of spore forming members of the order *Actinomycetales*. Int. J. Syst. Bacteriol. 33:173–180.

Takasawa, S., I. Kawamoto, I. Takahashi, M. Kohakura, R. Okachi, S. Sata, M. Yamamoto, and T. Nara. 1975. Platomycins A and B. I. Taxonomy of the producing strain and production, isolation and biological properties of platomycins. J. Antibiot. 28:656–661.

Tamura, A., I. Takeda, S. Naruto, and Y. Yoshimura. 1971. Chloramphenicol from *Streptosporangium viridogriseum* var. *kofuense*. J. Antibiot. 24:270.

Tamura, A., R. Furuta, S. Naruto, and H. Ishii. 1973. Actinotiocin, a new sulfur-containing peptide antibiotic from *Actinomadura pusilla*. J. Antibiot. 26:343–350.

Terekhova, L. P., O. A. Galatenko, and T. P. Preobrazhenskaya. 1982. *Actinomadura fulvescens* sp. nov. and *Actinomadura turkmeniaca* sp. nov. and their antagonistic properties. [In Russian]. Antibiotiki 27:87–92.

Thiemann, J. E. 1970. Studies of some genera and species of the *Actinoplanaceae*, p. 245–257. In: H. Prauser (ed.), The *Actinomycetales*. Gustav Fischer Verlag, Jena, Germany.

Thiemann, J. E., 1974a. Genus *Planomonospora* Thiemann, Pagani and Beretta, p. 719–720. In: R. E. Buchanan, and N. E. Gibbons (ed.), Bergey's manual of determinative bacteriology, 8th ed. Williams and Wilkins, Baltimore.

Thiemann, J. E. 1974. Genus *Planobispora* Thiemann and Beretta, p. 720–721. In: R. E. Buchanan, and N. E. Gibbons (ed.), Bergey's manual of determinative bacteriology, 8th ed. Williams and Wilkins, Baltimore.

Thiemann, J. E., and G. Beretta. 1968. A new genus of the *Actinoplanaceae: Planobispora* gen. nov. Arch. Microbiol. 62:157–166.

Thiemann, J. E., H. Pagani, and G. Beretta. 1967. A new genus of the *Actinoplanaceae: Planomonospora* gen. nov. Giorn. Microbiol. 15:27–38.

Thiemann, J. E., H. Pagani, and G. Beretta. 1968. A new genus of *Actinomycetales: Microtetraspora* gen. nov. J. Gen. Microbiol. 50:295–303.

Uchida, K., and K. Aida. 1977. Acyl type of bacterial cell wall: its simple identification by colormetric method. J. Gen. Appl. Microbiol. 23:249–260.

Umezawa, I., K. Kamiyama, H. Takeshita, J. Awaya, and S. Omura. 1976. A new antitumour antibiotic, PO-357. J. Antibiot. 29:1249–1251.

Van Brummelen, J., and J. C. Went. 1957. *Streptosporangium* isolated from forest litter in the Netherlands. Antonie van Leeuwenhoek J. Microbiol. 23:385–392.

Vobis, G. 1985. Spore development in sporangia-forming actinomycetes, p. 443–452. In: G. Szabó, S. Bíró, and M. Goodfellow (ed.), Biological, biochemical and biomedical aspects of actinomycetes. Akadémiai Kiadó, Budapest.

Vobis, G. 1989a. Genus *Planomonospora* Thiemann, Pagani and Beretta 1967, 29[AL], p. 2539–2543. In: S. T. Williams, M. E. Sharpe, and J. G. Holt (ed.), Bergey's manual of systematic bacteriology, vol. 4. Williams and Wilkins, Baltimore.

Vobis, G. 1989b. Genus *Planobispora* Thiemann and Beretta 1968, 157[AL], p. 2536–2539. In: S. T. Williams, M. E. Sharpe, and J. G. Holt (ed.), Bergey's manual of systematic bacteriology, vol. 4. Williams and Wilkins, Baltimore.

Vobis, G., and H. W. Kothe. 1985. Sporogenesis in sporangiate actinomycetes. Front. Appl. Microbiol. 1:25–47.

Vobis, G., and H. W. Kothe. 1989. Genus *Spirillospora* 1963, 61[AL], p. 2543–2545. In: S. T. Williams, M. E. Sharpe, and J. G. Holt (ed.), Bergey's manual of systematic bacteriology, vol. 4. Williams and Wilkins, Baltimore.

Waksman, S. A. 1950. The actinomycetes: Their nature, occurrence, activities and importance. Ann. Crypt. Phytopath. 9:1–230.

Waksman, S. A. 1961. The actinomycetes, volume 2: classification, identification and descriptions of genera and species. Bailliere, Tindall and Cox Limited, London.

Waldron, Jr., C. R., C. A. Becker-Vallone, and D. E. Eveleigh. 1986. Isolation and characterization of a cellulolytic actinomycete *Microbispora bispora*. Appl. Microbiol. Biotechnol. 24:477–486.

Wellington, E. M. H., and S. T. Williams. 1978. Preservation of actinomycete inoculum in frozen glycerol. Microbios Lett. 6:151–159.

Weyland, H. 1969. Actinomycetes in North Sea and Atlantic Ocean sediments. Nature (London) 223:858.

Weyland, H., E. Helmke, K. Weber, and T. Richter. 1982. Madurose in a LL-DAP containing actinomycete. Proceedings of the 5th International Symposium on Actinomycete Biology, Mexico.

Whitham, T. S. 1988. Selective isolation, characterisation and identification of streptosporangia. Ph.D. Thesis, University of Newcastle upon Tyne, UK.

Williams, S. T. 1970. Further investigations of actinomycetes by scanning electron microscopy. J. Gen. Microbiol. 62:67–73.

Williams, S. T., and G. P. Sharples. 1976. *Streptosporangium corrugatum* sp. nov., an actinomycete with some unusual morphological features. Int. J. Syst. Bacteriol. 26:45–52.

Williams, S. T., and E. M. H. Wellington. 1980. Micromorphology and fine structure of actinomycetes, p. 139–165. In: M. Goodfellow, and R. G. Board (ed.), Microbiological classification and identification. Academic Press, London.

Willoughby, L. G. 1969a. A study of aquatic actinomycetes. The allochthonous leaf component. Nova Hedwigia. 18:45–113.

Willoughby, L. G. 1969b. A study of the aquatic actinomycetes of Blenham Tarn. Hydrobiologiya 34:465–483.

Yablonsky, M. D., K. O. Elliston, and D. E. Eveleigh. 1989. The relationship between the endoglucanase MbcelA of *Microbispora bispora* and the cellulases of *Cellulomonas fimi,* p. 73–83. In: M. P. Coughlan (ed.), Enzyme systems for lignocellulose degradation. Elsevier Applied Science, London.

Zimmermann, W. 1989. Hemicellulolytic enzyme systems from actinomycetes, p. 167–181. In: M. P. Coughlan (ed.), Enzyme systems for lignocellulose degradation. Elsevier Applied Science, London.

The Genus *Nocardiopsis*

REINER MICHAEL KROPPENSTEDT

History

Today, the genus *Nocardiopsis* harbors five species and one subspecies: *N. alborubida* (ATCC 23612[T], DSM 40465[T]), *N. alba* subsp. *alba* (DSM 43377[T]), *N. alba* subsp. *prasina* (ATCC 35940[T], DSM 43845[T]), *N. antarcticus* (VKM A-836[T], DSM 43884[T]), *N. dassonvillei* (ATCC 23218[T], DSM 43111[T]), and *N. listeri* (ATCC 27442[T], DSM 40297[T]). For a long time, the classification of the *Nocardiopsis* species was very difficult, and their designations changed frequently (Table 1). This confusion occurred because of the deficiency of good discrimination markers. Originally, morphology and physiology were the only criteria used to differentiate *Nocardiopsis* from other mesophilic aerobic actinomycete genera such as *Streptomyces* and

Nocardia, which also produce conidia or arthrospores on an aerial mycelium. Although morphology is a very valuable marker for preliminary identification of aerobic sporeforming actinomycetes, this character has only a limited value in detailed classification, because the same morphological structures are found in different actinomycete genera, and a high degree of morphological overlap exists between phylogenetically unrelated taxa (Gordon and Horan, 1968; Goodfellow and Cross, 1984; Stackebrandt, 1986; Stackebrandt et al., 1983). Table 2 lists the most important morphological characters of the major genera of sporoactinomycetes.

The taxonomy of the *Actinomycetales* first acquired a solid basis with the introduction of chemical markers. The first chemotaxonomic

Table 1. References for the history of the genus *Nocardiopsis*.[a]

DSM number	Original assignment	Current designation Bergey's Manual (1989) (Meyer, 1989)	Approved lists of bacterial names (1990)
43748[T]	*A. africana* (1,2)[b]	*N. africana* (3)	*M. africana* (4)
40465[T]	"*Act. alborubidus*" (5)		*N. alborubida* (6)
43377[T]	*A. dassonvillei* (7)	*N. dassonvillei* (8)	*N. alba* subsp. *alba* (6)
43845[T]		*N. dass.* subsp. *prasina* (9)	*N. alba* subsp. *prasina* (6)
43884[T]		*N. antarcticus*	*N. antarctica* (10)
	St. dassonvillei (11)		
	Noc. dassonvillei (12)		
43111[T]	*A. dassonvillei* (7)	*N. dass.* subsp. *dass.* (9)	*N. dassonvillei* (6)
40297[T]	*Act. listeri* (13)	"*Sm. listeri*" (14)	*N. listeri* (6)
43679[T]	*A. coeruleofusca* (1)	*N. coeruleofusca* (3)	*S. coeruleofusca* (15)
43885[T]	*A. flava* (1)	*N. flava* (3)	*S. flava* (15)
43749[T]	*A. longispora* (1)	*N. longispora* (3)	*S. longispora* (15)
43853[T]	—	*N. mutabilis* (16)	*S. mutabilis* (17)
43886[T]	—	*N. syringae* (18)	*S. syringae* (15)

DSM, Deutsche Sammlung von Mikroorganismen und Zellkulturen, Braunschweig, FRG; A, *Actinomadura*; Act, *Actinomyces*; M, *Microtetraspora*; N, *Nocardiopsis*; Noc, *Nocardia*; S, *Saccharothrix*; Sm, *Streptomyces*; St, *Streptothrix*.
[a]Numbers in parentheses correspond to the following references: (1) Preobrazhenskaya and Sveshnikova, 1974; (2) Greiner-Mai et al., 1987; (3) Preobrazhenskaya et al., 1982; (4) Kroppenstedt et al., 1990; (5) Kudrina, 1957; (6) Grund and Kroppenstedt, 1990; (7) H. A. Lechevalier and M. P. Lechevalier, 1970; (8) Meyer, 1976; (9) Miyashita et al., 1984; (10) Abyzov et al., 1983; (11) Brocq-Rousseau, 1904; (12) Liegard and Landrieu, 1911; (13) Erikson, 1935; (14) Waksman and Henrici, 1948; (15) Grund and Kroppenstedt, 1989; (16) Shearer et al., 1983; (17) Labeda and Lechevalier, 1989; (18) Gauze et al., 1974.

Table 2. Morphological characters of *Nocardiopsis* and 11 other genera of sporoactinomycetes.

	Spore chains on special aerial hyphae	AM[a] disintegrate into spores	Multiple cell division	Spore chains, 50 or more spores	Fragmentation of substrate mycelium	Sporangia or pseudosporangia	Motile spores
Nocardiopsis	−	+	−	+	+/−	−	−
Saccharothrix	−	+	−	+	−	−	−
Glycomyces	−	+	−	+	−	−	−
Actinomadura	+	−	−	−	−	−	−
Microtetraspora	+	−	−	−	−	−[b]	−
Streptomyces	+	−	−	−	−	−	−
Amycolatopsis	−	−	−	+	+	−	−
Kibdellosporangium	+	−	−	+	+/−	+	−
Saccharopolyspora	−	+	−	+	+/−	−	−
Pseudonocardia	−	+	+	+	+	−	−
Amycolata	−	+	+	+	+	−	−
Nocardia	−	+	−	V	+	−	−

Symbols: +, present; −, absent; +/−, character inconsistent within a genus.
[a]AM, aerial mycelium.
[b]Pseudosporangia are found in *M. carminata, M. roseolaviolacea, M. helvata,* and *M. pusilla,* however.
Adapted from Goodfellow (1989); Kothe et al. (1989); Kroppenstedt et al. (1990); Labeda et al (1984); Labeda et al. (1985); Lechevalier et al. (1986); Meyer (1976); Preobrazhenskaya et al. (1982); Shearer et al. (1986); Williams and Wellington (1981).

marker used in classification was cell wall chemistry, based on the analysis of amino acids from purified cell walls (Cummins and Harris, 1958). This work was continued by H. A. Lechevalier and M. P. Lechevalier (1970), who analyzed the cell walls of more than 600 strains of actinomycetes. The results of this study led to the concept of "cell wall chemotypes." M. P. Lechevalier and H. A. Lechevalier (1970a) used specific differences of amino acid composition in the cell wall to differentiate aerobic actinomycetes into four groups. Based on the chemical data and on some specific morphological characters, they created the new actinomycete genus *Actinomadura.* This genus harbored three species, *A. madura, A. pelletieri, A. dassonvillei* (H. A. Lechevalier and M. P. Lechevalier, 1970). The two unifying characters of these three aerobic actinomyces species were: their ability to produce spores on aerial mycelium, and a cell wall composition of N-acetyl-muramic acid, N-acetyl-glucosamine, alanine, glutamic acid, and *meso*-2,6-diaminopimelic acid, i.e., cell wall chemotype III. Table 3 lists the cell wall chemotypes of *Nocardropsis* and other actinomycetes that produce conidia or arthrospores. Cell wall chemotype III, corresponds to the peptidoglycan type A1γ (Schleifer and Kandler, 1972). Using cell wall chemistry, it was possible to separate these three *Actinomadura* species from morphologically related taxa showing a different cell wall composition, e.g., type I *Streptomyces,*

II *Glycomyces,* or IV *Nocardia* and *Pseudonocardiacea* (Table 3). However, using only morphology and this limited set of chemical markers, the genus *Actinomadura* quickly became a dumping ground for many unrelated chemotype III isolates. Further progress in the classification of these organisms was then achieved by the introduction of an additional chemotaxonomical marker into taxonomy, the "whole cell sugar pattern" (M. P. Lechevalier and H. A. Lechevalier, 1970a, 1970b; Lechevalier et al., 1971). This made possible the separation of the cell chemotype III organisms into those having the diagnostic sugar madurose (3–O-methyl-D-galactose) in the whole cell hydrolysates, i.e., *Actinomadura,* and those lacking this sugar, i.e., *Nocardiopsis.* Based on the characteristic development of spores, including the specific zigzag formation of aerial hyphae before spore delimitation (Williams et al., 1974) and the lack of madurose (M. P. Lechevalier and H. A. Lechevalier, 1970a), Meyer (1976) created a new genus for *A. dassonvillei.* The name *Nocardiopsis* for this new genus was selected because of the nocardia-like appearance of *N. dassonvillei.* Like *Nocardia, N. dassonvillei* does not produce special spore-forming hyphae. In *N. dassonvillei,* the whole aerial mycelium disintegrates completely into smooth long spores (Williams et al., 1974). The results of the fatty acid and menaquinone analyses, as well as the molecular genetic data, confirmed the separation of *No-*

Table 3. Chemical markers used to differentiate *Nocardiopsis* from other actinomyces that produce conidia or arthrospores on aerial mycelia.

	Cell wall type[a]	Muramic acid type[b]	Diagnostic sugar	Sugar type[c]	Mycolic acid[d]	GC content (mol%)	Diagnostic phospholipid[g]	Phospholipid type[h]	Principal menaquinone[i]	Menaquinone type[j]	Fatty acid type[k]
Nocardiopsis	III	Acetyl	None	C	None	64–69	PC, PME	III	10/4, 10/6	4c2	3d
Saccharothrix	III	Acetyl	Rhamnose, galactose	E	None	70–76	OH-PE, l-PE	II	9/4	2D	3f
Streptoalloteichus[e]	III	Acetyl	Rhamnose, galactose	E	None	–	PE, PI, DPG	II	9/6, 10/6	3C	nd
Actinosynnema[f]	III	Acetyl	Rhamnose, galactose	E	None	71	PE, OH-PE	II	9/4, 10/4	3B	3f
Actinomadura	III	Acetyl	Madurose, galactose	B	None	66–70	PI, PIM, DPG	I	9/4, 9/6, 9/8	4B2	3a
Microtetraspora	III	Acetyl	Madurose, galactose	B	None	66–69	PGN, OH-PE	IV	9/0, 9/2, 9/4	4A2	3c
Glycomyces	II	Glycolyl	Xylose, arabinose, galactose	D	None	71–73	PI, PIM, DPG	I	9/4, 10/4	3B	2c
Streptomyces	I	Acetyl	Various	n.c.	None	66–75	PE, OH-PE var.	II	9/4, 9/6, 9/8	4B1	2c
Kitasatosporia	I	Acetyl	Rhamnose, galactose	E	None	67–73	PE	II	9/4, 9/6, 9/8	4B1	2c
Sporithya	I	Acetyl	None	n.c.	None	–		n.d.	9/6	4B1	3a
Nocardioides	I	Actyl	None	n.c.	None	66	PG, OH-PG, DPG	I	8/4	2B	3c
Amycolatopsis	IV	Acetyl	Arabinose, galactose	A	None	66–69	PE, OH-PE	II	9/2, 9/4	4A1	3f
Kibdellosporangium[e]	IV	Acetyl	Arabinose, galactose	A	None	66	PE, OH-PE	II		nd	3f
Pseudoamycolata	IV	Acetyl	Arabinose, galactose	A	None	72	PE, PME	II	8/4	2B	3e

continued

Table 3. Continued

	Cell wall type[a]	Muramic acid type[b]	Diagnostic sugar	Sugar type[c]	Mycolic acid[d]	GC content (mol%)	Diagnostic phospholipids[g]	Phospholipid type[h]	Principal menaquinone[i]	Menaquinone type[j]	Fatty acid type[k]
Actinopolyspora	IV	Acetyl	Arabinose, galactose	A	None	64	PC, PE	III	9/4, 9/6	4B2	2e
Saccharopolyspora	IV	Acetyl	Arabinose, galactose	A	None	77	PC, 1-PE, OH-PE, PME	III	9/2, 9/4, 10/4	3B/4A	3e
Pseudonocardia	IV	Acetyl	Arabinose, galactose	A	None	79	PC, 1-PE, OH-PE	III	8/4	2B	3f
Amycolata	IV	Acetyl	Arabinose, galactose	A	None	68–71	PC, PME	III	8/4	2B	3e
Nocardia	IV	Glycolyl	Arabinose, galactose	A	Nocardomycolic	64–72	PE	II	8/4c	2E	1b

[a]Cell wall types are as given in H. A. Lechevalier and M. P. Lechevalier (1970); cell wall types correspond to the following peptidoglycan types (Schleifer and Kandler, 1972): I, A3γ; and II, III, and IV, A1γ.

[b]Taken from Uchida and Aida (1977).

[c]Whole cell sugar types are as given in M. P. Lechevalier and H. A. Lechevalier (1970). n.c., not characteristic; diagnostic sugars may be present, but have no taxonomic implication in this cell wall type.

[d]Taken from Minnikin et al. (1975).

[e]Sporangiospores are found in addition.

[f]Flagellated spores are released in aqueous environment.

[g]Abbreviations for phospholipids: PC, phosphatidylcholine; PE, phosphatidylethanolamine; OH-PE, hydroxy-PE; 1-PE, lyso-PE; PME, phosphatidylmethylethanolamine. In addition, other phospholipids that are not diagnostic may also be found: DPG, diphosphatylglycerol; PG, phosphatidylglycerol; PI, phosphatidylinositol; and PIM, phosphatidylmannosides. OH-PE var., hydroxy-PE, is not present in all streptomycetes species; the occurrence of hydroxy-PE, is diagnostic for some *Streptomyces* species, i.e., *S. coelicolor*, *S. rimosus*, *S. violaceoniger* (Kawanami et al., 1969; R. M. Kroppenstedt, unpublished observations).

[h]Phospholipid types according to Lechevalier et al., (1977).

[i]10/4, MK-10(H₄), i.e., a menaquinone having two of the 10 isoprene units saturated; 8/6c, MK-8(H₄ ω cylic). From Collins et al. (1988); Howard et al. (1986).

[j]Menaquinone types according to Kroppenstedt (1985) are exemplified by the following: type 1, no hydrogenation of the isoprenoid chain, e.g., MK-9; type 2, only one menaquinone present, di- or tetrahydrogenated, e.g., MK-9(H₄); type 3, a tetrahydrogenated multiprenyl menaquinone, e.g., MK-9(H₄) and MK-10(H₄); type 4, menaquinones with the same isoprenoid chain length but increasing degree of hydrogenation, e.g., MK-9(H₄), MK-9(H₆), and MK-9(H₈); positions of double bonds in isoprenoid chain is diagnostic, e.g., MK-9(II,III-H₄) vs. MK-9(III,VIII-H₄) 4A1 vs. 4A2; and MK-9(II,III,IX-H₆) vs. MK-9(II,III,VIII-H₆) 4B1 vs. 4B2. From Batrakov and Bergelson (1978); Collins et al. (1988); Yamada et al. (1972a, 1972b).

[k]Fatty acid types according to Kroppenstedt (1985); for further information, see Table 4.

cardiopsis dassonvillei and *Actinomadura* into two genera. (Athalye et al., 1984; Alderson et al., 1985; Fischer et al., 1983). Preobrazhenskaya et al. (1982) transferred four additional *Actinomadura* species into the genus *Nocardiopsis*, i.e., *A. africana, A. coeruleofusca, A. flava,* and *A. longispora.* However, subsequent chemotaxonomic studies and partial oligonucleotide sequencing of ribosomal RNA from *Nocardiopsis* and *Actinomadura* revealed that these species did not belong either to *Nocardiopsis* or *Actinomadura* and that both genera are heterogeneous (Goodfellow et al., 1988; Grund and Kroppenstedt, 1990; Poschner et al., 1985). The heterogeneity of these taxa was underlined by the results of numerical taxonomic studies that were based on phenotypic characters (Alderson et al., 1984; Athalye et al., 1984; Goodfellow and Alderson, 1979).

In the meantime, Labeda et al. (1984) described a nocardioform actinomycete which had been isolated from a soil sample collected in Australia. Because of its unique combination of chemical markers, they created for this strain a new genus, *Saccharothrix.* Although this isolate resembled *Nocardiopsis dassonvillei* morphologically and in some chemotaxonomic characters, i.e., it is cell wall chemotype III/C, Labeda et al. (1984) did not classify this isolate in the genus *Nocardiopsis* because it has different phospholipids (i.e., type II) and because it had the menaquinones MK-9 (H_4) + MK-10 (H_4). In addition they detected the sugar rhamnose in whole cell hydrolysates of all *Saccharothrix* strains. This sugar was never found in *Nocardiopsis* (Fischer et al., 1983; Grund and Kroppenstedt, 1989, 1990) but is present in *Streptoalloteichus hindustanicus* (Goodfellow, 1989). This separation of *S. australiensis* the type species of the genus *Saccharothrix* from *N. dassonvillei* has been validated by molecular genetic data. It has been shown by 16S rRNA sequencing results that *Nocardiopsis* occupies a phylogenetic position between *Microtetraspora* and *Streptomyces* (Goodfellow et al., 1988; Kroppenstedt et al., 1990), whereas *S. australiensis* is closely related to members of the family Pseudonocardiaceae (Bowen et al., 1989). The results of chemotaxonomic study on nearly all the type strains of *Nocardiopsis* and related genera showed that six of the eight *Nocardiopsis* species were misclassified (Grund and Kroppenstedt, 1989, 1990). Five of the species, *N. coeruleofusca, N. flava, N. longispora, N. mutabilis,* and *N. syringae,* belong to the genus *Saccharothrix,* and one, *N. africana,* to the genus *Microtetraspora* (Table 1; Goodfellow et al. 1988; Kroppenstedt et al. 1990; Poschner et al., 1985). Phenetic data indicated that the remaining single member species, *N. antarcticus,* is identical to the type species for *N. dassonvillei* (Meyer, 1989). DNA-DNA hybridization data and numerical analyses using fatty acid results and phenetic data showed that all tested strains of *N. dassonvillei* belong to one genus but should be divided into two species (Fisher et al., 1983; Grund, 1987). Based on these results and some additional studies, a new species, *N. alba,* was created for those strains which fell in a separate cluster apart from *N. dassonvillei.* This study showed in addition that *N. dassonvillei* subsp. *prasina* was actually closer to *N. alba.* It was transferred to *N. alba,* with the revision of its name to *N. alba* subsp. *prasina.* An investigation of the A_2 pm ("meso-DAP") streptomycetes (Pridham and Lyons, 1968; Shirling and Gottlieb, 1972) revealed that two members of this group, *"Streptomyces alborubidus"* and *"Streptomyces listeri,"* belonged to *Nocardiopsis* and they were therefore reclassified into this genus, as *N. alborubidus* and *N. listeri* (Grund and Kroppenstedt, 1990).

Habitat

The natural habitat of *Nocardiopsis* is the soil. In screening programs searching for new antimicrobial products or other metabolites, members of this genus are frequently isolated from soil together with streptomycetes and some other actinomycetes (Mishra et al., 1987a). The isolation of *N. dassonvillei* from other sources is reported from time to time. The first description of *Nocardiopsis dassonvillei (Streptothrix dassonvillei)* by Brocq-Rousseau (1904) was made on strains isolated from mildewed grain and fodder. Later, Liegard and Landrieu (1911) reported a case of ocular conjunctivitis from which they could isolate a microorganism matching the description of Brocq-Rousseau's strain. They called this strain *Nocardia dassonvillei* because of its nocardia-like appearance. Lacey (1977) detected *N. dassonvillei* in cotton waste and occasionally in hay. In a study on *Nocardiopsis* and related organisms, Gordon and Horan (1968) found that 15 of their 26 strains were clinical isolates; only nine of these strains were collected from soil samples. Although most of the cultures originated from human or animal infections, Gordon and Horan (1968) speculated that such cultures might usually be discarded in medical laboratories as contaminants because of their macroscopic resemblance to cultures of *Streptomyces griseus.* This opinion was supported by the findings of Mishra et al. (1988) who could isolate 22 *Nocardiopsis* strains from soil samples. Abyzov et al. (1983) isolated a single *Nocardiopsis* strain from an Antarctic glacier at a depth horizon of 85 m, which corresponded to an age of over 2,000 years. This isolate was named *N. antarcticus* reflecting its origin. Although this culture was isolated from a cold region, it is a mesophilic species with an optimal growth temperature of 37°C and an inability to grow at 4°C.

In a screening program for alkalophilic bacteria from soils, Mikami et al. (1982) isolated 20 actinomycetes that grew at pH 11.5. Nine of these isolates resembled *N. dassonvillei* morphologically and had the same III/C cell wall chemotype. These strains were later classified as *N. dassonvillei* subsp. *prasina* (Miyashita et al., 1984). At the same time, Mikami et al. (1985) screened the 420 International Streptomycetes Project (ISP) streptomycetes for alkalophilic strains and found that only six of the 420 strains were able to grow at a pH of 11.5 or above. Cell wall analyses of the alkalophilic strains revealed that half of these strains had been misclassified. One strain (ISP 5011[T], DSM 40011[T]) was later classified as *Amycolata autotrophica* (Lechevalier et al., 1986), another (ISP 5465[T], DSM 40465[T]) as *Nocardiopsis alborubida* (Grund and Kroppenstedt, 1989), and another strain *Streptomyces caeruleus* (ISP 5103[T], DSM 40103[T]) showed a close relationship in its chemical markers to *Saccharothrix* (Grund, 1987). Other alkalophilic soil isolates that grew at pH 10 were identified by Tsujibo et al. (1988) as *N. dassonvillei*. Based on these findings, it can be concluded that soil is the natural habitat of *Nocardiopsis* and that mildly alkaline conditions (pH 8) are best for optimal growth. For some *N. dassonvillei* strains (Tsujibo et al., 1988) and other species of this genus, i.e., *N. alba* subsp. *prasina* (Miyashita et al., 1984) and *N. alborubida* (Mikami et al., 1982, 1985), the pH has to be even higher (pH 10) to get acceptable growth rates. When alkalophilic strains are inoculated on agar plates at the lowest possible pH for their growth, they begin to grow slowly by changing the pH of the plate until it reaches their optimal pH (Horikoshi and Akiba, 1982; Mikami et al., 1985). There is strong evidence that strains isolated under alkalophilic conditions belong to the genus *Nocardiopsis*.

Cultivation

Nocardiopsis dassonvillei grows very well at 28°C on most agar media used in the International Streptomyces Project (ISP) (Shirling and Gottlieb, 1966). After 14–21 days of incubation under optimal conditions, the color of the white aerial mycelium will change to yellow or gray, indicating the production of spore chains. For optimal growth of *N. alba* subsp. *alba* and *N. antarctica,* the pH of the ISP media should be adjusted to pH 8.5 with sterile sodium carbonate after autoclaving. Although *N. listeri* grows quite well on the ISP media, no aerial mycelium is produced. Hickey-Tresner agar is the only medium where sparse white aerial mycelia are seen. The truly alkalophilic species, i.e. *N. alborubida* and *N. alba* subsp. *prasina* need an even higher pH of 10 to produce aerial mycelium and spores on yeast extract-malt extract agar. The pH of 10 and above has to be adjusted with sodium hydroxide after autoclaving. The following media are used for cultivating most species of the genus *Nocardiopsis:*

ISP Medium 2: Yeast Extract-Malt Extract Agar (Pridham et al., 1956–1957)

Yeast extract (Difco)	4.0 g
Malt extract (Difco)	10.0 g
Dextrose	4.0 g
Distilled water	1.0 liter
Agar	20.0 g

Adjust to pH 7.3.

ISP Medium 3: Oatmeal Agar (Küster, 1959a)

Oatmeal	20.0 g
Agar	18.0 g

Cook or steam 20 g oatmeal in 1 liter of distilled water for 20 min. Filter through cheese cloth. Add distilled water to restore volume of filtrate to 1 liter. Add trace salt solution. Adjust to pH 7.5 with NaOH. Add 18 g agar; liquefy by steaming at 100°C for 15–20 min.

Trace salt solution:

$FeSO_4 \cdot 7H_2O$	0.1 g
$MnCl_2 \cdot 4H_2O$	0.1 g
$ZnSo_4 \cdot 7H_2O$	0.1 g
Distilled water	100.0 ml

ISP Medium 4: Inorganic Salts-Starch Agar (Küster, 1959b)

Solution I: Difco soluble starch, 10.0 g. Make a paste of the starch with a small amount of cold distilled water and bring to a volume of 500 ml.

Solution II:

K_2HPO_4 (anhydrous)	1.0 g
$MgSO_4 \cdot 7H_2O$	1.0 g
NaCl	1.0 g
$(NH_4)_2SO_4$	2.0 g
$CaCO_3$	2.0 g
Distilled water	500 ml
Trace salt solution	1.0 ml

Adjust to pH 7.5. Mix solutions I and II together. Add 20 g of agar. Liquify agar by steaming at 100°C for 10–20 minutes.

ISP Medium 5: Glycerol-Asparagine Agar
(Pridham and Lyons, 1961)

L-Asparagine (anhydrous)	1.0 g
Glycerol	10.0 g
K₂HPO₄ (anhydrous)	1.0 g
Distilled water	1.0 liter
Trace salt solution	1.0 ml

Adjust to pH 7.4. Add 20 g of agar.

ISP Medium 7: Tyrosine Agar Medium (Shinobu, 1958)

Glycerol	15.0 g
L-Tyrosine	0.5 g
L-Asparagine	1.0 g
K₂HPO₄ (anhydrous)	0.5 g
MgSO₄ · 7H₂O	0.5 g
NaCl	0.5 g
FeSO₄ · 7H₂O	0.01 g
Distilled water	1.0 liter
Trace salt solution	1.0 ml

Adjust to pH 7.5. Add 20 g of agar.

Hickey-Tresner Agar Medium (Hickey and Tresner, 1952)

Dextrin	10.0 g
Tryptone	2.0 g
Meat extract	1.0 g
Yeast extract	1.0 g
CoCl₂	2.0 g
Distilled water	1.0 liter
Agar	15.0 g

Preservation

For short-term preservation, sporulated *Nocardiopsis* strains may be kept on suitable agar slants at 4°C and transferred every four months. To prevent the agar from drying, the tubes are tightly sealed with silicone stoppers. For longer preservation, spore suspensions or homogenized mycelia are mixed with glycerol to a final concentration of 25% and kept at −25°C (Wellington and Williams, 1978). Zippel and Neigenfind (1988) tested the viability, stability of auxotrophic markers, and antibiotic production of strains kept at 25°C in 45% glycerol. No drastic changes of the markers could be noticed after one year. Good results have been obtained for nonsporulating *Nocardiopsis* strains by this method because no freezing of the suspension occurs at −25°C using the high concentration of glycerol. For long-term preservation, lyophilization of spores and mycelia suspended in 10% skim milk is a convenient method. Alternatively, a very simple, reliable, and time-saving method is N₂ cryopreservation of living cells in small polyvinyl chloride (PVC) tubes ("straws") at −196°C. This method has been described by Hoffmann (1989a, 1989b) for long-term preservation of *Basidiomycetes*. The pro-cedure has been tested for actinomycetes, including *Nocardiopsis*. This method proved to be very useful for actinomycetes, including *Nocardiopsis* and will therefore be described briefly. *Nocardiopsis* strains are harvested from well-sporulated cultures grown on suitable agar media in petri dishes. A 2 × 25mm piece of sterile PVC tube is pressed into the mycelia and the agar and carefully raised, excising the agar plug. This is repeated until the tube is filled with agar. The filled tubes are placed in a sterile cryo vial with the screw cap marked with the strain accession number. With a given volume of 1.8 ml, one vial will hold up to 13 tubes. Two vials are prepared for each strain and then fixed to a metal clamp for freezing in the gas phase of the liquid nitrogen container. After 10–15 minutes, the temperature falls below −130°C, and the clamp can be positioned into the liquid phase at −196°C. A container with a capacity of 250 liters will hold at least 8,000 vials or 4,000 strains.

For viability testing, one tube is removed from the vial within the gas atmosphere of the container and placed directly on a suitable agar medium for thawing. After a few days incubation, mycelium will be visible. For those strains that do not produce abundant mycelium, the plugs may be pushed out of the tubes by a sterile needle.

Isolation

For the isolation and enrichment of *Nocardiopsis* strains, no specific procedures have been described (Meyer, 1989). Usually the same methods can be used as those employed for the isolation of streptomycetes (Kutzner, 1981; see also Chapter 41). Good results have been obtained using the GAC-agar medium of Nonomura and Ohara (1971). The recipe for this medium is given below. The following special method for the isolation of alkalophilic *Nocardiopsis* strains has been reported by Horikoshi (1971): A small amount of soil is suspended in 1 ml of sterilized water. 100 μl aliquots of the suspension are spread on dry agar plates (recipe given below). The plates are incubated for 7 to 14 days at 27°C. Spores or mycelia of isolated colonies are picked up on a needle and streaked on various cultivation media for color determination and for macroscopic, microscopic, and chemotaxonomic examination.

GAC-Agar Medium (Nonomura and Ohara, 1971)

Solution A:	
Glucose	1.0 g
L-Asparagine	1.0 g

K₂HPO₄	0.3 g

K_2HPO_4 — 0.3 g
$MgSO_4 \cdot 7 H_2O$ — 0.3 g
NaCl — 0.3 g
Distilled water — 800 ml
Trace salts — 1.0 ml
Agar — 15.0 g

Solution B:
Casamino Acids   0.5 g/l

Antibiotics (cycloheximide, nystatin, 50µg/ml; poly-myxin B, 4µg/ml; penicillin, 0.8µg/ml) may be added for isolation. Four ml of solution B is poured over 15 ml of hardened solution A in a plate.

Trace salt solution:
$FeSO_4 \cdot 7H_2O$ — 10 mg/ml
$MnSO_4 \cdot 7H_2O$ — 1 mg/ml
$CuSO_4 \cdot 5H_2O$ — 1 mg/ml
$ZnSO_4 \cdot 7H_2O$ — 1 mg/ml

Adjust to pH 8.

Horikoshi Agar for Alkalophilic *Nocardiopsis* Strains (Horikoshi, 1971)

Glucose — 10.0 g
Peptone — 5.0 g
Yeast extract — 5.0 g
K_2HPO_4 — 1.0 g
$MgSO_4$ — 0.2 g
$NaCO_3$ — 10.0 g
Distilled water — 1.0 liter
Agar — 20.0 g

Adjust to pH 8. Sodium carbonate is sterilized separately and added to the autoclaved medium.

Identification

Macroscopic and microscopic features of *Nocardiopsis* strains may be used for presumptive identification. However, the final classification has to be proven by chemical markers. Most *Nocardiopsis* strains produce a sparse-to-abundant white aerial mycelium that in most species becomes yellow-gray (griseus-colored) during spore formation. The aerial hyphae are long, moderately branched, straight and flexuous (recti flexibilis). They show a typical zig-zag formation before spore delimitation. The elongated spores are smooth and differ in length. They are enclosed within a fibrillar sheath and have thickened polar walls.

Although *Nocardiopsis* strains resemble those of *Streptomyces griseus* macroscopically (Gordan and Horan, 1968), the two taxa can be differentiated by their differing types of spore formation and spores see (Table 2). In contrast to *Streptomyces* where spores of the same shape are delimited almost simultaneously, in *N. dassonvillii* the cross walls are formed in a relatively uncoordinated manner resulting in spores of various length. Another genus that has often

been confused with *Nocardiopsis* is *Actinomadura* (see Table 1). However, both genera differ in morphology and in their specific mode of spore formation. In *Actinomadura*, special spore-forming hyphae form on a sterile aerial mycelium, either singly or in clusters, producing chains or spores showing spirals with 1.5 to 2 turns, whereas in *Nocardiopsis*, special spore-forming hyphae are absent. In this latter genus, the straight, spiralled aerial mycelium disintegrates completely into fragmentation spores (Table 2). However, it is more difficult to differentiate *Nocardiopsis* from *Glycomyces* on morphological characters alone. Slight differences may be found in the formation of substrate and aerial mycelium between species of these two genera. Labeda et al. (1985) reported stable substrate mycelia and scant aerial mycelia carrying short chains of square-ended conidia in *Glycomyces*, whereas in many *Nocardiopsis* strains the substrate mycelium fragments very early depending on strain and culture conditions. In addition most strains of the latter one develop abundant aerial mycelium with long hyphae that fragment into chains of conidiaspores completely (Meyer, 1989). No dissimilarities could be found in morphological characters to *Saccharothrix* (Labeda et al., 1984).

For a definite identification of *Nocardiopsis* strains, the determination of chemotaxonomic markers is essential. The combination of markers that are listed in Table 3 are diagnostic for this taxon. *Nocardiopsis* has the cell wall chemotype III/C, i.e., it has meso-diaminopimelic acid, alanine, and glutamic acid in its cell walls. In whole cell hydrolysates, glucose and galactose are also detected (type C). No diagnostic sugars like arabinose, madurose, or rhamnose have been found (Grund and Kroppenstedt, 1990; Lechevalier and Lechevalier, 1970a). The muramic acid of the cell wall is acetylated (Kroppenstedt, 1987). No mycolic acids are present in whole cell methanolysates (Mordarska et al., 1972; Minnikin et al., 1975).

The best diagnostic character for the differentiation of *Nocardiopsis* from all other actinomycetes taxa is their specific phospholipid pattern. The phospholipid pattern is composed of phosphatidylcholine (cardiolipin), phosphatidylmethylethanolamine (PME), phosphatidylglycerol (PG), phosphatidylinositol (PI), and small amounts of diphosphatidylglycerol (DPG). In addition, two spots of unknown glycolipids and 2–4 unknown phospholipids that run above DPG are detected on two-dimensional thin-layer chromatogram (TLC) plates (Fig. 1; Grund and Kroppenstedt, 1990; Mordarska et al., 1983; Minnikin et al., 1977). The

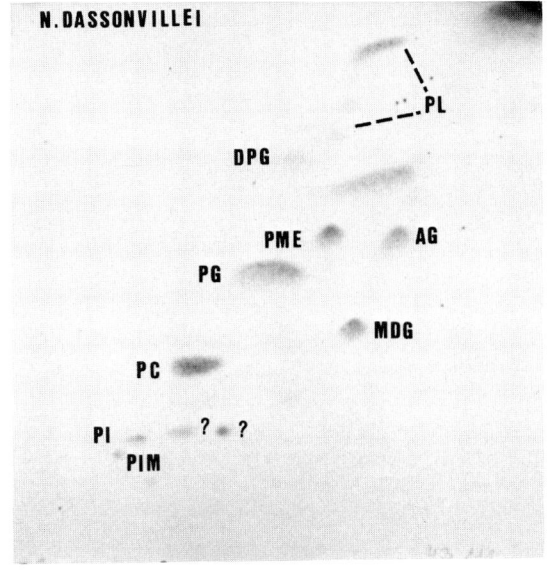

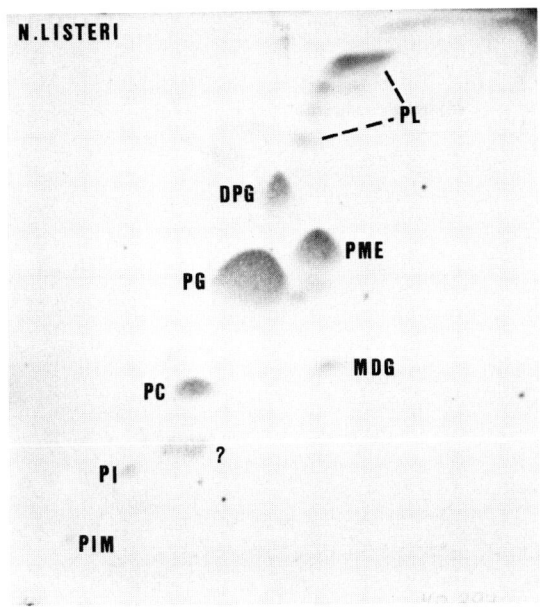

Fig. 1. Two-dimensional thin-layer chromatograms of polar lipids of *Nocardiopsis dassonvillei* DSM 43111 and *N. listeri* DSM 40297. Abbreviations: DPG, diphosphatidylglycerol; PME, phosphatidylmethylethanolamine; PG, phosphatidylglycerol; PC, phosphatidylcholine; PI, phosphatidylinositol; PIM, phosphatidylinositolmannosides; AG, monoacetylated glucose; MDG, monomannosyl diglyceride; PL, unknown phospholipids specific for *Nocardiopsis*; ?, lipids of unknown structure.

occurrence of phosphatidylcholine in lipid extracts of *Nocardiopsis* corresponds with phospholipid type III, according to Lechevalier et al. (1977). *Nocardiopsis* synthesizes menaquinones with an isoprene side chain of 10 units and depending on species up to four of these isoprene units are saturated MK-10 (H_0) to MK-10 (H_8). Small amounts of the MK-9 series are also

found (Fig. 2) (Fischer et al., 1983; Grund and Kroppenstedt, 1990; Minnikin et al., 1978).

Nocardiopsis synthesizes terminally branched and 10-methyl-branched fatty acids showing a chain length from 14 to 18 carbon atoms. Hydroxy fatty acids have never been detected in *Nocardiopsis* strains (Grund and Kroppenstedt, 1990). Among the terminally branched fatty acids, 14-methyl-heptadecanoic acid (iso-16:0; 21–35%) and 14-methyl-hexadecanoic acid (anteiso-17:0; 7–24%) are the main components. Substantial amounts (6–18%) of the 10-methyl branched tuberculostearic acid i.e. 10-methyl-octadecanoic acid (10-methyl-18:0) and its precursor the unsaturated cis 9,10 octadecenoic acid (18:1 cis; 2–19%) has been found in addition (Grund and Kroppenstedt, 1990). According to Kroppenstedt (1985) this fatty acid pattern belongs to the fatty acid type 3d. Table 4 lists the fatty acid types found among *Actinomycetales*. They can be combined into 3 main types and 14 subtypes (diagnostic fatty acids of the main types are surrounded by bold lines those of subtypes by thin lines). Type 1 showing unbranched saturated and unsaturated fatty acids and those fatty acids for which unsaturated fatty acids are precursors, i.e., cyclopropane and 10-methyl-branched fatty acids. Type 2 is composed of iso- and anteiso-branched fatty acids. In type 3 both types, the terminally branched (iso/anteiso) and the 10-methyl-branched fatty acids are present. The subtypes of types 2 and 3 are mainly based on quantitative differences of the fatty acids and the occurrence of hydroxy fatty acids.

Based on these markers (see Table 3), *Nocardiopsis* species are easily differentiated from other actinomycetes and may be identified at the genus level. The differences in fatty acid patterns of *A. madura* and *N. dassonvillei* were first reported by Agre et al. (1975) and Kroppenstedt and Kutzner (1976, 1978).

For rapid identification of *Nocardiopsis* strains, lipid analysis is very useful. The specificity of these markers is very high and may be used solely to differentiate species of this taxon from all other actinomycetes. Although *Nocardiopsis* species share phospholipid pattern III with four other actinomycete genera, i.e., *Amycolata*, *Faenia*, *Pseudonocardia*, and *Saccharopolyspora*, *Nocardiopsis* strains are easily differentiated from these genera because *Nocardiopsis* strains synthesize characteristic phospho- and glycolipids not found in the other taxa. Separating the lipid extract on TLC plates using the solvent system of Minnikin et al. (1984), 3–4 additional phospholipids with high R_f-value above DPG and two glycolipids right below PME can be detected on TLC plates (Fig.

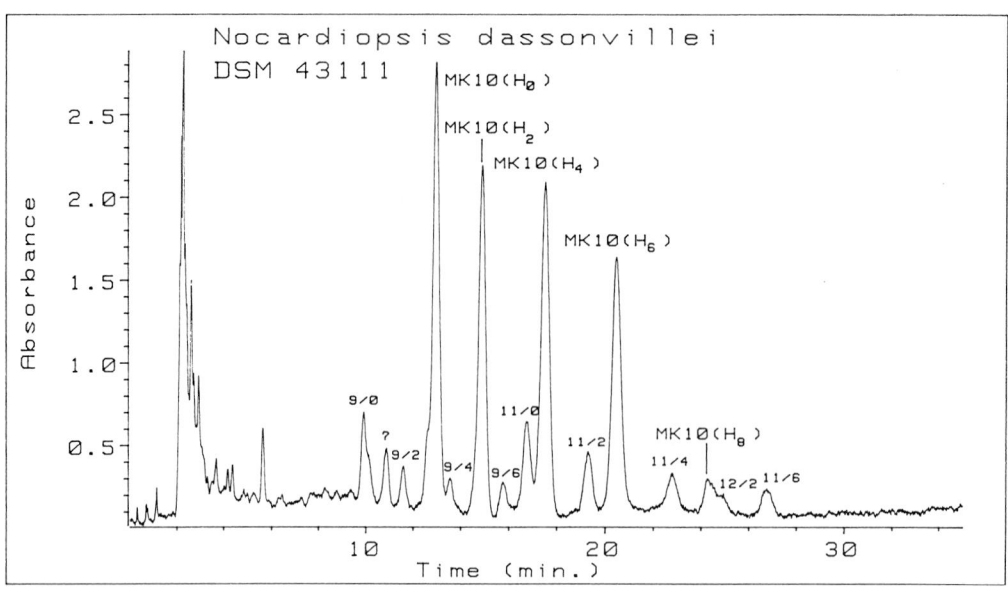

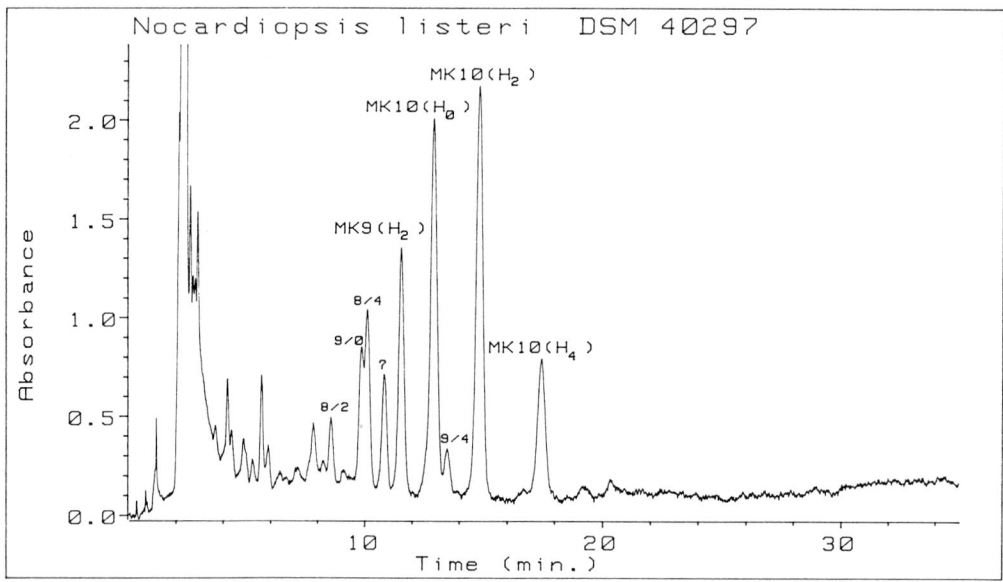

Fig. 2. Menaquinone profiles of *Nocardiopsis dassonvillei* (top) and *N. listeri* (bottom). The extent of hydrogenation of the 10 isoprene units is shown by the subscript of the abbreviation. For instance, MK-10(H$_8$) is a menaquinone with four hydrogenated isoprene units. Injection volume 5 μl; 5% of the extract prepared by the method of Minnikin et al. (1984) without further purification.

1; Grund, 1987; Minnikin et al., 1977). The two glycolipids were identified by Mordarska et al. (1983) as monomannosyl diglyceride and monoacetylated glucose respectively. The structures of the unknown phospholipids (PL) are still unidentified (Fig. 1). In addition, the genera with the above-listed phospholipid type, type III, may be further differentiated from *Nocardiopsis* by looking for specific phospholipids containing primary or secondary amino groups, i.e., phosphotidylethanolamine (PE) and PME in *Faenia;* lyso-PE, hydroxy-PE, and PE in *Pseudonocar-*dia; and lyso-PE, PME, and PE in *Saccharopolyspora*. In *Nocardiopsis,* only one ninhydrin-positive spot, PME, is present.

Nocardiopsis species show complicated menaquinone patterns containing menaquinones from MK-10 to MK-10 (H$_8$) (Fig. 2). Small amounts of the MK-9 and/or MK-11 series are also found. In *N. alborubidus* and *N. listeri* the highly saturated MK-10 (H$_6$) and MK-(H$_8$) are usually missing. They synthesize higher amounts of MK-9 (H$_4$) and can be separated from the other *Nocardiopsis* strains by this dif-

Table 4. The composition of the fatty acid types among the species of Actinomycetales.

Actinomycetales species	Type	Iso-15:0	Anteiso-15:0	Iso-16:0	16:0	10-Methyl 16:0	Iso-17:0	Anteiso-17:0	17:0	17:1	10-Methyl 17:0	Iso-18:0	18:0	18:1	10-Methyl 18:0	OH2	Cyclo 19
Corynebacterium[b]	1 a	−	−	−	+++	−	−	−	−	−	−	−	+	+++	−	−	−
Nocardia[c]	1 b	−	−	−	+++	++	−	−	−	−	+	−	+	++	+++	−	−
Actinomyces israelii	1 c	−	−	−	+++	−	−	−	−	−	−	−	+	++	−	−	++
Saccharomonospora	2 a	+	−	+++	++	−	+	+	+	+	−	+	−	−	−	++	−
Thermoactinomyces	2 b	+++	++	++	+	−	+++	++	−	−	−	−	−	−	−	(V)	−
Streptomyces[d]	2 c	++	+++	++	+	−	+	++	−	−	−	+	+	++	−	−	−
Actinoplanes	2 d	+++	++	+++	++	−	++	++	+	+	−	+	+	++	−	−	−
Actinomadura	3 a	−	−	++	+++	+	−	−	+	+	+	+	+	+	++	−	−
Micromonospora	3 b	+++	+	++	−	−	++	++	++	++	++	−	−	+	−	−	−
Microtetraspora	3 c	++	+	+++	+	+	+	+	++	+	++	+	+	+	+++	++	−
Nocardiopsis	3 d	+	+	+++	+	−	+	+++	+	+	+	+	+	++	+	−	−
Amycolata	3 e	++	−	+++	+	−	+	+	−	−	+	−	−	−	−	−	−
Amycolatopsis	3 f	++	+	+++	+	−	+	+	++	++	+	−	−	−	−	+	−
Saccharothrix	3 g	++	+	+++	+	−	+	++	+	+	+	−	−	−	−	+	−

[a] Diagnostic fatty acids of main types are surrounded by bold lines, subtypes by thin lines. Symbols: +, 1–5%; ++, 5–15%; +++, 15–25%; ++++, >25%; (V), variable, usually less than 2% for one component.

[b] Except C. ammoniagenes and C. bovis which belong to type 1b.

[c] Including Gordona, Mycobacterium, Rhodococcus, and Tsukamurella.

[d] Occurrence of hydroxy fatty acids is species specific among Streptomyces. Branched 2-hydroxy fatty acids could be found in all tested strains of S. coelicolor (30 strains), S. rimosus (14 strains), S. violaceusniger (18 strains) and in 20 of 27 strains of S. hygroscopicus. No hydroxy fatty acids could be detected in S. albus (33 strains), S. fradiae (25 strains), S. glaucescens (8 strains), S. griseus (22 strains), S. violaceoruber (16 strains), S. viridochromogenes (25 strains), Y. Zhang and R. M. Kroppenstedt, unpublished observations.

ference (Fig. 2). Although menaquinone MK-10 (H_4) is found in some other sporoactinomycetes, the latter can be distinguished because here MK-10 (H_4) always occurs in a different combination, either together with MK-9 (H_4), as in *Faenia* and *Saccharothrix* or in combination with high amounts of MK-10 (H_8) and MK-11 (H_6), as in some *Thermomonospora* strains.

The fatty acid patterns of *Nocardiopsis* provide a large amount of taxonomic information. *Nocardiopsis* strains are easily separated from all other actinomycetes on the basis of their specific fatty acid patterns (Table 4; Fischer et al., 1983; Poschner et al., 1985). For differentiation from other taxa, both qualitative and quantitative differences of the fatty acid patterns are useful (Grund and Kroppenstedt, 1989, 1990). To compare the huge amount of information present in the fatty acid patterns, numerical methods and the aid of computers have to be applied. For comparison of chemotaxonomic data, principal component analyses can be applied, using the data of the matrix of Euclidian distances of the fatty acids. Figure 3 shows a two-dimensional plot based on principal component analyses of fatty acid results of 58 species belonging to the genera *Nocardiopsis, Actinomadura, Microtetraspora* and *Saccharothrix*. A clear-cut separation of all four genera is obtained based solely on the differences in fatty acids pattern (R.M. Kroppenstedt, unpublished observations).

Nocardiopsis species may be readily identified by comparing the color of the aerial and substrate mycelium, the growth at different pH, the pattern of carbon-source utilization, and some other physiological tests (Table 5). Most *Nocardiopsis* species produce white aerial mycelium, which becomes yellowish or grayish in most but not all strains after two to three weeks. *N. alba* subsp. *prasina* develops, on mineral starch agar and oatmeal agar, a leek-green aerial mycelium, whereas isolated colonies of *N. antarctica* and *N. alborubida* become gray after three to four weeks. *N. alborubida* is easily distinguished from all other *Nocardiopsis* strains by the red-orange color of its substrate mycelium, whereas all other strains show a nonspecific yellow-brown substrate mycelium. *N. alba* subsp. *prasina* and *N. alborubida* are alkalophilic strains that grow up to pH 12 but not at pH 6. Their optimal pH for growth is pH 10. The other *Nocardiopsis* species are alkali tolerant with a lower pH optimum for growth (Table 5). *N. alborubida* occupies an isolated position in physiological tests. This species may be differentiated from all other *Nocardiopsis* species by its ability to degrade quinate, ben-

zoate, and hydroxybenzoate, but not cellobiose, adonitol, inositol, or xanthine. The strains of the type species *N. dassonvillei* grow on malonate and xylose. The ability to degrade xylose is shared by *N. antarcticus* and *N. listeri*. *N. listeri* and *N. antarcticus* are differentiated because the former can grow on rhamnose and does not produce aerial mycelium on all ISP media, whereas *N. antarcticus* does not grow on rhamnose and produces abundant aerial mycelium on nearly all media. The subspecies *N. alba* subsp. *prasina* is distinguished from *N. alba* subsp. *alba* by its higher pH optimum for growth, the absence of an alkaline reaction when grown on malonate, and the inability to use sucrose as a sole source of carbon.

Properties of Soil Isolates

When microorganisms are isolated from soil using alkaline conditions a high percentage of *Nocardiopsis* strains are usually found. Miyashita et al. (1984) reported that under these conditions half of the actinomycetes strains belong to the genus *Nocardiopsis*. Table 5 shows that this taxon harbors many alkalophilic or at least alkalitolerant species. As expected, the alkalophilic strains synthesize many different alkaline-tolerant enzymes (Horikoshi and Akiba, 1982). Mikami et al. (1985) examined alkalophilic amylases from these strains and found that they all had a pH optimum of about 10. In contrast, the neutrophilic *N. dassonvillei* strains, which grow better at pH 7.5, synthesize amylases with a pH optimum of 8.

In a screening program searching for microbial metabolites with interesting properties, Mishra et al. (1987a) isolated 942 microorganisms from soil of which 832 were aerobic actinomyces, most of them (302) *Streptomyces*. The residual 530 isolates were distributed among 17 different actinomycete genera, including 22 that were classified as *Nocardiopsis*. All isolates were tested for the production of specific metabolites. The *Nocardiopsis* strains showed no effect against the tested algae or higher plants (Mishra et al., 1987a) but six of the 22 *Nocardiopsis* isolates showed specific herbicidal properties. Four strains inhibited the germination of cress seeds and two effected seeds of barnyard grass (Mishra et al., 1988). Bioassays for insecticidal properties from these strains were performed using mosquito larvae of *Aedes aegypti* and for nematicidal activities using the free-living nematode *Penagrellus revidivus*. One *Nocardiopsis* strain showed insecticidal properties but no nematicidical proper-

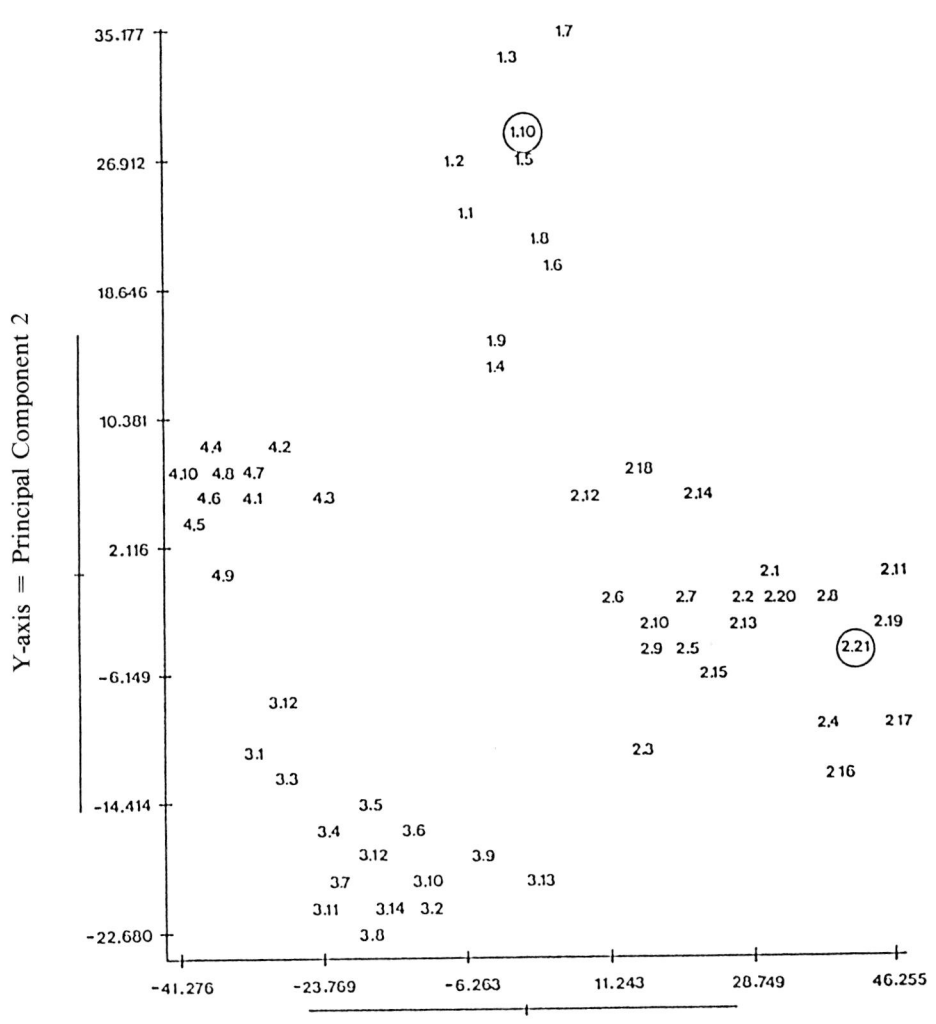

X-axis = Principal Component 1

Fig. 3. Two-dimensional plot based on principal component analyses, using the Euclidean distances of the fatty acids of 58 species from the genera *Nocardiopsis, Actinomadura, Microtetraspora,* and *Saccharothrix*. The numbers on the graph indicate the fatty acid group and the names of the various species are as follows: Mean − X = −0.064. Mean − Y = 0.193. Standard deviation − X = 26.045. Standard deviation − Y = 15.206. **1. Nocardiopsis.** 1.1 N. alborubida, 1.2 N. listeri, 1.3 N. dassonvillei 43111, 1.4 N. dassonvillei 43120, 1.5 N. dassonvillei 43379, 1.6 N. dassonvillei 43119, 1.7 N. dassonvillei 43235, 1.8 N. dassonvillei 43377, 1.9 N. dassonvillei 43378, 1.10 N. antarcticus / N. alba ssp. prasina. **2. Actinomadura.** 2.1 A. madurae, 2.2 A. malachitica, 2.3 A. livida, 2.4 A. macra, 2.5 A. atramentaria, 2.6 A. echinospora, 2.7 A. libanotica, 2.8 A. coerulea, 2.9 A. kijaniata, 2.10 A. verrucosospora, 2.11 A. vinacea, 2.12 A. umbrina, 2.13 A. oligospora, 2.14 A. aurantiaca, 2.15 A. luteofluorescens, 2.16 A. yumaensis, 2.17 A. fulvescens, 2.18 A. pelletieri, 2.19 A. cremea 43936, 2.20 A. viridis, 2.21 A. cremea 43676 / A. citrea. **3. Microtetraspora.** 3.1 M. niveoalba, 3.2 M. africana, 3.3 M. fusca, 3.4 M. salmonea, 3.5 M. rubra, 3.6 M. helvata, 3.7 M. glauca, 3.8 M. pusilla, 3.9 M. roseoviolacea, 3.10 M. ferruginea, 3.11 M. spiralis, 3.12 M. fastidiosa, 3.13 M. angiospora, 3.14 M. aurantiaca, 3.15 M. recticatena. **4. Saccharothrix.** 4.1 S. flava, 4.2 S. coeruleofusca, 4.3 S. longispora, 4.4 S. syringae, 4.5 S. coerulea, 4.6 S. flava, 4.7 S. aerocolonigenes 40034, 4.8 S. aerocolonigenes 43385, 4.9 S. mutabilis, 4.10 S. australiensis. (Methods described by Eorolen and Lehtonen, 1988, and O'Donnell, 1985.)

ties could be found among any of these *Nocardiopsis* isolates (Mishra et al., 1987b). The growth of the fungus *Fusarium oxysporum* f. sp. *albedensis,* a pathogen of date palm, was inhibited by a *Nocardiopsis* strain (Sabou and Bounaga, 1987). The growth of some other soil fungi was not affected by *Nocardiopsis dassonvillei* metabolites.

Kase et al. (1986) isolated, from Japanese soil, a *Nocardiopsis* strain, K-252, which produced a potent new protein kinase C inhibitor designated K-252a (Kase et al., 1986). Three

Table 5. Differentiation characteristics of *Nocardiopsis* species.[a]

Character	Species					
	N. dassonvillei	*N. alborubida*	*N. alba* subsp. *alba*	*N. alba* subsp. *prasina*	*N. antarcticus*	*N. listeri*
Color of aerial mycelium[a]	White to yellowish grayish (griseus)	Gray	White to yellowish grayish (griseus)	White to yellowish grayish (griseus)	Gray	White on HT agar
Color of substrate mycelium	Yellowish to brownish	Yellow to orange	Yellowish to brownish	Yellow	Dark brown	Yellow
Presence of melanin	V	−	−	−	+[b]	−
pH for optimal spore formation	7	10	9	10	8	8
Decarboxylation of:						
Quinate	−	+	−	−	NT	−
Malonate	+	+	−	−	NT	−
Degradation of:						
Benzoate	−	+	−	−	NT	−
p-Hydroxybenzoate	−	+	−	−	NT	−
Decomposition of:						
Hypoxanthine	+	−	+	+	NT	−
Ca oxalate	(+)	+	−	−	NT	−
Xanthine	+	−	+	+	NT	+
Growth on:						
Xylose	+	+	−	−	+	+
Melezitose	−	+	−	−	−	−
Cellobiose	+	−	+	+	NT	+
Rhamnose	+	+	−	−	−	+
Sucrose	+	+	+	−	+	+
Adonitole	−	+	−	−	−	−
Inositole	−	+	−	−	−	−

V, variable results; NT, not tested; (+), most but not all of the tested strains show positive results.

[a]The typical color of the aerial mycelium (spores) is only obtained under optimal pH conditions.

[b]Abyzov et al. (1983); no melanin production on peptone-iron agar and tyrosine-agar (R. M. Kroppenstedt, unpublished observations).

Taken from Athalye et al. (1985); Grund and Kroppenstedt (1989, 1990, unpublished); Mikami et al. (1985); Miyashita et al. (1984), Shirling and Gottlieb (1972).

structurally related compounds, K-252b, c, and d, were later found in the culture broth of another *Nocardiopsis* isolate, K-290 (Nakanishi et al., 1986). The antiinflammatory and antiallergic effects of the metabolite K-252a were described by Ohmori et al. (1988). In addition, the metabolite K-252a inhibits the light chain kinase of smooth muscle myosin (Nakanishi et al., 1988) and the action of nerve growth factor on PC12 cells (Koizumi et al., 1988; Matsuda and Fukuda, 1988).

Phenazine antibiotics could be extracted from the mycelium of an alkalophilic *Nocardiopsis* strain, OPC-15, (Tsujibo et al., 1988). *N. dassonvilleii* strain OPC produced different phenazine antibiotics under different culture conditions (incubation period, incubation temperature, and temperature shift). 1,6-Dihydroxyphenazine (compound III) was obtained from the mycelium after incubation for 6–8 days at 27°C, while 1,6-dihydroxyphenazine 5,10 dioxide (compound I, iodine) was isolated after incubation for 6 days at 27°C followed by further incubation for two days at 4°C. The production of a new indole alkaloid, pendolmycin, was reported from a *Nocardiopsis* strain, SA 1715, isolated from soil collected in a river near Shanghai (Yamashita et al., 1988). Pendolmycin is an inhibitor of phosphatidylinositol turnover induced in human epidermoid carcinoma by epidermal growth factor. The structurally related metabolites teleocidin B and lyngbyatoxin have been isolated from a *Streptomyces* strain and from lipid extracts of an Hawaiian shallow-water variety of the cyanobacterium *Lyngbya majuscula,* respectively. Thus, the same secondary metabolite may be synthesized in different taxa, indicating that the occurrence of a metabolite may not have any taxonomic relevance.

Current Taxonomic Position
of *Nocardiopsis*

After nearly 100 years of searching for a home for *Nocardiopsis dassonvillei* strains, the taxonomic position of this taxon has now been elucidated. Based on chemotaxonomy, numerical taxonomy using phenetic data, DNA-DNA reassociation studies, and 16S rRNA analyses, many actinomycete taxa had to be reclassified (Athalye et al., 1983; Fischer et al., 1983; Goodfellow et al., 1988; Grund and Kroppenstedt, 1989, 1990; Kroppenstedt et al, 1990; Poschner et al., 1985; Stackebrandt, 1986). The six currently validly described *Nocardiopsis* species, *N. alba* subsp. *alba*, *N. alba* subsp. *prasina*, *N. alborubida*, *N. antarcticus*, *N. dassonvillei*, and *N. listeri* form a homogeneous cluster among the other taxa of the order Actinomycetales. The *Nocardiopsis* species are clearly separated from all other actinomycetes by their unique combination of chemical markers (Kroppenstedt, 1987). In the phylogenetic tree of Actinomycetales, which was constructed by comparison of the 16S rRNA sequences, the 16S rRNA cataloging data, and the DNA-DNA hybridization results, the genus *Nocardiopsis* occupies a position between *Streptomyces* and *Microtetraspora* and related taxa (Goodfellow et al., 1988; Kroppenstedt et al., 1990; Stackebrandt et al., 1983; Stackebrandt, 1986). To maintain the integrity of the genus *Nocardiopsis* for the future, only those actinomycete isolates should be included into this genus whose position in the genus *Nocardiopsis* has been proved by DNA-DNA homology studies or which show the typical chemotaxonomy of *Nocardiopsis* (Table 3).

Literature Cited

Abyzov, S. S., S. N. Philipova, and V. D. Kuznetsov. 1983. *Nocardiopsis antarcticus,* a new species of actinomycetes, isolated from the ice sheet of the central antarctic glacier. Izv. Akad. Nauk SSSR Ser. Biol. 4:559–568.

Agre, N. S., T. P. Efimova, and L. N. Guzeva. 1975. Heterogeneity of the genus *Actinomadura* Lechevalier and Lechevalier. Microbiology (English translation of Mikrobiologiya) 44:200–233.

Akimov, V. K., L. I. Evtushenko, and S. V. Dobritsa. 1989. *Pseudoamycolata halophobica* gen. nov., sp. nov. Int. J. Syst. Bacteriol. 39:457–461.

Alderson, G., M. Athalye, and R. P. White. 1984. Numerical methods in the taxonomy of sporoactinomycetes, p. 597–615. In: L. Ortiz-Ortiz, L. F. Bojalil, and V. Yakoleff (ed.), Biological, biochemical and biomedical aspects of actinomycetes. Academic Press, London.

Alderson, G., M. Goodfellow, and D. E. Minnikin. 1985. Menaquinone composition in classification of *Strep-*

tomyces and other sporoactinomycetes. J. Gen. Microbiol. 131:1671–1679.

Athalye, M., M. Goodfellow, J. Lacey, and R. P. White. 1985. Numerical classification of *Actinomadura* and *Nocardiopsis.* Int. J. Syst. Bacteriol. 35:86–98.

Athalye, M., M. Goodfellow, and D. E. Minnikin. 1984. Menaquinone composition in the classification of *Actinomadura* and related taxa. J. Gen. Microbiol. 130:817–823.

Batrakov, S. G., and L. D. Bergelson. 1978. Lipids of the streptomycetes. Structural investigation and biological interrelation. Chemistry and Physics. 21:1–29.

Becker, B., M. P. Lechevalier, and H. A. Lechevalier. 1964. Rapid differentiation between Nocardia and Streptomyces by paper chromatography of whole-cell hydrolysates. Appl. Microbiol. 12:421–423.

Becker, B., M. P. Lechevalier, and H. A. Lechevalier. 1965. Chemical composition of cell wall preparations from strains of various form genera of aerobic actinomycetes. Appl. Microbiol. 13:236–243.

Bowen, T., E. Stackebrandt, M. Dorsch, and T. M. Embley. 1989. The phylogeny of *Amycolata autotrophica, Kibdelosporangium aridum* and *Saccharothrix australiensis.* J. Gen. Microbiol. 135:2529–2536.

Brocq-Rousseau, D. 1904. Sur un *Streptothrix.* References Generale Botanique 16:219–230.

Collins, M. D., M. Faulkner, and R. M. Keddie. 1984. Menaquinone composition of some sporeforming actinomycetes. System. Appl. Microbiol. 5:20–29.

Collins, M. D., and D. Jones. 1981. Distribution of isoprenoid quinone structural types in bacteria and their taxonomic implications. Microbiol. Rev. 45:316–354.

Collins, M. D., R. M. Keddie, and R. M. Kroppenstedt. 1983. Lipid composition of *Arthrobacter simplex, Arthrobacter tumescens,* and possibly related taxa. System. Appl. Microbiol. 4:18–26.

Collins, M. D., and R. M. Kroppenstedt. 1987. Structures of the partially saturated menaquinones of *Glycomyces rutgersensis.* FEMS Microbiol. Lett. 44:215–219.

Collins, M. D., R. M. Kroppenstedt, J. Tamaoka, K. Komagata, and T. Kinoshita. 1988. Structures of the tetrahydrogenated menaquinones from *Actinomadura angiospora, Faenia rectivirgula,* and *Saccharothrix australiensis.* Curr. Microbiol. 17:275–279.

Cummins, C. S., and H. Harris. 1958. Studies on cell-wall composition and taxonomy of *Actinomycetales* and related groups. J. Gen. Microbiol. 18:173–189.

Deutsche Sammlung von Mikroorganismen und Zellkulturen, (DSM) Catalogue of Strains. 1989, 4th. ed., German Collection of Microorganisms and Cell Cultures, Braunschweig, FRG.

Eerolen, E., and O-P. Lehtonen. 1988. Optimal data processing proceedure for automating bacterial identification by gas-liquid chromatography of cellular fatty acids. J. Clin. Microbiol. 26:1745–1753.

Embley, T. M., R. Wait, G. Dobson, and M. Goodfellow. 1987. Fatty acid composition in the classification of *Saccharopolyspora hirsuta.* FEMS Microbiol. Lett. 41:131–135.

Embley, T. M., A. G. O'Donnell, J. Rostron, and M. Goodfellow. 1988. Chemotaxonomy of wall type IV actinomycetes which lack mycolic acids. J. Gen. Microbiol. 134:953–960.

Erikson, D. 1935. The pathogenic aerobic organisms of the actinomyces group. Med. Res. Counc. (G.B.) Spec. Rep. Ser. 203:1–61.

Fischer A., R. M. Kroppenstedt, and E. Stackebrandt. 1983. Molecular-genetic and chemotaxonomic studies on *Actinomadura* and *Nocardiopsis*. J. Gen. Microbiol. 129:3433–3446.

Gauze, G. F., T. S. Maksimova, O. L. Olkhovotova, M. A. Sveshnikova, G. V. Kochetkova, G. B. Ilchenko. 1974. Production of madumycin, an antibacterial antibiotic, by *Actinomadura flava* sp. nov. Antibiotiki 9:771–775.

Goodfellow, M. 1989. The actinomycetes I, Supragenetic classification of actinomycetes, p. 2333–2339. In: S. T. Williams, M. E. Sharpe, and J. G. Holt (ed.), Bergey's manual of systematic bacteriology, vol. 4. Williams and Wilkins, Baltimore.

Goodfellow, M., and Alderson, G. 1979. Numerical taxonomy of *Actinomadura* and related actinomycetes. J. Gen. Microbiol. 112:95–111.

Goodfellow, M., and T. Cross. 1984. Classification, p. 7–164. In: M. Goodfellow, M. Modarski, and S. T. Williams (ed.), The biology of actinomycetes. Academic Press, London.

Goodfellow, M., E. Stackebrandt, and R. M. Kroppenstedt. 1988. Chemotaxonomy and actinomycete systematics. In: Y. Okami, T. Beppu, and H. Ogawara (ed.), Biology of actinomycetes 88. Japan Scientific Societies Press, Tokyo.

Gordon, R., and A. Horan. 1968. *Nocardia dassonvillei*, a macroscopic replica of *Streptomyces griseus*. J. Gen. Microbiol. 50:235–240.

Greiner-Mai, E., R. M. Kroppenstedt, F. Korn-Wendisch, and H. J. Kutzner. 1987. Morphological and biochemical characterization and emended descriptions of thermophilic actinomycetes species. Syst. Appl. Microbiol. 9:97–109.

Grund, E. 1987. Untersuchungen zur Chemotaxonomie einiger Actinomyceten und coryneformer Bakterien. Ph.D. Thesis, Technische Hochschule Darmstadt, FRG.

Grund, E. and R. M. Kroppenstedt. 1989. Transfer of five nocardiopsis species to the genus *Saccharothrix* Labeda et al. 1984. J. Syst. Appl. Microbiol. 12:267–274.

Grund, E., and R. M. Kroppenstedt. 1990. Chemotaxonomy and numerical taxonomy of the genus *Nocardiopsis* Meyer 1976. Int. J. Syst. Bact. 40:5–11.

Hickey, R. J. and H. D. Tresner. 1952. A cobald containing medium for sporulation of *Streptomyces* species. J. Bacteriol. 64:891–892.

Hoffmann, P. 1989a. Cryopreservation of fungi. World Federation for Culture Collections. Technical information sheet No. 5. UNESCO/WFCC/Education Committee. Braunschweig, FRG.

Hoffmann, P. 1989b. Cryopreservation of basidiomycete cultures. Mushroom Science XII (Part I), Proceeding of the 12th International Congress on the Science and Cultivation of Edible Fungi (1987). Braunschweig, FRG.

Hofheinz, W., and H. Grisebach. 1965. Die Fettsäuren von *Streptomyces erythreus* und *Streptomyces halstedii*. Z. Naturforsch. 20:43–53.

Horikoshi, K. 1971. Production of alkaline enzymes by alkalophilic microorganisms. Part I. Alkaline protease produced by *Bacillus* no. 221. Agric. Biol. Chem. 35:1407–1414.

Horikoshi, K., and T. Akiba. 1982. Alkalophilic microorganisms: A new world. Japan Scientific Press, Tokyo, Japan.

Howard, O. W, E. Grund, R. M. Kroppenstedt, and M. D. Collins. 1986. Structural determination of a novel naturally occurring cyclic vitamin K. Biochem. Biophys. Res. Commun. 140:916–923.

Itoh, T., T. Kudo, F. Parenti, and A. Seino. 1989. Amended description of the genus *Kineosporia* based on chemotaxonomic and morphological studies. Int. J. Syst. Bact. 9:168–173.

Kase, H. K. Iwahashi, and Y. Matsuda. 1986. K-252a, a potent inhibitor of protein kinase c from microbial origin. J. Antibiot. 39:1059–1065.

Kawanami, J., A. Kimura, Y. Nakagawa, and H. Otsuka. 1969. Lipids of *Streptomyces sioyaensis*, V: On the 2-hydroxy-13-methyl-tetradecanoic acid from phosphatidylethanolamine. Chem. Phys. Lipids. 3:29–38.

Koizumi S., M. L. Contreras, Y. Matsuda, T. Hama, P. Lazarovici, and G. Guroff. 1988. K-252a: A specific inhibitor of the action of nerve growth factor on PC 12 cells. J. Neurosc. 8:715–721.

Korn-Wendisch, F., A. Kempf, E. Grund, R. M. Kroppenstedt, and H. J. Kutzner. 1989. Transfer of *Faenia rectivirgula* Kurup and Agre 1983 to the genus *Saccharopolyspora* Lacey and Goodfellow 1975, elevation of *Saccharopolyspora hirsuta* subsp. taberi Labeda 1987 to species level, and emended description of the genus *Saccharopolyspora*. Int. J. Syst. Bacteriol. 39:430–441.

Kothe, H. W., G. Vobis, R. M. Kroppenstedt, and A. Henssen. 1989. A taxonomic study of mycolateless, wall chemotype IV actinomycetes. Syst. Appl. Microbiol. 12:61–69.

Kroppenstedt, R. M. 1979. Chromatographische Identifizierung von Mikroorganismen dargestellt am Beispiel der Actinomycten. Kontakte Merck. 2:12–21.

Kroppenstedt, R. M. 1982a. Anwendung von Dünnschicht- und Gas-Chromatographie in der Bakterientaxonomie. GIT-Chromatographie. Suppl. 1982:34–40.

Kroppenstedt, R. M. 1982b. Separation of bacterial menaquinones by HPLC using reverse phase (RP-18) and a silver loaded ion exchanger. J. Liquid Chromat. 5:2359–2367.

Kroppenstedt, R. M. 1982c. Anwendung chromatographischer HP-Verfahren (HPTLC und HPLC) in der Bakterien-Taxonomie. GIT-Labor-Medizin. 5:266–275.

Kroppenstedt, R. M. 1985. Fatty acid and menaquinone analysis of actinomycetes and related organisms, p. 173–199. In: M. Goodfellow, and D. E. Minnikin (ed.), Chemical methods in bacterial systematics No. 20. SAB Technical Series, Academic Press, New York.

Kroppenstedt, R. M. 1987. Chemische Untersuchungen an *Actinomycetales* und verwandten Taxa, Korrelation von Chemosystematik und Phylogenie. Habilitationsschrift. Technische Hochschule Darmstadt. Darmstadt, FRG.

Kroppenstedt, R. M., and H. J. Kutzner. 1976. Biochemical markers in the taxonomy of Actinomycetales. Experientia 32:319–320.

Kroppenstedt, R. M., and H. J. Kutzner. 1978. Biochemical taxonomy of some problem actinomycetes. Zentralbl. Bakteriol. Parasitenkd. Infektionskr. Hyg. Abt. 1 Orig. Reihe C Suppl. 6:125–133.

Kroppenstedt, R. M., F. Korn-Wendisch, V. J. Fowler, and E. Stackebrandt. 1981. Biochemical and molecular genetic evidence for transfer of *Actinoplanes armeniacus* into the family *Streptomycetaceae.* Zbl. Bakt. Hyg., Abt. Orig. C2:254–262.

Kroppenstedt, R. M., E. Stackebrandt, and M. Goodfellow. 1990. Taxonomic revision of the actinomycete genera *Actinomadura* and *Microtetraspora.* J. Syst. Appl. Bact. Microbiol. 13:148–160.

Kudrina, E. S. 1957. p. 109–111. In: G. F. Gauze, T. P. Preobrashinskaya, E. S. Nudrina, N. O. Blinoo, J. D. Ryabova, and M. A. Sveshnikova. (ed.), Problems of classification of actinomycetes-antagonists. Government Publishing House of Medical Literature, Moscow.

Küster, E. 1959a. Outline of comparative study of criteria used in characterization of the actinomycetes. Int. Bull. Bact. Nomen. and Taxon. 9:98–104.

Küster, E. 1959b. Introductory remarks and welcome. "Round table conference on streptomycetes." 7th International Congress of Microbiology, Stockholm, August 4–5, 1958. Int. Bull Bact. Nomen. and Taxon. 9:57–61.

Kutzner, H. J. 1981. The Family Streptomycetaceae, p. 2029–2090. In: Starr, M. P., Stolp, H., Trüper, H. G., Balows, A. and Schlegel, H. G. (ed.), The prokaryotes: a handbook on habitats, isolation, and identification of bacteria. Springer-Verlag, Berlin.

Labeda, D. P., and M. P. Lechevalier. 1989. Amendment of the genus *Saccharothrix* Labeda et al. 1984 and descriptions of *Saccharothrix espanensis* sp. nov., *Saccharothrix cryophilis* sp. nov., and *Saccharothrix mutabilis* comb. nov. Int. J. Syst. Bacteriol. 39:429–423.

Labeda, D. P., R. T. Testa, M. P. Lechevalier, and H. A. Lechevalier. 1984. Saccharothrix: A new genus of the Actinomycetales related to *Nocardiopsis.* Int. J. Syst. Bacteriol. 34:426–431.

Labeda, D. P., R. T. Testa, M. P. Lechevalier, and H. A. Lechevalier. 1985. *Glycomyces,* a new genus of the Actinomycetales. Int. J. Syst. Bacteriol. 35:417–421.

Lacey, J. 1977. The ecology of actinomycetes in fodders and related substrates, Proceeding of the Warsaw Symposium on *Streptomyces* and *Nocardia.* Zentralbl. Bakteriol. Parasitenkd. Infektionskr. Hyg. Abt. 1, Suppl. 6:161–170.

Lechevalier, H. A., and M. P. Lechevalier. 1970. A critical evaluation of the genera of aerobic actinomycetes, p. 393–405. In: H. Prauser (ed.), The Actinomycetales. Gustav Fischer Verlag, Jena, Germany.

Lechevalier, H. A., M. P. Lechevalier, and N. N. Gerber. 1971. Chemical composition as a criterion in the classification of actinomycetes. Adv. Appl. Microbiol. 14:47–72.

Lechevalier, M. P., and H. A. Lechevalier. 1970a. Composition of whole-cell hydrolysates as a criterion in the classification of aerobic actinomycetes, p. 311–316. In: H. Prauser (ed.), The Actinomycetales. VEB-Fischer, Jena, Germany.

Lechevalier, M. P., and H. A. Lechevalier. 1970b. Chemical composition as a criterion in the classification of aerobic actinomycetes. Int. J. Syst. Bacteriol. 20:435–443.

Lechevalier, M. P., C. de Bievre, and H. A. Lechevalier. 1977. Chemotaxonomy of aerobic actinomycetes: Phospholipid composition. Biochem. Ecol. Syst. 5:249–260.

Lechevalier, M. P., H. Prauser, D. P. Labeda, and J. S. Ruan. 1986. Two new genera of nocardioform actinomycetes: *Amycolata* gen. nov. and *Amycolatopsis* gen. nov. Int. J. Syst. Bacteriol. 36:29–37.

Liegard, H., and Landrieu. 1911. Un cas de mycose conjunctivale. Ann. Ocul. 146:418–426.

Matsuda, Y., and J. Fukuda. 1988. Inhibition by K-252a, a new inhibitor of protein kinase, of nerve growth factor-induced neurite outgrowth of chick embryo dorsal root ganglion cells. Neurosc. Lett. 87:11–17.

Meyer, J. 1976. *Nocardiopsis,* a new genus of the order Actinomycetales. Int. J. Syst. Bacteriol. 26:487–493.

Meyer, J. 1989. Genus *Nocardiopsis* 1976, p. 2562–2569. In: S. T. Williams, M. E. Sharpe, and J. G. Holt (ed.), Bergey's manual of systematic bacteriology, vol. 4. Williams and Wilkins, Baltimore.

Mikami, Y., K. Miyashita, and T. Arai. 1982. Diaminopimelic acid profiles of alkalophilic and alkali-resistant strains of actinomycetes. J. Gen. Microbiol. 128:1709–1712.

Mikami, Y., K. Miyashita, and T. Arai. 1985. Alkalophilic actinomycetes. Actinomycetes 19:176–191.

Minnikin, D., and A. G. O'Donnell. 1984. Actinomycete envelope lipid and peptidoglycan conmposition, p. 337–388. In: M. Goodfellow, M. Mordarski and S. T. Williams (ed.), Biology of the actinomycetes. Academic Press, London.

Minnikin, D. E., L. Alshamaony, and M. Goodfellow. 1975. Differentiation of *Mycobacterium, Nocardia,* and related taxa by thin-layer chromatographic analysis of whole-organism methanolysates. J. Gen. Microbiol. 88:200–204.

Minnikin, D. E., M. D. Collins, and M. Goodfellow. 1978. Menaquinone patterns in the classification of nocardioform and related taxa. Zentralbl. Bakteriol. Parasitenkd. Infektionskr. Hyg. Abt. 1 Orig. Reihe C Suppl. 6:85–90

Minnikin, D. E., A. G. O'Donnell, M. Goodfellow, G. Alderson, M. Athalye, K. P. Schaal, and J. H. Parlett, J. H. 1984. An integrated procedure for extraction of bacterial isoprenoid quinones and polar lipids. J. Microbiol. Methods. 2:233–241.

Minnikin, D. E., T. Pirouz, and M. Goodfellow. 1977. Polar lipid composition in the classification of some *Actinomadura* species. Int. J. Syst. Bacteriol. 27:118–121.

Mishra, S. K., J. E. Keller, J. R. Miller, R. M. Heisey, M. G. Nair, and A. R. Putnam. 1987a. Insecticidal and nematicidal properties of microbial metabolites. J. Industr. Microbiol. 2:267–276.

Mishra, S. K., W. H. Taft, A. R. Putnam, and S. K. Ries. 1987b. Plant growth regulatory metabolites from novel actinomycetes. J. Plant Growth Regul. 6:75–84.

Mishra, S. K., C. J. Whitenack, and A. R. Putnam. 1988. Herbicidal properties of metabolites from several genera of soil microorganisms. Weed Science 36:122–126.

Miyashita, M., Y. Mikami, and T. Arai. 1984. Alkalophilic actinomycete, *Nocardiopsis dassonvillei* subsp. *prasina* subsp. nov. isolated from soil. Int. J. Syst. Bacteriol. 34:405–409.

Mordarska, H., A. Gamian, and J. Carrasco. 1983. Sugar-containing lipids in the classification of representative *Actinomadura* and *Nocardiopsis* species. Arch. Immunol. Exp. 31:135–143.

Mordarska, H., M. Mordarski, and M. Goodfellow. 1972. Chemotaxonomic characteristics and classification of of some nocardioform bacteria. J. Gen. Microbiol. 73:77–86.

Nakanishi, S., Y. Matsuda, K. Iwahashi, and H. Kase. 1986. K-252b, c and d, potent inhibitors of protein kinase c from microbial origin. J. Antibiot. 39:1066–1071.

Nakanishi, S., K. Yamada, H. Kase, S. Nakamura, and Y. Nonomura. 1988. K-252a, a novel microbial product, inhibits smooth muscle myosin light chain kinase. J. Biol. Chem. 263:6215–6219.

Nonomura, H., and Y. Ohara. 1971. Distribution of actinomycetes in soil. J. Ferm. Technol. 49:904–912.

O'Donnell, A. G. 1985 Numerical analysis of chemotaxonomic data, p. 403–414. In: M. Goodfellow, D. Jones, and F. G. Priest (ed.), Computer assisted bacterial systematics. Academic Press, London.

Ohmori, K., H. Ishii, H. Manabe, H. Satoh, T. Tamura, and H. Kase. 1988. Antiflammatory and antiallergic effects of a novel metabolite of Nocardiopsis sp. as a potent protein kinase C inhibitor from microbial origin. Drug Res. 38:809–814.

Poschner, J., R. M. Kroppenstedt, A. Fischer, and E. Stackebrandt. 1985. DNA-DNA-reassociation and chemotaxonomic studies on Actinomadura, Microbispora, Microtetrapora, Micropolyspora, and Nocardiopsis. Syst. Appl. Microbiol. 6:264–270.

Preobrazhenskaya, T. P., and M. A. Sveshnikova. 1974. New species of the genus Actinomadura. Mikrobiologiya 43:864–868.

Preobrazhenskaya, T. P., M. A. Sveshnikova, and G. F. Gauze. 1982. On the transfer of certain species of the genus Actinomadura Lechevalier et Lechevalier 1970 to the genus Nocardiopsis Meyer 1976. Mikrobiologiya 51:111–113.

Pridham, T. G., P. Anderson, C. Foley, L. A. Lindenfelser, C. W. Hesseltine, and R. G. Benedict. 1956–1957. A selection of media for maintainance and taxonomic study of streptomycetes. Antibiotics Ann. 1956/1957:947–953.

Pridham, T. G., and A. J. Lyons, Jr. 1968. Streptomyces albus (Rossi Doria) Waksman et Henricic: Taxonomic study of strains labeled Streptomyces albus. J. Bacteriol. 81:431–441.

Sabaou, N., and N. Bounaga. 1987. Actinomycetes parasite de champignons: étude des especes, specificite de l'action parasitaire au genre Fusarium et antagonist dans le sol envers Fusarium oxysporum f.sp. albedinis (Kilian et Maire) Gordon. Can. J. Microbiol. 33:445–451.

Schleifer, K. H., and O. Kandler. 1972. Peptidoglycan types of bacterial cell walls and their taxonomic implications. Bacteriol. Rev. 36:407–477.

Shearer, M. C., P. M. Colman, R. M. Ferrin, L. J. Nisbet, and C. H. Nash III. 1986. New genus of the Actinomycetales: Kibdelosporangium aridum gen. nov., sp. nov. Int. J. Syst. Bacteriol. 36:47–54.

Shearer, M. C., P. M. Colman, and C. H. Nash III. 1983. Nocardiopsis mutabilis, a new species of nocardioform bacteria isolated from soil. Int. J. Syst. Bacteriol. 33:369–374.

Shinobu, R. 1958. Physiological and cultural study for the identification of soil actinoycete species. Mem. Osaka Univ. B. Nat. Sci. 7:1–76.

Shirling, E. B., and D. Gottlieb. 1966. Methods for the characterization of Streptomycetes species. Int. J. Syst. Bacteriol. 16:313–340.

Shirling, E. B., and D. Gottlieb. 1972. Cooperative description of type cultures of Streptomyces. V. Additional species descriptions from first and second studies. Int. J. Syst. Bacteriol. 18:279–392.

Skerman, V. B. D., V. McGowan, and P. H. A. Sneath. 1980. Approved lists of bacterial names. Int. J. Syst. Bacteriol. 30:225–420

Stackebrandt, E. 1986. The significance of "Wall Types" in phylogenetically based taxonomic studies on actinomycetes, p. 497–506. In: G. Szabo, S. Biro, and M. Goodfellow (ed.), Biological, biochemical and biomedical aspects of actinomycetes. Akademiai Kiado, Budapest.

Stackebrandt, E., W. Ludwig, E. Seewald, and K. H. Schleifer. 1983. Phylogeny of sporeforming members of the order Actinomycetales. Int. J. Syst. Bacteriol. 33:173–180.

Suzuki, K. 1988. Cellular fatty acid analysis in actinomycete taxonomy, p. 251–256. In: Y. Okami, T. Beppu, and H. Ogawara, (ed.), Biology of actinomycetes 88. Japan Scientific Societies Press, Tokyo.

Takahashi, Y., and S. Omura. 1987. Kitasatosporia, a genus of the order Actinomycetales. Kitasato Arch. Exp. Med. 60:1–14

Tsujibo, H., T. Sato, M. Inui, H. Yamamoto, and Y. Inamori. 1988. Intracellular accumulation of phenazine antibiotics produced by an alkalophilic actinomycete. I. Taxonomy, isolation and identification of the phenazine antibiotics. Agr. Biol. Chem. 52:301–306.

Uchida, K., and K. Aida. 1977. Acyl type of bacterial cell wall: Its simple identification by colorimetric method. J. Gen. Appl. Microbiol. 23:249–260.

Waksman, S. A. and A. T. Henrici. 1948. Family Actinomycetaceae Buchanan and family Streptomycetaceae Waksman and Henrici, p. 961. In: R. S. Breed, E. G. D. Murray, and A. P. Hitchens (ed.), Bergey's manual of determinative bacteriology, 6th ed. Williams & Wilkins Co., Baltimore.

Wellington, E. M. H., and S. T. Williams. 1978. Preservation of actinomycete inoculum in frozen glycerol. Microbios Lett. 6:151–157.

Williams, S. T., G. P. Sharpels, and R. M. Bradshaw. 1974. Spore formation in Actinomadura dassonvillei (Brocq-Rousseu) Lechevalier and Lechevalier. J. Gen. Microbiol. 84:415–419.

Williams, S. T., and E. M. H. Wellington. 1981. The genera Actinomadura, Actinopolyspora, Excellospora, Microbispora, Micropolyspora, Microtetraspora, Nocardiopsis, Saccharopolyspora, and Pseudonocardia. In: M. P. Starr, H. Stolp, H. G. Trüper, A. Balows, and H. G. Schlegel (ed.), The prokaryotes: a handbook of habitats, isolation, and identification of bacteria. Springer-Verlag, Berlin.

Yamada, Y., K. Aoki, and Y. Tahara. 1982a. The structure of the hexahydrogenated isoprenoid side-chain menaquinone with nine isoprene units isolated from Actinomadura madurae. J. Gen. Microbiol. 28:321–429.

Yamada, Y., C. F. Hou, J. Sasaki, Y. Tahara, and H. Yoshioka. 1982b. The structure of the octahydrogenated isoprenoid side-chain menaquinone with nine isoprene units isolated from Streptomyces albus. J. Gen. Appl. Microbiol. 28:519–529.

Yamashita, T., M. Imoto, K. Isshiki, T. Sawa, H. Naganawa, S. Kurasawa, B.-Q. Zhu, and K. Umezawa. 1988. Isolation of a new indole alkaloid, pendolmycin, from Nocardiopsis. J. Nat. Prod. 51:1184–1187.

Yano, I., Y. Furukawa, and M. Kusunose. 1969. Phospholipids of Nocardia coeliaca. J. Bacteriol. 98:124–130.

Zippel, M., and M. Neigenfind. 1988. Preservation of streptomycetes. J. Gen. Microbiol. 34:7–14.

CHAPTER 49

The Genus *Corynebacterium*—Nonmedical

WOLFGANG LIEBL

Introduction

Corynebacteria belong to the large group of Gram-positive bacteria with a high GC content, which constitutes the Actinomycetes subdivision of the Gram-positive eubacteria, as grouped by 16S rRNA cataloging (Stackebrandt and Woese, 1981). Studies of W. Ludwig and K. H. Schleifer (personal communication) using comparative 23S rRNA sequence homology are in agreement with this suggestion and confirm the phylogenetic position of the genus *Corynebacterium*.

Chemotaxonomic studies comparing the cell wall composition (peptidoglycan structure, occurrence of mycolic acids) and lipid profiles suggest that the genera *Mycobacterium, Nocardia,* and *Rhodococcus* are the closest relatives of *Corynebacterium,* and the four genera are combined in the "CMN group" (Barksdale, 1970). It has been suggested that it might be suitable to place these genera, together with *Caseobacter* (later transferred to *Corynebacterium,* Collins et al., 1989), into a family called Mycobacteriaceae (Barksdale, 1981; Jones and Collins, 1986). On the other hand, there is also a different suggestion to classify the mycolate-containing, cell-wall-type IV actinomycetes genera *Gordona, Nocardia, Rhodococcus,* and *Tsukamurella* in the family Nocardiaceae (see the following Chapter 51).

The genus *Corynebacterium* was originally created by Lehmann and Neumann (1896). In the course of the following decades, many bacterial isolates were assigned to the genus, often for reasons that are not justifiable from a modern taxonomic point of view. Therefore, the genus *Corynebacterium* traditionally comprised an extremely diverse collection of microorganisms brought together in one group mainly on the basis of their cell morphology, staining properties, and respiratory metabolism, but the use of these criteria alone obviously did not result in a homogenous group of organisms. Chemotaxonomical markers are now available that define the genus *Corynebacterium* well enough to redraw the borderline of this taxon. This trend is documented in several recent publications proposing the transfer of various species from *Corynebacterium* to other genera as well as the inclusion of species previously placed elsewhere to the genus (see Collins and Cummins, 1986; Collins and Bradbury, 1986; Collins, 1987a, 1987b; Collins et al., 1988). However, it should be kept in mind that the genus *Corynebacterium* is still very diverse and contains medically important species (see preceding chapter) as well as saprophytic and nonpathogenic species. None of the "plant pathogenic coryneform bacteria" listed in the eighth edition of *Bergey's Manual of Determinative Bacteriology* (Cummins et al., 1974), which were treated as members of the genus *Corynebacterium* for many years, is a true *Corynebacterium* sensu stricto (see Collins and Bradbury, 1986). This notion was supported mainly by the accumulation of chemotaxonomic data (i.e., cell wall composition, mycolic acid content, GC content), arguing against the retention of phytopathogenic "corynebacteria" in the genus, which finally led to the reclassification of these bacteria in other genera, mainly in *Curtobacterium* and *Clavibacter* (Collins and Jones, 1983; Goodfellow, 1984a, 1984b; Collins et al., 1981, 1982b; Collins and Jones, 1982; Davis et al., 1984; see also Chapter 62 by Collins and Bradbury).

In *Bergey's Manual of Systematic Bacteriology,* 16 "legitimate" species of *Corynebacterium* are listed (Collins and Cummins, 1986). Most of these are considered to be of medical significance and are therefore treated in Chapter 50, while the following will be treated here: *C. callunae, C. flavescens, C. glutamicum,* and *C. vitarumen.* Additionally, the following species are included in this chapter for the reasons given below: *C. ammoniagenes,* formerly *Brevibacterium ammoniagenes,* has been transferred to the genus *Corynebacterium* (Collins, 1987b); *C. amycolatum,* a relatively new species (Collins et al., 1988); *C. variabilis,* formerly *Ar-*

throbacter variabilis; was transferred to the genus *Corynebacterium* (Collins, 1987a); and *Caseobacter polymorphus* was proposed to be a subjective synonym of *C. variabilis* (Collins et al., 1989).

In addition to the species listed above, certain bacteria frequently encountered in the relevant literature will be discussed. Their systematic classification has not been clarified but numerous data exist (Abe et al., 1967; Suzuki et al., 1981; Minnikin et al., 1978; W. Liebl, unpublished observations) indicating their close relatedness, if not identity, with *C. glutamicum*: *C. lilium*, *Brevibacterium flavum*, *B. lactofermentum*, and *B. divaricatum*.

Of the nomenclatural species *B. flavum*, *B. lactofermentum*, *B. divaricatum*, only *B. divaricatum* is included in the *Approved Lists of Bacterial Names* (Skerman et al., 1980), and none is a true member of the genus *Brevibacterium*. Therefore, data obtained with these species will be included with the discussion of the properties of *C. glutamicum*.

The characteristic features of the genus *Corynebacterium*, as outlined by Collins and Cummins (1986), are also true for the bacteria described in this section: Gram-positive (sometimes unevenly stained), nonsporing, nonmotile, not acid-fast, straight or slightly curved rods, ovals, or clubs; often with metachromic granules; often typical V-shaped arrangements of cells; facultatively anaerobic to aerobic; catalase positive; chemoorganotrophic; peptidoglycan directly cross-linked of the type Alγ (Schleifer and Kandler, 1972), with *meso*-diaminopimelic acid (*meso*-A$_2$pm) as the cross-linking amino acid; predominant cell wall sugars are arabinose and galactose; mycolic acids (corynemycolic acids = short-chain α-substituted-β-hydroxy acids with 22–36 carbon atoms) present (except in *C. amycolatum*); straight-chain saturated or monounsaturated fatty acids, 10-methyl-branched chain acids often present; eight and/or nine isoprene units in dihydrogenated menaquinones. Including the reclassified and new species of *Corynebacterium*, the GC content of the DNA of the genus covers the range of approximately 51–65 mol%.

The use of the term "coryneform bacteria" (the origin of this expression and its introduction into the literature was examined by Keddie, 1978) as a supergeneric group designation is troublesome since it traditionally has been applied to a diverse collection of bacteria of various genera (e.g., *Arthrobacter*, *Cellulomonas*, *Mycobacterium*, and *Brevibacterium*, to name just a few; see Keddie, 1978; Goodfellow and Minnikin, 1981). The expression is used in a vague sense even in the recent literature. The rather confusing situation is best documented by the structure of the chapter entitled "Coryneform Group of Bacteria" in the eighth edition of *Bergey's Manual of Determinative Bacteriology* (Rogosa et al., 1974), where, as Barksdale (1981) points out, *Corynebacterium* is placed, "along with a polytypic array of bacteria whose common property would seem to be pleomorphism." Crombach (1986) even refers to the coryneforms as a "catch-all" group. The use of the expression "coryneform bacteria" may previously have been justified due to the lack of a meaningful taxonomic concept using characteristics other than morphology. However, the classification methods available now are much more advanced and have led to a more sophisticated picture of the interrelationships among the "coryneform" organisms. Thus, we feel that the term "coryneform" should be avoided where possible since it is taxonomically unsatisfactory. It may nevertheless be a useful expression if used as a purely descriptive term of bacterial morphology.

Habitats

The bacteria discussed in this chapter were originally isolated from a broad variety of habitats, including soil, feces, diary products, animal skin, and plant material (Table 1). *Arthrobacter variabilis* and *Caseobacter polymorphus*, both now accommodated in the new species *Corynebacterium variabilis* (Collins, 1987a; Collins et al., 1989), were originally isolated from animal fodder and cheese, respectively (Müller, 1961; Crombach, 1978). In a numerical taxonomic study of cheese-smear coryneform bacteria isolated from the rind of different cheese varieties, Seiler (1986) found that about 20% of the white and yellow-colored coryneform isolates clustered in a phenon closely resembling the "*Brevibacterium ammoniagenes* group" of Seiler (1983). Similar strains were also isolated from piggery wastes (Seiler and Hennlich, 1983). Interestingly, significant numbers of bacteria displaying properties characteristic of the genus *Corynebacterium* have also been isolated from marine samples (Bousfield, 1978). However, these isolates were not differentiated to the species level. Nevertheless, it seems that nonmedical corynebacteria are widely disseminated in nature, although there are no data available concerning the numbers present in different habitats.

Table 1. Differential characteristics of saprophytic *Corynebacterium* species.

	C. flavescens	*C. callunae*	*C. glutamicum*	*C. vitarumen*	*C. variabilis*	*C. ammoniagenes*	*C. amycolatum*
Acid from:							
Glucose	+	+	+	+	−	ND	+
Galactose	+	−	−	ND	ND	ND	−
Lactose	−	−	−	−	−	ND	−
Maltose	−	+	+	+	ND	ND	+
Sucrose	−	+	+	+	−	ND	V
Trehalose	−	+	+	+	ND	ND	V
Tyrosine hydrolysis	ND	ND	−	ND	−	+	−
Esculin hydrolysis	ND	−	−	+	ND	ND	−
Hippurate hydrolysis	−	+	+	−	ND	+	V
Nitrate reduced to nitrite	−	−	+	+	ND	+	ND
Starch hydrolysis	−	−	−	−	−	−	+
Urease	−	+	+	+	ND	+	V
Voges-Proskauer	+	ND	−	+	ND	ND	+
10-Methyl-octadecanoic acid	−	−	−	−	+	+	ND
Mycolic acids	+	+	+	+	+	+	−
GC content (mol%)	58.3	51.2	55.5–57.5	64.8	65.0	53.7–55.8	61.0
Habitat	Dairy products	Heather	Soil, animal feces, vegetables, fruits	Rumen of cow	Animal fodder, cheese	Feces of infants, piggery waste	Human skin

Symbols: +, most or all strains display the property; −, most or all strains do not display the property; V, variable; ND, no data available or conflicting data.

Data compiled from Abe et al., 1967; Barksdale et al., 1979; Collins, 1987a, 1987b; Collins and Cummins, 1986; Lanéelle et al., 1980; Seiler, 1983; Yamada and Komagata, 1972.

Isolation, Cultivation, and Preservation

Nonmedical corynebacteria can be isolated from a variety of different habitats (soil, water, plant material, animals; see Table 1). However, no selective media or enrichment procedures have apparently been described which are specific for this group of organisms.

Most strains normally grow well if cultivated aerobically at 30°C in standard peptone-yeast extract media like the *Corynebacterium* medium (1989 catalog of strains, medium no. 53, Deutsche Sammlung von Mikroorganismen) shown below, although growth on very rich media such as Brain heart infusion (Difco) is generally faster and more abundant.

Corynebacterium Medium

Casein peptone, tryptic digest   10 g

Yeast extract	5 g
Glucose	5 g
NaCl	5 g
Agar (optional)	15 g
Distilled water	1 liter

Adjust the pH to 7.2–7.4.

For *C. ammoniagenes,* a chemically defined medium was described by Nara et al. (1969a). Synthetic media, e.g., BMCG broth (Liebl et al., 1989a, see recipe below), have been designed for *C. glutamicum;* they support abundant growth of this species. The addition of certain substances such as 0.1% citrate or low concentrations (10^{-5} M) of some dihydroxyphenolic compounds (catechol, protocatechuate) has been shown to greatly stimulate growth of *C. glutamicum* in synthetic broth, presumably by assisting in the assimilation of iron (Von der Osten et al., 1989; Liebl et al., 1989a).

BMCG Synthetic Broth

(NH₄)₂SO₄	7 g
Distilled water	850 ml
10× M9 solution	100 ml

Autoclave for 20 min at 121°C; then add the following aseptically:

200× Salt solution	5 ml
Trace element solution	2 ml
CaCl₂(1M)	0.05 ml
Vitamin stock solution	1 ml
Glucose (20%)	50 ml
Catechol (10 mM)	1 ml

10× M9 solution: 60 g Na₂HPO₄, 30 g KH₂PO₄, 5 g NaCl, 10 g NH₄Cl, 1 liter distilled water, pH 7.3.
200× Salt solution: 80 g MgSO₄·7H₂O, 4 g FeSO₄·7H₂O, 0.4 g MnSO₄·H₂O, 5 g NaCl, 1 liter distilled water.
Trace element solution: 88 mg Na₂B₄O₇·10H₂O, 40 mg (NH₄)₆Mo₇O₂₄·4H₂O, 10 mg ZnSO₄·7H₂O, 270 mg CuSO₄·5H₂O, 7.2 mg MnCl₂·4H₂O, 870 mg FeCl₃·6H₂O.
Vitamin stock solution: 1 mg biotin, 10 mg thiamine HCl per 1 ml H₂O.
200× Salt solution, trace element solution, and 1 M CaCl₂ are autoclaved separately; vitamin and glucose solutions are filter sterilized; the catechol stock is adjusted to a neutral pH, sterilized by filtration, stored in aliquots at −20°C, and added aseptically to the medium just before inoculation.

Corynebacteria are generally exacting in their nutritional requirements. It is noteworthy that all strains of glutamic acid-producing corynebacteria (*C. glutamicum* and similar strains) are dependent upon the presence of biotin in the growth medium, and some also require thiamine or *p*-aminobenzoic acid (Abe et al., 1967).

Strains of *Corynebacterium* grown on agar slants or plates can be kept at 4°C for at least 4–6 weeks. Collins and Cummins (1986) reported that cultures grown in chopped meat medium (without glucose) will remain viable for many months if stored in the dark at room temperature. Long-term storage can be achieved by adding 20% glycerol or 7% dimethylsulfoxide/ (DMSO) to liquid cultures and freezing at −70°C. Lyophilization is also a reliable means of long-term preservation. For this purpose, 10 ml of overnight cultures grown in the *Corynebacterium* medium described above, can be harvested, resuspended in 1 ml of 10% skim milk (Difco) and, upon lyophilization of 0.2-ml aliquots, stored over silica gel in sealed glass tubes.

Identification

Morphology

For *C. glutamicum*, we have not encountered problems in recognizing the morphological features typical of *Corynebacterium* with cells

grown in various media and taken after different incubation periods, i.e., somewhat irregular rod-shaped cells, often arranged in V-formations, etc. However, there are morphological differences depending on the media used and the culture age (see Fig. 1). Therefore, for the investigation of the species described in this chapter and in order to have reproducible culture conditions for reasons of comparison, it may be useful to employ EYGA medium when examining morphological features, as recommended by Cure and Keddie (1973).

EYGA Medium

K₂HPO₄	1.1 g
KH₂PO₄	0.86 g
CaCl₂	0.025 g
MgSO₄·7H₂O	0.2 g
NaCl	0.1 g
(NH₄)₂SO₄	0.5 g
Yeast extract	1 g
Glucose	1 g
Vitamin B₁₂	0.002 mg
Trace element solution	3 ml
Agar	12 g
Distilled water	1 liter

Adjust the pH to 6.8.

Trace element solution: 5 g EDTA, 2.2 g ZnSO₄·7H₂O, 0.57 g MnSO₄·4H₂O, 0.5 g FeSO₄·7H₂O, 0.161 g CoCl₂·6H₂O, 0.157 g CuSO₄·5H₂O, 0.151 g Na₂MoO₄·2H₂O, and 600 ml distilled water.

Identification and Differentiation from Other Genera

A bacterial isolate which fulfills the following criteria can be considered to be a member of *Corynebacterium* sensu stricto: facultatively anaerobic (although some may be strictly aerobic); coryneform morphology; cell wall containing *meso*-A₂pm, arabinose, and galactose; short-chain mycolic acids with 22–36 carbon atoms; GC content of 51–65 mol%. Using these criteria only, certain rhodococci may be confused with *Corynebacterium*, although their GC content (approximate range for *Rhodococcus* is 60–69 mol%) is generally higher than that found for most corynebacteria. Also, rhodococci are considered to be aerobic while most representatives of *Corynebacterium* are facultatively anaerobic.

Mycolic acids (2-alkyl-branched-3-hydroxy acids) are considered to be valuable chemotaxonomic markers for *Corynebacterium* and related taxa (Minnikin et al., 1978). They are unique lipophilic components of the cell envelopes of the genera *Corynebacterium*, *Gordona*, *Mycobacterium*, *Nocardia*, *Rhodococcus*, and *Tsukamurella*. The most convenient method for the determination of mycolic acids (see Keddie and Jones, 1981, for an evaluation of the methods available) is the acid methanolysis of bacterial cells and subsequent thin-layer-chromatographic analysis of the resulting mycolic acid methyl esters (Minnikin et al., 1975, 1978).

The presence of mycolic acids is a criterion which allows the genera mentioned above to be separated from all other bacteria of similar ("coryneform") morphology. Also, mycolic acid-containing genera can be differentiated to a certain degree by examining the size of these compounds (see Collins and Cummins, 1986). However, part of the genus *Rhodococcus* overlaps with *Corynebacterium* with respect to the mycolate sizes (Collins et al., 1982a), which may pose problems. With the exception of *C. amycolatum,* which does not contain mycolic acids (Collins et al., 1988), all bacterial species discussed in this chapter have mycolic acids with 26–38 carbon atoms (Minnikin et al., 1978; Lanéelle et al., 1980).

Differentiation at the Species Level

Biochemical characteristics which may be useful for differentiating the saprophytic *Corynebacterium* species are listed in Table 1.

Genetics

The biotechnological importance of amino acid-producing corynebacteria has resulted in the development of cloning systems for these bacteria. There are no data available concerning the genetics of the other corynebacteria discussed in this chapter. As a prerequisite for the construction of cloning vectors, numerous corynebacterial strains were screened for the presence of plasmids. Plasmids were detected in strains of *C. glutamicum, B. lactofermentum,* and *C. callunae,* with sizes ranging from 2.7 to 95 kb (Miwa et al., 1984; Katsumata et al., 1984; Sandoval et al., 1984; Sandoval et al., 1985; Yoshihama et al., 1985). Apart from a 29-kb *C. glutamicum* plasmid carrying streptomycin- and spectinomycin-resistance determinants (Katsumata et al., 1984), all plasmids are cryptic. A multitude of cloning vectors for amino acid-producing corynebacteria have been constructed, most of which are shuttle vectors based on cryptic corynebacterial plasmids (Martin et al., 1987; Archer et al., 1989).

Several methods for the introduction of recombinant plasmid molecules in corynebacteria are available. Protoplast transformation, a method established for various other Gram-positive host organisms, was developed for amino acid-producing corynebacteria simultaneously by different groups (Katsumata et al., 1984; Santamaria et al., 1984; Yoshihama et al., 1985). In these transformation systems, genuine (spherical) protoplasts are not used, but osmotic-sensitive cells which retain rod-shaped morphology. Electroporation was shown to be a more convenient and much more efficient method for transformation of *C. glutamicum* (Liebl et al., 1989c; Haynes and Britz, 1989; Dunican and Shivnan, 1989; Bonamy et al., 1990). A further method for the introduction of recombinant DNA into corynebacteria has also been described: Schäfer et al. (1990) reported the conjugal transfer of mobilizable shuttle plasmids from Gram-negative *Escherichia coli* to various corynebacterial strains.

Polyethylene glycol (PEG)-induced protoplast fusion has been used to obtain hybrid organisms from mutant strains of amino acid-producing corynebacteria (Karasawa et al., 1986) and may have potential in the breeding of improved industrial amino acid producers. However, since it is a rather nonspecific method, its usefulness in basic research is limited.

Phages of several amino acid-producing corynebacteria *(C. glutamicum, C. lilium, B. lactofermentum, B. flavum)* have been described (e.g., Oki and Ogata, 1968; Patek et al., 1985; Trautwetter et al., 1987; Sonnen et al., 1990). Certain phages have been used in transfection experiments with protoplasts of amino acid-producing corynebacteria in order to optimize protoplast transformation protocols (Katsumata et al., 1984; Yeh et al., 1985; Sanchez et al., 1986). Phage-mediated transduction in *B. flavum* was reported by Momose et al. (1976). A cosmid vector for glutamic acid-producing bacteria has been constructed which was transduced at low frequency through phage infection (Miwa et al., 1985). However, too little is now known about the biology of the bacteriophages of saprophytic corynebacteria to make possible their use as tools in genetic research.

Numerous observations made during transformation experiments suggest the wide dissemination of restriction/modification systems in corynebacteria (e.g., Katsumata et al., 1984; Follettie and Sinskey, 1986; Bonamy et al., 1990). The problems posed by these systems for cloning experiments can be overcome by the use of restriction-deficient host strains (Liebl et al., 1989c; Liebl and Schein, 1990).

An interesting feature of the molecular biology of *C. glutamicum* is the broad acceptance of heterologous expression signals by this organism. Heterologous antibiotic-resistance genes of Gram-negative *(Escherichia coli)* and Gram-positive origin *(Staphylococcus aureus, Enterococcus faecalis, Streptomyces acrimycini,*

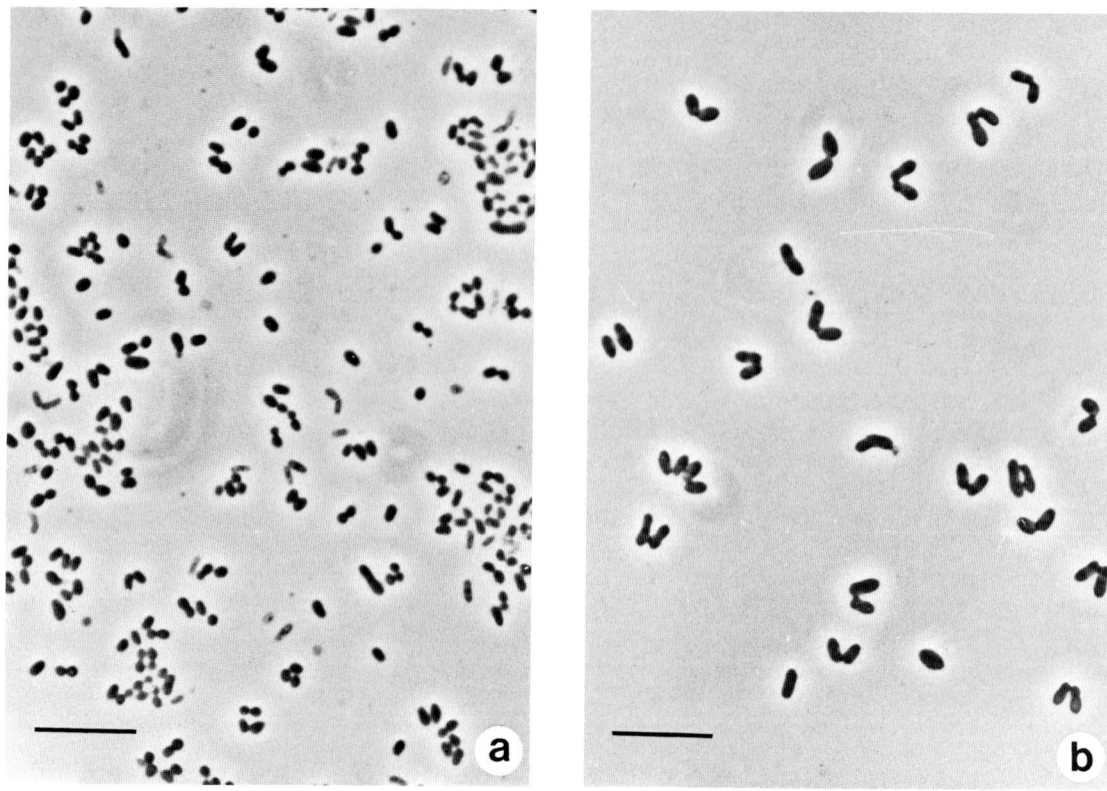

Fig. 1. Phase contrast photomicrographs of cells of *C. glutamicum* from cultures grown on (a) *Corynebacterium* agar without glucose for 48 h; (b) BMCG synthetic broth for 16 h. Note the cell size difference between the two photos. Bar = 10 μm.

Streptomyces hygroscopicus) were observed to function in *C. glutamicum (B. lactofermentum)* (Ozaki et al., 1984; Batt et al., 1985; Yoshihama et al., 1985; Santamaria et al., 1984, 1987). Additionally, a number of other heterologous genes have been expressed in these bacteria, e.g., threonine biosynthetic genes from *Escherichia coli* (Patek et al., 1989), an α-amylase gene from *Bacillus amyloliquefaciens* (Smith et al., 1986), exoglucanase and endoglucanase A genes from *Cellulomonas fimi* (Paradis et al., 1987), and the genes encoding thermonuclease from *Staphylococcus aureus* and lipase from *Staphylococcus hyicus* (Liebl et al., 1989b). The extracellular enzymes encoded by the *Cellulomonas* and *Staphylococcus* genes are efficiently excreted to the culture medium by the corynebacterial hosts (Paradis et al., 1987; Liebl et al., 1989b). Furthermore, the *E. coli lac*UV5, *tac,* and *trp* promoters are functional in *B. lactofermentum* (Morinaga et al., 1987). The broad recognition of heterologous expression signals by *C. glutamicum/B. lactofermentum* contrasts with the situation found with other Gram-positive cloning hosts, e.g., *Bacillus subtilis,* where genes of Gram-negative origin are generally not effi-

ciently expressed due to the more stringent requirements of the this organisms transcription/translation apparatus (McLaughlin et al., 1981; Moran et al., 1982; Graves and Rabinowitz, 1986). Most genes introduced into amino acid-producing corynebacteria by recombinant DNA methods are homologous genes from these bacteria themselves, encoding enzymes involved in amino acid biosynthesis or deregulated derivatives thereof. Using this strategy it may be possible to obtain improved production strains by raising the activity of rate-limiting enzymes and thus removing some bottlenecks of amino acid biosynthesis.

Physiology

There have been very few studies of the metabolism of saprophytic corynebacteria. Only amino acid-producing corynebacteria, particularly *Brevibacterium flavum* and *C. glutamicum,* have been studied in detail. In *B. flavum,* the uptake of glucose and fructose seems to be catalyzed by specific phosphoenolpyruvate:sugar phosphotransferase (PTS) systems

(Mori and Shiio, 1987a, 1987b), although conflicting data have been presented by Marauska et al. (1981). The main route for hexose breakdown in amino acid-producing corynebacteria is the Embden-Meyerhof pathway, followed by further oxidation through the TCA cycle. Additionally, the key enzymes of the glyoxylate pathway, which are necessary for the assimilation of acetate, have been demonstrated in these bacteria (Shiio et al., 1959). Maltose and sucrose are cleaved intracellularly in *B. flavum* by maltase and invertase, respectively. The catabolism of mannitol is initiated by mannitol dehydrogenase. Ribose and gluconate are assimilated via the pentose phosphate pathway (Mori and Shiio, 1987a). Furthermore, amino acid-producing corynebacteria are able to assimilate various organic acids and ethanol (Yamada and Komagata, 1972; Oki et al., 1968).

C. glutamicum and similar bacteria are used for the fermentative production of amino acids, with L-glutamic acid and L-lysine being the most important as far as quantity is concerned (see "Applications," this chapter). Therefore, some aspects of the pathways and the regulation of the biosynthesis of these amino acids will be mentioned here. Since the biosynthetic pathways in *C. glutamicum* and similar bacteria are basically the same as those found in other organisms, the question arises: what pecularities make these bacteria superior amino acid excreters? In the case of the production of L-glutamic acid, which is synthesized via the reductive amination of α-ketoglutarate by action of NADP⁺-specific glutamate dehydrogenase (Fig. 2), it was repeatedly suggested that an apparent lack of α-ketoglutarate dehydrogenase, resulting

in an incomplete citric acid cycle, was an important prerequisite for overproduction of the amino acid (Shiio et al., 1961; Kinoshita and Tanaka, 1972). This view still persists in some recently published biotechnology textbooks. However, the presence of α-ketoglutarate dehydrogenase has been demonstrated in *B. flavum*, but its unstable nature makes its detection and isolation difficult (Shiio and Ujigawa-Takeda, 1980). Nevertheless, this point may still be of importance for glutamic acid production since the activity of α-ketoglutarate dehydrogenase seems to be substantially lower than the activity of glutamate dehydrogenase (Shiio and Ujigawa-Takeda, 1980).

The generally accepted explanation for high-level glutamate accumulation in biotin-limited fermentations is that the exit of this amino acid from biotin-starved cells leads to a decreased internal concentration of glutamate, thus relieving a feedback control mechanism and resulting in further glutamate synthesis. Under biotin-excess conditions, on the other hand, no exit of glutamate occurs and therefore glutamate synthesis ceases when a certain intracellular level is reached. A high intracellular glutamate concentration results in the inhibition of citrate synthase, thus leading to the accumulation of aspartate, which in turn inhibits PEP carboxylase activity (see Kinoshita, 1985). It is noteworthy in this context that in "short-time fermentations," *C. glutamicum* cells have a relatively high internal steady state concentration of glutamate (130–200 mM) under biotin-limited conditions (producer cells) as well as under biotin-excess conditions and maintained a high internal level (100 mM) even in the ab-

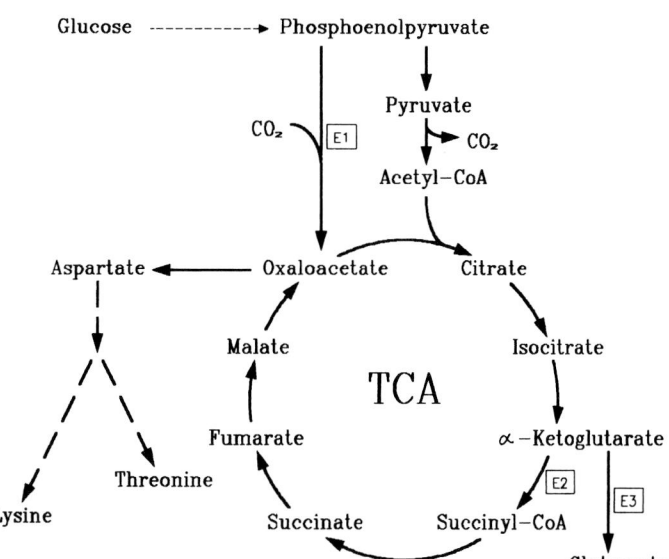

Fig. 2. Metabolic pathways in glutamic acid-producing corynebacteria. Abbreviations: E1, phosphoenolpyruvate carboxylase; E2, α-ketoglutarate dehydrogenase; E3, glutamate dehydrogenase.

sence of a carbon source (Hoischen and Krämer, 1989).

In order to understand the reasons why amino acid-producing corynebacteria are so well suited for the overproduction of L-lysine, an amino acid of the aspartic acid family, it is important to be aware of the relative simplicity of the regulation of this multi-branched biosynthetic pathway when compared to that found in other bacteria like *E. coli* (Fig. 3). In *C. glutamicum* the synthesis of lysine uses a different pathway from the one used in *E. coli*, *C. glutamicum* uses the diaminopimelate dehydrogenase reaction (Misono et al., 1979). There is no regulation of the synthesis of enzymes of the lysine pathway in *C. glutamicum* (see Yeh et al., 1988). Homoserine dehydrogenase, the enzyme located at the branching point leading to threonine, isoleucine, and methionine, is repressed by methionine (Nara et al., 1961; Cremer et al., 1988). Inhibition of enzyme activity in the lysine biosynthetic sequence was only found for aspartate kinase (inhibited by lysine plus threonine; Nakayama et al., 1966; Cremer et al., 1988). Inhibition of this enzyme by the con-

certed action of lysine plus threonine occurs at the first step of the pathway. This type of regulation represents a means for overall control of the synthesis of the amino acids of both major branches of the pathway. The flow of aspartate semialdehyde to either the diaminopimelate/lysine or the methionine/threonine/isoleucine branch is regulated solely at the level of homoserine dehydrogenase by repression of the enzyme by methionine and sensitive feedback inhibition by threonine. Additionally, homoserine dehydrogenase activity is inhibited 50% by L-isoleucine (Cremer et al., 1988). There is no regulation of dihydrodipicolinate synthase, the other enzyme at the branching point competing for aspartate semialdehyde and first enzyme of the diaminopimelate/lysine-specific part of the pathway. Accumulation of threonine efficiently shuts down the methionine/threonine/isoleucine branch of the pathway by sensitive feedback inhibition of homoserine dehydrogenase, directing the flow of aspartate semialdehyde into the lysine branch. Excess lysine in turn can lead to the decrease of carbon flow into the whole biosynthetic pathway via

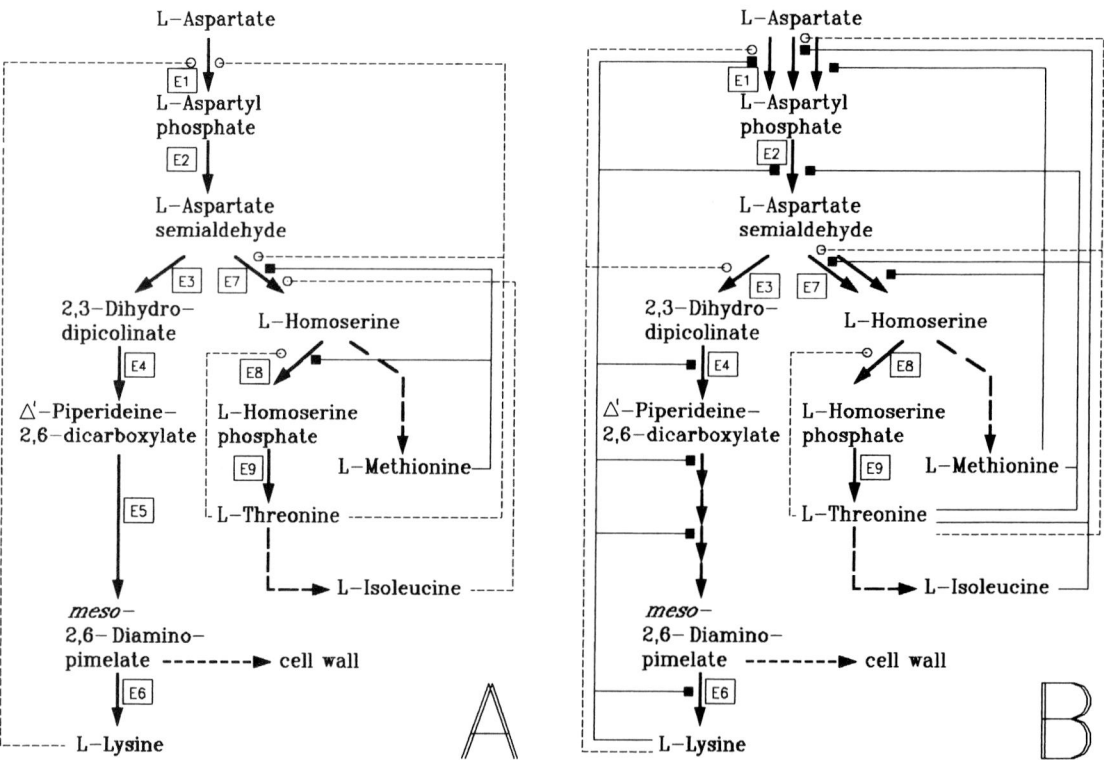

Fig. 3. Regulation of the enzymes involved in the biosynthesis of L-lysine and L-threonine from aspartic acid in (A) amino acid-producing corynebacteria and (B) *Escherichia coli*. Thin lines: repression. Dashed lines: inhibition. Regulation of the L-methionine and L-isoleucine-specific sequences are not shown. Abbreviations: E1, aspartate kinase; E2, aspartate semialdehyde dehydrogenase; E3, dihydrodipicolinate synthase; E4, dihydrodipicolinate reductase; E5, diaminopimelate dehydrogenase; E6, diaminopimelate decarboxylase; E7, homoserine dehydrogenase; E8, homoserine kinase; E9, threonine synthase.

the multivalent inhibition of aspartate kinase by lysine plus threonine. In contrast to the situation found with amino acid-producing corynebacteria, the regulation of amino acid biosynthesis in other bacteria is often more complex. For example, *E. coli* has three distinct and separately regulated aspartate kinase isoenzymes, two of which are bifunctional proteins displaying aspartate kinase plus homoserine dehydrogenase activity (see Cohen, 1983). The rather simple regulation strategies found in amino acid-producing corynebacteria may explain the relative ease of obtaining high-level lysine producers. Single mutations relieving the feedback inhibition of aspartate kinase or leading to decreased homoserine dehydrogenase activity have resulted in the isolation of lysine-overproducing strains of *B. flavum* (Sano and Shiio, 1970; Shiio and Sano, 1969). Another factor of importance for lysine overproduction is the replenishment of intermediates of the citric acid cycle withdrawn for amino acid synthesis. The crucial anaplerotic enzyme in amino acid-producing corynebacteria during growth on glucose is phosphoenolpyruvate (PEP) carboxylase (Ozaki and Shiio, 1969). Amplification of cloned PEP carboxylase (Sano et al., 1987; Eikmanns et al., 1989) in a lysine production strain (deregulated aspartate kinase) has resulted in a 15–20% increase in the rate of production of lysine by *C. glutamicum* (B. J. Eikmanns, personal communication). The process of lysine excretion by corynebacterial cells, which has not been studied in detail to date, may also prove to be of utmost importance for fermentative lysine production (L. Eggeling, personal communication).

Applications

Certain saprophytic corynebacteria have a long tradition as industrial microorganisms in biotechnological production processes. The most prominent example is the production of amino acids by *C. glutamicum* and other similar bacteria. *C. glutamicum* (synonym *Micrococcus glutamicus*) was first isolated in 1957 by Kinoshita while searching for an efficient glutamic acid-producing microorganism among various bacterial isolates (see Kinoshita and Nakayama, 1978). Subsequently, a large number of glutamic acid-producers were isolated and classified in different genera as various different species, e.g., *Brevibacterium flavum, B. lactofermentum, B. roseum, B. divaricatum, C. lilium, C. herculis, Microbacterium ammoniaphilum*, etc. However, the separate species status of these organisms is in serious doubt (see "Introduction," this chapter).

Glutamic acid is produced on a large scale by fermentation. Worldwide production of monosodium glutamate (MSG), which is used mainly as a flavor enhancer, was estimated at approximately 270 million kg in 1988 (Hepner et al., 1988). Under certain culture conditions, *C. glutamicum* excretes remarkably large amounts of glutamic acid into the growth medium (>100 g per liter). In typical fermentations, the molar conversion of sugar to glutamic acid by resting cells is 50–75% (Kinoshita and Nakayama, 1978). A critical factor in glutamic acid fermentations is the concentration of biotin in the culture broth. On one hand, biotin is an essential growth factor for *C. glutamicum* and similar bacteria. On the other hand, a limiting biotin concentration is necessary for good glutamic acid production. The limiting concentration of biotin for optimal glutamate accumulation is approximately 1 μg/l (Kinoshita and Tanaka, 1972; Hoischen and Krämer, 1989). It is, however, not clear how biotin starvation actually triggers glutamate excretion. Biotin limitation leads to suboptimal growth and gives rise to alterations in the lipid content and composition of the cell membrane by interfering with fatty acid synthesis (Takinami et al., 1968; Hoischen and Krämer, 1990). However, in contrast to previous assumptions (Shiio et al., 1963; Kimura, 1963; Demain and Birnbaum, 1968), this does not result in glutamate excretion due to mere "leakiness" of the cells via relaxed membrane permeability (Hoischen and Krämer, 1989). Also, glutamate excretion by producing *C. glutamicum* cells is probably not achieved by the reverse action of the glutamate uptake system present in this organism, as had been suggested by Clement et al. (1984) and Clement and Lanéelle (1986). Instead, Hoischen and Krämer (1989) have provided convincing evidence that a specific, energy-dependent secretion system is responsible for glutamate excretion by *C. glutamicum* under biotin limitation. This efficient active excretion system for glutamate is also likely to be an important participant in glutamic acid overproduction. With an experimental setup which employs monitoring the lipid composition and glutamate excretion activity upon adding biotin to biotin-limited (glutamate producing) cells, it was observed that changes in glutamate excretion can be provoked on a much shorter time scale than the one in which significant alterations in the lipid content and composition take place (Hoischen and Krämer, 1990). Therefore, it appears that the decreased lipid content of the membrane, although a necessary prerequisite, is not sufficient for high glutamate efflux activity (Hoischen and Krämer, 1990).

In order to use cheap substrates rich in biotin, (e.g., cane molasses) for glutamic acid fermentation, other processes have been developed which do not require biotin limitation. The addition of penicillin during the logarithmic growth phase leads to effective glutamate excretion. This treatment does not provoke cell lysis but causes a rapid, 97–99.5% decrease in cell viability (Kinoshita, 1985). Alternative methods include the addition of fatty acid derivatives such as polyoxyethylene sorbitan monooleate (Tween 60) during growth or the use of oleic acid or glycerol-requiring auxotrophs which are kept under oleic acid or glycerol limitation, respectively, during fermentation (see Kinoshita and Nakayama, 1978).

Another amino acid which is produced fermentatively by C. glutamicum and similar bacteria is L-lysine (1988 estimate: 70 million kg per year; Hepner et al., 1988). This essential amino acid is used mainly as a food supplement. While glutamic acid production by C. glutamicum can be achieved simply by adjusting the critical culture parameters (biotin limitation, addition of penicillin, etc.), the high-yield fermentation of lysine by this organism requires the use of metabolically mutated strains. Kinoshita et al. (1958) isolated both homoserine mutants and methionine- plus threonine-requiring mutants that produced L-lysine from carbohydrate. A different strategy employs the use of regulatory mutants which are selected for their increased resistance to antimetabolites. Isolation of B. flavum mutants resistant to the lysine analogue S-(β-aminoethyl)-L-cysteine (AEC) led to strains with increased lysine production due to the release of feedback inhibition of the first step of lysine synthesis from aspartate (Sano and Shiio, 1970). These mutants contain a deregulated aspartate kinase which is no longer subject to the concerted feedback inhibition by lysine plus threonine. Industrial strains used for the production of amino acids generally have combinations of auxotrophic and regulatory mutations.

Similarly, the production of threonine by corynebacteria has been brought about with α-amino-β-hydroxyvaleric acid (AHV, a L-threonine analogue)-resistant mutants or methionine-auxotrophic, AHV-resistant double mutants (Shiio and Nakamori, 1970; Nakamori and Shiio, 1972). Various other amino acids, e.g., isoleucine, valine, tryptophane, phenylalanine, and others, can also be produced by fermentation processes employing appropriate strains of C. glutamicum or similar bacteria (see Kinoshita and Nakayama, 1978; Kinoshita, 1985).

Traditionally, strains suited for amino acid production have been obtained by multiple rounds of classical random mutagenesis and screening procedures. Present efforts are directed toward the application of gene cloning techniques and transformation systems now available for corynebacteria to the specific design of improved amino acid producers, with special emphasis on the construction of production strains for "high-cost" amino acids such as threonine and tryptophane.

Saprophytic corynebacteria are also used in the fermentative production of nucleotides, which are of interest primarily as flavor-enhancing additives in foods (see Komata, 1976). Mutant strains of C. glutamicum have been isolated which excrete the purine ribonucleoside 5′-monophosphates 5′-inosinic acid (IMP), 5′-xanthylic acid (XMP), and 5′-guanylic acid (GMP) (see Demain, 1978). Similarly, adenine-requiring mutants of C. ammoniagenes (previously Brevibacterium ammoniagenes, recently transferred to the genus Corynebacterium by Collins, 1987b) were found to accumulate IMP in fermentations (Nara et al., 1967). Remarkably, intact wild-type cells of C. ammoniagenes are able to convert the purine bases hypoxanthine, guanine, and adenine to the respective nucleoside monophosphate IMP (for hypoxanthine) or mono-, di-, and triphosphates GMP, GDP, GTP (for guanine), and AMP, ADP, ATP (for adenine), respectively (Nara et al., 1967, 1968a; Tanaka et al., 1968). In this case, "salvage synthesis" reactions occur which are extraordinary in that they take place extracellularly, provided certain requirements are fulfilled, i.e., high phosphate and magnesium concentrations and the supplementation of thiamine and pantothenic acid as well as growth-limiting amounts of manganese ions (Nara et al., 1968b). The conditions employed result in the exit of ribose-5′-phosphate, phosphoribosyl pyrophosphate (PRPP), and probably also ATP from the cells (see Demain, 1978). Additionally, the salvage enzymes 5′-phosphoribose pyrophosphokinase and purine nucleotide pyrophosphorylases appear extracellularly, which are thought to catalyze the conversion of an added base to its nucleotide according to the reaction sequence illustrated in Fig. 4. The effects caused by the nutritional changes used in these types of fermentation, in particular by manganese limitation, are drastic: unbalanced growth ensues, the cellular lipid composition is altered, DNA synthesis ceases, cell viability drops, and cell morphology is affected (Furuya et al., 1970; Oka et al., 1968; Thaler and Diek-

Fig. 4. Extracellular salvage synthesis of purine nucleotides by *C. ammoniagenes*. Abbreviations: E1, 5′-phosphoribose pyrophosphokinase; E2, purine nucleotide pyrophosphorylase.

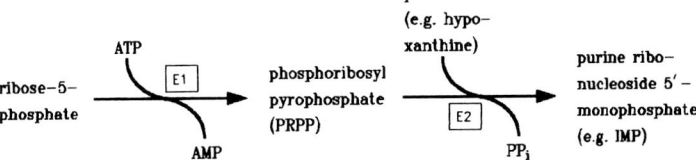

mann, 1979; Auling et al., 1980). It has been proposed that one of the main targets of manganese starvation in *C. ammoniagenes* is the process of ribonucleotide reduction, which is catalyzed by a manganese-dependent ribonucleotide reductase in this organism (Auling et al., 1980; Willing et al., 1988). In the presence of nonlimiting amounts of manganese, nucleotide production can be achieved by the addition of certain antibiotics or surfactants or the use of mutant strains which accumulate IMP under excess manganese conditions (Nara et al., 1969b; Furuya et al., 1969). In addition to the production of nucleotides from purine bases, *C. ammoniagenes* was found to be suited for the fermentative production of nicotinamide adenine dinucleotide (NAD) when cultured in medium containing adenine and nicotinic acid or nicotinamide. Nicotinic acid mononucleotide was produced if only nicotinic acid or nicotinamide was present (Nakayama et al., 1968).

Literature Cited

Abe, S., K. Takayama, and S. Kinoshita. 1967. Taxonomical studies on glutamic acid-producing bacteria. J. Gen. Appl. Microbiol. 13:279–301.

Archer, J. A. C., M. T. Follettie, and A. J. Sinskey. 1989. Biology of *Corynebacterium glutamicum*: a molecular approach, p. 27–33. In: C. L. Hershberger, S. W. Queener, and G. Hegeman (ed.), Genetics and molecular biology of industrial microorganisms. American Society for Microbiology, Washington, DC.

Auling, G., M. Thaler, and H. Diekmann. 1980. Parameters of unbalanced growth and reversible inhibition of deoxyribonucleic acid synthesis in *Brevibacterium ammoniagenes* ATCC 6872 induced by depletion of Mn²⁺. Inhibitor studies on the reversibility of deoxyribonucleic acid synthesis. Arch. Microbiol. 127:105–114.

Barksdale, L. 1970. *Corynebacterium diphtheriae* and its relatives. Bacteriol. Rev. 34:378–422.

Barksdale, L. 1981. The genus *Corynebacterium*, p. 1827–1837. In: M. P. Starr, H. Stolp, H. G. Trüper, A. Balows and H. G. Schlegel (ed.), The prokaryotes, vol. 2. Springer-Verlag, Berlin.

Barksdale, L., M. A. Lanéelle, M. C. Pollice, J. Asselineau, M. Welby, and M. V. Norgard. 1979. Biological and chemical basis for the reclassification of *Microbacterium flavum* Orla-Jensen as *Corynebacterium flavescens* nom. nov. Int. J. Syst. Bacteriol. 29:222–233.

Batt, C. A., W. S. Shanabruch, and A. J. Sinskey. 1985. Expression of pAMα1 tetracycline resistance gene in

Corynebacterium glutamicum: Segregation of antibiotic resistance due to intramolecular recombination. Biotechnol. Lett. 7:717–722.

Bonamy, C., A. Guyonvarch, O. Reyes, F. David, and G. Leblon. 1990. Interspecies electro-transformation in corynebacteria. FEMS MIcrobiol. Lett. 66:263–270.

Bousfield, I. J. 1978. The taxonomy of coryneform bacteria from the marine environment, p. 217–233. In: I. J. Bousfield and A. G. Callely (ed.), Coryneform bacteria. Academic Press, London.

Clement, Y., and G. Lanéelle. 1986. Glutamate excretion mechanism in *Corynebacterium glutamicum*: triggering by biotin starvation or by surfactant addition. J. Gen. Microbiol. 132:925–929.

Clement, Y., B. Escoffier, M. C. Trombe, and G. Lanéelle. 1984. Is glutamate excreted by its uptake system in *Corynebacterium glutamicum?* A working hypothesis. J. Gen Microbiol. 130:2589–2594.

Cohen, G. N. 1983. The common pathway to lysine, methionine, and threonine, p. 147–171. In: K. M. Herrmann and R. L. Somerville (ed.), Amino acids: Biosynthesis and genetic regulation. Addison-Wesley Publishing Co., Reading, PA.

Collins, M. D. 1987a. Transfer of *Arthrobacter variabilis* (Müller) to the genus *Corynebacterium*, as *Corynebacterium variabilis* comb. nov. Int. J. Syst. Bacteriol. 37:287–288.

Collins, M. D. 1987b. Transfer of *Brevibacterium ammoniagenes* (Cooke and Keith) to the genus *Corynebacterium*, as *Corynebacterium ammoniagenes* comb. nov. Int. J. Syst. Bacteriol. 37:442–443.

Collins, M. D., and J. F. Bradbury. 1986. Plant pathogenic species of *Corynebacterium*, p. 1276–1283. In: P. H. A. Sneath, N. S. Mair, M. E. Sharpe, and J. G. Holt (ed.), Bergey's manual of systematic bacteriology, vol. 2. Williams and Wilkins, Baltimore.

Collins, M. D., and C. S. Cummins. 1986. Genus *Corynebacterium* Lehmann and Neumann 1896, p. 1266–1276. In: P. H. A. Sneath, N. S. Mair, M. E. Sharpe, and J. G. Holt (ed.), Bergey's manual of systematic bacteriology, vol. 2. Williams and Wilkins, Baltimore.

Collins, M. D., and D. Jones. 1982. Taxonomic studies on *Corynebacterium beticola* (Abdou). J. Appl. Bacteriol. 52:229–233.

Collins, M. D., and D. Jones. 1983. Reclassification of *Corynebacterium flaccumfaciens*, *Corynebacterium betae*, *Corynebacterium oortii* and *Corynebacterium poinsettiae* in the genus *Curtobacterium*, as *Curtobacterium flaccumfaciens* comb. nov. J. Gen Microbiol. 129:3545–3548.

Collins, M. D., D. Jones, and R. M. Kroppenstedt. 1981. Reclassification of *Corynebacterium ilicis* (Mandel, Guba and Litsky) in the genus *Arthrobacter*, as *Arthrobacter ilicis* comb. nov. Zentralbl. Bakteriol. Mikrobiol. Hyg. Abt. I. Orig. C 2:318–323.

Collins, M. D., M. Goodfellow, and D. E. Minnikin. 1982a. A survey of the structure of mycolic acids in *Corynebacterium* and related taxa. J. Gen. Microbiol. 128:129–149.

Collins, M. D., D. Jones, and R. M. Kroppenstedt. 1982b. In Validation of the publication of new names and combinations previously effectively published outside the IJSB. List No. 9. Int. J. Syst. Bacteriol. 32:384–385.

Collins, M. D., R. A. Burton, and D. Jones. 1988. *Corynebacterium amycolatum*, sp. nov. a new mycolic acidless *Corynebacterium* species from human skin. FEMS Microbiol. Lett. 49:349–352.

Collins, M. D., J. Smida, and E. Stackebrandt. 1989. Phylogenetic evidence for the transfer of *Caseobacter polymorphus* (Crombach) to the genus *Corynebacterium*. Int J. Syst. Bacteriol. 39:7–9.

Cremer, J., C. Treptow, L. Eggeling, and H. Sahm. 1988. Regulation of enzymes of lysine biosynthesis in *Corynebacterium glutamicum*. J. Gen. Microbiol. 134:3221–3229.

Crombach, W. H. J. 1978. *Caseobacter polymorphus* gen. nov., sp. nov., a coryneform bacterium from cheese. Int. J. Syst. Bacteriol. 28:354–366.

Crombach, W. H. J. 1986. Genus *Caseobacter* 1978, p. 1318–1319. In: P. H. A. Sneath, N. S. Mair, M. E. Sharpe, and J. G. Holt (ed.), Bergey's manual of systematic bacteriology, vol. 2. Williams and Wilkins, Baltimore.

Cummins, C. S., R. A. Lelliott, and M. Rogosa. 1974. Genus I. *Corynebacterium* Lehmann and Neumann 1896, 350, p. 602–617. In: R. E. Buchanan, and N. E. Gibbons (ed.), Bergey's manual of determinative bacteriology, 8th ed. Williams and Wilkins, Baltimore.

Cure, G. L., and R. M. Keddie. 1973. Methods for the morphological examination of aerobic coryneform bacteria, p. 123–135. In: R. G. Board, and D. N. Lovelock (ed.), Sampling—Microbiological monitoring of environments. Society for Applied Bacteriology Technical Series 7. Academic Press, New York.

Davis, M. J., A. G. Gillespie, Jr., A. K. Vidaver, and R. W. Harris. 1984. *Clavibacter*: a new genus containing some phytopathogenic coryneform bacteria, including *Clavibacter xyli* subsp. *xyli* sp. nov., subsp. nov. and *Clavibacter xyli* subsp. *cynodontis* subsp. nov., pathogens that cause ratoon stunting disease of sugar cane and Bermudagrass stunting disease. Int. J. Syst. Bacteriol. 34:107–117.

Demain, A. L. 1978. Production of nucleotides by microorganisms, p. 187–208. In: A. H. Rose (ed.), Primary products of metabolism. Academic Press, London.

Demain, A. L., and J. Birnbaum. 1968. Alteration of permeability for the release of metabolites from the microbial cell. Curr. Top. Microbiol. Immunol. 46:1–25.

Dunican, L. K., and E. Shivnan. 1989. High frequency transformation of whole cells of amino acid producing coryneform bacteria using high voltage electroporation. Bio/Technology. 7:1067–1070.

Eikmanns, B. J., M. T. Follettie, M. U. Griot, and A. J. Sinskey. 1989. The phosphoenolpyruvate carboxylase gene of *Corynebacterium glutamicum*: Molecular cloning, nucleotide sequence, and expression. Mol. Gen. Genet. 218:330–339.

Follettie, M. T., and A. J. Sinskey. 1986. Recombinant DNA technology for *Corynebacterium glutamicum*. Food Technol. 40:88–94.

Furuya, A., S. Abe, and S. Kinoshita. 1969. Accumulation of 5'inosinic acid by a manganese-insensitive mutant of *Brevibacterium ammoniagenes*. Appl. Microbiol. 18:977–984.

Furuya, A., S. Abe, and S. Kinoshita. 1970. Effects of manganese and adenine on 5' inosinic acid accumulation by a mutant of *Brevibacterium ammoniagenes*. Agr. Biol. Chem. 34:210–221.

Goodfellow, M. 1984a. Reclassification of *Corynebacterium fascians* (Tilford) Dowson in the genus *Rhodococcus*, as *Rhodococcus fascians* comb. nov. Syst. Appl. Microbiol. 5:225–229.

Goodfellow, M. 1984b. In validation of the publication of new names and combinations previously effectively published outside the IJSB. List No. 10. Int. J. Syst. Bacteriol. 34:503–504.

Goodfellow, M., and D. E. Minnikin. 1981. Introduction to the coryneform bacteria, p. 1811–1826. In: M. P. Starr, H. Stolp, H. G. Trüper, A. Balows and H. G. Schlegel (ed.), The prokaryotes, vol. 2. Springer-Verlag, Berlin.

Graves, M. C., and J. C. Rabinowitz. 1986. In vivo and in vitro transcription of the *Clostridium pasteurianum* ferredoxin gene. J. Biol. Chem. 261:11409–11415.

Haynes, J. A., and M. L. Britz. 1989. Electrotransformation of *Brevibacterium lactofermentum* and *Corynebacterium glutamicum*: growth in tween 80 increases transformation frequencies. FEMS Microbiol. Lett. 61:329–334).

Hepner, L., and Associates. Ltd. 1988. Amino acids—markets and opportunities 1987/88–1994. Tavistock House North, London.

Hoischen, C., and R. Krämer. 1989. Evidence for an efflux carrier system involved in the secretion of glutamate by *Corynebacterium glutamicum*. Arch. Microbiol. 151:342–347.

Hoischen, C., and R. Krämer. 1990. Membrane alteration is necessary but not sufficient for effective glutamate secretion in *Corynebacterium glutamicum*. J. Bacteriol. 172:3409–3416.

Jones, D., and M. D. Collins. 1986. Irregular, nonsporing Gram-positive rods, p. 1261–1266. In: P. H. A. Sneath, N. S. Mair, M. E. Sharpe, and J. G. Holt (ed.), Bergey's manual of systematic bacteriology, vol. 2. Williams and Wilkins, Baltimore.

Karasawa, M., O. Tosaka, S. Ikeda, and H. Yoshii. 1986. Application of protoplast fusion to the development of L-threonine and L-lysine producers. Agr. Biol. Chem. 50:339–346.

Katsumata, R., A. Ozaki, T. Oka, and A. Furuya. 1984. Protoplast transformation of glutamate-producing bacteria with plasmid DNA. J. Bacteriol. 159:306–311.

Keddie, R. M. 1978. What do we mean by coryneform bacteria? p. 1–12. In: I. J. Bousfield and A. G. Callely (ed.), Coryneform bacteria. Academic Press, London.

Keddie, R. M., and G. L. Cure. 1977. The cell wall composition and distribution of free mycolic acids in named strains of coryneform bacteria and in isolates from various natural sources. J. Appl. Bacteriol. 42:229–252.

Keddie, R. M., and D. Jones. 1981. Saprophytic, aerobic coryneform bacteria, p. 1838–1878. In: M. P. Starr, H. Stolp, H. G. Trüper, A. Balows and H. G. Schlegel (ed.), The prokaryotes, vol. 2. Springer-Verlag, Berlin.

Kimura, M. 1963. The effect of biotin on the amino acid biosynthesis by *Micrococcus glutamicus*. J. Gen. Appl. Microbiol. 9:205–212.

Kinoshita, S. 1985. Glutamic acid bacteria, p. 115–142. In: A. L. Demain, and N. A. Solomon (ed.), Biology of industrial microorganisms. The Benjamin/Cummins Publishing Company, London.

Kinoshita, S., and K. Tanaka. 1972. Glutamic acid, p. 263–324. In: K. Yamada, S. Kinoshita, T. Tsunoda, and K. Aida (ed.), The microbial production of amino acids. Kodansha Ltd., Tokyo; John Wiley and Sons, New York.

Kinoshita, S., and K. Nakayama. 1978. Amino acids. p. 209–261. In A. H. Rose (ed.), Primary products of metabolism. Academic Press, London.

Kinoshita, S., K. Nakayama, and S. Kitada. 1958. L-Lysine production using microbial auxotroph. J. Gen. Appl. Microbiol. 4:128–129.

Komata, Y. 1976. Utilization in foods, p. 299–319. In: K. Ogata, S. Kinoshita, T. Tsunoda, and K. Aida (ed.), Microbial production of nucleic acid-related substances. Kodansha Ltd., Tokyo; John Wiley and Sons, New York.

Lanéelle, M. A., J. Asselineau, M. Welby, M. V. Norgard, T. Imaeda, M. C. Pollice, and L. Barksdale. 1980. Biological and chemical basis for the reclassification of *Brevibacterium vitarumen* (Bechdel et al.) Breed (Approved lists, 1980) as *Corynebacterium vitarumen* (Bechdel et al.) comb. nov. and *Brevibacterium liquefaciens* Okabayashi and Musuo (Approved lists, 1980) as *Corynebacterium liquefaciens* (Okabayashi and Masuo) comb. nov. Int. J. Syst. Bacteriol. 30:539–546.

Lehmann, K. B., and R. Neumann. 1896. Atlas und Grundriss der Bakteriologie und Lehrbuch der speciellen bakteriologischen Diagnostik, 1st ed. Munich: Lehmann.

Liebl, W., and B. Schein. 1990. Isolation of restriction deficient mutants of *Corynebacterium glutamicum*. Proceedings of "Dechema-Biotechnology-Conferences," vol. 4. (in press.)

Liebl, W., R. Klamer, and K. H. Schleifer. 1989a. Requirement of chelating compounds for the growth of *Corynebacterium glutamicum* in synthetic media. Appl. Microbiol. Biotechnol. 32:205–210.

Liebl, W., K. H. Schleifer, and A. J. Sinskey. 1989b. Secretion of heterologous proteins by *Corynebacterium glutamicum*, p. 553–559. In: L. O. Butler, C. Harwood, and B. E. B. Moseley (ed.), Genetic transformation and expression. Intercept Ltd., Andover, England.

Liebl, W., A. Bayerl, B. Schein, U. Stillner, and K. H. Schleifer. 1989c. High efficiency electroporation of intact *Corynebacterium glutamicum* cells. FEMS Microbiol. Lett. 65:299–304.

Marauska, D. F., M. P. Ruklish, and N. I. Galynina. 1981. Energy-dependence of glucose transport in *Brevibacterium flavum*. Mikrobiologiya 50:763–768.

Martin, J. F. 1989. Molecular genetics of amino acid-producing corynebacteria, p. 25–59. In: S. Banmberg, I. Hunter, and M. Rhodes (ed.), Society for general microbiology symposium, 44. Cambridge University Press, Cambridge.

Martin, J. F., R. Santamaria, H. Sandoval, G. del Real, L. M. Mateos, J. A. Gil, and A. Aguilar. 1987. Cloning systems in amino acid-producing corynebacteria. Bio/Technology 5:137–146.

McLaughlin, J. R., C. L. Murray, and J. C. Rabinowitz. 1981. Unique features in the ribosome binding site sequence of the Gram-positive *Staphylococcus aureus* β-lactamase gene. J. Biol. Chem. 256:11283–11291.

Minnikin, D. E., L. Alshamaony, and M. Goodfellow. 1975. Differentiation of *Mycobacterium, Nocardia* and related taxa by thin-layer chromatographic analysis of whole cell methanolysates. J. Gen. Microbiol. 88:200–204.

Minnikin, D. E., M. Goodfellow, and M. D. Collins. 1978. Lipid composition in the classification and identification of coryneform and related taxa, p. 85–160. In: I. J. Bousfield and A. G. Callely (ed.), Coryneform bacteria. Academic Press, London.

Misono, H., H. Togawa, T. Yamamoto, and K. Soda. 1979. *Meso-α,ε*-diaminopimelate D-dehydrogenase: distribution and the reaction product. J. Bacteriol. 137:22–27.

Miwa, K., H. Matsui, M. Terabe, S. Nakamori, K. Sano, and H. Momose. 1984. Cryptic plasmids in glutamic acid producing bacteria. Agr. Biol. Chem. 48:2901–2903.

Miwa, K., H. Matsui, M. Terabe, K. Ito, K. Ishida, H. Takagi, S. Nakamori, and K. Sano. 1985. Construction of novel shuttle vectors and a cosmid vector for the glutamic acid-producing bacteria *Brevibacterium lactofermentum* and *Corynebacterium glutamicum*. Gene 39:281–286.

Momose, H., S. Miyashiro, and M. Oba. 1976. On the transducing phages in glutamic acid producing bacteria. J. Gen. Appl. Microbiol. 22:119–129.

Moran, C. P. Jr., N. Lang, S. F. J. LeGrice, G. Lee, M. Stephens, A. L. Sonnenshein, J. Pero, and R. Losick. 1982. Nucleotide sequences that signal the initiation of transcription and translation in *Bacillus subtilis*. Mol. Gen. Genet. 186:339–346.

Mori, M., and I. Shiio. 1987a. Pyruvate formation and sugar metabolism in an amino acid-producing bacterium, *Brevibacterium flavum*. Agr. Biol. Chem. 51:129–138.

Mori, M., and I. Shiio. 1987b. Phosphoenolpyruvate: sugar phosphotransferase systems and sugar metabolism in *Brevibacterium flavum*. Agr. Biol. Chem. 51:2671–2678.

Morinaga, Y., M. Tsuchiya, K. Miwa, and K. Sano. 1987. Expression of *Escherichia coli* promoters in *Brevibacterium lactofermentum* using the shuttle vector pEB003. J. Biotechnol. 5:305–312.

Müller, G. 1961. Mikrobiologische Untersuchungen über die "Futterverpilzung durch Selbsterhitzung." III. Mitteilung: Ausführliche Beschreibung neuer Bakterienspecies. Zentralbl.Bakteriol. Parasitenkd. Infektionskr. Hyg. Abt. 2 Orig. Reihe A. 114:520–537.

Nakamori, S., and I. Shiio. 1972. Microbial production of L-threonine. Part III. Production by methionine and lysine auxotrophs derived from α-amino-β-hydroxyvaleric acid resistant mutants of *Brevibacterium flavum*. Agr. Biol. Chem. 36:1209–1216.

Nakayama, K., H. Tanaka, H. Hagino, and S. Kinoshita. 1966. Studies on lysine fermentaion. V. Concerted feedback inhibition of aspartokinase and the absence of lysine inhibition on aspartic semialdehyde-pyruvate condensation in *Micrococcus glutamicus*. Agr. Biol. Chem. 30:611–616.

Nakayama, K., Z. Sato, H. Tanaka, and S. Kinoshita. 1968. Production of NAD and nicotinic acid mononucleotide with *Brevibacterium ammoniagenes*. Agr. Biol. Chem. 32:1331–1336.

Nara, T., H. Samejima, C. Fujita, M. Ito, K. Nakayama, and S. Kinoshita. 1961. L-Homoserine fermentation. VI. Effect of threonine and methionine on L-homoserine in dehydrogenase in *Micrococcus glutamicus* 534-Co147. Agr. Biol. Chem. 25:532–541.

Nara, T., M. Misawa, and S. Kinoshita. 1967. Production of 5'-Inosinic acid by an adenine auxotroph of *Brevibacterium ammoniagenes*. Agr. Biol. Chem. 31:1351–1356.

Nara, T., M. Misawa, and S. Kinoshita. 1968a. Fermentative production of 5'-purine ribonucleotides by *Brevibacterium ammoniagenes*. Agr. Biol. Chem. 32:561–567.

Nara, T., M. Misawa, and S. Kinoshita. 1968b. Pantothenate, thiamine and manganese in 5'-purine ribonucleotide production by *Brevibacterium ammoniagenes*. Agr. Biol. Chem. 32:1153–1161.

Nara, T., T. Komuro, M. Misawa, and S. Kinoshita. 1969a. Growth responses of *Brevibacterium ammoniagenes*. Agr. Biol. Chem. 33:1030–1036.

Nara, T., M. Misawa, T. Komuro, and S. Kinoshita. 1969b. Effect of antibiotics and surface-active agents on 5'-purinenucleotide production by *Brevibacterium ammoniagenes*. Agr. Biol. Chem. 33:1198–1204.

Oka, T., K. Udagawa, and S. Kinoshita. 1968. Unbalanced growth death due to depletion of Mn^{2+} in *Brevibacterium ammoniagenes*. J. Bacteriol. 96:1760–1767.

Oki, T., and K. Ogata. 1968. Size and morphology of *Brevibacterium* phages. Agr. Biol. Chem. 32:241–248.

Oki, T., Y. Sayama, Y. Nishimura, and A. Ozaki. 1968. L-Glutamic acid formation by microorganisms from ethanol. Agr. Biol. Chem. 32:119–120.

Ozaki, H., and I. Shiio. 1969. Regulation of the TCA and glyoxylate cycles in *Brevibacterium flavum*. J. Biochem. 66:297–311.

Ozaki, A., R. Katsumata, T. Oka, and A. Furuya. 1984. Functional expression of the genes of *Escherichia coli* in Gram-positive *Corynebacterium glutamicum*. Mol. Gen. Genet. 196:175–178.

Paradis, F. W., R. A. I. Warren, D. G. Kilburn, and R. C. Miller Jr. 1987. The expression of *Cellulomonas fimi* cellulase genes in *Brevibacterium lactofermentum*. Gene 61:199–206.

Patek, M., J. Ludvik, O. Benada, J. Hochmannova, J. Nesvera, V. Krumphanzl, and M. Bucko. 1985. New bacteriophage-like particles in *Corynebacterium glutamicum*. Virology 140:360–363.

Patek, M., O. Navratil, J. Hochmannova, J. Nesvera, and J. Hubacek. 1989. Expression of the threonine operon from *Escherichia coli* in *Brevibacterium flavum* and *Corynebacterium glutamicum*. Biotechnol. Lett. 11:231–236.

Rogosa, M., C. S. Cummins, R. A. Lelliott, and R. M. Keddie. 1974. p. 599–632. Coryneform group of bacteria. In: R. E. Buchanan, and N. E. Gibbons (ed.), Bergey's manual of determinative bacteriology, 8th ed. Williams and Wilkins, Baltimore.

Sanchez, F., M. A. Penalva, C. Patino, and V. Rubio. 1986. An efficient method for the introduction of viral DNA into *Brevibacterium lactofermentum* protoplasts. J. Gen. Microbiol. 132:1767–1770.

Sandoval, H., A. Aguilar, D. Paniagua, and J. F. Martin. 1984. Isolation and physical characterization of plasmid pCCl from *Corynebacterium callunae* and con-

struction of hybrid derivatives. Appl. Microbiol. Biotechnol. 19:409–413.

Sandoval, H., G. Del Real, L. M. Mateos, A. Aguilar, and J. F. Martin. 1985. Screening of plasmids in non pathogenic corynebacteria. FEMS Microbiol. Lett. 27:93–98.

Sano, K., and I. Shiio. 1970. Microbial production of L-lysine. III. Production by mutants resistant to S-(2-aminoethyl)-L-cysteine. J. Gen. Appl. Microbiol. 16:373–391.

Sano, K., K. Ito, K. Miwa, and S. Nakamori. 1987. Amplification of the phosphoenol pyruvate carboxylase gene of *Brevibacterium lactofermentum* to improve amino acid production. Agr. Biol. Chem. 51:597–599.

Santamaria, R., J. A. Gil, J. M. Mesas, and J. F. Martin. 1984. Characterization of an endogenous plasmid and development of cloning vectors and a transformation system in *Brevibacterium lactofermentum*. J. Gen. Microbiol. 130:2237–2246.

Santamaria, R. I., J. F. Martin, and J. A. Gil. 1987. Identification of a promoter sequence in the plasmid pUL340 of *Brevibacteriium lactofermentum* and construction of new cloning vectors for corynebacteria containing two selectable markers. Gene 56:199–208.

Schäfer, A., J. Kalinowski, R. Simon, A. H. Seep-Feldhaus, and A. Pühler. 1990. High-frequency conjugal transfer from Gram-negative *Escherichia coli* to various Gram-positive coryneform bacteria. J. Bacteriol. 172:1663–1666.

Schleifer, K. H., and O. Kandler. 1972. Peptidoglycan types of bacterial cell walls and their taxonomic implications. Bacteriol. Rev. 35:407–477.

Seiler, H. 1983. Identification key for coryneform bacteria derived by numerical taxonomic studies. J. Gen. Microbiol. 129:1433–1471.

Seiler, H. 1986. Identification of cheese-smear coryneform bacteria. J. Dairy Res. 53:439–449.

Seiler, H., and W. Hennlich. 1983. Characterization of coryneform bacteria in piggery wastes. System. Appl. Microbiol. 4:132–140.

Shiio, I., and S. Nakamori. 1970. Microbial production of L-threonine. II. Production by α-amino-β-hydroxyvaleric acid resistant mutants of glutamate producing bacteria. Agr. Biol. Chem. 34:448–456.

Shiio, I., and K. Sano. 1969. Microbial production of L-lysine. II. Production by mutants sensitive to threonine or methionine. J. Gen. Appl. Microbiol. 15:267–287.

Shiio, I., and K. Ujigawa-Takeda. 1980. Presence and regulation of α-ketoglutarate dehydrogenase complex in a glutamate-producing bacterium, *Brevibacterium flavum*. Agr. Biol. Chem. 44:1897–1904.

Shiio, I., S. I. Otsuka, and T. Tsunoda. 1959. Glutamic acid formation from glucose by bacteria. I. Enzymes of the Embden-Meyerhof-Parnas pathway, the Krebs-cycle, and the glyoxylate bypass in cell extracts of *Brevibacterium flavum* No. 2247. J. Biochem. 46:1303–1311.

Shiio, I., S. I. Otsuka, and N. Katsuya. 1963. Cellular permeability and extracellular formation of glutamic acid in *Brevibacterium flavum*. J. Biochem. 53:333–340.

Shiio, I., S. I. Otsuka, and M. Takahashi. 1961. Significance of α-ketoglutaric dehydrogenase on the glutamic acid formation in *Brevibacterium flavum*. J. Biochem. 51:164–165.

Skerman, V. B. D., V. McGowan, and P. H. A. Sneath. 1980. Approved lists of bacterial names. Int. J. Syst. Bacteriol. 30:225–420.

Smith, M. D., J. L. Flickinger, D. W. Lineberger, and B. Schmidt. 1986. Protoplast transformation in coryneform bacteria and introduction of an α-amylase gene from *Bacillus amyloliquefaciens* into *Brevibacterium lactofermentum*. Appl. Environ. Microbiol. 51:634–639.

Sonnen, H., J. Schneider, and H. J. Kutzner. 1990. Characterization of φGA1, an inducible phage particle from *Brevibacterium flavum*. J. Gen. Microbiol. 136:567–571.

Stackebrandt, E., and C. R. Woese. 1981. The evolution of prokaryotes, p. 1–31. In: M. J. Carlile, J. F. Collins, and B. E. B. Moseley (ed.), Molecular and cellular aspects of microbial evolution. Cambridge University Press, Cambridge.

Stanek, J. L., and G. D. Roberts. 1974. Simplified approach to identification of aerobic actinomycetes by thin-layer chromatography. Appl. Microbiol. 28:226.231.

Suzuki, K., T. Kaneko, and K. Komagata. 1981. Deoxyribonucleic acid homologies among coryneform bacteria. Int. J. Syst. Bacteriol. 31:131–138.

Takinami, K., H. Yoshii, Y. Yamada, H. Okada, and K. Kinoshita. 1968. Control of L-glutamic acid fermentation by biotin and fatty acid. Amino Acid Nucl. Acid 18:120–160.

Tanaka, H., Z. Sato, K. Nakayama, and S. Kinoshita. 1968. Formation of ATP, GTP, and their related substances by *Brevibacterium ammoniagenes*. Agr. Biol. Chem. 32:721–726.

Thaler, M., and H. Diekmann. 1979. The effect of manganese deficiency on lipid content and composition in *Brevibacterium ammoniagenes*. Eur. J. Appl. Microbiol. Biotechnol. 6:379–387.

Trautwetter, A., C. Blanco, and A. M. Sicard. 1987. Structural characteristics of the *Corynebacterium lilium* bacteriophage CL31. J. Virol. 61:1540–1545.

Von der Osten, C. H., C. Gioannetti, and A. J. Sinskey. 1989. Design of a defined medium for growth of *Corynebacterium glutamicum* in which citrate facilitates iron uptake. Biotechnol. Lett. 11:11–16.

Willing, A., H. Follmann, and G. Auling. 1988. Ribonucleotide reductase of *Brevibacterium ammoniagenes* is a manganese enzyme. Eur. J. Biochem. 170:603–611.

Yamada, K., and K. Komagata, 1972. Taxonomic studies on coryneform bacteria. IV. Morphological, cultural, biochemical, and physiological characteristics. J. Gen. Appl. Microbiol. 18:399–416.

Yeh, P., J. Oreglia, and A. M. Sicard. 1985. Transfection of *Corynebacterium lilium* protoplasts. J. Gen. Microbiol. 131:3179–3183.

Yeh, P., A. M. Sicard, and A. J. Sinskey. 1988. General organization of the genes specifically involved in the diaminopimelate-lysine biosynthetic pathway of *Corynebacterium glutamicum*. Mol. Gen. Genet. 212:105–111.

Yoshihama, M., K. Higashiro, E. A. Rao, M. Akedo, W. G. Shanabruch, M. T. Follettie, G. C. Walker, and A. J. Sinskey. 1985. Cloning vector system for *Corynebacterium glutamicum*. J. Bacteriol. 162:591–597.

CHAPTER 50

The Genus *Corynebacterium*—Medical

ALEXANDER VON GRAEVENITZ and THOMAS KRECH

The Taxonomic Framework

Medically important corynebacteria share the general characteristics of the genus *Corynebacterium* (Collins and Cummins, 1986). Thus, they are straight or slightly curved rods that may be club-shaped or ellipsoidal. Their ends may be tapered or may (in *C. matruchotii*) show a whip handle. Pleomorphism within a culture depends on species and medium. Arrangement of the cells may be angular ("Chinese letters," V forms), or palisade-like. They are Gram-positive but staining may be uneven, particularly in older cultures. Resistance to decolorization is species-dependent. Metachromatic granules (volutin or Babes-Ernst bodies) are frequently seen. These granules consist of polymetaphosphate and may number from one to several; if there are two, they are most often located on each pole. The granules stain reddish purple with methylene blue and retain stains. Corynebacteria show no acid fastness and mycelia, capsules, and spores are absent. Some species show piliation but no flagella; there is no motility. Both aerobic and facultatively anaerobic species are known. Corynebacteria have catalase and use organic carbon and energy sources. The cell wall contains meso-diamino-pimelic acid, and at least arabinose and galactose; in addition, corynemycolic acids, which are short-chained mycolic acids (22–36 C atoms). Cellular fatty acids are straight-chain saturated and mono-unsaturated; some species (for example, *C. bovis*) also contain 10-methyloctadecanoic acid (tuberculostearic acid). The respiratory quinones are dihydrogenated menaquinones with eight to nine isoprene units. The phospholipids are diphosphatidyl glycerol, phosphatidyl inositol, and monoacylated phosphatidyl mannoside. In some species, glycolipids are also found. The GC content is 51–63 mol%.

Nutritional Requirements

Most species are nutritionally exacting, requiring amino acids and/or vitamins, and purine/pyrimidine. Some will grow on nutrient agar and many require the addition of animal protein (serum, blood). The term lipophilic is used fairly loosely in the literature and refers both to species or strains whose growth is stimulated in vitro by lipids (egg yolk, oleic acid, or Tween 80) as well as to those that are lipid dependent (Ersgaard and Justesen, 1984; McGinley et al., 1985; Smith, 1969a; Somerville, 1973). For diagnostic purposes, both kinds of organisms may be called lipophilic. Lipid(s) may be supplied in the medium from blood or serum. Growth in broth is rarely abundant; pellicle formation serves to characterize certain types (e.g., in *C. diphtheriae*). Growth of obligately aerobic species is, of course, inhibited under anaerobic conditions. The optimum temperature for medical corynebacteria is 35–37°C; the range is 15–42°C.

Chemotaxonomic Features

Table 1 lists relevant criteria for chemotaxonomy of the corynebacteria.

Certain taxa have been included in this chapter although their taxonomic status is unclear at present: 1) *C. ulcerans*. This is not recognized as a species in *Bergey's Manual* (Collins and Cummins, 1986), but DNA homology studies (Groman et al., 1984) hint at a distinct species. 2) *C. jeikeium*. This species has been proposed recently for organisms formerly named CDC Group JK after Johnson and Kaye (Jackman et al., 1987) and has been validated in 1988 (List No. 24, Int. J. Syst. Bacteriol. 38:136–137). 3) *C. kutscheri*. This species is still listed in the genus *Corynebacterium* although its GC content is below the accepted range (Pitcher, 1983) and its cell wall contains rhamnose (Collins and Cummins, 1986). 4) *C. bovis*. This species is called "not a member of the genus *Corynebacterium*" in *Bergey's Manual* (Collins and Cummins, 1986) but is listed in the addendum to that genus. The GC content is outside the accepted range for corynebacteria (Pitcher, 1983), and the species contains short-chain mycolic

Table 1. Chemotaxonomic data of medically relevant corynebacteria.

Species	Additional fatty acids[a]	Mycolic acids (chain length)	Additional cell wall sugars[b]	GC content[c] (mol%)	Facultative/ obligate aerobe[d]	Major menaquinones[i]	Additional data
C. diphtheriae	C16:1ω9>C18:2>C18:0	C28–C36	–	52–55	F	MK-8(H₂)	Type intermedius may be lipophilic
C. ulcerans	-"-	C26–C36	–	54	F	-"-	
C. pseudotuberculosis	-"-	C26–C36	glu, mann	52–53	F	-"-	
C. xerosis	C18:0	C28–C36	mann	55–59	F	MK-8(H₂)	
C. renale	C18:0	C26–C36	glu, mann	53–58	F	MK-8(H₂)	
C. kutscheri	C18:0	C26–C36	mann, rha	46	F	ND[e]	
C. striatum	C18:0	ND[e]		58	F	MK-8(H₂)	
C. pseudodiphtheriticum	C18:0	C28–C36	glu	55–57	O	MK-8(H₂)	
C. minutissimum	C18:0>TBSA[a]	C26–C36	–	57–59	F	MK-8,9(H₂)	Lipophilic
C. jeikeium	C18:0>C18:2	C32–C36	mann, glu[h]	58–61	f	ND	Lipophilic
CDC group F	C18:0>C18:2	ND	ND	ND	F	ND	
CDC group G	C18:0>C18:2	ND	ND	ND	F	ND	
CDC group I	C18:0>C18:2	ND	ND	ND	F	ND	
CDC group D-2	C18:0>C18:2>TBSA	ND	ND	ND	O	ND	Lipophilic
CDC group 1	C16:0>C18:1	ND	ND	ND	F	ND	
CDC group 2	C16:0>C18:1	ND	ND	ND	F	ND	Catalase negative
C. bovis	C18:0>C18:2>TBSA	C22–C32	–	67–69	F	MK-9(H₂)	Lipophilic
CDC group ANF	ND	ND	ND	0	ND	ND	
C. matruchotii[g]	C18:0	C28–C36	–	55–58	F	MK-8,9(H₂)	

[a] Main fatty acids in addition to C16:0 and C18:1ω9. TBSA, tuberculostearic acid.
[b] In addition to galactose and arabinose: glu, glucose; mann, mannose; rha, rhamnose.
[c] Rounded up/off.
[d] F, facultatively anaerobic; O, obligately aerobic.
[e] ND, no data available.
[f] No anaerobic growth.
[g] Formerly *Leptotrichia buccalis, L. dentium.*
[h] According to Young et al. (1981).
[i] The numbers indicate the number of isoprenoid units in the menaquinone. H₂ indicates that two of the isoprenoid units are saturated.
References: Alshamaony et al. (1977); Collins and Cummins (1986); Collins et al. (1977, 1979, 1982a, 1982b, 1986); Coyle et al. (1985); Hollis and Weaver (1981); Na'Was et al. (1987); von Graevenitz et al. (1990).

acids (Table 1). Biochemical tests, however, differ substantially between reference strains and strains investigated in two series from the United Kingdom (Harrigan, 1966; Hollis and Weaver, 1981; Jayne-Williams and Skerman, 1966). 5) CDC Groups D-2, F, G, I, ANF, 1 and 2 (Hollis and Weaver, 1981; Na'Was et al., 1987). These groups have not been assigned to the genus *Corynebacterium* but biochemical, morphological, and fatty acid data (von Graevenitz et al., 1990) would support their inclusion although GC content, mycolic acids, menaquinones, and cell wall sugars are unknown thus far.

On the other hand, certain taxa have been omitted. They are either medically unimportant, poorly defined, or only morphologically similar to corynebacteria but fail to fulfill essential criteria that would make them members of the genus *Corynebacterium*: 1) *C. pilosum* and *C. cystitidis*. These were once thought to be biotypes of *C. renale* but have been assigned species status in *Bergey's Manual* (Collins and Cummins, 1986) because of biochemical, serological, and GC differences; furthermore, they show only a low degree of hybridization with *C. renale* ATCC 10848. In contrast to *C. renale, C. pilosum* and *C. cystitidis* hydrolyze starch and fail to hydrolyze casein. Only *C. cystitidis* produces acid from xylose, and only *C. pilosum* reduces nitrate (Collins and Cummins, 1986). 2) *C. genitalium* and *C. pseudogenitalium* (Evangelista et al., 1984). Morphologically, chemotaxonomically, and biochemically, these proposed species are close to *C. jeikeium; C. genitalium* ATCC No. 33798 is indistinguishable from CDC group D-2. *C. genitalium* is described as oxidative and always fructose- and sucrose-negative, while *C. pseudogenitalium* is largely fermentative, always fructose- and occasionally sucrose-positive (Evangelista et al., 1984). The first CDC group JK strains described were sucrose-negative but fructose-variable (Riley et al., 1979); the proposed species *C. jeikeium* is fructose- and sucrose-negative. DNA-DNA homology and protein electrophoretic studies of *C. genitalium* ATCC 33030 did not support inclusion of *C. genitalium* into *C. jeikeium* (Jackman et al., 1987). No similar studies exist for *C. pseudogenitalium*. 3) *C. aquaticum* and the CDC groups A-1 and A-2 (*Oerskovia xanthineolytica, O. turbata*), A-3, A-4, and A-5 (Hollis and Weaver, 1981). These are motile and possess mainly branched fatty acids that exclude them from the genus *Corynebacterium* (von Graevenitz et al., 1990). 4) CDC groups B-1 and B-3 (Hollis and Weaver, 1981). While they are nonmotile, they also mainly possess branched fatty acids (von Graevenitz et al.,

1990). 5) CDC group E (Hollis and Weaver, 1981). This organism is catalase-negative, and is probably an aerotolerant *Bifidobacterium adolescentis*. 6) Unspeciated *"lipophilic diphtheroids."* This is an ill-defined group of corynebacteria from the human skin which has been investigated by some authors who were mostly unable to speciate them (Smith, 1969a; Somerville, 1973). Common to them was either lipid dependence or growth stimulation by lipids (see definition of lipid dependence in "Nutritional Requirements," this chapter). Some of them may have been *C. jeikeium, C. xerosis, C. minutissimum,* or CDC group D-2, some may have belonged to as yet undefined taxa. 7) Species formerly counted among the corynebacteria which have now been assigned to other genera, i.e., *Arcanobacterium haemolyticum* (formerly *C. haemolyticum*), *Actinomyces pyogenes* (formerly *C. pyogenes*), *Gardnerella vaginalis* (formerly *C. vaginale*) and *Rhodococcus equi* (formerly *C. equi*).

Isolation

Corynebacteria are frequently isolated from human or animal specimens. They occur either as true pathogens or as cutaneous or mucocutaneous contaminants. The latter are particularly bothersome in specimens that are normally sterile, such as blood or body fluids.

Maintenance of bacteria is assured in ordinary semisolid transport media. Medically important corynebacteria appear within 24 to 48 h on sheep blood agar incubated aerobically or in a CO_2 atmosphere. They do not grow on enteric media. Several species (see Table 1) do not grow under anaerobic conditions. Lipids and vitamins are generally supplied through blood (or serum in the case of broths), but addition of 0.3–3% Tween 80 may be necessary for optimal growth. Chocolatized blood agar can be inhibitory to some strains, e.g., of *C. diphtheriae* (Glass, 1939) or *C. jeikeium* (A. von Graevenitz, unpublished observation).

Selection of corynebacteria is accomplished in the clinical laboratory by the use of media inhibitory to Gram-negative bacteria, such as colistin-nalidixic acid blood agar (as used for streptococci). Also, a fosfomycin disk may be placed on blood agar to select coryneforms in the surrounding inhibition zone (Wirsing von König et al., 1988). Furoxone (50–100 mg/1), added to trypticase-soy yeast-extract agar with lipids, selects corynebacteria as well (Smith, 1969b).

Special mention must be made of *C. diphtheriae*. Since the prevalence of diphtheria today is low in North America and Europe, addition of a selective medium for *C. diphtheriae* (as outlined below) is necessary only in suspicious cases or for epidemiological purposes, i.e., in the search for carriers. Appropriate specimens would be swabbings from the nose, the throat, and particluarly from skin lesions; rarely from the conjunctiva or the vagina. If rapid processing of such specimens is not possible, a silica gel-transport medium (Kim-Farley et al., 1987) or a specially designed liquid enrichment medium for *C. diphtheriae* (Calalb et al., 1961) may be used.

Liquid-Enrichment Medium for *C. diphtheriae*
(Calalb et al., 1961, modified as personally
communicated by P. Maximescu)

Proteose peptone no. 3 (Difco)	9.0 g
Meat extract	9.0 g
NaCl	2.7 g
Na$_2$HPO$_4$·10H$_2$O	1.8 g
Glucose	1.8 g
Distilled water	900.0 ml
Potassium tellurite, 2% (w/v)	75.0 ml

The thoroughly mixed solution is adjusted to pH 7.4 and Seitz-filtered. The following sterile items are added aseptically:

Bovine serum	100.0 ml
Nystatine (1,000 units/ml)	1.15 ml
Di-Sodium-Fosfomycin (Boehringer)	200.0 mg
Glucose-6-phosphate (Boehringer Mannheim)	33.64 mg
Egg yolk suspension (chicken)	from 10 eggs
L-cystine 1% (w/v)	1.0 ml

The entire medium is mixed well and placed in tubes (2–3 ml per tube). It can be stored at 4°C for 1 to 2 months.

The liquid enrichment may give better isolation rates in antibiotically pretreated patients, since inhibiting substances present in the specimen are diluted. This medium is also recommended for epidemiological purposes, i.e., in the search for carriers. It is kept either at room temperature or at 4°C. Addition of fosfomycin (132 mg/l) and nystatin (860,000 I.U./l) are recommended to prevent overgrowth by staphylococci and fungi. When using silica gel for transportation, it is essential that desiccated swabs be incubated overnight in a broth supplemented with plasma or blood before they are plated on agar media (Kim-Farley et al., 1987). Likewise, swabs in liquid enrichment medium are incubated overnight at 35–37°C before subcultivation on solid media.

Media suitable for the isolation of *C. diphtheriae* are Loeffler or Pai slants. These are nonselective but favor growth of *C. diphtheriae* and

its characteristic microscopic morphology (see "Identification and Typing," this chapter).

Loeffler Slants (MacFaddin, 1985)

To 100 ml peptone broth containing 1% glucose, adjusted to pH 7.6, add 300 ml of sterile normal bovine serum. After thorough mixing, distribute in 3 ml aliquots into tubes and inspissate for 30 min in a slanted position. Sterilize on 3 successive days at 75–85°C for 2 h.

Pai Slants (MacFaddin, 1985)

Whole chicken eggs	
Glucose	5.0 g
Distilled water	1 l
	500.0 ml

Mix gently and filter through two layers of gauze. Add 120.0 ml of glycerol and mix gently to avoid formation of bubbles. Dispense aliquots of 5 ml into tubes and autoclave in a slanted position. Avoid foaming by controlled air escape and influx. The medium can be stored at 4°C for 6 to 8 weeks.

Also available are various selective media such as Tinsdale agar (Tinsdale, 1947; Moore and Parsons, 1958), cystine tellurite blood agar (Frobisher, 1937), and the previously described liquid enrichment medium (Calalb et al., 1961). They rely on the capacity of *C. diphtheriae* and other coryneform bacteria to reduce tellurium salts (Conradi and Trosch, 1912). Tellurium crystals accumulate in corynebacteria (Morton and Anderson, 1941; Tucker et al., 1962), rendering colonies black.

Tinsdale Agar (MacFaddin, 1985; modified as personally communicated by P. Maximescu)

Proteose peptone no. 3 (Difco)	20.0 g
NaCl	5.0 g
Agar (Difco)	20.0 g
Distilled water	1 l

The molten medium is adjusted to pH 7.4 and autoclaved for 15 min at 121°C. This basic medium can be stored at 4°C for 3 to 4 months. After cooling down to 55°C, the following sterile solutions are aseptically added:

Formalinized sheep blood (100 ml blood + 0.125 ml formalin)	3.0 ml
0.1 N NaOH	60.0 ml
L-cystine 0.4% (0.24 g L-cystine dissolved in 60 ml 0.1 N HCl at 60 to 70°C)	60.0 ml
Potassium tellurite, 1% (w/v) aqueous	30.0 ml
Sodium thiosulfate (2.5 g dissolved in 100 ml distilled water at 60°C)	17.0
Bovine serum (sterile)	100.0–150.0 ml

The medium is poured into petri dishes and stored at 4°C for no more than 4 days. Each lot must be checked with a reference strain for satisfactory halo production.

Cystine Tellurite Blood Agar (MacFaddin, 1985)

Heart infusion agar, 2%, sterile, cooled to 50°C	100.0 ml
K_2TeO_3 (potassium tellurite), 0.3%, autoclaved	15.0 ml
Sheep blood, sterile	5.0 ml
Mix well and add:	
L-cystine powder, sterile	0.005 g

Pour into petri dishes while shaking steadily. The shelf life at 4°C is about 1 month.

The growth of most noncoryneform bacteria is inhibited by tellurite. On Tinsdale agar, *C. diphtheriae, C. ulcerans,* and *C. pseudotuberculosis* produce not only black colonies, but also a brown to gray halo whose color depends on the kind of medium used. Thus, the halo allows a preliminary differentiation. Some other species, however, primarily *Staphylococcus aureus* and streptococci, may produce black colonies and, rarely, halos. Since some strains of *C. diphtheriae* may not grow on some older batches of Tinsdale agar, cystine tellurite blood agar (on which no halo will form) has been recommended throughout (Coyle et al., 1985).

Two selective media for *C. jeikeium* have also been devised. One, a trypticase soy broth to determine carrier rates, contains 1% yeast, 0.5% Tween 80, and 10 mg/l gentamicin plus 1 mg/l cephapirin (Tompkins et al., 1982). The other one is lecithin-Tween medium (Wichmann et al., 1984), a tryptose agar supplemented with 3% Tween 80, 0.5% phosphatidyl choline (lecithin), 0.1% histidine, 0.5% sodium thiosulfate, 0.3% glycerol, 100 mg/l ticarcillin, 200 mg/l 5-fluorocytosine, and 100 mg/l fosfomycin (with 30 mM glucose-6-phosphate).

Preservation of Cultures

Well-grown cultures of corynebacteria may be kept on Loeffler or Pai slants at room temperature but transfer of cultures should be done every 6 months. Alternatively, corynebacteria can be stored for years when lyophilized. Any preservation medium can be used for lyophilization. The following protocol has been used successfully in our laboratory:

A 24-h broth is centrifuged at 3,000 rpm for 15 min, the supernatant decanted, and the sediment resuspended in the same volume of 0.9% NaCl containing 2% bovine serum albumin. Lyophilization in aliquots of 1 ml follows. For regrowth, the freeze-dried bacteria are resuspended in 1 ml of nutrient broth and streaked onto blood agar.

Identification and Typing

Stains

Gram staining of direct material or cultures from routine media reveals the features listed in Table 2. The diagnostic value of stained slides is limited since other Gram-positive rods (e.g., *Arcanobacterium, Rhodococcus*) may show similar morphologies. Furthermore, direct smears of suspected diphtheritic material are of low sensitivity.

Differential stains from special media, however, are of greater diagnostic value, particularly for *C. diphtheriae*:

Loeffler's Methylene Blue Stain (MacFaddin, 1985)

Methylene blue chloride	1.0 g
Ethanol 96%	10.0 ml

Mix and keep for 7 days at 37°C. Prepare a 10% aqueous solution and add 10 ml of a 0.1% (w/v) aqueous KOH solution. After incubation at 37°C for 1 week, the solution is filtered and ready for use. Smears are heat fixed, stained for 10 min, and washed with tap water.

Neisser's Stain

Solution I:	Methylene blue	1.0 g
	Ethanol 96%	20.0 ml
	Distilled water	950.0 ml
	CH_3COOH (> 95%) (glacial acetic acid)	50.0 ml
Solution II:	Crystal violet	1.0 g
	Ethanol 96%	10.0 ml
	Distilled water	300.0 ml

Mix freshly two parts of solution I with one part of solution II before use.

Solution III:	Lugol's solution	100.0 ml
	Lactic acid, conc.	1.0 ml
Counter stain:	Chrysoidin	2.0 g
	Distilled water	300.0 ml
Dissolve by heating to 100°C.		

The heat-fixed smears are stained with the mixture of solution I and II for 20 to 30 s. After washing with tap water, solution III is applied for 5 s. Counterstaining is done for 5 min.

A presumptive identification of *C. diphtheriae* can be made from such stains of cultures on Loeffler or Pai slants (morphologic characteristics are optimally expressed on Loeffler slants). Gram-stained smears of *C. diphtheriae* show slender, pleomorphic rods whose ends may be slightly swollen, producing a club shape. Cells in V- and L-shaped arrangements creating "Chinese characters" are observed as well. When Neisser's stain is used, intense brownish

Table 2. Morphology of corynebacteria on sheep blood agar.

Species	Colonies at 24–48 h	Microscopic morphology
C. diphtheriae	[a]	[a]
C. ulcerans	Matt, move en masse when touched	Similar to *C. diphtheriae* type gravis
C. pseudotuberculosis	Small, yellowish white	Short, irregular staining, clubs, metachromatic granules
C. xerosis	R[b] or S[b], slightly yellowish to tan	Short to long, irregular staining, barred, occasionally clubs
C. renale	White to buff, dry	Coccoid to long, irregular staining, occasionally pointed ends
C. kutscheri	Irregular edges, gray to yellow	Beaded slender rods, occasionally clubs and pointed ends, metachromatic granules
C. striatum	S, white, small, some yellowish	Short, barred, clubs, dumbbell shapes
C. pseudodiphtheriticum	S, white to buff	Short to coccoid, evenly staining (median septum may be unstained)
C. minutissimum[c]	S, small, circular	Short, some irregular staining, palisading, V forms
C. jeikeium	S, small, grayish white	Short, pleomorphic, clubs, V forms, granules, palisades
CDC group F[d]	S, small	Medium to large, pleomorphic
CDC group G[d]	S, small	Coccoid to medium, barred, clubs
CDC group D-2[d]	S, whitish, opaque	Coccoid to short, palisades, V forms
C. bovis[d]	S, white to cream, circular	Coccoid to medium, barred, clubs
CDC group ANF	S, medium	Uncharacteristic
C. matruchotii	R, opaque, tough, adherent, three colonial types	Pleomorphic, filamentous, branching, "whip handles"
CDC group 1 and 2	S, opaque, circular, white to cream-colored	Uncharacteristic

[a]See text.
[b]R, rough; S, smooth.
[c]If grown on Mueller-Hinton agar (Cohen and Nickolai, 1969) or on media with 20% bovine fetal serum, red to orange fluorescence under UV light.
[d]Slow growth (colonies visible ≥24 h).

black metachromatic granules at one or both ends or in the middle of yellow-stained cells can be seen. The V- and L-shaped arrangement together with the brown polar bodies gives the impression of scattered matches. With Loeffler's methylene blue stain the metachromatic granules appear reddish against the blue background of the cell body. On all other media (e.g., blood agar, cystine tellurite blood agar, Tinsdale agar) metachromatic granules form poorly.

On Loeffler slants of *C. ulcerans,* pleomorphic coccoid forms predominate, and few metachromatic granules are seen. *C. pseudotuberculosis* yields coccoid forms as well.

Cultures

Colonial morphologies of the various *Corynebacterium* species are listed on Table 2. None are so characteristic as to yield a tentative diagnosis. Three cultural types of *C. diphtheriae,* commonly referred to as biotypes, can be distinguished (McLeod, 1943): *gravis, mitis,* and *intermedius.* The cultural aspects of the biotypes are best expressed, but not always easily recognized, on cystine tellurite blood agar. The *gravis* type forms large grey-black colonies with

a granular, radially striated surface and crenated margins (daisyheadlike colonies). The middle-sized colonies of the *mitis* type are intense black with regular edges and a convex, glossy surface. The *intermedius* type grows in small to middlesized colonies with a flat or convex granular surface and a raised black enter. On Loeffler slants, *C. diphtheriae* develops cream-colored, smooth, or granular colonies. *C. ulcerans* grows similarly but more luxuriantly. *C. pseudotuberculosis* grows on Loeffler slants in umbonate, opaque, friable, and slightly yellowish colonies.

On blood agar, *C. diphtheriae* shows pearl-grayish colonies. Weak beta hemolysis is observed with most *mitis* strains and a few *gravis* strains, but rarely with *intermedius* strains. In broth, *gravis* forms a pellicle, *intermedius* granular and *mitis,* diffuse turbidity. *C. ulcerans* shows a narrow zone of diffuse lysis and resembles *C. diphtheriae* type *gravis* on blood agar. Growth of *C. pseudotuberculosis* on blood agar is poor. Colonies are surrounded by a narrow zone of diffuse lysis.

A presumptive diagnosis of *C. diphtheriae* can also be made if typical colonies on Tinsdale agar show coryneform bacteria in the Gram stain (which may take up to 48 h).

Biochemical Characteristics

Biochemical reactions provide for a final species diagnosis. In a fair percentage of all corynebacterial isolates, however, a correct identification is impossible (Smith et al., 1973), and some strains diagnosed as corynebacteria by morphological criteria turn out to belong to the genera *Actinomyces, Lactobacillus, Arcanobacterium, Propionibacterium, Listeria, Nocardia, Erysipelothrix, Mycobacterium, Gardnerella,* and *Streptococcus* (Clarridge, 1986). Table 3 lists only those test results that are agreed upon by various authors (Collins and Cummins, 1986 Coyle et al., 1985; Estrangin et al., 1987; Harrigan, 1966; Hollis and Weaver, 1981; Jayne-Williams et al., 1966), and have differential diagnostic value. Contradictory results have been reported for hippurate hydrolyis, gelatinase, growth in 6.5% and 9% NaCl, Voges-Proskauer reaction, hydrolysis of various Tweens and of casein, β-galactosidase, DNase, and formation of propionic acid (Reddy and Kao, 1978; Estrangin et al., 1987). For the determination of fermentative vs. nonfermentative carbohydrate utilization, Triple Sugar Iron agar (TSI) or Kligler's Iron Agar (KIA) largely suffices; it may have to be supplemented with a few drops of serum. For the enteric fermentative base with Andrade's indicator and carbohydrates, addition of serum may also be necessary. In our hands, cystine trypticase agar (as used for fastidious Gram-negative rods) has also worked well. Hydrogen sulfide production in *C. diphtheriae, C. ulcerans,* and *C. pseudotuberculosis* can be detected on Tinsdale agar (see "Isolation," this chapter) or in tubes with Pisu's medium:

Pisu's Medium (Saragea et al., 1979, modified after T. Krech)

Meat extract (Oxoid)	4.0 g
Proteose peptone no. 3 (Difco)	20.0 g
NaCl	5.0 g
Agar (Bacto agar Difco)	7.0 g
Distilled water	1 l

Dissolve in steam and sterilize by filtration. Adjust pH to 7.5. Distribute in aliquots of 80 ml in 200 ml Erlenmeyer flasks and heat for one in steam. Cool down to 60°C and add the following substances to each flask:

Horse serum, sterile	30.0 ml
L-cystine, 1% (w/v)	15.0 ml
Lead acetate, 10% (w/v), sterilized	1.0 ml

Distribute in aliquots of 2 to 3 ml in small tubes and leave vertically on table to solidify. The medium can be stored at 4°C for up to 1 month. Inoculation is made by stabbing. High-cystinase producers render the agar diffuse black. A black colorization of the surface or the stabbing canal only is interpreted as negative.

TSI or KIA are not sensitive enough to indicate H_2S formation, and lead acetate paper is too sensitive (Hollis and Weaver, 1981).

For urease, nitratase, and esculin hydrolysis (the latter negative for all species listed except *C. kutscheri* and some *C. matruchotii* strains) traditional identification media can be used. The CAMP test was found positive in *C. renale* and in a minority of *C. xerosis* (Zaki, 1965). The phospholipase D of *C. ulcerans* and *C. pseudotuberculosis* has the effect of inhibiting the enhanced zone of hemolysis formed in the CAMP test at the junction of β-staphylococcal and β-streptococcal hemolysins, and eliciting hemolysis in conjunction with *R. equi* (Barksdale et al., 1981). The two species as well as *C. diphtheriae,* in contrast to all other species, do not form pyrazinamidase (Table 3).

Pyrazinamidase Test (Sulea et al., 1980) Using Pyrazinamide Agar

Dubos broth base	6.5 g in 1 l of distilled H_2O
Pyrazinamide	0.1 g
Na pyruvate	2.0 g
Agar	15.0 g

Dispense in 5-ml tubes, autoclave 15 lb 15 min. Slant. Inoculate slants with bacteria from overnight growth on chocolate agar slants. Incubate at 37°C for 24 h. A pink band in the agar (pyrazinoic acid) indicates the presence of pyrazinamidase.

Tests for lipid requirement have to start from a lipid-free medium which can be obtained by adding 0.1% activated charcoal (Ersgaard and Justesen, 1984) to the agar base.

There are also rapid methods for the diagnosis of corynebacteria: Minitek (BBL Microbiology Systems, Baltimore, MD) (Slifkin et al., 1986); API 20S (Analytab Products, Plainview, NY) (Kelly et al., 1984); API 20 Strep (API Systems, La Balme-les Grottes, France) (Tillotson et al., 1988); RIM series (Austin Biological Laboratories Inc., Austin TX) (Grasmick and Bruckner, 1987) and special carbohydrate media (Thompson et al., 1983; Hollis et al., 1980) have been used for various, but not for all, *Corynebacterium* spp. mentioned in Tables 1 through 3.

Detection of Toxin

Toxigenicity testing should be performed on all isolates of *C. diphtheriae, C. ulcerans,* and *C. pseudotuberculosis.* It is important to use at least 10 colonies, since toxigenic and nontoxigenic colonies may be found in the same strain (Simmons et al., 1980). Toxin can be detected in vivo, i.e., in the guinea pig test, or in vitro, i.e., in the agar immunodiffusion test (Elek,

Table 3. The most important biochemical characteristics of medical corynebacteria.

Species/Taxon	Action on sheep blood	Phospholipase D	Utilization of glucose	Nitrate reduction	Urease	Acid from: Glucose	Sucrose	Maltose	Starch	Others
C. diphtheriae	β/−	−	F	+[a]	−	+	−	+	V[b]	Halo on Tinsdale positive, pyrazinamidase negative
C. ulcerans	ly/−	v	F	−	+	+	V[D]	+/+[D]	+	
C. pseudotuberculosis	ly/−	v	F	v	+	+	V[D]	+[D]/+	−	
C. xerosis	ly/−	−	F	+	−	+	+/+[D]	+/+[D]	+	
C. renale	−/ly	−	F	−	+	+	−	−	−	Esculin positive[g]
C. kutscheri	ly/−	−	F	+	+	+	+	+	v	
C. striatum	−/α	−	F	+	−	+	+	−	v	
C. pseudodiphtheriticum	−	−	NF	+	+	−	−	−	−	
C. minutissimum	−	−	F	−	−	+	V	+	−	
C. jeikeium	−	−	F?(ND)	−	−	+[D]/+	−	V[D]	−	[d]
CDC group F	−	ND	F	v	−	+/+[D]	d	+/+[D]	ND	[d]
CDC group G	−	ND	F	d	−	+/+[D]	+/+[D]	V	ND	[d,f]
CDC group I	−/ly	ND	F	+	−	+	−	d	ND	Pyrazinamidase positive[a]
CDC group D-2	−/ly	−	NF	−	+	−	−	−	−	
C. bovis	−	−	F	−	+	+	−	+	ND	[d]
CDC group ANF	−/ly	ND	NF	d	−	−	−	−	ND	
C. matruchotii	ly/α	−	F	+/(−)	−	+	+/+[D]	V	V	Esculin v[g]
CDC group 1	α	ND	F	+	−	+/+[D]	+/+[D]	+	+	
CDC group 2	−	ND	F	−	−	+/+[D]	−	+/+[D]	+	

α, alpha hemolysis; β, beta hemolysis (see text for *C. diphtheriae*); +, positive; −, negative; V, variable; /, or (more frequent reactions listed first); ( ), occasional strain; ND, no data available; d, different biotypes in CDC groups F, G, I, ANF; D, delayed beyond 48 h; F, fermentative; ly, diffuse lysis; NF, nonfermentative.

[a] *C. diphtheriae* biotype *mitis* var. *belfanti* is nitrate negative.

[b] *C. diphtheriae* biotype *mitis* and *intermedius* are starch negative, biotype *gravis* is starch positive.

[c] *C. renale* sensu stricto. For *C. pilosum* and *C. cystitidis*, see text.

[d] Needs serum for growth.

[e] Important as differential characteristic vs. *C. diphtheriae*.

[f] Indistinguishable from *C. matruchotii* except for its serum requirement and microscopic morphology.

[g] Only *C. kutscheri* and some *C. matruchotii* are esculin positive.

1949). The guinea pig test is a reliable test for diphtheria toxin detection even in inexperienced hands. Two guinea pigs are required. The inoculum is prepared either from the entire growth of a 24-h Loeffler or Pai slant suspended in 12 ml of sterile broth or from a heavy 24- to 48-h broth culture (saline cannot be used since C. diphtheriae rapidly loses viability in saline suspensions). The inoculum should have at least a McFarland No. 3 density. One animal is injected intraperitoneally with 1,000–2,000 I.U. of diphtheria antitoxin. One to 3 h later, this animal and an unprotected animal are inoculated subcutaneously with 2 to 3 ml of the culture suspension. A toxigenic strain will usually cause death in the unprotected animal within 1 to 3 days, whereas the protected (control) animal will survive. Characteristically, the toxin causes swelling and hemorrhages of the adrenal glands. The same reaction may be obtained by a diphtheria toxin-producing strain of C. ulcerans. The second C. ulcerans toxin, identical with the toxin of C. pseudotuberculosis (phospholipase D), is also lethal for guinea pigs but is not neutralized by diphtheria toxin and does not affect the adrenal glands. It can be shown by inhibition of the CAMP test (see above).

In the Elek test, the test strain is streaked at a right angle to an antitoxin-saturated filter paper strip laid on Elek agar. Toxin production is indicated by development of a precipitation line between the culture streak and the filter paper within 16–48 h of incubation at 35–37°C. The specificity of the test greatly depends on the quality of the antiserum used. The numerous problems encountered with this test have been described by several authors (Bickham and Jones, 1972; Snell et al., 1984). Phospholipase D does not interfere with the test.

Elek Agar (MacFaddin, 1985)

Solution I:	Proteose peptone no. 3	
	(Difco)	3.0 g
	Maltose	0.6 g
	Distilled water	100.0 ml

Adjust solution to pH 7.8 and sterilize by filtration.

Solution II:	Agar no. 1 (Oxoid)	3.0 g
	NaCl	1.0 g
	Distilled water	100.0 ml

Adjust to pH 7.8 and dissolve in steam for 1 h.

The solutions are mixed and distributed in aliquots of 10 ml into tubes and heated in steam for 15 min on 3 successive days. After cooling to 40°C, 2–4 ml of sterile bovine serum is added to each tube and the contents are poured into petri dishes. Immediately after solidification, antitoxin filter paper strips are gently pressed with sterile forceps onto the agar surface. Instead of serum, K-L enrichment broth (MacFaddin, 1985) (Difco Laboratories, Detroit, MI) may be used.

Preparation of the antitoxin filter paper strips:
Sterile filter paper strips of 6 by 1 cm are soaked with 150–200 µl of diphtheria antitoxin diluted to about 500 I.U./ml. The excess fluid is removed and the strips are dried for 1 h at 37°C. They can be stored in a closed tube at 4°C for several months.

The test must never be performed without a positive and negative control strain. Optimally, one single precipitation line is produced with the positive control and positive test strains.

Diphtheria toxin can also be detected by inoculation of embryonated eggs (Knothe, 1955), in tissue culture cells (Gabliks and Solotorovsky, 1962), or by immunoassays (Holmes and Perlow, 1975; Krech and Wittelsburger, 1987). Some of these tests have been used for direct toxin detection in body fluids, but have not been evaluated on a large scale.

Detection of Epidemiological Markers

To analyze epidemiological interrelationships of C. diphtheriae isolates, phage typing has been used successfully (Saragea et al., 1979). Recently, polypeptide analysis by sodium dodecyl sulfate polyacrylamide gel electrophoresis (SDS-PAGE) and DNA fingerprinting have been added. The results compared favorably with phage typing (Krech et al., 1988; Rappuoli et al., 1988). Plasmid analysis (Khabbaz et al., 1986; Kerry-Williams and Noble, 1986) and protein analysis by SDS-PAGE (Jackman et al., 1987) have also been used for C. jeikeium typing.

Habitat, Medical Significance, Pathogenicity, and Antibiotic Susceptibility

Corynebacterium diphtheriae, C. ulcerans, and C. pseudotuberculosis

C. diphtheriae occurs primarily in humans, but it has been isolated from animals, albeit very rarely, e.g., from teat ulcers of otherwise healthy cows (Greathead and Bisschop, 1963). It may, however, have been transmitted from sores on the hands of milkers. C. diphtheriae is found not only in diseased but also in healthy individuals (carriers) (Frost et al., 1936). Spread is from person to person via droplets or contact. Diphtheric skin ulcers are an important source of infection in developing countries (Gunatillake and Taylor, 1968). In a recent resurgence of diphtheria in an improverished district of Seattle, cutaneous manifestations played an important role (Harnisch et al., 1989).

The full clinical picture of diphtheria is caused by toxin-producing strains, but nontoxigenic bacteria can most likely cause local symptoms of mild diphtheria such as sore throat and enlarged tonsils (Edward and Allison, 1951). However, before the diagnosis of diphtheria caused by nontoxigenic strains is established, other possibilities have to be ruled out. First, toxin detection in the Elek test is not always reliable, as pointed out above (Snell et al., 1984; Bickham and Jones, 1972). Second, toxigenic and nontoxigenic variants of a single strain can be found in the same patient (Simmons et al., 1980); and third, other pathogens such as Epstein-Barr virus or hemolytic streptococci have to be excluded. Nontoxigenic strains have also produced endocarditis (McCloskey, 1985; Namnyak et al., 1987) and arthritis (Guran et al., 1979). The toxin, however, is the principal pathogenic factor in causing organ complications in severe diphtheria (Pappenheimer, 1982). Diphtheria toxin, a heat-labile polypeptide, is produced by bacteria replicating at the site of infection, and the toxin is then transported by the bloodstream throughout the body. The toxin causes damage of heart, kidney, and peripheral nerves by interference with cell protein synthesis (Pappenheimer, 1982). The fragment A of the toxin inhibits polypeptide chain elongation by inactivating the elongation factor EF-2 which is necessary for translocation of polypeptidyl-transfer RNA from the acceptor to the donor site on the ribosome. The mechanism of EF-2 inactivation involves catalysis of a reaction that yields free nicotinamide plus an inactive ADP-ribose-EF-2 complex. Fragment B is required for transport of fragment A into the cell.

Inflammation at the entry site (usually the oropharynx) is characterized by a diphtheric pseudomembrane consisting of fibrin, bacteria, and necrotic cells of the mucous membrane. The term "pseudomembrane" means that the membrane is not a normal host constituent. It is tightly attached to the underlying tissue and bleeding occurs if attempts are made to pull it off. The presence of local pseudomembrances and normal or moderately elevated body temperature in combination with general weakness or even toxic appearance, and occasional organ complications (myocarditis, nephritis, and neuritis) are indications of diphtheria. Bacteremia is rare. A detailed description of the manifestations of diphtheria is given by McCloskey (1985). Active immunization with diphtheria toxoid protects from severe disease and acts as an immunologic barrier against epidemic spread of diphtheria. Furthermore, the carrier rate decreases in parallel with mass-immunization (Centers for Disease Control, 1985).

C. ulcerans was first isolated in humans with sore throats (Gilbert and Stewart, 1927). It can also be cultivated from the nasopharynx of human carriers (Cook and Jebb, 1952) although, unlike *C. diphtheriae,* spread from person to person has not been shown so far. The pathogen rarely causes mastitis in cows (Higgs et al., 1967) and has been isolated in a high percentage from the respiratory tract of horses (Maximescu et al., 1974). Human infection is usually associated with consumption of raw milk and generally is seen clinically as a mild sore throat; however, rare infections indistinguishable from diphtheria with formation of pseudomembranes have been seen (Henriksen and Grelland, 1952; Meers, 1979). *C. ulcerans* strains may produce phospholipase D identical to the *C. pseudotuberculosis* toxin and a protein identical to diphtheria toxin (see "Detection of Toxin," this chapter) (Maximescu et al., 1974; Wong and Groman, 1984; Carne and Onon, 1982). Diphtheria toxin producers as well as nonproducers can cause human disease.

The natural habitats of *C. pseudotuberculosis* (formerly called "*C. ovis*") are animals such as sheep and horses. The organism causes ulcerative lymphangitis of the limbs in horses with abscesses of internal organs and caseous (pseudotuberculous) lymphadenitis in sheep. Lymphadenitis in cattle, goats, and deer has also been reported. Infection in humans occurs rarely, either by ingestion of milk or by direct contact (Goldberger et al., 1981). It usually is seen as granulomatous, necrotizing lymphadenitis of the axilla, groin, or cervical region. One case of eosinophilic pneumonia has been reported (Keslin et al., 1979). The organism has been found most frequently in Australia, but human infections have also been described from the United States and France (Peloux et al., 1980). Although *C. pseudotuberculosis* is capable of phage-induced diphtheria toxin production (Maximescu et al., 1974), not one among the human isolates reported to date has been shown to produce this toxin. Rather, the toxin it produces is phospholipase D (Onon, 1979).

Antibiotics have no effect on the action of diphtheria toxin but stop multiplication of the organisms in diseased persons and carriers, thereby limiting the size of an outbreak. Penicillin and erythromycin are largely effective agents in vitro. Erythromycin has been more effective against the carrier state in some epidemics (Zalma et al., 1970) but resistance has been observed (Coyle et al., 1979).

Corynebacterium xerosis

C. xerosis is a member of the normal flora of the skin and occasionally of mucocutaneous membranes (nasopharynx). It has caused a variety of infections in compromised hosts, such as endocarditis on normal or prosthetic valves, septicemia, empyema, wound infection, pneumonia, and CAPD peritonitis (Bennet et al., 1986; Eliakim et al., 1983; Lipsky et al., 1982; Porschen et al., 1977; Valenstein et al., 1988) Antibiotic susceptibility is variable; strains resistant to penicillin, aminoglycosides, and clindamycin are not uncommon (Porschen et al., 1977).

The *Corynebacterium renale* group

C. renale is a rare agent of urinary infection in cattle (Collins and Cummins, 1986), and has been involved in a human breast abscess (Peloux et al., 1981). *C. pilosum* and *C. cystitidis* have so far, with one exception (pus from an anal abscess; Chatelain et al., 1980), not been seen in humans but have been isolated from the urogenital tract of cows (*C. pilosum*), the prepuce of steer, and, rarely, the urine of cows (*C. cystitidis*). Both can cause cystitis and pyelonephritis. Toxins have not been found.

Corynebacterium kutscheri

Apparently, *C. kutscheri* is a common commensal in mice, rats, and voles that develop pseudotuberculous lesions if body defenses are altered. Experimental (parenteral and oral) infection is also possible. The possibility of an exotoxin has been discussed (Wilson, 1984). There is one report of chorioamnionitis in humans due to *C. kutscheri* (Fitter et al., 1979) and one of septic arthritis (Messina et al., 1989).

Corynebacterium striatum

This organism is a normal inhabitant of the human nasal mucosa (Collins and Cummins, 1986). It has been isolated from respiratory specimens and from the blood in two patients with compromised defense mechanisms (Barr and Murphy, 1986; Bowstead and Santiago, 1980). Strains from cows with mastitis may not refer to the same organisms (Collins and Cummins, 1986).

Corynebacterium pseudodiphtheriticum

This organism is part of the normal pharyngeal flora. Disease is rare: endocarditis (Johnson and Kaye, 1970; Rubler et al., 1982), pulmonary infection (Donaghy and Cohen, 1983; Miller et al., 1986), urinary infection (Nathan et al.,

1982) and lymphadenitis (LaRocco et al., 1987) have been reported. Most of the patients involved were compromised hosts.

Corynebacterium minutissimum

This organism was first isolated in 1961 from human cases of erythrasma (Sarkany et al., 1961), a superficial skin infection characterized by pruritic, reddish brown macular patches mainly in intertriginous areas. Apparently, it can also be part of the normal flora of the skin (Coyle et al., 1985). Rare infections reported are septicemia in a leukemic (Guarderas et al., 1986), endocarditis (Brian and Brucker, 1985), and breast abscesses (Berger et al., 1984).

Corynebacterium jeikeium

This bacterium, first reported in 1976 as "CDC group JK" (Hande et al., 1976), is now recognized as a frequent colonizer of the inguinal, perirectal, and axillary skin in hospitalized patients (Larson et al., 1986; Young et al., 1981). Normal individuals harbor them more rarely and in lower counts than patients, and males more frequently than females (Gill et al., 1981; Kerry-Williams and Noble, 1987; Stamm et al., 1979). Risk factors for colonization and infection other than prolonged hospitalization are neutropenia, antibiotic treatment, and indwelling plastic devices (Lipsky et al., 1982; Riebel et al., 1986). Infections, almost all of them nosocomial and occurring in approximately 20% of colonized individuals (Riebel et al., 1986), are: septicemia (with and without skin lesions) and other intravenous catheter infections (Gill et al., 1981), endocarditis on native or prosthetic valves (Vanbosterhaut et al., 1989), meningitis (Hoffmann et al., 1983), shunt infections (Keren et al., 1988), peritonitis after CAPD (Altwegg et al., 1984), pneumonia (McNaughton et al., 1988), and purulent infections elsewhere (Gronemeyer et al., 1980). Environmental contamination may be extensive (Quinn et al., 1984). Molecular methods have led to different conclusions regarding the question whether *C. jeikeium* is transmitted by cross-infection, or exclusively selected by antibiotic treatment (Kerry-Williams and Noble, 1986; Khabbaz et al., 1986). Use of selective media (see "Isolation," this chapter) may increase the number of positive specimens up to 10-fold (Wichmann et al., 1984). Some suspect that *C. jeikeium* is a lipophilic diphtheroid of the skin which was selected by antibiotic treatment (McGinley et al., 1985). It is generally resistant to β-lactam antibiotics, aminoglycosides, and trimethoprim-sulfamethoxazol, less frequently also to chloramphenicol, tetracycline, clindamycin, and erythromycin. It has always been found susceptible to vancomycin and to the newer quinolones (Riley et al., 1979; Kerry-Williams and Noble, 1987). There are strains, however, that are susceptible to more antibiotics. The mean thickness of the surface layer (exterior to the cell wall) of resistant *C. jeikeium* cells was significantly greater than on susceptible cells (Blom and Heltberg, 1986). Mice are resistant against infection except when pretreated with cyclophosphamide and type-2 carrageenan (Traub, 1985).

   C. genitalium and *C. pseudogenitalium* (see "Chemotaxonomic Features," this chapter) have been observed by one group of authors as agents of urethritis (Evangelista et al., 1984).

CDC Groups F, G, I, ANF, 1, and 2

Group F corynebacteria have been cultured mainly from the human genitourinary tract, group G from the eye, and groups I, ANF, and 1 and 2 from several different sites (Hollis and Weaver, 1981; Na'Was et al., 1987). There is one report each of endocarditis with the groups G-2 and I-1 (Austin and Hill, 1983; Farrer, 1987), and one of a sternal wound infection with group I-2 (Riche et al., 1989).

CDC Group D-2

This organism commonly colonizes the skin, particularly the groin, of hospitalized patients. As a pathogen, it is mainly involved in urinary tract infections associated with an alkaline urine and struvite (ammonium magnesium phosphate) stones ("alkaline encrusted cystitis"). Risk factors are urologic procedures, previous use of antibiotics, age over 65 years, and previous urinary tract infections (which have given rise to inflammatory foci) (Nadal et al., 1988; Soriano and Fernandez-Roblas, 1988). The organism has rarely caused other diseases such as pneumonia (Jacobs and Perlino, 1979), endocarditis (Langs et al., 1988), bacteremia, and peritonitis associated with continuous ambulatory peritoneal dialysis (Vanbosterhaut et al., 1987). Multiple resistance to antibiotics (like in *C. jeikeium*) is the rule (Soriano et al., 1987).

Corynebacterium bovis

This organism is a normal inhabitant of the bovine udder where it can also cause mastitis. Musculoskeletal abscesses may follow horning or other trauma (Lipsky et al., 1982). *C. bovis* has been isolated from very few human infections (endocarditis, meningitis, chronic otitis media, a leg ulcer, and bacteremia after a ventriculojugular shunt procedure) (Bolton et al., 1975; Vale and Scott, 1977).

Corynebacterium matruchotii

This organism is found in the oral cavity of humans and primates, in dental calculi, and plaques. It has been isolated from three cases of ocular infections, and possibly from a maxillary abscess (Wilhelmus et al., 1979).

Literature Cited

Alshamaony, L., M. Goodfellow, D. E. Minnikin, G. H. Bowden, and J. M. Hardie. 1977. Fatty and mycolic acid composition of *Bacterionema matruchotii* and related organisms. J. Gen. Microbiol. 98:205–213.

Altwegg, M., K. Zaruba, and A. von Graevenitz. 1984. Corynebacterium group JK peritonitis in patients on continuous ambulatory peritoneal dialysis. Klin. Wochenschr. 62:793–794.

Austin, G. E., and E. O. Hill. 1983. Endocarditis due to *Corynebacterium* CDC group G2. J. Infect. Dis. 147:1106.

Barksdale, L., R. Linder, I. T. Sulea, and M. Pollice. 1981. Phospholipase D activity of *Corynebacterium pseudotuberculosis* (*Corynebacterium ovis*) and *Corynebacterium ulcerans*, a distinctive marker within the genus *Corynebacterium*. J. Clin. Microbiol. 13:335–343.

Barr, J. G., and G. P. Murphy. 1986. *Corynebacterium striatum*: an unusual organism isolated in pure culture from sputum. J. Infect. 13:297–298.

Bennet, W. M., W. K. Stewart, and A. C. Scott. 1986. Recurrent *Corynebacterium* CAPD peritonitis treated without catheter removal. Perit. Dial. Bull. 6:43–44.

Berger, S. A., A. Gorea, J. Stadler, M. Dan, and M. Zilberman. 1984. Recurrent breast abscesses caused by *Corynebacterium minutissimum*. J. Clin. Microbiol. 20:1219–1220.

Bickham, S. T., and W. L. Jones. 1972. Problems in the use of the in vitro toxigenicity test for *Corynebacterium diphtheriae*. Am. J. Clin. Pathol. 57:244–246.

Blom, J., and O. Heltberg. 1986. The ultrastructure of antibiotic-susceptible and multi-resistant strains of group JK diphtheroid rods isolated from clinical specimens. Acta Pathol. Microbiol. Immunol. Scand. Sect. B 94:301–308.

Bolton, W. K., M. A. Sande, D. E. Normansell, B. C. Sturgill, and F. B. Westervelt 1975. Ventriculojugular shunt nephritis with *Corynebacterium bovis*. Am. J. Med. 59:417–423.

Bowstead, T. T., and S. M., Santiago 1980. Pleuropulmonary infection due to *Corynebacterium striatum*. Br. J. Dis. Chest 74:198–200.

Brian, H. J., and A. J. Brucker. 1985. Embolic retinopathy due to *Corynebacterium minutissimum* endocarditis. Br. J. Ophthalmol. 69:29–31.

Calalb, G., A. Saragea, P. Maximescu, N. Cioroianu, A. Popescu, S. Popa, and A. Mihailescu. 1961. Recherches sur un milieu liquide d'enrichissement pour le diagnostic bactériologie de la diphthérie. Arch. Roum. Pathol. Exp. 20:95–101.

Carne, H. R., and E. O. Onon. 1982. The exotoxins of *Corynebacterium ulcerans.* J. Hyg. Camb. 88:173–191.

Centers for Disease Control. 1985. Diphtheria, tetanus, and pertussis: guidelines for vaccine prophylaxis and other preventive measures. MMWR 34:405–426.

Chatelain, R., E. P. Espaze, A. L. Courtieu, and H. H. Mollaret. 1980. Isolement de *Corynebacterium pilosum* dans un pus d'origine humaine. Méd. Mal. Infect. 10:361–363.

Clarridge, J. E. 1986. When, why, and how far should coryneforms be identified? Clin. Microbiol. Newsl. 8:32–34.

Cohen, S. N., and D. Nickolai. 1969. Simple medium for pigment production by the erythrasma diphtheroid. Appl. Microbiol. 17:479–480.

Collins, M. D., and C. S. Cummins. 1986. Genus *Corynebacterium* Lehmann and Neumann 1896, 350^AL, p. 1266–1276. In:P. H. A. Sneath (ed.), Bergey's manual of systematic bacteriology, vol. 2. Williams & Wilkins, Baltimore.

Collins, M. D., M. Goodfellow, and D. E. Minnikin. 1979. Isoprenoid quinones in the classification of coryneform and related bacteria. J. Gen. Microbiol. 110:127–136.

Collins, M. D., M. Goodfellow, and D. E. Minnikin. 1982a. A survey of the structures of mycolic acids in *Corynebacterium* and related taxa. J. Gen. Microbiol. 128:129–149.

Collins, M. D., M. Goodfellow, and D. E. Minnikin. 1982b. Fatty acid composition of some mycolic acid-containing coryneform bacteria. J. Gen. Microbiol. 128:2503–2509.

Collins, M. D., T. Pirouz, and M. Goodfellow. 1977. Distribution of menaquinones in actinomycetes and corynebacteria. J. Gen. Microbiol. 100:221–230.

Conradi, H., and P. Trosch. 1912. Ein Verfahren zum Nachweis der Diphtheriebazillen. Münch. Med. Wschr. 59:1652–1653.

Cook, G. T., and W. H. H. Jebb. 1952. Starch-fermenting, gelatin-liquefying corynebacteria and their differentiation from *C. diphtheriae gravis.* J. Clin. Pathol. 5:161–164.

Coyle, M. B., D. G. Hollis, and N. B. Groman. 1985. *Corynebacterium* spp. and other coryneform organisms, p. 193–204. In: E. H. Lennette et al., (ed.), Manual of clinical microbiology. American Society for Microbiology, Washington, DC.

Coyle, M. B., B. H. Minshew, J. A. Bland, and P. C. Hsu. 1979. Erythromycin and clindamycin resistance in *Corynebacterium diphtheriae* from skin lesions. Antimicrob. Agents Chemother. 16:525–527.

Donaghy, M., and J. Cohen. 1983. Pulmonary infection with *Corynebacterium hofmannii* complicating systemic lupus erythematosus. J. Infect. Dis. 147:962.

Edward, D. G., and V. D. Allison. 1951. Diphtheria in the immunized, with observation on a diphtheria-like disease associated with nontoxigenic strains of *Corynebacterium diphtheriae.* J. Hyg. Camb. 49:205–219.

Elek, S. D. 1949. The plate virulence test for diphtheria. J. Clin. Pathol. 2:250–258.

Eliakim, R., P. Silkoff, G. Lugassy, and J. Michel. 1983. *Corynebacterium xerosis* endocarditis. Arch. Intern. Med. 143:1995.

Ersgaard, H., and T. Justesen. 1984. Multiresistant lipophilic corynebacteria from clinical specimens. Acta Pathol. Microbiol. Immunol. Scand. Sect. B 92:39–43.

Estrangin, E., B. Thiers, and Y. Peloux. 1987. Apport des microméthodes et de l'analyse en chromatographie en phase gazeuse des acides carboxyliques issus de la fermentation du glucose dans l'identification des corynébactéries. Ann. Biol. Clin. 45:285–289.

Evangelista, A. T., K. M. Coppola, and G. Furness. 1984. Relationship between group JK corynebacteria and the biotypes of *Corynebacterium genitalium* and *Corynebacterium pseudogenitalium.* Can. J. Microbiol. 30:1052–1057.

Farrer, W. 1987. Four-valve endocarditis caused by *Corynebacterium* CDC group Il. South. Med. J. 80:923–925.

Fitter, W. F., D. J. Se Sa, and R. Richardson. 1979. Chorioamnionitis and funisitis due to *Corynebacterium kutscheri.* Arch. Dis. Child. 55:710–712.

Frobisher, M., Jr. 1937. Cystine-tellurite agar for *C. diphtheriae.* J. Infect. Dis. 10:99–105.

Frost, H. W., M. Frobisher, F. A. van Volkenburgh, and M. L. Levin. 1936. Diphtheria in Baltimore. A comparative study of morbidity, carrier prevalence and antitoxic immunity in 1921–24 and 1933–36. Am. J. Hyg. 24:568–586.

Gabliks, J., and M. Solotorovsky. 1962. Cell culture reactivity to diphtheria, staphylococcus, tetanus and *Escherichia coli* toxins. J. Immunol. 88:505–512.

Gilbert, R., and F. C. Stewart. 1927. *Corynebacterium ulcerans:* A pathogenic microorganism resembling *C. diphtheriae.* J. Lab. Clin. Med. 12:756–761.

Gill, V. J., C. Manning, M. Lamson, P. Woltering, and P. A. Pizzo. 1981. Antibiotic-resistant group JK bacteria in hospitals. J. Clin. Microbiol. 13:472–477.

Glass, V. 1939. The effect of blood digest and heme on the growth of *C. diphtheriae.* J. Pathol. Bacteriol. 49:549–561.

Goldberger, A. C., B. A. Lipsky, and J. J. Plorde. 1981. Suppurative granulomatous lymphadenitis caused by *Corynebacterium ovis (pseudotuberculosis).* Am. J. Clin. Pathol. 76:486–490.

Grasmick, A. E., and D. A. Bruckner. 1987. Comparison of rapid identification method and conventional substrates for identification of *Corynebacterium* group JK isolates. J. Clin. Microbiol. 25:1111–1112.

Greathead, M. M., and P. J. N. R. Bisschop. 1963. A report on the occurrence of *C. diphtheriae* in dairy cattle. S. Afr. Med. J. 37:1261–1262.

Groman, N., J. Schiller, and J. Russell. 1984. *Corynebacterium ulcerans* and *Corynebacterium pseudotuberculosis* responses to DNA probes derived from corynephage beta and *Corynebacterium diphtheriae.* Infect. Immun. 45:511–517.

Gronemeyer, P. S., A. S. Weissfeld, and A. C. Sonnenwirth. 1980. Corynebacterium group JK bacterial infection in a patient with an epicardial pacemaker. Am. J. Clin. Pathol. 74:838–842.

Guarderas, J., A. Karnad, S. Alvarez, and S. L. Berk. 1986. *Corynebacterium minutissimum* bacteremia in a patient with chronic myeloid leukemia in blast crisis. Diag. Microbiol. Infect. Dis. 5:327–330.

Gunatillake, P. D., and T. Taylor. 1968. The role of cutaneous diphtheria in the acquisition of immunity. J. Hyg. Camb. 66:83–88.

Guran, Ph., H.-H. Mollaret, R. Chatelain, M. Gropman, F. Prigent, and G. Béal. 1979. Arthrite purulente à bacille diphtérique atoxinogène. Arch. Franc. Pédiat. 36:926–929.

Hande, K. R., F. G. Witebsky, M. S. Brown, C. B. Shulman, S. E. Anderson, Jr., A. S. Levine, J. D. MacLowry, and B. A. Chabner. 1976. Sepsis with a new species of *Corynebacterium*. Ann. Intern. Med. 85:423–426.

Harnisch, J. P., E. Tronca, Ch. M. Nolan, M. Turck, and K. K. Holmes. 1989. Diphtheria among alcoholic urban adults. A decade of experience in Seattle. Ann. Intern. Med. 111:71–82.

Harrigan, W. F. 1966. The nutritional requirements and biochemical reactions of *Corynebacterium bovis*. J. Appl. Bacteriol. 29:380–391.

Henriksen, S. D., and R. Grelland. 1952. Toxigenicity, serological reactions and relationships of the diphtheria-like organism. J. Pathol. Bacteriol. 64:503–511.

Higgs, T. M., A. Smith, L. M. Cleverly, and F. K. Neave. 1967. *Corynebacterium ulcerans* infections in a dairy herd. Vet. Rec. 81:34–35.

Hoffmann, S., H. Ersgaard, T. Justesen, and H. Friis. 1983. Fatal meningitis with group JK *Corynebacterium* in a leucopenic patient. Eur. J. Clin. Microbiol. 2:213–215.

Hollis, D. G., F. O. Sottnek, W. J. Brown, and R. E. Weaver. 1980. Use of the rapid fermentation test in determining carbohydrate reactions of fastidious bacteria in clinical laboratories. J. Clin. Microbiol. 12:620–623.

Hollis, D. G., and R. E. Weaver. 1981. Gram-positive organisms: A guide to identification. Centers for Disease Control, Special Bacteriology Section, Atlanta.

Holmes, R. K., and R. B. Perlow. 1975. Quantitative assay of diphtheria toxin and of immunologically crossreacting proteins by reverse passive hemagglutination. Infect. Immun. 12:1392–1400.

Jackman, P. J. H., D. G. Pitcher, S. Pelczynska, and P. Borman. 1987. Classification of corynebacteria associated with endocarditis (group JK) as *Corynebacterium jeikeium* sp. nov. System. Appl. Microbiol. 9:83–90.

Jacobs, Jr., N. F., and C. A. Perlino. 1979. "Diphtheroid" pneumonia. South. Med. J. 72:475–476.

Jayne-Williams, D. J. and T. M. Skerman. 1966. Comparative studies on coryneform bacteria from milk and dairy sources. J. Appl. Bacteriol. 29:72–92.

Johnson, W. D., and D. Kaye. 1970. Serious infections caused by diphtheroids. Ann. N.Y. Acad. Sci. 174:568–576.

Kelly, M. C., I. D. Smith, R. J. Anstey, J. H. Thornley, and R. P. Rennie. 1984. Rapid identification of antibiotic-resistant corynebacteria with the API 2OS system. J. Clin. Microbiol. 19:245–247.

Keren, G., T. Geva, B. Bogokovsky, and E. Rubinstein. 1988. *Corynebacterium* group JK pathogen in cerebrospinal fluid shunt infection. J. Neurosurg. 68:648–650.

Kerry-Williams, S. M., and W. C. Noble. 1986. Plasmids in group JK coryneform bacteria isolated in a single hospital. J. Hyg. Camb. 97:255–263.

Kerry-Williams, S. M., and W. C. Noble. 1987. Group JK coryneform bacteria. J. Hosp. Infect. 9:4–10.

Keslin, M. H., E. L. McCoy, J. J. McCusker, and J. S. Lutch. 1979. *Corynebacterium pseudotuberculosis*. A new case of infectious eosinophilic pneumonia. Am. J. Med. 67:228–231.

Khabbaz, R. F., J. B. Kaper, M. R. Moody, S. C. Schimpff, and J. H. Tenney. 1986. Molecular epidemiology of group JK *Corynebacterium* on a cancer ward: lack of evidence for patient-to-patient transmission. J. Infect. Dis. 154:95–99.

Kim-Farley, R. J., T. I. Soewarso, S. Rejeki, S. Soeharto, A. Karyadi, and S. Nurhayati. 1987. Silica gel as transport medium for *Corynebacterium diphtheriae* under tropical conditions (Indonesia). J. Clin. Microbiol. 25:964–965.

Knothe, H. 1955. Zum Nachweis des Diphtherietoxins bei Diphtheriestämmen von Diphtheriebakterienträgern im bebrüteten Hühnerei. Dtsch. Med. Wschr. 80:785–787.

Krech, T., J. de Chastonay, and E. Falsen. 1988. Epidemiology of diphtheria: Polypeptide and restriction enzyme analysis in comparison with conventional phage typing. Eur. J. Clin. Microbiol. Infect. Dis. 7:232–237.

Krech, T., and C. Wittelsbürger. 1987. Immunologische Methoden zum Nachweis von Diphtherie-Toxin (Passive Hämagglutination und ELISA zum Toxinnachweis aus Kulturen und im Serum). Zentralbl. Bakteriol. A 265:124–135.

Langs, J. C., D. de Briel, C. Sauvage, J. F. Blickle, and H. Akel. 1988. Endocardite à *Corynebacterium* du groupe D2, à point de départ urinaire. Méd. Mal. Infect. 5:293–295.

LaRocco, M., C. Robinson, and A. Robinson. 1987. *Corynebacterium pseudodiphtheriticum* associated with suppurative lymphadenitis. Eur. J. Clin. Microbiol. 6:79.

Larson, E. L., K. J. McGinley, J. J. Leyden, M. E. Cooley, and G. H. Talbot. 1986. Skin colonization with antibiotic-resistant (JK group) and antibiotic-sensitive lipophilic diphtheroids in hospitalized and normal adults. J. Infect. Dis. 153:701–706.

Lipsky, B. A., A. C. Goldberger, L. S. Tompkins, and J. J. Plorde. 1982. Infections caused by nondiphtheria corynebacteria. Rev. Infect. Dis. 4:1220–1235.

MacFaddin, J. F. 1985. Media for isolation-cultivation-identification-maintenance of medical bacteria. Williams & Wilkins, Baltimore.

Maximescu, P., A. Oprisan, A. Pop, and E. Potorac. 1974. Further studies on *Corynebacterium* species capable of producing diphtheria toxin (*C. diphtheriae, C. ulcerans, C. ovis*). J. Gen. Microbiol. 82:49–56.

McCloskey, R. V. 1985. *Corynebacterium diphtheriae* (Diphtheria), p. 1171–1174. In: G. L. Mandell, R. G. Douglas, and J. E. Bennett (ed.), Principles and practice of infectious diseases, 2nd ed. John Wiley & Sons, New York.

McGinley, K. J., J. N. Labows, J. M. Zeckman, K. M. Nordstrom, G. F. Webster, and J. J. Leyden. 1985. Pathogenic JK group corynebacteria and their similarity to human cutaneous lipophilic diphtheroids. J. Infect. Dis. 152:801–806.

McLeod, J. W. 1943. The types *mitis, intermedius* and *gravis* of *Corynebacterium diphtheriae*. Bacteriol. Rev. 7:1–41.

McNaughton, R. D., R. R. Villamieva, R. Donnelly, J. Freedman, and R. Nawrot. 1988. Cavitating pneumonia caused by *Corynebacterium* group JK. J. Clin. Microbiol. 26:2216–2217.

Meers, P. D. 1979. A case of classical diphtheria, and other infections due to *Corynebacterium ulcerans*. J. Infect. 1:139:–142.

Messina, O. D., J. A. Maldonado-Cocco, A. Pescio, A. Farinati, and O. Garcia-Morteo. 1989. *Corynebacterium kutscheri* septic arthritis. Arthr. Rheum. 32:1053.

Miller, R. A., A. Rompalo, and M. B. Coyle. 1986. *Corynebacterium pseudodiphtheriticum* pneumonia in an immunologically intact host. Diag. Microbiol. Infect. Dis. 4:165–171.

Moore, M. S., and E. I. Parsons. 1958. A study of modified Tinsdale's medium for the primary isolation of *Corynebacterium diphtheriae*. J. Infect. Dis. 102:88–93.

Morton, H. E., and T. F. Anderson. 1941. Electron microscopic studies of biological reactions. I. Reduction of potassium tellurite by *Corynebacterium diphtheriae*. Proc. Soc. Exp. Biol. Med. 46:272–276.

Nadal, D., M. Schwöbel, and A. von Graevenitz. 1988. *Corynebacterium* group D2 and urolithiasis in a boy with megacalycosis. Infection 16:245–247.

Namnyak, S. S., R. P. Bhat, A. Al-Jama, and S. E. Fathalla. 1987. Prosthetic valve endocarditis caused by *Corynebacterium diphtheriae* in a patient with pemphigus vulgaris. J. Clin. Microbiol. 25:1330–1332.

Nathan, A. W., D. R. Turner, C. Aubrey, J. S. Cameron, D. G. Williams, C. S. Ogg, and M. Bewick. 1982. *Corynebacterium hofmannii* infection after renal transplantation. Clin. Nephrol. 6:315–318.

Na'Was, T. E., D. G. Hollis, C. W. Moss, and R. E. Weaver. 1987. Comparison of biochemical, morphologic, and chemical characteristics of Centers for Disease Control fermentative coryneform groups 1, 2, and A-4. J. Clin. Microbiol. 25:1354–1358.

Onon, E. O. 1979. Purification and partial characterization of the exotoxin of *Corynebacterium ovis*. Biochem. J. 177:181–186.

Pappenheimer, A. M. 1982. Diphtheria: Studies on the biology of an infectious disease. Harvey Lect. 76:45–73.

Peloux, Y., C. Maresca, and J.-H. Oddou. 1980. La lymphadénite suppurée provoquée par *Corynebacterium pseudotuberculosis* à propos d'un cas observé chez un berger alpin. Mediterr. Méd. 234:7–12.

Peloux, Y., R. Chatelain, and R. Erny. 1981. Abcès du sein provoqué par une corynebactérie du group "renale". Pathol. Biol. 29:299–300.

Pitcher, D. G. 1983. Deoxyribonucleic acid base composition of *Corynebacterium diphtheriae* and other corynebacteria with cell wall type IV. FEMS Microbiol. Lett. 16:291–295.

Porschen, R. K., Z. Goodman, and B. Rafai. 1977. Isolation of *Corynebacterium xerosis* from clinical specimens. Am. J. Clin. Pathol. 68:290–293.

Quinn, J. P., P. M. Arnow, D. Weil, and J. Rosenbluth. 1984. Outbreak of JK diphtheroid infections associated with environmental contamination. J. Clin. Microbiol. 19:668–671.

Rappuoli, R., D. M. Perugini, and E. Falsen. 1988. Molecular epidemiology of the 1984–1986 outbreak of diphtheria in Sweden. N. Engl. J. Med. 318:12–14.

Reddy, C. A., and M. Kao. 1978. Value of acid metabolic products in identification of certain corynebacteria. J. Clin. Microbiol. 7:428–433.

Riche, O., V. Vernet, C. Rouger, and V. Erhardt. 1989. Suppuration à *Corynebacterium* I₂. La Presse Méd. 18:1033–1034.

Riebel, W., N. Frantz, D. Adelstein, and P. J. Spagnuolo. 1986. *Corynebacterium* JK: A cause of nosocomial device-related infection. Rev. Infect. Dis. 8:42–49.

Riley, P. S., D. G. Hollis, G. B. Utter, R. E. Weaver, and C. N. Baker. 1979. Characterization and identification of 95 diphtheroid (group JK) cultures isolated from clinical specimens. J. Clin. Microbiol. 9:418–424.

Rubler, S., L. Harvey, A. Avitabile, and T. Abenavoli. 1982. Mitral valve obstruction in a case of bacterial endocarditis due to *Corynebacterium hofmanii*. N.Y. State J. Med. 82:1590–1594.

Saragea, A., P. Maximescu, and E. Meitert. 1979. *Corynebacterium diphtheriae*: Microbiological methods used in clinical and epidemiological investigations, p. 61–176. In: T. Bergan and J. R. Norris (ed.), Methods in microbiology, vol. 13. Academic Press, New York.

Sarkany, I., D. Taphu, and H. Blank. 1961. The etiology and treatment of erythrasma. J. Invest. Dermatol. 37:283–290.

Simmons, L. E., J. D. Abbott, M. E. Macaulay, A. E. Jones, A. G. Ironside, B. K. Mandal, T. N. Stanbridge, and P. Maximescu. 1980. Diphtheria carriers in Manchester: Simultaneous infection with toxigenic and nontoxigenic mitis strains. Lancet I:304–305.

Slifkin, M., G. M. Gil, and C. Engwall. 1986. Rapid identification of group JK and other corynebacteria with the Minitek system. J. Clin. Microbiol. 24:177–180.

Smith, R. F. 1969a. Characterization of human cutaneous lipophilic diphtheroids. J. Gen. Microbiol. 55:433–443.

Smith, R. F. 1969b. A medium for the study of the ecology of human cutaneous diphtheroids. J. Gen. Microbiol. 57:411–417.

Smith, R. F., D. Blasi, and S. L. Dayton. 1973. Isolation and characterization of corynebacteria from burned children. Appl. Microbiol. 26:554–559.

Snell, J. J. S., J. V. Demello, P. S. Gardner, W. Kwantes, and R. Brooks. 1984. Detection of toxin production by *Corynebacterium diphtheriae*: results of a trial organised as part of the United Kingdom National External Microbiological Quality Assessment Scheme. J. Clin. Pathol. 37:796–799.

Somerville, D. A. 1973. A taxonomic scheme for aerobic diphtheroids from human skin. J. Med. Microbiol. 6:215–224.

Soriano, F., and R. Fernandez-Roblas. 1988. Infections caused by antibiotic-resistant *Corynebacterium* group D2. Eur. J. Clin. Microbiol. Infect. Dis. 7:337–341.

Soriano, F., C. Ponte, M. Santamaria, A. Torres, and R. Fernandez-Roblas. 1987. Susceptibility of urinary isolates of *Corynebacterium* group D2 to fifteen antimicrobials and acetohydroxamic acid. J. Antimicrob. Chemother. 20:349–355.

Stamm, W. E., L. S. Tompkins, F. K. Wagner, G. W. Counts, E. D. Thomas, and J. D. Meyers. 1979. Infection due to *Corynebacterium* species in marrow transplant patients. Ann. Intern. Med. 91:167–173.

Sulea, I. T., M. C. Pollice, and L. Barksdale. 1980. Pyrazine carboxylamidase activity in *Corynebacterium*. Int. J. Syst. Bacteriol. 30:466–472.

Thompson, J. S., D. R. Gates-Davis, and D. C. T. Yong. 1983. Rapid microbiochemical identification of *Corynebacterium diphtheriae* and other medically important corynebacteria. J. Clin. Microbiol. 18:926–929.

Tillotson, G., M. Arora, M. Robbins, and J. Holton. 1988. Identification of *Corynebacterium jeikeium* and *Corynebacterium* CDC group D2 with the API 20 Strep system. Eur. J. Clin. Microbiol. Infect. Dis. 7:675–678.

Tinsdale, G. F. W. 1947. A new medium for the isolation and identification of *C. diphtheriae* based on the production of hydrogen sulfide. J. Pathol. Bacteriol. 59:461–466.

Tompkins, L. S., F. Juffali, and W. E. Stamm. 1982. Use of selective broth enrichment to determine the prevalence of multiply resistant JK corynebacteria on skin. J. Clin. Microbiol. 15:350–351.

Traub, W. H. 1985. Multiple drug-resistant *Corynebacteriaceae:* in vitro and in vivo (murine) studies. Chemotherapy 31:372–382.

Tucker, F. L., J. W. Walper, M. D. Appleman, and J. Donahue. 1962. Complete reduction of tellurite to pure tellurium metal by microorganisms. J. Bacteriol. 83:1313–1314.

Vale, J. A., and G. W. Scott. 1977. *Corynebacterium bovis* as a cause of human disease. Lancet II:682–684.

Valenstein, P., A. Klein, C. Ballow, and W. Greene. 1988. *Corynebacterium xerosis* septic arthritis. Am. J. Clin. Pathol. 89:569–571.

Vanbosterhaut, B., G. Claeys, J. Gigi, and G., Wauters. 1987. Isolation of *Corynebacterium* group D2 from clinical specimens. Eur. J. Clin. Microbiol. 7:418–419.

Vanbosterhaut, B., I. Surmont, J. Vandeven, G. Wauters, and J. Vandepitte. 1989. *Corynebacterium jeikeium* (group JK diphtheroids) endocarditis. A report of five cases. Diag. Microbiol. Infect. Dis. 12:265–268.

von Graevenitz, A., G. Osterhout, and J. Dick. 1990. Grouping of some Gram-positive rods by automated fatty acid analysis: Diagnostic implications. (In press).

Wichmann, S., C. H. Wirsing von Koenig, E. Becker-Boost, and H. Finger. 1984. Isolation of *Corynebacterium* group JK from clinical specimens with a semiselective medium. J. Clin. Microbiol. 19:204–206.

Wilhelmus, K. R., N. M. Robinson, and D. B. Jones. 1979. *Bacterionema matruchotii* ocular infections. Am. J. Ophthalmol. 87:143–147.

Wilson, G. 1984. *Corynebacterium* and other coryneform organisms, p. 94–113. In: M. T. Parker (ed.), Topley and Wilson's principles of bacteriology, virology, and immunity, 7th ed., vol. 2. E. Arnold Ltd., London.

Wirsing von König, C. H., T. Krech, H. Finger, and M. Bergmann. 1988. Use of fosfomycin disks for isolation of diphtheroids. Eur. J. Clin. Microbiol. Infect. Dis. 7:190–193.

Wong, T. P., and N. Groman. 1984. Production of diphtheria toxin by selected isolates of *Corynebacterium ulcerans* and *Corynebacterium pseudotuberculosis.* Infect. Immun. 43:1114–1116.

Young, V. M., W. F. Meyers, M. R. Moody, and S. C. Schimpff. 1981. The emergence of coryneform bacteria as a cause of nosocomial infections in compromised hosts. Am. J. Med. 70:646–650.

Zaki, M. M. 1965. Relations between staphylococcal beta-lysin and different corynebacteria. Vet. Rec. 77:941.

Zalma, V. M., J. J. Older, and G. F. Brooks. 1970. The Austin, Texas, diphtheria outbreak. Clinical and epidemiological aspects. JAMA 211:2125–2129.

CHAPTER 51

The Family Nocardiaceae

MICHAEL GOODFELLOW

Actinomycetes with *meso*-diaminopimelic acid (*meso*-A$_2$pm), arabinose, and galactose in the wall peptidoglycan (wall chemotype IV sensu Lechevalier and Lechevalier, 1970a) fall into two markedly distinct aggregate groups (Goodfellow and Lechevalier, 1989; Goodfellow and Minnikin, 1984). Wall chemotype IV actinomycetes containing mycolic acids, (high-molecular-weight 3-hydroxy fatty acids with a long alkyl branch in the two position) are classified in the genera *Corynebacterium, Gordona, Mycobacterium, Nocardia, Rhodococcus,* and *Tsukamurella;* their mycolateless counterparts are in the family Pseudonocardiaceae (see Chapter 42). The mycolic acid-containing strains have many properties in common (Goodfellow and Cross, 1984; Goodfellow and Wayne, 1982) and form a recognizable suprageneric group (Mordarski et al., 1980a; Stackebrandt and Woese, 1981; Stackebrandt et al., 1983). The genera *Corynebacterium* (Chapters 49 and 50) and *Mycobacterium* (Chapter 52 and 53), however, are often considered separately from the other mycolic acid-containing taxa which constitute the family Nocardiaceae Castellani and Chalmers 1919.

The family Nocardiaceae can now largely be defined on the basis of chemotaxonomic criteria. The taxon should be restricted to actinomycetes with the following characteristics: 1) a peptidoglycan composed of *N*-acetylglucosamine, D-alanine, L-alanine, and D-glutamic acid with *meso*-A$_2$ pm as the diamino acid and muramic acid in the *N*-glycolated form (Uchida and Aida, 1979); 2) a polysaccharide fraction of the wall rich in arabinose and galactose (whole-organism sugar pattern type A sensu Lechevalier and Lechevalier, 1970b); 3) a phospholipid pattern consisting of diphosphatidylglycerol, phosphatidylethanolamine (taxonomically significant nitrogenous phospholipid), phosphatidylinositol, and phosphatidylinositol mannosides without phosphatidylcholine or phospholipids containing glucosamine (i.e., phospholipid type 2, Lechevalier et al., 1977); 4) a fatty acid profile showing major amounts of straight chain, unsaturated, and tuberculostearic acids (Kroppenstedt, 1985); 5) mycolic acids with 48 to 78 carbons (Alshamaony et al., 1976a, 1976b; Goodfellow et al., 1978, 1982a); and 6) GC content of the DNA within the range 66 to 74 mol%. The type genus is *Nocardia* Trevisan 1889, 9[AL].

Some of the major properties of the mycolic acid-containing genera are shown in Table 1. Representatives of these taxa have also been found to be closely related using a variety of serological techniques. The most comprehensive serological studies have used immunodiffusion techniques, and common precipitinogens have been detected among corynebacteria, gordonae, mycobacteria, nocardiae, and rhodococci (Lind and Ridell, 1976; Lind et al., 1980). The suprageneric relationships established between representative mycolic acid-containing actinomycetes are shown in Fig. 1.

Classification

Nocardia Trevisan 1889, 9[AL]

Nocardiae are aerobic, catalase-positive actinomycetes that form rudimentary to extensively branched, substrate hyphae that often fragment in situ, or on mechanical disruption, into rod-shaped to coccoid, nonmotile elements. Aerial hyphae, at times visible only microscopically, are almost always present. Short-to-long chains of well-to-poorly differentiated conidia may occasionally be found on the aerial hyphae and, more rarely, on both aerial and substrate mycelia. Nocardiae are chemoorganotrophic, having an oxidative type of metabolism. In addition to the chemical properties already mentioned, the nocardial peptidoglycan is of the A1γ type (Schleifer and Kandler, 1972). The organism contains mycolic acids with 40 to 60 carbon atoms and up to three double bonds; the fatty acid esters released on pyrolysis gas chromatography of mycolic esters contain 12 to 18 carbon atoms and may be saturated or unsaturated. The predominant menaquinone in most nocardiae corresponds to a hexahydrogenated menaquinone with eight isoprene units in which the end two units are cyclized (i.e., II, III-tetrahydro-ω-[2,6,6-trimethylcyclohex-2-enylmethyl] menaquinone-6); *N. amarae* contains dihydrogenated menaquinones with nine isoprene units as the major isoprenologue. The GC content of the DNA is 64 to 72 mol%.

Type species: *Nocardia asteroides* (Eppinger) Blanchard 1896, 856[AL], Opinion 58, Judicial Commission 1985, 538. Type strain: ATCC 19247.

The genus *Nocardia* currently encompasses 11 validly described species: *N. amarae, N. asteroides, N. brasiliensis, N. farcinica* (group Kyoto-1 sensu Tsukamura, 1969; *N. asteroides* subgroup B, Schaal and Reutersberg, 1978), *N. otitidis-caviarum* (formerly *N. caviae), N. pinensis* (Blackall et al., 1989), and *N. seriola* (Kudo et al., 1988), and the less extensively studied *N. brevicatena, N. carnea, N. transvalensis,* and *N. vaccinii* (Goodfellow and Lechevalier, 1989; Gordon et al., 1978).

The continued inclusion of *N. amarae* in the genus is open to doubt. Strains of *N. amarae* formed a homogeneous cluster on the edge of an aggregate group corresponding to *Nocardia* (Goodfellow et al., 1982a), have dihydrogenated menaquinones with nine isoprene units as major isoprenolog, release C_{16} and C_{18} monounsaturated esters on pyrolysis of methyl mycolates, and are unable to grow in lysozyme broth (Goodfellow et al., 1982a; Lechevalier and Lechevalier, 1974). Phages that specifically lyse *N. asteroides* (Andrzejewski et al., 1978; Prauser, 1976, 1981; Pulverer et al., 1975), *N. brasiliensis* (Pulverer et al., 1975), *N. carnea* (Williams et al., 1980), *N. otitidis-caviarum* and *N. vaccinii* (Prauser, 1976) do not lyse the type strain of *N. amarae* (Williams et al., 1980).

Nocardia amarae, N. brasiliensis, N. farcinica, N. otitidis-caviarum, N. pinensis, and *N. seriolae* are homogeneous species (Blackall et al., 1989; Goodfellow, 1971; Goodfellow et al., 1982a; Kudo et al., 1988), but *N. asteroides* has been found to be heterogeneous on the basis of DNA homology (Bradley et al., 1978; Mordarski et al., 1977, 1978), numerical taxonomy (Goodfellow, 1971; Orchard and Goodfellow, 1980; Schaal and Reutersberg, 1978; Tsukamura, 1969), phage sensitivity (Pulverer et al., 1975) and immunology (Kurup and Schribner, 1981; Magnusson, 1976; Pier and Fichtner, 1971; Ridell, 1981). However, using more sensitive immunological techniques, Kurup et al. (1983) concluded that the seven immunotypes previously recognized should be retained within the species *N. asteroides,* rather than be assigned to separate species as previously proposed by Pier and Fichtner. Additional representatives of the remaining species need to be examined for them to be defined with precision.

Gordona (Tsukamura, 1971) Stackebrandt et al. 1988, 345[AL]

The genus *Gordona* (Tsukamura, 1971) was proposed for some slightly acid-fast actinomycetes isolated from soil and the sputa of patients with pulmonary disease. The three original species, *G. bronchialis, G. rubra,* and *G. terrae,* were subsequently reclassified in the revised and redescribed genus *Rhodococcus* (Tsukamura, 1974a; Goodfellow and Alderson, 1977). Rhodococcal species were subsequently assigned to two aggregate groups based primarily on chemical and serological properties (Goodfellow, 1986). All species originally classified in the genus *Gordona* contained mycolic acids with between 48 and 66 carbon atoms and major amounts of dihydrogenated menaquinones with nine isoprene units, whereas the remaining strains were characterized by shorter chain mycolic acids (34 to 52 carbon atoms) and dihydrogenated menaquinones with eight isoprene units as the predominant isoprenolog (Alshamaony et al., 1976b; Collins et al., 1977, 1985). The two aggregate groups were also recognized by their antibiotic sensitivity profiles (Goodfellow and Orchard, 1974), delayed skin reaction on sensitized guinea pigs, and polyacrylamide gel electrophoresis of cell extracts (Hyman and Chaparas, 1977).

The discovery that the two aggregate groups were phylogenetically distinct led Stackebrandt et al. (1988) to revive the genus *Gordona* Tsukamura 1971 for organisms classified as *R. bronchialis, R. rubropertinctus, R. sputi,* and *R. terrae;* DNA homology data indicate that *R. obuensis* Tsukamura 1982 should be considered as a subjective synonym of *R. sputi* (Zakrzewska-Czerwińska et al., 1988). Gordonae are catalase positive, arylsulfatase negative, and have an oxidative type of metabolism. In addition to the aforementioned chemical properties they have an A1α type of peptidoglycan (Schleifer and Kandler, 1972). The fatty acid esters released on pyrolysis gas chromatography of mycolic esters contain 16 to 18 carbon atoms. The GC content of the DNA is 63 to 69 mol%.

Type species: *Gordona bronchialis* (Tsukamura, 1971) Stackebrandt et al. 1988, 345[AL]. Type strain: ATCC 25592.

Rhodococcus Zopf 1891, 28[AL]

Rhodococci have a long and checkered taxonomic history (Bousfield and Goodfellow, 1976;

Table 1. Characteristics of genera encompassing wall chemotype IV actinomycetes containing mycolic acids.[a]

Characteristic	Gordona	Family Nocardiaceae				Corynebacterium	Mycobacterium
		Nocardia	Rhodococcus	Tsukamurella			
Morphology	Rods and cocci	Mycelium, later fragmenting into rods and cocci; usually some aerial mycelium	Rods to extensively branched mycelium; latter fragments into irregular rods and cocci; usually no aerial hyphae	Rods occurring singly, in pairs or masses; coccobacillary forms produced		Pleomorphic rods, often club-shaped; commonly in angular and palisade arrangement	Rods, occasionally branched filaments; no aerial mycelium[c]
Time for visible colonies	1–3 days	1–5 days	1–3 days	1–3 days		1–2 days	2–40 days
Degree of acid-fastness (not necessarily also alcohol-fastness)	Often partially acid-fast	Often partially acid-fast	Often partially acid-fast	Weak to strongly acid-fast		Sometimes weakly acid-fast	Usually strongly acid-fast
Strictly aerobic	+	+	+	+		−	+
Arylsulfatase produced	−	(−)	−	−		−	+
Peptidoglycan type[d]	A1γ	A1γ	A1γ	A1γ		A1γ	A1γ
Acyl group of muramic acid[e]	N-glycolated	N-glycolated	N-glycolated	N-glycolated		N-acetylated	N-glycolated
Fatty acid types[f]	S,U,T	S,U,T	S,U,T	S,U,T		S,U,(T)[g]	S,U,T[h]
Mycolic acid type:[i]							
Overall size (number of carbons)	48–66	46–60	34–52	64–78		22–38	60–90
Number of double bonds	1–4	0–3	0–2	1–6		0–2	1–3
Fatty acid esters released on pyrolysis (number of carbons)	16–18	12–18	12–16	20–22		8–18	22–26
Phospholipid type[j]	2	2	2	2		1	2
Predominant menaquinone(s)[k]	MK-9(H$_2$)	MK-8(H$_{4\,\omega\text{-cycl}}$)	MK-8(H$_2$)	MK-9		MK-8(H$_2$), -9(H$_2$)	MK-9(H$_2$)
GC content (mol%)	63–69	64–72	67–73	∽7–68		51–59	62–70

(continued)

[a]Adapted from Collins et al. (1989), Goodfellow (1989), and Stackebrandt et al. (1988).

[b]Symbols: +, positive; −, negative; (−), usually negative.

[c]*Mycobacterium farcinogenes* and *M. xenopi* may occasionally produce aerial mycelium.

[d]A, cross-linkage between positions 3 and 4 of adjacent peptide subunits; 1, peptide bridge absent; γ, *meso*-A_2pm at position 3 of the tetrapeptide subunits (Schleifer and Kandler, 1972).

[e]Acyl group detected using a simple glycolate test (Uchida and Aida, 1979).

[f]Abbreviations: S, straight chain; U, monounsaturated; T, tuberculostearic (10-methyloctadecanoic) acid; bracket indicates variable occurrence. See footnote to Table 2 for reference to methods.

[g]*Corynebacterium bovis* contains tuberculostearic acid (Collins et al., 1982a; Lechevalier et al., 1977).

[h]*Mycobacterium gordonae* lacks substantial amounts of tuberculostearic acid (Minnikin et al., 1985; Tisdall et al., 1979).

[i]In mycobacterial mycolic acids, double bonds may be converted to cyclopropane rings; methyl branches and other oxygen functions may be present (Dobson et al., 1985; Minnikin et al., 1984a). See footnote to Table 2 for reference to methods.

[j]Phospholipid types: 1, phosphatidylglycerol (variable) and phosphatidylinositol; 2, phosphatidylethanolamine (Lechevalier et al., 1977, 1981). See footnote to Table 2 for reference to methods.

[k]Abbreviations exemplified by MK-8(H_2), menaquinones having two of the eight isoprene units hydrogenated.

[l]Nocardiae were originally reported to have predominant amounts of MK-8(H_4). However, the major component was later shown to correspond to a hexahydrogenated menaquinone with eight isoprene units in which the end of two units of the multiprenyl side chain were cyclized (Collins et al., 1987; Howarth et al., 1986). The predominant isoprenolog of *N. amarae* is MK-9(H_2) (Goodfellow et al., 1982a).

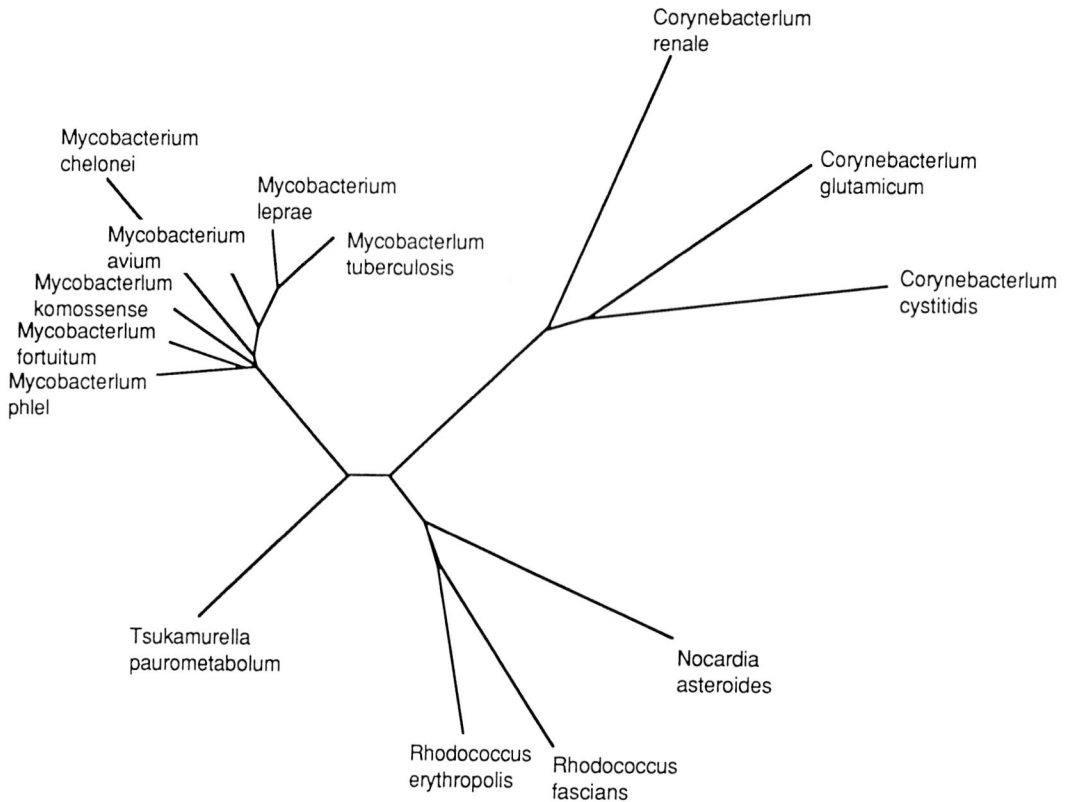

Fig. 1. Unrooted phylogenetic tree of mycolic acid-containing taxa. (From Collins et al., 1988, with permission.)

Goodfellow and Cross, 1984). The epithet *rhodochrous* (Zopf, 1891) was resurrected by Gordon and Mihm (1957) for actinomycetes bearing a multiplicity of generic and specific names but with many properties in common with both mycobacteria and nocardiae. The taxon was provisionally assigned to the genus *Mycobacterium* but was later considered to merit generic status. The genus *Rhodococcus* was eventually reintroduced (Goodfellow and Alderson, 1977; Tsukamura, 1974a) and encompasses 16 species in the current edition of *Bergey's Manual of Systematic Bacteriology* (Goodfellow, 1989). Most of these were circumscribed in numerical taxonomic surveys (Goodfellow and Alderson, 1977; Goodfellow et al., 1982b, 1982c; Rowbotham and Cross, 1977a; Tsukamura, 1974a) and later shown to be homogeneous on chemical and molecular systematic grounds (Minnikin and Goodfellow, 1980; Mordarski et al., 1980b; Zakrzewska-Czerwińska et al., 1988).

Given the reintroduction of the genus *Gordona*, the genus *Rhodococcus* is left as a homogeneous taxon encompassing 13 species (see Table 4). The redefined genus contains aerobic, Gram-positive, partially acid-fast, nonmotile actinomycetes which form rods to extensively branched substrate mycelium. The growth cycle starts with the coccus or short rod stage, different organisms then showing a more or less complex series of morphological stages: cocci may germinate only into short rods, or form filaments with side projections, or show elementary branching, or in the most differentiated forms produce branched hyphae. The next generation of cocci or short rods is produced by fragmentation of the rods, filaments, and hyphae. Some strains produce sparse, microscopically visible, aerial hyphae that may be branched or aerial synnemata consisting of unbranched filaments that coalesce and project upwards.

Rhodococci grow well on standard laboratory media at 30°C although some strains require thiamine. Colonies may be rough, smooth, or mucoid; and pigmented buff, cream, yellow, orange, or red, though colorless variants occur. Rhodococci are catalase positive, sensitive to lysozyme, arylsulfatase negative, and produce acid from glucose oxidatively. In addition to the chemical properties mentioned earlier, the rhodococcal peptidoglycan belongs to the A1γ type (Schleifer and Kandler, 1972). The fatty acid esters released on pyrolysis gas chromatography of mycolic esters contain 12 to 16 carbon atoms. The GC content of the DNA ranges from 67 to 73 mol%.

Type species: *Rhodococcus rhodochrous* (Zopf, 1891) Tsukamura 1974, 43[AL]. Type strain: ATCC 13808.

Genus *Tsukamurella* Collins et al. 1989, 387[AL]

This taxon was introduced to accommodate organisms classified as *Corynebacterium paurometabolum* and *Rhodococcus aurantiacus*. *Corynebacterium paurometabolum* was proposed by Steinhaus (1941) for bacteria isolated from the mycetome and ovaries of the bedbug *(Cimex lectularis)* but its placement in the genus *Corynebacterium* was questioned (Collins and Jones, 1982; Jones, 1975). The organism has an A1γ peptidoglycan (Cummins, 1971; Schleifer and Kandler, 1972) but was distinguished from corynebacteria by the presence of long, highly unsaturated, mycolic acids (Collins and Jones, 1982). A similar series of unsaturated mycolic acids were detected in *Rhodococcus aurantiacus* (Goodfellow et al., 1978; Tomiyasu and Yano, 1984), the generic position of which was also equivocal (Goodfellow, 1986). This species, first described as *Gordona aurantiaca* (Tsukamura and Mizuno, 1971), was subsequently reclassified in the genus *Rhodococcus* (Tsukamura, 1974a; Tsukamura and Yano, 1985). In contrast, Goodfellow et al. (1978) considered that *aurantiaca* strains were worthy of generic status as they formed a numerically circumscribed taxon equivalent in rank to the genera *Corynebacterium*, *Mycobacterium*, *Nocardia*, and *Rhodococcus* and contained characteristic mycolic acids and unsaturated menaquinones with nine isoprene units. *Corynebacterium paurometabolum* was also found to have menaquinones containing unsaturated multiprenyl side chains (Collins and Jones, 1982).

Corynebacterium paurometabolum and *Rhodococcus aurantiacus* were reduced to a single species and reclassified in the genus *Tsukamurella* using 16S rRNA sequence data and the results of experiments outlined above (Collins et al., 1989). The genus accommodates aerobic, Gram-positive, weakly acid-fast, nonmotile actinomycetes forming straight to slightly curved rods that occur singly, in pairs or in masses. Coccobacillary elements also occur. The organism forms white/creamy to orange, convex colonies that are dryish but easily emulsified. Tsukamurellas are catalase positive, arylsulfatase negative, and produce acid from glucose oxidatively. In addition to having a wall chemotype IV, the organism contains mycolic acids with 62 to 78 carbon atoms and one to six double bonds. The fatty acid esters released on pyrolysis gas chromatography of mycolic esters have 20 to 22

carbon atoms. Menaquinones are the sole respiratory quinones with unsaturated components with nine isoprene units predominating. The GC content of the DNA is 67 to 68 mol%.

Type species: *Tsukamurella paurometabolum* (Steinhaus, 1941) Collins et al. 1989. Type strain: ATCC 8368.

Habitats

Members of the family Nocardiaceae are common and widely distributed in terrestrial habitats but some are opportunistic pathogens for humans and animals. Nocardiaceae strains are probably involved in the turnover of plant material in soil but little is known of their role within this habitat. However, developments in selective isolation procedures can be expected to yield information on the occurrence, distribution, numbers, and activity of actinomycetes in natural habitats (Goodfellow and O'Donnell, 1989).

Nocardiae as Animal Pathogens

Nocardiae cause a variety of suppurative infections in humans and animals (Beaman, 1984; Pulverer and Schaal, 1978; Schaal and Beaman, 1984). Human infections may be divided clinically into superficial, pulmonary, and systemic nocardiosis. They can be recognized with certainty only by laboratory means and their occurrence is almost certainly underestimated. Most infections of the internal organs in nontropical countries are caused by *N. asteroides* or *N. farcinica* and only a few by *N. brasiliensis* or *N. otitidis-caviarum* (Beaman and Sugar, 1983; Beaman et al., 1976). *Nocardia farcinica* shows a greater degree of virulence than *N. asteroides* (Pulverer and Schaal, 1978).

Nocardiosis develops as an opportunistic infection that complicates chronic debilitating primary disease such as leukemia, lymphoma, and other neoplasms, or in patients undergoing immunosuppressive, therapeutic procedures (Schaal and Beaman, 1984). Increases in the reported incidence of nocardiosis can be attributed to the extensive use of immunosuppressive drugs, improved isolation procedures, and increased awareness among clinical microbiologists. Recent estimates indicate that considerably more than 1,000 cases of nocardiosis are diagnosed each year in the United States (Beaman, 1988). Some 20 to 50% of all heart transplant patients develop nocardial infection, and 0.5 to 1.0% of all AIDS patients are said to contract systemic nocardiosis. Most major hospitals in the USA have between 2 and 15 cases a year, and the disease is recognized in all 50 states (Beaman and Sugar, 1983). The pathogenicity of *Nocardia* strains is highly variable and is influenced by factors that include the age of the culture of the pathogen, rate of growth, route of infection, and by the immune status of the host.

In systemic nocardiosis, the primary lesion is usually the lungs, but secondary and often fatal infections may develop in the central nervous system and less frequently, in other internal organs. Nocardiae appear to invade the host either by direct inhalation of contaminated dust particles or through soil contaminated wounds. Host-parasite interactions may be studied in murine models (Beaman, 1984). Localized cutaneous or subcutaneous infections are usually primary infections caused either by *N. asteroides* or *N. brasiliensis*.

Actinomycete mycetomas are localized, chronic, progressive infections of the skin and subcutaneous tissue (Mahgoub and Murray, 1973), which are endemic in many tropical and subtropical regions. They are characterized by subcutaneous granulomata and abscesses and by areas of induration. Sinus tracts, often multiple, may discharge granules consisting of colonies of the causal organism. Similar fistulous lesions are occasionally seen in patients from temperate countries (Serrano et al., 1986). Infections are frequently through the foot, especially in localities where people cannot afford footwear. However, many extrapedal cases are seen in other parts of the body that come into contact with soil. *Nocardia brasiliensis* is the main causal agent in many tropical areas, such as the sugar plantations of Mexico. *Nocardia asteroides* and *N. otitidis-caviarum* occasionally cause mycetomas. Human mycetoma has been simulated in a mouse model (González-Ochoa, 1973), which is now used to study agents of host-parasite relationships (Ortiz-Ortiz et al., 1984).

Nocardiae are also responsible for a wide range of animal diseases (Beaman and Sugar, 1983). *Nocardia asteroides* is the most common agent, followed by *N. brasiliensis* and *N. otitidis-caviarum*. Diseases include abortion, mycetoma, as well as pulmonary and systemic nocardiosis. In dairy cattle, the predominant infection is mastitis, and epizootes have also been reported. Other animals susceptible to nocardial infection include birds, dogs, fish, goats, horses, and swine. *Nocardia seriolae* is a pathogen of cultured fish (Kudo et al., 1988).

Nocardiosis can be considered a communicable disease. Animal-to-animal transmission occurs in dairy cows (Bushnell et al., 1979). Good evidence of spread from person to person

was obtained in a hospital in which six of nine immunosuppressed patients undergoing renal transplantation, who occupied the same room at various times over a six-month period, developed disease caused by the same strain of *N. asteroides* (Houang et al., 1980).

Nocardiae as Saprophytes, Symbionts, and Plant Pathogens

Nocardiae are common in soil (Cross et al., 1976); populations up to $7.3 \times 10_4$/g dry weight have been found in environmental samples from tropical and temperate regions (Orchard et al., 1977). They also form mutualistic associations with blood-sucking arthropods and occur in aquatic habitats (Cross et al., 1976), where they have been implicated in the biodeterioration of natural rubber joints in water and sewage pipes (Hutchinson et al., 1975). Most isolates from these habitats have been identified as *N. asteroides,* but the apparent predominance of these species may merely reflect the isolation methods used. Large populations of *N. amarae* and *N. pinensis,* and smaller numbers of *N. asteroides* and *N. otitidis-caviarum,* have been isolated from foams occasionally formed in sewage treatment plants of the activated-sludge type (Blackall et al., 1988, 1989; Lechevalier and Lechevalier, 1974; Lechevalier et al., 1977). *Nocardia amarae* produces a surfactant and cells that are very hydrophobic (Blackall and Marshall, 1989). In the context of froth flotation theory, these factors are essential for foam production and transport of cells from aqueous to bubble phase. The addition of montmorillonitic clay to cultures prior to foaming prevents foam stabilization. *Nocardia vaccinii,* the only well-documented nocardial plant pathogen, causes galls in blueberry (Demaree and Smith, 1952). Little is known about the ecology of the remaining *Nocardia* species.

Gordonae, Rhodococci, and Tsukamurellas as Saprophytes, Symbionts, and Pathogens

Gordonae and rhodococci are widely distributed in nature but only rarely encountered as primary pathogens in healthy individuals. They have been isolated from soil, aquatic habitats, and from gut contents of blood-sucking arthropods with which they may form a mutualistic association (Cross et al., 1976; Goodfellow and Aubert, 1980). *Gordona terrae, Rhodococcus erythropolis, R. rhodochrous,* and *R. ruber* have been isolated from soil (Goodfellow and Williams, 1983), *R. chlorophenolicus* from a pentachlorophenol enrichment culture inoculated from lake sediment (Apajalahti et al., 1986),

R. marinonascens from marine sediments (Helmke and Weyland, 1984), and *R. rhodochrous* from foam on the surface of aeration tanks in activated sludge plants (Lemmer and Kroppenstedt, 1984).

Gordona bronchialis is associated with the sputa of patients suffering from cavitary pulmonary tuberculosis and bronchiectasis (Tsukamura, 1971), and *G. sputi* may cause tuberculosis-like mesenteric lymphadentis in swine (Tsukamura et al., 1988a). *Rhodococcus equi* is an important veterinary pathogen that causes bronchopneumonia, ulcerative enteritis and lymphadenitis in foals (Barton and Hughes, 1980). In Ontario it was reported to cause 10% of deaths of foals under six months of age (Zink et al., 1986). The organism infects other domesticated animals is being isolated with increasing frequency from human patients, especially from those with generally decreased resistance to infections (Eberzole and Paturzo, 1988; Van Etta et al., 1983). Oral immunization can protect foals against severe challenge with *R. equi* (Chirino-Trejo et al., 1987b). However, it seems that the humoral response to *R. equi* whole-cell antigens is not important in protection against disease (Chirino-Trejo and Prescott, 1986). This observation is consistent with the behavior of the organism as a facultative intracellular pathogen (Zink et al., 1985).

Rhodococcus coprophilus has been the subject of extensive ecological studies (Rowbotham and Cross, 1977b). Although this organism grows on herbivore dung, high numbers have also been recorded from grazed pastures and from rivers, streams, and lake muds that receive runoff from land devoted to dairy farming. It seems that the coccal survival stage contaminates grass in pastures or hay used during the winter months for fodder, and remains viable after injection and passage through the rumen. The significant correlation found between the numbers of *R. coprophilus* and fecal streptococci (enterococci) in polluted water led Al-Diwany and Cross (1978) to the view that the organism be considered as an indicator of farm animal effluent. *Rhodococcus coprophilus* is also seen to have potential in water and diary bacteriology as a specific indicator organism of fecal pollution due to farm animal wastes (Mara and Oragui, 1981; Oragui and Mara 1985).

Tsukamurella paurometabolum has been isolated from sputa of patients with tuberculosis or tuberculosis-like lung disease (Tsukamura and Mizuno, 1971) as well as from the ovaries of the bedbug (*Cimex lectularis:* Steinhaus, 1941). The organism has also been reported to cause lung disease (Tsukamura and Kawakami, 1982), lethal meningitis (Prinz et al., 1985), and necrotizing tenosynovitis (Tsukamura et al., 1988b).

Isolation and Cultivation

Classical Methods of Selective Isolation for Nocardiae and Rhodococci

Classical isolation procedures have depended on the capacity of nocardiae and rhodococci to use hydrocarbons as sole sources of carbon for energy and growth (Cross et al., 1976; Tárnok, 1976). Modifications of Söhngen's (1913) paraffin baiting method have often been used to isolate *Nocardia* and *Rhodococcus* strains from soil (Portaels, 1976; Schaal and Bickenbach, 1978), but are of little value in quantitative studies because they merely indicate the presence or absence of organisms in samples. Many other bacteria and fungi that metabolize paraffin wax frequently outgrow nocardiae and rhodococci on coated glass rods and render isolation impossible. Alternative isolation methods, such as inoculating guinea pig or hamster testicles with soil suspensions supplemented with penicillin and streptomycin (Conti-Diaz et al., 1971), or plating suspensions onto media that contain sodium azide and cholesterol acetate (Farmer, 1962), underestimate populations of nocardiae and rhodococci in natural habitats.

Selective Isolation of Nocardiae

Large numbers of nocardiae have been isolated from soil by plating out soil suspensions onto diagnostic sensitivity test (DST) agar supplemented with antifungal and various combinations of antibacterial antibiotics (Orchard and Goodfellow, 1974; Orchard et al., 1977). Inoculated plates are incubated for up to 21 days at 25°C. Colonies with pink to red stroma, covered to a greater or lesser extent with white aerial hyphae, are characteristic of nocardiae. To date, most isolates have been classified as *N. asteroides,* although laboratory strains of *N. brasiliensis* and *N. otitidis-caviarum* do grow on DST media (Orchard et al., 1977). It seems likely that at least some strains of *N. asteroides* will be inhibited by the antibacterial antibiotics in DST agar (Schaal and Heimerzheim, 1974). A selective medium free of such antibiotics has been used to isolate *N. asteroides* from clinical material (Schaal, 1972).

Sabouraud dextrose agar supplemented with chloramphenicol has been recommended for the recovery of *Nocardia* species from sputum specimens (Ajello and Roberts, 1981), but many nocardiae are inhibited by chloramphenicol (Gutmann et al., 1983). There is encouraging evidence that chemically defined media containing paraffin agar may be useful for the selective isolation of nocardiae from clinical specimens (Shawar et al., 1990). Nonselective media such as brain heart infusion and Sabouraud dextrose agars have also been recommended for the isolation of *Nocardia* species from clinical material (Schaal, 1977). Decontamination of respiratory specimens has also been shown to be toxic for nocardiae (Murray *et al.*, 1987).

Nocardia amarae, N. asteroides, and *N. otitidis-caviarum* have been isolated by plating sewage foam onto either Czapek's agar supplemented with yeast extract (0.2%) (Higgins and Lechevalier, 1969) or glycerol agar (Gordon and Smith, 1953) and incubating plates at 28°C for 5 to 7 days. Using nitrite medium (Winogradsky, 1949), Cross et al. (1976) have reported high numbers of nocardiae from soils in the USSR, but it seems likely that many of their isolates were rhodococci. Isolation of *N. pinensis* strains was achieved by micromanipulation using the Skerman micromanipulator (Skerman, 1968) and either yeast extract-glucose agar or tryptone-yeast extract agar (Blackall et al., 1989).

Nocardiae grow well on most laboratory media, including brain heart infusion (Difco 0418), nutrient, Sabouraud dextrose, Bennett's (Jones, 1949), yeast extract-glucose (Waksman, 1950), modified Sauton's (Mordarska et al., 1972) and yeast-malt extract (ISP medium 2; Difco 0770) agars. Media should be incubated aerobically at 25 to 37°C for up to 3 weeks.

Selective Isolation of Gordonae and Tsukamurellas

A pretreatment combined with a selective medium has been developed for the isolation of gordonae and tsukamurellas from sputa and soils (Tsukamura, 1971; Tsukamura et al., 1988b). Samples and specimens are added to an equal volume of NaOH (4%) and liquefied, either by incubation at 37°C for 30 minutes or by shaking at room temperature for 15 to 30 minutes. The digests are inoculated onto Ogawa egg medium (Tsukamura, 1972) and incubated for 4 to 8 weeks.

For soil samples, soil (5 g) is suspended in distilled water (25 ml) and shaken vigorously in a 300-ml Erlenmeyer flask at room temperature on a reciprocal shaker (stroke: 10 cm, 60 cpm) for 30 min. The suspension is allowed to settle for 10 min, 15 ml of the supernatant is added to an equal volume of NaOH (8%), and the mixture is shaken for 10 min before centrifugation at 500 × g for 15 min. The residue is suspended in 10 ml of a 1% NaH_2PO_4 solu-

tion, and 0.002 ml is added to Ogawa egg medium slants, which are incubated for 4 to 8 weeks.

Colonies growing on the Ogawa egg medium are transferred to fresh slants supplemented with sodium salicylate (0.5 mg/ml), which inhibits *Mycobacterium tuberculosis* (Tsukamura, 1962). Smears, prepared from colonies growing on supplemented slants after 3 weeks of incubation at 37°C, are stained by the Ziehl-Neelson method and observed under the light microscope. Slightly acid-fast, rod-shaped bacteria are typical of gordonae; those strongly acid-fast are mycobacteria. Gordonae produce rough, reddish or pinkish colonies on Ogawa egg slants that have been plugged with cotton wool.

A special semisolid medium, the main nutritive constituents of which were proteose peptone, rabbit serum, gelatin, mineral rabbit kidney, and carbohydrates, were used for the original isolation of *Tsukamurella paurometabolum* (Steinhaus, 1941). The organism grew in 24 to 48 hours when subsequently transferred to beef infusion agar. Gordonae and tsukamurellas grow well on most standard laboratory media, including Bennett's (Jones, 1949), brain heart infusion (Difco 0418), modified Sauton's supplemented with thiamine (Mordarska et al., 1972), and glucose yeast extract (Waksman, 1950) agars, and Lowenstein-Jensen's medium (BBL 20908).

Selective Isolation of Rhodococci

A heart-pretreatment method combined with a selective medium has been devised to isolate *Rhodococcus coprophilus* from both aquatic and terrestrial habitats (Rowbotham and Cross, 1977b). Thus, environmental samples (2 ml) of cream, milk, or water in 100- × 12-mm glass tubes sealed with silicon rubber bungs are heated in a water bath for 6 min at 55°C prior to further dilution or plating out. Water samples are stored at 4°C before heating. Suspensions of dung, grass, soil (1:10), or hay (1:50) are homogenized in 1/4-strength Ringer's solution containing gelatin (0.01%), pH 7.0, before heat pretreatment. Immediately after pretreatment, samples are shaken on a Vortex mixer and portions (0.2 ml) spread onto M3 agar plates (see "Media for Isolation and Cultivation") which are incubated at 30°C for 7 days.

Rhodococcus chlorophenolicus was isolated from a pentachlorophenol enrichment culture (Apajalahti et al., 1986); *R. luteus* and *R. maris* from soil and the skin and intestinal contents of carp *(Cyprinus carpier)* on mineral salts agar enriched with *n*-alkanes and incubated at 28°C (Nesterenko et al., 1982); *R. marinonascens* from marine sediments using a number of rich media supplemented with seawater and incubated for 8 to 12 weeks at 18°C (Weyland, 1969), and *R. erythropolis* and *R. rhodochrous* on mineral salts media supplemented with *m*-cresol or phenol (Gray and Thornton, 1928). Similarly, *R. equi* was isolated from soil, feces, lymph nodes, and the intestinal contents of several animal species using a selective medium, (NANAT medium) supplemented with cycloheximide, nalidixic acid, novobiocin, and potassium tellurite (Mutimer and Woolcock, 1980; Woolcock et al., 1979). A selective enrichment broth (TANT broth) containing cycloheximide, nalidixic acid, penicillin, and potassium tellurite incubated at 30°C and used in conjunction with Tinsdale medium (Oxoid) and modified M3 medium has also been used to isolate *R. equi* (Barton and Hughes, 1981).

Rhodococci have been isolated from soil using Czapek's agar (Higgins and Lechevalier, 1969), glycerol agar (Gordon and Smith, 1953), and Winogradsky's nitrate medium (Winogradsky, 1949), and from diseased sweet peas using potato dextrose agar (Tilford, 1936). The organism grows well on standard laboratory media such as Bennett's (Jones, 1949), brain heart infusion (Difco 0418), glucose yeast extract (Waksman, 1950), modified Sauton's plus thiamine (Mordarska et al., 1972), and nutrient agars, and Lowenstein-Jensen's medium (BBL 20908).

Media for Isolation and Cultivation

The following are the recipes for those media mentioned in the preceding section that are not commercially available:

Isolation Media

Chitin Agar (Weyland, 1969)

Chitin, hydrolyzed and precipitated	10 g
Peptone	0.5 g
Yeast extract	0.1 g
FePO$_4$·H$_2$O	0.01 g
Seawater	750 ml
Distilled water	250 ml
Agar	15 g

Adjust to pH 7.5.

Czapek-Dox Agar (Weyland, 1969)

Sucrose	15 g
NaNO$_3$	2 g
FePO$_4$·H$_2$O	0.01 g
Magnesium glycerophosphate	0.5 g
Seawater	750 ml
Distilled water	250 ml

Agar	15 g

Adjust to pH 7.6–7.8.

DST Agar (Orchard and Goodfellow, 1974)

Diagnostic sensitivity (DST) agar (Oxoid, CM261) is supplemented with cycloheximide and nystatin (antifungal antibiotics), and various concentrations of chlortetracycline HCl, demethylchlortetracycline HCl, and methacycline HCl (antibacterial antibiotics). The DST agar and cycloheximide are autoclaved separately at 15 psi (1.013 × 10$_5$ Pa) for 20 min, and the remaining antibiotics are filtered. The individual antibiotics are pipetted separately into petri dishes and the basal medium is added to give the following concentrations (μg/ml) for four different media:

Medium 1: demethylchlortetracycline HCl (demeclocycline HCl), 5; actidione, 50; and mycostatin, 50.

Medium 2: methacycline HCl, 10; actidione, 50; and mycostatin, 50.

Medium 3: chlortetracycline HCl (aureomycin), 45; demethylchlortetracycline HCl, 5; actidione, 50; and mycostatin, 50.

Medium 4: chlortetracycline HCl, 45; methacycline HCl, 10; actidione, 50; and mycostatin, 50.

Media 1 and 2 generally give higher counts of *Nocardia* because of the satisfactory inhibition of other soil bacteria. However, when large mixed populations of bacteria occur, the more selective formulations, media 3 and 4, may be required. The number of unwanted bacteria can also be reduced by heating soil suspensions (2 ml) in a water bath at 55°C for 6 minutes.

Glycerol Agar (Gordon and Smith, 1953)

Beef extract	3 g
Glycerol	7%
Peptone	5 g
Agar	15 g
Soil extract (see below)	1 liter

Adjust to pH 7.0.

Soil extract is prepared by autoclaving 1,000 g of air-dried soil (that has been sifted through a no. 9 mesh screen) with 2.4 liter of tap water at 121°C for 60 minutes. The preparation is then decanted and filtered through paper pulp.

M3 Agar (Rowbotham and Cross, 1977b)

KH$_2$PO$_4$	0.466 g
Na$_2$HPO$_4$	0.732 g
KNO$_3$	0.01 g
NaCl	0.29 g
MgSO$_4$·7H$_2$O	0.1 g
CaCO$_3$	0.02 g
FeSO$_4$·7H$_2$O	200 μg
ZnSO$_4$·7H$_2$O	180 μg
MnSO$_4$·7H$_2$O	20 μg
Sodium propionate	0.2 g
Agar	18 g
Distilled water	1 liter

Adjust to pH 7.0.

Cycloheximide and thiamine HCl, sterilized by membrane filtration, are added to the autoclaved and cooled medium to give final concentrations of 50 mg/l and 4.0 mg/l, respectively.

Mineral Salts Medium (Gray and Thornton, 1928)

K_2HPO_4	1.0 g
$MgSO_4 \cdot 7H_2O$	0.2 g
NaCl	0.1 g
$(NH_4)_2SO_4$ or KNO_3	0.5–1.0 g
$CaCl_2$	0.1 g
$FeCl_3$	0.02 g
Distilled water	1 liter

Soil (0.5–1g) is inoculated into 100 ml of mineral salts solution supplemented, (after autoclaving) with m-cresol (0.05%) or phenol (0.05–1.0%). Subcultures are made into fresh flasks containing the same medium, and isolates plated onto appropriate agar media containing m-cresol or phenol.

Mineral Salts Agar (Nesterenko et al., 1982)

KH_2PO_4	0.14 g
KNO_3	1.0 g
$MgSO_4$	0.1 g
NaCl	1.0 g
Na_2HPO_4	0.6 g
Mixture of n-alkanes (C-12 to C-22)	20.0 g
Distilled water	500 ml
Tap water	500 ml

Rhodococci are isolated on this medium using the method of Yamada et al. (1963).

Mineral Salts Paraffin Agar (Shawar et al., 1990)

KH_2PO_4	3 g
K_2HPO_4	1 g
NH_4Cl	5 g
NH_4NO_3	1 g
$FeSO_4$	0.05 g
$MgSO_2 \cdot 7H_2O$	0.05 g
$MnSO_4$	0.05 g
$ZnSO_4$	0.05 g
Bacto agar (Difco)	17 g
Distilled water	1 liter

Adjust to pH 7.2. Mineral salts paraffin agar is prepared by mixing 9 parts of the carbon-free agar medium with 1 part paraffin oil (Soybolt viscosity at 37.8°C equals 345–355; J. T. Baker Chemical Co., Phillipsburg, NJ).

Modified Czapek's Agar (Higgins and Lechevalier, 1969)

$NaNO_3$	2 g
K_2HPO_4	1 g
$MgSO_4 \cdot 7H_2O$	0.5 g
KCl	0.5 g
$FeSO_4$	0.01 g
Sucrose	30.0 g
Yeast extract	2 g
Agar	15 g
Distilled water	1 liter
Adjust to pH 7.2	

Münz Paraffin Agar (Nesterenko et al., 1978b)

KH_2PO_4	0.14 g
KNO_3	1.0 g
$MgSO_4 \cdot 7H_2O$	0.1 g
NaCl	1.0 g
Na_2HPO_4	0.6 g

Agar	15 g
Distilled water	1 liter
Adjust to pH 7.2.	

10 ml of liquid paraffin is added to the carbon-free medium after sterilization.

NANAT Medium (Woolcock et al., 1979)

Tryptone soya broth (Oxoid)	30 g
Yeast extract (Oxoid)	1 g
Agar (Difco)	15 g
Distilled water	1 liter

Sterilize at 121°C for 15 min. When cool add:

Cycloheximide	40 μg/ml
Nalidixic acid	20 μg/ml
Novobiocin	25 μg/ml
Potassium tellurite to 0.005%.	

Nitrite Medium (Winogradsky, 1949)

$NaNO_2$	2 g
Na_2CO_4 (anhydrous)	1 g
K_2HPO_4	0.5 g
Agar	15 g
Distilled water	1 liter

Ogawa Egg Medium (Tsukamura, 1972)

Whole chicken eggs	200 ml
Glycerol	6 ml
Malachite green (2%)	6 ml
Sodium glutamate (1%) plus KH_2PO_4 (1%)	100 ml
Adjust to pH 6.8, and sterilize at 90°C for 60 min.	

Tellurite-Actidione-Nalidixic Acid-Penicillin (TANT) Selective Enrichment Broth (Barton and Hughes, 1981)

Trypticase soy broth	30 g
Actidione (cycloheximide)	50 μg/ml
Nalidixic acid	20 μg/ml
Potassium tellurite	0.05 g
Penicillin	10 units/ml
Distilled water	1 liter

TANT broth (10 ml) inoculated with a 1-g environmental sample (e.g., soil or feces) is incubated at 30°C for 6 to 7 days. The broth is then subcultured onto Tinsdale (Oxoid) agar supplemented with 50 μg/ml cycloheximide and onto M3 agar (Rowbotham and Cross, 1977b) modified by the addition of 0.005% potassium tellurite. Colonies of *Rhodococcus equi* appear on these selective media after 4 to 5 days at 30°C.

Tryptone Yeast Extract Agar (Blackall et al., 1989)

Glucose	5 g
Tryptone	3 g
Yeast extract	5 g
Distilled water	1 liter
Adjust to pH 7.0 and sterilize at 121°C for 15 min.	

Yeast Glucose Agar (Blackall et al., 1989)

Glucose	10 g
Yeast extract	10 g
Agar	15 g

Distilled water   1 liter
Adjust to pH 7.0 and sterilize at 121°C for 15 min.

Cultivation Media

Bennett's Agar (Jones, 1949)

Beef extract	1 g
Glucose	10 g
N-Z amine A (enzymatic digest of casein)	2 g
Yeast extract	1 g
Agar	15 g

Adjust to pH 7.3 with NaOH.

Modified Bennett's agar can also be used; in it, glucose is replaced by glycerol (10 g), and N-Z amine A by Bacto-casitone (2 g, Difco).

Glucose-Yeast Extract Agar (Waksman, 1950)

Glucose	10 g
Yeast extract (50% solution)	20 ml
Agar	15 g

Adjust to pH 6.8.

Modified Sauton's Medium (Mordarska et al., 1972)

Asparagine	5.0 g
Casein hydrolysate	2.0 g
Glucose	15.0 g
Sodium citrate	1.5 g
KH_2PO_4	5.0 g
$MgSO_4 \cdot 7H_2O$	0.5 g
K_2SO_4	0.5 g
Ferric ammonium citrate	Trace
Distilled water	1 liter

Adjust to pH 7.2. The glucose should be sterilized separately. For the cultivation of *Rhodococcus* strains, the medium should be supplemented with thiamine (50 mg/l).

Preservation of Cultures

Serial transfer from agar slants of appropriate media (see above) every two months, with storage at 4°C between transfers, is the most convenient method for shortterm storage. Lyophilization, storage in liquid nitrogen, or as a frozen glycerol suspension can be used for long-term preservation. For lyophilization, cell masses are suspended in a suitable fluid, e.g., 7.5% (w/v) glucose serum, or skimmed milk plus 7.5% (w/v) glucose. For storage in liquid nitrogen, the microorganisms are inoculated into small test tubes containing the appropriate medium and incubated until satisfactory growth is visible. The tubes are then closed with cotton wool plugs, dipped in liquid paraffin wax, and placed in a liquid nitrogen container.

Glycerol suspensions are prepared by scraping growth from heavily inoculated agar plates and making heavy suspensions in 3 ml of aqueous glycerol in bijoux bottles which are stored at −20°C (Wellington and Williams, 1978). The frozen glycerol suspensions serve both as a convenient means of long-term preservation and a quick source of inoculum. Working inocula are obtained by thawing suspensions at room temperature prior to treating as for broth cultures. After use, glycerol suspensions are promptly frozen and stored again at −20°C.

Identification

Mycolic acid-containing actinomycetes can be difficult to distinguish both from one another and from other bacteria using conventional staining and morphological properties. Thus, *Nocardia* strains that lack aerial hyphae cannot always be separated from mycobacteria, rhodococci, and "bald" streptomycetes, whereas those producing abundant aerial hyphae can be confused with members of the family Pseudonocardiaceae. Reliable differentiation at the genus level depends upon the application of chemotaxonomic techniques. Simplified procedures have been devised for a number of chemical tests.

Identification at the Generic Level

Examination of whole-organism methanolysates for the presence of mycolic acids is the first stage in the chemical procedure. Mycolic acids vary considerably in structure, ranging from relatively simple mixtures of saturated and unsaturated acids found in corynebacteria to very complex mixtures characteristic of mycobacteria (Minnikin and Goodfellow, 1980). Qualitative evaluation of mycolic acids can be easily and quickly achieved using a thin-layer chromatographic technique devised by Minnikin et al. (1975). Methanolysates of mycobacteria give a multispot pattern of mycolates, and those of corynebacteria, gordonae, rhodococci, and tsukamurellas produce a single spot value of which reflects the chain length and structure of the mycolic acids (Goodfellow et al., 1976; Minnikin et al., 1980, 1984). Mycolic esters can be positively identified on thin-layer chromatograms by their characteristic immobility when plates are subsequently washed in methanol-water (5:2, v/v) (Minnikin et al., 1975). The discontinuous distribution of mycolic acid types can also be exploited for the identification of mycobacterial species (Minnikin et al., 1984a, 1985b).

The mycolic acids from mycobacteria are precipitated from ether solution by addition of an equal (Kanetsuna and Bartoli, 1972) or double (Hecht and Causey, 1976) volume of ethanol. Mycolic acids from corynebacteria, gordonae, nocardiae, and rhodococci are not precipitated by such a procedure, but Hecht and Causey (1976) detected their presence by thin-layer chromatography of the supernatant. The behavior of the mycolic acids of *Tsukamurella* strains in these systems has still to be determined.

When the presence of mycolic acids has been detected, their esters can be isolated and studied further by pyrolysis gas chromatography, by mass spectrometry of the intact esters, by gas chromatography of trimethylsilyl ether derivatives, and by high-performance liquid chromatography (HPLC) of bromophenacyl derivatives. On pyrolysis, mycolic acid methyl esters yield long-chain methyl esters and aldehydes which can be analyzed directly by use of gas pyrolysis, whereas the remaining mycolic acid-containing organisms produce esters with fewer carbon atoms (Table 2). Mycolic esters from *N. amarae* release unsaturated C_{16} and C_{18} major components (Lechevalier and Lechevalier, 1974) while production of unsaturated C_{20} and C_{22} esters characterize the mycolates from *Tsukamurella parometabolum* (Goodfellow et al., 1978). Mycolic acids with unsaturation in the chain in the 2-position have also been observed in two nocardial strains, *N. carnea* and *N. vaccinii,* but no details have been reported (Lechevalier and Lechevalier, 1974).

The pyrolysis of mycolates can also be observed by mass spectrometry, where the highest peaks in the spectra correspond to anhydromycolates formed by loss of water from the parent molecule (Etémadi, 1967). The overall size of mycolates, their degree of unsaturation, and the nature of both long-alkyl chains may be determined by mass spectrometry. However, the complex mixtures of homologs usually present and the competing fragmentation pathways make interpretation difficult in some cases (Alshamaony et al., 1976a, 1976b; Maurice et al., 1971). The analysis can be taken a stage further by using gas chromatography-mass spectrometry of trimethylsilyl and tertbutyldimethylsilyl derivatives of mycolic acids (Athalye et al., 1984; Pommier and Michel, 1985; Tomiyasu et al., 1981; Yano et al., 1978). This procedure separates mycolic ester derivatives into their homologous components each of which can be analysed by mass spectrometry. Detailed mycolic acid analyses may eventually help in the identification of all mycolic acid-containing species.

Different classes of mycobacterial mycolic acids have been separated by HPLC (Qureshi et al., 1978; Steck et al., 1978). This separation is based on the chain lengths, the degree of unsaturation, and other functional groups found in these fatty acids (Svensson et al., 1982). Re-

verse-phase HPLC of *p*-bromophenacyl esters of mycolic acids may provide a rapid way of distinguishing between mycolic acid-containing genera (Butler et al., 1986).

Simple chromatographic analysis of whole-organism methanolysates provide a simple and reliable means of separating corynebacteria, mycobacteria, and nocardiae (Minnikin et al., 1975, 1980). Mycobacteria and nocardiae may also be distinguished from corynebacteria as the former contain DNA that is relatively rich in GC (62–72 mol%) and *N*-glycolyl residues in the glycan moiety of their peptidoglycan (Table 1). In contrast, it can be difficult to separate corynebacteria from some rhodococci, and to a lesser extent gordonae, nocardiae, rhodococci, and tsukamurellas. Lipid markers can, however, be used to differentiate all six genera that accommodate mycolic acid-containing actinomycetes (Table 2).

Members of the genus *Rhodococcus* generally contain longer mycolic acids (34–52 carbon atoms) than those of *Corynebacterium* (22–38 carbon atoms) but an unambiguous distinction between the two genera cannot presently be made by analysis of mycolic acids alone (Collins et al., 1982b). Several rhodococci have mycolic acids that overlap in size with those of corynebacteria (e.g., *R. equi;* Collins et al., 1982b; Barton et al., 1989). Rhodococci are described as aerobic (Goodfellow and Alderson, 1977); although the vast majority of corynebacteria are undoubtedly facultative, obligate aerobic strains do occur which cannot be assigned with confidence to one genus or the other. Criteria which may prove to be of value in the identification of such "difficult strains" include information from DNA base composition and peptidoglycan analyses. Representatives of the genus *Rhodococcus* generally have DNA with a higher GC content (67–73 mol%) than *Corynebacterium* (51–59 mol%), and rhodococci contain *N*-glycolyl residues in the glycan moiety of their peptidoglycan (Table 1). Similarly, the presence of 10-methyloctadecanoic (tuberculostearic) acid and phosphatidylethanolamine can help in the assignment of strains between the two genera. Members of the genus *Rhodococcus* contain tuberculostearic acid whereas corynebacteria, with the exception of *C. bovis,* lack this fatty acid (Collins et al., 1982a; Lechevalier et al., 1977). Indeed, all other mycolic acid-containing actinomycetes, apart from *M. gordonae,* contain substantial amounts of tuberculostearic acid (Minnikin et al., 1985a; Tisdall et al., 1979). With the exception of corynebacteria, all mycolic acid-containing strains also contain phosphatidylethanolamine, in addition to diphosphatidylglycerol and the characteristic

Table 2. Lipid markers that are useful in the identification of wall chemotype IV actinomycetes containing mycolic acids.[a]

Genus or species	Fatty acid[b]	Mycolic acid pattern[c]	Solubility of mycolic acids[d]	Mycolate size (number of carbons)[e]	Esters released on pyrolysis of mycolate[f]	Major menaquinone[g]	Diagnostic phospholipid[h]
Mycobacterium spp.	Straight-chain, unsaturated, and tuberculostearic acids	Multispot	Insoluble	60–90	22:0–26:0	$MK-9(H_2)$	PE
Tsukamurella paurometabolum		Single spot	ND	64–78	20:0–22:0	MK-9	PE
Gordona spp.			Soluble	48–66	16:0–18:0	$MK-9(H_2)$	PE
Nocardia spp.				46–60	12:0–18:0	$MK-8(H_4 \, \omega\text{-cycl})$	PE
N. amarae				46–54	16:1–18:1	$MK-9(H_2)$	PE
Rhodococcus spp.				34–52	12:0–16:0	$MK-8(H_2)$	PE
Corynebacterium bovis				22–32	8:0–10:0	$MK-9(H_2)$	–
Corynebacterium spp.	Straight-chain and unsaturated acids only	Single spot	Soluble	26–38	14:0–18:0 (14:1–18:1)	$MK-8(H_2), -9(H_2)$	–

[a]Adapted from Collins et al. (1989); Goodfellow (1989); and Stackebrandt et al. (1988).

[b]Determined by gas chromatography (Collins et al., 1982a; Kroppenstedt, 1985; Lechevalier et al., 1977).

[c]Established by thin-layer chromatography (Goodfellow et al., 1976; Minnikin et al., 1975, 1980, 1984b).

[d]In ethanol-diethyl ether (Hecht and Causey, 1976; Kanetsuna and Bartoli, 1972).

[e]Detected by mass spectrometry (Alshamaony et al. 1976a, 1976b; Collins et al., 1982b), gas chromatography-mass spectrometry (Athalye et al., 1984; Tomiyasu et al., 1981; Yano et al., 1978) or high-performance liquid chromatography (Butler et al., 1986).

[f]Esters detected by pyrolysis gas chromatography (Collins et al., 1982b; Goodfellow et al., 1978; Lechevalier et al., 1971).

[g]Detected by chromatographic or physicochemical analysis (Collins, 1985; Collins et al., 1977; Kroppenstedt, 1985; Minnikin et al., 1984c). See footnotes to Table 1 for explanation of abbreviations.

[h]PE, phosphatidylethanolamine—a diagnostic phospholipid that can occur in addition to diphosphatidylglycerol and phosphatidylinositol mannosides. Polar lipid patterns are determined by thin layer chromatography and chemical analysis (Lechevalier et al., 1977, 1981; Minnikin et al., 1977).

phosphatidylinositol mannosides (Lechevalier et al., 1977, 1981; Minnikin et al., 1977).

Gordonae, nocardiae, rhodococci, and tsukamurellas can be separated solely on the basis of their predominant menaquinones (Table 2). The presence of fully unsaturated menaquinones with nine isoprene units (Collins and Jones, 1982; Goodfellow et al., 1978) serves to distinguish *Tsukamurella paurometabola* from all other mycolic acid-containing actinomycetes. Similarly, *Nocardia* strains, apart from *N. amarae,* contain a pseudotetrahydrogenated menaquinone with eight isoprene units where the end of the multiprenyl side chain is cyclized (Blackall et al., 1989; Collins et al., 1987; Howarth et al., 1986). Gordonae and rhodococci can also be distinguished from one another, and from most related strains, as the former contain dihydrogenated menaquinones with nine isoprene units as the predominant isoprenologue and the latter the corresponding menaquinone with eight isoprene units (Collins et al., 1977, 1985). Detailed analytical procedures for determining menaquinone composition have been described (Collins, 1985; Collins et al., 1987; Kroppenstedt, 1985; Minnikin et al., 1984c). A small-scale integrated chemotaxonomic procedure is available for the detection of wall and lipid markers (O'Donnell et al., 1985).

Susceptibility to bleomycin (2.5 μg/ml), 5-fluorouracil (20 μg/ml), mitomycin (10 μg/ml), and β-galactosidase activity may prove useful in the differentiation of mycolic acid-containing genera (Tsukamura, 1974b, 1981a, 1981b, 1982b). There is also evidence that rhodococci are more resistant than mycobacteria to prothionamide (Ridell, 1983), and that they can be distinguished from the latter on the basis of the acid-fast stain, arylsulfatase activity, capacity to use sucrose as a carbon source, and inability to metabolize trimethylamine as a simultaneous carbon and nitrogen source (Tsukamura, 1971). It is also possible to separate *Rhodococcus* from *Mycobacterium* and *Nocardia* (Ridell and Norlin, 1973), and from *Corynebacterium* (Ridell, 1977), using diagnostic precipitinogens.

Identification at the Subgeneric Level

Nocardia species can be difficult to identify, though several diagnostic schemes are available for this purpose (Goodfellow and Lechevalier, 1989; Gordon et al., 1978). Several species can be recognized using a number of degradative, nutritional, and physiological tests (Table 3) but little confidence can be placed in the predictiveness of these properties with poorly circumscribed species such as *N. brevicatena, N. carnea, N. transvalensis,* and *N. vaccinii.* However,

the improvements in nocardial classification provide a good base in the search for new tests that can be weighted for diagnostic purposes. Laboratory specialists should be aware of the occasional occurrence of atypical strains of *N. asteroides* (Boiron et al., 1990).

Serological tests for the diagnosis of *Nocardia* infections have been investigated as alternatives to current microbiological laboratory procedures. However, most of the serological tests that have been developed have failed to reach the desired sensitivity or specificity. Cross-reactions of *Nocardia* antigens derived from culture filtrates or whole-organism extracts with sera from cases of tuberculosis or leprosy have been reported (Blumer and Kaufman, 1979; Shainhaus et al., 1978), as has a lack of reactivity in sera from documented cases of *Nocardia* infection (Blumer and Kaufman, 1979). Monoclonal antibodies developed for use in the localization, purification, and characterization of *Nocardia* antigens may possibly serve as diagnostic reagents (Jimenez et al., 1990). A 55-kDa protein present in *N. asteroides, N. brasiliensis,* and *N. otitidis-caviarum* has been tested in serological assays yielding promising results (Angeles and Sugar, 1987; Sugar and Angeles, 1987).

DNA probes are now available for the rapid and specific identification of *N. asteroides* (Brownell and Belcher, 1990). The advantage of DNA probes over most other approaches to the identification of nocardial pathogens is that, when combined with target amplification or polymerase chain reaction, the need to culture the target organisms is eliminated.

Rhodococci can be assigned to 13 established taxospecies (Goodfellow et al., 1990) most of which have been shown to correspond with genomic species (Zakrzewska-Czerwińska et al., 1988). Nevertheless, it remains difficult to identify rhodococci partly due to the incomplete sample of strains included in comparative studies. This problem is compounded given the inconsistencies found between the results of biochemical, degradative, nutritional, and physiological tests in different studies. Chromogenic and fluorogenic substrates have been shown to be highly reactive and sensitive enough to demonstrate a range of putatively diagnostic-enzyme activities in suspensions of nongrowing rhodococcal cultures (Goodfellow et al., 1987, 1991; Mutimer and Woolcock, 1982). A number of properties of tentative diagnostic value for the identification of rhodococci are shown in Table 4. Similar features can be provisionally weighted for the identification of *Gordona* and *Tsukamurella* species (Table 5).

Table 3. Characteristics differentiating the species of *Nocardia*.[a,b,c]

Character	N. amarae	N. asteriodes	N. brasiliensis	N. brevicatena	N. carnea	N. farcinica	N. otitidis-caviarum	N. pinensis	N. seriola	N. transvalensis	N. vaccinii
Decomposition of:											
Esculin	+	+	+	+	+	+	+	−	+	+	+
Casein	−	−	+	−	−	−	−	−	−	−	−
Elastin	−	−	+	−	−	−	−	ND	ND	−	−
Hypoxanthine	−	−	+	−	−	−	+	−	−	+	−
Testosterone	−	+	+	+	+	+	−	ND	ND	−	−
Tyrosine	−	−	+	−	−	−	−	−	−	−	−
Xanthine	−	−	−	−	−	−	+	−	−	v	−
Acid from:											
Adonitol	−	−	−	−	−	−	−	ND	−	+	−
Arabinose	−	−	−	−	−	−	v	ND	−	−	+
Erythritol	−	−	−	−	−	−	−	ND	−	+	−
Glucose	+	+	+	−	+	+	+	+	+	+	+
Sorbitol	−	−	−	−	+	−	−	ND	−	v	v
Utilization of:											
Inositol	+	−	+	+	−	−	+	ND	ND	−	+
Mannitol	+	−	+	+	−	−	+	ND	ND	−	+
Rhamnose	+	−	−	+	−	+	−	ND	ND	−	+
Production of:											
Nitrate reductase	+	+	+	−	+	+	+	−	ND	+	+
Urease	+	+	+	−	−	+	+	+	−	+	+
Resistance to lysozyme	−	+	+	+	+	+	+	ND	+	+	+
Survival at 50°C/8 h	−	+	−	+	+	+	+	ND	−	+	+

[a]Adapted from Blackall et al. (1989); Goodfellow and Lechevalier (1989); and Kudo et al. (1988).

[b]Symbols: +, ≥90% of strains positive; −, 90% of strains negative; v, variable; ND, not determined.

[c]For details of test procedures, see Goodfellow (1971), Goodfellow and Alderson (1977), Goodfellow and Pirouz (1982), Gordon (1967), and Lacey and Goodfellow (1975).

[d]*Nocardia asteroides* and *N. farcinica* can also be distinguished by the ability of the latter to use *iso*-amyl alcohol; 2,3-butylene glycol; and 1,2-propylene glycerol as sole carbon sources (Schaal, 1977).

Comprehensive comparative taxonomic studies designed to highlight markers that can be weighted for diagnostic purposes are now possible given the improvements in the classification of *Gordona, Rhodococcus,* and *Tsukamurella* strains. Serological techniques may prove to be of value in distinguishing between species assigned to these genera (Hyman and Chaparas, 1977). There is also evidence that whole cell protein patterns may be useful in the identification of *R. equi* (Chirino-Trejo and Prescott, 1987).

Metabolism and Genetics

The earlier, and some of the later, work on the genetics and metabolism of Nocardiaceae strains is difficult to interpret, given the uncertain taxonomic status of the organisms examined. Many investigations reported on "nocardiae" can be attributed to strains now classified either in the genus *Rhodococcus* (Bradley, 1978; Brownell and Denniston, 1984; Peczyńska-Czoch and Mordarski, 1988) or the family Pseudonocardiaceae (Hütter and Eckhardt, 1988; Matsushima et al., 1987; Schupp et al., 1975). Despite the difficulties, certain general deductions can be drawn from the literature.

The catabolic potential of nocardiae and rhodococci, and perhaps gordonae and tsukamurellas, not only includes the capacity to assimilate carbohydrates and proteins but also unusual compounds such as aliphatic hydrocarbons, nitroaromatic compounds, bicyclic and polycyclic hydrocarbons, aniline, pyridine, and sterols (Cain, 1981; Janke et al., 1986; Peczyńska-Czoch and Mordarski, 1988; Tarnok, 1976; van Ginkel et al., 1987; Williams et al., 1989). Detergents and pesticides, including warfarin, are also modified (Goodfellow and Williams, 1983), and 1,2-epoxyalkanes produced from 7-alkanes (Fukuhashi et al., 1981). The ability to synthesize complex lipids (Minnikin,

Table 4. Characters differentiating the species of *Rhodococcus*.[a]

	R. aichiensis	R. chlorophenolicus	R. coprophilus	R. equi	R. erythropolis	R. fascians	R. globerulus	R. luteus	R. marinonascens	R. maris	R. rhodnii	R. rhodochrous	R. ruber
Morphogenetic sequence	R-C	R-C	H-R-C	R-C	EB-R-C	H-R-C	EB-R-C	EB-R-C	H-R-C	R-C	EB-R-C	EB-R-C	H-R-C
Decomposition of:													
Adenine	ND	ND	−	+	+	+	−	ND	−	ND	−	v	v
Tyrosine	ND	ND	−	−	v	+	−	+	v	−	+	+	+
Enzymatic tests													
a. APIZYM:													
Cysteine arylamidase	−	−	−	v	−	+	−	+	−	−	−	−	−
α-Glucosidase	+	+	−	v	+	+	+	+	−	+	−	−	+
Valine arylamidase	+	−	−	v	−	+	−	−	−	−	+	−	−
b. Cleavage of 7-amino-4-methylcoumarin (7AMC) substrates:													
β-Arginine-7AMC	−	−	+	+	+	−	v	+	+	+	+	+	+
CBZ-phenylalanine-arginine-7AMC	+	+	+	−	−	+	+	+	+	+	+	−	−
Pyroglutarate-7AMC	−	−	−	−	−	+	−	+	−	−	−	−	−
c. Cleavage of 4-methyl-umbelliferone (4-MU)-7AMC:													
4MU-β-d-Galactopyranoside	+	−	−	+	−	−	+	−	−	−	−	−	−
4MU-β-d-Glucopyranoside	+	+	+	+	+	+	+	−	+	−	−	+	+
4MU-Pyrophosphate	+	+	v	+	+	−	+	+	−	−	−	+	+
4MU-β-d-Xylopyranoside	+	+	−	+	+	−	+	−	−	−	−	+	+
Growth on sole carbon sources (%):													
Lactose (1.0)	+	−	−	−	+	−	−	+	−	−	−	+	+
Maltose (1.0)	ND	−	+	+	+	+	+	+	−	−	−	+	+
Mannose (1.0)	+	+	+	+	−	+	−	+	+	+	−	+	+
Androsterone (0.1)	+	−	−	v	+	+	+	+	−	+	−	+	+
Butan-2-3-diol (0.1)	+	+	+	+	+	+	−	+	+	+	−	+	+
Citraconic acid (0.1)	+	+	−	−	+	+	+	+	−	+	−	+	+
p-Hydroxybenzoic acid (0.1)	+	−	−	v	+	+	−	+	−	+	−	−	−
d-Mandelic acid (0.1)	−	−	−	−	+	+	+	+	−	+	−	+	+
Pimelic acid (0.1)	+	+	−	v	+	−	−	−	−	−	−	+	+
Sodium glutamate (0.1)	−	−	−	+	+	+	+	+	−	+	−	+	+
Sodium citrate (0.1)	+	+	+	+	+	−	+	+	−	−	−	+	+
Testosterone (0.1)	+	+	+	+	+	−	−	+	−	+	−	+	+

[a]Data and test procedures from Goodfellow (1989); Goodfellow and Alderson (1977); and Goodfellow et al. (1991). For symbols, see Table 3.

Table 5. Characteristics differentiating the species of *Gordona* and *Tsukamurella paurometabolum*.[a]

	G. bronchialis	G. rubropertincta	G. sputi	G. terrae	T. paurometabolum
Enzymatic tests:					
a. APIZYM:					
Cysteine arylamidase	+	+	−	−	+
α-Glucosidase	+	−	+	+	+
Phosphatase alkaline	+	−	−	−	−
b. Cleavage of 7-amino-4-methylcoumarin substrates (7AMC):					
D-Alanine-7AMC	+	+	−	+	−
CBZ-glycine-proline-7AMC[a]	−	+	−	+	−
Iso-leucine-7AMC	−	−	+	−	+
Pyroglutamate-7AMC	−	−	−	+	+
Valine-7AMC	−	+	+	+	−
c. Cleavage of 4-methylumbelliferone substrates (4MU-):					
4MU-β-D-galactopyranoside	+	−	+	−	+
4MU-β-D-mannopyranoside	+	−	+	−	−
Utilization as sole carbon source (%):					
Cellobiose (1.0)	+	−	+	+	−
Lactose (1.0)	−	−	+	+	−
Meso-inositol (1.0)	+	−	−	−	+
Raffinose (1.0)	−	−	+	+	−
Butan-2,3-diol (0.1)	−	−	+	+	+
Pimelic acid (0.1)	−	+	−	+	+
Testosterone (0.1)	−	+	−	+	−

[a]Symbols: as for Table 3; CBZ, the carbobenzoxy-*N*-protecting group of amino acids.
Data and test procedures taken from Goodfellow et al. (1990).

1982; Ioneda, 1988), including mycolic acids, distinguishes members of the family Nocardiaceae from all other bacteria apart from corynebacteria and mycobacteria.

Rhodococci have frequently been isolated from soils polluted with petroleum (Nesterenko et al., 1978a, 1978b) and have been implicated in the degradation of humic acid (Cross et al., 1976) and lignin-related compounds (Eggeling and Sahm, 1980, 1981; Rast et al., 1980). They also produce enzymes that mediate transformation of xenobiotics (Peczyńska-Czoch and Mordarski, 1988), including nitrile (Nagasawa and Yamada, 1989) and chlorinated phenolic compounds (Häggblom, 1988). *Rhodococcus* strains have been shown to mineralize pentachlorophenol and to degrade several other polychlorinated phenols, guaiacols, and syringols (Apajalahti and Salkinoja-Salonen, 1986, 1987; Häggblom et al., 1988). The degradation of polychlorinated phenols is initiated by hydroxylation to chlorinated p-hydroquinones. The reaction sequence for complete dechlorination also involves three reductive dechlorinations, producing 1,2,4-trihydroxybenzene (Häggblom et al., 1989). Thus, all of the chlorine moieties are removed from the polychlorinated compound prior to ring cleavage. Phenol degradation by rhodococci has been examined in batch

and continuous culture (Hensel and Straube, 1990; Straube et al., 1990). Oligocarbophily has been reported for a rhodococcal strain labelled "*Nocardia corallina*" (Tárnok, 1976).

The microbial degradation of nitriles proceeds through two enzymatic pathways. Nitrilase catalyzes the direct cleavage of nitriles to yield the corresponding acids and ammonia, whereas nitrile hydratase catalyzes the hydration of nitriles to amides. These enzymes have been detected in diverse bacteria, including rhodococci (Harper, 1977, 1985; Nagasawa et al., 1988a). Nagasawa and his colleagues demonstrated the occurrence of a cobalt induced and cobalt-containing nitrile hydratase in *Rhodococcus rhodochrous* J1; the cells of this organism also produce a nitrilase (Nagasawa et al., 1988b) that has been purified and characterized (Kobayashi et al., 1989a). A novel nitrilase that preferentially catalyzes the hydrolysis of aliphatic nitriles has been found in *R. rhodochrous* K22, a facultative crotononitrile-utilizing actinomycete (Kobayashi et al., 1990). This enzyme is remarkable in its broad substrate specificity for aliphatic nitriles. Aliphatic olefinic nitriles, especially low-molecular-mass ones with 2 to 5 carbon atoms, were especially good substrates for the enzymes. The nucleotide sequence of the nitrile hydratase gene of *R. rhodochrous* J1 has been determined (Kobayashi et al., 1990).

Genetic recombination and plasmids have been reported in *N. asteroides* (Kasweck and Little, 1982; Kasweck et al., 1982) and nocardiophages for *N. asteroides* (Andrzejewski et al., 1978; Prauser, 1976, 1981a; Pulverer et al., 1975), *N. brasiliensis* (Pulverer et al., 1975), *N. carnea* (Williams et al., 1980), *N. otitidis-caviarum,* and *N. vaccinii* (Prauser, 1976). Members of the family Nocardiaceae are generally susceptible to nocardiophages whereas strains classified in the families Mycobacteriaceae and Pseudonocardiaceae are not (Prauser, 1981b; Williams et al., 1980). Conversely, phages which lyse *Amycolatopsis mediterranei* are inactive against nocardiae (Thiemann et al., 1964).

Developments in rhodococcal genetics have been reviewed by Brownell and Denniston (1984). Recombination was reported in the model organism *R. erythropolis* by Adams and Bradley (1963). Over 60 genetic traits have been used in the development of a *R. erythropolis* linkage map, and temperate phages are available as cloning vectors for establishing a gene cloning system. Genetic recombination also occurs between rhodococcal strains labelled *"Nocardia opaca"* and *"Nocardia restricta."* A transferable plasmid described for a strain of the former carries traits allowing for chemolithoautotrophic growth (*aut⁺*) including genes for hydrogenase, phosphoribulokinase, and ribulose-bisphosphate-carboxylase production (Reh, 1981; Reh and Schlegel, 1981). The *"N. opaca"* strain is able to transfer the *aut⁺* trait to related organisms and to *R. erythropolis;* in the presence of CO_2 and H_2 it has a generation time of 7 hours.

A miniplasmid has been characterized for a rhodococcal strain labelled *"R. corallina"* (Kirby and Usdin, 1985) and *Rhodococcus* genes encoding pigment production cloned and expressed in *Escherichia ecoli* (Hill et al., 1989). Plasmid-determined transformation of *cis*-abienol and schareol has been achieved in *"Nocardia restricta* JTS-162" (Hieda et al., 1982). A generalized transducing bacteriophage is available for *Rhodococcus erythropolis* (Dabs, 1987).

Applications

Nocardiaceae strains, notably rhodococci, produce antibiotics (Okami and Hotta, 1988), perform useful chemical modifications of complex compounds (Peczyńska-Czoch and Mordarski, 1988), and degrade diverse pollutants (Lacey, 1988), including phthalate esters (Kurane et al.,

1979, 1980, 1984), nitriles (Kobayashi et al., 1990) and polychlorinated phenols (Apajalahti and Salkinoja-Salonen, 1986; Häggblom et al., 1988, 1989). Consequently, mycolic acid-containing actinomycetes may have a role to play in bioremediation and detoxification of contaminated soils. Rhodococci may also prove to be a valuable source of biosurfactants (Gerson and Zajic, 1977). Biosurfactants from a strain labelled *"Mycobacterium rhodochrous"* were found to enhance oil recovery using the water flooding technique (Wagner et al., 1976). It is also possible that the cloning and expression of rhodococcal genes in *E. coli* (Hill et al., 1989) will facilitate the production of large amounts of pigment for chemical characterization and industrial purposes, and permit the generation of chromogenic vectors which do not require expensive substrates.

Nitrile-converting enzymes from rhodococci are attracting increasing attention as catalysts for processing organic chemicals, since they can convert nitriles to the corresponding higher value acids or amides, notably acrylamide (Nagasawa and Yamada, 1989). Optimal conditions have been established for the production of nicotinic acid from 3-cyanopyridine (Mathew et al., 1988), and p-aminobenzoic acid from p-aminobenzonitrile (Kobayashi et al., 1989b) with resting cells of *R. rhodochrous* J1. Furthermore, the development of stable activated sludge systems with *Rhodococcus* would represent a major advance in the treatment of toxic nitrile industrial wastes (Kobayashi et al., 1990) as conventional activated sludge systems are readily incapacitated by such wastes.

Literature Cited

Adams, J. N., and S. G. Bradley. 1963. Recombination events in the bacterial genus *Nocardia.* Science 140:1392–1394.

Ajello, L., and G. D. Roberts. 1981. Mycetomas, p. 1033. In: W. J. Hausler (ed.), Diagnostic procedures for bacterial, mycotic and parasitic infections. American Public Health Association, Washington.

Al-Diwany, L. J., and Cross, T. 1978. Ecological studies on nocardioforms and other actinomycetes in aquatic habitats. Zbl. Bakt. Suppl. 6:153–160.

Alshamaony, L., M. Goodfellow, and D. E. Minnikin. 1976a. Free mycolic acids as criteria in the classification of *Nocardia* and the *'rhodochrous'* complex. J. Gen. Microbiol. 92:188–199.

Alshamaony, L., M. Goodfellow, D. E. Minnikin, and H. Mordarska. 1976b. Free mycolic acids as criteria in the classification of *Gordona* and the *'rhodochrous'* complex. J. Gen. Microbiol. 92:183–187.

Angeles, A. M., and A. M. Sugar. 1987. Rapid diagnosis of nocardiosis with enzyme immunoassay. J. Infect. Dis. 155:292–296.

Andrzejewski, J., G. Müller, E. Röhrscheidt, and D. Pietkiewicz. 1978. Isolation, characterization and classification of *Nocardia asteroides* bacteriophage. Zbl. Bakt. Suppl. 6:319–326.

Apajalahti, J. H. A., and M. S. Salkinoja-Salonen. 1986. Degradation of polychlorinated phenols by *Rhodococcus chlorophenolicus*. Appl. Microbiol. Biotech. 25:62–67.

Apajalahti, J. H. A., and M. S. Salkinoja-Salonen. 1987. Dechlorination and *para*-hydroxylation of polychlorinated phenols by *Rhodococcus chlorophenolicus*. Bacteriol. 169:675–681.

Apajalahti, J. H. A., P. Kärpänoja, and M. S. Salkinoja-Salonen. 1986. *Rhodococcus chlorophenolicus* sp. nov., a chlorophenol-mineralising actinomycete. Int. J. Syst. Bacteriol. 36:246–251.

Athalye, M., W. C. Noble, A. I. Mallet, and D. E. Minnikin. 1984. Gas-chromatography-mass spectrometry of mycolic acids as a tool in the identification of medically important coryneform bacteria. J. Gen. Microbiol. 130:513–519.

Barton, M. D., and K. L. Hughes. 1980. *Corynebacterium equi*: A review. Vet. Bull. 50:65–80.

Barton, M. D., and K. L. Hughes. 1981. Comparison of three techniques for the isolation of *Rhodococcus (Corynebacterium) equi* from contaminated sources. J. Clin. Microbiol. 13:219–221.

Barton, M. D., M. Goodfellow, and D. E. Minnikin. 1989. Lipid composition in the classification of *Rhodococcus equi*. Zbl. Bakt. 272:154–170.

Beaman, B. L. 1984. Actinomycete pathogenesis, p. 457–479. In: M. Goodfellow, M. Mordarski, and S. T. Williams (ed.), The biology of the actinomycetes. Academic Press, London.

Beaman, B. L. 1988. *Nocardia* in the etiology of nocardiosis, p. 3–4. In: Actinomycetes as opportunistic pathogens (abstracts). 3rd International Symposium of the Research Center for Pathogenic Fungi and Microbial Toxicoses, Chiba University, Japan.

Beaman, B. L., and A. M. Sugar. 1983. Interaction of *Nocardia* in naturally acquired infections in animals. J. Hyg. 91:393–419.

Beaman, B. L., J. Burnside, B. Edwards, and W. Causey. 1976. Nocardial infections in the United States, 1972–1974. J. Infect. Dis. 134:285–289.

Blackall, L. L., A. E. Harbers, P. F. Greenfield, and A. C. Hayward. 1988. Actinomycete scum problems in Australian activated sludge plants. Wat. Sci. Tech. 20:493–495.

Blackall, L. L., and K. C. Marshall. 1989. The mechanism of stabilization of actinomycete foams and the prevention of foaming under laboratory conditions. J. ind. Microbiol. 4:181–188.

Blackall, L. L., J. H. Parlett, A. C. Hayward, D. E. Minnikin, P. F. Greenfield, and A. E. Harbers. 1989. *Nocardia pinensis* sp. nov., an actinomycete found in activated sludge foams in Australia. J. Gen. Microbiol. 135:1547–1558.

Blanchard, R. 1896. Parasites végétaux á l'exclusion des bactéries, p. 811–932. In: C. Bouchard (ed.), Traité de pathologie générale, volume 2. G. Masson, Paris.

Blumer, S. O., and L. Kaufman. 1979. Microimmunodiffusion test for nocardiosis. J. Clin. Microbiol. 10:308–312.

Boiron, P., C. Lafaurie, A. Rabbache, J. Brown, R. Carteret, and J. Petit. 1990. Urease-negative *Nocardia asteroides* causing cutaneous nocardiosis. J. Clin. Microbiol. 28:801–802.

Bousfield, I. J., and Goodfellow, M. 1976. The 'rhodochrous' complex and its relationships with allied taxa, p. 39–65. In: M. Goodfellow, G. H. Brownell and J. A. Serrano (ed.), The biology of the nocardiae. Academic Press, London.

Bradley, S. G. 1978. Physiological aspects of nocardiae. Zbl. Bakt. Suppl. 6:287–302.

Bradley, S. G., L. W. Enquist, and H. E. Scribner, III. 1978. Heterogeneity among deoxyribonucleotide sequences of actinomycetales, p. 207–224. In: E. Freerksen, I. Tárnok, and J. H. Thumin (ed.), Genetics of the Actinomycetales. Gustav Fischer Verlag, Stuttgart.

Brownell, G. H., and K. E. Belcher. 1990. DNA probes for the identification of *Nocardia asteroides*. J. Clin. Microbiol. 28:2082–2086.

Brownell, G. H., and K. Denniston. 1984. Genetics of nocardioform bacteria, p. 201–228. In: M. Goodfellow, M. Mordarski, and S. T. Williams (ed.), The biology of the actinomycetes. Academic Press, London.

Bushnell, R. B., A. C. Pier, R. E. Fichtner, B. L. Beaman, H. A. Boos, and M. D. Salman. 1979. Clinical and diagnostic aspects of herd problems with nocardial and mycobacterial mastitis. Am. Assoc. Vet. Lab. Diag. 22:1–12.

Butler, W. R., D. G. Ahearn, and J. O. Kilburn. 1986. High performance liquid chromatography of mycolic acids as a tool in the identification of *Corynebacterium, Nocardia, Rhodococcus,* and *Mycobacterium* species. J. Clin. Microbiol. 23:182–185.

Cain, R. B. 1981. Regulation of aromatic and hydroaromatic catabolic pathways in nocardioform actinomycetes. Zbl. Bakt. Suppl. 11:335–354.

Castellani, A., and A. J. Chambers. 1917. Manual of tropical medicine, 3rd ed. Williams, Wood and Co., New York.

Chirino-Trejo, J. M., and J. F. Prescott. 1987a. Polyacrylamide gel electrophoresis of whole cell preparations of *Rhodococcus equi*. Can. J. Jet. Res. 51:297–300.

Chirino-Trejo, J. M., and J. F. Prescott. 1987b. Antibody response of horses to *Rhodococcus equi* antigens. Can. J. Jet. Res 51:301–305.

Chirino-Trejo, J. M., J. F. Prescott, and J. A. Yager. 1987. Protection of foals against experimental *Rhodococcus equi* pneumonia by oral immunization. Can. J. Vet. Res. 51:444–447.

Collins, M. D. 1985. Isoprenoid quinone analyses in bacterial classification and identification, p. 267–287. I: M. Goodfellow, and D. E. Minnikin (ed.), Chemical methods in bacterial systematics. Academic Press, London.

Collins, M. D. and D. Jones. 1982. Lipid composition of *Corynebacterium paurometabolum* (Steinhaus). FEMS Microbiol. Lett. 13:13–16.

Collins, M. D., T. Pirouz, M. Goodfellow, and D. E. Minnikin. 1977. Distribution of menaquinones in actinomycetes and corynebacteria. J. Gen. Microbiol. 100:221–230.

Collins, M. D., M. Goodfellow, and D. E. Minnikin. 1982a. Fatty acid composition of some mycolic acid-containing coryneform bacteria. J. Gen. Microbiol. 128:2503–2509.

Collins, M. D., M. Goodfellow, and D. E. Minnikin. 1982b. A survey of the structure of mycolic acids in *Coryne-*

bacterium and related taxa. J. gen. Microbiol. 128:129–149.

Collins, M. D., M. Goodfellow, D. E. Minnikin, and G. Alderson. 1985. Menaquinone composition of mycolic acid-containing actinomycetes and some sporoactinomycetes. J. Appl. Bacteriol. 58:77–86.

Collins, M. D., O. W. Howarth, E. Grund, and R. M. Kroppenstedt. 1987. Isolation and structural determination of new members of the vitamin K series in *Nocardia brasiliensis*. FEMS Microbiol. Lett. 41:35–39.

Collins, M. D., J. Smida, M. Dorsch, and E. Stackebrandt. 1989. *Tsukamurella* gen. nov. harboring *Corynebacterium paurometabolum* and *Rhodococcus aurantiacus*. Int. J. Syst. Bacteriol. 38:385–391.

Conti-Diaz, I. A., E. Gezuele, E. Civila, and J. E. Mackinnon. 1971. Termotolerancia y acción patógena de cepas de *Nocardia asteroides* aislades de fuentes naturales. Revta Urug. Patol. Clin. Microbiol. 9:232–241.

Cross, T., T. J. Rowbotham, E. N. Mishustin, E. Z. Tepper, F. Antoine-Portaels, K. P. Schaal, and H. Bickenbach. 1976. The ecology of nocardioform actinomycetes, p. 337–371. In: M. Goodfellow, G. H. Brownell, and J. A. Serrano (ed.), The biology of the nocardiae. Academic Press, London.

Cummins, C. S. 1971. Cell wall composition in *Corynebacterium bovis* and some other corynebacteria. J. Bacteriol. 105:1227–1228.

Dabs, E. R. 1987. A generalized transducing bacteriophage for *Rhodococais erythropolis*. Mol. Gen. Genet. 206:116–120.

Demaree, J. B., and N. R. Smith. 1952. *Nocardia vaccinii* n. sp. causing galls on blueberry plants. Phytopathology 42:249–252.

Dobson, G., D. E. Minnikin, S. M. Minnikin, J. H. Parlett, M. Goodfellow, M. Ridell, and M. Magnusson. 1985. Systematic analysis of complex mycobacterial lipids, p. 237–265. In: M. Goodfellow, and D. E. Minnikin, Chemical methods in bacterial systematics. Academic Press, London.

Eberzole, L. L., and J. L. Paturzo. 1988. Endophthalinitis caused by *Rhodococcus equi* Prescott serotype 4. J. Clin. Microbiol. 26:1221–1222.

Eggeling, J. and H. Sahm. 1980. Degradation of coniferyl alcohol and other lignin-related aromatic compounds by *Nocardia* sp. DSM 1069. Arch. Mikrobiol. 126:141–148.

Eggeling, L., and H. Sahm. 1981. Degradation of lignin-related aromatic compounds by *Nocardia* spec. DSM 1069 and the specificity of demethylation. Zbl. Bakt. Suppl. 11:361–366.

Etémadi, A. H. 1967. The use of pyrolysis gas chromatography and mass spectrometry in the study of the structure of mycolic acids. J. Gas Chromat. 5:447–456.

Farmer, R. 1962. Influence of various chemicals in the isolation of Nocardia from soil. Proc. Acad. Sci. 43:254–256.

Fukuhashi, K., A. Taoka, S. Uchida, I. Karube, and S. Suzuki. 1981. Production of 1,2-epoxyalkanes from 1-alkanes by *Nocardia corallina* B-276. Eur. J. Appl. Microbiol. Biotechnol. 12:39–45.

Gerson, D. F., and J. E. Zajic. 1977. Bitumen extraction from tar sands with microbial surfactants. The Oil Sands of Canada-Venezuela. CIM Special Publication No. 17:705–710.

Gonzáles-Ochoa, A. 1973. Virulence of nocardiae. Can J. Microbiol. 19:901–904.

Goodfellow, M. 1971. Numerical taxonomy of some nocardioform bacteria. J. Gen. Microbiol. 69:33–80.

Goodfellow, M. 1986. Genus *Rhodococcus* Zopf 1891, 28[AL], p. 1472–1481. In: P. H. A. Sneath, N. S. Mair, M. E. Sharpe, and J. G. Holt (ed.), Bergey's manual of systematic bacteriology, vol. 2. Williams and Wilkins, Baltimore.

Goodfellow, M. 1989. Genus *Rhodococcus* Zopf 1891, 28[AL], p. 2362–2371. In: S. T. Williams, M. E. Sharpe, and J. G. Holt (ed.), Bergey's manual of systematic bacteriology, vol. 4. Williams and Wilkins, Baltimore.

Goodfellow, M., and G. Alderson. 1977. The actinomycete genus *Rhodococcus:* a home for the *'rhodochrous'* complex. J. Gen. Microbiol. 100:99–122.

Goodfellow, M., and E. Aubert. 1980. Characterization of rhodococci from the intestinal tract of *Rapa Nui* cockroaches, p. 231–240. In: G. L. Nogrady (ed.), Microbiology of Easter Island, vol. 2. Sovereign Press, Oakville, Canada.

Goodfellow, M., and T. Cross. 1984. Classification, p. 7–164. In: M. Goodfellow, M. Mordarski and S. T. Williams (ed), The biology of the actinomycetes. Academic Press, London.

Goodfellow, M., and M. P. Lechevalier. 1989. Genus *Nocardia* Trevisan 1889, 9[AL], p. 2350–2361. In: S. T. Williams, M. E. Sharpe, and J. G. Holt (ed.), Bergey's manual of systematic bacteriology, vol. 4. Williams and Wilkins, Baltimore.

Goodfellow, M., and D. E. Minnikin. 1984. A critical evaluation of *Nocardia* and related taxa, p. 588–596. In: L. Ortiz-Ortiz, L. F. Bojalil, and V. Yakoleff (ed.), Biological, biochemical and biomedical aspects of actinomycetes. Academic Press, Orlando.

Goodfellow, M. and A. G. O'Donnell. 1989. Search and discovery of industrially significant actinomycetes, p. 343–383. In: S. Baumberg, M. Rhodes, and I. Hunter (ed.) Microbial products : new approaches. Cambridge University Press, Cambridge.

Goodfellow, M., and V. A. Orchard. 1974. Antibiotic sensitivity of some nocardioform bacteria and its value as a criterion for taxonomy. J. Gen. Microbiol. 83:375–387.

Goodfellow, M., and T. Pirouz. 1982. Numerical classification of sporoactinomycetes containing *meso*-diaminopimelic acid in the cell wall. J. Gen. Microbiol. 128:503–527.

Goodfellow, M., and L. G. Wayne. 1982. Taxonomy and nomenclature, p. 471–521. In: C. Ratledge, and J. L. Stanford (ed.), The biology of the mycobacteria, vol. 1. Academic Press, London.

Goodfellow, M., and S. T. Williams. 1983. Ecology of actinomycetes. Ann. Rev. Microbiol. 37:189–216.

Goodfellow, M., M. D. Collins, and D. E. Minnikin. 1976. Thin-layer chromatographic analysis of mycolic acid and other long-chain components in whole-organism methanolysates of coryneform and related taxa. J. Gen. Microbiol. 96:351–358.

Goodfellow, M., P. A. B. Orlean, M. D. Collins, L. Alshamaony, and D. E. Minnikin. 1978. Chemical and numerical taxonomy of strains received as *Gordona aurantiaca*. J. Gen. Microbiol. 109:57–68.

Goodfellow, M., D. E. Minnikin, C. Todd, G. Alderson, S. M. Minnikin, and M. D. Collins. 1982a. Numerical and chemical classification of *Nocardia amarae*. J. Gen. Microbiol. 128:1283–1297.

Goodfellow, M., Beckham, A. R., and M. D. Barton. 1982b. Numerical classification of *Rhodococcus equi* and related actinomycetes. J. Appl. Bacteriol. 53:199–207.

Goodfellow, M., C. R. Weaver, and D. E. Minnikin. 1982c. Numerical classification of some rhodococci, corynebacteria and related organisms. J. Gen. Microbiol. 128:731–735.

Goodfellow, M., E. G. Thomas, and A. L. James. 1987. Characterisation of rhodococci using peptide hydrolase substrates based on 7-amino-4-methylcoumarin. FEMS Microbiol. Lett. 44:349–355.

Goodfellow, M., E. G. Thomas, A. C. Ward, and A. L. James. 1991. Classification of *Rhodococcus*. Zbl. Bakt. 298:(in press).

Gordon, R. E. 1967. The taxonomy of bacteria, p. 293–321. In: T. R. G. Gray, and D. Parkinson (ed.), The ecology of soil bacteria. Liverpool University Press, Liverpool.

Gordon, R. E., and J. E. Mihm. 1957. A comparative study of some new strains received as nocardiae. J. Bacteriol. 73:15–27.

Gordon, R. E., and M. M. Smith. 1953. Rapidly growing acid fast bacteria. I. Species description of *Mycobacterium phlei* Lehmann and Neumann and *Mycobacterium smegmatis* (Trevisan) Lehmann and Neumann. J. Bacteriol. 66:41–48.

Gordon, R. E., S. K. Mishra, and D. A. Barnett. 1978. Some bits and pieces of the genus *Nocardia; N. carnea, N. vaccinii, N. transvalensis, N. orientalis* and *N. aerocolonigenes*. J. Gen. Microbiol. 108:69–78.

Gray, P. H. H., and H. G. Thornton. 1928. Soil bacteria that decompose certain aromatic compounds. Zbl. Bakt. Parasitkde. Abt. II 73:74–96.

Gutmann, L., F. W. Goldstein, M. D. Kitzis, B. Hautefort, C. Darmon, and J. F. Acar. 1983. Susceptibility of *Nocardia asteroides* to 46 antibiotics, including 22 β-lactams. Antimicrob. Agents Chemother. 23:248–251.

Häggblom, M. 1988. Degradation and transformation of chlorinated phenolic compounds by strains of *Rhodococcus* and *Mycobacterium*. Ph.D. Thesis, University of Helsinki.

Häggblom, M. M., D. Jänke, and M. S. Salkinoja-Salonen. 1989. Hydroxylation and dechlorination of tetrachlorohydroquinone by *Rhodococcus* sp. strain CP-2 cell extracts. Appl. Environ. Microbiol. 55:516–519.

Häggblom, M. M., L. J. Nohynek, and M. S. Salkinoja-Salonen. 1988. Degradation and O-methylation of polychlorinated phenol compounds by *Rhodococcus* and *Mycobacterium* strains. Appl. Environ. Microbiol. 54:3043–3052.

Harper, D. B. 1977. Microbiol metabolism of aromatic nitriles. Biochem. J. 165:309–319.

Harper, D. B. 1985. Characterization of a nitrilase from *Nocardia* sp. (rhodochrous group) NCIB 11215, using p-hydroxybenzonitrile as sole carbon source. Int. J. Biochem. 17:677–683.

Hecht, S. T., and W. A. Causey. 1976. Rapid methods for the detection and identification of mycolic acids in aerobic actinomycetes and related bacteria. J. Clin. Microbiol. 4:284–287.

Helmke, E., and H. Weyland. 1984. *Rhodococcus marinonascens* sp. nov., an actinomycete from the sea. Int. J. Syst. Bacteriol. 34:127–138.

Hensel, J., and G. Straube. 1990. Kinetic studies of phenol degradation by *Rhodococcus* sp. II. Continuous cultivation. Antonie van Leeuwenhoek 57:33–36.

Hieda, T., Y. Mikami, Y. Obi, and T. Kisaka. 1982. Plasmid-determined transformation of *cis*-abienol and sclareol in *Nocardia restricta* JTS-162. Agr. Biol. Chem. 46:305–306.

Higgins, M. L., and M. P. Lechevalier. 1969. Poorly lytic bacteriophage from *Dactylosporangium thailandensis* (Actinomycetales). J. Virol. 3:210–216.

Hill, R., S. Hart, N. Illing, R. Kirby, and D. R. Woods. 1989. Cloning and expression of *Rhodococcus* genes encoding pigment production in *Escherichia coli*. J. Gen. Microbiol. 135:1507–1513.

Houang, E. T., I. S. Lovett, F. D. Thompson, A. R. Harrison, A. M. Joekes, and M. Goodfellow. 1980. *Nocardia asteroides*—A transmissible disease. J. Hosp. Infect. 1:31–40.

Howarth, O. W., E. Grund, R. M. Kroppenstedt, and M. D. Collins. 1986. Structural determination of a new naturally occurring cyclic vitamin K. Biochem. Biophys. Res. Commun. 140:916–923.

Hutchinson, M., J. W. Ridgway, and T. Cross. 1975. Biodeterioration of rubber in contact with water, sewage and soil, p. 187–202. In: D. W. Lovelock, and R. J. Gilbert (ed.), Microbial aspects of the deterioration of materials. Academic Press, London.

Hütter, R. A., and T. Eckhardt. 1988. Genetic manipulation, p. 89–184. In: M. Goodfellow, S. T. Williams, and M. Mordarski (ed.), Actinomycetes in biotechnology. Academic Press, London.

Hyman, I. S., and S. D. Chaparas. 1977. A comparative study of the 'rhodochrous' complex and related taxa by delayed type skin reactions on guinea pigs and by polyacrylamide gel electrophoresis. J. Gen. Microbiol. 100:363–371.

Ioneda, T., 1988. Biochemical and physiological aspects of mycolic acids and mycolyl derivatives, p. 463–468. In: Y. Okami, T. Beppu, and H. Ogawara (ed.), Biology of actinomycetes '88. Japan Scientific Societies Press, Tokyo.

Janke, D., B. Schukat, and H. Prauser, 1986. Screening among nocardioform bacteria strains able to degrade aniline and monochloranilines. J. Basic Microbiol. 6:341–350.

Jones, D. 1975. A numerical taxonomic study of coryneform and related bacteria. J. Gen. Microbiol. 87:52–96.

Jones, K. L. 1949. Fresh isolates of actinomycetes in which the presence of sporangeous aerial mycelia is a fluctuating characteristic. J. Bacteriol. 57:141–145.

Judicial Commission. 1985. Opinion 58. Confirmation of the types in the Approved Lists as nomenclatural types including recognition of *Nocardia asteroides* (Eppinger, 1891) Blanchard 1896 and *Pasteuria multicida* (Lehmann and Neumann, 1899) Rosenbusch and Merchant 1939 as the respective type species of the genera *Nocardia* and *Pasteurella* and the rejection of the species name *Pasteurella gallicida* (Burrill 1883) Buchanan 1925. Int. J. Syst. Bacteriol. 35:538.

Jiménez, T., A. M. Diaz, and H. Zlotnik. 1990. Monoclonal antibodies to *N. asteroides* and *N. brasiliensis* antigens. J. Clin. Microbiol. 28:87–91.

Kanetsuna, F., and A. Bartoli. 1972. A simple chemical method to differentiate *Mycobacterium* from *Nocardia*. J. Gen. Microbiol. 70:209–212.

Kasweck, K. L., and M. L. Little. 1982. Genetic recombination in *Nocardia asteroides*. J. Bacteriol. 149:403–406.

Kasweck, K. L., M. L. Little, and S. G. Bradley. 1981. Characteristics of plasmids in *Nocardia asteroides*. Actin. Rel. Org. 16:57–63.

Kasweck, K. L., M. L. Little, and S. G. Bradley. 1982. Plasmids in mating strains of *Nocardia asteroides*. Dev. Ind. Microbiol. 23:279–286.

Kirby, R., and K. Usdin. 1985. The isolation and restriction mapping of a miniplasmid from the actinomycete *Nocardia corallina*. FEMS Microbiol. Lett. 27:57–59.

Kobayashi, M., T. Nagasawa, and H. Yamada. 1989a. Nitrilase of *Rhodococcus rhodochrous* J1: purification and characterization. Eur. J. Biochem. 182:349–356.

Kobayashi, M., T. Nagasawa, N. Yanaka, and H. Yamada. 1989b. Nitrilase-catalyzed production of p-aminobenzoic acid from p-aminobenzonitrile with *Rhodococcus rhodochrous* J1. Biotechnol. Lett. 11:27–30.

Kobayashi, M., N. Yanaka, T. Nagasawa, and H. Yamada. 1990. Purification and characterization of a novel nitrilase of *Rhodococcus rhodochrous* K22 that acts on aliphatic nitriles. J. Bacteriol. 4807–4815.

Kroppenstedt, R. M. 1985. Fatty acid and menaquinone analysis of actinomycetes and related organisms, p. 173–199. In: M. Goodfellow, and D. E. Minnikin (ed.), Chemical methods in bacterial systematics. Academic Press, London.

Kudo, T., K. Hatai, and A. Seino. 1988. *Nocardia seriolae* sp. nov. causing nocardiosis of cultured fish. Int. J. Syst. Bacteriol. 38:173–178.

Kurane, R., T. Suzuki, and Y. Takahara. 1979. Removal of phthalate esters by activated sludge inoculated with a strain of *Nocardia erythropolis*. Agr. Biol. Chem. 43:421–427.

Kurane, R., T. Suzuki, and Y. Takahara. 1980. Metabolic pathway of phthalate esters by *Nocardia erythropolis*. Agric. Biol. Chem. 44:523–527.

Kurane, R., T. Suzuki, and S. Fukuoka. 1984. Purification and some properties of a phthalate ester hydrolyzing enzyme from *Nocardia erythropolis*. Appl. Microbiol. Biotech. 20:378–383.

Kurup, V. P., and G. H. Schribner. 1981. Antigenic relationship among *Nocardia asteroides* immunotypes. Microbios 31:25–30.

Kurup, V. P., J. E. Peichura, E. Y. Ting, and J. A. Orlowski. 1983. Immunochemical characterization of *Nocardia asteroides* antigens: Support for a single species concept. Can. J. Microbiol. 29:425–432.

Lacey, J. 1988. Actinomycetes as biodeteriogens and pollutants in the environment, p. 359–432. In: M. Goodfellow, S. T. Williams, and M. Mordarski (ed.), Actinomycete biology. Academic Press, London.

Lacey, J., and M. Goodfellow. 1975. A novel actinomycete from sugar-cane bagasse, *Saccharopolyspora hirsuta*, gen. et sp. nov. J. Gen. Microbiol. 88:75–85.

Lechevalier, H. A., and M. P. Lechevalier. 1970a. Chemical composition as a criterion in the classification of aerobic actinomycetes. Int. J. Syst. Bacteriol. 20:435–443.

Lechevalier, H. A., M. P. Lechevalier, P. E. Wyszkowski, and F. Mariat. 1977. Actinomycetes found in sewage-treatment plants of the activated sludge type, p. 227–247. In: T. Arai (ed.), Actinomycetes: The boundary microorganisms. Toppan, Tokyo.

Lechevalier, M. P., and H. A. Lechevalier. 1970b. A critical evaluation of the genera of aerobic actinomycetes, p. 393–405. In: H. Prauser (ed.), The Actinomycetales. Gustav Fischer Verlag, Jena.

Lechevalier, M. P., and H. A. Lechevalier. 1974. *Nocardia amarae* sp. nov., an actinomycete common in foaming activated sludge. Int. J. Syst. Bacteriol. 24:278–288.

Lechevalier, M. P., A. C. Horan, and H. A. Lechevalier. 1971. Lipid composition in the classification of nocardiae and mycobacteria. J. Bacteriol. 105:313–318.

Lechevalier, M. P., C. De Biévre, and H. A. Lechevalier. 1977. Chemotaxonomy of aerobic actinomycetes: phospholipid composition. Biochem. Syst. Ecol. 5:249–260.

Lechevalier, M. P., A. E. Stern, and H. A. Lechevalier. 1981. Phospholipids in the taxonomy of actinomycetes. Zbl. Bakt. Suppl. 11:111–116.

Lemmer, H., and R. M. Kroppenstedt. 1984. Chemotaxonomy and physiology of some actinomycetes isolated from scumming activated sludge. Syst. Appl. Microbiol. 5:124–135.

Lind, A., and M. Ridell. 1976. Serological relationships betwen *Nocardia*, *Mycobacterium*, *Corynebacterium* and the 'rhodochrous' taxon, p. 220–235. In: M. Goodfellow, G. H. Brownell, and J. A. Serrano (ed.), The biology of the nocardiae. Academic Press, London.

Lind, A., and M. Ridell. 1983. Immunological classification: Immunodiffusion and immunoelectrophoresis, p. 67–82. In: G. Kubica, and L. G. Wayne (ed.), The mycobacteria: a sourcebook. Marcel Dekker, New York.

Lind, A., Ö. Ouchterlony, and M. Ridell. 1980. Mycobacterial antigens, p. 275–303. In: G. Meissner, and Schimiedel (ed.), Infektionskrankheiten und ihre Erreger, Bd. 4, Mycobakterien und mykobakterielle Krankheiten. Fischer Verlag, Jena, Germany.

Magnusson, M. 1976. Sensitin tests in *Nocardia* taxonomy, p. 236–265. In: M. Goodfellow, G. H. Brownell, and J. A. Serrano (ed.), The biology of the nocardiae. Academic Press, London.

Mahgoub, E. S., and I. G. Murray. 1973. Mycetoma. Heinemann, London.

Mara, D. D., and J. I. Oragui. 1981. Occurrence of *Rhodococcus coprophilus* and associated actinomycetes in faeces, sewage, and fresh water. Appl. Environ. Microbiol. 42:1037–1042.

Mathew, C. D., T. Nagasawa, M. Kobayashi, and H. Yamada. 1988. Nitrilase-catalyzed production of nicotinic acid from 3-cyanopyridine in Rhodococcus rhodochrous J1. Appl. Environ. Microbiol. 54:1030–1032.

Matsushima, P., M. A. McHenney, and R. H. Baltz. 1987. Efficient transformation of *Amycolatopsis orientalis* (*Nocardia orientalis*) protoplasts by *Streptomyces* plasmids. J. Bacteriol. 169:2298–2300.

Maurice, M. T., M. J. Vacheron, and G. Michel. 1971. Isolément d'acides nocardiques de plusieurs espèces de *Nocardia*. Chem. Phys. Lipids 7:9–18.

Minnikin, D. E. 1982. Lipids: Complex lipids, their chemistry, biosynthesis and roles, p. 95–184. In: C. Ratledge and J. L. Stanford (ed.), The biology of the mycobacteria, vol. 1. Academic Press, London.

Minnikin, D. E. and M. Goodfellow. 1980. Lipid composition in the classification and identification of acid-fast bacteria, p. 189–256. In: M. Goodfellow, and R. G. Board (ed.), Microbiological classification and identification. Academic Press, London.

Minnikin, D. E., L. Alshamaony, and M. Goodfellow. 1975. Differentiation of *Mycobacterium, Nocardia* and related taxa by thin-layer chromatographic analysis of whole-organism methanolysates. J. Gen. Microbiol. 88:200–204.

Minnikin, D. E., P. V. Patel, L. Alshamaony, and M. Goodfellow. 1977. Polar lipid composition in the classification of *Nocardia* and related bacteria. Int. J. Syst. Bacteriol. 27:104–117.

Minnikin, D. E., I. G. Hutchinson, A. B. Caldicott, and M. Goodfellow. 1980. Thin-layer chromatography of methanolysates of mycolic acid-containing bacteria. J. Chromat. 188:221–233.

Minnikin, D. E., S. M. Minnikin, J. H. Parlett, M. Goodfellow, and M. Magnusson. 1984a. Mycolic acid patterns of some species of *Mycobacterium*. Arch. Microbiol. 139:225–231.

Minnikin, D. E., S. M. Minnikin, A. G. O'Donnell, and M. Goodfellow. 1984b. Extraction of mycobacterial mycolic acids and other long-chain compounds by an alkaline methanolysis procedure. J. Microbiol. Meth. 2:243–249.

Minnikin, D. E., A. G. O'Donnell, M. Goodfellow, G. Alderson, M. Athalye, A. Schaal, and J. H. Parlett. 1984c. An integrated procedure for the extraction of isoprenoid quinones and polar lipids. J. Microbiol. Meth. 2:233–241.

Minnikin, D. E., G. Dobson, M. Goodfellow, P. Draper, and M. Magnusson. 1985a. Quantitative comparison of the mycolic and fatty acid composition of *Mycobacterium leprae* and *Mycobacterium gordonae*. J. Gen. Microbiol. 131:2013–2021.

Minnikin, D. E., S. M. Minnikin, J. H. Parlett, and M. Goodfellow. 1985b. Mycolic acid patterns of some rapidly-growing species of *Mycobacterium*. Zbl. Bakt. Hyg. A259:446–460.

Mordarska, H., M. Mordarski, and M. Goodfellow. 1972. Chemotaxonomic characters and classification of some nocardioform bacteria. J. Gen. Microbiol. 71:77–86.

Mordarski, M., K. P. Schaal, K. Szyba, G. Pulverer, and A. Tkacz. 1977. Interrelation of *Nocardia asteroides* and related taxa as indicated by deoxyribonucleic acid reassociation. Int. J. Syst. Bacteriol. 27:66–70.

Mordarski, M., K. P. Schaal, A. Tkacz, G. Pulverer, K. Szyba, and M. Goodfellow. 1978. Deoxyribonucleic acid base composition and homology studies on *Nocardia*. Zbl. Bakt. Suppl. 6:91–97.

Mordarski, M., M. Goodfellow, A. Tkacz, G. Pulverer, and K. P. Schaal. 1980a. Ribosomal ribonucleic acid similarities in the classification of *Rhodococcus* and related taxa. J. Gen. Microbiol. 118:313–319.

Mordarski, M., M. Goodfellow, K. Szyba, A. Tkacz, G. Pulverer, and K. P. Schaal. 1980b. Deoxyribonucleic acid reassociation in the classification of the genus *Rhodococcus*. Int. J. Syst. Bacteriol. 30:521–527.

Murray, P. R., R. L. Heeren, and A. C. Niles. 1987. Effect of decontamination procedures on recovery of *Nocardia* spp. J. Clin. Microbiol. 25:2010–2011.

Mutimer, M. D., and J. B. Woolcock. 1980. *Corynebacterium equi* in cattle and pigs. Vet. Q 2:25–27.

Mutimer, M. D., and J. B. Woolcock. 1982. APIZYM for identification of *Corynebacterium equi*. Zbl. Bakt. Hyg. I. Abt. Orig. C3:410–415.

Nagasawa, T., M. Kobayashi, and H. Yamada. 1988. Optimum culture conditions for the production of ben-

zonitrilase by *Rhodococcus rhodochrous* J1. Arch. Microbiol. 150:89–94.

Nagasawa, T., K. Takeuchi, and H. Yamada. 1988a. Occurrence of a cobalt-induced and cobalt-containing nitrile hydratase in *Rhodococcus rhodochrous* J1. Biochem. Biophys. Res. Commun 155:1008–1016.

Nagasawa, T., and H. Yamada. 1989. Microbial transformations of nitriles. TIBTECH 7:153–158.

Nesterenko, O. A., S. A. Kasumova, and S. I. Kvasnikov. 1978a. Microorganisms of the *Nocardia* genus and the 'rhodochrous' group in soils of the Ukranian S.S.R. Mikrobiologija 47:866–870.

Nesterenko, O. A., S. A. Kasumova, and S. I. Kvasnikov. 1978b. Microorganisms of the genus *Nocardia* and the 'rhodochrous' group in the soils of the Ukranian S.S.R., USSR. Microbiology 47:699–703.

Nesterenko, O. A., T. M. Nogina, S. A. Kasumova, E. I. Kvasnikov, and S. G. Batrakov. 1982. *Rhodococcus luteus* nom. nov. and *Rhodococcus maris* nom. nov. Int. J. Syst. Bacteriol. 32:1–14.

O'Donnell, A. G., D. E. Minnikin, and M. Goodfellow. 1985. Integrated lipid and wall analysis of actinomycetes, p. 131–143. In: M. Goodfellow, and D. E. Minnikin (ed.). Chemical methods in bacterial systematics. Academic Press, London.

Okami, Y., and K. Hotta. 1988. Search and discovery of new antibiotics, p. 33–67. In: M. Goodfellow, S. T. Williams, and M. Mordarski (ed.), Actinomycete biology. Academic Press, London.

Oraqui, J. L., and D. D. Mara. 1985. Fecal streptococci, *Rhodococcus coprophilus* and bifidobacteria as specific indicator organisms of fecal pollution. J. Appl. Bact. 59:v–vi.

Orchard, V. A., and M. Goodfellow. 1974. The selective isolation of Nocardia from soil using antibiotics. J. Gen. Microbiol. 85:160–162.

Orchard, V. A., and M. Goodfellow. 1980. Numerical classification of some named strains of *Nocardia asteroides* and related isolates from soil. J. Gen. Microbiol. 118:295–312.

Orchard, V. A., M. Goodfellow, and S. T. Williams. 1977. Selective isolation and occurrence of nocardiae in soil. Soil Biol. Biochem. 9:233–238.

Ortiz-Ortiz, L., E. I. Melendro, and C. Conde. 1984. Host-parasite relationship in infections due to *Nocardia brasiliensis*, p. 119–133. In: L. Ortiz-Ortiz, L. F. Bojalil, and V. Yakoleff (ed.), Biological, biochemical and biomedical aspects of actinomycetes. Academic Press, Orlando.

Peczyńska-Czoch, W., and M. Mordarski. 1988. Actinomycete enzymes, p. 219–283. In: M. Goodfellow, S. T. Williams, and M. Mordarski (ed.), Actinomycetes in biotechnology. Academic Press, London.

Pier, A. C., and R. E. Fichtner. 1971. Serologic typing of *Nocardia asteroides* by immunodiffusion. Am. Rev. Resp. Dis. 103:698–707.

Pommier, M. T., and G. Michel. 1985. Occurrence of corynomycolic acids in strains of *Nocardia otitidis-caviarum*. J. Gen. Microbiol. 131:2637–2641.

Portaels, F. 1976. Isolation and distribution of nocardiae in the Bas-Zaire. Annls Soc. Belge Méd. Trop. 56:73–83.

Prauser, H. 1976. Host-phage relationships in nocardioform organisms, p. 266–284. In: M. Goodfellow, G. H. Brownell, and J. A. Serrano (ed.), The biology of the nocardiae. Academic Press, London.

Prauser, H. 1981a. Taxon specificity of lytic actinophages that do not multiply in the cells affected. Zbl. Bakt. Suppl. 11:87–92.

Prauser, H. 1981b. Nocardioform organisms: General characterisation and taxonomic relationships. Zbl. Bakt. Suppl. 11:17–24.

Prinz, G., E. Bán, S. Fekete, and Z. Szabó. 1985. Meningitis caused by *Gordona aurantiaca (Rhodococcus aurantiacus)*. J. Clin. Microbiol. 22:472–474.

Pulverer, G. H., and K. P. Schaal. 1978. Pathogenicity and medical importance of aerobic and anaerobic actinomycetes. Zbl. Bakt. Suppl. 6:417–427.

Pulverer, G., H. Schütt-Gerowitt, and K. P. Schaal. 1975. Bacteriophages of *Nocardia asteroides*. Med. Microbiol. Immunol. 161:113–122.

Qureshi, N., K. Takayama, H. C. Jordi, and H. K. Schnoes. 1978. Characterisation of the purified components of a new homologous series of α-mycolic acids from *Mycobacterium tuberculosis* H37 Ra. J. Biol. Chem. 253:5411–5417.

Rast, H. G., G. Engelhardt, W. Diegler, and P. R. Wallhoffer. 1980. Bacterial degradation of model compounds for lignin and chlorophenol derived lignin bound residues. FEMS Microbiol. Lett. 8:259–263.

Reh, M. 1981. Chemolithoautotrophy as an autonomous and transferable property of *Nocardia opaca* lb. Zbl. Bakt. Suppl. 11:577–583.

Reh, M., and H. G. Schlegel. 1981. Hydrogen autotrophy as a transferable genetic character of *Nocardia opaca* lb. J. Gen. Microbiol. 126:327–336.

Ridell, M. 1977. Studies on corynebacterial precipitinogens common to mycobacteria, nocardiae and rhodococci. Int. Arch. Allergy Appl. Immunol. 55:468–475.

Ridell, M. 1981. Immunodiffusion studies of some *Nocardia* strains. J. Gen. Microbiol. 123:69–74.

Ridell, M. 1983. Sensitivity to capreomycin and prothionamide in strains of *Mycobacterium, Nocardia, Rhodococcus* and related taxa for taxonomical purposes. Zbl. Bakt. Hyg. A255:309–316.

Ridell, M., and M. Norlin. 1973. Serological study of *Nocardia* by using mycobacterial precipitation reference systems. J. Bacteriol. 113:1–7.

Rowbotham, T. J., and T. Cross. 1977a. *Rhodococcus coprophilus* sp. nov.: An aerobic nocardioform actinomycete belonging to the *'rhodochrous'* complex. J. Gen. Microbiol. 100:123–138.

Rowbotham, T. J., and T. Cross. 1977b. Ecology of *Rhodococcus coprophilus* and associated actinomycetes in freshwater and agricultural habitats. J. Gen. Microbiol. 100:231–240.

Schaal, K. P. 1972. Zur mikrobiologischen Diagnostik der Nocardiose. Zbl. Bakt. Parasitkde. (Abt I) Orig. Reihe A 220:242–246.

Schaal, K. P. 1977. *Nocardia, Actinomadura* and *Streptomyces*, p. 131–144. In: A. von Graevenitz (ed.), Chemical microbiology, vol. 1. CRC Handbook Series in Chemical Laboratory Sciences, Section E. CRC Press, Cleveland.

Schaal, K. P., and B. L. Beaman. 1984. Chemical significance of actinomycetes, p. 389–424. In: M. Goodfellow, M. Mordarski, and S. T. Williams (ed.), The biology of actinomycetes. Academic Press, London.

Schaal, K. P., and H. Bickenbach. 1978. Soil occurrence of pathogenic nocardiae. Zbl. Bakt. Suppl. 6:429–434.

Schaal, K. P., and H. Heimerzheim. 1974. Mikrobiologische Diagnose und Therapie der Lungennocardiose. Mykosen 17:313–319.

Schaal, K. P., and H. Reutersberg. 1978. Numerical taxonomy of *Nocardia asteroides*. Zbl. Bakt. Suppl. 6:53–62.

Schleifer, K. H., and O. Kandler. 1972. Peptidoglycan types of bacterial cell walls and their taxonomic implications. Bacteriol. Rev. 36:407–477.

Schupp, T., R. Hütter, and D. A. Hopwood. 1975. Genetic recombination in *Nocardia mediterranei*. J. Bacteriol. 121:128–135.

Serrano, J. A., B. L. Beaman, J. E. Viloria, M. A. Mejia, and R. Zamora. 1986. Histological and ultrastructural studies of human actinomycetomas, p. 647–662. In: G. Szabó, S. Biró, and M. Goodfellow (ed.), Biological, biochemical and biomedical aspects of actinomycetes. Akadémiai Kiadó, Budapest.

Shainhaus, J. Z., A. C. Pier, and D. A. Stevens. 1978. Complement fixation antibody test for human nocardiosis. J. Clin. Microbiol. 8:516–519.

Shawar, R. M., D. G. Moore, and M. T. Larocco. 1990. Cultivation of *Nocardia* spp. on chemically defined media for selective recovery of isolates from clinical specimens. J. Clin. Microbiol. 28:508–512.

Skerman, V. B. D. 1968. A new type of micromanipulator and microforge. J. Gen. Microbiol. 54:287–297.

Söhngen, N. L. 1913. Benzin, Petroleum, Paraffinöl und Paraffin als Kohlenstoff und Energiequelle für Mikroben. Centbl. Bakt. Parasitkde, Abt. 2 37:595–609.

Stackebrandt, E., and C. R. Woese. 1981. Towards a phylogeny of the actinomycetes and related organisms. Curr. Microbiol. 5:197–202.

Stackebrandt, E., W. Ludwik, E. Seewaldt, and K. H. Schleifer. 1983. Phylogeny of sporeforming members of the order Actinomycetales. Int. J. Syst. Bacteriol. 33:173–180.

Stackebrandt, E., J. Smida, and M. D. Collins. 1988. Evidence of phylogenetic heterogeneity within the genus *Rhodococcus*: Revival of the genus *Gordona* (Tsukamura). J. Gen. Appl. Microbiol. 34:341–348.

Steck, P. A., B. A. Schwartz, M. S. Rosendahl, and G. R. Gray. 1978. Mycolic acids—a reinvestigation. J. Biol. Chem. 253:5625–5629.

Steinhaus, E. A. 1941. A study of the bacteria associated with thirty species of insects. J. Bacteriol. 42:757–790.

Straube, G., J. Hensel, C. Niedan, and E. Straube. 1990. Kinetic studies of phenol degradation by *Rhodococcus* sp P1. Antonie van Leeuwenhoek 57:29–32.

Sugar, A. M., and Angeles, A. M. 1987. Identification of a common immunodominant protein in culture filtrate of three *Nocardia* species and use in etiologic diagnosis of mycetoma. J. Clin. Microbiol. 25:2278–2280.

Svensson, L., L. Sisfontes, G. Nyborg, and R. Blanstrand. 1982. High performance liquid chromatography and glass capillary gas chromatography of geometric and positional isomers of long chain monounsaturated fatty acids. Lipids 17:50–59.

Tárnok, I. 1976. Metabolism in nocardiae and related bacteria, p. 451–500. In: M. Goodfellow, G. H. Brownell, and J. A. Serrano (ed.), The biology of the nocardiae. Academic Press, London.

Thiemann, J. E., C. Hengeller, A. Virgilio, O. Buelli, and G. Licciardello. 1964. Rifampicin 33. Isolation of actinophages active on *Streptomyces mediterranei* and characterisation of phage-resistant strains. Appl. Microbiol. 12:261–268.

Tilford, P. E. 1936. Fasciation of sweet peas caused by *Phytomonas fascians* n. sp. J. Agr. Res. 53:383–394.

Tisdall, P. A., G. D. Roberts, and J. P. Anhalt. 1979. Identification of clinical isolates with gas-liquid chromatography alone. J. Clin. Microbiol. 10:506–514.

Tomiyasu, I., and I. Yano. 1984. Separation and analysis of novel polyunsaturated mycolic acids from a psychrophilic acid-fast bacterium, *Gordona aurantiaca*. Eur. J. Biochem. 139:173–180.

Tomiyasu, I., S. Toriyama, I. Yano, and M. Masui. 1981. Changes in molecular species composition of nocardomycolic acids in *Nocardia rubra* by the growth temperature. Chem. Phys. Lipids 28:41–54.

Trevisan, V. 1989. Generiele Specie delle Batteriacee. Zanaboni and Gabuzzi, Milano.

Tsukamura, M. 1962. Differentiation of *Mycobacterium tuberculosis* from other mycobacteria by sodium salicylate susceptibility. Am. Rev. resp. Dis. 86:81–83.

Tsukamura, M. 1969. Numerical taxonomy of the genus *Nocardia*. J. Gen. Microbiol. 56:265–287.

Tsukamura, M. 1971. Proposal of a new genus, *Gordona*, for slightly acid-fast organisms occurring in sputa of patients with pulmonary disease and in soil. J. Gen. Microbiol. 68:15–26.

Tsukamura, M. 1972. An improved selective medium for atypical mycobacteria. Jap. J. Microbiol. 16:243–246.

Tsukamura, M. 1974a. A further numerical taxonomic study of the *rhodochrous* group. Jap. Microbiol. 18:37–44.

Tsukamura, M. 1974b. Differentiation of the *Mycobacterium rhodochrous* group from nocardiae by β-galactosidase activity. J. Gen. Microbiol. 80:553–555.

Tsukamura, M. 1981a. Tests from susceptibility to mitomycin C as aids in differentiating the genus *Rhodococcus* from the genus *Nocardia* and for differentiating *Mycobacterium fortuitum* and *Mycobacterium chelonei* from other rapidly growing mycobacteria. Microbiol. Immunol. 25:1197–1199.

Tsukamura, M. 1981b. Differentiation between the genera *Mycobacterium*, *Rhodococcus* and *Nocardia* by susceptibility to 5-fluorouracil. J. Gen. Microbiol. 125:205–208.

Tsukamura, M. 1982a. Numerical analysis of the taxonomy of nocardiae and rhodococci. Division of *Nocardia asteroides sensu stricto* into two species and descriptions of *Nocardia paratuberculosis* sp.nov. Tsukamura (formerly Kyoto-1 group of Tsukamura), *Nocardia nova* sp. nov. Tsukamura, *Rhodococcus chubuensis* sp.nov. Tsukamura and *Rhodococcus obuensis* sp.nov. Tsukamura. Microbiol. Immunol. 26:1101–1119.

Tsukamura, M. 1982b. Differentiation between the genera *Rhodococcus* and *Nocardia* and between species of the genus *Mycobacterium* by susceptibility to bleomycin. J. Gen. Microbiol. 128:2385–2388.

Tsukamura, M., and K. Kawakami. 1982. Lung infection caused by *Gordona aurantiaca (Rhodococcus aurantiacus)*. J. Clin. Microbiol. 16:604–607.

Tsukamura, M., and S. Mizuno. 1971. A new species *Gordona aurantiaca* occurring in sputa of patients with pulmonary disease. Kekkaku 46:93–98. (in Japanese).

Tsukamura, M., and I. Yano. 1985. *Rhodococcus sputi* sp. nov., nom.rev., and *Rhodococcus aurantiacus* sp.nov., nom. rev. Int. J. Syst. Bacteriol. 35:364–368.

Tsukamura, M., C. Komatsuzaki, R. Sakai, K. Kaneda, T. Kudo, and A. Seino. 1988a. Mesenteric lymphadenitis of swine caused by *Rhodococcus sputi*. J. Clin. Microbiol. 26:155–157.

Tsukamura, M., K. Hikosaka, K. Nishimura, and S. Hara. 1988b. Severe progressive subcutaneous abscesses and necrotizing tenosynovitis caused by *Rhodococcus aurantiacus*. J. Clin. Microbiol. 26:201–205.

Uchida, K., and K. Aida. 1979. Taxonomic significance of cell-wall acyl type in *Corynebacterium-Mycobacterium-Nocardia* group by a glycolate test. J. Gen. Appl. Microbiol. 25:169–183.

Van Etta, L. L., G. A. Filice, R. M. Ferguson, and D. N. Gerding. 1983. *Corynebacterium equi* : A review of 12 cases of human infection. Rev. Infect. Dis. 5:1012–1018.

Van Ginkel, C. G., H. G. J. Welten, S. Hartmans, and J. A. M. de Bont. 1987. Metabolism of trans-2-butene and butane in *Nocardia* TB1. J. Gen. Microbiol. 133:1713–1720.

Wagner, F., W. Lindorfer, and W. Schultz. 1976. Verfahren zur Verbesserung des Ausbeute bei Gewinnung von Erdol durch Wasserfluten. German patent No. 24:19–267.

Waksman, S. A. 1950. The actinomycetes: Their nature, occurrence, activities and importance. Ann. Crypt. Phytopath. 9:1–230.

Wellington, E. M. H., and S. T. Williams. 1978. Preservation of actinomycete inoculum in frozen glycerol. Microbios Lett. 6:151–159.

Weyland, H. 1969. Actinomycetes in North Sea and Atlantic Ocean sediments. Nature 233:851.

Williams, D. R., P. W. Trudgill, and D. G. Taylor. 1989. Metabolism of 1,8-cineole by a *Rhodococcus* species: Ring cleavage reactions. J. Gen. Microbiol. 135:1957–1967.

Williams, S. T., E. M. H. Wellington, and L. S. Tipler. 1980. The taxonomic implications of the reactions of representative *Nocardia* strains to actinophage. J. Gen. Microbiol. 119:173–178.

Winogradsky, S. 1949. Microbiologie du sol. Masson et Cie, Paris.

Woolcock, J. B., A. M. T. Farmer, and M. D. Mutimer. 1979. Selective medium for *Corynebacterium equi* isolation. J. Clin. Microbiol. 9:640–642.

Yamada, K., J. Takahashi, K. Kobayashi, and Y. Imada. 1963. Studies on utilisation of hydrocarbons by microorganisms. Agric. Biol. Chem. 27:390–395.

Yano, I., K. Kageyama, Y. Ohno, M. Masui, E. Kusunose, M. Kusunose, and N. Akinori. 1978. Separation and analysis of molecular species of mycolic acids in *Nocardia* and related taxa by gas chromatography mass spectrometry. Biomed. Mass Spectrometry 5:14–24.

Zakrzewska-Czerwinska, J., M. Mordarski, and M. Goodfellow. 1988. DNA base composition and homology values in the classification of some *Rhodococcus* species. J. Gen. Microbiol. 134:2807–2813.

Zink, M. C., J. A. Yager, J. F. Prescott, and B. N. Wilkie. 1985. In vitro phagocytosis and killing of *Corynebacterium equi* by alveolar macrophages of foals. Am. J. Vet. Res. 46:2171–2174.

Zink, M. C., J. A. Yager, and W. L. Smart. 1986. *Corynebacterium equi* infection in horses 1958–1984: A review of 131 cases. Can. Vet. J. 27:213–217.

Zopf, W. 1891. Uber Ausscheidung von Fellfarbstoffen (Lipochromen) seitens gewisser Spattpilze. Ber. Deut. Bot. Ges. 9:22–28.

The Genus *Mycobacterium*—Nonmedical

SYBE HARTMANS and JAN A. M. DE BONT

Introduction

Mycobacteria are aerobic, acid-fast actinomycetes that usually form slightly curved or straight nonmotile rods (0.2–0.6 × 1.0–10 μm). Branching and mycelium-like growth may take place with fragmentation into rods and coccoid elements. Many species form whitish or cream-colored colonies but especially among the rapid growers, there are also many bright yellow or orange species containing carotenoid pigments (David, 1984). In some cases, the pigments are only formed in response to light (photochromogenic species) but most pigmented species also form these pigments in the dark (scotochromogenic species).

Mycobacterium is the only genus listed in the Family Mycobacteriaceae in *Bergey's Manual of Systematic Bacteriology* (Wayne and Kubica, 1986), but the genus is considered to be closely related to the other mycolic acid-containing genera of cell wall chemotype IV: *Caseobacter, Corynebacterium, Nocardia,* and *Rhodococcus* (Goodfellow and Cross, 1984).

However, based on 16S rRNA studies, it has recently been proposed that the Family Mycobacteriaceae encompasses the genera *Mycobacterium, Nocardia,* and *Rhodococcus,* with the genus *Corynebacterium* in the separate Family Corynebacteriaceae (Fox and Stackebrandt, 1987). Chemical differentiation of mycobacteria from the other mycolic acid-containing genera is possible by analysis of the fatty acid esters formed upon pyrolysis of the mycolic acid esters, in combination with the identification of the major menaquinone present in the plasma membrane (Table 1). The Ziehl-Neelsen stain for acid fastness, however, remains the most obvious method to quickly identify mycobacteria (Barksdale and Kim, 1977).

Mycobacteria are the causal agents of two important diseases, tuberculosis and leprosy, and there has thus been a significant clinical interest in the two responsible species. This started with the work of Koch (1882), who detected the tubercle bacillus in stained infected tissues. The generic name *Mycobacterium* was introduced by Lehmann and Neumann (1896) to include the tubercle and leprosy bacilli. For many years after this work, isolates other than *M. tuberculosis,* but resembling it in staining characteristics, were described as "atypical mycobacteria." After the discovery in the early 1950s that several of these "atypical mycobacteria" could also produce disease in humans (see Kubica, 1978), it was recognized that identification of these strains was required. The classification of Runyon, separating the mycobacteria into four groups (photochromogens, scotochromogens, nonphotochromogens, and rapid growers) was introduced in the late 1950s as a systematic base for the description of the "atypical mycobacteria" (Wayne, 1984). This division, based on pigmentation and rate of growth, is still of use to the clinical mycobacteriologist (see Chapter 53).

The separation of the genus into two major groups based on the growth rate of the individual species forms the basis of mycobacterial taxonomy. Although not exactly following this division, many slow growers are either associated with, or are the causal agents of diseases of humans or other animals. Most rapid growers are not known to be associated with human diseases and consequently are often considered as nonpathogens, although in a strict sense, "nonpathogens" may not really exist (Tsukamura, 1984). Rapidly growing mycobacteria are common saprophytes in natural habitats but have received much less attention than the clinically more relevant slow-growing species.

More than 300 species names for mycobacteria have been published over the years but as a result of the ongoing efforts of the International Working Group on Mycobacterial Taxonomy (IWGMT), the 1977 list of approved species contained only 21 slowly and 14 rapidly growing species (Kubica, 1978). However, the 1980 list of approved species (Skerman et al., 1980) listed 25 slowly growing species and 16 rapidly growing species of mycobacteria, illustrating the dynamics of mycobacterial taxon-

Table 1. Differential characteristics of the mycolic acid-containing genera of wall chemotype IV.

Genus	Mycolic acids		Predominant menaquinone	N-glycolyl in glycan moiety of cell wall
	Overall size (number of carbons)	Ester pyrolysis products (number of carbons in chain)		
Caseobacter	30–36	14–18	MK-8(H$_2$), MK-9(H$_2$)[a]	−
Corynebacterium	22–36	8–18	MK-8(H$_2$), MK-9(H$_2$)	−
Mycobacterium	60–90	22–26	MK-9(H$_2$)	+
Nocardia	44–60	12–18	MK-8(H$_4$)	+
Rhodococcus	34–64	12–18	MK-8(H$_2$), MK-9(H$_2$)	+
Tuskamurella	64–78	20–22	MK-9	+

[a]MK-8(H$_2$) indicates a menaquinone with one of the eight isoprenoid units hydrogenated.
Adapted from Collins et al. (1988); Goodfellow and Cross (1984).

omy. In the following four years, 13 new species were recognized (Moore et al., 1985), 12 of which were rapidly growing species (Wayne and Kubica, 1986). The rapidly growing species were isolated predominantly from soil. Subsequently two more species of rapidly growing mycobacteria have been proposed: *M. moriokaense* sp. nov. isolated from soil (Tsukamura et al., 1986) and *M. poriferae* sp. nov. isolated from a marine sponge (Padgitt and Moshier, 1987) increasing the total number of recognized species to 56 (Table 2).

Habitats

Mycobacteria have been isolated from a very diverse array of biotopes including material of both mammalian and nonmammalian origin, as for instance fresh and salt water, soil, and dust. Some species of the nonpathogenic saprophytes may also occur as opportunistic pathogens. The pathogenic species and their habits are covered in Chapter 53 and will only be discussed in this chapter in relation to nonmammalian environments.

Although many *Mycobacterium* species have been isolated from environmental samples, this does not necessarily imply that these strains can all grow under these conditions. Essential in assigning a species to the natural flora of a specific environment is that it should be capable of multiplying actively in these environments (Kazda, 1983). If it lacks this property then it should be regarded as a contaminant. It is very difficult to discriminate between these two possibilities because the chance of isolating "contaminating" mycobacteria from environmental samples is quite large as a result of the ability of mycobacteria to survive for very long periods under nongrowth conditions. *M. paratuberculosis,* for example, was reported to survive for 252 days in a soil-water slurry (Jorgensen, 1977; cited by Kazda, 1983).

Table 2. Classification of the species of the genus *Mycobacterium*.[a]

Slow growers	Rapid growers
M. africanum	*M. agri*
M. asiaticum	*M. aichiense*
M. avium	*M. aurum*
M. bovis	*M. austroafricanum*
M. farcinogenes	*M. chelonae*
M. gastri	subsp. *abscessus*
M. gordonae	subsp. *chelonae*
M. haemophilum	*M. chitae*
M. intracellulare	*M. chubuense*
M. kansasii	*M. diernhoferi*
M. leprae	*M. duvalii*
M. lepraemurium	*M. fallax*
M. malmoense	*M. flavescens*
M. marinum	*M. fortuitum*
M. microti	*M. gadium*
M. nonchromogenicum	*M. gilvum*
M. paratuberculosis	*M. komossense*
M. scrofulaceum	*M. moriokaense*
M. shimoidei	*M. neoaurum*
M. simiae	*M. obuense*
M. szulgai	*M. parafortuitum*
M. terrae	*M. phlei*
M. triviale	*M. porcinum*
M. tuberculosis	*M. poriferae*
M. ulcerans	*M. pulveris*
M. xenopi	*M. rhodesiae*
	M. senegalense
	M. smegmatis
	M. sphagni
	M. thermoresistibile
	M. tokaiense
	M. vaccae

[a]All these species are listed in *Bergey's Manual of Systematic Bacteriology* (Wayne and Kubica, 1986), with the exception of *M. moriokaense* (Tsukamura et al., 1986) and *M. poriferae* (Padgitt and Moshier, 1987).

The most recent reviews on the ecology of mycobacteria are those of Kazda (1983), Collins et al. (1984), and Tsukamura (1984). Kazda (1983) classified the mycobacteria into four groups on the basis of ecologically relevant properties. He distinguished obligate pathogenic, facultative pathogenic, potentially pathogenic, and saprophytic species.

Included in the group of obligate pathogens, which presumably are unable to multiply outside living beings, are the species *M. tuberculosis, M. bovis, M. africanum, M. asiaticum, M. malmoense, M. microti, M. simiae, M. szulgai,* and *M. haemophilum.* Of these species, *M. tuberculosis* and *M. bovis* have been isolated from waste water (Kazda, 1983). *M. farcinogenes* and *M. shimodei* can also be included in the group of obligate pathogens as they have been isolated from mammalian sources only (Wayne and Kubica, 1986; Tsukamura, 1982).

Kazda assigned *M. leprae, M. paratuberculosis,* and *M. ulcerans* to the provisional group of the facultative pathogens as their incidence in natural biotopes has not been investigated in detail. Future research may result in the assignment of these species to either the potential or obligate pathogens (Kazda, 1983).

Examples of potential pathogens are *M. avium, M. chelonae, M. fortuitum, M. intracellulare, M. kansasii, M. marinum, M. senegalense, M. scrofulaceum,* and *M. xenopi* (Kazda, 1983). These species can grow in natural biotopes without losing their pathogenic properties, and they have been isolated from a variety of biotopes (Kazda, 1983; Tsukamura, 1984).

The saprophytic mycobacteria include the slowly growing species *M. gordonae, M. nonchromogenicum, M. triviale, M. terrae,* and *M. gastri,* and all the rapidly growing species (Table 2), with the exception of *M. chelonae, M. fortuitum,* and *M. senegalense,* which are classified as potential pathogens (Kazda, 1983). The saprophytic and potentially pathogenic groups (Kazda, 1983) are found in many environments (Tsukamura, 1984) and only occasionally occur as an accompanying flora in pathogenic processes. The ability of organisms from these groups to utilize many different growth substrates and their capacity to survive and multiply under a wide range of environmental conditions apparently enable them to compete successfully with other organisms in many biotopes.

The abundance of the fast-growing species in soils was already demonstrated by Jones and Jenkins (1965). In screening soil samples, 101 acid-fast strains were isolated. Of the 93 isolates studied, five were scotochromogens and all but three of the nonphotochromogens were rapidly growing mycobacteria, forming colonies within 7 days at 25°C.

The most commonly occurring *Mycobacterium* in the environment is probably *M. fortuitum,* as it is often the major *Mycobacterium* species (40–80%) isolated from soil samples (Tsukamura, 1984). Strains of the slowly growing *M. nonchromogenicum* complex *(M. nonchromogenicum, M. terrae,* and *M. triviale)* and the rapidly growing species *M. aurum, M. smegmatis,* and *M. agri* have also been isolated frequently from soil samples (Tsukamura, 1984). Isolates from hospital dust were also predominantly *M. fortuitum* (43%), with significant numbers of strains of the *M. nonchromogenicum* complex (25%) and *M. gordonae* (18%), whereas from house dust the majority of the strains isolated belonged to the MAIS *(M. avium-intracellulare-scrofulaceum)* complex (55%), and significant numbers of the *M. nonchromogenicum* complex (23%), and also *M. gordonae* (11%) (Tsukamura, 1984).

The "tap water scotochromogen," *M. gordonae,* is the strain most often isolated from water samples of various origins (Collins et al., 1984). Besides *M. gordonae,* strains of *M. terrae, M. phlei,* and *M. fortuitum* have also been isolated at a high frequency from various nonmarine aquatic environments (Viallier and Viallier, 1973). Analysis of municipal water supplies and water supplies in hemodialysis centers showed strains of *M. fortuitum, M. chelonae, M. scrofulaceum, M. gordonae,* and of the *M. avium* and *M. terrae* complexes to be present (Carson et al., 1988). The average number of mycobacteria detected in the municipal water supplies was 74 ± 42 per 100 ml. Marine water samples, usually from coastal waters, predominantly yielded strains of the MAIS complex, although strains of *M. gordonae* and *M. terrae* were also isolated (Collins et al., 1984).

It should, however, be emphasized when considering studies on the isolation of various *Mycobacterium* species from the environment, that environmental samples are routinely decontaminated, and that the distribution of the species isolated need not necessarily reflect the distribution of these species in the original sample. It may well be that certain species, or cells in a particular physiological state, exhibit a higher- or lower-than-average survival rate during the decontamination procedure. Application of the decontamination procedure therefore may result in an over- or underestimate of the prevalence of a particular species.

Another consequence of routinely applied decontamination procedures is that no reliable data can be obtained on the numbers of viable mycobacteria present in natural environments. The actual numbers present in soil samples may also be considerably higher than anticipated from viable counts of nondecontaminated samples due to the hydrophobic characteristics of the mycobacterial cell wall, which may result in significant bacterial adhesion to surfaces.

Isolation

Mycobacteria are not always readily isolated from natural samples. They grow relatively slowly and are therefore easily overgrown by faster-growing organisms. However, taking advantage of the resistance of mycobacteria to adverse conditions, decontamination procedures and selective media have been developed to increase the efficiency of isolation procedures (Kubica and Good, 1981). Most of the decontamination methods were developed for the isolation of mycobacteria from specimens originating from diseased humans or animals and exploit the resistance of the acid-fast mycobacteria to alkaline and acidic conditions. The most commonly applied decontamination procedures involve treatment of the samples with NaOH (2–4%) for 15–30 min at room temperature or at 37°C (Jenkins et al., 1982). Methods for the selective isolation of mycobacteria from environmental samples have been reviewed by Songer (1981), who, besides reviewing decontamination methods, also discussed numerous selective growth media.

It should be realized, however, that no information is available on the number and type(s) of mycobacteria that are lost as a result of the decontamination procedure. Furthermore, the selectivity of the growth conditions (medium composition and incubation temperature) also affects the number and species of mycobacteria that eventually are isolated.

Portaels et al. (1988) have recently compared different decontamination procedures and selective media for the isolation of mycobacteria from soil samples. The best results (low contamination in combination with high positivity rates) were obtained using the following procedure:

Isolation of Mycobacteria from Soil (Portaels et al., 1988)

Add 0.5 g of soil (wet weight) to 5 ml sterile trypticase soy broth, shake vigorously, and incubate at 37°C for 5 h. After sedimentation of the soil particles, add 5 ml malachite green (0.2%), 1 ml cycloheximide (500 μg per ml), and 5 ml NaOH (1 M) to the supernatant. After 30 min at room temperature, neutralize the mixture with HCl (1 M) and centrifuge at 2,000 g for 20 min. Inoculate the pellet on Ogawa egg medium (see below) containing cycloheximide (500 μg per ml). Incubate at 30°C and inspect every 2 weeks until growth is observed.

Ogawa Egg Medium (Tsukamura et al., 1986)

Add to 100 ml water containing sodium glutamate (1%) and KH$_2$PO$_4$ (1%), 200 ml of whole eggs, 6 ml glycerol, and 6 ml of a malachite green solution (2%). The medium is made up as slopes by heating for 60 min at 90°C.

Using the above method, mycobacteria were isolated from 91% of the soil samples with only 4% of the tubes being contaminated, i.e., exhibiting growth of non-acid-fast microorganisms within 6 months (Portaels et al., 1988). Omission of cycloheximide, malachite green, or both compounds from the egg medium resulted in contamination rates of 14%, 49%, and 63%, respectively.

The decontamination method and incubation conditions may even select for a specific mycobacterial species. *M. moriokaense*, for instance, is the predominant species isolated from soil samples using the following procedure:

Isolation of *M. moriokaense* (Tsukamura et al., 1986)

Shake soil sample (20 g) with 100 ml of 0.9% NaCl for 30 min and allow to settle for 15 min. Filter the supernatant and centrifuge the filtrate for 20 min at 700 g. Mix the precipitate in the centrifuge tube with 3 ml of KOH (1%). After 5 min, inoculate 0.02 ml portions of this suspension on slopes of Ogawa egg medium. Incubate the inoculated slopes at 42°C for 7–10 days and subculture the colonies.

The nonpigmented acid-fast colonies isolated using the above procedure were characterized as *M. moriokaense* (80%) and as *M. fortuitum* (20%). The strong selectivity of this procedure is probably due to the combination of the relatively high incubation temperature and the short incubation time. Unfortunately, no mention was made of the numbers of any pigmented acid-fast colonies which might also have grown on the slopes (Tsukamura et al., 1986).

Another approach for selectively isolating mycobacteria is to use specific carbon sources in simple mineral media. Many years ago Söhngen (1913) demonstrated that paraffin could be used to enrich mycobacteria from soil and water samples. Likewise, enrichment cultures with ethene as the sole carbon source also exclusively led to the isolation of mycobacteria (Hartmans et al., 1989). Other genera, for instance *Xanthobacter* and *Nocardia* species, may also grow on ethene (van Ginkel et al., 1987a). However, the very low growth rates of these species with ethene may explain why enrichment cultures with ethene result in the isolation of the faster-growing mycobacteria.

Morpholine (1-oxa-4-azacyclohexane), which for many years was considered nonbiodegradable due to its persistence in waste-water-treatment plants, has recently been demonstrated to be degraded by *Mycobacterium* isolates (Knapp and Brown, 1988; Cech et al., 1988). All strains isolated on morpholine from municipal activated sludge systems and river water were identified as mycobacteria (Knapp and Brown, 1988).

Another approach combines the selectivity of a single carbon source with the selectivity of an antibiotic in enrichment cultures. Gram-positive methanol-utilizing bacteria were isolated from different soil samples with methanol as carbon source in the presence of the antibiotic polymyxin B (Urakami and Yano, 1989). The 14 isolates were all very similar and were identified as mycobacteria closely resembling *M. fortuitum*. The growth rates with methanol, though, varied quite significantly. Enrichment procedures on methanol in the absence of polymyxin B usually result in the isolation of Gram-negative genera.

Although mycobacteria are isolated quite easily from many different environments, they are difficult to quantify, especially in heterogeneous matrices such as soil. As mentioned by Songer (1981), the attachment of mycobacteria to surfaces has not yet been investigated. Considering the hydrophobic character of mycobacteria (van Loosdrecht et al., 1987), it is conceivable that a large percentage of the mycobacteria present in soil adhere to the surfaces of solids. Biofouling of cellulose diacetate membranes used in reverse-osmosis water purification plants is an example of this characteristic. It has been proposed that the initial step in this process is the attachment of mycobacteria (Ridgway et al., 1984). It is therefore possible that many of the mycobacteria present in environmental samples are overlooked, especially if interfaces are present.

Cultivation

Mycobacteria can utilize a wide range of carbon compounds. Very often glycerol is designated the preferred carbon since it is utilized as sole source of carbon and energy by all cultivable mycobacteria (Ratledge, 1982b). For clinical isolates, an absolute requirement for carbon dioxide has been reported (Ratledge, 1982b). It has been observed repeatedly that the lag phase of rapidly growing mycobacteria is shortened considerably when liquid cultures are incubated stationary rather than shaken. This effect may be due to a stimulatory effect of carbon dioxide accumulating in the medium of the stationary culture.

Mycobacteria can use a variety of nitrogen sources, including amino acids and ammonium. In many of the frequently used media for the cultivation of mycobacteria, asparagine is the nitrogen source (Jenkins et al., 1982). Many species can also reduce nitrate (Table 3) and use it as a source of nitrogen. Mycobacteria cannot fix nitrogen. The *"M. flavum"* 301 capable of fixing nitrogen (Ratledge, 1982b), is in fact a *Xanthobacter flavus* strain (see Chapter 119). Table 3 lists a number of characteristics of rapidly growing species that are useful in identification.

Most mycobacteria do not require any specific growth factors or vitamins in the growth medium. Exceptions are *M. haemophilum*, which requires hemin, and *M. paratuberculosis*, which requires mycobactin, and of course *M. leprae*, which has until now not been cultivated in vitro (Goodfellow and Wayne, 1982).

Problematic in the cultivation of many strains of mycobacteria in liquid culture is the tendency of these organisms to form aggregates and to adhere to the surfaces of growth vessels, probably as a result of the hydrophobic nature of the cell wall. Detergents such as Tween 80 are therefore often added to the growth medium to reduce lumping and to stimulate growth (Ratledge, 1982b). Due to this tendency of most mycobacteria to form clumps and to adhere to the surfaces of laboratory fermentors, there have been very few studies of mycobacteria using continuous cultures. One exception was reported by Lowrie et al. (1979). Using a special fermentor in which all stainless steel had been replaced by either glass, Teflon, or titanium, *M. bovis* BCG and *M. microti* were grown with 0.08% Tween 80 in the growth medium and by aerating with air containing 5% carbon dioxide. The air supply was introduced in the gas phase of the fermentor, and the stirring rate was kept low to prevent the formation of bubbles.

Some species, however, may be grown in chemostat cultures without having to take any specific precautions. *M. phlei* can be grown dispersed in a fermentor (Girbal et al., 1989) without the addition of a detergent, and consequently it has been used quite often for metabolic studies (Ratledge, 1982b). Two other examples of mycobacteria cultivated in continuous culture without detergents are *M. aurum* strain L1 growing on vinyl chloride (Hartmans et al., 1985), and the ethene-utilizing strain E3 growing on ethene (van Ginkel et al., 1987b).

The mineral salts medium (Wiegant and de Bont, 1980) used in the chemostat studies of *M. aurum* L1 contained 0.2 ml of the Vishniac and Santer (1957) trace element solution per liter. Subsequent chemostat studies using this me-

Table 3. Identifying characteristics of the rapidly growing species of mycobacteria.

Species	Characteristic[a]															
	1	2	3	4	5	6	7	8	9	10	11	12	13	14	15	16
"M. acetamidolyticum"	N	+	−	−	+	−	+	+	+	+	−	−	−	−	−	−
M. agri	N	+	+	+	−	+	+	+	v	+	+	−	+	−	−	−
M. aichiense	S		−	+	+	−	v	−	−	−	−	−	−	v	+	−
M. aurum	S	−	−	+	v	−	+	v	v	v	v	−	+	v	+	−
M. austroafricanum	S	−	−	+	+	v	+	+	−	+	−	−	+	−	+	−
M. chelonae																
Subsp. *abscessus*	N	+	−	+	+	v	+	−	−	v	−	−	−	v	−	−
Subsp. *chelonae*	N	−	−	−	+	v	+	−	−	v	−	−	−	v	−	−
M. chitae	N	−	−	+	−	+	+	+	+	+	−	−	−	−	−	−
M. chubuense	S	−	−	+	−	−	+	+	−	+	−	−	+	v	v	v
M. diernhoferi	N	−	−	+	−	−	+	+	+	+	v	−	−	+	−	−
M. duvalii	S	−	−	−	−	+	+	+	−	−	−	−	−	+	−	
M. fallax	N	−	−	−	−	+	+	+	−							
M. flavescens	S	v	−	−	v	+	+	+	−	v	−	−	−	−	v	v
M. fortuitum	N	v	−	+	+	v	v	+	+	v	+	−	−	v	v	−
M. gadium	S	−	−	−					+							
M. gilvum	S	−		v	−	v	+	−	+	v	+				+	
M. komossense	S	−	−	+	−	+	+	+	−	−	−	+	−	+	+	+
M. moriokaense	N	+	−	+	+	+	+	+	−	−	v	−	−	v	+	+
M. neoaurum	S	v	−	+	+	−	+	v	+	+	+	−	v	+	+	−
M. obuense	S	−	−	+	+	v	+	−	−	+	+	−	−	v	+	v
M. parafortuitum	P	v	−	+	v	−	v	v	v	+	−	−	v	−	+	−
M. phlei	S	+	+	+	−	+	+	+	v	+	v	−	v	v	+	v
M. porcinum	N	+	−	+	+	v	+	−	+	+	+	+	−	+	+	−
M. poriferae	S	−	−	+	−			−	−	v	−		+	+	+	+
M. pulveris	N	+	−	v	−	+	+	+	−	+	−	−	−	−	−	−
M. rhodesiae	S		+	+	+	−							+		+	
M. senegalense	N							+	+	+	+	+			+	
M. smegmatis	N	+	+	+	+	−	−	+	+	−	+	+	+	+	+	+
M. sphagni	S	−	−	−	−	−	+	−	−	−	−	−	−	v	+	−
M. thermoresistibile	S	+	+	−	−	+	+	+	−	−	−	−	−	−	−	−
M. tokaiense	S		−	+	+	v	+	−	+	+	+	+	+	v	+	+
M. vaccae	P	+	−	+	v	−	v	v	+	+	+	+	v	v	+	−

Symbols: N, nonchromogenic; S, scotochromogenic; P, photochromogenic; +, >85% positive; v, variable; −, <15% positive.

[a]Numbers correspond to the following characteristics: 1, pigmentation; 2, growth at 42°C; 3, growth at 45°C; 4, tolerates 0.2% picric acid; 5, arylsulfatase (3 days); 6, α-esterase; 7, β-esterase; 8, nitrate reduction (24 h); 9, acetamidase; 10, nicotinamidase; 11, allantoinase; 12, succinamidase; 13, xylose utilized; 14, trehalose utilized; 15, mannitol utilized; 16, sorbitol utilized.

Data adapted from Kubica et al. (1972); Saito et al. (1977); Tsukamura (1981); Tsukamura et al. (1981, 1983c, 1986); Tsukamura and Ichiyama (1986); Wayne (1984); Wayne and Kubica (1986); and Padgitt and Moshier (1987).

dium (S. Hartmans and J. A. M. de Bont, unpublished observations) with vinyl chloride as the carbon source, revealed that the culture was iron-limited, rather than carbon-limited. An improved mineral salts medium with an increased iron content was therefore formulated:

Basal Mineral Salts Medium for the Cultivation of Mycobacteria

  The medium contains the following ingredients:

K_2HPO_4	1.55 g
$NaH_2PO_4 \cdot 2H_2O$	0.85 g
$(NH_4)_2SO_4$	2.0 g
$MgCl_2 \cdot 6H_2O$	0.1 g
EDTA	10 mg
$ZnSO_4 \cdot 7H_2O$	2 mg
$CaCl_2 \cdot 2H_2O$	1 mg
$FeSO_4 \cdot 7H_2O$	5 mg
$Na_2MoO_4 \cdot 2H_2O$	0.2 mg
$CuSO_4 \cdot 5H_2O$	0.2 mg
$CoCl_2 \cdot 6H_2O$	0.4 mg
$MnCl_2 \cdot 2H2O$	1 mg
Deionized water	1 liter

Identification

The simplest test for discriminating mycobacteria from other prokaryotes is the Ziehl-Neelsen acid-fast stain. Mycobacteria stained with carbol fuchsin resist decolorization with hydrochloric acid-alcohol. Barksdale and Kim (1977) have termed this property "mycobacterial acid-fastness" or acid-alcohol-fastness, in contrast to the acid-fastness of several other mycolic acid-containing genera, which resist decolorization with dilute mineral acids but which can be decolorized with hydrochloric acid-alcohol. A detailed discussion or the acid-fast stain is presented in the excellent review of Barksdale and Kim (1977).

An alternative method to reliably distinguish mycobacteria from other genera containing mycolic acids (Table 1) is to analyze the mycolic acid methyl esters by thin-layer chromatography (TLC) (Minnikin et al., 1980; Daffé et al., 1983) and to characterize the major menaquinone (Collins et al., 1977). Alternatively, the fatty acid methyl esters formed during pyrolysis gas liquid chromatography of mycolic acid methyl esters can also be analyzed (Kusaka and Mori, 1986).

Identification schemes for mycobacteria are based on a clear distinction between slowly growing and rapidly growing species. Species forming colonies from dilute inocula within 7 days under optimal conditions are classified as rapid growers. Those requiring more than 7 days are designated slow growers. The difference in growth rates is actually quite distinct, with the slow growers usually requiring 10 days or more to form colonies, whereas all rapidly growing species, including the relatively slowly growing *M. flavescens* and *M. thermoresistibile,* form colonies within 4–5 days (Jenkins et al., 1982).

It has been argued that the importance attributed to this characteristic may lead to erroneous results in the identification of mycobacteria, as isolates are usually only compared with known species of the same growth rate category. However, evidence that the rapidly growing and slowly growing mycobacteria are also clearly separated by 16S RNA studies (Stahl and Urbance, 1990) support the significance usually attributed to this characteristic in identifying mycobacteria.

The identification of slowly growing strains is described in Chapter 53 as well as in other reviews (Jenkins et al., 1982; Wayne, 1984; Wayne and Kubica, 1986).

Identification of Rapidly Growing Mycobacteria

Identification of rapidly growing mycobacteria can be based on a number of characteristics. Chromogenicity is often used as an initial criterion, primarily of course, due to its obvious nature. Numerical studies based on a large

number of biochemical and physiological properties have been used to define the presently accepted species, and consequently many of these characteristics can be used in the identification of new isolates. Chemical analysis of cell constituents such as mycolic acids and mycobactins are also very powerful tools in the identification of mycobacteria. DNA analysis including hybridization studies with specific probes and the identification of specific sequences are the most recently developed techniques in mycobacterial identification.

Chromogenicity and Morphology

Chromogenicity is often used as a convenient criterion for identifying or classifying mycobacterial species, but some caution should be taken, as the significance of pigment production in the physiology of mycobacterial species and its taxonomic value are not very well studied (Kaneda et al., 1988). Thus, novel isolates should be compared with both the pigmented as well as with the nonpigmented species (e.g., Tsukamura et al., 1986). When determining chromogenic properties, the effect of light on pigmentation or even on growth itself should also be considered (de Bont et al., 1980).

A description of the morphological characteristics of the different species has been given by Wayne and Kubica (1986), although, according to Tsukamura et al. (1981), colonial morphology is not very useful in differentiating mycobacterial species.

Biochemical and Physiological Characteristics

A probability matrix for the identification of the slowly growing mycobacteria has been developed on the basis of a large pool of collected data (Wayne et al., 1980). Although many species of rapidly growing mycobacteria also have been subjected to numerical analysis of a large number of biochemical characteristics (Saito et al., 1977; Tsukamura, 1981; Tsukamura et al., 1981, 1983c), such a probability matrix has not yet been developed for the rapidly growing species (Wayne, 1984). Identification of the many rapidly growing mycobacterial species that have been proposed during the last 10 years was, however, usually based on studies involving numerical analysis (Kazda, 1980; Lévy-Frébault et al., 1983; Tsukamura, 1981, 1982; Tsukamura and Ichiyama, 1986; Tsukamura et al., 1981, 1983a, 1983b, 1983c, 1986; Padgitt and Moshier, 1987). Besides the numerical taxonomic data, additional criteria such as DNA hybridization experiments are generally required to warrant the proposal of a new species (Wayne, 1984).

Based on a number of these numerical analysis studies of a large number of strains (Kubica et al., 1972; Saito et al., 1977; Tsukamura, 1981; Tsukamura et al., 1981, 1983c; Tsukamura and Ichiyama, 1986), we have listed several physiological and biochemical characteristics which should be of help in the identification of newly isolated rapidly growing species (Table 3). A detailed description of the methods is given by Vestal (1975). The diagnostic table for rapidly growing species published by Wayne and Kubica (1986) and the references cited therein are also very useful.

Generally, additional tests, such as mycolic acid analysis and DNA-hybridization experiments, are required before an exact identification can be made. Identification of strains of the "*M. parafortuitum* complex," which includes *M. parafortuitum, M. diernhoferi, M. aurum* and *M. neoaurum,* for example, is sometimes rather difficult. Several strains that clustered with the type strain of *M. neoaurum* in the numerical study of Saito et al. (1977) clustered with the *M. aurum* type strain in a subsequent study by Tsukamura et al. (1983c). In the second study, the above species and the novel species *M. austroafricanum* could be separated satisfactorily. Nevertheless, in view of the similarities observed, the preservation of the *M. parafortuitum* complex containing the five species mentioned above was considered to be justified (Tsukamura et al., 1983c).

The identification of *M. fortuitum* is another example showing that a small number of tests is not sufficient for identifying strains of rapidly growing mycobacteria. This clinically most important rapidly growing species, is often identified solely on the basis of its positive threeday arylsulfatase test. To distinguish it from *M. chelonae,* NaCl tolerance, nitrate reduction, and Fe uptake are often also determined (Kubica and Good, 1981). However, as can be seen in Table 3, several other nonphotochromogenic species are also positive in the 3-day arylsulfatase test.

residues are *N*-glycolated, rather than *N*-acetylated as in most other *meso*-A$_2$pm containing prokaryotes (Table 1) (Brennan, 1988). Also very characteristic for mycobacteria is the lipidrich thick cell envelope containing very longchain mycolic acids (Table 1) (Draper, 1982; Minnikin and O'Donnell, 1984; Rastogi et al., 1986; Mamadou et al., 1989). The characteristic mycolic acids have received much attention in relation to the identification of mycobacteria, especially at the species level.

Mycolic acids are high-molecular-weight, 3-hydroxy fatty acids substituted with an aliphatic side chain at the C_2 position (Fig. 1). The molecular mass of mycobacterial mycolic acids varies from C_{60} to C_{90} (Table 1). Pyrolysis of the methyl esters of mycobacterial mycolic acids yields long-chain meroaldehydes and fatty acid methyl esters (Fig. 2) with chain lengths of 22–26 carbon atoms (Table 1). This is a very useful characteristic in separating mycobacteria from the other mycolic acid-containing genera, although discrimination between the genera *Tsukamurella* and *Mycobacterium* also requires identification of the major menaquinone (Table 1).

Methanolysates of mycobacterial mycolic acids can be resolved into several classes based on the presence or absence of different functional groups in the longer carbon chain, i.e., R_1 in Fig. 1, of the mycolic acid molecule. Both one-dimensional (Daffé et al., 1983) and two-dimensional (Minnikin et al., 1980) TLC methods have been used. The least polar compounds are termed α-mycolic acids and α'-mycolic acids. They do not contain any oxygen functions apart from the 3-hydroxy and carboxy functions. In the more polar mycolic acids, the longer chain is substituted with methoxy, keto, epoxy, or carboxy functions (Minnikin, 1982). The various epoxy mycolic acids produce differing characteristic TLC patterns, depending on the method by which the methanolysate (acidic or basic methanolysis) was prepared (Minnikin et al., 1984a).

The predominant α-mycolic acids generally contain 74–82 carbon atoms whereas the α'-my-

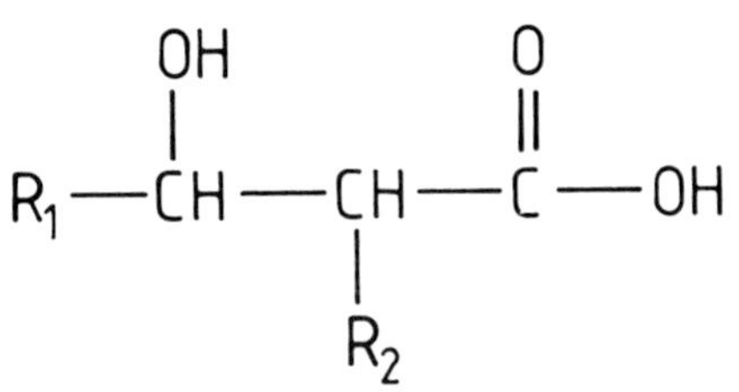

Chemical Analysis of the Mycobacterial Cell

Cell wall and fatty acid analysis has had much impact on the classification of those Gram-positive prokaryotes that have wall chemotype IV (Goodfellow and Cross, 1984), i.e., prokaryotes with a cell wall containing *meso*-diaminopimelic acid (*meso*-A$_2$pm), arabinose, and galactose (Lechevalier and Lechevalier, 1970). In the mycobacterial peptidoglycan, the peptidoglycan

Fig. 1. General formula for mycolic acids. R_1 and R_2 represent variable side chains.

Fig. 2. Rearrangement of the mycolic acid methyl ester during pyrolysis to the meroaldehyde (containing R_1) and the fatty acid methyl ester (containing R_2).

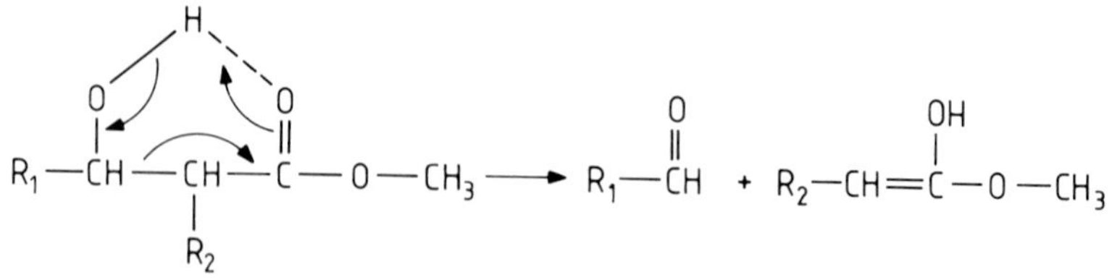

$$R_1\text{—CH—CH—C—O—CH}_3 \longrightarrow R_1\text{—CH} + R_2\text{—CH}=\text{C—O—CH}_3$$

colic acids are of a lower molecular weight, with the predominant type containing 60–68 carbon atoms (Kaneda et al., 1988). Five different types of long-chain α-mycolic acids have been separated and identified using argentation TLC, gas chromatography, and mass spectrometry (Minnikin et al., 1984b; Kaneda et al., 1988). These types are as follows:

I. dicyclopropanoyl;
II. monocyclopropanoyl monoenoic;
II. methylated monocyclopropanoyl monoenoic;
III. dienoic; and
III. methylated dienoic.

The α'-mycolic acids generally contain only one or two double bonds without any cyclopropanoyl functions (Kaneda et al., 1988). Very often (e.g., Collins et al., 1988) the α-mycolic acids of mycobacteria are described as always containing one or two double bonds. This is probably due to the fact that mycolic acids containing a cyclopropanoyl ring and mycolic acids containing a double bond are not separated by TLC. In fact, many species contain only dicyclopropanoyl α-mycolic acids without any double bonds as shown in Table 4, which lists the mycolic acid content by species (Kaneda et al., 1988).

For a number of species, e.g., *M. fortuitum, M. smegmatis* (Minnikin et al., 1984a), *M. phlei, M. thermoresistibile* (Lévy-Frébault et al., 1986b), and *M. chitae* (Minnikin et al., 1985), the mycolic acids of large numbers of strains have been analyzed, all resulting in a characteristic pattern for each species. Analysis of seven *M. aurum* strains, however, revealed two different mycolic acid patterns (Table 4), with two strains also containing α'-mycolic acids. This discrepancy was also observed by Lanéelle et al. (1988).

The composition of the α- and α'-mycolic acids of a number of strains has been further analyzed by Kaneda et al. (1988) using gas chromatography and mass spectrometry. These authors separated the rapid growers into three groups based on the types of α-mycolic acids present. Group A only contained dienoic α-mycolic acids, group B contained α-mycolic acids containing double bonds and cyclopropanoyl

rings, and group C only contained dicyclopropanoyl α-mycolic acids. Group A could be further divided into two groups based on the number of double bonds in the α'-mycolic acids (Table 4). No α'-mycolic acids were detected in *M. gilvum* by Kaneda et al. (1988), whereas several other authors reported α'-mycolic acids to be present in this species based on TLC studies. Unfortunately, *M. fallax* was not included in the study of Kaneda et al. (1988), as *M. fallax* was previously reported to contain unique α-mycolic acids containing three double bonds (Lévy-Frébault et al., 1983). Comparison of the data from these authors, however, also reveals a discrepancy in the reported composition of α-mycolic acids of *M. triviale*. Lévy-Frébault et al., (1983) could not detect unsaturated α-mycolic acids in the type strain of *M. triviale,* whereas Kaneda et al. (1988) detected only dienoic α-mycolic acids in the *M. triviale* strain they analyzed.

The chain lengths of the fatty acid esters released upon pyrolysis of mycolic acid methyl esters have also been analyzed for a number of strains (Table 4). Usually only the major chain length is reported. With the exception of *M. aurum* and *M. diernhoferi* (Table 4), no contradictory results were reported. *M. aurum* ATCC 25793 contained almost equal amounts of C_{22} and C_{24} (Kusaka and Mori, 1986), whereas *M. aurum* ATCC 25797 mainly contained C_{24} (Valero-Guillén and Martín-Luengo, 1986) while the type strain *M. aurum* ATCC 23366 only contained C_{22} (Valero-Guillén and Martín-Luengo, 1986; Lanéelle et al., 1988) although it had previously been reported to contain mainly C_{24} (Daffé et al., 1983).

As the analysis of mycobacterial mycolic acid methyl esters by TLC provides a sensitive and relatively easy method to determine mycolic acid patterns, the technique clearly can be of great use in the identification of mycobacteria and in mycobacterial systematics. For a number of species, however, a larger number of strains has to be analyzed to confirm the consistency of this characteristic within the species. Especially the situation observed within the *M. aurum* species requires further research, possibly resulting in the reassignment of several strains presently recognized as belonging to this species

Table 4. Classification of the mycolic acids and pyrolysis products of the rapidly growing species of mycobacteria.

Species	Mycolic acids[a]	α-Mycolic acids[b]	α'-Mycolic acids[b]	Ester pyrolysis products[c] 1	2	3
"*M. acetamidolyticum*"		III, III'	III	24		
M. agri	αα'm	I	IV	24	24	
M. aichiense	αkc				24	
M. aurum	αα'kc/αkc	I/III[d]	IV	22, 24	22/24	22B
M. austroafricanum						
M. chelonae	αα'	III, III'	IV		24	24B
M. chitae	αα'e	III, III'	IV	24		24B
M. chubuense	αα'kc				22, 24	
M. diernhoferi	αkc	I, II, III		22	24	22A
M. duvalii	αα'kc	I	IV	22, 24	22, 24	
M. fallax	α					
M. flavescens	αkc				24	24B
M. fortuitum						
subsp. *fortuitum*	αα'e	III, III'	III	24	24	24B
subsp. *peregrinum*	αα'e	III, III'	III	24	24	24B
M. gadium	αkc					
M. gilvum	αα'kc	I		22, 24	22	
M. komossense	αkmc/αkc					
M. moriokaense	αkc					
M. neoaurum	αkc				22	22A
M. obuense	αα'kc				22	
M. parafortuitum	αα'kc	I, II, II', III, III'	IV	22	22	22A
M. phlei	αkc	I, II, II'		22, 24	24	24A
M. porcinum	αα'e/αe	III, III'	III	24		
M. poriferae	αkc					
M. pulveris		I	IV	22, 24		
M. rhodesiae	αkc	I, II, II', III, III'		22, 24	24	24A
M. senegalense	αα'e			24[e]		
M. smegmatis	αα'e	III, III'	IV	24	24	24B
M. sphagni	αkc				24	
M. thermoresistibile	αα'km	I	IV	24		24B
M. tokaiense	αkc				24	
M. vaccae	αα'kc	I	IV	22, 24	22	22B

[a]Abbreviations: m, methoxy-; k, keto-; c, dicarboxy mycolic acids. Data adapted from Daffé et al. (1983); Minnikin et al. (1984b, 1985); Valero-Guillén and Martin-Luengo (1986); Lévy-Frébault et al. (1986b); Tsukamura et al. (1986); Luquin et al. (1987); Padgitt and Moshier (1987).

[b]Abbreviations: I, dicyclopropanoyl; II, monocyclopropanoyl monoenoic; II', methylated monocyclopropanoyl monoenoic; III, dienoic; III', methylated dienoic; IV, monounsaturated α'-mycolic acids. Data adapted from Kaneda et al. (1988).

[c]Data adapted from: 1, Kaneda et al. (1988); 2, Valero-Guillén and Martin-Luengo (1986); 3, Kusaka and Mori (1986). Chain lengths of the fatty acid methyl esters formed upon pyrolysis: 22A, $C_{22} > 80\%$; 22B, $C_{22} > 50\%$ and $C_{24} > 20\%$; 24A, $C_{24} > 50\%$ and $C_{22} > 20\%$; 24B, $C_{24} > 80\%$.

[d]Type strain data adapted from Lanéelle et al. (1988).

[e]Data adapted from Daffé et al. (1983).

to other (sub) species. These studies should be combined with the data from biochemical numerical studies. The chain length of the fatty acid methyl ester formed during pyrolysis may be of additional use in characterizing new isolates. The detailed characterization of the α-mycolic acids, as was reported by Kaneda et al. (1988), would seem to be less practical in view of the required analytical equipment.

The use of mycobactins (lipid-soluble intracellular siderophores of mycobacteria) as chemotaxonomic markers was proposed in the past. Their use was, however, often hampered

by the difficulty of acquiring sufficient amounts of material for analysis. Mycobactin yields of 1–8% (w/w) were reported for *M. smegmatis* using a simple glycerol/asparagine/phosphate medium solidified with agar (Hall and Ratledge, 1982). Subsequently, the same method was used to analyze the mycobactins of a number of different rapidly growing mycobacteria using TLC and high-performance liquid chromatography (HPLC) (Hall and Ratledge, 1984). With most strains tested, mycobactin yields of 4–6% (w/w) were obtained. *M. aurum*, *M. parafortuitum*, and *M. thermoresistibile* did not form detectable amounts of mycobactin under the standard conditions, though more than two-thirds of these strains did yield mycobactins (3–5%) when grown on glucose/yeast extract/agar. The *M. vaccae, M. chelonae* subsp. *chelonae,* and *M. komossense* strains tested did not produce detectable amounts of mycobactin on either growth medium. TLC analysis of 32 strains of 15 species of rapidly growing mycobacteria using two different solvent systems revealed that, with the exception of *M. flavescens,* the R_f values of the different strains within a single species were generally the same.

TLC analysis of mycobactins is a relatively simple technique, and based on the data presented by Hall and Ratledge (1984), it appears to be a very useful chemotaxonomic character with a high discriminatory power. However, just as in the case of mycolic acid patterns, much larger numbers of strains have to be analyzed to confirm the chemotaxonomic utility of the mycobactin R_f value in identifying mycobacterial species.

DNA Hybridization Studies

DNA hybridization studies of the genomic DNA have been performed with a number of mycobacteria (Bradley, 1973; Baess, 1982; Estrada-Garcia et al., 1989 Lévy-Frébault et al., 1984, 1986a; Garcia and Taberés, 1986; Imaeda et al., 1988; Hurley et al., 1988) and have generally confirmed the species status already derived from numerical taxonomy studies (Wayne, 1984) (also see "Phylogeny and Taxonomy" section). More discriminative techniques involving specific DNA probes hybridizing with total DNA (Zainuddin and Dale, 1989; McFadden et al., 1990) or with DNA amplified with the polymerase chain reaction (PCR) (Hartskeerl et al., 1989) have been used, but they have focussed mainly on the slowly growing species. Sequences of 16S rRNA have also been obtained by direct sequencing of DNA amplified by the PCR method. They were demonstrated to be useful in the differentiation of mycobacteria at the species level (Rogall et al., 1990). Clearly, the PCR technique will have a significant impact on the identification of mycobacteria in the near future (also see Chapter 53).

Physiology

The physiology of mycobacteria will not be discussed in an exhaustive manner in this chapter as this subject has been dealt with in excellent reviews by Ratledge (1976, 1982b) and also in the reviews of Ramakrishnan et al. (1972) and Masood et al. (1985). Furthermore, contrary to what was sometimes suspected in the past, mycobacterial physiology differs from that of other aerobic saprophytes only in minor aspects. This section will focus on the areas of mycobacterial physiology that are typical for the genus. The metabolism of unsaturated gaseous hydrocarbons is one area where the role of the mycobacteria is especially important. Another area to be discussed is fatty acid biosynthesis, because mycobacteria clearly differ in this respect from most other prokaryotes. Reserve materials, fatty acid composition, and iron uptake are also discussed.

Catabolic Activities

Mycobacteria are metabolically versatile organisms. They not only grow on common substrates such as sugars, alcohols, and organic acids, but also on a large variety of hydrocarbons including branched-chain, unsaturated, aromatic, and cyclic hydrocarbons (Söhngen, 1913; Lukins and Foster, 1963). Mycobacteria also degrade polycyclic aromatic hydrocarbons, such as pyrene (Heitkamp et al., 1988a, 1988b) and phenanthrene (Guerin and Jones, 1988). Some mycobacteria grow on the simple one-carbon compounds methanol and methylamines (Kato et al., 1988; Urakami and Yano, 1989). In one strain, which was identified as a *M. gastri* species, 3-hexulose-6-phosphate synthase was present, indicating formaldehyde incorporation via the ribulose-monophosphate pathway in this strain (Kato et al., 1988). Autotrophic growth on carbon dioxide and hydrogen gas of several strains of *M. smegmatis, M. marinum,* and *M. fortuitum* was already reported by Lukins and Foster in 1963 (20% of the strains tested). The propene-utilizing strain called *Mycobacterium* Py1 (de Bont et al., 1980) also grew autotrophically. Under these conditions, ribulose-1,5-bisphosphate carboxylase and a membrane-bound hydrogenase were induced (C. G. van Ginkel and J. A. M. de Bont, unpublished observations). Reports in the older literature on mycobacteria growing on methane should be treated skeptically since such isolates were either lost before a rigorous identification was performed, or no longer grew on methane when investigated by others later.

Metabolism of Gaseous Hydrocarbons

In the past, one of the incentives for studying microorganisms degrading gaseous hydrocarbons was the suspected relation between the numbers of these organisms present in soil and the presence of fossil fuel reserves. Such a relationship has, however, not been substantiated (Brisbane and Ladd, 1972).

Growth of mycobacteria with the C_2 hydrocarbons ethane and ethene (ethylene) has been described several times. Using ethane as carbon source, Davis et al. (1956) isolated several types of ethane-utilizing bacteria with mycobacteria predominating. The isolated mycobacteria could be divided into two groups based on their capacity to grow on complex media. About half of the isolates did not grow on nutrient agar or with glycerol as sole carbon source and were placed in a separate novel species, *"M. paraffinicum."* The assignment of *"M. paraffinicum"* to the genus *Mycobacterium* is, however, questionable, as it was shown to contain trehalose mycolates of a relatively low molecular weight (Minnikin and Goodfellow, 1980). Using ethene as carbon source, several mycobacteria were isolated (de Bont, 1976). *Mycobacterium* E20 also grew with ethane. Ethane metabolism was via acetate and ethene metabolism via epoxyethane (de Bont and Harder, 1978). Epoxyethane was further degraded to acetyl-CoA in a CoA-and NAD-dependent reaction (de Bont and Harder, 1978). In contrast to the alkene monooxygenase induced by growth on ethene, the ethane hydroxylase activity of *Mycobacterium* E20 could not be detected in cell-free extracts of ethane-grown cells (de Bont et al., 1979).

Of the short-chain hydrocarbons, the metabolism of propane and its derivatives has received the most attention in the past. The initial step in propane degradation has been the subject of speculation and study for quite some time. Both oxidation of the primary and secondary carbon atoms of propane has been shown to occur (Perry, 1980). *M. vaccae* JOB5 has been used in several cases to study three-carbon metabolism (see Perry, 1980). In this strain, acetone and acetol as well as acetate have been detected as intermediates in propane metabolism. These observations contrast with acetone metabolism in several unidentified Gram-positive bacteria that degraded acetone via pyruvate (Taylor et al., 1980).

Mycobacterium Py1, which was isolated with propene as carbon source (de Bont et al., 1980), did not grow on propane or acetone. However, it utilized acetol, which was oxidatively transformed into acetate and formaldehyde by acetol monooxygenase through a Bayer-Villiger type of reaction (Hartmans and de Bont, 1986). Propene metabolism in *Mycobacterium* Py1 proceeds via an initial oxidation to epoxypropane which is subsequently carboxylated, presumably to acetoacetate.

Fatty Acid Biosynthesis

Mycobacterial lipids have been subject of numerous studies, often with an emphasis on taxonomic aspects (Ratledge, 1982a; Minnikin, 1982; Brennan, 1988). Besides the characteristic lipid-rich cell wall, mycobacteria also differ from most other prokaryotes in aspects of fatty acid biosynthesis. Fatty acid synthetases are generally divided into two types: the type I "eukaryotic" system and the type II "prokaryotic" system. The type II system, which readily dissociates into separate proteins with discrete catalytic activities, has been detected in all bacteria studied, with the exception of some mycobacteria and corynebacteria. The type I system present in animals and eukaryotic microorganisms, has also been found in *M. smegmatis* and *Corynebacterium diphtheriae*.

Fatty acid biosynthesis in *M. smegmatis* has been studied in detail by Bloch and coworkers (Bloch, 1977). They described two fatty acid synthetase (FAS) activities in this species. One is the extensively studied multienzyme complex called FAS-I, which is probably very much like the FAS of eukaryotes. It is however, unique in that it produces both very long (C_{24} and C_{26}) as well as the more common (C_{16} and C_{18}) saturated CoA esters of fatty acids, and its activity is stimulated by certain polysaccharides. The 3-*O*-methylmannose polysaccharide (MMP) and 6-*O*-methylglucose polysaccharide (MGLP) affect the K_m for both acetyl-CoA and malonyl-CoA and can bind palmitoyl-CoA, thus restricting further chain elongation and consequently influencing the bimodal product distribution. The biosynthesis of these polysaccharides has recently been studied by Ballou and coworkers (Weisman and Ballou, 1984a, 1984b; Kamisango et al., 1987).

The second FAS in *M. smegmatis* studied by Bloch and coworkers is similar to the type II FAS found in other prokaryotes in its requirement for an acyl-carrier protein (ACP). De novo fatty acid synthesis is, however, not observed with the type II FAS of *M. smegmatis*, so that it should actually be regarded as a fatty-acid-elongating system, elongating acyl-CoA esters of C_{16} to C_{28} (Odriozola et al., 1977). The polysaccharides which affect the type I FAS do not affect the type II elongating system of *M. smegmatis*, and FAS II activity is not inhibited by palmitoyl-CoA.

Another fatty-acid-elongation system (FES I), isolated from *M. smegmatis*, requires acetyl-CoA and is apparently ACP-independent (Shimakata et al., 1977). It exhibits optimal activity with the C_8 and C_{10} acetyl-CoA esters. The enzymatic activities of this fatty acid-elongation system may however, under physiological conditions be involved in the β-oxidation of fatty acids based on the stereospecificity of the 3-oxoacyl-CoA reductase which forms L-hydroxyacyl-CoA esters (Shimakata et al., 1979). L-hydroxyacyl-CoA esters normally are intermediates in the degradation of fatty acids.

Subsequently, a third elongation system has been described which was apparently also ACP-independent, but which required malonyl-CoA instead of acetyl-CoA. This FES exhibits activity with C_{10} to C_{24} acetyl-CoA esters with an optimum for stearyl-CoA (Kikuchi and Kusaka, 1982). Also, a very long-chain fatty acid-elongation system was isolated from *M. avium* (Kikuchi et al., 1989), which possibly is involved in the synthesis of mycolic acids. It differs from the ACP-independent, malonyl-CoA-incorporating-elongation system of *M. smegmatis* (Kikuchi and Kusaka, 1982) in its cofactor requirements and its sensitivity towards isoniazid. The authors suggest that isoniazid possibly affects the 3-oxoacyl-CoA and enoyl-CoA reductase activities. This would explain the previously observed effect of isoniazid specifically inhibiting the synthesis of mycolic acids (Winder, 1982).

In contrast to the common bacterial pathway of unsaturated fatty acid formation by elongation of decenoyl-CoA, resulting in the formation of palmitoleic and *cis*-vaccenic ($C_{18:1}\Delta^{11}$) acid, biosynthesis of unsaturated fatty acids in mycobacteria is accomplished by desaturation of stearyl-CoA to oleoyl-CoA, and to a lesser extent palmitoyl-CoA to palmitoleoyl-CoA, by the particulate Δ^9 desaturase (Ratledge, 1982a). A very long-chain, soluble, fatty acid Δ^{15} desaturase was isolated from *M. smegmatis*, which exhibited optimal activity with lignoceroyl-CoA ($C_{24:0}$) (Kikuchi and Kusaka, 1986).

The most common mycobacterial branched-chain fatty acid is tuberculostearic acid (D-10-methylstearic acid). It is formed by methylation of oleic acid residues already esterified in phospholipids (Ratledge, 1982a). The methyl group is derived from *S*-adenosylmethionine.

The various different fatty acid synthases and elongating systems that have been found in mycobacteria are of course a reflection of the many different fatty acids present in these organisms. The different constituents will not be discussed in detail here, as excellent reviews are available (Minnikin, 1982; Brennan, 1988).

Reserve Materials

Besides the mycolic acid composition (see "Identification"), other lipid components of mycobacteria have also been studied. Mycobacteria can contain considerable amounts of triacylglycerols, especially in glycerol-grown cells, resulting in the formation of fat bodies (lipid vacuoles) (Brennan, 1988). These lipids may be utilized as a reserve material, though glycogen and trehalose have also been suggested as reserve materials (Ratledge, 1982b). In nitrogen-limited batch cultures of *M. phlei*, both lipid and glycogen accumulation were observed (Antoine and Tepper, 1969). Exchange of these cells to a medium with a high nitrogen content without carbon resulted in restoration of growth and a decrease of the lipid and glycogen content. Similar experiments with *M. smegmatis* (Elbein and Mitchell, 1973) focussed on glycogen and trehalose levels. *M. smegmatis* grown under nitrogen-limiting conditions had an increased glycogen content, which was rapidly utilized when these cells were transferred to a medium containing sufficient carbon and nitrogen. The trehalose levels were more or less the same under all growth conditions, indicating that under these conditions trehalose is not a reserve material, although the turnover rates were very high (Elbein and Mitchell, 1973).

Starvation experiments performed with *Mycobacterium* sp. strain E3, a "*M. parafortuitum* complex" species, grown under nitrogen or carbon limitation in chemostat cultures have been performed (de Haan et al., 1991). Cells grown under nitrogen limitation contained 10 times more glycogen and only 50% more trehalose and lipid than cells grown under carbon limitation. The reserve materials of these cells were monitored during a 2-day incubation in buffer in the absence of carbon or nitrogen sources. Trehalose levels fell from 2–3% to less than 0.5% within 8 h. The lipid/protein ratio remained constant during the 2-day starvation experiment. Glycogen was only consumed in the cells which had been grown under nitrogen limitation. The 2% glycogen present in cells grown under carbon limitation was not consumed during the 2-day-starvation experiment. Experiments in which the NADH-dependent oxidation of propene to 1,2-epoxypropane by starving cells was monitored showed that nitrogen-limited-grown cells, containing the higher glycogen levels, produced more of the epoxide (de Haan et al., 1991).

Lipid Composition

The phospholipids of *M. smegmatis* and *M. phlei* have been studied in some detail (Dhariwal et al., 1976). Major phospholipids present in mycobacteria are phosphatidylinositol, phosphatidylethanolamine, diphosphatidylglycerol (cardiolipin), and the phosphatidylinositol mannosides. The phosphatidylinositol mannosides are highly characteristic for actinomycetes and coryneform bacteria (Brennan, 1988).

The phospholipid and fatty acid composition are to a large extent dependent on the culture conditions. The growth temperature and the carbon source as well as the ratio of carbon and nitrogen sources affect the lipid composition (Dhariwal et al., 1977; King and Perry, 1975). Dhariwal et al. (1976) also reported the effect of culture age on the fatty acid composition. The major change they observed was an increase in tuberculostearic acid accompanied by a decrease in oleic acid (18:1) content. This composition is probably a reflection of the biosynthesis of tuberculostearic acid, which is formed by methylation of esterified oleic acid residues. This perhaps also explains why relatively high tuberculostearic acid contents have often been reported, as, very often, cultures were analyzed which were already in the stationary growth phase.

An important aspect which should be emphasized is that these studies were all performed with batch-grown cultures. This implies that the growth conditions and growth rates were rarely constant, thus making it difficult to ascribe the observed changes in fatty acid composition to a specific factor.

More reliable experimental data require the application of chemostat cultures. As discussed in the section on cultivation, the continuous culture of mycobacteria in chemostats has been demonstrated several times. Therefore the regulation of lipid composition, or indeed many other aspects of mycobacterial physiology, should be studied using chemostat cultures.

Iron Uptake

Mycobacterial iron metabolism, especially iron uptake, has been extensively studied by the group of Ratledge (Ratledge, 1982b, 1984). Mycobacteria appear to be unique in producing two different siderophores (exochelins and mycobactins), probably necessitated by the thick lipoidal nature of the cell envelope. Exochelins are extracellular siderophores, which until now have only been poorly characterized. Two types of exochelins, differing in their solubility in organic solvents, have been described (Ratledge, 1982b). The very hydrophobic mycobactins, which are located within the cell envelope, have been studied to a greater extent. The potential for using the mycobactin composition in the identification of rapidly growing mycobacteria (Hall and Ratledge, 1984) is discussed in the section on identification. Recent work has focussed on the regulation of the biosynthesis of the different siderophores and other iron-regu-

lated proteins (Sritharan and Ratledge, 1989). It was shown for *M. neoaurum* that these components were coordinately expressed in the presence of low iron concentrations (< 0.2 mg/1). Increasing the iron concentration to 0.5 mg/1 or more resulted in repression of the synthesis of all three components (Sritharan and Ratledge, 1989).

Phylogeny and Taxonomy

Taxonomic studies traditionally have relied heavily on morphological characteristics, but subsequently numerical taxonomy and chemotaxonomic characteristics have played an important role in determining taxonomic relationships. Cell wall and mycolic acid analyses have proved to be of great value in separating the actinomycete genera, while numerical taxonomy has had the most impact on the subgeneric level (Goodfellow and Wayne, 1982). DNA hybridization of genomic DNA and, more recently, 16S rRNA similarity data are of course important tools in determining phylogenetic relations both on the generic and subgeneric levels.

Based on chemotaxonomic studies, the genus *Mycobacterium* was placed within the CNM *(Corynebacterium-Nocardia-Mycobacterium)* complex. Together with the genera *Caseobacter, Rhodococcus,* and *Tsukamurella,* these genera all have cell wall chemotype IV (Lechevalier and Lechevalier, 1970) and contain mycolic acids (Collins et al., 1988).

In the most recent proposal for a hierarchical structure for the Gram-positive bacteria, based on 16S rRNA similarity data, the Family Mycobacteriaceae includes the genera *Mycobacterium, Nocardia,* and *Rhodococcus,* whereas the genus *Corynebacterium* is in a separate Family Corynebacteriaceae (Fox and Stackebrandt, 1987). Both families are placed within the Order Actinomycetales, which is one of the three orders within the Class Actinomycetes, i.e., Gram-positive bacteria with a high (>55%) GC content. The genus *Tsukamurella* (Collins et al., 1988) should probably also be placed within the Family Mycobacteriaceae.

Interestingly, 16S rRNA similarity studies confirm the traditional division of the genus *Mycobacterium* into rapidly and slowly growing species (Stahl and Urbance, 1990). This study, however, also suggests that the type strain of *M. chitae* is incorrectly classified within the genus *Mycobacterium,* although the mycolic acid composition of *M. chitae* (Ausina et al., 1985, Lévy-Frébault et al., 1986b; Kaneda et al., 1988) and also numerical taxonomy data (Saito et al.,

1977) indicate *M. chitae* to be an acceptable species of the genus.

The complete nucleotide sequence of the 16S rRNA gene of *M. bovis* BCG has been determined (Suzuki et al., 1988) as well as the sequence of the 3'-end of the 16S rRNA of *M. smegmatis, M. vaccae, M. tuberculosis* and *M. leprae* (Estrada-Garcia et al., 1989). Comparison of these five sequences revealed at least 98.5% similarity of helix 39 of the 16S rRNA, indicating this sequence is highly conserved within the genus *Mycobacterium*. Comparison of this sequence with those of other genera revealed that *Mycobacterium* shares most similarity with *Streptomyces lividans* (93%) and significantly less with *Bacillus subtilis* (74%) and *E. coli* (58%) (Estrada-Garcia et al., 1989). Larger sequences of about 1 kb of the 5'-end of 16S rRNA, determined by direct sequencing of DNA amplified by the polymerase chain reaction, have been reported (Rogall et al., 1990). With the exception of the species of the *M. tuberculosis* complex and of *M. gastri,* which could not be distinguished from *M. kansasii,* all analyzed species could be distinguished based on nucleotide sequences of the 200–400 region (using the *E. coli* numbering system). An extended helix was found between positions 451 and 482 (*E. coli* numbering) for most slowly growing mycobacteria. In the rapidly growing species, a much smaller helix is present (Stahl and Urbance, 1990). *M. triviale* (Stahl and Urbance, 1990) and *M. simiae* (Rogall et al., 1990) were, however, exceptions, having the smaller helix which is characteristic for the rapidly growing species.

DNA-DNA homology studies can be used to discriminate between more closely related taxons, e.g., the different species of a genus. It should be noted, however, that DNA-DNA hybridization data from different laboratories can only be compared when reference strains are included. DNA homology studies with selected strains of *M. tuberculosis, M. bovis, M. bovis.* BCG, *M. microti,* and *M. africanum* revealed that all strains exhibited more than 90% DNA relatedness, whereas DNA relatedness between *M. tuberculosis* and other slowly growing mycobacteria ranged from 9–53% (Imaeda, 1985). Although *M. microti, M. tuberculosis, M. bovis,* and *M. africanum* can be distinguished phenotypically, numerical taxonomy places them all in a cluster distinct from other slowly growing mycobacteria (Wayne and Kubica, 1986). 16S rRNA sequencing data of 14 *M. tuberculosis,* two *M. bovis,* and two *M. africanum* isolates was also identical (Rogall et al., 1990). Considering the above, it is anticipated that a proposal will be made to reduce these four species to one species of *M. tuberculosis,* which possibly will be subdivided into subspecies (Wayne and Kubica, 1986).

DNA-DNA hybridization studies have also led to a better understanding of the relationship between *M. paratuberculosis* and the MAIS *(M. avium–M. intracellulare–M. scrofulaceum)* complex. The MAIS complex strains have been grouped together based on phenotypic characteristics (Wayne, 1984), but DNA hybridization studies indicate that *M. scrofulaceum* shows little DNA similarity with *M. avium* and *M. intracellulare* (Hurley et al., 1988). Based on these studies it was suggested that *M. avium, M. intracellulare,* and *M. paratuberculosis* should be grouped together as biovars of a single species (Hurley et al., 1988). The 16S rRNA sequences can, however, be used to differentiate between representatives of each of these species (Rogall et al., 1990).

DNA homology studies with rapidly growing mycobacteria have also been performed (Baess, 1982; Lévy-Frébault et al., 1984, 1986a; Garcia and Taberés, 1986). In the study of Baess (1982) *M. chelonae* subsp. *chelonae* and *M. chelonae* subsp. *abscessus* could not be distinguished using the spectrophotometric method, but with the more sensitive S1 nuclease method, the two subspecies were clearly distinct (Lévy-Frébault et al., 1986a). Within the subspecies, homology values were higher than 73% and ΔT_m less than 2°C, whereas homology values between the *M. chelonae* groups were 26 to 52% with ΔT_m values of more than 8°C. Both studies indicate that the unrecognized species *"M. perigrinum,"* often grouped within the *M. fortuitum* complex, should be revived as an independent species.

Genetics

Our knowledge of the genetics and the molecular biology of mycobacteria lags behind that of thoroughly characterized species such as *Escherichia coli* or *Bacillus subtilis*. This is partly due to the low growth rate of mycobacteria but also due to the ineffectiveness of many of the standard molecular biology techniques when applied to the mycobacteria. Recently, however, significant progress has been made, part of which has already been reviewed (Grange, 1982; Hopwood et al., 1988; Konicek et al., 1988).

Organization of Genetic Information

The estimated genome sizes of mycobacteria vary from $3–5.5 \times 10^9$ daltons (Baess and Mansa, 1978). As a comparison, the genome size of *E. coli* is 2.5×10^9 daltons. Mycobacterial DNA has a high GC content, varying from 66 to 71 mol% for the strains examined by Baess and Mansa (1978).

The presence of extrachromosomal DNA (plasmids) has been demonstrated conclusively in several species. Most of the strains studied were, however, slow-growing pathogens (Hopwood et al., 1988). Plasmids have also been described in fast-growing species although no selective markers could be attributed to them (Labidi et al., 1984). A 173-kb plasmid isolated from a *M. scrofulaceum* species isolated from the environment encoded mercury and copper resistance (Meissner and Falkinham, 1984; Erardi et al., 1987). This strain contained a total of four plasmids varying in size from 15 to 300 kb. Two smaller mycobacterial plasmids have been studied in more detail: Plasmid pLR7 (15.3 kb) from *M. intracellulare* has been mapped (Crawford and Bates, 1984), and the complete nucleotide sequence of pAL5000, a 4,837-bp plasmid from *M. fortuitum,* has been determined (Rauzier et al., 1988).

The organization and sequences of mycobacterial genes coding for ribosomal RNA (rRNA) have been studied quite intensively. The rRNA can be easily isolated and sequenced, and these sequences have potential as taxonomic markers (Cox and Katoch, 1986) and are used in determining phylogenic relationships (Woese, 1987) (also see "Identification" and "Phylogeny and Taxonomy").

Using rRNA probes derived from *E. coli*, hybridization with DNA from *M. phlei* and *M. smegmatis* revealed that these fast-growing strains contain two rRNA operons, whereas the slow-growing *M. tuberculosis* and *M. intracellulare* appeared to possess only one rRNA operon (Bercovier et al., 1986). The slow-growing *M. bovis* BCG also contains only one set of rRNA genes (Suzuki et al., 1987). It is tempting to speculate that the number of rRNA operons present in the genome forms the genetic basis for the difference in growth rate between the rapidly and slowly growing mycobacteria. *E. coli* grows much faster than the fast-growing mycobacteria and possesses seven rRNA operons (Bercovier et al., 1986).

The rRNA genes of *M. smegmatis* have been studied by restriction analysis (Bercovier et al., 1989), and, like the *M. bovis* BCG rRNA genes (Suzuki et al., 1987), the genes coding for the different rRNAs are organized in operons in the order 16S–23S–5S, as in other eubacteria (Woese, 1987; Clark-Curtiss, 1990).

Genetic Recombination

Papers concerning the transfer of mycobacterial DNA through transduction and conjugation published in the 1970s have been reviewed (Konicek et al., 1988; Grange, 1982) and will not be discussed here. Genetic recombination of mycobacteria by spheroplast fusion to produce genetically modified strains for sterol transformation has also been reported (Jekkel et al., 1989).

It is, however, essential that other methods of genetic recombination are available for a better genetic characterization of mycobacteria. Methods to efficiently transfer DNA between mycobacteria and *E. coli* and vice versa are consequently an important tool. The approach of Jacobs et al. (1987) to attack this problem was to construct a vector which replicates as a plasmid in *E. coli* and as a phage in mycobacteria. This was achieved by introducing an *E. coli* plasmid replicon into a nonessential region of mycobacteriophage TM4, a temperate phage of *M. avium*. The resulting "phasmid" phAE1 grows as a lytic phage in *M. smegmatis* and replicates as a plasmid in *E. coli*. DNA transfer into *M. smegmatis* was achieved by transfection of protoplasts and introduction into *E. coli* was done by in vitro packaging with lambda proteins.

Subsequently, efforts were undertaken to use the same approach to construct lysogenic phasmids, that would allow the introduction and maintenance of DNA in growing mycobacteria. Shuttle "phasmids" were constructed in a similar manner as described above using the temperate phage L1 which stably lysogenizes *M. smegmatis* by integrating in the chromosome (Snapper et al., 1988). With one of these phasmids (phAE19) it was possible to lysogenize *M. smegmatis* protoplasts and generate kanamycin- and chloramphenicol-resistant colonies, thus illustrating the possibility of introducing and expressing foreign genes in mycobacteria (Snapper et al., 1988).

A much larger stimulation of research of mycobacterial genetics is expected from the construction of plasmids capable of replicating in both *E. coli* and mycobacteria. Snapper et al. (1988) have constructed such hybrid shuttle plasmids by randomly inserting the *E. coli* plasmid pIJ666, containing an origin of replication and the genes for kanamycin and chloramphenicol resistance in pAL5000 from *M. fortuitum.* Transformation of the pIJ666::pAL5000 library into *M. smegmatis* protoplasts was not successful. Subsequently the high voltage electroporation technique was applied. This method had previously been demonstrated to be useful in the transformation of other Gram-positive bacteria (Chassy and Flickinger, 1987). Electroporation conditions were optimized for the uptake of lytic D29 phage DNA by intact

M. smegmatis cells resulting in $> 5 \times 10^3$ plaque-forming units per μg DNA. Electroporation of *M. smegmatis* using the optimized procedure with the pIJ666::pAL5000 recombinant library yielded 1–10 kanamycin resistant transformants per μg DNA. Plasmids isolated from the transformants were used in retransformation experiments yielding kanamycin resistant *E. coli* and *M. smegmatis* colonies.

Gicquel-Sanzey et al. (1989) used a similar approach to construct the 9.2-kb vector pAL8 by combining pAL5000 with an *E. coli* plasmid and a gene coding for kanamycin resistance. However, transformation of spheroplasts with the pAL8 vector was not successful using conditions under which transformation with the lytic mycobacteriophage D29 resulted in transformation efficiencies of 10^4 to 10^5 per μg DNA. Using the electroporation technique, Gicquel-Sanzey et al. (1989) reported transformation frequencies for *M. smegmatis* of 10 per μg pAL8 DNA. Electroporation of *M. smegmatis* with phage D29 resulted in transformation frequencies of 10^3 per μg DNA, similar to the rate reported by Snapper et al. (1988). Both groups report much higher transformation efficiencies (10^3 per μg DNA) for *M. bovis* BCG. Recently though, high-efficiency-transforming mutants of *M. smegmatis* have been isolated, yielding more than 10^5 transformants per μg DNA (Snapper et al., 1990b). Interestingly, these mutants do not show enhanced transformation frequencies with an integrating vector that recombines into the *M. smegmatis* chromosome. It is suggested that the mutation possibly affects plasmid replication and maintenance in *M. smegmatis*. Using these mutants, an essential replication region of pAL5000 was mapped and the gene coding for the 65-kDa stress-protein antigen of *M. leprae* was expressed (Snapper et al., 1990a).

Deletion experiments with pAL8 have resulted in the construction of a smaller shuttle plasmid of 6.6 kb (pRR3) incorporating only 2.58 kb of pAL5000 (Ranes et al., 1990). The transformation rates of *M. bovis* BCG and a high-transforming mutant of *M. smegmatis* (Snapper et al., 1990b) using the electroporation technique with this plasmid were 10^4 per μg pRR3 DNA.

Transformation by electroporation of *M. aurum* and *M. smegmatis* with the broad-host-range, Gram-negative vector pJRD215 has also been demonstrated (Hermans et al., 1991). In contrast to the constructs described above, this cosmid vector does not contain mycobacterial DNA, i.e., a mycobacterial origin of replication. An advantage of pJRD215 is that it contains the phage lambda *cos* site, allowing the cloning of relatively large DNA fragments in *E. coli* and thereby facilitating the construction of genomic libraries. Expression of pJRD215 in mycobacteria should allow the screening of such libraries by the complementation of mutants. The kanamycin- and streptomycin-resistance genes carried by pJRD215 were both expressed in *M. aurum*. The transformation efficiency as determined by screening for kanamycin resistance, was rather low at 2×10^2 transformants per μg DNA. Enhancement of the transformation frequency by pretreatment of the cells with isoniazid, as was demonstrated for the transformation of *M. aurum* with pAL8 (Hermans et al., 1990), might also be applied successfully for the transformation of mycobacteria with pJRD215.

Another approach to stably introduce DNA in mycobacteria, using a vector without a mycobacterial origin of replication, requires homologous recombination to take place. This was demonstrated with a shuttle vector that can replicate autonomously in *E. coli* but must integrate into homologous DNA for survival in *M. smegmatis* (Husson et al., 1990). The vector, pY6002, contained an *E. coli* origin of replication and the *pyrF* gene of *M. smegmatis* with the kanamycin-resistance gene *aph* as an insert. The *pyrF*-negative mutants are both uracil auxotrophic and fluorouracil resistant. The transformation frequency of *M. smegmatis* by electroporation with this vector ranged from 10 to 500 transformants per μg DNA. Integration of pY6002 in the chromosomal DNA of the prototrophic "wild-type" *M. smegmatis* gave two types of kanamycin-resistant recombinants resulting from either a single or double recombination event. Class I recombinants resulting from a single recombination event contained the entire plasmid as well as a functional *pyrF* gene, whereas class II recombinants were uracil auxotrophic and fluorouracil resistant. In class II recombinants, the plasmid had apparently integrated in the chromosome at the *pyrF* locus, and replaced it. Class II transformants could be retransformed with a plasmid containing the intact *pyrF* gene without an *aph* insert (pY6001). Selection for uracil prototrophs gave both classes of transformants. Class I, resulting from a single recombination event, contained both the disrupted and the wild-type *pyrF* gene, while class II kanamycin-sensitive transformants only contained the wild-type *pyrF* gene. Using this technique, the 65-kDa stress-protein antigen of *M. leprae* was expressed in *M. smegmatis* at detectable levels. Two vectors were used with the *M. leprae* gene on a 3.6-kb fragment inserted at different sites of pY6002. The transformation frequencies were comparable to those of pY6002. The two transformants expressing the 65-kDa antigen were kanamycin resistant and fluorouracil sensitive, indicating them to be class I transformants (Husson et al., 1990).

The methods and tools allowing the transformation and molecular genetic manipulation of mycobacteria described above (Jacobs et al., 1987; Snapper et al., 1988, 1990a; Husson et al., 1990; Hermans et al., 1991) should result in major advances in mycobacterial research in the near future.

Applications

The most obvious applications of mycobacteria are of course associated with the disease tuberculosis. Examples are the BCG (Bacillus Calmette-Guérin) vaccine, which is derived from the *Mycobacterium bovis* BCG Strain, and the production of tuberculin, which is extracted from *Mycobacterium* cultures. Tuberculin is used diagnostically in a delayed hypersensitivity type of skin test for the detection of a current or previous infection with *M. tuberculosis.*

In this section we will focus on nonmedical applications. Mycobacteria are used mainly in the area of biocatalysis in order to perform specific transformation reactions. The justification for using microorganisms or enzymes for a particular transformation is usually the selectivity and specificity (regio- or stereospecificity) exhibited by the biocatalyst (Meijer et al., 1985).

One area in biotechnology that has a large number of commercial applications utilizing mycobacteria is the biotransformation of steriods. These processes involve either the modification of the steroid nucleus of natural or synthetic sterols or the selective degradation of the side chain of naturally occurring sterols such as cholesterol and β-sitosterol (Martin, 1984). The products formed can subsequently be chemically further transformed to pharmacologically active sterols (Martin, 1984). Cholesterol degradation by mycobacteria has been known for a long time (Söhngen, 1913) and usually involves simultaneous degradation of both the side chain and the steroid nucleus. Selective modification of only the nucleus or the side chain can be achieved by employing inhibitors in the biotransformation process or by using mutants (Martin, 1984). Wovcha et al. (1978) for example obtained *M. fortuitum* mutants blocked in various steps of sitosterol degradation. With these mutants, intermediates of the steroid-degradation pathway can be produced. Some of these intermediates can be used as substrates for the production of medically useful steroids. Most commercial processes for the se-

lective side-chain cleavage of sterols employ mutant strains of *M. fortuitum* or *M. parafortuitum,* although other *Mycobacterium* species and bacteria of related genera are also used (Martin, 1984).

Many biocatalytic processes comprise only one enzymatic step, using enzyme preparations or whole (permeabilized) cells. An example is the utilization of *M. neoaurum* ATCC 25795 containing a L-specific aminopeptidase with a very high stereospecificity and a broad substrate specificity. Using permeabilized *M. neoaurum* cells, a wide range of L- or D-α-methyl-substituted amino acids can be produced by stereoselective hydrolysis of racemic mixtures of the corresponding amides (Kamphuis, 1987).

Another potential application of mycobacteria in the field of biocatalysis is the production of optically active epoxides from alkenes (Habets-Crützen et al., 1985; Hartmans et al., 1989). Optically pure epoxides form versatile starting materials for the chemical synthesis of optically active pharmaceutical compounds. Screening a number of bacteria from different genera revealed that alkene-grown mycobacteria produce the epoxides examined (1,2-epoxypropane, 1,2-epoxybutane, and 2,3-epoxy-1-chloropropane) in the highest enantiomeric excess (Weijers et al., 1988). However, substantial research efforts will be required to increase the specific activity and the operational stability of the biocatalysts before commercial production of epoxides by the alkene-utilizing mycobacteria can be realized. In this respect, product toxicity and cofactor regeneration have been studied for the production of 1,2-epoxypropane with the ethene-utilizing *Mycobacterium* strain E3 (Habets-Crützen and de Bont, 1985, 1987).

One aspect that has received some attention is the use of organic solvents and immobilization of mycobacteria for application in continuously operated processes (Steinert et al., 1986; Flygare and Larsson, 1987). The advantages of immobilization, in terms of increased reactor productivity or a better operational stability, were, however, not very evident. One advantage that immobilization of cells can offer is that organic solvents, which can function as a reservoir for substrates and products with a low water solubility, can be used without the problems of cell aggregation and biocatalyst/solvent separation which are associated with the use of organic solvents with free cells (Linko and Linko, 1985). However, immobilization does not protect the cells against the adverse effects many solvents have on mycobacteria (Brink and Tramper, 1985; Steinert et al., 1986). In the study of Brink and Tramper (1985), complete retention of activity was observed in the presence of dioctyl and didecyl phthalate, but most solvents resulted in decreased biocatalytic activity.

The effect of water-immiscible organic solvents on growth has also been tested with the rapidly growing strain *Mycobacterium* E3 and several other bacteria from different genera (Rezessy-Szabó et al., 1986). From these experiments it appeared that *Mycobacterium* E3 was not particularly resistant to organic solvents compared to other bacteria, as for instance *Pseudomonas* species.

Another area of research concerning the potential application of mycobacteria is in environmental biotechnology. Removal of traces of the plant hormone ethene from storage facilities for fruit using immobilized ethene-utilizing mycobacteria is one example (van Ginkel et al., 1986). Unfortunately, the activity of the ethene-utilizing mycobacteria was very low at the ethene levels (often less than 1 ppm) prevailing in the fruit storage facilities. Another example is the application of the vinyl chloride-utilizing *Mycobacterium* strain L1 to remove the carcinogenic vinyl chloride from industrial waste gases (Hartmans et al., 1985). The compound is metabolized by an initial oxidation step to the corresponding epoxide by alkene monooxygenase. Accumulation even of very low concentrations of this reactive intermediate, which can occur due to fluctuations in the vinyl chloride supply, irreversibly inhibited the alkene monooxygenase. As such fluctuations will probably frequently occur, practical application of this strain does not seem realistic.

The use of mycobacteria in the bioremediation of contaminated sediments has also been suggested. Addition of a pyrene-degrading *Mycobacterium* strain to sediments resulted in an enhancement of the mineralization rates of several polycyclic aromatic hydrocarbons. Further investigations were suggested to assess the potential of this strain in the bioremediation of contaminated sediments (Heitkamp and Cerniglia, 1989).

A disadvantage of mycobacteria in general is their relatively low growth rate and hence their low catalytic activities. However, mycobacteria are likely to play an increasingly important role as a source of interesting biocatalytic capacities now that the development of the molecular genetic tools required to increase the expression of desired activities, either in mycobacteria or other hosts, is well under way.

Literature Cited

Antoine, A. D., and B. S. Tepper. 1969. Environmental control of glycogen and lipid content of *Mycobacterium phlei*. J. Gen. Microbiol. 55:217–226.

Ausina, V., M. Luquin, and L. Margarit. 1985. Mycolic acids of *Mycobacterium chitae*. J. Gen. Microbiol. 131:2237–2239.

Baess, I. 1982. Deoxyribonucleic acid relatedness among species of rapidly growing mycobacteria. Acta Path. Microbiol. Immunol. Scand. Sect. B 90:371–175.

Baess, I., and B. Mansa. 1978. Determination of genome size and base ratio on deoxyribonucleic acid from mycobacteria. Acta Path. Microbiol. Immunol. Scand. Sect. B 86:309–312.

Barksdale, L., and K.-S. Kim. 1977. *Mycobacterium*. Bacteriol. Rev. 41:217–372.

Bercovier, H., O. Kafri, and S. Sela. 1986. Mycobacteria possess a surprisingly small number of ribosomal RNA genes in relation to the size of their genome. Biochem. Biophys. Res. Comm. 136:1136–1141.

Bercovier, H., O. Kafri, D. Kornitzer, and S. Sela. 1989. Cloning and restriction analysis of ribosomal RNA genes from *Mycobacterium smegmatis*. FEMS Microbiol. Lett. 57:125–128.

Bloch, K. 1977. Control mechanisms for fatty acid synthesis in *Mycobacterium smegmatis*. Adv. Enzymol. 45:1–84.

Bradley, S. G. 1973. Relationships among mycobacteria and nocardiae based upon deoxyribonucleic acid reassociation. J. Bacteriol. 113:645–651.

Brennan, P. J. 1988. *Mycobacterium* and other actinomycetes, p. 203–298. In: C. Ratledge and S. G. Wilkinson (ed.), Microbial lipids, vol. 1. Academic Press, London.

Brink, L. E. S., and J. Tramper. 1985. Optimization of organic solvent in multiphase biocatalysis. Biotechnol. Bioeng. 27:1258–1269.

Brisbane, P. G., and J. N. Ladd. 1972. Growth of *Mycobacterium paraffinicum* on low concentrations of ethane in soils. J. Appl. Bacteriol. 35:659–665.

Carson, L. A., L. A. Bland, L. B. Cusick, M. S. Favero, G. A. Bolan, A. L. Reingold, and R. B. Good. 1988. Prevalence of nontuberculous mycobacteria in water supplies of hemodialysis centers. Appl. Environ. Microbiol. 54:3122–3125.

Cech, J. S., P. Hartman, M. Slosarek, and J. Chudoba. 1988. Isolation and identification of a morpholine-degrading bacterium. Appl. Environ. Microbiol. 54:619–621.

Chassy, B. M., and J. L. Flickinger. 1987. Transformation of *Lactobacillus casei* by electroporation. FEMS Microbiol. Lett. 44:173–177.

Clark-Curtiss, J. E. 1990. Genome structure of mycobacteria, p. 77–96. In: J. J. McFadden (ed.), Molecular biology of the mycobacteria. Academic Press, London.

Collins, C. H., J. M. Grange and M. D. Yates. 1984. Mycobacteria in water. J. Appl. Bacteriol. 57:193–211.

Collins, M. D., T. Pirouz, M. Goodfellow, and D. E. Minnikin. 1977. Distribution of menaquinones in actinomycetes and corynebacteria. J. Gen. Microbiol. 100:221–230.

Collins, M. D., J. Smida, M. Dorsch, and E. Stackebrandt. 1988. *Tsukamurella* gen. nov. harboring *Corynebacterium paurometabolum* and *Rhodococcus aurantiacus*. Int. J. Syst. Bacteriol. 38:385–391.

Cox, R. A., and V. M. Katoch. 1986. Evidence for genetic divergence in ribosomal RNA genes in mycobacteria. FEBS Lett. 195:194–198.

Crawford, J. T., and J. H. Bates. 1984. Restriction endonuclease mapping and cloning of *Mycobacterium intracellulare* plasmid pLR7. Gene 27:331–333.

Daffé, M., M. A. Lanéelle, C. Asselineau, V. Lévy-Frébault, and H. David. 1983. Intérêt taxonomique des acides gras des mycobactéries: proposition d'une méthode d'analyse. Ann. Microbiol. (Paris) 134B:241–256.

David, H. L. 1984. Carotenoid pigments of the mycobacteria, p. 537–545. In: G. P. Kubica, and L. G. Wayne (ed.), The mycobacteria: a source book, Part A. Marcel Dekker, New York.

Davis, J. B., H. H. Chase, and R. L. Raymond. 1956. *Mycobacterium paraffinicum* n. sp., a bacterium isolated from soil. Appl. Microbiol. 4:310–315.

de Bont, J. A. M. 1976. Oxidation of ethylene by soil bacteria. Antonie van Leeuwenhoek 42:59–71.

de Bont, J. A. M., and W. Harder. 1978. Metabolism of ethylene by *Mycobacterium* E20. FEMS Microbiol. Lett. 3:89–93.

de Bont, J. A. M., M. M. Attwood, S. B. Primrose, and W. Harder. 1979. Epoxidation of short-chain alkenes in *Mycobacterium* E20: the involvement of a specific mono-oxygenase. FEMS Microbiol. Lett. 6:183–188.

de Bont, J. A. M., S. B. Primrose, M. D. Collins, and W. Harder. 1980. Chemical studies on some bacteria which utilize gaseous unsaturated hydrocarbons. J. Gen. Microbiol. 117:97–102.

de Haan, A., M. R. Smith, W. Voorhorst, and J. A. M. de Bont. 1991. Co-factor regeneration in the production of 1,2-epoxypropane by *Mycobacterium* sp. E3. Submitted for publication.

Dhariwal, K. R., A. Chander, and T. A. Venkitasubramanian. 1976. Alterations in lipid constituents during growth of *Mycobacterium smegmatis* CDC 46 and *Mycobacterium phlei* ATCC 354. Microbios 16:169–182.

Dhariwal, K. R., A. Chander, and T. A. Venkitasubramanian. 1977. Environmental effects on lipids of *Mycobacterium phlei* ATCC 354. Can. J. Microbiol. 23:7–19.

Draper, P. 1982. The anatomy of mycobacteria, p. 9–52. In: C. Ratledge, and J. Stanford (ed.), The biology of the mycobacteria, vol. 1. Physiology, identification and classification. Academic Press, London.

Elbein, A. D., and M. Mitchell. 1973. Levels of glycogen and trehalose in *Mycobacterium smegmatis* and the purification and properties of the glycogen synthetase. J. Bacteriol. 113:863–873.

Erardi, F. X., M. L. Failla, and J. O. Falkinham III. 1987. Plasmid-encoded copper resistance and precipitation by *Mycobacterium scrofulaceum*. Appl. Environ. Microbiol. 53:1951–1954.

Estrada-Garcia, I. C. E., M. Joseph Colston, and R. A. Cox. 1989. Determination and evolutionary significance of nucleotide sequences near to the 3′-end of 16S ribosomal RNA of mycobacteria. FEMS Microbiol. Lett. 62:285–290.

Flygare, S., and P.-O. Larsson. 1987. Steroid transformation using magnetically immobilized *Mycobacterium* sp. Enzyme Microb. Technol. 9:494–499.

Fox, G. E., and E. Stackebrandt. 1987. The application of 16S rRNA cataloging and 5S rRNA sequencing in bacterial systematics, p. 405–458. In: R. R. Colwell and R. Grigorova (ed.), Methods in microbiology, vol. 19. Current methods for classification and identification of microorganisms. Academic Press, London.

Garcia, M. J., and E. Tabarés. 1986. Separation of *Mycobacterium gadium* from other rapidly growing mycobacteria on the basis of DNA homology and restriction endonuclease analysis. J. Gen. Microbiol. 132:2265–2269.

Gicquel-Sanzey, B., J. Moniz-Pereira, M. Gheorghiu, and J. Rauzier. 1989. Structure of pAL5000, a plasmid from *M. fortuitum* and its utilization in transformation of mycobacteria. Acta Leprol. 7:208–211.

Girbal, E., R. A. Binot, and P. F. Monsan. 1989. Production, purification, properties and kinetic studies of free and immobilized polyphosphate: glucose-6-phosphotransferase from *Mycobacterium phlei*. Enzyme Microb. Technol. 11:518–527.

Goodfellow M., and T. Cross. 1984. Classification, p. 8–164. In: M. Goodfellow, M. Mordarski, and S. T. Williams (ed.), The biology of the actinomycetes. Academic Press, London.

Goodfellow, M., and L. G. Wayne. 1982. Taxonomy and nomenclature, p. 471–521. In: C. Ratledge, and J. Stanford (ed.), The biology of the mycobacteria, vol. 1. Physiology, identification and classification. Academic Press, London.

Grange, J. M. 1982. The genetics of mycobacteria and mycobacteriophages, p. 309–351. In: C. Ratledge, and J. Stanford (ed.), The biology of the mycobacteria, vol. 1. Physiology, identification and classification. Academic Press, London.

Guerin, W. F., and G. E. Jones. 1988. Mineralization of phenanthrene by a *Mycobacterium* sp. Appl. Environ. Microbiol. 54:937–944.

Habets-Crützen, A. Q. H., S. J. N. Carlier, J. A. M. de Bont, D. Wistuba, V. Schurig, S. Hartmans, and J. Tramper. 1985. Stereospecific formation of 1,2-epoxypropane, 1,2-epoxybutane and 1-chloro-2,3-epoxypropane by alkene-utilizing bacteria. Enzyme Microbiol. Technol. 7:17–21.

Habets-Crützen, A. Q. H., and J. A. M. de Bont. 1985. Inactivation of alkene oxidation by epoxides in alkene- and alkane-grown bacteria. Appl. Microbiol. Biotechnol. 22:428–433.

Habets-Crützen, A. Q. H., and J. A. M. de Bont. 1987. Effect of co-substrates on 1,2-epoxypropane formation from propene by ethene-utilizing mycobacteria. Appl. Microbiol. Biotechnol. 26:434–438.

Hall, R. M., and C. Ratledge. 1982. A simple method for the production of mycobactin, the lipid-soluble siderophore, from mycobacteria. FEMS Microbiol. Lett. 15:133–136.

Hall, R. M., and C. Ratledge. 1984. Mycobactins as chemotaxonomic characters for some rapidly growing mycobacteria. J. Gen. Microbiol. 130:1883–1892.

Hartmans, S., J. A. M. de Bont, J. Tramper, and K.Ch.A. M. Luyben. 1985. Bacterial degradation of vinyl chloride. Biotechnol. Lett. 7:383–388.

Hartmans, S., and J. A. M. de Bont. 1986. Acetol monooxygenase from *Mycobacterium* Py1 cleaves acetol into acetate and formaldehyde. FEMS Microbiol. Lett. 36:155–158.

Hartmans, S., J. A. M. de Bont, and W. Harder. 1989. Microbial metabolism of short-chain unsaturated hydrocarbons. FEMS Microbiol. Rev. 63:235–264.

Hartskeerl, R. A., M. Y. L. de Wit, and P. R. Klatser. 1989. Polymerase chain reaction for the detection of *Mycobacterium leprae*. J. Gen. Microbiol. 135:2357–2364.

Heitkamp, M. A., and C. E. Cerniglia. 1989. Polycyclic aromatic hydrocarbon degradation by a *Mycobacterium* sp. in microcosms containing sediment and water from a pristine ecosystem. Appl. Environ. Microbiol. 55:1968–1973.

Heitkamp, M. A., W. Franklin, and C. E. Cerniglia. 1988a. Microbial metabolism of polycyclic aromatic hydrocarbons: isolation and characterization of a pyrene-degrading bacterium. Appl. Environ. Microbiol. 54:2549–2555.

Heitkamp, M. A., J. P. Freeman, D. W. Miller, and C. E. Cerniglia. 1988b. Pyrene degradation by a *Mycobacterium* sp.: Identification of ring oxidation and ring fission products. Appl. Environ. Microbiol. 54:2556–2565.

Hermans, J., J. G. Boschloo, and J. A. M. de Bont. 1990. Transformation of *Mycobacterium aurum* by electroporation: the use of glycine, lysozyme and isonicotinic acid hydrazide in enhancing transformation efficiency. FEMS Microbiol. Lett. 72:221–224.

Hermans, J., C. Martin, G. N. M. Huijberts, T. Goosen, and J. A. M. de Bont. 1991. Transformation of *Mycobacterium aurum* and *Mycobacterium smegmatis* with the broad-host-range Gram-negative cosmid vector pJRD125. Molec. Microbiol. (in press).

Hopwood, D. A., T. Kieser, M. J. Colston, and F. I. Lamb. 1988. Molecular biology of mycobacteria. Brit. Med. Bull. 44:528–546.

Hurley, S. S., G. A. Splitter, and R. A. Welch. 1988. Deoxyribonucleic acid relatedness of *Mycobacterium paratuberculosis* to other members of the family Mycobacteriaceae. Int. J. Syst. Bacteriol. 38:143–146.

Husson, R. N., B. E. James, and R. A. Young. 1990. Gene replacement and expression of foreign DNA in mycobacteria. J. Bacteriol. 172:519–524.

Imaeda, T. 1985. Deoxyribonucleic acid relatedness among selected strains of *Mycobacterium tuberculosis, Mycobacterium bovis, Mycobacterium bovis* BCG, *Mycobacterium microti,* and *Mycobacterium africanum.* Int. J. Syst. Bacteriol. 35:147–150.

Imaeda, T., G. Broslawski, and S. Imaeda. 1988. Genomic relatedness among mycobacterial species by nonisotopic blot hybridization. Int. J. Syst. Bacteriol. 38:151–156.

Jacobs, W. R., M. Tuckman, and B. R. Bloom. 1987. Introduction of foreign DNA into mycobacteria using a shuttle phasmid. Nature 327:532–534.

Jekkel, A., E. Csajági, E. Ilköy, and G. Ambrus. 1989. Genetic recombination by spheroplast fusion of sterol-transforming *Mycobacterium* strains. J. Gen. Microbiol. 135:1727–1733.

Jenkins, P. A., S. R. Pattyn, and F. Portaels. 1982. Diagnostic bacteriology, p. 441–470. In: C. Ratledge, and J. Stanford (ed.), The biology of the mycobacteria, vol. 1. Physiology, identification and classification. Academic Press, London.

Jones, R. J., and D. E. Jenkins. 1965. Mycobacteria isolated from soil. Can. J. Microbiol. 11:127–133.

Kamisango, K.-I., A. Dell, and C. E. Ballou. 1987. Biosynthesis of the mycobacterial *O*-methylglucose lipopolysaccharide. J. Bacteriol. 262:4580–4586.

Kamphuis, J., C. H. M. Schepers, W. H. J. Boesten, M. J. A. Roberts, P. J. H. Peters, J. A. M. van Balken, E. M. Meijer, and H. E. Schoemaker. 1987. Stereoselective enzymatic hydrolysis of α-disubstituted amino acid amides with an aminopeptidase from *Mycobacterium neoaurum*, p. 164. In: O. M. Neijssel, R. R. van der Meer, and K. Ch. A. M. Luyben (ed.), Proceedings of the 4th European Congress on Biotechnology, vol. 2. Elsevier Science Publishers, Amsterdam.

Kaneda, K., S. Imaizumi, S. Mizuno, T. Baba, M. Tsukamura, and I. Yano. 1988. Structure and molecular species composition of three homologous series of α-mycolic acids from *Mycobacterium* spp. J. Gen. Microbiol. 134:2213–2229.

Kato, N., N. Miyamoto, M. Shimao, and C. Sakazawa. 1988. 3-Hexulose phosphate synthase from a new facultative methylotroph *Mycobacterium gastri* MB19. Agric. Biol. Chem. 52:2659–2661.

Kazda, J. F. 1980. *Mycobacterium sphagni* sp. nov. Int. J. Syst. Bacteriol. 30:77–81.

Kazda, J. F. 1983. The principles of the ecology of mycobacteria, p. 323–415. In: C. Ratledge, and J. Stanford (ed.), The biology of the mycobacteria, vol. 2. Immunological and environmental aspects. Academic Press, London.

Kikuchi, S., and T. Kusaka. 1982. New malonyl-CoA-dependent fatty acid elongation system in *Mycobacterium smegmatis*. J. Biochem. 92:839–944.

Kikuchi, S., and T. Kusaka. 1986. Isolation and partial characterization of a very long-chain fatty acid desaturation system from the cytosol of *Mycobacterium smegmatis*. J. Biochem. 99:723–731.

Kikuchi, S., T. Takeuchi, M. Yasui, T. Kusaka, and P. E. Kolattukudy. 1989. A very long-chain fatty acid elongation system in *Mycobacterium avium* and a possible mode of action of isoniazid on the system. Agric. Biol. Chem. 53:1689–1698.

King, D. H., and J. J. Perry. 1975. The origin of fatty acids in the hydrocarbon-utilizing microorganism *Mycobacterium vaccae*. Can. J. Microbiol. 21:85–89.

Knapp, J. S., and V. R. Brown. 1988. Morpholine degradation. Int. Biodet. 24:299–306.

Koch, R. 1882. Die Aetiologie der Tuberkulose. Berl. Klin. Wochenschr. 19:221–230.

Konicek, J., M. Konickova-Radochova, G. Y. Daraselia, and M. Slosarek. 1988. Mycobacteria in the light of modern genetics development. Folia Microbiol. 33:71–19.

Kubica, G. P. 1978. Classification and nomenclature of the mycobacteria. Ann. Microbiol. (Inst. Pasteur) 129A:7–12.

Kubica, G. P., I. Baess, R. E. Gordon, P. A. Jenkins, J. B. G. Kwapinski, C. McDurmont, S. R. Pattyn, H. Saito, V. Silcox, J. L. Stanford, K. Takeya, and M. Tsukamura. 1972. A co-operative numerical analysis of the rapidly growing mycobacteria. J. Gen. Microbiol. 73:55–70.

Kubica, G. P., and R. C. Good. 1981. The genus *Mycobacterium* (except *M. leprae*), p. 1962–1984. In: M. P. Starr, H. Stolp, H. G. Trüper, A. Balows, and H. G. Schlegel (ed.), The prokaryotes: a handbook of habitats, isolation and identification of bacteria, vol. 2. Springer Verlag, Berlin.

Kusaka, T., and T. Mori. 1986. Pyrolysis gas chromatography-mass spectrometry of mycobacterial mycolic acid methyl esters and its application to the identification of *Mycobacterium leprae*. J. Gen. Microbiol. 132:3403–3406.

Labidi, A., C. Dauguet, K. S. Goh, and H. L. David. 1984. Plasmid profiles of *Mycobacterium fortuitum* complex isolates. Curr. Microbiol. 11:235–340.

Laneelle, M.-A., C. Lacave, M. Daffé, and G. Lanéelle. 1988. Mycolic acids of *Mycobacterium aurum*. Eur. J. Biochem. 177:631–635.

Lechevalier, M. P., and H. A. Lechevalier. 1970. Chemical composition as a criterion in the classification of aerobic actinomycetes. Int. J. Syst. Bacteriol. 20:435–443.

Lehmann, K. B., and R. Neumann. 1896. Atlas und Grundriss der Bakteriologie und Lehrbuch der speziellen bakteriologischen Diagnostik. J. F. Lehmann, München.

Lévy-Frébault, V., M. Daffé, E. Restrepo, F. Grimont, P. A. D. Grimont, and H. L. David. 1986b. Differentiation of *Mycobacterium thermoresistible* from *Mycobacterium phlei* and other rapidly growing mycobacteria. Ann. Microbiol. (Paris) 137A:143–151.

Lévy-Frébault, V., F. Grimont, P. A. D. Grimont, and H. L. David. 1984. Deoxyribonucleic acid relatedness study of *Mycobacterium fallax.* Int. J. Syst. Bacteriol. 34:423–425.

Lévy-Frébault, V., F. Grimont, P. A. D. Grimont, and H. L. David. 1986a. Deoxyribonucleic acid relatedness study of the *Mycobacterium fortuitum-Mycobacterium chelonae* complex. Int. J. Syst. Bacteriol. 36:458–460.

Lévy-Frébault, V., E. Rafidinarivo, J.-C. Promé. J. Grandry, H. Boisvert, and H. L. David. 1983. *Mycobacterium fallax* sp. nov. Int. J. Syst. Bacteriol. 33:336–343.

Linko, Y.-Y., and P. Linko. 1985. Immobilized biocatalysts in organic synthesis and chemical production, p. 159–178. In: J. Tramper. H. C. van der Plas, and P. Linko (ed.), Biocatalysis in organic syntheses. Elsevier Science Publishers, Amsterdam.

Lowrie, D. B., V. R. Aber, and P. S. Jackett. 1979. Phagosome-lysosome fusion and cyclic adenosine 3':5'-monophosphate in macrophages infected with *Mycobacterium microti, Mycobacterium bovis* BCG or *Mycobacterium lepraemurium.* J. Gen. Microbiol. 110:431–441.

Lukins, H. B., and J. W. Foster. 1963. Utilization of hydrocarbons and hydrogen by mycobacteria. Z. Allg. Mikrobiol. 3:251–264.

Luquin, M., L. Margarit, M. J. Condom, and V. Ausina. 1987. Mycolic acids of *Mycobacterium porcinum.* Int. J. Syst. Bacteriol. 37:75–77.

Mamadou, D., M.-A. Dupont, and N. Gas. 1989. The cell envelope of *Mycobacterium smegmatis:* cytochemistry and architectural implications. FEMS Microbiol. Lett. 61:89–94.

Martin, C. K. A. 1984. Sterols, p. 79–95. In: K. Kieslich (ed.), Biotechnology, vol. 6A. Biotransformations. Chemie Verlag, Weinheim.

Masood, R., Y. K. Sharma, and T. A. Venkitasubramanian. 1985. Metabolism of mycobacteria. J. Biosci. 7:421–431.

McFadden, J. J., Z. Kunze, and P. Seechurn. 1990. DNA probes for detection and identification, p. 139–172. In: J. J. McFadden (ed.), Molecular biology of the mycobacteria. Academic Press, London.

Meijer, E. M., W. H. J. Boesten, H. E. Schoemaker, and J. A. M. van Balken. 1985. Use of biocatalysts in the industrial production of specialty chemicals, p. 135–156. In: J. Tramper. H. C. van der Plas, and P. Linko (ed.), Biocatalysis in organic syntheses. Elsevier Science Publishers, Amsterdam.

Meissner, P. S., and J. O. Falkinham III. 1984. Plasmid-encoded mercuric reductase in *Mycobacterium scrofulaceum.* J. Bacteriol. 157:669–672.

Minnikin, D. E. 1982. Lipids: Complex lipids, their chemistry, biosynthesis, and roles, p. 95–184. In: C. Ratledge, and J. Stanford (ed.), The biology of the mycobacteria, vol. 1. Physiology, identification and classification. Academic Press, London.

Minnikin, D. E., and M. Goodfellow. 1980. Lipid composition in the classification and identification of acid-fast bacteria, p. 189–256. In: M. Goodfellow, and R. G. Board (ed.), Microbiological classification and identification. Academic Press, London.

Minnikin, D. E., I. G. Hutchinson, A. B. Caldicott, and M. Goodfellow. 1980. Thin-layer chromatography of methanolysates of mycolic acid-containing bacteria. J. Chromatogr. 188:221–233.

Minnikin, D. E., S. M. Minnikin, I. G. Hutchinson, M. Goodfellow, and J. M. Grange. 1984a. Mycolic acid patterns of representative strains of *Mycobacterium fortuitum, 'Mycobacterium peregrinum'* and *Mycobacterium smegmatis.* J. Gen. Microbiol. 130:363–367.

Minnikin, D. E., S. M. Minnikin, J. H. Parlett, M. Goodfellow, and M. Magnusson. 1984b. Mycolic acid patterns of some species of *Mycobacterium.* Arch. Microbiol. 139:225–231.

Minnikin, D. E., S. M. Minnikin, J. H. Parlett, and M. Goodfellow. 1985. Mycolic acid patterns of some rapidly-growing species of *Mycobacterium.* Zentralbl. Bakteriol. Mikrobiol. Hyg., Ser A 259:446–460.

Minnikin D. E., and A. G. O'Donnell. 1984. Actinomycete envelope lipid and peptidoglycan composition, p. 337–388. In: M. Goodfellow, M. Mordarski, and S. T. Williams (ed.), The biology of the actinomycetes. Academic Press, London.

Moore, W. E. C., E. P. Cato, and L. V. H. Moore. 1985. Index of the bacterial and yeast nomenclature changes published in the international journal of systematic bacteriology since the 1980 approved lists of bacterial names (1 January 1980 to 1 January 1985). Int. J. Syst. Bacteriol. 35:382–407.

Odriozola, J. M., J. A. Ramos, and K. Bloch. 1977. Fatty acid synthetase activity in *Mycobacterium smegmatis*—characterization of the acyl carrier protein-dependent elongating system. Biochim. Biophys. Acta 488:207–217.

Padgitt, P. J., and S. E. Moshier. 1987. *Mycobacterium poriferae* sp. nov., a scotochromogenic, rapidly growing species isolated from a marine sponge. Int. J. Syst. Bacteriol. 37:186–191.

Perry, J. J. 1980. Propane utilization by microorganisms. Adv. Appl. Microbiol. 26:89–115.

Portaels, F., A. de Muynck, and M. P. Sylla. 1988. Selective isolation of mycobacteria from soil: a statistical analysis approach. J. Gen. Microbiol. 134:849–855.

Ramakrishnan, T., P. Suryanarayana Murthy, and K. P. Gopinathan. 1972. Intermediary metabolism of mycobacteria. Bacteriol. Rev. 36:65–108.

Ranes, M. G., J. Rauzier, M. Lagranderie, M. Gheorghiu, and B. Gicquel. 1990. Functional analysis of pAL5000, a plasmid from *Mycobacterium fortuitum:* construction of a "mini" *Mycobacterium-Escherichia coli* shuttle vector. J. Bacteriol. 172:2793–2797.

Rastogi, N., C. Frehel, and H. L. David. 1986. Triple-layered structure of mycobacterial cell wall: evidence for the existence of a polysaccharide-rich outer layer in 18 mycobacterial species. Curr. Microbiol. 13:237–242.

Ratledge, C. 1976. The physiology of the mycobacteria. Adv. Microbiol. Physiol. 13:115–244.

Ratledge, C. 1982a. Lipids: cell composition, fatty acid bio-syntheses, p. 53–93. In: C. Ratledge, and J. Stanford (ed.), The biology of the mycobacteria, vol. 1. Physiology, identification and classification. Academic Press, London.

Ratledge, C. 1982b. Nutrition, growth and metabolism, p. 185–271. In: C. Ratledge, and J. Stanford (ed.), The biology of the mycobacteria, vol. 1. Physiology, identification and classification. Academic Press, London.

Ratledge, C. 1984. Metabolism of iron and other metals by mycobacteria, p. 603–627. In: G. P. Kubica, and L. G. Wayne (ed.), The mycobacteria: a source book, part A. Marcel Dekker, New York.

Rauzier, J., J. Moniz-Pereira, J., and B. Gicquel-Sanzey. 1988. Complete nucleotide sequence of pAL5000, a plasmid from Mycobacterium fortuitum. Gene 71:315–321.

Rezessy-Szabó, J. M., G. N. M. Huijberts, and J. A. M. de Bont. 1986. Potential of organic solvents in cultivating micro-organisms on toxic water-insoluble compounds, p. 295–302. In: C. Laane, J. Tramper. and M. D. Lilly (ed.), Biocatalysis in organic media. Elsevier Science Publishers, Amsterdam.

Ridgway, H. F., M. G. Rigby, and D. G. Argo. 1984. Adhesion of a Mycobacterium sp. to cellulose diacetate membranes used in reverse osmosis. Appl. Environ. Microbiol. 47:61–67.

Rogall, T., T. Flohr, and C. Böttger. 1990. Differentiation of Mycobacterium species by direct sequencing of amplified DNA. J. Gen. Microbiol. 136:1915–1920.

Saito, H., R. E. Gordon, I. Juhlin, W. Käppler, J. B. G. Kwapinski, C. McDurmont, S. R. Pattyn, E. H. Runyon, J. L. Stanford, I. Tarnok, H. Tasaka, M. Tsukamura, and J. Weiszfeiler. 1977. Cooperative numerical analysis of rapidly growing mycobacteria. Int. J. Syst. Bacteriol. 27:75–85.

Shimakata, T., Y. Fujita, and T. Kusaka. 1977. Acetyl-CoA-dependent elongation of fatty acids in Mycobacterium smegmatis. J. Biochem. 82:725–732.

Shimakata, T., Y. Fujita, and T. Kusaka. 1979. Purification and characterization of 3-hydroxyacyl-CoA dehydrogenase from Mycobacterium smegmatis. J. Biochem. 86:1191–1198.

Skerman, V. B. D., V. McGowan, and P. H. A. Sneath. 1980. Approved lists of bacterial names. Int. J. Syst. Bacteriol. 30:225–420.

Snapper, S. B., B. R. Bloom, and W. R. Jacobs Jr. 1990a. Molecular genetic approaches to mycobacterial investigation, p. 199–218. In: J. J. McFadden (ed.), Molecular biology of the mycobacteria. Academic Press, London.

Snapper, S. B., L. Lugosi, A. Jekkel, R. E. Melton, T. Kieser, B. R. Bloom, and W. R. Jacobs, Jr. 1988. Lysogeny and transformation of mycobacteria: Stable expression of foreign genes. Proc. Natl. Acad. Sci. USA 85:6987–6991.

Snapper, S. B., R. E. Melton, S. Mustafa, T. Kieser, and W. R. Jacobs Jr. 1990b. Isolation and characterization of efficient plasmid transformation mutants of Mycobacterium smegmatis. Molec. Microbiol. 4:1911–1919.

Söhngen, N. L. 1913. Benzin, Petroleum, Paraffinöl und Paraffin als Kohlenstoff- und Energiequelle für Mikroben. Zentralblatt für Bakteriologische Parasitenkunde, Abt. II 37:595–609.

Songer, J. G. 1981. Methods for selective isolation of mycobacteria from the environment. Can. J. Microbiol. 27:1–7.

Sritharan, M., and C. Ratledge. 1989. Co-ordinated expression of the components of iron transport (mycobactin, exochelin and envelope proteins) in Mycobacterium smegmatis. FEMS Microbiol. Lett. 60:183–186.

Stahl, D. A., and J. W. Urbance. 1990. The division between fast- and slow-growing species corresponds to natural relationships among the mycobacteria. J. Bacteriol. 172:116–124.

Steinert, H.-J., K. D. Vorlop, and J. Klein. 1986. Steroid side chain cleavage with immobilized living cells in organic solvents, p. 51–63. In: C. Laane, J. Tramper, and M. D. Lilly (ed.), Biocatalysis in organic media. Elsevier Science Publishers, Amsterdam.

Suzuki, Y., A. Nagata, Y. Ono, and T. Yamada. 1988. Complete nucleotide sequence of the 16S rRNA gene of Mycobacterium bovis BCG. J. Bacteriol. 170:2886–2889.

Suzuki, Y., K. Yoshinaga, Y. Ono, A. Nagata, and T. Yamada. 1987. Organization of rRNA genes in Mycobacterium bovis BCG. J. Bacteriol. 169:839–843.

Taylor, D. G., P. W. Trudgill, R. E. Cripps, and P. R. Harris. 1980. The microbial metabolism of acetone. J. Gen. Microbiol. 118:159–170.

Tsukamura, M. 1981. Numerical analysis of rapidly growing, nonphotochromogenic mycobacteria, including Mycobacterium agri (Tsukamura 1972) Tsukamura sp. nov., nom. rev. Int. J. Syst. Bacteriol. 31:247–258.

Tsukamura, M. 1982. Mycobacterium shimoidei sp. nov., nom. rev., a lung pathogen. Int. J. Syst. Bacteriol. 32:67–69.

Tsukamura, M. 1984. The "non-pathogenic" mycobacteria—Their distribution and ecology in non-living reservoirs, p. 1339–1359. In: G. P. Kubica, and L. G. Wayne (ed.), The mycobacteria: a source book, part B. Marcel Dekker, New York.

Tsukamura, M., and S. Ichiyama. 1986. Numerical classification of rapidly growing nonphotochromogenic mycobacteria. Microbiol. Immunol. 30:863–882.

Tsukamura, M., S. Mizuno, and S. Tsukamura. 1981. Numerical analysis of rapidly growing, scotochromogenic mycobacteria, including Mycobacterium obuense sp. nov., nom. rev., Mycobacterium rhodesiae, sp. nov., nom. rev., Mycobacterium aichiense sp. nov., nom. rev., Mycobacterium chubuense, sp. nov., nom. rev., and Mycobacterium tokaiense sp. nov., nom. rev. Int. J. Syst. Bacteriol. 31:263–275.

Tsukamura, M., S. Mizuno, and H. Toyama. 1983a. Mycobacterium pulveris sp. nov., a nonphotochromogenic Mycobacterium with an intermediate growth rate. Int. J. Syst. Bacteriol. 33:811–815.

Tsukamura, M., H. Nemoto, and H. Yugi. 1983b. Mycobacterium porcinum sp. nov., a porcine pathogen. Int. J. Syst. Bacteriol. 33:162–165.

Tsukamura, M., H. J. van der Meulen, and W. O. K. Grabow. 1983c. Numerical taxonomy of rapidly growing, scotochromogenic mycobacteria of the Mycobacterium parafortuitum complex: Mycobacterium austroafricanum sp. nov. and Mycobacterium diernhoferi sp. nov., nom. rev. Int. J. Syst. Bacteriol. 33:460–469.

Tsukamura, M., I. Yano, and T. Imaeda. 1986. Mycobacterium moriokaense sp. nov., a rapidly growing, nonphotochromogenic Mycobacterium. Int. J. Syst. Bacteriol. 36:333–338.

Urakami, T., and I. Yano. 1989. Methanol-utilizing *Mycobacterium* strains isolated from soil. J. Gen. Appl. Microbiol. 35:125–133.

Valero-Guillén, P. L., and F. Martín-Luengo. 1986. 1-Tetradecanol, a new alcohol found in the cell wall of some rapidly growing chromogenic mycobacteria. FEMS Microbiol. Lett. 35:59–63.

van Ginkel, C. G., A. Q. H. Habets-Crützen, A. R. M. van der Last and J. A. M. de Bont. 1987b. A description of microbial growth on gaseous alkenes in a chemostat culture. Biotechnol. Bioeng. 30:799–804.

van Ginkel, C. G., H. G. J. Welten, and J. A. M. de Bont. 1987a. Oxidation of gaseous and volatile hydrocarbons by selected alkene-utilizing bacteria. Appl. Environ. Microbiol. 53:2903–2907.

van Ginkel, C. G., H. G. J. Welten, J. A. M. de Bont, and H. A. M. Boerrigter. 1986. Removal of ethene to very low concentrations by immobilized *Mycobacterium* E3. J. Chem. Technol. Biotechnol. 36:593–598.

van Loosdrecht, M. C. M., J. Lyklema, W. Norde, G. Schraa, and A. J. B. Zehnder. 1987. The role of bacterial cell wall hydrophobicity in adhesion. Appl. Environ. Microbiol. 53:1893–1897.

Vestal, A. L. 1975. Procedures for the isolation and identification of mycobacteria. U.S. Department of Health, Education and Welfare, Publication no. (CDC) 76–8230. Center for Disease Control, Atlanta.

Viallier, J., and G. Viallier. 1973. Inventaire des mycobacteries de la nature. Ann. Soc. Belge Med. Trop. 53:361–371.

Vishniac, W., and M. Santer. 1957. The thiobacilli. Bacteriol. Rev. 21:195–213.

Wayne, L. G. 1984. Mycobacterial speciation, p. 25–65. In: G. P. Kubica, and L. G. Wayne (ed.), The mycobacteria: a source book, part A. Marcel Dekker, New York.

Wayne, L. G., E. J. Krichevsky, L. L. Love, R. Johnson, and M. I. Krichevsky. 1980. Taxonomic probability matrix for use with slowly growing mycobacteria. Int. J. Syst. Bacteriol. 30:528–538.

Wayne, L. G., and G. P. Kubica. 1986. Genus *Mycobacterium* Lehmann and Neumann 1896, 363, p. 1436–1457. In: P. H. A. Sneath, N. S. Mair, M. E. Sharpe, and J. G. Holt (ed.), Bergey's manual of systematic bacteriology, vol. 2. The Williams and Wilkins Co., Baltimore.

Weijers, C. A. G. M., C. G. van Ginkel, and J. A. M. de Bont. 1988. Enantiomeric composition of lower epoxyalkanes produced by methane-, alkane- and alkene-utilizing bacteria. Enzyme Microbiol. Technol. 10:214–218.

Weisman, L. S., and C. E. Ballou. 1984a. Biosynthesis of the mycobacterial methylmannose polysaccharide, Identification of an $\alpha 1 \rightarrow 4$-mannosyltransferase. J. Biol. Chem. 259:3457–3463.

Weisman, L. S., and C. E. Ballou. 1984b. Biosynthesis of the mycobacterial methylmannose polysaccharide, Identification of a 3-*O*-methyltransferase. J. Biol. Chem. 259:3464–3469.

Wiegant, W. M., and J. A. M. de Bont. 1980. A new route for ethylene glycol metabolism in *Mycobacterium* E44. J. Gen. Microbiol. 120:325–331.

Winder, F. G. 1982. Mode of action of the antimycobacterial agents and associated aspects of the molecular biology of the mycobacteria, p. 353–438. In: C. Ratledge, and J. Stanford (ed.), The biology of the mycobacteria, vol. 1. Physiology, identification and classification. Academic Press, London.

Woese, C. R. 1987. Bacterial evolution. Microbiol. Rev. 51:221–271.

Wovcha, M. G., F. J. Antosz, J. C. Knight, L. A. Kominek, and T. R. Pyke. 1978. Bioconversion of sitosterol to useful steroidal intermediates by mutants of *Mycobacterium fortuitum*. Biochim. Biophys. Acta 531:308–321.

Zainuddin, Z. F., and J. W. Dale. 1989. Polymorphic repetitive DNA sequences in *Mycobacterium tuberculosis* detected with a gene probe from a *Mycobacterium fortuitum* plasmid. J. Gen. Microbiol. 135:2347–2355.

CHAPTER 53

The Genus *Mycobacterium*—Medical

ROBERT C. GOOD

For many years, the taxonomy of the family Mycobacteriaceae changed very little, but in the 1970s and 1980s, application of the techniques of Adamsonian classification and DNA technology provided better methods for recognition and classification of the many species within the single genus *Mycobacterium*. Tuberculosis and leprosy, the most significant diseases caused by species in the genus, have been recognized throughout recorded history. Leprosy and *Mycobacterium leprae* are discussed in Chapter 54; therefore, they are considered only briefly here. Tuberculosis is still a common disease throughout the world and is discussed in detail in this chapter, which is an extension of material presented in the first edition of *The Prokaryotes* (Kubica and Good, 1981).

Fifty-four species make up the genus *Mycobacterium* as it appears in the current edition of *Bergey's Manual of Systematic Bacteriology* (Wayne and Kubica, 1986). Of the species of medical importance, there are 10 strictly human pathogens, i.e., species that have been isolated only from human specimens but not from environmental sources (Table 1), and an additional nine species are considered to be opportunistic pathogens. The six described species of animal pathogens are also listed in Table 1.

Mycobacterium tuberculosis, often called the tubercle bacillus, was stained in infected tissue and cultured on an inspissated serum medium in the classic studies of Koch (1882); he was able to isolate the bacterium on a laboratory medium, to produce characteristic disease in experimental animals by the injection of bacilli from a pure culture, to observe the bacilli in infected tissues, and to recover the inoculated bacterium from experimentally infected animals by culture. These observations linking the etiologic agent to an infectious disease are referred to as Koch's postulates; they have formed the basis of strategies for the treatment and control of tuberculosis and have been widely accepted for defining the etiology of all infectious diseases.

Table 1. *Mycobacterium* species associated with human and animal diseases.

Strict pathogens[a]	Nonphotochromogens
M. africanum[b]	*M. avium*[b]
M. asiaticum[c]	*M. intracellulare*[b]
M. bovis[b]	Rapid growers
M. haemophilum[d]	*M. fortuitum*[b]
M. leprae[b]	*M. chelonae*[b]
M. malmoense[d]	Animal pathogens—slow growing
M. shimoidei[d]	*M. farcinogenes*
M. simiae[b,c]	*M. lepraemurium*
M. tuberculosis[b]	*M. microti*
M. ulcerans[b]	*M. paratuberculosis*
Photochromogens	Animal pathogens—fast growing
M. kansasii[b]	*M. porcinum*
M. marinum[b]	*M. senegalense*
Scotochromogens	
M. scrofulaceum[b]	
M. szulgai[b]	
M. xenopi[b]	

[a]These species are not associated with an environmental source.
[b]Classified as a pathogen by Runyon (1974).
[c]Species usually classified as a photochromogen.
[d]Species usually classified as a nonphotochromogen.

Koch (1890) found that subcutaneous inoculation of normal guinea pigs with tubercle bacilli resulted in slow development of an ulcerated lesion that did not heal. On the other hand, when the bacilli were injected subcutaneously into previously infected guinea pigs, a localized lesion rapidly developed, ulcerated, and then resolved. This reaction, which can also be induced by the injection of cell-free culture filtrates, is called the "Koch phenomenon" and is an expression of delayed-type hypersensitivity that develops as the result of the primary infection.

Early investigators considered the similar human and bovine infections to be of common etiology. However, Smith (1898) was able to identify two separate species on the basis of pathogenicity for animals. Strains (*M. tuberculosis*) isolated from humans produced progressive disease when injected into guinea pigs, but only limited disease when injected into rabbits and cattle. On the other hand, strains *(M. bovis)* of bovine origin produced progressive disease in all three animal species. In further studies, Smith (1904–1905) differentiated the bacterial species in vitro on the basis of the final acidity achieved in broth cultures. This report provided the first evidence that mycobacteria gave reproducible responses under in vitro culture conditions, and refinements in procedures have led to the development of standardized, highly reproducible tests for the recognition of species of *Mycobacterium* (Wayne et al., 1974, 1976). Development of the complex media required for the growth of *M. tuberculosis* has had a major impact on procedures in the clinical laboratory. Until recently, laboratory confirmation of a clinical diagnosis of tuberculosis by culture techniques still required from 4 to 8 weeks.

Bacteria similar to tubercle bacilli were found in diseases of animals in the three decades following isolation of the tubercle bacillus. *M. avium* was associated with a tuberculosis-like disease in fowl by Strauss and Gameleïa (1891); *M. paratuberculosis* was isolated from cattle and sheep with chronic enteritis by Johne and Frothingham (1895); *M. lepraemurium* was isolated from rats by Dean (1903), Rabinowitsch (1903), and Stefansky (1903). Since these early descriptions, other species of acid-fast bacilli have been isolated and characterized from human and animal diseases as well as from environmental sources. Characteristics of these strains are included in the description of currently accepted species in *Bergey's Manual* (Wayne and Kubica, 1986).

All species in the genus *Mycobacterium* are acid-fast, i.e., once stained with a basic dye such as fuchsin, they resist decolorization with mineral acids or with acidified organic solvents. Lipids in the cell wall limit the movement of dye so that heat or other agents must be used in procedures to stain the bacilli (Goren et al., 1978). When stained by an acid-fast procedure, the stain is taken up irregularly in most instances, so that the cells appear beaded or granular with heavily stained areas separated by nonstained segments. Barksdale and Kim (1977) review the theoretical and practical details of the acid-fast characteristic and discuss the selection of dyes for light and fluorescent microscopy. Practical application and standardization of acid-fast staining and microscopy have been described by Smithwick (1976). The Gram stain does not include special agents or involve methods to allow dyes to easily penetrate mycobacterial cells, so tubercle bacilli appear to be only weakly Gram positive.

Mycobacterial Diseases

Tuberculosis

ETIOLOGY. *M. tuberculosis, M. bovis, M. africanum,* and *M. microti* are all grouped into the "*M. tuberculosis* complex," on the basis of studies of DNA homology (Baess, 1979; Imaeda, 1985). However, they are discussed here as if they were individual species. *M. microti,* the vole bacillus, is of such limited pathogenicity for humans that it has been used as a vaccine against tuberculosis. The other three species in the complex do cause human disease, tuberculosis. Differentiation of the species is important for epidemiologic purposes, but all three species that cause tuberculosis are significant to public health control procedures. Confirmation of the clinical diagnosis of tuberculosis is dependent on the isolation and identification of *M. tuberculosis* from an appropriate specimen.

EPIDEMIOLOGY. Of all of the species of *Mycobacterium, M. tuberculosis* is encountered most frequently in the laboratory (Good, 1980; Good and Snider, 1982) and is the cultivable species most often responsible for disease. Despite many studies aimed at determining optimal chemotherapy, preventive therapy, use of vaccination, control measures, and the enhancement of public awareness, it is estimated that each year throughout the world, 8 million new cases occur and 3 million persons die of tuberculosis (Styblo and Rouillon, 1981). This disease has affected all segments of the population throughout history and has exacted a high toll. It has been quite accurately called "the white plague" (Dubos and Dubos, 1952) because of the human suffering it has caused.

From the reports published by Frost (1937) and Styblo (1980), Murray et al. (1990) conclude that mortality due to tuberculosis has declined steadily in developed countries since at least the turn of the century. However, they present data showing that the estimated incidence of all forms of tuberculosis in 1990 ranges from 120 to 229 cases per 100,000 population in developing countries. Data in this study suggest that more than 2.5 million people in developing countries die from tuberculosis each year, probably more than from any other single infectious agent. In addition, the annual incidence of new cases of all forms of tuberculosis is in excess of 7.1 million in the developing world. All age groups can be afflicted, but the greatest incidence and mortality are in the adult population (15 to 59 years of age). Whereas new cases in developed countries occur largely through activation of an earlier infection (and in an over-50 age group), cases in developing countries occur as the result of person-to-person contact in a younger age group. However, populations in both developing and developed countries are subject to increasing rates of infection with the human immunodeficiency virus (HIV) and subsequent development of acquired immunodeficiency syndrome (AIDS), which increases susceptibility to tuberculosis (Chretien, 1990). Certain behaviors, such as intravenous drug use, also lead to an increased risk of acquisition of tuberculosis.

When the use of effective chemotherapy became widespread in the United States in 1953, there were 84,304 cases of tuberculosis reported annually. In 1987, 22,517 cases were reported. During this same period, the case rate had dropped from 53.0 to 9.3 per 100,000 population (Bloch et al., 1989). However, the downward trend occurring between 1953 and 1984 changed markedly in 1985 when the number of reported cases was only 0.2% below that for 1984, and the numbers in 1986 and 1987 increased over those for 1984. Using the average annual decline of 6.7% for 1981 through 1984, Bloch et al. (1989) and Rieder et al. (1989) estimated that 9,226 excess cases had occurred though 1987. Greater increases in cases occurred in males between 15 and 44 years of age who were black or Hispanic or were from Puerto Rico or a foreign country. HIV-positive persons have a high risk of developing tuberculosis (Chaisson et al. 1987; Colebunders et al., 1989; Mann et al., 1986; Pitchenik et al., 1984, 1988; Sunderam et al., 1986). Rieder et al. (1989) attribute the lack of a continuing fall in the reported number of new cases of tuberculosis in the United States to the immunosuppressive effects of HIV infection. This can then lead to activation of a site of dormant tuberculous infection or make the person more susceptible to primary infection (see below).

DISEASE PROGRESSION. *M. tuberculosis* is spread from person to person by droplet nuclei, i.e., by droplets of 1 to 5 μm, which are generated during coughing, singing, sneezing, or similar activities where breath is expelled rapidly by a person with pulmonary tuberculosis. Because of their small size, droplets may be suspended for some time and constitute a continuing hazard. Control of the biohazard posed by aerosols involves use of high-efficiency filters, irradiation of room air with ultraviolet light at a germicidal wave length, dilution of the bacilli in air, and directing the air exhaust from a contaminated area away from occupied space.

Tubercle bacilli are inhaled, reach the alveoli, and are retained. As described by Youmans (1979), the bacilli are phagocytized, and then multiply slowly in neutrophils and alveolar macrophages. An inflammatory cellular reaction develops that does not significantly limit multiplication of the bacilli, which disseminate via the lymphatics to the blood and then to other organs. The lungs are also reinoculated, but in immunocompetent persons, the bacilli are controlled, i.e., do not continue to multiply, so that the remaining type of lesion is a small calcified nodule and hilar lymph nodes are enlarged. This residuum of infection is called the Ghon complex. Two results of infection can be detected at this time: hypersensitivity to tuberculin and inhibition of intracellular multiplication of bacilli in macrophages.

Youmans (1979) describes reinfection tuberculosis, whether initiated by reinoculation from an active case or by activation of an old infection, as being a circumscribed lesion with evidence of necrosis. Development of necrosis in the lesion is the result of hypersensitivity to tuberculin that developed in the course of the primary infection. The pathogenesis of tuberculosis is described by Dannenberg (1980, 1982), who has detailed the contribution of the immune reaction to progress of the disease. The relationship between delayed-type hypersensitivity and cell-mediated immunity fluctuates during different stages of infection and disease so that when low levels of antigen are present, macrophages accumulate, become activated, and destroy the bacilli, but when high levels of antigen are present as in progressive disease, tissue necrosis occurs (Dannenberg, 1989). Des Prez and Heim (1990) describe the characteristics of tuberculosis as it occurs in different age groups and in various tissues and organs.

TUBERCULIN HYPERSENSITIVITY. Delayed-type hypersensitivity develops following infection with tubercle bacilli. In this reaction, an area of induration reaches maximum diameter 48 to 72h following the intradermal injection of tuberculin. "Old tuberculin" (OT) is a concentrated culture filtrate of *M. tuberculosis* that has been heat sterilized and preserved with glycerol. A purified protein derivative (PPD), that was free of sensitizing proteins, was prepared by Seibert (1934) for use in place of the complex culture filtrate mixture. A standard preparation (PPD-S), first made available in 1941 (Seibert and Glenn, 1941), is still being used to standardize commercial preparations of tuberculin. The injection of 0.1 μg of PPD-S produces a 12-mm (mean) zone of erythema in sensitized guinea pigs, which is the standard delayed-type-hypersensitivity reaction in this species. Unfortunately, this well-characterized supply of PPD-S that has been used in so many studies is getting low, but a new standard is being prepared.

The standard test dose to determine human sensitivity to tuberculoprotein is five tuberculin units (TU; one TU is equivalent to the biological activity of 0.00002 mg of PPD-S) inoculated intradermally, usually into the volar aspect of the forearm. This is the standard Mantoux test procedure.

The multiple puncture test is an alternative procedure using a device with four or more prongs that have been coated either with OT or with Tuberculin, PPD. The device is pressed into the skin of the forearm so that the tines penetrate the skin layer. However, the amount of protein on the tines varies and release from the tines into the tissue is not uniform. Therefore, a standard dose of antigen is not delivered (Reichman, 1979).

Hypersensitivity can be detected 4 to 6 weeks after infection with tubercle bacilli, and the reaction that develops is influenced by a number of factors (Snider, 1982), including the method of test administration (Chaparas et al., 1985), co-infection with viral agents (Mori and Shiozawa, 1985), age and immune status of the patient (Capewell and Leitch, 1986; Perez-Stable et al., 1988), and medication (Hsu, 1983; Sipka et al., 1983; Tager et al., 1985). The distribution of reaction sizes elicited in culture-positive persons by PPD-S produces a symmetric curve with a mean value of 16 mm.

The site of injection of tuberculin (Mantoux test) is examined after 48 to 72 h under a good light. Only the zone of induration as identified by touch is measured in millimeters, and the area of erythema is ignored (American Thoracic Society, 1981b; Reichman, 1979). Individuals who are recent converters are those whose tuberculin skin test shows ≥10-mm increase within a 2-year period for those <35 years old and ≥15-mm increase for those ≥35 years old (Centers for Disease Control, 1990). If persons are exposed to a tuberculous patient, they are considered tuberculin positive when the reaction is ≥5 mm. Persons who reside in areas where infection with opportunistic *Mycobacterium* species is common are not considered positive unless the reaction is ≥15 mm.

Results of tuberculin testing are used as criteria to determine need for preventive therapy (CDC, 1990), since a positive tuberculin reaction indicates past or present infection with tubercle bacilli, but it is not prima facie evidence of active disease. Also, induction of hypersensitivity does not imply development of specific acquired resistance, i.e., protective immunity, to challenge infection (Orme, 1988; Youmans, 1979), but many investigators in the field believe hypersensitivity and protective immunity are related and tend to use the terms interchangeably. Tuberculin hypersensitivity, rather than protective immunity, is used routinely as a measure of the quality of vaccination with BCG, the attenuated strain of *M. bovis* accepted as a viable vaccine against tuberculosis (Centers for Disease Control, 1988). As stated above, infection with *Mycobacterium* species other than the *M. tuberculosis* complex can induce delayed-type hypersensitivity that will be elicited also by tuberculin (Edwards and Palmer, 1968; Palmer and Edwards, 1966, 1967), so the reaction resulting from the injection of tuberculin must be interpreted in light of all available clinical and epidemiologic data about the patient (American Thoracic Society, 1981a). PPDs prepared from other mycobacterial species were tested for their ability to elicit delayed-type hypersensitivity reactions in a study of naval recruits from all geographic areas of the United States (Palmer and Edwards, 1967); however, these results cannot be compared directly with the response to PPD-S since the former were standardized only on the basis of weight, not on biologic activity.

Hypersensitivity is used to identify newly infected persons, but this technique is complicated because of the "booster phenomenon" (Thompson et al., 1979). It is characterized as follows: the response to tuberculin in a person with mycobacterial infection may weaken over time, and even become negative, but it can be stimulated by introduction of the antigen, and the person may respond as positive when retested with PPD or OT after a week or longer. Therefore, testing adults in two steps can reduce the probability of misidentifying a person as uninfected.

Surveys of tuberculin hypersensitivity in residents in the United States have been summarized by Snider (1982), who reported that the specificity and sensitivity of the tuberculin test are 90%, and the positive predictive value in a population with high prevalence of tuberculosis is 0.99 while it is 0.08 in a population of low prevalence.

IMMUNOLOGY AND SEROLOGY. Mycobacterial antigens are often complexes consisting of proteins, lipids, and polysaccharides. In general, humoral responses to tuberculous infection and disease have not been consistent and a reliable serologic method for diagnosis of tuberculosis has not been developed. Identification of more specific antigens, improvement in methods for detection of antigen-antibody reaction, and more selective interpretation of the results suggest that soon serologic diagnosis may be successful (Daniel, 1988; Good, 1989; Ridell, 1988). The status of partially characterized antigens for the performance of serodiagnostic tests has been evaluated (Daniel and Debanne, 1987; Grange, 1984). Major problems in the interpretation of tests with such antigens result from rather poor sensitivity and specificity values. Some progress has been made in improving the reliability of the tests, primarily by using monoclonal antibodies directed against species-specific antigenic determinants (Ivanyi et al., 1988). Competition immunoassays have been quite reliable in smear-positive tuberculosis, but antibody levels in smear-negative, extrapulmonary, or childhood tuberculosis were too low to be used for a reliable diagnosis.

TUBERCULOSIS AND AIDS. HIV-induced immunosuppression produced by AIDS or HIV infection is a predisposing factor for the development of tuberculosis and other mycobacterial diseases (American Thoracic Society, 1987). Diagnosis of tuberculosis usually precedes diagnosis of AIDS, but this is not always the case. The clinical picture of tuberculosis is not typical in patients with AIDS, and, at presentation, infiltrates may be seen in any area of the lung, often associated with mediastinal or hilar lymphadenopathy or both. Pulmonary disease is often not cavitary and may not appear in the apical areas of the lung. Extrapulmonary disease is common, occurring in 60 to 70% of patients, and disseminated disease is quite frequently seen in patients with HIV infection. Modilevsky et al. (1989) reported that the diagnosis of tuberculosis preceded that of AIDS in 26 (67%) of 39 cases, while diagnosis of disease due to the *M. avium* complex (MAC) never preceded the recognition of AIDS in 55 cases. Pulmonary changes were seen in roentgenographs of 83% of 29 AIDS patients with pulmonary tuberculosis, compared with 25% of 28 patients with MAC. Sputum smears are more likely to be positive, and pleuritis and lymphadenitis are more likely to develop in patients with tuberculosis than in patients with disease due to MAC. Mycobacteremia is more frequent in infections with MAC than in tuberculosis.

Laboratory diagnosis of tuberculosis in patients with HIV infection requires that a number of specimens be examined including respiratory secretions, bronchial washings, lung tissue, pleural fluid, lymph node tissue, bone marrow, blood, urine, stool, brain biopsy tissues, and cerebrospinal fluid. Granulomas may not form in infected tissues, and acid-fast bacilli may not be seen on microscopic examination.

Due to immunosuppression, the tuberculin test may be interpreted as negative in a patient with disease. However, every person in whom the diagnosis of AIDS is suspected should be tuberculin tested, and specimens should be examined for the presence of mycobacteria (American Thoracic Society, 1987).

CHEMOTHERAPY. Adequate treatment of tuberculosis is dependent on the use of multiple drugs for an extended period of time. Routinely, specimens are taken for drug susceptibility tests, but empiric therapy is usually initiated before results of these tests are known. Most strains of *M. tuberculosis* in the United States are susceptible to the primary drugs (Good et al., 1985; Kopanoff et al., 1978). Therefore, the recommended treatment is a 6-month regimen that includes isoniazid (300 mg/kg maximal daily dose [MDD]), rifampin (600 mg/kg MDD), and pyrazinamide (20 to 30 mg/kg, but not to exceed 2 g MDD) given for 2 months followed by a maintenance dose of isoniazid and rifampin for 4 months (American Thoracic Society, 1986). Should resistance to isoniazid be suspected, ethambutol is included in the regimen, at least until the results of drug susceptibility tests are available. Therapy should be changed if results of susceptibility testing indicate resistance to one or more of the drugs. Selection of alternative therapy is based on drug susceptibility test results and clinical information about the patient.

The search for new drugs continues so that less toxic, more efficient regimens can be formulated, and dosage schedules can be reduced in length and complexity (O'Brien and Snider, 1985). Characteristics of the drugs with antituberculosis activity have been reviewed by Gangadharam and Iseman (1987), and mode of action can be used to formulate an active regimen. The quinolones have shown activity against a number of *Mycobacterium* species in vitro (Crowle et al., 1988; Gay et al., 1984; Grosset et al., 1988; Saito et al., 1985), and the studies of Tsukamura (1985) have indicated that ofloxacin is effective in treatment of tuberculosis.

Preventive therapy is used to hinder development of disease in infected persons, particularly persons who have just become tuberculin positive. The American Thoracic Society (1986) recommends treatment of such high-risk persons with the MDD of 300 mg of isoniazid for 6 to 12 months. During the course of this treatment, individuals must be carefully monitored for adverse effects. Rifampin is given in those instances where exposure has been to an isoniazid-resistant strain.

Other Mycobacterioses

Although *M. tuberculosis* has been the species most often associated with human disease in the past, the role of other mycobacteria in human disease was shown clearly in the 1950s (Buhler and Pollak, 1953; Crow et al., 1957; Timpe and Runyon, 1954). Timpe and Runyon (1954) found that isolates from 88 diseased patients could be differentiated from human and bovine tuberculosis strains and classified into three groups on the basis of characteristic colonial morphology and virulence for mice and guinea pigs. Runyon (1959, 1965) simplified classification of such isolates and placed them into one of four groups on the basis of two easily determined cultural characteristics: speed of growth and colony pigmentation. Therefore, these four groups are identified as slowly growing strains that are either photochromogenic, scotochromogenic, or nonphotochromogenic, or as rapidly growing strains. Recognition of these characteristics has made identification somewhat easier and has provided clearer differentiation of *M. tuberculosis* from other species of *Mycobacterium*.

Laboratory procedures are accurate and reproducible so that most mycobacterial isolates can be identified at the species level; however, several undesirable terms have crept into mycobacterial nomenclature. They were used in the period between 1955 and 1980 when identification tests for species differentiation were not well standardized. As early as 1974, Runyon (1974) recognized the problem with nomenclature and published guidelines for standard terminology. Therefore, *M. avium* complex (MAC) is now used to include both *M. avium* and *M. intracellulare,* and *M. fortuitum* complex (MFC) to include both *M. fortuitum* and *M. chelonae.* This agrees with Runyon's suggestion that each complex is made up of two or more species whose distinction is of no medical importance. With MAC species, the concept generally applies, but identification to species and subspecies level has become important in recent years as a way to follow the epidemiology of disease in special population groups and as a guide to appropriate therapy.

The term *M. avium-intracellulare* has no taxonomic status, so it and its acronym MAI should not be used. Also, "MAIS complex" was suggested by Hawkins (1977) to identify a group of pigmented strains with unusual biochemical reactions. The term is jargon in most of the laboratories that use it today since it is used to refer to either MAC or *M. scrofulaceum,* and they can be differentiated easily. As stated previously, the *M. tuberculosis* complex was recognized because four biotypes, previously considered to be distinct species, could not be differentiated on the basis of similarity of their deoxyribonucleic acids (DNA). However, recognition of the biotypes is important in many instances, particularly for determining epidemiology and evaluating public health significance. Since these biotypes can be differentiated phenotypically, the definition of species should be reexamined to consider results both of DNA homology and determination of phenotypic characteristics.

Characteristics of mycobacterioses associated with the various opportunistic species have been reviewed in some detail (Fadda, 1988; Good, 1985; Sanders and Horowitz, 1990; Wolinsky, 1979, 1984; Woods and Washington, 1987; Wayne, 1979). These diseases, with the exception of progressive infections with MAC, are discussed only briefly here.

Mycobacteriosis caused by species other than those in the *M. tuberculosis* complex is not the same disease as tuberculosis. The etiology of mycobacterioses cannot be determined on the basis of clinical signs, histologic testing, or even the way the bacilli look in culture. Since management of the cases will differ, depending on the etiologic agent, the *Mycobacterium* species causing disease must be identified. Since these are not reportable diseases, the magnitude of the problem due to any one species is difficult to determine.

A total of 5,469 patients with infections due to nontuberculous mycobacteria were identified in a national survey conducted for the years 1981 and 1982 (O'Brien et al., 1987). Isolates of MAC were most frequent, followed by *M. kansasii, M. fortuitum, M. scrofulaceum,* and *M. chelonae.* Since almost all of these organisms were isolated from sputum, temporary colonization cannot be ruled out; however, the investigators established rigid criteria to assure that the isolates were associated with progressive disease. While isolation of *M. tuberculosis* from a clinical specimen is diagnostic of disease, isolation of many other species cannot be accepted as an assignment of etiology since they are widespread in the environment and may represent contamination or temporary colonization. Strict diagnostic criteria must be applied to interpretation of these opportunistic mycobacterial pathogens from patient specimens. Criteria applied by the American Thoracic Society (1981a) include clinical evidence of disease of unknown etiology and either repeated isolation of the same *Mycobacterium* species, usually in large numbers, or isolation of the mycobacteria from a closed lesion that has been sampled using aseptic techniques. Diagnosis and management of cases due to *Mycobacterium* species other than *M. tuberculosis* are reviewed by Davidson (1989).

MYCOBACTERIUM AVIUM COMPLEX (MAC). *M. avium* and *M. intracellulare* cannot be easily differentiated on the basis of biochemical reactions, and often strains of *M. scrofulaceum* give similar patterns, particularly in those instances when the MAC isolates appear scotochromogenic. The MAC are grouped with nonphotochromogens, not because they are nonpigmented, but because induction of pigment is not influenced by exposure to light. Using methods detailed by Schaefer (1979), species in the complex have been identified by seroagglutination reactions (Schaefer, 1965; Wolinsky and Schaefer, 1973). Brennan (1984) reviewed studies of the typing antigens on the cell surface and concluded that those for MAC have a common lipopeptidyl-*O*-(3,4-di-*O*-methyl-rhamnose) core. In this formulation, the serologic specificity is conveyed by the outer one or two sugar residues, and the individual compounds responsible can be detected by thin-layer chromotography.

DNA homology studies have indicated that the assignment of MAC isolates to a species on the basis of the serovar may be incorrect (Baess, 1983). Serovars of MAC and their associations with species are shown in Table 2. Before this species relationship was known, Meissner and Anz (1977) and Jorgensen (1978) concluded that chickens and wild fowls were the natural reservoir of serovars 1, 2, and 3; that serovars 4, 6, and 8 through 11 were associated with insects and sawdust; and that serovars 7 and 12 through 21 were associated with environmental reservoirs. They also believed that serovars 1, 2, and 3 from fowls were a major source of human infection, but the same serotypes from swine and cattle were not. Serovar 8 was isolated from immunocompetent hosts most frequently in the period 1976 to mid-1978 by McClatchy (1981), and other serovars were 16, 4, 19, 9, 42, and 1 in decreasing order of frequency. For the period from 1979 through 1982, 448 isolates were serotyped at the Centers for Disease Control, and again serovar 8 was seen most often,

followed by serovars 1, 14, 4, 16, 9, 42, and 6 in order of frequency (Good, 1985).

As described by Wolinsky (1979, 1984), the average case of pulmonary disease due to *M. avium* occurs in a 48-year-old white male with long-standing chronic obstructive lung disease (COPD), silicosis, or a similar condition; there is not a history of tuberculosis in family members, and a source for tuberculosis is not identified among contacts; when first seen, the average patient has a 3-month history of increasingly productive cough, night sweats, low-grade fever, and moderate weight loss. Fibrosis and thin-walled cavities often can be seen in chest roentgenographs. Repeated cultures of sputum will be positive for acid-fast bacilli.

A different view is presented by a series of patients admitted to a large medical center over a 10-year period who did not have preexisting lung disease or immunosuppression, but whose disease was diagnosed as due to MAC (Prince et al., 1989). The diagnosis in 21 patients without predisposing factors was on the basis of pulmonary symptoms, abnormalities on chest roentgenographs, and in 14 patients, evidence of invasive disease in biopsy specimens. On presentation, most patients were females (86%) with a mean age of 66 years (range from 46 to 84 years) who did not report a history of fever or weight loss, but who did have a persistent cough of 25 weeks' duration with the production of purulent sputum (86%) before the correct diagnosis was made. Radiographs indicated slowly progressive nodular opacities in 71% of patients, but only 24% had cavitary disease.

The authors concluded from their study of this series of patients that pulmonary disease due to MAC can occur in persons without predisposing conditions, and that the disease may not be recognized for some time because of its slow progression. After recognition of disease in patients such as these, a reason for infection in a normal host must be considered, whether the source is unique or the infecting strain is of unusual virulence (Iseman, 1989). Since four of the patients in this study died of the MAC infection (Prince et al., 1989), the idea that chemotherapy is needed only in a few patients to control the indolent disease may be in error (Iseman, 1989).

MAC is the most frequent mycobacterial isolate from patients with AIDS (American Thoracic Society, 1987), and between 17% and 28% of patients are found to have disseminated infections before death (Macher, 1984; Murray et al., 1984). Only 37 cases of disseminated MAC infections were reported prior to the beginning of the AIDS epidemic (Horsburgh et al., 1985). The rate of disseminated infections with MAC

Table 2. Serotypes of *M. avium, M. intracellulare,* and *M. scrofulaceum.*

Species designation	Serotype number
M. avium	1 through 6; 8 through 11
M. intracellulare	7; 12 through 20
M. scrofulaceum	41 through 43
Not assigned to species	21 through 28

in patients with AIDS is 5.5% (Horsburgh and Selik, 1989). This value is based on primary surveillance data collected at the time AIDS was reported, so the actual number may be much larger, as indicated by reports describing later events in the course of the disease. Between 50% and 56% of AIDS patients have had MAC infection diagnosed from specimens taken at autopsy (Armstrong et al., 1985; Macher, 1984). Young et al. (1986) isolated mycobacteria from 104 of more than 300 AIDS patients at the University of California, Los Angeles (UCLA) from 1980 through 1985, and 89 were MAC, including those from four patients whose cultures also contained *M. gordonae* and from one patient who had a dual infection with *M. kansasii*. Only two patients' cultures were identified as *M. tuberculosis;* four were *M. kansasii,* four were *M. gordonae,* two were *M. fortuitum,* two were *M. smegmatis,* and one was *M. scrofulaceum.*

Isolates from 212 AIDS patients examined in a reference laboratory consisted of 177 MAC, 20 *M. tuberculosis,* 4 *M. scrofulaceum,* 2 *M. szulgai,* 2 *M. kansasii,* and 1 each of *M. asiaticum, M. flavescens, M. fortuitum, M. gordonae, M. malmoense,* an unidentified scotochromogen, and a rapid grower that was not in the *M. fortuitum* complex (Good, 1985). In another study of 1,984 cases of disseminated nontuberculous mycobacterial disease in AIDS patients, MAC accounted for 96.1%, *M. kansasii* for 3%, *M. gordonae* for 0.6%, and *M. fortuitum* and *M. chelonae* for 0.3% each (Horsburgh and Selik, 1989). MAC occurs as a late opportunistic infection in patients with AIDS (Modilevsky et al., 1989; Jacobson, 1988; Flepp et al., 1988).

In the first study of MAC as a cause of disease in patients with AIDS, all strains were serotype 8 and some cross-reacted with serotype 4 (Greene et al., 1982). Of the strains submitted to the Centers for Disease Control as isolates from 146 patients with AIDS, 43% were type 4, 15% were type 8, 10% were type 1, 9% were type 6, and 3% cross-reacted with antisera prepared against types 4 and 8 (Good, 1985). This distribution is different from that seen in patients without AIDS since serotype 4 predominated rather than serotype 8.

The study has now extended to include isolates from 727 patients with AIDS (Yakrus and Good, 1990). Of the strains submitted from medical centers across the United States, 630 (87%) were typable and almost all of the isolates were *M. avium* (serotypes 1 through 6 and 8 through 11). Observations made earlier were still valid; serotype 4 predominated, accounting for 40% of the strains, serotype 8 for 17%, and serotype 1 for 9%. A geographic distribution was found in this series of isolates in that serotype 4 accounted for 49% and 42% of isolates in New York City and San Francisco, respectively, but for only 14% of isolates in Los Angeles. On the other hand, 36% of the isolates in Los Angeles were serotype 8, while 8% and 16% of isolates were serotype 8 in New York City and San Francisco, respectively. Strains submitted for serotyping were primarily from blood, and 42% of both serotype 4 and 8 isolates were from this source. Isolates from tissue, sputum, and stool specimens also accounted for significant numbers of specimens. However, serotype distribution did not vary on the basis of source of specimen. Eleven of the 22 isolates of *M. intracellulare* were from sputum, suggesting that isolation of the species from this source may represent temporary colonization rather than a forewarning of the development of progressive disease.

In a retrospective study at the Mycobacteria Reference Laboratory, Boston, Massachusetts, 1,953 patients were identified whose cultures had been positive for MAC in the years 1972 through 1983 (du Moulin et al., 1985). Isolates of MAC from normally sterile sites increased fivefold during the period of study. Most were from persons in areas of high density population and in communities where water was transported from long distances rather than being obtained from a local source. In a series of reference cultures received by the Georgia State Laboratory in a 12-month period, the most frequent serotypes were 1, 16, 14, and 13 in decreasing order (Hitz, 1986). These represented all isolates including sputum and bronchial wash. Among 26 isolates from specimens indicative of progressive disease, serotype 4 predominated and accounted for 23% of isolates from these sites, followed by types 8, 14, and 16, each representing 12% of the isolates.

du Moulin et al. (1988) recovered *M. avium* from 14 of 34 water specimens taken from sites in a hospital that included 18 cold water taps and 16 hot water taps. *M. avium* was recovered more frequently and in greater numbers from the hot water sources, and seven of 11 strains tested were serotype 4. A chlorine level of 1 mg/liter in the water was suggested as being inadequate to control the growth and persistence of *M. avium* in the water supply since in a separate study, the authors found mycobacteria to be resistant to this level of chlorine (Pelletier et al., 1988).

Specific agglutinating antibody has been demonstrated in the sera of five immunocompetent patients with pulmonary disease due to *M. intracellulare* (Dawson et al., 1974). Data from another study to detect antibody to predominant MAC were interpreted as an absence of antibody to mycobacteria in AIDS patients who had confirmed pulmonary infections (Wayne et al., 1986). The failure to detect high antibody titers in patients whose blood cultures were positive for MAC suggested to the authors that the mycobacterial infection occurred after paralysis of the immune system so that response to new antigens could not be initiated. This supports the conclusion that MAC infections are a late event in the time span of progressive AIDS.

Large plasmids have been found in strains of *M. avium* isolated from AIDS patients (Crawford and Bates, 1986). These same strains were found to be virulent in an animal model and to be associated with increased catalase activity and decreased oxygen metabolite release from macrophages (Gangadharam et al., 1988, 1989; Pethel and Falkinham, 1989).

Patients with AIDS who develop infections with MAC usually present with disseminated disease and a history of progressive weight loss, intermittent fever, chills, night sweats, and diarrhea that may extend over weeks or months (Young, 1988). Occasionally the patient presents with pulmonary infiltrate or lymphadenopathy, and cultures of the pulmonary secretions are positive. Young (1988) states that *M. avium* infection usually follows episodes with one of the other opportunistic pathogens and indicates a deteriorating clinical status, with low numbers (<100) of CD4-positive lymphocytes. *M. avium* recovered from blood is positive evidence of disseminated disease, and detection of granulomas in bone marrow is strongly suggestive of infection, while a positive culture is as significant as isolation from blood (Young, 1988). Correlation with clinical signs is necessary to interpret positive stains and cultures from specimens of the respiratory tract, but these results suggest that follow-up specimens of blood, bone marrow, and stool should be cultured. Isolates from specimens of the gastrointestinal tract have a high positive predictive value for the development of disseminated disease. The decision to treat infection on the basis of isolation of MAC from any of these sites is based on the physician's sense of whether it can provide the patient with an improved quality of life (Young, 1988).

Treatment with multiple drugs did not result in clinical improvement of AIDS patients with disseminated MAC infections, and mycobacteria were never eliminated from all tissues (Hawkins et al., 1986); mycobacteremia persisted in 24 of 26 treated patients, but the numbers of bacilli recovered by quantitative cultures of blood were reduced in three patients. In this study, if infection with MAC was found before death, disseminated infection with the organism was found at autopsy. Prince et al. (1989) also found that infections due to MAC did not respond well to therapy since 8 of 21 immunocompetent patients in their study relapsed after cessation of therapy, and four died of progressive disease. In vitro tests have shown that strains of MAC are not usually susceptible to antituberculosis drugs (Good et al, 1985; Heifets and Lindholm-Levy, 1987, 1989; Heifets et al., 1987). Rifabutin is one of the new drugs that have been tested for activity against MAC (Woodley and Kilburn, 1982), and in spite of its activity in vitro, a firm conclusion regarding its efficacy against disease due to MAC has not been reached (O'Brien et al., 1987).

M. KANSASII. *M. kansasii* is a slow-growing photochromogen that is distributed through tap water (Bailey et al., 1970; McSwiggan and Collins, 1974), and its contamination of a metropolitan water distribution system (Steadham, 1980) has been reported. When the organisms are present in the water supply, there is an increase in the number of isolations in a hospital setting (Wright et al., 1985), and these may or may not be associated with disease. Biofilms present in water systems may be the habitat for mycobacteria associated with water, i.e., *M. avium, M. xenopi, M. marinum, M. gordonae,* and *M. kansasii.* They also may be the site for constant renewal of the organisms in the water supply (Schulze-Robbecke and Fischeder, 1989).

M. kansasii was the third most common, potentially pathogenic *Mycobacterium* species isolated in state public health laboratories (Good and Snider, 1982). A range of clinical manifestations from fatal infection to colonization was reported by Schraufnagel et al. (1986), who described patients with this disease as principally males with COPD. The disease is usually considered to occur at a rather constant rate in the population, but a significant increase has been noted in Japan (Tsukamura et al., 1988). Trehalose-containing lipooligosaccharides occur on the surface of variants with rough colony morphology, but not on variants with smooth morphology, which may be a factor in the pathogenesis of certain strains (Belisle and Brennan, 1989). The trehalose-containing antigens occur independently from the phenolic glycolipid surface antigen that is specific for *M. kansasii.*

M. MARINUM. Skin abrasions may become infected with *M. marinum,* a photochromogenic species with an optimal growth temperature of 25 to 35°C. The organism is freeliving and causes disease in fish, but humans may be infected when swimming, fishing, or cleaning a tropical fish tank.

In a review of the disease in humans, Collins et al. (1985) described the lesions occurring on the elbows, knees, hands, and feet as resembling those of sporotrichosis. Lesions can resolve with no treatment other than excision and drainage, but a long period of healing may be required. Antimicrobial agents that have been used for treatment include rifampin, ethambutol, trimethoprim, sulfonamides, and tetracyclines.

M. PARATUBERCULOSIS AND CHRONIC INFLAMMATORY BOWEL DISEASE (CROHN'S DISEASE).

M. paratuberculosis is the etiologic agent of Johne's disease, a chronic intestinal disease in ruminants. It has also been suspected as a cause of human intestinal disease, particularly Crohn's disease, because of the characteristics of the lesion and the appearance of sections stained by the acid-fast procedure. Recent evidence linking *M. paratuberculosis* to this disease has been summarized by Chiodini (1989). Discovery of the etiology of Crohn's disease is very important because of the improvements that can be made in prevention, case management, and treatment. But evidence is not totally in support of the mycobacterial association (Cho et al., 1986; Gitnick et al., 1989; Graham et al., 1987; Kobayashi et al., 1989). There is some evidence indicating that the organisms may occur as spheroplasts in intestinal lesions, and culture of this cell-wall-free organism and the requirement for mycobactin can influence the success of routine culture procedures (Markesich et al., 1988). Also, *M. paratuberculosis, M. avium,* and strains isolated from wood pigeons shared species homology in studies of DNA (Levy-Frebault et al., 1989). Serologic detection of antibodies in patients with Crohn's disease (Blaauwgeers, et al., 1989; Prantera et al., 1989) and the response of some, but not all, of these patients to antituberculosis drugs further blurs the final conclusion.

RAPIDLY GROWING MYCOBACTERIA.

M. fortuitum and *M. chelonae* are the two rapidly growing species in the genus that are responsible for human disease. *M. fortuitum* consists of at least three biovariants, *fortuitum, perigrinum,* and up to three unnamed biovariants, and *M. chelonae* consists of two subspecies, *chelonae* and *abscessus,* and an *M. chelonae*-like organism referred to as MCLO (Silcox et al., 1981). Infections with rapidly growing mycobacteria have been found following surgical procedures or accidental, penetrating trauma (Wallace et al., 1983); however, disease due to these organisms is not limited to cutaneous infection and may

appear in disseminated form or be more localized, e.g., as cervical lymphadenitis, keratitis, or endocarditis associated with implantation of a contaminated prosthetic heart valve. In one series of cases of human disease, both species of rapidly growing mycobacteria were encountered with equal frequency, but 80% of the isolates of *M. chelonae* were subspecies *abscessus,* and 83% of the isolates of *M. fortuitum* were biovariant *fortuitum* (Wallace et al., 1983). Treatment for the different infections is difficult and varies (Wallace et al., 1983), but selection of a chemotherapeutic regimen should be based on minimal inhibitory concentrations determined for the particular strain by in vitro susceptibility tests (Swenson et al., 1982).

These two rapidly growing *Mycobacterium* species of medical importance were combined as the *M. fortuitum* complex. However, strains of rapidly growing mycobacteria isolated from outbreaks of infection in sternal wounds following cardiac surgery (Hoffman et al., 1981) were found to differ by specific laboratory tests (Silcox et al., 1981). Because of the epidemiologic considerations and differences in the antimicrobial susceptibility patterns between the two species, they are now routinely differentiated. Strains isolated from 45 sporadic cases of sternal wound infection were more likely to be *M. fortuitum* or *M. smegmatis* than *M. chelonae,* and isolates from individual outbreaks were genotypically distinct, but more than one strain was involved in five of eight outbreaks investigated (Wallace et al., 1989). A direct source for the contaminating organisms in sternal wound outbreaks was not found by careful epidemiologic investigations in the initial outbreaks, but a pail of nonsterile ice used in the operating arena was identified as the source of *M. chelonae* subsp. *abscessus* in one study (Kuritsky et al., 1983). The relationship of the isolate from the sternal wound and from instruments used in the operating room was shown by determination of the resistance of the strains to heavy metals and by plasmid analysis (Wallace et al., 1989).

The occurrence of bacteremia, soft tissue infection, access-graft infection, and widely disseminated disease due to *M. chelonae* in patients with chronic renal failure were all traced to units processed with water in a hemodialysis center (Bolan, et al., 1985). Dialyzers were treated with 2% formaldehyde solution for disinfection, but as shown by Carson et al. (1978), isolates of *M. chelonae* and *M. fortuitum* are resistant to chemical disinfectants. In an extension of this study, water samples collected at dialysis centers throughout the United States showed that viable mycobacteria were present in the municipal water supplies, in water treated at the centers for preparation of dialysis fluids, and in water used to rinse dialyzers and prepare disinfectant solutions. Mycobacteria isolated in this study were selected in part by 1% formaldehyde treatment and were identified as *M. fortuitum, M. chelonae, M. chelonae*-like organisms, *M. scrofulaceum, M. gordonae, M. avium* complex, and *M. terrae* complex, with rapidly growing species making up about 65% of the isolates. Occurrence of *M. fortuitum* in the ice machine of a hospital was responsible for colonization of patients and for putting immunocompromised patients at risk, but none developed disease during the period of investigation (Laussucq et al., 1988).

M. smegmatis is usually associated with secretions of the normal genitalia, but Wallace et al. (1988) have identified 19 isolates that were associated with skin or soft tissue infections. The authors suggest that failure to identify the species more often is due to the phenotypic characteristics that it shares with *M. fortuitum*. *M. thermoresistibile,* a saprophytic species, has also been associated with human infection (Weitzman et al., 1981), so presumably under the proper circumstances, any *Mycobacterium* species can invade the human host and cause progressive pathologic changes.

Habitats

Mycobacterium species may be obligate parasites, opportunistic pathogens, or free-living saprophytes. *M. tuberculosis* is well recognized as the cause of tuberculosis, and the reservoir of this organism is humans, since it does not regularly produce disease in other animal species and has not been isolated from environmental sources. Other species, such as *M. avium* complex and *M. paratuberculosis,* are more closely associated with diseases in animals, but they represent etiologic agents of opportunistic infection in the human population.

M. tuberculosis represents over 50% of the total mycobacteria isolated in state laboratories in the United States (Good and Snider, 1982). *M. avium* complex, an opportunistic human pathogen, represents 17%, and *M. gordonae,* a slowly growing saprophytic scotochromogen, represents 15%. *M. fortuitum* and *M. chelonae,* rapidly growing species, are opportunistic pathogens that represent, collectively, about 6% of the species identified in state laboratories. Other species listed in Table 1 as a cause of human disease represent <2% of isolates, except for *M. kansasii* and *M. terrae* complex, which repre-

sented 2.74% and 2.15% of isolates, respectively.

Tubercle bacilli may infect farm animals and household pets such as cats, dogs, and birds (Ackerman et al., 1974; Feldman, 1960; Snider, 1971; Snider et al., 1971). *M. bovis,* the etiologic agent of bovine tuberculosis, may be transmitted to humans, primarily through ingestion of raw milk from an infected cow. Disease due to this organism has essentially been eradicated in the United States as the result of a program to tuberculin-test all cattle and slaughter all reactors. Therefore, isolates of the species are uncommon in state laboratories (Good, 1980; Good and Snider, 1982), and identification of *M. bovis* as a cause of human disease is reason for a major effort by public health programs to find the infected herd of cattle. The risk of contamination of the food supply led to the establishment of pasteurization temperatures based on the thermal death times for tubercle bacilli.

Contaminated meat fed to cats was the source of *M. bovis* infection in animal quarters of a laboratory (Isaac et al., 1983). In this same outbreak, other cats and an opossum were infected by the inhalation of bacilli that were shed from the sinus of one of the animals. *M. bovis* is endemic in the badger population of the British Isles, and disease in this animal is reported to be a source of infection in cattle (Cheeseman et al., 1989). Many aspects of the bovine tubercle bacillus have been reviewed by Collins and Grange (1983), who discuss the bacteriology, clinical manifestations of disease, epidemiology, and eradication programs. Others have also reported the epidemiologic features of the disease in cattle herds of Great Britain (Wilesmith, 1983) and in the human population of southeast England, where disease due to *M. bovis* accounts for about 1% of all tuberculosis cases. *M. tuberculosis* can be transmitted from the respiratory and urinary tracts of humans to cattle (Yates and Grange, 1988), which develop disease that needs to be differentiated from bovine tuberculosis.

M. africanum did not separate by numerical taxonomy into a cluster apart from *M. tuberculosis* and *M. bovis* (David et al., 1978) and was shown to have 90% homology with these species in DNA hybridization studies (Imaeda, 1985). *M. africanum* was originally isolated in certain parts of Africa (Castets et al., 1963, 1969; Pattyn et al., 1970), and its occurrence appears to be limited to that geographic area, or as a cause of disease in former residents who have resettled in other areas.

Mycobacterium species have been isolated from water, milk, and soil samples (Carson et al., 1988; Collins et al., 1984; Dunn and Hodgson, 1982; Goslee and Wolinsky, 1976; Kazda, 1973a, 1973b; Kazda and Hoyte, 1972; Kubica et al., 1963; Laussucq et al., 1988; Matthews et al., 1976; Reznikov and Leggo, 1974; Hitz, 1986). Reznikov et al. (1971) related the serologic type of MAC isolated from house dust to human infection and thereby suggested that the source of the organism was the environment of the patient.

Throughout the last decade, investigators in New York and Virginia have joined to study the occurrence of MAC and *M. scrofulaceum* in the environment and to relate their occurrence to the epidemiology of infections (Brooks et al., 1984; Falkinham et al., 1980, 1989; Fry et al., 1986; George et al., 1980; Gruft et al., 1981; Martin et al., 1987; Parker et al., 1983; Wendt et al., 1980). From these studies, it appears that mycobacteria are aerosolized from ocean, lake and river water, and they can be isolated from air samples and rain water. Geographically, MAC and *M. scrofulaceum* tend to occur more frequently in the southeastern United States, and statistical analysis shows a highly significant correlation between the recovery of large numbers of bacilli and high acidity of the soil.

Rapidly growing species in the environment were discussed in the preceding section.

Isolation Techniques

Collection of Specimens

Specimens submitted to a clinical laboratory for the isolation of *Mycobacterium* species should be collected in clean, sterile containers before treatment is started. If containers are reused, they must be cleaned with dichromate-sulfuric acid solution or a comparable cleaning solution to prevent carryover of bacilli. Specimens should be delivered to the laboratory expeditiously, and if they are to be sent through the mail, they should be packaged with extreme care to prevent breakage and to comply with national or international postal requirements (Kent and Kubica, 1985).

Most of the specimens processed for isolation of mycobacteria are sputum specimens collected because of suspected pulmonary disease. A series of five or six early-morning sputum specimens collected on succeeding days is usually adequate for bacteriologic confirmation of disease (Blair et al., 1976; Krasnow and Wayne, 1969). Patients should be instructed to rinse the mouth to clear it of food, mouthwash, or drugs. Sample collection will be more successful if the patient is instructed on the definition of sputum and on how to best produce a good specimen (Kent and Kubica, 1985). If the patient is unable to produce a good specimen, sputum may be induced by the inhalation of aerosolized, warm, hypertonic (5% to 10%) saline solution. Such induced specimens are usually thin and watery, so they must be labeled as induced specimens so that the laboratory will not discard them as saliva and inadequate. A 24- to 48-h pooled specimen may be collected from those patients who have difficulty producing sputum. Specimens should be collected in well-ventilated areas by personnel protected with high-efficiency masks since aerosols of tubercle bacilli are generated, particularly during collection of induced specimens.

Gastric washings are collected when patients cannot provide a sputum specimen, when sputum cultures are negative but roentgenography suggests tuberculosis, or under other special circumstances. Gastric lavage specimens should be collected only from hospitalized patients and in the early morning before breakfast, preferably before the patient arises. Twenty to 50 ml of sterile, distilled water are introduced into the stomach through a disposable, plastic gastric tube, and the liquid is withdrawn with a 50-ml syringe. The material should be processed rapidly since mycobacteria die quickly in aspirated gastric washings. If the specimen must be held, it should be neutralized by the addition of 1.5 ml of 40% disodium phosphate.

Bronchial washings are collected in many instances rather than gastric washings. Lavage fluid or transtrachial aspirates should be delivered to the laboratory rapidly, but these specimens do not need to be neutralized.

Spinal fluid, pleural fluid, blood, bone marrow, and other normally sterile specimens are collected aseptically by aspiration or surgical procedures. Heparin (0.2 mg per ml) or sterile potassium oxalate (0.01 to 0.02 ml of 10% neutral oxalate solution per ml) should be added to specimens that may clot (Kent and Kubica, 1985). Isolation of mycobacteria from normally sterile body fluids that may contain few bacilli is more successful if they are inoculated at the bedside into liquid medium such as Middlebrook 7H9, Dubos Tween albumin broth, or Proskauer-Beck at a ratio of 1 part fluid specimen to 5–10 parts liquid medium. Cultures should be incubated in the dark at 35 to 37°C in an atmosphere of 10% CO_2 and gently shaken daily. When acid-fast bacilli are seen in the cultures at weekly smear examination, subcultures are made to solid egg- or agar-based media or both. If the broth is still negative after 4 to 6 weeks, the entire culture is centrifuged at 3,000 $\times$ g or greater, and the sediment is streaked to one or more solid media as above.

Blood is routinely taken in cases of suspected disseminated disease, particularly from patients with AIDS. Three methods have been proposed by Kiehn et al. (1985) for culturing blood. Blood collected in a Vacutainer tube (100 × 16 mm) containing 5.95 mg of sodium polyanethole sulfonate (Becton Dickinson and Co., Paramus, NJ) is centrifuged at 1,500 × g for 30 minutes, and 0.1 ml of the concentrated sediment is placed onto both the surface of 7H11 agar in a petri dish and a slant of Lowenstein-Jensen medium. If high-grade bacteremia is suspected, blood in the Vacutainer tube is diluted 1- and 100-fold with 7H9 broth before centifugation. The second method is treatment of the blood with the Isolator-10 lysis-centrifugation system (Du Pont Co., Wilmington, DE). Ten ml of blood are collected in the tube which contains an anticoagulant and a lytic agent, saponin. The tube is inverted several times and delivered to the laboratory. After lysis, the tube is centrifuged at 3,000 × g for 30 min, and the concentrated sediment of 1.5 ml is placed in equal portions onto the surface of 7H11 agar on four plates. The third method for isolation from blood is with the Isolator-1.5 lysis tube. Only 1.5 ml of blood need be collected into the tube, which is then cultured without centrifugation. If high-grade bacteremia is expected, the lysates are diluted before culture onto plates of medium.

Pierce et al. (1983) found that the lysis centrifugation system (Isolator) could be used for isolation of mycobacteria from blood, and although comparisons were not made, a radiometric detection system (BACTEC, now manufactured by Becton Dickinson Diagnostic Instrument Systems, Towson, MD) was also effective. Witebsky et al. (1988) made direct comparisons of the 7.5-ml Isolator and BACTEC procedure with 13A medium using the same blood specimens and found that recovery was about the same with either method, that 12 to 14 days were needed for detection, and that even though the BACTEC 13A medium required less manipulation of the specimens, it did not provide quantitation of the bacilli. Other normally sterile body fluids may be processed by either of these methods if it is necessary to free the bacilli ingested by white cells.

Urine specimens should be collected in the early morning as a midstream or total voided specimen (Kent and Kubica, 1985). Excessive contamination of the specimen may be a problem if it is not refrigerated or processed rapidly.

Stool specimens from AIDS patients are regularly cultured to confirm dissemination of MAC. The methodology presented by Kiehn et al. (1985) suggests that when an acid-fast stain of unprocessed fecal material is positive, a suspension consisting of 1 g of feces in 5 ml of 7H9 broth should be treated by the sputum decontamination method (see below) and cultured on selective media.

Specimens should be rapidly transmitted to the laboratory in a safe manner. In some instances they must be transported by mail. Shipment of diagnostic specimens such as urine, sputum, and tissue, and of cultures of etiologic agents which include all *Mycobacterium* species, is subject to the packaging and labeling requirements of the Interstate Quarantine Regulations (Federal Register, Title 42, Chapter 1, Part 72, revised July 30, 1972). Diagnostic specimens must be packaged to withstand leakage of contents, shocks, pressure changes, and other conditions incident to ordinary handling in transportation. Cultures of *Mycobacterium* species require additional procedures as described in Section 72.25. Details of shipping procedures are given by Kent and Kubica (1985).

Microscopic Examination of Specimens

Detection of acid-fast bacilli by microscopic examination of a specimen is used in many parts of the world as the only confirmation of a diagnosis of tuberculosis. In developed countries that depend on isolation and identification of an etiologic agent, the smear is helpful to evaluate a specimen and to follow the effectiveness of therapy.

New clean slides should be used for all tests with specimens. Caseous, bloody, or purulent particles should be selected from sputum specimen for direct smears (Smithwick, 1976). The selected part of the specimen, approximately 0.01 ml, is spread over an area 1 × 2 cm and allowed to air-dry. The smear is heat-fixed by passing it through the flame of a Bunsen burner three or four times or by heating on an electric slide warmer at 65–75°C for 2 or more h. Whether the specimen has been concentrated by one of the methods described below or is sampled directly, it is treated the same way during the staining procedure.

If the specimen is for microscopy only, add a volume of 5% sodium hypochlorite (commercial bleach) equal to the volume of the specimen, close the container securely, and mix on a test-tube mixer or other shaking machine until liquefied. Allow to stand at least 10 but no more than 15 min. Dilute with sterile water and centrifuge at 2,000 × g for 15 min. Decant the supernatant and smear the sediment onto the microscope slide as described above.

Hypochlorite solution will kill the mycobacteria, and if it is allowed to act too long, the bacilli will disintegrate. Therefore, the time limitations given in the procedure must be carefully followed.

Staining Procedures (Smithwick, 1976)

Fluorochrome staining procedures are becoming widely used in developed countries, but the older procedures based on staining with fuchsin are still important in laboratories without fluorescent microscopes.

Ziehl-Neelsen Acid-Fast Stain

Reagents:
1. Fuchsin-phenol: Dissolve 0.3 g basic fuchsin in 10 ml 90–95% ethanol or methylated spirits. In a separate container, dissolve 5 g of phenol crystals in 95 ml distilled water. Add 90 ml of phenol solution to the fuchsin solution and mix well.
2. Acid alcohol: Add 3 ml concentrated hydrochloric acid to 97 ml 90–95% ethanol or methylated spirits.
3. Methylene blue: Dissolve 0.3 g methylene blue chloride in 100 ml distilled water.

Procedure:
1. Prepare and heat-fix smears.
2. Cover smear with absorbent paper and add approximately 5 drops of fuchsin-phenol solution.
3. Heat the bottom of the slide with a Bunsen burner or electric heater until the stain begins to steam. Heat for 5 min, but do not boil or allow to dry.
4. Remove paper with forceps, rinse smears with tap water, and drain.
5. Flood smear with acid alcohol and destain for at least 2 min, then rinse in tap water and drain.
6. Flood smear with methylene blue and counterstain for 1–2 min.
7. Rinse, drain, and air dry.

Kinyoun Acid-Fast Stain

Reagents:
1. Fuchsin-phenol: Dissolve 4 g basic fuchsin in 20 ml 95% ethanol, and add a solution of 8 g crystalline phenol in 100 ml distilled water.
2. Acid alcohol: 3 ml concentrated hydrochloric acid in 97 ml 95% ethanol.
3. Methylene blue: Dissolve 0.3 g methylene blue chloride in 100 ml distilled water.

Procedure:
1. Prepare and heat-fix smears.
2. Cover the smear with absorbent paper, add 5 drops of stain solution, and stain for 5 minutes.
3. Remove paper with forceps, rinse smear with tap water, and drain.
4. Flood smear with acid alcohol and destain for at least 2 min.
5. Flood smear with methylene blue and counterstain for 1–2 min. Rinse, drain, and air dry.

Blair Fluorescence Acid-Fast Stain

Reagents:
1. Auramine O-phenol: Dissolve 0.1 g auramine 0 in 10 ml 95% ethanol. Mix with a solution prepared by dissolving 3 g phenol crystal in 87 ml of distilled water. Store at room temperature protected from light.
2. Acid alcohol: Add 0.5 ml concentrated hydrochloric acid to 100 ml 70% ethanol.
3. Potassium permanganate: Dissolve 0.5 g potassium permanganate ($KMnO_4$) in 100 ml distilled water.

Procedure:
1. Prepare and heat-fix smear.
2. Flood smear with auramine O-phenol solution and allow to remain for 15 min.
3. Rinse with chlorine-free water and drain.
4. Flood with acid alcohol and destain for 2 minutes.
5. Rinse and drain.
6. Flood with potassium permanganate and allow to stand for 2 min.
7. Rinse, drain, and air dry.

Truant Fluorescence Acid-Fast Stain

Reagents:
1. Auramine O-rhodamine B-phenol: Dissolve 1.5 g auramine 0 and 0.75 g rhodamine B in a solution of 75 ml glycerol, 10 ml heat-liquified phenol crystals, and 50 ml distilled water. Filter the solution through glass wool. Store at room temperature protected from light.
2. Acid alcohol: Add 0.5 ml concentrated hydrochloric acid to 100 ml 70% ethanol.
3. Potassium permanganate: Dissolve 0.5 g potassium permanganate ($KMnO_4$) in 100 ml distilled water.

Procedure:
1. Prepare and heat-fix smear.
2. Flood smear with auramine-rhodamine solution and allow to stand for 15 min.
3. Rinse with chlorine-free water and drain.
4. Flood with acid alcohol and destain for 2 minutes.
5. Rinse and drain.
6. Flood with potassium permanganate and allow to stay for 2 min.
7. Rinse, drain, and air dry.

Reading Smears

Smears should be examined microscopically by making three longitudinal sweeps of the stained area parallel to the length of the slide (Smithwick, 1976). About 100 fields can be viewed in a single sweep if magnification is 1,000-fold, as it usually is for fuchsin-stained smears. Therefore, a total of 300 fields will be observed. The recommended minimal number of fields to search depends on the magnification and ranges from 30 fields at 250× magnification (with fluorescence) to 300 at 1,000× (with light microscopy). Time required to scan a slide with fluorescence microscopy is 1.5 min, as opposed to the 15 min required to scan the 300 fields for light microscopy. Smithwick (1976) has proposed the method shown in Table 3 for reporting results of smear examination. The smear is not reported as positive unless ≥3 acid-fast bacilli are in the 300 fields viewed by the fuchsin-staining methods or an equivalent number on

Table 3. Reporting scheme for acid-fast smears.[a]

Number of AFB[b] seen	Report
None in 3 sweeps (300 fields)	Negative
1–2 per 300 fields	Doubtful; request repeat specimen
1–9 per 100 fields	1+
1–9 per 10 fields	2+
1–9 per field	3+
>9 per field	4+

[a]When smear examinations are made at other than 800–1,000×, compensation must be made for the lower magnification used. See Smithwick, 1976, for discussion of the "conversion factors."
[b]Acid-fast bacteria.

the basis of the magnification used for the fluorochrome stain.

Digestion—Decontamination of Specimens

Specimens sent to clinical laboratories often are contaminated, and mucus that entraps mycobacteria is usually present. Digestion with a mucolytic agent frees the entrapped mycobacteria and makes the specimen easier to process. Treatment with a toxic agent kills unwanted microbial flora in the specimen. The N-acetyl-L-cysteine-sodium hydroxide (NALC) method (Kent and Kubica, 1985; Kubica et al., 1963; Sommers and Good, 1985) is one of the most widely used procedures in the United States (see description below). Acetyl-cysteine, the mucolytic agent, loses activity rapidly once it is put into solution, so the solution must be prepared daily. High pH is used for decontamination, but many of the mycobacteria will also be killed, particularly if the final concentration of NaOH exceeds 1% or the temperature is permitted to rise above 30°C (Kent and Kubica, 1985). Care must be taken to follow directions for decontamination precisely in order to recover the maximal numbers of bacilli.

The cetylpyridinium chloride-sodium chloride (CPC) method (see below) was proposed by Smithwick et al. (1975) to digest and decontaminate sputa in transit, so that upon arrival in the laboratory the specimen can be concentrated by centrifugation and placed directly onto media. Advantages of this method are that contaminants do not overgrow the mycobacteria before the specimen is processed; less laboratory time is needed because the digestion-decontamination process occurs during mailing; more positive cultures are detected because the procedure is less mycobactericidal than the NALC method; and the overall contamination rate of inoculated medium is lower. The principal disadvantage of the method is that cetyl-

pyridinium chloride is bacteriostatic for mycobacteria and must be neutralized with phospholipid or diluted so that the mycobacteria will grow well in culture. Phospholipid in egg yolk added to the egg-based media for mycobacteria is adequate for this purpose. Although not reported, it should be adequate to resuspend the centrifuged specimen in sterile buffer or water containing lecithin prior to culture. The CPC method is particularly good for processing environmental samples (du Moulin and Stottmeier, 1978; du Moulin et al., 1988) because highly toxic chemicals are not used, and there is a greater chance of recovering mycobacteria from those specimens that contain few bacilli.

Other digestion-decontamination methods have been described that use strong acid or alkali, but recovery of mycobacteria is generally not as good as it is with the NALC and CPC methods. Details of the methods are described by, among others, Collins et al. (1985), Kent and Kubica (1985), Kubica and Good (1981), and Sommers and Good (1985).

Large-volume specimens, such as gastric lavage or urine, should be centrifuged at 2,000 × g to concentrate the sediment, which is then processed as for sputum. Stool specimens proved to contain acid-fast bacilli by examination of stained smears are suspended at the rate of 1 g of feces per 5 ml 7H9 broth, and the suspension is treated by the sputum decontamination method (Kiehn et al., 1985).

All specimen decontamination methods are toxic for mycobateria as well as for the contaminating flora, and overtreatment, either in time, temperature, or concentration can markedly reduce the numbers of surviving acid-fast bacilli detected in culture. Therefore, the digestion—decontamination procedure should be as gentle as possible and should be compatible with an overall contamination rate that does not exceed 5% (Kent and Kubica, 1985).

NALC Method for Digestion—Decontamination (Kubica et al., 1963; Kent and Kubica, 1985)

Prepare in advance 100 ml of 4% sodium hydroxide (4 g NaOH in 100 ml distilled water) and 100 ml of 2.9% sodium citrate (2.9 g anhydrous sodium citrate in 100 ml distilled water); these solutions may be sterilized by autoclaving and stored on the shelf in screw-top flasks until needed. Prior to use, mix the NaOH and citrate solutions together (200 ml final volume) and add powdered N-acetyl-L-cysteine to a concentration of 0.5% (1.0 g NALC to 200 ml). This final solution should be used within 18–24 h.

Transfer 10 ml (or less) of sputum (or other clinical specimen) to sterile, 50-ml, screw-cap centrifuge tubes (preferably plastic). Smaller tubes may be used, but in such cases the volume of the specimen should not exceed

1/5 the capacity of the tube. Add an equal volume of the NALC digestant, then replace and securely tighten the screw cap on the tube. Mix the contents of the tube by swirling on a vortex-type test-tube mixer until the specimen is liquefied (usually 5–15 seconds). Let stand for 15 min at room temperature to effect decontamination.

Fill tubes to the 50-ml mark (about 12 mm from the top) with sterile, distilled water or pH 6.8 phosphate buffer (0.067 M). Dilution minimizes the continuing action of NaOH and lowers the specific gravity of the suspension so that centrifugation is more efficient and the neutralization of alkali is not necessary. Tighten screw caps and mix contents of tubes by swirling. Centrifuge at 2,000–3,000 × g for 15 min using aerosol-free, sealed centrifuge cups. Pour off the supernatant fluid and resuspend the sediment in a small volume (1 or 2 ml) of sterile, 0.2% bovine albumin fraction V (prepare in 0.85% sodium chloride, adjust pH to 6.8–7.0, and sterilize by Seitz or membrane filtration).

Place the resuspended sediment (both undiluted and after a 1:10 dilution in sterile saline) onto the surface of at least two media of different basic composition (e.g., an egg base such as Lowenstein-Jensen, and an agar base such as 7H11).

CPC Method for Digestion—Decontamination (Smithwick et al., 1975)

Dissolve 10 g of cetylpyridinium chloride and 20 g of sodium chloride in 1 liter of sterile distilled water. Store at room temperature. The solution is self-sterilizing (except for mycobacteria) and remains stable if tightly capped and protected from excess heat and light.

Sputa to be shipped should be collected in 50-ml, screw-cap centrifuge tubes. Add an equal volume of the CPC reagent, tighten the cap, and shake by hand until the specimen becomes homogenized. Package the specimens in a watertight, double mailing container (Kent and Kubica, 1985) and send to the laboratory. The specimen is decontaminated in transit without any appreciable reduction in numbers of mycobacteria that may be present. Upon receipt, fill each tube to within 12 mm of the top with sterile, distilled water, recap, and centrifuge at 2,000–3,000 × g for 20 min. Decant the supernatant fluid, resuspend the sediment in 1 ml of sterile water or saline, and inoculate to egg-based medium.

Primary Culture Media

Recovery of mycobacteria from clinical specimens is dependent on providing the conditions that are necessary for optimal growth, and all identification tests for species depend on a well-grown primary culture. Most mycobacteria are not very fastidious, but until strains have adapted to an in vitro growth environment, the temperature, medium, atmospheric gas concentrations, and other conditions conducive to good growth should be optimal. Many different media have been proposed for the primary cultivation of acid-fast bacteria, and most of these are egg-potato base or serum-agar base formulations. Three media used worldwide for primary isolation, drug susceptibility testing, in vitro identification tests, and determination of colonial morphology will be discussed in detail. The BACTEC method of primary culture will also be covered in this section.

Modified Lowenstein-Jensen Egg Medium (Kent and Kubica, 1985)

Clean fresh eggs, not over 1 week old, in soap solution by scrubbing with a hand brush. After scrubbing, let the eggs stand in the soap solution for 30 min. Rinse thoroughly in running water, then soak eggs in 70% ethanol for 15 min. Break the eggs into a sterile flask and homogenize by hand shaking. Filter the eggs through four layers of sterile gauze into a sterile graduated cylinder. Eighteen to 24 eggs (depending upon size) are needed to provide the required 1 liter of homogenized whole egg. Set the homogenized eggs aside and prepare the basal salt solution.

Place 600 ml distilled water in a 2-liter flask and dissolve in order:

KH_2PO_4 (anhydrous)	2.4 g
$MgSO_4 \cdot 7H_2O$	0.24 g
Magnesium citrate	0.6 g
L-Asparagine	3.6 g
Glycerol	12.0 ml
Potato flour	30.0 g

Autoclave at 121°C for 30 min. Cool to room temperature and add 20 ml of a freshly prepared 2% aqueous solution of malachite green and the 1 liter of homogenized whole eggs. Mix the medium well and pour into a sterile aspirator bottle or funnel assembled with a bell attachment (test-tube filling device). Dispense medium into appropriate containers (e.g., 8 ml in 20 × 150-mm, sterile, screw-cap test tubes) in desired volumes. Slant tubes and coagulate by inspissation (moist heat) at 85°C for 50 min. Incubate the finished medium at 37°C for 48–72 h to determine sterility. Medium may be stored for several months in the refrigerator (4–6°C) if caps are tightened securely to prevent drying.

Lowenstein-Jensen egg medium can be made selective for mycobacteria (Gruft, 1971; Petran and Vera, 1971). This and other egg media may be inoculated with bacteriological loops, pipettes (capillary or serologic), or cotton swabs. The usual inoculum is 0.1 ml or 1 to 2 loopfuls of the digested, decontaminated clinical material, or the appropriate dilution of a suspension from a liquid or solid culture of the organism. Spread the inoculum to cover the surface of the slant. Tubes should be incubated at 35–37°C in the slanted position for at least 7 days to permit even distribution of the inoculum over the entire surface of the medium. After this time, tubes may be incubated upright. Most incubation is continued for a least 8 weeks at 35–37°C before the tubes are discarded as negative. Some mycobacteria, especially those recovered from superficial lesions (e.g., *M. marinum, M. ulcerans*), grow better at 30–33°C on primary isolation; others, like *M. xenopi,* may prefer 42–43°C.

Agar-Based Media (Kent and Kubica, 1985)

The most used agar-based media are Dubos oleic acid-albumin medium and Middlebrook 7H10 and 7H11 media (see below). The 7H11 medium is 7H10 agar that has been enriched by the addition of enzymatic digest of casein. Drug susceptibility determinations are usually done in Middlebrook 7H10 medium, and the 7H11 modification of the medium should not be used unless results are controlled and restandardized to establish minimal inhibitory concentrations that are reproducible and that can be interpreted in the same manner as tests on 7H10 medium.

After preparation, the media should be protected from direct light and heating. Store plates in clean plastic bags and tighten caps on tubes to prevent drying. Medium should not be used if it has dried, appears contaminated, or changes color from the greenish hue to a yellowish hue.

Middlebrook 7H10 Agar Medium

Middlebrook 7H10 agar-powder base and Middlebrook OADC enrichment, both from commercial suppliers, are completely satisfactory. The recommended method is to prepare the medium in quantities of 200–400 ml; preparation in larger volumes requires longer heating to solubilize all of the agar, and this generally results in a medium of inferior quality.

Suspend 3.8 g of Middlebrook 7H10 agar base in 180 ml of freshly distilled water in a 400-ml flask and add 1 ml glycerol. Swirl the powered agar base into suspension without heating, and sterilize in the autoclave at 121°C for 10 min. As soon as the pressure reaches ambient, remove the flask of medium from the autoclave, swirl gently to mix, and place in a 52–56°C water bath. When the medium has cooled to 54–56°C, add 20 ml OADC enrichment, swirl carefully to mix (avoid bubbles), and pour into plates or tubes as soon as possible (within 1 h after autoclaving).

Allow to solidify at room temperature, and store in a plastic bag shielded from light at room temperature. Inoculate plates with 0.1 ml of treated specimen by dropping from a capillary pipette or by spreading the inoculum over the surface of the plate. After inoculum has absorbed into the medium, place plates singly into clear polyethylene bags and seal with tape or heat. The plates can be examined directly through the plastic bag. Incubate in an atmosphere supplemented with 10% CO_2 for optimal growth on primary culture.

Middlebrook 7H11 Agar Medium

Suspend 3.78 g of Middlebrook 7H11 agar base in 180 ml of distilled water containing 1 ml of glycerol. Autoclave for 15 min at 121°C, then cool to 50–52°C in a water bath. Add 20 ml of OADC enrichment that has been warmed to room temperature and swirl to mix.

Dispense within 1 h of autoclaving. Inoculate and incubate as directed with 7H10 medium.

SELECTIVE AGAR MEDIUM. Either the 7H10 or 7H11 medium can be made selective for mycobacteria by the aseptic addition of antimicrobial agents at 50–52°C prior to pouring the medium into final containers. Drug concentrations recommended by Mitchison et al. (1972), as modified by McClatchy et al. (1976), are (per ml final concentration): polymyxin B, 200 units; amphotericin B, 10 μg; carbenicillin, 50 μg; and trimethoprim, 10 or 20 μg.

Middlebrook 7H9 Liquid Medium

This medium is recommended for subculturing stock strains, preparing inocula for drug susceptibility tests, and making some dilutions. The medium is prepared from commercially available powdered base and is supplemented with Middlebrook albumin-dextrose-catalase (ADC) enrichment after sterilization and cooling. A final concentration of 0.22% glycerol or 0.05% Tween 80 is used in the medium, but both agents are not added to the same lot of medium.

Suspend 4.7 g of the dehydrated base in 900 ml of distilled water containing 2 ml glycerol or 0.5 g Tween 80. Autoclave at 121°C for 15 minutes, cool to 45°C in a water bath, and aseptically add 100 ml ADC enrichment. Dispense 5-ml amounts of the complete, sterile medium into 20 × 150 mm screwcap tubes and tighten the caps to prevent evaporation.

The medium can be stored for several months without losing the ability to support growth of mycobacteria.

Radiometric Methodology (BACTEC) for Primary Culture

Middlebrook et al. (1977) described an automatable radiometric detection procedure for *M. tuberculosis* in selective medium. The 7H12 medium designed for the procedure is commercial 7H9 broth base supplemented with 0.1% enzymatic hydrolysate of casein. After sterilization, the antimicrobial agents previously described (Mitchison et al., 1972; McClatchy et al., 1976) were added to the medium along with 3 units of catalase per ml final concentration as follows: amphotericin B, 5 μg/ml; polymyxin B, 50 μg/ml; carbenicillin, 25 μg/ml; and trimethoprim, 2.5 μg/ml. Palmitic-1-^{14}C acid (1 mCi) was dissolved in 2 ml of 0.05 N sodium hydroxide, which was then diluted with 5% bovine serum albumin to use concentration of 10 μCi/ml. The acid-albumin solution was added to the medium to a final concentration of 0.5% albumin and 1 μCi radioactivity/ml, and the final pH was adjusted to 6.5. Both glycerol and glucose were omitted so that the medium was more selective for mycobacteria. $^{14}CO_2$ released by the metabolism of palmitic acid was collected in a closed vial and quantitated in an automatable ion chamber system currently marketed as BACTEC 460 with the BACTEC TB Hood (Becton Dickinson Diagnostic Instrument Systems, Towson, MD).

There are several advantages to this procedure: mycobacteria grow more rapidly in liquid media; detection of enzymatic activity is more sensitive than reading turbidity in the medium; and the method is adaptable to a number of analyses including detection, identification, and determination of drug susceptibility. Roberts et al. (1983) found that *M. tuberculosis* grew from smear-positive specimens in the Middlebrook 7H12 medium in an average of 8.3 days, and other *Mycobacterium* species could be detected in 5.2 days. Detection by the conventional method was 19.4 and 17.8 days, respectively. The mean detection time from smear-negative specimens was 13.7 days by the BACTEC procedure and 26.3 days by conventional techniques. Salfinger et al. (1988) reported that the recovery time of *M. avium* complex from blood specimens cultured on 7H11 agar plates was 7 to 21 days (mean 9.6 days), and in 7H12 medium, it was 3–12 days (mean 6.6 days). Studies by many investigators have demonstrated the advantages of BACTEC methodology directly or with specific improvements for detection and isolation of *Mycobacterium* species from primary cultures (Cutler et al., 1987; Heifets, 1986; Hoffner, 1988; Kirihara et al., 1985; Kononov et al., 1988; Morgan et al., 1985; Raja et al., 1988; Siddiqi et al., 1988; Strand et al., 1989; Takahashi and Foster, 1983; Witebsky et al., 1988). Occasional problems have been reported with the BACTEC system (Conville and Witebsky, 1989; Salfinger et al., 1988), but these have not been significant compared to conventional detection methods for mycobacteria.

BACTEC Method (Siddiqi, 1988)

Digest and decontaminate specimens as directed above. If the specimen exceeds 10 ml, concentrate by centrifuging at 2,000 to 3,000 × *g* for 15 to 20 min, then resuspend concentrates in 0.067 M phosphate buffer, pH 6.8. Clean the septum of the BACTEC vial with a disinfectant and rinse with 70% isopropyl alcohol. Then, pretest vials on the BACTEC 460 instrument to establish an atmosphere enriched with 5% CO_2. Combine concentrates and inject 0.5 ml into the vial using a separate, new, sterile tuberculin syringe with attached needle for each specimen. Sterilize the septum with disinfectant again and rinse with 70% alcohol. For primary cultures, PANTA supplement (final concentration per ml is polymyxin B, 50 µg; amphotericin B, 5 µg; nalidixic acid, 20 µg, trimethoprim, 5 µg; and azlocillin, 10 µg) must be added to medium in BACTEC vials. After all of the additions have been made, incubate vials at 37°C and test on the BACTEC 460 according to manufacturer's instructions.

The schedule for testing vials may be 2 to 3 times a week for 3 weeks and then once a week after that for up to 6 weeks. At each test, radioactivity measured in the gas in the head-space is recorded on a scale ranging from 0 to 999 where 100 units, designated Growth Index (GI) units are equal to 0.27 µCi Once the culture is positive, i.e., the Growth Index (GI) is ≥10, read daily and make an acid-fast smear when the GI reaches 100. When the GI is ≥300, use as an inoculum for drug susceptibility tests.

Identification

Mycobacteria are chemically complex, and immunologic responses to the bacilli are equally complex, as emphasized in the review by Barksdale and Kim (1977). This complexity has been used in many instances to design specific methods of detection and identification. Mycobacteria grow slowly, yet rapid detection is often critical for institution of appropriate treatment and public health measures. Therefore, basic research into the activities of cellular components of these bacteria has led inevitably to development of new diagnostic procedures apart from the standard biochemical tests. Research studies in this area, very active in the past decade, continue to result in new techniques so that procedures to establish taxonomic position or to improve clinical laboratory diagnosis now may be selected rather than mandated. Newer techniques in lipid analysis (Butler et al., 1986; Bulter and Kilburn, 1988; Lambert et al., 1986; Minnikin and Goodfellow, 1980, Minnikin et al., 1983, 1984a, 1984b), serologic typing (Brennan, 1981; Good, 1985; Good and Beam, 1984; McIntyre and Stanford, 1986; Tsang et al., 1983, 1984; Yakrus and Good, 1990), phage typing (Jones, 1988; Jones and Woodley, 1983; Jones et al., 1982; Snider et al., 1984), DNA homology (Baess, 1983; Eisenach et al., 1986, 1988; Imaeda, 1985; Levy-Frebault et al., 1986; McFadden, et al., 1987a; Picken et al., 1988), differentiation by seroprecipitation of catalases (Wayne and Diaz, 1985, 1986), and restriction fragment testing (Collins and DeLisle, 1985; Crawford and Bates, 1984; McFadden et al., 1987b; Shoemaker et al., 1986; Whipple et al., 1987, 1989) are available only in special laboratories, but they have aided in genus and species assignment and in the development of markers to be used in epidemiology. Development of many of these new methods will make diagnosis of mycobacterial diseases possible by detection of their chemical imprint, whether it is a specific antigen, a unique antibody, or a particular nucleotide sequence. Methods may soon be available that have sensitivity and specificity never before achieved, that are less expensive than standard methods, and that allow reports of analysis within a day of specimen receipt (Crawford et al., 1989).

IDENTIFICATION TECHNIQUES. Routine procedures used by various investigators for identification of species of medical importance in the genus *Mycobacterium* may not be identical, so published studies must be consulted for precise techniques. Techniques referred to here for performance of biochemical tests, determination of chromogenicity and rate of growth, and description of colonial morphology are those developed in or recommended by the Mycobacteriology Laboratory, Centers for Disease Control (Kent and Kubica, 1985; Kubica, 1984; Sommers and Good, 1985; Vestal, 1975). When possible, the tests should be performed using standardized procedures recommended by the International Working Group on Mycobacterial Taxonomy (Wayne et al., 1974, 1976). Procedures using mycolic acid patterns, tuberculostearic acid detection, radiometric techniques, and DNA probes have improved the ability of the clinical laboratory to rapidly detect and identify mycobacteria responsible for human disease or to diagnose the condition.

IDENTIFICATION OF *MYCOBACTERIUM* SPECIES BY MYCOLIC ACID PATTERN ANALYSIS (BULTER ET AL., 1986; BUTLER AND KILBURN, 1988). Lipids detected in mycolic acid-containing genera of bacteria by thin-layer chromatography have been used as taxonomic criteria for classification (Minnikin et al., 1978, 1980; Minnikin and Goodfellow, 1980). High-performance liquid chromatography (HPLC) of bromophenacyl esters of mycolic acids was used by Butler et al. (1986) to identify representative mycolic acid patterns for species of *Corynebacterium, Rhodococcus, Nocardia,* and *Mycobacterium.* Also, gas-liquid chromatography (GLC) of the methyl esters of mycolic acid cleavage products has been used as the basis of an identification scheme for *Mycobacterium* species (Lambert et al. 1986; Maliwan et al. 1988). Mycolic acids are β-hydroxy-α-branched fatty acids with the general formula:

$$R_1\text{-CH(OH)-CH-COOH}$$
$$|$$
$$R_2$$

where R_2 is an alkyl chain (C_nH_{2n+1}). R_1 and R_2 carbon chains vary in length in the different genera so that mycolic acids in corynebacteria are C_{22} to C_{38}, *Nocardia* are C_{44} through C_{60} and mycobacteria are C_{60} through C_{90}. Mycolic acids

in other genera, such as *Rhodococcus* and *Tsukamurella,* are of intermediate chain lengths.

Method for Mycolic Acid Analysis (Butler and Kilburn, 1988)

> Add at least 10^7 bacilli from culture to a tube and saponify in a 25% potassium hydroxide solution (in 50% aqueous solution of methanol) by autoclaving at 121°C for 1 h. Cool the suspension, adjust to pH 2, and extract with chloroform. Dry the extract, redissolve in potassium bicarbonate solution, and dry again. Dissolve the mycolic acid residue in chloroform and derivatize to the *p*-bromophenacyl ester of the mycolic acid.

Ultraviolet-absorbing characteristics of the ester are used to develop a pattern of mycolic acids eluted from a HPLC column. HPLC absorption patterns differentiate genera containing mycolic acids and provide unique patterns for identification of species of *Mycobacterium,* as well as of *Nocardia* and *Rhodococcus* (Butler et al., 1987). Elution patterns of derivatized mycolic acids are standardized by inclusion of a high-molecular-weight standard in each run, and computer-assisted identification of species is being developed. The procedure is rapid in that cultures can be identified in about 2 h, and most species of *Mycobacterium* can be identified with little problem. Original published reports should be consulted for the presentation of representative patterns and the details of test performance.

IDENTIFICATION OF *MYCOBACTERIUM* SPECIES BY NAP TEST (BACTEC) (GROSS AND HAWKINS, 1985; LASZLO AND SIDDIQI, 1984; LASZLO AND HANDZEL, 1986). Species in the *M. tuberculosis* complex are selectively inhibited by NAP (*p*-nitro-α-acetylamino-β-hydroxy-propiophenone) a compound that is an intermediate in the synthesis of chloramphenicol. Growth of tubercle bacilli is inhibited by NAP, as is the enzymatic activity of the cell, so that evolution of $^{14}CO_2$ is minimal or absent. When the Growth Index (GI) of cultures growing in 12B medium reaches 50 to 100, the purity of the culture is confirmed by Gram staining, and 1 ml of the suspension is transferred to a new vial that contains a disc impregnated with 5 µg NAP in 2 ml of 7H12 medium. GI in the vial containing NAP is measured over the next 2 to 6 days and is compared with the original (control) vial that provided the inoculum. If GI in the control vial continues to increase, but there is no change or a decrease in the vial containing NAP, the organisms in culture are in the *M. tuberculosis* complex. This determination can usually be made in about 4 days, because mycobacteria not in the *M. tuberculosis* complex will have a GI above 500 in 4 to 6 days, depending on the species. The NAP test can also be performed with cultures on a solid medium if colonies are suspended in 0.2% bovine albumin at a turbidity equivalent to a McFarland #1 standard, and 0.1 ml of the suspension is added to a vial of 7H12 medium that is then treated as described above. Differential identification of the *M. tuberculosis* complex by this procedure is accomplished with a specificity of >99% (Morgan et al., 1985). Susceptibility to NAP can also be tested in any other liquid or solid medium and the result interpreted as turbidity or colony growth (Laszlo and Eidus, 1978).

IDENTIFICATION OF *M. TUBERCULOSIS* COMPLEX, *M. AVIUM*, AND *M. INTRACELLULARE* WITH SPECIFIC ^{125}I-DNA PROBES. Nucleic acid probes can be selected that will detect either organisms with highly specific characteristics or broader groups that share one or more characteristics (Kohne, 1989). Therefore, probes can be made specific to the genus, to the species, or to one or more species that share sequences. Probes complementary to RNA sequences of all *Mycobacterium* species, as well as to only *M. tuberculosis, M. avium,* and *M. intracellulare,* have been developed and are available commercially (Gen-Probe, Inc., San Diego, CA). In evaluation of ^{125}I-labeled probes, only 1 of 185 strains of *Mycobacterium* species failed to react, and none of the strains in other genera tested reacted (Woodley et al., 1989). Therefore, sensitivity of the test was 99.5% and specificity was 100%. Specificity and sensitivity, respectively, for the other three probes were 100% and 99.1% for *M. tuberculosis,* 100% and 100% for *M. avium,* and 100% and 100% for *M. intracellulare.* Failure of the standard methods to adequately identify strains of the *M. avium* complex was a problem in this evaluation, but rechecking tests and methods led to the conclusion that results with the Gen-Probes were correct, and the biochemical tests were not accurate.

Identification of *Mycobacterium* species in clinical laboratories with Gen-Probe kits has been very successful (Drake et al., 1987; Kiehn and Edwards, 1987; Musial et al., 1988), and DNA species probes have been generally accepted for this purpose. In an evaluation of the *M. avium* and *M. intracellulare* probes for accuracy and applicability to a clinical laboratory setting, all isolates were correctly identified, with no false positives recorded among 66 other cultures of mycobacteria and 8 of *Nocardia* (Drake et al., 1987). Using 9% hybridization as the limit for identification, the range with 94 isolates of *M. avium* was 11.5% to 72.7%, and with 40 strains of *M. intracellulare,* it was 22.7% to 60.7%. The method was judged simple to perform in this study and could be completed in as little as 2 h.

Ellner et al. (1988) collected mycobacterial cells from BACTEC vials with a GI ≥80 and prepared a suspension equivalent to a McFarland #1 standard for tests with Gen-Probe kits. A total of 176 isolates of *M. tuberculosis,* 110 of *M. avium,* and 5 of *M. intracellulare* were detected. Since 66% of isolates were identified within 2 weeks of setting up cultures and 100% within 4 weeks, time until reports are issued from the clinical laboratory can be reduced by 3 to 5 weeks.

It is necessary to have 10^5 to 10^7 cells available for the Gen-Probe Rapid Diagnostic System, and positive and negative controls must be tested also. Conville et al. (1989) concentrated mycobacterial growth in BACTEC vials by three different methods: centrifugation of one ml. of culture from the BACTEC vial for 5 min. at 3,400 × *g* followed by resuspension in sterile distilled water; treatment of the culture with Lysing Reagent (Gen-Probe), centrifugation, resuspension in Wash Reagent (Gen-Probe), centrifugation, and resuspension in water; and treatment with the Lysing Reagent only. The highest number of specimens was positive when both reagents were used prior to probe analysis. Therefore, a valid result can be obtained with the probe only if the test is performed as specified in package inserts that accompany the kits. All measurements must be precise, and all steps must be followed as closely as possible in order to obtain an accurate result.

DIAGNOSIS OF TUBERCULOUS MENINGITIS BY DETECTION OF TUBERCULOSTEARIC ACID IN SPINAL FLUID. Tuberculostearic acid was first isolated from *M. tuberculosis* and purified by Anderson and Chargaff (1929). This acid is a major component found by GLC following pyrolysis of mycobacterial cells (Larsson et al., 1979, 1980; Lambert et al., 1986). Detection of the acid in spinal fluid is diagnostic for tuberculous meningitis (Brooks et al., 1987; Elias et al., 1989; French et al., 1987a; Mardh et al., 1983).

Meningitis due to infection with *M. tuberculosis* is a very serious, life-threatening disease that must be diagnosed quickly so that effective therapy can be initiated (Tandon et al., 1988). Detection of tuberculostearic acid in spinal fluid can provide a better means of diagnosing this disease when cultures require weeks to report and the bacterium cannot always be recovered. Equipment for performance of the test varies from expensive to very expensive, since a mass spectrometer should be integrated into the GLC system. A method for the detection of femtomole quantities of the acid as the basis for diagnosis has been proposed (Brooks et al., 1987), but assays incorporating selective ion monitoring can be more precise. Detection of tuberculostearic acid in sputum has been proposed as diagnostic for pulmonary tuberculosis (French et al., 1987b; Larsson et al., 1981, 1987), but it is not yet clear whether other *Mycobacterium* species, including those that establish transient colonization, will give a positive test result also.

ROUTINE PROCEDURES FOR IDENTIFICATION OF *MYCOBACTERIUM* SPECIES (JENKINS ET AL., 1982; KENT AND KUBICA, 1985; KUBICA, 1984; SOMMERS AND GOOD, 1985; VESTAL, 1975). As stated in the introduction to this section, methods for performing routine tests for identification will not be described here. Results of the tests for identification are taken mostly from Good (1985), Good et al. (1985), and Kent and Kubica (1985).

IDENTIFICATION OF *MYCOBACTERIUM* SPECIES BY ROUTINE TEST PROCEDURES. Mycobacteria are grouped, on the basis of growth within 7 days, into rapid growers and slow growers as shown in Table 1. Pigmented, rapidly growing species are generally not associated with human infections, but the nonpigmented species are categorized on the basis of key reactions as shown in Table 4. Growth on MacConkey agar without crystal violet and lack of pigmentation may be used to separate the clinically significant, rapidly growing species from saprophytic species. Only subgroups of *M. chelonae* and *M. fortuitum* are positive in the 3-day arylsulfatase test. Identification to the particular subgroup is important because of differences in response to drugs (Wallace et al., 1983, 1985, 1988).

Testing for niacin production is important for the identification of *M. tuberculosis,* but some strains of *M. simiae,* which also produces pulmonary disease, may produce niacin as well. Differentiation of the two species is important because of the public health significance associated with identification of a patient with tuberculosis and because of the difference in therapeutic regimen required for treatment of the disease. Reactions for differentiation of the two species are shown in Table 5. Determination of NAP susceptibility by the BACTEC procedure and testing for percent homogenization with the DNA probe to *M. tuberculosis* (Gen-Probe) will readily differentiate the strains. Among routine differential tests, nitrate reduction and the two measures of catalase activity are most helpful. Not all strains of *M. simiae* are photochromogenic, but this is a valuable differential characteristic.

Differential reactions for nonphotochromogenic *Mycobacterium* species that may be seen in the clinical laboratory are given in Table 6. The NAP test and DNA probes are particularly useful in providing rapid identification of the more commonly encountered species, *M. avium* and *M. intracellulare.* *M. tuberculosis* is also a nonphotochromogenic species but is not included in the table since it is usually identified with the first niacin test run on rough, nonpigmented slow growers. The 10-day, Tween hydrolysis test distinguishes the potentially pathogenic nonphotochromogens, which are negative, from the commonly saprophytic species which are Tween positive. The major exception is that *M. malmoense* is Tween positive. Other reactions listed in Table 6 are used to identify individual nonphotochromogenic species.

MAC represents the second-most-frequently identified mycobacterial group in state laboratories (Good and Snider, 1982), but other nonphotochromogens are encountered only rarely. Therefore, it is usually not necessary for the

Table 4. Key tests for identification of rapidly growing mycobacteria of medical significance.

Test	*M. fortuitum* biovar			*M. chelonae* subsp.		*M. chelonae*-like	*M. smegmatis*
	fortuitum	*peregrinum*	3	*chelonae*	*abscessus*		
Arylsulfatase, 3 days	+	+	+	+	+	+	−
Nitrate reduction	+	+	+	−	−	−	+
Iron uptake	+	+	+	−	−	+[a]	+
NaCl tolerance, 28°C	+	+	+	−	+	−	+
Growth on:							
MacConkey agar without crystal violet	+	+	+	+	+	+	+
Sodium citrate	−	−	−	+	−	+	−
Mannitol	−	+	+	−	−	+	+
Inositol	−	−	+	−	−	−	+

[a]Reaction is tan rather than a deep rust color.
Adapted from Silcox et al. (1981) and Wallace et al. (1988).

Table 5. Key tests for differentiation of the niacin-producing nonphotochromogenic *Mycobacterium* species.[a]

Test	*M. tuberculosis*	*M. simiae*
Nitrate reduction	+	−
Catalase, room temperature	<45 mm	>45 mm
Catalase, 68°C	−	+
Pigment, photochromogenic	−	+
Temperature range for growth	33–39°C	22–41°C
Inhibition by NAP (BACTEC)[a]	Yes	No
M. tuberculosis DNA probe (Gen-Probe)[a]	+	−

[a]See text for explanation of tests.
Adapted from Kent and Kubica, 1985.

clinical microbiologist to subdivide *M. terrae* complex or to identify the saprophytic *M. gastri.* However, two potentially pathogenic species, *M. simiae* and *M. malmoense,* may be misidentified as MAC. Probes now available for the MAC species will make misidentification in this group less likely. *M. simiae* is included in a description of the nonphotochromogens since it often requires at least 8 h of exposure to light to stimulate pigment production, and then takes 24 h or more to display the photoactivated pigment. Identification of *M. xenopi* cultures, which may or may not be pigmented, depends on growth at 41 to 43°C and appearance of colonies on an agar medium, i.e., branching and filamentous extensions that can be observed around a circular core. This "bird's nest" appearance can be best seen with low-power microscopic observation of young colonies.

M. kansasii is the most frequently encountered photochromogenic species. Separation from *M. marinum* and *M. simiae,* other less frequently encountered photochromogenic species, can best be accomplished with Tween hydrolysis, nitrate reduction, pyrazinamidase, and 2-week arylsulfatase, as indicated in Table 7. The mycolic acid patterns of identity will also differentiate the species.

Key tests for identification of scotochromogenic species are shown in Table 8. Clinical laboratories must be able to identify *M. gordonae,* a saprophytic species, because it is the third-most-frequent mycobacterial isolate reported by state laboratories (Good and Snider, 1982). The organism is widely distributed in the environment and regularly appears as a contaminant in clinical specimens, but it has been shown to be the etiologic agent in infection only in exceedingly rare instances. Species that are scotochromogenic can be identified on the basis of mycolic acid patterns (Butler and Kilburn, 1988). Tests for Tween hydrolysis and urease

may also be important, particularly since many of the strains of *M. gordonae* are urease positive. *M. scrofulaceum* is not isolated with great frequency in clinical laboratories in the United States, but it is still found as a cause of cervical lymphadenopathy in many parts of the world.

As shown in Table 9, species in the *M. tuberculosis* complex can be differentiated on the basis of five tests (Collins and Yates, 1984; Collins et al., 1985). Oxygen preference is determined in the following way: deliver 0.2 ml of a homogenous suspension of bacilli in phosphate buffer about 1 cm below the surface of 5 to 10 ml of semisolid Middlebrook or Kirchner medium contained in a tube; mix contents carefully to avoid aeration; incubate for 18 days. If the culture is aerobic, growth will appear at or near the surface, but if it is microaerophilic, growth will be seen as a band 10 to 20 mm below the surface. These reactions are important in identifying many of the strains that react with the DNA probe, fail to grow in the presence of NAP, but do not appear to be typical *M. tuberculosis* strains.

Overview of Identification

Identification of *Mycobacterium* species is time consuming and requires experience in interpreting the results of various biochemical tests. Certainly, variation is found in responses to all of the tests, e.g., not all strains of *M. tuberculosis* are niacin positive, and not all strains of *M. gordonae* fail to produce urease. The new techniques of NAP susceptibility and reactivity with DNA probes make differentiation of *Mycobacterium* species much easier, but the most significant improvement is apt to be identification by the unique mycolic acid patterns. A culture of any of the 54 species can be identified in a 2- to 4-h time span, a single run through the HPLC. Another important method is the detection of tuberculostearic acid in patient specimens without the need for prolonged incubation. Further developments are expected using the new techniques of polymerase chain reaction that require only brief incubation periods, if any at all. Also, as selection of antigens and monoclonal antibodies improves, enzyme-linked immunoassays will become more sensitive and specific.

The changing patterns of mycobacterial disease emphasize the importance of laboratory support for the diagnosis. Control of tuberculosis and the other mycobacterial diseases is dependent on a definite diagnosis in a short period of time, perhaps even while the patient is still in the physician's office. It is conceivable that, within the next decade, such diagnoses will be

Table 6. Key tests for identification of nonphotochromogenic *Mycobacterium* species.

Test	M. bovis	M. avium	M. intracellulare	M. xenopi	M. gastri	M. terrae complex	M. simiae	M. malmoense	M. haemophilum
Catalase, room temperature	<45 mm	<45 mm	<45 mm	<45 mm	<45 mm	<45 mm	<45 mm	<45 mm	<45 mm
Catalase, 68°C	–	+	+	+	+	+	+	+	–
Tween hydrolysis	–	–	–	–	–	–	+	+	–
Tellurite reduction	–	+	+	±	–	+	+	+	–
Nitrate reduction	+	–	–	–	–	+	–	–	–
Inhibition by TCH	+	–	–	–	–	–	–	–	ND
NaCl tolerance	–	–	–	–	–	–	–	–	–
Urease	+	–	–	–	+	–	+	–	ND
Pyrazinamidase (4 day/7 day)	–/–	+/+	+/+	+/+	–/+	+/+	+/+	+/+	+/+
Pigment, photochromogenic	–	–	–	–	–	–	+	–	–
Niacin	–	–	–	–	–	–	+	–	–
Inhibition by NAP (BACTEC)	+	–	–	–	–	–	–	–	–
DNA probe (Gen-Probe):									
M. tuberculosis	+	–	–	–	–	–	–	–	–
M. avium	–	+	–	–	–	–	–	–	–
M. intracellulare	–	–	+	–	–	–	–	–	–

Adapted from Kent and Kubica, 1985.

Table 7. Key tests for identification of photochromogenic *Mycobacterium* species.

Test	M. kansasii	M. marinum	M. simiae
Tween hydrolysis	+	+	−
Nitrate reduction	+	−	−
Catalase, room temperature	>45 mm	>45 mm	>45 mm
Niacin	−	−	+
Pyrazinamidase, 4 days	−	+	+
Arylsulfatase, 2 weeks	+	+	−

Table 8. Key tests for identification of scotochromogenic *Mycobacterium* species.

Test	M. scrofulaceum	M. szulgai	M. xenopi	M. gordonae	M. flavescens
Tween hydrolysis	−	±	−	+	+
Nitrate hydrolysis	−	+	−	−	+
Catalase, room temperature	>45 mm	>45 mm	<45 mm	>45 mm	>45 mm
NaCl tolerance	−	−	−	−	±
Urease[a]	+	+	−	−	+
Arylsulfatase, 2 weeks	∓	∓	+	±	+

[a]Using urease disks.
Adapted from Kent and Kubica, 1985.

Table 9. Key tests for subdivision of *M. tuberculosis* complex.

	M. tuberculosis		M. africanum		M. bovis	
	Classic	Asian	I	II	Classic	BCG
Nitrate reduction	+	+	−	+	−	−
Oxygen preference	A	A	M	M	M	A
Susceptible to:						
TCH	−	+	+	+	+	+
PZA	+	+	+	+	−	−
CS	+	+	+	+	+	−
Inhibition by NAP (BACTEC)	+	+	+	+	+	+
M. tuberculosis DNA probe (Gen-Probe)	+	+	+	+	+	+

Abbreviations: A, aerobic; M, microaerophilic; TCH, thiopen-2-carboxylic acid hydrazide; PZA, pyrazinamide; CS, cycloserine.
Adapted from Collins and Yates, 1984; Collins et al., 1985.

made from urine or serum specimens. However, mycobacteriologists in clinical laboratories will be challenged to maintain the working knowledge and experience necessary to evaluate new procedures as they are introduced.

Laboratory Proficiency

Three levels of laboratory service are recognized in the United States. This concept reflects the frequency and species of mycobacteria isolated in clinical laboratories and the tests required for identification and antimicrobic susceptibility. (American Thoracic Society, 1983a, 1983b). Diagnostic laboratory personnel need to maintain proficiency to the level of service they provide and a smooth referral service needs to be established for higher levels of diagnostic tests. It is essential that every mycobacteriology labo-

ratory be equipped with proper space, hoods, and centrifuges to prevent laboratory infections, and that all personnel be trained in methods for isolation and identification of mycobacteria and in biosafety procedures (Kent and Kubica 1985; Richardson and Barkley, 1988).

Literature Cited

Ackerman, L. J., S. C. Benbrook, and B. C. Walton. 1974. *Mycobacterium tuberculosis* infection in a parrot (*Amazona farinosa*). Am. Rev. Respir. Dis. 109:388–390.
American Thoracic Society. 1981a. Diagnostic standards and classification of tuberculosis and other mycobacterial diseases, 14th ed. Am. Rev. Respir. Dis. 123:343–358.
American Thoracic Society. 1981b. The tuberculin skin test. Am. Rev. Respir. Dis. 124:356–363.

American Thoracic Society. 1983a. Levels of laboratory services for mycobacterial diseases. Am. Rev. Respir Dis. 128:1.

American Thoracic Society. 1983b. The levels of service concept in mycobacteriology, p. 19–25. In: Am. Thorac. Soc. News, Summer 1983.

American Thoracic Society. 1986. Treatment of tuberculosis and tuberculosis infection in adults and children. Am. Rev. Respir. Dis. 134:355–363.

American Thoracic Society. 1987. Mycobacterioses and the acquired immunodeficiency syndrome. Am. Rev. Respir. Dis. 136:492–496.

Anderson, R. J., and E. Chargaff. 1929. The chemistry of the lipoids of tubercle bacilli. VI. Concerning tuberculostearic acid and phthioic acid from the acetone-soluble fat. J. Biol. Chem. 85:77–88

Armstrong, D., J. W. M. Gold, J. Dryjanski, E. Whimbey, B. Polsky, C. Hawkins, A. E. Brown, E. Bernard, and T. Kiehn. 1985. Treatment of infections in patients with the acquired immunodeficiency syndrome. Ann. Intern. Med. 103:738–743.

Baess, I. 1979. Deoxyribonucleic acid relatedness among species of slowly-growing mycobacteria. Acta Pathol. Microbiol. Scand. 87:221–226.

Baess, I. 1983. Deoxyribonucleic acid relationships between different serovars of M. avium, M. intracellulare, and M. scrofulaceum. Acta Pathol. Microbiol. Scand. 91:201–203.

Bailey, R. K., S. Wyles, M. Dingley, F. Hesse, and G. W. Kent. 1970. The isolation of high catalase Mycobacterium kansasii from tap water. Am. Rev. Respir. Dis. 101:430–431.

Barksdale, L., and K.-S. Kim. 1977. Mycobacterium. Bacteriol. Rev. 41:217–372.

Belisle, J. T., and P. J. Brennan. 1989. Chemical basis of rough and smooth variation in mycobacteria. J. Bacteriol. 171:3465–3470.

Blaauwgeers, J. L., P. K. Das, A. W. Slob, and H. J. Houthoff. 1989. Human gut wall reactivity to monoclonal antibodies against M. avium glycolipid in relation to Crohn's disease. Acta Leprol. 7:138–140.

Blair, E. B., G. L. Brown, and A. H. Tull. 1976. Computer files and analyses of laboratory data from tuberculous patients. II. Analyses of six years' data on sputum specimens. Am. Rev. Respir. Dis. 113:427–432.

Bloch, A. B., H. L. Rieder, G. D. Kelly, S. M. Cauthen, C. H. Hayden, and D. E. Snider, Jr. 1989. The epidemiology of tuberculosis in the United States: implications for diagnosis and treatment. Clin. Chest Med. 10:297–313.

Bolan, G., A. L. Reingold, L. A. Carson, V. A. Silcox, C. L. Woodley, P. S. Hayes, A. W. Hightower, L. McFarland, J. W. Brown III, N. J. Petersen, M. S. Favero, R. C. Good, and C. V. Broome. 1985. Infections with Mycobacterium chelonei in patients receiving dialysis and using processed hemodialyzers. J. Infect. Dis. 152:1013–1019.

Brennan, P. J. 1981. Structures of the typing antigens of atypical mycobacteria: a brief review of present knowledge. Rev. Infect. Dis. 3:905–913.

Brennan, P. J. 1984. Antigenic peptidoglycolipids, phospholipids and glycolipids, p. 467–489. In: G. P. Kubica and L. G. Wayne (ed.), The mycobacteria, a sourcebook. Marcel Dekker, New York.

Brooks, J. B., M. I. Daneshvar, D. M. Fast., and R. C. Good. 1987. Selective procedures for detecting femtomole quantities of tuberculostearic acid in serum and cerebrospinal fluid by frequency-pulsed electron-capture gas-liquid chromatography. J. Clin. Microbiol. 25:1201–1206.

Brooks, J. B., M. I. Daneshvar, R. L. Haberberger, and I. A. Mikhail. 1990. Rapid diagnosis of tuberculous meningitis by frequency-pulsed electron-capture gas-liquid chromatography detection of carboxylic acids in cerebrospinal fluid. J. Clin. Microbiol. 28:989–997.

Brooks, R. W, B. C. Parker, and J. O. Falkinham III. 1984. Epidemiology of infection by nontuberculous mycobacteria. V. Numbers in eastern United States soils and correlation with soil characteristics. Am. Rev. Respir. Dis. 130:630–633.

Buhler, V. B., and A. Pollak. 1953. Human infection with atypical acid-fast organisms. Am. J. Clin. Pathol. 23:363–374.

Butler, W. R., D. G. Ahearn, and J. O. Kilburn. 1986. High-performance liquid chromatography of mycolic acids as a tool in the identification of Corynebacterium, Nocardia, Rhodococcus, and Mycobacterium species. J. Clin. Microbiol. 23:182–185.

Butler, W. R., and J. O. Kilburn. 1988. Identification of major slowly growing pathogenic mycobacteria and Mycobacterium gordonae by high-performance liquid chromatography of their mycolic acids. J. Clin. Microbiol. 26:50–53.

Butler, W. R., J. O. Kilburn, and G. P. Kubica. 1987. High-performance liquid chromatography analysis of mycolic acids as an aid in laboratory identification of Rhodococcus and Nocardia species. J. Clin. Microbiol. 25:2126–2131.

Capewell, S., and A. G. Leitch. 1986. Tuberculin reactivity in a chest clinic: the effects of age and prior BCG vaccination. Br. J. Dis. Chest. 80:37–44.

Carson, L. A., L. A. Bland, L. B. Cusick, M. S. Favero, G. A. Bolan, A. L. Reingold, and R. C. Good. 1988. Prevalence of nontuberculous mycobacteria in water supplies of hemodialysis centers. Appl. Environ. Microbiol. 54:3122–3125.

Carson, L. A., N. J. Petersen, M. S. Favero, and S. M. Aguero. 1978. Growth characteristics of atypical mycobacteria in water and their comparative resistance to disinfectants. Appl. Environ. Microbiol. 36:839–846.

Castets, M., H. Boisvert, F. Grumbach, M. Brunel, and N. Rist. 1963. Les bacilles tuberculeux du type africain. Rev. Tuberc. Pneumol. 32:179–184.

Castets, M., N. Rist, and H. Boisvert. 1969. La variété africane du bacille tuberculeux humain. Méd. Afr. Noire. 16:321–322.

Centers for Disease Control. 1988. Use of BCG vaccines in the control of tuberculosis: a joint statement by the ACIP and the Advisory Committee for Elimination of Tuberculosis. Morbid. Mortal. Weekly Rep. 37:663–675.

Centers for Disease Control. 1990. The use of preventive therapy for tuberculous infection in the United States. Morbid. Mortal. Weekly Rep. 39 (No. RR-8):9–12.

Chaisson, R. E., G. F. Schecter, C. Theuer, G. W. Rutherford, and D. Echenberg. 1987. Tuberculosis in patients with the acquired immunodeficiency syndrome. 1987. Am. Rev. Respir. Dis. 136:570–574.

Chaparas, S. D., H. MacVandiviere, I. Melvin, G. Koch, and C. Becker. 1985. Tuberculin test—variability with the Mantoux procedure. Am. Rev. Respir. Dis. 132:175–177.

Cheeseman, C. G., J. W. Wilesmith, and F. A. Stuary. 1989. Tuberculolsis: the disease and its epidemiology in the badger, a review. Epidemiol. Inf. 103:113–125

Chiodini, R. J. 1989. Crohn's disease and the mycobacterioses: a review and comparison of two disease entities. Clin. Microbiol. Rev. 2:90–117.

Cho, S. N., P. J. Brennan, H. H. Yoshimura, B. I. Korelitz, and D. Y. Graham. 1986. Mycobacterial etiology of Crohn's disease: serologic study using common mycobacterial antigens and a species-specific glycolipid antigen from *Mycobacterium paratuberculosis*. Gut 27:1353–1356.

Chretien, J. 1990. The cursed duet. Bull. Int. Union Tuberc. Lung Dis. 65:25–28.

Colebunders, R. L., R. W. Ryder, N. Nzilambi, K. Dikilu, J. C. Willame, M. Kaboto, N. Bagala, J. Jeugmans, K. Muepu, H. L. Francis, J. M. Mann, and T. C. Quinn. 1989. HIV infection in patients with tuberculosis in Kinshasa, Zaire. Am. Rev. Respir. Dis. 139:1082–1085.

Collins, C. H., and J. M. Grange. 1983. The bovine tubercle bacillus. J. Appl. Bacteriol. 55:13–29.

Collins, C. H., J. M. Grange, and M. D. Yates. 1984. Mycobacteria in water. J. Appl. Bacteriol. 57:193–211.

Collins, C. H., J. M. Grange, and M. D. Yates. 1985. Organization and practice in tuberculosis bacteriology. Butterworths, London.

Collins, C. H., and M. D. Yates. 1984. *Mycobacterium africanum* and the 'african' tubercle bacilli. Med. Lab. Sci. 41:410–413.

Collins, D. M., and G. W. DeLisle. 1985. DNA restriction endonuclease analysis of *M. bovis* and other members of the tuberculosis complex. J. Clin. Microbiol. 21:562–564.

Conville, P. S., J. F. Keiser, and F. G. Witebsky. 1989. Comparison of three techniques for concentrating positive BACTEC 13A bottles for mycobacterial DNA probe analysis. Diagn. Microbiol. Infect. Dis. 12:309–314.

Conville, P. S., and F. G. Witebsky. 1989. Inter-bottle transfer of mycobacteria by the BACTEC 460. Diagn. Microbiol. Infect. Dis. 12:401–406.

Crawford, J. T., and J. H. Bates. 1984. Restriction endonuclease mapping and cloning of *Mycobacterium intracellulare* plasmid pLR7. Gene. 27:331–333.

Crawford, J. T., and J. H. Bates. 1986. Analysis of plasmids in *Mycobacterium avium-intracellulare* isolates from persons with acquired immunodeficiency syndrome. Am. Rev. Respir. Dis. 134:659–661.

Crawford, J. T., K. D. Eisenach, and J. H. Bates. 1989. Diagnosis of tuberculosis: present and future. Semin. Respir. Infect. 4:171–181.

Crow, H. E., C. T. King, C. E. Smith, R. F. Corpe, and I. Stergus. 1957. A limited clinical, pathologic, and epidemiologic study of patients with pulmonary lesions associated with atypical acid-fast bacilli in the sputum. Am. Rev. Tuberc. 75:199–222.

Crowle, A. J., N. Elkins, and M. H. May. 1988. Effectiveness of ofloxacin against *Mycobacterium tuberculosis* and *Mycobacterium avium,* and rifampin against *M. tuberculosis* in cultured human macrophages. Am. Rev. Respir. Dis. 137:1141–1146.

Cutler, R. R., P. Wilson, and F. V. Clarke. 1987. The effect of polyoxyethylene stearate (POES) on the growth of mycobacteria in radiometric 7H12 Middlebrook TB medium. Tubercle 68:209–220.

Daniel, T. M. 1988. Antibody and antigen detection for the immunodiagnosis of tuberculosis: Why not? What more is needed? Where do we stand today? J. Infect. Dis. 158:678–680.

Daniel, T. M., and S. M. Debanne. 1987. The serodiagnosis of tuberculosis and other mycobacterial diseases by enzyme-linked immunosorbent assay. Am. Rev. Respir. Dis. 135:1137–1151.

Dannenberg, A. M., Jr. 1980. Pathogenesis of tuberculosis, p. 1264–1281. In: A. P. Fishman (ed.), Pulmonary diseases and disorders. McGraw-Hill Book Company, New York.

Dannenberg, A. M., Jr. 1982. Pathogenesis of pulmonary tuberculosis. Am. Rev. Respir. Dis. 125 (Suppl.):25–30.

Dannenberg, A. M., Jr. 1989. Immune mechanisms in the pathogenesis of pulmonary tuberculosis. Rev. Infect. Dis. 11 (Suppl.):369–378.

David, H. L., M.-T. Jahan, A. Jumin, J. Grandry, and E. H. Lehman. 1978. Numerical taxonomy analysis of *Mycobacterium africanum*. Int. J. Syst. Bacteriol. 28:467–472.

Davidson, P. T. 1989. The diagnosis and management of disease caused by *M. avium* complex, *M. kansasii,* and other mycobacteria. Clin. Chest Med. 10:431–433.

Dawson, D. J., E. W. Abrahams, and Z. M. Blacklock. 1974. Serum agglutinins in disease caused by *Mycobacterium intracellulare*. Appl. Microbiol. 27:1164–1166.

Dean, G. 1903. A disease of the rat caused by an acid-fast bacillus. Zentralb. Bakteriol. Parasitenkd. Abt. 1 34:222–224.

Des Prez, R. M., and C. R. Heim. 1990. *Mycobacterium tuberculosis,* p. 1877–1906. In: G. L. Mandell, R. G. Douglas, Jr., and J. E. Bennett (ed.), Principles and practice of infectious diseases, 3rd ed. Churchill Livingstone, New York.

Drake, T. A., J. A. Hindler, O. G. W. Berlin, and D. A. Bruckner. 1987. Rapid identification of *Mycobacterium avium* complex in culture using DNA probes. J. Clin. Microbiol. 25:1442–1445.

Dubos, R., and J. Dubos. 1952. The white plague. Tuberculosis, man and society. Little, Brown and Co., Boston.

du Moulin, G. C., I. H. Sherman, D. C. Hoaglin, and K. D. Stottmeier. 1985. *Mycobacterium avium* complex, an emerging pathogen in Massachusetts. J. Clin. Microbiol. 22:9–12.

du Moulin, G. C., and K. D. Stottmier. 1978. Use of cetylpyridinium chloride in the decontamination of water for culture of mycobacteria. Appl. Environ. Microbiol. 36:771–773.

du Moulin, G. C., K. D. Stottmeier, P. A. Pelletier, A. Y. Tsang, and J. Hedley-Whyte. 1988. Concentration of *Mycobacterium avium* by hospital hot water systems. J. Am. Med. Assoc. 260:1599–1601.

Dunn, B. L., and D. J. Hodgson. 1982. 'Atypical' mycobacteria in milk. J. Appl. Bacteriol. 52:373–376.

Edwards, L. B., and C. E. Palmer. 1968. Identification of the tuberculous infected by skin tests. Ann. N. Y. Acad. Sci. 154:140–148.

Eisenach, K. D., J. T. Crawford, and J. H. Bates. 1986. Genetic relatedness among strains of the *Mycobacterium tuberculosis* complex. Am. Rev. Respir. Dis. 133:1065–1068.

Eisenach, K. D., J. T. Crawford, and J. H. Bates. 1988. Repetitive DNA sequences as probes for *Mycobacterium tuberculosis*. J. Clin. Microbiol. 26:2240–2245.

Elias, J., J. P. DeConing, S. A. Vorster, and H. F. Joubert. 1989. The rapid and sensitive diagnosis of tuberculous meningitis by the detection of tuberculostearic acid in cerebrospinal fluid using gas chromatography-mass spectrometry with selective ion monitoring. Clin. Biochem. 22:463–467.

Ellner, P. D., T. E. Kiehn, R. Cammarata, and M. Hosmer. 1988. Rapid detection and identification of pathogenic mycobacteria by combining radiometric and nucleic acid probe methods. J. Clin. Microbiol. 26:1349–1352

Fadda, G. 1988. Mycobacterioses excluding tuberculosis, p. 382–395. In: A. Balows, W. J. Hausler, Jr., and E. H. Lennette (ed.), Laboratory diagnosis of infectious diseases: principles and practice. Springer-Verlag, New York.

Falkinham, J. O., III., K. L. George, and B. C. Parker. 1989. Epidemiology of infection by nontuberculous mycobacteria. VII. Absence of mycobacteria in chicken litter. Am. Rev. Respir. Dis. 139:1347–1349.

Falkinham, J. O., III, B. C. Parker, and H. Gruft. 1980. Epidemiology of infection by nontuberculous mycobacteria. I. Geographic distribution in the eastern United States. Am. Rev. Respir. Dis. 121:931–937.

Feldman, W. H. 1960. Avian tubercle bacilli and other mycobacteria, their significance in the eradication of bovine tuberculosis. Am. Rev. Respir. Dis. 81:666–673.

Flepp, M., K. Rhyner, R. Luethy, P. Greminger, U. Vurma-Rapp, C. Wolfisberg, A. Burnens, and W. Siegenthaler. 1988. Mycobacterial infection in patients with HIV infection. Dtsch. Med. Wochenschr. 113:711–718.

French, G L., C. Y. Chan, S. W. Cheung, and K. T. Oo. 1987b. Diagnosis of pulmonary tuberculosis by detection of tuberculostearic acid in sputum by using gas chromatography-mass spectrometry with selected ion monitoring. J. Infect. Dis. 156:356–362.

French, G. L., C. Y. Chan, S. W. Cheung, R. Teoh., M. J. Humphries, G. O'Mahony. 1987a. Diagnosis of tuberculous meningitis by detection of tuberculostearic acid in cerebrospinal fluid. Lancet i:117–119.

Frost, W. H. 1937. How much control of tuberculosis? Am. J. Public Health 27:759–766.

Fry, K. L., P. S. Meissner, and J. O. Falkinham, III. 1986. Epidemiology of infection by nontuberculous mycobacteria. VI. Identification and use of epidemiologic markers for studies of *Mycobacterium avium, M. intracellulare,* and *M. scrofulaceum.* Am. Rev. Respir. Dis. 134:39–43.

Gangadharam, P. R. J., and M. D. Iseman. 1987. Antimycobacterial drugs. Antimicrob. Agents Ann. 2:14–35.

Gangadharam, P. R. J., V. K. Perumal, J. T. Crawford, and J. H. Bates. 1988. Association of plasmids and virulence of *Mycobacterium avium* complex. Am. Rev. Respir. Dis. 137:212–214.

Gangadharam, P. R. J., V. K. Perumal, B. T. Jairam, N. R. Podapati, R. B. Taylor, and J. F. LaBrecque. 1989. Virulence of *Mycobacterium avium* complex strains from acquired immune deficiency syndrome patients: relationship with characteristics of the parasite and host. Microb. Path. 7:263–278.

Gay, J. D., D. R. DeYoung, and G. D. Roberts. 1984. In vitro activities of norfloxacin and ciprofloxacin against *Mycobacterium tuberculosis, M. avium* complex, *M. chelonei, M. fortuitum,* and *M. kanasasii.* Antimicrob. Agents Chemother. 26(1):94–96.

George, K. L., B. C. Parker, H. Gruft, and J. O. Falkinham, III. 1980. Epidemiology of infection by nontuberculous mycobacteria. II. Growth and survival in natural waters. Am. Rev. Respir. Dis. 122:89–94.

Gitnick, G., J. Collins, B. Beaman, D. Brooks, M. Arthur, T. Imaeda, and M. Palieschesky. 1989. Preliminary report on isolation of mycobacteria from patients with Crohn's disease. Dig. Dis. Sci. 34:925–932.

Good, R. C. 1980. Isolation of nontuberculous mycobacteria in the United States, 1979. J. Infect. Dis. 142:779–783.

Good, R. C. 1985. Opportunistic pathogens in the genus *Mycobacterium.* Ann. Rev. Microbiol. 39:347–369.

Good, R. C. 1988. Tuberculosis, p. 504–518. In: A. Balows, W. J. Hausler, Jr., and E. H. Lennette (ed.), Laboratory diagnosis of infectious diseases: principles and practice. Springer-Verlag, New York.

Good, R. C. 1989. Serologic methods for diagnosing tuberculosis. Ann. Intern. Med. 110:97–98.

Good, R. C., and R. E. Beam. 1984. Seroagglutination, p. 105–122. In: G. P. Kubica and L. G. Wayne (ed.), The mycobacteria: A sourcebook. Marcel Dekker, Inc., New York.

Good, R. C., and T. D. Mastro. 1989. The modern mycobacteriology laboratory: How it can help the clinician. Clin. Chest Med. 10:315–322.

Good, R. C., V. A. Silcox, and J. O. Kilburn. 1985. Identification and drug susceptibility test results for *Mycobacterium* spp. Clin. Microbiol. Nwsltr. 7:133–136.

Good, R. C., and D. E. Snider, Jr. 1982. Isolation of nontuberculous mycobacteria in the United States, 1980. J. Infect. Dis. 146:829–833.

Goren, M. B., M. Cernich, and O. Brokl. 1978. Some observations on mycobacterial acid-fastness. Am. Rev. Respir. Dis. 118:151–154.

Goslee, S., and E. Wolinsky. 1976. Water as a source of potentially pathogenic mycobacteria. Am. Rev. Respir. Dis. 113:287–292.

Graham, D. Y., D. C. Markesich, and H. H. Yoshimura. 1987. Mycobacteria and inflammatory bowel disease. Results of culture. Gastroenterology 92:436–442.

Greene, J. B., S. G. Sidhu, S. Lewin, J. F. Levine, H. Masur, M. S. Simberkoff, P. Nicholas, R. C. Good, S. B. Zolla-Pazner, A. A. Pollock, M. L. Tapper, and R. S. Holzman. 1982. *Mycobacterium avium-intracellulare:* a cause of disseminated life-threatening infection in homosexuals and drug abusers. Ann. Intern. Med. 97:539–546.

Grange, J. M. 1984. The humoral immune response in tuberculosis, its nature, biological role and diagnostic usefulness. Adv. Tuberc. Res. 21:1–78.

Gross, W. M., and J. E. Hawkins. 1985. Radiometric selective inhibition tests for differentiation of *Mycobacterium tuberculosis, Mycobacterium bovis,* and other mycobacteria. J. Clin. Microbiol. 21:565–568.

Grosset, H. H. C. C. Guelpa-Lauras, E. G. Perani, and C. Beoletto. 1988. Activity of ofloxacin against *Mycobacterium leprae* in the mouse. Int. J. Lepr. 56:259–264.

Gruft, H. 1971. Isolation of acid-fast bacilli from contaminated specimens. Health Lab. Sci. 8:79–82.

Gruft, H., J. O. Falkinham III, and B. C. Parker. 1981. Recent experience in the epidemiology of disease caused by atypical mycobacteria. Rev. Infect. Dis. 3:990–996.

Hawkins, C. C., J. W. M. Gold, E. Whimbey, T. E. Kiehn, P. Brannon, R. Cammarata, A. E. Brown, and D. Armstrong. 1986. *Mycobacterium avium* complex infections in patients with the acquired immunodeficiency syndrome. Ann. Intern. Med. 105:184–188

Hawkins, J. E. 1977. Scotochromogenic mycobacteria which appear intermediate between *Mycobacterium avium-intracellulare* and *Mycobacterium scrofulaceum*. Am. Rev. Respir. Dis. 116:963–964.

Heifets, L. B. 1986. Rapid automated methods (BACTEC system) in clinical mycobacteriology. Semin. Respir. Infect. 1:242–249.

Heifets, L. B., M. D. Iseman, and P. J. Lindholm-Levy. 1987. Determination of MICs of conventional and experimental drugs in liquid medium by the radiometric method against *Mycobacterium avium* complex. Drugs Exp. Clin. Res. 13:529–538.

Heifets, L. B., and P. J. Lindholm-Levy. 1987. Bacteriostatic and bactericidal activity of ciprofloxacin and ofloxacin against *Mycobacterium tuberculosis* and *Mycobacterium avium* complex. Tubercle 68:267–276.

Heifets, L. B., and P. J. Lindholm-Levy. 1989. Comparison of bactericidal activities of streptomycin, amikacin, kanamycin, and capreomycin against *Mycobacterium avium* and *M. tuberculosis*. Antimicrob. Agents Chemother. 33:1298–1301.

Hitz, J. R. 1986. D.P.H. dissertation. University of North Carolina, Chapel Hill.

Hoffman, P. C., D. W. Fraser, F. Robicsek, P. R. O'Bar, and C. U. Mauney. 1981. Two outbreaks of sternal wound infections due to organisms of the *Mycobacterium fortuitum* complex. J. Infect. Dis. 143:533–542.

Hoffner, S. E. 1988. Improved detection of *Mycobacterium avium* complex with the BACTEC radiometric system. Diagn. Microbiol. Infect. Dis. 10:1–6.

Horsburgh, C. R., U. G. Mason, D. C. Farhi, and M. D. Iseman. 1985. Disseminated infection wit-*Mycobacterium avium-intracellulare*. Medicine 64:39–48.

Horsburgh, C. R., and R. M. Selik. 1989. The epidemiology of disseminated nontuberculous mycobacterial infection in the acquired immunodeficiency syndrome (AIDS). Am. Rev. Respir. Dis. 139:4–7.

Hsu, K. H. K. 1983. Tuberculin reaction in children treated with isoniazid. Am. J. Dis. Child. 137:1090–1092.

Imaeda, T. 1985. Deoxyribonucleic acid relatedness among selected strains of *Mycobacterium tuberculosis*, *Mycobacterium bovis*, *Mycobacterium bovis* BCG, *Mycobacterium microti*, and *Mycobacterium africanum*. Int. J. Syst. Bacteriol. 35:147–150.

Isaac, J., J. Whitehead, J. W. Adams, M. D. Barton, and P. Coloe. 1983. An outbreak of *Mycobacterium bovis* infection in cats an an animal house. Aust. Vet. J. 60:243–245.

Iseman, M. D. 1989. *Mycobacterium avium* complex and the normal host: the other side of the coin. N. Engl. J. Med. 321:896–898.

Ivanyi, J., G. H. Bothamley, and P. S. Jackett. 1988. Immunodiagnostic assays for tuberculosis and leprosy. Brit. Med. Bull. 44:635–649.

Jenkins, P. A., S. R. Pattyn, and F. Portaels. 1982. Diagnostic bacteriology, p. 441–470. In: C. Ratledge and J. Stanford (ed.), The biology of the mycobacteria. Academic Press, London.

Jacobson, M. A. 1988. Mycobacterial diseases. Tuberculosis and *Mycobacterium avium* complex. Infect. Dis. Clin. North Am. 2:465–474.

Johne, H. A., and L. Frothingham. 1895. Ein eigenthumlicher fall von tuberculose beim rinde. Dtsch. Zeitschr. Thiermed. Vergleich. Pathol. 21:438–454.

Jones, W. D., Jr. 1988. Bacteriophage typing of *Mycobacterium tuberculosis* cultures from incidents of suspected laboratory cross-contamination. Tubercle 69:43–46.

Jones, W. D., Jr., R. C. Good, N. J. Thompson, and G. D. Kelly. 1982. Bacteriophage types of *Mycobacterium tuberculosis* in the United States. Am. Rev. Respir. Dis. 125:640–643.

Jones, W. D., Jr., and C. L. Woodley. 1983. Phage-type patterns of *Mycobacterium tuberculosis* from southeast Asian immigrants. Am. Rev. Respir. Dis. 127:348–349.

Jorgensen, J. B. 1978. Serologic investigation of strains of *Mycobacterium avium* and *Mycobacterium intracellulare* isolated from animal and non-animal sources. Nordisk Veterinaer Medicin 30:155–162.

Kazda, J. 1973a. Die Bedeutung von Wasser für die Verbreitung von potentiell pathogen Mykobacterien. I. Mögleichkeiten für eine Vehmehrung von Mykobakterien. Zentralb. Bakteriol. Parasitenkd. Infektionskr. Hyg. Abt. 1 Orig., Rhihe A 158:161–169.

Kazda, J. 1973b. Die Bedeutung von Wasser für die verbreitung von potentiell pathogenen Mykobakterien. II. Vehmehrung der Mykobakterien in Gewässermodellen. Zentralb. Bakteriol. Parasitenkd. Infektionskr. Hyg. Abt. 1 Orig., Rhihe A 158:170–176.

Kazda, J., and R. Hoyte. 1972. Zur Ökologie von *Mycobacterium intracellulare* serotype Davis. Zentralb. Bakteriol. Parasitenkd. Infektionskr. Hyg. Abt. 1 Orig., Rhihe A 222:506–509.

Kent, P. T., and G. P. Kubica. 1985. Public health mycobacteriology: a guide for the level III laboratory. U. S. Department of Health and Human Services, Centers for Disease Control, Atlanta.

Kiehn, T. E., and F. F. Edwards. 1987. Rapid identification using a specific DNA probe of *Mycobacterium avium* complex from patients with acquired immunodeficiency syndrome. J. Clin. Microbiol. 25:1551–1552.

Kiehn, T. E., F. F. Edwards, P. Brannon, A. Y. Tsang, M. Maio, J. W. M. Gold, E. Whimbey, B. Wong, J. K. McClatchy, and D. Armstrong. 1985. Infections caused by *Mycobacterium avium* complex in immunocompromised patients: diagnosis by blood culture and fecal examination, antimicrobial susceptibility tests, and morphological and seroagglutination characteristics. J. Clin. Microbiol. 21:168–173.

Kirihara, M. J., S. L. Hillier, and M. B. Coyle. 1985. Improved detection times for *Mycobacterium avium* complex and *Mycobacterium tuberculosis* with the BACTEC radiometric system. J. Clin. Microbiol. 22:841–845.

Kobayashi, K., M. J. Blaser, and W. R. Brown. 1989. Immunohistochemical examination for mycobacteria in intestinal tissues from patients with Crohn's disease. Gastroenterology 96:1009–1015.

Koch, R. 1882. Die aetiologie der tuberculose. Berl. Klin. Wochenschr. 19:221–30.

Koch, R. 1890. Weitere Mittheilungen über ein Heilmitted gegen Tuberculose. Deutsche medizinische Wochenschrift 16:756–757.

Kohne, D. E. 1989. The use of DNA probes to detect and identify microorganisms, p. 11–35. In: B. Kleger, D. Jungkind, E. Hinks, and L. A. Miller (ed.), Rapid methods in clinical microbiology. Plenum Press, New York.

Kononov, Y., K. D. Ta, and L. Heifets. 1988. Effect of egg yolk on growth of *Mycobacterium tuberculosis* in 7H12 liquid medium. J. Clin. Microbiol. 26:1395–1397.

Kopanoff, D. E., J. O. Kilburn, J. L. Glassroth, D. E. Snider, Jr., L. S. Farrer, and R. C. Good. 1978. A continuing survey of tuberculosis primary drug resistance in the United States: March 1975 to November 1977. Am Rev. Respir. Dis. 118:835–842.

Krasnow, I., and L. G. Wayne. 1969. Comparison of methods for tuberculosis bacteriology. Appl. Microbiol. 18:915–917.

Kubica, G. P. 1984. Clinical microbiology, p. 133–175. In: G. P. Kubica and L. G. Wayne (ed.), The mycobacteria: a sourcebook. Marcel Dekker, Inc., New York.

Kubica, G. P., R. E. Beam, and J. W. Palmer. 1963. A method for the isolation of unclassified acid-fast bacilli from soil and water. Am. Rev. Respir. Dis. 88:718–720.

Kubica, G. P., W. E. Dye, M. L. Cohn, and G. Middlebrook. 1963. Sputum digestion and decontamination with N-acetyl-L-cysteine-sodium hydroxide for culture of mycobacteria. Am. Rev. Respir. Dis. 87:775–779.

Kubica, G. P., and R. C. Good. 1981. The genus *Mycobacterium* (except *M. leprae*), p. 1962–1984. In: M. P. Starr, H. Stolp, H. G. Truper, A. Balows, and H. G. Schlegel (ed.), The prokaryotes: A handbook on habitats, isolation, and identification of bacteria. Springer-Verlag, Berlin.

Kuritsky, J. N., M. G. Bullen, C. V. Broome, V. A. Silcox, R. C. Good, and R. J. Wallace, Jr. 1983. Sternal wound infections and endocarditis due to organisms of the *M. fortuitum* complex. Ann. Intern. Med. 98:938–939.

Lambert, M. A., C. W. Moss, V. A. Silcox, and R. C. Good, 1986. Analysis of mycolic acid cleavage products and cellular fatty acids of *Mycobacterium* species by capillary gas chromatography. J. Clin. Microbiol. 23:731–736.

Larsson, L., P.-A. Mardh, and G. Odham. 1979. Detection of tuberculostearic acid in mycobacteria and nocardiae by gas chromatography and mass spectrometry using selected ion monitoring. J. Chromatogr. 163:221–224.

Larsson, L., P.-A. Mardh, G. Odham, G. Westerdahl. 1980. Detection of tuberculostearic acid in biological specimens by means of glass capillary gas chromatography-electron and chemical ionization mass spectrometry, utilizing selected ion monitoring. J. Chromatogr. 182:402–408.

Larsson, L., P.-A. Mardh, G. Odham, and G. Westerdahl. 1981. Use of selected ion monitoring for detection of tuberculostearic and C_{32} mycocerosic acid in mycobacteria and in five-day-old cultures of sputum specimens from patients with pulmonary tuberculosis. Acta Pathol. Microbiol. Scand., Sect. B. 89:245–251.

Larsson, L., G. Odham, G. Westerdahl, and B. Olsson. 1987. Diagnosis of pulmonary tuberculosis by selected-ion monitoring: improved analysis of tuberculostearate in sputum using negative-ion mass spectrometry. J. Clin. Microbiol. 25:893–896.

Laszlo, A., and L. Eidus. 1978. Test for differentiation of *M. tuberculosis* and *M. bovis* from other mycobacteria. Can. J. Microbiol. 24:754–756.

Laszlo, A., and V. Handzel. 1986. Radiometric diagnosis of mycobacteria. Eur. J. Clin. Microbiol. 5:152–155.

Laszlo, A., and S. H. Siddiqi. 1984. Evaluation of a rapid radiometric differentiation test for the *Mycobacterium tuberculosis* complex by selective inhibition with p-nitro-α-acetylamino-β-hydroxy-propiophenone. J. Clin. Microbiol. 19:694–695.

Laussucq, S., A. L. Baltch, R. P. Smith, R. P. Smith, R. W. Smithwick, B. J. Davis, E. K. Desjardin, V. A. Silcox, A. B. Spellacy, R. T. Zeimis, H. M. Gruft, R. C. Good, and M. L. Cohen. 1988. Nosocomial *Mycobacterium fortuitum* colonization from a contaminated ice machine. Am. Rev. Respir. Dis. 138:891–894.

Levy-Frebault, V. V., M. F. Thorel, A. Varnerot, and B. Gicquel. 1989. DNA polymorphism in *Mycobacterium paratuberculosis,* "wood pigeon mycobacteria," and related mycobacteria analyzed by field inversion gel electrophoresis. J. Clin. Microbiol. 27:2823–2826.

Levy-Frebault, V., F. Grimont, P. A. D. Grimont, and H. L. David. 1986. Deoxyribonucleic acid relatedness study of the *Mycobacterium fortuitum-Mycobacterium chelonae* complex. Int. J. Syst. Bacteriol. 36:458–460.

Macher, A. M. 1984. Acquired immunodeficiency syndrome: epidemiologic, clinical, immunologic, and therapeutic considerations. Ann. Intern. Med. 100:92–106.

Maliwan, N., R. W. Reid, S. R. Pliska, T. J. Bird, and J. R. Zvetina. 1988. Identifying *Mycobacterium tuberculosis* cultures by gas-liquid chromatography and a computer-aided pattern recognition model. J. Clin. Microbiol. 26:182–187.

Mann, J., D. E. Snider, Jr., H. Francis, T. C. Quinn, R. L. Colebunders, P. Piot, J. W. Curran, N. Nzilambi, N. Bosenge, M. Malonga, D. Kalunga, M. M. Nzingg, and N. Bagala. 1986. Association between HTLV-III/LAV infection and tuberculosis in Zaire. J. Am. Med. Assoc. 256:346.

Mardh, P.-A., L. Larsson, N. Holby, H. C. Engback, and G. Odham. 1983. Tuberculostearic acid as a diagnostic marker in tuberculous meningitis. Lancet i:367.

Markesich, D. C., D. Y. Graham, and H. H. Yoshimura. 1988. Progress in culture and subculture of spheroplasts and fastidious acid-fast bacilli isolated from intestinal tissues. J. Clin. Microbiol. 26:1600–1603.

Martin, E. C., B. C. Parker, and J. O. Falkinham III. 1987. Epidemiology of infection by nontuberculous mycobacteria. VII. Absence of mycobacteria in southeastern groundwaters. Am. Rev. Respir. Dis. 136:344–348.

Matthews, P. R. J., P. Collins, and P. W. Jones. 1976. Isolation of mycobacteria from dairy creamery effluent sludge. J. Hyg. 76:407–413.

McClatchy, J. K. 1981. The seroagglutination test in the study of nontuberculous mycobacteria. Rev. Infect. Dis. 3:867–870.

McClatchy, J. K., R. F. Waggoner, W. Kanes, M. S. Cernich, and T. L. Bolton. 1976. Isolation of mycobacteria from clinical specimens by use of a selective 7H-11 medium. Am. J. Clin. Pathol. 65:412–415.

McFadden, J. J., P. D. Butcher, R. Chiodini, and J. Hermon-Taylor. 1987a. Crohn's disease-isolated mycobacteria are identical to *Mycobacterium paratuberculosis,* as determined by DNA probes that distinguish between mycobacterial species. J. Clin. Microbiol. 25:796–801.

McFadden, J. J., P. D. Butcher, J. Thompson, R. Chiodini, and J. Hermon-Taylor. 1987b. The use of DNA probes identifying restriction-fragment-length polymorphisms to examine the *Mycobacterium avium* complex. Mol. Microbiol. 1:283–291.

McIntyre, G., and J. L. Stanford. 1986. The relationship between immunodiffusion and agglutination serotypes of *Mycobacterium avium* and *Mycobacterium intracellulare*. Eur. J. Respir. Dis. 69:135–141.

McSwiggan, D. A., and C. H. Collins. 1974. The isolation of *M. kansasii* and *M. xenopi* from water systems. Tubercle 55:291–297.

Meissner, G., and W. Anz. 1977. Sources of *Mycobacterium avium* complex infection resulting in human disease. Am. Rev. Respir. Dis. 116:1057–1064.

Middlebrook G., Z. Reggiardo, and W. D. Tigertt. 1977. Automatable radiometric detection of growth of *Mycobacterium tuberculosis* in selective media. Am. Rev. Respir. Dis. 115:1066–1069.

Minnikin, D. E., G. Dobson, and I. G. Hutchinson. 1983. Characterization of phthiocerol dimycocerosates from *Mycobacterium tuberculosis*. Biochim. Biophys. Acta 753:445–449.

Minnikin, D. E., and M. Goodfellow. 1980. Lipid composition in the classification and identification of acid-fast bacteria, p. 189–256. In: M. Goodfellow and R. G. Board (ed.), Microbiological classification and identification. Academic Press, London.

Minnikin, D. E., M. Goodfellow, and M. D. Collins. 1978. Lipid composition in the classification and identification of coryneform and related taxa, p. 85–160. In: I. J. Bousfield and A. G. Callely (ed.), Coryneform bacteria. Academic Press, London.

Minnikin, D. E., I. G. Hutchinson, A. B. Caldicott, and M. Goodfellow. 1980. Thin-layer chromatography of methanolysates of mycolic acid-containing bacteria. J. Chromatogr. 188:221–233.

Minnikin, D. E., S. M. Minnikin, J. H. Parlett, M. Goodfellow, and M. Magnusson. 1984a. Mycolic acid patterns of some species of *Mycobacterium*. Arch. Microbiol. 139:225–231.

Minnikin, D. E., J. H. Parlett, M. Magnusson, M. Ridell, and A. Lind. 1984b. Mycolic acid patterns of representatives of *Mycobacterium bovis* BCG. J. Gen. Microbiol. 130:2733–2736.

Mitchison, D. A., B. W. Allen, L. Carol, J. M. Dickinson, and V. R. Aber. 1972. A selective oleic acid albumin agar medium for tubercle bacilli. J. Med. Microbiol. 5:165–175.

Modilevsky, T, F. R. Sattler, and P. F. Barnes. 1989. Mycobacterial disease in patients with human immunodeficiency virus infection. Arch. Intern. Med. 149:2201–2205.

Morgan, M., C. D. Horstmeier, D. R. DeYoung, and G. D. Roberts. 1983. Comparison of a radiometric method (BACTEC) and conventional culture media for recovery of mycobacteria from smear-negative specimens. J. Clin. Microbiol. 18:384–388.

Morgan, M. A., K. A. Doerr, H. O. Hempel, N. L. K. Goodman, and G. D. Roberts. 1985. Evaluation of the *p*-nitro-acetylamino-β-hydroxypropiophenone differential test for identification of *Mycobacterium tuberculosis* complex. J. Clin. Microbiol. 21:634–635.

Mori, T., and K. Shiozawa. 1985. Suppression of tuberculin hypersensitivity caused by rubella infection. Am. Rev. Respir. Dis. 131:886–888.

Murray, C. J. L., K. Styblo, and A. Rouillon. 1990. Tuberculosis in developing countries: burden, intervention and cost. Bull. Int. Union Tuberc. Lung Dis. 65:6–24.

Murray, J. F., C. P. Felton, S. M. Garay, M. S. Gottlieb, P. C. Hopewell, D. E. Stover, and A. S. Teirstein. 1984. Pulmonary complications of the acquired immunodeficiency syndrome. N. Eng. J. Med. 310:1682–1688.

Musial, C. E., L. S. Tice, L. Stockman, and G. D. Roberts. 1988. Identification of mycobacteria from culture by using Gen-Probe Rapid Diagnostic System for *Mycobacterium avium* complex and *Mycobacterium tuberculosis* complex. J. Clin. Microbiol. 26:2120–2123.

O'Brien, R. J., L. J. Geiter, and D. E. Snider, Jr. 1987. The epidemiology of nontuberculous mycobacterial diseases in the United States: results from a national survey. Am. Rev. Respir. Dis. 135:1007–1014.

O'Brien, R. J., M. A. Lyle, and D. E. Snider, Jr. 1987. Rifabutin (ansamycin LM 427): a new rifamycin-S derivative for the treatment of mycobacterial diseases. Rev. Infect. Dis. 9:519–530.

O'Brien, R. J., and D. E. Snider, Jr. 1985. Tuberculosis drugs—old and new. Am. Rev. Respir. Dis. 131:309–311.

Orme, I. M. 1988. Induction of nonspecific acquired resistance and delayed-type hypersensitivity, but not specific acquired resistance in mice inoculated with killed mycobacterial vaccines. Infect. Immun. 56:3310–3312.

Palmer, C. E., and L. B. Edwards. 1966. Sensitivity to mycobacterial PPD antigens with some laboratory evidence of its significance. Tuberkuloza. 18:193–200.

Palmer, C. E., and L. B. Edwards. 1967. Tuberculin test in retrospect and prospect. Arch. Environ. Health. 15:793–808.

Parker, B. C., M. A. Ford, H. Gruft, and J. O. Falkinham III. 1983. Epidemiology of infection by nontuberculous mycobacteria. IV. Preferential aerosolization of *Mycobacterium intracellulare* from natural waters. Am. Rev. Respir. Dis. 128:652–656.

Pattyn, S. R., F. Portaels, L. Spanoghe, and J. Magos. 1970. Further studies on African strains of *Mycobacterium tuberculosis*. Comparison with *Mycobacterium bovis* and *Mycobacterium microti*. Ann. Soc. Belg. Méd. Trop. Parasitol. Mycol. 50:211–228.

Pelletier, P. A., G. C. du Moulin, and K. D. Stottmeier. 1988. Mycobacteria in public water supplies: comparative resistance to chlorine. Microbiol. Sci. 5:147–148.

Perez-Stable, E. J., D. Flaherty, G. Schecter, G. Slutkin, and P. C. Hopewell. 1988. Conversion and reversion of tuberculin reactions in nursing home residents. Am. Rev. Respir. Dis. 137:801–804.

Pethel, M. L., and J. O. Falkinham. 1989. Plasmid-influenced changes in Mycobacterium avium catalase activity. Infect. Immun. 57:1714–1718.

Petran, E., and H. D. Vera. 1971. Media for selective isolation of mycobacteria. Health Lab. Sci. 8:225–230.

Picken, R. N., S. J. Plotch, Z. Wang, B. C. Lin, J. J. Donegan, and H. L. Yang. 1988. DNA probes for mycobacteria. Mol. Cell. Probes. 2:111–124.

Pierce, P. F., D. R. DeYoung, and G. D. Roberts. 1983. Mycobacteremia and the new blood culture systems. Ann. Intern. Med. 99:786–789.

Pitchenik, A. E., C. Cole, B. W. Russell, M. A. Fischli, T. J. Spira, and D. E. Snider, Jr. 1984. Tuberculosis, atypical mycobacteriosis, and the acquired immunodeficiency syndrome among non-Haitian patients in south Florida. Ann. Intern. Med. 101:641–645.

Pitchenik A. E., D. Fertel, and A. B. Bloch. 1988. Mycobacterial disease: epidemiology, diagnosis, treatment, and prevention. Clin. Chest Med. 9:425–441.

Prantera, C., Bothamley G., S. Levenstein, R. Mangiarotti, and R. Argentieri. 1989. Crohn's disease and mycobacteria: two cases of Crohn's disease with high antimycobacterial antibody levels cured by dapsone therapy. Biomed. Pharmacother. 43:295–299.

Prince, D. S., D. D. Peterson, R. M. Steiner, J. E. Gottlieb, R. Scott, H. L. Israel, W. G. Figueroa, and J. E. Fish. 1989. Infection with *Mycobacterium avium* complex in patients without predisposing conditions. N. Engl. J. Med. 321:863–868.

Rabinowitsch, L. 1903. Ueber eine durch säure-fest bakterien hervorgerufene hauterkrankung der ratten. Zentralbl. Bakteriol. Parasitenkd. Abt. 1, 33:577–580.

Raja, A., A. R. Machicao, A. B. Morrissey, M. R. Jacobs, and T. M. Daniel. 1988. Specific detection of *Mycobacterium tuberculosis* in radiometric cultures by using an immunoassay for antigen 5. J. Infect. Dis. 158:468–470.

Reichman, L. B. 1979. Tuberculin skin testing: the state of the art. Chest 76 (Suppl.):764–770.

Reznikov, M., and J. H. Leggo. 1974. Examination of soil in the Brisbane area for organisms of the *Mycobacterium avium-intracellulare-scrofulaceum* complex. Pathology. 6:269–273.

Reznikov, M., J. H. Leggo, and D. J. Dawson. 1971. Investigation of seroagglutination of strains of the *Mycobacterium intracellulare-M. scrofulaceum* group from house dusts and sputum in southeastern Queensland. Am. Rev. Respir. Dis. 104:951–953.

Richardson, J. H., and W. E. Barkley (ed.). 1988. Biosafety in microbiological and biomedical laboratories, 2nd ed, HHS Publication No. (NIH) 88–8395. U. S. Department of Health and Human Services, Centers for Disease Control, Atlanta, and National Institutes of Health, Bethesda, MD.

Ridell, M. 1988. Serodiagnosis of tuberculosis. Eur. Respir. J. 1:587–588.

Rieder, H. L., G. M. Cauthen, G. D. Kelly, A. B. Bloch, and D. E. Snider, Jr. 1989. Tuberculosis in the United States. J. Am. Med. Assoc. 262:385–389.

Roberts, G. D., N. L. Goodman, L. Heifets, H. W. Larsh, T. H. Lindner, J. K. McClatchy, M. R. McGinnis, S. H. Siddiqi, and P. Wright. 1983. Evaluation of the BACTEC radiometric method for recovery of mycobacteria and drug susceptibility testing of *Mycobacterium tuberculosis* from acid-fast smear-positive specimens. J. Clin. Microbiol. 18:689–696.

Runyon, E. H. 1959. Anonymous mycobacteria in pulmonary disease. Med. Clin. North Am. 43:273–290.

Runyon, E. H. 1965. Pathogenic mycobacteria. Adv. Tuberc. Res. 14:235–287.

Runyon, E. H. 1974. Ten mycobacterial pathogens. Tuberc. 55:235–240.

Saito, H., K. Sato, T. Watanabe, and H. Tomioka. 1985. In vitro and in vivo susceptibilities of *Mycobacterium fortuitum* to ofloxacin, p. 319–320. In: J. Ishigami (ed.), Recent advances in chemotherapy: proceedings of the 14th International Congress of Chemotherapy, Kyoto, 1985. University of Tokyo Press, Tokyo.

Salfinger, M., E. W. Stool, D. Piot, and L. Heifets. 1988. Comparison of three methods for recovery of *Mycobacterium avium* complex from blood specimens. J. Clin. Microbiol. 26:1225–1226.

Sanders, W. E., Jr. and E. A. Horowitz. 1990. Other *Mycobacterium* species, p. 1914–1926. In: G. L. Mandell,

R. G. Douglas, Jr., and J. E. Bennett (ed.), Principles and practice of infectious diseases, 3rd ed. Churchill Livingstone, New York.

Schaefer, W. B. 1965. Serologic identification and classification of the atypical mycobacteria by their agglutination. Am. Rev. Respir. Dis. 92 (Suppl.):85–93.

Schaefer, W. B. 1979. Serological identification of atypical mycobacteria, p. 323–343. In: T. Bergan and J. R. Norris (ed.), Methods in microbiology. Academic Press, London.

Schraufnagel, D. E., J. A. Leech, and B. Pollak. 1986. *Mycobacterium kansasii:* colonization and disease. Br. J. Dis. Chest 80:131–137.

Schulze-Robbecke R, and R. Fischeder. 1989. Mycobacteria in biofilms. Zentralbl. Bakteriol. Mikrobiol. Hyg. Abt 1, Orig. B, Hyg. Umwelthyg. Krankenhaushyg. Arbeitshyg. Praev. Med. 188:385–390.

Seibert, F. B. 1934. The isolation and properties of the purified protein derivative of tuberculin. Am. Rev. Tuberc. 30 (Suppl.):713–720.

Seibert, F. B., and J. T. Glenn. 1941. Tuberculin purified protein derivative. Preparation and analyses of a large quantity for standard. Am. Rev. Tuberc. 44:9–25.

Shoemaker, S. T., J. H. Fisher, W. D. Jones, Jr., and C. H. Scoggin. 1986. Restriction fragment analysis of chromosomal DNA defines different strains of *Mycobacterium tuberculosis*. Am. Rev. Respir. Dis. 134:210–213.

Siddiqi, S. H. 1988. BACTEC TB System: product and procedure manual. Becton Dickinson Diagnostic Instrument Systems, Towson, MD.

Siddiqi, S. H., J. P. Libonati, M. E. Carter, N. M. Hooper, J. F. Baker, C. C. Hwangbo and L. E. Warfel. 1988. Enhancement of mycobacterial growth in Middlebrook 7H12 medium by polyoxyethylene stearate. Curr. Microbiol. 17:105–110.

Silcox, V. A., R. C. Good, and M. M. Floyd. 1981. Identification of clinically significant *Mycobacterium fortuitum* complex isolates. J. Clin. Microbiol. 14:686–691.

Sipka, S., G. Abel, L. Czirják, K. Dankó, T. Szilágyi, and J. Fachet. 1983. Inhibition of tuberculin reaction by suramin in guinea pigs. Ann. Immunol. Hung. 23:187–190.

Smith, T. 1898. A comparative study of bovine tubercle bacilli and of human bacilli from sputum. J. Exp. Med. 3:451–511.

Smith, T. 1904–1905. Studies in mammalian tubercle bacilli. III. Description of a bovine bacillus from the human body. A culture test for distinguishing the human from the bovine type of bacilli. J. Med. Res. 13:253–300.

Smithwick, R. W. 1976. Laboratory manual for acid-fast microscopy, 2nd ed. U. S. Department of Health, Education and Welfare. Centers for Disease Control, Atlanta.

Smithwick, R. W., C. B. Stratigos, and H. L. David. 1975. Use of cetylpyridinium chloride and sodium chloride for the decontamination of sputum specimens that are transported to the laboratory for the isolation of *Mycobacterium tuberculosis*. J. Clin. Microbiol. 1:411–413.

Snider, D. E., Jr. 1982. The tuberculin skin test. Am. Rev. Respir. Dis. 125:108–118.

Snider, D. E. Jr., W. D. Jones, and R. C. Good. 1984. The usefulness of phage typing *Mycobacterium tuberculosis* isolates. Am. Rev. Respir. Dis. 130:1095–1099.

Snider, W. R. 1971. Tuberculosis in canine and feline populations. Am. Rev. Respir. Dis. 104:877–887.

Snider, W. R., D. Cohen, J. S. Reif, S. C. Stein, J. E. Pier. 1971. Tuberculosis in canine and feline populations. Study of high risk populations in Pennsylvania, 1966–1968. Am. Rev. Respir. Dis. 104:866–876.

Sommers, H. M., and R. C. Good. 1985. *Mycobacterium*, p. 216–248. In: E. H. Lennette, A. Balows, W. J. Hausler, Jr., and H. J. Shadomy (ed.), Manual of clinical microbiology, 4th ed. American Society for Microbiology, Washington, DC.

Steadham, J. E. 1980. High catalase strains of *Mycobacterium kansasii* isolated from water in Texas, USA. J. Clin. Microbiol. 11:496–498.

Stefansky, W. K. 1903. Eine lepreaähnlich erkrankung der haut und der lymphdrusen bei wanderratten. Zentralbl. Bakteriol. Parasitenkund., Abt. 1, 33:481–487.

Strand, C. L., C. Epstein, S. Verzosa, E. Effatt, P. Hormozi, and S. H. Siddiqi. 1989. Evaluation of a new blood culture medium for mycobacteria. Am. J. Clin. Pathol. 91:316–318.

Strauss, I., and N. Gameleïa. 1891. Recherches expèrimentales sur la tuberculose; la tuberculose humaine; sa distinction de la tuberculose des oiseaux. Arch. Med. Exper. Anat. Pathol. 3:457–484.

Styblo, K. 1980. Recent advances in epidemiological research in tuberculosis. Adv. Tuberc. Res. 20:1–63.

Styblo, K., and A. Rouillon. 1981. Tuberculosis in the world. II. Estimated global incidence of smear-positive pulmonary tuberculosis. Unreliability of officially reported figures on tuberculosis. Bull. Int. Union Tuberc. 56:118–126.

Sunderam, G., R. McDonald, and T. Maniatis. 1986. Tuberculosis as a manifestation of the acquired immunodeficiency syndrome (AIDS). J. Am. Med. Assoc. 256:362–366.

Swenson, J. M., C. Thornsberry, and V. A. Silcox. 1982. Rapidly growing mycobacteria: Testing of susceptibility to 34 antimicrobial agents by broth microdilution. Antimicrob. Agents Chemother. 22:186–192.

Tager, I. B., R. Kalaidjian, L. Baldini, and R. E. Rocklin. 1985. Variability in the intradermal and in vitro lymphocyte responses to PPD in patients receiving isoniazid chemoprophylaxis. Am. Rev. Respir. Dis. 131:214–220.

Takahashi, H., and V. Foster. 1983. Detection and recovery of mycobacteria by a radiometric procedure. J. Clin. Microbiol. 17:380–381.

Tandon, P. N., R. Bhatia, and S. Bhargava. 1988. Tuberculos meningitis. Handb. Clin. Neurol. 8:195–226.

Thompson, N. J., J. L. Glassroth, and D. E. Snider, Jr. 1979. The booster phenomenon in serial tuberculin testing. Am. Rev. Respir Dis. 119:587–597.

Timpe, A., and E. H. Runyon. 1954. The relationship of "atypical" acid-fast bacteria to human disease. J. Lab. Clin. Med. 44:202–209.

Tsang, A. Y., V. L. Barr, J. K. McClathy, M. Goldberg, I. Drupa, and P. J. Brennan. 1984. Antigenic relationships of the *Mycobacterium fortuitum-Mycobacterium chelonae* complex. Int. J. Syst. Bacteriol. 34:35–44.

Tsang, A. Y., I. Drupa, M. Goldberg, J. K. McClatchy, and P. J. Brennan. 1983. Use of serology and thin-layer chromatography for the assembly of an authenticated collection of serovars within the *Mycobacterium avium-Mycobacterium intracellulare-Mycobacterium scrofulaceum* complex. Int. J. Syst. Bacteriol. 33:285–292.

Tsukamura, M., N. Kita, H. Shimoide, H. Arakawa, and A. Kuze. 1988. Studies on the epidemiology of nontuberculous mycobacteriosis in Japan. Am. Rev. Respir. Dis. 137:1280–1284.

Tsukamura, M., E. Nakamura, S. Yoshii, and H. Amano. 1985. Therapeutic effect of a new anibacterial substance ofloxacin (DL-8280) on pulmonary tuberculosis. Am. Rev. Respir. Dis. 131:352–356.

Vestal, A. 1975. Procedures for the isolation and identification of mycobacteria. U. S. Department of Health, Education and Welfare publication no. (CDC) 76–8230. Centers for Disease Control, Atlanta.

Wallace, R. J., Jr., J. M. Musser, S. I. Hull, V. A. Silcox, L. C. Steele, G. D. Forrester, A. Labidi, and R. K. Selander. 1989. Diversity and sources of rapidly growing mycobacteria associated with infections following cardiac surgery. J. Infect. Dis. 159:708–716.

Wallace, R. J., Jr., D. R. Nash, M. Tsukamura, Z. M. Blacklock, and V. A. Silcox. 1988. Human disease due to *Mycobacterium smegmatis*. J. Infect. Dis. 158:52–59.

Wallace, R. J., Jr., J. M. Swenson, V. A. Silcox, and M. G. Bullen. 1985. Treatment of nonpulmonary infections due to *Mycobacterium fortuitum* and *Mycobacterium chelonei* on the basis of in vitro susceptibilities. J. Infect. Dis. 152:500–514.

Wallace, R. J., Jr., J. M. Swenson, V. A. Silcox, R. C. Good, J. A. Tschen, and M. S. Stone. 1983. Spectrum of disease due to rapidly growing mycobacteria. Rev. Infect. Dis. 5:657–679.

Wayne, L. G. 1979. The "atypical" mycobacteria: recognition and disease association. Crit. Rev. Microbiol. 12:185–222

Wayne, L. G., and G. A. Diaz. 1985. Identification of mycobacteria by specific precipitation of catalase with absorbed sera. J. Clin. Microbiol. 21:721–725.

Wayne, L. G., and G. A. Diaz. 1986. Differentiation between T-catalases derived from *Mycobacterium avium* and *Mycobacterium intracellulare* by a solid-phase immunosorbent assay. Int. J. Syst. Bacteriol. 36:363–367.

Wayne, L. G., H. C. Engbaek, H. W. B. Engle, S. Froman, W. Gross, J. Hawkins, W. Käppler, A. G. Karlson, H. H. Kleeberg, I. Krasnow, G. P. Kubica, C. McDurmont, E. E. Nel, S. R. Pattyn, K. H. Schröder, S. Showalter, I. Tarnok, M. Tsukamura, B. Vergmann, and E. Wolinsky. 1974. Highly reproducible techniques for use in systematic bacteriology in the genus *Mycobacterium:* Tests for pigment, urease, resistance to sodium chloride, hydrolysis of Tween 80, and β-galactasidase. Int. J. Syst. Bacteriol. 24:412–419.

Wayne, L. G., H. W. B. Engle, C. Grassi, W. Gross, J. Hawkins, P. A. Jenkins, W. Käppler, H. H. Kleeberg, I. Krasnow, E. E. Nel, S. R. Pattyn, P. A. Richards, S. Showalter, M. Slosarek, I. Szabo, I. Tarnok, M. Tsukamura, B. Vergmann, and E. Wolinsky. 1976. Highly reproducible techniques for use in systematic bacteriology in the genus *Mycobacterium:* Tests for niacin and catalase and for resistance to isoniazid, thiophene-2-carboxylic acid hydraziade, hydroxylamine, and *p*-nitrobenzoate. Int. J. Syst. Bacteriol. 26:311–318.

Wayne, L. G., and G. P. Kubica. 1986. Genus *Mycobacterium* Lehmann and Neumann 1896, p. 1436–1457. In: P. H. A. Sneath, N. S. Mair, M. E. Sharpe, and J. G.

Holt (ed.), Bergey's manual of systematic bacteriology. Williams and Wilkins, Baltimore.

Wayne, L. G., L. S. Young, and M. Bertram. 1986. Absence of mycobacterial antibody in patients with acquired immune deficiency syndrome. Eur. J. Clin. Microbiol. 5:363–365.

Weitzman, I., D. Osadczyi, M. L. Corrado, and D. Karp. 1981. *Mycobacterium thermoresistibile:* a new pathogen for humans. J. Clin. Microbiol. 14:593–595.

Wendt, S. L., K. L. George, B. C. Parker, H. Gruft, J. O. Falkinham, III. 1980. Epidemiology of infection by nontuberculous mycobacteria. III. Isolation of potentially pathogenic mycobacteria from aerosols. Am. Rev. Respir. Dis. 122:259–263.

Whipple D., P. Kapke, and R. Andrews. 1989. Analysis of restriction endonuclease fragment patterns of DNA from *Mycobacterium paratuberculosis.* Vet. Microbiol. 19:189–194.

Whipple, D. L., R. B. Febvre., R. E. Andrews, and A. B. Thiermann. 1987. Isolation and analysis of restriction endonuclease digestive patterns of chromosomal DNA from *Mycobacterium paratuberculosis* and other *Mycobacterium* species. J. Clin. Microbiol. 25:1511–1515.

Wilesmith, J. W. 1983. Epidemiological features of bovine tuberculosis in cattle herds in Great Britain. J. Hyg., Camb. 90:159–176.

Witebsky, F. G., J. F. Keiser, P. S. Conville, R. Bryan, C. H. Park, R. Walker, and S. H. Siddiqi. 1988. Comparison of BACTEC 13A medium and Du Pont isolator for detection of mycobacteremia. J. Clin. Microbiol. 26:1501–1505.

Wolinsky, E. 1979. Nontuberculous mycobacteria and associated diseases. Am. Rev. Respir. Dis. 119:107–159.

Wolinsky, E. 1984. Nontuberculous mycobacteria and associated diseases, p. 1141–1207. In: G. P. Kubica and L. G. Wayne (ed.), The mycobacteria, a sourcebook. Marcel Dekker, New York.

Wolinsky, E., and W. B. Schaefer. 1973. Proposed numbering scheme for mycobacterial serotypes by agglutination. Int. J. Syst. Bacteriol. 23:182–183.

Woodley, C. L., and J. O. Kilburn. 1982. In vitro susceptibility of *Mycobacterium avium* complex and *Mycobacterium tuberculosis* strains to a spiro-piperidyl rifamycin. Am. Rev. Respir. Dis. 126:586–587.

Woodley, C. L., V. A. Silcox, M. M. Floyd, and G. P. Kubica. 1989. The use of DNA probes for rapidly identifying cultures of *Mycobacterium,* p. 51–56. In: B. Kleger, D. Jungkind, E. Hinks, and L. A. Miller (ed.), Rapid methods in clinical microbiology. Plenum Press, New York.

Woods, G. L., and J. A. Washington II. 1987. Mycobacteria other than *Mycobacterium tuberculosis:* review of microbiologic and clinical aspects. Rev. Infect. Dis. 9:275–294.

Wright, E. P., C. H. Collins, and M. D. Yates. 1985. *Mycobacterium xenopi* and *Mycobacterium kansasii* in a hospital water supply. J. Hosp. Infect. 6:175–178.

Yakrus, M. A., and R. C. Good. 1990. Geographic distribution, frequency, and specimen source of *Mycobacterium avium* complex serotypes isolated from patients with acquired immunodeficiency syndrome. J. Clin. Microbiol. 28:926–929.

Yates, M. D., and J. M. Grange. 1988. Incidence and nature of human tuberculosis due to bovine tubercle bacilli in south-east England: 1977–1987. Epidemiol. Infect. 101:225–229.

Youmans, G. P. 1979. Pathogenesis of tuberculosis, p. 317–326. In: G. P. Youmans (ed.), Tuberculosis. W. B. Saunders, Philadelphia.

Young, L. S. 1988. *Mycobacterium avium* complex infection. J. Infect. Dis. 157:863–867.

Young, L. S., C. B. Inderlied, O. G. Berlin, and M. S. Gottlieb. 1986. Mycobacterial infections in AIDS patients, with an emphasis on the *Mycobacterium avium* complex. Rev. Infect. Dis. 8:1024–1033.

CHAPTER 54

Mycobacterium leprae

THOMAS M. SHINNICK

Although *Mycobacterium leprae* was one of the first bacterial pathogens of humans to be described (Hansen, 1874), progress on understanding the basic biology and pathogenicity of this organism has been greatly hampered by the inability to find a conventional laboratory medium or tissue culture system that can support its growth. Consequently, the only means of propagating this organism at present is by using experimental animals. Furthermore, it has been found that the nine-banded armadillo can be used to produce large numbers of bacilli (Kirchheimer and Storrs, 1971; Storrs, 1971). Relatively little is known therefore, about the taxonomy, genetics, and biochemistry of this species. A corollary of this is that much of what we do know about *M. leprae* has come from studies of the disease it causes (leprosy or Hansen's disease) and from the animal models. As such, this chapter emphasizes the characteristics and behavior of *M. leprae* in experimental animal model systems and in humans. Several excellent reviews on the clinical aspects, epidemiology, immunology, and pathology of leprosy, and on the biochemistry and immunochemistry of *M. leprae* have been published recently, and the reader is referred to these for additional information (Bloom and Godal, 1983; Bloom and Mehra, 1984; Fine, 1982; Gaylord and Brennan, 1987; Hastings, 1986; Hastings and Franzblau, 1988; Jopling and McDougall, 1988; Kaplan and Cohn, 1986; Stewart-Tull, 1982).

Although *M. leprae* can occasionally be found in the body extracellularly, the bacillis appears to be able to replicate only within cells of the host, most commonly in macrophages and Schwann cells (Bloom and Godal, 1983; Kaplan and Cohn, 1986). Hence, *M. leprae* is considered to be an obligate intracellular pathogen. In host cells, the bacilli are found singly or in clumps referred to as *globi* (Cowdry, 1940). The bacilli are straight or slightly curved, Gram-positive, acid-fast, alcohol-fast, nonmotile rods ranging from 1 to 8 μm in length and 0.2 to 0.5 μm in width (Draper, 1983). Acid- and alcohol-fastness refers to the ability of the bacillus to retain the color of certain dyes, usually carbol fuchsin, following treatment with mild acid and alcohol, respectively.

M. leprae has been placed in the genus *Mycobacterium* in the family Actinomycetales based mainly on cell structure, staining properties and chemical composition as well as on the basis of the presence of mycolic acids, antigens characteristic of mycobacteria, and a lipid-rich cell envelope (Draper, 1976; Harboe et al., 1977; Stanford et al., 1975). For example, the *M. leprae* bacillus closely resembles *M. tuberculosis* bacilli in size, morphology, and staining characteristics, although it does stain a little more deeply with carbol fuchsin and the staining is somewhat less acid fast. Recently, analyses of the ribosomal RNA sequences of armadillo-grown *M. leprae* by nucleic acid hybridization techniques (Sela et al., 1989) and by ribosomal RNA sequence comparisons (Smida et al., 1988) revealed that *M. leprae* is closely related to the corynebacteria, nocardia, and mycobacteria, especially to the two slowly growing *Mycobacterium* species *M. avium* and *M. tuberculosis*. Although these observations indicate that *M. leprae* should be classified in the genus *Mycobacterium*, several features distinguish *M. leprae* from other members of the *Mycobacterium* genus. These are: 1) loss of acid-fastness upon extraction with pyridine, although *M. smegmatis, M. vaccae,* and *M. phlei* do lose acid fastness after prolonged exposure to pyridine (Fisher and Barksdale, 1971; McCormick and Sanchez, 1979; Skinsnes et al., 1975); 2) ability to oxidize 3,4-dihydroxyphenylalanine (Prabhakaran and Kirchheimer, 1966); 3) replacement of L-alanine with glycine in the linking peptide of peptidoglycan (Draper, 1976); 4) 56% GC content as compared with 65–70% for most *Mycobacterium* species (Clark-Curtiss et al., 1985; Imaeda et al., 1982; Wayne and Gross, 1968); and 5) lack of substantial genomic DNA homology with other *Mycobacterium* species (quite in contrast with the ribosomal rRNA results) (Athwal et al.,

1984; Grosskinski et al., 1989). Although these differences are not sufficient to exclude *M. leprae* from the genus *Mycobacterium,* they are sufficient to render the final taxonomic position of *M. leprae* somewhat in doubt.

The Disease

Hansen's disease (leprosy) is a chronic infectious granulomatous disease that primarily affects the peripheral nervous system, skin, and mucous membranes, especially nasal mucosa. In advanced cases, other tissues—including muscle, testes, capillary endothelium, liver, spleen, and bone marrow—can be affected (Desikan and Job, 1968). Although leprosy is not usually fatal in and of itself, 20–30% of patients with untreated or neglected infections develop crippling deformities of hands and feet. A sequela of these deformities is the social stigma that has been historically associated with leprosy. An addition consequence is that suicide is a common cause of death among infected individuals, particularly during episodes of exacerbation of the lesions (erythema nodosum leprosum).

Leprosy has been estimated to afflict 10–12 million individuals world-wide, with most cases being found in tropical and subtropical regions (Sixth Report of the WHO Expert Committee on Leprosy, 1988). Climatic factors do not seem to have a significant impact on the disease, however, since leprosy was widespread in Europe, particularly Norway, in the past centuries (Browne, 1975), and cases have been reported from above the Arctic Circle (Sansarricq, 1981). Within endemic areas, cases of leprosy appear to cluster geographically, with the prevalence of disease exceeding 10 cases per 1000 population in the high prevalence areas (Bloom and Godal, 1983). Although this disease has been cited as the "least infectious" of communicable diseases (McWhirter, 1981), epidemics of leprosy have occurred on some Pacific Islands in which as much as 35% of the population developed the disease over a 20-year period (Wade and Ledowski, 1952).

A feature of the disease that has intrigued clinicians and immunologists is the variety of clinical manifestations that can be found. These range from a single lesion with no detectable bacilli to multiple lesions containing large numbers of bacilli. Bacterial counts up to 5×10^9 organisms per gram of tissue have been found (Collaborative effort, 1975). The variety of disease symptoms is not related to the genetics of the bacterium but rather to the immune responsiveness of the host (Bloom and Mehra,

1984; Kaplan and Cohn, 1986). A clinically and experimentally useful categorization of this disease spectrum is the Ridley-Jopling classification scheme which is based on immunopathologic features of the disease (Ridley and Jopling, 1966). Tuberculoid leprosy (TT) is at one pole of the spectrum and is characterized by one or a few localized skin or nerve lesions, a strong cellular immune response, and a weak humoral immune response. Histologically, one finds well-organized epithelioid cell granulomas with multinucleated giant cells and abundant lymphocytes in TT lesions, but bacilli are usually absent ($<10^5$ bacilli/gram of tissue). At the other extreme is lepromatous leprosy (LL), characterized by numerous small, bilaterally symmetrical skin lesions, a weak or absent cellular immune response, and high antibody titers. Histologically, one observes a foamy macrophage granuloma with few lymphocytes in LL lesions, and the macrophages contain numerous bacilli ($>10^8$ bacilli/gram). Within these extremes are other conditions, including borderline lepromatous (BL; numerous skin lesions, numerous bacilli, granulomas with undifferentiated macrophages and histiocytes), borderline (BB; numerous skin lesions, few acid-fast bacilli, nerve involvement, epithelioid cell granulomas with no giant cells), and borderline tuberculoid (BT; multiple skin lesions, occasional bacilli, nerve involvement, diffuse epithelioid cell granulomas). One additional category, called indeterminant leprosy, often found when a patient first seeks treatment, usually consists of a single small hypopigmented plaque. Such a lesion may remain indeterminant, regress spontaneously, or progress into a lesion that falls into one of the above categories.

It is generally accepted that *Mycobacterium leprae* is the etiologic agent of leprosy. However, not all of Koch's postulates have been fulfilled for identifying *M. leprae* as the causative agent of human leprosy. That is, a pure culture of *M. leprae* has not been developed from a single bacillus and shown to cause disease. The inability to grow the organism outside of animal hosts is the major stumbling block to completing Koch's postulates. Nonetheless, much evidence has accumulated in support of the role of *M. leprae* as the agent of the disease. The evidence includes the following: 1) *M. leprae* is found in leprosy patients and not in nonleprosy patients; 2) *M. leprae* bacilli are invariably present in lepromatous lesions; 3) leprosy patients display characteristic immune responses to *M. leprae* antigens (Mitsuda and Fernandez reactions; see Jopling and McDougall; 1988); 4) antibodies and T-cells reactive with antigens or epitopes uniquely expressed by *M. leprae* can be isolated from leprosy patients (reviewed in Gaylord and Brennan, 1987); 5) a phenolic glycolipid (PGL-I; Hunter and Brennan, 1981; Hunter et al., 1982) uniquely found in *M. leprae* is also uniquely found in lesions, sera, and urines from leprosy patients (Cho et al., 1983; Cho et al., 1986; Koster et al., 1987; Vemuri et al., 1985; Young et al., 1985); 6) *M. leprae*-specific nucleotide sequences can be found in leprosy lesions (Clark-Curtiss and Docherty, 1989); and 7) drug susceptibility of *M. leprae* in animal models parallels drug efficacy in patients (reviewed in Shepard, 1986).

Several properties of the *M. leprae* bacillus contribute to the features of the disease:

1. *M. leprae* has a generation time of 11–13 days in experimental animals (Levy, 1976; Shepard and McRae, 1971b). Such slow growth might influence the length of time from infection to disease (median interval 2–8 years; Fine, 1982) and the chronic nature of the infection.

2. *M. leprae* grows best at 30°C, which may explain its affinity for cooler parts of the body including skin and nasal mucosa (Shepard, 1965).

3. *M. leprae* infects Schwann cells and is the only bacterial pathogen capable of entering peripheral nerves (Job, 1971; Ridley et al., 1987; Stoner, 1979). These facets play a role in the loss of nerve function and the generation of crippling deformities.

4. *M. leprae* can supress the cellular immune response, perhaps through induction of suppressor T-cells (Bloom and Mehra, 1984; Mehra et al., 1984) and/or rendering the infected macrophage defective for activation or antigen processing (Desai et al., 1989; Prasad et al., 1987; Sibley and Krahenbuhl, 1988). This may allow the host to tolerate the large bacterial load seen in lepromatous leprosy patients (up to 10^{13} organisms).

5. The unique phenolic glycolipid of *M. leprae*, called PGL-I, is a major component of the cell envelope (2% of total bacterial mass; Hunter and Brennan, 1981), can scavenge hydroxyl radicals and superoxide anions in vitro (Chan et al., 1989) and can inhibit the oxidative response in macrophages that have ingested *M. leprae* (Vachula et al., 1989). Hence, PGL-I may play a role in the ability of the bacteria to survive within macrophages and may also play a role in immunosuppression (Mehra et al., 1984; Prasad et al., 1987).

6. *M. leprae* can survive and multiply within macrophages and Schwann cells. Three possible strategies of intracellular survival have been proposed for *M. leprae*. First, *M. leprae* bacilli might avoid the bactericidal activities of the phagocytic cells by escaping from the phagosome (phagolysosome?) and multiplying in the cytoplasm of the infected cells (Mor, 1983). Second, viable *M. leprae* have been reported to prevent the fusion of phagosomes and lysosomes in murine macrophages (Frehel and Rastogi, 1987; Sibley et al., 1987), but not in Schwann cells (Steinhoff et al., 1989). A similar prevention of phagosome-lysosome fusion is an intracellular survival strategy used by *M. tuberculosis* (d'Arcy Hart, 1982). Third, inside phagosomes *M. leprae* generates around the bacillus characteristic electron microscope image called an "electron transparent zone (ETZ)" (Draper and Rees, 1970). The ETZ might represent a physical barrier to prevent degradative or bactericidal proteins of the lysosome from reaching the bacterial surface. Regardless of the precise strategy used, one result of evading the bactericidal and degradative activities of the phagocytic cells might be that the *M. leprae* antigens would be less likely to be processed and presented to the immune system. If so, the bacilli might thereby prevent or escape a protective immune response.

Habitat

Leprosy and *M. leprae* have historically been considered to be confined to humans. Indeed, humans are the major host and reservoir of the leprosy bacillus. Humans, however, are not the only possible habitat for *M. leprae*. Naturally acquired, leprosy-like infections have been observed in armadillos (*Dasypus novemcinctus;* Walsh et al., 1975), chimpanzees (*Pan troglodytes;* Donham and Leininger, 1977), and Mangabey monkeys (*Cercocebus torquatas atys;* Meyers et al, 1985). Since the bacilli infecting these animal species are identical to the human pathogen, based on a variety of DNA homology and biochemical studies (Athwal et al., 1984; Clark-Curtiss and Walsh, 1989; Meyers et al., 1985), leprosy should be considered a zoonotic disease (Walsh et al., 1981). Furthermore, the possibility of the transmission of *M. leprae* from animals to humans has been raised by the observation that five armadillo handlers in Texas developed leprosy in the absence of any known contact with a human source of *M. leprae* (Lumpkin et al., 1983).

The isolation of *M. leprae* from soil in Bombay and other environmental sources has also been reported (Kazda, 1981a; Kazda et al., 1986). It is unclear, however, if these reports identify an environmental niche for *M. leprae* or if the isolated bacilli represent "environmental contamination" from *M. leprae*-infected individuals. For example, one possible source of "environmental contamination" might be nasal secretions of lepromatous leprosy patients (see below). Finally, free-living amoebae have been suggested as a potential reservoir for *M. leprae* (Grange and Rowbotham, 1987), although no convincing evidence for the presence or multiplication of *M. leprae* in amoebae has been published.

The mode of transmission of the leprosy bacillus remains unknown. Evidence is accumulating that infection occurs predominantly by way of the respiratory route, although other routes may be responsible for some cases (reviewed in Pallen and McDermott, 1986). Some of the evidence for respiratory transmission includes: 1) the epidemiology of transmission is consistent with spread by a respiratory route (Barton, 1974; Davey and Rees, 1974). 2) The major portal of exit of the leprosy bacillus is the nose. Lepromatous leprosy patients can shed up to 10^8 bacilli per day in nasal discharges, and the bacilli can survive for several days in dried secretions (Davey and Rees, 1974; Shepard, 1962). 3) Bacilli can be aerosolized by coughing and sneezing. 4) The nose has been suggested as a possible site of initial infection in humans (Pallen and McDermott, 1986), and immunodeficient mice can be infected by exposing them to an aerosol containing *M. leprae* (Rees and McDougall, 1976; Chehl et al., 1985). 5) As originally discussed by Koch (1897) and Schaffer (1898), the similarities between tuberculosis and leprosy (e.g., the similarity between the large numbers of *M. leprae* in nasal secretions and the large numbers of *M. tuberculosis* in sputa) is suggestive of an analogous route of transmission.

At present, however, one can not exclude transmission by skin-to-skin contact or by insect vectors. For example, ulcerating skin lesions could be a source of bacilli that might enter a susceptible host through a skin abrasion. Similarly, broken skin might be a portal of entry for bacilli that were aerosolized or otherwise deposited in the environment. With respect to insect vectors, *M. leprae* can be found in flies that have fed on nasal secretions from lepromatous leprosy patients (Greater, 1975; Kirchheimer, 1976), and the transmission of *M. leprae* from humans to mouse footpad by a mosquito vector (*Aedes aegypti*) has been reported (Narayanan et al., 1977). Another possible route is via breast milk since *M. leprae* bacilli can be found in the breast milk of lepromatous patients (Pedley, 1968). The role, if any, of the gastrointestinal tract in the transmission of leprosy is not known. Finally, occasional cases of leprosy may result from accidental inoculation of susceptible individuals by needle prick (Wade, 1948).

Isolation and Propagation

The currently available methods for the isolation and propagation of *M. leprae* are greatly constrained by the lack of an axenic or cell-culture cultivation system. In the laboratory, *M. leprae* has been propagated in both immunocompetent and immunodeficient mice and rats (Colston and Hilson, 1976; Fieldsteel and Levy, 1976; Hilson, 1965; Rees, 1966; Shepard, 1960), hamsters (Binford, 1959), armadillos (Kirchheimer and Storrs, 1971), and in Mangabey, Rhesus, and African green monkeys (Wolf et al., 1985). The features of the growth and pathology of *M. leprae* in each species is somewhat different, and the choice of animal model often depends on exactly what experimental question is being asked (reviewed in Shepard, 1986).

Staining Methods

One consequence of having only animal systems in which to propagate *M. leprae* is that much of the work depends on the enumeration of acid-fast bacilli by direct counting in microscopic fields. The standard staining method is a modified Ziehl-Neelsen technique (e.g., see Jenkins et al., 1982, or Jopling and McDougall, 1988). Briefly, bacilli on a glass slide are covered with a carbol-fuchsin solution and left at room temperature for 20 minutes (cold Ziehl-Neelsen) or heated for 15 minutes over a boiling water bath (hot Ziehl-Neelsen). The slide is washed with water, destained with 1% hydrochloric acid in 70% ethanol, and washed again with water. The sample is then counterstained with 1% methylene blue, washed in water, and air dried. Several methods for the actual counting of bacilli in the microscope and for using the counts to determine the concentration of bacilli in a sample have been described (Hanks et al., 1964; Shepard and McRae, 1968). The uniformity of staining exhibited by a bacillus is used as an indication of the viability of the bacillus (Hansen and Looft, 1895; Shepard and McRae, 1965a; Waters and Rees, 1962). Rods showing uniform staining are considered to be viable, while bacilli displaying fragmented or granular staining are considered non viable. A "morphologic index" is calculated as the percentage of uniformly staining bacilli and has been used to follow the progress of chemotherapy (Waters and Rees, 1962).

Propagation in Mice

The most frequently used propagation system is growth in the footpad of immunocompetent mice, as was originally described in the landmark paper of Shepard (1960). In this animal model, one typically propagates *M. leprae* by injecting mice subcutaneously with about 5×10^3 bacilli in one of the hind footpads, although as few as 1–10 bacilli are sufficient to establish an infection (Shepard and McRae, 1965a). A typical growth curve displays an initial lag phase followed by logarithmic growth with a doubling time of 11–13 days and a stationary phase with a plateau level of about $1–2 \times 10^6$ bacilli per footpad. The plateau level is thought to be due to development of a cell-mediated immune response to the bacillus. As a consequence, the viability of the *M. leprae* in the footpad decreases dramatically during the plateau phase. Thus, to propagate a strain one should harvest bacilli prior to or early in the plateau phase.

A typical growth curve is accompanied by a typical histopathologic picture. Early in infection one finds single bacilli or small clumps of 2–5 bacilli aligned in a parallel fashion ("cigar packs") in tissue macrophages, muscle, and tendon cells. Later, a cellular infiltrate mainly of macrophages and a few lymphocytes appears, followed by development of a microscopic granuloma with a few large, "foamy" macrophages in the center surrounded by a loose collection of macrophages and lymphocytes. The large macrophages contain numerous bacilli. In late stages, one occasionally observes invasion of the nerves.

In the absence of a T-cell response, multiplication continues to much higher levels. Thus, immunodeficient animals such as thymectomized-irradiated mice (Rees, 1966), neonatally thymectomized Lewis rats (Fieldsteel and Levy; 1976), and nude mice (*nu/nu*) (Colston and Hilson, 1976; Chehl et al., 1983) have been used to propagate *M. leprae*. The minimum infectious dose and generation time during logarithmic growth appear to be the same in the immunodeficient animals as in immunocompetent animals. The bacilli, however, can multiply to up to about 10^9 cells per foot pad in the neonatally thymectomized Lewis rat and about 5×10^{10} cells per footpad in the nude mouse. In the nude mouse one frequently observes spread of the *M. leprae* to other cool sites in the body (e.g., the ears) and invasion into nerves. Overall, the histopathology of the disease in the immunodeficient mice resembles that of human lepromatous leprosy. One drawback to the use of these animals on a routine basis is that they are particularly susceptible to a variety of other infections, and hence, special precautions need to be taken to ensure the survival of the animals for the duration of the experiment (often greater than 1 year).

Propagation in the Armadillo

A second major advance in leprosy research was the development of the nine-banded armadillo as an animal model system for the propagation of *M. leprae* (Kirchheimer and Storrs; 1971; Storrs, 1971). The key here is that in the armadillo the disease resembles human lepromatous leprosy and one can isolate very large numbers of bacilli from each infected armadillo. For example, the bacillary load in lesions can reach 10^{10} organisms per gram of tissue, and the total bacillary load can exceed 10^{12} bacilli per armadillo. Also, the liver and spleen are heavily infected and are easily manipulated sources of bacilli. The availability of the large amounts of *M. leprae* from armadillos opened the way for biochemical and molecular biologic analyses of the bacillus as well as providing a way to produce sufficient *M. leprae* for use in vaccine trials. One technical problem with the use of armadillos is that the *M. leprae* infection is fatal for the armadillo. Hence, to maximize bacillary yield and avoid postmortem contamination of tissues with other bacteria, care must be taken to sacrifice the animal and harvest bacilli just prior to the animal succumbing to the *M. leprae* infection itself. Finally, one must be aware that armadillos may harbor other naturally acquired, fastidious, acid-fast mycobacteria (Kazda, 1981b) and may harbor naturally acquired *M. leprae* (Walsh et al., 1975).

Primate Infection

Naturally occurring *M. leprae* infections have been observed in several primates, and *M. leprae* has been experimentally transmitted to three species of monkeys (Wolf et al., 1985). While the disease in these animals seems to parallel closely disease in humans, their use in the routine isolation and propagation of *M. leprae* is deterred by economic and humane concerns. Nonetheless, the primates may prove to be important model systems for studies on pathology, immunology, and vaccine development.

Maintenance of Viable Cells

Finally, a concern of those working with *M. leprae* is that the bacillus does not store well. Good viability of *M. leprae* in suspensions or in biopsy specimens is maintained for 7–10 days at 0–4°C (Shepard and McRae, 1965b).

For long-term storage, one can store the bacilli in liquid nitrogen. Three important points here are: 1) the sample should contain 10% dimethyl sulphoxide or 10% glycerol; 2) slow freezing is required ($<1°C$/min); and 3) the bacilli rapidly lose viability upon repeated freezing and thawing (Colston and Hilson, 1979; Portaels et al., 1988). For routine laboratory work one usually maintains several stocks of *M. leprae* in continuous serial passage. Of course, one is concerned that serial passaging of *M. leprae,* which typically involves transfer of 10^3 to 10^4 bacilli from mouse to mouse, might lead to changes in the bacillus, e.g., outgrowth of variants. However, some strains have been serially passaged in the Hansen's Disease Laboratory at the Centers for Disease Control (CDC) for over 25 years without any detectable changes in growth pattern (rate and plateau level) or histopathology of infection. Furthermore, by restriction-fragment-length-polymorphism (RFLP) studies, Clark-Curtiss and Walsh (1989) have found that the *M. leprae* bacilli isolated from leprosy patients in India, armadillos in Louisiana, and a naturally infected Mangabey monkey were virtually identical. This exceptional conservation of nucleotide sequence might be related to the long generation time of *M. leprae* or to the presence of a very efficient DNA-damage-repair system. Overall, these observations suggest that *M. leprae* changes very slowly, if at all, during serial passaging.

Isolation from Clinical Specimens

With respect to the isolation of *M. leprae* from a clinical specimen (e.g., biopsy of an LL lesion), the first steps are usually to purify the bacilli and to inject them into the footpad of an immunocompetent mouse such as the BALB/c or CFW strain (reviewed in Shepard, 1986). To purify the bacilli, the usual procedure is to release the bacilli from their intracellular location within a biopsy of infected tissue by mincing the sample, suspending it in Hank's Balanced Salt solution containing 0.1% BSA, and homogenizing in the presence of 2 to 3-mm-diameter glass beads. To reduce potential contamination with other bacteria, the sample can be treated with 0.5M NaOH for 15 minutes and then neutralized with HCl. Following this treatment, the cells are usually harvested by centrifugation and resuspended in Hank's Balanced Salt solution containing 0.1% BSA. Samples of the bacterial suspension are stained and examined microscopically to determine the number of uniformly staining, acid-fast bacilli. One then injects portions of the samples containing 10^4 or fewer bacilli into the footpads of

mice. Tissues from the mice are harvested periodically and examined for the presence and numbers of acid-fast bacilli and the histopathology of the site of infection.

These initial steps allow one to produce sufficient bacilli to begin the process of determining if the isolated organism is *M. leprae* (described below). In addition, the rate of growth in the footpad and the histopathology of the infection in the mouse footpad may be useful in distinguishing *M. leprae* from other mycobacteria (Shepard, 1986). That is, only *M. leprae, M. lepraemurium, M. marinum,* and *M. ulcerans* of the *Mycobacterium* species, grow in the mouse footpad, and each displays a characteristic growth rate and histopathology. For example, *M. marinum* infections appear in about 2 weeks and often cause ulceration at the site of inoculation.

Identification

The lack of an in vitro system for the cultivation of *M. leprae* makes positive identification particularly difficult. Often a presumptive identification of an organism as *M. leprae* is made based simply on two observations—staining characteristics and lack of growth on conventional laboratory media. For example, with clinical specimens, any noncultivable, acid-fast, rod-shaped bacillus isolated from a site displaying a histopathology characteristic of a leprosy lesion is presumed to be *M. leprae*. Although such a definition is adequate for decisions on patient management, it is not particularly useful if one is trying to identify an organism isolated from a source devoid of histopathologic information, such as nasal secretions or the environment. Also, such a definition is clearly insufficient to prove an organism is *M. leprae*. Undoubtedly, the imprecise nature of the definition has contributed to the numerous erroneous claims for the cultivation of *M. leprae* in vitro.

Until recently, only a few rather labor-intensive, time-consuming, and technically difficult tests, such as growth in mouse footpads or presence of *M. leprae*-specific components, were available to confirm the presumptive identification. Fortunately, a variety of relatively simple biochemical, immunological, and molecular-biological tests are now becoming available that should allow a rapid and definitive identification of *M. leprae*. Some of the potentially useful identification tests are: 1) morphology and staining characteristics; 2) failure to grow on bacteriologic media; 3) growth and histopathology in animal models; 4) drug susceptibility patterns; 5) presence of unique biochemical components or activities; 6) presence of characteristic antigens; and 7) presence of specific nucleotide sequences.

The identification process still starts with an examination of the staining characteristics of the bacilli. *M. leprae* bacilli isolated from lesions are Gram-positive rods. The carbol fuchsin stain is not extractable with acid or alcohol but is extractable with pyridine (Fisher and Barksdale, 1971). Of the other mycobacteria, only *M. smegmatis, M. vaccae,* and *M. phlei* display somewhat similar pyridine-extractable acid-fast characteristics (McCormick and Sanchez, 1979; Skinsnes et al., 1975). An important caveat here is that absence of acid-fastness can not be taken as an indication that the organism is not *M. leprae,* since a variety of mycobacteria lose the acid-fast characteristic at various times during their growth.

The second routinely used test is a determination of growth on any of a variety of media used for the propagation of mycobacteria, such as Lowenstein-Jensen, Middlebrook 7H9, or Dubos medium. *M. leprae* is the only *Mycobacterium* species that does *not* grow on at least one type of axenic medium. Of course, one needs to include a control test, such as inoculation into mouse footpads, to ensure that the original preparation did indeed contain viable bacteria.

M. leprae grows in mice, rats, armadillos, and monkeys and produces a characteristic pattern of proliferation and histopathologic changes in each animal (for details, see Shepard, 1986). Hence, growth characteristics can be used for identification purposes. For practical and economic reasons, one usually starts by analyzing growth in mice and then proceeds onto other animal models only as warranted. A key step in this test is careful observation of histopathologic changes. For example, *M. leprae* is the only *Mycobacterium* species to invade peripheral nerves and such invasion is prominent in the later stages of an *M. leprae* infection in the thymectomized-irradiated mouse and in the armadillo. Hence, observation of acid-fast bacilli in nerves is a good indication that the organism is *M. leprae.*

Biochemical Tests

Among the mycobacteria, *M. leprae* has a unique pattern of drug susceptibilities (Shepard, 1971; Shepard et al., 1983; Hastings and Franzblau, 1988). Thus, one could confirm a presumptive identification by measuring drug susceptibility in: 1) inhibition of in vivo growth in the mouse footpad (Shepard, 1971); 2) in-hibition of growth in macrophage culture (Mittal et al., 1983; Ramasesh et al., 1987); or 3) effect of the drug on PGL-I synthesis, palmitate oxidation, or ATP generation in bacilli maintained in a synthetic medium (Franzblau, 1988; Franzblau and Hastings, 1987; Franzblau et al., 1987). Note that the bacilli do not multiply in the systems used by Franzblau and colleagues.

Several identification tests are based on the observation that *M. leprae* contains components, structures, enzymes, and antigens that apparently are not present in any of the other *Mycobacterium* species. For example, enzymatic activities that are apparently unique to *M. leprae* (among the mycobacteria) include the ability to oxidize 3,4-dihydroxyphenylalanine (DOPA) and the synthesis of phenolic glycolipid I (PGL-I). DOPA-oxidase activity can be measured in vitro by following the conversion of DOPA to the pigmented product indole-5,6-quinone (Prabhakaran and Kirchheimer, 1966). This test is not entirely specific for *M. leprae,* since some non-acid-fast bacteria and, more importantly, some mammalian tissues have similar oxidase (phenolase) activities (Prabhakaran, 1967). PGL-I biosynthesis can be followed by measuring the incorporation of ^{14}C-palmitate into PGL-I in an *M. leprae*/macrophage co-culture system (Ramasesh et al., 1987) or in bacilli maintained in a synthetic medium (Franzblau and Hastings, 1987). Since PGL-I is unique to *M. leprae,* this assay is particularly useful. Also, characterization of the susceptibility of the in vitro reaction to various drugs can help confirm the identification.

Characteristic biochemical components of *M. leprae* include: 1) peptidoglycan that contains glycine instead of L-alanine in the linking peptide (Draper, 1976); 2) a GC content of 56 mol% as opposed to GC contents of 65–70 mol% for the other *Mycobacterium* species (Clark-Curtiss et al., 1985; Imaeda et al., 1982; Wayne and Gross, 1968); 3) PGL-I. However, other mycobacteria (e.g., *M. tuberculosis*) have structurally similar phenol-phthiocerol triglycosides (Brennan, 1983; Chatterjee et al., 1989; Daffe et al., 1988; Hunter and Brennan; 1981); and 4) specific mycolic acids (Draper, 1976). Analysis of the mycolic acids may be particularly informative since the high pressure liquid chromatography (HPLC) pattern of mycolic acids produced following alkaline methanolysis seems to be unique for each species of *Mycobacterium.* Indeed, Butler and Kilburn (1988) have devised a relatively simple and rapid scheme based on such HPLC patterns for identifying many of the *Mycobacterium* species.

Immunological Tests

Several antigens (or epitopes?) also appear to be uniquely present in *M. leprae* (reviewed in Gaylord and Brennan, 1987). Particularly useful reagents are monoclonal antibodies directed against the *M. leprae*-specific oligosaccharide of PGL-I (Mehra et al., 1984; Young et al., 1984). Such antibodies have been shown to react only with *M. leprae* and have been used in the serodiagnosis of leprosy (Cho et al., 1986; Koster et al., 1987; Sanchez et al., 1986; Young et al., 1985). Several other monoclonal antibodies are also available that react with other targets that appear to be *M. leprae*-specific (reviewed in Gaylord and Brennan, 1987). One could also assay the reactivity of the unknown with T-cell clones that are directed against *M. leprae*-specific components. Other potential immunological tests include fluorescent-antibody stains (Abe et al., 1980), immunodiffusion assays (Payne et al., 1982), immunoprecipitation assays (Harboe et al., 1978), and the ability to induce a delayed type hypersensitivity reaction in infected individuals similar to that induced by authentic *M. leprae* preparations, e.g. lepromin (Shepard and Guinto, 1963). In any immunological test, one should be aware that *M. leprae* does contain many antigens and epitopes that are cross-reactive with antigens found in one or more of the other *Mycobacterium* species as well as antigens that are cross-reactive with organisms that are evolutionary very distant, including humans.

Genetic Tests

One can measure the relatedness of two organisms by determining the degree of hybridization between genomic DNAs from the two organisms (Brenner, 1989). Cells from the same species have DNA homologies of greater than 70% and a difference in melting temperature of < 5°C. Such genomic DNA:DNA hybridization studies have been reported for *M. leprae* and a variety of other *Mycobacterium* species. *M. leprae* shows little if any homology with the other mycobacteria (Athwal et al., 1984; Grosskinsky et al., 1989). Thus, this sort of test might be useful in identifying *M. leprae*. Unfortunately, such genomic DNA hybridization studies require fairly large amounts of DNA (10–100 μg, 10^9–10^{10} bacilli), not easily obtained for a non-axenically cultivable bacterium.

One way to circumvent the need for such large amounts of DNA is to use a labelled (e.g., radioactive) nucleic acid probe to assess DNA homologies (reviewed in Tenover, 1989). Using labelled probes, one can reduce the number of organisms needed for analysis to 10^5 to 10^6. Potential hybridization probes (nucleotide sequences that hybridize only with *M. leprae* nucleic acids) for *M. leprae* have been identified and cloned into *E. coli*, which allows easy production of large amounts of the probes. Of particular interest is the work of Clark-Curtiss and colleagues (Clark-Curtiss and Docherty, 1989; Grosskinsky et al., 1989) who have identified a sequence that is present in about 19 copies in the *M. leprae* genome and is not present in other mycobacteria. Using this sequence as a probe and purified DNA, they were able to detect as few as 4×10^3 genome equivalents of *M. leprae* DNA. Furthermore, they were able to detect homologous sequences in skin biopsy samples from lepromatous leprosy patients, which is of potential clinical importance. Incidentally, the *M. leprae* genome appears to be very stable with respect to this sequence in that all *M. leprae* isolates (from patients, armadillos, and monkeys) examined to date have identical hybridization patterns, with the exception of a single difference found in only one of the armadillo isolates (Clark-Curtiss and Walsh, 1989). This observation suggests that the probe should be capable of detecting all *M. leprae* isolates.

Several research groups are currently exploring the potential use of gene amplification techniques such as the polymerase chain reaction (PCR) to detect and identify *M. leprae*. In this technique, one uses oligonucleotide primers to direct the amplification of a particular nucleotide sequence via a bidirectional polymerase cascade reaction (Guatelli et al., 1989; Mullis and Faloona, 1987). The specificity of the reaction is determined by the choice of primers. In some preliminary studies, we have used oligonucleotide primers corresponding to various regions of the gene encoding the 65-kilodalton(kDa) antigen (B. B. Plikaytis and T. M. Shinnick, unpublished observations). Primers representing sequences unique to *M. leprae* direct the amplification of the gene from *M. leprae* but not from the other *Mycobacterium* species tested. The potential power of this technique can be demonstrated by two results. First, using these primers, we can specifically detect as few as 10 copies of the *M. leprae* 65-kDa antigen gene in a sample containing 10^6 copies of the *M. tuberculosis* 65-kDa antigen gene. Second, using crude lysates of *M. leprae*, grown in mouse footpads we can detect a positive signal in a sample containing as few as 20 bacilli.

These results indicate that the gene amplification technology can be used to amplify specifically sequences from *M. leprae*. Importantly, without optimizing the system, we were able to detect the equivalent of as few as 20 bacilli in a sample—a sensitivity much better than the currently available clinical tests. However, much work needs to be done before we can truly assess the potential clinical impact of this technology for leprosy. For example, among other things the system must be optimized with respect to: 1) target sequence and primers; 2) recovery of bacilli and isolation of DNA; 3) reaction conditions; and 4) detection systems. Nonetheless, the degree of sensitivity and specificity displayed by the primers described above is quite encouraging for the potential use of this technology in the rapid detection of small numbers of mycobacteria in clinical specimens and the positive identification of the organism as *M. leprae*.

Summary

Although much has been learned about the structure and composition of the *M. leprae* bacillus, the challenge for the identification and characterization of *M. leprae* still lies with the inability to propagate the organism outside of animal hosts. Until a suitable axenic culture system is developed, one must rely on animal models, hence making identification and characterization quite laborious. Fortunately, there are methods and procedures on the horizon for the analysis of very small numbers of bacilli (PCR) and for the analysis of the metabolic activity of bacilli (Franzblau and Hastings, 1987). These advances will probably facilitate the identification of organisms as *M. leprae* and the characterization of relevant biochemical traits.

Literature Cited

Abe, M., F. Ninagowa, Y. Yoshino, T. Ozawa, K. Saikawa, and T. Saito. 1980. Int. J. Lepr. 48:109.

Athwal, R. S., S. S. Deo, and T. Imaeda. 1984. Deoxyribonucleic acid relatedness among *Mycobacterium leprae, Mycobacterium tuberculosis,* and selected bacteria by dot blot and spectrophotometric deoxyribonucleic acid hybridization assays. Int. J. Syst. Bacteriol. 34:1136–1141.

Barton, R. F. E. 1974. A clinical study of the nose in leprosy. Lepr. Rev. 45:135–144.

Binford, C. E. 1959. Histiocytic granulomatous mycobacterial lesions produced in the golden hamster (*Cricetus auratus*) inoculated with human leprosy—negative results using other animals. Lab. Invest. 8:901–924.

Bloom, B. R., and V. Mehra. 1984. Immunological unresponsiveness in leprosy. Immunol. Rev. 80:5–28.

Bloom, B. R., and T. Godal. 1983. Selective primary health care: Strategies for control of disease in the developing world. V. Leprosy. Rev. Infect. Dis. 5:765–780.

Brennan, P. J. 1983. The phthiocerol-containing surface lipids of *Mycobacterium leprae:* a perspective of past and present work. Int. J. Lepr. 51:387–396.

Brenner, D. J. 1989. DNA hybridization for characterization, classification, and identification of bacteria. pp. 75–103. In: B. Swaminathan and G. Prakash, (ed.), Nucleic Acid and Monoclonal Antibody Probes: Applications in Diagnostic Microbiology. Marcel Dekker, Inc., New York.

Browne, S. G. 1975. Some aspects of the history of leprosy: the leprosy of yesteryear. Proc. R. Soc. Med. 68:485–493.

Butler, W. R., and J. O. Kilburn. 1988. Identification of major slow-growing pathogenic mycobacteria and *Mycobacterium gordonae* by high-performance liquid chromatography of their mycolic acids. J. Clin. Microbiol. 26:50–53.

Chan, J., T. Fujiwara, P. Brennan, M. McNeil, S. J. Turco, J.-C. Sibille, M. Snapper, P. Aisen, and B. R. Bloom. 1989. Microbial glycolipids: Possible virulence factors that scavenge oxygen radicals. Proc. Natl. Acad. Sci. USA 86:2453–2457.

Chatterjee, D., C. M. Bozic, C. Kinsley, S.-N., and P. J. Brennan. 1989. Phenolic glycolipids of *Mycobacterium bovis:* New structures and synthesis of a corresponding seroreactive neoglycoprotein. Infect. Immun. 57:322–330.

Chehl, S., C. K. Job, and R. C. Hastings. 1985. The nose: a site for the transmission of leprosy in nude mice. Am. J. Trop. Med. Hyg. 34:1161–1166.

Chehl, S., J. Ruby, C. K. Job, and R. C. Hastings. 1983. The growth of *Mycobacterium leprae* in nude mice. Lepr. Rev. 54:283–304.

Cho, S., S. W. Hunter, R. H. Gelber, T. H. Rea, and P. J. Brennan. 1986. Quantitation of the phenolic glycolipid of *Mycobacterium leprae* and relevance to glycolipid antigenemia in leprosy. J. Infect. Dis. 153:560–569.

Cho, S., D. L. Yanagihara, S. W. Hunter, R. H. Gelber, and P. J. Brennan. 1983. Serological specificity of phenolic glycolipid I from *Mycobacterium leprae* and use in serodiagnosis of leprosy. Infect. Immun. 41:1077–1083.

Clark-Curtis, J. E., and M. A. Docherty. 1989. A species-specific repetitive sequence in *Mycobacterium leprae* DNA. J. Infect. Dis. 159:7–15.

Clark-Curtiss, J. E., W. R. Jacobs, M. A. Docherty, L. R. Ritchie, and R. Curtiss III. 1985. Molecular analysis of DNA and construction of genomic libraries of *Mycobacterium leprae*. J. Bacteriol. 161:1093–1102.

Clark-Curtiss, J. E., and G. P. Walsh. 1989. Conservation of genomic sequences among isolates of *Mycobacterium leprae*. J. Bacteriol. 171:4844–4851.

Collaborative effort of the U.S. Leprosy Panel of the U.S.-Japan Cooperative Medical Sciences Program and the Leonard Wood Memorial. 1975. Rifampin therapy of lepromatous leprosy. Am. J. Trop. Med. Hyg. 24:475–484.

Colston, M. J., and G. R. F. Hilson. 1976. Growth of *Mycobacterium leprae* and *M. marinum* in congenitally athymic (nude) mice. Nature 262:399–401.

Colston, M. J., and G. R. F. Hilson. 1979. The effect of freezing and storage in liquid nitrogen on the viability and growth of *Mycobacterium leprae*. J. Med. Micro. 12:137–142.

Cowdry, E. V. 1940. Cytologic Studies on globi in leprosy. Am. J. Pathol. 16:103–136.

Daffe, M., M. A. Laneelle, C. Lacave, and G. Laneelle. 1988. Monoglycosyldiacylphenol-pthiocerol of *Mycobacte-*

rium tuberculosis and *Mycobacterium bovis*. Biochim. Biophys. Acta 958:443–449.

d'Arcy Hart, P. 1982. Lysosome fusion responses of macrophages to infection: behaviour and significance, p. 437–447. In: M. L. Karnovsky and L. Bolis (ed.), Phagocytosis: past and future. Academic Press, New York.

Davey, T. F., and R. J. W. Rees. 1974. The nasal discharge in leprosy: clinical and bacteriological aspects. Lepr. Rev. 45:121–134.

Desai, S. D., T. J. Birdi, and N. H. Antia. 1989. Correlation between macrophage activation and bactericidal function and *Mycobacterium leprae* antigen presentation in macrophages of leprosy patients and normal individuals. Infect. Immun. 57:1311–1317.

Desikan, K. V., and C. K. Job. 1968. A review of postmortem findings in 37 cases of leprosy. Int. J. Lepr. 36:31–44.

Donham, K. J., and J. R. Leininger. 1977. Spontaneous leprosy-like disease in a chimpanzee. J. Infect. Dis. 136:132–136.

Draper, P. 1976. The cell walls of *Mycobacterium leprae*. Int. J. Lepr. 44:95–98.

Draper, P. 1983. The bacteriology of *Mycobacterium leprae* Tubercle 64:43–56.

Draper, P., and R. J. W. Rees. 1970. Electron-transparent zone of mycobacterium may be a defence mechanism. Nature 228:860–861.

Fieldsteel, A. H., and L. Levy. 1976. Neonatally thymectomized Lewis rats infected with *Mycobacterium leprae:* response to primary infection, secondary challenge, and large inocula. Infect. Immun. 14:736–741.

Fine, P. E. M. 1982. Leprosy: The epidemiology of a slow-growing bacterium. Epidemiol. Rev. 4:161–188.

Fisher, C. A., and L. Barksdale. 1971. Elimination of the acid-fastness but not the gram positivity of leprosy bacilli after extraction with pyridine. J. Bacteriol. 106:707–708.

Franzblau, S. G. 1988. Oxidation of palmitic acid by *Mycobacterium leprae* in an axenic medium. J. Clin. Microbiol. 26:18–21.

Franzblau, S. G., E. B. Harris, and R. C. Hastings. 1987. Axenic incorporation of [U-¹⁴C] palmitic acid into phenolic glycolipid I of *Mycobacterium leprae*. FEMS Microbiol. Lett. 48:107–114.

Franzblau, S. G., and R. C. Hastings. 1987. Rapid in vitro metabolic screen for antileprosy compounds. Antimicrob. Agents Chemother. 31:780–783.

Frehel, C., and N. Rastogi. 1987. *Mycobacterium leprae* surface components intervene in the early phagosome-lysosome fusion inhibition event. Infect. Immun. 55:2916–2921.

Gaylord, H., and P. J. Brennan. 1987. Leprosy and the leprosy bacillus: recent developments in characterization of antigens and immunology of the disease. Annu. Rev. Microbiol. 41:645–675.

Grange, J. M., and T. J. Rowbotham. 1987. Microbe dependence of *Mycobacterium leprae:* A possible intracellular relationship with protozoa. Int. J. Lepr. 55:565–566.

Greater, J. G. 1975. The fly as potential vector in the transmission of leprosy. Lepr. Rev. 46:279–286.

Grosskinsky, C. M., W. R. Jr., Jacobs, J. E. Clark-Curtiss, and B. R. Bloom. 1989. Genetic relationships between *Mycobacterium leprae, Mycobacterium tuberculosis,* and candidate leprosy vaccine strains by DNA hybrid-ization: Identification of an *M. leprae*-specific repetitive sequence. Infect. Immun. 57:1535–1541.

Guatelli, J. C., T. R. Gingeras, and D. D. Richman. 1989. Nucleic acid amplification in vitro: Detection of sequences with low copy numbers and application to diagnosis of human immunodeficiency virus type 1 infection. Clin. Microbiol. Rev. 2:217–226.

Hanks, J. H., B. R. Chatterjee, and M. F. Lechat. 1964. A guide to the counting of mycobacteria in clinical and experimental materials. Int. J. Lepr. 32:156–167.

Hansen, G. A. 1874. Causes of Leprosy. Norsk. Mag. for Laegervidenskaben 4:1–88.

Hansen, G. A., and C. Looft. 1895. Leprosy: in its clinical and pathological aspects. Reprinted by John Wright, Bristol, 1973.

Harboe, M., O. Closs, B. Bjorvatn, G. Kronvall, and N. H. Axelsen. 1977. Antibody response in rabbits to immunization with *Mycobacterium leprae*. Infect. Immun. 18:792–805.

Hastings, R. C. (ed.). 1986. Leprosy. Churchill Livingstone, Edinburgh/New York.

Hastings, R. C., and Franzblau, S. G. 1988. Chemotherapy of leprosy. Annu. Rev. Toxicol. 28:231–45.

Hilson, G. R. F. 1965. Observations on the inoculation of *M. leprae* in the foot pad of the white rat. Int. J. Lepr. 33:662–665.

Hunter, S. W., and P. J. Brennan. 1981. A novel phenolic glycolipid from *Mycobacterium leprae* possibly involved in immunogenicity and pathogenicity. J. Bacteriol. 147:728–735.

Hunter, S. W., T. Fujiwara, and P. J. Brennan. 1982. Structure and antigenicity of the major specific glycolipid antigen of *Mycobacterium leprae*. J. Biol. Chem. 257:15072–15078.

Imaeda, T., W. F. Kirchheimer, and L. Barksdale. 1982. DNA isolated from *Mycobacterium leprae:* genome size, base ratio, and homology with other related bacteria as determined by optical DNA-DNA reassociation. J. Bacteriol. 150:414–417.

Jenkins, P. A., S. R. Pattyn, and F. Portaels. 1982. Diagnostic Bacteriology. p. 441–470. In: C. Ratledge and J. Stanford (ed.), The Biology of the Mycobacteria, vol. 1. Academic Press, London.

Job, C. K. 1971. Pathology of peripheral nerve lesions in lepromatous leprosy. A light and electron microscopic study. Int. J. Lepr. 39:251–268.

Jopling, W. H., and A. C. McDougall. 1988. Handbook of Leprosy. Heinemann Professional Publishing Ltd, Oxford.

Kaplan, G., and Z. A. Cohn. 1986. The immunobiology of leprosy. Int. Rev. Exp. Pathol. 28:45–78.

Kazda, J. 1981a. Occurrence of non-cultivatable acid-fast bacilli in the environment and their relationship to *M. leprae*. Lepr. Rev. 52 (Suppl. 1):85–92.

Kazda, J. 1981b. Nine-banded armadillos in captivity: prevention of losses due to parasitic diseases. Some remarks on mycobacteria-free maintenance. Int. J. Lepr. 49:345–346.

Kazda, J., R. Ganapati, C. Revankar, T. M. Buchanan, D. B. Young, and L. M. Irgens. 1986. Isolation of environment-derived *Mycobacterium leprae* from the soil in Bombay. Lepr. Rev. 57 (Suppl. 3):201–208.

Kirchheimer, W. F. 1976. The role of arthropods in the transmission of leprosy. Int. J. Lepr. 44:104–107.

Kirchheimer, W. F., and E. E. Storrs. 1971. Attempts to establish the armadillo (*Dasypus novemcinctus* Linn.) as a model for the study of leprosy. 1. Report of lepromatoid leprosy in an experimentally infected armadillo. Int. J. Lepr. 39:693–702.

Koch, R. 1897. Die Lepraerkrankungen in Kreise Memel. Abstracted in Baumgartens Jahresbericht 14:428.

Koster, F. T., D. M. Scollard, E. T. Umland, D. B. Fishbein, W. C. Hanley, P. J. Brennan, and K. E. Nelson. 1987. Cellular and humoral immune response to a phenolic glycolipid antigen (PGL-I) in patients with leprosy. J. Clin. Microbiol. 25:551–556.

Levy, L. 1976. Studies on the mouse foot pad technique for cultivation of *Mycobacterium leprae*. 3. Doubling time during logarithmic multiplication. Lepr. Rev. 47:103–106.

Lumpkin, L. R., G. F. Fox, and J. E. Wolf. 1983. Leprosy in five armadillo handlers. J. Am. Acad. Dermatol. 9:899–903.

McCormick, G. T., and R. M. Sanchez. 1979. Pyridine extractability of acid-fastness of *M. leprae*. Int. J. Lepr. 47:495–499.

McWhirter, N. (ed.). 1981. Guiness Book of Records. 28th ed. Trowbridge: Redwood Burn Ltd.

Mehra, V., P. J. Brennan, E. Rada, J. Convit, and B. R. Bloom. 1984. Lymphocyte suppression in leprosy induced by unique *M. leprae* glycolipid. Nature 308:194–196.

Meyers, W. M., G. P. Walsh, H. L. Brown, C. H. Binford, G. D, Jr., Imes, T. L. Hadfield, C. J. Schlagel, Y. Fukunishi, P. J. Gerone, R. H. Wolf, B. J. Gormus, L. N. Martin, M. Harboe, and T. Imaeda. 1985. Leprosy in a Mangabey monkey—naturally acquired infection. Int. J. Lepr. 53:1–14.

Mittal, A., M. Sathish, P. S. Seshadri, and I. Nath. 1983. Rapid radiolabeled-microculture method that uses macrophages for in vitro evaluation of *Mycobacterium leprae* viability and drug susceptibility. J. Clin. Microbiol. 17:704–707.

Mor, N. 1983. Intracellular location of *Mycobacterium leprae* in macrophages of normal and immunodeficient mice and effect of rifampicin. Infect. Immun. 42:802–811.

Mullis, K. B., and F. A. Faloona. 1987. Specific synthesis of DNA in vitro via a polymerase-catalyzed chain reaction. Methods in Enzymol. 155:335–350.

Narayanan, E., Sreevetsa, W. F. Kirchheimer, and B. M. S. Bedi. 1977. Transfer of leprosy bacilli from patients to mouse foot pads by *Aedes aegypti*. Leprosy in India 49:181–186.

Pallen, M. J., and R. D. McDermott. 1986. How might *Mycobacterium leprae* enter the body. Lepr. Rev. 57:289–297.

Payne, S. N., P. Draper, and R. J. W. Rees. 1982. Serological activity of purified glycolipid from *Mycobacterium leprae*. Int. J. Lepr. 50:220–221.

Pedley, J. C. 1968. The presence of *M. leprae* in the lumina of the female mammary gland. Lepr. Rev. 39:201.

Portaels, F., K. Fissette, K. De Ridder, P. M. Macedo, A. De Muynck, and M. T. Silva. 1988. Effects of freezing and thawing on the viability and the ultrastructure of in vivo grown mycobacteria. Int. J. Lepr. 56:580–587.

Prabhakaran, K. 1967. Oxidation of 3,4-dihydroxyphenylalanine (DOPA) by *Mycobacterium leprae*. Int. J. Lepr. 35:42–51.

Prabhakaran, K., and W. F. Kirchheimer. 1966. Use of 3,4-dihydroxyphenylalanine oxidation in the identification of *Mycobacterium leprae*. J. Bacteriol. 92:1267–1268.

Prasad, H. K., R. S. Mishra, and I. Nath. 1987. Phenolic glycolipid I of *Mycobacterium leprae* induces general suppression of in vitro concanavalin A responses unrelated to leprosy type. J. Exp. Med. 165:239.

Ramasesh, N., R. C. Hastings, and J. L. Krahenbuhl. 1987. Metabolism of *Mycobacterium leprae* in macrophages. Infect. Immun. 55:1203–1206.

Rastogi, N., and H. L. David. 1988. Mechanisms of pathogenicity in mycobacteria. Biochemie 70:1101–1120.

Rees, R. J. W. 1966. Enhanced susceptibility of thymectomized and irradicated mice to infection with *Mycobacterium leprae*. Nature 211:657–658.

Rees, R. J. W., and A. C. McDougall. 1976. Airborne infection with *Mycobacterium leprae* in mice. Int. J. Lepr. 44:99–103

Ridley, D. S., and W. H. Jopling. 1966. Classification of leprosy according to immunity. A five-group system. Int. J. Lepr. 34:255–273.

Ridley, M. J., M. F. R. Waters, and D. S. Ridley. 1987. Events surrounding the recognition of *Mycobacterium leprae* in nerves. Int. J. Lepr. 55:99–108.

Sanchez, G. A., A. Malik, C. Tougne, P. H. Lambert, and H. D. Engers. 1986. Simplification and standardization of serodiagnostic tests based on phenolic glycolipid I (PGL-I) antigen. Lepr. Rev. 57:(Suppl. 2):83–93.

Sansarricq, H. 1981. Leprosy in the world today. Lepr. Rev. 51 (Suppl. 1):15–31.

Schaffer, X. Ueber der Verbreitung der Leprabacillen von den oberen Luftwegen aus. 1898. Arch. Dermatol. Syphil. 44:159–174.

Sela, S., J. E. Clark-Curtiss, H. Bercovier. 1989. Characterization and taxonomic implications of the rRNA genes of *Mycobacterium leprae*. J. Bacteriol. 171:70–73.

Shepard, C. C. 1960. The experimental disease that follows the injection of human leprosy bacilli into the footpads of mice. J. Exp. Med. 12:445–454.

Shepard, C. C. 1962. The nasal excretion of *Mycobacterium leprae* in leprosy. Int. J. Lepr. 30:10–18.

Shepard, C. C. 1965. Stability of *Mycobacterium leprae* and temperature optimum for growth. Int. J. Lepr. 33:541–547.

Shepard, C. C. 1971. A survey of drugs with activity against *M. leprae* in mice. Int. J. Lepr. 39:340–348.

Shepard, C. C. 1986. Experimental Leprosy. p. 269–286. In: R. C. Hastings (ed.), Leprosy. Churchill Livingstone, Edinburgh.

Shepard, C. C., and R. S. Guinto. 1963. Immunological identification of foot-pad isolates as *Mycobacterium leprae* by lepromin reactivity in leprosy patients. J. Exp. Med. 118:195–204.

Shepard, C. C., and D. H. McRae. 1965a. *Mycobacterium leprae* in mice: Minimal infectious dose, relationship between staining and infectivity, and effect of cortisone. J. Bacteriol. 89:365–372.

Shepard, C. C., and D. H. McRae. 1965b. *Mycobacterium leprae*: Viability at O°C, 31°C, and during freezing. Int. J. Lepr. 33:316–323.

Shepard, C. C., and D. H. McRae. 1968. A method for counting acid-fast bacteria. Int. J. Lepr. 36:78–82.

Shepard, C. C., and D. H. McRae. 1971. Hereditary characteristics that varies among isolates of *Mycobacterium leprae*. Infect. Immun. 3:121–126.

Shepard, C. C., R. M. Vanlandingham, and L. L. Walker. 1983. Recent studies of antileprosy drugs. Lepr. Rev. 54:23S–30S.

Sibley, L. D., S. G. Franzblau, and J. L. Krahenbuhl. 1987. Intracellular fate of *Mycobacterium leprae* in normal and activated macrophages. Infect. Immun. 55:680–685.

Sibley, L. D., and J. L. Krahenbuhl. 1988. Defective activation of granuloma macrophages from *Mycobacterium leprae*-infected nude mice. J. Leuk. Biol. 43:60–66.

Sixth Report of the WHO Expert Committee on Leprosy. 1988. World Health Organization Technical Report Series No. 768. Geneva.

Skinsnes, O. K., P. H. C. Chang, and E. Matsuo. 1975. Acid-fast properties and pyridine extraction of *M. leprae*. Int. J. Lepr. 43:392–398.

Smida, J., J. Kazda, and E. Stackebrandt. 1988. Molecular-genetic evidence for the relationship of *Mycobacterium leprae* to slow-growing pathogenic mycobacteria. Int. J. Lepr. 56:449–454.

Stanford, J. L., G. A. W. Rook, J. Convit, T. Godal, G. Kronvall, R. J. W. Rees, and G. P. Walsh. 1975. Preliminary taxonomic studies of the leprosy bacillus. Brit. J. Exp. Path. 56:579–585.

Steinhoff, U., J. R. Golecki, J. Kazda, and S. H. E. Kaufmann. 1989. Evidence for phagosome-lysosome fusion in *Mycobacterium leprae*-infected murine Schwann cells. Infect. Immun. 57:1008–1010.

Stewart-Tull, D. E. S. 1982. *Mycobacterium leprae*—The bacteriologists' enigma, p. 273–307. Inc: C. Ratledge and J. Stanford, (ed.), The biology of mycobacteria, vol. 1. Academic Press, London.

Stoner, G. L. 1979. Importance of the neural predilection of *Mycobacterium leprae* in leprosy. Lancet ii:994–997.

Storrs, E. E. 1971. The nine-banded armadillo: a model for leprosy and other biomedical research. Int. J. Lepr. 39:703–714.

Tenover, F. C. (ed.). 1989. DNA probes for infectious diseases. CRC Press, Boca Raton, FL.

Vachula, M., T. J. Holzer, and B. R. Andersen. 1989. Suppression of monocyte oxidative response by phenolic glycolipid I of *Mycobacterium leprae*. J. Immunol. 142:1696–1701.

Vemuri, N., L. Kandke, P. R. Mahadevan, S. W. Hunter, and P. J. Brennan. 1985. Isolation of phenolic glycolipid from human lepromatous nodules. Int. J. Lepr. 53:489.

Wade, H. W. 1948. The Michigan inoculation cases. Int. J. Lepr. 16:465–475.

Wade, H. W., and V. Ledowski. 1952. The leprosy epidemic at Naura: a review with data on the status since 1937. Int. J. Lepr. 31:34–45.

Walsh, G. P., W. M. Meyers, C. H. Binford, P. J. Gerone, R. H. Wolf, and J. R. Leininger. 1981. Leprosy—a zoonosis. Lepr. Rev. 52 (Suppl. 1):77–83.

Walsh, G. P., E. E. Storrs, H. P. Burchfield, W. M. Meyers, and C. H. Binford. 1975. Leprosy-like disease occurring naturally in armadillos. J. Reticuloendothel. Soc. 18:347–351.

Waters, M. F. R., and R. J. W. Rees. 1962. Changes in the morphology of *Mycobacterium leprae* in patients under treatment. Int. J. Lepr. 30:266–277.

Wayne, L. G., and W. M. Gross. 1968. Base composition of deoxyribonucleic acid isolated from mycobacteria. J. Bacteriol. 95:1481–1482.

Wolf, R. H., B. J. Gormus, L. N. Martin, G. B. Baskin, G. P. Walsh, W. M. Meyers, and C. H. Binford. 1985. Experimental leprosy in three species of monkeys. Science 227:529–531.

Young, D. B., J. P. Harnisch, J. Knight, and T. M. Buchanan. 1985. Detection of phenolic glycolipid I in sera of patients with lepromatous leprosy. J. Infect. Dis. 152:1078.

Young, D. B., S. R. Khanolkar, L. L. Barg, and T. M. Buchanan. 1984. Generation and characterization of monoclonal antibodies to the phenolic glycolipid of *Mycobacterium leprae*. Infect. Immun. 43:183–188.

CHAPTER 55

The Genus *Arthrobacter*

DOROTHY JONES and R. M. KEDDIE

Conn (1928) described a group of bacteria, extremely numerous in certain soils, which were unusual in that they appeared as Gram-negative rods in young cultures and as Gram-positive cocci in older cultures. For these bacteria, Conn (1928) created the species *Bacterium globiforme*, which, as *Arthrobacter globiformis*, was later to become the type species of the genus *Arthrobacter*. The abundance in soil of bacteria similar to Conn's organism, and of other coryneform bacteria, was confirmed later by Jensen (1933, 1934) and Topping (1937, 1938), who, however, referred to them as soil corynebacteria, and by Taylor and Lochhead (1937), who used the name *Bacterium globiforme*. Jensen (1934) considered that these soil bacteria should be classified in the genus *Corynebacterium* because of their morphological resemblance to corynebacteria of animal origin. However, Conn (1947) vigorously opposed this view and created the genus *Arthrobacter* (by reviving an old name), with *A. globiformis* as the type species and with two of Jensen's soil corynebacteria as additional species (Conn and Dimmick, 1947).

In addition to their characteristic morphology and staining reactions, members of the genus *Arthrobacter* were originally described as being highly aerobic, nutritionally nonexacting, and capable of liquefying gelatin slowly (Conn and Dimmick, 1947). These features were chosen mainly to distinguish *Arthrobacter* from *Corynebacterium* as represented by *C. diphtheriae* and similar animal parasitic species. However, because of its poor circumscription (see Gibson, 1953; Jensen, 1952), the genus *Arthrobacter* was not widely accepted until it was included as a member of the family Corynebacteriaceae in the seventh edition of *Bergey's Manual of Determinative Bacteriology* (Breed et al., 1957). But by that time, the genus had been extended to include the two nutritionally exacting species *A. terregens* (Lochhead and Burton, 1953) and *A. citreus* (Sacks, 1954), and shortly afterwards two others were added (Lochhead, 1958a). Indeed, one of Conn's strains of *A. globiformis*

was shown subsequently to require biotin for growth (Chan and Stevenson, 1962; Morris, 1960). Thus the concept had developed of *Arthrobacter* as a genus of soil bacteria whose major distinguishing feature was a growth cycle in which the irregular rods in young cultures were replaced by coccoid forms in older cultures; these coccoid forms, when transferred to fresh medium, produced outgrowths ("germinated") to give irregular rods again, and so the cycle was repeated (Fig. 1).

This dependence on morphological features and habitat in the circumscription led to a great deal of confusion in the classification of the genus *Arthrobacter* and thus created considerable problems in the identification of new isolates as arthrobacters. Thus, isolates from soil and, more especially, those from other habitats have frequently been referred to in the literature as arthrobacters on the basis of morphological features alone, even though they were not necessarily similar to *A. globiformis* in other respects.

It was not surprising that when representatives of the genus were examined by more modern taxonomic methods such as numerical taxonomy (Jones, 1978), various chemotaxonomic techniques (Bowie et al., 1972; Keddie and Cure, 1977, 1978; Minnikin et al., 1978b; Schleifer and Kandler, 1972), and determinations of DNA base ratios (see Skyring and Quadling, 1970; Skyring et al., 1971), it was found to be heterogeneous.

The genus *Arthrobacter* as defined in the eighth edition of *Bergey's Manual* (Keddie, 1974) was heterogeneous, as was noted by Keddie and Jones (1981) in the first edition of *The Prokaryotes*. They referred to *Arthrobacter* in this broad sense as *Arthrobacter* sensu lato (Keddie and Jones, 1981). In *Bergey's Manual of Systematic Bacteriology* (Keddie et al., 1986), the genus was limited to those species which, like the type, *A. globiformis*, contain lysine as the cell wall diamino acid, i.e., *Arthrobacter* sensu stricto (Keddie and Jones, 1981). Thus some species formerly considered to be arthrobacters (Keddie, 1974) have now been removed from the genus.

The two species formerly named *A. terregens* and *A. flavescens*, which contain ornithine as the cell wall diamino acid (Schleifer and Kandler, 1972; Keddie and Cure, 1977), have been transferred to the genus *Aureobacterium* as *Aur. terregens* and *Aur. flavescens* (Collins et al., 1983). The species *Arthrobacter radiotolerans* has now been transferred to the new genus *Rubrobacter* as *R. radiotolerans* (Suzuki et al., 1988).

Although now resolved, the position of the two species *A. simplex* and *A. tumescens* has been more problematical. They were shown by Cummins and Harris (1959) to differ from *A. globiformis* in containing LL-diaminopimelic acid (LL-A$_2$pm) as the cell wall diamino acid, and many other taxonomic differences were detected subsequently (see Keddie et al., 1986, for further details). In the case of *A. simplex,* 16S rRNA cataloging studies showed this species to be only distantly related to *A. globiformis* (Stackebrandt et al., 1980). Conversely, 5S rRNA sequencing indicated that *Pimelobacter simplex* (*A. simplex*—see below) clearly belonged to an *"Arthrobacter—Micrococcus—Cellulomonas"* subgroup of the coryneform bacteria (Park et al., 1987). While there was general agreement that *A. simplex* and *A. tumescens* should be removed from *Arthrobacter,* there was disagreement about where they should be accommodated. Suzuki and Komagata (1983) created the genus *Pimelobacter* for the LL-A$_2$pm-containing coryneform bacteria and distinguished three species by use of DNA-DNA base-pairing techniques. The first species, *Pimelobacter simplex,* contained most strains of *A. simplex* (and also strains named *"Brevibacterium lipolyticum"*), the second, *P. tumescens,* for strains formerly called *A. tumescens,* and a third species, *P. jensenii,* was created for a single strain originally identified as an *A. simplex* strain (Gundersen and Jensen, 1956). The same authors also concluded from their DNA homology studies that *A. simplex* and *A. tumescens* were only distantly related to *Nocardioides albus,* a nocardioform organism. However, O'Donnell et al. (1982) considered that *A. simplex* (but not *A. tumescens*) closely resembled *Nocardioides* species in chemotaxonomic (particularly lipid) characters and proposed that it be transferred to that genus as *Nocardioides simplex.* This view received support from other studies (reviewed by Keddie et al., 1986). In an attempt to resolve the problem, Collins et al. (1989) made a comparative study of the 16S rRNA from the type strains of *Nocardioides albus, N. luteus, Pimelobacter simplex, P. tumes-*

cens, and *P. jensenii* by reverse transcriptase sequencing and compared the results with those from 18 previously studied actinomycetes from 14 different genera. The study confirmed the reclassification of *P. (Arthrobacter) simplex* in the genus *Nocardioides* as *N. simplex* as proposed by O'Donnell et al. (1982) and also showed that *P. jensenii* should be transferred to that genus as *N. jensenii* (Collins et al., 1989). However, *P. (Arthrobacter) tumescens* was so distinct from all the other actinomycete taxa that Collins et al. (1989) proposed its reclassification in a new genus, *Terrabacter,* as *T. tumescens.* The taxonomic status of the single strain of the species *A. duodecadis* remains unresolved. Details of this and some other species named *Arthrobacter,* but now excluded from the genus, are given by Keddie et al. (1986) and for *A. siderocapsulatus* and *A. viscosus* by Collins (1986).

However, Collins (1986) has shown that the lipid composition of *"A. sialophilus"* (Tanenbaum and Flashner, 1977) and the phytopathogen *Agrobacterium pseudotsugae* is consistent with their being members of *Arthrobacter* sensu stricto (Keddie and Jones, 1981), but no formal proposal has been made to include them in the genus.

As presently circumscribed, the genus *Arthrobacter* contains two "groups of species" referred to as the *A. globiformis*/*A. citreus* group and the *A. nicotianae* group. These groups differ in their peptidoglycan structure, teichoic acid content, and lipid composition (see "Further Identification of Arthrobacters" and Tables 2 and 3 for details). It has been suggested that the genus should be restricted to those bacteria that exhibit the characteristics of the *A. globiformis*/*A. citreus* group (Minnikin et al., 1978a; Collins and Kroppenstedt, 1983). However, on the basis of DNA-DNA homology studies of representative arthrobacters (Stackebrandt and Fiedler, 1979) and later 16S rRNA cataloging studies, Stackebrandt et al. (1983) concluded that the genus *Arthrobacter* contained two "nuclei," one represented by the *A. globiformis*/*A. citreus* group of species and the other by the *A. nicotianae* group. This view was adopted by Keddie et al. (1986) and is the one accepted here.

The results of 16S rRNA cataloging studies (Stackebrandt et al., 1980; Stackebrandt and Woese, 1981) indicate that the genus *Arthrobacter* is related to the other coryneform genera, *Aureobacterium, Cellulomonas, Curtobacterium,* and *Microbacterium,* and is more distantly related to *Brevibacterium.* All of these genera are members of the high GC "actinomycete" branch of the Gram-positive eubacteria (Stackebrandt and Woese, 1981). The studies of Stackebrandt et al. (1980) also showed that on a phylogenetic basis the *Arthrobacter* species could not be separated from members of the genus *Micrococcus*. While accepting that the genera *Arthrobacter* and *Micrococcus* are very closely related phylogenetically, we treat them here as distinct taxa for practical purposes.

Habitats of *Arthrobacter*

In many ecological studies, isolates have been identified as arthrobacters or described as "arthrobacter-like" simply because they showed the rod-coccus growth cycle and the staining reactions characteristic of the genus. Although the sequence of morphological changes that occurs during the growth cycle is an important distinguishing feature of the genus, it does not occur exclusively in arthrobacters. For example, it is also seen in the genus *Brevibacterium* and in at least some members of the genus *Rhodococcus*. Accordingly, some of the strains described as arthrobacters in the literature cited below may belong to other morphologically similar taxa, especially if they are isolates from habitats other than soil.

Soil and Similar Habitats

Many studies have shown that bacteria of the genus *Arthrobacter* form a numerically important fraction of the indigenous bacterial flora of soils from different parts of the world; they are sometimes the most numerous, single bacterial group recorded in aerobic plate counts (Hagedorn and Holt, 1975b; Holm and Jensen, 1972; Lowe and Gray; 1972; Mulder and Antheunisse; 1963; Skyring and Quadling, 1969; Soumare and Blondeau, 1972). However, both the numbers and the proportions of arthrobacters in "total" counts decrease with increasing soil acidity (Hagedorn and Holt, 1975b; Lowe and Gray, 1972). Among the explanations advanced for their numerical predominance are their extreme resistance to drying (Boylen, 1973; Chen and Alexander, 1973; Labeda et al., 1976; Mulder and Antheunisse, 1963; Robinson et al., 1965) and to starvation (Boylen and Ensign, 1970; Boylen and Mulks, 1978; Zevenhuizen, 1966), factors important in the survival of microorganisms in soil (Gray, 1976). However, the nutritional versatility of the commonly occurring species undoubtedly also plays a part (see below).

Both psychrophilic and psychrotrophic strains of the genus were reported to be the most abundant and active bacteria in subterranean cave silts (Gounot, 1967), and they also occur in glacier silts (Moiroud and Gounot, 1969). The genus was also represented among isolates from oil brines raised from soil layers some 200–700 m deep (Iizuka and Komagata, 1965). Arthrobacters capable of dissolving aluminium silicates were reported to be common on "karst" rocks, and most were considered to be capable of dinitrogen fixation (Smyk, 1970; Smyk and Ettlinger, 1963). Members of the genus have also been implicated in the growth of manganese nodules in the sea (Ehrlich, 1963, 1968).

There is now much evidence that the predominant soil arthrobacters can use a wide and diverse range of organic substrates as sole or principal sources of carbon and energy (Hagedorn and Holt, 1975a; Keddie, 1974). This nutritional versatility is characteristic of those species which do not require vitamins or other organic growth factors or which require only biotin. Most species now recognized, including *A. globiformis* and new isolates from soil belong to this nutritional category, with *A. citreus* (type strain) being a notable exception. The detailed carbon nutrition of representatives of a number of species of soil arthrobacters together with 26 unnamed isolates, mainly from soil, was examined by J. D. Owens and R. M. Keddie and G. L. Cure and R. M. Keddie (unpublished observations quoted in Keddie et al., 1986). The following substrates were utilized by some 90% or more of the strains tested: D-xylose, D-glucose, D-mannose, D-galactose, D-fructose, cellobiose, maltose, trehalose, sucrose, raffinose, melezitose, D-gluconolactone, salicin, acetate, propionate, pentanoate, heptanoate, succinate, fumarate, DL-lactate, DL-malate, citrate, pyruvate, oxaloacetate, glycerol, mannitol, *m*-hydroxybenzoate, *p*-hydroxybenzoate, uric acid, glycine, L-α-alanine, D-α-alanine, L-isoleucine, L-threonine, L-lysine, L-arginine, L-aspartate, L-glutamate, L-phenylalanine, L-tyrosine, L-proline, L-histidine, 1,4-butanediamine, agmatine, tyramine, betaine, sarcosine, and creatine. *A. citreus* (type strain) is less nutritionally versatile than the species mentioned above (Keddie, 1974). Other studies have demonstrated the ability of arthrobacters to utilize aromatic compounds (Stevenson, 1967) and nucleic acids and their degradation products (Antheunisse, 1972). The frequent recovery of *Arthrobacter* species from enrichment cultures in which various diverse, organic compounds are supplied as sole carbon sources in simple, mineral media is further evidence of this nutritional versatility. Such compounds include nicotine (Giovanozzi-Sermanni, 1959; Keddie et al., 1966; Sguros, 1955), puromycin amino-nucleoside (Greenberg and Barker, 1962), 2-hydroxypyridine (Ensign and Rittenberg, 1963; Kolenbrander et al., 1976), *n*-alkanes (Klein et al., 1968), lower alcohols (Akiba et al., 1970), choline (Kortstee, 1970), picolinic acid (Tate and Ensign, 1974), and squalene (Yamada et al. 1975).

Arthrobacters have also been shown to degrade herbicides such as disodium endoxohexahydrophthalate ("Endothal") (Jensen, 1964), and 2,4-dichlorophenoxyacetate (Cacciari et al., 1971; Loos et al., 1967; Sharpee et al., 1973). Others degrade pesticides, usually by cometabolism, e.g., "diazinon" (Sethunathan and Pathak, 1971) and *m*-chlorobenzoate, the central molecule in many pesticides (Horvath and Alexander, 1970).

Fish and Other Similar Habitats

Coryneform bacteria appear to be common on fish (both marine and freshwater) and on some other seafoods. Many are morphologically similar to arthrobacters (and to *Brevibacterium*) and are frequently referred to by that name. Thus "arthrobacters" have been reported to occur in shark spoilage (referred to as *"Corynebacterium" globiformis* and *C. helvolum;* Wood, 1950), eviscerated freshwater fish (Roth and Wheaton, 1962), fish-pen slime (Chai and Levin, 1975), and Pacific shrimp (Lee and Pfeifer, 1977). Sieburth (1964) reported the isolation of an *Arthrobacter* species from sea water.

However, in the few cases in which coryneform isolates from fish have been examined by suitable modern methods, their relationship to *Arthrobacter* has proved to be much more remote than has been inferred from their morphological similarity, and most, if not all of them belong to other taxa (see Bousfield, 1978; Crombach, 1974a, 1974b).

Sewage and Similar Habitats

Bacteria identified as *Arthrobacter* species have been isolated from sewage (Nand and Rao, 1972) and from "brewery sewage" (Kaneko et al., 1969). Arthrobacters physiologically similar to those from soil were common in dairy-waste activated sludge (Mulder and Antheunisse, 1963) and were considered to play an important role in the process (Adamse, 1968). A number of the activated-sludge strains were reported by the above authors to decompose phenol. However, subsequent examination of four of these phenol-decomposing strains using chemotaxonomic methods revealed that they were members of the "rhodochrous" taxon *(Rhodococcus)* (Keddie and Cure, 1977). Similarly, although

Schefferle (1966) considered that the predominant coryneform bacteria in poultry deep litter could be placed in the genus *Arthrobacter,* only one of seven of these litter strains subsequently examined by Keddie and Cure (1977) was considered to be a legitimate *Arthrobacter* species. Although many coryneform isolates from aerated, animal-manure slurries are similar in morphology to *Arthrobacter,* only 1 of 16 such isolates examined was identified as *A. globiformis;* most of the remainder were considered to be *"rhodochrous"* strains, and a few were similar to *Brevibacterium linens* (Keddie and Cure, 1977).

Such examples clearly illustrate that new isolates should not be assigned to the genus *Arthrobacter* on the basis of morphology and conventional features alone.

Other Habitats

Coryneform bacteria seem to be relatively common on the aerial surfaces of plants (Austin et al., 1978; Keddie et al., 1966; Mulder et al., 1966) but few have been shown to be legitimate arthrobacters. However, "arthrobacter-like" organisms have been isolated from frozen vegetables (Splittstoesser et al., 1967). Also, two of a number of strains isolated from cauliflower by Lund (1969) were later identified as *A. globiformis* strains by Keddie and Cure (1977), but it is possible that they were contaminants from soil rather than indigenous plant bacteria. However, the organism formerly called *Corynebacterium ilicis,* a pathogen of American holly (Mandel et al., 1961), has been shown to be a legitimate *Arthrobacter* species and has now been transferred to that genus as *A. ilicis* (Collins et al., 1981). The type strain of *A. protophormiae* (formerly *Brevibacterium protophormiae;* Lysenko, 1959) was isolated from an insect, *Protophormia terraenovae,* but other strains were isolated from soil (see Keddie et al., 1986).

Arthrobacters were reported to be numerous in commercial preparations of liquid eggs, probably as a result of contamination from the shells, but they were rarely isolated from turkey giblets (Kraft et al., 1966). Arthrobacters do not appear to have been isolated from clinical sources.

Isolation of Arthrobacters

Arthrobacters have normally been isolated from soil and similar habitats by plating on suitable nonselective ("total count") media, and then picking from a large, random selection of colonies, and identifying as arthrobacters those isolates that show a rod-coccus growth cycle (Holm and Jensen, 1972; Lowe and Gray, 1972; Skyring and Quadling, 1969). This method is only suitable for habitats such as soil in which arthrobacters form an appreciable proportion of the aerobic, cultivable population. A further extension of this technique was introduced by Mulder and Antheunisse (1963), who devised what was essentially a method for screening isolates picked from a nonselective medium for those that showed the typical rod-coccus growth cycle of arthrobacters (see also Mulder et al., 1966; Veldkamp, 1965). Hagedorn and Holt (1975b) devised a selective medium said to be suitable for the enumeration of arthrobacters in soil. Thus, when nine different soils were plated on this medium, an average of 74% of the colonies that developed were identified as arthrobacters, and the numbers were similar to those estimated from counts on a nonselective medium (Hagedorn and Holt, 1975b).

However, if identification of the isolates is based only, or largely, on the provision of a rod-coccus growth cycle as was the case in the studies quoted, then bacteria from other genera such as *Brevibacterium* and *Rhodococcus* may be mistaken for arthrobacters. If, however, isolates are screened for the presence of lysine in the cell wall by using a rapid method of cell wall analysis, then this problem is overcome.

As noted above, a number of *Arthrobacter* species have been isolated from enrichment cultures using a variety of organic substrates as sole carbon and energy sources in mineral salts media. However, the primary purpose of such enrichments was not to isolate arthrobacters but to obtain isolates capable of utilizing the particular substrates studied. Accordingly, it is not possible to assess the value of these enrichment methods for the isolation of particular strains of arthrobacters.

Nonselective Media for Isolation of Arthrobacters

The aim of this approach is to use media and conditions of incubation which give the maximum possible counts of soil bacteria capable of growth under aerobic conditions. The media used must therefore contain sufficient amounts of all the organic growth factors and mineral constituents required to provide the diverse nutritional requirements of the indigenous soil bacteria (see Lochhead, 1958b). But, at the same time, media must be sufficiently poor in carbon and energy sources to limit the size of colonies, thereby minimizing antagonistic effects be-

tween the components of the population. Various modifications of soil extract agar have been the media most widely used. Lochhead, who pioneered the use of such media, strongly advocated the use of soil extract agar without other additions (Lochhead and Burton, 1956), but for some soils, addition of low concentrations of yeast extract and glucose can give higher counts (Jensen, 1968). Examples of both kinds of soil extract agars are given below. If necessary, the growth of fungi may be suppressed by incorporating the antibiotics nystatin (50 μg/ml) and cycloheximide (50 μg/ml) in the medium (Williams and Davies, 1965). Other factors important in plating soil samples have been discussed by Jensen (1968). The more important of these are: 1) the soils should be examined within a few hours of sampling; 2) the primary dilution should be dispersed by using a laboratory blender but avoiding heating; 3) a suitable diluent should be used (see below); and 4) plates should be incubated at 25°C for a minimum of 2 weeks.

Jensen (1968) recommends Winogradsky's standard salt solution as a diluent. It has the following composition (Holm and Jensen, 1972):

Winogradsky's Standard Salt Solution

K$_2$HPO$_4$	0.25 g
MgSO$_4$	0.125 g
NaCl	0.125 g
Fe$_2$(SO$_4$)$_3$	0.0025 g
MnSO$_4$	0.0025 g
Deionized water	1 liter
Adjust pH to 6.5–6.7.	

Soil extract and dilute peptone solutions (0.05–0.1%, wt/vol) have also been used successfully but one-fourth-strength Ringer's solution, physiological saline, and tap water are unsuitable (Jensen, 1968). In this context, Owens and Keddie (1969) noted that a chelated mineral salts solution which they devised, but without (NH$_4$)$_2$SO$_4$ (mineral base E-N), was a suitable diluent for coryneform bacteria. Mineral base E-N gave slightly better survival of *Arthrobacter globiformis* than a simple salts solution and was markedly superior to traditional diluents such as one-fourth-strength Ringer's solution or physiological saline. The preparation of mineral base E-N was described by Cure and Keddie (1973).

Isolation of Arthrobacters Using Soil Extract Agar

Soil Extract (Lochhead and Burton, 1957)

To prepare the soil extract, add 1 kg of soil to 1 liter of tap water and autoclave at 121°C for 20 min; filter and restore volume to 1 liter with tap water. A fertile garden soil usually gives the best results (Jensen, 1968).

Soil Extract Agar for Isolating Arthrobacters (Lochhead and Burton, 1957)

Soil extract (see above), 1 liter; K$_2$HPO$_4$, 0.2 g; agar, 15 g; final pH 6.8. Autoclave at 121°C for 20 min. Colonies are picked into tubes of soil extract semisolid medium: soil extract, 1 liter; K$_2$HPO$_4$, 0.2 g; yeast extract, 1 g; agar, 3 g; final pH 6.8. Autoclave at 121°C for 20 min.

Soil Extract Agar for Isolating Arthrobacters (Holm and Jensen, 1972)

Soil extract (see above)	400 ml
Tap water	600 ml
Glucose	1 g
Peptone	1 g
Yeast extract	1 g
K$_2$HPO$_4$	1 g
Agar	20 g

Adjust pH to 6.5–6.7.

After sterilization and immediately before use, a filter-sterilized solution of cycloheximide is added to the medium to give a final concentration of 40 mg/liter. Before pouring plates, a layer of sterile agar is poured in the bottom of the plates and allowed to solidify. This prevents colonies from spreading between the agar and the bottom of the petri dishes. Colonies are picked onto slants of the same medium.

Isolation and Enumeration of Arthrobacters from Soil (Mulder and Antheunisse, 1963)

Dilutions are prepared and plates poured using the following "poor" medium (g/liter tap water):

Ca(H$_2$PO$_4$)$_2$	0.25
K$_2$HPO$_4$	1.0
MgSO$_4$·7H$_2$O	0.25
(NH$_4$)$_2$SO$_4$	0.25
Casein	1.0
Yeast extract	0.7
Glucose	1.0
Agar	10.0

Adjust pH to 6.9–7.0.

After incubation for 5 days at 25°C, colonies are counted and a large number are transferred to agar slants of the same composition. The slants are incubated for 7 days at 25°C and then examined microscopically. Colonies that consist of coccoid cells are then transferred to a "rich" medium containing (% wt/vol): yeast extract, 0.7; glucose, 1.0; agar, 1.0. The cultures are examined in the exponential phase of growth (usually not more than 24 h), and those showing "germinating" cocci and irregular rods are considered to be "arthrobacters."

Obviously, in the sense used by Mulder and Antheunisse (1963), the term "arthrobacters" refers to all bacteria that show the rod-coccus growth cycle of the genus *Arthrobacter*.

Isolation of Arthrobacters by Selective Media

Selective Medium of Hagedorn and Holt (1975b)

In this method, plate counts are made by spreading 0.1-ml amounts of suitable dilutions over the surface of sterile medium in petri dishes. Peptone solution (0.5%, wt/vol) was used as the diluent by Hagedorn and Holt (1975b). The selective medium has the following composition:

Trypticase soy agar (BBL)	0.4%
Yeast extract (Difco)	0.2%
NaCl	2.0%
Cycloheximide	0.01%
Methyl red (Harleco)	150 μg/ml
Agar	1.5%

The methyl red is filter-sterilized and added aseptically to the autoclaved, cooled medium. The medium is adjusted to the pH of the particular soil being examined.

The selective properties of the medium are said to be unaffected by pH values in the range 5.0–8.5. After incubation for 10 days at 25°C, the plates are counted. Colonies are transferred to slants of trypticase soy agar containing 0.2% yeast extract and examined microscopically for the possession of a morphological growth cycle as described in *Bergey's Manual* (Keddie, 1974). From the results obtained, the authors concluded that 78% of the counts on the selective medium was a suitable approximation of the arthrobacter counts for the soils studied. The authors state that:

The combination of actidione [cycloheximide] at 0.01% and NaCl at 2.0% effectively inhibited all fungi and most streptomycetes, nocardia, and Gram-negative bacteria. The methyl red at 150 μg/ml inhibited other Gram-positive bacteria (bacilli and micrococci) but did not affect the arthrobacters. The pH of the medium, between 5.0 and 8.5, did not affect its selectivity, and the combination of trypticase soy agar at 0.4% and yeast extract at 0.2% gave the highest yield of arthrobacters with the addition of the selective ingredients over the other basal media [tested].

The selective medium gave arthrobacter counts several times higher than those on the nutritionally "poor" medium of Mulder and Antheunisse (1963) for the four soils examined (Hagedorn and Holt, 1975b).

Preservation of Cultures

Stab cultures by loop in TSX semisolid medium (Keddie et al., 1966) or in the soil extract semisolid medium of Lochhead and Burton (1957) (see above) will remain viable for at least three months at room temperature (ca. 20°C) provided they are not allowed to dry out. Cultures may be preserved for longer periods (at least 10 years) by freezing in glass beads at $-70°C$ (Jones et al., 1984). For long-term storage, lyophilization is suitable.

Identification of Arthrobacters

Members of the genus *Arthrobacter* can be recognized by examining the following characters: morphology and staining reactions, oxygen relations, acid production from glucose, and cell wall composition.

The most distinctive feature of arthrobacters is the marked change of form that occurs during the growth cycle on complex media. Stationary phase cultures (usually 2–7 days) are composed entirely or largely of coccoid cells (Fig. 1d) which, on transfer to fresh complex medium, produce one and sometimes two (or occasionally more) outgrowths that give rise to the irregular rods characteristic of exponential phase cultures (Fig. 1a-c). Some of the cells are arranged in V-formations but more complex angular arrangements may also occur. Cells may show primary branching, but true mycelia (showing secondary branching) are not produced. As growth proceeds, the rods become shorter and are eventually replaced by the coccoid forms characteristic of stationary phase

cultures (Fig. 1d). For more detailed accounts of *Arthrobacter* morphology and morphogenesis, see Keddie (1974), Luscombe and Gray (1971), Clark (1972), and Duxbury and Gray (1977). Both rod and coccoid forms are Gram positive, but may decolorize readily, and are not acid-fast. The rods are nonmotile or motile by one subpolar or a few lateral flagella. They are obligate aerobes. The mode of metabolism is respiratory, never fermentative; little or no acid is formed from sugars in peptone media. The cell wall peptidoglycan contains lysine as diamino acid.

Additional features include the following: they are catalase-positive. The optimum temperature for growth is 25–30°C, and most *Arthrobacter* species grow in the range about 10°C–35°C; many strains also grow at 5°C, and a few grow at 37°C. Some obligately psychrophilic isolates with a growth range of about $-5°C$ to 20°C that are considered to be a new species of *Arthrobacter* ("A. glacialis") were described by Moiroud and Gounot (1969). They do not survive heating at 63°C for 30 min in skim milk. They do not hydrolyze cellulose. DNase is produced and gelatin is usually liquefied (Keddie et al., 1986). The GC content of the DNA of most species is in the range 59–66 mol% but that of *A. atrocyaneus* is higher, at

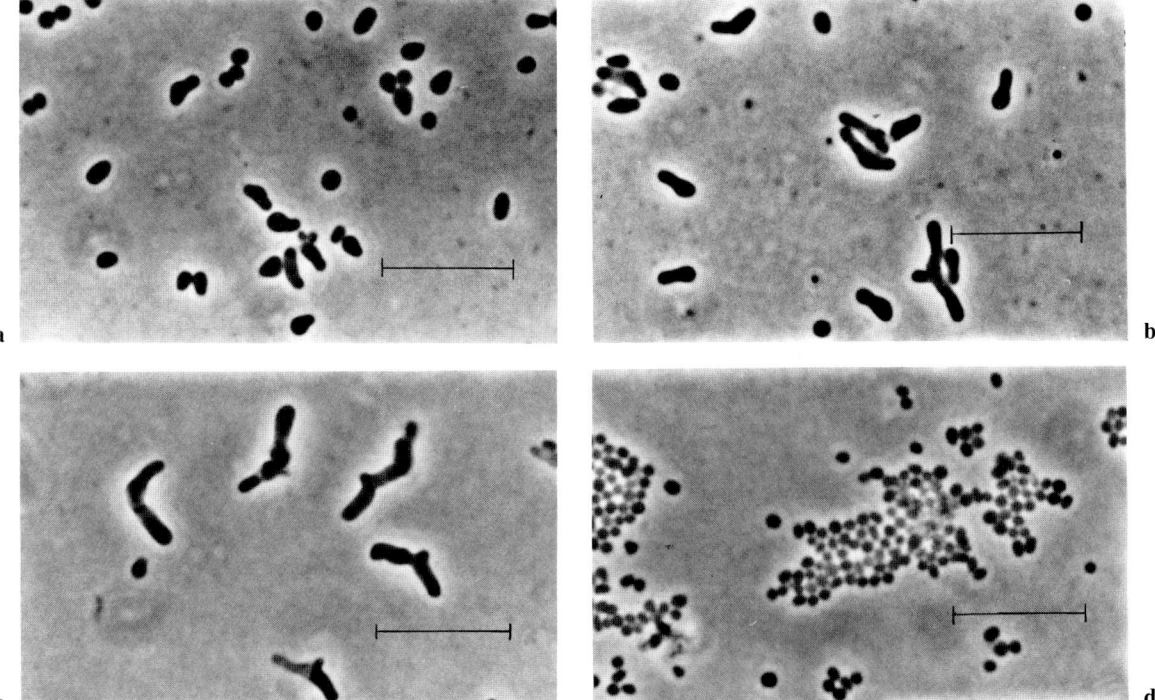

Fig. 1. *Arthrobacter globiformis* (ATCC 8010) grown on medium EYGA at 25°C; the inoculum was of coccoid cells as shown in (d). (a) After 6 h, showing outgrowth of rods from coccoid cells. (b) After 12 h. (c) After 24 h. (d) After 3 days. Bars = 10 μm.

about 70 mol% (see Keddie et al., 1986, for references).

The extent to which the morphology of arthrobacters changes during the growth cycle is markedly influenced by the nutritional status of the medium (see Clark, 1972; Ensign and Wolfe, 1964; Luscombe and Gray, 1971; Veldkamp et al., 1963); therefore, the medium used for morphological studies must be chosen with care (see Cure and Keddie, 1973). Many different media have been used for this purpose and most are based on soil extract (e.g., see Holm and Jensen, 1972; Lochhead and Burton, 1957). However, such media may give inconsistent results because of the variable nature of soil extract and because some mineral components may be precipitated to different extents during preparation. Medium EYGA (Cure and Keddie, 1973) was devised to overcome such problems and has proved to be satisfactory for examining the morphology of arthrobacters.

Cell wall composition may be determined by one of the rapid methods described by Bousfield et al. (1985) in which cell wall material is prepared by alkali treatment of whole cells.

Further Identification of Arthrobacters

Fifteen species of *Arthrobacter* are recognized in *Bergey's Manual of Systematic Bacteriology* (Keddie et al., 1986). Many of these species were created for single strains that possessed some unusual feature, such as a requirement for a particular growth factor, production of an unusual pigment, the ability to utilize a particular substrate, and so on. Accordingly, such species do not necessarily represent the commonly occurring arthrobacters in the habitats from which they were isolated originally and this may in part explain the common experience that many new soil isolates show little resemblance to the named strains used as reference cultures (see Hagedorn and Holt, 1975a; Keddie et al., 1966; Skyring and Quadling, 1969; Seiler et al., 1980). Also, because most species descriptions are based on single strains, the range of variation within the species is not known. Therefore some of the phenotypic features described for a particular species may be strain rather than species characteristics.

Most *Arthrobacter* species now recognized closely resemble the type species *A. globiformis* in a large number of phenotypic (largely nutritional) characters. For this reason they were referred to as the "globiformis" group of arthrobacters in the first edition of *The Prokaryotes* (Keddie and Jones, 1981) and included the species *A. atrocyaneus, A. aurescens, A. crystallopoietes, A. histidinolovorans, A. nicotianae, A.*

oxydans, A. pascens, A. polychromogenes (considered a subspecies of *A. oxydans* by Stackebrandt and Fiedler, 1979), *A. ramosus,* and *A. ureafaciens,* together with the species *Brevibacterium sulfureum* (now *A. sulfureum*), and the plant pathogen *Corynebacterium ilicis* (now *A. ilicis*) (see Keddie et al., 1986). A list of substrates utilized by 90% or more of single representatives of these species is given in the section on "Soil and Similar Habitats" above. Detailed nutritional data are not available for the two remaining species now recognized, *A. protophormiae* and *A. uratoxydans* (see Keddie et al., 1986). The remaining species, *A. citreus* (type strain only), is readily distinguished from the other species. It has a complex nutrition and requires biotin, thiamin, nicotinic acid, tyrosine, methionine, cystine, and a siderophore such as ferrichrome or mycobactin for growth (Seidman and Chan, 1969). *A. citreus* also utilizes a much more limited range of compounds as sole or major carbon and energy sources than the other species (Keddie et al., 1986). A number of other strains are listed as *A. citreus* in culture collections but do not have the characteristics of the type strain (ATCC 11624). They were considered to be strains of *A. protophormiae* and *A. uratoxydans* by Stackebrandt et al. (1983) (see Keddie et al., 1986).

However, despite the close phenotypic similarity of most species of *Arthrobacter* (*A. citreus* excepted), two groups within the genus can be distinguished according to their peptidoglycan structure, lipid composition, and the presence or absence of teichoic acids in the cell walls.

The peptidoglycans of all *Arthrobacter* species contain lysine as diamino acid but more detailed analysis revealed considerable heterogeneity in the cell wall peptidoglycans of the different species. All contain a group A peptidoglycan, i.e., one in which the cross-linkage is between positions 3 and 4 of the peptide subunits. However, the peptide subunits are linked by a number of different interpeptide bridges, depending on species (Schleifer and Kandler, 1972). Within these numerous different peptidoglycan types, two groups occur, which are referred to as the A3α and A4α variations. In the A3α variation, found in *A. globiformis, A. citreus,* and most other species (the "*A. globiformis/A. citreus*" group of Keddie et al., 1986), the interpeptide bridge contains only monocarboxylic acids and/or glycine. In the A4α variation, however, found in *A. nicotianae* and three other species (the "*A. nicotianae*" group of Keddie et al., 1986), the interpeptide bridge always contains a dicarboxylic acid and in most strains also contains alanine (Schleifer and Kandler, 1972). Thus members of the *A. globiformis/A.*

citreus group have peptidoglycans with the A3α variation (Schleifer and Kandler, 1972) and contain dihydrogenated menaquinones with nine isoprene units [MK-9(H$_2$)] as their major isoprenoid quinones (Yamada et al., 1976; Collins et al., 1979, 1981; Collins and Jones, 1981). They do not contain teichoic acids in the cell wall (Fiedler and Schäffler, 1987). In contrast, species of the *A. nicotianae* group have peptidoglycans with the A4α variation (Schleifer and Kandler, 1972; Stackebrandt et al., 1983) and, with the exception of *A. sulfureus,* have unsaturated menaquinones with eight isoprene units (MK-8) as major components, although they also contain substantial amounts of unsaturated menaquinones with nine isoprene units (MK-9) as well (Yamada et al., 1976; Collins and Jones, 1981; Collins and Kroppenstedt, 1983). *A. sulfureus* either contains MK-9 as the major menaquinone (Yamada et al., 1976; Collins et al., 1979) or comparable amounts of MK-9 and MK-10 (Collins and Kroppenstedt, 1983). *A. nicotianae* group strains also differ from *A. globiformis/A. citreus* group strains in possessing cell wall teichoic acids of the poly(glycero- phosphate) type (Fiedler and Schäffler, 1987). Further differences occur between the two groups in the types of polar lipids they contain. All *A. globiformis/A. citreus* group strains contain diphosphatidylglycerol, phosphatidylglycerol, and phosphatidylinositol. With the exception of *A. citreus,* they also contain several glycolipids (Shaw and Stead, 1971; Kostiw et al., 1972; Collins et al., 1981); *A. citreus* contains fewer glycolipids than the other species (Collins and Kroppenstedt, 1983).

The major characteristics of the 11 species of the *A. globiformis/A. citreus* group are listed in Table 1. The characteristics of the four species of the *A. nicotianae* group are listed in Table 2. To distinguish the two groups, the peptidoglycan "variation" may be determined by the rapid screening method described by Schleifer and Kandler (1972) (see also Schleifer and Seidl, 1985). Menaquinone composition may be determined by reverse phase partition thin layer chromatography (TLC) or high performance liquid chromatography (HPLC) (Collins et al., 1980, 1983). Polar lipid patterns may be obtained by two-dimensional TLC analysis of free lipid extracts (Collins et al., 1983). Fiedler and Schäffler (1987) state that *A. nicotianae* group strains may be recognized by testing for the presence of lysine, phosphorus, and glycerol in the cell wall. Further differentiation of the species within the *A. globiformis/A. citreus* group and the *A. nicotianae* group depends in large measure on DNA-DNA hybridization techniques (Fiedler and Stackebrandt, 1979; Stacke-

brandt et al., 1983). *A. citreus* can readily be distinguished but the remaining species share a high degree of phenotypic similarity, and it is not possible to differentiate them by using phenotypic characters. Those characters which may be of value in differentiation are listed in Tables 1 and 2.

REFERENCE STRAINS. *A. globiformis,* ATCC 8010 (NCIMB 8907). *A. nicotianae,* ATCC 15236 (NCIMB 9458). (Note: NCIMB was formerly NCIB.)

Biochemical and Physiological Properties

All species are catalase positive and probably contain cytochromes, though detailed information on the cytochrome content is not available for all species. The cytochrome composition of *A. globiformis* ATCC 4336 (NCIMB 8602) is *bcaa$_3$o* when the strain is in the exponential phase of growth but changes to *bcaa$_3$od* in the stationary phase of growth after the rod-shaped cells become oxygen limited and lose their ability to retain the Gram stain, and coccoid forms appear (Meyer and Jones, 1973; Jones, 1980). A similar observation was made by Faller and Schleifer (1981) when studying the apparent correlation between morphology and cytochrome content in *A. crystallopoietes.* They noted that rod-shaped cells in exponential phase cultures contained large amounts of cytochrome *aa$_3$* but only traces of cytochrome *d,* whereas coccoid cells in early stationary phase cultures contained relatively larger amounts of cytochrome *d* with lower amounts of cytochrome *aa$_3$.* However, in keeping with the observations of Meyer and Jones (1973) and Jones (1980), they showed that the increase in cytochrome *d* was not correlated with a change in cell morphology but resulted from the oxygen limitation that occurs in late exponential/early stationary phase cultures.

The fragmentary information available on metabolic pathways in *Arthrobacter* species has been reviewed by Krulwich and Pelliccione (1979). In summary, those few species which have been studied fall into two groups with respect to the pathways of carbohydrate dissimilation utilized. In *A. globiformis, A. ureafaciens, A. crystallopoietes,* and *"A. pyridinolis,"* the primary pathways are the Embden-Meyerhof-Parnas (EMP) pathway and, to a smaller extent, the hexose monophosphate (HMP) pathway. In contrast, *A. pascens* and *A. atrocyaneus*

Table 1. Some characteristics[a] of the *Arthrobacter* species having peptidoglycans of the A3α variation[b] and MK-9 (H₂)[c] as major menaquinones—the *A. globiformis*/*A. citreus* group.[d,e]

Characteristic	A. globiformis[f]	A. crystallopoietes	A. pascens	A. ramosus	A. aurescens	A. histidinolovorans	A. ilicis	A. ureafaciens	A. atrocyaneus	A. oxydans	A. citreus
Peptidoglycan type[b]	Lys-Ala₃	Lys-Ala	Lys-Ala₂	Lys-Ala₄	Lys-Ala-Thr-Ala	Lys-Ala-Thr-Ala	Lys-Ala-Thr-Ala	Lys-Ala-Thr-Ala	Lys-Ser-Ala₂₋₃	Lys-Ser-Thr-Ala	Lys-Thr-Ala₂
Cell wall sugars[g]	Gal, Glu	Gal, Glu	Gal, Glu	Gal, Rha, Man	Gal, (Man)	Gal, Glu	Gal, Rha, Man	Gal, (Man)	Gal, Glu, (Man)	Gal, Glu	Gal
Vitamin requirement; none or biotin only	+	+	+	+	+	+	+[h]	+	+	+	−
Nicotine utilization	−	−	−	−	−	−	−	−	−	+	−
Starch hydrolysis	+	−	+	−	+	−	−	−	+	+	−
Motility	−	−	−	+	−	−	+	−	+	−	+

[a] Most species are represented only by their type strain; therefore, the range of variation within the species is not known. For this reason, data on carbon source utilization tests are not included although studies on a wider range of strains may prove them to be useful. See section on "Soil and Similar Habitats" for a list of near-universal substrates. There are conflicting reports in the literature about the responses of the type strains of some species in some common tests such as nitrate reduction, urea hydrolysis, etc., and therefore these data have been omitted. Other data are from original and/or revised descriptions of species, and from Keddie et al. (1966), Yamada and Komagata (1972), and Cure and Keddie, and Robertson and Keddie, unpublished observations.

[b] From Schleifer and Kandler (1972); and Stackebrandt and Fiedler (1979).

[c] From Collins and Jones (1981).

[d] Species are arranged according to peptidoglycan type (Stackebrandt and Fiedler, 1979).

[e] Symbols: +, 90% or more of the strains are positive; −, 90% or more of strains are negative.

[f] Lys, L-lysine; Ala, L-alanine; Thr, L-threonine; Ser, L-serine; Gal, galactose; Glu, glucose; Rha, rhamnose; Man, mannose; ( ), conflicting reports on occurrence.

[g] From Keddie and Cure (1978).

[h] *A. ilicis* grows in a mineral salts-glucose medium only when provided with casamino acids (S. Robertson and R. M. Keddie, unpublished observations).

Table 2. Characters[a] most useful in differentiating the *Arthrobacter* species which have peptidoglycans of the A4α variation[b] and MK-8[c] (or MK-9[d]) as major menaquinones—the *A. nicotianae* group.[e,f]

Characteristics	*A. nicotianae*	*A. protophormiae*	*A. uratoxydans*	*A. sulfureus*
Number of strains studied	6	6	2	3
Peptidoglycan type[b]	Lys-Ala-Glu[g]	Lys-Ala-Glu	Lys-Ala-Glu	Lys-Glu
Major wall sugars[h]	Gal, Glc (one strain)	ND	ND	Gal, Glc (one strain)
Hydrolysis of:				
Starch	+	−	−	−
Casein	+	d	+	−
Utilization of:				
4-Hydroxybenzoate	+	+ (5/6)[i]	−	+
Glyoxylate	d	d	+	−
L-Asparagine	+	+ (5/6)	−	+
L-Arginine	d	+	−	+
L-Histidine	d	+	−	+
D-Xylose	+	− (5/6)	−	−
D-Ribose	+	d	−	d
L-Arabinose	+	+ (5/6)	−	−
D-Galactose	+	+ (5/6)	−	d
L-Rhamnose	−	− (5/6)	+	d
2,3-Butylene glycol	+ (5/6)	−	−	d
Glycerol	+	+	−	−

[a] Adapted from Table 5 in Stackebrandt et al. (1983).

[b] From Schleifer and Kandler (1972); and Stackebrandt et al. (1983).

[c] From Collins and Jones (1981); and Collins and Kroppenstedt (1983).

[d] *A. sulfureus* ("Brevibacterium sulfureum") ATCC 19098 (NCIMB 10355) contains MK-9 as major menaquinone (Collins and Kroppenstedt, 1983); a second strain designated *A. sulfureus* by Stackebrandt et al. (1983), ATCC 15170 (formerly named *A. citreus*), contains similar amounts of MK-9 and MK-10 as major menaquinones (Collins and Kroppenstedt, 1983).

[e] All *A. nicotianae* group strains which have been examined contain teichoic acids (Fiedler and Schäffler, 1987) and contain the polar lipids diphosphatidylglycerol and phosphatidyglycerol, but not phosphatidylinositol (Collins and Kroppenstedt, 1983). According to Stackebrandt et al. (1983) more than 90% of *A. nicotianae* group strains utilize: acetate, propionate, valerate, capronate, heptanoate, caprylate, succinate, DL-malate, citrate, DL-lactate, fumarate, D-gluconate, glycine, L-proline, L-threonine, and L-aspartate; none utilizes adipate, levulinate, and acetamide.

[f] Symbols: see Table 1, footnote [e]. Also: d, 11–89% of strains are positive; ND, no data.

[g] Lys, L-lysine; Ala, L-alanine; Glu, L-glutamic acid; Gal, galactose; Glc, glucose.

[h] From Keddie and Cure (1978).

[i] Fraction of strains giving indicated reaction.

use the Entner-Doudoroff and HMP pathways (Krulwich and Pelliccione, 1979). A list of carbon compounds utilized by a number of species as sole or major carbon sources is given in the section on "Soil and Similar Habitats."

Details of the peptidoglycan structure, lipid composition, menaquinone content, and the presence or absence of teichoic acids of the *Arthrobacter* species now recognized are given in the section "Further Identification of Arthrobacters."

Einck et al. (1973) have summarized the data available on bacteriophages active on *Arthrobacter* species.

Arthrobacters may be distinguished from other coryneforms and morphologically similar bacteria which have lysine in the cell wall or which show a rod-coccus growth cycle by the characters listed in Table 3. The differentiation from *Arthrobacter* of some former species now removed from the genus is also given in Table 3.

Biotechnological Potential

Undoubtedly, soil is the most important habitat of bacteria of the genus *Arthrobacter*. Their abundance in soils of various types and in different geographical locations has been amply demonstrated by many investigators. Their numerical predominance, coupled with the nutritional versatility of the commonly occurring species (Hagedorn and Holt, 1975a; Keddie, 1974), suggests that they may be important agents of mineralization in soil and possibly also in some other habitats. Among the compounds reported to be degraded by arthrobacters are certain herbicides and pesticides, as well as a wide range of naturally occurring and synthetic molecules of various degrees of complexity. Other roles that have been ascribed to at least some *Arthrobacter* species are phytohormone production (Barea et al., 1976; Katznelson and Cole, 1965; Rivière, 1963) and dinitro-

Table 3. Characters[a] differentiating *Arthrobacter* from similar genera which either have a rod-coccus growth cycle or which have lysine as the cell wall diamino acid.[b]

| Genus or species | Mycelium produced | Rod-coccus cycle[c] | Oxygen requirement | Acid from glucose[d] | Cell wall | | Major menaquinone[f] |
					Diamino acid[e]	Glycine present	
Arthrobacter	−	+	Aerobic	−	Lysine	−	MK-9 (H$_2$) or MK-8 and/ or MK-9
Aureobacterium terregens/ *Aur. flavescens*[g]	−	+	Aerobic	W	Ornithine	+	MK-12 and/ or MK-13
Nocardioides simplex[h]	−	+	Aerobic	−	LL-A$_2$pm	+	MK-8 (H$_4$)
Terrabacter tumescens[i]	−	+	Aerobic	−	LL-A$_2$pm	+	MK-8 (H$_4$)
Microbacterium	−	−	Equivocal	+	Lysine	+	MK-11, MK-12
Renibacterium	−	−	Aerobic	−	Lysine	+	MK-9
Oerskovia	+	−	Facultative	+	Lysine	−	MK-9 (H$_4$)
Brevibacterium	−	+	Aerobic	−	*meso*-A$_2$pm	−	MK-8 (H$_2$)
Rhodococcus	D	D	Aerobic	−	*meso*-A$_2$pm	−	MK-8 (H$_2$) or MK-9 (H$_2$)

[a] Data from Keddie and Cure, 1977, 1978; Keddie and Bousfield, 1980; Sanders and Fryer, 1980; Collins and Jones, 1981; Keddie and Jones, 1981; Lechevalier and Lechevalier, 1981; Collins, 1982; Collins et al., 1989.

[b] Symbols: +, 90% or more of strains are positive; −, 90% or more of strains are negative; W, weak; D, only a proportion of strains studied have the feature cited.

[c] Similar to that in *A. globiformis*.

[d] In peptone-based media.

[e] A$_2$pm, diaminopimelic acid.

[f] MK-8, MK-9, etc., indicates the number of isoprene units in the menaquinone; (H$_2$), (H$_4$), etc., indicates the number of double bonds hydrogenated.

[g] Formerly *Arthrobacter terregens* and *A. flavescens*.

[h] Formerly *A. simplex*.

[i] Formerly *A. tumescens*.

gen fixation (Cacciari et al., 1971; Smyk, 1970; Smyk and Ettlinger, 1963). Putative *Arthrobacter* species have also been reported to lyse yeast cells (Kitamura et al., 1972) and mycelium of *Fusarium roseum,* a carnation-root pathogen (Morrisey et al., 1976; Szajer and Koths, 1973). In the latter case, the *Arthrobacter* strain investigated produced a chitinase and was considered a possible means of biological control of *Fusarium* diseases.

Products of actual or potential commercial importance obtained from *Arthrobacter* species include glutamic acid (Tanaka and Kimura, 1972; Veldkamp et al., 1963), and α-ketoglutaric acid (Tanaka and Kimura, 1972), although it is likely that in some of the examples mentioned the bacteria concerned were not legitimate *Arthrobacter* species (see Keddie and Cure, 1977, 1978). Also, Veldkamp et al. (1966) described a strain of *A. globiformis* that produced large amounts of riboflavin.

In some species of *Arthrobacter,* the morphological changes that occur during the growth cycle have been shown to be subject to nutritional control, and such species have proved to be useful in the study of bacterial morphogenesis (see review by Clark, 1972).

Literature Cited

Adamse, A. D. 1968. Formation and final composition of the bacterial flora of a dairy waste activated sludge. Water Research 2:665–671.

Akiba, T., Ueyama, H., Seki, M., and Fukimbara, T. 1970. Identifications of lower alcohol-utilizing bacteria. Journal of Fermentation Technology 48:323–328.

Antheunisse, J. 1972. Decomposition of nucleic acids and some of their degradation products by microorganisms. Antonie van Leeuwenhoek Journal of Microbiology and Serology 38:311–327.

Austin, B., Goodfellow, M., and Dickinson, C. H. 1978. Numerical taxonomy of phylloplane bacteria from *Lolium perenne.* Journal of General Microbiology 104:139–155.

Barea, J. M., Navarro, E., and Montoya, E. 1976. Production of plant growth regulators by rhizosphere phosphate-solubilizing bacteria. Journal of Applied Bacteriology 40:129–134.

Bousfield, I. J. 1978. The taxonomy of coryneform bacteria from the marine environment, p. 217–233. In: Bousfield, I. J., and Callely, A. G. (ed.), Special publications of the Society for General Microbiology I. Coryneform bacteria. London: Academic Press.

Bousfield, I. J., Keddie, R. M., Dando, T. R., and Shaw, S. 1985. Simple rapid methods of cell wall analysis as an aid in the identification of aerobic coryneform bacteria, p. 221–236. In: Goodfellow, M., and Minnikin, D. E. (ed.), Chemical methods in bacterial systematics. Society for Applied Bacteriology, Technical Series No. 20. London: Academic Press.

Bowie, I. S., Grigor, M. R., Dunckley, G. G., Loutit, M. W., and Loutit, J. S. 1972. The DNA base composition and fatty acid constitution of some Gram-positive pleomorphic soil bacteria. Soil Biology and Biochemistry 4:397–412.

Boylen, C. W. 1973. Survival of *Arthrobacter crystallopoietes* during prolonged periods of extreme desiccation. Journal of Bacteriology 113:33–37.

Boylen, C. W., and Ensign, J. C. 1970. Long-term starvation survival of rod and spherical cells of *Arthrobacter crystallopoietes*. Journal of Bacteriology 103:569–577.

Boylen, C. W., and Mulks, M. H. 1978. The survival of coryneform bacteria during periods of prolonged nutrient starvation. Journal of General Microbiology 105:323–334.

Breed, R. S., Murray, E. G. D., and Smith, N. R. (ed.). 1957. Bergey's manual of determinative bacteriology, 7th ed. Baltimore: Williams & Wilkins.

Cacciari, I., Giovannozzi-Sermanni, G., Grappelli, A., and Lippi, D. 1971. Nitrogen fixation by *Arthrobacter* sp. I-Taxonomic study and evidence of nitrogenase activity of two new strains. Annali di Microbiologia ed Enzymologia 21:97–105.

Chai, T. J., and Levin, R. E. 1975. Characteristics of heavily mucoid bacterial isolates from fish pen slime. Applied Microbiology 30:450–455.

Chan, E. C. S., and Stevenson, I. L. 1962. On the biotin requirement of *Arthrobacter globiformis*. Canadian Journal of Microbiology 8:403–405.

Chen, M., and Alexander, M. 1973. Survival of soil bacteria during prolonged desiccation. Soil Biology and Biochemistry 5:213–221.

Clark, J. B. 1972. Morphogenesis in the genus *Arthrobacter*. CRC Critical Reviews in Microbiology 1:521–544.

Collins, M. D. 1982. Lipid composition of *Renibacterium salmoninarum* (Sanders and Fryer). FEMS Microbiology Letters 13:295–297.

Collins, M. D. 1986. Lipid composition of *Arthrobacter siderocapsulatus, A. viscosus, "A. oxamicetus," "A. sialophilus," "A. stabilis,"* and *"Agrobacterium pseudotsugene."* Systematic and Applied Microbiology 8:1–7.

Collins, M. D., Dorsch, M., and Stackebrandt, E. 1989. Transfer of *Pimelobacter tumescens* to *Terrabacter* gen. nov. as *Terrabacter tumescens* comb. nov. and of *Pimelobacter jensenii* to *Nocardioides* as *Nocardioides jensenii* comb. nov. International Journal of Systematic Bacteriology 39:1–6.

Collins, M. D., Goodfellow, M., and Minnikin, D. E. 1979. Isoprenoid quinones in the classification of coryneform and related bacteria. Journal of General Microbiology 110:127–136.

Collins, M. D., and Jones, D. 1981. The distribution of isoprenoid quinone structural types in bacteria and their taxonomic implications. Microbiological Reviews 45:316–354.

Collins, M. D., Jones, D., Keddie, R. M., Kroppenstedt, R. M., and Schleifer, K. H. 1983. Classification of some coryneform bacteria in a new genus *Aureobacterium*. Systematic and Applied Microbiology 4:236–252.

Collins, M. D., Jones, D., and Kroppenstedt, R. M. 1981. Reclassification of *Corynebacterium ilicis* (Mandel, Guba and Litsky) in the genus *Arthrobacter* as *Arthrobacter ilicis* comb. nov. Zentralblatt für Bakteriologie, Mikrobiologie und Hygiene, Abt. I., Orig. C. 2:318–323.

Collins, M. D., and Kroppenstedt, R. M. 1983. Lipid composition as a guide to the classification of some coryneform bacteria containing an A4α type peptidoglycan (Schleifer and Kandler). Systematic and Applied Microbiology 4:95–104.

Collins, M. D., Shah, H. N., and Minnikin, D. E. 1980. A note on the separation of natural mixtures of bacterial menaquinones using reverse-phase partition thin-layer chromatography. Journal of Applied Bacteriology 48:277–282.

Conn, H. J. 1928. A type of bacteria abundant in productive soils, but apparently lacking in certain soils of low productivity. New York State Agricultural Experimental Station Technical Bulletin No. 138:3–26.

Conn, H. J. 1947. A protest against the misuse of the generic name *Corynebacterium*. Journal of Bacteriology 54:10.

Conn, H. J., and Dimmick, I. 1947. Soil bacteria similar in morphology to *Mycobacterium* and *Corynebacterium*. Journal of Bacteriology 54:291–303.

Crombach, W. H. J. 1974a. Relationships among coryneform bacteria from soil, cheese and sea fish. Antonie van Leeuwenhoek Journal of Microbiology and Serology 40:347–359.

Crombach, W. H. J. 1974b. Morphology and physiology of coryneform bacteria. Antonie van Leeuwenhoek Journal of Microbiology and Serology 40:361–376.

Cummins, C. S., and Harris, H. 1959. Taxonomic position of *Arthrobacter*. Nature 184:831–832.

Cure, G. L., and Keddie, R. M. 1973. Methods for the morphological examination of aerobic coryneform bacteria, p. 123–135. In: Board, R. G., and Lovelock, D. N. (ed.), Sampling-microbiological monitoring of environments. Society for Applied Bacteriology Technical Series 7. New York: Academic Press.

Duxbury, T., and Gray, T. R. G. 1977. A microcultural study of the growth of cystites, cocci and rods of *Arthrobacter globiformis*. Journal of General Microbiology 103:101–106.

Ehrlich, H. C. 1963. Bacteriology of manganese nodules. I. Bacterial action on manganese in nodule enrichments. Applied Microbiology 11:15–19.

Ehrlich, H. C. 1968. Bacteriology of manganese nodules. II. Manganese oxidation by cell-free extract from a manganese nodule bacterium. Applied Microbiology 16:197–202.

Einck, K. H., Pattee, P. A., Holt, J. G., Hagedorn, C., Miller, J. A., and Berryhill, D. L. 1973. Isolation and characterisation of a bacteriophage of *Arthrobacter globiformis*. Journal of Virology, 12:1031–1033.

Ensign, J. C., and Rittenberg, S. C. 1963. A crystalline pigment produced from 2-hydroxypyridine by *Arthrobac-*

ter crystallopoietes n. sp. Archiv für Mikrobiologie 47:137–153.

Ensign, J. C., and Wolfe, R. S. 1964. Nutritional control of morphogenesis in *Arthrobacter crystallopoietes.* Journal of Bacteriology 87:924–932.

Faller, A. H., and Schleifer, K. H. 1981. Effects of growth phase and oxygen supply on the cytochrome composition and morphology of *Arthrobacter crystallopoietes.* Current Microbiology 6:253–258.

Fiedler, F., and Schäffler, M. J. 1987. Teichoic acids in cell walls of strains of the *"nicotianae"* group of *Arthrobacter:* a chemotaxonomic marker. Systematic and Applied Microbiology 9:16–21.

Gibson, T. 1953. The taxonomy of the genus *Corynebacterium.* Atti del VI Congresso Internazionale di Microbiologia, Roma 1:16–20.

Giovanozzi-Sermanni, G. 1959. Una nuova specie di *Arthrobacter* determinante la degradazione della nicotina: *Arthrobacter nicotianae.* II Tabacco 63:83–86.

Gounot, A. M. 1967. Role biologique des *Arthrobacter* dans les limons souterrains. Annales de l'Institut Pasteur 113:923–945.

Gray, T. R. G. 1976. Survival of vegetative microbes in soil. Symposium of the Society for General Microbiology 26:327–364.

Greenberg, J., and Barker, H. A. 1962. A ferrichrome-requiring arthrobacter which decomposes puromycin aminonucleoside. Journal of Bacteriology 83:1163–1164.

Gundersen, K., and Jensen, H. L. 1956. A soil bacterium decomposing organic nitro-compounds. Acta Agriculturae Scandinavica 6:100–114.

Hagedorn, C., and Holt, J. G. 1975a. A nutritional and taxonomic survey of *Arthrobacter* soil isolates. Canadian Journal of Microbiology 21:353–361.

Hagedorn, C., and Holt, J. G. 1975b. Ecology of soil arthrobacters in Clarion-Webster toposequences of Iowa. Applied Microbiology 29:211–218.

Holm, E., and Jensen, V. 1972. Aerobic chemoorganotrophic bacteria of a Danish beech forest. Oikos 23:248–260.

Horvath, R. S., and Alexander, M. 1970. Cometabolism of *m*-chlorobenzoate by an *Arthrobacter.* Applied Microbiology 20:254–258.

Iizuka, H., and Komagata, K. 1965. Microbiological studies on petroleum and natural gas. III. Determination of *Brevibacterium, Arthrobacter, Micrococcus, Sarcina, Alcaligenes,* and *Achromobacter* isolated from oil-brines in Japan. Journal of General and Applied Microbiology 11:1–14.

Jensen, H. L. 1933. Corynebacteria as an important group of soil microorganisms. Proceedings of the Linnean Society of New South Wales 58:181–185.

Jensen, H. L. 1934. Studies on saprophytic mycobacteria and corynebacteria. Proceedings of the Linnean Society of New South Wales 59:19–61.

Jensen, H. L. 1952. The coryneform bacteria. Annual Review of Microbiology 6:77–90.

Jensen, H. L. 1964. Studies on soil bacteria *(Arthrobacter globiformis)* capable of decomposing the herbicide Endothal. Acta Agriculturae Scandinavica 14:193–207.

Jensen, V. 1968. The plate count technique, p. 158–170. In: Gray, T. R. G., and Parkinson, D. (ed.), The ecology of soil bacteria. Liverpool: Liverpool University Press.

Jones, C. W. 1980. Cytochrome patterns in classification and identification including their relevance to the oxidase test, p. 127–138. In: Goodfellow, M., and Board, R. G. (ed.). Microbial classification and identification. Society for Applied Bacteriology Symposium Series 8. New York Academic Press.

Jones, D. 1978. An evaluation of the contribution of numerical taxonomy to the classification of the coryneform bacteria, p. 13–46. In: Bousfield, I. J., and Callely, A. G. (ed.), Coryneform bacteria. Special Publications of the Society for General Microbiology I. London: Academic Press.

Jones, D., Pell, P. A., and Sneath, P. H. A. 1984. Maintenance of bacteria on glass beads at −60°C to −76°C, p. 35–40. In: Kirsop, B. E., and Snell, J. J. S. (ed.). Maintenance of microorganisms: A manual of laboratory methods. London: Academic Press.

Kaneko, T., Kitamura, K., and Yamamoto, Y. 1969. *Arthrobacter luteus* nov. sp. isolated from brewery sewage. Journal of General and Applied Microbiology 15:317–326.

Katznelson, H., and Cole, S. E. 1965. Production of gibberellin-like substances by bacteria and actinomycetes. Canadian Journal of Microbiology 11:733–741.

Keddie, R. M. 1974. *Arthrobacter,* p. 618–625. In: Buchanan, R. E., and Gibbons, N. E. (ed.), Bergey's manual of determinative bacteriology, 8th ed. Baltimore: Williams & Wilkins.

Keddie, R. M., and Bousfield, I. J. 1980. Cell wall composition in the classification and identification of coryneform bacteria, pp. 167–188. In: Goodfellow, M., and Board, R. G. (ed.), Microbiological classification and identification. Society for Applied Bacteriology Symposium Series No. 8. New York: Academic Press.

Keddie, R. M., Collins, M. D., and Jones, D. 1986. Genus *Arthrobacter* p. 1288–1301. In: Sneath, P. H. A., Mair, N. S., Sharpe, M. E., and Holt, J. G. (ed), Bergey's manual of systematic bacteriology, vol. 2. Baltimore: Williams & Wilkins.

Keddie, R. M., and Cure, G. L. 1977. The cell wall composition and distribution of free mycolic acids in named strains of coryneform bacteria and in isolates from various natural sources. Journal of Applied Bacteriology 42:229–252.

Keddie, R. M., and Cure, G. L. 1978. Cell wall composition of coryneform bacteria, p. 47–84. In: Bousfield, I. J., and Callely, A. G. (ed.), Coryneform bacteria. Special Publications of the Society for General Microbiology I. London: Academic Press.

Keddie, R. M., and Jones, D. 1981. Saprophytic, aerobic coryneform bacteria, p. 1838–1878. In: Starr, M. P., Stolp, H., Trüper, H. G., Balows, A., and Schlegel, H. G. (ed.). The prokaryotes: A handbook on habitats, isolation and identification of bacteria. Berlin: Springer-Verlag.

Keddie, R. M., Leask, B. G. S., and Grainger, J. M. 1966. A comparison of coryneform bacteria from soil and herbage: Cell wall composition and nutrition. Journal of Applied Bacteriology 29:17–43.

Kitamura, K., Kaneko, T., and Yamamoto, Y. 1972. Lysis of viable yeast cells by enzymes of *Arthrobacter luteus.* I. Isolation of lytic strain and studies of its lytic activity. Journal of Applied and General Microbiology 18:57–71.

Klein, D. A., Davis, J. A., and Casida, L. E. Jr. 1968. Oxidation of *n*-alkanes to ketones by an *Arthrobacter* species. Antonie van Leeuwenhoek Journal of Microbiology and Serology 34:495–503.

Kolenbrander, P. E., Lotong, N., and Ensign, J. C. 1976. Growth and pigment production by *Arthrobacter pyridinolis* n. sp. Archives of Microbiology 110:239–245.

Kortstee, G. J. J. 1970. The aerobic decomposition of choline by microorganisms. I. The ability of aerobic organisms, particularly coryneform bacteria, to utilize choline as the sole carbon and nitrogen source. Archiv für Mikrobiologie 71:235–244.

Kostiw, L. L., Boylen, C. W., and Tyson, B. J. 1972. Lipid composition of growing and starving cells of *Arthrobacter crystallopoeites*. Journal of Bacteriology 111:103–111.

Kraft, A. A., Ayres, J. C., Torrey, G. S., Salzer, R. H., and da Silva, G. A. N. 1966. Coryneform bacteria in poultry, eggs and meat. Journal of Applied Bacteriology 29:161–166.

Krulwich, T. A., and Pelliccione, N. J. 1979. Catabolic pathways of coryneforms, nocardias and mycobacteria. Annual Review of Microbiology 33:95–111.

Labeda, D. P., Liu, K. C., and Casida, L. E. Jr. 1976. Colonization of soil by *Arthrobacter* and *Pseudomonas* under varying conditions of water and nutrient availability as studied by plate counts and transmission electron microscopy. Applied and Environmental Microbiology 31:551–561.

Lechevalier, H. A., and Lechevalier, M. P. 1981. Actinomycete genera "in search of a family," p. 2118–2123. In: Starr, M. P., Stolp, H., Trüper, H. G., Balows, A., and Schlegel, H. G. (ed.) The prokaryotes: A handbook on habitats, isolation and identification of bacteria. Berlin: Springer-Verlag.

Lee, J. S., and Pfeifer, D. K. 1977. Microbiological characteristics of Pacific shrimp (*Pandalus jordani*). Applied and Environmental Microbiology 33:853–859.

Lochhead, A. G. 1958a. Two new species of *Arthrobacter* requiring respectively vitamin B_{12} and the terregens factor. Archiv für Mikrobiologie 31:163–170.

Lochhead, A. G. 1958b. Soil bacteria and growth-promoting substances. Bacteriological Reviews 22:145–153.

Lochhead, A. G., and Burton, M. O. 1953. An essential bacterial growth factor produced by microbial synthesis. Canadian Journal of Botany 31:7–22.

Lochhead, A. G., and Burton, M. O. 1956. Importance of soil extract for the enumeration and study of soil bacteria, p. 157–161. Transactions of the 6th International Congress of Soil Science, Paris.

Lochhead, A. G., and Burton, M. O. 1957. Qualitative studies of soil micro-organisms. XIV. Specific vitamin requirements of the predominant bacterial flora. Canadian Journal of Microbiology 3:35–42.

Loos, M. A., Roberts, R. N., and Alexander, M. 1967. Phenols as intermediates in the decomposition of phenoxyacetates by an *Arthrobacter* species. Canadian Journal of Microbiology 13:679–690.

Lowe, W. E., and Gray, T. R. G. 1972. Ecological studies on coccoid bacteria in a pine forest soil. I. Classification. Soil Biology and Biochemistry 4:459–468.

Lund, B. M. 1969. Properties of some pectolytic, yellow pigmented, Gram-negative bacteria isolated from fresh cauliflowers. Journal of Applied Bacteriology 32:60–67.

Luscombe, B. M., and Gray, T. R. G. 1971. Effect of varying growth rate on the morphology of *Arthrobacter*. Journal of General Microbiology 69:433–434.

Lysenko, O. 1959. The occurrence of species of the genus *Brevibacterium* in insects. Journal of Insect Pathology 1:34–42.

Mandel, M., Guba, E. F., and Litsky, W. 1961. The causal agent of bacterial blight of American holly p. 61. Bacteriological Proceedings.

Meyer, D. J., and Jones, C. W. 1973. Distribution of cytochromes in bacteria: relationship to general physiology. International Journal of Systematic Bacteriology 23:459–467.

Minnikin, D. E., Collins, M. D., and Goodfellow, M. 1978a. Menaquinone patterns in the classification of nocardioform and related bacteria. Zentralblatt für Bakteriologie, Parasitenkunde, Infektionskranheit und Hygiene Abt. 1, Suppl. 6:85–90.

Minnikin, D. E., Goodfellow, M., and Collins, M. D. 1978b. Lipid composition in the classification and identification of coryneform and related taxa, p. 85–160. In: Bousfield, I. J., and Callely, A. G. (ed.), Special publications of the Society for General Microbiology. I. Coryneform bacteria. London: Academic Press.

Moiroud, A., and Gounot, A. M. 1969. Sur une bactérie psychrophile obligatoire isolée de limons glaciaires. Comptes Rendus Hebdomadaires des Seances de l'Academie des Sciences, Serie D 269:2150–2152.

Morris, J. G. 1960. Studies on the metabolism of *Arthrobacter globiformis*. Journal of General Microbiology 22:564–582.

Morrisey, R. F., Dugan, E. P., and Koths, J. S. 1976. Chitinase production by an *Arthrobacter sp.* lysing cells of *Fusarium roseum*. Soil Biology and Biochemistry 8:23–28.

Mulder, E. G., Adamse, A. D., Antheunisse, J., Deinema, M. H., Woldendorp, J. W., and Zevenhuizen, L. P. T. M. 1966. The relationship between *Brevibacterium linens* and bacteria of the genus *Arthrobacter*. Journal of Applied Bacteriology 29:44–71.

Mulder, E. G., and Antheunisse, J. 1963. Morphologie, physiologie et écologie des *Arthrobacter*. Annales de l'Institut Pasteur 105:46–74.

Nand, K., and Rao, D. V. 1972. *Arthrobacter mysorens*—a new species excreting L-glutamic acid. Zentralblatt für Bakteriologie, Parasitenkunde, Infektionskrankheiten und Hygiene, Abt. 2, Orig. 127:324–331.

O'Donnell, A. G., Goodfellow, M., and Minnikin, D. E. 1982. Lipids in the classification of *Nocardioides*: reclassification of *Arthrobacter simplex* (Jensen) Lochhead in the genus *Nocardioides* (Prauser) emend. O'Donnell *et al.* as *Nocardioides simplex* comb. nov. Archive für Mikrobiologie 133:323–329.

Owens, J. D., and Keddie, R. M. 1969. The nitrogen nutrition of soil and herbage coryneform bacteria. Journal of Applied Bacteriology 32:338–347.

Park, Y.-H., Hori, H., Suzuki, K.-I., Osawa, S., and Komagata, K. 1987. Phylogenetic analysis of the coryneform bacteria by 5S rRNA sequences. Journal of Bacteriology 169:1801–1806.

Rivière, J. 1963. Action des microorganismes de la rhizosphère sur la croissance du blé. II. Isolement et caractérisation des bactéries produisant des phytohormones. Annales de L'Institut Pasteur 105:303–314.

Robinson, J. B., Salonius, P. O., and Chase, F. E. 1965. A note on the differential response of *Arthrobacter* spp. and *Pseudomonas* spp. to drying in soil. Canadian Journal of Microbiology 11:746–748.

Roth, N. G., and Wheaton, R. B. 1962. Continuity of psychrophilic and mesophilic growth characteristics in the genus *Arthrobacter*. Journal of Bacteriology 83:551–555.

Sacks, L. E. 1954. Observations on the morphogenesis of *Arthrobacter citreus,* spec. nov. Journal of Bacteriology 67:342–345.

Sanders, J. E., and Fryer, J. L. 1980. *Renibacterium salmoninarum* gen. nov. spec. nov., the causative agent of bacterial kidney disease in salmonid fishes. International Journal of Systematic Bacteriology 30:496–502.

Schefferle, H. E. 1966. Coryneform bacteria in poultry deep litter. Journal of Applied Bacteriology 29:147–160.

Schleifer, K. H., and Kandler, O. 1972. Peptidoglycan types of bacterial cell walls and their taxonomic implications. Bacteriological Reviews 36:407–477.

Schleifer, K. H., and Seidl, P. H. 1985. Chemical composition and structure of murein. P. 201–219. In: Goodfellow, M. and Minnikin, D. E. (ed.), Chemical methods in bacterial systematics. Society for Applied Bacteriology Technical series no. 20. London: Academic Press.

Seidman, P., and Chan, E. C. S. 1969. Growth of *Arthrobacter citreus* in a chemically-defined medium and its requirement for chelating agents with schizokinen activity. Journal of General Microbiology 58: v.

Seiler, H., Braatz, R., and Ohmeyer, G. 1980. Numerical cluster analysis of coryneform bacteria from activated sludge. Zentralblatt für Bakteriologie, Parazitenkunde, Infektionskrankheiten und Hygiene Abt 1. Orig C. 1:357–375.

Sethunathan, N., and Pathak, M. D. 1971. Development of a diazinon-degrading bacterium in paddy water after repeated application of diazinon. Canadian Journal of Microbiology 17:699–702.

Sguros, P. L. 1955. Microbial transformations of the tobacco alkaloids. I. Cultural and morphological characteristics of a nicotinophile. Journal of Bacteriology 69:28–37.

Sharpee, K. W., Duxbury, J. M., and Alexander, M. 1973. 2,4-Dichlorophenoxyacetate metabolism by *Arthrobacter* sp.: Accumulation of a chlorobutenolide. Applied Microbiology 26:445–447.

Shaw, N., and Stead, A. 1971. Lipid composition of some species of *Arthrobacter*. Journal of Bacteriology 107:130–133.

Sieburth, J. McN. 1964. Polymorphism of a marine bacterium *(Arthrobacter)* as a function of multiple temperature optima and nutrition. Proceedings of the Symposium on Experimental Marine Ecology. Occasional Publication No. 2:11–16.

Skyring, G. W., and Quadling, C. 1969. Soil bacteria: Comparisons of rhizosphere and nonrhizosphere populations. Canadian Journal of Microbiology 15:473–488.

Skyring, G. W., and Quadling, C. 1970. Soil bacteria: A principal component analysis and guanine-cytosine contents of some arthrobacter-coryneform soil isolates and of some named cultures. Canadian Journal of Microbiology 16:95–106.

Skyring, G. W., Quadling, C., and Rouatt, J. W. 1971. Soil bacteria: Principal component analysis of physiological descriptions of some named cultures of *Agrobacterium,*

Arthrobacter, and *Rhizobium.* Canadian Journal of Microbiology 17:1299–1311.

Smyk, B. 1970. Fixation of atmospheric nitrogen by the strains of *Arthrobacter*. Zentralblatt für Bakteriologie, Parasitenkunde, Infektionskrankheiten and Hygiene, Abt. 2 Orig. 124:231–237.

Smyk, B. N., and Ettlinger, L. 1963. Recherches sur quelque espèces d'arthrobacter fixatrices d'azote isolées des roches karstiques alpines. Annales de L'Institut Pasteur 105:341–348.

Soumare, S., and Blondeau, R. 1972. Caractéristiques microbiologiques des sol de la région du nord de la France: Importance des 'Arthrobacters.' Annales de L'Institut Pasteur 123:239–249.

Splittstoesser, D. F., Wexler, M., White, J., and Colwell, R. R. 1967. Numerical taxonomy of Gram-positive and catalase-positive rods isolated from frozen vegetables. Applied Microbiology 15:158–162.

Stackebrandt, E., and Fiedler, F. 1979. DNA-DNA homology studies among *Arthrobacter* and *Brevibacterium.* Archives of Microbiology 120:289–295.

Stackebrandt, E., Fowler, V. J., Fiedler, F., and Seiler, H. 1983. Taxonomic studies on *Arthrobacter nicotianae* and related taxa. Description of *Arthrobacter uratoxydans* sp. nov. and *Arthrobacter sulfureus* sp. nov. and reclassification of *Brevibacterium protophormiae* as *Arthrobacter protophormiae* comb. nov. Systematic and Applied Microbiology 4:470–486.

Stackebrandt, E., Lewis, B. J., and Woese, C. R. 1980. The phylogenetic structure of the coryneform group of bacteria. Zentralblatt für Bakteriologie, Parasitenkunde, Infektionskrankheiten und Hygiene, Abt. 2, Orig. C 1:137–149.

Stackebrandt, E., and Woese, C. R. 1981. The evolution of the prokaryotes, p. 1–3. In: Carlile, M. J., Collins, J. F., and Moseley, B. E. B. (ed.), Molecular and cellular aspects of microbial evolution. Symposium of the Society for General Microbiology 32. Cambridge: Cambridge University Press.

Stevenson, I. L. 1967. Utilization of aromatic hydrocarbons by *Arthrobacter spp.* Canadian Journal of Microbiology 13:205–211.

Suzuki, K., Collins, M. D., Iijima, E., and Komagata, K. 1988. Chemotaxonomic characterization of a radiotolerant bacterium *Arthrobacter radiotolerans:* description of *Rubrobacter radiotolerans* gen. nov., comb. nov. FEMS Microbiology Letters 52:33–40.

Suzuki, K., and Komagata, K. 1983. *Pimelobacter* gen. nov.—a new genus of coryneform bacteria with LL-diaminopimelic acid in the cell wall. Journal of General and Applied Microbiology 29:59–71.

Szajer, C., and Koths, J. S. 1973. Physiological properties and enzymatic activity of an *Arthrobacter* capable of lysing *Fusarium sp.* Acta Microbiologica Polonica Series B 5:81–86.

Tanaka, K., and Kimura, K. 1972. Process for producing L-glutamic acid and alpha-ketoglutaric acid. United States Patent No. 3,642,576.

Tanenbaum, S. W., and Flashner, M. 1977. *Arthrobacter sialophilus* sp. nov. a neuramidase-producing coryneform. Canadian Journal of Microbiology 23:1568–1572.

Tate, R. L., and Ensign, J. C. 1974. A new species of *Arthrobacter* which degrades picolinic acid. Canadian Journal of Microbiology 20:691–694.

Taylor, C. B., and Lochhead, A. G. 1937. A study of *Bacterium globiforme* Conn in soils differing in fertility. Canadian Journal of Research C 15:340–347.

Topping, L. E. 1937. The predominant micro-organisms in soils. I. Description and classification of the organisms. Zentralblatt für Bakteriologie, Parasitenkunde, Infektionskrankheiten und Hygiene, Abt. 2 Orig. 97:289–304.

Topping, L. E. 1938. The predominant micro-organisms in soils. II. The relative abundance of the different types of organisms obtained by plating, and the relation of plate to total counts. Zentralblatt für Bakteriologie, Parasitenkunde, Infektionskrankheiten und Hygiene, Abt. 2 Orig. 98:193–201.

Veldkamp, H. 1965. The isolation of *Arthrobacter*. Zentralblatt für Bakteriologie, Parasitenkunde, Infektionskrankheiten und Hygiene, Abt. 1 Orig. Suppl. 1:265–269.

Veldkamp, H., van den Berg, G., and Zevenhuizen, L. P. T. M. 1963. Glutamic acid production by *Arthrobacter globiformis*. Antonie van Leeuwenhoek Journal of Microbiology and Serology 29:35–51.

Veldkamp, H., Venema, P. A. A., Harder, W., and Konings, W. N. 1966. Production of riboflavin by *Arthrobacter globiformis*. Journal of Applied Bacteriology 29:107–113.

Williams, S. T., and Davies, F. L. 1965. Use of antibiotics for selective isolation and enumeration of actinomycetes in soil. Journal of General Microbiology 38:251–261.

Wood, E. J. F. 1950. The bacteriology of shark spoilage. Australian Journal of Marine and Freshwater Research 1:129–138.

Yamada, K., and Komagata, K. 1972. Taxonomic studies on coryneform bacteria. IV. Morphological, cultural, biochemical and physiological characteristics. Journal of General and Applied Microbiology 18:399–416.

Yamada, Y., Motoi, H., Kinoshita, S., Takada, N., and Okada, H. 1975. Oxidative degradation of squalene by *Arthrobacter* species. Applied Microbiology 29:400–404.

Yamada, Y., Inouye, G., Tahara, Y., and Kondo, K. 1976. The menaquinone system in the classification of coryneform and nocardioform bacteria and related organisms. Journal of General and Applied Microbiology 22:203–214.

Zevenhuizen, L. P. T. M. 1966. Formation and function of the glycogen-like polysaccharide of *Arthrobacter*. Antonie van Leeuwenhoek Journal of Microbiology and Serology 32:356–372.

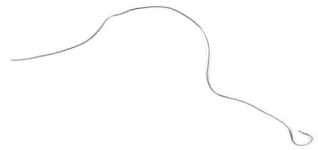

CHAPTER 56

The Genus *Micrococcus*

MILOSLAV KOCUR, WESLEY E. KLOOS, and KARL-HEINZ SCHLEIFER

The genus *Micrococcus* consists of Gram-positive spheres occurring in tetrads and in irregular clusters that are usually nonmotile and nonsporeforming. They are catalase positive and usually aerobic with strictly respiratory metabolism. Most species produce carotenoid pigments. The GC content of the DNA ranges from 65 to 75 mol%. There are nine species recognized in the genus (see later, Table 2). The data on GC content of the DNA, chemical cell wall analysis, and a comparative analysis of 16S rRNA sequences indicate that the genus *Micrococcus* is more closely related to the genus *Arthrobacter* than it is to other coccoid genera such as *Staphylococcus* and *Planococcus* (Keddie, 1974; Kloos et al., 1974; Kocur et al., 1971; Stackebrandt and Woese, 1979). For these reasons it cannot be included with the genera *Staphylococcus* and *Planococcus* in the same family Micrococcaceae. Therefore, both the genus *Micrococcus* and the genus *Arthrobacter* should be regarded as closely related, but separate genera.

Habitats

Mammalian skin is now considered as primary habitat of micrococci. Micrococci are found more consistently and usually in larger populations on mammalian skin than from other sources (Carr and Kloos, 1977; Glass, 1973; Marples, 1965, Noble and Sommerville, 1974; Kloos et al., 1974, 1976).

Human Skin

Human skin is a rich source of micrococci. It was shown that 96% of 115 people living in 18 different states in the USA carried cutaneous populations of micrococci (Kloos et al., 1974). The percentages of individuals carrying various *Micrococcus* species were as follows: *M. luteus,* 90%; *M. varians,* 75%; *M. lylae,* 33%; *M. nishinomiyaensis,* 28%; *M. kristinae,* 25%; *M. roseus,* 15%; *M. sedentarius,* 13%; *M. agilis,* 4%.

Populations of *M. luteus* were usually relatively large and this organism was isolated on the average, from 51% of the different skin sites sampled of individuals carrying this species. In a temporal study by Kloos and Musselwhite (1975), it was found that micrococci usually constituted from 1 to 20% of the total aerobic bacteria isolated from the skin of the head, legs, and arms, but less than 1% of those isolated from the high bacterial density areas of the nares and axillae. In the same temporal study, 80% of the individuals carried micrococci on the head, legs, and arms for at least 1 year. Where strain (or clonal) populations could be resolved, it was determined that certain strains of *M. luteus, M. varians, M. kristinae,* and *M. sedentarius* persisted on specific individuals up to 1 year, which strongly suggested a resident status. Several strains of *M. luteus* that were monitored for longer periods persisted for up to 2 1/2 years. Most strains appeared to be more transient and were isolated once or more for periods up to 6 months. In another temporal study of the micrococci of infant skin, Carr and Kloos (1977) reported an increase in the occurrence of micrococci with increasing age up to 10 weeks. Micrococci were rarely isolated from infants less than 1 week of age.

Animal Skin

Micrococci have been isolated from the skin of a variety of mammals, including squirrels of the genus *Sciurus*, rats, raccoons, opossums, horses, swine, cattle, dogs, and various primates (Kloos et al., 1976). The predominant species found on nonhuman mammals studied to date was *M. varians. M. luteus* was rarely isolated from nonhuman mammals. Other *Micrococcus* species found on the skin of humans appear to have a narrow host range, as their populations have not yet been discovered on other mammals. Most strains of these specialized species required one or more amino acids and vitamins for growth (Farrior and Kloos, 1975, 1976). The distribution of micrococci on other animals (e.g., birds, reptiles, amphibians, and fish) has not yet been adequately determined.

There are only few data on the distribution of micrococci on fish. Micrococci appear to be numerically diverse on the skin of marine and fresh-water fish, often comprising 1 to 25% of the bacterial population. Similar percentages of micrococci have been found within alimentary tract of different fish (Horsley, 1977). Micrococci predominate in the bacterial population of marine fish captured on the south Australian coasts, which indicates that this genus is most numerous within these waters (Gillespie and Macrae, 1975). It has been found that micrococci and coryneforms predominate in the bacterial population of sharks (Venkataraman and Sreenivasan, 1953, 1955). The high incidence of micrococci reported on certain fish may be due to their population density in the water or contamination from landing nets or excessive handling by the angler (Horsley, 1977). Micrococci and coryneforms are found among the major groups of bacteria of crustacean shellfish. The percentage composition varies. Williams et al. (1952) found the main groups in the whole Gulf shrimps to be *Acinetobacter (Achromobacter), Micrococcus, Pseudomonas,* and *Bacillus,* whereas Sreenivasan (1959) found that prawns in India carried predominately micrococci and coryneforms. The bacterial flora of the Ontario fresh-water fish examined was similar to that reported from marine fish, in which species *Pseudomonas* and *Micrococcus* were encountered frequently (Evelyn and McDermott, 1961).

Other Sources

With the exception of animal and dairy products, which may be considered as secondary sources of micrococci, most of the sources, such as soil, estuarine mud, marine and fresh water, plants, fomites, dust, and air contain small isolated populations of micrococci (Aaronson, 1955; Abd-el-Malek and Gibson, 1948; Baird-Parker, 1962; Doeringer and Dugan, 1973; Kitchell, 1962; Pohja, 1960; Streby-Andrews and Kloos, 1971; ZoBell and Upham, 1944).

Soil is not the primary source of micrococci as had been originally assumed. *Micrococcus luteus* cells died relatively quickly when they were added to natural soil. Microscopic observation showed that the cells were being physically destroyed by bacterial predators in the soil. One of the predators was *Streptoverticillium* sp. (Casida, 1980a, 1980b).

Micrococci formed only a limited part of bacterial population of sea water and therefore they have not been studied thoroughly in marine en-

vironments (Anderson, 1962; Brisou, 1955; Wood, 1952; ZoBell and Upham, 1944). It has been found that sand taken from public swimming areas at several North Carolina ocean beaches often contained small populations of *M. luteus,* whereas sand taken from remote ocean island beaches, frequented by small numbers of indigenous animals and a few anglers, only occasionally contained populations of *M. varians* and never *M. luteus,* a species common on human skin, but rare on the skin of other animals.

Some micrococci may well survive in beer. Dickscheit (1961) isolated 33 strains of micrococci from samples of beer and also "yellow sarcinas" which were probably strains of *M. kristinae, M. varians,* or *M. luteus.* Also Back (1980) isolated several strains from beer which produced a fruity odor. *M. kristinae* formed sediment in beer bottles and changed the flavor of beer.

Isolation

Direct Isolation from Skin Under Nonselective Conditions

Various semiquantitative procedures have been described for isolating aerobic bacteria from human skin (Kloos and Musselwhite, 1975; Pachtman et al., 1954; Smith, 1970; Williamson, 1965; Williamson and Kligman, 1965). The procedure described by Kloos and Musselwhite (1975) has been used for the isolation of micrococci and is suitable for use with human as well as other mammalian skin.

Isolation of Micrococci from Human and Other Mammalian Skin (Kloos and Musselwhite, 1975)

> Sterile cotton swabs were moistened with a detergent containing 0.1% Triton X-100 (Packard) in 0.075M phosphate buffer, pH 7.9 (Williamson, 1965), and rubbed vigorously, with rotation, over approximately 8-cm² sites. Swabbing was performed for 5 s on sites of the forehead, cheek, chin, nares, and axillae that usually contained large populations of bacteria and for 15 s on sites of the arms and legs that usually contained relatively small populations. Swabs taken from the forehead, cheek, chin, external naris, arms, and legs were immediately applied directly on agar media (standard agar plate, 100-mm diameter) by rubbing, with rotation, over the entire surface for two consecutive times. Swabs taken from the anterior nares and axillae were immediately rinsed once in 5 ml of detergent, and the rinse was applied to the surface of agar media. Later, during the course of the study, we observed that adults often contain populations of bacteria on the forehead, cheek, chin, and external nares that were too large to be analyzed by inoculating swabs directly onto media. In these instances, samples taken from a single swab rinse proved

to be more satisfactory and produced well isolated colonies.

The Isolating Medium (P agar) of Naylor and Burgi (1956)

Peptone (Difco)	5.0 g
Sodium chloride	5.0 g
Glucose	1.0 g
Agar (Difco)	15.0 g
Distilled water	1 liter

Inoculated agar media are incubated under aerobic conditions at 34°C for 4 days, at which time colonies are counted and recorded according to morphology and pigment. Subcultures (of selected colonies showing *Micrococcus* morphology) are stored at 4°C. For convenience, the original isolation plates could be stored at 4°C for 2 to 3 weeks prior to the isolation of cultures.

This procedure can be modified for use with other mammalian skin as follows.

Isolation of Micrococci from Nonhuman Mammalian Skin (Kloos, et al., 1976)

Sites on the body were exposed for swabbing by parting the pelage away from the site area. Swabbing was performed for 5 s on sites in the anterior naris and ventral pouch (opossums only) that usually contained large populations of bacteria and for 15 s on sites of the forehead, forelimbs, hindlimbs, abdomen, and back that usually contained relatively small populations. Swabs taken from the forehead, forelimbs, hindlimbs, abdomen, and back were immediately applied directly on agar media by rubbing, with rotation, over the entire surface for two consecutive times. Swabs taken from the anterior nares and ventral pouch were immediately rinsed once in 5 ml of detergent and then applied to the surface of agar media. (The authors of this chapter have found that rinsing in detergent is also necessary for swabs taken from the face, back, abdomen, and perineum of many Old World monkeys and great apes.)

Direct Isolation from Skin Under Selective Conditions

For shipping agar plates between collecting points or for use with nonhuman mammals, P agar was supplemented with the mold inhibitor cycloheximide (50 μg/ml) (Kloos et al., 1974; Kloos et al., 1976), Many mammals, because of their close contact with soil and foliage, carry large populations of fungi on their skin. If allowed to grow, fungal contaminants may rapidly cover the surface of unsupplemented P agar plates, making the isolation and characterization of bacteria very difficult. If *Bacillus* species are numerous on skin, it may be necessary to supplement P agar with 7% Nace to inhibit their large or spreading colonies. However, such a high amount of sodium chloride will interfere with distinctive *Micrococcus* species colony morphology and *Micrococcus* colonies will not

be easily distinguished from *Staphylococcus* colonies.

Selective medium for isolating micrococci (and corynebacteria) in the presence of populations of staphylococci has been described by Curry and Borovian (1976). The nitrofuran-containing medium, known as FTO agar, permits the growth of micrococci and prevents the growth of staphylococci. It is particularly useful for sampling areas of the skin such as the nares, axillae, and perineum where *Staphylococcus* populations are usually very large.

FTO Agar Medium for Direct, Selective Isolation of Micrococci from Skin (Curry and Borovian, 1976)

Tryptic soy agar (Difco) or Trypticase soy agar (BBL)	40.0 g
Yeast extract	1.0 g
Tween 80	5.0 g
Distilled water	1 liter

After autoclaving and cooling to 48°C, 0.1% of a 0.5 acetone stock of the dye Oil Red 0 and 10% of a 0.05% acetone stock of nitrofuran (Furoxone) is added. To prevent precipitation, the latter is added slowly from a 100-ml graduate into swirling agar. Flasks are then left open or loosely covered in the water bath, to allow acetone volatilization prior to pouring and hardening plates on a level surface. The plates can be incubated at 34 to 37°C for 3–4 days for adequate development of *Micrococcus* colonies. To reduce competition and crowding by lipophilic corynebacteria, Tween 80 should be omitted from the formula. FTO agar is conveniently prepared in 700-ml amounts in 1-liter Erlenmeyer flasks.

Cultivation

Most of micrococci grow well on nutrient agar or on P agar (Naylor and Burgi, 1956) at 37°C. The only exception is *M. agilis,* which is psychrotrophic and grows best at 22 to 25°C, and *M. halobius,* which requires 5% NaCl in cultivating medium such as nutrient agar and P agar.

Preservation

Cultures of micrococci may be stored on nutrient agar in a refrigerator (5°C) for 3 to 5 months if they are in perfectly sealed tubes. They may also be stored on nutrient agar under liquid paraffin in a refrigerator (5°C) for 1 to 2 years. The most reliable method for long-term preservation is lyophilization, using standard methods (Kirsop and Snell, 1984) or in liquid nitrogen.

Identification

Separation of Staphylococci from Micrococci

The characters that separate staphylococci from micrococci are listed in Table 1. Marked differences in the DNA base and cell wall compositions, fatty acid, and menaquinone patterns permit an accurate separation of the members of the two genera. However, these characters cannot be determined in the routine laboratory; therefore, other characters that can be easily analyzed are also listed in Table 1.

A simple test for the separation of staphylococci from micrococci was described by Schleifer and Kloos (1975). It is based on the ability of staphylococci to produce acid aerobically from glycerol in the presence of 0.4 μ/ml of erythromycin and on their sensitivity to lysostaphin. Recent studies on the isolation of micrococci and staphylococci from dry sausage have indicated that the addition of erythromycin to the glycerol medium is not absolutely necessary (Fischer and Schleifer, 1980). Only strains of *Micrococcus kristinae* and a few strains of *M. roseus* produce small amounts of acid aerobically from glycerol, but these organisms can be easily distinguished from staphylococci by their convex colony profile and characteristic colony pigment (Kloos et al., 1974). Furazolidone agar proved to be a suitable medium for the separation of micrococci from staphylococci (Rheinbaben and Hadlok, 1981).

Table 1. Separation of staphylococci from micrococci.

Character	*Staphylococcus*	*Micrococcus*	Reference
Anaerobic fermentation of glucose	+ ($\pm$, −)	− ($\pm$, +)	Evans and Kloos, 1972
FP agar	No growth	Growth	Rheinbaben and Hadlok, 1981
Bacitracin disk diffusion test	Resistant	Susceptible	Falk and Guering, 1983
Acid from glycerol-erythromycin medium	+ (−)	− (+)	Schleifer and Kloos, 1975
Selective medium containing thiocyanate plus azide	Growth	No growth	Schleifer and Krämer, 1980
Susceptibility to the vibriostatic agent 0/129 (0.5 mg/disc)	Growth	No growth	Bouvet et al., 1982
Resistance to lysostaphin	−	+	Klesius and Schuhardt, 1968; Schleifer and Kloos, 1975
Oxidase test	− (+)	+ (−)	Faller and Schleifer, 1981
Modified benzidine test	− (+)	+ (−)	Faller and Schleifer, 1981
Fructose-1,6-biphosphate aldolase (FBP)	Class I[a]	Class II	Götz et al., 1979; Fischer et al., 1982
DNA base composition (mol% GC)	30–38	66–73	Kocur et al., 1971
Major menaquinones	MK-6 to MK-8	Partially hydrogenated MK-7(H$_2$) to MK-9(H$_2$)	Collins and Jones, 1981
Long-chain unbranched fatty acids (C-18:0, C-20:0)	+	−	Schleifer and Kroppenstedt, personal communication
Cell wall composition: Peptidoglycan			
Position 1	Ala	Ala	Schleifer, 1986
Position 3	Lys	Lys	
Main interpeptide bridge	Gly$_{5-6}$, Ala-Gly$_4$, or Gly$_{3-5}$, Ser$_{1-2}$	Peptide subunit Ala$_{3-4}$, Asp-Thr-Ala$_3$, or Ser$_2$-D-Glu	
Teichoic acid	+	−	Schleifer, 1973; Schleifer and Kloos, 1975

+, positive; $\pm$, weak; −, negative. Symbols in parentheses denote a character frequency of less than 30%.
[a]*S. chromogenes, S. hyicus,* and *S. intermedius* contain both class I and II FBP-aldolases; *S. caseolyticus* contains only class II.

FP Agar for Separating *Micrococcus* from
Staphylococcus (Rheinbaben and Hadlok, 1981)

Peptone	10.0 g
Yeast extract	5.0 g
NaCl	5.0 g
Glucose	1.0 g
Agar	12.0 g
Distilled water	1 liter
pH 7.0	

After autoclaving and cooling to 48°C, 100 ml of a 0.02%
acetone solution of furazolidone are mixed under slow
stirring with the basal medium. Before pouring the
plates, the flasks are left open or loosely covered in a
water bath for 3 to min to allow evaporation of acetone.

Micrococci are susceptible to vibriostatic
agent 0/129 (0,5 mg/disc) while staphylococci
are resistant to it which enables to separate both
genera. The recommended medium for 0/129
susceptibility testing is Mueller-Hinton agar
(Bouvet et al., 1982).

Differentiation of *Micrococcus* species

Micrococci may be differentiated into nine species by means of tests listed in Table 2. Their
pigment production and colony morphology
may be used as a simple test for their presumptive identification (Kloos et al., 1974).
Certain difficulties may occur in the differentiation of *M. luteus* and *M. lylae*, as both species
have several features in common. However, *M.
lylae* can be distinguished from *M. luteus* by
cream-white or unpigmented colonies, lack of
growth on inorganic nitrogen agar, lysozyme resistance, and cell wall peptidoglycan. The most
common yellow pigmented species, *M. luteus*
and *M. varians,* differ in acid production from
glucose, nitrate reduction, lysozyme susceptibility, growth on inorganic nitrogen agar, and
on Simmons' citrate agar, and oxidase reaction.
M. roseus differs from other species in having
pink colonies, nitrate reduction, and an inability to hydrolyze gelatin.

Micrococcus agilis differs significantly from
other micrococci in several features. It possesses
flagella, is psychrophilic, and exhibits β-galactosidase activity. *M. kristinae* is a clearly separated species, too. It produces acid from glucose under anaerobic conditions, From glycerol
aerobically, it forms acetoin and hydrolyses esculin. *M. kristinae* produces unique wrinkled
growth on purple agar (Difco) containing 1%
maltose. The orange-pigmented *M. nishinomiyaensis* may be further distinguished from
M. kristinae, which forms very pale orange colonies, by growth on 7.5% NaCl agar, acetoin
production, and esculin hydrolysis. *M. sedentarius* differs from other *Micrococcus* species by
being resistant to penicillin and methicillin,

producing often water-soluble exopigment,
growing very slowly, and reacting positively to
the arginine dihydrolase test. *M. halobius* can
be easy separated from other species, as it requires at least 5% NaCl for growth.

Physiological and Biochemical Properties

Detailed studies on physiological and biochemical properties of micrococci are rather scarce.
Most of these studies have been performed before a clear separation of micrococci from
staphylococci was possible. Therefore, only
those studies will be taken into consideration
where a classification as a *Micrococcus* is guar-
anteed.

Growth Requirements

Chemically defined media for the growth of *M.
luteus* and for the growth and the pigmentation
of *M. roseus* have been devised (Wolin and Naylor, 1957; Grula et al., 1961; Cooney and
Thierry, 1966). Strains of *M. luteus* may grow
in a defined medium containing pyruvate or
glutamate as carbon and energy source, biotin,
and mineral salts (Perry and Evans, 1966; Salton, 1964). Initiation of growth of *M. luteus* in
a defined medium depends upon the presence
of an iron-binding compound, such as phenolic
compounds (e.g., catechol) or ferrichrome (Salton, 1964; Walsh et al., 1971).

Metabolic Properties

Most micrococci are strictly aerobic. Carbon-
containing compounds, oxidized to carbon
dioxide and water, include acetate, lactate, pyruvate, succinate, fructose, galactose, glucose,
glycerol, maltose, and sucrose. Variable oxidation of mannitol, sorbitol, arabinose, rhamnose,
ribose, xylose, and starch. Dulcitol is not oxidized (Saz and Krampitz, 1954; Rosypal and
Kocur, 1963; Perry and Evans, 1960, 1966).
Glucose is metabolized by fructose-1, 6-bi-
phosphate and hexose monophosphate pathways and citric acid enzymes; (Dawes and
Holmes, 1958; Perry and Evans, 1966; Blevins
et al., 1969). They possess, like most eubacteria,
a class II D-fructose-1,6-biphosphate adolase
(Götz et al., 1979). However, strains belonging
to *M. varians* and *M. kristinae* can also grow
facultatively anaerobic and produce L-lactic
acid from glucose. An NAD-dependent L-lactate dehydrogenase could be found in representatives of both species (Hartinger and Schleifer,
unpublished observations). *M. luteus* possesses

Table 2. Abbreviated scheme for the differentiation of species of the genus *Micrococcus*.[a]

Species	Major pigment[b]	Water-soluble exopigment	Growth on Simmons citrate agar	Growth on inorganic nitrogen agar	Acetoin	Nitrate reduction	Oxidase	Aerobic acid from		Lysozyme susceptibility[c]	β-Galactosidase	Arginine dihydrolase	Growth at 37°C	Peptidoglycan type	Amino sugar in cell wall polysaccharide
								Glucose	Glycerol						
M. luteus	Y>CW	–	–	+>±,–	–	–>+	++	–	–	S	–	–	+	L-Lys-peptide subunit	Mannosamineuronic acid
M. lylae	CW, U	–	–	–	–>±	–>+	+,±	–	–	SR	–	–	+	L-Lys-Asp	Galactosamine
M. varians	Y	–	+>±,–	–>±	±,–	+>±	–±	+	–	R	–	–	+	L-Lys-L-Ala$_{3-4}$	Galactosamine
M. roseus	PR>OR	–	–>±	–	±>–	+>±	–±	+,±	–	SR-R	–	–	+	L-Lys-L-Ala$_{3-4}$	Galactosamine
M. agilis	R	–	ND	ND	–	–	+	–	–	R	+	–	–	L-Lys-Thr-L-Ala$_3$	Glucosamine
M. kristinae	PO	–	–	–	+	–>±	+,±	++	++	R	–+	–	+	L-Lys-L-Ala$_3$	Glucosamine
M. nishinomiyaensis	O	–>+	–	±,–	–>±	+,±,–	+,±	–>±	–	SR-R	–	–	+	L-Lys-L-Ser-D-Glu	Galactosamine
M. sedentarius	CW>BY	+,±	–	–	–	–	–	–>±	–	S-SR	–	+	+	Uncertain	–
M. halobius	U	–	ND	ND	–	–	+	+	–	R	+	–	+	ND	ND

+ +, strong positive; +, positive; ±, weak; –, negative; ND, not determined.

[a]A single listed symbol denotes a character frequency of about 70–100%; the notation > denotes "a frequency greater than"; a comma between symbols denotes nearly equal frequency.

[b]CW, cream-white; U, unpigmented; PR, pastel red; OR, orange red; R, red; PO, pale orange; O, orange; BY, buttercup yellow; Y, yellow.

[c]S, susceptible (minimal inhibitory concentration, MIC: below 5 μg/ml); SR, slightly resistant (MIC: 5–50 μg/ml); R, resistant (MIC: above 100 μg/ml).

1305

an NAD-dependent and an NAD-independent malic dehydrogenase which can oxidize malic acid to oxalacetic acid (Cohn, 1956). A bacterial tyramine oxidase which catalyzes the oxidation of tyramine and dopamine was for the first time isolated from *M. luteus* (Sarcina lutea) by Yamada et al. (1967). A phosphoglucomutase, which catalyzes the interconversion of glucose-l-phosphate and glucose-6-phosphate, was purified from *M. luteus* (Hanabusa et al., 1966). The metabolism of exogeneous pyrimidine bases and nucleosides was also investigated in *M. luteus* (Auling et al., 1982; Auling and Moss, 1984). The presence of thymidine kinase and phosphorylase could be demonstrated, whereas uridine phosphorylase occurred in a rather weak activity and uridine kinase was not detectable. In contrast to *E. coli*, the pyrimidine nucleotide phosphorylases are not inducible.

Membrane composition

The cytoplasmic membrane composition, in particular that of *M. luteus,* has been thoroughly studied (Salton, 1987). The major fatty acid component of *Micrococcus* strains is a methyl-branched C_{15}-saturated acid (Girard, 1971; Onishi and Kamekura, 1972; Thirkell and Gray, 1974; Jantzen et al., 1974; Brooks et al., 1980). Micrococci contain relatively high amounts of long-chain aliphatic hydrocarbons in the range C_{22} to C_{33} (Tornabene et al., 1970; Morrison et al., 1971; Kloos et al., 1974). Cardiolipin and phosphatidylglycerol are the major phospholipids. Unidentified phospholipids a and b were described by Komura et al. (1975). *Micrococcus* species partially contain hydrogenated menaquinones of the type MK-7, MK-8, and MK-9 (Jeffries, 1969; Jeffries et al., 1969; Yamada et al., 1967).

All studied micrococci contain cytochromes of *a-*, *b-*, *c-*, and *d-* types (Faller et al., 1980). Besides cytochromes, menaquinones, and carotenoids, NADH, malate, lactate, and succinate dehydrogenases are also part of the electron-transport system of *M. luteus* (Erickson and Parker, 1969; Crowe and Owen, 1983a). The succinate dehydrogenase was purified and cross reactions with specific antisera directed to the membrane-bound enzyme of other micrococci could be shown, but no cross reaction occurred with the enzyme from a wide variety of Gram-positive and Gram-negative bacteria (Crowe and Owen, 1983b). The proton-translocating ATP synthase of *M. luteus* has been thoroughly studied (Salton and Schor, 1974; Schmitt et al., 1978).

It consists of the typical F_0F_1 complex. The cytoplasmic membrane is also the site of en-

zymes involved in phospholipid, peptidoglycan, and teichuronic biosynthesis (De Siervo and Salton, 1971; Park and Matsuhashi, 1984; Traxler et al., 1982). Membranes of *M. luteus* are unusually rich in mannose; much of this is bound in succinylated lipomannan (Owen and Salton, 1975). Part of the mannose is also found in membrane glycoproteins (Doherty et al., 1982). The occurrence of glycoproteins is a rather unusual feature for eubacteria.

Genetics Properties

DNA Relationships

Transformation studies by Kloos (1969a) and DNA-DNA hybridization studies by Ogasawara-Fujita and Sakaguchi (1976) showed that there is no close genetic or DNA relationship among members of the species *M. luteus, M. roseus,* and *M. varians.* Schleifer et al. (1979) demonstrated that DNA homology values between the type strains of *M. luteus* and *M. lylae* were in the range of 40 to 50% at optimal reassociation conditions, whereas, values of only 10 to 18% were obtained between these species and *M. kristinae* or *M. varians.* Kloos et al. (1974) showed a significant genetic relationship between *M. luteus* and *M. lylae* in transformation studies. Genetics exchange occurred at about 1 to 5% of homologous values using DNA isolated from *M. lylae* with *M. luteus* auxotrophic recipients. An epigenetic relationship between *M. luteus* and *M. lylae* has been demonstrated by comparative immunological studies of catalase using both double-immunodiffusion tests and quantitative micro-complement fixation assays (Rupprecht and Schleifer, 1977). The nucleotide sequence of a 23S rRNA gene of *M. luteus* was determined (Regensburger et al., 1988a). From sequence comparison to 23S rRNA genes from other bacteria, specific probes could be constructed (Regensburger et al., 1988b). One probe (pAR28) reacted only with *M. luteus* and *M. lylae,* another probe (pAR27) reacted with all micrococci and arthrobacters.

Transformation

Genetic exchange in the genus *Micrococcus* was first demonstrated by transformation of Fleming's *M. luteus* strain (formerly *M. lysodeikticus*) (Kloos, 1968; Kloos and Schultes, 1969; Mahler and Grossmann, 1968; Okubo and Nakayama, 1968). Following the determination of optimal conditions for transformation (Kloos, 1969b), a procedure for transformation was developed for use in genetic mapping (Kloos and

Rose, 1970). Transformation has not been reported in other micrococcal species.

Transformation of *Micrococcus luteus* ATCC 27141 Auxotrophs (Kloos and Rose, 1970)

An 18-h P agar slant culture of the recipient strain was suspended in 1 ml saline and diluted 1/100 in saline. Aliquots of 0.1 ml [about 5×10^6 colony-forming units (CFU)] from the diluted suspension were added to duplicate tubes containing 1 ml defined broth supplemented with 20 μg/ml of the required metabolite (e.g., L-tryptophan, L-histidine, adenosine, etc.) (Kloos, 1969b). Cultures were shaken in a 34°C water bath with a Burrell Wrist-Action Shaker at a setting of 4 for 24 h. Following growth, cells (about $1-3 \times 10^8$ CFU/ml) were centrifuged and resuspended in 1 ml transformation buffer: 50 mM Tris (hydroxymethyl) aminomethane + 1-mM $SrCl_2$, pH 7.0. Donor DNA, 0.1 μg in crosses or 10 μg for the construction of double auxotrophic mutants for three-point crosses, in 0.1 ml saline was added to each tube and the mixtures were shaken in a 30°C water bath at a setting of 4 for 30 min. DNA uptake was terminated by the addition of deoxyribonuclease (5μg/ml) and 5 mM $MgSO_4$ with incubation at 37°C for 15 min. After treatment, cells were centrifuged and resuspended in 1 ml saline. Aliquots of 0.1 ml were taken from an appropriate saline dilution of the original suspension and spread on duplicate defined agar plates. Prototrophs were scored after incubation at 34°C for 72 h. Donor-type cotransformants (on appropriate supplemented defined agar plates) were scored after incubation for 96 h. Using the above procedure transformation frequencies of 0.1–0.7 percent of recipient CFU could be obtained with prototroph donor DNA.

Competence in transformation has been demonstrated in *M. luteus* strains ATCC 27141 and its relatives (ATCC 15801, PU, UM, WRU) and strains ATCC 540 and ATCC 8673.

Chromosomal Gene Arrangements

Tryptophan and histidine biosynthesis genes have been mapped in *M. luteus* by using two-point "best-fit" and donor-types (contransformation) analyses and three-point transformation (Kane-Falce and Kloos, 1975; Kloos and Rose, 1970). These studies have indicated that at least two histidine genes are closely linked, probably contiguous, to the tryptophan gene cluster containing trpE, trpC, trpB, and trpA genes, in that order. Tryptophan genes are located in the main cluster mentioned above and at least in one other unlinked site, containing trpD and trpF. Histidine genes are distributed in at least four regions of the chromosome. The largest cluster contains four genes: his (EAH or F), hisB, hisC, and his (EAH or F). Mapping purine biosynthesis genes by the two-point best-fit method has indicated significant separation of the purine genes purE, purJ, and purC and

no linkage of these genes to purH (Mohapatra and Kloos, 1974, 1975).

Plasmid Composition

Plasmids have been detected in a small to moderate percentage (7–55%) of strains in most species of micrococci (Mathis and Kloos, 1984). They have not yet been reported in the species *M. lylae* and *M. sedentarius*. Most strains carrying plasmids exhibit only one or two types, ranging in size from 1 to 100SMDa. Plasmid patterns appear to be slightly more complex in *M. nishinomiyaensis*. Thirty-six percent of strains in this species carried two to three different plasmids. To date, most micrococcal plasmids remain cryptic. Several small plasmids identified in *M. nishinomiyaensis*, *M. roseus*, *M. varians*, and *M. luteus* have been studied with restriction endonucleases and may be suitable for use as cloning vectors following some modification.

Pathogenicity for Humans

The data about possible pathogenicity of micrococci for humans are very poor and controversial. Micrococci, however, may be considered as opportunistic pathogens, particularly in view of the increasing number of immunocompromised patients. Micrococci have been reported to be associated with various infections, especially those of urinary tract (Meers, et al., 1975; Roberts, 1967; Sellin et al., 1975; Telander and Wallmark, 1975). However, these and similar reports cannot be evaluated because the discussed strains were either not sufficiently taxonomically described or incorrectly identified. As a result of the present classification, all the strains from such infections tested to date have been shown to be staphylococci.

Therefore, only the papers of the last 10 to 15 years describing human infections caused by micrococci can be taken into consideration. A case of septic shock caused by *Micrococcus luteus* was described by Albertson et al. (1978). *M. luteus* was reported in a case of cavitating pneumonia of an immunosuppressed patient (Souhami et al., 1979) and in case of meningitidis (Fosse et al., 1986). *M. kristinae* has been isolated in pure culture from several types of infections in the U.S.A. *M. sedentarius* has been found to be associated with pitted keratolysis (Nordstrom et al., 1987). Micrococci have been isolated from blood and surgical specimens of patients associated with heart diseases and septic complications following cardiac surgery. Some of the isolates were probably *M. lylae* strains. There is no clear proof that micrococci are pathogenic for nonimmunocompromised humans. Pathogenicity of micrococci for plants has not yet been described.

Application

Micrococci have been reported to be used in the processing of fermented meat products to improve their color, aroma, flavor, and keeping quality (Kitchell, 1962; Niinivaara and Pohja, 1957; Pohja, 1960). However, based on an updated classification of these organisms, most strains used in processing have proved to be staphylococci, in particular, *S. carnosus,* and occasionally *S. xylosus* (W.E. Kloos, unpublished observations). The true micrococci that have been useful in meat processing have proved to be *Micrococcus varians.* Studies by Fischer and Schleifer (1980) showed that besides *M. varians, M. kristinae* could also be isolated from fermented sausage. Species such as *Micrococcus luteus* and *M. varians* have been widely used in industries as assay organisms to test various antibiotics in body fluids, feeds, milk, and pharmaceutical preparations (Bowman, 1957; Coates and Argoudelis, 1971; Grove and Randall, 1955; Kirshbaum and Arret, 1959; Simon and Yin, 1970); the vitamin biotin (Aaronson, 1955); and the enzyme lysozyme (Dickman and Proctor, 1952).

Micrococci are among the few bacteria that synthetize long-chain (C_{21}-C_{34}) aliphatic hydrocarbons. These hydrocarbons may have potential economic value if they can be produced and extracted economically and then processed into useful lubricating oils or other petroleum substitutes (Kloos et al., 1974; Morrison et al. 1971; Tornabene et al., 1970).

Literature Cited

Aaronson, S. 1955. Biotin assay with a coccus, *Micrococcus sodonensis,* nov. sp. J. Bacteriol. 69:67–70.

Abd-el-Malek, Y., and Gibson, T. 1948. Studies in the bacteriology of milk. II. The staphylococci and micrococci of milk. J. Dairy Res. 15:249–260.

Albertson, D., Natsions, G A., and Gleckman, R. 1978. Septic shock with *Micrococcus luteus.* Arch. Intern. Med. 138:487–488.

Anderson, J. I. W. 1962. Studies on micrococci isolated from the North Sea. J. Appl. Bacteriol. 25:362–368.

Auling, G., and Moss, B. 1984. Metabolism of pyrimidine bases and nucleosides in the coryneform bacteria *Brevibacterium ammoniagenes* and *Micrococcus luteus.* J. Bacteriol. 158:733–736.

Auling, G., Prelle, H., and Diekmann, H. 1982. Incorporation of deoxyribonucleosides into DNA of coryneform bacteria and the relevance of deoxyribonucleoside kinases. Eur. J. Biochem. 121:365–370.

Back, W. 1980. Taxonomische Untersuchungen an beerschädlichen Bakterien. Habilitationschrift. Technische Universität, Munich.

Baird-Parker, A. C. 1962. The occurrence and enumeration, according to a new classification, of micrococci and staphylococci in bacon and on human and pig skin. J. Appl. Bacteriol. 25:352–361.

Blevins, W. T., Perry J. J., and Evans, J. B. 1969. Growth and macromolecular biosynthesis by *Micrococcus sodonensis* during the utilization of glucose and lactate. Can. J. Microbiol. 15:383–388.

Bouvet, P., Chatelain, R., and Riou, J. Y. 1982. Intéret du composé vibriostatique 0/129 pour différencier les genres *Staphylococcus* et *Micrococcus.* Ann. Inst. Pasteur 113 B: 449–453.

Bowman, F. W. 1957. Test organisms for antibiotic microbial assays. Antibiot. Chemother. 7:639–640.

Brisou, J. 1955. La microbiologie de milieu marin. Editions Médicales Flammarion, Paris.

Brooks, B. W., Murray, R. G. E., Johnson, J. L., Stackebrandt, E., Woese, C. R., and Fox, G. E. 1980. Redpigmented micrococci: A basis for taxonomy. Int. J. Syst. Bacteriol. 30:627–646.

Carr, D. L., and Kloos, W. E. 1977. Temporal study of the staphylococci and micrococci of normal infant skin. Appl. Environ. Microbiol. 34:673–680.

Casida, L. E., Jr. 1980a. Death of *Micrococcus luteus* in soil. Appl. Environ. Microbiol. 39:1031–1034.

Casida, L. E., Jr. 1980b. Bacterial predators of *Micrococcus luteus* in soil. Appl. Environ. Microbiol. 39:1035–1041.

Coates, J. H., and Argoudelis, A. D. 1971. Microbial transformation of antibiotics: Phosphorylation of clindamycin by *Streptomyces coelicolor* Müller. J. Bacteriol. 108:459–464.

Cohn, D. V. 1956. The oxidation of malic acid by *Micrococcus lysodeikticus.* J. Biol. Chem. 221:413–420.

Collins, M. D., and Jones, D. 1981. The distribution of isoprenoid quinone structural types in bacteria and their taxonomic implications. Microbiol. Rev. 45:316–354.

Cooney, J. J., and Thierry, O. C. 1966. A defined medium for growth and pigment synthesis of *Micrococcus roseus.* Can. J. Microbiol. 12:83–89.

Crowe, B. A., and Owen, P. 1983a. Immunochemical analysis of respiratory-chain components of *Micrococcus luteus (lysodeikticus).* J. Bacteriol. 153:498–505.

Crowe, B. A., and Owen, P. 1983b. Molecular properties of succinate dehydrogenase isolated from *Micrococcus luteus (lysodeikticus).* J. Bacteriol. 153:1493–1501.

Curry, J. C., and Borovian, G. E. 1976. Selective medium for distinguishing micrococci from staphylococci in the clinical laboratory. J. Clin. Microbiol. 4:455–457.

Dawes, E. A., and Holmes, W. H. 1958. On the quantitative evaluation of routes of glucose metabolism by the use of radioactive glucose. Biophys. Acta 34:551–552.

De Siervo, A. J., and Salton, M. R. J. 1971. Biosynthesis of cardiolipin in the membranes of *Micrococcus lysodeikticus.* Biochim. Biophys. Acta 239:280–292.

Dickman, S. R., and Proctor, C. M. 1952. Factors affecting the activity of egg white lysozyme. Arch. Biochem. Biophys. 40:364–372.

Dickscheit, R. 1961. Beiträge zur Physiologie und Systematik der Pediokokken des Bieres. Zentralbl. Bakteriol. Parasitenkd. II. Abt 114:270–284, 458–471.

Doeringer, R. H., and Dugan, P. R. 1973. Growth relationship between the blue-green alga *Anacystis nidulans* and *Sarcina flava* in mixed culture. Abst. Ann. Meet. Am. Soc. Microbiol. 1973:45.

Doherty, H., Condon, C., and Owen, P. 1982. Resolution and in vitro glycosylation of membrane glycoproteins in *Micrococcus luteus (lysodeikticus)*. FEMS Microbiol. Lett. 15:331–336.

Erickson, S. K., and Parker, G. L. 1969. The electron-transport system of *Micrococcus luteus (Sarcina lutea)*. Biochim. Biophys. Acta 180:56–62.

Evans, J. B., and Kloos, W. E. 1972. Use of shake cultures in a semisolid thioglycolate medium for differentiating staphylococci from micrococci. Appl. Microbiol. 23:326–331.

Evelyn, T. P. T., and McDermott, L. A. 1961. Bacteriological studies of fresh-water fish. I. Isolation of aerobic bacteria from several species of Ontario fish. Can. J. Microbiol. 7:375–382.

Falk, D., and Guering, S. J. 1983. Differentiation of *Staphylococcus* and *Micrococcus* spp. with the taxo A bacitracin disk. J. Clin. Microbiol. 18:719–721.

Faller, A., Götz, F., and Schleifer, K. H. 1980. Cytochrome patterns of staphylococci and micrococci and their taxonomic implications. Zbl. Bakteriol. I. Abt. Orig. C1:26–39

Faller, A., Schleifer, and K. H. 1981. Modified oxidase and benzidine tests for separation of staphylococci from micrococci. J. Clin. Microbiol. 13:1031–1035.

Farrior, J. W., and Kloos, W. E. 1975. Amino acid and vitamin requirements of *Micrococcus* species isolated from human skin. Int. J. Syst. Bacteriol. 25:80–82.

Farrior, J. W., and Kloos, W. E. 1976. Sulfur amino acid auxotrophy in *Micrococcus* species isolated from human skin. Can. J. Microbiol. 22:1680–1690.

Fischer, S., Luczak, H., and Schleifer, K. H. 1982. Improved methods for the detection of class I and class II fructose-1, 6-biphosphate aldolases in bacteria. FEMS Microbiol. Lett. 15:103–108.

Fischer, U., and Schleifer, K. H. 1980. Zum Verkommen der Gram-positiven, katalase-positiven Kokken in Rohwurst. Fleischwirtschaft. 60:1046–1051.

Fosse, T., Peloux, Y., Granthil, C., Toga, B., Bertrando, J., and Sethian, M. 1986. Meningitis due to *Micrococcus luteus*. Eur. J. Clin. Study Treat. Infect. 13:280–281.

Gillespie, N. C., and Macrae, I. C. 1975. The bacterial flora of some Queensland fish and its ability to cause spoilage. J. Appl. Bacteriol. 39:91–100.

Girard, A. E. 1971. A comparative study of the fatty acids of some micrococci. Can. J. Microbiol. 17:1503–1508.

Glass, M. 1973. *Sarcina* species on the skin of the human forearm. Trans. St. John's Hosp. Dermatol. Soc. 59:56–60.

Götz, F., Nürnberger, E., and Schleifer, K. H. 1979. Distribution of class I and class II D-fructose-1, 6-biphosphate aldolase in various Gram-positive bacteria. FEMS Microbiol. Lett. 5:253–257.

Grove, D. C. and Randall, W. A. 1955. Assay Methods of Antibiotics. Antibiotics Monographs No. 2. Medical Encyclopedia, New York.

Grula, E. A., Luk, S-K., and Chu, Y-C. 1961. Chemically defined medium for growth of *Micrococcus lysodeikticus*. Can. J. Microbiol. 7:27–32.

Hanabusa, K., Dougherty, H. W., Del Rio, C., Hashimoto, T., and Handler, P. 1966. Phosphoglucomutase. II. Preparation and properties of phosphoglucomutases from *Micrococcus lysodeikticus* and *Bacillus cereus*. J. Biol. Chem. 241:3930–3939.

Horsley, R. W. 1977. A review of the bacterial flora of teleosts and elasmobranchs, including methods for its analysis. J. Fish. Biol. 10:529–553.

Jantzen, E., Bergan, T., and Bøvre, K. 1974. Gas chromatography of bacterial whole cell methanolysates. VI. Fatty acid composition of strains within *Micrococcaceae*. Acta Pathol. Microbiol. Scand. Sec. B 82:785–798.

Jeffries, L. 1968. Sensitivity to novobiocin and lysozyme in the classification of *Micrococcaceae*. J. Appl. Bacteriol. 31:436–442.

Jeffries, L. 1969. Menaquinone in the classification of *Micrococcaceae* with observations on the application of lysozyme and novobiocin sensitivity tests. Int. J. Syst. Bacteriol. 19:183–187.

Jeffries, L., Cawthorne, M. A., Harris, M., Cook, B., and Diplock, A. T. 1969. Menaquinone determination in the taxonomy of *Micrococcaceae*. J. Gen. Microbiol. 54:365–380.

Kane, C. M., and Kloos, W. E. 1970. Transformation of *Sarcina flava* and *Micrococcus flavocyaneus*. Genet. Res., Camb. 15:339–343.

Kane-Falce, C. M., and Kloos, W. E. 1975. A genetic and biochemical study of histidine biosynthesis in *Micrococcus luteus*. Genetics 79:361–376.

Keddie, R. M. 1974. *Arthrobacter*, p. 618–625. In: R. E. Buchanan and N. E. Gibbons (ed.), Bergey's manual of determinative bacteriology, 8th ed. Williams and Wilkins, Baltimore.

Kirshbaum, A., and Arret, B. 1959. Outline of details for assaying the commonly used antibiotics. Ant. Chem. 9:613–617.

Kirsop, B. E., and Snell J. J. S. (ed.). 1984. Maintenance of micro-organisms. Academic Press, London.

Kitchell, A. G. 1962. Micrococci and coagulase negative staphylococci in cured meats and meat products. J. Appl. Bacteriol. 25:416–431.

Klesius, P. H., and Schuhardt, V. T. 1968. Use of lysostaphin in the isolation of highly polymerized deoxyribonucleic acid and in the taxonomy of aerobic *Micrococcaceae*. J. Bacteriol. 95:739–743.

Kloos, W. E. 1968. Evidence of genetic exchange in *Micrococcus lysodeikticus*. Bacteriol. Proc. 1968:55.

Kloos, W. E. 1969a. Transformation of *Micrococcus lysodeikticus* by various members of the family *Micrococcaceae*. J. Gen. Microbiol. 59:247–255.

Kloos, W. E. 1969b. Factors affecting transformation of *Micrococcus lysodeikticus*. J. Bacteriol 98:1397–1399.

Kloos, W. E., and Musselwhite, M. S. 1975. Distribution and persistence of *Staphylococcus* and *Micrococcus* species and other aerobic bacteria on human skin. Appl. Microbiol. 30:381–395.

Kloos, W. E., and Rose, N. E. 1970. Transformation mapping of tryptophan loci in *Micrococcus luteus*. Genetics 66:595–605.

Kloos, W. E., and Schultes, L. M. 1969. Transformation in *Micrococcus lysodeikticus*. J. Gen. Microbiol. 55:307–317.

Kloos, W. E., Tornabene, T. G., and Schleifer, K. H. 1974. Isolation and characterization of micrococci from human skin, including two new species: *Micrococcus lylae* and *Micrococcus kristinae*. Int. J. Syst. Bacteriol. 24:79–101.

Kloos, W. E., Zimmerman, R. J., and Smith, R. F. 1976. Preliminary studies on the characterization and distribution of *Staphylococcus* and *Micrococcus* species on animal skin. Appl. Environ. Microbiol. 31:53–59.

Kocur, M., Bergan, T., and Mortensen, N. 1971. DNA base composition of Gram-positive cocci. J. Gen. Microbiol. 69:167–183.

Komura, I., Yamada, K., and Komagata, K. 1975. Taxonomic significance of phospholipid composition in aerobic Gram-positive cocci. J. Gen. Appl. Microbiol. 21:97–107.

Leifson, E. 1963. Determination of carbohydrate metabolism of marine bacteria. J. Bacteriol. 85:1183–1184.

Mahler, I., and Grossman, L. 1968. Transformation of radiation sensitive strains of *Micrococcus lysodeikticus*. Biochem. Biophys. Res. Commun. 32:776–781.

Marples, M. J. 1965. The ecology of the human skin. Charles C. Thomas, Springfield, Illinois.

Mathis, J. N., and Kloos, W. E. 1984. Isolation and characterization of *Micrococcus* plasmids. Curr. Microbiol. 10:163–171.

Meers, P. D., Whyte, W., and Sandys, G. 1975. Coagulase-negative staphylococci and micrococci in urinary tract infections. J. Clin. Pathol. 28:270–273.

Mohapatra, N., and Kloos, W. E. 1974. Biochemical and genetic studies of laboratory purine auxotrophic strains *Micrococcus luteus*. Can. J. Microbiol. 20:1751–1754.

Mohapatra, N., and Kloos, W. E. 1975. Biochemical characterization and genetic mapping of purine genes in *Micrococcus luteus*. Genet. Res., Camb. 26:163–171.

Morrison, S. J., Tornabene, T. G., and Kloos, W. E. 1971. Neutral lipids in the study of the relationships of members of the family *Micrococcaceae*. J. Bacteriol. 108:353–358.

Naylor, H. B., and Burgi, E. 1956. Observations on abortive infections of *Micrococcus lysodeikticus* with bacteriophage. Virology 2:577–593.

Niinivaara, F. P., and Pohja, M. S. 1957. Erfahrungen über die Herstellung von Rohwurst mittels einer Bakterienreinkultur. Fleischwirtschaft 9:789–790.

Noble, W. C., and Somerville, D. A. 1974. Microbiology of human skin. W. B. Saunders, London.

Nordstrom., K. M., McGinley, K. J., Zechman, J. M., and Leyden, J. J. 1987. Similarities between *Dermatophilus congolensis* and *Micrococcus sedentarius*: Identity of the etiologic agent of pitted keratolysis. Abstr. Ann. Meet. Amer. Soc. Microbiol. 244.

Ogasawara-Fujita, N., and Sakahuchi, K. 1976. Classification of micrococci on the basis of deoxyribonucleic acid homology. J. Gen. Microbiol. 94:97–106.

Okubo, S., and Nakayama, H. 1968. Evidence of transformation in *Micrococcus lysodeikticus*. Biochem. Biophys. Res. Commun. 32:825–830.

Onishi, H., and Kamekura, H. 1972. *Micrococcus halobius* sp. n. Int. J. Syst. Bacteriol. 22:233–236.

Owen, P., and Salton, M. R. J. 1975. A succinylated mannan in the membrane system of *Micrococcus lysodeikticus*. Biochem. Biophys. Res. Comm. 63:875–880.

Pachtman, E. A., Vicher, E. E., and Brunner, M. J. 1954. The bacteriologic flora in seborrhoeic dermatitis. J. Invest. Dermatol. 22:389–397.

Park, W., and Matsuhashi, M. 1984. *Staphylococcus aureus* and *Micrococcus luteus* peptidoglycan transglycosylases that are not penicillin-binding proteins. J. Bacteriol. 157:538–544.

Perry, J. J., and Evans, J. B. 1960. Oxidative metabolism of lactate and acetate by *Micrococcus sodonensis*. J. Bacteriol. 79:113–118.

Perry, J. ., and Evans, J. B. 1966. Oxidation and assimilation of carbohydrates by *Micrococcus sodonensis*. J. Bacteriol. 91:33–38.

Pohja, M. S. 1960. Micrococci in fermented meat products. Classification and description of 171 different strains. Acta Agralia Fennica 96:1–80.

Regensburger, A., Ludwig, W., Frank, R., Blöcker, H., and Schleifer, K. -H. 1988a. Complete nucleotide sequence of a 23S ribosomal RNA gene from *Micrococcus luteus*. Nucl. Acids, Res. 16:2344.

Regensburger, A., Ludwig, W., and Schleifer, K. -H. 1988b. DNA probes with different specificities from a cloned 23S rRNA gene of *Micrococcus luteus*. J. Gen. Microbiol. 134:1197–1204.

Rheinbaben, K. E. v. and Hadlok, R. M. 1981. Rapid distinction between micrococci and staphylococci with furazolidone agars. Antonie van Leeuwenhoek 47:41–51.

Roberts, A. P. 1967. *Micrococcaceae* from the urinary tract in pregnancy. J. Clin. Pathol. 20:631–632.

Rosypal, S., and Kocur, M. 1963. The taxonomic significance of the oxidation of carbon compounds by different strains of *Micrococcus luteus*. Antonie van Leeuwenhoek 29:313–318.

Rupprecht, M., and Schleifer, K. H. 1977. Comparative immunological study of catalases in the genus *Micrococcus*. Arch. Microbiol. 114:61–66.

Salton, M. R. J. 1964. Requirements of dehydroxyphenols for the growth of *Micrococcus lysodeikticus* in synthetic media. Biochim. Biophys. Acta 86:421–422.

Salton, M. R. J. 1987. Bacterial membrane proteins. Microbiol. Sciences 4:100–105.

Salton, M. R. J., and Schor, M. T. 1974. Release and purification of *Micrococcus lysodeikticus* ATPase from membranes extracted with n-butanol. Biochim. Biophys. Acta 345:74–82.

Saz, H. J., and Krampitz, L. O. 1954. The oxidation of acetate by *Micrococcus lysodeikticus*. J. Bacteriol. 67:409–418.

Schleifer, K. H. 1973. Chemical composition of staphylococcal cell walls, p. 13–23. In: Jeljaszewicz (ed.), Staphylococci and staphylococcal infections. Recent progress. S. Karger, Basel.

Schleifer, K. H. 1986. Section 12. Gram-positive cocci, p. 999–1003. In: P. H. A. Sneath, N. S. Mair, M. E. Sharpe, J. G. Holt (ed.), Bergey's manual of systematic bacteriology, vol. 2. Williams and Wilkins, Baltimore.

Schleifer, K. H., Heise, W., and Meyer, S. A. 1979. Deoxyribonucleic acid hybridization studies among soms micrococci. FEMS Lett. 6:33–36.

Schleifer, K. H., and Kloos, W. E. 1975. A simple test system for the separation of staphylococci from micrococci. J. Clin. Microbiol. 1:337.

Schleifer, K. H., Kloos, W. E., and Kocur M. 1981. The genus *Micrococcus*, p. 1539–1547. In: M. P. Starr, H. Stolp, H. G. Trüper, A. Balows and H. C. Schlegel (ed.), The prokaryotes: a handbook on habitats, isolation and identification of bacteria. Springer-Verlag, Berlin.

Schleifer, K. H., and Krämer, E. 1980. Selective medium for isolating staphylococci. Zentralbl. Bakteriol. Mikrobiol. Hyg. Abt. I Orig. Cl:270–280.

Schmitt, M., Rittinghaus, K., Scheurich, P., Schwulera, U., and Dose, K. 1978. Immunological properties of membrane-bound adenosine triphosphatiase. Biochim. Biophys. Acta 509:410–418.

Sellin, M. A., Cooke, D. I., Gillespie, W. A., Sylvester, D. G. H., and Anderson, J. D. 1975. Micrococcal urinary tract infections in young women. Lancet ii:570–572.

Simon, H. J., and Yin, E. J. 1970. Microbioassay of antimicrobial agents. Appl. Microbiol. 19:573–579.

Smith, R. F. 1970. Comparative enumeration of lipophilic and nonlipophilic cutaneous diphtheroids and cocci. Appl. Microbiol. 19:254–258.

Souhami. L., Feld, R., Tuffnell, P. G., and Feller, T. 1979. *Micrococcus luteus* pneumonia: A case report and review of the literature. Med. Pediatr. Oncol. 7:309–314.

Sreenivasan, A. 1959. A note on the bacteriology of prawns and their preservation by freezing. J. Sci. Ind. Res. 18C:119.

Stackebrandt, E., and Woese, C. R. 1979. A phylogenetic dissection of the family *Micrococcaceae*. Curr. Microbiol. 2:317–322.

Streby-Andrews, M. E., and Kloos, W. E. 1971. Amino acid auxotrophy in natural strains of *Micrococcus luteus*. Bacteriol. Proc. 1971:27.

Telander, B., and Wallmark, G. 1975. *Micrococcus* subgroup 3-a common cause of acute urinary tract infection in women. Lakartidningen 72:1967.

Thirkell, D., and Gray, E. M. 1974. Variation in the lipid fatty acid composition in purified membrane fractions from *Sarcina aurantiaca* in relation to growth phase. Antonie van Leeuwenhoek. J. Microbiol. Serol. 40:71–78.

Tornabene, T. G., Morrison, S. J., and Kloos, W. E. 1970. Aliphatic hydrocarbon contents of various members of the family *Micrococcaceae*. Lipids 5:929–937.

Traxler, C. I. Goustin, A. S., and Anderson, J. S. 1982. Elongation of teichuronic acid chains by a wall-membrane preparation from *Micrococcus luteus*. J. Bacteriol. 150:649–656.

Venkataraman, R., and Sreenivasan, A. 1953. The bacteriology of freshwater fish. Ind. J. Med. Res. 41:385–392.

Venkataraman, R., and Sreenivasan, A. 1955. Bacterial flora of fresh shark. Curr. Sci. 11:380–381.

Walsh, B. L., O'Dor, J., and Warren, R. A. J. 1971. Chelating agents and the growth of *Micrococcus lysodeikticus*. Can. J. Microbiol. 17:593–597.

Williams, O. B., Rees, H. B., and Campbell, L. L. 1952. The bacteriology of Gulf Coast shrimp. I. Experimental procedures and quantitative results. Texas J. Sci. 4:49–52.

Williamson, P. 1965. Quantitative estimation of cutaneous bacteria, p. 3–11. In: Maibach, H. I., and Hildick-Smith, G. (ed.), Skin bacteria and their role in infection. McGraw-Hill, New York.

Williamson, P., and Kligman, A. M. 1965. A new method for the quantitative investigation of cutaneous bacteria. J. Invest. Dermatol. 45:498–503.

Wolin, H. L., and Naylor, H. B. 1957. Basic nutritional requirements of *Micrococcus lysodeikticus*. J. Bacteriol. 74:163–167.

Wood, E. J. F. 1952. The micrococci in marine environments. J. Gen. Microbiol. 6:205.

Yamada, H., Uwajima, T., Kumagai, H., Watanabe, M., and Ogata, K. 1967. Crystalline tyramine oxidase from *Sarcina lutea*. Biochem. Biophys. Res. Commun. 27:350–355.

ZoBell, C. E., and Upham, C. 1944. A list of marine bacteria including descriptions of sixty new species. Bulletin of Scripps Institute of Oceanography, University of California 5:239–292.

The Genus *Renibacterium*

T. MARTIN EMBLEY

The genus *Renibacterium* was proposed by Sanders and Fryer (1980) to accommodate the single species, *Renibacterium salmoninarum*, the causal agent of bacterial kidney disease (BKD) of salmonid fish. This disease causes significant mortalities in salmonid populations in Europe (Fryer and Sanders, 1981; Ghittino et al., 1977; Pfeil-Putzien et al., 1985), Japan (Fryer and Sanders, 1981), Canada, and North America (Bell et al., 1984; Fryer and Sanders, 1981). BKD was first reported among wild *Salmo salar* on the rivers Dee and Spey in Scotland (Mackie et al., 1933). However, the casual agent was not cultivated on laboratory media until much later (Earp, 1950). The difficulties in growing *R. salmoninarum* led to problems in its classification because workers tended to rely on cell shape and staining properties. Reports of "Chinese letter" formation, pleomorphism, and the occurrence of metachromatic granules led to the bacterium being classified as a member of the genus *Corynebacterium* (Ordal and Earp, 1956; Smith, 1964). However, Goodfellow et al. (1976) analyzed the lipids of a single BKD isolate and were unable to detect mycolic acids, which are a characteristic feature of the genus *Corynebacterium*. Sanders and Fryer (1980) later demonstrated that the cell wall of BKD strains contained lysine and that three strains had an average GC content of 53 mol%. Taken together, these data strongly supported the formation of a new genus (Sanders and Fryer, 1980). Additional investigations of the lipid composition of BKD isolates supported this proposal (Collins, 1982; Embley et al., 1983). Thus, strains of *Renibacterium salmoninarum* contain unsaturated menaquinones with Mk-9 as major component, and monomethyl branched fatty acids comprising 12-methyltetradecanoic (*ai*-15:0) and 14-methylhexadecanoic (*ai*-17:0) acids. They contain diphosphatidylglycerol and phosphatidylglycerol, and an unusually large number of glycolipids. Further work on the cell wall structure revealed that the peptidoglycan is cross-linked by a glycyl-alanine interpeptide bridge (type A3 alpha; Fiedler and

Draxl, 1986; Kusser and Fiedler, 1983). Moreover, the α-carboxyl group of D-glutamic acid in position two of the peptide subunit is substituted through a D-alanine amide. An unusual cell wall polysaccharide containing galactose, rhamnose, *N*-acetylglycosamine and *N*-acetylfucosamine is also present (Fiedler and Draxl, 1986). The analysis of the 16S rRNA catalog from *R. salmoninarum* showed it to be a distinct genus within the Gram-positive bacteria (Stackebrandt et al., 1988), its nearest neighbours being members of the genus *Arthrobacter*. This was considered a surprising result since *Arthrobacter* contains organisms with relatively high GC content (>60 mol%), mainly saturated menaquinones (e.g., MK-8H$_2$, Mk-9H$_2$), and far fewer glycolipids (Stackebrandt et al., 1988). Similarities between renibacteria and arthrobacters include the possession of iso and anteiso fatty acids and a similar peptidoglycan structure. Further work is needed to confirm the precise phylogeny of *Renibacterium*.

Habitat

R. salmoninarum appears to be an obligate pathogen of members of the family Salmonidae, i.e., trout, salmon, and char. It has a broad host range in the family and has been reported (Fryer and Sanders, 1981) in sockeye salmon *(Oncorhynchus nerka)*, coho salmon *(O. kisutch)*, chinook salmon *(O. tshawytscha)*, pink salmon *(O. gorbuscha)*, cherry salmon *(O. masou)*, Atlantic salmon *(Salmo salar)*, rainbow trout *(Salmo gairdneri)*, brown trout *(S. trutta)*, cut-throat trout *(S. clarki)*, brook trout *(Salvelinus fontinalis)*, and lake trout *(S. namaycush)*. There is as yet no evidence that *Renibacterium salmoninarum* is capable of a free-living existence in an aquatic environment (Austin and Rayment, 1985). However, there are currently no effective methods for isolating small numbers of renibacteria from samples containing large numbers of heterotrophic bacteria. The slow growth of *R. salmoninarum* often means that selective isolation plates inoculated with water or sediment samples are overgrown before renibacteria can form macroscopic colonies. As a consequence, there has been no systematic attempt to isolate renibacteria from the environment. *R. salmoninarum* has been detected in the feces of infected fish using a selective isolation medium (Austin et al., 1983) and by fluorescent antibody methods (Mitchum et al., 1979). Furthermore, viable renibacteria have been isolated from sediment in tanks containing experimentally infected fish (Austin and Rayment, 1985). In these experiments, renibacteria were not isolated from water samples taken from the same tanks. However, the cell surface of *R. salmoninarum* is hydrophobic and it has been suggested (Daly and Stevenson, 1987) that this may cause it to leave the water column and attach to surfaces and organic material.

Epizootiology

As discussed above, it is difficult to obtain evidence for water-born transmission of BKD. Nevertheless, horizontal transmission is considered to be a significant source of infection within the high population densities found in modern fish farms and hatcheries (Bell et al., 1984; Fryer and Sanders, 1981). Carrier fish, i.e., fish showing no overt disease symptoms, have been detected using culture and serological methods and may constitute a source of infection (Austin and Rayment, 1985; Austin et al., 1983; Fryer and Sanders, 1981; Paterson et al., 1979). The disease has also been detected in wild salmonid populations in fresh and salt water (Banner et al., 1983, 1986; Ellis et al. 1978; Fryer and Sanders, 1981; Pippy, 1969). In at least one case, disease-free hatchery trout were infected with *R. salmoninarum* after release into a stream where BKD had previously occurred in the wild brook trout population (Fryer and Sanders, 1981; Mitchum et al., 1979). There do not appear to be any other hosts for *R. salmoninarum* but no systematic survey has been carried out (Bell and Traxler, 1986). Reports that Pacific herring can be infected with *R. salmoninarum* (Traxler and Bell, 1988), and a report of a bacterium in eels that reacted with antisera raised to renibacteria (Dutil and Lallier, 1984), need to be substantiated.

Vertical transmission via eggs from infected brood fish is now believed to be a major route of infection with BKD (Allison, 1958; Bullock et al., 1978). Eggs in brood fish are probably infected through the micropyle from contact with infected coelomic fluid (Evelyn et al., 1984, 1986a). In one study (Bruno and Munro, 1986a), bacteria with the staining characteristics of *R. salmoninarum* were also observed in developing oogonia. Infected eggs may subsequently give rise to infected progeny (Bruno and Munro, 1986a; Evelyn et al., 1986a). Treatment of eggs with iodine or other agents may reduce, but does not appear to prevent, vertical transmission (Bullock et al., 1978; Bruno and Munro, 1986; Evelyn et al., 1984, 1986b). This is probably due to the fact that once inside the egg, the bacterium is sheltered from the effects of surface disinfectants (Bruno and Munro, 1986; Evelyn et al., 1984). Injection of brood fish with erythromycin prior to spawning, has been reported to facilitate the entry of the antibiotic into the egg and reduce the chance of infection (Armstrong et al., 1989; Bullock and Leek, 1986; Evelyn et al., 1986b).

Diagnosis

BKD is classically diagnosed by observation of clinical signs (Fryer and Sanders, 1981). However, the external pathology of BKD is variable and normally observed in the terminal stages of the disease (Belding and Merrill, 1935; Bell, 1961; Earp, 1950; Hoffmann et al., 1984; Mackie et al., 1933; Rucker et al., 1951, 1954; Smith, 1964; Snieszko and Griffon, 1955; Wood and Wallis, 1955; Wood and Yasutake, 1956; Young and Chapman, 1978). Exophthalmos, abdominal distension, superficial blebs or blisters, hemorrhagic areas, and abscesses on the exterior body surface have all been reported. Internally, BKD is a systemic infection with a predilection for kidney tissue although small gray lesions may also occur on the liver and spleen. These lesions are often filled with tissue debris and viable bacteria (Fryer and Sanders, 1981). In advanced cases, the entire kidney is swollen and necrotic. Petechial hemorrhages of the muscle lining the peritoneum are reported to be characteristic lesions in Atlantic salmon (Smith, 1964). The bacterium is frequently intracellular in infected tissues (Ordal and Earp, 1956; Snieszko and Griffon, 1955; Wood and Yasutake, 1956; Yamamoto, 1975). Phagocytosed bacteria appear to retain their characteristic morphology and may continue cell division (Young and Chapman, 1978). Histologically the disease can be characterized as a systemic diffuse granulomatous (histiocytic) inflammation (Bruno, 1986; Snieszko and Griffon, 1955; Wolke, 1975; Wood and Yasutake, 1956; Young and Chapman, 1978). Data from studies of the physiology and haematology of infected fish reflect the observed pathological changes (Aldrin et al., 1978; Bruno, 1986; Bruno and Munro, 1986c; Fryer and Sanders, 1981; Iwama et al., 1986).

As mentioned above, the appearance of clinical signs often occurs late in the disease when fish are already dying. As such, these features are of little use in detecting the disease prior to it becoming widespread within fish stocks. In addition, *R. salmoninarum* has been isolated from fish exhibiting no overt disease symptoms (Austin et al., 1983). Consequently there has been considerable interest in developing sensitive and rapid methods for the detection of renibacteria. These include fluorescent antibody tests (Bullock and Stuckey, 1975; Bullock et al., 1980; Elliott and Barila, 1987; Laidler, 1980; Mitchum et al., 1979; Paterson et al., 1979, 1980), latex agglutination (Dixon, 1987b), staphylococcal coagglutination (Kimura, 1978), immunodiffusion (Chen et al., 1974; Kimura et al., 1978), the peroxidase-antiperoxidase procedure (Sakai et al., 1987), and enzyme-linked-immunosorbent assay (ELISA) (Dixon, 1987a; Pascho and Mulcahy, 1987; Pascho et al., 1987). A number of studies have compared the various methods (Austin and Rayment, 1985; Cipriano et al., 1985; Dixon, 1987b; Evelyn et al., 1981; Hoffmann et al., 1989; Lee and Gordon, 1987; Pascho et al., 1987; Sakai et al., 1987; Yoshimizu et al., 1988). While each method has its adherents, there is no general agreement as to which is the best. All suffer from disadvantages, such as lack of sensitivity, reproducibility, cross reactions, and false positives. The development of monoclonal antibodies to specific renibacterial surface antigens may eventually provide more sensitive and reliable immunodiagnostics (Arakawa et al., 1987; Weins and Kaattari, 1989). At the moment, the most widely used methods for survey work and presumptive diagnosis are the direct and indirect fluorescent antibody methods, and the ELISA techniques. However, in cases concerning the culling or the movement of stock to new areas, they should be confirmed using an additional method, preferably by culture (Evelyn et al., 1981; Fryer and Sanders, 1981). For maximum sensitivity with culture, precautions should be taken to remove inhibitory substances from the sample and antibiotics should be incorporated into the isolation medium (Austin et al., 1983; Daly and Stevenson, 1988; Evelyn et al., 1981).

Pathogenic Mechanisms

Bacterial kidney disease is a chronic infection in which invasiveness (Bell et al., 1988) and host tissue responses lead to death of the host (Wolke, 1975). Unfortunately, it is difficult to induce reproducible experimental infection other than by injecting large numbers of bacteria, a situation unlikely to occur in nature. A number of studies have demonstrated the hydrophobic and auto-agglutinating properties of the cell surface of *R. salmoninarum* (Bandin et al., 1989; Bruno, 1988; Daly and Stevenson, 1987). It has been suggested that hydrophobicity contributes to invasiveness by facilitating adherence to host tissues and uptake and survival of *R. salmoninarum* in macrophages (Bandin et al., 1989; Daly and Stevenson, 1987). In a preliminary study (Bruno, 1988), strains that showed high hydrophobicity were shown to be more virulent for experimentally infected rainbow trout than for weakly hydrophobic strains. No differences in the gross pathology of infected fish were observed irrespective of the strain injected. Hydrophobicity is lost following heating in water at 50 to 62°C (Bruno, 1988; Daly and Stevenson, 1987; Evelyn et al., 1973), and may also be lost upon repeated passage on laboratory media (Bandin et al., 1989; Bruno, 1988). Strains of *R. salmoninarum* possess hemagglutinin activity against erythrocytes from a variety of homiotherms (Bandin et al., 1989; Daly and Stevenson, 1987). In contrast, only a few strains show hemagglutinin activity against fish erythrocytes (Bandin et al., 1989; Daly and Stevenson, 1987). The hemagglutinins can be removed by washing cells with water (Daly and Stevenson, 1987). *Renibacterium* secretes a hemolysin on media supplemented with blood and has low levels of proteolytic activity on substrates such as casein and gelatin (Bruno and Munro, 1986b; Goodfellow et al., 1985). No role for these activities has been assigned for in vivo pathogenicity. A recent report of an extracellular toxin (Shieh, 1988) needs to be confirmed and its role in the disease process investigated.

Treatment

There are currently no completely effective methods of treating fish infected with *Renibacterium salmoninarum*. Some of the problems are probably due to the fact that *R. salmoninarum* occurs intracellularly and can persist at low levels in asymptomatic fish (Fryer and Sanders, 1981; Young and Chapman, 1978). Thus, the beneficial effects of drug regimes have often lasted only as long as treatment continued. The main method for controlling or preventing the spread of BKD has been the use of erythromycin, either injected or fed (Bell et al., 1988; Bullock et al., 1975; Fryer and Sanders, 1981; Wolf and Dunbar, 1959). However, the demonstration of resistance to erythromycin in a laboratory strain of *R. salmoninarum* suggests that this drug should be used with care (Austin, 1985; Bell et al., 1988). There has been considerable interest in the use of vaccination against *R. salmoninarum* (Fryer and Sanders, 1981; McCarthy et al., 1984; Paterson et al., 1980, 1981a, 1981b). However, the results of vaccine trials have been inconclusive and as yet there is no effective vaccine on the market. A number of studies have demonstrated that dietary supplements may influence the occurrence or severity of artificially induced BKD (Bell et al., 1984; Bowser et al., 1988; Fryer and Sanders, 1981; Wedemyer and Ross, 1973). Studies have also shown that stressing fish may increase the severity of BKD, or precipitate the onset of the disease (Fryer and Sanders, 1981). It seems likely, therefore, that good husbandry will remain an important factor in the control of BKD.

Isolation

The most common method for the isolation of *R. salmoninarum* is to streak kidney material or material that has discharged from lesions onto kidney disease medium 2 (KDM2) agar (Evelyn, 1977) (see below). Incubation is normally at 15 to 17°C and plates should be placed in plastic bags to preserve moisture. The selectivity of KDM2 can be improved by the addition of cycloheximide (50 μg/ml), D-cycloserine (12.5 μg/ml), oxolinic acid (2.5 μg/ml), and polymixin B sulphate (25 μg/ml) as filter-sterilized solutions (Austin et al., 1983). KDM2 supplemented with polymyxin B sulfate (15 μg/ml), furazolidone (4 μg/ml), and amphotericin (20 μg/ml) has also been used for selective isolation. Colonies of *R. salmoninarum* may take up to 6 weeks to appear, depending on the density of cells within the sample. The colonies are white and convex and often have a sticky elastic texture. A number of studies have demonstrated that salmonid tissue may contain a substance that inhibits the growth of *R. salmoninarum* on isolation plates (Daly and Stevenson, 1988; Evelyn et al., 1981). The inhibition may be prevented (Evelyn et al., 1981) by first washing homogenized tissue in peptone-saline (0.1% wt/vol; 0.85%, wt/vol, respectively).

Kidney Disease Medium 2 (Evelyn, 1977)

Contains per liter:

L-Cysteine·HCl	1 g
Tryptone	10 g
Yeast extract	0.5 g
Agar	15 g

The pH is adjusted to 6.5 to 6.8.

The ingredients are dissolved in 800 ml of distilled water, the pH is adjusted to 6.5 to 6.8, and the medium is autoclaved at 121°C for 15 min. After cooling to approximately 55°C, 200 ml of sterile calf serum is added.

Cultivation

The first attempts to culture *Renibacterium salmoninarum* were upon media incorporating ingredients such as fish extract and minced chick embryo (Earp, 1950). Growth upon these media was extremely sparse and unreliable. Subsequently, Ordal and Earp (1956) demonstrated that cysteine was required for growth of *R. salmoninarum* and devised a cysteine blood agar that gave good growth from heavy inoculum within 10 days at 17°C. This formulation was later modified (Evelyn, 1977; Evelyn et al., 1973) to produce KDM2 (see above), which is widely used for the cultivation of *R. salmoninarum*. Incubation is at 15°C. If a heavy inoculum is used, then growth may appear within 7 days. With a dilute inoculum, macroscopic colonies may take up to 6 weeks to appear. The role of calf serum in KDM2 has been the subject of much investigation. It may act by binding toxic substances that would otherwise prevent or retard the growth of *R. salmoninarum,* especially from single cells. This has led to attempts to replace serum with components such as starch (McCarthy et al., 1984) and activated charcoal (Daly and Stevenson, 1985; Shieh, 1988). Wolf and Dunbar (1959) used Mueller-Hinton medium (which contains starch) supplemented with cysteine for routine maintenance of renibacteria. We have found that this medium is less reliable than KDM2 for growth from dilute inocula. Daly and Stevenson (1985) replaced serum in KDM2 with 0.1% (wt/vol) activated charcoal (Sigma) and were able to grow single colonies from dilute inoculum. This suggests that KDM2 incorporating charcoal may be a cheaper and more easily prepared alternative to KDM2. There have been two unsuccesful attempts to define the growth requirements of *R. salmoninarum*. Embley et al. (1982) devised a medium in which the only nondefined component was tryptone at 1% (wt/vol). Although this medium could be used for routine subculture, growth from dilute inoculum was poor. More recently, Shieh (1988) devised a complete medium containing charcoal in which the nondefined components are peptone and yeast extract. Growth in this medium was reported to compare favorably to growth in KDM2 broth formulated with charcoal or serum (Shieh, 1988).

Preservation of Cultures

Strains of *R. salmoninarum* can be preserved as suspensions prepared from young plate cultures in 15 to 20% (wt/vol) sterile aqueous glycerol stored at $-20°C$ (Wellington and Williams, 1979). In our experience, such suspensions retain viability for up to 1 year. Renibacteria can also be lyophilised using standard procedures.

Identification

R. salmoninarum can be easily differentiated from other Gram-positive taxa using the chemotaxonomic features described in the introduction. Strains are aerobic, strongly Gram-positive, asporogenous, non-acid-fast, nonencapsulated, nonmotile, short rods (0.3 to 1.0 by 1.0 to 1.5μm), often occurring in pairs. Growth is very slow at 5 and 22°C, and absent at 30°C (Smith, 1964). Strains of *R. salmoninarum* are remarkably uniform in their biochemical reactions (Bruno and Munro, 1986b; Goodfellow et al., 1985). All strains examined are catalase positive and cytochrome oxidase negative. Using the API-ZYM strip (API laboratory products) most strains produce acid and alkaline phosphatase, α-D-mannosidase, C8 esterase (caprylate), and trypsin-like activity. Since the API-ZYM strip requires metabolic activity but not growth, these tests can be used to give a quick tentative identification without long incubation periods (Austin et al., 1983; Goodfellow et al., 1985). Freshly isolated renibacteria produce zones of β hemolysis on media containing blood. The majority of strains degrade casein, DNA, gelatin, and Tweens 20, 40, 60, and 80 (Bruno and Munro, 1986b; Goodfellow et al., 1985). Most of the serological methods mentioned in the section on diagnosis can also be used to identify strains from plate cultures.

Physiology, Immunology, and Genetics

The cell surface antigens of strains of *Renibacterium salmoninarum* have been the subject of a number of investigations (Fiedler and Draxl, 1986; Weins and Kaatari, 1989). Fiedler and Draxl (1986) isolated a cell wall polysaccharide containing galactose, rhamnose, *N*-acetylglucosamine, and *N*-acetylfucosamine. They were subsequently able to detect an immune response to the polysaccharide in rabbits vaccinated with whole renibacteria. Most studies have concentrated on protein antigens (Arakawa et al., 1987; Getchell et al., 1985; Turaga et al., 1987; Weins and Kaattari, 1989). The major surface protein antigen appears to be a 57 kDa protein, named antigen F (Getchell et al., 1985).

Monoclonal antibodies have been raised to this protein and shown to be highly specific for renibacteria (Weins and Kaatari, 1989).

There have been no investigations of the physiology of renibacteria. The type strain of *R. salmoninarum* does not appear to contain any plasmids (Taranzo et al., 1983).

Literature Cited

Aldrin, J. F., M. Meval, J. Y. Robert, M. Vigneulle, and F. Baudin-Laurencon. 1978. Incidences metaboliques de la corynebacteriose experimentale chez le saumon coho *(Oncorhynchus kisutch).* Bull. Soc. Sci. Vet. Med. Comp. Lyon 80:79–90.

Allison, L. N. 1958. Multiple sulfa therapy of kidney disease among brook trout. Prog. Fish Cult. 20:66–68.

Arakawa, C. K., J. E. Sanders, and J. L. Fryer. 1987. Production of monoclonal antibodies against *Renibacterium salmoninarum.* J. Fish Dis. 10:249–253.

Armstrong, R. D., T. P. T. Evelyn, S. W. Martin, W. Dorward, and H. W. Ferguson. 1989. Erythromycin levels within eggs and alevins derived from spawning broodstock chinook salmon *Oncorhynchus tshawytscha* injected with the drug. Dis. Aquat. Org. 6:33–36.

Austin, B. A. 1985. Chemotherapy of bacterial fish diseases, p. 19–26. In: A. E. Ellis (ed.), Fish and shellfish pathology. Academic Press, London.

Austin, B. A., and J. N. Rayment. 1985. Epizootiology of *Renibacterium salmoninarum* the causal agent of bacterial kidney disease in salmonid fish. J. Fish Dis. 8:505–509.

Austin, B., T. M. Embley, and M. Goodfellow. 1983. Selective isolation of *Renibacterium salmoninarum.* FEMS Microbiol. Lett. 17:111–114.

Bandin, I., Y. Santos, J. L. Barja, and A. E. Toranzo. 1989. Influence of the growth conditions on the hydrophobicity of *Renibacterium salmoninarum* evaluated by different methods. FEMS Microbiol. Lett. 60:71–78.

Banner, C. R., J. S. Rohovec, and J. L. Fryer. 1983. *Renibacterium salmoninarum* as a cause of mortality among chinook salmon in salt water. J. World Maricult. Soc. 14:236–239.

Banner, C. R., J. J. Long, J. L. Fryer, and J. S. Rohovec. 1986. Occurrence of salmonid fish infected with *Renibacterium salmoninarum* in the Pacific Ocean. J. Fish Dis. 9:273–275.

Belding, D. L. and B. Merril. 1935. A preliminary report on a hatchery disease of the *Salmonidae.* Trans. Am. Fish Soc. 65:76–84.

Bell, G. R. 1961. Two epidemics of apparent kidney disease in cultured pink salmon *Oncorhynchus gorbuscha.* J. Fish. Res. Brd. Can. 18:559–562.

Bell, G. R., and Traxler, G. S. 1986. Resistance of the Pacific lamprey, *Lampetra tridenata* (Gairdner), to challenge by *Renibacterium salmoninarum,* the causative agent of kidney disease in salmonids. J. Fish Dis. 9:277–279.

Bell, G. R., D. A. Higgs, and G. S. Traxler. 1984. The effect of dietary ascorbate, zinc, and manganese on the development of experimentally induced bacterial kidney disease in sockeye salmon *(Onchorhynchus nerka).* Aquaculture 36:293–311.

Bell, G. R., G. S. Traxler, and C. Dworschak. 1988. Development in vitro and pathogenicity of an erythromycin resistant strain of *Renibacterium salmoninarum*, the causative agent of bacterial kidney disease in salmonids. Dis. Aquat. Org. 4:19–25.

Bowser, P. R., R. B. Landy, G. A. Wooster, and J. G. Babish. 1988. Efficacy of elevated dietary fluoride for the control of *Renibacterium salmoninarum* infection in rainbow trout *Salmo gairdneri*. J. World Aquacult. Soc. 19:1–7.

Bruno, D. W. 1986. Histopathology of bacterial kidney disease in laboratory infected rainbow trout, *Salmo gairdneri* Richardson, and Atlantic salmon, *Salmo salar* L., with reference to naturally infected fish. J. Fish Dis. 9:523–537.

Bruno, D. W. 1988. The relationship between auto-agglutination, cell surface hydrophobicity and virulence of the fish pathogen *Renibacterium salmoninarum*. FEMS Microbiol. Lett. 51:135–140.

Bruno, D. W., and A. L. S. Munro. 1986a. Observations on *Renibacterium salmoninarum* and the salmonid egg. Dis. Aquat. Org. 1:83–87.

Bruno, D. W., and A. L. S. Munro. 1986b. Uniformity in the biochemical properties of *Renibacterium salmoninarum* isolates from several sources. FEMS Microbiol. Lett. 33:247–250.

Bruno, d. W., and A. L. S. Munro. 1986c. Haematological assessment of rainbow trout, *Salmo gairdneri* Richardson, and Atlantic salmon, *Salmo salar* L., infected with *Renibacterium salmoninarum*. J. Fish Dis. 9:195–204.

Bullock, G. L., and S. L. Leek. 1986. Use of erythromycin in reducing vertical transmission of bacterial kidney disease. Vet. Hum. Toxicol. 28 (suppl. 1):18–20.

Bullock, G. L., and H. M. Stuckey. 1975. Fluorescent antibody identification and detection of the *Corynebacterium* causing kidney disease in salmonids. J. Fish. Res. Brd. Can. 32:2224–2227.

Bullock, G. L., H. M. Stuckey, and D. Mulcahy. 1978. Corynebacterial kidney disease: egg transmission following iodophore disinfection. Fish Hlth. News 7:51–52.

Bullock, G. L., B. R. Griffon, and H. M. Stuckey. 1980. Detection of *Corynebacterium salmoninus* by direct fluorescent antibody test. Can. J. Fish. Aquat. Sci. 37:719–721.

Chen, P. K., G. L. Bullock, H. M. Stuckey, and A. C. Bullock. 1974. Serological diagnosis of corynebacterial kidney disease of salmonids. J. Fish. Res. Brd. Can. 31:1939–1940.

Cipriano, R. C., C. E. Starliper, and J. H. Schachte. 1985. Comparative sensitivities of diagnostic procedures used to detect bacterial kidney disease in salmonid fishes. J. Wildl. Dis. 21:144–148

Collins, M. D. 1982. Lipid composition of *Renibacterium salmoninarum* (Sanders and Fryer). FEMS Microbiol. Lett. 13:295–297.

Daly, J., and R. W. Stevenson. 1985. Charcoal agar, a new growth medium for the fish disease bacterium *Renibacterium salmoninarum*. Appl. Environ. Microbiol. 50:868–871.

Daly, J., and R. M. Stevenson. 1987. Hydrophobic and haemagglutinating properties of *Renibacterium salmoninarum*. J. Gen. Microbiol. 133:3575–3580.

Daly, J., and R. W. Stevenson. 1988. Inhibitory effects of salmonid tissue on the growth of *Renibacterium salmoninarum*. Dis. Aquat. Org. 4:169–171.

Dixon, P. F. 1987a. Detection of *Renibacterium salmoninarum* by the enzyme-linked immunosorbant assay (ELISA). J. Appl. Ichthyol. 3:77–78.

Dixon, P. F. 1987b. Comparison of serological techniques for the identification of *Renibacterium salmoninarum*. J. Appl. Ichthyol. 3:131–138.

Dutil, J. D., and R. Lattier. 1984. Testing bacterial infection as a factor involved in the mortality of catadromous eels *(Anguilla rostrata)* migrating down the St. Laurence estuary (Canada). Naturaliste Can. (Rev. Ecol. Syst.) 111:395–400.

Earp, B. J. 1950. Kidney disease in young salmon. MS. Thesis, University of Washington.

Elliot, D. G, and T. Y. Barilla. 1987. Membrane filtration-fluorescent antibody straining procedure for detecting and quantifying *Renibacterium salmoninarum* in coelomic fluid of chinook salmon *(Oncorhynchus tshawytscha)*. Can. J. Fish. Aquat. Sci. 44:206–210.

Ellis, R. W., A. J. Novotny, and L. W. Harrell. 1978. Case report of kidney disease in a wild chinook salmon *(Oncorhynchus tshawytscha)* in the sea. J. Wildlife Dis. 14:121–123.

Embley, T. M., M. Goodfellow, and B. Austin. 1982. A semi-defined growth medium for *Renibacterium salmoninarum*. FEMS Microbiol. Lett. 14:299–301.

Embley, T. M., M. Goodfellow, D. E. Minnikin, and B. Austin. 1983. Fatty acid, isoprenoid quinone and polar lipid composition of *Renibacterium salmoninarum* J. Appl. Bacteriol. 55:31–37.

Evelyn, T. P. T. 1977. An improved growth medium for the kidney disease bacterium and some notes on using the medium. Bull. Off. Int. Epizoot. 87:511–513.

Evelyn, T. P. T., G. E. Hoskins, and G. R. Bell. 1973. First record of bacterial kidney disease in an apparently wild salmonid in British Columbia. J. Fish. Res. Brd. Can. 30:1578–1579.

Evelyn, T. P., J. E. Ketcheson, and L. Prosperi-Porta. 1981. The clinical significance of immunofluorescence-based diagnoses of the bacterial kidney disease carrier. Fish Pathol. 15:293–300.

Evelyn, T. P., J. E. Ketcheson, and L. Prosperi-Porta. 1984. Further evidence for the presence of *Renibacterium salmoninarum* in salmonid eggs and for the failure of providone-iodine to reduce the intra-ovum infection rate in water hardened eggs. J. Fish Dis. 7:173–182.

Evelyn, T. P. T., L. Prosperi-Porta, and J. E. Ketcheson. 1986a. Experimental intra-ovum infection of salmonid eggs with *Renibacterium salmoninarum* and vertical transmission of the pathogen with such eggs despite their treatment with erythromycin. Dis. Aquat. Org. 1:197–202.

Evelyn, T. P. T., J. E. Ketcheson, and L. Prosperi-Porta. 1986b. Use of erythromycin as a means of preventing vertical transmission of *Renibacterium salmoninarum*. Dis. Aquat. Org. 2:7–11.

Fiedler, F., and R. Draxl. 1986. Biochemical and immunochemical properties of the cell surface of *Renibacterium salmoninarum*. J. Bacteriol. 168:799–804.

Fryer, J. L., and Sanders, J. E. 1981. Bacterial kidney disease of salmonid fish. Ann. Rev. Microbiol. 35:273–298.

Getchell, R. G., J. S. Rohovec, and J. L. Fryer. 1985. Comparison of *Renibacterium salmoninarum* isolates by antigenic analysis. Fish Pathol. 20:149–159.

Ghittino, P., S. Andruetto, and S. Viglianni. 1977. Recent findings in trout kidney pathology in Italy. Bull. Off. Int. Epiz. 87:491–493.

Goodfellow, M., M. D. Collins, and D. E. Minnikin. 1976. Thin-layer chromatographic analysis of mycolic acid and other long chain components in whole organism methanolysates of coryneform and related taxa. J. Gen. Microbiol. 96:351–358.

Goodfellow, M., T. M. Embley, and B. Austin. 1985. Numerical taxonomy and emended description of *Renibacterium salmoninarum*. J. Gen. Microbiol. 131:2739–2752.

Hoffmann, R., W. Popp, and S. van der Graff. 1984. Atypical BKD predominantly causing occular and skin lesions. Bull. Eur. Assoc. Fish Pathol. 4:7–9.

Hoffmann, R. W., G. R. Bell, C. Pfeil-Putzein, and M. Ogawa. 1989. Detection of *Renibacterium salmoninarum* in tissue sections by different methods—a comparative study with special regard to the indirect immunohistochemical peroxidase technique. Fish Pathol. 24:101–104.

Iwama, G. K., G. L. Greer, and D. J. Randall. 1986. Changes in selected haematological parameters in juvenile chinook salmon subjected to a bacterial challenge and a toxicant. J. Fish Biol. 28:563–572.

Kimura, T. 1978. Bacterial kidney disease of salmonids. Fish Pathol. 13:43–52.

Kimura, T., Y. Ezura, K. Tajima, and M. Yoshimizu. 1978. Serological diagnosis of bacterial kidney disease of salmonids (BKD): immunodiffusion test by heat stable antigen extracted from infected kidney. Fish Pathol. 13:103–108.

Kusser, W., and F. Fiedler. 1983. Murein type and polysaccharide composition of cell walls from *Renibacterium salmoninarum*. FEMS Microbiol. Lett. 20:391–394.

Laidler, T. 1980. Detection and identification of the bacterial kidney disease (BKD) organism by the indirect fluorescent antibody technique. J. Fish Dis. 3:67–69.

Lee, E., and M. R. Gordon. 1987. Immunofluorescence screening of *Renibacterium salmoninarum* in the tissues and eggs of farmed chinook salmon spawners. Aquaculture 65:7–14.

Mackie, T. J., J. A. Arkwright, T. E. Pryce-Tannatt, J. C. Mottram, W. D. Johnston, and W. J. M. Menzies. 1933. The second interim report of the furunculosis committee. Her Majesty's Stationery Office, Edinburgh.

McCarthy, D. H., T. R. Croy, and D. F. Amend. 1984. Immunisation of rainbow trout, *Salmo gairdneri* Richardson, against bacterial kidney disease: preliminary efficacy evaluation. J. Fish Dis. 7:65–71.

Mitchum, P. L., L. E. Sherman, and G. T. Baxter. 1979. Bacterial kidney disease in feral populations of brook trout *(Salvelinus fontinalis)*, brown trout *(Salmo trutta)*, and rainbow trout *(Salmo gairdneri)*. J. Fish. Res. Brd. Can. 36:1370–1376.

Ordal, E. J., and B. J. Earp. 1956. Cultivation and transmission of the aetiological agent of kidney disease in salmonid fishes. Proc. Soc. Exp. Biol. Med. 92:85–88.

Pascho, R. J., and D. Mulcahy. 1987. Enzyme-linked immunosorbant assay for a soluble antigen of *Renibacterium salmoninarum*, the causative agent of salmonid bacterial kidney disease. Can. J. Fish. Aquat. Sci. 44:183–191.

Pascho, R. J., D. G. Elliott, R. W. Mallett, and D. Mulcahy. 1987. Comparison of five techniques for the detection of *Renibacterium salmoninarum* in adult coho salmon. Trans. Am. Fish. Soc. 116:882–890.

Paterson, W. D., C. Gallant, and D. Desautels. 1979. Detection of bacterial kidney disease in wild salmonids in the Margaree river system and adjacent waters using an indirect fluorescent antibody technique. J. Fish. Res. Brd. Can. 36:1464–1468.

Paterson, W. D., S. P. Lall, and D. Desautels. 1980. Bacterial kidney disease in Atlantic salmon (*Salmo salar*) in Nova Scotia, Canada. Fish Path. 15:283–292.

Paterson, W. D., S. P. Lall, and D. Desautels. 1981a. Studies on bacterial kidney disease in Atlantic salmon (*Salmo salar*) in Canada. Fish Pathol. 15:283–292.

Paterson, W. D., D. Desautels, and J. Weber. 1981b. The immune response of Atlantic salmon *(Salmo salar)* to the causative agent of bacterial kidney disease *(Renibacterium salmoninarum)*. J. Fish Dis. 4:99–111.

Pfiel-Putzien, C. von., R. Hoffman, W. Popp, and M. Schaurer. 1985. Zur Verbreitung der bacterial kidney disease (BKD) der Salmoniden in der Bundersrepublik Deutschland. Zbl. Vet. Med. B. 32:541–547.

Pippy, J. H. C. 1969. Kidney disease in juvenile Atlantic salmon *(Salmo salar)* in the Margaree River. J. Fish. Res. Brd. Can. 26:2535–2537.

Rucker, R. R., Bernier, A. F., W. J. Whipple, and R. E. Burrows. 1951. Sulfadiazine for kidney disease. Prog. Fish Cult. 13:135–137.

Rucker, R. R., B. J. Earp, and E. J. Ordal. 1954. Infectious diseases of Pacific salmon. Trans. Am. Fish. Soc. 83:297–312.

Sakai, M., G. Koyama, S. Atsuta, and M. Kobayashi. 1987. Detection of *Renibacterium salmoninarum* by a modified peroxidase-antiperoxidase (PAP) procedure. Fish Pathol. 22:1–5.

Sanders, J. E., and J. L. Fryer. 1980. *Renibacterium salmoninarum* gen. nov. sp. nov. the causative agent of bacterial kidney disease. Int. J. System. Bacteriol. 30:496–502.

Shieh, H. S. 1988. Blood free media for the cultivation of the fish kidney disease bacterium, *Renibacterium salmoninarum*. Microbios Lett. 37:141–145.

Smith, I. W. 1964. The occurrence and pathology of Dee disease. Fresh water and fisheries research 34. Her Majesty's Stationery Office, Edinburgh.

Snieszko, S. F., and P. J. Griffon. 1955. Kidney disease in brook trout and its treatment. Prog. Fish Cult. 17:3–13.

Stackebrandt, E., U. Wehmeyer, H. Nader, and F. Fiedler. 1988. Phylogenetic relationship of the fish pathogenic *Renibacterium salmoninarum* to *Arthrobacter*, *Micrococcus* and related taxa. FEMS Microbiol. Lett. 50:117–120.

Taranzo, A. E., J. L. Barja., R. R. Colwell, and F. M. Hetrick. 1983. Characterisation of plasmids in bacterial fish pathogens. Infect. Immun. 39:184–192.

Traxler, G. S., and G. R. Bell. 1988. Pathogens associated with impounded Pacific herring *Clupea harengus pallasi*, with emphasis on viral erythrocytic necrosis (VEN) and atypical *Aeromonas salmonicida*. Dis. Aquat. Org. 5:93–100.

Turaga, P. S. D., G. D. Weins, and S. L. Kaattari. 1987. Analysis of *Renibacterium salmoninarum* antigen production *in situ*. Fish Pathol. 22:209–214.

Wedemyer, G. A., and A. J. Ross. 1973. Nutritional factors in the biochemical pathology of corynebacterial kidney disease in the coho salmon *(Oncorhynchus kisutch)*. J. Fish. Res. Brd. Can. 30:296–298.

Weins, G. D., and S. L. Kaattari. 1989. Monoclonal antibody analysis of common surface proteins of *Renibacterium salmoninarum*. Fish Pathol. 24:1–7.

Wellington, E., and S. T. Williams. 1979. Preservation of actinomycete inoculum in frozen glycerol. Microbios Lett. 6:151–157.

Wolf, K., and C. E. Dunbar. 1959. Test of 34 therapeutic agents for control of kidney disease in trout. Trans. Am. Fish. Soc. 88:117–129.

Wolke, R. E. 1975. The pathology of bacterial and fungal diseases affecting fish, p. 76–78. In: W. E. Ribelin and G. Migaki (ed.), The pathology of fishes. The University of Wisconsin Press, Wisconsin.

Wood, E. M., and W. T. Yasutake. 1956. Histopathology of kidney disease in fish. Am. J. Pathol. 32:845–857.

Wood, J. W., and J. Wallis. 1955. Kidney disease in adult chinook salmon and its transmission by feeding to young chinook salmon. Research Briefs of the Fisheries Commision of Oregon 6:32–40.

Yamamoto, T. 1975. Infectious pancreatic necrosis virus and bacterial kidney disease appearing concurrently in populations of *Salmo gairdneri* and *Salvelinus fontinalis*. J. Fish. Res. Brd. Can. 32:92–95.

Yoshimizu, M., R. Ji, M. Sami, and T. Kimura. 1988. Comparison of FITC conjugate avidin-biotin complex ABC method and indirect FAT for the detection rate of *Renibacterium salmoninarum* antigen in carrier fish in BKD. Fish Pathol. 23:171–174.

Young, C. L., and G. B. Chapman. 1978. Ultrastructural aspects of the causative agent and histopathology of bacterial kidney disease in brook trout *(Salvelinus fontinalis)*. J. Fish. Res. Brd. Can. 35:1234–1248.

CHAPTER 58

The Genus *Stomatococcus*

ERKO STACKEBRANDT

The genus *Stomatococcus,* with *S. mucilaginosus* as its only species, was first described after significant phylogenetic and biochemical differences to otherwise morphologically similar organisms were detected (Bergan and Kocur, 1982). *S. mucilaginosus* ("*Staphylococcus salivarius,*" Andrewes and Gordon, 1907, or "*Micrococcus mucilaginosus*" Bergan et al., 1970) is defined to contain organisms which are Gram-positive, nonmotile, nonsporeforming, encapsulated, spherical cells, and which show either a weakly positive or a negative catalase reaction. Cells are arranged mostly in clusters, and occasionally in pairs or tetrads. Following the tradition of defining higher taxa by morphology, *Stomatococcus*—together with the genera *Micrococcus, Staphylococcus,* and *Planococcus*—were described to constitute the family Micrococcaceae (Bergan and Kocur, 1982; Schleifer, 1986), although a dissection of this family on phylogenetic grounds had already been proposed by Stackebrandt and Woese (1979). Chemical and phylogenetic evidence suggest that the species *S. mucilaginosus* constitutes an individual line of descent within the order Actinomycetales belonging to a yet-to-be described family that in addition contains the genera *Micrococcus, Arthrobacter, Renibacterium,* and *Jonesia* (Stackebrandt et al., 1983, 1988; Rocourt et al., 1987). Earlier reports on 16S rRNA analyses had shown *S. mucilaginosus* to branch off even deeper than the other member of the family (Stackebrandt et al., 1983). The shift of the branching point was due to the inclusion of more representatives from other genera of this phylogenetic cluster, embracing *Arthrobacter, Brevibacterium, Dermatophilus, Microbacterium, Cellulomonas,* and their respective related taxa.

Habitats

Stomatococci are normal inhabitants of the mouth and the upper respiratory tract of humans (Bergan et al., 1970), comprising about 3.5% of the predominant cultivable aerobic microflora of the human oral cavity. *S. mucilaginosus* is indigenous to the human tongue and pharynx but it has also been isolated from the nasopharynx, bronchial secretions, blood (Gordon, 1967; Bergan et al., 1970; Rubin et al., 1978; Pinsky et al., 1989), and from smears of human dental plaque (Bowden, 1969). It may also be among the early colonizers of human teeth (Nyvad and Kilian, 1987).

Isolation

Swab samples from the human tongue or other parts of the oral cavity are inoculated onto blood agar or trypticase soy agar which is incubated at 30–37°C. Colonies are usually mucoid, transparent, or whitish and adherent to agar surface.

Media and Cultivation

Under aerobic conditions, stomatoccoci show good growth within 24–30 h at 30°C on complex media such as brain heart infusion agar (Difco 0418) or casein peptone (tryptic digest)–glucose–yeast extract–NaCl (1:0.5:0.5:0.5%, w/v) broth, pH 7.2–7.4. No growth occurs in media prepared with 5% NaCl or 40% bile (Baird-Parker, 1974).

Preservation

Stock cultures are routinely maintained on agar slants using one of the media indicated above and stored at 4°C. Cultures are transferred to fresh medium every 4 weeks. Liquid cultures can be frozen at −12°C and survive several months. Long-term preservation is done by lyophilization or storage in liquid nitrogen. No information on the viability has been reported.

Identification

The same morphological and biochemical tests as are used for staphylococci and micrococci are recommended (Bergan and Kocur, 1986). Cells are nonmotile and spherical, arranged mostly in clusters, occasionally in pairs or tetrads. A copious capsule is formed (Silva et al., 1977). Oxidase and benzidine tests are negative (Schleifer, 1986). The catalase reaction is negative in about one-half of the strains investigated (Gordon, 1967; Bergan and Kocur, 1986). Cells are resistant to lysostaphin (Schleifer and Kloos, 1975). A variety of sugars are fermented anaerobically. Acid but no gas is produced, although no information on end products is available.

S. *mucilaginosus* is susceptible to micrococcal bacteriophages, most of which are also specifically active on *Micrococcus luteus* (Bauske et al., 1978). *Stomatococcus*-specific phages have as yet not been found. Cytochrome patterns (Stackebrandt et al., 1983) and fatty-acid and polar lipid patterns (Jantzen et al., 1974; Amadi et al., 1988) distinguish stomatococci from micrococci and staphylococci (Faller et al., 1980; Lennarz and Talamo, 1966; Nahaie et al., 1984). With respect to the presence of unsaturated menaquinones, stomatococci resemble staphylococci and *M. luteus,* but not the other *Micrococcus* species (Amadi et al., 1988). The peptidoglycan type differs among strains of *S. mucilaginosus,* in that the interpeptide bridge consists of either L-alanine, D-serine, or glycine (Schleifer and Kandler, 1972; Stackebrandt et al., 1983).

Despite the variation in peptidoglycan types and the GC content of 56–60.5 mol%, all 10 strains of *S. mucilaginosus* included in DNA hybridization studies revealed such a high degree of relatedness (more than 75%) that their allocation to a single species seems justified.

The 16S rRNA catalog of the type strain (Ludwig et al., 1981) contains a high fraction of *Stomatococcus*-specific oligonucleotides so that new isolates can be allocated unambiguously to this species.

The type strains of *S. mucilaginosus* is ATCC 25256 (CCM 2417, NCTC 10663).

Pathogenicity

The isolation of strains of *S. mucilaginosus* has recently been reported from the blood cultures of patients with endocarditis, in patients who were intravenous drug abusers (Relman et al., 1987; Coudron et al., 1987) and in patients with preexisting valvular heart disease (Rubin et al., 1978; Prag et al., 1985), bacteremia (Barlow et al., 1986), and recurrent peritonitis during chronic ambulatory peritoneal dialysis (Ragnaud et al., 1981). The presence of *S. mucilaginosus* as a member of the indigenous microflora of the upper respiratory tract in humans makes it unlikely that this organism is pathogenic. However, the increasing number of reports on the presence of *S. mucilaginosus* in clinical cases can not be ignored. Further studies should show whether this species should be considered an opportunistic pathogen.

Literature Cited

Amadi, E. N., Alderson, G., and D. E. Minnikin. 1988. Lipids in the classification of the genus *Stomatococcus.* System. Appl. Microbiol. 10:111–115.

Andrewes, F. W., and M. H. Gordon. 1907. Report on the biological characters of the staphylococci pathogenic for man. Ann. Rep. Med. Offr. Loc. Govt. 35: 543–560.

Baird-Parker, A. C. 1974. *Staphylococcus* Rosenbach, p. 483–489. In: R. E. Buchanan and N. E. Gibbons (ed.), Bergey's manual of determinative bacteriology, 8th ed. Williams and Wilkins, Baltimore.

Barlow, J. F., K. A. Vogele, and P. F. Dzintars. 1986. Septicemia with *Stomatococcus mucilaginosus.* Clin. Microbiol. Newsletter 8: 22.

Bauske, R., Peters, G., and G. Pulverer. 1978. Activity spectrum of micrococcal and staphylococcal phages. Zbl. Bakt. Hyg., Abt. Orig. A 241: 24–29.

Bergan, T., and M. Kocur. 1982. *Stomatococcus mucilaginosus* gen. nov. spec. nov., ep. rev., a member of the family *Micrococcaceae.* Int. J. Syst. Bacteriol. 32: 374–377.

Bergan, T., and M. Kocur. 1986. *Stomatoccocus* Bergan and Kocur, p. 1008–1010. In: P. H. A. Sneath, N. S. Mair, M. E. Sharpe, J. G. Holt (ed.), Bergey's manual of systematic bacteriology, vol. 2. Williams and Wilkins, Baltimore.

Bergan, T., Bøvre, K., and B. Hovig 1970. Priority of *Micrococcus mucilaginosus* Migula 1900 over *Staphylococcus salivarius* Andrewes and Gordon 1907 with proposal of a neotype strain. Int. J. Syst. Bacteriol. 20: 107–113.

Bowden, G. H. 1969. The components of cells walls and extracellular slime of four strains of *Staphylococcus salivarius* isolated from human dental plaque. Arch. Oral. Biol. 14: 685–697.

Coudron, P. E., Markowitz, S. M., Mohanty, L. B., Schatzki, P. F., and J. M. Payne, 1987. Isolation of *Stomatococcus mucilaginosus* from drug user with endocarditis. J. Clin. Microbiol. 25: 1359–1363.

Faller, A. H., F. Götz, and K. H. Schleifer. 1980. Cytochrome patterns of staphylococci and micrococci and their taxonomic implications. Zbl. Bakt. Hyg., I. Abt. Orig. C1: 26–39.

Gordon, D. F. 1967. Reisolation of *Staphylococcus salivarius* from the human oral cavity. J. Bacteriol. 94: 1281–1286.

Jantzen, E., T. Bergan, and K. Bovre. 1974. Gas chromatography of bacterial whole cell methanolysates. VI. Fatty acid composition of strains within Micrococcaceae. Acta Path. Microbiol. Scand Sect. B 82: 785–798.

Lennarz, W. J., and B. Talamo. 1966. The chemical characterization and enzymatic synthesis of mannolipids in Micrococcus lysodeikticus. J. Biol. Chem. 241: 2702–2719.

Ludwig, W., K.-H. Schleifer, G. E. Fox., E. Seewaldt, and E. Stackebrandt. 1981. A phylogenetic analysis of staphylococci, Peptococcus saccharolyticus and Micrococcus mucilaginosus. Gen. Microbiol. 125: 357–366.

Nahaie, M. R., M. Goodfellow, D. E. Minnikin, and V. Hajek. 1984. Polar lipid and isoprenoid quinone composition in the classification of Staphylococcus. J. Gen. Microbiol. 130: 2427–2437.

Nyvad, B., and M. Kilian. 1987. Microbiology of the early colonization of human enamel and root surfaces in vivo. Scand. J. Dent. Res. 95: 369–380.

Pinsky, R. L., Piscitelli, V., and J. E. Petterson. 1989. Endocarditis caused by relatively penicillin-resistant Stomatococcus mucilaginosus. J. Clin. Microbiol. 27: 215–216

Prag, J., E. Kjoller, and F. Espersen. 1985. Stomatococcus mucilaginosus endocarditis. Eur. J. Clin. Microbiol. 4: 422–424.

Ragnaud, J.-M., Marceau, C., Roche-Bezain, M. B., and C. Wone. 1981. Peritonite a 'rechute a Stomatococcus mucilaginosus chez une malade traitee par dialyse peritoneale continue ambulatoire. Press. Med. 14: 2063.

Relman, D. A., K. Rouff, and M. J. Ferraro. 1987. Stomatococcus mucilaginosus endocarditis in an intravenous drug abuser. J. Infect. Dis. 155: 1080–1082.

Rocourt, J., U. Wehmeyer, and E. Stackebrandt. 1987. Transfer of Listeria denitrificans to a new genus Jonesia gen. nov. as Jonesia denitrificans comb. nov. Int. J. Syst. Bacteriol. 37: 266–270.

Rubin, S. J., Lyons, R. W., and A. J. Murcia. 1978. Endocarditis associated with cardial catheterization due to a Gram-positive coccus designated Micrococcus mucilaginosus incertae sedis. J. Clin. Microbiol. 7: 546–549.

Schleifer, K.-H. 1986. Micrococcaceae, p. 1003–1035. In: P .H. A. Sneath, N. S. Mair, M. E. Sharpe, J. G. Holt, (ed.), Bergey's manual of systematic bacteriology, vol. 2. Williams and Wilkins, Baltimore.

Schleifer, K.-H., and O. Kandler. 1972. Peptidoglycan types of bacterial cell walls and their taxonomic implications. Bacteriol. Rev. 36: 407–477.

Schleifer, K. H., and W. E. Kloos. 1975. A simple test for the separation of staphylococci from micrococci. J. Clin. Microbiol. 1:337–338.

Silva, M. T., J. J. Polonia, and M. Kocur. 1977. The fine structure of Micrococcus mucilaginosus. J. Submicrosc. Cytol. 9: 53–66.

Stackebrandt. E., and C. R. Woese. 1979. A phylogenetic dissection of the family Micrococcaceae. Curr. Microbiol. 2: 317–322.

Stackebrandt, E., C. Scheuerlein, and K.-H. Schleifer. 1983. Phylogenetic and biochemical studies on Stomatococcus mucilaginosus. System. Appl. Microbiol. 4:207–217.

Stackebrandt, E., U. Wehmeyer, H. Nader, and F. Fiedler. 1988. Phylogenetic relationship of the fish pathogenic Renibacterium salmoninarum to Arthrobacter, Micrococcus and related taxa. FEMS Microbiol. Lett. 50: 117–120.

CHAPTER 59

The Family Cellulomonadaceae

ERKO STACKEBRANDT and HELMUT PRAUSER

Although not yet validly described, the family Cellulomonadaceae can be defined to include the genera *Cellulomonas, Oerskovia, Promicromonospora,* and *Jonesia.* The rational for the establishment of this family is based mainly on phylogenetic grounds. Comparative 16S rRNA cataloging of members of these genera indicates that they form an individual line of descent within the order Actinomycetales (Stackebrandt et al., 1980a; 1983). Phylogenetically neighboring groups to the cellulomonads are *Arthrobacter, Renibacterium, Micrococcus, Stomatococcus, Dermatophilus,* and *Microbacterium* and its related genera. *Cellulomonas, Arthrobacter,* and *Micrococcus* also formed a phylogenetic tight group by 5S rRNA sequences (Park et al. 1987). Members of the Cellulomonadaceae are characterized by a combination of phenotypic characters that distinguish them from the neighboring taxa: from *Microbacterium* and related taxa by the type A crosslinking of the peptidoglycan moities, from *Dermatophilus* and *Brevibacterium* by the lack of *meso*-diaminopimelic acid in the peptidoglycan and from all neighbors by the presence of phospholipid type 5 (see below).

The intrafamily structure shows that *Cellulomonas* and *Oerskovia* form a closely related pair of genera, while *Promicromonospora* and *Jonesia* are more distantly related, not only to each other but to the former pair of genera as well. Depending on the number and taxonomic position of outside reference organisms, *Promicromonospora* and *Jonesia* may be separated from the other two genera of the family to form a phylogenetically separate sub-branch (Rocourt et al., 1987a). Within the family, *Jonesia* is the most diverse taxon in that the GC content of its DNA is about 13 mol% lower than the contents reported for the other three members and its menaquinones are unhydrogenated. Morphological and chemotaxonomic characteristics for the differentiation of members of the family Cellumonadaceae and related genera are given in Table 1. The morphologies of members

of the family Cellulomonadaceae are given in Figures 1–12.

Prior to the recognition that these morphologically diverse organisms share a substantial degree of genetic similarity *Cellulomonas* was included in the "coryneform group of organisms" (Keddie, 1974), *Oerskovia* and *Promicromonospora* were considered "genera incertae sedis" of the family Nocardiaceae (McClung, 1974) and Micromonosporaceae (Küster, 1974), respectively, while *Jonesia denitrificans* was a member of the genus *Listeria* (Seeliger and Welshimer, 1974) with doubtful taxonomic status. Taxonomic studies on members of *Cellulomonas* included *Cellulomonas cartae* (subsequently placed to *C. cellulans* (Stackebrandt and Keddie, 1986), which in DNA-DNA hybridization studies revealed a low degree of relationship to other species of the genus. Representatives of *Oerskovia,* however, were not included in this study. When in a later phylogenetic survey by 16S rRNA representatives of many actinomycete genera were investigated, the relationship between cellulomonads and oerskoviae became obvious. The degree of similarity between strains of both genera was as high as that separating the most unrelated species of other genera, e.g., *Arthrobacter* and *Micrococcus.* Moreover, *Cellulomonas cellulans* was much more closely related to *Oerskovia xanthineolytica* then to any of the other *Cellulomonas* species. Despite genus-specific differences in morphological (Prauser, 1986) and chemotaxonomic properties, e.g., peptidoglycan type (Seidl et al., 1980; Stackebrandt et al., 1980b), in fatty acid composition (Minnikin et al., 1979) and in susceptibility to phages (Prauser, 1986), the genera *Cellulomonas* and *Oerskovia* were united under *Cellulomonas* (Stackebrandt et al., 1982). Numerical taxonomic studies gave contradictory results. While Alderson and Amadi (1985) supported the proposal to reduce *Oerskovia* to a synonym of *Cellulomonas,* Bousfield et al. (1983) found cellulomonads and oerskoviae to group in different clusters. As pointed out by Prauser (1986), dis-

Table 1. Differential properties of members of the family Cellulomonadaceae and certain phylogenetically and morphologically related genera.

Character	Cellulomonas	Oerskovia	Promicromonospora	Jonesia	Dermatophilus	Arthrobacter	Renibacterium	Agromyces	Microbacterium	Brevibacterium	Nocardia	Actinomadura	Nocardioides
Branched rods	+	−	−	+	−	+	−	−	+	+	−	−	−
Branching mycelium	−	P,T	P	−	T	−	−	T	−	−	P,T	P	P
Fragmentation of mycelium in older cultures	−	+	+	−	−	−	−	+	−	−	d	−	+
Aerial mycelium produced	−	−	+	−	+	−	−	−	−	−	+°	d	+
Motility or motile elements produced	+	+	−	+	+	d	−	−	d	−	−	−	−
Strict aerobe	−	−	+	−	−	+	+	−	−	+	+	+	+
Peptidoglycan type	A4β	A4α	A4α	A4α	A1γ	A3α, A4α	A3α	B2α	B1β	A1γ	A1γ	A1γ	A3γ
Cell wall type	VIII	VI	VI	VI	III	VI	VI	VII	VI	III	IV	III	I
Phospholipid type	PV	PV	PV	ND	PI	PI	PI	PI	PI	PI	PII	PI	PI
Fatty acid pattern	2c	2c	ND	2c	1a	2c	2c	2c	2c	2c	1b	3a	3a
Whole sugar pattern	−	−	−	−	B	−	−	−	−	C	A	B	−
Menaquinone (MK) type	9(H₄)	9(H₄)	9(H₄)	9	8(H₄)	9(H₂)	9	11, 12, 13	11, 12	8(H₂)	8(H₄), 9(H₂)	9(H₆)	8(H₄)
GC content (mol%)	72–76	70–75	70–75	56–58	57–59	59–70	53–54	71–77	69–75	60–67	64–72	66–69	66–73

Abbreviations: P, persistent; T, transient; °, lacking in some strains; d, 11–89% of the strains are positive; ND, no data.
Chemotaxonomic properties were abbreviated as indicated: Peptidoglycan type (Schleifer and Kandler, 1972); phospholipid type, fatty acid pattern and whole cell pattern (Lechevalier and Lechevalier, 1981); menaquinone type (Collins and Jones, 1981).
Adapted from Lechevalier and Lechevalier (1981a, 1981b); Goodfellow (1989); and Lechevalier (1989).

tinct differences in peptidoglycan types and low DNA homologies between members of the two genera, which in addition are placed in two different sublines of descent, are convincing arguments to keep *Cellulomonas* and *Oerskovia* as separate taxonomic entities.

The Genus *Cellulomonas*

The genus *Cellulomonas* is represented by eight validly described species (Stackebrandt and Kandler, 1979; Bagnara et al., 1985). However, *C. cellulans* (previously *Nocardia cellulans* and *Cellulomonas cartae,* respectively) is today recognized as a member of *Oerskovia* (Bousfield et al., 1983; Keddie and Jones, 1981; Prauser, 1986). Although a formal proposal for the transfer has as yet not been made, this species will be dealt with in the chapter on *Oerskovia.*

Habitat

The main habitat of cellulomonads appears to be the soil, from which the original cultures were isolated (Kellerman et al., 1913; Kauri and Kushner, 1985). In addition to the habitats indicated previously (Stackebrandt and Keddie, 1986), recent emphasis placed on the cellulolytic activity of these organisms has resulted in the successful isolation of *Cellulomonas* spp. from rumen (Lee and Lee, 1986), activated sludge (Ramasamy et al., 1981), and cellulose-enriched environments such as bark and wood (Deschamps, 1982; Przybyl, 1979), and sugar fields (de Leon and Joson, 1980).

Media

Moderate growth occurs on meat extract, peptone agar, or media based on yeast extract or peptone at around neutral pH. Growth-promoting factors in yeast extract are, in part, thiamine and biotin. These factors can be supplemented by adding a few drops of a sterile commercially available multivitamin solution (e.g., Multibionta, Merck).

Enrichment

As described by Stackebrandt and Keddie (1984), a suitable procedure is to enrich cellulomonads in mineral-based medium containing a low concentration (0.05–0.1%) of yeast extract to which filter paper is added as principal carbon source. Isolates are tested on plates containing Avicel, Solka floc, CF11 cellulose, carboxymethyl cellulose, or phosphoric acid-treated cellulose (Kauri and Kushner, 1985) (see also "Physiological Properties"). Cellulose deg-

radation can be visualized due to the formation of clearing zones. This method is not selective for *Cellulomonas* and isolates must be screened for typical coryneform morphology. Cells are routinely grown in shake flasks at 30°C.

Preservation

For short-term preservation, stab cultures in semisolid medium should remain viable for several months at room temperature. For long-term preservation, lyophilization is recommended.

Identification

Cellulomonads are slender irregular rods (ca. 0.4–0.8 μm in diameter) which may vary considerably in length and which may be arranged in V-formation (Fig. 1). On agar media they may undergo a more or less expressed morphological growth cycle. In this case colonies resulting from single cells appear nocardioform, i.e. they show mycelia-like fringes (Fig. 2). The filaments, which may be branching, fragment subsequently into rods (Fig. 3). Nocardioform mycelia-like structures may also appear, when filaments penetrate into the agar medium (Fig. 4). Week-old cultures are composed mainly of short rods but a proportion of the cells may be coccoid. Except for *C. fermentans* which has been reported to be nonmotile (Bagnara et al. 1985), all species possess polar multitrichous flagella (Thayer, 1984).

Cellulomonas species have been extensively characterized with respect to chemotaxonomic properties. The diagnostic amino acid in position 3 of the peptide subunit of the peptidoglycan is ornithine with the interpeptride bridge containing either D-aspartic acid (as in *C. flavigena*) or D-glutamic acid (all other species) (Fiedler and Kandler, 1973); rhamnose is the dominant cell wall sugar in most strains (Stackebrandt and Kandler, 1979); branched 13-methyltetradecanoic (i-15) and 12-methyltetradecanoic (ai-15) and straight-chain pentadecanoic (C_{15}) are the dominant fatty acids (Minnikin et al., 1979; Suzuki and Komagata, 1983); phosphatidylglycerol (Lechevalier et al., 1981c), diphosphatidylglycerol and a phosphoglycolipid are the major polar lipids (Minnikin et al., 1979); menaquinones of the MK9 (H_4) type are the predominant isoprenoid quinones (Collins and Jones, 1981). Neither genus-specific nor species-specific phages have as yet been detected. The individual species can be differentiated by a set of characters which require determination of peptidoglycan, cell wall sugars, endproducts of glucose degradation, and the

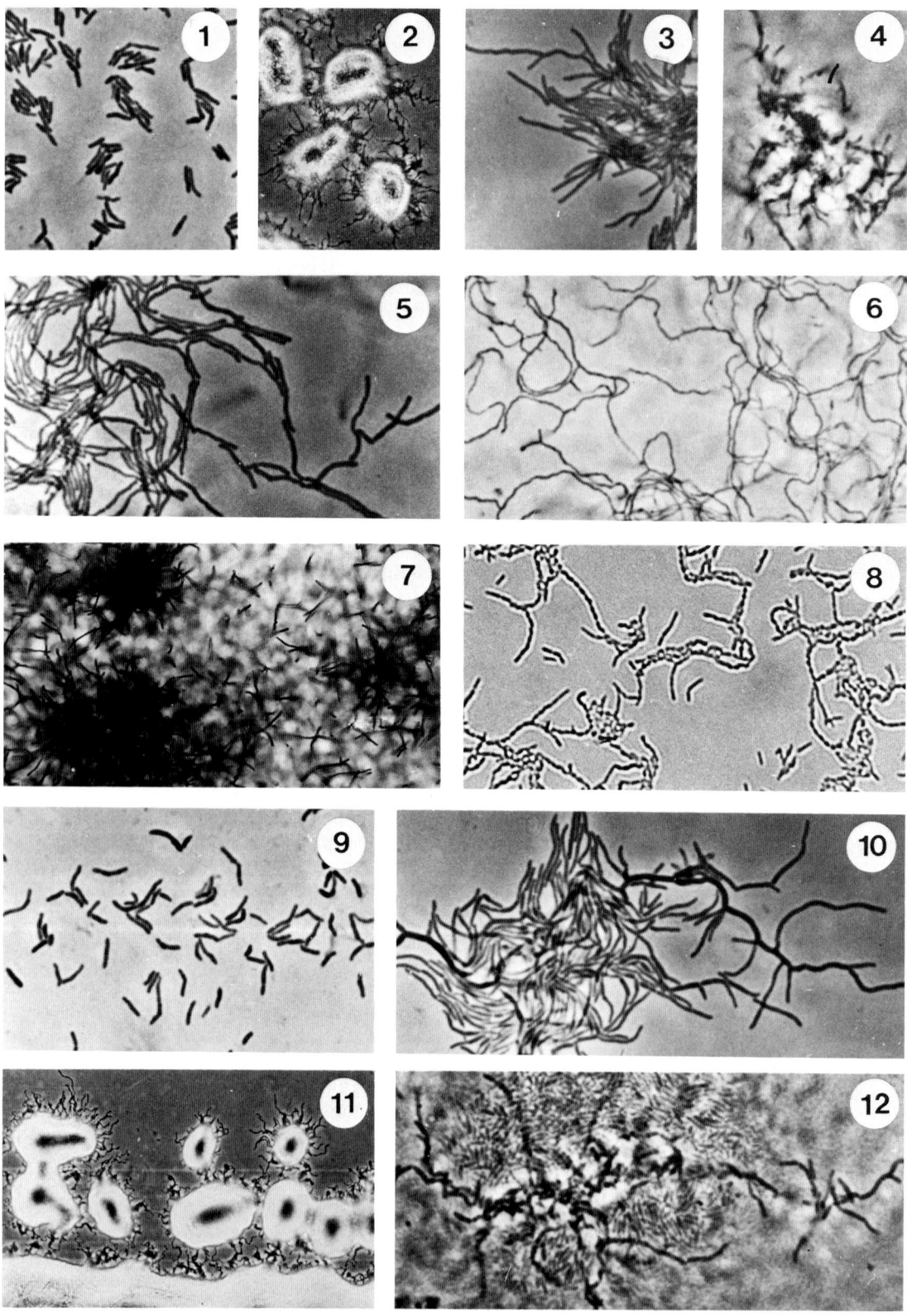

Fig. 1. *Cellulomonas flavigena* CCM 1926 (ATCC 482). Coryneform rods. 28 h slide culture on Bennett's saccharose agar. Phase contrast. Fig. 2–4. *Cellulomonas biazotea* NCIB 8077. Fig. 2. Colonies with nocardioform fringes. 28 h mannose glycerol agar plate. Direct phase contrast micrograph without coverslip. Fig. 3. Nocardioform microcolony. 28 h slide culture on Bennett's saccharose agar. Phase contrast. Fig. 4. Nocardioform fragmentation of hyphae penetrating into the agar. 28 h slide culture on Bennett's saccharose agar. Phase contrast. Fig. 5. *Oerskovia turbata* NCIB 10587. Nocardioform

utilization of certain sugars and acids (Stackebrandt and Kandler, 1979).

The GC content of the DNA has been determined in several laboratories using different methods. Values ranging between 71.0 and 76 mol% do not allow differentiation of species (cf. Stackebrandt and Kandler, 1979; Bagnara et al., 1985). DNA hybridization studies (Stackebrandt and Kandler, 1979; Bagnara et al., 1985; Prauser, 1986) were in most cases performed with a single strain of a species only. Data from different laboratories are in excellent agreement.

With binary DNA homology values of less than 60% all species can be considered genetically well defined (Stackebrandt and Kandler, 1979). Only the type species *C. flavigena* had been included in sequencing studies of 16S (Stackebrandt et al., 1980a) and 5S rRNA (Park et al., 1987). Both studies came to the same conclusion, indicating that cellulomonads are more closely related to arthrobacters and micrococci than to members of the genus *Corynebacterium*. The currently described species are not easy to differentiate by morphological and physiological properties. Compilations of biochemical reactions are given by Stackebrandt and Kandler (1979), Bagnara et al. (1985), and McHan and Cox (1987). The last authors propose a simple identification key for most species; for the characterization of new isolates, it is recommended that comparisons should be made to the type strains of the species which are deposited in various culture collections. The type strains are: *C. biazotea* (ATCC 486, DSM 20112, NCIB 8077); *C. flavigena* (ATCC 482, DSM 20109, NCIB 8073); *C. cellasea* (ATCC 487, DSM 20118, NCIB 8078); *C. fimi* (ATCC 484, DSM 20113, NCIB 8980); *C. gelida* (ATCC 488, DSM 20111, NCIB 8076); *C. fermentans* (DSM 3133); and *C. uda* (ATCC 491, DSM 20107, NCIB 8200).

Physiological Properties

All cellulomonads are able to grow under aerobic and microaerophilic conditions and *C. fermentans* (Bagnara et al., 1985) and *C. uda* ATCC 21399 (Dermoun et al., 1988) grow under strictly anaerobic conditions as well. Anaerobically, resting cells catabolize glucose to various end products. Radiorespirometric studies indicated that glucose is catabolized mainly via the Embden-Meyerhof (EM) pathway and, to some extent, through the hexose monophosphate (HMP) shunt (Stackebrandt and Kandler, 1979). This finding is supported by determination of respective key enzyme activities of these pathways in *C. flavigena* (Kim, 1987) and *C. uda* (Marschoun et al., 1987). As shown with *C. flavigena,* gluconate is catabolized via the Entner-Doudoroff (ED) pathway and HMP shunt (Kim, 1987). End products of aerobic glucose dissimilation are mainly CO_2 and either acetate or acetate and L-lactate (Stackebrandt and Kandler, 1980a). Under anaerobic conditions, resting cells produce mainly CO_2, acetate, lactate, and ethanol but succinate and formate are also formed. Similar products, although in varying amounts, were reported for *C. fermentans* (Bagnara et al., 1987) and *C. uda* grown under strictly anaerobic conditions (Marschoun et al., 1987).

Resting cultures of all *Cellulomonas* strains investigated show a highly effective symmetric interchange and an asymmetric redistribution of carbon atoms within the hexose molecule (Stackebrandt and Kandler, 1980a). Under energy excess conditions, *C. uda* accumulates glycogen and trehalose which in turn are degraded at different rates during carbon starvation (Schimz and Overhoff, 1987a, 1987b).

The most salient feature of cellulomonads is their ability to attack cellulose. Different kinds of cellulose fibers have been tested, including amorphous (Dermoun and Belaich, 1985; de Conick-Chosson, 1988), swollen, phosphoric-acid-treated (Kauri and Kushner, 1985), microcrystalline (Vladut-Talor et al., 1986; Dermoun and Belaich, 1988), and carboxymethyl cellulose (CMC). With one exception (Thayer et al., 1984), studies have been performed with a single *Cellulomonas* strain only. As reported by

---

fragmentation of hyphae on the agar surface. 28 h slide culture on Bennett's saccharose agar. Phase contrast. Fig. 6. *Promicromonospora citrea* RIA 562 (ATCC 15908). Mycelium growing on the agar surface and penetrating into the agar. 28 h slide culture on Bennett's saccharose agar. Phase contrast. Fig. 7 and 8. *Promicromonospora citrea* IMET 7536. Fig. 7. Aerial mycelium. 28 h Bennett's saccharose agar plate. Direct micrograph without coverslip. Fig. 8. Aerial hyphae fragmented into regular spore-like elements. Obtained by pressing a coverslip onto the aerial mycelium shown in Fig. 7, followed by the transfer of the slip onto a slide. Micrograph without adding water. Fig. 9–12. *Jonesia denitrificans* DSM 20603. Fig. 9. Coryneform rods from 28 h slant culture on Bennett's saccharose agar. Phase contrast. Fig. 10. Fragmenting nocardioform mycelium. 20 h slide culture on complex organic agar M 79. Phase contrast. Fig. 11. Colonies with nocardioform fringes. 24 h agar plate culture on R-medium. Direct phase contrast micrograph without coverslip. Fig. 12. Nocardioform fragmentation of hyphae penetrating into the agar. Blurred: rods resulting from fragmentation of hyphae on the agar surface. 28 h slide culture on Bennett's saccharose agar. Phase contrast. Magnifications: Figures 1, 3–6, 9, 10, 12: 1300 ×. Figures 2: 250 ×, 7: 380 ×, 8: 800 ×, 11: 160 ×.

Kauri and Kushner (1985), degradation of cellulose does not depend on cell-to-fiber contact, but cellulases from three *Cellulomonas* strains were active even when they were physically separated from the fibers. Microcalorimetric studies with aerobically grown *C. uda* ATCC 21399 revealed that microcrystalline cellulose (Avicel pH 101, Cellulose MN300, Whatman cc41) is less efficiently attacked (Dermoun et al., 1984) than amorphous cellulose (phosphoric-acid-treated Whatman cc41 cellulose) (Dermoun and Belaich, 1985). Under anaerobic conditions, celluloses with varying degrees of crystallinity were metabolized by *C. uda* ATCC 21399 with the same efficiency as by aerobically grown cells. However, as compared to the catabolic capacities obtained under aerobic conditions, the growth yield was markedly reduced (Dermoun et al., 1988). Optimal liquefaction of carboxymethyl cellulose (CMC) gels occurred/ in a synthetic medium at 40°C at pH of 7.0–7.5 (Thayer et al., 1984). According to their cellulolytic activities, cellulomonads could be divided into two distinct phena which correlated with results of serological (Braden and Thayer, 1976) and DNA hybridization studies (Stackebrandt and Kandler, 1979).

Glucose and cellobiose have a strongly inhibitory effect on cellulose breakdown. In strain NRCC 2406 1% glucose prevented decomposition of different kinds of celluloses (Solka floc, Avicel, CFII) (Vladut-Talor et al., 1986), and cellulases were repressed in all *Cellulomonas* strains tested by the addition of 0.5% glucose or cellobiose (Thayer et al., 1984). To reduce catabolic repression of cellulase formation, it was suggested that end products should be removed (Kauri and Kushner, 1985), but contradictory results were reported for the effect of glucose on the growth of cellulose-grown cells. While stimulation was found with strains cultivated on CMC (Thayer et al., 1984), strain NRCC 2406 showed a sharp decline in viable counts when cultivated in the presence of both Solka flocs and glucose.

The cellulase system has been studied mainly in *C. fimi* (Whittle et al., 1982; Gilkes et al., 1984; Langsford et al., 1984; Warren et al., 1987a; Greenberg et al., 1987a; Owolabi et al., 1988; Guo et al., 1988) and *C. uda* (Nakamura et al., 1986a, 1986b, 1986c; Han and Kim, 1987; Poulsen and Petersen, 1987). Supernatants from *C. fimi* contained up to 10 compounds with CM-cellulase activity, three of which showed a substrate-bound activity (Langsford et al., 1984). Cloned components of the cellulase system of a *C. fimi* strain in *Escherichia coli* revealed the presence of at least three classes of β-1,4-glucanases, β-1,4-endoglucanase (Eng, en-

coded by *cenA* genes) (EC 3.2.1.4), at least one β-1,4-exoglucanase (Exg, encoded by *cex* genes) (EC 3.2.1.91), and a strictly intracellular β-glucosidase (EC 3.2.1.21) which hydrolyzes cellobiose to glucose (Gilkes et al., 1984; Wakarchuk et al., 1984). The first two enzymes were characterized by molecular genetic methods, including expression and variation of the level, regulation, initiation, termination, and sequence analysis of *cenA* and, partly, or *cenB* (Greenberg et al., 1987b; Owolabi et al., 1988; Guo et al., 1988). Both enzymes are subject to regulation by the carbon source provided. The nucleotide sequence of a *cenA* gene of *C. fimi* has been determined. It is 1350-bp long and encodes a polypeptide of 449 amino acids. The *N*-terminal amino acids of the endoglucanase constitute a leader peptide which is functional in secretion of the enzyme to the periplasma of *E. coli*. The nucleotide sequence of the *cex* gene of *C. fimi* consists of 484 amino acids, the leader sequence is 41 amino acids long (O'Neill et al., 1986a) and allows secretion of the Eng protein into the periplasm (O'Neill et al., 1986b). The binding domains of cellulose of the *cex* and *cenA* genes have been determined (Gilkes et al., 1988). Warren et al. (1987b) were able to construct a fusion between the *cex* and the *cenA* gene of *C. fimi*. The chimeric gene was expressed in *E. coli* to give a product which exhibited both endo- and exoglucanase activities. The protein, which was exported into the periplasm, however, did not bind to microcrystalline cellulose. Molecular studies have been performed on a β-glucosidase (EC 3.2.1.21) from *C. uda* CB4 which were found to be excreted by *E. coli* C600 (Nakamura et al., 1986a, 1986b). The sequence of the endoglucanase (CMCase) has been determined (Nakamura et al., 1986c). For *C. uda* ATCC 21399, nine components with carboxymethyl cellulase activity were recovered, and one of them was purified. As already reported for *Cellulomonas* sp. IIbc (Beguin and Eisen, 1978) and *C. fimi* (Langsford et al., 1984), the cellulase (endoglucanase) is a glycoprotein (Han and Kim, 1987; Poulsen and Petersen, 1987). Heterologous hybridization with DNA fragments carrying *Clostridium thermocellum* endoglucanases A *(celA)* and B *(celB)* genes as probes revealed a weak but distinct signal between the *celA* probe and DNA from *C. uda* ATCC 20108. No homology was detected with the *celB* probe (Petre et al., 1985). Cellobiose phosphorylase (EC 2.4.1.20) has been identified in the soluble cytosol fraction of strain DSM 20108 (Schimz and Decker, 1985). A biochemical survey on the cellulase system of *C. fermentans* (Bagnara et al., 1986) indicates the presence of two endoglucanases (EC 3.2.1.4) and possibly an exglucanase (EC 3.2.1.91) as also reported for *C. fimi*. six different endoglucanases have been reported in a new *Cellulomonas* isolate (Prasertsan and Doelle, 1986).

Xylan-degrading enzymes, such as a cell-associated β-xylanase and a cell-bound β-xylosidase, have been characterized in *C. uda* (Rapp and Wagner, 1986). Substrate-bound xylanase activity has also been detected on sugarcane bagasse pith cultured with *Cellulomonas* sp. IIbc (Rodriguez et al. 1985). Amylase activities have been reported for *C. flavigena* (McCarthy and Pembroke, 1988).

Applications

Like other cellulose- and hemicellulose-degrading organisms, *Cellulomonas* strains have been considered potential candidates for waste disposal (Ramasamy et al., 1981; Dunlap and Callihan, 1974), and composting of bagasse (Richard and Peiris, 1981), pith, leaves of sugar canes (Diaz and Guirola, 1983; Richard and Peiris, 1981; Rajoka and Malik, 1986), dried palm oil mill effluent (Agamuthu and Tan, 1985), shredded newspapers (Rapp et al. 1984), or even as producers of chemicals from low-cost substrates.

Mixed cultures consisting of *Cellulomonas* sp. ATCC 21399, *Desulfovibrio vulgaris* strain J.J., and *Methanosarcina barkeri* 227 were highly efficient in converting xylan to methane via hydrolysis and acidogenesis (strain ATCC 21399), acetogenesis (strain J.J.), and methanogenesis (strain 227) (Guyot, 1986). The same *Cellulomonas* strain has also been used in mixed cultures with *Rhodopseudomonas capsulata* to photoevolve molecular hydrogen by the nitrogenase system of the phototrophic strain with cellulose as the sole carbon source (Odom and Wall, 1983).

The ability of cellulomonads to attack cellulose and wheat straw under microaerobic or even anaerobic conditions has been used in mixed cultures to provide nitrogen-fixing strains of *Bacillus macerans* and *Azospirillum brasilense* with energy-yielding products (Halsall and Gibson, 1985, 1986; Halsall and Goodchild, 1986). Good nitrogen-fixing rates have been reported for the pair *A. brasilense* ATCC 29145 and *Cellulomonas* strain CS117. The latter is a mutant strain selected for its increased production of cellulase and reduced sensitivity to inhibition or repression by accumulated cellobiose and glucose (Haggatt et al., 1978).

The level of expression by *E. coli* of a structural gene coding for an endonuclease from *C. fimi* was increased by replacing transcriptional and translational regulatory elements with those of the *E. coli lac* operon (Owolabi et al., 1988). Using an *E. coli/Brevibacterium lactofermentans* shuttle vector, Paradis et al. (1987) were able to demonstrate expression of high levels of *cex* and *cenA* genes from *C. fimi* in *B. lactofermentans*. Expression in the heterologous Gram-positive organism was five times higher than in *E. coli*.

The gene for *C. fimi* β-1,4-exoglucanase was successfully inserted into the galactosidase peptide gene of *Saccharomyces cerevisiae* 284 to produce a fused exoglucanase/galactosidase signal peptide. *S. cerevisiae* cells thus transformed produced active exonuclease which was able to hydrolyze xylan, CMC, 4-methylumbelliferyl-β-D-cellobiose, and *p*-nitrophenyl-β-D-cellobiose. As compared to the native *C. fimi* enzyme, the yeast-mediated version was more thermostable (Curry et al., 1988). These experiments are encouraging in that they indicate that purified redesigned cellulases as well as heterologous organisms harboring genes of the cellulase system from cellulomonads may play an increasingly important role in the above-mentioned fields.

Initial experiments leading to the development of a genetic system for cellulomonads include the formation of protoplasts of *C. flavigena* NCIB 12901 (Bae and Lee, 1986) and strain CS 11 (Lee and Bae, 1986) and protoplast fusion to obtain recombinants of strain CS 11 (Kim and Lee, 1985). The isolation of the restriction enzyme Cfl I from *C. flavigena* IFO 3753, cleaving the sequence 5′CTGCAG3′, has been reported (Hiraoka et al., 1985).

The Genus *Oerskovia*

The organisms allocated or to be allocated to the genus *Oerskovia* have been known for decades but were classified in different genera of coryneform and nocardioform bacteria. The genus *Oerskovia* itself has to be emended, and the misplaced strains are still in need of formalization of their nomenclatural status. Although this will not be done here, the underlying facts will be given.

The genus was united with *Cellulomonas* (Stackebrandt et al., 1982). The rational for this union was based on an overestimation of phylogenetic relationships between members of these two genera while, at the same time, neglecting the presence of genus-specific chemotaxonomic properties. Today, *Oerskovia* is considered to be well separated from, though closely related to, *Cellulomonas* (Prauser, 1986). Supporting arguments are DNA-DNA similarities between 20 and 26% (Prauser, 1986; Stackebrandt et al., 1980b), 16S rRNA similarity coefficients, (S_{AB}) of about 0.65 (Stackebrandt et al., 1988), lysine versus ornithine in the peptidoglycan of cellulomonads (Prauser, 1966; Fiedler and Kandler, 1973), unsusceptibility of cellulomonads to *Oerskovia* phages (Prauser, 1986), differences in the fatty acid patterns (Minnikin et al., 1979), and differences in the formation of mycelia.

Two species have been validly published: the type species *Oerskovia turbata* (Prauser et al., 1970) and *O. xanthineolytica* Lechevalier (1972). The species *"O. citrea"* was invalidly published in a patent (Nunokawa et al., 1975).

Motile nocardioforms were reported by Topping (1937), Ørskov (1938), Jensen (1953), and others. Erikson (1954) described "Oerskov's motile nocardia, strain 27" as *Nocardia turbata*. On the basis of a few shared physiological characters, Jones and Bradley (1964) suggested a close relationship between Ørskov's motile nocardia and *Cellulomonas biazotea*. Prauser (1967b) demonstrated phenetic similarities among 12 isolates from different soils and the type strain of *N. turbata*. However, *meso*-diaminopimelic acid, the diagnostic amino acid of the peptidoglycan of nocardiae, was not found in the *N. turbata* strains. Although not all strains were motile, in most of the strains, motility by flagellation of fragmented hyphal parts was observed, a property not otherwise found in true nocardiae. On the basis of lysine as a diagnostic constituent of the peptidoglycan and the motility of "spores" (Higgins et al., 1967), i.e., rodlike hyphal fragments (Sukapure et al., 1970), *N. turbata* was combined in a new genus as *Oerskovia turbata* (Prauser et al., 1970). The genus was emended (Lechevalier, 1972) due to the capability of oerskoviae to grow anaerobically on trypticase-soy agar as one of 15 media tested, to attack glucose oxidatively and fermentatively, and to be catalase-negative when grown under anaerobic conditions.

The paper introducing the second species, *O. xanthineolytica,* also contained information on "nonmotile *Oerskovia*-like strains" ("NMOs"). These organisms were obviously similar to the nonmotile and motile strains of Prauser (1967b) as demonstrated among other criteria by their susceptibility to *Oerskovia* phages (Prauser, 1984b, 1986; Prauser and Falta, 1968). The definition of the genus *Oerskovia* to include only motile organisms turned out to be a drawback in subsequent taxonomic studies. As a result, the misplaced nonmotile strains constituting the species *Nocardia cellulans* (Metcalf and Brown, 1957), *"Corynebacterium manihot"* (Collard, 1963), *Brevibacterium lyticum* (Takayama et al., 1960), *Brevibacterium fermentans* (Chatelain and Second, 1976), *"Arthrobacter lu-*

teus" (Kaneko et al., 1969), *Cellulomonas cartae* (Stackebrandt and Kandler., 1980), as well as the NMOs (Lechevalier, 1972) and *Promicromonospora enterophila* (Jager et al., 1983) are still in need of their formal inclusion in the genus *Oerskovia* which is to be amended to include nonmotile organisms, too. Most probably, the organisms mentioned, except for *P. enterophila,* are members of *O. xanthineolytica.* The present classificatory and nomenclatural situation of *O. xanthineolytica* needs some comment. Stackebrandt et al. (1982) combined the type species of *Oerskovia*, *O. turbata,* as *Cellulomonas turbata,* i.e. they united *Cellulomonas* and *Oerskovia*. As a result, *O. xanthineolytica* lost its home. Previously at the occasion of describing *C. cartae* (ex *"C. cartalyticum"*), Stackebrandt and Kandler (1980) proposed *O. xanthineolytica* and *N. cellulans* among the other species mentioned above as subjective synonyms of *C. cartae.* Subsequently Stackebrandt and Keddie (1986) formally combined *N. cellulans* as *C. cellulans* and consequently included the strains hitherto known as *O. xanthineolytica* in *C. cellulans.* Since the union of *Cellulomonas* and *Oerskovia* is no longer maintained (Prauser, 1986; the present paper), *O. xanthineolytica* is at present the appropriate home for all of the synonymous species, but needs to be renamed *O. cellulans* for the reason of priority, as soon as possible. In order to avoid confusion in the further paragraphs of this article the intermediate combination of *C. cellulans* will be neglected here, i.e., the respective organisms will be presented by their original names. The organisms were studied in various combinations and subjected to various methods. Hence, the evidence of identity or close relationship, as suggested by Bousfield (1972), Jones (1975), Keddie and Cure (1977), Collins et al. (1979a), Minnikin et al. (1979), Stackebrandt et al. (1980b), Stackebrandt and Kandler (1980), and Prauser (1986) will be reviewed in a complex and somewhat simplified way below.

Those organisms that are identical or close to *O. xanthineolytica* share the peptidoglycan type L-Lys-D-Ser-β-D-Asp (Seidl et al., 1980; for additional references, see below). They are susceptible to *Oerskovia* phages, particularly to those which are active against *O. xanthineolytica* (Prauser, 1984b, 1986). Numerical taxonomic studies revealed high phenetic similarity (Bousfield, 1972; Jones, 1975; Seiler, 1983; Seiler et al., 1977). DNA-DNA similarities obtained for *B. fermentans, C. cellulans, "Corynebacterium manihot,"* and *O. xanthineolytica* ranged from 60 to 68 mol% (Prauser, 1986; Stackebrandt et al., 1980b), indicating relationships nearly at the species level. Comparative 16S rRNA cataloging revealed a high homology level (0.93b S_{AB}) for *Nocardia cellulans* and *C. cartae* (Stackebrandt et al., 1980b). *C. cartae* was transferred to N. cellulans because of high 16S rRNA similarities (Stackebrandt and Woese, 1981), and *N. cellulans* was transferred to *Cellulomonas* as *C. cellulans.*

On the basis of the occurrence of spores, nonmotile strains isolated from feces of a millipede were placed in the genus *Promicromonospora* as *P. enterophila* (Jáger et al., 1983). However, the spores, which were observed in only 6 of 54 strains of the new species were killed by heating at 80°C for 5 min (Jáger et al., 1983) and could not be detected in the type strain (H. Prauser, unpublished observations). Since Lechevalier (1972), Kuimova et al. (1983), Prauser (1986), and Takahashi et al. (1987) were unable to detect spores even in *Promicromonospora citrea*, the occurrence of spores is very doubtful in both species. With respect to the physiological characters (Jager et al., 1983), *P. enterophila* is closer to oerskoviae than to *P. citrea*. The exclusive susceptibility of *P. enterophila* strains to *Oerskovia* phages showed them to be definitely members of this genus (Prauser, 1986). Since strains of *P. enterophila* and *O. turbata* share not only the susceptibility to species-specific *O. turbata* phages (Prauser, 1986) but also the lack of hydrolysis of xanthine and the presence of lysine and threonine as diagnostic peptidoglycan constituents, strains of these two species differ only in motility, a character of doubtful taxonomic value. Thus, strains of *P. enterophila* should be classified in future as nonmotile strains of *O. turbata*.

Habitat

Oerskoviae were randomly and rarely isolated from various types of soils from different geographical regions, composts, brewery sewage (Kaneko et al., 1969), decaying plant materials, aluminum hydroxide gel antacid (Lechevalier, 1972), and occasionally from various clinical specimens (Cruickshank et al., 1979; Reller et al., 1975; Sottnek et al., 1977).

Results from the group of I. M. Szabó throw new light on the ecology of oerskoviae. Strains of the misplaced species *P. enterophila* constitute the major part of the actinomycete microflora of the intestines and feces of litter-inhabiting millipedes, e.g., *Chromatoiulus projectus* (Dzingov et al., 1982; Jager et al., 1983; Szabó et al., 1983, 1986). The species occurs also in the feces of the cave-inhabiting blind isopode *Mesoniscus graniger* (Bodnar et al., 1989). A large homogenous population of facultatively anaerobic *Oerskovia*-type nocardioforms was also found in the gut contents of adult specimens of the common earthworm *Lumbricus polyphemus* (Ravasz et al., 1987; Szabó et al., 1986). None of these oerskoviae could be isolated from the surrounding feeding habitats of the animals. Moreover, the inability of the gut and feces nocardioforma to survive in the natural soil and litter habitat could be demonstrated (Márialigeti et al., 1985). Untreated and sterilized samples were inoculated with suspensions of isolated oerskoviae from the feces. They disappeared within 14 days from the untreated samples but could be reisolated from the reinfected sterilized soil over a one-year period at continuously increasing rates. It might be speculated, that phages are involved in the disappearance of these nocardioforms. This can be deduced from the finding that it is easier to isolate *Oerskovia* phages from soils and composts than the organisms themselves (Prauser, 1976).

Isolation

Selective media have not been reported for the isolation of strains of the genus. Oerskoviae grow on a wide range of media as do many other soil bacteria. The main problem in their isolation is the exclusion of organisms that cover large areas of the isolation plates, e.g. swarming bacilli, pseudomonads, and hyphal fungi, as well as the suppression of the numerous streptomycetes, which may be confused at first sight with young stages on the isolation plates and which may possibly antagonize the oerskoviae. The procedures and media discussed in the section on the genus *Promicromonospora* below, can also be used for *Oerskovia*. The possible use of tap water agar (1.5% crude agar in tap water; Lechevalier and Lechevalier, 1989) may also be used to isolate oerskoviae.

Identification

PHAGE TYPING. Members of the genus *Oerskovia,* including all nonmotile species and the NMOs of Lechevalier (1972) discussed above, are most easily identified by their susceptibility to at least one phage of a genus-specific set of phages (Prauser, 1984b, 1986; Prauser and Falta, 1968). The set consists at present of 5, out of 13 available phages (02 = IMET 5022 [IMET = Institute for Microbiology and Experimental Therapy, Jena], 03 = IMET 5026, 05 = IMET 5025, 06 = IMET 5024, 013 = IMET 5027; Prauser et al., 1987). Besides true lytic effects, the occurrence of "clearing effects" only, i.e., phage-induced lysis without phage propagation, was reported for two strains (Prauser, 1986). In general, such effects were also shown to indicate affiliation of a strain under study to the respective genus (Prauser, 1981b). The phenomenon of a clearing effect caused by the *Oerskovia* phage 02 on the type strain of *O. xanthineolytica* has been reported by Klein et al. (1981). Recently isolated phages, which cause true lysis with the strains hitherto affected by clearing effects only, will probably complete the set of *Oerskovia* phages subsequently to comprehensive studies of their host ranges.

LIFE CYCLE, MORPHOLOGICAL, AND COLONY CHARACTERS. Oerskoviae are typical nocardioforms but may show a coryneform appearance depending on the particular strain, the age of the culture, and the external growth conditions. During the exponential growth phase of the type strain of *O. xanthineolytica,* five morphological types, ranging from true mycelia to irregular coryneform rods, may occur, the actual type depending on the influence of external factors such as the kind of the culture (liquid, chemostat, agar), the degree of aeration, and the kind of carbon and nitrogen sources (Kretschmer, 1981). The morphological variability together with certain difficulties in demonstrating motility have contributed to the problems in the taxonomy of oerskoviae.

In general, on agar media, hyphae develop that branch and grow on the surface or penetrate into the agar to form an extensive mycelium. Beginning in the oldest parts of the mycelium, the hyphae fragment into sections of different size and more or less irregular shape (Fig. 5). Fragmentation may continue to yield rodlike cells of coryneform appearance and arrangement. During the process of fragmentation, flagella may develop and motility may be seen simultaneously in a more or less large fraction of the cells. In Perfilieff capillaries, active motion of large mycelial parts has been observed (H. Prauser, unpublished observations). In order to facilitate the observation of motility or to state its definitive absence, agar plates are inoculated by streaking the strain under study onto the surface using an old inoculation loop to produce scratches in the agar surface. After overnight incubation, single colonies are wetted with a loopful of buffer and covered by a small cover slip (0.5 × 0.5 cm). In positive cases, motile cells will be seen by phase contrast microscopy in the scratches near the colonies.

Different kinds of flagellation have been reported, i.e., subpolar tufts of one to three flagella (Higgins et al., 1967) and monotrichous flagella (for short cellular elements) and peritrichous flagella (for longer cells) (Sukapure et al., 1970). All elements, motile and nonmotile, resulting from fragmentation may give rise to new my-

celia independently of their size. Aerial mycelium and single microscopical aerial hyphae are lacking.

Colonies are lemon-yellow to whitish. Their consistency is smooth and the surface is glistening with a tendency to dull. Edges show mycelial or at least hyphal character, resembling those of other nocardioforms. Similar effects have also been reported for two *Cellulomonas* species (Prauser, 1986).

Physiological Properties

Oerskoviae are aerobic to facultatively anaerobic and produce acid from a variety of carbohydrates (Stackebrandt and Kandler, 1980a; Lechevalier and Lechevalier, 1981b). More extensive studies on glucose metabolism have only been done with *C. cartae* and *N. cellulans.* Acetic acid is the main acidic intermediary product of aerobic glucose dissimilation by *C. cartae,* acidifying peptone-yeast extract-glucose medium to pH 4.8 within 48–72 h. L-Lactic acid is produced in much smaller amounts. After exhaustion of glucose, the acids are then oxidized, resulting in an increase of the pH of the medium (to pH 8.8 after 8 days). Resting cells ferment glucose to CO_2, acetic acid, and L-lactic acid (Stackebrandt and Kandler, 1980a). The electrophoretic mobility of the L-lactate dehydrogenase (LDH) is 1.33 (using rabbit muscle L-LDH isoenzyme I as a standard of 1.0). The motility of the L-LDH is therefore significantly higher than those of cellulomonads, which ranged between 1.04 and 1.07 (Stackebrandt and Kandler, 1979). Determination of enzyme activities in *N. cellulans* indicates that glucose and gluconate (after reduction to glucose) are both metabolized via the EM pathway and the HMP shunt (Kim, 1987).

N. cellulans and NMOs of type A and B possess cellulolytic activities but, in contrast to the well-characterized cellulose-degrading system in *Cellulomonas* spp., almost no information is available for oerskoviae. In *C. cartae,* cellobiose is hydrolyzed by a β-glucosidase (EC 3.2.1.21) to give two molecules of glucose. The mechanism is different from the phosphorolytic cleavage of cellobiose of genuine cellulomonads, where one mole of α-glucose-1-phosphate and one mole of glucose is obtained (Schimz et al. 1983).

Oerskoviae, including NMOs, are producers of a variety of extracellular enzymes including chitinase (Mann et al., 1978), dextranase (Hayward and Sly, 1976), keratinase (Goodfellow, 1971), and hydrolases which degrade the walls of both live and dead yeast cells (Obata et al., 1977a, 1977b; Jeffries et al., 1977; Mann et al.,

1978; Scott and Schekman, 1980; Jeffries and Macmillan, 1981). In *O. xanthineolytica,* four different β-(1,3)-glucanases were identified which showed distinct action patterns in batch cultures against walls of viable yeast (Jeffries and Macmillan, 1981). Enzyme I was identified as an endoglucanase (EC 3.2.1.6) which displayed high activity against laminarin and lichenan but not against viable yeast cells; enzyme II was also an endoglucanase (EC 3.2.1.29) active against both viable yeast cells and laminarin but with restricted activity against lichenan. The other two enzymes (IIIa and b) were identified as β-D-glucan-pentaose hydrolases which were more active on yeast glucan than on laminarin. One component of yeast glucan hydrolases showed proteolytic activities (Obata et al., 1977a, 1977b) and, as reported by Scott and Schekman (1980), complete lysis of yeast cells required the synergistic enzyme activities of a "lyticase," consisting of glucanases, and an alkaline protease. Synthesis and regulation of these enzymes isolated from *O. xanthineolytica* LLG109 grown in continuous culture has been described by Andrews and Asenjo (1987). This system is induced by yeast glucan and subject to catabolite repression by glucose.

Chemotaxonomic Characters

Lysine is the main diagnostic amino acid of the peptidoglycan (Higgins et al., 1967; Prauser, 1966), corresponding to cell wall type VI (Lechevalier and Lechevalier, 1981a). More precise analysis revealed two cell wall types for *O. turbata,* namely L-Lys-L-Thr-β-D-Asp and L-Lys-L-Thr-β-D-Glu, while *O. xanthineolytica* exhibits the type L-Lys-D-Ser-β-D-Asp (Seidl et al., 1980). The latter type was also shown to be present in those oerskoviae which were erroneously classified in coryneform genera, i.e., in *"Corynebacterium manihot"* (Fiedler et al., 1970), *C. cartalyticum* (Stackebrandt et al., 1978), *B. fermentans,* and *N. cellulans* (Stackebrandt et al., 1980).

Fatty acid patterns of both *Oerskovia* species are similar and correspond to the branched chain iso(i) and anteiso(ai) type with a clear preference for ai-15 followed by i-15 and with the almost complete absence of straight-chain fatty acids. Cellulomonads, on the other hand exhibit a rather high amount of the straight-chain fatty acids (Minnikin et al., 1979). Concerning the phospholipids, the results of Komura et al. (1975) and Lechevalier et al. (1977) correspond well, if it is conceded that phosphatidyl-inositol mannosides could not be completely separated from unknown glucosamine-containing phospholipids. According to the

classification of Lechevalier et al. (1977), oerskoviae, as well as cellulomonads (Lechevalier et al., 1981c) and promicromonosporae, contain type PV phospholipids. The type is characterized by the occurrence of unknown glucosamine-containing phospholipids and phosphatidyl glycerol and by the lack of phosphatidyl choline. Phosphatidyl inositol and diphosphatidyl glycerol, widely distributed phospholipids, were also present. Mycolic acids are lacking (Aristarkova et al., 1981; Mordarska et al., 1972). Tetrahydromenaquinones with nine isoprenoid units—MK-9(H$_4$)—predominate. Menaquinones of the types MK-9 (H$_2$), MK-8 (H$_4$), and MK-7 are minor components (Collins et al., 1977, 1979a; Yamada et al., 1976). Teichoic acids are lacking (Evtushenko et al., 1984a). The cytochromes belong to the *a, b,* and *c* types (Seidl et al., 1980).

The GC content of the DNA of oerskoviae as defined in this chapter ranges between 70 to 75 mol% (Tm) (Lechevalier, 1972; Jones and Bradley 1964; Prauser, 1966).

Oerskoviae as Pathogens

The pathogenic potential of oerskoviae was first described by Reller et al. (1975). Strains identified as *O. turbata* were frequently isolated from the blood taken from a patient suffering from endocarditis after homograft replacement of the aortic valve. The source of *O. turbata* remained speculative although contamination of harvested heart valves with *O. turbata* have been reported (Reller et al., 1975). The second documented instance of opportunistic infection of a human by a member of *Oerskovia* was reported by Cruikshank et al. (1979). In this case, a nonmotile strain of *Oerskovia* was found to cause pyonephritis. The occurrence of oerskoviae in various clinical sources was documented (Sottnek et al., 1977) when a large number of motile, Gram-positive, nonsporeforming, yellow-pigmented organisms could be allocated to the two described *Oerskovia* species. Thirty-five of 57 clinical isolates, collected by the Bacteriology Division, Centers for Disease Control (CDC) over a period of 20 years, were identified as either *O. turbata* (9 strains), or *O. xanthineolytica* (26 strains). The source of the *O. turbata* isolates was heart tissues, heart valves, blood, and tissues, while the sources of the *O. xanthineolytica* isolates varied more widely, including blood, urine, sputum, cerebrospinal fluid, and wounds. Since no satisfactory case histories for any of the isolates was available, their clinical significance could not be elucidated.

Two cases have also been reported in which *O. xanthineolytica* was described as the causative agent for endophthalmitis (Hussain et al., 1987) and meningitis (Kailath et al., 1988). As in previous cases, the clinical significance was beyond doubt but the source of infection remained speculative. Information on infections with oerskoviae in animals are even rare.

Applications

The ability of *O. xanthineolytica* to produce inducible extracellular enzymes which rapidly degrade walls of various yeasts may be of biotechnological significance (Macmillan et al., 1974), such as release of single-cell protein from yeast or cellulolytic fungi (Mann et al., 1978) or production of heterologous proteins expressed by transformed yeast cells. Yeast spheroplasts are a prerequisite for molecular genetic studies, and under appropriate osmotic conditions can be produced with the *O. xanthineolytica* enzyme. In the absence of osmotic support and with sodium dodecyl sulfate as detergent, yeast cells lyse, and the cell contents are gently released, allowing the isolation of undegraded high-molecular-weight nucleic acids (De Kloet, 1984).

The Genus *Promicromonospora*

The genus *Promicromonospora* encompasses two species, the type species *P. citrea* (Krassilnikov et al., 1961) and *P. sukumoe* (Takahashi et al., 1987), which is represented by a single strain only. The recognized differences between the two species are of minor taxonomic importance, particularly since *P. citrea* was shown by numerical taxonomic studies (Kalakoutskii et al., 1989; N. S. Agre, personal communication) and by DNA hybridization studies (Prauser, 1986; H. Prauser, unpublished observations) to be highly inhomogenic. Neither the DNA hybridization cluster nor the phenetic cluster could as yet be defined by descriptive taxonomic characters. Thus, possible synonymy of *P. sukumoe* to *P. citrea* should be taken into consideration. A third species, *P. enterophila* (Jáger et al., 1983), was shown to actually be a member of the genus *Oerskovia* (see above).

The genus *Promicromonospora* was established to harbor an organism that was understood to connect the genera *"Proactinomyces" (Nocardia)* and *Micromonospora* from a genealogical point of view (Krassilnikov et al., 1961). However, the presence of lysine as the diagnostic diamino acid of the peptidoglycan (Yamaguchi, 1965), fatty acids of the branched chain iso/anteiso type (Andreyev et al., 1983), and menaquinone of the MK-9(H$_4$) type,

(Collins and Jones, 1981) excluded a close relationship to either genus. Lechevalier and Lechevalier (1981b) included the genus among those "in search of a family." Based on comparative 16S rRNA cataloging, *Promicromonospora citrea* clustered with the genera *Oerskovia, Cellulomonas, Micrococcus,* and *Arthrobacter* (Stackebrandt et al., 1983). Consequently, Stackebrandt and Schleifer (1984) placed *Promicromonospora* together with these other genera in the tentatively described family Arthrobacteraceae (Fox and Stackebrandt, 1987). Within this family, *Promicromonospora* occupied an isolated position. Results of numerical phenetic studies reflected the same situation (Alderson and Amadi, 1986). In a subsequent phylogenetic study, the misclassified organism *Listeria denitrificans,* consequently described in a new genus as *Jonesia denitrificans,* was found to be the nearest neighbor of *P. citrea* (Rocourt et al., 1987a). The 16S rRNA cataloging similarity (S$_{AB}$) for this pair of species was 0.65, whereas the pair joined the *Cellulomonas-Oerskovia* group, as well as arthrobacteria, micrococci, and *Dermatophilus congolense* at a level of S$_{AB}$ values of 0.60 (Rocourt et al., 1987a; Stackebrandt et al., 1988). *Promicromonospora* shares about 20% DNA hybridization values with oerskoviae and cellulomonads, but less than 15% with arthrobacteria and micrococci (Prauser, 1986; H. Prauser, unpublished observations). Based upon the RNA and DNA studies, as well as upon chemotaxonomic data, the genus *Promicromonospora* appears to be a true member of the family Cellulomonadaceae.

Habitat

Strains of promicromonosporae have been isolated from different types of soils collected from several geographical localities. Strains could also be obtained from compost and from aluminum hydroxide gel antacid (Lechevalier, 1972). Lacey (1988) refers to the isolation of these organisms from humidifiers in hot-air central heating systems.

Since promicromonosporae do not require special growth conditions, the small number of known isolates seems strange in view of their wide distribution. Possibly the fraction of these organisms in microbial ecosystems is really small, or the actual niche of promicromonosporae has as yet not been elucidated. On the other hand, these organisms may have frequently been neglected because they lack striking characters.

Interestingly, five of the six known genus-specific phages have been isolated from compost, though most of the samples surveyed for the isolation of *Promicromonospora* phages were from soil and mud (Prauser et al., 1987; H. Prauser, unpublished observations). Disregarding the improbable possibility that these phages might propagate in hitherto unknown hosts in the respective ecosystems, this observation indicates that promicromonosporae are most frequently distributed in substrates which undergo intensive degradation of a variety of organic matter.

Isolation

Selective isolation methods have not been reported for promicromonosporae. The strains were isolated by chance during more general isolation programs. Thus, various media and methods were used in the isolation. Since promicromonosporae (Alferova et al. 1989) as well as most of the coryneform bacteria (Wakisaka et al., 1982) are not susceptible to nalidixic acid and cycloheximide, these compounds can be successfully used to suppress most of the contaminating bacilli and hyphal fungi, which both spread quickly on the isolation plates. Concentrations of 20 to 50 mg/l of nalidixic acid and 50 mg/1 of cycloheximide, respectively, are recommended. The initial solution of nalidixic acid (small amounts of sodium hydroxide help to make an aqueous solution) should be sterilized by filtration prior to the addition to the molten agar media for the basal as well as for the upper layer of double-layer agar plates. Cycloheximide withstands sterilization in the autoclave. Vitamins and yeast extract supported growth of some strains on synthetic media (Kalakoutskii et al., 1989).

In our experience, the use of double-layer agar plates is superior to the direct spreading of suspensions of soil samples onto the agar surface. For double-layer agar plates, drying of the basal layers at ca. 37°C is essential. Prior to a large-scale isolation study of a given sample, a test should be performed to define the most appropriate dilution among the suspected successful ones (usually three serial dilutions at steps of 10^{-1} according to previous experience). Tests should also be done to determine the necessary measures to reduce the number and growth of contaminant bacteria and fungi. To the molten agar medium for the upper layer (6 ml for petri dishes, 100 mm in diameter), 1 ml of the chosen suspension and the calculated amount of the nalidixic acid stock solution are added at a temperature of 48°C. The procedure must be performed quickly in order to distribute both additives homogenously in the medium and to avoid its solidification prior to pouring it onto the basal layer.

The following media were successfully used for the isolation of these actinomycetes in our laboratory:

Complex Organic Medium (Medium M 79)
(Prauser and Falta, 1968)

Dextrose	10 g
Peptone	10 g
Casamino acids (Difco)	2 g
Yeast extract	2 g
NaCl	6 g
Distilled water	1 liter
Agar	15 g
Adjust to pH 7.0	

Soy-Yeast-Soil Extract Agar (M 137)

Tryptic soy broth (Difco)	1.0 g
Yeast extract	1.0 g
Soil extract	0.2 liter
Distilled water	0.8 liter
Agar	15.0 g
Adjust to pH 7.0	

Recipe for soil extract

Soil	1000 g
Distilled water	1000 ml
Autoclave for 20 min.	
Add 10 g Ca CO_3, mix, and filter.	

Addition of nalidixic acid and/or cycloheximide at concentrations appropriate for the particular sample (see above) is recommended for both media. For the upper layers of double-layer agar plates, the concentration of agar is 0.6 % to 0.8% depending on the gelling properties of the agar used.

Since these media differ fundamentally in their composition, it is obvious that many other media may be used, as well. Unfortunately, agar M 137 is also highly suitable for the isolation of sporoactinomycetes, particularly for the frequent and widely distributed streptomycetes. The plates should therefore be inspected by use of a stereomicroscope at low magnification (5 to 20×). The observer will quickly learn to identify streptomycetes and other sporoactinomycetes and to exclude them from isolation. The isolation of promicromonosporae and other actinomycetes is accomplished by use of a sharp inoculation needle (kept sharp by cutting it occasionally at an angle with scissors) under a stereomicroscope at low magnification.

Cultivation

Promicromonosporae may be cultivated at about 28° to 30°C on various media in the form of agar slants or of submerged culture. Most suitable are media like the complex organic medium M 79, given above for isolation; however, antagonizing substances should be omitted.

Other suitable media are the Rich (R) medium (See below) (Yamada and Komagata, 1972), peptone-corn extract agar (Agre, 1964), and oatmeal agar (Shirling and Gottlieb, 1966). For agar slant cultures, serial transfers at three-month intervals are recommended. Submerged cultures on rotary shakers develop within one to two days, agar slant cultures within two to three days.

Rich (R) Medium (Yamada and Komagata, 1972)

Bacto-peptone (Difco)	10 g
Yeast extract (Difco)	5 g
Casamino acids (Difco)	5 g
Beef extract (Difco)	2 g
Malt extract (Difco)	5 g
Glycerol	2 g
MgSO$_4$ · 7H$_2$O	1 g
Tween 80	0.05 g
Distilled water	1 liter
Bacto agar (Difco)	20 g
Adjust to pH 7.2.	

Preservation

Freeze-dried samples can be obtained by suspending the growth of an agar slant culture in 2 ml of medium M 79, transferring 0.2 ml of this suspension into 0.5 ml skim milk in a 2-ml ampoule, followed by one of the conventional freeze-drying procedures. For preservation in or above liquid nitrogen, suspensions of shaken submerged cultures are mixed with dimethyl sulfoxide (DMSO) at a final concentration of 5%. Aliquots of 0.5 to 0.8 ml are placed in 2-ml ampoules, which are directly transferred into vessels containing the liquid nitrogen. Using the method of Prauser (1984a), serial removals within a few days and even shifts at temperatures above 20°C are possible without loss of viability. In this case, minicultures (0.5 ml of appropriate agar medium in cotton-plugged tubes, ca. 8 × 80 mm, inoculated by use of an inoculation loop) are cultivated until sufficient growth occurs. The protruding part of the cotton plug is then cut off, and the top of the tube dipped into molten paraffin. In our experience, these minicultures may be kept alive with liquid nitrogen for more than 10 years.

Identification

PHAGE TYPING. At the genus level, promicromonosporae may be identified most quickly and easily by phage typing. After the first report on the activity of taxon-specific phages was published (Prauser, 1967a), Prauser and Falta (1968) described the study of phage-host ranges as an effective tool for stimulating and supporting taxonomic studies in the field of nocardioforms and sporoactinomycetes. In addition to true lysis, the taxonomic implication of so-called clearing effects, i.e., phage-dependent lysis in lawns of bacteria without propagation of the phage, could also be shown (Prauser, 1981b). Reviews on phage typing, including the applied methods, were given by Prauser (1976, 1984b). At least one of five *Promicromonospora* phages, preferably phage IMET 5030, caused true lysis of all of the 26 strains of *P. citrea* obtained from M. P. Lechevalier or N. S. Agre, or isolated in our own laboratory (Prauser, 1986). In order to obtain reliable identification of promicromonosporae, a set of three phages (P2 = IMET 5029, P3 = IMET 5030, and P4 = IMET 5031) was selected (Prauser et al., 1987). The type strain of *P. sukumoe* was also found susceptible to two phages of this set (H. Prauser, unpublished observations). The taxon specificity of phage IMET 5030 was studied and confirmed by Williams et al. (1980). *Promicromonospora* phages do not attack *Jonesia denitrificans,* oerskoviae, or cellulomonads. Vice versa, promicromonosporae are not susceptible to any of the available phages that are specific for strains of other actinomycete genera.

LIFE CYCLE AND MORPHOLOGICAL AND COLONY CHARACTERS. Promicromonosporae display a nocardioform life cycle which most closely resembles the cycles known for the genera *Nocardioides* and *Nocardia.* The particular life cycle, the implicated morphological characters, and the resulting appearance of the colonies depend primarily on the individual strains (Kuimova et al., 1983) and, to some degree, on the composition of the growth medium, temperature, and age of the culture. The organisms grow between 10 and 42°C and have their optimal growth temperature at about 30°C. In general, the irregularly septated hyphae, 0.5–1.0 µm in diameter, of an extended substrate mycelium grow readily on the surface and penetrate into the agar (Fig. 6). They subsequently fragment into nonmotile elements of different size and shape. On media rich in organic nitrogen, the colonies tend to a "bacteroid" habit (Kalakoutskii et al., 1989) while on poor media, they exhibit a leathery appearance. Fragmentation is promoted by higher temperature and retarded at lower temperatures. Colonies are smooth and shiny to dull and wrinkled. Depending on the particular strains, they are intensely lemon-yellow, yellowish, or whitish, displaying all intermediate shadings. Exposure to daylight intensifies the yellow color.

Hyphal fragmentation results in the formation of irregularly curved, slender sections of different length, in more or less regular rods of coryneform appearance, including V- and Y-forms, and even of coccoid cells. All of these nonmotile elements represent stages of the life cycle and may give rise to new mycelia. Lateral and terminal sessile of short-stalked single spores (Krassilnikov et al., 1961; Luedemann, 1974) or spore-shaped elements (Evtushenko et al., 1982; Kuimova et al., 1983); Kalakoutskii et al., 1989) were reported. Lechevalier (1972), Kuimova et al. (1983), Prauser (1986), and Takahashi et al. (1987) could not confirm the presence of spores. Evtushenko et al. (1982) described nine morphological types as part of the life cycle as well as the polymorphism of *P. citrea,* including the morphotypes in submerged culture. Since structures like the chlamydospores and giant cells described above also are found in other sporoactinomycetes, nocardioforms, and coryneforms, and since no biological functions could be hitherto attributed to them, they may be provisionally interpreted as involution forms. The ultrastructural characters of the promicromonospora (Evtushenko et al., 1982) completely resemble those of *Nocardioides albus* (Prauser, 1981a).

On several media—such as peptone-corn extract agar (Agre, 1964), oatmeal agar (Shirling and Gottlieb, 1966), Bennett's saccharose agar, and some synthetic media, e.g., inorganic salts-starch agar (Shirling and Gottlieb, 1966)—aerial mycelium may be formed within 1 to 5 days (Fig. 7). In general, it is discernible solely by microscopic inspection, particularly with strains continuously propagated by serial transfers. The aerial mycelium rarely covers the whole colony as a whitish mat; it may occur only in patches or at the edges of the colonies. Originally, the curved or irregular, occasionally branched aerial hyphae were recorded to be sterile (Krassilnikov et al., 1961; Evtushenko et al., 1982). However, using a simple technique, fragmentation of these hyphae could be demonstrated. By gently pressing a dry coverslip onto aerial mycelia, the fragmented hyphae adhere to the slip. The slip will be either transferred directly onto a slide or onto a film of agar spread on a slide. Observation will be performed by bright field or by phase contrast microscopy, respectively. In most cases chains of hyphal fragments, rods or nearly coccoid spore-like cells will be seen depending on the developmental stage of the aerial mycelium (Fig. 8) (Prauser, 1986).

Physiological Characters

Promicromonosporae are Gram-positive, non-acid-fast, catalase-positive, and urease-positive. They reduce nitrate, are aerobic, and metabolize glucose oxidatively. For a few strains, very slow anaerobic growth was observed on trypticase-soy agar. Many carbon and nitrogen sources are utilized. As the sole carbon source L-arabinose, D-galactose, cellobiose, D-fructose, D-maltose, mannose, raffinose, sucrose, trehalose, and D-xylose were found for the vast majority of the *P. citrea* strains tested (Kalakoutskii et al., 1989). Esculin, casein, gelatin, and starch were hydrolyzed. Key enzymes of the ED pathway were absent (Kersters and De Ley, 1968).

To our knowledge, application of *P. citrea* strains in biotechnological processes or implications in pathogenic processes have not yet been reported. The type strain of *P. sukumoe* produces a new antibiotic called 7-hydro-8-methylpteroylglutamylglutamic acid (Murata et al., 1987).

Chemotaxonomic Characters

Lysine was found as a diagnostic amino acid of the cell wall as early as 1965 (Yamaguchi, 1965). According to the classification of Lechevalier and Lechevalier (1981a), the cell wall is of type VI. Aspartic acid is lacking (Yamaguchi, 1965; Evtushenko et al., 1984b), while galactose is usually present (Yamaguchi, 1965). Differing results were reported for the molecular ratio of the amino acids of peptidoglycan. Stackebrandt et al. (1983) report the ratio of alanine to glutamic acid to lysine to be 4:1:1. Deducing the structure from these components, the interpeptide bridge would consist of two alanine residues (variation Lys-Ala$_2$, corresponding to type A3α of Schleifer and Kandler, 1972). However, Evtushenko et al. (1984b) reported that the interpeptide bridge contained a glutamic acid residue which replaces one alanine residue (variation L-Lys-Ala-Glu, type A4α). This is in accord with the ratio of alanine to glutamic acid to lysine as 3.01:2.12:1.00, i.e., 3:2:1, given as a preliminary personal communication of O. Kandler in 1975 for the type strain of *P. citrea.*

Mycolic acids (Kalakoutskii et al., 1989) as well as teichoic acids (Evtushenko et al., 1984a; Takahashi et al., 1987) are lacking. Among the isoprenoid quinones, tetrahydrogenated menaquinones with nine isoprenoid units, MK-9(H$_4$), predominate (Collins and Jones, 1981; Takahashi et al., 1987). Major fatty acids are the 12- and 13-methyltetradecanoic acids (ai-15:0, i-15:0) (Andreyev et al., 1983).

According to Lechevalier et al. (1977), diagnostic phospholipids are composed of phosphatidyl glycerol, acyl-phosphatidyl glycerol, and unknown glucosamine-containing phospholipids (type PV). Takahashi et al. (1987), who did not detect phosphatidyl glycerol in either of the two described species, suggest the types PI or PIV for the genus.

The GC content of the DNA ranges from 70 to 75 mol% (T_m) (Evtushenko et al., 1984b; Tsyganov et al., 1966; Yamaguchi 1967).

The Genus *Jonesia*

Jonesia denitrificans has recently been described for the misclassified species *Listeria denitrificans* (Rocourt et al., 1987a). This organism was originally placed by Sohier et al. (1948) in this genus without species status and later named *L. denitrificans* by Prévot (1961). The low degree of similarity between this species and genuine members of *Listeria* had been recognized for more than 20 years, and there was general agreement that this species should be removed from the genus. Evidence came from numerical analysis (Chatelain and Second, 1976; Jones, 1975; Stuart and Pease, 1972; Stuart and Welshimer, 1974), serology (Welshimer and Meredith, 1971), chemotaxonomy (Fiedler and Seger, 1983; Fiedler et al., 1984; Stuart and Welshimer, 1974; Collins et al., 1979b, 1983), and the determination of intra- and intergeneric relationships (Stuart and Welshimer, 1974; Rocourt et al., 1987a). In 1977, the Subcommittee on the Taxonomy of *Listeria* and Related Bacteria suggested that *L. denitrificans* be designated as nomen generum perplexum (Int. J. Syst. Bacteriol. 28:443). The phylogenetic study of Rocourt et al. (1987a) then clearly demonstrated the relationship of *L. denitrificans* to members of the order Actinomycetales (sensu Stackebrandt, 1981), while the genuine members of *Listeria* clustered with lactobacilli, streptococci, and bacilli (Ludwig et al., 1984; Rocourt et al., 1987b).

The membership of *J. denitrificans* to the family Cellulomonadaceae is not clearcut. Before sequence information on 16S rRNA oligonucleotides became available for this organism, cellulomonads, oerskovias, and *P. citrea* formed a loose but distinct sub-group within the *Arthrobacter/Brevibacterium/Cellulomonas* subline of descent. When the 16S rRNA catalog was included in the analysis, *Promicromonospora* and *Jonesia* formed an individual line which was separated from the *Cellulomonas/Oerskovia* line (Rocourt et al., 1987a). *J. denitrificans* differs from members of the other three genera of this family in its lower GC content and its composition of peptidoglycan and isoprenoid quinone. The rational for keeping *Jonesia* within the family is primarily based on the findings that it is a phylogenetic neighbor of *Promicromonospora,* that it forms vegetative hyphae and branched rods on solid media, and that in many biochemical reactions, including

cellulolytic activity, it behaves like NMOs of group A (Sottnek et al., 1977). However, after more representatives of the four genera of the family been subjected to phylogenetic analysis, *Jonesia* may be removed from the Cellulomonadaceae.

Habitats

The only known strain of *J. denitrificans* was isolated from boiled ox blood (Sohier et al. 1948) but the natural habitat is not known. Schefferle (1966) isolated three coryneform strains from poultry deep litter which clustered at a similarity of about 80% with *J. (L.) denitrificans* in a numerical phenetic study (Jones, 1975) but these isolates were not further characterized.

Media and Cultivation

Since *J. denitrificans* was treated as a *Listeria* strain, cultivation was mostly done in the more complex and rich media necessary for growing genuine *Listeria* strains. Cultures were grown in either brain heart infusion broth (Difco); tryptose phosphate broth (TD Difco, 2%; NaC1, 1%; glucose, 0.2%, Na_2HPO_4, 0.35%); nutrient broth (neutralized bacteriological peptone [Oxoid L 34], 1%; Lab Lemco, 0.8%; NaC1, 0.5%; pH 7.4); or nutrient broth No. 2 (Oxoid), supplemented with glucose, 0.4%. Cells are routinely grown in shake flasks at 30°C.

Media for cultivation on agar plates were tryptose blood agar base (Difco) supplemented with 1% glucose, or heart infusion agar (HIA, Difco), or trypticase soy agar (TSA, BBL).

Preservation of Cultures

The bacteria may be preserved for some months by stab inoculation into nutrient agar (Nutrient Agar [Oxoid], or other similar media) in small screw-capped containers. After incubation overnight at 30°C, the caps should be screwed tightly to prevent evaporation and the containers stored in the dark, preferably at refrigerator temperatures. Alternatively, tubes with paraffin-waxed sealed stoppers may be used (D. Jones, personal communication) Longer term preservation (over 10 years) may be achieved by storing bacteria on glass beads at −70°C (Feltham et al., 1978; Jones et al., 1984). The bacteria can also be preserved by lyophilization.

Identification

On agar media the type strain of *Jonesia denitrificans* DSM 20603 displays a distinct morphological growth cycle. Cells transferred from agar or submerged cultures, respectively, are typically coryneform, i.e. they are straight to irregularly bent rods of variable length. V-formations are frequent (Fig. 9). These cells produce hyphal or even mycelia-like structures which undergo fragmentation resulting in rods as described above (Fig. 10). Colonies show mycelial fringes (Fig. 11), and filaments may even penetrate into the agar, may branch there, and may subsequently fragment into rods (Fig. 12). *Jonesia denitrificans* shares the property of producing a morphological growth cycle with representatives of other genera of the family Cellulomonadaceae (Fig. 1–5) and with other nocardioform bacteria. In stained smears, *J. denitrificans* is Gram positive but many of the older cells stain Gram negative. Cells are motile by peritrichous flagella. Colonies on nutrient agar are grayish, and translucent to opaque, becoming yellowish after 2 to 3 weeks.

Information on chemotaxonomic properties is absolutely necessary to unambiguously identify this species. The most useful properties are: the rather low GC content of about 57 mol% (Seeliger and Jones, 1986); peptidoglycan of the Lys-Ser-D-Glu-type (Fiedler and Seger, 1983); unsaturated menaquinone of the MK9 type (Collins et al., 1979b); long-chain fatty acids and polar lipids (Collins et al., 1983); and the presence of an *N*-acetyl galactosamine-substituted poly (ribitol phosphate) teichoic acid (Fiedler et al., 1984).

For comparison, the type strain of *J. denitrificans* is available as ATCC 14870 and CIP 55134.

Physiological Properties

Like other members of the family, *J. denitrificans* grows under aerobic conditions and forms acid from a large variety of sugars, polysaccharides, and other compounds in fermentation broth base. No information is available on the mode and quality of end products formed. Cellulose, starch, DNA, and RNA are hydrolyzed. Other useful information on simple biochemical reactions is compiled by Sottnek et al. (1977) and Seeliger and Jones (1986).

Literature Cited

Agamuthu, P., and E. L. Tan. 1985. Digestion of dried palm oil mill effluent by *Cellulomonas* species. Microbios Letters 30:109 113.

Agre, N. S. 1964. A contribution to the technique of isolation and cultivation of thermophilic actinomycetes (in Russian). Mikrobiologiya 33:913–917.

Alderson, G., and E. N. Amadi. 1986. Numerical taxonomic study of *Cellulomonas* and possibly related genera, p. 597. In: G. Szabó, S. Biro and M. Goodfellow (ed.),

Biological, biochemical, and biomedical aspects of actinomycetes, Part B. Akademiai Kiado, Budapest.

Alferova, I. V., L. P. Terekova, and H. Prauser. 1989. Selective medium with nalidixic acid for the isolation of antibiotic-producing actinomycetes (in Russian). Antibiotiki i Khimoterapiya 34:344–348.

Andrews, B. A., and J. A. Asenjo. 1987. Continuous-culture studies of synthesis and regulation of extracellular β(1–3) glucanase and protease enzymes from *Oerskovia xanthineolytica*. Biotech. Bioeng. 30:628–637.

Andreyev, L. V., L. I. Evtushenko, and N. S. Agre. 1983. Fatty acid composition of *Promicromonospora citrea*. Mikrobiologiya 52:58–63.

Aristarkhova, V. I. and E. G. Rodionova. 1981. Chemotaxonomic properties of *Nocardia*-like organisms isolated from compost. Mikrobiologiya 50:844–848.

Bae, M., and E. J. Lee. 1986. Studies on the protoplast formation of *Cellulomonas flavigena* and its observations under scanning electron microscope. Kor. J. Appl. Microbiol. Bioeng. 2:175–179.

Bagnara, C., C. Gaudin, and J. P. Belaich. 1986. Purification and partial characterization of two extracellular endoglucanases from *Cellulomonas fermentans*. Biochem. Biophys. Res. Comm. 140:219–229.

Bagnara, C., C. Gaudin, and J. P. Belaich. 1987. Physiological properties of *Cellulomonas fermentans*, a mesophilic cellulolytic bacterium. Appl. Microbiol. Biotechnol. 26:170–176.

Bagnara, C., R. Toci, C. Gaudin, and J. P. Belaich. 1985. Isolation and characterization of a cellulolytic microorganism, *Cellulomonas fermentans* sp. nov. Int. J. Syst. Bacteriol. 35:502–507

Beguin, P., and H. Eisen, 1978. Purification and partial characterization of three extracellular cellulases from *Cellulomonas* sp. Eur. J. Biochim. 87:525–531.

Bodnar, G., I. M. Szabó, and A. Zicsi. 1989. Untersuchungen über die intestinalen Actinomyceten-Gemeinschaften von *Mesoniscus graniger* Friv./Isopoda. Memoires de Biospeologie 17:131–136.

Bousfield, I. J. 1972. A taxonomic study of some coryneform bacteria. J. Gen. Microbiol. 71:441–445.

Bousfield, I. J., G. L. Smith, T. R. Dando, and G. Hobbs. 1983. Numerical analysis of total fatty acid profiles in the identification of coryneform, nocardioform and some other bacteria. J. Gen. Microbiol. 129:37–5394.

Braden, A. R., and D. W. Thayer. 1976. Serological study of *Cellulomonas*. Int. J. Syst. Bacteriol. 26:123–126.

Chatelain, R., and L. Second. 1966. Taxonomie numerique de quelques *Brevibacterium*. Ann. Inst. Pasteur 111:630–644.

Collard, P. 1963. A species isolated from fermenting cassava roots. J. Appl. Bacteriol. 26:115–116.

Collins, M. D., T. Pirouz, M. Goodfellow, and D. E. Minnikin. 1977. Distribution of menaquinones in actinomycetes and corynebacteria. J. Gen. Microbiol. 100:221–230.

Collins, M. D., M. Goodfellow, and D. E. Minnikin. 1979a. Isoprenoid quinones in the classification of coryneform and related bacteria. J. Gen. Microbiol. 110:127–136.

Collins, M. D., D. Jones, M. Goodfellow, and D. E. Minnikin. 1979b. Isoprenoid quinone composition as a guide to the classification of *Listeria*, *Brochothrix*, *Erysipelothrix* and *Caryophanon*. J. Gen. Microbiol. 11:453–457.

Collins, M. D., M. Goodfellow, D. E. Minnikin, and G. Alderson. 1985. Menaquinone composition of mycolic acid-containing actinomycetes and some sporoactinomycetes. J. Appl. Bacteriol. 58:77–86.

Collins, M. D., and D. Jones. 1981. Distribution of isoprenoid quinone structural types in bacteria and their taxonomic implications. Microbiol. Rev. 45:316–354.

Collins, M. D., S. Feresu, and D. Jones. 1983. Cell wall, DNA base composition and lipid studies on *Listeria denitrificans* (Prévot). FEMS Microbiol. Lett. 18:131–134.

Cruickshank, J. G., A. H. Gawler, and C. Shaldon. 1979. *Oerskovia* species. Rare opportunistic pathogens. J. Med. Microbiol. 12:513–515.

Curry, C., N. Gilkes, G. O'Neill, R. C. Miller, Jr., and N. Skipper. 1988. Expression and secretion of a *Cellulomonas fimi* exoglucanase in *Saccharomyces cerevisiae*. Appl. Environ. Microbiol. 54:476–484.

De Coninck-Chosson, J. 1988. Aerobic degradation of cellulose and adsorption properties of cellulases in *Cellulomonas uda* JC3: effects of crystallinity of substrate. Biotech. Bioeng. 31:495–501.

De Kloet, S. R. 1984. Rapid procedure for the isolation of high-molecular weight RNA and DNA from yeast using lyticase and sodium dodecyl sulfate. J. Microbiol. Meth. 2:189–196.

De Leon, C. A., and L. M. Joson. 1980. Conversion of celluloses to protein. Acta Manilana Ser. Natl. Appl. Sci. 19:75–77.

Dermoun, Z., and J. P. Belaich. 1985. Microcalorimetric study of cellulose degradation by *Cellulomonas uda* ATCC 21399. Biotech. Bioeng. 27:1005–1011.

Dermoun, Z., and J. P. Belaich. 1988. Crystalline index change in cellulose during aerobic and anaerobic *Cellulomonas uda* growth. Appl. Microbiol. Biotechnol. 27:399–404.

Dermoun, Z., C. Gaudin, and J. P. Belaich. 1984. Microcalorimetric study of aerobic growth of *Cellulomonas* sp. 21399 on various carbohydrates. Appl. Microbiol. Biotechnol. 19:281–287.

Dermoun, Z., C. Gaudin, and J. P. Belaich. 1988. Effects of end-product inhibition of *Cellulomonas uda* anaerobic growth on cellobiose chemostat culture. J. Bact. 170:2827–2831.

Deschamps, A. M. 1982. Nutritional capacities of bark and wood decaying bacteria with particular emphasis on condensed tannin degrading strains. Eur. J. Path. 12:252–257.

Diaz, P. L., and H. A. Guirola. 1983. Fermentation study of cellulosic materials of sugarcane by species of the genus *Cellulomonas*. Rev. Cienc. Biol. 14:283–298.

Dunlap, C. E., and C. D. Callihan. 1974. Single cell protein production from cellulosic waste, p. 335–347. In: H. Yen (ed.), Recycling and disposal of solid wastes: industrial, agricultural, domestic. Ann Harbor Sci. Publ. Inc., Ann Arbor, MI, USA.

Dzingov, A., K. Márialigeti, K. Jáger, E. Contreras, L. Kondics, and I. M. Szabó. 1982. Studies on the microflora of millipedes *(Diplopoda)*. I. A comparison of actinomycetes isolated from surface structures of the exoskeleton and the digestive tract. Pedobiologia 24:1–7.

Erikson, D. 1954. Factors promoting cell division in a "soft" mycelial type of *Nocardia: Nocardia turbata* n. sp. J. Gen. Microbiol. 11:198–208.

Evtushenko, L. I., N. A. Janushkene, G. M. Streshinskaya, I. B. A. Naumova, and N. S. Agre. 1984a. Occurrence of teichoic acids in representatives of the order Actinomycetales. Dokl. Akad. Nauk SSSR 278:237–239.

Evtushenko, L. I., G. F. Levanova, and N. S. Agre. 1984b. Nucleotide composition of DNA and amino acid composition of A4 peptidoglycan in *Promicromonospora citrea*. Mikrobiologiya 53:519–520.

Evtushenko, L. I., D. T. Pataraya, N. S. Agre, and L. V. Kalakoutskii. 1982. Polymorphism of *Promicromonospora citrea* cells. Mikrobiologiya 51:466–471.

Feltham, R. K. A., A. K. Power, P. A. Pell, and P. H. A. Sneath. 1978. A simple method for storage of bacteria at −76°C. J. Appl. Bacteriol. 44:313–316.

Fiedler, F., and O. Kandler. 1973. Die Mureintypen der Gattung *Cellulomonas*. Bergey et al. Arch. Mikrobiol. 89:4150.

Fiedler, F., K. H. Schleifer, B. Cziharz, E. Interschick, and O. Kandler. 1970. Murein types in arthrobacter, brevibacteria, corynebacteria and microbacteria. Publ. Fac. Sci. Univ. Brno 509–514:111–122.

Fiedler, F., and J. Seger. 1983. The murein types of *Listeria grayi*, *Listeria murrayi*, and *Listeria denitrificans*. System. Appl. Microbiol. 4:444–450.

Fiedler, F., J. Seger, A. Schrettenbrunner, and H. P. R. Seeliger. 1984. The biochemistry of murein and cell wall teichoic acids in the genus *Listeria*. Syst. Appl. Microbiol. 5:52–96.

Fox, G. E., and E. Stackebrandt. 1987. The application of 16S rRNA cataloguing and 5S rRNA sequencing in bacterial systematics, p. 405–458. In: R. R. Colwell and R. Grigorova (ed.), Methods in microbiology, vol. 19. Academic Press, London.

Gilkes, N. R., D. G. Kilburn, M. L. Langsford, R. C. Miller, W. W. Wakarchuk, R. A. J. Warren, D. J. Whittle, and K. R. Wong. 1984. Isolation and characterisation of *Escherichia coli* clones expressing cellulase genes from *Cellulomonas fimi*. J. Gen. Microbiol. 130:1377–1384.

Gilkes, N. R., R. A. J. Warren, R. C. Miller, Jr., and D. G. Kilburn. 1988. Precise excision of the cellulose binding domains from two *Cellulomonas fimi* cellulases by a homologous protease and the effect of catalysis. J. Biol. Chem. 263:10401–10407

Goodfellow, M. 1971. Numerical taxonomy of some nocardioform bacteria. J. Gen. Microbiol. 69:33–80.

Goodfellow, M. 1989. Suprageneric classification of actinomycetes, p. 2333–2337. In: S. T. Williams, M. E. Sharpe, and J. G. Holz (ed.) Bergey's manual of systematic bacteriology, vol. 4. Williams & Wilkins, Baltimore, MD.

Greenberg, N. M., R. A. J. Warren, D. G. Kilburn, and R. C. Miller, Jr. 1987a. Regulation, initiation, and termination of the cenA and cex transcripts of *Cellulomonas fimi*. J. Bacteriol. 169:646–653.

Greenberg, N. M., R. A. J. Warren, D. G. Kilburn and R. C. Miller, Jr. 1987b. Regulation and initiation of cenB transcripts of *Cellulomonas fimi*. J. Gen. Microbiol. 169:4674–4677.

Guo, Z., N. Arfman, E. Ong, N. R. Gilkes, D. G. Kilburn, R. A. J. Warren, and R. C. Miller, Jr. 1988. Leakage of *Cellulomonas fimi* cellulases from *Escherichia coli*. FEMS Microbiol. Lett. 49:279–283.

Guyot, J. P. 1986. Role of formate in methanogenesis from xylane by *Cellulomonas* sp. associated with methano-

Table 2. Differential characteristics of species of the genus *Brevibacterium*.

Characteristic	*B. linens*	*B. casei*	*B. epidermidis*	*B. iodinum*
Colony color	Yellow-orange	Gray-white	Gray-white	Gray-white
Crystals of iodinin	−	−	−	+
Oxidase	−	−	−	+
Survival at 60°C for 30 min	−	+	+	−
GC content (mol%)	60–64	66–67	63–64	61–63

+, present; −, absent.

ium from other actinomycete and coryneform genera are shown in Table 1.

Differentiation of *Brevibacterium* Species

Brevibacterium linens may be distinguished from other species of the genus in producing yellow to deep orange colonies (Mulder et al., 1966; Jones et al., 1973). In contrast, the colonies of *B. iodinum* are characterized by the production of purple extracellular crystals of iodinin (Davis, 1939; Clemo and Daglish, 1950; Sneath, 1960). *B. casei* and *B. epidermidis* produce gray-white colonies. Other tests useful in differentiating species are shown in Table 2. *B. casei* and *B. epidermidis* can only be distinguished at present by DNA base composition and DNA-DNA hybridization tests.

Applications

B. linens is a resident of the exterior of surface-ripened soft cheeses of the Limburger variety, where it is considered to contribute to the surface color and ripening (Albert et al., 1944). *B. linens* produces methanethiol from L-methionine (Sharpe et al., 1976, 1978). Methanethiol is an important constitutent of the aroma of cheddar cheese, and it is thought that the production of this compound by *B. linens* may contribute to the aroma and flavor of surface-ripened cheeses such as Limburger and Romadour.

Literature Cited

Albert, J. O., H. F. Long, and B. W. Hammer, 1944. Classification of the organisms important in dairy products IV. *Bacterium linens*. Agricultural Experimental Station, Iowa Research Bulletin 328:234–259.

Anderton, W. J., and S. G. Wilkinson, 1980. Evidence for the presence of a new class of teichoic acid in the cell wall of bacterium NCTC9742. J. Gen. Microbiol. 118:343–351.

Anderton, W. J., and S. G. Wilkinson, 1985. Structural studies on a mannitol teichoic acid from the cell wall of bacterium N.C.T.C.9742. Biochem. J. 226:587–599.

Bousefield, I. J., 1978. The taxonomy of coryneform bacteria from the marine environment, p. 217–233. In: Bousefield, I. J., Callely, A. G. (ed.), Coryneform bacteria. Academic Press, London.

Bousefield, I. J., G. L. Smith, T. R. Dando, and G. Hobbs, 1983. Numerical analysis of total fatty acid profiles in the identification of coryneform, nocardioform and some other bacteria. J. Gen. Microbiol. 129:375–394.

Breed, R. S., 1953. The *Brevibacteriaceae* fam. nov. of order Eubacteriales. Riassunti delle Communicazione VI Congresso Internazionale di Microbiologia, Roma 1:13–1.

Clemo, G. R., and A. F. Daglish, 1950. The phenazine series. Part VIII. The constitution of the pigment of *Chromobacterium iodinum*. J. Chem. Soc. 1481–1485.

Collins, M. D., J. A. E. Farrow, M. Goodfellow, and D. E. Minnikin, 1983. *Brevibacterium casei* sp. nov. and *Brevibacterium epidermidis* sp. nov. Syst. Appl. Microbiol. 4:388–395.

Collins, M. D., D. Jones, R. M. Keddie, and P. H. A. Sneath, 1980. Reclassification of *Chromobacterium iodinum* (Davis) in a redefined genus *Brevibacterium* (Breed) as *Brevibacterium iodinum* nom. rev.; comb. nov. J. Gen. Microbiol. 120:1–10.

Colwell, R. R., R. V. Citarella, I. Ryman, and G. B. Chapman, 1969. Properties of *Pseudomonas iodinum*. Can. J. Microbiol. 15:851–857.

Crombach, W. H. J., 1974. Relationships among coryneform bacteria from soil, cheese and sea fish. Antonie van Leeuwenhoek J. Microbiol. Serol. 40:347–359.

Davis, J. G., 1939. *Chromobacterium iodinum* (n.sp.). Zentralbl. Bakteriol. Parasitenkd. Infektionskr. Hyg. Abt. II 100:273–276.

Fiedler, F., M. J. Schäffler, and E. Stackebrandt, 1981. Biochemical and nucleic acid hybridization studies on *Brevibacterium linens* and related strains. Arch. Microbiol. 129:85–93.

Jones, D., and R. M. Keddie, 1985. GENUS *Brevibacterium* p. 1301–1313. Sneath, P. H. A., N. A. Mair, M. E. Sharpe and J. G. Holt (ed.), In-Bergey's manual of systematic bacteriology, vol 2. Williams and Wilkins, Baltimore.

Jones, D., J. Watkins, and S. K. Erickson, 1973. Taxonomically significant colour changes in *Brevibacterium linens* probably associated with a carotenoid-like pigment. J. Gen. Microbiol. 77:145–150.

Keddie, R. M., and G. L. Cure, 1977. The cell wall composition and distribution of free mycolic acids in named strains of coryneform bacteria and in isolates from various natural sources. J. Appl. Bacterial. 42:229–252.

Mohan, K., 1981. *Brevibacterium* sp. from poultry. Antonie van Leeuwenhoek. J. Microbiol. Serol. 47:449–453.

Mulder, E. G., A. D. Adamse, J. Antheunisse, M. H. De-
    inema, J. W. Woldendrop, and L.P.T.M. Zevenhuizen,
    1966. The relationship between *Brevibacterium linens*
    and bacteria of the genus *Arthrobacter*. J. Appl. Bac-
    teriol. 29:44–71.

Pitcher, D. G., and W. C. Noble, 1978. Aerobic diphtheroids
    of human skin. In: Bousefield, I. J., Callely, A. G. (ed.),
    Coryneform bacteria. Academic Press, London.

Schefferle, H. E., 1966. Coryneform bacteria in poultry deep
    litter. J. Appl. Bacteriol. 29:147–160.

Schleifer, K. H., and O. Kandler, 1972. Peptidoglycan types
    of bacterial cell walls and their taxonomic implications.
    Bacteriol. Rev. 36:407–477.

Sharpe, M. E., B. A. Law, and B. A. Phillips, 1976. Cory-
    neform bacteria producing methanethiol. J. Gen. Mi-
    crobiol. 94:430–435.

Sharpe, M. E., B. A. Law, B. A. Phillips, and D. G. Pitcher,
    1977. Methanethiol production by coryneform bacte-
    ria: strains from dairy and human skin sources and
    Brevibacterium linens. J. Gen. Microbiol. 101:345–349.

Sneath, P. H. A., 1960. A study of the genus *Chromobac-
    terium*. Iowa State J. Sci. 34:243–500.

Stackebrandt, E., B. J. Lewis, and C. R. Woese, 1980. The
    phylogenetic structure of the coryneform group of bac-
    teria. Zentralbl. Bakteriol. Mikrobiol. Hyg. Abt. II Orig.
    C 1:137–149.

CHAPTER 62

The Genera *Agromyces, Aureobacterium, Clavibacter, Curtobacterium,* and *Microbacterium*

MATTHEW D. COLLINS and JOHN F. BRADBURY

Introduction

This chapter deals with a number of saprophytic and phytopathogenic irregularly rod-shaped bacteria, whose mode of metabolism is entirely (or primarily) respiratory and which have cell walls containing the unusual group B type peptidoglycan. This peptidoglycan is characterized by a cross-linkage between the α-carboxyl group of D-glutamic acid in position 2 of the peptide subunit and the C-terminal D-alanine of an adjacent subunit (Schleifer and Kandler, 1972). The bacteria include members of the genera *Microbacterium, Aureobacterium, Curtobacterium,* and *Clavibacter*. The "microaerophilic" filamentous actinomycete species *Agromyces ramosus* also possesses a group B type peptidoglycan and shares a number of important chemical criteria with the above genera. Other chemical features common to this group include: absence of mycolic acids; and presence of major amounts of *anteiso-* and *iso-*methyl branched fatty acids, unsaturated menaquinones, high GC content of the DNA (about 68 to 75 mol%), and diphosphatidylglycerol, phosphatidylglycerol, and a variety of glycolipids (Collins, 1982; Collins and Jones, 1980; Collins et al., 1980, 1983a, 1983b; Döpfer et al., 1982). Support for the close relationship of these genera also comes from rRNA cistron similarity studies (Döpfer et al., 1982). DNA-rRNA hybridization studies clearly demonstrate that all of the coryneform taxa (including *Agromyces*) containing the group B peptidoglycan are phylogenetically closely related and belong to a single rRNA cistron cluster (differences in the $T_{m(e)}$ values of DNA-rRNA heteroduplexes cover a relatively narrow range of about. 8°C) (Döpfer et al., 1982).

Orla-Jensen (1919) first established the genus *Microbacterium* to accommodate a diverse collection of Gram-positive nonsporeforming rods isolated during studies on lactic acid-producing bacteria. The genus originally contained four species, *M. lacticum, M. flavum, M. mesentericum,* and *M. liquefaciens*. Much later, two fur-

ther species, *M. thermosphactum* (McLean and Sulzbacher, 1953) and the glutamic acid-producing "patent" strain *M. ammoniaphilum* (Abe et al., 1967) were included in the genus. It had always been recognized that the genus *Microbacterium* was very heterogeneous, but it has only been during the past 10 or so years that the systematic relationships of these various species have been resolved. *Microbacterium flavum* and *M. ammoniaphilum* have been shown to be authentic corynebacteria (Barksdale et al., 1979; Collins et al., 1982), whereas *M. liquefaciens* has been reclassified in a new genus, *Aureobacterium* (Collins et al. 1983b). The species *M. thermosphactum* has been assigned to the genus *Brochothrix* (Sneath and Jones, 1976) and is phylogenetically more closely related to *Listeria* and other low-GC Gram-positive bacteria than to actinomycetes. As a result of systematic chemotaxonomic investigations, the genus *Microbacterium* has been redefined largely in chemical terms and is now restricted to those organisms which contain a group B type peptidoglycan based upon L-lysine, predominantly methyl-branched fatty acids, long unsaturated menaquinones (MK-10 to MK-12), and a GC content of 69 to 75 mol% (Collins et al., 1983a). Defined in this way, the genus *Microbacterium* includes *M. lacticum* (the type species) and the species *M. imperiale* (formerly *Brevibacterium imperiale*) and *M. laevaniformans* (formerly *Corynebacterium laevaniformans*) (Collins et al., 1983a). A fourth species, *M. arborescens* (formerly *Flavobacterium arborescens*), which conforms to the above definition has also recently been added to the genus (Imai et al., 1984).

The genus *Curtobacterium* was established by Yamada and Komagata (1972) for some former *Brevibacterium* species and plant pathogenic corynebacteria that contained D-ornithine in their cell peptidoglycan, had a high GC content (66 to 71 mol%), and showed rather weak acid production from some carbohydrates. Originally included in the new genus were *Corynebacterium flaccumfaciens, Brevibacterium al-*

bidum, B. citreum, B. helvolum, B. insectiphilum, B. luteum, B. pusillum, B. saperdae, and *B. testaceum* (Yamada and Komagata, 1972). It became evident during subsequent chemotaxonomic investigations that *B. testaceum* and *B. saperdae* were quite distinct from other *Curtobacterium* species in possessing unusually long menaquinones (MK-11, MK-12) (Collins et al., 1980, 1983b; Collins and Jones, 1981), whereas other members of the genus *Curtobacterium* invariably synthesize MK-9 (Collins et al., 1980). *B. testaceum* and *B. saperdae* also differ from other curtobacteria in containing *N*-glycoly residues in the glycan moiety of their walls (Uchida and Aida, 1977) and in containing an additional glycine residue in the interpeptide bridge of their peptidoglycan (Schleifer and Kandler, 1972; Collins et al., 1983b). These chemical differences led Collins et al. (1983b) to reclassify these two species in a new genus, *Aureobacterium,* with the species *A. liquefaciens* (formerly *M. liquefaciens*), *A. barkeri* (formerly *Corynebacterium barkeri*), *A. flavescens* (formerly *Arthrobacter flavescens*), and *A. terregens* (formerly *Arthrobacter terregens*). When the *Approved Lists of Bacterial Names* were compiled (Skerman et al., 1980), *B. helvolum* and *B. insectiphilum* were excluded because of doubts over the authenticity of the strains, and the plant pathogens were included in *Corynebacterium,* but not in *Curtobacterium.* Collins and Jones (1983) formally proposed that the plant pathogens *C. flaccumfaciens, C. betae, C. oortii,* and *C. poinsettiae* be included in the genus *Curtobacterium* as four pathovars of the single species *Curtobacterium flaccumfaciens.* This proposal was consistent with the high overall phenetic similarity (Dye and Kemp, 1977) and genomic relatedness (Döpfer et al., 1982) of the four phytopathogens. Hence, in *Bergey's Manual of Systematic Bacteriology* (Komagata and Suzuki, 1986), five species were included in the genus: *Cu. albidum, Cu. citreum, Cu. flaccumfaciens* (and pathovars), *Cu. luteum,* and *Cu. pusillum.* Dunleavy (1989) recently described a sixth species, the saprophyte *Cu. plantarum.*

The genus *Clavibacter* was established by Davis et al. (1984) to accommodate those plant pathogenic corynebacteria whose cell wall peptidoglycan contains 2,4-diaminobutyric acid (Dab) as the dibasic cross-linking acid (type B2γ; Schleifer and Kandler, 1972). *Clavibacter* currently contains five species: *Cl. michiganensis* (including five subspecies), *Cl. iranicum, Cl. rathayi, Cl. tritici,* and *Cl. xyli* (including two subspecies *xyli* and *cynodontis*). The limited DNA-DNA homology studies performed to date (Starr et al., 1975; Döpfer et al., 1982) are somewhat conflicting but they indicate—when taken together with biochemical criteria (Dye and Kemp, 1977), soluble protein patterns (Carlson and Vidaver, 1982), lipid composition (Collins and Jones, 1980; Collins, 1983), and enzyme polymorphism (Riley et al., 1988)—that *Cl. michiganensis* subspecies *michiganensis, insidiosus, nebraskensis,* and *sepedonicus* all warrant separate specific status (see Collins and Bradbury, 1986). Two other bacteria merit inclusion in the genus *Clavibacter* as separate species; the bacterium which causes toxicity in annual ryegrass and the organism designated *"Corynebacterium agropyri"* (Riley et al., 1988). The branching, highly filamentous actinomycete *Agromyces ramosus* also contains a peptidoglycan based on 2,4-diaminobutyric acid and in spite of its morphology is phylogenetically closely related to *Clavibacter* and the other group B genera (Collins, 1982; Döpfer et al., 1982). Characters useful in distinguishing the genera *Agromyces, Aureobacterium, Clavibacter, Curtobacterium,* and *Microbacterium* are shown in Table 1.

The Genus *Agromyces*

The genus *Agromyces* contains one species, *Agromyces ramosus,* young cultures of which are composed of branched, filamentous elements which, as growth progresses, undergo septation and fragmentation to yield coccoid and irregular (diphtheroid) forms. Cells are nonmotile and stain Gram-positive with segments of the mycelium becoming Gram-negative during fragmentation. *Agromyces* is microaerobic to aerobic, has an oxidative metabolism, and is catalase and oxidase negative. Although benzidine negative, it produces cytochromes *b, c* and *aa*$_3$ (Jones et al., 1970). Flavoproteins are present and menaquinones (MK-11, MK-12, and MK-13) are the sole respiratory quinones (Collins, 1982). An alcohol-soluble yellow carotenoid is produced, which is particularly noticeable when cells are grown in the presence of blood or catalase. The cell wall peptidoglycan contains 2,4-diaminobutyric acid as the diamino acid (type B2γ of Schleifer and Kandler, 1972). The long-chain cellular fatty acids are primarily of the *anteiso*-methyl branched types with 12-methyltetradecanoic (*anteiso*-C$_{15:0}$) and 14-methylhexadecanoic (*anteiso*-C$_{17:0}$) acids predominating. The membrane polar lipids comprise diphosphatidylglycerol, phosphatidylglycerol, and two glycolipids (Collins, 1982). The GC content of the DNA is 71(Bd) to 76.7(T$_m$) mol%.

Table 1. Characters useful in distinguishing *Agromyces, Aureobacterium, Clavibacter, Curtobacterium,* and *Microbacterium*.

	Diamino acid in peptidoglycan	Menaquinone	Mycelium produced	Oxygen requirement
Agromyces	Dab	MK-11, MK-12	+	Aerobic to microaerobic
Aureobacterium	D-Orn	MK-11, MK-12	−	Aerobic
Clavibacter	Dab	MK-9, MK-10	−	Aerobic
Curtobacterium	D-Orn	MK-9	−	Aerobic
Microbacterium	Lys	MK-11, MK-12	−	Aerobic

Dab, diaminobutyric acid; D-Orn, D-ornithine; Lys, L-lysine.

Habitats

The natural habitat of *Agromyces ramosus* is soil. It has been isolated from a wide range of soil types ranging from fertile meadow to barren desert soils. In these soils it is reported to be 10- to 100-fold more numerous than the total colony-forming microflora (Gledhill and Casida, 1969).

Isolation and Cultivation

Although *Agromyces* occurs in large numbers in soil, isolation is often difficult due to problems with the organism adjusting to conditions of laboratory cultivation. Once adapted, however, it usually grows without problem on nutritionally rich media. *Agromyces* may be isolated from soil using the dilution frequency procedure of Casida (1965), which allows fastidious organisms to adapt to laboratory media without competition from other species. Heart infusion agar at 30°C is a suitable medium for growth (Gledhill and Casida, 1969).

Heart Infusion Agar

Heart infusion broth (Difco)	25 g
Casitone (Difco)	4 g
Yeast extract	5 g

pH 7 to 7.4.

Agromyces is probably microaerophilic because it lacks catalase activity since the growth of this oxidative organism is best under full oxygen tension, provided that accumulating H_2O_2 is destroyed. This destruction (and therefore enhanced growth) can be achieved by the addition of catalase, fresh horse blood, or MnO_2 (Jones et al., 1970).

Identification

Agromyces ramosus can be readily distinguished from other coryneforms and related actinomycetes by its characteristic cellular morphology, by being catalase, benzidine, and oxidase negative, and by simultaneously possessing an oxidative metabolism. It can also be readily identified on the basis of its cell wall and membrane composition (Schleifer and Kandler, 1972; Collins, 1982). A cell wall peptidoglycan containing diaminobutyric acid in combination with the presence of predominantly branched fatty acids and MK-12 serves to distinguish *Agromyces ramosus* from all actinomycete taxa described to date.

The Genus *Aureobacterium*

Young cultures of aureobacteria are composed of irregular, short, slender rods (about 0.4 to 0.6 μm diam and 0.8 to 3 μm length) occurring singly or in groups. Cells from older cultures (3 to 7 d) are generally shorter. V-formations are common. No mycelium is produced. Cells are Gram-positive, and either motile or nonmotile. A variety of pigments are produced (yellow, orange, or red; depending on species). Nutritional requirements are complex. Some species (*A. flavescens, A. terregens*) require the terregens factor (this may be replaced by coprogen or ferrichrome) for growth. Aureobacteria are obligately aerobic, catalase positive, and may show slow and weak oxidative acid production from some carbohydrates. The cell wall peptidoglycan is based on D-ornithine (type B2β) (Schleifer and Kandler, 1972) with a glycine residue within the interpeptide bridge. All species examined to date possess high levels of N-glycolyl residues in the glycan moiety of the cell wall, indicating the presence of N-glycolyl muramic acid instead of the more usual N-acetyl form (Uchida and Aida, 1977; Collins et al., 1983b). A variety of sugars (e.g., galactose, glucose, rhamnose, and 6-deoxytalose) occur in the cell walls; composition varies and depends on strain and species (Keddie and Cure, 1977). Mycolic acids are not produced. The long-chain cellular fatty acids are mainly of the *anteiso*-methyl branched types, although *iso*-branched acids are also present in substantial amounts. *Anteiso*-$C_{15:0}$ and *anteiso*-$C_{17:0}$ are the major fatty acids. All aureobacteria contain diphosphatidylglycerol, phosphatidylglycerol, and a single diglycosyldiacylglycerol (Collins et al., 1983b). The latter is different from the glycolipid of microbacteria (Collins et al., 1983b). The report of the presence of phosphatidylinositolmannosides in *A. testaceum* is incorrect (Komura et al., 1975). Menaquinones are the sole respiratory quinones present. All strains contain very long, unsaturated prenologs, such as MK-11, MK-12, and MK-13 (Collins et al., 1980, 1983b). The GC content reported for aureobacteria are generally within the range of 67 to 70 mol%. Significantly higher values (73 to 76 mol%) have been reported by Döpfer et al. (1982).

Habitats

Little is known about the natural habitats of aureobacteria, although they have been isolated from a variety of sources. *A. liquefaciens* has been isolated from milk, cheese, dairy products, and dairy equipment. Organisms which phenotypically resemble *A. liquefaciens* have also been isolated from soil (Topping, 1937) and frozen vegetables (Splittstoesser et al., 1967). *A. flavescens* and *A. terregens* are found in soil, whereas *A. barkeri* has been isolated from raw domestic sewage (Dias et al., 1962). The only known source of *A. testaceum* is rice (Komagata and Iizuka, 1964). *A. saperdae* was originally isolated from the body cavity of the insect *Saperda caracharias* (Lysenko, 1959).

Isolation and Cultivation

We are not aware of any selective media for *Aureobacterium* species. All aureobacteria are nutritionally exacting. The species *A. terregens* and *A. flavescens* have a requirement for the terregens factor (Lochhead, 1958; Lochhead and Burton, 1953). These species can be cultivated on a variety of conventional media supplemented with soil extract. A suitable medium is soil extract nutrient agar.

Soil Extract Nutrient Agar

Lab-Lemco beef extract	1 g
Yeast extract	2 g
Peptone	5 g
NaCl	5 g
Agar	15 g
Soil extract (see below)	1 liter

Adjust to pH 7.0. Soil extract is prepared by suspending 500 to 1000 g soil in 1 liter of tap water and autoclaving at 121°C for 30 mins. Clear supernatant is obtained by filtration or centrifugation.

A good general-purpose medium for the cultivation of other *Aureobacterium* species is coryneform agar.

Coryneform Agar

Casein-peptone-tryptic digest	10 g
Yeast extract	5 g
Glucose	5 g
NaCl	5 g
Agar	15 g
Water	1 liter

Adjust to pH 7.2 to 7.4.

Identification

There are currently no simple biochemical/physiological tests for the identification of the genus. Although cell shape is a useful initial test, many other coryneform taxa (e.g., *Curtobacterium*, *Cellulomonas*, *Corynebacterium*, *Microbacterium*) have a similar morphology. Several chemical tests must, therefore, be performed to distinguish aureobacteria from these taxa. The demonstration of ornithine in the cell wall serves to distinguish aureobacteria from a host of other actinomycete and coryneform taxa (including most of those genera with a group B peptidoglycan). Cellulomonads also contain walls based upon ornithine (group A peptidoglycan) but differ from aureobacteria in being cellulolytic. Aureobacteria may also be readily distinguished from cellulomonas species by analysis of their polar lipids (Minnikin et al., 1979; Collins et al., 1983b) and menaquinones (Collins and Jones, 1981; Collins et al., 1983b). Aureobacteria may be distinguished from curtobacteria by their longer menaquinones (MK-11, MK-12).

The Genus *Clavibacter*

Cells are irregular rods, Gram-positive, nonmotile, nonsporeforming, and not acid-fast. They do not produce mycelium nor show any marked rod- to-coccus cycle of development and primary branching is rare, but arrangement of cells in V-formations and palisades is often seen. Colonies are white or yellow to orange, occasionally pink, depending on species, strain, and medium, and usually smooth, entire, and slow growing. The pigments of *Clavibacter michiganensis* subsp. *michiganensis,* which is most variable in color, are carotenoids (Saperstein et al., 1954). Most strains of *Cl. michiganensis* subsp. *insidiosus,* when grown on YDC medium (Dye and Kemp, 1977), also produce a dark blue-gray insoluble pigment (the bipyridyl pigment indigoidine) within their colonies (Starr, 1958; Kuhn et al., 1965). Nutritional requirements are complex and incompletely known. Probably all species need B vitamins, some are known to require amino acids, and *Cl. xyli* needs additions of hemin chloride and bovine serum albumin (Collins and Bradbury, 1986; Davis et al., 1984). All species are obligately aerobic, catalase positive, and produce acid rather weakly from some carbohydrates, but not from rhamnose or ribose. Kovacs' oxidase, nitrate reductase, tyrosinase, urease, and lipase tests are negative.

The cell wall contains Dab and rhamnose, but not mycolic acids or arabinose. The peptidoglycan type is B2γ[L-Dab]D-Glu-D-Dab (Schleifer and Kandler, 1972). The predominant fatty acids are *anteiso*-$C_{15:0}$, *anteiso*-$C_{17:0}$, and *iso*-$C_{16:0}$ (Collins and Jones, 1980; Bousfield et al., 1983; Stead, 1988,). Menaquinones are the sole respiratory quinones, the predominant forms being unsaturated menaquinones with 9 or 10 isoprene units (Collins and Jones, 1980; Collins, 1983). Polar lipids are diphosphatidylglycerol, phosphatidylglycerol, and some characteristic glycosyldiacylglycerols, the composition of which varies between species (Collins and Jones, 1980; Collins 1983). Reports of the GC content vary within the range 67–78 mol%, with values near 70–72 mol% being most frequently reported.

Habitats and Pathogenicity

All known species of *Clavibacter* are plant pathogens, and their main habitats are considered to be their respective plant hosts. Each pathogen has been found in one or a few usually related genera of plants (Bradbury, 1986), and they are thought to show considerable host specificity. Thus *Cl. iranicus* and *Cl. tritici* occur in wheat, where they cause gumming diseases, particulary affecting influorescences. Another as-yet-unnamed species causes gumming in *Lolium rigidum* and sometimes in other grasses in the same pasture (Bird and Stynes, 1977; Riley, 1987), and produces a neurotoxin that can be fatal to grazing animals (Riley, 1987). These three species and probably *Cl. rathayi*, which causes a similar disease in *Dactylis glomerata,* require nematodes of the genus *Anguina* as vectors (Sabet, 1954; Gupta and Swarup, 1972). They all show specific adhesion to *Anguina* species (Riley et al., 1988), and the bacteria apparently are then carried on the nematode surface into the ovaries of the grasses. *"Corynebacterium agropyri,"* which was isolated from 30- to 40-year-old herbarium material of an *Agropyron* sp. (reference cultures having been lost before the *Approved Lists* were compiled), also shows specific adhesion to *Anguina* species but pathogenicity could not be shown (Riley et al., 1988).

Clavibacter michiganensis subsp. *insidiosus* causes wilting and stunting in alfalfa, lucerne,

and a few other pasture legumes. *Cl. michiganensis* subsp. *michiganensis* causes wilt and canker in tomato and has been found in three *Solanum* species. *Cl. michiganensis* subsp. *nebraskensis* occurs only in *Zea mays,* where it causes leafspot, blight, and wilt. *Cl. michiganensis* subsp. *sepedonicus* causes wilt and tuber rot in potato, and *Cl. michiganensis* subsp. *tessellarius* causes a mosaic-like syndrome on wheat. *Cl. xyli* subsp. *cynodontis* and *Cl. xyli* subsp. *xyli* occur in *Cynodon dactylon* and *Saccharum officinarum* respectively, where they invade the vascular systems and cause stunting diseases.

Apart from diseased host plants, all species of *Clavibacter* can live in plant debris, which may serve as a source of inoculum, but survival in natural soils without host debris is poor (Schuster and Coyne, 1974). With the exception of the two subspecies of *Clavibacter xyli, Cl. michiganensis* subsp. *sepedonicus,* and *Cl. michiganensis* subsp. *tessellarius,* there is evidence that all clavibacters can survive on or in seed (Richardson, 1979, 1983). *Cl. michiganensis* subsp. *sepedonicus* can survive for long periods on farm machinery, sacks, and bins (Richardson, 1957), and can survive in potato and sugar beet without causing symptoms (Hayward, 1974; Bugbee et al., 1987). Such latent survival and survival in the absence of the host are important in the epidemiology and control of the disease.

Isolation

These slow-growing bacteria may be isolated from diseased plants if certain general rules are followed. The choice of suitable diseased material is of prime importance and excision and handling should utilize careful aseptic technique to avoid unecessary contamination with faster-growing bacteria. Finally a suitable medium and incubation temperature must be used.

The distribution of a bacterial pathogen in a diseased plant is highly localized, and for successful isolation one must select the parts which contain bacteria. In a wilting plant this will usually be a section of the vascular tissue at the base of, or below, the wilting part. It will usually show some discoloration and bacterial ooze, or streaming will be seen coming from the cut ends of vessels placed in water. Flaccid or scorched-looking leaves are likely to result from the water shortage caused by vascular infection lower in the plant, and such leaves are not likely to contain the pathogen, although they may be infected by secondary bacteria. With cankers and necrotic lesions, an area where the bacteria are active, usually at the edge, is often characterized by a water-soaked or a dull greasy appearance of the tissue. When such areas run into the healthy tissue without a definite boundary, the lesion is actively growing and it is here that the best material for isolation will be found. A sharply defined boundary between old, necrotic tissue and healthy tissue usually indicates that the pathogen is dormant or has died out and isolation is then difficult or impossible. When symptoms are more general in the plant, such as stunting and chlorosis in lucerne *(Medicago sativa)* caused by *Cl. michiganensis* subsp. *insidiosus,* or the mosaic in wheat caused by *Cl. michiganensis* subsp. *tessellarius,* a vascular infection should be suspected and the vessels examined for the presence of bacteria. Old necrotic areas, rotten parts, or parts showing advanced stages of disease are likely to contain many saprophytes and these will rapidly overgrow the very slow-growing *Clavibacter* species, making direct isolation impossible from such material.

The chosen piece of tissue should be carefully excised with instruments sterilized by flaming, placed in a little sterile water, and teased apart to release the bacteria. It is often an advantage to allow the comminuted tissue to stand for a period of a few minutes to an hour to allow the bacteria to swim or diffuse into the water. The resulting suspension is then streaked out onto a suitable agar medium. Serial dilutions may be streaked, the aim being to obtain areas of single colonies on at least some of the plates.

Isolations on nonselective media can give information on the numbers of different bacteria present in the chosen tissue, and sometimes result in the isolation of an unexpected or new pathogen. Suitable nonselective media for growth of most clavibacters include nutrient agar to which 1% glucose or 5% sucrose has been added (Lelliott and Stead, 1987; Collins and Bradbury, 1986), and glucose-yeast-calcium (GYCA) medium (Moffett et al., 1983).

GYCA Medium

Glucose	5 g
Yeast extract (Difco)	5 g
Finely ground CaCO₃	40 g
Agar	15 g
Water	1 liter

A finely ground commercial grade of $CaCO_3$ is usually best, and the powder should be dispersed in the agar just before setting. Rapid setting, e.g., in a cold water bath, is advantageous for keeping the carbonate suspended.

Better growth of some strains can be obtained with Doepel's medium (Lelliott and Stead,

1987) or with nutrient broth-yeast (NBY) medium (Vidaver, 1967; Vidaver and Davis, 1988).

Doepel's Medium

Glucose	10 g
Casein hydrolysate (not vitamin free)	8 g
Yeast extract	8 g
K₂HPO₄	2 g
MgSO₄·7H₂O	0.3 g
Agar	12 g
Water	1 liter

NBY Medium

Nutrient broth	8 g
Yeast extract	2 g
K₂HPO₄	2 g
KH₂PO₄	0.5 g
Agar	15 g
Water	1 liter

50 ml 10% glucose and 1.0 ml 1M $MgSO_4 \cdot 7H_2O$ are both added after separate sterilization.

The more complex SC medium must be used for growing *Cl. xyli.*

SC Medium

Cornmeal agar	17 g
Papaic digest of soy meal (or phytone peptone or soytone)	8 g
K₂HPO₄	1 g
KH₂PO₄	1 g
MgSO₄·7H₂O	0.2 g
Water	1 liter
Bovine hemin chloride (15 ml of 0.1% solution in 0.05 N NaOH)	15 mg
Bovine serum albumin, fraction V (10 ml of 20% solution)	2 g
Glucose (1 ml of 50% solution)	0.5 g
Cysteine (10 ml of 10% solution)	1 g

The last three solutions should be filter sterilized and added to the molten medium after it has been sterilized and cooled to 50°C. The pH is then adjusted to 6.6.

Colonies of *Cl. xyli* on SC medium are 0.1 to 0.3 mm in diameter, circular, convex, entire, and nonpigmented after 2 weeks incubation at 30°C (Davis et al., 1980).

Selective Media

Selective media are often useful for isolation of these slow-growing bacteria. These include CNS medium, which is based on NBY medium and used for isolation of *Cl. michiganensis* subsp. *nebraskensis* (Gross and Vidaver, 1979). For the original recipe, one liter of NBY medium (see above) is made, with the addition of 10 g lithium chloride. To this sterile, molten medium cooled to 50°C are added: nalidixic acid, 25 mg (2.5 ml freshly made 1% solution in 0.1 M NaOH); polymyxin B sulfate, 32 mg (3.2 ml of

1% aqueous solution of 8,000 USP units/mg, or a total of 256,000 units); cycloheximide, 40 mg (4 ml of 1% aqueous solution); and tetrachloroisophthalonitrile, 0.66 mg (0.625 ml of Bravo 6F diluted 1:50). A slightly different formula has also been published (Vidaver and Davis, 1988). Smidt and Vidaver (1984) found that one particular preparation of polymyxin B sulfate did not allow growth of single colonies, even at one-tenth of the usual concentration. This was overcome either by using a different preparation of antibiotic or omitting the lithium chloride. It is thus advisable to test new batches of antibiotic before use. The same authors later reported that lithium chloride is transiently toxic to newly isolated *Cl. michiganensis* subsp. *nebraskensis* (Smidt and Vidaver, 1986). This can be overcome either by omitting it from the medium, or by allowing the plant suspension to stand in a buffer containing sodium for 1–2 h before streaking out.

CNS agar medium is also useful for isolating *Cl. michiganensis* subsp. *michiganensis* and *Cl. tritici,* but gives variable or no growth with other clavibacters (Gross and Vidaver, 1979). This is the medium of choice for those species and subspecies which grow, as the appearance of the colonies is similar to that on NBY and is of diagnostic value.

For the isolation of *Cl. xyli* subsp. *cynodontis,* a selective medium (SCMS) was developed from SC medium (Davis and Augustin, 1984). To prepare it, 50 mg of cycloheximide, 50 mg of 3-hydroxy-5-methylisoxazole (hymexazole, Tachigaren; Sankyo Co., Japan), 10 mg of colistin methanesulfonate, and 50 mg of polymyxin B sulfate were dissolved in 10 ml of water, filter sterilized, and added to 1 liter of molten SC agar at 50°C. On this medium, 7–14 d are required for clearly visible colonies, rather than 5–7 d on the nonselective medium SC. Colonies are round, convex, entire, and yellow orange. This medium was not used for *Cl. xyli* subsp. *xyli,* but may well be selective for this organism also, as it is closely related. If selective media cannot be used, successful isolation of these two pathogens requires surface sterilization of the plant material and attention to aseptic technique to reduce the growth of saprophytes to the absolute minimum.

Medium D2 (Kado and Heskett, 1970) is reported to be selective for various coryneforms.

Medium D2

Glucose	10 g
Casein hydrolysate	4 g
Yeast extract	2 g
NH$_4$Cl	1 g
LiCl	5 g
MgSO$_4$ · 7H$_2$O	0.3 g
Tris HCl buffer	1.2 g
Agar	15 g
Water	1 liter

The pH is adjusted to 7.8 before autoclaving. The molten medium is cooled to 50°C and polymyxin B sulfate, 40 mg, and sodium azide, 2 mg, are added immediately before dispensing. The original publication specifies "(300 units)" after polymyxin B sulfate, 40 mg. In practice, we find that 40 mg is effective. This weight of commercially available preparations represents about 300,000 USP units.

This medium not only allows growth of various coryneforms, including species of *Clavibacter* and *Curtobacterium,* but also human and animal pathogenic coryneforms, Gram-positive cocci, some Gram-negative bacteria, and some fungi. The coryneforms are reported to be distinguishable by their small, light-yellow, circular, convex, glistening colonies.

Identification

The identification of *Clavibacter* species can be approached in a number of different ways. Normally a plant pathologist will need only a presumptive identification to confirm the cause of a disease observed in the field. In that case it is usually sufficient to isolate slow-growing, non-motile, Gram-positive rods showing a tendency to coryneform morphology and agreeing in colonial characteristics (color, texture, consistency, etc.) with the expected pathogen. Confirmation may then be possible by inoculation into a healthy host plant to produce the characteristic symptoms. However, a successful pathogenicity test may prove difficult if the right plants at the right stage of growth are not available, or if the cereal and grass pathogens *Cl. rathayi, Cl. tritici,* and *Cl. iranicus* are involved, as these have an infection route via nematodes.

If information on origin, host plant, and symptoms is lacking, a definitive identification will require a number of additional tests. An organism that is obligately aerobic, catalase positive, oxidase negative, contains Dab in the cell wall peptidoglycan, and whose the predominant menaquinones has 9 or 10 isoprene units is distinguished from other coryneforms and belongs to the genus *Clavibacter.* Phenotypic tests to distinguish species and subspecies of *Clavibacter* are somewhat equivocal; Davis et al. (1984) obtained many results at variance with those of Dye and Kemp (1977) using the same strains. However, different methods were used, and if the methods are carefully followed, the results should be meaningful. Some useful phenotypic distinctions are outlined in Table 2.

Table 2. Differential characteristics of the species and subspecies of *Clavibacter*.

Characteristic	Cl. iranicus	Cl. rathayi	Cl. tritici	Cl. m. insidiosus	Cl. m. michiganensis	Cl. m. nebraskensis	Cl. m. sepedonicus	Cl. m. tessellarius	Cl. x. cynodontis	Cl. x. xyli
Fastidious, requires SC or similar medium	−	−	−	−	−	−	−	−	+	+
Color of colonies on NBY or SC medium	Y	Y	Y	Y	Y[a]	O[b]	W	Y	Y	W
Indigoidine (dark blue, slightly diffusible pigment)	−	−	−	(+)	−[c]	−	−	−	−	−
Growth on CNS agar	−	v	+	−	+	+	−	+	−	−
Growth on TTC agar[d]	ND	ND	ND	+	+	−	−	+	−	−
Acid production from:										
Sorbitol	−	−	−	−	−	+[e]	+[e]	+	−	−
Inulin	+[e]	−	+	−	−	−	−	−	−	−
Utilization of acetate	−	−	+	−	−[e]	+	+	−	−	−
Isoprene units of major menaquinone[f]	10	10	10	9	9	9	9	9	ND	ND

Symbols: +, positive; −, negative; (+), most isolates positive; Y, yellow; O, orange; W, white; ND, no data.
[a]Rarely orange, red, pink, or colorless strains are seen.
[b]Sometimes yellow strains are seen.
[c]Occasionally an isolate produces indigoidine.
[d]TTC agar contains (g/l) glucose, 5; peptone, 10; casein hydrolysate, 1; agar, 17; 2,3,5-triphenyltetrazolium chloride, 0.05 (5 ml 1% aqueous solution filter sterilized and added to the molten medium cooled to 50°C).
[e]The opposite result was reported by Dye and Kemp (1977) using different methods.
[f]The method is given in Collins (1985).
Adapted from Davis et al. (1984), Vidaver and Davis (1988), and Collins and Bradbury (1986).

A number of rapid methods useful for identification or confirmation have been developed, and some look promising. Most, however, require special apparatus or facilities that may not be available. Serological methods are very sensitive and particularly useful for detection. Immunofluorescence allows direct observation of cells in the plant host and is useful for detection of *Cl. michiganensis* subspp. *insidiosus, michiganensis,* and *sepedonicus* (Lelliott and Stead, 1987). Agglutination, double gel diffusion, and indirect fluorescent antibody staining have all been used for *Cl. michiganensis* subsp. *sepedonicus,* and various cross-reactions have been reported to interfere (Slack et al., 1979; De Boer, 1982). Unfortunately, even with the use of monoclonal antibodies, cross-reactions occur, not only between clavibacters, but also with saprophytic coryneforms (De Boer and Wieczorek, 1984). Analysis of soluble proteins by polyacrylamide gel electrophoresis is also rapid and can give repeatable patterns (Carlson and Vidaver, 1982). DNA hybridization probes have recently been used for *Cl. michiganensis* subsp. *sepedonicus* (Johansen et al., 1989).

The Genus *Curtobacterium*

Cells are irregular, unbranched rods tending to become shorter with age; Gram-positive, nonsporeforming, and multiplying by a bending type of cell division. Usually motile with sparse peritrichous flagella, catalase positive, obligately aerobic, chemoorganotrophic and capable of growth on nutrient agar. Acid is produced rather weakly from glucose and fructose and usually from xylose, rhamnose, and some other carbohydrates. Pyruvic acid is used as a sole carbon source, and most strains also use acetic, fumaric, glycolic, and succinic acids. Acid is not produced from lactose, trehalose, sorbitol, salicin, dextrin, inulin, or starch. DNAse is produced, but not urease. Gelatin is hydrolyzed by most strains. Growth usually occurs in 5% NaCl, but only one species will grow in 10%. The cell wall peptidoglycan contains D-ornithine as the diamino acid of the peptide bridge and is of type B2β, [L-homoserine]-D-Glu-D-Orn of Schleifer and Kandler (1972). Mycolic acids are not present. The predominant fatty acids are *anteiso*-C$_{15:0}$, *aneteiso*-C$_{17:0}$ with substantial amounts of *iso*-C$_{16:0}$, and some others (Collins et al., 1980; Stead, 1988). Polar lipids are diphosphatidylglycerol, phosphatidylglycerol, and some glycosyldiacylglycerols (Collins et. al., 1980). Menaquinones are the sole respiratory quinones; unsaturated menaquinones with nine isoprene units predominating. The

GC content of the DNA is within the range of 68.3–75.2 mol%.

Habitats

Five of the six known species have been isolated from plant material, and *Curtobacterium pusillum* was isolated from oil brine in a Japanese oil field (Iizuka and Komagata, 1965). *Cu. albidum, Cu. citreum,* and *Cu. luteum* were all isolated from rice grain (Komagata and Iizuka, 1964) but are not known to cause any disease on rice. The main occurrence of *Cu. flaccumfaciens* is in diseased plants (its four pathovars causing diseases in their respective host plants). The pathovar *flaccumfaciens* causes a wilt of beans *(Phaseolus* species and *Glycine max);* pathovar *poinsettiae* causes stem canker and leaf spot of poinsettia; pathovar *betae* causes a wilt and leaf spot of red beet; pathovar *oortii* causes a disease of tulip. All are able to infect their hosts systemically, and all are transmitted in or on seed (Richardson, 1979, 1983). *Curtobacterium plantarum* was isolated from all 200 species of plant leaves examined (Dunleavy, 1989). A total of 62 plant families were represented, including herbs, shrubs, and trees from a widespread area in the state of Iowa (USA). The isolation method used involved no surface sterilization, so it is not possible to deduce whether the bacteria were carried within the leaf tissues or closely adherent to the surfaces. In seed, however, the bacteria were apparently carried within the testa as they were not eliminated by rigorous surface sterilization.

Isolation

To isolate plant pathogens, the principles outlined in the section on the isolation of *Clavibacter* species should be followed, even though the plant pathogenic curtobacteria grow more rapidly. Nutrient agar to which 1% glucose has been added is a satisfactory medium for isolation and growth (Lelliott and Stead, 1987). Kado and Heskett (1970) reported that their selective medium D2 (see above) gave good growth of one strain each of *Curtobacterium flaccumfaciens* pv. *flaccumfaciens* and *Cu. flaccumfaciens* pv. *poinsettiae.* Vidaver and Starr (1981) report that growth occurs on CNS medium (see above) with *Cu. flaccumfaciens* pvs. *oortii* and *betae,* but that strains of pvs. *flaccumfaciens* and *poinsettiae* vary; some grow well, others not at all. Later, Vidaver and Davis (1988) reported that all four pathovars grow on CNS medium. No selective media or special methods for enrichment have been developed for the non-plant pathogens, which may be isolated by dilution techniques on media based on peptone, yeast extract, and glucose (Komagata and Suzuki, 1986), and on tryptic soy agar (Dunleavy, 1989).

Identification

For phytopathogenic strains it is probably simplest to examine morphology and staining reactions (Gram-positive with at least some tendency to pleomorphism or to angular and palisade cell arrangements, but no strong tendency for a rod-to-coccus developmental cycle) and then base the species determination on host and symptoms, confirmed by a pathogenicity test. For saprophytes, soil isolates, strains of unknown origin, or suspected new species, identification to genus level requires additional chemotaxonomic and laboratory testing. A cell wall analysis to demonstrate the presence of ornithine is essential (see Schleifer and Kandler, 1972; Bousfield et al., 1985). Acid production and good growth with glucose, fructose, or other sugar should be shown to occur in the aerobic tube only, indicating the strictly aerobic pattern of metabolism, unlike that of *Cellulomonas,* which also contains ornithine but is a facultative anaerobe. Cellulose hydrolysis, which is negative for *Curtobacterium,* can be tested by inoculating a strip of filter paper standing half-submerged in 0.5% peptone water and testing for breakdown of the paper (Skerman, 1967). This may take several weeks. The menaquinone composition should also be examined (see Collins, 1985 for methods); *Curtobacterium* contains largely MK-9, whereas *Aureobacterium* contains longer prenologs (mostly MK-11 to MK-13), and *Cellulomonas* contains the tetrahydrogenated MK-9 (H_4). Polar lipid analyses also readily distinguish curtobacteria and cellulomonads (Minnikin et al., 1979; Collins et al., 1980). Some simple additional tests that might also help to confirm identification are catalase (positive), urease (negative), and motility (usually positive).

Further identification to species level in the laboratory is likely to involve a considerable number of tests or use of specialized techniques, such as serology, phage typing, protein analysis by PAGE, or DNA hybridization, to compare with strains of known identity.

The Genus *Microbacterium*

In young cultures, cells of microbacteria are irregular short slender rods (ca. 0.4 to 0.8 μm in diameter and 1 to 4μm in length) occurring singly or in groups. Some cells are arranged at angles to each other to give V-forms. Primary branching is uncommon. In older cultures (3 to 7 d) the cells become shorter (a proportion may be coccoid), but a marked rod-coccus growth cycle does not occur. Both rods and coccoid forms are Gram-positive, but older cultures may fail to retain the Gram stain. Cells may be nonmotile or motile. Pigments are produced (gray-white to yellow-orange depending on species). Nutritional requirements are complex, all require B vitamins, and some require amino acids also. Metabolism is primarily respiratory, but may also be fermentative. Some workers have reported that growth does not occur under strictly anaerobic conditions (Abd-el-Malek and Gibson, 1952; Orla-Jensen, 1919), whereas others that weak anaerobic growth occurs (Robinson, 1966; Jones, 1975). The cell wall peptidoglycan contains lysine as the diamino acid of type B1α, [L-lysine]-D-glutamic acid-glycine-L-lysine *(M. lacticum* and *M. laevaniformans)* or of type B1β, [L-homoserine]-D-glutamic acid-glycine$_2$-L-lysine *(M. imperiale* and *M. arborescens)* (Schleifer and Kandler, 1972). The glycan moiety of the cell wall contains both *N*-glycolyl and *N*-acetyl residues (Uchida and Aida, 1977). Mycolic acids are not produced. The long-chain cellular fatty acids are mainly of the *anteiso-* and *iso*-methyl branched types with *anteiso*-$C_{15:0}$, *iso*-$C_{16:0}$, and *anteiso*-$C_{17:0}$ acids predominating. All microbacteria contain diphosphatidylglycerol, phosphatidylglycerol, and dimannosyldiacylglycerol as their major polar lipids. Minor amounts of monomannosyldiacylglycerol and a phosphoglycolipid may also occur (Collins et al., 1983a). Menaquinones are the sole respiratory quinones present. Unsaturated menaquinones with 10, 11 and 12 isoprene units predominate. The GC content is within 69 to 75 mol%.

Habitats

Microbacteria have been isolated from a variety of habitats. *M. lacticum* is usually found in milk, powdered milk, cheese, and on dairy equipment, and it is considered to represent a major part of the thermoduric, coryneform bacterial population of such sources (Jayne-Williams and Skerman, 1966; Thomas et al., 1967; Gillies, 1971). *M. laevaniformans* has been isolated from raw sewage and activated sludge (Dias and Bhat, 1962; Dias, 1963). The only known report of the isolation of *M. imperiale* is from the alimentary canal of the imperial moth *Eacles imperialis* (Steinhaus, 1941). Organisms which phenotypically resemble microbacteria have also been isolated from fresh beef, poultry giblets, and raw and pasteurized eggs (Kraft et al., 1966; Splittstoesser et al., 1967).

Table 3. Differential characteristics of the species of the genus *Microbacterium.*[a]

Characteristic	M. arborescens	M. imperiale	M. lacticum	M. laevaniformans
Colony color	Red to orange-red	Red to orange	Gray-white to yellow	Yellow
Nitrate reduction	−	−	+	−
H₂S production	+	−	−	+
Hydrolysis of:				
Gelatin	+	+[d]	+[w]	+
Starch	−	+[d]	+[w]	+
Acid from:				
Arabinose	+[w]	+	−	−
Xylose	+[w]	+[d]	−	−
Sucrose	+[w]	+	−	+
Peptidoglycan type	B1β	B1β	B1α	B1α

Symbols: +, positive; −, negative; +[d], delayed reaction; +[w], weak positive.
Physiological data are from Collins et al. (1983a), Dias and Bhat (1962), Imai et al. (1984), Jones (1975), and Steinhaus (1941).

The identity of these organisms, however, cannot be certain.

Isolation and Cultivation

Since most strains of *M. lacticum* are thermoduric (survive 63°C for 30 min or 72°C for 15 min in skim milk) (Abd-el-Malek and Gibson, 1952; Jayne-Williams and Skerman, 1966) the usual procedure for isolating this organism from dairy products involves plating laboratory-pasteurized (63°C for 30 min) samples onto a suitable, nonselective medium. However, the pasteurization step should be omitted if nonthermoduric strains are also required. A suitable nonselective medium is yeast extract milk agar (YMA; Harrigan and McCance, 1976).

An enrichment procedure has been described by Dias and Bhat (1962) for the isolation of *M. laevaniformans* from raw sewage and activated sludge. Bacteria other than *M. laevaniformans* are, however, also enriched by this method. A good general-purpose medium for the cultivation of all microbacteria is coryneform agar (see "The Genus *Aureobacterium*"). Strains should be incubated aerobically at 30°C.

Identification

Identification of microbacteria requires an examination of cellular morphology and a number of chemotaxonomic tests. Members of the genus *Microbacterium* contain lysine as the diamino acid and thus may be differentiated from other coryneform bacteria with a group B peptidoglycan. Indeed, the combination of a cell wall based on lysine and the presence of long unsaturated menaquinones (MK-10 to MK-12) serves to distinguish microbacteria from all the other actinomycete and coryneform taxa described to date. Polar lipid analyses are also useful in distinguishing microbacteria from members of the genus *Aureobacterium* (Collins et al., 1983b).

It is currently not easy to select phenetic characters that allow the definitive identification of the four *Microbacterium* species. However, characters selected from the literature which may be of value in differentiating the species are shown in Table 3. It should be noted that the recently described *M. arborescens* (Imai et al., 1984) is similar to *M. imperiale* in possessing the unusual B1β type murein. These taxa are also biochemically very similar, and nucleic acid studies are necessary determine whether or not they are, in fact, separate species.

Literature Cited

Abd-el-Malek, Y., and Gibson, T. 1952. Studies in the bacteriology of milk III. The corynebacteria of milk. J. Dairy Res. 19:153–159.

Abe, S., Takayama, K., and Kinoshita, S. 1967. Taxonomical studies on glutamic acid-producing bacteria. J. Gen. Appl. Microbiol. 13:279–301.

Barksdale, L. Laneelle, M. A., Pollice, M. C., Asselineau, J., Welby, M., and Norgard, M. V. 1979. Biological and chemical basis for the reclassification of *Microbacterium flavum* Orla-Jensen as *Corynebacterium flavescens* nom.nov. Int. J. Syst. Bacteriol. 29:222–233.

Bird, A. F., and Stynes, B. A. 1977. The morphology of a *Corynebacterium* sp. parasitic on annual rye grass. Phytopathology 67:828–830.

Bousfield, I. J., Keddie, R. M., Dando, T. R., and Shaw, S. 1985. Simple rapid methods of cell wall analysis as an aid in the identification of aerobic coryneform bacteria, p. 221–236. In: M. Goodfellow and D. E. Minnikin (ed.), Chemical methods in bacterial systematics. Academic Press, London.

Bousfield, I. J., Smith, G. L., Dando, T. R., and Hobbs, G. 1983. Numerical analysis of total fatty acid profiles in the identification of coryneform, nocardioform and some other bacteria. J. Gen. Microbiol. 129:375–394.

Bradbury, J. F. 1986. Guide to plant pathogenic bacteria. Commonwealth Agricultural Bureau (CAB) International Mycological Institute, Kew, UK.

Bugbee, W. M., Gudmestad, N. C., Secor, G. A., and Nolte, P. 1987. Sugar beet as a symptomless host for *Corynebacterium sepedonicum*. Phytopathology 77:765–770.

Carlson, R. R., and Vidaver, A. K. 1982. Taxonomy of *Corynebacterium* plant pathogens, including a new pathogen of wheat, based on polyacrylamide gel electrophoresis of cellular proteins. Int. J. Syst. Bacteriol. 32:315–326.

Casida, L. E. 1965. Abundant microorganisms in soil. Appl. Microbiol. 13:327–334

Collins, M. D. 1982. Lipid composition of *Agromyces ramosus*. FEMS Microbiol. Lett. 14:187–189.

Collins, M. D. 1983. Cell wall peptidoglycan and lipid composition of the phytopathogen *Corynebacterium rathayi* (Smith). Syst. Appl. Microbiol. 4:193–195.

Collins, M. D. 1985. Isoprenoid quinone analyses in bacterial classification and identification, p. 267–287. In: M. Goodfellow and D. E. Minnikin (ed.), Chemical methods in bacterial systematics. Academic Press, London.

Collins, M. D., and Bradbury, J. F. 1986. Plant pathogenic species of *Corynebacterium*, p. 1276–1283. In: P. H. A. Sneath, N. A. Mair, M. E. Sharpe, and J. G. Holt (ed.), Bergey's manual of systematic bacteriology, vol. 2. Williams and Wilkins, Baltimore.

Collins, M. D., Goodfellow, M., and Minnikin, D. E. 1980. Fatty acid, isoprenoid quinone and polar lipid composition in the classification of *Curtobacterium* and related taxa. J. Gen. Microbiol. 118:29–37.

Collins, M. D., Goodfellow, M., and Minnikin, D. E. 1982. A survey of the structure of mycolic acids in *Corynebacterium* and related taxa. J. Gen. Microbiol. 128:129–149.

Collins, M. D., and Jones, D. 1980. Lipids in the classification and identification of coryneform bacteria containing peptidoglycans based on 2,4-diaminobutyric acid. J. Appl. Bacteriol. 48:459–470.

Collins, M. D. and Jones, D. 1981. The distribution of isoprenoid quinone structural types in bacteria and their taxonomic implications. Microbiol. Rev. 45:316–354.

Collins, M. D. and Jones, D. 1983. Reclassification of *Corynebacterium flaccumfaciens, Corynebacterium betae, Corynebacterium oortii* and *Corynebacterium poinsettiae* in the genus *Curtobacterium*, as *Curtobacterium flaccumfaciens* comb.nov. J. Gen. Microbiol. 129:3545–3548.

Collins, M. D., Jones, D., Keddie, R. M., Kroppenstedt, R. M., and Schleifer, K. H. 1983b. Classification of some coryneform bacteria in a new genus *Aureobacterium*. Syst. Appl. Microbiol. 4:236–252.

Collins, M. D., Jones, D., and Kroppenstedt, R. M. 1983a. Reclassification of *Brevibacterium imperiale* (Steinhaus) and *"Corynebacterium laevaniformans"* (Dias and Bhat) in a redefined genus *Microbacterium* (Orla-Jensen), as *Microbacterium imperiale* comb.nov. and *Microbacterium laevaniformans* nom.rev.; comb.nov. System. Appl. Microbiol. 4:65–78.

Davis, M. J., and Augustin, B. J. 1984. Occurrence in Florida of the bacterium that causes bermuda grass stunting disease. Plant Disease 68:1095–1097.

Davis, M. J., Gillespie, A. G., Jr., Harris, R. W., and Lawson, R. H. 1980. Ratoon stunting disease of sugarcane: Isolation of the causal bacterium. Science 210:1365–1367.

Davis, M. J., Gillespie, A. G., Jr., Vidaver, A. K., and Harris, R. W. 1984. *Clavibacter:* a new genus containing some phytopathogenic coryneform bacteria, including *Clavibacter xyli* subsp. *xyli* sp.nov., subsp.nov. and *Clavibacter xyli* subsp. *cynodontis* subsp. nov., pathogens that cause ratoon stunting disease of sugar-cane and Bermudagrass stunting disease. Int. J. Syst. Bacteriol. 34:107–117.

De Boer, S. H. 1982. Cross reaction of *Corynebacterium sepedonicum* antisera with *C. insidiosum, C. michiganense,* and an unidentified coryneform bacterium. Phytopathology 72:1474–1481.

De Boer, S. H., and Wieczorek, A. 1984. Production of monoclonal antibodies to *Corynebacterium sepedonicum*. Phytopathology 74:1431–1434.

De Boer, S. H., Wieczorek, A., and Kumer, A. 1988. An ELISA test for bacterial ring rot of potato with a new monoclonal antibody. Plant Disease 72:874–878.

Dias, F. F. 1963. Studies in the bacteriology of sewage. J. Indian Inst. Sci. 45:36–48.

Dias, F., and Bhat, J. V. 1962. A new levan producing bacterium, *Corynebacterium laevaniformans* nov. spec. Antonie van Leeuwenhoek J. Microbiol. Serology 28:63–72.

Dias, F. F., Bilimoria, M. H., and Bhat, J. V. 1962. *Corynebacterium barkeri* nov. spec., a pectinolytic bacterium exhibiting a biotin-folic acid inter-relationship. J. Ind. Inst. Sci. 44:59–67.

Döpfer, H., Stackebrandt, E., and Fiedler, F. 1982. Nucleic acid hybridization studies on *Microbacterium, Curtobacterium, Agromyces* and related taxa. J. Gen. Microbiol. 128:1697–1708.

Dunleavy, J. M. 1989. *Curtobacterium plantarum* sp. nov. is ubiquitous in plant leaves and is seed transmitted in soybean and corn. Int. J. Syst. Bacteriol. 39:240–249.

Dye, D. W., and Kemp, W. J. 1977. A taxonomic study of plant pathogenic *Corynebacterium* species. N.Z. J. Agric. Res. 20:563–582.

Gillies, A. J. 1971. Significance of thermoduric organisms in Queensland Cheddar cheese. Australian J. Dairy Technol. 26:145–149.

Gledhill, W. E., and Casida, L. E. 1969. Predominant catalase-negative soil bacteria III. *Agromyces*, gen.n., microorganisms intermediary to *Actinomyces* and *Nocardia*. Appl. Microbiol. 18:340–349

Gross, D. C., and Vidaver, A. K. 1979. A selective medium for the isolation of *Corynebacterium nebraskense* from soil and plant parts. Phytopathology 69:82–87.

Gupta, P., and Swarup, G. 1972. Ear-cockle and yellow ear rot disease of wheat II. Nematode bacterial association. Nematologia 18:320–324.

Harrigan, W. F., and McCance, M. E. 1976. Laboratory methods in food and dairy microbiology, revised ed. Academic Press, London.

Hayward, A. C. 1974. Latent infections by bacteria. Annual Review of Phytopathlogy 12:87–97.

Iizuka, H., Komagata, K. 1965. Microbiological studies on petroleum and natural gas. III. Determination of *Brevibacterium, Arthrobacter, Micrococcus, Sarcina, Alcaligenes* and *Achromobacter* isolated from oil brines in Japan. J. Gen. Appl. Microbio. 11:1–14.

Imai, K., Takeuchi, M., and Banno, I. 1984. Reclassification of *"Flavobacterium arborescens"* (Frankland and Frankland) Bergey et al. in the genus *Microbacterium* (Orla-Jensen) Collins et al., as *Microbacterium arborescens* comb.nov., nom.rev. Curr. Microbiol. 11:281–284.

Jayne-Williams, D. J., and Skerman, T. M. 1966. Comparative studies on coryneform bacteria from milk and diary sources. J. Appl. Bacteriol. 29:72–92.

Johansen, I. E., Rasmussen, O. F., and Heide, M. 1989. Specific identification of *Clavibacter michiganese* subsp. *sepedonicum* by DNA-hybridization probes. Phytopathology 79:1019–1023.

Jones, D. 1975. A numerical taxonomic study of coryneform and related bacteria. J. Gen. Microbiol. 87:52–96.

Jones, D., Watkins, J., and Meyer, D. J. 1970. Cytochrome composition and effect of catalase or growth of *Agromyces ramnosus*. Nature 226:1249–1250.

Kado, C. I., and Heskett, M. G. 1970. Selective media for isolation of *Agrobacterum, Corynebacterium, Erwinia, Pseudomonas* and *Xanthomonas*. Phytopathology 60:969–976.

Keddie, R. M., and Cure, G. L. 1977. The cell wall composition and distribution of free mycolic acids in named strains of coryneform bacteria and in isolates from various natural sources. J. Appl. Bacteriol. 42:229–252

Komagata, K., and Iizuka, H. 1964. New species of *Brevibacterium* isolated from rice (Studies on the microorganisms of cereal grains. Part VII). [In Japanese] J. Agri. Chem. Soc. Japan 38:496–502

Komagata, K. and Suzuki, K. 1986. Genus *Curtobacterium*, p. 1313–1317. P. H. A. Sneath, N. A. Mair, M. E. Sharpe, and J. G. Holt (ed.), Bergey's manual of systematic bacteriology, vol. 2. Williams and Wilkins, Baltimore.

Komura, I., Yámada, K., Otsuka, S., and Komagata, K. 1975. Taxonomic significance of phospholipids in coryneform and nocardioform bacteria. J. Gen. Appl. Microbiol. 21:251–261.

Kraft, A. A., Ayres, J. C., Torrey, G. S., Salzer, R. H., and da Silva, G. A. N. 1966. Coryneform bacteria in poultry, eggs and meat. J. Appl. Bacteriol. 29:161–166.

Kuhn, R., Starr, M. P., Kuhn, D. A., Bauer, H., and Knackmuss, H. J. 1965. Indigoidine and other bacterial pigments related to 3,3'-bipyridyl. Archiv. Mikrobiologie 51:71–84.

Lelliott, R. A., and Stead, D. E. 1987. Methods for the diagnosis of bacterial diseases of plants. 216pp. Blackwell Scientific, Oxford, UK.

Lochhead, A. G. 1958. Two new species of *Arthrobacter* requiring respectively vitamin B12 and the terregens factor. Archiv. Mikrobiologie 31:163–170.

Lochhead, A. G., and Burton, M. O. 1953. An essential bacterial growth factor produced by microbial synthesis. Canadian J. Bot. 31:7–22.

Lysenko, O. 1959. The occurrence of species of the genus *Brevibacterium* in insects. J. Insect Pathol. 1:34–42.

McLean, R. A., and Sulzbacher, W. L. 1953. *Microbacterium thermosphactum,* spec.nov.; a nonheat resistant bacterium from fresh pork sausage. J. Bacteriol. 65:428–433.

Minnikin, D. E., Collins, M. D., and Goodfellow, M. 1979. Fatty acid and polar lipid composition in the classification of *Cellulomonas, Oerskovia* and related taxa. J. Appl. Bacteriol. 47:87–95.

Moffett, M. L., Fahy, P. C., and Cartwright, D. 1983. *Corynebacterium,* p. 45–65. In: P. C. Fahy and G. J. Persley (ed.), Plant bacterial diseases; a diagnostic guide. Academic Press, Sydney, Australia.

Orla-Jensen, S. 1919. The lactic acid bacteria. Host & Son, Copenhagen.

Richardson, L. T. 1957. Quantitative determination of viability of potato ring rot bacteria following storage, heat, and gas treatment. Canadian J. Botany 35:647–656.

Richardson, M. J. 1979. An annotated list of seed-borne diseases, third edition. Phytopathological Paper 23. Commonwealth Agricultural Bureau (CAB) International Mycological Institute, Kew, UK.

Richardson, M. J. 1983. An annotated list of seed-borne diseases, 3rd ed., Suppl. 2. International Seed Testing Association, Zurich.

Riley, I. T. 1987. Serological relationships between strains of coryneform bacteria responsible for annual ryegrass toxicity and other plant pathogenic corynebacteria. Int. J. Syst. Bacteriol. 35:153–159.

Riley, I. T., Reardon, T. B., and McKay, A. C. 1988. Genetic analysis of plant pathogenic bacteria in the genus *Clavibacter* using allozyme electrophoresis. J. Gen. Microbiol. 134:3025–3030.

Robinson, K. 1966. Some observations on the taxonomy of the genus *Microbacterium*. I. Cultural and physiological reactions and heat resistance. J. Appl. Bacteriol. 29:607–615

Sabet, K. A. 1954. On the host range and systematic position of the bacteria responsible for the yellow slime diseases of wheat (*Triticum vulgare* Vill.) and cocksfoot grass (*Dactylis glomerata* L.). Annals Appl. Biol. 41:606–611.

Saperstein, S., Starr, M. P., and Filfus, J. A. 1954. Alterations in carotenoid synthesis accompanying mutation in *Corynebacterium michiganense*. J. Gen. Microbiol. 10:85–92

Schleifer, K. H., and Kandler, O. 1972. Peptidoglycan types of bacterial cell walls and their taxonomic implications. Bacteriol. Rev. 36:407–477.

Schuster, M. L., and Coyne, D. P. 1974. Survival mechanisms of phytopathogenic bacteria. Ann. Rev. Phytopathology 12:199–221.

Skerman, V. B. D. 1967. A guide to the identification of the genera of bacteria, 2 ed. Williams & Wilkins, Baltimore.

Skerman, V. B. D., McGowan, V., and Sneath, P. H. A. 1980. Approved lists of bacterial names. Int. J. Syst. Bacteriol. 30:225–420.

Slack, S. Kelman, A., and Perry, J. 1979. Comparison of three serodiagnostic assays for the detection of *Corynebacterium sepedonicum*. Phytopathology 69:186–189.

Smidt, M. L., and Vidaver, A. K. 1984. Inhibition of *Corynebacterium michiganense* subsp. *nebraskense* by certain lots of polymyxin used in the selective medium. Plant Disease 68:536.

Smidt, M. L., and Vidaver, A. K. 1986. Differential effects of lithium chloride on in vitro *Clavibacter michiganense* subsp. *nebraskense* depending upon inoculum source. Appl. Environ. Microbiol. 52:591–593.

Sneath, P. H. A., and Jones, D. 1976. *Brochothrix,* a new genus tentatively placed in the family Lactobacillaceae. Int. J. Syst. Bacteriol. 26:102–104

Splittstoesser, D. F., Wexler, M., White, J., and Colwell, R. R. 1967. Numerical taxonomy of Gram-positive and catalase-positive rods isolated from frozen vegetables. Appl. Microbiol. 15:158–162.

Starr, M. P. 1958. The blue pigment of *Corynebacterium insidiosum.* Archiv. Mikrobiologie 30:325–334.

Starr, M. P., Mandel, M., and Murata, N. 1975. The phytopathogenic coryneform bacteria in the light of DNA base composition and DNA-DNA segmental homology. J. Gen. Appl. Microbiol. 21:13–26.

Stead, D. E. 1988. Identification of bacteria by computer assisted fatty acid profiling. Acta Horticulturae 225:39–46.

Steinhaus, E. A. 1941. A study of the bacteria associated with thirty species of insects. J. Bacteriol. 42:757–790.

Thomas, S. B., Druce, R. G., Peters, G. J., and Griffiths, D. G. 1967. Incidence and significance of thermoduric bacteria in farm milk samples: A reappraisal and review. J. Appl. Bacteriol. 30:265–298.

Topping, L. E. 1937. The predominant micro organisms in soils. I. Description and classification of the organisms. Zent. Bakt. Parasit, Inf. Hyg. Abt. 2 Orig. 97:289–304.

Uchida, K., and Aida, K. 1977. Acyl type of bacterial cell wall: its simple identification by colorimetric method. J. Gen. Appl. Microbiol. 23:249–260

Vidaver, A. K. 1967. Synthetic and complex media for rapid detection of fluorescence of phytopathogenic pseudomonads: effect of the carbon source. Appl. Microbiol. 15:1523–1524.

Vidaver, A. K., and Davis, M. J. 1988. Coryneform plant pathogens, p. 104–113. In: N. W. Schaad (ed.), Laboratory guide for identification of plant pathogenic bacteria, 2nd ed. APS Press, St. Paul, MN, USA.

Vidaver, A. K., and Starr, M. P. 1981. Phytopathogenic coryneform and related bacteria, p. 1879–1887. In: M. P. Starr, H. Stolp, H. G. Truper, A. Balows, and H. G. Schlegel (ed.), The prokaryotes, vol 2. Springer-Verlag, Berlin.

Yamada, K., and Komagata, K. 1972. Taxonomic studies on coryneform bacteria. V. Classification of coryneform bacteria. J. Gen. Appl. Microbiol. 18:417–43.

The Genus *Staphylococcus*

WESLEY E. KLOOS, KARL-HEINZ SCHLEIFER, and FRIEDRICH GÖTZ

Ogston (1883) introduced the name *Staphylococcus* (σταφυλή , a bunch of grapes) for the group-micrococci causing inflammation and suppuration. He was the first to differentiate two kinds of pyogenic cocci or micrococci: one arranged in groups or masses was called *Staphylococcus* and another arranged in chains was named Billroth's *Streptococcus*. Although Ogston's reference to *Staphylococcus* was only in the descriptive sense, Rosenbach (1884) used the term in the taxonomic sense and provided a formal description of the genus *Staphylococcus,* dividing the genus into the two species *Staphylococcus aureus* and *S. albus*. In the same publication, Rosenbach also used the trinomials *S. pyogenes aureus* and *S. pyogenes albus* to denote these species. Passet (1885) added a third species, *S. citreus*. At this point in time, cell morphology and type of cell aggregation (or division) provided the criteria for genus classification, and colony pigment (or color) was the sole criterion for staphylococcal species classification. The interest in pigment was due to Verneuil's (1880) observations on the color of pus, and to Ogston's (1882) connecting the yellow or orange-yellow appearance of pus with the color of growth of the infecting micrococci. Later, it was shown that pigment should not be the only basis for classification, since it was not a stable character in many strains.

Prior to Ogston's investigations, staphylococci were included together with other spherical bacteria under various classifications and names. The ancient traces of staphylococci, as indicated by the descriptions of diseases which they are now known to produce, were noted by the early Jewish, Egyptian, Greek, and Roman writers and also suggested by their customs (reviewed by Bulloch, 1938; Hare, 1954; Hooten, 1930; Ruffer, 1914). In his treatise *De Contagione, Contagiosis Morbis et eorum Curatione,* Fracastorius (1546) presented the theory of *seminaria contagiones* and introduced the term *seminaria* to denote the "seeds" or "germs" of disease. "His work contained the first scientific statement of the true nature of contagion, of infection, of disease germs and the modes of transmission of infectious diseases.... These germs he describes as particles too small to be apprehended by our senses, but which in appropriate media, are capable of reproduction and thus of infecting the surrounding tissues." (Garrison, 1910). In Book II of the treatise, Fracastorius describes various infections including furuncles (boils), many of which were probably caused by staphylococci. It was Leeuwenhoek (1684, 1695) who first observed spherical and other bacteria under the microscope. He called them animalcules and affectionately spoke of them as "beasties" (beejes) or "little creatures" (cleijne Schepsels). In Leeuwenhoek's figures of bacteria from the human mouth (Letter 39, 17 Sept. 1683), some in Figure E may have been micrococci or staphylococci. Linnaeus (1767) was unable to classify the animalcules discovered by Leeuwenhoek and others following him. In his *Systema naturae,* he grouped them under "Vermes" and in a class with six species which he called "Chaos". The species *Chaos infusorium* contained various bacteria. The first attempt to classify "infusoria" was made by Müller (1773–1774). In his work entitled *Vermium terrestrium et fluviatilium, seu animalium infusoriorum ... succinta historia,* he introduced the genera *Monas* and *Vibrio,* which contained bacterial forms as well as those of protozoa and algae. The spherical and near-spherical forms were placed in the genus *Monas*. Ehrenberg (1838) expanded Müller's classification of the infusoria but again confused the bacteria with lower animals, due partly to the heterogeneous nature of the material with which they worked and to limitations of their microscopes. Ehrenberg coined the name bacteria and applied it to the genus *Bacterium*. Microorganisms which we now recognize as bacteria in the broader sense were distributed by Ehrenberg into three families: Monadina, Cryptomonadina, and Vibrionia. He placed the spherical bacteria (Kugelmonaden) in the genus *Monas* in the family Monadina.

The greatest advances in the early systematics of bacteria were made by Cohn (1872). In his classical *Untersuchungen über Bacterien* he formulated the groundwork for modern ideas in bacteriology. He regarded his genera as natural groups, but his species as largely provisional. He recognized that morphology alone was insufficient for classification, i.e., bacteria similar in form could differ from one another in physiological characters and in their products. He described four tribes and placed the spherical types in the tribe Sphaerobacteria (Kugelbacterien). This tribe contained the one genus *Micrococcus*. At about the same time, spherical bacteria found in pus, abscesses, and blood of people with pyemia were commonly called micrococci (Birch-Hirschfeldt, 1872; Koch, 1878; Von Recklinghausen, 1871), a term taken from Hallier's "micrococcus" which was a nucleated yeast (1867). Klebs (1872) called these bacteria *Microsporon septicum;* Hueter (1872) and Vogt (referred to by Koch, 1878) named them monads. Many of these bacteria were probably staphylococci and/or streptococci. Pasteur (1880), unknown to Ogston, described small spherical bacteria isolated from pus obtained from furuncles and osteomyelitis and considered them to be pathogenic. He was describing Ogston's staphylococci. It was Ogston (1880, 1882, 1883), however, who demonstrated in a convincing manner that staphylococci produce inflammation and suppuration. The main criticism brought against his teaching was that cocci were not the only bacteria or agent capable of producing suppuration.

Near the end of the 19th century, all spherical bacterial were included in the family Coccaceae (Zopf, 1885). Zopf and later Migula (1897) placed the mass-forming staphylococci and tetrad-forming micrococci in the genus *Micrococcus*. Flügge (1886) rearranged the cocci and maintained the genus *Staphylococcus* separate from *Micrococcus*. He differentiated the two genera primarily on the basis of their action on gelatin and on relation to their hosts. Micrococci were variable in their action on gelatin and were saprophytic whereas staphylococci liquefied gelatin and were parasitic and/or pathogenic. The genera *Staphylococcus, Micrococcus,* and *Planococcus,* containing Gram-positive, catalase-positive cocci, were later placed in the family Micrococcaceae (Prevot, 1961; Pribram, 1929), where they have remained to date. From the early 1900s to the mid-1950s, taxonomists added confusion to the classification by lumping staphylococci and micrococci together, first in the genus *Micrococcus* (Hucker, 1924; 1948) and then later in the genus *Staphylococcus* (Shaw et al., 1951). Evans et al. (1955) proposed

separating staphylococci from micrococci on the basis of their relation to oxygen. The facultative cocci were placed in the genus *Staphylococcus* and the obligate aerobes were placed in the genus *Micrococcus*. Although we now know that there are some exceptions to this proposal, the views of Evans et al. (1955) placed staphylococcal systematics once again on a fruitful course.

By the mid-1960s, a clear distinction could be made between staphylococci and micrococci on the basis of their DNA base composition (Rosypal et al., 1966; Silvestri and Hill, 1965). Members of the genus *Staphylococcus* have a GC content of their DNA of 30 to 39 mol%. Members of the genus *Micrococcus* have a GC content within the range of 63 to 73 mol%. The wide divergence in DNA base composition shows that these two genera are not closely related. More recent systematic studies have distinguished staphylococci from micrococci and other cocci on the basis of the chemical composition of cell walls (Endl et al., 1983; Schleifer, 1986; Schleifer and Kandler, 1972), composition of cytochrome (Faller et al., 1980) and of menaquinone (Collins and Jones, 1981), susceptibility to lysostaphin and erythromycin (Schleifer and Kloos, 1975a), bacitracin (Falk and Guering, 1983), and furazolidone (Baker, 1984) and its related compound nitrofuran (Curry and Borovian, 1976), hybridization of DNA-DNA (Kloos and Schleifer, 1986; Schleifer et al., 1979) and of DNA-rRNA (Kilpper et al., 1980), comparative oligonucleotide cataloging of 16S rRNA (Ludwig et al., 1981), and comparative immunology of catalases (Schleifer, 1986). Genetic mapping has indicated that the composition and arrangements of the histidine and tryptophan gene clusters (or operons) are very different in the *Staphylococcus* type species *S. aureus* (Kloos and Pattee, 1965; Proctor and Kloos, 1970) and the *Micrococcus* type species *M. luteus* (Kane-Falce and Kloos, 1975; Kloos and Rose, 1970). In *S. aureus,* each of these amino acid gene groups is located in a single operon, though the operons are not closely linked. In *M. luteus,* the corresponding gene groups are each located in two or more separate gene clusters, and one of the histidine gene clusters is very closely linked to a tryptophan gene cluster. These observations suggest a fundamental difference in the genetic organization of the two different type species and probably also of the two genera. The wide variety of genetic and epigenetic criteria indicates that the genus *Staphylococcus* forms a coherent and well-defined natural group that is widely divergent from the genus *Micrococcus*. The genus *Staphylococcus* belongs to the broad *Bacillus-Lacto-*

bacillus-Streptococcus cluster. More specifically, the closest extant relatives of staphylococci appear to be the planococci, enterococci, bacilli, and *Brochothrix thermosphacta* (Ludwig et al., 1985). The closest relatives of *Micrococcus* are the arthrobacters (Kloos, et al., 1974; Stackebrandt and Woese, 1979). (See also Chapter 55).

In the early 1960s, Baird-Parker (1962, 1963, 1965) conducted several large taxonomic studies of staphylococci and micrococci isolated from various sources. He classified these organisms on the basis of several easily determined characters, including cell arrangements, colony pigment, carbohydrate reactions, and enzyme and physiological tests. Two of his groups, though redefined today by newer systematics, became the species *S. epidermidis* and *S. saprophyticus*. Systematic bacteriologists began to look deeper into the molecular structure, immunology, genetics, and ecology of staphylococci, leading to the introduction of new species and subspecies. Currently, 28 species are recognized in the genus *Staphylococcus*. These include *S. aureus* (Rosenbach, 1884), *S. epidermidis, S. haemolyticus, S. saprophyticus, S. cohnii,* and *S. xylosus* (Schleifer and Kloos, 1975b), *S. capitis, S. warneri, S. hominis,* and *S. simulans* (Kloos and Schleifer, 1975a), *S. sciuri* and *S. lentus* (Kloos et al., 1976a; Schleifer et al., 1983), *S. intermedius* (Hájek, 1976), *S. hyicus* and *S. chromogenes* (Devriese et al., 1978; Hájek et al., 1986), *S. saccharolyticus* (Kilpper-Bälz and Schleifer, 1981), *S. caseolyticus* (Schleifer et al., 1982), *S. carnosus* (Schleifer and Fischer, 1982), *S. auricularis* (Kloos and Schleifer, 1983), *S. caprae* and *S. gallinarum* (Devriese et al., 1983), *S. arlettae, S. equorum,* and *S. kloosii* (Schleifer et al., 1984), *S. lugdunensis* and *S. schleiferi* (Freney et al., 1988), *S. delphini* (Veraldo et al., 1988), and *S. felis* (Igimi et al., 1989). Seven subspecies have also been described, three of which have been given names: *S. aureus* subsp. *anaerobius* (De La Fuente et al., 1985), *S. capitis* subsp. *ureolyticus* (Bannerman and Kloos, 1991), *S. cohnii* subsp. *urealyticum* and *S. cohnii* subsp. 3 (Kloos and Wolfshohl, 1983, and unpublished observation), and *S. haemolyticus* subsp. 2 and *S. warneri* subsp. 2 from nonhuman primates (Kloos and Wolfshohl, 1979). The nonhuman primate subspecies *S. auricularis* subsp. 2 has been given only a brief mention (Kloos, 1986a). It is expected that more *Staphylococcus* species and subspecies will be discovered as additional host species and environments are examined. In the future, more of the rarer species and subspecies should be encountered, providing a more complete view of the evolutionary relationships within this large and highly successful genus.

Ecology

Skin as a Habitat

Staphylococci are widespread in nature, though they are found more consistently and in denser populations on the skin, skin glands, and mucous membranes of mammals and birds. They are sometimes found in the pharynx, mouth, blood, mammary glands, and intestinal, genitourinary, and upper respiratory tracts of these hosts. The largest populations of staphylococci are found in regions of the skin supplied with large numbers of pilosebaceous units and sweat glands and on the skin and mucous membranes surrounding openings to the body surface. Staphylococcal communities may be found living in the follicular canals, the openings to sweat glands, the capacious lumen of sebaceous follicles, and on the surface of and beneath desquamating epithelial scales (Noble and Pitcher, 1978; Noble and Somerville, 1974). Most knowledge concerning the ecology of staphylococci has been obtained by collectively sampling the surface, hair, and epidermal invaginations of skin using scrubbing or swabbing methods (Kloos and Musselwhite, 1975; Noble and Somerville, 1974; Williamson and Kligman, 1965). The body regions most commonly studied on humans include the scalp, face, external auditory meatus, anterior nares, axillae, arms, palms, inguinal and perineal area, legs, and interdigital spaces of the foot (Kloos, 1986a, 1986b; Kloos and Musselwhite, 1975; Kloos and Schleifer, 1975a; Marples, 1965; Noble and Somerville, 1974). A variety of habitats are present in the human cutaneous ecosystem and they can be distinguished by differences in the density and structure of the microbial communities inhabiting them, as well as by their anatomical and physiological properties.

The body regions studied on nonhuman primates include the head, anterior nares, external auditory meatus, axillae, chest, abdomen, back, arms, legs, and, when present, specialized scent glands (Kloos and Wolfshohl, 1979, 1983). The cutaneous habitats of lower primates are significantly different from those of humans, especially with respect to the number, distribution, and anatomy of hair follicles and of glands (Montagna and Ellis, 1963) and lipid composition (Nicolaides, 1974).

A variety of cutaneous habitats have been sampled in other mammals but the flora has not been studied as quantitatively as it has in primates. Most of the investigations have focused on the staphylococci of domestic animals including cattle, pigs, horses, sheep, goats, dogs, and cats (Amtsberg, 1978; Baba et al., 1980;

Devriese, 1979, 1986; Devriese and Adegoke, 1985; Devriese and Derycke, 1979; Devriese and Oeding, 1976; Devriese et al., 1984; Hájek, 1976; Hájek and Marsalek, 1976; Igimi et al., 1989; Kloos et al., 1976b; Krogh and Kristensen, 1976; Langlois et al., 1983; Oeding et al., 1974; Phillips and Kloos, 1981; Phillips et al., 1980; Poutrel, 1984; Raus and Love, 1983; Sompolinsky, 1950; Watts et al., 1984). Several studies have described natural populations of staphylococci living on the skin of wild carnivores, rodents, lagamorphs, and marsupials (Hájek and Marsalek, 1971, 1976; Kloos and Schleifer, 1981; Kloos et al., 1976a, 1976b; Oeding et al., 1973). Several reports have been made on the staphylococci of domestic fowl (Devriese and Oeding, 1975, 1976; Devriese et al., 1978, 1983; Evans et al., 1983; Hájek and Marsalek, 1969, 1971; Hummel and Witte, 1985; Smith and Crabb, 1960; Wise, 1971; Witte et al., 1977), and one major report has been made on wild birds, including gulls, ducks, pheasants, buzzards, goshawks, buntings, warblers, martins, and swallows (Akatov et al., 1985).

Flies of the genera *Musca, Fannia,* and *Stomoxys,* commonly found in human and/or animal habitations, can carry populations of staphylococci and appear to be significant vectors of these organisms in an epizootiological chain (Hájek and Balusek, 1985).

Host Range

Because staphylococci are so widespread in nature, one must use rather stringent criteria based on temporal studies and population size to estimate residency status and host range (Kloos and Musselwhite, 1975; Noble and Somerville, 1974; Price, 1938). The organisms can be divided into two classes based on residency status; residents are bacteria that produce relatively stable populations and increase in numbers mainly by multiplication of those already present; they are indigenous to the host. Transients, on the other hand, are relatively rare on unexposed skin, but occur more frequently on exposed skin and may be easily removed. They are contaminating organisms that do not multiply significantly and come from extraneous sources. Cross-contamination of staphylococci can occur readily where different host species come in contact with one another, but where host specificity is high, the transient organisms will usually be eliminated within several hours or days, unless the normal defense barriers are compromised. Ideally, determination of natural host range should be made with host species that are relatively isolated in nature. Most of the ecological studies reported fall

short of this ideal situation, but, nevertheless, have indicated some clear patterns of host and niche preferences for certain *Staphylococcus* species.

Species of the *S. epidermidis* species group (with the possible exception of *S. caprae*) have a preference for primate hosts (Kloos, 1980; Kloos and Musselwhite, 1975; Kloos and Schleifer, 1981). *S. epidermidis* is the most prevalent and persistent *Staphylococcus* species on human skin. It is found over much of the body surface and produces the largest populations where moisture content and nutrition are high, such as in the anterior nares, axillae, inguinal and perineal area, and toewebs. This species may be found occasionally on other hosts, such as domestic animals, but it is presumably transferred there from human sources. *S. hominis* is also prevalent on human skin. Its population size is usually second or equal to *S. epidermidis* on skin sites where apocrine glands are numerous (e.g., axillae and inguinal and perineal area). It can also colonize the drier regions of skin (e.g., on the extremities) more successfully than other species. *S. haemolyticus* shares many of the habitats of *S. hominis,* but it is usually found in smaller populations. Some individuals may carry unusually large populations of *S. haemolyticus.* A different subspecies of *S. haemolyticus* is found living on nonhuman primate skin (e.g., *Pan, Macaca, Cercocebus, Erythrocebus, Microcebus,* and *Lemur*) that can be distinguished from the human-adapted subspecies on the basis of DNA-DNA hybridization (Kloos and Wolfshohl, 1979). Since it is difficult to distinguish colonies of the nonhuman primate *S. haemolyticus* subspecies from a sibling species, provisionally designated *S. "simians",* found also on nonhuman primate skin, adequate enumeration of this subspecies is not yet possible. *S. warneri* is found usually in small numbers on human skin, though a few individuals may carry unusually large populations. Like the situation for *S. haemolyticus,* a different subspecies of *S. warneri* is found living on nonhuman primate skin. *S. warneri* is a major species on nonhuman primates, especially on the more advanced Cercopithecoidea and Pongidae. Occasionally, small transient populations of *S. haemolyticus* or *S. warneri* may be isolated from domestic animals. *S. capitis* produces large populations on the human scalp following puberty, and in climax communities it is usually the predominant species of this region. It is also found on other regions of the adult head, e.g., forehead, face, eyebrows, and external auditory meatus in moderate-sized to large populations. The largest populations are found in areas where sebaceous glands are numerous and well-developed. *S. capitis* subsp. *ureolyticus* is present on regions of the head in rather small populations and, like *S. capitis* (above), may be found only occasionally on other body sites (Bannerman and Kloos, 1991). This subspecies has been isolated from both human and nonhuman primate skin (e.g., *Pan*). *S. caprae* is an unusual member of the *S. epidermidis* species group in that it is principally found on the skin of domestic goats or in their milk (Devriese et al., 1983; Poutrel, 1984).

S. auricularis is one of the major species found living in the adult human, external auditory meatus and demonstrates a strong preference for this niche (Kloos and Schleifer, 1983). A different subspecies of *S. auricularis* is found in the ear and specialized scent (or marking) glands of nonhuman primates (e.g., *Pan, Pongo, Cercopithecus, Lemur, Galago,* and *Microcebus*) (Kloos, 1985, 1986a).

S. aureus is a major species of primates, though specific ecovars or biotypes can be found occasionally living on different domestic animals or birds (Kloos, 1980; Meyer, 1967). This species is found infrequently on nonprimate wild animals. On humans, *S. aureus* demonstrates a niche preference for the anterior nares, especially in the adult. Here it can exist as a resident or as a transient member of the normal flora. *S. aureus* demonstrates a selective adherence to nasal epithelial (mucosal) cells (Aly et al., 1981). Nasal carrier rates range from less than 10% to more than 40% in normal adult human populations residing outside of the hospital (Noble and Somerville, 1974). The nasal adherence of *S. aureus* is significantly greater for carriers of this species than for noncarriers. *S. aureus* subsp. *anaerobius* is found living on sheep (De la Fuente et al., 1985).

Species of the *S. saprophyticus* species group demonstrate a variety of host ranges from humans to lower mammals and birds (Devriese, 1986; Kloos, 1980). As a group, these staphylococci are most prevalent on lower primates and mammals. Those species found most frequently on primates include *S. saprophyticus, S. cohnii,* and *S. xylosus. S. saprophyticus* is found usually in small, transient populations on the skin of humans or other primates. This species possesses surface properties that allow it to adhere readily to urogenital cells (Colleen et al., 1979), which ultimately may lead to urinary tract infections (Anderson et al., 1981; Marrie et al., 1982). It is also sometimes isolated from lower mammals and environmental sources. *S. cohnii* is found as a temporary resident or transient on human skin. *S. cohnii* subsp. *urealyticum* is sometimes found on human skin, but it is often one of the major species and subspecies of nonhuman primates, especially the lower primates. The largest populations of this subspecies are found living on the Tupaiidae, Prosimii, and Ceboidea (Kloos and Wolfshohl, 1983). A third subspecies is also found on the Ceboidea. *S. xylosus* is often found as a transient on the skin of lower primates and other mammals, and occasionally on birds (Akatov et al., 1985; Devriese and Adegoke, 1985; Kloos et al., 1976b). The related species *S. kloosii* has been found living on a variety of lower mammals including wild marsupials, rodents, and carnivores, and less frequently on domestic animals (Kloos, 1980; Schleifer et al., 1984). *S. arlettae* has been isolated from poultry and goats, *S. equorum* from horses, and *S. gallinarum* from poultry (Devriese et al., 1983; Schleifer et al., 1984).

S. intermedius is a major species of the domestic dog (Hájek, 1976; Krogh and Kristensen, 1976). This species can be found in relatively large populations on canine skin and can on occasion be transferred to the skin of human handlers (Kloos et al., 1976). *S. intermedius* appears to be also indigenous to a variety of other carnivores, including the mink *(Mustela)* (Hájek and Marsalek, 1976; Hájek et al., 1972; Oeding et al., 1973), fox *(Vulpes)* (Hájek and Marsalek, 1976), and raccoon *(Procyon)* (Kloos et al., 1976b). It has also been isolated from horses and pigeons (Hájek and Marsalek, 1971, 1976). *S. felis* is one of the major species of the domestic cat (Igimi et al., 1989). *S. hyicus* and *S. chromogenes* are found predominantly on domestic ungulates such as pigs, cattle, and horses (Devriese, 1986; Devriese and Derycke, 1979; Devriese et al., 1983; Phillips et al., 1980). *S. lentus* has been isolated in large populations from domestic sheep and goats (Kloos et al., 1976a, 1976b) and occasionally from other domestic animals (Devriese and Adegoke, 1985).

Staphylococci have been isolated sporadically from a wide variety of environmental sources such as soil, beach sand, seawater, fresh water, plant surfaces and products, feeds, meat and poultry, dairy products, and on the surfaces of cooking ware, utensils, furniture, clothing, blankets, carpets, linens, paper currency, and dust and air in various inhabited areas. With the exception of the animal products, most environmental sources contain small, transient populations of staphylococci, many of which are probably contaminants disseminated by human, animal, or bird host carriers. It is possible that certain of the species, e.g., *S. sciuri* and *S. xylosus,* which are capable of growing in habitats containing only an inorganic nitrogen source, might be more free-living than other staphylococci (Emmett and Kloos, 1975, 1979).

These species have been isolated in small numbers from beach sand, natural waters, and marsh grass (Kloos and Schleifer, 1981) and also from plant products (Bucher et al., 1980; Pioch et al., 1988).

Opportunistic Pathogens

The coagulase-positive species *S. aureus, S. intermedius,* and *S. delphini,* and the coagulase-variable species *S. hyicus* are regarded as potentially serious pathogens. *S. aureus,* since its early discovery as an opportunistic pathogen, continues to be a major cause of mortality and is responsible for a variety of infections. In the late 1950s and early 1960s, *S. aureus* caused considerable morbidity and mortality as a nosocomial pathogen of hospitalized patients. Methicillin-resistant *S. aureus* (MRSA) strains have emerged in the 1980s as a major clinical and epidemiological problem in hospitals. Among the major human infections caused by this species are furuncles, carbuncles, impetigo, toxic epidermal necrolysis (scalded skin syndrome), pneumonia, osteomyelitis, acute endocarditis, myocarditis, pericarditis, enterocolitis, mastitis, cystitis, prostatitis, cervicitis, cerebritis, meningitis, bacteremia, toxic shock syndrome, and abscesses of the muscle, skin, urogenital tract, central nervous system, and various intraabdominal organs. Food poisoning is often attributed to staphylococcal enterotoxin. Comprehensive reviews on the nature of human infections caused by *S. aureus* can be found in the following texts: *Staphylococcus Pyogenes and Its Relation to Disease* (Elek, 1959), *The Staphylococci* (Cohen, 1972), *Staphylococci and Staphylococcal Infections: Recent Progress* (Jeljaszewicz, 1973, 1976, 1981, 1985), *Microbiology of Human Skin* (Noble and Somerville, 1974), and *Staphylococci and Staphylococcal Infections,* Vol. 1, *Clinical and Epidemiological Aspects* and Vol. 2, *The Organism In Vivo and In Vitro* (Easmon and Adlam, 1983).

Staphylococcus aureus is also capable of producing infections in a variety of other mammals and birds. The more common natural infections include mastitis, synovitis, arthritis, endometritis, furuncles, suppurative dermatitis, pyemia, and septicemia. The association of staphylococci with veterinary medicine began in the late 1880s with the isolation of *S. aureus* from mastitis in sheep (Nocard and Mollereau, 1887) and cattle (Guillebeau, 1890). Staphylococcal mastitis in either a clinical or subclinical form may have considerable economic consequences in the dairy industry. *S. aureus* subsp. *anaerobius* is the etiologic agent of an abscess disease in sheep, symptomatically similar to caseous

lymphadenitis (De la Fuente et al., 1985). Reference may be made to the above-mentioned texts and, in addition, to *Infectious Diseases of Animals: Diseases Due to Bacteria* (Stableforth and Galloway, 1959), *Diseases of the Mammary Glands of Domestic Animals* (Heidrich and Renk, 1967), and *Hagan and Bruner's Microbiology and Infectious Diseases of Domestic Animals* (Timoney et al., 1988) for information on the nature of *S. aureus* infections in animals and birds.

S. intermedius is a serious opportunistic pathogen of dogs and may cause otitis externa, pyoderma, abscesses, reproductive tract infections, mastitis, and purulent wound infections (Devriese and Hájek, 1980; Devriese and Oeding, 1976; Hájek, 1976; Krogh and Kristensen, 1976; Phillips and Kloos, 1981; Raus and Love, 1983; Szynkiewicz et al., 1985). *S. hyicus* has been implicated as the etiologic agent of infectious exudative epidermitis (greasy pig disease) and septic polyarthritis of pigs (Devriese, 1977; Devriese and Hájek, 1980; Phillips and Kloos, 1981; Phillips et al., 1980; Sompolinsky, 1953). It can also cause skin lesions in cattle and horses (Devriese and Derycke, 1979; Devriese et al., 1983), osteomyelitis in poultry and cattle (Carnaghan, 1966; De Kesel and Devriese, 1982; Wise, 1971), and occasionally has been associated with mastitis in cattle (Devriese and De Keyser, 1980). *S. delphini* has been implicated in purulent skin lesions of dolphins (Varaldo et al., 1988).

The coagulase-negative staphylococcal species constitute a major component of the normal microflora of the human, and for this reason have generally been regarded as saprophytes or organisms with no or very low virulence. However, over the last two decades there has been an increase in the documentation of infections due to coagulase-negative staphylococci, especially with the species *S. epidermidis.* The increase in infections by these organisms has been correlated with the wide medical use of prosthetic and indwelling devices and the growing number of immunocompromised patients in hospitals (Pulverer, 1985). Infectious processes may result from the introduction of endogenous staphylococci beyond the normal integumentary barriers. Of the 13 coagulase-negative species recognized from humans, *S. epidermidis* appears to have the greatest pathogenic potential and adaptive diversity. This species has been implicated in bacteremia, native and prosthetic valve endocarditis, osteomyelitis, pyoarthritis, peritonitis during continuous ambulatory dialysis, mediastinitis, infections of permanent pacemakers, vascular grafts, cerebrospinal fluid shunts, prosthetic joints, and a variety of orthopedic devices, and urinary tract infections including cystitis, urethritis, and pyelonephritis (Archer, 1984; Archer et al., 1980; Baddour et al., 1986; Baumgart et al., 1983; Bor et al., 1983; Brause, 1986; Choo et al., 1981; Clarke, 1979; Forse et al., 1979; George et al., 1979; Hamory and Parisi, 1987; John et al., 1978; Kraus and Spector, 1983; Leighton and Little, 1986; Lowy and Hammer, 1983; Martin et al., 1989; Morris et al., 1986; Paley et al., 1986; Peters et al., 1982; 1984; Pfaller and Herwaldt, 1988; Ponce de Leon et al., 1986; Richardson et al., 1978; Rubin et al., 1980; Winston et al., 1983). Reviews on the nature of human infections caused by *S. epidermidis* and other coagulase-negative species can be found in the texts: *The Staphylococci: Proceedings of the Vth International Symposium on Staphylococci and Staphylococcal Infections* (Jeljaszewicz, 1985), *Coagulase-Negative Staphylococci* (Mårdh and Schleifer, 1986), and *Pathogenicity and Clinical Significance of Coagulase-Negative Staphylococci* (Pulverer et al., 1987.) Nosocomial methicillin-resistant *S. epidermidis* (MRSE) strains have become a serious clinical problem in the 1980s, especially in patients with prosthetic heart valves or who have undergone other forms of cardiac surgery (Archer and Tenenbaum, 1980; Karchmer et al., 1983). *S. epidermidis* has also been occasionally associated with mastitis in cattle (Baba et al., 1980; Devriese and De Keyser, 1980; Holmberg, 1986).

Certain other coagulase-negative species have been associated with infections in humans and animals. *S. haemolyticus* is the second most frequently encountered species of this group found in human clinical infections. It has been implicated in native valve endocarditis, septicemia, peritonitis and urinary tract infections, and is occasionally associated with wound, bone, and joint infections (Caputo et al., 1985; Fleurette et al., 1987; Gill et al., 1983; John et al., 1987; Kristinsson et al., 1986; Leighton and Little, 1986; Marsik and Brake, 1982; Martin et al., 1989; Nord et al., 1976; Ponce de Leon et al., 1986; Sewell et al., 1982). *S. haemolyticus* has been occasionally associated with mastitis in cattle (Baba et al., 1980). *S. lugdunensis* appears to be a significant opportunistic pathogen in man. This species was identified only two years ago (Freney et al., 1988), but is now being commonly recognized in clinical infections in a number of countries. Cultures of this species were first obtained from infections in the early 1980s but many were misidentified as *S. warneri* or *S. hominis*. *S. lugdunensis* has been implicated in native and prosthetic valve endocarditis, septicemia, brain abscess, and chronic osteoarthritis and infections of soft tissues,

bone, peritoneal fluid, and catheters, especially in patients with underlying diseases (Etienne et al., 1989; Fleurette et al., 1989; Freney et al., 1988; Herchline and Ayers, 1989). *S. schleiferi,* also recently characterized as a new species, has been implicated in human brain empyema, osteoarthritis, bacteremia, wound infections, and infections associated with a cranial drain and jugular catheter (Fleurette et al., 1989; Freney et al., 1988; Jean-Pierre et al., 1989). This species occurs less frequently than *S. lugdunensis* in the hospital environment and human infections. *S. saprophyticus* is an important opportunistic pathogen in human urinary tract infections, especially in young, sexually active females (Anderson et al., 1981; Gillespie et al., 1978; Hovelius, 1986; Jordan et al., 1980; Marrie et al., 1982; Shrestha and Darrell, 1979; Wallmark et al., 1978). It is considered to be the second most common cause of urinary tract infections, such as acute cystitis or pyelonephritis, in these patients. This species can also produce urinary tract infections in men, but in contrast to the infection among women, it is most commonly seen in elderly patients with predisposing diseases of the urinary tract (Hovelius, 1986; Hovelius et al., 1984; Marrie et al., 1982). *S. saprophyticus* has occasionally been isolated from wound infections and septicemia (Fleurette et al., 1987; Marsik and Brake, 1982).

Several other coagulase-negative species have been implicated at low incidence in a variety of human infections. In most cases, patients with these infections had predisposing or underlying diseases that caused intensive changes in the immune system and had also experienced surgery or intravascular manipulations. *S. warneri* has been on occasion the etiologic agent of vertebral osteomyelitis, native valve endocarditis, and urinary tract infections in males and females (Caputo et al., 1985; Dan et al., 1984; Karthigasu et al., 1986; Leighton and Little, 1986). This species has been associated with mastitis in cattle (Devriese, 1979; Devriese and De Keyser, 1980). *S. simulans* has been associated with human chronic osteomyelitis and pyarthrosis (Males et al., 1985) and bovine mastitis (Devriese and De Keyser, 1980; Watts et al., 1984). *S. felis,* a relative of *S. simulans,* has been isolated from clinical infections in cats, including external ear otitis, cystitis, abscesses, wounds, and other skin infections (Igimi et al., 1989). *S. hominis* has been associated with human endocarditis, peritonitis, septicemia, and arthritis (Bowman and Buck, 1984; Fleurette et al., 1987). Some of the earlier reports indicating an association of this species with infections were in error, due to the confusion of this species with phosphatase-negative strains of *S. ep-*

idermidis, S. chromogenes, a close relative of *S. hyicus,* is commonly isolated from the milk of cows suffering from mastitis, although its role as an etiologic agent is questionable (Devriese and De Keyser, 1980; Langlois et al., 1983; Watts et al., 1984).

Mechanisms of Pathogenicity

The mechanisms of staphylococcal pathogenicity have been studied most extensively in the species *S. aureus.* This aggressive and invasive species has at its disposal a variety of extracellular enzymes and toxins and antiphagocytic components of the cell wall to use against the host. Some of the principal toxins include the α- and δ-hemolysins and leukocidin, and perhaps of less significance, the β- and γ-hemolysins. Certain strains of *S. aureus* are capable of producing other damaging toxins such as enterotoxin, exfoliatin, or the toxin responsible for toxic shock syndrome. The enzyme coagulase has been regarded as a virulence factor in that it may inhibit the bactericidal activity of normal serum and phagocytosis. *S. aureus* lipase may influence the nonspecific primary defense system by increasing the chemotactic response of polymorphonuclear leukocytes. The role of other extracellular enzymes in pathogenicity is less understood. Protein A is a cell wall component that impairs chemotaxis and opsonization, probably because immunoglobulin IgG attaches to this protein, making the Fc portion of the antibody unavailable for attachment to a specific receptor on the phagocytic cell membrane. Cell wall peptidoglycan can also interfere with serum opsonization and phagocytosis. In addition to these virulence factors, some strains produce an extracellular capsule that presumably protects these organisms from phagocytosis by polymorphonuclear leukocytes. Descriptions of the above toxins, enzymes, cell wall components, and capsules are given in a following section of this chapter, "Exported Compounds in Staphylococci." The ability of *S. aureus* to bind to fibronectin and collagen exposed in open wounds and microtrauma allows this organism to readily establish itself in traumatized tissues (Wadstrom et al., 1987). Binding to laminin allows this species to localize in basal membranes in various body sites. Staphylococcal pathogenicity also depends upon the effectiveness of the host defense systems, which include the natural barriers of skin and mucous membranes and phagocytic and humural immune responses. During the past decade it has become increasingly apparent that patients with deficient numbers of phagocytic cells or disorders in their function have increased suscepti-

bility to staphylococcal infection (Quie et al., 1983).

S. epidermidis pathogenesis has received some attention in the last decade. This species can elaborate a variety of extracellular enzymes and toxins in varying amounts, but their role in pathogenicity has not yet been elucidated (Gemmel, 1987). Most studies on mechanisms of pathogenicity have focused on the adherence of this organism to foreign bodies or devices and slime production. The first step in the pathogenesis of foreign body infections is probably the adherence of cells to the surface of the biomaterial. Once adherent, the cells multiply and produce extracellular substances (e.g., slime). Continued multiplication leads to the formation of microcolonies that become embedded in a matrix of slime (glycocalyx), which may subsequently contribute to the persistence of bacteria and infection (Christensen et al., 1982; Peters, 1985; Peters et al., 1987). *S. epidermidis* slime may also play a role in foreign body infections by interfering with host defenses (Gray et al., 1987; Johnson et al., 1987). Slime has been shown to inhibit the functions of polymorphonuclear granulocytes as well as lymphocytes. Adherence to traumatized host tissues may include an association of these organisms with fibronectin and collagen (Wadström and Rozgonyi, 1986; Wadström et al., 1987). Furthermore, capsule production by some strains of *S. epidermidis* may enhance the ability of this species to colonize traumatized vascular walls and the endocardium (Baddour et al., 1984). These virulence properties of *S. epidermidis* are also shared by some related species of the *S. epidermidis* species group, though these species are associated with infections much less frequently. The other species are present in significantly smaller populations on their human hosts, which may be a factor in reducing their availability for initiating infection; however, it would be expected that the different species would also have somewhat different capabilities to produce infection. This difference is not clearly understood yet.

Studies of *S. saprophyticus* pathogenesis have suggested some possible mechanisms for pathogenicity by this species (Mårdh, 1986). However, information on suspected virulence properties is quite limited and still inconclusive. *S. saprophyticus* can adhere to a variety of epithelial cells. Colleen et al. (1979) have reported that this species adhered more readily to human urethelium and periurethral cells than to human buccal and skin cells. Kasprowicz et al. (1987) were unable to demonstrate a selective binding affinity for urinary tract isolates of *S. saprophyticus,* but did find a tendency of respiratory isolates to adhere in higher numbers to oral and respiratory epithelia. *S. saprophyticus* can cause direct hemagglutination of sheep erythrocytes, in contrast to other staphylococcal species and most Gram-positive bacteria; however, it does not agglutinate human erythrocytes (Hovelius and Mårdh, 1979). It has been suggested that the receptor responsible for hemagglutination of *S. saprophyticus* is also likely to be responsible for the attachment of this organism to urothelium (Mårdh, 1986). This species demonstrates unique surface projections, though additional studies must be made to see whether these structures are related to the adherence phenomena (Christiansen and Mårdh, 1986). Reviews of mechanisms of pathogenicity in the staphylococci can be found in the various texts mentioned in the previous section, "Opportunistic Pathogens."

Isolation Techniques

Isolation of *Staphylococcus aureus* from Foods

S. aureus has been confirmed to be the causative agent of many cases of severe food poisoning (reviewed by Bergdoll, 1972; Bergdoll and Bennett, 1976; Minor and Marth, 1972). Its presence in foods, therefore, is of major concern. *S. aureus* is very susceptible to heat treatment and most sanitizing agents. Hence, when it or its enterotoxins are found in processed foods, poor sanitation is usually indicated. Detailed procedures for preparing food samples for analysis, isolating and enumerating *S. aureus,* and detecting staphylococcal enterotoxins in foods can be found in the following texts: *Compendium of Methods for the Microbiological Examination of Foods* (Speck, 1984), *Association of Official Analytical Chemists (AOAC) Official Methods of Analysis,* 14th and 15th Editions (AOAC, 1984, 1990), *Bacteriological Analytical Manual,* 6th Edition (Center for Food Safety and Applied Nutrition, U.S. Food and Drug Administration, 1984). In addition to the recommended optimum sensitivity and microslide methods (Bergdoll and Bennett, 1984), it is now possible to detect enterotoxins in culture and in contaminated foods directly using the following rapid tests: radioimmunoassay (RIA) (Miller et al., 1978), enzyme-linked immunosorbent assay (ELISA) (Freed et al., 1982), and reverse passive latex agglutination (RPLA) (Igarashi et al., 1985). (An ELISA kit is available from Toxin Technology, Inc., Madison, WI, and an RPLA kit (SET-RPLA) is available from Oxoid, USA, Columbia, MD.) We will describe here several of the procedures recommended for isolating *S. aureus* from foods.

NONSELECTIVE ENRICHMENT PROCEDURES. It is often necessary to use nonselective enrichment procedures for the detection of *S. aureus* in processed foods, especially when it is suspected that the food contains a small number of cells that may have been injured, e.g., as a result of heating, freezing, desiccation, or storage, and whose growth could be inhibited by toxic components of a selective enrichment media. The following nonselective (repair) enrichment procedure of Tatini et al. (1984) is appropriate for this use:

Nonselective Enrichment of *Staphylococcus aureus* (Tatini et al., 1984)

> A 50-ml aliquot of a 1:10 dilution of the food sample homogenate is transferred in 50 ml of double-strength trypticase soy broth (TSB). Incubate the preparation for 3 h at 35 to 37°C. Then add 100 ml of a single-strength TSB containing 20% NaCl. Incubate for 24 ± 2 h at 35 to 37°C. Transfer 0.1-ml aliquots of the culture to each of duplicate Baird-Parker agar plates, and spread the inoculum so as to obtain isolated colonies. Incubate the inoculated plates for 46 ± 2 h at 35 to 37°C. Select two or more colonies suspected to be *S. aureus* from each plate. *S. aureus* colonies are usually ≥ 1.5 mm in diameter, jet-black to dark gray, smooth, convex, have entire margins and off-white edge, and may show an opaque zone and/or a clear halo extending beyond the opaque zone. Test selected colonies for coagulase activity. Results should be reported as *S. aureus* present or absent in 5 g of food, following the results of coagulase testing.

The coagulase test procedure recommended by the Association of Official Analytical Chemists (AOAC) for the identification of *S. aureus* from foods is given below, but slight modifications of this procedure are proposed for use with clinical isolates (Kloos and Lambe, 1990).

Coagulase Test for the Identification of *Staphylococcus aureus* Isolated from Food (Association of Official Analytical Chemists, 1984)

> "To brain heart infusion (BHI) broth cultures add 0.5 ml reconstituted coagulase plasma with ethylenediaminetetraacetic acid (EDTA) [Reconstitute plasma according to manufacturer's directions. If not available, reconstitute desiccated coagulase plasma (rabbit) and add disodium dihydrate EDTA to final concentration of 0.1% in reconstituted plasma.] and mix thoroughly. Incubate at 35–37°C and examine periodically over 6-h intervals for clot formation. Any degree of clot formation is considered a positive reaction. Small or poorly organized clots may be observed by gently tipping tube so that liquid portion of reaction mixture approaches lip of tube; clots will protrude above liquid surface. Coagulase-positive cultures are considered to be *S. aureus.* Test positive and negative controls simultaneously with

cultures of unknown coagulase reactivity. Recheck doubtful coagulase test results on BHI cultures which have been incubated at 35–37°C for >18 but ≤48 h."

With this procedure, false-positive tests may occur with mixed cultures, but this will probably be avoided if only well-isolated colonies typical of *S. aureus* are chosen. On rare occasions, coagulase-negative mutants of *S. aureus* may be present in foods and overlooked by the above procedures, and the presence of other coagulase-positive staphylococci, such as *S. intermedius* and certain strains of *S. hyicus,* may be misrepresented as *S. aureus* using the coagulase test alone.

Baird-Parker agar base when supplemented with egg yolk tellurite enrichment is recommended for the detection and enumeration of coagulase-positive staphylococci in foods. Baird-Parker Agar for Enumeration of *Staphylococcus aureus* (Schwab et al., 1984)

Basal medium:
Tryptone	10.0 g
Beef extract	5.0 g
Yeast extract	1.0 g
Glycine	12.0 g
Lithium chloride·6H$_2$0	5.0 g
Agar	20.0 g

This basal medium may be special-ordered from Difco Laboratories, Detroit, MI. Suspend ingredients in 950 ml distilled water. Boil to dissolve completely. Dispense 95.0 ml portions in screw capped bottles. Autoclave 15 minutes at 121°C. Adjust final pH to 6.8–7.2 at 25°C.

Egg yolk tellurite enrichment:
Soak eggs in aqueous mercuric chloride 1:1000 for not less than one min. Rinse in sterile water and dry with a sterile cloth. Aseptically crack eggs and separate whites and yolks. Blend yolk and sterile physiological saline solution (3 + 7 v/v) in high-speed sterile blender for 5 seconds. Mix 50.0 ml blended egg yolk to 10.0 ml of filter-sterilized 1% potassium tellurite. Mix and store at 2 to 8°C. Bacto egg-tellurite enrichment is a commercial preparation available from DIFCO.

Preparation of plates:
Add 5.0 ml prewarmed (45 to 50°C) enrichment to 95 ml melted basal medium, which has been adjusted to 45 to 50°C. Mix well (avoiding bubbles), and pour 15.0 to 18.0 ml into sterile 15 × 100 mm petri dishes. Plates can be stored at 2 to 8°C in plastic bags for 4 weeks. Immediately prior to use, spread 0.5 ml per plate of 20% solution of [membrane] filter-sterilized sodium pyruvate and dry plates at 50°C for 2 h or 4 h at 35°C with agar surface uppermost.

SELECTIVE ENRICHMENT PROCEDURES. Selective enrichment is recommended for raw food ingredients and unprocessed foods expected to contain <100 *S. aureus* cells/g and a large population of competing species. The recom-

mended procedure of the AOAC is widely accepted and uses the most probable number technique.

Most Probable Number (MPN) Technique for the Selective Enrichment of *Staphylococcus aureus* (Association of Official Analytical Chemists, 1987).

"Inoculate 3 tubes of trypticase soy broth with 10% NaCl and 1% sodium pyruvate at each test dilution with 1 ml aliquots of decimal dilutions of sample. Maximum dilution of sample must be high enough to yield negative end point. Incubate 48 h at 35°. Using 3 mm loop, transfer 1 loopful from each growth-positive tube to dried Baird-Parker medium plates. Vortex-mix tubes before streaking if growth is visible only on bottom or sides of tubes. Streak so as to obtain isolated colonies. Incubate 48 h at 35–37°C."

The interpretation of colonial growth and identification of *S. aureus* on Baird-Parker agar given by the AOAC is very similar to that given above by Tatini et al. (1984).

"For each plate showing growth, pick≥ 1 colony suspected to be *S. aureus*. With sterile needle transfer colonies to tubes containing 0.2ml brain heart infusion (BHI) broth and to agar slants containing any suitable maintenance medium, e.g., trypticase soy agar, standard plate count agar, etc. Incubate BHI culture suspensions and slants 18–24 h at 35°C."

The BHI culture suspensions are used as inocula for the coagulase test (described above), and the slant cultures are used for ancillary tests or repeats of the coagulase test, if results are questionable. Report the most probable number (MPN) of *S. aureus*/g from tables of MPN values (Table 46:01, Association of Official Analytical Chemists, 1984).

DIRECT SURFACE PLATING PROCEDURES. These procedures are sometimes preferred over the MPN technique for the detection of *S. aureus* in raw or unprocessed foods as they are more rapid and are regarded by some investigators to be more accurate than MPN.

Surface Plating Procedure for the Enumeration of *Staphylococcus aureus* (Tatini et al., 1984; adapted from Association of Official Analytical Chemists, 1980)

"The sensitivity of this procedure may be increased by using larger volumes >1 ml) distributed over >3 replicate plates. Plating of two or more decimal dilutions may be required to obtain plates with the desired number of colonies per plate. For each dilution to be plated, aseptically transfer 1 ml of sample suspension to triplicate plates of Baird-Parker agar and distribute the 1 ml inoculum equally over the triplicate plates (e.g., 0.4, 0.3, and 0.3 ml). Spread the inoculum over the surface of the agar using sterile, bent glass spreading rods. Avoid the extreme edges of the plate. Retain the plates in an upright position until the inoculum is adsorbed by the medium (about 10 min on properly dried plates). If the inoculum is not readily adsorbed, plates may be placed in an incubator in an upright position for about 1 h before inverting. Invert plates and incubate 45 to 48 h at 35 to 37°C. Select plates containing 20 to 200 colonies unless plates at only lower dilutions (>200 colonies) have colonies with the typical appearance of *S. aureus*. If several types of colonies are observed which appear to be *S. aureus,* count the number of colonies of each type and record counts separately. When plates at the lowest dilution plated contain <20 colonies, they may be used. If plates containing >200 colonies have colonies with the typical appearance of *S. aureus* and typical colonies do not appear on plates at higher dilutions, use these plates for enumeration of *S. aureus,* but do not count nontypical colonies. Select one or more colonies of each type counted and test for coagulase production. Coagulase positive cultures may be considered to be *S. aureus.* Add the number of colonies on triplicate plates represented by colonies giving a positive coagulase test, and multiply the total by the same dilution factor. Report this number as *S. aureus* per gram of product tested."

Direct enumeration of coagulase-positive *S. aureus* can be made on Baird-Parker agar containing rabbit plasma-fibrinogen tellurite (Boothby et al., 1979) or Baird-Parker agar without egg yolk to which a tempered pork plasma-fibrinogen overlay agar has been added (Hauschild et al., 1979). In place of Baird-Parker agar, some laboratories have reported the successful use of tellurite polymyxin egg yolk agar (Crisley et al., 1964), Kalium-Rhodanid (= potassium thiocyanate)-Actidione-Natriumazid (= sodium azide)-Egg yolk-Pyruvate (KRANEP) agar (Sinell and Baumgart, 1966), and Schleifer-Krämer (SK) agar (Schleifer and Krämer, 1980) for the selective isolation and enumeration of staphylococci from foods.

SK Agar for Selective Isolation of Staphylococci
(Schleifer and Krämer, 1980)

Basal medium:	
Tryptone or peptone from casein	10.0 g
Beef extract	5.0 g
Yeast extract	3.0 g
Glycerol	10.0 g
Sodium pyruvate	10.0 g
Glycine	0.5 g
KSCN	2.25 g
$NaH_2PO_4 \cdot H_2O$	0.6 g
$Na_2HPO_4 \cdot 2H_2O$	0.9 g
LiCl	2.0 g
Agar	13.0 g
Distilled H_2O	1 liter

Adjust pH to 7.2. Autoclave at 121°C for 15 min, cool down in water bath to 45°C and add 10 ml of a 0.45% sterile-filtered solution of sodium azide. Mix medium thoroughly and pour immediately into petri dishes. The medium can be stored at 4°C for at least one week. Staphylococci can be detected in various foods at levels as low as 100 colony-forming-units (CFU)/g of food. As in the case of KRANEP agar, the addition of egg yolk or pork plasma to the basal medium can provide a basis for distinguishing *S. aureus* from the coagulase-negative staphylococci.

The recovery of some animal species (e.g., *S. caprae* and *S. chromogenes*) can be improved by adding 5% sheep blood and/or reducing the level of sodium azide in SK agar from 45 mg/1 to 15 mg/1 (Harvey and Gilmour, 1988.)

Isolation of Staphylococci from Clinical Specimens

The isolation and enumeration of staphylococci from clinical specimens are routine operations in the hospital and veterinary clinical laboratory. Procedures for handling specimens and isolating and enumerating staphylococci are described in the American Society for Microbiology (ASM) *Manual of Clinical Microbiology,* fourth and fifth editions (Lennett et al., 1985; Balows et al., 1991) and the American Public Health Association (APHA) *Diagnostic Procedures for Bacterial Infections,* seventh edition (Wentworth, 1987).

PREPARATION OF BLOOD CULTURES. Blood cultures are usually indicated when there is a sudden increase in the pulse rate and temperature of the patient, a change in sensorium, and the onset of chills, prostration, and hypotension. Another indication is prolonged, mild, and intermittent fever in association with a heart murmur. Timing of collection is usually not critical when bacteremia is expected to be continuous, such as with endocarditis, endarteritis, typhoid fever, and brucellosis. Bacteremia is usually intermittent in most other infections. In these cases, timing may be very important, and bacteremia may precede the onset of fever or chills by as much as 1 h. Staphylococci are one of the major groups of bacteria that can produce a serious bacteremia.

Collecting Blood Cultures for Isolation of Bacteria
(Isenberg el al., 1985)

"In patients with suspected bacterial endocarditis, three blood cultures are sufficient to isolate the etiological agent in nearly all instances. These should be collected separately and, the condition of the patient permitting, at no less than hourly intervals within a 24-h period. In intermittent bacteremias, three separate blood cultures within 24 to 48 h are usually sufficient to isolate the etiological agent. ... In patients who have received antimicrobial agents before blood collection, a total of four to six separate blood cultures may be necessary to isolate the etiological agent, or consideration should be given to using special techniques to adsorb or inactivate the antibiotic(s) present in the blood. It is essential that blood for culture be collected aseptically, first by cleaning the skin with 80 to 95% alcohol and then by applying 2% iodine [or iodophore] in concentric fashion to the venipuncture site.... the iodine or iodophore should remain intact on the skin for at least 1 min. The intended venipuncture site should not be touched ... After the venipuncture, any residual iodine should be removed with an alcohol sponge or pad. It is recommended that 10 to 20 ml of blood [in adults] be collected for each culture.... In infants and children, collection of 1 to 5 ml appears to be satisfactory."

Needham (1987) recommends collecting 20 ml of blood from adults, 5 to 10 ml from children, and 1 to 2 ml from neonatal patients. Blood may be inoculated directly into culture media at the bedside of the patient or transported to the laboratory in a sterile, evacuated tube containing sodium polyanetholesulfonate (SPS) and then inoculated into culture media. Blood should be inoculated into culture media at a 10% volume to counteract the normal bactericidal activities. A variety of commercially available broth media are suitable for the culture of blood, e.g., tryptic soy broth (Difco Laboratories), trpticase soy broth (BBL Microbiology Systems, Cockeysville, MD), Columbia broth, and brain heart infusion (BHI) broth. One culture or bottle should be incubated anaerobically (e.g., with CO_2) and the other incubated aerobically. Cultures should be incubated at 35°C and examined later on the same day and daily thereafter for up to 7 days for evidence of growth. Longer incubation may be preferable for specimens from seriously ill patients who appear unresponsive to antibiotic therapy. Gram-stained smears and anaerobic subcultures of suspected positive cultures should be prepared immediately.

PREPARATION OF BODY FLUID CULTURES. In addition to blood, staphylococci may invade a variety of other body fluids, such as cerebrospinal fluid (CSF) and joint, intraocular, pericardial, peritoneal, and pleural space fluids. It usually is easiest to establish the etiological agent in infections of normally sterile body sites, provided puncture and handling of the specimen are performed under conditions of strict asepsis. The skin should be disinfected with providone-iodine or a 1–2% solution of tincture of iodine. The specimen should be injected immediately into a sterile (screw-cap) tube or bottle. Since infection of spaces containing body fluids may be due to anaerobes, it is recommended that fluid or pus be collected with a sterile syringe and needle (with air expelled) and the material be injected into an anaerobic transport tube or bottle containing CO_2. Staphylococci will transport satisfactorily under these conditions. Since there may be only a small number of microorganisms present in clear or slightly cloudy fluids, these specimens should be centrifuged to concentrate the organisms. The supernatant fluid should be removed for chemical or serological studies, and the sediment used for both smear and culture purposes. Very purulent material should be smeared directly and stained for the presence of bacteria.

PREPARATION OF URINE CULTURES. Urinary tract infections in humans due to staphylococci are commonly caused by the species *S. saprophyticus, S. epidermidis,* and *S. aureus* and usually involve the upper urinary tract. Guidelines for the diagnosis of urinary tract infections may be obtained from the text *Detection, Prevention, and Management of Urinary Tract Infections* (Kunin, 1979). The reliability of a culture of a single, clean-voided urine specimen is about 80% in the female and about 100% in the adult male (if he is circumcised or has carefully retracted foreskin and cleansed glans). The reliability in females can be increased to 90% with the collection of two specimens and nearly 100% with three, if the specimens contain the same organism. In asymptomatic patients, two or three specimens should be collected to document bacteriuria. Suprapubic aspiration may be indicated in patients who have low bacterial counts in clean-voided specimens and also in neonates and young infants. Urine should be transported in a sterile, screw-capped container or tube. It should be cultured within 1 h of collection or stored in a refrigerator for no more than one or two days prior to culturing. Most references to diagnostic criteria state that a significant bacteriuria occurs when there are 100,000 cells or more per ml in a clean-voided, midstream specimen obtained from asymptomatic patients. With acute dysuria and frequency in young, sexually active females, a colony count as low as 100 per ml may be a useful criterion. Many significant urinary tract infections due to *S. saprophyticus* are associated with only 100 to 10,000 colonies per ml. If significant, these low counts may be substantiated by repetition of the procedure. Among other methods, significant bacteriuria may also be determined by microscopic examination of a Gram-stained smear of uncentrifuged urine. The presence of at least two bacteria per 1000-× microscopic field of the Gram-stained smear is approximately equal to 100,000 or more cells per ml (Pollock, 1983).

PROCEDURES FOR THE ISOLATION AND CULTURE OF STAPHYLOCOCCI. Staphylococci from a variety of clinical specimens are usually isolated in primary culture on blood agar and in a fluid medium such as thioglycolate broth. A general discussion of the preparation of specimens for primary culturing and inoculation of media and colony isolation methodology can be found in the ASM *Manual of Clinical Microbiology* (1985, 1991).

Isolation and Culture of Staphylococci from Clinical Specimens (Kloos and Lambe, 1991)

"... [B]lood agar (preferably sheep blood agar) and a fluid medium such as thioglycolate broth should be inoculated. On blood agar, abundant growth of most staphylococcal species occurs within 18 to 24 h. By this time, colonies will be 1 to 3 mm in diameter, and they are usually circular, smooth, and raised, with a butyrous consistency. Only individual colonies should be picked for preliminary identification testing at this time (e.g., from acute infections). Since most species cannot be distinguished from one another on the basis of colony morphology within a 24-h incubation period, colonies should be allowed to grow for at least an additional two days before the primary isolation plate is confirmed for species and/or strain composition (Kloos and Schleifer, 1975a). This is particularly important if it is necessary to sample more than one colony to obtain sufficient inocula and for determining the predominant organism or a pure culture. Failure to hold plates for 72 h, can result in (i) selection of more than one species or strain if two or more colonies are sampled to produce an inoculum, (ii) selection of an organism(s) not producing the infection, if the specimen contains two or more different species or strains, and (iii) incorrectly labeling a mixed culture as a pure culture. Colonies should be Gram-stained, subcultured, and tested for genus, species, and, when applicable, strain properties. Most staphylococci of major medical interest produce growth in the upper as well as the lower anaerobic portions of thioglycolate broth or semisolid agar (Kloos and Schleifer, 1975b)."

Fecal specimens suspected of containing infecting staphylococci (e.g., associated with staphylococcal enterocolitis) and other specimens from potentially heavily contaminated sources should also be inoculated on a selective medium such as SK agar (described above), Columbia CNA agar, lipase-salt-mannitol agar (LSM) (Remel, Lenexa, KS, USA), tellurite glycine agar, phenylethyl alcohol agar, or mannitol salt agar. These media inhibit the growth of Gram-negative bacteria in addition to some other contaminating species. Incubation of these cultures should be for at least 48 to 72 h for discernable colony development.

Preparation of Blood Agar (Phillips and Nash, 1985)

Beef heart muscle, infusion form	375.0 g
Tryptose or Thiotone peptic digest of animal tissue USP	10.0 g
Sodium chloride	5.0 g
Agar	15.0 g
Distilled or demineralized water	1 liter

Adjust to pH 7.4. Autoclave at 121°C for 15 min. Cool to 50°C, and add 5% sterile defibrinated rabbit blood.

Sheep blood is commonly used as a substitute for rabbit blood. Bovine blood may also be used for determining hemolysis characteristics of the various species. Blood agar base may be obtained commercially. Other media suitable for the culture of staphylococci such as tryptic or trypticase soy agar or P agar (see below) may be substituted for the blood agar base.

Isolation of Staphylococci from Skin and Mucous Membranes

Several basic methods are available for isolating staphylococci and other aerobic bacteria from skin and the adjacent mucous membranes (reviewed by Noble and Somerville, 1974). Washing or swabbing methods disperse cutaneous bacteria to provide samples of uniform composition. They break up large aggregates or microcolonies on skin into smaller colony-forming units (CFU) and in some cases single cells. Impression methods estimate the number of microcolonies or aggregates of bacteria on the skin surface. Biopsy methods can determine the location of bacteria in microniches on skin. Most of the sampling of aerobic bacteria on skin and mucous membranes has been performed using scrubbing and swabbing methods. The swab technique described by Kloos and Musselwhite (1975) is suitable for use with human as well as other mammalian skin. The medium most widely used for the isolation and culture of natural populations of staphylococci is P agar (Kloos et al., 1974), which is prepared as follows:

P agar for the Isolation and Identification of Staphylococci (Phillips and Nash, 1985)

Peptone	10.0 g
Yeast extract	5.0 g
Sodium chloride	5.0 g
Glucose	1.0 g
Agar	15.0 g
Distilled water	1 liter

Adjust pH to 7.5 before autoclaving. Autoclave at 121°C for 15 min.

From each inoculated swab, a series of dilutions are prepared and plated on standard-size (15× 100mm) P agar plates, in an attempt to obtain 50 to 300 isolated colonies on a plate for identification and enumeration. Inoculated plates are incubated at 34 to 35°C for 3 to 4 days and then held at room temperature for an additional 2 days. Each colony type is enumerated and one or two representatives of each type per plate are examined further for distinguishing genus, species, and strain characteristics. Colony morphology can be a useful supplementary character in the identification of species and strains. Selective media may be used in addition to the nonselective P agar if bacterial and/or fungal populations are very large (e.g., from the human inguinal and perineal area or from the skin of certain animals) and/or if the cutaneous flora contains species producing large, spreading colonies. Pigment production of colonies may be enhanced by the addition of milk, fat, glycerol monoacetate, or soaps to P agar or heart infusion agar (Willis et al., 1966).

Isolation of Staphylococci from Water

The presence of potentially pathogenic staphylococci in recreational waters, swimming pools, water that might be added to foods, and hydrotherapy pools poses a threat to human health (reviewed by Evans, 1977). Although several of the staphylococcal species are serious opportunistic pathogens, most attention is focused on the presence of *S. aureus* in water. Staphylococci are somewhat resistant to halogen disinfectants. For this reason, significant numbers of these organisms can remain viable for extended periods of time in inadequately treated bathing places. Methods for the recovery and enumeration of *S. aureus* from water using a membrane filter technique and a multiple-tube procedure are described in the manual jointly published by the APHA, American Water Works Association (AWWA), and Water Pollution Control Federation (WPCF) entitled *Standard Methods for the Examination of Water and Wastewater*, 16th Edition (Greenberg et al., 1985).

Identification

Members of the genus *Staphylococcus* are Gram-positive cocci (0.5 to 1.5µm in diameter) that occur singly, in pairs, tetrads, short chains (three or four cells), and irregular "grape-like" clusters. They are nonmotile, nonsporeforming, and usually are unencapsulated or have limited capsule formation. Most species are facultative anaerobes and they are positive for the catalase and benzidine test. With the exception of *S. saccharolyticus* and *S. aureus* subsp. *anaerobius,* growth is more rapid and abundant under aerobic conditions. These exceptional staphylococci are also catalase-negative. Staphylococcal cell walls contain teichoic acid and peptidoglycan. The peptidoglycans contain more than

2 moles of glycine per mole of lysine. Most species contain *a*- and *b*-type cytochromes. The exceptional species *S. caseolyticus, S. lentus,* and *S. sciuri* contain *a-, b-,* and two *c*-type cytochromes. Menaquinones are unsaturated (normal). The GC composition is in the range of 30–39 mol%, Staphylococci are generally susceptible to lysostaphin (some species more than others), furazolidone, and nitrofuran, and resistant to erythromycin and bacitracin at low levels. In the routine laboratory, rapid distinction of staphylococci from micrococci can be made by demonstrating the susceptibility of staphylococci to 200µg of lysostaphin per ml and resistance to erythromycin at 0.04µg per ml, plus the production of acid from glycerol (Schleifer and Kloos, 1975a) or, alternatively, demonstrating susceptibility of staphylococci to a 100µg furazolidone disk and resistance to a 0.04-unit bacitracin disk (Baker, 1984). Furthermore, staphylococci, with the exceptions of *S. caseolyticus, S. lentus,* and *S. sciuri,* exhibit a negative reaction with the rapid modified oxidase test; whereas, micrococci are positive for this test (Faller and Schleifer, 1981).

Differentiation of Species Groups

Staphylococcal species can be grouped on the basis of their natural or genomic relationships as determined by DNA-DNA hybridization and the thermal stability of DNA heteroduplexes (Kloos and Schleifer, 1986; Kloos and Wolfshohl, 1979, 1983; Schleifer, 1986; Schleifer et al., 1979). DNA relationships of the various species groups, species, and subspecies are shown in the dendrogram (Fig. 1). These relationships have been supported by extensive phenotypic character analyses. Currently, the following six species groups can be recognized:

1. *S. epidermidis* species group: *S. epidermidis, S. capitis, S. caprae, S. saccharolyticus, S. warneri, S. haemolyticus,* and *S. hominis. S. lugdunensis* has diverged somewhat from this group on the basis of DNA-DNA hybridization criteria, but shares many of the phenotypic characteristics of the group. This species group is coagulase-negative and novobiocin-susceptible. The interpeptide bridge of the peptidoglycan contains glycine and significant quantitities of L-serine. Members of this group are slightly resistant to lysostaphin (MIC ≥ 100 µg/ml). They usually require a relatively large number (5–12) of amino acids for growth.

2. *S. saprophyticus* species group: *S. saprophyticus, S. cohnii,* and *S. xylosus. S. equorum, S. arlettae, S. kloosii,* and *S. gallinarum* are phenotypically at the margin of this group.

% DNA - DNA homology

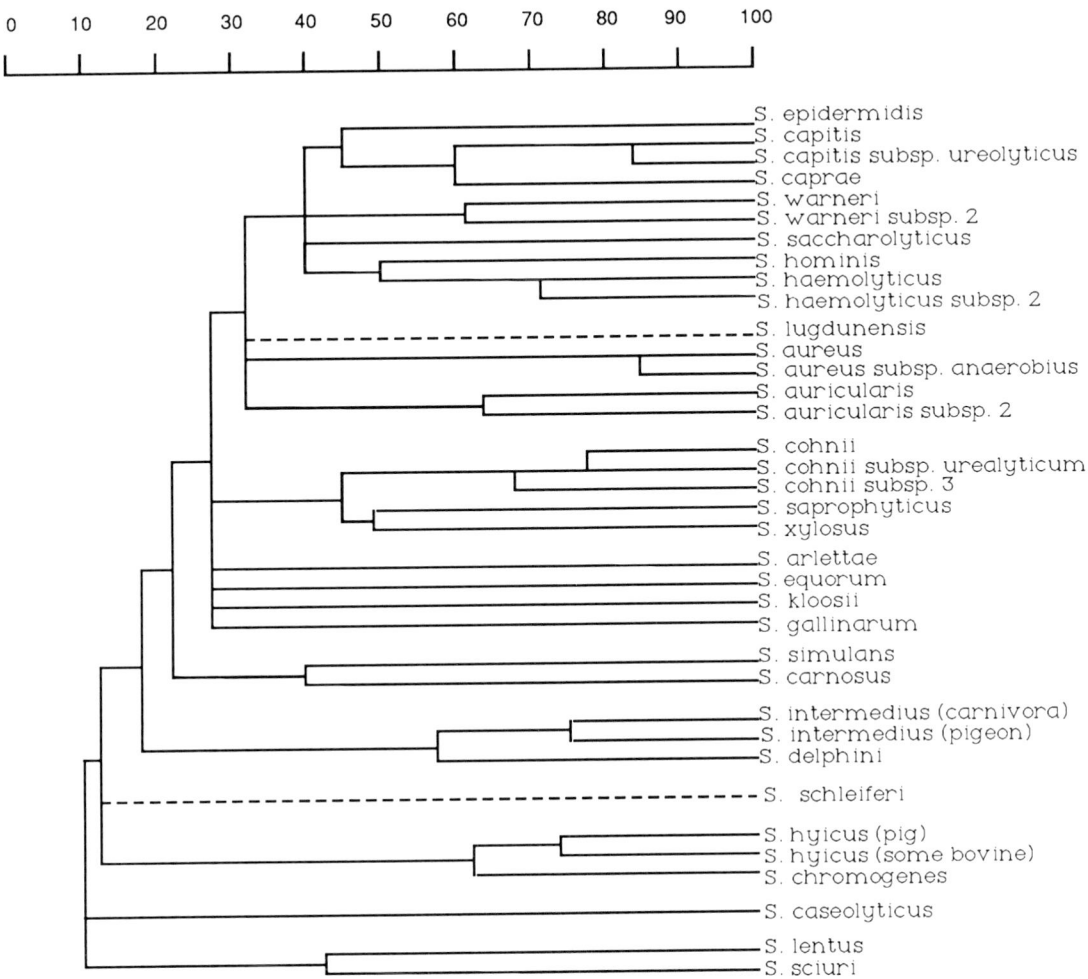

Fig. 1. Dendrogram of the DNA relationships of the *Staphylococcus* species and subspecies. The dashed lines indicate a tentative position based on DNA-DNA hybridization with one reference (*S. epidermidis* species group).

This species group is coagulase-negative and novobiocin-resistant.

3. *S. simulans* species group: *S. simulans* and *S. carnosus*. *S. felis* is phenotypically at the margin of this group. This species group is coagulase-negative and novobiocin-susceptible. In contrast to the *S. epidermidis* species group, members of the *S. simulans* species group demonstrate β-galactosidase activity.

4. *S. intermedius* species group: *S. intermedius* and *S. delphini*. This species group is coagulase-positive and novobiocin-susceptible. The group (on the basis of information from *S. intermedius*) can be distinguished from the coagulase-positive, novobiocin-susceptible species *S. aureus* by its β-galactosidase activity and degradation of galactose via the Leloir pathway. *S. aureus* degrades galactose via the tagatose-6-phosphate pathway. *S. schleiferi* is at the margin of this group.

5. *S. hyicus* species group: *S. hyicus* and *S. chromogenes*. The species group is coagulase-variable *(S. hyicus)* or coagulase-negative *(S. chromogenes)* and novobiocin-susceptible. Members of this group do not produce acetylmethyl carbinol (acetoin) and do not exhibit β-galactosidase activity.

6. *S. sciuri* species group: *S. sciuri* and *S. lentus*. This species group is coagulase-negative and novobiocin-resistant. The interpeptide bridge of the peptidoglycan contains glycine and L-

alanine. Members of this group are positive for the modified oxidase test mentioned above (they contain *c*- type cytochromes).

Differentiation of Species and Subspecies

The classification of species and subspecies of staphylococci was based on a variety of phenotypic character analyses and DNA-DNA (genomic) relationships. DNA homology is the most reliable criterion available today for determining species boundaries and has been used in the formal classification of staphylococcal species and most subspecies (Fig. 1). Selected phenotypic characters were also useful in classification because they had a high predictive value in identifying DNA homology groups. Characters studied at the cellular and population levels, including morphological and physiological properties, enzyme reactions, and intrinsic resistance to certain antibiotics, have been included in practical identification schemes, such as the one given in Table 1. Molecular studies of phenotypic characters have also provided a basis for determining epigenetic relationships.

CONVENTIONAL METHODS. Conventional methods for the determination of phenotypic characters at the cellular and population levels were developed first and then examined for their correlation to DNA homology (reviewed by Kloos and Schleifer, 1986; Schleifer, 1986). Key characters now used for species and subspecies identification include the following: colony morphology, oxygen requirements, coagulase, clumping factor, heat-stable nuclease (thermonuclease), hemolysins, catalase, oxidase, alkaline phosphatase, urease, ornithine decarboxylase, pyrrolidonyl arylamidase, β-galactosidase, acetoin production, nitrate reduction, esculin hydrolysis, aerobic acid production from a variety of carbohydrates including D-trehalose, D-mannitol, D-mannose, D-turanose, D-xylose, D-cellobiose, L-arabinose, maltose, α-lactose, sucrose, and raffinose, and intrinsic resistance to novobiocin and polymyxin B (reviewed by Kloos and Lambe, 1990). Some conventional methods may require up to three to five days before a final result can be obtained, while others only require several hours for interpretation. They are usually quite reliable and have served as a reference for more recent studies aimed at simplifying and expediting character analyses.

RAPID IDENTIFICATION SYSTEM. To facilitate identification in the routine or clinical laboratory, several manufacturers have developed rapid species identification kits or automated systems requiring only a few hours to one day for the completion of tests. Identification of a number of the *Staphylococcus* species can be made with an accuracy of 70 to >90% (Almeida et al., 1983; Crouch et al., 1987; Giger et al., 1984; Hussain et al., 1986; Kloos and Wolfshohl, 1982) using the commercial systems. After their introduction, these systems have been improved and expanded to include more species. It is expected that their reliability will continue to increase as the result of a growing data base and the addition of more discriminating tests. For some systems, reliability depends upon additional testing as suggested by the manufacturer and/or by published evaluations of the product. The major companies and their products include: 1) API Analytab Products, Plainview, NY—API STAPH-IDENT and the dms STAPH Trac Kits; 2) API SYSTEM S.A., La Balme-les-Grottes, France—STAPH Trac and ATB 32 STAPH Kits; 3) Baxter Healthcare Corporation, MicroScan Division, West Sacramento, CA—MicroScan Pos Combo Type 6 Panel and MicroScan RAPID POS ID Panel, for use with the automated autoSCAN-W/A system; 4) Becton Dickinson Diagnostic Instrument Systems, Towson, MD—Sceptor Staphylococcus MIC/ID Panel; 5) Becton Dickinson Microbiology Systems, Cockeysville, MD-Minitek Gram-Positive Set; 6) Chemapol Co. Ltd., Praha, Czechoslovakia—STAPHY test kit; and 7) Vitek Systems, Hazelwood, MO—Gram-Positive Identification Card GPI for use with the automated VITEK System. Depending on the system, from 8 to 32 tests (or character determinations) are included for the identification of *Staphylococcus* species. The automated systems produced by Baxter-MicroScan and Vitek incubate inoculated trays or cards, respectively, read and interpret results, and, with the aid of their programmed computer, determine identification of the organism. Some of the rapid identification systems have introduced new key characters for the identification of staphylococcal species and subspecies such as the pyrrolidonyl arylamidase, ornithine decarboxylase, urease, β-glucosidase, β-glucuronidase, and β-galactosidase enzyme tests.

The presumptive identification of *S. aureus* can be made also by using rapid commercial slide tests for detecting hemagglutination by clumping factor (BBL Microbiology Systems, Cockeysville, MD; BioMerieux, Marcy l'Etiole, France) or latex agglutination tests that detect both clumping factor and protein A (Scott Laboratories, Inc., Fiskeville, R.K., I-M, Inc., Operating Unit of American MicroScan, Lexington, KY; Carr-Scarborough Microbiologicals,

Inc., Stone Mountain, GA; Wellcome Diagnostics, Dartford, England).

DIFFERENTIATION ON THE BASIS OF BIOCHEMICAL AND CHEMICAL PROPERTIES. The most important biochemical and chemical properties found useful for identifying *Staphylococcus* species and some subspecies have been the cell wall composition, configuration of lactate produced from glucose fermentation, occurrence of NAD-dependent L- and/or D-lactate dehydrogenase, pathways for galactose (and lactose) degradation, class of fructose-1, 6-bisphosphate aldolases, and types of cytochromes (reviewed by Schleifer, 1986). These properties will be discussed in subsequent sections of this chapter.

Durham and Kloos (1978) reported that certain *Staphylococcus* species could be distinguished on the basis of their cellular fatty acid composition. More recently, Microbial ID, Inc., Newark, DE, has developed a Microbial Identification System (MIS) which automates the identification of various bacteria, including staphylococcal species and certain subspecies, by combining cellular fatty acid analysis with computerized high-resolution gas chromatography. In general, the fatty acid profiles correlated with DNA homology data.

Cell Wall Composition

The ultrastructure and chemical composition of the cell wall of staphylococci is similar to that of other Gram-positive bacteria. It consists of a thick (usually 60–80 nm), rather homogeneous, and not very electron-dense layer. It is made up of peptidoglycan, teichoic acid, and protein (Schleifer, 1983).

A characteristic feature of the peptidoglycan of staphylococci is the occurrence of glycine-rich interpeptide bridges. Penta- and hexaglycine interpeptide bridges are found in about half of the staphylococcal species (peptidoglycan type: Lys-Gly$_{5-6}$). In the other half, a minor part of the glycine residues can be replaced by L-serine (peptidoglycan type: Lys-Gly$_4$, Ser). In two species *(S. sciuri, S. lentus)*, an L-alanine instead of a glycine residue is bound to lysine of the peptide subunit (peptidoglycan type: Lys-Ala-Gly$_4$).

Cell wall teichoic acids are water-soluble polymers containing repeating phosphodiester groups that are covalently linked to peptidoglycan. They consist of polyol (glycerol, ribitol), sugars and/or N-acetylamino sugars. Most staphylococci contain glycerol or ribitol teichoic acids (Table 2). The teichoic acids consist of polymerized polyol phosphates that are substituted with various combinations of sugars and/

or N-acetylamino sugar residues, and also ester-linked D-alanine residues. In some cases N-acetylamino sugar residues can also form an integral part of the polymer chain (Endl et al., 1983, 1984). The occurrence of the same major components does not always mean that the structure of the teichoic acid is identical; for example, the teichoic acids of *S. capitis* and *S. hyicus* show a similar composition but their structures are quite different.

Peptidoglycan as well as cell wall teichoic acids not only have structural functions but also exhibit biological activities. Endotoxin-like properties, other pathogenic or toxic properties e.g., dermonecrotic reaction and inhibition of leucocyte migration) and nonpathogenic properties (e.g., adjuvant and mitogenic activities) have been reported for peptidoglycan preparations. Cell wall teichoic acids play an important role for phage adsorption (Chatterjee, 1969; Schleifer and Steber, 1974) and for the binding of divalent cations (Archibald, 1974).

The best-studied cell wall protein from staphylococci is protein A. It is found in over 90% of all *S. aureus* strains. It is covalently linked to the peptidoglycan. Each of its five repetitive, homologous domains binds to the Fc region of immunoglobulins from many mammals (Moks et al., 1986). Protein A is not only used commercially for various immunological methods, but also for preparative and analytical purposes. Other cell surface proteins play an important role as receptors to adhesion proteins of eukaryotes, such as fibronectin, fibrinogen, laminin, and collagen (Wadström and Rozgonyi, 1986). This binding to adhesion proteins represents a mechanism of bacterial attachment to the tissues. *S. aureus* grown on solid surface selectively enhances or expresses cell surface proteins in comparison to cells grown under same conditions in a liquid medium (Cheung and Fischetti, 1988). Infecting bacteria are often surface associated and their cell envelope proteins should therefore be more similar to that of bacteria grown on a solid than in liquid medium.

Metabolism

Phosphotransferase System for Lactose and Other Sugars

Carbohydrate uptake can principally occur in two ways: either the carbohydrate is taken up and accumulates inside the cell in an unchanged form, or it is covalently modified during transport. The mechanism for the first type of carbohydrate uptake has not been well studied in staphylococci. In the second type of carbohy-

Table 1. Differentiation of *Staphylococcus* species and subspecies.

Character	S. epidermidis	S. capitis	S. capitis subsp. ureolyticus	S. caprae	S. saccharolyticus	S. warneri	S. haemolyticus	S. hominis	S. lugdunensis	S. aureus	S. aureus subsp. anaerobius	S. schleiferi	S. auricularis
Colony size ≥ 6mm	−	−	−	d	−	d	+	−	d	+	−	−	−
Colony pigment	−	−	(d)	−	−	d	d	d	d	+	−	−	−
Anaerobic growth	+	(+)	(+)	(+)	+	+	(+)	(±)	+	+	(+)	+	(+)
Aerobic growth	+	+	+	+	(±)	+	+	+	+	+	(±)	+	(+)
Staphylocoagulase	−	−	−	−	−	−	−	−	−	+	+	−	−
Clumping factor	−	−	−	−	−	−	−	−	(+)	+	−	+	−
Thermonuclease	−	−	−	−	−	−	−	−	−	+	+	+	−
Hemolysis	(d)	(d)	(d)	(d)	−	(d)	(+)	−	(+)	+	+	(+)	−
Catalase	+	+	+	+	−	+	+	+	+	+	−	+	+
Modified oxidase	−	−	−	−	−	−	−	−	−	−	−	−	−
Alkaline phosphatase	+	−	−	(+)	d	−	−	−	−	+	+	+	−
Pyrrolidonyl arylamidase	−	−	(d)	d	ND	−	+	−	+	−	ND	+	d
Ornithine decarboxylase	(d)	−	−	−	ND	−	−	−	+	−	ND	−	−
Urease	+	−	+	+	ND	+	−	+	d	d	ND	−	−
β-Glucosidase	(d)	−	−	−	ND	+	d	−	+	+	−	−	−
β-Glucuronidase	−	−	−	−	ND	d	d	−	−	−	−	−	−
β-Galactosidase	−	−	−	−	ND	−	−	−	−	−	−	(+)	(d)
Arginine dihydrolase	d	d	+	+	+	d	+	d	−	+	ND	+	d
Acetoin production	+	d	d	+	ND	+	+	d	+	+	−	+	−
Nitrate reduction	+	d	+	+	+	d	+	d	+	+	−	+	(d)
Esculin hydrolysis	−	−	−	−	ND	−	−	−	−	−	−	−	−
Novobiocin resistance	−	−	−	−	−	−	−	−	−	−	−	−	−
Acid (aerobically) from:													
D-Trehalose	−	−	−	(+)	−	+	+	d	+	+	−	d	(+)
D-Mannitol	−	+	+	d	−	d	d	−	−	+	ND	−	−
D-Mannose	(+)	+	+	+	(+)	−	−	−	+	+	−	+	−
D-Turanose	(d)	−	−	−	ND	(d)	(d)	+	(d)	+	ND	−	(d)
D-Xylose	−	−	−	−	−	−	−	−	−	−	−	−	−
D-Cellobiose	−	−	−	−	−	−	−	−	−	−	−	−	−
L-Arabinose	−	−	−	−	−	−	−	−	−	−	−	−	−
Maltose	+	−	+	(d)	−	(+)	+	+	+	+	+	−	(+)
Sucrose	+	(+)	+	−	−	+	+	(+)	+	+	+	−	d
N-Acetylglucosamine	−	−	−	−	ND	−	+	d	+	+	−	(+)	−
Raffinose	−	−	−	−	−	−	−	−	−	−	−	−	−

Symbols: +, 90% or more strains positive; ±, 90% or more strains weak-positive; −, 90% or more strains negative; d, 11 to 89% of strains positive; ND, not determined. Parentheses indicate a delayed reaction.
Adapted from Kloos (1990); Kloos and Lambe (1991); Kloos and Schleifer (1986); Schleifer (1986).

drate transport, the phosphoenol-pyruvate (PEP):carbohydrate phosphotransferase system (PTS), the carbohydrate is phosphorylated during the transport process (Fig. 2). The PTS-mediated group translocation delivers exogenous carbohydrates as phosphate esters to the cell cytoplasm.

In *Staphylococcus,* glucose, mannose, glucosamine, fructose, *N*- acetylglucosamine, lactose, galactose, mannitol, and β-glucosides are taken up by the PTS (Reizer et al., 1988). The lactose-specific PTS of *S. aureus* is one of the best-studied systems (Egan and Morse, 1966; Hengstenberg et al., 1967). It is comprised of four enzymes: Enzyme I (EI), HPr, Enzyme III (EIII)^{lac}, and Enzyme II (EII)^{lac}.

EI and HPr are the general proteins of the PTS and are involved in most PTS-mediated sugar transports; both enzymes are soluble. *S. aureus* mutants which are defective in one or the other enzyme reveal a pleiotropic phenotype (Egan and Morse, 1965). The EI of *S. aureus* was partially purified and has a molecular weight of about 80,000. It appears to be mon-

	S. saprophyticus	*S. cohnii*	*S. cohnii* subsp. *urealyticum*	*S. xylosus*	*S. kloosii*	*S. equorum*	*S. arlettae*	*S. gallinarum*	*S. simulans*	*S. carnosus*	*S. felis*	*S. intermedius*	*S. delphini*	*S. hyicus*	*S. chromogenes*	*S. caseolyticus*	*S. sciuri*	*S. lentus*
	+	d	+	+	d	−	d	+	+	+	+	+	+	+	+	−	+	−
	d	−	d	d	d	−	+	d	−	−	−	−	−	−	+	d	d	d
	(+)	d	(+)	d	−	−	−	(+)	+	+	+	(+)	(+)	+	+	(±)	(+)	(±)
	+	+	+	+	+	(+)	+	+	+	+	+	+	+	+	+	+	+	(+)
	−	−	−	−	−	−	−	−	−	−	−	+	+	d	−	−	−	−
	−	−	−	−	−	−	−	−	−	−	−	d	−	−	−	−	−	−
	−	−	−	−	−	−	−	−	−	−	−	+	−	+	−	ND	−	−
	−	(d)	(d)	−	(d)	(d)	−	(d)	(d)	−	(d)	d	+	−	−	−	−	−
	+	+	+	+	+	+	+	+	+	+	+	+	+	+	+	+	+	+
	−	−	−	−	−	−	−	−	−	−	−	−	−	−	−	+	+	+
	−	−	+	d	d	(+)	(+)	(+)	(d)	+	+	+	+	+	+	−	+	(±)
	−	−	d	d	d	−	−	−	+	+	ND	+	ND	−	d	+	−	−
	−	−	−	−	−	−	−	−	−	−	ND	−	ND	−	−	−	−	−
	+	−	+	+	d	+	−	+	+	−	+	+	+	d	+	−	−	−
	d	−	−	+	d	ND	ND	+	−	−	−	d	ND	d	d	−	+	+
	−	−	+	+	d	+	+	d	d	−	−	−	ND	+	−	−	−	−
	+	−	+	+	d	d	d	d	+	+	+	+	ND	−	−	−	−	−
	−	−	−	−	−	−	−	+	+	+	d	+	+	+	+	d	−	−
	+	d	d	d	d	−	−	−	d	+	−	−	−	−	−	−	−	−
	−	−	−	d	−	+	−	+	+	+	+	+	+	+	+	+	+	+
	−	−	−	d	d	d	−	+	−	−	ND	−	ND	−	−	ND	+	+
	+	+	+	+	+	+	+	+	−	−	−	−	−	−	−	−	+	+
	+	+	+	+	+	+	+	+	d	d	+	+	−	+	+	d	+	+
	d	d	+	+	+	+	+	+	+	+	+	(d)	(+)	−	d	−	+	+
	−	(d)	+	+	−	+	+	+	d	+	+	+	+	+	+	−	(d)	(+)
	+	−	−	d	−	d	+	+	−	−	ND	d	ND	−	d	−	(±)	(±)
	−	−	−	+	(d)	+	+	+	−	−	−	−	−	−	−	−	(d)	(±)
	−	−	−	−	−	(d)	−	+	−	−	−	−	ND	−	−	−	+	+
	−	−	−	d	d	+	+	+	−	−	−	−	−	−	−	−	d	d
	+	(d)	(+)	+	d	d	+	+	(±)	−	−	(±)	+	−	d	+	(d)	d
	+	−	−	+	(±)	+	+	+	+	−	d	+	+	+	+	d	+	+
	d	−	d	+	−	d	−	+	+	ND	+	+	ND	+	d	ND	d	d
	−	−	−	−	−	−	+	+	−	−	−	−	ND	−	−	ND	−	+

omeric. EI is phosphorylated by PEP, a reaction which requires Mg^{2+}. The phosphorylation occurs at a histidine residue in the *N*3 position. HPr, the second protein of the PTS, is a small protein with a molecular weight of 8,300; its amino acid sequence has been determined (Beyreuther et al. 1977). HPr is phosphorylated by EI-P in position *N*1 of His-15 (Schrecker et al. 1975). A comparison of the HPr primary structures of *S. aureus*, *Enterococcus faecalis*, *Bacillus subtilis*, and the Gram-negative *Escherichia coli* revealed that the HPrs from the Gram-positive organisms are very similar in size and that they contain three highly conserved centers: the active center around His-15, an adjacent region around Tyr-37, and a region around Ser-46 (Reizer et al., 1988). Ser-46 may be a site for ATP-dependent phosphorylation in Gram-positive bacteria.

Depending on the specific sugar phosphorylated, HPr (HPr-P) can transfer its phosphate either directly to EII, or via EIII to EII. In the lactose-specific PTS of *S. aureus*, EIII[lac] is involved. It is a 103-residue protein whose amino

Table 2. Classification of different types of cell wall teichoic acid.

Teichoic acid class	Teichoic acid type	Main substituent[a]	Species
Poly(polyolphosphate)	Poly(glycerolphosphate)	Glc, GlcNAc	S. caprae, S. cohnii, S. epidermidis, S. gallinarum, S. intermedius, S. saccharolyticus, S. warneri
		GlcNAc	S. haemolyticus, S. hominis, S. capitis, S. intermedius
		GalNAc	Some strains of S. aureus
		Glc, GalNAc	S. carnosus
		GalNAc, GlcNAc	S. simulans
	Poly(ribitolphosphate)	GlcNAc	S. aureus
	Poly(glycerolphosphate) + Poly(ribitolphosphate)	GlcNAc	S. arlettae, S. equorum, S. kloosii
		GalNAc	S. saprophyticus, S. xylosus Some strains of S. kloosii
Poly(glycerolphosphate-glycosylphosphate)	Poly(glycerolphosphate-N-acetylglucosaminylphosphate)	—	S. hyicus, S. sciuri
Poly(glycosylphosphate)	Poly(N-acetylglucosaminylphosphate)	—	S. auricularis, S. caseolyticus

[a]Glc, glucose; GlcNAc, N-acetylglucosamine; GalNAc, N-acetylgalactosamine.
From Endl et al. (1983) and Schleifer (1986).

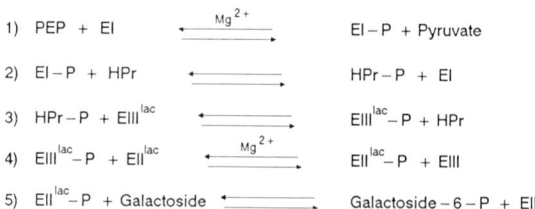

Fig. 2. Reaction scheme for the lactose-specific PTS of *S. aureus*. Enzyme I (EI) and HPr, synthesized constitutively by the cell, are also necessary for the transport of other carbohydrates. Enzyme III[lac] (EIII[lac]) is probably membrane-associated and is an inducible component of the lac operon. Enzyme II[lac] (EII[lac]) is an inducible galactoside-specific integral membrane protein.

acid sequence has been determined (Stuber et al., 1985). The native form of EIII[lac] consists of three identical monomers with a total molecular weight of 36,000. Each of the subunits is phosphorylated via HPr-P at the $N3$ position of His-82. With an altered EIII[lac], in which Gly-18 is replaced by a glutamyl residue, the interaction with EII[lac] is affected, but not the phosphorylation by Hpr-P, indicating that the hydrophobic N- terminal region of EIII[lac] is necessary for the interaction with the membrane-bound EII[lac] (Sobek et al. 1984). EII was partially purified and has a molecular weight of about 55,000. It is suggested that the PTS is composed of a membrane-bound multi-enzyme complex. Evidence for a specific interaction of HPr and EI-II[lac] of *S. aureus* stems from NMR measurements (Kalbitzer et al., 1981).

Some of the PTS-sugars such as glucose are taken up in preference to others (e.g., lactose). An explanation for this effect could be that the various EIII enzymes have different affinities to HPr-P. Not all sugars in staphylococci are taken up via the PTS. Examples of non-PTS carbohydrates are the pentoses D-ribose, D-xylose, and L-arabinose and the corresponding pentitols. These carbohydrates are taken up unsubstituted in *S. xylosus* and *S. saprophyticus* (Lehmer and Schleifer, 1980). In *S. aureus*, maltose is not taken up by the PTS, but maltose uptake may be connected to the PTS, because EI mutants of *S. aureus* are unable to utilize maltose (Button et al. 1973).

Glucose Metabolism

Staphylococci are facultatively anaerobic microorganisms. The Embden-Meyerhof-Parnas (EMP; glycolytic) and the oxidative hexose monophosphate (HMP) pathways are the two central routes of glucose metabolism (Fig. 3). There is no evidence for the existence of the Entner-Doudoroff pathway. Earlier studies of glucose metabolism, mainly confined to the oxidative aspects in *S. aureus*, are reviewed by Blumenthal (1972).

The major end product of anaerobic glucose metabolism in *S. aureus* is lactate (73–94%); acetate (4–7%) and traces of pyruvate are also formed (Theodore and Schade, 1965). Under aerobic growth conditions, acetate and CO_2 are the predominant end products; only 5–10% of the glucose carbon appears as lactate (Strasters

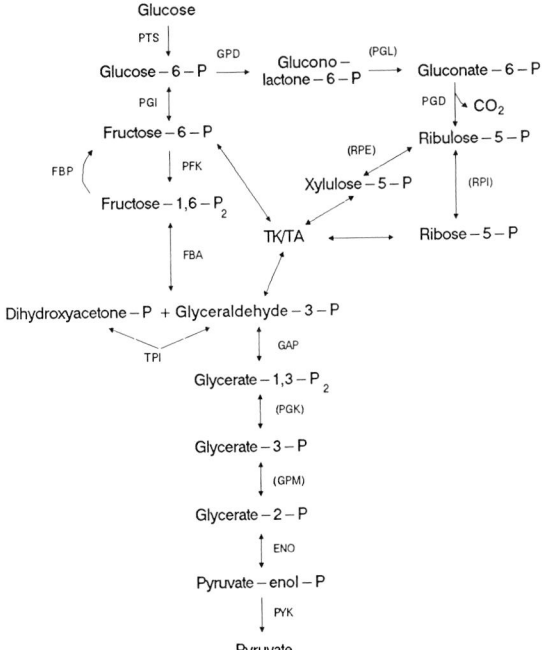

Fig. 3. Schematic representation of glycolysis (EMP) and the pentose phosphate (HMP) pathway in *Staphylococcus*. Most of the enzyme reactions have been confirmed; those which are tentative are in parentheses. Symbols: EMP enzymes: PTS, phosphotransferase system; PGI, phosphoglucose isomerase; PFK, phosphofructosekinase; FBP, fructose-1,6-bisphosphatase; FBA, fructose-1,6-bisphosphate aldolase; TPI, triose-phosphate isomerase; GAP, glyceraldehyde-3-phosphate dehydrogenase; PGK, phosphoglycerate kinase; GPM, phosphoglycerate mutase; ENO, enolase; PYK, pyruvate kinase. HMP enzymes: glucose-6-phosphate dehydrogenase; PGL, 6-phosphogluconolactonase; PGD, 6-phosphogluconate dehydrogenase; RPE, ribulose-5-phosphate epimerase; RPI, ribose-5-phosphate isomerase; TK, transketolase; TA, transaldolase.

and Winkler, 1963). In the presence of glucose, glycolysis is enhanced and many enzymes of the HMP pathway and the citric acid cycle are suppressed; furthermore, the oxidation of pyruvate and the cytochrome content is decreased in glucose-grown *S. aureus* cells. Glucose-mediated catabolite repression is markedly pronounced in staphylococci. The addition of glucose to aerobically grown *S. aureus* cells reduces both glucose degradation via the HMP pathway and the subsequent oxidation of pyruvate via the TCA cycle. An investigation of the specific activity of various HMP enzymes revealed no marked differences in extracts of *S. aureus* grown with or without glucose in nutrient broth. However, the specific activities of two enzymes of the EMP pathway (glyceraldehyde-3-phosphate dehydrogenase and lactate dehydrogenase) were markedly increased in the presence of glucose. Furthermore, in the presence of glucose the spe-

cific activities of the TCA cycle enzymes (succinate dehydrogenase and fumarase) were markedly decreased, and fumarase activity was not detectable. The reduced activities of the TCA cycle enzymes are very likely due to a repression of their biosynthesis; this phenomenon is referred to as the "glucose effect." The glucose-mediated inhibition of the TCA cycle activity could be triggered by the observed increase in the ATP pool. Citrate synthases of staphylococci are not affected by NADH, but are severely inhibited by ATP (Hoo et al., 1971). In the presence of glucose, the NAD-dependent lactate dehydrogenase of *S. aureus* has considerable activity under aerobic growth conditions; although its activity is about 10 times less than that under anaerobic conditions. It is therefore not surprising that in the presence of glucose, some lactic acid is produced even under aerobic conditions (Garrard and Lascelles, 1968). Under aerobic growth conditions in the absence of glucose, an oxidation of acetate, succinate, and malate by resting *S. aureus* cell suspensions is observed manometrically (Collins and Lascelles, 1962; Strasters and Winkler, 1963). However, when glucose or galactose (0.04 M) is present in the growth medium, the oxidation of these substrates is abolished. Growth in the presence of glucose also results in a 40-fold decrease in the cytochrome content (Strasters and Winkler, 1963). The presence of a pyruvate dismutation system in staphylococci was first described by Krebs (1937). Pyruvate dehydrogenase activity was demonstrated both in *S. aureus* and *S. epidermidis* (Sivakanesan and Dawes, 1980).

Regulation of Lactate Dehydrogenase Activity

Anaerobically grown *S. epidermidis* cells ferment glucose with the production of lactate and trace amounts of acetate, formate, and CO_2. As for *S. aureus*, glucose is metabolized principally via glycolysis and to a limited extent by the HMP oxidative pathway (Sivakanesan and Dawes, 1980). However, certain staphylococcal species, such as *S. epidermidis* and *S. intermedius*, are distinguished by a fructose-1, 6-bisphosphate (FBP)-activated, NAD-dependent L-lactate dehydrogenase (LDH) (Götz and Schleifer, 1975, 1976). The enzyme has a total molecular weight of 130,000 and is composed of four subunits. Physiological studies with *S. epidermidis* showed that the intracellular concentration of FBP influences the LDH activity. Under anaerobic growth conditions and in the presence of glucose, a high intracellular concentration of FBP is reached, resulting in maximal LDH activity. In cells grown in a glucose-limited medium under anaerobic conditions, the FBP pool is exhausted because of glucose limitation. In this case, LDH is not fully saturated with FBP and its activity is not maximal. Similar results to this were obtained under aerobic conditions in a glucose-excess medium (Götz and Schleifer, 1978).

Class I Aldolase

Aldolase is one of the key enzymes of the glycolytic pathway. There are two forms of fructose-1, 6-bisphosphate aldolase which can be differentiated by their catalytic and structural properties. Class I aldolases function via the formation of a Schiff base intermediate between the substrate and the amino group of a lysine residue of the enzyme. Class II aldolases do not form a Schiff base intermediate but contain an essential divalent cation, such as Zn^{2+}, Ca^{2+}, or Fe^{2+}, and can be inhibited by EDTA. Since class I aldolases are typical for higher animals and plants and are only found in a few bacteria, it is surprising that nearly all staphylococcal species possess a class I aldolase (Götz et al., 1979). The only exceptions are *S. intermedius* and *S. hyicus,* which possess both classes of aldolases, and *S. caseolyticus,* which possesses only a class II aldolase (Fischer et al., 1982). The class I aldolase of *S. aureus* strain ATCC 12600 was purified and appears to be a monomer with a molecular weight of 33,000 (Götz et al., 1980).

Galactose Metabolism

In many microorganisms, galactose is usually metabolized via the Leloir pathway (Fig. 4A). In *S. aureus,* galactose is converted to D-galactose-6-P, which is further metabolized

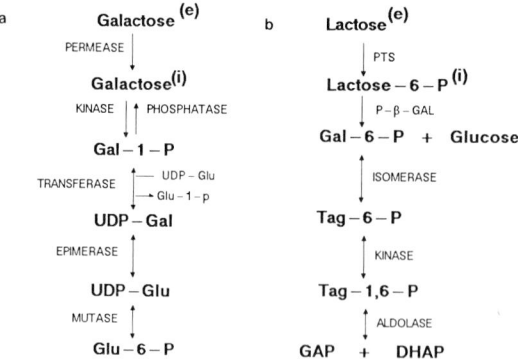

Fig. 4. Galactose catabolism: (a) Leloir pathway. (b) Tagatose-6-P pathway. Symbols: UDP-Gal, uridine diphosphate galactose; UDP-Glu, uridine diphosphate glucose; GAP, glyceraldehyde-3-phosphate; DHAP, dihydroxyacetone phosphate.

through tagatose derivatives (Fig. 4B) (Bissett and Anderson, 1973, 1974). A study of the distribution of the tagatose-6-P pathway in various staphylococcal species revealed that the key enzyme of this pathway (tagatose-6-P kinase) is found in *S. aureus, S. epidermidis* and *S. hominis.* These species do not possess enzymes of the Leloir pathway. The tagatose-6-P kinase is inducible by growth on galactose or lactose. In contrast, *S. intermedius, S. saprophyticus,* and *S. xylosus* use galactose only via the Leloir pathway and not via the tagatose-6-P pathway (Schleifer et al., 1978). The cloned lactose operon of *S. aureus* contains not only the phospho-β-galactosidase gene and genes involved in lactose uptake, but also the genes of the tagatose-6-P pathway (Breidt and Stewart, 1987; Oskouian and Stewart, 1987).

Pentose Metabolism

The uptake of pentoses and pentitols is PTS-independent (Lehmer and Schleifer, 1980). Many staphylococcal species are unable to ferment pentoses, such as D-ribose, L-arabinose, or D-xylose, but are capable of pentose uptake. For example, anaerobically grown *S. epidermidis* cells are unable to ferment ribose, mannitol, and lactose, but transport measurements confirm that ribose is taken up (Sivakanesan and Dawes, 1980). The genes encoding the enzymes for ribose utilization in *S. hyicus* have been cloned in *S. carnosus,* which thereby obtained the ability to utilize D-ribose (Keller et al., 1984). The *S. hyicus* DNA donor strain possessed D-ribokinase and D-ribose-5-P isomerase, which were absent in *S. carnosus.* Although the parent *S. carnosus* strain was unable to utilize ribose, the strain possessed a D-ribose-inducible uptake system which was severely repressed by the addition of glucose.

Respiratory System

Taber and Morrison (1964) identified three cytochromes in *S. aureus:* cytochrome *a*-602 and cytochromes *b*-555 and *b*-557. The two *b*-type cytochromes have different reactives. Cytochrome *b*-557 is reduced in the presence of 2-heptyl-4-hydroxyquinoline-N-oxide (HOQNO), while all the other hemoproteins of the respiratory chain remain oxidized. In contrast to cytochromes *a*-602 and *b*-555, cytochrome *b*-557 is not reduced by ascorbate-dichlorophenolindophenol, suggesting that it has a lower reduction potential than the other cytochromes. The most significant difference, however, is that cytochrome *b*-557, unlike cytochromes *b*-555 and *a*-602, does not react with carbon monoxide. Cytochrome *b*-557 appears to be an intermediate electron carrier of the cytochrome *b* or *b₁*-type. Cytochrome *b*-555, which is also found in *S. epidermidis,* binds carbon monoxide and probably is the major terminal oxidase; it is therefore now referred to as cytochrome *o*. There are some indications that the prosthetic groups of cytochromes *o* and *b*-557 are protohemins. The role of cytochrome *a*-602 as a terminal oxidase is less clear. It is speculated that it may be involved with nitrate reductase activity. In particulate enzyme preparations from *S. aureus,* succinate oxidase, NAD-linked ethanol oxidase, and NADH-oxidase activities are detectable. Succinate oxidase activity is inhibited by UV light (340 nm) and, surprisingly, by amytal (barbiturate A), which inhibits electron transfer from NADH to ubiquinone in higher organisms. The two oxidation-reduction dyes, methylene blue and 2,6-dichlorophenol-indophenol, interact with the electron transfer system. The reduced form of 2,6-dichlorophenol-indophenol reduces cytochrome *a*-602 and cytochrome *o* without reducing cytochrome *b*-557.

Our limited current knowledge about the overall electron transfer sequence is summarized in Fig. 5. The NAD-linked ethanol oxidase could not be definitely localized. The scheme is incomplete since the respiratory chains of *S. aureus* and other staphylococcal species are composed of two more minor cytochromes of the *b*-type in addition to cytochromes *a*-602, *b*-557 and *o*-555 (Faller et al. 1980). Cytochrome *b*-552 is found in all staphylococcal species except *S. sciuri.* Cytochromes *b*-560 and *b*-566 are widely distributed in staphylococci. Cytochromes of the *c* type (e.g., *c*-549 and *c*-554) are found in *S. sciuri, S. lentus,* and *S. caseolyticus.*

Staphylococci possess menaquinones (MK, vitamin K₂) as their sole isoprenoid quinones. There is a species-specific variation with regard to the length of the isoprenoid side chains. The two principal menaquinones have seven to eight isoprene units (Collins and Jones, 1981). Menaquinones are located in the cytoplasmic membrane (White and Frerman, 1967) and play important roles in electron transport and oxidative phosphorylation.

Nitrate Reductase Activity and Hemin Biosynthesis

Early investigations have suggested that some staphylococci have no anaerobic nitrate respiration (Jacobs et al., 1963). Under strictly anaerobic or aerobic growth conditions, the staphylococci examined did not reduce nitrate, although membrane-associate nitrate reductase activity is present in anaerobically grown cells. However, under semi-aerobic growth conditions, nitrate was readily reduced to nitrite, suggesting that O_2 is involved, either in its function as an electron acceptor or as an inducer of hemin biosynthesis. Nitrite accumulated in cells and was not further metabolized. Under anaerobic growth conditions, nitrate could be reduced only in the presence of hemin. This result indicates that an electron carrier required for nitrate reduction is not present and that the protein involved requires heme for activity.

Jacobs and Conti (1965) investigated the cytochrome content and respiratory capacity of a coagulase-negative *Staphylococcus* strain under anaerobic and aerobic conditions. In anaerobically grown cells, the quantity of all cytochromes was markedly reduced. When anaerobically grown cells were cultivated in the presence of hemin, the cytochrome *o*-555 and *b*-557 content and the respiratory rate in resting cell suspensions were nearly as high as that with aerobically grown cells. Cytochrome *a* was not detectable under these conditions, suggesting that it does not play a major role in respiration

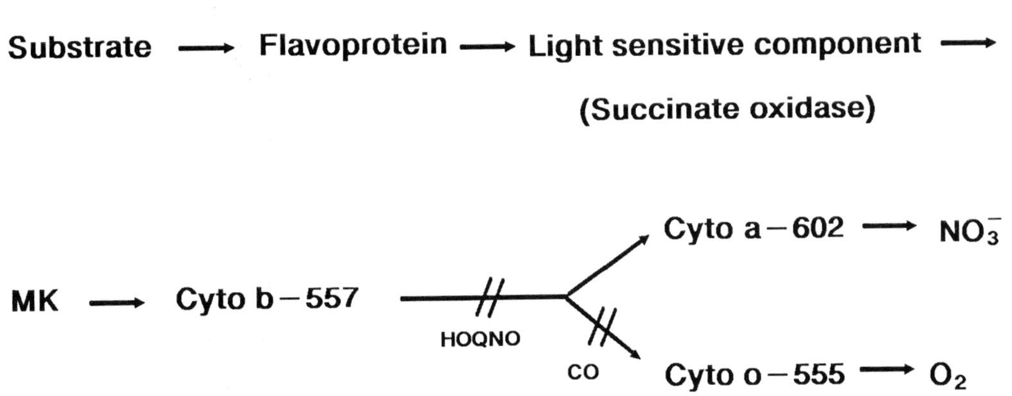

Substrate ⟶ **Flavoprotein** ⟶ **Light sensitive component** ⟶

(Succinate oxidase)

MK ⟶ **Cyto b − 557** ⟶ ╱╱ → **Cyto a − 602** ⟶ **NO₃⁻**

HOQNO   ╱╱

CO   **Cyto o − 555** ⟶ **O₂**

Fig. 5. Postulated electron transfer sequence according to Taber and Morrison (1964). The position of the menaquinones (MK) is not definitely known.

and that protoheme (hemin) is not converted to heme *a*. Anaerobic growth leads to an impaired formation of the respiratory system as indicated by the decrease (10%) of the respiratory rate with resting cell suspensions. Since this inhibition can be overcome by the addition of hemin, it is very likely that in *S. epidermidis* and some other staphylococcal species, such as *S. carnosus* and *S. xylosus* (F. Götz, unpublished observations), the biosynthesis of heme is blocked under anaerobic conditions. The major role of O_2 in the formation of the respiratory system in *S. epidermidis* is its involvement in heme biosynthesis. The biosynthesis of the cytochrome apoenzymes appears to be constitutive. The alternate electron acceptor, NO_3^-, cannot substitute for O_2 even though the nitrate reductase is present under anaerobic conditions. One of the oxidative steps in heme biosynthesis, possibly the oxygenase reaction, may exhibit a requirement for molecular oxygen, and perhaps no other electron acceptor can substitute.

Some staphylococci accumulate large amounts of coproporphyrin during anaerobic growth. In the presence of hemin, the accumulation is markedly reduced, indicating that hemin causes a negative feedback inhibition of porphyrin biosynthesis. No accumulation of coproporphyrin is observed under aerobic conditions (Heady et al., 1964). Anaerobically grown cells contain only about 10 to 15% of the amount of protoheme found in aerobically grown cells (Jacobs et al., 1969). In the presence of S-aminolevulinic acid, the level of coproporphyrin under anaerobic incubation is eight fold higher than under aerobic incubation, suggesting that S-aminolevulinic acid is a precursor in heme synthesis, as it is in mammals, and that all the enzymes for heme synthesis are present in anaerobically grown cells. The addition of NO_3^- to anaerobically grown cells does not cause an increase in the protoheme content.

S. aureus subsp. *anaerobius* grows only anaerobically, is respiration deficient, and catalase- and benzidine-negative. A mutant was isolated which is able to grow aerobically (De la Fuente et al., 1986). Both the wildtype and the mutant produced an intense magenta red color upon illumination with UV light (366 nm), excreted coproporphyrin (mainly type III) into the medium under anaerobic growth conditions, and accumulated uroporphyrin (mainly type I) intracellularly. The mutant did not fluoresce when grown aerobically. None of the strains accumulated protoporphyrin, which is in agreement with the results described above for some other staphylococci. Catalase activity was inversely proportional to coproporphyrin accumulation; under anaerobic conditions, the level

of catalase activity was only 5% of that found during aerobic growth. In a heme-requiring *S. aureus* mutant, inactive apocatalase was also synthesized in the absence of hemin (Jensen and Hyde, 1963). If hemin (0.5 μg/ml) was added to the growth medium of a coagulase-negative staphylococcus under anaerobic growth conditions, catalase activity was nearly as high as under aerobic growth (Heady et al., 1964). It appears that here too, the low level of catalase activity under anaerobic growth is a result of the impaired heme synthesis rather than a repressed synthesis of apocatalase.

Amino Acid and Vitamin Requirements

S. aureus requires the amino acids arginine, proline, and cysteine, and the vitamins niacin (or nicotinic acid) and thiamine for growth on minimal media (Emmett and Kloos, 1975, 1979; Tschäpe, 1973). Some strains also require one or more of the amino acids threonine, methionine, valine, isoleucine, glycine, phenylalanine, tyrosine, tryptophan, histidine, or alanine. Biotin is an essential growth factor of *S. aureus* when glutamic acid instead of glucose is used as a carbon source in a minimal medium (Mah et al., 1967).

The coagulase-negative species *S. epidermidis, S. hominis, S. haemolyticus, S. warneri,* and *S. capitis,* all members of the *S. epidermidis* species group, have requirements for arginine and proline and often for valine, leucine, histidine, and tryptophan (Emmett and Kloos, 1975, 1979). Some strains also require one or more of the amino acids threonine, methionine, cysteine, isoleucine, glycine, serine, phenylalanine, tyrosine, and alanine. *S. capitis* generally requires more different amino acids than the other species mentioned. The vitamins required by this group include nicotinic acid and thiamine and often biotin and/or pantothenic acid (Gretler et al., 1955). Species of the *S. saprophyticus* species group generally require fewer amino acids than those of the *S. epidermidis* species group. Most *S. saprophyticus* strains require proline and/or valine, whereas most *S. cohnii* strains require proline and/or arginine. Only a small percentage (5 to 20%) of *S. xylosus* strains require proline or any other amino acid. Many strains of this species and the distantly related species *S. sciuri* can grow on an inorganic nitrogen source [e.g., $(NH_4)_2SO_4$] as a sole source of nitrogen. Many strains of *S. saprophyticus, S. cohnii,* and *S. xylosus* require nicotinic acid, thiamine, and/or pantothenic acid (Cove et al., 1983; M. Emmett, unpublished observations).

Exported Proteins

Coagulase and Clumping Factor

Coagulase and clumping factor are extracellular and cell-bound proteins; their biochemical properties were reviewed by Jeljaszewicz et al. (1983). Coagulase stimulates the clotting reaction in plasma by forming a stoichiometric, noncovalent complex with prothrombin. In contrast to the prothrombin activation to form thrombin performed by blood-clotting factor Xa, the formation of staphylothrombin does not involve proteolytic cleavage (Hemker et al., 1975). The molecular weight of coagulase differs in various *S. aureus* strains but is normally in the range of 40,000–64,000. Coagulase is regarded as a virulence factor because coagulase-deficient *S. aureus* mutants are less virulent against mice (Jonsson et al., 1985). Recently the gene for coagulase *(coa)* has been cloned (Phonimdaeng et al., 1988).

Clumping factor is also a typical protein of *S. aureus*. It is a cell-wall component with a molecular weight of 21,000. Clumping factor causes an aggregation of cells in the presence of an appropriate animal plasma or fibrinogen. The clumping of cells involves adsorption of fibrinogen to the cells. Clumping factor cannot be clotted by thrombin (Lipinski et al., 1967).

Lipase

It has been known for a long time that certain staphylococci produce extracellular lipases (Weld et al., 1963). Various lipases from different *S. aureus* strains have been isolated and analyzed biochemically (Rollof et al., 1987; Tysky et al., 1983). Most of the staphylococcal lipases are distinguished by a wide substrate specificity. They hydrolyze both water-soluble and water-insoluble triglycerides as well as oxyethylene-sorbitan fatty acids (Tweens). The lipase genes of *S. hyicus* (Götz et al., 1985) and of *S. aureus* (Lee and Iandolo, 1986) have been sequenced. The *S. hyicus* lipase preprotein is composed of 641 amino acids with a predicted molecular weight of 71,382. The sequences responsible for the secretion of the lipase gene have been used to secrete foreign proteins in *S. carnosus* (Liebl and Götz, 1986). Biochemical studies with the *S. hyicus* lipase revealed that triglycerides were fully hydrolyzed to free fatty acids and glycerol and that the enzyme also hydrolyzed naturally occurring phosphatidylcholines and lysophospholipids to free fatty acids and water-soluble products. The *S. hyicus* lipase is extracellularly cleaved by a specific protease to a 46-kDa active lipase form (van Oort et al., 1989).

Lysostaphin

Lysostaphin is an extracellular enzyme of *S. simulans* biovar *staphylolyticus* that lyses staphylococcal cells. Purification and characterization of lysostaphin revealed that the cell wall lytic activity is due to a glycylglycine endopeptidase (Schindler and Schuhardt, 1965). The endopeptidase is a zinc-metalloenzyme (Browder et al., 1965; Trayer and Buckley, 1970). The main target of the endopeptidase is the staphylococcal interpeptide bridge, which in *S. aureus* and some other staphylococcal species is composed of five glycine residues. If two or more glycine residues of the interpeptide bridge are replaced by serine, as in *S. epidermidis*, the cells become less susceptible to lysostaphin (Robinson et al., 1979; Schleifer and Kloos, 1975a). Lysostaphin is important for studies of staphylococcal genetics, for it is used for DNA isolation and protoplast formation. The lysostaphin precursor protein (whose gene has been cloned and sequenced) appears to be organized as a preprolysostaphin with many repetitive sequences in the propeptide protion (Heinrich et al., 1987; Recsei et al., 1987).

Nuclease

The staphylococcal nuclease is biochemically and physically one of the best-studied enzymes. It is also referred to as "micrococcal" nuclease and is an extracellular phosphodiesterase produced by *S. aureus*. The nuclease hydrolyzes RNA and DNA to 3′-phosphomononucleotides (Anfinsen et al., 1971). The secreted form of the nuclease is composed of only 149 amino acids and lacks disulfide bonds. Ca^{2+} ions are required for catalytic activity. The enzyme is distinguished by its thermoresistance and is therefore sometimes referred to as thermonuclease. For a review of the numerous biochemical and physical chemical studies, see Tucker et al. (1978, 1979). The nuclease *(nuc)* gene from *S. aureus* (Foggi) has been cloned and sequenced (Shortle, 1983). Gene expression studies revealed that there are two forms of the enzyme; nuclease A is the secreted, 149-amino acid, classical nuclease, while nuclease B is a membrane-bound precursor of nuclease A found in exponentially growing cells (Davis et al., 1977). The calcium liganding and solvent structure in the active site were determined from the crystal structure of the ternary complex composed of the enzyme, Ca^{2+}, and the inhibitor pdTp (deoxythymidine 3′, 5′-bisphosphate) (Loll and Lattmann, 1989).

Penicillinase

Resistance of *S. aureus* to β-lactam antibiotics is mainly the result of the production of an extracellular β-lactamase, which hydrolyzes the C–N bond in the β-lactam ring. In most strains of *S. aureus,* the β-lactamase genes are encoded on 20–30-kb multi-resistance plasmids (Shalita et al., 1980). Some of the genes are inducible by penicillin or its analogs. The plasmid-encoded penicillinase of *S. aureus* (pPC1) is composed of 257 amino acid residues and has a molecular weight of 28,823, as calculated from the amino acid sequence (Ambler, 1975). The mature and secreted enzyme contains no cysteine or tryptophan. The signal peptide contains a cysteine residue in position 17, which is involved in the membrane attachment by a thioether modification (Nielsen and Lampen, 1982). The plasmid-encoded β-lactamase genes of several strains of *S. aureus* have been cloned and sequenced. The derived amino acid sequence of these strains revealed a striking similarity, having only eight or less amino acid exchanges (East and Dyke, 1989).

Phosphatases

Two types of phosphatases in *S. aureus* have been described: an alkaline and an acid phosphatase (Abramson, 1972; Arvidson, 1983; Malveaux and San Clemente, 1969). Approximately 50% of the acid phosphatase is released into the culture medium. The residual activity appears to be loosely bound to the cell surface. This enzyme can be released from cells with 1 M KCl at pH 8.5. It has a molecular weight of 58,000 and a pH optimum of 5.2. The alkaline phosphatase is firmly bound to the cell surface and has a pH optimum at 10.8. Alkaline phosphatase production is inhibited when *S. aureus* is grown in the presence of inorganic phosphate.

Protein A

Protein A is probably the best-studied cell surface protein in Gram-positive bacteria. It is associated with the cell surface of most *S. aureus* strains. Protein A interacts with human, rabbit, and guinea-pig IgG in a pseudo-immune Fc reaction. The structure, binding properties, biosynthesis, and applications of protein A have been reviewed by Forsgren et al. (1983). The complete nucleotide sequence of the protein A gene *(spa)* reveals extensive internal repetitive sequences (Uhlen et al., 1984a); a 58-amino acid unit, responsible for IgG binding, is repeated 5 times; and an 8-amino acid unit, possibly responsible for the binding to the cell wall of *S. aureus,* is repeated 12 times. Deduced from the nucleotide sequence, the preprotein of protein A has a molecular weight of 58,703. The *spa* gene can be expressed in certain transformed coagulase-negative staphylococcal species and

is similarly localized (Uhlen et al., 1984b). Several secretion vectors were constructed using the signal peptide and the flanking IgG-binding domain (Nilsson et al., 1985a).

Proteinases

Many *S. aureus* strains are proteolytic. Three different types of proteinases have been studied in detail (see review by Arvidson, 1983): 1) **Serine-proteinase.** The enzyme is an endoprotease isolated from *S. aureus* V8 and cleaves peptide bonds at the carboxy-terminal side of glutamic acid and aspartic acid (Drapeau et al., 1972). Its molecular weight is 27,000 and its pH optimum is between 4.0 and 7.8. The enzyme is inhibited by diisopropylfluorophosphate. 2) **Metallo-proteinase.** The enzyme is inhibited by EDTA and *o*-phenanthroline (Arvidson, 1973) and cleaves peptide bonds at the amino-terminal side of hydrophobic residues. Its pH optimum is 7.0 and its molecular weight is 28,000. 3) **Thiol-proteinase.** The enzyme exhibits extensive proteolytic activity with no apparent preferential cleavage site. The pH optimum is between 8.0 and 8.8. It is a small protein of 12,500 molecular weight. It is inhibited by oxidizing agents and is active only in the reduced form.

Staphylokinase

Staphylokinase (SAK) is mainly produced by coagulase-positive staphylococci and exhibits fibrinolytic activity. Fibrinolysis is caused by binding of SAK to plasminogen, which becomes activated to plasmin upon SAK-binding. SAK does not react with fibrin itself but is a nonproteolytic plasminogen activator like the streptokinase produced by *Streptococcus pyogenes.* In this respect, SAK differs from the mammalian urokinase, a proteolytic enzyme which splits plasminogen to form plasmin. Staphylococcal strains producing SAK are frequently found to be lysogenic for phage belonging to serogroup B or F (Winkler et al., 1965). The *sak* gene was cloned and sequenced by two groups (Behnke and Gerlach, 1987; Sako and Tsuchida, 1983). The molecular weight of SAK deduced from the nucleotide sequence is 18,490.

Membrane-Damaging Toxins

S. aureus produces various membrane-damaging and cytotoxic toxins. These toxins damage the cytoplasmic membrane of eukaryotic cells. Each toxin differs in its protein characteristics and mode of interaction with membranes (Mollby, 1983; Telestam, 1983; Wadström, 1983).

The α-toxin is cytotoxic and cytolytic to a wide variety of cell types. It is especially hemolytic to rabbit erythrocytes and is dermonecrotic, neurotoxic, and lethal to experimental animals when injected intravenously at a low dose. Because of its broad biological effect, it has been suggested that α-toxin should be called a membrane-damaging toxin rather than a hemolysin (McCartney and Arbuthnott, 1978). The DNA sequence of the α-toxin gene from *S. aureus* strain Wood 46 has been determined (Gray and Kehoe, 1984). The α-toxin is secreted as a water-soluble, 33-kDa protein that contains three short, highly hydrophobic regions in addition to a number of short, weakly hydrophobic regions. The precursor protein contains a 26-amino acid signal peptide. The α-toxin oligomerizes in the target membranes, forming amphiphilic ring-shaped hexamers (Reichwein et al., 1987). These hexamers become partially imbedded within the lipid bilayer to create aqueous transmembrane pores with an effective diameter of 2 to 3 nm (Bhakdi et al., 1984). The channel-forming α-toxin also stimulates leukotriene formation in rabbit polymorphonuclear leukocytes, and it acts as Ca^{2+} gates allowing passive Ca^{2+} influx into the cell (Suttorp and Habben, 1988).

The β-toxin is a 30-kDa protein (Jeljaszewicz, 1972; Wiseman, 1975). It is the only membrane-damaging toxin with a known enzymatic mode of action; it has sphingomyelinase C (phospholipase C) activity. In addition to sphingomyelin, lysophosphatidylcholine is the only other phospholipid which is enzymatically degraded by this enzyme. It is cytotoxic to a wide variety of cell types, causing significant degradation of membrane sphingomyelin. The β-toxin is also called "hot-cold hemolysin." The name refers to the following phenomenon: sheep or ox erythrocytes do not lyse when treated with β-toxin at 37°C in the presence of Mg^{2+}, unless they are then chilled for a short period at 4°C (Bernheimer et al., 1974). Sheep and ox erythrocytes are especially rich in sphingomyelin, which is located in the outer region of the membrane. It is suggested that by hydrolyzing the sphingomyelin of intact erythrocytes, β-toxin could induce the formation of lipid monolayers in regions of the membrane. The hot-cold hemolysis process could be explained if the lipid monolayer was sufficiently stable at 37°C but not below 10°C. The β-toxin requires Mg^{2+} for activity; Ca^{2+} and Zn^{2+} are inhibitors.

The γ-toxin lyses rabbit erythrocytes. The toxin also lyses human leucocytes and is cytotoxic to human lymphoblast cells (Fackrell and Wiseman, 1976). Very little is known about the mode of action of γ-toxin. Taylor and Bernheimer (1974) showed that the γ-toxin of *S. aureus* strain Smith 5R is composed of two components. The closely linked γ-toxin genes (*hlg*A and *hlg*B) encode proteins of 32 and 36 kDa respectively (Cooney et al., 1988).

The δ-toxin is a small 26-amino acid peptide with a calculated molecular weight of 2,977 (Fitton et al., 1980). It is distinguished by its thermostability, apparent nonantigenicity, high surface activity, solubility in hydrophobic solvents, inhibition by phospholipids, relatively broad activity spectrum on erythrocytes and other cells, and similarities to the surface-active bee mellitin polypeptide and the detergent Triton X-100 (Möllby, 1983). Its affinity for phospholipids could explain its primary role of effecting or facilitating the extrusion or insertion of larger polypeptides through or into the cytoplasmic membrane (Fitton et al., 1980).

Staphylococcal Enterotoxins

The staphylococcal enterotoxins (SE) are protein toxins that cause the typical staphylococcal food poisoning syndrome when ingested (Bergdoll, 1979, 1983). However, the SEs also have other biological activities, including pyrogenicity, T-cell mitogenicity, and induction of human immune interferon. SEs are distinguished by their high resistance to proteolytic digestion, high temperature stability, and broad pH stability.

SEs can be classified into five serological groups, A,B,C,D, and E, referred to as SEA, SEB, SEC, SED, and SEE, respectively. SEC is further subdivided into SEC-1, SEC-2, and SEC-3, based on differences in minor epitopes (Bergoll, 1979). The SEA gene (*ent*A) is encoded on a staphylococcal phage. The mature protein is composed of 233 amino acid residues (molecular weight 27,100); the signal peptide is composed of 24 amino acids (Betley and Mekalanos, 1988). The mature SEA and SEB are transiently associated with the cell wall before being released extracellularly (Christianson et al., 1985). The nucleotide sequence of the chromosomally linked SEB gene (*ent*B) has been determined by Jones and Khan (1986). The mature SEB consists of 239 amino acids (molecular weight 28,336). There is evidence that the *ent*B gene is structurally unstable and possibly part of a mobile genetic element such as a phage or transposon (Johns and Khan, 1988). The nucleotide sequence of the SEC-1 gene (*ent*C-1) was determined by Bohach and Schlievert (1987). The mature protein is composed of 239 amino acids (molecular weight 27,496); the signal peptide consists of 27 amino acids. The *ent*C-1 gene may be encoded on the chromosome or a plasmid. The nucleotide sequence of the SEE gene (*ent*E) was determined by Couch et al. (1988). The mature extracellular form is composed of 230 amino acids (molecular weight 26,425); the signal peptide consists of 27 amino acids.

The sequenced staphylococcal enterotoxins share extensive sequence and structural homology at the protein level. SEA and SEE mature proteins have a rather high (82%) sequence homology. The homology between the other SEs is in the range of 50 to 75%. The staphylococcal enterotoxins also share homology with the streptococcal pyrogenic exotoxin A, the scarlet fever toxin (Johnson et al., 1986).

Exfoliative Toxin A and B

Infection of human neonates with certain *S. aureus* strains can cause Ritter's disease, also known as the staphylococcal scalded skin syndrome (SSSS) (Arbuthnott, 1983; Rogolsky, 1979). The clinical symptoms of SSSS include a wide-spread erythema, followed by loosening of the skin, which then peels in places, leaving extensive raw areas. The molecular basis of the action of the two responsible exfoliative toxins (ETA and ETB) remains unknown. Both ETA and ETB cause the same histological changes in the skin of humans and mice. The nucleotide sequences of the ET genes, *eta* and *etb*, have been determined by several groups (Lee, C.Y. et al., 1987: O'Tool and Foster, 1987; Sakurai et al., 1988). The DNA sequence data reveal proteins of 242 amino acids for the mature ETA (molecular weight 26,951) and 246 amino acids for the mature ETB (molecular weight 27,318). ETA and ETB possess signal peptides of 38 and 31 amino acids, respectively. At the DNA level, the genes have more than 50% homology; at the protein level, homology is found at the N-termini, the middle and the C-termini of the toxins. A comparison of the relative hydrophilicity and hydrophobicity (hydropathicity) indicates highly conserved domains. The *eta* gene is chromosomally located, and the *etb* gene is usually plasmid-born. Despite the in similarities in size and structure, ETA differs from ETB by its heat stability and its Cu^{2+}-dependence for activity (Rogolsky, 1979).

Toxic Shock Syndrome Toxin (TSST-1)

Staphylococcal toxic shock syndrome (TSS) is characterized by fever, hypotension or dizziness, erythematous rash, desquamation of skin upon recovery, and rapid onset of clinical shock (Bergdoll, 1983; Reiser et al., 1983). The illness is frequently associated with the use of tampons during menstruation by young adult women (Davis et al., 1980). According to the nucleotide sequence, the mature protein of TSST-1 is composed of 194 amino acids (molecular weight 22,049); the signal peptide is composed of 40 amino acids (Blomster-Hautamaa et al., 1986). Computer analyses of the amino acid sequence show that TSST-1 has little or no sequence homology with biologically related toxins, streptococcal pyrogenic exotoxin A, and staphylococcal enterotoxins B and C.

Exopeptides with Antibiotic Activity

It has been known for some time that certain staphylococci produce lantibiotics; small polypeptides of 19 to 34 amino acids, with antibacterial activity. A main feature of lantibiotics is the presence of several sulfide rings consisting of two unusual amino acids, *meso*- lanthionine (Lan) and 3-methyllanthionine (Me-Lan); hence the name "lantibiotic" (lanthionine-containing antibiotic) (Schnell et al., 1988). Lantibiotics are produced by a wide range of microorganisms including *Staphylococcus, Lactococcus, Bacillus,* and *Streptomyces*. In staphylococci, three lantibiotics have been described and their chemical structure elucidated: epidermin (Allgaier et al., 1986), gallidermin (Kellner et al., 1988), and Pep5 (Kellner et al., 1989). Epidermin and Pep5 are produced by *S. epidermidis* strains and gallidermin is isolated from *S. gallinarum*. The fermentation of *S. epidermidis* and the epidermin isolation procedure is described by Hörner et al. (1989) and Fiedler et al. (1987). Lantibiotic activity is mainly directed against Gram-positive bacteria. Some of the lantibiotics have high antimicrobial activity against pathogenic Gram-positive bacteria such as *Propionibacterium acne,* staphylococci, and streptococci. It is therefore of interest to determine whether some lantibiotics might be of therapeutic value in topical treatment of certain forms of acne. Recently, it was found that the investigated lantibiotics are synthesized ribosomally and not via multienzyme complexes as are certain cyclic peptide antibiotics (Schnell et al., 1988, 1989). According to the DNA sequence of the epidermin and gallidermin structural genes, both lantibiotics are organized as 52-amino acid prepeptides, which are processed and posttranslationally modified to tetracyclic 21-peptide amide antibiotics. The mature sequence of epidermin corresponds to the C-terminal 22-peptide segment. The cationic lantibiotics epidermin, gallidermin, Pep5, and nisin (from *Streptococcus lactis*) depolarize bacterial and planar lipid membranes in a voltage-dependent manner (Kordel et al., 1988; Ruhr and Sahl, 1985). These lantibiotics are also able to form pores (up to 1 nm in diameter) in the membrane.

Certain staphylococci isolated from the urogenital flora release small peptides which can inhibit the growth of gonococci. The anti-gonococcal substance produced by *S. haemolyticus* has a broad anti-gonococcal spectrum and a narrow antibacterial spectrum (Bisaillon et al., 1981; Frenette et al., 1984). Three small antigonococcal peptides have been isolated and sequenced (Watson et al., 1988). All of the peptides are 44 amino acids long, and they share 65 to 75% sequence homology. They contain a positively charged residue (lysine) at the third position, followed by a core of hydrophobic residues. This suggests that the three peptides are possible signal sequences of one or more secreted or membrane-associated proteins.

Siderophores (Staphyloferrin A and B)

The growth of *S. aureus* in human serum depends, in part, on the iron concentration (Schade, 1963). The production of a staphylococcal iron chelator has been reported by Marcelis et al. (1978) and Maskell (1980). Also, certain membrane proteins of *S. aureus* (Domingue et al., 1989) and *S. epidermidis* (William et al., 1988) are preferentially expressed under iron limitation. These proteins are possibly part of a siderophore-mediated iron uptake system. A highly hydrophilic compound has been isolated from low-iron culture broth of an *S. hyicus* strain. The compound exhibits siderophore activity to the producer and to 37 other staphylococcal strains. The previously unknown metabolite was designated staphyloferrin A and consists of two molecules of citric acid, each linked to D-ornithine by an amide bond (Meiwes et al., 1989). Staphyloferrin A and a second iron-regulating compound, staphyloferrin B, are found in the culture fluid of various staphylococcal species. The structure of staphyloferrin A is shown in Fig. 6.

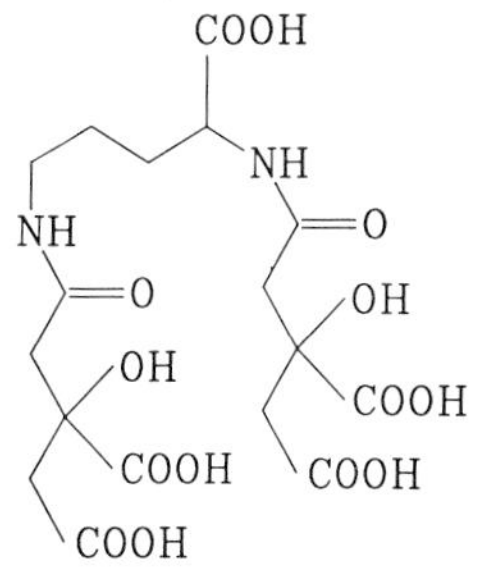

Capsules and Slime

A mucoid colony morphology on agar is an indication of bacterial encapsulation. A capsule is a covering layer outside the cell wall, demonstrable by light microscopy, and having a definite external surface. Its detection by light microscopy implies a thickness of greater than 200 nm. Capsules thinner than 200 nm exist, which are designated as microcapsules. Slime is an exopolysaccharide, but is not as firmly attached to the surface as a capsule and its production is often dependent on specific nutritional conditions. The isolation, chemical composition, and the biological and clinical significance of

Fig. 6. Structure of staphyloferrin A. (From Meiwes et al., 1989.)

staphylococcal capsules and slime were reviewed by Wilkinson (1983).

The three *S. aureus* strains whose capsules have been best characterized biologically and biochemically are strains M, T, and Smith diffuse. Strain M produces a capsule composed of 2-acetamido-2-deoxy-D-galacturonic acid, 2-acetoamido-2-deoxy-D-fucose, and taurin (Liau and Hash, 1977). Strain T contains 2-acetamido-2-deoxy-D-mannuronic acid and 2-acetamido-2-deoxy-D-fucose (Wu and Park, 1971). The Smith diffuse capsule consists of 2-acetamido-2-deoxy-D-glucuronic acid and 2-acetamido-2-deoxy-L-alanyl glucuronic acid (Hanessian and Haskell, 1964). The major constituents of the capsule of the mucoid strain SA1, a toxic shock isolate of *S. aureus,* are 2-acetamido-2-deoxy-α-galacturonic acid (4-O linked), 2-acetamido-2-deoxy-α-fucose (3-O linked), and taurin (Lee, J. C., et al., 1987). The chemical composition of slime is somewhat different from that of capsules. The main components are galacturonic acid, glucuronic acid, galactose, various neutral sugars, and amino acids (Wilkinson, 1983).

A comparison of the biological properties of encapsulated and unencapsulated *S. aureus* strains reveals that encapsulated strains are more virulent against mice (Wiley and Maverakis, 1974), much less susceptible to phagocytosis (Peterson et al., 1978), not phage-typable and negative in clumping factor (Yoshida and Ekstedt, 1968). Because of the strong adherence properties of slime, this component may be a major pathogenic factor of *S. epidermidis* (Christensen et al., 1987). Tojo et al. (1988) have described a "capsular polysaccharide adhesin" (CPA), rich in galactose and glucosamine, that may play a role in the initial attachment of this species to smooth surfaces. This factor may be separate from slime or certain of its components.

Genetics

Most of the studies on staphylococcal genetics have been carried out with *Staphylococcus aureus,* mainly because of its importance in clinical microbiology. Comparatively few studies have been done with *S. epidermidis.* In recent years, a new host-vector system has been developed with *S. carnosus.* Strains of this species are used as starter cultures in food industry.

Transformation

With staphylococci DNA, transformation was first described in *S. aureus* strain NCTC 8325 (Lindberg et al., 1972). This strain is lysogenic, harboring the prophages $\phi 11$, $\phi 12$, and $\phi 13$ (Novick, 1967). It was demonstrated that optimal transformation was only achieved when the cells were harvested in the early exponential growth phase. Transformation is dependent on the presence of Ca^{2+} ions, which are most effective at a concentration of 0.1 M. The transformation rate is also influenced by the purity of the isolated DNA. In *S. aureus,* the optimal pH and temperature for transformation are 6.8 and 30°C, respectively (Rudin et al., 1973). The ability of *S. aureus* to become competent for transformation and also for transfection is not only dependent on the presence of high concentrations of Ca^{2+} ions, but also on the presence of certain bacteriophages such as $\phi 11$, $\phi 14$, or $\phi 83a$ (Rudin et al., 1973; Sjöström and Philipson, 1974; Sjöström et al, 1973). These three phages are antigenically related and belong to the serological phage group B. The involvement of bacteriophage in conferring competence to *S. aureus* is due to an interaction of externally supplied phage components with the surface of the cell. This component is inactivated by pronase and is inhibited by antiserum prepared against purified particles of serological group B phage (Thompson and Pattee, 1981). The competence-conferring activity in crude lysates of staphylococcal bacteriophage 80 was concentrated and purified (Birmingham and Pattee, 1981). The purified preparation exhibited lytic activity against cells of *S. aureus* 8325-4 and was composed of a large number of headless phage tails and what may have been a large number of the phage adsorption organelles. The competence-conferring activity thus appears to be a unit morphological precursor of the $\phi 80$ virion that may mediate transfection and transformation in the presence of 0.1 M $CaCl_2$, in a manner related to the mechanism whereby normal phage particles initiate infection by introduction of the viral DNA into the host cell. Transformation played a major role in determining the chromosomal map of *S. aureus* (Pattee et al., 1984; Pattee, 1990).

Competent *S. aureus* cells can also be transformed with isolated and purified plasmid DNA (Lindberg and Novick, 1973; Lindberg et

al., 1972). The transformation frequency, however, is very low and normally on the order of 10^3 transformants per μg plasmid DNA. The transformation frequency of competent *S. aureus* cells is markedly influenced by the restriction and modification system of the recipient cell. With a plasmid isolated from a nonhomologous *S. aureus* strain, another species, or even another genus such as *Bacillus subtilis,* the transformation frequency is reduced by two orders of magnitude (Sjöström et al., 1979).

Transduction

Ritz and Baldwin (1958) were the first to report transduction in *S. aureus,* using phage 52 to transduce penicillin resistance into a penicillin-susceptible strain. With the aid of transduction, various chromosomal markers, such as resistance to novobiocin, production of coagulase, or the ability to ferment various sugars can be introduced into recipient cells. Transduction was used in the fine-structure mapping of amino acid biosynthesis genes and their clusters (or operons) in *S. aureus* (Kloos and Pattee, 1965; Pattee et al., 1974; Proctor and Kloos, 1970; Smith and Pattee, 1967).

A set of phages is used to differentiate various *S. aureus* strains (International Phage Typing System) (Rosenblum and Dowell, 1960). According to sensitivity to these phages, *S. aureus* strains could be divided into four groups (Stobberingh et al., 1979). Not all bacteriophages used in the phage typing of *S. aureus* conferred general transduction. Only those phages belonging to serological group B possessed transducing capacities (Pattee and Baldwin, 1961). It is thus remarkable that this is the same phage group which confers competence to transformation in *S. aureus.* Transducing bacteriophages pack chromosomal DNA instead of phage DNA into the phage head. The size of the packed DNA is similar to that of the phage DNA (Beryhill and Pattee, 1969). Transducing DNA fragments, which include at least the histidine biosynthesis genes, are highly uniform in size and their nature is not influenced much by the choice of phage or donor strain (Kloos and Pattee, 1965; Pattee et al., 1968).

Plasmid DNA much smaller than the phage DNA can also be transduced. These plasmids are present in high copy number (10–40 copies/cell), and there is evidence that more than one plasmid is packed (Ubelaker and Rosenblum, 1978). The frequency of transduction of plasmids increases with decreasing size of the plasmid. In *S. aureus,* small plasmids such as pT181 are transduced in the form of linear concatamers usually containing only plasmid DNA. Insertion of any phage DNA into the plasmid causes an approximately 10^5-fold increase in the transduction frequency (Novick et al., 1986). In the recipient strain, the plasmid multimers are then processed by a replicative mechanism to the monomeric form. Transduction of plasmids, like transformation, is affected by the host-specific restriction-modification system. One can assume that the host-specific restriction-modification system influences the distribution of resistance plasmids under natural conditions.

Cotransduction of plasmids occurs via a transitory cointegration between plasmid DNAs which is resolved in the recipient strain. A systematic investigation of cointegration between the staphylococcal plasmids pC194, pE194, pUB110, pT181, pSN2, and pS194 (Table 3) revealed a new type of site-specific recombination (Novick et al., 1981). Two different recombination sites (RS) were found, which were named RS_A and RS_B. RS_B is present once on each of the six plasmids. The plasmids pT181 and pE194 carry in addition the recombination site RS_A. DNA sequencing revealed that the 70-bp RS_A sequences are homologous to each other, as the 30-bp RS_B sequences are to each other (Novick et al., 1984). Recombination occurs between the homologous recombination sites of RS_A and RS_B, respectively. Some staphylococcal plasmids, such as pT181, encode a site-specific recombination function called *pre* (Gennaro et al., 1987). This plasmid recombinase appears to be specific for short homologous DNA regions and to be involved in recombination at RS_A and RS_B sites.

Conjugation

As the use of gentamicin in medicine increased, a rapid spread of gentamicin-resistance genes in *S. epidermidis* and *S. aureus* was observed in hospitals (Bint et al., 1977). Transmission of this resistance can occur within the species *S. aureus,* and also from *S. aureus* to *S. hominis* or *S. epidermidis* on human and mouse skin (Naidoo and Noble, 1978, 1981). Gentamicin resistance is frequently associated with kanamycin, tobramycin, and sisomicin resistance. The resistance genes for these antibiotics are normally located on plasmids (Jaffe et al., 1982; Wood et al., 1977). Studies of the transmission of gentamicin resistance from *S. epidermidis* to *S. aureus* revealed that the basis of transmission is very likely a conjugal transfer process (Forbes and Schaberg, 1983). Most of the self-transmissible (Tra$^+$) gentamicin-resistance plasmids in *S. aureus* are larger than 30 kb (Archer and Johnston, 1983; Forbes and Schaberg, 1983).

For example, the plasmid pCRG1600 is a 52.9-kb self-transmissible plasmid encoding resistance to aminoglycoside and β-lactam antibiotics in *S. aureus*. In this plasmid, the *tra* function is located in a 14.6-kb region (Asch et al., 1984). As we know from the Enterobacteriaciae, conjugation is a highly effective method of spreading antibiotic-resistance markers. We now know this occurs within the genus *Staphylococcus* and also between the genera *Enterococcus* and *Staphylococcus*. It is expected that in the near future we will learn more about conjugation and the genes involved in transmission.

Protoplast Fusion

The experimental conditions for plasmid transfer and genetic recombination in *S. aureus* and some coagulase-negative staphylococci by protoplast fusion have been described by Götz et al. (1981). Protoplasts are prepared by treatment with lysostaphin and lysozyme in a buffered medium with 0.7 to 0.8 M sucrose. Regeneration of cell walls is accomplished on a hypertonic agar medium containing succinate and bovine serum albumin. Transfer of plasmids occurs after treatment of the protoplast mixtures with polyethylene glycol (molecular weight 6,000) between strains of the same or different species, although at approximately 100-fold lower frequency in the latter case. Recombination of the chromosomal genes in fused protoplasts requires simultaneous treatment of the mixed protoplasts with polyethylene glycol and $CaCl_2$. Antibiotic-resistance plasmids can be introduced into the parental strains and used as primary markers to detect protoplast fusion. The method can be applied in the construction of strains with multiple mutant characters. It allows for the isolation of recombinants after fusion of mutants and recombination can occur at a high frequency.

Staphylococcal Plasmids

Plasmids have played an important role as transmitters of antibiotic resistance. In 1946, six years after penicillin was first introduced into medicine, 14% of the *S. aureus* strains isolated in hospitals were already penicillin-resistant. In 1947 and 1949, 38% and 59%, respectively, of clinical *S. aureus* strains were penicillin-resistant (Hardy, 1981). Most of the penicillin-resistant staphylococcal strains produced a penicillin-inducible β-lactamase. It could be demonstrated that cultivated penicillin-resistant staphylococcal strains very frequently lose the penicillin-resistance marker, giving the first hint of a plasmid-encoded function. It is typical for staphylococcal penicillin-resistance plasmids to also contain, in addition to the penicillin-resistance marker (Pc), heavy metal-resistance markers conferring resistance to lead (Pb), mercury (Hg), arsenate (Asa), arsenite (Asi), or cadmium (Cd). The genetic basis of antimicrobial resistance of *S. aureus* was reviewed by Lyon and Skurray (1987). Table 3 summarizes some of the more thoroughly characterized staphylococcal resistance plasmids. The plasmids are divided into 13 different incompatibility groups (Iordanescu, 1979; Iordanescu and Surdeanu, 1980; Iordanescu et al., 1978). The incompatibility marker is one of the most reliable characteristics for the classification of plasmids (Novick, 1987). With respect to size and copy number of the various resistance plasmids in staphylococci, there are essentially two groups. The small plasmids (3–6 kb) normally possess a higher copy number (20–80 copies per bacterial chromosome) and carry one resistance marker. The larger resistance plasmids (20–30 kb) are normally present in a lower copy number (8–12 copies per bacterial chromosome) and sometimes carry more than one resistance marker. A few plasmids are distinguished by a relaxation complex (RC) (Novick, 1976). These plasmids form a DNA-protein complex at a particular site. In the presence of high salt concentration or SDS, the plasmid is nicked at this site, thus transforming the supercoiled form of the plasmid into an open circle. Several of the prototype plasmids which have been studied extensively will be discussed in the following section.

TETRACYCLINE-RESISTANCE PLASMID pT181. Most studies on plasmid replication have been carried out with plasmid pT181 by Novick and co-workers. pT181 is a small tetracycline-resistance plasmid that was originally isolated from *S. aureus* (Iordanescu, 1976). The plasmid has a copy number of about 20 and belongs to incompatibility group 3 (Inc 3) (Iordanescu et al., 1978). The DNA of pT181 has been completely sequenced and has a size of 4,437 bp (Kahn and Novick, 1983). The nucleotide sequence revealed four open reading frames (A–D), all transcribed in the same orientation. Gene A represents *repC,* which is necessary for replication of pT181. Genes B and D are involved in tetracycline resistance. Gene C *(pre)* encodes a site-specific recombinase (Gennaro et al., 1987; Novick et al., 1984). Tetracycline resistance due to this system displays a specific exit (efflux) mechanism preventing the antibiotic from reaching its target in the cell. The role of the RepC protein was investigated, mainly using in vitro studies (Kahn et al., 1981; Kahn and Novick, 1983). Mutations which led

to an overproduction of RepC in *S. aureus* caused an increase in the copy number of pT181 (Novick et al., 1982). The RepC protein may be the pacemaker for the replication of this plasmid. Mutations which affect the copy number of the plasmid are located within the *repC* leader region in proximity to the *ori*. The origin of pT181 is located at the beginning of the *repC* coding sequence (Kahn et al., 1982). The replication of pT181 occurs unidirectionally following the mechanism of a rolling circle. The expression of *repC* is regulated by means of anti-sense RNAs (counter transcripts that block expression of *repC*). The counter transcripts induce premature termination (attenuation) of the initiator mRNA by promoting the formation of a termination-causing hairpin that is just 5′ to the *repC* start codon (Carleton et al., 1984; Novick et al., 1989). In the absence of the counter transcripts (i.e., in deletion mutants), an upstream sequence, the preemptor, pairs with the proximal arm of the terminator hairpin, preventing termination and permitting transcription of *repC*. This system differs from classical attenuators in that attenuation is driven by anti-sense RNAs rather than by tRNA-induced stalling of ribosomes. Originally it was believed that the *repC* Shine-Dalgarno sequence and start codon were sequestered by base pairing with the anti-sense RNA (Novick et al., 1985). The two anti-sense RNA molecules, I and II are active in trans and control expression of *repC* in a negative manner. The Inc3A incompatibility of pT181 is also influenced by the two RNA molecules I and II. They are active in trans and inhibit the replication of a different incoming plasmid belonging to the same incompatibility group (Inc3A).

CHLORAMPHENICOL-RESISTANCE PLASMID pC194. Plasmid pC194 is a small chloramphenicol-resistance plasmid originally isolated from *S. aureus* (Iordanescu, 1975). Horinouchi and Weisblum (1982b) determined the nucleotide sequence of pC194, which was later modified by Dagert et al. (1984). The plasmid has a size of 2,910 bp and the sequence revealed four open reading frames. The *cat* gene (ORF B) encodes a chloramphenicol acetyl transferase. pC194 is a broad host-range plasmid. Ehrlich (1977) showed that pC194 replicates in *Bacillus subtilis,* and the chloramphenicol-resistance marker is expressed. Since then, several other staphylococcal plasmid vectors have been constructed for *B. subtilis* (Gryczan et al., 1978). At first glance, it is surprising that staphylococcal genes are expressed in such phenotypically different microorganisms as *B. subtilis.* However, 16S rRNA homology studies indicate

Table 3. Characteristics of the plasmids found in *Staphylococcus*.

Incompatibility group	Origin	Plasmid	Phenotype	Size (kb)	Relaxation complex	Copy number/chromosome	Reference
Inc 1	*S. aureus*	pI258	Pc, Asa, Asi, Hg, Cd, Pb, Bi, Em	28.2	RC⁻	8–12	Peyru et al., 1969
Inc 2	*S. aureus*	pII147	Pc, Asa, Hg, Cd, Pb	31.8	RC⁻	8–12	Novick and Richmond, 1965
Inc 3	*S. aureus*	pT127	Tc	4.4	RC⁻	80–100	Iordanescu et al., 1978
	S. aureus	pT169	Tc	4.4		80–100	Novick and Bouanchaud, 1971
	S. aureus	pT181	Tc, Rep (ts)	4.437	RC⁻	20	Iordanescu, 1975
Inc 4	*S. aureus*	pC221	Cm	4.4	RC⁺	90–100	Wilson et al., 1981
Inc 5	*S. aureus*	pS177	Sm	4.4	RC⁺	80–100	Novick, 1976
	S. aureus	pS194	Sm				Iordanescu et al., 1978
	S. aureus	pS169	Sm				Novick, 1976
	S. aureus	pUB109	Sm				Novick, 1976
Inc 6	*S. aureus*	pK545	Km/Nm	22.7	RC⁻	2–3	Ruby and Novick, 1975
	S. aureus	pSH2	Km/Nm	13.6			Stiffler et al., 1974
Inc 7	*S. aureus*	pUB101	Pc, Cd, Fa	22.1	RC⁻	8–12	Chopra et al., 1973
Inc 8	*S. aureus*	pC194	Cm	2.910	RC⁻		Iordanescu, 1976
Inc 9	*S. aureus*	pUB112	Cm	4.5	RC⁻	23	Chopra et al., 1973
Inc 10	*S. aureus*	pC223	Cm	4.5	RC⁺		Novick, 1976
Inc 11	*S. aureus*	pE194	Em	3.728	RC⁻		Horinouchi and Weisblum, 1982a
Inc 12	*S. aureus*	pE1764	Em	2.8			Iordanescu and Surdeanu, 1980
	S. aureus	pE5	Em	2.473			Projan et al., 1987
	S. aureus	pE2222	Em	2.6		10	Iordanescu and Surdeanu, 1980
	S. epidermidis	pNE131	Em	2.355			Lampson and Parisi, 1986a, 1986b
Inc 13	*S. aureus*	pUB110	Km/Nm/Tm	4.5		9–16	Gryczan et al., 1978
Uncertain	*S. xylosus*	pSX267	Asa, Asi	29.5		Multicopy	Götz et al., 1983
	S. simulans	pMK148	Tc	4.5			Kreutz and Götz, 1984
	S. aureus	pSH8	Gm, Pm, EtBr, conjugative	44		Multicopy	Keller et al., 1983; McDonnell et al., 1983
Plasmids of other microorganisms that replicate in staphylococci	*S. haemolyticus*	pIP855	Lm	2.5			Leclercq et al., 1985
	Enterococcus faecalis	pAMβ1	Em, conjugative	25		2 (in *Enterococcus*)	Clewell et al., 1974; Vescovo et al., 1983
	E. faecalis	pR1405	Em, conjugative	25			Engel et al., 1980
	Streptococcus agalactiae	pIP501	Em, Cm, conjugative	30			Horodniceanu et al., 1979; Schaberg et al., 1982

that the genus *Bacillus* is clearly related to *Staphylococcus*. pC194 also replicates in *Streptococcus pneumoniae* (Barany et al., 1982) and in *Saccharomyces cerevisiae* (Goursot et al., 1982). The chloramphenicol resistance of pC194, and of another chloramphenicol resistance plasmid pC221, is inducible by chloramphenicol (Shaw, 1983). The regulation of the inducible chloramphenicol acetyltransferase gene of the *S. aureus* plasmid pUB112 was studied by Brückner and Matzura (1985).

ERYTHROMYCIN-RESISTANCE PLASMIDS pE194 AND pNE131. The plasmid pE194 is a small plasmid originally isolated from *S. aureus* by Iordanescu and Surdeanu (1980). The plasmid confers erythromycin-inducible resistance to macrolide and lincosamide antibiotics and streptogramin B, which are referred to as MLS antibiotics. The plasmid has a size of 3,728 bp (Horinouchi and Weisblum, 1982a). The erythromycin-resistance gene *(ermC)* encodes a 29-kDa 23S rRNA methyltransferase (adenine methylase). The erythromycin resistance is due to N^6, N^{6-} dimethylation of an adenine residue at position 2058 of the 23S rRNA (Weisblum, 1975). Although the host plasmid pE194 has been studied extensively, it is uncommon in natural populations of staphylococci.

pNE131 is a small plasmid isolated from *S. epidermidis* by Parisi et al. (1981). It is similar to the *S. aureus* plasmid pE2222 (Table 3) and is in the same incompatibility group. The plasmid confers erythromycin-constitutive resistance to macrolide and lincosamide antibiotics and streptogramin B. The plasmid has been sequenced and has a size of 2,355 bp (Lampson and Parisi, 1986b). The erythromycin-resistance-gene region carried by this plasmid is almost identical to the inducible *ermC* gene region of pE194, except that a 107-bp deletion occurs, which removes the mRNA leader sequence required for inducible expression (Lampson and Parisi, 1986a). pNE131 contains some other regions of homology with pE194 and also some homology with the cryptic plasmid pSN2 of *S. aureus*. pNE131 is also related to the *S. aureus* plasmids pE1764 and pE5 (Table 3), which contains an intact mRNA leader sequence for inducible *ermC* expression. These and other plasmids of the Inc12 incompatibility group are the most common host plasmids for *ermC* among the staphylococci (Thakker-Varia et al., 1987).

KANAMYCIN-RESISTANCE PLASMID pUB110. pUB110 is a 4.5-kb plasmid conferring resistance to kanamycin that was originally isolated from *S. aureus* (Chopra, et al., 1973). pUB110

is often used as a vector, especially in *B. subtilis* (Gryczan et al., 1980). Interestingly, pUB110 shares a high DNA homology with plasmid pBC16 of *B. subtilis* and plasmid pAMαl, a tetracycline-resistance plasmid from *Enterococcus faecalis* (Perkins and Youngman, 1983). This study suggests that within the Gram-positive organisms *Bacillus, Enterococcus,* and *Staphylococcus,* certain plasmids may be exchangeable. The kanamycin-resistance gene *(kan)* encodes a kanamycin nucleotidyl transferase. This enzyme was purified and characterized by Sadaie et al. (1980). The nucleotide sequence of the nucleotidyl transferase gene was determined by Matsumura et al. (1984).

Staphylococcal Transposons

TRANSPOSON TN*551*. Tn*551* is a 5.3-kb transposon which occurs naturally on the staphylococcal plasmid pI258. The transposon carries an erythromycin-resistance gene *(ermB),* which is constitutively expressed (Novick et al., 1979). Tn*551* possesses inverted repeats at the ends with a length of 14 nucleotides and during transposition creates 5-bp-long direct repeats at the ends (Kahn and Novick, 1980). Based on this organization and the extensive nucleotide homology of the inverted repeats, Tn*551* belongs to the Tn*3* transposon family. All members of the Tn*3* family transpose by a two-step mechanism (Heffron, 1983). Tn*551* can integrate at various positions on the *S. aureus* chromosome and some insertions lead to insertional inactivation of markers. Thus, Tn*551* is a valuable tool in transposon mutagenesis (Luchansky and Pattee, 1984). The frequency of Tn*551* translocation was estimated to be 10^{-4}. For transposon mutagenesis in staphylococci, a temperature-sensitive mutant of pI258 carrying Tn*551* is used. At elevated temperatures, the plasmid is lost, and clones into which the transposon has inserted into the chromosome can be selected with erythromycin (Pattee, 1981; Berger-Bächi, 1983).

TRANSPOSON TN*554*. Tn*554* is a 6,691-bp transposon which was originally found in *S. aureus* and confers resistance to MLS antibiotics and spectinomycin (Murphy et al., 1985; Phillips and Novick, 1979). It is integrated at a specific position on the chromosome and is rarely found in plasmids (Krolewski et al., 1981; Murphy et al., 1981). The transposon contains six open reading frames designated: *tnpA, tnpB,* and *tnpC,* which encode transposition functions; *ermA,* for MLS resistance; *spc,* for spectinomycin resistance; and ORF, for an unknown function (Murphy et al., 1985). It has several unique features distinguishing it from other transposable elements. For example, its ends are asymmetric and lack inverted or direct terminal repeats, it does not generate a duplication of a target sequence upon transposition, and it is extremely site-specific. Transposition to the chromosome occurs in only one orientation. The target site contains a 6-bp sequence that corresponds to the terminal 6 bp of the right end of Tn*554*.

TRANSPOSON TN*4001*. Tn*4001* is 4.5-kb transposon which confers resistance to gentamicin, tobramycin, and kanamycin (Lyon et al., 1984). Tn*4001* possesses approximately 1.3-kb-long inverted repeats at the ends which probably constitute IS elements. The frequency of transposition is similar to that of Tn*551*, namely 10^{-4}. Tn*4001* was found on the 27-kb aminoglycoside-resistance plasmid pSK1 in *S. aureus* (Lyon et al., 1983). The plasmid also determines resistance to ethidium bromide and quaternary ammonium substances.

Chromosomal Map of *S. aureus*

The chromosomal map of *S. aureus* NCTC 8325 was determined with the aid of transformation and transduction. The distances between the various genetic markers are relative to the cotransducibility of the markers. The map is presented in Fig. 7 (Pattee et al., 1984; Pattee, 1990). Based on the protoplast fusion technique of staphylococci, which allows the isolation of recombinants (Götz et al., 1981), the linkage map could be further extended. Stahl and Pattee (1983a, 1983b) were able to determine the orientation of the three known linkage groups. The markers used to construct the map are shown on the outside of the circle and are defined in the figure legend. There are only two places where the continuity of the circular map has not been established (Fig. 7, gaps in the circle). The length of the *S. aureus* chromosome was determined by pulsed-field agarose gel electrophoresis of *Sma*I-digested chromosomal DNA. The estimated size of the *S. aureus* chromosome is 2,780 kb, as shown in the center of Fig. 7.

Applied Genetics of *S. carnosus*

It has been known for a long time that in the ripening process of dry sausages, Gram-positive and catalase-positive cocci play an important role. The predominant microorganism in fermented meat is *Staphylococcus carnosus,* which in early literature was regarded as a *Micrococcus* (Lerche and Sinell, 1955; Niinivaara and Pohja, 1956). However, based on modern taxonomic criteria, it was determined that *S. carnosus* is a new staphylococcal species whose name is derived from its occurrence in meat (Schleifer and Fischer, 1982). For more than 40 years, *S. carnosus* has been used alone or in combination with other microorganisms, such as pediococci or lactobacilli, as a starter culture for the production of raw sausage.

S. carnosus has only a low DNA homology with *S. aureus* and other staphylococcal species and produces no toxins, hemolysins, protein A, coagulase, or clumping factor, markers which are typical for many *S. aureus* strains. Thus, there exists within the genus *Staphylococcus,* a similar situation as for *Streptococcus,* where the genus is comprised both of potentially pathogenic and of food-grade species. Since *S. carnosus* is considered to be food grade, it was particularly worthwhile to develop a gene cloning system for this microorganism.

Many staphylococcal plasmids can be transferred into *S. carnosus* by the protoplast transformation method (Götz et al., 1983). The efficiency of transformation is about 10^6 transformants per μg plasmid DNA. Storage of the protoplasts in deep-frozen (at $-70°C$) form for several months can be made without affecting their transforming capability (Götz and Schumacher, 1987). Recently, optimal conditions for electroporation were determined for *S. carnosus* and other staphylococcal species (Augustin and Götz, 1990). With this method, plasmids can be transformed with an efficiency up to 3×10^5 transformants per μg pC194 plasmid DNA.

In order to facilitate cloning of DNA in *S. carnosus,* several plasmid vectors have been constructed (Keller et al., 1983). The constructed chimeric plasmids pCT20 and pCE10 are both stable in *S. aureus* and *S. carnosus* (Keller et al., 1983). Another suitable plasmid vector, pCA43, is composed of the replicon pC194 and the arsenate/arsenite-resistance operon derived from pSX267 (Kreuz and Götz, 1984). The 29.5-kb plasmid pSX267 originated from *Staphylococcus xylosus* DSM 20267 and encodes arsenate, arsenite, and antimony III resistance. By heteroduplex analysis and Southern blotting, it was found that the region for arsenate, arsenite, and antimony III resistance shares a high degree of homology with the same resistance region of the multiple resistance plasmid pI258 of *S. aureus* (Götz et al., 1983).

Secretion Vectors for *S. carnosus, S. aureus,* and *Escherichia coli*

Two staphylococcal exoproteins, lipase and protein A, have been used in gene fusion experi-

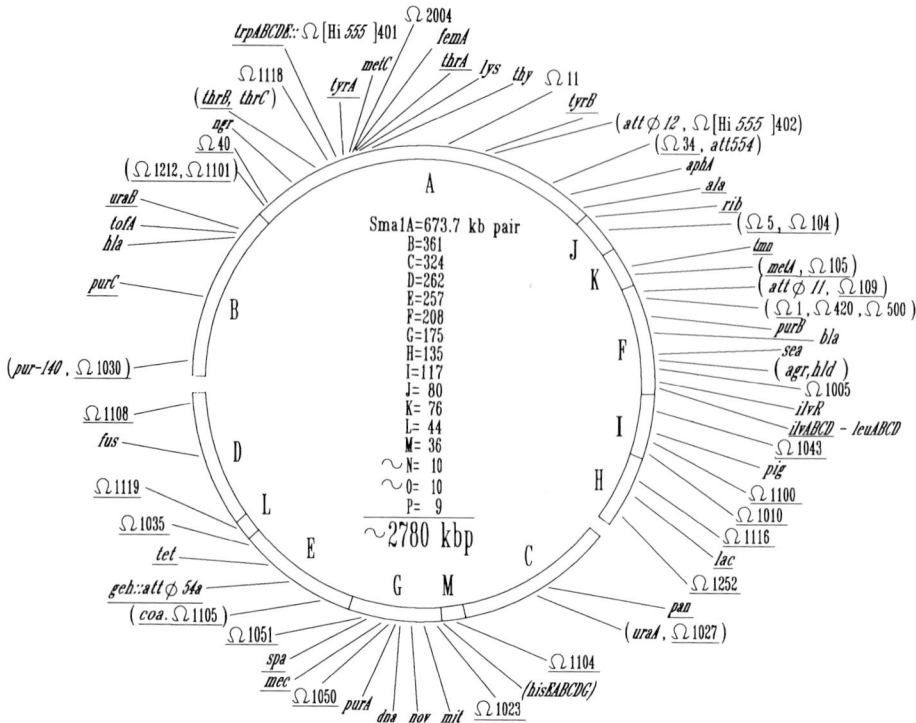

Fig. 7. Chromosome map of *Staphylococcus aureus* NCTC 8325. Markers used to construct the map are shown on the outside of the circle, and they are defined as follows: *agr (hla),* accessory gene regulator (of several extracellular toxins); *ala,* L-alanine requirement; *aphA,* aminoglycoside phosphotransferase (APH3′III); *attø11,* prophage ø11 integration site; *attø12,* prophage ø12 integration site; *attøL54a,* prophage øL54a integration site (within the *geh* gene); *att554,* primary insertion site for Tn*554; bla (pen),* β-lactamase production; *coa,* coagulase; *dna,* temperature-sensitive DNA synthesis; *fus,* fusidic acid resistance; *geh,* glycerol ester hydrolase (lipase); *his,* L-histidine requirement (*hisE, A, B, C, D, G*); *hla (hly),* α-toxin structural gene; *hlb,* β-hemolysin; *hld,* δ-hemolysin; *hlg,* γ-hemolysin; *ilv-leu,* requirement for L-isoleucine, L-valine, and L-leucine (*ilvA, B, C, D; leuA, B, C, D*); *ilvR,* resistance to D-leucine; *lac,* phospho-β-galactosidase; *lip,* lipoic acid requirement; *lys,* L-lysine requirement (probably *lysO, A, B, F, G*); *mec* (Tn*4291*), methicillin resistance; *metA,* Met⁻; *metC,* Met⁻, β-cystathionase; *mit,* mitomycin C, nitrosoguanidine, and UV sensitivity; *ngr,* deficient in apurinic endonuclease; *nov,* resistance to novobiocin; *pan,* panthothenate requirement; *pig,* absence of golden-yellow pigment; *purA,* adenine requirement; *purB,* adenine + guanine requirement; *purC,* purine requirement; *purD,* guanine requirement; *rib,* riboflavin requirement; *rif,* resistance to rifampicin; *sea (entA),* enterotoxin A production; *spa,* staphylococcal protein A; *tet,* resistance to tetracycline; *thrA,* requirement for L-lysine, L-methionine, and L-threonine; and failure to convert aspartate to aspartate-β-semialdehyde; *thrB,* Thr⁻, homoserine kinase; *thrC,* Thr⁻, threonine synthetase; *thy,* thymine requirement; *tmn,* resistance to tetracycline and minocycline; *tofA,* temperature-sensitive, osmotically remedial cell wall synthesis; *trp,* L-tryptophan requirement (*trpABFCDE* operon); *tyrA,* L-tyrosine requirement; *tyrB,* L-tyrosine requirement; *uraA,* uracil requirement; *uraB,* uracil requirement. The order of the markers shown in parentheses is unknown. Continuity of the circular map has not been established for the region between Ω1108 and Ω1030, or for the region between *pan* and Ω1252. The symbol Ω denotes insertion sites of transposons or IS-like elements. (From Pattee, 1990.)

ments to secrete heterologous proteins. The lipase gene of *Staphylococcus hyicus* was cloned in *S. carnosus* and sequenced (Götz et al., 1985). The signal sequence of the lipase comprises the first 38 N-terminal amino acids (van Oort et al., 1989). However, gene fusion studies between the lipase and the *E. coli* β-lactamase in *S. carnosus* revealed that the lipase possesses adjacent to the signal peptide, a peptide domain of approximately 200 amino acids that is essential for the secretion of hybrid proteins. For heterologous protein secretion, the secretion vector pALβ8 is routinely used (Liebl and Götz, 1986).

The protein A gene of *S. aureus* is especially well suited for use in gene fusions since the interaction between protein A and the immunoglobulin IgG permits the efficient and rapid purification of the fusion protein by IgG-Sepharose affinity chromatography. The protein A gene was sequenced by Uhlen et al. (1984a). Two secretion vectors using the protein A gene are available, pRIT2T and pRIT5 (Nilsson et al., 1985a). pRIT2T is designed for inducible expression of intracellular fusion proteins in *E. coli.* The protein A gene is under control of the phage lambda P_r promoter. pRIT5 contains the protein A signal sequence and the IgG binding domain. Heterologous proteins fused after the IgG binding domain are transported to the periplasmic space of *E. coli* or, in Gram-positive cells, the fusion proteins are translocated across the membrane into the growth medium. pRIT5 represents a shuttle vector between *E. coli* and *S. aureus*. The signals for protein A secretion have been analyzed by Abrahmsén et al. (1985). Fusion of proteins after the IgG binding part of protein A led to efficient secretion in both *E. coli* and *S. aureus*. With the aid of protein A fusion vectors, efficient secretion and purification was also achieved with the human insulin-like growth factor I (Nilsson et al., 1985b).

With the biochemical and genetic technology now available, it will be possible to determine the gene, protein, and regulatory structures and functions for various components of metabolic processes and virulence properties of the staphylococci. Many laboratories are currently engaged in such a pursuit. The future for staphylococcal research should be exciting and characterized by new views of ecology, pathogenesis, genetic and population interactions, and industrial applications at the molecular level.

Literature Cited

Abrahmsén, L., T. Moks, B. Nilsson, U. Hellman, and M. Uhlén. 1985. Analysis of signals for secretion in the staphylococcal protein A gene. EMBO J. 4:3901–3906.

Abramson, C. 1972. Staphylococcal enzymes, p. 187–248. In: J. O. Cohen (ed.), The staphylococci. Wiley Interscience, New York.

Akatov, A. K., V. Hájek, T. M. Samsonova, and J. Balusek. 1985. Classification and drug resistance of coagulase-negative staphylococci isolated from wild birds, p. 125–127. In: J. Jeljaszewicz (ed.), The staphylococci: proceedings of the Vth international symposium on staphylococci and staphylococcal infections. Gustav Fischer Verlag, Stuttgart.

Allgaier, H., G. Jung, R. G. Werner, U. Schneider, and H. Zähner. 1986. Epidermin: sequencing of a heterodet tetracyclic 21-peptide amide antibiotic. Eur. J. Biochem. 160:9–22.

Almeida, R. J., J. H. Jorgensen, and J. E. Johnson. 1983. Evaluation of the AutoMicrobic System Gram-Positive Identification Card for species identification of coagulase-negative staphylococci. J. Clin. Microbiol. 18:438–439.

Aly, R., H. R. Shinefield, and H. I. Maibach. 1981. *Staphylococcus aureus* adherence to nasal epithelial cells: studies of some parameters, p. 171–179. In: H. I. Maibach and R. Aly (ed.), Skin microbiology: relevance to clinical infection. Springer-Verlag, New York.

Ambler, R. P. 1975. Amino-acid sequence of *Staphylococcus aureus* penicillinase. Biochem. J. 151:197–218.

Amtsberg, G. 1978. Nachweis von Exfoliation auslösenden Substanzen in Kulturen von *Staphylococcus hyicus* des

Schweines und *Staphylococcus epidermidis* Biotyp 2 des Rindes. Zentralbl. Vet. Med. B 26:257–272.

Anderson, J. D., A. M. Clarke, M. E. Anderson, J. L. Isaac-Renton, and M. G. McLoughlin. 1981. Urinary tract infections due to *Staphylococcus saprophyticus* biotype 3. Can. Med. Assoc. J. 124:415–418.

Anfinsen, C. B., P. Cuatrecasas, and H. Taniuchi. 1971. Staphylococcal nuclease, chemical properties and catalysis, p. 177–204. In: P. Boyer (ed.), The enzymes, vol. 4. Academic Press, New York.

Arbuthnott, J. P. 1983. Epidermolytic toxins, p. 599–617. In: C. S. F. Easmon and C. Adlam (ed.), Staphylococci and staphylococcal infections, vol. 2. Academic Press, London.

Archer, G. L. 1984. *Staphylococcus epidermidis*: the organism, its diseases and treatment, p. 25–48. In: J. S. Remington and M. N. Swartz (ed.), Current clinical topics in infectious diseases, vol. 5. McGraw-Hill, New York.

Archer, G. L., and J. L. Johnston. 1983. Self-transmissible plasmids in staphylococci that encode resistance to aminoglycosides. Antimicrob. Agents Chemother. 24:70–77.

Archer, G. L., and M. J. Tenenbaum. 1980. Antibiotic-resistant *Staphylococcus epidermidis* in patients undergoing cardiac surgery. Antimicrob. Agents Chemother. 17:269–272.

Archer, G. L., G. J. Vazquez, and J. L. Johnson. 1980. Antibiotic prophylaxis of experimental endocarditis due to methicillin-resistant *Staphylococcus epidermidis*. J. Infect. Dis. 142:725–731.

Archibald, A. R. 1974. The structure, biosynthesis and function of teichoic acid. Adv. Microb. Physiol. 11:53–95.

Arvidson, S. O. 1973. Studies on extracellular proteolytic enzymes from *Staphylococcus aureus*. II. Isolation and characterization of an EDTA-sensitive protease. Biochim. Biophys. Acta 301:149–157.

Arvidson, S. O. 1983. Extracellular enzymes from *Staphylococcus aureus*, p. 745–808. In: C. S. F. Easmon and C. Adlam (ed.), Staphylococci and staphylococcal infections, vol. 2. Academic Press, London.

Asch, D. K., R. V. Goering, and E. A. Ruff. 1984. Isolation and preliminary characterization of a plasmid mutant derepressed for conjugal transfer in *Staphylococcus aureus*. Plasmid 12:197–202.

Association of Official Analytical Chemists. 1980. Official methods of analysis, 13th ed, p. 826–827. Association of Official Analytical Chemists, Washington, D.C.

Association of Official Analytical Chemists. 1984. Official methods of analysis, 14th ed, p. 941–976. Association of Official Analytical Chemists, Arlington, VA.

Association of Official Analytical Chemists. 1987. Changes in official methods of analysis, 14th ed., 3rd suppl. J. Assoc. Off. Anal. Chem. 70:393–394.

Association of Official Analytical Chemists. 1990. Official methods of analysis, 15th ed, p.425–497. Association of Official Analytical Chemists, Arlington, VA.

Augustin, J., and F. Götz. 1990. Transformation of *Staphylococcus epidermidis* and other staphylococcal species with plasmid DNA by electroporation. FEMS Microbiol. Lett. 66:203–208.

Baba, E., T. Fukata, and H. Matsumoto. 1980. Ecological studies on coagulase-negative staphylococci in and around udder. Bull. Univ. Osaka Pref., Ser. B 32:70–75.

Baddour, L. M., G. D. Christensen, M. G. Hester, and A. L. Bisno. 1984. Production of experimental endocarditis by coagulase-negative staphylococci: variability in species virulence. J. Infect. Dis. 150:721–727.

Baddour, L. M., T. N. Phillips, and A. L. Bisno. 1986. Coagulase-negative staphylococcal endocarditis: occurrence in patients with mitral valve prolapse. Arch. Intern. Med. 146:119–121.

Baird-Parker, A. C. 1962. The occurrence and enumeration, according to a new classification, of micrococci and staphylococci in bacon and on human and pig skin. J. Appl. Bacteriol. 25:352–361.

Baird-Parker, A. C. 1963. A classification of micrococci and staphylococci based on physiological and biochemical tests. J. Gen. Microbiol. 30:409–427.

Baird-Parker, A. C. 1965. The classification of staphylococci and micrococci from world-wide sources. J. Gen. Microbiol. 38:363–387.

Baker, J. S. 1984. Comparison of various methods for differentiation of staphylococci and micrococci. J. Clin. Microbiol. 19:875–879.

Balows, A., W. J. Hausler, Jr., K. L. Herrmann, H. D. Isenberg, and H. J. Shadomy (ed.) 1991. Manual of clinical microbiology, 5th ed. American Society for Microbiology, Washington, D.C. (in press).

Bannerman, T. L., and W. E. Kloos. 1991. *Staphylococcus capitis* subsp. *ureolyticus* subsp. Nov. from human stain. Int. J. Syt. Bacteriol. 41. (in press).

Barany, F., J. D. Boeke, and A. Tomasz. 1982. Staphylococcal plasmids that replicate and express erythromycin resistance in both *Streptococcus pneumoniae* and *Escherichia coli*. Proc. Natl. Acad. Sci. USA 79:2991–2995.

Baumgart, S., S. E. Hall, J. M. Campos, and R. A. Polin. 1983. Sepsis with coagulase-negative staphylococci in critically ill newborns. Am. J. Dis. Child. 137:461–463.

Behnke, D., and D. Gerlach. 1987. Cloning and expression in *Escherichia coli, Bacillus subtilis,* and *Streptococcus sanguis* of a gene for staphylokinase—a bacterial plasminogen activator. Mol. Gen. Genet. 210:528–534.

Bergdoll, M. S. 1972. The enterotoxins, p. 301–331. In: J. O. Cohen (ed.), The staphylococci. John Wiley and Sons, New York.

Bergdoll, M. S. 1979. Staphylococcal intoxications, p. 443–493. In: H. Riemann and F. L. Bryan (ed.), Foodborne infections and intoxications. Academic Press, New York.

Bergdoll, M. S. 1983. Enterotoxin, p. 559–598. In: C. S. F. Easmon and C. Adlam (ed.), Staphylococci and staphylococcal infections, vol. 2. Academic Press, London.

Bergdoll, M. S., and R. W. Bennett. 1976. Staphylococcal enterotoxins, p. 387–416. In: M. L. Speck (ed.), Compendium of methods for the microbiological examination of foods. American Public Health Association, Washington, D.C.

Bergdoll, M. S., and R. W. Bennett. 1984. Staphylococcal enterotoxins, p. 428–457. In: M. L. Speck (ed.), Compendium of methods for the microbiological examination of foods, 2nd ed. American Public Health Association, Washington, D.C.

Berger-Bächi, B. 1983. Insertional inactivation of staphylococcal methicillin resistance by Tn551. J. Bacteriol. 154:479–487.

Bernheimer, A. W., L. S. Avigad, and K. S. Kim. 1974. Staphylococcal sphingomyelinase (beta-hemolysin). Ann. N.Y. Acad. Sci. 236:292–306.

Beryhill, D. L., and P. A. Pattee. 1969. Bouyant density analysis of staphylococcal bacteriophage 80 transducing particles. J. Virol. 4:804–806.

Betley, M. J., and J. J. Mekalanos. 1988. Nucleotide sequence of the type A staphylococcal enterotoxin gene. J. Bacteriol. 170:34–41.

Beyreuther, K., H. Raufuss, O. Schrecker, and W. Hengstenberg. 1977. The phosphoenolpyruvate-dependent phosphotransferase system of *Staphylococcus aureus.* I. Amino acid sequence of the phosphocarrier protein HPr. Eur. J. Biochem. 75:275–286.

Bhakdi, S., M. Muhly, and R. Füssle. 1984. Correlation between toxin binding and hemolytic activity in membrane damage by staphylococcal alpha-toxin. Infect. Immun. 46:318–323.

Bint, A. J., R. H. George, D. E. Healing, R. Wise, and M. Davies. 1977. An outbreak of infection caused by a gentamicin resistant *Staphylococcus aureus.* J. Clin. Pathol. 30:165–167.

Birch-Hirschfeldt, F. V. 1872. Die neuern pathologish-anatomischen Untersuchungen über Krankmachende Schmarotzerpilze. Schmidt's Jahrbb. 115:97–109.

Birmingham, V. A., and P. A. Pattee. 1981. Genetic transformation in *Staphylococcus aureus:* isolation and characterization of a competence-conferring factor from bacteriophage 80 alpha lysates. J. Bacteriol. 148:301–307.

Bisaillon, J. G., R. Beaudet, S. Lafond, S. A. Saheb, and M. Sylvester. 1981. Antigonococcal and antibacterial spectra of some bacterial isolates of the urogenital flora. Rev. Can. Biol. 40:215–227.

Bissett, D. L., and R. L. Anderson. 1973. Lactose and D-galactose metabolism in *Staphylococcus aureus:* pathway of D-galactose 6-phosphate degradation. Biochem. Biophys. Res. Commun. 52:651–647.

Bissett, D. L., and R. L. Anderson. 1974. Lactose and D-galactose metabolism in group N streptococci: presence of enzymes for both the D-galactose 1-phosphate and D-tagatose 6-phosphate pathways. J. Bacteriol. 117:318–320.

Blomster-Hautamaa, D., B. N. Kreiswirth, J. S. Kornblum, R. P. Novick, and P. M. Schlievert. 1986. The nucleotide and partial amino acid sequence of toxic shock syndrome toxin-1. J. Biol. Chem. 261:15783–15786.

Blumenthal, H. J. 1972. Glucose catabolism in staphylococci, p. 111–135. In: J. O. Cohen (ed.), The staphylococci. Wiley Interscience, New York.

Bohach, G. A., and P. M. Schlievert, 1987. Nucleotide sequence of staphylococcal enterotoxin C-1 gene and relatedness to other pyrogenic toxins. Mol. Gen. Genet. 209:15–20.

Boothby, J., C. Genigeorgis, and M. H. Fanelli. 1979. Tandem coagulase/thermonuclease agar method for the detection of *Staphylococcus aureus.* Appl. Environ. Microbiol. 37:298–302.

Bor, D. H., R. M. Rose, J. F. Modlin, R. Weintraub, and G. H. Friedland. 1983. Mediastinitis after cardiovascular surgery. Rev. Infect. Dis. 5:885–897.

Bowman, R. A., and M. Buck. 1984. *Staphylococcus hominis* septicaemia in patients with cancer. Med. J. Aust. 140:26–27.

Brause, B. D. 1986. Infections associated with prosthetic joints. Clin. Rheum. Dis. 12:523–535.

Breit, F. Jr. and G. C. Stewart. 1987. Nucleotide and deduced amino acid sequences of the *Staphylococcus au-*

reus phospho-β-galactose gene. Appl. Environm. Microbiol. 53:969–973.

Browder, H. P., W. A. Zygmunt, J. R. Young, and P. A. Tavormina. 1965. Lysostaphin: enzymatic mode of action. Biochem. Biophys. Res. Commun. 19:383–389.

Brückner, R., and H. Matzura. 1985. Regulation of the inducible chloramphenicol acetyltransferase gene of the *Staphylococcus aureus* plasmid pUB112. EMBO J. 4:2295–2300.

Bucher, E., G. Beck, and K. H. Schleifer. 1980. Vorkommen und Verteilung von Staphylokokken und Mikrokokken in Sojaextraktionsschroten. Zentralbl. Bakteriol. Hyg. Abt. 1 Orig. C 1:320–329.

Bulloch, W. 1938. The history of bacteriology. Oxford University Press, London.

Button, D. K., J. B. Egan, W. Hengstenberg, and M. L. Morse. 1973. Carbohydrate transport in *Staphylococcus aureus*. IV. Maltose accumulation and metabolism. Biochem. Biophys. Res. Commun. 52:850–855.

Caputo, G. M., G. L. Archer, S. B. Calderwood, and A. W. Karchmer. 1985. Native valve endocarditis due to coagulase-negative staphylococci. Abstr. 25th Inter. Sci. Conf. Antimicrob. Agents Chemother. 1985:898.

Carleton, S., S. J. Projan, S. K. Highlander, S. M. Moghazeh, and R. P. Novick. 1984. Control of pT181 replication II. Mutational analysis. EMBO J. 3:2407–2414.

Carnaghan, R. B. A. 1966. Spinal cord compression in fowls due to spondylitis caused by *Staphylococcus pyogenes*. J. Comp. Pathol. 76:9–14.

Chatterjee, A. W. 1969. Use of bacteriophage-resistant mutants to study the nature of the bacteriophage receptor site of *Staphylococcus aureus*. J. Bacteriol. 98:519–527.

Cheung, A. L., and V. A. Fischetti. 1988. Variation on the expression of cell wall proteins of *Staphylococcus aureus* grown on solid and liquid media. Infect. Immun. 56:1061–1065.

Choo, M. H., D. R. Holmes, B. J. Gersch, J. D. Maloney, J. Meredith, J. R. Pluth, and J. Trusty. 1981. Permanent pacemaker infections: characterization and management. Am. J. Cardiol. 48:559–564.

Chopra, I., P. M. Bennett, and R. W. Lacey. 1973. A variety of staphylococcal plasmids present as multiple copies. J. Gen. Microbiol. 79:343–345.

Christensen, G. D., L. M. Baddour, and W. A. Simpson. 1987. Phenotypic variation of *Staphylococcus epidermidis* slime production in vitro and in vivo. Infect. Immun. 55:2870–2877.

Christensen, G. D., W. A. Simpson, A. L. Bisno, and E. H. Beachey. 1982. Adherence of slime-producing strains of *Staphylococcus epidermidis* to smooth surfaces. Infect. Immun. 37:318–326.

Christiansen, G., and P.-A. Mårdh. 1986. Electron microscopy of negative-stained *Staphylococcus saprophyticus* reveals filamentous surface protrusions. In: P. -A. Mårdh and K. H. Schleifer (ed.), Coagulase-negative staphylococci. Almqvist and Wiksell International, Stockholm.

Christianson, K. K., R. K. Tweten, and J. J. Iandolo. 1985. Transport and processing of staphylococcal enterotoxin A. Appl. Environm. Microbiol. 50:696–697.

Clarke, A. M. 1979. Prophylactic antibiotics for total hip arthroplasty—the significance of *Staphylococcus epidermidis*. J. Antimicrob. Chemother. 5:493–502.

Clewell, D. B., Y. Yagi, G. Dunny, and S. K. Schultz. 1974. Characterization of three plasmid deoxyribonucleic acid molecules in a strain of *Streptococcus faecalis*: identification of a plasmid determining erythromycin resistance. J. Bacteriol. 117:283–289.

Cohen, J. O. (ed.) 1972. The staphylococci. John Wiley and Sons, New York.

Cohn, F. 1872. Untersuchungen über Bacterien. Beitr. Biol. Pflanz. Bd. 1, Heft 2:127–224.

Colleen, S., B. Hovelius, Å. Wieslander, and P.-A. Mårdh. 1979. Surface properties of *Staphylococcus saprophyticus* and *Staphylococcus epidermidis* as studied by adherence tests and two-polymer, aqueous phase systems. Acta Pathol. Microbiol. Immunol. Scand. Sect. B 87:321–328.

Collins, F. M., and J. Lascelles. 1962. The effect of growth conditions on oxidation and dehydrogenase activity in *Staphylococcus aureus*. J. Gen. Microbiol. 29:531–535.

Collins, M. D., and D. Jones. 1981. Distribution of isoprenoid quinone structural types and their taxonomic implications. Microbiol. Rev. 45:316–354.

Cooney, J., M. Mulvey, J. P. Arbuthnott, and T. J. Foster. 1988. Molecular cloning and genetic analysis of the determinant for gamma-lysin, a two-component toxin of *Staphylococcus aureus*. J. Gen. Microbiol. 134:2179–2188.

Couch, J. L., M. T. Soltis, and M. J. Betley. 1988. Cloning and nucleotide sequence of the type E staphylococcal enterotoxin gene. J. Bacteriol. 170:2954–2960.

Cove, J. H., J. N. Keareney, K. T. Holland, and W. J. Cundliffe. 1983. The vitamin requirements of *Staphylococcus cohnii*. J. Appl. Bacteriol. 54:203–208.

Crisley, F. D., R. Angelotti, and M. J. Foter. 1964. Multiplication of *Staphylococcus aureus* in synthetic cream fillings and pies. Pub. Health Reports 79:369–376.

Crouch, S. F., T. A. Pearson, and D. M. Parham. 1987. Comparison of modified Minitek System with Staph-Ident System for species identification of coagulase-negative staphylococci. J. Clin. Microbiol. 25:1626–1628.

Curry, J. C., and G. E. Borovian. 1976. Selective medium for distinguishing micrococci from staphylococci in the clinical laboratory. J. Clin. Microbiol. 4:455–457.

Dagert, M., I. Jones, A. Goze, S. Romac, B. Niaudet, and S. D. Ehrlich. 1984. Replication functions of pC194 are necessary for efficient transduction by M13 phage. EMBO J. 3:81–86.

Dan, M., G. J. R. Marien, and G. Goldsand. 1984. Endocarditis caused by *Staphylococcus warneri* on a normal aortic valve following vasectomy. Can. Med. Assoc. J. 131:211–213.

Davis, A., I. B. Moore, D. S. Parker, and H. Taniuchi. 1977. Nuclease B, a possible precursor of nuclease A, an extracellular nuclease of *Staphylococcus aureus*. J. Biol. Chem. 252:6544–6553.

Davis, J. P., P. J. Chesney, P. J. Wand, M. Laventure, and the investigation and laboratory team. 1980. Toxic-shock syndrome. Epidemiologic features, recurrence, risk factors, and prevention. N. Engl. J. Med. 303:1429–1435.

De Kesel, A., and L. Devriese. 1982. Successful treatment of a case of osteomyelitis caused by a *Staphylococcus hyicus* in a heifer. Vlaams Diergeneesk. Tijdschr. 51:222–225.

De la Fuente, R., K. H. Schleifer, F. Götz, and H.-P. Köst. 1986. Accumulation of porphyrins and pyrrole pigments by *Staphylococcus aureus* subsp. *anaerobius* and its aerobic mutant. FEMS Microbiol. Lett. 35:183–188.

De la Fuente, R., G. Suarez, and K. H. Schleifer. 1985. *Staphylococcus* subsp. *anaerobius* subsp. nov., the causal agent of abscess disease in sheep. Int. J. Syst. Bacteriol. 35:99–102.

Devriese, L. A. 1977. Isolation and identification of *Staphylococcus hyicus*. Am. J. Vet. Res. 38:787–792.

Devriese, L. A. 1979. Identification of clumping-factor-negative staphylococci isolated from cows' udders. Res. Vet. Sci. 26:313–320.

Devriese, L. A. 1986. Coagulase-negative staphylococci in animals, p. 51–57. In: P.-A. Mårdh and K. H. Schleifer (ed.), Coagulase-negative staphylococci. Almqvist and Wiksell International, Stockholm.

Devriese, L. A., and G. O. Adegoke. 1985. Identification of coagulase-negative staphylococci from farm animals. J. Appl. Bacteriol. 55:45–55.

Devriese, L. A., and H. De Keyser. 1980. Prevalence of different species of coagulase-negative staphylococci on teats and in milk samples from dairy cows. J. Dairy Res. 47:155–158.

Devriese, L. A., and J. Derycke. 1979. *Staphylococcus hyicus* in cattle. Res. Vet. Sci. 26:356–358.

Devriese, L. A., and V. Hájek. 1980. Identification of pathogenic staphylococci isolated from animals and foods derived from animals. J. Appl. Bacteriol. 49:1–11.

Devriese, L. A., V. Hájek, P. Oeding, S. A. Meyer, and K. H. Schleifer. 1978. *Staphylococcus hyicus* (Sompolinsky 1953) comb. nov. and *Staphylococcus hyicus* subsp. *chromogenes* subsp. nov. Int. J. Syst. Bacteriol. 28:482–490.

Devriese, L. A., D. Nzuambe, and C. Godard. 1984. Identification and characterization of staphylococci isolated from cats. Vet. Microbiol. 9:279–285.

Devriese, L. A., and P. Oeding. 1975. Coagulase and heat-resistant nuclease producing *Staphylococcus epidermidis* strains from animals. J. Appl. Bacteriol. 39:197–207.

Devriese, L. A., and P. Oeding. 1976. Characteristics of *Staphylococcus aureus* strains isolated from different animal species. Res. Vet. Sci. 21:284–291.

Devriese, L. A., B. Poutrel, R. Kilpper-Bälz, and K. H. Schleifer. 1983. *Staphylococcus gallinarum* and *Staphylococcus caprae*, two new species from animals. Int. J. Syst. Bacteriol. 33:480–486.

Devriese, L. A., K. Vlaminck, J. Nuytten, and Ph. De Keersmaecker. 1983. *Staphylococcus hyicus* in skin lesions of horses. Equine Vet. J. 15:263–265.

Domingue, P. A. G., P. A. Lambert, and M. R. W. Brown. 1989. Iron depletion alters surface-associated properties of *Staphylococcus aureus* and its association to human neutrophils in chemiluminescence. FEMS Microbiol. Lett. 59:265–268.

Drapeau, G. R., Y. Baily, and J. Houmard. 1972. Purification and properties of an extracellular protease of *Staphylococcus aureus*. J. Biol. Chem. 247:6720–6726.

Durham, D. R., and W. E. Kloos. 1978. A comparative study of the total cellular fatty acids of *Staphylococcus* species of human origin. Int. J. Syst. Bacteriol. 28:223–228.

Easmon, C. S. F., and C. Adlam (ed.) 1983. Staphylococci and staphylococcal infections, vol. 1 and 2. Academic Press, London.

East, A. K., and K. G. H. Dyke. 1989. Cloning and sequence determination of six *Staphylococcus aureus* β-lactamases and their expression in *Escherichia coli* and *Staphylococcus aureus*. J. Gen. Microbiol. 135:1001–1015.

Egan, J. B., and M. L. Morse. 1965. Carbohydrate transport in *Staphylococcus aureus:* characterization of the defect of a pleiotropic transport mutant. Biochem. Biophys. Acta 109:172–183.

Egan, J. B., and M. L. Morse. 1966. Carbohydrate transport in *Staphylococcus aureus.* III. Studies of the transport process. Biochem. Biophys. Acta 112:63–73.

Ehrenberg, C. G. 1838. Die Infusionsthierchen als vollkommene Organismen. Leipzig.

Ehrlich, S. D. 1977. Replication and expression of plasmids from *Staphylococcus aureus* in *Bacillus subtilis*. Proc. Natl. Acad. Sci. USA 74:1680–1682.

Elek, S. D. 1959. *Staphylococcus pyogenes* and its relation to disease. E. and S. Livingstone Ltd., Edinburgh.

Emmett, M., and W. E. Kloos. 1975. Amino acid requirements of staphylococci isolated from human skin. Can. J. Microbiol. 21:729–733.

Emmett, M., and W. E. Kloos. 1979. The nature of arginine auxotrophy in cutaneous populations of staphylococci. J. Gen. Microbiol. 110:305–314.

Endl, J., P. H. Seidl, F. Fiedler, and K. H. Schleifer. 1983. Chemical composition and structure of cell wall teichoic acids of staphylococci. Arch. Microbiol. 135:215–223.

Endl, J., P. H. Seidl, F. Fiedler, and K. H. Schleifer. 1984. Determination of cell wall teichoic acid structures of staphylococci by rapid chemical and serological screening methods. Arch. Microbiol. 137:272–280.

Engel, H. W. B., N. Soedirman, J. A. Rost, W. J. Van-Leeuwen, and J. D. A. VanEmbden. 1980. Transferability of macrolide, lincomycin, and streptogramin resistances between group A, B, and D streptococci, *Streptococcus pneumoniae,* and *Staphylococcus aureus.* J. Bacteriol. 142:407–413.

Etienne, J., B. Pangon, C. Leport, M. Wolff, B. Clair, C. Perrone, Y. Brun, and A. Bure. 1989. *Staphylococcus lugdunensis* endocarditis. Lancet i:390.

Evans, J. B. 1977. Coagulase-positive staphylococci as indicators of potential health hazards from water, p. 126–130. In: A. W. Hoadley and B. J. Dutka (ed.), Bacterial indicators and health hazards associated with water. American Society for Testing and Materials, Philadelphia.

Evans, J. B., G. A. Ananaba, C. A. Pate, and M. S. Bergdoll. 1983. Enterotoxin production by atypical *Staphylococcus aureus* from poultry. J. Appl. Bacteriol. 54:257–261.

Evans, J. B., W. L. Bradford, Jr., and C. F. Niven. 1955. Comments concerning the taxonomy of the genera *Micrococcus* and *Staphylococcus*. Int. Bull. Bacteriol. Nomencl. Taxon. 5:61–66.

Fackrell, H. B., and G. M. Wiseman. 1976. Properties of the gamma haemolysin of *Staphylococcus aureus* "Smith 5R". J. Gen. Microbiol. 92:11–24.

Falk, D., and S. J. Guering. 1983. Differentiation of *Staphylococcus* and *Micrococcus* spp. with the taxo A Bacitracin disk. J. Clin. Microbiol. 18:719–721.

Faller, A. H., F. Götz, and K. H. Schleifer. 1980. Cytochrome patterns of staphylococci and micrococci and their taxonomic implications. Zentralbl. Bakteriol. Hyg. Abt. 1 Orig C 1:26–39.

Faller, A. H., and K. H. Schleifer. 1981. Modified oxidase and benzidine tests for separation of staphylococci from micrococci. J. Clin. Microbiol. 13:1021–1035.

Fiedler, H. P., T. Hörner, and A. Wörn. 1987. Separation of polypeptide antibiotics by reversed-phase high-performance liquid chromatography. Chromatographia 24:433–438.

Fischer, S., H. Luczak, and K. H. Schleifer. 1982. Improved methods for the detection of class I and class II fructose-1, 6-bis-phosphate aldolases in bacteria. FEMS Microbiol. Lett. 15:103–108.

Fitton, J. E., A. Dell, and W. V. Shaw. 1980. The amino acid sequence of the delta haemolysin of *Staphylococcus aureus*. FEBS Lett. 115:209–212.

Fleurette, J., M. Bes, Y. Brun, J. Freney, F. Forey, M. Coulet, M. E. Reverdy, and J. Etienne. 1989. Clinical isolates of *Staphylococcus lugdunensis* and *S. schleiferi*: bacteriological characteristics and susceptibility to antimicrobial agents. Res. Microbiol. 140:107–118.

Fleurette, J., Y. Brun, M. Bes, M. Coulet, and F. Forey. 1987. Infections caused by coagulase-negative staphylococci other than *S. epidermidis* and *S. saprophyticus*, p. 195–208. In: G. Pulverer, P. G. Qui, and G. Peters (ed.), Pathogenicity and clinical significance of coagulase-negative staphylococci. Gustav Fischer Verlag, Stuttgart.

Flügge, C. 1886. Die Mikroorganismen. F.C.W. Vogel, Leipzig.

Forbes, B. A., and D. R. Schaberg. 1983. Transfer of resistance plasmids from *Staphylococcus epidermidis* to *Staphylococcus aureus*: evidence for conjugative exchange of resistance. J. Bacteriol. 153:627–634.

Forse, R. A., C. Dixon, K. Bernard, L. Martinez, P. H. McLean, and J. L. Meakins. 1979. *Staphylococcus epidermidis*: an important pathogen. Surgery 86:507–514.

Forsgren, A., V. Ghetie, R. Lindmark, and J. Sjöquist. 1983. Protein A and its exploitation, p. 429–480. In: C. S. F. Easmon and C. Adlam (ed.), Staphylococci and staphylococcal infections, vol. 2. Academic Press, London.

Fracastorius, H. 1546. *De Contagione, Contagiosis Morbis et eorum Curatione*, Lib I-III. Venice.

Freed, R. C., M. L. Evenson, R. F. Reiser, and M. S. Bergdoll. 1982. Enzyme-linked immunosorbent assay for detection of staphylococcal enterotoxins in foods. Appl. Environ. Microbiol. 44:1349–1355.

Frenette, M., R. Beaudet, J.-G. Bisaillon, M. Sylvestre, and V. Portelance. 1984. Chemical and biological characterization of a gonococcal growth inhibitor produced by *Staphylococcus haemolyticus* isolated from urogenital flora. Infect. Immun. 46:340–345.

Freney, J., Y. Brun, M. Bes, H. Meugneir, F. Grimont, P. A. D. Grimont, C. Nervi, and J. Fleurette. 1988. *Staphylococcus lugdunensis* sp. nov. and *Staphylococcus schleiferi* sp. nov., two species from human clinical specimens. Int. J. Syst. Bacteriol. 38:168–172.

Garrard, W., and J. Lascelles. 1968. Regulation of *Staphylococcus aureus* lactate dehydrogenase. J. Bacteriol. 95:152–156.

Garrison, F. H. 1910. Fracastorius, Athanasius Kircher and the germ theory of disease. Science 31:500–502.

Gennaro, M. L., J. Kornblum, and R. P. Novick. 1987. A site specific recombination function in *Staphylococcus aureus* plasmids. J. Bacteriol. 169:2601–2610.

George, R., L. Leibrock, and M. Epstein. 1979. Long term analysis of cerebrospinal fluid shunt infections: a 25-year experience. J. Neurosurg. 51:804–811.

Giger, O., C. C. Charilaou, and K. R. Cundy. 1984. Comparison of API Staph-Ident and DMS Staph-Trac systems with conventional methods used for the identification of coagulase-negative staphylococci. J. Clin. Microbiol. 19:68–72.

Gill, V. J., S. T. Selepak, and E. C. Williams. 1983. Species identification and antibiotic susceptibilities of coagulase-negative staphylococci isolated from clinical specimens. J. Clin. Microbiol. 18:1314–1319.

Gillespie, W. A., M. A. Sellin, P. Gill, M. Stephens, L. A. Tuckwell, and A. L. Hilton. 1978. Urinary tract infection in young women with special reference to *Staphylococcus saprophyticus*. J. Clin. Pathol. 31:348–350.

Götz, F., S. Ahrne, and M. Lindberg. 1981. Plasmid transfer and genetic recombination by protoplast fusion in staphylococci. J. Bacteriol. 145:74–81.

Götz, F., S. Fischer, and K. H. Schleifer. 1980. Purification and characterization of an unusually heat-stable and acid/base stable class I fructose-1, 6-bisphosphate aldolase from *Staphylococcus aureus*. Europ. J. Biochem. 108:295–301.

Götz, F., B. Kreuz, and K. H. Schleifer. 1983. Protoplast transformation of *Staphylococcus carnosus* by plasmid DNA. Mol. Gen. Genet. 189:340–342.

Götz, F., E. Nürnberger, and K. H. Schleifer. 1979. Distribution of class I and class II D-fructose-one, 6-bisphosphate aldolases in various staphylococci, peptococci and micrococci. FEMS Microbiol. Lett. 5:253–257.

Götz, F., F. Popp, and K. H. Schleifer. 1985. Complete nucleotide sequence of the lipase gene from *Staphylococcus hyicus* cloned in *Staphylococcus carnosus*. Nucl. Acids Res. 13:5895–5906.

Götz, F., and K. H. Schleifer. 1975. Purification and properties of fructose-1, 6-diphosphate activated L-lactate dehydrogenase from *Staphylococcus epidermidis*. Arch. Microbiol. 105:303–312.

Götz, F., and K. H. Schleifer. 1976. Comparative biochemistry of lactate dehydrogenases from staphylococci, p. 245–252. In: J. Jeljaszewicz (ed.), Staphylococci and staphylococcal diseases. Gustav Fischer Verlag, Stuttgart.

Götz, F., and K. H. Schleifer. 1978. Biochemical properties and the physiological role of the fructose-1, 6-bisphosphate activated L-lactate dehydrogenase from *Staphylococcus epidermis*. Europ. J. Biochem. 90:555–561.

Götz, F., and B. Schumacher. 1987. Improvements of protoplast transformation in *Staphylococcus carnosus*. FEMS Microbiol. Lett. 40:285–288.

Goursot, R., A. Goze, B. Niaudet, and S. D. Ehrlich. 1982. Plasmids from *Staphylococcus aureus* replicate in yeast *Saccharomyces cerevisiae*. Nature 298:488–490.

Gray, E. D., W. E. Regelmann, and G. Peters. 1987. Staphylococcal slime and host defenses: effects on lymphocytes and immune function, p. 45–54. In: G. Pulverer, P. G. Quie, and G. Peters (ed.), Pathogenicity and clinical significance of coagulase-negative staphylococci. Gustav Fischer Verlag, Stuttgart.

Gray, G. S., and M. Kehoe. 1984. Primary sequence of the α-toxin gene from *Staphylococcus aureus* Wood 46. Infect. Immun. 46:615–618.

Greenberg, A. E., R. R. Trussell, and L. S. Clesceri (ed.). 1985. Standard methods for the examination of water and wastewater, 16th ed. American Public Health Association, Washington, D.C.

Gretler, A. C., P. Muccolo, J. B. Evans, and C. F. Nivan, Jr. 1955. Vitamin nutrition of the staphylococci with special reference to their biotin requirements. J. Bacteriol. 70:44–49.

Gryczan, T. J., S. Contente, and D. Dubnau. 1978. Characterization of *Staphylococcus aureus* plasmids introduced by transformation into *Bacillus subtilis*. J. Bacteriol. 134:318–329.

Gryczan, T. J., A. G., Shivakumar, and D. Dubnau. 1980. Characterization of chimeric plasmid cloning vehicles in *Bacillus subtilis*. J. Bacteriol. 141:246–253.

Guillebeau, A. 1890. Studien über Milchfehler und Euterentzundungen bei Rindern und Ziegen. I. Über Ursachen der Euterentzundung. Landwirt Jahrb. Schweiz 4:27–44.

Hájek, V. 1976. *Staphylococcus intermedius,* a new species isolated from animals. Int. J. Syst. Bacteriol. 26:401–408.

Hájek, V., and J. Balusek. 1985. Staphylococci from flies of different environments, p. 129–133. In: J. Jeljaszewicz (ed.), The staphylococci: Proceedings of the Vth international symposium on staphylococci and staphylococcal infections. Gustav Fischer Verlag, Stuttgart.

Hájek, V., L. A. Devriese, M. Mordarski, M. Goodfellow, G. Pulverer, and P. E. Veraldo. 1986. Elevation of *Staphylococcus hyicus* subsp. chromogenes (Devriese et al., 1978) to species status: *Staphylococcus chromogenes* (Devriese et al., 1978) comb. nov. Syst. Appl. Microbiol. 8:169–173.

Hájek, V., and E. Marsalek. 1969. A study of staphylococci isolated from the upper respiratory tract of different animal species. II. Biochemical properties of *Staphylococcus aureus* strains of pigeon origin. Zentralbl. Bakteriol. Parasitenkd. Infektionskr. Hyg. Abt. 1 Orig. Reihe A 212:67–73.

Hájek, V., and E. Marsalek. 1971. A study of staphylococci isolated from the upper respiratory tract of different animal species. IV. Physiological properties of *Staphylococcus aureus* strains of hare origin. Zentralbl. Bakteriol. Parasitenkd. Infektionskr. Hyg. Abt. 1 Orig. Reihe A 216:168–174.

Hájek, V., and E. Marsalek. 1976. Evaluation of classificatory criteria for staphylococci, p. 11–21. In: J. Jeljaszewicz (ed.), Staphylococci and staphylococcal diseases. Gustav Fischer Verlag, Stuttgart.

Hájek, V., E. Marsalek, and J. Hubacek. 1972. A study of staphylococci isolated from the upper respiratory tract of different animal species. V. Physiological properties of *Staphylococcus aureus* strains isolated from mink. Zentralbl. Bakteriol. Parasitenkd. Infektionskr. Hyg. Abt. 1 Orig. Reihe A 222:194–199.

Hallier, E. 1867. Gärungserscheinungen, Untersuchungen über Gärung, Fäulnis und Verwesung mit Berüchsichtigung der Miasmen und Kontagien, Sowie der Desinfektion. Leipzig.

Hamory, B. H., and J. T. Parisi. 1987. *Staphylococcus epidermidis:* a significant nosocomial pathogen. Am. J. Infect. Control 15:59–74.

Hanessian, S., and T. H. Haskell. 1964. Structural studies on staphylococcal polysaccharide antigen. J. Biol. Chem. 239:2758–2764.

Hardy, K. 1981. Bacterial plasmids. Nelson and Sons Ltd., Nairobi.

Hare, R. 1954. Pomp and pestilence. Gollancz, London.

Harvey, J., and A. Gilmour. 1988. Isolation and characterization of staphylococci from goats milk produced in Northern Ireland. Lett. Appl. Microbiol. 7:79–82.

Hauschild, A. H. W., C. E. Park, and R. Hilheimer. 1979. A modified pork plasma agar for the enumeration of *Staphylococcus aureus* in foods. Can. J. Microbiol. 25:1052–1057.

Heady, R. E., N. J. Jacobs, and R. H. Deibel. 1964. Effect of hemin supplementation on porphyrin accumulation and catalase synthesis during anaerobic growth of *Staphylococcus*. Nature 203:1285–1286.

Heffron, F. 1983. Tn3 and its relatives, p. 223–260. In: J. A. Shapiro (ed.), Mobile genetic elements. Academic Press, New York.

Heidrich, H. J., and W. Renk. 1967. Diseases of the mammary glands of domestic animals (translated by L. W. Van Den Heever). W. B. Saunders, Philadelphia.

Heinrich, P., R. Rosenstein, M. Böhmer, P. Sonner, and F. Götz. 1987. The molecular organization of the lysostaphin gene and its sequences repeated in tandem. Mol. Gen. Genet. 209:563–569.

Hemker, H. C., B. M. Bas, and A. D. Muller. 1975. Activation of a pro-enzyme by a stoichiometric reaction with another protein. The reaction between prothrombin and staphylocoagulase. Biochim. Biophys. Acta 379:180–188.

Hengstenberg, W., J. B. Egan, and M. L. Morse. 1967. Carbohydrate transport in *Staphylococcus aureus*. V. The accumulation of phosphorylated carbohydrate derivatives, and evidence for a new enzyme splitting lactose phosphate. Proc. Nat. Acad. Sci. USA. 58:274–279.

Herchline, T. E., and L. Ayers. 1989. Clinical characteristics of *Staphylococcus lugdunensis* infection. Abstr. Annu. Meet. Am. Soc. Microbiol. 1989:419.

Holmberg, O. 1986. Coagulase-negative staphylococci in bovine mastitis, p. 203–211. In: P.-A. Mårdh and K. H. Schleifer (ed.), Coagulase-negative staphylococci. Almqvist and Wiksell International, Stockholm.

Hoo, R. L., M. Wadke, and H. J. Blumenthal. 1971. Control of the hexosemonophosphate (HMP) pathway by NAD in *Staphylococcus aureus*. Bacteriological Proc. 133.

Hooten, E. A. 1930. The indians of Pecos Pueblo. Yale University Press, New Haven.

Horinouchi, S., and B. Weisblum. 1982a. Nucleotide sequence and functional map of pE194, a plasmid that specifies inducible resistance to macrolide, lincosamide, and streptogramin type B antibiotics. J. Bacteriol. 150:804–814.

Horinouchi, S., and B. Weisblum. 1982b. Nucleotide sequence and functional map of pC194, a plasmid that specifies inducible chloramphenicol resistance. J. Bacteriol. 150:815–825.

Hörner, T., H. Zähner, R. Kellner, and G. Jung. 1989. Fermentation and isolation of epidermin, a lanthionine containing polypeptide antibiotic from *Staphylococcus epidermidis*. Appl. Microbiol. Biotechnol. 30:219–225.

Horodniceanu, T., L. Bougueleret, N. El-Sohl, D. H. Bouanchaud, and Y. A. Chabbert. 1979. Conjugative R plasmids in *Streptococcus agalactiae* (group B). Plasmid 2:197–206.

Hovelius, B. 1986. Epidemiological and clinical aspects of urinary tract infections caused by *Staphylococcus saprophyticus,* p. 195–202. In: P.-A. Mårdh and K. H. Schleifer (ed.), Coagulase-negative staphylococci. Almqvist and Wiksell International, Stockholm.

Hovelius, B., S. Colleen, and P.-A. Mårdh. 1984. Urinary tract infections in men caused by *Staphylococcus saprophyticus*. Scand. J. Infect. Dis. 16:37–41.

Hovelius, B., and P.-A. Mårdh. 1979. Haemagglutination by *Staphylococcus saprophyticus* and other staphylococcal species. Acta Pathol. Microbiol. Immunol. Scand. Sect. B 87:45–50.

Hucker, G. J. 1924. Studies on the Coccaceae. IV. The classification of the genus *Micrococcus* Cohn. Tech. Bull. N.Y. St. Agric. Exp. Sta. No. 102. Cornell University, Geneva.

Hucker, G. J. 1948. *Micrococcaceae*, p. 235–294. In: R. S. Breed, E. G. D. Murray, and A. P. Hitchens (ed.), Bergey's manual of determinative bacteriology, 6th ed. Williams and Wilkins, Baltimore.

Hueter, C. 1872. Zur Aetiologie and Therapie der metastasirenden Pyämie. Dtsch. Zeitschr. Chir. 1:91–125.

Hummel, R., and W. Witte. 1985. Ecological, epidemiological and experimental results on the question of a mutual transfer of drug resistance between *Staphylococcus aureus* strains from man and from animals, p. 613–616. In: J. Jeljaszewicz (ed.), The staphylococci: proceedings of the Vth international symposium on staphylococci and staphylococcal infections. Gustav Fischer Verlag, Stuttgart.

Hussain, Z., L. Stoakes, D. L. Stevens, B. C. Schieven, R. Lannigan, and C. Jones. 1986. Comparison of the MicroScan system with the API Staph-Ident system for species identification of coagulase-negative staphylococci. J. Clin. Microbiol. 23:126–128.

Igarashi, H., M. Shingaki, H. Fujikawa, H. Ushioda, T. Terayama. 1985. Detection of staphylococcal enterotoxins in food poisoning outbreaks by reversed passive latex agglutination, p. 255–257. In: J. Jeljaszewicz (ed.), The staphylococci: proceedings of the Vth international symposium on staphylococci and staphylococcal infections. Gustav Fischer Verlag, Stuttgart.

Igimi, S., S. Kawamura, E. Takahashi, and T. Mitsuoka. 1989. *Staphylococcus felis*, a new species from clinical specimens from cats. Int. J. Syst. Bacteriol. 39:373–377.

Iordanescu, S. 1975. Recombinant plasmid obtained from two different, compatible staphylococcal plasmids. J. Bacteriol. 124:597–601.

Iordanescu, S. 1976. Temperature-sensitive mutant of a tetracycline resistance staphylococcal plasmid. Arch. Roum. Path. Exp. Microbiol. 35:257–264.

Iordanescu, S. 1979. Incompatibility-deficient derivatives of a small staphylococcal plasmid. Plasmid 2:207–215.

Iordanescu, S., and M. Surdeanu. 1980. New incompatibility groups for *Staphylococcus aureus* plasmids. Plasmid 4:256–260.

Iordanescu, S., M. Surdeanu, P. Della Latta, and R. P. Novick. 1978. Incompatibility and molecular relationships between small staphylococcal plasmids carrying the same resistance marker. Plasmid. 1:468–479.

Isenberg, H., J. A. Washington II, A. Balows, and A. C. Sonnenwirth. 1985. Collection, handling, and processing of specimens, p. 73–98. In: E. H. Lennette, A. Balows, W. J. Hausler, Jr., and H. J. Shadomy (ed.), Manual of clinical microbiology, 4th ed. American Society for Microbiology, Washington, D.C.

Jacobs, N. J., and S. F. Conti. 1965. Effect of hemin on the formation of the cytochrome system of anaerobically grown *Staphylococcus epidermidis*. J. Bacteriol. 89:675–679.

Jacobs, N. J., M. J. Jacobs, and G. S. Sheng. 1969. Effect of oxygen on heme and porphyrin accumulation from δ-aminolevulinic acid by suspensions of anaerobically grown *Staphylococcus epidermidis*. J. Bacteriol. 99:37–41.

Jacobs, N. J., J. Johantges, and H. Deibel. 1963. Effect of anaerobic growth on nitrate reduction by *Staphylococcus epidermidis*. J. Bacteriol. 85:782–787.

Jaffe, H. W., H. M. Sweeney, R. A. Weinstein, S. A. Kabins, C. Nathan, and S. Cohen. 1982. Structural and phenotypic varieties of gentamicin resistance plasmids in hospital strains of *Staphylococcus aureus* and coagulase-negative staphylococci. Antimicrob. Agents Chemother. 21:773–779.

Jean-Pierre, H., H. Darbas, A. Jean-Roussenq, and G. Boyer. 1989. Pathogenicity in two cases of *Staphylococcus schleiferi*, a recently described species. J. Clin. Microbiol. 27:2110–2111.

Jeljaszewicz, J. 1972. Toxins (hemolysins), p. 249–280. In: J. O. Cohen (ed.), The staphylococci. Wiley-International, New York.

Jeljaszewicz, J. (ed.). 1973. Staphylococci and staphylococcal infections. Polish Medical Publishers, Warsaw.

Jeljaszewicz, J. (ed.). 1976. Staphylococci and staphylococcal diseases. Gustav Fischer Verlag, Stuttgart.

Jeljaszewicz, J. (ed.). 1981. Staphylococci and staphylococcal infections. Gustav Fischer Verlag, Stuttgart.

Jeljaszewicz, J. (ed.) 1985. The staphylococci: proceedings of the Vth international symposium on staphylococci and staphylococcal infections. Gustav Fischer Verlag, Stuttgart.

Jeljaszewicz, J., L. M. Switalski, and C. Adlam. 1983. Staphylocoagulase and clumping factor, p. 525–557. In: C. S. F. Easmon and C. Adlam (ed.), Staphylococci and staphylococcal infections, vol. 2. Academic Press, London.

Jensen, J., and M. O. Hyde. 1963. "Apocatalase" of catalase-negative staphylococci. Science 141:45–46.

John, J. F., P. K. Gramling, and N. M. O'Dell. 1978. Species identification of coagulase-negative staphylococci from urinary tract infections. J. Clin. Microbiol. 8:435–437.

Johns, M. B., and S. A. Khan. 1988. Staphylococcal enterotoxin B gene is associated with a discrete genetic element. J. Bacteriol. 170:4033–4039.

Johnson, G. M., W. E. Regelmann, E. D. Gray, G. Peters, and P. G. Quie. 1987. Staphylococcal slime and host defenses: effects on polymorphonuclear granulocytes, p. 33–44. In: G. Pulverer, P. G. Quie, and G. Peters (ed.), Pathogenicity and clinical significance of coagulase-negative staphylococci. Gustav Fischer Verlag, Stuttgart.

Johnson, L. P., J. J. L'Italien, and P. A. Schlievert. 1986. Streptococcal pyrogenic exotoxin A (scarlet fever toxin) is related to *Staphylococcus aureus* enterotoxin B. Mol. Gen. Genet. 203:354–356.

Jones, C. L., and S. A. Khan. 1986. Nucleotide sequence of the enterotoxin B gene from *Staphylococcus aureus*. J. Bacteriol. 166:29–33.

Jonsson, P., M. Lindberg, I. Haraldsson, and T. Wadström. 1985. Virulence in *Staphylococcus aureus* in a mouse mastitis model: studies of alpha hemolysin, coagulase, and protein A as possible virulence determinants with protoplast fusion and gene cloning. Infect. Immun. 49:765–769.

Jordan, P. A., A. Iravani, G. A. Richard, and H. Baer. 1980. Urinary tract infections caused by *Staphylococcus saprophyticus*. J. Infect. Dis. 142:510–515.

Kalbitzer, H. R., J. Deutscher, W. Hengstenberg, and P. Rösch. 1981. Phosphoenolpyruvate-dependent phosphotransferase system of *Staphylococcus aureus*. ¹H-NMR studies of phosphorylated and unphosphorylated Factor III^lac and its interaction with the phospho-carrier protein HPr. Biochemistry 21:6178–6185.

Kane-Falce, C., and W. E. Kloos. 1975. A genetic and biochemical study of histidine biosynthesis in *Micrococcus luteus*. Genetics 79:361–376.

Karchmer, A. W., G. L. Archer, and W. E. Dismukes. 1983. *Staphylococcus epidermidis* causing prosthetic valve endocarditis: microbiologic and clinical observations as guides to therapy. Ann. Intern. Med. 98:447–455.

Karthigasu, K. T., R. A. Bowman, and D. I. Grove. 1986. Vertebral osteomyelitis due to *Staphylococcus warneri*. Ann. Rheum. Dis. 45:1029–1030.

Kasprowicz, A., A. Bialecka, and P. B. Heczko. 1987. Surface properties of *Staphylococcus saprophyticus* strains isolated from various sources, p. 77–81. In: G. Pulverer, P. G. Quie, and G. Peters (ed.), Pathogenicity and clinical significance of coagulase-negative staphylococci. Gustav Fischer Verlag, Stuttgart.

Keller, G., K. H. Schleifer, and F. Götz. 1983. Construction and characterization of plasmid vectors for cloning in *Staphylococcus aureus* and *Staphylococcus carnosus*. Plasmid 10:270–278.

Keller, G., K. H. Schleifer, and F. Götz. 1984. Cloning of the ribokinase gene of *Staphylococcus hyicus* subsp. *hyicus* in *Staphylococcus carnosus*. Arch. Microbiol. 140:218–224.

Kellner, R., G. Jung, T. Hörner, H. Zähner, N. Schnell, K.-D. Entian, and F. Götz. 1988. Gallidermin: a new lanthionine-containing polypeptide antibiotic. Eur. J. Biochem. 177:53–59.

Kellner, R., G. Jung, M. Josten, C. Kaletta, K.-D. Entian, and H.-G. Sahl. 1989. Pep5: structure elucidation of a large lantibiotic. Angew. Chem. Int. Ed. Engl. 28:616–619.

Khan, S. A., G. K. Alder, and R. P. Novick. 1982. Functional origin of replication of pT181 plasmid DNA is contained within a 168-basepair segment. Proc. Natl. Acad. Sci. USA 79:4580–4584.

Khan, S. A., S. M. Carleton, and R. P. Novick. 1981. Replication of plasmid pT181 DNA *in vitro*: requirement for a plasmid-encoded product. Proc. Natl. Acad. Sci. USA 78:4902–4906.

Khan, S. A., and R. P. Novick. 1980. Terminal nucleotide sequences of Tn551, a transposon specifying erythromycin resistance in *Staphylococcus aureus*: homology with Tn3. Plasmid 4:148–154.

Khan, S. A., and R. P. Novick. 1983. Complete nucleotide sequence of pT181, a tetracycline-resistance plasmid from *Staphylococcus aureus*. Plasmid 10:251–259.

Kilpper, R., U. Buhl, and K. H. Schleifer. 1980. Nucleic acid homology studies between *Peptococcus saccharolyticus* and various anaerobic and facultative anaerobic Gram-positive cocci. FEMS Microbiol. Lett. 8:205–210.

Kilpper-Bälz, R., and K. H. Schleifer. 1981. Transfer of *Peptococcus saccharolyticus* Foubert and Douglas to the genus *Staphylococcus*: *Staphylococcus saccharolyticus*

(Foubert and Douglas) comb. nov. Zentralbl. Bakteriol. Mikrobiol. Hyg. Abt. 1 Orig. C 2:324–331.

Klebs, E. 1872. Beiträge zur pathologischen Anatomie der Schusswunden. Vogel, Leipzig.

Kligman, A. M. 1975. The bacteriology of normal skin, p. 13–31. In: H. I. Maibach and G. Hildick-Smith (ed.), Skin bacteria and their role in infection. McGraw-Hill Book Co., New York.

Kloos, W. E. 1980. Natural populations of the genus *Staphylococcus*. Annu. Rev. Microbiol. 34:559–592.

Kloos, W. E. 1985. Staphylococcus, p. 431–434. In: S. P. Parker (ed.), McGraw-Hill 1986 Yearbook of Science and Technology. McGraw-Hill Book Co., New York.

Kloos, W. E. 1986a. Ecology of human skin, p. 37–50. In: P.-A. Mårdh and K. H. Schleifer (ed.), Coagulase-negative staphylococci. Almqvist and Wiksell International, Stockholm.

Kloos, W. E. 1986b. Community structure of coagulase-negative staphylococci in humans, p. 132–138. In: L. Leive (ed.), Microbiology 1986. American Society for Microbiology, Washington, D.C.

Kloos, W. E., and D. W. Lambe, Jr. 1991. Staphylococcus. In: A. Balows, W. J. Hausler, Jr., K. Herrmann, H. Isenberg, and H. J. Shadomy (ed.), Manual of Clinical Microbiology, 5th ed. American Society for Microbiology, Washington, D.C. (in press).

Kloos, W. E. and M. S. Musselwhite. 1975. Distribution and persistence of *Staphylococcus* and *Micrococcus* species and other aerobic bacteria on human skin. Appl. Microbiol. 30:381–395.

Kloos, W. E. and P. A. Pattee. 1965. Transduction analysis of the histidine region in *Staphylococcus aureus*. J. Gen. Microbiol. 39:195–207.

Kloos, W. E., and N. E. Rose. 1970. Transformation mapping of tryptophan loci in *Micrococcus luteus*. Genetics 66:595–605.

Kloos, W. E., and K. H. Schleifer. 1975a. Isolation and characterization of staphylococci from human skin. II. Description of four new species: *Staphylococcus warneri, Staphylococcus capitis, Staphylococcus hominis,* and *Staphylococcus simulans*. Int. J. Syst. Bacteriol. 25:62–79.

Kloos, W. E., and K. H. Schleifer. 1975b. Simplified scheme for routine identification of human *Staphylococcus* species. J. Clin. Microbiol. 1:82–88.

Kloos, W. E., and K. H. Schleifer. 1981. The genus *Staphylococcus*, p. 1548–1569. In: M. P. Starr, H. Stolp, H. G. Trüper, A. Balows, and H. G. Schegel (ed.), The prokaryotes: a handbook on habitats, isolation, and identification of bacteria. Springer-Verlag, New York.

Kloos, W. E., and K. H. Schleifer. 1983. *Staphylococcus auricularis* sp. nov.: an inhabitant of the human external ear. Int. J. Syst. Bacteriol. 33:9–14.

Kloos, W. E., and K. H. Schleifer. 1986. Genus IV. *Staphylococcus* Rosenbach 1884, p. 1013–1035. In: J. G. Holt, P. H. A. Sneath, N. S. Mair, and M. S. Sharpe (ed.), Bergey's manual of systematic bacteriology, Vol. 2. Williams and Wilkins, Baltimore.

Kloos, W. E., K. H. Schleifer, and R. F. Smith. 1976a. Characterization of *Staphylococcus sciuri* sp. nov. and its subspecies. Int. J. Syst. Bacteriol. 26:22–37.

Kloos, W. E., T. G. Tornabene, and K. H. Schleifer. 1974. Isolation and characterization of micrococci from human skin, including two new species: *Micrococcus lylae* and *Micrococcus kristinae*. Int. J. Syst. Bacteriol. 24:79–101.

Kloos, W. E., and J. F. Wolfshohl. 1979. Evidence for deoxyribonucleotide sequence divergence between staphylococci living on human and other primate skin. Curr. Microbiol. 3:167–172.

Kloos, W. E., and J. F. Wolfshohl. 1982. Identification of *Staphylococcus* species with the API STAPH-IDENT system. J. Clin. Microbiol. 16:509–516.

Kloos, W. E., and J. F. Wolfshohl. 1983. Deoxyribonucleotide sequence divergence between *Staphylococcus cohnii* subspecies populations living on primate skin. Curr. Microbiol. 8:115–121.

Kloos, W. E., R. J. Zimmerman, and R. F. Smith. 1976b. Preliminary studies on the characterization and distribution of *Staphylococcus* and *Micrococcus* species on animal skin. Appl. Environ. Microbiol. 31:53–59.

Koch, R. 1878. Untersuchungen über die Aetiologie der Wundinfections-krankheiten. F.C.W. Vogel, Leipzig.

Kordel, M., R. Benz, and H.-G. Sahl. 1988. Mode of action of the staphylococcin-like peptide Pep5: voltage-dependent depolarization of bacterial and artificial membranes. J. Bacteriol. 170:84–88.

Kraus, E. S., and D. A. Spector. 1983. Characteristics and sequelae of peritonitis in diabetics and nondiabetics receiving chronic intermittent peritoneal dialysis. Medicine 62:52–57.

Krebs, H. A. 1937. Dismutation of pyruvic acid in *Gonococcus* and *Staphylococcus*. Biochem. J. 31:661–671.

Kreuz, B., and F. Götz. 1984. Construction of *Staphylococcus* plasmid vector pCA43 conferring resistance to chloramphenicol, arsenate, arsenite and antimony. Gene 31:310–304.

Kristinsson, K. G., R. C. Spencer, and C. B. Brown. 1986. Clinical importance of production of slime by coagulase-negative staphylococci in chronic ambulatory peritoneal dialysis. J. Clin. Pathol. 39:117–118.

Krogh, H. V., and S. Kristensen. 1976. A study of skin diseases in dogs and cats. II. Microflora of the normal skin of dogs and cats. Nord. Vet. Med. 28:459–463.

Krolewski, J. J., E. Murphy, R. P. Novick, and M. Rush. 1981. Site-specificity of the chromosomal insertion of *Staphylococcus aureus* transposon Tn554. J. Mol. Biol. 152:19–33.

Kunin, C. M. 1979. Detection, prevention and management of urinary tract infections, 3rd ed. Lea and Febiger, Philadelphia.

Lampson, B. C., and J. T. Parisi. 1986a. Naturally occurring *Staphylococcus epidermidis* plasmid expressing constitutive macrolide-lincosamide-streptogramin B resistance contains a deleted attenuator. J. Bacteriol. 166:479–483.

Lampson, B. C., and J. T. Parisi. 1986b. Nucleotide sequence of the constitutive macrolide-lincosamide-streptogramin B resistance plasmid pNE131 from *Staphylococcus epidermidis* and homologies with *Staphylococcus aureus* plasmids pE194 and pSN2. J. Bacteriol. 167:888–892.

Leclercq, R., C. Carlier, J. Duval, and P. Courvalin. 1985. Plasmid-mediated resistance to lincomycin by inactivation in *Staphylococcus haemolyticus*. Antimicrob. Agents Chemother. 28:421–424.

Langlois, B. E., R. J. Harmon, and K. Akers. 1983. Identification of *Staphylococcus* species of bovine origin

with the API Staph-Ident system. J. Clin. Microbiol. 18:1212–1219.

Lee, C. Y., and J. J. Iandolo. 1986. Lysogenic conversion of staphylococcal lipase is caused by insertion of the bacteriphage L54a genome into the lipase structural gene. J. Bacteriol. 166:385–391.

Lee, C. Y., J. J. Schmidt, A. D. Johnson-Winegar, L. Spero, and J. J. Iandolo. 1987. Sequence determination and comparison of the exfoliative toxin A and toxin B genes from *Staphylococcus aureus*. J. Bacteriol. 169:3904–3909.

Lee, J. C., F. Michon, N. E. Perez, C. A. Hopkins, and G. B. Pier. 1987. Chemical characterization and immunogenicity of capsular polysaccharide isolated from mucoid *Staphylococcus aureus*. Infect. Immun. 55:2191–2197.

Leeuwenhoek, A. 1684. Letter 39, dated Sep. 17. 1683, containing some microscopical observations, about animals in the scurf of the teeth, the substance called worms in the nose, the cuticula consisting of scales. Philos. Trans. 14:568–574.

Leeuwenhoek, A. 1695. Letter 75, dated 16 Sept. 1692. Arc. Nat. Det., Lugd. Bat. 1695:334–339.

Lehmer, A., and K. H. Schleifer. 1980. Metabolism of pentose and pentitols by *Staphylococcus xylosus* and *Staphylococcus saprophyticus*. Zentralbl. Bakteriol. Mikrobiol. Hyg. Abt. 1 Orig. C1:109–123.

Leighton, P. M., and J. A. Little. 1986. Identification of coagulase-negative staphylococci isolated from urinary tract infections. Am. J. Clin. Pathol. 85:92–95.

Lennette, E. H., A. Balows, W. J. Hausler, Jr., and H. J. Shadomy (ed.) 1985. Manual of clinical microbiology, 4th ed. American Society for Microbiology, Washington, D.C.

Lerche, M., and H. J. Sinell. 1955. Zum Vorkommen von Kokken in Rohwürsten. Archiv Lebensmittelhygiene 6:194–198.

Liau, D. F., and J. H. Hash. 1977. Structural analysis of the surface polysaccharide of *Staphylococcus aureus* M. J. Bacteriol. 131:194–200.

Liebl, W., and F. Götz. 1986. Lipase directed export of *Escherichia coli* β-lactamase in *Staphylococcus carnosus*. Mol. Gen. Genet. 204:166–173.

Lindberg, M., and R. P. Novick. 1973. Plasmid-specific transformation in *Staphylococcus aureus*. J. Bacteriol. 115:139–145.

Lindberg, M., J. E. Sjöström, and T. Johansson. 1972. Transformation of chromosomal and plasmid characters in *Staphylococcus aureus*. J. Bacteriol. 109:844–847.

Linnaeus, C. 1767. Sytema naturae, 12th ed. Holmiae.

Lipinski, B., J. Hawiger, and J. Jeljaszewicz. 1967. Staphylococcal clumping with soluble fibrin monomer complexes. J. Exp. Med. 126:979–988.

Loll, P. J., and E. E. Lattman. 1989. The crystal structure of the ternary complex of staphylococcal nuclease, Ca²⁺, and the inhibitor pdTp, refined at 1,65 A°. Proteins, Structure, Function and Genetics 5:183–201.

Lowy, F. D., and S. M. Hammer. 1983. *Staphylococcus epidermidis* infections. Ann. Intern. Med. 99:834–839.

Luchansky, J. B., and P. A. Pattee. 1984. Isolation of transposon Tn551 insertions near chromosomal markers of interest in *Staphylococcus aureus*. J. Bacteriol. 159:894–899.

Ludwig, W., K. H. Schleifer, G. E. Fox, E. Seewaldt, and E. Stackebrandt. 1981. A phylogenetic analysis of staphylococci, Peptococcus saccharolyticus and Micrococcus mucilaginosus, J. Gen. Microbiol. 125:357–366.

Ludwig, W., E. Seewaldt, R. Kilpper-Bälz, K. H. Schleifer, L. Magrum, C. R. Woese, G. F. Fox, and E. Stackebrandt. 1985. The phylogenetic position of Streptococcus and Enterococcus. J. Gen. Microbiol. 131:543–551.

Lyon, B. R., J. W. May, and R. A. Skurray. 1983. Analysis of plasmids in nosocomial strains of multiple-antibiotic-resistant Staphylococcus aureus. Antimicrob. Agents Chemother. 23:817–826.

Lyon, B. R., J. W. May, and R. A. Skurray. 1984. Tn4001: a gentamicin and kanamycin resistance transposon in Staphylococcus aureus. Mol. Gen. Genet. 193:554–556.

Lyon, B. R., and R. A. Skurray. 1987. Antimicrobial resistance of Staphylococcus aureus: genetic basis. Microbiol. Rev. 51:88–134.

Mah, R. A., D. Y. C. Fung, and S. A. Morse. 1967. Nutritional requirements of Staphylococcus aureus S-6. Appl. Microbiol. 15:866–870.

Males, B. M., W. R. Bartholomew, and D. Amsterdam. 1985. Staphylococcus simulans septicemia in a patient with chronic osteomyelitis and pyarthrosis. J. Clin. Microbiol. 21:255–257.

Malveaux, F. J., and C. L. San Clemente. 1969. Staphylococcal acid phosphatase: Preliminary physical and chemical characterization of the loosely bound enzyme. J. Bacteriol. 97:1215–1219.

Marcelis, J. H., J. H. den Daas-slagt, and J. A. A. Hoogkamp-Korstanje. 1978. Iron requirement and chelator production of staphylococci, Streptococcus faecalis and Enterobacteriaceae. Antonie van Leeuwenhoek J. Microbiol. 44:257–267.

Mårdh, P.-A. 1986. Pathogenicity of Staphylococcus saprophyticus. A review, p. 131–138. In: P.-A. Mårdh and K. H. Schleifer (ed.), Coagulase-negative staphylococci. Almqvist and Wiksell International, Stockholm.

Mårdh, P.-A., and K. H. Schleifer (ed.) 1986. Coagulase-negative staphylococci. Almqvist and Wiksell International, Stockholm.

Marples, M. J. 1965. The ecology of the human skin. Charles C. Thomas, Springfield.

Marrie, T. J., C. Kwan, M. A. Noble, A. West, and L. Duffield. 1982. Staphylococcus saprophyticus as a cause of urinary tract infections. J. Clin. Microbiol. 16:427–431.

Marsik, F. J., and S. Brake. 1982. Species identification and susceptibility to 17 antibiotics of coagulase-negative staphylococci isolated from clinical specimens. J. Clin. Microbiol. 15:640–645.

Martin, M. A., M. A. Pfaller, and R. P. Wenzel. 1989. Coagulase-negative staphylococcal bacteremia. Ann. Intern. Med. 110:9–16.

Maskell, J. P. 1980. The function interchangeability of enterobacterial and staphylococcal iron chelators. Antonie van Leeuwenhoek J. Microbiol. 46:343–351.

Matsumura, M., Y. Katakura, T. Imanaka, and S. Aiba. 1984. Enzymatic and nucleotide sequence studies of a kanamycin-inactivating enzyme encoded by a plasmid from thermophilic bacilli in comparison with that encoded by plasmid pUB110. J. Bacteriol. 160:413–420.

McCartney, C. and J. P. Arbuthnott. 1978. Mode of action of membrane damaging toxins produced by staphylococci, p. 89–127. In: J. Jeljaszewicz and T. Wadström

(ed.), Bacterial toxins and cell membranes. Academic Press, New York.

McDonnell, R. W., H. M. Sweeney, and S. Cohen. 1983. Conjugational transfer of gentamicin resistance plasmids intra- and inter- specifically in Staphylococcus aureus and Staphylococcus epidermidis. Antimicrob. Agents Chemother. 23:151–160.

Meiwes, J., H. -P. Fiedler, H. Haag, H. Zähner, S. Konetschny-Rapp, and G. Jung. 1990. Isolation and characterization of staphyloferrin A, a compound with siderophore activity from Staphylococcus hyicus DSM 20459. FEMS Microbiol. Lett. 67:201–206.

Meyer, W. 1967. A proposal for subdividing the species Staphylococcus aureus. Int. J. Syst. Bacteriol. 17:387–389.

Migula, W. 1897. System der Bakterien: Handbuch der Morphologie, Entwickelungsgeschichte und Systematik der Bakterien. Gustav Fischer Verlag, Jena.

Miller, B. A., R. F. Reiser, and M. S. Bergdoll. 1978. Detection of staphylococcal enterotoxin A, B, C, D, and E in foods by radioimmunoassay, using staphylococcal cells containing protein A as immunoadsorbent. Appl. Environ. Microbiol. 36:421–426.

Minor, T. E., and E. H. Marth. 1972. Staphylococcus aureus and staphylococcal food intoxications. A review. IV. Staphylococci in meat, bakery products, and other foods. J. Milk Food Techn. 35:228–241.

Moks, T., L. Abrahamsen, B. Nilsson, U. Hellman, J. Sjöquist, and M. Uhlen. 1986. Staphylococcal protein A consists of five IgG-binding domains. Eur. J. Biochem. 156:637–643.

Möllby, R. 1983. Isolation and properties of membrane damaging toxins, p. 619–669. In: C. S. F. Easmon and C. Adlam (ed.), Staphylococci and staphylococcal infections, vol. 2. Academic Press, London.

Montagna, W., and R. A. Ellis. 1963. New approaches to the study of the skin of primates, p. 179–196. In: J. Buettner-Janusch (ed.), Evolutionary and genetic biology of primates, vol. 1. Academic Press, New York.

Morris, I. M., P. C. Mattingly, and B. E. Gostelow. 1986. Coagulase-negative Staphylococcus as a cause of joint infection. Brit. J. Rheum. 25:414–415.

Müller, O. F. 1773–1774. Vermium Terrestrium et Fluviatilium, seu Animalium Infusoriorum, Helminthicorum et Testaceorum, Non Marinorum, Succincta Historia, 2 vol. Havniae et Lipsiae.

Murphy, E., L. Huwyler, and M. d.C. de Freire Bastos. 1985. Transposon Tn554: complete nucleotide sequence and isolation of transposition-defective and antibiotic-sensitive mutants. EMBO J. 4:3357–3365.

Murphy, E., S. Phillips, I. Edelman, and R. P. Novick. 1981. Tn554: isolation and characterization of plasmid insertions. Plasmid 5:292–305.

Naidoo, J., and W. C. Noble. 1978. Transfer of gentamicin resistance between strains of Staphylococcus aureus on skin. J. Gen. Microbiol. 107:391–393.

Naidoo, J., and W. C. Noble. 1981. Transfer of gentamicin resistance between coagulase-negative and coagulase-positive staphylococci on skin. J. Hyg. Camb. 86:183–187.

Needham, C. A. 1987. Specimen collection, p. 3–26. In: B. B. Wentworth (ed.), Diagnostic procedures for bacterial infections, 7th ed. American Public Health Association, Washington, D.C.

Nicolaides, N. 1974. Skin lipids: their biochemical uniqueness. Science 186:19–26.

Nielsen, J. B. K., and J. O. Lampen. 1982. Membrane bound penicillinases in Gram-positive bacteria. J. Biol. Chem. 257:4490–4495.

Nilsson, B., L. Abrahmsen, and M. Uhlen. 1985a. Immobilisation and purification of enzymes with staphylococcal protein A gene fusion vectors. EMBO J. 4:1075–1080.

Nilsson, B., E. Holmgren, S. Josephson, S. Gatenback, L. Philipson, and M. Uhlen. 1985b. Efficient secretion and purification of human insulin-like growth factor I with a gene fusion vector in staphylococci. Nucl. Acids. Res. 13:1151–1162.

Niinivaara, F. P., and M. S. Pohja. 1956. Über die Reifung der Rohwurst. I. Mitt: Die Veäanderung der Bakterienflora während der Reifung. Zeitschritt für Lebensmittel Untersuchung und Forschung 104:413–422.

Noble, W. C., and D. G. Pitcher. 1978. Microbial ecology of the human skin. Adv. Microb. Ecol. 2:245–289.

Noble, W. C., and D. A. Somerville. 1974. Microbiology of human skin. W. B. Saunders, London.

Nocard, E., and A. Mollereau. 1887. Sur une mammite contagieuse des vaches laitières. Ann. Inst. Pasteur (Paris) 1:109–126.

Nord, C.-E., S. Holta-Oie, Å. Ljungh, and T. Wadström. 1976. Characterization of coagulase-negative staphylococcal species from human infections, p. 105–111. In J. Jeljaszewicz (ed.), Staphylococci and staphylococcal diseases. Gustav Fischer Verlag, Stuttgart.

Novick, R. 1967. Properties of a cryptic high-frequency transducing phage in *Staphylococcus aureus.* Virology 33:155–166.

Novick, R. 1976. Plasmid-protein relaxation complexes in *Staphylococcus aureus.* J. Bacteriol. 127:1177–1187.

Novick, R. P. 1987. Plasmid incompatibility. Microbiol. Rev. 51:381–395.

Novick, R. P., G. K. Adler, S. Majumder, S. A. Khan, S. Carleton, W. D. Rosenblum, and S. Iordanescu. 1982. Coding sequence for the pT181 *repC* product: a plasmid-coded protein uniquely required for replication. Proc. Natl. Acad. Sci. USA 79:4108–4112.

Novick, R. P., and D. Bouanchaud. 1971. Extrachromosomal nature of drug resistance in *Staphylococcus aureus.* Ann. N.Y. Acad. Sci. 182:279–294.

Novick, R. P., J. Edelman, and S. Lofdahl. 1986. Small *Staphylococcus aureus* plasmids are transduced as linear multimers which are formed and resolved by replicative processes. J. Mol. Biol. 192:209–220.

Novick, R. P., I. Edelman, M. D. Schwesinger, A. D. Gruss, E. C. Swanson, and P. A. Pattee. 1979. Genetic translocation in *Staphylococcus aureus.* Proc. Natl. Acad. Sci. USA 76:400–404.

Novick, R. P., S. Iordanescu, S. J. Projan, J. Kornblum, and I. Edelman. 1989. pT181 plasmid replication is regulated by a countertranscript-driven transcriptional attenuator. Cell 59:395–404.

Novick, R. P., S. Iordanescu, M. Surdeanu, and I. Edelman. 1981. Transduction related cointegrate formation between staphylococcal plasmids: a new type of site-specific recombination. Plasmid 6:159–172.

Novick, R. P., S. J. Projan, C. Kumar, S. Carleton, S. Gruss, S. K. Highlander, and J. Kornblum. 1985. Replication control for pT181, an indirectly regulated plasmid, p. 299–320. In: D. R. Helinski, S. N. Cohen, D. B. Clew-

ell, D. A. Jackson and A. Hollaender (ed.), Plasmids in bacteria. Plenum Press, New York.

Novick, R. P., S. J. Projan, W. Rosenblum, and I. Edelman. 1984. Staphylococcal plasmid cointegrates are formed by host- and phage-mediated general *rec* systems that act on short regions of homology. Mol. Gen. Genet. 195:374–377.

Novick, R. P., and M. H. Richmond. 1965. Nature and interactions of the genetic elements governing penicillinase synthesis in *Staphylococcus aureus.* J. Bacteriol. 90:467–480.

Oeding, P., V. Hájek, and E. Marsalek. 1973. A comparison of antigenic structure and phage pattern with biochemical properties of *Staphylococcus aureus* strains isolated from hares and mink. Acta Pathol. Microbiol. Scand. Sec. B 81:567–570.

Oeding, P., V. Hájek, and E. Marsalek. 1974. A comparison of antigenic structure and phage pattern with biochemical properties of *Staphylococcus aureus* strains isolated from horses. Acta Pathol. Microbiol. Scand. Sec. B 82:899–903.

Ogston, A. 1880. Ueber Abscesse. Arch. Klin. Chir. 25:588–600.

Ogston, A. 1882. Micrococcus poisoning. J. Anat. Physiol. (London) 16:526–567.

Ogston, A. 1883. Micrococcus poisoning. J. Anat. Physiol. (London) 17:24–58.

Oort, van M. G., A. M. T. J. Deveer, R. Dijkman, M. L. Tjeenk, H. M. Verheij, G. H. de Haas, E. Wenzig, and F. Götz. 1989. Purification and substrate specificity of *Staphylococcus hyicus* lipase. Biochem. 28:9278–9285.

Oskouian, B., and G. C. Stewart. 1987. Cloning and characterization of the repressor gene of the *Staphylococcus aureus* lactose operon. J. Bacteriol. 169:5459–5465.

O'Tool, P. W. and T. J. Foster. 1987. Nucleotide sequence of the epidermolytic toxin A gene of *Staphylococcus aureus.* J. Bacteriol. 169:3910–3915.

Paley, D., C. F. Moseley, P. Armstrong, and C. G. Prober. 1986. Primary osteomyelitis caused by coagulase-negative staphylococci. J. Ped. Orthop. 6:622–626.

Parisi, J. T., J. Robbins, B. C. Lampson, and D. W. Hecht. 1981. Characterization of a macrolide, lincosamide, and streptogramin resistance plasmid in *Straphylococcus epidermidis.* J. Bacteriol. 148:559–564.

Passet, J. 1885. Ueber Mikroorganismen der eiterigen Zellgewebsentzündung des Menschen. Fortschr. Med. 3:33–43, 68–73.

Pasteur, L. 1880. De L'extension de la théorie des germes à l'étiologie de quelques maladies communes. C.R. Acad. Sci. 90:1033–1044.

Pattee, P. A. 1981. Distribution of Tn551 insertion sites responsible for auxotrophy on the *Staphylococcus aureus* chromosome. J. Bacteriol. 145:479–488.

Pattee, P. A. 1990. *Staphylococcus aureus,* p 2.23–2.27. In: S. J. O'Brien (ed.), Genetic maps: locus maps of complex genomes, 5th ed. Cold Spring Harbor Laboratory, Cold Spring Harbor, NY.

Pattee, P. A., and J. N. Baldwin. 1961. Transduction of resistance to chlortetracycline and novobiocin in *Staphylococcus aureus.* J. Bacteriol. 82:875–881.

Pattee, P. A., J. M. Jones, and S. E. Yost. 1984. Chromosome map of *Staphylococcus aureus* p. 126–130. In: S. J. O'Brien (ed.), Genetic maps 1984: a compilation of linkage and restriction maps of genetically studied organisms, vol. 3. Cold Spring Harbor Laboratory, Cold Spring Harbor, NY.

Pattee, P. A., W. E. Kloos, J. B. Bodensteiner, and A. Zara. 1968. Homogeneity in a *Staphylococcus* transducing fragment. J. Virol. 2:652–654.

Pattee, P. A., T. Schutzlank, D. H. Kay, and M. H. Laughlin. 1974. Genetic analysis of the leucine biosynthetic genes and their relationship to the *ilv* gene cluster. Ann. N.Y. Acad. Sci. 236:175–186.

Pattee, P. A., N. W. Thompson, D. Haubrich, and R. P. Novick. 1977. Chromosomal map locations of integrated plasmids and related elements in *Staphylococcus aureus.* Plasmid. 1:38–51.

Perkins, J. B., and P. J. Youngman. 1983. *Streptococcus* plasmid pAMα1 is a composite of two separable replicons, one of which is closely related to *Bacillus* plasmid pBC16. J. Bacteriol. 155:607–615.

Peters, G. 1985. Staphylococcal "plastic" foreign body infections: evidence and pathogenesis, p. 515–524. In: J. Jeljaszewicz (ed.), The staphylococci: proceedings of the Vth international symposium on staphylococci and staphylococcal infections. Gustav Fischer Verlag, Stuttgart.

Peters, G., R. Locci, and G. Pulverer. 1982. Adherence and growth of coagulase-negative staphylococci on surfaces of intravenous catheters. J. Infect. Dis. 146:479–482.

Peters, G., F. Schumacher-Perdreau, B. Jansen, M. Bey, and G. Pulverer. 1987. Biology of *S. epidermidis* extracellular slime, p. 15–32. In: G. Pulverer, P. G. Quie, and G. Peters (ed.), Pathogenicity and clinical significance of coagulase-negative staphylococci. Gustav Fischer Verlag, Stuttgart.

Peters, G., F. Saborowski, R. Locci, and G. Pulverer. 1984. Investigations on staphylococcal infection of intravenous endocardial pacemaker-electrodes. Am. Heart J. 108:359–364.

Peterson, P. K., B. J. Wilkinson, Y. Kim, D. Schmeling, and P. G. Quie. 1978. Influence of encapsulation on staphylococcal opsonization and phagocytosis by human polymorphonuclear leukocytes. Infect. Immun. 19:943–949.

Peyru, G., L. F. Wexler, and R. P. Novick. 1969. Naturally occurring penicillinase plasmids in *Staphylococcus aureus.* J. Bacteriol. 98:215–221.

Pfaller, M. A., and L. A. Herwaldt. 1988. Laboratory, clinical, and epidemiological aspects of coagulase-negative staphylococci. Clin. Microbiol. Rev. 1:281–299.

Phillips, E., and P. Nash. 1985. Culture media, p. 1051–1092. In: E. H. Lennette, A. Balows, W. J. Hausler, Jr., and H. J. Shadomy (ed.), Manual of clinical microbiology, 4th ed. American Society for Microbiology, Washington, D.C.

Phillips, S., and R. P. Novick. 1979. A site-specific repressor-controlled transposon in *Staphylococcus aureus.* Nature 278:476–478.

Phillips, W. E., Jr., R. E. King, and W. E. Kloos. 1980. Isolation of *Staphylococcus hyicus* subsp. *hyicus* from a pig with septic polyarthritis. Am. J. Vet. Res. 41:274–276.

Phillips, W. E., Jr., and W. E. Kloos. 1981. Identification of coagulase-positive *Staphylococcus intermedius* and *Staphylococcus hyicus* subsp. *hyicus* isolates from veterinary clinical specimens. J. Clin. Microbiol. 14:671–673.

Phonimdaeng, P., M. O'Reilly, P. W. O'Tool, and T. Foster. 1988. Molecular cloning and expression of the coagulase gene of *Staphylococcus aureus* 8325-4. Gen. Microbiol. 134:75–83.

Pioch, G., H. Heyne, and W. Witte. 1988. Koagulase-negative Staphylokokken-Spezies in Mischfuttermitteln und auf Getreide. Zentralbl. Mikrobiol. 143:157–171.

Pollock, H. M. 1983. Laboratory techniques for detection of urinary tract infection and assessment of value. Am. J. Med. 75 (Suppl. 1B):79–84.

Ponce de Leon, S., S. H. Guenther, and R. P. Wenzel. 1986. Microbiologic studies of coagulase-negative staphylococci isolated from patients with nosocomial bacteraemias. J. Hosp. Infect. 7:121–129.

Poutrel, B. 1984. Udder infection of goats by coagulase-negative staphylococci. Vet. Microbiol. 9:131–137.

Prévot, A. R. 1961. Traité de systématique bactérienne, vol. 2. Dunod, Paris.

Pribram, E. 1929. A contribution to the classification of microorganisms. J. Bacteriol. 18:361–394.

Price, P. B. 1938. The bacteriology of normal skin: a new quantitative test applied to a study of the bacterial flora and the disinfectant action of mechanical cleansing. J. Infect. Dis. 63:301–318.

Proctor, A. R., and W. E. Kloos. 1970. The tryptophan gene cluster of *Staphylococcus aureus.* J. Gen. Microbiol. 64:319–327.

Projan, S. J., M. Monod, C. S. Narayanan, and D. Dubnau. 1987. Replication properties of pIM13, a naturally occurring plasmid found in *Bacillus subtilis,* and of its close relative pE5, a plasmid native to *Staphylococcus aureus.* J. Bacteriol. 169:5131–5139.

Pulverer, G. 1985. On the pathogenicity of coagulase-negative staphylococci, p. 1–9. In: J. Jeljaszewicz (ed.), The staphylococci: Proceedings of the Vth international symposium on staphylococci and staphylococcal infections. Gustav Fischer Verlag, Stuttgart.

Pulverer, G., P. G. Quie, and G. Peters (ed.). 1987. Pathogenicity and clinical significance of coagulase-negative staphylococci. Gustav Fischer Verlag, Stuttgart.

Quie, P. G., E. L. Mills, K. L. Gates, J. S. Abramson, W. E. Regelmann, and P. K. Peterson. 1983. Functional defects of granulocytes and susceptibility to staphylococcal disease, p. 243–273. In: C. S. F. Easmon and C. Adlam (ed.), Staphylococci and staphylococcal infections, vol. 1. Academic Press, London.

Raus, J., and D. N. Love. 1983. Characterization of coagulase-positive *Staphylococcus intermedius* and *Staphylococcus aureus* isolated from veterinary clinical specimens. J. Clin. Microbiol. 18:789–792.

Recsei, P. A., A. D. Gruss, and R. P. Novick. 1987. Cloning, sequence, and expression of the lysostaphin gene from *Staphylococcus simulans.* Proc. Natl. Acad. Sci. USA 84:1127–1131.

Reichwein, J., F. Hugo, M. Roth, A. Sinner, and S. Bhakdi. 1987. Quantitative analysis of the binding and oligomerization of staphylococcal alpha-toxin in target erythrocyte membranes. Infect. Immun. 55:2940–2944.

Reiser, R. E., R. N. Robbins, G. P. Khoe, and M. S. Bergdoll. 1983. Purification and some physicochemical properties of toxic-shock toxin. Biochemistry 22:3907–3912.

Reizer, J., M. H. Saier, Jr., J. Deutscher, F. Grenier, J. Thompson, and W. Hengstenberg. 1988. The phosphoenolpyruvate: sugar phosphotransferase system in

Gram-positive bacteria: properties, mechanism, and regulation. Crit. Rev. Microbiol. 15:297–338.

Richardson, J. V., R. B. Karp, J. W. Kirklin, and W. E. Dismukes. 1978. Treatment of infective endocarditis: a 10 year comparative analysis. Circulation 58:589–597.

Ritz, H. L. and J. N. Baldwin. 1958. Induction of penicillinase production in staphylococci by bacteriophage. Bacteriol. Proc. 1958:40.

Robinson, J. M., J. K. Hardman, and G. L. Sloan. 1979. Relationship between lysostaphin endopeptidase production and cell wall composition of *Staphylococcus staphylolyticus*. J. Bacteriol. 137:1158–1164.

Rogolsky, M. 1979. Nonenteric toxins of *Staphylococcus aureus*. Microbiol. Rev. 43:320–360.

Rollof, J., S. A. Hedström, and P. Nilsson-Ehle. 1987. Purification and characterization of a lipase from *Staphylococcus aureus*. Biochem. Biophys Acta 921:364–369.

Rosenbach, F. J. 1884. Mikro-organismen bei den Wund-Infections-Krankheiten des Menschen. J. F. Bergmann, Wiesbaden.

Rosenblum, E. D., and C. E. Dowell. 1960. Lysogeny and bacteriophage typing in coagulase positive staphylococci. J. Infect. Dis. 106:297–303.

Rosypal, S., A. Rosypalová, and J. Horejš. 1966. The classification of micrococci and staphylococci based on their DNA base composition and Adansonian analysis. J. Gen. Microbiol. 44:281–292.

Rubin, J., W. A. Rogers, H. M. Taylor, E. D. Everett, B. F. Prowant, L. V. Fruto, and K. D. Nolph. 1980. Peritonitis during continuous ambulatory peritoneal dialysis. Ann. Intern. Med. 92:7–13.

Ruby, C., and R. P. Novick. 1975. Plasmid interactions in *Staphylococcus aureus*: non-additivity of compatible plasmid DNA pools. Proc. Natl. Acad. Sci. USA 72:5031–5035.

Rudin, L., J. E. Sjöström, M. Lindberg, and L. Philipson. 1973. Factors affecting competence for transformation in *Staphylococcus aureus*. J. Bacteriol. 118:155–164.

Ruffer, M. A. 1914. Pathological notes on the royal mummies of the Cairo museum. Mittheil. Gesch. Med. Naturwissensch. 13:239–253.

Ruhr, E., and H. G. Sahl. 1985. Mode of action of the peptide antibiotic nisin and influence on cytoplasmic and artificial membrane vesicles. Antimicrob. Agents Chemother. 27:841–845.

Sadaie, Y., K. C. Burtis, and R. H. Doi. 1980. Purification and characterization of a kanamycin nucleotidyltransferase from plasmid pUB110-carrying cells of *Bacillus subtilis*. J. Bacteriol. 141:1178–1182.

Sako, T., and N. Tsuchida. 1983. Nucleotide sequence of the staphylokinase gene from *Staphylococcus aureus*. Nucl. Acids Res. 11:7679–7693.

Sakurai, S., H. Suzuki, and I. Kondo. 1988. DNA sequencing of the *eta* gene coding for staphylococcal exfoliative toxin serotype A. J. Gen. Microbiol. 134:711–717.

Schaberg, D. R., D. B. Clewell, and L. Glatzer. 1982. Conjugative transfer of R-plasmids from *Streptococcus faecalis* to *Staphylococcus aureus*. Antimicrob. Agents Chemother. 22:204–207.

Schade, A. L. 1963. Significance of serum iron for the growth, biological characteristics and metabolism of *Staphylococcus aureus*. Biochem. Z. 338:140–148.

Schindler, C. A., and V. T. Schuhardt. 1965. Purification and properties of lysostaphin: a lytic agent for *Staphylococcus aureus*. Biochim. Biophys. Acta 97:242–250.

Schleifer, K. H. 1983. The cell envelope, p. 358–428. In: C. S. F. Easmon and C. Adlam (ed.), Staphylococci and staphylococcal infections, vol. 2. Academic Press, London.

Schleifer, K. H. 1986. Taxonomy of coagulase-negative staphylococci, p. 11–26. In:P.-A. Mårdh and K. H. Schleifer (ed.), Coagulase-negative staphylococci. Almqvist and Wiksell International, Stockholm.

Schleifer, K. H., and U. Fischer. 1982. Description of a new species of the genus *Staphylococcus*: *Staphylococcus carnosus*. Int. J. Syst. Bacteriol. 32:153–156.

Schleifer, K. H., U. Geyer, R. Kilpper-Bälz, and L. A. Devriese. 1983. Elevation of *Staphylococcus sciuri* subsp. *lentus* (Kloos et al.) to species status: *Staphylococcus lentus* (Kloos et al.) comb. nov. Syst. Appl. Microbiol. 4:382–387.

Schleifer, K. H., A. Hartinger, and F. Götz. 1978. Occurrence of D-tagatose-6-phosphate pathway of D-galactose metabolism among staphylococci. FEMS Microbiol. Lett. 3:9–11.

Schleifer, K. H., and O. Kandler. 1972. Peptidoglycan types of bacterial cell walls and their taxonomic implications. Bacteriol. Rev. 36:407–477.

Schleifer, K. H., R. Kilpper-Bälz, and L. A. Devriese. 1984. *Staphylococcus arlettae* sp. nov., *S. equorum* sp. nov., and *S. kloosii* sp. nov.: three new coagulase-negative staphylococci from animals. Syst. Appl. Microbiol. 5:501–509.

Schleifer, K., R. Kilpper-Bälz, U. Fischer, A. Faller, and J. Endl. 1982. Identification of 'Micrococcus candidus' ATCC 14852 as a strain of *Staphylococcus epidermidis* and of 'Micrococcus caseolyticus' ATCC 13548 and Micrococcus varians ATCC 29750 as members of a new species, *Staphylococcus caseolyticus*. Int. J. Syst. Bacteriol. 32:15–20.

Schleifer, K. H., and W. E. Kloos. 1975a. A simple test system for the separation of staphylococci from micrococci. J. Clin. Microbiol. 1:327–338.

Schleifer, K. H., and W. E. Kloos. 1975b. Isolation and characterization of staphylococci from human skin. I. Amended descriptions of *Staphylococcus epidermidis* and *Staphylococcus saprophyticus* and descriptions of three new species: *Staphylococcus cohnii*, *Staphylococcus haemolyticus*, and *Staphylococcus xylosus*. Int. J. Syst. Bacteriol. 25:50–61.

Schleifer, K. H., and E. Krämer. 1980. Selective medium for isolating staphylococci. Zentralbl. Bakteriol. Hyg. Abt. 1 Orig. C 1:270–280.

Schleifer, K. H., S. A. Meyer, and M. Rupprecht. 1979. Relatedness among coagulase-negative staphylococci: deoxyribonucleic acid reassociation and comparative immunological studies. Arch. Microbiol. 122:93–101.

Schleifer, K. H., and J. Steber. 1974. Chemische Untersuchungen am Phagenrezeptor von *Staphylococcus epidermidis*. Arch. Microbiol. 98:251–270.

Schnell, N., K.-D. Entian, F. Götz, T. Hörner, R. Kellner, and G. Jung. 1989. Structural gene isolation and prepeptide sequence of gallidermin, a new lanthionine containing antibiotic. FEMS Microbiol. Lett. 58:263–268.

Schnell, N., K.-D. Entian, U. Schneider, F. Götz, H. Zähner, R. Kellner, and G. Jung. 1988. Prepeptide sequence of epidermin, a ribosomally synthesized antibiotic with four sulphide-rings. Nature 333:276–278.

Schrecker, O., R. Stein, W. Hengstenberg, M. Gassner, and D. Stehlik. 1975. The staphylococcal PEP dependent phosphotransferase system, proton magnetic resonance (PMR) studies on the phosphoryl carrier protein HPr. Evidence for a phosphohistidine residue in the intact phospho-HPr molecule. FEBS Lett. 51:309–311.

Schwab, A. H., H. V. Leininger, and E. M. Powers. 1984. Media, reagents, and stains, p. 788–897. In:M. L. Speck (ed.), Compendium of methods for the microbiological examination of foods, 2nd ed. American Public Health Association, Washington, D.C.

Sewell, C. M., J. E. Clarridge, E. J. Young, and R. K. Guthrie. 1982. Clinical significance of coagulase-negative staphylococci. J. Clin. Microbiol. 16:236–239.

Shalita, Z., E. Murphy, and R. P. Novick. 1980. Penicillinase plasmids of Staphylococcus aureus: structural and evolutionary relationships. Plasmid 3:291–311.

Shaw, C., J. M. Stitt, and S. T. Cowan. 1951. Staphylococci and their classification. J. Gen. Microbiol. 5:1010–1023.

Shaw, W. V. 1983. Chloramphenicol acetyltransferase: enzymology and molecular biology. Crit. Rev. Biochem. 4:1–43.

Shortle, D. 1983. A genetic system for analysis of staphylococcal nuclease. Gene 22:181–189.

Shrestha, T. L., and J. H. Darrell. 1979. Urinary infection with coagulase-negative staphylococci in a teaching hospital. J. Clin. Pathol. 32:299–302.

Silvestri, L. G., and L. R. Hill. 1965. Agreement between deoxyribonucleic acid base composition and taxonomic classification of Gram-positive cocci. J. Bacteriol. 90:136–140.

Sinell, H. J., and J. Baumgart. 1966. Selektivnährboden zur Isolierung von Staphylokokken aus Lebensmitteln. Zentralbl. Bakteriol. Abt. 1 Orig. 197:447–461.

Sivakanesan, R. and E. A. Dawes. 1980. Anaerobic glucose and serine metabolism in Staphylococcus epidermidis. J. Gen. Microbiol. 118:143–157.

Sjöström, J. E., M. Lindberg, and L. Philpson. 1973. Competence for transfection in Staphylococcus aureus. J. Bacteriol. 113:576–585.

Sjöström, J. E., S. Löfdahl, and L. Philipson. 1979. Transformation of Staphylococcus aureus by heterologous plasmids. Plasmid 2:529–535.

Sjöström, J. E., and L. Philipson. 1974. Role of the ϕ11 phage genome in competence of Staphylococcus aureus. J. Bacteriol. 119:19–32.

Smith, C. D., and P. A. Pattee. 1967. Biochemical and genetic analysis of isoleucine and valine biosynthesis in Staphylococcus aureus. J. Bacteriol. 93:1832–1838.

Smith, H. W., and W. E. Crab. 1960. The effect of diets containing tetracyclines and penicillin on the Staphylococcus aureus flora of the nose and skin of pigs and chickens and their human attendants. J. Pathol. Bacteriol. 79:243–249.

Sobek, H. M., K. Stüber, K. Beyreuther, W. Hengstenberg, and J. Deutscher. 1984. Staphylococcal phosphoenolpyruvate-dependent phosphotransferase system: purification and characterization of a defective lactose-specific Factor III protein. Biochemistry 23:4460–4464.

Sompolinsky, D. 1950. Impetigo contagiosa suis. Maanedsskrift for Dyrlaeger 61:401–453.

Sompolinsky, D. 1953. De l'impétigo contagiosa suis et du Micrococcus hyicus n. sp. Schweiz. Arch. Tierneilk. 95:302–309.

Speck, M. L. (ed.) 1984. Compendium of methods for the microbiological examination of foods, 2nd ed. American Public Health Association, Washington, D.C.

Stableforth, A. W., and I. A. Galloway (ed.) 1959. Infectious diseases of animals, vol. 2, Diseases due to bacteria. Academic Press, London.

Stackebrant, E., and C. R. Woese. 1979. A phylogenetic dissection of the family Micrococcaceae. Curr. Microbiol. 2:317–322.

Stahl, M. L., and P. A. Pattee. 1983a. Computer-assisted chromosome mapping by protoplast fusion in Staphylococcus aureus. J. Bacteriol. 154:395–405.

Stahl, M. L. and P. A. Pattee. 1983b. Confirmation of protoplast fusion-driven linkages in Staphylococcus aureus transformation with protoplast DNA. J. Bacteriol. 154:406–412.

Stiffler, P. W., H. M. Sweeney, and S. Cohen. 1974. Cotransduction of plasmids mediating resistance to tetracycline and chloramphenicol in Staphylococcus aureus. J. Bacteriol. 120:934–944.

Stobberingh, E. E., J. A. Meijers, and J. H. Van Kats-Renand. 1979. The sensitivity of phage DNA for a restriction enzyme from Staphylococcus aureus. Antonie van Leeuwenhoek J. Microbiol. 35:19–23.

Strasters, K. C., and K. C. Winkler. 1963. Carbohydrate metabolism of Staphylococcus aureus. J. Gen. Microbiol. 33:213–229.

Stüber, K., J. Deutscher, H. M. Sobek, W. Hengstenberg, and K. Beyreuther. 1985. Amino acid sequence of the amphiphilic phosphocarrier protein Factor III^{lac} of the lactose-specific phosphotransferase system of Staphylococcus aureus. Biochemistry 24:1164–1168.

Suttorp, N., and E. Habben. 1988. Effect of staphylococcal alpha-toxin on intracellular Ca^{2+} in polymorphonuclear leukocytes. Infect. Immun. 56:2228–2234.

Szynkiewicz, Z. M., S. Kryński, M. N. Bánbura, M. Binek, and E. Becla. 1985. Staphylococcus intermedius infection in dogs in the years 1978 to 1983 and characteristics of isolated strains, p. 641–645. In: J. Jeljaszewicz (ed.), The staphylococci: Proceedings of the Vth international symposium on staphylococci and staphylococcal infections. Gustav Fischer Verlag, Stuttgart.

Taber, H. W., and M. Morrison. 1964. Electron transport in staphylococci: properties of a particle preparation from exponential phase Staphylococcus aureus. Arch. Biochem. Biophys. 105:367–379.

Tatini, S. R., D. G. Hoover, and R. V. F. Lachica. 1984. Methods for the isolation and enumeration of Staphylococcus aureus, p. 411–427. In:M. L. Speck (ed.), Compendium of methods for the microbiological examination of foods. American Public Health Association, Washington, D.C.

Tayler, A. G., and A. W. Bernheimer. 1974. Further characterization of staphylococcal gamma hemolysin. Infect. Immun. 10:54–59.

Telestam, M. 1983. Modes of membrane damaging action of staphylococcal toxins, p. 705–744. In: C. S. F. Easmon and C. Adlam (ed.), Staphylococci and staphylococcal infections, vol. 2. Academic Press, London.

Theodore, T. S. and A. L. Schade. 1965. Carbohydrate metabolism of iron-rich and iron-poor Staphylococcus aureus. J. Gen. Microbiol. 40:385–395.

Thompson, N. E., and P. A. Pattee. 1981. Genetic transformation in Staphylococcus aureus: demonstration of a competence-conferring factor of bacteriophage origin in bacteriophage 80 alpha lysates. J. Bacteriol. 148:294–300.

Timoney, J. F., J. H. Gillespie, F. W. Scott, and J. E. Barlough (ed.) 1988. Hagan and Bruner's microbiology and infectious diseases of domestic animals, 8th ed. Comstock Publishing, Ithaca, New York.

Tojo, M., N. Yamashita, D. A. Goldmann, and G. B. Pier. 1988. Isolation and characterization of a capsular polysaccharide adhesin from *Staphylococcus epidermidis*. J. Infect. Dis. 157:713–722.

Trayer, H. R., and C. E. Buckley. 1970. Molecular properties of lysostaphin, a bacteriolytic agent specific for *Staphylococcus aureus*. J. Biol. Chem. 245:4842–4846.

Tschäpe, H. 1973. Genetic studies on nutrient markers and their taxonomic importance, p. 57–62. In: J. Jeljaszewicz (ed.), Staphylococci and staphylococcal infections. Polish Medical Publishers, Warszawa.

Tucker, P. W., E. E. Hazen Jr., and F. A. Cotten. 1978. Staphylococcal nuclease reviewed: a prototypic study in contemporary enzymology. I. Isolation, physical and enzymatic properties. Mol. Cell. Biochem. 22:67–77.

Tucker, P. W., E. E. Hazen Jr., and F. A. Cotten. 1979. Staphylococcal nuclease reviewed: a prototypic study in contemporary enzymology. IV. The nuclease as a model for protein folding. Mol. Cell. Biochem. 23:131–142.

Tyski, S., W. Hryniewicz, and J. Jeljaszewicz. 1983. Purification and some properties of the staphylococcal extracellular lipase. Biochim. Biophys. Acta 749:312–317.

Ubelaker, M. H., and E. D. Rosenblum. 1978. Transduction of plasmid determinants in *Staphylococcus aureus* and *Escherichia coli*. J. Bacteriol. 133:699–707.

Uhlen, M., B. Guss, B. Nilsson, S. Gatenbeck, L. Philipson, and M. Lindberg. 1984a. Complete sequence of the staphylococcal gene encoding Protein A: a gene evolved through multiple duplications. J. Biol. Chem. 259:1695–1702.

Uhlen, M., B. Guss, B. Nilsson, F. Götz, and M. Lindberg. 1984b. Expression of the gene encoding Protein A in *Staphylococcus aureus* and coagulase-negative staphylococci. J. Bacteriol. 159:713–719.

United States Food and Drug Administration. 1984. Bacteriological analytical manual, 6th ed. Center for Food Safety and Applied Nutrition, U.S. Food and Drug Administration, Washington, D.C.

Varaldo, P. E., R. Kilpper-Bälz, F. Biavasco, G. Satta, and K. H. Schleifer. 1988. *Staphylococcus delphini* sp. nov., a coagulase-positive species isolated from dolphins. Int. J. Syst. Bacteriol. 38:436–439.

Verneuil. 1880. De la suppuration orangée. Arch. Gén. Méd. 146:641–654.

Vescovo, M., L. Morelli, V. Bottazzi, and M. J. Gasson. 1983. Conjugal transfer of broad-host-range plasmid pAMβl into enteric species of lactic acid bacteria. Appl. Environm. Microbiol. 46:753–755.

Von Recklinghausen, N. D. 1871. Quoted by Von Lingelsheim, W. (1900). Atiologie und Therapie der Staphylokokkeninfektionen. Urban and Schwarzenberg, Berlin.

Wadström, T. 1983. Biological effects of cell damaging toxins, p. 671–704. In: C. S. F. Easmon and C. Adlam (ed.), Staphylococci and staphylococcal infections, vol. 2. Academic Press, London.

Wadström, T., P. Speziale, F. Rozgonyi, Å. Ljungh, I. Maxe, and C. Ryden. 1987. Interactions of coagulase-negative staphylococci with fibronectin and collagen as possible first step of tissue colonization in wounds and other tissue trauma, p. 83–91. In: G. Pulverer, P. G. Quie, and G. Peters (ed.), Pathogenicity and clinical significance of coagulase-negative staphylococci. Gustav Fischer Verlag, Stuttgart.

Wadström, T., and F. Rozgonyi. 1986. Virulence determinants of coagulase-negative staphylococci, p. 123–130. In: P.-A. Mårdh and K. H. Schleifer (ed.), Coagulase-negative staphylococci. Almqvist and Wiksell International, Stockholm.

Wallmark, G. I., I. Anemark, and B. Telander. 1978. *Staphylococcus saprophyticus*: a frequent cause of urinary tract infections among female outpatients. J. Infect. Dis. 138:791–797.

Watson, D. C., M. Yaguchi, J.-G. Bisaillon, R. Beaudet, and R. Morosoli. 1988. The amino acid sequence of a gonococcal growth inhibitor from *Staphylococcus haemolyticus*. Biochem. J. 252:87–93.

Watts, J. L., J. W. Pankey, and S. C. Nickerson. 1984. Evaluation of the Staph-Ident and STAPHase systems for identification of staphylococci from bovine intrammary infections. J. Clin. Microbiol. 20:448–452.

Weisblum, B. 1975. Altered methylation of ribosomal ribonucleic acid in erythromycin-resistant *Staphylococcus aureus*, p. 199–206. In: D. Schlessinger (ed.), Microbiology 1974. American Society for Microbiology, Washington, D.C.

Weld, J. T., B. H. Kean, and W. M. O'Leary. 1963. Production of octadecenoic acid in plasma by *Staphylococcus aureus*. Proc. Soc. Exp. Biol. Med. 112:448–451.

Wentworth, B. B. (ed.) 1987. Diagnostic procedures for bacterial infections, 7th ed. American Public Health Association, Washington, D.C.

White, D. C., and F. E. Frerman. 1967. Extraction, characterization, and cellular localization of the lipids of *Staphylococcus aureus*. J. Bacteriol. 94:1854–1867.

Wiley, B. B., and N. H. Maverakis. 1974. Capsule production and virulence among strains of *Staphylococcus aureus*. Ann. N.Y. Acad. Sci. 236:221–232.

Wilkinson, B. J. 1983. Staphylococcal capsules and slime, p. 481–523. In: C. S. F. Easmon and C. Adlan (ed.), Staphylococci and staphylococcal infection, vol. 2. Academic Press, London.

William, P., S. P. Denyer, and R. G. Finsch. 1988. Protein antigens of *Staphylococcus epidermidis* grown under iron-restricted conditions in human peritoneal dialysate. FEMS Microbiol. Lett. 50:29–33.

Williamson, P., and A. M. Kligman. 1965. A new method for the quantitative investigation of cutaneous bacteria. J. Invest. Dermatol. 45:498–503.

Willis, A. T., J. J. O'Connor, and J. A. Smith. 1966. Colonial pigmentation of *Staphylococcus aureus*. J. Pathol. Bacteriol. 92:97–106.

Wilson, C. R., S. E. Skinner, and W. V. Shaw. 1981. Analysis of two chloramphenicol resistance plasmids from *Staphylococcus aureus*: insertional inactivation of Cm resistance, mapping of restriction sites and construction of cloning vehicles. Plasmid 5:245–258.

Winkler, K. C., J. DeWaart, C. Grootsen, B. J. M. Zegers, N. F. Tellier, and C. D. Vertegt. 1965. Lysogenic conversion of staphylococci to loss of betatoxin. J. Gen. Microbiol. 39:321–333.

Winston, D. J., D. V. Dudnick, M. Chapin, W. G. Ho, R. P. Gale, and W. J. Martin. 1983. Coagulase-negative staphylococcal bacteremia in patients receiving immunosuppressive therapy. Arch. Intern. Med. 143:32–36.

Wise, D. R. 1971. Staphylococcal osteomyelitis of the avian vertebral column. Res. Vet. Sci. 12:169–171.

Wiseman, G. M. 1975. The hemolysins of *Staphylococcus aureus.* Bacteriol. Rev. 39:317–344.

Witte, W., R. Hummel, W. Meyer, H. Exner, and R. Wundrack. 1977. Ecology of *Staphylococcus aureus:* characterization of strains from chicken. Z. Allg. Mikrobiol. 17:639–646.

Wood, D. O., M. J. Carter, and G. K. Best. 1977. Plasmid-mediated resistance to gentamicin in *Staphylococcus aureus.* Antimicrob. Agents Chemother. 12:513–517.

Wu, T. C. M., and J. T. Park. 1971. Chemical characterization of a new surface antigenic polysaccharide from a mutant of *Staphylococcus aureus.* J. Bacteriol. 108:874–884.

Yoshida, K., and R. D. Ekstedt. 1968. Regulation of mucoid growth of *Staphylococcus aureus* to clumping factor reaction, morphology in serum-soft agar, and virulence. J. Bacteriol. 96:902–908.

Zopf, W. 1885. Die Spaltpilze, 3 Aufl. E. Trewendt, Breslau.

CHAPTER 64

The Genus *Streptococcus*—Oral

JEREMY M. HARDIE and ROBERT A. WHILEY

The classification of the genus *Streptococcus* has undergone major revisions in recent years as a result of the application of molecular and chemotaxonomic approaches (Schleifer and Kilpper-Bälz, 1987). This has resulted in the creation of two new genera, *Enterococcus* (Chapter 66) and *Lactococcus* (Chapter 67) to include species previously allocated to the "enterococcus" (or "fecal") and the "lactic" groups of streptococci, respectively (Hardie, 1986a; Jones, 1978; Sherman, 1937). In addition, some other species previously considered to be streptococci, such as the anaerobes *S. hansenii, S. morbillorum, S. parvulus,* and *S. pleomorphus,* should now be excluded from the genus (Schleifer and Kilpper-Bälz, 1987).

The species which remain within the genus *Streptococcus* following these taxonomic revisions include those which Sherman (1937) described as the "viridans" groups, and for which others have preferred the term "oral streptococci" (Hardie, 1986b; Jones, 1978). The type of hemolysis produced by these streptococci when grown on blood agar varies considerably, both between and within species, so that strains may display either complete (beta), partial (alpha), or no (gamma) hemolysis. In view of this variation, the term viridans would seem to be rather misleading for the group as a whole. Since these streptococci are found predominantly, but not exclusively, in the mouth and upper respiratory tract, the continued use of the term oral streptococci, as suggested in *Bergey's Manual of Systematic Bacteriology* (Hardie, 1986a, 1986b) is perhaps justified and convenient for descriptive purposes, although it has no particular taxonomic validity.

In addition to comprising part of the normal body flora, the oral streptococci are also involved in a number of human diseases, usually as opportunistic pathogens. As described below, members of the "*S. mutans* group" are particularly associated with dental caries, members of the "*S. milleri* group" occur frequently in purulent infections in various parts of the body, and almost all species have been reported as etiological agents in infective endocarditis.

Identification of the oral streptococci to species level has been problematic in the past, because of taxonomic and nomenclatural uncertainties. However, in the light of current knowledge, it should be possible to identify most isolates with a reasonable degree of confidence. Only by proper characterization of these organisms will it be possible to elucidate fully their ecological distribution, transmission, and role in infectious processes.

In this chapter consideration is given to the habitat, distribution, isolation, and taxonomy of the oral streptococci, and to their role in human disease. Although *S. pneumoniae* should be regarded as one of the oral streptococci on the basis of molecular and numerical taxonomic studies, rather than belonging to the pyogenic group (Schleifer and Kilpper-Bälz, 1987), this species is not considered in detail here. This important pathogen is discussed in the chapter on medical aspects of the genus (Chapter 65).

Habitat and Distribution

The oral or viridans streptococci form an important component of the normal microbial flora of the mouth and the upper respiratory tract of humans (Hardie and Marsh, 1978a; Jones, 1978). These sites appear to be their main habitats, although some species can be isolated from other body sites and from feces (Van Houte et al., 1971; Unsworth, 1980) and have also been found in soil (Gledhill and Casida, 1969). Relatively little is known about the distribution of these streptococci in other animals, although several of the species found in humans have also been isolated from the dental plaque of animals (Dent et al., 1978).

Within the oral cavity, a variety of surfaces and ecological niches are available for colonization, and it is well established that different species of streptococci (and other bacteria) preferentially become established at different sites

(Hardie and Bowden, 1974a; Marsh and Martin, 1984). Thus, for example, *S. salivarius* is found in relatively high numbers on the dorsal surface of the tongue and in saliva but not in dental plaque, whereas *S. sanguis* preferentially colonizes the tooth surface, and *S. vestibularis* appears to favor the vestibular mucosa.

The naturally occurring proportions of different species on particular types of surface in the mouth, including the nonshedding, hard surfaces of the teeth and the various types of mucous membrane, have been shown to correlate with their experimentally observed adherence to such surfaces (Gibbons and Van Houte, 1975; Gibbons, 1984). The mechanisms by which these bacteria adhere to surfaces have been the subject of much study, and several symposia have been devoted to this subject (Beachey, 1980; Berkeley et al., 1980; ten Cate et al., 1984; Ellwood et al., 1979). It is thought that bacteria have specific surface ligands or adhesins on their surfaces which enable them to bind to complementary host tissue components. The adhesins often possess lectinlike or hydrophobic properties and may be present on filamentous surface appendages like pili or fimbriae (Gibbons, 1984; Mergenhagen et al., 1987; Weerkamp et al., 1984). The relationship of surface structures to adhesion, coaggregation with other bacteria, and hydrophobicity have been studied in detail in some species, including *S. salivarius* (Handley et al., 1984, 1987; Weerkamp and Jacobs, 1982) and *S. sanguis* (Handley et al., 1985), and the ultrastructural characteristics of the complex of different surface appendages defined. Growth of the strain called *S. salivarius* (HB) and of four adhesion-deficient mutants under different conditions in a chemostat indicated that the ability to adhere to buccal epithelial cells and to coaggregate with *Veillonella parvula* did not vary with growth rate, whereas cell surface thickness and hydrophobicity and cell surface proteins were phenotypically variable characteristics (Harty and Handley, 1989).

Both sucrose-independent and sucrose-dependent mechanisms are involved in the adherence of *S. mutans* to tooth surfaces (Gibbons, 1984; Koga et al., 1986). Extracellular glucan synthesis is not essential for initial attachment, as once believed, but is important in subsequent retention and accumulation of plaque deposits. As described below, mutant strains which lack the enzymes responsible for production of these polymers have reduced caries-inducing potential in experimental animals (Koga et al., 1986; Loesche, 1986).

Isolation and Cultivation

Nutritional Requirements

The nutritional requirements of oral streptococci, as with all facultative anaerobic streptococci, include amino acids, peptides and proteins, a carbohydrate source, fatty acids, vitamins, and purines and pyrimidines, in addition to inorganic ions. These requirements necessitate the use of complex media that often contain meat extract. In addition, an elevated CO_2 level (typically 5%) during incubation is essential for the growth of several species, including *S. mutans,* strains of the *S. milleri* group, and *S. pneumoniae*. Strains referred to as nutritionally variant streptococci (NVS) also require the addition to culture media of approximately 0.001% pyridoxal HCL or pyridoxamine diHCl for growth.

Sampling

Oral streptococci may be isolated from almost any type of clinical specimen, including blood cultures, pus, wound or surface swabs, body fluids, and biopsies. When taking material from the mouth and upper respiratory tract, soft tissue surfaces can be sampled with a cotton swab and whole saliva, possibly stimulated by chewing a piece of sterile paraffin wax, by dribbling into a sterile bottle or tube. As an alternative to collecting saliva, particularly in infants, quantitatively similar results can be obtained by taking a sample from the dorsal surface of the tongue with a standard disposable plastic loop (Beighton, 1986).

For collection of samples of dental plaque, it is usually necessary to use a rigid instrument of some kind in order to scrape the adherent material from the tooth surface, although dental floss can be used for interproximal areas between the teeth. A variety of techniques and instruments have been used by different investigators, including dental probes, scalers, curettes, abrasive strips, hyperdermic needles, wires, wood sticks, and paper points (Hardie and Bowden, 1976b).

Transport Medium

If immediate laboratory processing of a sample is not possible, a suitable transport medium is required. The reduced transport fluid (RTF) of Syed and Loesche (1972) is often used and is suitable for holding a clinical sample or streptococcal population at room temperature.

Reduced Transport Fluid (RTF) (in ml/100 ml RTF)

Na$_2$CO$_3$ (8% solution; filter-sterilized)	0.5
0.1M Ethylenediamine tetraacetate (EDTA) 1%	1.0
DL-Dithiothreitol (Clelands reagent) (filter-sterilized)	2.0
Solution a: 0.6% K$_2$HPO$_4$	7.5
Solution b: 0.6% KH$_2$PO$_4$, 1.2% NaCl, 1.2% (NH$_4$)$_2$SO$_4$, 0.24% MgSO$_4$	7.5

Routine Cultivation

Oral streptococci can be maintained in the Laboratory by culture on a variety of blood-containing agar media, such as oxoid blood agar no. 2 (Oxoid Ltd., Hants, U.K.) with 5% horse blood. For liquid culture, commercially available Todd-Hewitt broth (Oxoid Ltd., Hants, U.K.), brain-heart infusion broth (Difco Laboratories, Michigan, USA), and other formulations may be used. The following broth medium (strep. base plus 0.5% glucose) has been used for many years in the author's laboratory (Hardie and Bowden, 1974b; Whiley and Hardie, 1989):

Strep Base (in g/l)

Proteose peptone (Oxoid)	20
Yeast extract (Difco)	5
NaCl	5
Na$_2$HPO$_4$	1
Glucose	5

Dissolve ingredients in distilled water; adjust pH to 7.6. Autoclave at 121°C for 15 min.

Isolation of Extracellular Polysaccharide-Producing Streptococci

Two of the most commonly used selective isolation media for oral streptococci contain sucrose, and these allow some species to produce characteristic colonies as a result of extracellular polysaccharide formation from this substrate. The recipes for these agar media (TYC and MS) are given below.

Trypticase-Yeast Extract-Cystine (TYC) 5% Sucrose Agar (De Stoppelaar et al., 1967) (in g/l)

Trypticase (BBL)	15.0
Yeast extract (Difco)	5.0
L-Cystine	0.2
Na$_2$SO$_3$	0.1
NaCl	1.0
NaHCO$_3$	0.1
NaCl	1.0
Na$_2$HPO$_4$·12H$_2$O	2.0
NaHCO$_3$	2.0
Sodium acetate (3H$_2$O)	20.0
Sucrose	50.0
Agar	12.0

Adjust pH to 7.3. Autoclave at 121°C for 15 min.

Commercial TYC is marketed by Lab M (London Analytical and Bacteriological Media Ltd., Lanc., U.K.), but substituting trypticase (BBL) and yeast extract (Difco) with Lab M tryptone and yeast extract, respectively.

Colonial Morphology on TYC Agar

Several species of oral streptococci give rise to characteristic colonial morphologies on TYC agar that may be useful as an aid to identification (Hardie and Marsh, 1978b). However, it must be emphasized that the presumptive identification of a strain should always be backed up by additional biochemical criteria.

Descriptions of the colonial morphology of some oral streptococci on TYC agar are given below:

S. mutans: Rough, heaped, irregular colonies, resembling frosted glass. Mostly crumbly, although whole colonies can be picked off the agar. White, gray or yellow in color, 0.5–2 mm in diameter. May produce a drop of liquid (water-soluble glucan) on top of the colony or a puddle of polysaccharide around the colony.

S. sobrinus: Rough, irregular colonies, less heaped than *S. mutans*. White, 0.5–mm in diameter. Produce a white "halo" or milky zone in the agar surrounding the colony.

S. sanguis: Smooth or rough, hard and rubbery colonies. Colonies adhere strongly to the agar making them difficult to remove with a loop. Gray, white, or colorless, 1–3 mm in diameter. Some strains do not produce extracellular polysaccharide.

S. oralis: Two types produced: 1) a hard type similar to *S. sanguis* and 2) a soft, smooth, nonadherent type. Gray, white, or colorless, 0.5–2.0 mm in diameter. Polysaccharide production is a variable characteristic of this species.

S. anginosus and related taxa: Rough, dry, crumbly, or smooth and soft colonies. White or gray, 1–2 mm in diameter. No polysaccharide produced.

S. vestibularis: Smooth, soft, white colonies, 1–2 mm in diameter. No extracellular polysaccharide is produced.

S. gordonii: Colonies of polysaccharide-producing strains resemble those of *S. sanguis*. Non-polysaccharide producing strains are also encountered.

S. mitis and *"S. parasanguis"*: These taxa consist of non-polysaccharide-producing strains. Colonial morphology is similar to soft *S. oralis* strains, with no particular distinctive features.

S. salivarius: Colonies domed, smooth, and mucoid. Older colonies may become hard, pitting the agar. White or gray, 2–6 mm diameter.

MITIS SALIVARIUS (MS) AGAR. This widely used agar is for the selection of oral streptococci and contains 5% sucrose, trypan blue, crystal violet, and potassium tellurite as selective agents. Commercial sources include BBL and Difco. One recipe (Difco) is detailed below:

Mitis Salivarius (MS) Agar (in g/l)

Bacto tryptose (Difco)	10
Proteose peptone no. 3 (Difco)	5
Proteose peptone (Difco)	5
Bacto dextrose (Difco)	1
Bacto saccharose (Difco)	50
K_2HPO_4	4
Trypan blue	0.075
Bacto crystal violet (Difco)	0.0008
Bacto agar (Difco)	15

Dissolve by heating in deionized water to boiling. Sterilize at 121°C for 15 minutes. Cool to 50–55°C. Add 1 ml of 3.5% potassium tellurite per liter of medium.

When grown on MS agar, strains of *S. mutans* produce rough colonies that often look like frosted glass in appearance. *S. salivarius* produces either large (2–5 mm diameter), mucoid, smooth colonies resembling "gum drops" or rough, irregular colonies; *S. sanguis* can form hard, rubbery, adherent colonies (called zooglea) of less than 2 mm in diameter. *S. vestibularis* strains grown anaerobically on MSA produce 2–3 mm in diameter, matte, colonies with undulate edges.

SELECTIVE MEDIA FOR *S. MUTANS*. Most media that have been developed for selection of mutans streptococci have been based on either MS agar or TYC agar, but with increased amounts of sucrose and the addition of bacitracin.

Two frequently used examples (one MSA-based and the other TYC-based) are:

Mitis salivarius sucrose bacitracin (MSB) (Gold et al., 1973): MSA plus 15% sucrose plus 0.2 units/ml bacitracin.

Trypticase, yeast extract, cystine (TYC) agar plus 15% sucrose plus 0.1 unit/ml bacitracin (= TYCSB) (Van Palenstein Helderman et al., 1983). A variation has been described (Wade et al., 1986) on which optimum recovery/growth of both *S. mutans* (strain National Collection of Type Cultures (NCTC) 10449) and *S. sobrinus* (strain 6715) was achieved. This medium consisted of TYC agar plus 15% (w/v) additional sucrose and 0.2 units of bacitracin per ml (compared to 0.1 unit per ml in the original TYCSB formulation).

Although TYCSB was developed in response to the shortcomings of MS-based media (Van Palenstein Helderman et al., 1983), the increase recovery of *S. mutans* initially reported has not always been verified in other studies (Beighton et al., 1989).

Other selective media for *S. mutans* that have been described include: MS agar plus a final concentration of 40% sucrose (MS40S) (Ikeda and Sandham, 1972); mannitol-sorbitol-fuchsin-azide agar (MSFA) (Linke 1977); and glucose-sucrose-tellurite-bacitracin agar (GSTB) (Tanzer et al., 1983).

Media for Chromophore Detection in Nutritionally Variant Streptococci (NVS)

As stated previously, the NVS can be grown using routine complex media supplemented with pyridoxal HCl or pyridoxamine diHCl (0.001%).

Complex Medium for NVS

A liquid medium has been developed which increases the detection rate of the red chromophore that is used as a diagnostic marker for NVS strains (Stein and Libertin, 1989a). The medium is called THBP + YE and consists of: Todd Hewitt broth (Difco) plus 0.8% (w/v) yeast extract (Difco) plus 10 g/ml pyridoxal HCl (Sigma).

These authors also reported the use of a semidefined medium (CDMT) based on the defined medium of Van de Rijn and Kessler (1980).

CHROMOPHORE ASSAY. For detection of the red chromophose the following procedure is used (Bouvet et al., 1981; Stein and Libertin, 1989a): strains are grown overnight in 30 ml of THBP + YE medium. Cells are harvested by centrifugation at 3,000 ×g for 15 min and resuspended in a volume of 2N HCl equal to the pellet and heated at 100°C. The appearance of a pink color within 5 min is taken as a positive result.

Classification and Identification

The classification and identification of the oral or viridans streptococci has long been regarded as a difficult area of streptococcal taxonomy (Hardie and Marsh, 1978a). Early attempts at classification on the basis of hemolysis and the possession of group-specific polysaccharide (Lancefield) antigens, which proved so useful with some streptococci (Lancefield, 1933), were unsuccessful (Lancefield, 1925a, 1925b). Subsequent efforts to classify the viridans streptococci by serological methods often resulted in the formation of groups exhibiting considerable biochemical heterogeneity and did not produce practically useful schemes (Ball, 1985; Selbie et al., 1949; Solowey, 1942; Williamson, 1964).

The use of biochemical and physiological tests in the division of the oral streptococci has proved more successful. From the earliest applications of this approach at the turn of the century (Andrews and Horder, 1906; Gordon, 1905) to the more recent numerical taxonomic studies (Bridge and Sneath, 1983; Carlsson, 1968; Colman, 1968; Hardie et al., 1982), the phenotypic characterization of strains has formed the basis of the majority of taxonomic studies of these streptococci (for reviews of the literature see: Coykendall, 1989a; Hardie and Bowden, 1976a; Hardie and Marsh, 1978a). Significant advances were achieved by combining data from cell wall analysis (Colman and Williams, 1965), genetic transformation experiments (Colman, 1969), and a large numerical taxonomic investigation (Colman, 1968). From these studies, six distinct species were recognized among the viridans streptococci: *Streptococcus salivarius, S. mitior,* (now called *S. oralis*) *S. milleri, S. sanguis, S. mutans* and *S. pneumoniae* (Colman, 1976; Colman and Williams, 1972).

A large number of publications appeared during the 1960s and 1970s on the taxonomy of the oral streptococci, but there were some disagreements between the proposed classification schemes (e. g., Carlson, 1968; Colman and Williams, 1972—see Hardie and Bowden, 1976a) and the nomenclatural systems used (Colman and Williams, 1972; Facklam, 1977). The increased use of genotypic criteria which followed the early transformation experiments of Colman (1969) included GC content determinations and DNA-DNA hybridization experiments. Initially, these studies were directed toward *Streptococcus mutans* (Coykendall, 1970, 1971, 1977), but more recently these techniques have also been used to clarify the taxonomy of the majority of the oral streptococci and thus provide a more reliable classification on which identification schemes can be based. Organisms for which these genetic approaches have been used include *S. mutans* and "mutans-like" streptococci (Beighton et al., 1984; Coykendall, 1977, 1983; Coykendall et al., 1976; Schleifer et al., 1984; Whiley et al., 1988), *S. salivarius* and close relatives (Coykendall and Gustafson, 1985; Farrow and Collins, 1984a; Kilpper-Bälz et al., 1982; Whiley and Hardie, 1988), the *S. milleri* group (Coykendall et al., 1987; Ezaki et al., 1986; Farrow and Collins, 1984b; Kilpper-Bälz et al., 1984; Knight and Shlaes, 1988; Welborn et al., 1983; Whiley and Hardie, 1989), and *S. sanguis, S. oralis (S. mitior), S. mitis,* and closely related taxa (Coykendall and Munzenmaier, 1978; Coykendall and Specht, 1975; Coykendall and Wesbecher,

1990; Kilpper-Bälz et al., 1985; Schmidhuber et al., 1987; Welborn et al., 1983; Whiley and Hardie, 1990).

Other chemotaxonomic and molecular techniques that have been applied to the study of oral streptococci include cell wall (peptidoglycan) analyses (Schleifer and Kandler, 1972), multilocus enzyme electrophoresis (Gilmour et al., 1987), DNA-rRNA hybridization, and 16S rRNA sequence analyses. (Ludwig et al., 1985). These approaches have shed light on the natural relationships between streptococcal species and, as mentioned earlier, have also led to the division of the original genus *Streptococcus* into three genera: *Streptococcus, Enterococcus,* and *Lactococcus* (Schleifer and Kilpper-Bälz, 1984; Schleifer et al., 1985). Phylogenetic relationships between these and other Gram-positive bacteria can also be determined by such approaches (Ludwig et al., 1985).

At the time of writing, there are 19 recognized and named species of oral streptococci, although this number is likely to increase as new species are described. Data from rRNA cataloging and nucleic acid hybridization studies indicate that these can be divided into three main species groups: the *S. oralis* group, *S. mutans* group, and *S. salivarius* group (Schleifer and Kilpper-Bälz, 1987), and that these groups can be further divided into "subgroups" of closely related species (Table 1).

A brief outline of each of the recognized species of oral streptococci is given below, together with identification tables.

Streptococcus mutans Group

Although the species *Streptococcus mutans* was not included in the eighth edition of *Bergey's Manual of Determinative Bacteriology* (Buchanan and Gibbons, 1974), subsequent rapid taxonomic developments have established a group that now consists of seven distinct species: *S. mutans, S. sobrinus, S. cricetus, S. rattus, S. ferus, S. macacae* and *S. downei.*

The streptococci originally designated *S. mutans* were isolated from carious human teeth by Clarke (1924). Despite Clarke's observations and the reported isolation of *S. mutans* from bacterial endocarditis (Abercrombie and Scott, 1928), little attention was directed towards this species until the early 1960s. The demonstration that caries could be experimentally induced and transmitted in animals (Fitzgerald and Keyes, 1960; Keyes, 1960) and that similar, "caries-inducing" streptococci were present in human plaque (Krasse, 1966; Zinner et al., 1965), stimulated renewed interest in strains resembling *S. mutans.* The biochemical charac-

Table 1. Species groups and subgroups within the oral streptococci.

Main group[a]	Subgroup	Species	GC content (mol%)	Peptidoglycan type
S. mutans	A	S. mutans	36–38	Lys-Ala$_{2-3}$
		S. rattus	41–43	Lys-Ala$_{2-3}$
	B	S. sobrinus	44–46	Lys-Thr-Ala
		S. downei	41–42	Lys-Thr-Ala[b]
		S. cricetus	42–44	Lys-Thr-Ala
	Ungrouped	S. ferus	43–45	Lys-Ala$_{2-3}$
	Not known	S. macacae	35–36	ND
S. oralis	A	S. oralis	38–42	Lys-direct
		S. mitis	40–41	Lys-direct
		(S. pneumoniae)	36–37	Lys-Ala$_2$(Ser)
	B	S. sanguis	40–46	Lys-Ala$_{1-3}$
		S. gordonii	38–42	Lys-Ala$_{1-3}$
		S. parasanguis	41–43	ND
		"Tufted fibril group"	ND	ND
	S. milleri group	S. anginosus	37–39.5	Lys-Ala$_{1-3}$
		S. constellatus	36–38.5	Lys-Ala$_{1-3}$
		S. intermedius	37–38.5	Lys-Ala$_{1-3}$
S. salivarius		S. salivarius	37–40	Lys-Thr-Gly
		S. vestibularis	38–41	Lys-Ala$_{1-3}$[b]
		S. thermophilus	37–40	Lys-Ala$_{2-3}$
Nutritionally variant streptococci (NVS)	Not known	S. adjacens	36–37	ND
	Not known	S. defectivus	46–47	ND

[a]Based on Schleifer and Kilpper-Bälz (1987).
[b]Data of K. H. Schleifer (personal communication).
Based on data from: Beighton et al. (1984); Bouvet et al. (1989); Coykendall (1974, 1976, 1977, 1983); Coykendall and Gustafson (1985); Coykendall and Munzenmaier (1978); Coykendall and Specht (1975); Coykendall and Wesbecher (1990); Coykendall et al. (1987); Ezaki et al. (1986); Farrow and Collins (1984a, 1984b); Kilian et al. (1989a); Kilpper-Bälz et al. (1982, 1984, 1985); Knight and Shlaes (1988); Ludwig et al. (1985); Schleifer and Kilpper-Bälz (1987); Schleifer et al. (1984); Schmidhuber et al. (1987); Welborn et al. (1983); Whiley and Hardie (1988); Whiley et al. (1988); Whiley and Hardie (1989); Whiley et al. (1990a).

teristics of this "species" enabled several authors to recognize clusters of strains to which it corresponded (Carlsson, 1968; Colman and Williams, 1972; Drucker and Melville, 1971). However, in contrast to the initially perceived phenotypic homogeneity of these strains, subsequent studies showed considerable heterogeneity among the mutans-like streptococci. Eight serovars, designated a through h, have been demonstrated (Bratthall, 1970; Perch et al., 1974; Beighton et al., 1981), and variation on the basis of biochemical biotypes, cell wall studies (Hardie and Bowden, 1974b), electrophoretic separation of membrane proteins, comparative enzymes studies, intracellular proteins, and whole-cell-derived proteins has been fully documented (see Hamada and Slade, 1980, and Hamada et al., 1986, for reviews of the literature).

The division of the S. mutans strains into five distinct species designated S. mutans, S. sobrinus, S. rattus, S. cricetus, and S. ferus resulted from a series of genotypically based studies (Coykendall, 1974, 1977, 1983; Coykendall et al., 1974, 1976). More recent studies have in-

creased the number of mutans-like species to seven, with the addition of S. macacae (Beighton et al., 1984) and S. downei (Whiley et al., 1988). The taxonomic position of one of the species placed within the S. mutans group, S. ferus, is presently less certain because this species is included only on the basis of DNA homology (Schleifer et al., 1984) and appears to be more closely related to the S. oralis group, as measured by multilocus enzyme electrophoresis (Gilmour et al., 1987). The characteristics of the currently recognized species within the group are shown in Table 2.

Streptococcus salivarius

The name Streptococcus salvarius was originally given by Andrewes and Horder (1906) to a relatively easily recognized streptococcus that was common in human saliva, present in the intestine, and also isolated occasionally from patients with endocarditis, terminal septicemia, and peritonitis. These streptococci characteristically produced short chains in broth, caused clotting of milk, reduced neutral red, produced

Table 2. Biochemical properties of species within the *Streptococcus mutans* group.

Characteristic	S. mutans	S. sobrinus	S. cricetus	S. rattus	S. macacae	S. downei	S. ferus
Acid produced from:							
Mannitol	+	+	+	+	+	+	+
Sorbitol	+	d	+	+	+	−	+
Melibiose	+	d	+	+	−	−	−
Raffinose	+	d	+	+	+	−	−
Inulin	+	−	+	+	−	+	+
Dextrin	−	−	−	−	−	ND	+
Starch	−	−	−	−	−	−	+
Hydrolysis of:							
Arginine	−	−	−	+	−	−	−
Esculin	+	d	d	+	+	−	+
Glycogen	−	ND	ND	ND	−	−	+
Production of:							
Acetoin	+	+	+	+	+	+	+
H_2O_2	−	+	−	−	−	−	−
Resistance to bacitracin	+	+	−	+	−	−	−
Cell wall carbohydrates	Glucose, rhamnose	Glucose, galactose, rhamnose	Glucose, galactose, rhamnose	Galactose, rhamnose	Glucose, rhamnose	Glucose, galactose, rhamnose	Glucose, rhamnose
Serotype	c,e,f,	d,g	a	b	c	h	c

Symbols: +, 90% or more of strains are positive; −, 90% or more of strains are negative; d, 11–89% of strains are positive; ND, not determined.
Data from: Beighton et al. (1981); Bratthall (1970); Coykendall (1970); Hamada and Slade (1980); Hardie (1986b); Kral and Daneo-Moore (1981); Perch et al. (1974); Schleifer et al. (1984); Whiley et al. (1988).

acid from sucrose, lactose, and raffinose (usually), rarely fermented inulin, never fermented mannitol, and were unable to grow on gelatin at 20°C.

Later workers tended to use fewer tests in assigning strains to *S. salivarius,* thus obscuring the distinction between this and the much-more heterogeneous species *S. mitis.* Unfortunately, in addition, many nonhemolytic streptococci were allocated to the poorly defined "species" *S. mitis,* and this has led to some confusion ever since (see below).

Further characterization of *Streptococcus salivarius* on the basis of physiological properties (Sherman et al., 1943), extracellular polysaccharide (fructan/levan) production from sucrose (Niven et al., 1941a), nutritional requirements (Niven and Smiley, 1942), and serology (Farmer, 1953; Montague and Knox, 1968; Sherman et al., 1943; Williams, 1956) enabled this species to be identified relatively easily in subsequent taxonomic studies (Bridge and Sneath, 1983; Carlsson, 1968; Colman and Williams, 1972; Facklam, 1977; Hardie et al., 1982). Strains of *S. salivarius* are typically nonhemolytic, although some beta hemolytic strains have been described (Saunders and Ball, 1980), produce acid from inulin, lactose, raffinose, salicin, and trehalose but not from mannitol, sorbitol, or melibiose, can hydrolyze esculin but not arginine, produce acetoin, and

frequently produce urease. Most strains produce levan as an extracellular polysaccharide from sucrose, and they may react with Lancefield group K antiserum.

DNA-DNA hybridization experiments have demonstrated a close relationship between strains of *S. salivarius* and *S. thermophilus* (Kilpper-Bälz et al., 1982), and it was later proposed that these be redesignated *S. salivarius* subspecies *salivarius* and *S. salivarius* subspecies *thermophilus* respectively (Farrow and Collins, 1984a). However, more extensive hybridization studies have shown that the actual level of DNA homology shared between these two, albeit closely related, species and also with the newly described *Streptococcus vestibularis* does warrant separate species status for these organisms (Schleifer and Kilpper-Bälz, 1987; Whiley and Hardie, 1988).

The different phenotypic characteristics of *S. salivarius* and related species are shown on Table 3.

Streptococcus vestibularis

Streptococcus vestibularis (Whiley and Hardie, 1988) is the name given to a group of alpha-hemolytic streptococci that had been isolated mainly from the vestibular mucosa of the human mouth and that included several unidentified strains from a earlier numerical taxo-

Table 3. Differential characteristics of *S. salivarius* and closely related species.

	S. salivarius	*S. vestibularis*	*S. thermophilus*
Acid from:			
N-Acetylglucosamine	+	+	−
Amygdalin	+	d	−
Cellobiose	+	d	−
Inulin	d	−	−
Maltose	+	+	−
Melezitose	−	−	d
Melibiose	−	−	d
Ribose	−	−	d
Trehalose	d	d	−
Hydrolysis of Esculin	+	+	−
Production of:			
Acetoin (Voges-Proskauer)	d	d	+
PAL[a]	d	−[b]	−
ecp[a]	+	−	−
α-Galactosidase	d	−[b]	−
β-Galactosidase	d	+[b]	+
H_2O_2	−	+	−
Urease	d	+	−
Growth at 45°C	−	−	+
Growth in 0.0004% crystal violet	+	+	−

Symbols: see Table 2.

[a]ecp: extracellular polysaccharide; PAL, alkaline phosphatase.

[b]From R. A. Whiley (unpublished observations).

Data from: Bridge and Sneath (1983); Carlsson (1968); Colman and Ball (1984); Farrow and Collins (1984); Hardie (1986b); Whiley and Hardie (1988).

nomic study (Carlsson, 1968). These strains (designated group IV) were distinguished by being able to produce acid from lactose, salicin, and cellobiose but not mannitol, sorbitol, inulin, or raffinose, and hydrolyzed starch and esculin but not arginine, produced hydrogen peroxide, urease and, usually, acetoin, but were unable to produce extracellular polysaccharide from sucrose.

Chemotaxonomic data indicated a close relationship with *S. salivarius* from the presence of eicosenoic ($C_{20:1}$) acids by capillary gas-liquid chromatography and from whole-cell-derived polypeptide patterns by SDS-PAGE. DNA-DNA hybridization studies confirmed that *S. vestibularis* strains represented one of a group of three closely related species that also included *S. salivarius* and *S. thermophilus*.

The distinguishing characteristics of these three species are shown in Table 3.

Streptococcus anginosus (S. milleri Group)

The background of the taxonomy and nomenclature of the streptococci presently classified as *S. anginosus* (Andrewes and Horder, 1906; Coykendall et al., 1987) is confused. Much of our current understanding of these streptococci rests on the work of Colman and Williams (1972) who recognized the overall similarity among the strains variously referred to in the literature as *"Streptococcus MG"* (Mirick et al., 1944), the hemolytic and non-hemolytic streptococci of Lancefield group F (Ottens and Winkler, 1962), the minute-colony-forming streptococci of Lancefield groups F and G (Bliss, 1937; Long and Bliss, 1934), and those streptococci given the name *S. milleri* by Guthof (1956). Guthof had first proposed the name *Streptococcus milleri* for a group of non-hemolytic streptococci from purulent oral infections which were able to hydrolyze esculin and arginine, were able to grow on 40% bile agar and at 45°, but could not ferment mannitol or sorbitol.

Facklam (1977) also noted a close similarity among several species previously described by different authors as *S. anginosus* (Andrews and Horder, 1906; Deibel and Seeley, 1974), *S. constellatus* (Holdeman and Moore, 1974), *S. intermedius* (Holdeman and Moore, 1974), and *Streptococcus MG* (Mirick et al., 1944). Rather than include all these together within *S. milleri* as Colman and Williams had suggested, Facklam divided them into two groups, designated *S. MG-intermedius* (lactose fermenters) and *S. anginosus-constellatus* (lactose nonfermenters).

In practice, most British and other European microbiologists tended to include all such strains within *S. milleri* according to the Colman and Williams approach while many in the USA followed Facklam's system. In an attempt to reduce the confusion caused by the use of two conflicting nomenclatures, Facklam (1984) suggested that nonhemolytic strains should continue to be divided on the basis of lactose fermentation into *S. constellatus* (lactose nonfermenters) and *S. intermedius* (lactose fermenters) with beta-hemolytic strains (in possession of a Lancefield group antigen A, C, F, or G, or remaining ungroupable), being identified as *S. anginosus*.

In addition to nomenclatural problems, results from taxonomic studies on the *S. milleri* group have been contradictory. While several investigations have indicated that these strains are closely related, sharing a high degree of phenotypic similarity (Coykendall et al., 1987; French et al., 1989; Labbe et al., 1985; Lütticken et al., 1978; Mejaré, 1975; Mejaré and Edwardsson, 1975), others have demonstrated heterogeneity on the basis of long-chain fatty acid analyses (Cookson et al., 1989; Drucker and Lee, 1981), serological data (Colman and Williams, 1972; Lütticken et al., 1978; Yakushiji et al., 1988), GC content (Drucker and Lee, 1983), and fermentation patterns (Drucker and Lee, 1981; Poole and Wilson, 1979; Ruoff and Kunz, 1982).

Conflicting data have also been obtained from the several published DNA-DNA hybridization studies. Some have concluded that strains within the *Streptococcus milleri* group should be regarded as belonging to a single species (Coykendall et al., 1987; Ezaki et al., 1986; Farrow and Collins, 1984b; Welborn et al., 1983;) whereas others have detected several centers of taxonomic variation among these streptococci (Kilpper-Bälz et al., 1984; Knight and Shlaes, 1988; Whiley and Hardie, 1989). Some of these differences may have been due to the respective stringencies of the methods employed, as suggested by several of the research groups working in this area (Coykendall et al., 1987; Knight and Shlaes, 1988; Whiley and Hardie, 1989). In addition, in some cases, the selection of strains for study on the basis of the type of hemolysis produced on blood agar and on lactose fermentation may have produced data that were unrepresentative of the milleri group as a whole (Ezaki et al., 1986; Knight and Shlaes, 1988). Studies using DNA-DNA hybridization (S1-nuclease method), whole-cell-derived polypeptide patterns by SDS-PAGE, and phenotypic tests have confirmed the presence of three distinct, albeit closely related, taxa that include the reference strains of *S. constellatus* (National Collection of Dairy Organisms [NCDO] 2226), *S.*

intermedius (NCDO 2227), and *S. anginosus* (NCTC 10713), respectively (Whiley and Hardie, 1989; Whiley et al., 1990b).

Differential characteristics of the three species within the *S. milleri* group are listed in Table 4.

Streptococcus sanguis

The taxonomic history of *S. sanguis* provides an excellent example of the progress that has resulted from a shift away from a predominantly serological approach to the use of biochemical and physiological data, cell wall studies, and finally to the application of genotypic criteria.

Streptococcus sanguis was the name given by White and Niven (1946) to the alpha-hemolytic, dextran-forming streptococci isolated from the blood and heart vegetations of patients with bacterial endocarditis. It had previously been referred to as *Streptococcus* s.b.e. (Loewe et al., 1946). The early literature was largely concerned with the serological/antigenic analyses of strains variously called *Streptococcus* s.b.e., *S. sanguis*, and the Lancefield group H streptococci. The latter strains were originally isolated from human throats and described by Hare (1935), but unfortunately the strain used to raise the original group H antiserum was not recorded. Later investigators reporting the presence of "H" antigen in some strains of *S. sanguis* used different strains to raise their group H antisera (Dodd, 1949; Porterfield, 1950; Farmer, 1954) with the result that the identity of the group H antigen and the taxonomic position of these strains remained the subject of much confusion (Cole et al., 1976; Hardie and Bowden, 1976c). Further serological analysis of strains identified as *S. sanguis* led to the description of antigens a–e, of which antigen a, characterized as a glycerol teichoic acid, was considered to be the group H antigen (Rosan 1973, 1976; Rosan and Argenbright, 1982). However, a recent report has shown that several of the strains examined in Rosan's studies fall into the newly described species *S. gordonii* (Kilian et al., 1989a). Thus it appears that there is a considerable degree of overlap of these antigens, including the group H antigen, among different taxa.

Although the name *S. sanguis* had originally been used to describe dextran-producing strains, Colman and Williams (1972) considered that both dextran-positive and dextran negative strains should be included within this species. *S. sanguis* was characterized as being: usually alpha-hemolytic, producing hydrogen peroxide, able to hydrolyze arginine and escu-

Table 4. Differential characteristics of strains of *S. anginosus, S. constellatus,* and *S. intermedius* (*S. milleri* group).

	S. anginosus	*S. constellatus*	*S. intermedius*
Acid from:			
Amygdalin	+	−	d
Lactose	+	d	+
Production of:			
β-N-Acetylglucosaminidase	−	−	+
β-N-Acetylgalactosaminidase	−	−	+
α-Galactosidase	d	−	−
β-Galactosidase	−	−	+
α-Glucosidase	d	+	+
β-Glucosidase	+	−	d
β-D-Fucosidase[a]	−	−	+
Sialidase	−	−	+
Hyaluronidase	−	+	+
H_2O_2	d	−	−
Hemolysis	α,β,γ	α,β,γ	α,γ,β
Lancefield groups	−,A,C,F,G	−,F	−,F

Symbols: see Table 2.
[a]From R. A. Whiley (unpublished observations).
Based on data from: Kilpper-Balz et al (1984); Knight and Shlaes (1988); Whiley and Hardie (1989); Whiley et al. (1990b).

lin, able to produce acid from trehalose and salicin, infrequently fermenting raffinose, and unable to produce acid from mannitol, sorbitol, arabinose, and glycerol. These authors excluded those strains previously designated as *S. sanguis* serotype II (Washburn et al., 1946) which lacked rhamnose in their cell walls and failed to hydrolyze arginine or esculin, preferring to give these isolates the name *"S. mitior"* (Schottmüller, 1903).

An alternative nomenclature was proposed by Facklam (1977) who suggested dividing those strains resembling the *S. mitior* of Colman and Williams (1972) into two groups on the basis of raffinose fermentation, thus creating *S. sanguis* biotype II (raffinose fermenters) and *S. mitis* (raffinose non-fermenters). In Facklam's system, strains of *S. sanguis,* as defined by Colman and Williams, were designated *S. sanguis* biotype I. The use of different nomenclatural systems for these streptococci has given rise to considerable confusion in the literature, which persists to this day.

DNA-based studies (Coykendall and Munzenmaier, 1978; Coykendall and Specht, 1975; Coykendall and Wesbecher, 1990; Schmidhuber et al., 1987; Welborn et al., 1983; Whiley and Hardie, 1990) have revealed the existence of four DNA homology groups within strains designated *S. sanguis* (Colman, 1976; Colman and Williams, 1972) or *S. sanguis* biotype I (Facklam, 1977). Such studies have also unequivocally separated these streptococci from strains corresponding to *S. mitior* (Colman, 1976; Colman and Williams, 1972) or *S. sanguis* biotype

II/*S. mitis* (Facklam, 1977). Two of these four genetic groups are represented by the validity published species *S. sanguis* (type strain NCTC 7863 = ATCC 10556) and *S. gordonii* (type strain NCTC 12261) (Kilian et al., 1989a). Within the latter group is included strain NCTC 3165, which was previously given as the type strain of *S. mitis* in the *Approved Lists of Bacterial Names* (Skerman et al., 1980). Phenotypically, this strain does not closely match the original description of *S. mitis* and, as discussed below, an alternative type strain has now been proposed (Coykendall, 1989b; Kilian et al., 1989b).

The third genetic group among the sanguis-like streptococci has recently been described as a new species, for which the name *S. parasanguis* has been proposed (Whiley et al., 1990a). The fourth known group has not yet been fully described and is known at present simply as the "tufted fibril group" (see Table 1) (Handley et al., 1985).

Differential characteristics of *S. sanguis* and related species are shown in Table 5.

Streptococcus gordonii

Streptococcus gordonii is the name given by Kilian et al. (1989a) to the group of strains that closely resemble *S. sanguis* and to which the former type strain of *S. mitis* (NCTC 3165) has been shown to belong in DNA-DNA hybridization studies (Coykendall and Specht, 1975; Schmidhuber et al., 1987; Whiley and Hardie, 1990).

Table 5. Differential characteristics of *S. sanguis, S. oralis,* and closely related taxa.[a]

	S. sanguis	S. gordonii	S. parasanguis	Tufted fibril group	S. oralis	S. mitis
Acid from:						
Amygdalin	−	+	d	−	−	−
Arbutin	+	+	ND	ND	−	−
Inulin	+	+	−	+	−	−
Sorbitol	d	−	−	−	−	d
Hydrolysis of:						
Esculin	d	+	d(wk)[a]	+(wk)[a]	−	−
Arginine	+	+	+	+	−	d
Production of:						
PAL[a]	−	+	d	−	d	d
α-L-Fucosidase	−	+	ND	ND	−	−
β-D-Fucosidase	d	−	+	ND	+	d
β-D-Glucosaminidase	−	+	ND	ND	+	d
Neuraminidase	−	−	ND	ND	+	d
IgA1 protease	+	−	ND	ND	+	−
ecp[a]	+	+	−	d	d	−
Ability to bind salivary α-amylase	−	+	d	+	−	+

Symbols: see Table 2.

[a]ecp, extracellular polysaccharide; PAL, alkaline phosphatase; (wk), weak positive result.

Data from: Coykendall (1989); Coykendall and Munzenmaier (1978); Coykendall and Specht (1975); Coykendall and Wesbecher (1990); Douglas et al (1990); Handley et al. (1985); Kilian et al. (1989); Kilpper-Bälz et al. (1985); Knight and Shlaes (1988); Schmidhuber et al. (1987); Whiley et al. (1990a).

The phenotypic similarity of these strains to *S. sanguis* stems from their ability to hydrolyze arginine and esculin and to produce extracellular polysaccharide from sucrose, but they differ in their ability to ferment amygdalin, and in possessing β-glucosaminidase, β-mannosidase, α-L-fucosidase, and a strong alkaline phosphatase activity (Kilian et al., 1989a). They also exhibit lower GC content (38–39 mol%) compared to *S. sanguis* (40–43 mol%) and are able to bind salivary α-amylase (Douglas et al., 1990).

Although strain NCTC 3165 is now included in the species *S. gordonii*, it is not phenotypically typical since it lacks the ability to produce extracellular polysaccharide but does possess β-galactosidase activity. Not surprisingly, therefore, it has not been suggested as the type strain. Properties of *S. gordonii* are shown in Table 5.

Streptococcus parasanguis

The name *Streptococcus parasanguis* has been proposed for a group of strains isolated from clinical specimens (throats, blood and urine). These alpha hemolytic streptococci have been shown to be more closely related to *S. sanguis* than to other oral streptococci on the basis of DNA hybridization studies (Whiley et al., 1990a). Some of the strains included in this species had fallen into unnamed DNA homology groups in previous studies, namely the "MGH group" (Coykendall and Wesbecher, 1990), *S.*

intermedius DNA homology group III (Knight and Shlaes, 1988) and DNA homology group IV (Whiley and Hardie, 1989).

Biochemical and physiological characteristics useful in differentiating strains of *S. parasanguis* from other oral and viridans streptococci are shown in Table 5.

Tufted Fibril Group

Evidence of another, as yet unnamed, genetic group within strains that resemble *S. sanguis* has been reported on the basis of DNA-DNA hybridization experiments (Coykendall, 1989a). These strains were characterized by the presence of tufts and fibrils on their cell walls (Handley et al., 1985), were able to hydrolyze arginine, gave weak esculin hydrolysis, did not produce acid from raffinose, or produce alkaline phosphatase.

Streptococcus mitis

Andrewes and Horder (1906) used the name *Streptococcus mitis* to describe a saprophytic streptococcus, found mainly in human saliva and feces, that was short chained, grew well at 20°C on gelatin, did not clot milk but often reduced neutral red, and nearly always fermented lactose and sucrose and sometimes also the glucosides salicin and coniferin.

Due to the small number of tests later used to identify strains as *S. mitis* (e.g., Holman, 1916), this species was often poorly differentiated from *S. salivarius*. Sherman (1937) emphasized the unsatisfactory description of *S. mitis* in his review and used the term *S. mitis* group for what had become a "rather ill defined and heterogeneous" collection of strains that were characterized mainly on the basis of negative characters. They were described as producing marked greening (alpha hemolysis), being unable to ferment arabinose, glycerol, inulin, mannitol, sorbitol, and xylose, and unable to hydrolyze sodium hippurate, or produce polysaccharide from sucrose. Most strains fermented salicin, and were sometimes able to ferment raffinose, hydrolyze arginine and esculin, and grow at 45°C, but were seldom able to ferment trehalose or grow on 10% bile agar. The production of ammonia from arginine by some strains was taken to indicate that more than one species was present in this group. Streptococci identified as *S. mitis* according to the description of Sherman et al. (1943) were isolated in a number of studies from the human throat, the oral cavity, and from blood, including cases of subacute bacterial endocarditis (Carlsson, 1967; Farmer, 1953; Guggenheim, 1968; Loewe et al., 1946; Morris, 1954; Niven and White, 1946). Failure to demonstrate a group antigen (Sherman et al., 1943) was followed by studies that showed the *S. mitis* group to be serologically heterogeneous (Farmer 1953; Williamson, 1964). Some studies also reported the production of dextran from strains considered to be *S. mitis* (Guggenheim, 1968; Hehre and Neill, 1946).

Despite the emphasis placed in the eighth edition of *Bergey's Manual of Determinative Bacteriology* (Deibel and Seeley, 1974) on the poor definition of this species, the name *S. mitis* was included in the *Approved Lists of Bacterial Names* (Skerman et al., 1980) with strain NCTC 3165 assigned as type strain. Unfortunately, this strain was inappropriate since it more closely resembled *S. sanguis* (*S. sanguis* biotype I). Streptococcal strains considered by many bacteriologists to belong to *S. mitis* were those otherwise referred to as *S. sanguis* biotype II or *S. mitior* (Cole et al., 1976; Colman and Williams, 1972; Kilian et al., 1986; Rosan, 1973). DNA-DNA hybridization studies (Coykendall and Wesbecher, 1990; Kilpper-Bälz et al., 1985; Schleifer and Kilpper-Bälz, 1987; Schmidhuber et al., 1987; Welborn et al., 1983; Whiley and Hardie, 1990) have clearly demonstrated that strain NCTC 3165 belongs to a distinct taxon for which the name *Streptococcus gordonii* has been proposed (Kilian et al., 1989a), and this species has been described in more detail above.

There have been two recent proposals for the rejection of strain NCTC 3165 as the type species of *S. mitis* (Coykendall, 1989b; Kilian et al., 1989b). Pending a decision on this by the Judicial Commission of the International Committee on Systematic Bacteriology, Kilian et al., (1989b) have proposed a new type strain for *S. mitis*, NS51 (= NCTC 12261). These authors also proposed to assign to the species *S. mitis* those streptococci that share the same peptidoglycan type (lysine direct), lack both glycerol teichoic acid and significant amounts of rhamnose, but which contain a ribitol teichoic acid in their cell walls (Colman and Williams, 1965; 1972; Kilpper-Bälz et al., 1985; Rosan, 1976). These strains sometimes hydrolyze arginine, do not produce extracellular polysaccharide from sucrose, and produce IgA1 protease less frequently than do strains of *S. oralis*. Several of the strains included by Kilian et al., (1989a) in their description of *S. mitis* were previously found to belong to a distinct DNA homology group by Coykendall and Munzenmaier (1978). This group showed between 30 to 35% DNA homology with strains ultimately shown to belong to *S. oralis* (Bridge and Sneath, 1982; Kilpper-Bälz et al., 1985). Some of the distinguishing characteristics of *S. mitis* as now recognized are listed in Table 5.

Streptococcus oralis

Streptococcus oralis was the name given by Bridge and Sneath (1982, 1983) to a group of isolates originally from the human oral cavity (Carlsson, 1967) that included strains corresponding to both *S. sanguis* and *S. mitior* (Colman and Williams, 1972).

In a later study incorporating cell wall analysis, physiological data, and nucleic acid hybridization, those strains which resembled *S. sanguis* were excluded to give an emended description of *S. oralis* (Kilpper-Bälz et al., 1985). Cell walls of the redefined species contained ribitol and choline, but lacked rhamnose, and possessed directly cross-linked peptidoglycan with lysine as the diamino acid. These studies also indicated that *S. oralis* strains were relatively closely related to strains of *S. pneumoniae*.

The valid publication of the description of *S. oralis* (Bridge and Sneath, 1982; Kilpper-Bälz et al., 1985) means that this is now the approved name for strains which, in the past, have been identified as *S. mitior* (Colman and Williams, 1972), *S. sanguis* biotype II, or *S. mitis* (Facklam, 1977).

In addition to the cell wall characteristics described above, strains of *S. oralis* are usually α-hemolytic on blood agar, produce acid from *N*-acetyl glucosamine, lactose, and sucrose, with some strains able to produce acid from arbutin, dextrin, cellobiose, glycerol, glycogen, *meso*-inositol, melezitose, melibiose, raffinose, salicin, and starch. These bacteria are infrequently able to ferment amygdalin, arabinose, erythritol, inulin, mannitol, rhamnose, sorbitol, and sorbose, do not hydrolyse esculin or arginine, but sometimes produce extracellular glucan from sucrose. Some of the characters of *S. oralis* are shown in Table 5.

Nutritionally Variant Streptococci

Nutritionally variant streptococci (NVS) (alternatively referred to in the literature as satelliting, thiol-requiring, vitamin B$_6$-dependent, pyridoxal-dependent, or nutritionally deficient streptococci) were originally recognized by Frenkel and Hirsch (1961) as ungroupable viridans streptococci that grew as satellite colonies around colonies of staphylococci and required sulfhydryl compounds for normal growth. These are fastidious bacteria, which form part of the normal flora of the human throat as well as the urogenital and intestinal tracts, and require the addition of cysteine or one of the active forms of vitamin B$_6$, such as pyridoxal hydrochloride or pyridoxamine dihydrochloride, for growth on complex media (Carey et al., 1975). As discussed later, they are of particular interest to the clinical microbiologist because of their role in infective endocarditis and other conditions such as otitis, abscesses of the brain and pancreas, wounds, pneumonia, osteomyelitis, and cancer (Bouvet et al., 1989).

The taxonomic position of the NVS has remained uncertain until very recently. Initially, these streptococci were considered to be closely related to *S. oralis* (i.e., *S. mitior* (Colman and Williams, 1972), or *S. sanguis* biotype II and *S. mitis* (Facklam, 1977), on the basis of their physiological characteristics (Bouvet et al., 1981; Carey et al., 1975; Cooksey et al., 1979; Roberts et al., 1979; Schiller and Roberts, 1982; Van de Rijn and Bouvet, 1984). This belief was reinforced by the demonstration of a characteristic, pH-dependent, red chromophore in the cell walls of both the NVS and *S. mitis* strains, as well as by the absence of rhamnose and the presence of ribitol teichoic acid in their walls (Bouvet et al., 1981; Van de Rijn and Bouvet, 1984). However, studies on the penicillin-binding protein (PBP) patterns and enzymatic activities distinguished the NVS from strains resembling *S. oralis* and divided the former into two groups: group I contained strains with PBP pattern I and biotype I, while group II contained strains of PBP pattern II, and included biotypes 2 and 3 (Bouvet et al., 1985; Gutmann et al., 1982). Using DNA-DNA hybridization, Bouvet et al. (1989) have shown that these groups within the NVS represent two new species called *Streptococcus defectivus* (group I strains) and *S. adjacens* (group II strains). The biochemical properties of these new species and of *S. oralis* are shown in Table 6.

Identification

Despite the extensive taxonomic and nomenclatural changes that are taking place within the

Table 6. Differential characteristics of NVS (*S. defectivus* and *S. adjacens*) and *S. oralis* (*S. mitior*).

	S. defectivus	*S. adjacens*	*S. oralis*
Acid from:			
Glycogen	−	−	d
Inulin	−	d	−
Lactose	d	−	+
Raffinose	d	−	d
Ribose	−	−	+
Starch	+	−	d
Trehalose	+	−	d
Production of:			
Acetoin (VP)	d	d	−
PAL	−	−	d
α-Galactosidase	+	−	d
β-Galactosidase	+	−	+
β-Glucosidase	−	d	−
Pyrrolidonylarylamidase	+	+	−
ecp	−	−	d
GC content (mol%)	46–46.6	36.6–37.4	37.6–40.5

Symbols: see Table 2.
Data from: Bouvet et al. (1985, 1989); Colman and Ball (1984); Kilian et al. (1989a); Kilpper-Bälz et al. (1985); Schmidhuber et al. (1987).

oral streptococci, there is still a need for reliable phenotypic tests for the routine identification of freshly isolated strains. Several identification schemes have been proposed in the past, with some forming the bases of currently available commercial identification kits (Colman and Ball, 1984; Facklam, 1977; Facklam et al., 1984; Fertally and Facklam, 1987; French et al., 1989; Peterson et al., 1988). These commercial kits are based on classification schemes which now need to be revised in order to accommodate the several new species of oral streptococci that have been described and to take account of the nomenclatural changes that have occurred.

Several authors have highlighted potentially useful tests for an identification scheme for the oral streptococci. These tests have been brought together in Table 7.

Oral Streptococci and Disease

In addition to comprising a significant proportion of the normal microbial flora of the mouth in healthy individuals, oral streptococci are involved in several different types of disease (Hardie and Marsh, 1978a). These include conditions which occur locally in the mouth, such as dental caries, dental abscesses, and possibly periodontal disease, as well as pathological processes in other parts of the body, including infective endocarditis and abscesses in various organs. Pyogenic infections arising in and around the oral cavity, which frequently involve streptococci together with a variety of anaerobes, may spread to other regions by the blood stream, the lymphatics, or by direct extension along fascial planes (Schlossberg, 1987). Such infections may be extremely severe and life-threatening, particularly when they cause obstruction of the airway or spread to the brain.

Dental Caries

Although, theoretically, any acidogenic bacteria in dental plaque may contribute to the demineralization of enamel which is responsible for the initiation of dental caries, most studies in recent years have been focused primarily on *Streptococcus mutans* or "the mutans streptococci." The extensive literature on *S. mutans* since the original description of the species by Clark (1924) has been fully reviewed by others (Hamada and Slade, 1980; Hamada et al., 1986; Loesche, 1986) and no attempt will be made here to cite all the many hundreds of publications that have appeared in recent years.

ANIMAL STUDIES The most direct evidence for the cariogenic potential of different species of streptococci and some other genera has come from experimental infections in germ-free or gnotobiotic animals. Since the early studies (Fitzgerald and Keyes, 1960; Orland et al., 1955), which showed that monoinfections of rats and hamsters could induce caries and that the disease was transmissible from one animal to another (Keyes, 1960), many different bacterial strains have been tested in a variety of animal model systems (Fitzgerald, 1968; Tanzer, 1981).

When cariogenicity was first demonstrated in animals, the taxonomy and nomenclature of the streptococci tested was less well defined than it is now. It is clear that members of the "mutans group," and *S. mutans* in particular, have the ability to induce caries in animals, but several other species have also been shown to have some cariogenic potential, including certain strains of "*S. milleri*" (Drucker and Green, 1978), *S. salivarius* (Drucker et al., 1984a), and *S. oralis* (Willcox et al., 1987). In contrast, strains of *S. sanguis* and *S. mitis* are reported to produce little or no caries in gnotobiotic rats (Drucker and Green, 1978; Drucker et al., 1984b; Fitzgerald, 1968).

Although some differences are evident, due to variations in the types of animals used, bacterial strains tested, and experimental conditions employed, it appears that most strains of *S. mutans* are able to produce high levels of caries, usually affecting different types of tooth surface (e.g., fissures and smooth surfaces). With other species, not all strains are cariogenic, the overall level of caries is generally less than that produced by *S. mutans,* and the lesions may be restricted to particular tooth surfaces. Thus, there is a spectrum of caries-inducing potential among the oral streptococci, with *S. mutans* being most active and other species, such as *S. sanguis* and *S. mitis* displaying negligible cariogenic activity.

HUMAN STUDIES. Many studies have been undertaken in order to establish the relationship between specific oral bacteria and dental caries in various human populations, most of which have concentrated on *S. mutans* and lactobacilli as the putative cariogenic organisms (Loesche, 1986). In addition to examining different groups of subjects of different ages, country of origin, disease status, and other variables, such studies have varied by being either cross-sectional or longitudinal in design. Since dental caries develops in a person over a period of time and is difficult to diagnose in the initial stages, especially in the depths of occlusal fissures and on

Table 7. Some useful differential characteristics of oral streptococci.

	S. sanguis	*S. gordonii*	*S. parasanguis*	*S. oralis*	*S. mitis*	*S. salivarius*	*S. vestibularis*	*S. anginosus*	*S. constellatus*	*S. intermedius*	*S. mutans*	*S. sobrinus*	*S. adjacens*	*S. defectivus*
Acid from:														
Amygdalin	-	+	d	-	-	+	d	+	-	d	d	-	ND	ND
Arbutin	+	+	ND	-	-	+	ND	ND	ND	ND	+	-	ND	ND
Inulin	+	+	-	-	-	d	-	-	d	-	+	-	d	-
Lactose	+	-	+	+	+	d	+	+	-	+	+	+	-	d
Mannitol	-	d	d	+(wk)	-	-	-	d	d	-	+	+	-	-
Melibiose	d	d	d	d	d	d	-	d	d	d	+	d	-	d
Raffinose	d	-	-	-	d	d	-	-	-	d	+	d	-	-
Sorbitol	d		-		d	-	-			-	+	d	-	-
Hydrolysis of:														
Arginine	+	+	+	-	d	-	-	+	+	+	-	-	-	-
Esculin	d	+	d(wk)	-	-	+	+	+	d	d	+	d	-	-
Production of:														
Acetoin (VP)	-	-	-	-	-	d	d	+	+	+	+	+	d	d
PAL[a]	-	+	d	d	d	d	-	+	+	+	-	-	-	-
ecp[a]	+	+	-	d	-	+	-	+	+	+	+	+	-	-
α-L-Fucosidase	-	+	ND	-	d	-	ND	-	ND	-	-	ND	ND	ND
β-D-Fucosidase	d	-	+	+	d	d	ND	-	-	ND	+	ND	ND	ND
α-Galactosidase	d	d	d	d	d	d	-	d	-	+	-	+	-	+
β-Glucosaminidase	-	+	ND	+	+	-	ND	d	-	-	+	ND	ND	ND
H₂O₂	+	+	+	+	ND	-	+	d	+	ND	-	+	ND	ND
Hyaluronidase	-	ND	ND	d	-	-	ND	-	+	d	-	-	ND	ND
IgA₁ protease	+	-	ND	+	d	-	ND	-	ND	+	-	-	ND	ND
Neuraminidase	-	-	ND	+	-	-	ND	-	ND	ND	-	-	ND	ND
Urease	-	-	ND	-	-	d	+	-	ND	ND	-	-	ND	ND
α-Amylase binding	-	+	d	-	+	d	-	-	-	ND	-	-	ND	ND
Lancefield group	H*[b], -	H*, -	F,G,C,B,-	-,K	-,O	-,K	ND	-,A,C,F,G	-,F	-,F	-(E)	-	-	-(H)
Hemolysis	α	α	α	α	α	γ,α,β	α	α,γ,β	β,α,γ	α,γ	α,γ,β	α,γ	α	α
pH-dependent chromophore	-	-	ND	d	ND	-	ND	-	-	-	-	ND	+	+

Symbols: see Table 2.

[a] Abbreviations: ecp, extracellular polysaccharide; PAL, alkaline phosphatase; H*, results vary according to antiserum used.

Data from: Bouvet et al. (1985, 1989); Colman and Williams (1972); Coykendall (1989); Coykendall and Munzenmaier (1978); Coykendall and Specht (1975); Coykendall and Wesbecher (1990); Douglas (1983); Douglas et al. (1990); Handley et al. (1985); Hardie (1986b); Hardie et al. (1984, 1985); Kilian et al. (1989); Kilpper-Bälz et al. (1984, 1985); Knight and Shlaes (1988); Kral and Daneo-Moore (1981); Perch et al. (1974); Schmidhuber et al. (1987); Whiley and Hardie (1988, 1989); Whiley et al. (1988, 1990a).

approximal surfaces between the teeth, the longitudinal approach is preferable when attempting to demonstrate cause and effect relationships.

Notwithstanding problems associated with experimental design and techniques, many studies have reported a strong association between mutans streptococci, particularly *S. mutans,* and human caries, using either dental plaque or saliva as samples (see Krasse, 1988; Loesche, 1986; MacFarlane, 1989, for reviews). A similar, albeit less strong, association has also been recorded between lactobacilli and human caries, but it is not entirely clear whether these bacteria are involved in initiation or subsequent progression of the lesions. Although the weight of the published evidence supports the concept of a key role for *S. mutans* in most types of human caries (e.g., Kristoffersson et al., 1985; Loesche et al., 1984), some longitudinal studies have failed to demonstrate a clear association between this species and the initiation of caries at specific sites (Hardie et al., 1977; Mikkelsen and Poulsen, 1976; Mikkelsen et al., 1981), thus giving some credence to the possibility that other organisms may, on occasions, be responsible. This would be in accord with the evidence from animal experiments and with the suggestion that the flora associated with dental caries may vary at different stages of disease progression (Marsh et al., 1989).

CARIES PREDICTION. In order to identify individuals who are at greatest risk of developing dental caries, and thus allow expensive and time-consuming caries preventive measures to be concentrated on those who need them most, dentists have long been in search of convenient predictive, or "caries-activity," tests (Bibby and Shern, 1978). The salivary lactobacillus count has been used for many years for this purpose, but more recently attention has been given to the predictive value of *S. mutans* levels in saliva, either alone or in combination with lactobacillus counts (Crossner, 1981; Klock and Krasse, 1979; Krasse, 1988).

Several studies, mostly from Scandinavia, have shown that subjects with high salivary *S. mutans* levels, the threshold value being set between 10^5–10^6 colony-forming-units (CFU) per ml of saliva, are more likely to develop new carious lesions than those with low levels (Emilson and Krasse, 1985; Köhler et al., 1981, 1988; Zickert et al., 1983). Early establishment of mutans streptococci in the mouths of young children appears to be particularly significant to their subsequent risk of developing caries (Alaluusua and Renkonen, 1983), and preventive measures directed towards children and their

mothers may prevent or delay such colonization (Köhler et al., 1983; Zickert et al., 1982).

The clinical significance of particular levels of mutans streptococci in saliva varies from population to population. For example, in England (Beighton et al., 1987), Kenya (Beighton et al., 1989), and Sweden (Emilson and Krasse, 1985), high levels of these streptococci are associated with high caries scores, whereas in some other countries such as Sudan (Carlsson et al., 1987; El Tayeb et al., 1985), Mozambique (Carlsson et al., 1985), and Tanzania (Matee et al., 1985), high levels of *S. mutans* are accompanied by apparently low caries experience. Such observations underline the importance of other factors, such as diet, in addition to the presence or absence of particular microorganisms (Beighton et al., 1989). In the industrialized countries, where caries levels in the child population as a whole fell during the 1980s, it is particularly important to be able to identify "high risk" groups or individuals within the community.

It appears from the available information that a combination of *S. mutans* and lactobacillus counts in saliva may give a better prediction of caries risk than either parameter alone (Krasse, 1988; Stecksen-Blicks, 1985). Dip-slide tests for determining the levels of these organisms in saliva are commercially available (Alaluusua et al., 1984; Emilsson and Krasse, 1985; Larmas, 1975) and should help to simplify the use of microbiological techniques for assessment of caries risk.

VIRULENCE FACTORS. The cariogenicity of the mutans streptococci is due to their ability to colonize the tooth surface, forming a constituent of dental plaque, and to produce acid. It has long been known that dietary sucrose plays an important etiological role in dental caries, and a considerable amount of research has been carried out on the metabolism of sucrose by the mutans streptococci (Hamada and Slade, 1980; Loesche, 1986). In addition to producing acid from sucrose, *S. mutans* and *S. sobrinus* also utilize this substrate for the formation of extracellular glucans and fructans. Various structural forms of these polymers are produced which may be cell-associated or released into the surrounding environment, and their formation depends upon the activities of different glucosyl- or fructosyl-transferases. Depending on the specific linkages present, these extracellular polysaccharides may be water soluble or insoluble. One particular water-insoluble α-(1–3)-glucan of *S. sobrinus* (originally referred to as *S. mutans*) is known as mutan (Guggenheim, 1970). In addition to extracellular polysaccharides, *S.*

mutans strains produce an amylopectin-like intracellular polysaccharide which can be metabolized when external sources of carbohydrate are unavailable.

Mutant strains of *S. sobrinus* and *S. mutans* which have a reduced ability to produce extracellular polysaccharides have been shown to have lower caries-inducing potential in experimental animals (de Stoppelaar et al., 1971; Freedman et al., 1981, 1983; Johnson et al. 1977; Michalek et al., 1975; Otake et al., 1978; Tanzer et al., 1974). Similarly, mutants of *S. mutans* with decreased acid production (Mao and Rosen, 1980), decreased aciduricity (de Stoppelaar et al., 1971), or decreased intracellular polysaccharide production (Tanzer et al., 1976) are also less cariogenic in animals. The absence of lactate dehydrogenase (LDH) in mutants of *S. rattus* is also associated with markedly reduced caries-inducing ability (Johnson et al., 1980). More recently, modern genetic analysis and cloning techniques have been applied to the study of virulence determinants of the mutans streptococci (Curtiss et al., 1986; Russell, 1990).

TRANSMISSION OF *S. MUTANS*. Biotyping and serotyping methods have been employed to distinguish between strains of mutans streptococci but, as explained elsewhere, the different types recognized by such approaches are now mostly considered to be distinct species. In order to demonstrate transmission of strains within families or other groups of people, it is necessary to have a more precise typing system. Bacteriocin typing has been used with some success for this purpose, and studies indicate that the mother is usually the primary source from whom infants acquire *S. mutans* (Berkowitz and Jones, 1985; Berkowitz and Jordan, 1975; Rogers, 1977, 1981). More recently, the close similarity between strains of *S. mutans* isolated from mothers and their children has clearly been shown by plasmid-DNA analysis (Caufield et al., 1988) and chromosomal DNA restriction fragment analysis (Caufield and Walker, 1989).

CARIES VACCINE. Since the 1970s, work has been in progress in a number of laboratories to develop an anti-caries vaccine based on *S. mutans*. A variety of preparations have been tested in rodents and primates, including vaccines based on whole cells, crude cell wall preparations, glucosyltransferases, and several purified protein antigens, but, to date, no human clinical trials have been reported (for reviews see Cohen et al., 1983; Curtiss, 1986; Hamada et al., 1986; Klein and Scholler, 1988; Lehner et al., 1981; McGhee and Michalek, 1981; Russell and John-son, 1987; Russell and Mestecky, 1986). Reservations have been expressed about the desirability of such a vaccine, partly because of concern about safety and the possibility of inducing antibodies against heart tissue, (with which *S. mutans* may cross-react) and partly because of the decrease in caries in many industralized countries and the availability of other effective and totally safe caries-preventive measures. However, it is arguable that a caries vaccine might be extremely valuable in countries with an increasing caries problem and limited dental manpower resources.

The mechanisms by which *S. mutans* vaccines exert their caries-preventive effect in animals are not fully understood, and both systemic (IgG) and local (IgA) antibody responses have been considered to be significant in protection. The use of molecular approaches to identify the protective antigens should help to clarify the situation (Curtiss, 1986; Hamada et al., 1986). The gene coding for the 190-kDa surface protein antigen (PAc) of a serotype c *S. mutans* (strain MT8148) has been cloned (Okahasi et al., 1989a), and the complete nucleotide sequence of the gene has been determined (Okahashi et al., 1989b).

Periodontal Diseases

There is very little evidence to suggest that the oral streptococci play a significant role in the various types of periodontal disease. Most of the extensive recent literature on the microbial etiology of periodontal diseases points towards Gram-negative anaerobes *(Porphyromonas* and *Bacteroides)* as the most likely causative agents, together with *Actinobacillus actinomycetemcomitans, Fusobacterium,* and *Treponema* species. However, as pointed out by Gossling (1988) in her review of the pathogenicity of the *S. milleri* group, a few studies have indicated a possible relationship between members of this group of streptococci and periodontal disease. Strains identified as *S. intermedius* were found to be present in higher numbers in diseased sites than in healthy sites in one series of studies (Moore et al., 1984, 1985), and in another (Haffajee et al., 1985), the presence of *S. intermedius* appeared to be correlated with a poor response to periodontal treatment. Although the data are far from conclusive at present, the possible involvement of members of the *S. milleri* group in some forms of periodontal disease cannot be entirely discounted and further investigation would seem to be indicated.

Purulent Infections

Purulent infections in various parts of the body are frequently found to be associated with a mixed bacterial flora, often involving streptococci together with one or more obligate anaerobes. This is particularly evident with dentoalveolar abscesses where up to eight isolates per specimen may be recovered, the mean number usually being between three and four per abscess (Heimdahl et al., 1985; Lewis et al., 1986).

In a large survey of streptococci from systemic infections (Parker and Ball, 1976), the species most often isolated from purulent lesions was *"S. milleri."* In this particular series, it was recovered from brain abscesses, meningitis, pleural empyema, and various intraabdominal abscesses, and accounted for about a third of the streptococci isolated from these cases. However, other streptococci, including *S. pyogenes, S. agalactiae,* and small numbers of other species (including *S. salivarius, S. sanguis, S. mitior, S. faecalis,* and *S. bovis*), were also reported.

The association of the *S. milleri* group with purulent infections has been noted by many workers since their occurrence in dental abscesses was first described by Guthof (1956). The extensive literature on the clinical significance, occurrence, and pathogenicity of these streptococci has recently been reviewed by several authors (Gossling, 1988; Ruoff, 1988; Van der Auwera, 1985), and the list of sites and conditions from which they have been reported is impressive (Table 8). Case reports and reviews of particular types of infection caused by these streptococci continue to appear regularly (e.g., Admon et al., 1987; Chua et al., 1989; Cimolai et al., 1988; Kambal, 1987; Rabe et al., 1988; Ruoff et al., 1985).

The confusion which has surrounded the taxonomy and nomenclature of the *S. milleri* group, as described above, means that some of the earlier epidemiological and ecological data are difficult to interpret since the precise identity of the streptococci isolated is not always clear. For example, where the isolates have been described simply as *S. milleri,* it would be preferable in the light of current knowledge to determine which of the genetic groups (i.e., *S. anginosus, S. constellatus,* or *S. intermedius*) they resemble. Unfortunately, even when these specific epithets have been used, the identifications may not always be reliable if based solely on conventional biochemical and physiological characters of the strains. In other cases, different designations such as "group F streptococci" have been used to describe streptococci which belong to the *S. milleri* group (Libertin et al., 1985; Shlaes et al., 1981).

Infections with the *S. milleri* group are generally thought to be of endogenous origin, since these organisms form part of the commensal

Table 8. Isolation of members of the *S. milleri* group from infections at different sites (excluding bacteremia and infective endocarditis).

Site or clinical condition	Selected references[a]
Dental abscesses	1,2,4,5,9
Maxillary sinusitis	1,2,3
Brain abscess	1,2,3,6,7,9,11
Meningitis	1,2,7,9
Pharyngitis	1,2,15,16
Lung abscess	1,2,3,11
Pleural empyema	1,2,3,7,9,11
Liver abscess	1,2,8,9,11
Intraabdominal abscess	1,2,3,7,9,12,16
Peritonitis	1,2,3
Appendicitis	1,2,12,13,16
Female genital tract	1,2,14,16
Neonatal sepsis	1,2
Spinal epidural abscess	10
Bone and joint infections	1,2,9
Pace-maker infection	3
Vascular graft infection	3
Others (miscellaneous)	1,2,3,9,12,16

[a]References cited:

1. Gossling, 1988	9. Shlaes et al., 1981
2. Ruoff, 1988	10. Ghosh et al., 1988
3. Van der Auwera, 1985	11. Admon et al., 1987
4. Heimdahl et al., 1985	12. Kambal, 1987
5. Lewis et al., 1986	13. Poole and Wilson, 1977
6. De Louvois et al., 1977	14. Rabe et al., 1988
7. Parker and Ball, 1976	15. Cimolai et al., 1988
8. Chua et al., 1989	16. Poole and Wilson, 1976

flora of the mouth and upper respiratory tract, the gastrointestinal tract, and the female urogenital tract. However, the exact site of origin of the infecting organism may not always be obvious, especially when the pathogen has reached the affected organ via the bloodstream. Some patients with debilitating conditions such as malignancies, blood dyscrasias, immunosuppressive therapy, or diabetes mellitus, develop purulent infections with these streptococci (as well as other organisms), although such predisposing conditions are not essential for infections with the *S. milleri* group.

It has been reported that prophylaxis with antimicrobial agents such as metronidazole and aminoglycoside antibiotics prior to abdominal surgery can predispose to suppurative, postoperative infections with the *S. milleri* group (Tresadern et al., 1983), and this may occur following appendectomy in children (Madden and Hart, 1985). Such streptococcal infections may be facilitated by the suppression of other components of the gut flora, especially the anaerobes, by these antimicrobial agents.

PATHOGENICITY OF *S. MILLERI* GROUP. Experimental infections in animals with milleri group streptococci have been reported by several investigators, using either pure or mixed cultures (see Gossling, 1988, for review). Relatively large inocula (10^8–10^{10} cells) have generally been required to produce suppurative infections (Brook and Walker, 1984). Experimental infections in mice with bacteria from dental abscesses indicated that *S. milleri* group strains were less virulent than the anaerobic Gram-negative species tested, although they nevertheless did induce abscess formation (Lewis et al., 1988).

Several possible virulence factors have been described, including hyaluronidase (Kilpper-Bälz et al., 1984), gelatinase and collagenase (Steffen and Hentges, 1981), and DNase and RNase (Marshall and Kaufman, 1981; Pullian et al., 1980). A recent taxonomic study indicated that hyaluronidase activity may be a constant characteristic of strains in the same DNA homology groups as *S. intermedius* and *S. constellatus,* but this enzyme was not found in any of the 12 strains of *S. anginosus* examined (Whiley and Hardie, 1989). An immunosuppressive fraction from a strain of *S. intermedius* which strongly suppresses lymphocyte and fibroblast proliferation, has also been described (Arala-Chaves et al., 1981).

The only potential virulence factor so far clearly shown to be of significance in experimental infections with these streptococci is the possession of a polysaccharide capsule. Strains with >50% of cells encapsulated were found to be capable of inducing abscesses in pure culture, whereas a nonencapsulated strain of *S. intermedius* was ineffective (Brook and Walker, 1985).

It is apparent that further studies are required in order to establish whether all members of the *S. milleri* group are equally pathogenic, and to elucidate further their virulence determinants.

Infective Endocarditis

The oral streptococci have long been recognized as important etiological agents in endocarditis. Bacteremias arising from the mouth, as during tooth extraction and other dental operative procedures, invariably involve oral streptococci, and these may settle on the endocardium in "at risk" patients with preexisting heart valve lesions (Hardie and Marsh, 1978a). However, it should be noted that the oral cavity is not the only source of such transient bacteremias, and, in a significant proportion of cases of endocarditis, the organisms may enter the bloodstream from the gut, genitourinary tract, or other body sites (Bayliss et al., 1983).

In reports of the microorganisms found in infective endocarditis, a decrease in the proportion of cases caused by oral (viridans) streptococci has been noted over the last 20 years or so (Bouvet and Acar, 1984; Roberts et al., 1979). While streptococci and enterococci together still comprise by far the most frequently reported organisms, the proportions of Lancefield group D streptococci (including *Enterococcus* species and *S. bovis*) appear to have increased, together with those of staphylococci and other microorganisms (Table 9). However, the oral streptococci as a whole still account for the largest number of cases of endocarditis and remain an important cause of morbidity and mortality.

The identification of streptococcal species amongst strains isolated from infective endocarditis has been reported by several groups (e.g., Bayliss et al., 1983; Bouvet and Acar, 1984; Facklam, 1977; Horaud and Delbos, 1984; Lowes et al., 1980; Moulsedale et al., 1980; Parker and Ball, 1976). Direct comparisons between studies are complicated by variations in the taxonomic systems employed by different authors, but an attempt is made to summarize the reported bacteriological findings in Table 10, using currently accepted nomenclature. Future studies should include some of the more recently recognized species whose role, if any, in infective endocarditis is not currently known.

As illustrated in Table 10, *S. sanguis* and *S. oralis* (the latter is usually reported as *S. mitis,* "*S. mitior,*" or *S. sanguis* II) have been isolated most often in these series, but significant numbers of *S. mutans* and the "*S. milleri*" group have also been recorded. The exact prevalence of each species is difficult to determine from published reports, since some studies are based on a series of cases of endocarditis at a particular center over a given period while others depend on analyses of isolates sent to a reference laboratory for identification.

The nutritionally variant streptococci have been estimated as causing 5–6% of cases of endocarditis due to viridans streptococci (Roberts et al., 1979). Although not all authors have reported the occurrence of these streptococci in

Table 9. Microorganisms isolated from infective endocarditis.

Bacterial type	Isolation frequency	
	Before 1960	Since 1970
Viridans streptococci	60%	40–50%
Group D streptococci	8–10%	10–20%
Staphylococcus aureus	<10%	10–15%
Staphylococcus epidermidis	<5%	5–10%
Gram-negative bacilli	<5%	6–8%
Other microorganisms	1%	2–5%
Negative cultures	15%	8–12%

Adapted from Bouvet and Acar (1984).

Table 10. Species of streptococci isolated from infective endocarditis.

Species[a]	Isolation frequency (% of streptococcal species)
Streptococcus sanguis (including *S. sanguis* I)[b]	7–29%
S. oralis (including "*S. mitior*," *S. mitis,* and *S. sanguis* II)	7–30%
S. mutans	3–18%
S. salivarius	1– 4%
S. milleri group (including *S. anginosus, S. constellatus,* and *S. intermedius*)	3–15%
Nutritionally variant streptococci (NVS)	8–17%
All oral (viridans) streptococci	66–80%
S. bovis (including *S. bovis* I and II)	5–17%
Enterococci	5–10%

[a]Species names are given using currently accepted nomenclature.
[b]Strains reported as *S. sanguis* may include representatives of *S. gordonii* (Kilian et al., 1989a).
Summary of data from: Bayliss et al. (1983); Bouvet and Acar (1984); Facklam (1977); Horaud and Delbos (1984); Lowes et al. (1980); Moulsdale et al. (1980); Parker and Ball (1976); Roberts et al. (1979).

their series, it is possible that they may account for at least some of the culture-negative cases that are recorded. The potential therapeutic problems posed by these streptococci are well recognized, and a combination of penicillin and aminoglycoside therapy is usually recommended (Stein and Nelson, 1987; Stein and Libertin, 1989b). The taxonomic status of the nutritionally variant streptococci has, until recently, been very uncertain, but the proposal of two new species, *S. defectivus* and *S. adjacens,* to include such strains now provides a firmer basis for future studies (Bouvet et al., 1989).

The importance of distinguishing between *S. bovis* biotypes I and II, and *S. salivarius,* when identifying strains from bacteremias has been pointed out by Ruoff et al. (1989). These authors noted a striking association between *S. bovis* I bacteremia and underlying endocarditis, and also confirmed the previously reported correlation between this organism and the presence of colonic neoplasms (Klein et al., 1979). The role in endocarditis, if any, of the recently described species *S. vestibularis* (Whiley and Hardie, 1988), remains to be determined, but at least some isolates identified in the past as *S. sanguis* may have been representatives of the recently described *S. gordonii* (Kilian et al., 1989a).

The diagnosis of infective endocarditis is normally confirmed in the laboratory by isolation of the causative organism from blood cultures, but in up to 10% of cases such cultures may be negative (Bayliss et al., 1983). The use of serological tests, such as immunoblotting, provides an alternative method for diagnosis of such cases (Burnie et al., 1987).

Antibiotic Susceptibility of Oral Streptococci

The species of streptococci considered in this chapter are generally thought to be fully susceptible to penicillin, although the existence of resistant strains has been known for many years (Garrod and Waterworth, 1962; Naiman and Barrow, 1963; Phillips et al., 1976). Examination of one series of strains from infective endocarditis, excluding enterococci, showed that 91% of strains were sensitive to benzylpenicillin (Etienne et al., 1984). The penicillin-resistant strains were identified in this study as *S. sanguis* I, *S. sanguis* II, and *S. mitis.* Resistance to erythromycin was found in 16% of strains. In another study of endocarditis isolates (Horaud and Delbos, 1984), 96% of strains were sensitive to penicillin, but 20% were resistant to tetracycline, and 8% were found to be multiply resistant.

Serious, life-threatening infections due to penicillin-resistant viridans streptococci, including *S. intermedius* and *S. mitis,* have been reported, and penicillin resistance may be associated with altered penicillin-binding proteins (Quinn et al., 1988). Tolerance to penicillin has also been observed amongst viridans streptococci, in which the lethal (bactericidal) effect of the antibiotic is greatly reduced (Catto et al., 1987; Handwerger and Tomasz, 1985; Powley et al., 1989; Slater and Greenwood, 1983). Thus, as pointed out by Quinn and colleagues (1988), it is desirable to report antibiotic-sensitivity results on all isolates of viridans streptococci from sites that are normally sterile.

Molecular Studies

As reviewed in other sections, the application of molecular and genetic techniques, such as nucleic acid hybridization and RNA sequencing, has had a considerable impact on our current understanding of the taxonomy of the oral streptococci (Schleifer and Kilpper-Bälz, 1987). The development of DNA probes for identification of streptococci, as described for *S. oralis* (Schmidhuber et al., 1988), is likely to be extended to cover other species. For epidemiological studies, the technique of DNA fingerprinting by restriction pattern analysis provides another potentially useful tool (Skjold et al., 1987), and this has already been applied in a study on transmission of *S. mutans* (Caufield and Walker, 1989).

Streptococcus mutans has probably received more attention than any other oral microorganism, with many studies on various aspects of the molecular biology of sucrose metabolism, surface antigens, plasmids, and the development of vaccines (Hamada and Slade, 1980; Hamada et al., 1986; Loesche, 1986). Genetic approaches to the study of virulence have been extensively used in research on *S. mutans* (Curtiss, 1986; Russell and Gilpin, 1987), and during the next few years it is likely that other oral streptococci will receive similarly detailed examination and analysis of putative virulence factors (Russell, 1990). Such studies should lead to a greatly enhanced understanding of the mechanisms by which oral streptococci may cause disease. With the anticipated development of specific probes against these streptococci, it will also become possible to undertake far more sophisticated studies on their ecology, epidemiology, and role in various disease processes.

Literature Cited

Abercrombie, G. F. and W. M. Scott. 1928. A case of infective endocarditis due to *Streptococcus mutans*. Lancet ii:697–699.

Admon, D., M. A. Ephros, D. Gavish, and R. Raz. 1987. Infection with *Streptococcus milleri*. J. Infect. 14:55–60.

Alaluusua, S. and O-V. Renkonen. 1983. *Streptococcus mutans* establishment and dental caries experience in children from 2 to 4 years old. Scand. J. Dent. Res. 91:453–457

Alaluusua, S., J. Savdainen, H. Tuompo, and L. Gronroos. 1984. Slide-scoring method for estimation of *Streptococcus mutans* levels in saliva. Scand. J. Dent. Res. 92:127–133.

Andrewes, F. W. and J. Horder. 1906. A study of the streptococci pathogenic for man. Lancet ii 708–713.

Arala-Chaves, M. P., M. T. Porto, P. Arnaud, M. J. Saraiva, H. Geada, C. C. Patrick, and H. H. Fudenberg. 1981. Fractionation and characterization of the immuno-suppressive substance in crude extracellular products released by *Streptococcus intermedius*. J. Clin. Invest. 68:294–302.

Ball, L. C. 1985. Serological identification of *Streptococcus sanguis* and *Streptococcus mitior*. J. Clin. Pathol. 38:452–454.

Bayliss, R., C. Clarke, C. M. Oakley, W. Somerville, A. G. W. Whitfield, and S. E. J. Young. 1983. The microbiology and pathogenesis of infective endocarditis. Br. Heart J. 50:513–519.

Beachey, E. H. (ed.) 1980. Bacterial adherance. Receptors and Recognition, Series B, vol. 6. Chapman and Hall, London.

Beighton, D. 1986. A simplified procedure for estimating the level of *Streptococcus mutans* in the mouth. Brit. Dent. J. 160:329–330.

Beighton, D., H. Hayday, R. R. B. Russell, and R. A. Whiley. 1984. *Streptococcus macacae* sp. nov. from dental plaque of monkeys *(Macaca fascicularis)*. Int. J. Syst. Bacteriol. 34:332–335.

Beighton, D., F. Manji, V. Baelum, O. Fejerskov, N. W. Johnson, and J. M. A. Wilton. 1989. Association between salivary levels of *Streptococcus mutans, Streptococcus sobrinus*, lactobacilli and caries experience in Kenyan adolescents. J. Dent. Res. 68:1242–1246.

Beighton, D., H. R. Rippon, and H. E. C. Thomas. 1987. The distribution of *Streptococcus mutans* serotypes and dental caries in a group of 5- to 8-year-old Hampshire schoolchildren. Brit. Dent. J. 162:103–106.

Beighton, D., R. R. B. Russell, and H. Hayday. 1981. The isolation and characterization of *Streptococcus mutans* serotype h from dental plaque of monkeys (Macaca fascicularis) J. Gen. Microbiol. 124:271–279.

Berkeley, R. C. W., J. M. Lynch, J. Melling, P. R. Rutter, and B. Vicent (ed.). 1980. Microbial Adhesion to Surfaces. Ellis Horwood Ltd., Chichester.

Berkowitz, R. J., and P. Jones. 1985. Mouth-to-mouth transmission of the bacterium *Streptococcus mutans* between mother and child. Arch. Oral Biol. 30:377–379.

Berkowitz, R. J., and H. V. Jordan. 1975. Similarity of bacteriocins of *Streptococcus mutans* from mother and infant. Archives of Oral Biology. 20:725–730.

Bibby, B. G., and R. J. Shern. (ed.). 1978. Methods of Caries Prediction. Information Retrieval Inc., Washington, D.C.

Bliss, E. A. 1937. Studies upon minute streptococci III. Serological differentiation. J. Bacteriol. 33:625–642.

Bouvet, A., and J. F. Acar. 1984. New bacteriological aspects of infective endocarditis. Eur. Heart J. 5 (Suppl. C):45–48.

Bouvet, A., F. Grimont, and P. A. D. Grimont. 1989. *Streptococcus defectivus* sp.nov. and *Streptococcus adjacens* sp.nov., nutritionally variant streptococci from human clinical specimens. Int. J. Syst. Bacteriol. 39:290–294.

Bouvet, A., I. van de Rijn, and M. McCarty. 1981. Nutritionally variant streptococci from patients with endocarditis: growth parameters in a semisynthetic medium and demonstration of a chromophore. J. Bacteriol. 146:1075–1082

Bouvet, A., F. Villeroy, F. Cheng, C. Lamesch, R. Williamson, and L. Gutman. 1985. Characterization of nutritionally variant streptococci by biochemical tests and penicillin-binding proteins. J. Clin. Microbiol. 22:1030–1034.

Bratthall, D. 1970. Demonstration of five serological groups of streptococcal strains resembling *Streptococcus mutans*. Odont. Revy 21:143–152.

Bridge, P. D., and P. H. A. Sneath. 1982. *Streptococcus gallinarum* sp.nov. and *Streptococcus oralis* sp.nov. Int. J. Syst. Bacteriol. 32:410–415.

Bridge, P. D., and P. H. A. Sneath. 1983. Numerical taxonomy of *Streptococcus*. J. Gen. Microbiol. 129:565–597.

Brook, I., and R. I. Walker. 1984. Pathogenicity of anaerobic Gram-positive cocci. Infect. Immun. 45:320–324.

Brook, I., and R. Walker. 1985. The role of encapsulation in the pathogenesis of anaerobic Gram-positive cocci. Can. J. Microbiol. 31:176–180.

Buchanan, R. E., and N. E. Gibbons. 1974. Bergey's manual of determinative bacteriology, 8th ed. Williams and Wilkins, Baltimore.

Burnie, J. P., M. Holland, R. C. Matthews, and W. Lees. 1987. Role of immunoblotting in the diagnosis of culture negative and enterococcal endocarditis. J. Clin. Pathol. 40:1149–1158.

Carey, R. A., K. C. Gross, and R. B. Roberts. 1975. Vitamin B₆-dependent *Streptococcus mitior* (mitis) isolated form patients with systemic infections. J. Infect. Dis. 131:722–726.

Carlsson, J. 1967. Presence of various types of non-hemolytic streptococci in dental plaque and in other sites of the oral cavity in man. Odontologisk Revy 18:55–74.

Carlsson, J. 1968. A numerical taxonomic study of human oral streptococci. Odontologisk Revy 19:137–160.

Carlsson, P., L. A. Gandour, B. Olsson, B. Rickardsson, and K. Abbas. 1987. High prevalence of mutans streptococci in a population with extremely low prevalence of dental caries. Oral Microbiol. Immunol. 2:121–124.

Carlsson, P., B. Olsson, and D. Bratthall. 1985. The relationship between the bacterium *Streptococcus mutans* in saliva and dental caries in children in Mozambique. Arch. Oral. Biol. 29:385–393.

Catto, B. A., M. R. Jacobs, and D. M. Shlaes. 1987. *Streptococcus mitis*. A cause of serious infection in adults. Arch. Intern. Med. 147:885–888.

Caufield, P. W., K. Ratanapridakul, D. N. Allen, and G. R. Cutter. 1988. Plasmid-containing strains of *Streptococcus mutans* cluster within family and racial cohorts: implications for natural transmission. Infect. Immunol. 56:3216–3220.

Caufield, P. W., and T. M. Walker. 1989. Genetic diversity within *Streptococcus mutans* evident from chromosomal DNA restriction fragment polymorphisms. J. Clin. Microbiol. 27:274–278.

Chapman, G. H. 1944. The isolation of streptococci from mixed cultures. J. Bacteriol. 48:113–114.

Chua, D., H. H. Reinhart, and J. D. Sobel. 1989. Liver abscess caused by *Streptococcus milleri*. Rev. Infect. Dis. 11:197–202.

Cimolai, N., R. W. Elford, L. Bryan, C. Anand, and P. Berger. 1988. Do the β-hemolytic non-group A streptococci cause pharyngitis? Rev. Infect. Dis. 10:587–601.

Clarke, J. K. 1924. On the bacterial factor in the aetiology of dental caries. Brit. J. Exp. Path. 5:141–147.

Cohen, B., S. L. Peach, and R. R. B. Russell. 1983. Immunization against dental caries, p. 255–294. In: C. S. F. Easmon and J. Jeljaszewicz (ed.), Medical Microbiology, vol. 2. Immunisation against bacterial disease. Academic Press, London.

Cole, R. M., G. B. Calandra, E. Huff, and K. N. Nugent. 1976. Attributes of potential utility in differentiating among "Group H" streptococci or *Streptococcus sanguis*. J. Dent. Res. 55:A142–A153.

Colman, G. 1968. The application of computers to the classification of streptococci. J. Gen. Microbiol. 50:149–158.

Colman, G. 1969. Transformation of viridans-like streptococci. J. Gen. Microbiol. 57:247–255.

Colman, G. 1976. The viridans streptococci, p. 179–198. In: De Louvois (ed.), Selected topics in clinical bacteriology. Balliere Tindal, London.

Colman, G., and L. C. Ball. 1984. Identification of streptococci in a medical laboratory. J. Appl. Bacteriol. 57:1–14.

Colman, G., and R. E. O. Williams. 1965. The cell walls of streptococci. J. Gen. Microbiol. 41:375–387.

Colman, G., and R. E. O. Williams. 1972. Taxonomy of some human viridans streptococci, p. 281–299. In: L. W. Wannamaker and J. M. Matsen (ed.), Streptococci and Streptococcal Disease. Academic Press, London.

Cooksey, R. C., F. S. Thompson, and R. R. Facklam. 1979. Physiological characterization of nutritionally variant streptococci. J. Clin. Microbiol. 10:326–330.

Cookson, B., H. Talsania, S. Chinn, and I. Phillips. 1989. A qualitative and quantitative study of the cellular fatty acids of '*Streptococcus milleri*' with capillary gas chromatography. J. Gen. Microbiol. 135:831–838.

Coykendall, A. L. 1970. Base composition of deoxyribonucleic acid isolated from cariogenic streptococci. Arch. Oral Biol. 15:365–368.

Coykendall, A. L. 1971. Genetic heterogeneity in *Streptococcus mutans*. J. Bacteriol. 106:192–196.

Coykendall, A. L. 1974. Four types of *Streptococcus mutans* based on their genetic, antigenic and biochemical characteristics. J. Gen. Microbiol. 83:327–338.

Coykendall, A. L. 1977. Proposal to elevate the subspecies of *Streptococcus mutans* to species status based on their molecular composition. Int. J. Syst. Bacteriol. 27:26–30.

Coykendall, A. L. 1983. *Streptococcus sobrinus* nom. rev. and *Streptococcus ferus* nom. rev.: habitat of these and other mutans streptococci. Int. J. Syst. Bacteriol. 33:883–885.

Coykendall, A. L. 1989a. Classification and identification of the viridans streptococci. Clin. Microbiol. Rev. 2:315–328.

Coykendall, A. L. 1989b. Rejection of the type strain of *Streptococcus mitis* (Andrewes and Horder 1906). Request for an opinion. Int. J. System. Bacteriol. 39:207–209.

Coykendall, A. L., D. Bratthall, K. O'Connor, and R. A. Dvarskas. 1976. Serological and genetic examination of some nontypical *Streptococcus mutans* strains. Infect. and Immun. 14:667–670.

Coykendall, A. L., and K. B. Gustafson. 1985. Deoxyribonucleic acid hybridisations among strains of *Streptococcus salivarius* and *Streptococcus bovis*. Int. J. Syst. Bacteriol. 35:274–280.

Coykendall, A. L., and A. J. Munzenmaier. 1978. Deoxyribonucleic acid base sequence studies on glucan-producing and glucan-negative strains of *Streptococcus mitior*. Int. J. Syst. Bacteriol. 28:511–515.

Coykendall, A. L., and P. A. Specht. 1975. DNA base sequence homologies among strains of *Streptococcus sanguis*. J. Gen. Microbiol. 91:92–98.

Coykendall, A. L., P. A. Specht, and H. H. Samol. 1974. *Streptococcus mutans* in a wild, sucrose-eating rat population. Infect. and Immun. 10:216–219.

Coykendall, A. L., and P. M. Wesbecher. 1990. Genetic relationships among some "viridans" streptococci. In: R. Lütticken (ed.), 10th Lancefield International Symposium on Streptococci and Streptococcal Diseases (in press).

Coykendall, A. L., P. M. Wesbecher, and K. B. Gustafson. 1987. '*Streptococcus milleri*,' *Streptococcus constellatus* and *Streptococcus intermedius* are later synonyms of *Streptococcus anginosus*. Int. J. Syst. Bacteriol. 37:222–228.

Crossner, G-G. 1981. Salivary lactobacillus counts in the prediction of caries activity. Commun. Dent. Oral Epidemiol. 9:182:190.

Curtiss, R., III. 1986. Genetic analysis of *Streptococcus mutans* virulence and proposects for an anticaries vaccine. J. Dent. Res. 65:1034–1045.

Curtiss, R., III., R. Goldsmidst, J. Barrett, W. Jacobs, R. Pastian, M. Lyons, S. M. Michalek, and J. Mestecky. 1986. Cloning virulence determinants from *Streptococcus mutans* and the use of recombinant clones to construct bivalent oral vaccine strains to confer protective immunity against *S. mutans*-induced dental caries, p. 173–180. In: S. Hamada, S. Michalek, H. Kyiono, L. Menaker and J. R. McGhee (ed.), Molecular microbiology and immunobiology of *S. mutans*. Elsevier Science Publishers, Amsterdam.

Deibel, R. H., and H. W. J. Seeley. 1974. Genus I. Streptococcus Rosenbach 1884,22, p. 490–509. In: R. E. Buchanan and N. E. Gibbons (ed.), Bergey's manual of determinative bacteriology, 8th ed. Williams and Wilkins, Baltimore, MD.

De Louvois, J., P. Gortvai, and R. Hurley. 1977. Bacteriology of abscesses of the central nervous system: a multicentre prospective study. Brit. Med. J. 2:981–984.

Dent, V. E., J. M. Hardie, and G. H. Bowden. 1978. Streptococci isolated from dental plaque of animals. J. Appl. Bacteriol. 44:249–258.

De Stoppelaar, J. D., J. Van Houte, and C. E. De Moor. 1967. The presence of dextran forming bacteria resembling *Streptococcus bovis* and *Streptococcus sanguis* in human dental plaque. Arch. Oral Biol. 12:1199–1201.

De Stoppelaar, J. D., K. Konig, A. Plasschaert, and J. Van der Hoeven. 1971. Decreased cariogenicity of a mutant of *Streptococcus mutans*. Arch. Oral Biol. 16:971–975.

Dodd, R. 1949. Serologic relationship between *Streptococcus* group and *Streptococcus sanguis*. Proc. Soc. Exp. Biol. Med. 70:598–599.

Douglas, C. W. I. 1983. The binding of human salivary-amylase by oral strains of streptococcal bacteria. Arch. Oral Biol. 28:567–573.

Douglas, C. W. I., A. A. Pease, and R. A. Whiley. 1990 Amylase binding as a discriminator among oral streptococci. FEMS Microbiol. Lett. 66:193–198.

Drucker, D. B., and R. M. Green. 1978. The relative cariogenicities of *Streptococcus milleri* and other viridans group streptococci in gnotobiotic hooded rats. Arch. Oral Biol. 23:183–187

Drucker, D. B., and S. M. Lee. 1981. Fatty acid fingerprints of 'Streptococcus milleri" *Streptococcus mitis,* and related species. Int. J. Syst. Bacteriol. 31:219–225.

Drucker, D. B., and S. M. Lee. 1983. Possible heterogeneity of *Streptococcus milleri* determined by DNA mol% (guanine plus cytosine) measurement and physiological characterization. Microbios 38:151–157.

Drucker, D. B., and T. H. Melville. 1971. The classification of some oral streptococci of human or rat origin. Archs. Oral Biol. 16:845–853.

Drucker, D. B., A. P. Shakespeare, and R. M. Green. 1984a. The production of dental plaque and caries by the bacterium *Streptococcus salivarius* in gnotobiotic WAG/RIJ rats. Arch. Oral Biol. 29:437–443.

Drucker, D. B., A. P. Shakespeare, and R. M. Green. 1984b. In vivo dental plaque-forming ability and relative cariogenicity of the bacteria *Streptococcus mitis* and *Streptococcus sanguis* I and II in mono-infected gnotobiotic rats. Arch. Oral Biol. 29:1023–1031.

El Tayeb, Y., D. Bratthall, and P. Carlsson. 1985. Dental caries and *Streptococcus mutans* in Sudanese schoolchildren. Odontostomatol. Trop. 8:77–80.

Ellwood, D. C., J. Melling, and P. Rutter. (ed.). 1979. Adhesion of microorganisms to surfaces. Academic Press, London.

Emilson, C-G, and B. Krasse. 1985. Support for and implications of the Specific Plaque Hypothesis. Scand. J. Dent. Res. 93:96–104.

Emilson, C-G., and B. Krasse. 1986. Comparison between a dip-slide test and plate count for determination of *Streptococcus mutans* infection. Scand. J. Dent. Res. 94:500–506.

Etienne, J., L. D. Gruer, and J. Fleurette. 1984. Antibiotic susceptibility of streptococcal strains associated with infective endocarditis. Eur. Heart J. 5 (Suppl. C):33–37.

Ezaki, T., R. Facklam, N. Takeuchi, and E. Yabuuchi. 1986. Genetic relatedness between the type strain of *Streptococcus anginosus* and minute-colony-forming beta-hemolytic streptococci carrying different Lancefield grouping antigens. Int. J. Syst. Bacteriol. 36:345–347.

Facklam, R. R. 1977. Physiological differentiation of viridans streptococci. J. Clin. Microbiol. 5:184–201.

Facklam, R. R. 1984. The major differences in the American and British *Streptococcus* taxonomy schemes with special reference to *Streptococcus milleri*. Eur. J. Clin. Microbiol. 3:91–93.

Facklam, R. R., D. L. Rhoden, and P. B. Smith. 1984. Evaluation of the Rapid Strep System for the identification of clinical isolates of *Streptococcus* species. J. Clin. Microbiol. 20:894–898.

Farmer, E. D. 1953. Streptococci of the mouth and their relationship to subacute bacterial endocarditis. Proc. Soc. Royal Soc. Med. 46:201–208.

Farmer, E. D. 1954. Serological subdivisions among the Lancefield group H streptococci. J. Gen. Microbiol. 11:131–138.

Farrow, J. A. E., and M. D. Collins. 1984a. DNA base composition, DNA/DNA homology and long-chain fatty acid studies on *Streptococcus thermophilus* and *Streptococcus salivarius*. J. Gen. Micrbiol. 130:357–362.

Farrow, J. A. E., and M. D. Collins. 1984b. Taxonomic studies on streptococci of serological groups C, G and L and possibly related tax. System. App. Microbiol. 5:483–493.

Fertally, S. S., and R. Facklam. 1987. Comparison of physiologic tests used to identify non-beta-hemolytic aerococci, enterococci and streptococci. J. Clin. Mircobiol. 25:1845–1850.

Fitzgerald, R. J. 1968. Dental caries research in gnotobiotic animals. Caries Res. 2:139–146.

Fitzgerald, R. J., and P. H. Keyes. 1960. Demonstration of the etiologic role of streptococci in experimental caries in the hamster. J. Am. Dent. Assoc. 61:9–19.

Freedman, M. L., J. M. Tanzer, and A. L. Coykendall. 1981. The use of genetic variants in the study of dental caries, p. 247–269. In: J. M. Tanzer (ed.), Animal models in cariology. Information Retrieval, Washington, D.C.

Freedman, M. L., J. M. Tanzer, E. Swayne, and G. Allenspzch-Petrzilka. 1983. Colonization and virulence of *Streptococcus sobrinus:* the roles of glucan-associated phenomena revealed by the use of mutants, p. 39–50. In: R. J. Doyle and J. E. Ciardi (ed.), Glucosyl-transferases, glucans, sucrose and dental caries. A Special Supplement to Chemical Senses. Information Retrieval Ltd., Washington, D.C.

French, G. L., H. Talsania, J. R. H. Charlton, and I. Phillips. 1989. A physiological classification of viridans streptococci by use of the API-20 STREP System. J. Med. Microbiol. 28:275–286.

Frenkel, A., and W. Hirsch. 1961. Spontaneous development of L forms of streptococci requiring secretions of other bacteria or sulphydryl compounds for normal growth. Nature 191:728–730.

Garrod, L. P., and P. M. Waterworth. 1962. The risks of dental extraction during penicillin treatment. Br. Heart J. 24:39–46.

Ghosh, K., R. Ducan, and P. G. E. Kennedy. 1988. Acute spinal epidural abscess caused by *Streptococcus milleri.* J. Infect. 16:303–304.

Gibbons, R. J. 1984. Adherent interactions which may affect microbial ecology in the mouth. J. Dent. Res. 63:378–385.

Gibbons, R. J., and J. van Houte. 1975. Bacterial adherence in oral microbial ecology. Ann. Revs. Microbiol. 29:19–44.

Gilmour, M. N., T. S. Whittam, M. Kilian, and R. K. Selander. 1987. Genetic relationships among the oral streptococci. J. Bacteriol. 169:5247–5257.

Gledhill, W. E., and L. E. Casida. 1969. Predominant catalase-negative soil bacteria. 1. Streptococcal population indigenous to soil. Appl. Microbiol. 17:208–213.

Gold, O. G., H. V. Jordan, and J. van Houte. 1973. A selective medium for *Streptococcus mutans.* Arch. Oral Biol. 18:1357–1364.

Gordon, M. H. 1905. A ready method of differentiating streptococci and some results already obtained by its application. Lancet ii:1400–1403.

Gossling, J. 1988. Occurrence and pathogenicity of the *Streptococcus milleri* Group. Rev. Infect. Dis. 10:257–285.

Guggenheim, B. 1968. Streptococci of dental plaques. Caries Res. 2:147–163.

Guggenheim, B. 1970. Enzymatic hydrolysis and structure of water-insoluble glucan produced by glucosyl-transferases from a strain of *Streptococcus mutans.* Helv. Odont. Acta. 14:Suppl. V:89–108.

Guthof, O. 1956. Ueber pathogene 'Vergrünede Streptokokken'; Streptokokken Befunde bei dentogenen abszessen und Infiltraten im Bereich der Mundhöhle. Zentbl. Bakt. Parasit. Abt. 1. 166:533–564.

Gutmann, L., A. Bouvet, and J. F. Acar. 1982. Pencillin-binding proteins of nutritionally variant streptococci. FEMS Microbiol. Lett. 14:11–14.

Haffajee, A. D., S. S. Socransky, and J. L. Ebersole. 1985. Survival analysis of periodontal sites before and after periodontal therapy. J. Clin. Periodontol. 12:553–567.

Hamada, S., S. M. Michalek, H. Kiyono, L. Menaker, and J. R. McGhee. 1986. Molecular microbiology and immunobiology of *Streptococcus mutans.* Elsevier, Amsterdam.

Hamada, S., and H. D. Slade. 1980. Biology, immunology and cariogenicity of *Streptococcus mutans.* Microbiol. Revs. 44:331–384.

Handley, P. S., P. L. Carter, and J. Fielding. 1984. *Streptococcus salivarius* strains carry either fibrils or fimbriae on the cell surface. J. Bacteriol. 157:64–72.

Handley, P. S., P. L. Carter, J. E. Wyatt, and L. M. Hesketh. 1985. Surface structures (peritrichous fibrils and tufts of fibrils) found on *Streptococcus sanguis* strains may be related to their ability to coaggregate with other oral genera. Infect. Immun. 47:217–227.

Handley, P. S., D. W. S. Harty, J. E. Wyatt, C. R. Brown, J. P. Doran, and A. C. C. Gibbs. 1987. A comparison of the adhesion, coaggregation and cell-surface hydrophobicity of fibrillar and fimbrate strains of *Streptococcus salivarius.* J. Gen. Microbiol. 133:3207–3217.

Handwerger, S., and A. Tomasz. 1985. Antibiotic tolerance among clinical isolates of bacteria. Rev. Infect. Dis. 7:368–386.

Hardie, J. M. 1986a. Genus *Streptococcus* Rosenbach 1884, p. 1043–1047. In: P. H. A. Sneath, N. S. Mair and M. E. Sharpe (ed.), Bergey's manual of systematic bacteriology, vol. 2. Williams and Wilkins, Baltimore.

Hardie, J. M. 1986b. Oral Streptococci, p. 1054–1063. In: P. H. A. Sneath, N. S. Mair and M. E. Sharpe (ed.), Bergey's manual of systematic bacteriology, vol. 2. Williams and Wilkins, Baltimore.

Hardie, J. M., and G. H. Bowden. 1974a. The normal microbial flora of the mouth, p. 47–83. In: F. A. Skinner and J. G. Carr (ed.), The normal microbial flora of man. Academic Press, London.

Hardie, J. M., and G. H. Bowden. 1974b. Cell wall and serological studies on *Streptococcus mutans.* Caries Research 8:301–316.

Hardie, J. M., and G. H. Bowden. 1976a. Physiological classification of oral viridans streptococci. J. Dent. Res. 55:A166-A176.

Hardie, J. M., and G. H. Bowden. 1976b. The microbial flora of dental plaque: bacterial succession and isolation considerations, p. 63–87. In: H. M. Stiles, W. J. Loesche and T. C. O'Brien (ed.), Microbial aspects of dental caries, vol. 1. Information Retrieval Inc., Washington, D.C.

Hardie, J. M., and G. H. Bowden. 1976c. Some serological cross-reactions between *Streptococcus mutans, S. sanguis,* and other dental plaque streptococci. J. Dent. Res. 55 (Special Issue C):C50-C58.

Hardie, J. M., and P. D. Marsh. 1978a. Streptococci and the human oral flora, p. 157–206. In: F. A. Skinner and L. B. Quesnel (ed.), Streptococci. Academic Press, London.

Hardie, J. M., and P. D. Marsh. 1978b. Isolation media for streptococci. Oral Streptococci, p. 380–383. In: F. A. Skinner and L. B. Quesnel (ed.), Streptococci. Academic Press, London.

Hardie, J. M., P. L. Thomson, R. J. South, P. D. Marsh, G. H. Bowden, A. S. McKee, E. D. Fillery, and G. L. Slack. 1977. A longitudinal epidemiological study on dental plaque and the development of dental caries—interim results after two years. J. Dent. Res. 56:90C-98C.

Hardie, J. M., R. A. Whiley, and M. Sackin. 1982. A numerical taxonomic study of oral streptococci, p. 59–60. In: S. E. Holm and P. Christensen (ed.), Basic concepts of streptococci and streptococcal diseases. Reed Books Ltd, Chertsey, UK.

Hare, R. 1935. The classification of haemolytic streptococci from the nose and throat of normal human beings by means of precipitin and biochemical tests. J. Pathol. Bacteriol. 41:499–512.

Harty, D. W. S., and P. S. Handley. 1989. Expression of the surface properties of the fibrillar *Streptococcus salivarius* HB and its adhesion deficient mutants grown in continuous culture under glucose limitation. J. Gen. Microbiol. 135:2611–2621.

Hehre, E. J., and J. M. Neill. 1946. Formation of serologically reactive dextrans by streptococci from subacute bacterial endocarditis. J. Exper. Med. 83:147–162.

Heimdahl, A., L. von Konow, T. Satoh, and C. E. Nord. 1985. Clinical appearance of orofacial infections of oral origin in relation to microbiological findings. J. Clin. Microbiol. 22:299–302.

Holdeman, L. V., and W. E. C. Moore. 1974. New genus, *Coprococcus,* twelve new species emended descriptions of four previously described species of bacteria from human feces. Int. J. System. Bacteriol. 24:260–277.

Holman, W. L. 1916. The classification of streptococci. J. Med. Res. 34:377–443.

Horaud, T., and F. Delbos. 1984. Viridans streptococci in infective endocarditis: species distribution and susceptibility to antibiotics. Eur. Heart J. 5 (Suppl. C.):39–44.

Ikeda, T., and H. J. Sandham. 1972. A medium for the recognition and enumeration of *Streptococcus mutans.* Arch. Oral Biol. 17:601–604.

Johnson, M. C., J. T. Bozzola, and I. L. Shechmeister. 1977. Biochemical study of the relationship of extracellular glucan to adherence and cariogenicity in *Streptococcus mutans* and an extracellular polysaccharide mutant. J. Bacteriol. 129:351–357.

Johnson, C. P., S. M. Gross, and J. D. Hillman. 1980. Cariogenic potential in vitro in man and in vivo in the rat of lactate dehydrogenase mutants of *Streptococcus mutans.* Arch. Oral Biol. 25:707–713.

Jones, D. 1978. Composition and differentiation of the genus *Streptococcus,* p. 1–49. In: F. A. Skinner and L. B. Quesnel (ed.), Streptococci. Society for Applied Bacteriology Symposium Series no. 7. Academic Press, London.

Kambal, A. M. 1987. Isolation of *Streptococcus milleri* from clinical specimens. J. Infect. 14:217–223.

Keyes, P. H. 1960. The infectious and transmissable nature of experimental dental caries. Arch. Oral Biol. 1:304–320.

Kilian, M., L. Mikkelsen, and J. Henrichsen. 1989a. Taxonomic study of viridans streptococci description of *Streptococcus gordonii* sp.nov. and emended descriptions of *Streptococcus sanguis* (White and Niven, 1946), *Streptococcus oralis* (Bridge and Sneath, 1982) and *Streptococcus mitis* (Andrewes and Horder, 1906). Int. J. System. Bacteriol. 39:471–484.

Kilian, M., L. Mikkelsen, and J. Henrichsen. 1989b. Replacement of the type strain of *Streptococcus mitis.* Request for an opinion. Int. J. System. Bacteriol. 39:498–499.

Kilian, M., B. Nyvad, and L. Mikkelsen. 1986. Taxonomic and ecological aspects of some oral streptococci, p. 391–400. In: S. Hamada, S. M. Michalek, H. Kiyono, L. Menaker and J. R. McGhee (ed.), Molecular microbiology and immunobiology of *Streptococcus mutans.* Elsevier Science Publishers, Amsterdam.

Kilpper-Bälz, R., G. Fischer, and K. H. Schleifer. 1982. Nucleic acid hybridization of Group N and Group D streptococci. Current Microbiol. 7:245–250.

Kilpper-Bälz, R., B. L. Williams, R. Lutticken, and K. H. Schleifer. 1984. Relatedness of *"Streptococcus milleri"* with *Streptococcus anginosus* and *Streptococcus constellatus.* System. Appl. Microbiol. 5:494–500.

Kilpper-Bälz, R., P. Wenzig, and K. H. Schleifer. 1985. Molecular relationships and classification of some viridans streptococci as *Streptococcus oralis* and emended description of *Streptococcus oralis* (Bridge and Sneath, 1982). Int. J. System. Bacteriol. 35:482–488.

Klein, J. P., and M. Scholler. 1988. Recent advances in the development of a *Streptococcus mutans* vaccine. Eur. J. Epidemiol. 4:419–425.

Klein, R. S., M. T. Catalano, S. C. Edberg, J. I. Casey, and N. H. Steigbigel. 1979. *Streptococcus bovis* septicemia and carcinoma of the colon. Ann. Intern. Med. 91:560–562.

Klock, B., and B. Krasse. 1979. A comparison between different methods for prediction of caries activity. Scand. J. Dent. Res. 87:129–139.

Knight, R. G., and D. M. Shlaes. 1988. Physiological characteristics and deoxyribonucleic acid relatedness of *Streptococcus intermedius* strains. Int. J. Syst. Bacteriol. 38:19–24.

Koga, T., N. Okahashi, H. Asakawa, and S. Hamada. 1986. Adherence of *Streptococcus mutans* to tooth surfaces, p. 111–120. In: S. Hamada, S. M. Michalek, H. Kiyono, L. Menaker and J. R. McGhee (ed.), Molecular microbiology and immunobiology of *Streptococcus mutans.* Elsever Science Publishers, Amsterdam.

Köhler, B., B. M. Pettersson, and D. Bratthall. 1981. *Streptococcus mutans* in plaque and saliva and the development of caries. Scand. J. Dent. Res. 89:19–25.

Köhler, B., D. Bratthall, and B. Krasse. 1983. Preventive measures in mothers influence the establishment of the bacterium *Streptococcus mutans* in their infants. Arch. Oral Biol. 28:225–231.

Köhler, B., I. Andreen, and B. Jonsson. 1988. The earlier the colonization of mutans streptococci, the higher the caries prevalence at 4 years of age. Oral Microbiol. Immun. 3:14–17.

Krasse, B. 1966. Human streptococci and experimental caries in hamsters. Arch. Oral Biol. 11:429–436.

Krasse, B. 1988. Biological factors as indicators of future caries. Int. Dent. J. 38:219–225.

Kristoffersson, K., J. G. Grondahl, and D. Bratthall. 1985. The more *Streptococcus mutans,* the more caries on approximal surfaces. J. Dent. Res. 64:58–61.

Labbe, M., P. Van Der Auwera, Y. Glupczński, F. Crockaert, and E. Yourassowsky. 1985. Fatty acid composition of *Streptococcus milleri.* Eur. J. Clin. Microbiol. 4:391–393.

Lancefield, R. C. 1925a. The immunological relationships of *Streptococcus viridans* and certain of its chemical fractions. I. Serological reactions obtained with antibacterial sera. J. Exp. Med. 42:377–395.

Lancefield, R. C. 1925b. The immunological relationships of *Streptococcus viridans* and certain of its chemical factions. II. Serological reactions obtained with antinucleo-protein sera. J. Exp. Med. 42:397–412.

Lancefield, R. C. 1933. The serological differentiation of human and other groups of hemolytic streptococci. J. Exper. Med. 57:571–595.

Larmas, H. 1975. A new dip-slide method for the counting of salivary lactobacilli. Proc. Finn-Dent. Soc. 71:31–35.

Lehner, T., M. W. Russel, J. Caldwell, and R. Smith. 1981. Immunization with purified protein antigens from *Streptococcus mutans* against dental caries in rhesus monkeys. Infect. Immun. 34:407–415.

Lewis, M. A. O., T. W. McFarlane, and D. A. McGowan. 1986. Quantitative bacteriology of acute dento-alveolar abscess. J. Med. Microbiol. 21:101–104.

Lewis, M. A. O., T. W. McFarlane, D. A. McGowan, and D. G. MacDonald. 1988. Assessment of the pathogenicity of bacterial species isolated from acute dento-alveolar abscesses. J. Med. Microbiol. 27:109–116.

Libertin, C. R., P. E. Hermans, and J. A. Washington II. 1985. Beta-hemolytic group F streptococcal bacteremia: a study and review of the literature. Rev. Infect. Dis. 7:498–503.

Linke, H. A. B. 1977. A new medium for the isolation of *Streptococcus mutans* and its differentiation from other oral streptococci. J. Clin. Microbiol. 5:604–609.

Loesche, W. J. 1986. Role of *Streptococcus mutans* in human dental decay. Microbiol. Rev. 50:353–380.

Loesche, W. J., S. Eklund, R. Earnest, and B. Burt. 1984. Longitudinal investigation of bacteriology of human fissure decay: epidemiological studies in molars shortly after eruption. Infect. Immun. 46:765–772.

Loewe, L., N. Plummer, C. F. Niven, and J. M. Sherman. 1946. *Streptococcus s.b.e.* in subacute bacterial endocarditis. J. Amer. Med. Assoc. 130:257.

Long, P. H., and E. A. Bliss. 1934. Studies upon minute hemolytic streptococci I. The isolation and cultural characteristics of minute β-hemolytic streptococci. J. Exp. Med. 60:619–631.

Lowes, J. A., G. Williams, S. Tabaqchali, I. M. Hill, J. Hamer, E. Houang, E. J. Shaw, and G. M. Rees. 1980. 10 years of infective endocarditis at St. Bartholomews's Hospital: analysis of clinical features and treatment in relation to prognoisis and mortality. Lancet. i:133–136.

Ludwig, W., E. Seewaldt, R. Kilpper-Bälz, K. H. Schleifer, L. Magrum, C. R. Woese, G. E. Fox, and E. Stackebrandt. 1985. The phylogenetic position of *Streptococcus* and *Enterococcus*. J. Gen. Microbiol. 131:543–551.

Lütticken, R., U. Wendorff, D. Lütticken, E. A. Johnson, and L. W. Wannamaker. 1978. Studies on streptococci resembling *Streptococcus milleri* and on an associated surface-protein antigen. J. Med. Microbiol. 11:419–431.

McFarlane, T. W. 1989. Plaque-related infections. J. Med. Microbiol. 29:161–170.

Madden, N. P., and C. A. Hart. 1985. *Streptococcus milleri* in appendicitis in children. J. Pediatr. Surg. 20:6–7.

Mao, S. W., and S. Rosen. 1980. Cariogenicity of *Streptococcus mutans*. J. Dent. Res. 59:1620–1626.

Marsh, P. D., and M. V. Martin. 1984. Oral Microbiology, 2nd ed. Van Nostrand Reinhold (UK) Co. Ltd., Wokingham.

Marsh, P. D., A. Featherstone, A. S. McKee, A. S. Hallsworth, C. Robinson, J. A. Weatherall, H. N. Newman, and A. F. V. Pitter. 1989. A microbiological study of early caries in approximal surfaces in schoolchildren. J. Dent. Res. 68:1151–1154.

Marshall, R., and A. K. Kaufman. 1981. Production of deoyribonuclease, ribonuclease, coagulase and hemolysins by anaerobic Gram-positive cocci. J. Clin. Microbiol. 13:787–788.

Matee, M. I., F. H. M. Mikx, J. E. F. N. Frencken, G. J. Truin, and H. M. H. M. Ruiken. 1985. Selection of a micromethod and its use in the estimation of salivary *Streptococcus mutans* and Lactobacillus counts in relation to dental caries in Tanzanian children. Caries Res. 19:497–506.

McGhee, J. R., and S. M. Michalek. 1981. Immunobiology of dental caries: microbial aspect and local immunity. Ann. Rev. Microbiol. 35:595–638.

Mejaré, B. 1975. Characteristics of *Streptococcus milleri* and *Streptococcus mitior* from infected dental root canals. Odont. Revy 26:291–308.

Mejaré, B., and S. Edwardsson. 1975. *Streptococcus milleri* (Guthof); an indigenous organism of the human oral cavity. Arch. Oral Biol. 20:757–762.

Mergenhagen, S. E., A. L. Sandberg, B. M. Chassy, M. J. Brennan, M. K. Yeung, J. A. Donkersloot, and J. O. Cisar. 1987. Molecular basis of bacterial adhesion in the oral cavity. Rev. Infect. Dis. 9 (Suppl. 5):S467–S474.

Michalek, S. M., J. R. McGhee, and J. M. Navia. 1975. Virulence of *Streptococcus mutans*: a sensitive method for evaluating cariogenicity in young gnotobiotic rats. Infect. Immun. 12:69–75.

Mikkelsen, L., S. B. Jensen, and J. Jakobsen. 1981. Microbiol studies on plaque from carious and caries-free proximal tooth surfaces in population with high caries experience. Caries Res. 15:451–452.

Mikkelsen, L., and S. Poulsen. 1976. Microbiological studies on plaque in relation to development of dental caries in man. Caries Res. 10:178–188.

Mirick, G. S. L. Thomas, E. C. Curnen, and F. L. Horsfall. 1944. Studies on a non-hemolytic streptococcus isolated from the respiratory tract of human beings. I. Biological characteristics of *Streptococcus MG*. J. Exp. Med. 80:391–406.

Montague, E. A., and Knox K. W. 1968. Antigenic components of the cell wall of *Streptococcus salivarius*. J. Gen. Microbiol. 54:237–246.

Moore, W. E. C., L. V. Holdeman, E. P. Cato, I. J. Good, E. P. Smith, R. R. Ranney, and K. G. Pulcanis. 1984. Variation in periodontal floras. Infect. Immun. 46:720–726.

Moore, W. E. C., L. V. Holdeman, E. P. Cato, R. M. Smibert, J. A. Burmeister, K. G. Pulcanis, and R. R. Ranney. 1985. Comparative bacteriology of juvenile periodontitis. Infect. Immun. 48:507–519.

Morris, E. O. 1954. The bacteriology of the oral cavity. III. *Streptococcus*. Brit. Dent. J. 96:95–108.

Moulsdale, M. T., S. J. Eykyn, and I. Phillips. 1980. Infective endocarditis, 1970–1979. A study of culture-positive cases in St. Thomas' Hospital. Quart. J. Med. 49:315–328.

Naiman, R. A., and J. G. Barrow. 1963. Penicillin-resistant bacteria in the mouths and throats of children receiving continuous prophylaxis against rheumatic fever. Ann. Intern. Med. 58:768–772.

Niven, C. F. Jr., and K. L. Smiley. 1942. The nutritive requirements of the 'viridans' streptococci with special reference to *Streptococcus salivarius*. (abstract). J. Bacteriol. 44:260.

Niven, C. F., Jr., K. L. Smiley, and J. M. Sherman. 1941. The production of large amounts of a polysaccharide by *Streptococcus salivarius*. J. Bacteriol. 41:479–484.

Niven, C. F., Jr., and J. C. White. 1946. A study of streptococci associated with subacute bacterial endocarditis. J. Bacteriol. 51:790.

Okahashi, N., C. Sasakawa, M. Yoshikawa, S. Hamada, and T. Koga. 1989a. Cloning of a surface protein antigen gene from serotype c *Streptococcus mutans*. Mol. Microbiol. 3:221–228.

Okahashi, N., C. Sasakawa, M. Yoshikawa, S. Hamada, and T. Koga. 1989b. Molecular characteristics of a surface protein antigen gene from serotype c *Streptococcus mutans*, implicated in dental caries. Mol. Microbiol. 3:673–678.

Orland, F. J., J. R. Blayney, R. W. Harrison, J. A. Reyniers, P. C. Trexler, R. F. Ervin, H. A. Gordon, and M. Wagner. 1955. Experimental caries in germ-free rats inoculated with enterococci. J. Am. Dent. Assoc. 50:259–272.

Otake, S., J. R. McGhee, M. Hirasawa, K. Williams, R. R. Arnold, J. L. Babbi, H. Kiyono, S. Cox, S. M. Michalek, T. Shiota, T. Ikeda, and K. Ochiai. 1978. Use of mutants in the elucidation of virulence of *Streptoccus mutans*. Adv. Exp. Med. Biol. 107:673–683.

Ottens, H., and K. C. Winkler. 1962. Indifferent and haemolytic streptococci possessing Group-antigen F. J. Gen. Microbiol. 28:181–191.

Parker, M. T., and L. C. Ball. 1976. Streptococci and aerococci associated with systemic infection in man. J. Med. Microbiol. 9:275–302.

Perch, B., E. Kjems, and T. Ravn. 1974. Biochemical and serological properties of *Streptococcus mutans* from various human and animal sources. Acta path. et Microbiol. Scand.(B). 82:357–370.

Peterson, E. M. J. T. Shigei, A. Woolard, and L. M. De La Maza. 1988. Identification of viridans streptococci by three commercial systems. Amer. J. Clin. Pathol. 90:87–91.

Phillips, I., C. Warren, J. M. Harrison, P. Sharples, L. C. Ball, and M. T. Parker. 1976. Antibiotic susceptibilities of streptococci from the mouth and blood of patients treated with penicillin or lincomycin and clindamycin. J. Med. Microbiol. 9:393–404.

Poole, P. M., and G. Wilson. 1976. Infection with minute-colony-forming β-haemolytic streptococci. J. Clin. Path. 29:740–745.

Poole, P. M., and G. Wilson. 1977. *Streptococcus milleri* in the appendix. J. Clin. Pathol. 30:937–942.

Poole, P. M., and G. Wilson. 1979. Occurrence and cultural features of *Streptococcus milleri* in various body sites. J. Clin. Path. 32:764–768.

Porterfield, J. S. 1950. Classification of the streptococci of subacute bacterial endocarditis. J. Gen. Microbiol. 4:92–101.

Powley, L., J. Meeson, and D. Greenwood. 1989. Tolerance to penicillin in streptococci of viridans group. J. Clin. Pathol. 42:77–80.

Pullian, L., R. K. Porschen, and W. K. Hadley. 1980. Biochemical properties of CO_2-dependent streptococci. J. Clin. Microbiol. 12:27–31.

Quinn, J. P., C. A. DiVincenzo, D. A. Lucks, R. L. Luskin, K. L. Shatzer, and S. A. Lerner. 1988. Serious infections due to penicillin-resistant strains of viridans streptococci with altered penicillin-binding proteins. J. Infect. Dis. 157:764–769.

Rabe, L. K., K. K. Winterscheid, and S. L. Hillier. 1988. Association of viridans group streptococci from pregnant women with bacterial vaginosis and upper genital tract infection. J. Clin. Microbiol. 26:1156–1160.

Roberts, R. B., A. G. Krieger, N. L. Schiller, and K. C. Gross. 1979. Viridans streptococcal endocarditis: the role of various species, including pyridoxal-dependent streptococci. Rev. Infect. Dis. 1:955–966.

Rogers, A. H. 1977. Evidence for the transmissibility of human dental caries. Aust. Dent. J. 22:53–56.

Rogers, A. H. 1981. The source of infection in the intrafamilial transfer of *Streptococcus mutans*. Caries Res. 15:26–31.

Rosan, B. 1973. Antigens of *Streptococcus sanguis*. Infect. and Immun. 7:205–211.

Rosan, B. 1976. Relationship of the cell wall composition of Group H streptococci and *Streptococcus sanguis* to their serological properties. Infect. Immun. 13:1144–1153.

Rosan, B., and L. Argenbright. 1982. Antigenic determinant of the Lancefield Group H antigen of *Streptococcus sanguis*. Infect. Immun. 38:925–931.

Ruoff, K. L. 1988. *Streptococcus anginosus ("Streptococcus milleri")*: the unrecognized pathogen. Clin. Microbiol. Rev. 1:102–108.

Ruoff, K. L., and L. Kunz. 1982. Identification of viridans streptococci isolated form clinical specimens. J. Clin. Microbiol. 15:920–925.

Ruoff, K. L., L. Kunz, and M. J. Ferraro. 1985. Occurrence of *Streptococcus milleri* among beta-hemolytic streptococci isolated from clinical specimens. J. Clin. Microbiol. 22:149–151.

Ruoff, K. L., S. I. Miller, C. V. Garner, M. J. Ferraro, and S. B. Calderwood. 1989. Bacteremia with *Streptococcus bovis* and *Streptococcus salivarius*: clinical correlates of more accurate identification of isolates. J. Clin. Microbiol. 27:305–308.

Russell, M. W., and J. Mestecky. 1986. Potential for immunological intervention against dental caries. J. Biol. Buccale 14:159–175.

Russell, R. R. B. 1990. Genetic analysis and genetic probes for oral bacteria. In: D. B. Ferguson (ed.), Molecular genetics of the oral cavity. Frontiers in Oral Physiology 8: (in press) Karger, Basel.

Russell, R. R. B., and M. L. Gilpin. 1987. Identification of virulence components of mutans streptococci, p. 201–204. In: J. J. Ferretti and R. Curtiss (ed.), Streptococcal genetics. American Society for Microbiology, Washington, D.C.

Russell, R. R. B., and N. W. Johnson. 1987. The propects for vaccination against dental caries. Brit. Dent. J. 162:29–34.

Saunders, K. A., and L. C. Ball. 1980. The influence of the composition of blood agar on beta-haemolysis by *Streptococcus salivarius*. Med. Lab. Sci. 37:341–345.

Schiller, N. L., and R. B. Roberts. 1982. Vitamin B_6 requirements of nutritionally variant *Streptococcus mitior*. J. Clin. Microbiol. 15:740–743.

Schleifer, K. H., and O. Kandler. 1972. Peptidoglycan types of bacteria cell walls and their taxonomic implications. Bacteriol. Rev. 36:407–477.

Schleifer, K. H., and R. Kilpper-Bälz. 1984. Transfer of *Streptococcus faecalis* and *Streptococcus faecium* to the genus *Enterococcus* nom.rev. as *Enterococcus faecalis* comb.nov. and *Enterococcus faecium* comb.nov. Internat. J. System. Bacteriol. 34:31–34.

Schleifer, K. H., and R. Kilpper-Bälz. 1987. Molecular and chemotaxonomic approaches to the classification of streptococci, enterococci and lactococci: a review. System. Appl. Microbiol. 10:1–19.

Schleifer, K. H., R. Kilpper-Bälz, J. Kraus, and F. Gehring. 1984. Relatedness and classification of *Streptococcus mutans* and "mutans-like" streptococci. J. Dent. Res. 63:1047–1050.

Schleifer, K. H., J. Kraus, C. Dvorak, R. Kilpper-Bälz, M. D. Collins, and W. Fischer. 1985. Transfer of *Streptococcus lactis* and related streptococci to the genus *Lactococcus* gen.nov. Syst. Appl. Microbiol. 6:183–195.

Schlossberg, D. (ed.). 1987. Infections of the head and neck. Springer-Verlag, New York.

Schmidhuber, S., R. Kilpper-Bälz, and K. H. Schleifer. 1987. A taxonomic study of *Streptococcus mitis, S. oralis* and *S. sanguis.* System. Appl. Microbiol. 10:74–77.

Schmidhuber, S., W. Ludwig, and K. H. Schleifer. 1988. Construction of a DNA probe for the specific identification of *Streptococcus oralis.* J. Clin. Microbiol. 26:1042–1044.

Schottmüller, H., 1903. Die Artunterscheidung der für den Menschen pathogenen Streptokokken durch Blutagar. Münchener Medizinische Wochenschrift 50:849–853; 909–912.

Selbie, F. R., R. D. Simon, and R. H. H. Robinson. 1949. Serological classification of viridans streptococci for subacute endocarditis, teeth and throats. Brit. Med. J. 2:667–672.

Sherman, J. M. 1937. The streptococci. Bacteriol. Revs. 1:3–97.

Sherman, J. M. C. F. Niven, and K. L. Smiley. 1943. *Streptococcus salivarius* and other non-hemolytic streptococci of the human throat. J. Bacteriol. 45:249–263.

Shlaes, D. M., P. I. Lerner, E. Wolinsky, and K. V. Gopalakrishna. 1981. Infections due to Lancefield Group F and related streptococci *(S. milleri, S. anginosus).* Medicine 60:197–207.

Skerman, V. B. D., V. McGowan, and P. H. A. Sneath. 1980. Approved lists of bacterial names. Int. J. System. Bacteriol. 30:225–420.

Skjold, S. A., P. G. Quie, L. A. Fries, M. Barnham, and P. P. Cleary. 1987. DNA fingerprinting of *Streptococcus zooepidemicus* (Lancefield group C) as an aid to epidemiological study. J. Infect. Dis. 155:1145–1150.

Slater, G. J., and D. Greenwood. 1983. Detection of penicillin tolerance in streptococci. J. Clin. Pathol. 36:1353–1356.

Solowey, M. 1942. A serological classification of viridans streptococci with special reference to those isolated from subacute bacterial endocarditis. J. Exp. Med. 76:109–126.

Stecksen-Blicks, C. 1985. Salivary counts of lactobacilli and *Streptococcus mutans* in caries prediction. Scand. J. Dent. Res. 93:204–212.

Steffen, E. K., and D. J. Hentges. 1981. Hydrolytic enzymes of anaerobic bacteria isolated from human infections. J. Clin. Microbiol. 14:153–156.

Stein, D. S., and C. R. Libertin. 1989a. A double-blinded comparative evaluation of three media for chromophore testing with viridans and nutritionally variant (deficient) streptococci. Amer. J. Clin. Pathol. 91:589–593.

Stein, D. S., and C. R. Libertin. 1989b. Molecular analysis of viridans and nutritionally deficient (variant) streptococci causing sequential episodes of endocarditis in a patient. Amer. J. Clin. Pathol. 91:602–624.

Stein, D. S., and K. E. Nelson. 1987. Endocarditis due to nutritionally deficient streptococci: therapeutic dilemma. Rev. Infect. Dis. 9:908–916.

Syed, S. A., and W. J. Loesche. 1972. Survival of human dental plaque flora in various transport media. Appl. Microbiol. 24:638–644.

Tanzer, J. M. (ed.). 1981. Animal models in cariology. Special Supplement, Microbiology abstracts. Information Retrieval Inc., Washington, D.C.

Tanzer, J. M., A. C. Borjesson, L. Kurasz, L. Laskoski, M. Testa, and B. Krasse. 1983. GSBT, an alternative to MSB agar. J. Dent. Res. 62:241 (Abstract).

Tanzer, J. M., M. L. Freedman, R. J. Fitzgerald, and R. H. Larson. 1974. Diminished virulence of glucan synthesis-defective mutants of *Streptococcus mutans.* Infect. Immun. 10:197–203.

Tanzer, J. M., M. L. Freedman, F. N. Woodiel, R. L. Eifert, and L. A. Rinehimer. 1976. Association of *Streptococcus mutans* virulence with synthesis of intracellular polysaccharide, p. 597–616. In: H. M. Stiles, W. J. Loesche, and T. C. O'Brien (ed.), Proceedings: Microbial aspects of dental caries (a special supplement to Microbiology Abstracts), vol. 3. Information Retrieval, Inc., Washington, D.C.

Ten Cate, J. M., F. A. Leach, and J. Arends. 1984. Bacterial Adhesion and Preventative Dentistry. IRL Press, Oxford.

Tresadern, J. C., R. J. Farrand, and M. H. Irving. 1983. *Streptococcus milleri* and surgical sepsis. Ann. R. Coll. Surg. Engl. 65:78–79.

Unsworth, P. F. 1980. The isolation of streptococci from human faeces. J. Hyg. (Camb.) 85:153–164.

Van der Auwera, P. 1985. Clinical significance of *Streptococcus milleri.* Eur. J. Clin. Microbiol. 4:386–390.

Van de Rijn, I., and A. Bouvet. 1984. Characterization of a pH-dependent chromophore from nutritionally variant streptococci. Infect. Immun. 43:28–31.

Van de Rijn, I., and R. E. Kessler. 1980. Growth characteristics of group A streptococci in a new chemically defined medium. Infect. Immun. 27:444–448.

Van Houte, J., H. V. Jordan, and S. Bellack. 1971. Proportions of *Streptococcus sanguis,* an organism associated with subacute bacterial endocarditis, in human feces and dental plaque. Infect. and Immun. 4:658–659.

Van Palenstein Helderman, W. H., M. Ijsseldijk, and J. H. Huis in't Veld. 1983. A selective medium for the two major subgroups of the bacterium *Streptococcus mutans* isolated from human dental plaque and saliva. Arch. Oral Biol. 28:599–603.

Wade, W. G., M. J. Aldred, and D. M. Walker. 1986. An improved medium for isolation of *Streptococcus mutans.* J. Med. Microbiol. 22:319–323.

Washburn, M. R., J. C. White, and C. F. Niven. Jr. 1946. *Streptococcus S.B.E.*: immunological characteristics. J. Bacteriol. 51:723–729.

Weerkamp, A. H., and T. Jacobs. 1982. Cell wall-associated protein antigens of *Streptococcus salivarius*: purification, properties and function in adherence. Infect. Immun. 38:233–242.

Weerkamp, A. H., H. C. Van der Mei, D. P. E. Engelen, and C. E. A. de Windt. 1984. Adhesion receptors (adhesins) of oral streptococci, p. 85–97. In: J. M. ten Cate, F. A. Leach and J. Arends, (ed.), Adhesion and preventative dentistry. IRL Press, Oxford.

Welborn, P. P., W. K. Hadley, E. Newbrun, and D. M. Yajko. 1983. Characterization of strains of viridans streptococci by deoxyribonucleic acid hybridization and physiological tests. Int. J. Syst. Bacteriol. 33:293–299.

Whiley, R. A., H. Y. Fraser, C. W. I. Douglas, J. M. Hardie, A. M. Williams, and M. D. Collins. 1990a. *Streptococcus parasanguis* sp. nov., an atypical viridans streptococcus from human clinical specimens. FEMS Microbiol. Lett. 68:115–122.

Whiley, R. A., H. Y. Fraser, J. M. Hardie and D. Beighton. 1990b. Phenotypic differentiation of *Streptococcus intermedius*, *Streptococcus constellatus*, and *Streptococcus anginosus* strains within the "*Streptococcus milleri* group." J. Clin. Microbiol. 28:1497–1501.

Whiley, R. A., and J. M. Hardie. 1988. *Streptococcus vestibularis* sp.nov. from the human oral cavity. Int. J. System. Bacteriol. 38:335–339.

Whiley, R. A., and J. M. Hardie. 1989. DNA-DNA hybridization studies and phenotypic characteristics of strains within the "*Streptococcus milleri* group". J. Gen. Microbiol. 135:2623–2633.

Whiley, R. A., and J. M. Hardie. 1990. DNA-DNA hybridization studies on strains of *Streptococcus sanguis*. In: R. Lütticken (ed.), 10th Lancefield International Symposium on Streptococci and Streptococcal Diseases (in press).

Whiley, R. A., R. R. B. Russell, J. M. Hardie, and D. Beighton. 1988. *Streptococcus downei* sp.nov. for strains pre-viously described as *Streptococcus mutans* serotype h. Int. J. System. Bacteriol. 38:25–29.

White, J. C., and C. F. Niven. Jr. 1946. *Streptococcus S.B.E.*: a *Streptococcus* associated with subacute bacterial endocarditis. J. Bacteriol. 51:711–722.

Willcox, M. D. P., D. B. Drucker, and R. M. Green. 1987. Relative cariogenicity and in vivo plaque-forming ability of the bacterium *Streptococcus oralis* in gnotobiotic WAG/RIJ rats. Arch. Oral Biol. 32:455–457.

Williams, R. E. O. 1956. *Streptococcus salivarius* (vel. *hominis*) and its relations to Lancefield's Group K. J. Path. Bacteriol. 72:15–25.

Williamson, C. K. 1964. Serological classification of viridans streptococci from the respiratory tract of man, p. 607–622. In: C. A. Leone (ed.), Taxonomic biochemistry and serology. Ronald Press Co., New York.

Yakushiji, T., R. Konagawa, M. Oda, and M. Inoue. 1988. Serological variation in oral *Streptococcus milleri*. J. Med. Microbiol. 27:145–151.

Zickert, I., C. G. Emilson, and B. Krasse. 1982. Effect of caries preventive measures in children highly infectd with the bacterium *Streptococcus mutans*. Arch. Oral Biol. 27:861–868.

Zickert, I., C. G. Emilson, and B. Krasse. 1983. Correlation of level and duration of *Streptococcus mutans* infection with incidence of dental caries. Infect. Immunol. 39:982–985.

Zinner, D. D., J. M. Jablon, A. P. Aran, and M. S. Saslaw. 1965. Experimental caries induced in animals by streptococci of human origin. Proc. Soc. Exp. Bio. Med. 118:766–770.

CHAPTER 65

The Genus *Streptococcus*—Medical

KATHRYN L. RUOFF

Recent taxonomic studies have engendered numerous changes in the genus *Streptococcus*. What was once considered a single genus of catalase-negative Gram-positive coccoid bacteria is now divided into the genera *Streptococcus, Enterococcus,* and *Lactococcus* (Schleifer and Kilpper-Balz, 1987). As a result of genetic and physiological studies, some of the species remaining in the genus *Streptococcus* have been redefined and renamed, and our concepts of streptococcal classification have been altered.

Bacteria now included in the genus *Streptococcus* are nonsporeforming chemoorganotrophs forming coccoid or coccobacillary cells arranged in pairs or chains. These nutritionally fastidious organisms ferment carbohydrates with the production of lactic acid as the major end product. Although streptococci are incapable of using oxygen metabolically, they grow in its presence. Thus, they have been referred to as aerotolerant anaerobes. Some capnophilic (CO_2-requiring) strains have also been described. Organisms previously referred to as obligately anaerobic streptococci (Hardie, 1986) appear to be genetically unrelated to streptococci (Ludwig et al., 1988). Occasional strains of streptococci synthesize peroxidase enzymes that produce false-positive catalase reactions when hydrogen peroxide is used as the assay reagent. None of the streptococci are capable of synthesizing heme groups, and are consequently negative in the benzidine test (Deibel and Evans, 1960) and the porphyrin test described by Wong (1987).

Differentiation of streptococci from facultatively anaerobic members of the family Micrococcaceae (*Staphylococcus, Stomatococcus*) is usually easily accomplished on the basis of the catalase test and cellular morphology. Streptococci may, however, be easily confused with facultatively anaerobic Gram-positive cocci of the genera *Enterococcus* (see Chapter 66), *Aerococcus* (Chapter 68), *Leuconostoc* (Chapter 69), *Pediococcus* (Chapter 68), *Gemella* (Chapter 74), and *Lactococcus* (Chapter 67). At present, it seems that lactococci are extremely rare in clinical specimens. The other five genera may be encountered in clinical specimens, and can be differentiated from streptococci by the characteristics noted in Table 1 (Facklam et al., 1989). In addition to production of the enzyme pyrrolidonyl arylamidase, enterococci are distinguished from the other genera in Table 1 by their relative resistance to antibiotics. *Leuconostoc* and *Pediococcus,* only recently recognized in clinical specimens, are unusual in their resistance to vancomycin, a cell wall synthesis inhibiting antibiotic active against virtually all Gram-positive bacteria. Aside from cellular morphology and arrangement, the inability to grow well under anaerobic conditions distinguishes the genus *Aerococcus* from streptococci and from the other bacteria listed in Table 1 (Evans 1986). These nonstreptococcal genera are covered in detail elsewhere in this volume (see Chapters 66–69).

In the past, the characterization of medically important streptococci was restricted to a few easily determined traits, such as the hemolytic reactions originally described by Brown (1919) and the antigens discovered by Lancefield (1933). While the contributions of these streptococcologists are invaluable, reliance solely on hemolytic and serological reactions for characterizing streptococci resulted in an inadequate classification system. The more recent application of comprehensive physiological testing and genetic homology studies has led to the creation of an updated classification system based on currently accepted taxonomic criteria. Table 2 presents the streptococci of medical importance examined in this chapter, notes their distinguishing characteristics, and comments on their nomenclature and taxonomic standing. Other streptococci are covered in Chapter 64.

Habitats

Streptococci of medical importance inhabit the skin and mucous membranes of the respiratory, alimentary, and urogenital tracts. Table 3 lists

Table 1. Differential characteristics of catalase-negative, benzidine-negative, facultatively anaerobic Gram-positive cocci.

Characteristic	*Streptococcus*	*Enterococcus*	*Leuconostoc*	*Pediococcus*	*Aerococcus*	*Gemella*
Cellular morphology and arrangement	Spherical or ovoid cells in pairs, chains	Spherical or ovoid cells in pairs, chains	Spherical or ovoid cells in pairs, chains	Spheres in tetrads, pairs	Spheres in tetrads, pairs	Cocci in pairs; adjacent sides flattened; short chains
Good growth under anaerobic conditions	+	+	+	+	−	−[a]
Gas produced from glucose	−	−	+	−	−	−
Resistance to vancomycin	−	−	+	+	−	−
Pyrrolidonyl arylamidase[b]	V[c]	+	−	−	+	V

+, positive; −, negative; V, variable.

[a]Weak growth under anaerobic conditions is characteristic of *Gemella hemolysans* (Reyn, 1986). The proposal by Kilpper-Balz and Schleiffer (1988) to include *Streptococcus morbillorum* in the genus *Gemella* as *Gemella morbillorum* would negate the usefulness of this characteristic for genus identification, since *G. morbillorum* grows well anaerobically.

[b]Methods for assay of this enzyme appear in the "Identification" section, this chapter.

[c]*Streptococcus pyogenes* (beta-hemolytic, Lancefield group A) produces this enzyme; only occasional strains of other streptococci resembling enterococci are positive, thus this is a useful characteristic for differentiating enterococci from non-beta-hemolytic streptococci.

Table 2. Clinically important streptococci isolated from humans: nomenclature and basic characteristics.

Species or division	Synonyms	Hemolysis[a]	Antigens	Comments
S. pyogenes		β	Lancefield group A	Forms large colonies in comparison to beta-hemolytic group A strains of *S. anginosus* (Hardie and Whiley, Chapter 64).
S. agalactiae		β, γ	Lancefield group B	
Group C and G streptococci (large-colony)	Group C: *S. equisimilis* *S. equi* *S. zooepidemicus* *S. dysgalactiae*	β	Lancefield group C or G	Human strains of large-colony-forming group C and G streptococci are closely related. *S. dysgalactiae* has been proposed as a species name for these organisms (Farrow and Collins, 1984). Not related to small-colony-forming beta-hemolytic group C and G *S. anginosus* (Hardie and Whiley, Chapter 64).
S. pneumoniae		α	Capsular polysaccharide antigens	
S. bovis		γ, α	Lancefield group D	Previously referred to as a "non-enterococcal" group D streptococcus. Enterococci also have the group D antigen.

[a]α, alpha hemolysis; β, beta hemolysis; γ, gamma hemolysis (no reaction on blood agar).

the habitats and summarizes the important pathogenic activities of some of the major streptococcal groups. *S. pyogenes* (group A) is a well-described pathogen capable of elaborating an impressive array of erythrogenic and cytolytic (hemolytic) toxins (Wannamaker, 1983). Non-suppurative sequelae of *S. pyogenes* infection include rheumatic fever and acute glomerulonephritis. Acute rheumatic fever, after declining in incidence in developed countries, appears to

Table 3. Habitats and pathogenicity of clinically important streptococci.

Species or group	Habitat	Pathogenicity	References
S. pyogenes	Throat	Pharyngitis	Bisno, 1990a, 1990b
	Skin	Respiratory tract infection	Wannamaker, 1979
	Rectum	Skin and soft tissue infection	
		Rheumatic fever	
		Glomerulonephritis	
		Endocarditis	
S. agalactiae	Genital tract	Neonatal infection	Edwards and Baker, 1990
	Gastrointestinal tract	Urogenital infection	Gallagher and Watanakunakorn, 1985
	Throat	Skin and soft tissue infection	Gallagher and Watanakunakorn, 1986
	Skin	Endocarditis	Patterson and Hafeez, 1976
		Respiratory tract infection	
Group C and G	Throat	Respiratory tract infection	Barnham et al., 1983
streptococci	Genital tract	Pharyngitis	Benjamin and Perriello, 1976
(large-colony	Gastrointestinal tract	Endocarditis	Chung, 1982
forms)	Skin	Skin and soft tissue infection	Downing and Spirazza, 1986
		Meningitis	Duca et al., 1969
		Bone and joint infection	Gaunt and Seal, 1987
		Nephritis	Gnann et al., 1987
			Goldmann and Breton, 1978
			Mohr et al., 1979
			Smyth et al., 1988
			Stamm and Cobbs, 1980
S. pneumoniae	Throat	Respiratory tract infection	Burman et al., 1985
		Ear and eye infection	Jette et al., 1989
		Septicemia	Mufson, 1990
S. bovis	Gastrointestinal tract	Endocarditis	Klein et al., 1987
		Bacteremia associated with colonic cancer	

be once again on the increase in the United States. The 1980s brought reports of a resurgence of this disease (Congeni et al., 1987; Hosier et al., 1987; Veasey et al., 1987). Although the reasons for this observed increase in acute rheumatic fever are not fully understood, one suggestion has been that mucoid (due to a thick hyaluronic acid capsule) type M-18 *S. pyogenes* may be involved (Marcon et al., 1988). Others (Kaplan et al., 1989) have implicated multiple serotypes of *S. pyogenes* in documented cases of rheumatic fever. In spite of decades of research into the pathogenic propensities of *S. pyogenes,* there is still much to be learned about this clinically important species.

Group B streptococci (*S. agalactiae),* originally recognized as bovine pathogens, have become increasingly important in human infections. Heavy colonization of the maternal genital tract is correlated with a high risk of infection in newborns (Jones et al., 1984). Early onset (within the first few days of life) and late onset (within 1 week–4 months after birth) syndromes of group B infection are overwhelming diseases characterized by septicemia and meningitis. Intrapartum antibiotic prophylaxis of colonized mothers (Boyer and Gotoff, 1986) and immunization of pregnant women (Baker

et al., 1988) have been suggested as strategies to prevent neonatal group B infection. Serious infection by *S. agalactiae* also occurs in adults and usually coincides with serious underlying disease (Gallagher and Watanakunakorn, 1985).

Beta-hemolytic large-colony-forming streptococci with Lancefield's group C or G antigen are isolated from the human throat, skin, respiratory, and gastrointestinal tracts, and are responsible for a variety of infections. Streptococci with the same Lancefield antigens are also important animal pathogens. Group C strains were formerly divided into several species on the basis of their fermentative reactions (*S. equi, S. equisimilis, S. zooepidemicus,* and *S. dysgalactiae.,*).) Recent work suggests that *S. equisimilis* and *S. dysgalactiae,* along with other human isolates producing group G or L Lancefield antigens, form one species. The name *S. dysgalactiae* has been proposed for this group. Animal isolates of *S. zooepidemicus* and *S. equi* appear to be closely related and *S. zooepidemicus* has been proposed as a subspecies of *S. equi* (Farrow and Collins, 1984). The name *S. canis* has been suggested for animal, but not human, strains of group G streptococci (De Vriese et al., 1986). It should be noted that large-colony-forming beta-hemolytic group C and G streptococci are not related to *S. anginosus* strains (small colony formers) displaying beta-hemolysis and group C or G antigen (see Chapter 64).

Pneumococci (*S. pneumoniae*) reside in the upper respiratory tract in 5–70% of healthy adults (Hendley et al., 1975). These streptococci are important agents of pneumonia, bacteremia, otitis media, and meningitis. During the past two decades an increasing number of reports of penicillin-resistant pneumococcal isolates have appeared in the literature (Appelbaum, 1987). These pathogens produce polysaccharide capsules, of which more than 80 different serotypes have been described. Surveillance of the prevailing serotypes causing infection is important for formulating capsular vaccines which have been available since the late 1970s.

Streptococcus bovis is currently the sole Lancefield group D species of medical importance. Enterococci, now considered as a separate genus (Schleifer and Kilpper-Balz, 1984 see also Chapter 66) also produce the group D antigen and were previously regarded as streptococcal species which could be differentiated from *S. bovis* on the basis of their ability to grow in the presence of 6.5% sodium chloride. Both enterococci and *S. bovis* can hydrolyze the glycoside esculin in the presence of 40% bile; this activity was previously used to separate these two groups of organisms from other streptococci. *S. bovis* inhabits the gastrointestinal tract and when isolated from the bloodstream, may indicate the presence of colonic cancer (Klein et al., 1987). Based on physiological characteristics, *S. bovis* can be separated into two biotypes (Facklam, 1972; Parker and Ball, 1976), one of which is able to produce a polysaccharide when grown in the presence of sucrose. The polysaccharide-producing biotype is more often associated with colonic neoplasms and endocarditis than its nonpolysaccharide-producing counterpart (Ruoff et al., 1989).

Isolation

Streptococci are responsible for many different infections and are consequently isolated from numerous types of clinical specimens. If recommended practices for specimen collection and transport (Isenberg et al., 1985) are adhered to, streptococci present in clinical specimens should remain viable. Throat swabs for culture of group A (*S. pyogenes*) streptococci may be transported directly to the laboratory in a sterile container if processing will occur within 2 h. If a longer interval between collection and processing is anticipated, a transport medium should be used. For extended transport times (longer than 24 h) commercially available silica gel or filter paper transport systems are appropriate (Facklam and Carey, 1985).

In spite of the nutritionally fastidious nature of streptococci, many commonly used laboratory media support their growth. Rich agar-containing media which have little or no reducing sugar (tryptic soy, heart infusion, and brucella) and are supplemented with 5% animal blood (sheep or horse) are excellent for cultivation of streptococci and determination of hemolysis. Selective media for streptococci include media selective for Gram-positive bacteria in general (e.g., phenylethyl alcohol or Columbia colistin-nalidixic acid agars) as well as media developed especially for the isolation of specific groups of streptococci. Many studies have investigated the utility of selective media for the culture of *S. pyogenes* from throat swabs. Black and Van Buskirk (1973) advocated the addition of gentamicin to Columbia blood agar for selection of beta-hemolytic streptococci, but Murray and colleagues (1976) found that gentamicin supplementation of trypticase soy blood agar restricted growth of *S. pyogenes*. Gunn and co-workers (1977) suggested that the addition of sulfamethoxazole and trimethoprim to tryptic soy blood agar would enhance recovery of group A and B streptococci. Their findings were corroborated by Kurzynski and Meise (1979), but other workers (Dykstra et al., 1979; Libertin et al., 1983) found no advantage for recovery of *S. pyogenes* with sulfamethoxazole and trimethoprim-supplemented media. A commercially available medium employing sulfamethoxazole, trimethoprim, colistin, and crystal violet as selective agents for group A streptococci was described by Carlson and co-workers (1985). Selective media for group B beta-hemolytic streptococci have also been described (de la Rosa et al., 1983; Lim et al., 1987).

While most streptococci grow in ambient atmospheres, anaerobic incubation enhances growth and the expression of beta-hemolysis, due to the functioning of both oxygen-stable and oxygen-labile hemolysins in the absence of oxygen. Incubation in the presence of elevated CO_2 concentrations (CO_2 incubator or candle jar) is optimal for recovery of capnophilic streptococcal strains and should be employed when streptococci other than *S. pyogenes* are sought or when normally sterile body fluids or wound specimens are cultured. Incubation in air or elevated CO_2 concentrations interferes with the detection of beta-hemolysis, but in the case of throat cultures, adherence to the inoculation procedures described below obviate these atmosphere-induced problems of hemolysis determination.

Throat Cultures

Sheep blood-supplemented agars are recommended for throat cultures because *Haemophilus haemolyticus*, a beta-hemolytic throat inhabitant, fails to grow on these media, due to the low levels of NAD in sheep blood. This results in the reduction of the number of beta-hemolytic colonies to be investigated as suspected streptococci. Throat culture swabs may be directly inoculated onto blood agar plates, or may be used to make pour plates as explained below.

Streak-Stab Plates (Facklam and Carey, 1985)

Approximately one-sixth of the plate is inoculated by rolling the swab across the agar surface (Fig. 1). The plate is then streaked with a wire loop to produce isolated colonies. After streaking, the agar is stabbed as shown in Fig. 1, with the loop held at a 90° angle to the agar surface to introduce as little oxygen as possible into the stab. The relatively anaerobic environment within the agar stab will enhance hemolysis and obviate the need for anaerobic incubation of the plates, which would otherwise be necessary for the accurate determination of hemolysis of surface colonies.

Blood Agar Pour Plates (Facklam and Carey, 1985)

Figure 2 illustrates the preparation of blood agar pour plates for beta-hemolytic streptococcal isolation. The throat swab is first introduced into 1 ml of broth (Todd-Hewitt, tryptic soy, heart infusion) and incubated for 2–24 h at 35°C. For specimens received promptly (within 2–4 h after collection), a 2 h incubation is sufficient. If the transit time of the specimen is between 4 and 8 h, incubation for a minimum of 2 h is

recommended, while specimens that have been in transit for more than 8 h should be incubated for 4 to 5 h. Swabs may also be incubated in broth overnight. After incubation, the swab is rolled against the inside of the tube to express as much broth as possible. The swab is then removed to a sterile tube. A loopful of broth is transferred to a tube containing 15 to 20 ml of blood agar base which has been melted, cooled to 50°C, and supplemented with 0.8 to 1.0 ml of sterile defibrinated sheep blood. Broths that have been incubated overnight should be diluted in sterile saline (one loopful of broth in 15 ml saline) before being added to the melted blood agar (one loopful of the saline dilution is added to the agar). The contents of the inoculated melted blood agar tube are mixed thoroughly before delivery to a sterile petri dish. The original specimen swab can be used to inoculate one-half of the surface of the solidified agar. This will allow for both surface and subsurface growth. Pour plates can be incubated in any atmosphere. Subsurface colonies should be examined for hemolysis reactions, as described below.

Identification

Determination of Hemolytic Reactions

As mentioned previously, an anaerobic environment is essential for the accurate determination of hemolytic reactions. Under these conditions, both streptolysin O and streptolysin S are functional and there is no interference from peroxide, which may be produced during aero-

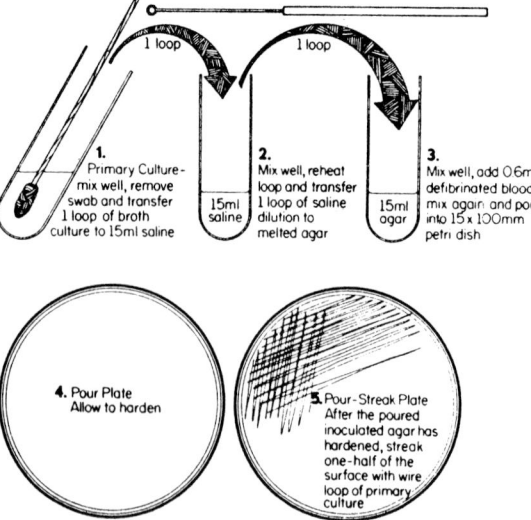

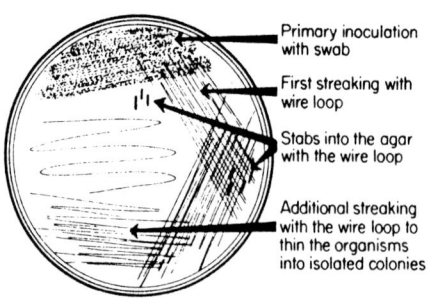

Fig. 1. The blood agar streak-stab method for isolation of beta-hemolytic streptococci.

Fig. 2. Method for preparing blood agar pour plates used in the isolation of beta-hemolytic streptococci.

bic growth of the beta-hemolytic streptococcus being examined or by neighboring alpha-hemolytic streptococci on the surface of a streak plate. Figure 3 illustrates the microscopic appearance (approximately 60× magnification) of hemolytic reactions of subsurface streptococcal colonies in pour plates and of streptococcal growth surrounding stabs in the agar. Beta-hemolysis is characterized by complete lysis of red blood cells. In alpha-hemolysis, partial lysis of blood cells occurs. Some intact cells remain, and the agar surrounding streptococcal growth appears greenish on macroscopic inspection. A zone of partial lysis surrounded by a zone of complete lysis is referred to as alpha prime hemolysis. The term "gamma-hemolysis" refers to a lack of action on blood agar, evidenced by nonhemolytic organisms. While accurate determination of hemolytic reactions requires microscopic examination, plates are more often examined only macroscopically in the clinical laboratory. For macroscopic examination, plates must be viewed with a light source behind them. Under these conditions beta-hemolysis appears as complete clearing of the agar surrounding a stab.

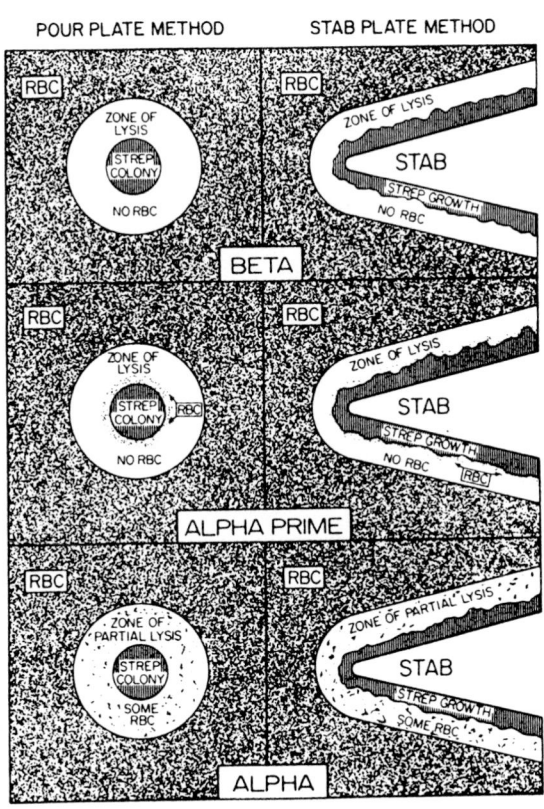

Fig. 3. Microscopic appearance of streptococcal hemolytic reactions.

Serological Determination

Lancefield's descriptions of streptococcal antigens in the 1930s fostered intense interest in the serological characteristics of streptococci. For close to half a century, Lancefield serology was considered as perhaps the most important characteristic for identification of beta-hemolytic streptococci, based on the premise that possession of a given Lancefield antigen predicted an isolate's relatedness with other isolates producing the identical antigen. Although more recent studies of streptococcal physiology and nucleic acid relatedness have engendered movement away from strict adherence to classification based on Lancefield serology, determination of streptococcal group antigens is still an important identification technique.

Prior to the 1980s, serological techniques were somewhat cumbersome and few laboratories employed them routinely. Streptococci to be serologically grouped were first grown overnight in Todd-Hewitt or another suitable broth. After centrifugation of the culture, the streptococcal cell pellet was subjected to a procedure designed to extract the Lancefield group antigen from the cell wall. Table 4 lists the methods commonly employed for antigen extraction. The extracted antigens were then identified via precipitin testing, using antisera specific for the various group antigens. A micromethod for precipitin testing, performed in capillary tubes, can be employed to conserve antisera (Facklam and Carey, 1985).

More recently, commercially available products have made streptococcal grouping feasible as a routine procedure in the clinical laboratory. These products usually employ enzymes or nitrous acid as extraction agents. Solubilized antigens are mixed with a suspension of particles that have been coated with antiserum specific for a given Lancefield antigen. The particles employed are either latex beads (latex agglutination techniques) or killed *Staphylococcus aureus* cells (coagglutination techniques). When streptococcal antigen is mixed with particles coated with the homologous antiserum, a macroscop-

Table 4. Methods for extraction of Lancefield antigens.

Extraction agent	Reference
Hot hydrochloric acid	Lancefield, 1933
Hot formamide	Fuller, 1938
Streptomyces albus enzymes	Maxted, 1948
Heat (autoclave)	Rantz and Randall, 1955
Pronase B	Ederer et al., 1972
Nitrous acid	El Kholy et al., 1974
Streptomyces albus enzymes and lysozyme	Watson et al., 1975

ically visible clumping of the particles occurs. Detailed instructions for performing these tests vary depending on the product, and are supplied by the manufacturer. These products offer relatively rapid and simple methods for Lancefield grouping in the clinical laboratory. Daly and Seskin (1988) have evaluated two of the newer products and cite publications evaluating numerous other products.

Useful Biochemical Tests

Before the advent of commercially available products for serological characterization, much effort was expended on developing simple biochemical tests which could presumptively identify or predict the Lancefield group of a given streptococcal isolate. These tests were easily performed in the clinical laboratory and were viewed as acceptable alternatives to the rigors of serological studies. These tests are still useful for streptococcal identification and are still widely employed. Table 5 summarizes the use of the tests described below for identification of streptococci.

BACITRACIN AND SXT SUSCEPTIBILITY TESTS. Bacitracin susceptibility is a characteristic of group A beta-hemolytic streptococci (S. pyogenes). A few colonies of the isolate to be tested are streaked for confluent growth onto onefourth of a sheep blood-supplemented trypticase soy agar plate. A 0.04 unit bacitracin disk is placed on the plate, which is then incubated overnight at 35°C. Any visible zone of inhibition of growth surrounding the disk is interpreted as susceptibility to bacitracin. SXT disks (23.75 μg sulfamethoxazole plus 1.25 μg trimethoprim) are used with beta-hemolytic streptococci in the same manner as bacitracin disks and afford more complete characterization when used in conjunction with bacitracin. S.

pyogenes is usually bacitracin-susceptible and SXT-resistant; group B streptococci are resistant to the agents in both disks; and group C and G large-colony-forming streptococci are bacitracin-resistant and SXT-susceptible (Facklam and Carey, 1985).

PYR TEST. This test determines the presence or absence of the enzyme pyrrolidonyl arylamidase (PYR), elaborated by S. pyogenes and members of the genera Aerococcus and Enterococcus (Table 1). Facklam and co-workers (1982) found the PYR test to be equally as sensitive as, and more specific than, bacitracin susceptibility for S. pyogenes identification. Bosley et al. (1983) described a test method employing commercially available PYR broth (Todd-Hewitt broth supplemented with a 0.01% L-pyrrolidonyl-beta-naphthylamide). A small amount (0.2 ml) of this broth is inoculated with a loopful of growth from a blood agar medium, and incubated at 36°C for 4 h. After addition of a drop of PYR reagent (N,N-dimethylaminocinnamaldehyde), a positive test is indicated by development of a deep red color, due to reaction of liberated β-naphthylamine with the PYR reagent. Any other color development, including orange, is interpreted as a negative result. Rapid tests for determination of pyrrolidonyl arylamidase are also available (Ellner et al., 1985; Wasilauskas and Hampton, 1984). While the PYR test is extremely useful for streptocococcal and enterococcal identification, users should remember that in addition to S. pyogenes and enterococci, strains of Aerococcus, Gemella (Facklam et al., 1989), nutritionally variant streptococci (Bouvet et al., 1985), and staphylococci (Oberhofer, 1986) may give positive PYR reactions.

CAMP TEST. Most group B streptococci produce a diffusible extracellular protein referred to as the CAMP factor (Facklam et al., 1979).

Table 5. Differentiation of streptococci with the use of biochemical tests.

| | Susceptibility to: | | PYR | CAMP test | Hydrolysis of hippurate | Bile esculin | Growth in 6.5% NaCl | Optochin and bile susceptibility |
	Bacitracin	SXT						
S. pyogenes (group A)	+	−	+	−	−	−	−	−
S. agalactiae (group B)	−[a]	−	−	+	+	−	+[a]	−
Large colony (group C and G)	−[a]	+	−	−	−	−	−	−
S. pneumoniae			−	−	−	−	−	+
S. bovis	−	+[a]	−	−	−	+	−	−

+, positive; −, negative, SXT, sulfamethoxazole and trimethoprim; PYR, pyrrolidonyl arylamidase; CAMP test, test for enhancement of hemolysis by Staphylococcus aureus beta lysin.
[a]Exceptions occur occasionally.
Adapted from Facklam and Carey (1985) and other sources.

This protein, named for its discoverers, Christie, Atkins and Munch-Peterson, acts in synergy with staphylococcal beta lysin, producing complete lysis of red blood cells. The streptococcus to be tested is used to make a single straight streak perpendicular to and 1–2 mm away from a similar streak of a beta-lysin-producing staphylococcus. Trypticase soy agar containing 5% sheep blood is the recommended medium. After overnight incubation at 35°C, a positive CAMP test is evidenced by the production of an arrowhead-shaped zone of complete hemolysis in the area into which both CAMP factor and beta lysin have diffused. Modifications of the CAMP test employing disks containing beta lysin (Wilkinson, 1977) or clostridial alpha-toxin (Smith and Ngui-Yen, 1980) have also been described.

HIPPURATE HYDROLYSIS TEST. Virtually all group B beta-hemolytic streptococci (*S. agalactiae*) are capable of hydrolyzing hippurate. In Hwang and Ederer's (1975) rapid hippurate hydrolysis test, a loopful of streptococcal growth from a blood agar plate is suspended in 0.4 ml of an aqueous 1% sodium hippurate solution. After incubation for 2 h at 37°C, 0.2 ml of a ninhydrin solution (3.5% ninhydrin in a 1:1 mixture of butanol and acetone) is added to the reaction tube. After a 10-min reincubation, development of a deep purple color, caused by reaction of ninhydrin with glycine formed from hippurate hydrolysis, signifies a positive reaction.

OPTOCHIN SUSCEPTIBILITY AND BILE SUSCEPTIBILITY TESTS. These two tests are useful for identification of pneumococci, which are inhibited by optochin (ethylhydrocupreine) and lyse rapidly in the presence of bile. Optochin disks are placed onto a quadrant of a blood agar plate inoculated with two to three colonies of the suspected pneumococcus. After overnight incubation at 35°C in a candle jar or CO_2 incubator, optochin susceptibility is indicated by an inhibition zone of greater than 14 mm with a 6-mm disk, or greater than 16 mm with a 10-mm disk (Facklam and Carey, 1985). The bile solubility test can be performed by placing a few drops of 10% sodium deoxycholate onto the suspected pneumococcal colony. After incubation of the plate until the detergent solution is absorbed into the agar, pneumococcal colonies will have disappeared due to lysis (Finegold and Baron, 1986).

BILE ESCULIN AND SALT TOLERANCE TESTS. Streptococci of Lancefield's group D (*S. bovis*) are able to hydrolyze esculin in the presence of 40% bile, but are unable to grow in the presence of 6.5% NaCl. This differentiates them from enterococci (bile esculin-positive, salt tolerant) and other streptococci (Table 5). A positive bile esculin test is indicated by blackening of more than one-half of a bile esculin agar slant (Phillips and Nash, 1985) which has been lightly inoculated (two to three colonies) with the organism to be tested and incubated for 24 to 48 h at 35°C. Growth on bile esculin agar may occur without blackening; this is interpreted as a negative reaction (Facklam and Carey, 1985). Salt tolerance is indicated by growth in heart infusion broth containing 6.5% NaCl after incubation at 35°C for 24 to 48 h.

RAPID VP TEST. Use of a rapid Voges-Proskauer (VP) test for detection of acetoin production has been advocated by Bucher and von Graevenitz (1984) for differentiation of beta-hemolytic *S. anginosus* (group C or G) from large-colony-forming group C and G streptococci. A turbid suspension of the organism to be tested is made in 0.2 ml of MR-VP broth. After a 5-h incubation at 37°C, one drop each of 0.5% creatine, alpha naphthol solution (0.8 g in 12 ml 95% ethanol), and 40% KOH is added to the tube. The contents of the tube are shaken thoroughly and observed for up to 30 min for development of a pink to red color, indicating a positive test. *S. anginosus* strains (beta-hemolytic group A, C, or G) are VP-positive, in contrast to large colony group C and G beta-hemolytic streptococci and *S. pyogenes*.

Tests for Identification of *S. bovis*

The supplemental tests described below can be used for identification of *S. bovis* and the differentiation of this species from the "viridans" streptococci described by Hardie and Whiley in Chapter 64. Alternatively, commercially available systems (see next section) may be used for species identification of these streptococci. Table 6 summarizes the characteristics of *S. bovis* and its variant biotype.

CARBOHYDRATE FERMENTATION TESTS (FACKLAM, 1972). To 900 ml of heart infusion broth, add 100 ml of a 10% solution of the carbohydrate to be tested and 1 ml of indicator solution (1.6 g of bromcresol purple in 100 ml of 95% ethanol). Dispense the medium in 3 ml amounts into screw-capped tubes and autoclave for 10 min at 15 psi. Inoculate each tube with one to two drops of an overnight broth culture (Todd-Hewitt, or heart infusion broth with glucose) and incubate at 35°C for up to 72 h. A positive reaction (acid production) causes the

Table 6. Identification of biotypes of *S. bovis*.

Characteristic	*S. bovis* (biotype I)	*S. bovis* variant (biotype II)
Bile esculin	+	+
Growth in 6.5% NaCl	−	−
Arginine hydrolysis	−	−
Starch hydrolysis	+	−[a]
Esculin hydrolysis	+	+
Acidification of:		
Lactose	+	+
Mannitol	+	−
Sorbitol	−	−
Inulin	+[a]	−
Raffinose	+[a]	v
Polysaccharide production	+, levan	−[a]

+, positive; −, negative; v, variable.
[a]Exceptions occur occasionally.
Adapted from Facklam and Carey (1985).

indicator to change in color from purple to yellow.

ARGININE HYDROLYIS TEST (PHILLIPS AND NASH, 1985). Moeller decarboxylase medium with arginine may be purchased or prepared by adding the following ingredients to 1 liter of distilled water: peptone, 5 g; beef extract, 5 g; pyridoxal, 5 mg; L-arginine, 10 g; glucose, 0.5 g; bromcresol purple, 0.625 ml; and cresol red solution (0.5 g cresol red dissolved in 26.2 ml of 0.01 N NaOH and diluted to 250 ml), 2.5 ml. Adjust the pH to between 6.0 and 6.5, dispense in 3 ml amounts and autoclave for 10 min at 15 psi. Inoculate with one or two drops of an overnight broth culture of the organism to be tested and cover the medium with sterile mineral oil. Incubate at 35°C for 72 h. A negative reaction appears as a yellow color due to acid production from glucose. If arginine is attacked, the liberation of basic products elevates the pH and is indicated by a purple color (positive reaction).

ESCULIN HYDROLYSIS TEST (PHILLIPS AND NASH, 1985). Esculin agar contains 1.0 g esculin, 0.5 g ferric citrate, and 40.0 g heart infusion agar per liter of distilled water. The medium is heated to dissolve the ingredients, cooled to 55°C, and then the pH is adjusted to 7.0. After dispensing into tubes and autoclaving at 15 psi for 15 min, the medium is cooled in a slanted position. Solidified slants are inoculated with one or two drops of an overnight broth culture of the organism to be tested, and incubated at 35°C for 72 h. Esculin hydrolysis is indicated by blackening of the medium.

STARCH HYDROLYSIS TEST (FACKLAM, 1972). Heart infusion agar is supplemented with soluble starch (2%) before sterilizing (15 psi, 15 min), cooling, and pouring into plates. After inoculation and incubation of the plates for 48 h, they are flooded with Gram's iodine. Starch hydrolysis appears as a clear area surrounding the streptococcal growth, in contrast to the purple color produced by reaction of iodine with unhydrolyzed starch.

PRODUCTION OF POLYSACCHARIDE FROM SUCROSE (FACKLAM, 1972). Heart infusion agar is supplemented with 5% sucrose prior to sterilization (15 psi, 15 min), cooling, and dispensing into plates. The medium is inoculated and incubated for up to 72 h at 35°C. Levan-producing streptococci form large gummy colonies on this medium. The production of dextran, another type of polysaccharide synthesized from sucrose, is indicated by firm adherence of colonies to the medium.

Commercially Available Identification Systems Employing Biochemical Tests

Beginning in the 1980s, Several identification products for streptococci and other Gram-positive bacteria became commercially available. These systems identify streptococci on the basis of their physiological reactions (carbohydrate fermentations, enzyme activities) and at present are most useful for characterizing viridans streptococci, since beta-hemolytic streptococci can be presumptively identified by simple, more cost-effective methods (Table 5). Full instructions for use of these systems are supplied with the products. They all employ micromethods, are simple to use, and generally provide results in less time than conventional methods. Numerous evaluations of these products have been published (Appelbaum et al., 1984; Appelbaum et al., 1986; Facklam et al., 1984; Facklam et al., 1985; Ruoff and Kunz, 1983).

Direct Detection of Streptococci in Clinical Specimens

Several serological methods can be employed for the direct detection of streptococcal antigens in clinical specimens. Table 7 lists these methods and summarizes their applications. Most of the methods may also be used for identification of cultures of streptococci. The Quellung test for pneumococcal detection is perhaps the oldest of these techniques and is performed by mixing a drop of the specimen to be tested with a loopful of methylene blue stain and a loopful of antiserum (omniserum, a polyvalent antiserum reactive with more than 80 types of pneumococcal capsular polysaccharide). This mixture is examined microscopically for pneumococci with accentuated capsules (Fig. 4). Re-

Table 7. Methods employed for direct detection of streptococci in clinical specimens.

Method	Streptococcus detected/specimen type	References
Agglutination	Group A/throat swabs	Facklam, 1987
	Group B/urogenital, blood, urine, CSF[a]	Ingram et al., 1982
		Jones et al., 1984
		Slifkin et al., 1982
		Webb et al., 1980
	Pneumococci/CSF, sputum	Drow et al., 1983
		Edward and Coonrod, 1980
	Group A,B,C,D,G/blood	Wellstood, 1982
Enzyme immunoassay	Group A/throat swabs	Schwabe et al., 1987
		Patel et al., 1987
Quellung test	Pneumococci/sputum, CSF	Merrill et al., 1973
Immunofluorescence	Group A/throat swabs	Moody et al., 1963
	Group B/urogenital swabs	Romero and Wilkinson, 1974
	Pneumococci/sputum	Wicher et al., 1982
Counterimmunoelectrophoresis	Pneumococci/sputum, CSF, blood, urine	Kenny, 1983
	Group B/CSF	Webb et al., 1980

[a]CSF, cerebrospinal fluid.

action of antiserum with the pneumococcal capsule enhances its visualization (Facklam and Carey, 1985).

The technical simplicity of agglutination techniques has resulted in their widespread use. Several commercially available products for detection of streptococcal antigens have appeared recently, including products for direct detection of group A streptococci in throat swabs. In these methods, the entire swab (containing streptococci) is subjected to an antigen extraction procedure. The swab is removed from the extraction medium (enzyme or nitrous acid solution) and any antigens in the extraction fluid are detected with agglutination reagents. Facklam (1987) summarized the findings of numerous evaluations of these products. While most products are highly specific, the observed sensitivities are quite variable. False-negative results are often associated with the presence of low numbers of streptococci in the specimen. The rapidity with which results are generated by the direct detection tests make them attractive for use in a medical office or clinic setting, since results may be obtained while the patient waits. However, the variable sensitivities shown by these products suggest that all negative test results should be followed up by culture.

Pathogenicity and Virulence

Extracellular Products

Exotoxin production represents a major virulence factor of *Streptococcus pyogenes* and other streptococci. The hemolysins streptolysin O (oxygen labile) and streptolysin S (oxygen stable) are capable of damaging the membranes of many types of cells and organelles. These cytolysins are also produced by large-colony-forming group C and G streptococci (Wannamaker, 1983). Streptolysin O is immunogenic, and the presence of elevated host antibodies is indicative of recent streptococcal infection. The deleterious effects of streptolysins have been hypothesized to play a role in the pathogenesis of acute rheumatic fever (Bisno, 1990a).

Other extracellular products of *S. pyogenes* include hyaluronidase, DNAse enzymes A, B, C, and D, streptokinase (a promoter of fibrin clot breakdown), and NADase. The first three products mentioned above may facilitate the spread of infection, while all four are antigenic. Presence of elevated host antibody to these products is, as in the case of anti-streptolysin

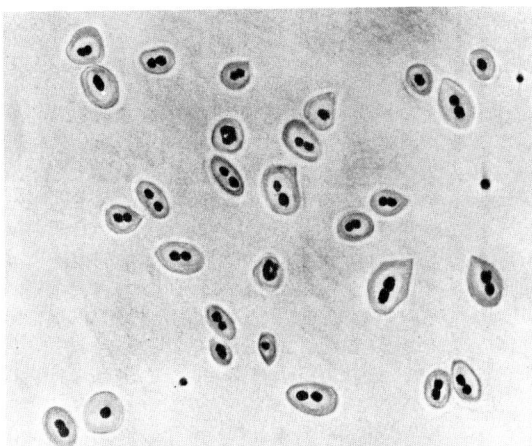

Fig. 4. The Quellung reaction for identification of pneumococci.

O antibodies, taken as documentation of streptococcal infection.

Three (A, B, and C) streptococcal pyrogenic
(erythrogenic) toxins have been described as extracellular products of *S. pyogenes.* In addition
to causing the characteristic rash of scarlet fever,
pyrogenic toxins display cytotoxicity and cause
enhancement of susceptibility to endotoxin.
Toxin production is associated with infection of
streptococci by temperate bacteriophages
(Bisno, 1990b; Wannamaker, 1983). Recent reports of severe group A streptococcal infections
characterized by a toxic shock-like syndrome
have implicated pyrogenic toxin A-producing
streptococci (Stevens et al., 1989).

Cell-associated Factors

Capsules function as virulence factors in several
streptococci, usually by virtue of their antiphagocytic properties. In *S. pyogenes,* the capsule is composed of hyaluronic acid, while various type-specific carbohydrates constitute the
capsules of group B streptococci and *S. pneumoniae.* In the pneumococcus, capsules play an
important role in virulence, since nonencapsulated (rough) strains are avirulent. Yeung and
Mattingly (1984) found that increased levels of
capsule in serotype III group B streptococci correlated with increased virulence.

Streptococcal M proteins have been extensively studied in *S. pyogenes,* and serve as serological typing molecules within this Lancefield group. *S. pyogenes* M proteins have
antiphagocytic properties and elicit the production of protective host antibodies. Proteins similar to the M protein of group A strains have
been described in group C and G streptococci.
M proteins from *S. pyogenes* display elongated
coiled structures with the C-terminal ends an-
chored in the cell membrane and the N-terminal ends extending from the cell surface to
form projections. The amino acid sequence of
the N-terminal region of the molecule seems to
be variable and responsible for the numerous
(more than 80) existing M serotypes (Fischetti,
1989).

M-protein-producing streptococci also contain M-associated proteins (MAP), which are
divided into two serologically distinct classes,
MAP I and MAP II. Production of opacity factor (OF), a lipoproteinase that produces opacity
in serum, is correlated with the presence of
MAP II antigen. MAP I antigen is found in the
majority of group A strains associated with
rheumatic fever. Recent evidence suggests that
the MAP I antigen may actually represent part
of the M protein itself (Fischetti, 1989).

Streptococcal lipoteichoic acid is considered
here as a virulence factor because it functions
as an adhesin, promoting attachment of streptococci to mucosal surfaces. Attachment and
subsequent colonization are requisite steps in
the process of infection. Extensive studies on
the adherence of *S. pyogenes* to epithelial cells,
recently reviewed by Beachey and Courtney
(1987), suggest that lipoteichoic acid molecules
form complexes with M or other surface proteins resulting in orientation of the lipid end of
the lipoteichoic acid molecule towards the outer
surface of the bacterial cell. The lipid moiety of
lipoteichoic acid then binds with the N-terminal region of fibronectin molecules on the host
cell surface, mediating adherence. Myhre and
Kuusela (1983) demonstrated that group A, C,
and G streptococci bound purified fibronectin
in an in vitro assay, suggesting a similar adherence mechanism among all three Lancefield
groups.

In group B streptococci, lipoteichoic acid also
appears to play a role in virulence. Nealon and
Mattingly (1983) found that isolates from diseased neonates had higher lipoteichoic acid levels than strains isolated from asymptomatic infants. Group B streptococcal lipoteichoic acid
is cytotoxic for human cells in culture (Goldschmidt and Panos, 1984), and mediates adherence of the streptococci to human embryonic, fetal, and adult oral epithelial cells
(Nealon and Mattingly, 1984). Although these
observations suggest that the adherence process
in group B streptococci is similar to that described for *S. pyogenes,* Butler and colleagues
(1987) demonstrated the inability of group B
streptococci to bind purified fibronectin in an
in vitro assay system. Further research is
needed to clarify the role of lipoteichoic acid in
adherence of group B streptococci.

Antimicrobial Susceptibility

Beta-hemolytic streptococci are usually extremely susceptible to penicillin. Even after decades of use of this antibiotic for treating streptococcal infections, it remains the drug of
choice, followed by first-generation cephalosporins, erythromycin, or vancomycin as alternate
therapy (Moellering, 1990). In general, group A
streptococci seem to be more sensitive to penicillin than group B, C, or G streptococci (Baker
et al., 1981; Bourbeau and Campos, 1982; Rolston et al., 1982). While penicillin is still an
effective drug for treating pneumococcal infections, penicillin-resistant pneumococci have
been isolated since the late 1960s. Strains with
high level and intermediate level resistance have
been described. Alterations in penicillin-binding proteins resulting in their decreased affinity
for penicillin have been found in resistant
strains (Appelbaum, 1987).

Antibiotic tolerance, first described in a pneumococcal strain, also occurs in other streptococci. Tolerance to an antibiotic is usually measured by the magnitude of the ratio of the minumum bactericidal concentration (MBC) to the minimal inhibitory concentration (MIC). If the MBC/MIC ratio is greater than 32, the organism is considered tolerant to the drug in question. Compared with the antibiotic concentration required to inhibit growth, a much higher concentration is required for killing of a tolerant organism: presumably this can affect the efficacy of therapy (Handwerger and Tomasz, 1985). Tolerance to penicillin has been described in group A (Dagan et al., 1987; Krasinski et al., 1986), group B (Kim and Anthony, 1981), and group C and G streptococci (Rolston et al., 1982; Noble et al., 1980).

Streptococcal Vaccines

Pneumococcal vaccines, consisting of purified capsular material, have been available since the 1970s. These vaccines contain capsular polysaccharides of the most prevalent serotypes isolated from infections, and induce antibody production in normal adults. Unfortunately, the vaccines are often not effective in those with poor immunological response (very young children, the elderly, immunocompromized patients). In spite of the variable efficacy of pneumococcal vaccine, it is still useful for preventing disease in certain segments of the population (Mufson, 1990). Vaccines containing pneumococcal polysaccharides conjugated to proteins may prove useful for eliciting protective antibodies in young children (Austrian, 1989).

The prevention of devastating perinatal group B streptococcal infections has been attempted with intrapartum antibiotic administration (Boyer and Gotoff, 1986) and more recently with polysaccharide vaccine. In a trial employing purified type III capsular polysaccharide, about two-thirds of the pregnant women vaccinated developed immunoglobulin G (IgG) antibodies which can cross the placenta. Future vaccines with increased immunogenicity should prove to be promising for prevention of neonatal group B streptococcal disease (Baker et al., 1988).

Literature Cited

Appelbaum, P. C. 1987. World-wide development of antibiotic resistance in pneumococci. Eur. J. Clin. Microbiol. 6:367–377.

Appelbaum, P. C., M. R. Jacobs, J. I. Heald, W. M. Palko, A. Duffet, R. Crist, and P. A. Naugle. 1984. Comparative evaluation of the API 20S system and the AutoMicrobic system gram-positive identification card for species identification of streptococci. J. Clin. Microbiol. 19:164–168.

Appelbaum, P. C., M. R. Jacobs, W. M. Palko, E. E. Frauenhoffer, and A. Duffett. 1986. Accuracy and reproducibility of the IDS Rapid STR system for species identification of streptococci. J. Clin. Microbiol. 23:843–846.

Austrian, R. 1989. Pneumococcal polysaccharide vaccines. Rev. Infect. Dis. 11:S598–S602.

Baker, C. J., M. A. Rench, M. S. Edwards, R. J. Carpenter, B. M. Hays, and D. L. Kasper. 1988. Immunization of pregnant women with a polysaccharide vaccine of group B streptococcus. New Engl. J. Med. 319:1180–1185.

Baker, C. N., C. Thornsberry, and R. R. Facklam. 1981. Synergism, killing kinetics, and antimicrobial susceptibility of group A and B streptococci. Antimicrob. Ag. Chemother. 19:716–725.

Barnham, M., T. J. Thornton, and K. Lange. 1983. Nephritis caused by *Streptococcus zooepidemicus* (Lancefield group C). Lancet i:945–948.

Beachey, E. H., and H. S. Courtney. 1987. Bacterial adherence: the attachment of group A streptococci to mucosal surfaces. Rev. Infect. Dis. 9 (supplement 5):S475–S481.

Benjamin, J. T., and V. A. Perriello. 1976. Pharyngitis due to group C hemolytic streptococci in children. J. Pediatr. 89:254–256.

Bisno, A. L. 1990a. Nonsuppurative streptococcal sequelae: rheumatic fever and glomerulonephritis, p. 1528–1539. In: G. L. Mandell, R. G. Douglas, Jr., and J. E. Bennett (ed.), Principles and practice of infectious diseases. John Wiley and Sons, New York.

Bisno, A. L. 1990b. *Streptococcus pyogenes,* p. 1519–1528. In: G. L. Mandell, R. G. Douglas, Jr., and J. E. Bennett (ed.), Principles and practice of infectious diseases. John Wiley and Sons, New York.

Black, W. A., and F. Van Buskirk. 1973. Gentamicin as a selective agent for the isolation of beta hemolytic streptococci. J. Clin. Pathol. 26:154–156.

Bosley, G. S., R. R. Facklam, and D. Grossman. 1983. Rapid identification of enterococci. J. Clin. Microbiol. 18:1275–1277.

Bourbeau, P., and J. M. Campos. 1982. Current antibiotic susceptibility of group A β-hemolytic streptococci. J. Inf. Dis. 145:916.

Bouvet, A., F. Villeroy, F. Cheng, C. Lamesch, R. Williamson, and L. Gutmann. 1985. Characterization of nutritionally variant streptococci by biochemical tests and penicillin binding proteins. J. Clin. Microbiol. 22:1030–1034.

Boyer, K. M., and S. P. Gotoff. 1986. Prevention of early-onset neonatal group B streptococcal disease with selective intrapartum chemoprophylaxis. N. Engl. J. Med. 314:1665–1669.

Brown, J. H. 1919. The use of blood agar for the study of streptococci. Monographs of the Rockefeller Institute for Medical Research, no. 9. The Rockefeller Institute for Medical Research, New York.

Bucher, C., and A. von Graevenitz. 1984. Differentiation in throat cultures of group C and G streptococci from *Streptococcus milleri* with identical antigens. Eur. J. Clin. Microbiol. 3:44–45.

Burman, L. A., R. Norrby, and B. Trollfors. 1985. Invasive pneumococcal infections: incidence, predisposing factors, and prognosis. Rev. Infect. Dis. 7:133–142.

Butler, K. M., C. J. Baker, and M. S. Edwards. 1987. Interaction of soluble fibronectin with group B streptococci. Infect. Immun. 55:2404–2408.

Carlson, J. R., W. G. Merz, B. E. Hansen, S. Ruth, and D. G. Moore. 1985. Improved recovery of group A beta-hemolytic streptococci with a new selective medium. J. Clin. Microbiol. 21:307–309.

Chung, S. J. 1982. Meningitis caused by *Streptococcus equisimilis* (Group C). S. Med. J. 75:769.

Congeni, B., C. Rizzo, J. Congeni, and V. V. Sreenivasan. 1987. Outbreak of acute rheumatic fever in northeast Ohio. J. Pediatr. 111:176–179.

Dagan, R., M. Ferne, M. Sheinis, M. Alkan, and E. Katzenelson. 1987. An epidemic of penicillin- tolerant group A streptococcal pharyngitis in children living in a closed community: mass treatment with erythromycin. J. Inf. Dis. 156:514–516.

Daly, J. A., and K. C. Seskin. 1988. Evaluation of rapid, commercial latex techniques for serogrouping beta-hemolytic streptococci. J. Clin. Microbiol. 26:2429–2431.

Deibel, R. H., and J. B. Evans. 1960. Modified benzidine test for the detection of cytochrome-containing respiratory systems in microorganisms. J. Bacteriol. 79:356–360.

de la Rosa, M., R. Villareal, D. Vega, C. Miranda, and A. Martinezbrocal. 1983. Granada medium for detection and identification of group B streptococci. J. Clin. Microbiol. 18:779–785.

DeVriese, L. A., J. Hommez, R. Kilpper-Balz, and K. H. Schleifer. 1986. *Streptococcus canis* sp. nov.: a species of group G streptococci from animals. Int. J. Syst. Bacteriol. 36:422–425.

Dowing, G. J., and C. Spirazza. 1986. Group C beta-hemolytic streptococcal endocarditis. Pediatr. Infect. Dis. 5:703–704.

Drow, D. L., D. F. Welch, D. Hensel, K. Eisenbach, E. Long, and M. Slifkin. 1983. Evaluation of the Phadebact CSF test for detection of the four most common causes of bacterial meningitis. J. Clin. Microbiol. 18:1358–1361.

Duca, E., T. C. Radu, A. Vita, P. Talasman-Niculescu, E. Bernescu, C. Feldi and V. Rosca. 1969. A new nephritogenic streptococcus. J. Hyg. Camb. 67:691–698.

Dykstra, M. A., J. C. McLaughlin, and R. C. Bartlett. 1979. Comparison of media and techniques for detection of group A streptococci in throat swab specimens. J. Clin. Microbiol. 9:236–238.

Ederer, G. M., M. M. Hermann, R. Bruce, J. M. Matsen, and S. S. Chapman. 1972. Rapid extraction method with pronase B for grouping beta-hemolytic streptococci. Appl. Microbiol. 23:285–288.

Edward, E. A., and J. D. Coonrod. 1980. Coagglutination and counterimmunoelectrophoresis for detection of pneumococcal antigens in sputum of pneumonia patients. J. Clin. Microbiol. 5:488–491.

Edwards, M. S., and C. J. Baker. 1990. *Streptococcus agalactiae* (group B streptococcus), p. 1554–1563. In: G. L. Mandell, R. G. Douglas, Jr., and J. E. Bennett (ed.), Principles and practice of infectious diseases. John Wiley and Sons, New York.

El Kholy, A., L. W. Wannamaker, and R. M. Krause. 1974. Simplified extraction procedure for serological groupings of beta-hemolytic streptococci. Appl. Microbiol. 28:863–869.

Ellner, P. D., D. A. Williams, M. E. Hosmer, and M. A. Cohenford. 1985. Preliminary evaluation of a rapid colorimetric method for the presumptive identification of group A streptococci and enterococci. J. Clin. Microbiol. 22:880–881.

Evans, J. B. 1986. Genus *Aerococcus* Williams, Hirch, and Cowan 1953, 475[AL] p. 1080. In: P. H. A. Sneath, N. S. Mair, M. E. Sharpe, and J. G. Holt (ed.), Bergey's manual of systematic bacteriology, vol. 2. Williams and Wilkins, Baltimore.

Facklam, R. R. 1972. Recognition of group D streptococcal species of human origin by biochemical and physiological tests. Appl. Microbiol. 23:1131–1139.

Facklam, R. R. 1987. Specificity study of kits for detection of group A streptococci directly from throat swabs. J. Clin. Microbiol. 25:504–508.

Facklam, R., G. S. Bosley, D. Rhoden, A. R. Franklin, N. Weaver, and R. Schulman. 1985. Comparative evaluation of the API 20s and Auto Microbic Gram-Positive Identification systems for non-beta-hemolytic streptococci and aerococci. J. Clin. Microbiol. 21:535–541.

Facklam, R. R., and R. B. Carey. 1985. Streptococci and aerococci, p. 154–175. In: E. H. Lennette, A. Balows, W. J. Hausler, Jr., and H. J. Shadomy (ed.), Manual of clinical microbiology, 4th ed. American Society for Microbiology Washington, D.C.

Facklam, R., D. Hollis, and M. D. Collins. 1989. Identification of gram-positive coccal and coccobacillary vancomycin-resistant bacteria. J. Clin. Microbiol. 27:724–730.

Facklam, R. R., J. F. Padula, E. C. Wortham, R. C. Cooksey, and H. A. Rountree. 1979. Presumptive identification of group A, B and D streptococci on agar plate media. J. Clin. Microbiol. 9:665–672.

Facklam, R. R., D. L. Rhoden, and P. B. Smith. 1984. Evaluation of the Rapid Strep system for the identification of clinical isolates of *Streptococcus* species. J. Clin. Microbiol. 20:894–898.

Facklam, R. R., L. G. Thacker, B. Fox, and L. Eriquez. 1982. Presumptive identification of streptococci with a new test system. J. Clin. Microbiol. 20:894–898.

Farrow, J. A. E., and M. D. Collins. 1984. Taxonomic studies on streptococci of serological groups C, G and L and possibly related taxa. Syst. Appl. Microbiol. 5:483–493.

Finegold, S. M., and E. J. Baron. 1986. Bailey and Scott's diagnostic microbiology, 7th ed, p. 113. C. V. Mosby, St. Louis.

Fischetti, V. A. 1989. Streptococcal M protein: molecular design and biological behavior. Clin. Microbiol. Rev. 2:285–314.

Fuller, A. T. 1938. The formamide method for the extraction of polysaccharides from beta haemolytic streptococci. Brit. J. Exp. Pathol. 19:130–139.

Gallagher, P. G., and C. Watanakunakorn. 1985. Group B streptococcal bacteremia in a community teaching hospital. Am. J. Med. 78:795–800.

Gallagher, P. G., and C. Watanakunakorn. 1986. Group B streptococcal endocarditis: report of seven cases and review of the literature, 1962–1985. Rev. Infect. Dis. 8:175–188.

Gaunt, P. N., and D. V. Seal. 1987. Group G streptococcal infections. J. Infect. 15:5–20.

Gnann, J. W., B. M. Gray, F. M. Griffin, Jr., and W. E. Dismukes. 1987. Acute glomerulonephritis following group G streptococcal infection. J. Infect. Dis. 156:411–412.

Goldmann, D. A., and S. J. Breton. 1978. Group C streptococcal surgical wound infections transmitted by an anorectal and nasal carrier. Pediatr. 61:235–237.

Goldschmidt, J. C. Jr., and C. Panos. 1984. Teichoic acids of *Streptococcus agalactiae*: chemistry, cytotoxicity, and effect on bacterial adherence to human cells in tissue culture. Infect. Immun. 43:670–677.

Gunn, B. A., D. K. Ohashi, C. A. Gaydos, and E. S. Holt. 1977. Selective and enhanced recovery of group A and B streptococci from throat cultures with sheep blood agar containing sulfamethoxazole and trimethoprim. J. Clin. Microbiol. 5:650–655.

Handwerger, S., and A. Tomasz. 1985. Antibiotic tolerance among clinical isolates of bacteria. Rev. Infect. Dis. 7:368–386.

Hardie, J. M. 1986. Genus *Streptococcus* Rosenbach 1884, 22[AL] Oral streptococci, p. 1054–1063. In:P. H. A. Sneath, N. S. Mair, M. E. Sharpe, and J. G. Holt (ed.), Bergey's manual of systematic bacteriology, vol. 2. Williams and Wilkins, Baltimore.

Hendley, J. O., M. A. Sande, P. M. Stewart, and J. M. Gwaltney. 1975. Spread of *Streptococcus pneumoniae* in families. I. Carriage rates and distribution of types. J. Infect. Dis. 132:55–61.

Hosier, D. M., J. M. Craenen, D. W. Teske, and J. J. Wheller. 1987. Resurgence of acute rheumatic fever. Am. J. Dis. Child. 141:730–733.

Hwang, M., and G. M. Ederer. 1975. Rapid hippurate hydrolysis method for presumptive identification of group B streptococci. J. Clin. Microbiol. 1:114–115.

Ingram, D. L., D. M. Suggs, and A. W. Pearson. 1982. Detection of group B streptococcal antigen in early onset and late onset group B streptococcal disease with the Wellcogen Strep B latex agglutination test. J. Clin. Microbiol. 16:656–658.

Isenberg, H. D., J. A. Washington II, A. Balows, and A. C. Sonnenwirth. 1985. Collection, handling, and processing of specimens, p. 73–98. In:E. H. Lennette, A. Balows, W. J. microbiology, 4th ed. American Society for Microbiology, Washington, D.C.

Jette, L. P., F. Lamothe, and the Pneumococcus Study Group. 1989. Surveillance of invasive *Streptococcus pneumoniae* infection in Quebec, Canada, from 1984 to 1986: serotype distribution, antimicrobial susceptibility, and clinical characteristics. J. Clin. Microbiol. 27:1–5.

Jones, D. E., K. S. Kanarek, and D. V. Lim. 1984. Group B streptococcal colonization patterns in mothers and their infants. J. Clin. Microbiol. 20:438–440.

Kaplan, E. L., D. R. Johnson, and P. P. Cleary. 1989. Group A streptococcal serotypes isolated from patients and sibling contacts during the resurgence of rheumatic fever in the United States in the mid-1980's. J. Infect. Dis. 159:101–103.

Kenny, G. E. 1983. Counterimmunoelectrophoresis for the diagnosis of pneumococcal respiratory and other infections, p. 105–111. In: J. D. Coonrod, L. J. Kunz, and M. J. Ferraro (ed.), The direct detection of microorganisms in clinical samples. Academic Press, New York.

Kilpper-Balz, R., and K. H. Schleifer. 1988. Transfer of *Streptococcus morbillorum* to the genus *Gemella* as *Gemella morbillorum* comb. nov. Int. J. Syst. Bacteriol. 38:442–443.

Kim, K. S., and B. S. Anthony. 1981. Penicillin tolerance in group B streptococci isolated from infected neonates. J. Infect. Dis. 144:411–419.

Klein, R. S., S. W. Warman, G. G. Knachmuhs, S. C. Edberg, and N. H. Steigbigel. 1987. Lack of association of *Streptococcus bovis* with noncolonic gastrointestinal carcinoma. Am. J. Gastroenterol. 82:540–543.

Krasinski, K., B. Hanna, P. LaRussa, and D. Desiderio. 1986. Penicillin tolerant group A streptococci. Diag. Microbiol. Infect. Dis. 4:291–297.

Kurzynski, T. A., and C. K. Meise. 1979. Evaluation of sulfamethoxazole-trimethoprim blood agar plates for recovery of group A streptococci from throat cultures. J. Clin. Microbiol. 9:189–193.

Lancefield, R. C. 1933. Serological differentiation of human and other groups of hemolytic streptococci. J. Exp. Med. 57:571–595.

Libertin, C. R., A. D. Wold, and J. A. Washington II. 1983. Effects of trimethoprim-sulfamethoxazole and incubation atmosphere on isolation of group A streptococci. J. Clin. Microbiol. 18:680–682.

Lim, D. V., W. J. Morales, and A. F. Walsh. 1987. Lim group B strep broth and coagglutination for rapid identification of group B streptococci in preterm pregnant women. J. Clin. Microbiol. 25:452–453.

Ludwig, W., M. Weizenegger, R. Kilpper-Balz, and K. H. Schleifer. 1988. Phylogenetic relationships of anaerobic streptococci. Int. J. Syst. Bacteriol. 38:15–18.

Marcon, M. J., M. M. Hribar, D. M. Hosier, D. A. Powell, M. T. Brady, A. C. Hamoudi, and E. L. Kaplan. 1988. Occurrence of mucoid M-18 *Streptococcus pyogenes* in a central Ohio pediatric population. J. Clin. Microbiol. 26:1539–1542.

Maxted, W. R., 1948. Preparation of streptococcal extracts for Lancefield grouping. Lancet ii:255–256.

Merrill, C. W., J. M. Gwaltney, Jr., J. W. Hendley, and M. A. Sande. 1973. Rapid identification of pneumococci, Gram stain vs. the Quellung reaction. N. Engl. J. Med. 288:510–512.

Moellering, R. C. Jr. 1990. Principles of anti-infective therapy, p. 206–218. In: G. L. Mandell, R. G. Douglas, Jr., and J. E. Bennett (ed.), Principles and practices of infectious diseases. John Wiley and Sons, New York.

Mohr, D. N., D. J. Feist, J. A. Washington II, and P. E. Hermans. 1979. Infections due to group C streptococci in man. Am. J. Med. 66:450–456.

Moody, M. D., A. C. Siegel, B. Pittman, and C. C. Winter. 1963. Fluorescent-antibody identification of group A streptococci from throat swabs. Am. J. Public Health 53:1083–1092.

Mufson, M. 1990. *Streptococcus pneumoniae*, p. 1539–1550. In: G. L. Mandell, R. G. Douglas, Jr., and J. E. Bennett (ed.), Principles and practice of infectious diseases. John Wiley and Sons, New York.

Murray, P. R., A. D. Wold, C. A. Schreck, and J. A. Washington II. 1976. Effects of selective media and atmosphere of incubation on the isolation of group A streptococci. J. Clin. Microbiol. 4:54–59.

Myhre, E. B., and P. Kuusela. 1983. Binding of human fibronectin to group A, C, and G streptococci. Infect. Immun. 40:29–34.

Nealon, T. J., and S. J. Mattingly. 1983. Association of elevated levels of cellular lipoteichoic acids of group B streptococci with human neonatal disease. Infect. Immun. 39:1243–1251.

Nealon, T. J., and S. J. Mattingly. 1984. Role of cellular lipoteichoic acids in mediating adherence of serotype III strains of group B streptococci to human embryonic, fetal, and adult epithelial cells. Infect. Immun. 43:523–530.

Noble, J. T., M. B. Tyburski, M. Berman, J. Greenspan, and M. J. Tenenbaum. 1980. Antibiotic tolerance in group G streptococci. Lancet ii:982.

Oberhofer, T. R. 1986. Value of the L-pyrrolidonyl-β- naphthylamide hydrolysis test for identification of select gram-positive cocci. Diag. Microbiol. Infect. Dis. 4:43–47.

Parker, M. T., and L. C. Ball. 1976. Streptococci and aerococci associated with systemic infections in man. J. Med. Microbiol. 9:275–302.

Patel, K., A. L. Chittom, R. Toshniwal, and F. E. Kocka. 1987. Rapid commercial test for direct detection of group A streptococci in throat swabs. Eur. J. Clin. Microbiol. 6:193–194.

Patterson, M. J., and A. E. B. Hafeez. 1976. Group B streptococci in human disease. Bacteriol. Rev. 40:774–792.

Phillips, E., and P. Nash. 1985. Culture media, p. 1051–1092. In: E. H. Lennette, A. Balows, W. J. Hausler, Jr., and H. J. Shadomy (ed.), Manual of clinical microbiology, 4th ed. American Society for Microbiology, Washington, D.C.

Rantz, L. A., and E. Randall. 1955. Use of autoclaved extracts of hemolytic streptococci for serological grouping. Stanford Med. Bull. 13:290–291.

Reimer, L. G., and L. B. Reller. 1981. Growth of nutritionally variant streptococci on common laboratory and 10 commercial blood culture media. J. Clin. Microbiol. 14:329–332.

Reyn, A. 1986. Genus Gemella Berger 1986, 253^AL, p. 1081–1082. In: P. H. A. Sneath, N. S. Mair, M. E. Sharpe, and J. G. Holt (ed.), Bergey's manual of systematic bacteriology, vol. 2. Williams and Wilkins, Baltimore.

Rolston, K. V. I., J. L. LeFrock, and R. F. Schell. 1982. Activity of nine antimicrobial agents against Lancefield group C and G streptococci. Antimicrob. Ag. Chemother. 22:930–932.

Romero, R., and H. W. Wilkinson. 1974. Identification of group B streptococci by immunofluorescence staining. Appl. Microbiol. 28:199–204.

Ruoff, K. L., and L. J. Kunz. 1983. Use of the Rapid STREP system for identification of viridans streptococcal species. J. Clin. Microbiol. 18:1138–1140.

Ruoff, K. L., S. I. Miller, C. V. Garner, M. J. Ferraro, and S. B. Calderwood. 1989. Bacteremia with Streptococcus bovis and Streptococcus salivarius: clinical correlates of more accurate identification of isolates. J. Clin. Microbiol. 27:305–308.

Schleifer, K. H., and R. Kilpper-Balz. 1984. Transfer of Streptococcus faecalis and Streptococcus faecium to the genus Enterococcus nom. rev. as Enterococcus faecalis comb. nov. and Enterococcus faecium comb. nov. Int. J. Syst. Bacteriol. 34:31–34.

Schleifer, K. H., and R. Kilpper-Balz. 1987. Molecular and chemotaxonomic approaches to the classification of streptococci, enterococci and lactococci: a review. Syst. Appl. Microbiol. 10:1–9.

Schwabe, L. D., M. T. Small, and E. L. Randall. 1987. Comparison of Test Pack Strep A test kit with culture technique for detection of group A streptococci. J. Clin. Microbiol. 25:309–311.

Slifkin, M., D. Freedel, and G. M. Gil. 1982. Direct serogrouping of group B streptococci from urogenital and gastric swabs with nitrous acid extraction and the Phad

ebact streptococcus test. Am. J. Clin. Pathol. 78:850–853.

Smith, J. A., and J. H. Ngui-Yen. 1980. Evaluation of a clostridial alpha- toxin disk test for rapid presumptive identification of group B streptococci. J. Clin. Microbiol. 12:18–21.

Smyth, E. G., A. P. Pallet, and R. N. Davidson. 1988. Group G streptococcal endocarditis: two case reports, a review of the literature and recommendations for treatment. J. Infect. 161:169–176.

Stamm, A. M., and C. G. Cobbs. 1980. Group C streptococcal pneumonia: report of a fatal case and review of the literature. Rev. Infect. Dis. 2:889–898.

Stevens, D. L., M. H. Tanner, J. Winship, R. Swarts, K. M. Ries P. M. Schlievert, and E. Kaplan. 1989. Severe group A streptococcal infections associated with a severe toxic shock-like syndrome and scarlet fever toxin A. New Engl. J. Med. 321:1–7.

Veasy, L. G., S. E. Wiedmeier, G. S. Orsmond, H. D. Ruttenberg, M. M. Boucek, S. J. Roth, V. F. Tait, J. A. Thompson, J. A. Daly, E. L. Kaplan, and H. H. Hill. 1987. Resurgence of acute rheumatic fever in the intermountain area of the United States. N. Engl. J. Med. 316:421–427.

Wannamaker, L. W. 1979. Changes and changing concepts in the biology of group A streptococci and in the epidemiology of streptococcal infections. Rev. Infect. Dis. 1:967–973.

Wannamaker, L. W. 1983. Streptococcal toxins. Rev. Infect. Dis. 5 (supplement 4):S723-S732.

Wasilauskas, B. L., and K. D. Hampton. 1984. Evaluation of the Strep-A-Fluor identification method for group A streptococci. J. Clin. Microbiol. 20:1205–1206.

Watson, B. K., R. C. Moellering, Jr., and L. J. Kunz. 1975. Identification of streptococci: use of lysozyme and Streptomyces albus filtrate in the preparation of extracts for Lancefield grouping. J. Clin. Microbiol. 1:274–278.

Webb, B. J., M. S. Edwards, and C. J. Baker. 1980. Comparison of slide coagglutination test and countercurrent immunoelectrophoresis for detection of group B streptococcal antigen in cerebrospinal fluid from infants with meningitis. J. Clin. Microbiol. 11:263–265.

Wellstood, S. 1982. Evaluation of Phadebact and Streptex Kits for rapid grouping of streptococci directly from blood cultures. J. Clin. Microbiol. 15:226–230.

Wicher, K., C. Kalinka, P. Mlodozeniec, and N. Rose. 1982. Fluorescent antibody technic used for identification and typing of Streptococcus pneumoniae. Am. J. Clin. Pathol. 77:72–77.

Wilkinson, H. W. 1977. CAMP-disk test for presumptive identification of group B streptococci. J. Clin. Microbiol. 6:42–45.

Wong, J. D. 1987. Porphyrin test as an alternative to benzidine test for detecting cytochromes in catalase-negative gram-positive cocci. J. Clin. Microbiol. 25:2006–2007.

Yeung, M. K., and S. J. Mattingly. 1984. Biosynthetic capacity for type-specific antigen synthesis determines the virulence of serotype III strains of group B streptococci. Infect. Immun. 44:217–221.

CHAPTER 66

The Genus *Enterococcus*

LUC A. DEVRIESE, MATTHEW D. COLLINS, and REINHARD WIRTH

The genus *Enterococcus* was first proposed by Schleifer and Kilpper-Bälz (1984) to accommodate the Lancefield group D fecal streptococci species *Streptococcus faecalis* and *S. faecium*, now called *Enterococcus faecalis* and *E. faecium*. Chemotaxonomic studies have now resulted in the assignment of several other group D streptococci to the genus: *Enterococcus avium, E. casseliflavus, E. durans, E. gallinarum,* and *E. malodoratus* (Collins et al., 1984); five new species, *E. hirae, E. mundtii, E. pseudoavium, E. raffinosus,* and *E. solitarius,* have also been described (Collins et al., 1986, 1989b; Farrow and Collins, 1985; Williams et al., 1989).

The enterococci are Gram-positive cocci which occur mostly in pairs or in short chains. A few species are pigmented *(E. mundtii, E. casseliflavus)* or motile *(E. casseliflavus, E. gallinarum)*. The group D antigen is not confined solely to enterococci. A number of other streptococcal species *(S. alactolyticus, S. bovis, S. equinus, S. suis)* possess this antigen. *S. bovis* and *S. equinus* have historically been associated with the enterococci due to their similar habitat and serological reaction (Sherman, 1937; Jones, 1978). However these latter species are members of *Streptococcus* sensu strictu and this assignment demonstrates the overemphasis placed on serology in streptococcal classification. In fact, *Streptococcus cecorum* (Devriese et al., 1983), which lacks the group D antigen, has recently been reclassified in the genus *Enterococcus* (Williams et al., 1989).

Hybridization between ribosomal RNA (rRNA) and DNA and the determination of sequences of 16S rRNA Kilpper-Bälz et al. 1982; Ludwig et al., 1985; Schleifer and Kilpper-Bälz, 1984, 1987) have shown that enterococci form a phylogenetically coherent group equivalent in rank to *Streptococcus* and *Lactococcus*. Recent comparative analyses of full 16S rRNA sequences using reverse transcriptase indicate that the enterococci are phylogenetically more closely related to some motile group N strains and to strains of *Listeria* than to *Streptococcus* or *Lactococcus* (Collins et al., 1989a). A list of accepted species of enterococci is given in Table 1.

Habitats

Enterococci are predominantly inhabitants of the intestines of humans and other animals. Certain host specificities may be noted: *E. durans* was found in humans and in poultry but not in several other farm animal species; *E. gallinarum* occurs in poultry but *E. avium* is often found in mammals (Devriese et al., 1987). Although the latter species was first described as being typically associated with poultry (Nowlan and Deibel, 1967), it appears to be rare in modern poultry stock. *E. faecium* is the most frequently occurring enterococcal species in healthy chickens (L. A. Devriese, unpublished observations) and in pigs (Molitoris et al., 1986). According to Medrek and Barnes (1962), the enterococci are not present in large numbers in the gut of cattle but they are more frequent in sheep.

In humans, *E. faecalis* and *E. faecium* are the most frequent species. In countries such as England and the United States, *E. faecalis* tends to predominate in the human gut, but in others such as India, Japan, and Uganda, the incidence

Table 1. The accepted species of the genus *Enterococcus*.

Species	Reference
E. faecalis	Schleifer and Kilpper-Bälz, 1984
E. faecium	Schleifer and Kilpper-Bälz, 1984
E. hirae	Farrow and Collins, 1985
E. durans	Collins et al., 1984
E. gallinarum	Collins et al., 1984
E. avium	Collins et al., 1984
E. mundtii	Collins et al., 1986
E. casseliflavus	Collins et al., 1984
E. malodoratus	Collins et al., 1984
E. pseudoavium	Collins et al., 1989b
E. solitarius	Collins et al., 1989b
E. raffinosus	Collins et al., 1989b
E. cecorum	Williams et al., 1989

of *E. faecium* is equal to or greater than that of *E. faecalis* (Hill et al., 1971). Diet or other factors may influence this (Finegold et al., 1974), but not much is known with certainty about this topic.

The enterococci are among the predominant flora of the intestine in the first 2–3 days of life in many animal host species and they decline to markedly lower levels after 2 or more weeks (Smith and Crabb, 1961). The proportions in the gut of distinct species may also vary with age. Chicks carry and excrete large numbers of *E. faecalis* when a few days old, but afterwards this organism decreases sharply, while *E. faecium* remains constant and *E. gallinarum* increases (Kaukas et al., 1986). Watanabe et al. (1981) reported frequent isolations of *E. avium* from children, this species being absent in samples from adults.

E. faecalis and *E. faecium* can be found associated with plants, as are the pigmented *E. mundtii* and *E. casseliflavus*. The latter two species are rare in the intestines of mammals and birds (Martin and Mundt, 1972). Strains from plants and insects more actively digest starch than strains from mammals (Geldreich et al., 1964; Mundt, 1973). In moderate climates, the enterococci disappear from the plant world during the winter, reappearing in the spring and becoming more and more frequent as the plants grow and flourish. It is thought that insects play an important role in this seasonal variation (Martin and Mundt, 1972). This facet of the ecology of enterococci needs further study, especially in the light of the new classification of the genus. The possibility that enterococci associated with insects or plants are different from strains occurring in mammals, merits further investigation.

Isolation, Cultivation, and Preservation

The enterococci are chemoorganotrophs, and their complex nutritional requirements are normally satisfied by the media commonly used containing peptone or similar products (Niven and Sherman, 1944: Deibel, 1964). They can be cultured for various purposes in brain heart infusion and other rich media. Most strains of *E. cecorum* require >3% CO_2 in the atmosphere for growth (Devriese et al., 1983).

The ability of enterococci to grow under adverse conditions is used widely in its selective isolation and in the differentiation of these bacteria from streptococci (see below). They are resistant or relatively insensitive to many anti-

biotics active against Gram-positive bacteria, for instance tetracyclines, aminoglycosides, sulfonamides, several penicillins, and the lincosamides.

More than 60 different selective media have been described, most of which also allow growth of certain streptococci. The commonly available selective media have been designed to support the growth of *Streptococcus bovis* as well as the enterococci. This can be a disadvantage and may lead to confusion, because *S. bovis* may behave very differently from the enterococci (see "Applications"). Comparisons of media have been published by Barnes (1959), Mallmann and Seligmann (1950), Pavlova et al. (1972), Sabaj et al. (1971), and Switzer and Evans (1974).

Sodium azide is the most widely used selective agent. Also used have been growth in the presence of bile, thallium acetate, ethyl violet, crystal violet, potassium tellurite, potassium thiocyanate, 2,3,5-triphenyl tetrazolium chloride (TTC), kanamycin, nalidixic acid, oxolonic acid, polymyxin, or colistin, but though useful, these agents are less selective than sodium azide.

Most media contain between 0.2 and 0.5% sodium azide; SF broth of Hajna and Perry (1943); BAGG (buffered azide glucose glycerol) of Hajna (1951); EVA (ethyl violet azide) broth of Litsky et al. (1953); KF broth of Kenner et al. (1961); selective medium of Levin et al. (1975) with TTC and cycloheximide for use with sea water; and M2 agar of Rutkowski and Sjogren (1987) with TTC and nalidixic acid for use in the quality control of drinking water.

Other media rely mainly on thallium acetate as a selective substance: TITG (thallous tetrazolium glucose) agar of Barnes (1956); and TSTA (tyrosine sorbitol thallous agar) of Mead (1963).

Resistance to 20 μg/ml of the aminoglycoside antibiotic kanamycin in combination with sodium azide offers a useful means for isolation, and Mossel et al. (1978) have described a kanamycin esculin azide broth and agar for use in enrichment and isolation:

Kanamycin Esculin Broth

Tryptone	20 g/l
Yeast extract	5 g/l
Sodium chloride	5 g/l
Sodium citrate	1 g/l
Esculin	1 g/l
Ferric ammonium citrate	0.5 g/l
Sodium azide	0.15 g/l
Kanamycin sulfate	0.02 g/l

Several media have been designed for use with membrane filters which allow the examination of relatively large volumes of water, but they can also be used for other purposes. An example of these media has been devised by Slanetz and Bartley (1957):

Tryptose	20 g/l
Yeast extract	5 g/l
Glucose	2 g/l
Disodium phosphate	4 g/l
Sodium azide	0.4 g/l
TTC	0.1 g/l
Agar	12 to 15 g/l

Media such as SF broth and KF agar and broth, which contain 0.04 to 0.05% sodium azide, inhibit *S. bovis* to some extent while allowing good recovery of enterococci (Switzer and Evans, 1974). *E. cecorum*, however, is also inhibited by these and even lower concentrations of sodium azide (Devriese et al., 1983). KF broth and agar have the following composition:

Proteose peptone	10 g/l
Yeast extract	10 g/l
Sodium chloride	5 g/l
Sodium glycerophosphate	10 g/l
Maltose	20 g/l
Lactose	1 g/l
Sodium azide	0.4 g/l
Bromcresol purple	0.015 g/l

Enterococcus strains are usually preserved by lyophilization. They can also be kept for 2 years at $-20°C$ in a lyophilization medium containing nutrient broth, inactivated horse serum, and glucose without actual lyophilization. Litmus milk plus chalk at 4°C is suitable for short-term preservation (3 months).

Identification

With the exception of *E. cecorum,* all enterococcal species possess the Lancefield group D antigen, and the group D reaction is, therefore, a useful presumptive test. Strains of some species *(E. avium, E. raffinosus),* however, give a weak or negative reaction. Even more serious, the nonenterococcal species *S. alactolyticus, S. bovis* and *S. equinus* are group D positive, as are some *Leuconostoc* and most *Pediococcus* strains (Facklam and Collins, 1989).

Growth in the presence of 6.5% NaCl is one of the best identification characters of the genus, although *E. cecorum* and some *E. avium* strains do not grow at this NaCl concentration, and some *Aerococcus viridans* are resistant. The lactococci can be differentiated by their group N streptococcal antigen and the leuconostocs by

their resistance to vancomycin, by gas production from glucose, and by a negative pyrrolidonyl arylamidase reaction (Facklam and Collins 1989). The last character also differentiates to some extent the enterococci (all positive, except *E. cecorum*) from the streptococci, which are mostly negative. Growth and hydrolysis of esculin in the presence of 40% bile can also be used (Facklam and Moody, 1970): all known enterococci are positive but *S. bovis* and some strains of *S. suis* react similarly. Other characters that are generally useful indicators for the recognition of enterococci include growth at 10°C and at pH 9.6, and survival when heated at 60°C for 30 min. However, these latter tests are difficult to standardize, and the same species that cause difficulties in the other tests mentioned above react atypically.

It can be concluded that the well-known species which are most frequently associated with humans are easily recognized as enterococci, but difficulties may arise with strains from animals and environmental sources. Species identification of enterococci is usually conducted with identification sets commercially available for identification of streptococci (Devriese et al., 1987; Facklam and Collins, 1989), often in combination with a few laboratory-made test media. Tests useful in the identification of species are shown in Table 2. Certain characters are especially valuable: *E. gallinarum* and *E. casseliflavus* are motile, and *E. mundtii* and *E. casseliflavus* are pigmented. The new species *E. cecorum, E. raffinosus,* and *E. pseudoavium,* as well as *E. avium* and *E. malodoratus,* do not hydrolyze arginine. *E. hirae* can be differentiated from *E. durans,* the species with which it was originally confused, by its positive sucrose and melibiose reaction. Also acid production from mannitol, sorbitol, and sorbose are particularly helpful, and they can be used, together with arginine hydrolysis, as key tests for the identification of human strains (Facklam and Collins, 1989).

Penicillin-binding protein patterns give excellent correlation with the results of DNA-DNA hybridizations, and their determination affords a precise method for species identification (and the recognition of new species) (Williamson et al., 1986; Collins et al., 1989b).

Phage typing with species-specific phages or pooled phages has been used for identification (Pleceas and Brandis, 1974), and phage typing systems have been described for use in epidemiological investigations (Caprioli et al., 1975; Kühnen et al., 1988; Smyth et al., 1987). Serotyping of *E. faecalis* (Sharpe and Shattock, 1952), biotyping, and especially bacteriocine (enterococcin) typing combined with phage typ-

Table 2. Characteristics useful for the differentiation of the enterococci.

Test	E. faecalis	E. faecium	E. hirae	E. durans	E. gallinarum	E. avium	E. casseliflavus	E. mundtii	E. malodoratus	E. pseudoavium	E. solitarius	E. raffinosus	E. cecorum
Acid from:													
L-arabinose	−	+	−	−	+	+	+	+	−	−	−	+	−
Mannitol	+	+	−	−	+	+	+	+	+	+	+	+	−
Sorbitol	+	−	−	−	−	+	−	D[a]	+	+	+	+	−
Raffinose	−	D[b]	D	−	+	−	+	+	+	−	−	+	+
Saccharose	+	D+	+	−	+	+	+	+	+	−	+	+	+
Melibiose	D−	+	+	−	+	D−	+	+	+	−	−	+	+
Motility	−	−	−	−	+[c]	−	+	−	−	−	−	−	−
Yellow pigment	−	−	−	−	−	−	+	+	−	−	−	−	−
Growth in 6.5% NaCl	+	+	+	+	+	D	+	+	+	+	+	+	−
Pyrrolidonyl-arylaminidase	+	+	+	+	+	+	+	+	+	+	+	+	−

[a]D, different; D−, usually negative; D+, usually positive.

[b]Most *E. faecium* strains in poultry are raffinose positive; strains associated with other hosts are negative.

[c]*E. gallinarum* was incorrectly described as nonmotile in the species description of Bridge and Sneath (1982).

ing (Kühnen et al., 1988) can be used to study hospital epidemiology.

Physiology

As is the case for the genetics of enterococci, only limited data are available concerning the biochemical and physiological properties of members of this genus. Since most species were described within the last few years, nearly all data concern *E. faecalis* and *E. faecium*, which have been known for a long time. Two factors appear to be responsible for our limited knowledge of enterococcal physiology: 1) there is no readily available defined medium for enterococci; 2) enterococci perform a simple type of fermentation, namely a homofermentative lactic acid fermentation and therefore, the study of the physiology of enterococci has not attracted much attention.

Despite various attempts to develop a defined medium for enterococci, there have been no reports of such a medium being generally applicable for growth of all isolates of *E. faecalis*. The members of this genus are, therefore, generally referred to as having complex and variable nutritional requirements. Indeed, only rich media like Todd-Hewitt Broth or Brain Heart Infusion Broth support a rapid growth to high densities. Even LB medium (1.0% tryptone, 0.5% yeast extract), which is an excellent complex medium for *E. coli*, yields slow and poor growth. RPMI 1640 medium (Gibco/BRL), which is a defined medium developed for the cultivation of eukaryotic cells (composed of 7 inorganic salts, all 20 L-amino acids, and 11 vitamins, and glucose as carbon source) does not support the growth of the *E. faecalis* strains tested so far, whereas other fastidious bacteria such as lactococci will grow in this medium.

Growth Requirements

According to Niven and Sherman (1944), enterococci require the vitamins biotin, nicotinate, panthothenate, riboflavin, and pyridoxine. In addition, *E. faecium*, but not *E. faecalis*, requires folic acid. Lipoate is absolutely necessary for the growth of many *E. faecalis* strains, but only if pyruvate is used as energy source (a detailed description of the relevant observations is given by Deibel, 1964). This absolute requirement can even be used for quantitative determinations of lipoate; it should be noted that, at higher oxygen tensions, up to 20 times more lipoate is required by *E. faecalis* compared to the amount required for anaerobic growth. Interestingly, under aerobic conditions, in *E. faecalis*, an H_2O-forming NADH oxidase is found (Hoskins et al., 1962). This enzyme has been purified to homogeneity (Schmidt et al., 1986); however, as yet no data on the possible involvement of a lipoate derivative in the reactions catalyzed are available.

The characteristics of the requirement for amino acids is still not clear; at least 10 amino acids are absolutely necessary for growth, with others showing an additional stimulatory effect (McCoy and Wender, 1953; Niven and Sherman, 1944). It seems that single L-amino acids support growth better than acid or enzymatic hydrolysates of casein; α-keto derivatives of amino acids and D-amino acids can be used only in exceptional cases (Deibel, 1964).

Metabolism

Energy may be obtained by the degradation of some amino acids. *E. faecalis* uses arginine as an energy source by a sequence of reactions that involves deamination to citrulline in a first step. Phosphorolysis of citrulline leads in a second step to the formation of carbamoyl phosphate and ornithine. The latter is excreted by the cell and the energy-rich bond of carbamoyl phosphate is transferred to ADP, yielding ATP, CO_2, and NH_3. The decarboxylated derivative of arginine, agmatine, can also be used by *E. faecalis* in a similiar way. In this case, the end products are the excreted putrescine, ATP, CO_2, and NH_3. Since only one mole of ATP is generated in this way by the degradation of one mole of arginine or agmatine, it is not surprising that both an arginine/ornithine and an agmatine/putrescine antiporter could be identified in *E. faecalis* (Poolman et al., 1987; Driessen et al., 1988). These systems allow the concurrent influx of arginine or agmatine and the efflux of ornithine or putrescine, respectively, without energy consumption. If canavanine is added to an *E. faecalis* culture using arginine as energy source, a stimulation of growth was observed (Himmel and Zimmermann, 1963). Probably canavanine inhibits the arginine dihydrolase, thereby increasing the amount of arginine available for the ATP-generating reactions described earlier. *E. faecalis*, but not *E. faecium*, can use serine as an energy source via deamination to pyruvate.

The main route for energy production is the homofermentative lactic acid pathway, yielding mainly L-lactic acid from glucose. In their study on numerical taxonomy of streptococci, Bridge and Sneath (1983) reported that the final pH is below 4.25 for all strains belonging to *Enterococcus* phenons if they are grown in liquid culture under fermentative conditions. In contrast to earlier data (Deibel, 1964), Bridge and Sneath (1983) reported that pentoses are fermented by 30 to 100% of all strains tested. The two reports agree that ribose is fermented by all enterococci. The production of acid from a great variety of carbon/energy sources varies greatly between different enterococci, and therefore this trait can be used for taxonomic purposes. All reports agree that erythritol cannot be used by any strain of enterococci; differences as to the acid production from *meso*-inositol were reported, however.

The end products of glucose fermentation are mainly ethanol, formic acid, and acetic acid, if the pH is kept above neutral (Gunsalus and Niven, 1942). As noted above, lactic acid is the main product of glucose fermentation if a drop in the pH to values below 5 is allowed. Under aerobic conditions (without added heme—see below) glucose is converted to acetic acid, acetoin, and CO_2 (London and Appleman, 1962). This latter type of glucose conversion needs vigorous agitation and results in a 40% higher cell yield than anaerobic glucose conversion. Two explanations were offered for the higher yield under aerobic conditions: 1) aerobically, the pyruvate formed from glucose will be oxidized to acetic acid and CO_2, yielding additional energy (as mentioned above, lipoate is absolutely necessary for this reaction); 2) it was argued that the production of acetoin will result in slower lowering of pH, meaning less growth inhibition. In this connection, it should be stressed again that in all these experiments complex media were used, which might well have contained unidentified carbon/energy sources. It was also argued that the H_2O-forming NADH oxidase of *E. faecalis* might allow a better growth yield under aerobic conditions. At a first glance, the use of two moles of $NADH_2$ and one mole of O_2 to generate two moles of H_2O might appear energetically wasteful. On the other hand, bacteria not possessing a respiratory chain could make use of substrate-chain phosphorylations and dispose of excess electrons without the need to produce energetically wasteful H_2. Therefore, the NADH oxidase might enable *E. faecalis* to use substrates that this bacterium could not use otherwise because it lacks hydrogen acceptors.

The distribution of NADH oxidases, NADH peroxidases, L-lactate oxidases, pyruvate oxidase, and superoxide dismutase was studied by Zitzelsberger et al. (1984). These authors tested 23 different strains of "streptococci", three of which (*E. avium, E. faecalis,* and *E. faecium*) belonged to the genus *Enterococcus*. All 23 strains possessed superoxide dismutase and NADH oxidase activity. NADH peroxidase activity was found in 7 of the 23 strains, including *E. avium* and *E. faecalis*, whereas L-lactate oxidase activity was detected only in 4 of the 23 strains, including *E. avium* and *E. faecium*. Pyruvate oxidase activity was found only in *E. faecalis*. These data indicate that many streptococci are able to cope with aerobic conditions, and that at least some enterococci in addition can show increased growth under aerobic conditions.

This is especially true if heme is added to aerobically grown enterococci, especially *E. faecalis*. It was reported (Sijpensteijn, 1970;

Ritchey and Seeley, 1976) that the great majority of *E. faecalis* strains are not able to synthesize heme on their own, but possess the ability to establish a "cytochrome-like" respiratory chain if supplemented with heme. Therefore, at least *E. faecalis* can be viewed as a facultative anaerobic species with the potential to perform aerobic metabolism via a respiratory chain if supplied with heme. This potential dual nature of *E. faecalis* metabolism has been discussed in detail by Whittenbury (1978). It should be noted that *E. faecalis* encounters aerobic conditions only under special circumstances, e.g., when present in blood (*E. faecalis* is a causative agent of endocarditis), and that in *E. faecalis* conjugative plasmids carrying hemolysin determinants are not uncommon (see "Genetics").

The principal routes of glycerol metabolism under both aerobic and anaerobic conditions are known (Jacobs and VanDenmark, 1960). Aerobically, glycerol is phosphorylated in the first step yielding glycerol-3-phosphate. An FAD-linked dehydrogenase then converts glycerol-3-phosphate to dihydroxyacetone phosphate. The latter compound is also formed under anaerobic conditions and finally leads to the formation of lactate. The anaerobic use of glycerol in the presence of fumarate as election acceptor has attracted much attention, because only *E. faecalis* is able to use this substance directly. Here, glycerol is dehydrated in a first step to yield dihydroxyacetone, which is further phosphorylated to dihydroxyacetone phosphate. The dehydrogenase is NAD-linked, and the reducing equivalents are used for the reduction of fumarate.

The use of gluconate as an energy source also merits attention, because this compound is transported via a phosphotransferase (PTS) system into *E. faecalis* (Bernsmann et al., 1982). This is a unique situation, in as much as PTS-dependent transport was thought to be restricted to uncharged carbohydrates. The intracellular gluconate-6-phosphate probably is metabolized further via the pentose phosphate route. It was reported that *E. faecalis* contains a 6-phosphogluconate dehydrogenase that can convert gluconate-6-phosphate to ribulose-5-phosphate and CO_2 (Bernsmann et al., 1982).

The existence of at least one other PEP-dependent PTS mechanism in *E. faecalis,* namely for lactose, was demonstrated by Heller and Röschenthaler (1978). These authors identified an enzyme that they called β-D-phosphogalactosidase. It splits the intracellular lactose phosphate (which is accumulated in the cell by the lactose-PTS system) into glucose and galactose-6-phosphate. The latter is converted to dihydroxyacetone phosphate and glyceraldehyde-3-

phosphate via the tagatose route. The β-D-phosphogalactosidase was reported to be induced by lactose and galactose, but not by glucose, mannose, and maltose. It has also been reported that "lactose metabolic genes" from *"Streptococcus cremoris, lactis, S. mutans, S. sanguis,* and *S. faecalis"* are homologous to each other (Inamine et al., 1986). The DNA probe used in these experiments contained β-D-phosphogalactosidase- *and* lactose-PTS genes, and therefore it still remains unclear if the β-D-phosphogalactosidase gene of *E. faecalis* is homologous to that of the other species. However, it appears that lactose can be imported by *E. faecalis* cells also via a second route, namely by a lactose permease. Bridge and Sneath (1983) reported the existence of β-galactosidase in 90% of their strains of *E. faecalis.* Very interestingly, a gene has been sequenced that shows high homology to the β-galactosidase genes from *Escherichia coli* and *Streptococcus thermophilus* (R. Wirth, unpublished observations).

Chemical Composition

The cell wall of enterococci contains peptidoglycan of the type lys-ala$_{2-3}$ *(E. faecalis)* or lys-D-asp *(E. faecium, E. avium, E. casseliflavus, E. durans, E. gallinarum,* and *E. malodoratus).* The type D antigen seems to consist of an unusual glycerol teichoic acid (Wicken and Baddiley, 1963; Wicken et al., 1963); as with all other teichoic acids, the backbone consists of glycerol molecules that are linked through phosphodiester groups at positions 1 and 3. Position 2 of glycerol is esterified with either the glucose disaccharide kojibiose *(E. faecalis)* or the glucose trisaccharide kojitriose *(E. faecium).* The unusual nature of this polymer stems from the fact that either the amino acid D-alanine *(E. faecalis)* or the amino acids D-alanine and L-lysine *(E. faecium)* are in an ester linkage to the hydroxyl groups of glucose rather than to the polyol. This antigen seems to be "hidden" in the cell wall; sometimes group D antigen is considered to have an intracellular location, which might explain the difficulties incurred in serological grouping of enterococci (for a detailed discussion, see Deibel, 1964).

Enterococci normally are considered to be facultatively anaerobic organisms having a preference for anaerobic conditions. They are not able to synthesize porphyrins, and therefore cannot produce cytochromes. However, it was reported that *E. faecalis* is able to synthesize cytochromes if the medium contains hemin (Sijpensteijn, 1970; Ritchey and Seeley, 1976). It was of interest, therefore, to look for non-cytochrome electron carriers in enterococci.

Collins and Jones (1979) determined the content of isoprenoid quinones in "streptococci" of groups D and N. *E. faecalis* contained demethylmenaquinones with nine isoprenoid units; *E. casseliflavus* had menaquinones with eight isoprenoid units, while neither menaquinones nor ubiquinones were detected in *E. avium, E. durans, E. faecium,* and *E. malodoratus.*

The composition of the major fatty acids found in enterococci is also of taxonomic value. According to Schleifer and Kilpper-Bälz (1984) and Collins et al. (1984), the main compounds are: hexadecanoic, octodecanoic, and *cis*-11,12-methylenoctadecanoic acids for *E. faecalis* and *E. faecium;* the same fatty acids are present in *E. durans, E. gallinarum,* and *E. casseliflavus,* however, with decreasing amounts of the last compound. *E. avium* contains substantial amounts of tetradecanoic and hexadecanoic acid, with minute amounts of octadecanoic and *cis*-11,12-methylenoctadecanoic acid. *E. malodoratus* seems to possess the same spectrum as *E. avium,* with a somewhat higher molar ratio of the latter two compounds.

Genetics

E. faecalis is the only species of the genus *Enterococcus* that has been genetically characterized to any extent. However, the genetic characterization has not been performed on the chromosomal DNA. Rather, the special interest in *E. faecalis* arose when it became clear that this species contains many plasmids and that at least some of them can be transferred conjugatively to other bacteria. It should be noted also, that it was in this species of the genus *Enterococcus* that two new genetic systems—namely conjugative transposons and sex pheromone plasmids—were discovered.

Transposons and Plasmids

As early as 1964, the conjugal transfer of chloramphenicol resistance was reported, and in 1973 a conjugative hemolysin determinant was identified. The first proof for the existence of plasmid DNA in group D streptococci was provided in 1972 (Courvalin et al., 1972). The first clear demonstration of plasmid involvement in conjugation was provided by Jacob and Hobbs in 1974 and within seven years (Clewell, 1981), a total of 47 plasmids had been identified in *E. faecalis;* of these 29 were classified as conjugative, 4 as nonconjugative and 14 were not defined. The same report cited a total of 38 plasmids known in eight other species of streptococci. Because many of the conjugative

E. faecalis plasmids seem to have a rather broad host range, *E. faecalis* sometimes is regarded as functioning like a "plasmid reservoir" for other bacterial species found in its natural habitat. As an example for this, one can cite plasmid pIP501 (originally isolated from *Streptococcus agalactiae*), which can be conjugatively transferred from *E. faecalis* to at least nine different species of streptococci, as well as to bacilli, lactobacilli, pediococci, listeria, and clostridia (Krah and Macrina, 1989).

The rapid acquisition and distribution of drug resistance determinants in *E. faecalis* can be attributed to plasmids (see below) and transposons. Tn917 seems to be the prototype for nonconjugative transposons in *E. faecalis*. It carries a macrolide-lincosamide-streptogramin (MLS) resistance determinant, and its transposition frequency can be enhanced by exposure to low concentrations of erythromycin (Tomich et al., 1980). This induction phenomenon probably is linked to the expression of resistance-related genes (Shaw and Clewell, 1985). Tn917-like transposons (which may or may not show erythromycin-enhanced induction of transposition) seem to be widespread in *E. faecalis* (Rollins et al., 1985). Tn917 was the first transposon of a Gram-positive bacterium to be completely sequenced (Shaw and Clewell, 1985) and used for the construction of useful derivatives. One of the most useful Tn917-derivatives allows the in vivo construction of transcriptional gene fusions using the *lacZ* system (Perkins and Youngman, 1986). The sequence data confirmed the hypothesis that Tn917 belongs to the Tn3 family, due to the observed sequence similarities of the inverted repeats, the generation of a five base-pair duplication upon insertion, and sequence similarities between the resolvase genes enabling resolution of inserted transposons.

A completely new type of genetic element, namely the conjugative transposon was identified when it became clear that some antibiotic-resistance determinants of streptococci are transfered in an unusual way. These elements behave like transposons with respect to integration into chromosomal or plasmid DNA. Depending on the bacterial host, integration can be at random or at certain hot spots of the target. The transfer of these elements between different cells (intra- or interspecies specific) takes place in a conjugation-like, DNase-resistant manner. In no case, however, could a corresponding plasmid be identified.

The prototype of this class of genetic element is Tn916, which was first identified in *E. faecalis* (Franke and Clewell, 1981; Gawron-Burke and Clewell, 1982). Tn916 is just one representative of this type of genetic element; Tn*918* and Tn*919* are very closely related to Tn*916*, but exhibit a different host range. Tn*1545* and Tn*3701* were isolated originally from *Streptococcus pyogenes* and *S. pneumoniae*, respectively, and show similarity to Tn*916* (LeBouguenec et al., 1988; Courvalin and Carlier, 1986). All the above-mentioned transposons encode at least one drug-resistance determinant, *tetM*. The existence of other conjugative transposons in bacteria belonging to Lancefield groups A,B,C,D,F, and G, and especially in *S. pneumoniae, S. agalactiae, S. pyogenes,* and *S. sanguis* was reviewed by Clewell and Gawron-Burke (1986). The *tetM* determinant is by far the most common in conjugative transposons; it was argued, therefore, that the *tet* determinant was the first to have acquired conjugative ability, and that conjugative transposons play(ed) a major role in the spread of antibiotic resistance in streptococci (Clewell and Gawron-Burke, 1986). Various conjugative transposons may have different host ranges; Tn*916* can be transferred conjugatively to *S. mutans, S. agalactiae, S. pyogenes, Lactococcus lactis, Bacillus thuringiensis, Staphylococcus aureus, Listeria monocytogenes,* and *Mycoplasma hominis*. Introduction of Tn*916* into *Streptococcus sanguis, S. pneumoniae, Bacillus subtilis, Haemophilus influenzae,* and *Escherichia coli* by transformation has also been reported. It is noteworthy that Tn*916* excises at a very high frequency under nonselective conditions in the Gram-negative host *E. coli*. It seems that excision is exact under such conditions, and therefore the following cloning strategy was developed for genes of Gram-positive bacteria: a mutant is obtained in the desired gene by insertion of Tn*916;* the corresponding DNA fragment is subcloned into *E. coli*. The original gene can be recovered under nonselective conditions (i.e., in the absence of tetracycline) in *E. coli* after excision of Tn*916*.

It was found that *Enterococcus faecalis* can harbor a special class of conjugative plasmids that are known as sex pheromone plasmids. A comparison of "normal" conjugative plasmids and sex pheromone plasmids is given in Table 3. In general, normal conjugative plasmids carry drug resistance determinants; these are found in sex pheromone plasmids in only exceptional cases. Very often, both types of plasmids are found in one strain. Therefore, the transfer frequency of the normal conjugative plasmids can increase indirectly in such a situation to numbers typical for sex pheromone plasmids. This increase is due to the fact that the sex pheromone plasmids ensure a tight physical contact, over a time long enough for

conjugation, between the nonmotile donor strains and the recipient strains (see below).

Normal conjugative plasmids have a broad host range. Conjugal transfer from *E. faecalis* to *Lactobacillus plantarum, Streptococcus sanguis,* and *Clostridium acetobutylicum* has been reported. Conjugal transfer of a pBR322/pAMβ1 shuttle vector into *Escherichia coli* was achieved, albeit at a low frequency (5×10^{-9} per donor cell) by Trieu-Cuot et al. (1988). It was argued that shuttle vectors, using the pIP501 or pAMβ1 replicon combined with an *E. coli* replicon, could be very useful for genetic manipulations of a great variety of Gram-positive bacteria. Such shuttle vectors have now been constructed by different groups and include pSA3, pAM401, and pWM401 (Dao and Ferretti, 1985; Wirth et al., 1986; Wirth et al., 1987).

Sex pheromone plasmids seem to be restricted to *E. faecalis,* and only in this bacterial species has the sex pheromone system been detected (Dunny et al., 1978). A working model for its function was first given by Dunny et al. (1979); an updated version is given in Fig. 1.

The working model for the sex pheromone system states the following: plasmid-free strains of *E. faecalis* can excrete small, linear peptides that act as pheromones. Such pheromones indicate to a potential donor cell (which carries the corresponding sex pheromone plasmid) the presence of a recipient strain. In response to the presence of sex pheromone, the donor synthesizes "aggregation substance." This "adhesive" protein enables donor and recipient to form tight aggregates, as shown in Fig. 2; the physical contact enables the nonmotile mating partners to stay in close contact for a period of time long enough to ensure conjugative transfer of the sex pheromone plasmid. Donor strains do not excrete the sex pheromone corresponding to the plasmid they carry (at the moment it is uncertain if they excrete a modified form of this peptide called mc1 in Fig. 1). Instead they excrete an inhibitor peptide that counteracts the effect of the corresponding sex pheromone. Donor strains do, however, excrete sex pheromones not related to the sex pheromone plasmid(s) they carry. Therefore, a strain that does not carry all the sex pheromone plasmids can still act as a donor for a sex pheromone plasmid it harbors, and as a recipient for a sex pheromone plasmid it does not harbor.

The existence of such a system explains the difference in conjugation frequency observed for normal conjugative and sex pheromone plasmids (see Table 3). In liquid culture, the sex pheromone system allows the physical contact needed for conjugation to be formed; if cells are

Table 3. Comparison of the characteristics of "normal" conjugative plasmids and sex pheromone plasmids of *E. faecalis*.

	"Normal" conjugative plasmids	Sex pheromone plasmids
Response to sex pheromones	No	Yes
Size (kb)	22 to 30	> 50 (pAM373: 37)
Host range	Broad	Limited to *E. faecalis*
Transfer frequency in liquid culture[a]	$\leqslant 10^{-6}$	10^{-1} to 10^{-3}
Transfer frequency on solid medium[b]	10^{-2} to 10^{-4}	10^{-1} to 10^{-3}
Drug resistance determinants	Often observed; especially MLS; also multiple resistances observed	Only observed in 3 of 11 cases

[a]Per donor cell; after a 4 hours incubation in liquid medium.
[b]Per donor cell; on a filter placed on a semisolid medium.

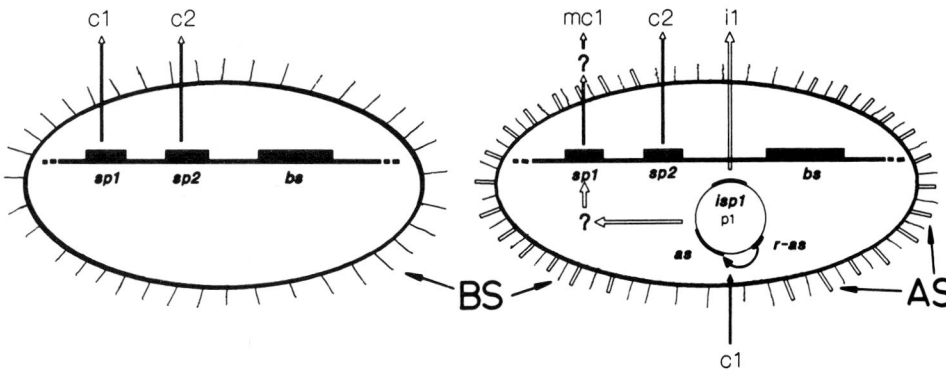

Fig. 1. Working model for the function of the sex pheromone system of *E. faecalis*. A plasmid-free recipient cell of *E. faecalis* is shown on the left and a donor cell harboring the hypothetical sex pheromone plasmid p1 is shown on the right. The horizontal line represents chromosomal DNA, and the thin circle represents the sex pheromone plasmid p1. *sp1, sp2*, and *bs* denote the genes leading to the production and secretion of the sex pheromones c1 and c2, and to the synthesis of binding substance (BS). The sex pheromone plasmid p1 carries the genes *isp1, as,* and *r-as*. The first two bring about the synthesis of an inhibitor peptide i1 and an aggregation substance (AS), and *r-as* is a postulated regulatory function for the expression of *as*. mc1 represents a postulated modified form of the sex pheromone c1. Note: only in the case of *as* a structural gene was characterized up to now. Therefore, the use of gene symbols does not necessarily imply the existence of single structural genes.

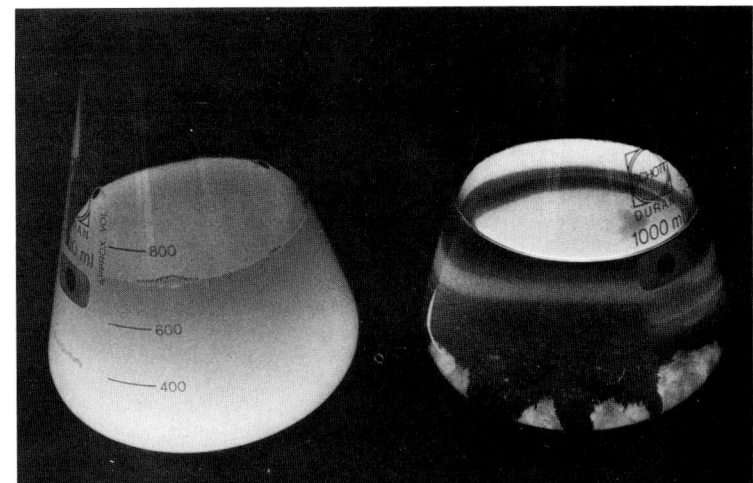

Fig. 2. Demonstration of the clumping phenomenon induced by the sex pheromone. The left flask contains a culture of *E. faecalis* which harbors a sex pheromone plasmid but has not been induced by sex pheromone. The culture in the right flask, which also carries a sex pheromone plasmid, was induced by the corresponding sex pheromone. Note the even distribution of the culture on the left, and the large clumps in the culture on the right.

brought together on solid surfaces, the difference in conjugation frequency is much less pronounced. Furthermore, the existence of such a system explains why *E. faecalis* is "loaded" with plasmids: this bacterial species possesses a unique plasmid collection mechanism.

Up to now, 11 different types of sex pheromone plasmids have been described (Clewell and Weaver, 1989). Their main characteristics are given in Table 4. From these data, it is evident that there is no common phenotype for all these plasmids. On the basis of DNA hybridizations, it could be demonstrated that all these plasmids (with the exception of pAM373) share one DNA region (R. Wirth, unpublished observations). Interestingly, data from Ehrenfeld and Clewell (1987) indicated that this region might code for the aggregation substance. Sequencing data (see below) subsequently has confirmed this (R. Wirth, unpublished observations). Therefore, the only common feature for all the sex pheromone plasmids is their ability to allow cells to express aggregation substance after induction by sex pheromones.

Up to now, the sequences of four of the sex pheromone peptides and of two inhibitor peptides are known (Mori et al., 1984; Suzuki et al., 1984; Mori et al, 1986; Mori et al., 1988). In all cases, linear peptides with no unusual amino acids were found (for details, see Clewell and Weaver, 1989). Sex pheromones are named after the corresponding plasmid: cAD1 (where c stands for clumping) induces the conjugation of the sex pheromone plasmid pAD1, etc. The postulated chromosomal location of the structural gene(s) encoding these peptides is supported by the fact that plasmid-free strains excrete sex pheromones. A direct proof that sex pheromones are translation products (i.e., they are synthesized by ribosomes) and are not assembled enzymatically like the peptide antibiotics of bacilli, is lacking.

Data of Ehrenfeldt et al. (1986) indicate that the receptor for aggregation substance is lipoteichoic acid; this, however, has yet to be confirmed.

Earlier work from the same laboratory indicated that aggregation substance coats donor cells after induction by sex pheromones as an evenly distributed fuzzy layer (Yagi et al., 1983). Subsequent work using a combination of immunogold-labelling and high-resolution scanning electron microscopic techniques showed that the aggregation substance is composed of "hairy" structures (Galli et al., 1989), which, as is shown in Fig. 3, are incorporated only into "old" cell walls (Wanner et al., 1989). Such a type of incorporation would allow a rapid dissociation of donor cells from mating aggregates: after transfer of the plasmid division of the newly created donor can then give rise to cells free of aggregation substance.

The postulated location of the structural gene for the aggregation substance on the corresponding sex pheromone plasmid was recently confirmed (Christie et al., 1988; Galli et al., 1990). Christie et al, reported the cloning and expression of pCF10-specific aggregation substance in *Escherichia coli*. The second group sequenced the structural gene for pAD1-specific aggregation substance. No amino acid sequence similarities to other surface proteins of Gram-positive bacteria were found, with the exception of an amino terminal signal sequence and a carboxy terminal cell wall region followed by a membrane anchor. In particular, there were no

Table 4. Known sex pheromone plasmids of *E. faecalis*.

Plasmid	Size (kb)	Encoded phenotype[a]	Known sex pheromone
pAM373	37	Unknown	cAM373
pAD1	58	Bac; Hly; Uvr	cAD1
pAMλ1		Identical to pAD1	cAD1
pJH2	58	Bac; Hly	cAD1
pAM322[b]	71	Pen^r; Gm^r; Km^r; Tm^r	cAD1
pAM323	63	Em^r	
pCF10	65	Tc^r	
pAMλ3	54	Unknown	
pOB1	68	Bac; Hly	
pAMλ2	54	Bac	
pPD1	56	Bac	cPD1
pAM324	56	Unknown	

[a]Bac, production of bacteriocin; Hly, production of hemolysin; Uvr, ultraviolet-light resistance; Em^r, resistance to erythromycin; Gm^r, resistance to gentamicin; Km^r, resistance to kanamycin; Pen^r, resistance to penicillin; Tc^r, resistance to tetracycline; Tm^r, resistance to tobramycin.
[b]This plasmid has been renamed pBEM10.

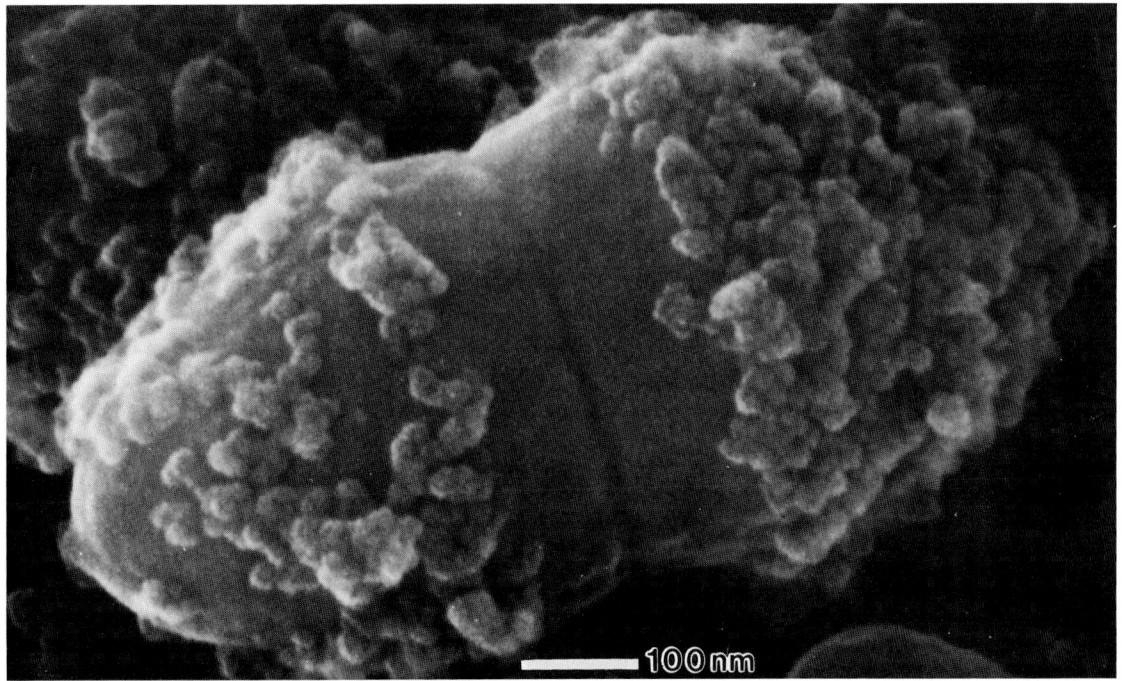

Fig. 3. Distribution of the aggregation substance on the cell surface of *E. faecalis*. The aggregation substance on the surface of *E. faecalis* was reacted with a specific antibody. A secondary antibody that was conjugated to gold particles was used for visualization of the surface protein by high-resolution scanning electron microscopy. It is evident that the aggregation substance is not located in the septal region, which is defined as the only region in which new cell wall material is incorporated.

protein internal repeats, as is known for example for *Streptococcus pneumoniae* M proteins, or for proteins A and G of *S. aureus* and a Lancefield group C streptococcus, respectively. Galli et al. (1990) reported the existence of two possible regions in the aggregation substance (a rather large protein of 142 kDa) that might mediate adhesion to eukaryotic cells; no data on this are available at the moment, however. Such a possible binding to eukaryotic cells seems very attractive: it could explain why the opportunistic pathogen *Enterococcus faecalis* evolved a mechanism to "collect" the corresponding plasmids.

The sex pheromone system of *E. faecalis* is unique; it is, unclear, however, if the production of the small linear peptides is unique. This stems from the fact that bacterial species other than *E. faecalis* have been found to excrete sex pheromone-like activities. They were identified in the course of studies aimed at screening for the possible existence of a sex pheromone system in bacterial species other than *E. faecalis* (Clewell et al., 1985). The respective species are: *Enterococcus faecium*, *Streptococcus sanguis*, and *Staphylococcus aureus* (the number of positive strains were: 2 of 2 strains tested, 2 of 12 strains tested, and 23 of 23 strains tested, re-

spectively). In all these cases, the same activity was observed; a culture supernatant of the strains induced formation of aggregation substance in an *E. faecalis* strain harboring the sex pheromone plasmid pAM373. In no case, however, could pAM373 be transferred conjugatively to the other bacterial species; very probably this plasmid does not replicate in them. The structure of the sex pheromone cAM373 excreted by *E. faecalis* is known and the activities excreted by the other species are certainly brought about by small, linear peptides, but their structure is unknown.

Transformation

No natural transformation systems in the genus *Enterococcus* are known, despite efforts to find one. Early genetic experiments used a two-step procedure. First, the plasmid of interest was transformed into *Streptococcus sanguis* (strain Challis), and in a second step, transferred from this species into *E. faecalis* by conjugative mobilization. However, this method was cumbersome, because of the time needed and the fact that the *S. sanguis* transformation procedure is prone to the formation of deletions in the plasmids. The establishment of a reliable transformation procedure for *E. faecalis* was of great importance for genetic experiments. A first protocol for polyethylene glycol-induced protoplast transformation (Smith, 1985) was soon followed by an improved version that produced transformation frequencies of up to 10^5 transformants per μg of plasmid DNA (Wirth et al., 1986). An electroporation method is also available that is at least as efficient (Fiedler and Wirth, unpublished observations). Because nonprotoplasted cells can be used, this new method reduces the time for recovery of transformants from 72 to 24 hours.

Pathogenicity

E. faecalis is by far the most common member of the enterococci to cause infections in humans (Kaye, 1982). *E. faecium* accounts for less than 20% of clinical isolates from the various organ systems, except those from the respiratory tract (Hahn. 1981) and the incidence of other enterococci is very low.

In humans, enterococci are frequently associated with intraabdominal and pelvic infections, but usually other organisms are also present. Only in endocarditis and in urinary infections are they commonly the only organism involved.

The pathogenic significance of enterococci in polymicrobial infections is controversial. Fass et al. (1973) reported that polymicrobial abdominal and pelvic infections with aerobic and anaerobic bacteria, including enterococci, responded to treatment with antibiotics that were not active against the enterococci. Matlow et al. (1989) found, however, that *E. faecalis* contributed to the severity of a rat model of fecal peritonitis.

Enterococci, usually *E. faecalis,* are the cause of 5 to 15% of human endocarditis cases, most often in elderly men with prostatic disease or in younger women with pregnancy-related problems (Kaye, 1982). Also, *E. avium, E. casseliflavus, E. durans, E. faecium, E. gallinarum,* and *E. raffinosus* have been isolated from clinical endocarditis (Facklam and Collins, 1989). The organisms causing endocarditis usually originate from the genito-urinary tract (Mandell et al., 1970). Enterococcal urinary infections are mostly asymptomatic (Horvitz and von Graevenitz, 1977) and related to instrumentation and indwelling catheters. In recent years the enterococci have become more and more important in nosocomial infections (Murray, 1989).

The pathogenicity mechanisms of the enterococci are unknown. Strains from lesions in humans often produce a hemolysin that may be a virulence factor (Ike et al., 1987). Such strains were formerly considered to constitute a separate variety or even a subspecies of *E. faecalis,* named *zymogenes.* Hemolysin genes seem to be plasmid coded (Clewell, 1981); Table 4 lists those located on sex pheromone plasmids. The hemolysin-bacteriocin determinant of pAD1 enhances virulence in intraperitoneal infections in mice (Ike et al., 1987).

Protein antigens produced by serum-grown *E. faecalis* cells and present in the sera of patients with *E. faecalis* endocarditis are possibly useful in the serodiagnosis of this infection (Aitchison et al., 1987; Shorrock et al., 1990). Information about surface components that may be of importance in adhesion, pathogenicity, and immunity, is lacking.

As stated in the section on "Isolation," the enterococci are resistant to many antibiotics. Compared to the streptococci, they are relatively resistant to the penicillins, *E. faecium* even more so than *E. faecalis.* Ampicillin, piperacillin, inempenem, and vancomycin are among the most active in vitro (Kaye, 1982; Kim et al., 1987). In endocarditis and other serious infections for which bactericidal antibiotic activity is needed, combinations with aminoglycoside antibiotics, usually gentamicin, are indicated (Kaye, 1982). When used alone, the aminoglycosides are not active. Different types of high-level resistance may impair their synergistic action with penicillins. One important mechanism is based on inactivation of the antibiotics by adenylating or phosphorylating enzymes, mediated by plasmids that are transferable by conjugation (Courvalin et al., 1978). Others involve the ribosomal target site of the aminoglycosides (Zimmerman et al., 1971) and cell penetration of the antibiotics (Moellering et al., 1980). A comprehensive review of the epidemiology of enterococcal antibiotic resistance in the hospital environment is given by George and Uttley (1990) and by Murray (1989).

In animals, the enterococci are most often present in polymicrobial infections of the same types as those found in humans. Polymicrobial infection with *E. faecalis* is frequent in both chronic and recurrent otitis externa of dogs. Monoinfections include septicemia and endocarditis in poultry and ducks (Domermuth and Gross, 1969; Sandhu, 1988), tracheitis and airsac infection in canaries, septicemia and arthritis in hypogammaglobulemic newborn calves, iatrogenic and asympomatic bacteriuria in dogs (L. A. Devriese, unpublished observations). Other conditions have been described in various animal species, but the literature is vague (Hahn, 1980). Experimental infections in chickens showed that *E. faecalis* was the most pathogenic species, but strains identified as *E. faecium* and *E. durans* also caused endocarditis (Domermuth and Gross, 1969)

An as-yet-unexplained and possibly important phenomenon concerns specific attachment of enterococci to the brush border of small intestinal villi of animals, as has been described and reproduced with *E. hirae* in suckling rats (Etheridge et al., 1988) and in foals (Tzipori et al., 1984). The strain studied by Tzipori and coworkers, which also produced enteritis in gnotobiotic piglets and which was identified as *E. durans,* is in fact a strain of *E. hirae* (L. A. Devriese, unpublished observations). Adherence of *E. durans* (also identified later as *E. hirae,* L. A. Devriese, unpublished observations) to small intestinal enterocytes has been described also in dog pups suffering from diarrhea (Collins et al., 1988). In this context, the role of enterococci as causes of growth depression in chicks merits attention. This was first described by Eyssen and De Somer (1967) and later confirmed by Fuller et al. (1979) and Houghton et al. (1981), with strains which were identified as *E. hirae* by Farrow and Collins (1985). The so-called nutritional antibiotics that improve growth and feed conversion in farm animals may be active against these bacteria, but this possible explanation of their mode of action has not been proved or disproved yet.

The resistance of the enterococci to adverse conditions makes them good indicators of unsanitary conditions in certain types of dried and frozen foods. This is, however, not always a simple matter, as is illustrated by the following: frozen vegetables may contain high numbers (> 1000 colony-forming units (CFU)/g) of enterococci, while dried vegetables and spices usually contain fewer than 100 CFU/g. In a study of streptococci and enterococci from such foods, Mundt (1976) found that only 3% of the strains isolated resembled animal or human *E. faecalis* and 12% *E. faecium*. Mundt called other strains "plant type *'faecalis'*" or "casseliflavus biotype" (some of which were later renamed *E. mundtii*) (Collins et al., 1986), and nearly 50% of strains remained unidentified. The majority of these showed characteristics suggestive of enterococci.

Artificial contamination of the gut of humans (Lewenstein et al., 1979) and farm animals (Gudding and Larssen, 1985; Underdahl et al., 1982; Ushe and Nagy, 1985) with enterococci, mainly *E. faecium,* has received much attention as a means to prevent enteric disease of humans and animals or to improve the growth of animals (Roth and Kirchgessner, 1986). The findings concerning pathogenic *E. coli* indicate that the treatments may lower enteral colonization of *E. coli,* but the effects on disease and mortality are less clear.

Applications

The enterococci are commonly monitored as indicators of fecal pollution in the purification of drinking water and certain types of foods. According to some workers (Burman, 1978), enterococci die more slowly in water than *Escherichia coli* and other Enterobacteriaceae do, and this makes them useful indicator organisms in water subject to remote or intermittent fecal pollution. This has been demonstrated most convincingly by Geldreich and Kenner (1969) with *E. faecalis*. The same authors also found that *Streptococcus bovis* dies at a much faster rate than *E. coli*. Confusion arises because *S. bovis* is often isolated and counted together with the enterococci (Anonymous, 1985).

Meats and meat products are principally contaminated with enterococci in packing plants, rather than at the retail level (Stiles et al., 1978). *E. faecalis* was found to be the predominant *Enterococcus* species in beef and pork cuts, while *E. faecium* and *E. durans* (possibly *E. hirae,* according to the more recent classification) was most frequently isolated from processed meat.

Literature Cited

Anonymous. 1985. Tests for the fecal *Streptococcus* group p. 902–910. In: M. A. H. Franson (ed.), Standard methods for the examination of water and wastewater. Publications Office Am. Publ. Health Ass., Washington, D.C.

Aitchison, E. J., P. A. Lambert, E. G. Smith, and I. D. Farrell. 1987. Serodiagnosis of *Streptococcus faecalis* endocarditis by immunoblotting of surface protein antigens. J. Clin. Microbiol. 25:211–215.

Barnes, E. M. 1956. Methods for the isolation of faecal streptococci (Lancefield group D) from bacon factories. J. Appl. Bacteriol. 19:193–203.

Barnes, E. M. 1959. Differential and selective media for the faecal streptococci. J. Sci. Food Agric. 12:656–662.

Bernsmann, P., C. A. Alppert, P. Muss, J. Deutscher, and W. Hengstenberg. 1982. The bacterial PEP-dependent phosphotrasferase system mechanism of gluconate phosphorylation in *Streptococcus faecalis*. FEBS Letters 138:101–103.

Bridge, P. D., and P. H. A. Sneath. 1982. Numerical taxonomy of *Streptococcus*. J. Gen. Microbiol. 114:27–33.

Burman, N. P. 1961. Some observations on coli-aerogenes bacteria and streptococci in water. J. Appl. Bacteriol. 24:368–376.

Caprioli, T., F. Zaccour, and S. S. Kasatiya. 1975. Phage typing scheme for group D streptococci isolated from the human urogenital tract. J. Clin. Microbiol. 2:311–317.

Christie, P. J., M. S. Kao, J. C. Adsit, and G. M. Dunny. 1988. Cloning and expression of genes encoding pheromone-inducible antigens of *Enterococcus (Streptococcus) faecalis*. J. Bacteriol. 170:5161–5168.

Clewell, D. B. 1981. Plasmids, drug resistance, and gene transfer in the genus *Streptococcus*. Microbiol. Rev. 45:409–436.

Clewell, D. B., F. Y. An, B. A. White, and C. Gawron-Burke. 1985. *Streptococcus faecalis* sex pheromone (cAM373) also produced by *Staphylococcus aureus* and identification of a conjugative transposon (Tn918). J. Bacteriol. 162:1212–1220.

Clewell, D. B., and C. Gawron-Burke. 1986. Conjugative transposons and the dissemination of antibiotic resistance in streptococci. Ann. Rev. Microbiol. 40:635–659.

Clewell, D. B., and K. E. Weaver. 1989. Sex pheromones and plasmid transfer in *Enterococcus faecalis*. Plasmid 21:175–184.

Collins, J. E., M. E. Bergeland, C. J. Lindeman, and J. R. Duimstra. 1988. *Enterococcus (Streptococcus) durans* adherence in the intestine of a diarrheic pup. Vet. Pathol. 25:396–398.

Collins, M. D., C. Ash, J. A. E. Farrow, S. Wallbanks, and A. M. Williams. 1989a. 16S Ribosomal ribonucleic acid sequence analysis of lactococci and related taxa. Description of *Vagococcus fluvialis* gen.nov., sp. nov. FEMS Microbiol. Letters. 57:283–288.

Collins, M. D., R. R. Facklam, J. A. E. Farrow, and R. Williamson. 1989b. *Enterococcus raffinosus* sp. nov., *Enterococcus solitarius* sp. nov. and *Enterococcus pseudoavium* sp. nov. FEMS Microbiol. Lett. 57:283–288.

Collins, M. D., J. A. E. Farrow, and D. Jones 1986. *Enterococcus mundtii* sp. nov. Int. J. Syst. Bacteriol. 36:8–12.

Collins, M. D., and D. Jones. 1979. The distribution of isoprenoid quinones in streptococci of serological groups D and N. J. Gen. Microbiol. 114:27–33.

Collins, M. D., D. Jones, J. A. E. Farrow, R. Kilpper-Bälz, and K. H. Schleifer. 1984. *Enterococcus avium* nom. rev., comb. nov.; *E. casseliflavus* nom. rev., comb. nov.; *E. durans* nom. rev., comb. nov.; *E. gallinarum* comb. nov.; and *E. malodoratus* sp. nov. Int. J. Syst. Bacteriol. 34:220–223.

Courvalin, P. M., C. Carlier. 1986. Transposable multiple resistance in *Streptococcus pneumoniae*. Molec. Gen. Genet. 205:291–297.

Courvalin, P. M., C Carlier, and Y. A. Chabbert. 1972. Plasmid-linked tetracycline and erythromycin resistance in group D "Streptococcus". Ann. Inst. Pasteur (Paris) 123:755–759.

Courvalin, P. M., W. V. Shaw, and A. E. Jacob. 1978. Plasmid mediated mechanisms of resistance to aminoglycoside antibiotics and to chloramphenicol in group D streptococci. Antimicrob. Ag. Chemother. 13:716–725.

Dao, M. L., and J. J. Feretti. 1985. *Streptococcus-Escherichia coli* shuttle vector pSA3 and its use in the cloning of streptococcal genes. Appl. Env. Microbiol. 49:111–119.

Deibel, R. H. 1964. The group D streptococci. Bacteriol. Rev. 28:330–366.

Devriese, L. A., G. N. Dutta, J. A. E. Farrow, A. Van De Kerckhove, and B. A. Phillips. 1983. *Streptococcus ce-*

corum, a new species isolated from chickens. Int. J. Syst. Bacteriol. 33:772–776.

Devriese, L. A., A. Van De Kerckhove, R. Kilpper-Bälz, and K. H. Schleifer. 1987. Characterization and identification of *Enterococcus* species isolated from animals. Int. J. Syst. Bacteriol. 37:257–259.

Domermuth, C. H., and W. B. Gross. 1969. A medium for isolation and tentative identification of fecal streptococci, and their role as avian pathogens. Avian Dis. 13:394–399.

Driessen, A. J. M., E. J. Smid, and W. N. Konings. 1988. Transport of diamines by *Enterococcus faecalis* is mediated by an agmatine-putrescine antiporter. J. Bacteriol. 170:4522–4527.

Dunny, G. M., B. L. Brown, and D. B. Clewell. 1978. Induced cell aggregation and mating in *Streptococcus faecalis:* evidence for a bacterial sex pheromone. Proc. Natl. Acad. Sci. (USA) 75:3479–3483.

Dunny, G. M., R. A. Craig, R. L. Carron, and D. B. Clewell. 1979. Plasmid transfer in *Streptococcus faecalis:* production of multiple sex pheromones by recipients. Plasmid 2:454–465.

Ehrenfeldt, E. E., R. E. Kessler, and D. B. Clewell. 1986. Identification of pheromone-induced surface proteins in *Streptococcus faecalis* and evidence of a role for lipoteichoic acid in formation of mating aggregates. J. Bacteriol. 168:6–12.

Ehrenfeldt, E. E., and D. B. Clewell. 1987. Transfer functions of the *Streptococcus faecalis* plasmid pAD1: organization of plasmid DNA encoding response to sex pheromone. J. Bacteriol. 169:3473–3481.

Etheridge, M., R. H. Yolken, and S. L. Vonderfecht. 1988. *Enterococcus hirae* implicated as a cause of diarrhea in suckling rats. J. Clin. Microbiol. 26:1741–1744.

Eyssen, H., and P. De Somer. 1967. Effects of *Streptococcus faecalis* and a filterable agent on growth and nutrient absorption in gnotobiotic chicks. Poultry Sci. 46:323–333.

Facklam, R. R., and M. D. Collins. 1989. Identification of *Enterococcus* species isolated from human infections by a conventional test scheme. J. Clin. Microbiol. 27:731–734.

Facklam, R. R., and M. D. Moody. 1970. Presumptive identification of group D streptococci: the bile-esculin test. Appl. Microbiol. 20:245–250.

Farrow, J. A. E., and M. D. Collins. 1985; *Enterococcus hirae,* a new species that includes amino acid assay strain NCDO 1258 and strains causing growth depression in young chickens. Int. J. Syst. Bacteriol. 35:73–75.

Fass, R. J., J. F. Scholand, and J. R. Hodges. 1973. Clindamycin in the treatment of serious anaerobic infections. Ann. Intern. Med. 78:853–859.

Finegold, S. M., H. R. Attebery, and V. L. Sutter. 1974. Effect of diet on human faecal flora: comparison of Japanese and American diets. Am. J. Clin. Nutrition. 27:1456–1469.

Franke, A. E., and D. B. Clewell. 1981. Evidence for conjugal transfer of a *Streptococcus faecalis* transposon (Tn916) from a chromosomal site in the absence of plasmid DNA. Cold Spring Harbor Symp. Quant. Biol. 45:77–80.

Fuller, R., M. E. Coates, and G. F. Harrison. 1979. The influence of specific bacteria and a filterable agent on the growth of gnotobiotic chicks. J. Appl. Bacteriol. 46:335–342.

Galli, D., F. Lottspeich, and R. Wirth. 1990. Sequence analysis of pAD1-encoded aggregation substance of *Enterococcus faecalis*. Molec. Microbiol. 4:895–904.

Galli, D., R. Wirth, and G. Wanner. 1989. Identification of aggregation substances of *Enterococcus faecalis* cells after induction by sex pheromones. Arch. Microbiol. 151:486–490.

Gawron-Burke, C., and D. B. Clewell. 1982. A transposon in *Streptococcus faecalis* with fertility properties. Nature 300:281–284.

Geldreich, E. E., and B. A. Kenner. 1969. Concepts of fecal streptococci in stream pollution. J. Water Poll. Control Fed. 41:R336-R352.

Geldreich, E. E., B. A. Kenner. and P. W. Kabler. 1964. Occurrence of coliforms, fecal coliforms, and streptococci in vegetation and insects. Appl. Microbiol. 12:63–69.

George, R. C., and A. H. C. Uttley. 1990. Susceptibility of enterococci and epidemiology of enterococcal infection in the 1980s. Epidem. Inf. 103:403–413.

Gudding, R., and R. B. Larssen. 1985. Some effects of *Streptococcus faecium* M 74 in piglets. Nord. Vet. Med. 37:48–49.

Gunsalus, I. C., and C. F. Niven Jr. 1942. The effect of pH on the lactic acid fermentation. J. Biol. Chem. 145:131–136.

Hahn, G. 1980. Enterokokken (serologische Gruppe D), p. 235–242. In: H. Blobel, and Th. Schliesser (ed.), Handbuch der Bakteriëllen Infektionen bei Tieren. G. Fisher, Jena, Germany.

Hahn, G. 1981. Ergebnisse aus der Streptokokken-Zentrale in Kiel von 1965 bis 1980: Gruppe D-Streptokokken (Enterokokken). Zbl. Bakt. Hyg. I. Abt. Orig. A 250:431–445.

Hajna, A. A. 1951. A buffered azide glucose-glycerol broth for the presumptive and confirmative tests for faecal streptococci. Public Health Lab. 9:80–81.

Hajna, A. A., and C. A. Parry. 1943. Comparative study of presumptive and confirmative identification for bacteria of the coliform group and for faecal streptococci. Am. J. Public Health. 33:550–556.

Heller, K., and R. Röschenthaler. 1978. Beta-D-phosphogalactosidase-galactohydrolase of *Streptococcus faecalis* and the inhibition of its synthesis by glucose. Can. J. Microbiol. 24:512–515.

Hill, R. J., B. S. Drasar, V. Aries, J. S. Crowther, G. Hawksworth, and R. E. O. Williams. 1971. Bacteria and aetiology of cancer of the large bowel. Lancet 1:95–100.

Himmel, J. M., and L. N. Zimmerman. 1963. Growth stimulation of *Streptococcus faecalis* var. *liquifaciens* by canavanine. J. Bacteriol. 86:490–493.

Horvitz, R. A., and A. von Graevenitz. 1977. A clinical study of the role of enterococci as sole agents of wound and tissue infection. Yale J. Biol. Med. 50:391–395.

Hoskins, D. D., H. R. Whiteley, and B. Mackler. 1962. The reduced diphosphopyridine nucleotide oxidase of *Streptococcus faecalis*: purification and properties. J. Biol. Chem. 237:2647–2651.

Houghton, S. B., R. Fuller, and M. E. Coates. 1981. Correlation of growth depression of chicks with the presence of *Streptococcus faecium* in the gut. J. Appl. Bacteriol. 51:113–120.

Ike, Y., H. Hashimoto, and D. B. Clewell. 1987. High incidence of hemolysin production by *Enterococcus (Streptococcus) faecalis* strains associated with human parenteral infections. J. Clin. Microbiol. 25:1524–1528.

Inamine, J. M., L. N. Lee, and D. J. LeBlanc. 1986. Molecular and genetic characterization of lactose-metabolic genes of *Streptococcus cremoris*. J. Bacteriol. 167:855–862.

Jacobs, N. J., and P. J. VanDenmark. 1960. Comparison of the mechanism of glycerol oxidation in aerobically and anaerobically grown *Streptococcus faecalis*. J. Bacteriol. 79:532–538.

Jones, D. 1978. Composition and differentiation of the genus *Streptococcus*, p. 1–49. In: F. A. Skinner, and L. B. Quesnel (ed.), Streptococci. Academic Press, London.

Kaukas, A., M. Hinton, and A. H. Linton. 1986. Changes in the faecal enterococcal population of young chickens and its effect on the incidence of resistance to certain antibiotics. Lett. Appl. Microbiol. 2:5–8.

Kaye, D. 1982 Enterococci: biologic and epidemiologic characteristics and in vitro susceptibility. Arch. Int. Med. 142:2006–2009.

Kenner, B. A., H. F. Clark, and P. W. Kabler. 1961. Faecal streptococci. I. Cultivation and enumeration of streptococci in surface waters. Appl. Microbiol. 9:15–20.

Kilpper-Bälz, R., G. Fischer, and K. H. Schleifer. 1982. Nucleic acid hybridization of group N and group D streptococci. Curr. Microbiol. 7:245–250.

Kim, M. J., M. Weisser, S. Gottschall, and E. L. Randall. 1987. Identification of *Streptococcus faecalis* and *Streptococcus faecium* and susceptibility studies with newly developed antimicrobial agents. J. Clin. Microbiol. 25:787–790.

Krah, E. R., III and F. L. Macrina. 1989. Genetic analysis of the conjugal transfer determinants encoded by the streptococcal broad-host-range plasmid pIP501. J. Bacteriol. 171:6005–6012.

Kühnen, E., F. Richter, K. Richter, and L. Andries. 1988. Establishment of a typing system for group D streptococci. Zbl. Bakt. Hyg. A 267:322–330.

LeBougenec, C., G. de Cepedes, and T. Houraud. 1988. Molecular analysis of a composite chromosomal conjugative element (Tn3701) of *Streptococcus pyogenes*. J. Bacteriol. 170:3930–3936.

Levin, M. A., J. R. Fischer, and V. J. Cabelli. 1975. Membrane filter technique for enumeration of enterococci in marine waters. Appl. Microbiol. 30:66–71.

Lewenstein, A., G. Frigerio, and M. Monori. 1979. Biological properties of SF 68, a new approach in the treatment of diarrheal diseases. Curr. Therap. Res. 26:967–981.

Litsky, W., W. L. Mallmann, and C. W. Filfield. 1953. A new medium for the detection of enterococci in water. Amer. J. Publ. Health 43:873–879.

London, J., and M. D. Appleman. 1962. Oxidative and glycerol metabolism of two species of enterococci. J. Bacteriol. 84:597–598.

Ludwig, W., E. Seewaldt, R. Kilpper-Bälz, K. H. Schleifer, L. Magrum, C. R. Woese, G. E. Fox, and E. Stackebrandt. 1985. The phylogenetic position of *Streptococcus* and *Enterococcus*. J. Gen. Microbiol. 131:543–551.

Mallmann, W. L., and E. B. Seligmann Jr. 1950. A comparative study of media for the detection of streptococci in water and sewage. Amer. J. Public Health 40:286–289.

Mandell, G. L., D. Kaye, M. E. Levison. 1970. Enterococcal endocarditis: an analysis of 38 patients observed at the New York Hospital-Cornell Medical Center. Arch. Intern. Med. 125:258–264.

Martin, J. D., and J. O. Mundt. 1972. Enterococci in insects. Appl. Microbiol. 24:575–580.

Matlow, A. G., J. M. A. Bohnen, C. Nohr, N. Christou, and J. Meakins. 1989. Pathogenicity of enterococci in a rat model. J. Inf. Dis. 160:142–145.

McCoy, T. A., and S. H. Wender. 1953. Some factors affecting the nutritional requirements of *Streptococcus faecalis*. J. Bacteriol. 65:660–665.

Mead, G. C. 1963. A medium for the isolation of *Streptococcus faecalis*, sensu strictu. Nature 3:1323–1324.

Medrek, T. F., and E. M. Barnes. 1962. The distribution of group D streptococci in cattle and sheep. J. Appl. Bacteriol. 25:159–168.

Moellering Jr, R. C., B. E. Murray, and S. C. Schoenbaum. 1980. A novel mechanism of resistance to penicillin-gentamycin synergism in *Streptococcus faecalis*. J. Infect. Dis. 141:81–86.

Molitoris, E., M. I. Krichevski, D. J. Fagerberg, and C. L. Quarles. 1986. Effects of dietary chlortetracycline on the antimicrobial resistance of porcine faecal streptococcaceae. J. Appl. Bacteriol. 60:111–120.

Mori, M., A. Isogai, Y. Sakagami, M. Fujino, C. Kitada, D. B. Clewell, and A. Suzuki. 1986. Isolation and structure of the *Streptococcus faecalis* sex pheromone inhibitor, iAD1, that is excreted by the donor strain harboring plasmid pAD1. Agric. Biol. Chem. 50:539–541.

Mori, M., Y. Sakagami, Y. Ishii, A. Isogai, C. Kitada, M. Fujino, J. D. Adsit, G. M. Dunny, and A. Suzuki. 1988. Structure of cCF10, a peptide sex pheromone which induces conjugative transfer of the *Streptococcus faecalis* tetracycline resistance plasmid, pCF10. J. Biol. Chem. 263:14574–14578.

Mori, M., Y. Sakagami, M. Narita, A. Isogai, M. Fujino, C. Kitada, R. A. Craig, D. B. Clewell, and A. Suzuki, 1984. Isolation and structure of the bacterial sex pheromone, cAD1, that induces plasmid transfer in *Streptococcus faecalis*. FEBS Letters 178:97–100.

Mossel, D. A. A., P. G. H. Bijker, and I. Eelderink. 1978. Streptococci of Lancefield groups A,B and D and those of buccal origin in foods: their public health significance, monitoring and control, p. 315–333. In: F. A. Skinner, and L. B. Quesnel (ed.), Streptococci. Academic Press, London.

Mundt, J. O. 1973. Litmus milk reaction as a distinguishing feature between *Streptococcus faecalis* of human and non-human origins. J. Milk Food Technol. 36:364–367.

Mundt, J. O. 1976. Streptococci in dried and frozen foods. J. Milk Food Technol. 39:413–416.

Niven, C. F., and J. M. Sherman. 1944. Nutrition of the enterococci. J. Bacteriol. 47:335–342.

Murray, B. E. 1989. The life and times of the enterococcus. Clin. Microbiol. Rev. 3:40–65.

Nowlan, S. S., and R. H. Deibel. 1967. Group Q streptococci. J. Bacteriol. 94:291–196.

Pavlova, M. T., F. T. Brezenski, and W. Litsky. 1972. Evaluation of various media for isolation, enumeration and identification of faecal streptococci from natural sources. Health Lab. Sci. 9:289–298.

Perkins, J. B., and P. J. Youngman. 1986. Construction and properties of Tn917-lac, a transposon derivative that mediates transcriptional gene fusions in *Bacillus subtilis*. Proc. Natl. Acad. Sci. USA 83:140–145.

Pleceas, P., and H. Brandis. 1974. Differentiation des principales espèces de streptocoques du groupe D par des

mélanges de bacteriophages spécifiques. Ann. Microbiol. (Inst. Pasteur) 125:463–470.

Poolman, A. J. M. Driessen, and W. N. Konings. 1987. Regulation of arginine-ornithine exchange and the arginine deiminase pathway in *Streptococcus lactis*. J. Bacteriol. 169:5597–5604.

Ritchey, T. W., and H. W. Seeley, 1976. Distribution of cytochrome-like respiration in streptococci. J. Gen. Microbiol. 93:195–203.

Rollins, L. D., L. N. Lee, and D. J. LeBlanc. 1985. Evidence for disseminated erythromycin resistance determinant mediated by Tn917-like sequences among group D streptococci isolated from pigs, chickens, and humans. Antimicrob. Ag. Chemother. 27:439–444.

Roth, F. X., and M. Kirchgessner. 1986. Zur nutritiven Wirksamkeit von *Streptococcus faecium* (Stamm M 74) in der Kükenmast. Arch. Geflügelk. 50:225–228.

Rutkowski, A. A., and R. E. Sjogren. 1987. Streptococcal population profiles as indicators of water quality. Water Air Soil Poll. 34:273–284.

Sabaj, J., V. L. Sutter, and S. M. Finegold. 1971. Comparison of selective media for isolation of presumptive group D streptococci from human faeces. Appl. Microbiol. 22:1008–1011.

Sandhu, T. S. 1988. Fecal streptococcal infection in commercial white pekin ducklings. Avian Dis. 32:570–583.

Schleifer, K. H., and R. Kilpper-Bälz. 1984. Transfer of *Streptococcus faecalis* and *Streptococcus faecium* to the genus *Enterococcus* nom. rev. as *Enterococcus faecalis* comb. nov. and *Enterococcus faecium* comb. nov. Int. J. Syst. Bacteriol. 34:31–34.

Schleifer, K. H., and R. Kilpper-Bälz. 1987. Molecular and chemotaxonomic approaches to the classification of streptococci, enterococci and lactococci: a review. Syst. Appl. Microbiol. 10:1–19.

Schmidt, H. L., W. Stöcklein, J. Danzer, P. Kirch, and B. Limbach. 1986. Isolation and properties of an H2O-forming NADH oxidase from *Streptococcus faecalis*. Eur. J. Biochem. 156:149–155.

Sharpe, M. E., and P. M. F. Shattock. 1952. The serological typing of group D streptococci associated with outbreaks of neonatal diarrhea. J. Gen. Microbiol. 6:150–165.

Shaw, J. H., and D. B. Clewell. 1985. Complete nucleotide sequence of macrolide-lincosamide-streptogramin resistance transposon Tn917 in *Streptococcus faecalis*. J. Bacteriol. 164:782–796.

Sherman, J. M. 1937. The streptococci. Bacteriol. Rev. 1:3–97.

Shorrock, P. J., P. A. Lambert, E. J. Aitchison, E. G. Smith, E. D. Farrell, and E. Gutschik. 1990. Serological response in *Enterococcus faecalis* endocarditis determined by enzyme-linked immunosorbent assay. J. Clin. Microbiol. 28:195–200.

Sijpesteijn, A. K. 1970. Induction of cytochrome formation and stimulation of oxidative dissimilation by haemin in *Streptococcus faecalis* strain 10C1. J. Bacteriol. 96:1595–1600.

Slanetz, L. W., and C. H. Bartley. 1957. Numbers of enterococci in water, sewage and faeces determined by the membrane filter technique with an improved medium. J. Bacteriol. 74:591–595.

Smith, H. W. and W. E. Crabb. 1961. The faecal bacterial flora of animals and man: its development in the young. J. Pathol. Bacteriol. 82:53–66.

Smith, M. D. 1985. Transformation and fusion of *Streptococcus faecalis* protoplasts. J. Bacteriol. 162:92–97.

Smyth, C. J., H. Matthews, M. K. Halpenny, H. Brandis, and G. Colman. 1987. Biotyping, serotyping and phage typing of *Streptococcus faecalis* isolated from dental plaque in the human mouth. J. Med. Microbiol. 23:45–54.

Stiles, M. E., N. W. Ramji, L. K. Ng, and D. C. Paradis. 1978. Incidence and relationship of group D streptococci with other indicator organisms in meats. Can. J. Microbiol. 24:1502–1508.

Suzuki, A., M. Mori, Y. Sakagami, A. Isogai, M. Fujino, C. Kitada, R. A. Craig, and D. B. Clewell. 1984. Isolation and structure of bacterial sex pheromone, cPD1. Science 226:849–850.

Switzer, R. E., and J. B. Evans. 1974. Evaluation of selective media for enumeration of group D streptococci in bovine faeces. Appl. Microbiol. 28:1086–1087.

Tomich, P. K., F. A. An, and D. B. Clewell. 1980. Properties of erythromycin-inducible Tn917 in *Streptococcus faecalis*. J. Bacteriol. 151:1217–1221.

Trieu-Cuot, P., C. Carlier and P. Courvalin. 1988. Conjugative plasmid transfer from *Enterococcus faecalis* to *Escherichia coli*. J. Bacteriol. 170:4388–4391.

Tzipori, S., J. Hayes, L. Sims, and M. Withers. 1984. *Streptococcus durans*: an unexpected enteropathogen of foals. J. Inf. Dis. 150:589–593.

Underdahl, N. R., A. Torres-Medina, and A. R. Doster. 1982. Effect of *Streptococcus faecium* C-68 in control of *Escherichia coli*-induced diarrhea in gnotobiotic pigs. Am. J. Vet. Res. 43:2227–2232.

Ushe, T. C., and B. Nagy. 1985. Inhibition of small intestinal colonization of enterotoxigenic *Escherichia coli* by *Streptococcus faecium* M74 in pigs. Zbl. Bakt. Hyg. I. Abt. Orig. B 181:374–382.

Wanner, G., H. Formanek, D. Galli, and R. Wirth. 1989. Localization of aggregation substances of *Enterococcus faecalis* after induction by sex pheromones. Arch. Microbiol. 151:491–497.

Watanabe, T., H. Shimohashi, Y. Kawai, and M. Mutai. 1981. Studies on streptococci. I. Distribution of faecal streptococci in man. Microbiol. Immunol. 25:257–269.

Whittenbury, R. 1978. Biochemical characteristics of *Streptococcus* species, p. 51–69. In: F. A. Skinner, and L. B. Quesnel (ed.), Streptococci. Academic Press, London.

Wicken, A. J., and J. Baddiley. 1963. Structure of intracellular teichoic acids from group D streptococci. Biochem. J. 87:54–62.

Wicken, A. J., S. D. Elliott, and D. Baddiley. 1963. The identity of streptococcal group D antigen with teichoic acid. J. Gen. Microbiol. 31:231–239.

Williams, A. M., J. A. E. Farrow, and M. D. Collins. 1989. Reverse transcriptase sequencing of 16S ribosomal RNA from *Streptococcus cecorum*. Lett. Appl. Microbiol. 8:185–189.

Williamson, R., L. Gutman, R., T. Horand, F. Delbros and J. F. Acar. 1986. Use of penicillin-binding proteins for the identification of enterococci. J. Gen. Microbiol. 131:1927–1937.

Wirth, R., F. Y. An, and D. B. Clewell. 1986. Highly efficient protoplast transformation system for *Streptococcus faecalis* and a new *Escherichia coli-S. faecalis* shuttle vector. J. Bacteriol. 165:831–836.

Wirth, R., F. Y. An, and D. B. Clewell. 1987. Highly efficient cloning system for *Streptococcus faecalis*: protoplast transformation, shuttle vectors, and applications, p. 25–27. In: J. J. Feretti, and R. Curtiss (ed.), Streptococcal genetics. ASM Publ. Washington, D.C.

Yagi, Y., R. E. Kessler, J. H. Shaw, D. E. Lopatin, F. Y. An, and D. B. Clewell. 1983. Plasmid content of *Streptococcus faecalis* strain 39–5 and identification of a pheromone (cPD1)-induced surface antigen. J. Gen. Microbiol. 129:1207–1215.

Zimmerman, R. A., R. C. Moellering Jr., and A. F. Weinberg. 1971. Mechanism of resistance to antibiotic synergism in enterococci. J. Bacteriol. 105:873–879.

Zitzelberger, W., F. Götz, and K. H. Schleifer. 1984. Distribution of superoxide dismutases, oxidases, and NADH peroxidase in various streptococci. FEMS Microbiol. Letters 21:243–246.

The Genus *Lactococcus*

MICHAEL TEUBER, ARNOLD GEIS, and HORST NEVE

The lactococci are distinguished by important significance for basic microbiology, genetics, and molecular biology, and general microbial biochemistry as well as for food science and biotechnology. The present-day significance is in the large-scale use of the lactococci for industrial fermentations, especially dairy products.

The first studies of the lactococci were initiated by Joseph Lister (1873), who was attempting to prove Pasteur's germ theory of fermentative changes. In his experiments with boiled milk as a nutrient medium, he obtained by chance the first pure bacterial culture. It is worthwhile to recall in the context of this handbook his original discussion of this discovery, marking the dawn of bacterial taxonomy:

Admitting then that we had here to deal with only one bacterium, it presents such peculiarities both morphologically and physiologically as to justify us, I think, in regarding it a definite and recognizable species for which I venture to suggest the name *Bacterium lactis*. This I do with diffidence, believing that up to this time no bacterium has been defined by reliable characters. Whether this is the only bacterium that can occasion the lactic acid fermentation, I am not prepared to say.

This bacterium was later renamed *Streptococcus lactis* (Löhnis, 1909; Orla-Jensen, 1919). On the basis of exhaustive reinvestigations, Schleifer et al. (1985) proposed that the N streptococci (Lancefield, 1933) be separated from the oral *streptococci,* the enterococci, and the hemolytic streptococci and suggested the new genus name *Lactococcus*. The position of the lactococci within the clostridium branch of the evolutionary phylogenetic tree is shown in Fig. 1. The lactococci are clearly separated from pathogenic genera of streptococci (Stackebrandt and Teuber, 1988). Lactococci are generally recognized as safe (GRAS) for human consumption, and they have never been identified as causes of infectious disease. This is also evident by their large scale, deliberate use in the dairy industry as starter cultures for many different products (see Table 1 and 2). This development was initiated at the turn of the century when H. Weigmann, V. Storch, and H. W. Conn identified lactococci as the essential components of the mesophilic microflora in spontaneously fermented cream and milk. This finding led to the introduction of pure starter cultures of lactic acid bacteria to the dairy field for use in the

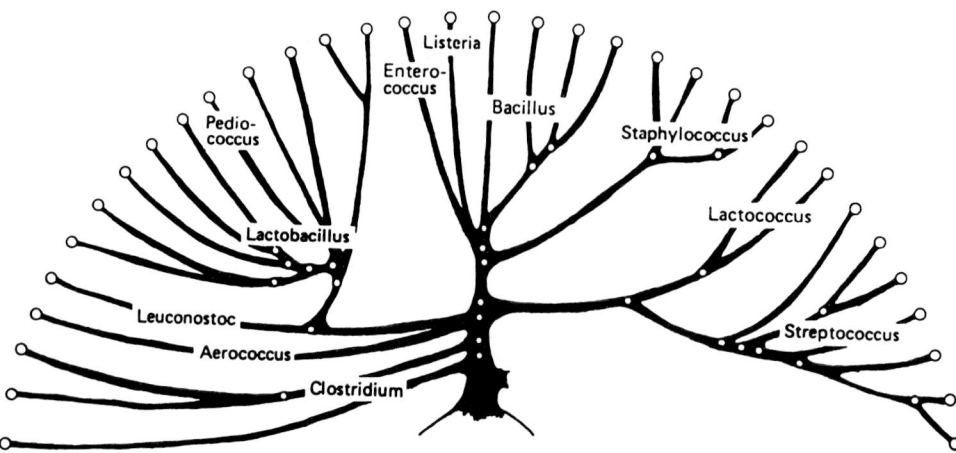

Fig. 1. Evolutionary tree of the *Clostridium* branch of Gram-positive bacteria showing the position of *Lactococcus lactis* and *L. raffinolactis* as an offshoot of the *Streptococcus* branch. The data are based on 16S rRNA oligonucleotide sequence similarities. (From Stackebrandt and Teuber, 1988.)

fermentation and ripening of milk, cream, and cheese (Weigmann, 1905–1908).

In recent years, the physiology, biochemistry, genetics, and molecular biology of the lactococci have gained much attention due to the great economic importance of these organisms.

It is the purpose of this chapter to outline these modern developments, in addition to providing the basic facts on the habitats, enrichment, isolation, and differentiation of the lactococci.

Habitats

The lactococci comprise the species *Lactococcus lactis, L. garviae, L. plantarum,* and *L. raffinolactis* (Table 3). *L. lactis* subsp. *lactis* and *L. lactis* subsp. *"diacetylactis"** have commonly been detected directly or following enrichment in plant material, including fresh and frozen corn, corn silks, navy beans, cabbage, lettuce, peas, wheat middlings, grass, clover, potatoes, cucumbers, and cantaloupe (Sandine et al., 1972). Lactococci are usually not found in fecal material or soil. Only small numbers occur on the surface of the cow and in its saliva. Since raw cow's milk consistently contains *L. lactis* subsp. *lactis,* and to a much lesser extent *L. lactis* subsp. *"diacetylactis"* and *L. lactis* subsp. *cremoris,* it is tempting to suggest that lactococci enter the milk from the exterior of the udder during milking and from the feed, which may be the primary source of inoculation. *L. lactis* subsp. *cremoris* has hitherto not been isolated with certainty from habitats other than milk, fermented milk, cheese, and starter cultures.

Other Lactococci

Out of 34,935 streptococcal specimens from humans and animals, which were sent to the Streptokokkenzentrale (Bundesanstalt für Milchforschung, FRG) Kiel, for identification during the period from 1965 to 1977, 1,433 strains were classified as *L. lactis* subsp. *lactis,* 40 as *L. lactis* subsp. *cremoris,* and 22 as *L. lactis* subsp. *"diacetylactis"* (Hahn and Tolle, 1979). This low count of lactococci within a large body of pathogenic and fecal streptococci reflects the accepted view that lactococci are apathogenic to humans and animals and that their incidence

---

*According to Schleifer et al. (1985), the subspecific status *"diacetylactis"* is no longer valid. Because of the importance of diacetyl-forming lactococci, especially in dairy fermentations, we still distinguish, for practical reasons, between *L. lactis* subsp. *lactis* and *L. lactis* subsp. *"diacetylactis."*

Table 1. Lactococci as components of starter cultures for fermented dairy products.[a]

Type of product	Composition of starter culture
1. Cheese type without eye formation (Cheddar, Camembert, Tilsit)	*Lactococcus lactis* subsp. *cremoris,* 95 to 98%; *Lactococcus lactis* subsp. *lactis,* 2 to 5%
2. Cottage cheese, quarg, fermented milks, cheese types with few or small eyes (e.g., Edam)	*Lactococcus lactis* subsp. *cremoris,* 95%; *Leuconostoc mesenteroides* subsp. *cremoris,* 5%; or *Lactococcus lactis* subsp. *cremoris,* 85 to 90%; *Lactococcus lactis* subsp. *lactis,* 3%; *Leuconostoc mesenteroides* subsp. *cremoris,* 5%
3. Cultured butter, fermented milk, buttermilk, cheese types with round eyes (e.g., Gouda)	*Lactococcus lactis* subsp. *cremoris,* 70 to 75%; *Lactococcus lactis* subsp. *"diacetylactis,"* 15 to 20%; *Leuconostoc mesenteroides* subsp. *cremoris,* 2 to 5%
4. Taette (Scandinavian ropy milk)	*Lactococcus lactis* subsp. *cremoris* (ropy strain)
5. Viili (Finnish ropy milk)	*Oidium lactis* (covers surface); *Lactococcus lactis* subsp. *cremoris* (ropy strain)
6. Casein	*Lactococcus lactis* subsp. *cremoris*
7. Kefir	Kefir grains containing lactose-fermenting yeasts (e.g., *Candida kefir*), *Lactobacillus kefir, Lactobacillus kefiranofacians, Lactococcus lactis* subsp. *lactis*

[a]The quantitative composition has been taken from the culture catalog of a major worldwide supplier.

may be due to contamination from food and feed, respectively.

Information on the habitats of the species *L. garviae, L. plantarum,* and *L. raffinolactis* is scarce. Only a few strains have been isolated from samples of cows with mastitis. Koch's postulates for *L. garviae* as infective agent of mastitis have not been fulfilled.

In the horse stomach, where an active lactic fermentation takes place, 3×10^8 lactococci have been counted per gram of content (Giesecke and Henderickx, 1973).

The most important habitats for lactococci, however, are in the dairy industry, and are shown in Table 1.

Table 2. Geographical distribution of the production of cheese and butter in 1987.[a]

	Production (metric tons)	
Location	Cheese	Butter and ghee
Developed market economies		
North America	3,114,470	604,322
Western Europe	5,276,384	2,088,500
Oceania	290,469	374,650
Others	169,543	80,682
Developing market economies		
Africa	43,295	56,604
Latin America	667,327	209,783
Near East	753,555	320,654
Far East	39,429	1,013,562
Others	–	1,253
Centrally planned economies		
Asia	135,301	65,800
Eastern Europe and USSR	3,220,029	2,578,767

Adapted from the FAO Production Yearbook, vol. 41, 1987.

Enrichment and Isolation

Lactic acid bacteria are nutritionally fastidious. They all require complex media for optimal growth. In synthetic media, all strains of lactococci require amino acids such as isoleucine, valine, leucine, histidine, methionine, arginine, and proline, and vitamins (niacin, Ca-pantothenate, and biotin) (Anderson and Elliker, 1953).

Isolation from Plant Material

Plant material like grass and herbages is the natural source of lactic acid bacteria. Ensilage allows the enrichment of lactococci, leuconostocs, and pediococci. The isolation procedure takes advantage of the sequential growth of the above-mentioned genera.

Ensilage Enrichment of Lactococci, Leuconostocs, and Pediococci (Whittenbury, 1965)

Grasses and other plant material are collected and cut in pieces as aseptically as possible. The prepared material is then placed in sterile glass tubes and compressed. Fifty grams of material are enough to fill a 3 × 20 cm tube. The tubes are sealed in a way that permits pressurized gas to escape but prevents the entry of oxygen. A number of silages are prepared and incubated at 30°C. Tubes are opened beginning on the second day. The silage is removed and placed in flasks of sterile water, which are vigorously shaken. These suspensions can then either be streaked directly onto agar plates or diluted and pour-plated. The 2- to 3-day-old silages are the best sources for lactococci and leuconostocs; 4- to 7-day-old silages are best for pediococci.

A modification of the above method has been reported by Weiler and Radler (1970). Grape leaves are homogenized with the same amount of acetate buffer (pH 5.4; 0.2 M). The homogenate is placed in sterile tubes. The tubes are sealed and incubated at 30°C.

Isolation from Dairy Products

The main problem in isolating lactic acid bacteria from dairy products is a proper dissolution or dispersion of the solid or semisolid fat-containing material.

Isolation of Lactic Acid Bacteria from Cheese Other than Cottage Cheese (Olson et al., 1978)

Using aseptic technique, thoroughly comminute or mix each sample until representative portions can be removed. Heat 99-ml dilution blanks of sterile, freshly prepared (less than 7 days old), aqueous 2% sodium citrate to 40°C. Aseptically transfer 11 g of cheese to a sterile blender container previously warmed to 40°C and add the warmed sodium citrate blank. Mix for 2 min at a speed sufficient to emulsify the sample properly, invert the container to rinse particles from the interior walls, and remix for approximately 10 s. Inadvertent heating to temperatures in excess of 40°C from friction in agitation may occur with some mechanical blenders. This should be determined before use of the blender so corrective action can be taken, if needed. If heating is unavoidable with equipment available, mixing periods of less than 2 min should be used provided that complete emulsification is obtained. The 1:10 dilution should be plated or further diluted immediately, great care being taken to avoid air bubbles or foam.

As an alternative method, 1 g ± 10 mg is rapidly weighed into a presterilized 177-ml (6 oz) Whirl-Pak bag or its equivalent. Close the bag, transfer it to a flat surface, and macerate the contents into a fine paste by rolling a 15 × 125 mm test tube or similar cylindrical object over the bag. The sample should not be forced into the corners or the ties seal area of the bag. Open the bag and add 9 ml of 2% sodium citrate at 40°C. Reclose the bag and roll the contents, as described above, to form a fine emulsion and proceed with plating immediately. Enumeration of bacterial species, such as lactic streptococci, that form chains may not be feasible with this method. However, this problem was not evident in a collaborative study involving analysis of Cheddar and Romano cheeses in nine laboratories in the USA.

Isolation of Lactic Acid Bacteria from Cottage Cheese (Olson et al., 1978)

Place the sterile blender container on a proper balance and tare. With a sterile spatula, mix contents of the cottage cheese container; or if in a tightly closed plastic sample pouch, gently knead and mix the enclosed curd. Aseptically remove the cover of the blender and place it upside down on the balance beside the container. Weigh 11 g of cottage cheese into the sterile container. Add 99 ml of warmed (40°C) sterile 2% sodium citrate solution as described in (1) (Isolation from Cheese Other than Cottage Cheese) to disperse and dilute the cottage cheese curd. Proceed as in (1) and plate appropriate

Table 3. Characteristics differentiating species and subspecies of the genus *Lactococcus*.

Species and subspecies	Source	Peptidoglycan type[a]	Major menaquinones[b]	Acid production from							Hydrolysis of arginine
				Galactose[c]	Lactose	Maltose	Melibiose	Melizitose	Raffinose	Ribose	
L. lactis subsp. *lactis*	Raw milk and dairy products	Lys-D-Asp	MK-9,MK-8	+	+	+	−	−	−	+	+
L. lactis subsp. *cremoris*	Raw milk and dairy products	Lys-D-Asp	MK-9,MK-8	+	+	−	−	−	−	−	−
L. lactis subsp. *hordniae*	Leaf hopper	Lys-D-Asp	MK-8,MK-9	−	−	−	−	−	−	−	+
L. garviae	Bovine samples	Lys-Ala-Gly-Ala	MK-9,MK-8	+	+	V	V	−	−	+	+
L. plantarum	Frozen peas	Lys-Ser-Ala	−	−	−	+	−	+	−	−	−
L. raffinolactis	Raw milk	Lys-Thr-Ala	−	+	+	+	+	V	+	V	V

[a]Abbreviations according to Schleifer and Kandler (1972). Asp, aspartic acid; gly, glycine; lys, lysine; ser, serine; thr, threonine.
[b]Abbreviations according to Collins and Jones (1981). MK-8, menaquinone with n=8 isoprene units; MK-9, menaquinone with n=9 isoprene units.
[c]+, positive; −, negative, V, variable.
Adapted from Schleifer (1987).

dilutions immediately. The alternative bag method described in (1) may also be used.

Isolation of Lactic Acid Bacteria from Cultured Milk, Cultured Cream, Yogurt, Acidophilus Milk, Bulgarian Buttermilk, and Similar Cultured or Acidified Semifluid Products (Olson et al., 1978)

After thoroughly mixing the sample, weigh 11 g of product into a sterile wide-mouth container, add 99 ml of sterile buffered distilled water (40°C), shake until a homogeneous dispersion is obtained, and withdraw appropriate amounts of this 1:10 dilution for plating or further dilution. The sample may be dispersed in 2% sodium citrate with a mechanical blender as described in (2) (Isolation from Cottage Cheese).

Unfortunately, no satisfactory selective medium is available for the isolation of lactococci. Two media, both commercially available, are generally accepted to give reliable growth of these organisms. The medium proposed by Elliker et al. (1956) is widely used for the isolation and enumeration of lactococci. M17 medium (Terzaghi and Sandine, 1975), a complex medium supplemented by 1.9% β-disodium glycerophosphate, resulted in improved growth of lactococci.

Elliker Agar Medium for Isolation of Lactococci (Elliker et al., 1956)

Medium contains, per liter:

Tryptone	20.0 g
Yeast extract	5.0 g
Gelatin	2.5 g
Dextrose	5.0 g
Lactose	5.0 g
Sucrose	5.0 g
Sodium chloride	4.0 g
Sodium acetate	1.5 g
Ascorbic acid	0.5 g
Agar	15.0 g

The medium has a pH of 6.8 before autoclaving.

This medium is probably the most cited for the isolation and growth of lactococci, although it is unbuffered. This disadvantage can be overcome by the addition of suitable buffer substances. Addition of 0.4% (wt/vol) of diammonium phosphate improves the enumeration of lactic streptococci on Elliker agar. Colony counts were up to about eight times greater due to improved buffering capacity (Barach, 1979).

M17 Medium for Isolation of Lactococci (Terzaghi and Sandine, 1975)

Phytone peptone	5.0 g
Polypeptone	5.0 g
Yeast extract	5.0 g
Beef extract	2.5 g
Lactose	5.0 g

Ascorbic acid	0.5 g
β-disodium glycerophosphate	19 g
1.0 M $MgSO_4 \cdot 7H_2O$	1.0 ml
Glass-distilled water	1 l

The medium is sterilized at 121°C for 15 min. The pH of the broth is 7.1. Solid medium contains 10 g agar/l of medium.

This medium is useful for the isolation of all strains of *L. lactis* subsp. *cremoris*, *L. lactis* subsp. *lactis*, *L. lactis* subsp. *"diacetylactis,"* and *Streptococcus salivarius* subsp. *thermophilus* and mutants of those strains lacking the ability to ferment lactose.

Addition of a pH indicator dye (bromcresol purple) and reduction of β-disodium glycerophosphate (5 g/l) allows an easy differentiation between lactose fermenting (large yellow colonies) and nonfermenting (small white colonies) strains (Kondo and McKay, 1984).

Shankar and Davies (1977) demonstrated the suppression of *Lactobacillus delbrueckii* subsp. *bulgaricus* in M17 medium. The majority of these *Lactobacillus* strains failed to grow in this medium adjusted to pH 6.8. Since M17 medium supported good growth of *Streptococcus salivarius* subsp. *thermophilus*, it can be used for the selective isolation of this microorganism from yogurt.

In recent years, this medium has become the standard for genetic investigations of lactococci (see below).

Lactococcal bacteriophages can be efficiently demonstrated and distinguished on M17 agar. Plaques larger than 6 mm in diameter could be observed as well as turbid plaques, indicating lysogeny (Terzaghi and Sandine, 1975).

ENUMERATION OF CITRATE-FERMENTING BACTERIA IN LACTIC STARTER CULTURES AND DAIRY PRODUCTS. To control gas and aroma (diacetyl) production in the fermentation of various dairy products, it is important to know the quantitative composition of the used starter cultures. *Leuconostoc* species and *Lactococcus lactis* subsp. *"diacetylactis"* are components of many mesophilic starter cultures (Table 1). These organisms are able to ferment citrate with concomitant production of CO_2 and diacetyl.

For the collective enumeration of leuconostocs and *L. lactis* subsp. *"diacetylactis"* in starters and fermented dairy products, a whey agar containing calcium lactate and casamino acids (WACCA) has been introduced by Galesloot et al. (1961).

WACCA 0.5% Medium for Enumeration of *Leuconostoc* and *Lactococcus lactis* subsp. *"diacetylactis"* (Galesloot et al., 1961)

Composition of the medium:
Dissolve in 1 l of whey, 5 g Ca-lactate · 5 H_2O, 7 g casitone, and 0.5% yeast extract. Adjust to pH 7.3 with

Ca(OH)$_2$-suspension. Steam for 30 min, filter and adjust to pH 7.1 with a NaOH solution. Add 1 ml MnSO$_4$ solution (40 mg MnSO$_4$ · 4 H$_2$O/100 ml). Dissolve 15 g agar, clarify with 5 g albumin, and sterilize for 15 min at 110°C (15 ml/tube).

Preparation of whey:
Add to 1 l of high-temperature-short-time (HTST)-pasteurized fresh skim milk at 30°C, 0.3 ml 35% CaCl$_2$ solution and 0.3 ml commercial rennet (strength 1:10,800). Cut coagulum after 30 min at 30°C. Filtrate after 2 h at 45°C.

Preparation of Ca-citrate suspension:
Suspend 28 g Ca-citrate (Merck) in a 100-ml 1.5% carboxymethyl cellulose solution prepared at 45°C. Allow to precipitate for 2 h at 45°C. The supernatant is steamed for 30 min.

Application:
Add 0.3–0.7 ml Ca-citrate suspension per 15-ml WACCA 0.5% (48°C). The amount to be added has to be adapted to the type of starter under investigation. Incubate for 5 days at 25°C. Count after 2, 3, and 5 days.

A different medium developed by Nickels and Leesment (1964) for the same purpose gives comparable results. A modified medium based on the different action of the lactose analogue 5-bromo-4-chloro-3-indolyl-β-D-galactopyranoside (Xgal) has been recently suggested (Vogensen et al., 1987).

Production and Preservation of Starter Cultures

Liquid, dried, and frozen starter cultures are in use (Jespersen, 1977; Wigley, 1977). Starter cultures must have a high survival rate of microorganisms coupled with optimum activity for the desired technological performance; e.g., the fermentation of lactose to lactate, controlled proteolysis of casein, and production of aroma compounds like diacetyl. Since the genes for lactose and citrate fermentation as well as for certain proteases are located on plasmids (Efstathiou and McKay, 1976; Kempler and McKay, 1979a, 1979b), continuous culture has not been successful because fermentation-defective variants easily develop (Lawrence et al., 1976). In most instances, pasteurized or sterilized skim milk is the basic nutrient medium for the large-scale production of starter cultures because it ensures that only lactococci fully adapted to the complex medium milk will develop. For liquid starter cultures, the basic milk medium may be supplemented with yeast extract, glucose, lactose, and calcium carbonate. To obtain optimum activity and survival, it may be necessary to neutralize the lactic acid that is produced by addition of sodium or ammonium hydroxide. Since many strains of lactic streptococci produce hydrogen peroxide during growth under microaerophilic conditions, it has been beneficial to add catalase to the growth media, thus leading to cell densities of more than 10^{10} viable units per milliliter of culture (Stanley, 1977).

Of course, many important details of the art of producing starter cultures are not being published for the protection of natural commercial interests. One example for the manufacture of freeze-dried yogurt starters has been described by Tamine and Robinson (1976). In contrast to this procedure, which uses reconstituted skim milk, the media for the production of concentrated starters are clarified by proteolytic digestion of skim milk with papain or bacterial enzymes to avoid precipitation of casein in the separators used to collect the streptococcal biomass (Stanley, 1977).

Batch Fermentation for Concentrated Starter Cultures (Stanley, 1977)

The (pilot) plant consists of 40-gal and 100-gal batch pasteurization tanks for use in medium preparation. Medium is pumped via a single, centrally placed peristaltic pump through a pasteurizer and into the 40 gal fermenter. This fermenter is fitted with pH control equipment, a stirrer for efficient mixing and hot and cold water jackets for temperature control. The pasteurized medium is aseptically inoculated with appropriate starter and incubated for 16–18 h at 22°C. A positive air pressure is maintained in the fermenter throughout this time. When grown, the culture is rapidly cooled in the fermenter, and harvested in a selfcleaning clarifier. The concentrated cells are automatically reconstituted by the desludging action of the separator and are pumped into the CCV (cell collection vessel) where gentle agitation ensures homogeneity of the product and a cooling device maintains a refrigeration temperature. From the CCV the concentrate passes to the microflow cabinet where it is packed semiautomatically in syringes via a syringe filler unit. These syringes are then frozen under vapor phase liquid nitrogen and stored until ready for use.

Further valuable details have been disclosed in the symposium "The Manufacture and Use of Starters for the Dairy Industry" (Cox, 1977; Gordon and Shapton, 1977; Jespersen, 1977; Lewis, 1977; Osborne, 1977; Wigley, 1977).

Identification

Lactococci are Gram-positive, microaerophilic cocci which lack the cytochromes of the respiratory chain (Hardie, 1986). They can be simply differentiated from pediococci and leuconostocs by the main fermentation products from glucose (see Table 4).

The common morphology consists of spherical or ovoid cells, 0.5 to 1 μm in diameter, in pairs or more-or-less long chains (Figs. 2–5). The lactic streptococci are in practice differentiated by their growth behavior at different temperatures (Table 5) into the mesophilic species *Lactococcus lactis* (Schleifer et al., 1985) and the thermophilic species *Streptococcus sal-*

Table 4. Differentiation scheme for lactococci, pediococci, and leuconostocs.

Fermentation products of glucose	Genus
L(+)-Lactic acid	*Lactococcus*
D(−)-Lactic acid, CO_2, acetic acid, ethanol	*Leuconostoc*
DL-Lactic acid	*Pediococcus*

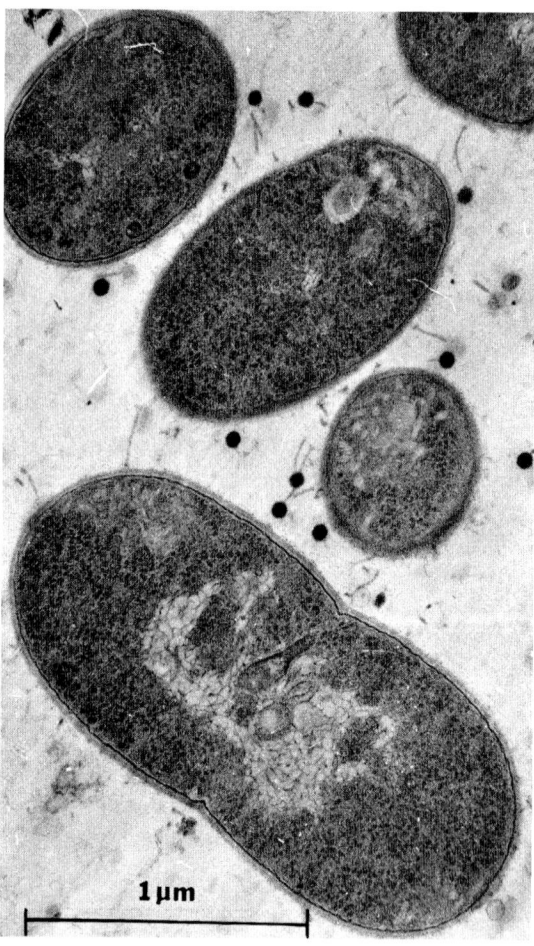

Fig. 2. Electron micrograph of thin sections of *Lactotoccus lactis* subsp. *lactis* at an early stage after infection with a specific bacteriophage. The large cell shows the typical features of a Gram-positive bacterium. It clearly demonstrates the almost rodlike appearance of a cell just starting to divide (compare to Figs. 3, 4, and 5). (Courtesy of J. Lembke.)

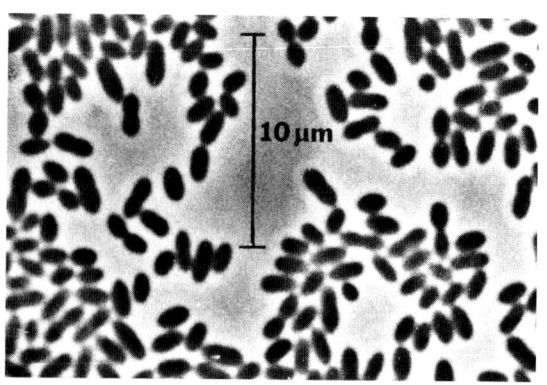

Fig. 3. Phase-contrast micrograph of *Lactococcus lactis* subsp. *lactis*. This strain is forming pairs of ovoid cells.

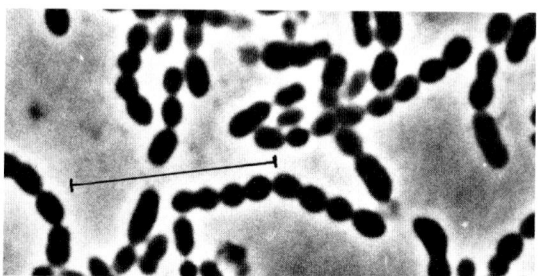

Fig. 4. Phase-contrast micrograph of *Lactococcus lactis* subsp. *lactis*. This strain is forming chains. Bar = 10 μm.

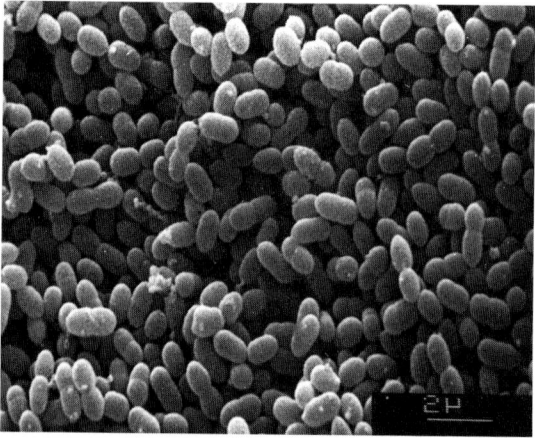

Fig. 5. Scanning electron micrograph of *Lactococcus lactis* subsp. *"diacetylactis"* growing in pairs of ovoid cells. Bar = 2 μm.

ivarius subsp. *thermophilus*. *S. salivarius* subsp. *thermophilus* is differentiated at first glance from the enterococci by its inability to grow in the presence of 6.5% NaCl. Problems may arise in the identification of *L. lactis* subsp. *"diacetylactis"* if the plasmid-coded fermentation of citrate is lost. The GC content of DNA ranges from 34 to 43 mol% (Schleifer et al., 1985). The

Table 5. Physiological and other properties of dairy lactococci used for identification and differentiation.

Properties	L. lactis subsp. lactis	L. lactis subsp. "diacetylactis"	L. lactis subsp. cremoris
Growth at 10°C	+	+	+
Growth at 40°C	+	+	−
Growth at 45°C	−	−	−
Growth in 4% NaCl	+	+	−
Growth in 6.5% NaCl	−	−	−
Growth at pH 9.2	+	+	−
Growth with methylene blue (0.1% milk)	+	+	−
Growth in presence of bile (40%)	+	+	+
NH_3 from arginine	+	+	−
CO_2 from citrate	−	+	−
Diacetyl and acetoin	−	+	−
Fermentation of maltose	+	+	Rarely
Hydrolysis of starch	−	−	−
Heat resistance (30 min at 60°C)	V	V	V
Serological group[a]	N	N	N
GC content of DNA[b] (mol%)	33.8–36.9	33.6–34.8	35.0–36.2

+, positive; −, negative; V, variable.

[a]Lancefield, 1933.

[b]Knittel, 1965.

genome sizes of different lactococci were estimated to be 2,300 to 2,600 kb (LeBourgeois et al., 1989).

Since the mesophilic dairy species differ only in a few properties (Table 5), the GC data are meaningless for this differentiation. That the three mesophilic dairy species are closely related is also implicated by the observation that many bacteriophages cross the "subspecies" line and attack strains of all three species (Lembke et al., 1980). Also, the plasmid patterns investigated so far do not allow a species differentiation (see below), but do allow the reidentification of strains. Another approach is differentiation on the basis of protein patterns after gel electrophoresis of soluble cell extracts (Jarvis and Wolff, 1979). By this method, classification of closely related strains seems possible. Collins and Jones (1979) have shown that the lactococci contain menaquinones with nine isoprene units as the major component, in contrast to dimethylmenaquinones with nine isoprene units and menaquinones with eight isoprene units in enterococci. At the moment, it is not possible to assess the number of different lactococci strains existing in dairies and starter cultures throughout the world.

Genetics

Genetic interest in the lactococci has been mainly focused in recent years on several properties of these bacteria important for dairy fermentation. These include, among others, the ability for lactose fermentation, casein breakdown, production of diacetyl from citrate, and resistance against the attack by bacteriophages. The relative instability of these properties is linked to the localization of relevant genes on extrachromosomal elements which may be lost in the fermentation process.

Lactococci possess an unusually large complement of plasmids, ranging in molecular weights from about 2 to more than 100 kb. The number of different plasmids in strains isolated from dairy cultures varies from one to 12 (Andresen et al., 1984; Cords et al., 1974; Davies et al., 1981; LeBlanc and Lee, 1979; Otto et al., 1982; Pechmann und Teuber, 1980; Yu et al., 1982) (Fig. 6). Plasmid-free strains were only obtained after plasmid curing by treatment with acridine dyes, ethidium bromide, nalidixic acid, growth at elevated temperature, or protoplast regeneration (Gasson, 1983; Geis et al., 1983; Kuhl et al., 1979; Pechmann and Teuber, 1980). Plasmid curing has provided the first evidence for plasmid linkage of a variety of properties (Table 6, columns 2 and 3). In some cases, the function of the plasmid was confirmed by plasmid transfer (see below).

Mechanisms of Gene Transfer

Plasmids can be transferred in and among lactococci by several mechanisms like conjugation, transduction, protoplast fusion, and in recent time by several transformation methods:

1. Transduction. Gene transfer by transduction was performed with virulent and temperate bacteriophages. Although transduction is severely limited by the narrow host ranges of

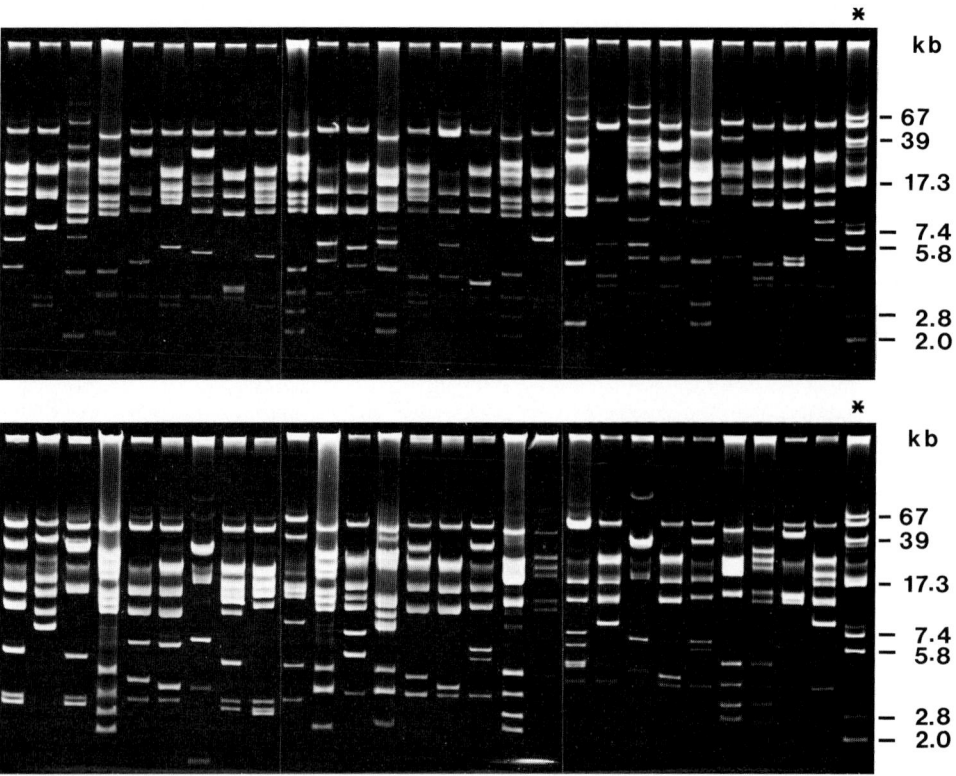

Fig. 6. Plasmid profiles of 54 *Lactococcus lactis* subsp. *lactis* strains isolated from an undefined mixed strain starter culture. Plasmid DNA from *L. lactis* subsp. *cremoris* AC1 (Neve et al., 1984) was used as size standard (indicated by asterisks on the right lanes). Electrophoresis was done on 0.8% agarose gels. (Adapted from Andresen et al., 1984.)

the transducing phages—for most of the temperate phages no host strains are available— several genes were transferred by this mechanism, including genes necessary for lactose, maltose, and mannose metabolism (McKay et al., 1973), protease activity (McKay and Baldwin, 1974), streptomycin resistance, and tryptophan biosynthesis (Allen et al., 1963; Sandine et al., 1962). Transductional transfer of the MLS-antibiotic resistance plasmid pAMβ and of small lactococcal vectors was also demonstrated (Gasson and Davies, 1980a).

2. Conjugation. Transfer of plasmids by conjugation is a rather common phenomenon in lactococci (Neve et al., 1987). A variety of lactococcal plasmids encodes for their own conjugal transfer. Most plasmids, however, are nonconjugative but many of them can be transferred when mobilized by a conjugative plasmid. Plasmids encoding genes for lactose-metabolism (Gasson and Davies, 1980b; McKay et al., 1980) and sucrose (Gasson, 1984; Gonzales and Kunka, 1985), proteinase activities (Gasson and Davies, 1980b; Kuhl et al., 1979), the production of antagonistic components and the resistance against these

substances (Davey, 1984; Neve et al., 1984; Scherwitz et al., 1983), for bacteriophage resistance (see below), citrate metabolism (Neve et al., 1987) and "slime polymer" synthesis (Vedamuthu and Neville, 1986) were transferred directly or by cotransfer with a conjugative plasmid. The conjugative transposons Tn916, Tn919, and Tn1545, originating from *Streptococcus sanguis* FC1, *Enterococcus faecalis* DS16, and *S. pneumoniae* BM4200, were transferred to a number of lactococcal strains. The application of transposons will become a powerful tool in lactococcal genetics, especially for the analysis of chromosomally encoded genes (Courvalin and Carlier, 1986; Hill et al., 1985, 1987; Renault and Heslot, 1987; Renault et al., 1989).

3. Protoplast fusion. Using polyethylene-glycol (PEG) as the fusogenic agent, protoplasts of several strains of lactococci were fused, resulting in recombinants that contained genetic markers of both fusion partners (Gasson, 1980; Okamoto et al., 1983). Intergeneric protoplast fusion involving lactococci and *Bacillus subtilis* or lactobacilli was also performed (Cocconcelli et al., 1986; van der Lelie et al., 1988; van der Vossen et al., 1988).

Table 6. Cloned genes from lactococci.

Gene	Gene product or phenotype	Origin	Cloning vector	Host	Reference
*lac*G	P-β-Gal EC 3.2.1.85	*L. lactis* Z268	pUC	*E. coli*	Boizet et al., 1988
		L. lactis 712 (pLP712)	pAT153 pNZ12	*E. coli* *L. lactis* MG1363	Maeda and Gasson, 1986 de Vos and Simons, 1988 de Vos and Gasson, 1989
		L. lactis W67 (pTD1)	pBR322, pUC7 pGKV210	*E. coli, L. lactis* Bu2-60, *L. lactis* MG1363	Geis, 1984
		L. lactis LMO232 (pLM2001)	pDB101	*Streptococcus sanguis*	Harlander et al., 1984
		L. lactis subsp. *cremoris* H2 (pJI70)	pUC18/19	*E. coli*	Inamine et al. 1986
*prt*P	Cell-wall-associated proteinase	*L. lactis* 712 (pLP712)	pCK1	*L. lactis* MG1363 *B. subtilis*	Gasson et al. 1987
		L. lactis subsp. *cremoris* WG2 (pWV05)	pGKV2	*L. lactis* MG1363 *B. subtilis*	Kok et al., 1985
		L. lactis subsp. *cremoris* SK11 (pSK111)	pNZ121	*L. lactis* MG1363 *B. subtilis*	de Vos, 1986
*prt*M	Proteinase maturation protein	pWV05 pSK111	pGKV2 pIL253	*L. lactis* MG1363 *L. lactis* MG1363	Haandrikman et al., 1989 Vos et al., 1989
—	Tagatose-1,6-biphosphate aldolase	*L. lactis* H1(pD31)	pACYC184	*E. coli*	Limsowtin et al., 1986
—	α-Acetolactate decarboxylase	*L. lactis* DSM20384 (chromosomal DNA)	pBR322	*E. coli*	Goelling and Stahl, 1988
—	Bacteriocin	*L. lactis* WM4 (pNP2)	pGB301	*L. lactis* LM0230	Sherwitz-Harmon and McKay, 1987
—	Bacteriocin resistance	*L. lactis* subsp. *cremoris* 9B4 (p9B4-6)	PGKV210 pGV1 pMG24	*L. lactis* 1403	van Belkum et al., 1989
*spa*N	Nisin	*L. lactis* ATCC11454	AJ1, pUC9	*E. coli*	Buchman et al., 1988
*nis*A	Nisin	*L. lactis* 6F3	pUC19	*E. coli*	Kaletta and Entian, 1989
—	NisinR	*L. lactis* DRC3 (pNP40)	pSA3	*L. lactis*	Froseth et al., 1988
—	UV-resistance	*L. lactis* 594 (pIL7)	pHV1301	*L. lactis* IL1403	Chopin et al., 1986
—	R+/M+	*L. lactis* subsp. *lactis, cremoris*	pIL1403	*L. lactis* IL1403	Gautier et al., 1987
—	Phage resistance (Hsp+)	*L. lactis* subsp. *cremoris* R1 (pTR2030)	EMBL3 pSA3	*L. lactis* MG1363	Hill et al., 1989
—	Phage resistance	*L. lactis* KR2 (pKR223)	pGB301	*L. lactis* LM230	Laible et al., 1987

4. Protoplast transformation. Genetic analysis of lactococci and the application of recombinant DNA technology to these bacteria was, until recently, severely impeded by the lack of an efficient transformation method. By modifying a transformation protocol for *B. subtilis* protoplast transformation (Chang and Cohen, 1979), a method, albeit rather inefficient, was developed for the transformation of lactococci with plasmid and bacteriophage DNA (Geis, 1982; Kondo and McKay, 1982). Although the original protocols for the PEG-induced protoplast transformation were subsequently optimized

(Kondo and McKay, 1984; Simon et al., 1986; van der Vossen et al., 1988; von Wright et al., 1985; Woskow and Kondo, 1987), only a limited number of strains can be transformed efficiently (10^4–10^7 transformants/μg DNA). Probably due to poor protoplast regeneration, many strains are still not transformable by these methods.

5. Electroporation. Transformation of lactococci has also been achieved by use of electroporation. The main advantage of this method seems to be its general applicability without the need to prepare and regenerate protoplasts. Transfer efficiencies, however, are strain specific and optimal conditions have to be determined (Harlander, 1987; Powell et al., 1988; van der Lelie et al., 1988).

Gene Cloning

The availability of efficient DNA transformation systems was a prerequisite for the application of recombinant DNA technology to lactococci. In the last few years a number of plasmid-based cloning vectors have been developed and suitable host vector systems are now available which have been used to clone several homologous and heterologous genes into lactococci and hosts of other genera (Tables 6 and 7).

Applications

Lactococci are employed in single and mixed cultures for the production of different kinds of cheeses, fermented milks, cultured butter, and casein (see Table 1).

The biochemical and technological functions of lactococci necessary for milk fermentation and cheese production can be summarized as follows:

1. Formation of lactic acid from lactose. This early function in fermentation starts with the phosphorylation of lactose during transport via a specific, membrane bound phosphotransferase system. The lactose-6-phosphate is split by a phospho-β-galactosidase into glucose and galactose-6-phosphate, which is further metabolized by the tagatose-6-phosphate pathway. L(+)-lactate is the main end-product of glycolysis. The resulting lowered pH values (4.0 to 5.6) as compared to milk (6.6 to 6.7) prevent or retard growth of spoilage bacteria, especially clostridia, staphylococci, Enterobacteriaceae, and psychrophilic Gram-negative-like *Pseudomonas*. If the isoelectric point of casein at pH 4.6 to 4.8 is approached, casein is precipitated. This effect is used to curdle milk in the production of cottage cheese, quarg, sour milk, yogurt, and casein. Starter bacteria for this purpose are *Lactococcus lactis* subsp. *lactis* and *L. lactis* subsp. *cremoris* (besides lactobacilli and *Streptococcus salivarius* subsp. *thermophilus*).

2. Formation of diacetyl from citrate. Diacetyl is the most characteristic aroma compound provided by *Lactococcus lactis* subsp. *"diacetylactis."* It is derived from the citrate of milk (about 0.1%) present in solution and in the casein micelles as a casein-citrate-calcium-phosphate complex. The pathway goes through oxaloacetate, pyruvate, and α-acetolactate with a coupled release of carbon dioxide, which induces eye formation in cheese but also unwanted floating of the curd in the manufacture of cottage cheese or quarg if an unbalanced mixed starter is used (see Table 1).

Table 7. Heterologous genes cloned in lactococci.

Gene	Gene product or relevant phenotype	Origin	Cloning vector	Reference
cat	Chloramphenicol acetyl-transferase	*S. aureus* (pC194)	pWV01[a]	Kok et al., 1984
cat86		*B. pumilis* (pPL608)	pSH71[a]	van der Vossen et al., 1985
erm C	Ribosome methylase	*S. aureus* (pE194 cop6)	pWV01[a]	Kok et al., 1984
—	3'5"-Aminoglycoside-phosphotransferase (Km^R)	*S. aureus* (pUB110)	pSH71[a]	Gasson and Anderson, 1985
tet M	Tc^R	*E. faecalis* (pAD1)	pGV1	Hill et al., 1988
tet M	Tc^R	*E. faecalis* JHZ2 (pAM180)	pIL204	Loureiro dos Santos and Chopin, 1987
lac Z	β-Galactosidase	*E. coli*	pNZ17	de Vos and Simons, 1988
hel	Lysozyme	Chicken	pMG36	van den Guchte et al., 1989
—	Chymosin	Bovine	pNZ18	Simons et al., 1988
—	Thaumatin	*Thaumatococcus daniellii*		Novak and Batt, 1987

[a]Cryptic plasmids from lactococci.

3. Limited proteolysis during cheese ripening. Many strains of all lactococcal species possess a β-casein specific, cell-wall-associated protease, together with a complement of peptidases including an aminopeptidase, an X-prolyl-dipeptidyl aminopeptidase, and a dipeptidase, which are necessary for growth in milk and may also function during cheese ripening (Kiefer-Partsch et al., 1989; Kok and Venema, 1988; Thomas and Pritchard, 1987).

The lactococci and lactobacilli needed by the dairy and meat industries are supplied by firms and institutes that have specialized in the production of starter cultures. In the FRG, for example, at least 12 local, national, and international commercial enterprises produced and distributed the cultures for the manufacture in 1988 of about 1 million tons of cheese, 500,000 tons of fermented milk products, and 400,000 tons of lactic butter.

The cultures are typically provided in either of the three following forms: 1) fresh, fluid cultures containing about 0.5×10^9 viable lactococci per milliliter; 2) freeze-dried (lyophilized) cultures containing about 2×10^9 viable lactococci per gram; and 3) deep-frozen, concentrated cultures containing about 10^{11} to 10^{12} viable lactococci per milliliter.

Fresh cultures deteriorate during improper shipment and storage. The lyophilized material is convenient to ship and may be kept at 5 to 10°C for at least 6 months. Deep-frozen, concentrated cultures can be shipped and stored indefinitely in liquid nitrogen. After shipment in dry ice and storage at -45°C, keeping time is at least 1 month (Jespersen, 1977), but much longer at -60°C.

Fluid and freeze-dried cultures are the seed material for the bulk starters to be prepared in the factory, whereas deep-frozen, concentrated cultures are preferred for direct vat inoculation but are more expensive than the conventional cultures.

Traditional cheese production proceeds in open vats. The milk needed for cheese-making—e.g., raw milk for Emmental or Grana, pasteurized for most others—cannot be a sterile environment. It may contain a variable number (up to more than 10^6/ml) of a variety of Gram-positive, acid-producing bacteria and increasingly Gram-negative, psychrophilic microorganisms due to an increase of refrigeration time during production, transport, and workup of milk. Usually, starter cultures overgrow the endogenous microflora of milk which may, however, considerably contribute to the final quality, taste, and aroma of cheese (Stadhouders, 1986).

The main microbiological problem is the common infection with bacteriophages, which leads to a slowdown or in severe cases to a breakdown of the lactic fermentation and the ripening process (see below).

The use of deep-frozen starter cultures would eliminate the possibility that important technological functions coded for by plasmid DNA (lactose metabolism, protease activity) are being lost during the preparation of mother cultures and bulk starters by microbiologically untrained personnel. Although a large body of knowledge exists about cheese starters, the microbial ecology of cheese production and ripening is still not fully understood (Lawrence et al., 1976; Stadhouders, 1975). The economical significance of lactococci is fundamental for the food industry. The world production of cheese in 1987 was estimated by the FAO at 13,709,802 tons, that of butter and ghee at 7,394,576 tons (Food and Agriculture Organization of the United Nations, 1987). Table 2 demonstrates that production is highest in the technologically developed parts of the world. Together with the amounts of fermented milks for which no accurate estimates were available, we may assume for 1987 a total of 2×10^{10} kg of dairy products manufactured with the aid of lactococci and lactobacilli. Since the numbers of lactococci easily reach 10^9 viable units per gram of cheese or sour milk (Stadhouders, 1975), the bacterial biomass handled annually by the dairy industry can be calculated to be on the scale of 10^5 tons if we include the biomass present in typical by-products such as cheese whey and buttermilk and assume a weight of 10^{-12} g per viable bacterium. Based on the wholesale prices for butter (\$310 to \$350/100 kg) and cheese (\$270 to \$350/100 kg Cheddar and Gouda, respectively) as recorded by the FAO for 1987 (Food and Agriculture Organization of the United Nations, 1987) a value of the world output of fermented milk products of about \$50 billion (5×10^{10}) can be implicated.

Bacteriophages

Phages attacking dairy starter cultures were reported in 1935 (Whitehead and Cox, 1935), and these viruses are still the main cause for slow or incomplete fermentations. Dairy processes are performed under nonsterile conditions, since 1. the substrate milk cannot be sterilized prior to inoculation; and 2. open tanks are used. In addition, milk fermentations have been scaled up to 50,000 l and due to economic reasons, short-time fermentations with more than one filling per day are common (Daly, 1983;

Lawrence et al., 1976; Teuber and Lembke, 1983). Therefore, once having been introduced in a dairy plant, bacteriophages specific for the used lactic acid bacteria easily spread and high titers of 10^{10} phages/ml of cheese whey may result within a few days. Bacteriophages are easily distributed in dairy plants by air (in particular, whey aerosols), personnel, equipment, and the milk itself. Whey powder may be infected with bacteriophages since phages survive pasteurization and spray-drying (Chopin, 1980).

A variety of concepts has been elaborated in order to minimize the risk of a bacteriophage contamination.

1. The necessity to work as far as possible under strictly hygienic conditions (e.g., by using closed fermentation utilities with filter systems that retain phage particles). The dairy equipment has to be subjected to continuous daily cold disinfection. Active chlorine and peracetic acid are suitable to totally inactivate bacteriophages (Lembke and Teuber, 1981a).
2. Propagation of mother and bulk cultures prior to the batch fermentations are frequently done in phage-inhibitory media (Sandine, 1977), which usually contain phosphate and/or citrate chelating divalent cations (in particular Ca^{2+}), which are essential for lytic bacteriophage development (Collins et al., 1950; Potter and Nelson, 1952).
3. By using deep-frozen (Gilliland and Speck, 1974) or lyophilized (Stadhouders et al., 1969) concentrated starter cultures, which are inoculated directly into the culture tank, prolonged exposure of the mother culture and the bulk culture to bacteriophages in the cheese factory can be avoided efficiently.
4. Starter bacteria exhibiting different phage spectra are used as single strains or are blended for a defined multiple starter culture within a rotation scheme. By continous phage monitoring in the plant, phage-sensitive strains are immediately replaced by alternative cultures (Huggins, 1984; Hull, 1977).
5. However, the use of starter bacteria with total or at least high phage insensitivity is desirable.

Phage-resistant mutants can easily be isolated after prolonged challenge of a culture with bacteriophages (King et al., 1983; Limsowtin and Terzaghi, 1976; Möller and Teuber, 1988). These mutants have to be examined very carefully for their usefulness in dairy fermentations, since they frequently reveal low acid production or protease activity (slow variants) (Jarvis, 1981; Lawrence, 1978). By use of a laboratory activity test with a starter culture (Heap and Lawrence, 1976), bacterial strains are continuously challenged with high-titer phage lysates for at least seven passages under conditions simulating the commercial cheese fermentation process with a 38°C cheese cooking step.

Traditionally used starter cultures with unknown strain composition (undefined mixed strain starters) are a good reservoir for the selection of *phage-insensitive strains* (Chopin et al., 1976; Heap and Lawrence, 1976; Stadhouders, 1975). These starter bacteria have been used in dairy fermentations for many years, giving rise to natural selection of phage-resistant strains. Plasmid-encoded phage resistance mechanisms have been detected frequently in these strains (see below, this chapter).

Virulent Bacteriophages

Virulent bacteriophages specific for lactococci have been found in cheese factories throughout the world. The majority of these phages were classified into three morphological groups (small isometric-, large isometric-, and prolate-headed phages), but phages with an unusual peculiar morphology were detected occasionally (Keogh and Shimmin, 1974; Lembke et al., 1980; Terzaghi, 1976). The small isometric- and prolate-headed phages were most common. Phages of the same morphology could be further differentiated by immunological techniques (phage neutralization test (Jarvis, 1977), the ELISA test (Lembke and Teuber, 1979), Western blotting (Relano et al., 1987), and immuno electron microscopy (Lembke and Teuber, 1981b). Phages with different morphologies revealed no DNA-DNA homology (Coveney et al., 1987; Jarvis, 1984a; Lautier and Novel, 1987). Seven distinct genetic groups of phages were found, of which the type phages are shown in Fig. 7 (Braun, Jr. et al., 1989). The genomes of all investigated lactococcal bacteriophages are composed of double-stranded DNA with genome sizes ranging from 18.1 to 54.5 kb (Braun, Jr. et al., 1989). Physical maps of the genomes of the most common small isometric- and prolate-headed phages have been constructed (Braun, Jr. et al., 1989; Coveney et al., 1987; Jarvis and Meyer, 1986; Loof et al., 1983; Powell and Davidson, 1985). These phages contain linear genomes with cohesive ends and an unusually low number of restriction enzyme recognition sites (Loof et al., 1983; Powell and Davidson, 1986). Heteroduplex analysis with DNAs of related small isometric-headed phages allowed the exact localization of homologous DNA sequences on the phage genomes (Coveney et al., 1987; Jarvis and Meyer, 1986; Loof and Teuber, 1986).

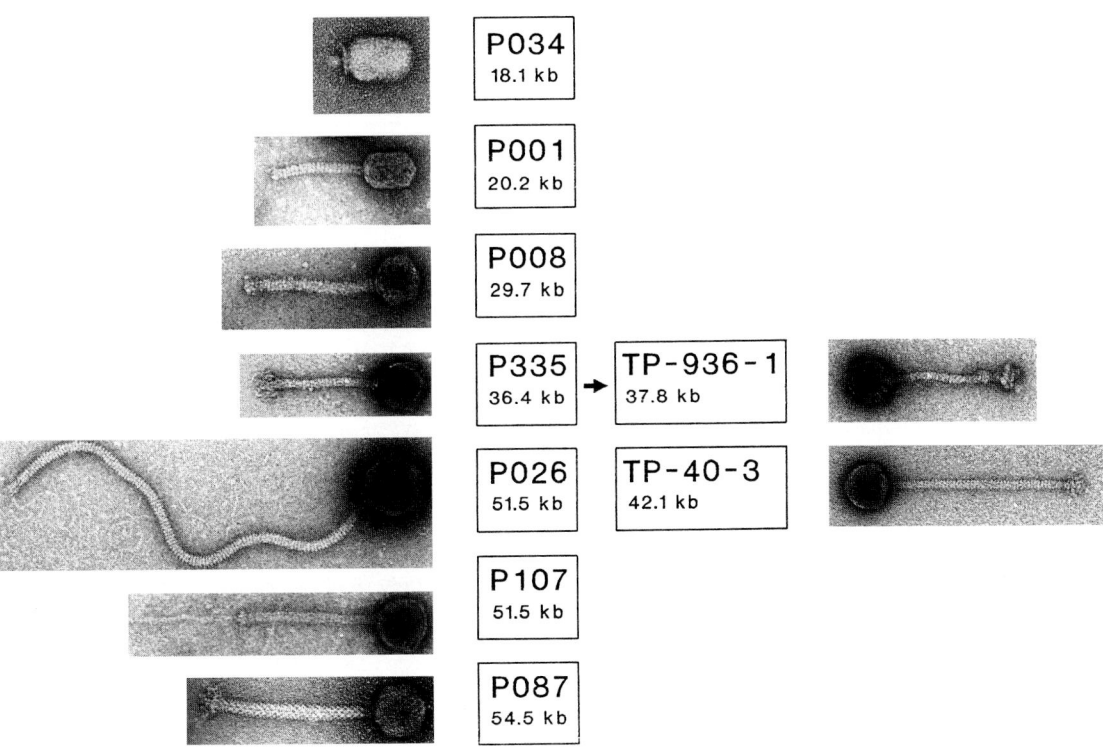

Fig. 7. Electron micrographs of virulent and temperate lactococcal-type phages with their corresponding genome sizes. The arrow indicates genetic relationship between virulent and temperate phages. (Adapted from Braun, Jr. et al., 1989.)

Temperate Bacteriophages

Most lactococcal starter strains are lysogenic, harboring one or more prophages that can be liberated by induction with mitomycin C, UV-light irradiation, or even spontaneously (Park and McKay, 1975; Reiter, 1949; Terzaghi and Sandine, 1981; Teuber and Lembke, 1983). The prophages are isometric-headed with linear DNA molecules of genome sizes of 37.8 to 46 kb (Braun, Jr. et al., 1989), which are frequently redundant and circularly permuted (Lakshmidevi et al., 1988). These temperature phages can be divided into at least three taxonomic groups with no significant DNA homology. In general, no DNA homology exists between the temperate and the majority of virulent phages (Jarvis, 1984b). Only one group of virulent phages is genetically closely related to the temperate phages (Braun, Jr. et al., 1989; Lautier and Novel, 1987; Relano et al., 1987). It is still uncertain whether spontaneously released temperate phages may be the origin of severe virulent bacteriophage infections, since indicator strains for lytic propagation of temperate phages have rarely been isolated from multiple-strain starter cultures (Lawrence et al., 1976; Teuber and Lembke, 1983). Prophage-curing of lysogenic strains has been performed, and formal evidence for the lysogenic state has been demonstrated by relysogenization of the prophage-cured hosts (Gasson and Davies, 1980c; Georghiou et al., 1981).

Bacteriophage Resistance Mechanisms in Lactococci

In the 1980s, the interests of dairy microbiologists have focused on plasmid-directed bacteriophage resistance mechanisms which include: 1. blocking of phage-specific adsorption to the cell envelope (de Vos et al., 1984; Sanders and Klaenhammer, 1983); 2. multiple restriction/modification (R/M) systems, (Chopin et al., 1984; Sanders and Klaenhammer, 1981); and 3. abortive infections (Baumgartner et al., 1986; Klaenhammer and Sanozky, 1985; Laible et al., 1987; McKay and Baldwin, 1984; Murphy et al., 1988).

More than one phage insensitivity mechanism may be present in individual strains (Froseth et al., 1987; Gautier and Chopin, 1987;

Sanders and Klaenhammer, 1983). However, phage insensitivity is frequently restricted to a limited group of phages (Jarvis and Klaenhammer, 1986).

Adsorption of phages to phage receptors localized in the cell envelope is a highly specific event (Budde-Niekiel and Teuber, 1987). Phage receptors have been localized in the cell wall and the rudimentary work described so far indicated that a proteinaceous and/or a carbohydrate component may be part of the receptor structure (Keogh and Pettingill, 1983; Oram and Reiter, 1968). Localization of the phage receptor in the lipoprotein fraction of the cytoplasmic membrane also has been reported (Oram, 1971).

A variety of R/M systems have been detected in the lactococcal starters, and multiple systems can occur in individual strains (Boussemaer et al., 1980). Activity of a type II restriction enzyme has been shown in a lactococcal strain (Fitzgerald et al., 1982).

Phage insensitivity mechanisms, which are not due to blocking of phage adsorption or R/M-systems, are classified as *abortive infection* and are poorly understood. Phage resistance based on R/M systems or abortive infection is frequently lost at elevated temperatures (37 to 40°C), which are used in dairy processing (Jarvis, 1988; McKay and Baldwin, 1984; Pearce, 1978; Sanders and Klaenhammer, 1980).

Genome sizes of plasmids encoding phage insensitivity vary from 5.1 to 106 kb (Gautier and Chopin, 1987; Jarvis, 1988). Phage-insensitivity plasmids could be transferred by conjugation (Baumgartner et al., 1986; Chopin et al., 1984; Froseth et al., 1988; Higgins et al., 1988; Jarvis, 1988; Wetzel et al., 1986) or transformation (Gautier and Chopin, 1987; Josephsen and Vogensen, 1989; Simon et al., 1985) into phage-sensitive strains, and successful industrial use of phage-resistant transconjugants has been reported (Sanders et al., 1986; Sing and Klaenhammer, 1986). Additive effects by transferring more than one phage-insensitivity plasmid into one recipient strain have been demonstrated (Chopin et al., 1984; Daly and Fitzgerald, 1987).

Plasmid-borne phage resistance determinants have been cloned (Hill et al., 1989; Laible et al., 1984) and may result in new strains with "super" phage-insensitivity.

Literature Cited

Allen, L. K., W. E. Sandine, and P. R. Elliker. 1963. Transduction in *Streptococcus lactis*. J. Dairy Res. 30:351–357.

Anderson. A. W., and P. R. Elliker. 1953. The nutritional requirements of lactic streptococci isolated from starter cultures. I. Growth in a synthetic medium. J. Dairy Sci. 36:161–167.

Andresen, A., A. Geis, U. Krusch, and M. Teuber. 1984. Plasmid profiles of mesophilic dairy starter cultures. Milchwissenschaft 39:140–143.

Barach, J. T. 1979. Improved enumeration of lactic acid streptococci on Elliker agar containing phosphate. Appl. Environ. Microbiol. 38:173–174.

Baumgartner, A., M. Murphy, C. Daly, and G. F. Fitzgerald. 1986. Conjugative co-transfer of lactose and bacteriophage resistance plasmid from *Streptococcus cremoris* UC653. FEMS Microbiol. Lett. 35:233–237.

Boizet, B., D. Villeval, P. Slos, M. Novel, and A. Mercenier. 1988. Isolation and structural analysis of the phospho-β-galactosidase gene from *Streptococcus lactis* Z268. Gene 62:249–261.

Boussemaer, J. P., P. P. Schrauwen, J. L. Sourrouille, and P. Guy. 1980. Multiple modification/restriction systems in lactic streptococci and their significance in defining a phage-typing system. J. Dairy Res. 47:401–409.

Braun, Jr., V., S. Hertwig, H. Neve, A. Geis, and M. Teuber. 1989. Taxonomic differentiation of bacteriophages of *Lactococcus lactis* by electron microscopy, DNA-DNA hybridization, and protein profiles. J. Gen. Microbiol. 135:2551–2560.

Buchman, G. W., S. Banerjee, and J. N. Hansen. 1988. Structure, expression, and evolution of a gene encoding the precursor of nisin, a small protein antibiotic. J. Biol. Chem. 263:16260–16266.

Budde-Niekiel, A., and M. Teuber. 1987. Electron microscopy of the adsorption of bacteriophages to lactic acid streptococci. Milchwissenschaft 42:551–554.

Chang, S., and S. N. Cohen. 1979. High frequency transformation of *Bacillus subtilis* protoplasts by plasmid DNA. Mol. Gen. Genet. 168:111–115.

Chopin, A., M.-C. Chopin, A. Moillo-Batt, and P. Langella. 1984. Two plasmid-determined restriction and modification systems in *Streptococcus lactis*. Plasmid 11:260–263.

Chopin, M.-C. 1980. Resistance of 17 mesophilic lactic *Streptococcus* bacteriophages to pasteurization and spray-drying. J. Dairy Res. 47:131–139.

Chopin, M.-C., A. Chopin, A. Rouault, and D. Simon. 1986. Cloning in *Streptococcus lactis* of plasmid-mediated UV resistance and effect on prophage stability. Appl. Environ. Microbiol. 51:233–237.

Chopin, M.-C., A. Chopin, and C. Roux. 1976. Definition of bacteriophage groups according to their lytic action on mesophilic lactic streptococci. Appl. Environ. Microbiol. 32:741–746.

Cocconcelli, P. S., L. Morelli, M. Vescovo, and V. Bottazzi. 1986. Intergeneric protoplast fusion in lactic acid bacteria. FEMS Microbiol. Lett. 35:211–214.

Collins, E. B., F. E. Nelson, and C. E. Parmelee. 1950. The relation of calcium and other constituents of a defined medium to proliferation of lactic *Streptococcus* bacteriophage. J. Bacteriol. 60:533–542.

Collins, M. D., and D. Jones. 1979. The distribution of isoprenoid quinones in streptococci of serological groups D and N. J. Gen. Microbiol. 114:27–33.

Collins, M. D., and D. Jones. 1981. Distribution of isoprenoid quinone structural types in bacteria and their taxonomic implications. Microbiol. Rev. 45:316–345.

Cords, B. R., L. L. McKay, and P. Guerry. 1974. Extrachromosomal elements in group N streptococci. J. Bacteriol. 117:1149–1152.

Courvalin, P., and C. Carlier. 1986. Transposable multiple antibiotic resistance in *Streptococcus pneumoniae.* Mol. Gen. Genet. 205:291–297.

Coveney, J. A., G. F. Fitzgerald, and C. Daly. 1987. Detailed characterization and comparison of four lactic streptococcal bacteriophages based on the morphology, restriction mapping, DNA homology, and structural protein analysis. Appl. Environ. Microbiol. 53:1439–1447.

Cox. W. A. 1977. Characteristics and use of starter cultures in the manufacture of hard pressed cheese. J. Soc. Dairy Technol. 30:5–15.

Daly, C. 1983. The use of mesophilic cultures in the dairy industry. Antonie van Leeuwenhoek 49:297–312.

Daly, C., and G. F. Fitzgerald. 1987. Mechanisms of bacteriophage insensitivity in the lactic streptococci, p. 259–268. In: J. J. Ferretti and R. E. Curtiss III (ed.), Streptococcal genetics. American Society for Microbiology, Washington, D.C.

Davey, G. P. 1984. Plasmid associated with diplococcin production in *Streptococcus cremoris.* Appl. Environ. Microbiol. 48:895–896.

Davies, F. L., H. M. Underwood, and M. J. Gasson. 1981. The value of plasmid profiles for strain identification in lactic streptococci and the relationship between *Streptococcus lactis* 712, ML3 and C2. J. Appl. Bacteriol. 51:325–337.

de Vos, M. W. 1986. Genetic improvement of starter-streptococci by the cloning and expression of a gene coding for a non-bitter proteinase, p.465–472. In: E. Magnien (ed.), Biomolecular engineering in the European community. Martinus Nijhoff Publisher, Dordrecht, The Netherlands.

de Vos, W. M., and M. J. Gasson. 1989. Structure and expression of the *Lactococcus lactis* gene for phospho-β-galactosidase (lacG) in *Escherichia coli* and *Lactococcus lactis.* J. Gen. Microbiol. 135:1833–1846.

de Vos, W. M., and G. Simons. 1988. Molecular cloning of lactose genes in dairy lactic streptococci: The phospho-β-galactosidase and β-galactosidase genes and their expression. Biochimie 70:461–473.

de Vos, W. M., H. M. Underwood, and F. L. Davies. 1984. Plasmid encoded bacteriophage resistance in *Streptococcus cremoris* SK11. FEMS Microbiol. Lett. 23:175–178.

Efstathiou, J. D., and L. L. McKay. 1976. Plasmids in *Streptococcus lactis:* Evidence that lactose metabolism and proteinase activity are plasmid linked. Appl. Environ. Microbiol. 32:38–44.

Elliker, P. R., A. W. Anderson, and G. Hannesson. 1956. An agar medium for lactic acid streptococci and lactobacilli. J. Dairy Sci. 39:1611–1612.

Fitzgerald, G. F., C. Daly, L. R. Brown, and T. R. Gingeras. 1982. ScrFI: a new sequence specific endonuclease from *Streptococcus cremoris.* Nucl. Acids Res. 10:8171–8179.

Food and Agriculture Organization of the United Nations. 1987. 1986 FAO production yearbook, vol. 41. FAO Statistics Series No. 82. Food and Agriculture Organization, Rome.

Froseth, B. R., S. K. Harlander, and L. L. McKay. 1988. Plasmid-mediated reduced phage sensitivity in *Streptococcus lactis* KR5. J. Dairy Sci. 71:275–284.

Gasson, M. J. 1980. Production, regeneration and fusion of protoplasts in lactic streptococci. FEMS Microbiol. Lett. 9:99–102.

Gasson, M. J. 1983. Plasmid complements of *Streptococcus lactis* NCDO 712 and other lactic streptococci after protoplast-induced curing. J. Bacteriol. 154:1–9.

Gasson, M. J. 1984. Transfer of sucrose fermenting ability, nisin resistance and nisin production into *Streptococcus lactis* 712. FEMS Microbiol. Lett. 21:7–10.

Gasson, M. J., and P. H. Anderson. 1985. High copy number plasmid vectors for the use in lactic streptococci. FEMS Microbiol. Lett. 30:193–196.

Gasson, M. J., and F. L. Davies. 1980a. Conjugal transfer of the drug resistance plasmid pAMβ in the lactic streptococci. FEMS Microbiol. Lett. 7:51–53.

Gasson, M. J., and F. L. Davies. 1980b. High-frequency conjugation associated with *Streptococcus lactis* donor cell aggregation. J. Bacteriol. 143:1260–1264.

Gasson, M. J., and F. L. Davies. 1980c. Prophage-cured derivatives of *Streptococcus lactis* and *Streptococcus cremoris.* Appl. Environ. Microbiol. 40:964–966.

Gasson, M. J., S. H. A. Hill, and P. H. Anderson. 1987. Molecular genetics of metabolic traits in lactic streptococci. p. 242–245. In: J. J. Ferretti, and R. E. Curtiss III (ed.), Streptococcal genetics. American Society for Microbiology, Washington, D.C.

Galesloot, T. E., F. Hassing, and J. Stadhouders. 1961. Agar media voor het isoleren en tellen van aromabacterien in zuursels. Neth. Milk Dairy J. 15:127–150.

Gautier, M., and M.-C. Chopin. 1987. Plasmid-determined systems for restriction and modification activity and abortive infection in *Streptococcus cremoris.* Appl. Environ. Microbiol. 53:923–927.

Gautier, M., M. Veaux, and M.-C. Chopin. 1987. Cloning of three plasmid-determined systems for restriction/modification and abortive infection. FEMS Microbiol. Rev. 46:P44.

Geis, A. 1982. Transfection of protoplasts of *Streptococcus lactis* subsp. *diacetylactis.* FEMS Microbiol. Lett. 15:119–122.

Geis. A. 1984. Molecular cloning of plasmid DNA from lactic acid streptococci and bacteriophage DNA in *E. coli.* Bundesanstalt fur Milchforschung, Annual Report 1984, B 32.

Geis, A., J. Singh, and M. Teuber. 1983. Potential of lactic streptococci to produce bacteriocin. Appl. Environ. Microbiol. 45:205–211.

Georghiou, D., S. H. Phua, and E. Terzaghi. 1981. Curing of a lysogenic strain of *Streptococcus cremoris* and characterization of the temperate bacteriophage. J. Gen. Microbiol. 122:295–303.

Giesecke, D., and H. K. Henderickx. 1973. Biologie und Biochemie der mikrobiellen Verdauung, p. 294. BLV Verlagsgesellschaft, Bern, Vienna.

Gilliland, S. E., and M. L. Speck. 1974. Frozen concentrated cultures of lactic starter bacteria. A review. J. Milk Food Technol. 37:107–111.

Goelling, D., and U. Stahl. 1988. Cloning and expression of an α-acetolactate decarboxylase gene from *Streptococcus lactis* subsp. *diacetylactis* in *Escherichia coli.* Appl. Environ. Microbiol. 54:1889–1891.

Gonzales, C. F., and B. S. Kunka. 1985. Transfer of sucrose-fermenting ability and nisin production phenotype among lactic streptococci. Appl. Environ. Microbiol. 49:627–633.

Gordon, J. F., and N. Shapton. 1977. Characteristics and use of starters for the manufacture of yoghurt, cottage cheese, cultured buttermilk and other fermented products. J. Soc. Dairy Technol. 30:15–22.

Haandrikman, A. J. J. Kok, H. Laan, S. Soemitro, A. M. Ledeboer, W. N. Konings, and G. Venema. 1989. Identification of a gene required for maturation of an extracellular lactococcal serin proteinase. J. Bacteriol. 171:2789–2794.

Hahn, G., and A. Tolle. 1979. Ergebnisse aus der Strepto-kokken-Zentrale in Kiel von 1965 bis 1977—Ein Über-blick. Zentralbl. Bakteriol., Parasitenk., Infektionskr., Hyg., Abt. 1 Orig., Reihe A 244:427–438.

Hardie, J. M. 1986. Genus Streptococcus Rosenbach 1884, 22, p. 1043–1071. In: Sneath, P. H. A., N. S. Mair, M. E. Sharpe, and J. G. Holt (ed.), Bergey's manual of sys-tematic bacteriology, vol. 2. Williams & Wilkins, Bal-timore.

Harlander, S. K. 1987. Transformation of Streptococcus lac-tis by electroporation, p.229–233. In: J. J. Ferretti and R. Curtiss III (ed.), Streptococcal genetics. American Society for Microbiology, Washington, D.C.

Harlander, S. K., L. L. McKay, and C. F. Schachtele. 1984. Molecular cloning of the lactose-melabolizing genes from Streptococcus lactis. Appl. Environ. Microbiol. 48:347–351.

Heap, H. A., and R. C. Lawrence. 1976. The selection of starter strains for cheesemaking. N.Z.J. Dairy Sci. Technol. 11:16–20.

Higgins, D. L., R. B. Sanozky-Dawes, and T. R. Klaenham-mer. 1988. Restriction and modification activities from Streptococcus lactis ME2 are encoded by a self-trans-missible plasmid, pTN20, that forms cointegrates dur-ing mobilization of lactose-fermenting ability. J. Bac-teriol. 170:3435–3442.

Hill, C., C. Daly, and G. F. Fitzgerald. 1985. Conjugative transfer of transposon Tn919 to lactic acid bacteria. FEMS Microbiol. Lett. 30:115–119.

Hill, C., C. Daly, and G. F. Fitzgerald. 1987. Development of high-frequency delivery system for transposon Tn919 in lactic streptococci: random insertion in Strep-tococcus lactis subsp. diacetylactis 18–16. Appl. Envi-ron. Microbiol. 53:74–78.

Hill, C., D. A. Romero, D. S. McKenney, K. R. Finer, and T. R. Klaenhammer. 1989. Localization, cloning, and expression of genetic determinants for bacteriophage resistance (Hsp) from the conjugative plasmid pTR2030. Appl. Environ. Microbiol. 55:1684–1689.

Hill, C., G. Venema, C. Daly, and G. F. Fitzgerald. 1988. Cloning and characterization of the tetracycline resist-ance determinant of and several promoters from within the conjugative transposon Tn919. Appl. Environ. Mi-crobiol. 54:1230–1236.

Huggins, A. R. 1984. Progress in dairy starter culture tech-nology. Food. Technol. 38:41–50.

Hull, R. R. 1977. Methods for monitoring bacteriophage in cheese factories. Aust. J. Dairy Technol. 32:63–64.

Inamine, J. M., L. N. Lee, and D. . LeBlanc. 1986. Molec-ular and genetic characterization of lactose-metabolic genes of Streptococcus cremoris. J. Bacteriol. 167:855–862.

Jarvis, A. W. 1977. The serological differentiation of lactic streptococcal bacteriophages. N.Z.J. Dairy Sci. Tech-nol. 12:176–181.

Jarvis, A. W. 1981. The use of whey-derived phage-resist-ance starter strains in New Zealand cheese plants. N.Z.J. Dairy Sci. Technol. 16:25–31.

Jarvis, A. W. 1984a. Differentiation of lactic streptococcal phages into phage species by DNA-DNA homology. Appl. Environ. Microbiol. 47:343–349.

Jarvis, A. W. 1984b. DNA-DNA homology between lactic streptococci and their temperate and lytic phages. Appl. Environ. Microbiol 47:1031–1038.

Jarvis, A. W. 1988. Conjugal transfer in lactic streptococci of plasmid-encoded insensitivity to prolate- and small isometric-headed bacteriophages. Appl. Environ. Mi-crobiol. 54:777–783.

Jarvis, A. W., and T. R. Klaenhammer. 1986. Bacteriophage resistance conferred on lactic streptococci by the con-jugative plasmid pTR2030: effects on small isometric-, large isometric-, and prolate-headed phages. Appl. En-viron. Microbiol. 51:1272–1277.

Jarvis, A. W., and J. Meyer. 1986. Electron microscopic het-eroduplex study and restriction endonuclease cleavage analysis of the DNA genomes of three lactic strepto-coccal bacteriophages. Appl. Environ. Microbiol. 51:566–571.

Jarvis, A. W., and J. M. Wolff. 1979. Grouping of lactic streptococci by gel electrophoresis of soluble cell ex-tracts. Appl. Environ. Microbiol. 37:391–398.

Jespersen, N. J. T. 1977. The use of commercially available concentrated starters. J. Soc. Dairy Technol. 30:47–51.

Josephsen, J., and F. K. Vogensen. 1989. Identification of three different plasmid-encoded restriction/modifica-tion systems in Streptococcus lactis subsp. cremoris W56. FEMS Microbiol. Lett. 59:161–166.

Kaletta, C., and K.-D. Entian. 1989. Nisin, a peptide anti-biotic: Cloning and sequencing of the nisA gene and posttranslational processing of its peptide product. J. Bacteriol. 171:1597–1601.

Kempler, G. M., and L. L. McKay. 1979a. Characterization of plasmid deoxyribonucleic acid in Streptococcus lactis subsp. diacetylactis: evidence for plasmid-linked citrate utilization. Appl. Environ. Microbiol. 37:316–323.

Kempler, G. M., and L. L. McKay. 1979b. Genetic evidence for plasmid-linked lactose metabolism in Streptococcus lactis subsp. diacetylactis. Appl. Environ. Microbiol. 37:104–1043.

Keogh, B. P., and G. Pettingill. 1983. Adsorption of bac-teriophage eb7 on Streptococcus cremoris EB7. Appl. Environ. Microbiol. 45:1946–1948.

Keogh, B. P., and P. D. Shimmin. 1974. Morphology of the bacteriophages of lactic streptococci. Appl. Microbiol. 27:411–415.

Kiefer-Partsch, B., W. Bockelmann, A. Geis, and M. Teuber. 1989. Purification of an X-prolyl-dipeptidyl aminopep-tidase from the cell wall proteolytic system of Lacto-coccus lactis subsp. cremoris. Appl. Microbiol. Bio-technol. 31:75–78.

King, W. R., E. B. Collins, and E. L. Barrett. 1983. Fre-quencies of bacteriophage-resistant and slow acid-pro-ducing variants of Streptococcus cremoris. Appl. En-viron. Microbiol. 45:1481–1485.

Klaenhammer, T. R., and R. B. Sanozky. 1985. Conjugal transfer from Streptococcus lactis ME2 of plasmid en-coding phage resistance, nisin resistance and lactose fermenting ability: evidence for a high-frequency con-jugative plasmid responsible for abortive infection of virulent bacteriophage. J. Gen. Microbiol. 131:1531–1541.

Knittel, M. D. 1965. Genetic homology and exchange in lactic acid streptococci. Ph.D. Thesis. Oregon State University.

Kok, J., J. M. B. M. van der Vossen, and G. Venema. 1984. Construction of plasmid cloning vectors for lactic strep-tococci which also replicate in Bacillus subtilis and Escherichia coli. Appl. Environ. Microbiol. 48:726–731.

Kok, J., J. M. van Dijl, J. M. B. M. van der Vossen, and G. Venema. 1985. Cloning and expression of a Strepto-

coccus cremoris proteinase in *Bacillus subtilis* and *Streptococcus lactis*. Appl. Environ. Microbiol. 50:94–101.

Kok, J., and G. Venema. 1988. Genetics of proteinase of lactic acid bacteria. Biochimie 70:475–488.

Kondo, J. K., and L. L. McKay. 1982. Transformation of *Streptococcus lactis* protoplasts by plasmid DNA. Appl. Environ. Microbiol. 43:1213–1215.

Kondo, J. K., and L. L. McKay. 1984. Plasmid transformation of *Streptococcus lactis* protoplasts: Optimization and use in molecular cloning. Appl. Environ. Microbiol. 48:252–259.

Kuhl, S. A., L. D. Larsen, and L. L. McKay. 1979. Plasmid profiles of lactose-negative and proteinase-deficient mutants of *Streptococcus lactis* C10, ML3, and M18. Appl. Environ. Microbiol. 37:1193–1195.

Laible, N. J., P. L. Rule, S. K. Harlander, and L. L. McKay. 1987. Identification and cloning of plasmid deoxyribonucleic acid coding for abortive phage infection from *Streptococcus lactis* subsp. *diacetylactis* KR2. J. Dairy Sci. 70:2211–2219.

Lakshmidevi, G., B. E. Davidson, and A. J. Hillier. 1988. Circular permutation of the genome of a temperate bacteriophage from *Streptococcus cremoris* BK5. Appl. Environ. Microbiol. 54:1039–1045.

Lancefield, R. C. 1933. A serological differentiation of human and other groups of hemolytic streptococci. J. Exp. Medicine 57:571–595.

Lautier, M., and G. Novel. 1987. DNA-DNA hybridizations among lactic streptococcal temperate and virulent phages belonging to distinct lytic groups. J. Industr. Microbiol. 2:151–158.

Lawrence, R. C. 1978. Action of bacteriophage on lactic acid bacteria: consequences and protection. N.Z.J. Dairy Sci. Technol. 13:129–136.

Lawrence, R. C., T. D. Thomas, and B. E. Terzaghi. 1976. Reviews of the progress of dairy science: cheese starters. J. Dairy Res. 43:141–193.

LeBlanc, D. J., and L. N. Lee. 1979. Rapid screening procedure for detection of plasmids in streptococci. J. Bacteriol. 140:1112–1115.

LeBourgeois, P., M. Mata, and Paul Ritzenthaler. 1989. Genome comparison of *Lactococcus* strains by pulsed-field gel electrophoresis. FEMS Microbiology Letters 59:65–70.

Lembke, J., U. Krusch, A. Lompe, and M. Teuber. 1980. Isolation and ultrastructure of bacteriophages of group N (lactic) streptococci. Zentralbl. Bakteriol. Hyg. I Abt. Orig. C1:79–91.

Lembke, J., and M. Teuber. 1979. Detection of bacteriophages in whey by an enzyme-linked immunosorbent assey (ELISA). Milchwissenschaft 34:457–458.

Lembke, J., and M. Teuber. 1981a. Inaktivierung von Bakteriophagen durch Desinfektionsmittel. Deutsche Molkerei-Zeitung 102:2–6.

Lembke, J., and M. Teuber. 1981b. Serotyping of morphologically identical bacteriophages of lactic streptococci by immunoelectronmicroscopy. Milchwissenschaft 36:10–12.

Lewis, J. E. 1977. Starter manufacture at individual cheese factories. J. Soc. Dairy Technol. 30:32–35.

Limsowtin, G. K. Y., V. L. Grow, and L. E. Pearce. 1986. Molecular cloning and expression of the *Streptococcus lactis* tagatose 1,6 bisphosphate aldolase gene in *Escherichia coli*. FEMS Microbiol. Lett. 33:79–83.

Limsowtin, G. K. Y., and B. E. Terzaghi. 1976. Phage resistant mutants: their selection and use in cheese factories. N.Z.J. Dairy Sci. Technol. 11:251–256.

Lister, J. 1873. A further contribution to the natural history of bacteria and the germ theory of fermentative changes. Quart. Microbiol. Sci. 13:380–408.

Löhnis, F. 1909. Die Benennung der Milchsäurebakterien. Zentralbl. Bakteriol. Parasitenk. Infektionskr. und Hyg., Abt. 2 22:553–555.

Loof, M., J. Lembke, and M. Teuber. 1983. Characterization of the genome of the *Streptococcus lactis* subsp. *diacetylactis* bacteriophage P008 wide-spread in German cheese factories. Syst. Appl. Microbiol. 4:413–423.

Loof, M., and M. Teuber. 1986. Heteroduplex analysis of the genomes of *Streptococcus lactis* "subsp. *diacetylactis*" bacteriophages of the P008-type isolated from German cheese factories. Syst. Appl. Microbiol. 8:226–229.

Loureiro dos Santos, A. L., and A. Chopin. 1987. Shotgun cloning in *Streptococcus lactis*. FEMS Microbiol. Lett. 42:209–212.

Maeda, S., and M. J. Gasson. 1986. Cloning, expression and localization of the *Streptococcus lactis* gene for phospho-β-D-galactosidase. J. Gen. Microbiol. 132:331–340.

McKay, L. L., and K. A. Baldwin. 1974. Simultaneous loss of proteinase- and lactose-utilizing enzyme activities in *Streptococcus lactis* and reversal of loss by transduction. Appl. Microbiol. 28:342–346.

McKay, L. L., and K. A. Baldwin. 1984. Conjugative 40-megadalton plasmid in *Streptococcus lactis* subsp. *diacetylactis* DRC3 is associated with resistance to nisin and bacteriophage. Appl. Environ. Microbiol. 47:68–74.

McKay, L. L., K. A. Baldwin, and P. M. Walsh. 1980. Conjugal transfer of genetic information in group N streptococci. Appl. Environ. Microbiol. 40:84–91.

McKay, L. L., B. R. Cords, and K. A. Baldwin. 1973. Transduction of lactose metabolism in *Streptococcus lactis* C2. J. Bacteriol. 115:810–815.

Möller, V., and M. Teuber. 1988. Selection and characterization of phage-resistant mesophilic lactococci from mixed-strain dairy starter cultures. Milchwissenschaft 43:482–486.

Murphy, M. C., J. L. Steele, C. Daly, and L. L. McKay. 1988. Concomitant conjugal transfer of reduced-bacteriophage-sensitivity mechanisms with lactose- and sucrose-fermenting ability in lactic streptococci. Appl. Environ. Microbiol. 54:1951–1956.

Neve, H., A. Geis, and M. Teuber. 1984. Conjugal transfer and characterization of bacteriocin plasmids in group N (lactic acid) streptococci. J. Bacteriol. 157:833–838.

Neve, H., A. Geis, and M. Teuber. 1987. Conjugation, a common plasmid transfer mechanism in lactic acid streptococci of dairy starter cultures. Syst. Appl. Microbiol. 9:151–157.

Nickels, C., and H. Leesment. 1964. Methode zur Differenzierung und quantitativen Bestimmung von Säureweckerbakterien. Milchwissenschaft 19:374–378.

Novak, S. R., and C. A. Batt. 1987. Expression of the intensely sweet protein thaumatin in *Streptococcus lactis*. FEMS Microbiol. Rev. 46:P15.

Okamoto, T., Y. Fujita, and R. Irie. 1983. Fusion of protoplasts of *Streptococcus lactis*. Agric. Biol. Chem. 47:2675–2676.

Olson, N. F., R. F. Anderson., and R. Sellars. 1978. Microbiological methods for cheese and other cultured products, p. 161–164. In: Marth, E. H. (ed.), Standard methods for the examination of dairy products, 14th ed. American Public Health Association, Washington, D.C.

Oram, J. D. 1971. Isolation and properties of a phage receptor substance from the plasma membrane of *Streptococcus lactis* ML3. J. Gen. Virol. 13:59–71.

Oram, J. D., and B. Reiter. 1968. The adsorption of phage to group N streptococci. The specificity of adsorption and the location of phage receptor substances in cell-wall and plasma-membrane fractions. J. Gen. Virol. 3:103–119.

Orla-Jensen, S. 1919. The lactic acid bacteria. Host & Son, Copenhagen.

Osborne, R. J. W. 1977. Production of frozen concentrated cheese starters by diffusion cultures. J. Soc. Dairy Techn. 30:40–44.

Otto, R., W. M. de Vos, and J. Gavrieli. 1982. Plasmid DNA in *Streptococcus cremoris* Wg2: influence of pH on selection in chemostats of a variant lacking a protease plasmid. Appl. Environ. Microbiol. 43:1272–1277.

Park, C., and L. L. McKay. 1975. Induction of prophage in lactic streptococci isolated from commercial dairy starter cultures. J. Milk Food Technol. 38:594–597.

Pearce, L. E. 1978. The effect of host-controlled modification on the replication rate of a lactic streptococcal bacteriophage. N. Z. J. Dairy Sci. Technol. 13:166–171.

Pechmann, H., and M. Teuber. 1980. Plasmid pattern of group N (lactic) streptococci. Zentralbl. Bakteriol. Internat. J. of Mikrobiol. and Hyg., Abt. 1 Orig., Reihe C 1:133–136.

Potter, N. N., and F. E. Nelson. 1952. Effects of calcium on proliferation of lactic streptococcus bacteriophage. I. Studies on plaque formation with modified plating technique. J. Bacteriol. 64:105–111.

Powell, I. B., M. G. Achen, A. J. Hillier, and B. E. Davidson. 1988. A simple and rapid method for genetic transformation of lactic streptococci by electroporation. Appl. Environ. Microbiol. 54:655–660.

Powell, I. B., and B. E. Davidson. 1985. Characterization of streptococcal bacteriophage c6A. J. Gen. Virol. 66:2737–2741.

Powell, I. B., and B. E. Davidson. 1986. Resistance to in vitro restriction of DNA from lactic streptococcal bacteriophage c6A. Appl. Environ. Microbiol. 51:1358–1360.

Reiter, B. 1949. Lysogenic strains of lactic streptococci. Nature 164:667–668.

Relano, P., M. Mata, M. Bonneau, and P. Ritzenthaler. 1987. Molecular characterization and comparison of 38 virulent and temperate bacteriophages of *Streptococcus lactis*. J. Gen. Microbiol. 133:3053–3063.

Renault, P., C. Gaillardin, and H. Heslot. 1989. Product of the *Lactococcus lactis* gene required for malolactic fermentation is homologous to a family of positive regulators. J. Bacteriol. 171:3108–3114.

Renault, P., and H. Heslot. 1987. Selection of *Streptococcus lactis* mutants defective in malolactic fermentation. Appl. Environ. Microbiol. 53:320–324.

Sanders, M. E., and T. R. Klaenhammer. 1980. Restriction and modification in group N streptococci: Effect of heat on development of modified lytic bacteriophage. Appl. Environ. Microbiol. 40:500–506.

Sanders, M. E., and T. R. Klaenhammer. 1981. Evidence for plasmid linkage of restriction and modification in *Streptococcus cremoris* KH. Appl. Environ. Microbiol. 42:944–950.

Sanders, M. E., and T. R. Klaenhammer. 1983. Characterization of phage-sensitive mutants from a phage-insensitive strain of *Streptococcus lactis*: evidence for a plasmid determinant that prevents phage adsorption. Appl. Environ. Microbiol. 46:1125–1133.

Sanders, M. E., P. J. Leonhard, W. D. Sing, and T. R. Klaenhammer. 1986. Conjugal strategy for construction of fast acid-producing, bacteriophage-resistant lactic streptococci for use in dairy fermentations. Appl. Environ. Microbiol. 52:1001–1007.

Sandine, W. E. 1977. New techniques in handling lactic cultures to enhance their performance. J. Dairy Sci. 60:822–828.

Sandine, W. E., P. R. Elliker, L. K. Allen, and W. C. Brown. 1962. Genetic exchange and variability in lactic *Streptococcus* starter organisms. J. Dairy Sci. 45:1266–1271.

Sandine, W. E., P. C. Radich, and P. R. Elliker. 1972. Ecology of the lactic streptococci. A review J. Milk Food Technol. 35:176–184.

Scherwitz, K. M., K. A. Baldwin, and L. L. McKay. 1983. Plasmid linkage of a bacteriocin-like substance in *Streptococcus lactis* subsp. *diacetylactis* WM4: transferability to *Streptococcus lactis*. Appl. Environ. Microbiol. 45:1506–1512.

Scherwitz-Harmon, K., and L. L. McKay. 1987. Restriction enzyme analysis of lactose and bacteriocin plasmids from *Streptococcus lactis* subsp. *diacetylactis* WM4 and cloning of *Bcl*I fragments coding for bacteriocin production. Appl. Environ. Microbiol. 53:1171–1174.

Schleifer, K. H. 1987. Recent changes in the taxonomy of lactic acid bacteria. FEMS Microbiol. Rev. 46:201–203.

Schleifer, K. H., and O. Kandler. 1972. Peptidoglycan types of bacterial cell walls and their taxonomic implications. Bacteriol. Rev. 36:407–477.

Schleifer, K. H., J. Kraus, G. Dvorak, R. Kilpper-Bälz, M. D. Collins, and W. Fischer. 1985. Transfer of *Streptococcus lactis* and related streptococci to the genus *Lactococcus* gen. nov. Syst. Appl. Microbiol. 6:183–195.

Shankar, P. A., and F. L. Davies. 1977. A note on the suppression of *Lactobacillus bulgaricus* in media containing β-glycerophosphate and application of such media to selective isolation of *Streptococcus thermophilus* from yoghurt. J. Soc. Dairy Technol. 30:28–30.

Simon, D., A. Rouault, and M.-C. Chopin. 1985. Protoplast transformation of group N streptococci with cryptic plasmids. FEMS Microbiol. Lett. 26:239–241.

Simon, D., A. Rouault, and M.-C. Chopin. 1986. High-efficiency transformation of *Streptococcus lactis* protoplasts by plasmid DNA. Appl. Environ. Microbiol. 52:394–395.

Simons, G., G. Rutten, M. Hornes, and W. de Vos. 1988. Production of bovine prochymosin by lactic acid bacteria. Proc. 2nd Netherlands Biotechnol. Congress 1988:183–187.

Singh, W. D., and T. R. Klaenhammer. 1986. Conjugal transfer of bacteriophage resistance determinants on pTR2030 into *Streptococcus cremoris* strains. Appl. Environ. Microbiol. 51:1264–1271.

Stackebrandt, E., and M. Teuber. 1988. Molecular taxonomy and phylogenetic position of lactic acid bacteria. Biochimie 70:317–324.

Stadhouders, J. 1975. Microbes in milk and dairy products. An ecological approach. Neth. Milk Dairy J. 29:104–126.

Stadhouders, J. 1986. The control of cheese starter activity. Neth. Milk Dairy J.40:155–173.

Stadhouders, J., L. A. Jansen, and G. Hup. 1969. Preservation of starters and mass production of starter bacteria. Neth. Milk Dairy J.23:182–199.

Stanley, G. 1977. The manufacture of starters by batch fermentation and centrifugation to produce concentrates. J. Soc. Dairy Technol. 30:36–39.

Tamime, A. Y., and R. K. Robinson. 1976. Recent developments in the production and preservation of starter cultures for yogurt. Dairy Industries International 41:408–411.

Terzaghi, B. E. 1976. Morphologies and host sensitivities of lactic streptococcal phages from cheese factories. N.Z.J. Dairy Sci. Technol. 11:155–163.

Terzaghi, B. E., and W. E. Sandine. 1975. Improved medium for lactic streptococci and their bacteriophages. Appl. Microbiol. 29:807–813.

Terzaghi, B. E., and W. E. Sandine. 1981. Bacteriophage production following exposure of lactic streptococci to ultraviolet radiation. J. Gen. Microbiol. 122:305–311.

Teuber, M., and J. Lembke. 1983. The bacteriophages of lactic acid bacteria with emphasis on genetic aspects of group N lactic acid streptococci. Antonie van Leeuwenhoek 49:283–295.

Thomas, T. D., and G. G. Pritchard. 1987. Proteolytic enzymes of dairy starter cultures. FEMS Microbiol. Rev. 46:245–268.

van Belkum, M. J., B. J. Hayema, A. Geis, J. Kok, and G. Venema. 1989. Cloning of two bacteriocin genes from a lactococcal bacteriocin plasmid. Appl. Environ. Microbiol. 55:1187–1191.

van der Guchte, M., J. M. B. M. van der Vossen, J. Kok, and G. Venema. 1989. Construction of a lactococcal expression vector: expression of hen egg white lysozyme in *Lactococcus lactis* subsp. *lactis*. Appl. Environ. Microbiol. 55:224–228.

van der Lelie, D., J. M. B. M. van der Vossen, and G. Venema. 1988. Effect of plasmid incompatibility on DNA transfer to *Streptococcus cremoris*. Appl. Environ. Microbiol. 54:865–871.

van der Vossen, J. M. B. M., J. Kok, and G. Venema. 1985. Construction of cloning, promoter-screening and terminator-screening shuttle vectors for *Bacillus subtilis* and *Streptococcus lactis*. Appl. Environ Microbiol. 50:540–542.

van der Vossen, J. M. B. M., J. Kok, D. van der Lelie, and G. Venema. 1988. Liposome-enhanced transformation of *Streptococcus lactis* and plasmid transfer by intergeneric protoplast fusion of *Streptococcus lactis* and *Bacillus subtilis*. FEMS Microbiol. Lett. 49:323–329.

Vedamuthu, E. R., and J. M. Neville. 1986. Involvement of a plasmid in production of ropiness (mucoidness) in milk cultures by *Streptococcus cremoris*. Appl. Environ. Microbiol. 51:677–682.

Vogensen, F. K., T. Karst, J. J. Larsen, B. Kringelum, D. Ellekjaer, and E. Waagner Nielsen. 1987. Improved direct differentiation between *Leuconostoc cremoris*, *Streptococcus lactis* and *Streptococcus cremoris/Streptococcus lactis* on agar. Milchwissenschaft 42:646–648.

von Wright, A., A.-M. Taimisto, and S. Sivela. 1985. Effect of Ca^{2+} ions on plasmid transformation of *Streptococcus lactis* protoplasts. Appl. Environ. Microbiol. 50:1100–1102.

Vos, P., M. van Asseldonk, F. van Joveren, R. Siezen, G. Simon, and W. M. deVos. 1989. A maturation protein is essential for production of active forms of *Lactococcus lactis* SK11 serin proteinase located in or secreted from the cell envelope. J. Bacteriol. 171:2795–2802.

Weigmann, H. 1905–1908. Das Reinzuchtsystem in der Butterbereitung und in der Käserei, p. 293–309. In: Lafar, F. (ed), Handbuch der Technischen Mykologie, vol. 2. Mykologie der Nahrungsmittelgewerbe. Gustav Fischer Verlag, Jena, Germany.

Weiler, H. G., and F. Radler. 1970. Milchsäurebakterien aus Wein und von Rebenblättern, Zentralbl. Bakteriol., Parasitenk., Infektionskr. Hyg., Abt. 2 Orig. 124:707–732.

Wetzel, A., H. Neve, A. Geis, and M. Teuber. 1986. Transfer of plasmid-mediated phage resistance in lactic acid streptococci. Chem. Mikrobiol. Technol. Lebensm. 10:86–89.

Whitehead, H. R., and G. A. Cox. 1935. The occurrence of bacteriophage in cultures of lactic streptococci. N.Z.J. Sci. Technol. 16:319–320.

Whittenbury, R. 1965. The enrichment and isolation of lactic acid bacteria from plant material. Zentralbl. Bakteriol., Parasitenk., Infektionskr. Hyg., Abt. 1 Suppl. 1:395–398.

Wigley, R. C. 1977. The use of commercially available concentrated starters. J. Soc. Dairy Technol. 30:45–47.

Woskow, S. A., and J. K. Kondo. 1987. Effect of proteolytic enzymes on transfection and transformation of *Streptococcus lactis* protoplasts. Appl. Environ. Microbiol. 53:2583–2587.

Yu, R. S. T., T. V. Hung, and A. A. Azad. 1982. Rapid screening of highly purified plasmids in lactic streptococci. Austr. J. Dairy Technol. 37:99–103.

The Genera *Pediococcus* and *Aerococcus*

NORBERT WEISS

In the first edition of *The Prokaryotes,* the two genera *Pediococcus* and *Aerococcus* were treated in two different chapters: "The Family Streptococcaceae (Nonmedical Aspects)" (Teuber and Geis, 1981) and "The Family Streptococcaceae (Medical Aspects)" (Facklam and Wilkinson, 1981), respectively. However, studies of 16s rRNA cataloging and sequencing have shown that *Pediococcus pentosaceus* (Stackebrandt et al., 1983) and its nearest relative *P. acidilactici* (D. Yang and C.R. Woese, personal communication; see Fig. 5 in Chapter 69) should be placed in the phylogenetic cluster formed by the lactobacilli and the leuconostocs. This conclusion was confirmed by Collins et al. (1990), who determined nucleotide sequences of the type strains of all the presently recognized species of the genus *Pediococcus* except *Pediococcus inopinatus.* Collins et al. (1990) also demonstrated that all aciduric, facultatively anaerobic species form a phylogenetically tight group, with only P. *dextrinicus* being somewhat peripheral, as shown in Table 1). Because of its relatively high DNA-DNA homology to both *P. damnosus,* the type species of the genus, and *P. parvulus* (Back and Stackebrandt, 1978), one can assume that *P. inopinatus* also belongs to this group. The nonaciduric, microaerophilic species *P. urinaeequi* turned out to be phylogenetically almost identical with *Aerococcus viridans,* as had already been assumed by many workers on the basis of the biochemical-cultural similarities summarized by Bergan et al. (1984). *A. viridans* (synonyms: "*Gaffkya homari,*" "*Pediococcus homari*"), the only recognized species within the genus, is not specifically related to any of the lactic-acid-producing genera and represents a single line of descent (Stackebrandt and Teuber, 1988; Collins et al., 1990). For the salt-requiring, nonaciduric, facultatively anaerobic species *P. halophilus* (synonym: "*P. soyae*"), no close phylogenetic relatives could be detected. Its reclassification in the new genus *Tetragenococcus* therefore has been proposed (Collins et al., 1990).

In the near future, the redefined genus *Pediococcus* should be restricted to aciduric, facultatively anaerobic species, which should provisionally be called the "true" pediococci. For the time being and for practical reasons, it seems justified to discuss all the Gram-positive, tetrad-forming, catalase-negative, lactic-acid-producing cocci within one chapter of this handbook.

The Genus Pediococcus

Habitats

Pediococci, mainly of the species *P. pentosaceus* and *P. acidilactici,* occur on a great variety of

Table 1. Examples of homology values for a 1,340-nucleotide region of 16s rRNAs of pediococci, *Aerococcus viridans,* and *Lactobacillus pentosus.*

	1	2	3	4	5	6	7	8	9
1. *P. damnosus*	100								
2. *P. acidilactici*	96.6	100							
3. *P. pentosaceus*	96.5	98.3	100						
4. *P. parvulus*	98.7	97.0	96.7	100					
5. *P. dextrinicus*	94.0	93.8	93.2	94.5	100				
6. *L. pentosus*	94.6	94.8	94.4	94.6	93.5	100			
7. *A. viridans*	89.9	89.3	89.0	89.6	89.6	89.3	100		
8. *P. urinaeequi*	89.3	90.3	89.6	89.8	90.5	90.2	99.9	100	
9. *P. halophilus*	88.7	89.7	88.3	87.4	88.6	88.8	89.7	90.4	100

Adapted from M. D. Collins, personal communication.

plants and fruits, although only in small numbers (Mundt et al., 1969; Back, 1978). In fermenting plant materials, such as silage, cucumbers, or olives, the pediococci often multiply rapidly, becoming a major component of the lactic acid flora primarily in the early stages of the fermentation process (Vaughn, 1975; Dellaglio and Torriani, 1986). *P. parvulus, P. inopinatus,* and *P. dextrinicus* have been isolated from silage, sauerkraut, and fermented beans; the latter two species have also occasionally been found in beer (Back, 1978). Pedioccocci occur also in proteinaceous foods as such as fresh and cured meat, raw sausages (Reuter, 1970), fresh and marinated fish (Blood, 1975), and cheese (Dacre, 1958). *P. damnosus* is found mainly in beer and in the brewery environment, but it has occasionally also been isolated from wine and cider. Growth in beer causes cloudiness and other faults. In modern brewing, spoilage problems with pediococci have been overcome mainly by advanced contamination control measures. The salt-tolerant *P. halophilus* is found in soy sauce and pickling brines (Back, 1978). Within the last few years, pediococci have been given more attention in clinical microbiology (Colman and Efstratiou, 1987; Ruoff et al., 1988; Facklam et al., 1989; Riebel and Washington, 1990). Their role as possible pathogens, however remains questionable (Riebel and Washington, 1990).

Isolation

Because of variations in cultural characteristics, no general isolation procedure can be recommended. Pediococci from plants and fermenting plant materials can be isolated on MRS agar (de Man et al., 1960) or SL acetate agar (Rogosa et al., 1951) can be used.

MRS Agar for Isolating and Propagating
Pediococci (de Man et al., 1960)

For 1 liter of medium, dissolve 15 g agar in distilled water by steaming. Add:

Oxoid peptone	10.00 g
Meat extract	10.00 g
Yeast extract	5.00 g
KH_2PO_4	2.00 g
Diammonium citrate	2.00 g
Glucose	20.00 g
Tween 80	1.00 ml
Na acetate·$3H_2O$	5.00 g
$MgSO_4$·$7H_2O$	0.50 g
$MnSO_4$·H_2O	0.25 g
Distilled water	1 liter

Adjust pH to 6.2 to 6.4. Sterilize at 121°C for 15 min.

Selective SL Medium for Isolating Pediococci
(Rogosa et al., 1951)

For 1 liter of medium:

Trypticase	10.00 g
Yeast extract	5.00 g
KH_2PO_4	6.00 g
Diammonium citrate	2.00 g
$MgSO_4$·$7H_2O$	0.58 g
$MnSO_4$·H_2O	0.15 g
$FeSO_4$·$7H_2O$	0.03 g
Glucose	20.00 g
Tween 80	1.00 ml
Na acetate·$3H_2O$	25.00 g
Agar	15.00 g

Dissolve the agar in 500 ml water by boiling. Dissolve all other ingredients in 500 ml water, adjust pH to 5.4 using glacial acetic acid and mix with the melted agar. Boil for a further 5 min and pour plates or distribute the hot medium in convenient amounts in sterile screw-capped bottles; no further sterilization is given. Avoid repeated melting and cooling.

Pediococci from beer or brewery habitats can be isolated on MRS agar adjusted to pH 5.5. Back (1978) recommends a 1:1 mixture of MRS medium and beer for fastidious beer-spoiling strains. Plates should be incubated in an atmosphere of 90% N_2 + 10% CO_2 at 22°C. For the isolation of halophilic pediococci from soy sauce and brines, the use of MRS agar adjusted to pH 7.0 and supplemented with 4 to 6% NaCl is recommended. Cycloheximide can be added to all media to suppress yeast growth.

Maintenance

Cultures for storage and conservation should not be allowed to overgrow. All pediococci can be kept as stab cultures in the appropriate medium at 4°C with monthly transfers. An excellent procedure for medium-term conservation (at least one year) is obtained by mixing a well-developed but not overgrown culture with an equal volume of sterile glycerol in a small screw-capped vial and storage at −20°C. Loops of inoculum can thus be taken without exposing the cells to the killing effect of the freezing/thawing procedure. For long-term conservation, lyophilization by usual procedures or storage in liquid nitrogen is advisable.

Identification

Pediococci are Gram-positive, never-elongated cocci dividing alternately in two planes at right angles to form tetrads. Tetrads, however, may not always easily be detectable and often only pairs and clusters of cells can be seen. In the author's experience, for microscopic examinations, cultures grown in MRS broth are better than cultures grown in the media commonly used in most clinical laboratories; also, phase contrast observation is superior to stained preparations. Motility, spore formation, and nitrate reduction are never detected. Cytochromes are absent. Pediococci are catalase-negative, and the benzidine test (Deibel and Evans, 1960) always gives negative results. Strains of some species, however, display pseudocatalase activity especially when grown on media with low carbohydrate content. Glucose is fermented, probably by the Embden-Meyerhof pathway, to DL-, or in the case of *P. dextrinicus* and *P. halophilus,* to L(+)-lactic acid. Gas is not produced from glucose. These characters allow the separation of pediococci from all other Gram-positive cocci with high reliability. Characteristics useful to distinguish pediococci from the most similar aerococci are summarized in Table 2.

Identification of species of the genus *Pediococcus* is best achieved by determining the range of temperature, pH, and NaCl concentration at which growth occurs and by measuring physiological characteristics such as carbohydrate fermentation, hydrolysis of arginine, and the isomer(s) of lactic acid produced. Characteristics that have been proven most useful for the identification key are presented in Table 3. However, since strains which are highly related by DNA-DNA homology may reveal a broad variability within their phenotypic characters (Dellaglio and Torriani, 1986), no single characteristic can be taken as absolutely discriminatory. Thus, one should compare results with all the data compiled in Table 4.

Plasmids

Plasmids have been detected in many strains of *P. pentosaceus* isolated from ensiled high-moisture corn grain (Dellaglio et al., 1984). The number of plasmids in the 16 strains studied varied from none to three, and their molecular

Table 2. Key characteristics for differentiating "true" pediococci from aerococci.

Characteristic	Pediococci	Aerococci
Growth at pH 5	+	−
Growth at pH 9	−	+
Facultative anaerobic	+	−
Microaerophilic	−	+
Peptidoglycan type	Lys-D-Asp	Lys-direct
Major $C_{16:1\omega9cis}$ CFA[a]	−	+
Major $C_{18:1\omega9cis}$ CFA[a]	−	+
Vancomycin (30-μg disc)[b]	Resistant	Sensitive
Pyrrolidonyl-arylase[a,b]	+	−
Leucine aminopeptidase[a]	−	+

Symbols: +, positive or present; −, negative or absent.
[a]CFA, cellular fatty acid; from Bosley et al. (1990).
[b]From Facklam et al. (1989).

weight ranged from 1.3 to 30 MDa. Curing studies with novobiocin gave evidence that the production of a bacteriocin-like substance by "*Pediococcus cerevisiae*" FBB63 may be linked to a 10.5-MDa plasmid (Graham and McKay, 1985). Daeschel and Klaenhammer (1985) reported that production of a bacteriocin named pediocin A was associated with a plasmid in two strains of *P. pentosaceus* isolated from cucumber fermentation. Evidence was obtained that both bacteriocin immunity and bacteriocin production are encoded by the plasmid. The authors also proposed the use of pediocin-producing strains in food fermentations. The abilities to ferment raffinose, melibiose, and sucrose by three strain of *P. pentosaceus* were found to be encoded by plasmids (Gonzalez and Kunka, 1986). Hoover et al. (1986; cited in Raccach, 1987) showed that lactose fermentation in a strain of *P. pentosaceus* is also associated with a plasmid, as is sucrose utilization in a strain of *P. acidilactici.*

Applications

Pediococcal strains, mainly *P. pentosaceus* and *P. acidilactici,* are used for several biotechnological applications such as the processing and preservation of meat and plant food. These organisms play an important role as components of some commercial starter cultures for raw sausages. Biotechnological aspects of pediococci have recently been reviewed with much competence by Raccach (1987).

The Genus *Aerococcus*

The genus *Aerococcus* consists of only one species, *A. viridans.*

Habitats

Aerococci are isolated from sources quite different than those of pediococci, such as from the air in hospital environments (Kerbaugh and Evans, 1968), from vegetation, outdoor air, and dust (Deibel and Niven, 1960), from human clinical specimens (Bosley et al., 1990), and as a marine organism causing a fatal disease of lobsters (Hitchner and Snieszko, 1947; Wiik et al., 1986). The taxonomic kinship of these strains from different sources has been documented repeatedly by physiological, serological, chemotaxonomic, and DNA homology studies, most recently by Bosley et al. (1990).

The medical importance of aerococci as causative agents of different diseases such as osteomyelitis, endocarditis, and septic arthritis is

Table 3. Key for presumptive identification of pediococci.

I. a) ribose fermented, NH$_3$ from arginine	II
I. b) ribose not fermented, no NH$_3$ from arginine	IV
II. a) growth in 15% NaCl, L(+)-lactate produced	*P. halophilus*
II. b) no growth in 15% NaCl, DL-lactate produced	III
III. a) growth at 50°C, maltose not fermented	*P. acidilactici*
III. b) no growth at 50°C, maltose fermented	*P. pentosaceus*
IV. a) starch fermented, L(+)-lactate produced	*P. dextrinicus*
IV. b) starch not fermented, DL-lactate produced	V
V. a) growth at 35°C	VI
V. b) no growth at 35°C	*P. damnosus*
VI. a) lactose fermented	*P. inopinatus*
VI. b) lactose not fermented	*P. parvulus*

Table 4. Differential characteristics of the species of the genus *Pediococcus*.

	1. P. dam-nosus	2. P. parvu-lus	3. P. inopi-natus	4. P. dex-trinicus	5. P. pen-tosaceus	6. P. acidi-lactici	7. P. halo-philus	8. P. uri-naeequi
Growth at pH 4.5	+	+	ND	−	+	+	−	−
Growth at pH 7.0	−	+	+	+	+	+	+	+
Growth at 45°C	−	−	−	−	d	+	−	−
Growth in 10% NaCl	−	−	−	−	d	−	+	−
Pseudocatalase	−	−	ND	−	+	+	−	−
NH$_3$ from arginine	−	−	−	−	+	+	−	−
Acid produced from[a]								
Ribose	−	−	−	−	+	+	+	ND
Arabinose	−	−	−	−	+	d	+	d
Xylose	−	−	−	−	d	+	−	d
Mannitol	−	−	−	−	−	−	−	d
Lactose	−	−	+	d	d	d	−	d
Maltose	d	+	+	+	+	−	+	+
Sucrose	d	−	d	d	−	−	+	+
Trehalose	+	d	+	−	+	d	+	+
Melezitose	d	−	−	−	−	−	+	−
Dextrin	−	−	d	+	−	−	−	+
Starch	−	−	−	+	−	−	−	−
Lactate isomer	DL	DL	DL	L(+)	DL	DL	L(+)	L(+)

Symbols: +, 90% or more strains positive; −, 90% or more strains negative; d, 11–89% strains positive; ND, no data.
[a]Most strains ferment fructose, galactose, glucose, mannose, cellobiose, amygdalin, esculin, and salicin. Most strains do not ferment sorbitol, glycerol, sorbose, inulin, dulcitol, and inositol. Melibiose and raffinose may be fermented by strains of *L. pentosaceus, P. acidilactici,* and *P. urinaeequi.*

documented by many case reports (see Facklam and Wilkinson, 1981; Janosek et al., 1980; Taylor and Trueblood, 1985), but not fully understood.

Isolation and Maintenance of Aerococci

According to Evans (1986), *Aerococcus* strains can be isolated on a wide range of selective and nonselective media generally employed for other Gram-positive cocci such as staphylococci and enterococci. Because of the comparatively slower and weaker growth of aerococci, they may be overlooked in many isolation procedures. Blood agar containing 0.001% potassium tellurite and 0.00025% crystal violet is rec-

ommended as a selective medium. APT or tryptic soy broth (TSB) are satisfactory nonselective media. *Aerococcus* strains can be maintained on APT or TSB agar slants. For medium- and long-term preservation, the procedures recommended above for pediococci can also be used.

Identification

Most criteria that are helpful for separating members of the genus *Aerococcus* from other similar, Gram-positive, catalase-negative cocci have already been discussed above (Table 2). The microaerophilic nature is best detected in agar shake cultures; whereas surface growth is very weak, a heavy band of discrete colonies is produced just beneath the surface. Anaerobic growth is mostly absent or very delayed and consists of only a few single colonies. Distinct tetrads can be observed most reliably in young cultures grown in rich broth media. A weak pseudocatalase activity is often detected. Acetyl-methylcarbinol is not produced. Arginine is not hydrolyzed, and nitrates are not reduced to nitrites. Growth is not prevented by 40% bile, 10% NaCl, 0.01% potassium tellurite, or a pH of 9.6. Acid but no gas is produced from glucose, fructose, galactose, mannose, maltose, and sucrose and usually also from lactose, trehalose, and mannitol. Final pH in glucose broth does not drop below 5.0. L(+)-lactic acid is the main fermentation product from glucose. The peptidoglycan of cell walls is of the L-lysine-direct type.

Literature Cited

Back, W. 1978. Zur Taxonomie der Gattung Pediococcus. Phänotypische und genotypische Abgrenzung der bisher bekannten Arten sowie Beschreibung einer neuen bierschädlichen Art: Pediococcus inopinatus. Brauwiss. 31:237–250, 312–320, 336–343.

Back, W., and E. Stackebrandt. 1978. DNS/DNS-Homologiestudien innerhalb der Gattung *Pediococcus*. Arch. Microbiol. 118:79–85.

Bergan, T., R. Solberg, and O. Solberg. 1984. Fatty acid and carbohydrate cell composition in pediococci and aerococci, and identification of related species, p. 179–211. In: T. Bergan (ed.), Methods in microbiology, vol. 16. Academic Press, London.

Blood, R. M. 1975. Lactic acid bacteria in marinated herring, p. 195–208. In: J. G. Carr, C. V. Cutting, and G. C. Whiting (ed.), Lactic acid bacteria in beverages and food. Academic Press, London.

Bosley, G. S., P. L. Wallace, C. W. Moss, A. G. Steigerwalt, D. J. Brenner, J. M. Swenson, G. A. Hebert, and R. R. Facklam. 1990. Phenotypic characterization, cellular fatty acid composition, and DNA relatedness of aerococci and comparison to related genera. J. Clin. Microbiol. 28:416–421.

Collins, M. D., A. M. Williams, and S. Wallbanks. 1990. The phylogeny of *Aerococcus* and *Pediococcus* as determined by 16s rRNA sequence analysis: description of *Tetragenococcus* gen. nov. FEMS Microbiol. Lett. 70:255–262.

Colman, G., and A. Efstratiou. 1987. Vancomycin-resistent leuconostocs, lactobacilli and now pediococci. J. Hosp. Infect. 10:1–3.

Dacre, J. C. 1958. A note on the pediococci in New Zealand cheddar cheese. J. Dairy Res. 25:414–417.

Daeschel, M. A., and T. R. Klaenhammer. 1985. Association of a 13.6-megadalton plasmid in *Pediococcus pentosaceus* with bacteriocin activity. Appl. Environ. Microbiol. 50:1538–1541.

Deibel, R. H., and J. B. Evans. 1960. Modified benzidine test for the detection of cytochrome-containing respiratory systems in microorganisms. J. Bacteriol. 79:356–360.

Deibel, R. H., and C. F. Niven, Jr. 1960. Comparative study of *Gaffkya homari, Aerococcus viridans,* tetrad-forming cocci from meat curing brines, and the genus *Pediococcus.* J. Bacteriol. 79:175–180.

Dellaglio, F., and S. Torriani. 1986. DNA-DNA homology, physiological characteristics and distribution of lactic acid bacteria isolated from maize silage. J. Appl. Bacteriol. 60:83–92.

Dellaglio, F., M. Vescovo, L. Morelli, and S. Torriani. 1984. Lactic acid bacteria in ensiled high-moisture corn grain: physiological and genetic characterization. System. Appl. Microbiol. 5:534–544.

De Man, J. C., M. Rogosa, and M. E. Sharpe. 1960. A medium for the cultivation of lactobacilli. J. Appl. Bacteriol. 23:130–135.

Evans, J. B. 1986. Genus *Aerococcus,* p. 1080. In: P. H. A. Sneath, N. Mair, M. E. Sharpe, and J. G. Holt (ed.), Bergey's manual of systematic bacteriology, vol. 2. William and Wilkins, Baltimore.

Facklam, R., D. Hollis, and M. D. Collins. 1989. Identification of Gram-positive coccal and coccobacillary vancomycin-resistant bacteria. J. Clin. Microbiol. 27:724–730.

Facklam, R., and H. W. Wilkinson. 1981. The family Streptococcaceae (Medical aspects), p. 1572–1597. In: M. P. Starr, H. Stolp, H. G. Trüper, A. Balows, and H. G. Schlegel (ed.), The prokaryotes. Springer-Verlag, Berlin.

Gonzalez, C. F., and B. S. Kunka. 1986. Evidence for plasmid linkage of raffinose utilization associated α-galactosidase and sucrose hydrolase activity in *Pediococcus pentosaceus.* Appl. Environ. Microbiol. 51:105–109.

Graham, D. C., and L. L. McKay. 1985. Plasmid DNA in strains of *Pediococcus cerevisiae* and *Pediococcus pentosaceus.* Appl. Environ. Microbiol. 50:532–534.

Kerbaugh, M. A., and J. B. Evans. 1968. *Aerococcus viridans* in the hospital environment. Appl. Microbiol. 16:519–523.

Mundt, J. O., W. G. Beattie, and F. R. Wieland. 1969. Pediococci residing on plants. J. Bacteriol. 98:938–942.

Raccach, M. 1987. Pediococci and biotechnology. Crit. Rev. Microbiol. 14:291–309.

Reuter, G. 1970. Laktobazillen und eng verwandte Mikroorganismen in Fleisch und Fleischerzeugnissen. 4. Mitteilung: Die Ökologie von Laktobazillen, Leuconostoc-Spezies und Pediokokken. Fleischwirtschaft 50:1397–1399.

Riebel, W. J., and J. A. Washington. 1990. Clinical and microbiologic characteristics of pediococci. J. Clin. Microbiol. 28:1348–1355.

Rogosa, J., J. A. Mitchell, and R. F. Wiseman. 1951. A selective medium for isolation and enumeration of oral and fecal lactobacilli. J. Bacteriol. 62:132–133.

Ruoff, K. L., D. R. Kuritzkes, J. S. Wolfson, and M. J. Ferraro. 1988. Vancomycin-resistant Gram-positive bacteria isolated from human sources. J. Clin. Microbiol. 26:2064–2068.

Stackebrandt, E., V. J. Fowler, and C. R. Woese. 1983. A phylogenetic analysis of lactobacilli, *Pediococcus pentosaceus* and *Leuconostoc mesenteroides.* System. Appl. Microbiol. 4:326–337.

Stackebrandt, E., and M. Teuber. 1988. Molecular taxonomy and phylogenetic position of lactic acid bacteria. Biochimie 70:317–324.

Taylor, P. V., and M. C. Trueblood. 1985. Septic arthritis due to *Aerococcus viridans*. J. Rheumatol. 12:1604–1605.

Teuber, M., and A. Geis. 1981. The family Streptococcaceae (Nonmedical aspects), p. 1614–1630. In: M. P. Starr, H. Stolp, H. G. Trüper, A. Balows, and H. G. Schlegel (ed.), The prokaryotes. Springer-Verlag, Berlin.

Vaughn, R. H. 1975. Lactic acid fermentation of olives with special reference to California conditions, p. 307–323. In: J. G. Carr, C. V. Cutting, and G. C. Whiting (ed.), Lactic acid bacteria in beverages and food. Academic Press, London.

CHAPTER 69

The Genus *Leuconostoc*

WILHELM H. HOLZAPFEL and ULRICH SCHILLINGER

The genus *Leuconostoc* (Lc.) has a physiology similar to that of the genera *Lactobacillus* (Lb.), *Pediococcus,* and other lactic acid bacteria. Phylogenetically, these three genera are considered intermixed (Stackebrandt and Teuber, 1988; Stackebrandt et al., 1983), and the 16S rRNA sequences indicate that the leuconostocs and lactobacilli form a common phylogenetic unit (Stackebrandt and Woese, 1981). These genera are separated phenotypically mainly by differences in their morphological features (Garvie, 1976, 1986; Kandler and Weiss, 1986). Although morphologically similar to the lactococci and other streptococci, leuconostocs have more features in common with the heterofermentative lactobacilli than with any other group of lactic acid bacteria. Moreover, morphology is considered an unreliable indicator of bacterial relationships (Fox et al., 1980).

The main criteria for the physiological differentiation of the lactic acid bacteria genera of concern, are presented in Table 4 and are discussed elsewhere.

Apart from the common metabolic features of leuconostocs and heterofermentative lactobacilli, some taxonomically exceptional species of the heterofermentative lactobacilli (e.g., *Lb. fructosus, Lb. viridescens,* and *Lb. minor*) share key characteristics with the leuconostocs. These include the absence of arginine dihydrolase, the production of predominantly D(−) lactate from glucose, and a peptidoglycan type similar to that of the leuconostocs (Kandler, 1970; Kandler and Weiss, 1986; Schleifer and Kandler, 1972). Moreover, the *Leuconostoc* species do not form a homogeneous group among themselves, and 23S rRNA similarity studies have shown that *Lc. mesenteroides* subsp. *mesenteroides* and *Lc. paramesenteroides* belong to two widely separated clusters (Schillinger et al., 1989). The taxonomically exceptional species *Lb. fructosus* was included in the larger leuconostoc cluster, while *Lb. confusus, Lb. halotolerans, Lb. kandleri, Lb. minor,* and *Lb. viridescens* form part of the *Lc. paramesenteroides* cluster (see Fig. 5). The acidophilic species, *Lc. oenos,* does not belong to either cluster (Schillinger et al., 1989). A recent comparative study of the 16S rRNA sequences (Yang and Woese, 1989) confirmed these findings. Moreover, Yang and Woese (1989) showed *Lb. confusus, Lb. halotolerans, Lb. kandleri, Lb. minor,* and *Lb. viridescens* to form a natural phylogenetic group together with the leuconostocs which they called the "leuconostoc branch" of the lactobacilli. These relationships are illustrated by the phylogenetic tree in Fig. 1, suggested by Yang and Woese (1989).

Presently, the genus *Leuconostoc* comprises nine species, of which five are relatively newly described.

Lately, leuconostocs have increasingly been associated with the spoilage of meat and meat products, and two new species, *Lc. carnosum* and *Lc. gelidum,* have been reported for this habitat (Shaw and Harding, 1989).

Awareness of the essential contribution of wine leuconostocs (*Lc. oenos*) to wine quality, and to malo-lactic fermentation, has stimulated research on the "acidophilic" leuconostocs in recent years.

In this chapter, relevant aspects concerning the genus *Leuconostoc,* as was presented by Teuber and Geis (1981) have been incorporated.

Habitats

Leuconostocs are fastidious chemoorganotrophs and share numerous natural and artificial habitats with the lactobacilli; in these habitats, amino acids and peptides, carbohydrates, fatty acids, nucleic acids and other growth factors, including vitamins, meet the nutritional requirements. All strains need biotin, nicotine, thiamine, and pantothenic acid or its derivatives. The typical energy-generating mechanism is carbohydrate fermentation (both by the hexoses monophosphate and phosphoketolase pathways) in combination with substrate-level phosphorylation. The end products of glucose fermentation are CO_2, ethanol, and D(−) lactate, although some strains may form acetate

Fig. 1. Phylogenetic tree for the "leuconostoc branch" lactobacilli, as suggested by Yang and Woese (1989). Bar
= 5% distance of relative evolutionary relationship.

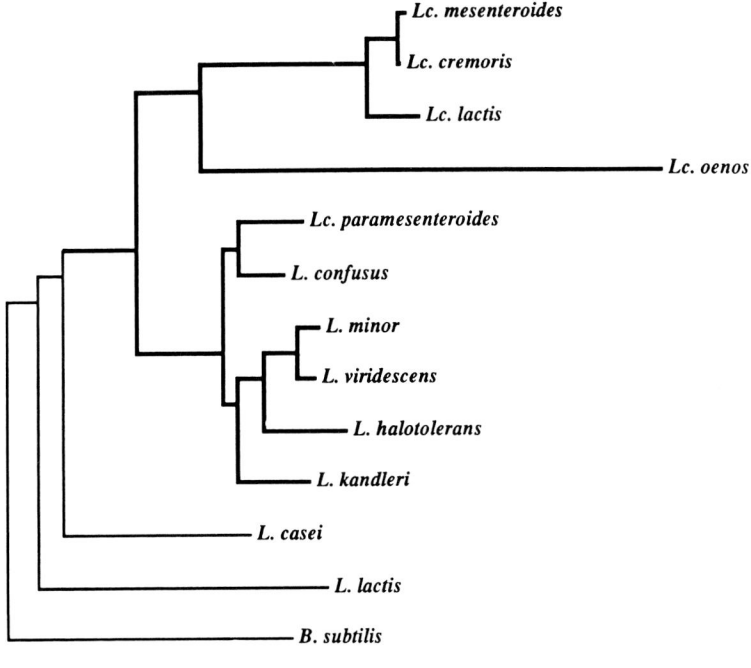

5%

instead of ethanol. The L(+) isomer of lactate
is produced only when malate is fermented, and
other organic acids (e.g., citric acid) provide
D(−) lactate (Kandler, 1982). Leuconostocs
prefer neutral to slightly acidic media, and, with
the exception of acidophilic *Lc. oenos,* are less
aciduric than the lactobacilli. Best growth has
been found within the temperature range of 20
to 30°C, with a minimum around 5°C. Exceptions are the "meat" leuconostocs, which are
able to grow at 1°C (Shaw and Harding, 1989).
Most strains are aerotolerant, but growth is usually more prolific under reduced atmosphere.

Beverages

WINE AND CIDER. *Lc. mesenteroides* subsp.
mesenteroides, Lc. oenos, and some other lactic
acid bacteria populate the surface of grapes and
vine leaves (Wibowo et al., 1985) and are present at low levels in wine must (Weiler and Radler, 1970), but only *Lc. oenos* is able to survive
the alcoholic fermentation of the must (Davis
et al., 1986; Fleet et al., 1984; Lafon-Lafourcade
et al., 1983). *Lc. oenos* is the organism mainly
responsible for the so-called malo-lactic fermentation (Beelman and Gallander, 1979)
which normally follows in red wines the completion of the alcoholic fermentation accom-

plished by yeasts. Lafon-Lafourcade et al.
(1983) suggest that *Lc. oenos* for the most part
originates from winery equipment and materials. Depending on the type of wine, the region
and the technique of processing, the malo-lactic
fermentation may be regarded as favorable or
detrimental (Radler, 1975). Conversion of
L(+)-malic acid to L(+)-lactic acid plus carbon
dioxide leads to a natural reduction of acidity,
ensures biological stability of the wine before
bottling, and influences the flavor and aroma of
wine (Davis et al., 1985b; Kunkee, 1967; Pilone
et al., 1966). The malo-lactic fermentation, on
the other hand, may be detrimental to wines
with high pH, as some Australian wines, resulting in a wine of lower quality and growth of
spoilage bacteria (Rankine, 1977). Thus, the
control of the malo-lactic fermentation is of
great importance to the wine-making industry.
Commercial strains of *Lc. oenos* are available
to induce the malo-lactic fermentation in wines
and modern processes may involve the use of
immobilized cells (Crapisi et al., 1987; Spettoli
et al., 1982). Growth of *Lc. oenos* and the rate
of malo-lactic fermentation are mainly affected
by wine pH, the concentrations of ethanol and
sulfur dioxide, and by the yeast strains used for
the alcoholic fermentation (Wibowo et al.,
1988). *Lc. oenos* is well adapted to a high-al-

cohol content and a low pH and is the only bacterial species isolated from wines at pH below 3.5 (Costello et al., 1983; Davis et al., 1986). Fleet et al. (1984) observed that several different strains of *Lc. oenos* may evolve during the malo-lactic fermentation in a single wine. Similarily, in cider, leuconostocs are associated with the malo-lactic fermentation (Carr and Whiting, 1971). According to Salih et al. (1988), *Lc. oenos* was predominant in cider at the end of the alcoholic fermentation and during the malo-lactic fermentation.

Lc. oenos may also be involved in the spoilage of fermented fruit juices such as wines, ciders, and perries (fermented pear juice) (Splittstoesser, 1982). Its growth may lead to haze and sediment, gassiness due to CO_2, and an increase in metabolic compounds that are responsible for off-flavor.

OTHER ALCOHOLIC BEVERAGES. Bryan-Jones (1975) and Priest and Pleasants (1988) report that leuconostocs are usually involved in both grain and malt whiskey distillery fermentations. *Lc. mesenteroides* subsp. *mesenteroides* belongs to the microbial population responsible for fermentation and may contribute to the final flavor of the product (Priest and Pleasants, 1988). In pulque, the product of spontanous fermentation of agave juice, leuconostocs producing polysaccharides or dextrans are regarded as responsible for the characteristic viscosity of the beverage, and leuconostocs also increase the acidity of the juice very rapidly, inhibiting the growth of other less-desirable bacteria (Sanchez-Marroquin, 1953; Steinkraus, 1983). Exceptionally acid-tolerant strains of *Lc. mesenteroides* subsp. *mesenteroides* able to grow at pH 3.3 were isolated from fruit mashes (Heinzl and Hammes, 1986). Leuconostocs also contribute to the fermentation of palm sap (Atputharajah et al., 1986; Faparusi, 1971; Shamala and Sreekantiah, 1988) and are associated with traditional African-fermented beverages (Sanni and Oso, 1988).

NONALCOHOLIC BEVERAGES. In fruit juices, leuconostocs may contribute to flavor deterioration as a result of increased levels of diacetyl and acetoin produced as metabolites (Hatcher et al., 1984). *Lc. mesenteroides* subsp. *mesenteroides* and *Lc. paramesenteroides* were isolated from orange juices (Sodeko et al., 1987) and citrus molasses (Parish and Higgins, 1988). *Lc. oenos* was found as a spoilage organism in canned mango juice (Ethiraj and Suresh, 1985).

Meat and Meat Products

Rapid developments in vacuum and modified-gas packaging and in refrigeration technologies have influenced the distribution pattern of fresh meat during recent years, resulting in the establishment of a "novel" food ecosystem, the refrigerated, vacuum-packaged meat product. In this system, a spoilage pattern concomitant with a dominant psychrotolerant population of lactobacilli and leuconostocs is reported with increasing frequency. Often considered as one of the minor groups among the "meat lactic" bacteria (Allen and Foster, 1960; Christopher et al., 1979; Egan and Shay, 1982; Hitchener et al., 1982; Kitchell and Shaw, 1975; Mol et al., 1971; Reuter, 1970b, 1975, 1981; Seideman et al., 1976), leuconostocs have been reported more recently as a major group on refrigerated, vacuum, or modified-atmosphere packaged meats (Hanna et al., 1981, 1983; Savell et al., 1981, 1986; Schillinger and Lücke, 1986; Shaw and Harding, 1984), livers and kidneys (Hanna et al., 1982), and processed meat products (Borch and Molin, 1988; Korkeala et al., 1988; Korkeala and Màkelà, 1989; Von Holy and Holzapfel, 1989). Domination of leuconostocs, in addition to meat lactobacilli, was also recorded for dark, firm, and dry (DFD) and normal beef steaks displayed for 3 to 6 days after wrapping in high oxygen-barrier film (Vanderzant et al., 1983).

Meat and processed, cured meat products apparently provide a favorable growth substrate for some leuconostocs. Refrigeration (2–4°C) in combination with vacuum packaging or reduced (modified) atmospheric conditions, favor the proliferation and eventual domination of leuconostocs and "atypical" lactobacilli (*Lb. curvatus* and *Lb. sake*) (Holzapfel and Gerber, 1986; Reuter, 1981; Von Holy and Holzapfel, 1989). Pure-culture inoculation of vacuum-packaged beef followed by storage at 5°C revealed that growth of a meat-associated leuconostoc was comparable to that of meat lactobacilli, with both strains reaching population levels of $>10^8/cm^2$ after 7 days (Egan and Shay, 1982). The leuconostocs produced a slightly more severe off-flavor than the lactobacilli. These results were confirmed only in part by Schillinger and Lücke (1986) who reported that in vacuum-packaged beef stored at 2°C the growth of a meat *Leuconostoc* strain used for inoculation was not comparable to that of other lactic acid bacteria. However, in non-inoculated meat, *Leuconostoc* strains have very often been found to dominate the population after 4 weeks. It may be concluded that refrigerated storage at 2°C (as compared to 5°C) favors the progressive domination of leuconostocs and typical meat lactobacilli in vacuum-packaged meat (see Table 1) (Schillinger and Lücke, 1986). Inoculation studies on lean and fat tissue of beef, pork, and

Table 1. Lactic acid bacteria predominating on vacuum-packaged meat.[a]

Experiment number	Cutting room	Meat	Storage temperature	Dominating population after:	
				13–16 days	23–30 days
3	A	Pork[b]	5°C	Acetic acid sensitive[c]	Acetic acid sensitive[c] *Lb. sake* (in part)
4	B	Pork	5°C	Acetic acid sensitive[c]	Acetic acid sensitive[c]
5,6,7	C	Pork	2°C	*Lactobacillus (Lb.) sake* *Carnobacterium (Cb.) divergens*	*Lb. sake*
8,9,10	C	Pork	2°C	*Cb. divergens* *Leuconostoc*	*Leuconostoc* *Lb. sake*
11,12,13	C	Beef	2°C	*Lb. sake* *Leuconostoc*	*Lb. sake* *Leuconostoc*
14,15,16	C	Beef	2°C	*Cb. divergens* *Leuconostoc*	*Leuconostoc* *Lb. sake*
17,18,19	D	Beef	2°C	Mixed population, acetic acid sensitive	*Lactococcus* *Leuconostoc*

[a]Oxygen permeability of the laminates was less than 25 ml/day $\times$ m^2 $\times$ bar; pH 24h after slaughtering = 5.8.
[b]Hot deboned pork was used in experiment 3.
[c]Not identified.
Adapted from Schillinger and Lücke (1986).

lamb have shown that proliferation of *Lc. mesenteroides* subsp. *mesenteroides* on lean samples is comparable or slightly better than on fat (Vanderzant et al., 1986). In model studies on the spoilage of commercially marketed meat, Toji (1979) has found the availability of glycogen, other sugars, and protein breakdown products to be key factors determining the growth of leuconostocs and other meat bacteria. Collins-Thompson and Lopez (1982) observed that *Brochothrix thermosphacta* (a typical meat-spoilage bacterium) was able to compete successfully with *Lc. mesenteroides* subsp. *mesenteroides* in vacuum-packed bologna at 5°C. In a special case of spoilage, refrigerated (1–2°C) beef strip loins, stored either vacuum packaged or in different combinations of O$_2$-CO$_2$-N$_2$ atmospheres, exhibited a spoilage association that was dominated up to 100% by leuconostocs and was associated with defects such as surface discoloration, general unattractiveness, and off-odors (Savell et al., 1981). For vacuum-packaged beef, domination by *Leuconostoc* spp. representing 45% of the total isolates, has been reported by Patterson and Baird (1977). Leuconostocs have also been found as the predominant group, in addition to the *Lb. curvatus/Lb. sake* group, throughout the manufacture and prolonged storage of vacuum-packaged smoked "vienna-type" sausages, as is shown in Table 2 (Von Holy and Holzapfel, 1989). Out of 466 representative isolates from spoiled products (with total bacterial numbers of 10^8–10^9/g), 36.3% were characterized as *Leuconostoc* spp.; their association with the sausage surface and their relative contribution to the spoilage population was studied by scanning electron microscopy (Fig. 2 and 3) (Von Holy and Holzapfel, 1989). In addition to souring and slight (though normally not adverse aroma and taste) defects, a leuconostoc-dominated spoilage pattern of vacuum-packaged processed sausages was often accompanied by gas production (blowing) and slime formation. The impact of this spoilage problem is reflected by an average direct loss of 3.5%, based on turnover, that can be directly related to lactic acid-bacterial metabolic activities (Von Holy and Holzapfel, 1989).

A fermented, sour aroma and taste were observed as the main sensory defects for vacuum-packed cooked ring sausages (Korkeala et al., 1985), for which leuconostocs were implicated as one of the major microbial groups involved in spoilage (Korkeala and Mäkelä, 1989). Korkeala et al. (1988) reported frequent occurrence of ropy slime at the surface of Finnish vacuum-packed cooked meat products and they found *Lc. mesenteroides* subsp. *mesenteroides* among the lactic acid bacteria responsible for this defect. The isolated strains differed from the known leuconostocs in producing ropy slime without sucrose.

Based on the inhibition of Gram-negative meat spoilage bacteria, leuconostocs were reported only to show moderate potential for ground meat preservation (Dubois et al., 1979). Attempts to utilize leuconostocs in mixed starter cultures for summer sausages produced results inferior of those with *Lactobacillus* cultures alone (Burrowes et al., 1986). Lactic acid bacteria have been shown to reduce nitrite in cured meat products, with the highest reduction rate found for *Lc. mesenteroides* subsp. *mes-*

Table 2. Changes in predominant lactic-acid bacteria populations from vacuum-packaged, smoked vienna sausages with increasing product age.

Sample type	Percentage isolates per biogroup[a]							Number of isolates characterized
	IA+B	II	III	IVA	IVB	V	VI	
Freshly manufactured samples	9.2	36.9	9.3	20.0	16.9	0	7.7	65
Fresh (4–10 days postmanufacture)	30.2	43.1	21.5	2.6	2.6	0	0	74
Returned (terminally spoiled)	3.8	36.3	58.0	0	1.9	0	0	466

[a]Identification of biogroups: IA = heterofermentative lactobacilli, arginine positive; IB = heterofermentative lactobacilli, arginine negative; II = *Leuconostoc* spp.; III = homofermentative lactobacilli; IVA = pedicocci; IVB = streptococci/enterococci; V = *Carnobacterium*; VI = "thermophilic" lactobacilli.
From Von Holy and Holzapfel (1989).

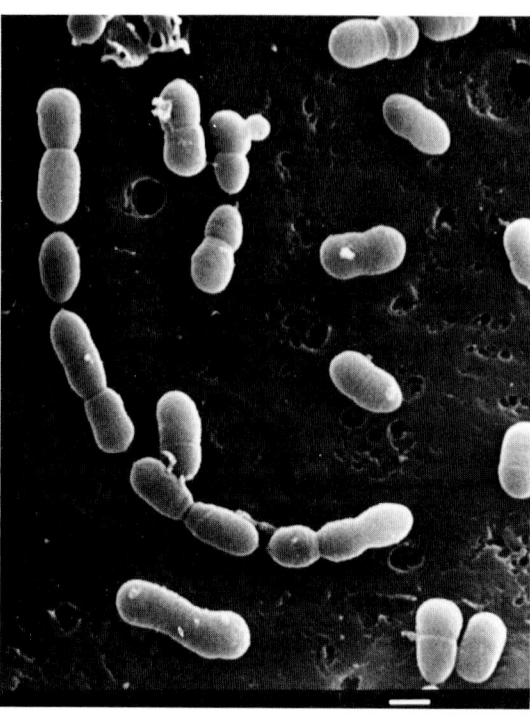

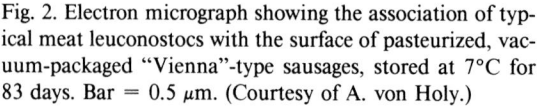

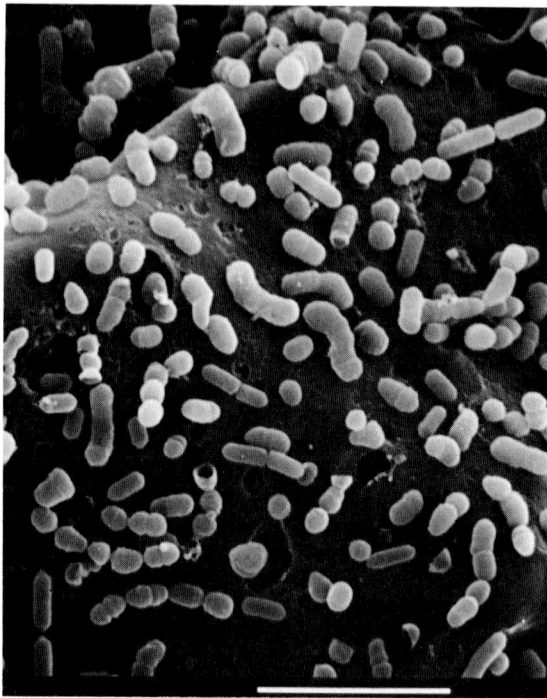

Fig. 2. Electron micrograph showing the association of typical meat leuconostocs with the surface of pasteurized, vacuum-packaged "Vienna"-type sausages, stored at 7°C for 83 days. Bar = 0.5 μm. (Courtesy of A. von Holy.)

Fig. 3. Electron micrograph showing a mixed population of leuconostocs (ovoid cells) and lactobacilli (straight to curved rods) associated with the surface of spoiled vacuum-packaged, smoked, "Vienna"-type sausages. Bar = 5 μm. (Courtesy of A. von Holy.)

enteroides (Collins-Thompson and Rodriguez-Lopez, 1981).

In previous reports, meat-associated leuconostocs were often identified as *Lc. mesenteroides* subsp. *mesenteroides* or *Lc. paramesenteroides* (Collins-Thompson and Rodriguez-Lopez, 1981; Hitchener et al., 1982; Holzapfel and Gerber, 1986; Reuter, 1970a; Savell et al., 1981). Their resemblance to acknowledged *Leuconostoc* spp., however, has been doubted (Shaw and Harding, 1984) and in a numerical taxonomic study Borch and Molin (1988) found them phenotypically distant from type strains of *Leuconostoc* species. In a recent study, Shaw

and Harding (1989) used numerical taxonomy and DNA homology to classify *Leuconostoc* isolates from chill-stored meats and recognized two new species, *Lc. carnosum* and *Lc. gelidum*. The taxonomic features of these species are discussed elsewhere in this chapter. The ability of most strains to grow at 1°C but not at 37°C (Shaw and Harding, 1989) is of practical importance and indicates their typical psychrotrophic character. This is also supported by the data of Savell et al. (1981) and Schillinger and Lücke (1986) (discussed in this section), pointing at the competitive behavior of meat leuconostocs at low temperatures of 1 to 2°C. All strains isolated from processed meat products by Borch and Molin (1988) grew at 5°C but not at 45°C, and may have similar psychrotrophic character as the Shaw and Harding (1989) strains; their relatedness to the new species, however, is not clear.

Milk and Dairy Products

Often associated with milk and milk products, the leuconostocs may bring about undesirable effects in fresh milk; selected strains, however, are of special value in starter cultures for a variety of fermented milk products.

MILK. Although leuconostocs are typical contaminants of the dairy environment, they have only limited growth potential in refrigerated milk. Previously often confused with the morphologically similar streptococci (genus *Lactococcus*) (Schleifer et al., 1985) which are better adapted to the milk substrate, they can be clearly distinguished by simple phenotypical criteria such as heterofermentation and production of the D(−) lactate isomer.

The association of leuconostoc contaminants with raw and pasteurized milk is generally recognized, but relatively little is known about their involvement in and contribution to the spoilage association of this product. Earlier reports on the slow development of leuconostocs in refrigerated milk (Juffs and Babel, 1975) were supported by observations by Galli et al. (1983), who report that *Leuconostoc* spp. (together with lactococci) predominated only when milk was stored at 12°C and not at 5°C. Studies on 182 representative strains of lactic acid bacteria associated with raw milk in Brazil (Antunes and De Oliveira, 1986), showed *Lc. mesenteroides* subsp. *cremoris* as a minor group, representing only 1.1% of the total. The predominant bacteria involved in the spoilage of pasteurized milk were studied by Keller et al. (1987), who reported *Leuconostoc* spp. to constitute 9.6% of 258 representative isolates investigated. Using phenotypic criteria, including the cell wall peptidoglycan type, and determination of the GC content of the DNA, Keller (1987) classified 13 out of 23 strains as *Lc. mesenteroides* subsp. *mesenteroides*. The remaining strains were represented by *Lc. lactis* (5), *Lc. amelibiosum* (4), and *Lc. paramesenteroides* (1) (Keller, 1987). Two parameters important in expressing the thermal sensitivity of an organism are D_{10} (decimal reduction time) and Z values. D_{10} is the time necessary for a 10-fold reduction in viable count at a given temperature. The Z value is the increase in temperature necessary to reduce

the D_{10} by a factor of 10. Decimal reduction (D_{10}) values for these strains, determined at 72°C in full-cream milk ranged from 4.54 to 5.52. Using different temperatures for determing heat destruction, z-values of 4.39 to 4.86 were calculated (Keller, 1987). Generation times of 22 h at 5°C determined in milk (Keller, 1987) indicated a nonpsychrotrophic nature of the strains of *Lc. mesenteroides* subsp. *mesenteroides* tested, when compared to the criterion of 8 h at 4°C suggested by Law and Mabbit (1983). According to Sadovski et al. (1980) *Lc. mesenteroides* subsp. *cremoris* showed the slowest growth of starter organisms under psychrotrophic conditions.

In a study of 1760 psychrotrophic bacterial isolates from raw ewe's milk, stored at 4 and 7°C, Nunez et al. (1984) have found the Gram-negative bacteria to predominate, yet considered *Lc. mesenteroides* subsp. *dextranicum* as the main Gram-positive "psychrotrophic" species. However, no conclusive evidence on the psychrotrophic nature of milk-associated leuconostocs has been provided yet. Leuconostocs (represented by *Lc. paramesenteroides*) were reported to represent a minor group among lactic acid bacteria isolated from salted raw milk, incubated at 30°C for 4 to 21 d (El-Gendy et al., 1983).

FERMENTED MILK PRODUCTS. In contrast to the lactococci, leuconostocs are not competitive growers or important producers of lactic acid in milk. The ability of certain strains to produce the flavor compound diacetyl, however, has led to their frequent incorporation into starter cultures for buttermilk, butter, quarg (cream cheese), and cheese (Collins, 1972; Collins and Speckman, 1974; Garvie, 1984; Quist et al., 1987; Sandine and Elliker, 1970). Their use in formulated starter cultures for kefir production has also been reported (Duitschaever et al., 1987; Marshall and Cole, 1985). *Lc. lactis* and *Lc. mesenteroides* subsp. *cremoris* contribute to diacetyl production by carbohydrate heterofermentation in associative growth with the lactococci (Garvie, 1984), and their involvement in flavor-production even of spontaneously fermented Moroccan dairy products (Iben and Smen) has been documented (Tantaoui-Elaraki and El Marrakchi, 1987). *Lc. paramesenteroides* was found to be the dominant species of lactic acid bacteria in "dadih," a traditional fermented milk in Indonesia (Hosono et al., 1989). Increased flavor production is stimulated at reduced pH, but especially associated with citrate-lyase-positive strains in the presence of citrate (Cogan, 1975; Collins and Speckman, 1974; Speckman and Collins, 1968). According to

Garvie (1984), all strains of *Lc. mesenteroides* subsp. *cremoris* examined used citrate. Walker and Gilliland (1987), however, reported considerable variations in the amount of diacetyl produced among strains of this subspecies. More information on the genetic basis of stability and expression of this feature is needed. Further aspects of the microbiology, handling, and propagation of starter cultures are discussed in Chapter 67 on *Lactococcus* and were also reviewed by Daly (1983).

Plant Materials

HERBAGE. Among lactic acid bacteria found on plants, *Leuconostoc* appears to be predominant (Mundt et al., 1967). Out of 400 lactic isolates studied by Stirling and Whittenbury (1963), 80% were leuconostocs, 10% pediococci, and the remainder lactobacilli. Species of *Leuconostoc* other than *Lc. mesenteroides* subsp. *mesenteroides* are only rarely isolated from plant materials (Mundt, 1970). On living undamaged plant tissue, leuconostocs occur in relatively low numbers. They have to compete with less fastidious, mainly Gram-negative bacteria and aerobic sporeformers. The presence of leuconostocs is apparently associated with the release of nutrients from damaged or decaying plant material (Stirling and Whittenbury, 1963). Their numbers increase with the degree of plant maturity (Daeschel et al., 1987) and during harvesting and ensiling. In vegetables such as peas and beans prepared for freezing, *Lc. mesenteroides* subsp. *mesenteroides* may multiply and cause sourness, discoloration, or off-flavor (Sharpe and Pettipher, 1983). *Lc. mesenteroides* subsp. *mesenteroides* dominated among the isolates of lactic acid bacteria responsible for spoilage of fresh ready-to-use grated carrots (Carlin et al., 1989).

FERMENTED FOODS OF PLANT ORIGIN. *Lc. mesenteroides* subsp. *mesenteroides* plays an important role in the fermentation of vegetables such as sauerkraut and cucumbers. Although not the dominant species present on cabbage at the time of shredding, *Lc. mesenteroides* subsp. *mesenteroides* initiates the fermentation of sauerkraut and is then succeeded by the more acid-tolerant lactobacilli (Pederson, 1930; Stamer, 1975). At elevated fermentation temperature or increased salt concentration, *Pediococcus cerevisiae* (now named *P. pentosaceus*) may also develop (Pederson and Albury, 1969). The same microbial succession was observed during fermentation of cucumbers or other pickles as well as olives (Vaughn, 1985). *Lc. mesenteroides* subsp. *mesenteroides* is the first to appear and

is quickly succeeded by *Lactobacillus brevis* and *Lactobacillus plantarum,* which then dominate and complete the fermentation. The sequence of lactic acid bacteria is dependent upon the initial load, growth rates, and salt- and acid-tolerances (Daeschel et al., 1987). *Lc. mesenteroides* subsp. *mesenteroides* is apparently better adapted to plant materials and initiates growth more rapidly than other lactic acid bacteria and produces the acid and carbon dioxide essential for the inhibition of the aerobic mesophilic organisms that might destroy crispness in the cabbage or cucumbers (Steinkraus, 1983). Carbon dioxide produced by leuconostocs replaces the air and creates an anaerobic atmosphere favorable for the stabilization of ascorbic acid and for preservation of the natural flavor of the vegetable (Pederson, 1930). *Lc. mesenteroides* subsp. *mesenteroides* is able to grow at 7.5°C (Vaughn, 1985). *Lc. mesenteroides* subsp. *mesenteroides* is less salt tolerant than the other lactic acid bacteria involved in vegetable fermentation. In salt stock pickles, the initial salt concentration is two- to three-fold higher than that employed in sauerkraut and *Lc. mesenteroides* subsp. *mesenteroides* therefore plays a less-active role in pickle fermentations (Stamer, 1988). On the other hand, slimy or ropy kraut may be the result of dextran formation caused by *Lc. mesenteroides* subsp. *mesenteroides* (Vaughn, 1985). *Lc. mesenteroides* subsp. *mesenteroides* has also been found to be involved in the preservation of small-sized ripe tomatoes by lactic acid fermentation (Beltran-Edeza and Hernandez-Sanchez, 1989). *Lc. mesenteroides* subsp. *dextranicum* may be part of sour-dough microbial population and distinctively influences the bread taste (Lönner and Preve-Akesson, 1989).

Lc. mesenteroides subsp. *mesenteroides* is also predominant and responsible for initiating the fermentation of many traditional lactic acid fermented foods in the tropics. High numbers of *Lc. mesenteroides* subsp. *mesenteroides* were isolated from vegetable products like the Indonesian Sayur-Asin prepared from mustard cabbage (Puspito and Fleet, 1985) and from starchy products like cassava (Okafor, 1977) or kocho, an African acidic fermented product from *Ensete ventricosa* (Gashe, 1987). Strains of *Lc. mesenteroides* subsp. *mesenteroides* have been found to produce a highly active linamarase which hydrolyses the cyanogenic glucoside linamarin present in cassava (Okafor and Ejiofor, 1985). In acidic, leavened breads and similar pancake-like products such as the Indian idli or dosa and the Ethiopian tef (Gashe, 1985) *Lc. mesenteroides* subsp. *mesenteroides* is responsible for souring, strengthening, and leavening the bread (Mukherjee et al., 1965; Steinkraus, 1983; Soni et al., 1986). *Lc. mesenteroides* subsp. *mesenteroides* is also involved in the fermentation of seeds of the African oil bean tree (Antai et al., 1986), of vanilla beans, coffee (Frank and Dela Cruz, 1964), and cocoa (Ostovar and Keeney, 1973; Passos et al., 1984).

SILAGE. In silage, lactic acid bacteria are responsible for the primary acid fermentation which is essential to prevent multiplication of clostridia (Whittenbury et al., 1967). Fermentation is usually initiated by *Enterococcus faecalis* and *Lc. mesenteroides* subsp. *mesenteroides* which replace the aerobic Gram-negative microbial population initially present on herbage intended for silage (Daeschel et al., 1987). Eventually leuconostocs are in turn superseded by aciduric lactobacilli (Langston et al., 1962; Woolford, 1985). Investigating the microbial population of high-moisture corn grain, Dellaglio et al. (1984) have found that at ensiling, 67% of the lactic acid bacteria consisted of leuconostocs and heterofermentative lactobacilli and that after 120 days, 84% of the isolates belonged to the homofermentative lactobacilli. Leuconostocs isolated by Dellaglio and Torriani (1986) and Dellaglio et al. (1984) from ensiled high-moisture corn grain and maize silage were identified as *Lc. paramesenteroides* by using DNA-DNA hybridization techniques. Grazia and Suzzi (1984) report the occasional occurrence of leuconostocs in Italian maize and alfalfa silage.

SUGAR. Leuconostocs are known to be responsible for deterioration effects in the sugar industry, and are one of the first bacterial groups studied for their causative role in commercial losses (Van Tieghem, 1878). During harvesting, sugar cane is contaminated with *Lc. mesenteroides* subsp. *mesenteroides* (previously *Lc. mesenteroides*), which is able to grow within the cut stalks and cause souring of the cane juice (Tilbury, 1975). The high sugar content (about 15%) and the initial pH of 5.0 to 5.5 render cane juice a suitable substrate for lactic acid bacteria. In hot, humid climates, growth of *Lc. mesenteroides* subsp. *mesenteroides* in harvested cane may result in a loss of 1 to 5% of total sugar for each day between harvesting and processing (Tilbury, 1975). Moreover, large amounts of dextran may be synthesized from sucrose by *Lc. mesenteroides* subsp. *mesenteroides* causing undesirable complication of the refining process (Pivnick, 1980). In beet sugar plants, similar spoilage is caused by slime-producing strains of *Lc. mesenteroides* subsp. *mesenteroides* and heterofermentative lactobacilli (Sharpe et al.,

1972). The ability of *Lc. mesenteroides* subsp. *mesenteroides* to produce dextrans from sucrose by a dextransucrase has been exploited for the production of commercially valuable dextrans on an industrial scale and is mentioned elsewhere in this chapter.

Leuconostocs as Pathogens

According to several recent reports, vancomycin-resistant leuconostocs were isolated from clinical sources (Coovadia et al., 1987, 1988; Hardy et al., 1988; Horowitz et al., 1987; Isenberg et al., 1988; Luetticken and Kunstmann, 1988; Rubin et al., 1988; Ruoff et al., 1988; Wenocur et al., 1988). Some of these were associated with infections such as purulent meningitis (Coovadia et al., 1987, 1988) and an odontogenic infection (Wenocur et al., 1988). Classification of these vancomycin-resistant, Gram-positive cocci with leuconostocs has been supported by comparative taxonomic studies involving DNA homology (Farrow et al., 1989). These studies resulted in the classification of some of these clinical isolates into two new species, *Lc. citreum* and *Lc. pseudomesenteroides* (Farrow et al., 1989). Whether these vancomycin-resistant leuconostocs are indeed true pathogens is doubtful, and Koch's postulates have still to be demonstrated.

Enrichment and Isolation

Knowledge of the physiology and typical environmental conditions favoring growth of specific strains are important prerequisites for the successful isolation of *Leuconostoc* species. Leuconostocs are typically associated with mixed populations of lactic acid bacteria in natural and artificial plant and food environments. The complex growth requirements and general physiological features of most leuconostocs are comparable to those of the lactobacilli, pediococci, and other lactic acid bacteria. These aspects explain the difficulty of obtaining pure leuconostoc cultures in a one-step selective operation.

Information on the composition of general media for the cultivation, maintenance, and semi-selective isolation of leuconostocs has been compiled by Garvie (1984), and is summarized in Table 3.

Isolation from Plant Material

Plant materials such as grass, herbages, and vegetables are the natural habitat of several types of lactic acid bacteria. Ensilage allows the enrichment of most *Leuconostoc* spp. in addition

Table 3. Ingredients of the various media used in the identification of species of the genus *Leuconostoc*.[a]

Ingredient	Medium[b]						
	YGPB	MRS	Medium 1	Medium 2	ATB	CMB	DTB
Glucose	1.0	2.0	1.0	1.0	1.0	1.0	1.0
Peptone	1.0	1.0	1.0	1.0	1.0	1.0	0.75
Meat extract	0.8	0.8					
Yeast extract	0.3	0.5	0.5	0.5	0.5	0.5	0.25
NaCl	0.5						
KH_2PO_4	0.25		0.5	0.5		0.25	0.25
K_2HPO_4	0.25	0.2					
$MgSO_4 \cdot 7H_2O$	0.2	0.2	0.2	0.2	0.2	0.2	0.2
$MnSO_4 \cdot 4H_2O$	0.005	0.005	0.005	0.005	0.005	0.005	0.005
Ammonium citrate		0.2	0.5				0.1
Citric acid				0.5		0.25	
DL-Malic acid						0.25	
Sodium acetate		0.5	0.25	0.25			0.25
Tween 80		0.1	0.1	0.1		0.1	0.1
Tomato juice				10.0	25.0		10.0
Usual pH	6.8	6.2	6.5	4.8	4.8	4.8	6.5

[a]Values are given in percent (w/v) of each ingredient. 1.5% agar is added when solid media are required. 0.05% cysteine hydrochloride is added to all media when required. Sterilization is normally at 15 lb for 15 min. Special techniques used with different media are mentioned in the text.
[b]YGPB, yeast glucose phosphate broth (Garvie, 1976); MRS, (DeMan et al., 1960); media 1 and 2: (Garvie, 1969); ATB, acid tomato broth (Garvie and Mabbitt, 1967); CMB, citrate-malate broth (Garvie and Mabbitt, 1967); DTB, dilute tomato broth (Garvie and Mabbitt, 1967).
From Garvie (1984).

to pediococci, lactococci, and lactobacilli. Favorable conditions for the initial dominance by leuconostocs exist in "spontaneous" vegetable fermentation processes, involving 2% brine. In these environments, leuconostocs are selectively enriched during the early stages of fermentation. The choice of semi- or non-selective media, such as MRS (De Man et al., 1960) or APT (Evans and Niven, 1951) for isolation, often depends on conditions where leuconostocs predominate.

Ensilage Enrichment of Lactoccocci, Streptococci, Leuconostocs, and Pediococci (Whittenbury, 1965a)

Grasses and other plant material are collected and cut in pieces as aseptically as possible. The prepared material is then placed into sterile glass tubes and compressed. Fifty grams of material are enough to fill a 3-×20-cm tube. The tubes are sealed in a way that permits gas under pressure to escape but prevents the entry of oxygen. A number of silages is prepared and incubated at 30°C. Tubes are opened beginning on the second day. The silage is removed and placed into flasks containing sterile water, which are vigorously shaken. Then, these suspensions can either be spread directly onto agar plates or diluted and pour-plated. The 2- to 3-day-old silages are the best sources for lactococci, streptococci, and leuconostocs.

A modification of the above method was reported by Weiler and Radler (1970). Grape leaves are homogenized with the same amount of acetate buffer (pH 5.4; 0.2 M). The homogenate is placed into sterile tubes. The tubes are sealed and incubated at 30°C.

Tetrazolium-sucrose (TS) Agar for Isolation of Leuconostocs and Streptococci (mainly lactococci) from Thawed Frozen Peas and Vegetables (Cavett et al., 1965)

The basal medium contains per liter of distilled water:

Evans peptone	10 g
Lab-lemco	10 g
Sucrose	50 g
New Zealand agar	14 g
pH 6.0	

Lots of 100 ml each are sterilized at 121°C for 20 min in 100-ml quantities, and 1 ml of a 1% solution (w/v) of filter-sterilized 2,3,5-triphenyltetrazolium chloride (TTC) is added.
The leuconostocs produce translucent, glassy, or watery colonies, containing gummy polysaccharide.

Thallous-acetate-tetrazolium-sucrose (TTS) Agar for Isolation of Leuconostocs from Plant Materials (Cavett et al., 1965)

Satisfactory results of leuconostoc isolation were obtained by modification of the Barnes (1956) medium. Glucose was substituted by sucrose and 1 ml of filter-sterilized 10% solution of thallous acetate was added to 100 ml of the TS medium (Cavett et al., 1965).

Differential Medium for the Enumeration of Homofermentative and Heterofermentative Lactic

Acid Bacteria from Fermented Vegetables (HHD-medium) (McDonald et al., 1987)

For 1 liter of medium:

Fructose	2.5 g
KH$_2$PO$_4$	2.5 g
Trypticase peptone	10.0 g
Phytone peptone	1.5 g
Casamino acids	3.0 g
Yeast extract	1.0 g
Tween 80	1.0 g
Bromocresol green[a]	20.0 ml
Agar (if applicable)	20.0 g

The pH of the dissolved medium is adjusted to 7.0 ± 0.02; and the medium is autoclaved at 121°C for 15 min.

[a]The bromocresol green is prepared as a stock solution by dissolving 0.1 g bromocresol green in 30 ml of 0.01 N NaOH. On the agar medium, homofermentative colonies are blue to green, while heterofermentative colonies remain white (McDonald et al., 1987).

Glucose-Yeast Extract Agar for Isolation of *Leuconostoc* and *Pediococcus* (Whittenbury, 1965a)

Medium contains, per liter:

Glucose	5.0 g
Yeast extract	5.0 g
Peptone	5.0 g
Meat extract	5.0 g
Agar	15 g
pH 6.5	

Acetate Agar for Isolation of *Leuconostoc* and *Pediococcus* (Whittenbury, 1965b)

This agar is a modification of the medium proposed by Keddie (1951) as being selective for lactobacilli.

Meat extract	50 g
Peptone	5.0 g
Yeast extract	5.0 g
Glucose	10 g
Tween 80	0.5 ml
Tap water	900 ml

The pH is adjusted to 5.4. Medium is autoclaved at 121°C for 15 min. Before plating, 100 ml of sterile 2 M acetic acid-sodium acetate buffer (pH 5.4) is added.

Yeast Extract-Glucose-Citrate (YGC) Broth for Isolation of *Leuconostoc* (Garvie, 1967a)

YGC has been suggested for the isolation and cultivation of leuconostocs from different sources. The medium is prepared by adding the following compounds to 1 liter of distilled water:

Peptone	10.0 g
Lemco	10.0 g
Yeastrel	5.0 g
Glucose	10.0 g
Triammonium citrate	5.0 g
Sodium acetate	2.0 g
MgSO$_4$·7H$_2$O	0.2 g

MnSO$_4$·4H$_2$O	0.05 g
Tween 80	0.05 ml

The pH is adjusted to 6.7 and the medium is autoclaved at 121°C for 15 min.

Medium for Isolation of *Lc. mesenteroides* subsp. *mesenteroides* from Plant Materials (Vrbaski et al., 1988)

The medium contains per liter:

Sucrose	100.0 g
Yeast extract	2.5 g
K$_2$HPO$_4$	5.0 g
Ammonium sulfate	0.2 g
MgSO$_4$·7H$_2$O	0.2 g
NaCl	0.6 g
Agar (if necessary)	20.0 g

The pH is adjusted to 7.8 before sterilization at 114°C for 20 min. Growth of *Lc. mesenteroides* subsp. *mesenteroides* results in a highly viscous broth medium within 24 h (Vrbaski et al., 1988).

Isolation from Meat and Meat Products

MRS-agar, adjusted to pH 5.7 and with 0.2% of potassium sorbate added, allows the selective isolation of leuconostocs and lactobacilli from meat products. No direct differentiation between lactobacilli and leuconostocs is possible on this medium. Reuter (1970b) recommended the use of thallium acetate medium according to Barnes (1956) for the selective isolation of leuconostocs from meat in the presence of a predominating lactobacillus population. Carnobacteria have been found resistant to thallium acetate, and may grow on this medium, especially if adjusted to pH >7.0 (W. H. Holzapfel, unpublished observations).

Isolation from Milk and Dairy Products

Most semi-selective media for leuconostocs also support growth of lactobacilli and pediococci and do not allow differentiation between these groups. These include media such as MRS (De Man et al., 1960) or Rogosa SL-medium (Rogosa et al., 1951); colonies on these media need to be further identified. General aspects concerning these and related media for lactobacilli, pediococci, and leuconostocs are discussed in Chapter 70.

To control gas and aroma (diacetyl) production in the fermentation of various diary products, it is important to know the quantitative composition of the starter cultures used. *Leuconostoc* species and *Lactococcus lactis* subsp. *diacetylactis* are components of many mesophilic starter cultures. These organisms are able to ferment citrate with concomitant production of CO$_2$ and diacetyl.

For the collective enumeration of leuconostocs and *Lactococcus lactis* subsp. *diacetylactis* in starters and fermented dairy products, a whey agar containing calcium lactate and casamino acids (WACCA) has been introduced by Galesloot et al. (1961).

WACCA 0.5% Medium for Enumeration of *Leuconostoc* and *Lactococcus lactis* subsp. *diacetylactis* (Galesloot et al., 1961)

Composition of the medium:

Dissolve 5 g Ca-lactate·5H$_2$O, 7 g Casitone, and 0.5% yeast extract in 1 liter of whey. Adjust to pH 7.3 with Ca(OH)$_2$-suspension. Steam for 30 min. Filtrate to pH 7.1 with a NaOH solution. Add 1 ml MnSO$_4$ solution (40 mg MnSO$_4$·4H$_2$O/100 ml). Dissolve 15 g agar. Clarify with 5 g albumin. Sterilize for 15 min 110°C (15 ml/tube).

Preparation of whey:

Add 0.3 ml 35% CaCl$_2$ solution and 0.3 ml commercial rennet (strength 1:10.800) to 1 liter of high-temperature-short-time (HTST)-pasteurized fresh skim milk at 30°C. Cut coagulum after 30 min at 30°C. Filtrate after 2 h at 45°C.

Preparation of Ca-citrate suspension:

Suspend 28 g Ca-citrate (Merck) in a 100-ml 1.5% carboxymethylcellulose solution, prepared at 45°C. Allow to precipitate for 2 h at 45°C. The supernatant is steamed for 30 min.

Application:

Add 0.3 to 0.7 ml Ca-citrate suspension per 15 ml WACCA 0.5% (48°C). The amount to be added has to be adapted to the type of starter under investigation. Incubate for 5 days at 25°C. Count after 2, 3, and 5 days.

A different medium (KCA) developed by Nickels and Leesment (1964) for the same purpose yields comparable results.

For the selective isolation and enumeration of *Leuconostoc* strains from mixed strain starter cultures, the following medium has been proposed.

HP Medium for Enumeration of *Leuconostoc* (Pearce and Halligan, 1978)

Composition per liter:

Phytone	20.0 g
Yeast extract	6.0 g
Beef extract	10.0 g
Tween 80	0.5 g
Ammonium citrate	5.0 g
FeSO$_4$·7H$_2$O	0.04 g
MgSO$_4$·7H$_2$O	0.2 g
MnSO$_4$·4H$_2$O	0.05 g
Glucose (sterilized separately)	10.0 g

The addition of tetracycline (0.12 μg/ml) to the medium selectively inhibits the growth of streptococci, and, therefore, allows the direct enumeration of *Leuconostoc* species in mixed-strain cultures.

A rapid enzymatic method has been found suitable for the quantitative enumeration of *Lc. lactis* in mixed cultures (Boquien et al., 1989). This method relies on the linear correlation that was shown to exist between population numbers of *Lc. lactis* (strain CNRZ 1091) and the activities of α-galactosidase and citrate lyase. It enables the estimation of a bacterial population within 2 h (Boquien et al., 1989). The method may have limited application for undefined mixed populations since some lactic acid bacteria other than leuconostocs may produce similar enzymes. Other aspects concerning starter cultures, including their enumeration, production, and preservation, are discussed by Teuber et al. (see Chapter 67).

Isolation from Wine

The "wine"-leuconostocs differ in some physiological properties from other leuconostocs. Media for their selective enumeration rely on their acidophilic nature (Garvie and Farrow, 1980) alcohol tolerance and adaption to the wine environment. Several acidic media have been employed for the isolation of leuconostocs (*Lc. oenos*) and of pediococci from wine (Garvie, 1967a; Weiler and Radler, 1970).

Acidic Tomato Broth (ATB) for Isolation of *Leuconostoc* and *Pediococcus* (Garvie, 1967a)

Medium contains, per liter:

Peptone	10.0 g
Yeastrel	5.0 g
Glucose	10.0 g
MgSO$_4$·7H$_2$O	0.2 g
MnSO$_4$·4H$_2$O	0.05 g
Tomato juice	25% (v/v)

The pH is 4.8 and the medium is autoclaved at 121°C for 15 min. Before use, a solution of cysteine hydrochloride sterilized by filtration is added to a final concentration of 0.05% (w/v).

Isolation of Lactic Acid Bacteria from Wine (Weiler and Radler, 1970) (not completely selective for *Lc. oenos*)

For 1 liter of medium:

Peptone	5.0 g
Yeast extract	5.0 g
Glucose	10.0 g
Diammonium hydrogencitrate	2.0 g
Sodium acetate·3H$_2$O	5.0 g
Tween 80	1.0 g
KH$_2$PO$_4$	5.0 g
MgSO$_4$·7H$_2$O	0.5 g
MnSo$_4$·4H$_2$o	0.2 g
FeSO$_4$·7H$_2$O	0.05 g
Agar	15.0 g

The pH is 5.3 to 5.4. Medium is sterilized 15 min at 121°C. To inhibit growth of yeasts, the medium is supplemented with sorbic acid at a final concentration of 0.05%.

Davis et al. (1985a) recommended the use of either MRS-agar (De Man et al., 1960) or tomato-juice-agar (Ingraham et al., 1960), adjusted to pH 5.5 and supplemented with 50 μg of cycloheximide per milliliter for the direct isolation of *Lc. oenos* from wine.

Supplementation of general medium with 40 to 80% wine, has been found to enhance growth of *Lc. oenos* (Davis et al., 1985b).

Fructose and Tween 80 (FT)-Medium for *Lc. oenos* (Cavin et al., 1988)

Composition per liter:

Casamino acids	5 g
Yeast extract	4 g
KH$_2$PO$_4$	0.6 g
KCl	0.45 g
CaCl$_2$·2H$_2$O	0.13 g
MgSO$_4$·7H$_2$O	0.13 g
MnSO$_4$·H$_2$O	0.003 g
Tween 80	1 ml
L(−)-malic acid	10 g
Agar (as desired)	15 g
D(+)-fructose	35 g
D(+)-glucose	5.0 g

The pH of the medium was adjusted with 10 N NaOH to 5.2, and then autoclaved at 121°C for 15 min.

Malo-lactic Differential (MLD) Medium for Screening *Lc. oenos* Strains Defective in Malo-Lactic Fermentation (Cavin et al., 1989)

This medium is based on the FT medium (Cavin et al., 1988) (discussed before) which is supplemented with 100 g of cellulose MN 300 and 0.1 g of bromocresol green per liter. Lower concentrations of fructose (4 g/liter) and glucose (1 g/liter) substitute those used for the FT medium, and were added to the sterilized medium as filter-sterilized stock solutions consisting of 8% fructose and 2% glucose (Cavin et al., 1989). Defective malo-lactic colonies show an acid reaction and remain yellow-green, while the malo-lactic positive colonies turn blue.

Dicks (1989) Recommended Acidic Grape Broth for Cultivation and Maintenance of *Lc. oenos* Strains

For 1 liter of medium:

Glucose	10 g
Peptone	10 g
Yeast extract	5 g
MgSO$_4$·7H$_2$O	0.2 g
MnSO$_4$·H$_2$O	0.05 g
Grape juice	250 ml

The pH is adjusted to 4.8 with 10 N NaOH.

The reader is also referred to procedures used by Nonomura et al. (1965, 1967) for studying malo-lactic bacteria from wine.

Media for Physiological Studies

1. "Basal" MRS-media

MRS-broth without citrate, acetate, beef extract, or glucose (pH 6.5) and with 0.004% chlorophenol red indicator added, can be used for determination of the sugar fermentation pattern. In another modification, MRS-broth without acetate or citrate (BM-MRS-broth) was recommended (Shaw and Harding, 1989; Wilkinson and Jones, 1977).

2. "Sugar" Basal Broth (Garvie, 1984) for Sugar Fermentation

Composition per liter:

Oxoid peptone	10.0 g
Yeast extract	2.5 g
Tween 80	0.1 g
Bromocresol purple (1.6% solution in ethanol)	1.0 ml
pH 6.8	

Fill 5-ml amounts into test tubes, and autoclave at 121°C for 15 min.

3. "Sugar" Basal Broth for Sugar Fermentation Pattern of *Lc. oenos* (Garvie, 1984)

The normal "sugar basal broth," mentioned before can be modified for *Lc. oenos* by adjusting the pH to 5.2, including 0.5% agar, and substituting bromocresol purple by 0.004% bromocresol green (Garvie, 1984).

4. Gas Production from Glucose

MRS-broth without citrate, using Durham tubes (Holzapfel and Gerber, 1983).

Gibson's medium (Gibson and Abdel-Malek, 1945) may give better results for slow gas producers (Briggs, 1953), and was recommended by Borch and Molin (1988).

Milk agar (Garvie, 1984) with the following composition:

a. Litmus milk	800 ml
Glucose	5.5 g
Yeast extract	2.0 g
b. Tomato juice (Oxoid)	100 ml
c. Nutrient agar	200 ml

To prepare, mix a and b, adjust to pH 6.8, heat to 45°C and add c melted. Fill volumes of 10 ml each into vials, and autoclave 115°C for 10 min. Cool in water. For use, melt, cool to 45°C, and inoculate with 0.25 ml of culture, solidify in cold water, and pour 4.0 ml of nutrient agar as top layer (Garvie, 1984).

5. Dextran Production

MRS-agar, modified by addition of 10% (w/v) sucrose (Sharpe, 1962).

Sucrose Agar for Dextran Production (Garvie, 1984) Composition per liter:

Tryptone	10.0 g
Yeast extract	5.0 g
K$_2$HPO$_4$	5.0 g
Diammonium citrate	5.0 g
Sucrose	50.0 g

Agar   15.0 g
pH 7.0

Autoclave at 121°C for 15 min.

General Growth Conditions

Most leuconostocs are relatively insensitive to oxygen, although more prolific growth is often observed under reduced atmospheres. Especially for the isolation of leuconostocs from vacuum-packaged meats, and for *Lc. oenos* from wine, microaerophilic to anaerobic conditions are recommended. Gas mixtures of N_2, H_2, or CO_2 or anaerobic "gas-generating kits," such as GasPak (Oxoid), or Anaerocult A (Merck), provide more favorable conditions for surface colony growth. Cysteine hydrochloride (0.05%–0.1%) may be added to broth media.

Incubation temperatures should be 20 to 25°C with the lower range being more favorable for *Lc. oenos.* Depending on the strain and other growth factors, an incubation period from 48 h up to 10 days may be necessary.

Cultivation, Maintenance, and Conservation of Cultures

The general procedures of cultivation, maintenance, and preservation as used for lactobacilli (see Chapter 70) are applicable to most leuconostocs.

MRS broth or agar is generally used for axenic cultivation, and cultures may be stored as stab cultures (MRS-agar) at 4°C for 1 to 2 weeks. Vitality may be retained in yeast glucose litmus milk + calcium carbonate for several months (Sharpe, 1981). Information on general cultivation and maintenance of strains has been summarized in Table 3. Juven (1979) recommended the preservation of stock cultures, partly dehydrated, on granular pumice stone for 6 to 24 months at room temperature. Lyophilization of *Lc. mesenteroides* subsp. *mesenteroides* in a solution of native dextran resulted in good long-term survival and retention of the dextran-forming ability (Valakhanovich et al., 1975). *Lc. lactis* and *Lc. mesenteroides* subsp. *cremoris,* important in starter cultures for their diacetyl-producing ability, were stored at −30°C for 3 months without loss of viability (Oberman et al., 1986). Initial freezing was at −70°C, and sterile milk or cream (18%) gave equally good results as protecting agents.

For short-term maintenance of acidophilic leuconostocs, e.g., *Lc. oenos,* stab cultures in Acidic Tomato Medium at pH 4.8 (Garvie and Mabbitt, 1967) may be used.

For lyophilization, harvesting of *Lc. oenos* in the mid-log phase gave maximum viability (Kole et al., 1982). Kole and Altosaar (1984) reported the increase in viability and bile resistance of bile resistant *Lc. oenos* strain 44.40 upon lyophilization.

Physiology

Metabolism

As known for other lactic acid bacteria, growth of leuconostocs is dependent on the presence of a fermentable sugar. Glucose is fermented by all species, but fructose is generally preferred. Leuconostocs do not possess an FDP-aldolase. Hexoses are degraded by a combination of the hexoses monophosphate system and the phosphoketolase pathway, yielding equimolar amounts of D(−)-lactate, ethanol, and CO_2 (De Moss et al., 1951; Gunsalus and Gibbs, 1952). However, in *Lc. oenos* this metabolic pathway has not been fully confirmed. Glucose-6-phosphate dehydrogenase and xylulose-5-phosphoketolase are the key enzymes present in all species (Garvie, 1986). In most species, both NAD and NADP may serve as coenzymes of the glucose-6-phosphate dehydrogenase, but in *Lc. oenos,* only NADP is used (Garvie, 1975). Glucose is phosphorylated and then oxidized to 6-phosphogluconate followed by decarboxylation. The resulting pentose is converted into lactic acid and ethanol. Some strains with an oxidative mechanism produce acetate instead of ethanol and pentose fermentation yields equimolar quantities of D(−)-lactate and acetate. Polysaccharides and most alcohols are not attacked. Malate is converted into L(+)-lactate and CO_2 by strains of *Lc. oenos* and *Lc. mesenteroides* subsp. *mesenteroides.* Citrate is metabolized to acetate and lactate. At low pH, however, diacetyl and acetoin may be produced from citrate (Cogan et al., 1981). Acetate and tartrate are not used. Little is known about the production of biogenic amines by leuconostocs. No tyramine formation was detected in strains of *Leuconostoc* isolated from fresh and vacuum-packaged meat (Edwards et al., 1987) or from *Lc. oenos* strains isolated from wine (Cilliers and van Wyk, 1985). Choudhury et al. (unpublished results) observed production of tyramine from tyrosine by a strain of *Lc. oenos* (DSM 20206).

Nutritional Requirements

Nutritional requirements have been found to be variable among different species and also among different strains of the same species. Like other lactic acid bacteria, leuconostocs need a complex medium containing vitamins, nucleotide bases, and amino acids. All species require nicotinic acid, thiamine, biotin, and pantothenic acid. *Lc. oenos* prefers a gluco-derivative of pantothenic acid (Amachi et al., 1971) which is also known as tomato juice factor (TJF) (Garvie and Mabbitt, 1967), but the degree of dependence varies with different strains and growth conditions. Folic acid is not required for growth of *Lc. lactis. Lc. mesenteroides* subsp. *mesenteroides* requires only glutamic acid and valine whilst other subspecies and species are dependent upon a variety of amino acids (Garvie, 1967b).

Identification

Differentiation of the Genus *Leuconostoc* from Other Lactic Acid Bacteria

Leuconostoc species are differentiated from other lactic acid bacteria on the basis of phenotypical criteria such as coccoid appearance, formation of gas, inability of arginine hydrolysis, and production of the D(−)-lactate isomer from glucose (Table 4), but the formation of dextran from sucrose and cell wall composition may also be helpful for identification. Morphologically, it is often difficult to separate leuconostocs from streptococci, lactococci, and lactobacilli, because leuconostocs, lactococci, and streptococci may form ovoid or even rod-shaped cells (in older cultures or under stress); and lactobacilli (e. g., *Lb. coryniformis* or *Lb. sake*) may produce very short rods or ellipsoid cells. This explains why, in the past, when identification of leuconostocs was still based mainly on morphology and slime production, dextran-

forming strains of *Lb. confusus* were frequently misidentified as *Lc. mesenteroides* subsp. *mesenteroides* (Holzapfel and Kandler, 1969). Sharpe et al. (1972) report that many strains of this *Lactobacillus* sent to the National Collection of Dairy Organisms were erroneously labelled as *Lc. mesenteroides* subsp. *mesenteroides*. Gas formation is an easily determinable characteristic that allows separation of homofermentative lactobacilli, pediococci, lactococci, enterococci, and streptococci from leuconostocs. Most gas-forming heterofermentative lactobacilli can be distinguished by their ability to hydrolyze arginine and by the formation of DL-lactate, yet some heterofermentative *Lactobacillus* species are difficult to separate because of their phenotypic similarities with *Leuconostoc*. For instance, *Lb. viridescens* and *Lb. fructosus* are not able to produce ammonia from arginine and form mainly D(−)-lactate from glucose (Kandler and Weiss, 1986). However, cells of these two species usually are more elongated than those of leuconostocs.

Species Differentiation

The list of recognized *Leuconostoc* species presented in *Bergey's Manual of Systematic Bacteriology,* Vol. 2 (Garvie, 1986) comprises only four species: *Lc. mesenteroides* with the three subspecies *dextranicum, cremoris,* and *mesenteroides; Lc. paramesenteroides; Lc. lactis;* and *Lc. oenos.* In 1989, however, five new *Leuconostoc* species were described: *Lc. amelibiosum* (Schillinger et al., 1989), *Lc. carnosum* (Shaw and Harding, 1989), *Lc. citreum* (Farrow et al., 1989), *Lc. gelidum* (Shaw and Harding, 1989), and *Lc. pseudomesenteroides* (Farrow et al., 1989). The differentiation characteristics of all *Leuconostoc* species are given in Table 5. *Lc.*

Table 4. Differentiation of leuconostocs from other lactic acid bacteria.

| | Leuconostocs | Lactobacilli | | Streptococci (including lactococci and enterococci) | Pediococci | Carnobacteria |
		Hetero-fermentative	Homo-fermentative			
Morphology	Cocci or coccobacilli	Coccobacilli or rods	Rods	Cocci (coccobacilli)	Cocci in tetrads	Rods
Gas from glucose	+	+	−	−	−	(+)
Hydrolysis of arginine	−	±	∓	− or +	− or +	+
Dextran from sucrose	− or +	∓	∓	∓	−	−
Type of lactic acid	D(−)	DL	D(−), DL, or L(+)	L(+)	DL or L(+)	L(+)

+, positive reaction; (+), weak positive reaction; ±, mostly positive, only a few species negative; ∓, mostly negative, only a few species positive.
Derived from Garvie (1984) and Shaw and Harding (1984).

Table 5. Differentiation within the genus *Leuconostoc*.

Leuconostoc species	Growth at 1°C	Growth at 37°C	Growth at pH 4.8	Growth in 10% ethanol	Esculin hydrolysis	Acid formed from: L-Arabinose	Cellobiose	Fructose	Galactose	Lactose	Maltose	Mannitol	Mannose	Melibiose	Raffinose	Ribose	Salicin	Sucrose	Trehalose	D-Xylose	Dextran from sucrose	Lemon-yellow pigment	Mol% GC	Peptidoglycan type
Lc. amelibiosum	+	ND	−	ND	+	+	+	+	ND	−	+	−	ND	ND	−	−	+	+	+	−	+	−	42	Lys-Ala$_2$
Lc. carnosum	+	∓	ND	ND	d	−	d	+	∓	−	∓	d	d	d	−	d	d	+	+	−	∓	−	39	ND
Lc. citreum	ND	d	ND	ND	∓	∓	∓	+	∓	−	+	d	+	∓	−	−	∓	+	+	∓	ND	∓	38–40	ND
Lc. gelidum	+	∓	ND	ND	+	+	+	+	+	+	d	−	+	+	+	∓	+	+	+	+	∓	−	37	ND
Lc. lactis	ND	+	−	−	−	−	−	+	+	+	+	−	d	d	d	∓	d	+	−	−	∓	−	43–45	Lys-Ala$_2$
Lc. mesenteroides subsp. *cremoris*	ND	−	−	−	−	−	−	−	+	+	−	−	−	−	−	−	−	−	−	−	−	∓	38–40	Lys-Ser-Ala$_2$
Lc. mesenteroides subsp. *dextranicum*	ND	+	−	−	d	−	∓	±	d	d	±	∓	d	d	d	ND	∓	±	+	d	+	−	37–40	Lys-Ser-Ala$_2$
Lc. mesenteroides subsp. *mesenteroides*	ND	d	−	−	±	+	d	+	±	d	±	d	±	±	d	±	±	+	+	d	+	−	37–39	Lys-Ser-Ala$_2$
Lc. oenos	ND	d	+	+	+	±	±	+	d	−	−	−	d	d	−	d	d	−	+	d	−	−	37–39	Lys-Ser$_2$, Lys-Ala-Ser
Lc. paramesenteroides	ND	d	∓	−	d	±	d	+	+	d	+	∓	+	+	d	+	−	±	+	∓	−	−	37–38	Lys-Ala$_2$
Lc. pseudomesenteroides	ND	+	ND	ND	d	±	±	+	±	d	+	±	+	±	±	+	d	±	+	+	ND	∓	38–41	Lys-Ser-Ala$_2$[a]

+, positive reaction; −, negative reaction; ±, most strains positive; ∓, most strains negative; d, variable reaction; ND, no data.
[a] Deduced from information on the cell wall of strain DSM 20193 (Schillinger et al., 1989).
Data from Farrow et al. (1989), Garvie (1984), Schillinger et al. (1989), and Shaw and Harding (1989).

oenos is recognized the most easily because of its growth at a low pH (4.2–4.8 or even lower) and in the presence of 10% ethanol. All other recognized *Leuconostoc* species do not grow in media with an initial pH value of 4.2 or less (Garvie, 1967a). These so-called nonacidophilic leuconostocs are more difficult to distinguish from each other because of their similar phenotypic properties. *Lc. mesenteroides* subsp. *cremoris* shows a stronger tendency to chain formation than other species (Fig. 4). Production of a lemon yellow pigment (*Lc. citreum*) and dextran formation from sucrose may be helpful for species differentiation.

SUGAR FERMENTATION PATTERN. The sugar fermentation pattern is of limited value for identification of the nonacidophilic leuconostocs because of the considerable variation among different strains of the same species (Table 5) on the one hand, and because of the similarity of the range of fermented sugars of different species on the other. Carbohydrate utilization

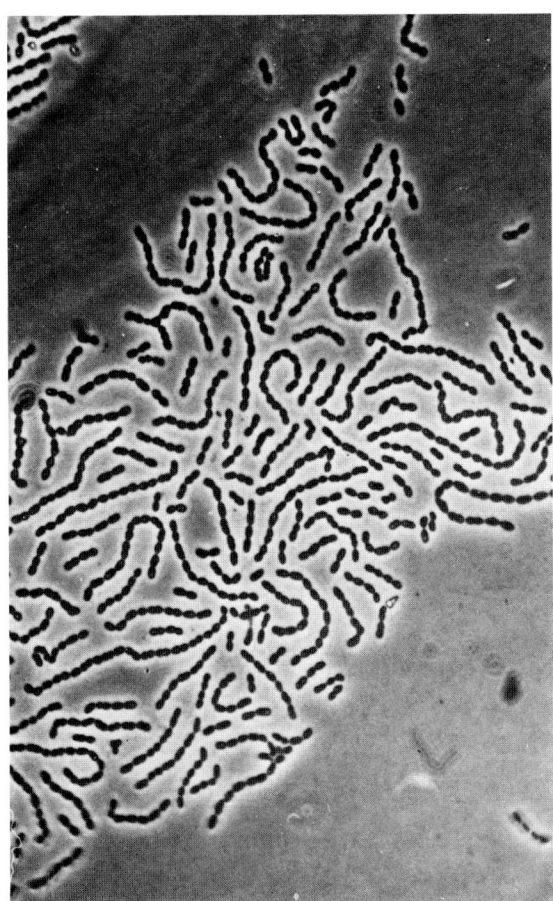

Fig. 4. Photomicrograph showing cellular arrangement typical of *Leuconostoc mesenteroides* subsp. *cremoris* (strain DSM 20346).

patterns of strains belonging to different species may be identical. Out of 15 different sugars, 13 may be fermented by strains of *Lc. mesenteroides* subsp. *dextranicum* (Table 5) and there are also many differences in the fermentation pattern among different strains of *Lc. mesenteroides* subsp. *mesenteroides*. Only *Lc. mesenteroides* subsp. *cremoris* is easily distinguished from the other leuconostocs by its very limited range of fermented sugars, consisting of glucose, galactose, and lactose. Sugars most helpful for differentiation of the other species seem to be arabinose, melibiose, trehalose, and xylose. Determination of the volatile metabolites by means of Freon 11 extraction may also be useful for species differentiation (Tracey and Britz, 1989a). There are also many differences in carbohydrate fermentation patterns among strains of *Lc. oenos,* which has therefore been regarded as a phenotypically heterogeneous species (Garvie, 1967a; Garvie and Farrow, 1980; Tracey and Britz, 1987, 1989b). Accordingly, some authors (Peynaud and Domerc, 1968; Rossi et al., 1978) suggested dividing *Lc. oenos* into different species on the basis of carbohydrate utilization. Studies of the genetic relatedness of 53 strains of *Lc. oenos* (Dicks and Van Vuuren, 1989) have shown, however, that this species is genetically homogeneous.

ELECTROPHORETIC MOBILITY AND IMMUNOLOGICAL RELATIONSHIPS OF ENZYMES. Within the genus *Lactobacillus,* there is a considerable variation in the electrophoretic mobility of lactate dehydrogenases (LDHs) among different species; the electrophoretic pattern of these enzymes therefore is a useful taxonomic marker for species differentiation (Gasser, 1970). In contrast to this, the electrophoretic behavior of the NAD-dependent D-LDHs of the nonacidophilic species of *Leuconostoc* is remarkably uniform (Garvie, 1969). The D-LDHs of *Lc. mesenteroides* and its subspecies, as well as *Lc. lactis* and *Lc. paramesenteroides,* have an identical electrophoretic mobility suggesting that these species possess a common lactate dehydrogenase, whereas the acidophilic species, *Lc. oenos* has a distinct D-LDH (Garvie, 1969). Similar results were obtained for the dehydrogenases of glucose-6-phosphate and 6-phosphogluconate (Garvie, 1975). It is not possible to use electrophoretic patterns of any of these dehydrogenases to distinguish the nonacidophilic species from each other. Immunological studies of Hontebeyrie and Gasser (1975), however, revealed differences among these electrophoretically identical enzymes. Cross-reactions obtained by double diffusion with antisera prepared against D-LDH and glucose-6-phosphate dehydrogenase from *Lc. lactis* allowed the leuconostocs to be divided into seven groups of identical specificity.

NUMERICAL ANALYSIS OF TOTAL SOLUBLE CELL PROTEIN PATTERNS. Dicks and van Vuuren (1989) demonstrated that electrophoresis of total cell-soluble proteins may be applied to differentiate among *Leuconostoc* species. The groupings obtained by numerical analysis of total cell-soluble proteins corresponded well to the groupings obtained by DNA-DNA hybridization.

CELLULAR FATTY ACID ANALYSIS. Fatty acid profiles obtained by gas chromatography allowed differentiation between *Leuconostoc* species (Schmitt et al., 1989). Shaw and Harding (1989) found complete agreement of DNA-DNA hybridization results and the classification of meat leuconostocs as determined by cellular fatty acid analysis.

CELL WALL PEPTIDOGLYCAN. Schleifer and Kandler (1972), who studied the cell wall structure of bacteria including lactic acid bacteria, have found that the amino acid sequences of the interpeptide bridge which is covalently linked to two adjacent peptidoglycans are useful characteristics for classification. The methods for detection of peptidoglycan types are given by these authors. All species of *Leuconostoc* have a similar type of peptidoglycan that is also present in some heterofermentative lactobacilli and that contains alanine, or serine and alanine, as typical constitutents of the interpeptide bridge but there are also some species-specific differences that may be used for species differentiation. Strains belonging to *Lc. mesenteroides* and its subspecies and *Lc. pseudomesenteroides* are characterized by the Lys-Ser-Ala$_2$ type, whereas the peptidoglycan of *Lc. paramesenteroides, Lc. lactis,* and *Lc. amelibiosum* lacks serine (Lys-Ala$_2$ type); *Lc. oenos* contains the Lys-Ala-Ser or Lys-Ser-Ser type (Table 5).

DNA-BASE COMPOSITION. Most leuconostocs have a similar GC content of DNA, namely about 37 to 40 mol%. In *Lc. lactis* and *Lc. amelibiosum,* it is higher (about 42–45 mol%). Therefore the determination of the mol% GC content of DNA is of limited value for identification.

Genotypic Relationships

DNA-DNA HYBRIDIZATIONS. Different authors have used DNA-DNA hybridization techniques to clarify the genetic relationships of the *Leu-*conostoc species. Hontebeyrie and Gasser (1977) employed the hydroxylapatite method (Brenner, 1973), whereas Garvie (1976) and Farrow et al. (1989) applied the membrane filter method of Denhardt (1966), and Schillinger et al. (1989) used both the filter method and the optical method of determination from denaturation rates (De Ley et al., 1970). Shaw and Harding used the S1 nuclease procedure (Crosa et al., 1973) for their hybridizations. The results of these different hybridization techniques were in good agreement. They are summarized in Table 6.

A very close relatedness between *Lc. mesenteroides* and the leuconostocs previously classified as *Lc. dextranicum* and *Lc. cremoris* had been suggested by the immunological similarities that were found among the D($-$)-lactate dehydrogenases and among the glucose-6-phosphate dehydrogenases of these leuconostocs (Gasser and Hontebeyrie, 1977; Hontebeyrie and Gasser, 1975). DNA-DNA hybridization revealed that *Lc. mesenteroides, Lc. cremoris,* and *Lc. dextranicum* indeed belong to a single deoxyribonucleic acid homology group, and accordingly *Lc. cremoris* and *Lc. dextranicum* were included as subspecies into the species *Lc. mesenteroides* (Garvie, 1983). The subspecies status was chosen because of the great phenotypical differences between *Lc. cremoris* and the other two species of importance to the dairy industry. *Lc. mesenteroides* subsp. *cremoris* seems to represent a biotype adapted to the special environment of creameries. The DNA-DNA hybridization studies of Garvie (1976), Hontebeyrie and Gasser (1977), and Schillinger et al. (1989) have shown that not all strains classified as *Lc. mesenteroides* subsp. *mesenteroides* are highly homologous to the type strain of this species.

Farrow et al. (1989) who included three of these strains into their DNA-DNA hybridization studies of vancomycin-resistant clinical *Leuconostoc* isolates, have found that they belong to two groups of high DNA homology genetically distinct from each other and from all other *Leuconostoc* species. For these two homology groups, the new species *Lc. citreum* and *Lc. pseudomesenteroides* have been proposed. DNA-DNA hybridizations were also useful in clarifying the taxonomic status of strains which had tentatively been designated as *Lc. mesenteroides* subsp. *amelibiosum* (Holzapfel, 1969; Kandler, 1970) because of their phenotypic similarity with *Lc. mesenteroides*. Since no high genetic relationship to *Lc. mesenteroides* and any other species of *Leuconostoc* has been found, the conclusion was drawn that these strains represent a separate species which has been called

Table 6. Levels of DNA-DNA homology for leuconostocs.

| Species or subspecies | % Homology with ³[H]DNA from | | | | |
	Lc. mesenteroides subsp. *mesenter.*	*Lc. pseudomesenteroides*	*Lc. paramesenteroides*	*Lc. lactis*	*Lc. carnosum*
Lc. mesenteroides subsp. mesenteroides	73–108	24–63	7–18	7–38	1–17
Lc. mesenteroides subsp. cremoris	66–106	20–50	5–10	5–35	4–10
Lc. mesenteroides subsp. dextranicum	84–110	38–49	5–19	6–35	5–8
Lc. pseudomesent.	18–48	70–100	9–22	13–36	ND
Lc. paramesent.	6–19	7–30	82–100	8–14	0–3
Lc. lactis	16–49	23–61	0–25	74–100	7–22
Lc. amelibiosum	39	34	22	32	78–116
Lc. carnosum	19–32	ND	0–6	0–25	3–21
Lc. gelidum	9–31	ND	0–3	0–10	ND
Lc. citreum	21–30	27–56	10–20	23	ND
Lc. oenos	8–15	5–10	5–7	3–11	ND

ND, no data.
Data compiled from Farrow et al. (1989), Garvie (1976), Hontebeyrie and Gasser (1977), Schillinger et al. (1989), and Shaw and Harding (1989).

Lc. amelibiosum (Schillinger et al., 1989). DNA-homology studies of *Leuconostoc* strains isolated from chill-stored meats (Shaw and Harding, 1989) revealed the existence of two other new species for which the names *Lc. carnosum* and *Lc. gelidum* were proposed. The DNA of *Lc. paramesenteroides* have been found to show only a low degree of hybridization with all other species of *Leuconostoc* (Table 6). Even lower DNA-homology values have been found between *Lc. oenos* and the remaining leuconostocs, indicating a very distant genomic relationship between the acidophilic and non-acidophilic species of *Leuconostoc*.

DNA-rRNA HYBRIDIZATIONS AND COMPARATIVE ANALYSIS OF 16S rRNA SEQUENCES. A specific relationship between leuconostocs and some heterofermentative lactobacilli possessing a similar type of peptidoglycan is suggested by the common phenotypic properties. Sharpe et al. (1972) have found that the D(−)-LDH common to the nonacidophilic species of *Leuconostoc* was electrophoretically the same as the D(−)-LDH of *Lb. confusus* and *Lb. viridescens*. The high similarity of these lactate dehydrogenases was confirmed by enzyme hybridization studies (Garvie, 1975). Garvie (1975) concluded from these results that these lactic-acid bacteria may belong to a single genus. Nucleic acid hybridization studies may be used to clarify the relationships on a genetic level. In contrast to DNA-DNA hybridizations which allow detection only of closely related organisms DNA-rRNA hybridization using 23S ribosomal RNA cistrons which are a more conserved part of the genome (De Ley and De Smedt, 1975), can be expected to yield information about the genetic relationships between less closely related organisms. In such DNA-rRNA hybridizations, the degree of relatedness is not expressed as percentage homology, but in terms of thermal stability of the duplexes formed between DNA and rRNA. Melting points (Tm$_{(e)}$) of the hybrids of DNA and RNA are determined and transferred into rRNA similarity maps (De Smedt and De Ley, 1977) to highlight the interrelationships among species which have undergone DNA-rRNA hybridizations.

Only a few data of DNA-rRNA hybridizations exist for lactic-acid bacteria. Results of DNA-rRNA hybridizations of Garvie (1981) and Schillinger et al. (1989) confirmed the distant relationship of the acidophilic species *Lc. oenos* to the remaining leuconostocs. Schillinger et al. (1989) have found the nonacidophilic leuconostocs to belong to two different clusters of rRNA similarity (Fig. 5). One of them comprises almost all nonacidophilic species except for *Lc. paramesenteroides* which showed a higher degree of rRNA similarity to *Lb. confusus*, *Lb. viridescens*, and other heterofermentative lactobacilli which share many phenotypic characteristics with the leuconostocs.

These results are in good agreement with the findings of Yang and Woese (1989) who applied comparative analysis of 16S rRNA sequences to clarify the phylogenetic relationships of these bacteria. The results demonstrated that leuconostocs and lactobacilli of the *Lb. confusus/Lb. viridesces* type form a natural grouping which the authors called "leuconostoc branch" of lac-

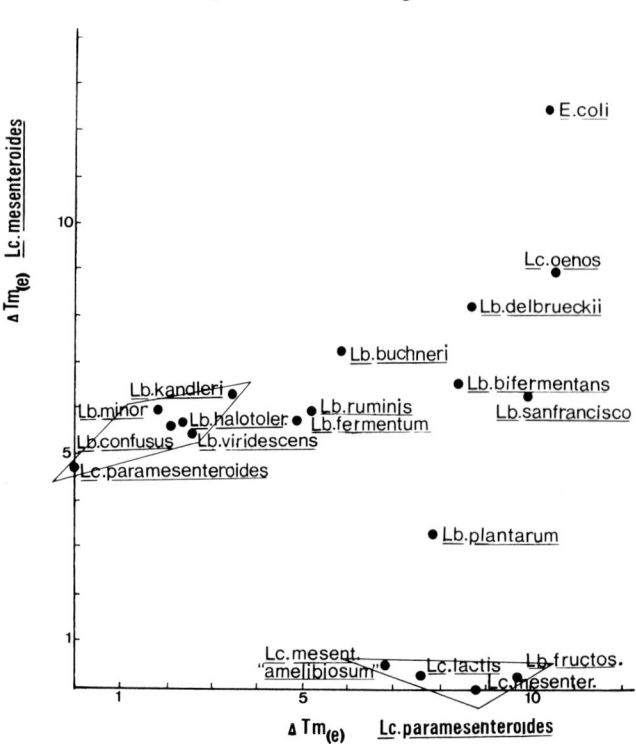

Fig. 5. Similarity map of hybrids between 23S rRNA of *Leuconostoc mesenteroides* subsp. *mesenteroides* DSM 20343ᵀ and *Leuconostoc paramesenteroides* DSM 20288ᵀ and DNA of various leuconostocs and lactobacilli. Organisms sharing high rRNA similarity are surrounded by a line (Schillinger et al., 1989).

tobacilli; and which they subdivided into two subgroups, one comprising leuconostocs exclusively and the other containing *Lc. paramesenteroides* and the above-mentioned lactobacilli. Comparative analysis of 16S rRNA sequences revealed also that the genome of *Lc. oenos* evolves more rapidly than that of all other lactic-acid bacteria; this explains why a very distant relationship between *Lc. oenos* and the nonacidophilic leuconostocs has been found by DNA-DNA and DNA-rRNA hybridizations.

Genetics

Plasmids

Information on extrachromosomal elements of leuconostocs is rare. O'Sullivan and Daly (1982) were the first who reported the occurrence of plasmids in *Leuconostoc*. All of the 10 strains examined harbored at least one plasmid species in the 2.5 to 40 Md range. Orberg and Sandine (1984) who investigated 18 *Leuconostoc* strains (including four strains of *Lc. oenos*) have found plasmids varying in molecular mass from about 1.9 to about 76 Md. Janse et al. (1987) detected plasmid DNA in 11 out of 45 strains of *Lc. oenos* and Dellaglio et al. (1984) demonstrated the presence of plasmids (range from 1.2–35 Md) in all of 15 *Lc. paramesenteroides* strains investigated. Elements of lactose and citrate metabolism may be encoded by plasmids (David

and De Vos, 1987; O'Sullivan and Daly, 1982); in contrast to this, vancomycin resistance seems to be chromosomally determined (Horowitz et al., 1989). Plasmid DNA encoding for nisin production (Tsai and Sandine, 1987a), lactose-fermenting ability (Tsai and Sandine, 1987b), protease activity (Caplice et al., 1987), and antibiotic resistance (Pucci et al., 1988) was transferred conjugatively from *Lactococcus, Enterococcus,* or *Streptococcus* species to *Leuconostoc*. Recently, the transformation of a strain of *Lc. paramesenteroides* by electroporation was demonstrated by David et al. (1989).

Bacteriophages

The presence of bacteriophages of *Lc. oenos* was demonstrated in several Swiss wines (Sozzi et al., 1982), Australian wines (Davis et al., 1985a), and South African wines (Nel et al., 1987). Bacteriophages of other *Leuconostoc* species were isolated from dairy products (Saxelin et al., 1986; Shin, 1983; Shin and Sato, 1979; Sozzi et al., 1978). Shin and Sato (1980) found lysogenic strains in all three subspecies of *Lc. mesenteroides* which delivered phage-like particles after treatment with mitomycin C. *Lc. oenos* phages can attack *Lc. oenos* starter cultures and delay or prevent malo-lactic fermentation (Henick-Kling et al., 1986a). Phage activity is inhibited by ethanol at concentrations above 5% (v/v) (Davis et al., 1985). Henick-Kling et al. (1986b) have found that phage inhibition was strongest when phage infection occurred at the beginning of the bacterial growth phase, at low temperatures and at low pH. *Lc. oenos* strains of different phage sensitivity can be selected and used in rotating starter cultures for controlled fermentation.

Applications and Biotechnology

The contribution of leuconostocs to fermentation processes of vegetable products (sauerkraut and cucumbers), and to flavor production by starter cultures for fermented milk products, has been well studied and documented. Increasing awareness of the importance of malo-lactic fermentation in wine has stimulated research into the physiology and metabolism of the "wine leuconostocs" (*Lc. oenos*) These aspects are discussed in the section "habitats." The presently available information on physiology, substrate-related growth characteristics, and genetics of these bacteria allows new approaches to industrial use.

With the increasing information on the genetic systems of leuconostocs, methods of genetic modification will be available for the improvement of technologically important strains. For instance, the gene for the malo-lactic enzyme of *Lc. oenos* has been cloned in *Escherichia coli* (Lautensach and Subden, 1984). Selected strains of *Lc. mesenteroides* subsp. *cremoris* and *Lc. lactis* are now being optimized for accelerated flavor production in fermented milk products, while selected strains of *Lc. mesenteroides* subsp. *dextranicum* and *Lc. mesenteroides* subsp. *mesenteroides* seem to improve initial fermentation conditions for sauerkraut and cucumbers. The use of bacteriocinogenic *Leuconostoc* strains in selected starter cultures and concomitant inhibition of undesired naturally occurring lactic acid bacteria in the fermentation of vegetable products has been suggested (Daeschel, 1989). These prospects seem promising in view of reports on the production of inhibitory substances active against other leuconostocs and lactococci, by *Lc. mesenteroides* subsp. *dextranicum* and two unidentified *Leuconostoc* strains (Orberg and Sandine, 1984). The antimicrobial effect of leuconostocs against psychrotrophic spoilage bacteria and pathogens in milk and dairy products was reported frequently (Branen et al., 1975; Fernandez Escartin et al., 1984; Juffs and Babel, 1975; Ross, 1981; Speck, 1972). The use of slowly growing antagonistic leuconostoc strains for "protection" of refrigerated milk was suggested by Juffs and Babel (1975), and as antagonists against *Pseudomonas fluorescens* causing spoil-

age of ground beef by Branen et al. (1975) and Reddy et al. (1970). Strains of *Lc. mesenteroides* subsp. *mesenteroides* are industrially used to produce dextrans. Dextrans and their derivatives have been used by the pharmaceutical and chemical industry as plasma extenders, anticoagulants (sulfated dextrans), and adsorbents (DEAE dextran) (Jeanes, 1978, cited by Garvie, 1984; Sinskey et al., 1986). In the medical field, modified dextrans are used as adjuvants for interferon induction, carriers of insulin and vitamin B12, and enzyme stabilization (Garvie, 1984). Starter cultures containing leuconostocs were also developed for the desugarization of egg white by fermentation (Bean et al., 1963; Galluzzo et al., 1974).

In the wine industry, phages that destroy *Lc. oenos* activity may cause serious problems and require the development of strategies for their control. One approach would be to select phage-resistant strains that could form the basis of a culture bank of *Lc. oenos* strains. It should allow the combination of suitable organisms for multiple strain starter cultures, based on different phage-sensitivity patterns and fermentation characteristics (Henick-Kling et al., 1986b).

Literature Cited

Allen, J. R., and E. M. Foster. 1960. Spoilage of vacuum packaged sliced processed meat during refrigerated storage. Food Res. 25:19–25.

Amachi, T., S. Imamoto, and H. Yoshizumi. 1971. A growth factor for malo-lactic fermentation bacteria. II. Structure and synthesis of a novel pantothenic acid derivative isolated from tomato juice. Agric. Biol. Chem. 35:1222–1230.

Antai, S. P., and M. H. Ibrahim. 1986. Microorganisms associated with African locust bean (Parkia-Filicoidea) fermentation for "dawadawa" production. J. Appl. Bacteriol. 61:145–148.

Antunes, L. A. F., and J. S. De Oliveira. 1986. Lactic-acid bacteria of raw milk: frequency and characterization (Port.) Arg. Biol. Tecnol. (Curitiba), 29:505–514.

Atputharaja, J. D., S. Widanapathirana, and U. Samarajeewa. 1986. Microbiology and biochemistry of natural fermentation of coconut palm sap. Food Microbiol. 3:273–280.

Barnes, E. M. 1956. Methods for the isolation of fecal streptococci (Lancefield Group D) from bacon factories. J. Appl. Bacteriol. 19:193–203.

Bean, M. L., K. Jjichi, T. F. Sugihara, J. J. Meehan, and L. Kline. 1963. Effects of modified processing techniques on the performance of egg white solids. Cereal Sci. Today 8:127.

Beelman, R. B., and J. F. Gallander. 1979. Wine deacidification. Adv. Food Res. 25:1–53.

Beltran-Edeza, L. M., and H. Hernandez-Sanchez. 1989. Preservation of ripe tomatoes by lactic acid fermentation. Lebensm.-Wiss. Technol. 22:65–67.

Borch, E., and C. Molin. 1988. Numerical taxonomy of psychrotrophic lactic acid bacteria from prepared meat and meat products. Antonie van Leeuwenhoek 54:301–323.

Boquien, C. Y., Corrieu, G., and M. J. Desmazeaud. 1989. Enzymatic methods for determining populations of *Streptococcus cremoris* AM2 and *Leuconostoc lactis* CNRZ 1091 in pure and mixed cultures. Appl. Microbiol. Biotechnol. 30:402–407.

Branen, A. L., H. C. Go, and R. P. Genske. 1975. Purification and properties of antimicrobial substances produced by *Streptococcus diacetilactis* and *Leuconostoc citrovorum*. J. Food Sci. 40:446–450.

Brenner, D. J. 1973. Deoxyribonucleic acid reassociation in the taxonomy of enteric bacteria. Int. J. Syst. Bacteriol. 23:298–307.

Briggs, M. 1953. The classification of lactobacilli by means of physiological tests. J. Gen. Microbiol. 9:234–248.

Bryan-Jones, G. 1975. Lactic acid bacteria in distillery fermentations, p. 165–175. In: J. C. Carr, C. V. Cutting, and G. C. Whiting (ed.), Lactic acid bacteria in beverages and foods. Academic Press, London.

Burrowes, J., F. H. Schmidt, K. L. Smith, and J. V. Chambers. 1986. Evaluation of summer sausage manufactured using mixed lactobacillus and leuconostoc starter culture. J. Food Prot. 49:280–281.

Caplice, E., G. F. Fitzgerald, and C. Daly. 1987. Conjugative transfer of a recombinant proteinase plasmid from *Streptococcus lactis* to *Leuconostoc mesenteroides*. FEMS Microbiol. Rev. 46:P35.

Carlin, F., C. Nguyen-The, P. Cudennec, and M. Reich. 1989. Microbiological spoilage of fresh ready-to-use grated carrots. Sci. Aliments 9:371–386.

Carr, J. G., and G. C. Whiting. 1971. Microbiological aspects of production and spoilage of cider. J. Appl. Bacteriol. 34:81–93.

Cavett, J. J., G. J. Dring, and A. W. Knight. 1965. Bacterial spoilage of thawed frozen peas. J. Appl. Bacteriol. 28:241–251.

Cavin, J. F., P. Schmitt, A. Arias, J. Lin, and C. Divies. 1988. Plasmid profiles in *Leuconostoc* species. Microbiol. Alim. Nutr. 6:55–62.

Cavin, J. F., H. Prevost, J. Lin, P. Schmitt, and C. Divies. 1989. Medium for screening *Leuconostoc oenos* strains defective in malo-lactic fermentation. Appl. Environ. Microbiol. 55:751–753.

Christopher, F. M., S. C. Seideman, Z. L. Carpenter, G. C. Smith, and C. Vanderzant. 1979. Microbiology of beef packaged in various gas atmospheres. J. Food Prot. 42:240–244.

Cilliers, J. D., and C. J. van Wyk. 1985. Histamine and tyramine content of South African wine. South African J. Enol. Vitic. 6:35–40.

Cogan, T. M. 1975. Citrate utilization in milk by *Leuconostoc cremoris* and *Streptococcus diacetilactis*. J. Dairy Res. 42:139–146.

Cogan, T. M., M. O'Dowd, and D. Mellerick. 1981. Effects of pH and sugar on acetoin production from citrate by *Leuconostoc lactis*. Appl. Environ. Microbiol. 41:1–8.

Collins, E. B. 1972. Biosynthesis of flavor compounds by microorganisms. J. Dairy Sci. 55:1022–1028.

Collins, E. B., and R. A. Speckman. 1974. Influence of acetaldehyde on growth and acetoin production by *Leuconostoc citrovorum*. J. Dairy Sci. 57:1428–1431.

Collins-Thompson, D. L., and G. Rodriguez-Lopez. 1981. Depletion of sodium nitrite by lactic-acid bacteria isolated from vacuum packed bologna. J. Food Prot. 44:593–595.

Collins-Thompson, D. L., and G. Rodriguez-Lopez. 1982. Control of *Brochothrix thermosphacta* by *Lactobacillus* spp. in vacuum-packed bologna. Can. Inst. Food Sci. Technol. J. 15:307–309.

Coovadia, Y. M., Z. Solwa, and J. Van Den Ende. 1987. Meningitis caused by vancomycin-resistant *Leuconostoc* sp. J. Clin. Microbiol. 25:1784–1785.

Coovadia, Y. M., Z. Solwa, and J. Van Den Ende. 1988. Potential pathogenicity of *Leuconostoc*. Lancet i:3–7.

Costello, P. J., G. J. Morrison, T. H. Lee, and G. H. Fleet. 1983. Numbers and species of lactic acid bacteria in wines during vinification. Food Technol. Austr. 35:14–18.

Crapisi, A., P. Spettoli, M. P. Nuti, and P. Zamorani. 1987. Comparative traits of *Lactobacillus brevis*, *Lactobacillus fructivorans*, and *Leuconostoc oenos* immobilized cells for the control of malo-lactic fermentation in wine. J. Appl. Bacteriol. 63:513–522.

Crosa, J. H., D. J. Brenner, and S. Falkow. 1973. Use of a single-strand specific nuclease for analysis of bacterial and plasmid deoxyribonucleic acid homo- and heteroduplexes. J. Bacteriol. 115:904–911.

Daeschel, M. A. 1989. Antimicrobial substances from lactic acid bacteria for use as food preservatives. Food Techn. 43:164–167.

Daeschel, M. A., R. E. Andersson, and H. P. Fleming. 1987. Microbial ecology of fermenting plant materials. FEMS Microbiol. Rev. 46:357–367.

Daly, C. 1983. Use of mesophilic cultures in the dairy industry. Antonie van Leeuwenhoek. 49:297–312.

David, S., and W. M. De Vos. 1987. Genetic characterization of *Leuconostoc* spp. from dutch starter cultures. FEMS Microbiol. Rev. 46:P26.

David, S., G. Simons, and W. M. De Vos. 1989. Plasmid transformation by electroporation of *Leuconostoc paramesenteroides* and its use in molecular cloning. Appl. Environ. Microbiol. 55:1483–1489.

Davis, C. R., N. F. A. Silveira, and G. H. Fleet. 1985a. Occurrence and properties of bacteriophages of *Leuconostoc oenos* in Australian wines. Appl. Environ. Microbiol. 50:872–876.

Davis, C. R., D. Wibowo, R. Eschenbruch, T. H. Lee, and G. H. Fleet. 1985b. Practical implications of malo-lactic fermentation: a review. Am. J. Enol. Vitic. 36:290–301.

Davis, C. R., D. J. Wibowo, T. H. Lee, and G. H. Fleet. 1986. Growth and metabolism of lactic-acid bacteria during and after malo-lactic fermentation of wines at different pH. Appl. Environ. Microbiol. 51:539–545.

De Ley, J., H. Cattoir, and A. Reynaerts. 1970. The quantitative measurement of DNA hybridization from renaturation rates. Eur. J. Biochem. 12:133–142.

De Ley, J., and J. De Smedt. 1975. Improvements of the membrane filter method for DNA/rRNA hybridization. Antonie van Leeuwenhoek J. Microbiol. 41:287–307.

Dellaglio, F., and S. Torriani. 1986. DNA-DNA homology, physiological characteristics and distribution of lactic-acid bacteria isolated from maize silage. J. Appl. Bacteriol. 60:83–92.

Dellaglio, F., M. Vescovo, L. Morelli, and S. Torriani. 1984. Lactic acid bacteria in ensiled high-moisture corn grain:

physiological and genetic characterization. System. Appl. Microbiol. 5:534–544.

De Man, J. C., M. Rogosa, and M. E. Sharpe. 1960. A medium for the cultivation of lactobacilli. J. Appl. Bacteriol. 23:130–135.

De Moss, R. D., R. C. Bard, and I. C. Gunsalus. 1951. The mechanism of the hetero-lactic fermentation: a new route of ethanol formation. J. Bacteriol. 62:499–511.

Denhardt, D. T. 1966. A membrane-filter technique for the detection of complementary DNA. Biochem. Biophys. Res. Commun. 23:641–646.

De Smedt, J., and J. De Ley. 1977. Intra- and intergeneric similarities of *Agrobacterium* ribosomal ribonucleic acid cistrons. Int. J. Syst. Bacteriol. 27:220–240.

Dicks, L. M. T. 1989. A taxonomic study of *Leuconostoc oenos*. Ph.D. manuscript University of Stellenbosch; South Africa.

Dicks, L. M. T., and H. J. J. van Vuuren. 1989. Relatedness of *Leuconostoc oenos* revealed by numerical analysis of total soluble cell proteins, deoxyribonucleic acid (DNA) base compositions, and DNA-DNA hybridizations. Int. J. Syst. Bacteriol. (In press)

Dubois, G., H. Beaumier, and H. Charbonneau. 1979. Inhibition of bacteria isolated from ground meat by Streptococcaceae and Lactobacillaceae. J. Food Sci. 44:1649–1652.

Duitschaever, C. L., N. Kemp, and E. Emmons. 1987. Pure culture formulation and procedure for the production of kefir. Milchwissenschaft 42:80–82.

Edwards, R. A., R. H. Dainty, C. M. Hibbard, and S. V. Ramantanis. 1987. Amines in fresh beef of normal pH and the role of bacteria in changes in concentration observed during storage in vacuum packs at chill temperatures. J. Appl. Bacteriol. 63:427–434.

Egan, A. F., and B. J. Shay. 1982. Significance of lactobacilli and film permeability in the spoilage of vacuum-packaged beef. J. Food Sci. 47:1119–1122 and 1126.

El-Gendy, S. M., H. Abdel-Galil, Y. Shahin, and F. Z. Hegazi. 1983. Characteristics of salt tolerant lactic-acid bacteria, in particular lactobacilli leuconostocs and pediococci isolated from salted raw milk. J. Food Prot. 46:429–433.

Ethiraj, S., and E. R. Suresh. 1985. A note on the occurrence of *Leuconostoc oenos* as a spoilage organism in canned mango juice. J. Appl. Bacteriol. 59:239–242.

Evans, J. B., and C. F. Niven. 1951. Nutrition of the heterofermentative lactobacilli that cause greening of cured meat products. J. Bacteriol. 62:599–603.

Faparusi, S. I. 1971. Microflora of fermenting palm sap. J. Food Sci. Technol. 8:206.

Farrow, J. A. E., R. R. Facklam, and M. D. Collins. 1989. Nucleic acid homologies of some vancomycin-resistant leuconostocs and description of *Leuconostoc citreum* sp. nov. and *Leuconostoc pseudomesenteroides*. Int. J. Syst. Bacteriol. 39:279–283.

Fernandez Escartin, E., R. Torres Vitela, and A. Castillo, Ayala. 1984. Antagonism of *Streptococcus, Lactobacillus* and *Leuconostoc* strains isolated from fresh raw cheeses against some enteropathogenic bacteria (Span.). Rev. Latinoam Microbiol. 26:47–52.

Fleet, G. H., S. Lafon-Lafourcade, and P. Ribereau-Gayon. 1984. Evolution of yeasts and lactic acid bacteria during fermentation and storage of Bordeaux wines. Appl. Environ. Microbiol. 48:1034–1038.

Fox, G. E., E. Stackebrandt, R. B. Hespell, J. Gibson, J. Maniloff, T. A. Dyer, R. S. Wolfe, W. E. Balch, R. Tanner, L. Magrum, L. B. Zablen, R. Blakemore, R. Gupta, L. Bohnen, B. J. Lewis, D. A. Stahl, K. R. Luehrsen, K. N. Chen, and C. R. Woese. 1980. The phylogeny of prokaryotes. Science 209:457–463.

Frank, H. A., and A. S. Dela Cruz. 1964. Role of incidental microflora in natural decomposition of mucilage layer in Kona coffee. Food Sci. 29:850–853.

Galesloot, Th. E., F. Hassing, and J. Stadhouders. 1961. Agar media voor het isoleren en tellen van aromabacterien in zuursels. Neth. Milk Dairy J. 15:127–150.

Galli, A., G. Ottogalli, and L. Franzetti. 1983. Studies on the psychrotrophic microflora of raw milk. Ann. Microbiol. Enzymol. 33:99–108.

Galluzzo, S. J., O. J. Cotterill, and R. T. Marshall. 1974. Fermentation of whole egg by heterofermentative streptococci. Poultry Sci. 53:1575–1584.

Garvie, E. I. 1967a. *Leuconostoc oenos* sp. nov. J. Gen. Microbiol. 48:431–438.

Garvie, E. I. 1967b. The growth factor and amino acid requirements of species of the genus *Leuconostoc*, including *Leuconostoc paramesenteroides* (sp. nov.) and *Leuconostoc oenos*. J. Gen. Microbiol. 48:439–447.

Garvie, E. I. 1969. Lactic dehydrogenases of strains of the genus *Leuconostoc*. J. Gen. Microbiol. 58:85–94.

Garvie, E. I. 1975. Some properties of gasforming lactic-acid bacteria and their significance in classification, p. 339–349. In: J. G. Carr, C. V. Cutting, and G. C. Whiting (ed.), Lactic-acid bacteria in beverages and food. Academic Press, London.

Garvie, E. I. 1976. Hybridization between the deoxyribonucleic acids of some strains of heterofermentative lactic-acid bacteria. Int. J. Syst. Bacteriol. 26:116–122.

Garvie, E. I. 1981. Sub-divisions within the genus *Leuconostoc* as shown by RNA-DNA-hybridization. J. Gen. Microbiol. 127:209–212.

Garvie, E. I. 1983. *Leuconostoc mesenteroides* subsp. *cremoris* (Knudsen and Sørensen) comb. nov. and *Leuconostoc mesenteroides* subsp. *dextranicum* (Beijerinck) comb. nov. Int. J. Syst. Bacteriol. 33:118–119.

Garvie, E. I. 1984. Separation of species of the genus *Leuconostoc* and differentiation of the leuconostocs from other lactic acid bacteria, p.147–178. In: T. Bergan (ed.), Methods in microbiology, vol. 16. Academic Press, London.

Garvie, E. I. 1986. Genus *Leuconostoc*, p.1071–1075. In: P. H. A. Sneath, N. S. Mair, M. E. Sharpe, and J. G. Holt (ed.), Bergey's manual of systematic bacteriology, vol. 2. Williams and Wilkins, Baltimore.

Garvie, E. I. and J. P. E. Farrow. 1980. The differentiation of *Leuconostoc oenos* from non-acidophilic species of *Leuconostoc* and the identification of five strains from the American Type Culture Collection. Am. J. Enol. Vitic. 31:154–157.

Garvie, E. J. and L. A. Mabbitt. 1967. Stimulation of the growth of *Leuconostoc oenos* by tomato juice. Arch. Microbiol. 55:398–407.

Gashe, B. A. 1985. Involvement of lactic-acid bacteria in the fermentation of Tef eragrostis-Tef an Ethiopian fermented food. J. Food Sci. 50:800–801.

Gashe, B. A. 1987. Kocho fermentation. J. Appl. Bacteriol. 62:473–478.

Gasser, F. 1970. Electrophoretic characterization of lactic dehydrogenases in the genus *Lactobacillus*. J. Gen. Microbiol. 62:223–239.

Gasser, F., and M. Hontebeyrie. 1977. Immunological relationships of glucose-6-phosphate dehydrogenase of *Leuconostoc mesenteroides* NCDO 768 (=ATCC 12291). Int. J. Syst. Bacteriol. 27:6–8.

Gibson, T., and Y. Abdel-Malek. 1945. The formation of carbon dioxide by lactic acid bacteria and *Bacillus licheniformis* and a cultural method of detecting the process. J. Dairy Res. 14:35–44.

Grazia, L., and G. Suzzi. 1984. A survey of lactic acid bacteria in Italian silage. J. Appl. Bacteriol. 56:373–379.

Gunsalus, I. C., and M. Gibbs. 1952. The heterolactic fermentation: position of C^{14} in the products of glucose dissimilation by *Leuconostoc mesenteroides*. J. Biol. Chem. 194:871–875.

Hanna, M. O., C. Vanderzant, G. C. Smith, and J. W. Savell. 1981. Packaging of beef loin steaks in 75% CO_2 + 25% O_2. 2. Microbiological properties. J. Food Prot. 44:928–933.

Hanna, M. O., G. C. Smith, J. W. Savell, F. K. McKeith, and C. Vanderzant. 1982. Effects of packaging methods on the microbial flora of livers and kidneys from beef or pork. J. Food Prot. 45:74–81.

Hanna, M. O., J. W. Savell, G. C. Smith, D. E. Purser, F. A. Gardner, and C. Vanderzant. 1983. Effect of growth of individual meat bacteria on pH, color, and odor of aseptically prepared vacuum-packaged round steaks. J. Food Prot. 46:216–221, 225.

Hardy, S., K. L. Ruoff, E. A. Catlin, and J. J. Santos. 1988. Catheter-associated infection with a vancomycin-resistant Gram-positive coccus of the *Leuconostoc*-sp. Pediatr. Infect. Dis. J. 7:519–520.

Hatcher, W. S., Jr., E. C. Hill, D. F. Splittstoesser, and J. L. Weihe. 1984. Fruit beverages, ch. 47, p. 645. In: M. L. Speck (ed.), Compendium of methods for the microbiological examination of foods. Am. Pub. Health Assoc., Washington, D.C.

Heinzl, H., and W. P. Hammes. 1986. Die mikroaerophile Bakterienflora von Obstmaischen. Chem. Mikrobiol. Technol. Lebensm. 10:106–109.

Henick-Kling. T., T. H. Lee, and D. J. D. Nicholas. 1986a. Inhibition of bacterial growth and malo-lactic fermentation in wine by bacteriophage. J. Appl. Bacteriol. 61:287–294.

Henick-Kling. T., T. H. Lee, and D. J. D. Nicholas. 1986b. Characterization of the lytic activity of bacteriophages of *Leuconostoc oenos* isolated from wine. J. Appl. Bacteriol. 61:525–534.

Hitchener, B. J., A. F. Egan, and P. J. Rogers. 1982. Characteristics of lactic acid bacteria isolated from vacuum-packaged beef. J. Appl. Bacteriol. 52:31–37.

Holzapfel, W. H. 1969. Aminosäuresequenz des Mureins und Taxonomie der Gattung *Leuconostoc*. Ph.D. dissertation, University of Munich, Germany.

Holzapfel, W. H., and E. S. Gerber. 1983. *Lactobacillus divergens* sp. nov., a new heterofermentative *Lactobacillus* species producing L(+)-lactate. Syst. Appl. Microbiol. 4:522–534.

Holzapfel, W. H., and E. S. Gerber. 1986. Predominance of *Lactobacillus curvatus* and *Lactobacillus sake* in the spoilage association of vacuum-packaged meat products. Prov. 32nd Europ. Meeting. Meat Res. Workers, p.26, 4:3, Ghent, Belgium, Aug. 24–29, 1986.

Holzapfel, W., and O. Kandler. 1969. Zur Taxonomie der Gattung *Lactobacillus* Beijerinck. VI. *Lactobacillus coprophilus* subsp. *confusus* nov. subsp., eine neue Unterart der Untergattung Betabacterium. Zentralbl. Bakteriol. Parasitenkd. Infektionskr. Hyg. Abt. 2,123:657–666.

Hontebeyrie, M., and F. Gasser. 1975. Comparative immunological relationships of two distinct sets of isofunctional dehydrogenases in the genus *Leuconostoc*. Int. J. Syst. Bacteriol. 25:1–6.

Hontebeyrie, M., and F. Gasser. 1977. Deoxyribonucleic acid homologies in the genus *Leuconostoc*. Int. J. Syst. Bacteriol. 27:9–14.

Horowitz, H. W., S. Handwerger, K. G. Van Horn, and G. P. Wormser. 1987. *Leuconostoc*, an emerging vancomycin-resistant pathogen. Lancet ii:1329–1330.

Horowitz, H., A. Kolokathis, and S. Handwerger. 1989. Genetics of vancomycin resistance in *Leuconostoc* spp. Abstr. Annu. Meet. Am. Soc. Microbiol. A-90:16.

Hosono, A., R. Wardojo, and H. Otani. 1989. Microbial flora in dadih, a traditional fermented milk in Indonesia. Lebensm. Wiss. Technol. 22:20–24.

Ingraham, J. L., R. H. Vaughn, and G. M. Cooke. 1960. Studies of the malolactic organisms of Californian wines. Am. J. Enol. Vitic. 11:1–4.

Isenberg, H. D., E. M. Vellozzi, J. Shapiro, and L. G. Rubin. 1988. Clinical laboratory challenges in the recognition of *Leuconostoc* spp. J. Clin. Microbiol. 26:479–483.

Janse, B. J. H., B. D. Wingfield, I. S. Pretorius, and H. J. J. van Vuuren. 1987. Plasmids in *Leuconostoc oenos*. Plasmid 17:173–175.

Juffs, H. S., and F. J. Babel. 1975. Inhibition of psychrotrophic bacteria by lactic cultures in milk stored at low temperatures. J. Dairy Sci. 58:1612–1619.

Juven, B. J. 1979. A simple method for long-term preservation of stock cultures of lactic acid bacteria. J. Appl. Bacteriol. 47:379–382.

Kandler, O. 1970. Amino acid sequence of the murein and taxonomy of the genera *Lactobacillus, Bifidobacterium, Leuconostoc* and *Pediococcus*. Int. J. Syst. Bacteriol. 20:491–507.

Kandler, O. 1982. Gärungsmechanismen bei Milchsäurebakterien. Forum Mikrobiologie 5:16–22.

Kandler, O., and N. Weiss. 1986. Genus *Lactobacillus*, p. 1209–1234. In: P. H. Sneath, N. S. Mair, and M. E. Sharpe (ed.), Bergey's manual of systematic bacteriology, vol. 2. Williams and Wilkins, Baltimore.

Keddie, R. M. 1951. The enumeration of lactobacilli on grass and silage. Proceed. Soc. Appl. Bacteriol. 14:157–160.

Keller, J. J. 1987. The microbiological spoilage of pasteurized milk in the retail trade, with special reference to the Gram-positive coccoid organisms (Afr.) M.Sc. (Agric.) Thesis, University of Pretoria, South Africa.

Keller, J. J., W. H. Holzapfel, and M. A. Steinmann. 1987. The microbiological population differences between pasteurized and spoiled pasteurized milk (Afr.). S. Afr. J. Dairy Sci. 19:85–95.

Kitchell, A. G., and B. G. Shaw. 1975. Lactic acid bacteria in fresh and cured meat, p.209–220. In: J. G. Carr, C. V. Cutting, and G. C. Whiting (ed.). Lactic acid bacteria in beverages and food. Academic Press, London.

Kole, M., J. Altosaar. 1984. Bovine bile resistance increases *Leuconostoc oenos* 44.40 viability upon lyophilization. Appl. Environ. Microbiol. 47:1150–1153.

Kole, M., J. Altosaar, and P. Duck. 1982. Pilot scale production and preservation of a new malolactic culture

Leuconostoc oenos 44.40 for use in secondary wine fermentation. Biotechnol. Lett. 44:695–700.

Korkeala, H., and P. Mäkelä. 1989. Characterization of lactic acid bacteria isolated from vacuum-packed cooked ring sausages. Int. J. Food Microbiol. 9:33–43.

Korkeala, H., S. Lindroth, M. Suihko, A. Kuhmonen, and P.-L. Penttilä. 1985. Microbiological and sensory quality changes in blood pancakes and cooked ring sausage during storage. Int. J. Food Microbiol. 2:279–291.

Korkeala, H., T. Suortti, and P. Mäkelä. 1988. Ropy slime formation in vacuum-packed cooked meat products caused by homofermentative lactobacilli and a *Leuconostoc* species. Int. J. Food Microbiol. 7:339–347.

Kunkee, R. E. 1967. Malo-lactic fermentation. Adv. Appl. Microbiol. 9:235–279.

Lafon-Lafourcade, S., E. Carre, and P. Ribereau-Gayon. 1983. Occurrence of lactic acid bacteria during the different stages of vinification and conservation of wines. Appl. Environ. Microbiol. 46:874–880.

Langston, C. W., C. Bouma, and R. M. Conner. 1962. Chemical and bacteriological changes in grass silage during the early stages of fermentation. 1. Bacteriological changes. J. Dairy Sci. 45:618–624.

Lautensach, A., and R. E. Subden. 1984. Cloning of malic-acid assimilating activity from *Leuconostoc oenos* in *Escherichia coli*. Microbios 39:29–40.

Law, B. A., and L. A. Mabbit. 1983. New methods for controlling the spoilage of milk and milk products, p. 131–148. In: T. A. Roberts, and F. A. Skinner, Food Microbiology: Advances and Prospects. Academic Press, London.

Lönner, C., and K. Preve-Akesson. 1989. Effects of lactic acid bacteria on the properties of sour dough bread. Food Microbiol. 6:19–35.

Luetticken, R., and G. Kunstmann. 1988. Vancomycin-resistant Streptococcaceae from clinical material. Zentralbl. Bakteriol. Mikrobiol. Hyg. Ser. A. 267:379–382.

Marshall, V. M., and W. M. Cole. 1985. Methods for making kefir and fermented milks based on kefir. J. Dairy Res. 52:451–456.

McDonald, L. C., R. F. McFeeters, M. A. Daeschel, and H. P. Fleming. 1987. A differential medium for the enumeration of homofermentative and heterofermentative lactic acid bacteria. Appl. Environ. Microbiol. 53:1382–1384.

Mol, J. H. H., J. A. E. Hietbrink, H. W. M. Mollen, and J. Van Tinteren. 1971. Observations on the microflora of vacuum packed sliced cooked meat products. J. Appl. Bacteriol. 34:377–397.

Mukherjee, S. K., M. N. Albury, C. S. Pederson, A. G. Van Veen, and K. H. Steinkraus. 1965. Role of *Leuconostoc mesenteroides* in leavening the batter of idli, a fermented food of India. Appl. Microbiol. 13:227–231.

Mundt, J. O. 1970. Lactic acid bacteria associated with raw plant food material. J. Milk Food Technol. 33:550–553.

Mundt, J. O., W. F. Graham, and I. E. McCarty. 1967. Spherical lactic acid-producing bacteria of southern-grown raw and processed vegetables. Appl. Microbiol. 15:1303–1308.

Nel, L., B. D. Wingfield, L. J. Van der Meer, and H. J. J. Van Vuuren. 1987. Isolation and characterization of *Leuconostoc oenos* bacteriophages from wine and sugarcane. FEMS Microbiol. Lett. 44:63–68.

Nickels, C., and H. Leesment. 1964. Methode zur Differenzierung und quantitativen Bestimmung von Säureweckerbakterien. Milchwissenschaft 19:374–378.

Nonomura, H., T. Yamazaki, and Y. Ohara. 1965. Die Äpfelsäure- Milchsäure-Bakterien, welche aus japanischen Weinen isoliert wurden. Mitt. Klosterneuburg 15A:241–254.

Nonomura, H., T. Yamazaki, and Y. Ohara. 1967. Die Äpfelsäure-Milchsäure-Bakterien welche aus französischen und spanischen Weinen isoliert wurden. Mitt. Klosterneuburg 17:345–351.

Nunez, J. A., F. J. Chavarri, and M. Nunez. 1984. Psychrotrophic bacterial flora of raw ewe milk with particular reference to Gram-negative rods. J. Appl. Bacteriol. 57:23–30.

Oberman, H., Z. Libudzisz, and A. Piatkiewicz. 1986. Physiological activity of deep-frozen concentrates of *Leuconostoc* strains. Nahrung 30:147–154.

Okafor, N. 1977. Microorganisms associated with cassava fermentation for garri production. J. Appl. Bacteriol. 42:279–284.

Okafor, N., and M. A. N. Ejiofor. 1985. The linamarase of *Leuconostoc mesenteroides*, production, isolation and some properties. J. Sci. Food Agric. 36:669–678.

Orberg, P. K., and W. E. Sandine. 1984. Common occurrence of plasmid DNA and vancomycin resistance in *Leuconostoc* spp. Appl. Environ. Microbiol. 48:1129–1133.

Ostovar, K., and P. G. Keeney. 1973. Isolation and characterization of microorganisms involved in the fermentation of Trinidad's cacao beans. J. Food Sci. 38:611–617.

O'Sullivan, T., and C. Daly. 1982. Plasmid DNA in *Leuconostoc* species. Ir. J. Food Sci. Technol. 6:206.

Parish, M., and D. Higgins. 1988. Isolation and identification of lactic acid bacteria from samples of citrus molasses and unpasteurized orange juice. J. Food Sci. 53:645–646.

Passos, F. M. L., D. O. Silva, A. Lopez, C. L. L. F. Ferreira, and W. V. Guimaraes. 1984. Characterization and distribution of lactic-acid bacteria from traditional cocoa-bean fermentations in Bahia, Brazil. J. Food Sci. 49:205–208.

Patterson, J. T., and K. J. Baird. 1977. Some findings on the microbiology of vacuum-packaged beef. J. Sci. Food Agric. 28:1043–1044.

Pearce, L. E., and A. C. Halligan. 1978. Cultural characteristics of *Leuconostoc* strains from cheese starters, p.520–521. In: Congrilait 20th International Dairy Congress, Paris, France.

Pederson, C. S. 1930. Floral changes in the fermentation of sauerkraut. N. Y. S. Agric. Exp. Sta. Techn. Bull. 168:1–37.

Pederson, C. S., and M. N. Albury. 1969. The sauerkraut fermentation. N. Y. S. Agric. Exp. Sta. Bull. 824, Geneva, NY.

Peynaud, E., and S. Domercq. 1968. Etude de quatre cents souches de conques heterolactiques isolés de vins. Ann. Inst. Pasteur 19:159–169.

Pilone, G. J., P. E. Kunkee, and A. D. Webb. 1966. Chemical characterisation of wines fermented with various malolactic bacteria. Appl. Microbiol. 14:608–615.

Pivnick, H. 1980. Sugar, cocoa, chocolate, and confectionieries, p.778–821. In: J. H. Silliker, R. P. Elliott, A. C. Baird-Parker, F. L. Bryan, J. H. B. Christian, D. S. Clark, J. C. Olson, Jr., and T. A. Roberts (ed.), Microbial ecology of foods, vol. 2. Academic Press, New York.

Priest, F. G., and J. G. Pleasants. 1988. Numerical taxonomy of some leuconostocs and related bacteria isolated from Scotch whisky distilleries. J. Appl. Bacteriol. 64:379–388.

Pucci, M. J., M. E. Monteschio, and C. L. Kemker. 1988. Intergeneric and intrageneric conjugal transfer of plasmid-encoded antibiotic resistance determinants in *Leuconostoc* spp. Appl. Environ. Microbiol. 54:281–287.

Puspito, H., and G. H. Fleet. 1985. Microbiology of Sayur-Asin fermentation. Appl. Microbiol. Biotechnol. 22:442–445.

Quist, K. B., D. Thomsen, and E. Hoier. 1987. Effect of ultrafiltered milk and use of different starters on the manufacture, fermentation and ripening of Havarti cheese. J. Dairy Res. 54:437–446.

Radler, F. 1975. The metabolism of organic acids by lactic-acid bacteria. In: J. G. Carr, C. V. Cutting, and G. C. Whiting (ed.), Lactic-acid bacteria in beverages and food. Academic Press, London.

Rankine, B. C. 1977. Developments in malo lactic fermentation of Australian red table wines. Am. J. Enol. Vitic. 28:27–33.

Reddy, S. G., R. L. Henrickson, and H. C. Olson. 1970. The influence of lactic cultures on ground beef quality. J. Food Sci. 35:787–791.

Reuter, G. 1970a. Laktobazillen und eng verwandte Mikroorganismen in Fleisch und Fleischwaren. 3. Mitteilung: Abgrenzung der einzelnen Keimgruppen. Fleischwirtschaft 50:1081–1084.

Reuter, G. 1970b. Laktobazillen und eng verwandte Mikroorganismen in Fleisch und Fleischwaren. 4. Mitteilung: Die Ökologie von Laktobazillen, *Leuconostoc* species und Pediokokken. Fleischwirtsch. 50:1397–1399.

Reuter, G. 1975. Classification, ecology and some biochemical activities of lactobacilli of meat products, p.221–229. In: J. G. Carr, C. V. Cutting, and G. C. Whiting (ed.), Lactic acid bacteria in beverages and food. Academic Press, London.

Reuter, G. 1981. Psychrotrophic lactobacilli in meat products, p.253–258. In: T. A. Roberts, G. Hobbs, J. H. B. Christian, and N. Skovgaard (ed.), Psychrotrophic microorganisms in spoilage and pathogenicity. Academic Press, London.

Rogosa, M., J. A. Mitchell, and R. F. Wiseman. 1951. A selective medium for the isolation and enumeration of oral and faecal lactobacilli. J. Bacteriol. 62:132–133.

Ross, G. D. 1981. The inhibition of growth of spoilage microorganisms in milk by *Streptococcus lactis* ssp. *diacetylactis, Leuconostoc cremoris* and *Leuconostoc dextranicum.* Aust. J. Dairy Technol. 36:147–152.

Rossi, J., L. Costamagna, and F. Cleventi. 1978. La flora malolactica in alcuni vini dell' Italia centrole. Annali Facolta di Agraria dell' Universita di Perugia 33:187–196.

Rubin, L. G., E. Vellozzi, J. Shapiro, and H. D. Isenberg. 1988. Infection with vancomycin-resistant streptococci due to *Leuconostoc* sp. J. Infect. Dis. 157:216.

Ruoff, K. L., D. R. Kuritzkes, J. S. Wolfson, and M. J. Ferraro. 1988. Vancomycin-resistant Gram-positive bacteria isolated from human sources. J. Clin. Microbiol. 26:2064–2068.

Sadovski, A. Y., S. Gordin, and J. Foreman. 1980. Psychrotrophic growth microorganisms in a cultured milk product. J. Food Prot. 43:765–768.

Salih, A. G., J. F. Drilleau, F. F. Cavin, C. Divies, and C. M. Bourgeois. 1988. A survey of microbiological aspects of cider making. Int. Inst. Brew. 94:5–8.

Sanchez-Marroquin, A., and P. H. Hope. 1953. Agave juice: Fermentation and chemical composition studies of some species. J. Agric. Food Chem. 1:246–249.

Sandine, W. E., and P. R. Elliker. 1970. Microbiologically induced flavors and fermented foods flavor in fermented dairy products. J. Agr. Food Chem. 18:557–562.

Sanni, A. I., and B. A. Oso. 1988. The production of agad-agidi a Nigerian fermented beverage. Nahrung 32:319–326.

Savell, J. W., M. O. Hanna, C. Vanderzant, and G. C. Smith. 1981. An incident of predominance of *Leuconostoc* sp. in profile of steaks stored in O_2-CO_2-N_2 atmospheres. J. Food Prot. 44:742–745.

Savell, J. W., D. B. Griffin, C. W. Dill, G. R. Acuff, and C. Vanderzant. 1986. Effect of film oxygen transmission rate on lean color and microbiological characteristics of vacuum-packaged beef knuckles. J. Food Prot. 49:917–919.

Saxelin, M. L., E. L. Nurmiaho-Lassila, V. T. Merilainen, and R. I. Forsen. 1986. Ultrastructure and host specificity of bacteriophages of *Streptococcus cremoris, Streptococcus lactis* ssp. *diacetylactis* and *Leuconostoc cremoris* from Finnish fermented milk viili. Appl. Environ. Microbiol. 52:771–777.

Schillinger, U., and F.-K. Lücke. 1986. Milchsäurebakterien-Flora auf vakuumverpacktem Fleisch und ihr Einfluss auf die Haltbarkeit. Fleischwirtschaft 66:1515–1520.

Schillinger, U., W. Holzapfel, and O. Kandler. 1989. Nucleic acid hybridization studies on *Leuconostoc* and heterofermentative lactobacilli and description of *Leuconostoc amelibiosum* sp. nov. System. Appl. Microbiol. 12:48–55.

Schleifer, K. H., and O. Kandler. 1972. Peptidoglycan types of bacterial cell walls and their taxonomic implications. Bacteriol. Rev. 36:407–477.

Schleifer, K. H., J. Kraus, C. Dvorak, R. Kilpper-Bälz, M. D. Collins, and W. Fischer. 1985. Transfer of *Streptococcus lactis* and related streptococci to the genus *Lactococcus* gen. nov. Syst. Appl. Microbiol. 6:183–195.

Schmitt, P., A. G. Mathot, and C. Davies. 1989. Fatty acid composition of the genus *Leuconostoc.* Milchwissenschaft 44:556–559.

Seideman, S. C., C. Vanderzant, M. D. Hanna, Z. L. Carpenter, and G. C. Smith. 1976. Effect of various types of vacuum packages and length of storage on the microbial flora of wholesale and retail cuts of beef. J. Milk Food Technol. 39:745–753.

Shamala, T. R., and K. R. Sreekantiah. 1988. Microbial and biochemical studies on traditional Indian palm wine fermentation. Food Microbiol. 5:157–162.

Sharpe, M. E. 1962. Lactobacilli in meat products. Food Manuf. 37:582–584.

Sharpe, M. E. 1981. The genus *Lactobacillus,* p. 1653–1679. In: M. Starr, M. Stolp, H. Trüper, A. Balows, and K. Schlegel (ed.), The prokaryotes. Springer-Verlag, New York.

Sharpe, M. E., and G. L. Pettipher. 1983. Food spoilage by lactic acid bacteria. Economic Microbiology 8:199–223.

Sharpe, M. E., E. I. Garvie, and R. H. Tilbury. 1972. Some slime forming heterofermentative species of the genus *Lactobacillus.* Appl. Microbiol. 23:389–397.

Shaw, B. G., and C. D. Harding. 1984. A numerical taxonomy study of lactic acid bacteria from vacuum packed beef, pork, lamb and bacon. J. Appl. Bacteriol. 56:25–40.

Shaw, B. G., and C. H. Harding. 1989. *Leuconostoc gelidum* sp. nov. and *Leuconostoc carnosum* sp. nov. from chill-stored meats. Int. J. Syst. Bacteriol. 39:217–223.

Shin, J. 1983. Some characteristics of *Leuconostoc cremoris* bacteriophage isolated from blue cheese. Jpn. J. Zootech. Sci. 54:481–486.

Shin, C., and Y. Sato. 1979. Isolation of *Leuconostoc* bacteriophages from dairy products. Jpn. J. Zootech. Sci. 50:419–422.

Shin, C., and Y. Sato. 1980. Lysogeny in leuconostocs. Jpn. J. Zootech. Sci. 51:478–484.

Sinskey, A., S. Jamas, D. Easson Jr., and C. Rha. 1986. Biopolymers and modified polysaccharides, p. 73–114. In: S. K. Harlander, and T. P. Labuza (ed.), Biotechnology in food processing. Noyes Publications, Park Ridge, New Jersey.

Sodeko, O. O., Y. S. Jzuagbe, and M. E. Ukhun. 1987. Effect of different preservative treatments on the microbial population of Nigerian orange juice. Microbios. 51:133–144.

Soni, S. K., D. K. Sandhu, K. S. Vilkhu, and N. Kamra. 1986. Microbiological studies on *Dosa* fermentation. Food Microbiol. 3:45–54.

Sozzi, T., J. M. Poulin, R. Maret, and R. Pousaz. 1978. Isolation of a bacteriophage of *Leuconostoc mesenteroides* from dairy products. J. Appl. Bacteriol. 44:159–161.

Sozzi, T., F. Gnaegi, N. D'Amico, and H. Hose. 1982. Difficultées de fermentation malolactique du vin dues à des bacteriophages de *Leuconostoc oenos*. Revue Suisse de Viticulture Arboriculture et Horticulture 14:17–23.

Speck, M. L. 1972. Control of food-borne pathogens by starter cultures. J. Dairy Sci. 55:1019–1022.

Speckman, R. A., and E. B. Collins. 1968. Diacetyl biosynthesis in *Streptococcus diacetilactis* and *Leuconostoc citrovorum*. J. Bacteriol. 95:174–180.

Spettoli, P., A. Bottacin, M. P. Nuti, and A. Zamorani. 1982. Immobilization of *Leuconostoc oenos* ML-34 in calcium alginate gels and its application to wine technology. Am. J. Enol. Vitic. 33:1–5.

Splittstoesser, D. F. 1982. Microorganisms involved in the spoilage of fermented fruit juices. J. Food Prot. 45:874–877.

Stackebrandt, E., and C. R. Woese. 1981. The evolution of prokaryotes, p.1–31. In: M. J. Carlile, J. F. Collins, and B. E. B. Moseley (ed.), Molecular and cellular aspects of microbial evolution. Society of General Microbiology Ltd., Symposium 32, Cambridge University Press, Cambridge, UK.

Stackebrandt, E., and M. Teuber. 1988. Molecular taxonomy and phylogenetic position of lactic acid bacteria. Biochimie 70:317–324.

Stackebrandt, E., V. J. Fowler, and C. R. Woese. 1983. A phylogenetic analysis of lactobacilli, *Pediococcus pentosaceus* and *Leuconostoc mesenteroides*. Syst. Appl. Microbiol. 4:326–337.

Stamer, J. R. 1975. Recent developments in the fermentation of sauerkraut, p. 267–280. In: J. G. Carr, C. V. Cutting, and G. C. Whiting (ed.), Lactic acid bacteria in beverages and food. Academic Press, London.

Stamer, J. R. 1988. Lactic acid bacteria in fermented vegetables, p. 67–85. In: R. K. Robinson (ed.), Developments in food microbiology-3. Elsevier Applied Science, London.

Steinkraus, K. H. 1983. Lactic acid fermentation in the production of foods from vegetables, cereals and legumes. Antonie van Leeuwenhoek J. Microbiol. 49:337–348.

Stirling, A. C., and R. Whittenbury. 1963. Sources of the lactic acid bacteria occurring in silage. J. Appl. Bacteriol. 26:86–90.

Tantaoui-Elaraki, A., and A. El Marrakchi. 1987. Study of Moroccan dairy products Iben and Smed. Mircen J. Appl. Microbiol. Biotechnol. 3:211–222.

Teuber, M., and A. Geis. 1981. The family Streptococcaceae (nonmedical aspects), p.1614–1630. In: M. P. Starr, H. Stolp, H. G. Trüper, A. Balows, and H. G. Schlegel (ed.), The prokaryotes. Springer Verlag, Berlin.

Tilbury, R. H. 1975. Occurrence and effects of lactic acid bacteria in the sugar industry, p.177–191. In: J. G. Carr, C. V. Cutting, and G. C. Whiting (ed.), Lactic acid bacteria in beverages and food. Academic Press, London.

Toji, Y. 1979. A hygienic ecological study on meat spoilage: Meat spoilage model (Jap.). Bull. Azabu Vet. Coll. 4:135–154.

Tracey, R. P., and T. J. Britz. 1987. A numerical taxonomic study of *Leuconostoc oenos* strains from wine. J. Appl. Bacteriol. 63:523–532.

Tracey, R. P., and T. J. Britz. 1989a. Freon 11 extraction of volatile metabolites formed by certain lactic acid bacteria. Appl. Environ. Microbiol. 55:1617–1623.

Tracey, R. P., and T. J. Britz. 1989b. Cellular fatty acid composition of *Leuconostoc oenos*. J. Appl. Bacteriol. 66:445–456.

Tsai, H. J., and W. E. Sandine. 1987a. Conjugal transfer of nisin plasmid genes from *Streptococcus lactis* 7962 to *Leuconostoc dextranicum* 181. Appl. Environ. Microbiol. 53:352–357.

Tsai, H. J., and W. E. Sandine. 1987b. Conjugal transfer of lactose-fermenting ability from *Streptococcus lactis* C2 to *Leuconostoc cremoris* CAF7 yields *Leuconostoc* that ferment lactose and produce diacetyl. J. Ind. Microbiol. 2:25–34.

Valakhanovich, A. J., L. A. Vitkovskaya, R. V. Zaretskaya, O. F. Balatnikova, T. V. Polushina, and L. J. Mitskevich. 1975. Use of lyophilization for preserving a commercial strain of *Leuconostoc-mesenteroides* SF-4: Khim-Farm ZH. 9:39–43.

Vanderzant, C., L. K. Chesser, J. W. Savell, F. A. Gardner, and G. C. Smith. 1983. Effect of addition of glucose citrate and citrate lactic-acid on microbiological and sensory characteristics of steaks from normal and dark firm and dry beef carcasses displayed in polyvinyl chloride film and in vacuum packages. J. Food Prot. 46:775–780.

Vanderzant, C., J. W. Savell, M. O. Hanna, and V. Potluri. 1986. A comparison of growth of individual meat bacteria on the lean and fatty tissue of beef, pork and lamb. J. Food Sci. 51:5–8,11.

Van Tieghem, P. E. L. 1878. Sur la gomme de sucrerie. Ann. d. Sci. Nat. Bot. 6e Ser. 6 7:180–202.

Vaughn, R. H. 1985. The microbiology of vegetable fermentations, p.49–109. In: B. J. B. Wood (ed.), Microbiology of fermented foods, vol. 2. Elsevier, New York.

Von Holy, A., and W. H. Holzapfel. 1989. Spoilage of vacuum packaged processed meats by lactic acid bacteria, and economic consequences. Proc. Xth WAFVH International Symposium, Stockholm, Sweden, 6–9 July 1989.

Vrbaski, L., B. Pekic, and M. Hauk. 1988. An improved medium for isolation of *Leuconostoc mesenteroides.* Mikrobiologija (Belgr.) 25:79–90.

Walker, D. K., and S. E. Gilliland. 1987. Buttermilk manufacture using a combination of direct acidification and citrate fermentation by *Leuconostoc cremoris.* J. Dairy Sci. 70:2055–2062.

Weiler, H. G., and F. Radler. 1970. Milchsäurebakterien aus Wein und von Rebenblättern. Zentralbl. Bakteriol. Parasitenkd. Infektionskr. Hyg. Abt. 2 Orig. 124:707–732.

Wenocur, H. S., M. A. Smith, E. M. Vellozzi, J. Shapiro, and H. D. Isenberg. 1988. Odontogenic infection secondary to *Leuconostoc* spp. J. Clin. Microbiol. 26:1893–1894.

Whittenbury, R. 1965a. The enrichment and isolation of lactic acid bacteria from plant material. Zentralbl. Bakteriol. Parasitenkd. Infektionskr. Hyg. Abt. 1 Suppl. 1:395–398.

Whittenbury, R. 1965b. A study of some pediococci and their relationship to *Aerococcus viridans* and the enterococci. J. Gen. Microbiol. 40:97–106.

Whittenbury, R., P. McDonald, and D. G. Brian-Jones. 1967. A short review on some biochemical and microbiological aspects of ensilage. J. Sci. Food Agric. 18:441–444.

Wibowo, D., R. Eschenbruch, C. R. Davis, G. H. Fleet, and T. H. Lee. 1985. Occurrence and growth of lactic-acid bacteria in wine: a review. Am. J. Enol. Vitic. 36:302–313.

Wibowo, D., G. H. Fleet, T. H. Lee, and R. E. Eschenbruch. 1988. Factors affecting the induction of malo-lactic fermentation in red wines with *Leuconostoc oenos.* J. Appl. Bacteriol. 64:421–428.

Wilkinson, B. J., and D. Jones. 1977. A numerical taxonomic survey of *Listeria* and related bacteria. J. Gen. Microbiol. 98:399–421.

Woolford, M. K. 1985. The silage fermentation, p.85–112. In: B. J. B. Wood (ed.), Microbiology of fermented foods. Elsevier, New York.

Yang, D., and C. R. Woese. 1989. Phylogenetic structure of the "leuconostocs": an interesting case of rapidly evolving organism. System. Appl. Microbiol. 12(2):145–149.

The Genera *Lactobacillus* and *Carnobacterium*

WALTER P. HAMMES, NORBERT WEISS, and WILHELM HOLZAPFEL

Lactobacilli are characterized as Gram-positive, nonsporeforming rods, are catalase-negative when growing without blood, usually nonmotile, do not usually reduce nitrate, and utilize glucose fermentatively (Kandler and Weiss, 1986). They may be either homofermentative, producing more than 85% lactic acid from glucose, or heterofermentative, producing lactic acid, CO_2, ethanol, and/or acetic acid in equimolar amounts. The genus *Lactobacillus* presently comprises more than 50 recognized species. These are compiled in Table 1 with the most recent changes in nomenclature indicated by arrows. Based on the results of nucleic acid hybridization and sequencing studies, the genus is now a well-defined group of organisms. The type species is *Lactobacillus delbrueckii* Leichmann 1896 (Beijerinck, 1901). Until recently, members of the genus *Carnobacterium* were also included in the genus *Lactobacillus* (Collins et al., 1987), with the type species *Carnobacterium divergens*. These organisms share some habitats but differ in certain physiological properties, the most important of which are compiled in Table 2. One species, *C. piscicola*, even contains strains pathogenic to fish (Cone, 1982). Thus, the separation of *Carnobacterium* from *Lactobacillus* renders the lactobacilli a homogeneous group of nonpathogenic bacteria which are useful to humans in several respects: They are indispensable agents of the fermentation of foods and feed and exert probiotic effects in humans and animals. On the other hand, *Carnobacterium* is known exclusively for its adverse effects as an agent of food spoilage or fish disease. With regard to these special properties, it is justified to treat the two genera within one chapter. In the first edition of *The Prokaryotes*, the genus *Lactobacillus* was excellently treated by M. E. Sharpe (1981), so it is the intention of the authors to concentrate on new results and points of interest in this chapter.

The Genus *Lactobacillus*

Habitats

Lactobacilli are strictly fermentative and have complex nutritional requirements, needing to be supplied with carbohydrates, amino acids, peptides, fatty acid esters, salts, nucleic acid derivatives, and vitamins. Large amounts of lactic acid and small amounts of other compounds are the products of their carbohydrate metabolism. Lactobacilli grow in a variety of habitats, wherever high levels of soluble carbohydrate, protein breakdown products, vitamins, and a low oxygen tension occur. They are aciduric or acidophilic, different species having adapted themselves to grow under widely different environmental conditions, and their production of high levels of lactic acid lowers the pH of the substrate and suppresses many other bacteria; these factors account for the wide distribution of lactobacilli and their successful establishment in many markedly different habitats (Sharpe, 1981).

Humans and Animals

ORAL CAVITY. The oral cavity of humans and animals represents a mucosal membrane region that is constantly exposed to exogenous factors, especially during intake of food and breathing. Gram-negative anaerobes and streptococci predominate in this mainly anaerobic environment, while the lactobacilli constitute only about 1% of the total microbes in the healthy human mouth (London, 1976). The composition and density of the microbial population is determined mainly by three ecological determinants, nutrition, redox potential, and adhesion. In addition, several unspecific factors are operative, including (Knoke and Bernhardt, 1986): 1) the saliva (flow rate, redox potential, buffer capacity and pH, concentration of ions, salts and enzymes); 2) the swallowing process;

Table 1. List of the species of the genus *Lactobacillus*.[a]

L. acetotolerans[VP] (Entani et al., 1986)
L. acidophilus[AL]
L. agilis[VP] (Weiss et al., 1981)
L. alimentarius[VP] (Reuter, 1983)
L. amylophilus[VP] (Nakamura and Crowell, 1979)
L. amylovorus[VP] (Nakamura, 1981)
L. animalis[VP] (Dent and Williams, 1982)
L. aviarius subsp. *araffinosus*[VP] (Fujisawa et al., 1984)
L. aviarius subsp. *aviarius*[VP] (Fujisawa et al., 1984)
L. bavaricus[VP] (Stetter and Stetter, 1980)
L. bifermentans[VP] (Kandler et al., 1983a)
L. brevis[AL]
L. buchneri[AL]
L. bulgaricus → *L. delbrueckii* subsp. *bulgaricus*
L. carnis synonym *Carnobacterium piscicola*
L. casei[AL]
L. casei subsp. *alactosus* → *L. paracasei* subsp. *paracasei*
L. casei subsp. *pseudoplantarum* → *L. paracasei* subsp. *paracasei*
L. casei subsp. *rhamnosus* → *L. rhamnosus*
L. casei subsp. *tolerans* → *L. paracasei* subsp. *tolerans*
L. catenaformis[AL]
L. cellobiosus synonym *L. fermentum*
L. collinoides[AL]
L. confusus[AL]
L. coryniformis subsp. *coryniformis*[AL]
L. coryniformis subsp. *torquens*[AL]
L. crispatus[AL]
L. curvatus[AL]
L. delbrueckii subsp. *bulgaricus*[VP] (Weiss et al., 1983)
L. delbrueckii subsp. *delbrueckii*[AL]
L. delbrueckii subsp. *lactis*[VP] (Weiss et al., 1983)
L. divergens → *Carnobacterium divergens*
L. farciminis[VP] (Reuter, 1983)
L. fermentum[AL]
L. fructivorans[AL]
L. fructosus[AL]
L. gasseri[VP] (Lauer and Kandler, 1980)
L. graminis[VP] (Beck et al., 1988)
L. halotolerans[VP] (Kandler et al., 1983b)
L. hamsteri[VP] (Mitsuoka and Fujisawa, 1987)
L. helveticus[AL]
L. heterohiochii synonym *L. fructivorans*
L. hilgardii[AL]
L. homohiochii[AL]
L. jensenii[AL]
L. kandleri[VP] (Holzapfel and van Wyk, 1982)
L. kefir[VP] (Kandler and Kunath, 1983)
L. kefiranofaciens[VP] (Fujisawa et al., 1988)
L. lactis → *L. delbrueckii* subsp. *lactis*
L. leichmannii synonym *L. delbrueckii* subsp. *lactis*
L. malefermentans[VP] (Farrow et al., 1988)
L. mali[AL]
L. maltaromicus[AL]
L. minor[VP] (Kandler et al., 1983b)
L. minutus[AL]
L. murinus[VP] (Hemme et al., 1980)
L. oris[VP] (Farrow and Collins, 1988)
L. parabuchneri[VP] (Farrow et al., 1988)
L. paracasei subsp. *paracasei*[VP] (Collins et al., 1989)
L. paracasei subsp. *tolerans*[VP] (Collins et al., 1989)
L. pentosus[VP] (Zanoni et al., 1987)

(*continued*)

Table 1. *Continued*

L. piscicola → *Carnobacterium*
L. plantarum[AL]
L. reuteri[VP] (Kandler et al., 1982)
L. rhamnosus[VP] (Collins et al., 1989)
L. rogosae[AL]
L. ruminis[AL]
L. sake[AL]
L. salivarius subsp. *salicinius*[AL]
L. salivarius subsp. *salivarius*[AL]
L. sanfrancisco[VP] (Weiss and Schillinger, 1984)
L. sharpeae[VP] (Weiss et al., 1981)
L. suebicus[VP] (Kleynmans et al., 1989)
L. trichodes synonym *L. fructivorans*
L. vaccinostercus[VP] (Okada et al., 1979)
L. vaginalis[VP] (Embley et al., 1989)
L. viridescens[AL]
L. vitulinus[AL]
L. xylosus → *Lactococcus lactis* subsp. *lactis*
L. yamanashiensis synonym *L. mali*

[a]The superscript AL indicates that the name was included on the *Approved Lists of Bacterial Names* (Skerman et al., 1980). The superscript VP indicates that the name, although not on the *Approved Lists of Bacterial Names,* was subsequently validly published in the *International Journal of Systematic Bacteriology.* In the latter case, the literature concerning the description of the new species or the new combination is given. The most recent changes in nomenclature are indicated by arrows.

3) the diet (composition, consistency, frequency of meals); 4) microbial interactions; 5) mouth hygiene; and 6) factors such as phagocytosis and leukodiapedesis.

Streptococci represent the major part of the microbial population at all sites of the human mouth. The anatomical features, in combination with the previously mentioned factors will, however, determine the presence and relative proportions of other bacterial groups, including lactobacilli. In general, lactobacilli have been found to constitute <0.1% of the populations of either the cheek and tongue, <0.005% of intragingival plaque, and <1% each of the saliva and the gingival crevice (Marsh and Martin, 1984). Studies of tooth surfaces of infants and young children (Carlsson and Gothefors, 1975) have shown lactobacilli to be present only in very small numbers or as transients in the mouth. Lactobacilli developing in the mouth of 2- to 5-year-old children consisted mostly of *L. casei* and *L. rhamnosus,* and occasionally of *L. acidophilus* and *L. fermentum. L. casei* was found especially in children with carious lesions. Rogosa et al. (1953), in their classical paper on the identification of oral lactobacilli, identified 500 strains isolated from saliva specimens of 130 school children. *L. casei* and *L. fermentum* were the predominant species present in 59% and 45% of the samples, respectively,

Table 2. Key characteristics for differentiating lactobacilli from carnobacteria.

Characteristic	Lactobacillus	Carnobacterium
Growth on acetate agar (pH 5.4)	+	−
Growth at pH 4.5	+	−
Growth at pH 9.0[a]	−	+
Lactic acid isomers produced	L(+), D(−), DL	L(+)
Type of diamino acid in peptidoglycan[b]	Lys, mDpm, Orn[b]	mDpm[b]
Major $C^{18:1}$-cellular fatty acid[c]	cis-Vaccenic acid(11,12)	Oleic acid(9,10)
Fermentation type (glucose)	Homo or Hetero	Atypical homo[d]
GC content (mol%)	32–55[e]	33.0–37.2[c]

[a]In modified MRS-broth with 2% sucrose.

[b]Abbreviations based on Schleifer and Kandler (1972).

[c]Collins et al. (1987). The numbers give the position of the double bond.

[d]Glycolysis with minor and retarded CO_2 production (De Bruyn et al., 1988). Homo, predominant fermentation product is lactic acid; hetero, CO_2, ethanol, and acetic acid produced in addition to lactic acid.

[e]Kandler and Weiss (1986).

L. acidophilus in 22%, and L. brevis in 17%. L. buchneri, L. salivarius, L. plantarum, and L. cellobiosus occurred less frequently. L. salivarius, although representing only about 2% of the total oral lactobacillus population in humans, has its almost unique habitat here, and in hamsters, where it represents the dominant part of the homofermentative lactobacilli (Rogosa et al., 1953). Those findings have been confirmed by other studies on children and adults (London, 1976) in which similar species were isolated, and L. casei and L. fermentum were found to predominate. All strains of lactobacilli isolated from deep dental plaque by Shovell and Gillis (1972) were L. casei, and other work has confirmed that this species is the prevalent lactobacillus in plaque (Basson and Van Wyk, 1982; Depaola, 1989; Hahn et al., 1989; Wijeyeweera and Kleihberg, 1989). Kneist et al. (1988) found that L. casei subsp. rhamnosus is the dominant lactobacillus in softened and hard carious dentin of 125 deciduous molars. In relative importance, this species was followed by L. plantarum, L. casei subsp. casei, L. curvatus, L. xylosus, and L. coryniformis. Although present in carious lesions, the lactobacilli are not considered to be actively involved in carious progression. Kneist et al. (1988) and Russel and Ahmed (1978) have shown that neither L. acidophilus, nor L. casei or L. fermentum is able to form plaque alone (i.e., without the participation of Streptococus mutans or S. sanguis). Yet, lactobacilli were found to appear at the site of 85% of progressive lesions before clinical diagnosis of carious progression was made (Boyar and Bowden, 1985). The association of heterofermentative lactobacilli with dental caries is not often reported; however, Tina (1987) identified 12 heterofermentors among 46 lactobacillus isolates from carious lesions, of which 7

were classified as L. cellobiosus and 5 as L. fermentum.

Heterofermentative Lactobacillus strains isolated from human saliva and classified as L. brevis (Hayward, 1957; Hayward and Davis, 1956), were found not to be genetically or biochemically related to this or other recognized heterofermentative species, and were classified into a new species, Lactobacillus oris (Farrow and Collins, 1988).

Using a modified Rogosa SL agar medium (with melezitose as the only sugar), Claesson and Crossner (1985) found L. casei to be the most common oral Lactobacillus species, representing a higher proportion of the salivary lactobacillus population in children than in adults.

It is recognized that the microbial population of the saliva is not representative of different areas of the mouth, and although it has been suggested that dental plaque contributes little to this population (Hakgudener, 1985; Hardie and Bowden, 1974; Matee et al., 1985), recent reports indicate a definite relationship between the incidence of caries and the lactobacillus population of the saliva (Alaluusua et al., 1989; Bjarnason, 1989; Krasse, 1988; Minah et al., 1985).

Denepitiya and Kleinberg (1982) found that dental plaque was a significant although not the major source for salivary sediment. Due to fluctuations in numbers of lactobacilli and pH of the saliva, Sullivan and Schroeder (1989) concluded that lactobacilli had only low caries-predictive ability for children from 5 to 7 years of age. Data from other workers, however, contradicted this report, and specifically suggest the salivary lactobacillus population to be a useful tool for early diagnosis of caries (Crossner and Unell, 1986; Krasse, 1988; Vanderas, 1986; Wikner, 1986). Neither the use of fluoride (gel) toothpaste, fluoride-containing mouth rinse (Brown et al., 1983; Etemadzeh et al., 1989), nor fluoride-containing chewing gum (Ekstrand et al., 1985), or even chlorhexidine treatments (Lundstrom and Krasse, 1987) were found to decrease oral lactobacilli numbers significantly. By contrast, *S. mutans* numbers were reduced by chlorhexidine treatment (Lundstrom and Krasse, 1987).

Detailed studies of dental plaque show that it is initiated largely by organisms capable of adhering to tooth surfaces, including *Streptococcus mutans,* and consists partly of extracellular glucose polymers produced by the oral streptococci. Only at this stage may lactobacilli multiply within this matrix. Surveys of noncarious dental plaque show that lactobacilli are present either only in small numbers of 5–12% or may not be isolated at all (Hardie and Bowden, 1974). The presence of unrestricted carious cavities considerably increases the lactobacillus count, and *L. casei* in particular appears to be associated with such lesions (Kneist et al., 1988). However, as the low pH (initiated by the streptococci) found in carious cavities favors lactobacilli, the high count may at least in part be the result of caries and not the cause (Alaluusa et al., 1987; Hardie and Bowden, 1974; Wijeyeweera and Kleinberg, 1989).

INTESTINAL TRACT. Conditions influencing and controlling the intestinal microbial population of homeothermic and poikilothermic animals are discussed by Clarke and Bauchop (1977), of humans in particular by Drasar and Hill (1974), Hentges (1983), Irrgang and Sonnenborn (1988) and Knoke and Bernhardt (1986), and of the gastrointestinal (GI) system, by Goldin (1986), Grubb et al. (1989), and Simon and Gorbach (1984).

With a total population of about 10^{14} microorganisms, the human GI tract harbors about 400 to 500 different bacterial species and subspecies, of which the largest part (10^{10}–10^{12}/g) is found in the colon (Goldin, 1986; Irrgang and Sonnenborn, 1988). Especially for the lower parts of the GI tract, anaerobic techniques are important for obtaining a reliable picture of the population size and distribution. The assessment of the lactobacilli, proportional to the total in vivo microbial population of the intestinal tract of humans, is difficult, and modern concepts of colonization and population ratios are largely based on data obtained from monogastric animals. Moreover, in several studies using conventional anaerobic techniques for enumeration and isolation, no distinction was made between anaerobic lactobacilli and bifidobacteria (Sharpe, 1981). The genus *Bifidobacter-*

ium, a major bacterial group of the gastrointestinal tract, resembles *Lactobacillus* in several physiological properties. However, being the most ancient group of the Actinomycetes subdivision of the Gram-positive eubacteria, it is only distantly related to the lactobacilli (Stackebrandt and Teuber, 1988).

In the neonatal mammal—such as the human infant, piglet, and calf—the intestinal tract is sterile at birth. It is rapidly colonized from the mouth, breast, hands, and rectum, usually initially with the mother's vaginal and perianal population (Knoke and Bernhardt, 1986). In detailed studies of a wide variety of young animals, Smith (1965a, 1971) showed that at birth the GI tract becomes flooded with multiplying bacteria, including coliforms, clostridia, and other anaerobes, followed rapidly by lactobacilli. The balance of the microbial population is then controlled by acid secretion in the stomach (although this is buffered to some extent by the milk imbibed) and ingestion of immunoglobulins and other protective factors in the mother's milk (reviewed by Reiter, 1978), so that lactobacilli soon become the dominant organisms and other groups rapidly decline (Knoke and Bernhardt, 1986). Lactobacilli predominate in the stomach and small intestine, but in the jejunum and large intestine, strict anaerobes such as bacteroides predominate, and lactobacilli constitute only 0.07–1% of the total bacteria. This desirable predominance of lactobacili in the upper intestine, which is established on suckling, helps to prevent the potentially lethal diarrhea or scouring that occurs in young animals when enteropathogenic coliforms proliferate in the upper GI tract.

The production of antimicrobial substances by intestinal lactobacilli may also contribute to this "protective" effect (Fernandes et al., 1987; Mehta et al., 1983). Such "active" products are formed by (among others) *L. acidophilus*—probably the most important *Lactobacillus* species of the GI tract—and some have been identified as bacteriocins. Examples are lactacin B (Barefoot and Klaenhammer, 1983), lactacin F (Muriana and Klaenhammer, 1987), acidophilin (Shahani et al., 1977), and an unidentified broad-spectrum bacteriocin product by lactobacilli isolated from the murine GI tract (McCormick and Savage, 1983). Possible mechanisms by which lactobacilli suppress the rest of the bacterial population are not fully understood yet, and are discussed under "Antimicrobial Activities."

Comparing the fecal microbial population of breast-fed and bottle-fed infants, significantly higher numbers of *L. acidophilus* were found in the feces of the latter (Mitsuoka and Kaneuchi,

1977) with *Bifidobacterium breve* dominating in both groups (Benno et al., 1984). As with other young animals, the mother's milk protects the infant against neonatal coliform infections and promotes the development of high numbers of bifidobacteria and low numbers of lactobacili. Various aspects of the influence of feeding on the microflora are discussed by Reiter (1978). In the feces of 66 infants aged 3–220 days, including breast- and bottle-fed babies, Mitsuoka et al. (1975) found *L. acidophilus, L. fermentum,* and *L. salivarius* to be present at levels varying from 10^3 to 10^{10}/g feces. These they regarded as the indigenous species while *L. fermentum* was isolated from the feces of 6 out of 10 breast-fed infants, all less than 7 days old (Sharpe, 1981). Most strains, previously identified as *L. fermentum,* are probably representatives of *L. reuteri.* This aspect is discussed elsewhere in this section. In the young calf, Contrepois and Gouet (1973) confirmed the findings of Smith (1965a) of the dominance of lactobacili in the stomach and upper GI tract, although they also observed varying numbers of strict anaerobes to be present from the stomach onwards. From the abdominal fluid of the milk-fed calf, isolates have been identified as *Lactobacillus lactis* and *L. fermentum,* while, in addition to these two species, *L. lactis* was isolated from the dried stomach of the young calf, indicating that the milk-fed calf stomach is their natural habitat (Sharpe, 1981). This observation deserves special attention in view of their practical importance and occurrence in hard cheese of the Emmental and Parmesan types (see below), where the natural calf stomach is used as a source of rennin. In the rumen of young calves, *L. acidophilus* was the dominant lactobacillus, with *L. fermentum* present in lower numbers (Marounek et al., 1988). The latter, however, was found to be the major lactobacillus adhering to columnar epithelial cells of calves (Mäyrä-Mäkinen et al., 1983). Lactobacilli are indigenous to the stomach of rodents and pigs and to the crop of chickens, thickly colonizing the surface of stratified squamous epithelium in the oesophagus, crop, or stomach of these neonatal and young animals (Savage, 1977; Tannock et al., 1982, 1987). This adherent population, which is continuously shedding organisms from epithelial surfaces, is instrumental in controlling colonization and composition of the indigenous lactobacilli of the GI tract. Colonization is host-specific, i.e., only strains of lactobacilli isolated from birds will adhere to the chicken crop (Fuller, 1973) and only strains isolated from the rat will adhere to the rat stomach epithelium (Kawai and Shegara, 1977). In the neonatal piglet, Fuller et al. (1978) found *L.*

fermentum, L. acidophilus, and *L. salivarius,* sometimes associated with streptococci, to be colonizing the stomach epithelial cells; these species were also present in the duodenal contents, at levels of 10^7 to 10^8/g (Barrow et al., 1977). Epithelial association appeared to be an important factor in the maintenance of populations of lactobacilli in the porcine tract (Pedersen and Tannock, 1989). Using biotin-labelled plasmid DNA probes, Tannock (1989) showed an association of *L. delbrueckii* and *L. reuteri* with the epithelium of the murine forestomach. In the pig intestine, *L. acidophilus* and *L. reuteri* were found to be the dominant *Lactobacillus* species.

In the chicken crop, biotypes of *L. salivarius* (Sarra et al., 1985) and *L. fermentum* were the species adhering to the epithelium (Fuller, 1973); some *L. fermentum* strains may have been synonymous with *L. reuteri* (Kandler et al., 1980), for which a chromosome-located determinant of crop adhesion was suggested (Sarra et al., 1986).

In the alimentary tract of chickens, *L. salivarius* and *L. acidophilus* appear to be the major species of homofermentative lactobacilli (Kvasnikov et al., 1982; Sarra et al., 1985) and *L. reuteri* the major heterofermenter (Sarra et al., 1985). Strains of *L. fermentum* reported by Kvasnikov et al. (1981) were probably identical to *L. reuteri.*

L. reuteri, probably the dominant heterofermentative lactobacillus of the GI tract of humans and most animals (Kandler et al., 1980; Kandler and Weiss, 1986; Reuter, 1965; Sarra et al., 1985), shows close physiological resemblance, but no genetic relationship, with *L. fermentum* (Kandler and Weiss, 1986). According to Talarico et al. (1988), its ability to produce a potent, broad-spectrum antimicrobial agent (reuterin) from glycerol, may be unique among the lactobacilli.

The production of reuterin and other bacteriocins by lactobacilli is discussed elsewhere in this section. During the first year of life, the intestinal microbial population gradually stabilizes and soon approaches the typical composition of adults (Stark and Lee, 1982).

In the stomach of homeothermic animals and humans, the number and dominant types of microorganisms are largely influenced by the pH. At a normal stomach pH of 3.0 (range 2.2 to 4.2), most organisms ingested with nutrients and saliva are killed by hydrochloric acid in the stomach, leaving a population of about 10^3/ml that is dominated by lactobacilli and streptococci (Goldin, 1986; Irrgang and Sonnenborn, 1988). The acid-tolerant and typically nonresident population of the stomach and duodenum may be influenced by different antimicrobial factors (acid, peristalsis, antibodies), of which the pH is the most important. At pH <3.0 a dysbiosis that favors lactobacilli and yeasts may be expected, while dysbiosis at pH >3.0 may cause the increase of the lactobacillus (and total) population to $>10^6$/ml of gastric juice (Knoke and Bernhardt, 1986). In the normal human stomach, *L. acidophilus* ("type I") and *L. fermentum* appear to be the main representatives of the lactobacilli (Knoke and Bernhardt, 1986).

With a pH range of 4.3 to 6.4 (for different animal species), the anterior part of the stomach favors the development of increasing numbers of lactobacilli. The population of the jejunum as a transitional zone between the stomach and the ileum is still dominated by lactobacilli and streptococci, representing about 90% of the total population of maximally 10^5/g. In the ileum, with up to 10^9 microorganisms/g, the relative proportion of the lactobacilli drops and numbers of up to 10^5/g are typical (Simon and Gorbach, 1984), with the strictly anaerobic bacteria as the dominant group. The largest part and the highest concentration of the GI microbial population is harbored in the colon, with total numbers ranging from 10^{10} to 10^{12}/g, and the lactobacilli only account for numbers between 10^6 and 10^{10}/g (Drasar and Hill, 1974; Goldin, 1986; Simon and Gorbach, 1984). At least one-third of the feces content is accounted for by bacterial mass (Stephens and Cummings, 1980), of which $\geq$95% is contributed by nonsporeforming Gram-negative and Gram-positive bacteria (Irrgang and Sonnenborn, 1988). In contrast to previous assumptions, it has been shown that the composition of the fecal microbial population is not completely representative of the situation in the lower intestines (Irrgang and Sonnenborn, 1988). Lactobacilli isolated from the feces, usually at levels of between 10^4 and 10^9/g, consist of *L. acidophilus, L. fermentum, L. salivarius,* and, more irregularly, *L. lactis, L. casei, L. plantarum, L. brevis,* and *L. buchneri* (Reuter, 1965). By feeding people diets high in lactobacilli, tracing these last five species to the feces, and then excluding lactobacilli from the diet, Reuter (1965) was able to show that these five species were all transients. These results, together with those of other workers (Bianchi-Salvadori et al., 1984; Fernandes et al., 1987; Finegold et al., 1977; Kandler and Weiss, 1986; Mitsuoka et al., 1975; Moore and Holdeman 1974; Sarra and Dellaglio, 1984; Sarra et al., 1989) indicate that *L. acidophilus, L. gasseri, L. fermentum, L. reuteri,* and *L. salivarius* are the predominant lactobacilli in the lower (human) intestine and feces. In one study, Dellaglio and Sarra (1984) found *L. lactis* as the dominant *Lactobacillus* species in the human intestine. Different biotypes of these species may be present and are not the same as those found in animals (Mitsuoka et al., 1975), and host-specific cryptic plasmids are thought to be involved in their adhesion (Lin and Savage, 1985). On the other hand, *Lactobacillus* strains fed to the same species from which they were isolated do not necessarily survive the passage through the stomach and small intestine (Jonsson et al., 1985). In addition to the representative *Lactobacillus* species mentioned, a number of species, probably of lesser importance, and some strictly anaerobic ones were found infrequently associated with the GI tract, including *L. animalis* (Dent and Williams, 1982), *L. crispatus* (Kandler and Weiss, 1986), *L. murinus* (from mice and rats) (Hemme et al., 1980; Moore and Holdeman, 1974), and *L. ruminis* (isolated from the human GI tract) (Sharpe et al., 1973b; 1973c). A new species, *Lactobacillus hamsteri,* was recently isolated from the intestine of hamsters (Mitsuoka and Fujisawa, 1987). *L. catenaforme* (isolated from human feces) and *L. rogosae* were found not to be phylogenetically related to the lactobacilli (Kandler and Weiss, 1986). Anaerobic lactobacilli may at times be part of the majority anaerobic population of the feces, present at levels of 10^{10}/g feces (Mitsuoka and Ohno, 1977). In the large intestine, however, simple sugars are not available, and organisms must rely on sugars derived from GI-tract mucins and plant material (Salyers et al., 1977), and it seems likely that lactobacilli must rely on other organisms degrading these complex carbohydrates to obtain their energy sources.

The relative contribution of *L. acidophilus* to the *Lactobacillus* population is generally recognized (Lidbeck et al., 1987; Luchansky et al., 1989) although substantial proof of the constancy of population numbers has been lacking thus far. Fecal samples from two pigs, kept on a constant diet and analyzed at weekly intervals over 3 months, contained a *Lactobacillus* population of 10^8 to 10^{10}/g, to which *L. acidophilus,* on average, contributed a relatively constant figure of 10% (Von Aulock and Holzapfel, 1987). Out of 49 strains of heterofermentative lactobacilli isolated as a minor group in this study, 42 were identified as *L. reuteri,* 4 as *L. fermentum,* while 3 remained unidentified (Tina, 1987). The identity of most *L. acidophilus* strains mentioned in the literature may be questioned in view of recent reports on *L. gasseri,* which is found in similar habitats as *L. acidophilus* and cannot be distinguished from this species, or even from *L. crispatus* or *L. amylovorus*, by simple phenotypic criteria (Kandler and Weiss, 1986). To some extent, therefore, the same nomenclatural discrepancies as for *L. reuteri* and *L. fermentum* may exist here.

Detailed studies show that in the adult pig, *L. fermentum, L. acidophilus, L. cellobiosus,* and *L. salivarius* predominate in the feces (Fuller et al., 1960; Mitsuoka, 1969). Studies of the ileum, cecum, and colon epithelial tissues of beagle dogs (Davis et al., 1977) showed that they harbor a wide variety of lactobacilli: *L. acidophilus* in the highest numbers, then *L. leichmannii, L. plantarum,* and *L. fermentum.* The much smaller numbers of many other species isolated may have been transients. Mitsuoka et al. (1976) isolated *L. acidophilus, L. salivarius,* and *L. fermentum,* including a number of different biotypes, from the feces of dogs. In the rat, *L. acidophilus* and *L. salivarius* predominated (Raibaud et al., 1973), some thermobacteria (for definition see "Grouping of Species of Lactobacilli," this chapter) being able to ferment pentoses and one isolate being proteolytic (Moreau et al., 1976). Kawai and Shegara (1977) isolated a strain of *L. fermentum* that could colonize rat stomach epithelial cells, and in the mouse, Roach et al. (1977) isolated several different groups of unclassified thermobacteria, some also able to ferment ribose, which were the indigenous lactobacilli colonizing the stomach. Pentose-fermenting, thermobacteria lactobacilli, resembling *L. acidophilus* and *L. gasseri,* were also isolated from the feces of pigs (Von Aulock, 1981). These strains were found to produce lactate and acetate in equimolar ratios from ribose and L-arabinose (Radmore et al., 1984).

Relatively little is still known about the *lactobacillus* population of the ruminant and especially of the rumen of adult animals. The rumen begins to function when a solid diet is fed. Anaerobic lactobacilli have been isolated as part of the majority population from ruminating calves (Bryant et al., 1958), and similar organisms have been isolated from older animals and adult cattle (Hungate, 1966; Jayne-Williams, 1979; Sharpe et al., 1973c). Several *Lactobacillus* strains associated with the rumen show resistance to antibiotics normally used in animal husbandry (Dutta and Devries, 1981). In a taxonomic study on strains isolated by Stewart and Duncan (1985) from the rumen of sheep treated with avoparcin, Stewart et al. (1988) found them to be related to *L. fermentum.* These strains may be identical to *L. reuteri.*

The level of lactobacilli and bifidobacteria in the rumen, often counted as one group, is also influenced by the diet. On high-fiber diets, lactobacilli may constitute 1–2% of the total population whereas on high-carbohydrate diets they may constitute 15–20% of the total (Latham et al., 1971).

Several external factors may influence the composition and numbers of the intestinal microbial population. The influence of stress on GI microorganisms has been reviewed by Tannock (1983) and Suzuki et al. (1989). Crowding and continued heat stress (31°C) caused a slight (though not significant) increase in the GI *Lactobacillus* population of rats. Under continued heat stress (35°C), however, the lactobacillus population in the duodenum, jejunum, and cecum of chickens decreased slightly (Suzuki et al., 1989).

Although dietary stress may cause population shifts in the GI tract (Tannock, 1983), Moore and Holdeman (1975) could not detect any significant GI population shift in persons changing from an omnivorous to a vegetarian diet. The findings of Maier et al. (1974), however, contradicted the conclusions of Moore and Holdeman (1975). Takahashi et al. (1982) reported a significant reduction in the *Lactobacillus* population in the cecum of chicken fed a diet formulated to be low in lysine, as compared to a control diet. Most reports, however, indicate that the type of nutrition only has a minor influence on the microbial population of the GI tract (Drasar and Barrow, 1985; Knoke and Bernhardt, 1986). The prospective therapeutic and dietetic potential of intestinal lactobacilli has evoked increased interest in their metabolic activities and beneficial physiological properties relevant to the GI environment. The presence of lactobacilli in the feces could not serve as proof of their colonization in, and adhesion to, the GI tract. Continued therapeutic and nutritional benefits from fermented milk and other food products, as well as from food and feed supplements and adjuncts containing viable lactobacilli, can only result when these organisms survive and proliferate in the GI tract.

Although it is accepted that lactobacilli such as *L. acidophilus* will establish in the GI tract (Watkins and Miller, 1983), the mechanism of colonization of the epithelial habitat and of adhesion to squamous epithelia is not yet fully understood (Savage, 1983). The GI tract represents an extremely complex, "steady-state" ecosystem in which numerous niches (Savage, 1977) are colonized by microcolonies, surrounded by glycocalyx (Costerton et al., 1983). This enables the GI microbial population to become established as a dense biofilm covering undigested particles, as well as the mucosal layer and the intestinal epithelial surface (Costerton et al., 1983). The complexity of this ecological system contributes to its relative stability, with the "autochtonous" microbial population, including the lactobacilli, being responsible for vital functions in the host organism (Savage, 1977). During studies on the colonization of the mouse gastric epithelium by lactobacilli, Sherman and Savage (1986) showed lipoteichoic acids to be involved in the adhesive interaction with keratinized epithelial cells. Lipoteichoic acid was also indicated as a mediating factor for the adhesion of *Bifidobacterium bifidum* subsp. *pennsylvanicum* to human colonic epithelial cells (Op den Camp et al., 1985). *L. bulgaricus* was found to colonize the large intestine, rather than the small intestine, of germ-free mice, by adherence to the mucosal epithelium (Bianchi-Salvadori et al., 1984). Observing the adhesion of *L. acidophilus* and *L. bulgaricus* to both human and pig intestinal cells, Conway et al. (1987) suggested that a non-specific mechanism may be involved. Conway and Kjelleberg (1989), however, found adhesion of *L. fermentum* to mouse stomach squamous epithelium to be mediated by a protein. Electron microscopy revealed an extracellular polysaccharide to be involved in the strong adherence of a *L. acidophilus* strain to human intestinal tissue cells (Hood and Zottola, 1987). Lorenz et al. (1982a) suggested a mechanism by which adherence to the epithelial top cell layer of the rat stomach by lactobacilli is mediated by piliform appendages (PA). These PA were shown to contain capsular acidic mucopolysaccharide material that had an affinity for special receptor sites on the bacterial cell wall and the epithelial mucus.

Sarra et al. (1986) provided evidence for a chromosomal determinant involved in the adherence of one *L. reuteri* strain to chicken crop. In colonization studies of the human intestine, Sarra and Dellaglio (1984) found that only two out of four different genotypes of *L. acidophilus* (one isolated from calf feces and one from human origin) were able to settle in the GI tract. Using a human fetal intestinal cell line as an in vitro system, Kleeman and Klaenhammer (1982) found two adherence mechanisms to be involved, one of which required calcium and was nonspecific, thereby allowing all the test strains to adhere. A calcium-independent mechanism was operative in four human strains of *L. acidophilus,* and Kleeman and Klaenhammer (1982) concluded that a combination of in vitro and in vivo adherence studies may enable the screening of candidate organisms for microbiotic supplements. Another approach is the use of biotin-labeled plasmid DNA probes by which the association of strains of *L. delbrueckii*

and *L. reuteri* to murine forestomach membranes could be detected (Tannock, 1989).

Among the diverse metabolic activities accounted for by the GI microbial population, the anticholesteremic and bile-salt hydrolytic activities of lactobacilli have received special attention in recent years. Other therapeutic benefits attributed to the lactobacilli include the alleviation of lactose intolerance, protection against intestinal infections, and even antitumor activities (Fernandes et al., 1987; Fuller, 1989).

Biss et al. (1971) (quoted by Fernandes et al., 1987) observed the low serum cholesterol level and low incidence of ischemic heart disease of the Masai people, and Mann and Spoerry (1974) (quoted by Gilliland et al., 1985) related this phenomenon to the consumption of large amounts of milk fermented with a wild *Lactobacillus* strain. Observations were also made that the consumption of yogurt decreased the serum cholesterol level of rabbits (Kiyosawa et al., 1984), humans (Hepner et al., 1979), and mature boars (Danielson et al., 1989). In the latter study, yogurt fermented with *L. acidophilus* strains isolated from the feces of mature boars caused a reduction in serum cholesterol and low-density lipoproteins, but not of serum triglycerides or high-density lipoproteins (Danielson et al., 1989). The intake of milk fermented with *L. acidophilus* was shown to reduce serum cholesterol levels in rats (Grunewald, 1982), laying hens (Tortuero et al., 1975), and infants (Harrison and Peat, 1975).

Discrepancies in data on the reduction of the serum cholesterol level have been related to the strain specificity of *L. acidophilus* and *L. casei* (Nielsen and Gilliland, 1985). Evidence of the direct assimilation of cholesterol by *L. acidophilus* strains was given by Gilliland et al. (1985), at the same time showing this activity to be strain-specific, and to be dependent on bile and anaerobic conditions.

The ability to deconjugate bile salts appears to be another strain-specific property of intestinal lactobacilli, and has been shown by Gilliland and Speck (1977) to occur only in an anaerobic environment. Studying the deconjugation ability of bacteria isolated from the small intestines of liver cirrhosis patients, Shindo et al. (1989) found all strains of *L. acidophilus, L. buchneri, L. bulgaricus, L. cellobiosus,* and *L. plantarum* negative for this property. Tannock et al. (1989), however, found lactobacilli to be the main contributors to total bile-salt hydrolase activity in the murine intestinal tract.

In view of its potential value as a dietary adjunct and therapeutic agent, *L. acidophilus* has been shown to improve lactose digestion in the intestinal tract (Gilliland and Kim, 1982; Kim and Gilliland, 1983). Several *Lactobacillus* species, including strains of *L. acidophilus* and *L. ruminis,* were found to synthesize group B vitamins in the digestive tract of commercially raised chickens (Kvasnikov et al., 1985).

Antitumor activities have been related to lactic fermented foods as well as intestinal lactobacilli (Friend and Shahani, 1984). Although the mechanisms involved are not fully understood yet, at least three aspects may contribute to this phenomenon: 1) destruction of carcinogens such as nitrosamines or elimination of procarcinogens (e.g., nitrite); 2) inhibition of procarcinogenic activation enzymes (e.g., azoreductase) in other bacteria producing these enzymes; and 3) tumor suppression (reviewed by Fernandes et al., 1987; Goldin, 1986; Groeneveld and Leitzmann, 1987).

Intestinal lactobacilli appear to stimulate the immunoregulatory system, and increased phagocytic activities were observed when *L. casei* and *L. plantarum* were administered parenterally (Fuller, 1989). De Simone et al. (1987) found *L. acidophilus* and bifidobacteria to influence the regulation of gamma-interferon production by human peripheral blood lymphocytes in vitro. The feeding of fermented milks containing *L. acidophilus* and *L. casei* resulted in the activation of the immune system of mice, as shown by an increase in both the phagocytic and lymphocytic activity (Perdigon et al., 1988). It has been postulated that stimulation of the immunoregulatory system involves the translocation of GI lactobacilli to organs (Berg, 1983) such as the liver, lungs, and spleen (Fuller, 1989).

HUMAN VAGINA. In the healthy adult woman, the pH of the vagina is normally ≤4.5. The *Lactobacillus* species predominant in the vaginas of normal women are believed to maintain this low pH through their fermentative activities and to protect against invasion of undesirable microorganisms (Hill et al., 1985). Rogosa and Sharpe (1960), who cite the earlier literature, identified isolates from normal, nonpregnant women as *L. acidophilus* (67%), *L. fermentum* (19%), *L. casei* subsp. *rhamnosus* (10%), and *L. cellobiosus* (4%). Other studies confirmed the predominance of *L. acidophilus* and *L. fermentum* and reported the occasional isolation of *L. casei, L. plantarum, L. brevis, L. delbrueckii, L. lactis, L. bulgaricus, L. leichmanii,* and *L. salivarius* (Lenzner, 1966; Wylie and Henderson, 1969). After the description of *L. jensenii* isolated from vaginal discharges by Gasser et al. (1970), that species was found as part of the normal flora of many women (Carlsson and Gothefors, 1975; J. Narvus, cited in

Sharpe, 1981; Eschenbach et al., 1989). A new *Lactobacillus* species, *L. vaginalis,* has also been isolated from the vagina of patients suffering from trichomoniasis (Embley et al., 1989). Whereas in most cases only one species represents the vaginal flora, some women have more than one species present (Carlsson and Gothefors, 1975; Levison et al., 1977; Eschenbach et al., 1989; N. Weiss, unpublished observations). There are few data on the numbers present in the vagina. Levison et al. (1977) found that they occurred at more than 10^5/ml in normal vaginal secretion, and Eschenbach et al. (1989) reported a mean of 10^7/ml.

In one of our laboratories (N. Weiss), a study on a large number of lactobacilli isolated from the vagina is in progress to identify the strains using chemotaxonomical methods. For this study, it has become obvious that about 60% of the isolates identified by the classical phenotypic characteristics as *L. acidophilus* actually belong to *L. gasseri* or *L. crispatus*. This is in good agreement with the report of Johnson et al. (1980) that a relatively high proportion of the strains within homology group B-1—today labeled *L. gasseri* (Lauer et al., 1980)—and homology group A-2—now identified as *L. crispatus* (Cato et al., 1983)—originated from vaginal samples. Some of the isolates, indistinguishable from *L. fermentum* by physiological criteria only, were identified as *L. reuteri*.

Classically, glycogen is regarded as the only source of carbohydrate present in the vagina. However, Rogosa and Sharpe (1960) found that only some of their isolates fermented glycogen, and Wylie and Henderson (1969) found only 3 of their 42 isolates to be positive. In the study of N. Weiss, about 25% of the isolates under investigation have been able to ferment glycogen. All positive strains belong to *L. acidophilus* or *L. crispatus*. Stewart-Tull's finding (1964) that vaginal strains of *L. acidophilus* could ferment glycogen only in the presence of normal human serum (containing a glycogenase) suggested that the majority of vaginal lactobacilli are likely to obtain available carbohydrate from enzymatic breakdown of the polysaccharide by tissues or possibly by other organisms.

Eschenbach et al. (1989) observed that 96% of healthy women harbored hydrogen peroxide-producing facultative anaerobic *Lactobacillus* species in their vagina but only 6% of women suffering from unspecific bacterial vaginosis. They concluded that hydrogen peroxide production by lactobacilli may represent a nonspecific antimicrobial defense mechanism of the normal vaginal ecosystem.

LACTOBACILLI OF INSECTS. Although insects could be interesting plant vectors, little is known about the lactobacilli in these animals. Kvasnikov, Kotljar, and Vasileva (cited by London, 1976) isolated strains of *L. casei* and *L. cellobiosus* from the honeybee, silkworm moth, and from fruit flies; Ruiz-Arguesco and Rodriguez-Navarro (1975) found *L. viridescens* in the stomach of the honeybee. Shrivastava (1982) suggested a new species, *Lactobacillus eurydice,* to be typically associated with honey bees and bumble bees; this species name, however, has not been validated yet. The favorable effect of *L. acidophilus* and *L. bulgaricus,* administered in combination with propionibacteria via mulberry leaves to silk worms and their intestinal microbial population, was reported by Rizvanov et al. (1982). In a study of 44 *Lactobacillus* strains isolated from the intestines of insects (mainly termites), Tina (1987) grouped 31 strains as homofermentative, 10 as *L. brevis,* and 3 as *L. cellobiosus.* These studies on heterofermentative lactobacilli in the environment revealed *L. brevis, L. confusus,* and *L. fermentum* to be associated with frog feces (Tina, 1987).

LACTOBACILLI AND PATHOGENICITY. Some lactobacilli, particularly *L. rhamnosus,* have been isolated from diseased humans suffering from subacute bacterial endocarditis, systemic septicemia, and abscesses. Other species sometimes found under such conditions are *L. acidophilus, L. plantarum,* and occasionally *L. salivarius.* Bayer et al. (1978), Berger (1974), and Sharpe et al. (1973a) describe a number of cases and discuss pathogenic aspects. It is considered that the site of invasion may often be the oral cavity, and that in some instances the lactobacilli may be secondary invaders. Bourne et al. (1978) reviewed the literature on 21 cases of bacteremia where lactobacilli were one of three genera implicated. *L. brevis* was associated with lung infections, complicated by lung cancer, indicating an opportunistic behavior (Niido, 1982). In one case of chorioamnionitis, an unclassified *Lactobacillus* species was implicated as the sole causative organism (Lorenz et al., 1982b). *L. buchneri* and *L. fermentum* isolated from feces were reported as the causative agents of D-lactic acidosis in two patients with short bowel syndrome (Satoh et al., 1982). Since the report refers to intoxication, these lactobacilli cannot be regarded as pathogens. Dickgiesser et al. (1984) related *L. gasseri* to a case of urosepsis. In one case of subacute endocarditis, the causative role of *L. rhamnosus* was clearly established (Naudé et al., 1988). In no case has a true pathogenicity been established by the fulfillment of Koch's postulates.

Plants and Materials of Plant Origin

The scanty data available on the occurrence of lactobacilli on plants suggest that only small numbers are present on intact plant material (Keddie, 1959). Lactic acid bacteria can be recovered from the surfaces of the leafy parts at numbers of 10–1,000 per g where they represent 0.01–1% of the total microbial flora (Daeschel et al., 1987). Stirling and Whittenbury (1963) found that leuconostocs constituted 80% of the isolates, and lactobacilli only 10%. Species isolated included homofermenters and heterofermenters, represented mainly by *Lactobacillus plantarum* and *L. fermentum,* respectively. Lactobacilli were also reported for a wide variety of plants in a subtropical area (Mundt and Hammer, 1968), on which small numbers of *L. brevis* were found, and occasionally *L. casei, L. viridescens, L. cellobiosus,* and *L. salivarius.* The numbers of lactobacilli vary with the climate and the season, being highest in warm and humid environments and during flowering (Kvasnikov et al., 1983). In addition, flowers, fruiting structures, and plant lesions with sap release are higher in numbers. After the harvest, lesions provide more nutrients, leading to increased numbers. Lactic acid bacteria of the rhizosphere have been reviewed by Kvasnikov et al. (1983). *L. plantarum, L. brevis,* and *L. fermentum* have been found most frequently. The widespread presence of lactobacilli, though in low numbers, supports the hypothesis that plants are a primary habitat for some *Lactobacillus* species (Daeschel et al., 1987).

Little is known about the interdependence of plants and lactobacilli. A protective effect of lactobacilli on plants was discussed by Daeschel et al. (1987) since certain lactic acid bacteria produce antagonistic compounds (Visser et al., 1986), which contribute to an inhibition of *Xanthomonas campestris, Erwinia carotovora* and *Pseudomonas syringae.* Furthermore, acid is formed which lowers the pH at injured parts and acts additionally in preventing growth of opportunistic phytopathogens. On the other hand, phytoncides (antimicrobials from plants) probably do not exert strong effects on lactic acid bacteria, as is indicated by their excellent growth during fermentation of vegetables. A well-known exception is oleuropein and its breakdown products, which inhibit lactobacilli, pediococci, and leuconostocs (Etchells et al., 1975).

Part of our information on lactobacilli occurring on plants is derived from microbiological studies of fermentation processes. Thus, the microbial population at the time of initiation of the process is known for cabbage (Buckenhüskes et al., 1986), silage raw materials (Langston and Bouma, 1960a, 1960b), carrots, and beets (Andersson, 1984), and fruits like grapes (Weiller and Radler, 1970) and pears (Heinzl and Hammes, 1986). Although the numbers detected at this stage do not reflect those of the living plant, they indicate what type of organisms may potentially perform the fermentation process or might become agents of food spoilage.

Soil, Water, Sewage and Manure

The presence of lactobacilli in soil and water depends on the content of fermentable substrates. Thus, they are more frequent in soils (Kvasnikov et al., 1983) in which plants grow and constitute a part of the bacterial plant rhizosphere or are washed off from the phyllosphere. They are further involved in the breakdown of decaying matter. Correspondingly, lactobacilli are not found in fresh or marine waters but occur in sewage. From aquatic sources a multitude of heterofermentative and homofermentative lactobacilli have been isolated (Weiss et al., 1981). Heterofermenters were 25% of the total lactic acid bacteria and comprised strains of *L. fermentum, L. reuteri, L. brevis,* and a minor group of *L. confusus* and *Leuconostoc* species. The major part of the homofermenters consisted of *L. plantarum* and *L. ruminis.* Also found were *L. casei, L. acidophilus, L. farciminis, L. curvatus, L. sake, L. lactis, L. salivarius,* and *L. coryniformis.* Two new species were isolated from this environment, *L. sharpeae* and *L. agilis.* The lactic acid bacteria in sewage are present at numbers of 10^4–10^5/ml and originate from other partially unknown habitats. The same holds true for manure, from which *L. curvatus* and *L. coryniformis* were first isolated (Abo-Elnaga and Kandler, 1965). Similarly, *L. vaccinostercus* has been isolated from cow dung exclusively (Okada et al., 1979).

Food Fermentations

RAW MATERIALS OF PLANT ORIGIN. After harvest, plant materials undergo lactic acid fermentation when the content of sugar or starch is high, neutral or weakly acid conditions prevail, and access of oxygen is prevented. Lactic acid bacteria are further associated with alcoholic fermentations performed by yeasts, with which they commonly share the habitat.

VEGETABLES AND FRUITS. The lactic acid fermentation of vegetables and fruits are traditional processes to protect plant raw materials from spoilage. Products and microorganisms involved have been reviewed recently (Buckenhüskes and Hammes, 1990; Steinkraus, 1983; Wood, 1985). Most information on lactobacilli on these products are available for sauerkraut, pickles, and olives, because these are the products of main economic importance in the western world. Besides these foods, a vast multitude of products from various raw materials are fermented which, however, have only regional meaning or are bought by special consumer groups. The processes are commonly performed by spontaneous fermentation. The microbial population of fresh raw materials is dominated by aerobic bacteria and yeasts, whereas lactic acid bacteria represent only a minor component (Kandler et al., 1986; Pederson, 1979). During fermentation, under the influence of osmotically active salt, anaerobiosis, death of cells, increasing availability of nutrients derived from the plant contents, and drop in pH and redox potential, lactic acid bacteria gain dominance and undergo characteristic qualitative successions. With sauerkraut and pickles as examples, the role of lactobacilli in vegetable fermentation can be described.

For sauerkraut, there is general agreement (Brunkow et al., 1925; Buckenhüskes et al., 1986; Daeschel et al., 1987; Kandler, 1983; Kandler et al., 1986; Pederson, 1979; Stamer et al., 1975) that the succession proceeds via the growth of *Leuconostoc mesenteroides,* which is followed by "betabacteria," mainly *Lactobacillus brevis.* Homofermentative bacteria become dominant thereafter for a period whose length depends on temperature. They consist mainly of *L. plantarum* (old synonym: *L. cucumeris*) (Brunkow et al., 1925; Pederson, 1969), *L. curvatus,* and *L. sake.* Other lactobacilli of minor importance resemble *L. paracasei* (formerly *L. casei* subsp. *pseudoplantarum*) and *L. bavaricus,* which is characterized by formation of L(+) lactic acid, but comprises strains that should be allotted to *L. sake* or *L. curvatus* (Int. J. Syst. Bacteriol., 1987). Lactococci, enterococci, and pediococci may also be found but their numbers are commonly low, usually below 10% (Kandler et al., 1986) of total lactic acid bacteria (LAB).

For the fermentation of olives and cucumbers, whole fruits or vegetables in brine are used. Therefore, the nutrients are not as readily available as in sauerkraut and silage, where they are released by shredding and chopping, respectively. As shown by Daeschel et al. (1987), lactic acid bacteria can grow not only in the brine, but after brining also within the plant tissue, where they likely enter via the stomata of the epidermis. The succession of lactic acid bacteria in cucumber fermentation resembles that of sauerkraut (Pederson, 1979). Etchells et al. (1975) observed growth of lactic acid bacteria in the following order of increasing prevalence: *Leuconostoc mesenteroides, Enterococcus faecalis, Pediococcus cerevisiae, Lactobacillus brevis,* and *L. plantarum.* Kandler (1983) has found that besides *L. plantarum,* also *L. curvatus* and *L. sake* are important species in cucumber fermentation.

The fermentation of juices of fruits and vegetables is a new method to obtain beverages of appealing flavor. The juices are pasteurized and fermented with the aid of starter cultures (see below).

CEREAL PRODUCTS. The production of bread requires leavening of the dough. Before yeasts were available for bakeries, sourdough was the only biological leavening agent (Spicher, 1983). Today, breads from wheat may be leavened with yeast exclusively but sourdough or liquid preferments containing lactobacilli (*L. fermentum, L. plantarum* and others) (Miller and Johnson, 1958) are also in use. In addition to leavening, the application of sourdough improves the flavor and shelf life of bread. For Italian sweet baked goods of the panettone type, the leavening is the major desired effect of the sourdough. Breads made from rye flour or mixtures of wheat and more than 20% of rye require acidification of the dough, which is traditionally performed by the addition of sourdough. Chemical acidification in combination with yeast as leavening agent is alternatively practiced. Traditionally, sourdough contains yeasts and lactobacilli which have formed a mixture of adapted organisms in the course of their continuous propagation. The role of lactic acid bacteria in sourdough was first recognized by Holliger (1902). The description of the organisms reflected the state of taxonomy and isolation techniques of that time. The importance of heterofermentative lactobacilli was suggested by Henneberg (1909). Investigation of Danish sourdoughs led Knudsen (1924) to the conclusion that "Betabacteriumγ" is a highly adapted organism which is the true sourdough bacterium. From his description, it can be deduced that "Betabacteriumγ" is identical with *L. sanfrancisco,* first described in San Francisco sourdough bread by Kline and Sugihara (1971). More recent studies of the microorganisms in sourdough revealed the presence of yeasts and of homofermentative and heterofermentative lactobacilli: *L. delbrueckii, L. acidophilus, L.*

plantarum, L. casei, L. farciminis, L. homohiochii, L. brevis, L. buchneri, L. fermentum, L. hilgardii, L. sanfrancisco, and *L. viridescens* (Spicher and Loenner, 1985; Spicher, 1987). The microbiology of sourdough and its application has been reviewed by Spicher (1983) and Sugihara (1985).

For the production of soda crackers in the USA, a lactobacillus-yeast fermentation of the dough is used. Sugihara (1978) detected *L. plantarum* as the dominant species in the dough and *L. delbrueckii* in significant numbers. From these species, a pure starter preparation has been developed for improved control of the fermentation process.

Similar fermentation processes take place in numerous doughs or batters of indigenous fermented foods made from cereals or other starch containing raw materials which are not baked but simply boiled or steamed or even consumed without any heating. *Leuconostoc* and *Enterococcus* are usually important fermentation agents in addition to lactobacilli. These products are discussed by Steinkraus (1983), Wood and Hedge (1985), and Holzapfel (1989).

SILAGE. As for the fermentation of foods, the production of silage has a tradition dating back to ancient times. The main agents of the fermentation are lactobacilli, which are therefore of considerable economic importance. Silage is made from various raw materials, of which grass, hay, and maize play the major role. Depending on the quality of the raw material, the dry matter content, and the technology of ensilation, populations of lactic acid bacteria develop which determine the final quality of silage. Since clostridia can multiply at pH values above 4.2, it is important to decrease the pH below that value (Groß and Riebe, 1974). This limit depends, however, on water activity and therefore may be higher when the dry matter content is high. With poorly acidified silages *Listeria monocytogenes* may also cause hygienic problems (Woolford, 1984). Only after streptococci and leuconostocs have multiplied, do lactobacilli dominate the silage microflora, together with pediococci (Langston et al., 1962). Species chiefly isolated have been *Lactobacillus plantarum, L. casei, L. brevis, L. buchneri,* unclassified "streptobacteria," *L. coryniformis, L. curvatus, L. casei, L. fermentum, L. acidophilus,* and *L. salivarius* (Abo-Elnaga and Kandler, 1965; Azeezullah et al., 1973; Keddie, 1959; Langston and Bouma, 1960a, 1960b; Grazia and Suzzi, 1984; Dellaglio and Torriani, 1986). Thorough investigation of the lactic acid bacteria involved in fermentation of grass and red clover, ensiled as fresh or prewilted material, was performed by Beck et al. (1987). 612 isolates were characterized and their role in the course of fermentation was evaluated. Leuconostocs and *"Lactobacillus coprophilus"* were the dominant bacteria at the initial phase, after which other species gained importance. Pediococci, *L. plantarum,* and *L. graminis,* together with unclassified heterofermentative lactobacilli, persisted for up to 90 days, whereas the organisms of the initial phase died off. There is increasing use of starter cultures for production of silage (Pahlow and Honig, 1986; Seale, 1986).

RAW MATERIALS OF ANIMAL ORIGIN. In modern technologies, the fermentation of raw materials of animal origin is characterized by a widely accepted application of starter cultures. Therefore, these aspects are included under "Technical Applications."

FERMENTED MILKS. A multitude of different types of fermented milks is produced worldwide (reviewed by Oberman, 1985). Products containing lactobacilli, produced with the aid of starter cultures and known to the western world, are yogurt, kefir, and some dietary products, of which acidophilus milk is most representative. Starter cultures for yogurt production contain *Streptococcus salivarius* subsp. *thermophilus* and *Lactobacillus delbrueckii* subsp. *bulgaricus.* These bacteria multiply in the milk during incubation at 40–42°C in a symbiotic association (Marshall and Law, 1984; Pette and Lolkema, 1950; Robinson and Tamime, 1981). The *Lactobacillus* species has stronger proteolytic activity, and the amino acids and dipeptides produced by it stimulate the growth of the *Streptococcus* species (Shankar and Davies, 1978), which in its turn reduces the redox potential and forms formic acid that is stimulatory for the *Lactobacillus* cells (Galesloot et al., 1968). Optimum yogurt flavor is also produced in association (Bottazzi and Vescovo, 1969), with *Lactobacillus* being the primary producer. The major flavor impact is derived from acetaldehyde but acetone and diacetyl also contribute to the flavor. Threonine was found to be the precursor for acetaldehyde production by *L. delbrueckii* subsp. *bulgaricus* (Lees and Jago, 1978).

Kefir is a product of combined alcoholic and lactic acid fermentation, containing 0.9–1.1% lactic acid and 0.5–1% alcohol. The fermentation is initiated by the addition of kefir grains or starter cultures derived from them. These grains have been found to contain yeast (mainly *Candida kefir*) and lactic acid bacteria. Lactobacilli are the dominant group in the grains and in addition leuconostocs and lactococci were

also identified (Kunath, 1983). *Lactobacillus acidophilus, L. kefir* (Kandler and Kunath, 1983), and *L. kefiranofaciens* (Fujasawa et al., 1988) are the most characteristic lactobacilli, the latter organism being responsible for production of kefiran, a polymer which is the matrix of the grains (Toba et al., 1986; La Rivière et al., 1967). The composition of the microbial population changes from the grains to the ready-to-drink kefir. In the final product, lactococci *(Streptococcus lactis)* prevail and the ratio of *L. acidophilus/L. kefir* changes from 0.89/0.11 to 0.2/0.8 (Kunath, 1983).

Acidophilus milk is consumed mainly for its therapeutic effect (Fernandes et al., 1987; Robinson and Tamime, 1981). The taste of the product is rather astringent. *L. acidophilus* is known to produce antagonistic compounds, to exert probiotic effects (see above), and to grow poorly in milk. Special care has to be taken to keep the numbers of lactobacilli at a high level since viable cells are considered essential for the therapeutic activity. Thus, the shelf life is usually restricted to around one week.

CHEESE. Two groups of lactobacilli can be differentiated based on the time of their growth during the process of cheese making. These are the thermophilic lactobacilli, added deliberately as components of starter cultures, and the mesophilic lactobacilli which are derived from the environment and emerge during the ripening process. The thermophilic starter cultures are applied for the production of cheese types which require elevated temperatures during the process of curd preparation. For example, the production of Emmental cheese includes cooking the curd grains at 53–57°C. The thermophilic lactic acid bacteria grow best at about 45°C. They survive this step and multiply during the cooling-down period (Auclair and Accolas, 1983). Additional examples of cheese types produced with thermophilic cultures are Gruyere, Gorgonzola, Mozarella, Cacciocavallo, and Provolone (Bottazzi et al., 1973). Species included in starter cultures are *L. helveticus, L. delbrueckii* subsp. *lactis,* and subsp. *bulgaricus,* in combination with *Streptococcus thermophilus.* Mesophilic lactobacilli grow in virtually all types of cheese. For example, in cheddar cheese, numbers of 10^6–10^8 are found after 10 to 60 days and are maintained for 4–6 months. Species involved include *L. casei, L. plantarum, L. brevis,* and *L. buchneri* (Chapman and Sharpe, 1981). For the production of white brined cheese, starter cultures were applied successfully containing lactobacilli in combination with lactococci, enterococci, leuconostocs, or *S. thermophilus.* Species employed were *L. casei, L.*

plantarum, L. helveticus, and *L. delbrueckii* sub. *bulgaricus* (for review, see Haddadin, 1986).

FERMENTED SAUSAGES. Fermented sausages are made from raw meat and gain their characteristic texture, color, flavor and, above all, their resistance to spoilage, during a ripening process in which lactobacilli are the essential fermentation agents. Fungi, yeasts, and micrococci or staphylococci may also contribute to the special character of the product (Hammes, 1986; Liepe, 1983; Lücke, 1985). In the traditional way, sugar is added as a substrate for lactobacilli which convert it mainly to lactic acid, thereby decreasing the pH to values which prevent the growth of food spoiling or poisoning organisms. Decrease of water activity and addition of nitrite (or nitrate) support this effect (Meisel et al., 1989). Formerly designated as atypical lactobacilli, *L. sake* and *L. curvatus* are the dominant lactobacilli, but *L. plantarum, L. alimentarius,* and *L. farciminis* have been also isolated (Reuter, 1970, 1975). Starter cultures containing *L. plantarum, L. sake, L. curvatus,* or pediococci, frequently combined with micrococci or staphylococci, have been developed and are widely applied (Hammes et al., 1985).

Food Spoilage

ACID FOODS. Lactobacilli play an important role in the spoilage of processed foods. They may exert their effect during processing or especially in the final product. Food products with pH values below 4.1 (ICMSF, 1980; Buckenhüskes et al., 1988) are commonly not sterilized but only pasteurized since the out-growth of endospores of Bacillaceae is not of concern. Further factors that may protect low-acid foods and especially juices and juice-containing soft drinks are the presence of essential oils or benzoic acid, absence of specific nutrients or growth factors, and, in carbonated drinks, CO_2. In the case of underprocessing or leakages of food preserves, juices and juice-containing beverages, lactobacilli may grow and cause formation of slime, gas, off flavors, turbidity, and changes in acidity. Species involved are mainly *Leuconostoc mesenteroides* and less commonly *Lactobacillus confusus, L. buchneri, L. casei,* and *L. plantarum* (Back, 1981). In citrus juices, *L. brevis* and *L. plantarum* can multiply at a pH of less than 3.5 and at a temperature of 10°C (Juven, 1976; Murdock and Hatcher, 1975).

LACTIC-ACID-FERMENTED FOODS. Lactobacilli are also agents of spoilage of fermented foods. In products obtained by lactic acid fermentation, the effect is commonly exerted during the

fermentation process. As a result of the growth of inappropriate strains, the fermentation product may become unacceptable. Examples were found in the formation of red color in sauerkraut caused at elevated pH by *L. brevis* (Stamer, 1975) and formation of bloaters in cucumber fermentation as a result of gas formation by heterofermenters and homofermenters, which form CO_2 from malate (McFeeters et al., 1984).

In cheese, the citrate-utilizing species *L. casei* and the heterofermentative *L. brevis* may produce excessive CO_2, giving rise to unwanted gas pockets in cheese and blowing of packaged cheeses (Fryer et al., 1970; Keller and Jaarsman, 1975). Undesired small cracks ("Boekelscheuren") are formed in Gouda and Edam cheese by *L. bifermentans,* an organism which produces CO_2 and free H_2 (Pette and van Beynum, 1943). Slime-forming strains of *L. plantarum* can multiply in cheese pickling brines, causing ropiness, and salt-tolerant streptobacteria multiplying in rennet cause serious texture and flavor defects in Dutch cheese (Stadhouders and Veringa, 1967). Orange-pigmented strains *(L. plantarum* subsp. *rudensis* or *L. brevis* subsp. *rudensis)* may multiply in hard cheese (Sharpe, 1962) and in white brined cheese (Chomakov, 1962). Formation of biogenic amines has been observed in cheese, and lactobacilli have been identified as causative agents. *L. buchneri* was isolated from high-level-histamine Swiss cheese (Sumner et al., 1985). From Gouda cheese (Joosten and Northolt, 1987), strains of *L. buchneri* and *L. brevis* were isolated which produced histamine and tyramine, respectively. A potential for formation of tyramine and histamine in dairy-related bacteria has been observed also for *L. casei* (histamine) and lactococci (tyramine) (Voigt and Eitenmiller, 1978).

Sensory defects of raw fermented sausages have been attributed to the growth of inappropriate species of lactobacilli during the ripening process (Corretti, 1958). These defects may be traced back to mistakes in technology or formulation and are greatly reduced by the use of starter culture preparations. The sensory aberrations included defects in color (loss of color stability, gray or green core, green surface spots) and flavor. Corretti (1958) isolated, from spoiled sausages, strains of *L. plantarum, L. delbrueckii, L. brevis, L. buchneri,* lactococci, and *Leuconostoc mesenteroides.* In sausages, the activities of these bacteria may be the same as observed for lactic acid bacteria involved in meat spoilage (see below).

FERMENTED BEVERAGES. In fermented beverages obtained by alcoholic fermentation, lactobacilli may contribute to the quality of the product but may also cause spoilage.

WINE. Lactobacilli commonly occur in many types of wine, despite the high level of ethanol, the low pH of 3.2–3.8, and the added SO_2 present. They can exert profound effects on the quality of wine. When they are present in high numbers, they produce a combination of the various typical wine defects, the prevalence of which differs, depending on the strains present (Dittrich, 1977). Their ability to metabolize organic acids such as malic, citric, and tartaric acid may affect the final product in either a desirable (see Chapter 69) or an undesirable way. In low-acid wines, this process is detrimental and must be controlled. Although most lactobacilli from wines decompose malate, the source of malolactic bacteria is uncertain (Kunkee, 1967). They have been detected only sporadically and in small numbers on grapes and grape leaves and may well be part of the established microbial population of the winery itself. A variety of malolactic-fermenting species have been isolated, including *Lactobacillus plantarum, L. casei,* unclassified "streptobacteria," and the heterofermentative species *L. brevis, L. buchneri, L. hilgardii, L. trichodes, L. fructivorans, L. desidiosus,* and *L. mali* (Barre, 1969; Chalfan et al., 1977; Maret and Sozzi, 1977, 1979; Maret et al., 1979; Nonomura and Ohara, 1967; Peynaud and Domercq, 1967, 1970; Pilone et al., 1966; Weiller and Radler, 1970). The homofermentative species disappear during alcoholic fermentation in favor of pediococci and heterofermentative lactic acid bacteria. The same species of lactobacilli are found in French, Spanish, German, Australian, Californian, and Japanese wines. From high-temperature (40–43°C) fermenting grape musts, however, Barre (1978) has isolated pentose-fermenting "thermobacteria," some of which closely resemble *L. acidophilus.*

The decomposition of tartaric acid in wine is usually connected with severe spoilage of wine. Only a few strains of homofermentative and heterofermentative lactobacilli perform this reaction (Krumperman and Vaughn, 1966). Radler and Yannisses (1972) observed that the tartrate-decomposing systems of *L. plantarum* and *L. brevis* differ in metabolic pathway (CO_2, lactate, and acetate are formed by both organisms, but succinate is an additional product of *L. brevis*) and several other criteria.

Other effects of lactobacilli in wine are due to the production of diacetyl from citric acid, a substance which enhances flavor when present in traces, but causes spoilage if present in excess; spoilage such as bitterness occurring together with excess formation of mannitol from fermentation of fructose; and occasional flocculent growth of *L. trichodes* (Amerine and Kunkee, 1968).

APPLE CIDER. The indigenous microbial population of an apple cider factory is composed partly of lactobacilli. As with wine lactobacilli, only strains that are adapted to survive a low pH, low levels of C and N compounds, and presence of increasing ethanol will form part of the permanent population. Many of these selected strains, particularly the heterofermentive species, can metabolize malic and citric acid, as well as quinic acid, a substance present at relatively high levels in apple cider (Carr, 1959). Lactobacilli may multiply in stored ciders, where they bring about several changes. They may take part in the malolactic fermentation, which will often be beneficial to the flavor of the cider; they may also metabolize citrate and pyruvate, yielding acetate, lactate, and acetoin. The metabolizing of fructose and quinic acid to acetate, CO_2 and dihydroxyshikimate is of special interest (Whiting, 1975), since the acetate formed is detrimental to flavor (Sharpe, 1981).

Heterofermentative isolates from cider are usually *Lactobacillus brevis.* They have an optimum pH of 4.0–5.0 and metabolize actively only at these low pH values. Fructose is preferentially utilized, glucose often only weakly (Carr, 1959). Slime-forming strains cause ropiness by production of polysaccharide from glucose, fructose, or maltose but not from sucrose (Millis, 1951). Homofermentative species are mainly strains of *L. plantarum* and also *L. yamanashiensis (L. mali)* (Carr et al., 1977), which does not, however, utilize quinic acid.

BEER. In the brewery, lactobacilli may cause spoilage but are also applied for useful purposes. Heterofermentative strains are prevalent in spoiled beer and preferentially ferment maltose. They grow poorly on glucose unless an arginine supplement is present as an additional energy source (Rainbow, 1975). They ferment a narrow range of carbohydrates, have complex nutritional requirements, and are able to tolerate pH values of 3.8–4.3 found in this environment. Their tolerance to the hop resin humolene, unusual for Gram-positive organisms, and the rapidity with which such resistance is acquired (Richards and Macrae, 1964), suggests that they have become adapted to the brewery as their natural habitat (Sharpe, 1981).

Back (1981) obtained about 1,000 strains of bacteria from contaminated beer. These could be alloted to 13 Gram-positive species belonging to five genera and were proven to be obligately or potentially beer-spoiling organisms. The most frequently occurring species were *L. brevis* (28%), *Pediococcus damnosus* (27%), *L. casei* (11%), *L. lindneri* (9%), and *L. coryniformis* (6%). In addition, *L. curvatus, L. plan-*

tarum, L. buchneri and two atypical groups of heterofermentative lactobacilli, one of which contained the new species *L. brevissimile,* were isolated (Back, 1987). The growth of the lactobacilli causes a silky turbidity accompanied by acidity and off flavors because of diacetyl (Scherrer, 1972), while slime-forming strains cause ropiness (Williamson, 1959). As discussed by Back (1987), *L. lindneri* and *L. brevissimile* may pass through filters because of their small size and contaminate the beer in the final container. The latter organism does not cause severe spoilage and is not easy to culture. These species die off in the bottle rather quickly and, therefore, escaped detection, until recently.

FRUIT MASHES. For the production of fruit brandies, mashes of fruits, such as apples, pears, plums, and cherries, are fermented and distilled. To ensure a "clean" yeast fermentation of the mashes, sulfuric acid is added to decrease the pH to about 3.0. In this environment, highly acidophilic and alcohol-resistant lactic acid bacteria grow and may lower the yield of alcohol or add undesired flavor-active compounds (Heinzl and Hammes, 1986; Kleynmans et al., 1989). Some strains of *L. plantarum* and *L. suebicus* grow at pH 2.5 in the presence of 12 and 14% ethanol, respectively. *L. suebicus* grows during and, above all, after the alcoholic fermentation and was found in pear and apple mashes, exclusively. Other lactobacilli present at this fermentation phase and exceeding numbers of 10^6 colony-forming-units (CFU)/ml were *L. plantarum, L. brevis,* and *L. hilgardii.*

GRAIN MASHES. During the manufacture of malt whiskey, lactobacilli may multiply and reach high numbers during the fermentation process itself. In contrast to brewing, the malt is not boiled and lactobacilli are often present in the malted barley. Thermophilic strains may multiply during fermentation when a rich supply of nutrients is readily available. This rapidly depresses the pH to such an extent that the activity of the debranching enzymes and residual amylases of the yeast are inhibited, fermentation is not completed, and a much lower yield of alcohol results (MacKenzie and Kenny, 1965; Simpson, 1968). In addition, detrimental flavor compounds such as hydrogen sulfide may be produced (Geddes, 1986). Species of lactobacilli isolated from distillery fermentations include *L. fermentum, L. brevis, L. casei, L. delbrueckii,* and *L. plantarum* (Bryan-Jones, 1975).

FOODS PRESERVED BY ACETIC ACID. The application of vinegar as a food preservative is a traditional method of preventing spoilage. Acetic acid is an effective acidulant because it exists in acid foods largely in the undissociated form (pK 4.75) which can pass the cell membrane (Baird-Parker, 1980). Mayonnaise, dressings, and salads are examples of foods whose microbial stability (and taste) is mainly affected by acetic acid. When access to oxygen is prevented, the spoilage is most commonly caused by yeasts and lactobacilli with exceptional high tolerance to acetic acid. Highest resistance was found for *L. acetotolerance* (Entani et al., 1986), which grows in fermenting rice vinegar broth and can tolerate 4–5% acetic acid at pH 3.5. In mayonnaise and salad dressings, *L. fructivorans* can cause spoilage (Charlton et al., 1934; Kurtzman et al., 1971) at pH 3.7–3.8. Salads composed of mayonnaise or dressings and ingredients like potatoes, meat, marinated fish, eggs, vegetables, etc. have been investigated by Baumgart et al., 1983). From 81 samples, six species of lactobacilli were isolated (in order of prevalence): *L. plantarum, L. buchneri, L. brevis, L. delbrueckii, L. casei, L. fructivorans.* These organisms are commonly tolerant to the added benzoic and sorbic acid and form gas and off flavors in the product.

The spoilage of marinated fish (mainly herring) involves tolerance not only to acetic acid but also to salt (Meyer, 1965). Lactic acid bacteria can grow and cause sensory defects and swelling of the containers due to gas formation. CO_2 may originate from fermented carbohydrates or, where these are missing, from decarboxylation of amino acids (Meyer, 1956a). The biogenic amines formed from the respective amino acids are γ-aminobutyric acid, cadaverine, tyramine, and histamine (reviewed by Blood, 1975). The species involved in spoilage are *L. brevis, L. buchneri, L. fermentum, L. pastorianum, L. delbrueckii, L. plantarum,* and *L. casei* (Blood, 1970; Kreuzer, 1957; Lerche, 1960; Meyer, 1956a, 1956b; Reuter, 1965; Yurtyeri, 1963). *L. plantarum* and *L. casei* have been also identified as causatives of ropiness in cooked marinades (Priebe, 1970).

SUGAR PROCESSING. Of major interest is the effect of lactic acid bacteria during sugar production (Sharpe, 1981). They may cause losses in yield and, due to excretion of dextran, interfere with the efficiency of the production process and with crystal formation. With cane sugar, most spoilage is caused by leuconostocs. However, sugar-tolerant, acidophilic strains of lactobacilli (able to multiply in 15% sucrose), consisting mainly of *L. confusus* (Sharpe et al., 1972) and occasionally of *L. plantarum* and *L. casei,* multiply in cane juice, causing souring and deterioration of canes. Most of these strains, including *L. plantarum* and *L. casei,* produce large amounts of dextran from the sucrose (Tilbury, 1975). They come from the cane itself and from contaminated equipment. Similar spoilage occurs with beet sugar production (Tilbury, 1975), where strains isolated include *L. casei, L. plantarum, L. cellobiosus,* and *L. fermentum* (Kvasnikov et al., 1976).

MILK. Aseptically drawn raw milk contains no lactobacilli when it leaves the udder, but contamination with these organisms rapidly occurs, from the dairy utensils, dust, grass, silage, and other feed stuffs. Milk is an ideal substrate for bacterial growth, but conditions that allow contamination and multiplication favor other organisms, and lactobacilli are usually outgrown. In the United Kingdom, single-herd milks produced under good hygienic conditions contain small numbers of lactobacilli, of >1 to 50/ml, whereas bulked herd market milks usually contain about 10^3/ml (Sharpe, 1981). Species present include *L. casei, L. plantarum, L. brevis, L. coryniformis, L. curvatus,* and occasionally *L. buchneri, L. lactis,* and *L. fermentum* (reviewed by Abo-Elnaga and Kandler, 1965; Sharpe, 1962). Raw ewe's milk contains the same species (Chomakov and Kirov, 1975). Pasteurization of the milk (HTST, 71.7°C/15 sec) usually destroys all lactobacilli present, except for heat-resistant *L. paracasei* subsp. *tolerans* (Abo-Elnaga and Kandler, 1965) Such heat-treated milks, when used for processing, rapidly become recontaminated from the creamery environment with the same species of lactobacilli (reviewed by Sharpe, 1962). Thereafter, lactobacilli are so universally present in milk and dairy products that generally only strains having unusual characteristics cause spoilage. In liquid milk, slime-producing strains of *Lactobacillus casei, L. brevis, L. bulgaricus,* and *L. acidophilus* occasionally produce ropiness, and *L. maltaromicus* may produce a malty flavor (Miller et al., 1974).

MEAT AND MEAT PRODUCTS. Technologies developed for extending the shelf life of meat and meat products include curing, smoking, and packaging in films of low gas permeability together with applying vacuum or controlled atmosphere and refrigeration. These technologies depress the growth of putrefactive microorganisms but create more or less selective conditions for the growth of lactobacilli, leuconostocs (see Chapter 69), carnobacteria (see below), and *Brochothrix thermosphacta* (Egan, 1983; Kitchell and Shaw, 1975). Their growth results in souring, slime formation, off-odor, and greening (Reuter, 1975). Properties of lactobacilli which

lead to their preponderance in meat are: psychrotrophy and tolerance to high CO_2 tensions, to nitrite, to salt, and to low pH values. The numbers of lactobacilli on fresh meat and meat products are usually below 10^3/g and less than 10/cm² of the meat surface can grow under the selective conditions present in vacuum packages (Egan, 1983). During cold storage, these "atypical" lactobacilli usually become the dominant group, and their identity has been investigated in many studies (Allen and Foster, 1960; Cavett, 1963; Gardner, 1968; Hitchener et al., 1982; von Holy and Holzapfel, 1989; Holzapfel and Gerber, 1986; Mol et al., 1971; Schillinger and Lücke, 1986). It has become evident that meat and meat products are a habitat of lactic acid bacteria hitherto not detected elsewhere. Studies of Reuter (1975) and Shaw and Harding (1984) revealed that lactobacilli dominate the competitive leuconostocs or carnobacteria when acidification and/or reduction of the water activity by drying (or smoking) or addition of salt and nitrite are applied, e.g., in dry raw sausages and bacon.

A thorough investigation of lactobacilli in meat and meat products performed by Reuter (1975) showed that "atypical streptobacteria" are the dominant group in all types of meat products. These have been alloted to *L. curvatus* and *L. sake* by Kagermeier (1981). In addition *L. plantarum, L. casei, L. farciminis, L. alimentarius, L. brevis,* and *L. halotolerans* have been found. The latter has been transferred to the rank of species, as *L. halotolerans,* by Kandler et al. (1983b) since it is not related to *L. viridescens* isolated by Niven and Evans (1957) as a causative agency of greening. The greening defect originates from the formation of H_2O_2, which reacts with myoglobin to form the green pigment choleglobin. Sulfmyoglobin is another green pigment which is formed by the reaction of myoglobin with H_2S. Lactobacilli which form H_2S have been isolated from meat and contribute in this way to spoilage of vacuum-packed meat. The genetic information for H_2S production is plasmid-encoded (Shay and Egan, 1981, Shay et al., 1988). Formation of hydrogen sulfide and methyl mercaptan by lactobacilli was also described for strains isolated from spoiled Parma hams (Cantoni et al., 1969). The numbers of lactobacilli in spoiled meats may exceed 10^7 or even 10^8/g in products with added sugar, e.g.) raw sausages and vacuum-packed cooked sausages (Reuter, 1975).

The animal source of vacuum-packed meat influences the composition of the microbial spoilage association. Shaw and Harding (1984) observed a preponderance of "non-aciduric streptobacteria" (carnobacteria) in pork and lamb, as opposed to beef where lactobacilli prevail. This observation is in partial accordance with earlier studies reviewed by Kitchell and Shaw (1975), which indicated a preponderance of *Brochothrix thermosphacta* on vacuum-packed lamb. Furthermore, lactobacilli do not belong to the spoilage population of refrigerated chicken meat. This is mainly due to the packaging techniques, which consist usually of wrapping without application of vacuum or controlled atmosphere. In addition, when controlled atmosphere is applied, carnobacteria predominate and lactobacilli are rarely found (see "The Genus Carnobacterium," this chapter).

Technical Applications

Lactobacilli are frequently used as technical and analytical tools. Of supreme importance is their use as starter cultures, while their application as probiotics (see section "Probiotics," this chapter) receives growing interest (see below). The large-scale production of lactic acid with the aid of *L. leichmannii (L. delbrueckii)* is one of the oldest biotechnical processes for producing chemicals by pure cultures (Buchta, 1983). Finally, still in use is the determination of growth factors (vitamins and amino acids) in complex mixtures using auxotrophic strains of lactobacilli.

Starter Cultures

The application of starter cultures for the production of fermented foods of plant origin has still not been very successful in practice. According to Fleming et al. (1985), starter cultures in combination with the traditional technology do not influence the fermentation process in a way that superior products or an improvement of the economy of the process can be obtained. A limited application has been found for *L. bavaricus,* which is used to produce" L(+)-sauerkraut." This product is sold in health stores because of its content of L(+)-lactic acid. On the other hand, for the production of fermented vegetable juices, several starter cultures are in use. In this case, it is possible to pasteurize the raw material and to initiate the fermentation without any competing indigenous flora. Starter organisms for fermentation of vegetable juices have been reviewed by Buckenhüskes and Hammes (1990) and include *L. acidophilus, L. bavaricus, L. brevis, L. casei, L. delbrueckii, L. helveticus, L. plantarum, L. salivarius.* For production of juices containing L(+)-lactic acid, *L. bavaricus* and *L. casei* are in use.

Fruit juices have been fermented with the aid of *L. casei* (Wiesenberger et al., 1986). In this process, the flavor was changed due to degradation of malic acid and formation of L(+)-lactic acid.

The application of lactic acid bacteria in wine production is now receiving commercial application (Krieger, 1989). The main aim is to reduce the content of malate, which is converted to L(+)-lactic acid and CO_2 with a decrease in acidity. In addition, both the stability and the flavor of the wine are improved. Most work has been performed with *Leuconostoc oenos,* but lactobacilli have also been applied. Organisms that are sensitive to the acid-and alcohol-containing environment are added to the must before fermentation (e.g., *L. plantarum*) while more tolerant cultures containing *L. brevis* or *L. casei* have also been added to the wine.

Lactobacilli are also used in breweries. In Germany, a sour wort is fermented at 48–50°C, which is added to the mash at 1–2%, thereby decreasing the pH by 0.2 to 0.4 pH units. This causes the following desired effects: higher enzyme activity in the mash and better separation of protein in wort boiling, followed by a quick fermentation. The sensory quality of the beer is also improved since it becomes more mellow in taste, lighter in color, and the foam has a higher stability. Finally, the microbial stability is positively affected. Maltose-fermenting strains of *L. delbrueckii* subsp. *delbrueckii* are most suitable. The cultures are propagated in the brewery and contain *L. delbrueckii* subsp. *lactis, L. amylovorus, L. fermentum,* and *L. rhamnosus,* and, occasionally, *L. helveticus* (Back, 1988).

In addition, lactic-acid-fermented types of beer are well known, e.g., Berliner Weisse beer, which is produced with *L. brevis* (K. Wackerbauer, personal communication) and sorghum beer (Haggblade and Holzapfel, 1989), in which the souring process is performed at 48–50°C with a thermophilic *Lactobacillus* strain (probably not *L. delbrueckii,* W. Holzapfel, unpublished observations) as the main fermenting organism. Similarly, the Russian drink called kwas is a sour beer made from bread, which obtains acidity from fermentation by homofermentative and heterofermentative lactobacilli.

Modern techniques of breadmaking make extensive use of starter culture preparations, the majority of which are mixed strain cultures. The application of starter cultures in modern baking technology requires the additional use of bakers yeast. The starter cultures commonly contain the organisms present in the sourdough but at varying ratios (Spicher, 1984). Böcker and Hammes (1990) observed that in a so-called "pure culture sour," more than 99% of the lactobacilli consisted of two strains of *L. sanfrancisco* and one strain of *L. brevis*. This sourdough culture preparation has been used for over 60 years, and its composition has remained remarkably unchanged in various batches of starter in spite of the use of a nonaseptic production process, which is similar to the traditional method of sourdough propagation. Starter preparations consisting of defined strains are also available. Single strain cultures contain *L. brevis, L. sanfrancisco, L. delbrueckii,* or *L. plantarum,* while multiple strain cultures consist of combinations of *L. plantarum, L. sanfrancisco,* and *L. fructivorans,* or *L. brevis,* and *Saccharomyces cerevisiae* (Budolfsen-Hansen, 1988; Spicher, 1987; Sugihara, 1985).

For silage fermentation, the most frequently applied lactic acid bacteria in starter culture preparations were *L. plantarum* (61%), *Lactococcus lactis* (31%), *Lactobacillus acidophilus* (23%), *Pediococcus acidilactici* (19%), *Lactobacillus brevis* (16%), *Enterococcus faecium* (11%), *Lactobacillus lactis* (6%), and *L. bulgaricus* (5%) (Pahlow and Honig, 1986). The cultures should ensure quick acidification to pH values below 4.0 and should not be of the heterofermentative type, because of possible losses in nutritional value. Their application should further ensure that spoilage microorganisms do not affect the quality, even after the opening of the silo when oxygen gains access. Present knowledge suggests that the cultures should include fast-growing strains of *L. plantarum* and *P. acidilactici,* one of which possesses cellulase activity (Seale, 1986).

Probiotics

Originally referring to substances from one protozoan stimulating another, and later describing animal feed supplements beneficial to the host by mediation of its GI microbial population, probiotics are now defined as "a live microbial feed supplement which beneficially affects the host animal by improving its intestinal microbial balance" (Fuller, 1989). Probiotics as dietary and therapeutic adjuncts for humans and animal nutrition are now well established in the market and may provide effective and highly acceptable alternatives to conventional growth promotors and therapeutic agents. Their favorable effect on growth and general health is most probably the result of a combination of factors discussed before in this section. In earlier considerations, Gilliland (1979) has listed four desirable properties of an organism to be used as dietary adjunct: 1) normal inhabitant of the GI tract; 2) resistance to inhibitory systems in the digestive tract; 3) potential for producing beneficial effects; and 4) retention of viability during storage as dietary preparations.

The present status of probiotics, though still in the early stages of application, has been excellently and comprehensively reviewed by Fuller (1989). Referring to the complexity of the underlying mechanism, he emphasized the vital need for additional (basic) information on antagonism, growth rate in the intestine, and attachment to gut epithelial cells (Fuller, 1989). The species most widely used in probiotic products are mainly intestinal strains of *L. acidophilus, L. casei, L. helveticus, L. lactis, L. plantarum, L. salivarius,* as well as *Enterococcus faecium, Enterococcus faecalis, Bifidobacterium* species, and *Escherichia coli*. Special attention has been given recently to possible anti-mutagenic effects induced by viable lactic acid bacteria. The clastogenic effects induced by a strong chemical mutagen have been inhibited to about 80% by lyophilized cultures of *L. acidophilus* and *Bifidobacterium* in experiments with small laboratory animals (H. Renner, personal communication). Beneficial effects have also been claimed for nonintestinal strains of *L. bulgaricus, Streptococcus thermophilus* (Fuller, 1989), and a combination of mesophilic lactobacilli, including *L. curvatus* and *L. sake*. The "Nurmi-concept," related in principle to the probiotic approach, involves the introduction of the heterogenous (more or less undefined) microbial cecum population of adult poultry to newly hatched chicks, thereby rendering them immediately resistant to 10^3 to 10^6 infectious doses of *Salmonella* (Pivnick and Nurmi, 1982). Aspects of probiotics in poultry nutrition have been reviewed by Jernigan et al. (1985).

Isolation

Media for the isolation of lactobacilli must take into account the aciduric or acidophilic nature of these organisms and their complex nutritional requirements. In some cases, species have adapted to extreme environmental conditions and can only grow on media that simulate their natural habitat. This includes in some cases even strictly anaerobic growth conditions. All media must contain adequate growth factors, usually with yeast extract as a source of vitamins, as well as peptone, manganese, acetate, and the stimulatory Tween 80. A low pH, ranging between 4.5 and 6.2, favors growth. In some habitats, particular spoilage situations, lactobacilli may constitute the only organisms present; more often, they occur together with other organisms, which may include other lactic acid bacteria and yeasts.

Media

When lactobacilli are the majority population, MRS agar (see below) (de Man et al., 1960) can often be used for isolation. This medium is discussed in detail by Sharpe and Fryer (1965) and compared with the somewhat-similar APT medium (Evans and Niven, 1951), which is commonly used for isolating *Lactobacillus viridescens* as well as other lactobacilli and carnobacteria from meat products. The growth of especially fastidious lactobacilli, mainly obligately heterofermentative species, has been found to be supported best by the modified Homohiochii medium described by Kleynmans et al. (1989). These media are generally used for cultivation of lactobacilli and other lactic acid bacteria, and may be regarded as semiselective.

MRS Agar for Isolating and Propagating Lactobacilli (de Man et al., 1960)

For 1 liter of medium, dissolve 15 g agar in distilled water by steaming. Add:

Oxoid peptone	10.00 g
Meat extract	10.00 g
Yeast extract	5.00 g
K_2HPO_4	2.00 g
Diammonium citrate	2.00 g
Glucose	20.00 g
Tween 80	1.00 ml
Na acetate	5.00 g
$MgSO_4 \cdot 7H_2O$	0.58 g
$MnSO_4 \cdot 4H_2O$	0.25 g

Adjust pH to 6.2 to 6.4. Sterilize at 121°C for 15 min.

APT Medium for Isolating and Propagating Lactobacilli (Evans and Niven, 1951)

For 1 liter of medium, dissolve 15 g agar in distilled water by steaming. Add:

Tryptone	10.00 g
Yeast extract	5.00 g
K_2HPO_4	5.00 g
Na citrate	5.00 g
NaCl	5.00 g
Glucose	10.00 g
Tween 80	1.00 ml
$MgSO_4 \cdot 7H_2O$	0.80 g
$MnCl_2 \cdot 4H_2O$	0.14 g
$FeSO_4 \cdot 7H_2O$	0.04 g

Adjust pH to 6.7 to 7.0. Sterilize at 121°C for 15 min.

Modified Homohiochii Medium for Isolating and Propagating Obligately Heterofermentative Lactobacilli (Kleynmans et al., 1989)

For 1 liter of medium, dissolve 15 g agar in distilled water by steaming. Add:

Tryptone	10.00 g
Yeast extract	7.00 g
Meat extract	2.00 g
Glucose	5.00 g
Fructose	5.00 g
Maltose	2.00 g
Na gluconate	2.00 g
Diammonium citrate	2.00 g
Na acetate	5.00 g
Tween 80	1.00 ml
$MgSO_4 \cdot 7H_2O$	0.20 g
$MnSO_4 \cdot 4H_2O$	0.05 g
$FeSO_4 \cdot 7H_2O$	0.01 g
Mevalonic acid lactone	0.03 g
Cysteine hydrochloride	0.50 g

Adjust pH to 5.4. Sterilize at 121°C for 15 min and add 40 ml ethanol per liter.

When lactobacilli occur only as part of a complex population, selective media are required. The most widely used of these in the past is the acetate (SL) medium of Rogosa et al. (1953). In this and other similar media, the selective action is based on a low pH of 5.4, a high concentration of acetate ions (inhibitory to many other organisms), and the presence of the growth stimulatory substance Tween 80. These media are further described and discussed by Sharpe (1960). Care has to be taken in preparing SL medium, particularly with regard to final pH since, if this is higher than 5.4, streptococci may not be inhibited. Dehydrated preparations can be purchased commercially, but the final pH should be checked.

Selective SL Medium for Isolating Lactobacilli (Rogosa et al., 1953)

For 1 liter of medium:

Trypticase	10.00 g
Yeast extract	5.00 g
KH_2PO_4	6.00 g
Diammonium citrate	2.00 g
$MgSO_4 \cdot 7H_2O$	0.58 g
$MnSO_4 \cdot 4H_2O$	0.15 g
$FeSO_4 \cdot 7H_2O$	0.03 g
Glucose	20.00 g
Tween 80	1.00 ml
Na acetate $\cdot 3H_2O$	25.00 g
Agar	15.00 g

Dissolve the agar in 500 ml water by boiling. Dissolve all other ingredients in 500 ml water, adjust pH to 5.4 using glacial acetic acid and mix with the melted agar. Boil for a further 5 min, and pour plates or distribute the hot medium in convenient amounts in sterile screw-capped bottles; no further sterilization is given. Avoid repeated melting and cooling.

This SL medium is recommended for isolation of a wide range of lactobacilli. However, the common meat-spoiling species *L. viridescens* and other species adapted to very acidic environments will not grow. Streptococci, carnobacteria, and other organisms are inhibited, but most pediococci and leuconostocs (from dairy and fermented vegetable sources), some enterococci and bifidobacteria (from intestinal sources), and yeasts may grow. As these pediococci and leuconostocs have metabolic characteristics in common with lactobacilli and many causes similar changes in a product, their detection may be useful (Sharpe, 1962). Growth of yeasts may be eliminated by addition of cycloheximide at a concentration of 10 mg/liter.

ORAL CAVITY, INTESTINE, AND VAGINA. SL medium was designated initially for selective isolation of lactobacilli from oral and intestinal sources (Rogosa et al., 1951) and has remained the medium of choice. Some bifidobacteria and occasional enterococci may also grow, and colonies may have to be further identified. Dashkevicz and Feighner (1989) suggested a medium for the detection of bile-salt hydrolase active lactobacilli based on SL or MRS medium supplemented with taurocholic or taurochenodeoxycholic acid. Care should be taken to ensure anaerobic growth conditions. Best results are obtained by application of the Hungate technique.

MILK AND DAIRY PRODUCTS. SL medium is used for isolation of lactobacilli from milk, cheese, and fermented milks. Cheese starter lactococci are completely suppressed when enumerating cheese samples. Leuconostocs and pediococci, often found in milk and cheese, are not inhibited and colonies may have to be further identified. SL may not be optimum for some "thermophilic" lactobacilli from dairy sources. Therefore, SL supplemented with 0.5% meat extract is recommended. For selective isolation of *L. delbrueckii* subsp. *bulgaricus* from yogurt, M16 agar (Terzaghi and Sandine, 1975), with pH adjusted to 5.6 with 1.0 M acetic acid, has been used successfully (Davies et al., 1977). The isolation and cultivation of *L. kefiranofaciens* from kefir requires the use of KPL medium (Toba et al. 1986).

KPL Medium for Isolating *L. kefiranofaciens* (Toba et al. 1986)

For 1 liter of medium:

Lactic acid whey (see below)	930.0 ml
White table wine (see below)	70.0 ml
Glucose	10.0 g
Galactose	10.0 g
Tween 80	1.0 ml
Agar	15.0 g

Mix all ingredients except the wine, adjust pH to 5.5, and boil to dissolve the agar. Sterilize at 121°C for 15 min.

Lactic acid whey: Adjust 10% skim milk to pH 5.5 using lactic acid, and boil for 30 min. Remove precipitate by filtration.

The table wine (preferentially SO$_2$-free) is filter-sterilized and added to the autoclaved medium.

For broth medium, use deproteinized lactic acid whey prepared as follows: Adjust 10% skim milk to pH 5.5 using lactic acid and boil for 30 min. Remove precipitate by filtration. Adjust filtrate to pH 7.0 with 2N NaOH and boil for 30 min. Remove precipitate by filtration and readjust pH to 5.5 using 2N HCl.

MEAT AND MEAT PRODUCTS. The use of APT rather than MRS for isolation of *L. viridescens* and other lactobacilli from meat is probably traditional rather than necessary, as these organisms also grow profusely on MRS. Kitchell and Shaw (1975), discussing media for isolation from meats, suggest MRS in addition to APT, incorporating 0.1% thallous acetate with the pH adjusted to 5.5, or SL adjusted to pH 5.8. A sorbic acid medium was recommended by Reuter (1968) for the selective enumeration of lactobacilli from meat and meat products. Enterococci, leuconostocs, and pediococci may be observed as small (pin-point) colonies and thus can be distinguished from the more prolifically growing lactobacilli.

Selective Sorbic Acid Medium for Isolation of Lactobacilli from Meat and Meat Products (Reuter, 1968)

For 1 liter of medium:

Trypticase	10.00 g
Meat extract	10.00 g
Yeast extract	5.00 g
Glucose	20.00 g
Tween 80	1.00 ml
Na acetate	5.00 g
Na citrate	3.00 g
MgSO$_4$·7H$_2$O	0.20 g
MnSO$_4$·4H$_2$O	0.05 g
Agar	20.00 g

0.4 g sorbic acid is added as slightly alkaline solution in water. After dissolving the ingredients at 100°C, the pH is adjusted to 5.0 at 50°C. The medium is heated to 100°C, filtered through cotton wool, and boiled again for 30 min.

A modification of the sorbic acid medium, using 0.2% potassium sorbate and a higher pH of 5.7, was found to be especially favorable for the recovery of large numbers of lactobacilli from meat products (Holzapfel and Gerber, 1986). This medium is based on MRS. The commercially available product can be used, to which the 0.2% potassium sorbate is added after boiling and adjusting the pH. Autoclaving is optional since boiling for 30 min is sufficient.

FERMENTED VEGETABLES AND SILAGE. For silage isolates, SL medium is used. For the low-pH, fermented vegetable processes (using cucumber, sauerkraut, olives, etc.), SL medium and the modified Homohiochii medium are suggested.

FRUIT JUICES. For spoilage organisms from orange juice, Juven (1976) recommends APT medium. Murdock et al. (1952), for the same purpose, recommend a medium containing orange serum.

Orange-Serum Medium for Isolation of Spoilage Organisms from Fruit Juices (Murdock et al., 1952)

For 1 liter of medium dissolve:

Trypticase	10.0 g
Yeast extract	3.0 g
Orange extract	5.0 g
Glucose	4.0 g
K$_2$HPO$_4$	3.0 g
Agar	17.0 g

Adjust pH to 5.5. Sterilize at 115°C for 15 min, avoiding overheating.

FERMENTED BEVERAGES. Isolations from wine, beer, and fermented grain mashes, where lactobacilli have adapted to extremely specialized environments, require quite different types of media. It may be necessary to include some of the natural substrate to provide any unknown growth factors essential for strains which have become particularly adapted to their environments. Often, tomato juice can replace these specific growth factors. It may be necessary to suppress such aciduric organisms as yeasts, molds, and acetobacters (early work cited by Sharpe, 1960) with inhibitory agents.

WINE. For isolation of the slow-growing lactobacilli from wines—both those taking part in the malolactic fermentation and spoilage strains—tomato juice and yeast extract are highly stimulatory and should be included in the medium. The pH should not exceed 5.0. An addition of 4 to 5% ethanol to all media is also highly recommended. Yoshizumi (1975) suggests the following medium:

Tomato Juice Medium for Isolating Lactobacilli from Wine (Yoshizumi, 1975)

For 1 liter of medium, dissolve in distilled water:

Glucose	10.00g
Yeast extract	5.00 g
Polypeptone	5.00 g
KH$_2$PO$_4$	0.50 g
KCl	0.12 g
CaCl$_2$·2H$_2$O	0.12 g
NaCl	0.12 g
MgSO$_4$·7H$_2$O	0.12 g
MnSO$_4$·4H$_2$O	0.03 g
Bromcresol green	0.03 g
Agar	15.00 g
Canned tomato juice	150.00 ml

Steam the agar in the water to dissolve it first. Adjust the final pH to 5.0. Sterilize at 121°C for 15 min. A fungistat should be added to inhibit the growth of yeasts. The author suggests adding Eurocidin or Kabicidin (100 mg/liter). If these cannot be obtained, cycloheximide (100 mg/liter) or sorbic acid (1.2 g/liter) have also been used (Chalfan et al., 1977). It is essential to cultivate under anaerobic conditions for isolation from later stage of fermentation (Yoshizumi, 1975).

For the isolation of lactobacilli from some wines, it is necessary to use a grape-based medium with added yeast extract and a pH of 3.2 to 4.5, depending on the wine being examined (Castino et al., 1975; Chalfan et al., 1977). The modified Homohiochii medium has been found to support excellent growth of practically all typical lactobacilli associated with wine.

CIDER. An apple juice-based medium is recommended (Carr and Davies, 1970) consisting of apple juice plus 1% yeast extract, with the specific gravity adjusted to 1.040, and the pH value to 4.8 with NaOH. The addition of 3% agar is necessary to ensure a firm gel.

BEER. For isolation of spoilage-causing brewery lactobacilli, many selective media have been used (Hsu and Taparowsky, 1977). Boatwright and Kirsop (1976) described a sucrose medium and confirmed the usefulness of cycloheximide, polymyxin B, and phenyl ethanol in suppressing yeasts and Gram-negative bacteria. This sucrose agar compared favorably with other brewery media and could be used to cultivate a wide range of lactobacilli, but pediococci and leuconostocs also grew.

Sucrose Agar for Brewery Isolates (Boatwright and Kirsop, 1976)

For 1 liter of medium, dissolve in distilled water:

Sucrose	50.0 g
Oxoid peptone	10.0 g
Yeast extract	5.0 g
NaCl	5.0 g
MnSO$_4$·4H$_2$O	0.5 g
MgSO$_4$·7H$_2$O	0.5 g
Tween 80	0.1 g
CaCO$_3$	3.0 g
Bromcresol green	20.0 mg

Agar	20.0 g

Steam the agar first to dissolve it. Final pH is adjusted to 6.2. Sterilize at 121°C for 15 min after distributing in convenient amounts. Microbial inhibitors are added to the molten agar just before pouring plates—cycloheximide as a filter-sterilized solution to give a final concentration of 10 mg/ml, and 2-phenyl ethanol without dilution or sterilization, final concentration 0.3%

A double-concentrated MRS medium adjusted with beer to normal concentration before autoclaving can also be used for cultivation of typical beer lactobacilli. Excellent results for recovery of typical beer lactic acid bacteria are obtained with NBB medium (Back, 1980). The preparation is rather complicated and tedious. A commercial ready-to-use product, however, is available.

GRAIN MASHES. MRS agar was found to be unsatisfactory for isolations from grain mash, and a medium based on a mixture of filter-sterilized malt extract and yeast autolysate has been developed (Bryan-Jones, 1975). Further improvement can be achieved by the addition of wheat flour or bran. For the isolation and propagation of lactobacilli from mageu, a traditional sour maize beverage of southern Africa, the following special medium is recommended (Holzapfel, 1989).

Medium for Isolation of Typical Lactobacilli from Mageu (Holzapfel, 1989)

For 1 liter of medium:

Filtrate of 5% maize meal porridge	1.00 l
Tween 80	4.00 ml
Whey powder	20.00 g
Wheat flour	10.00 g
$MgSO_4 \cdot 7H_2O$	0.10 g
$MnSO_4 \cdot 4H_2O$	0.05 g
Triammonium citrate	4.00 g
Sucrose	20.00 g
Cysteine hydrochloride	1.00 g
Agar	15.00 g

Adjust pH before sterilization to 6.0. Sterilize at 121°C for 15 min. Incubate at 50°C for 3–5 days under reduced atmosphere. Inclusion of triphenyl tetrazolium chloride facilitates the enumeration of single colonies against the white medium.

SAKE. For isolation of spoilage lactobacilli from rice wines, the modified Homohiochii medium can be recommended.

SOURDOUGH. Highest recovery of lactic acid bacteria from sourdough was obtained with modified Homohiochii medium which included the following supplements per liter (W. P. Hammes, unpublished observations): 21

g of bakers' yeast presuspended in 100 ml deionized water and 50 g wheat bran. After sterilizing in centrifuge beakers, the medium is centrifuged (15,000 × g for 15 min) and the supernatant, slightly turbid medium is aseptically distributed into appropriate containers. For preparing agar plates, mix double concentrated broth with equal amounts of sterile, hot 3% water agar.

Gaseous Environment

Most lactobacilli grow better either anaerobically or in the presence of increased CO_2 tension, particularly on first isolation. Agar plates should be incubated in an atmosphere of 90% N_2 + 10% CO_2. Surface plating is recommended so that different colonial types can be observed if present, often indicating the presence of more than one species or biotype. For isolation of oxygen-sensitive intestinal lactobacilli, poured, dried plates must be prereduced by overnight incubation, preferably in an anaerobic jar equipped with catalyst and gas-generating kit. Good anaerobic conditions for isolations should generally also be applied where thermophilic lactobacilli can be expected (raw milk and dairy products, fermentations at elevated temperatures). When selective media are used, particularly the SL medium, care must be taken not to dry the plates too long or the concentrated acetate at the agar surface may inhibit growth. Isolates from humans, animals, and some dairy products are incubated at 37°C; from other habitats at 30°C; and at 22°C from low-temperature sources.

Culture Maintenance

Once isolated, unless there are special growth requirements, most species of lactobacilli can be cultured in MRS broth (de Man et al., 1960) or maintained for short periods in MRS agar stabs. For anaerobic lactobacilli, 0.05% cysteine hydrochloride or 0.1% ascorbic acid should be added, the broth steamed just before use, and organisms cultured under 90% N_2 + 10% CO_2. For some strains or species of lactobacilli, particularly heterofermentative ones, a carbohydrate other than glucose, such as maltose, fructose, or a pentose, gives better growth. For fastidious strains, supplements applied for their isolation are required for their culturing, e.g., freshly prepared yeast autolysate and wheat bran for *L. sanfrancisco*. For several other species, the modified Homohiochii medium (Kleynmans et al., 1989) has given satisfactory results.

CONSERVATION OF CULTURES. For preservation for 3–6 months, strains can be stored in yeast glucose litmus milk (YGLM) + calcium carbonate (Bryan-Jones, 1975; Sharpe and Fryer, 1965). To reconstituted skim milk powder or fresh skim milk, add litmus at a final concentration of 0.01%; yeast extract, 0.2%; glucose, 1%; liver extract, 0.25%; and calcium carbonate, 5%. Divide into 10-ml amounts, sterilize at 121°C for 10 min. Before use, tubes should be incubated for 1 week to check sterility. This medium is not suitable for the acidophilic heterofermentative isolates from wines and cider. Stab cultures in tomato juice agar at pH 5.0 are preferred. Optimum conditions for medium-term conservation (at least one year) are obtained by mixing a well-grown culture with an equal volume of sterile glycerol and storing at −20°C. Loops of inoculum can thus be taken without exposing the cells to the killing effect of the freezing/thawing procedure.

For long-term preservation, lyophilization is an excellent method for maintaining lactobacilli. Using the method described by Phillips et al. (1975), centrifuged packed cells from a vigorously growing broth cultures are resuspended in sterile horse serum containing 7.5% glucose and freeze-dried using the standard technique of Lapage et al. (1970). Similar results are obtained by suspending the cells in 10% skim milk. Ampules are sealed under vacuum and stored at 5–8°C. Both methods generally result in good recovery of cultures even after 20 years (N. Weiss, unpublished observations).

Highest numbers of survivors after conservation are required for the preparation of freeze-dried starter cultures. The general principles rely on the use of cryoprotectants and the prevention of access of oxygen and moisture during preparation and storage. Cryoprotectants have been reviewed by Bousfield and MacKenzie (1976). The composition of the growth medium and the time of harvest of the culture exert profound effects on the competence of cells for freezing. There are no methods that can be generally applied to all strains, so optimum conditions have to be determined for each strain.

Most optimum results are generally obtained by preservation in liquid nitrogen. 5% DMSO or 10% glycerol are routinely added as cryoprotectants before freezing. This method is applied especially for maintaining patent strains.

Identification

Organisms sharing habitats with lactobacilli and often growing on the same selective media are leuconostocs, pediococci, bifidobacteria, and occasionally carnobacteria (see later), lactococci, streptococci, and enterococci. In most cases, these organisms should be all morphologically distinguishable from lactobacilli, but sometimes there are difficulties. For example, cells of some freshly isolated strains of *L. sake* (Fig. 1) from meat form very short rods and only partially return to bacillary growth after subculturing. In addition, some heterofermentative lactobacilli grow as coccobacilli and are usually differentiated from morphologically similar leuconostocs only by their production of ammonia from arginine (exceptions: *L. viridescens*, *L. fructosus*, *L. sanfrancisco*, and *L. vaccinostercus*) and by forming DL- and not D(−)-lactic acid from glucose. Some homofermentative anaerobic lactobacilli from intestinal sources may be morphologically similar to certain bifidobacteria. They can be differentiated by determination of their end products from glucose fermentation. Whereas the homofermentative lactobacilli produce mainly lactic acid, bifidobacteria are heterofermentative and form acetic acid and lactic acid in a ratio of about 2:1. Curved cells are typical for *L. curvatus*, which may even assume a spiral shape (Fig. 2) . Similar morphology was also described for *L. delbrueckii* subsp. *bulgaricus* (Bottazzi, 1988). Pediococci, also isolated on most selective media for lactobacilli, can be differentiated morphologically, as can normally lactococci, streptococci, and enterococci. Carnobacteria

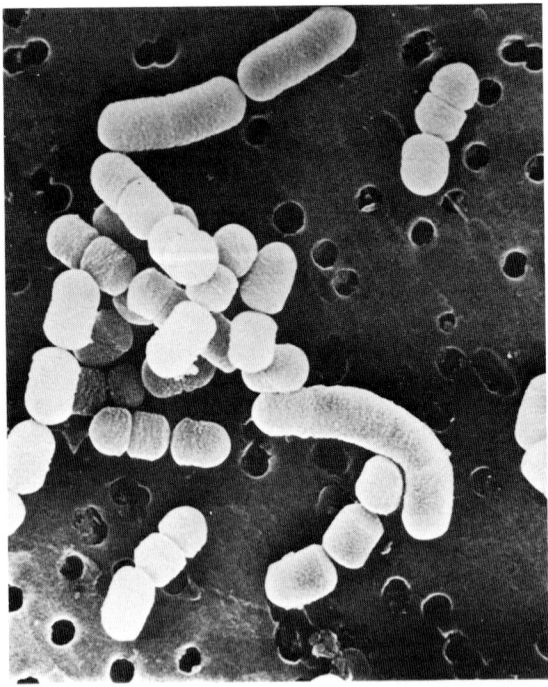

Fig. 1. Electron micrograph showing typical morphology of a *Lactobacillus sake* isolate from meat, with strong tendency to coccoidal rod shape. Bar = 0.5 μm. (Courtesy of Elizabeth S. Visser.)

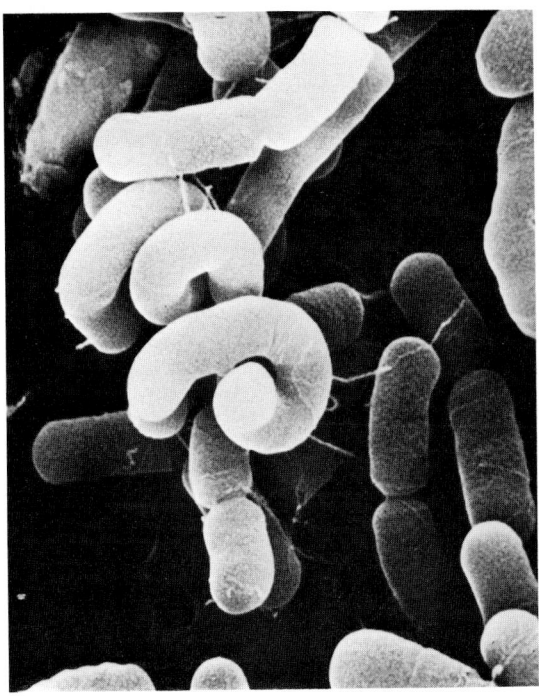

Fig. 2. Electron micrograph of *Lactobacillus curvatus* isolated from spoiled processed meat, showing typical spiral and curved shapes. Bar = 0.5 μm. (Courtesy of Elizabeth S. Visser.)

can be separated by the characteristics listed in Table 2.

Grouping of Species of Lactobacilli

Comparative analysis of 16S rRNA sequences (Yang and Woese, 1989) have been performed with lactobacilli by D. Yang and C. R. Woese (unpublished observations). Examples of these investigations are shown in Figs. 3, 4, and 5. These represent the survey of the "full lacto group" and, as branches thereof, the "casei group" and the "lactis group." From these data, it becomes evident that the genus *Carnobacterium* (Fig. 3) is only distantly related to lactobacilli, whereas the relationships of pediococci (Fig. 5) and leuconostocs (Fig. 4) to lactobacilli are rather close. It is of major importance that the type of carbohydrate fermentation, as the traditional basis of grouping of lactobacilli, is not strictly related to the evolution of the organisms. Thus, obligately heterofermentative lactobacilli are not grouped in one single evolutionary branch but are found at various branches. The same is true for the obligate homofermenters. Nevertheless, several branches in evolution can be recognized, which have in common the type of carbohydrate fermentation. These groups might be the basis of

future grouping based on evolutionary relationships.

To handle the large number of species, a grouping based on biochemical characteristics is still advisable. The most advanced system, as proposed by Kandler and Weiss (1986), is a modification of the division of the genus *Lactobacillus* into three subgenera by Orla-Jensen (1919, 1943). These former subgenera, *Thermobacterium, Streptobacterium,* and *Betabacterium,* are related to the following groups respectively: "obligate homofermenters," "facultative heterofermenters," and "obligate heterofermenters." The species belonging to these three groups are compiled in Tables 3, 4, and 5 with reference to their key physiological characteristics. The basic metabolic background of the new grouping relies on the carbohydrate metabolism, which is treated in more details in this chapter under "Physiology" (see below). The obligate homofermenters (group I) produce more than 85% lactic acid from glucose, as do the facultative heterofermenters. Whereas obligate homofermenters are unable to ferment pentose or gluconate, facultative heterofermenters (group II) ferment at least some of these carbohydrates. From glucose, the obligate heterofermenters (group III) produce equimolar amounts of CO_2, lactic acid, and acetic acid and/or ethanol.

Determination of Special Characteristics

Identification of species of lactobacilli requires, in many cases, the determination of many physiological and biochemical characteristics. Carbohydrate fermentation patterns may be determined either by the conventional test tube method or by commercially available miniaturized rapid systems. Results may not always be identical and sometimes vary from laboratory to laboratory. In addition, some species exhibit a great strain- to-strain variability in these tests which may in part be explained by the encoding of specific properties on plasmids (see below). In our experience, a reliable identification requires the determination of additional and less variable characteristics. These tests are described in more detail by Kandler and Weiss (1986) and are summarized as follows: the determination of the isomers of lactic acid produced from glucose can easily be achieved enzymatically with commercial test kits. The presence or absence of *meso*-diaminopimelic acid (*m*-DAP) in the cell wall can also be checked with a minimum of effort for a large number of strains. The determination of the peptidoglycan type is very helpful for the identification of some obligately heterofermentative

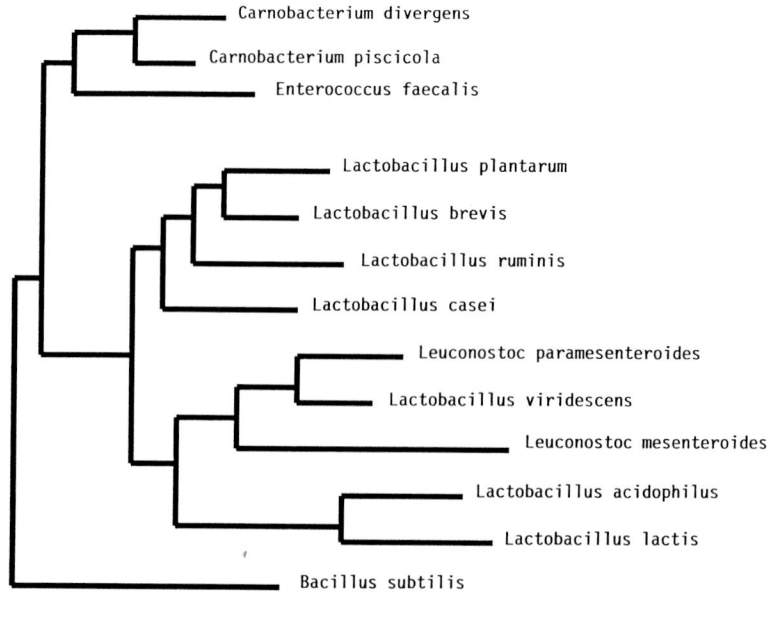

Fig. 3. Phylogenetic tree for the "full lacto group" of the lactobacilli, determined according to the procedure described by Yang and Woese (1989). Bar represents a relative evolutionary distance of 10%. (D. Yang and C. R. Woese, personal communication.)

species (see Table 5), but requires much more time expenditure and experience.

Comparison of total soluble protein patterns by polyacrylamide gel electrophoresis (PAGE) has proven to be a good tool for the separation of species which can hardly be differentiated by any other method (Dicks and van Vuuren, 1987, 1988). The accuracy of the method is demonstrated by the fact that *L. brevis* strain ATCC 8007 could not be allotted to any other cluster containing *L. brevis* (Dicks and van Vuuren, 1988), consistent with the observation of Kandler and Kunath (1983) that this strain is in fact *L. kefir*. In addition, *L. fermentum* ATCC 23272 and *L. reuteri* DSM 20016, both derived from G. Reuter's strain F275 and both identified as *L. reuteri* by Kandler et al. (1980), were well separated from *L. fermentum* by PAGE but not by carbohydrate fermentations. Reliable results are also obtained by determination of the electrophoretic mobility of lactic acid dehydrogenases (LDH) in polyacrylamide gels, as described by Hensel et al. (1977). Relative migration distances of species have been compiled by Kandler and Weiss (1986). This method is especially useful in the separation of the otherwise phenotypically very similar group consisting of *L. acidophilus, L. crispatus,* and *L. gasseri.*

Studies on antigenic determinants have not contributed to an improved identification system for lactobacilli within the last decade. For the older literature, a survey is presented by Sharpe (1981).

Information on DNA base composition is known for all the *Lactobacillus* species described presently (see Tables 3, 4, and 5). DNA-DNA homology studies have been performed for nearly all species of the genus and are the main basis for the present species concept.

Identification of Species

For the presumptive identification of a *Lactobacillus* strain, dichotomous keys are presented in Tables 6 and 7. Group I and group II species were combined in one key because a definite allocation of some species to one group only is not possible. As with any simplified system, some special strains may not be allocated correctly, and therefore the results should be reconfirmed by comparison with the characteristics compiled in Tables 3, 4, and 5. Among the species listed in Table 1, *L. catenaformis* and *L. minutus* have not been considered because they are no longer regarded as lactobacilli (Stackebrandt and Teuber, 1988); their affilia-

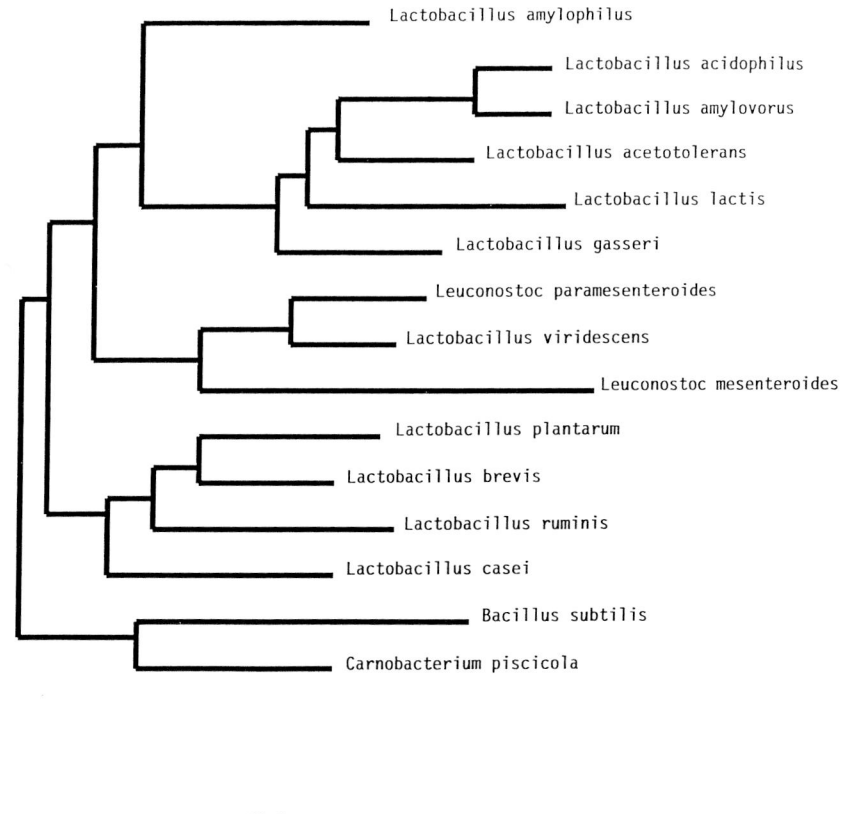

Fig. 4. Phylogenetic tree for the "lactis group" ("*delbrueckii* subsp. *lactis* group") of the lactobacilli, determined according to the procedure described by Yang and Woese (1989). Bar represents a relative evolutionary distance of 10%.

tion is presently uncertain. The same is true for *L. rogosae* (Kandler and Weiss, 1986). The incorporation into the key of some species (*L. acetotolerans, L. aviarius, L. kefiranofaciens,* and *L. vaginalis)* was not feasible. *L. acetotolerans* and *L. kefiranofaciens* are best characterized by their unusual growth requirements. *L. aviarius* is characterized by its unusual coccoidal morphology and anaerobic nature. *L. vaginalis* can be distinguished from the phenotypically similar *L. fermentum* and *L. reuteri* by its delayed fermentation of ribose.

Physiology

The metabolism of lactobacilli appears simple because of their complex nutritional requirements and their fermentative nature. There are, however, certain traits which are characteristic for special groups, species, or even strains that contribute in a unique manner not only to their adaptation to but also their effect on a substrate. Thus, using the various substrates, quite distinct species may be revealed which affect the sensory quality of foods during both the desired

fermentation process and the detrimental spoilage process.

Carbohydrate metabolism leads to the production of lactic acid or of lactic acid, CO_2, and acetate and/or ethanol. The homolactic species utilize hexoses by the glycolytic pathway, and the heterofermentative species by the 6-phosphogluconate pathway (Kandler, 1983). The latter pathway requires phosphoketolase, an enzyme which is absent in the obligate homofermenters but present in facultative heterofermentative species. Unusual pathways for carbohydrate metabolism appear to be present in certain groups of lactobacilli. For example, Barre (1978) isolated thermophilic lactobacilli from fermenting grape must that utilized pentoses and performed a homolactic fermentation. The same was observed for isolates from the intestine of rats related to *L. salivarius* (Raibaud et al., 1973) and for an unknown thermophilic species (Fukui et al, 1957). In the latter case, a new fermentative pathway was suggested because sedoheptulose was transported but not fermented by these organisms, which should be the case when the pentoses were degraded via the pentose-phosphate pathway. The ferment-

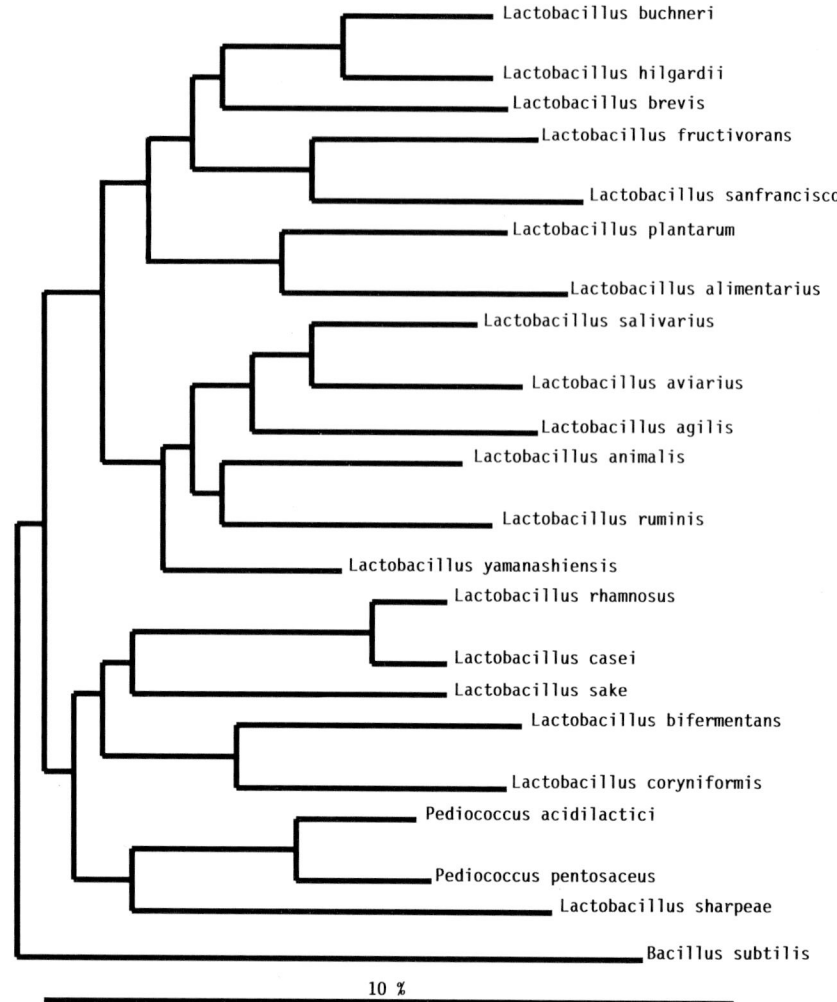

Fig. 5. Phylogenetic tree for the "casei group" of the lactobacilli, determined according to the procedure described by Yang and Woese (1989). Bar represents a relative evolutionary distance of 10%.

able carbohydrates and the products formed both depend on the environment. Thus, glucose represses the activity of ketolases in the facultative heterofermenters, and the pH also exerts effects on the type of fermentable carbohydrates used (Carr, 1987).

The nature of the end products of carbohydrate metabolism are strongly affected by the presence of oxidants (Condon, 1983, 1987). In the presence of oxygen, lactobacilli transfer electrons to this molecule, leading to the formation of superoxide (O_2^-), H_2O_2, or H_2O. Known enzymatic activities involved are NADH:H_2O_2 oxidase, NADH:H_2O oxidase, pyruvate oxidase, α-glycerophosphate oxidase, and NADH peroxidase. Superoxide dismutase activity has not been found in lactobacilli, although the dismutation of superoxide is catalyzed by internally accumulated manganese (Archibald and Fridovich, 1981; Götz et al.,

1980). Hydrogen peroxide can be removed either by NADH peroxidases or by catalases. Catalase activity in lactobacilli depends either on the presence of hematin (Whittenbury, 1964; Wolf and Hammes, 1988) or on a manganese-containing pseudocatalase. True catalase activity has been found in *L. plantarum, L. pentosus, L. delbrueckii, L. sake, L. brevis, L. buchneri, L. fermentum* (Whittenbury, 1964; Wolf and Hammes, 1988), and pseudocatalase in *L. plantarum, P. pentosaceus, E. faecalis,* a *Leuconostoc* species (Johnston and Delwiche, 1962), and *L. mali* (W. P. Hammes, unpublished observations). Another group of oxidants affecting the metabolism of lactic acid bacteria are nitrate and nitrite (Dodds and Collins-Thompson, 1985; Wolf and Hammes, 1988; Wolf et al., 1990). Nitrate reductases have been found in *L. plantarum, L. pentosus, L. fermentum,* and *L. casei* (Costilow and Humphreys, 1955; Langs-

Table 3. Key physiological characteristics of the obligately homofermentative *Lactobacillus* species.

Species	GC content (mol%)	Peptidoglycan type	Lactic acid isomer(s)	Growth at 15°C	NH$_3$ from arginine	Carbohydrate fermented											
						Amygdalin	Cellobiose	Galactose	Lactose	Maltose	Mannitol	Mannose	Melibiose	Raffinose	Salicin	Sucrose	Trehalose
L. delbrueckii subsp. *delbrueckii*	49–51	Lys-D-Asp	D	−	d	−	d	−	−	d	−	+	−	−	−	+	d
L. delbrueckii subsp. *lactis*	49–51	Lys-D-Asp	D	−	d	+	d	d	+	+	−	+	−	−	+	+	+
L. delbrueckii subsp. *bulgaricus*	49–51	Lys-D-Asp	D	−	−	−	−	−	+	−	−	−	−	−	−	−	−
L. acidophilus	34–37	Lys-D-Asp	DL	−	−	+	+	+	+	+	−	+	d	d	+	+	d
L. amylophilus	44–46	Lys-D-Asp	L	+	ND	−	−	+	−	+	−	+	−	−	−	−	−
L. amylovorus	40–41	Lys-D-Asp	DL	−	ND	+	+	+	−	+	−	+	+	+	+	+	+
L. animalis	41–44	Lys-D-Asp	L	−	−	d	+	+	+	+	−	+	+	−	d	+	−
L. aviarius subsp. *araffinosus*	39–43	Lys-D-Asp	L(D)	−	ND	d	d	−	−	+	−	+	d	−	d	+	+
L. aviarius subsp. *aviarius*	39–43	Lys-D-Asp	DL	−	ND	d	+	d	d	+	−	+	d	+	+	+	+
L. crispatus	35–38	Lys-D-Asp	DL	−	−	+	+	+	+	+	−	+	−	−	+	+	−
L. farciminis	34–36	Lys-D-Asp	L(D)	+	+	+	+	+	+	+	−	+	−	−	+	+	+
L. gasseri	33–35	Lys-D-Asp	DL	−	−	+	+	+	d	d	−	+	d	d	+	+	d
L. hamsteri	33–35	Lys-D-Asp	DL	−	ND	+	+	+	+	+	−	d	+	+	+	+	d
L. helveticus	38–40	Lys-D-Asp	DL	−	−	−	−	+	+	d	−	d	−	−	−	−	d
L. jensenii	35–37	Lys-D-Asp	D	−	+	+	+	+	−	d	−	+	−	+	+	+	+
L. kefranofaciens	34–35	ND	D(L)	−	ND	−	−	+	+	+	d	ND	+	+	−	+	−
L. mali	32–34	DAP	L	+	−	+	d	d	−	−	−	+	−	−	+	+	+
L. ruminis	44–47	DAP	L	−	−	+	+	+	d	+	−	−	+	+	−	+	−
L. salivarius	34–36	Lys-D-Asp	L	−	−	−	−	+	+	+	+	+	+	+	d	+	+
L. sharpeae	53	DAP	L	+	−	+	+	+	+	+	−	+	+	+	+	+	−
L. vitulinus	34–37	DAP	D	−	−	+	+	+	+	+	−	+	+	+	+	+	d

Symbols: +, 90% or more of strains are positive; −, 90% or more of strains are negative; d, 11–89% of strains are positive; ND, no data available; DAP, diaminopimelic acid. Parenthesized isomers indicate <15% of total lactic acid.

Table 4. Key physiological characteristics of the facultatively heterofermentative *Lactobacillus* species.

Species	GC content (mol%)	Peptidoglycan type	Lactic acid isomer(s)	Growth at 15°C	Amygdalin	Arabinose	Cellobiose	Esculin	Gluconate	Mannitol	Melezitose	Melibiose	Raffinose	Ribose	Sorbitol	Sucrose	Xylose
L. acetotolerans I	35–36	Lys-D-Asp	DL	−	−	−	−	w	−	+	−	−	−	+	−	−	−
L. acetotolerans II	35.5–36.5	Lys-D-Asp	D(L)	−	−	−	+	+	−	−	−	−	−	−	−	−	−
L. agilis	43–44	DAP	L	−	+	−	+	+	−	+	+	+	+	+	d	+	−
L. alimentarius	36–37	Lys-D-Asp	L(D)	+	ND	d	+	+	+	−	+	−	−	+	−	+	−
L. bavaricus	41–43	Lys-D-Asp	L	+	+	−	+	+	+	−	−	+	−	+	−	+	−
L. casei	45–47	Lys-D-Asp	L	+	+	−	+	+	+	+	+	−	−	+	+	+	−
L. coryniformis subsp. *coryniformis*	45	Lys-D-Asp	D(L)	+	−	−	−	d	+	+	−	d	d	−	d	+	−
L. coryniformis subsp. *torquens*	45	Lys-D-Asp	D	+	−	−	−	−	+	−	−	−	−	−	−	+	−
L. curvatus	42–44	Lys-D-Asp	DL	+	−	−	+	+	+	−	−	−	−	+	−	d	−
L. graminis	41–43	Lys-D-Asp	DL	+	+	−	+	+	−	−	−	−	−	−	−	+	+
L. homohiochii	35–38	Lys-D-Asp	DL	+	−	−	d	ND	−	d	−	−	−	d	−	−	−
L. maltaromicus[a]	36	DAP	L	+	+	−	+	ND	ND	+	−	+	−	+	+	+	−
L. murinus	43–44	Lys-D-Asp	L	−	d	+	+	+	−	d	−	+	+	+	−	+	−
L. paracasei subsp. *paracasei*	45–47	Lys-D-Asp	L	+	+	−	+	+	+	+	+	+	−	+	d	+	−
L. paracasei subsp. *tolerans*	45–47	Lys-D-Asp	L	+	−	−	−	−	w	−	−	−	−	−	−	−	−
L. pentosus	46–47	DAP	DL	+	+	+	+	ND	+	+	d	+	+	+	+	+	+
L. plantarum	44–46	DAP	DL	+	+	d	+	+	+	+	+	+	+	+	+	+	d
L. rhamnosus	45–47	Lys-D-Asp	L	+	+	d	+	+	+	+	+	−	−	+	+	+	−
L. sake	42–44	Lys-D-Asp	DL	+	+	+	+	+	+	+	−	+	−	+	−	+	−

For symbols, see Table 3; w = weak positive reaction.

[a]On the basis of ribosome sequence analysis and the pattern of fatty acids, this species should be grouped with *Carnobacterium* (C. Woese, personal communication; N. Weiss, unpublished observations).

Table 5. Key physiological characteristics of the obligately heterofermentative *Lactobacillus* species.[a]

Species	GC content (mol%)	Peptidoglycan type	Growth at 15°C	NH₃ from arginine	Arabinose	Cellobiose	Esculin	Galactose	Maltose	Mannose	Melezitose	Melibiose	Raffinose	Ribose	Sucrose	Trehalose	Xylose
L. bifermentans	45	Lys-D-Asp	+	−	−	−	−	+	+	+	−	−	−	+	−	−	−
L. brevis	44–47	Lys-D-Asp	+	+	+	−	d	d	+	−	−	+	d	+	d	−	d
L. buchneri	44–46	Lys-D-Asp	+	+	+	−	d	d	+	−	+	+	d	+	d	−	d
L. collinoides	46	Lys-D-Asp	+	+	+	−	+	+	+	+	+	+	+	+	−	−	+
L. confusus	45–47	Lys-Ala	+	+	−	+	+	+	+	+	−	−	−	+	+	−	+
L. fermentum	52–54	Orn-D-Asp	−	+	d	d	−	+	+	w	−	+	+	+	+	d	d
L. fructivorans	38–41	Lys-D-Asp	+	+	−	−	−	−	d	−	−	−	−	−	d	−	−
L. fructosus	47	Lys-Ala	+	−	−	−	ND	−	−	+	−	−	−	w	−	+	−
L. halotolerans	45	Lys-Ala-Ser	+	+	−	−	−	−	+	+	−	−	−	+	−	−	−
L. hilgardii	39–41	Lys-D-Asp	+	+	−	−	−	d	+	−	d	−	−	+	d	−	+
L. kandleri	39	Lys-Ala-Gly-Ala₂	+	+	−	−	−	+	−	−	−	−	−	+	−	−	−
L. kefir	41–42	Lys-D-Asp	+	+	d	−	−	−	+	−	−	+	−	+	−	−	−
L. malefermentans	41–42	Lys-D-Asp	+	+	−	−	−	−	+	−	−	−	−	+	−	−	−
L. minor	44	Lys-Ser-Ala₂	+	+	−	+	+	+	+	+	−	+	+	+	+	+	+
L. oris	49–51	Lys-D-Asp	−	−	+	d	d	+	+	d	+	+	+	+	+	d	+
L. parabuchneri	44	Lys-D-Asp	+	+	+	−	−	+	+	ND	−	+	+	+	+	−	−
L. reuteri	40–42	Lys-D-Asp	−	+	+	−	ND	+	+	−	+	+	+	+	+	−	−
L. sanfrancisco	36–38	Lys-Ala	+	−	−	−	ND	+	+	−	−	−	−	+	−	−	−
L. suebicus	40,4	DAP	+	ND	+	d	−	+	+	ND	−	d	−	+	d	−	+
L. vaginalis	38–41	ND	−	ND	−	−	d	+	+	+	−	+	+	d	+	−	+
L. vaccinostercus	36	DAP	−	−	+	w	−	w	+	−	−	−	−	+	−	−	+
L. viridescens	41–44	Lys-Ala-Ser	+	−	−	w	−	w	+	+	−	−	−	−	d	d	−

For symbols, see Tables 3 and 4.

[a] The following sugars are generally fermented: fructose (exceptions: *L. malefermentans*, *L. sanfrancisco*, and *L. vaccinostercus*) and glucose. The following sugars are generally not fermented: amygdalin (exceptions: *L. confusus* and *L. oris*), mannitol (exceptions: *L. bifermentans* and *L. kandleri*), rhamnose (exception: *L. bifermentans*), and sorbitol.

Table 6. Key for presumptive identification of obligately homofermentative and facultatively heterofermentative species of the genus *Lactobacillus*.

I. *meso*-A$_2$pm present in cell hydrolysates
  A. Ribose fermented
    1. Growth at 15°C
      a. DL-Lactic acid produced
        aa. Glycerol and xylose fermented ...... *L. pentosus*
        ab. Glycerol and xylose not
          fermented .... *L. plantarum*
      b. L(+)-Lactic acid produced .... *L. maltaromicus*
  B. No growth at 15°C .... *L. agilis*
    1. Ribose not fermented
      a. Growth at 15°C, L(+)-lactic acid produced
        aa. Maltose fermented, Sucrose not
          fermented .... *L. sharpeae*
        ab. Maltose not fermented, Sucrose
          fermented .... *L. mali*
      b. No growth at 15°C
        ba. L(+)-Lactic acid produced .... *L. ruminis*
        bb. D(−)-Lactic acid produced .... *L. vitulinus*
II. *meso*-A$_2$pm not present in cell hydrolysates
  A. Ribose fermented
    1. Growth at 15°C
      a. DL-Lactic acid produced
        aa. Sucrose fermented
          aaa. Mannitol fermented .... *L. paracasei*
          aab. Mannitol not fermented
            aaba. Melibiose fermented .... *L. sake*
            aabb. Melibiose not
              fermented .... *L. curvatus*
        ab. Sucrose not fermented .... *L. homohiochii*
      b. L(+)-Lactic acid produced
        ba. Mannitol fermented
          baa. Rhamnose fermented, growth at
            45°C .... *L. rhamnosus*
          bab. Rhamnose not acid, no growth at
            45°C .... *L. paracasei*
        bb. Mannitol not fermented
          bba. Lactose fermented, trehalose not
            fermented .... *L. bavaricus*
          bbb. Lactose not fermented, trehalose
            fermented .... *L. alimentarius*
    2. No growth at 15°C
      a. DL-Lactic acid produced .... *L. hamsteri*
      ab. Optionally active lactic acid produced
      b. L(+)-Lactic acid produced .... *L. murinus*
      c. D(−)-Lactic acid produced .... *L. jensenii*
  B. Ribose not fermented
    1. Growth at 15°C
      a. DL-Lactic acid produced .... *L. graminis*
        aa. Optically active lactic acid produced
      b. D(−)-Lactic acid produced .... *L. coryniformis*
      c. L(+)-Lactic acid produced
        ca. Mannitol fermented .... *L. casei*
        cb. Mannitol not fermented
          cba. Starch hydrolyzed .... *L. amylophilus*
          cbb. Starch not hydrolyzed .... *L. farciminis*
    2. No growth at 15°C
      a. DL-Lactic acid produced

(continued)

Table 6. *Continued*

      aa. Sucrose fermented
        aaa. Starch hydrolyzed .... *L. amylovorus*
        aab. Starch not hydrolyzed
          aaba. Cell wall teichoic acid
            present .... *L. acidophilus* and
                            L. crispatus
          aabb. Cell wall teichoic acid
            absent .... *L. gasseri*
      ab. Sucrose not fermented .... *L. helveticus*
      b. L(+)-Lactic acid produced
        ba. Mannitol fermented .... *L. salivarius*
        bb. Mannitol not fermented .... *L. animalis*
      c. D(−)-Lactic acid
        produced .... *L. delbrueckii* and *L. jensenii*

ton and Bouma, 1960a; 1960b) and nitrite reductase in *L. lactis, L. leichmannii, L. buchneri, L. plantarum, L. acidophilus, L. viridescens, L. sake, L. farciminis, L. pentosus, L. brevis,* and *L. suebicus* (Wolf et al., 1990; Fournaud et al., 1964; Collins-Thompson and Lopez, 1981). Two types of nitrite reductases are known, those depending on the presence of hematin *(L. plantarum* and *L. pentosus)* (Wolf and Hammes, 1988; Wolf et al., 1990) and heme-independent enzymes *(L. delbrueckii* subsp. *lactis, L. sake, L. farciminis, L. brevis, L. buchneri* and *L. suebicus)* (Dodds and Collins-Thompson, 1985; Wolf et al., 1990). In the former type of reaction, ammonia is produced, whereas in the latter type, NO and N$_2$O are the products of nitrite reduction.

The presence of an oxidant leads to the formation of end products of carbohydrate metabolism which are more oxidized than lactic acid, such as CO$_2$ plus acetate, acetoin, or diacetyl (Condon, 1987). A most instructive example of the effect of oxygen on carbohydrate metabolism was provided by Dirar and Collins (1973), who observed that at low galactose concentration (1–6 mmol/1), the carbohydrate was degraded by *L. plantarum* to almost exclusively acetic acid (93%) in addition to carbon dioxide. The formation of acetate catalyzed by acetate kinase increases the energy yield in this organism. The same mechanism is active in heterofermenters where acetylphosphate is derived from the reaction catalyzed by phosphoketolase. In the presence of oxygen, the formation of ethanol is reduced in favor of acetate (Kandler, 1983). In organisms devoid of phosphate acetyl transferase and alcohol dehydrogenase (Condon, 1987) (e.g., *L. brevis, L. buchneri*), fructose is used as oxidant and is reduced to mannitol. Alternatively, quinic acid is reduced to dihydroshikimic acid by a *L. collinoides* strain isolated from cider, and a similar mechanism was

Table 7. Key for presumptive identification of obligately heterofermentative species of the genus *Lactobacillus*.

I. CO_2 + H_2 produced from lactate .... *L. bifermentans*

II. CO_2 + H_2 not produced from lactate

  A. *meso*-A_2pm present in cell hydrolysates

    1. Growth at pH 3.3 and 12% of ethanol .... *L. suebicus*

    2. No growth under the above conditions .... *L. vaccinostercus*

  B. *meso*-A_2pm not present in cell hydrolysates

    1. Ribose fermented

      a. Growth at 15°C

        aa. Cellobiose fermented

          aaa. Slime capsule, xylose fermented .... *L. confusus*

          aab. No slime capsule, xylose not fermented .... *L. minor*

        ab. Cellobiose not fermented

          aba. Mannitol fermented .... *L. kandleri*

          abb. Mannitol not fermented

            abba. Arabinose fermented

              abbaa. Melezitose fermented

                abbaaa. Xylose fermented .... *L. buchneri*

                abbabb. Xylose not fermented ..... *L. parabuchneri*

              abbab. Melezitose not fermented

                abbaba. Amygdalin and gentiobiose fermented .... *L. oris*

                abbabb. Amygdalin and gentiobiose not fermented .... *L. brevis, L. collinoides,* and *L. kefir*

            abbb. Arabinose not fermented

              abbba. Melibiose fermented .... *L. kefir*

              abbbb. —Melibiose not fermented

                —Xylose fermented .... *L. hilgardii*

                —Xylose not fermented

                —Fructose fermented

                —Mannose fermented .... *L. halotolerans*

                —Mannose not fermented .... *L. fructivorans*

                —Fructose not fermented .... *L. malefermentans*

      b. No growth at 15°C, growth at 45°C

        ba. Peptidoglycan contains ornithine .... *L. fermentum*

        bb. Peptidoglycan contains lysine .... *L. reuteri*

    2. Ribose not fermented

      a. Sucrose fermented .... *L. viridescens*

      b. Sucrose not fermented

        ba. Maltose fermented .... *L. sanfranciseo*

        bb. Maltose not fermented .... *L. fructosus*

---

also detected in *L. plantarum* (Whiting and Coggins, 1974). Increased carbohydrate utilization and acetate formation was also observed for *L. brevis* and *L. buchneri* with glycerol and glucose as substrates. Glycerol was reduced to propanediol-1–3 via 3-hydroxypropanal as an intermediate (Schütz and Radler, 1984).

The synthesis of carbohydrate polymers by certain lactobacilli strongly affects the physical properties and visual appearance of a substrate. Thus, slime formation during sugar processing (see above), on surfaces of meat and meat products, in beer (Lawrence, 1988), wine, cider, and soft drinks is a detrimental property of lactobacilli (Carr and Davies, 1970; Williamson, 1959; Dunican and Seeley, 1965). On the other hand, in film jölk (Sharpe, 1979) and yogurt (Davis, 1975), slime formation by lactic acid bacteria is a desired property. The composition of the slimes has been only poorly investigated. The major product is dextran (Sharpe et al., 1972) but heteropolymers are also formed, as, for example, kefiran (La Riviere, 1967).

The metabolism of carbohydrates is the main source of energy and involves substrate-level phosphorylation. In addition to the ATP synthesized during glycolysis, ATP can be derived from acetylphosphate which is formed in reactions catalyzed by pyruvate oxidase or pyruvate decarboxylase. A further ATP-generating step involves the breakdown of arginine via the arginine deiminase (dihydrolase) pathway, whereby carbamate kinase synthesizes ATP from carbamoylphosphate with the formation of ammonia and CO_2. This mechanism is present in heterofermentative lactobacilli and was studied in *L. buchneri* (Manca de Nadra et al., 1988). A final energy-generating mechanism was proposed by Michels et al. (1979). Their "energy recycling model" postulates that cells can excrete end products together with protons, resulting in generation of a proton motive force which can be used for the production of metabolic energy. The functioning of such a mechanism in *Lactococcus cremoris,* as shown by ten Brink and Konings (1982) and Renault et al. (1988), provided evidence that the performance of the malolactic fermentation serves additionally as a mechanism for building up a proton motive force.

The transport of sugar and its regulation in lactic acid bacteria, in particular in the homofermentative species, have recently attracted interest because of their special advantages as models for studying these metabolic traits (Thompson, 1988). Unfortunately, the majority of investigations have been performed with lactococci, and only few lactobacilli were thoroughly studied. Glucose (and other saccharides) is transported with the aid of permease and is phosphorylated in the cytoplasm. The homofermentative lactobacilli *L. casei* and *L. plantarum* additionally contain a phosphoenol pyruvate (PEP): glucose phosphotransferase system which is missing in heterofermentative species (Romano et al., 1979). The presence of the specific PEP-dependent sugar: phosphotransferase system (PTS) in *L. casei* was detected for ribitol and xylitol (London and Hausman, 1982) as well as lactose (Chassy and Thompson, 1983). In lactose-fermenting lactobacilli, both permease and the PTS appear to be present (Premi et al., 1972), except for *L. casei,* since in five strains of this species, no β-galactosidase but only β-D-phosphogalactosidase could be detected. A plasmid-encoded β-galactosidase was, however, detected in *Lactobacillus casei* ATCC 393 (Chassy, 1987).

The utilization of proteins and peptides by lactobacilli again is better known for lactococci *(Streptococcus lactis)* than for lactobacilli. The main interest in these metabolic characteristics is derived from the role of lactic acid bacteria in milk and dairy products. Since milk is low in its content of amino acids and peptides, these bacteria have to hydrolyze milk proteins and to transport the resulting peptides. In addition, proteolysis is important in cheese ripening as it affects the texture and flavor in a desired (but sometimes faulty) way (reviewed by Thomas and Pritchard, 1987). Proteinases and peptidases of lactobacilli are not released into the medium but are bound to the cell wall. They may also act on substrates after lysis of the cells and release of cytoplasmic activities. The specifities of proteinases are usually determined on the basis of the site of their action on the various milk protein fractions. For example, the cell-wall-associated proteinase of *L. helveticus* hydrolyzed both κ-casein and β-casein (Ezzat et al., 1985), and the enzyme from *L. bulgaricus* was active on all types of casein (Chandan et al., 1982). For further degradation, peptidases of various types are active: dipeptidase, tripeptidases, aminopeptidases, carboxypeptidase, and arylamidases. The degradation products of proteins are transported into the cell by peptide transport systems, peptidase-coupled transport, and amino acid transport. Extra-cellular peptidase and transport systems of lactobacilli have been poorly characterized. Finally, intracellular peptidases degrade the peptides to amino acids. Generally, the proteinase and peptidase activities of thermophilic starter lactobacilli are higher than those of *S. thermophilus* (Hemme et al., 1981; Shankar and Davies, 1978). For the production of Emmenthal cheese, however, strains with low proteinase activity are required

and are selected accordingly for use in starter cultures (Steffen, 1976, 1979).

Antimicrobial Activities

Antagonistic activity of *Lactobacillus* cultures is a widely observed and frequently reported phenomenon. Early reports suggested the production of "antibiotic-like" substances by different lactobacilli (Wheater et al., 1951), *L. acidophilus* (Fernandes et al., 1987; Sabine, 1963; Tramer, 1966; Vincent et al., 1959), *L. helveticus* (Vincent et al., 1959; Wheater et al., 1951), and *L. lactis* (Reiter et al., 1980; Wheater et al., 1952). In part, the antimicrobial effect could be related to the production of lactic acid (mainly responsible for pH reduction and the preservative effect in lactic fermented foods) (Wood, 1985) and of hydrogen peroxide, produced by *L. acidophilus* (Collins and Aramaki, 1980) and *L. bulgaricus* and *L. lactis* (Dahiya and Speck, 1968), and shown to inhibit *Staphylococcus aureus* (Dahiya and Speck, 1968; Wheater et al., 1952) and *Pseudomonas* spp. (Price and Lee, 1970).

In vivo growth inhibition of *Escherichi coli* in the GI tract was attributed to the activation of the lactoperoxidase antibacterial system by H_2O_2, produced by *L. lactis* (Reiter et al., 1980). This system has been reviewed by Reiter and Härnulv (1984). Although lactobacillin, one suggested "antibiotic-like" substance of *L. lactis,* was shown to actually be hydrogen peroxide (Wheater et al., 1952), substances with typical antibiotic properties were indeed found to contribute to specific antimicrobial activities of lactobacilli. The role and relevance of "natural antibiotics" have only been studied in a few *Lactobacillus* strains, thus far. The bacteriocins have received special attention in recent years and have been reviewed by Daeschel (1989), Geis (1989), and Klaenhammer (1988). The present knowledge of bacteriocins of lactobacilli is summarized in Table 8. They can be clearly distinguished from other nonproteinaceous and broad-spectrum antibiotics of lactobacilli, and are defined as proteinaceous macromolecules that exert bactericidal activity against a limited range of organisms relatively closely related to the producer (Tagg et al., 1976). Factors such as specific bacteriocin receptors of sensitive cells, and plasmids as genetic determinants of bacteriocin production and immunity, are also involved (Daeschel, 1989). According to these criteria, only some of the previously mentioned "natural antibiotics" can be classified as bacteriocins. As a typical example, lactacin B from *L. acidophilus* inhibits a "narrow" spectrum of

Table 8. Properties of bacteriocins from lactobacilli.

Bacteriocin type	Producer organism	Molecular weight (KDa)	Resistance to heat	Susceptibility to proteases	Maximum production (growth phase)	Inhibitory spectrum (susceptible organisms)	Evidence for plasmid involvement Production	Immunity	Plasmid size (MDa)	Reference
Lactacin B	*L. acidophilus*	6–6.5	+ (100°C, 60 min)	+: protease, proteinase K	Early stationary phase	Lactobacilli	−	−	−	Barefoot and Klaenhammer (1983, 1984)
Lactacin F	*L. acidophilus*	2.5	+ (99°C, 20 min; 121°C, 15 min)	+: ficin, protease K, trypsin, *B. subtilis* protease	ND	Lactobacilli, enterococci	+	+	68; 52	Muriana and Klaenhammer (1987, 1989)
Unnamed	*L. brevis*	ND	+ (121°C, 60 min)	+: pronase E, trypsin	Late log phase	Lactobacilli, leuconostocs	ND	ND	ND	Rammelsberg (1988)
Unnamed	*L. casei*	37–39	− (50°C, 90 min)	−: pronase E, trypsin	ND	Lactobacilli	ND	ND	ND	Rammelsberg (1988)
Unnamed	*L. fermentum*	ND	+ (96°C, 30 min)	+: trypsin, pepsin	ND	Lactobacilli	ND	ND	ND	DeKlerk and Smit (1967)
Helveticin J	*L. helveticus*	37	− (100°C, 30 min)	+: trypsin, pepsin, ficin, proteinase K, pronase, subtilisin	Late log phase	Lactobacilli	−	−	−	Joerger and Klaenhammer (1986)
Lactocin 27	*L. helveticus*	12	+ (100°C, 60 min)	+: trypsin, pronase −: ficin	Early stationary phase	Lactobacilli	ND	ND	ND	Upreti and Hinsdill (1975)
Plantaricin A	*L. plantarum*	>8	+ (100°C, 30 min)	+: protease	Mid-log phase	Lactic acid bacteria	−	−	−	Daeschel et al. (1986)
Sakacin A	*L. sake*	ND	+ (100°C, 20 min)	+: trypsin, pepsin	Late log phase	Lactobacilli, leuconostocs, enterococci, listerias	+	+	18	Schillinger and Lücke (1989)

Adapted from U. Schillinger, personal communication.

closely related organisms, such as *L. bulgaricus, L. helveticus, L. lactis,* and *L. leichmannii* (Barefoot and Klaenhammer, 1984). Isolated as a thermostable macromolecular protein lipopolysaccharide complex of ca. 100 kDa (Barefoot and Klaenhammer, 1983), lactacin B was purified to resolve an active protein component of 6,000 to 6,500 Da that showed sensitivity to proteinase K (Barefoot and Klaenhammer, 1984). The frequency of bacteriocin production appears to be high for some species, including *L. acidophilus,* for which Barefoot and Klaenhammer (1983) reported 63% of the investigated strains positive, and showing inhibition only of closely related organisms. Mehta et al. (1983) reported a 5,400-Da inhibitory protein from *L. acidophilus* AC_1 that showed broad activity spectrum in vitro against various pathogenic organisms, including *Salmonella typhi, Shigella flexnerii, Pseudomonas aeruginosa,* and *Staphylococcus aureus.* The bacteriocinic nature of this substance has not been finally established. It appears that for typical bacteriocins from lactobacilli, antimicrobial activities have been reported only for closely related Gram-positive bacteria (Klaenhammer, 1988).

Several broad-spectrum antibiotic-like substances, such as acidophilin and lactocidin (from *L. acidophilus*) and bulgarican (from *L. bulgaricus*) are presumably not bacteriocins in nature (Reddy et al., 1984; Shahani et al., 1977; Vincent et al., 1959) but have not been chemically characterized. Showing broad-spectrum inhibition of nonrelated bacteria, and especially of Gram-negative intestinal pathogens, these "antibiotics" and their producer strains have been suggested to influence the microecology of the GI tract favorably (Fernandes et al., 1987; Reddy et al., 1984; Shahani et al., 1977; Whitt and Savage, 1987). Following an early suggestion by Winkelstein (cited by Vincent et al., 1959), the implantation of *L. acidophilus* and other intestinal lactobacilli in the digestive tract has received increased attention in recent years (Gilliland, 1979; Klaenhammer, 1982; Nahaishi, 1986; Sandine, 1979; Shahani and Ayebo, 1980). The suggested "probiotic" effects of certain lactobacilli in the GI tract may be brought about by a combination of different factors (including bacteriocins), referred to in this section.

In spite of increased research interest in bacteriocins and "antibiotics" of lactobacilli, relatively little is known about their in vivo activities in habitats such as the GI tract. Such information is important in view of the use of *Lactobacillus* preparations as dietary and therapeutic supplements and as "protective" cultures for biological food preservation. Probiot-

ics are discussed elsewhere in this section. In vitro studies have shown several factors to influence the production and activity of bacteriocins. The bacteriocin-producing ability of some *L. acidophilus* strains was found to be related to the type of growth medium and was inhibited by bile salts (Fernandes et al., 1988; Shahani et al., 1976). Bactericidal activity of lactacin B from *acidophilus* (Barefoot and Klaenhammer, 1983) and plantacin B from *L. plantarum* (West and Warner, 1988) could only be demonstrated on solid media and not in liquid cultures. Optimal production of helveticin J (Joerger and Klaenhammer, 1986), plantaricin A (Daeschel et al., 1987), and sakacin A (Schillinger and Lücke, 1989) was observed during the mid- to late log phase of growth. For lactacin B (Barefoot and Klaenhammer, 1983) and lactocin 27 (Upreti and Hinsdill, 1975), however, maximal concentrations were detected during the early stationary phase. Bacteriocin production and its stability is likewise influenced by the initial pH of the growth medium, and Muriana and Klaenhammer (1987) reported maximum yield of lactacin F when the pH of MRS broth was maintained at 7.0 during cultivation of *L. acidophilus* 88. Joergen and Klaenhammer (1986), however, observed the highest accumulation of helveticin J when *L. helveticus* was cultivated at pH 5.5. Lactobacillus bacteriocins generally show a bactericidal mode of action specifically against closely related organisms, excluding all Gram-negative bacteria (Daeschel, 1989; Klaenhammer, 1988). In contrast to the lactococci, the involvement of plasmid DNA could not be demonstrated for the majority of bacteriocins produced by lactobacilli, the exceptions being lactacin F (for *L. acidophilus*) (Muriana and Klaenhammer, 1987) and sakacin A (for *L. sake*) (Schillinger and Lücke, 1989). Using electroporation and conjugation techniques, various plasmids have been transferred to *L. acidophilus* strain ADH, while transduction of plasmid DNA mediated by temperate bacteriophage (O adh), could also be demonstrated (Luchansky et al., 1989). Reuterin, a potent, broad-spectrum antibiotic active against several Gram-positive and Gram-negative bacteria, fungi, yeasts, and protozoa, is produced by *L. reuteri* when glycerol is present in the medium (Talarico et al., 1988). The antimicrobial agent was identified as an equilibrium mixture of "monomeric, hydrated monomeric, and cyclic dimeric forms of β-hydroxypropionaldehyde" (Talarico and Dobrogosz, 1989). The production of these and related substances does not appear to be restricted to *L. reuteri,* but was also reported for *Streptococcus lactis* var. *maltigenes* by Morgan et al. (1966), and for *L. buch-*

neri and *L. brevis* by Schütz and Radler (1984). Being nonproteinaceous and showing extremely broad-spectrum antimicrobial activity, it cannot be classified as a bacteriocin. Results obtained from in vitro studies suggest that reuterin is effectively produced under pH, temperature, and relative anaerobic conditions resembling those in GI tract regions inhabited by *L. reuteri* (Chung et al., 1989; Dobrogosz et al., 1989). The presence of a variety of different microorganisms was shown to stimulate the "heterologous" production of reuterin (Chung et al., 1989) while several strains of lactobacilli inhabiting the GI tract were found insensitive to reuterin (Dobrogosz et al., 1989). These properties have been discussed with reference to a high potential of *L. reuteri* for use as dietary adjunct, or even of purified reuterin for food preservation (Chung et al., 1989; Daeschel, 1989; Dobrogosz et al., 1989).

Genetics

Only limited data are available on the genetics of lactobacilli. With regard to their multiple nutritional requirements, Morishita et al. (1974, 1981) could show that the genes coding for the synthesis of most amino acids and vitamins are not absent but exist as "silent genes." With *L. casei, L. plantarum, L. helveticus,* and *L. acidophilus* at a frequency of 10^{-4} to 10^{-8} per survivor, mutants could be obtained which had reverted to prototrophy for an amino acid or vitamin. For *L. casei,* quintuple mutants were created which had lost the requirement for serine, aspartic acid, leucine, isoleucine, and tyrosine. Thus, the requirement for specific nutrients in lactobacilli may originate from some minor defects in a specific structural gene, the promotor, or a regulatory element. The number of the eliminated pathways and the degree of mutation leading to this effect depends on the environment to which a species is adapted (Morishita et al., 1981).

Lactobacilli contain plasmids, as was first shown by Chassy et al. (1976). Most are cryptic but the following metabolic functions have been found to be encoded by plasmids: lactose metabolism (Chassy and Alpert, 1989; Shimizu-Kadota, 1988), drug resistance (Axelsson et al., 1988; Gibson et al., 1979; Ishiwa and Iwata, 1980; Lin and Savage, 1986; Vescovo et al., 1982), maltose utilization (Liu et al., 1988), protein hydrolysis (Morelli et al., 1986), cysteine metabolism (Shay et al., 1988), fermentation of *N*-acetyl-D-glucosamine (Smiley and Fryder, 1978), and bacteriocin production and immunity (Schillinger and Lücke, 1989; Muriana and Klaenhammer, 1987).

Conjugation (Gibson et al., 1979; Langella and Chopin, 1989; Shrago et al., 1986; Thompson and Collins, 1988; Vescovo et al., 1983) and transduction (Raya et al., 1989) have been shown to function as mechanisms of recombination in lactobacilli. Transfer of a nonconjugative vector by means of conjugative mobilization has been performed with *L. plantarum* (Shrago and Dobrogosz, 1988). No evidence for transformation or natural competence is available. On the other hand, in vitro transformation has been performed with protoplasts. This method yields low transfer rates only (Lin and Savage, 1986; Morelli et al., 1987). Higher transfer rates were obtained by electroporation (Chassy and Flickinger, 1987). In addition, intrageneric (Iwata et al., 1986) and intergeneric (Cocconcelli et al., 1986) protoplast fusion and transfection (Boizet et al., 1988; Cosby et al., 1988; Shimizu-Kadota and Kudo, 1984) have been successfully applied.

For genetic engineering experiments, *E. coli*–*Lactobacillus* shuttle vectors were constructed from endogenous plasmids of lactobacilli (Leer et al., 1987; Josson et al., 1989). Vectors derived from replicons of other Gram-positive organisms, mostly lactococci, have been constructed and successfully transferred into lactobacilli (Kok et al., 1984; de Vos, 1987). For stable maintenance of new properties, DNA can be integrated into the *Lactobacillus* chromosome (Scheirlinck et al., 1989). Transposons have also been introduced into lactobacilli (Knauf et al., 1989; Aukrust and Nes, 1988). Heterologous genes have been expressed in lactobacilli coding for α-amylase, endoglucanase (Bates et al., 1989; Scheirlinck et al., 1989), and antibiotic resistance.

Bacteriophages

Lactobacillus phages, which can be isolated from the common habitats of lactobacilli, were reviewed by Sechaud et al. (1988) and Sozzi et al. (1981). The importance of these phages becomes obvious in food fermentations, where they interfere with the process in a way that may result in downgrading of the product, or in a time delay or even complete breakdown of the fermentation process. As the use of starter cultures is most developed in the dairy field, the effect of phages is best known for lactococci. Phages of thermophilic lactobacilli have been found responsible for difficulties in acidification encountered in the preparation of cheese and yogurt (Peake and Stanley, 1978; Accolas and

Spillmann, 1979). Virulent phages specific for strains of *L. lactis, L. bulgaricus,* and *L. helveticus* used for the production of Swiss and Emmental cheese or yogurt have been reported (Kiuru and Tybeck, 1955; Sozzi and Maret, 1975; Alatossava and Pythilä, 1980; Trautwetter et al., 1986; Peake and Stanley, 1978). Watanabe et al. (1970) and Shimizu-Kadota and Sakurai (1982) describe virulent phages active against *L. casei,* isolated from yakult, which is a Japanese dietetic fermented milk product.

Despite the fact that the field of application for lactobacilli is much broader than that of lactococci, no reports of phage problems in products such as fermented sausages, fermented vegetables, sourdough, or silage are known. This can be explained by the fact that the fermentation of these products depends less on starters or takes place in a nearly solid matrix. In addition, it appears likely that in cases where the fermentation process was partially inhibited, phages might have been the causative agent, but have just not been looked for. There are few reports of studies of the interrelationship between phage and lactobacilli in meat products (Biewald, 1968; Nes and Sorheim, 1984; Trevors et al., 1983, 1984; Heidel and Hammes, 1990). From these studies it was concluded that phages are present in meat products or even in starter culture preparations but they cause only a time delay in fermentation of model substrates.

Reports of phage-carrying bacteria involved in the malolactic fermentation suggest a potential threat of bacteriophage to the wine industry. Lee (1978) demonstrated the existence of temperate phages in *L. casei* and *L. hilgardii* strains isolated from wine. Virulent phages from abnormal malolactic fermentations were also isolated, which were, however, active against *Leuconostoc oenos* (Sozzi et al., 1982; Henick-Kling et al., 1986).

Lysogeny in lactobacilli was first demonstrated by Coetzee and de Klerk (1962) in two strains of *L. fermentum.* Several studies have now established that lysogeny is widespread among lactobacilli (Yokokura et al., 1974; Cluzel et al., 1987; Shimizu-Kadota and Sakurai, 1982; Tohyama et al., 1972; Sakurai et al., 1970; McArthur and Barefoot, 1986; Raya et al., 1989; Stetter et al., 1977; Mata et al., 1986). Evidence for lysogeny was generally limited to detection of cell lysis following induction with mitomycin C or ultraviolet radiation and visualization of phage-like particles under the electron microscope. Plaque formation or lytic activity were detected only in a few cases in which indicator strains were available (Cluzel et al., 1987; Yokokura et al., 1974).

Temperate phages are a potential source of virulent phages, as it was shown by the example of *L. casei* phage ϕ FSW. This temperate phage converts into a virulent phage either by point mutation or by transposition of a mobile genetic element in the phage genome, arising from the bacterial chromosome (Shimizu-Kadota et al., 1983, 1985; Shimizu-Kadota and Tsuchida, 1984). This mobile genetic element, called ISL 1, is the first transposable element described in lactobacilli.

Studies of Sozzi et al. (1976) suggest that MRS is the optimal medium for the formation of plaques by bacteriophages of thermophilic lactobacilli. The addition of calcium chloride sometimes favors the formation of plaques and is therefore recommended. Investigations on the influence of incubation temperature on the development of lactic acid bacteria and their phages showed that the phages have mostly the same temperature minimum, maximum, and optimum as their hosts. Thus, temperature regulation cannot be used to prevent phage attacks in industrial fermentations (Sozzi et al., 1978). Bacteriophage-inactivation effects of basic amino acids (Murata et al., 1974), thiol reducing agents (Murata et al., 1972), and ascorbic acid (Murata et al., 1971) have been reported, suggesting that the target attacked was not protein, but DNA. All *Lactobacillus* phages investigated have a double-stranded linear DNA, but with mostly cohesive ends which enable circularization of the DNA. The genome size ranges from 40 kbp found for the phage PL-1 of *L. casei* to 73 kbp found for the phage B2 of *L. plantarum* (Nes et al., 1988). There may be a correlation between the large genome size of phage B2 and its small burst size of only 12–14 phages per cell. Phage PL-1 has a size of 200–600 phages per cell. Phage IL-1 of *L. casei* has also a genome size of 40 kbp and a burst size of 200 phages per cell.

Electron microscopic examination showed the tail-first orientation of phage adsorption (Watanabe et al., 1984). Studies of receptor sites have shown that D-galactose and L-rhamnose comprise a phage receptor for a strain of *L. casei,* serological group B. Addition of these sugars inactivated the phage, and a phage-resistant mutant lacked galactosamine in its surface component (Yokokura, 1977). This effect is interesting, as this species of *Lactobacillus* has no teichoic acid in its cell wall, indicating that wall teichoic acid is not involved here in absorption as in some other Gram-positive bacteria such as staphylococci. Using isolated cell wall fragments of host strain *L. casei* PL-1, Watanabe et al. (1977) have shown that cell walls of *L. casei* may have incomplete receptors. The mode of action of a defective phage of *L. salivarius,* able to kill both lysogenic and nonlysogenic strains of *L. salivarius,* was similar to that of colicin (Tohyama, 1973) and these defective phages may be similar to bacteriocins.

Morphology of *Lactobacillus* phages was, in the absence of further information, the classification criterion used in accordance with the system of Bradley. Most phages have isometric heads and long contractile or noncontractile tails, belonging therefore to Bradley's group A or B. *Lactobacillus* bacteriophages so far reported were classified in Bradley groups by Sozzi et al. (1981).

The most valid criterion in studies of phage taxonomy is DNA homology, demonstrated by DNA-DNA hybridization. A close relatedness between virulent and temperate phages of lactobacilli has also been demonstrated by DNA-homology studies. Mata and Ritzenthaler (1988) and Lahbib-Mansais et al. (1988) compared 18 virulent and 4 temperate phages of *L. delbrueckii* subsp. *bulgaricus* and subsp. *lactis* by DNA-DNA hybridization. All temperate and 15 virulent phages could be alloted to homology group a and group b contained the remaining three virulent phages. Five virulent phages of *L. helveticus* were also investigated and were shown to be related to one another, but unrelated to the phages of group a and b. Cluzel et al. (1987) investigated 10 temperate and 16 virulent *L. delbrueckii* subsp. *bulgaricus* and subsp. *lactis* phages and could assign 7 temperate and 12 lytic phages to the previously described group a. Phages differing completely from those of group a and b were assigned to two new unrelated groups, called c and d. The phages of groups a, c, and d share the same host range.

The Genus *Carnobacterium*

Certain "atypical" lactobacilli that were repeatedly isolated from refrigerated vacuum-packaged beef (Von Holy, 1983) were thought to be heterofermentative (Holzapfel and Gerber, 1983) and clearly differed from the atypical streptobacteria of meat and meat products described by Reuter (1970, 1975). Because of the differences from all other taxonomic groups and species of lactobacilli described to that stage, these bacteria were incorporated in a new species, *Lactobacillus divergens* (Holzapfel and Gerber, 1983). Properties such as CO_2 and $L(+)$-lactic acid production from glucose, and their relative inability to grow on acetate agar (Holzapfel and Gerber, 1983; Shaw and Harding, 1984, 1985), indicated a relationship to

"atypical" lactobacilli isolated from chicken meat (Thornley, 1957; Thornley and Sharpe, 1959; Barnes, 1976) and vacuum-packaged meat (Hitchener et al., 1982; Shaw and Harding, 1984). From the latter, another new species, *Lactobacillus carnis,* was isolated (Shaw and Harding, 1985), which, however, was found to be homologous to representatives of a salmonid fish pathogen (Cone, 1982), *Lactobacillus piscicola* (Hiu et al., 1984), with which *L. carnis* was reduced to synonymy (Collins et al., 1987). In an attempt to clarify their taxonomic position, Collins et al. (1987) applied physiological and biochemical criteria to classify these lactobacillus-like organisms into a new genus, *Carnobacterium.* The majority of the poultry strains were allocated to *Carnobacterium divergens* and *Carnobacterium piscicola,* while the remainder were incorporated into two new species, *Carnobacterium gallinarum* and *Carnobacterium mobile.*

Habitats of Carnobacteria

The close phenotypical resemblance and comparable nutritional requirements of lactobacilli and carnobacteria suggest that they may share common habitats. This is true for all ecosystems where carnobacteria have thus far been found. These facts may explain why carnobacteria, although isolated and studied before, were not recognized as a separate taxon earlier.

Thornley (1957) isolated atypical lactobacilli from irradiated chicken meat, in which numbers of 10^8/g were reached during storage in a nitrogen atmosphere at 5°C. In what was probably the first report on strains corresponding to carnobacteria, mention was made of failure to grow on acetate-containing media (Thornley, 1957). In a more detailed study of isolates from chicken and minced beef, Thornley and Sharpe (1959) identified three groups of atypical lactobacilli, all of which contained diaminopimelic acid in the cell wall. Feresu and Jones (1988) included these strains in a comprehensive numerical taxonomic study and showed group 2 of Thornley and Sharpe (1959) to form three clusters, two of which were found to resemble *Carnobacterium carnis (piscicola)* (phenon 9) and *C. divergens* (phenon 11), respectively. Collins et al. (1987) assigned the majority of the strains, originally isolated from chicken (Thornley, 1957; Thornley and Sharpe, 1959), to these two species and to *Brochothrix thermosphacta,* while the remainder were designated to two new species, *C. gallinarum* and *C. mobile.* One strain (NCDO 1230) identified as *C. piscicola* was isolated from human plasma (Collins et al., 1987).

The association of atypical lactobacilli (probably resembling carnobacteria) with meat and meat products has been mentioned by several workers. Among 140 strains of streptobacteria isolated from bacon, 11 failed to grow on acetate agar at pH 5.8 (Kitchell and Shaw, 1975). Reference to "atypical" streptobacteria as one of the major groups on vacuum-packaged bacon was made by Tonge et al. (1964), Kitchell (1964), and Kitchell and Shaw (1975). This and other groups of lactobacilli reach numbers of 10^8/g on vacuum-packaged Wiltshire bacon, both under refrigeration and at room temperature. "Sour spoilage" resulting from their metabolic activities is not considered a less adverse defect than that caused by proteolytic bacteria. The question of the real identity of these "atypical" streptobacteria remains open. The majority of these strains isolated from bacon probably resemble *Lactobacillus curvatus* and *Lactobacillus sake;* these species, however, do not represent genotypically and phenotypically homogeneous taxa (Holzapfel and Gerber, 1986; Hastings and Holzapfel, 1987a, 1987b; Reuter, 1981). On the other hand, they can be clearly distinguished from the carnobacteria, which share some common phenetic properties not found among the lactobacilli (see Table 2).

Blickstad and Molin (1983) reported the isolation of "unidentified" lactobacilli from smoked pork loin and frankfurter sausages, and Borch and Molin (1988) identified *C. divergens* as one of the dominating psychrotrophic bacterial groups on cured, CO_2-packed pork and sliced, vacuum-packed ham. Contrasting with its nonaciduric nature (Shaw and Harding, 1984, 1985) *C. piscicola* was also isolated from fermented sausages (Schillinger and Lücke, 1987).

The foremost and hitherto best-recognized habitat of *C. divergens* and *C. piscicola* appears to be unprocessed, refrigerated red meat. Vacuum packaging of the specific nutrient-rich substrate and refrigeration seem to create a favorable ecosystem in which these carnobacteria may dominate. Thus far, these two *Carnobacterium* species have been isolated from minced meat (Holzapfel and Gerber, 1983; Shaw and Harding, 1984, 1985), refrigerated, vacuum-packaged unprocessed beef, pork and lamb (Borch and Molin, 1988; Hitchener et al., 1982; Schillinger and Lücke, 1986, 1987; Shaw and Harding, 1984, 1985), and CO_2-packed pork (Borch and Molin, 1988).

Some members of *C. piscicola* appear to be fish pathogens, and have been isolated from diseased rainbow trout *(Salmo gairdneri),* cut-throat trout *(Salmo clarki),* and chinook salmon *(Oncorhynchus tshawytscha)* (Cone, 1982; Hiu et al., 1984). These bacteria have been isolated over several years from hatcheries in the State of Oregon, and were most frequently associated with infected fish of one year and older and especially under conditions of stress associated with handling and spawning (Evelyn and McDermott, 1961; Hiu et al., 1984; Ross and Toth, 1974). The infections caused pathological lesions and blisters on and under the skin, and different signs in the internal organs. According to Hiu et al. (1984) only *C. piscicola* (and no typical lactobacilli) have thus far been found to cause septicemia ("lactobacillosis") in salmonid and other fish species. More recently, *C. piscicola* was found to be associated with clinical and subclinical peritonitis of Australian salmonids (Humphrey et al., 1987).

Isolation of Carnobacteria

Apart from their nonaciduric nature, carnobacteria possess the relatively complex nutritional requirements of the lactobacilli. Hiu et al. (1984) have found that *C. piscicola* isolates from diseased fish requires folic acid, riboflavin, pantothenate, and niacin for growth, but not vitamin B_{12}, biotin, thiamine, or pyridoxal. *C. divergens* and *C. piscicola* isolates from meat have also been found to not require thiamine (De Bruyn, 1987). A neutral to high pH, ranging from 6.8 to 9.0, favors growth and the range 8.0 to 9.0 (Fig. 6) may serve to selectively inhibit lactobacilli that are often found in association with carnobacteria. The omission of acetate from conventional lactobacillus media such as MRS (de Man et al., 1960) also favors the growth of carnobacteria (see Fig. 6).

Nonselective and Semiselective Isolation

When the microbial population is dominated by carnobacteria (which may be the case for vacuum packaged, refrigerated meat, and poultry, and for diseased fresh water fish), isolation is possible by direct streaking or plating onto some nonselective "universal" media.

From diseased fish, direct streakings were made onto brain heart infusion agar or tryptic soy agar (Hiu et al., 1984) that are commercially available. For the nonselective recovery of carnobacteria from refrigerated, vacuum-packaged meat, Standard-I-Agar (pH 7.2 to 7.5) (Merck) has been found to be useful (Von Holy, 1983); aerobic incubation was either at 25°C for 3 days or at 7°C for 10 days. Alternatively, CASO-medium (Merck) or tryptic soy agar (Difco) with

Fig. 6. Growth yield (absorbance) of *Carnobacterium divergens* and *Lactobacillus sake* at different pH values in MRS broth with and without acetate, after 24 h at 30°C. *L. sake,* without acetate: ———; *L. sake,* with acetate: — — —; *C. divergens,* without acetate: - - - - -; *C. divergens,* with acetate: — · — · —.

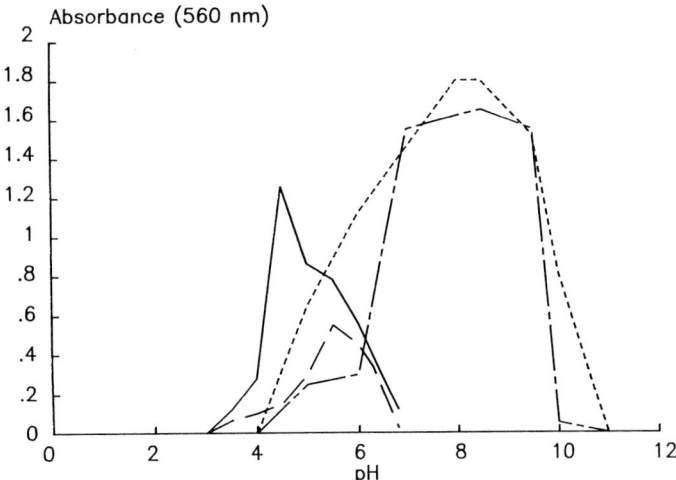

0.3% added yeast extract may be used for this purpose.

After adjusting the pH to 8.0 and substituting sucrose for glucose, Standard-I-Agar may be applied for the semiselective enumeration of carnobacteria in the presence of a dominating *Lactobacillus* population. The identity of typical catalase-negative colonies, however, should be verified by microscopy and by use of some of the criteria in Table 2.

Borch and Molin (1988) used tryptone glucose extract agar (TGE, Oxoid) (pH 7.0) for determining the total bacterial count of refrigerated, prepacked meat and meat products and recovered *C. piscicola* and *C. divergens* after incubation at 28°C for 3 days.

Prior to their recognition as separate group, the carnobacteria were probably most frequently isolated on MRS and APT agar as part of the *Lactobacillus* population of meat and poultry (Hitchener et al., 1982; Schillinger and Lücke, 1986; 1987; Shaw and Harding, 1984; Thornley and Sharpe, 1959). Since MRS agar was found to inhibit the growth of some "*L. divergens*" isolates, it was modified (as "D-MRS") by increasing the pH to 8.5, omitting acetate, and by substituting glucose by sucrose (Bosch and Holzapfel, 1985, unpublished observations; De Bruyn, 1987; Holzapfel and Long, 1984, unpublished observations). The substitution by sucrose was necessary to reduce Maillard reactions during autoclaving.

D-MRS for Cultivation and Semiselective Enumeration of Carnobacteria (amounts per liter)

Universal peptone	10 g
Yeast extract	5 g
Beef extract	5 g
Sucrose	20 g
Tween 80	1 g
K₂HPO₄	2 g
Triammonium citrate	2 g
MgSO₄·7H₂O	0.1 g
MnSO₄·4H;2O	0.05 g

K_2HPO_4, Triammonium citrate, $MgSO_4 \cdot 7H_2O$, $MnSO_4 \cdot 4H;2O$

For 1 liter of medium, the ingredients are added after dissolving 15 g of agar in distilled water. The pH is adjusted to 8.5 using NaOH, before sterilising at 121°C for 10 to 15 min.

Increasing the $MnSO_4 \cdot 7H_2O$ concentration to 0.4% may stimulate colony growth of *C. divergens* and *C. piscicola* strains (Bosch and Holzapfel, 1985, unpublished observations).

Incubation is at 25°C for 3 days, either aerobically or using a reduced gas atmosphere (e.g., Anaerocult, Merck).

Selective Isolation Medium

Selective Cresol Red Thallium Acetate Sucrose (CTAS) agar (WPCM, 1989) for Isolation and Enumeration of *C. divergens* and *C. piscicola*

Peptone from casein	10.0 g
Yeast extract	10.0 g
Sucrose	20.0 g
Tween 80	1.0 g
Sodium citrate	15.0 g
MnSO₄·4H₂O	4.0 g
K₂HPO₄	2.0 g
Thallium acetate	1.0 g
Nalidixic acid	0.04 g
Cresol red	0.004 g
Triphenyl tetrazolium chloride	0.01 g
Agar	15.0 g
water	1 liter

Add all the components except triphenyl tetrazolium chloride to 990 ml of water and bring to a boil to dissolve completely, cool to 55°C, and adjust the pH of the medium to 9.1 with 1 N NaOH. Autoclave at 121°C, and add 1 ml of a 1% solution of triphenyl tetrazolium chloride. Pour about 15 ml of medium per petri dish. After surface spreading or direct plating, the plates are incubated aerobically at 30°C for 24–48 h or at 25°C for 3–4 days.

Substitution of sucrose by 2% inulin may be used to distinguish *C. piscicola* colonies from enterococci. The latter may grow on CTAS medium, but do not ferment inulin. In general, *C. divergens* shows sparser growth than *C. piscicola* and may appear as pin-point sized colonies, often without changing the red color of the medium to yellow, and showing a metallic-bronze shine.

Typical colony morphology, being convex, shiny, and ranging from yellowish to pinkish, can be best observed on CTAS agar when triphenyl tetrazolium chloride is omitted. When sucrose is substituted by inosine, *C. piscicola* strains especially show dry, umbonate colony growth, comparable to "*β*-colonies" (Mitsuoka, 1969), often observed for *Lactobacillus acidophilus*.

Cultivation, Maintenance, and Conservation if Carnobacteria

Cultivation

Although MRS (broth or agar) can be used for the cultivation of axenic cultures, growth is often poor, while some strains are unable to sustain growth on subcultivation on this medium. APT medium (Evans and Niven, 1951) at pH 7.0, CASO-medium (Merck) with 0.3% added yeast extract, or even Standard-I-Medium (Merck) at pH 7.2 provide better growth for practically all *Carnobacterium* strains. For general cultivation purposes, and especially for fastidious strains, D-MRS (pH 8.0–8.5) is recommended. Incubation is at 25°C for 24–48 h, either in air or under slightly reduced atmosphere.

Maintenance

General maintenance of cultures is as for lactobacilli, taking into account the nonaciduric nature of the carnobacteria and selecting APT, CASO, or D-MRS media in the pH range 7.0 to 8.5. Stab cultures in D-MRS agar (pH 8.0 to 8.5) should be kept at 1–4°C and transplanted every 2–3 weeks. Addition of 5% calcium carbonate to cultivation broth (e.g., APT) may serve to protect vitality of stock cultures over several weeks at 1–4°C.

Conservation

Lyophilization as used for lactobacilli gives satisfactory results, provided cryoprotective agents (milk solids, lactose, or horse serum) is added to the cell suspension, and ampules are sealed under high vacuum and stored at 8 to 12°C. Superior results were obtained by cryopreservation at −80°C, using glycerol-peptone protective broth for suspending late logarithmic cells, harvested by centrifugation or by rinsing surface growth from agar media. Borch and Molin (1988) recommended the storage of strains as dense cultures in APT broth at −20°C.

Identification

The general description of the carnobacteria corresponds to that of the lactobacilli, in as far as the former are Gram-positive, catalase-negative, nonsporeforming rods that are usually nonmotile, do not reduce nitrate, and have a fermentative metabolism (Collins et al., 1987; Holzapfel and Gerber, 1983; Shaw and Harding, 1985). More precisely, the carnobacteria may be described as short to medium length, straight, slender rods with rounded ends, occuring singly, in pairs, or sometimes in short chains. Initially described as "atypical" heterofermenters (Holzapfel and Gerber, 1983), metabolic studies on *L. divergens* (syn. *C. divergens* DSM 20623) have shown this organism to ferment glucose via the glycolytic pathway (De Bruyn et al., 1987, 1988). Only some secondary decarboxylation/dissimilation of pyruvate/lactate to acetate, formate, and CO_2 was observed, explaining the slow and often minimal production of CO_2 in glucose broth media. Using D-[U-^{14}C] glucose, De Bruyn et al. (1988) showed that about 75% of the lactate and less than 10% of each of the formate and acetate were produced from glucose. It was postulated that the remainder of these products may be derived from endogenous non-glucose sources. Lactate and small amounts of ethanol and acetate are produced from ribose (Holzapfel and Gerber, 1983).

Other properties that are shared with most lactobacilli include the inability to produce gelatinase, indole, or H_2S. The peptidoglycan is of the *meso*-diaminopimelic acid-direct type (described by Schleifer and Kandler, 1972). More than 95% of the lactic acid is produced as the L(+) isomer. Most strains grow at 0–2°C, especially in typical meat habitats under vacuum packaging. No growth occurs at 45°C, in presence of 8% NaCl, or at pH 3.9 (Collins et al., 1987; Holzapfel and Gerber, 1983).

Using D-MRS broth (W. H. Holzapfel and M. Long, 1984, unpublished observations), optimum growth of *C. divergens* was observed within a pH range of 8.0 to 9.5. Addition of 1.5% acetate raised the minimum pH for growth from about 4.8 to 6.0 (see Fig. 5).

The major properties for differentiating the carnobacteria from the lactobacilli are shown in

Table 2. During isolation and identification procedures, the acetate sensitivity (as can be demonstrated by failure to grow on acetate agar, Rogosa et al., 1951) and the ability to grow at pH 8.5 to 9.0, may serve as routine tests for the recognition of carnobacteria among the rod-shaped lactic acid bacteria. The predominant $C_{18:1}$ fatty acid of carnobacteria, identified as oleic acid (9, 10) differs from that of the lactobacilli (*cis*-vaccenic acid [11,12]; moreover,

carnobacteria do not possess lactobacillic acid (Collins et al., 1987; Shaw and Harding, 1985).

Differentiation of Species of Carnobacteria

Four species are presently recognized to comprise the genus *Carnobacterium: C. divergens, C. gallinarum, C. mobile,* and *C. piscicola* (Collins et al., 1987). Although these species rep-

Table 9. Selected physiological characteristics of species of *Carnobacterium.*

Characteristic	*C. divergens*	*C. gallinarum*	*C. mobile*	*C. piscicola*
Growth at:				
0°C	+	+	+	+
45°C[a]	−	−	−	−
Motility	−	−	+	−
Voges-Proskauer test	+/−	+	−/+	+
Acid produced from:				
Amygdalin	+	−	+	+
L-Arabinose	−	−	−	−
Galactose	−	+	+	+
Gluconate	+	+	−	+
Inulin[b]	−	−	+	+
Lactose	−	−	+	+
Mannitol	−	−	−	+
Melezitose	+/−	+	−	+/−
Melibiose	−	−	−	+
α-Methyl-D-glucoside	−	+	−	+
α-Methyl-D-mannoside	−	−	−	+
Raffinose	−	−	−	−/+
Sorbitol	−/+[c]	−	−	−/+
D-Tagatose	−	+	−/+	−
D-Turanose	−	+	−	+/−
D-Xylose	−	+	−	−
Growth in presence of (%):				
Cd acetate (0.1)	−/w	ND	ND	+
Co nitrate (0.5)	−	ND	ND	+
Cu sulfate (0.01)	−	ND	ND	+
Mercury(ll) chloride (0.5)	−	ND	ND	w/−
Mn sulfate-7-hydrate (3.0)	−	ND	ND	+
Na selenite (0.5)	w	ND	ND	w
K tellurite	+	ND	ND	+
Pb acetate	−/w	ND	ND	+
Sensitivity to (in μg):				
Erythromycin (15)	+	ND	ND	+
Gentamycin (10)	+/−	ND	ND	−
Kanamycin (30)	−/+	ND	ND	−
Methicillin (10)	−	ND	ND	−
Nalidixic acid (30)	−	ND	ND	−
Tetracycline (50)	+	ND	ND	+
GC content (mol%)	33.0–36.4	34.3–36.4	35.5–37.2	33.7–36.4

Symbols: +/−, occasional strain negative; −/+, occasional strain positive; ND, no data; w, weak positive reaction.
[a]According to Feresu and Jones (1988), more than 80% of their isolates resembling *C. divergens* and *C. piscicola* grew at 45°C!
[b]Hiu et al. (1984) reported that all 17 *C. piscicola* strains were negative for inulin fermentation.
[c]Some strains were positive according to Feresu and Jones (1988).
Compiled from Borch and Molin, 1988; W. Bosch and W. H. Holzapfel, 1985, unpublished observations; Collins et al., 1987; Feresu and Jones, 1988; Hiu et al., 1984; Holzapfel and Gerber, 1983; Shaw and Harding, 1985.

resent four separate DNA homology groups, they have been found to belong to the same ribosomal RNA homology cluster (Champomier et al., 1989; Collins et al., 1987). Polyacrylamide gel electrophoresis of the whole cell proteins confirmed the homology of *L. carnis* with *C. piscicola,* and the status of *C. divergens* as a separate species (F. Venter, P. L. Steyn, E. Van Zyl and W. H. Holzapfel, unpublished observations). The key physiological criteria for their differentiation are given in Table 9. Simple physiological tests may be relied on for species differentiation and are described by Borch and Molin (1988), Collins et al. (1987), Holzapfel and Gerber (1983), and Shaw and Harding (1985). Sugar fermentation patterns (see Table 9) can especially be used for routine differentiation of *Carnobacterium* species. The key sugars for this purpose are amygdalin, galactose, gluconate, inulin, mannitol, melibiose, α-methyl-D-glucoside, D-turanose, and D-xylose. Application of the API 10E and API 50 CH systems (Analytab Products), as was practiced by Collins et al. (1987), or the Minitek System (BBL) (Borch and Molin, 1988), contributes to a simplification of the working procedure. These commercialized "rapid" test systems seem to produce similar results to conventional sugar fermentation tests in the broth used by Holzapfel and Gerber (1983) or on the agar as used by Shaw and Harding (1985). Shaw and Harding (1985) added 0.5% of a filter-sterilized sugar solution to the following basal agar medium after sterilization: yeast extract (Difco), 0.6%; trypticase (BBL), 1.5%; cysteine HCl, 0.02%; Tween 80, 0.1%; chlorophenol red, 0.004%; $MgSO_4 \cdot 7H_2O$, 0.02%; $MnSO_4 \cdot 4H_2O$, 0.005%; agar, 1.2%; pH 6.4.

Taking into account the nonaciduric nature of the carnobacteria, adjustment of the pH of basal media to neutral and even higher may be advizable. Care should be taken that a suitable indicator (e.g., bromcresol purple or cresol red) is chosen according to the adjusted pH value. Most tests used for lactobacilli (described elsewhere in this chapter) may be applied to the carnobacteria provided a higher pH is used. For motility testing, stab inoculation into YGPB medium, modified to contain 0.1% glucose, 0.1% lactose, and 0.2% agar (Oxoid no. 3) (Collins et al., 1987) is recommended.

All *Carnobacterium* strains seem to share the following common properties (Borch and Molin, 1988; Collins et al., 1987): acid production from cellobiose, D-fructose, D-glucose, maltose, D-mannose, ribose, salicin, and sucrose; acid not produced from arabinose, arabitol, dulcitol, erythritol, fucose, glycogen, inositol, raffinose, L-rhamnose, L-sorbose, xylitol, and L-

xylose. All strains investigated by Collins et al. (1987) produced arginine deiminase (dihydrolase) and β-galactosidase.

Carnobacteria appear to be facultative aerobes, as are most lactobacilli. Although cultivation in air does not seem to reduce colony growth of most strains on agar plate surfaces significantly, *C. divergens* strains especially show more prolific growth in a reducing environment. *C. piscicola,* on the other hand, generally shows higher resistance to adverse conditions and stress factors than *C. divergens.* Examples of differences in heavy metal resistance are shown in Table 9. Furthermore, *C. piscicola* has been found to grow well in D-MRS broth with 10% diammonium citrate added, as compared to about 6% citrate as the highest concentration tolerated by *C. divergens* (W. Bosch and W. H. Holzapfel, 1985, unpublished observations). Their relatively high resistance to thallium acetate and sodium azide enable most carnobacteria to grow well on media selective, for *Enterococcus,* such as Barnes agar (Barnes, 1959) and CATC agar (Burkwall and Hartman, 1964; Reuter, 1968). *C. piscicola* produces slightly better growth than *C. divergens* on these media, but their colonies appear relatively similar to those of typical enterococci.

Investigations into the construction of selective media for carnobacteria revealed their more or less similar heavy metal resistance as strains of *Enterococcus faecium* and *E. faecalis* (Bosch and Holzapfel, 1985, unpublished observations). Moreover, using about 60 different antibiotic sensitivity test discs, practically identical antibiotic resistance spectra have been found for *C. divergens* and enterococci.

Edwards et al. (1987) used pure culture meat inoculation experiments for studying amine production in vacuum-packaged beef stored at 1°C. Tyramine production has been shown to be restricted to lactobacilli and strains of *C. piscicola* and *C. divergens;* strains of leuconostocs and *Brochothrix thermosphacta* were negative.

Literature Cited

Abo-Elnaga, I. G., and O. Kandler. 1965. Zur Taxonomie der Gattung *Lactobacillus* Beijerinck. I. Das Subgenus *Streptobacterium* Orla-Jensen. Zentralbl. Bakteriol. Parasitenkd. Infektionskr. Hyg. Abt. II 119:1–36.

Accolas, J.-P., and H. Spillmann. 1979. The morphology of six bacteriophages of *Streptococcus thermophilus.* J. Appl. Bacteriol. 47:135–144.

Alaluusua, S., E. Kleemola-Kujula, M. Nystrom, M. Evalahti, and L. Gronroos. 1987. Caries in the primary teeth and salivary *Streptococcus mutans* and *Lactobacillus* levels as indicators of caries in permanent teeth. Pediatr. Dent. 9:126–130.

Alaluusua, S., M. Nyström, L. Grönroos, and L. Peck. 1989. Caries-related microbiological findings in a group of teenagers and their parents. Caries Res. 23:49–54

Alatossava, T., and M. J. Pythilä. 1980. Characterization of a new *Lactobacillus lactis* bacteriophage. IRCS Med. Sci. Biochem. 8:297–298.

Allen, J. R., and E. M. Foster. 1960. Spoilage of vacuum-packed sliced processed meats during refrigerated storage. Food Res. 25:19–25.

Amerine, M. A., and R. E. Kunkee. 1968. Microbiology of wine-making. Ann. Rev. Microbiol. 22:323.

Andersson, R. 1984. Characteristics of the bacterial flora isolated during spontaneous lactic acid fermentation of carrots and red beets. Lebensm. Wiss. u. Technol. 17:282–286.

Archibald, F. S., and I. Fridovich. 1981. Manganese and defenses against oxygen toxicity in *Lactobacillus plantarum*. J. Bacteriol. 145:442–451.

Auclair, J., and J.-P. Accolas. 1983. Use of thermophilic lactic starters in the dairy industry. Antonie van Leeuwenhoek 49:313–326.

Aukrust, T., and I. F. Nes. 1988. Transformation of *Lactobacillus plantarum* with the plasmid pTV1 by electroporation. FEMS Microbiol. Lett. 52:127–132.

Axelsson, L. T., S. E. I. Ahrne, M. C. Andersson, and S. R. Stahl. 1988. Identification and cloning of a plasmid-encoded erythromycin resistance determinant of *Lactobacillus reuteri*. Plasmid 20:171–174.

Axelsson, L. T., T. C. Chung, W. J. Dobrogosz, and S. E. Lindgren. 1989. Production of a broad spectrum antimicrobial substance by *Lactobacillus reuteri*. Microb. Ecol. Health Dis. 2:131–136.

Azeezullah, M., A. K. Sharma, K. C. Landon, and R. A. Srinirasan. 1973. Microflora of oats and berseem silages. Indian J. Dairy Sci. 26:232–236.

Back, W. 1980. Bierschädliche Bakterien. Nachweis und Kultivierung bierschädlicher Bakterien im Betriebslabor. Brauwelt 120:1562–1569.

Back, W. 1981. Schädliche Mikroorganismen in Fruchtsäften, Fruchtnektaren und süssen, alkoholfreien Erfrischungsgetränken. Brauwelt 121:43–48.

Back, W. 1987. Neubeschreibung einer bierschädlichen Laktobazillen-Art: *Lactobacillus brevisimilis* spec. nov. Monatsschr. f. Brauwiss. 40:484–488.

Back, W. 1988. Biologische Säuerung. Monatsschr. f. Brauwiss. 41:152–156.

Baird-Parker, A. C. 1980. Organic acids, p. 126–135. In: ICMSF, Microbial ecology of foods. Academic Press, New York.

Barefoot, S. F., and T. R. Klaenhammer. 1983. Detection and activity of lactacin B, a bacteriocin produced by *Lactobacillus acidophilus*. Appl. Environ. Microbiol. 45:1808–1815.

Barefoot, S. F., and T. R. Klaenhammer. 1984. Purification and characterization of the *Lactobacillus acidophilus* bacteriocin lactacin B. Antimicr. Agents Chemother. 26:328–334.

Barnes, E. M. 1959. Differential and selective media for faecal streptococci. J. Sci. Food Agric. 10:656–662.

Barnes, E. M. 1976. Microbiological problems of poultry at refrigerator temperatures—a review. J. Sci. Food Agric. 27:777–782.

Barre, P. 1969. Taxonomie numérique de lactobacilles isolates du vin. Archiv Mikrobiol. 68:74–86.

Barre, P. 1978. Identification of thermobacteria and homofermentative, thermophilic pentose-utilizing lactobacilli from high temperature fermenting grape musts. J. Appl. Bacteriol. 44:125–129.

Barrow, P. A., R. Fuller, and A. J. Newport. 1977. Changes in the microflora and physiology of the anterior intestinal tract of pigs weaned at 2 days with special reference to the pathogenesis of diarrhea. Infection and Immunity 18:586–595.

Basson, N., and C. Van Wyk. 1982. Colonization of the tooth surface by lactobacilli. J. Dent. Res. 61:603.

Bates, E. E. M., H. J. Gilbert, G. P. Hazelwood, J. Huckle, J. I. Laurie, and S. P. Mann. 1989. Expression of *Clostridium thermocellum* endoglucanase gene in *Lactobacillus plantarum*. Appl. Environ. Microbiol. 55:2095–2097.

Baumgart, J., B. Weber, and B. Hanekamp. 1983. Mikrobiologische Stabilität von Feinkosterzeugnissen. Fleischwirtschaft 63:93–94.

Bayer, S. S., A. W. Chow, D. Betts, and L. B. Guze. 1978. Lactobacillemia—report of nine cases. Important clinical and therapeutic considerations. Am. J. Med. 64:808–813.

Beck, R., F. Groß, and T. Beck. 1987. Untersuchungen zur Kenntnis der Gärfuttermikroflora. Zeitschrift "Das wirtschaftseigene Futter" 33:13–33.

Beck, R., N. Weiss, and J. Winter. 1988. *Lactobacillus graminis* sp. nov., a new species of facultative heterofermentative lactobacilli surviving at low pH in grass silage. System. Appl. Microbiol. 10:279–283.

Benno, Y., K. Sawada, and T. Mitsuoka. 1984. The intestinal microflora of infants. Composition of fecal flora in breast-fed and bottle-fed infants. Microbiol. Immunol. 28:975–986.

Berg, R. D. 1983. Translocation of indigenous bacteria from the intestinal tract, p. 335–352. In: D. J. Hentges (ed.), Human intestinal microflora in health and disease. Academic Press, London.

Berger, U. 1974. Pathogenicity of lactobacilli. Deutsche Medizinische Wochenschrift 99:1200–1203.

Beurne, K. A., J. L. Beebe, Y. A. Lue, and P. D. Ellner. 1978. Bacteremia due to *Bifidobacterium*, *Eubacterium* or *Lactobacillus*. Twenty-one cases and review of the literature. Yale J. Biol. Med. 51:505–512.

Bianchi-Salvadori, B., P. Camaschella, and E. Bazzigaluppi. 1984. Distribution and adherence of *Lactobacillus bulgaricus* in the gastrointestinal tract of germ-free animals. Milchwissenschaft 39:387–391.

Biewald, U. 1968. Untersuchungen über *Lactobacillus*-Phagen aus Rohwürsten. Monatsh. Veterinaermed. 23:489–492.

Bjarnason, S. 1989. On dental health in Icelandic children. Observations during a clinical dentifrice trial. Swed. Dent. J. Supp. 57:1–40.

Blickstad, E., and G. Molin. 1983. The microbial flora of smoked pork loin and frankfurter sausage stored in different gas atmospheres at 4°C. J. Appl. Bacteriol. 54:45–56.

Blood, R. M. 1970. The microbiology of semi-preserved fish products. M. Sc. Thesis, University of London.

Blood, R. M. 1975. Lactic acid bacteria in marinated herring, p. 195–208. In: J. G. Carr, C. V. Cutting, and G. C. Whitting (ed.), Lactic acid bacteria in beverages and food. Academic Press, London, New York, San Francisco.

Boatwright, J., and B. H. Kirsop. 1976. Sucrose agar—a growth medium for spoilage organisms. J. Inst. Brewing 82:343–346.

Böcker, G., and W. P. Hammes. 1990. *Lactobacillus sanfrancisco* in a commercial sour dough starter preparation. Food Biotechnol. 4:475.

Boizet, B., I. L. Flickinger, and B. M. Chassy. 1988. Transfection *Lactobacillus bulgaricus* protoplasts by bacteriophage DNA. Appl. Environ. Microbiol. 54:3014–3018.

Borch, E., and G. Molin. 1988. Numerical taxonomy of psychrotrophic lactic acid bacteria from prepacked meat and meat products. Antonie van Leeuwenhoek 54:301–323.

Bottazzi, V. 1988. An introduction to rod-shaped lactic acid bacteria. Biochimie 70:303–315.

Bottazzi, V., and M. Vescovo. 1969. Carbonyl compounds produced by yoghurt bacteria. Neth. Milk Dairy J. 23:71.

Bottazzi, V., M. Vescovo, and F. Dellaglio. 1973. Microbiology of Grana cheese. IX. Characteristics and distribution of *Lactobacillus helveticus* biotypes in natural whey cheese starter. Scienza e Technica Lattierocasearia 24:23–39.

Bourne, K. A., J. L. Beebe, Y. A. Lue, and P. D. Ellner. 1978. Bacteremia due to *Bifidobacterium*, *Eubacterium* or *Lactobacillus*: twenty-one cases and review of the literature. Yale J. Biol. Med. 51:505–512.

Bousfield, I. J. and A. R. McKenzie. 1976. Inactivation of bacteria by freeze drying, p. 329–344. In: F. A. Skinner and W. B. Hugo (ed.), Inhibition and inactivation of vegetative microbes. Academic Press, London, New York, San Francisco.

Boyar, R. M., and G. H. Bowden. 1985. The microflora associated with the progression of incipient carious lesions in teeth of children living in a fluoridated water area. Caries Res. 19:298–306.

Brown, L. R., J. O. White, J. M. Horton, S. Dreizen, and J. L. Streckfuss. 1983. Effect of continuous fluoride gel use on plaque fluoride retention and microbial activity. J. Dent. Res. 62:746–751.

Brunkow, O. R., W. H. Peterson, and E. B. Fred. 1925. A study of the influence of inoculation upon the fermentation of sauerkraut. J. Agr. Res. 30:955–960.

Bryan-Jones, G. 1975. Lactic acid bacteria in distillery fermentations, p. 165–175. In: J. G. Carr, C. V. Cutting, and G. C. Whiting (ed.), Lactic acid bacteria in beverages and food. Academic Press, London.

Bryant, M. P., N. Small, C. Bouma, and I. Robinson. 1958. Studies on the composition of the ruminal flora and fauna of young calves. J. of Dairy Sci. 41:1747–1767.

Budolfsen-Hansen, G. 1988. Freeze-dried starter cultures for sour-dough bread. ZFL 7:597–602.

Buchta, K. 1983. Lactic acid, p. 409–417. In: H. J. Rehm and G. Reed (ed.), Biotechnology, vol. 3c. Verlag Chemie, Weinheim, Germany.

Buckenhüskes, H., K. Gierscher, and W. P. Hammes. 1988. Optimierung des Pasteurisierungseffektes bei Sauergemüse. Indust. Obst- und Gemüseverwertung. 73:315–322.

Buckenhüskes, H., and W. P. Hammes. 1990. Einsatz von Starterkulturen bei der Verarbeitung von Obst und Gemüse. Bioengineering. 2:34–42.

Buckenhüskes, H., M. Schneider, and W. P. Hammes. 1986. Die milchsaure Vergärung pflanzlicher Rohware unter besonderer Berücksichtigung der Herstellung von Sauerkraut. Chem. Mikrobiol. Technol. Lebensm. 10:42–53.

Burkwall, M. K., and P. A. Hartman. 1964. Comparision of direct plating media for the isolation and enumeration of enterococci in certain frozen foods. Appl. Microbiol. 12:18–23.

Cantoni, C., M. A. Bianchi, P. Renon, and S. Daubert. 1969. Ricerche sulla putrefazione del prosciutto crudo. Archivo Veterinario Italiano 20:354–370.

Carlsson, J. 1989. Microbial aspects of frequent intake of products with high sugar concentrations. Scand. J. Dent. Res. 97:110–114.

Carlsson, J. and L. Gothefors. 1975. Transmission of *Lactobacillus jensenii* and *Lactobacillus acidophilus* from mother to child at the time of delivery. J. Clin. Microbiol. 1:124–128.

Carr, J. G. 1959. Some special characteristics of the cider lactobacilli. J. Appl. Bact. 22:377–383.

Carr, J. G. 1987. Microbiology of ciders and wines, p. 291–307. In: J. R. Norris, and G. L. Pettipher (ed.), Essays in agricultural and food microbiology. John Wiley & Sons, Chichester, UK.

Carr, J. G., and P. A. Davies. 1970. Homofermentative lactobacilli of ciders including *Lactobacillus mali* nov. spec. J. Appl. Bacteriol. 33:768–774.

Carr, J. G., P. A. Davies, F. Dellaglio, M. Vescovo, and R. A. D. Williams. 1977. The relationship between *Lactobacillus mali* from cider and *Lactobacillus yamanashiensis* from wine. J. Appl. Bacteriol. 42:219–228.

Castino, M., L. Usseglio-Tomasset, and A. Gandini. 1975. Factors which affect the spontaneous inclination of the malo-lactic fermentation in wines. The possibility of transmission by inoculation and its effect on organoleptic properties, p. 139–148. In: J. G. Carr, C. V. Cutting, and G. C. Whiting (ed.), Lactic acid bacteria in beverages and food. Academic Press, London.

Cato, E. P., W. E. C. Moore, and J. L. Johnson. 1983. Synonymy of strains of *"Lactobacillus acidophilus"* groups A-2 (Johnson et al. 1980) with the type strain of *Lactobacillus crispatus* (Brygoo and Aladame 1953) Moore and Holdeman 1970. Int. J. Syst. Bacteriol. 33:426–428.

Cavett, J. J. 1963. A diagnostic key for identifying the lactic acid bacteria of vacuum-packed bacon. J. Appl. Bacteriol. 26:453–470.

Chalfan, Y., I. Goldberg, and R. I. Mateles. 1977. Isolation and characterization of malo-lactic bacteria from Israeli red wine. J. Food Science 42:939–968.

Champomier, M. C., M. C. Montel, and R. Talon. 1989. Nucleic acid relatedness studies on the genus *Carnobacterium* and related taxa. J. Gen. Microbiol. 135:1391–1394.

Chandan, R. C., P. J. Argyle, and G. E. Mathison. 1982. Action of *Lactobacillus bulgaricus* proteinase preparations on milk proteins. J. Dairy Sci. 65:1408–1413.

Chapman, H. R., and M. E. Sharpe. 1981. Microbiology of cheese, p. 157–243. In: R. K. Robinson (ed.), Dairy microbiology, vol. 2. Appl. Sci. Publishers, London.

Charlton, D. B., M. E. Nelson, and C. H. Werkman. 1934. Physiology of *Lactobacillus fructivorans* spec. nov. isolated from spoiled salad dressing. Iowa State J. Sci. 9:1–11.

Chassy, B. M. 1987. Prospects for the genetic manipulation of lactobacilli. FEMS Microbiol. Rev. 46:297–312.

Chassy, B. M., and C. A. Alpert. 1989. Molecular characterization of the plasmid encoded lactose-PTS of *Lactobacillus casei*. FEMS Microbiol. Rev. 63:157–166.

Chassy, B. M., and J. L. Flickinger. 1987. Transformation of *Lactobacillus casei* by electroporation. FEMS Microbiol. Lett. 44:173–177.

Chassy, B. M., E. Gibson, and A. Giuffrida. 1976. Evidence for extrachromosomal elements in *Lactobacillus*. J. Bacteriol. 127:1576–1578.

Chassy, B. M., and J. Thompson. 1983. Regulation of lactose-phosphoenolpyruvate-dependent phosphotransferase system and β-D-phosphogalactoside galactohydrolase activities in *Lactobacillus casei*. J. Bacteriol. 154:1195–1203.

Chomakov, H. V. 1962. Rusty spots in brined white cheese. Mikrobiologiya 31:726–730.

Chomakov, H. V., and N. Kirov. 1975. Lactic acid bacteria in raw milk and white pickled cheese. Dairy Sci. Abstr. 37:134.

Chung, T. C., L. Axelsson, S. E. Lindgren, and W. E. Dobrogosz. 1989. In vitro studies on reuterin synthesis by *Lactobacillus reuteri*. Microb. Ecol. Health Dis. 2:137–144.

Claesson, R., and C. G. Crossner. 1985. Presence of *Lactobacillus casei* in saliva from children and adults using a new medium. Scand. J. Dent. Res. 93:17–22.

Clarke, R. T. J., and T. Bauchop (ed.). 1977. Microbial ecology of the gut. Academic Press, London.

Cluzel, P.-J., J. Serio, and J. P. Accolas. 1987. Interactions of *Lactobacillus bulgaricus* temperature bacteriophage 0448 with host strains. Appl. Environ. Microbiol. 53:1850–1854.

Cocconcelli, P. S., L. Morelli, M. Vescovo, and V. Bottazzi. 1986. Intergeneric protoplast fusion in lactic acid bacteria. FEMS Microbiol. Lett. 35:211–214.

Coetzee, J. N., and H. C. de Klerk. 1962. Lysogeny in the genus *Lactobacillus*. Nature 187:348–349.

Collins, E. B., and K. Aramaki. 1980. Production of hydrogen peroxide by *Lactobacillus acidophilus*. J. Dairy Sci. 63:353–357.

Collins, M. D., J. A. E. Farrow, B. A. Philips, S. Ferusu, and D. Jones. 1987. Classification of *Lactobacillus divergens, Lactobacillus pisciola,* and some catalase-negative, asporogenous, rod-shaped bacteria from poultry in a new genus, *Carnobacterium*. Int. J. Syst. Bacteriol. 37:310–316.

Collins, M. D., B. A. Phillips, and P. Zanoni. 1989. Deoxyribonucleic acid homology studies of *Lactobacillus casei, Lactobacillus paracasei* sp. nov., subsp. *paracasei* and subsp. *tolerans,* and *Lactobacillus rhamnosus* sp. nov., comb. nov. Int. J. Syst. Bacteriol. 39:105–108.

Collins-Thompson, D. L., and G. R. Lopez. 1981. Depletion of sodium nitrite by lactic acid bacteria isolated from vacuum-packed Bologna. J. Food Prot. 44:593–595.

Condon, S. 1983. Aerobic metabolism of lactic acid bacteria. Irish J. Food Sci. Technol. 7:15–25.

Condon, S. 1987. Responses of lactic acid bacteria to oxygen. FEMS Microbiol. Rev. 46:269–280.

Cone, D. K., 1982. A *Lactobacillus* sp. from diseased female rainbow trout, *Salmo gairdneri* Richardson, in Newfoundland, Canada. J. Fish Diseases 5:479–485.

Contrepois, M., and Ph. Gouet. 1973. La microflore du tube digestif du jeune veau péruminant: Dénombrement de quelques groupes bacteriens à différents niveaux du tube digestif. Annales de Recherche Vétérinaire 4:161–170.

Conway, P. L., S. L. Gorbach, and B. R. Goldin. 1987. Survival of lactic acid bacteria in the human stomach and adhesion to intestinal cells. J. Dairy Sci. 70:1–12.

Conway, P. L., and S. Kjelleberg. 1989. Protein-mediated adhesion of *Lactobacillus fermentum* strain 737 to mouse stomach squamous epithelium. J. Gen. Microbiol. 135:1175–1186.

Coretti, K. 1958. Rohwurstfehlfabrikate durch Laktobazillen. Die Fleischwirtschaft 4:218–225.

Cosby, W. M., J. A. Casas, and W. J. Dobrogosz. 1988. Formation, regeneration and transfection of *Lactobacillus plantarum* protoplasts. Appl. Environ. Microbiol. 54:2599–2602.

Costerton, J. W., K. R. Rozee, and K. J. Cheng. 1983. Colonisation of particulates, mucons, and intestinal tissue. Prog. Nutr. Food Sci. 7:91–105.

Costilow, R. N., and T. W. Humphreys. 1955. Nitrate reduction by certain strains of *Lactobacillus plantarum*. Science 121:168.

Crossner, C. G., and L. Unell. 1986. Salivary lactobacillus counts as a diagnostic and didactic tool in caries prevention. Community Dent. Oral Epidemiol. 14:156–160.

Daeschel, M. A. 1989. Antimicrobial substances from lactic acid bacteria for use as food preservatives. Food Technol. 43:164–167.

Daeschel, M. A., R. E. Andersson, and H. P. Fleming. 1987. Microbial ecology of fermenting plant materials. FEMS Microbial. Rev. 46:357–367.

Daeschel, M. A., M. C. McKenney, and L. C. McDonald. 1986. Characterization of a bacteriocin from *Lactobacillus plantarum*. Abstr. Proc. 86th ASM-Meeting. H9: p. 146, Washington, D.C.

Dahiya, R. S. and M. L. Speck. 1968. Hydrogen peroxide formation by lactobacilli and its effect on *Staphylococcus aureus*. J. Dairy Sci. 51:1568

Danielson, A. D., E. R. Peo Jr., K. M. Shahani, A. J. Lewis, P. J. Whalen, and M. A. Amer. 1989. Anticholesteremic property of *Lactobacillus acidophilus* fed to mature boars. J. Anim. Sci. 67:966–974.

Dashkevicz, M. P., and S. D. Feighner. 1989. Development of a differential medium for bile salt hydrolase active *Lactobacillus* spp. Appl. Environ. Microbiol. 55:11–16.

Davies, F. L., P. A. Shankar, and H. M. Underwood. 1977. The use of milk concentrated by reverse osmosis for the manufacture of yoghurt. J. Soc. Dairy Technol. 30:23–27.

Davis, J. G. 1975. The microbiology of yoghourt, p. 245–263. In: J. G. Carr, C. V. Cutting, and G. C. Whiting (ed.), Lactic acid bacteria in beverages and food. Academic Press, London.

Davis, C. P., D. Cleven, E. Balish, and C. E. Yale. 1977. Bacterial association in the gastrointestinal tract of Beagle dogs. Appl. Environ. Microbiol. 34:194–206.

De Bruyn, I. N. 1987. Glucose metabolism of *Lactobacillus divergens*. M.Sc. (Microbiol.) Thesis, Univ. of Pretoria.

De Bruyn, I. N., W. H. Holzapfel, L. Visser, and A. J. Louw. 1988. Glucose Metabolism by *Lactobacillus divergens*. J. Gen. Microbiol. 134:2103–2109.

De Bruyn, I. N., A. I. Louw, L. Visser, and W. H. Holzapfel. 1987. *Lactobacillus divergens* is a homofermentative organism. System. Appl. Microbiol. 9:173–175.

De Klerk, H. C., and J. N. Coetzee. 1967. Bacteriocinogeny in *Lactobacillus fermenti*. Nature 214:609.

De Klerk, H. C., and J. A. Smit. 1967. Properties of a *Lactobacillus fermenti* bacteriocin. J. Gen. Microbiol. 48:309.

De Man, J. C., M. Rogosa, and M. E. Sharpe. 1960. A medium for the cultivation of lactobacilli. J. Appl. Bacteriol. 23:130–135.

De Simone, G., M. Ferrazzi, M. Di Seri, F. Mongio, L. Baldinelli, and S. Di Fabio. 1987. The immunoregulation of the intestinal flora. Bifidobacteria and lactobacilli modulate the production of gamma-IFN induced by pathogenic bacteria. Int. J. Immunother. 3:151–158.

De Vos, W. M. 1987. Gene cloning and expression in lactic streptococci. FEMS Microbiol. Rev. 46:281–295.

Deeth, H. C. 1984. Yogurt and cultured products. Austr. J. Dairy Technol. 39:111–113.

Deibel, R. H., and P. A. Hartman. 1976. The enterococci, p. 370–373. In: M. L. Speck (ed.), Compendium of methods for the microbiological examination of foods. Am. Public Health Assoc., Washington, D. C.

Dellaglio, F., and P. G. Sarra. 1984. *Lactobacillus lactis* as dominant lactobacilli flora in human intestine. Microbiologica (Bologna) 7:381–384.

Dellaglio, F., and S. Torriani. 1986. DNA-DNA homology, physiological characteristics and distribution of lactic acid bacteria isolated from maize silage. J. Appl. Bacteriol. 60:83–92.

Denepitiya, L., and J. Kleinberg. 1982. A comparison of the microbial compositions of pooled human dental plaque and salivary sediment. Arch. Oral Biol. 27:739–746.

Dent, V. E., and R. A. Williams. 1982. *Lactobacillus animalis* sp. nov., a new species of *Lactobacillus* from the alimentary canal of animals. Zentralbl. Bakteriol. Mikrobiol. Hyg. I. Abt. Orig. Reihe C 3:377–386.

Depaola, P. F. 1989. Integration of host and environmental factors associated with root caries. J. Dent. Res. 68 (spec. issue):178.

Dickgiesser, U., N. Weiss, and D. Fritsche. 1984. *Lactobacillus gasseri* as the cause of septic urinary infections. Infection 12:14–16.

Dicks, L. M. T., and H. J. J. van Vuuren. 1987. Relatedness of heterofermentative *Lactobacillus* species revealed by numerical analysis of total soluble cell protein patterns. J. Syst. Bacteriol. 37:437–440.

Dicks, L. M. T., and H. J. J. van Vuuren. 1988. Identification and physiological characteristics of heterofermentative strains of *Lactobacillus* from South African red wines. J. Appl. Bacteriol. 64:505–513.

Dirar, H., and E. B. Collins. 1973. Aerobic utilisation of low concentrations of galactose by *Lactobacillus plantarum*. J. Gen. Microbiol. 78:211–215.

Dittrich, H. H. 1977. Mikrobiologie des Weines. Handbuch der Getränketechnologie. Eugen Ulmer Verlag, Stuttgart.

Dobrogosz, W. J., I. A. Casas, G. C. Pagano, T. L. Talarico, and B. M. Sjoberg. 1989. *Lactobacillus reuteri* and the enteric microbiota, p. 283–292. In: R. Grubb, T. Midtvedt, and E. Norin (ed.), The regulatory and protective role of the normal microflora. Wenner-Gren International Symposium Series, vol. 52. MacMillan Press, Basingstoke, UK.

Dodds, K. L., and D. L. Collins-Thompson. 1985. Characteristics of nitrite reductase activity in *Lactobacillus lactis* Ts4. Can. J. Microbiol. 31:558–562.

Drasar, B. S., and P. A. Barrow. 1985. Intestinal Microbiology. In: J. A. Cole, C. J. Knowles, and D. Schlessinger (ed.), Aspects of microbiology, vol. 10. Am. Soc. of Microbiology, Washington, D.C.

Drasar, B. S., and M. J. Hill. 1974. Human intestinal flora. Academic Press, London.

Dunican, L. K., and H. W. Seeley. 1965. Extracellular polysaccharide synthesis by members of the genus *Lactobacillus*: conditions for formation and accumulation. J. Gen. Microbiol. 40:297–308.

Dutta, G. A., and L. A. Devries. 1981. Sensitivity and resistance to growth promoting agents in animal lactobacilli. J. Appl. Bacteriol. 51:283–288.

Edwards, R. A., R. H. Dainty, C. M. Hibbard, and S. V. Ramantanis. 1987. Amines in fresh beef of normal pH, and the role of bacteria in changes in concentration observed during storage in vacuum packs at chill temperatures. J. Appl. Bacteriol. 63:427–434.

Egan, A. F. 1983. Lactic acid bacteria of meat and meat products. Antonie van Leeuwenhoek 49:327–336.

Ekstrand, J., D. Birkhed, L. E. Lindgren, A. Oliveby, S. Edwardsson, and G. Frostell. 1985. The effect of repeated intake of a sugar free fluoride-containing chewing gum on acidogenicity and microbial composition of dental plaque. Scand. J. Dent. Res. 93:309–314.

Embley, T. M., N. Faquir, W. Bossart, and M. D. Collins. 1989. *Lactobacillus vaginalis* sp. nov. from the human vagina. Int. J. Syst. Bacteriol. 39:367–370.

Entani, E., H. Masai, and K.-I. Suzuki. 1986. *Lactobacillus acetotolerans*, a new species from fermented vinegar broth. Int. J. Syst. Bacteriol. 36:544–549.

Eschenbach, D. A., P. R. Davick, B. L. Williams, S. J. Klebanoff, K. Young-Smith, C. M. Critchlow, and K. K. Holmes. 1989. Prevalence of hydrogen peroxide-producing *Lactobacillus* species in normal women and women with bacterial vaginosis. J. Clin. Microbiol. 27:251–256.

Etchells, J. L., R. N. Costilow, T. A. Bell, and H. A. Rutherford. 1961. Influence of gamma radiation on the microflora of cucumber fruit and blossoms. Appl. Microbiol. 9:145–149.

Etchells, J. L., H. P. Fleming, and T. A. Bell. 1975. Factors influencing the development of lactic acid bacteria during fermentation of brined cucumbers and olives, p. 281–305. In: J. G. Carr, C. V. Cutting, and G. C. Whiting (ed.), Lactic acid bacteria in beverages and food. Academic Press, London.

Etemadzeh, H., J. H. Meurman, H. Murtomaa, H. Torkko, L. Lappi, and M. Roos. 1989. Effect on plaque growth and salivary microorganisms of amine fluoride-stannous fluoride and chlorhexidine-containing mouth rinses. J. Clin. Periodontol. 16:175–178.

Evans, J. B., and C. F. Niven. 1951. Nutrition of the heterofermentative lactobacilli that causes greening of cured meat products. J. Bacteriol. 62:599–603.

Evelyn, T. P. T., and L. A. McDermott. 1961. Bacteriological studies of fresh water fish. Can. J. Microbiol. 7:375–382.

Ezzat, N., M. El Soda, C. Bouillanne, C. Zevaco, and P. Blanchard. 1985. Cell wall associated proteinases in *Lactobacillus helveticus*, *Lactobacillus bulgaricus* and *Lactobacillus lactis*. Milchwissenschaft 40:140–143.

Farrow, J. A. E., and M. D. Collins. 1988. *Lactobacillus oris* sp. nov. from the human oral cavity. Int. J. Syst. Bacteriol. 38:116–118.

Farrow, J. A. E., B. A. Phillips, and M. D. Collins. 1988. Nucleic acid studies on some heterofermentative lactobacilli: Description of *Lactobacillus malefermentans* sp. nov. and *Lactobacillus parabuchneri* sp. nov. FEMS Microbiol. Lett. 55:163–168.

Feresu, S. B., and D. Jones. 1988. Taxonomic studies on *Brochothrix, Erysipelothrix, Listeria* and atypical Lactobacilli. J. Gen. Microbiol. 134:1165–1183.

Fernandes, C. F., K. M. Shahani, and M. A. Amer. 1987. Therapeutic role of dietary lactobacilli and lactobacillic fermented dairy products. FEMS Microbiol. Rev. 46:343–356.

Fernandes, C. F., K. M. Shahani, and M. A. Amer. 1988. Effect of nutrient media and bile salts on growth and antimicrobial activity of *Lactobacillus acidophilus*. J. Dairy Sci. 71:3222–3229.

Finegold, S. M., V. L. Sutter, P. T. Sugihara, H. A. Elder, S. M. Lehmann, and R. L. Phillips. 1977. Fecal microbial flora in Seventh Day Adventist population and control subjects. Amer. J. Clin. Nutr. 30:1781–1792.

Fleming, H. P., R. F. McFeeters, and M. A. Daeschel. 1985. The Lactobacilli, Pediococcus and Leuconostocs: vegetable products, p. 97–118. In S. E. Gilliland (ed.), Bacterial starter cultures for foods. CRC Press, Inc., Boca Rota, FL.

Fournaud, J., P. Raibaud, and G. Mocquot. 1964. Etude de la réduction des nitrites par une souche de *Lactobacillus lactis*. Mise en évidence de ce métabolisme chez d'autres bactéries du genre *Lactobacillus*. Ann. Inst. Pasteur Lille 15:213–224.

Friend, B. A., and K. M. Shahani. 1984. Antitumour properties of lactobacilli and dairy products fermented by lactobacilli. J. Food Prot. 47:717–723.

Fryer, T. F., M. E. Sharpe, and B. Reiter. 1970. Utilization of milk citrate by lactic acid bacteria and "blowing" of film wrapped cheese. J. Dairy Res. 37:17–28.

Fujisawa, T., S. Adachi, T. Toba, K. Arihara, and T. Mitsuoka. 1988. *Lactobacillus kefiranofaciens* sp. nov. isolated from Kefir grains. Int. J. Syst. Bacteriol. 38:12–14.

Fujisawa, T., S. Shirasaka, J. Watanabe, and T. Mitsuoka. 1984. *Lactobacillus aviarius* sp. nov.: a new species isolated from the intestine of chickens. System. Appl. Microbiol. 5:414–420.

Fukui, S., A. Oi, A. Obayashi, and K. Kitahara. 1957. Studies on the pentose metabolism by microorganisms. I. A new type-lactic acid fermentation of pentoses by lactic acid bacteria. J. Gen. Appl. Microbiol. 3:258–268.

Fuller, R. 1973. Ecological studies on the lactobacillus flora associated with the crop epithelium of the fowl. J. Appl. Bacteriol. 36:131–139.

Fuller, R. 1989. Probiotics in man and animals. A review. J. Appl. Bacteriol. 66:365–378.

Fuller, R., P. A. Barrow, and B. E. Brooker. 1978. Bacteria associated with the gastric epithelium of neonatal pigs. Appl. Environ. Microbiol. 35:582–591.

Fuller, R., L. G. M. Newland, C. A. E. Briggs, R. Braude, and K. G. Mitchell. 1960. The normal intestinal flora of the pig. IV. The effect of dietary supplements of penicillin, chlortetracycline or copper sulphate on the faecal flora. J. Appl. Bacteriol. 23:195–205.

Galesloot, T. E., F. Hassing, and H. A. Verenga. 1968. Symbiosis in yoghurt. I. Stimulation of *Lactobacillus bulgaricus* by a factor produced by *Streptococcus thermophilus*. Neth. Milk Dairy J. 22:50.

Gardner, G. A. 1968. Effects of pasteurization or added sulphite on the microbiology of stored vaccum-packed bacon-burgers. J. Appl. Bacteriol. 31:462–478.

Gasser, F., M. Mandel, and M. Rogosa. 1970. *Lactobacillus jensenii* sp. nov., a new representative of the subgenus *Thermobacterium*. J. Gen. Microbiol. 62:219–222.

Geddes, P. A. 1986. The production of hydrogen sulphide by *Lactobacillus* spp. in fermenting wort, p. 364–370. In: J. Campbell, and F. G. Priest (ed.), Proceedings of the second Avremore conference on malting, brewing, and distilling. Institute of Brewing, London.

Geis, A. 1989. Antagonistic compounds produced by lactic acid bacteria. Kieler Milchwirtsch. Forsch. Ber. 41:97–104.

Gibson, E. M., N. M. Chase, S. B. London, and J. London. 1979. Transfer of plasmid-mediated antibiotic resistance from streptoccoci to lactobacilli. J. Bacteriol. 137:614–619.

Gilliland, S. E. 1979. Beneficial interrelationships between certain microoganisms and humans: candidate organisms for use as dietary adjuncts. J. Food Prot. 42:164–167.

Gilliland, S. E., and H. S. Kim. 1982. Reduction of lactose mal-absorption in humans by consumption of milk containing different numbers of *Lactobacillus acidophilus*. J. Dairy Sci. 65 (Suppl. 1):220.

Gilliland, S. E., C. R. Nelson, and C. Maxwell. 1985. Assimilation of cholesterol by *Lactobacillus acidophilus*. Appl. Environ. Microbiol. 49:377–381.

Gilliland, S. E., and M. L. Speck. 1977. Deconjugation of bile acids by intestinal lactobacilli. Appl. Environ. Microbiol. 33:15–18.

Goldin, B. 1986. In situ bacterial metabolism and colon mutagens. Annu. Rev. Microbiol. 40:467–493.

Götz, F., E. F. Elstner, B. Sedewitz, and E. Lengfelder. 1980. Oxygen utilization by *Lactobacillus plantarum*. II. Superoxide and superoxide dismutation. Arch. Microbiol. 125:215–220.

Grazia, L., and G. Suzzi. 1984. A survey of lactic acid bacteria in Italian silage. J. Appl. Bacteriol. 56:373–379.

Groeneveld, M., and C. Leitzmann. 1987. Zum Vorkommen antikanzerogener Substanzen in Lebensmitteln, speziell in milchsauren Produkten. Aktuelle Ernährungsmedizin 12:202–204.

Groß, F., and K. Riebe. 1974. Gärfutter. Verlag Eugen Ulmer, Stuttgart.

Grubb, R., T. Midtvedt, and E. Norin (ed.). 1989. The regulatory and protective role of the normal microflora. Wenner-Gren International Symposium Series, Vol. 52. MacMillan Press, Basingstoke, UK.

Grunewald, K. K. 1982. Serum cholesterol levels in rats fed skim milk fermented by *Lactobacillus acidophilus*. J. Food Sci. 47:2078–2079.

Grunewald, K. K. 1985. Influence of bacterial starter cultures on nutritional value of foods: effects of *Lactobacillus acidophilus* fermented milk on growth and serum cholesterol in laboratory animals. Cult. Dairy Prod. J. 20:26–27.

Haddadin, M. S. Y. 1986. Microbiology of white-brined cheeses, p. 67–89. In: R. K. Robinson (ed.), Developments in food microbiology-2. Elsevier Applied Science Publishers, London.

Haggblade, S., and W. H. Holzapfel. 1989. Industrialization of Africa's indigenous beer brewing, p. 191–283. In: K. H. Steinkraus (ed.), Industrialization of indigenous fermented foods. Marcel Dekker Inc., New York.

Hahn, C. L., W. A. Falkler Jr., G. E. Minah, and J. Palmer. 1989. Clinical and bacteriological studies of dental caries. J. Dent. Res. 68 (spec. issue):214.

Hakgudener, Y. 1985. Lactobacilli in the saliva of persons with carious and noncarious teeth. Mikrobiyol. Bul. 19:29–35.

Hamdan, I. Y., and E. M. Mikolajcik. 1974. Acidolin: an antibiotic produced by *Lactobacillus acidophilus*. J. Antibiot. 27:631–636.

Hammes, W. P. 1986. Starterkulturen in der Fleischwirtschaft. Chem. Mikrobiol. Technol. Lebensmittel 9:131–143.

Hammes, W. P., I. Rölz, and A. Bantleon. 1985. Mikrobiologische Untersuchung der auf dem deutschen Markt vorhandenen Starterkulturpräparate für die Rohwurstbereitung. Fleischwirtschaft 65:5/1–6; 6/1–6.

Hansen, G. B. 1988. Freeze-dried starter cultures for sourdough bread. ZFL 7:597–602.

Hardie, J. M., and G. H. Bowden. 1974. The normal microflora of the mouth, p. 47–83. In: Skinner, F. A., J. G. Carr (ed.), The normal microflora of man. Academic Press, London.

Harrison, V. C., and G. Peat. 1975. Serum cholesterol and bowel flora in the newborn. Am. J. Clin. Nutr. 28:1351–1355.

Hastings, J. W., and W. H. Holzapfel. 1987a. Numerical taxonomy of lactobacilli surviving radurization of meat. Int. J. Food Microbiol. 4:33–49.

Hastings, J. W., and W. H. Holzapfel. 1987b. Conventional taxonomy of lactobacilli surviving radurization of meat. J. Appl. Bacteriol. 62:209–216.

Hayward, A. C. 1957. A comparison of *Lactobacillus* species from human saliva with those from other natural sources. Br. Dent. J. 102:450–451.

Hayward, A. C. and G. H. G. Davis. 1956. The isolation and classification of *Lactobacillus* strains from Italian saliva samples. Br. Dent. J. 101:43–46.

Heidel, M., and W. P. Hammes. 1990. Bacteriophages in lactobacilli. Food Biotechnology 4:552.

Heinzl, H., and W. P. Hammes. 1986. Die mikroaerophile Bakterienflora von Obstmaischen. Chem. Mikrobiol. Technol. Lebensmittel 10:106–109.

Hemme, D., P. Raibaud, R. Ducluzeau, J. V. Galpin, Ph. Sicard, and J. Van Heijekoort. 1980. *Lactobacillus murinus* n. sp., une nouvelle espèce de la flore dominante autochtone du tube digestif du rat et de la souris. Ann. Microbiol. (Inst. Pasteur) 131A:297–308.

Hemme, D., V. Schmal, and J. Auclair. 1981. Effect of the addition of extracts of thermophilic lactobacilli on acid production by *Streptococcus thermophilus* in milk. J. Dairy Res. 48:139–148.

Henick-Kling, T., T. H. Lee, and D. J. D. Nicholas. 1986. Inhibition of bacterial growth and malolactic fermentation in wine by bacteriophage. J. Appl. Bacteriol. 61:287–293.

Henneberg, W. 1909. Gärungsbakteriologisches Praktikum, Betriebsuntersuchungen und Pilzkunde. Paul Parey, Berlin.

Hensel, R., U. Mayr, K. O. Stetter, and O. Kandler. 1977. Comperative studies of lactic acid dehydrogenases in lactic acid bacteria. I. Purification and kinetics of the allosteric L-lactic acid dehydrogenase from *Lactobacillus casei* subsp. *casei* and *Lactobacillus curvatus*. Arch. Microbiol. 112:81–93.

Hentges, D. J. (ed.), 1983. Human intestinal microflora in health and disease. Academic Press, New York.

Hepner, G., R. Fried, S. St. Jeor, L. Fusetti, and R. Morin. 1979. Hypocholesterolemic effect of yogurt and milk. Am. J. Clin. Nutr. 32:19–24.

Hill, G. B., D. A. Eschenbach, and K. K. Holmes. 1985. Bacteriology of the vagina. Scand. J. Urol. Nephrol. Suppl. 86:23–29.

Hitchener, B. J., A. F. Egan, and P. J. Rogers. 1982. Characteristics of lactic acid bacteria isolated from vacuum-packaged beef. J. Appl. Bacteriol. 52:31–37.

Hiu, S. R., R. A. Holt, N. Sriranganathan, R. J. Seidler, and J. L. Fryer. 1984. *Lactobacillus piscicola*, a new species from salmonid fish. Int. J. Syst. Bacteriol. 34:393–400.

Holliger, W. 1902. Bakteriologische Untersuchungen über Mehlteiggärung. Zentralbl. Bakteriol. Parasitenkd. Infektionskr. Hyg. Abt. II 9:305–312; 361–371; 395–425; 473–483; 521–537.

Holzapfel, W. H. 1989. Industrialization of Mageu fermentation in South Africa, p. 285–328. In: K. H. Steinkraus (ed.), Industrialization of indigenous fermented foods. Marcel Dekker Inc., New York.

Holzapfel, W. H., and E. S. Gerber. 1983. *Lactobacillus divergens* sp. nov., a new heterofermentative *Lactobacillus* species producing L(+)-lactate. Syst. Appl. Microbiol. 4:522–534.

Holzapfel, W. H., and E. S. Gerber. 1986. Predominance of *Lactobacillus curvatus* and *Lactobacillus sake* in the spoilage association of vacuum-packed meat products. Prov. 32nd Europ. Meeting. Meat Res. Workers, p. 26, 4:3, Ghent, Belgium, Aug. 24–29, 1986.

Holzapfel, W. H., and E. P. van Wyk. 1982. *Latobacillus kandleri* sp. nov., a new species of the subgenus *Betabacterium*, with glycine in the peptidoglycan. Zentralbl. Bacteriol. Microbiol. Hyg. Abt. I Orig. C 3:495–502.

Hood, S. K., and E. A. Zottola. 1987. Electron microscopic study of the adherence properties of *Lactobacillus acidophilus*. J. Food Sci. 52:791–792, 805.

Hsu, W. P., and J. A. Taparowsky. 1977. Growth response of common brewery bacteria to different media. Brewers Digest 52:48–53.

Humphrey, J. D., C. E. Lancaster, N. Gudkovs, and J. W. Copland. 1987. J. Fish. Dis. 10:403–410.

Hungate, R. E. 1966. The rumen and its microbes. Academic Press, New York.

ICMSF. 1980. Microbial ecology of foods, vol. 2, food commodities. The International Commission on Microbiological Specifications for Foods. Academic Press, New York.

International Journal of Systematic Bacteriology. 1987. International committee on systematic bacteriology. Subcommittee on the taxonomy of the lactobacilli, bifidobacteria, and related organisms. Int. J. Syst. Bacteriol. 37:469–470.

Irrgang, K., and U. Sonnenborn (ed.). 1988. Beziehungen zwischen Wirtsorganismus und Darmflora. Schattauer Verlag, Stuttgart.

Ishiwa, H., and S. Iwata. 1980. Drug resistance plasmids in *Lactobacillus fermentum*. J. Gen. Microbiol. 26:71–74.

Iwata, M., M. Mada, and H. Ishiwa. 1986. Protoplast fusion of *Lactobacillus fermentum*. Appl. Environ. Microbiol. 52:392–393.

Jayne-Williams, D. J. 1979. The bacterial flora of healthy and bloating calves. J. Appl. Bacteriol. 47:271–284.

Jernigan, M. A., R. D. Miles, and A. S. Arafa. 1985. Probiotics in poultry nutrition. A review. World's Poult. Sci. J. 41:99–107.

Joerger, M. C., and T. R. Klaenhammer. 1986. Characterization and purification of helveticin J, and evidence for a chromosomally determined bacteriocin produced by *Lactobacillus helveticus* 481. J. Bacteriol. 167:439–446.

Johnson, J. L., C. F. Phelps, C. S. Cummins, J. London, and F. Gasser. 1980. Taxonomy of the *Lactobacillus acidophilus* group. Int. J. Syst. Bacteriol. 30:53–68.

Johnston, M. A., and E. A. Delwiche. 1962. Catalase of the *Lactobacillaceae*. J. Bacteriol. 83:936–938.

Jonsson, E., L. Bjorck, and C. O. Claesson. 1985. Survival of orally administered *Lactobacillus* strains in the gut of cannulated pigs. Livestick Prod. Sci. 12:279–285.

Joosten, H. M. L. J., and M. D. Northolt. 1987. Conditions allowing the formation of biogenic amines in cheese. 2. Decarboxylative properties of some non-starter bacteria. Neth. Milk Dairy 41:259–280.

Josson, K., T. Scheirlinck, F. Michiels, C. Platteeuw, P. Stanssens, H. Joos, P. Dhaese, Zabeau and J. Mahillon. 1989. Characterization of a Gram-positive broad-host-range plasmid isolated from *Lactobacillus hilgardii*. Plasmid 21:9–20.

Juven, B. J. 1976. Bacterial spoilage of citrus products at pH lower than 3.5. J. Food Prot. 39:819–822.

Kagermeier, A. 1981. Taxonomie und Vorkommen von Milchsäurebakterien in Fleischprodukten. Thesis, Ludwig-Maximilian University Munich, West Germany.

Kandler, O. 1983. Carbohydrate metabolism in lactic acid bacteria. Antonie van Leeuwenhoek 49:209–224.

Kandler, O., W. P. Hammes, M. Schneider, and K. O. Stetter. 1986. Microbial interaction in sauerkraut fermentation. Proc. IV. Int. Symp. Microb. Ecol., p. 302–308.

Kandler, O., and P. Kunath. 1983. *Lactobacillus kefir* sp. nov., a component of the microflora of kefir. System. Appl. Microbiol. 4:286–294.

Kandler, O., U. Schillinger, and N. Weiss. 1983a. *Lactobacillus bifermentans* sp. nov., nom. rev., an organism forming CO_2 and H_2 from lactic acid. System. Appl. Microbiol. 4:408–412

Kandler, O., U. Schillinger, and N. Weiss. 1983b. *Lactobacillus halotolerans* sp. nov., nom. rev. and *Lactobacillus minor* sp. nov., nom. rev. Syst. Appl. Microbiol. 4:280–285.

Kandler, O., K. O. Stetter, and R. Köhl. 1980. *Lactobacillus reuteri* sp. nov., a new species of heterofermentative lactobacilli. Zentralbl. Bakteriol. Mikrobiol. Hyg. Abt.1. Orig. Reihe C 1:264–269.

Kandler, O., and N. Weiss. 1986. Regular, non-sporing Gram-positive rods, p. 1208–1234. In: P. H. A. Sneath, N. Mair, M. E. Sharpe, and J. G. Holt (ed.), Bergey's manual of systematic bacteriology, vol. 2. William and Wilkins, Baltimore.

Karjalainen, S., M. Hamalainen, L. Karhuvaara, and E. Soderling. 1987. Effect of variations in sucrose consumption on salivary lactobacillus count and sucrose activity in man. Acta Odontol. Scand. 45:289–296.

Kawai, Y., and N. Shegara. 1977. Specific adhesion of lactobacilli to keratinized epithelial cells of the rat stomach. Amer. J. Clin. Nutr. 30:1777–1780.

Keddie, R. M. 1959. The properties and classification of lactobacilli isolated from grass and silage. J. Appl. Bacteriol. 22:403–416.

Keller, J. J., and J. Jaarsman. 1975. Lactobacilli and gas formation of film wrapped Cheddar cheese. South African J. Dairy Technol. 7:183–185.

Kim, H. S., and S. E. Gilliland. 1983. *Lactobacillus acidophilus* as a dietary adjunct for milk to acid lactose digestion in humans. J. Dairy Sci. 66:959–966.

Kitahara, K., T. Kaneko, and O. Goto. 1957. Taxonomic studies on the hiochi-bacteria, specific saprophytes of sake, identification of hiochi-bacteria. J. Gen. Appl. Microbiol. 3:102–110

Kitchell, A. G. 1964. Meat Research. 7. Microbiology Ann. Rep. Low-Temp. Res. Sta., Cambridge, p. 26. London:H.M.S.O.

Kitchell, A. G., and B. G. Shaw. 1975. Lactic acid bacteria in fresh and cured meat, p. 209–220. In: J. G. Carr, C. V. Cutting, and G. C. Whiting (ed.), Lactic acid bacteria in beverages and food. Academic Press, London.

Kiuru, U. I. T., and E. Tybeck. 1955. Characteristics of bacteriophage active against lactic acid bacteria in Swiss cheese. Suomen Kemistileht 28:57–62.

Kiyosawa, H., C. Sugawara, N. Sugawara, and H. Miyake. 1984. Effect of skim milk and yogurt on serum lipids and development of sudanophilic lesions in cholesterol fed rabbits. Am. J. Clin. Nutr. 40:479–484.

Klaenhammer, T. R. 1982. Microbiological considerations in selection and preparation of *Lactobacillus* strains for use in dietary adjuncts. J. Dairy Sci. 65:1339–1349.

Klaenhammer, T. R. 1988. Bacteriocins of lactic acid bacteria. Biochimie 79:337–349.

Kleeman, E. G., and T. R. Klaenhammer. 1982. Adherence of *Lactobacillus* spp. to human fetal intestinal cells. J. Dairy Sci. 65:2063–2069.

Kleynmans, U., H. Heinzl, and W. P. Hammes. 1989. *Lactobacillus suebicus* spec. nov. an obligately heterofermentative *Lactobacillus* species isolated from fruit mashes. Syst. Appl. Microbiol. 11:267–271.

Kline, L., and T. F. Sugihara. 1971. Microorganisms of the San Francisco Sour Dough Bread Process. II. Isolation and characterization of undescribed bacterial species responsible for the souring activity. Appl. Microbiol. 21:459–465.

Knauf, H. J., R. F. Vogel, and W. P. Hammes. 1989. Introduction of the transposon Tn919 into *Lactobacillus curvatus* Lc2-c. FEMS Microbiol. Lett. 65:101–104.

Kneist, S., R. Heinrich, and W. Kuenzel. 1988. Zum Vorkommen von Laktobazillen im karioesen Dentin. Zahn-, Mund-, Kieferheilkd. 76:123–127.

Knoke, M. and H. Bernhardt (ed.). 1986. Mikroökologie des Menschen. Mikroflora bei Gesunden und Kranken. VCH Verlagsgesellschaft, Weinheim.

Knudsen, S. 1924. Über die Milchsäurebakterien des Sauerteiges und ihre Bedeutung für die Sauerteiggärung. Den. Kgl. Veterinaer-og Landbohoskoles Arsskrift.

Kok, J., J. M. B. M. van der Vossen, and G. Venema. 1984. Construction of plasmid cloning vectors for lactic streptococci which also replicate in *Bacillus subtilis* and *Escherichia coli*. Appl. Environ. Microbiol. 48:726–731.

Konings, W. N. 1985. Generation of metabolic energy by end-product efflux. Trends Biochem. Sci. 10:317–319.

Krasse, B. 1988. Biological factors as indicator of future caries. Int. Dent. J. 38:219–225.

Kreuzer, R. 1957. Untersuchungen über den biologisch bedingten Verderb von Fischwaren und seine Verhinderung. I. Kaltmarinaden-Organismen und Milieu. Arch. Fisch Wiss. 8:104–159.

Krieger, S. A. 1989. Optimierung des biologischen Säureabbaus in Wein mit Starterkulturen. Thesis. University of Hohenheim, West Germany.

Kristoffersson, K., and D. Birkhed. 1987. Effects of partial sugar restriction for 6 weeks on numbers of *Streptococcus mutans* in saliva and interdental plaque in man. Caries Res. 21:79–86.

Krumperman, P. H., and R. H. Vaughn. 1966. Some lactobacilli associated with decomposition of tartaric acid in wine. Am. J. Enol. Vitic. 17:185–190.

Kunath, P. 1983. Die Mikroflora von Kefir. Thesis, Ludwig-Maximilians-University, Munich.

Kunkee, R. E. 1967. Malo-lactic fermentation. Adv. Appl. Microbiol. 9:235–279.

Kurtzman, C. P., R. Rogers, and C. W. Hesseltine. 1971. Microbiological spoilage of mayonnaise and salad dressings. Appl. Microbiol. 21:870–874.

Kvasnikov, E. I., G. S. Eliseeva, N. K. Kovalenko, and T. N. Shishlevskaya. 1985. Synthesis of group B vitamins by lactic acid bacteria in the digestive tract of commercially raised chickens (Russ.). Mikrobiol. Zh. (Kiev) 47:26–30.

Kvasnikov, E. I., A. N. Kotljar, and Z. A. Vasileva. 1976. Species composition and certain peculiarities in physiology of bacteria isolated in sugar production. Mikrobiologichnii Zhurnal 38:434–438.

Kvasnikov, E. I., N. K. Kovalenko, and O. A. Nesterenko. 1983. Lactic acid bacteria in nature and the national economy. Appl. Biochem. and Microbiol. 18:665–676.

Kvasnikov, E. I., T. N. Shishlevskaya, and N. K. Kovalenko. 1982. Lactic acid bacteria in the digestive tract of chickens grown under commercial conditions (Russ.). Mikrobiol. Zh. (Kiev) 43:703–708.

La Riviere, J. W. M., P. Kooiman, and K. Schmidt. 1967. Kefiran, a novel polysaccharide produced in the kefir grain by *Lactobacillus brevis*. Arch. Microbiol. 59:269–278.

Lahbib-Mansais, Y., M. Mata, and P. Ritzenthaler. 1988. Molecular taxonomy of *Lactobacillus* phages. Biochimie 70:429–435.

Langella, P., and A. Chopin. 1989. Conjugal transfer of plasmid pIP501 from *Lactococcus lactis* to *Lactobacillus delbrueckii* subsp. *bulgaricus* and *Lactobacillus helveticus*. FEMS Microbiol. Lett. 60:149–153.

Langston, C. W., and C. Bouma. 1960a. Types and sequence change of bacteria in orchard grass and alfalfa silages. J. Dairy Sci. 43:1575–1584.

Langston, C. W., and C. Bouma. 1960b. A study of the microorganisms from grass silage. II. The lactobacilli. Appl. Microbiol. 8:223–234.

Langston, C. W., C. Bouma, and R. M. Conner. 1962. Chemical and bacteriological changes in grass silage during the early stages of fermentation. II. Bacteriological changes. J. Dairy Sci. 45:618–624.

Lapage, S. P., J. E. Shelton, T. G. Mitchell, and A. R. Mackenzie. 1970. Culture collections and the preservation of bacteria, p.135–228. In: J. R. Norris, and D. M. Rib-

bons (ed.), Methods in microbiology vol. 3A. Academic Press, London.

Latham, M. J., M. E. Sharpe, and J. D. Sutton. 1971. The microbial flora of the rumen of cows fed hay and high cereal rations and its relationships to the rumen fermentation. J. Appl. Bacteriol. 34:425–434.

Lauer, E., C. Helming, and O. Kandler. 1980. Heterogeneity of the species *Lactobacillus acidophilus* (Moro) Hansen and Moquot as revealed by biochemical characteristics and DNA-DNA hybridisation. Zentralbl. Bakteriol. Hyg., Abt. I Orig. C 1:150–168.

Lauer, E., and O. Kandler. 1980. *Lactobacillus gasseri* sp. nov., a new species of the subgenus *Thermobacterium*. Zentralbl. Bakteriol. Mikrobiol. Hyg. Abt. I Orig. C 1:75–78.

Lawrence, D. R. 1988. Spoilage organisms in beer, p. 1–48. In: R. K. Robinson (ed.), Developments in food microbiology. 3. Elsevier Applied Science Publishers, London.

Lee, A. 1978. Bacteriophages associated with lactobacilli isolated from wine. 5th Int. Oenolog. Symp., Auckland 1978, p. 287–295. Aust. N. Z. Ass. Advance Sci., 1978.

Leer, R. J., M. Posno, J. M. M. van Rijn, B. L. Lokman, and P. H. Pouwells. 1987. Transformation of *Lactobacillus plantarum* by plasmid DNA. FEMS Microbiol. Rev. 46:P20, A32.

Lees, G. J., and J. R. Jago. 1978. Role of acetaldehyde in metabolism: a review. II. The metabolism of acetaldehyde in cultured dairy products. J. Dairy Sci. 61:1216.

Lenzner, A. A. 1966. Some results of the investigation of lactobacilli of human microflora. Survey of Research in Medicine, Tartu State University, 1940–1965 191:69–75.

Lerche, M. 1960. Bombage-Ursache in Fischpräserven. Berl. Münch. tierärztl. Wochenschrift 75:12–14.

Levison, M. E., L. C. Corman, E. R. Carrington, and D. Kaye. 1977. Quantitative microflora of the vagina. Am. J. Obstetrics Gynaecol. 127:80–85.

Lidbeck, A., J. A. Gustafsson, and N. E. Nord. 1987. Impact of *Lactobacillus acidophilus* supplements on the human oropharyngeal and intestinal microflora. Scand. J. Infect. Dis. 19:531.

Liepe, H.-U. 1983. Starter cultures in meat production, p. 400–424. In: H.-J. Rehm, and G. Reed (ed.), Biotechnology, vol. 5. Verlag Chemie, Weinheim.

Lin, J. H., and D. C. Savage. 1985. Cryptic plasmids in *Lactobacillus* strains isolated from the murine gastrointestinal tract. Appl. Environ. Microbiol. 49:1004–1006.

Lin, J. H., and D. C. Savage. 1986. Genetic transformation of rifampicin resistance in *Lactobacillus acidophilus*. J. Gen. Microbiol. 132:2107–2111.

Liu, M. L., J. K. Kondo, M. B. Barnes, and D. T. Bartholomew. 1988. Plasmid linked maltose utilization in *Lactobacillus* ssp. Biochimie 70:351–355.

London, J. 1976. The ecology and taxonomic status of the lactobacilli. Annu. Rev. Microbiol. 30:279–301.

London, J., and S. Hausman. 1982. Xylitol-mediated transient inhibition of ribitol utilization by *Lactobacillus casei*. J. Bacteriol. 150:657–661.

Lorenz, A., F. K. Gruette, H. Haenel, and U. Sanders. 1982a. A mechanism of association of lactobacilli with the rat stomach epithelium. Zentralbl. Bakteriol. Mikrobiol. Hyg. Abt. I Orig. Reihe A 252:9–16.

Lorenz, R. P., P. C. Applebaum, R. M. Ward, and J. J. Botti. 1982b. Chorioamniotis and possible neonatal infection associated with *Lactobacillus* species. J. Clin. Microbiol. 16:558–561.

Luchansky, J. B., E. G. Kleeman, R. R. Raya, and T. R. Klaenhammer. 1989. Genetic transfer systems for delivery of plasmid deoxyribonucleic acid to *Lactobacillus acidophilus* ADH: conjugation, electroporation and transduction. J. Dairy Sci. 72:1408–1417.

Lücke, F.-K. 1985. Fermented sausages, p. 41–83. In: B. J. B. Wood (ed.), Microbiology of fermented foods, vol. 2. Elsevier Applied Science Publishers, London.

Lundstrom, F., and B. Krasse. 1987. *Streptococcus mutans* and lactobacilli frequency in orthodontic patients; the effect of chlorhexidine treatments. Eur. J. Orthod. 9:109–116.

Lyons, T. P., and A. H. Rose. 1977. Whisky, p. 635–692. In: A. H. Rose (ed.), Economic microbiology, vol. 1: Alcoholic beverages. Academic Press, London.

MacKenzie, K. G., and M. C. Kenny. 1965. Non-volatile organic acid and pH changes during the fermentations of distiller's wort. J. Inst. Brewing 71:160–165.

Maier, B. B., M. A. Flynn, G. G. Burton, R. K. Tsutakawa, and D. J. Hentges. 1974. Effects of a high-beef diet on bowel flora. A preliminary report. Am. J. Clin. Nutr. 27:1470–1474.

Manca de Nadra, M. C., A. A. Pesce De Ruiz Holgado, and G. Oliver. 1988. Arginine dihydrolase pathway in *Lactobacillus buchneri*: a review. Biochimie 70:367–374.

Maret, R. and T. Sozzi. 1977. Flore malolactique de mouts et de vins du Canton du Valais (Suisse). I. Lactobacilles et pediocoques. Ann. Technol. Agric. 27:255–273.

Maret, R. and T. Sozzi. 1979. Flore malolactique de mouts et de vins du Canton du Valais (Siusse). II. Evolution des populations de lactobacilles et de pediocoques au cours de la vinification d'un vin blanc (un Fendant) et d'un vin rouge (une Dole). Ann. Technol. Agric. 28:31–40.

Maret, R., T. Sozzi, and D. Schellenberger. 1979. Flore malolactique de mouts et de vins du Canton Valais (Suisse). III.- Les Leuconostocs. Ann. Technol. Agric. 28:41–55.

Marounek, M., K. Jehlickova, and V. Kmet. 1988. Metabolism and some characteristics of lactobacilli isolated from the rumen of young calves. J. Appl. Bacteriol. 65:43–48.

Marsh, P., and M. Martin, 1984. Oral microbiology, 2nd ed. Series: Aspects of microbiology 1. Van Nostrand Reinhold (UK) Co., Ltd.

Marshall, V. M. E., and B. A. Law. 1984. The physiology and growth of diary lactic acid bacteria, p. 67–98. In: F. L. Davies, and B. A. Law (ed.), Advances in the microbiology and biochemistry of cheese and fermented milk. Elsevier Applied Science Publishers, London.

Marshall, V. M., S. M. Phillips, and A. Turvey. 1982. Isolation of a hydrogen peroxide-producing strain of *Lactobacillus* from calf gut. Res. Vet. Sci. 32:259–260.

Mata, M., and P. Ritzenthaler. 1988. Present state of lactic acid bacteria phage taxonomy. Biochimie 70:395–399.

Mata, M., A. Trautwetter, G. Luthaud, and P. Ritzenthaler. 1986. Thirteen virulent and temperate bacteriophages of *Lactobacillus bulgaricus* and *Lactobacillus lactis* belong to a single DNA homology group. Appl. Environ. Microbiol. 52:812–818.

Matee, M. I., F. Mikx, J. E. Frencken, G. J. Fruin, and H. M. Ruiken. 1985. Selection of a micromethod and its use in the estimation of salivary *Streptococcus mutans* and *Lactobacillus* counts in relation to dental caries in Tanzanian children. Caries Res. 19:497–506.

Mäyrä-Mäkinen, A., M. Manninen, and H. Glyllenberg. 1983. The adherence of lactic acid bacteria to the columnar epithelium cells of pigs and calves. J. Appl. Bacteriol. 55:241–246.

McArthur, J. L., and S. F. Barefoot. 1986. Isolation and characterization of temperate bacteriophage in *Lactobacillus acidophilus*. Abstr. Ann. Meeting: 236.

McCormick, E. L., and D. C. Savage. 1983. Characterization of *Lactobacillus* sp. strain 100–37 from the murine gastrointestinal tract: ecology, plasmid content, and antagonistic activity toward *Chlostridium ramosum* H1. Appl. Environ. Microbiol. 46:1103–1112.

McFeeters, R. F., H. P. Fleming, and M. A. Daeschel. 1984. Malic acid degradation and brined cucumber bloating. J. Food Sci. 49:999.

Mehta, A. M., K. A. Patel, and P. J. Dave. 1983. Purification and properties of the inhibitory protein isolated from *Lactobacillus acidophilus* ACl. Microbios. 38:73–81.

Meisel, C., K. H. Gehlen, A. Fischer, and W. P. Hammes. 1989. Inhibition of the growth of *Staphylococcus aureus* in dry sausages by *Lactobacillus curvatus*, *Micrococcus varians*, and *Debaryomyces hansenii*. Food Biotechnol. 3:145–168.

Meyer, V. 1956a. Die Bestimmung der Bombage-Arten bei Fischkonserven. Fischwirtschaft 8:212–214.

Meyer, V. 1956b. Probleme des Verderbens von Fischkonserven in Dosen. II. Aminosäuredecarboxylase durch Organismen der *Betabacterium Buchneri*- Gruppe als Ursache bombierter Marinaden. Veröff. Inst. Meeresforsch. Bremerh. 4:1–16.

Meyer, V. 1965. Marinades, p. 221. In: G. Borgström (ed.), Fish as food. Academic Press, London.

Michels, P. A. M., J. P. J. Michels, J. Boonstra, and W. N. Konings. 1979. Generation of an electrochemical proton gradient in bacteria by the excretion of metabolic end products. FEMS Microbiol. Lett. 5:357–364.

Miller, A., III, M. E. Morgan, and L. M. Libbey. 1974. *Lactobacillus maltaromicus*, a new species producing a malty aroma. Int. J. Syst. Bacteriol. 24:346–354.

Miller, B. S., and J. A. Johnson. 1958. Present knowledge concerning baking processes employing pre-ferments. Wallerstein. Lab. Commun. 21:115–137.

Millis, N. F. 1951. Some bacterial fermentations in cider. Thesis. University of Bristol, Bristol, England.

Minah, G. E., E. S. Solomon, and K. Chu. 1985. The association between dietary sucrose consumption and microbial population shifts at 6 oral sites in man. Arch. Oral Biol. 30:397–402.

Mitsuoka, T. 1969. Vergleichende Untersuchungen über die Lactobazillen aus den Faeces von Menschen, Schweinen und Hühnern. Zentralbl. Bakteriol., Parasitenk., Infektionskr. Hyg. Abt. 1 orig. 210:32–51.

Mitsouka, T., and T. Fujisawa. 1987. *Lactobacillus hamsteri*, a new species from the intestine of hamsters. Proc. Jap. Acad. Ser. B, Phys. Biol. Sci. 63:269–272.

Mitsuoka, T., K. Hayakawa, and N. Kimura. 1975. The fecal flora of man. III Communication: The composition of lactobacillus flora of different age groups. Zentralbl. Bakteriol. Parasitenk. Infektionskr. Hyg. Abt. 1 Orig. Reihe A 232:499–511.

Mitsuoka, T., and C. Kaneuchi. 1977. Ecology of the bifidobacteria. Am. J. Clin. Nutr. 30:1799–1810.

Mitsuoka, T., N. Kimura, and A. Kobayashi. 1976. Studies on the composition of the fecal flora of healthy dogs with the special references of lactobacillus flora and bifidobacterium flora. Zentralbl. Bakteriol. Parasitenk. Infektionskr. Hyg. Abt. 1 Orig. Reihe A 235:485–493.

Mitsuoka, T., and K. Ohno. 1977. Fecal flora of man. V. Communication: The fluctuations of the fecal flora of the healthy adult. Zentr. f. Bakteriol., Abt. 1., Orig. Reihe A 238:228–236.

Mol, J. H. H., J. E. A. Hietbrink, H. W. M. Mollen, and J. Van Tinteren. 1971. Observations on the microflora of vacuum packed sliced cooked meat products. J. Appl. Bacteriol. 34:377–397.

Moore, W. E. C., and L. V. Holdeman. 1974. Human fecal flora: The normal flora of 20 Japanese-Hawaiians. Appl. Microbiol. 27:961–979.

Moore, W. E. C., and L. V. Holdeman. 1975. Discussion of current bacteriological investigations of the relationships between intestinal flora, diet and colon cancer. Cancer Res. 35:3418–3420.

Moreau, M.-C., R. Ducluzeau, and P. Raibaud. 1976. Hydrolysis of urea in the gastrointestinal tract of "monaxenic" rats: Effect of immunization with strains of ureolytic bacteria. Infect. and Immun. 13:9–15.

Morelli, L., P. S. Cocconcelli, V. Bottazzi, G. Damiani, L. Ferretti, and V. Sgaramella. 1987. Lactobacillus protoplast transformation. Plasmid 17:73–75.

Morelli, L., M. Vescovo, P. S. Cocconcelli, and V. Bottazzi. 1986. Fast and slow milk coagulating variants of Lactobacillus helveticus HLM-1. Can. J. Microbiol. 32:758–760.

Morgan, M. E., R. C. Lindsay, L. M. Libbey, and R. L. Pereira. 1966. Identity of additional aroma constituents in milk cultures of Streptococcus lactis var. maltigenes. J. Dairy Sci. 49:15–18.

Morishita, T., Y. Deguchi, M. Yajima, T. Sakurai, and T. Yura. 1981. Multiple nutritional requirements of lactobacilli: genetic lesions affecting amino acid biosynthetic pathways. J. Bacteriol. 148:64–71.

Morishita, T., T. Fukada, M. Shirota, and T. Yura. 1974. Genetic bases of nutritional requirements in Lactobacillus casei. J. Bacteriol. 120:1078–1084.

Mundt, J. O., W. F. Graham, and I. E. McCarty. 1967. Spherical lactic acid-producing bacteria of southern-grown raw and processed vegetables. Appl. Microbiol. 15:1303–1308.

Mundt, J. O., and J. L. Hammer. 1968. Lactobacilli on plants. Appl. Microbiol. 16:1326–1330.

Murata, A., K. Kitagawa, and R. Saruno. 1971. Inactivation of bacteriophages by ascorbic acid. Agr. Biol. Chem. 35:294–296.

Murata, A., K. Kitagawa, H. Inmaru, and R. Saruno. 1972. Inactivation of bacteriophages by thiol reducing agents. Agr. Biol. Chem. 36:1065–1067.

Murata, A., M. Odaka, and S. Mukuno. 1974. The bacteriophage-inactivating effect of basic amino acids; arginine, histidine and lysine. Agr. Biol. Chem. 38:477–478.

Murdock, D. I., J. F. Folinazzo, and V. S. Troy. 1952. Evaluation of plating media for citrus concentrates. Food Technol. 6:181–185.

Murdock, D. I., and W. S. Hatcher. 1975. Growth of microorganisms in chilled orange juice. J. Milk and Food Technol. 38:393–396.

Muriana, P. M., and T. R. Klaenhammer. 1987. Conjugal transfer of plasmid-encoded determinants for bacteriocin production and immunity in Lactobacillus acidophilus 88. Appl. Environ. Microbiol. 53:553–560.

Muriana, P. M., and T. R. Klaenhammer. 1989. Purification and physical characterization of a bacteriocin (lactacin F) produced by Lactobacillus acidophilus. Abstr. Annu. Meet. Am. Soc. Microbiol. 43:326. New Orleans.

Nahaisi, M. H. 1986. Lactobacillus acidophilus: therapeutic properties, products and enumeration, p. 153–178. In: R. K. Robinson (ed.), Developments in food microbiology, 2. Elsevier Applied Science Publishers, London.

Nakamura, L. K. 1981. Lactobacillus amylovorus, a new starch-hydrolizing species from cattle waste-corn fermentation. Int. J. Syst. Bacteriol. 31:56–63.

Nakamura, L. K., and C. D. Crowell. 1979. Lactobacillus amylophilus, a new starch-hydrolyzing species from swine waste-corn fermentation. Develop. Ind. Microbiol. 20:531–540.

Naudé, W. D. T., A. Swanepoel, R. H. Böhmer, and E. Bolding. 1988. Endocarditis caused by Lactobacillus casei ssp. rhamnosus: a case report. S. Afr. Med. J. 73:612–614.

Nes, I. F., J. Brendehaug, and K. O. von Husby. 1988. Characterization of the bacteriophage B2 of Lactobacillus plantarum ATCC 8014. Biochimie 70:423–427.

Nes, J. F., and O. Sorheim. 1984. Effect of infection of bacteriophage in a starter culture during the production of salami dry sausage; a model study. J. Food Sci. 49:337–340.

Nielsen, J. W., and S. E. Gilliland. 1985. Variations in cholesterol assimilation by individual strains of Lactobacillus acidophilus and Lactobacillus casei from human intestine. J. Dairy Sci. 68 (Suppl. 1): 83.

Niido, S. 1982. The bacterial flora of the lower respiratory tract. 2. A comparison of the types of lactobacilli detected in lavage water through the fibre optic bronchoscope and in the feces (Jap.). Yokohama Med. J. 33:243–246.

Niven, C. F., and J. B. Evans. 1957. Lactobacillus viridescens nov. spec. A heterofermentative species that produces a green discoloration of cured meat pigments. J. Bacteriol. 73:758–759.

Nonomura, H., and Y. Ohara. 1967. Die Klassifikation der Apfelsäure-Milchsäure-Bakterien. Mitt. Klosterneuburg 17:449–465.

Oberman, H. 1985. Fermented milks, p. 167–195. In: B. J. B. Wood (ed.), Microbiology of fermented foods. Elsevier Applied Science Publishers, London.

Okada, S., Y. Suzuki, and M. Kozaki. 1979. A new heterofermentative Lactobacillus species with meso-diaminopimelic acid in peptidoglycan, Lactobacillus vaccinostercus Kozaki and Okada sp. nov. J. Gen. Appl. Microbiol. 25:215–221.

Op den Camp, H. J. M., A. Oosterhof, and J. H. Veerkamp. 1985. Interaction of bifidobacterial lipoteichoic acid with human intestinal epithelial cells. Infect. Immun. 47:332–334.

Orla-Jensen, S. 1919. The lactic acid bacteria. Høst and Son, Copenhagen.

Orla-Jensen, S. 1943. Die echten Michsäurebakterien. Ejnar Munksgaard, Copenhagen

Pahlow, G., and H. Honig. 1986. Wirkungsweise und Einsatzgrenzen von Silage-Impfkulturen aus Milchsäurebakterien. Zeitschrift "Das wirtschaftseigene Futter" 32:20–35.

Peake, S. E., and G. Stanley. 1978. Partial characterization of bacteriophage *Lactobacillus bulgaricus* isolated from yoghurt. J. Appl. Bacteriol. 44:321–323.

Pedersen, K., and G. W. Tannock. 1989. Colonization of the porcine gastrointestinal tract by lactobacilli. Appl. Environ. Microbiol. 55:279–283.

Pederson, C. S. 1969. Sauerkraut. Adv. Food Res. 10:233–291.

Pederson, C. S. 1979. Microbiology of food fermentations, 2nd ed. The AVI Publ. Comp., Westport, CT.

Perdigon, G., M. E. N. De Macias, S. Alvarez, G. Oliver, and A. P. De Ruiz Holgado. 1988. Systemic augmentation of the immune response in mice by feeding fermented milks with *Lactobacillus casei* and *Lactobacillus acidophilus*. Immunology 63:17–24.

Pette, J. W., and J. van Beynum. 1943. Boekelscheurbacterien. Rijkslandbauproefstation te hoorn. Versl. Landbouwkd. Onderz. 490:315–346.

Pette, J. W., and H. Lolkema. 1950. Yoghurt. I. Symbiosis and antibiosis in mixed cultures of *Lactobacillus bulgaricus* and *Streptococcus thermophilus*. II. Growth-stimulating factors for *Streptococcus thermophilus*. Neth. Milk Dairy J. 4:197–224.

Peynaud, E., and S. Domercq. 1967. Etude de quelques bacilles homolactiques isolés de vins. Arch. Microbiol. 70:255.

Peynaud, E., and S. Domercq. 1970. Etude de deux cent-cinquante souches de bacilles hétérolactiques isolés de vins. Arch. Microbiol. 70:348.

Phillips, A. B., M. J. Latham, and M. E. Sharpe. 1975. A method for freeze drying rumen bacteria and other strict anaerobes. J. Appl. Bacteriol. 38:319–322.

Pilone, G. J., R. E. Kunkee, and A. D. Webb. 1966. Chemical characterization of wines fermented with various malolactic bacteria. Appl. Microbiol. 14:608–615.

Pivnick, H., and E. Nurmi. 1982. The Nurmi concept and its role in the control of salmonellae in poultry, p. 41–70. In: R. Davies (ed.), Developments in food microbiology, 1. Applied Science Publishers, London.

Premi, L., W. E. Sandine, and P. R. Elliker. 1972. Lactose-hydrolyzing enzymes of *Lactobacillus* species. Appl. Microbiol. 24:51–57.

Price, R. J., and J. S. Lee. 1970. Inhibition of *Pseudomonas* species by hydrogen peroxide producing lactobacilli. J. Milk Food Technol. 33:13.

Priebe, K. 1970. Untersuchungen zur Ursache und zur Vermeidung des Auftretens von Fadenziehen bei Bratheringsmarinaden. Arch. Lebensmittelhyg. 2:1–23.

Radler, F., and C. Yanisses. 1972. Weinsäureabbau bei Milchsäurebakterien. Arch. Microbiol. 82:219–239.

Radmore, K., W. H. Holzapfel, and P. L. Steyn. 1984. Physiology of a pentose-utilizing thermophilic *Lactobacillus*. FEMS Microbiol. Lett. 23:191–194.

Raibaud, P., J. V. Galpin, R. Ducluzeau, G. Mocquot, and G. Oliver. 1973. Le genre *Lactobacillus* dans le tube digestif du rat. I. Charactères des souches homofermentaires isolées de rats holo- et gnotoxeniques. Annales de Microbiologie 124A:83–109.

Rainbow, C. 1975. Beer spoilage lactic acid bacteria, p. 149–158. In: J. G., Carr, C. V. Cutting, and G. C. Whiting (ed.), Lactic acid bacteria in beverages and food. Academic Press, London.

Rammelsberg, M. 1988. Antagonistische Beziehungen bei Milchsäurebakterien. Ph.D. Dissertation, University of Mainz.

Raya, R. R., E. G. Kleeman, J. B. Luchansky, and T. R. Klaenhammer. 1989. Characterization of the temperate bacteriophage ADH and plasmid transduction in *Lactobacillus acidophilus* ADH. Appl. Environ. Microbiol. 55:2206–2213.

Reddy, G. V., K. M. Shahani, B. A. Friend, and R. C. Chandan. 1984. Natural antibiotic activity of *Lactobacillus acidophilus* and *bulgaricus*. III. Production and partial purification of bulgarican from *Lactobacillus bulgaricus*. Cult. Dairy Prod. J. 19:7–11.

Reiter, B. 1978. Antimicrobial systems in milk. J. Dairy Res. 45:131–147.

Reiter, B., and B. G. Härnulv. 1984. Lactoperoxidase antibacterial system. Natural occurrence, biological functions and practical applications. J. Food Prot. 47:724–732.

Reiter, B., V. M. Marshall, and S. M. Phillips. 1980. The antibiotic activity of the lactoperoxidase-thiocyanate-hydrogen peroxide system in the calf abomasum. Res. Vet. Sci. 28:116–122.

Renault, P., C. Gaillardin, and H. Heslot. 1988. Role of malolactic fermentation in lactic acid bacteria. Biochimie 70:375–379.

Reuter, G. 1965. Das Vorkommen von Laktobazillen in Lebensmitteln und ihr Verhalten im menschlichen Intestinaltrakt. Zentralbl. Bakteriol., Parasitenk., Infektionskr. Hyg., Abt. 1 Orig. 197:468–487.

Reuter, G. 1968. Erfahrungen mit Nährböden für die selektive mikrobiologische Analyse von Fleischerzeugnissen. Arch. Lebensmittelhyg. 19:53–57; 84–89.

Reuter, G. 1970. Laktobazillen und eng verwandte Mikroorganismen in Fleisch und Fleischwaren. II. Die Charakterisierung der isolierten Laktobazillenstämme. Fleischwirtschaft 50:954–962.

Reuter, G. 1975. Classification problems, ecology and some biochemical activities of lactobacilli of meat products, p. 221–229. In: J. G. Carr, C. V. Cutting, and G. C. Whiting (ed.), Lactic acid bacteria in beverages and food. Academic Press, London.

Reuter, G. 1981. Psychrotrophic lactobacilli in meat products, p. 253–258. In: T. A. Roberts, G. Hobbs, J. H. B. Christian, and N. Skovgaard (ed.), Psychrotrophic microorganisms in spoilage and pathogenicity. Academic Press, London.

Reuter, G. 1983. *Lactobacillus alimentarius* sp. nov., nom. rev. and *Lactobacillus farciminis* sp. nov., nom. rev.. System. Appl. Microbiol. 4:277–279.

Richards, M., and R. M. Macrae. 1964. The significance of the use of hops in regard to the biological stability of beer. II. The development of resistance to hop resins by strains of lactobacilli. J. Inst. Brewing 70:484–488.

Rizvanov, K., G. Kanarev, N. K. Kh"ng, and J. Tsirkov. 1982. Some strains of lactic acid and propionic acid bacteria as they affect development and productivity of silkworms (Bulg.). Zhivotnov'd Nauki 19:122–128.

Roach, S., D. C. Savage, and G. W. Tannock. 1977. Lactobacilli isolated from the stomach of conventional mice. Appl. Environ. Microbiol. 33:1197–1203.

Robinson, R. K., and A. Y. Tamime. 1981. Microbiology of fermented milks, p. 245–278. In: R. K. Robinson (ed.), Dairy microbiology, vol. 2: The microbiology of milk products. Applied Science Publishers, London.

Rogosa, M. and M. E. Sharpe. 1960. Species differentiation of human vaginal lactobacilli. J. Gen. Microbiol. 23:197–201.

Rogosa, M. J. A. Mitchell, and R. F. Wiseman. 1951. A selective medium for the isolation and enumeration of oral and fecal lactobacilli. J. Bacteriol. 62:132–133.

Romano, A. H., J. D. Trifone, and M. Brustolon. 1979. Distribution of the phosphoenolpyruvatw:glucose phosphotransferase system in fermentative bacteria. J. Bacteriol. 139:93–97.

Ross, A. J., and R. J. Toth. 1974. *Lactobacillus*—a new fish pathogen? Prog. Fish Cult. 36:191.

Russell, C., and F. I. K. Ahmed. 1978. Interrelationships between lactobacilli and streptococci in plaque formation on a tooth in an artificial mouth. J. Appl. Bacteriol. 45:373–382.

Sabine, D. B. 1963. Antibiotic effect of *Lactobacillus acidophilus*. Nature 199:811.

Sakurai, T., T. Takahashi, and H. Arai. 1970. The temperate phages of *Lactobacillus salivarius* and *Lactobacillus casei*. Jap. J. Microbiol. 14:333–336.

Salyers, A. A., S. E. H. West, J. R. Vercellotti, and A. T. D. Wilkins. 1977. Fermentation of mucins and plant polysaccharides by anaerobic bacteria from the human colon. Appl. Environ. Microbiol. 34:529–533.

Sandine, W. E. 1979. Roles of *Lactobacillus* in the intestinal tract. J. Food Prot. 42:259–262.

Sandine, W. E., K. S. Muralidhara, P. R. Elliker, and D. C. England. 1972. Lactic acid bacteria in food and health: A review with special reference to entero-pathogenic *Escherichia coli* as well as certain enteric diseases and their treatment with antibiotics and lactobacilli. J. Milk Food Technol. 35:691–702.

Sarra, P. G., and F. Dellaglio. 1984. Colonization of a human intestine by 4 different genotypes of *Lactobacillus acidophilus*. Microbiologica (Bologna) 7:331–340.

Sarra, P. G., F. Dellaglio, and V. Bottazzi. 1985. Taxonomy of lactobacilli isolated from the alimentary tract of chickens. Syst. Appl. Microbiol. 6:86–89.

Sarra, P. G., M. Vescovo, and M. Fulgoni. 1986. Study on crop adhesion genetic determinant in *Lactobacillus reuteri*. Microbiologica (Bologna) 9:279–286.

Sarra, P. G., M. Vescovo, and V. Bottazzi. 1989. Antagonism and adhesion among isogenic strains of *Lactobacillus reuteri* in the cecum of gnotobiotic mice. Microbiologica (Pavia) 12:69–74.

Satoh, T., K. Narisawa, T. Konno, T. Katoh, J. Fujiyama, A. Tomoe, K. Metoki, K. Hayasaka, and K. Toda. 1982. D-Lactic acidosis in 2 patients with short bowel syndrome. Bacteriological analyses of the fecal flora. Eur. J. Pediatr. 138:324–326.

Savage, D. C. 1977. Microbial ecology of the gastrointestinal tract. Annu. Rev. Microbiol. 31:107–133.

Savage, D. C. 1983. Mechanisms by which indigenous microorganisms colonize gastrointestinal epithelial surfaces. Prog. Food. Nutr. Sci. 7:65–74.

Savage, D. C. 1987. Factors influencing biocontrol of bacterial pathogens in the intestine. Food Technol. 41:82.

Scheirlinck, T., J. Mahillon, H. Joos, H., P. Dhaese, and F. Michiels. 1989. Integration and expression of amylase and endoglucanase genes in the *Lactobacillus plantarum* chromosome. Appl. Environ. Microbiol. 55:2130–2137.

Scherrer, A. 1972. Formation and analysis of diacetyl, 2,3-pentanedione, acetoin and 2,3-butanediol in wort and beer. Wallerstein Lab. Com. 35:5–33.

Schillinger, U., and F.-K. Lücke. 1986. Milchsäurebakterien-Flora auf vakuumverpacktem Fleisch und ihr Einfluss auf die Haltbarkeit. Fleischwirtschaft 66:1515–1520.

Schillinger, U., and F.-K. Lücke. 1987. Identification of lactobacilli from meat and meat products. Food Microbiol. 4:199–208.

Schillinger, U., and F.-K. Lücke. 1989. Antibacterial activity of *Lactobacillus sake* isolated from meat. Appl. Environ. Microbiol. 55:1901–1906.

Schleifer, K. H., and O. Kandler. 1972. Peptidoglycan types of bacterial cell walls and their taxonomic implications. Bacteriol. Rev. 36:407–477.

Schütz, H., and F. Radler. 1984. Anaerobic reduction of glycerol to propanediol-1,3 by *Lactobacillus brevis* and *Lactobacillus buchneri*. Syst. Appl. Microbiol. 5:169–178.

Seale, D. R. 1986. Bacterial inoculants as silage additives. J. Appl. Bacteriol. Symp. Suppl.:9S-26S.

Sechaud, L., P.-J. Cluzel, M. Rousseau, M., A. Baumgartner, and J.-P. Accolas. 1988. Bacteriophages of lactobacilli. Biochimie 70:401–410.

Shahani, K. M., and A. D. Ayebo. 1980. Role of dietary lactobacilli in gastrointestinal microecology. Am. J. Clin. Nutr. 33:2448–2457.

Shahani, K. M., J. R. Vakil, and A. Kilara. 1976. Natural antibiotic activity of *Lactobacillus acidophilus* and *bulgaricus*. I. Cultural conditions for the production of antibiosis. Cult. Dairy Prod. J. 11:14–17.

Shahani, K. M., J. R. Vakil, and A. Kilara. 1977. Natural antibiotic activity of *Lactobacillus acidophilus* and *bulgaricus*. II. Isolation of Acidophilin from *L. acidophilus*. Cult. Dairy Prod. J. 12:8–11.

Shankar, P. A., and F. L. Davies. 1978. Amino acid and peptide utilization by *Streptococcus thermophilus* in relation to yogurt manufacture. Streptococci. Soc. Appl. Bacteriol. Symp. Series 7:402–403.

Sharpe, M. E. 1960. Selective media for the isolation and enumeration of lactobacilli. Lab. Practice 9:223–227.

Sharpe, M. E. 1962. Enumeration and studies of lactobacilli in food products. Dairy Sci. Abstr. 24:165–171.

Sharpe, M. E. 1979. Lactic acid bacteria in the dairy industry. J. Soc. Dairy Tech. 32:9–18.

Sharpe, M. E. 1981. The genus *Lactobacillus*, p. 1653–1679. In: M. P. Starr, H. Stolp, H. G. Trüper, A. Balows, and H. G. Schlegel (ed.), The prokaryotes. Springer-Verlag, Berlin.

Sharpe, M. E., J. H. Brock, K. W. Knox, and A. J. Wicken. 1973b. Glycerol teichoic acid as a common antigenic factor in lactobacilli and some other Gram-positive organisms. J. Gen. Microbiol. 74:119–126.

Sharpe, M. E., and T. F. Fryer. 1965. Media for lactic acid bacteria. Lab. Practice 14:697–701.

Sharpe, M. E., E. I. Garvie, and R. H. Tilbury. 1972. Some slime-forming heterofermentative species of the genus *Lactobacillus*. Appl. Microbiol. 23:389–397.

Sharpe, M. E., L. R. Hill, and S. P. Lapage. 1973a. Pathogenic lactobacilli. J. Med. Microbiol. 6:281–286.

Sharpe, M. E., M. J. Latham, E. I. Garvie, J. Zirngibl, and O. Kandler. 1973c. Two new species of *Lactobacillus* isolated from the bovine rumen, *Lactobacillus ruminis* sp. nov. and *Lactobacillus vitulinus* sp. nov. J. Gen. Microbiol. 77:37–49.

Shaw, B. G., and C. D. Harding. 1984. A numerical taxonomic study of lactic acid bacteria form vacuum-packed beef, pork, lamb and bacon. J. Appl. Bacteriol. 56:25–40.

Shaw, B. G., and C. D. Harding. 1985. Atypical lactobacilli from vacuum-packaged meats: comparison by DNA hybridization, cell composition and biochemical tests with a description of *Lactobacillus carnis* sp. nov. Syst. App. Microbiol. 6:291–297.

Shay, B. J., and A. F. Egan. 1981. Hydrogen sulphide production and spoilage of vacuum-packed beef by a *Lactobacillus*, p. 241–251. In: T. A. Roberts, G. Hobbs, J. H. B. Christian, and N. Skovgaard (ed.), Psychrotrophic microorganisms in spoilage and pathogenicity. Academic Press, London.

Shay, B. J., A. F. Egan, M. Wright, and P. J. Rogers. 1988. Cysteine metabolism in an isolate of *Lactobacillus sake:* Plasmid composition and cysteine transport. FEMS Microbiol. Lett. 56:183–188.

Sherman, L. A., and D. C. Savage. 1986. Lipoteichoic acids in *Lactobacillus* strains that colonize the mouse gastric epithelium. Appl. Environ. Microbiol. 52:302–304.

Shimizu-Kadota, M. 1988. Cloning and expression of the phospho-β-galactosidase genes on the lactose plasmid and the chromosome of *Lactobacillus* C257 in *Escherichia coli*. Biochimie 70:523–529.

Shimizu-Kadota, M., M. Kiwaki, H. Hirokawa, and N. Tsuchida. 1985. ISL1: a new transposable element in *Lactobacillus casei*. Mol. Gen. Genet. 200:193–198.

Shimizu-Kadota, M. and S. Kudo. 1984. Liposome-mediated transfection of *Lactobacillus casei* spheroplasts. Agric. Biol. Chem. 48:1105–1107.

Shimizu-Kadota, M., and T. Sakurai. 1982. Prophage curing in *Lactobacillus casei* by isolation of a thermoinducible mutant. Appl. Environ. Microbiol. 43:1284–1287.

Shimizu-Kadota, M., T. Sakurai, and N. Tsuchida. 1983. Prophage origin of a virulent phage appearing on fermentations of *Lactobacillus casei* S-1. Appl. Environ. Microbiol. 45:669–674.

Shimizu-Kadota, M., and N. Tsuchida. 1984. Physical mapping of the virion and the prophage DNAs of a temperate *Lactobacillus* phage FSW. J. Gen. Microbiol. 130:423–430.

Shindo, K., T. Mizuno, H. Shionoiri, and K. Tarao. 1989. Deconjugation of bile acids by aerobic bacteria isolated from the small intestine of cirrhotic patients. Curr. Ther. Res. Clin. Exp. 45:955–961.

Shovell, F. E., and R. E. Gillis. 1972. Biochemical and antigenic studies of lactobacilli isolated from deep dentinal caries. II. Antigenic aspects. J. Dent. Res. 51:583–587.

Shrago, A. W., B. W. Chassy, and W. J. Dobrogosz. 1986. Conjugal plasmid transfer (pAMβ1) in *Lactobacillus plantarum*. Appl. Environ. Microbiol. 52:574–576.

Shrago, A. W., and W. J. Dobrogosz. 1988. Conjugal transfer of group B streptococcal plasmids and comobilization of *Escherichia coli-Streptococcus* shuttle plasmids to *Lactobacillus plantarum*. Appl. Environ. Microbiol. 54:824–826.

Shrivastava, K. P. 1982. *Bacterium eurydice* strains from bumble bees. J. Invertebr. Pathol. 40:180–185.

Simon, G. L., and S. L. Gorbach. 1984. Intestinal flora in Health and Disease. Gastroenterology 86:174–193.

Simpson, A. C. 1968. Manufacture of Scotch malt whisky. Process Biochem. 3:9–12.

Skerman, V. B. D., V. McGowan, and P. H. A. Sneath. 1980. Approved list of bacterial names. Int. J. Syst. Bacteriol. 30:225–420.

Smiley, M. B., and V. Fryder. 1978. Plasmids, lactic acid production and *N*-acetyl-D-glucosamine fermentation in *Lactobacillus helveticus* subsp. *jugurti*. Appl. Environ. Microbiol. 35:777–781.

Smith, H. W. 1965a. The development of the flora of the alimentary tract in young animals. J. Pathol. Bacteriol. 90:495–513.

Smith, H. W. 1965b. Observations on the flora of the alimentary tract of animals and factors affecting its composition. J. Pathol. Bacteriol. 89:95–122.

Smith, H. W. 1971. The bacteriology of the alimentary tract of domestic animals suffering from *Escherichia coli* infection. Ann. of the New York Academy of Sciences. 176:110–125.

Sozzi, T., H. Hose, D. Amico, and F. Gnaegi. 1982. Bacteriophage of *Leuconostoc oenos*. Internat. Union Microbiol. Soc., 13th Intern. Congr. Microbiol., Boston, Massachusetts, 174, Abstr. P86:2.

Sozzi, T., and R. Maret. 1975. Isolation and characterization of *Streptococcus thermophilus* and *Lactobacillus helveticus* phages from Emmenthal starters. Lait 55:269–288.

Sozzi, T., R. Maret, and J. M. Poulin. 1976. Study of plating efficiency of bacteriophages of thermophilic lactic acid bacteria on different media. Appl. Environ. Microbiol. 32:131–137.

Sozzi, T., J.-M. Poulin, and R. Maret. 1978. Effect of incubation temperature on the development of lactic acid bacteria and their phages. J. Dairy Res. 45:259–265.

Sozzi, T., K. Watanabe, K. Stetter, and M. Smiley. 1981. Bacteriophages of the genus *Lactobacillus*. Intervir. 16:129–135.

Spicher, G. 1983. Baked goods. In: G. Reed (ed.), Food and feed production with microorganisms: Biotechnology vol. 5. Verlag Chemie, Weinheim, Germany.

Spicher, G. 1984. Die Mikroflora des Sauerteiges. XVII. Mitteilung: Weitere Untersuchungen über die Variabilität der Mikroflora handelsüblicher Sauerteig-Starter. Z. Lebensm. Unters. Forsch. 178:106–109.

Spicher, G. 1987. Die Mikroflora des Sauerteiges. XXII. Mitteilung: Die in Weizensauerteigen vorkommenden Lactobacillen. Z. Lebensm. Unters. Forsch. 184:300–303.

Spicher, G., and C. Loenner. 1985. Die Mikroflora des Sauerteiges. XXI. Mitteilung: Die in Sauerteigen schwedischer Bäckereien vorkommenden Lactobacillen. Z. Lebensm. Unters. Forsch. 181:9–13.

Spicher, G., and H. Stephan. 1987. Handbuch Sauerteig: Biologie, Biochemie, Technologie. Behr's Verlag, Hamburg.

Stackebrandt, E., and M. Teuber. 1988. Molecular taxonomy and phylogenetic position of lactic acid bacteria. Biochimie 70:317–324.

Stadhouders, J., and H. A. Veringa. 1967. Texture and flavor defects in cheese caused by bacteria from contaminated rennet. Neth. Milk Dairy J. 21:192–207.

Stamer, J. R. 1975. Recent developments in the fermentation of sauerkraut, p. 267–280. In: J. G. Carr, C. V., Cutting, and G. C. Whiting (ed.), Lactic acid bacteria in beverages and food. Academic Press, London.

Stark, P. L., and A. Lee. 1982. The microbial ecology of the large bowel of breast-fed and formulated infants during the first year of life. J. Med. Microbiol. 15:189–203.

Steffen, C. 1976. Praktische Überlegungen zu den Ursachen der Nachgärung und deren Bekämpfung im Emmentalerkäse. SMF 5:43–50.

Steffen, C. 1979. Vergleichende Untersuchungen in Emmentalerkäsen mit und ohne Nachgärung. VI. Schlußfolgerungen. SMF 8:44–48.

Steinkraus, K. H. 1983. Handbook of indigenous fermented foods. Marcel Dekker Inc., New York.

Stephens, A. M., and J. H. Cummings. 1980. The microbial contribution to human fecal mass. J. Med. Microbiol. 13:45–56.

Stetter, K. O. 1977. Evidence for frequent lysogeny in lactobacilli: temperate bacteriophages within the subgenus *Streptobacterium*. J. Virol. 24:685–689.

Stetter, H., and K. O. Stetter. 1980. *Lactobacillus bavaricus* sp. nov., a new species of the subgenus *Streptobacterium*. Zentralbl. Bakteriol. Mikrobiol. Hyg. Abt. I Orig. C 1:70–74

Stewart, C. S., and S. H. Duncan. 1985. The effect of avoparein on cellulolytic bacteria of the ovine rumen. J. Gen. Microbiol. 131:427–435.

Stewart, C. S., S. H. Duncan, and H. J. Flint. 1988. Characteristics of isolates of *Lactobacillus fermentum* from the rumen of sheep. Lett. Appl. Microbiol. 6:133–135.

Stewart-Tull, D. E. S. 1964. Evidence that vaginal lactobacilli do not ferment glycogen. Am. J. Obstet. Gynaecol. 88:676–679.

Stirling, A. C., and R. Whittenbury. 1963. Sources of the lactic acid bacteria occurring in silage. J. Appl. Bacteriol. 26:86–90.

Sugihara, T. F. 1978. Microbiology of the soda cracker process. I. Isolation and identification of microflora. J. Food Prot. 41:977–979.

Sugihara, T. F. 1985. Microbiology of breadmaking, p. 249–263. In B. J. B. Wood (ed.), Microbiology of fermented foods. Elsevier Applied Science Publishers, London.

Sullivan, A., and U. Schroeder. 1989. Systematic analysis of gingival state and salivary variables as predictors of caries from 5 to 7 years of age. Scand. J. Dent. Res. 97:25–32.

Sumner, S. S., W. Speckhard, E. B. Somers, and S. L. Taylor. 1985. Isolation of histamine-producing *Lactobacillus buchneri* from Swiss cheese implicated in a food poisoning outbreak. Appl. Environ. Microbiol. 50:1094–1096.

Suzuki, K., Y. Kodama, and T. Mitsuoka. 1989. Stress and intestinal flora. Bifidobacterium and Microflora 8:23–38.

Tagg, J. R., A. S. Dajanii, and L. W. Wannamaker. 1976. Bacteriocins of Gram-positive bacteria. Bacteriol. Rev. 40:722–756.

Takahashi, M., M. Kametaka, and T. Mitsuoka. 1982. Influence of diets low in protein or lysine on the intestinal flora of chicks with reference to cecal contents. J. Nutr. Sci. Vitaminol. 28:501–510.

Talarico, T. L., I. A. Casas, T. C. Chung, and W. J. Dobrogosz. 1988. Production and isolation of Reuterin, a growth inhibitor produced by *Lactobacillus reuteri*. Antimicrob. Agents Chemother. 32:1854–1858.

Talarico, T. L., and W. J. Dobrogosz. 1989. Chemical characterization of an antimicrobial substance produced by *Lactobacillus reuteri*. Antimicrob. Agents Chemother. 33:674–679.

Tannock, G. W. 1983. Effect of dietary and environmental stress on the gastrointestinal microbiota, p. 517–539. In: D. J. Hentges (ed.), Human intestinal microflora in health and disease. Academic Press, New York.

Tannock, G. W. 1989. Biotin-labeled plasmid DNA-probes for detection of epithelium-associated strains of lactobacilli. Appl. Environ. Microbiol. 55:461–464.

Tannock, G. W., R. Blumershine, and R. Archibald. 1987. Demonstration of epithelium-associated microbes in the oesophagus of pigs, cattle, rats and deer. FEMS Microbiol. Ecol. 45:199–203.

Tannock, G. W., M. P. Dashkevicz, and S. D. Feighner. 1989. Lactobacilli and bile salt hydrolase in the murine intestinal tract. Appl. Environ. Microbiol. 55:1848–1895.

Tannock, G. W., O. Szylit, Y. Duval, and P. Raibaud. 1982. Colonization of tissue surfaces in the gastrointestinal tract of gnotobiotic animals by lactobacillus strains. Can. J. Microbiol. 28:1196–1198.

Ten Brink, B., and W. N. Konings. 1982. The electrochemical proton gradient and lactate concentration gradient in *Streptococcus cremoris* grown in batch culture. J. Bacteriol. 152:682–686.

Terzaghi, B. E., and W. E. Sandine. 1975. Improved medium for lactic streptococci and their bacteriophages. Appl. Microbiol. 29:807–813.

Thomas, S. B., R. G. Druce, G. J. Peters, and D. G. Griffiths. 1967. Incidence and significance of thermoduric bacteria in farm milk supplies: a reappraisal and review. J. Appl. Bacteriol. 30:265–298.

Thomas, S. B., J. W. Egdell, L. F. L. Clegg, and W. A. Cuthbert. 1950. Thermoduric organisms in milk. I. A review of the literature. Proc. Soc. Appl. Bacteriol. 13:27–64.

Thomas, T. D., and G. G. Pritchard. 1987. Proteolytic enzymes of dairy starter cultures. FEMS Microbiol. Rev. 46:245–268.

Thompson, J. 1988. Lactic acid bacteria: model systems for in vivo studies of sugar transport and metabolism in Gram-positive organisms. Biochimie 70:325–336.

Thompson, J. K., and M. A. Collins. 1988. Evidence for the conjugal transfer of the broad host range plasmid pIP501 into strains of *Lactobacillus helveticus*. J. Appl. Bacteriol. 65:309–319.

Thornley, M. J. 1957. Observations on the microflora of minced chicken meat irradiated with 4MeV cathode rays. J. Appl. Bacteriol. 20:286–298.

Thornley, M. J., and M. E. Sharpe. 1959. Microorganisms from chicken meat related to both lactobacilli and aerobic sporeformers. J. Appl. Bacteriol. 22:368–376.

Tilbury, R. H. 1975. Occurrence and effect of lactic acid bacteria in the sugar industry, p. 177–191. In: J. G. Carr, C. V. Cutting, and G. C. Whiting (ed.), Lactic acid bacteria in beverages and food. Academic Press, London.

Tina, R. 1987. A taxonomic evaluation of *Betabacterium* Orla-Jensen. M.Sc. (Microbiol.) Thesis, University of Pretoria, South Africa.

Toba, T., S. Abe, K. Arihara, and S. Adachi. 1986. A medium for the isolation of capsular bacteria from kefir grains. Agric. Biol. Chem. 50:2673–2674.

Tohyama, K. 1973. Studies on temperate phages of *Lactobacillus salivarius*. II. Mode of action of defective phage 208 on nonlysogenic and homologous lysogenic *Lactobacillus salivarius* strains. Jap. J. Microbiol. 17:173–180.

Tohyama, K., T. Sakurai, H. Arai, and A. Oda. 1972. Studies on temperate phages of *Lactobacillus salivarius*. Jap. J. Microbiol. 16:385–395.

Tonge, R. J., A. C. Baird-Parker, and J. J. Cavett. 1964. Chemical and microbiological changes during storage of vacuum packed sliced bacon. J. Appl. Bacteriol. 27:252–264.

Tortuero, F., A. Brenes, and J. Rioperez. 1975. The influence of intestinal (cecal) flora on serum cholesterol. Am. J. Clin. Nutr. 36:1106–1111.

Tramer, J. 1966. Inhibitory effect of *Lactobacillus acidophilus.* Nature 211:204–205.

Trautwetter, A., P. Ritzenthaler, T. Alatossava, and M. Mata-Gilsinger. 1986. Physical and genetic characterization of the genome of *Lactobacillus lactis* bacteriophage LL-H. J. Virol. 59:551–555.

Trevors, K. E., R. A. Holley, and A. G. Kempton. 1983. Isolation and characterization of a *Lactobacillus plantarum* bacteriophage isolated from a meat starter culture. J. Appl. Bacteriol. 54:281–288.

Trevors, K. E., R. A. Holley, and A. G. Kempton. 1984. Effect of bacteriophage on the activity of lactic acid starter cultures used in the production of fermented sausage. J. Food Sci. 49:650–653.

Upreti, G. C., and R. D. Hinsdill. 1975. Production and mode of action of Lactocin 27: Bacteriocin from a homofermentative lactobacillus. Antimicrob. Agents Chemother. 7:139–145.

Vanderas, A. P. 1986. Bacteriologic and nonbacteriologic criteria for identifying individuals at high risk of developing dental caries: a review. J. Public Health Dent. 46:106–113.

Vaughn, R. H. 1985. The microbiology of vegetable fermentations, p. 49–110. In: B. J. B. Wood (ed.), Microbiology of fermented foods, vol. 1. Elsevier Applied Science Publishers, London.

Vescovo, M., L. Morelli, and V. Bottazzi, 1982. Drug resistance plasmids in *Lactobacillus acidophilus* and *Lactobacillus reuteri.* Appl. Environ. Microbiol. 43:50–56.

Vescovo, M., L. Morelli, V. Bottazzi, and M. J. Gasson. 1983. Conjugal transfer of broad host range plasmid pAMβ1 into enteric species of lactic acid bacteria. Appl. Environ. Microbiol. 46:753–755.

Vincent, J. G., R. C. Veomett, and R. F. Riley. 1959. Antibacterial activity associated with *Lactobacillus acidophilus.* J. Bacteriol. 78:477–484.

Visser, R., W. H. Holzapfel, J. J. Bezuidenhout, and J. M. Kotze. 1986. Antagonism of lactic acid bacteria against phytopathogenic bacteria. Appl. Environ. Microbiol. 52:552–555.

Voigt, M. N., and R. R. Eitenmiller. 1978. Role of histidine and tyrosine decarboxylases and mono- and diamine oxidases in amine buildup in cheese. J. Food Prot. 41:182–186.

Von Aulock, M. H. M. 1981. Voorkoms es spesic-eienskappe van *Lactobacillus acidophilus* in die spysverteringskanaal van die vark. M.Sc. (Microbiol.) Thesis, University of Pretoria, South Africa.

Von Aulock, M. H. M., and W. H. Holzapfel. 1987. Frequency and occurrence of *Lactobacillus acidophilus* in the gut of the pig, as indicated by its presence in the feces. Onderstepoort J. vet. Res. 54:581–584.

Von Holy, A. 1983. Bacteriological studies on the extension of the shelf life of raw minced beef. M.Sc. (Microbiol.) Thesis, University of Pretoria, South Africa.

Von Holy, A., and W. H. Holzapfel. 1989. Spoilage of vacuum packaged processed meats by lactic acid bacteria, and economic consequences. Proc. Xth WAFVH Intern. Symp., 6–9 July 1989, Stockholm, Sweden.

Watanabe, K., K. Ishibashi, Y. Nakashima, and T. Sakurai. 1984. A phage-resistant mutant of *Lactobacillus casei* which permits phage adsorption but not genome injection. J. Gen. Virol. 65:981–986.

Watanabe, K., S. Takesue, K. Jin-Nai, and T. Yoshikawa. 1970. Bacteriophage active against the lactic acid beverage-producing bacterium *Lactobacillus casei.* Appl. Microbiol. 20:409–415.

Watanabe, K., S. Takesue, and K. Ishibashi. 1977. Reversibility of the adsorption of bacteriophage P1-1 to the cell walls isolated from *Lactobacillus casei.* J. Gen. Virol. 34:189–194.

Watkins, B. A. and B. F. Miller. 1983. Colonization of *Lactobacillus acidophilus* in gnotobiotic chicks. Poultry Sci. 62:2152–2157.

Weiller, H. G., and F. Radler. 1970. Milchsäurebakterien aus Wein und von Rebenblättern. Zentralbl. Bakteriol. Parasitenkd. Infektionskr. Hyg. Abt. II 124:707–732.

Weiss, N., U. Schillinger, M. Laternser, and O. Kandler. 1981. *Lactobacillus sharpeae* sp. nov. and *Lactobacillus agilis* sp. nov., two new species of homofermentative, *meso*-diaminopimelic acid-containing lactobacilli isolated from sewage. Zentralbl. Bakteriol. Hyg., Abt. I Orig. C 2:242–253.

Weiss, N., U. Schillinger, and O. Kandler. 1983. *Lactobacillus lactis, Lactobacillus leichmannii* and *Lactobacillus bulgaricus,* subjective synonyms of *Lactobacillus delbrueckii* subsp. *lactis* comb. nov. and *Lactobacillus delbrueckii* subsp. *bulgaricus* comb. nov. System. Appl. Microbiol. 4:552–557.

Weiss, N., and U. Schillinger. 1984. *Lactobacillus sanfrancisco* sp. nov., nom. rev. System. Appl. Microbiol. 5:230–232.

West, C. A., and P. J. Warner. 1988. Plantacin B, a bacteriocin produced by *Lactobacillus plantarum* NCDO 1193. FEMS Microbiol. Lett. 49:163–165.

Wheater, D. M., A. Hirsch, and A. T. R. Mattick. 1951. "Lactobacillin," an antibiotic from lactobacilli. Nature 168:659.

Wheater, D. M., A. Hirsch, and A. T. R. Mattick. 1952. Possible identity of "Lactobacillin" with hydrogen peroxide produced by lactobacilli. Nature 170:623–624.

Whiting, G. C. 1975. Some biochemical and flavour aspects of lactic acid bacteria in ciders and other alcoholic beverages, p. 69–85. In: J. G. Carr, C. V. Cutting, and G. C. Whiting (ed.), Lactic acid bacteria in beverage and food. Academic Press, London.

Whiting, G. C., and R. A. Coggins. 1974. A new nicotinamide-adenine dinucleotide dependent hydro-aromatic dehydrogenase of *Lactobacillus plantarum* and its role in formation of (−) t-3,t-4-dihydrocyclohexane-c-carboxylate. Biochem. J. 141:35–42.

Whitt, D. D., and D. C. Savage. 1987. Lactobacilli as effectors of host functions: no influence on the activities of enzymes in enterocytes of mice. App. Environ. Microbiol. 53:325–330.

Whittenbury, R. 1964. Hydrogen peroxide formation and catalase activity in the lactic acid bacteria. J. Gen. Microbiol. 35:13–26.

Wiesenberger, A., E. Kolb, J. A. Schildmann, and H.-M. Dechent. 1986. Die Lactofermentation natürlicher Substrate mit niedrigen pH-Werten. Chem. Mikrobiol. Technol. Lebensm. 10:32–36.

Corynebacteriaceae. In the eighth edition of *Bergey's Manual of Determinative Bacteriology* (Buchanan and Gibbons, 1974), the genus *Listeria*, now containing four species *(L. monocytogenes, L. denitrificans, L. grayi,* and *L. murrayi)* was treated as a "genus of uncertain affiliation" in the section that contained the family Lactobacillaceae. Further numerical, taxonomic studies (Jones, 1975; Wilkinson and Jones, 1977; Feresu and Jones, 1988) and chemical studies (Collins and Jones, 1981; Kamisango et al., 1982; Fiedler and Seger, 1983, Fiedler et al., 1984; Ruhland and Fiedler, 1987; Feresu and Jones, 1988) all reinforced the distinctness of *Listeria* from the corynebacteria and showed that phenotypically *Listeria* resembled the genus *Brochothrix.* The suprageneric position of the genus remained unclear, however, until the 16S rRNA cataloging studies of Stackebrandt et al. (1983) demonstrated that *Listeria* (but not *L. denitrificans)* was a distinct taxon within the *Clostridium-Lactobacillus-Bacillus* branch of the bacterial phylogeny constructed by Stackebrandt and Woese (1981). Within this phylogeny, *Listeria* is most closely related to the genus *Brochothrix* (Ludwig et al., 1984).

The intrageneric composition of the genus has also been problematical. On the basis of growth characteristics, biochemical reactions, and chemical studies, including DNA composition and DNA-DNA hybridization, it was generally agreed that the species *L. denitrificans,* which was first isolated from cooked blood by Sohier et al., (1948), was misclassified [see Seeliger and Jones (1986) for detailed discussion]. Seeliger and Jones (1986) treated the taxon with the genus *Listeria* as incertae sedis, only because it could not be placed with confidence in any other described taxonomic grouping. However, 16S rRNA cataloging studies have resolved the problem, and *Listeria denitrificans* is now classified in a new genus, *Jonesia,* as *J. denitrificans* (Rocourt et al., 1987a).

Seeliger and Jones (1986) also treated *L. grayi* and *L. murrayi* as incertae sedis because of the conflict between the interpretation of numerical taxonomic and chemical studies on the one hand and those of DNA-DNA hybridization studies on the other. While the results of the former studies all indicated that these species are members of the genus *Listeria,* the results of the DNA-DNA hybridization studies supported the suggestion of Stuart and Welshimer (1974) that *L. grayi* and *L. murrayi* be reclassified as a new genus for which they proposed the name *"Murraya"* [see Seeliger and Jones (1986) for references and fuller discussion]. The problem has been resolved by the results of 16S

rRNA cataloging studies that showed that *L. murrayi* is closely related (similarity coefficient (S_{AB}) value, 0.73) to *L. monocytogenes* (Rocourt et al., 1987b).

It is only since the DNA-DNA hybridization studies of Rocourt et al., (1982a) that the taxonomy of some listeriae allocated over the years to the species *L. monocytogenes* has been resolved. Seeliger and Welshimer (1974) and Welshimer (1981) noted the unsatisfactory classification within *L. monocytogenes* of strains that differed from the original description of the species (Murray et al., 1926); some exhibited pronounced β-hemolysis while others, isolated mainly from fecal and environmental sources, were nonhemolytic and also nonpathogenic for mice. Stuart and Welshimer (1973, 1974) demonstrated two DNA-DNA hybridization groups among strains labelled *L. monocytogenes* but could not distinguish the groups by any other criteria. Seeliger (1981) proposed that some nonhemolytic and nonpathogenic strains be recognized as a new species, *L. innocua,* a name originally used for these strains by Seeliger and Schoofs (1979). The comprehensive DNA-DNA hybridization study by Rocourt et al. (1982a) of a large collection of hemolytic and nonhemolytic strains, all designated *L. monocytogenes,* indicated the existence of five DNA-DNA homology groups. One of these groups contained the type strain of *L. monocytogenes;* a second contained strains previously designated *L. innocua;* and a third group contained strains first described by Ivanow in 1962 (Ivanov, 1975) as exhibiting a pronounced β-hemolysis and now named *L. ivanovii* (Seeliger et al., 1984). The other two genomic groups were later named *L. seeligeri* and *L. welshimeri* (Rocourt and Grimont, 1983). Thus, as noted by Rocourt et al. (1987b), the genus *Listeria* currently contains seven species that represent two closely related but distinct lines of descent; one line includes *L. monocytogenes, L. innocua, L. ivanovii, L. seeligeri,* and *L. welshimeri;* the other *L. grayi* and *L. murrayi.* It should be noted that the separate species status of the last two species is doubtful. Some authors, notably Stuart and Welshimer (1974) and Feresu and Jones (1988), consider *L. murrayi* to be a subspecies of *L. grayi* but this reclassification has not been formally proposed.

Habitats

Until about 1960, *L. monocytogenes* was associated almost exclusively with infections in animals and less frequently in humans (Gray and Killinger, 1966; Seeliger, 1961; Welshimer,

As noted previously, there are many other published selective media for the detection of listeriae. That of Van Netten et al. (1988) has been found to be particularly useful in some laboratories but is not as widely used as the Oxford medium. Skalka and Smola (1983) recommend the incorporation of washed rabbit erythrocytes together with an extract of *Rhodococcus equi* (equi factor) into a medium containing acriflavin and nalidixic acid for the rapid selective detection of hemolytic (pathogenic) *Listeria* species. Blanco et al. (1989) have applied a "top layer" of agar containing washed sheep erythrocytes to listeriae cultured on a variety of selective media for 24–48 h. Before addition of the top layer containing the sheep cells, the plates were cooled at 4°C for 2 h; after addition, the plates were incubated at 30°C for 14 h. The authors report this method results in more distinct hemolysis with *L. monocytogenes* and *L. ivanovii* (Blanco et al., 1989).

Preservation of Cultures

Listeria cultures may be preserved for short periods (6 months to a year) by stab inoculation into nutrient agar in small, screw-capped bottles. After overnight incubation at 30°C, the caps should be screwed tightly to prevent evaporation and then stored in the dark at room, or preferably refrigerator, temperature. Longer-term preservation (over 10 years) may be achieved by freezing on glass beads at − 60 to −70°C (Jones et al., 1984). Cultures may also be preserved by freeze-drying (lyophilization).

Identification

Identification of new isolates as *Listeria* species requires examination of colonial and cellular morphology and staining reactions, growth at 37°C, relationships to oxygen and catalase production, together with a number of conventional biochemical tests such as acid production from various sugars and hydrolysis of esculin. Serological identification may be necessary. Chemical composition is also helpful but not necessary for routine identification. The features that are the most useful in differentiating listeriae from other Gram-positive, nonspore-forming, rod-shaped bacteria are listed in Table 1.

The colonial morphology of listeriae is similar for all species. Colonies on nutrient agar after incubation for 24–48 h at 37°C are 0.5 to 1.5 mm in diameter, round, translucent with a watery consistency, and low convex with a finely textured surface and entire margins. Colonies removed from a plate often leave an impression on the agar; although often sticky, they emulsify easily. When grown on a clear, solid medium, such as tryptose agar for 24–48 h and viewed with a dissecting microscope using obliquely transmitted light (Henry, 1933) as described by Gray (1957) (see "Isolation"), the colonies exhibit a distinctive blue-green sheen. Older cultures, particularly those of *L. grayi* and *L. murrayi,* may have an orange-yellow tinge, especially around the edge. With normal illumination, the colonies appear bluish gray on such media. Colonies of most strains show a tendency to dissociate on laboratory culture. A full description of colonial dissociation and the various colonial forms is given by Gray and Killinger (1966) and Seeliger (1961), but this is not usually a problem in identification of new isolates. Rough cultures do not exhibit a blue-green sheen when viewed by obliquely transmitted light.

The cellular morphology of listeriae is not particularly distinctive. The Gram-positive rods are short, 0.4–0.5 μm in diameter and 1–2 μm in length, with rounded ends. They are not acid-fast, do not form spores, and capsules are not present. The cells usually occur singly or in short chains but may also show palisade and Y-form arrangements that superficially resemble corynebacteria. Some cells from older cultures and from direct smears from partially treated clinical cases may stain Gram-negative. In older or rough cultures, filaments up to 20 μm in length can lead to confusion with *Erysipelothrix,* and the coccoid forms frequently seen in liquid cultures or in direct smears from infected tissue can be confused with enterococci or streptococci.

Catalase and motility tests are useful for initial screening. A positive catalase test on cultures grown at 37°C excludes enterococci, streptococci, the majority of lactobacilli (catalase production if present is very weak), *Carnobacterium, Erysipelothrix,* and *Brochothrix* (members of the last genus do not grow at 37°C and are catalase-positive only when grown at 25°C). A portion of growth on nutrient, not blood, agar, is removed and rubbed into a drop of H_2O_2 (3% concentration or greater) on a glass slide and observed for vigorous bubbling.

Motility excludes most species of small, Gram-positive rods. Motility tests should always be carried out at 20–25°C; at 37°C flagellar development is so poor that listeriae appear nonmotile. Two tubes of brain heart infusion (BHI) broth (Difco) or a similar medium are inoculated directly from isolated colonies and incubated at 37°C and 20–25°C.

Table 1. Features most useful in differentiating *Listeria* from morphologically similar genera.[a,b]

Property	*Listeria*	*Brochothrix*	*Carnobacterium*	*Lactobacillus*	*Erysipelothrix*	*Kurthia*
Growth at 35°C	+	−	+	+	+	+
Motility	+	−	−[c]	−[d]	−	+
Catalase	+	+	−	−[e]	−	+
Facultatively anaerobic	+	+	+	+	+	−
Acid from glucose	+	+	+	+	+	−
H₂S production	−[f]	−	−	−	+	−[f]
Major peptidoglycan diamino acid	*meso*-A₂pm[g]	*meso*-A₂pm	*meso*-A₂pm	L-Lysine, *meso*-A₂pm, or ornithine	L-Lysine	L-Lysine
Major menaquinone	MK-7	MK-7	−	−[h]	−	MK-7
Major fatty acid types[i]	S, A, I	S, A, I	S, U (C)	S, U (C)	S, A, I, U	S, A, I
GC content (mol%)	36–42	36–38	33–37	32–53	36–40	36–38

[a]Adapted from Collins et al., 1987; Feresu and Jones, 1988; Sneath and Jones, 1976; Talon et al., 1988.
[b]Symbols: +, 90% or more of strains positive; −, 90% or more of strains negative.
[c]Most strains are nonmotile, but *C. mobile* strains are motile (Collins et al., 1987).
[d]Most strains are nonmotile, but a few motile strains occur.
[e]Some strains give a positive catalase reaction.
[f]Weak production of H₂S by some strains of *Listeria grayi* and *Kurthia zopfii* has been reported.
[g]*meso*-A₂pm, *meso*-diaminopimelic acid.
[h]A very few lactobacilli contain menaquinones.
[i]S, straight-chain saturated; A, *anteiso*-methyl-branched; I, *iso*-methyl-branched; U, monounsaturated; C, cyclopropane-ring fatty acids. Those in parentheses may also be present.

Hanging drop preparations, prepared when turbidity becomes apparent, are observed microscopically. Few if any organisms will be motile at 37°C whereas many will be motile at the lower temperature, with a number of the rods exhibiting a characteristic tumbling motility. In doubtful cases, motility can be demonstrated by inoculating strains into 1/4-strength BHI agar in a U-shaped tube. The medium in one arm of the U tube is inoculated and incubated at 20–25°C. Progress of motile bacteria can be followed by a clouding of the medium just below the surface of the uninoculated arm of the U tube. Alternatively, the same medium in a deep tube may be lightly inoculated with a straight wire to the bottom of the tube; after about one week at 20°C, an "umbrella" layer of growth is seen just below the surface of the medium.

All listeriae are facultatively anaerobic and produce acid but no gas from glucose and some other sugars (features which distinguish *Listeria* species from the strictly aerobic genus *Kurthia*). Additional, useful, differential criteria are a positive Voges-Proskauer (VP) test, esculin hydrolysis (the basis of the black colony production on Oxford medium, see "Isolation"), alkaline phosphatase positive, urease negative, and oxidase negative. Detailed descriptions of biochemical activities are given by Seeliger and Jones (1986) and Rocourt and Catimel (1985).

Further Differentiation of *Listeria* Species

The characters most useful for distinguishing the species are hemolysis; production of acid from mannitol, rhamnose, D-xylose, and α-methyl-D-mannoside; and nitrate reduction. Table 2 lists the characters of the *Listeria* species.

Observation of hemolysis depends on the basal medium used and on the source of the blood (Skalka et al., 1982a; Pine et al., 1987). *Listeria monocytogenes*, *L. seeligeri* and *L. ivanovii* lyse a range of mammalilan erythrocytes but sheep or horse blood are recommended. Hemolysis is examined on sheep or horse blood agar plates by streaking or stabbing cultures on blood agar base (Difco) containing 5% (v/v) blood. On streak plates, colonies of *L. monocytogenes* produce narrow, clear zones of a β-type hemolysis that sometimes barely extend beyond the edge of the colony. Very weak β-type hemolysis is produced by strains of *L. seeligeri*. Hemolysis by strains of both these species is usually enhanced by stabbing the culture into the same medium because the listeriolysin is a sulfhydryl-activated (oxygen-sensitive) cytotoxin. Weak or doubtful hemolysis in both species can be potentiated by a modified CAMP (Christie et al., 1944) test. A suspected hemolytic strain is streaked at right angles (1–2 mm apart) to a β toxin producing strain of *Staphylococcus aureus* on a washed sheep blood (5% v/v) agar plate. After incubation for 24–48 h, hemolytic strains of *L. monocytogenes* and *L. seeligeri* exhibit enhanced hemolysis within the zone of toxin produced by the *S. aureus*. In contrast, *L. ivanovii* produces a wide zone or multiple zones of hemolysis on sheep blood (5%

Table 2. Differential features of the *Listeria* species.[a]

Property	L. monocytogenes	L. innocua	L. ivanovii	L. seeligeri	L. welshimeri	L. grayi	L. murrayi
β-Hemolysis[b,d]	+	−	+	+	−	−	−
CAMP test (*Staphylococcus aureus*)[d]	+	−	−	+	−	−	−
CAMP test (*Rhodococcus equi*)[d]	−	−	+	−	−	−	−
Acid production from:							
Gluconate[d]	−	−	−	−	−	+	+
Lactose	d	+	+	+	+	+	+
D-Lyxose	d	d	−	d	d	−	+
Mannitol[d]	−	−	−	−	−	+	+
Melezitose	d	+	d	d	d	−	−
α-Methyl-D-glucoside	+	+	+	+	+	+	−
α-Methyl-D-mannoside[d]	+	+	−	−	+	+	+
L-Rhamnose[d]	+	d	−	−	d	−	−
Ribose	d	d	+	d	d	+	+
Saccharose	d	d	d	d	+	−	−
D-Tagatose	d	d	−	d	d	−	−
D-Turanose	d	d	d	d	d	−	+
D-Xylose[d]	−	−	+	+	+	−	−
Hydrolysis of:							
Esculin	+	+	+	+	+	+	+
Hippurate	+	+	+	+	+	−	−
Phosphatase	+	+	+	+	+	+	+
Acid phosphatase (API ZYM)[c]	+	+	+	+	+	−	−
Phosphoamidase (API ZYM)	+	+	+	+	+	−	−
Reduction of nitrate to nitrite[d]	−	−	−	−	−	−	+
H₂S production	−	−	−	−	−	d	d
Pathogenicity for mice	+	−	+	−	−	−	−
GC content (mol%)	37–39	36–38	37–38	36	36	41–42	41–42.5

Symbols: +, 90% or more of strains positive; −, 90% or more strains negative; d, 11 to 89% of strains positive.
[a]Adapted from Seeliger and Jones (1986); Rocourt and Catimel (1985).
[b]Some strains of *L. monocytogenes* do not exhibit hemolysis on horse, sheep, and bovine blood; the type strain ATCC 15313 is nonhemolytic; *L. ivanovii* exhibits wide or multiple zones of hemolysis. *L. seeligeri* is very weakly hemolytic.
[c]API ZYM, API ZYM gallery; available from API La Balme les Grottes, Montalieu Vercieu, France.
[d]These features are especially useful for identification.

v/v) agar plates. No enhancement of *L. ivanovii* hemolysis is apparent when the modified CAMP test is performed with *S. aureus* but a characteristic "spadelike" enhancement of hemolysis occurs when *L. ivanovii* (but not *L. monocytogenes* or *L. seeligeri*) is plated adjacent to *Rhodococcus equi* (Table 2). Although not widely used, it should be noted that Skalka et al. (1982b) recommend the incorporation of the exosubstance of *Rhodococcus equi* in prepurified form into sheep blood agar plates for enhanced hemolysis of all hemolytic species of *Listeria*.

The four other species of *Listeria* are nonhemolytic but a weak direct contact hemolysis underneath the colonies may be apparent with nonhemolytic strains of *L. innocua* and the occasional nonhemolytic *L. monocytogenes*. Contact hemolysis, therefore, does not necessarily differentiate between strains of *L. innocua* and nonhemolytic *L. monocytogenes*.

Some workers prefer to examine hemolytic activity by the "well" hemolysis test. In this procedure, fresh sterile defibrinated sheep blood is held overnight at 2–4°C, the serum is carefully removed, and a 2% (v/v) suspension of the erythrocytes is made in phosphate-buffered saline (pH 7.4). 0.1 ml of the erythrocyte suspension plus 0.1 ml of the isolate to be investigated (cultured overnight in brain heart infusion broth at 37°C) are placed in the well of a microtiter plate (the wells must be V-shaped, not round) and incubated for 30–60 min at 37°C. Control wells containing uninoculated broth and *L. innocua* (negative controls) and *L. monocytogenes* (positive control) should always be included.

The composition of the basal medium and the pH indicator used are important for testing for the production of acid from carbohydrates. A suitable, reliable, basal medium is purple broth base (Difco) supplemented with 0.5 or 1% (w/v) of the carbohydrate. Some workers, however, prefer a peptone-water medium with phenol red as indicator, especially for the production of acid from L-rhamnose, D-xylose, and α-methyl-D-mannoside. Where possible, all carbohydrates should be sterilized by filtration and not by autoclaving. A more reliable but expensive method for testing for acid production from carbohydrates is the AP1 50 CH gallery (AP1, Montalieu Vercieu, France). Information on all the 49 test substrates in this system is not required for routine identification. Additional and differential tests for the seven species are listed in Tables 1 and 2.

The antigenic composition of listeriae is helpful but not essential for routine identification. The Seeliger-Donker-Voet scheme is based on that devised by Paterson (1940) who recognized four serovars on the basis of H and O antigens of strains designated *L. monocytogenes*. The scheme was later extended by Donker-Voet (1957), Seeliger (1953), Seeliger and Schoofs (1979), and Seeliger and Höhne (1979). Sixteen serovars are now recognized among the five antigenically related *Listeria* species. *L. grayi* and *L. murrayi* are antigenically different from the other species but closely related to each other (Table 3).

Some minor shared antigens result in serological cross-reactions between the somatic (0) antigens of listeriae and some Gram-positive bacteria such as staphylococci, enterococci, and even *Escherichia coli*. It is therefore important that only suitably absorbed antisera be used in serological tests (Seeliger, 1961); somatic and flagellar (H) antigens and antisera should be prepared as described by Seeliger and Höhne (1979). With the exception of the serovar 5 strains, all of which examined to date are members of the species *L. ivanovii*, there is no strict correlation between serovar and species. *Listeria innocua* and *L. welshimeri* can usually be differentiated serologically from *L. monocytogenes*, but differentiation between *L. monocytogenes* and *L. seeligeri* is not possible since both species share the same antigenic composition with several serovars each (Seeliger and Langer, 1989).

Serological methods alone are not useful for identification to species level (Table 4). The human and animal pathogenic species *L. monocytogenes* can be subdivided into at least 13 serovars but this serotyping scheme is of little value for practical epidemiological discrimination as a large proportion of strains from human and animal infections belong to one of a small number of serovars (Seeliger and Höhne, 1979; McLauchlin, 1987).

Sodium dodecyl sulfate-polyacrylamide gel electrophoresis (SDS-PAGE) studies of whole cell protein of all *Listeria* species indicate little potential for species discrimination. Lamont et al. (1986) found that the polypeptide profiles of representative strains of *L. monocytogenes, L. innocua, L. ivanovii, L. seeligeri,* and *L. welshimeri* were all very similar but distinct from those of *L. grayi,* which, not surprisingly, were very similar to those of *L. murrayi* (Lamont et al., 1986).

Table 3. Major serovars of the genus *Listeria*.[a]

Identification scheme												
Paterson	Seeliger-Donker-Voet	O antigens present										H antigens present
1	1/2a	I II (III)										A B
	1/2b	I II (III)										A B C
2	1/2c	I II (III)										B D
3	3a	II (III) IV										A B
	3b	II (III) IV					(XII XIII)					A B C
	3c	II (III) IV					(XII XIII)					B D
4	4a	(III)	(V)		VII	IX						A B C
	4ab	(III)	V VI	VII		IX X						A B C
	4b	(III)	V VI									A B C
	4c	(III)	V	VII								A B C
	4d	(III)	(V) VI		VIII							A B C
	4e	(III)	V VI	(VIII) (IX)								A B C
	5	(III)	(V) VI	(VIII)	X							A B C
	7?	(III)					XII XIII					A B C
	6a (4f)	(III)	V (VI) (VII)		(IX)			XV				A B C
	6b (4g)	(III)	(V) (VI) (VII)		IX X X1							A B C
L. grayi		(III)					XII XIV					E
L. murrayi		(III)					XII XIV					E

[a]Adapted from Seeliger and Höhne (1979). Other undesignated serovars are also known. Parentheses indicate that the antigen is not always present.

Multilocus enzyme electrophoresis studies with strains of *L. monocytogenes* from clinical and food sources indicate that this technique has excellent potential for epidemiological studies with this species. Piffaretti et al. (1989) analyzed a number of strains of *L. monocytogenes* electrophoretically for allelic variation at 16 genetic loci encoding metabolic enzymes. Forty-five distinct electrophoretic types (ETs) were distinguished. These could be divided into two clusters; one cluster contained *L. monocytogenes* of serovars 1/2b, 4a, and 4b, the other serovars 1/2a and 1/2c (Piffaretti et al., 1989). The authors suggested that the two clusters represented two primary phylogenetic divisions within the species *L. monocytogenes*. Very similar results were obtained by Bibb et al. (1990) but in their study one cluster contained all the *L. monocytogenes* strains of H antigen type a, the other all the *L. monocytogenes* strains of H antigen type b examined. While analysis of the ET subsets within the two clusters indicated no obvious trends in ET distribution within geographic areas, type of clinical presentation, or severity of disease, Bibb et al. (1990) considered that one ET (containing *L. monocytogenes* 4b isolates) represented a particularly virulent clone.

The type strain of *L. monocytogenes* is ATCC 15313. Unfortunately this strain is not β-hemolytic nor does it exhibit hemolysis in the CAMP test with *Staphylococcus aureus*. Further, it is difficult to determine its antigenic composition. A previously suggested reference strain ATCC 35152 (NCTC 7973) has been reported to contain a nonhemolytic variant (Pine et al., 1987). A suitable reference strain is ATCC 43248. The type strains of the other species are *L. innocua* (ATCC 33090), *L. ivanovii* (ATCC 19119), *L. seeligeri* (CIP 100100), *L. welshimeri* (CIP 8149), *L. grayi* (ATCC 19120), and *L. murrayi* (ATCC 25401).

General Properties

Chemical Composition

The chemical composition of the macromolecules of all species is very similar. The cell wall peptidoglycan contains *meso*-diaminopimelic acid (*meso*-A$_2$pm), glutamic acid, alanine, muramic acid, and glucosamine in a ratio that corresponds to the directly cross-linked *meso*-A$_2$pm containing peptidoglycan variation A1 γ of Schleifer and Kandler (1972); glucose, galactose, and rhamnose may or may not be present, and arabinose and galactose are never found (Fiedler and Seger, 1983; Fiedler et al.,

1984; Hether et al., 1983; Kamisango et al., 1982; Keeler and Gray, 1960; Schleifer and Kandler, 1972; Srivastava and Siddique, 1973; Ullmann and Cameron, 1969).

All strains examined contain ribitol teichoic acids; the structural diversity of these components correlates significantly with the different somatic antigenic properties of strains of the genus (Fiedler et al., 1984). Lipoteichoic acids are also present (Hether and Jackson, 1983; Ruhland and Fiedler, 1987), but probably do not contribute to the serological differentiation of *Listeria* strains since they are identical in the different serovars examined (Ruhland and Fiedler, 1987). In *L. grayi* and *L. murrayi,* which are serologically different from the five other species of the genus, the lipoteichoic acids are slightly modified in that they lack the galactosyl substituents in the glycerophosphate units of the hydrophilic chains (Ruhland and Fiedler, 1987). The failure of Wexler and Oppenheim (1979) to detect lipoteichoic acid in the membrane of a virulent strain of *L. monocytogenes* serovar 4b is difficult to explain.

Mycolic acids are absent. The detection of mycolic acidlike substances in *L. monocytogenes* (Mára and Michalec, 1977) has not been confirmed by other workers (Jones et al., 1979; Feresu and Jones, 1988).

The polar lipid composition of *L. monocytogenes* comprises phosphatidylglycerol, diphosphatidylglycerol, galactosyl-glucosyldiacylglycerol, and an uncharacterized glycophospholipid (Kosaric and Carroll, 1971; Shaw, 1974). The fatty acid composition of *L. monocytogenes, L. innocua,* and *L. ivanovii* is virtually identical. All contain predominantly straight-chain, saturated, anteiso-, and iso-methyl-branched chain types. The major fatty acids of all strains examined are 14-methylhexadecanoic (anteiso-C$_{17:0}$) and 12-methyl-tetradecanoic (anteiso-C$_{15:0}$) (Carroll et al., 1968; Tadayon and Carroll, 1971; Feresu and Jones, 1988). In their studies on the lipoteichoic acids of all *Listeria* species, Ruhland and Fiedler (1987) found the fatty acid residue to be comprised mainly of these two types.

All *Listeria* strains examined contain MK-7 as the major menaquinone component with MK-6 and MK-5 as minor components (Collins and Jones, 1981; Feresu and Jones, 1988).

Cytochromes a_1bdo are present when listeriae are grown on a suitable medium, e.g., brain heart infusion broth (Oxoid), with shaking (Feresu and Jones, 1988). Trivett and Meyer (1971), using a defined medium, failed to detect any cytochromes. In static culture, Meyer and Jones (1973) detected only cytochrome b.

Table 4. Distribution of serovars among some *Listeria* species.

Species	Serovars
L. monocytogenes	1/2a, 1/2b, 1/2c, 3a, 3b, 3c, 4a, 4ab, 4b, 4c, 4d, 4e, 7
L. innocua	3, 6a, 6b, 4ab, U/S[a]
L. ivanovii	5
L. seeligeri	1/2a, 1/2b, 1/2c, 4b, 4c, 4d, 6b, U/S[a]
L. welshimeri	1/2a, 4c, 6a, 6b, U/S[a]

[a]U/S, undesignated serovar.

[a]Adapted from Knorz and Hof (1986); Rocourt and Seeliger (1985).

The GC content of the DNA of the genus varies from 36–42 mol%. That of *L. monocytogenes,* 37–39 (Rocourt and Grimont, 1983; Stuart and Welshimer, 1974); *L. innocua,* 38; *L. ivanovii,* 38; *L. seeligeri,* 36; *L. welshimeri,* 36 (Rocourt and Grimont, 1983); *L. grayi,* 41; and *L. murrayi,* 42 mol% (Stuart and Welshimer, 1974).

Metabolism and Nutritional Requirements

Listeria species are facultatively anaerobic, though better growth of laboratory cultures is achieved in air. There is not a great deal of information on the metabolic pathways. Carbohydrate is essential for growth of *Listeria* strains. In studies with *L. monocytogenes,* Miller and Silverman (1959) reported that glucose could not be replaced as a carbon and energy source by gluconate, xylose, arabinose, or ribose. However, it is probable from their biochemical reactions that *L. innocua, L. monocytogenes,* and *L. seeligeri* can utilize xylose as a carbon source. Catabolism of glucose apparently proceeds by the Embden-Meyerhof (glycolysis) pathway both aerobically and anaerobically. Anaerobically, the end product is mainly lactic acid; aerobically, pyruvate, acetoin, lactic acid, and other end products are formed. There is no evidence for the Entner-Doudoroff pathway but glucose-6-phosphate dehydrogenase and 6-phosphogluconate dehydrogenase, key enzymes of the hexose-monophosphate pathway, have been detected in *L. monocytogenes* (Miller and Silverman, 1959). Pine et al. (1989) found that the formation of lactic and acetic acids in glucose casitone broth depended on the extent of aeration with an increase in acetic acid and decrease in lactic acid under strongly aerobic conditions. Recent chromatographic analyses of the aerobic fermentation products of glucose and amino acid metabolism of *Listeria* species indicated the production of acetic, isobutyric, butyric, isovaleric, phenylacetic, lactic, 2-hydroxybutyric, 2-hydroxyvaleric, and 2-hydroxyisocaproic acids. Propionic acid was not formed. Of the

alcohol and amino derivatives, only acetylmethylcarbinol, butylamine, and putrescine were identified (Daneshvar et al., 1989).

Exogenous citrate is not utilized. In a defined medium, Trivett and Meyer (1971) reported that neither pyruvate, acetate, citrate, isocitrate, α-ketoglutarate, succinate, fumarate, nor malate supported growth of *L. monocytogenes* in the absence of glucose, nor did they increase growth in the presence of glucose. Pyruvate, malate, succinate, and α-ketoglutarate have, however, been reported to be oxidized at low rates by *L. monocytogenes* (Friedman and Alm, 1962; Kolb and Seidel, 1960). In a complex medium, pyruvate is utilized as a carbon source by some strains. *Listeria monocytogenes* appears to possess a split noncyclic citrate pathway that has both an oxidative and a reductive portion. Trivett and Meyer (1971) speculated that the pathway is probably important in biosynthesis but not for a net gain of energy.

Biotin, riboflavin, thiamine, thioctic acid, and several amino acids, including cysteine, glutamine, isoleucine, leucine, and valine, are required for growth in a defined medium. Arginine, histidine, methionine, and tryptophan are reported to have a growth-stimulating effect on *L. monocytogenes* (for references, see Gray and Killinger, 1966; Ralovich, 1984; Siddiqi and Khan, 1989). In comparative studies with various minimal defined media, it has been found that the growth factors described by Friedman and Roessler (1961) and by Ralovich (1984) both allow growth of all seven *Listeria* species but better growth was obtained with those described by Friedman and Roessler (1961) (D. Jones, unpublished observations).

Iron is stimulatory for growth of *L. monocytogenes* in stationary and aerated cultures (Sword, 1966; Trivett and Meyer, 1971); aeration improves growth only in the presence of adequate iron (Trivett and Meyer, 1971).

All species produce catalase, and as noted previously, cytochromes, but the production of both is dependent on the medium and conditions of cultivation. Strains grown on media containing low levels of meat and yeast extract may be catalase-negative, and aeration and adequate iron are necessary for the detection of cytochromes (Feresu and Jones, 1988).

Temperature and pH

Listeriae grow between 1 and 45°C (Seeliger and Jones, 1986) but strain variation is evident at 45°C. The ability to grow at refrigeration temperatures is especially important both as a method of isolating *L. monocytogenes* (see "Isolation"), and because the numbers of this species present in foods stored at low temperatures can increase to potentially hazardous levels. At 4°C, generation times of 29–45 h in milk have been reported (Donnelly and Briggs, 1986; Rosenow and Marth, 1987).

Listeriae do not survive heating at 60°C for 30 min in laboratory media (Seeliger and Jones, 1986). The heat resistance of *L. monocytogenes* in foods, especially milk, has received a great deal of attention after an outbreak of listeriosis in Massachusetts was linked to the consumption of pasteurized milk (Fleming et al., 1985). Despite some reports to the contrary, it is now generally agreed that listeriae are killed at pasteurization temperatures (Bradshaw et al., 1985; Reed, 1986). A review of the various reported decimal reduction times (*D* values) in milk by Mackey and Bratchell (1989) indicates that reported differences in *D* values by various workers might be due to the heating method employed. With the sealed tube (ST) method, data indicate that vat pasteurization (62.8°C for 30 min) would achieve a 39-*D* reduction in numbers of listeriae, while those for higher-temperature short-time (HTST) pasteurization (71.7°C for 15 sec) by the slug-flow heat-exchanger (SF) predict a 5.2-*D* reduction. Thus the HTST processes provide a much smaller margin of safety. Mackey and Bratchell (1989) suggest that the SF model be used for calculating lethalities in milk at high temperatures (above 65°C) and the ST model for lower temperatures; based on the ST model, $D_{62.8} = 46$ sec. There is, as yet, no evidence to support the suggestion of Barza (1985) that *L. monocytogenes* may be afforded some protection from pasteurization by intracellular survival in bovine lymphocytes. The reported heat resistance of *L. monocytogenes* in different foods varies with the composition of a particular product and also with the method of subsequent detection; some methods are more favorable for recovery of sublethally damaged cells (see review of Mackey and Bratchell, 1989). There is evidence that preheating at 48°C for 1 h increases the heat resistance of *L. monocytogenes* subsequently exposed at 60°C (Fedio and Jackson, 1989).

Listeriae grow best at a neutral or alkaline pH. Some strains grow at pH 9.6 but are usually inhibited at a pH below 5.5 (Gray and Killinger, 1966; Seeliger, 1961). Viable transfers can rarely be made from cultures derived from fermentation studies where much acid is produced. Inhibition in such cultures may reflect the nature of the acid. Ahamad and Marth (1989) found that 0.1% of acetic, citric, and lactic acids inhibited the growth of *L. monocytogenes* in tryptose broth; the degree of inhibition increased as the temperature of incubation decreased. George et al. (1988) demonstrated growth of several strains of *L. monocytogenes* cultured at 20–30°C in a nutrient medium acidified with HCl to a pH of 4.4. The ability of the strains to grow at low pH values was influenced by temperature; at 7–10°C growth did not occur below pH 4.62 and at 4°C below pH 5.23. Similar results were obtained by McClure et al. (1989).

Irradiation

Listeriae are not particularly resistant to irradiation. Patterson (1989) found the sensitivity of four strains of *L. monocytogenes* comparable to that of salmonellae in both phosphate-buffered saline and poultry meat. As with temperature and pH, the susceptibility of the organisms to irradiation depends on the composition of the suspending material, and the recovery of the organisms after irradiation on the type of recovery medium used (Patterson, 1989).

Inhibitory Agents

All listeriae grow in complex media containing 6.5% NaCl, and the majority also grow at 10% NaCl when incubated at 30°C. Some strains of *L. monocytogenes* tolerate 20% NaCl, and Seeliger (1961) reported the recovery of a strain of this species after 1 year in 16% NaCl at pH 6.0. The ability of listeriae to grow in the presence of various concentrations of NaCl is dependent on the pH of the medium and the temperature of incubation (McClure et al., 1989; Conner et al., 1986).

All strains tested grow on 10% and 40% bile-containing agar. Wetzler et al. (1968) found generally better growth of *L. monocytogenes* on 40% (Bacto oxgall was used) than on 10% bile. Growth of all strains occurs on MacConkey agar.

Although all strains grow on media containing 0.025% thallous acetate, 3.75% potassium thiocyanate, 0.04% potassium tellurite, or 0.01% 2,3,5-triphenyltetrazolium chloride when incubated in a nutrient medium at 30–37°C, growth is partially inhibited by all these agents. Listeriae do not grow in the presence of 0.02%

sodium azide (Seeliger and Jones, 1986) or of potassium cyanide (Wetzler et al., 1968). No growth occurs on the medium devised for the isolation of *Brochothrix* (Gardner, 1966), and growth, if any, is poor on MRS medium devised by de Man et al. (1960) for the growth of lactobacilli.

An investigation by Best et al, (1990) on the efficacy of 14 different disinfectants against *L. innocua* and *L. monocytogenes* showed that strains of both species were more resistant to killing when dried on metal surfaces than when tested in suspension. Serum and especially milk interfered with the bactericidal properties of all the antiseptics. *Listeria monocytogenes* appeared to be more resistant to killing than did *L. innocua*. The most effective disinfectants for organisms dried on metal surfaces in the presence of serum were povidone-iodine, chlorhexidine gluconate, and glutaraldehyde, but none was effective in the presence of milk. Only one of the 14 disinfectants tested, sodium dichloroisocyanurate, was active against the listeriae dried on metal surfaces in the presence of milk. However, in suspension, all but four of the disinfectants tested were effective, regardless of the degree of the organic load, serum or milk (Best et al., 1990).

Listeriae are sensitive to a number of antibiotics. Sometimes, there are unexplained differences between the sensitivity of new isolates and the culture collection strains of *L. monocytogenes* (see Gray and Killinger, 1966; Wetzler et al., 1968). When tested in vitro by the agar diffusion method, listeriae are sensitive to ampicillin, carbenicillin, cephaloridine, chloramphenicol, erythromycin, furazolidone, neomycin, novobiocin, oleandomycin, ticarcillin, and azolocillin; they are less sensitive to chlortetracycline, oxytetracycline, tetracycline, gentamicin, kanamycin, nitrofurantoin, penicillin G, and streptomycin. Most strains are sensitive to methicillin. Listeriae are resistant to colistin sulfate, nalidixic acid, polymyxin B, and sulfonomides. It is important to recognize that this information is not necessarily relevant to the clinical effectiveness of the antibiotics in the treatment of listeriosis.

Currently, benzylpenicillin or ampicillin in combination with an aminoglycoside provide the treatment of choice for listeriosis; tetracycline, erythromycin, or chloramphenicol (alone or in combination) are alternative antibiotics for persons allergic to penicillins. However, antibiotic resistance has been recognized. An isolate of *L. monocytogenes* from a patient with meningoencephalitis was found to be resistant to chloramphenicol, erythromycin, streptomycin, and tetracycline. The multiple resistance

was plasmid borne, and evidence suggested that the replicon originated in "enterococci-streptococci" (Poyart-Salmeron et al., 1990).

Although most clinical isolates are uniformly susceptible to the above antibiotics, prognosis remains poor in the more severe forms of the disease. Boisivon et al. (1990) have therefore advocated consideration of antibiotics that are active inside macrophages. On the basis of in vitro tests, they recommended that a combination of trimethoprim-sulfamethoxazole and rifampicin merited evaluation in in vivo models of *L. monocytogenes* meningitis.

Larsen and Gundstrup (1965) reported the inhibition of *L. monocytogenes* by cultures of Gram-positive bacteria, notably *Bacillus* species, and by cell-free extracts of cultures of the same bacteria. Inhibition of *L. monocytogenes* by other microorganisms has also been reported (Ralovich, 1984). Interestingly, inhibition of listeriae by a heat-labile component of carrots has been noted (Beuchat and Brackett, 1990).

Bacteriophages, Bacteriocins, and Plasmids

Lysogeny is common in listeriae. Phages have been isolated from lysogenic strains with and without induction. While induction is useful for research studies on *Listeria* phages, it frequently results in the production of bacteriocins that may interfere with phage-typing patterns (Rocourt et al., 1982b). Bacteriophages have been isolated from many strains of *Listeria* species (Schultz, 1945; Sword and Pickett, 1961; Jasinska, 1964; Audurier et al., 1977, 1979a; Rocourt, 1986; Rocourt et al., 1982b, 1985b). In the last decade, phage-typing schemes of value for epidemiological purposes have been devised (Audurier et al., 1979b, 1984; Rocourt, 1986; Rocourt et al., 1985a; McLauchlin et al., 1986). Although phage typing is extremely useful, only 60–70% of *L. monocytogenes* isolates are typable (McLauchlin, 1987; McLauchlin et al., 1986). Strains of *L. ivanovii* appear to be particularly sensitive to phage lysis. Descriptions of the morphology of *Listeria* phages have been given by Sword and Pickett (1961); Jasinska (1964); Chiron et al. (1977); Ackermann et al. (1981); Rocourt et al. (1983); Rocourt (1986). The phages isolated so far are members of the Styloviridae (noncontractile tails) and Myroviridae (contractile tails) families of phages (Rocourt, 1986).

A high proportion of listeriae examined produce bacteriocins (Sword and Pickett, 1961; Hamon and Peron, 1962, 1963; Bradley and Dewar, 1966; Ortel, 1978; Rocourt, 1986). In addition to listeriae, these bacteriocins are active against staphyloccci and bacilli but not against Gram-negative bacteria. The general characteristics of the bacteriocins are given by Hamon and Peron (1962, 1963; Rocourt, 1986) and their morphological characteristics by Bradley and Dewar (1966). A bacteriocin-typing scheme useful for epidemiological purposes has not been developed. As noted by Rocourt et al. (1982b), bacteriocins are often produced when lysogenic listeria are induced to produce lytic phages.

Plasmids have been isolated from listerae (Perez-Diaz et al., 1981; Flamm et al., 1984; Fistrovici and Collins-Thompson, 1990). A plasmid coding for multiple antibiotic resistance was isolated by Poyart-Salmeron et al. (1990) from a clinical isolate of *L. monocytogenes*. Plasmid profiles have been used in environmental studies of *Listeria* species from raw milk (Fistrovici and Collins-Thompson, 1990).

Pathogenesis

Despite significant advances in recent years, knowledge of the pathogenesis of listeriosis remains imperfect. Listeriae are widely distributed in the environment, yet overt listeriosis is a rare disease. The pathogenic species *L. monocytogenes* (for humans and animals) and *L. ivanovii* (for animals, mainly sheep) are facultative intracellular pathogens. While the individuals who are most susceptible to listerial infection are the pregnant (with severe effects often seen in the fetus), the newborn, and stressed or immunocompromised subjects, infection also occurs in apparently healthy humans and animals (see "Habitats"). Moreover, the organisms have been recovered from the feces of asymptomatic subjects. Thus, exposure to pathogenic *Listeria* species does not always result in disease. The likelihood of systemic infection almost certainly depends on host susceptibility, infectious dose, and the virulence factor complement of the infecting organisms.

Most studies have been concerned with *L. monocytogenes*. The pioneering work of Mackaness (1962, 1969) and subsequent studies (see Chakraborty and Goebel, 1988) showed that initial survival within macrophages and other mammalian cells is essential for successful infection with the organism. The importance of T-cell-mediated immunity to listerial infection in the mouse model is well documented (Hahn and Kaufman, 1981; Gellin and Broome, 1989) and accords with the association of listeriosis with immunosuppressive therapy and pregnancy. The role of humoral defenses is less well understood. That opsonization may play a part in the immune response to infection is suggested by the enhanced susceptibility of neonates, attributed to the low levels of IgM and the diminished activity of the classic complement pathway during the neonatal period of life (see Gellin and Broome, 1989).

The size of the infectious dose necessary to initiate infection is unknown. Such information as is available is derived from studies in animals. Immunocompromised or pregnant mice succumb to smaller doses (see Gellin and Broome, 1989). The main portal of entry for the pathogen is also not known with certainty. However, evidence from animal models of listeriosis following feeding of the organism and from outbreaks of human disease associated with the ingestion of contaminated food points to the intestine as an important route. It has been speculated that gastric acid neutralization due to the use of antacids or cimetidene predisposes to infection. Damage to the intestinal mucosa due to a preexisting lesion or a coinfecting microbe has also been suggested as a predisposing factor (see Gellin and Broome, 1989). However, intestinal ulceration was noted by Pirie as early as 1927 as one of the main symptoms of listeriosis in gerbils following feeding with *L. monocytogenes*-contaminated food. In addition, a more recent animal model of listerial enteritis demonstrated invasion of an intact epithelial barrier (see Gellin and Broome, 1989). Further, in vitro studies have shown that *L. monocoytogenes* invades enterocytelike cells by inducing phagocytosis and that preexisting lesions are not required for invasion (Gaillard et al., 1987).

A number of products have been considered as virulence determinants in *L. monocytogenes*. These include hemolysin (listeriolysin O), phospholipase, catalase, superoxide dismutase, and other enzymes; the components referred to as monocytosis-producing activity (MPA), immunosuppressive activity (ISA), the delayed-type hypersensitivity (DTH) factor, and the protein p60 (Chakraborty and Goebel, 1988).

The application of modern genetic methods, in particular transposon mutagenesis and gene cloning, together with suitable in vitro and in vivo model systems have begun to clarify the basis of listerial pathogenesis. Protein p60 (Kuhn and Goebel, 1989; Köhler et al., 1990) may be involved in promoting adhesion and penetration into mammalian cells by inducing phagocytosis. Listeriolysin 0, a sulfhydryl-activated 59-kDa protein, is an essential virulence factor that plays a key role in the escape of the bacteria from the membrane-bound phagocytic vacuole and thus allows their subsequent multiplication in the cell cytoplasm. Nonhemolytic mutants demonstrate virtually no intracellular multiplication (Gaillard et al., 1986; Kathariou et al., 1987; Cossart et al., 1989).

Electron microscope studies have shown that once released into the cytoplasm, intracellular and cell-to-cell spread of *L. monocytogenes* involves encapsulation of the bacteria by short actin filaments and other actin-binding proteins (Dabiri et al., 1990; Mounier et al., 1990; Tilney and Portnoy, 1989). Once formed, this layer of actin filaments is reorganized, by an unknown mechanism, to form polar tails that appear to be associated with the movement of listeria inside the cytoplasm and in intercellular spread. A mutant strain of *L. monocytogenes* that induced actin polymerization but was not able to rearrange the actin capsule into polar tails and consequently remained entrapped in the actin cloud was shown to have much reduced virulence (Kuhn et al., 1990). Further studies on the 10 classes of mutants of *L. monocytogenes* constructed by transposon mutagenesis, which are all defective for intracellular growth and intercellular spread (Sun et al., 1990), should aid the identification of other factors involved in pathogenesis.

There has lately been renewed interest in early reports of an increase in virulence of *L. monocytogenes* cultured at refrigerator temperatures. Czuprynski et al. (1989) demonstrated that the virulence of *L. monocytogenes* is increased at 4°C for intravenously but not orally infected mice. This increased virulence is probably associated with stress proteins and reflects a highly artificial situation that requires direct entry of the bacterium into the blood stream. The same study does not support claims that hemolysin production is enhanced at 4°C. Thus there is no evidence to support the notion that *L. monocytogenes* in refrigerated products has enhanced virulence when ingested.

Far less work has been done on the pathogenesis of *L. ivanovii*. This species produces a sulfhydryl-activated listeriolysin highly homologous to that of *L. monocytogenes* (Leimeister-Wächter and Chakraborty, 1989) and in addition another cytolysin, sphingomyelinase C (Vazquez-Boland et al., 1989). The exact role of the sulfhydryl-activated listeriolysin (ivanolysin 0) and the sphingomyelinase in the pathogenesis of *L. ivanovii* is not known. *Listeria ivanovii* has a lower pathogenicity for mice and seems to selectively affect pregnant ewes (Vazquez-Boland et al., 1989).

Sequences homologous to the listeriolysin gene of *L. monocytogenes* have also been detected in the weakly hemolytic species *L. seeligeri* (Leimeister-Wächter and Chakraborty, 1989). This species is generally considered to be nonpathogenic; its inability to cause disease may be due to the absence of other factors. Gaillard et al. (1987) found that *L. seeligeri* is incapable of inducing its own phagocytosis in human enterocytes although it was capable of survival and multiplication following nonspecific uptake.

Literature Cited

Ackermann, H-W., Audurier, A., and Rocourt, J. 1981. Morphologie de bactériophages de *Listeria monocytogenes*. Annales de Virologie (Institut Pasteur) 132E:371–382.

Ahamad, N. and Marth, E. H. 1989. Behavior of *Listeria monocytogenes* at 7,13,21 and 35°C in tryptose broth acidified with acetic, citric or lactic acid. Journal of Food Protection 52:688–695.

Armstrong, D. 1985. *Listeria monocytogenes* p. 1177–1182. In: Mandell, G. L., Douglas, R. G., and Bennett, J. E. (ed.), Principles and practice of infectious diseases, 2nd ed. New York: John Wiley and Sons.

Audurier, A., Chatelain, B., Chalons, F., and Piéchaud, M. 1979a. Lysotypie de 823 souches de *Listeria monocytogenes* isolées en France de 1958 à 1978. Annales Microbiologie (Institut Pasteur) 130B:179–189.

Audurier, A., Rocourt, J., and Courtieu, A. L. 1977. Isolement et caractérisation de bactériophages de *Listeria monocytogenes*. Annales Microbiologie (Institut Pasteur) 128A:185–198.

Audurier, A. Rocourt, J., and Courtieu, A. L. 1979b. Phage typing system for *Listeria monocytogenes*, p. 108–121. In: Ivanov, I. (ed), Problems of listeriosis. National Agroindustrial Union, Center for Scientific Information, Sofia, Bulgaria.

Audurier, A., Taylor, A. G., Carbonelle, B., and McLauchlin, J. 1984. A phage typing system for *Listeria monocytogenes* and its use in epidemiological studies. Clinical and Investigative Medicine 7:229–232.

Barza, M. 1985. Listeriosis and milk. New England Journal of Medicine 312:438–440.

Best, M., Kennedy, M. E., and Coates, F. 1990. Efficacy of a variety of disinfectants against *Listeria* species. Applied and Environmental Microbiology 56:377–380.

Beuchat, L. R. and Brackett, R. E. 1990. Inhibitory effects of raw carrots on *Listeria monocytogenes*. Applied and Environmental Microbiology 56:1734–1742.

Bibb, W. F., Gellin, B. G., Weaver, R., Schwartz, B., Plikaytis, B. D., Reeves, M. W., Pinner, R. W., and Broome, C. V. 1990. Analysis of clinical and foodborne isolates of *Listeria monocytogenes* in the United States by multilocus enzyme electrophoresis and application of the method to epidemiologic investigations. Applied and Environmental Microbiology 56:2133–2141.

Bille, J. and Glauser, M-P. 1988. Listeriosis en Suisse. Bulletin des Bundesamtes fur Gesundheitswesen 3:28.

Blanco, M., Fernandez-Garayzabel, J. F., Dominguez, L., Briones, V., Vazquez-Boland, J. A., Blanco, J. L., Garcia, J. A., and Suarez, G. 1989. A technique for the direct identification of haemolytic-pathogenic listeria on selective plating media. Letters in Applied Microbiology 9:125–128.

Bockemühl, J., Feindt, E., Höhne, K., and Seeliger, H. P. R. 1974. Acridinfarbstoffe in Selektivnährböden zur Isolierung von *Listeria monocytogenes*. II Modifiziertes Stuart-Medium: Ein neues *Listeria*-Transport-Anreicherungsmedium. Medical Microbiology and Immunology 159:289–299.

Bockemühl, J., Seeliger, H. P. R., and Kathke, R. 1971. Acridinfarbstoffe in Selektivnährböden zur Isolierung von *Listeria monocytogenes*. Medical Microbiology and Immunology 157:84–95.

Boisivon, A., Guiomar, C., and Carbon, C. 1990. In vitro bactericidal activity of amoxicillin, gentamicin, rifampicin, ciprofloxacin and trimethoprim-sulfamethoxazole alone or in combination against *Listeria monocytogenes*. European Journal of Clinical Microbiology and Infectious Diseases 9:206–209.

Bradley, D. E. and Dewar, C. A. 1966. The structure of phage-like objects associated with non-induced bacteriocinogenic bacteria. Journal of General Microbiology 45:399–408.

Bradshaw, J. G., Peeler, J. T., Corwin, J. J., Hunt, J. M., Tierney, J. T., Larkin, E. P., and Twedt, R. M. 1985. Thermal resistance of *Listeria monocytogenes* in milk. Journal of Food Protection 48:743–745.

Breed, R S., Murray, E. G. D., and Hitchens, A. P. (ed.). 1948. Bergey's manual of determinative bacteriology, 6th ed. Baltimore: Williams & Wilkins.

Breed, R. S., Murray, E. G. D., and Smith, N. R. (ed.). 1957. Bergey's manual of determinative bacteriology, 7th ed. Baltimore: Williams & Wilkins.

Buchanan, R. E. and Gibbons, N. E. (ed.). 1974. Bergey's manual of determinative bacteriology, 8th ed. Baltimore: Williams & Wilkins.

Busch, L. A. 1971. Human listeriosis in the United States, 1967–1969. Journal of Infectious Disease 123:328–332.

Carroll, K. K., Cutts, J. H., and Murray, E. G. D. 1968. The lipids of *Listeria monocytogenes*. Canadian Journal of Biochemistry 46:899–904.

Chakraborty, T. and Goebel, W. 1988. Recent developments in the study of virulence in *Listeria monocytogenes*. Current Topics in Microbiology and Immunology 138:41–58.

Chiron, J. P., Maupas, Ph., and Denis, F. 1977. Ultrastructure des bactériophages de *Listeria monocytogenes*. Comptes Rend Biol. (Paris) 171:488–491.

Christie, R., Atkins, N. E., and Munch-Petersen, E. 1944. A note on the lytic phenomenon shown by group B streptococci. Australian Journal of Experimental Biology and Medical Sciences 22:197–200.

Colburn, K. G., Kaysner, C. A., Abeyta, C., and Wekell, M. M. 1990. *Listeria* species in a California coast esturine environment. Applied and Environmental Microbiology 56:2007–2011.

Collins, M. D. and Jones, D. 1981. The distribution of isoprenoid quinone structural types in bacteria and their taxonomic implications. Microbiological Reviews 45:316–354.

Collins, M. D., Farrow, J. A. E., Phillips, B. A., Feresu, S., and Jones, D. 1987. Classification of *Lactobacillus divergens, Lactobacillus piscicola*, and some catalase-negative, asporogenous, rod-shaped bacteria from poultry in a new genus *Carnobacterium*. International Journal of Systematic Bacteriology 37:310–316.

Conner, D. E., Brackett, R. E., and Beuchat, L. R. 1986. Effect of temperature, sodium chloride and pH on growth of *Listeria monocytogenes* in cabbage juice. Applied and Environmental Microbiology 52:59–63.

Cossart, P., Vincente, M. F., Mengaud, J., Baquero, F., Perez-Diaz, J. C., and Berche, P. 1989. Listeriolysin O is essential for virulence of *Listeria monocytogenes:* direct evidence obtained by gene complementation. Infection and Immunity 57:3629–3636.

Cummins, C. S., and Harris, H. 1956. The chemical composition of the cell wall in some Gram-positive bacteria and its possible value as a taxonomic character. Journal of General Microbiology 14:583–600.

Curtis, G. D. W., Mitchell, R. G., King, A. F., and Griffin, E. J. 1989a. A selective differential medium for the isolation of *Listeria monocytogenes*. Letters in Applied Microbiology 8:95–98.

Curtis, G. D. W., Nichols, W. W., and Falla, T. J. 1989b. Selective agents for listeria can inhibit their growth. Letters in Applied Microbiology 8:169–172.

Czuprynski, C. J., Brown, J. F., and Roll, J. T. 1989. Growth at reduced temperatures increases the virulence of *Listeria monocytogenes* for intravenously but not intragastrically inoculated mice. Microbial Pathogenesis 7:213–223.

Dabiri, G. A., Sanger, J. M., Portnoy, D. A., and Southwick, F. S. 1990. *Listeria monocytogenes* moves rapidly through the host cell cytoplasm by inducing directional actin assembly. Proceedings of the National Academy of Science USA 87:6068–6072.

Daneshvar, M. L., Brooks, J. B., Malcolm, G. B., and Pine, L. 1989. Analyses of fermentation products of *Listeria* species by frequency-pulsed electron-capture gas-liquid chromatography. Canadian Journal of Microbiology 35:786–793.

Davis, G. H. G., Fomin, L., Wilson, E., and Newton, K. G. 1969. Numerical taxonomy of *Listeria,* streptococci and possibly related bacteria. Journal of General Microbiology 57:333–348.

Davis, G. H. G. and Newton, K. G. 1969. Numerical taxonomy of some named coryneform bacteria. Journal of General Microbiology 56:195–214.

De Man, J. C., Rogosa, M., and Sharpe, M. E. 1960. A medium for the cultivation of lactobacilli. Journal of Applied Bacteriology 23:130–135.

Donker-Voet, J. 1957. Serological studies on some strains of *Listeria monocytogenes* Tijdschr. Diergeneesk 82:341–350.

Donnelly, C. W. and Baigent, G. J. 1986. Method for flow cytometric detection of *Listeria monocytogenes* in milk. Applied and Environmental Microbiology 52:689–695.

Donnelly, C. W. and Briggs, E. H. 1986. Psychrotrophic growth and thermal inactivation of *Listeria monocytogenes* as a function of milk composition. Journal of Food Protection 49:994–998.

Dumont, J. and Cotoni, L. 1921. Bacille semblable à celui de rouget de porc rencontre dans le L.C.R. d'un méningitique. Annales de l'Institut Pasteur 35:625–633.

Fedio, W. M. and Jackson, H. 1989. Effect of tempering on the heat resistance of *Listeria monocytogenes*. Letters in Applied Microbiology 9:157–160.

Fenlon, D. R. 1986a. Growth of naturally occurring *Listeria* spp. in silage: a comparative study of laboratory and farm ensiled grass. Grass and Forage Science 41:375–378.

Fenlon, D. R. 1986b. Rapid quantitative assessment of the distribution of *Listeria* in silage implicated in a suspected outbreak of listeriosis in calves. Veterinary Record 118:240–242.

Feresu, S. B. and Jones, D. 1988. Taxonomic studies on *Brochothrix, Erysipelothrix, Listeria* and atypical lactobacilli. Journal of General Microbiology 134:1165–1183.

Fiedler, F. and Seger, J. 1983. The murein types of *Listeria grayi, Listeria murrayi* and *Listeria denitrificans.* Systematic and Applied Microbiology 4:444–450.

Fiedler, F., Seger, J., Schrettenbrunner, A., and Seeliger, H. P. R. 1984. The biochemistry of murein and cell wall teichoic acids in the genus *Listeria.* Systematic and Applied Microbiology 5:360–376.

Fistrovici, E. and Collins-Thompson, D. L. 1990. Use of plasmid profiles and restriction endonuclease digest in environmental studies of *Listeria* spp. from raw milk. International Journal of Food Microbiology 10:43–50.

Flamm, R. K., Hinrichs, D. J., and Tomashaw, M. F. 1984. Introduction of pAM1 into *Listeria monocytogenes* by conjugation and homology between native *L. monocytogenes* plasmids. Infection and Immunity 44:157–161.

Fleming, D. W., Cochi, S. L., MacDonald, K. L., Brondrum, J., Hayes, P. S., Plikaytis, B. D., Holmes, M. B., Audurier, A., Broome, C. V., and Reingold, A. L. 1985. Pasteurized milk as a vehicle of infection in an outbreak of listeriosis. New England Journal of Medicine 312:404–407.

Friedman, M. E. and Alm, W. L. 1962. Effect of glucose concentration in the growth medium on some metabolic activities of *Listeria monocytogenes.* Journal of Bacteriology 84:375–376.

Friedman, M. E. and Roessler, W. G. 1961. Growth of *Listeria monocytogenes* in defined media. Journal of Bacteriology 82:528–533.

Gaillard, J. L., Berche, P., and Sansonetti, P. J. 1986. Transposon mutagenesis as a tool to study the role of hemolysin in the virulence of *Listeria monocytogenes.* Infection and Immunity 52:50–55.

Gaillard, J. L. Berche, P., Mounier, J., Richard, S., and Sansonetti, P. 1987. In vitro model of penetration and intracellular growth of *Listeria monocytogenes* in the human enterocyte-like cell line Caco-2. Infection and Immunity 55:2822–2829.

Gardner, G. A. 1966. A selective medium for the enumeration of *Microbacterium thermosphactum* in meat and meat products. Journal of Applied Bacteriology 29:455–460.

Gellin, B. G. and Broome, C. V. 1989. Listeriosis. Journal of the American Medical Association 261:1313–1320.

George, S. M., Lund, B. M. and Brocklehurst, T. F. 1988. The effect of pH and temperature on initiation of growth of *Listeria monocytogenes.* Letters in Applied Microbiology 6:153–156.

Gilbert, R. J., Hall, S. M. and Taylor, A. G. 1989. Listeriosis update. Public Health Laboratory Service Microbiology Digest 6:33–37.

Gitter, M. 1985. Listeriosis in farm animals in Great Britain, p. 191–200. In: Collins, C. H. (ed.), Isolation and identification of microorganisms of medical and veterinary importance. Symposium of the Society for Applied Bacteriology. Academic Press, London.

Gray, M. L. 1957. A rapid method for the detection of colonies of *Listeria monocytogenes.* Zentralblatt für Bakteriologie, Parasitenkunde, Infektionskrankeiten und Hygiene. Abt. 1. 169:373–377.

Gray, M. L. 1960. Isolation of *Listeria monocytogenes* from oat silage. Science 132:1767–1768.

Gray, M. L. and Killinger, A. H. 1966. *Listeria monocytogenes* and listeric infections. Bacteriological Reviews 30:309–382.

Gray, M. L., Stafseth, H. J., Thorp, F., Jr., Sholl, L. B., and Riley, W. F. 1948. A new technique for isolating listerellae from the bovine brain. Journal of Bacteriology 55:471–476.

Hahn, H. and Kaufman, S. H. E. 1981. The role of cell mediated immunity in bacterial infections. Reviews of Infectious Diseases 3:1121–1250.

Hamon, Y. and Peron, Y. 1962. Étude du pouvoir bactériocinogène dans le genre *Listeria.* I. Propriétes générales de ces bacteriocines. Annales de l'Institut Pasteur 103:876–889.

Hamon, Y. and Peron, Y. 1963. Étude du pouvoir bactériocinogène dans le genre *Listeria.* II. Individualité et classification des bactériocines en cause. Annales de l'Institut Pasteur 104:55–65.

Heisick, J. E., Wagner, D. E., Nierman, M. L., and Peeler, J. T. 1989. *Listeria* spp. found on fresh market produce. Applied and Environmental Microbiology 55:1925–1927.

Henry, B. S. 1933. Dissociation in the genus *Brucella.* Journal of Infectious Diseases 52:374–402.

Hether, N. W., Campbell, P. A., Baker, L. A., and Jackson, L. L. 1983. Chemical composition and biological functions of *Listeria monocytogenes* cell wall preparations. Infection and Immunity 39:1114–1121.

Hether, N. W. and Jackson, L. L. 1983. Lipoteichoic acid from *Listeria monocytogenes.* Journal of Bacteriology 156:809–817.

Hülphers, G. 1911. Lefvernekros hos kanin orsakad ef ej fovut bestrifuen bakterie. Svensk Veterinar Tidskrift. 16:265–273. Reprinted 1959 Medlemsbad for Sveriges Veterinarforbund. 11 (Suppl.).:10–16.

Ivanov, I. 1975. Establishment of non-motile strains of *Listeria monocytogenes* type 5, p. 18–26. In: Woodbine, M. (ed.), Problems of listeriosis. Leicester University Press, Leicester, England.

Ivanow, I. 1962. Untersuchungen uber die Listeriose der Schafe in Bulgarien. Monatsh. Veterinämed 17:729–736.

Jasinska, S. 1964. Bacteriophages of lysogenic strains of *Listeria monocytogenes.* Acta Microbiologica Polonica 13:29–44.

Jones, D. 1975. A numerical taxonomic study of coryneform and related bacteria. Journal of General Microbiology 87:52–96.

Jones, D., Collins, M. D., Goodfellow, M., and Minnikin, D. E. 1979. Chemical studies in the classification of the genus *Listeria* and probably related bacteria, p. 17–24. In: Ivanov, I. (ed.), Problems of Listeriosis. National Agroindustrial Union, Center for Scientific Information. Sofia, Bulgaria.

Jones, D., Pell, P. A., and Sneath, P. H. A. 1984. Maintenance of bacteria on glass beads at −60°C to −76°C, p. 35–40. In: Kirsop, B. E. and Snell, J. J. S. (ed.), Maintenance of microorganisms. A manual of laboratory methods. Academic Press, London.

Junttila, J. and Brander, M. 1989. *Listeria monocytogenes* septicaemia associated with the consumption of salted mushrooms. Scandanavian Journal of Infectious Diseases 21:339–342.

Kamisango, K., Saiki, I., Tanio, Y., Okmomura, H., Araki, Y., Sekikawa, I., Azuma, I., and Yamamura, Y. 1982. Structures and biological activities of peptidoglycans of *Listeria monocytogenes* and *Propionibacterium acnes*. Journal of Biochemistry 92:23–33.

Kampelmacher, E. H. and van Noorle Jansen, L. M. 1969. Isolation of *Listeria monocytogenes* from faeces of clinicially healthy humans and animals. Zentralblatt für Bakteriologie, Parasitenkunde, Infektionskrankheiten und Hygiene Abt 1 Orig. 211:353–359.

Kampelmacher, E. H. and van Noorle Jansen, L. M. 1975. Occurrence of *L. monocytogenes* in effluents, p. 66–70. In: Woodbine, M. (ed.), Problems of listeriosis. Leicester University Press, Leicester, England.

Kathariou, S., Metz, P., Hof, H., and Goebel, W. 1987. Tn*916*-induced mutations in the hemolysin determinant affecting virulence of *Listeria monocytogenes*. Journal of Bacteriology 169:1291–1297.

Keeler, R. F. and Gray, M. L. 1960. Antigenic and related biochemical properties of *Listeria monocytogenes*. I. Preparation and composition of cell wall material. Journal of Bacteriology 80:683–692.

Knorz, W. and Hof, H. 1986. Zur Pathogenität von Listerien. Immunität und Infektion 2:76–80.

Köhler, S., Leimeister-Wächter, M., Chakraborty, T., Lottspeich, F., and Goebel, W. 1990. The gene coding for protein p60 of *Listeria monocytogenes* and its use as a specific probe for *Listeria monocytogenes*. Infection and Immunity 58:1943–1950.

Kolb, E. and Seidel, H. 1960. Ein Beitrag zur Kenntnis des Stoffwechsels von *Listeria monocytogenes* (Typ 1) unter besonderer Berucksichtigung der Oxydation von Kohlenhydraten und Metaboliten des Tricarbonsaurecyclus und deren Beeinflussung durch Hemmstoffe. Zentralblatt für Veterinärmedicin 7:509–518.

Kosaric, N. and Carroll, K. K. 1971. Phospholipids of *Listeria monocytogenes*. Biochemica Biophysica Acta. 239:428–442.

Kuhn, M. and Goebel, W. 1989. Identification of an extracellular protein of *Listeria monocytogenes* possibly involved in intracellular uptake by mammalian cells. Infection and Immunity 57:55–61.

Kuhn, M., Prévost, C.-M., Mounier, J., and Sansonetti, P.J. 1990. A nonvirulent mutant of *Listeria monocytogenes* does not move intracellularly but still induces polymerization of actin. Infection and Immunity 58:3477–3486.

Lamont, R. J., Petrie, D. T., Melvin, W. T. and Postlethwaite, R. 1986. An investigation of the taxonomy of *Listeria* species by comparison of electrophoretic protein patterns, p 41–46. In: Courtieu, A-L., Espaze, E. P., and Raynaud, A. E. (ed.), Listeriose, listeria, listeriosis. Université de Nantes, Nantes, France.

Lamont, R. J., Postlethwaite, R., and MacGowan, A.P. 1988. *Listeria monocytogenes* and its role in human infection. Journal of Infection 17:7–28.

Larsen, H. E. and Gundstrup, A. 1965. The inhibitory action of various strains of *Bacillus* on *Listeria monocytogenes*. Nord Vet Med 17:336–341.

Leimeister-Wächter, M. and Chakraborty, T. 1989. Detection of listeriolysin, the thiol-dependent hemolysin in *Listeria monocytogenes*, *Listeria ivanovii* and *Listeria seeligeri*. Infection and Immunity 57:2350–2357.

Linnan, M. J., Mascola, L., Lou, X. D., Goulet, V., May, S., Salminen, C., Hird, D. W., Yonekura, M. L., Hayes, P., Weaver, R., Audurier, A., Plikaytis, B. D., Fannin, S. L., Kleeks, A., and Broome, C. V. 1988. Epidemic listeriosis associated with Mexican-style cheese. New England Journal of Medicine 319:823–828.

Lovett, J., Francis, D. W., and Hunt, J. M. 1987. *Listeria monocytogenes* in raw milk: detection, incidence and pathogenicity. Journal of Food Protection 50:188–192.

Ludwig, W., Schleifer, K. H., and Stackebrandt, E. 1984. 16S rRNA analysis of *Listeria monocytogenes* and *Brochothrix thermosphacta*. FEMS Microbiology Letters 25:199–204.

Lund, B. M. 1990. The prevention of foodborne listeriosis. British Food Journal 92:13–22.

Mackaness, G. B. 1962. Cellular resistance to infections. Journal of Experimental Medicine 116:381–406.

Mackaness, G. B. 1969. The influence of immunologically committed lymphoid cells on macrophage activity in vivo. Journal of Experimental Medicine 129:973–992.

Mackey, B. M. and Bratchell, N. 1989. A review: the heat resistance of *Listeria monocytogenes*. Letters in Applied Microbiology 9:89–94.

Mára, M. and Michalec, C. 1977. Chromatographic study of mycolic acid-like substances in lipids of *Listeria monocytogenes*. Journal of Chromatography 130:434–436.

McClure, P. J., Roberts, T. A., and Otto Oguru, P. 1989. Comparison of the effects of sodium chloride, pH and temperature on the growth of *Listeria monocytogenes* on gradient plates and in liquid medium. Letters in Applied Microbiology 9:95–100.

McLauchlin, J. 1987. A review. *Listeria monocytogenes*, recent advances in the taxonomy and epidemiology of listeriosis in humans. Journal of Bacteriology 63:1–11.

McLauchlin, J., Audurier, A., and Taylor, A. G. 1986. The evaluation of a phage-typing system for *Listeria monocytogenes* for use in epidemiological studies. Journal of Medical Microbiology 22:357–365.

Meyer, D. J. and Jones, C. W. 1973. Distribution of cytochromes in bacteria: relationship to general physiology. International Journal of Systematic Bacteriology 23:459–467.

Miller, I. L. and Silverman, S. J. 1959. Glucose metabolism of *Listeria monocytogenes*, p. 103. Bacteriological Proceedings. American Society for Microbiology.

Mounier, J., Ryter, A., Coquis-Rondon, M., and Sansonetti, P. J. 1990. Intracellular and cell-to-cell spread of *Listeria monocytogenes* involves interaction with F-actin in the enterocyte-like cell line Caco-2. Infection and Immunity 58:1048–1058.

Murray, E. G. D. 1953. The story of Listeria. Transactions of the Royal Society of Canada *XLVII* Series III:15–21.

Murray, E. G. D., Webb, R. A., and Swann, M. B. R. 1926. A disease of rabbits characterised by a large mononuclear leucocytosis, caused by a hitherto undescribed bacillus *Bacterium monocytogenes* (n. sp.) Journal of Pathology and Bacteriology 29:407–439.

Ortel, S. 1978. Untersuchungen über Monocine. Zentralblatt für Bakteriologie, Parasitenkunde, Infektionskrankheiten und Hygiene Abt 1. Orig A. 242:72–78.

Paterson, J. S. 1940. The antigenic structure of organisms of the genus *Listerella*. Journal of Pathology and Bacteriology 51:427–436.

Patterson, M. 1989. Sensitivity of *Listeria monocytogenes* to irradiation on poultry meat and in phosphate-buffered saline. Letters in Applied Microbiology 8:181–184.

Perez-Diaz, J. C., Vicente, M. F., and Baquero, F. 1981. Plasmids in *Listeria* p.33. In: Abstracts of VIIIth International Symposium on Problems of Listeriosis, Madrid. Ramon y Cajal, Madrid, Spain.

Piffaretti, J-C., Kressebuch, H., Aeschbacher, M., Bille, J., Bannerman, E., Musser, J. M., Selander, R. K., and Rocourt, J. 1989. Genetic characterization of clones of the bacterium *Listeria monocytogenes* causing epidemic disease. Proceedings of the National Academy of Sciences of the USA 86:3818–3822.

Pine, L., Malcolm, G. B., Brooks, J. B. and Daneshvar, M. L. 1989. Physiological studies on the growth and utilization of sugars by *Listeria* species. Canadian Journal of Microbiology 35:245–254.

Pine, L., Weaver, R. E., George, M. C., Pienta, P. A., Rocourt, J., Goebel, W., Kathariou, S., Bibb, W. F., and Malcolm, G. B. 1987. *Listeria monocytogenes* ATCC 35152 and NCTC 7973 contain a nonhemolytic, nonvirulent variant. Journal of Clinical Microbiology 25:2247–2251.

Pini, P. N. and Gilbert, R. J. 1988. The occurrence in the UK of *Listeria* species in raw chickens and soft cheeses. International Journal of Food Microbiology 6:317–326.

Pirie, J. H. H. 1927. A new disease of veld rodents "Tiger River Disease." Publication of the South African Institute of Research 3:163–186.

Pirie, J. H. H. 1940. *Listeria:* Change of name for a genus of bacteria. Nature 145:264.

Poyart-Salmeron, C., Carlier, C., Trieu-Cuot, P., Courtieu, A-L., and Courvalin, P. 1990. Transferable plasmid mediated antibiotic resistance in *Listeria monocytogenes*. Lancet 335:1422–1426.

Prentice, G. A. and Neaves, P. 1988. *Listeria monocytogenes* in food: its significance and methods for its detection. Bulletin of the International Dairy Federation 223:3–16.

Ralovich, B. 1975. Selective and enrichment media to isolate *Listeria*, p. 286–294. In: Woodbine, M. (ed.), Problems of listeriosis. Leicester University Press, Leicester, U.K.

Ralovich, B. 1984. Listeriosis research: present situation and perspective. Akademiai Kiado, Budapest.

Ralovich, B., Forray, A., Mero, E., Malovics, I., and Szazados, I. 1971. New selective medium for isolation of *L. monocytogenes*. Zentralblatt fur Bakteriologie, Mikrobiologie und Hygiene 1. Abt. Orig. 216:88–91.

Reed, P. 1986. "Listeria." University of Wisconsin Food Research Institute Annual Meeting. October, 1986. University of Wisconsin, Wisconsin, U.S.A.

Rocourt, J. 1986. Bacteriophages et bacteriocines du genre *Listeria*. Zentralblatt fur Bakteriologie Mikrobiologie und Hygiene A261:12–28.

Rocourt, J., Ackermann, H. W., Martin, M., Schrettenbrunner, A., and Selliger, H. P. R. 1983. Morphology of *Listeria innocua* bacteriophages. Annales de Virologie (Institut Pasteur) 134E:245–250.

Rocourt, J., Audurier, A., Courtieu, A-L., Durst, J., Ortel, S., Schrettenbrunner, A., and Taylor, A. G. 1985a. A multicentre study on phage typing of *Listeria monocytogenes*. Zentralblatt für Bakteriologie, Mikrobiologie und Hygiene A259:489–497.

Rocourt, J. and Catimel, B. 1985. Charactérisation biochimique des espèces du genre *Listeria*. Zentralblatt für Bakteriologie, Mikrobiologie und Hygiene A260:221–231.

Rocourt, J., Catimel, B., and Schrettenbrunner, A. 1985b. Isolement de bactériophages de *Listeria seeligeri* et *L. welshimeri*. Lysotypie de *L. monocytogenes, L. ivanovii, L. innocua, L. seeligeri* et *L. welshimeri*. Zentralblatt für Bakteriologie, Mikrobiologie und Hygiene A259:341–350.

Rocourt, J., Grimont, F., Grimont, P. A. D., and Seeliger, H. P. R. 1982a. DNA relatedness among serovars of *Listeria monocytogenes* sensu lato. Current Microbiology 7:383–388.

Rocourt, J. and Grimont, P. A. D. 1983. *Listeria welshimeri* sp. nov. and *Listeria seeligeri* sp. nov. International Journal of Systematic Bacteriology 33:866–869.

Rocourt, J., Schrettenbrunner, A., and Seeliger, H. P. R. 1982b. Isolation of bacteriophages from *Listeria monocytogenes* serovar 5 and *Listeria innocua*. Zentralblatt für Bakteriologie, Parasitenkunde, Infektionskrankheiten und Hygiene. Abt 1. 251:505–511.

Rocourt, J. and Seeliger, H. P. R. 1985. Distribution des espéces du genre *Listeria*. Zentralblatt für Bakteriologie, Mikrobiologie und Hygiene A259:317–330.

Rocourt, J., Wehmeyer, U., and Stackebrandt, E. 1987a. Transfer of *Listeria denitrificans* to a new genus *Jonesia* gen. nov. as *Jonesia denitrificans* comb. nov. International Journal of Systematic Bacteriology 37:266–270.

Rocourt, J., Wehmeyer, U., Cossart, P., and Stackebrandt, E. 1987b. Proposal to retain *Listeria murrayi* and *Listeria grayi* in the genus *Listeria*. International Journal of Systematic Bacteriology 37:298–300.

Rosenow, E. M. and Marth, E. H. 1987. Growth of *Listeria monocytogenes* in skim, whole and chocolate milk, and in whipping cream during incubation at 4, 8, 13, 21 and 35°C. Journal of Food Protection 50:452–459.

Ruhland, G. J. and Fiedler, F. 1987. Occurrence and biochemistry of lipoteichoic acids in the genus *Listeria*. Systematic and Applied Microbiology 9:40–46.

Schlech, W. F., Lavigne, P. M., Bortolussi, R. A., Allen, A. C., Haldane, E. V., Wort, A. J., Hightower, A. W., Johnson, S. E., King, S. H., Nicholls, E. S., and Broome, C. V. 1983. Epidemic listeriosis—evidence for transmission by food. New England Journal of Medicine 308:203–206.

Schleifer, K. H. and Kandler, O. 1972. Peptidoglycan types of bacterial cell walls and their taxonomic implications. Bacteriological Reviews 36:407–477.

Schultz, E. W. 1945. *Listerella* infections: a review. Stanford Medical Bulletin 3:135–151.

Seeliger, H. P. R. 1953. Zur Serodiagnostik der Listeriose mittels Agglutinations—und Komplementbindungsreaktion. Zeitschrift für Immun. Forsch. 110:252–264.

Seeliger, H. P. R. 1961. Listeriosis, 2nd ed. Karger, Basel.

Seeliger, H. P. R. 1981. Apathogene Listerien: *Listeria innocua* sp. n. (Seeliger et Schoofs 1977) Zentralblatt für Bakteriologie, Parasitenkunde Infektionskrankheiten und Hygiene. Abt 1. 249:487–493.

Seeliger, H. P. R. and Finger, H. 1985. Listeriosis, p. 264–289. In: Remington J. S., and Klein, J. O. (ed.), Infectious diseases of the fetus and new born infant, 2nd ed. W. B. Saunders, Philadelphia.

Seeliger, H. P. R. and Höhne, K. 1979. Serotyping of *Listeria monocytogenes* and related species, p. 31–49. In: Bergan, T. and Norris, J. R. (ed.), Methods in microbiology, vol 13. Academic Press, London, UK.

Seeliger, H. P. R. and Jones, D. 1986. The genus *Listeria,* p. 1235–1245. In: Sneath, P. H. A., Mair, N. S., and Sharpe, M. E. (ed.), Bergey's manual of systematic bacteriology, vol 2. Williams & Wilkins, Baltimore.

Seeliger, H. P. R. and Langer, B. 1989. Serological analysis of the genus *Listeria.* Its values and limitations. International Journal of Food Microbiology 8:245–248.

Seeliger, H. P. R., Rocourt, J., Schrettenbrunner, A., Grimont, P. A. D., and Jones, D. 1984. *Listeria ivanovii* sp. nov. International Journal of Systematic Bacteriology 34:336–337.

Seeliger, H. P. R. and Schoofs, M. 1979. Serological analysis of non-hemolyzing *Listeria* strains belonging to a species different from *Listeria monocytogenes,* p. 24–28. In: Ivanov, I. (ed), Problems of listeriosis. Proceedings of the VIIth International Symposium, Varna, 1977. National Agroindustrial Union, Center for Scientific Information, Sofia, Bulgaria.

Seeliger, H. P. R. and Welshimer, H. J. 1974. Genus *Listeria,* p 593–596. In: Buchanan, R. E. and Gibbons, N. E. (ed.), Bergey's manual of determinative bacteriology, 8th ed. Williams and Wilkins, Baltimore.

Shaw, N. 1974. Lipid composition as a guide to the classification of bacteria. Advances in Applied Microbiology 17:63–108.

Siddiqi, R. and Khan, M. A. 1989. Amino acid requirement of six strains of *Listeria monocytogenes.* Zentralblatt fur Bakteriologie. International Journal of Medical Microbiology 271:146–152.

Skalka, B. and Smola, J. 1983. Selective diagnostic medium for pathogenic *Listeria* spp. Journal of Clinical Microbiology 18:1432–1433.

Skalka, B., Smola, J., and Elischerová, K. 1982a. Different haemolytic activities of *Listeria monocytogenes* strains determined on erythrocytes of various sources and exploiting the synergism of the equi-factor. Zentralblatt Veterinärmed. Reihe B 29:642–649.

Skalka, B., Smola, J., and Elischerová, K. 1982b. Routine test for the in vitro differentiation of pathogenic and apathogenic *Listeria monocytogenes* strains. Journal of Clinical Microbiology 15:503–507.

Sneath, P. H. A., and Jones, D. 1976. *Brochothrix,* a new genus tentatively placed in the family Lactobacillaceae. International Journal of Systematic Bacteriology 26:102–104.

Sohier, R., Benazet F., and Piechaud, M. 1948. Sur un germe du genre *Listeria* apparemment non pathogene. Annales de l'Institut Pasteur 74:54–57.

Srivastava, K. K. and Siddique, I. H. 1973. Quantitative chemical composition of peptidoglycan of *Listeria monocytogenes.* Infection and Immunity 7:700–703.

Stackebrandt, E., Fowler, V. J., and Woese, C. R. 1983. A phylogenetic analysis of lactobacilli, *Pediococcus pentosaceus* and *Leuconostoc mesenteroides.* Systematic and Applied Microbiology 4:326–337.

Stackebrandt, E. and Woese, C. R. 1981. The evolution of prokaryotes, p. 1–31. In: Carlile, M. J., Collins, J. F., and Moseley, B. E. B. (ed.), Molecular and cellular aspects of evolution. Society for General Microbiology,

Symposium 32, Cambridge University Press, Cambridge.

Stuart, M. E. and Pease, P. E. 1972. A numerical study on the relationships of *Listeria* and *Erysipelothrix.* Journal of General Microbiology 73:551–565.

Stuart, R. D. 1959. Transport medium for specimens in Public Health bacteriology. Public Health Reports 74:431–438.

Stuart, S. E. and Welshimer, H. J. 1973. Intragenic relatedness of *Listeria* Pirie. International Journal of Systematic Bacteriology 23:8–14.

Stuart, S. E. and Welshimer, H. J. 1974. Taxonomic re-examination of *Listeria* Pirie and transfer of *Listeria grayi* and *Listeria murrayi* to a new genus *Murraya.* International Journal of Systematic Bacteriology 24:177–185.

Sun, A. N., Camilli, A., and Portnoy, D. A. 1990. Isolation of *Listeria monocytogenes* small-plaque mutants defective for intracellular growth and cell-to-cell spread. Infection and Immunity 58:3770–3778.

Sword, C. P. 1966. Mechanisms of pathogenesis in *Listeria monocytogenes* infection. 1. Influence of iron. Journal of Bacteriology 92:536–542.

Sword, C. P. and Pickett, M. J. 1961. The isolation and characterization of bacteriophages from *Listeria monocytogenes.* Journal of General Microbiology 25:241–248.

Tadayon, R. A. and Carroll, K. K. 1971. Effects of growth condition on the fatty acid composition of *Listeria monocytogenes* and comparison with the fatty acids of *Erysipelothrix* and *Corynebacterium.* Lipids 6:820–825.

Talon, R., Grimont, P. A. D., Grimont, F., Gasser, F., and Boeufgras, J. M. 1988. *Brochothrix campestris* sp. nov. International Journal of Systematic Bacteriology 38:99–102.

Tilney, L. G. and Portnoy, D. A. 1989. Actin filaments and growth, movement, and spread of the intracellular bacterial parasite, *Listeria monocytogenes.* The Journal of Cell Biology 109:1597–1608.

Trivett, T. L. and Meyer, E. A. 1971. Citrate cycle and related metabolism of *Listeria monocytogenes.* Journal of Bacteriology 107:770–779.

Ullmann, W. W. and Cameron, J. A. 1969. Immunochemistry of the cell walls of *Listeria monocytogenes.* Journal of Bacteriology 98:486–493.

Van Netten, P., Perales, I., and Mossel, D. A. A. 1988. An improved selective and diagnostic medium for isolation and counting of *Listeria* spp. in heavily contaminated foods. Letters in Applied Microbiology 7:17–21.

Vazquez-Boland, J-A., Dominguez, L., Rodriguez-Ferri, E. F., and Suarez, G. 1989. Purification and characterization of two *Listeria ivanovii* cytolysins, a sphingomyelinase C and a thiol-activated toxin (ivanolysin 0). Infection and Immunity 57:3928–3935.

Watkins, J. and Sleath, K. P. 1981. Isolation and enumeration of *Listeria monocytogenes* from sewage, sewage sludge, and river water. Journal of Applied Bacteriology 50:1–9.

Weiss, J. and Seeliger, H. P. R. 1975. Incidence of *Listeria monocytogenes* in nature. Applied Microbiology 30:29–32.

Welshimer, H. J. 1968. Isolation of *Listeria monocytogenes* from vegetation. Journal of Bacteriology 95:300–303.

Welshimer, H. J. 1981. The genus *Listeria* and related organisms, p. 1680–1687. In: Starr, M. P., Stolp, H., Trüper, H. G., Balows, A., and Schlegel, H. G. (ed.), The prokaryotes: a handbook on habitats, isolation, and identification of Bacteria. Springer-Verlag, Berlin.

Welshimer, H. J. and Donker-Voet, J. 1971. *Listeria monocytogenes* in nature. Applied Microbiology 21:516–519.

Wetzler, T. F., Freeman, N. R., French, M. L. V., Renkowski, L. A., Eveland, W. C. and Carver, O. J. 1968.

Biological characterisation of *Listeria monocytogenes*. Health Laboratory Science 5:46–62.

Wexler, H. and Oppenheim, J. D. 1979. Isolation, characterization and biological properties of an endotoxin-like material from the Gram-positive organism *Listeria monocytogenes*. Infection and Immunity 23:847–857.

Wilkinson, B. J. and Jones, D. 1977. A numerical taxonomic survey of *Listeria* and related bacteria. Journal of General Microbiology 98:399–421.

CHAPTER 72

The Genus *Brochothrix*

DOROTHY JONES

The genus *Brochothrix* contains Gram-positive, nonsporeforming, nonmotile, catalase-positive, facultatively anaerobic, regular, rod-shaped bacteria that show characteristic changes in cell morphology during growth. The genus was proposed by Sneath and Jones (1976) for some meat spoilage organisms, previously designated *Microbacterium thermosphactum* (McLean and Sulzbacher, 1953). The genus *Brochothrix* remained monospecific, containing only the type species *B. thermosphacta* (see Sneath and Jones, 1986), until Talon et al. (1988) described and named a second species, *B. campestris,* for isolates from soil and grass. Consequently, most of the information on the genus is derived from studies on *B. thermosphacta.*

Bacteria of the species *B. thermosphacta* have received a great deal of attention from taxonomists because of their former anomalous classification in the genus *Microbacterium* and from food microbiologists because of their association with off-odor development in meats, especially in prepacked products held at refrigeration temperatures. There is no evidence, however, that *Brochothrix* spp. are pathogenic to humans or animals. Neither species has been exploited in industrial processes.

History and Classification

The bacteria were first isolated and described by Sulzbacher and McLean (1951) during studies on pork sausage meat and were later allocated to the genus *Microbacterium* as a new species, *M. thermosphactum* (McLean and Sulzbacher, 1953). Their classification in the genus *Microbacterium* was due largely to the then poor circumscription of that genus (see Keddie and Jones, 1981). McLean and Sulzbacher (1953) noted the marked difference in cell morphology between *M. thermosphactum* and *M. lacticum,* the type species of the genus, and also commented on the close physiological resemblance between *M. thermosphactum* and the lactobacilli. However, at that time, both *Micro-*

bacterium and *Lactobacillus* were classified in the family Lactobacteriaceae and the main, distinction between the two genera was catalase production (Breed et al., 1948), therefore, McLean and Sulzbacher (1953) assigned their catalase-positive isolates to the genus *Microbacterium.*

Later workers confirmed and augmented the differences between *M. thermosphactum* and *M. lacticum,* not only in cell morphology (Davidson et al., 1968; Jones, 1975) but also in enzymology and protein profiles (Collins-Thompson et al., 1972; Robinson, 1966), in peptidoglycan structure (Schleifer, 1970; Schleifer and Kandler, 1972), and in DNA base composition (Collins-Thompson et al., 1972). In addition, numerical taxonomic studies showed that *M. thermosphactum* strains formed a relatively homogeneous taxon (intra-group similarity greater than 85%) quite distinct from *M. lacticum* (Davis and Newton, 1969; Davis et al., 1969; Jones, 1975; Wilkinson and Jones, 1977). The same studies indicated that the closest associates of *M. thermosphactum* were the genera *Listeria* and *Lactobacillus* and in one study, *Kurthia* (Davis and Newton, 1969). In none of the studies, however, was the similarity close enough to justify the inclusion of *M. thermosphactum* as a new species in any of these genera. Consequently, Sneath and Jones (1976) concluded that *M. thermosphactum* strains were sufficiently distinct from other Gram-positive bacteria to merit a separate genus and proposed that they be reclassified in a new genus *Brochothrix* as *B. thermosphacta*. These authors were aware that the taxonomic relatedness of *Brochothrix* to other Gram-positive bacteria at the suprageneric level was problematic. After evaluating all the data then available, including the presence of catalase and cytochromes in *B. thermosphacta* (Davidson and Hartree, 1968; Davidson et al., 1968) and the reported difference in fatty acid composition between *Brochothrix* and *Lactobacillus* (Shaw and Stead, 1970), they tentatively placed *Brochothrix* in the family Lactobacillaceae (Buchanan and

Gibbons, 1974); this family included the genus *Lactobacillus* and three other genera designated as having uncertain affiliation; *Listeria, Erysipelothrix,* and *Caryophanon.*

Subsequently, it became apparent that in phenotype, the members of the genus *Brochothrix* more closely resembled those of the genus *Listeria* than those of the genus *Lactobacillus* (see Sneath and Jones, 1986). Both *B. thermosphacta* and *Listeria* spp. possess catalase and cytochromes (Davidson and Hartree, 1968; Feresu and Jones, 1988). They also contain *meso*-diaminopimelic acid in the cell wall peptidoglycan (Schleifer and Kandler, 1972), possess menaquinones with seven isoprene units (MK-7) as the predominant isoprenoid quinone (Collins et al., 1979; Collins and Jones, 1981; Feresu and Jones, 1988) and contain predominantly methyl-branched chain fatty acids (Shaw, 1974; Feresu and Jones, 1988). This close phenotypic similarity between *B. thermosphacta* and *Listeria* spp. was confirmed by the results of the 16S rRNA oligonucleotide sequencing studies of Ludwig et al. (1984). These studies showed that *B. thermosphacta* and *L. monocytogenes* are phylogenetically very closely related and form one of the several sublines within the *Bacillus-Lactobacillus-Streptococcus* cluster of the clostridial subbranch of the Gram-positive eubacteria (Stackebrandt and Woese, 1981).

The species *B. campestris* was named by Talon et al. (1988) on the basis of numerical taxonomic and DNA hybridization studies of a number of strains of *Brochothrix* spp. isolated from a variety of sources, including grass and soil. As determined by the S1 nuclease method at 60°C, *B. campestris* strains exhibit about a 15% DNA relatedness to *B. thermosphacta* (Talon et al., 1988).

General Properties

Metabolism and Nutritional Requirements

Brochothrix spp. are aerobic and facultatively anaerobic (Sneath and Jones, 1986; Talon et al., 1988), but better growth is achieved by *B. thermosphacta* aerobically (see Gardner, 1981; Hitchener et al., 1979); no such information is yet available for *B. campestris.*

B. thermosphacta possesses enzymes for both the hexose-monophosphate and Embden-Meyerhof (glycolysis) pathways of glucose catabolism as well as a number of enzymes involved in pyruvate metabolism (Collins-Thompson et al., 1972; Grau, 1983). Fermentative metabolism of glucose always results in the production of L(+) lactic acid, but other end products depend on the growth conditions. McLean and Sulzbacher (1953) found only L(+) lactic acid present in sufficient quantities to detect. Davidson et al. (1968) detected small amounts of acetic and propionic acids in addition to L(+) lactic acid. Hitchener et al. (1979) found that in glucose-limited continuous culture under anaerobic conditions, L-lactate and ethanol were produced in the approximate ratio of 3:1. The studies of Grau (1983) showed the major end products of anaerobic glucose fermentation to be primarily L-lactate, acetate, formate, and ethanol, but that the ratios of these end products varied with conditions. Both the presence of acetate and formate and a pH below 6 increased L-lactate production from glucose. Of interest in this context is that the growth of *B. thermosphacta* is inhibited at a pH below 5 (Brownlie, 1966). Although McLean and Sulzbacher (1953) reported detectable CO_2 production (in Eldridge fermentation tubes), this has not been confirmed in any subsequent studies, but Hitchener et al. (1979) speculated that CO_2 was a likely end product in their studies on glucose-limited cultures.

The major end products of aerobic metabolism of glucose by *B. thermosphacta* growing on tryptone-based medium, on a minimal defined medium, or on meat are acetoin and acetic, isobutyric, isovaleric (3-methylbutyric), and 2-methylbutyric acids (Dainty and Hibbard, 1980 1983). In the minimal defined medium, glucose is believed to be the source of all the end products (Dainty and Hibbard, 1983); whereas in the complex medium and meat, only acetoin and acetic acid are derived from glucose; and isobutyric, isovaleric, and 2-methylbutyric acids are produced from valine, leucine, and isoleucine, respectively (Dainty and Hibbard, 1980, 1983). These compounds, or their derivatives, produce the sweet, sickly, malty odors that characterize the growth of *B. thermosphacta* (Dainty et al., 1985; Shaw and Dainty, 1985). The recent studies of Grau (1988) indicate that the substrates used by *B. thermosphacta* growing aerobically on meat include glucose, ribose, glycerol, glycerol-3-phosphate, and inosine; of these substrates only glucose and ribose are metabolized during anaerobic growth.

In cultures of *B. thermosphacta* grown on a tryptone-based medium, enzymes of the tricarboxylic acid (TCA) cycle are almost totally absent (Collins-Thompson et al., 1972). However, Grau (1979) has suggested, on the basis of studies with a defined medium, that the TCA-cycle enzymes may be sufficiently active to provide substrates for synthesis, but not to provide energy.

Collins-Thompson et al. (1971) demonstrated the presence of a glycerol ester hydrolase (lipase) in cell suspensions and cell-free extracts of *B. thermosphacta*. This lipase is active on tripropionin, tributyrin, tricaproin, tricaprylin and trilaurin but not on tripalmitin. Because the temperature optimum of the lipase is 35 to 37°C with little or no activity below 20°C, this enzyme is unlikely to be important in meat stored at refrigeration temperatures. The report that *B. thermospacta* cultures attack tributyrin (Sutherland et al., 1975) has not been confirmed by subsequent workers (Davis et al., 1969; Patterson and Gibbs, 1978).

B. thermosphacta requires cysteine, α-lipoate, nicotinate, pantothenate, *p*-aminobenzoate, biotin, and thiamin for growth (Grau, 1979). Thiamine can fulfill most, but not all, of the yeast extract requirements (Macaskie et al., 1981).

The production of cytochromes and catalase by *B. thermosphacta* depends on both the composition of the growth medium and the temperature of incubation (Davidson and Hartree, 1968; Davidson et al., 1968). Davidson et al. (1968) noted that *B. thermosphacta* strains grown on APT medium (BBL, Evans and Niven, 1951) incubated at 20°C were always catalase-positive but that weak or negative reactions were obtained on HIA (heart infusion agar) medium (Difco) incubated at the same temperature. The same authors reported that negative results were frequently obtained if the bacteria were grown on either medium at 30°C. Davidson and Hartree (1968) reported the same effects of growth medium and incubation temperature on the quantitative cytochrome content of the organism. On APT medium incubated at 20°C, *B. thermosphacta* contains cytochromes baa_3 (Davidson and Hartree, 1968). The temperature effect is difficult to explain; although not the optimum temperature for growth, *B. thermosphacta* grows well at 30°C. The effect of the composition of the culture medium may be related to the iron concentration. Davidson et al. (1968) noted that APT medium contains added iron (8.0 μg/ml), and more recently both Grau (1979) and Thomson and Collins-Thompson (1986) have noted a high ferric iron requirement for the aerobic growth of *B. thermosphacta* in defined media. More recently it was shown that manganese can partially replace the iron requirement of *B. thermosphacta* under iron-limiting conditions (Thomson and Collins-Thompson, 1988).

B. thermosphacta is oxidase negative (cytochrome *c* oxidase).

Temperature and pH

There is general agreement that the temperature limits of growth are between 0 and 30°C (see Gardner, 1981; Sneath and Jones, 1976; Talon et al., 1988). Limited growth of *B. thermosphacta* was noted at 35°C and one strain was reported to grow at 37 and 45°C (Gardner, 1981), but growth above 30°C has been rarely found. *B. campestris* does not grow at 37°C (Talon et al., 1988). The optimum temperature for growth is 20 to 25°C (see Gardner, 1981). The heat resistance of *B. thermosphacta* has received much attention because of its former classification in the genus *Microbacterium*, members of which are thermoduric. All workers agree that *B. thermosphacta* does not survive heating at 63°C for 5 min (see Gardner, 1981; Sneath and Jones, 1986).

The optimum pH for growth of *B. thermosphacta* is pH 7.0 but growth occurs within the range pH 5 to 9 (Brownlie, 1966).

Inhibitory Substances

The ability of *B. thermosphacta* to grow in the presence of NaCl has been examined by many workers (see Gardner, 1981). All strains grow at 6.5% NaCl and the majority can grow at NaCl concentrations up to 10% (Talon et al., 1988; Wilkinson and Jones, 1977). *B. campestris* strains do not grow in media with 8 or 10% NaCl and although there is no published information to indicate growth at 6.5% NaCl, circumstantial evidence indicates that *B. campestris* grows with 6.5% NaCl.

Growth of *B. thermosphacta* is inhibited by nitrite but the degree of inhibition is related to the pH of the medium and incubation temperature (Brownlie, 1966); low pH, low temperature, and high nitrite increase the inhibitory effect. The species does not appear to have a nitrite reductase system (Collins-Thompson and Rodriguez-Lopez, 1980). Dainty and Meredith (1972) found that 200-ppm nitrite at pH 5.5 stops RNA, DNA, and protein synthesis in *B. thermosphacta* but has no effect on membrane permeability. An interesting study on the combined effects of NaCl, $NaNO_2$, temperature, and pH on the growth of *B. thermosphacta* in broth cultures was done by Roberts et al. (1979).

In the United Kingdom, sulfur dioxide is permitted in sausage meat and *B. thermosphacta* tolerates SO_2 up to 500 ppm under both aerobic and anaerobic conditions (Dowdell and Board, 1971).

The studies of Macaskie (1982) indicate that in liquid culture palmitic acid is inhibitory to the growth of *B. thermosphacta*.

B. thermosphacta is more resistant to irradiation than common meat spoilage organisms such as *Pseudomonas.* The species has been frequently isolated from irradiated meat and poultry (see Gardner, 1981).

Serology

There do not appear to have been any systematic serological studies of *Brochothrix* spp. Wilkinson and Jones (1975) did not detect any serological relationship between *B. thermosphacta* and the genera *Erysipelothrix, Kurthia* or *Listeria* with antisera raised against representative strains of all four taxa.

Esterase Isoenzymes

Five different esterase isoenzymes have been detected by gel electrophoresis studies of *B. thermosphacta.* Different combinations of these enzymes resulted in the detection of seven groups among 26 strains examined. There was no correlation between source of isolation and groups based on esterase patterns (G. A. Gardner, personal communication).

Bacteriophages

Greer (1983) isolated bacteriophages active on *B. thermosphacta* from aqueous extracts of spoiled beef. Both phage plaque size and plating efficiency increased significantly when the incubation temperature was reduced from 25° to 1°C. The detection of 14 distinct phage lysotypes led Greer (1983) to suggest that phage typing may provide a rapid method of differentiating *B. thermosphacta* strains.

On the basis of their morphology as determined by electron microscopy, Ackermann et al. (1988) grouped these bacteriophages (and two isolated in France from broth cultures of lysogenic *B. thermosphacta* strains) into three viral species of the Myoviridae (species A19) or Siphoviridae (species NF5 and BL3). The bacteriophage species A19 are interesting because of their similarity to some bacteriophage species of the genera *Bacillus, Lactobacillus, Staphylococcus,* and *Streptococcus.* Surprisingly, none of the bacteriophages resembled those of the genus *Listeria* (Ackermann et al., 1988).

Plasmids

Dodd and Waites (1988) detected plasmids in all strains of *Brochothrix thermosphacta* isolated from sausages containing 450ppm sulfite, but not in the type strain (NCTC 10082). The plasmids ranged in size from 1,600 kDa to 43,000 kDa and plasmid numbers varied from one to eight per cell. Comparison of the plasmid profiles of the isolates showed that fewer profiles were present amongst the strains isolated from sausages stored for various periods of time at 4°C than were present amongst strains isolated from the same sausages on the day of manufacture. Dodd and Waites (1988) interpreted these observations as indicating that only some strains of *B. thermosphacta* were able to survive the conditions of storage.

Chemical Composition

Other than the presence of *meso*-diaminopimelic acid in the cell wall of *B. campestris* (Talon et al., 1988) and *B. thermosphacta* (Schleifer, 1970; Schleifer and Kandler, 1972), all the information on chemical composition is based on studies with *B. thermosphacta.* Arabinose and galactose are not present in the cell wall (see Schleifer and Kandler, 1972), nor are mycolic acids (Minnikin et al., 1978; Feresu and Jones, 1988). The major fatty acid is 12-methyltetradecanoic (*anteiso*-$C_{15:0}$ with substantial amounts of 14-methylhexadecanoic (*anteiso*-$C_{17:0}$) and 13-methyltetradecanoic (iso-$C_{15:0}$) acids (Feresu and Jones, 1988; Shaw and Stead, 1970). The major phospholipids are phosphatidylglycerol, diphosphatidylglycerol, and phosphatidylethanolamine. The glycolipid fraction contains an acylated glucose and small amounts of a glycosyl diglyceride tentatively identified as dimannosyl diglyceride (Shaw, 1974; Shaw and Stead, 1970).

Menaquinones are the sole respiratory quinones; MK-7 is the major component, and MK-6 and MK-5 are minor components (Collins et al., 1979; Collins and Jones, 1981; Feresu and Jones, 1988).

The GC content of the DNA of *B. thermosphacta* is 36–37 mol% (Collins et al., 1987; Collins-Thompson et al., 1972; Feresu and Jones, 1988) and of *B. campestris,* 38 mol% (Talon et al., 1988).

Habitats

The natural habitat of *Brochothrix* spp. is not known with certainty. This is mainly because for many years after the first reported isolation of the type species, *B. thermosphacta,* from finished pork sausage and its subsequent incrimination as an important psychrotophic spoilage organism of meat and meat products, most investigators concentrated only on these materials (see Gardner, 1981; Keddie and Jones, 1981). During studies on the development of a selective medium for the enumeration of *B. thermosphacta* from meat sources, Gardner (1966)

After incubation of appropriate samples on this medium at 20–22°C for 48 h, a large majority of the colonies are *B. thermosphacta;* the exceptions are a few pseudomonads.

Modified medium STAA for Selective Isolation of *Brochothrix* spp. from Grass, Soil, Feces, and Similar Material (Talon et al., 1988)

Prepare as for STAA medium but supplement with nalidixic acid (15 μg/ml) and oxacillin (5 μg/ml).

After incubation of appropriate samples on this medium at 20–22°C for 48 h, a large majority of the colonies are *Brochothrix* spp.

Enrichment of *Brochothrix* spp.

Enrichment of *Brochothrix* spp. is not usually performed, but holding meat or meat product samples under gas-permeable film at temperatures below 10°C can act as enrichment. Similar enrichment for other materials, grass or soil, could be useful but apparently has not been attempted.

Preservation of Cultures

Cultures may be preserved for short periods (6 months to 1 year) in nutrient agar (plus 0.1% glucose) stabs in screw-capped bottles stored in the dark at room or refrigeration temperature. Longer-term preservation (over 10 years with *B. thermosphacta*) may be achieved by freezing on glass beads at −60 to −70°C (Jones et al., 1984). Cultures may also be preserved by freeze drying (lyophilization).

Identification of *Brochothrix*

Identification of new isolates as *Brochothrix* spp. requires examination of cellular morphology and staining reactions, the maximum growth temperature, and relationships to oxygen and catalase production together with a number of conventional taxonomic tests, such as the ability to produce acid from various sugars. The chemical composition of the organisms also aids the identification of *Brochothrix* spp. but it is not usually necessary to perform the relevant analyses for routine identification. Features that differentiate the genus from other Gram-positive, nonsporeforming, rod-shaped bacteria are listed in Table 1.

The colonial morphology of *Brochothrix* spp. is not particularly diagnostic. After 24–48 h the colonies are circular, 0.75–1 mm in diameter, convex with entire margins, and not pigmented. In young cultures of *B. thermosphacta*, two types of colony varying in size and density may be present. These can be so distinct that the culture may appear to be contaminated (see Barlow and Kitchell, 1966). In older cultures of the same species, the edge of the colony breaks up and the center becomes raised to give a "fried egg" appearance. Neither of these two phenomena has been reported for *B. campestris* (Talon et al., 1988). *Brochothrix* spp. are non-hemolytic, but sometimes an area of weak greening is apparent around colonies on blood agar.

Gram-stains should be performed on 18–24 h and 2-day cultures grown on nutrient agar such as blood agar base (BAB no. 2, Oxoid) or APT medium incubated at 20–25°C. Exponential-phase cultures of *B. thermosphacta* show regular, unbranched rods that occur singly in pairs and short chains and in long, kinked, filamentous-like chains (Fig. 1a) that bend and loop to give characteristic knotted masses. This phenomenon has not yet been reported for *B. campestris;* in this species 24-h cultures are reported to consist of a mixture of long and short rods that usually occur singly or in pairs (Talon et al., 1988). In older cultures of *B. thermosphacta*, the rods give rise to coccoid forms (Fig. 1b), that when subcultured onto a suitable medium develop into rod forms. Both rod and coccal forms are Gram-positive, but a proportion may appear Gram-negative. They are nonmotile and do not form endospores or capsules. Talon et al. (1988) reported that about one-half of the *B. thermosphacta* strains they examined produced slime from sucrose in broth culture.

Brochothrix spp. are readily distinguishable from the morphologically similar members of the genus *Kurthia* by examining the oxygen requirements. *Brochothrix* is facultatively anaerobic whereas *Kurthia* is strictly aerobic. In addition, *Brochothrix* spp. produce acid from a wide variety of sugars. Most strains of *Kurthia* are highly motile.

The inability of *Brochothrix* spp. to grow at 35°C distinguishes them from members of the genera *Carnobacterium, Erysipelothrix, Lactobacillus,* and *Listeria.* Although facultatively anaerobic, *Brochothrix* spp. grow best aerobically. This feature, together with their ability to grow on unsupplemented nutrient agar, serves to distinguish *Brochothrix* spp. from members of the genera *Erysipelothrix* and *Lactobacillus* but not from *Carnobacterium* and *Listeria.*

The production of catalase distinguishes *Brochothrix* spp. from the genera *Carnobacterium, Erysipelothrix,* and the vast majority of *Lactobacillus* spp., but not from *Listeria.* But as noted previously, care must be taken in examining *Brochothrix* for catalase because its production depends on both the growth medium and the incubation temperature. Growth

Table 1. Features most useful in differentiating *Brochothrix* from morphologically similar genera.

	Brochothrix	*Listeria*	*Carnobacterium*	*Lactobacillus*	*Erysipelothrix*	*Kurthia*
Growth at 35°C	−	+	+	+	+	+
Motility	−	+	−a	−b	−	+
Catalase	+	+	−	−c	−	+
Facultatively anaerobic	+	+	+	+	+	−
Acid from glucose	+	+	+	+	+	−
H_2S production	−	−d	−	−	+	−d
Major peptidoglycan diamino acid	*meso*-A_2pm[e]	*meso*-A_2pm	*meso*-A_2pm	Lysine or *meso*-A_2pm or ornithine	L-lysine	L-lysine
Major menaquinone	MK-7[f]	MK-7	−	−g	−	MK-7
Major fatty acid types	S,A,I	S,A,I	S,U(C)	S,U(C)	S,A,I,U	S,A,I
GC mol%	36–38	36–42	33–37	32–53	36–40	36–38

Symbols: +, 90% or more of strains positive; −, 90% or more of strains negative.

a Most strains nonmotile, but *C. mobile* strains motile (Collins et al., 1987).

b Most strains nonmotile, but a few motile strains occur.

c Some strains give positive catalase reaction.

d Weak production of H_2S by some strains of *L. grayi* and *K. zopfii*.

e *meso*-A_2pm, *meso*-diaminopimelic acid (DAP).

f MK-7, menaquinone with 7 isoprenoid units.

g A very few lactobacilli contain menaquinones.

h S, straight chain saturated; U, monounsaturated; A, anteiso-methyl-branched; I, iso-methyl-branched; C, cyclopropane ring fatty acids. Those in parenthesis may be present.

Adapted from Collins et al., 1987; Feresu and Jones, 1988; Sneath and Jones, 1986; and Talon et al., 1988.

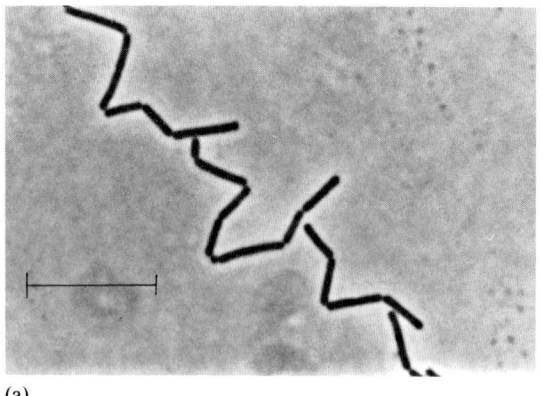
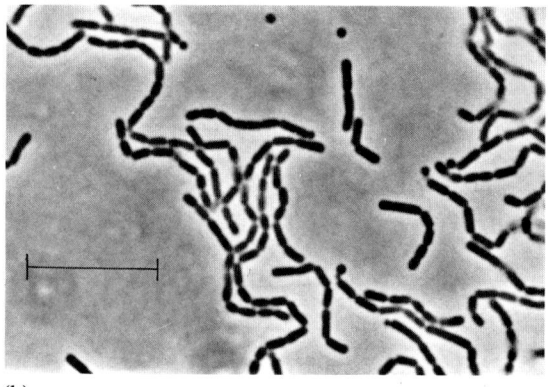

(a)   (b)

Fig. 1. *Brochothrix thermosphacta* (*Microbacterium thermosphactum*) (isolate) when grown on blood agar base no. 2 (Oxoid) at 25°C. (a) After 24 h, showing regular rods in chains. (b) After 48 h, showing development of coccoid forms. Bar = 10 μm.

on APT medium, BAB no. 2 or a similar medium for 24–48 h at 20–25°C is recommended; negative results are very frequently obtained if the bacteria are incubated at 30°C. Although catalase production by *Listeria* spp. is also dependent on the composition of the growth medium (Jones, 1975), all strains of *Listeria* are unequivocally catalase-positive at 35°C if the growth medium is suitable, whereas *Brochothrix* spp. do not grow at this temperature.

The genera *Brochothrix* and *Listeria* share many common characters. Members of both genera are facultatively anaerobic but grow better aerobically; produce acid from a variety of sugars; produce catalase, cytochromes, and menaquinones; possess a cell wall peptidoglycan with *meso*- diaminopimelic acid as the diamino acid; have similar polar lipid and fatty acid profiles and similar GC content (Table 1). Members of neither genus grow on acetate medium and only poorly on MRS (De Man et al., 1960) medium. However, they may be distinguished by their morphology, growth temperature, motility, and also serologically.

Colonies of the genus *Brochothrix* do not show the blue-green coloration exhibited by *Listeria* spp. when viewed by obliquely transmitted white light. The inability of *Brochothrix* spp. to

Table 2. Additional and differential features of *Brochothrix* spp.

	B. thermosphacta	B. campestris
Growth with:		
NaCl, 8% (2 days)[a]	+	−
NaCl, 10% (2–7 days)	d	−
Growth with and reduction of potassium tellurite, 0.05%[a]	+	−
Reduction of tetrazolium, 0.01%	d	+
Slime produced from sucrose	d	−
Hydrolysis of:		
Cellulose	−	ND
Hippurate[a]	−	+
Tween 20	−	ND
Tween 40	−	ND
Tween 60	d	ND
Tween 80	−	ND
Tyrosine	−	ND
Xanthine	−	ND
Production of:		
Phosphatase	+	ND
Sulfatase	−	ND
Acid from:		
Amygdalin	+	+
Arabinose	d	ND
Arbutin	d	+
Cellobiose	+	+
Dulcitol	+	ND
Galactose	+	ND
Gentobiose	+	d
Glycerol	+	+
Inositol	+	d
Inulin	d	ND
Lactose	+	ND
Mannitol	+	d
Melezitose	d	ND
Melibiose	d	ND
Raffinose	+	ND
Rhamnose[a]	−	+
Sorbitol	d	ND
Sorbose	−	ND
Starch	−	d
Sucrose	+	d
Tagatose	+	d
Xylose	d	ND
Resistance to furadoine	d	−

Symbols: +, 90% or more strains are positive; d, 11–89% of strains are positive; −, 90% or more of strains are negative; ND, no data.
[a]Features essential for identification.
Adapted from Feresu and Jones, 1988; Talon et al., 1988; and Wilkinson and Jones, 1977.

grow at 35°C and their lack of motility also distinguishes them from the genus *Listeria*. No serological cross-reactions have been demonstrated between *B. thermosphacta* and species of the genus *Listeria*.

In addition to the features already described, *B. thermosphacta* and *B. campestris* produce acid fermentatively from glucose, ribose, fructose, mannose, *N*-acetylglucosamine, salicin, maltose, and trehalose; are methyl-red and Voges-Proskauer positive. Esculin is hydrolyzed. Exogenous citrate and urea are not utilized. Indol and H_2S are not produced, nitrate is not reduced, gelatin is not liquefied, and arginine is not hydrolyzed (McLean and Sulzbacher, 1953; Sneath and Jones, 1976; Talon et al., 1988; Wilkinson and Jones, 1977).

With use of the API ZYM test system (API System, La Balme des Grottes, France) both species are reported to produce the following arylamidases: phenylalanine, histidine, glycylphenylalanine, seryl-tyrosine, glutamate, tryptophan, and histidyl-L-phenylalanine (Talon et al., 1988).

In the API esterase tests, Talon et al. (1988) noted that all five strains of *B. campestris,* but only 12 of 165 strains of *B. thermosphacta* tested, hydrolyzed naphthylbutyrate, naphthylvalerate, naphthylcaproate, naphthylcaprylate, naphthylnonanoate, and naphthylcaprate.

Both species are susceptible to a wide range of antibiotics. Using antibiotic disks on Mueller-Hinton agar, Talon et al. (1988) reported all strains of *B. thermosphacta* (104 tested) and *B. campestris* (5 tested) were sensitive to novobiocin, tetracycline (except for three strains of *B. thermosphacta*), amikacin, tobramycin, gentamicin, and ampicillin and were resistant to oxacillin, nalidixic acid, and colistin (except the type strain of both species and one other strain of *B. campestris*). A wider range of antibiotics was tested with *B. thermosphacta* (15 strains) on nutrient agar medium by Feresu and Jones (1988).

Further diagnostic information on both species is given in Table 2.

Sneath and Jones (1976) reported that *B. thermosphacta* does not produce deoxyribonuclease but later studies have detected both deoxyribonuclease and ribonuclease activity in strains of this species (Feresu and Jones, 1988; Wilkinson and Jones, 1977).

The results of many of the conventional tests can be influenced markedly by the media and methods used. Differences in methodology probably account for some of the discrepancies noted in the literature. A most reliable method for acid production from carbohydrates is the API 50CH test system. Media and methods such as those used by Talon et al. (1988) and Wilkinson and Jones (1977) are suitable for most tests and incubation should be at 20–25°C.

Differentiation of *Brochothrix* spp.

The most useful characters for differentiating the two species are growth in the presence of 8 and 10% NaCl, growth on BM medium (Wilkinson and Jones, 1977) containing 0.05% potassium tellurite, hippurate hydrolysis, and production of acid from rhamnose (see Table 2).

The type strain of *B. thermosphacta* is ATCC 11509 (NC1B 10018) and that of *B. campestris* is ATCC 43754 (CIP 10290 = S₃ Talon et al., 1988). Both are suitable reference strains, but it should be noted that *B. campestris* ATCC 43754 grows with 8% NaCl if incubated for 6 days (Talon et al., 1988).

Literature listed

Ackermann, H. W., Greer, G. G. and, Rocourt, J. 1988. Morphology of *Brochothrix thermosphacta* phages. Microbios 56:19–26.

Ayres, J. C. 1960. Temperature relationship and some other characteristics of the microbial flora developing on refrigerated beef. Food Research 25:1–18.

Bailey, J. S., Regan, J. O., Carpenter, J. A., Schuler, G. A., and Thompson, J. E. 1979. Types of bacteria and shelf life of evacuated carbon dioxide injected and ice-packed broilers. Journal of Food Protection 42:218–221.

Barlow, J. and, Kitchell, A. G. 1966. A note on the spoilage of prepacked lamb chops by *Microbacterium thermosphactum*. Journal of Applied Bacteriology 29:185–188.

Breed, R. S., Murray, E. G. D., and Hitchens, A. P. (ed.). 1948. Bergey's manual of determinative bacteriology, 6th ed. Williams & Wilkins, Baltimore.

Brownlie, L. E. 1966. Effects of some environmental factors on psychrophilic microbacteria. Journal of Applied Bacteriology 29:447–454.

Buchanan, R. E., and Gibbons, N. E. (ed.). 1974. Bergey's manual of determinative bacteriology, 8th ed. Williams & Wilkins, Baltimore.

Collins, M. D., Farrow, J. A. E., Phillips, B. A., Feresu, S., and Jones, D. 1987. Classification of *Lactobacillis divergens, Lactobacillus piscicola*, and some catalase-negative, asporogenous, rod-shaped bacteria from poultry in a new genus *Carnobacterium*. International Journal of Systematic Bacteriology 37:310–316.

Collins, M. D., Goodfellow, M., and Minnikin, D. E. 1979. Isoprenoid quinones in the classification of coryneform and related bacteria. Journal of General Microbiology 110:127–136.

Collins, M. D., and Jones, D. 1981. The distribution of isoprenoid quinone structural types in bacteria and their taxonomic implications. Microbiology Reviews 45:316–354.

Collins-Thompson, D. L., and Rodriguez-Lopez, G. 1980. Influence of sodium nitrite, temperature and lactic acid bacteria on the growth of *Brochothrix thermosphacta* under anaerobic conditions. Canadian Journal of Microbiology 26:1416–1421.

Collins-Thompson, D. L. Sorhaug, T., Witter, L. D., and Ordal, Z. J. 1971. Glycerol ester hydrolase activity in *Microbacterium thermosphactum*. Applied Microbiology 21:9–12.

Collins-Thompson, D. L., Sorhaug, T., Witter, L. D., and Ordal, Z. J. 1972. Taxonomic consideration of *Microbacterium lacticum, Microbacterium flavum* and *Microbacterium thermosphactum*. International Journal of Systematic Bacteriology 22:65–72.

Dainty, R. H., Edwards, R. A., and Hibbard, C. M. 1985. Time course of volatile compound formation during refrigerated storage of naturally contaminated beef in air. Journal of Applied Bacteriology 59:303–309.

Dainty, R. H., and Hibbard, C. M. 1980. Aerobic metabolism of *Brochothrix thermosphacta* growing on meat surfaces and in laboratory media. Journal of Applied Bacteriology 48:387–396.

Dainty, R. H., and Hibbard, C. M. 1983. Precursors of the major end-products of aerobic metabolism of *Brochothrix thermosphacta*. Journal of Applied Bacteriology 55:127–133.

Dainty, R. H., and Meredith, G. C. 1972. Mechanisms of inhibition of growth of bacteria by nitrite, p. 82. In: Meat Research Institute Annual Report 1972–73. HMSO, London.

Dainty, R. H., Shaw, B. G., Harding, C. D., and Michanie, S. 1979. The spoilage of vacuum-packed beef by cold tolerant bacteria, p. 83–100. In: Russell, A. D., Fuller, R. (ed.), Cold tolerant microbes in spoilage and the environment. Academic Press, London.

Davidson, C. M., and Hartree, E. F. 1968. Cytochrome as a guide to classifying bacteria: Taxonomy of *Microbacterium thermosphactum*. Nature 220:502–504.

Davidson, C. M., Mobbs, P., and Stubbs, J. M. 1968. Some morphological and physiological properties of *Microbacterium thermosphactum*. Journal of Applied Bacteriology 31:551–559.

Davis, G. H. G., Fomin, L., Wilson, E., and Newton, K. G. 1969. Numerical taxonomy of *Listeria,* streptococci and possibly related bacteria. Journal of General Microbiology 57:333–348.

Davis, G. H. G., and Newton, K. G. 1969. Numerical taxonomy of some named coryneform bacteria. Journal of General Microbiology 56:195–214.

De Man, J. C., Rogosa, M., and Sharpe, M. E. 1960. A medium for the cultivation of lactobacilli. Journal of Applied Bacteriology 23:130–135.

Dodd, C. E. R., and Waites, W. 1988. The use of plasmid profiling to determine strains of *Brochothrix thermosphacta* important in spoilage of sausages. Journal of Applied Bacteriology 65:XV.

Dowdell, M. J., and Board, R. G. 1971. The microbial associations in British fresh sausages. Journal of Applied Bacteriology 34:317–337.

Evans, J. B. and Niven, C. F., Jr. 1951. Nutrition of the heterofermentative lactobacilli that cause greening of cured meat products. Journal of Bacteriology 62:599–603.

Feresu, S. B., and Jones, D. 1988. Taxonomic studies on *Brochothrix, Erysipelothrix, Listeria* and atypical lactobacilli. Journal of General Microbiology 134:1165–1183.

Fournaud, J., Degas, T., Schmitt, O., and Sechet, J. 1980. Penetration des bacteries dans la viande, p.268–271. In: Proceedings of the 26th European Meeting of Meat Research Workers, 2, Colorado Springs, CO.

Gardner, G. A. 1966. A selective medium for the enumeration of *Microbacterium thermosphactum* in meat and meat products. Journal of Applied Bacteriology 29:455–460.

Gardner, G. A. 1981. *Brochothrix thermosphacta (Microbacterium thermosphactum)* in the spoilage of meats: a review, p.139–173. In: Roberts, T. A., Hobbs, G. A., Christian, J. H. B., Skovgaard, N. (ed.), Psychrotrophic microorganisms in spoilage and pathogenicity. Academic Press, London.

Gardner, G. A., Carson, A. W., and Patton, J. 1967. Bacteriology of prepacked pork with reference to the gas composition within the pack. Journal of Applied Bacteriology 30:321–333.

Gill, C. O., and Newton, K. G. 1977. The development of aerobic spoilage flora on meat stored at chill temperatures. Journal of Applied Bacteriology 43:189–195.

Gill, C. O., and Penney, N. 1977. Penetration of bacteria into meat. Applied and Environmental Microbiology 33:1284–1286.

Grau, F. H. 1979. Nutritional requirements of *Microbacterium thermosphactum*. Applied and Environmental Microbiology 38:818–820.

Grau, F. H. 1983. End products of glucose fermentation by *Brochothrix thermosphacta*. Applied and Environmental Microbiology 45:84–90.

Grau, F. H. 1988. Substrates used by *Brochothrix thermosphacta* when growing on meat. Journal of Food Protection 51:639–642.

Greer, G. G. 1983. Psychrotrophic *Brochothrix thermosphacta* bacteriophages isolated from beef. Applied and Environmental Microbiology 46:245–251.

Hitchener, B. J., Egan, A. F., and Rogers, P. J. 1979. Energetics of *Microbacterium thermosphactum* in glucose-limited continuous culture. Applied and Environmental Microbiology 37:1047–1052.

Holzapfel, W. H., and Hall, A. N. 1976. The microbiology of South African dried sausage. South African Journal of Animal Science 6:199–206.

Ingram, M., and Dainty, R. H. 1971. Changes caused by microbes in spoilage of meats. Journal of Applied Bacteriology 34:21–39.

Jones, D. 1975. A numerical taxonomic study of coryneform and related bacteria. Journal of General Microbiology 87:52–96.

Jones, D., Pell, P. A., and Sneath, P. H. A. 1984. Maintenance of bacteria on glass beads at −60°C to −76°C, p. 35–40. In: Kirsop, B. E., Snell, J. J. S. (ed.), Maintenance of microorganisms. A manual of laboratory methods. Academic Press, London.

Keddie, R. M., and Jones, D. 1981. Saprophytic, aerobic coryneform bacteria, p. 1838–1878. In: Starr, M. P., Stolp, H., Trüper, H. G., Balows, A., Schlegel, H. G. (ed.), The prokaryotes: a handbook on habitats, isolation and identification of bacteria. Springer-Verlag, Berlin.

Lannelongue, M., Hanna, M. O., Finne, G., Nickelson, R., and Vanderzant, C. 1982. Storage characteristics of finfish fillets (*Archosargus probatocephalus*) packaged in modified gas atmospheres containing carbon dioxide. Journal of Food Protection 45:440–444.

Ludwig, W., Schleifer, K. H., and Stackebrandt, E. 1984. 16S rRNA analysis of *Listeria monocytogenes* and *Brochothrix thermosphacta*. FEMS Microbiology Letters 25:199–204.

Macaskie, L. E. 1982. Inhibition of growth of *Brochothrix thermosphacta* by palmitic acid. Journal of Applied Bacteriology 52:339–343.

Macaskie, L. E., Dainty, R. H., and Henderson, P. J. F. 1981. The role of thiamin as a factor for the growth of *Brochothrix thermosphacta*. Journal of Applied Bacteriology 50:267–273.

McLean, R. A., and Sulzbacher, W. L. 1953. *Microbacterium thermosphactum*, spec. nov.; a nonheat resistant bacterium from fresh pork sausage. Journal of Bacteriology 65:428–433.

McMeekin, T. A. 1975. Spoilage association of chicken breast muscle. Applied Microbiology 29:44–47.

Minnikin, D. E., Goodfellow, M., and Collins, M. D. 1978. Lipid composition in the classification and identification of coryneform and related taxa. p. 85–160. In: Bousfield, I. J., Callely, A. G. (ed.), Coryneform bacteria. Special Publications of the Society for General Microbiology. Academic Press, London.

Mulder, S. J. 1978. *Microbacterium thermosphactum:* spoilage indicator of beef. Paper B4. In: Proceedings of the 24th European Meeting of Meat Research Workers, no. 1. Kulmbach, Germany.

Newton, K. G., Harrison, J. C. L., and Wauters, A. M. 1978. Sources of psychrotrophic bacteria on meat at the abattoir. Journal of Applied Bacteriology 45:75–82.

Nickelson, R., Finne, G., Hanna, M. O., and Vanderzant, C. 1980. Minced fish flesh from non-traditional Gulf of Mexico finfish species: bacteriology. Journal of Food Science 45:1321–1326.

Nychas, G. J., Dillon, V. M., and Board, R. G. 1988. Glucose, the key substrate in the microbiological changes occurring in meat and certain meat products. Biotechnology and Applied Biochemistry 10:203–231.

Paradis, D. C., and Stiles, M. E. 1978. A study of microbial quality of vacuum packaged sliced bologna. Journal of Food Protection 41:811–815.

Patterson, J. T., and Gibbs, P. A. 1977. Incidence and spoilage potential of isolates from vacuum-packed meat of high pH. Journal of Applied Bacteriology 43:25–38.

Patterson, J. T., and Gibbs, P. A. 1978. Sources and properties of some organisms isolated in two abattoirs. Meat Science 2:263–273.

Pierson, M. D., Collins-Thompson, D. L., and Ordal, Z. J. 1970. Microbiological, sensory and pigment changes of aerobically and anaerobically packaged beef. Food Technology 24:1171–1175.

Prior, B. A., and Casaleggio, C. 1978. The microbiology of polony. Journal of the South African Veterinary Association 49:115–119.

Roberts, T. A., Britton, C. R., and Shroff, N. N. 1979. The effect of pH, water activity, sodium nitrite and incubation temperature on growth of bacteria isolated from meats, p. 57–71. In: Jarvis, B., Christian, J. H. B., Michener, H. D. (ed.), Food microbiology and technology. Medecina Viva, Parma Italy.

Robinson, K. 1966. Some observations on the taxonomy of the genus *Microbacterium*. II. Cell wall analysis, gel electrophoresis and serology. Journal of Applied Bacteriology 29:616–624.

Rogosa, M., Mitchell, J. A., and Wiseman, R. F. 1951. A selective medium for the isolation and enumeration of oral and fecal lactobacilli. Journal of Bacteriology 62:132–133.

Schleifer, K. H. 1970. Die Mureintypen in der Gattung *Microbacterium*. Archiv fur Mikrobiologie 71:271–282.

Schleifer, K. H., and Kandler, O. 1972. Peptidoglycan types of bacterial cell walls and their taxonomic implications. Bacteriological Reviews 36:407–477.

Shaw, B. G., and Dainty, R. H. 1985. Microbiological and biochemical changes during spoilage of meat. Journal of the Science of Food and Agriculture 36:123–124.

Shaw, B. G., and Harding, C. D. 1978. The effect of nitrate and nitrite on the microbial flora of Wiltshire bacon after maturation and vacuum-packed storage. Journal of Applied Bacteriology 45:39–48.

Shaw, B. G., Harding, C. D., and Taylor, A. A. 1980. The microbiology and storage stability of vacuum packed lamb. Journal of Food Technology 15:397–405.

Shaw, N. 1974. Lipid composition as a guide to the classification of bacteria. Advances in Applied Microbiology 17:63–108.

Shaw, N., and Stead, D. 1970. A study of the lipid composition of *Microbacterium thermosphactum* as a guide to its taxonomy. Journal of Applied Bacteriology 33:470–473.

Sneath, P. H. A., and Jones, D. 1976. *Brochothrix,* a new genus tentatively placed in the family Lactobacillaceae. International Journal of Systematic Bacteriology 26:102–104.

Sneath, P. H. A., and Jones, D. 1986. The genus *Brochothrix,* p. 1249–1253. In: Sneath, P. H. A., Mair, N. S., and Sharpe, M. E. (ed.), Bergey's manual of systematic bacteriology, vol. 2. Williams & Wilkins, Baltimore.

Stackebrandt, E., and Woese, C. R. 1981. The evolution of prokaryotes, p. 1–31. In: Carlile, M. J., Collins, J. F., and Moseley, B. E. B. (ed.), Molecular and cellular aspects of evolution. Society for General Microbiology, Symposium 32. Cambridge University Press, Cambridge.

Sulzbacher, W. L., and McLean, R. A. 1951. The bacterial flora of fresh pork sausage. Food Technology 5:7–8.

Sutherland, J. P., Patterson, J. T., Gibbs, P. A., and Murray, J. G. 1975. Some metabolic and biochemical characteristics of representative microbial isolates from vacuum-packaged beef. Journal of Applied Bacteriology 39:239–249.

Talon, R., Grimont, P. A. D., Grimont, F., Gasser, F., and Boeufgras, J. M. 1988. *Brochothrix campestris* sp. nov. International Journal of Systematic Bacteriology 38:99–102.

Thomson, I. Q., and Collins-Thompson, D. L. 1986. Iron requirement of *Brochothrix thermosphacta.* International Journal of Food Microbiology 3:299–309.

Thomson, I. Q., and Collins-Thompson, D. L. 1988. Secondary role of manganese during iron deprivation of *Brochothrix thermosphacta.* Systematic and Applied Microbiology 10:195–199.

Thornley, M. J. 1957. Observations on the microflora of minced chicken meat irradiated with 4 MeV cathode rays. Journal of Applied Bacteriology 20:286–298.

Weidemann, J. F. 1965. A note on the microflora of beef muscle stored in nitrogen at 0°. Journal of Applied Bacteriology 29:365–367.

Wilkinson, B. J., and Jones, D. 1975. Some serological studies on *Listeria* and possibly related bacteria, p. 251–261. In: Woodbine, M. (ed.), Problems of listeriosis. Leicester University Press, Leicester, UK.

Wilkinson, B. J., and Jones, D. 1977. A numerical taxonomic survey of *Listeria* and related bacteria. Journal of General Microbiology 98:399–421.

Wolin, E. F., Evans, J. B., and Niven, C. F. 1957. The microbiology of fresh and irradiated beef. Food Research 22:682–686.

CHAPTER 73

The Genus *Erysipelothrix*

ANNETTE C. REBOLI and W. EDMUND FARRAR

The genus *Erysipelothrix* consists of a single species, *Erysipelothrix rhusiopathiae,* formerly known as *E. insidiosa.* In 1876, Koch first isolated this slender, pleomorphic, Gram-positive bacillus from the blood of mice that had been inoculated subcutaneously with blood from putrefied meat (Koch, 1878) and was designated *E. muriseptica.* In 1882, Loeffler observed a similar organism in the cutaneous blood vessels of a pig that had died of swine erysipelas and published the first good description of the organism (Loeffler, 1886). It is probable that a bacillus observed a few months previously by Pasteur and Dumas in pigs dying of rouget was the same organism as that described by Loeffler (Pasteur and Dumas, 1882). Trevisan proposed the name *E. insidiosa* in 1885. Rosenbach was the first to establish *Erysipelothrix* as a human pathogen. In 1909, he reported the isolation of the organism from a patient with localized cutaneous lesions and coined the term "erysipeloid" to distinguish these lesions from those of human erysipelas (Rosenbach, 1909). Subsequently, *Erysipelothrix* has been identified as the cause of infection in many animal species.

Rosenbach distinguished three species, *E. muriseptica, E. porci,* and *E. erysipeloides,* based on their murine, porcine, and human origins, respectively (Rosenbach, 1909). The name *Bacterium rhusiopathiae* (Migula, 1900) antedated the name *E. porci.* The combination *E. rhusiopathiae* was first proposed by Buchanan (1918). At least 36 names have appeared in the literature for species of this genus. With the appreciation that all strains belonged to a single species, the name *E. insidiosa* was proposed for *E. rhusiopathiae, E. muriseptica,* and *E. erysipeloides* (Langford and Hansen, 1953, 1954). In 1966, Shuman and Wellmann proposed that the name *E. insidiosa* be rejected in favor of *E. rhusiopathiae* which means literally "erysipelas thread of red disease."

Taxonomy

The genus *Erysipelothrix* is classified among the regular nonsporeforming Gram-positive rods (Jones, 1986). It contains only the type species, *E. rhusiopathiae,* but there is greater variation in serological, biochemical, and chemical properties of *E. rhusiopathiae* strains than is usually found within a single species (Erler, 1972; Feist, 1972; Flossman and Erler, 1972; White and Miritikani, 1976). Gel electrophoresis protein patterns also demonstrate much variation (White and Miritikani, 1976). Further studies are needed to determine whether the genus *Erysipelothrix* should contain more than one species.

The taxonomic position of *Erysipelothrix* with respect to other genera is uncertain. *Erysipelothrix* was once thought to be closely related to *Listeria* (Barber, 1939). Results of studies of cell wall peptidoglycan (Schleifer and Kandler, 1972), fatty acid patterns (Tadayon and Carroll, 1971), DNA hybridization studies (Stuart and Welshimer, 1974), and numerical taxonomic studies (Davis and Newton, 1969; Jones, 1975; Stuart and Pease, 1972; Wilkinson and Jones, 1977) do not support this relationship. There are no common antigens between strains of *Erysipelothrix* and *Listeria monocytogenes* as detected by immunodiffusion or passive hemagglutination tests (Pleszczynska, 1972). Differences between the two genera have been demonstrated in cell wall chemistry by chromatography and infrared spectrophotometry. Paper and thin-layer chromatography of acid hydrolysates of the purified cell wall show that *Erysipelothrix* is distinguishable from *Listeria. Erysipelothrix* has lysine and glycine in the cell wall (Mann, 1969); *Listeria* has meso-diaminopimelic acid.

Enzyme and DNA-base ratio studies reveal a closer relationship of *Erysipelothrix* to the family Lactobacillaceae than to Corynebacteriaceae (Flossmann and Erler, 1972). In a study of more than 200 strains of coryneform bacteria using 173 morphological, physiological, and biochemical tests and computer analysis, eight clusters were identified and *Erysipelothrix* was most closely related to *Streptococcus pyogenes* (Jones, 1975). In another study, the closest similarity of *Erysipelothrix* was to the genus *Gemella* (Wilkinson and Jones, 1977). In general, results of numerical taxonomic studies indicate that strains of *E. rhusiopathiae* form a distinct cluster that shows the closest similarity to the streptococci (Jones, 1986).

Results of 16S ribosomal RNA oligonucleotide studies indicate that *E. rhusiopathiae* is related to *Clostridium innocuum* (Jones, 1986). Both *E. rhusiopathiae* and *C. innocuum* contain lysine in the cell wall, but the detailed peptidoglycan structure of *C. innocuum* has not been elucidated. *C. innocuum* is a member of the RNA cluster which is part of the clostridial group and which contains the mycoplasmas (Stackebrandt and Woese, 1981; Woese et al., 1980). The mycoplasma cluster is related to the *Bacillus-Lactobacillus-Streptococcus* cluster. Nucleic acid hybridization studies should help clarify the taxonomic position of *Erysipelothrix*. The GC content of the DNA is 36–40 mol% (Flossman and Erler, 1972; White and Mirikitani, 1976).

Habitats

Occurrence in Animals

E. rhusiopathiae, and infections due to this organism, occur worldwide. It has been found as a commensal or a pathogen in a wide variety of vertebrate and invertebrate species, including "swine, sheep, cattle, horses, dogs, wild bears, kangaroos, reindeer, mice, wild rodents, seals, sea lions, cetaceans, mink, chipmunks, crustaceans, fresh and salt-water fish, crocodiles, caiman, stable flies, house flies, ticks, mites, mouse lice, turkeys, chickens, ducks, geese, guinea fowl, pigeons, sparrows, starlings, eagles, parrots, pheasants, peacocks, quail, parakeets, mud hens, canaries, finches, siskins, thrushes, blackbirds, turtledoves, and white storks" (Conklin and Steele, 1979; see also Creech, 1921; Gledhill, 1948; Hunter, 1974; Sneath et al., 1951; Wood, 1975; Woodbine, 1950). The major reservoir of *E. rhusiopathiae* is generally believed to be domestic swine, but rodents and birds are also frequently infected. The organism causes no known disease in fish but can grow and persist for long periods in the mucoid exterior slime of these animals (Wood, 1975).

The presence of *E. rhusiopathiae* in the external environment is probably secondary to contamination by infected animals; it may be found in sewage effluent from abattoirs and on ground contaminated with the feces of animals with apparent or inapparent infection. It now appears that, contrary to previous belief, the organism is not able to exist indefinitely in soil, but it may live long enough to cause infection in animals or humans weeks or months after initial soil contamination. Pathogenic strains of *E. rhusiopathiae* have been isolated from the feces of apparently healthy swine, and asymptomatic swine commonly harbor this organism in their tonsils and other lymphoid tissues (Wood, 1974). A cycle may thus occur in which asymptomatic carrier swine contaminate the soil of the swine pens, and organisms in the soil are transmitted to previously uninfected swine, some of which become asymptomatic carriers of *E. rhusiopathiae.*

Occurrence in Humans

The risk of human infection with *E. rhusiopathiae* is closely related to the opportunity for exposure to the organism (McGinnes and Spindle, 1934). Relation to age, sex, race, and socioeconomic status appears to reflect only opportunity for exposure. Most human cases are related to occupational exposure. Individuals at greatest risk for infection include butchers, fishermen, fish handlers, abattoir workers, veterinarians, and housewives (Berg, 1984; Conklin and Steele, 1979; Gorby and Peacock, 1988; Hillenbrand, 1953; Hunter, 1974; Klauder, 1938; Morrill, 1939; Wood, 1975; Woodbine, 1950), but erysipelothrix infection has been associated with a wide variety of occupations including "butchers, meat cutters, meat-processing workers, poultry-processing workers, meat inspectors, rendering plant workers, knackers, animal caretakers, farmers, fishermen, including lobster fishermen, fish and lobster handlers, fish-processing workers, crab and crayfish-processing workers, clam openers, veterinarians, including veterinary students, cooks, bakers, housewives, kitchen workers, food handlers, caterers, button makers (bone), game handlers, furriers, leather workers, soap makers, fertilizer workers, sewer workers, bacteriologists, laboratory workers, and stockyard workers" (Wood, 1975). Infection is especially common among individuals who handle fish. Hunter (1974) noted the occurrence of *fish-handlers' disease* among "fishermen, fish cleaners, gutters and picklers, fish porters, fish-box repairers, fish-lorry drivers, fish-meal workers, smoke driers, fish mongers, cooks and housewives who infect themselves through abrasions of the skin caused by the spines, fins, and bones of fish, especially skate." During World War II, outbreaks of the disease occurred in factories in Norway where fish was dried and tinned and cod heads made into fertilizer; the delay caused because the fishing boats had to sail in convoy apparently allowed the organisms to multiply, resulting in the high incidence of infections (Hunter, 1974). Diseases called *seal finger* and *whale finger* occur in those who capture these animals and scratch their hands on the steel ropes used in their work (Hillenbrand, 1953). Anglers may be infected through puncture wounds made by fish hooks or the teeth of fish, or by the claws of lobsters and crabs; in the series of 329 cases of erysipeloid described by Gilchrist, 323 were caused by injuries produced by crabs (Gilchrist, 1904). Abattoir workers, meat porters, butchers, and poulterers may become infected through small cuts from the knives they use or through abrasions caused by splinters of bone a condition called (*pork finger*). Veterinary pathologists may become accidentally inoculated by injury from knives or bone splinters during necropsy of infected animals, particularly poultry. Workers who peel potatoes and other root vegetables may become infected by earth contaminated with the manure of infected animals.

Most cases in humans and other animals probably occur via scratches or puncture wounds of the skin, but in some cases it appears that the organism has penetrated intact skin (McGinnes and Spindle, 1934). Human-to-human infection has not been documented. Although *E. rhusiopathiae* is killed by moist heat at 55°C for 15 minutes, it is resistant to many environmental influences, including salting, pickling, and smoking (Conklin and Steele, 1979). Meat and bacon may contain the organism after pickling for 170 days or after 30 days in a mixture of salt and potassium nitrate, and the organism has been recovered from ham after smoking. It may remain alive for 12 days in direct sunlight, and for many months in carcasses left to decay on the surface of the ground or buried as deep as 7 feet. It has also been found in city sewage containing drainage from abattoirs and stables.

Reporting of infections due to *E. rhusiopathiae* is not required by health authorities, so it is difficult to know whether the incidence of these infections is increasing or decreasing. Some technological changes in industries employing animal products have probably resulted in reduced contact between *E. rhusiopathiae* and humans. For example, nearly all buttons are now made of plastic, rather than bone. To the extent that such changes reduce occupa-tional exposure to the organism, the future incidence of erysipeloid, and more serious forms of infection with *E. rhusiopathiae*, will decline.

Diseases Due to *Erysipelothrix rhusiopathiae*

Disease in Animals

Swine erysipelas is worldwide in distribution. It was first recognized as an important disease in the United States during the 1930s, and by 1959 it had been reported from 44 states. The incidence then declined to a relatively low level by 1972, where it has remained. The present low prevalence of swine erysipelas, in spite of the widespread distribution of the organism, may be attributible largely to management practices such as maintenance of closed herds, use of confinement housing, improved waste disposal, vaccination, and use of antibiotics in feed (Wood, 1984).

Four clinical entities have been described in swine: 1) an acute septic form, in which the animals may die within a few days; 2) a subacute urticarial form marked by reddish purple rhomboid spots or "diamonds" in the skin (diamond skin disease); 3) a joint or arthritic form; and 4) a chronic cardiac form (endocarditis) (Gledhill, 1948; Klauder, 1944; Wood and Shuman, 1974; Woodbine, 1950).

In sheep, the typical form of the disease is polyarthritis, which begins when the lambs are 2 to 3 months old (Conklin and Steele, 1979). The gait becomes stiff, and the animals have difficulty rising and may fall down. The involved joints are usually swollen and the capsule is thickened. The organism can be cultured from the joint tissue. There are no lesions in visceral organs.

E. rhusiopathiae causes disease in many species of birds (Conklin and Steele, 1979). Adult male turkeys develop a disease called "bluecomb" because of the cyanotic skin color. The birds become droopy, develop diarrhea, and die. Hemorrhagic lesions are found in the breast and leg muscles, pericardium, liver, and spleen and the organism is easily isolated from these tissues. Large outbreaks have also occurred in domestic ducks, less commonly in chickens.

Although the greatest commercial impact of *E. rhusiopathiae* infection is due to disease in swine, infection of sheep, turkeys, and ducks is also of economic importance (Conklin and Steele, 1979). Both killed bacterin vaccines (formalin-killed whole culture adsorbed on aluminum hydroxide gel) and live avirulent vaccines have been available in the United States since the 1950s. Both are prepared from organisms of serovar 2, but provide cross-protection against many (but not all) other serovars (Sawada and Takahashi, 1987). Effectiveness of killed and live vaccines appears to be approximately equal (Shuman, 1959). Usage in the United States has roughly paralleled the incidence of swine erysipelas. Killed bacterins have also been used in turkeys, primarily during epizootics. An effective live avirulent vaccine that can be administered in drinking water has also been developed recently (Bricker and Saif, 1988).

Disease in Humans

The spectrum of disease seen in humans closely parallels that seen in swine (Grieco and Sheldon, 1970; Klauder, 1926; Phillips, 1986; Sikes, 1958). There are three well-defined clinical categories of human disease: 1) a localized cutaneous form, erysipeloid; 2) a generalized cutaneous form; and 3) a septicemic form which is often associated with endocarditis (Grieco and Sheldon, 1970).

Localized Cutaneous Form or Erysipeloid of Rosenbach

Erysipeloid is a localized skin infection that is actually a cellulitis. Because of its mode of acquisition, contact with infected animals, fish, or their products, with organisms gaining entrance via cuts or abrasions on the skin, lesions are usually confined to the fingers and hands. The patient complains of pain and swelling of the finger or part of the hand. The pain is often severe and may be described as a burning, throbbing, or itching sensation. A history may be elicited of a scratch or wound of the infected part by a bone or knife contaminated by animal secretions approximately 5 to 7 days or, at most, 2 weeks prior to the onset of symptoms (Price and Bennett, 1951). The infected area is swollen. The lesion consists of a well-defined, slightly elevated, violaceous zone that spreads peripherally as discoloration of the central area fades (King, 1946). Systemic effects are uncommon. Low-grade fever and arthralgias occur in approximately 10% of cases and lymphangitis and lymphadenopathy in approximately one-third (Nelson, 1955). There may be arthritis of an adjacent joint. Vesicles may be present but suppuration does not occur. The absence of suppuration along with the violaceous color, lack of pitting edema, and disproportionate pain help to distinguish erysipeloid from staphylococcal or streptococcal infection. Erysipeloid is a self-limited condition, the lesions usually resolving without therapy within 3 or 4 weeks.

Diffuse Cutaneous Form

This is a rare situation in which the violaceous cutaneous lesion progresses proximally from the site of inoculation or appears at remote areas (Ehrlich, 1946; Grieco and Sheldon, 1970). Bulla formation may occur. The patients often have systemic manifestations such as fever and joint pains, but blood cultures are negative. The clinical course is much more protracted than in the localized disease form, and recurrences are not uncommon. In one instance, a butcher who ate sausage from a pig slaughtered because of swine erysipelas developed widespread urticaria with the rhomboid pattern characteristic of swine erysipelas (Hunter, 1974).

Septicemia and Endocarditis

Systemic E. rhusiopathiae infection is uncommon. No cases of systemic disease were seen among the 500 cases of erysipeloid described by Nelson (1955) nor among the 329 cases reported by Gilchrist (1904). Fifty cases of E. rhusiopathiae infection have been reported with an extremely high incidence of endocarditis (90%) among these (Alexander and Goodwin, 1973; Baird and Benn, 1975; Berg, 1984; Blount, 1965; Borchardt et al., 1977; Fliegelman et al., 1985; Gransden and Eykyn, 1988; Grieco and Sheldon, 1970; Heggers et al., 1974; James et al., 1976; Klauder et al., 1943; Kramer et al., 1982; Lawes et al., 1952; Lerner and Weinstein, 1966, 1952; McCarty and Bornstein, 1960; McCracken et al., 1973; Mandal and Malloch, 1971; Morris et al., 1965; Muirhead and Reid, 1980; Normann and Kihlstrom, 1985; Park et al., 1976; Poretz, 1985; Procter, 1965; Russell and Lamb, 1940; Schiffman and Black, 1956; Scully et al., 1978; Silberstein, 1965; Simberkoff and Rahal, 1973; Townshend et al., 1973). All reported cases of endocarditis except one case of infection involving a Starr-Edwards prosthetic aortic valve (Gransden and Eykyn, 1988) have involved native valves. Only about one-third of patients had a history of an antecedent skin lesion or a concurrent characteristic skin lesion of erysipeloid (Gorby and Peacock, 1988). Gorby and Peacock (1988) compared clinical features of E. rhusiopathiae endocarditis with those of endocarditis caused by other bacteria (Kaye, 1976). They found a higher male-to-female ratio (which probably reflects occupational exposure), a greater propensity for involvement of the aortic valve, and a much higher mortality rate (38%) among patients with E. rhusiopathiae endocarditis. There was more prior heart disease among those with endocarditis caused by other organisms. In nearly 60% of patients, E. rhusiopathiae endocarditis apparently developed on previously normal heart valves. The clinical picture with respect to fever, peripheral skin stigmata of endocarditis, emboli, splenomegaly, hematuria, and mycotic aneurysm was similar for the two groups. Very few cases of endocarditis have occurred in immunocompromised patients, but a history of alcohol abuse was present in 33%. The presentation is most often subacute but may be acute (Gorby and Peacock, 1988; Schiffman and Black, 1956; Simberkoff and Rahal, 1973). The most common complication of endocarditis, congestive heart failure, was present in approximately 80% of patients (Fliegelman et al., 1985; Grieco and Sheldon, 1970; Heggers et al., 1974; Kramer et al., 1982; McCracken et al., 1973; Russell and Lamb, 1940). Myocardial abscesses and aortic valve perforation have been reported (Fowler et al., 1967; Heilman and Herrell, 1944; Kramer et al., 1982; Mandal and Malloch, 1971; Morris et al., 1965; Russell and Lamb, 1940). More than one-third of patients required valve replacement (Gorby and Peacock, 1988). Diffuse glomerular nephritis and meningitis have also been reported as complications (Silberstein, 1965; Simberkoff and Rahal, 1973). Ognibene et al. (1985) reported the first case of septic shock associated with this organism in a patient without convincing evidence of endocarditis.

Other Infections

Osseous necrosis of the thumb has been reported in a patient who developed fatal endocarditis (Klauder et al., 1943) and Torkildsen (1943) described a case of intracranial abscess. Chronic arthritis has been reported in a few cases in Europe (Ehrlich, 1946).

Treatment

The mainstay of treatment of infections caused by *E. rhusiopathiae* is antibiotic therapy. Although skin lesions usually heal spontaneously within 4 weeks, second attacks may occur and lesions may persist for months. Healing is hastened by antibiotic therapy. Susceptibility data are limited. Most strains are highly susceptible to penicillins, cephalosporins, erythromycin, and clindamycin (Gorby and Peacock, 1988; Heilman and Herrell, 1944; Poretz, 1985; Sneath et al., 1951; Stiles, 1947). Minimal inhibitory concentrations for penicillins have been reported to range from 0.0025 to 0.06 μg/ml with minimal bactericidal concentrations of 0.0025 to 0.75 μg/ml (Gorby and Peacock, 1988). Susceptibility to chloramphenicol and tetracycline is variable. Most strains are resistant to sulfonamides, trimethoprim-sulfameth-

oxazole, aminoglycosides, vancomycin, novobiocin, and polymyxins. Resistance to vancomycin is noteworthy because this agent is often used in empiric therapy of prosthetic valve endocarditis and in the treatment of native valve endocarditis due to Gram-positive organisms in individuals who are allergic to penicillins.

Penicillin G, in doses of 12 to 20 million units/day, is the drug of choice for serious infections caused by *E. rhusiopathiae* (Poretz, 1985). Recommended duration of therapy for endocarditis is 4 to 6 weeks, although shorter courses consisting of 2 weeks of intravenous therapy, followed by 2 to 4 weeks of oral therapy have been successful (Baird and Benn, 1975; Muirhead and Reid, 1980; Ognibene et al., 1985). There has been no reported experience in the treatment of penicillin-allergic patients. Cephalosporins are the most appropriate alternatives since both clindamycin and erythromycin are only bacteriostatic agents. Valve replacement has been necessary in about one-third of cases of endocarditis.

Pathogenesis and Pathology

The factors responsible for virulence in *E. rhusiopathiae* are not clearly defined. Most strains isolated from pigs with swine erysipelas belong to serotypes 1a, 1b, and 2. Serotype 1a is most common in septicemia. Ability to adhere to mammalian cells may be an important determinant of virulence. Virulent strains of *E. rhusiopathiae* adhere much more avidly than avirulent strains to porcine kidney cell lines (Takahashi et al., 1987), and strains isolated from swine with endocarditis or septicemia adhere more strongly to swine heart valve tissue in vitro than other isolates of *E. rhusiopathiae* (Bratberg, 1981). *E. rhusiopathiae* also produces a hyaluronidase and a neuraminidase, which cleaves α-glycosidic linkages of sialic acid, a mucopolysaccharide on the surface of mammalian cells. Neuraminidase activity is higher in virulent than in avirulent strains (Krasemann and Muller, 1975), and there is evidence that this enzyme may play a role in the pathogenesis of arthritis and thrombocytopenia in rats experimentally infected with *E. rhusiopathiae* (Nakato et al., 1986; Shinomiya and Nakato, 1985).

Intravenous injection of the organism into rabbits is fatal in 2 to 3 days. An erysipeloid rash develops in the injected ear, the lungs become hemorrhagic, and a pericardial exudate develops. Congestion of the viscera is noted with pinpoint focal necrosis in the liver and mononuclear cell infiltrates in the spleen (Smith, 1983). On histopathological examination, bacilli are scarce. Inoculation of the conjunctiva produces conjunctivitis which is often followed by fatal disseminated infection. Subcutaneous injections seldom cause death.

The rhomboidal skin lesions described in swine are the result of thrombotic vasculitis of end arterioles (Grieco and Sheldon, 1970). Injection of the organism into swine produces an inflammatory polyarthritis, lymphadenopathy, endocarditis, peripheral monocytosis, and focal necrosis of the liver and myocardium (Smith, 1983). The endocarditis in swine usually involves the mitral valve and there is a tendency for the vegetations to invade the mural endocardium (Russell and Lamb, 1940). Pathological changes in human cases of septicemia and endocarditis are indistinguishable from changes caused by other bacterial organisms.

Fatal septicemia occurred in a wild-caught opossum, 2 months after capture of the animal. *E. rhusiopathiae* was isolated from heart blood, liver, spleen, and lungs (Lonigro and LaRegina, 1988). Although no evidence of endocarditis was found, when this strain was inoculated intravenously into previously healthy opossums, all of three animals developed vegetations on the atrioventricular valve. Brown and Brenn stains revealed small Gram-positive rods within the vegetations, but the organism could not be grown from blood or vegetations of any of the animals (LaRegina et al., 1988).

Swine arthritis bears some similarities to human rheumatoid arthritis (Grieco and Sheldon, 1970; Phillips, 1986; Sikes, 1958). It is marked by pannus formation with destruction of cartilage at the site of pannus attachment, intraarticular fibrous adhesions and subchondral cellular reaction (Grieco and Sheldon, 1970; Sikes, 1958). Antigens of *E. rhusiopathiae* have been detected by electron microscopy and immunofluorescence in deep joint tissues of swine as long as 18 months after infection. It is not known whether development of chronic arthritis is due to specific immune reactions against *E. rhusiopathiae* or if autoimmune reactions are also involved (Wood, 1984).

Bacteriology

Morphology, Growth, and Biological Properties

The cell wall of *E. rhusipathiae* contains a peptidoglycan based on lysine (Feist, 1972). There are many sugars in the cell wall (arabinose, galactose, glucose, glucose-6-phosphate, galactose-6-phosphate, ribose, and xylose). Monosaccharide patterns show that there are three chemovars, which do not correlate with the serovars (Feist, 1972). The cell wall also contains 2- and 3-hydroxy and nonhydroxylated long chain fatty acids. The hydroxylated fatty acids are of the straight-chain saturated series; the nonhydroxylated fatty acids are predominantly of the straight-chain saturated and the monounsaturated acid series. Small amounts of iso- and anteiso-methyl-branched fatty acids are also present. Mycolic acids are not present. L-forms of *E. rhusiopathiae* have been described (Pachas and Currid, 1974; Stuart, 1972, Todorov, 1976).

E. rhusiopathiae is a straight or slightly curved, slender, rod-shaped organism with rounded ends. Organisms are arranged singly, in short chains, or in pairs in a "V" configuration or are grouped randomly. Filaments and long chains are sometimes seen. Nonencapsulated, nonsporulating, and nonmotile, it is Gram-positive but may appear Gram-negative because it decolorizes readily. It is not acid-fast. The exact growth requirements of the organism have not been determined. Several amino acids, riboflavin, and small amounts of oleic acid are required (Hunter, 1942) and growth is enhanced by tryptophan (Ewald, 1981). Growth occurs at temperatures ranging from 15 to 44°C, with an optimal temperature of 30 to 37°C, and at a pH of 7.2 to 7.6 with a range of 6.7 to 9.2 (Sneath et al., 1951). It is a facultative anaerobe. Growth is improved by 5 to 10% carbon dioxide. Heating at 60°C for 15 min is lethal. *E. rhusiopathiae* is able to grow in the presence of 0.2% phenol, 0.1% sodium azide, 0.001% crystal violet, 0.05% potassium tellurite, 0.02% thallous acetate, and 0.2% 2,3,5, triphenyl-tetrazolium chloride (Ewald, 1981; Sneath et al., 1951).

On blood agar it may be α-hemolytic but is never β-hemolytic (see Table 1). There is a dual colonial and microscopic appearance (Fig. 1 and 2). After growing for 24 to 48 h at 37°C on trypticase soy agar containing 5% sheep blood, colonies are very small (0.1 mm in diameter), convex, circular, and transparent with a smooth glistening surface and edge. These are smooth or S-forms. In older cultures, colonies are slightly larger and have opaque centers. Larger (0.2–0.4 mm diameter) flatter colonies with a matte surface and fimbriated edge are R-form or rough colonies. They may resemble miniature *Bacillus anthracis* colonies (Wilson and Miles, 1975). Both forms are usually light blue in color or sometimes green when viewed with oblique illumination. Intermediate forms are also seen (Barber, 1939; Ewald, 1981). S-form colonies dissociate to give rise to intermediate and R-form colonies. R-form colonies also give rise to S-forms (Ewald, 1962; Wawrzkiewicz, 1964). During these changes in morphology and

Table 1. Characteristics of *Erysipelothrix rhusiopathiae*.

Test	Reaction
Aerobic growth	+
Anaerobic growth	+
α-Hemolysis	+
β-Hemolysis	−
Catalase	−
Oxidase	−
Acid from glucose, fructose, galactose, and lactose	+
Gas from glucose, fructose, galactose, and lactose	−
Acid from maltose	−ᵃ
Acid from xylose, mannitol, and sucrose	−
Motility	−
H₂S in TSIᵇ	+
Nitrate reduction	−
Indole	−
Esculin	−
Voges-Proskauer	−
Methyl red	−
Methylene blue milk	−
Litmus milk	V
Gelation stab	Test-tube brush or "pipe cleaner" growth
Liquefaction of gelatin	−
Growth at 4°C	−
Susceptibility to NaCl (8.5%)	+
Susceptibility to neomycin	−

+, positive; −, negative; V, variable.
ᵃA few strains produce acid in 6 to 7 days.
ᵇTriple sugar iron agar.

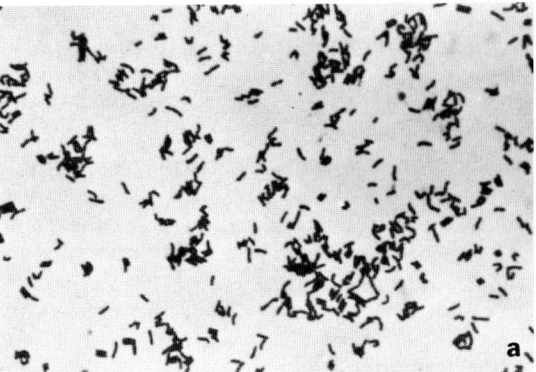

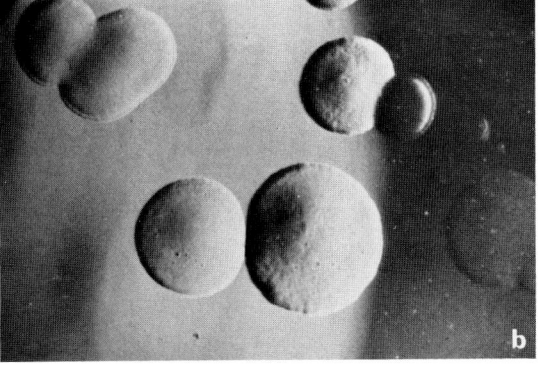

Fig. 1a,b. *Erysipelothrix rhusiopathiae*, smooth form.

cultural characteristics, there are also changes in virulence and antigenic properties. (The type-specific antigens remain unchanged.) In broth, S-form organisms cause a slight turbidity and a powdery deposit; R-forms have a tangled hairlike appearance. Microscopically, S-form organisms are the small, straight or slightly curved rods characteristic of the genus and measure 0.3 to 0.6 μm by 0.8 to 2.5 μm, while R-form organisms form long nonbranching filaments which may have a beaded appearance and which can be greater than 16 μm in length. Both forms stain evenly and may show deeply stained granules. Morphology varies with the medium, pH, and temperature of incubation. Acidic pH and temperature of 37°C favor R-forms (Wilson and Miles, 1975). Alkaline pH (7.6–8.2) and temperature of 30°C favor S-forms (Grieco and Sheldon, 1970). S-form organisms are seen in smears from blood and tissue and in acute forms of illness such as sepsis; R-forms or S-forms and R-forms are seen in more chronic conditions such as endocarditis or arthritis (Ewald, 1981).

E. rhusiopathiae is catalase and oxidase negative. Growth is improved by blood or 5 to 10%

serum, tryptophan, and glucose. The best growth occurs in 0.1% glucose broth or on 0.5% glucose agar. Larger amounts of glucose may be inhibitory. Glucose metabolism is via the Embden-Meyerhof-Parnas pathway, with a small amount by the hexose monophosphate shunt (Robertson and McCullough, 1968). The tricarboxylic acid cycle is relatively unimportant. Exogenous citrate is not used. Fermentative activity is weak (Smith, 1983). The fermentation pattern varies with the basal medium used (White and Shuman, 1961; Wood, 1970). Acid production from carbohydrates is usually poor or inconsistent when in 1% peptone water. It is recommended that 5 to 10% horse serum be added to the basal medium (Seeliger, 1974; White and Shuman, 1961; Wood, 1970), but since this is not always convenient, one may test for acid production in nutrient broth with 0.5 to 1% of the test carbohydrate and phenol red added as an indicator. In addition to lactic acid, small amounts of acetic acid, formic acid, ethyl alcohol, and CO₂ are produced from the fermentation of glucose. Acid without gas is produced within 48 h from glucose, lactose, fructose, and galactose (Karlson and Merchant, 1941). A few strains produce acid from maltose fermentation in 6 to 7 days. Xylose, mannitol, and sucrose are not fermented. This genus is indole and Voges-Proskauer negative and usually methyl red negative (Sneath et al., 1951). The methyl red test may be weakly positive when performed in a broth containing peptone, yeast extract, glucose, potassium phosphate, magnesium sulfate, and manganese sulfate (Wilkinson and Jones, 1977). There is no discoloration of methylene blue milk, little or no change in litmus milk, and no reduction of nitrate. Urea, esculin, sodium hippurate, starch, cellulose, and casein are not hydrolyzed. Xanthine and tyrosine are not degraded. The majority of strains produce hydrogen sulfide, but results can vary with the medium used (Ewald, 1964). Hydrogen sulfide production is an important test. It is best carried out on triple sugar iron agar (TSI) slants, in which hydrogen sulfide causes a blackened butt (Wood, 1970). An occasional old laboratory strain does not produce detectable hydrogen sulfide in TSI agar. Gelatin stab cultures yield a very characteristic pattern of growth described as a "test tube brush" or a "pipe cleaner" (Ewald, 1964; Jones, 1986; Weaver, 1985). After 24 h, growth is faint and hazy and limited to an area just below the surface. Within a few days, however, growth extends in a column to the bottom of the tube. There is no liquefaction of the gelatin. S-forms

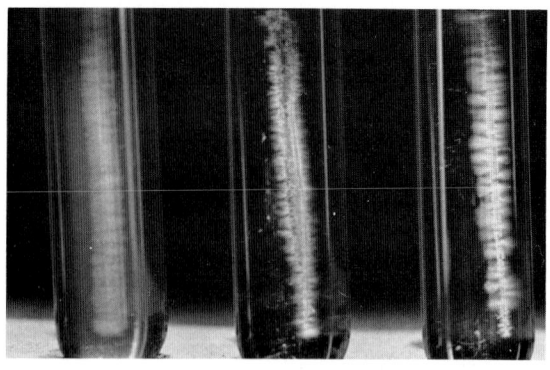

Fig. 3. Three *Erysipelothrix* strains on ferrochloride-gelatin medium after incubation of 48 h at 20°C. On the left, smooth strain; in the middle and on the right, dissociated strains.

produce fine horizontal outgrowths which extends only 2 to 3 mm from the stab in a typical "pipe-cleaner" growth (Ewald, 1964). R-forms extend further out and look like a "test-tube brush" (Fig. 3). This test is not convenient for most laboratories to do since the gelatin must be incubated at 22 to 25°C to maintain its solid state. Furthermore, this test is not required for identification.

Most known strains produce hyaluronidase (Ewald, 1957). There may be a correlation between production of this enzyme and virulence. Good producers of hyaluronidase usually belong to serovar 1 (Ewald, 1957, 1981). All strains produce neuraminidase but in various amounts. There is a good correlation with virulence (Krasemann and Müller, 1975; Müller and Krasemann, 1976; Müller and Seidler, 1975; Nikolov and Abrashev, 1976). There are also virulent strains that do not produce neuraminidase or which produce small amounts. Synthesis of this enzyme depends on the growth phase of the bacteria. There is maximum production at the end of the logarithmic phase of growth (Abrashev and Zamfirova, 1976).

Specimen Collection, Transport, and Maintenance

Routine blood culture techniques are adequate for specimen collection and organism growth in suspected cases of sepsis or endocarditis (Muirhead and Reid, 1980). Because organisms are located only in deeper parts of the skin in cases of erysipeloid, aspirates or biopsy specimens from the edge of the lesion are needed to obtain the organism (Price and Bennett, 1951). Biopsy should be of the entire thickness of the dermis. The organism may also be obtained from cases of erysipeloid by injecting saline into the edge

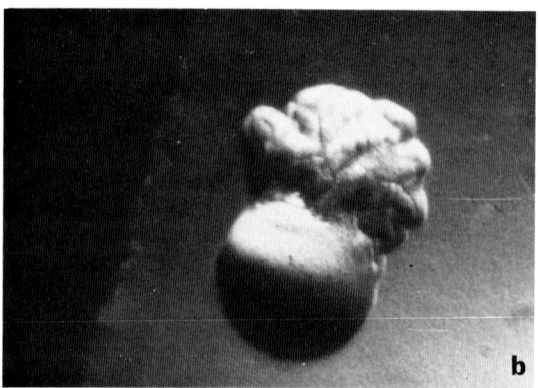

Fig. 2a,b. *Erysipelothrix rhusiopathiae,* smooth and rough forms.

of the lesion and aspirating some of the injected saline for culture (Smith, 1983).

Immediately after collection, the specimen should be put into an infusion broth of 1% glucose and kept at room temperature or refrigerated until it reaches the laboratory. Cultures can be maintained for several months by stab inoculation into tubes of nutrient agar (pH 7.4) (Jones, 1986). Freeze drying or freezing on glass beads at −70°C (Feltham, et al., 1978) is appropriate for long-term maintenance.

Isolation and Identification

Commercially available blood culture media are satisfactory for primary isolation from blood since *E. rhusiopathiae* is not particularly fastidious. Biopsy specimens or tissue aspirates from skin lesions should be put into infusion broth with 1% glucose and incubated in air or in 5 to 10% carbon dioxide at 35 to 37°C (Weaver, 1985). At 24-h intervals, subcultures to blood agar plates should be made. No visible growth occurs on potato or MacConkey's medium. Use of selective media is not necessary unless the specimen is heavily contaminated with other bacteria such as those from soil, manure, or animal tissues (Wood, 1965; Wood and Packer, 1972). Many selective media have been described, including medium ESB, a nutrient broth containing horse serum, kanamycin, neomycin, and vancomycin (Wood, 1965) and a tryptose blood agar containing crystal violet and sodium azide (Packer, 1943). Another selective liquid enrichment for *E. rhusiopathiae* contains kanamycin, crystal violet, sodium azide, and liquified phenol (Bohm, 1971). A solid medium which is a modification of this contains water blue and sucrose (Bohm, 1971). Since *E. rhusiopathiae* does not ferment sucrose, colonies appear colorless on the water blue sucrose agar. In cases of chronic infection in which the number of bacteria is small, enrichment by the addition of horse, calf, or swine serum in broth and incubation for longer than 10 days may be necessary.

Selective Media for Erysipelothrix Modified ESB Medium (Wood, 1965)

Nutrient broth No. 2 (Oxoid) dehydrated	25 g
Distilled Water	1 liter
Horse Serum	50 ml
Kanamycin	400 mg
Neomycin	50 mg
Vancomycin	25 mg

Specimen is placed in 10 ml of modified ESB medium and incubated overnight at 35°C. Five milliliters are removed, centrifuged at 1400 × g for 20 min, the sediment is resuspended in 1 to 2 ml of 0.8% saline, and a sample is plated on MBA medium. After incubation at 35°C for 24 to 48 h, the plate is examined for colonies.

MBA Medium (Harrington and Hulse, 1971)

Heart infusion agar (Difco)	40 g
Sodium azide	0.4 g
Distilled water	1 liter
Horse blood or	20 ml
Horse serum	50 ml

Packer's agar (Packer, 1943) is recommended for grossly contaminated specimens such as feces or soil because it is more selective for *E. rhusiopathiae* than is MBA medium (Jones, 1986).

Identification of the organism is based on the results of Gram-stain, lack of motility, hydrogen sulfide production, indole production, catalase activity, growth on agar containing potassium tellurite, and hemolysis on blood agar. The result of any one test is insufficient for identification. *E. rhusiopathiae* will need to be differentiated from other Gram-positive bacilli, in particular from *Actinomyces (Corynebacterium) pyogenes* and *Arcanobacterium (Corynebacterium) haemolyticum,* and from *Listeria monocytogenes.* These other organisms are β-hemolytic on blood agar and do not produce hydrogen sulfide in the butt of TSI slants. *Listeria monocytogenes* is catalase positive and motile. The neomycin susceptibility test can be used to distinguish *E. rhusiopathiae* from *Listeria monocytogenes,* the former being resistant to neomycin and the latter being susceptible (Fuzi, 1963). Some *Bacillus* species and streptococci form hydrogen sulfide but they can be differentiated from *E. rhusiopathiae* by formation of spores and by cellular morphology respectively. *E. rhusiopathiae* has occasionally been misidentified as a viridans streptococcus (Gorby and Peacock, 1988; Proctor, 1965). It has also been dismissed as a contaminant.

The mouse protection test is considered the best method for confirming an isolate as *E. rhusiopathiae* (Jones, 1986). In this test, a subcutaneous injection of 0.1 ml from an 18- to 24-h broth culture of the suspected *E. rhusiopathiae* is administered to mice along with a dose of commercial equine hyperimmune *E. rhusiopathiae* antiserum at another site. A control group of mice is injected with the broth culture but not the antiserum. If the organism is *E. rhusiopathiae,* the mice that did not receive antiserum die in 5 to 6 days, but those receiving antiserum are protected (Weaver, 1985). This test detects only those strains that are virulent for mice, but since most strains are virulent it is a good confirmatory test. One may also inject suspect clinical material subcutaneously into mice and isolate the organism from the kidneys or spleen when they die a few days later (Ewald, 1981).

E. rhusiopathiae has heat- and acid-stable, type-specific antigens and heat-labile, species-specific antigens (Grieco and Sheldon, 1970; Jones, 1986; Kalf and Grece, 1964). The type-specific antigens are polysaccharides. Strains may be identified serologically. In 1949, Dedié proposed that the two main serovars be designated as A and B and that strains which showed no reaction with A- or B-type-specific antiserum be designated as N. New serovars in group N were designated by consecutive letters of the alphabet. One of the problems with this system was that different serological methods were used. Kucsera (1973) recommended the use of a double agar-gel diffusion precipitin test using autoclaved antigens and type-specific antisera and a uniform system for designating serovars using arabic numbers. This numerical system is preferred over the older alphabetical system. Serovars 1 and 2 correspond to A and B of De-dié's system. These are the most common of the 22 serovars (Norrung, 1979; Wood et al., 1978). Strains of human and animal origin are antigenically alike (Sneath et al., 1951). Most virulent strains causing acute infection belong to serovar 1 or A. Strains of serovar 2 or B have been isolated primarily from chronic cases. They agglutinate chicken red blood cells that lyse when complement is added (Dinter, et al., 1976).

Direct and indirect fluorescent-antibody tests are also available in lieu of the mouse protection test to confirm identification of *E. rhusiopathiae* (Avilag et al., 1972; Heggers et al., 1974). They can be used to detect *E. rhusiopathiae* in tissues (Dacres and Groth, 1959; Seidler et al., 1971) and in enrichment broth (Harrington et al., 1974). In general, serological tests are not practical for routine use in a clinical laboratory for identification of the organism or for detection of antibody in patient sera. Bacteriophages active on *E. rhusiopathiae* strains have been isolated (Valerianov et al., 1976). A phage-typing system may prove to be useful.

Literature Cited

Abrashev, I., and Zamfirova, K. 1976. Dynamics in the accumulation of the enzyme neuraminidase depending on the growth phases of *Erysipelothrix insidiosa*. Acta Microbiologica, Virologica et Immunologica (Sofia) 4:27–32.

Alexander, W. D., and Goodwin, C. S. 1973. *Erysipelothrix* septicaemia. British Medical Journal 1:804.

Avilag, C., Unzucta, B. B. De., and Olguin, R. F. 1972. Indirect immunofluorescence test for the diagnosis of porcine erysipelas. Veterinaria (Mexico) 3:33–39.

Baird, P. J., and Benn, R. 1975. Reports of cases: *Erysipelothrix endocarditis*. Medical Journal of Australia 2:743–745.

Barber, M. 1939. A comparative study of *Listerella* and *Erysipelothrix*. Journal of Pathology and Bacteriology 48:11–23.

Berg, R. A. 1984. *Erysipelothrix rhusiopathiae*. Southern Medical Journal 77:1614.

Blount, J. G. 1965. Bacterial endocarditis. American Journal of Medicine 38:909–922.

Bohm, K. H. 1971. Neue Selektivnahrboden fur Rotlaufbakterien. Zentralblatt fur Bakteriologie, Parasitenkunde, Infektionskrankheiten und Hygiene. Abt. 1 Orig., Reihe A 218:330–334.

Borchardt, K. A., Sullivan, R. W., Blumberg, R. S., Gelber, R. H., Botch, V., Crull, S., and Ullyot, D. J. 1977. *Erysipelothrix rhusiopathiae* endocarditis. Western Journal of Medicine 127:149–151.

Bratberg, A. M. 1981. Selective adherence of *Erysipelothrix rhusiopathiae* to heart valves of swine investigated in an in vitro test. Acta Veterinaria Scandinavica 22:39–45.

Bricker, J. M., and Saif, Y. M. 1988. Use of a live oral vaccine to immunize turkeys against erysipelas. Avian Diseases 32:668–673.

Buchanan, R. E. 1918. Studies in the nomenclature and classification of the bacteria. Journal of Bacteriology 3:27–61.

Conklin, R. H., and Steele, J. H. 1979. *Erysipelothrix* infections, p. 327–337. In: J. H. Steele (ed.), CRC Handbook. Series in Zoonoses, vol. 1, (section A). CRC Press, Boca Raton, FL.

Creech, G. T. 1921. The bacillus of swine erysipelas isolated from urticarial lesions of swine in the United States. Journal of the American Veterinary Medical Association 59:139–150.

Dacres, W. G., and Groth, A. H., Jr. 1959. Identification of *Erysipelothrix insidiosa* with fluorescent antibody. Journal of Bacteriology 78:298–299.

Davis, G. H. G., and Newton, K. G. 1969. Numerical taxonomy of some named coryneform bacteria. Journal of General Microbiology 56:195–214.

Dedié, K. 1949. Die saureloslichen Antigene von *Erysipelothrix rhusiopathiae*. Monatshefte Veterinaermedizin. 4:7–10.

Dinter, Z., Diderholm, H., and Rockborn, G. 1976. Complement-dependent haemolysis following haemagglutination by *Erysipelothrix rhusiopathiae*. Zentralblatt fur Bakteriologie, Parasitenkunde, an Rotlaufbakterien. X. Die Differenzierung der Rotlaufbakterien nach chemischen Merkmalen. Archiv fur Experimentelle Veterinarmedizin 26:809–816.

Ehrlich, J. C. 1946. *Erysipelothrix rhusiopathiae* infection in man. Archives of Internal Medicine 78:565–577.

Erler, W. 1972. Serologisch, chemische und immunochemische Untersuchungen an Rotlaufbakterien. X. Die Differenzierung der Rotlaufbakterien nach chemischen Merkmalen. Arch. Exp. Veterinaermed. 26:809–816.

Ewald, F. W. 1957. Das Hyaluronidase-Bildungsvermogen von Rotlaufbakterien. Monatshefte fur Tierheilkunde 9:333–341.

Ewald, F. W. 1962. Uber die Dissoziation von *Erysipelothrix rhusiopathiae*. III. Mitteilung und Schluss: Die Virulenz dissoziierter Rotlaufbakterien. Monatsschrift für Tierheilkunde 14:260–267.

Ewald, F. W. 1964. Bakteriologie und Serologie humaner Rotlauferkrankungen unter besonderer Berucksichtigung des Ferrochlorid-Gelatine-Mediums nach Kaufmann. Arbeiten aus dem Paul-Ehrlich-Institut, Frankfurt am Main, Heft 61, 29–46.

Ewald, F. W. 1981. The genus *Erysipelothrix*, p. 1688–1700. In: M. P. Starr, H. Stolp, H. G. Truper, A. Balows and H. G. Schlegel (ed.), The prokaryotes: a handbook on habitats, isolation, and identification of bacteria. Springer-Verlag, New York.

Feist, H. 1972. Serologische, chemische und immunchemische Untersuchungen an Rotlaufbakterien. XII. Das Murein der Rotlaufbakterien. Archiv für Experimentelle Veterinarmedizin 26:825–834.

Feltham, R. K. A., Power, A. K., Pell, P. A., and Sneath, P. H. A. 1978. A simple method for storage of bacteria at −76°C. Journal of Applied Bacteriology 44:313–316.

Fliegelman, R. M., Cohen, R. S., and Zakhireh, B. 1985. *Erysipelothrix rhusiopathiae* endocarditis: report of a case and review of the literature. Journal of the American Osteopathic Association 85:39–42.

Flossmann, K. D., and Erler, W. 1972. Serologische, chemische und immunchemische Untersuchungen an Rothufbakterien. XI. Isolierung und Charakterisierung von Desoxyribonukleinsauren aus Rotlaufbaktereien. Archiv für Experimentelle Veterinarmedizin 26:817–824.

Fowler, N. O., Hamburger, M. H., and Bove K. E. 1967. Aortic valve perforation. American Journal of Medicine 42:539–546.

Fuzi, M. 1963. A neomycin sensitivity test for the rapid differentiation of *Listeria monocytogenes* and *Erysipelothrix rhusiopathiae*. Journal of Pathology and Bacteriology 85:524–525.

Gilchrist, T. C. 1904. Erysipeloid, with a record of 329 cases, of which 323 were caused by crab bites, or lesions produced by crabs. Journal of Cutaneous Diseases 22:507–519.

Gledhill, A. W. 1948. Discussion on swine erysipelas (*Erysipelothrix rhusiopathiae*) in man and animals. Proceedings of the Royal Society of Medicine 41:330–332.

Gorby, G. L., and Peacock, J. E., Jr. 1988. *Erysipelothrix rhusiopathiae* endocarditis: microbiologic, epidemiologic, and clinical features of an occupational disease. Reviews of Infectious Diseases 10:317–325.

Gransden, W. R., and Eykyn, S. J. 1988. *Erysipelothrix rhusiopathiae* endocarditis. Reviews of Infectious Diseases 10:317–325.

Grieco, M. H., and Sheldon, C. 1970. *Erysipelothrix rhusiopathiae*. Annals of the New York Academy of Sciences 174:523–532.

Harrington, R., Jr., and Hulse, D. C. 1971. Comparison of two plating media for the isolation of *Erysipelothrix rhusiopathiae* from enrichment broth culture. Applied Microbiology 22:141–142.

Harrington, R., Jr., Wood, R. L., and Hulse, D. C. 1974. Comparison of a fluorescent antibody technique and cultural method for the detection of *Erysipelothrix rhusiopathiae* in primary broth cultures. American Journal of Veterinary Research 35:461–462.

Heggers, I. P., Buddington, R. S., and McAllister, H. A. 1974. *Erysipelothrix* endocarditis diagnosis by fluorescence microscopy. Report of a case. American Journal of Clinical Pathology 62:803–806.

Heilman, F. R., and Herrell, W. E. 1944. Penicillin in the treatment of experimental infections due to *Erysipelothrix rhusiopathiae*. Proceedings of Staff Meetings, Mayo Clinic 19:340–345.

Hillenbrand, F. K. M. 1953. Whale finger and seal finger: their relation to erysipeloid. Lancet 1:680–681.

Hunter, D. 1974. The diseases of occupations, 5th ed., p. 709–712. The English Universities Press, Ltd., London.

Hunter, S. H. 1942. Some growth requirements of *Erysipelothrix* and *Listeria*. Journal of Bacteriology 43:629–640.

James, O., Morgan, O., Adam, M., and Schallibaum, E. 1976. *Erysipelothrix insidiosa* endocarditis: a report of two cases in Jamaica. West Indian Medical Journal 25:265–268.

Jones, D. 1975. A numerical taxonomic study of coryneform and related bacteria. Journal of General Microbiology 87:52–96.

Jones, D. 1986. Genus Erysipelothrix Rosenbach 367[al], p. 1245–1249. In: P. H. Sneath, N. S. Mair, and M. E. Sharpe (ed.), Bergey's manual of systematic bacteriology, vol. 2. Williams and Wilkins, Baltimore.

Kalf, G. F., and Grece, M. A. 1964. The antigenic components of *Erysipelothrix rhusiopathiae*. III. Purification of B- and C-antigens. Archives of Biochemistry and Biophysics 107:141–146.

Karlson, A. G., and Merchant, I. A., 1941. The cultural and biochemic properties of *Erysipelothrix rhusiopathiae*. American Journal of Veterinary Research 2:5–10.

Kaye, D. (ed.). 1976. Infective endocarditis. University Park Press, Baltimore.

King, P. F. 1946. Erysipeloid. Survey of 115 cases. Lancet 2:196–198.

Klauder, J. V. 1926. Erysipeloid and swine erysipelas in man. Journal of the American Medical Association 86:536–541.

Klauder, J. V. 1938. Erysipeloid as an occupational disease. Journal of the American Medical Association 111:1345–1348.

Klauder, J. V., Kramer, D. W., and Nicholas, L. 1943. *Erysipelothrix rhusiopathiae* septicemia; diagnosis and treatment. Report of fatal case of erysipeloid. Journal of the American Medical Association 122:938–943.

Klauder, J. V. 1944. *Erysipelothrix rhusiopathiae* infection in swine and in human beings: comparative study of the cutaneous lesions. Archives of Dermatology and Syphilology 50:151–159.

Koch, R. 1878. Untersuchungen uber die Atiologie der Wundinfektionskrankheiten. Vogel, Leipzig

Kramer, M. R., Gombert, M. E., Corrado, M. L., Ertgin, M. A., Burnett, V., and Ganguly, J. 1982. *Erysipelothrix rhusiopathiae* endocarditis. Southern Medical Journal 75:892.

Krasemann, C., and Müller, H. E. 1975. Die virulenz von *Erysipelothrix-rhusiopathiae*-stammen und neuraminidase-produktion. Zentralblatt für Bacteriologie, Parasitenkunde, Infektions-krankheiten und Hygiene. Abt. 1 Orig., Reihe A 231:206–213.

Kucsera, G. 1973. Proposal for standardization of the designations used for serotypes of *Erysipelothrix rhusio-*

pathiae (Migula) Buchanan. International Journal of Systematic Bacteriology 23:184–188.

Langford, G. C., and Hansen, P. A. 1953. *Erysipelothrix insidiosa*. Riass. Commun. VI. Congresso Internazionale di Microbiologia Roma 1:1–300.

Langford, G. C., and Hansen, P. A. 1954. The species of *Erysipelothrix*. Antonie van Leeuwenhoek Journal of Microbiology and Serology 20:87–92.

La Regina, M. C., Lonigro, J., Woods, L., Williams, G. A. and Vogler, G. A. 1988. Valvular endocarditis associated with experimental *Erysipelothrix rhusiopathiae* infection in the opossum (*Didelphis virginiana*). Laboratory Animal Science 38:159–161.

Lawes, F. A. E., Durie, E. G., Goldsworthy, N. E., and Spies, H. C. 1952. Subacute bacterial endocarditis caused by *Erysipelothrix rhusiopathiae*. Medical Journal of Australia 1:330–331.

Lerner, P. I., and Weinstein, L. 1966. Infective endocarditis in the antibiotic era. New England Journal of Medicine 274:199–206.

Loeffler, F. 1886. Experimentelle Untersuchungen uber Schweinerotlauf. Arbeiten aus dem Kaiserlichen Gesundheitsamt 1:46–55.

Lonigro, J. G., and La Regina, M. C. 1988. Characterization of *Erysipelothrix rhusiopathiae* isolated from an opossum (*Didelphis virginiana*) with septicemia. Journal of Wildlife Diseases 24(3):557–559.

McCarty, D., and Bornstein, S. 1960. *Erysipelothrix* endocarditis: report on a septicemic form of the erysipeloid of Rosenbach. American Journal of Clinical Pathology 33:39–42.

McCracken, A. W., Mauney, C. U., Huber, T. W., and McCloskey, R. V. 1973. American Journal of Clinical Pathology 59:219–222.

McGinnes, G. F., and Spindle, F. 1934. Erysipeloid condition among workers in a bone button factory due to the bacillus of swine erysipelas. American Journal of Public Health 24:32–35.

Mandal, B. N., and Malloch, J. A. 1971. Endocarditis due to *Erysipelothrix rhusiopathiae*. New Zealand Medical Journal 73:355–357.

Mann, S. 1969. Uber die Zellwandbausteine von *Listeria monocytogenes* und *Erysipelothrix rhusiopathiae*. Zentralblatt fur Bakteriologie, Parasitenkunde, Infektionskrankheiten und Hygiene, Abt. 1 Orig., Reihe A 209:510–518.

Migula, W. 1900. System der Bakterien. Handbuch der Morphologie, Entwicklungsgeschichte und Systematik der Bacterium. G. Fischer Verlag, Jena.

Morrill, C. C. 1939. Erysipeloid; occurence among veterinary students. Journal of Infectious Diseases 65:322–324.

Morris, C. A., Schwabacher, H., Lynch, P. G., Cross, C. D., and Dada, T. O. 1965. Two fatal cases of septicaemia due to *Erysipelothrix insidiosa*. Journal of Clinical Pathology 18:614–617.

Muirhead, N., and Reid, T. M. S., 1980. *Erysipelothrix rhusiopathiae* endocarditis. Journal of Infection 2:83–85.

Müller, H. E., and Krasemann, Ch. 1976. Immunitat gegen *Erysipelothrix rhusiopathiae*- Infektion durch aktive Immunisierung mit homologer Neuraminidase. Zeitschrift fur Imunitatsforschung 151:237–241.

Müller, H. E., and Seidler, D. 1975. Uber das Vorkommen Neuraminidase-Neutralisierender Antikorper bei: chronisch rotlaufkranden Schweinen. Zentralblatt fur Bakteriologie, Parasitenkunde. Infektionskrankheiten und Hygiene, Abt. 1 Orig., Reihe A 230:51–58.

Nakato, H., Shinomiya, K., and Mikawa, H. 1986. Possible role of neuraminidase in the pathogenesis of arteritis and thrombocytopenia induced in rats by *Erysipelothrix rhusiopathiae*. Pathology, Research and Practice 181:311–319.

Nelson, E. 1955. Five hundred cases of erysipeloid. Rocky Mountain Medical Journal 52:40–42.

Nikolov, P., and Abrashev, I. 1976. Comparative studies of the neuraminidase activity of *Erysipelothrix insidiosa*. Activity of virulent strains and avirulent variants of *Erysipelothrix insidiosa*. Acta Microbiologica, Virologica et Immunologica (Sofia) 3:28–31.

Normann, B., and Kihlstrom, E. 1985. *Erysipelothrix rhusiopathiae* septicaemia. Scandinavian Journal of Infectious Diseases 17:123–124.

Norrung, V. 1979. Two new serotypes of *Erysipelothrix rhusiopathiae*. Nordisk Veterinaermedicin 34:462–465.

Ognibene, F. P., Cunnion, R. E., Gill, V., Ambrus, J., Fauci, A. S., and Parrillo, J. E. 1985. *Erysipelothrix rhusiopathiae* bacteremia presenting as septic shock. American Journal of Medicine 78:861–864.

Pachas, W. N., and Currid, V. R. 1974. L-form induction, morphology, and development in two related strains of *Erysipelothrix rhusiopathiae*. Journal of Bacteriology 119:576–582.

Packer, R. A. 1943. The use of sodium azide and crystal violet in a selective medium for *Erysipelothrix rhusiopathiae* and *Streptococci*. Journal of Bacteriology 46:343–349.

Park, C. H., Poretz, D. M., and Goldenberg, R. 1976. *Erysipelothrix* endocarditis with cutaneous lesion. Southern Medical Journal 69:1101–1103.

Pasteur, L., and Dumas, M. 1882. Sur le rouget, ou mal rouge des porcs. Extrait d'une Lettre, Comptes Rendus Hebdomadaires des Seances de l'Academie des Sciences, Paris 95:1120–1121.

Phillips, P. E. 1986. Infectious agents in the pathogenesis of rheumatoid arthritis. Seminars in Arthritis and Rheumatism 16:1–10.

Pleszczynska, E. 1972. Comparative studies on *Listeria* and *Erysipelothrix*. I. Analysis of whole antigens. II. Analysis of antigen fractions. Polskie Archiwum Weterynaryjne 15, Fasc. 3:463–471, 473–481.

Poretz, D. M. 1985. *Erysipelothrix rhusiopathiae*, p. 1185–1186. In: G. L. Mandell, R. G. Douglas, Jr., J. E. Bennett, (ed.), Principles and practice of infectious diseases, 2nd ed. New York.

Price, J. E. L., and Bennett, W. E. J. 1951. The erysipeloid of Rosenbach. British Medical Journal 2:1060–1062.

Procter, W. I. 1965. Subacute bacterial endocarditis due to *Erysipelothrix rhusiopathiae*. American Journal of Medicine 38:820–824.

Robertson, D. C., and McCullough, W. G. 1968. Glucose catabolism of *Erysipelothrix rhusiopathiae*. Journal of Bacteriology 95:2112–2116.

Rosenbach, F. J. 1909. Experimentelle, morphologische und klinische Studien uber krankheitserregende Mikroorganismen des Schweinerotlaufs, des Erysipeloids und der Mausesepticamie. Zeitschrift fur Hygiene und Infektionskrankheiten 63:343–371.

Russell, W. O., and Lamb, M. E. 1940. *Erysipelothrix* endocarditis: a complication of erysipeloid. Journal of the American Medical Association 114:1045–1050.

Sawada, T., and Takahashi, T. 1987. Cross protection of mice and swine given live-organism vaccine against challenge exposure with strains of *Erysipelothrix rhusiopathiae* representing ten serovars. American Journal of Veterinary Research 48:81–84.

Schiffman, W. L., and Black, A. 1956. Acute bacterial endocarditis caused by *Erysipelothrix rhusiopathiae*. New England Journal of Medicine 255:1148–1150.

Schleifer, K. H., and Kandler, O. 1972. Peptidoglycan types of bacterial cell walls and their taxonomic implications. Bacteriological Reviews 36:407–477.

Scully, R. E., Galdabini, J., and McNeely, B. U. 1978. Weekly clinicopathological exercises. Case 16-1978. Case Records Massachusetts General Hospital. New England Journal of Medicine 298:957–962.

Seeliger, H. P. R. 1974. Genus *Erysipelothrix*, p. 597. In: Buchanan and Gibbons (ed.), Bergey's manual of determinative bacteriology, 8th ed. Williams and Wilkins, Baltimore.

Seidler, D., Trautwein, G., and Bohm, K. H. 1971. Nachweis von *Erysipelothrix insidiosa* mit fluoreszierenden Antikorpern. Zentralblatt fur Veterinarmedizin B, 18:280–292.

Shinomiya, K., and Nakato, H. 1985. An experimental model of arteritis: periarteritis induced by *Erysipelothrix rhusiopathiae* in young rats. International Journal of Tissue Reactions 7:267–271.

Shuman, R. D. 1959. Comparative experimental evaluation of swine erysipelas bacterins and vaccines in weanling pigs, with particular reference to the status of their dams. American Journal of Veterinary Research. 20:1002–1009.

Shuman, R. D., and Wellmann, G. 1966. Status of the species name *Erysipelothrix rhusiopathiae* with request for an opinion. International Journal of Systematic Bacteriology 16:195–196.

Sikes, D. 1958. A comparison of rheumatoid-like arthritis in swine and rheumatoid arthritis in man. Annals of the New York Academy of Science 70:717–725.

Silberstein, E. B. 1965. *Erysipelothrix* endocarditis: report of a case with cerebral manifestations. Journal of the American Medical Association 191:862–864.

Simberkoff, M. S., and Rahal, Jr., J. J. 1973. Acute and subacute endocarditis due to *Erysipelothrix rhusiopathiae*. American Journal of the Medical Sciences. 266:53–57.

Smith, G. 1983. Erysipelothrix and Listeria, p. 50–59. In: G. Wilson, A. Miles and M. T. Parker (ed.), Topley and Wilson's principles of bacteriology, virology and immunity. Williams and Wilkins, Baltimore.

Sneath, P. H. A., Abbott, J. D., and Cunliffe, A. C. 1951. The bacteriology of erysipeloid. British Medical Journal 2:1063–1066.

Stackebrandt, E., and Woese, C. R. 1981. Towards a phylogeny of the actinomycetes and related organisms. Current Microbiology 5:197–202.

Stiles, G. W. 1947. Chronic erysipeloid (swine erysipelas) in a man. The effect of treatment with penicillin. Journal of the American Medical Association 134:953–955.

Stuart, M. R. 1972. A note on the occurrence of core-like structures in association with *Erysipelothrix rhusiopathiae*. Journal of General Microbiology 73:571–572.

Stuart, M. R., and Pease, P. E. 1972. A numerical study on the relationships of *Listeria* and *Erysipelothrix*. Journal of General Microbiology. 73:551–565.

Stuart, S. E., and Welshimer, H. J. 1974. Taxonomic re-examination of *Listeria* Pirie and transfer of *Listeria grayi* and *Listeria murrayi* to a new genus *Murraya*. International Journal of Systematic Bacteriology 24:177–185.

Tadayon, R. A., and Carroll, K. K. 1971. Effect of growth conditions on the fatty acid composition of *Listeria monocytogenes* and comparison with the fatty acids of *Erysipelothrix* and *Corynebacterium*. Lipids 6:820–825.

Takahashi, T., Hirayama, N., Sawada, T., Tamura, Y., and Muramatsu, M. 1987. Correlation between adherence of *Erysipelothrix rhusiopathiae* strains of serovar 1A to tissue culture cells originated from porcine kidney and their pathogencity in mice and swine. Veterinary Microbiology 13:57–64.

Todorov, T. 1976. Induction of L-forms of *Erysipelothrix insidiosa* using antibiotics and lysozyme. Acta Microbiologica, Virologica et Immunologica (Sofia) 4:39–45.

Torkildsen, A. 1943. Intracranial *Erysipelothrix* abscess. Bulletin of Hygiene 18:1013–1015.

Townshend, R. H., Jephcott, A. E., and Yekta, M. H. 1973. *Erysipelothrix* septicemia without endocarditis. British Medical Journal 1:464.

Trevisan, V. Caratteri di alcuni nuovi generi di Batteriacee. Atti della Accademia Fisio-Medico-Statistica in Milano, Ser. 4, 3.92–107 (1885). International Bulletin of Bacteriological Nomenclature and Taxonomy 2:11–29.

Valerianov, T. S., Toschkoff, A., and Cholakova, S. 1976. Biological properties of *Erysipelothrix* phages isolated from lysogenic cultures. Acta Microbiologica, Virologica et Immunologica (Sofia) 3:32–38.

Wawrzkiewicz, K. 1964. Dissociation forms of *Erysipelothrix insidiosa*. (Poln.) Acta Microbiologica Polonica 13:45–54.

Weaver, R. E. 1985. Erysipelothrix, p. 209–210. In: E. H. Lennette, A. Balows, W. J. Hausler, Jr., and H. J. Shadomy (ed.), Manual of clinical microbiology, 4th ed. American Society for Microbiology, Washington, D.C.

White, T. G. and Mirikitani, F. K. 1976. Some biological and physical chemical properties of *Erysipelothrix rhysiopathiae*. Cornell Veterinarian 66:152–163.

White, T. G., and Shuman, R. D. 1961. Fermentation reactions of *Erysipelothrix rhusiopathiae*. Journal of Bacteriology 82:595–599.

Wilkinson, B. J., Jones, D. 1977 A numerical taxonomic survey of *Listeria* and related bacteria. Journal of General Microbiology 98:399–421.

Wilson, G. S., and Miles, A. A. 1975. Topley and Wilson's principles of bacteriology and immunity, 6th ed., vol. 1, *Erysipelothrix*, p. 554–558. Edward Arnold, London.

Woese, C. R., Maniloff, J., and Zablen, L. B. 1980. Phylogenetic analysis of the Mycoplasmas. Proceedings of National Academy of Sciences (USA) 77:494–498.

Wood, R. L. 1965. A selective liquid medium utilizing antibiotics for isolation of *Erysipelothrix insidiosa*. American Journal of Veterinary Research 26:1303–1308.

Wood, R. L. 1970. *Erysipelothrix*. In: Blair, Lennette, and Truant (ed.), Manual of clinical microbiology. American Society for Microbiology. Bethesda, Md.

Wood, R. L. 1974. Isolation of pathogenic *Erysipelothrix rhusiopathiae* from feces of apparently healthy swine. American Journal of Veterinary Research 35:41–43.

Wood, R. L. 1975. *Erysipelothrix* infection, p. 271–281. In: W. T. Hubbert, W. F. McCullough and P. R. Schnurrenberger (ed.), Diseases transmitted from animals to man, 6th ed. Thomas, Springfield, IL.

Wood, R. L. 1984. Swine erysipelas—a review of prevalence and research. Journal of the American Veterinary Medical Association 184:944–949.

Wood, R. L., Haubrich, D. R., and Harrington, R. 1978. Isolation of previously unreported serotypes of *Erysipelothrix rhusiopathiae* from swine. American Journal Veterinary Research 39:1958–1961.

Wood, R. L., and Packer, R. 1972. Isolation of *Erysipelothrix rhusiopathiae* from soil and manure of swine-raising premises. American Journal of Veterinary Research 33:1611–1620

Wood, R. L., and Shuman, R. D. 1974. Swine erysipelas, p. 565–620. In: H. W. Dunne, and A. D. Leman (ed.), Diseases of swine, 4th ed. Iowa State University Press, Ames, IA.

Woodbine, M. 1950. *Erysipelothrix rhusiopathiae*. Bacteriology and chemotherapy. Bacteriological Reviews 14:161–178.

CHAPTER 74

The Genus *Gemella*

ULRICH BERGER

The genus *Gemella* comprises two species, *G. haemolysans* and *G. morbillorum*. The cells are cocci which are either arranged in pairs, often with the adjacent sides flattened; or arranged in tetrads, short chains, or small irregular clusters. The cell wall is of the Gram-positive type. Gemella cells divide in two planes, generally at right angles to each other. Elongate and rod-shaped forms both occur. The organisms prefer an atmosphere rich in carbon dioxide, but growth, though more sparsely, can occur under both aerobic and strictly anaerobic conditions. Cytochrome oxidase, catalase, and peroxidase are absent, but superoxide dismutase is present. The GC content of the DNA is 33 mol% and 30 mol%, respectively.

Originally, *G. haemolysans* was described as a new *Neisseria* species by Thjötta and Böe (1938), who had isolated one strain from the sputum of a bronchitic patient. Later, some additional strains were isolated (Berger and Wezel, 1960). In contrast to representatives of the genus *Neisseria,* however, all strains of *Gemella* produced acid from glucose fermentatively (Berger, 1960b); lacked catalase, cytochrome oxidase, and peroxidase; possessed a cyanide-resistant respiratory system; formed small amounts of hydrogen peroxide on aerobic growth; and exhibited a reduced sensitivity to this compound (Berger, 1960a). Therefore, *N. haemolysans* was removed from the genus *Neisseria.* A new monospecific genus was created, which was named *Gemella* (Berger, 1960a, 1961). (*Gemellus* [Latin] is the diminutive form of *geminus* [twin].)

The high sensitivity of the genus to penicillins and its resistance to aminoglycosides and polymyxins were also consistent with its classification as a Gram-positive, streptococcus-like organism (Berger, 1960c). By gas chromatography, Yamakawa and Ueta (1964) demonstrated that whole cell extracts of *N. haemolysans* differed from those of the other pathogenic and commensal *Neisseria* spp. studied in the composition of both fatty acids and neutral sugars. This was essentially confirmed by Lambert et al. (1971) and Brooks et al. (1971, 1972). In addition, the latter authors as well as Morse et al. (1977) demonstrated differences between *Neisseria* and *Gemella* in metabolic products. Later, Chun et al. (1985) demonstrated by comparative enzyme electrophoresis of 19 species of the genus *Neisseria* and a few strains of other genera that *G. haemolysans* was genetically unrelated to any of the *Neisseria* spp. However, definite proof that *N. haemolysans* had to be removed from the genus *Neisseria* and that it represented a genus of its own had already been furnished by Reyn et al. (1966, 1970). They showed, with the electron microscope, that the cell wall of *G. haemolysans* is of the Gram-positive type. In addition, the mean GC content of its DNA, determined in five different isolates, was found to be 33.5 mol% (Reyn et al., 1970). This value is distinctly below those of the "true" *Neisseria* spp. (46.5–53.5 mol%) and of all genera of the family Neisseriaceae (38–55 mol%) (Bøvre, 1984). Facklam and Smith (1976) reported that *Gemella* strains morphologically and physiologically resembled the aerotolerant *Streptococcus morbillorum* (Facklam, 1977). Later, Facklam and Wilkinson (1981) regarded *G. haemolysans* and *S. morbillorum* as identical species. However, the more reliable method of comparative analysis of 16S rRNA revealed that *G. haemolysans* is part of the *"Clostridium"* sub-branch of Gram-positive bacteria which have a GC content of less than 55 mol% (Stackebrandt et al., 1982). The binary comparison of the similarity coefficients (S_{AB} values) showed *Gemella* to be even more closely related to the *Bacillus-Staphylococcus* group than to streptococci and lactobacilli.

G. morbillorum, the second species of the genus *Gemella,* was first described by Tunnicliff (1917). From the blood of 42 out of 50 patients in the preeruptive and early eruptive stages of measles, she isolated a Gram-positive, sometimes flattened diplococcus which occasionally exhibited rod-shaped morphology and formed short chains and small clusters. In primary cultures, the organism grew only under anaerobic conditions, but in most cases it could be adapted to aerobic growth by several transfers. On aerobic incubation it caused greening of blood agar. Some simple sugars were fermented. Later, she named this organism *Diplococcus rubeolae* (Tunnicliff, 1933). In the years following the first description, apparently the same organism was isolated from the same sources and under the same conditions by several authors from different countries (Degkwitz, 1927; Duval and Hibbard, 1927; Cary and Day, 1927; Kusama et al. 1930; Kodama, 1930; Prévot, 1933). Prévot (1933), being unaware of the name given by Tunnicliff (1933), designated the new species as *Diplococcus morbillorum*. The species names were based on the association of the organism with measles, but there is confusion here due to language differences; *rubeola* means "measles" only in British English but it refers to "German measles" in French and German medical terminology. In French- and German-speaking countries, the term for measles is *morbilli,* as it is, though not consistently, in the United States. To avoid misunderstandings arising from this terminological confusion, the name proposed by Prévot (1933) finally was adopted also by Tunnicliff (1936). In *Bergey's Manual of Determinative Bacteriology* (7th ed.), the organism was transferred to the genus *Peptostreptococcus* because of its primarily anaerobic nature (Smith, 1957). For the same reason and on the basis of its metabolic products, Holdeman and Moore (1975) initially assigned it to *Peptococcus.* However, after they realized that its major metabolic product was lactic acid, they transferred it to the genus *Streptococcus* as *S. morbillorum* (Holdeman and Moore, 1974; Holdeman et al. 1977).

As mentioned above, Facklam and Smith (1976) found that *S. morbillorum* and *G. haemolysans* were closely related organisms both morphologically and physiologically and, finally, stated that their names were mere synonyms (Facklam and Wilkinson, 1981; Facklam, 1984). The striking physiological similarity of *G. haemolysans* and *S. morbillorum* as observed by Facklam and his associates was substantiated by Berger and Pervanidis (1986). However, they detected a number of physiological and serological differences which enabled them to distinguish between these species. With isolates from pigs, Molitoris et al. (1986) also discriminated between *G. haemolysans* and *S. morbillorum* on the basis of biochemical and physiological tests. Since *S. morbillorum* (just as *G. haemolysans*) divides in two planes at right angles, Berger and Pervanidis (1986) concluded that this organism as defined could not be assigned to *Streptococcus.* They discussed its

transfer to the genus *Gemella* instead of conversely transferring *G. haemolysans* to the genus *Streptococcus.* By comparative analysis of 16S rRNA, it was demonstrated that *S. morbillorum* is related to *G. haemolysans* (Ludwig et al., 1988). Kilpper-Bälz and Schleifer (1988) proposed on the basis of DNA-DNA hybridization studies and certain physiological and chemotaxonomic properties the transfer of *S. morbillorum* to the genus *Gemella* as *G. morbillorum* comb. nov.

Habitats and Ecophysiology

Both *G. haemolysans* and *G. morbillorum* are commensals of the mucous membranes of humans and of certain warm-blooded animals.

In healthy people, *G. haemolysans* has only been isolated from the oral cavity and the upper respiratory tract. In two investigations of swabs from the nasopharyngeal mucosa, involving a total of 230 young adults, the organism was detected in 25% and 30% of the individuals, respectively (Berger and Wezel, 1960; Berger, 1985). Moreover, it was also found in dental plaque material of an individual examined by de Jong and van der Hoeven (1987). From the intestinal contents of breeding pigs and their litters, Molitoris et al. (1986) isolated 60 strains of *G. haemolysans* representing 5% of all catalase-negative isolates having streptococcal morphology.

For a long time, *G. morbillorum* has almost exclusively been associated with measles, where it was regularly isolated from blood and more-or-less frequently also from the throat, nose, eyes, and ears of the patients. Like *G. haemolysans, G. morbillorum* later was detected in the human oral cavity (Kannangara et al., 1981; Kolenbrander and Williams, 1983; Dzink et al., 1984). Unlike *G. haemolysans, G. morbillorum* is also found in the normal intestinal flora of humans (Holdeman et al., 1977). It represents about 0.1% of the total viable count (Holdeman and Moore, 1976). From the intestinal contents of breeding pigs, Molitoris et al. (1986) obtained 10 strains of *G. morbillorum.* Its proportion of all streptococcal isolates from this source amounted to less than 1%.

In the oral cavity, sterile saliva was found to support the growth of *G. haemolysans* in enrichment cultures from dental plaque matter, as well as in artificial mixtures of the dominant bacterial species from these cultures (de Jong and van der Hoeven, 1987). It is, however, not clear whether *G. haemolysans* itself secretes the necessary enzymes for breakdown of salivary glycoproteins or whether the accompanying microflora is providing the enzymes. Using the APIZYM system (API System S.A., Montalieu-Vercieu, France), only α-glucosidase was detected in *G. morbillorum,* and leucine aminopeptidase and chymotrypsin were detected in both *Gemella* species (Table 1) (Berger, 1985; Berger and Pervanidis, 1986).

Bacteria that produce poorly soluble exopolysaccharides are known to adhere to surfaces of teeth and epithelial cells. Production of dextran could not be demonstrated in *G. haemolysans* (Bridge and Sneath, 1983) and was not investigated in *G. morbillorum.* However, exopolysaccharides differing from those of streptococcal origin and not synthesized from sucrose were described by Reyn et al. (1970) and Mills et al. (1984). Their exact chemical structures are not yet understood, but apparently they are not dextrans, nor do they belong to the iodine-positive, starch-amylopectin-glycogen group produced by other oral bacteria (Berger and Wezel, 1960; Reyn et al., 1970; Facklam, 1977; Chatelain et al., 1982). In preparations for electron microscopy, the polysaccharides appear as fibrous material surrounding the cell, often radially arranged at its surface as a "gly-cocalyx." In *G. haemolysans,* this structure was interpreted as a narrow, slimy capsule (Reyn et al., 1970). In *G. morbillorum,* the polysaccharidic nature and low solubility in water were shown by Mills et al. (1984). That the exopolysaccharide from *G. morbillorum* constitutes an important adherence factor is shown by the endocarditis experiments cited below (Mills et al., 1984).

Another mechanism that might be responsible for the prevalence of *G. morbillorum* in dental plaque is based on the coaggregation phenomenon. The microflora of early plaque are characterized by a predominance of filamentous bacteria, such as *Actinomyces naeslundii* and *A. viscosus.* Most of the *A. naeslundii* strains were shown to coaggregate in vitro and in vivo with *G. morbillorum* strains, in a reaction that was inhibited by lactose or by previously heating the actinomycetes, but was not affected by saliva or by heating *G. morbillorum.* It was concluded that coaggregation is mediated by lectin-like structures on the surface of *A. naeslundii,* which bind to carbohydrate moieties on the surface of gemellae (Kolenbrander and Williams, 1983; Kolenbrander and Phucas, 1984).

In a comparison of *G. morbillorum* and five strains of different viridans streptococci, the amount of exopolysaccharide produced in the presence of glucose was shown to be the crucial factor for initiation of experimental bacterial endocarditis in rabbits. *G. morbillorum* which exhibited maximal exopolysaccharide production in vitro and in vivo also exhibited maximal adherence to artificially damaged heart valves, multiplied most rapidly, and formed the largest cardiac vegetations (Mills et al., 1984).

Both *Gemella* species are highly sensitive to the penicillins, cephalosporins, tetracyclines, chloramphenicol, and lincomycins. In addition, *G. haemolysans* is strongly inhibited by macrolide antibiotics (erythromycin, spiramycin, oleandomycin), vancomycin, ristocetin, novobiocin, and tyrothricin. A majority of strains are also inhibited by bacitracin and fusidic acid. *G. haemolysans* is moderately resistant to the aminoglycosides (streptomycin, kanamycin, gentamycin tobramycin, amikacin, and neomycin) and is fully resistant to polymyxin B and colistin. This resistance pattern is typical of Gram-positive organisms and resembles that of the streptococci. Sensitivity to chemotherapeutics such as sulfonamides, nitrofurazone, and nalidixic acid differs from one strain to another, but as a rule is low or lacking (Berger, 1960c; Reyn et al., 1970; Wilkinson and Jones, 1977; Chatelain et al., 1982; Buu-Hoi et al., 1982; Carles-Giraud et al., 1982; Bridge and Sneath,

Table 1. Biochemical reactions and enzymes common to both *G. haemolysans* and *G. morbillorum.*

Present	Absent[a]
Superoxide dismutase	Cytochrome oxidase
Acid from:	Catalase
Glucose	Acid from:
Fructose	Lactose
Galactose	Arabinose
Mannose	Rhamnose
Maltose	Inulin
Trehalose	Salicin
Sucrose	Polysaccharide
Starch	from sucrose
Glycogen	Production of:
Dextrin	Indole
N-Acetyl-glucosamine	Hydrogen sulfide[b]
C_4 esterase	Hydrolysis of:
C_8 esterase	Urea
Leucine aminopeptidase	Esculine
Chymotrypsin	Arginine
	Hippurate
	Reduction of nitrate

[a]In addition to the carbohydrates listed in the table, the following compounds were not attacked in the API 50 CH system: glycerol, erythritol, ribose, xylose, adonitol, sorbose, dulcitol, inositol, cellobiose, melibiose, melezitose, raffinose, fucose, and furanose. The following enzymes could not be demonstrated with the APIZYM system: C_{14} lipase, valine aminopeptidase, trypsin, α- and β-galactosidase, β-glucuronidase, β-glucosidase, β-glucosaminidase, α-mannosidase, and α-fucosidase.
[b]5% of *G. haemolysans* strains are weakly positive.
Modified from Berger and Pervanidis (1986).

1983; Thadepalli et al., 1983; Blin et al., 1984; Laudat et al., 1984; Chandrasekar et al., 1984; Mitchell and Teddy, 1985; Molitoris et al., 1986).

Isolation, Maintenance, and Preservation

Isolation

The methods for isolation depend on the material to be cultured and the purpose of investigation: The major difference concerns conditions required for growth. Whereas *G. haemolysans* should be incubated in the presence of free oxygen (with or without additional CO_2), for proliferation of *G. morbillorum*, oxygen has to be strictly excluded.

Oropharyngeal swabs are streaked onto blood agar plates for obtaining separate colonies. The expression of hemolysis, the crucial character for presumptive identification of *G. haemolysans*, depends on the choice of the "correct" blood species and the "correct" agar base. The organism consistently produces β-hemolysis only on rabbit blood agar (Berger and Wezel, 1960; Chatelain et al., 1982). On horse blood agar, complete (β) hemolysis was reported by Reyn et al. (1970) and Buu-Hoi et al. (1982), incomplete hemolysis by Chatelain et al. (1982), α-hemolysis by Bridge and Sneath (1983) and Mitchell and Teddy (1985), and no hemolysis at all by Wilkinson and Jones (1977), Carles-Giraud et al. (1982), Blin et al. (1984), and Laudat et al. (1984). As the agar base for the best hemolysin production (though not for best growth) Mueller-Hinton agar with 5% fresh or defibrinated rabbit blood is used. This is best suited for isolation and discrimination of colonies suspected of being *G. haemolysans* (Berger and Wezel, 1960). After inoculation, the plates are incubated aerobically in a humid atmosphere (closed jar, CO_2 incubator) with or without additional CO_2 (5–10%) at 35–37°C for 3–5 days. After two days of incubation, the plates are checked daily for minute hemolytic colonies from which pure cultures can be obtained by one or more transfers onto rabbit blood agar plates incubated as described above (Berger and Wezel, 1960; Berger, 1985).

For isolating *G. haemolysans* from septicemic infections, 5 ml of freshly drawn blood is transferred to any of the commercially available blood culture media and incubated aerobically or anaerobically. Incubation under aerobic conditions seems to yield somewhat better results than anaerobic cultures (Blin et al., 1984). Generally, the liquid medium becomes slightly turbid after three days of incubation. Then, 0.5 ml of culture fluid is transferred to 5% rabbit and sheep (or horse) blood agar and incubated aerobically under increased CO_2 pressure.

Isolation of *G. haemolysans* from cerebrospinal fluid (CSF) was performed by plating a small amount of CSF deposit on horse blood agar. Incubation for 2 days both aerobically, with or without additional CO_2, and anaerobically yielded scanty growth of small greyish colonies surrounded by a definite zone of α-hemolysis, which had to be differentiated from viridans streptococci (Mitchell and Teddy, 1985).

Isolation of *G. haemolysans* from dental plaque was achieved by de Jong and van der Hoeven (1987), who described the procedure in detail.

Thus, if the presence of *G. haemolysans* is suspected in any clinical specimen other than blood, it is recommended that both sheep (or horse or human) and rabbit blood agar plates be used and that the plates be incubated either in air or under increased CO_2 pressure. Smooth, minute colonies, which are visible to the naked eye after 2 or more days of incubation and are surrounded by a zone of β-hemolysis only on rabbit blood agar, should be regarded as *G. haemolysans* and should be investigated further.

G. morbillorum has been isolated from a much greater variety of clinical specimens than *G. haemolysans*. The specimens include blood; swabs from throat, tonsils, nose, and conjunctivae; dental plaque; intestinal contents of humans and pigs; samples from the genitourinary tract; and pus from a broad spectrum of suppurative infections ranging from brain abscess to madurafoot (Prévot et al., 1967; Facklam, 1977).

From swabs and pus, the organism has been isolated by streaking blood agar plates, generally containing 5% of defibrinated sheep blood. The minute colonies of *G. morbillorum* are either surrounded by a zone of greening (α-hemolysis) or leave the blood unchanged (Holdeman and Moore, 1974; Facklam, 1977). Cooksey et al. (1979) observed the same effects on rabbit blood agar. Facklam and Wilkinson (1981) failed to detect any influence of different blood sources on hemolysis due to *G. morbillorum*. On the other hand, Berger and Pervanidis (1986) observed strains causing β-hemolysis on rabbit blood agar, but α-hemolysis on sheep blood agar. Hemolysis was not only influenced by blood source, but also by the presence or absence of free oxygen, and by CO_2 pressure. Hence, hemolysis differs depending on strain and growth conditions; in *G. morbillorum* it is not a constant character, as it is in *G. haemolysans*.

From dental plaque, samples were taken with a sterile curette and immediately transferred to thioglycolate broth (Difco) containing 20% of beef infusion. Dilutions were made in the same medium, 0.1 ml portions were spread onto Columbia agar base (BBL) with blood added and incubated under an atmosphere containing H_2, CO_2, and N_2 (1:1:8), using the Gas-Pak apparatus (BBL). Any isolate that exhibited spherical morphology was subjected to further analysis, and *G. ("S.") morbillorum* was identified as described by Facklam (1977) (Kolenbrander and Williams, 1983).

Isolation of *G. morbillorum* from the intestinal tract of human male adults was achieved by Holdeman et al. (1976). The method employed is described by Moore and Holdeman (1974). The composition of the rumen-fluid-glucose-cellobiose agar (RGCA) used for culturing the anaerobic organisms is given in the VPI Anaerobe Laboratory Manual (Holdeman and Moore, 1975).

Maintenance

During experimental work, cultures of both *G. haemolysans* and *G. morbillorum* can be maintained on and in media of simple composition to which blood (5–7%), serum (5–10%) or ascitic fluid (10–25%) have been added. *G. haemolysans* can also be adapted to ordinary serum-free media, but growth is poor (Thjötta and Böe, 1938; Berger, 1961; Chatelain et al., 1982). Tween-80 (about 0.02%) and fermentable carbohydrates such as 0.2–0.5% glucose enhance growth of *G. morbillorum* in serum-free media (Prévot, 1933; Holdeman and Moore, 1974). For agar base, several commercial products such as heart infusion agar, Mueller-Hinton agar (Difco, BBL), GC agar base, trypticase soy agar (BBL), Columbia agar (Biomerieux) and blood agar base (Difco, Oxoid) have been used. Plates are inoculated and incubated aerobically as described. Freshly isolated strains of *G. morbillorum* not yet adapted to these conditions must be cultured anaerobically. Growth of this organism can be improved by addition of 0.001% pyridoxal-HCl to blood agar (Cooksey et al., 1979). Since on sheep blood agar *G. haemolysans* can not be distinguished from viridans streptococci, strains should be checked for purity after several transfers by streaking them onto rabbit blood agar, where they cause β-hemolysis.

Predominantly, heart infusion broth (Difco), brain heart infusion broth (Difco), and trypticase soy broth (BBL)—enriched with serum or ascitic fluid—served as liquid media. In these media, growth appears as scanty to gross, and either granular, flocculent, or smooth sediment with slightly turbid supernatant after one or more days of incubation. For *G. morbillorum,* Todd-Hewitt broth supplemented with 0.001% pyridoxal-HCl proved to be an excellent liquid medium (Cooksey et al., 1979), as did thioglycolate broth supplemented with 10% serum for *G. haemolysans* (Berger and Wezel, 1960). In these media formation of a copious deposit and dense turbidity was obtained. As serum-free fluid media used for working cultures of *G. morbillorum,* peptone-yeast extract solution (PY) with 0.02% Tween-80 and 0.2% glucose added was employed (Holdeman and Moore, 1974; Kolenbrander and Phucas, 1984), as well as a semisynthetic medium described by Ludwig et al. (1988). Protein-free Caso-bouillon (Stackebrandt et al., 1982) and a so-called basal medium broth (Bridge and Sneath, 1983) served for propagation of *G. haemolysans.*

Preservation

For short-term preservation (1 week to 1 month, approximately), *G. haemolysans* can be streaked onto heart infusion agar slants with 10% ascitic fluid or 10% inactivated serum added or stabbed into the same medium, incubated aerobically for one day, and then kept at room temperature (Berger and Wezel, 1960). In 50% sterile glycerol, *G. morbillorum* survived for 2 months (Prévot, 1933). Dzink et al. (1984) kept their isolates on trypticase soy agar plates with the addition of 5% sheep blood at 35–37°C in an atmosphere of N_2, H_2, and CO_2 (8:1:1) and transferred them weekly. Kannangara et al. (1981) kept stock strains of *G. morbillorum* in prereduced chopped meat-glucose broth at room temperature.

For long-term preservation (over one year) both *Gemella* spp. can be maintained in a lyophilized state or in media containing 15–20% glycerol at −70°C.

Identification

Morphology and Staining and Cultural Characteristics

The cells of both *Gemella* spp. are Gram-positive cocci. Form, arrangement, and mode of division are described above (see Introduction) and are illustrated in Figure 1. Though the cell wall is of the Gram-positive type, cells are easily decolorized and thus may appear Gram-variable or even Gram-negative. Both species exhibit a considerable pleomorphism which, in *G. hae-*

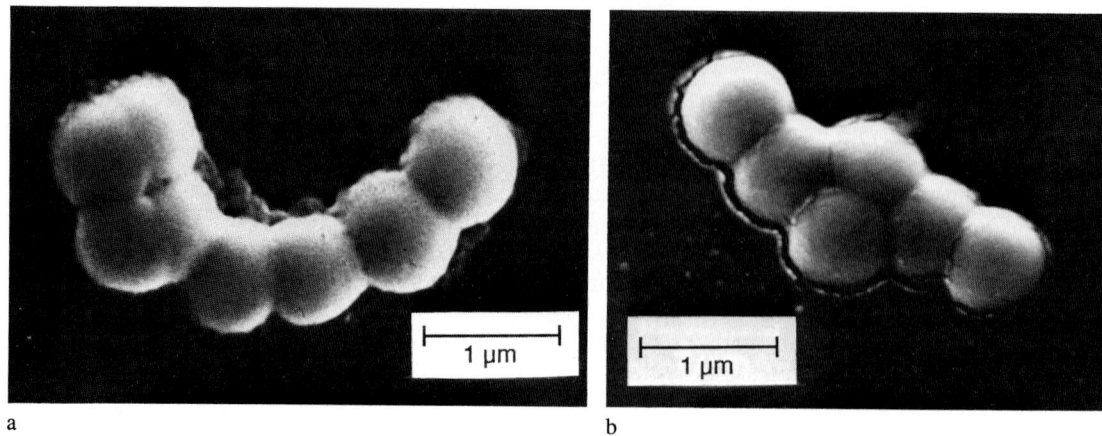

a   b

Fig. 1. Scanning electron micrographs of gold/palladium-coated *Gemella* spp. (a) *G. haemolysans* ATCC 10379. (b) *G. morbillorum* ATCC 27824.

molysans, is essentially confined to size, although elongate cells are occasionally seen (Berger, 1960a). Size of cells can vary from one strain to another (Berger, 1961) and from one culture to another (Reyn et al., 1970). The diameter varies from 0.5 μm to more than 1.0 μm, and "giant cells" have been observed (Berger, 1961; Chatelain et al., 1982; Reyn, 1986). The larger the cells, the more firmly the Gram stain is retained. In *G. morbillorum,* pleomorphism is more pronounced. Cells in pairs often differ considerably in size. Coccal forms frequently are elongate, occasionally with one end tapering. Rod-shaped cells occur and chains of considerable length, consisting of elongate cells, have been observed (Berger and Pervanidis, 1986). The average size of elongate cells is described as measuring 0.5 by 1.2 μm (range 0.3–0.8 by 0.5–1.4 μm) (Holdeman and Moore, 1974), but cells as long as 1.5 μm (Prévot, 1933) and even 2–3 μm (Kusama et al., 1930) apparently do occur. As in *G. haemolysans,* morphology of *G. morbillorum* varies with strain as with cultural conditions.

The Gram-variable character of *Gemella* may be the result of the variation in thickness of the cell wall. Though of the Gram-positive type, it is relatively thin and its thickness varies from about 10–20 nm (Reyn et al., 1970; Mills et al., 1984).

Electron microscopic studies revealed cells of *G. haemolysans* surrounded by a "corona" of floccular or fibrous material arranged radially at the cell surface (Reyn et al., 1966, 1970). An outer layer of very similar structure was also seen in *G. morbillorum* (Mills et al., 1984). From staining with a variety of polysaccharide stains it was concluded that the fibrous material in *G. morbillorum* is probably a polyanionic polysaccharide which is extensively cross-linked (Mills et al., 1984). The width of the capsule or glycocalyx is about 50–60 nm. The India ink-carbol fuchsin stain shows narrow halos surrounding the cells of both species, even under the light microscope (Berger, 1985; Berger and Pervanidis, 1986).

Round, whitish spots about 50 nm in diameter detected in the cytoplasma of *G. haemolysans* under the electron microscope were interpreted as lipid droplets (Reyn et al., 1970). By use of Sudan black stain (Burdon, 1946), these droplets were identified as poly-β-hydroxybutyrate granules, which were present in both *Gemella* spp. (U. Berger, unpublished observation).

Both species are nonmotile and do not form spores.

Growth of *Gemella* spp. is slow. Surface colonies are small, circular, entire, low-convex, translucent to opaque, smooth and occasionally somewhat mucoidal. The mucoidal colonies may exhibit confluent growth. No pigments are produced. After two days of incubation, colonies of *G. haemolysans* measure about 1 mm, and those of *G. morbillorum* are pinpoint to 0.5 mm in diameter. Differences in the size of colonies may also result from the kind and quality of culture media employed. From colonies of both organisms no homogenous suspensions for agglutination could be prepared (Berger and Wezel, 1960; Berger and Pervanidis, 1986).

After 2 days of incubation on rabbit blood agar plates, colonies of *G. haemolysans* are regularly surrounded by a distinct zone of β-hemolysis, whereas *G. morbillorum* produces greening (α-hemolysis) or fails to hemolyze; however, under certain conditions, exceptional strains of *G. morbillorum* may produce β-hemolysis as well (Berger and Pervanidis, 1986). Under strictly anaerobic conditions, β-hemolysis is less distinct and α-hemolysis is greatly reduced or lacking. For influence of blood source and agar base on hemolysis as well as for growth in liquid media see the Isolation section.

Serum, defibrinated or hemolysed blood, or ascitic fluid are indispensable components of ordinary solid and liquid culture media. Only *G. haemolysans* may be adapted to growth on media deficient in protein, but proliferation is scanty (Thjötta and Böe, 1938; Berger and Wezel, 1960). Though by addition of 5–10% CO$_2$ to the atmosphere, proliferation of both species is enhanced, gemellae are not strictly capnophilic organisms (Mitchell and Teddy, 1985; Berger and Pervanidis, 1986).

Optimum growth is at 35–37°C, but both species can develop over a wider range of temperatures. They cease to multiply at 10°C and 45°C (Reyn et al., 1970; Facklam, 1977; Wilkinson and Jones, 1977; Bridge and Sneath, 1983; Mitchell and Teddy, 1985; Molitoris et al., 1986).

Presence in culture media of 2% NaCl was tolerated by a minority of strains of *G. morbillorum*, but growth was completely inhibited by 4% NaCl (Holdeman and Moore, 1974; Facklam, 1977). According to Bridge and Sneath (1983), proliferation of *G. haemolysans* is inhibited by a concentration of 4% NaCl as well, but resistance to 6.5% and even 10% NaCl has been reported (Wilkinson and Jones, 1977; Molitoris et al., 1986). Resistance to optochin was demonstrated in both *G. haemolysans* (Reyn et al., 1970) and *G. morbillorum* (Facklam and Smith, 1976; Facklam, 1977).

Biochemical Properties

Both *Gemella* spp. have been reported to lack catalase, cytochrome oxidase and cytochromes (Table 1); in *G. haemolysans* lack of peroxidase was also observed (Berger, 1960a; Reyn et al., 1970; Facklam and Smith, 1976; Stackebrandt et al., 1982; Berger and Pervanidis, 1986; Kilpper-Bälz and Schleifer, 1988). However, cytochromes of types *b* and *d* were detected in cells of *G. haemolysans* incubated aerobically in the presence of hemin (Stackebrandt et al., 1982). In both species, superoxide dismutase was demonstrated (Stackebrandt et al., 1982; Kilpper-Bälz and Schleifer, 1988). Presence of some other enzymes important for aerobic growth, such as NADH- and L-lactate oxidases, were successfully assayed *G. haemolysans* (Stackebrandt et al., 1982). In cultures incubated aerobically, this organism apparently forms trace

amounts of hydrogen peroxide; it exhibited a moderate resistance to this compound. Its respiration is resistant to potassium cyanide (Berger, 1960a).

Glucose is degraded fermentatively by both *G. haemolysans* and *G. morbillorum* (Berger, 1960b; Wilkinson and Jones, 1977; Chatelain et al., 1982). Major end products of anaerobic glucose metabolism are L-lactic and acetic acids (Brooks et al., 1971; Holdeman and Moore, 1974; Stackebrandt et al., 1982). In both species, formic acid was detected as well (Brooks et al., 1971; Holdeman and Moore, 1974). As additional metabolites, isovaleric, α-hydroxyisocaproic, and α-ketoisocaproic acids, and *p*-hydroxyphenyl acetic and butyric acids were identified in cultures of *G. haemolysans* (Brooks et al., 1971, 1972), with trace amounts of succinic and pyruvic acids in those of *G. morbillorum* (Holdeman and Moore, 1974). Under aerobic conditions, end products of glucose metabolism formed by *G. haemolysans* were equimolar amounts of acetate and CO$_2$ (Stackebrandt et al., 1982). The presence of fructose-1,6-bisphosphate aldolase indicates that *G. haemolysans* degrades glucose by the Embden-Meyerhof pathway (Stackebrandt et al., 1982). Regularly, but in small amounts, acetyl methyl carbinol is formed by *G. haemolysans* (Berger and Wezel, 1960; Reyn et al., 1970; Morse et al., 1977; Berger, 1985), whereas in cultures of *G. morbillorum* this compound could be detected only exceptionally (Holdeman and Moore, 1974) or not at all (Holdeman et al., 1977; Berger and Pervanidis, 1986). Cadaverine and putrescine, common products of *Neisseria* spp., were not found in *G. haemolysans* (Morse et al., 1977).

G. haemolysans and *G. morbillorum* share a remarkable number of biochemical activities, as is to be expected in members of the same genus. Table 1 shows the reactions which in both species are either present or absent.

Acid production from some sugars is controversial. Only small amounts of acid are produced from galactose by *G. haemolysans* (Berger, 1985). In *G. morbillorum,* irregular acid production from galactose, was reported by Prévot (1933) and by Holdeman and Moore (1974). The API 50 CH system consistently yielded positive results. Acid production from trehalose is rather weak and is usually not detected by conventional methods (Berger and Pervanidis, 1986). Acid production from mannitol and sorbitol is a variable characteristic. By conventional methods only 9% of fresh isolates of *G. haemolysans* produced acid from these alcohols (Berger, 1985). On the other hand, the data base of the API 20 STREP system gives a value of

98% for positive strains. In this system, degradation of mannitol and sorbitol by *G. morbillorum* could not be detected.

Proteolytic activities, such as degradation of casein or liquefaction of gelatin or coagulated serum, were not detected in *Gemella* spp.

Though esterases were demonstrated in both of *Gemella* spp. (Table 1), hydrolysis of tributyrin and Tween-60 has only been found in *G. haemolysans* so far (Berger, 1961; Wilkinson and Jones, 1977). Also, in *G. haemolysans* lecithinase, RNase and DNase were detected by Wilkinson and Jones (1977); however, DNase activity could not be confirmed (Chatelain et al., 1982; Bridge and Sneath, 1983).

Nitrate is not reduced by *Gemella* spp. (Table 1). On the other hand, reduction of nitrite was proved to be a typical, regular characteristic in *G. haemolysans* but was not present in *G. morbillorum* and, therefore, it is a valuable character for species differentiation. For positive demonstration of this activity, low nitrite concentrations in liquid media (0.01% or even 0.001%) and prolonged incubation (5 days) are indispensable (Berger, 1985; Berger and Pervanidis, 1986). No gas formation was detected in nitrite broth.

Although there is a high degree of conformity of biochemical reactions in *G. haemolysans* and *G. morbillorum,* as shown in Table 2, the organisms can be differentiated by a limited number of laboratory tests. Indentification and differentation can also be achieved with the API 20 STREP rapid identification system (Appelbaum et al., 1984; Facklam et al., 1984; Laudat et al., 1984). With API 50 CH, APIZYM, and other miniaturized rapid systems, a considerable number of useful reactions can be performed (Bridge and Sneath, 1983; Laudat et al., 1984; Berger, 1985; Berger and Pervanidis, 1986). It must, however, be taken into account that in certain cases the results obtained with them may differ from those obtained with the conventional methods (Etienne et al., 1984; Berger, 1985).

The major fatty acids in cells of both *G. haemolysans* and *G. morbillorum* are palmitic (C_{16}) acid and stearic (C_{18}) acid. *G. haemolysans* differs from *G. morbillorum* only in its relatively high content of lauric acid (C_{12}). The fatty acid profiles do not allow a species differentiation within the genus *Gemella,* but they are helpful to distinguish *G. haemolysans* from *Neisseria* and *G. morbillorum* from viridans streptococci (Yamakawa and Ueta, 1964; Lambert et al., 1971; Labbé et al., 1985).

Serological Properties

Since most *Gemella* strains spontaneously agglutinate in physiological NaCl solution, the precipitation reaction was preferred in serological investigations. With the tube precipitation test, *G. haemolysans* could easily be separated from *Neisseria perflava* and *N. sicca,* the two *Neisseria* species which—biochemically and by producing hemolysis (*N. perflava* only)—resemble the *Gemella* species most closely (Berger and Wezel, 1960). In addition, the tube precipitation test was applied to prove the relationship to the

Table 2. Characteristics for differentiation of *G. haemolysans* and *G. morbillorum.*[a]

Characteristic	G. haemolysans		G. morbillorum	
	Type strain ATCC 10379	63 other strains	Type strain ATCC 27824	5 other strains
Morphology	Cocci, elongate cells rare		Cocci, elongate cells, rods	
Arrangement	Pairs, tetrads, clusters, short chains		Pairs, tetrads, clusters, more and longer chains	
Growth conditions	Aerobic better than anaerobic		Anaerobic better than aerobic	
β-Hemolysis (rabbit blood)	Optimum under 10% CO_2; in air, regularly and well marked		Very rare; if present, optimum under 10% CO_2; in air, absent or weak	
Reduction of nitrite	+	+	−	−
Acetoin from glucose	+	+	−	−
Alkaline phosphatase	+ +	+ +	−	20% +
Acid phosphatase	+ +	+ +	+	(+)/+
Pyrrolidone arylamidase	+	+/(+)	−	40% (+)
Acid from mannitol and sorbitol	−	9% (+)	(+)	40% (+)
α-Glucosidase	−	−	(+)	(+)/+
Alanine aminopeptidase	−	6% +	(+)	(+)/+
Leucine arylamidase	−	16% (+)	(+)	+

+, ≥95% positive; (+), weakly positive; + +, strongly positive. In the case of different reactions, percentage of positive strains is given.

[a]Modified from Berger and Pervanidis (1986).

type strain of a number of fresh isolates of *G. haemolysans*. Using antiserum raised against the type strain of *G. haemolysans* (ATCC 10379), all of 57 fresh isolates of this species were precipitated, though with differing intensities (Berger, 1985). The same antiserum did not react with extracts prepared from three strains of *G. morbillorum* (among them the type strain ATCC 27824), thus pointing to an antigenic nonidentity of the two *Gemella* species. On the other hand, an antiserum raised against *G. morbillorum* NCTC 11323 rapidly reacted not only with extracts prepared from three homologous strains (among them the type strain), but also, although more slowly, with those of about 70% of *G. haemolysans* isolates (among them the type strain ATCC 10379) (Berger and Pervanidis, 1986). Thus, a certain serological relationship between *G. haemolysans* and *G. morbillorum* exists.

Pathogenicity

Pathogenicity for Humans

Like virtually all commensal bacteria of the human microbiota, the gemellae, as "opportunistic pathogens," are able to cause severe localized and generalized infections predominantly in immunocompromized individuals.

From 1982 to 1984, five cases of subacute bacterial endocarditis and two cases of septicemia not involving the heart valves due to *G. haemolysans* have been reported in France. All patients recovered under antibiotic treatment (Chatelain et al., 1982; Buu-Hoi et al., 1982; Carles-Giraud et al., 1982; Laudat et al., 1984; Blin et al., 1984). One case of postoperative meningitis due to *G. haemolysans* occurred in a patient who had undergone thermolysis of the Gasserian ganglion for trigeminal neuralgia (Mitchell and Teddy, 1985).

Because of the remarkable regularity of its occurrence in the blood and throat of measles patients, *G. morbillorum* was initially believed to be the specific pathogen of this disease (Tunnicliff, 1917, 1933; Duval and Hibbard, 1927; Cary and Day, 1927). However, its presumptive role in measles was soon restricted to an auxiliary function in the propagation of the nonbacterial (viral) agent in vitro (Degkwitz, 1927) or in serving as its vector in vivo (Prévot, 1933). Much later, Prévot et al. (1967) reported on 12 strains isolated from various suppurative processes, such as dental abscess, tonsillitis, axillary hidroadenitis, pleuritis, lung abscess, peritonitis, subphrenic abscess, and pyelonephritis. Facklam (1977) described 46 isolates from hu-

man clinical specimens, predominantly from blood in endocarditis and septicemia, but in some instances also from the genitourinary and respiratory tract, from a brain abscess, and from purulent processes at other sites of the body. Apart from the blood isolates, whether they were sole or only concomitant causative agents is not clear. Holdeman et al. (1977) reported on 22 isolates from different human clinical specimens. The only source of more recent isolates of *G. morbillorum* was blood from patients with endocarditis, sepsis, or septicemia (Cooksey et al., 1979; Durack et al., 1983; Etienne et al., 1984; Mills et al., 1984; Coto and Berk, 1984). The strains described by Durack et al. (1983) represented 6% of viridans and 5% of all "streptococcal" isolates from 52 cases of endocarditis. According to Facklam and Wilkinson (1981), *G. morbillorum* constituted slightly more than 3% of the viridans streptococci isolated from all human specimens and less than 1% of the viridans isolates from endocarditis patients. Yet, one cannot be sure that all isolates labelled *G.("S.") morbillorum* really belonged to this species, since, in accordance with Facklam and Wilkinson (1981), many of the recent authors did not discriminate between *G.("S.") morbillorum* and *G. haemolysans*.

Experimental Pathogenicity

There is general agreement that in pure culture the gemellae are of little or no virulence for common laboratory animals.

With pure cultures of *G. haemolysans*, even with mucin or egg yolk added, no fatal infection could be produced in white mice when injected intraperitoneally (Thjötta and Böe, 1938; Berger, 1961).

G. morbillorum, injected intravenously in pure culture, caused pyrexia and leukopenia in rabbits and Japanese monkeys, but no weight loss was noticed (Hibbard and Duval, 1926; Duval and Hibbard, 1927; Cary and Day, 1927; Kusama et al., 1930; Kodama, 1930). Kodama (1930) explicitly stated that the organism was not markedly pathogenic for rabbits, guinea pigs, mice, and rats. Yet, if the virulence of *G. morbillorum* was increased by combining it with other organisms of low virulence or if resistance of the host was decreased by artificially damaging the heart valves, severe infections could be produced. In New Zealand white rabbits, Kannangara et al. (1981), Thadepalli et al. (1983) and Chandrasekar et al. (1984) could consistently produce lung abscesses when a mixture of pure cultures of *G. morbillorum, Fusobacterium nucleatum, Eubacterium lentum,* and *Bacteroides fragilis* was injected transtracheally. It was not possible to elicit a comparable infection by giving either a pure culture of *B. fragilis* alone or of *G. morbillorum, F. nucleatum,* and *E. lentum* in combination without *B. fragilis.* In some of the inoculated animals, *G. morbillorum* could be recovered from pus taken from the abscesses. Thus, this organism appears to play a definite role in the pathogenesis of experimental lung abscess, although by itself it is incapable of producing it.

In rabbits with artifically damaged heart valves, Mills et al. (1984) and Pulliam et al. (1985) could induce bacterial endocarditis by intravenous injection of 10^8–10^9 cfu of *G. morbillorum.* A persistent bacteremia developed, and in animals sacrificed 1–2 days after infection, the vegetations of heart valves were colonized by large numbers of gemellae. Since, in the experiments described, the animals were killed after a few days, nothing more is known about the "natural" course of the infection and its consequences.

Literature Cited

Appelbaum, P. C., P. S. Chaurushiya, M. R. Jacobs, and A. Duffett. 1984. Evaluation of the Rapid Strep System for species identification of streptococci. J. Clin. Microbiol. 19:588–591.

Berger, U. 1960a. *Neisseria haemolysans* (Thjötta und Böe, 1938). Untersuchungen zur Stellung im System. Zschr. f. Hyg. 146:253–259.

Berger, U. 1960b. Über den Kohlenhydrat-Stoffwechsel von *Neisseria* und *Gemella.* Zbl. Bakt. Parasitenk. Infektionskr. Hyg. Abt. I Orig. 180:147–149.

Berger, U. 1960c. Die Empfindlichkeit von *Gemella haemolysans* gegen einige gebräuchliche Antibiotica. Arch. Hyg. Bakt. 144:12–16.

Berger, U. 1961. A proposed new genus of Gram-negative cocci: *Gemella.* Internat. Bull. Bact. Nomencl. Taxon. 11:17–19.

Berger, U. 1985. Prevalence of *Gemella haemolysans* on the pharyngeal mucosa of man. Med. Microbiol. Immunol. 174:267–274.

Berger, U., and A. Pervanidis. 1986. Differentiation of *Gemella haemolysans* (Thjøtta and Bøe 1938) Berger 1960, from *Streptococcus morbillorum* (Prévot 1933) Holdeman and Moore 1974. Zbl. Bakt. Hyg. A 261:311–321.

Berger, U., and M. Wezel. 1960. Zur Frage der Identität hämolysierender saprophytischer Neisserien. Zschr. f. Hyg. 146:244–252.

Blin, C., V. Vialette, A. Tenaillon, D. Fischer, and G. Cosson. 1984. Septicémie à *Gemella haemolysans.* Méd. Mal. Infect. 14:163–165.

Bøvre, K. 1984. Family VIII. *Neisseriaceae* Prévot 1933, p. 288–290. In: N. R. Krieg and J. G. Holt (ed.), Bergey's manual of systematic bacteriology, vol. 1. Williams & Wilkins, Baltimore.

Bridge, P. D., and P. H. A. Sneath. 1983. Numerical taxonomy of *Streptococcus.* J. Gen. Microbiol. 129:565–597.

Brooks, J. B., D. S. Kellogg, L. Thacker, and E. M. Turner. 1971. Analysis by gas chromatography of fatty acids found in whole cultural extracts of *Neisseria* species. Can. J. Microbiol. 17:531–543.

Brooks, J. B., D. S. Kellogg, L. Thacker, and E. M. Turner. 1972. Analysis by gas chromatography of hydroxy acids produced by several species of *Neisseria.* Can. J. Microbiol. 18:157–168.

Burdon, K. L. 1946. Fatty material in bacteria and fungi revealed by staining dried, fixed slide preparations. J. Bact. 52:665–678.

Buu-Hoi, A., A. Sapoetra, C. Branger, and J. F. Acar. 1982. Antimicrobial susceptibility of *Gemella haemolysans* isolated from patients with subacute endocarditis. Eur. J. Clin. Microbiol. 1:102–106.

Carles-Giraud, D., P. Dellamonica, B. Monnier, and H. Duplay. 1982. Septicémie à *Gemella haemolysans.* A propos d'une observation. Méd. Mal. Infect. 12:255–256.

Cary, W. E., and L. A. Day. 1927. The etiology of measles. J. Amer. Med. Ass. 89:1206–1208.

Chandrasekar, P. H., K. V. I. Rolston, V. Chokkavelu, J. L. LeFrock, and D. W. Kannangara. 1984. Comparative efficacy of four antibiotics in anaerobic pulmonary infection. Chemotherapy 30:331–336.

Chatelain, R., J. Croize, P. Rouge, C. Massot, H. Dabernat, J. C. Auvergnat, A. Buu-Hoi, J. P. Stahl, and F. Bimet. 1982. Isolement de *Gemella haemolysans* dans trois cas d'endocardites bactériennes. Méd. Mal. Infect. 12:25–30.

Chun, P. K., G. F. Sensabaugh, and N. A. Vedros. 1985. Genetic relationships among *Neisseria* species assessed by comparative enzyme electrophoresis. J. Gen. Microbiol. 131:3105–3115.

Cooksey, R. C., F. S. Thompson, and R. R. Facklam. 1979. Physiological characterization of nutritionally variant streptococci. J. Clin. Microbiol. 10:326–330.

Coto, H., and S. L. Berk. 1984. Endocarditis caused by *Streptococcus morbillorum.* Amer. J. Med. Sci. 187:54–58.

Degkwitz, R. 1927. The etiology of measles. J. Inf. Dis. 41:304–316.

de Jong, M. H., and J. S. van der Hoeven. 1987. The growth of oral bacteria on saliva. J. Dent. Res. 66:498–505.

Durack, D. T., E. L. Kaplan, and A. L. Bisno. 1983. Apparent failures of endocarditis prophylaxis. Analysis of 52 cases submitted to a national registry. J. Amer. Med. Ass. 250:2318–2322.

Duval, C. W., and R. J. Hibbard. 1927. Further studies upon the etiology of measles. Proc. Soc. Exp. Biol. Med. 24:519–522.

Dzink, J. L., C. Smith, and S. S. Socransky. 1984. Semiautomated technique for identification of subgingival isolates. J. Clin. Microbiol. 19:599–605.

Etienne, J., M. E. Reverdy, L. D. Gruer, V. Delorme, and J. Fleurette. 1984. Evaluation of the API 20 STREP system for species identification of streptococci associated with infective endocarditis. Eur. Heart J. 5 (suppl. C):25–27.

Facklam, R. R. 1977. Physiological differentiation of viridans streptococci. J. Clin. Microbiol. 5:184–201.

Facklam, R. R. 1984. The major differences in the American and British *Streptococcus* taxonomy schemes with special reference to *Streptococcus milleri.* Eur. J. Clin. Microbiol. 3:91–93.

Facklam, R. R., D. L. Rhoden, and P. B. Smith. 1984. Evaluation of the Rapid Strep system for the identification of clinical isolates of *Streptococcus* species. J. Clin. Microbiol. 20:894–898.

Facklam, R. R., and P. B. Smith. 1976. The Gram-positive cocci. Human Pathology 7:187–194.

Fracklam, R. R., and H. W. Wilkinson. 1981. The family *Streptococcaceae* (medical aspects), p. 1572–1597. In: M. P. Starr, H. Stolp, H. G. Trüper, A. Balows, and H. G. Schlegel (ed.), The prokaryotes, vol. 2. Springer Verlag, Berlin.

Hibbard, R. J., and C. W. Duval. 1926. Studies upon the virus of measles. Proc. Soc. exp. Biol. Med. 23:853–856.

Holdeman, L. V., E. P. Cato, and W. E. C. Moore. 1977. Anaerobe Laboratory Manual, 4th ed., Blacksburg, Virginia.

Holdeman, L. V., I. J. Good, and W. E. C. Moore. 1976. Human fecal flora: Variation in bacterial composition within individuals and a possible effect of emotional stress. Appl. Environ. Microbiol. 31:359–375.

Holdeman, L. V., and W. E. C. Moore. 1974. New genus, *Coprococcus*, twelve new species, and emended descriptions of four previously described species of bacteria from human feces. Int. J. Syst. Bact. 24:260–277.

Holdeman, L. V., and W. E. C. Moore. 1975. Anaerobe Laboratory Manual, 3rd ed., Blacksburg, Virginia.

Kannangara, D. W., H. Thadepalli, V. T. Bach, and D. Webb. 1981. Animal model for anaerobic lung abscess. Infect. Immun. 31:592–597.

Kilpper-Bälz, R., and K. H. Schleifer. 1988. Transfer of *Streptococcus morbillorum* to the genus *Gemella* as *Gemella morbillorum* comb. nov. Int. J. Syst. Bact. 38:442–443.

Kodama, T. 1930. Bacteriological studies of measles, part 2. Further studies on Kusama's bacillus and cultivation of a diplococcus from measles patients. Kitasato Arch. Exp. Med. 7:226–234.

Kolenbrander, P. E., and C. S. Phucas. 1984. Effect of saliva on coaggregation of oral *Actinomyces* and *Streptococcus* species. Infect. Immun. 44:228–233.

Kolenbrander, P. E., and B. L. Williams. 1981. Lactose-reversible coaggregation between oral actinomycetes and *Streptococcus sanguis*. Infect. Immun. 33:95–102.

Kolenbrander, P. E., and B. L. Williams. 1983. Prevalence of viridans streptococci exhibiting lactose-inhibitable coaggregation with oral actinomycetes. Infect. Immun. 41:449–452.

Kusama, Yokoyama, and Ito. 1930. Bacteriological studies on the etiology of measles with Japanese monkeys. Kitasato Arch. Exp. Med. 7:220–225.

Labbé, M., P. van der Auwera, Y. Glupczynski, F. Crockaert, and E. Yourassowsky. 1985. Fatty acid composition of *Streptococcus milleri*. Eur. J. Clin. Microbiol. 4:391–393.

Lambert, M. A., D. G. Hollis, C. W. Moss, R. E. Weaver, and M. L. Thomas. 1971. Cellular fatty acids of nonpathogenic *Neisseria*. Can. J. Microbiol. 17:1491–1502.

Laudat, P., P. Cosnay, B. Icole, A. Raoult, P. Raynaud, and M. Brochier. 1984. Endocardite á *Gemella haemolysans*: une nouvelle observation. Méd. Mal. Infect. 14:159–161.

Ludwig, W., M. Weizenegger, R. Kilpper-Bälz, and K. H. Schleifer. 1988. Phylogenetic relationships of anaerobic streptococci. Int. J. Syst. Bact. 38:15–18.

Mills, J., L. Pulliam, L. Dall, J. Marzouk, W. Wilson, and J. W. Costerton. 1984. Exopolysaccharide production by viridans streptococci in experimental endocarditis. Infect. Immun. 43:359–367.

Mitchell, R. G., and P. J. Teddy. 1985. Meningitis due to *Gemella haemolysans* after radiofrequency trigeminal rhizotomy. J. Clin. Path. 38:558–560.

Molitoris, E., M. I. Kritchevsky, D. J. Fagersberg, and C. L. Quarles. 1986. Effects of dietary chlortetracycline on the antimicrobial resistance of porcine faecal *Streptococcaceae*. J. Appl. Bact. 60:111–120.

Moore, W. E. C., and L. V. Holdeman. 1974. Human fecal flora: The normal flora of 20 Japanese-Hawaiians. Appl. Microbiol. 27:961–979.

Morse, C. D., J. B. Brooks, and D. S. Kellogg. 1977. Identification of *Neisseria* by electron capture gas-liquid chromatography of metabolites in a chemically defined growth medium. J. Clin. Microbiol. 6:474–481.

Prévot, A. R. 1933. Etudes de systématique bactérienne. I. Lois générales.—II. Cocci anaérobies. Ann. Sci. Nat. Bot. 15:23–261.

Prévot, A. R., A. Turpin, and P. Kaiser. 1967. Les bactéries Anaérobies. Dunod, Paris.

Pulliam, L., L. Dall, S. Inokuchi, W. Wilson, W. K. Hadley, and J. Mills. 1985. Effects of exopolysaccharide production by viridans streptococci on penicillin therapy of experimental endocarditis. J. Inf. Dis. 151:153–156.

Reyn, A. 1986. Genus *Gemella* Berger 1960, p.1081–1082. In: P. H. A. Sneath, N. S. Mair, M. E. Sharpe, and J. G. Holt (ed.), Bergey's manual of systematic bacteriology, vol. 2. Williams & Wilkins, Baltimore.

Reyn, A., A. Birch-Andersen, and U. Berger. 1970. Fine structure and taxonomic position of *Neisseria haemolysans* (Thjøtta and Bøe 1938) or *Gemella haemolysans* (Berger 1960). Acta Path. Microbiol. Scand. B 78:375–389.

Reyn. A., A. Birch-Andersen, and S. P. Lapage. 1966. An electron microscope study of thin sections of *Haemophilus vaginalis* (Gardner and Dukes) and some possibly related species. Can. J. Microbiol. 12:1125–1136.

Smith, L. D. S. 1957. Genus V. *Peptostreptococcus*, p. 533–541. In: R. S. Breed, E. G. D. Murray, and N. R. Smith (ed.), Bergey's Manual of Determinative Bacteriology, 7th ed. Williams & Wilkins, Baltimore.

Stackebrandt, E., B. Wittek, E. Seewaldt, and K. H. Schleifer. 1982. Physiological, biochemical and phylogenetic studies on *Gemella haemolysans*. FEMS Microbiol. Letters 13:361–365.

Thadepalli, H., D. W. Kannangara, and V. T. Bach. 1983. Penicillin failure in the treatment of *Bacteroides fragilis* lung abscess. Chemotherapy 29:289–293.

Thjøtta, T., and J. Böe. 1938. *Neisseria hemolysans*. A hemolytic species of *Neisseria* Trevisan. Acta Path. Microbiol. Scand., Suppl.37, 527–531.

Tunnicliff, R. 1917. The cultivation of a *Micrococcus* from blood in pre-eruptive and eruptive stages of measles. J. Amer. Med. Ass. 68:1028–1030.

Tunnicliff, R. 1933. Colony formation of *Diplococcus rubeolae*. J. Inf. Dis. 52:39–53.

Tunnicliff, R. 1936. Opsonins for *Diplococcus morbillorum* and for *Streptococcus scarlatinae* in convalescent measles serum, convalescent scarlet fever serum and placental extract. J. Inf. Dis. 58:1–4.

Wilkinson, B. J., and D. Jones. 1977. A numerical taxonomic survey of *Listeria* and related bacteria. J. Gen. Microbiol. 98:399–421.

Yamakawa, T., and N. Ueta. 1964. Gaschromatographic studies of microbial components. I. Carbohydrate and fatty acid constitution of *Neisseria*. Japan. J. Exp. Med. 34:361–374.

CHAPTER 75

The Genus *Kurthia*

RONALD M. KEDDIE and DOROTHY JONES

In 1883, H. Kurth published a detailed description of a new bacterial species, *Bacterium zopfii,* which he had isolated from the intestinal contents of chickens (Kurth, 1883). Two years later, Trevisan (1885) created the genus *Kurthia* with *K. zopfii* as the type species. However, over the ensuing years Kurth's organism was given a variety of generic names, including *Zopfius* (Wenner and Rettger, 1919); the valid name *Kurthia* really came into general use only after the publication of the seventh edition of *Bergey's Manual of Determinative Bacteriology* (Breed et al., 1957). In the current edition of *Bergey's Manual,* the genus *Kurthia* contains two species, *K. zopfii* and *K. gibsonii* (Keddie and Shaw, 1986); a third species, *K. sibirica,* has subsequently been described (Belikova et al., 1986).

Kurth isolated the organism now called *K. zopfii* by streaking material from the intestinal contents of chickens onto nutrient gelatin plates. When the plates were incubated, the organism grew out from the original streak and through the gelatin in long, fine, apparently branched threads. For a number of years following the original isolation, *K. zopfii* aroused considerable interest because of its characteristic growth patterns in nutrient gelatin, and various attempts were made to explain the observed phenomena. Most characteristic is the appearance in a gelatin slant. If the slant is inoculated with a single central streak and incubated in the near vertical position, then the resultant growth resembles a bird's feather (Boyce and Evans, 1893; Jacobsen, 1907; Kufferath, 1911; Sergent, 1906, 1907; Zikes, 1903). The outgrowths, which appear to follow the lines of stress in the gelatin, are presumably a result of the organism's marked filament-forming ability coupled with an inability to hydrolyze gelatin; both motile and nonmotile strains exhibit this phenomenon (Keddie, 1949).

In the few decades following its original isolation, *K. zopfii* (or *K. zenkeri,* a synonym of *K. zopfii*) was isolated from a variety of sources, including fresh and putrefied meat (Günther, 1896; Jacobsen, 1907; Wenner and Rettger, 1919), waste water and air from abattoirs (Jacobsen, 1907), preserved sausage (Günther, 1896), and pus from a cat's ear (Boyce and Evans, 1893). It was also reported to occur in feces (Flügge, 1896), water (Flügge, 1896; Migula, 1900), and to be common in milk (Orla-Jensen, 1931). One strain of an organism considered to be *K. zopfii* was isolated from air at an altitude exceeding 3,000 m (10,000 ft) (Proctor, 1935)!

Following this early interest in *K. zopfii,* the genus *Kurthia* received scant attention for many years but the more recent rediscovery that *K. zopfii* and the organism now called *K. gibsonii* frequently occur as components of the aerobic flora of meats and meat products once again focused attention on the genus. However, bacteria of the genus *Kurthia* do not appear to have been implicated directly in the spoilage process of meats and meat products. There are also several scattered reports that organisms identified as "*Kurthia* spp." have been isolated from a variety of clinical materials but there is no evidence of pathogenicity in authentic members of the genus.

Bacteria of the genus *Kurthia* are strictly aerobic, Gram-positive rods that do not produce acid from glucose. Consequently, in earlier studies they were sometimes considered to be taxonomically related to one or more of the aerobic genera of coryneform bacteria. Indeed, some earlier numerical taxonomic surveys seemed to support this view (see Shaw and Keddie, 1983a). The "tentative" inclusion of *Kurthia* in the section on the "Coryneform Group of Bacteria" in the eighth edition of *Bergey's Manual of Determinative Bacteriology* unfortunately was interpreted as supporting this similarity (e.g. see Ludwig et al., 1981), and presumably led to the inclusion of the Chapter on "The Genus *Kurthia*" in the section on "The Coryneform Bacteria" in the first edition of *"The Prokaryotes"* (Starr et al., 1981). However, chemical studies of the peptidoglycan structure (Schleifer and Kandler, 1972), isoprenoid quinone composition (Collins et al., 1979), polar lipid composition (Goodfellow et al., 1980), and the GC content of the DNA (Belikova et al., 1980, 1986; Shaw and Keddie, 1984) distinguish *Kurthia* from members of the "coryneform group". Further, the results of rRNA oligonucleotide studies of one strain of *Kurthia* and various Gram-positive bacteria confirm that *Kurthia* is only very distantly related to the aerobic coryneform bacteria and is more closely related to members of the genera *Bacillus, Lactobacillus, Staphylococcus* and *Streptococcus* (Ludwig et al., 1981). This taxonomic placement is supported by the results of the more recent numerical taxonomic study of Shaw and Keddie (1983a), which showed a close phenotypic similarity between *Kurthia* and some aerobic *Bacillus* spp. but little similarity with various genera of aerobic coryneform bacteria.

Habitats

There have been few systematic studies of the occurrence of *Kurthia* spp. in natural materials. However, from the limited information available, it seems that the sources from which authentic *Kurthia* spp. are regularly isolated are meat and meat products and animal feces that have lain on soil, straw, etc., for a short time. There are also several reports of the isolation of *Kurthia* spp. from clinical materials and from other diverse sources, but in many of these cases the identification of these organisms as *Kurthia* spp. is open to considerable doubt. None belonged to any of the species now recognized.

Meat and Meat Products

Keddie (1949) isolated *K. zopfii* and the organism now called *K. gibsonii* from a variety of samples of fresh meat, fat, etc., and from meat that was allowed to putrefy at room temperature. These organisms were also readily isolated from meat, fat, waste water, etc. from an abattoir, thus confirming the early report of Jacobsen (1907). Such observations suggest that the immediate source of contamination of meat with *K. zopfii* and *K. gibsonii* is in the abattoir.

Ingram (1952) found that Gram-positive, nonsporeforming rods constituted about 10% of the aerobic flora of internally tainted, cured pork legs. Some of these rods were considered to be *Kurthia* spp. *K. zopfii* and *K. gibsonii* have been isolated from pork and pork products (Gardner et al., 1967; Shaw and Keddie, 1983a), various ground fresh meat products (Gardner, 1969; Shaw and Keddie, 1983a), from an eviscerated, polyethylene-wrapped chicken stored at 15°C until off-flavors developed, from an irradiated lamb carcass stored for about 7 weeks at 1°C (quoted by Gardner, 1969), and from spoiled British sausages (Dowdell and Board, 1971).

Information on the numbers of *Kurthia* in these various meat products is sparse. Gardner (1969) noted that although members of the genus were regularly isolated from fresh, comminuted meat products, they usually accounted for only up to 10% of the total aerobic count. On the other hand, in six samples of pork stored at 16°C for 5 days in gas-impermeable film, gas-permeable film, or with no film, *Kurthia* represented 12–44, 9–76, and 5–69%, respectively, of the total aerobic flora; but were not detected in similar samples stored at 2°C for 14 days (Gardner et al., 1967). From these and other observations (Gardner and Carson, 1967), Gardner (1969) concluded that *Kurthia* could compete favorably with other aerobic spoilage bacteria in meat stored at about 16°C, but not in meat stored at refrigeration temperatures. The two strains isolated by D'Aubert et al. (1975) from refrigerated, vacuum-packed meat and identified by them as *Kurthia* spp. did not have the characteristics of the genus.

What appeared to be the most unusual source was reported by Belikova et al. (1980), who isolated several strains of *Kurthia* from samples of the intestinal contents and stomach of the Magadan (Susuman) mammoth found preserved in the permafrost in East Siberia. Of 13 strains referred to by these authors as *K. zopfii*, four had the characters of that species, three appeared to be *K. gibsonii* and the remainder were psychrophilic strains that differed from both these species (see Keddie and Shaw, 1986). A new species, *K. sibirica,* was created for the psychrophilic strains (Belikova et al., 1986). While we obviously cannot comment on the ability of *Kurthia* to survive for several thousand years in permafrost, we are nevertheless certain that the highly aerobic *Kurthia* would be unable to multiply in the highly anaerobic environment of the animal gut. A solution to the problem became apparent when it was revealed that after the mammoth was removed from the permafrost it was kept in an unfrozen state for several days before examination (Belikova et al., 1986). Presumably the carcass became contaminated with *Kurthia* which then multiplied during storage. Indeed, more recent studies (see above) strongly suggest that Kurth's original isolates came from chicken intestines that had been stored at room temperature for a few days; like the mammoth, contamination with *Kurthia* (possibly from flies) would have been followed by multiplication during storage (see below).

Animal Feces in Contact with Soil, Straw, etc.

In our experience, *Kurthia* can regularly be isolated from the feces of certain domestic animals provided it has lain on soil, straw, etc., for a short time. Keddie (1949) was unable to isolate *Kurthia* from the cecal contents of chickens or from the freshly voided feces, but three of nine samples of chicken feces that had lain on the soil for some time yielded the organism. *K. zopfii* and *K. gibsonii* have been isolated from many different samples of soil-contaminated chicken feces and from various situations in poultry houses (Shaw and Keddie, 1983a). Strivastava et al. (1972) also reported the isolation of two presumptive *Kurthia* strains from a poultry house and a hatchery. Also, pig feces contaminated with straw or soil have yielded *K. zopfii* and *K. gibsonii,* as have similarly contaminated feces of horses and cows (Shaw and Keddie, 1983a).

Clinical Sources

Several strains of presumptive *Kurthia* spp. have been isolated from various clinical sources and most frequently from the feces of patients suffering from diarrhea. Some organisms were considered to be *"K. bessonii"* or *"K. variabilis"*, neither of which was included in The Approved Lists of Bacterial Names (Skerman et al., 1980) and are therefore not legitimate species.

Frequently, the descriptions of the isolates mentioned contain apparently contradictory statements (e.g., facultative anaerobes; do not ferment sugars) or are too limited to be able to judge whether or not the organisms were indeed legitimate kurthias.

Severi (1946) isolated what he considered to be a new species, *"K. variabilis"*, from feces in a case of mild food poisoning. Elston (1961) isolated three strains identified as *"K. bessonii"* from a pilonidal cyst, sputum, and a diarrheal stool, and Faoagali (1974) isolated a presumptive *Kurthia* sp. from a routine eye swab. Jarumilinta et al. (1976) isolated presumptive *Kurthia* spp. from the upper intestinal tract of six of 25 patients suffering from acute diarrhea, but only from one of 24 control patients. More recently Yang et al. (1985) isolated an organism they identified as *K. zopfii* from the blood of an infant with septicemia. However, in all of these cases, the connection between the occurrence of the presumptive *Kurthia* spp. and the clinical condition was, at most, tenuous.

Other Sources

The isolation of presumptive *Kurthia* spp. has been reported from sources as diverse as "sloughing spoilage" of ripe olives (*"K. bessonii"*; Patel and Vaughn, 1973), the gut of a crab (*"K. variabilis"*; Saha and Raychaudhuri, 1973), wet-stored wood (Berndt and Liese, 1973), and from dental plaque of beagle dogs (*Kurthia?;* Wunder et al. 1976); but like those from clinical sources, the accuracy of identification of many of these isolates is doubtful.

However, more recent evidence supports the early statement by Orla-Jensen (1931) that *K. zopfii* may occur in milk because this species was found at levels of about 10^4/ml in some samples of bulked, cold-stored (7°C) raw milks (C. M. Cousins, personal communication). *K. zopfii* has occasionally been isolated from soil and surface waters (Shaw and Keddie, 1983a) and a few observations indicate that organisms resembling *Kurthia* but distinct from the accepted species occur in certain peats; namely, low-moor peat (Janota-Bassalik, 1963) and Antarctic peat (Baker and Smith, 1972).

To our knowledge *Kurthia* spp. have not yet been exploited in any biotechnological process.

Isolation

K. zopfii and *K. gibsonii* may be isolated by using methods that exploit their unusual cultural properties. One of the most successful is a gelatin streak method similar to that used by Kurth and the early investigators (Keddie, 1949). However, a simple agar streak technique is a useful additional method (see Keddie, 1981).

These methods were not used for isolation of *K. sibirica* but the report that strains of this species produce a "bird's feather" growth on nutrient gelatin (Belikova et al., 1986) suggests that they might be successful if a suitably low incubation temperature is chosen.

Gelatin Streak Method for Isolating *Kurthia* (Keddie, 1949)

A nutrient gelatin medium (YNG) of the following composition is used. YNG medium contains per liter of distilled water:

Meat extract (Lab-Lemco powder, Oxoid)	4 g
Peptone (Difco)	5 g
Yeast extract (Difco)	2.5 g
NaCl	5 g
Gelatin (BDH)	60 g

The pH is adjusted to 7.0. For quantities up to 100 ml, sterilize at 115°C for 30 min. Medium YNB is medium YNG without gelatin.

Pour plates with about 20 ml of molten YNG medium and allow to solidify in the refrigerator. Inoculate plates heavily with a single central streak of the material to be examined (or with a suspension or macerate of solid material in a small amount of sterile water). Incubate plates at 20°C with the lids uppermost and examine daily. The gelatin is usually soon liquefied around the streak but in successful cultures, filamentous outgrowths appear beyond this zone in 2–3 days and after 3–4 days a tangled mass of filaments may completely permeate the solid part of the medium if it is not completely liquefied. To obtain a pure culture, a small piece of gelatin containing outgrowths is streaked on yeast extract-nutrient agar medium (YNA). YNA is similar to YNG but is solidified with agar instead of gelatin.

The composition of the nutrient gelatin medium used is important and the concentration and brand of gelatin used is particularly important; not all brands allow the typical outgrowths. The medium should be inoculated with a reference strain of *K. zopfii* (NCIB 9878) to test its ability to allow good outgrowth production. The gelatin manufactured by the BDH Chemical Co., Poole, England, is satisfactory but with some batches a higher concentration than that stated (up to 100 g/1) may be required. Although all *Kurthia* strains tested grow well in YNB, medium YNG prepared with some batches of gelatin has given poor growth. Dissolving the constituents of YNG in mineral base E (Owens and Keddie, 1969) to give MYNG (Shaw and Keddie, 1983a) overcomes this problem. When MYNG is autoclaved, a precipitate (which should be dispersed before pouring plates) is produced; the precipitate does not interfere with isolation. Overgrowth with fungi may be a problem with some materials. This can be prevented by adding nystatin (Squibb) to the molten YNG to a concentration of 10 units/ml before pouring plates.

Other bacteria that may produce outgrowths in gelatin similar to those of *Kurthia* are certain chain-forming *Bacillus* species, particularly *B. cereus* subsp. *mycoides,* although they may liquefy gelatin.

The agar streak method may be used in addition to that described above and can give successful results when rapid liquefaction of gelatin has prevented isolation of *K. zopfii* by the gelatin streak method.

Agar Streak Method for Isolating *Kurthia* (see Keddie, 1981)

The YNA medium consists of

Meat extract (Lab-Lemco powder, Oxoid)	4 g
Peptone (Difco)	5 g
Yeast extract (Difco)	2.5 g
NaCl	5 g
Agar	15 g

The pH is adjusted to 7.0. The YNA medium is inoculated with a single, central streak as described above.

At daily intervals, the edge of the streak is examined under low power (100×) of the microscope for the characteristic skeinlike outgrowths of *K. zopfii* grown on agar. Pure cultures are obtained by picking carefully from the edge of the outgrowths and plating on YNA as before.

A partial enrichment of *K. gibsonii* may be achieved by preparing a preliminary enrichment in YNB (YNG but without gelatin) incubated for 24 h at 45°C; a second subculture at 45°C is then made before inoculating and incubating YNG and YNA plates as described above. *Bacillus* spp. do not usually interfere in this modification of the method.

Isolation by Direct Plating

Kurthia spp. may be isolated by direct plating on YNA (or similar media) of materials such as meat and meat products if it forms a sufficiently high proportion of the population (Gardner, 1969).

Surface colonies of *Kurthia* spp. are recognized by their rhizoid form (Fig. 1a) and by the typical "medusa-head" appearance of young colonies when examined at low magnification. Such colonies have a skein-like structure which is resolved into whorls with whiplike outgrowths at the edge.

It should be noted that when pure cultures are streaked on YNA, a proportion of the colonies that develop may have a granular appearance instead of the typical rhizoid form (Keddie, 1949). Therefore, it is possible that when isolating *Kurthia* from natural materials, some may be missed because they produce nonrhizoid colonies. Isolates are then examined for the characteristic properties of *Kurthia;* useful screening tests are colony form, morphology and Gram reaction, production of "bird's feather" growth on nutrient gelatin slants (Fig. 1b), and aerobic growth in glucose nutrient agar shake cultures.

Preservation of Cultures

Cultures on YNA or nutrient agar slants should remain viable for at least 6 months when stored at room temperature (about 20°C), provided that they are not allowed to dry out. Longer-term preservation (over 10 years) may be achieved by freezing on glass beads at −60 to −70°C (Jones et al., 1984). Cultures can also be preserved by freeze drying (lyophilization).

Identification

Identification as a *Kurthia* sp. may be made on the basis of the following features. Gram-positive, regular, unbranched rods with rounded

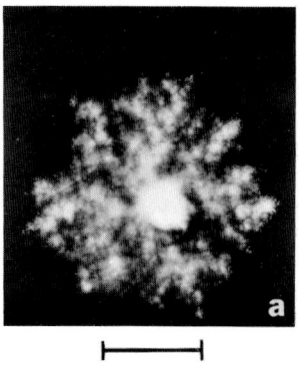

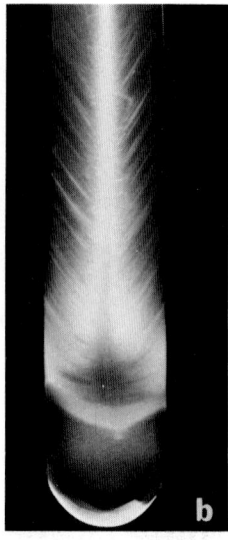

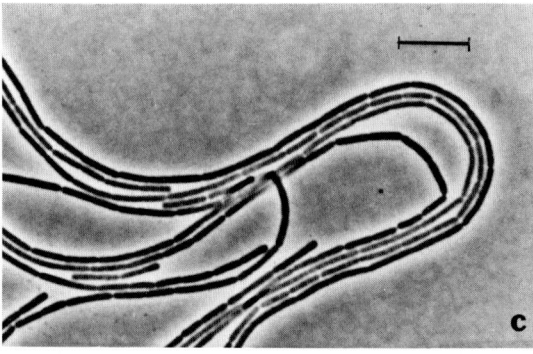

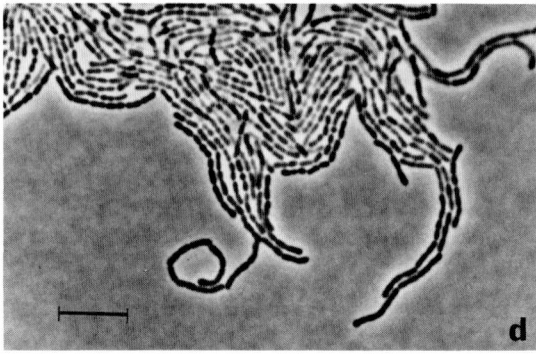

Fig. 1 (a–b) *Kurthia zopfii* (NCIB 9878). (a) Rhizoid colony on yeast nutrient agar after 4 days incubation at 25°C; bar = 10 mm. (b) Yeast nutrient gelatin slant showing "bird's feather" type of growth; incubated 5 days at 20°C. (c-d) *Kurthia zopfii* (isolate): edge of colony on yeast nutrient agar incubated at 25°C. (c) After 24 h, showing long filaments composed of rods. (d) After 3 days, showing development of coccoid forms. Bar = 10 μm.

ends, about 0.8×3–$8\mu m$ or longer, in long chains in exponential phase cultures (Fig. 1c) giving rise to cocoids, formed by fragmentation of the rods (Fig. 1d) or in some strains giving rise to short rods in stationary-phase cultures.

Strains which have been maintained in artificial culture for some time often give rise to short rods in stationary phase cultures (Shaw and Keddie, 1983a). The rods are usually motile by numerous peritrichous flagella, but nonmotile strains are known; they do not form endospores. Surface colonies on yeast nutrient agar are usually rhizoid (Fig. 1a) (but granular colonial variants occur) and have a "medusa-head" appearance under low magnification (100×). They are obligate aerobes and do not form acid from glucose or other carbohydrates in peptone media. In gelatin slant cultures (see medium YNG above) inoculated with a single central streak the growth resembles a bird's feather (Fig. 1b); the gelatin is not liquefied. They are catalase positive and grow best in the range of 25–30°C (*K. zopfii* and *K. gibsonii*) or 20–25°C (*K. sibirica*) at neutral pH.

Kurthia spp. give negative responses in most of the usual biochemical tests, e.g., indole, nitrate reduction; urease, lecithinase production; and hydrolysis of starch and esculin; however, some strains produce H_2S weakly (Jones, 1975; Shaw and Keddie, 1983a). The distinctive cell wall amino acids are lysine and aspartic acid (Belikova et al., 1980; Shaw and Keddie, 1984). In the two *K. zopfii* strains studied, the walls contain a group A peptidoglycan, type L-lysine-D-aspartic acid (Schleifer and Kandler, 1972). The major isoprenoid quinones in *K. zopfii* and *K. gibsonii* are unsaturated menaquinones with seven isoprene units (MK-7) (Collins et al., 1979). The major fatty acids in all three species are straight-chain saturated, anteiso- and isomethyl-branched chain acids: the major fatty acid is 12-methyltetradecanoic (anteiso-C15) acid in *K. zopfii* and *K. gibsonii* (Goodfellow et al., 1980), but with 13-methyltetradecanoic (iso-C15) acid in addition in *K. sibirica* (Belikova et al., 1986). The polar lipids in *K. zopfii* and *K. gibsonii* are diphosphatidylglycerol, phosphatidylglycerol, and phosphatidylethanolamine (Goodfellow et al., 1980). The GC content is remarkably constant among isolates of all three species and lies in the range of about 36–38 mol% (Belikova et al. 1980; Shaw and Keddie, 1984; Belikova et al., 1986). Glucose is not used as a carbon + energy source but acetate appears to be utilized by most strains of all three species (Shaw and Keddie, 1983a; Belikova et al., 1986). When supplied with a suitable source of amino acids, B vitamins are required (Shaw and Keddie, 1983b; Belikova et al., 1986). For other characters of *K. zopfii* and *K. gibsonii* see Table 1.

Bacteriophages isolated from *K. zopfii* are morphologically identical with the short-tailed, elongated head phages of the GA-1 species of *Bacillus* phages (Rocourt et al., 1983). The ac-

Table 1. Other features of *K. zopfii* and *K. gibsonii*.

Characteristics	*K. zopfii*	*K. gibsonii*
Acid from:		
Butan-1-ol	d	+
Propan-1-ol	+	+
Pentan-1-ol	d	d
Methanol	−	−
D-Mannitol	−	−
D-Sorbitol	−	−
m-Inositol	−	−
Ethanediol	+	+
Propanediol	−	−
2,3-Butanediol	−	−
m-Erythritol	−	−
D-Glucose	−	−
Hydrolysis of:		
Aesculin	−	−
Arginine	−	−
Tween 20	+	+
Tween 40	d	d
Tween 60	+	+
Tween 80	d	−
Tributyrin	d	d
Starch	−	−
Tyrosine	−	−
Xanthine	−	−
Hippurate	+	+
Cellulose	−	−
Uric acid	+	+
Casein	−	−
Chitin	−	−
Production of:		
H₂S	d	−
Acetoin	−	−
Dihydroxyacetone	−	−
Sulfatase	−	−
Urease	−	−
Acetamidase	−	−
Lecithinase	−	−
Gluconate oxidation	−	−
Indole	−	−
Growth at:		
40°C	−	+
50°C	−	−
Carbon + energy source:		
D-Ribose	d	d
D-Xylose	−	−
D-Glucose	−	−
D-Mannose	−	−
D-Fructose	−	d
Lactose	−	−
D-Raffinose	−	−
Glycogen	−	−
D-Glucuronate	−	−
N-Acetylglucosamine	+	d
Uridine	+ᵃ	+
Formate	−	d
Acetate	+ᵃ	+
Propionate	−	d
n-Butyrate	+	+
n-Pentanoate	−	−

continued

Table 1. *Continued*

Characteristics	*K. zopfii*	*K. gibsonii*
n-Hexanoate	d	d
n-Heptanoate	d	d
n-Octanoate	d	+
Isobutyrate	−	−
Succinate	+	+
Fumarate	+	+
Crotonate	d	+ᵃ
Tricarballylate	−	−
Aconitate	−	−
DL-Lactate	d	+ᵃ
L-Malate	+	+
DL-2-Hydroxybutyrate	−	−
DL-Glycerate	d	d
Citrate	−	d
Glyoxylate	−	−
Pyruvate	+	d
Oxalacetate	d	+
2-Oxoglutarate	d	−
Ethanediol	d	d
Propanediol	−	d
Glycerol	+	+
Adonitol	d	d
Ethanol	+	+
Propan-1-ol	d	d
Butan-1-ol	+	d
Acetaldehyde	d	d
Benzaldehyde	d	d
m-Hydroxybenzoate	−	−
Cinnamate	d	d
Phenylacetate	d	d
Uric acid	−	−
Glycine	+	d
L-Alanine	+	+ᵃ
L-Leucine	−	−
L-Isoleucine	−	d
L-Serine	+	+ᵃ
L-Threonine	+	d
L-Lysine	−	−
L-Citrulline	d	d
L-Arginine	+	d
L-Aspartate	+	+
L-Glutamate	+	+
L-Asparagine	+	+
L-Phenylalanine	−	−
L-Tyrosine	−	−
L-Histidine	d	d
L-Tryptophan	−	d
L-Proline	+	+
L-Hydroxyproline	d	d
DL-Homoserine	−	−
L-Cysteine	−	−
L-Methionine	−	−
Ethanolamine	−	−
Acetamide	−	−
Betaine	−	−
Creatinine	−	−
Hippurate	d	−
Allantoin	d	−
L-Ascorbate	−	−

continued

Table 1. *Continued*

Characteristics	K. zopfii	K. gibsonii
m-Erythrytol	−	−
Methanol	−	−
Crotonol	−	−
Cytosine	−	−
Xanthine	−	−

Symbols: −, 90% or more of strains are negative; +, 90% or more of strains are positive; and d, 11–89% of strains are positive.

[a]Eighty-eight percent of strains are positive.

Adapted from Shaw and Keddie (1983a). Reproduced with permission from *Bergey's Manual of Systematic Bacteriology*, vol. 2, Williams & Wilkins, Baltimore, 1986.

tivity of these phages on *K. gibsonii* and *K. sibirica* is not known.

There is, to our knowledge, no published information on the presence of plasmid DNA in *Kurthia* spp., but a recent study demonstrated plasmid DNA in the type strain *K. zopfii* (NCTC 10597). Further, the range of plasmids of this strain was identical to that of two isolates from sausages identified as *Brochothrix thermosphacta* (C. E. R. Dodd, personal communication). Although this observation suggests plasmid exchange between the genera, confirmation of such a phenomenon would require the demonstration of molecular identity between the plasmid complements.

Differentiation of *Kurthia* spp.

DNA-DNA hybridization studies of a few strains (some atypical) support the conclusion reached in the numerical taxonomic study of Shaw and Keddie (1983a) that *K. zopfii* and *K. gibsonii* are distinct species. They also show that the strains now included in the species *K. sibirica* are distinct from both species (Cherevach et al., 1983). The three species may be distinguished by using the characters listed in Table 2. It should be noted, however, that Shaw and Keddie (1983a, b, 1984) described a number of strains that had most or all of the characters of the genus *Kurthia* as defined by Keddie and Rogosa (1974), but could not be assigned to a species.

Those taxa most likely to be confused with *Kurthia* include *Brochothrix thermosphacta* (*Microbacterium thermosphactum,*) and certain *Bacillus* spp. Confusion with some aerobic, saprophytic coryneform bacteria is possible but is much less likely. *B. thermosphacta* (Sneath and Jones, 1976) is very similar to *Kurthia* in morphological features (see Davidson et al., 1968) but is nonmotile, facultatively anaerobic, and produces acid by fermentation from glucose and various other carbohydrates. On nutrient agar certain chain-forming *Bacillus* spp. may give colonies somewhat similar to *K. zopfii* and have a similar morphology in exponential-phase cultures. They may also give outgrowths in nutrient gelatin, although such growth is usually, if not always, followed by liquefaction. They may be distinguished by endospore formation (detected by a test of heat resistance following growth on nutrient agar supplemented with Mn^{2+}, $2\mu g/ml$, for at least 7 days). Those species most readily confused with *Kurthia,* e.g. *Bacillus cereus* subsp. *mycoides,* are facultatively anaerobic, produce acid from glucose, and liquefy gelatin.

Table 2. Differential characteristics of species of the genus *Kurthia*.

Characteristics	K. zopfii	K gibsonii	K. sibirica
Growth at 45°C	−	+	−
Survival at 55°C for 20 min[a]	−	+	ND
Acid from:			
Ethanol	+	−	−
Glycerol	−	+	+[b]
Colonies, yellow or cream	−	+	+
Deoxyribonuclease	−	+	ND
Ribonuclease	+	−	ND
Phosphatase	−	+	+
4-Amino-*n*-butyrate used as carbon and energy source	−	+	ND
Pantothenic acid required	+[c]	−[c]	+[d]
Nicotinic acid and pyridoxal-5-phosphate required	−[c]	−[c]	+[d]

Symbols: −, 12% or less of strains negative; +, 88% or more of strains positive; ND, no data.

[a]In skim milk.

[b]Slight acidity reported.

[c]When supplied with vitamin-free Casamino acids (Difco) and/or Casitone (Difco), biotin, and thiamine (Shaw and Keddie, 1983b).

[d]When supplied with casein hydrolysate (Difco), tryptophan, guanine, biotin, and thiamin. (Belikova et al., 1986).

Adapted from Keddie and Shaw (1986). Data on *K. sibirica* from Belikova et al. (1986).

Many aerobic coryneform bacteria (e.g., *Arthrobacter* spp.), like fresh *Kurthia* isolates, give coccoid cells in stationary-phase cultures, but the appearance in exponential phase cultures is quite distinct. In coryneform bacteria, the rods are irregular in form, may show rudimentary branching, and commonly occur in V formations but never in chains.

Reference strains: *K. zopfii*, NCIB 9878; *K. gibsonii*, NCIB 9758; *K. sibirica;* VKB B-1549.

Literature Cited

Baker, J. H., Smith, D. G. 1972. The bacteria in an Antarctic peat. Journal of Applied Bacteriology 35:589–596.

Belikova, V. L., Cherevach, N. V., Baryshnikova, L. M., Kalakutskii, L. V. 1980. Morphologic, physiologic and biochemical characteristics of *Kurthia zopfii*. Microbiology 49:51–55.

Belikova, V. L., Cherevach, N. V., Kalakutskii, L. V. 1986. A new species of bacteria of the genus *Kurthia*, *Kurthia sibirica* sp.nov. Mikrobiologiya 55:831–835.

Berndt, H., and Liese, W. 1973. Untersuchungen über das Vorkommen von Bakterien in wasserberieselten Buchenholzstammen. Zentralblatt für Bakteriologie, Parasitenkunde, Infektionskrankheiten und Hygiene, Abt. 2 128:578–594.

Boyce, R., and Evans, A. E. 1893. The action of gravity upon *Bacterium zopfii*. Proceedings of the Royal Society of London 54:300–312.

Breed, R. S., Murray, E. G. D., and Smith, N. R. (ed.). 1957. Bergey's manual of determinative bacteriology. 7th ed. Williams & Wilkins, Baltimore.

Cherevach, L. M., Tourova, T. P., and Belikova, V. L. 1983. DNA-DNA homology studies among strains of *Kurthia zopfii*. FEMS Microbiology Letters 19:243–245.

Collins, M. D., Goodfellow, M., Minnikin, D. E. 1979. Isoprenoid quinones in the classification of coryneform and related bacteria. Journal of General Microbiology 110:127–136.

D'Aubert, S., Cantoni, C., and Calcinardi, C. 1975. I corinebatteri nelle carni confezionate sottovuoto. Archivo Veterinario Italiano 26:65–70.

Davidson, C. M., Mobbs, P., and Stubbs, J. M. 1968. Some morphological and physiological properties of *Microbacterium thermosphactum*. Journal of Applied Bacteriology 31:551–559.

Dowdell, M. J., and Board, R. G. 1971. The microbial associations in British fresh sausages. Journal of Applied Bacteriology 34:317–337.

Elston, H. R. 1961. *Kurthia bessonii* isolated from clinical material. Journal of Pathology and Bacteriology 81:245–247.

Faoagali, J. L. 1974. *Kurthia*, an unusual isolate. American Journal of Clinical Pathology 62:604–606.

Flügge, C. 1896. Die Microorganismen, p. 277–278. vol.2. F. C. W. Vogel, Leipzig, Germany.

Gardner, G. A., 1969. Physiological and morphological characteristics of *Kurthia zopfii* isolated from meat products. Journal of Applied Bacteriology 32:371–380.

Gardner, G. A., and Carson, A. W. 1967. Relationship between carbon dioxide production and growth of pure strains of bacteria on porcine muscle. Journal of Applied Bacteriology 30:500–510.

Gardner, G. A., Carson, A. W., and Patton, J. 1967. Bacteriology of prepacked pork with reference to the gas composition within the pack. Journal of Applied Bacteriology 30:321–333.

Goodfellow, M., Collins, M. D., and Minnikin, D. E. 1980. Fatty acid and polar lipid composition in the classification of *Kurthia*. Journal of Applied Bacteriology 48:269–276.

Günther, C. 1896. Bakteriologische Untersuchungen in einem Falle von Fleischvergiftung. Archiv für Hygiene 28:153–158.

Ingram, M. 1952. Internal bacterial taints ('bone taint' or 'souring') of cured pork legs. Journal of Hygiene 50:165–181.

Jacobsen, H. C. 1907. Ueber einen richtenden Einfluss beim Wachstum gewisser Bakterien in Gelatine. Zentralblatt für Bakteriologie, Parasitenkunde, Infektionskrankheiten und Hygiene, Abt. 2 17:53–64.

Janota-Bassalik, L. 1963. Psychrophiles in low-moor peat. Acta Microbiologica Polonica 12:25–40.

Jarumilinta, R., Miranda, M., and Villarejos, V. M. 1976. A bacteriological study of the intestinal mucosa and luminal fluid of adults with acute diarrhoea. Annals of Tropical Medicine and Parasitology 70:165–179.

Jones, D. 1975. A numerical taxonomic study of coryneform and related bacteria. Journal of General Microbiology 87:52–96.

Jones, D., Pell, P. A., and Sneath, P. H. A. 1984. Maintenance of bacteria on glass beads at −60°C to −76°C. p.35–40. In: Kirsop, B. E., and Snell, J. J. S. (ed.), Maintenance of microorganisms. A manual of laboratory methods. Academic Press, London.

Keddie, R. M. 1949. A study of *Bacterium zopfii* Kurth. Dissertation. Edinburgh School of Agriculture, Edinburgh.

Keddie, R. M. 1981. The Genus *Kurthia*. p. 1888–1893. In: Starr, M. P., Stolp, H., Trüper, H. G., Balows, A., and Schlegel, H. G. (ed.), The prokaryotes: a handbook on habitats, isolation and identification of bacteria. Springer-Verlag, Berlin.

Keddie, R. M., and Rogosa, M. 1974. *Kurthia* p. 631–632. In: Buchanan, R. E., and Gibbons, N. E. (ed.), Bergey's manual of determinative bacteriology, 8th ed. Williams & Wilkins, Baltimore.

Keddie, R. M., and Shaw, S. 1986. Genus *Kurthia*. p.1255–1258. In: Sneath, P. H. A., Mair, N. S., Sharpe, M. E., and Holt, J. G. (ed.), Bergey's manual of systematic bacteriology, vol.2. Williams & Wilkins, Baltimore.

Kufferath, H. 1911. Note sur les tropismes du *Bacterium zopfii* "Kurth" Annales de l'Institut Pasteur 25:601–617.

Kurth, H. 1883. *Bacterium zopfii*. Ein Beitrag zur Kenntniss der Morphologie und Physiologie der Spaltpilze. Botanische Zeitung 41:369–386, 393–405, 409–420, 425–435.

Ludwig, W., Seewaldt, E., Schleifer, K. H., and Stackebrandt, E. 1981. The phylogenetic status of *Kurthia zopfii*. FEMS Microbiology Letters 10:193–197.

Migula, W. 1900. System der Bacterien, p. 815–816, vol.2. G. Fischer, Jena, Germany.

Orla-Jensen, S. 1931. Dairy bacteriology, p.51. 2nd English ed. J. & A. Churchill, London.

Owens, J. D., and Keddie, R. M. (1969). The nitrogen nutrition of soil and herbage coryneform bacteria. Journal of Applied Bacteriology 32:338–347.

Patel, I. B., and Vaughn, R. H. 1973. Cellulolytic bacteria associated with sloughing spoilage of California ripe olives. Applied Microbiology 25:62–69.

Proctor, B. E. 1935. The microbiology of the upper air II. Journal of Bacteriology 31:363–375.

Rocourt, J., Ackermann, H. W., and Brault, J. 1983. Isolement et morphologie de bactériophages de *Kurthia zopfii*. Annales Virologie (Institut Pasteur) 134E:557–567.

Rogosa, M., Cummins, C. S. Lelliott, R. A., and Keddie, R. M. 1974. Coryneform group of bacteria, p. 599–632. In: Buchanan, R. E.,. Gibbons, N. E. (ed.), Bergey's manual of determinative bacteriology. 8th ed. Williams & Wilkins, Baltimore.

Saha, N., and Raychaudhuri, D. N. 1973. A note of the bacterial flora in the gut of the crab. *Seylla serrata* (Forskal)(Crustacea: Decapoda), Science and Culture 39:361–363.

Schleifer, K. H., and Kandler, O. 1972. Peptidoglycan types of bacterial cell walls and their taxonomic implications. Bacteriological Reviews 36:407–477.

Sergent, E. 1906. Des tropismes du *Bacterium zopfii* Kurth, Première note. Annales de l'Institut Pasteur 20:1005–1017.

Sergent, E. 1907. Des tropismes du *Bacterium zopfii* Kurth, Deuxième note. Annales de l'Institut Pasteur 21:842–856.

Severi, R. 1946. L'azione patogena delle Kurthie e la loro sistematica. Una nuova specie: *"Kurthia variabilis"*. Giornale de Batteriologia e Immunologia 24:107–114.

Shaw, S., and Keddie, R. M. 1983a. A numerical taxonomic study of the genus *Kurthia* with a revised description of *Kurthia zopfii* and a description of *Kurthia gibsonii* sp.nov. Systematic and Applied Microbiology 4:253–276.

Shaw, S., and Keddie, R. M. 1983b. The vitamin requirements of *Kurthia zopfii* and *Kurthia gibsonii*. Systematic and Applied Microbiology 4:439–443.

Shaw, S., and Keddie R. M. 1984. The genus *Kurthia*: cell wall composition and DNA base content. Systematic and Applied Microbiology 5:220–224.

Skerman, V. B. D., McGowan, V., and Sneath, P. H. A. 1980. Approved Lists of Bacterial Names. International Journal of Systematic Bacteriology 30:225–420.

Sneath, P. H. A., and Jones, D. 1976. *Brochothrix*, a new genus tentatively placed in the family *Lactobacillaceae*. International Journal of Systematic Bacteriology 26:102–104.

Starr, M. P., Stolp, H., Trüper, H. G., Balows, A., and Schlegel, H. G. (ed.). 1981. The prokaryotes: a handbook on the habitats, isolation and identification of bacteria. Springer-Verlag, Berlin.

Strivastava, S. K., Singh, V. B., and Singh, N. P. 1972. Bacterial flora of poultry environment. Indian Journal of Microbiology 12:7–9.

Trevisan, V. 1885. Caretteri di alcuni nuovi generi di Batteriacee. Atti della Accademia Fisio-Medico-Statistica in Milano. Series 4 3:92–107.

Wenner, J. J., and Rettger, L. F. 1919. A systematic study of the Proteus group of bacteria. Journal of Bacteriology 4:331–353.

Wunder, J. A., Briner, W. W., and Calkins, G. P. 1976. Identification of the cultivable bacteria in dental plaque from the Beagle dog. Journal of Dental Research 55:1097–1102.

Yang, M., Sun, Y., Ge, P., Dong, Q., and Ma, Z. 1985. A case of infant septicemia caused by *Kurthia zopfii*. Chinese Journal of Microbiology and Immunology. 5:485.

Zikes, H. 1903. Die Wachstumserscheinungen von *Bacterium zopfii* auf Peptongelatine. Zentralblatt für Bakteriologie, Parasitenkunde, Infektionskrankheiten und Hygiene Abt. 2 11:59–61.

The Genus *Bacillus*—Nonmedical

RALPH A. SLEPECKY and H. ERNEST HEMPHILL

History

One of the earliest bacteria to be described was "Vibrio subtilis" by Ehrenberg in 1835. In 1872, Cohn renamed the organism *Bacillus subtilis* (Gordon, 1981). That organism was a charter member of a large and diverse genus, initiated by Cohn, that is part of the family Bacillaceae. This family's distinguishing feature is production of endospores, which are round, oval, or cylindrical highly refractile structures formed within bacterial cells. Spores were first described by Cohn in *subtilis* and later by Koch in the pathogen, *B. anthracis* (the only major pathogen of vertebrates in the genus). Cohn demonstrated the heat resistance of spores of *B. subtilis* and Koch first described in *B. anthracis* the developmental cycle of sporeformers, vegetative cell to spore and spore to vegetative cell (Keynan and Sandler, 1983). For the reasons of unusual spore resistance to chemical and physical agents; the developmental cycle; ubiquity of its members; and *B. anthracis* pathogenicity, the genus *Bacillus* attracted early interest which has continued since.

The endospore, either as the free spore or as the structure within the vegetative cell, in which case the whole entity is referred to as a sporangium, is readily detected using the phase contrast microscope (see Fig. 1.). This is because the spore at a point in the life cycle (to be detailed later) becomes highly refractile. Early workers used stains and special conditions (such as prolonged heating) to colorize the chemically impermeable spore (Doetsch, 1981). However, a Gram-stain is sufficient to determine the presence of spores because the spore remains unstainable while the vegetative cells or the vegetative part of the sporangia will stain. Because of this ease of microscopic detection of the spore and its heat resistance, many different endosporeformers can be easily found. Using any habitat—soil, water, food, etc—as the source, sporeformers can be readily isolated by suspending a sample in water and heating at 80°C for 10 to 30 min. Vegetative cells and other resting forms such as cysts and exospores are usually killed at that temperature. The heat-resistant endospore can then be plated on appropiate media and isolates recovered in 24 to 48 h. An idea of the kinds of habitats from which *Bacillus* species have been isolated can be obtained from Table 1. Heating the inoculum, when used in conjunction with cultivation at different temperatures, hydrogen ion concentrations, degrees of aeration, and substrates, has resulted in isolating many different species of endosporeformers. The media used for the isolation and cultivation of *Bacillus* species are listed in Table 2.

More often than not since the discovery of bacteria (and in every case since 1913), the possession of an endospore has been used as a premier characteristic in keys for the classification

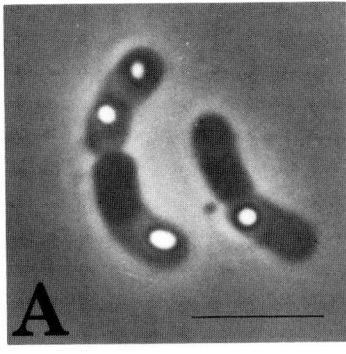 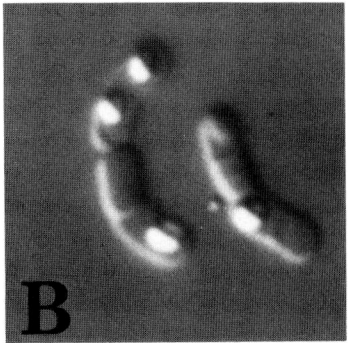

Fig. 1. *B. megaterium* sporeforming cells as seen in the phase contrast (A) or in the interference contrast (B) microscope. The refractile bodies in the center are the spores. Bar = 5 µm.

of bacteria. The family Bacillaceae was first formulated by Fisher in 1895 (Gordon, 1981). The features of the members of the genus *Bacillus* that distinguish it from other Bacillaceae (all endosporeformers) are their aerobic nature, which may be strict or facultative, rod shape, and catalase production. The other genera of sporeformers include *Sporolactobacillus,* which is microaerophilic and catalase-negative; *Clostridium,* anaerobic but does not reduce sulfate; *Desulfotomaculum,* anaerobic but does reduce sulfate; *Sporosarcina,* a coccus; and *Thermoactinomycetes,* which while forming endospores displays typical actinomycete characteristics.

General Taxonomic Considerations

Like the sirens of Greek mythology enticing the unsuspecting sailors, *Bacillus* species have captured the curiosity of many microbiologists. The first 107 years of the efforts to classify and identify members of the genus *Bacillus* is chronicled by R. E. Gordon (1981) who with her colleagues (Gordon et al., 1973; Smith et al., 1946, 1952) made many significant contributions on which the current classification (Claus and Berkeley, 1986) was built. The early attempts were "on rocky shoals" because a classification based on only the two characteristics of aerobic growth and endspore formation resulted in grouping together many bacteria possessing different kinds of physiology and occupying a variety of habitats. This heterogeneity in physiology, ecology, and genetics makes it difficult to categorize the genus or to make generalizations about it. The range of physiological life styles is impressive: degraders of most all substrates derived from plant and animal sources including cellulose, starch, proteins, agar, hydrocarbons, and others; antibiotic producers; heterotrophic nitrifiers; denitrifiers; nitrogen fixers; iron precipitators; selenium oxidizers; oxidizers and reducers of manganese; facultative chemolithotrophs; acidophiles; alkalophiles; psychrophiles, thermophiles and others (Slepecky, 1972; Norris et al., 1981; Claus and Berkeley, 1986) (see Table 2). Because of this vast diversity of physiological types, our knowledge of sporeformer ecology is slight (Slepecky, 1972; Norris et al., 1981; Slepecky and Leadbetter, 1977, 1984). In the main, sporeformers as part of the zymogenous flora of the soil are viewed as opportunists. Upon access to the proper germinants and substrates for subsequent outgrowth, they will actively contribute to and participate in the various microhabitats

Table 1. Origins of isolates of *Bacillus* species.

Name of *Bacillus* species	Habitats from which isolated
B. subtilis	Soil, water
B. acidocaldarius	Thermal acid water and soil
B. alcalophilus	pH 10 enrichment from soil
B. alvei	Soil, diseased bee larvae
B. amylolyticus	Soil
B. anthracis	Anthrax-diseased animals
B. azotoformans	Soil
B. badius	Feces, foods, marine sources
B. brevis	Soil, foods
B. cereus	Soil, foods
B. circulans	Soil
B. coagulans	Acid foods
B. fastidiosus	Soil, poultry litter
B. firmus	Soil, salt marshes
B. globisporus	Soil, water
B. insolitus	Soil
B. larvae	Diseased bee larvae
B. laterosporus	Soil, water
B. lautus	Soil, feces
B. lentimorbus	Diseased honeybee larvae
B. lentus	Soil, foods
B. licheniformis	Soil
B. macerans	Plant materials, food
B. macquariensis	Subantarctic soil
B. marinus	Marine sediment
B. megaterium	Soil
B. mycoides	Soil
B. pabuli	Soil, fodder
B. pantothenicus	Soil
B. pasteurii	Soil, water, sewage
B. popilliae	Diseased scarabid beetles
B. psychrophilus	Soil, water
B. pumilus	Soil
B. schlegelii	Lake sediment
B. sphaericus	Soil, water sediments, foods
B. stearothermophilus	Soil, hot spring, foods
B. thermoglucosidasius	Soil
B. thuringiensis	Soil, foods
B. validus	Soil

Based on Claus and Berkeley (1986).

which make up the soil's heterogeneous environment. Aerial distribution of the dormant spores may explain the occurrence of *Bacillus* species in most habitats examined. This diversity was apparent even with classical phenotypic characterizations based primarily on morphology (particularly size and position of the endospore within the vegetative cell), nutrition; growth characteristics; and various substrate utilization and physiological assessments.

At one time, 145 species made up the genus (Gordon, 1981). The understanding of the genus has been improved by augmenting the phenotypic characterizations with measurements of the DNA base composition and DNA-DNA hybridation Currently, there are listed in *Bergey's Manual of Systematic Bacteriology*

Table 2. Media used for the isolation and cultivation of *Bacillus* species.[a,b]

B. acidocaldarius	Part A: $(NH_4)_2SO_4$, 0.4; $MgSO_4$, 1.0; $CaCl_2\cdot2H_2O$, 0.5; KH_2PO_4, 6.0; distilled H_2O, 1 liter; pH adj. to 4.0
	Part B: glucose, 2.0; yeast ext., 2.0 distilled H_2O, 1 liter
	Combine A and B after sterilization
B. alcalophilus	Part A: glucose, 1.0; peptone, 5.0; yeast ext., 5.0; KH_2PO_4, 10.0 $MgSO_4\cdot7H_2O$, 0.2, distilled H_2O, 900 ml.
	Part B: $Na_2CO_3\cdot10H_2O$, 20; distilled H_2O, 100 ml.
	Combine A and B after sterilization (final pH = 10.5)
B. azotoformans	Peptone, 10.0; $Na_2HPO_4\cdot12H_2O$, 3.6; $MgSO_4\cdot7H_2O$, 0.03; $MnSO_4\cdot H_2O$, 0.05; KH_2PO_4, 1.0; NH_4Cl, 0.5; $CaCl_2\cdot2H_2O$, 0.1; distilled H_2O, 1 liter.
B. brevis	K_2HPO_4, 0.2; $MgSO_4\cdot7H_2O$, 0.02; NaCl, 0.02; $FeSO_4\cdot7H_2O$, 0.01; $MnSO_4\cdot H_2O$, 0.01; betaine, betaine·HCl or valine, 0.05M; agar, 16.0; distilled H_2O, 1 liter.
B. fastidiosus	K_2HPO_4, 0.8; KH_2PO_4, 0.2; $MgSO_4\cdot7H_2O$, 0.05; $CaCl_2\cdot2H_2O$, 0.05; $FeSO_4\cdot7H_2O$, 0.015; $MnSO_4\cdot H_2O$, 0.01; uric acid, 10.0; distilled H_2O, 1 liter.
B. lentus	Peptone, 10.0; meat ext., 10.0; agar, 15.0; distilled H_2O, 1 liter. adj. pH to 7.0–7.5; after sterilization, add 100 g urea, steam for 10 min.
B. licheniformis	Peptone, 5.0; meat ext., 3.0; KNO_3, 80.0; distilled H_2O, 1 liter; adj. pH to 7.0; fill glass-stoppered bottle to top for anaerobic conditions.
B. marinus	Peptone, 5.0; yeast ext., 1.0; $FePO_4\cdot4H_2O$, 0.01; agar, 15.0; aged sea water, 750 ml; distilled H_2O, 250 ml; adj. pH to 7.6.
B. pantothenicus	Nutrient broth + 4% (w/v) NaCl
B. pasteurii	Nutrient broth + 2% (w/v) urea
B. schlegelii	$Na_2HPO_4\cdot2H_2O$, 4.5; KH_2PO_4, 1.5; NH_4Cl, 1.0; $MgSO_4\cdot7H_2O$, 0.2; $CaCl_2\cdot2H_2O$, 0.01; ferric ammonium citrate, 0.005; $NaHCO_3$, 0.5; trace element soln, 5 ml ($ZnSO_4\cdot7H_2O$, 0.1; $MnCl_2\cdot4H_2O$, 0.03; H_3BO_3, 0.3; $CoCl_2\cdot6H_2O$, 0.02; $CuCl_2\cdot2H_2O$, 0.001; $NiCl_2\cdot6H_2O$, 0.02; $Na_2MoO_4\cdot2H_2O$, 0.03; distilled H_2O, 1 liter) Other: 65C, atmosphere of 0.05 atm. O_2 + 0.01 atm. CO_2 + 0.45 atm. H_2
B. stearothermophilus	Nutrient agar; incubate cultures at 55°C

[a]Numerical amounts are grams unless specified.
[b]Most other *Bacillus* cultures will grow on nutrient broth and nutrient agar.
Based on Claus and Berkeley (1986).

(BMSB) 40 recognized species (Claus and Berkeley, 1986), Table 4 lists these with their GC content. There are several validly published new species shown to be genetically and phenotypically distinct from other *Bacillus* species that have not been described in *Bergey's Manual*. These include *B. pulvifaciens* (Nakamura, 1984); *B. alginolyticus* and *B. chrondrotinus,* two alginate-degrading species, (Nakamura, 1987); *B. smithii* (Nakamura et al., 1988); *B. thermoleovorans.* an obligately thermophilic hydrocarbon-utilizing organism (Zarilla and Perry, 1987); *B. benzoevorans,* an aromatic acid and phenol degrader (Pichinoty et al., 1984); and *B. gordonae,* degrader of hydroxy aromatic compounds (Pichinoty et al., 1986).

There are more than 200 species of *Bacillus* in the category *"Species Incertae Sedis"* (Claus and Berkeley, 1986). These have been inadequately described or the orginal isolates have been lost. Presumably, these can be revived for listing in *Bergey's Manual* after reisolation and more detailed studies. For example, after extensive reconsideration of phenetic and molecular data it has been proposed that *B. flexus, B. fusiformis, B. kaustophilus, B. psychrosacchar-*

olyticus, B. simplex, (Priest et al., 1988), and *B. thiaminolyticus* (Nakumura, 1989) be recognized.

The literature contains many important experiments done with *Bacillus* isolates that have not yet been properly identified to species such as: *Bacillus* sp. strain SGI, a manganese-oxidizing and reducing organism (Johannes et al., 1986; Rosson and Nealson, 1982); *Bacillus* sp. strain C-59 and *Bacillus* sp. strain N-6, both alkalophilic organisms with unusual bioenergetic properties (Kitada and Horikoshi, 1987; Kitada et al., 1989); *Bacillus* sp. strain Gx6638, a novel alkaline and heat stable serine protease-secreting strain (Durham et al., 1987); and *Bacillus* sp. strain MGA3, a thermophilic methanol-utilizing species, mutants of which are capable of producing large amounts of lysine (Guettler and Hanson, 1988; Schendel et al., 1989).

An extensive list of phenetic characters of most members of the genus have been complied and procedures for the isolation and identification of individual species have been presented (Gordon et al., 1973; Berkeley and Goodfellow, 1981; Norris et al., 1981; Claus and Berkeley,

Table 3. Major distinguishing characteristics of some *Bacillus* species.

Thermophile	*B. stearothermophilus*
	B. thermodenitrificans
	B. caldotenax
Thermophilic acidophiles	*B. coagulans*
	B. acidocaldarius
Psychrophiles	*B. psychrophilus*
	B. macquariensis
	B. globisporus
	B. insolitus
	B. psychrosaccharolyticus
Alkalophile	*B. alcalophilus*
Facultative chemolithotroph	*B. schlegelii*
Nitrogen fixers	*B. polymyxa*
	B. macerans
	B. azotofixans
Alginate degraders	*B. alginolyticus*
	B. chrondrotinus
Aromatic acid and phenol degrader	*B. benzoevorans*
Hydroxy aromatic compounds degrader	*B. gordonae*
Nitrate reduction to N_2	*B. azotoformans*
	B. licheniformis
Thermophilic hydrocarbon utilizer	*B. thermoleovorans*
Growth restricted to uric acid, allantoin or allantoic acid	*B. fastidiosus*
Growth at high pH and NH_4Cl	*B. pasteurii*
Requires pantothenic acid	*B. pantothenticus*
Requires biotin	*B. pumilus*
Produces heat stable glucosidase	*B. thermoglucosidasius*
Insect pathogens[a]	*B. thuringiensis*
	B. larvae
	B. popilliae
	B. sphaericus
	B. lentimorbus
Human and animal pathogen	*B. anthracis*
	B. cereus

[a]See Chapter 77.

1986). Summaries of one such rendition are shown in Tables 5 and 6 (Norris et al., 1981).

The GC content (32–69 mol%) of the known *Bacillus* species as well as DNA hybridization experiments have revealed the heterogeneity of the genus (Priest, 1981; Fahmy et al., 1985) (See Table 4). Not only is there variation from species to species but there are differences in GC content within strains of a species identified on other bases. For example, the GC content of the *B. megaterium* group ranges from 36 to 45% (Hunger and Claus, 1981). It is thus understandable that Priest et al. (1981, 1988), who have conducted extensive numerical analysis of many unit characters in addition to the DNA studies, have proposed that the genus *Bacillus*

be split into multiple genera, since the intrageneric heterogeneity is as great as exists in most bacterial families. Priest et al. (1988) assigned 80 organisms of species rank to five or more cluster groups. Their studies reemphasized the heterogeneity of the *B. brevis, B circulans, B. coagulans, B. sphaericus,* and *B. stearothermophilus* groups.

A variety of techniques have been employed to find either a simple approach to *Bacillus* taxonomy or a quick and painless identification methodology. Assessment of lipid analyses (reviewed by Minnikin and Goodfellow, 1981) indicated that *Bacillus* could not be separated into discrete groups. On the other hand, some species could be delineated from others. For example, *B. acidocaldarius* could be characterized by its menaquinone (nine isoprenoid units, MK-9), cyclohexyl fatty acids, triterpenes, and complex lipids.

Using the API System (Analytab Products Incorporated) (a rapid identification system wherein many standardized biochemical assessments can be made on test strips) and some supplementary classical determinants, Logan and Berkeley (1981, 1984) have examined 1,075 *Bacillus* strains. They were able to show that the API System tests were more reproducible than the classical tests.

Pyrolysis gas-liquid chromatography has been applied to the problems of *Bacillus* taxonomy (O'Donnell and Norris, 1981; O'Donnell et al., 1988). Although there are still some problems with the technique, some promise for its use in classification and identification has been shown. For example, as with DNA-DNA hybridization studies, a separation has been made between *B. subtilis* and *B. amyloliquefaciens*. When gas-liquid chromatography was applied to examine the subgroups of *B. megaterium,* data were obtained that confirmed the heterogeneity of the group even though there was some difficulty in resolving relationships within the group.

Shute et al. (1984) have used Curie-point pyrolysis mass spectrometry as a taxonomic tool. *B. subtilis, B. pumilus, B. licheniformis,* and *B. amyloliquefaciens* could be separated using data obtained from nonsporulating cultures (those grown on nutrient agar); however, such was not the case with cultures sporulating on nutrient agar plus manganese.

Ribosomal RNA Sequencing

The most effective approach to *Bacillus* taxonomy may be analysis of 16S rRNA molecules by oligonucleotide sequencing (Fox et al., 1977; Stackebrandt and Woese, 1979). That technique

Table 4. DNA-base composition and sources of the type strains of *Bacillus* species.

Bacillus species	GC content (mol%)		Culture collection number				
	Tm[a]	BD[b]	ATCC[d]	DSM	NCIB	NCTC	NRRL
acidocaldarius	60.3	62.3	27009	446	11725		NRS1607
alcalophilus	37.0	36.7	27647	485	10436	4553	B14309
alvei	44.6	46.2	6344	29	9371	6352	B383
amylolyticus	ND[c]	53.0		3034			NRS290
anthracis	33.2	ND	14578		9388	10340	
azotoformans	ND	39.0	29788	1046			B14310
badius	43.8	43.5	14574	123	9364	10333	NRS663
brevis	47.3	47.4	8246	30	9372	2611	NRS604
cereus	35.7	36.2	14579	31	9373	2599	B3711
circulans	35.5	35.4	4513	11	9374	2610	B380
coagulans	47.1	44.5	7050	1	9365	10334	NRS609
fastidiosus	35.1	35.1	29604	91	11326		
firmus	41.4	40.7	14575	12	9366	10335	NRS613
globisporus	39.8	39.7	23301	4	11434		NRS1533
insolitus	35.9	36.1	23299	5	11433		
larvae	ND	50.0	9545				B2605
laterosporus	40.2	40.5	64	25	9367	6357	NRS314
lautus	ND	50–52		3035			NRS666
lentimorbus	37.7	ND	14707	2049	11202		B2522
lentus	36.3	36.4	10840	9	8773	4824	B396
licheniformis	46.4	44.7	14580	13	9375	10341	NRS1264
macerans	52.2	53.2	8244	24	9368	6355	B172
macquariensis	39.3	41.6	23464	2	9934	10419	B14306
marinus	37.6	38.0	29841	1297			B14321
megaterium	37.3	37.6	14581	32	9376	10342	B14308
mycoides	34.2	34.1	6462	2048			NRS273
pabuli	ND	48–50		3036			NRS924
pantothenticus	36.9	36.8	14576	26	8775	8162	NRS1321
pasteurii	38.5	38.4	11859	33	8841	4822	NRS673
polymyxa	44.3	45.6	842	36	8158	10383	NRS1105
popilliae	41.3	ND	14706	2047			B2309
psychrophilus	39.7	40.5	23304	3			NRS1530
pumilus	41.9	40.7	7061	27	9369	10337	NRS272
schlegelii	64.6	66.3	43741	2000			
sphaericus	37.3	37.1	14577	28	9370	10338	
stearothermophilus	51.9	51.5	12980	22	8923	10339	B1172
subtilis	42.9	43.1	6051	10	3610	3610	NRS744
thermoglucosidasius	45–46	ND	43742	2542			B14516
thuringiensis	33.8	34.3	10792	2046	9134		NRS996
validus	ND	53–54		3037			NRS1000

[a]Tm, GC content by thermal melting.
[b]BD, GC content by buoyant density.
[c]ND, not determined.
[d]ATCC, American Type Culture Collection; DSM, Deutsche Sammlung von Mikroorganismen; NCIB, National Collection of Industrial Bacteria; NCTC, National Collection of Type Cultures; NRRL, Northern Regional Research Laboratory.

holds much promise for leading microbial taxonomy into natural phylogenetic relationships. However, traditional taxonomists may be dismayed to find that *Bacillus* species show kinship with nonsporeforming species. Early studies with this powerful tool showed a close relationship among *Bacillus, Planococcus, Sporosarcina, Staphylococcus,* and *Thermoactinomycetes* (Stackebrandt et al., 1987; Stackebrandt and Woese, 1981). In a recent study 16S rRNA cataloging showed that *B. subtilis* and other ellipsoidal-sporeforming species, *B. cereus, B. megaterium,* and *B. pumilus,* formed a coherent cluster, while the round-sporeforming species, *B. sphaericus, B. globisporus,* and "*B. aminovorans*" did not cluster. Furthermore, the latter group were closer phylogenetically to nonsporeforming organisms as follows: *B. sphaericus* to *Caryophanon latum; B. globisporus* to *Filibacter limicola; B. pasteuri* to *Sporosarcina urea*

Table 5. Simplified key for the tentative identification of typical strains of *Bacillus* species.[a]

1. Catalase: positive 2	
negative 17	
2. Voges-Proskauer: positive 3	
negative 10	
3. Growth in anaerobic agar: positive 4	
negative 9	
4. Growth at 50°C: positive 5	
negative 6	
5. Growth in 7% NaCl: positive	*B. licheniformis*
negative	*B. coagulans*
6. Acid and gas from glucose (inorganic N): positive	*B. polymyxa*
negative 7	
7. Reduction of NO_3 to NO_2: positive 8	
negative	*B. alvei*
8. Parasporal body in sporangium: positive	*B. thuringiensis*
negative	*B. cereus*
9. Hydrolysis of starch: positive	*B. subtilis*
negative	*B. pumilus*
10. Growth at 65°C: positive	*B. stearothermophilus*
negative 11	
11. Hydrolysis of starch: positive 12	
negative 15	
12. Acid and gas from glucose (inorganic N): positive	*B. macerans*
negative 13	
13. Width of rod 1.0 μm or greater: positive	*B. megaterium*
negative 14	
14. pH in V-P broth $<$ 6.0: positive	*B. circulans*
negative	*B. firmus*
15. Growth in anaerobic agar: positive	*B. laterosporus*
negative 16	
16. Acid from glucose (inorganic N): positive	*B. brevis*
negative	*B. sphaericus*
17. Growth at 65°C: positive	*B. stearothermophilus*
negative 18	
18. Decomposition of casein: positive	*B. larvae*
negative 19	
19. Parasporal body in sporangium: positive	*B. popilliae*
negative	*B. lentimorbus*

[a]Numbers on the right indicate the number (on the left) of the next test to be applied until the right-hand number is replaced by a species name.
From Norris et al. (1981).

and *"B. aminovorans"* to *Planococcus citreus.* Cell wall composition agreed except with the last case. *B. stearothermophilus* fell outside the main *Bacillus* cluster and showed some relationship to *Thermoactinomycete vulgaris* (Stackebrandt et al., 1987).

In a more recent 16S rRNA sequencing survey, three major *Bacillus* taxonomic cluster groups were defined (Jurtshuk et al., 1989). This was accomplished by determining complete or partial sequences of 16S RNA on 35 recognized neotype reference strains or type species by the technique of Lane et al. (1985). The partial sequences analyzed typically exceeded 1,100 nucleotides. Phylogenetic analyses were performed using three different approaches (Sneath and Sokal, 1973; Fitch and Margoliash, 1967;

Saitou and Nei, 1987) which showed three major groupings of *Bacillus* spp., hereinafter referred to as clusters I, II, and III (see Table 7). The 16S rRNA *Bacillus* cluster groups were quite different from those previously noted by Stackebrandt et al. (1987). This is revealed by direct comparison to the commonly used morphological groupings (see Table 7). Except for morphological group II and the Unassigned Subgroup 2E, all strains sequenced fell into the *B. subtilis* cluster I grouping. *Bacillus* strains of morphological group II fell into all three 16S rRNA cluster groups and *B. macquariensis,* unlike other psychrophiles, fell into the *B. alvei* cluster II group.

Comparative 16S rRNA analyses on thermophilic and psychrophilic *Bacillus* strains

Table 6. Summary of the characters used in the simplified key for *Bacillus* species.

	Catalase	V-P reaction	Growth in anaerobic agar	Growth at 50°C	Growth in 7% NaCl	Acid and gas in glucose	NO₃ reduced to NO₂	Starch hydrolyzed	Growth at 65°C	Rods 1.0 μm wide or wider	pH in V-P medium <6.0	Acid from glucose	Hydrolysis of casein	Parasporal bodies
B. megaterium	+	−	−	−	+	−	V	+	−	+	V	+	+	−
B. cereus	+	+	+	−	+	−	+	+	−	+	+	+	+	V
B. thuringiensis	+	+	+	−	+	−	+	+	−	+	+	+	+	+
B. licheniformis	+	+	+	+	+	−	+	+	−	−	V	+	.+	
B. subtilis	+	+	−	+	+	−	+	+	−	−	V	+	+	−
B. pumilus	+	+	−	+	+	−	−	−	−	−	+	+	+	−
B. firmus	+	−	−	−	+	−	+	+	−	−	−	+	+	−
B. coagulans	+	+	+	+	−	−	V	+	−	V	+	+	V	
B. polymyxa	+	+	+	−	−	+	+	+	−	−	V	+	+	
B. macerans	+	−	+	+	−	+	+	+	−	−	−	+	−	
B. circulans	+	−	V	+	V	−	V	+	−	−	V	+	V	−
B. stearothermophilus	V	−	−	+	−	−	V	+	+	V	+	+	−	
B. alvei	+	+	+	−	−	−	−	+	−	V	+	+	+	
B. laterosporus	+	−	+	+	−	−	+	−	−	−	−	+	+	+
B. brevis	+	−	−	+	−	−	V	−	−	−	−	+	+	
B. larvae	−	−	+	−	+ᵃ	−	V	−	−	−	−	+	+	−
B. popilliae	−	−	+	−	+ᵃ	−	−	−	−	−	−	+	−	+
B. lentimorbus	−	−	+	−	−	−	−	−	−	−	−	+	−	−
B. sphaericus	+	−	−	−	V	−	−	−	−	V	−	−	V	−

+, Greater than 85% of strains examined by Gordon, Haynes, and Pang (1973) positive; −, greater than 85% of strains negative; V, variable character.
ᵃGrowth in 2% NaCl agar.

Table 7. *Bacillus* 16S rRNA cluster groups.

Morphological group	*B. subtilis* cluster I	*B. alvei* cluster II	*B. brevis* cluster III
I	*B. subtilis, B. cereus, B. licheniformis, B. pumilus, B. megaterium* strain Mohb, *B. coagulans, B. smithii*		
II	*B. circulans, B. larvae, B. stearothermophilus*	*B. alvei, B. polymyxa, B. macerans, B. azotofixans, B. pulvifaciens*	*B. brevis, B. laterosporus*
III	*B. sphaericus*		
Subgroup A	"*B. thiaminolyticus,*" *B. alcalophilus*		
Subgroup B	*B. lentus*		
Subgroup C	*B. freundenreichii,* "*B. aneurinolyticus*"		
Subgroup D	*B. pantothenticus*		
Subgroup E1	"*B. psychrophilus,*" *B. insolitus*		
Subgroup E2		*B. macquariensis*	

Table provided by Peter Jurtshuk.

(Wisotzkey et al., 1989) showed that the thermophiles, *B. stearothermophilus, B. thermodenitrificans* and *B. caldotenax* formed a subgroup within the *B. subtilis* cluster but separate from both the "thermotolerant" mesophilic, *B. subtilis* and *B. licheniformis* strains, and the moderate thermophile, *B. coagulans.* The psychrophilic strains, *B. psychrophilus* and *B. insolitus,* fell into cluster I while *B. macquariensis* fell into cluster II.

Because several species were included in the current study that had previously been examined by 16S rRNA oligonucleotide cataloging, it is possible to compare the two data sets directly. As a result, it is possible to augment the membership of cluster I to include *B. fastidiosus, B. firmus, B. badius* and *B. pasteurii* (C. B. Woese, personal communication). In addition, it is extremely likely that at least two nonspore-forming strains, *Planococcus citreus* (Stackebrandt and Woese, 1979) and *Filibacter limicola* (Clausen et al., 1985), as well as *Sporosarcina ureae* (Pechman et al., 1976), are properly regarded as members of cluster I.

Life Cycle; Sporulation and Germination as Models for Differentiation

Introduction

The processes of resting cell formation and the change back to the vegetative cell in a variety of prokaryotes (Losick and Shapiro, 1984) present excellent models for studying differentiation, with the added attendant advantages of microbial systems: ease of handling, use of large numbers of cells, fast growth, synchrony, and availability of mutants. The endospore models were recognized early and, therefore, more knowledge has been accumulated using them than with other prokaryotic systems. As the attempt to categorize the many species of *Bacillus* has had a long history, so have the efforts to unravel the many aspects of the life cycle of *Bacillus* (for a historical treatment see Keynan and Sandler, 1983). Because of the enormous literature in this area, the present treatment of the life cycle relies mainly on reviews of the subject and covers mainly highlights of germination and sporulation.

The cycle of germination, outgrowth, growth, and sporulation (shown schematically in Fig. 2) has been studied from many different aspects with many different species of spore-formers but because of the genetic versatility of *B. subtilis,* most work has focused on this species.

Germination and Outgrowth

Free spores usually must be activated for germination. Activation is a reversible process which conditions the spore for germination and increases the number of spores undergoing germination as well as the rate of germination. Spores can be activated by a variety of treatments, notably exposure to heat. During activation there is a loss of some coat protein, dipicolinic acid (DPA), and Zn^{2+} along with an increase in membrane fluidity. Germination, the breaking of the spore's highly dormant state, follows (recent reviews on germination include Setlow, 1983, and Foster and Johnstone, 1989). A series of degradative reactions is triggered in an unknown manner by simple compounds such as certain amino acids and ribosides or mixtures (no universal germinant has been described) or certain nonnutrient conditions, and can be monitored by the loss of spore refractility as seen in the phase-contrast microscope and by decrease in optical density. No metabolic activity can be detected during the first 2 min of germination of spores that require alaine or glucose for germination. Generation of ATP or production of known metabolic products of these initiators has not been found. Mutants deficient in key glycolytic pathway enzymes can germinate, thus ruling out glycolysis in the case of spores requiring glucose for germination. The same spores can be germinated by nonmetabolizable glucose analogs as well. (However, metabolism may play a role in the germination of *B. fastidiosus,* whose spores can only be germinated with uric acid, which is the main carbon source for these unique organisms (Aoki and Slepecky, 1973). One hypothesis suggests that germinants act on receptor proteins, possibly in the inner membrane, which then undergo conformational changes that alter permeability (Foster and Johnstone, 1989). This leads to an autocatalytic loss of heat resistance and to changes that initiate metabolism, leading to vegetative growth. Another view, based on the observation that inhibition of the electron transport system affects germination, postulates that respiration and ATP create a proton motive force, lending to the establishment of a proton gradient. The proton motive force is used for transport of ions from core to cortex to neutralize other ions (Gould, 1983).

Upon germination, the spores not only lose their resistance to heat but also resistance to radiation and injurious chemicals; their stainability also increases. Concomitantly with "phase" darkening, the spores swell, break out of their coats, and exude up to 30% of their dry weight; about one-half of the exudate consists of a calcium chelate of the spore-specific substance, DPA, and the remainder consists of peptidoglycan fragments (from the action of cortex lytic enzymes) and amino acids. The earliest measurable events are the loss of calcium, DPA, and heat resistance. This is followed by metabolic events using high-energy compounds pro-

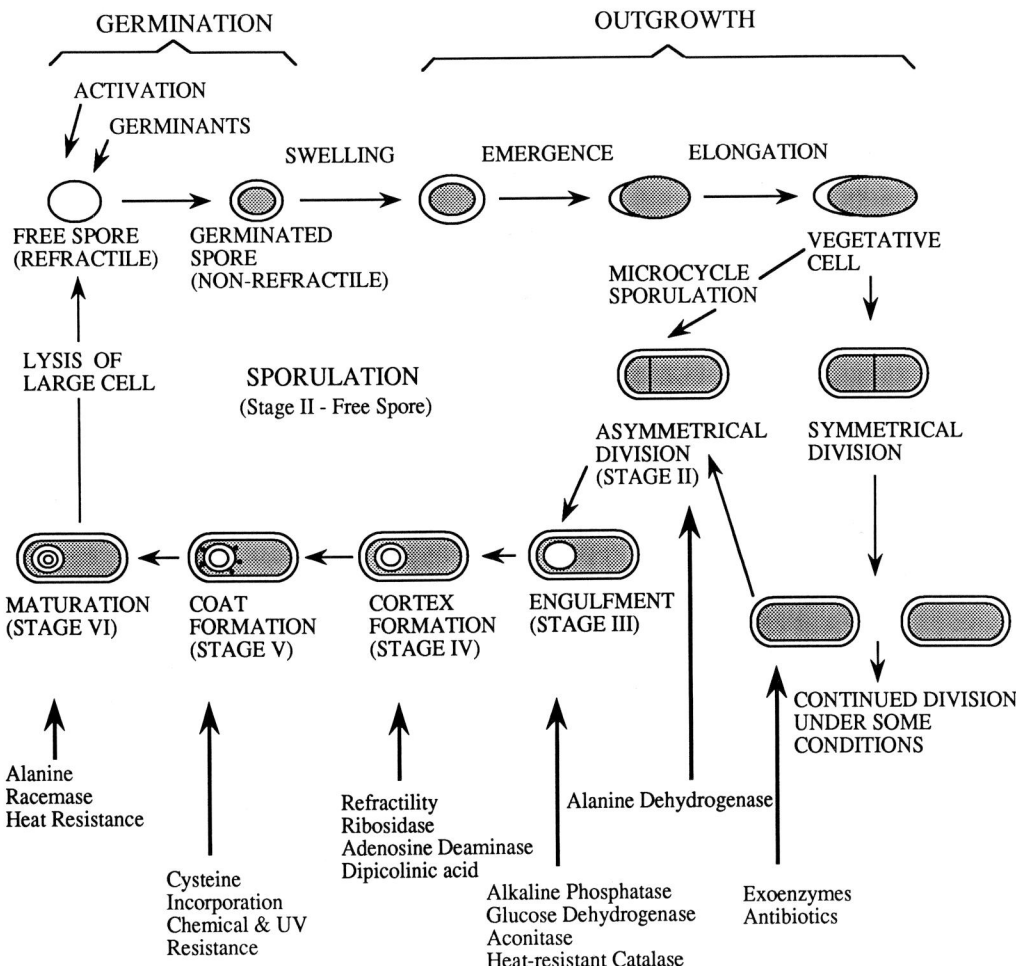

Fig. 2. Cycle of germination, outgrowth, and sporulation of a typical sporeforming bacterium. Also shown are some biochemical and physical events associated with various stages. (modification of a figure in Slepecky, R. A. (1978)

duced early in germination from energy reserves stored in the dormant spore.

RNA synthesis begins rapidly within 2 min of germination. The dormant spore lacks the ability to produce amino acids and amino acid biosynthesis is absent early in germination. During the first minutes of germination, 20% of the spore's protein is degraded, providing the source of amino acids for biosynthesis of new protein and small molecules (such as nucleotides) during outgrowth (reviewed by Setlow, 1988). The spore's enzymes are not degraded. Rather, a group of small acid-soluble proteins (SASP) are the source of the amino acids. These unique proteins, located in the core and sensitive to proteolysis, conprise 8 to 20% of the protein in the spore. Their molecular weight is low (5–11 kDa) and although they are not histones, they bind to the spore DNA. The proteins are degraded by a unique protease that has an absolute specificity for these proteins. They are

synthesized late in sporulation. Several of the five known SASP genes (referred to as *ssp*) have been cloned and in addition to coding proteins supplying amino acids in germination their products may have other roles. One has been shown to be involved in ultraviolet (UV) resistance.

Many germination mutants (abnormal germination phenotypes) have been described (reviewed by Moir et al., 1986; Foster and Johnstone, 1989). The *ger* (germination) genes are considered a subclass of *spo* (sporulation) loci, and are made up of several classes: I, structural genes for germination mechanism components; II, regulatory genes for class I; III, post-translation processing and assembly genes; and IV, genes for synthesizing spore structure (e.g., cortex) required for germination.

Outgrowth is the period during which the spore gradually becomes a vegetative cell and initiates new macromolecular synthesis. Genes associated with this period are *out* genes. DNA is replicated relatively late in outgrowth, just before division. The vegetative cell is then capable of undergoing various morphological and biochemical changes which lead either to a series of symmetric cell divisions if sufficient nutrients are present, whereas in stressful times (particularly nutrient limitation), subsequent spore formation or the production of a spore without intermittent cell division can occur. The latter is known as microcycle sporulation (Vinter and Slepecky, 1965).

Sporulation

Electron microscopical analyses of cells during sporulation has revealed seven stages (reviewed by Fitz-James and Young, 1969). The various stages are shown in Fig. 2. Vegetative cells are considered to be Stage 0. Upon induction input prior to actual sporulation, the nuclear material is in an axially disposed filament. This is stage I. However, since such a pattern does not appear to be unique to sporulating cells, current practice refers to cells in stages prior to stage II as preseptation cells. Segregation of the chromatin material to the poles of the cell occurs concomitantly with the invagination of the plasma membrane in an asymmetrical position on the cell which fuses to complete the spore septum. This is stage II. The mode of formation of this septum is similar to the formation of the transverse septum of symmetric vegetative cell division. (Indeed, it has been proposed that sporulation because of this and some other similarities is a modified prokaryotic division (Hitchins and Slepecky, 1969) and models for that view have been presented (Freese and Heinze, 1983). In sporulation, the division of the cell is not equal and subsequent proliferation of the larger cell's plasma membrane leads to complete engulfment of the "forespore" and liberation of the immature spore, surrounded now by a double unit membrane, into the cytoplasm of the larger cell. This completes stage III. This is a key step since this double membrane now has different transport properties owing to the opposing polarity of the two membranes.

During sporulation the vegetative cell is divided into two compartments each having a different fate and each displaying different patterns of gene expression (reviewed by Setlow, 1989). At this time, the cell is "committed" to complete the process of sporulation. The small cell eventually becomes the core of the spore while the large cell, the mother cell, goes on to produce the outer protective layers of the spore and then lyses to release the spore. Cortex material similar to vegetative cell peptidoglycan (but differing in the degree of cross-linking and other aspects) is laid down between the unit membranes, and its deposition corresponds in time to the accumulation in the core of DPA and calcium. Stage IV is now completed. Studies with cortex-less mutants show that the cortex is needed for refractility of the spore (when the spores become refractile, they can be seen in the phase-contrast microscope) and for accumulation of DPA. The cortex plays a fundamental role in the dehydration of the spore (reviewed by Gould, 1983).

During Stage V, protein coats are synthesized by the mother cell. There are about 10 major coat proteins in *B. subtilis* and they are encoded by *cot* genes, seven of which are known (Losick et al., 1986). The coat proteins are placed around the outside of the forespore. In some species, an additional protein layer called an exosporium is synthesized. Since the coat may play important roles in protection of the spore and its subsequent germination, it has been the subject of many investigations (reviewed by Aronson and Fitz-James, 1976; Losick et al., 1986). Electron microscopy reveals that *B. cereus* contains an outer coat showing a cross-patched pattern, an inner pitted layer, and a thin layer, the undercoat, while other species show distinct differences. *B. subtilis* possesses a very thick multilayered coat with an outer striped layer and *B. thuringiensis* has a coat deficient in the outer cross-patched layer. Differences show up as well within the major structural polypeptides and coat-associated proteases. These differences may be responsible for the variation found in germination and certain resistant properties (other than heat and UV resistance) of various species.

During vegetative growth and subsequent sporulation, a variety of proteases are produced (reviewed by Priest, 1977). There are six extracellular proteases and at least three major intracellular proteases—ISP, esterase A, and esterase B. They may be involved with turnover of intracellular proteins, the processing of protein precursors for spore coats, or inactivation of later sporulation enzymes, as well as other functions. As with other aspects of *B. subtilis* physiology, there are other considerations. Even though the genes *apr, npr, epr,* and *isp,* which code for the proteases alkaline (subtilisin), neutral (metallo-) "new" serine, and major intracellular serine, respectively, can be deleted, there still is some protease activity (Sloma et al., 1988). This finding suggests that there are other unindentified proteases.

As the spore matures (stage VI), it becomes resistant to heat and to a variety of organic solvents. Final lytic enzymes lyse the sporangial or mother cell liberating the free spore (stage VII). Figure 3 shows a cross section of a mature and heat-resistant spore. If the free spore is placed among the proper nutrients it will germinate, completing the cycle. The sporulation process takes 6 to 8 h at 37°C in *B. subtilis*.

Many other biochemical and physiological events occur during sporulation in addition to those indicated in Fig. 2 and those that can be surmised to be linked with the morphologically identified stages. Some vegetative enzyme activities disappear, some remain, others are modified, and new sporulation specific enzymes are synthesized. The number of sporulation-associated events is uncertain, but the genetic evidence suggests as many as 200 genes in 40 to 50 operons are involved (Losick et al., 1986; Mandelstam and Errington, 1987; Piggot, 1989; Youngman et al., 1989). Currently the genetic map, which includes sporulation and germi-

nation genes as well as all-known vegetative cell genes, contains 700 loci, more than 300 of which have been cloned and 180 of those sequenced (reviewed by Piggott, 1989). The ordered appearance of cytological and biochemical changes implies a sequential reading of the genome. The genetic data are consistent with there being a single linear dependent sequence to stage III with a more complex pattern of gene expression beyond that stage.

The genotype of mutants blocked at the various morphological stages in sporulation (as determined by electron microscopy) are designates *spoO, spoII, spoIII*, etc. For example, a *spoII* mutant would be arrested in stage II. Within each genotype, the designation *A, B, C*, etc. identifies different genetic locations; for example, *spoIIA, spoIIB* identify two separate mutations that cause arrest of developmental stage II. Eight loci have been shown to be concerned with stage 0, the five principal spoO genes being

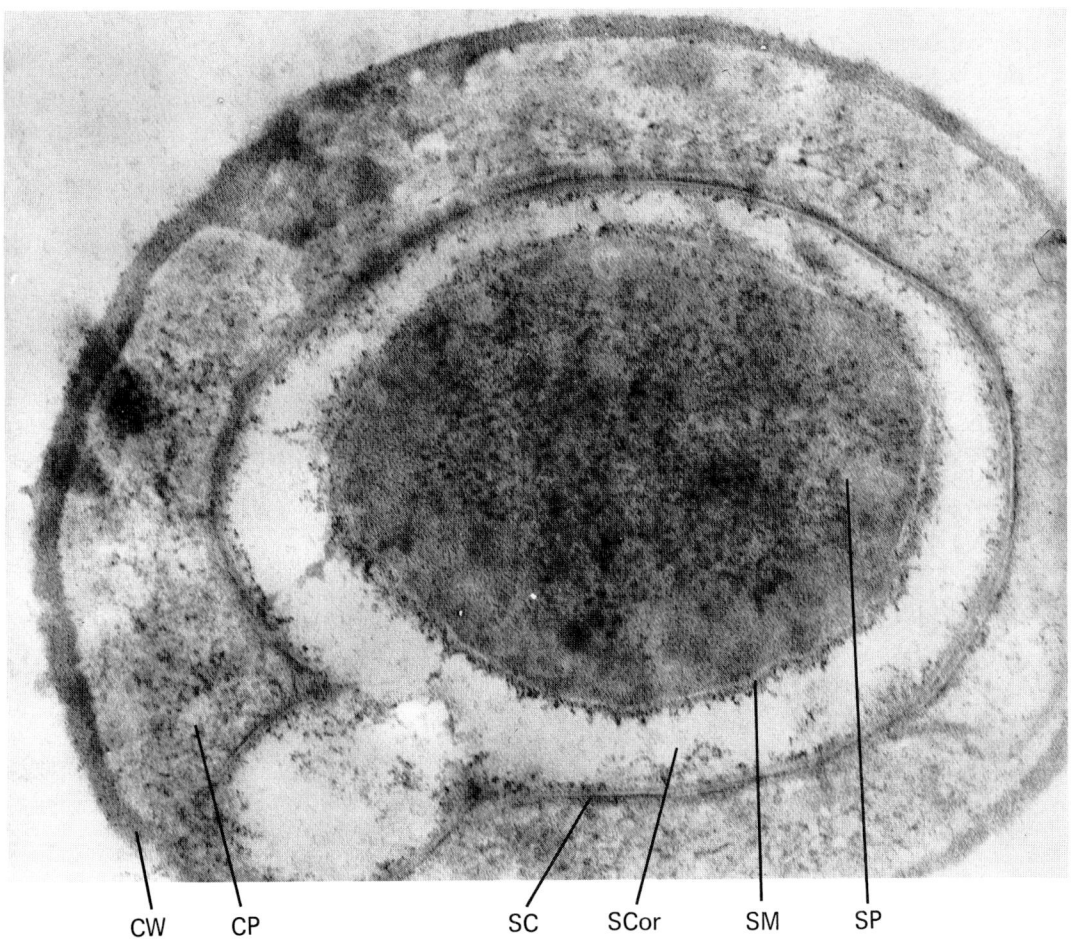

Fig. 3. Cross-section of *Bacillus megaterium* containing a spore and showing the sporangium (cell), cell protoplast (CP) and wall (CW), spore coat (SC), spore cortex (SCor), spore membrane (SM), and spore protoplast (SP). × 120,000. (Norris et al., 1981.)

spo0A, B, E, F, and *H.* These *spo0* mutants have pleiotropic effects on the phenotype, suggesting that their gene products regulate the expression of several genes. They do not make polar septa (the asymmetric division) or Stage 0 associated products—proteases, antibiotics and transformation competent cells. There are suppressors of spo0 mutations, i.e., suppressor mutations can restore sporulation in some spo0 mutants to the wild type level. For example, *sof-1* can suppress defects in *spo0F, B,* and *E* (Hoch et al., 1985). Others include *arbA* and *arbB* suppressors of *spo0A* mutations; *rvtA,* a *spo0* repressor; *ssa,* a reliever of suppression of alcohols; *crs,* an alleviator of glucose repression of sporulation; and *sapA* and *sapB.* suppressors involved with alkaline phosphate induction (Losick et al., 1986). Studies on these extragenic suppressors have given insights into the role of *spo0* mutants.

Sporulation, normally repressed at high growth rates in the presence of excess nutrients, will ensue at the beginning of stationary phase (arbitrarily defined as t_0) in batch culture upon limitation of carbon, nitrogen, or phosphate (reviewed by Smith, 1989, and Sonenshein, 1989). The initiating signal for sporulation is not known. However, sporulation initiation is associated with a decrease in the intracellular GTP pool (reviewed by Freese and Heinze, 1983). Decoynine, a drug that artificially reduces GTP level, induces sporulation even in the presence of excess nutrients. There may be a connection between GTP levels and a proposed sensing mechanism as follows: *spo0A, B,* and *F* share significant homology and *A* and *F* products show partial homology with proteins such as OmpR, NtrC, and SfrA which are "sensing proteins" involved in the transduction of various environmental signals (Trach et al., 1985; Nixon et al., 1986). Thus, *spo0A* and *spo0F* may play a role in sensing starvation and transferring a signal to other genes (see review by Smith, 1989, on the initiation of sporulation). A gene for a protein kinase that phosphorylates sporulation regulatory proteins Spo0A and Spo0F has been characterized (Pergo et al., 1989). The gene, *kinA,* previously known as *spo11J,* has been shown to code for a protein that is homologous to the transmitter class of proteins (Stragier, 1989). Defects in the methylation of a membrane-associated 40-kDa protein in some *B. subtilis spo0* mutants suggests that protein methylation also may be a part of the nutrient-sensing system (Golden and Bernlohr, 1989).

Multiple forms of RNA polymerase are of considerable importance in sporulation (reviewed by Doi and Wang, 1986; Losick et al.,

1986; Losick and Kroos, 1989; Moran, 1989; Setlow, 1989; Stragier, 1989). The holoenzyme is made up of an enzyme core comprised of subunits beta, beta' and alpha (the products of genes *rpoB, rpoC,* and *rpoA,* respectively) and attached sigma factors (the products of genes to be designated), regulatory proteins that determine the promoter specificity. Currently, there are nine known sigma factors, four associated with vegetative cells and five involved in sporulation (see Table 8). Losick and Pero (1981) suggested there is a succession of different sigma factors governing the transcription of the various temporal classes of sporulation genes.

The first acting sporulation sigma is σ^H which is required for the initiation of sporulation and it is involved with the transcription of a later gene, *spoVG.*

A key point in sporulation is stage II, which is the result of the asymmetric division event. stage II mutants and the associated gene products or functions are as follows: *spoIIAA* σ^E processing); *IIAC* (σ^E factor); *IIE* (pro σ^E); *IIGA*σ^E processing protease); *IIGB* (p^{31}; σ^E); *IIN* (Fts homology) and *IIJ* (ntrB protein kinase) (Leighton cited in Stragier (1989), and Stragier (1989)). That the *spoIIN* product has homology with *E. coli* Fts, a protein involved in symmetrical cell division, may suggest that the level of Fts controls the asymmetric division. In turn, the functions of the other *spoII* mutations listed above implies a rather complicated mechanism for the synthesis of σ^E. The sporulation septum (stage II), dependent on *spoIIAC* (σ^F) and *spoIIE* products, is required for *spoIIGA* processing activity, i.e., conversion of pro-σ^E by proteolytic cleavage to σ^E. The important finding by Stragier (1989) of the coupling of gene expression to morphogenesis may be a recurrent theme throughout other aspects of sporulation.

After Stage II, compartmentalization of sigma factors occurs. σ^E is active both in the mother cell and forespore compartment. σ^G acts only in the forespore compartment (Setlow, 1989) while σ^K, the product of *sig K,* a composite gene from *IVCB* plus *IIC* (due to a chromosome rearrangement requiring *SpoIIID* and a recombinase) acts only in the mother cell (Losick and Kroos, 1989). σ^K directs the transcription of *cotA* and *cotD,* spore coat genes.

This brief overview of some of the emerging information on regulation of gene expression during sporulation reflects the current view of a network of dependent pathways in which activation of developmental genes depends on the products of other developmental genes (Losick and Kroos, 1989).

Table 8. *Bacillus subtilis* sigma factors.

Sigma type	Previous designation	Gene	Physiological role	Time of action	Target genes	Cognate consensus −35	Promoter sequence −10
σ^A	σ^{55}, σ^{43}	*sig A* (*rpoD*)	Housekeeping	$<t_0$	Many	TTGACA	TATAAT
σ^B	σ^{37}	*sig B*	Unknown *ctc* transcription	$<t_0$	Many	AGGNTT	GGNATTGNT
σ^C	σ^{32}	*sig C³*	Unknown	$<t_0$	Many	AAATC	TANTGNTTNTA
σ^D	σ^{28}	*sig D*	Flagellar synthesis	$<t_0$	Flagellar genes	CTAA	CCGATAT
σ^E	σ^{24}	*sig E* *spoIIGB*	Engulfment; provides compartmentalization	$t_{1.5}$–$t_{3.5}$	*spoIID*	TTNAAA	CATATT
σ^F	$\sigma^{SpoIIAC}$	*sig F* *spoIIAC*	Asymmetric division (?)	About t_1–t_2	Unknown	Unknown but similar to σ^G	
σ^G	—	*sig G* *spoIIIG*	Forespore gene expression	t_3–t_5	*sspA-E-spoVA*	YGHATR	CAHWHTAH
σ^H	σ^{30}	*sig H* *spoOH*	Entry into stationary phase	About t_0	*spoVG*	CAGGA	GAATWWT
σ^k	σ^{27}	*spoIVCB* + *spoIIIC*	Mother cell gene expression	$t_{3.5}$–$t_{5.5}$	*cotA, cotD*	Unknown	

Modified from Helmann and Chamberlain (1988); Moran (1989); and Stragier (1989).

Surface Structures of *Bacillus*

S-layers

Crystalline surface layers of protein or glyco-protein subunits, called S-layers, are found in members of the genus *Bacillus* (reviewed by Sleytr and Messner, 1988). S-layers of individual strains of *Bacillus* have been shown to differ in molecular weight (40–200 kDa), the degree of glycosylation of the subunits, and the geometry of the S-layer lattice. For example, *B. stearothermophilus* contains one S-layer consisting of two types of glycan chains, one being a unique type of protein-carbohydrate linkage. On the other hand, *B. brevis* contains two S-layers, termed the outer wall protein (OWP) and the middle wall protein (MWP), external to the peptidoglycan layer. These form a hexagonal array in the cell wall. The nucleotide sequence of the entire MWP-OWP gene operon is known (Tsuboi et al., 1988). The gene encoding an S-layer protein of *B. sphaericus* has been cloned and sequenced (Bowditch et al., 1989). Not all *Bacillus* species contain S-layers and some strains within a species may lack such a layer. Furthermore, the type of lattice may vary from species to species and within strains of a species. *B. alvei, B. anthracis,* and *B. brevis* show a hexagonal array; *B. cereus, B. fastidiosus,* "*B. macroides,*" *B. megaterium, B. psychrophilus,* and *B. schlegelii* present a square lattice, while strains of *B. stearothermophilus* can be obtained that individually have one of the three types (Claus and Berkeley, 1986).

As with S-layers of other bacteria, their function in *Bacillus* is unknown. However, since it has been demonstrated that the S-layer can physically mask the negatively charged peptidoglycan sacculus in *B. stearothermophilus* and prevent autoagglutination, it has been postulated that the layer may play a key role in bacteria-metal interactions (Sleytr and Messner, 1988).

Capsules

The capsule (a homopolypeptide of D-glutamic acid) of *B. anthracis* as a virulence factor has been studied extensively (see Chapter 78). Other bacilli, such as *B. subtilis, B. megaterium,* and *B. licheniformis,* possess capsules containing the homopolypeptide of D- or L-glutamic acid as well (Makino et al., 1989). Some *Bacillus* species, e.g., *B. circulans, B. mycoides,* and *B. pumilus,* produce carbohydrate capsules. For example, *B. circulans* forms an extracellular polymer consisting of glucose and glucuronic acid (Claus and Berkeley, 1986). In the case of *B. megaterium* a heteropolysaccharide composed of D-glucose, D-xylose, D-galactose, and L-arabinose has been found in one strain (Cassity and Kolodziej, 1984).

Flagella

Most *Bacillus* species possess peritrichous flagella. Although some use has been made of H-antigens in setting up serotyping schemes, they have not been widely adopted (Claus and Berkeley, 1986). Chemotaxis has been studied extensively in *B. subtilis* (Ordal and Nettleton, 1985).

Cell Walls

Almost all *Bacillus* species tested have vegetative cell walls made up of peptidoglycan containing *meso*-diaminopimelic acid (*m*-DAP). The exceptions *(B. sphaericus* and the related species, *B. pasteurii* and *B. globisporus)* contain lysine instead (Bartlett and White, 1985). But even those species, as all others, contain *m*-DAP in the peptidoglycan of their spore cortex. Cell-wall turnover in Gram-positive bacteria, particularly *Bacillus* species which have been useful models, has been reviewed by Doyle et al. (1988). In addition to peptidoglycan in the cell wall, all *Bacillus* species contain large amounts of an anionic polymer, such as teichoic acid (a glycerol or ribitol-based polymer joined together by phosphodiester linkages to form a flexible linear strand) or teichuronic acid (uronic acid-based polymer) which are bonded to muramic acid residues. The type of this anionic polymer present depends on the levels of phosphate and magnesium in the growth medium. The glycerol teichoic acids vary a great deal between *Bacillus* species and within species. For example, *B. subtilis* can contain either glucosyl α or β (1→2) glycerol or glucosyl α (1→6) galactosyl α (1→1 or 3) glycerol), while *B. licheniformis* contains galactosyl α (1→2) glycerol. However, they are joined to the peptidoglycan through a common linkage disaccharide, acetylmannosaminyl(1→4)*N*-acetylglucosamine (Kaya et al., 1984).

As in other Gram-positive bacteria, lipoteichoic acids are found associated with most of the cell membranes of *Bacillus* species. These compounds are involved in the synthesis of wall teichoic acids as regulators of autholytic activity and as scavengers of bivalent ions. Those so tested in *Bacillus* fall into three groups based on the presence or absence of *N*-acetylglucosamine branches in the backbone chains of their lipoteichoic acids—hydrophilic poly (glycerolphosphate) chains and hydrophobic gentiobiosyldiacylglycerol anchors (Iwasaki et al., 1989). Group A is made up of strains of *B. subtilis, B. licheniformis,* and *B. pumilus;* group B, other *B. subtilis* strains and *B. cereus,* and group C, *B. polymyxa* and *B. circulans.*

Macrofibers

Macrofibers are multicellular and multistranded structures, hundreds of micrometers in length, produced by autocatalytic mutants of *B. subtilis* (Mendelson, 1978). These left- or right-handed helical structures have been used to study cell wall structure and growth. They are thought to reveal cell-surface molecular organization and force interactions in the cell wall not readily elucidated in the wild type, single-celled organism. The establishment and maintenance of macrofiber structure is influenced by both genetic and physiological factors (Briehl and Mendelson, 1987; Surana et al., 1988).

Membranes

The membranes of *Bacillus* species have been studied extensively because of their intrinsic interest; as a model of membrane structure in Gram-positive cells; with regard to their role in sporulation and germination; with respect to explanations for thermophily; and other reasons. For example, one explanation for the ability of thermophilic microorganisms such as *B. stearothermophilus* to grow at high temperatures is that the physical properties of the membrane are changed due to changes in the lipid composition in response to growth temperature (Gould, 1983).

There is great diversity in the range and type of lipids in *Bacillus* membranes (see review by Minnikin and Goodfellow, 1981) and wide variation in the fatty acids are found. The main phospholipids present are phosphatidylglycerol, diphosphatidylglycerol, and phosphatidylethanolamine; however, others are found as well. The major isoprenoid quinones are menaquinone, and most species contain menaquinones with seven isoprenoid units (MK-7). *B. acidocaldarius* is the exception and possesses MK-9 (Minniken and Goodfellow, 1981).

Two-dimensional polyacrylamide gel electrophoresis (PAGE) has been used to attempt to resolve all *B. subtilis* membrane polypeptides (see Shohayer and Chapra [1985] for one such study). Several membrane enzymes have been isolated and characterized, such as the lactate, malate, glycerate-3-phosphate, NADH, and succinate dehydrogenases.

Because of interest in synthesis and modification of the peptidoglycan layer, much attention has been paid to penicillin-binding proteins. Six have been found in *B. subtilis,* but one of the most penicillin-sensitive binding proteins, number 4, has been found to be absent from *B. subtilis* 168 (Buchanan, 1987).

Genetic Studies

The discovery of transformation in *B. subtilis* strain 168 by Spizizen (1958) was largely responsible for focusing attention on the genetics of the genus *Bacillus.* Strain 168 is thought to be a derivative of the type strain *B. subtilis* Marburg (see Hemphill and Whiteley, 1975), and is one of a relatively few bacilli in which competence for DNA uptake has been found to occur as a natural part of the life cycle. As a consequence of Spizizen's discovery and the later isolation of generalized transducing phages, our knowledge of the chromosomal organization of *B. subtilis* is second only to that of the enteric bacteria. (About one-half as many genetic markers are known in *B. subtilis* as in *Escherichia coli*.) Furthermore, the identification of numerous genes affecting sporulation in *B. subtilis* is providing a means for analyzing this complex developmental program, which is largely unique to the Gram-positives.

Transformation

The establishment of a competent state in broth culture is most efficiently brought about in a minimal salts medium (Anagnostopoulos and Spizizen, 1961; Bott and Wilson, 1967). As the bacteria enter stationary phase in this medium, a maximum of 20% become competent, with a 1 to 2% transformation frequency for a given marker. The development of competence is associated with a reduction in macromolecular synthesis that is initiated well before cells are transformable. Competent bacilli are relatively metabolically latent, and have a lower buoyant density compared to noncompetent bacteria (Dooley et al., 1971; Hadden and Nester, 1968). Although the period of competence overlaps the time in which sporulation is initiated, the two forms of physiological differentiation are thought not to be connected. Different media are preferred for the two processes and commitment to competence is reversible. Several lines of research suggest that the establishment of competence is coincidental with, and perhaps induces, a DNA-repair system analogous to the SOS regulon in *E. coli* (Love et al., 1985).

The development of the capacity of *B. subtilis* to take up DNA is associated with the appearance of several novel cellular proteins. Among these is a 75-kDa protein complex which has been isolated from the membrane of competent cells (Smith et al., 1985). The complex consists of two subunits of 18 kDa (polypeptide A) and 17 kDa (polypeptide B). Mutants lacking subunit A do not bind DNA to the cell surface whereas those deficient in B are defective in DNA entry. The 75-kDa protein complex also has nuclease activity that is probably associated with polypeptide B. A nuclease is expected, because it is known that one strand of transforming DNA is hydrolyzed in the process of entering the cell, resulting in a single-stranded

product on the cytoplasmic side of the membrane (Davidoff-Abelson and Dubnau, 1973a, 1973b). Competence factors that interact with the membrane have also been reported in *B. stearothermophilus* (Streips and Welker, 1971). *B. subtilis* cells show little specificity with regard to DNA uptake and may be transformed with homologous chromosomal DNA, plasmid DNA, or transfected with bacteriophage DNA.

There is evidence that strands of transforming DNA enter the cell at sites on the membrane where the chromosomal DNA is attached (teRiele et al., 1984). The penetrating DNA, now reduced to a single-strand form, is then brought in contact with the homologous region of the recipient chromosome. This step is probably mediated by the 45-kDa protein product of the *B. subtilis recE* gene and results in a complex in which the transforming DNA begins to displace one strand of the chromosomal DNA while hydrogen-bonding to the complementary strand. A continuation of the displacement reaction allows pairing and integration of several thousand bases of transforming DNA into the chromosome, while an equivalent amount of the homologous strand is removed and degraded.

Considerable thought has been given to the question of whether transformation is associated with genetic exchange in natural populations of bacilli (see Stewart and Carlson, 1986, for a review). The complexity of the transformation process with its requirement for unique competence factors appearing only at stationary phase suggests that the capacity to take-up exogenous DNA offers some selective advantage in the evolution of these bacteria. The fact that competence occurs only late in the growth cycle probably means the system is not designed to obtain DNA as a nutrient. Ephrati-Elizur (1968) found that *B. subtilis* cells excrete high-molecular-weight DNA into liquid culture as they grow. Under natural conditions this could be the source of donor DNA. Graham and Istock (1978) demonstrated that genetic exchange, thought to be mediated by transformation, occurs at high frequency between genetically labeled strains of *B. subtilis* in soil. Also, transformation frequencies in cultures in which the bacteria are allowed to attach to sand grains are much higher than in the standard liquid culture procedure (Lorenz et al., 1988).

Generalized Transduction

Bacteriophage capable of mediating generalized transduction have been reported in many species of *Bacillus* including *B. subtilis*, *B. cereus*, *B. megaterium*, *B. thuringiensis*, *B. anthracis*, and *B. stearothermophilus.* Thus, transduction offers an immediate advantage for genetic analysis over transformation in that it is applicable to more strains of these bacteria. In addition, some phages transduce fragments of DNA much larger than can be transferred via transformation, and this facilitates linking distant markers. PBS1, a bacteriophage that infects *B. subtilis* 168, can incorporate 5 to 10% of the bacterial chromosome in a single virion particle, and was instrumental in constructing the complete circular chromosomal map of *B. subtilis* (Lepesant-Kejzlarova et al., 1975). On the other hand, small DNA molecules such as those of some plasmids are not efficiently packaged in large phages, but can be transduced by a variety of small bacteriophages (Canosi et al., 1982). Some generalized transducing phages have relatively broad host ranges and can transfer plasmids between different species of bacilli (Ruhfel et al., 1984).

Little is known about the mechanism by which transducing particles are formed. PBS1 appears to package bacterial DNA randomly. There is little evidence for packaging sites as are indicated in *Salmonella* phage P22 (Jackson et al., 1978; Schmieger 1982, 1984). A *pac* site has been located in the genome of the small generalized transducing phage SPP1. However, its relevance, if any, to packaging transducing DNA is not clear (Deichelbohrer et al., 1982, 1985).

The process by which transducing DNA becomes incorporated into the recipient bacterial chromosome differs from that discussed earlier for transformation. In transduction, the donor DNA is thought to enter the infected cell as double-stranded DNA which then synapses with the homologous region of the recipient chromosome. Incorporation of the transducing fragment presumably results from a double crossover between the two DNA molecules. In support of this model, mutants of *B. subtilis* deficient in recombination functions have a reduced capacity to be transduced, while they can be transformed (Dodson and Hadden, 1980).

Other important *Bacillus* generalized transducing phages include CP15, a phage originally shown to transduce *B. cereus* (Thorne, 1968) but which can transfer plasmids between *B. cereus, B. thuringiensis,* and *B. anthracis* (Ruhfel et al., 1984). Bacteriophage MP13 has been important in mapping the chromosome of *B. megaterium* (Vary et al., 1982), and TP-13 and TP18 have been similarly used in genetic studies of *B. thuringiensis* (Barsomian et al., 1984.)

Specialized Transduction

Specialized transduction has been reported in several *B. subtilis* phages including φ105 (Shap-

iro et al., 1974), SPβ (Zahler et al., 1977), and φ3T (Odebralski and Zahler, 1982) and *B. amyloliquefaciens* phage H2 (Zahler et al., 1987). The most carefully examined of these is SPβ, a temperate bacteriophage that is normally carried as a prophage in *B. subtilis* 168 (Warner et al., 1977). SPβ can transduce markers proximal to its normal attachment site *attB* SPβ between the markers *ilvA* (threonine dehydratase) and *kauA* (ketoacid uptake) located near the terminus of chromosomal replication. In addition, SPβ prophage will insert at a variety of aberrant positions when it lysogenizes mutants of *B. subtilis* lacking *attB* SPβ. Induction of these lysogens gives rise to specialized transducing phages carrying genetic markers near the novel sites of integration (see Zahler, 1982, for a review). The gene order of SPβ prophage is a circular permutation of that present in the phage DNA (Spancake et al., 1984). Specialized transducing particles are thought to originate from errors in excision when the prophage is induced. The crossing-over event that leads to reformation of a circular phage DNA during induction may be displaced, resulting in removal of a portion of the bacterial DNA contiguous to the right or left boundaries of the prophage. Encapsidation of such hybrid DNA produces a specialized transducing particle.

Two types of transductants are recognized in the SPβ system (Zahler, 1982). The bacterial portion of the transducing phage DNA may undergo recombination with and replace the homologous region of the genome of the recipient. This so-called "replacement transduction" is similar to generalized transduction, except that only a limited number of genetic markers are involved. On the other hand, the infecting phage-bacterial DNA may incorporate as a prophage to produce a bacterium diploid for the *B. subtilis* genes associated with the prophage. Such "addition transductions" are most likely to occur if the recipient bacterium is already lysogenic for SPβ. The resulting merozygote may be used in complementation studies if different alleles are present on the prophage and bacterial chromosomes.

Conjugative Plasmids

Fertility plasmids capable of bringing about their own transfer from one bacterium to another have been described in several species of *Bacillus.* The capacity to produce the insecticidal delta toxin crystal protein in *B. thuringiensis* is encoded in large plasmids. Gonzalez et al. (1982) found that three strains of this bacterium transmitted the crystal-protein phenotype to *B. thuringiensis* variants which had lost

CHAPTER 76

the plasmid. Moreover, these plasmids could also be transferred to *B. cereus* and yielded transcipients that produce crystal protein. Battisti et al. (1985) reported the transfer of plasmids pXO11 and pXO12 from *B. thuringiensis* to *B. anthracis* and *B. cereus*. The transcipients, in turn, became effective donors, and in the case of those inheriting pXO12, also acquired the ability to produce parasporal crystals. Strains of *B. anthracis* that acquire plasmid pXO12 can subsequently mobilize and transfer nonconjugative plasmids present in the same cell. Using this system, the tetracycline resistance plasmid pBC16, the *B. anthracis* toxin plasmid pXO1, and the capsule plasmid pXO2 have been transmitted to *B. anthracis* and *B. cereus* recipients lacking these plasmids (Green et al. 1989). The small plasmid pBC16 is transferred at high frequency without direct interaction with pXO12; such transfer of a nonconjunctive plasmid is called donation. The large *B. anthracis* plasmids are apparently transferred by conduction. The latter involves formation of cointegrative molecules in the donor, and resolution of the cointegrates into pXO12 and the respective *B. anthracis* plasmid in the recipient. Cell-to-cell contact is necessary for plasmid transfer and is resistant to DNase, but little is known about the mechanisms or conjugative structures that may be involved.

A strain of *B. subtilis* (natto) has been found which carries a 55-kb self-transmissible plasmid (pLS20), which can be transferred to closely related strains and to restriction-deficient strains of *B. subtilis* (Koehler and Thorne, 1987). This plasmid also promotes the transfer of the tetracycline-resistance plasmid pBC16 from *B. subtilis* (natto) to a wide variety of *Bacillus* species including *B. anthracis*, *B. megaterium*, and *B. subtilis*. This is a much broader range of conjugative transmission than has been observed with the *B. thuringiensis* plamid. However, none of the conjugative plasmids have been found to mobilize and transfer chromosomal markers as is observed with the F plasmid of *E. coli*.

In addition to the naturally occurring transmissible plasmids of *Bacillus*, Christie et al. (1987) have identified a conjugative transposon (Tn925) which transfers from *Streptococcus faecalis* to *B. subtilis*.

Bacteriophages

Bacteriophages that infect *Bacillus* are common in soil. In addition, many strains of this genus are naturally lysogenic for one or more prophages. The most extensively studied *Bacillus* phages are those associated with *B. subtilis*, and these have been reviewed by Hemphill and Whiteley (1975), Rutberg (1982), and Zahler (1988).

With some exceptions, *Bacillus* phages have relatively narrow host ranges, probably at least in part because of restriction systems that make phage grown in one host incompatible with another strain (Ando et al., 1982). With the exception of the phages of *B. subtilis*, no scheme of classification has been adopted to organize the phages of this genus. Therefore, the bacteriophages described here will be grouped according to life cycle.

Temperate Bacteriophages

Most strains of bacilli that have been carefully examined have been found to release phage particles. These are of two types: defective phages that can kill but do not productively infect other strains (see "Defective Bacteriophages," this chapter), and those that grow on and lysogenize new host bacteria. *B. subtilis* 168 is lysogenic for phage SPβ and also releases defective phage PBSX. As an extreme, *B. thuringiensis* subsp. *aizawai* is polylysogenic for five unique temperate phages (Reynolds et al., 1988). Temperate phages are easily obtained from nature. If samples of soil are placed in broth and the mixture heated 10 min at 80°C, most free phage and nonsporeforming bacteria are destroyed. When the culture is allowed to incubate several hours and subsequently treated with mitomycin C, temperate phages are induced and released into the medium. (Some investigators inoculate the broth with the *Bacillus* strain of interest to enrich for phage, and then add mitomycin C.) The phages may then be recovered by filtering the solution and plating on an appropriate indicator.

Dean et al. (1978) have divided the temperate phages of *B. subtilis* and closely related species into four groups based upon serology, immunity, and physical characteristics (Table 9); the defective phages may be considered a fifth class. Several group III phages including *B. subtilis* phages SPβ and φ3T and *B. amyloliquefaciens* phage H2 can mediate specialized transduction. Group I phage φ105 also transduces genes close to its prophage attachment site (Shapiro et al., 1974). In addition, φ105, SPβ, ρ11, and others have been used as cloning vehicles, mostly for *B. subtilis* genes (see Zahler 1988, for a review).

Temperate bacteriophages often alter the biochemistry or phenotype of lysogenic bacteria and several examples of such prophage conversion have been observed in *B. subtilis*. Strains of this bacterium lysogenic for SPβ release a

Table 9. Major groups of temperate *B. subtilis* phages.

Group	Example	DNA size (kb)	Virion dimensions (nm)		Other members of group
			Head	Tail	
I	φ105	40	52 × 52	10 × 220	ρ14
II	SPO2	40	50 × 50	10 × 180	—
III	SPβ	126	72 × 82	12 × 358	φ3T, ρ11, Z, SPR
IV	SP6	53	61 × 61	12 × 192	—
V	PBSX	13	45 × 45	20 × 200	PBSZ

Modified from Dean et al. (1978) and Zahler (1988).

bacteriocin-like substance called betacin (Hemphill et al., 1980) which kills some *Bacillus* strains that do not harbor the SPβ prophage. Most group III bacteriophages including Φ3T, ρ11 and Z (but not SPβ) contain the structural gene for thymidylate synthetase, and express this gene continuously in lysogens. Stains of *B. subtilis* lysogenic for SP02 cannot be productively infected with virulent phage φ1 (Rettenmeir and Hemphill, 1974), and bacilli lysogenic for SPβ are protected by a similar interference system active against φ1*m* (Rettenmeir et al., 1979).

Defective Bacteriophages

Many species of *Bacillus* including *B. subtilis, B. amyloliquefaciens, B. pumilus,* and *B. laterosporus* release defective phages whose presence is revealed by their bactericidal activity against other strains or species of this genus (Hemphill and Whiteley, 1975; Steensma et al., 1978; Zahler 1988.) For example, *B. subtilis* strain 168 releases a defective phage called PBSX, which kills the cells of *B. subtilis* strain W23. Electron microscopic examination of the culture supernatant of strain 168 reveals typical phage particles. However, the PBSX virions cannot replicate and produce plaques; rather, they act much like a bacteriocin. *B. subtilis* strain W23, in turn, releases defective phage PBSZ, which has a bacteriocin activity against strain 168.

Despite being defective, these phages act much like other temperature viruses. Although small numbers are typically present in the culture supernatant, lysogens may be induced with UV light or mitomycin C and lyse and release large numbers of particles. (The situation is often confusing, however, because many bacilli including *B. subtilis* 168 are simultaneously lysogenic for one or more nondefective temperate phages.) Virions of PBSX are unmistakably phage-like in structure. They have a rather small head (45 nm in diameter) and a cylindrical, contractile tail 18 by 200 nm. The genes for structural proteins of PBSX have been mapped to a chromosomal position between the markers

metA and *metC* (Thurm and Garro, 1975). PBSX and presumably other phages of this group are defective for more than one reason. First, the particles do not contain an identifiable phage DNA. Instead, PBSX packages randomly selected 13-kb fragments of bacterial chromosomal DNA (Anderson and Bott, 1985). Second, although PBSX virions attach efficiently to susceptible strains and the tail appears to contract, the DNA in the phage heads is not injected (Okamoto et al., 1968).

Lytic Bacteriophages

Lytic bacteriophages that infect *Bacillus* are also common in soil and water. Again, the most detailed studies are of phages that infect *B. subtilis* or closely related species (see Hemphill and Whiteley, 1975, for a review.) Although many of these viruses are intrinsically interesting, two groups of subtilis phages have been the focus of studies on the regulation of viral transcription (see Geiduschek and Ito, 1982). The first class, which includes SP01 and SP82, is distinguished by the replacement of thymine by hydroxymethyluracil (hmU) in the viral DNA. The second class includes φ29 and relatives, which are very small, linear double-stranded DNA phages.

SP01 and SP82 have very complex temporal programs of transcription involving at least three phases of gene expression designated early, middle, and late. Early gene expression starts within 1 min of infection, and is thought to use the host RNA polymerase. Middle genes are activated about 4 min after infection, and the expression of late genes (which are turned on asynchronously) is first detected 10 min into the latent period. Initiation of transcription of the middle and late genes involves structural modifications of the *B. subtilis* RNA polymerase, such that the enzyme can recognize unique middle and late phage promoters that are not used by the unmodified polymerase (Tarkington and Pero, 1979). The product of at least one SP01/SP82 gene is thought to associate with the enzyme to activate middle genes (Hyde et al.,

1986), and at least two other phage-encoded polypeptides appear to modify the polymerase to synthesize late mRNAs (Fox, 1976 Tjian and Pero, 1976). Other viral genes associated with late transcription have been identified, but these may be required to initiate DNA replication, which is also involved in transcriptional control. Even as new mRNAs are made, transcription of at least some early and middle genes is repressed at characteristic times in the latent period of SP82 and SP01. Proteins that bind to phage DNA and block RNA synthesis have been isolated from infected cells, but the actual mechanism of repression is not clear. Although not directly related, the study of gene expression in these subtilis phages has influenced the investigation of somewhat analogous changes in RNA polymerase structure associated with sporulation.

Transcription of the ϕ29 genome also involves early and late mRNAs, which in this case are transcribed from different strands of the viral DNA. Early RNA is transcribed from the light (L) and late RNA from the heavy (H) strand. In vitro, *B. subtilis* RNA polymerase synthesizes early mRNA from as many as eight promoters (Mellado et al., 1988). At least a portion of the host RNA polymerase is also used for synthesis of late RNA, which is apparently initiated from a single promoter called A3. The latter has a -35 region that differs somewhat from the consensus -35 region of *B. subtilis* promoters (Vlcek and Paces, 1986). Transcription from A3 requires protein P4 encoded by ϕ29 gene 4 (Mellado et al., 1988).

Pseudotemperate Phages

Pseudotemperate bacteriophages are lytic phages that establish a relatively long-term association with bacteria that mimics true lysogeny (see Hemphill and Whiteley, 1975). Following adsorption and penetration, further advancement of the viral development cycle is often delayed; some infected cells continue to grow, divide, and even sporulate. A portion of the daughter cells may even be cured, and about one-half the spores produced from such cultures do not contain viral DNA. Pseudotemperate bacteriophages characteristically produce turbid plaques, and "pseudolysogens" obtained from these plaques may be subcultured. The pseudolysogens apparently consist of a balance between infected cells in a delayed latent period, bacteria containing maturing phage, and uninfected bacilli.

Included among the pseudotemperate phages are several of the best-known transducing phages of *Bacillus* such as *B. subtilis* phages PBS1, SP10, and SP15 and *B. thuringiensis* phage TP-13. The genomes of these viruses often contain modified DNA. For instance, the chromosome of PBS1 has a complete replacement of thymine by uracil (Takahashi and Marmur, 1963) and SP15 DNA substitutes 5-(4'-5'-dihydroxylpentyl)-uracil for thymine (Brandon et al., 1972).

Phage PBS1 (also referred to as PBS2, a clear-plaque mutant) and phage PMB12 carry out their entire development cycle in the presence of rifampin, an antibiotic that inhibits RNA synthesis by interacting with the bacterial RNA polymerase (Rima and Takahashi, 1974; Bramucci et al., 1977). In the case of PBS1 (PBS2), the resistance to rifampin is explained in part by the synthesis of a new phage-induced RNA polymerase that transcribes the viral late functions (Clark et al., 1974). Genetic evidence suggests that early RNA synthesis in PBS1 may be resistant to the drug because the virus, perhaps using a virion protein, induces a modification in the bacterial RNA polymerase that converts the latter to a resistant form (Osburne and Sonenshein, 1980).

One of the most intriguing phenomena associated with bacteriophage PMB12 is its capacity to convert pseudolysogens of certain sporulation-negative *(spo)* mutants of *B. subtilis* to a Spo+ phenotype (Kinney and Bramucci, 1981). Conversion has been observed in several stage O mutants, as well as in *B. subtilis* variants deficient in sporulation due to changes in bacterial RNA polymerase or 30S ribosomes. At least three PMB12 genes are involved in spore conversion. The products of these genes apparently interact with host-cell pathways at the earliest stages of sporulation. Conversion of spore-negative mutants has also been observed with *B. pumilis* phage PMB1 (Bramucci et al., 1977) and numerous other phages which infect this species (Keggins et al., 1978). Bacteriophage TP-13 similarly converts oligosporogenic, a-crystalliferous mutants of *B. thuringiensis* to spore-positive, crystal-positive phenotype (Perlak et al., 1979). This phage also mediates generalized transduction of large DNA fragments.

Plasmids

Plasmids are of widespread occurrence in the genus *Bacillus* and have been found in most species that have been screened. The vast majority are cryptic plasmids, that is, their presence has not been correlated with any unique property of the bacterial host. Although efforts have been made to analyze the relationship between different plasmids in a few species by comparing DNA-restriction fragment profiles, there is no accepted scheme for systematically classifying plasmids within the bacilli. Further confusing the study of plasmids in *Bacillus* is the fact that some of those used as vectors for gene cloning, such as pUB110, were actually derived from *Staphylococcus aureus* (Gryczan et al., 1978).

As shown in Table 10, some of the *Bacillus* plasmids do confer a recognizable phenotype to cells which carry them, such as antibiotic resistance, synthesis of toxins, unique metabolic activities, and transconjugation. Because of the medical and economic importance of their hosts, the plasmids conferring virulence in *B. anthracis* and *B. thuringiensis* have been of increasing interest. Virulence in *B. anthracis* requires the production of a capsule composed of poly-D-glutamic acid and an exotoxin consisting of three components: protective antigen, lethal factor, and edema factor. The capacity to produce the capsule is associated with plasmid pXO2 (Green et al., 1985), and the constituents of the toxin are encoded in plasmid pXO1 (Mikesell et al., 1983). Strains losing either plasmid are rendered avirulent. With the help of fertility plasmid pXO12 derived from *B. thuringiensis,* both pXO1 and pXO2 have been transferred to bacilli lacking these plasmids (Green et al., 1989). In the case of pXO1, cured strains of *B. anthracis* have been restored to toxin production (Thorne, 1985). The association of the toxin with plasmids is of historical interest. The original Pasteur vaccine strains of *B. anthracis* were produced by subculturing this bacterium at high temperatures which are now known to inhibit replication of pXO1. Indeed, two existing Pasteur strains lack this plasmid (Mikesell and Vodkin, 1985).

The genes for delta-exotoxin crystal proteins of *B. thuringiensis* strains are carried in large plasmids (see Chapter 77). These plasmids are somewhat divergent in size and the crystal proteins produced by *B. thuringiensis* strains differ in serology and toxicity to various insects (Höfte and Whiteley, 1989). Moreover, several of these plasmids, such as pXO12 of *B. thuringiensis* subsp. *thuringiensis* 4042, (Reddy et al., 1987; Green et al., 1989) have been shown to mediate conjugal transfer of themselves and other plasmids (see "Genetic Studies" this chapter).

The production of the parasporal body containing the crystal protein toxin is coordinated with sporulation; crystal protein gene expression begins about stage II (see Whiteley and Schnepf, 1986, for a review). Changes in transcription during sporulation are associated with modifications of RNA polymerase which enable the enzyme to recognize sporulation-specific promoters (see "Sporulation", this chapter). It has been suggested the activation of the crystal protein genes is also regulated by polymerase changes. DNA sequence analysis of the 133-kDa crystal protein gene from *B. thuringiensis* subsp. *kurstaki* HD-1-Dipel revealed two overlapping promoters, one used early in sporulation (BtI) and the other (BtII) activated midway through development (Wong et al., 1983). Comparison of these promoters with those of the carefully studied *B. subtilis* system show little sequence consensus with promoters that bind

Table 10. *Bacillus* plasmids.

Bacterium	Plasmid	DNA size (kb)	Phenotype associated with plasmid	Reference
B. anthracis	pXO1	168	Exotoxin (lethal factor, edema factor, protective antigen)	Mikesell et al., 1983 Tippetts and Robertson, 1988
	pXO2	85.6	Capsule	Green et al., 1985
B. cereus	pBC7	69	Bacteriocin	Bernhand et al., 1978
	pBC16	4.3	Tetracycline resistance	Bernhand et al., 1978
B. pumilus	pBL10	6.8	Bacteriocin	Lovett et al., 1976
B. subtilis	pIM13	2.2	Erythromycin resistance	Mahler and Halvorson, 1980
B subtilis (natto)	pLS19	5.4	Polyglutamate production	Hara et al., 1982
	pLS20	55	Self-transmissible plasmid, which also promotes transfer of other plasmids	Koehler and Thorne, 1987
B. thuringiensis	pXO12	112.5	Production of insecticidal crystal protein, and is a self-transmissible plasmid, which can co-transfer unrelated plasmids	Green et al., 1989
Bacillus species (thermophilic)	pTB19	26	Kanomycin and tetracycline resistance	Imanaka et al., 1981
	pTB20	4.3	Tetracycline resistance	Imanaka et al., 1981

vegetative RNA polymerase. Brown and Whiteley (1988) have isolated an RNA polymerase from strain HD-1-Dipel that directs transcription in vitro from the BtI promoter and also transcribes crystal protein genes from two other strains of *B. thuringiensis*. This polymerase contains a unique 35-kDa sigma subunit that is different from the major sigma subunit present in vegetative cells of *B. thuringiensis*. In addition, the presence of two regions of hyphenated dyad symmetry near the BtI and BtII promoters suggests that binding of other regulatory proteins in this region could be associated with gene expression.

Transcriptional Regulation in Biosynthetic and Catabolic Pathways

Bacillus species demonstrate great catabolic and biosynthetic versatility. It is only recently, however, that significant progress has been made in understanding how synthesis of enzymes is regulated. Again, most studies have concentrated on *B. subtilis* and closely related species.

Control of Biosynthetic Gene Expression

B. subtilis can be grown in minimal salts media with glucose as carbon source. Thus, this bacterium must have the capacity to synthesize de novo all amino acids, nucleotides, etc. As in the enteric bacteria, the structural genes for enzymes of common biosynthetic pathways are clustered on the *B. subtilis* chromosome (see Zalkin and Ebbole, 1988, for a review.) Such operon-like organization has been observed for the genes of the tryptophan, arginine, and iso-leucine-valine-leucine synthesis. In addition, the sequences encoding the enzymes for purine (*pur* genes) and pyrimidine (*pyr* genes) occur in multicistron clusters in *B. subtilis* (Ebbole and Zalkin, 1987; Lerner et al., 1987), which is not the case in *E. coli*. Gene clustering is associated with coordinate control of gene expression at the level of transcription by the products of the pathways. The *trp, arg,* and *ilv* operons are repressed, respectively, by tryptophan, arginine, and leucine. The *pur* genes are repressed by adenine and guanine nucleotides and *pyr* by pyrimidines.

It does not appear that regulation of the biosynthetic operons is mediated primarily through repressors, although the *arg* pathway may be an exception (Smith et al., 1986). Rather, the favored means of control appears to be termination/antitermination (attenuation)

mechanisms which affect the elongation of nascent mRNA molecules. In several operons studied, the 5'-end of the mRNA contains a leader of 100 to 300 nucleotides which does not encode part of the first structural gene. Within this leader is a region that can fold into two mutually exclusive secondary structures, one of which is a procaryote transcription termination signal. The other configuration is an antiterminator stem-and-loop (hairpin) which allows full-length message to be made. Control then resides in changing the ratio of termination vs. antitermination secondary structure.

In enteric bacteria, the leader sequence of amino acid biosynthetic gene clusters includes an open reading frame (ORF) for a short peptide, which is generally enriched in codons for the amino acid synthesized by that pathway. The formation of termination vs. antitermination hairpins is modulated by the rate at which the leader polypeptide is synthesized. In the *E. coli trp* operon, for example, the peptide is made slowly if intracellular levels of tryptophan are low, and the antitermination structure is generated. The converse is true if tryptophan levels are high. In *B. subtilis* those leader sequences that have been analyzed do not contain an ORF. Moreover, transcription of *pur* and *pyr* gene clusters is also thought to be controlled by termination/antitermination, and it is unlikely these operons could be regulated by a mechanism precisely analogous to the *E. coli* system.

In the *B. subtilis trp* operon, attenuation is regulated in part by a *trans*-acting factor thought to be the product of the methyltryptophan resistance locus *mtr* (Shimotsu et al., 1986). Mutations in this gene result in constitutive expression of the *trp* genes. It is likely that the *mtr* product, when activated by tryptophan, binds to the leader region of the nascent mRNA in such a manner as to prevent formation of the antiterminator structure, thus favoring termination. In addition, binding of tryptophan-activated *mtr* gene product to already completed mRNA molecules is thought to inhibit the initiation of translation, thereby further reducing the production of *trp* enzymes (Kuroda et al., 1988).

Transcriptional regulation of the *pur* operon of *B. subtilis* shows some similarities to that of the *trp* system, but may be even more complicated. Synthesis of inosine monophosphate (IMP) is repressed by both adenine and guanine nucleotides, but apparently by different mechanisms. The latter are thought to promote premature cessation of transcription at a rho-independent termination site in the leader sequence, while adenine nucleotides appear to repress transcription initiation (Zalkin and Ebbole, 1988). It is thought likely that when purine levels are high, a guanine nucleotide activates a regulatory molecule, which in turn binds to the leader region of the nascent mRNA and blocks formation of the antiterminator secondary structure.

The products of polycistronic mRNAs are sometimes subunits in a common enzyme complex. In these instances, the polypeptides are required in 1:1 stoichiometry. Sequence analysis of the *trp* and *pur* operons shows that both are characterized by many gene overlaps; that is, the 3'-end of one coding sequence overlaps with the 5'-end of a contiguous downstream coding sequence (Henner et al., 1984; Ebbole and Zalkin, 1987). The overlaps are thought to result in "translational coupling" in which synthesis of one polypeptide from a polycistronic mRNA is at least partially dependent on the translation of the contiguous upstream gene. In some instances, this coupled translation is thought to result in 1:1 stoichiometric synthesis of the products of the overlapping genes.

Control of Catabolic Gene Expression

Members of the family Bacillaceae catabolize a wide variety of simple and complex organic compounds, including mono-and disaccharides, and polysaccharides such as starch which are partially digested by extracellular enzymes. As in the enteric bacteria, the structural genes for enzymes involved in degradation of these substrates are often clustered on the *Bacillus* chromosome. All operons so far studied are transcribed from σ^A promoters, and control of gene expression involves repressors and termination/antitermination systems (see Klier and Rapport, 1988, for a review).

The pathway involved in gluconate catabolism in *B. subtilis* is apparently regulated by negative control mediated by a repressor. Two inducible enzymes are required: the transport protein gluconate permease and gluconate kinase which phosphorylates gluconate. The genes for these functions, *gntP* and *gntK*, respectively, are part of a four cistron operon *gntR-gntK-gntP-gntZ*. Mutations that inactivate *gntR* are constitutive. This gene encodes a 243 amino acid protein thought to inhibit transcription by binding near the promoter located 40 bp upstream from *gntR* (Fujita and Fujita, 1987). Curiously, there is a five base overlap between the coding sequence of *gntR* and *gntK* suggesting there may be a translational coupling between these two proteins (Fujita et al., 1986). The genes required for the growth of *B. subtilis* on xylose polymers are also organized in a five cistron regulon, in which one of the genes, *xylR*,

codes for a repressor that regulates expression of the entire gene cluster (Hastrup, 1988). Synthesis of the enzymes is induced by the presence of xylose in the medium. A repressor system has also been found to regulate the induction of penicillinase in *B. licheniformis* (Wittman and Wong, 1988).

Perhaps the most extensively studied catabolic system in the bacilli is that involved in sucrose catabolism. *B. subtilis* produces two β,D-fructofuranosidases, sucrase and levansucrase, after induction by sucrose. The former is an intracellular enzyme while the latter (which also catalyzes the formation of the fructose polymer levan) is secreted (see Klier and Rapoport, 1988). The structural genes for sucrase, *sacA,* and that for levansucrase, *sacB,* are widely separated on the *B. subtilis* chromosome. The *sacA* gene is tightly linked to *sacP*, which is thought to be a membrane-associated protein that transports and phosphorylates sucrose. The *sacB* gene and the *sacA-sacP* cistrons are linked to regulatory loci, *sacR* and *sacT*, respectively. Mutations in either of the regulatory loci results in constitutive synthesis of the corresponding enzyme. The most extensive studies of the mechanisms of control have been made in *sacR*.

The *sacR* locus contains the *sacB* promoter. Between the transcription start site and the *sacB* coding sequence is a 199-bp region that is the target of several regulatory effectors of expression of the levansucrase gene (Aymerich et al., 1986). Part of this control is thought to involve a region of dyad symmetry that could form a transcription termination signal in the nascent mRNA. Deletion of this perspective terminator results in constitutive synthesis of levansucrase, and previously isolated constitutive *sacR* mutants map to this putative stem-and-loop structure (Shimotsu and Henner, 1986; Steinmetz and Aymerich, 1986). Messenger RNA synthesis is thought to stop at this terminator unless an antiterminator protein, presumably regulated by sucrose, interacts with the nascent RNA to prevent formation or to allow bypass of the loop. One candidate for the antiterminator maps to *sacS,* a genetic locus that affects both the synthesis of levansucrase and sucrase. The *sacS* locus consists of two cistrons, *sacY* and *sacX* (Aymerich and Steinmetz, 1987; Steinmetz et al., 1988; Zukowski et al., 1988). The *sacY* gene is thought to encode a sucrose-dependent antiterminator protein whose target probably is the region of secondary structure in *sacR*, where it presumably acts to prevent formation of the stem-and-loop. Deletion of *sacY* abolishes levansucrase synthesis. Current data suggest the *sacX* gene directs the production of a negative regulator of *sacY;* mutants in this cistron produce levansucrase constitutively.

The products of two other regulatory genes, *sacU* and *sacQ,* have been implicated in regulation of levansucrase synthesis. The target of the products of these genes also appears to be the *sacR* region, and they may serve to modulate the levels of gene expression (Shimotsu and Henner, 1986; Zukowski and Miller, 1986). *sacQ* encodes a 46 amino acid polypeptide which appears to stimulate the transcription of several secreted enzymes. Some mutations in *sacQ* lead to higher levels of expression of target loci, and genes thought to be similar to *sacQ* have been cloned from *B. licheniformis* and *B. amyloliquefaciens.* It now appears that the *sacQ* polypeptide is one of many small polypeptides which somehow regulate the transcription of degradative enzymes in the family Bacillaceae. Others may include the products of the *sacV, sin,* and *ptrR* genes (see Klier and Rapoport, 1988).

The enzymes for the histidase pathway responsible for the degradation of histidine are encoded in a cluster of four genes *hutH-hutU-hutI-hutG* in *B. subtilis.* Expression of this *hut* operon is induced by histidine and is also subject to catabolite repression. Between the σ^A promoter and the first gene, *hutH,* is a regulatory locus, *hutR.* The latter has been cloned and sequenced and found to contain two regions of possible regulatory significance (Oda et al., 1988). One is an ORF (ORF1) for a 151 amino acid protein that acts *trans* in heterozygotes to regulate the *hut* operon. A mutant having an amino-acid substitution in this protein has been isolated and found to be uninducible for the histidase pathway enzymes. This suggests that the ORF1 protein is a positive regulator in expression of the *hut* operon, although the possibility that the mutant is a superrepressor cannot be entirely excluded. The second regulatory region located between ORF1 and the first structural gene *hutH* (ORF2) is a region of dyad symmetry which could form a rho-independent terminator. It is possible that the product of ORF1, presumably activated by histidine, acts as an antiterminator to allow complete transcription of the *hut* operon.

The picture emerging from these and other studies suggests that regulation of catabolic pathways in *B. subtilis* involves a combination of repressors and positive controls through antitermination. It should be noted that in the attenuation systems observed in the biosynthetic pathways the leader sequences appear to be able to form two mutually exclusive secondary structures, one of which is the terminator. In the several leader sequences studied in catabolic pathways, only the terminator stem-and-loop is indicated. Thus, the putative antiterminator proteins in catabolic systems probably act by blocking formation of or allowing bypass of the terminator, rather than acting to favor one secondary structure over the other.

Catabolite Repression

Catabolic pathways in *Bacillus* are subject to catabolite repression by high levels of glucose and a variety of other rapidly metabolized substrates (Nihashi and Fujita, 1984). This phenomenon is of special interest in bacteria of this genus because sporulation is also repressed by glucose. In the enteric bacteria, catabolite repression is mediated in part by the levels of cAMP and the interaction of this nucleotide with catabolite activator protein (CAP). However, cAMP apparently is not present in *Bacillus,* and thus could not be involved in catabolite repression in these organisms.

The promoters and associated control regions of many catabolic operons from *Bacillus* have been cloned and sequenced (Laoide et al., 1989; Melin et al., 1987; Oda et al., 1988). In some instances, these cloned regulatory systems, which for experimental purposes are sometimes fused to an indicator gene such as *E. coli lacZ,* continue to be subject to glucose suppression of transcription. Among the most thoroughly studied is the complex *amyR1-amyE,* the regulatory and structural genes for alpha-amylase of *B. subtilis,* and *amyR1-amyL,* the corresponding regulatory region and structural gene for alpha-amylase of *B. licheniformis.* Amylase synthesis is actually under two forms of regulation. One is a temporal control manifested by the fact that amylase genes are not fully induced until the onset of stationary phase. Temporal control may not occur in all *Bacillus* strains (Rothstein et al., 1986). In addition, in both *B. subtilis* and *B. licheniformis FDO2* production of alpha-amylase is repressed about 10-fold in glucose-containing cultures in early stationary phase, even when multiple copies of the complex are present on a plasmid (Laoide et al., 1989). Both forms of regulation are associated with the *amyR* regions. A plasmid-borne construct of *amyL* missing the promoter but including all transcribed sequences immediately adjacent and downstream from the promoter has been produced. When this fragment is attached at various distances from heterologous promoters, transcription continues to be subject to catabolite repression. This suggests the target of repression is not in the promoter itself or in regions upstream from it. One possibility is that a regulatory protein involved in mediating catabolite repression binds to a *cis*-acting site very close to the transcription start site, which in the case of *amyL* is only 29 to 31 nucleotides from the translation initiation codon (121 nucleotides for *amyE*). A search for consensus sequences downstream from several promoters of *B. subtilis* genes subject to catabolite repression has yielded a candidate sequence: 5′-ATTGTNA-3′ (Laoide et al., 1989). In the regulatory loci for *amyL* and *amyE*, this sequence is contained within an inverted repeat sequence that overlaps the transcription start site, and in the case of *amyL*, the translation initiation codon. A point mutation *(gra-10)* in this region of dyad symmetry relieves catabolite repression of *amyE* (Nicholson et al., 1987). In addition, Weickert and Chambliss (1989) have reported that deletion of DNA 3′ to the *amyR1* promoter does not impair temporal activation of chloramphenicol acetyltransferase in *amyR1-cat86* transcriptional fusions, but abolishes catabolite repression.

Mutants of *B. subtilis* that are resistant to at least some manifestations of catabolite repression have been isolated. Among mutants originally selected for resistance to glucose-mediated repression of sporulation (Takahashi, 1979) is *crysA43*, which maps to rpoD, the structural gene for σ^A (Price and Doi, 1985). The alteration in σ^A in this variant also relieves catabolite repression in an *amyRi=lacZ* fusion cloned in *B. subtilis* (Laoide and McConnell, 1989). This may suggest a connection between the mechanism of glucose repression affecting sporulation and catabolite repression of at least those genes activated in stationary phase. In addition, Sun and Takahashi (1984) localized another catabolite-resistant mutation, *crysE1*, to the *rpo* operon. This locus encodes the β and β' subunits of RNA polymerase. Taken together the two types of mutations occurring in subunits of RNA polymerase may suggest this enzyme is directly involved in catabolite repression or interacts with a protein, perhaps bound to the regulatory region, which mediates this phenomenon.

Resistance of Spores

The resting forms of bacteria are usually more resistant to various environmental stresses than their counterpart vegetative forms. The structure and composition of the resting (and dormant) form of endosporeforming bacteria, however, are quite different from other bacterial-resting forms (see Fig. 3). The core or protoplast containing the heat labile DNA, RNA, ribosomes, enzymes, and other proteins is surrounded by a primitive "germ cell wall." Moving toward the surface, there is a layer called the cortex which consists of peptidoglycan of a similar nature to that of the vegetative cell wall but with less cross linking in the peptides among other differences (Warth, 1978). A second cell membrane surrounds the cortex. The protoplast and cortex and their membranes are enclosed by layers of protein coat. A loose-fitting exosporium, appendages, and internal protein crystals may be found in some species.

Compared with vegetative cells, spores are more resistant to heat by a factor of 10^5 or more, to UV and ionizing radiation by 100-fold or more, and to desiccation, antibiotics, disinfectants, and other chemicals (reviewed by Roberts and Hitchins, 1969; Russell, 1982; Gould and Dring, 1974; Gould, 1983). Since spore resistance has been recognized for a long time and is important in food processing and sterilization considerations, it has been extensively studied particularly with the genus *Bacillus*. Spore resistance to physical and chemical agents other than heat and irradiation indicated above is also noteworthy. Vegetative cells are killed at 88,000 p.s.i. for 14 h hydrostatic pressure while their spores have been shown to require 176,000 p.s.i. for the same time period (Gould and Dring, 1974). Spores are about 10,000 times more resistant to hyperchlorites than are vegetative cells. Two hundred roentgens (r) $\times$ 10^3 of x-rays were required to kill 50% of treated *B. megaterium* spores, while 50% of treated *E. coli* cells were killed at 5.6×10^3 r. In every process designed for killing microorganisms, the spores are more resistant than vegetative cells and with few exceptions are the most resistant biological entities known. One exception is the nonsporeformer *Deinococcus radiodurans*, the most radiation-resistant organism yet discovered. Its resistance is presumably due to the possession of efficient repair mechanisms for radiation-induced changes in DNA.

The degree of heat resistance has been shown to depend not only on the species but also on the physiological environment in which the spores were formed. *B. stearothermophilus* spores are more heat resistant than *B. subtilis* spores which are more resistant than spores of *B. megaterium* (Roberts and Hitchins, 1969; Khoury et al., 1987). In addition, the spore resistance of each species depends on the temperature at which it was grown. For example, *B. subtilis* grown and sporulated at 20, 30, and 45°C produced spores having D_{90} values (the time required to kill 90% of the spores at 90°C) of 37, 78, and 99 min, respectively (Khoury et al., 1987). The spores were more temperature sensitive when formed in ethanol-supplemented media. Since temperature and ethanol are known to perturb the degree of order within membranes, this suggests that alteration of membrane function is an additional factor in the multifactorial nature of heat resistance. Other factors to consider in explaining heat resistance include protection of essential spore macromolecules (Murrell, 1981; Lindsay et al., 1985; Gerhardt and Marquis, 1989); specific effects of calcium dipicolinate (Lindsay and Murrell, 1986); mineralization (Marquis, 1989), and possibly foremost, dehydration (Gerhardt and Marquis, 1989). Current hypotheses on the heat resistance of endospores center on the dehydration of the protoplast (core) and the expandable cortex with its counterions (Gould, 1983). In the heat-resistant form, the spore coat is relatively impermeable to multivalent cations. The cortex (of high water content) contains an expanded electronegative peptidoglycan and mobile counterions exerting high osmotic pressure. The protoplast, of low water content, is osmotically dehydrated by the surrounding cortex and is, therefore, heat resistant. In the heat-sensitive form, there is a modified coat leaky to multivalent cations. The neutralized cortex, collapsed and free of counterions, exerts low osmotic pressure. The protoplast becomes partly hydrated and, therefore, heat sensitive. The theory of heat resistance, called the osmoregulatory expanded cortex theory, fits all the known facts but has yet to be proven or disproven. Germinated spores lose their heat resistance and yet, under special conditions, can be dehydrated to become both heat resistant and dormant once again (Gould, 1983). It is noteworthy that the heat resistance of some nonsporeformers can be increased in dehydrated cells. The expansion or contraction of the cortex is thought to account for the dehydrated state of the spore protoplast.

This reduced water content may also play a role in radiation resistance. It is thought that conformational differences between DNA in spore and vegetative cells may be associated with differences in hydration levels. The greater resistance of spores to UV radiation is also related to repair processes. The photoproduct formed during sporulation, 5-thyminyl-5,6-dehydrothymine (TDHT), is different from the thymine dimers formed in vegetative cells, both types of photoproducts being deemed as the cause of death. Dark repair mechanisms in germinating spores convert TDHT to thymine. The greater spore resistance to UV radiation is attributed to the more efficient removal of the unusual photoproduct (reviewed by Russell, 1982). However, as with the mechanism of spore heat resistance, the picture is complex and many factors are involved. For example, dipicolinic acid has been suggested as a protectant of spore DNA (Lindsay and Murrell, 1986) and it has been shown that the SASP may be involved (Setlow, 1988).

Likewise, explanations for ionizing radiation resistance of spores are not yet complete (Russell, 1982). Different spore DNA conformation (possibly due to dehydration), possession of coat layer radioprotectant substances, dipicolinic acid protection, and more efficient repair systems have all been implicated.

Spore coat layers, both coat protein and cortex, act as permeability barriers to toxic agents. At the present, this is the accepted explanation for the greater spore chemical resistance and resistance to lytic agents (Russell, 1982).

Spores as Biological Indicators

Endospores are often used as biological indicators. Known numbers of spores of various *Bacillus* species of predictable death rate can be placed on various solid substrates (usually strips of filter paper) and placed with the items to be sterilized. The strips are then checked for retention of viable organisms by immersion into culture medium (various strategies are employed) after the sterilization process is completed if growth occurs then survival of the spores has occurred and hence the procedure for processing was inadequate. A D-value (decimal reduction time or time to kill 90% of the population at a particular temperature or treatment) of 3.0 min for *B. subtilis* spores is required for ethylene oxide sterilization (600 mg ethylene oxide per liter at 50% relative humidity and 54°C). For moist heat (121°C) *B. stearothermophilus* spores with a D-value of 1.5 min and for dry heat (170°C) spores of *B. subtilis* with a D-value of 0.8 min are preferred. *B. pumilus* spores (D-value of 0.17 Mrad) are used when ionizing radiation is used for sterilization (Korczynski, 1981).

Production of Antibiotics

Members of the genus *Bacillus* are capable of producing antibiotics as secondary metabolites in the late logarithmic or early stationary phase of growth of batch cultures. As many as 169 of these secondary metabolites have been recorded; for example, various strains of *B. subtilis* have been shown to produce 68 antibiotics while *B. brevis* can produce 23 (Katz and Demain, 1977). A partial updated listing modified from that of Katz and Demain (1977) is presented (Table 11) to illustrate that there are other *Bacillus* antibiotic producers. Most of the antibiotics are active against Gram-positive organisms, although there are exceptions. The majority are peptide antibiotics but some belong to other chemical classes (e.g., butirosin is an aminoglycoside and protocin is a phosphorus-containing triene). Also indicated in Table 11 are those antibiotics whose structural genes have been mapped and the few whose genes have been cloned. The latter types of studies are important in their own right and serve as models for the study of expression of the genes of other antibiotics.

A controversy has existed for some time concerning the function of these antibiotics. Since they usually appear upon the onset of sporulation, it has been proposed that they may be important factors in the transition of vegetative cells to spores (reviewed by Katz and Demain, 1977). Support for this hypothesis that antibiotic production and sporulation may be regulated by the same or similar control mechanisms is now available. Transcription of the *B. subtilis tycA* gene (coding for tyrocidin synthe-

sis) is dependent on the products of certain Stage 0 sporulation regulatory genes (Marahiel et al., 1987). Thus it does appear that the two physiological events are partially coupled to regulatory events occurring at the onset of sporulation.

In addition, it has recently been demonstrated that gramicidin S functions as an inhibitor of outgrowth after germination (Daher et al., 1985).

Protein Secretion

Members of the genus *Bacillus* are able to secrete a wide variety of enzymes into the culture medium (reviewed by Priest, 1977; Mezes and Lampen, 1985). Every *Bacillus* species which has been checked produces at least one extracellular enzyme. These include many different carbohydrates, several kinds of proteases, penicillinases, nucleases, phosphatases, lipase, phospholipase C, thiaminase, and bacteriolytic enzymes. A vast literature exists on the use of *Bacillus* enzyme models for studying secretion mechanisms, cellular location, and regulation. For some time, there has been considerable in-

Table 11. Some *Bacillus* antibiotics.

Species	Antibiotic	Genes mapped	Genes cloned
B. subtilis	Subtilin[a]	+	+
	Surfactin[b]	+	−
	Bacilysin[a]	+	−
	Difficidin[c]	−	−
	Oxydifficidin[c]	−	−
	Bacillomycin F[d]	−	−
	Mycobacillin[d]	−	−
B. brevis	Gramicidin S[a]	+	+
	Linear Gramicidin[a]	−	−
	Tyrocidin[a]	+	+
B. licheniformis	Bacitracin[a]	+	+
	Proticin	−	−
B. pumilus	Pumilin[a]		
	Tetain[a]	Remaining antibiotics have neither been	
B. mesentericus	Esperin[a]	mapped nor cloned.	
B. polymyxa	Polymyxin[e]		
	Colistin[e]		
B. thiaminolyticus	Octopytin[a]		
	Baciphelacin[a]		
B. circulans	Circulin[e]		
	Butirosin[a]		
B. laterosporus	Laterospuramine[a]		
	Laterosporin[a]		
B. cereus	Biocerin[a]		
	Cerexin[a]		

[a]Anti-Gram-positive bacteria.
[b]Inhibitor of fibrin clotting.
[c]Broad spectrum antibiotic.
[d]Anti-fungal antibiotic.
[e]Anti-Gram-negative bacteria.

terest in producing large quantities of enzymes for industrial purposes—proteases for detergent supplementation, the brewing industry, various uses in the food industry and in leather manufacturing; and different amylases for brewery use, in bread making, and in the paper industry (see review by Debabov, 1982). Since many prokaryotic and eukaryotic genes can be fused to *B. subtilis*-derived regulatory regions and signal peptide sequences, such genetically manipulated organisms can be used for expressing and secreting many different heterologous proteins (for molecular cloning in *B. subtilis*, see Gryczan, 1982; Ganeson et al., 1982; Mezes and Lampen, 1985; Ganeson and Hoch, 1988). Interferon (Palva et al., 1983; Schien et al., 1986); human growth hormone (Honjo et al., 1986); and human interleukin-1 (Motley and Graham, 1988) are three such examples. However, yields of such proteins can be low, because extracellular neutral protease, subtilisin, esterases, and other proteases may degrade the secreted proteins. One strategy to overcome the problem and stabilize proteins has been to use strong glucose-insensitive promoters, protease deficient mutants, and catabolite repression of sporulation (Wong et al., 1986).

The widespread interest in protein secretion by *B. subtilis* has stimulated many studies on the genetics of secretion. There appears to be a general mechanism for regulating the synthesis of extracellular proteins. The activity of degradative enzymes can be increased by mutations at a number of loci that are unlinked to the structural genes for the affected enzymes. One family of genes, *senN*, *sacU*, *prtR* and *hpr*, codes for small regulatory proteins. For example, mutations at the *sacU* and *sacQ* loci can increase the expression of levan sucrase, alkaline protease, neutral protease, xylanase, beta-gluconase, alpha-amylase, and intracellular serine protease (Henner et al., 1988a). This type of stimulation of degradative enzymes is thought to be a global regulatory system turned on by a requirement for other carbon or nitrogen sources. The *SacU* product shares homology with other two-component sensor-regulator systems (Ronson et al., 1987; Henner et al., 1988; Kunst et al., 1988).

Other Considerations

Just as *Bacillus* species have been important as models for studying differentiation or secretion, they have been employed extensively to study other important biological problems. These include DNA replication (Winston and Sueoka, 1982) and repair (Yasbin, 1985); chemotaxis (Ordal and Nettleton, 1985); genetic transformation (Dubnau, 1982); and translation apparatus (Smith, 1982; Hager and Rabinowitz, 1985).

Literature Cited

Anagnostopoulos, C., and J. Spizizen. 1961. Requirements for transformation in *Bacillus subtilis*. J. Bacteriol. 81:741–746.

Anderson, L. M., and K. F. Bott. 1985. DNA packaging by the *Bacillus subtilis* defective bacteriophage PBSX. J. Virol. 54:773–780.

Ando, T., E. Hayase, S. Ikawa, and T. Shibata. 1982. Site-specific restriction endodeoxyribonucleases in Bacilli. p. 66–70. In: D. Schlessinger (ed.), Microbiology-1982. American Society of Microbiology, Washington, D.C.

Aoki, H, and R. A. Slepecky. 1973. Inducement of a heat-shock requirement for germination and production of increased heat resistance in *Bacilla fastidiosus* spores by manganous ins. J. Bacterial 114:137–143.

Aronson, A. I. and P. C. Fitz-James. 1976. Structure and morphogenesis of the bacterial spore coat. Bacteriol. Rev. 40:360–402.

Aymerich, S., G. Gonzy-Treboul, and M. Steinmetz. 1986. 5'-Noncoding region *sacR* is the target of all identified regulation affecting the levansucrase gene in *Bacillus subtilis*. J. Bacteriol. 166:993–998.

Aymerich, S. and M. Steinmetz. 1987. Cloning and preliminary characterization of the *sacS* locus from *Bacillus subtilis*, which controls the regulation of the exoenzyme levansucrase. Mol. Gen. Genet. 208:114–120.

Barsomian, G. D., N. J. Robillard, and C. B. Thorne. 1984. Chromosomal mapping of *Bacillus thuringiensis* by transduction. J. Bacteriol. 157:746–750.

Bartlett, A. T. M. and P. J. White. 1985. Species of *Bacillus* that make a vegetative peptidoglycan containing lysine lack diaminopimelate epimerase but have diaminopimelate dehydrogenase. J. Gen. Microbiol. 131:2145–2152.

Battisti, L., B. D. Green, and C. B. Thorne. 1985. Mating system for transfer of plasmids among *Bacillus anthracis*, *Bacillus cereus*, and *Bacillus thuringiensis*. J. Bacteriol. 162:543–550.

Berkeley, R. C. W. and M. Goodfellow (ed.). 1981. The Aerobic endosporeforming bacteria: classification and identification. Academic Press, London

Bernhard, K., H. Schrempf and W. Goebel. 1978. Bacteriocin and antibiotic resistance plasmids in *Bacillus cereus* and *Bacillus subtilis*. J. Bacteriol. 133:897–903.

Bott, K. F. and G. A. Wilson. 1967. Development of competence in the *Bacillus subtilis* transformation system. J. Bacteriol. 94:562–570.

Bowditch, R. D., P. Baumann, A. A. Yousten. 1989. Cloning and sequencing of the gene encoding a 125-Kilodalton surface-layer protein from *Bacillus sphaericus* 2362 and of a related cryptic gene. J. Bacteriol. 171:4178–4188.

Bramucci, M. G., K. Keggins, and P. S. Lovett. 1977. Bacteriophage conversion of spore-negative mutants to spore-positive in *Bacillus pumilis*. J. Virol. 22:194–202.

Brandon, C., P. M. Gallop, J. Marmur, H. Hayashi, and N. Nakanishi. 1972. Structure of a new pyrimidine from

Bacillus subtilis phage SP-15 nucleic acid. Nature New Biol 239:70–71.

Briehl, M. and N. H. Mendelson. 1987. Helix hand fidelity in *Bacillus subtilis* macrofibers after spheroplast regeneration. J. Bacteriol. 169:5838–5840.

Brown, K. L. and H. R. Whiteley. 1988. Isolation of a *Bacillus thuringiensis* RNA polymerase capable of transcribing crystal protein genes. Proc. Natl. Acad. Sci. USA. 85:4166–4170.

Buchanan, C. E. 1987. Absence of penicillin-binding protein 4 from an apparently normal strain of *Bacillus subtilis*. J. Bacteriol. 169:5301–5303.

Canosi, V., G. Luber, and T. A. Trautner. 1982. SPP1-mediated plasmid transduction. J. Virol. 44:431–436.

Cassity, R. R. and B. J. Kolodziej. 1984. Isolation partial characterization and utilization of a polysaccharide from *Bacillus megaterum* ATCC 19213. J. Gen. Microbiol. 130:535–539.

Christie, P. J., R. Z. Korman, S. A. Zahler, J. C. Adsit, and G. M. Dunny. 1987. Two conjugation systems associated with *Streptococcus faecalis* plasmid pCF10: identification of a conjugative transposon that transfers between *S. faecalis* and *B. subtilis*. J. Bacteriol. 169:2529–2536.

Clark, S., R. Losick, and J. Pero. 1974. New RNA polymerase from *Bacillus subtilis* infected with phage PBS2. Nature 252:21–24.

Claus, D. and R. C. W. Berkeley. 1986. The genus *Bacillus*, p. 1105–1139. In: P. H. A. Sneath (ed.), Bergey's manual of systematic bacteriology, vol. 2. Williams and Wilkins, Baltimore.

Clausen, V., J. G. Jones, and E. Stackebrandt. 1985. 16S ribosomal RNA analysis of *Filibacter limicola* indicates a close relationship to the genus *Bacillus*. J. Gen. Microbiol. 131:2659–2663.

Cowan, S. T. and K. J. Steel. 1974. Manual for the identification of medical bacteria, 2nd ed. Cambridge University Press, London.

Daher, E., E. Rosenberg, and A. L. Demain. 1985. Germination-initiated spores of *Bacillus brevis* nagano retain their resistance properties. J. Bacteriol. 161:47–50.

Davidoff-Abelson, R. and D. Dubnau. 1973a. Conditions affecting the isolation from transformed cells of *Bacillus subtilis* of high-molecular-weight single-stranded deoxyribonucleic acid of donor origin. J. Bacteriol. 116:146–153.

Davidoff-Abelson, R. and D. Dubnau. 1973b. Kinetic analysis of the products of donor deoxyribonucleate in transformed cells of *Bacillus subtilis*. J. Bacteriol. 116:154–162.

Dean, D. H., C. L. Fort and, J. A. Hoch. 1978. Characterization of temperate phages of *Bacillus subtilis*. Curr. Microbiol. 1:213–217.

Debabov, V. G. 1982. The industrial use of Bacilli, p. 331–370. In: D. A. Dubnau (ed.), The molecular biology of the bacilli. Academic Press, New York.

Diechelbohrer, I., J. C. Alonso, G. Luder, and T. Trautner. 1985. Plasmid transduction by *Bacillus subtilis* bacteriophage SPP1: Effects of DNA homology between plasmid and bacteriophage. J. Bacteriol. 162:1238–1243.

Diechelbohrer, I., W. Messer, and T. A. Trautner. 1982. Genome of *Bacillus subtilis* bacteriophage SPP1: structure and nucleotide sequence of *pac*, the origin of DNA packaging. J. Virol. 42:83–90

Dodson, L. A. and C. T. Hadden. 1980. Capacity for post-replication repair correlated with transducibility in Rec mutants of *Bacillus subtilis*. J. Bacteriol. 144:608–615.

Doetsch, R. N. 1981. Determinative methods of light microscopy, p. 21–33. In: P. Gerhart, (ed.), Manual of methods for general microbiology. American Society for Microbiology, Washington, D.C.

Doi, R. H. and L. F. Wang. 1986. Multiple procaryotic ribonucleic acid polymerase sigma factors. Microbiol. Rev. 50:227–243.

Dooley, D. C., C. T. Hadden, and E. W. Nester. 1971. Macromolecular synthesis in *Bacillus subtilis* during development of the competent state. J. Bacteriol. 108:668–679.

Doyle, R. J., J. Chaloupka and V. Vinter. 1988. Turnover of cell walls in microorganisms. Microbiol. Rev. 52:554–567.

Dubnau, D. A. 1982. Genetic transformation in *Bacillus subtilis*, p. 148–178. In: D. A. Dubnau, (ed.), The molecular biology of the bacilli, vol. 1. *Bacillus subtilis* Academic Press, New York.

Dubnau, D. A. (ed.). 1982. The molecular biology of the bacilli, vol. 1. *Bacillus subtilis*. Academic Press, New York.

Dubnau, D. A. (ed.). 1985. The molecular biology of the bacilli, vol. 2. Academic Press, New York.

Durham, D. R., D. B. Stewart, and E. J. Stellwag. 1987. Novel alkaline- and heat-stable serine proteases from alkalophilic *Bacillus* sp. Strain GX 6638. J. Bacteriol. 169:2762–2768.

Ebbole, D. J. and H. Zalkin. 1987. Cloning and characterization of a 12-gene cluster from *Bacillus subtilis* encoding nine enzymes for *de novo* purine nucleotide synthesis. J. Biol. Chem. 262:8274–8287.

Ephrati-Elizur, E. 1968. Spontaneous transformation in *Bacillus subtilis* Genet. Res. 11:83–96.

Fahmy, F., J. Flossdorf, and D. Claus. 1985. The DNA base composition of the type strains of the genus *Bacillus*. Syst. and Appl. Microbiol. 6:60–65.

Fitch, W. M. and E. Margoliash. 1967. Construction of phylogenetic trees: A method based on mutational distances as estimated from cytochrome c sequences is of general applicability. Science 155:279–284.

Fitz-James, P. C. and E. Young. 1969. Morphology of sporulation. p. 39–72. In: G. W. Gould and A. Hurst (ed.), The bacterial spore. Academic Press, New York.

Foster, S. J. and K. Johnstone. 1989. The trigger mechanism of bacterial germination, p. 89–108. In: I. Smith, R. A. Slepecky, and P. Setlow (ed.), Regulation of procaryotic development, structural and functional analysis of bacterial sporulation and germination. American Society for Microbiology, Washington, D.C.

Fox, G. E., K. R. Pechan, and C. R. Woese. 1977. Comparative cataloging of 16s ribosomal ribonucleic acid: molecular approach to prokaryotic systematics. Int. J. Syst. Bacteriol. 27:44–57.

Fox, T. D. 1976. Identification of phage SP01 proteins coded by regulatory genes *33* and *34*. Nature 262:748–753.

Freese, E. and J. Heinze. 1983. Metabolic and genetic control of bacterial sporulation, p. 101–172. In: A. Hurst and G. W. Gould (ed.), The bacterial spore, vol. 2. Academic Press, New York.

Freese, E. and J. Heinze, T. Mitani, and E. B. Freese. 1978. Limitation of nucleotides induces sporulations, p. 277–

285. In: G. Chambliss and J. C. Vary (ed.), Spores VII. American Society for Microbiology, Washington, D.C.

Fujita, Y. and T. Fujita. 1987. The gluconate operon *gnt* of *Bacillus subtilis* encodes its own transcriptional negative regulator. Proc. Natl. Acad. Sci. USA. 84:4524–4528.

Fujita, Y., T. Fujita, Y, Miwa, J. Nihashi, and Y. Aratani. 1986. Organization and transcription of the gluconate operon, *gnt*, of *Bacillus subtilis*. J. Biol. Chem. 261:13744–13753.

Ganesan, A. T., S. Chang, and J. A. Hoch (ed.). 1982. Molecular cloning and gene regulation in bacilli. Academic Press, New York.

Ganesan, A. T., and J. A. Hoch. (ed.). 1988. Genetics and biotechnology of bacilli, vol. 2. Academic Press, New York.

Geiduschek, E. P. and J. Ito. 1982. Regulatory mechanisms in the development of lytic bacteriophages in *Bacillus subtilis*, p. 203–245. In: D. A. Dubnau (ed.), The molecular biology of the bacilli, vol. 1. Academic Press, New York.

Gerhardt, P. and R. E. Marquis. 1989. Spore thermo-resistance mechanisms, p. 43–64. In: I. Smith, R. A. Slepecky, and P. Setlow (ed.), Regulation of procaryotic development structural and functional analysis of bacterial sporelation and generation. Am. Soc. Microbiol., Washington, D.C.

Golden, K. J. and R. W. Bernlohr. 1989. Defects in the nutrient-dependent methylation of a membrane-associate protein in *spo* mutants of *Bacillus subtilis*. Mol. and Gen. Genet 220:1–7.

Gonzalez, J. M., Jr., B. J. Brown, and B. C. Carlton. 1982. Transfer of *Bacillus thuringiensis* plasmids coding for delta-endotoxin among strains of *B. thuringiensis* and *B. cereus*. Proc. Natl. Acad. Sci. USA 79:6951–6955.

Gordon, R. E. 1981. One hundred and seven years of the genus *Bacillus*. In R. C. Berkeley, and M. Goodfellow (ed.), The aerobic endosporeforming bacteria. Academic Press, London.

Gordon, R. E., W. C. Haynes, and C. H.-N. Pang. 1973. The genus *Bacillus*. Handbook No. 427. U.S. Department of Agriculture, Washington, D.C.

Gould, G. W. 1983. Mechanisms of resistance and dormancy, p. 173–209. In: A. Hurst and G. W. Gould (ed.), The bacterial spore, vol. 2. Academic Press, New York.

Gould, G. W. and G. J. Dring. 1974. Mechanisms of spore heat resistance. Adv. Microbiol. Physiol. 2:137–161.

Graham, J. B. and C. A. Istock. 1978. Genetic exchange in *Bacillus subtilis* in soil. Mol. Gen. Genet. 166:287–290.

Green, B. D., L. Battisti, T. M. Koehler, C. B. Thorne, and B. E. Ivins. 1985. Demonstration of a capsule plasmid in *Bacillus anthracis*. Infect. Immun. 49:291–297.

Green, B. D., L. Battisti, and C. B. Thorne. 1989. Involvement of Tn *4430* in transfer of *Bacillus anthracis* plasmids mediated by *Bacillus thuringiensis* plasmid pXO12. J. Bacteriol. 171:104–113.

Gryczan, T. J. 1982. Molecular cloning in *Bacillus subtilis*, p. 307–330. In: D. A. Dubnau, (ed.), The molecular biology of the bacilli. vol. 1 Academic Press, New York.

Gryczan, T. J., S. Contente, and D. Dubnau. 1978. Characterization of *Staphylococcus aureus* plasmids introduced by transformation into *Bacillus subtilis*. J. Bacteriol. 134:318–329.

Guettler, M. and R. S. Hanson. 1988. Characterization of a methanol oxidizing member of the genus *Bacillus*. Abs. Ann. Mtg. Am. Soc. Microbiol, I-95, p.196.

Hadden, C. and E. W. Nester. 1968. Purification of competent cells in the *Bacillus subtilis* transformation system. J. Bacteriol. 95:876–885.

Hager, P. W. and J. C. Rabinowitz. 1985. Translation specificity in *Bacillus* subtilis, p. 1–32. In: D. A. Dubnau (ed.) The molecular biology of the bacilli, vol. 2. Academic Press, New York.

Hara, T., A. Aumayr, Y. Fujio, and S. Veda. 1982. Elimination of plasmid-linked polyglutamate production by *Bacillus subtilis* (natto) with acridine orange. Appl. Environ. Microbiol 44:1456–1458.

Hastrup, S. 1988. Analysis of the *Bacillus subtilis* xylose regulon, p. 79–83. In: A. T. Ganesan and J. A. Hoch (ed.), Genetics and biotechnology of bacilli, vol. 2. Academic Press, New York.

Helmann, J. D. and M. J. Chamberlain. 1988. Structure and function of bacterial sigma factors. Ann. Rev. Biochem. 57:839–872.

Hemphill, H. E., I. Gage, S. A. Zahler, and R. Korman. 1980. Prophage-mediated production of a bacteriocin-like substance by Spβ lysogens of *Bacillus subtilis*. Can. J. Microbiol. 23:45–51.

Hemphill, H. E. and H. R. Whiteley. 1975. Bacteriophages of *Bacillus subtilis*. Bacteriol. Rev. 39:257–315.

Henner, D. J., L. Band, and H. Shimotsu. 1984. Nucleotide sequence of the *Bacillus subtilis* tryptophon operon. Gene 34:169–177.

Henner, D. J., E. Ferrari, M. Perego, and J. A. Hoch. 1988a. Location of the targets of the *hpr-97, sacU32(hy)* and *sacQ 36(Hy)* mutations in upstream regions to the subtilis promoter. J. Bacteriol. 170:296–300.

Henner, D. J., M Yang, and E. Ferrari. 1988b. Localization of *B. subtilis sacU* (Hy) mutations to two linked genes with similarities to the conserved procaryotic family of two-component signalling systems. J. Bacteriol. 170:5102–5109.

Hitchins, A. D. and R. A. Slepecky. 1969. Bacterial sporulation as a modified procaryotic cell division. Nature (London) 223:804–807.

Hoch, J. A., K. Trach, I. Kawamura, and H. Saito. 1985. Identification of the transcriptional suppressor *SOF-1* as an alteration in the SpoOA protein, J. Bacteriol. 161:552–555.

Höfte, H. and H. R. Whiteley. 1989. Insecticidal crystal proteins of *Bacillus thuringiensis*. Microbiol. Rev. 53:242–255.

Honjo, M., A. Akaoka, A. Nakayama, and Y. Furutani. 1986. Secretion of human growth hormone in *B. subtilis* using prepropeptide coding region of *B. amyloliquefaciens* neutral protease gene. J. Biotechnol. 4:63–71.

Hunger, W. and D. Claus. 1981. Taxonomic studies on *Bacillus megaterium* and on agarolytic *Bacillus* strains, p. 217–239. In: R. C. Berkeley and M. Goodfellow (ed.), The aerobic endosporeforming bacteria. Academic Press, London.

Hyde, E. I., M. D. Hilton, and H. R. Whiteley. 1986. Interactions of *Bacillus subtilis* RNA polymerase with subunits determining the specificity of initiation. J. Biol. Chem. 261:16565–16570.

Imanaka, T., M. Fujii, and S. Aiba. 1981. Isolation and characterization of antibiotic resistance plamids from thermophilic bacilli and construction of deletion plasmids. J. Bacteriol. 146:1091–1097.

Iwasaki, H., A. Shimada, K. Yokoyama, and E. Iyo. 1989. Structure and glycosylation of lipoteichoic acids in *Bacillus* strains. J. Bacteriol. 171:424–429.

Jackson, E. N., D. A. Jackson, and R. J. Deans. 1978. *Eco* R1 analysis of bacteriophage P22 DNA packaging. J. Mol. Biol. 118:365–388.

Johannes, P. M. de V., F. G. Boogerd, and E. W. deV-de-Jong. 1986. Manganese reduction by a marine *Bacillus* species. J. Bacteriol. 167:30–34.

Jurtshuk, R. J., C. Lin, P. Candela, J. D. Wisotzkey, P. Jurtshuk, Jr., and G. E. Fox. 1989. 16S Ribosomal RNA sequencing studies on organisms of the *Bacillus* species. Abst. Ann. Meet. Am. Soc. Microbiol., R-10, p. 281.

Katz, E. and A. L. Demain. 1977. The peptide antibiotics of *Bacillus*: chemistry, biogenesis and possible functions. Bacteriol. Rev. 41:449–474.

Kaya, S., K. Yokoyama, Y. Araki, and E. Ito. 1984. N-*acetylmannos-aminyl* (14) N-acetylglucosamine, a linkage unit between glycerol teichoic acid and peptidoglycan in cell walls of several *Bacillus* strains. J. Bacteriol. 158:990–996.

Keggins, K. M., R. K. Nauman, and P. S. Lovett. 1978. Sporulation-converting bacteriophages for *Bacillus pumilis*. J. Virol. 27:819–822.

Keynan, A. and N. Sandler. 1983. Spore research in historical perspective, p. 1–48. In: A. Hurst and G. W. Gould (ed.), The bacterial spore, vol. 2. Academic Press, New York.

Khoury, P. H., S. L. Lombardi, and R. A. Slepecky. 1987. The perturbation of the heat resistance of bacterial spores by sporulation temperatures and ethanol. Current Microbiol. 15:15–19.

Kinney, D. M. and M. G. Bramucci. 1981. Analysis of *Bacillus subtilis* sporulation with spore-converting bacteriophage PMB12. J. Bacteriol. 145:1281–1285.

Kitada, M. and K. Horikoshi. 1987. Bioenergetic properties of alkalophilic *Bacillus* sp. strain C-59 on an alkaline medium containing K_2CO_3. J. Bacteriol 169:5761–5765.

Kitada, M., K. Onda, and K. Horikoshi. 1989. The sodium/proton antiport system in a newly isolated alkalophilic *Bacillus* sp. J. Bacteriol. 171:1879–1884.

Klier, A. F. and G. Rapoport. 1988. Genetics and regulation of carbohydrate catabolism in *Bacillus*. Ann. Rev. Microbiol. 42:65–95.

Koehler, T. M. and C. B. Thorne. 1987. *Bacillus subtilis* (natto) plasmid pLS20 mediates interspecies plasmid transport. J. Bacteriol. 169:5271–5278.

Korczynski, M. 1981. Sterilization, P. 476–486. In: P. Gerhardt (ed.), Manual of methods for general microbiology. American Society of Microbiology, Washington, D.C.

Kunst, F., M. Debarbouille, T. Msadek, M. Young, C. Mauel, D. Karomata, A. Klier, G. Rapoport, and R. Dedonder. 1988. Deduced polypeptides encoded by the *Bacillus subtilis sac U* locus share homology with two-component sensor-regulator systems. J. Bacteriol. 170:5093–5101.

Kuroda, M. I., D. Henner, and C. Yanofsky. 1988. *cis-* Acting sites in the transcript of the *Bacillus subtilis trp* operon regulate expression of the operon. J. Bacteriol. 170:3080–3088.

Lane, D. J., B. Pace, G. J. Olsen, D. A. Stahl, M. L. Sogin, and N. R. Pace. 1985. Rapid determination of 16S ribosomal RNA sequences for phylogenetic analysis. Proc. Natl. Acad. Sci. USA. 82:6955–6959.

Laoide, B., G. H. Chambliss, and D. J. McConnell. 1989. *Bacillus licheniformis* α-amylase gene, *amyL*, is subject

to promoter-independent calabotite repression in *Bacillus subtilis*. J. Bacteriol. 171:2435–2442.

Laoide, B. M. and D. J. McConnell. 1989. *cis* Sequences involved in modulating expression of *Bacillus licheniformis amyL* in *Bacillus subtilis*: effect of sporulation and catabolite repression resistance mutations on expression. J. Bacteriol. 171:2443–2450.

Lepesant-Kejzlarova, J., J.-A. Lepeseant, J. Walle, A. Billaut, and R. Dedonder. 1975. Revision of the linkage map of *Bacillus subtilis* 168: indications for circularity of the chromosome. J. Bacteriol. 121:823–834

Lerner, C. G., B. T. Stephenson, and R. L. Switzer. 1987. Structure of the *Bacillus subtilis* pyrimidine biosynthetic (Pyr) gene cluster. J. Bacteriol. 169:2202–2206.

Lindsay, J. A. and W. G. Murrell. 1986. Solution spectroscopy of dipicolinic acid interaction with nucleic acids: role in spore heat resistance. Curr. Microbiol. 13:255–259.

Lindsay, J. A., W. G. Murrell, and A. D. Warth. 1985. Spore resistance and the basic mechanism of heat-resistance, p. 162–186. In: L. E. Harris and A. J. Skopek (ed.), Sterilization of medical products, Vol. 3. Johnson & Johnson Pty., Ltd., Botany NSW, Australia.

Logan, N. A. and R. C. W. Berkeley. 1981. Classification and identification of members of the genus *Bacillus* using API tests, p. 105–140. In: R. C. W. Berkeley and M. Goodfellow (ed.), The aerobic endosporeforming bacteria: classification and identification." Academic Press, London.

Logan, N. A. and R. C. W. Berkeley. 1984. Identification of *Bacillus* strains using the API system. J. Gen. Microbiol. 130:1871–1882.

Lorenz, M. G., B. W. Aardema, and W. Wackernagel. 1988. Highly efficient genetic transformation of *Bacillus subtilis* attached to sand grains. J. Gen. Micro. 134:107–112.

Losick, R. and L. Kroos. 1989. Dependence pathways for the expression of genes involved in endospore formation in *Bacillus subtilis*, p. 223–242. In: I. Smith, R. A. Slepecky, and P. Setlow (ed.), Regulation of procaryotic development, structural and functional analysis on bacterial sporulation and germination. American Society for Microbiology, Washington, D.C.

Losick, R. and J. Pero. 1981. Cascades of sigma factors. Cell 25:582–584.

Losick, R. and L. Shapiro (ed.). 1984. Microbial development. Cold Spring Harbor Laboratory, NY.

Losick, R., P. Youngman, and P. J. Piggot. 1986. Genetics of endospore formation in *Bacillus subtilis*. Annv. Rev. Genet. 20:625–669.

Love, P. E., M. V. Lyle, and R. E. Yasbin. 1985. DNA-damage inducible *(din)* loci are transcriptionally activated in competent *Bacilus subtilis*. Proc. Natl. Acad. Sciences, 82:6201–6205.

Lovett, P. S., E. J. Duvall, and K. M. Keggins. 1976. *Bacillus pumilis* plasmed pPL10: Properties and insertion into *Bacillus subtilis* 168 by transformation. J. Bacteriol 127:817–828.

Mahler, I., and H. O. Halvorson. 1980. Two erythromycin resistance plasmids of diverse origin and their effect on sporulation in *Bacillus subtilis*. J. Gen. Microbiol. 120:259–263.

Makino, S-I, I. Uchida, N. Terakado, C. Sasakawa, and M. Yoshikawa. 1989. Molecular characterization and protein analysis of the cap region which is essential for encapsulation in *Bacillus anthracis*. J. Bacteriol. 171:722–730.

Mandelstam, J. and J. Errington. 1987. Dependent sequences of gene expression controlling spore formation in *Bacillus subtilis*. Microbiol. Sciences 4:238–244.

Marahiel, M. A., P. Zuber, G. Czekay, and R. Losick. 1987. Identification of the promoter for a peptide antibiotic biosynthesis gene from *Bacillus brevis* and its regulation in *Bacillus subtilis*. J. Bacteriol. 169:2215–2222.

Marquis, R. E. 1989. Minerals and bacterial spores, p. 147–161. In: T. J. Beveridge and R. J. Doyle (ed.) Bacterial interactions with metal ions. J. Wiley and Sons, New York.

Melin, L., K. Magnusson, and L. Rutberg. 1987. Identification of the promoter of the *Bacillus subtilis sdh* operon. J. Bacteriol. 169:3232–3236.

Mellado, R. P., I. Barthelemy, and M. Salas. 1988. Transcription initiation and termination signals of the *Bacillus subtilis* phage ϕ29 DNA, p. 215–219. In: A. T. Ganesan and J. A. Hoch (ed.), Genetics and biotechnology of bacilli, vol. 2. Academic Press, New York.

Mendelson, N. H. 1978 Helical *Bacillus subtilis* macrofibers: morphogenesis of a bacterial multicellular macroorganism. Proc. Natl. Acad. Sci. USA 75:2478–2482.

Mezes, P. S. and J. O. Lampen. 1985. Secretion of proteins by Bacilli, p. 151–185. In: D. A. Dubnau (ed.), The molecular biology of the bacilli, vol. 2. Academic Press, New York

Mikesell, P., B. E. Ivins, J. D. Ristroph, and T. M. Dreier. 1983. Evidence for plasmid-mediated toxin production in *Bacillus anthracis*. Infect. Immun. 39:371–376.

Mikesell, P. and M. Vodkin. 1985. Plasmids of *Bacillus anthracis*, p. 52–55. In: L. Leive (ed.), Microbiology-1985. American Society for Microbiology, Washington, D.C.

Minnikin, D. E. and M. Goodfellow. 1981. Lipids in the classification of *Bacillus* and related taxa, p. 59–103. In: Berkeley, R. C. and M. Goodfellow (ed.), The aerobic endosporeforming bacteria. Academic Press, London.

Moir, A., I. M. Feavers, and A. R. Zuberi. 1986. A spore germination operon in *Bacillus subtilis* 168, p. 183–194. In: A. T. Ganesan and J. A. Hoch (ed.), Bacillus molecular genetics and biotechnology applications. Academic Press, London.

Moran, C. P. 1989. Sigma factors and the regulation of transcription, p. 167–184. In: I. Smith, R. A. Slepecky and P. Setlow (ed.), Regulation of prokaryotic development, structural and functional analysis of bacterial sporulation and germination. American Society for Microbiology, Washington, D.C.

Motley, S. T. and S. Graham. 1988. Expression and secretion of human interleukin-1 in *Bacillus subtilis*, p. 371–376. In: A. T. Ganeson and J. A. Hoch (ed.), Genetics and biotechnology of bacilli, vol. 2. Academic Press, New York.

Murrell, W. G. 1981. Biophysical studies on the molecular mechanisms of spore heat resistance and dormancy, p. 64–77. In: H. S. Levinson, A. L. Sonenshein, and D. J. Tupper (ed.), Sporulation and germination. American Society for Microbiology, Washington, D.C.

Nakamura, L. K. 1984. *Bacillus pulvifaciens* sp. nov., nom. rev. Int. J. Syst. Bacteriol. 34:410–413.

Nakamura, L. K. 1987. *Bacillus alginolyticus* sp. nov. and *Bacillus chondritinus* sp. nov. Int. J. Syst. Bacteriol. 37:284–286.

Nakamura, L. K. 1989. *Bacillus thiaminolyticus* sp. nov., nom. rev. Abst. Annu. Mtg. Am. Soc. Microbiol. R-11, p. 282.

Nakamura, L. K., I. Blumenstock, and D. Claus. 1988. Taxonomic study of *Bacillus coagulans* Hammer 1915 with a proposal for *Bacillus smithii* sp. nov., Int. J. Syst. Bacteriol. 38:63–73.

Nicholson, W. L., Y-K, Paris, T. M. Henkin, M. Won, M. J. Weickert, J. A. Gaskell, and G. H. Chambliss. 1987. Catabolite repression-resistant mutations of the *Bacillus subtilis* alpha-amylase promoter affect transcription levels and are in an operator-like sequence. J. Mol. Biol. 198:609–618.

Nihashi, J-I. and Y. Fujita. 1984. Catabolite repression of inositol dehydrogenase and gluconate kinase synthesis in *Bacillus subtilis*. Biochimica et Biophysica Acta 798:88–95.

Nixon, B. T., C. W. Ronson, and F. M. Ausubel. 1986. Two component regulating systems responsive to environmental stimuli share strongly conserved domains with the nitrogen assimilation regulatory genes *ntrB* and *ntrC*. Proc. Natl. Acad. Sci. USA 83:7850–7854.

Norris, J. R., R. C. W. Berkeley, N. A. Logan, and A. G. O'Donnell. 1981. The genera *Bacillus* and *Sporolactobacillus*, p. 1711–1742. In: M. P. Starr, A. Stolp, A. G. Truper, A. Balows, and H. G. Schlegel (ed.), The prokaryotes, vol. 2. Springer-Verlag, Berlin.

Oda, M., A. Sogishita, and K. Furukawa. 1988. Cloning and nucleotide sequences of histidase and regulatory genes in the *Bacillus subtilis hut* operon and positive regulation of the operon. J. Bacteriol 170:3199–3205.

Odebralski, J. M. and S. A. Zahler. 1982. Specialized transduction of the *kauA* and *citK* genes of *Bacillus subtilis* by bacteriophage ϕ3T. Abstr. Am. Soc. Microbial. p. 130.

O'Donnell, A. G., H. J. H. Macfie, and J. R. Norris. 1988. An assessment of taxononic congruence between DNA-DNA hybridization and pyrolysis gas-liquid chromatographic classifications. J. Gen. Microbiol. 134:743–749.

O'Donnell, A. G. and J. R. Norris. 1981. Pyrolysis gas-liquid chromatographic studies in the genus *Bacillus* p. 141–179. In: R. C. W. Berkeley and M. Goodfellow (ed.), The aerobic endosporeforming bacteria, classification and identification. Academic Press, New York.

Okamoto, K., J. A. Mudd, J. Mangan, W. M. Huang, T. V. Subbaiah, and J. Marmur. 1968. Properties of the defective phage of *Bacillus subtilis*.. J. Mol. Biol. 34:413–428.

Ordal, G. W. and D. O. Nettleton. 1985. Chemotaxis in *Bacillus subtilis* p. 53–73. In: D. A. Dubnau (ed.), The molecular biology of the bacilli, vol. 2. Academic Press, New York.

Osburne, M., and A. L. Sonenshein. 1980. Inhibition by Lipiarmycin of bacteriophage growth in *Bacillus subtilis*. J. Virol. 33:945–953.

Palva, I., P. Lehtovaara, L. Kaariainen, M. Sibakov, L. Cantell, C. H. Schein, K. Kashiwagi, and C. Weismann. 1983. Secretion of interferon by *Bacillus subtilis*. Gene 22:229–235.

Pechman, K. J., B. J. Lewis, and C. R. Woese. 1976. Phylogenetic status of *Sporosarcina ureae*. Int. J. Syst. Bacteriol. 261:305–310.

Pergo, M., S. P. Cole, D. Burbulys, K. Trach, and J. A. Hoch. 1989. Characterization of the gene for a protein kinase which phosphorylates the sporulation-regulatory root-eins SpoOA and SpoOF of *Bacillus subtilis.* J. Bacteriol. 171:6187–6196.

Perlak, F. J., C. L. Mendelsohn, and C. B. Thorne. 1979. Converting bacteriophage for sporulation and crystal-formation in *Bacillus thuringiensis.* J. Bacteriol. 140:699–706.

Pichinoty, F., J. Asselineau, and M. Mandel. 1984. Characterisation biochimique de *Bacillus benzoevorans* sp. nov., une nouvelle espèce filamenteuse, engainée et mesophile, dégradant divers acides aromatiques et phenols. Ann. Microbiol. 135B:209–217.

Pichinoty, F., J. B. Waterbury, M. Mandel, and J. Asselineau. 1986. *Bacillus gordonae* sp. nov., Une nouvelle espèce appartenant au second groupe morphologique, dégradant divers composes aromatiques. Ann. Inst. Pasteur 137A:65–78.

Piggot, P. 1989. Revised genetic map of *B. subtilis* 168, p. 1–42. In: I. Smith, R. A. Slepecky, and P. Setlow (ed.), Regulation of procaryotic development, structural and functional analysis of bacterial sporulation and germination. American Society for Microbiology, Washington, D.C.

Potvin, B. W., R. J. Kelleher, Jr., and H. Gooder. 1975. Pyrimidine biosynthelic pathway of *Bacillus subtilis.* J. Bacteriol. 123:604–615.

Price, C. W. and R. H. Doi. 1985. Genetic mapping of *rpoD* implicates the major sigma factor of *Bacillus subtilis* RNA polymerase in sporulation intiation. Mol. Gen. Genet. 201:88–95.

Priest, F. G. 1977. Extracellular enzyme synthesis in the genus *Bacillus.* Bacterol. Rev. 41:711–753.

Priest, F. G. 1981. DNA homology in the genus *Bacillus,* p. 33–57. In: R. C. Berkeley and M. Goodfellow (ed.), The aerobic endosporeforming bacteria. Academic Press, London.

Priest, F. G., M. Goodfellow, and C. Todd. 1981. The genus *Bacillus:* A numerical analysis, p.91–103. In: R. C. Berkeley and M. Goodfellow (ed.) The aerobic endosporeforming bacteria. Academic Press, London.

Priest, F. G., M. Goodfellow, and C. Todd. 1988. A numerical classification of the genus *Bacillus.* J. Gen. Microbiol. 134:1847–1882.

Reddy, A., L. Battisti, and C. B. Thorne. 1987. Identification of self-transmissible plasmids in four *Bacillus thuringiensis* subspecies. J. Bacteriol. 169:5263–5270.

Rettenmier, C. W. and H. E. Hemphill. 1974. Abortive infection of lysogenic *Bacillus subtilis* 168(SPO2) by bacteriophage φ1. J. Virol. 13:870–880.

Rettenmeir, C. W., B. Gingell, and H. E. Hemphill. 1979. The role of temperate bacteriophage SPβ in prophage-mediated interference in *Bacillus subtilis.* Can. J. Microbiol. 25:1345–1351.

Reynolds, R. B., A. Reddy, and C. B. Thorne. 1988. Five unique temperate phages from a polylysogenic stain of *Bacillus thuringiensis* subsp. *aizawai.* J. Gen. Microbiol. 134:1577–1585.

Rima, B. K. and I. Takahashi. 1974. The synthesis of nucleic acids in *Bacillus subtilis* infected with phage PBSI. Can. J. Biochem. 51:1219–1224.

Roberts, T. A. and A. D. Hichins. 1969. Resistance of spores, p. 611–670. In: G. W. Gould and A. Hurst (ed.), The bacterial spore. Academic Press, London.

Ronson, C. W., B. T. Nixon, and F. Ausubel. 1987. Conserved domains in bacterial regulatory proteins that respond to environmental stimuli. Cell 49:579–581.

Rosson, A. R. and K. H. Nealson. 1982. Manganese binding and oxidation by spores of a marine *Bacillus.* J. Bacteriol. 151:1027–1034.

Rothstein, D. M., P. E. Devlin, and R. L. Cate. 1986. Expression of α-amylase in *Bacillus licheniformis.* J. Bacteriol 168:839–842.

Ruhfel, R. E., N. J. Robillard, and C. B. Thorne. 1984. Interspecies transduction of plasmids among *Bacillus anthracis, B. cereus,* and *B. thuringiensis.* J. Bacteriol. 157:708–711.

Russell, A. D. 1982. The bacterial spore, p. 1–24. In: A. D. Russell (ed.), The destruction of bacterial spores. Academic Press, London.

Rutberg, L. 1982. Temperate bacteriophages of *Bacillus subtilis.* p. 247–268. In: D. A. Dubnau (ed.), The molecular biology of the bacilli, vol. 1. Academic Press, New York.

Saitou, N. and M. Nei. 1987. The neighbor-joining method: a new method for reconstructing phylogenetic trees. Molec. Biol. and Evol. 4:406–425.

Schendel, F. J., C. E. Bremmon, M. G. Flickinger, M. Guettler, and R. S. Hanson. 1989. L-Lysine production from methanol at high cell densities of MGA3, a thermophilic *Bacillus.* Abst. Annu. Mtg. Am. Soc. Microbiol. p. 316.

Schien, C. H., K. Kashiwagi, A. Fujisawa, and C. Weissmann. 1986. Secretion of mature IFN-α2 and accumulation of uncleared precursor by *Bacillus subtilis* transformed with a hybrid α-amylase signal sequence-IFN-α2 gene Bio/Technology 4:719–725.

Schmieger, H. 1982. Packaging signals for phage P22 on the chromosome of *Salmonella typhimureim.* Mol. Gen. Genet. 187:516–518.

Schmieger, H. 1984. *Pac* sites are indispensible for in vivo packaging of DNA by P22. Mol. Gen. Genet. 195:252–255.

Setlow, P. 1983. Germination and outgrowth, p. 211–254. In: A. Hurst, and G. W. Gould (ed.), The bacterial spore, vol. 2. Academic Press, London.

Setlow, P. 1988. Small, acid-soluble spore proteins of *Bacillus* species: Structure, synthesis, genetics, function and degradation. Annv. Rev. Microbiol. 42:319–338.

Setlow, P. 1989. Forespore specific genes of *Bacillus subtilis:* function and regulation of expression, p. 211–222. In: I. Smith, R. A. Slepecky, and P. Setlow (ed.), Regulation of procaryotic development, structural and functional analysis of bacterial sporulation and germination. American Society for Microbiology, Washington, D.C.

Shapiro, J. M., D. H. Dean, and H. O. Halvorson. 1974. Low-frequency specialized transduction with *Bacillus subtilis* bacteriophage φ105. Virology 62:393–403.

Shimotsu, H., and D. J. Henner. 1986. Modulation of *Bacillus subtilis* levansucrase gene expression by sucrose and regulation of the steady state mRNA level by *sac* V and *sac* Q genes. J. Bacteriol. 168:380–388.

Shimotsu, H., M. I. Kuroda, C. Yanofsky, and D. J. Henner. 1986. Novel form of transcription attenuation regulates expression of the *Bacillus subtilis* tryptophan operon. J. Bacteriol. 166:461–471.

Shohayer, M. and I. Chapra. 1985. Composition of membranes from whole cells and minicells of *Bacillus subtilis.* J. Gen. Microbiol. 131:345–354.

Shute, L. A., C. S. Gutteridge, J. R. Norris and R. C. W. Berkeley. 1984. Curepoint pyrolysis mass spectrometry applied to characterization and identification of selected *Bacillus* species. J. Gen. Microbiol. 130:343–355.

Slepecky, R. A. 1972. Ecology of bacterial sporeformers, p. 297–313. In: H. G. Halvorson, R. Hanson, and L. L. Campbell (ed.), Spores V. American Society for Microbiology, Washington, D.C

Slepecky, R. A. 1978. Resistant forms, p. 14/1–14/31. In: J. R. Norris and M. H. Richmond (ed.), Essays in microbiology. John Wiley & Sons, New York.

Slepecky, R. A. and E. R. Leadbetter. 1977. The diversity of spore-forming bacteria: some ecological implications, p. 869–877. In: A. N. Barker, J. Wolf, D. J. Ellar, G. J. Dring, and G. W. Gould (ed.), Spore Research. 1976. Academic Press, London.

Slepecky, R. A. and E. R. Leadbetter. 1984. On the prevalence and roles of sporeforming bacteria and their spores in nature, p.79–99. In: A. Hurst and G. W. Gould (ed.), The bacterial spore, vol. 2. Academic Press, London.

Sleytr, U. B. and P. Messner. 1988. Crystalline surface layers in procaryotes. J. Bacteriol. 170:2891–2897.

Sloma, A., A. Ally, D. Ally, and J. Pero. 1988. Gene encoding a minor extracellular protease in *Bacillus subtilis*. J. Bacteriol. 170:5557–5563.

Smith, H., K. Wiersma, G. Venema, and S. Bron. 1985. Transformation in *Bacillus subtilis:* Further characterization of a 75,000-dalton protein complex involved in binding and entry of donor DNA. J. Bacteriol. 164:201–206.

Smith, I. 1982. The translational apparatus of *Bacillus subtilis,* p.111–147. In: D. A. Dubnau (ed.), The molecular biology of the bacilli, vol. 1. Academic Press, New York.

Smith, I. 1989. The initiation of sporulation, p. 185–210. In: I. Smith, R. A. Slepecky and P. Setlow (ed.), Regulation of procaryotic development, structural and functional analysis of bacterial sporulation and germination. American Society for Microbiology, Washington, D.C.

Smith, M. C. M., A Mountain, and S. Baumberg. 1986. Sequence analysis of the *Bacillus subtilis argC* promoter region. Gene 49:53–60.

Smith, N. R., R. E. Gordon, and F. E. Clark. 1946. Aerobic mesophilic sporeforming bacteria. US Dept. Agri. Misc. Publication 559. Washington, D.C.

Smith, N. R., R. E. Gordon, and F. E. Clark. 1952. Aerobic sporeforming bacteria. US Dept. Agr. Monograph 10. Washington, D.C.

Spancake, G. A., H. E. Hemphill, and P. S. Fink. 1984. Genome organization of SPβc2 bacteriophage carrying the thy P3 gene. J. Bacterial. 157:428–434.

Sneath, P. H. A., and R. R. Sokal. 1973. Numerical Taxonomy. W. H. Freeman, San Francisco.

Sonenshein, A. L. 1989. Metabolic regulation of sporulation and other stationary-phase phenomena, p. 109–130. In: I. Smith, R. A. Slepecky, and P Setlow (ed.), Regulation of procaryotic development, structural and functional analysis of bacerial sporulation and germination. American Society Microbiology, Washington, D.C.

Spizizen, J. 1958. Transformation of biochemically deficient strains of *Bacillus subtilis* by deoxyribonucleate. Proc. Natl. Acad. Sci. USA 44:1072–1078.

Stackebrandt, E. and C. R. Woese. 1979. A phylogenetic dissection of the family Micrococcaceae. Curr. Microbiol. 2:317–322.

Stackebrandt, E. and C. R. Woese. 1981. The evolution of prokaryotes, p. 1–31. In: M. J. Carlile, J. F. Collins, and B. E. B. Moseley (ed.), Molecular and cellular aspects of microbial evolution. Cambridge University Press, Cambridge.

Stackebrandt, E, W. Ludwig, M. Weizenegger, S. Dorn, T. J. McGill, G. E. Fox C. R. Woese, W. Schubert, and K-H. Schleifer. 1987. Comparative 16S RNA oligonucleotide analyses and murein types of round-sporeforming bacilli and nonsporeforming relatives. J. Gen. Microbiol. 133:2523–2529.

Steensma, H. Y., L. A. Robertson, and J. D. Van Elsas. 1978. The occurrence and taxonomic value of PBSX-like defective phages in the genus *Bacillus.* Antonie van Leeuwenhoock 44:353–366.

Steinmetz, M., and S. Aymerich. 1986. Analysis genetique de *sacR,* regulateur en cis de la synthese de la levane-saccharose de *Bacillus subtilis.* Ann. Microbiol. (Paris) 137A:3–14.

Steinmetz, M., S. Aymerich, G. Goney-Treboul, and D. LeCoq. 1988. Levansucrase induction by sucrose in *Bacillus subtilis* involves an antiterminator. Homology with the *Escherichia coli bgl* operon, p. 11–15. In: A. T. Ganesan and J. A. Hock (ed.), Genetics and biotechnology of bacilli, vol. 2. Academic Press, New York.

Stewart, G. J. and C. A. Carlson. 1986. The biology of natural transformation. Ann. Rev. Microbiol. 40:211–35.

Stragier, P. 1989. Temporal and spatial control of gene expression during sporulation: from facts to speculations, p. 243–254. In: I. Smith, R. A. Slepecky, and P. Setlow (ed.), Regulation of procaryotic development, structural and functional analysis of bacterial sporulation and germination. American Society for Microbiology Washington, D. C.

Streips, U. N. and N. E. Welker. 1971. Competence-inducing factor of *Bacillus stearothermophilus.* J. Bacteriol. 106:955–959.

Sun, D. and I. Takahashi. 1984. A catabolite-resistant mutation is localized in the *rpo* operon of *Bacillus subtilis.* Can. J. Microbiol. 30:423–429.

Surana, U., A. J. Wolfe, and N. H. Mendelson. 1988. Regulation of *Bacillus subtilis* macrofibe twist development by D-alanine. J. Bacteriol. 170:2328–2335.

Takahashi, I. 1979. Catabolite repression-resistant mutants of *Bacillus subtilis.* Can. J. Microbiol. 25:1283–1287.

Takahashi, I. and J. Marmur. 1963. Replacement of thymidylic acid by deoxyuridylic acid in the deoxyribonucleic acid of a transducing phage for *Bacillus subtilis.* Nature 197:794–795.

Tarkington, C. and J. Pero. 1979. Distinctive nucleotide sequences of promoters recognized by RNA polymerase containing a phage-coded "σ-like" protein. Proc. Nat. Acad. Sci. USA. 76:5465–5469.

teRiele, H. P. J. and G. Venema. 1984. Heterospecific transformation in *Bacillus subtilis:* protein composition of a membrane DNA complex containing a heterologous donor-recipient complex. Mol. Gen. Genetics. 197:478–485.

Thorne, C. B. 1968. Transducing bacteriophage for *Bacillus cereus.* J. Virol. 2:657–682.

Thorne, C. B. 1985. Genetics of *Bacillus anthracis,* p. 56–62. In: L. Leive, P. F. Bonventure, J. A. Morello, S. Schlesinger, S. D. Silver, and H. C. Wu (ed.), Microbiology-1985. American Society for Microbiology, Washington, D.C.

Thurm, P. and A. J. Garro. 1975. Isolation and characterization of prophage mutants of the defective *Bacillus subtilis* bacteriophage PBSX. J. Virol. 16:184–191.

Tippetts, M. T. and D. L. Robertson. 1988. Molecular cloning and expression of the *Bacillus anthrasis* edema factor toxin gene: a calmodulin-dependent adenylcyclase. J. Bacteriol. 170:2263–2266.

Tjian, R. and J. Pero. 1976. Bacteriophage SPO1 regulatory proteins directing late gene transcription in vitro. Nature 262:753–757.

Trach, K. A., J. W. Chapman, P. J. Piggot, and J. A. Hoch. 1985. Deduced product of the stage O sporulation gene *spoOF* shares homology with the SpoOA, OmpR and SfrA proteins. Proc. Natl. Acad. Sci. USA 82:7260–7264.

Tsuboi, A., R. Uchihi, T. Adachi, T. Sasaki, S. Hayakawa, H. Yamagata, N. Tsukagoshi, and S. Udaka. 1988. Characterization of the genes for the hexagonally arranged surface layer proteins in protein-producing *Bacillus brevis* 47: Complete nucleotide sequence of the middle wall protein gene. J. Bacteriol. 170:935–945.

Vary, P. S., J. C. Garbe, M. Franzen, and E. W. Frampton. 1982. MP13, a generalized transducing bacteriophage for *Bacillus megaterium.* J. Bacteriol. 149:112–119.

Vinter, V. and R. A. Slepecky. 1965. Direct transition of outgrowing bacterial spores to new sporangia without intermediate cell division. J. Bacteriol. 90:803–807.

Vlcek, C. and V. Paces. 1986. Nucleotide sequence of the late region of *Bacillus* phage Ø29 complete the 19285-bp sequence of Ø29 genome. Comparison with the homologous sequence of phage PZA. Gene 46:215–225.

Warner, F. D., G. A. Kitos, M. P. Romano, and H. E. Hemphill. 1977. Characterization of SPβ: a temperate bacteriophage from *Bacillus subtilis* 168M. Can. J. Microbiol. 23:45–51.

Warth, A. D. 1978. Molecular structure of the bacterial spore. Adv. Microb. Physiol. 17:1–45.

Weickert, M. J. and G. H. Chambliss. 1989. Genetic analysis of the promoter region of the *Bacillus subtilis* α-amylase gene. J. Bacteriol. 171:3656–3666.

Whiteley, H. R. and H. E. Schnepf. 1986. The molecular biology of parasporal crystal body formation in *Bacillus thuringiensis.* Ann. Rev. Microbiol. 40:549–576.

Winston, S. and N. Sueoka. 1982. DNA replication in *Bacillus subtilis,* p. 36–71. In: D. A. Dubnau (ed.), The molecular biology of the bacilli, vol. 1. Academic Press, New York.

Wisotzkey, J. D., P. Jurtshuk, Jr., and G. E. Fox. 1989. Comparative 16S rRNA analyses on thermophilic and psychrophilic bacillus species. Abst. Ann. Meet. Am. Soc. Microbiol., p. 281.

Wittman, V. and H. C. Wong. 1988. Regulation of the penicillinase genes of *Bacillus licheniformis:* Interaction of the *pen* repressor with its operators. J. Bacteriol. 170:3206–3212.

Wolf, J. and A. N. Barker. 1968. The genus *Bacillus:* aids to the identification of its species, p. 93–109. In: M. Gibbs and D. A. Shapton (ed.), Identification methods for microbiologists. Part B. Academic Press, London.

Wong, H. C., H. E. Schnepf, and H. R. Whiteley. 1983. Transcriptional and translational start sites for the *Bacillus thuringiensis* crystal protein gene. J. Biol. Chem. 258:1960–67.

Wong, S., F. Kawamura, and R. H. Doi. 1986. Use of *Bacillus subtilis* subtilisin signal peptide for efficient secretion of TEMB-lactamase during growth. J. Bacteriol. 168:1005–1009.

Yasbin, R. E. 1985. DNA repair in *Bacillus subtilis,* p. 33–52. In: Dubnau (ed.), The molecular biology of the bacilli, vol. 2. Academic Press, New York.

Yasbin, R. E., P. I. Fields, and B. J. Andersen. 1980. Properties of *Bacillus subtilis* 168 derivatives freed of their natural prophages. Gene 121:155–159.

Yasbin, R. E., G. A. Wilson, and F. E. Young. 1975. Transformation and transfection in lysogenic strains of *Bacillus subtilis:* evidence for selective induction of prophage in competent cells. J. Bacteriol. 121:296–304.

Youngman, P., H. Poth, B. Green, K. York, G. Olmedo, and K. Smith. 1989. Methods for genetic manipulation, cloning and functional analysis of sporulation genes in *Bacillus subtilis,* p. 65–88. In: I. Smith. R. A. Slepecky, and P. Setlow (ed.), Regulation of procaryotic development, structural and functional analysis of bacterial sporulation and germination. American Society for Microbiology, Washington, D.C.

Zahler, S. A. 1982. Specialized transduction in *Bacillus subtilis,* p 269–305. In: D. A. Dubnau (ed.), The molecular biology of the bacilli, vol. 1. Academic Press, New York.

Zahler, S. A. 1988. Temperate bacteriophages of *Bacillus subtilis,* p. 559–592. In: R. Calendar (ed.), The bacteriophages, vol. 1. Plenum Press, New York.

Zahler, S. A., R. Z. Korman, R. Rosenthal, and H. E. Hemphill. 1977. *Bacillus subtilis* bacteriophage SPβ: localization of the prophage attachment site and specialized transduction. J. Bacteriol 129:556–558.

Zahler, S. A., R. Z. Korman, C. Thomas, P. S. Fink, M. P. Weiner, and J. M. Odebralski. 1987. H2, a temperate bacteriophage isolated from *Bacillus amyloliquefaciens* strain H. J. Gen. Microbiol. 133:2937–2944.

Zalkin, H. and D. J. Ebbole. 1988. Organization and regulation of gene encoding biosynthetic enzymes in *Bacillus subtilis.* J. Biol. Chem. 263:1595–1598.

Zarilla, K. A. and J. J. Perry. 1987. *Bacillus thermoleovorans,* sp.nov., a species of obligately thermophilic hydrocarbon utilizing endosporeforming bacteria. System. App. Microbiol. 9:258–264.

Zukowski, M. M. and L. Miller. 1986. Hyperproduction of an intracellular heterologous protein in a *sac U^h* mutant of *Bacillus subtilis.* Gene 46:247–55.

Zukowski, M., L. Miller. P. Cogswell, and K. Chen. 1988. Inducible expression system based on sucrose metabolism genes of *Bacillus subtilis,* p. 17–22. In: A. T. Ganesan and J. A. Hoch (ed.), Genetics and biotechology of bacilli, vol. 2. Academic Press, New York.

CHAPTER 77

The Genus *Bacillus*—Insect Pathogens

DONALD P. STAHLY, ROBERT E. ANDREWS, and ALLAN A. YOUSTEN

Many different *Bacillus* species have been isolated from dead or dying insects. The *Bacillus* species commonly recognized as definite insect pathogens, however, are *B. popilliae, B. lentimorbus, B. larvae, B. thuringiensis,* and certain strains of *B. sphaericus* (de Barjac, 1981). These organisms, whose pathogenicity has been proved by fulfillment of Koch's postulates, are the subject of this review. The involvement of other *Bacillus* species in insect disease was reviewed by Krieg (1981) and several other authors in different chapters of the book *Microbial Control of Pests and Plant Diseases 1970–1980,* edited by H.D. Burges (1981).

Bacillus popilliae and Bacillus lentimorbus

B. popilliae and *B. lentimorbus* were named and described in 1940 by Dutky (1940). They are etiological agents of a naturally occurring disease of Japanese beetles, referred to as milky disease. The disease was so named because of the milky white appearance of the normally clear hemolymph of the diseased larvae (Hawley and White, 1935). The milkiness is due to the presence of very high numbers of spores of the bacteria; $2-5 \times 10^9$ spores per larva or $2-5 \times 10^{10}$ spores per ml (Dutky, 1940).

B. popilliae was named after the generic name of its host insect, the Japanese beetle, *Popillia japonica* Newman. The species *B. lentimorbus* was named by using the Latin adjective *lentus,* which means slow, and the Latin noun *morbus,* which means disease; literally, the slow disease (Dutky, 1940).

Both of these organisms have been used effectively for the biological control of the Japanese beetle (see Biotechnological Applications).

Habitats

Bacillus popilliae and *B. lentimorbus* are defined in nature as obligate pathogens according to the criteria of Bucher (1960). They are found only in association with a specific insect disease or in surrounding soil in the spore state. Both have a narrow host range, and in nature they probably grow only within the bodies of their host insects (Splittstoesser and Kawanishi, 1981). The host insects are all members of the order Coleoptera and the family Scarabaeidae.

Isolation, Growth, Maintenance, and Preservation of Cultures

These *Bacillus* species are isolated from hemolymph of infected beetle larvae where they occur as nearly pure cultures. Even when grubs of the Japanese beetle are co-infected with *B. popilliae* and *B. lentimorbus,* the diseased grubs prove to be infected with a single species, usually *B. popilliae* (Beard, 1946; Milner, 1981a).

To isolate, the grubs are first washed with water to remove most of the adhering soil, the surface is disinfected by immersion in 0.5% sodium hypochlorite (or 5% sodium hypochlorite [St. Julian et al., 1970]), the grub is rinsed, and the hemocoel is punctured with a dissecting needle. Hemolymph is then dripped into water. The spores can be freed of hemolymph by several alternating centrifugations and suspensions in water (Steinkraus, 1957a; Steinkraus and Tashiro, 1955).

Milner (1977) developed a useful, semiquantitative procedure for isolating a strain of *B. popilliae* (referred to as var. *rhopaea*) from soil. The procedure was based primarily on the observation that germination of *B. popilliae* spores was very slow in comparison to that of spores of other sporeformers. A soil suspension was made in a medium that promoted the germination of most spores except *B. popilliae.* The suspension was subjected to seven cycles of 40-min incubation followed by heating for 20 min at 70°C to kill germinated spores and vegetative cells. Further growth selectivity was provided by incubating the plated samples anaerobically (*B. popilliae* is facultative). Milner (1977) reported that about 15% of the *B. popilliae* spores in soil produced colonies and that a spore concentration of over 1.2×10^5 spores/g dry weight of soil could be quantified. Obviously, this procedure is very time consuming, and the possibility of quantitative error is great.

Both *B. popilliae* and *B. lentimorbus* are very fastidious. For example, neither will grow in nutrient broth, or if they grow, the growth is slight and ceases after three to four transfers following isolation from an insect host (Dutky, 1940). Spores or vegetative cells of both species can be grown on plates or in broth cultures of a variety of nutritionally rich, complex media. Two commonly used complex media are J-medium and MYPGP medium.

J-Medium for Growth of *B. popilliae* and *B. lentimorbus* (St. Julian et al., 1963)

Tryptone	5.0 g
Yeast extract	15.0 g
K₂HPO₄	3.0 g
Glucose (autoclaved separately)	2.0 g
Distilled water	to 1,000 ml

The pH is adjusted to 7.3–7.5. This medium may be solidified by inclusion of 20 g of agar.

MYPGP Medium for Growth of *B. popilliae* and *B. lentimorbus* (Costilow and Coulter, 1971)

Mueller-Hinton broth	10.0 g
Yeast extract	10.0 g
K₂HPO₄	3.0 g
Glucose (autoclaved separately)	0.5 g
Sodium pyruvate	1.0 g
Distilled water	to 1,000 ml

The pH is adjusted to 7.1. This medium may be solidified by inclusion of 20 g of agar.

Viable counts of vegetative cells or spores of *B. popilliae* and *B. lentimorbus* can be made by direct plating of suitable dilutions onto either of these media. The diluent should be either growth medium or 0.1% tryptone. Dilution in water or 0.85% NaCl leads to a rapid decline in viability (St. Julian et al., 1963). (If vegetative cells or spores are used to inject into grubs, the tryptone diluent is recommended because of the toxicity of growth medium [St. Julian et al., 1963].)

Inefficient germination and/or outgrowth complicates determination of spore numbers. It is common for 1 to 5% of microscopically visible spores to produce colonies. Aging of spores at 1–5°C (Splittstoesser and Farkas, 1966; Tashiro, 1957) or at room temperature (Milner, 1981a) provides some improvement. The other commonly used procedure is activation of germination by heating at 50°C for 15 min, although the effect is not dramatic. Sharpe et al. (1970) reported that 3.3% of microscopically visible spores of *B. popilliae* NRRL B-2309M

produced colonies compared to 5.7% after 15 min of heating at 50°C. Heating at 60 to 80°C resulted in fewer colonies (St. Julian and Hall, 1968). St. Julian et al. (1967) suspended spores of *B. popilliae* in various solutions prior to heating at 50°C for 20 min and plating. Adenosine, L-alanine, and glucose, alone and in combination, had no effect on the plating efficiency. Also, KCl, NaCl, MnCl₂, CaCl₂, MgCl₂, K₂HPO₄, and Na₂HPO₄ had little or no effect. Treatment of the spores with lysozyme, trypsin, and crude snail enzyme (from *Helix pomatia*) alone and in combination with various sugars had no effect. Splittstoesser and Farkas (1966) and Splittstoesser and Steinkraus (1962) reported that heating spores of *B. popilliae* at 60°C for 15 min in CaCl₂ (1 mM) at pH 7 caused about a 1,000-fold increase in colony count. This effect was prevented by KCl in concentrations as low as 10^{-6} M. There is no clear reason for the discrepancies between the results of the two groups with regard to heating in the presence of CaCl₂ and KCl. Perhaps the age of the spores was different. Splittstoesser and Farkas (1966) used spores freshly harvested from infected grubs. The age of the spores used by St. Julian et al. (1967) was not specified. Costilow and Coulter (1971) reported that spores of *B. popilliae* NRRL B-2309M heatshocked at 60°C for 15 min exhibited a plating efficiency of 60% when plated on the MYPGP medium as described above. Pyruvate was found to be important; when it was excluded from the medium, 7.7% plating efficiency occurred. Hrubant and Rhodes (1969) reported that treatment of spores of *B. popilliae* NRRL B-2309S with a crude extracellular enzyme produced by an unidentified *Bacillus* species (NRRL B-3425) increased the percentage germination from 8–16% to 36–43%. Splittstoesser et al. (1975) observed germination and outgrowth microscopically. They reported that more than 90% of spores of *B. popilliae* germinated and outgrew within 1 h when they were suspended in cabbage looper hemolymph and tyrosinase (quantity not specified), at an alkaline pH (unspecified) at 37°C.

Cells of *B. popilliae* and *B. lentimorbus* can be maintained in the vegetative state by transferring cultures from J-medium or MYPGP agar to fresh medium once every 7 to 10 days. The plates or tubes should be incubated at 30–33°C until pinpoint colonies are present, and then refrigerated. Gordon et al. (1973) employed monthly transfers of *B. popilliae* in tubes of J-medium without glucose and to which 0.1% agar was added and incubated at 25°C for the entire month. For *B. lentimorbus*, stock cultures were maintained in a biphasic medium (Haynes and Rhodes, 1963) on a shaker at 25°C. Equal volumes of 2× J-broth without glucose and 4% agar were sterilized separately. The agar was allowed to solidify and the broth was then added. Inoculum (5%) was transferred every 2 weeks.

The most convenient way to maintain cultures for long periods of time is to make use of the natural longevity of spores. Washed spore suspensions can be added to dry, sterile soil, smeared on glass microscope slides and dried, or kept as suspensions (in water or alcohol) (Milner, 1981a; St. Julian et al., 1978). Another way of maintaining cultures is to spread hemolymph from infected grubs directly (without washing the spores) onto slides and to air-dry them (Milner, 1981a). The problem with these approaches is that spores are formed reproducibly only by infection of their host larvae (see "Ecophysiology; Growth and Sporulation In Vitro",) this chapter.

Lyophilization has been used to preserve vegetative cells of *B. popilliae* and *B. lentimorbus.* Haynes et al. (1961) scraped cells of both species from agar-solidified medium and suspended them in bovine serum prior to lyophilization. Lingg et al. (1967) prepared cells for lyophilization by suspending them in a solution containing 5% sodium glutamate and 0.5% gum tragacanth.

Identification

No phylogenetic studies have been conducted on *B. popilliae* and *B. lentimorbus.* From the taxonomic studies that have been conducted, mainly based on phenotypic observations, these species do belong in the genus *Bacillus* (Claus and Berkely, 1986). They are rod-shaped, endosporeformers (with swollen sporangia) that have a typical Gram-positive cell wall structure (Black, 1968a, 1968b), the latter as observed by electron microscopy. (Actually, the Gram-staining reaction is reported to be negative except during the sporulation process.) They are facultative organisms, although they grow much better aerobically (Sharpe, 1966). One characteristic that makes these two species (and *B. larvae*) unusual members of the genus *Bacillus* is that they lack catalase. All other members of the genus, except *B. azotoformans* and some strains of *B. stearothermophilus,* are catalase-positive (Claus and Berkeley, 1986).

B. popilliae and *B. lentimorbus* are very similar, both causing milky disease in a restricted group of beetles belonging to the family Scarabaeidae. Of 42 phenotypic characteristics recommended by Claus and Berkeley (1986) for differentiation of *Bacillus* species, these two species share all but two; only *B. popilliae* produces a parasporal crystalline inclusion and is capable of growth in 2% NaCl. [The latter characteristic may be strain variable; Milner (1974) reported that a strain he refers to as *B. popilliae* var. *rhopaea* has a parasporal body but is incapable of growth in 2% NaCl.] Kaneda (1977) divided 19 *Bacillus* species into six groups based on fatty acid composition. *B. popilliae* and *B. lentimorbus* were placed in group B with *B. larvae* and *B. polymyxa* (see Chapter 76). This group's fatty acids consisted of less than 3% unsaturated fatty acids. The predominant (39–62%) fatty acid was the anteiso-C_{15} acid (range of chain length; 14–17) (Kaneda, 1969, 1977).

Dutky (1940) noted that there was a noticeable difference in the diseases caused by *B. popilliae* and *B. lentimorbus. B. popilliae* caused "type A milky disease," in which the infected grubs became increasingly white and opaque during the course of the infection. During the last stage of the disease, just before death, the grubs exhibited a slight brownish tinge. *B. lentimorbus* caused "type B milky disease." This disease state was superficially the same as type A in infected grubs found in late summer and fall. In grubs infected in the fall that survived until spring, however, a marked difference was noted, the color of these grubs turned a muddy brown. This darkening was due to extensive formation of blood clots, which are brown to black. Clots that accumulated in appendages blocked the insects' circulation, leading to a blackened, gangrenous condition of the affected parts. The biochemical basis of this clotting phenomenon is unknown.

Other differences between *B. popilliae* and *B. lentimorbus* have been reported. They differ in lipid composition (Bulla et al., 1970a; Kaneda, 1969), and they differ somewhat in the surface topography of their ridged spores (Bulla et al., 1969). Hrubant and Rhodes (1968) showed that antisera in rabbits presumably against surface antigens of B. popilliae and *B. lentimorbus* (the cells were incubated overnight at 4°C in 0.85% NaCl and 0.5% phenol before use as antigens). Antiserum against *B. popilliae* showed no cross-reaction in agglutination tests against *B. lentimorbus* cells and vice versa. Conversely, antisera made in rabbits against sonicated vegetative cells of both species showed significant cross-reaction in double diffusion assays and immunoelectrophoresis (Krywienczyk and Luthy, 1974).

Other data that are sometimes of taxonomic value relate to the type of peptidoglycan (Schleifer and Kandler, 1972) and the major type of quinone (Collins and Jones, 1981) possessed by cultures. Comparative data on *B. popilliae* and *B. lentimorbus,* however, are lacking. The *B. len-*

timorbus peptidoglycan is of the direct cross-linked *meso*-diaminopimelate type [from Ranftl (1972) as cited by Schleifer and Kandler (1972)]. No information exists on the peptidoglycan type of *B. popilliae*. The major quinone possessed by *B. popilliae* is a menaquinone with seven isoprene units (MK-7), but no information exists on the quinone(s) possessed by *B. lentimorbus* (Hess et al., 1979).

From the above and other data, it is understandable that differences of opinion have arisen concerning the taxonomy of these bacteria. Claus and Berkeley (1986), Gordon et al. (1973), and Krieg (1961) have preferred to consider these two as separate species, giving weight to the presence or absence of the parasporal body. Krywienczyk and Luthy (1974), Milner (1981a), and Wyss (1971) have given more weight to characteristics other than the parasporal body, and, thus, consider it more appropriate to refer to *"B. lentimorbus"* as *B. popilliae* var. *lentimorbus* and *"B. popilliae"* as *B. popilliae* var. *popilliae*.

Species names other than *B. popilliae* and *B. lentimorbus* have been proposed for some milky disease isolates. A bacterium was isolated in 1956 by Beard from an Australian Scarabaeidae, specifically *Heteronychus sanctae-helenae* Blanch, that had a milky disease. A very similar strain, referred to as *B. lentimorbus* var. Maryland, was studied by Tashiro and Steinkraus (1966) and Steinkraus and Tashiro (1967). Beard (1956) believed that his isolate differed sufficiently in morphology and in host specificity from *B. lentimorbus* that it deserved separate species status; he called it *B. eulomarahae*. The sizes of the vegetative cell rods and the spores of both Beard's isolate and *B. lentimorbus* var. Maryland are 0.3 × 3 μm and 0.2–0.4 μm in diameter, respectively. This compares to the corresponding dimensions of *B. lentimorbus* (Dutky, 1940) of 1 × 5 μm and 0.9 × 1.8 μm, respectively (Beard, 1956). These are rather significant differences. Studies on many more strains, including use of molecular taxonomic techniques, are required before this species designation should be officially recognized.

A separate species designation was also given to a culture isolated from a milky diseased common cockchafer grub (*Melolontha melolontha* Linneaus) in Switzerland by Wille (1956); i.e., *Bacillus fribourgensis*. Hurpin (1966) isolated another culture from a diseased grub of *M. melolontha* in France, and designated it as *B. popilliae* var. *melolontha*. These two cultures were reported to be essentially identical in appearance of the parasporal body, in their ability to grow and sporulate in tissue cultures of *Phyllophaga anxia* hemocytes (Lüthy et al., 1970),

in their antigenic composition based on double diffusion and complement fixation (Lüthy and Krywienczyk, 1972), and in their nutritional requirements (Wyss, 1971). They differed from the original *Bacillus popilliae* isolate of Dutky (1940) in host specificity, virulence for *M. melolontha* L. by different routes (Hurpin, 1967), nutritional requirements (Wyss, 1971), and a lower maximal growth temperature (33°C as compared to 37°C for *B. popilliae* Dutky) (Wyss, 1971). Most investigators agree that separate varietal status is probably more appropriate than separate species status.

Milner (1974) in Australia isolated a parasporal body-containing strain of *B. popilliae* from *Rhopaea verreauxi* that differed from both the Dutky (1940) strain (which he referred to as *B. popilliae* var. *popilliae*) and *B. popilliae* var. *melolontha*. It differed in host specificity (Milner, 1974, 1976), maximum growth temperature (32°C) (Milner, 1974), the ability to use carbohydrates and salicylic acid (it cannot use fructose and salicylic acid, compounds used by the other two varieties) (Milner, 1974; Wyss, 1971), and the inability to grow in 2% NaCl (Milner, 1974).

Other strains of milky disease-causing bacteria exist that possess parasporal bodies that should be investigated for possible varietal status in the species *B. popilliae*. Dumbleton (1945) in New Zealand isolated a culture from *Odontria zealandica* White that differs from *B. popilliae* var. *popilliae* in the shape of the parasporal body and in host specificity. White (1947) and later Adams (1949), Harris (1959), and Dutky (1963) studied milky disease in *Cyclocephala borealis,* the northern masked chafer. The *Bacillus popilliae* strain isolated from this species is unusual in host specificity and in having large and sometimes multiple parasporal bodies. Recently Boucias et al. (1986) found that milky disease of a serious sugarcane pest in Florida, *Cyclocephala parallela* Casey, was caused by *B. popilliae*. Fowler (1972, 1974) isolated a *B. popilliae* strain from the New Zealand grass grub *Costelytra zealandica* (White). Vyss (1986) in India isolated a strain of *B. popilliae* referred to by Milner as *B. popilliae* var. *holotrichia* from *Heterodera* (= *Holotrichia*) *consanguinea* (Blanchard), a grub that causes damage to sugarcane and other wet-season crops. Feng et al. (1982) in China studied a strain of *B. popilliae* designated SH-m5 that has pathogenicity for *Holotrichia oblita* Fald., *Anomala corpulenta* Mots., and *Exolontha serrulata* (Gyllenhal).

Probably different varieties also exist in the species *Bacillus lentimorbus*. For example, Beard (1956) in Australia referred to a culture that he isolated from milky diseased *Sericesthis pruinosa* (Dalm.) as *B. lentimorbus* var. *australis*. It differed from the originally isolated *B. lentimorbus* (Dutky, 1940) in host specificity. Other strains that should be studied include a strain lacking a parasporal body that was isolated from the New Zealand grass grub *Costelytra zealandica* (Fowler, 1972, 1974), strain RM17 isolated in Australia from *Aphodius tasmaniae* (Milner, 1981b; Milner and Beaton, 1981), and the strains isolated by Boucias (1986) in Florida from *Ligyrus subtropicus* Blatchley, the worst pest of Florida sugarcane. Investigations on taxonomy of *Bacillus lentimorbus* have definitely been hampered by the difficulty in growing the organism. Some strains, for example RM17 mentioned above (Milner, 1981b), have not yet been cultivated on artificial media.

In conclusion, considerable confusion exists concerning the appropriate nomenclature for milky disease bacteria. It is not clear whether *B. popilliae* and *B. lentimorbus* should be considered separate species. Varietal designations may or may not be appropriate. Phage typing would aid in these considerations but no phages active on these organisms have been discovered. Immunological techniques have only been used to a small extent. Molecular taxonomic approaches would be of great value with regard to these issues. For example, De Ley (1978) states that to belong to a single species, cultures should have GC contents that vary by no more than 2%. The reported GC content of *B. lentimorbus* and *B. popilliae* are 37.7 and 41.3mol%, respectively (Manichini et al., 1968), thus suggesting that separate species designation is appropriate. This conclusion is rendered less certain by the report of Fahmy et al. (1985) that the GC values for the same strain of some *Bacillus* species published by different investigators varied widely (up to 14%). Nucleic acid hybridization would be of even greater value than DNA base composition, but no data exist for *B. popilliae* and *B. lentimorbus*. It is perhaps appropriate to maintain separate species status, as has been done by the authors of the section on *Bacillus* species in the current edition of *Bergey's Manual of Systematic Bacteriology* (Claus and Berkeley, 1986), until more comparative data have been obtained.

Ecophysiology

NUTRITION. Both *B. popilliae* and *B. lentimorbus* are very fastidious. Most of the media in common use employ yeast extract, casein digest, and a carbohydrate, the latter reported to be essential for growth as an energy source (Bulla et al., 1978; Rhodes et al., 1966). Defined synthetic media have been developed (Sylvester and Costilow, 1964; Wyss, 1971). Thiamine is essential for growth, a discovery first made by Dutky (1947, 1963). Biotin, although not essential, is stimulatory for growth. Eleven amino acids are essential and three are stimulatory. Purines and pyrimidines are not required. Sylvester and Costilow (1964) reported that inclusion of barbituric acid is required for growth in their medium. The role of barbituric acid is unknown (Coulter and Costilow, 1970; Sylvester and Costilow, 1964).

Carbohydrates that can be used by all varieties of *B. popilliae* and *B. lentimorbus* tested include glucose, galactose, mannose, maltose, and trehalose (Milner, 1974; Steinkraus and Tashiro, 1967). All varieties tested except for *B. popilliae* var. *rhopaea* use fructose (Milner, 1974). No hydrolysis of starch, casein, or gelatin occurs (Claus and Berkeley, 1986).

These bacteria are facultative, but their growth is greatly enhanced by the presence of oxygen, as will be discussed below (see "Metabolism" this chapter).

GROWTH AND SPORULATION IN VIVO. The normal route of infection of these microbes is by feeding. St. Julian et al. (1970) described the infectious process as occurring in four phases: 1) during the first two days no bacteria are present in the hemolymph; 2) vegetative proliferation predominates from day 3 to day 5; 3) from day 5 to day 10 vegetative growth and sporulation occur concomitantly; and 4) the last phase, which usually terminates with insect death by day 14 to 21, is characterized by massive sporulation. As indicated previously, the number of spores at the time of death is as high as 5×10^{10}/ml. The insect grub continues feeding and appears normal until near death.

The most detailed description of how the bacteria migrate from the grub's intestinal tract into the hemolymph deals with infection by *Bacillus popilliae* var. *popilliae* of the European chafer, *Amphimallon majalis* (Splittstoesser et al., 1973; Splittstoesser et al., 1978). The spores germinate in the hindgut and vegetative cells are transported to the midgut by antiperistalsis. Here the vegetative cells penetrate the epithelial cells lining the midgut, mostly at the anterior end. This penetration seems to be by a phagocytosis-like event. Nothing is known about the mechanism of attachment that must precede phagocytosis. Perhaps the glucose- and mannose-containing capsular polysaccharide (Li et al., 1985) is involved. At the area of intracellular intrusion, hemocytes aggregate to form an inflammatory capsule on the hemocoel surface of the digestive tract. Considerable bacterial cell death occurs during this intracellular migration but a few viable cells pass into the hemolymph where they proliferate and sporulate. Once in the hemolymph, no phagocytosis by hemocytes occurs. Splittstoesser et al. (1973) indicated that bacilli tend to concentrate in connective tissue sheaths or in close contact with hemocytes during the sporulation process. Schwartz and Townshend (1968) reported that *B. popilliae* spores and vegetative cells in hemolymph of Japanese beetle larvae are also not phagocytized. They noticed no effect of the bacteria on hemolymph coagulation or on the number of hemocytes.

In the case of *B. thuringiensis,* the parasporal body alone causes most of the disease symptoms and may also aid in bacterial penetration to the hemocoel. With milky disease caused by *B. popilliae,* there is no such clear evidence. *B. lentimorbus* causes a disease that is almost identical to that caused by *B. popilliae,* and has no parasporal inclusion. Weiner (1978) isolated parasporal bodies from *B. popilliae* and determined that they were proteinaceous in nature. Whole or alkali-solubilized parasporal bodies fed to third-instar larvae of Japanese beetles were not toxic. Injection of solubilized parasporal bodies did cause death. Solubilized parasporal protein of *B. popilliae* var. *melolonthae* also causes loss of viability of *Melolontha melolontha* primary hemocyte cultures (Lüthy et al., 1976).

The precise cause of death is also not understood. Dutky (1963) observed that thiamine present in the hemolymph of healthy Japanese beetle larvae was absent in hemolymph of milky-diseased larvae. This observation led him to speculate that larval death was due to starvation for essential nutrients. Lüthy (1986) speculated that death is simply due to general exhaustion coured by the heavy infection. There is a possibility that one or more toxins may be involved since cell-free filtrates of cultures of *B. popilliae* are lethal upon injection (Dutky, 1963). The toxic component(s) is inactivated by heating at 50°C for 10 min. It is unknown whether this toxin(s) is elaborated in vivo. Sharpe and Detroy (1979) found that the fat bodies of Japanese beetle larvae infected with *B. popilliae* were reduced in weight about 75% from those of healthy larvae. They suggest that this depletion may cause death simply by preventing progress to the pupal stage.

One approach used to gain an understanding of conditions promoting growth and sporulation in the larval hemolymph of Japanese beetles was to determine changes that occurred in the hemolymph during the development of milky disease caused by *B. popilliae.* The results from a decade of research, including references to many publications, were published in a review by Bennett and Shotwell (1973), and will not be discussed in this review, except as they relate to the subsequent discussion of metabolism (see "Metabolism," this chapter).

GROWTH AND SPORULATION IN VITRO. Growth of *Bacillus popilliae* is exponential in most complex media used. After the maximum population is attained, there is usually a very rapid decline in cell viability with little or no sporulation occurring (Rhodes et al., 1966). The lack of efficient sporulation in vitro has been the major factor limiting the usefulness of these bacteria for insect control (see "Biotechnological Applications" this chapter). Only spores can be used for biological control, because of the rapid loss of viability in soil that occurs with vegetative cells.

Steinkraus and Tashiro (1955) were the first to obtain sporulation of *B. popilliae* and *B. lentimorbus* in vitro. Their strategy was based on the hypothesis that sporulation was induced by nutrient deficiency. They transferred cells growing on a complete solid medium to the surface of an agarsolidified starvation medium as a paste. Some sporulation occurred, but the spores fed to larvae were reduced in virulence. In a separate study, no significant sporulation of *B. popilliae* occurred in European chafer hemolymph, macerated larvae, or larval extracts (Steinkraus, 1957b). (Efficient sporulation occurs in the hemolymph of viable larvae of this insect host.)

Rhodes et al. (1965) obtained up to 0.3% sporulation of *B. popilliae* by the use of a solid medium that contained yeast extract, sodium acetate, potassium phosphate, and agar. The strain used was a derivative of strain NRRL B-2309, referred to as NRRL B-2309S, because it was isolated from a smooth colony on acetate agar. Addition of glucose or trehalose repressed sporulation. Sporulation frequency was unchanged by addition of soil extract, Japanese beetle hemolymph, or larval extract. It was later found that they could occasionally obtain between 10^5 and 10^7 spores/ml of this strain in liquid J-medium supplemented with 1.0% charcoal (Haynes and Rhodes, 1966; Haynes and Rhodes, 1969). The brand and specific batch of charcoal (Haynes et al., 1972), yeast extract (Haynes and Crowell, 1973), and tryptone (Haynes and Crowell, 1973) were crucial. Haynes and Whih (1972) reported that it was important to use spores from the previous culture of strain B-2309S as inoculum. Use of vegetative inocula resulted in loss of ability to sporulate. It is unknown whether spores of B-2309S are infective by feeding; they are by injection (St. Julian and Bulla, 1973).

Sharpe et al. (1970) isolated another strain of *B. popilliae*, referred to as NRRL B-2309M, from strain B-2309S. They developed an MYPT medium containing Mueller-Hinton broth, yeast extract, potassium phosphate, and trehalose that supported fair sporulation (up to 1.5 $\times$ 10^6 spores/ml). Sporulation was improved when the medium was filter-sterilized, and best when on solid MYPT medium (Sharpe et al., 1970; Sharpe and Rhodes, 1973). Asporogenic sectors frequently appeared in colonies. Although strain B-2309M sporulates fairly well in vitro, the spores are not commercially useful because they are infective for Japanese beetle larvae only by injection, not by feeding (Schwartz and Sharpe, 1970; Sharpe et al., 1970).

Costilow and Coulter (1971) developed MYPGP agar medium (see "Isolation, Growth, Maintenance, and Preservation of Cultures," this chapter), that supported 12% sporulation of *B. popilliae* NRRL B-2309M. The pyruvate in the medium was shown to be essential for sporulation.

Costilow et al. developed a broth medium and cultural conditions that resulted in the production of refractile bodies in cells of *B. popilliae* and *B. lentimorbus* that were not mature spores but could be premature, abortive spores (Costilow et al., 1966; Mitruka et al., 1967). Spores and the refractile bodies ("Costilow Bodies") of *B. popilliae* were similar in RNA, DNA, and protein content. Refractile bodies did not contain dipicolinic acid, whereas spores contained 1.9% dipicolinate. Catalase, although absent from vegetative cells, was present at low levels in both spores and refractile bodies (Mitruka et al., 1967).

Yousten et al. (1974) attempted, without success, to obtain sporulation of *B. popilliae* NRRL B-2309 by resuspension of postexponential phase cells of *B. popilliae* in glucose-supplemented spent broth prepared from a sporogenous strain of *B. subtilis*. Although unsuccessful, this approach was reasonable, particularly since the recent discovery of the extracellular factor(s) produced by *B. subtilis* that is required for efficient sporulation (Grossman and Losick, 1988).

Sporulation up to 50% was reported to occur when *B. popilliae* var. *melelontha* was grown in tissue culture medium with hemocytes of *Phyllophaga anxia* larvae (a June beetle) (Lüthy et al., 1970) or hemocytes of *Melolontha melolontha* larvae (Lüthy et al., 1976). The spores were infective by feeding. *Bacillus popilliae* NRRL

B-2309 (var. *popilliae*) grew but did not sporulate in either of these systems. Although the above results are promising, tissue culture techniques are too laborious and expensive to be useful for mass production of spores (Lüthy et al., 1976).

Results recently reported by Feng et al. (1982) are promising. They claim to have observed 50–60% sporulation when *B. popilliae* SH-m5 was cultured in J-broth supplemented with thiamine-HCl (0.01 g/l). Moreover, the spores produced in vitro were infective by feeding as well as by injection into larvae of three different species: *Holotrichia oblita* Fald., *Anomala corpulenta* Mots., and *Exolontha serrulata* (Gyllenhal).

METABOLISM. *Bacillus popilliae* and *B. lentimorbus* are facultative microbes but their growth is definitely enhanced in the presence of oxygen. Sharpe (1966) showed that the growth of *B. popilliae* in a fermenter was optimal at an aeration rate of 0.5 vol/vol/min (vvm), an airflow equivalent to an oxygen absorption rate of 0.5 mmol of oxygen/l/min. Weiner (see Rhodes, 1965) measured the dissolved oxygen concentration during growth of *B. popilliae* in a fermenter. As growth proceeded, a large drop in dissolved oxygen occurred. When growth was complete (and cell viability decreased) the dissolved oxygen concentration increased to a normal value. The hemolymph of healthy Japanese beetle third-instar larvae has a dissolved oxygen concentration of 0.078–0.1 μmol/ml (Weiner et al., 1966). When vegetative cells are abundant, a 36–53% drop in dissolved oxygen occurs, followed by a return to a normal level as sporulation proceeds (Weiner et al., 1966).

There are some indications in the literature that sporulation of *B. popilliae* is somewhat oxygen sensitive. The rapid loss in viability commonly observed (Rhodes et al., 1966) when growth of *B. popilliae* in liquid media is complete obviously precludes efficient sporulation. Costilow et al. (1966) reported that cessation of shaking at the end of the growth phase decreased the death rate. Sharpe et al. (1970) obtained optimal sporulation of *B. popilliae* NRRL B-2309S by incubating cells in MYPT broth with shaking for 3 days followed by no shaking for 10 to 15 days. The fact that sporulation occurs best on solid rather than in broth media may also be due to the reduced oxygen microenvironment, which is undoubtedly present when oxygen-metabolizing cells are so closely packed together. Oxygen is required for sporulation (Sharpe et al., 1970; Sharpe and Rhodes, 1973), but there may be a critical stage that requires reduced oxygen concentration.

B. popilliae, in addition to being catalase-negative as determined by the loop test, also lacks NADH peroxidase (Pepper and Costilow, 1965). Costilow and Coulter (1971) did detect some catalase in spores, but the activity was thousands-fold less than that in more common *Bacillus* species such as *B. subtilis* (Dingman and Stahly, 1984). Pepper and Costilow (1965) and Costilow et al. (1966) hypothesized that in *B. popilliae* the production of H_2O_2 during the stationary phase may be responsible for the rapid loss of viability and consequent absence of sporulation. Cell extracts of late exponential or early stationary phase cells produce H_2O_2 via a soluble NADH oxidation system (Pepper and Costilow, 1965). Attempts to prevent loss of viability by addition of catalase to cultures have been unsuccessful (Costilow et al., 1966; Steinkraus, 1957b). However, this finding does not rule out the possibility that intracellular H_2O_2 accumulation is responsible for the loss of viability. In larval hemolymph, the oxygen concentration may be poised perfectly so that the amount of available oxygen is adequate to provide the energy needed for sporulation but not enough to favor H_2O_2 accumulation. One other enzyme required for protection against oxygen toxicity is superoxide dismutase, the enzyme that catalyzes the conversion of superoxide (O_2^-) to H_2O_2 and O_2. Superoxide dismutase is produced by *B. popilliae* at an activity level approximately equivalent to that of *Escherichia coli* (Costilow and Keele, 1972). Only a single form of superoxide dismutase is synthesized in *B. popilliae* (Yousten and Nelson, 1976) and it is induced by aerobic growth (Yousten et al., 1973).

Both *B. popilliae* and *B. lentimorbus* oxidize glucose, primarily to acetate and lactate. Small amounts of glycerol and ethanol and traces of acetoin and acetaldehyde accumulate (Pepper and Costilow, 1964). The ratio of acetate to lactate produced by *B. popilliae* increases with increased aeration. Cells incubated anaerobically do not catabolize glucose. Glucose is catabolized by both *Bacillus* species solely by the Embden-Myerhof-Parnas (EMP) and the pentose phosphate (PP) pathways (Bulla et al., 1970b; Pepper and Costilow, 1964). Under normal aeration, glucose is catabolized by *B. popilliae* primarily (75–98%) by the EMP pathway (Bulla et al., 1970b; Pepper and Costilow, 1964). When the cells are aerated with 100% oxygen, 60% of the glucose is oxidized via the EMP pathway and 40% by the PP pathway (Pepper and Costilow, 1964). When *B. popilliae* cells were harvested from hemolymph of infected Japanese beetle larvae during growth and sporulation, they catabolized 75% of the glucose by the PP

pathway and 25% by the EMP pathway; a pattern opposite from that existing for cells grown in vitro (St. Julian et al., 1975). It is possible that the oxygen concentration in vivo is the factor that favors the PP pathway, but one or more different control mechanisms may be operative.

In larvae, glucose is not the major carbohydrate used by these bacteria. Trehalose (α-D-glucopyranosyl-α-D-glucopyranoside) is the major carbohydrate in Japanese beetle hemolymph. It is present at very high concentrations, about 0.7%, a concentration that inhibits vegetative growth of *B. popilliae* in vitro (Bennett and Shotwell, 1973). This fact led Bennett and Shotwell (1973) to speculate that trehalose might be present in hemolymph in a bound form to be released as needed by the pathogen. During sporulation in vivo, the concentration of trehalose is reduced to between 0.1 and 0.47% (Bennett and Shotwell, 1973). Thus, there is a significant concentration of trehalose present during sporulation. Although in vitro sporulation of *B. popilliae* NRRL B-2309S is repressed by trehalose (Rhodes et al., 1965), it is possible that trehalose is actually required for in vivo sporulation. Trehalose is transported into cells by a phosphotransferase system using phosphoenolpyruvate (PEP) as the phosphoryl donor (Bhumiratana et al., 1974). [The uptake mechanism used for glucose transport is unknown. No PEP-glucose phosphotransferase has been detected (St. Julian et al., 1975).] The intracellular transported form of trehalose, 6-0-phosphoryl-α-D-glucopyranosyl-α-D-glucopyranoside, is cleaved to glucose and glucose 6-phosphate by phosphotrehalase, a reaction first detected in *B. popilliae* (Bhumiratana et al., 1974). The intracellular level of PEP may regulate the uptake of trehalose. PEP is more likely to accumulate when the EMP pathway is used rather than the PP pathway. Thus, Bulla et al. (1978) proposed that perhaps the dependence on the EMP pathway by in vitro grown cells results in excessive accumulation of PEP, which, in turn, results in excessive intracellular accumulation of trehalose that represses sporulation.

Most *Bacillus* species oxidize glucose to pyruvate, lactate, or acetate during growth. When the glucose is gone, growth ceases, synthesis of enzymes of the tricarboxylic acid (TCA) cycle is induced, and the acids that accumulated during growth are oxidized to CO_2 and H_2O (Hanson et al., 1963, 1964; Nakata and Halvorson, 1960). *B. lentimorbus* NRRL B-2522 and *B. popilliae* B-2309 cells grown in vitro or in vivo (*B. popilliae* only) were not capable of oxidizing acetate (Bulla et al., 1971; McKay et al., 1971; St. Julian et al., 1975). Pepper and Costilow (1964) reported that a variant of strain B-2309, strain B-2309 PA, was able to oxidize acetate. This oxidation occurred at the end of the exponential growth phase. McKay et al. (1971) showed that several other variants of strain B-2309 also were able to oxidize acetate and they presented evidence in support of the existence of a complete TCA cycle. Although some of these strains are oligosporogenic in vitro, the strains that oxidize acetate most rapidly do not sporulate in vitro (McKay et al., 1971). It is possible that operation of a TCA cycle is not required for sporulation in vivo. The constant supply of trehalose during sporulation may serve as an adequate energy source (Bulla et al., 1978).

Genetics

A problem that complicates scientific study and commercial use of these organisms is the presence of genetic instability. The loss in ability to sporulate after prolonged cultivation on artificial media is a commonly observed phenomenon. Sharpe and Bulla (1978) reported that *Bacillus popilliae* NRRL B-2309 cells when plated yielded three different colonial variants, each of which exhibited a different generation time. Each colony type reverted to a mixture of all three types, but the tendency was to transform to the fastest growing variant, which proved to be noninfective when injected into Japanese beetle larvae. There seemed to be a correlation between slow growth rate and virulence.

Practically nothing is known about the genetics of these organisms. No conjugation or transduction system is known. In fact, no phages have been discovered that infect these bacteria. Bakhiet and Stahly (1986) developed a protoplast transformation system that was effective in permitting the transformation of plasmid pHV33 into *B. popilliae* NRRL B-2309S, but the frequency of transformation was so low that transformation attempts were often unsuccessful. One finding that facilitated production of protoplasts was that cell walls of *B. popilliae* are degraded by mutanolysin; they are resistant to lysozyme (Bakhiet and Stahly, 1985a). Development of a more efficient method of transformation will be a significant step toward gaining an understanding of the genetics of these organisms.

The only published information on plasmids present in these species is the report of Faust et al. (1979) that there are two plasmids present in the strain of *B. popilliae* isolated from spore powder produced by Fairfax Biological Laboratory, Inc., Clinton Corners, New York.

D. W. Dingman (personal communication) found that chromosomal DNA of *B. popilliae*

(nine strains) and *B. lentimorbus* (one strain) is resistant and sensitive to restriction by *Mbo*I and *Dpn*I, respectively. Chromosomal DNA of each of seven other *Bacillus* species is restricted by *Mbo*I but not *Dpn*I. This is evidence for the presence of ^{6}N-methyl adenine in the GATC sequences in DNA of *B. popilliae* and *B. lentimorbus*. Modification of this sequence is presumed to occur by a DNA methylase which functions like the *dam* methylase of *E. coli* (Marinus and Morris, 1973; Lacks and Greenberg, 1977).

Biotechnological Applications

B. popilliae and *B. lentimorbus* have been used effectively for the biological control of the Japanese beetle. The Japanese beetle was so named because it was first known to exist only in Japan. In 1916, Dickerson and Weiss of the New Jersey Department of Agriculture discovered several grubs (larvae) of the Japanese beetle in a nursery in the state of New Jersey, USA. It was speculated (Fleming, 1968) that they were imported in soil surrounding the rhizomes of Japanese iris. Since that time they have spread throughout almost all of the USA east of the Mississippi River. They have also "hitch-hiked" to the northwestern USA, Europe, and Australia.

Japanese beetle larvae eat the roots of grass (yards and pastures), ornamental plants, and garden or truck crops. Adult Japanese beetles eat leaves, flowers, and fruit of trees, ornamental and floral plants, citrus fruits, berry crops, garden vegetables, and farm crops (including corn, soybeans, clover, and alfalfa) (Klein, 1981). In 1982 the total cost for prevention, treatment, and replacement related to turfgrass damage alone in the USA was estimated to be $234 million (Ahmad et al., 1983). This cost does not include the extensive damage caused by the adult beetle, and omits some damage costs caused by the larval form of the beetle.

From 1939 to 1953 a massive program, involving USA federal and state governments, was carried on to accelerate the spread of *B. popilliae* throughout the area infested by the beetle (Fleming, 1968). Adequate biological control of the beetle was attained in much of the area at the time when the program ceased. Serious problems still occur today, however, and the spread of the Japanese beetle to new areas continues.

Spores for the government-subsidized program were produced in vivo; i.e., by the artificial infection of Japanese beetle larvae. Until recently this was also true for commercially available products. Fairfax Biological Laboratory, Inc. (Clinton Corners, NY) continues to obtain their spore preparations from spores generated in vivo. Ringer Corp. (Minneapolis, MN) is producing spores in vitro by use of a patented procedure (Ellis et al., 1989) purchased from Reuter Laboratories, Inc. (formerly of Haymarket, VA). However, their product, Grub Attack®, contains *Bacillus polymyxa* rather than *B. popilliae*. Also, all four strains deposited with the American Type Culture Collection (Rockville, MD) in connection with the patent are not *B. popilliae*. The preferred production strain, ATCC 53256, is *B. polymyxa* and the other strains (ATCC 53257, 53258, and 53259) are *Bacillus amylolyticus* (D. Stahly, unpublished observations).

Other areas of interest relate to the mechanism of pathogenicity and factors regulating host specificity of different strains (Klein, 1981). Research in these areas will undoubtedly aid and strengthen the biotechnological application of these bacteria.

Bacillus larvae

Bacillus larvae was named and described in 1906 by G. F. White, who provided evidence by experimental inoculation that this bacterium caused a fatal disease of honey bee larvae called American foulbrood (White, 1907). Maassen (1906) in Germany and Burri (1904) in Switzerland isolated cultures from larval remains that were named *B. brandenburgiensis* and *B. burri* (Cowan, 1911), respectively. These cultures later proved to be identical to White's isolate. Nomenclatural priority has been given to White. The only known host for *B. larvae* is the honey bee, *Apis mellifera* Linnaeus.

The use of the term "American" in "American foulbrood" is somewhat misleading because the disease occurs throughout the world. The term "foulbrood" is derived from the odor of decaying larvae (brood), described as resembling that of burned glue.

American foulbrood is a very serious disease in part because the bacteria sporulate in the larval remains so that the spores [more than 10^9 spores per larva (Holst, 1946; Shimanuki, 1980)], which are stable for at least 35 years (Haseman, 1961), are left to infect other larvae. When a colony is determined to be infected, government regulations require that the colony be destroyed (most commonly accomplished by burning). Shimanuki (1980) estimated that about 3% of all colonies inspected in the USA are infected with American foulbrood. The direct loss to U.S. apiaries from lost honey and beeswax (based on a total annual value of these products of $132 million [Levin, 1984]) is about

$4 million. This is a minor part of the total loss, however. It was estimated that in 1985 the value of the increased yield and quality of U.S. crops due to honey bee pollination was $9.3 billion (Robinson et al., 1989a, 1989b). Accordingly, the value of the lost pollination (3%) would be $279 million.

Habitats

The major source of *B. larvae* is the larval remains of honey bees in honeycombs. In a hive infected with American foulbrood, spores of *B. larvae* can also be isolated from the honey, wax, pollen, and hive walls (Gochnauer, 1981; Shimanuki and Knox, 1988).

Isolation, Growth, Maintenance, and Preservation of Cultures

Usually, *B. larvae* can be easily isolated from larval cadavers, since they commonly are present as almost pure cultures (Holst, 1945; Sturtevant, 1924). The relative lack of other organisms suggested to Holst (1945) that *B. larvae* might produce an antibiotic. Indeed, he found that one or more antibiotics were produced during sporulation that were effective against a variety of Gram-positive and Gram-negative species.

Sturtevant (1924) isolated pure cultures from diseased larval cadavers simply by suspending a small amount of the dried remains in water and then streaking a loop of the suspension onto a solid medium.

Rose (1969) developed a method for isolating *B. larvae* from a mixture of other bacteria, including spore-formers. His method is based on the fact that *B. larvae* spores, unlike those of most other sporeformers, do not readily germinate in nutrient broth (Rose, 1969; White, 1906). A few drops of an aqueous suspension of dried larval cadavers were added to a tube of nutrient broth and the tube was heated at 70°C for 15 min. Then it was incubated at 37°C for 12 h. Two additional heat treatments with an intermediate incubation were used before plating on a medium permitting germination of *B. larvae* spores and growth.

Shimanuki and Knox (1988) developed a method for determining the number of *B. larvae* spores in honey. Problems with honey are its high viscosity and the presence of antibacterial substances. Their procedure, in brief, involved heating the honey at 45°C to decrease its viscosity, diluting with water, dialysis, centrifugation, resuspension of the pellet in water, heating at 80°C for 10 min to kill vegetative cells, and plating on a solid medium. By use of this method they showed that *B. larvae* spores were present in 30% of beekeeper-packed honey and 100% of commercial-packed honey (blended from many sources).

Spores or vegetative cells of *B. larvae* can be grown on a variety of solid or liquid media. The medium used by Gordon et al. (1973) was the J-medium (St. Julian et al., 1963), described above for growth of *B. popilliae* and *B. lentimorbus*. Another medium that is satisfactory for the routine growth of *B. larvae* is "modified MYPGP medium", similar to the MYPGP medium described above.

Modified MYPGP Medium for Growth of *Bacillus larvae* (Dingman and Stahly, 1983)

Mueller-Hinton broth	10 g
Yeast extract	15 g
K₂HPO₄	3 g
Glucose (autoclaved separately)	2 g
Sodium pyruvate	1 g
Distilled water	to 1,000 ml

This medium may be solidified by inclusion of 20 g of agar.

When these or other solid media are used for determination of spore count, cultures are first heated at 65°C for 15 min to kill vegetative cells (Dingman and Stahly, 1983; Rose, 1969). A problem with spore counts is the low plating efficiency observed. Dingman and Stahly (1983) reported that heat-resistant counts were usually about 6% of the direct microscopic spore counts. The explanation for this discrepancy is unknown. Heat shocking of spores of *B. larvae* NRRL B-3650 at 60–80°C before plating promoted little to no increase in plating efficiency (Dingman, 1983).

B. larvae cultures can be maintained by growth on either of the above agar solidified media at 30–37°C for 2–3 days. Plates or slants are stored at 4°C and transferred every 2 weeks.

Frozen spore stocks are the most convenient for long-term culture preservation. Most *B. larvae* cultures exhibit low percentage sporulation when grown on J-agar or modified MYPGP agar at 37°C. Vegetative cells can be preserved by conventional lyophilization or by suspension in a freezing medium (e.g., 70-ml Hanks' 1× medium, 10-ml rabbit serum, 20 ml glycerol) and storage at −70°C.

Identification

No phylogenetic or molecular genetic studies have been performed on *B. larvae* other than an analysis of the nucleotide composition of its DNA; the type strain DNA has a GC content of 50 mol%, based on buoyant density (Nakamura, 1984).

Phenotypic observations indicate that *B. larvae* belongs in the genus *Bacillus* (Claus and Berkeley, 1986). The cells are rod shaped, and endospores are formed within swollen sporangia. No parasporal inclusions have been observed. The vegetative cells have a typical Gram-positive cell wall structure (Bakhiet and Stahly, 1985b). The endospores have a smooth surface, and the inner spore coat is somewhat unusual in that it consists of seven distinct lamellae (Bakhiet and Stahly, 1985b; Bulla et al., 1969).

Some strains are motile by peritrichous flagella. "Giant whips" are often microscopically observed in these strains (Gochnauer and L'Arrivee, 1969a; White, 1907). These were thought to be due to flagella actively discharged from the cells and in an aggregated state (Frank and Hoffman, 1968; Gochnauer and L'Arrivee, 1969a). Ludvik et al. (1983) observed by electron microscopy that in old sporulating cultures cell wall material separates from the cell surface in the form of thick filaments. They observed that the "giant whips" formed when free flagella aggregated with this material.

Other characteristics that are often of value in speciation are type of cell wall peptidoglycan (Schleifer and Kandler, 1972), cellular lipid composition (Kaneda, 1977), and major type of quinone present (Collins and Jones, 1981). Nothing is known about the type of cell wall peptidoglycan possessed by *B. larvae*. Based on lipid composition, Kaneda (1977) placed *B. larvae* in group B (out of six groups in the genus *Bacillus*) together with *B. popilliae*, *B. lentimorbus*, and *B. polymyxa*. The major quinone possessed by *B. larvae* is a menaquinone with seven isoprene units (MK-7). A minor amount of MK-3 is also present (Hess et al., 1979).

Bacillus larvae is similar to *B. popilliae* and *B. lentimorbus* in that it is fastidious. It does not survive serial transfer in nutrient broth. It is a facultative bacterium but it grows much better aerobically (Dingman and Stahly, 1984; Lochhead, 1928). Similar to *B. popilliae* and *B. lentimorbus*, *B. larvae* lacks catalase activity, as measured by observation of oxygen bubbles when a small drop of 10–30% H₂O₂ is applied to the colonies (Haynes, 1972). The catalase test is probably the most helpful diagnostic tool to use after initial isolation of the microbe, in addition to colony appearance.

The many other phenotypic characteristics of *B. larvae* that differentiate it from other *Bacillus* species are reviewed by Claus and Berkeley (1986) and will not be discussed here except as they relate to possible varietal diversity within the species. Jelinski (1985) conducted a study of biochemical reactions of 110 strains of *B.*

larvae from five sources. The strains were uniformly positive or negative for most tests. They were variable for nitrate reduction (58% of strains positive), acid production from mannitol (17% positive), and acid production from salicin (11% positive). If denitrification is actually occurring in the positive strains, then the difference between the positive and negative cultures might be sufficient to at least consider varietal differentiation. However, it has not been established that denitrification occurs in the positive strains. No published studies have indicated whether better growth occurs anaerobically in the presence or absence of nitrate. No studies have indicated whether nitrous oxide or nitrogen, products that normally result from denitrification, are produced by the positive strains. The other variable traits may or may not be significant from the perspective of a genetic relationship. Obviously, GC content and nucleic acid hybridization data are needed.

Immunological studies have been performed by several investigators. Antisera against *B. popilliae* and *B. lentimorbus* showed some cross-reaction in agglutination assays with *B. larvae* (Hrubant and Rhodes, 1968). Giauffret et al. (1970) demonstrated some cross-reaction in precipitation and immunofluorescence between *B. larvae* and *B. alvei*. Cross-reaction between *B. larvae* and *B. alvei* was not observed in immunodiffusion and immunofluorescence assays conducted by Peng and Peng (1979). Otte (1973) reported that antiserum against *B. larvae* produced little to no cross-reaction with 36 strains of other bacteria by immunofluorescence. Preliminary results of Giauffret et al. (1970) indicated that there are some important serological differences between strains of *B. larvae,* but more detailed investigations are needed to determine if varietal differentiation is justified.

Bacteriophage typing has been employed for varietal differentiation of *B. thuringiensis* and *B. sphaericus* strains. Several temperate phages of *B. larvae* have been isolated by different investigators (see "Genetics," this chapter). Some strains of *B. larvae* are sensitive, and some are immune (or resistant). Some of the strains are immune simply because they possess the same or related phages in the prophage state (Bakhiet and Stahly, 1988; Dingman et al., 1984; D. P. Stahly, unpublished observations). It is not clear, however, whether some of the cases of resistance are an indication of more basic cellular differences between *B. larvae* strains. More research is required before it can be determined whether phage typing will be of value for varietal differentiation.

Ecophysiology

NUTRITION. As stated previously, *Bacillus larvae* is fastidious, failing to grow on nutrient agar. This characteristic led early investigators to supplement their usual media with many different natural substances: e.g., macerated, healthy bee larvae (White, 1906); calf or pig brain extract (Maassen, 1908); egg yolk (White, 1919); egg yolk and yeast (Sturtevant, 1924); yeast extract and carrot extract (Lochhead, 1933); and minced chicken embryo (Tarr, 1937a). In 1942 Lochhead found that thiamine replaced all of these special requirements. His basic medium consisted of glucose, peptone, thiamine, K_2HPO_4, and various salts. Foster et al. (1950) found that poor growth occurred in the complex medium of Lochhead (1942) unless it was pretreated with charcoal, supplemented with starch, or a combination of the two. Their hypothesis is that anti-growth substances were removed or sequestered from the medium. Katznelson and Lochhead (1948) attempted to develop a synthetic medium for *B. larvae*. In addition to thiamine, a purine, either xanthine or guanine, was required. However, when a basal salts-glucose medium was supplemented with 18 amino acids, thiamine, and a purine, growth sufficient for serial transfer only occurred when the medium was made semisolid with agar.

B. larvae is capable of catabolizing the carbohydrates glucose, fructose, galactose, trehalose, glycerol, mannose, and ribose. Use of mannitol, xylose, sucrose, and salicin is strain variable (Azuma and Kitaoka, 1965; de Barjac, 1981; Jelinski, 1985; Lochhead, 1928).

B. larvae is facultative but grows much better in the presence of oxygen (Bailey, 1968). Bailey and Lee (1962) inoculated tubes containing a semisolid medium with spores of *B. larvae* and, after incubation, observed growth in a band (sometimes two bands) 5–10 mm below the surface. The authors concluded that the spores germinated and grew best under microaerophilic conditions. Dingman (1983) observed similar banded growth after inoculating semisolid medium with vegetative cells. The evidence of Bailey and Lee (1962) may not relate to the oxygen requirement for germination, as claimed, since it is possible that germination occurs at positions in the tubes other than where growth occurs.

GROWTH AND SPORULATION IN VIVO. Larvae of honey bees become infected by ingestion of spores of *B. larvae*. Spores are the only infective form (Tarr, 1937b), probably because vegetative cells are rapidly killed by royal jelly, the food material in which the larvae are immersed (Holst, 1946). When vegetative cells of *B. larvae* were suspended in royal jelly (10^7 cells/ml), < 10 viable cells/ml remained after 30 s at room temperature (N. Bakhiet and D. P. Stahly, unpublished observations). The bactericidal effect, due primarily to 10-hydroxy Δ^2-decenoic acid (Blum et al., 1959), is apparently diminished or absent in the midgut, where the spores germinate and grow.

The larvae are very susceptible to infection only up to an age of about 1.5 days. After 1.5–2 days of age they become totally resistant (Bamrick, 1964; Bamrick and Rothenbuhler, 1961; Hoage and Rothenbuhler, 1966). Davidson (1973) studied the histopathology of American foulbrood and in the process discovered a possible mechanism for the age-related development of resistance. She determined that vegetative cells in the midgut lumen invaded the peritrophic membrane by a phagocytosis-like process, traversed the epithelium, and entered the hemocoel, where they grew to high populations and caused a generalized septicemia. The first barrier, the peritrophic membrane, although present in larvae as young as 8 h, increases tremendously in thickness and "density of composition" with age (Davidson, 1970, 1973). Davidson suggested that this structural change may contribute to the age-related resistance. Nothing is known about the mechanism of attachment of *B. larvae* cells to the peritrophic membrane that must precede phagocytosis. Perhaps the glucose, mannose, and galactose-containing capsular polysaccharide (Li et al., 1985) is involved.

Death usually occurs at an age of 8 to 11 days; 2 days before or after the transition to the pupal stage. As stated above, death is preceded by growth in the hemolymph to high populations and a general invasion of most body tissues. Davidson (1973) reported heavy infection of the fat body and rupture of hemocytic membranes. The dead larvae rapidly decompose as *B. larvae* exhibits massive sporulation.

The question of whether a toxin or an enzyme of *B. larvae* is involved in pathogenicity is an open one. Aqueous extracts of cadavers of larvae killed by American foulbrood are toxic by feeding to honey bee larvae (Patel and Gochnauer, 1959). Such larval remains have a high level of protease activity (Holst and Sturtevant, 1940; Patel and Gochnauer, 1959), suggesting that protease activity and toxicity are related. Heating of the extract for 1 min at 100°C inactivated both protease activity and toxicity (Patel and Gochnauer, 1959). Holst and Sturtevant (1940) demonstrated protease activity only in sporulating cultures of *B. larvae*. Furthermore, certain asporogenic mutants did not produce protease. These facts led Bamrick (1964) to question involvement of proteases in the disease process; he suggested that their involvement was probably in postmortem decomposition, a process that occurred simultaneously with the appearance of spores. Gochnauer (1969) and Patel and Gochnauer (1972) separated multiple (two to three) proteases from cells at different stages of development. One enzyme, stated to be associated with "vegetative cells," was obtained from cells in cultures that were 3–10 days old; probably stationary-phase cells. Protease excretion by other *Bacillus* species occurs early in the sporulation process, at a time coincident with the beginning of the stationary phase (reviewed by Freese and Heinze, 1983). Since proteases are normally excreted early in sporulation, it is possible that proteases might be involved in late pathological changes in larvae, even before refractile spores are visible. The availability of a protease-minus, spore-plus mutant would aid in analyzing this question. Proteases are not involved in the initial, early penetration of *B. larvae* of the peritrophic membrane and epithelial cells. This conclusion is supported by the observation of Davidson (1973) that penetration of the peritrophic membrane is not accompanied by physical evidence of enzymatic digestion of the membrane.

SPORULATION IN VITRO. Whereas *Bacillus larvae* sporulates efficiently in the hemolymph and other tissues of bee larvae, most strains sporulate very poorly in or on artificial media. The study of American foulbrood has been slowed somewhat by the difficulty of obtaining large numbers of spores in vitro. Interest about in vitro sporulation is heightened by the possibility that what is learned about sporulation in *B. larvae* may, in part, be applicable to the other catalase-minus insect pathogens, *B. popilliae* and *B. lentimorbus*.

Sporulation has been reported to occur at a higher frequency in cells grown on solid media rather than in broth media. A variety of solid media has been used which promotes sporulation of *B. larvae* (Azuma and Kitaoka, 1965; Bailey and Lee, 1962; Dingman and Stahly, 1983; Foster et al., 1950; Gordon et al., 1973; Lochhead, 1928; Smith et al., 1949). Azuma and Kitaoka (1965) indicated that they obtained sporulation by growth on slants in tubes "closed tightly with rubber stoppers"; an indication that the sporulation process may be somewhat oxygen-sensitive.

Gochnauer (1969) and Gochnauer and L'Arrivee (1969a) used a biphasic system to obtain sporulation. The bottom phase consisted of

Difco brain-liver-heart agar. The top phase was a shallow layer of broth medium of the same components as the bottom phase (except agar) or a yeast extract, starch, and potassium phosphate medium (Bailey and Lee, 1962). Cultures were incubated at 35°C for 5–10 days without shaking.

There are only three reports of fair-to-good sporulation in single-phase, liquid media. St. Julian and Bulla (1971) obtained 2×10^7 spores/ml of the type strain, *B. larvae* NRRL B-2605, in liquid MD medium, which contains yeast extract, glucose, and K_2HPO_4. Gochnauer (1973) obtained 10^8 spores/ml after incubation with gentle shaking of *B. larvae* 34A in brain heart infusion broth plus thiamine hydrochloride. Dingman and Stahly (1983) obtained 5×10^8 spores/ml after incubation with gentle shaking of *B. larvae* NRRL B-3650 in TMYGP broth, which contains tris-maleate buffer, yeast extract, glucose, and sodium pyruvate. The level of aeration was critical to achieve sporulation of strain B-3650; aeration was required, but at a level much reduced from that used to obtain sporulation of more typical *Bacillus* species. This finding is consistent with the observation of Fitz-James and Young (1969) that sporulation of *B. larvae* was inhibited by excess shaking.

Lodesani et al. (1985) used a rather nontraditional method to obtain sporulation of strains that sporulate poorly if at all on solid media. A culture was grown in a broth medium, cells were concentrated by centrifugation and suspension in a small volume of phosphate buffer, and the suspension was placed in a sterile dialysis bag suspended in a broth medium containing yeast extract, potassium phosphate, and starch. The culture was incubated without shaking at 34°C for 30 days. The authors stated that, depending on the strain, 10–80% of the cells introduced into the dialysis bags sporulated.

METABOLISM. Aspects of metabolism discussed below deal with oxygen metabolism, protection against toxic forms of oxygen, pathways of glucose catabolism, and nutrient requirements for sporulation.

Bacillus larvae, when grown in the presence of air, consumes oxygen as would be expected (Dingman and Stahly, 1984). Most oxygen-utilizing bacteria possess superoxide dismutase to protect against superoxide (O_2^-) and catalase and/or peroxidase to protect against H_2O_2. *B. larvae* NRRL B-3650 resembles *B. popilliae* and *B. lentimorbus* in that it has superoxide dismutase levels approximately equal to those of *B. subtilis* 168, a "typical" *Bacillus* species, but lacks NADH peroxidase and has very low cat-

alase activity (Dingman and Stahly, 1984). Catalase activity is nondetectable during growth. It appears during the early stationary phase and increases during the time of appearance of refractile spores. The activities at these two stages are about 2,700 and 5,400 times less than those exhibited by *B. subtilis*. The relative absence of catalase (and NADH peroxidase) may be the cause of the apparent oxygen-sensitivity of sporulation (Dingman and Stahly, 1983). *B. larvae* does not accumulate H_2O_2 extracellularly, but intracellular accumulation remains a possibility (Dingman and Stahly, 1984).

Mylroie and Katznelson (1957) reported that *B. larvae* metabolizes glucose primarily to acetate during the exponential growth phase. They demonstrated the presence of enzymes of both the Embden-Meyerhoff-Parnas pathway and the pentose phosphate pathway. In an extension of this work, St. Julian and Bulla (1971) showed that key enzymes of the Entner-Doudoroff pathway, as well as those of the other two pathways, are present in *B. larvae* NRRL B-2605. This is rather unusual since the Entner-Doudoroff pathway has not been demonstrated in other *Bacillus* species, and the pathway rarely occurs in Gram-positive bacteria. Radiorespirometric analysis, however, indicates that the Entner-Doudoroff pathway is not being used to a significant extent; the pentose phosphate pathway is the primary pathway used.

No information exists on how glucose is transported into cells; no phosphoenolpyruvate: glucose phosphotranferase activity has been detected (St. Julian and Bulla, 1971). Also, no information exists on the mechanism of uptake and catabolism of trehalose, the main carbohydrate in honey bee larval hemolymph.

St. Julian and Bulla (1971) showed that glucose degradation by *B. larvae* NRRL B-2605 occurs during growth, resulting in a decline in pH due to acid (probably acetic acid) accumulation. When growth ends coincident with depletion of glucose from the medium, the pH increases. This pH increase is due to complete oxidation of acetate through the citric acid and glyoxylic acid cycles.

The pattern of glucose utilization by *B. larvae* NRRL B-3650 is surprisingly different from that described above for the type strain (NRRL B-2605) (Dingman and Stahly, 1983). Strain B-3650 does not degrade glucose during growth; glucose oxidation begins at the end of exponential growth. The only component of TMYGP broth required for growth of strain B-3650 is yeast extract. Acids were produced during growth, presumably from yeast extract components of the TMYGP broth (see "Ecophysiology; Sporulation In Vitro," this chapter).

Thus, glucose usage in this strain may be regulated by a type of catabolite repression. Another unusual property of strain B-3650 is that both glucose and pyruvate are required for sporulation. In most *Bacillus* species glucose represses sporulation. Glucose is needed by strain B-3650 for a late event(s) in sporulation; it could be added as late as 35 h (25 h after the end of exponential growth) without any decline in the number of spores formed. (Spores are first apparent at 45 h.) This requirement can be partially satisfied by substitution of glucosamine or potassium gluconate, but not by galactose, fructose, mannitol, or glycerol. Why pyruvate is required for sporulation is not readily apparent. Pyruvate is depleted from the medium before sporulation begins. Substitution of acetate, lactate, citrate, glutamate, or succinate does not satisfy this requirement (Dingman, 1983; Dingman and Stahly, 1983).

Genetics

Bacillus larvae exhibits instability with regard to sporulation ability and virulence. Tarr (1937b) reported that serial transfer on an artificial medium resulted in loss of ability to sporulate. Holst and Sturtevant (1940) observed frequent asporogenic sectors in colonies. Stahly and Livasy (unpublished observation) have made similar observations with *B. larvae* NRRL B-3650. Asporogenic colonies remained asporogenic; i.e., the change was not reversible. Shimanuki et al. (1965) indicated that virulence of *B. larvae* was increased by passage three times through honey bee larvae. To understand these and other phenomena, an understanding of the genetics of *B. larvae* is highly desirable.

Conditions have been developed for polyethylene glycol-mediated plasmid transformation and phage DNA transfection of protoplasts (Bakhiet and Stahly, 1985a; Chen and Yin, 1986). Although several phages have been isolated from *B. larvae* (see below), no reports of transduction exist. Also, there have been no reports of conjugation.

Three morphologically distinct temperate bacteriophages have been isolated from *B. larvae*. PBL1, originally isolated from *B. larvae* NRRL B-3553 by Gochnauer (1955, 1970), has been purified and characterized (Dingman et al., 1984; Gochnauer and L'Arrivee, 1969b). Based on comparative electron microscopic data, phage BLA, isolated by Drobnikova and Ludvik (1982), is a PBL1-like phage. PBL1-like phages are very common in *B. larvae* strains isolated from different geographical locations within the USA (D. P. Stahly, unpublished observations). Another temperate phage, PBL0.5,

was isolated from strain NRRL B-3558 (Dingman et al., 1984) and characterized (Bakhiet and Stahly, 1988). An apparently identical phage, PBL2, was isolated from strain NRRL B-3553, where it was found to coexist in the prophage state with PBL1 (Bakhiet and Stahly, 1988). It is probable that phage BL2, isolated by Benada et al. (1984b), is a PBL0.5-like phage; a conclusion based on comparative electron microscopic data. A morphologically distinct temperate phage, PBL3, was isolated from *B. larvae* GA and was partially characterized (Campana, N. Bakhiet, and D. P. Stahly, unpublished observations). It is hoped that these phages will be of value for phage typing and, perhaps, for development of a transduction system(s). None of the phages PBL1, PBL0.5, and PBL3 produced plaques on a variety of other *Bacillus* species, including *B. popilliae* (H.-W. Ackermann, personal communication; D. Takefman and D. P. Stahly, unpublished observations).

The only published information on plasmids in *B. larvae* is that of Benada et al. (1984a). They found cryptic plasmids in three of 12 strains tested.

Preliminary evidence suggests that a restriction-modification system exists in strain NRRL B-3555 (Bakhiet and Stahly, 1985).

Bacillus thuringiensis

Bacillus thuringiensis is a Gram-positive, facultative, sporeforming, rod-shaped bacterium that has been the subject of intense investigation in laboratories worldwide. As a member of the genus *Bacillus,* the organism shares with the other members of the taxon the ability to form endospores that are resistant to inactivation by heat, desiccation, and organic solvents. Biochemically, sporulation in *B. thuringiensis* is very similar to that in other sporeforming bacteria and the spores resemble those of other bacilli in morphology and composition. However, the formation of one or more parasporal crystalline bodies adjacent to the spore during stages III to V of sporulation distinguishes *B. thuringiensis* from most other *Bacillus* species (Andrews et al., 1982, 1987; Bulla et al., 1985).

B. thuringiensis is probably best known as an insect pathogen. Older cultures, which contain spores and parasporal crystals, are highly toxic to larvae of members of certain insect orders. This property has caused extensive interest in this organism because of the potential to use it to formulate more effective, specific, and environmentally safe insecticides (Bulla et al., 1985).

Although *B. thuringiensis* is capable of saprophytic existence in soil, it is also a frank insect pathogen. Not surprisingly, the initial discovery of *B. thuringiensis* was as an insect pathogen, since its pathogenicity made its existence in the environment more obvious. Probably the first description of *B. thuringiensis* was by Ishiwata, who, in 1901, described the etiological agent of "sotto disease" or "flacherie." Significantly, Ishiwata was also the first to describe the relationship between the culture's age and pathogenicity. Young cultures of the "sotto disease bacillus," as he termed his isolate, were not as pathogenic to larvae of the Japanese silk moth, *Bombyx mori,* as were older cultures (Bulla et al., 1980; Ishiwata, 1901).

Probably the first to suggest the use of *B. thuringiensis* for controlling insect pests was Berliner (1911, 1915), who, apparently independently of Ishiwata, described the causative agent of a disease of the Mediterranean meal moth, *Anagasta kuhniella.* Berliner, and later Mattes, described the crystalline inclusion body, or "Restokorper," within sporulated cells, and it was Berliner who first used the species name *thuringiensis,* deriving the name from the German province of Thuringia (Bulla et al., 1980).

As reviewed by Andrews et al. (1987), from the initial discovery of the microbe until the early 1950s, numerous studies appeared in the literature aimed at using *B. thuringiensis* in insect control without sufficient basic information to accomplish these goals. In the early 1950s, investigators began studying the basic biology of *B. thuringiensis* and considering some fundamental properties of its insect pathology. Steinhaus (1951), for example, published an electron micrograph of the crystal toxin, and in 1953 Hannay described the development of the crystal toxin in relation to the sporulation cycle. Hannay (1953) and Hannay and Fitz-James (1955) observed that the diamond-shaped crystals or parasporal bodies are composed primarily of protein and suggested that the crystals might be involved in the establishment of septicemia in insect larvae. Angus (1956a, 1956b) prepared toxic filtrates that included crystals alone and crystals plus spores and showed that the crystals were responsible for midgut paralysis and cessation of feeding, whereas the spores were required for septicemia.

Habitats

Bacillus thuringiensis is a member of a limited group of bacteria that occurs both naturally and can be added to an ecosystem to achieve insect control. A somewhat arbitrary distinction is made between "natural" and "artificial" habitats of *B. thuringiensis.* A habitat is defined as natural when *B. thuringiensis* can be isolated when there is no previous record of application of the organism to that ecosystem. A habitat is considered artificial, on the other hand, when there has been previous application of the organism for insect control. It is important to note that, in the latter case, artificial and natural may not be mutually exclusive; *B. thuringiensis* may occur naturally in environments where additional organisms have been applied.

Most *B. thuringiensis* cultures that have been studied were either isolated directly from infected insects or from soil associated with them. The spores of *B. thuringiensis* readily persist in soil and vegetative growth occurs when nutrients are available. All of the early isolates were pathogenic for insects and this pathogenicity became recognized as a differentiating characteristic of the species, in addition to possession of parasporal bodies. However, it is now apparent that *B. thuringiensis* isolates can be obtained from soil that has not been exposed to *B. thuringiensis*-derived insecticides and in which susceptible insects are not abundant. Moreover, many of these isolates are nontoxic for all insects tested. For example, Ohba and Aizawa (1986) isolated 189 cultures of *B. thuringiensis* from 136 soil samples collected from nonsericultural areas of Japan. The classification was based in part on possession of parasporal bodies (and not insect pathogenicity). According to Ohba et al. (1988), the majority of such *B. thuringiensis* isolates are nontoxic for a variety of insects.

De Lucca et al. (1981) conducted a similar survey of the presence of *B. thuringiensis* in soil in the USA. They found that *B. thuringiensis* represented between 0.5 and 0.005% of all *Bacillus* species isolated from the soil samples. Travers et al. (1987) screened soil samples collected from random locations in Montgomery County, MD, USA, for the presence of *B. thuringiensis.* It was found in almost all of the soil samples examined. Thus, it is obvious that *B. thuringiensis* is widespread in nature.

One might expect that epizootics caused by *B. thuringiensis* would be frequent, considering the widespread occurrence of the species in soil. However, epizootics are somewhat rare. Most epizootics have been limited to situations wherein the insect density is relatively high, providing better opportunity for establishing the disease within the population. In high density infestations of Douglas-fir tussock moth populations, for example, *B. thuringiensis* epizootics become a factor in reducing defoliation and eventual collapse of the population (Tunnock et al., 1974). Similarly, outbreaks of disease in insect populations have been reported in stored grain pests (Burges and Hurst, 1977; Vankova and Purrini, 1979) and in the European corn borer (Lynch et al., 1976). *B. thuringiensis* subsp. *israelensis,* which is a strain highly toxic to larvae of mosquitoes and some other hematophagous-insects (Klowden and Bulla, 1984; Tyrell et al., 1979), was first isolated from pond water in Israel wherein many diseased larvae were observed (Goldberg and Margalit, 1977; Margalit and Dean, 1985).

The artificial habitats are the locations where *B. thuringiensis* insecticides (usually a mixture of spores and crystals) are applied. Insecticides formulated with *B. thuringiensis* are currently being manufactured and used worldwide. Major producers in the USA, Europe, and the USSR make insecticides formulated with *B. thuringiensis*. Bulla et al. (1985) defined two groups of insecticides formulated with *B. thuringiensis.* Group I consists of products formulated with strains having toxicity against lepidopteran insects, whereas group II products are those with toxicities similar to *B. thuringiensis* subsp. *israelensis,* which are highly toxic to mosquitoes and certain other hematophagous dipteran larvae. With the discovery of strains such as subsp. *tenebrionis,* which are toxic to coleopteran larvae, a group III needs to be defined. Accordingly, *B. thuringiensis* may be applied as an insecticide on foliage, in soil, in water environments, and in food storage facilities (e.g., grain bins).

Isolation, Preservation, and Growth of Cultures

Isolation of *B. thuringiensis* from soil or other natural environments is greatly facilitated by use of selective techniques. Travers et al. (1987) took advantage of the observation that germination of spores in crystal-forming bacilli, including both *B. thuringiensis* and *B. sphaericus,* is inhibited by sodium acetate concentrations of approximately 0.25 M. Soil, which contained up to 10^9 bacteria/g, was inoculated into a nutrient medium that contained the sodium acetate. After a period of growth, the vegetative cells were eliminated by heat treatment and the remaining spores were isolated on a nutrient medium without acetate. The survivors from this treatment ranged from 20–96% *B. thuringiensis* and/or *B. sphaericus.* These two species are easily differentiated by observation of colonial and cellular morphology. Similarly, Saleh et al. (1969) exploited the observation that *B. thuringiensis* is relatively resistant to polymyxin B and penicillin G to isolate *B. thuringiensis* from soil.

Preservation of *B. thuringiensis* cultures is of special importance because of the common presence of multiple plasmids, some of which carry vital information; e.g., the crystal toxin gene (Faust et al., 1979; Gonzales et al., 1982). These plasmids may be lost during routine subculturing, and, therefore, crystal toxin production and, presumably, other plasmid-borne phenotypes may be simultaneously lost (Stahly et al., 1978). Investigators in R.E. Andrews' laboratory use three methods for culture preservation.

For long-term culture storage, cultures are grown until sporulation occurs (about 24 h) in liquid GYS medium (Nickerson and Bulla, 1974). After >90% of the cells in the culture contain mature, phase-bright spores, the cells are harvested by centrifugation and resuspended in 20% skim milk. This material is then lyophilized by standard techniques (Gherna, 1981). For routine laboratory operations, several agar slant cultures are prepared at the same time by inoculation with the *B. thuringiensis* culture to be preserved. These are incubated at 30°C for 4 days to allow extensive sporulation and then stored at 4°C for up to 6 months. When a culture is required for an experiment it can be started from one of these stored slants, which are used only once after they are removed from cold storage. The slant methodology has certain disadvantages. The inoculum contains frequently variable amounts of spores and the culture outgrowth can be somewhat erratic. For synchronized sporulation of *B. thuringiensis,* another method is used. The culture is grown to midlogarithmic phase in liquid modified GYS medium ($A_{600} = 0.5$) and then diluted 1:50 in modified GYS medium containing 50% (vol/vol) glycerol. This material is divided into 100-μl portions and stored in sterile Eppendorf test tubes at −70°C. Viability remains high for up to 1 year.

B. thuringiensis is not particularly fastidious and can be routinely cultured on many complex media; e.g., Difco nutrient agar, LB agar (Maniates et al., 1982), and Difco brain heart infusion agar. Although *B. thuringiensis* is facultative, it grows best aerobically. For growth in liquid media, cultures are usually aerated by vigorous shaking of cultures in Erlenmeyer or Fernbach flasks. Those in the R.E. Andrews' laboratory use 100 ml of medium in 300 ml Erlenmeyer flasks and 1 l of medium in 2.8-l Fernbach flasks. The use of flasks with baffled bottoms is not necessary, but may increase growth rates by about 10% (R. E. Andrews, unpublished observation).

Much of the understanding of sporulation and crystal toxin formation is derived from the observation that *B. thuringiensis* can be grown and induced to sporulate in near synchrony. That is, as the cells begin to sporulate, >99% of the cells will be in the same morphological stage. Bechtel and Bulla (1976), for example, used a medium that contained glucose, yeast extract, citric acid, and salts (modified GYS) to study sporulation and crystal toxin with the electron microscope. In these studies, the parasporal crystals formed in synchrony and allowed a detailed understanding of the morphological events related to sporulation. Later Andrews et al. (1981, 1982, 1985) used these same culture methods to show that crystal antigen first appeared in the cells at a defined time during the sporulation process, and that its synthesis was induced by de novo synthesis of crystal toxin specific mRNA. Moreover, this medium can be used to produce a nearly pure suspension of spores and crystals that can readily be used to purify spores and/or crystals for further use (Tyrell et al., 1981). The methodology works well with all subspecies tested to date (R. E. Andrews, unpublished observations).

Investigators in R. E. Andrews' laboratory currently use the following method to achieve synchronous sporulation without excessively long incubation times. A small volume (usually about 5 µl) of the vegetative cell culture preserved in 50% glycerol at −70° is inoculated into 100 ml of YEG broth (Tyrell et al., 1981). The culture is incubated overnight (about 15 h) at 30°C with vigorous shaking. Some calibration of the starting inoculum is required to insure that the overnight culture does not reach stationary phase. A transfer is made (10% inoculum) to modified GYS medium, and the cells are grown to mid-exponential phase (A_{600}, about 0.5). This culture serves as inoculum for the final experimental culture. Inoculum (10%) is transferred into modified GYS medium. Growth begins immediately with a generation time of 50–60 min, the cells enter stationary phase (the beginning of sporulation) after about 4 h, and sporulation is complete by 12 h. The preliminary transfers result in selection for rapidly growing cells, thus insuring development of physiological and morphological synchrony during sporulation.

Growth requirements for *B. thuringiensis* are relatively simple. Nickerson and Bulla (1974) and Kuznetsov and Khovrychev (1984) described growth and sporulation of several *B. thuringiensis* strains in synthetic media. In addition to basal salts these media, which were similar in composition, contained glucose and glutamate, although some strains required aspartate, citrate, alanine, and/or nicotinic acid.

Growth is slower than in complex media, with doubling times often in the range of 3–4 h, but such media should facilitate nutritional studies and selection of auxotrophic mutants.

The conditions used for commercial production of *B. thuringiensis* differ somewhat from those used in the laboratory. Fermentation media usually include unrefined substrates such as corn steep solids, molasses, corn starch, cotton seed flour, fish meal, hydrolyzed casein, groundnut cake, and soybean cake. Although sporulation is probably a prerequisite to crystal toxin, the goal of such fermentations is to optimize the production of crystal toxin rather than to maximize the formation of spores. Typically, fermentation begins with inoculation of a 15-l vessel from a seed culture. This 15-l culture is then used to inoculate larger vessels, and typical final fermentation volumes as large as 30,000 to 100,000 l are common (Andrews et al., 1987).

Identification

MORPHOLOGICAL CHARACTERISTICS. *Bacillus thuringiensis* cells are Gram-positive, sporeforming, and rod-shaped. The cells are straight rods that are typically 1.0–1.2 by 3–5 µm in size when grown in standard liquid media. Terminal to subterminal ellipsoidal spores are formed in sporangia that are not swollen. Probably the most characteristic distinguishing feature of *B. thuringiensis* is the presence of a parasporal crystal that forms adjacent to the spore, outside the exosporium during endospore formation. The presence of the parasporal crystal in cells and the production of insecticidal activity remain the best criteria available for differentiation of *B. thuringiensis* from the closely related species, *B. cereus* (Andrews et al., 1987; Baumann et al., 1984; Claus and Berkeley, 1986).

There is an interesting correlation between the spectrum of toxic action of a given strain and the shape of the crystal it produces. This is discussed in more detail elsewhere in this chapter. For the purpose of this discussion it is important to note, however, that the shape of the crystalline toxin can provide some important clues to the identity of the *B. thuringiensis* isolate in question. Subspecies *kurstaki*, for example, produces a single bipyrimidal crystal, whereas subsp. *israelensis* produces multiple, plemorphic, and globular crystals. If the task at hand is to differentiate between these two subspecies, the observation of crystals by use of a phase contrast microscope will provide meaningful information.

PHYSIOLOGICAL CHARACTERISTICS. Identification of *B. thuringiensis* by physiological characteristics is difficult. Aside from the presence of the crystal toxin, phenotypic characteristics to distinguish between *B. thuringiensis* and *B. cereus* are few. Claus and Berkeley (1986) compared approximately 40 phenotypic characteristics among the bacilli. These two species could not be differentiated by any of these physiological criteria. Baumann et al. (1984) and Lynch and Baumann (1985) examined 137 isolates of *B. thuringiensis* and 35 strains of *B. cereus* for the presence or absence of 99 phenotypic characteristics. Numerical analysis of the data showed that the various subspecies of *B. thuringiensis* (based on flagellar H antigen type) and *B. cereus* did not form two distinct clusters. On the contrary, *B. thuringiensis* and *B. cereus* strains were randomly dispersed throughout the dendrogram. Thus, although identification of an isolate as belonging to the *B. cereus-B. thuringiensis* group can be accomplished by using physiological criteria, other methods are required to distinguish between the two species. Also, a determination of physiological characteristics is of no value in differentiating between the *B. thuringiensis* subspecies.

IMMUNOLOGICAL CHARACTERISTICS. Immunological methods have also been used to identify *B. thuringiensis*. In addition, the basis for the subspecies classification of an isolate is largely based on antigen identification (de Barjac and Bonnefoi, 1962, 1973). Two antigenic targets have been used for immunological identification of *B. thuringiensis,* the flagellar antigens and the crystal toxin protein.

By immunologically analyzing strains for flagellar H antigens, more than 20 subspecies of *B. thuringiensis* have been identified (de Barjac and Bonnefoi, 1962, 1973; Krywienczyk et al., 1978). Although the major criterion for classification of an isolate as *B. thuringiensis* is the presence or absence of the crystal toxin, the use of H antigens has distinct advantages for the identification of subspecies. There is extensive evidence that the crystal toxin gene is frequently located on self transmissible plasmids (Battisti et al., 1985; Gonzales et al., 1982; Green et al., 1989; Klier et al., 1983; Ruhfel et al., 1984; Stahly et al., 1978). Moreover, the crystal toxin gene in many *B. thuringiensis* isolates has been shown to be surrounded by transposon-like DNA sequences (Bourgouin et al., 1988; Lereclus et al., 1984; Lereclus et al., 1986; Mahillon and Lereclus, 1988; Mahillon and Seurinck, 1988; Mahillon et al., 1985, 1987) and, therefore, may be transferable between strains. Thus, because the crystal toxin type produced by a *B. thuringiensis* isolate is probably a relatively inconstant characteristic, use of a more stable antigen as a target for identification is of greater

utility. Presumably the flagellar antigen would be of chromosomal origin and would therefore exhibit such stability.

Despite their comparatively unstable nature, crystal toxin antigens have been used for identification of *B. thuringiensis* subspecies. As taxonomic tools, the limitations of these methods have already been described (Krywienczyk, 1977; Krywienczyk et al., 1981). There are, however, some important applications for the use of these antigens; the observation that the crystal toxin presence is intimately correlated with insecticidal activity has been discussed previously. Thus, immunological determination of crystal toxin presence is a useful method for estimating toxic activity of insecticidal preparations. Rocket immune electrophoresis, for example, has been used to determine the toxin content in cultures of *B. thuringiensis* during growth and sporulation (Andrews et al., 1980, 1981, 1985). This method is rapid and simple, and it reliably estimates the toxic activity of cultures (Andrews et al., 1980). Immunological methods can also be used to quantify crystal toxin in the environment, and they are useful, therefore, for determination of residual insecticidal activity after pesticide application. Enzyme-linked-immunosorbent assay (ELISA), for example, can be used to determine crystal antigen with reasonable sensitivity even under field conditions (Wie et al., 1982, 1984).

In addition to quantification of insecticidal activity, there is another important use for immunological measurement of crystal toxin antigens. Immunological methods provide qualitative information about the crystal toxin. Wie et al. (1982) prepared antibody directed against the crystal toxin of an isolate of subsp. *kurstaki,* and then used heterologous crystal toxin proteins from several other subspecies to inhibit the homologous reaction in ELISAs. Of the five heterologous crystal types examined, four were toxic to lepidopteran larvae, and all four of these were highly inhibitory to the toxin antitoxin homologous reaction. As discussed below, DNA sequence data show that the toxin genes of lepidopteran-toxic strains of *B. thuringiensis* exhibit considerable homology. Tyrell et al. (1981) prepared sera against several crystal toxins from different subspecies and, by using Ouchterlony double diffusion assays, were unable to demonstrate cross-reactivity between the lepidopteran-toxic crystal antisera and the subsp. *israelensis* crystal toxin antigens. Interestingly, although no cross-reactivity was demonstrated with the Ouchterlony assays, Wie et al. (1982) found a low but significant (about 10%) inhibition of the subsp. *kurstaki* homologous reaction when the subsp. *israelensis* crystal protein was used in ELISA studies. The significance of this inhibition did not become clear until DNA sequencing studies revealed that these two toxins shared some similarity (Thorne et al., 1986). Therefore, immunological techniques employing crystal toxin antigens may have an important role in the regulatory process regarding pesticide registration. Under U.S. law, the Office of Pesticide Regulation of the Environmental Protection Agency has a responsibility to ensure that the toxin in a pesticide is the same as that contained therein at the time of the initial registration. Given the genetic exchange capacity of *B. thuringiensis,* these methods allow the registrant to characterize, in a standard manner, the active ingredient of their product, and to ensure that the current production strain expresses the same toxins and toxicity as when it was first registered.

GENETIC CHARACTERISTICS. Characterization of an isolate's DNA content is an important method for identifying *Bacillus thuringiensis.* The DNAs from *B. thuringiensis, B. anthracis,* and *B. cereus* are highly homologous, exhibiting >90% homology (Kaneko et al., 1978). Even though these other two species are common in soil, it should still be possible to use DNA probes to identify an isolate as *B. thuringiensis* by using a crystal toxin gene as a probe to detect homologous DNA target sequences in the test strain (Prefontaine et al., 1987). Use of such a DNA probe would obviate the need to detect the gene product, crystal toxin. As previously described, the major criterion for classification of an isolate as *B. thuringiensis* is the presence of crystal toxin. It is possible, however, that a crystal toxin gene may be present but not expressed. A mutation in the toxin gene could prevent expression. Klier and Lecadet (1976) first suggested and then Brown and Whiteley (1988) confirmed that a unique form of RNA polymerase is required for transcription of the crystal toxin gene in *B. thuringiensis.* Subsequently, Adams et al. (1989) have shown that a 20-kDa protein is required for expression of the crystal toxin gene. Mutations that effect expression of genes encoding one of these required components would create a strain that is unable to produce crystal toxin but contains an intact crystal toxin gene. It is unclear at present whether such mutants occur in nature, but their presence would certainly further confuse the taxonomic data.

Plasmid analysis can be used to identify specific isolates within the species *B. thuringiensis.* Most isolates of *B. thuringiensis* contain multiple plasmids ranging from as small as 1–2 kb in size to over 100 kb. Moreover, many isolates have unique plasmid size profiles that can be used in strain determination (Carlton and Gonzalez, 1985; Gonzales et al., 1982; Kronstad et al., 1983; Mahillon et al., 1988; Stahly et al., 1978; Stepanova and Azizbekian, 1987). Because of the potential loss of plasmids during extended culturing (Dean, 1984; Stahly et al., 1978) this method may be of somewhat limited utility as well.

Ecophysiology

Most evidence seems to indicate that the disease symptoms caused by *B. thuringiensis* are almost entirely due to the crystal toxin, often referred to as delta-endotoxin (Andrews et al., 1980; Fast, 1977). Andrews et al., (1987) have argued that the term "endotoxin" is inappropriate, because this term is commonly used to refer to toxic lipopolysaccharide components of cell walls from Gram-negative cells. (The toxin of *B. thuringiensis* shares some striking similarities to the toxins of *Clostridium botulinum* and *Clostridium perfringens* [Andrews and Bulla, 1981].) Spores alone are much less effective than either crystals alone or a mixture of the two. The relative ineffectiveness of spores alone is apparent in laboratory infectivity studies. Also, under field conditions, reinfection after application of insecticides is difficult to demonstrate (Couch and Ignoffo, 1981). In contrast, some evidence supports the view that other factors produced by *B. thuringiensis* are important in pathogenesis. As discussed in more detail below, β-exotoxin and certain proteases are produced by growing cells and are clearly active against insect systems. Some insects probably require viable spores for pathogenicity (Bulla et al., 1985). Finally, when an insect larva dies, the dead insect carcass usually contains relatively large quantities of spores and crystals (Aly, 1985; Aly et al., 1985; Prasertphon et al., 1974), indicating that septicemia typically follows the toxemia.

The observations that epizootics caused by *B. thuringiensis* occur relatively infrequently in nature, that the organism can be readily isolated from soils, that the crystal toxin is a normal component of the spore coat (Aronson et al., 1982; Somerville et al., 1968; Tyrell et al., 1981), and that *B. thuringiensis* is a relatively close relative of the common soil bacterium *B. cereus* (Andrews et al., 1987) imply a model for the microbe's ecology. This evidence suggests that the primary habitat of *B. thuringiensis* is soil. Somewhere over the course of evolution a common parent gained, either by mutation or genetic exchange, a spore coat protein that was toxic to insect larvae. By overproduction of this protein, the organism was able to kill an insect host and use the nutrients from that host in a saprophytic mode. This model would imply that the normal habitat for *B. thuringiensis* is soil, but that as the opportunity presents itself, the organism is able to use the insect as a temporary niche.

THE CRYSTAL TOXIN. As previously stated, *B. thuringiensis* is probably best known as a pathogen of insects. In nature, the disease is restricted to insect larvae; in the laboratory, toxicity to some adult insects of crystal protein has been reported when administered orally and/or rectally (Klowden and Bulla, 1984; Klowden et al., 1985). Moreover, the development stage of the larvae has a profound effect on its susceptibility to the toxin. Generally, younger larvae are more susceptible than older larvae (Andrews et al., 1987; Rasnitsyn et al., 1988). Because the route of entry of *B. thuringiensis* is through the oral cavity, ingestion of the toxin is mandatory, which requires insect feeding activity (Bulla et al., 1985).

Insects susceptible to the *B. thuringiensis* crystal toxin generally have alkaline midguts (pH ranges typically 10–12). At such pH the crystals solubilize and for some toxin types further activation may follow. Within the first minutes, a paralysis of the midgut and mouthparts ensues; these symptoms are accompanied by a drop in the pH of the insect gut and an increase in the hemolymph pH. The histological effects of the crystal toxin seem to be confined to the larval midgut epithelium and peritrophic membrane. Scanning electron micrographs of the larvae of *Manduca sexta* show that as soon as 1 h after oral administration of the toxin the microvilli in the midgut become shrunken, and after 4 h extensive midgut damage is observed. These histological changes are accompanied by a series of anomalies in the larvae's physiology. There is a severe restriction of potassium transport that is believed to be responsible for the maintenance of the high midgut pH. Active transport of potassium is inhibited about 78% within 10 min after ingestion, and, after several hours, all active transport ceases. Shortly after ingestion, the insect's peritrophic membrane becomes more permeable to particulate matter (Adang and Spence, 1983; Andrews et al., 1987; Gupta et al., 1985).

The mode of action of the crystal toxin has been reviewed (Andrews et al., 1987; Höfte and Whiteley, 1989). There is a marked similarity between the pathology of insects that have ingested the *B. thuringiensis* crystal toxin and that of humans suffering from cholera. Because of this similarity, much work has focused on the effect of the toxin on adenyl cyclase and intracellular cyclic AMP (cAMP) levels in diseased insects and in cultured cells that have been treated with toxin preparations. Knowles and Farndale (1988) found that treatment of cultured cabbage moth cells with *B. thuringiensis* toxin induced elevated cAMP levels and adenyl cyclase activity. However, when these cells were treated with the bee-venom toxin melittin, which also lyses the cells, they found similar effects. Therefore, the authors reasoned that the effect was secondary and was not directly related to the toxin's activity but rather consisted of a cellular response to lytic action.

Some evidence is beginning to accumulate relating to the mode of action of the toxin and its target or binding site. Knowles and Ellar (1987), for example, showed that insect cells cultured in the presence of ^{51}Cr, ^{86}Rb, and ^{3}H-uridine rapidly released those small molecules when treated with *B. thuringiensis* toxins that came from different subspecies. The receptor molecule may lie in the brush border membrane of epithelial cells. Sacchi et al. (1986) demonstrated that in midgut cells of the cabbage butterfly *Pieris brassicae*, K$^+$-gradient-dependent amino acid transport across the brush border membrane was inhibited by the *B. thuringiensis* toxin. Moreover, Hofmann et al. (1988a) demonstrated the presence of high affinity binding sites for the toxin of *B. thuringiensis* subsp. *thuringiensis* on the brush border membrane of midgut epithelial cells from *P. brassicae*. Several lines of evidence suggested that these were the natural binding sites for the *B. thuringiensis* toxin. I^{125}-labelled toxin readily bound to the target cells, whereas toxin binding to nontarget cells (rat small intestine epithelial cells) was observed with much lower affinity (Hofmann et al, 1988a). I^{127}-labelled, or unlabelled toxin competitively inhibited binding to the target cells but did not inhibit binding to the nontarget cells. The toxin from *B. thuringiensis* subsp. *thuringiensis*, although highly toxic for *P. brassicae*, is not toxic for larvae of the tobacco hornworm *Manduca sexta*. The toxin did not bind significantly to brush border membrane vesicles prepared from the larval midgut of *M. sexta*, a further indication that the binding to *P. brassicae* cells is specific (Hofmann et al., 1988b). Finally, Knowles et al. (1984) examined the effect of toxin on *Choristoneura fumiferana* CF1 cells in vitro. Preincubation of the toxin with N-acetylgalactosamine and N-acetylneuraminic acid specifically inhibited lysis of the target cells, and because N-acetylneuraminic acid has not been known to occur in insects, it was concluded that the toxin may recognize a specific plasma membrane glycoconjugate receptor with a terminal N-acetylgalactosamine residue. Considered togther these results suggest that the crystal toxin of many strains of *B. thuringiensis* binds to a glycoconjugate receptor on the brush border membrane of the insect midgut epithelium and then induces pore formation, resulting in leakage of the cells, followed by lysis.

OTHER PATHOGENIC FACTORS. Because of the importance of the crystal toxin to the activity of *Bacillus thuringiensis* insecticides, there has been a focus in research on this agent as a cause of insect pathology. It must be recognized, however, that other pathogenic factors are produced by certain *B. thuringiensis* isolates. Because of the nature of these factors, it is difficult to understand how they would function in ways other than to aid in an invasive propagation of *B. thuringiensis* cells in the insect.

One such factor is the so called β-exotoxin. The β-exotoxin, or fly factor, produced by some strains of *B. thuringiensis* is a heat stable nucleotide analog that is a potent inhibitor of RNA polymerase. This toxin is produced by β-exotoxin-positive cells during the stationary phase of growth and is broadly toxic to a variety of insects, both larvae and adults. Fly factor is an inhibitor of insect, mammalian, and bacterial RNA polymerases, and this activity is thought to be its mode of action. The heat stable properties and the toxic activity to *Musca domestica,* the house fly, are used to identify β-exotoxin activity in extracts (Beebee et al., 1972; Iandolo et al., 1976; Johnson, 1976, 1978; Johnson et al., 1975; Kim et al., 1972). Because of its toxicity and the possibility that it may be mutagenic, β-exotoxin-producing strains are currently banned in insecticidal formulations, but consideration has been given to the use of this factor itself as an insecticide (Bulla et al., 1985). As a final note, it is possible that the primary function of β-exotoxin is not as a pathogenic factor. Several kinds of observations lead to this conclusion. 1) β-Exotoxin is found primarily in spores, not in vegetative cells (Johnson, 1976) as would be expected for an invasive factor. 2) The RNA polymerases from sporulating cells of *B. thuringiensis* are up to 50% less sensitive to β-exotoxin than are those from vegetative cells, *E. coli,* or insects (Johnson, 1978). 3) Finally, there is no evidence to suggest that β-exotoxin-positive cultures are more pathogenic than are β-exotoxin-negative strains.

This evidence might lead to an alternative explanation for the presence of the toxic factor. Perhaps β-exotoxin is a normal regulatory factor important for control of sporulation in *B. thuringiensis;* the normal function might be to suppress expression of vegetative genes during sporulation. Under normal conditions, β-exotoxin is present at levels below normal detection in the so-called negative isolates. Strains that are termed positive simply have evolved a mechanism to overproduce the toxin, thus providing them a slight selective advantage in nature. Clearly, a better understanding of the role and occurrence of the β-exotoxin in *B. thuringiensis* is required to answer this question.

Many early investigators looked for specific immune systems analogous to the antibody response elicited by vertebrate animals when attacked by a bacterial invader. No specific agglutinating activity has been observed in insects. Insects do react with a defensive response during an intrusion by a bacterial invader, but the reaction is non-specific. The cecropins and attacins are two such classes of defensive proteins. These proteins, analogs of which have been identified in a wide variety of insects, are thought to function to lyse bacterial cells; the action is non-specific in that proteins induced by one bacterial species, for example, *E. coli* lyse cells from another species, *B. thuringiensis* (Hultmark et al., 1982, 1983; Hurlbert et al., 1985; Kaaya et al., 1987; Spies et al., 1986).

The finding that *B. thuringiensis* has been observed to produce a protease capable of specifically inactivating cecropins and attacins is particularly intriguing. This protein, inhibitor A, has been shown to attack and selectively destroy cecropins and attacins and thus to reduce the insect defense response. The protease is specific; although it does not seem to recognize a specific sequence, it attacks an open hydrophobic region near the C-terminus of the cecropin, and it does not attack globular proteins (Dalhammar and Steiner, 1984). The similarity between this pathogenic factor and an analogous factor produced by *Pseudomonas* species involved in infections in cystic fibrosis patients is striking. In the case of cystic fibrosis, it has been shown that several *Pseudomonas* isolates produce proteases that specifically degrade certain immunoglobins (Fick et al., 1984; Holder and Wheeler, 1984).

CRYSTAL TOXIN BIOCHEMISTRY. Despite the other pathogenic factors produced by *Bacillus thuringiensis,* by far the most research effort has been placed on the crystalline toxin, and this agent is therefore the best understood. There is a striking correlation between the shape of the parasporal crystal and the spectrum of toxicity it displays. The lepidopteran-toxic crystals are bipyrimidal in shape, the dipteran toxic crystals are pleomorphic, and coleopteran-toxic crystals are rectangular and flat. If toxicity truly resides in a discrete polypeptide subunit and if the individual polypeptide subunits have different ranges of toxicity to insects, then the molecular details of their structures must differ in significant ways and, hence, their crystals should have different topographical shapes. Therefore, it is possible that significant information about specificity of the proteins produced by a given isolate may be derived from a study of the crystal configuration.

Immunological comparison of the crystal toxin antigens has already been discussed. Antisera against crystals from several strains of pathotype I and from subsp. *israelensis* (pathotype II) were prepared by Tyrell et al. (1981). In Ouchterlony double diffusion assays, cross-reacting epitopes were not observed between pathotype I strains and the pathotype II strain, whereas an evidently homologous reaction was observed among the pathotype I strains. Using the more sensitive ELISA technique, however, differences among the pathotype I strains were noted and some similarity between these and the subsp. *israelensis* crystal proteins was apparent (Wie et al., 1982). Interestingly, crystal proteins from subsp. *tenebrionis* (pathotype III) did not serological cross-react with proteins from crystals of the other two pathotypes (Krieg et al., 1987b).

The crystals from all three pathotypes share some common properties. The most obvious of these is that they are all protein. Moreover, despite the antigenic diversity, the polypeptides contained in the crystals tend to have some common size ranges. For example, when crystal proteins from pathotypes I and II are separated on sodium dodecyl sulfate (SDS) polyacrylamide gels, major bands appear in the molecular weight (mol wt) range of 120,000–140,000 and a second band or group of bands in the mol wt range of 60,000–70,000. In addition, pathotype II crystals contain a third polypeptide (mol wt range of 23,000–30,000). Pathotype III crystals contain only proteins in the middle range. Under conditions of neutral and acid pH, the crystals are insoluble; in fact, the insolubility of the proteins and their tendency to re-form crystals and precipitate under such conditions remains a major problem in the laboratory. As previously discussed, insects sensitive to the crystal toxin of *B. thuringiensis* share the common property of having alkaline midguts. Not surprisingly, the crystals from all three pathotypes become soluble under alkaline conditions and in all cases the proteins retain toxicity. The crystal proteins from subsp. *israelensis* lose a substantial portion of their toxicity in insect bioassays when they are solubilized. However, this has been shown to result from a problem in the

insect assay procedure rather than from damage to the protein itself, because when the solubilized crystal proteins are adsorbed onto latex beads, the proteins regain most of their toxicity (Andrews et al., 1981; Bulla et al., 1981; Calabrese et al., 1980; Huber et al., 1981; Insell and Fitz-James, 1985; Lilley et al., 1980; Schnell et al., 1984; Tyrell et al., 1979).

PATHOTYPE I CRYSTAL PROTEINS. Crystal proteins from pathotype I isolates share a number of common features. The most obvious of these has already been pointed out; namely, they usually share a common crystal shape and contain proteins in a common size range (130,000–140,000). They are antigenically cross-reactive. The protein subunits are protoxin molecules that are converted to a toxic form after ingestion by a susceptible insect. Upon ingestion, the crystal becomes soluble in the insect midgut and is then activated, via proteolytic cleavage, to a toxin (mol wt = 68,000). The evidence suggests that the proteases responsible for toxin activation are insect-derived. Crystal proteins from subsp. *kurstaki* strain HD1, for example, are composed of repeating subunits of mol wt = 135,000 (Bulla et al., 1981). When the crystals are solubilized at pH 12, dialyzed against a pH 7.5 buffer, and allowed to stand at room temperature for several days, there is a conversion of some of the material to a smaller molecule (mol wt = 68,000) that is thought to be the toxin (Bulla et al., 1979). When these data appeared in the literature there were several items of controversy regarding this model. One area of concern was that the yield of toxic product was low relative to the quantity of starting crystal protein used, leading to some speculation that this was an artifact. Andrews et al. (1985) showed that under appropriate conditions the conversion of protoxin to toxin could be done in vitro using commercially available trypsin and that this treatment did not reduce the toxicity significantly. In these studies the soluble proteins from subsp. *kurstaki* were approximately four fold more toxic than were whole crystals. When these proteins were dialyzed against a dilute carbonate buffer and treated with limiting quantities of trypsin, nearly all the toxic activity (>90%) in the soluble crystal preparation was recovered, but the protein in the solution was now almost exclusively of a mol wt = 68,000. This means that the molar toxicity of the preparation remained constant, whereas the LC_{50} (concentration that is lethal to 50% of the treated insects) of the proteins had approximately doubled. Similar conversions were demonstrated using insect gut-derived proteases, confirming that enzymes from the larval midgut probably were responsible for the conversion in vivo (Tojo and Aizawa, 1983).

These data left one observation unexplained; crystals purified from all pathotype I strains examined before 1985 contained a molecule of approximately 68,000 mol wt (Tyrell et al., 1981). Since purified toxin comigrated with this smaller molecule, there was controversy over its origin. A better understanding of this problem came with the recognition that a strain of subsp. *kurstaki* (HD251) evidently did not contain, or contained very little, of the 68,000 mol wt protein in its crystals. Upon further examination, strain HD251 was shown to produce greatly reduced levels of intracellular proteases, but produced normal levels of toxicity. Subsequently, it was shown that the small quantity of 68,000 mol wt protein found in crystals of most pathotype I strains probably resulted from conversion of protoxin to toxin by intracellular proteases produced by *Bacillus thuringiensis* during the sporulation process (Andrews et al., 1985; Bibilos and Andrews, 1988).

Genetic observations confirmed conclusions derived from protein chemistry. From 1981 to 1989 at least 49 reports of cloning and/or sequencing of *B. thuringiensis* crystal toxin genes appeared in the literature (Table 1), and 30 of these reports were of genes encoding proteins with toxicity to Lepidoptera. Only two of these reports were of genes encoding proteins in the size range of 60,000–70,000 mol wt, whereas all of the others were of genes encoding proteins in the size range of 120,000–140,000 mol wt. Moreover, because both of the smaller proteins show a toxicity to both lepidopteran and dipteran larvae, it is unclear that these are true pathotype I toxins. Therefore, the DNA cloning and sequence data confirm that the crystal toxin subunit in pathotype I crystals are usually, if not always, composed of proteins of mol wt 120,000–140,000. DNA sequencing studies also confirmed the protoxin to toxin conversion hypothesis. Data from a number of laboratories showed that the toxin molecule was derived from the amino-terminal portion of the protoxin (Schnepf et al., 1985; Wabiko et al., 1985).

Crystals from some isolates contain multiple toxin proteins, each with a distinct host range of toxicity, and the total activity spectrum of the crystal represents a summation of the individual toxic spectra. For example, the crystals from subsp. *aizawai* contain at least two different toxin proteins, one of which is specific in its activity for *Pieris brassicae* and the other for *Spodoptera littoralis,* both of which are lepidopterans (Lecadet et al., 1988). The two types of subsp. *aizawai* genes have been cloned and

their distinct toxicities confirmed (Sanchis et al., 1988). Similar data were provided by Knowles and Ellar (1988), who observed that subsp. *aizawai* contained two proteins, one of which was toxic to *Choristoneura fumiferana* and the other to *S. frugiperda.* Multiple genes have been cloned from subsp. *entomocidus* (Visser et al., 1988), subsp. *morrisoni* (Granum et al., 1988), subsp. *thuringiensis* (Brizzard and Whiteley, 1988), and subsp. *kurstaki* (Widner and Whiteley, 1989).

There may be some clues in the literature regarding the significance of these different genes. In a report that is both interesting and alarming, McGaughey (1985) observed resistance to the *B. thuringiensis* toxin in an insect population. While trying to use *B. thuringiensis* to control *Plodia interpunctella* in grain bins, he observed that in some cases the populations were suppressed, whereas in others a similar concentration of insecticide did not achieve effective results. Upon further examination, it was observed that in as few as 15 generations of continual selection, the LC_{50} of the toxin increased nearly 100-fold. That a population can become resistant in such a short time raises obvious concerns about the use of *B. thuringiensis* insecticides. Interestingly, however, in later studies it was noted that insects were selectively resistant to certain toxin types while still sensitive to others (McGaughey and Johnson, 1987). The presence of resistance in a population may explain the presence of multiple toxin genes within a single isolate of *B. thuringiensis;* multiple genes may be a mechanism for overcoming the development of resistance. At the present time, resistance has only been observed in *P. interpunctella,* and it is unclear how widespread or universal this resistance will become.

PATHOTYPE II CRYSTAL PROTEINS. As previously indicated, one unique feature of the pathotype II isolates is that their crystals contain a polypeptide smaller than that observed in the other two pathotypes. Moreover, this smaller protein (mol wt = 23,000) is clearly the most abundant of the proteins in the crystal toxin. Also, the various polypeptides found in crystals of subsp. *israelensis* did not antigenically cross-react with each other, unlike the situation with pathotype I strains. Specifically, the proteins in the 120,000–140,000 mol wt range react with antibody prepared against the proteins in the 60,000–70,000 mol wt range, but the 23,000 mol wt proteins do not; nor do the larger proteins react with antibody prepared against the 23,000 mol wt protein (Pfannenstiel et al., 1986; Tyrell et al., 1981).

Table 1. Survey of reported *B. thuringiensis* crystal protein genes cloned between 1981 and 1989.

Source of the cloned gene (subspecies and/or strain)	Type of report (cloning and/or sequence)	Insect specificity of gene	Molecular weight of protein coded by gene	Reference
kurstaki HD1	Cloning	Lepidoptera	133,000	Schnepf and Whiteley, 1981
kurstaki HD1	Cloning	Lepidoptera	134,000	Held et al., 1982
thuringiensis 1715	Cloning	Lepidoptera	130,000	Klier et al., 1982
kurstaki HD73	Cloning	Lepidoptera	133,000	Kronstad and Whiteley, 1984
israelensis	Cloning	Cytolysin	26,000	Ward et al., 1984
kurstaki HD1	Cloning	Lepidoptera	133,000	Schnepf et al., 1985
kurstaki HD73	Cloning and sequence	Lepidoptera	133,000	Adang et al., 1985
aizawai	Cloning	Lepidoptera	133,000	Klier et al., 1985
kurstaki HD244	Cloning	Lepidoptera	140,000	McLinden et al., 1985
israelensis	Cloning	Diptera	Not reported	Sekar and Carlton, 1985
sotto	Cloning and sequence	Lepidoptera	144,000	Shibano et al., 1985
israelensis	Cloning and sequence	Cytolysin	28,000	Waalwijk et al., 1985
thuringiensis 1715	Cloning	Lepidoptera	140,000	Wabiko et al., 1985
israelensis	Cloning	Cytolysin	28,000	Bourgouin et al., 1986
		Diptera	130,000	
kurstaki HD1	Cloning and sequence	Lepidoptera	130,000	Geiser et al., 1986
san diego	Cloning	Coleoptera	65,000	Herrnstadt et al., 1986
thuringiensis	Cloning and sequence	Lepidoptera	130,000	Hofte et al., 1986
thuringiensis	Cloning	Lepidoptera	120,000	Honigman et al., 1986
kurstaki HD1	Cloning	Lepidoptera	130,000	Kronstad and Whiteley, 1986
thuringiensis HD2	Cloning	Lepidoptera	130,000	Kronstad and Whiteley, 1986
kurstaki	Cloning	Lepidoptera	135,000	Shivakumar et al., 1986
kurstaki	Cloning and sequence	Lepidoptera	131,000	Thorne et al., 1986
israelensis	Cloning and sequence	Diptera	58,000	Thorne et al., 1986
thuringiensis	Sequence	Lepidoptera	130,000	Wabiko et al., 1986
israelensis	Cloning	Diptera	130,000	Angsuthanasombat et al., 1987
aizawai HD133	Cloning	Lepidoptera	135,000	Chak and Ellar, 1987
kurstaki HD1	Sequence	Lepidoptera	135,000	Fischhoff et al., 1987
morrisoni	Cloning and sequence	Cytolysin	27,000	Galjart et al., 1987
aizawai IC1	Cloning	Lepidoptera and Diptera	130,000	Haider et al., 1987
san diego	Sequence	Coleoptera	65,000	Herrnstadt et al., 1987
tenebrionis	Cloning and sequence	Coleoptera	72,000	Hofte et al., 1987
aizawai IPL7	Cloning and sequence	Lepidoptera	131,000	Oeda et al., 1987
tenebrionis	Cloning and sequence	Coleoptera	72,000	Sekar et al., 1987
israelensis	Cloning and sequence	Diptera	130,000	Ward and Ellar, 1987
thuringiensis HD2	Cloning and sequence	Lepidoptera	130,000	Brizzard and Whiteley, 1988
israelensis	Sequence	Diptera	130,000	Chungjatupornchai et al., 1988
israelensis	Cloning and sequence	Cytolysin	28,000	Donovan et al., 1988a
		Diptera	72,000	
kurstaki HD263	Cloning and sequence	Diptera and Lepidoptera	66,000	Donovan et al., 1988b
morrisoni HD12	Cloning	Diptera and Lepidoptera	140,000	Granum et al., 1988
tenebrionis	Cloning and sequence	Coleoptera	73,000	McPherson et al., 1988
aizawai 7.29	Cloning	Lepidoptera	130,000	Sanchis et al., 1988
entomocidus	Cloning	Lepidoptera	130,000	Sanchis et al., 1989
israelensis	Cloning and sequence	Diptera	130,000	Sen et al., 1988
		Diptera	130,000	
israelensis	Sequence	Diptera	128,000	Tungpradubkul et al., 1988
entomocidus	Cloning	Lepidoptera	Not reported	Visser et al., 1988

continued

Table 1. *Continued*

Source of the cloned gene (subspecies and/or strain)	Type of report (cloning and/or sequence)	Insect specificity of gene	Molecular weight of protein coded by gene	Reference
israelensis	Cloning	Diptera	130,000	Ward and Ellar, 1988
		Diptera	130,000	
galleriae	Cloning	Diptera and Lepidoptera	61,000	Ahmad et al., 1989
entomocidus	Cloning and sequence	Lepidoptera	Not reported	Masson et al., 1989
		Lepidoptera		
kurstaki	Cloning and sequence	Lepidoptera	130,000	Widner and Whiteley,
		Lepidoptera	130,000	1989

Adapted from Hurley (1989).

Shortly after the initial observations regarding the polypeptide profiles in subsp. *israelensis,* a potent and highly nonspecific cytolytic activity was observed in these crystals. A number of investigators quickly purified the activity and ascribed it to the 23,000 mol wt protein (Armstrong et al., 1985; Chilcott and Ellar, 1988; Davidson and Yamamoto, 1984; Ibara and Federici, 1986; Insell and Fitz-James, 1985). Many of these investigators observed both cytolytic and toxic activity associated with this protein and identified it as the toxin molecule. The protein is active on cells from insects of several orders, on mammalian cells, and even on some bacterial cells. From a functional standpoint it was difficult to understand how such a nonspecific activity when tested on cells in vitro could have such a specific effect in vivo. Moreover, the toxicity of the purified cytolysin was approximately 100- to 1000-fold less than that of solubilized proteins of whole crystals. Accordingly, Hurley et al. (1985) first demonstrated that the activities could be separated, and then purified the 23,000 and 68,000 mol wt polypeptides (Hurley et al., 1987). The smaller protein, when purified, was highly cytolytic, whereas the larger protein was toxic to mosquito larvae (1000 times more toxic than the cytolysin). In addition to observing toxic activity in the 68,000 mol wt protein, high levels of toxicity were observed in the larger proteins as well. Hurley et al. (1987) attributed the disagreement between their data and that of the others to two factors: 1) In one report, a 65,000 mol wt molecule was purified and shown to be relatively nontoxic (Ibarra and Federici, 1986). However, these investigators used SDS to solubilize the toxin preparation; and at least with the lepidopteran toxins, this had previously been shown to eliminate activity (Huber et al., 1981). 2) When in purified form, the toxin was unstable and readily degraded to smaller nontoxic forms. It was unclear whether this instability was intrinsic or caused by contaminating

protease. Proteases associated with the crystal toxin of subsp. *israelensis* had already been reported (Chilcott et al., 1983). Although evidence from protein chemistry was becoming convincing, the ultimate proof would come from gene cloning experiments.

Bourgouin et al. (1986) constructed a library of *B. thuringiensis* subsp. *israelensis* plasmid DNA in an *E. coli* plasmid vector and identified three genes coding for proteins of 26, 73, and 135 kDa, respectively, located on a single 75-kb plasmid. Interestingly, the 73 and 135 kDa proteins produced from the cloned genes were both toxic, and the 73-kDa protein appeared to be a truncated version of the larger gene. The 23-kDa protein-encoding gene did not appear related to the other two, and the protein produced from the cloned gene was not toxic but was highly cytolytic. The surveyed gene cloning literature seems to support their conclusions. Of the 49 reports of cloned and/or sequenced crystal toxin genes from *B. thuringiensis* in the literature, five are of cytolysin genes, and all five genes encode proteins having molecular weights in the range of 26,000–28,000 (Table 1). There are 16 reports of dipteran toxic protein genes (including lepidopteran and dipteran toxic protein genes) from *B. thuringiensis* in the surveyed literature and of these four genes code for proteins in the 60–70 kDa range, whereas 11 code for proteins in the 120–140 kDa range, and 1 has no reported molecular weight (Table 1).

Since the isolation of subsp. *israelensis* there have been other isolations of pathotype II *B. thuringiensis* strains reported. These evidently belong to other subspecies based on their serotype and probably have proteins in their crystals similar to those in subsp. *israelensis* (Gill et al., 1987; Padua et al., 1980). Although the cytolysin is not the primary toxin, it is not without function. Evidence suggests that the cytolysin may act synergistically with the toxin (Hurley et al., 1987; Wu and Chang, 1985). There are evidently two forms of the toxin gene; however, their role is not understood.

PATHOTYPE III CRYSTAL PROTEINS. Owing to the relatively recent isolation of pathotype III strains, there is far less information regarding these crystal toxin proteins. It is interesting to note, however, that most of the controversy regarding the crystal proteins from the other two pathotypes was resolved not by protein chemistry but by gene cloning technology. Because pathotype III strains were isolated after gene cloning technology became available, it is not surprising that there has been much less controversy regarding the content of these crystals. The prototype strain for pathotype III is subsp. *tenebrionis*. The crystals from subsp. *tenebrionis* appear to contain two polypeptides of mol wt 68,000 and 72,000 (Krieg et al., 1987b). Five of the 49 cloned and/or sequenced genes reported in Table 1 are for pathotype III crystal proteins, and the data are in remarkable agreement with the protein chemistry data. The literature regarding pathotype III strains is not without controversy, however. There has been a report of a second subspecies of pathotype III, termed subsp. *san diego* (Ferro and Gelernter, 1989; Herrnstadt et al., 1986) but there are claims that this strain is identical to subsp. *tenebrionis* (Krieg et al., 1987a).

CRYSTAL TOXIN SYNTHESIS. There is substantial literature to support, at least with most subspecies of *B. thuringiensis*, that the production of crystal toxin is linked to sporulation. Several lines of evidence support this conclusion. First, the parasporal crystals appear in the cells in close proximity to the spore, and the time of their appearance when viewed by phase contrast and electron microscopy closely coincides with spore formation (Bechtel and Bulla, 1976). Fig. 1 shows a diagram of the relationship between sporulation and crystal toxin production in subsp. *kurstaki*. Moreover, crystal toxin antigen accumulates in the cells only between 4 and 6 h after the onset of sporulation (Andrews et al., 1981).

The synthesis of crystal toxin is evidently controlled at the level of transcription. Synthesis of mRNA specific for crystal toxin, as measured by in vitro translation (Andrews et al., 1982) and Northern blotting (Kronstad and Whiteley, 1984), correlates with the accumulation of crystal toxin antigen in sporulating cells. Moreover, there is evidence that formation of mRNA specific for crystal toxin requires a unique form of RNA polymerase (Brown and Whiteley, 1988). In subsp. *israelensis* the presence of a 27-kDa protein is required (Adams et al., 1989). There

is a notable exception to the association between crystal toxin production and sporulation, however. In subsp. *tenebrionis*, it appears that crystal toxin proteins are synthesized in vegetative cells (Sekar, 1988).

Genetics

The most extensive studies on the genetics of *B. thuringiensis* have focused on cloning and characterization of the crystal toxin genes. This information was presented in the previous section to aid in understanding the nature of the toxins and will not be reiterated in this section.

PLASMID BIOLOGY. *B. thuringiensis* isolates typically contain numerous plasmids, ranging in size from 2 to >200 kb. A given strain may carry up to 17 plasmids of distinct sizes (Gonzalez et al., 1981; Lereclus et al., 1982; Stahly et al., 1978). Several aspects of *B. thuringiensis* plasmid biology are unique and worthy of mention in this context.

In some strains with many plasmids (some of which are very large), up to 20% of the total DNA coding capacity may be plasmid-borne. Most of the plasmids in *B. thuringiensis* are cryptic in that functions have not been assigned to them; indeed, most of the plasmids from these strains can be cured with no observable change in phenotype (Gonzalez et al., 1981). It is unlikely, however, that an organism would maintain this amount of unproductive DNA, so it is likely that these cryptic plasmids serve some purpose.

Plasmid incompatibility is the inability for two plasmids to coexist in the absence of selective pressure. When two plasmids are of the same incompatibility group, they will not be maintained unless each contains a selective marker that can be used to assure that it is not lost. Because of the multiple plasmids found in *B. thuringiensis* isolates and because plasmid loss is not readily observed in these strains without some curing pressure, one must conclude that 1) there must be many incompatibility groups (at least 17) in *B. thuringiensis;* 2) that in cases where two or more of the plasmids are of the same incompatibility group, there must be some as of yet unrecognized selective pressure to maintain these plasmids; and/or 3) *B. thuringiensis* contains some mechanisms to defeat normal incompatibility groupings. In any case, further investigation of the multiple plasmids contained in *B. thuringiensis* should increase our understanding of plasmid incompatibility in bacteria in general.

Finally, *B. thuringiensis* plasmids have been shown to be transferred with relatively high fre-

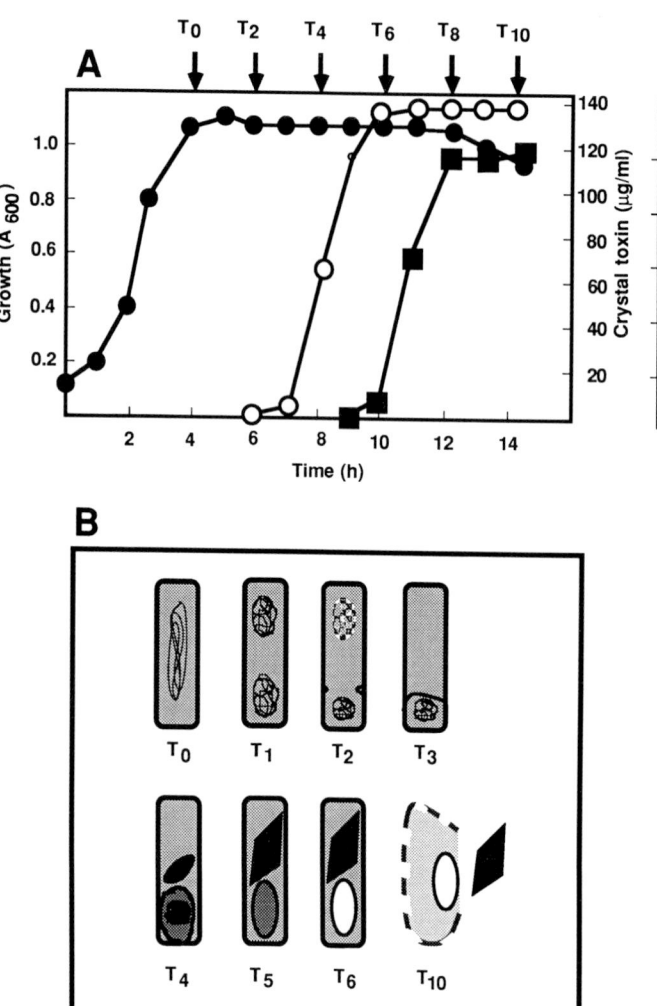

Fig. 1. Growth, sporulation, and crystal toxin formation in *B. thuringiensis*. (A) A typical growth curve of *B. thuringiensis* subsp. *kurstaki*. The solid circles represent growth, the open circles show the formation of crystal toxin antigen, and the closed squares show the percentage of cells that contain complete, phase-light spores. (B) An artist's conception of growth, sporulation, and crystal toxin production in *B. thuringiensis* subsp. *kurstaki*. The data are based on electron micrographs published by Bechtel and Bulla (1975) and on phase-contrast microscopic examination of the sporulating cultures shown in A. The cells were grown in liquid GYS medium as described in the text. Crystal toxin was determined as described by Andrews et al. (1981).

quency between many bacilli, including *B. thuringiensis, B. cereus, B. megaterium,* and *B. anthracis* (Battisti et al., 1985; Gonzalez et al., 1982; Green et al., 1989). Further understanding of this process will greatly extend the understanding of genetic exchange in Gram-positive bacteria in general, and further define safety parameters for genetically engineered microbes released into the environment.

TRANSPOSONS AND INSERTION SEQUENCES. Lereclus et al. (1983) described a unique 4.2-kb DNA sequence that was found on several large toxin-coding plasmids of *B. thuringiensis*. This particular sequence (Th-sequence) was found as an insertion in the *Streptococcus faecalis* plasmid, pAMβ1, which was introduced into *B. thuringiensis* subsp. *berliner* 1715. The 4.2-kb DNA sequence resembled an insertion sequence (IS element) and was found in close proximity to the protoxin genes from several *B. thuringiensis* subspecies. A 1.3-MDa segment of

DNA separated the Th-sequence from the subsp. *berliner* 1715 protoxin gene, and a similar DNA segment was located on the other extremity of the gene but in the opposite orientation. These DNA segments contained inverted repeat sequences and were referred to as IR*1*. When a plasmid carrying IR*1* and the Th-sequence was observed with an electron microscope, it appeared to be a transposon (Tn) (Lereclus et al., 1984). Therefore, the Th-sequence is now referred to as Tn*4430,* and is the first transposon isolated from the genus *Bacillus* (Lereclus et al., 1986).

This DNA segment is similar to the Tn3 family in structure and transpositional activity (Lereclus et al., 1986). In addition to its insertion properties, Tn*4430* has short terminal inverted repeats and promotes deletions adjacent to its insertion site similar to that observed with Class II elements (Lereclus et al., 1984, 1986). Nonetheless, the function of this transposon has not been identified. Similar inverted sequences,

IR*1750* and IR*2150*, flanking the plasmid-encoded 68-kDa protoxin gene of subsp. *kurstaki* HD-73, have been identified (Kronstad and Whiteley, 1984). Southern hybridization confirmed that these sequences were located in close proximity to the protoxin genes from 14 other strains of *B. thuringiensis,* and it was believed that IR*1750* and IR*2150* may be insertion sequences that mediate transposition. Nucleotide sequencing of one IR*1* element revealed all the characteristic features of an insertion sequence and was referred to as IS*231* (Mahillon et al., 1985). IS*231* (1656 bp) was isolated from subsp. *berliner* 1715 and found to be closely linked to the protoxin gene. It is delineated by two 20-bp inverted repeats flanked by two 11-bp direct repeats. IS*231* contains an open reading frame that spans almost the entire sequence, and DNA sequence homology has been observed between IS*231* and the *E. coli* IS*4* element.

Mahillon et al. (1987) sequenced two other insertion regions flanking the subsp. *berliner* 1715 crystal toxin gene and showed them to be variants of IS*231*. These sequences were referred to as iso-*231* elements. When the nucleotide sequences surrounding the iso-*231* elements were compared to IS*231*, it was shown that a structural association existed between these elements and the transposon Tn*4430*. It appeared that two IS*231* elements transposed into Tn*4430* where both the IS*231*s and the transposon Tn*4430* remained structurally intact. This structural association between IS*231* and Tn*4430* is similar to the organization of class I mobile genetic elements (i.e., Tn*5* or Tn*10*), and although the exact function of the IS*231* elements is unknown, it is thought they control Tn*4430* transposition (Mahillon et al., 1987).

In summary, the presence of repeated elements around crystal protein genes may provide two possible mechanisms for the dispersal of the toxin gene on different plasmids and for the integration of this gene into chromosomal DNA. The evidence for transposition of the Th-sequence into other plasmids and the arrangement of the repeated elements suggest that the toxin gene could undergo transposition. Alternatively, the presence of repeated elements on many different plasmids and on chromosomal DNA would also provide a mechanism for rearrangements of DNA. Further research is required to understand transposition in *B. thuringiensis* and to provide an increased understanding of the role of these repeated elements.

GENETIC TRANSFER SYSTEMS. Although there is substantial literature regarding the genetics of *Bacillus thuringiensis,* methodology for the introduction of foreign or cloned DNA into this organism is minimal. Three general mechanisms for introduction of DNA are usually described, transformation, conjugation, and transduction. All three forms of genetic exchange involving *B. thuringiensis* have been described in the literature.

Several investigators have reported transformation of *B. thuringiensis* protoplasts using methods similar to those used in *B. subtilis* (Chang and Cohen, 1979). Martin et al. (1981), for example, used a protoplast transformation protocol to transform *B. thuringiensis* with plasmid pC194. Unfortunately, they were unable to detect extrachromosomal pC194 in regenerated transformants and provided evidence that the plasmid integrated into the chromosome. Later, other workers physically identified the stable presence of pC194 in various transformants (Crawford et al., 1987; Fischer et al., 1984; Heierson et al., 1987). Moreover, other plasmids have been inserted into cells of *B. thuringiensis* by protoplast transformation (Alikhanian et al., 1981; Miteva et al., 1981). The methods used by different investigators for protoplast generation has varied. Fischer et al. (1984) used protoplasts generated by lysozyme treatment, whereas Heierson et al. (1987) and Crawford et al. (1987) generated protoplasts by using conditions that induced limited autolysin formation.

High voltage electroporation has been used in many systems, including plant cells, animal cells, and bacteria for introduction of foreign DNA. Recently, Bone and Ellar (1989) demonstrated use of this technique to transform several plasmids into *B. thuringiensis*. The method is rapid and effective, and yields transformation efficiencies in the range of 10^2–10^5/μg of DNA.

The naturally occurring conjugation systems in *B. thuringiensis* have already been discussed. Gonzalez et al. (1982), for example, described a natural plasmid transfer system that allowed them to more extensively analyze the relationship between crystal production and plasmid content. They noticed high frequency transfer of plasmids during logarithmic growth between parasporal crystal-positive (*cry*$^+$) strains and *cry*$^-$ strains. For instance, when *B. thuringiensis* subsp. *kurstaki cry*$^-$ recipients were mated with *cry*$^+$ *B. thuringiensis* subsp. *thuringiensis* donors, the *B. thuringiensis* subsp. *kurstaki* strains were converted to *cry*$^+$ phenotype. The size of the transmissible crystal-coding plasmid varied with the donor strain, and immunological analysis showed the *B. thuringiensis* subsp. *kur-*

staki cry⁺ transcipients to be hybrid strains, i.e., having the flagellar serotype of the recipient and crystals of the donor serotype.

Gonzalez and Carlton (1984), utilizing their high-frequency plasmid transfer system, genetically analyzed crystal toxin production in *B. thuringiensis* subsp. *israelensis*. They showed that *B. thuringiensis* subsp. *israelensis* contained eight plasmids ranging in size from 3.3–135 MDa. Curing studies showed that a 75-Mda plasmid was responsible for toxin production, because 15 B. thuringiensis subsp. *israelensis cry⁻* isolates were shown to lack this 75-MDa plasmid. This conclusion was supported by transfer of the 75-MDa plasmid into a *cry⁻* strain.

Transfer systems native to other bacteria can also be used for introduction of genetic information into *B. thuringiensis*. Franke and Clewell (1981) described a chromosome-borne tetracycline resistance element in *Streptococcus faecalis* strain DS16 that had the characteristics of a transposon. This 15–17 kb element, Tn*916*, was capable of transposition to several different conjugative plasmids containing hemolysin genes (pAD1 and pOB1) at frequencies of approximately 10⁻⁶. Moreover, Tn*916* was also capable of transposition at low frequency (10⁻⁸ per recipient) from plasmid-free derivatives of strain DS16 to plasmid-free *S. faecalis* recipients. This transfer was resistant to the action of DNase, did not require homologous host-mediated recombination functions, and occurred by a conjugation-like event that required direct contact between the donor and recipient. Tn*916* (16.4 kb) is classified as a Class IV transposon (Kleckner, 1981). Tn*916* has been transferred to a number of streptococcal species by the filter mating technique (Franke and Clewell, 1981; Nida and Cleary, 1983; Wanger and Dunny, 1985; Weiser and Rubens, 1987). Furthermore, Tn*916* has been transferred by filter matings to *Staphylococcus aureus, Listeria monocytogenes, Bacillus anthracis, Clostridium difficile,* and *Mycoplasma hominis* (Hachler et al., 1987; Ivins et al., 1987; Jones et al., 1987; Kathariou et al., 1987; Roberts and Kenny, 1987). The conjugative transfer of Tn*916* in the absence of plasmid DNA occurs at a frequency of 10⁻⁵–10⁻⁸ per donor, and DNA hybridization data has demonstrated that Tn*916* inserts into different sites on the recipient chromosome (Gawron-Burke and Clewell, 1982). Insertional mutagenesis using Tn*916* has been reported (Ivins et al., 1987; Kathariou et al., 1987; Nida and Cleary, 1983).

Naglich and Andrews (1988a) recently reported introduction of the *S. faecalis* transposon Tn*916* into *B. thuringiensis*. When introduced into *B. thuringiensis* by filter mating with *S. faecalis*, Tn*916* inserts into evidently random sites on the chromosome, as judged by Southern analysis; insertion into plasmid DNA has not been observed. In contrast, when Tn*916* was transferred from one strain of *B. thuringiensis* subsp. *israelensis* to an antibiotic-resistant derivative of the same strain, insertion occurred at one site, suggesting that in this case DNA probably integrated into the chromosomal DNA by homologous recombination.

To enhance the transfer frequency, Tn*916* was first introduced into *B. subtilis* and then transferred to an antibiotic-resistant strain of *B. thuringiensis* subsp. *israelensis*. With this method, transfer occurred at frequencies 10- to 100-fold greater than when transfer was from *S. faecalis*. As in the case of transfer from *S. faecalis*, Tn*916* inserted randomly into the chromosome of *B. thuringiensis,* but, as in the previous experiments, not into plasmid DNA (Naglich and Andrews, 1988b). Several lines of evidence support the conclusion that DNA transfer is a Tn*916*-dependent conjugal event. 1) Transfer of genetic material was only observed when the cells were impinged on a filter and then grown on solid medium; transfer was not observed in broth. 2) Incorporation of high levels of DNase into the medium upon which the filter was placed did not prevent the process. 3) Transfer was Tn*916*-dependent; no DNA transfer was observed in the absence of the transposon, which carries its own mobilization functions.

The *Staphylococcus aureus* plasmids, pC194 and pUB110, were introduced into *B. thuringiensis* subsp. *israelensis* by using Tn*916* as a mobilizing agent. Transfer of pC194 occurred only when *B. thuringiensis* subsp. *israelensis* was mated with a *B. subtilis* donor that contained both pC194 and Tn*916;* plasmid transfer was not observed in the absence of the transposon. *B. thuringiensis* transconjugants resistant to chloramphenicol (the plasmid marker, Cmʳ) and tetracycline (the transposon marker, Tetʳ) were detected at a frequency of 2 × 10⁻⁶ per recipient cell, whereas the Tetʳ phenotype but not the Cmʳ phenotype was observed at a frequency of 1 × 10⁻⁴. The converse, Cmʳ but not Tetʳ, was observed at a frequency of 3 × 10⁻⁵. Transconjugants were detected in filter matings only, not in broth. Southern hybridization data indicated that the transposon was integrated into the *B. thuringiensis* chromosome, whereas the pC194 probe hybridized to a covalently closed circle form of the same size as pC194. The Tetʳ phenotype was maintained during serial passage of *B. thuringiensis* without selection, whereas the Cmʳ phenotype was not.

The plasmid pUB110, also from *S. aureus,* can be transferred at similar frequencies, but mobilization of the plasmid pE194 has not been detected (Naglich and Andrews, 1988b).

Transduction in *B. thuringiensis* has also been reported. Lecadet et al. (1980), for example, demonstrated transfer of genetic markers in *B. thuringiensis* using the transducing phage CP54. A similar phage, CP51, has been shown to mediate generalized transduction in *B. thuringiensis,* as well as in *B. cereus* and *B. anthracis* (Thorne, 1978). CP51 transduces chromosomal markers at frequencies ranging from 10^{-6}–10^{-7} and can carry up to 60 kb of DNA (Yelton and Thorne, 1971). Also, phage CP51 induces movement of plasmid-borne DNA into *B. thuringiensis* (Ruhfel et al., 1984).

Biotechnological Applications

Despite the relatively large number of insect pathogens that have been proposed as insecticides, those products formulated with *B. thuringiensis* are the most extensively and widely used today. There have been recent reviews discussing the biotechnology of *B. thuringiensis* (Andrews et al., 1987; Bulla et al., 1985). Accordingly, only a few selected topics will be discussed herein.

Insecticides formulated with *B. thuringiensis* are extremely specific; they affect only a limited range of insects and are nontoxic to predatory insects, birds, and mammals, thus providing significant advantages over chemical insecticides. Because of their lack of toxicity to humans, these insecticides can be applied just prior to harvest, a time when an insect infestation can cause extreme damage to the crop in a short time. Because no human toxicity has been reported, there can be immediate field entry. Moreover, there is minimal risk of toxicity to those who apply the pesticide, and there is no significant risk to humans living near the application area. The latter is particularly important when mosquitoes must be controlled near residential areas.

There are, however, significant problems associated with the use of these insecticides. One is that they exhibit poor stability under field conditions. The insecticidal activity of pesticides formulated with *B. thuringiensis* has a half-life of 1–2 days (Beegle et al., 1981; Sorenson and Falcon, 1980). Therefore, frequent reapplications are required. At least in part because of this short field life, such insecticides are expensive to use when compared to the more widely used organic pesticides. Insecticidal activity is associated primarily with the crystal toxin rather than the spores (West et al.,

1984). In a soil environment, the low stability of crystal toxin is probably due to degradation by soil microflora (West, 1984). The lack of the toxin's stability on foliage is due at least in part to physical removal by precipitation. Research is needed to discover methods for stabilizing the crystal toxin.

Bulla et al., (1985) discussed three possible reasons that *B. thuringiensis* insecticides have not been more widely used. 1) For many insect control situations, broader-spectrum insecticides are preferred because of the multiple pests found on crops. Moreover, growers often believe that contact pesticides are more effective and have found that more cost-effective alternatives are available. Thus the marketplace is generally viewed as too small to attract widespread interest among agricultural companies. 2) Many agricultural companies are adapted to working with chemical pesticides and have difficulty working with a product of biological origin. 3) There has been little or no patent protection available to producers. The impact of the Chakrabarty court decision that allows patenting of genetically modified microbes remains unclear.

Despite these difficulties, *B. thuringiensis* has been the subject of intensive investigation using the tools of biotechnology. Because of the highly selective toxic activity associated with *B. thuringiensis,* biotechnologists see a potential to develop safe, effective, and economically feasible insect control agents. Research into new ways to use *B. thuringiensis* and/or its toxic products has focused in three general areas.

One area of interest centers on the problem of narrow host range. Although the advantages of such a narrow host range are clear, limited toxicity also restricts the use of *B. thuringiensis* as an insecticide. Wabiko et al. (1986), for example, compared the DNA and derived protein sequences of the crystal toxin genes from two strains of *B. thuringiensis* having different insect specificities. Interestingly, there were several areas of strong protein homology and three areas in which striking differences were clustered. The authors speculated that these three areas might be important for genetic determination of the insect specificity of the toxin. At the present time much effort is being exerted to understand the basis for insect specificity of the *B. thuringiensis* toxin in the belief that it may be possible to alter or expand the host range.

An alternative approach is that the toxin gene could be placed in another bacterium that would live longer in the environment, multiply, and, thus, remain in or on the insect's food supply for a longer period of time. In a recent review, Andrews et al. (1987) described three possible examples of this approach. One approach, for example, was to stabilize the subsp. *israelensis* toxin by cloning its gene into *Bacillus sphaericus,* an organism which is much more stable in water environments than *B. thuringiensis* var. *israelensis.* Another approach is to clone the toxin genes into pseudomonads, organisms known to adhere to certain parts of plants. Both living and killed strains containing toxin genes have been described. Also, it is possible to clone the subsp. *israelensis* toxin gene into cyanobacteria (Angsuthanasombat and Panyim, 1989). Such bacteria should both replicate in the water environment and serve to make the mosquito larval food supply toxic.

Finally, the *B. thuringiensis* toxin gene can be directly cloned into and expressed in plant systems. Indeed, insect-resistant transgenic plants containing the *B. thuringiensis* toxin gene have been reported (Fischhoff et al., 1987).

Bacillus sphaericus

Although aerobic bacilli that produce round spores were described early in this century, the isolation of a strain that was pathogenic to mosquitoes and matched the description of *Bacillus sphaericus* did not occur until 1964 (Kellen and Meyer, 1964). This strain and subsequent isolates all possess the phenotypic characteristics typical of the species (see Chapter 76). Both pathogenic and nonpathogenic strains are aerobic bacilli producing round, terminally located spores that swell the sporangium. The spores vary in size and somewhat in shape (round to slightly oval) depending upon growth conditions. These bacteria are easily distinguished microscopically and physiologically from the other well-known mosquito pathogen, *B. thuringiensis* subsp. *israelensis.* The cells and spores of *B. sphaericus* are smaller than those of *B. thuringiensis* and the former species is much less active in its metabolism of a variety of substrates than is the latter. *B. sphaericus* is similar to *B. thuringiensis* in that the insect pathogenicity of each is mediated by the production of toxin.

Habitats and Isolation

The first isolation of a *B. sphaericus* mosquito pathogen was made from moribund, fourth-instar larvae of *Culiseta incidens* collected in California (Kellen and Meyer, 1964; Kellen et al., 1965). Additional isolates have also been obtained from dead mosquito larvae (Singer, 1973). More recently, pathogenic isolates have been obtained from black flies (Weiser, 1984), from caterpillars and grasshoppers (Lysenko et al., 1985), and from snails (de Barjac et al., 1988). These isolates were shown to be pathogenic for mosquitoes but lacked pathogenicity for the insects from which they were obtained. There was no indication that the snail isolate had been tested for pathogenicity to snails. *B. sphaericus* mosquito pathogens have also been obtained from mud taken from pools, from the soil of a dried stream bed (Brownbridge and Margalit, 1987), and from a loamy soil (de Barjac et al., 1988). Since this bacterium is not an obligate parasite and since there is no indication of marked differences in growth rates or metabolism between pathogenic and nonpathogenic strains, it seems likely that pathogens may be recovered from the same soil and aquatic habitats that are occupied by the nonpathogens. However, it is known that the pathogenic strains can multiply in the cadavers of mosquito larvae that have been killed by the larvicidal toxin (Charles and Nicolas, 1986; Davidson et al., 1975, 1984). Thus, it is likely that larger populations of the pathogens may be found in aquatic habitats that contain dead or dying larvae.

Isolation of pathogenic *B. sphaericus* from dead larvae can be performed by crushing larvae in sterile water or buffer and streaking the homogenate onto a proteinaceous medium such as brain heart infusion agar or nutrient agar supplemented with 0.05% yeast extract. A medium containing amino acids or protein hydrolysate is required because *B. sphaericus* cannot utilize sugars as a carbon source (Russell et al., 1989). The inclusion of yeast extract in media is useful because many strains of *B. sphaericus* are auxotrophic for vitamins, frequently biotin and thiamine (White and Lotay, 1980). If isolations are made from mud or soil where the population of the pathogens might be lower than in larval cadavers, it is useful to heat the suspension to kill contaminating vegetative cells before streaking. *B. sphaericus* has been reported to be resistant to streptomycin, chloramphenicol, lincomycin, and bacitracin (Burke and McDonald, 1983; Kalfon et al., 1986; Yousten et al., 1985) and these antibiotics may be useful additions to selective media. A defined medium (BATS) utilizing arginine as the sole carbon and nitrogen source and streptomycin as a selective agent allowed growth of pathogenic strains of *B. sphaericus* but did not allow growth of 68% of the nonpathogenic strains tested. It also did not allow growth of several other aerobic, spore-forming bacteria.

BATS Defined Medium for Growth of *Bacillus sphaericus* Mosquito Pathogens (Yousten et al., 1985)

The medium contains per liter:

Na$_2$HPO$_4$	5.57 g
KH$_2$PO$_4$	2.4 g
MgSO$_4$·7H$_2$O	0.05 g
MnCl$_2$·4H$_2$O	0.004 g
FeSO$_4$·7H$_2$O	0.028 g
CaCl$_2$·2H$_2$O	0.002 g
L-Arginine	5.0 g
Thiamine	0.02 g
Biotin	2.0 μg
Streptomycin sulfate	0.1 g

The arginine, biotin, thiamine and streptomycin are prepared as a filter-sterilized stock solution. The Mg^{2+}, Mn^{2+}, Fe^{2+}, and Ca^{2+} salts are prepared as an acidified (0.03% [vol/vol] concentration of H$_2$SO$_4$), autoclaved stock solution. These two stock solutions are added to the autoclaved phosphate salts solution when the latter has cooled to 50°C.

At present there is no way to distinguish pathogenic strains from nonpathogenic strains of *B. sphaericus* by colony morphology or by any simple biochemical test. New isolates grown on solid media are examined microscopically for round spores and then grown in a liquid medium allowing good sporulation. This is followed by bioassay of the cell mass using susceptible mosquito larvae, e.g., *Culex quinquefasciatus*. Two rather similar media that have been widely used for obtaining good sporulation and concurrent toxic production are NYSM broth (Myers and Yousten, 1980) and MBS broth (Kalfon et al., 1983).

NYSM Broth (Myers and Yousten, 1980)

The medium contains per liter:

Nutrient broth	8.0 g
Yeast extract	0.5 g
MgCl$_2$·6H$_2$O	0.2 g
MnCl$_2$·4H$_2$O	10.0 mg
CaCl$_2$·2H$_2$O	0.1 g

Identification

THE PLACE OF THE PATHOGENIC STRAINS WITHIN THE SPECIES. Despite the phenotypic similarities and the same GC content (35–37 mol%) found in the pathogenic and nonpathogenic strains of *B. sphaericus,* large amount of genetic diversity appear to exist within this "species." The first indication of this diversity was given by Seki et al. (1978) and was greatly expanded to cover both pathogens and nonpathogens by Krych et al. (1980). Analysis using DNA hybridization revealed five major homology groups within the species. One of these groups (II) was subdivided into groups IIA and IIB. The former was composed of the seven mosquito pathogens included in the study, whereas all of the strains in the other homology groups were nonpathogenic. F. G. Priest (per-

sonal communication) used DNA hybridization to study additional pathogenic strains and obtained similar results. However, other than pathogenicity itself, no single phenotypic test allowed unambiguous differentiation of pathogens from nonpathogens. de Barjac et al. (1980) carried out numerical analysis of growth on 160 substrates to identify three groups; one of these contained nine pathogens and one nonpathogen and the other two groups contained 25 nonpathogens. Williston and Singer (1987) were able to separate six pathogens from the type strain of the species (a nonpathogen) and one other nonpathogen by electrophoretic analysis of aminopeptidase activity on four L-amino acid-beta naphthylamide substrates. F. G. Priest (personal communication) used a group of 13 phenotypic tests to successfully differentiate pathogenic strains from the closely related nonpathogenic strains of homology group IIB. He found that adenine utilization was the single test that most often correlated with pathogenicity. Although the DNA homology studies clearly indicate that the mosquito pathogens are not the same species as the type strain of *B. sphaericus* (ATCC 14577), it is not clear if the pathogens are a separate species from the nonpathogens of homology group IIB. It is not yet known if the gene for the toxin responsible for pathogenicity is located on the chromosome or on a plasmid. If it is located on a plasmid, speciation of the pathogens would leave one in the same unsatisfactory position as is presently the case with *B. thuringiensis* and *B. cereus.*

DIVERSITY AMONG THE MOSQUITO PATHOGENIC STRAINS. The pathogenic strains of *Bacillus sphaericus* have been subdivided by serotyping (H antigens), bacteriophage typing, serotyping based upon antigenicity of the cell surface protein (the S layer), and the level of toxicity to mosquito larvae. These methods have shown remarkable agreement in the establishment of groups of strains.

Flagellar serotyping (H antigens) has identified six groups among the pathogenic strains (de Barjac et al., 1980; 1985, 1988). Three of these serotypes (H5a,5b; H25; and H6) contain highly toxic strains, whereas the other three serotypes contain weakly toxic strains. Serotype H6 is unique in that it also contains some nontoxic strains (de Barjac et al., 1988).

Several lytic bacteriophages have been isolated for the *B. sphaericus* pathogens. These phages have been used to group the strains, and the groups identified by phage typing correspond closely to the groups identified by flagellar serotyping (Yousten, 1984; Yousten et al.,

1980). Phage group 3 contains the highly toxic strains of serotypes H5a,5b and H6.

The bacteriophages have been shown to attach to the bacterial cells via a high molecular weight glycoprotein (Lewis and Yousten, 1988). This protein is present in a fine, linear array (the S layer) on the cell surface of the mosquito pathogens (Lewis and Yousten, 1987). This is in contrast to the tetragonal arrangement of the S layer on the type strain, ATCC 14577. Immunological analysis of this protein revealed that it was distinct for each of the bacteriophage groups, although in several cases bacteriophages were neutralized by the protein from strains on which the phages failed to replicate. Thus, although the proteins are distinct for each phage group, they alone do not determine those groups. The gene encoding the S-layer protein from *B. sphaericus* 2362 has been cloned and sequenced (Bowditch et al., 1989). The nucleotide sequence indicated an open reading frame coding for a protein of 1,176 amino acids with a molecular weight of 125 kDa. A 30 amino acid leader peptide is cleaved from the 125-kDa protein during secretion resulting in a 122-kDa protein present on the cell surface. The 12 N-terminal amino acids of the 122-kDa S-layer protein and those of a 110-kDa larval toxin are identical. This larval toxin, which at one time was thought to be present in the parasporal body, thus appears to be derived from the S layer. Immunological studies also support this interpretation (Bowditch et al., 1989). Amino acid sequence comparisons of the 125-kDa protein with the S-layer proteins of other bacteria, revealed no sequence similarity to those of *Halobacterium halobium* or *Deinococcus radiodurans* but did show significant sequence similarity to the N-terminal portion of the "outer wall protein" of *Bacillus brevis* 47.

Ecophysiology

The major aspect of ecophysiology discussed below relates to the toxins produced by *B. sphaericus* strains that are mosquito pathogens. As indicated above, one of the mosquito larval toxins appears to be derived from the cell wall-associated S-layer glycoprotein. However, the toxin thought to be most important for pathogenicity is primarily located in a parasporal body or crystal (Davidson and Meyers, 1981). This inclusion body appears in the cell during the time of engulfment of the forespore by the forespore septum. Thin sections of the inclusions revealed a crystalline lattice structure having striations at a 6.3-nm interval (de Barjac and Charles, 1983; Yousten and Davidson, 1982). Unlike the paraspores of *B. thuringiensis*, which

form outside the exosporium, those of *B. sphaericus* are partially enclosed by an elongated exosporium. This results in many of the paraspores remaining attached to the endospore upon lysis of the mother cell at the completion of sporulation. When paraspores were fed to mosquito larvae, the protein matrix of the paraspores rapidly dissolved in the alkaline midgut leaving a meshlike envelope in the shape of the paraspore (Yousten and Davidson, 1982). The composition of the parasporal envelope is unknown.

The matrix of the parasporal body is composed of two proteins: a 41.9-kDa protein that is toxic to mosquito larvae and to cultured mosquito cells, and a 51.4-kDa protein that is toxic to neither. Genes encoding these proteins have been cloned and sequenced from three highly toxic strains (Baumann et al., 1988; Berry and Hindley, 1987; Hindley and Berry, 1987, 1988). No sequence similarity was found between the *B. sphaericus* toxin and the Diptera- and Lepidoptera-active toxins of *B. thuringiensis* subsp. *israelensis* and *kurstaki*, respectively (Baumann et al., 1988). The N-terminus of the toxin protein present in the paraspore is missing four amino acids as compared to the amino acid sequence deduced from the nucleotide sequence. It has been suggested that this deletion is produced by a *B. sphaericus* protease. Following ingestion of the 41.9-kDa toxin by mosquito larvae, the larval gut proteases remove an additional six amino acids from the N-terminus and approximately 20 amino acids from the C-terminus. This proteolysis enhanced the cytotoxicity of the protein for cultured mosquito cells (Broadwell and Baumann, 1987; Davidson, 1986; Davidson et al., 1987a). Interestingly, when the cloned 41.9-kDa toxin was produced in *E. coli*, it was not toxic to larvae unless it was combined with the 51.4-kDa (nontoxic) protein (Baumann et al., 1988).

In addition to the highly toxic strains, which produce a parasporal body containing the 41.9-kDa toxin, there are low-toxicity strains that do not produce a paraspore (Davidson and Myers, 1981). Probes that utilize the nucleotide sequence of the 41.9-kDa toxin did not detect these sequences in dot blots of the low toxicity strains (D. Wilcox, personal communication). The nature of the toxicity in these low toxicity strains is unknown and the possible relationship to the 110-kDa S layer-derived toxin remains to be determined.

The *B. sphaericus* toxin affects a variety of mosquito species from several genera but, unlike the toxin(s) from *B. thuringiensis* subsp. *israelensis*, it is not toxic to black fly larvae. It is much more toxic for mosquito larvae of the genus *Culex* than it is for *Aedes aegypti* (though certain other species of *Aedes* are quite sensitive). The degree of sensitivity to the toxin may be related to the ability of the toxin to bind to midgut cells in the larvae. Good binding of fluorescein-labeled toxin was observed in the gut of *Culex pipiens*, but there was no binding in the gut of *Aedes aegypti* (Davidson, 1989). The effects of the toxin have been studied both in mosquito larvae (Charles, 1987; Davidson, 1979; Singh and Gill, 1988) and in cultured mosquito cells (Davidson, 1986; Davidson and Titus, 1987; Davidson et al., 1987b). Although the pathological effects have been reported in detail for events occurring at the organ and cellular level, the mechanism of action of the toxin at the subcellular or molecular level is unknown.

Genetics

Although lytic bacteriophages have been isolated that use the mosquito pathogens as hosts, temperate phages capable of carrying out transduction have not been described. The phenomenon of lysogeny has not been demonstrated in these bacteria.

The procedures used to induce competence in *Bacillus subtilis* have not proven successful for carrying out transformation in *B. sphaericus*. However, plasmid-mediated transformation of polyethylene glycol (PEG)-treated protoplasts was performed by McDonald and Burke (1984). An improved PEG-protoplast transformation procedure using restriction-deficient strains of *B. sphaericus* 1593 produced about 1×10^3 transformants/μg of pUB110, pLT105, or pAMB1 DNA (Taylor and Burke, 1989). W. Burke (personal communication) also used electroporation for the introduction of plasmid DNA into *B. sphaericus*. Restriction and modification systems were suggested by the fact that in *B. sphaericus* 1593 the transformation efficiency was 10^4-fold greater when the plasmid DNA, pUB110, was isolated from *B. sphaericus* rather than *B. subtilis*. This was confirmed by agarose gel electrophoresis of pUB110 and PBC16 incubated with lysates from *B. sphaericus* and *B. subtilis*. The results indicated the restriction system is an isoschizomer of *Hae* III (McDonald and Burke, 1984; Burke, personal communication). Cloning vectors for the mosquito pathogenic strains are being constructed (Norton et al., 1985).

Conjugation between *B. sphaericus* 1593 and *Enterococcus (Streptococcus) faecalis* using the method of filter mating was carried out by Orzech and Burke (1984). The transfer process was more effective between *E. faecalis* and *B. sphaericus* than between strains of *B. sphaeri*-

cus. It was unaffected by either endogenous restriction enzyme activity or by the presence of exogenous DNase.

The reports of naturally occurring plasmids in the toxin-producing strains of *B. sphaericus* have provided somewhat conflicting information. Davidson et al. (1982) reported the presence of a single large plasmid in the highly toxic strains 1593 and 1881 but no plasmids in strains 1691 or 2362. Abe et al. (1983) found a single large plasmid in strain 1881 but no plasmids in strains 1593 or 1691. In the weakly pathogenic strain Kellen K, Davidson et al. (1982) found no plasmids but Abe et al. (1983) found five. Singer (1987) reported an approximately 75-MDa plasmid in the highly toxic strains 1593, 2362, and 2297 as well in the less toxic strain SSII-1. Strain 2297 also contained two smaller plasmids (3.4 and 3.2 MDa). A. A. Yousten (unpublished observations) obtained results similar to those of Singer. The approximately 75-MDa plasmid seems to be common among the highly toxic strains, but the plasmid is not restricted to them because it is also present in the much less toxic strain SSII-1. In *B. thuringiensis* the gene(s) for the toxin(s) are usually present on plasmids. However, this determination has not yet been made for the mosquito larval toxin of *B. sphaericus*.

Biotechnological Applications

The major biotechnological application of mosquito-pathogenic strains of *B. sphaericus* has been as a biological control agent.

Following ingestion of a mixture of spores and toxin-containing paraspores, the number of viable spores in the gut decreases until the death of the larva, which usually occurs about 24 h after ingestion of the toxin. At that time some unknown fraction of the spores germinate, grow vegetatively, and sporulate in the larval cadaver. This results in about a 10–100-fold increase in the number of spores compared to the number originally consumed (Charles and Nicolas, 1986; Davidson et al., 1984). Each larval cadaver seems capable of producing about 10^5–10^6 spores. This recycling of the spores in the cadaver may be related to the prolonged larval control reported for this bacterium compared to *B. thuringiensis* subsp. *israelensis* when sporulated cultures of these bacteria have been sprayed into water to control mosquito breeding (Singer, 1987). The attractiveness of *B. sphaericus* as a mosquito larvicide is related to its extended persistence in the water and to the unrelatedness of its toxin to that of *B. thuringiensis* subsp. *israelensis*. This latter characteristic might make it a useful alternative should resistance to the *B. thuringiensis* toxin appear. Efforts are being made to introduce toxin genes from *B. thuringiensis* subsp. *israelensis* into *B. sphaericus* with the goal of extending the lethal effects of the latter bacterium to *A. aegypti*, a species of mosquito quite resistant to the usual *B. sphaericus* toxin. This would also extend the environmental persistence of the *B. thuringiensis* toxin. An additional attempt to extend environmental persistence has been made by de Marsac et al. (1987), who cloned an 8.6-kb fragment of *B. sphaericus* 1593M DNA into the cyanobacterium *Anacystis nidulans R2*, using the shuttle vector pUC303. The resulting transformant was toxic to mosquito larvae, though at least 1000-fold less toxic than the *B. sphaericus* parent.

Literature Cited

Abe, K., R. Faust, and L. Bulla. 1983. Plasmid deoxyribonucleic acid in strains of *Bacillus sphaericus* and in *B. moritae*. J. Invertebr. Pathol. 41:328–335.

Adams, J. A. 1949. *Cyclocephala borealis* as a turf pest associated with the Japanese beetle in New York. J. Econ. Entomol. 42:626–628.

Adams, L. F., J. E. Visick, and H. R. Whiteley. 1989. A 20 kilodalton protein is required for efficient production of the *Bacillus thuringiensis* subsp. *israelensis* 27-kilodalton crystal protein in *Escherichia coli*. J. Bacteriol. 171:521–530.

Adang, M. J., and K. D. Spence. 1983. Permeability of the peritrophic membrane of the Douglas-fir tussock moth (*Orgyia pseudotsugata*). Comp. Biochem. Physiol. A. Comp. Physiol. 75:233–238.

Adang, M. J., M. J. Staver, T. A. Rocheleau, J. Leighton, R. F. Barker, and D. V. Thompson. 1985. Characterized full-length and truncated plasmid clones of the crystal protein of *Bacillus thuringiensis* subsp. *kurstaki* HD-73 and their toxicity to *Manduca sexta*. Gene 36:289–300.

Ahmad, S., H. T. Streu, and L. M. Vasvary. 1983. The Japanese beetle: A major pest of turfgrass. American Lawn Applicator April:4–10, 31.

Ahmad, W., C. Nicholls, and D. J. Ellar. 1989. Cloning and expression of an entomocidal protein gene from *Bacillus thuringiensis galleriae* toxic to both Lepidoptera and Diptera. FEMS Microbiol. Lett. 59:197–202.

Alikhanian, S. I., N. F. Ryabchenko, N. O. Bukanov, and V. A. Sakanyan. 1981. Transformation of *Bacillus thuringiensis* subsp. *galleria* protoplasts by plasmid pBC16. J. Bacteriol. 146:7–9.

Aly, C. 1985. Germination of *Bacillus thuringiensis* var. *israelensis* spores in the gut of *Aedes* larvae, Diptera: Culicidae. J. Invertebr. Pathol. 45:1–8.

Aly, C., M. S. Mulla, and B. A. Federici. 1985. Sporulation and toxin production by *Bacillus thuringiensis* var. *israelensis* in cadavers of mosquito larvae, Diptera: Culicidae. J. Invertebr. Pathol. 46:251–258.

Andrews, R. E., Jr., and L. A. Bulla, Jr. 1981. Toxins of sporeforming bacteria, p. 57–63. In: H. S. Levinson,
A. L. Sonenshein, and D. J. Tipper (ed.), Sporulation and germination. American Society for Microbiology, Washington, D.C.

Andrews, R. E., Jr., D. B. Bechtel, B. S. Campbell, L. I. Davidson, and L. A. Bulla, Jr. 1981. Solubility of parasporal crystals of *Bacillus thuringiensis* and presence of toxic protein during sporulation, germination, and outgrowth, p. 174–177. In: H. S. Levinson, A. L. Sonenshein, and D. J. Tipper (ed.), Sporulation and germination. American Society for Microbiology, Washington, D.C.

Andrews, R. E., Jr., M. M. Bibilos, and L. A. Bulla, Jr. 1985. Protease activation of the entomocidal protoxin of *Bacillus thuringiensis* subsp. *kurstaki*. Appl. Environ. Microbiol. 50:737–742.

Andrews, R. E., Jr., R. M. Faust, H. Wabiko, K. C. Raymond, and L. A. Bulla, Jr. 1987. The biotechnology of *Bacillus thuringiensis*. CRC Crit. Rev. Biotechnol. 6:163–232.

Andrews, R. E., Jr., J. J. Iandolo, B. S. Campbell, L. I. Davidson, and L. A. Bulla, Jr. 1980. Rocket immunoelectrophoresis of the entomocidal parasporal crystal of *Bacillus thuringiensis* subsp. *kurstaki*. Appl. Environ. Microbiol. 40:897–900.

Andrews, R. E., Jr., K. Kanda, and L. A. Bulla, Jr. 1982. In vitro translation of the entomocidal toxin of *B. thuringiensis*, p. 121–130. In: A. T. Ganesan, S. Chang, and J. A. Hoch (ed.), Molecular cloning and gene regulation in bacilli. Academic Press, Inc., New York.

Angsuthanasombat, C., W. Chungjatupornchai, S. Kertbundit, P. Lusananil, C. Settasatian, P. Wilairat, and S. Panyim. 1987. Cloning and expression of a 130-kDa mosquito-larvacidal delta-endotoxin of *Bacillus thuringiensis* var. *israelensis* in *Escherichia coli*. Mol. Gen. Genet. 208:384–389.

Angsuthanasombat, C., and S. Panyim. 1989. Biosynthesis of 130-kilodalton mosquito larvacide in the cyanobacterium *Agmenellum quadruplicatum* PR-6. Appl. Environ. Microbiol. 55:2428–2430.

Angus, T. A., 1956a. Association of toxicity with protein crystalline inclusions of *Bacillus sotto* Ishiwata. Can. J. Microbiol. 2:122–131.

Angus, T. A. 1956b. Extraction, purification and properties of *Bacillus sotto* toxin. Can. J. Microbiol. 2:416–426.

Armstrong, J. L., G. F. Rohrmann, and G. S. Beaudreau. 1985. Delta endotoxin of *Bacillus thuringiensis* subsp. *israelensis*. J. Bacteriol. 161:39–46.

Aronson, A. I., D. J. Tyrell, P. C. Fitz-James, and L. A. Bulla, Jr. 1982. Relationship of the syntheses of spore coat protein and parasporal crystal protein in *Bacillus thuringiensis*. J. Bacteriol. 151:399–410.

Azuma, R., and S. Kitaoka. 1965. Cultivation and properties of *Bacillus larvae* and decrease of its proteolytic activity on glucose media. Nat. Inst. Anim. Hlth. Quart. 5:138–145.

Bailey, L. 1968. Honey bee pathology. Annu. Rev. Entomol. 13:191–212.

Bailey, L., and D. C. Lee. 1962. *Bacillus larvae:* Its cultivation in vitro and its growth in vivo. J. Gen. Microbiol. 29:711–717.

Bakhiet, N., and D. P. Stahly. 1985a. Studies on transfection and transformation of protoplasts of *Bacillus larvae, Bacillus subtilis*, and *Bacillus popilliae*. Appl. Environ. Microbiol. 49:577–581.

Bakhiet, N., and D. P. Stahly. 1985b. Ultrastructure of sporulating *Bacillus larvae* in a broth medium. Appl. Environ. Microbiol. 50:690–692.

Bakhiet, N., and D. P. Stahly. 1986. Plasmid transformation of protoplasts of *Bacillus popilliae*. Abstr. Annual Meeting Amer. Soc. Microbiol. 1986:142.

Bakhiet, N., and D. P. Stahly. 1988. Properties of clear plaque mutants of the *Bacillus larvae* bacteriophages PBLO.5 and PBL2. J. Invertebr. Pathol. 52:78–83.

Bamrick, J. F. 1964. Resistance to American foulbrood in honey bees. V. Comparative pathogenesis in resistant and susceptible larvae. J. Insect Pathol. 6:284–304.

Bamrick, J. F., and W. C. Rothenbuhler. 1961. Resistance to American foulbrood in honey bees. IV. The relationship between larval age at inoculation and mortality in resistant and in a susceptible line. J. Insect Pathol. 3:381–390.

Battisti, L., B. D. Green, and C. B. Thorne. 1985. Mating system for transfer of plasmids among *Bacillus anthracis, Bacillus cereus,* and *Bacillus thuringiensis.* J. Bacteriol. 162:543–550.

Baumann, L., A. Broadwell, and P. Baumann. 1988. Sequence analysis of the mosquitocidal toxin genes encoding 51.4- and 41.9-kilodalton proteins from *Bacillus sphaericus* 2362 and 2297. J. Bacteriol. 170:2045–2050.

Baumann, L., K. Okamoto, B. M. Unterman, M. J. Lynch, and P. Baumann. 1984. Phenotypic characterization of *Bacillus thuringiensis* and *Bacillus cereus.* J. Invertebr. Pathol. 44:329–341.

Beard, R. L. 1946. Competition between two entomogenous bacteria. Science 103:371–372.

Beard, R. L. 1956. Two milky diseases of Australian Scarabaeidae. Can. Entomologist 88:640–647.

Bechtel, B. D., and L. A. Bulla, Jr. 1976. Electron microscope study of sporulation and parasporal crystal formation in *Bacillus thuringiensis.* J. Bacteriol. 127:1472–1481.

Beebee, T., A. Korner, and R. P. M. Bond. 1972. Differential inhibition of mammalian ribonucleic acid polymerases by an exotoxin from *Bacillus thuringiensis.* Biochem. J. 127:619–624.

Beegle, C. C., H. T. Dulmage, D. A. Wolfenbarger, and E. Martinez. 1981. Persistence of *Bacillus thuringiensis berliner* insecticidal activity on cotton foliage. Environ. Entomol. 10:201–206.

Benada, O., V. Drobniková, and J. Ludvik. 1984a. Plasmid DNA in *Bacillus larvae.* Folia Microbiologica 29:424.

Benada, O., J. Ludvík, and V. Drobniková. 1984b. Morphology of a new bacteriophage isolated from *Bacillus larvae.* Folia Microbiol. 29:520–521.

Bennett, G. A., and O. L. Shotwell. 1973. Isolation and characterization of constituents in hemolymph from healthy and diseased Japanese beetle larvae. Biotechnol. Bioeng. 15:1023–1037.

Berliner, E. 1911. Über die schalffsucht der mehlmottentaup. Z. Gesamte Getreidewes. 4:67–70.

Berliner, E. 1915. Über die schlaffsucht der mehlmottenraupe (*Ephestia kuhniella,* Zell.) und ihren erreger *Bacillus thuringiensis* n. sp. Z. Agnew. Entomol. 2:29–35.

Berry, C., and J. Hindley. 1987. *Bacillus sphaericus* strain 2362: Identification and nucleotide sequence of the 41.9 kDa toxin gene. Nucleic Acids Res. 15:5891.

Bhumiratana, A., R. L. Anderson, and R. N. Costilow. 1974. Trehalose metabolism by *Bacillus popilliae.* J. Bacteriol. 119:484–493.

Bibilos, M., and R. E. Andrews, Jr. 1988. Inhibition of *Bacillus thuringiensis* proteases and their effects on crystal toxin proteins and cell-free translations. Can. J. Microbiol. 34:740–747.

Black, S. H. 1968a. Cytology of milky disease bacteria. I. Morphogenesis of *Bacillus popilliae* in vivo. J. Invertebr. Pathol. 12:148–157.

Black, S. H. 1968b. Cytology of milky disease bacteria. II. Morphogenesis of *Bacillus popilliae* in vitro. J. Invertebr. Pathol. 12:158–167.

Blum, M. S., A. F. Novak, and S. Taber, III. 1959. 10-hydroxy-Δ2-decenoic acid, an antibiotic found in royal jelly. Science 130:452–453.

Bone, E. J., and D. J. Ellar. 1989. Transformation of *Bacillus thuringiensis* by electroporation. FEMS Microbiol. Lett. 58:171–178.

Boucias, D. G., R. H. Cherry, and D. L. Anderson. 1986. Incidence of *Bacillus popilliae* in *Ligyrus subtropicus* and *Cyclocephala parallela* (Coleoptera: Scarabaeidae) in Florida sugarcane fields. Environ. Entomol. 15:703–705.

Bourgouin, C., A. Delécluse, J. Ribier, A. Klier, and G. Rapoport. 1988. A *Bacillus thuringiensis* subsp. *israelensis* gene encoding a 125-kilodalton larvicidal polypeptide is associated with inverted repeat sequences. J. Bacteriol. 170:3575–3583.

Bourgouin, C., A. Klier, and G. Rapoport. 1986. Characterization of the genes encoding the haemolytic toxin and the mosquitocidal delta-endotoxin of *Bacillus thuringiensis israelensis.* Mol. Gen. Genet. 205:390–397.

Bowditch, R. D., P. Baumann, and A. A. Yousten. 1989. Cloning and sequencing of the gene encoding a 125-kilodalton surface-layer protein from *Bacillus sphaericus* 2362 and of a related cryptic gene. J. Bacteriol. 171:4178–4188.

Brizzard, B. L., and H. R. Whiteley. 1988. Nucleotide sequence of an additional crystal protein gene cloned from *Bacillus thuringiensis* subsp. *thuringiensis.* Nuc. Acids Res. 16:2723–2724.

Broadwell, A., and P. Baumann. 1987. Proteolysis in the gut of mosquito larvae results in further activation of the *Bacillus sphaericus* toxin. Appl. Environ. Microbiol. 53:1333–1337.

Brown, K. L., and H. R. Whiteley. 1988. Isolation of a *Bacillus thuringiensis* RNA polymerase capable of transcribing crystal protein genes. Proc. Natl. Acad. Sci. USA 85:4166–4170.

Brownbridge, M., and J. Margalit. 1987. Mosquito active strains of *Bacillus sphaericus* isolated from soil and mud samples collected in Israel. J. Invertebr. Pathol. 50:106–112.

Bucher, G. E. 1960. Potential bacterial pathogens of insects and their characteristics. J. Insect Pathol. 2:172–195.

Bulla, L. A., Jr., D. B. Bechtel, K. J. Kramer, Y. I. Shethna, A. I. Aronson, and P. C. Fitz-James. 1980. Ultrastructure, physiology and biochemistry of *Bacillus thuringiensis.* CRC Crit. Rev. Microbiol. 8:147–204.

Bulla, L. A., Jr., G. A. Bennett, and O. L. Shotwell. 1970a. Physiology of sporeforming bacteria associated with insects. II. Lipids of vegetative cells. J. Bacteriol. 104:1246–1253.

Bulla, L. A., Jr., R. N. Costilow, and E. S. Sharpe. 1978. Biology of *Bacillus popilliae.* Adv. Appl. Microbiol. 23:1–18.

Bulla, L. A., Jr., L. I. Davidson, K. J. Kramer, and B. L. Jones. 1979. Purification of the insecticidal toxin from the parasporal crystal of *Bacillus thuringiensis* subsp. *kurstaki*. Biochem. Biophys. Res. Commun. 91:1123–1130.

Bulla, L. A., Jr., R. M. Faust, R. Andrews, and N. Goodman. 1985. Insecticidal bacilli, p. 185–209. In: D. A. Dubnau (ed.), The molecular biology of the bacilli, vol. 2. Academic Press, Inc., New York.

Bulla, L. A., Jr., K. J. Kramer, D. J. Cox, B. L. Jones, L. I. Davidson, and G. L. Lookhart. 1981. Purification and characterization of the entomocidal protoxin of *Bacillus thuringiensis*. J. Biol. Chem. 256:3000–3004.

Bulla, L. A., Jr., G. St. Julian, and R. A. Rhodes. 1971. Physiology of sporeforming bacteria associated with insects. III. Radiorespirometry of pyruvate, acetate, succinate, and glutamate oxidation. Can. J. Microbiol. 17:1073–1079.

Bulla, L. A., G. St. Julian, R. A. Rhodes, and C. W. Hesseltine. 1969. Scanning electron and phase-contrast microscopy of bacterial spores. Appl. Microbiol. 18:490–495.

Bulla, L. A., G. St. Julian, R. A. Rhodes, and C. W. Hesseltine. 1970b. Physiology of sporeforming bacteria associated with insects. I. Glucose catabolism in vegetative cells. Can. J. Microbiol. 16:243–248.

Burges, H. D. 1981. Microbial control of pests and plant diseases 1970–1980. Academic Press, New York.

Burges, H. D., and J. A. Hurst. 1977. Ecology of *Bacillus thuringiensis* in storage moths. J. Invertebr. Pathol. 30:131–139.

Burke, W., and K. O. McDonald. 1983. Naturally occurring antibiotic resistance in *Bacillus sphaericus* and *Bacillus licheniformis*. Curr. Microbiol. 9:69–72.

Burri, R. 1904. Bakteriologische forschungen über die faulbrut. Schweiz. Bienenztg. 27:335–342, 360–365.

Calabrese, D. M., K. W. Nickerson, and L. C. Lane. 1980. A comparison of protein crystal subunit sizes in *Bacillus thuringiensis*. Can. J. Microbiol. 26:1006–1010.

Carlton, B. C., and J. M. Gonzalez, Jr. 1985. Plasmids and delta-endotoxin production in different subspecies of *Bacillus thuringiensis*, p. 246–252. In: J. A. Hoch and P. Setlow (ed.), Molecular biology of microbial differentiation. American Society for Microbiology, Washington, D.C.

Chak, K. F., and D. J. Ellar. 1987. Cloning and expression in *Escherichia coli* of an insecticidal crystal protein gene from *Bacillus thuringiensis* var. *aizawai* HD-133. J. Gen. Microbiol. 133:2921–2931.

Chang, S., and S. N. Cohen. 1979. High frequency transformation of *Bacillus subtilis* protoplasts by plasmid DNA. Mol. Gen. Genet. 168:111–115.

Charles, J. -F. 1987. Ultrastructural midgut events in culicidae larvae fed with *Bacillus sphaericus* 2297 spore/crystal complex. Ann. Inst. Pasteur/Microbiol. 138:471–484.

Charles, J. -F., and L. Nicolas. 1986. Recycling of *Bacillus sphaericus* 2362. Ann. Inst. Pasteur/Microbiol. 137B:101–111.

Chen, N., and X. Yin. 1986. Formation of protoplast and plasmid transformation in the genus *Bacillus*. Acta Microbiologica Sinica 26:134–142.

Chilcott, C. N., and D. J. Ellar. 1988. Comparative toxicity of *Bacillus thuringiensis* var. *israelensis* crystal proteins in vivo and in vitro. J. Gen. Microbiol. 134:2551–2558.

Chilcott, C. N., J. Kalmakoff, and J. S. Pillai. 1983. Characterization of proteolytic activity associated with *Bacillus thuringiensis* var. *israelensis* crystals. FEMS Microbiol. Lett. 18:37–41.

Chungjatupornchai, W., H. Höfte, J. Seurnick, C. Angsuthanasombat, and M. Vaeck. 1988. Common features of *Bacillus thuringiensis* toxins specific for Diptera and Lepidoptera. Eur. J. Biochem. 173:9–16.

Claus, D., and R. C. W. Berkeley. 1986. Genus *Bacillus* Cohn 1872, p. 1105–1207. In: P. H. A. Sneath, N. S. Mair, M. E. Sharpe, and J. G. Holt (ed.), Bergey's manual of systematic bacteriology, vol. 2. Williams and Wilkins, Baltimore.

Collins, M. D., and D. Jones. 1981. Distribution of isoprenoid quinone structural types in bacteria and their taxonomic implications. Microbiol. Rev. 45:316–354.

Costilow, R. N., and W. H. Coulter. 1971. Physiological studies of an oligosporogenous strain of *Bacillus popilliae*. Appl. Microbiol. 22:1076–1084.

Costilow, R. N., and B. B. Keele, Jr. 1972. Superoxide dismutase in *Bacillus popilliae*. J. Bacteriol. 111:628–630.

Costilow, R. N., C. J. Sylvester, and R. E. Pepper. 1966. Production and stabilization of cells of *Bacillus popilliae* and *Bacillus lentimorbus*. Appl. Microbiol. 14:161–169.

Couch, T. L., and C. M. Ignoffo. 1981. Formulation of insect pathogens, p. 621–634. In: H. D. Burges (ed.), Microbial control of pests and plant diseases, 1970–1980. Academic Press, NY.

Coulter, W. H., and R. N. Costilow. 1970. The role of barbituric acid in the nutrition of *Bacillus popilliae*. Can. J. Microbiol. 16:801–807.

Cowan, T. W. 1911. British bee-keeper's guide book to the management of bees in movable-comb hives. Madgwick, Houlston, and Co., Ltd., London.

Crawford, I. T., K. D. Greis, L. Parks, and U. N. Streips. 1987. Facile autoplast generation and transformation in *Bacillus thuringiensis* subsp. *kurstaki*. J. Bacteriol. 169:5423–5428.

Dalhammar, G., and H. Steiner. 1984. Characterization of inhibitor A, a protease from *Bacillus thuringiensis* which degrades attacins and cecropins, two classes of antibacterial proteins in insects. Eur. J. Biochem. 139:247–252.

Davidson, E. 1979. Ultrastructure of midgut events in the pathogenesis of *Bacillus sphaericus* strain SSII-1 infections of *Culex pipiens quinquefasciatus* larvae. Can. J. Microbiol. 25:178–184.

Davidson, E., and P. Myers. 1981. Parasporal inclusions in *Bacillus sphaericus*. FEMS Microbiol. Lett. 10:261–265.

Davidson, E., C. Shellabarger, M. Meyer, and A. Bieber. 1987b. Binding of the *Bacillus sphaericus* mosquito larvicidal toxin to cultured insect cells. Can. J. Microbiol. 33:982–989.

Davidson, E., S. Singer, and J. Briggs. 1975. Pathogenesis of *Bacillus sphaericus* strain SSII-1 infections in *Culex pipiens quinquefasciatus* larvae. J. Invertebr. Pathol. 25:179–184.

Davidson, E., J. Spizizen, and A. A. Yousten. 1982. Recent advances in the genetics of *Bacillus sphaericus*. Proc. Intl. Colloq. Invertebr. Pathol. Brighton, U.K. p. 14.

Davidson, E., and M. Titus. 1987. Ultrastructural effects of the *Bacillus sphaericus* mosquito larvicidal toxin on cultured insect cells. J. Invertebr. Pathol. 50:213–220.

Davidson, E. W. 1970. Ultrastructure of peritrophic membrane development in larvae of the worker honey bee (*Apis mellifera*). J. Invertebr. Pathol. 15:451–454.

Davidson, E. W. 1973. Ultrastructure of American foulbrood disease pathogenesis in larvae of the worker honey bee, *Apis mellifera.* J. Invertebr. Pathol. 21:53–61.

Davidson, E. W. 1986. Effects of *Bacillus sphaericus* 1593 and 2362 spore/crystal toxin on cultured mosquito cells. J. Invertebr. Pathol. 47:21–31.

Davidson, E. W. 1989. Variation in binding of *Bacillus sphaericus* toxin and wheat germ agglutinin to larval midgut cells of six species of mosquitoes. J. Invertebr. Pathol. 53:251–259.

Davidson, E. W., A. Bieber, M. Meyer, and C. Shellabarger. 1987a. Enzymatic activation of the *Bacillus sphaericus* mosquito larvicidal toxin. J. Invertebr. Pathol. 50:40–44.

Davidson, E. W., M. Urbina, J. Payne, M. Mulla, H. Darwazeh, H. Dulmage, and J. Correa. 1984. Fate of *Bacillus sphaericus* 1593 and 2362 spores used as larvicides in the aquatic environment. Appl. Environ. Microbiol. 47:125–129.

Davidson, E. W., and T. Yamamoto. 1984. Isolation and assay of the toxic component from the crystals of *Bacillus thuringiensis* var. *israelensis.* Curr. Microbiol. 11:171–174.

Dean, D. H. 1984. Biochemical genetics of the bacterial insect-control agent *Bacillus thuringiensis*: Basic principles and prospects for genetic engineering. Biotechnol. Genet. Eng. Rev. 2:341–363.

de Barjac, H. 1981. Insect pathogens in the genus *Bacillus,* p. 241–250. In: R. C. W. Berkeley and M. Goodfellow (ed.), The aerobic endosporeforming bacteria: Classification and identification. Academic Press, New York.

de Bariac, H., and A. Bonnefoi. 1962. Essai de classification biochimique et sérologique de 24 souches de *Bacillus* du type *thuringiensis.* Entomophaga 7:5–31.

de Barjac, H., and A. Bonnefoi. 1973. Mise au point sur la classification des *Bacillus thuringiensis.* Entomophaga 18:5–17.

de Barjac, H., and J. -F. Charles, 1983. Une nouvelle toxine active sur les moustiques presente dans des inclusions crystalline produites par *Bacillus sphaericus.* C. R. Acad. Sci. Paris. 296:905–910.

de Barjac, H., I. Larget-Thiery, V. Cosmao-Dumanoir, and H. Ripouteau. 1985. Serological classification of *Bacillus sphaericus* strains on the basis of toxicity to mosquito larvae. Appl. Microbiol. Biotechnol. 21:85–90.

de Barjac, H., I. Thiery, V. Cosmao-Dumanoir, E. Frachon, P. Laurent, J. -F. Charles, S. Hamon, and J. Ofori. 1988. Another *Bacillus sphaericus* serotype harboring strains very toxic to mosquito larvae: serotype H6. Ann. Inst. Pasteur/Microbiol. 139:363–377.

de Barjac, H., M. Veron, and V. Cosmao-Dumanoir. 1980. Caracterisation biochemique et serologique de souches de *Bacillus sphaericus* pathogenes ou non pour les moustiques. Ann. Microbiol. (Paris). 131B:191–201.

De Ley, J. 1978. Modern molecular methods in bacterial taxonomy: Evaluation, applications, prospects. Proc. 4th Int. Conf. Plant Path. Bact. Angers 1978, 347–357.

de Marsac, N. T., F. de la Torre, and J. Szulmajster. 1987. Expression of the larvicidal gene of *Bacillus sphaericus* 1593M in the cyanobacterium *Anacystis nidulans* R2. Mol. Gen. Genet. 209:396–398.

DeLucca, A. J., II, J. G. Simonson, and A. D. Larson. 1981. *Bacillus thuringiensis* distribution in soils of the United States. Can. J. Microbiol. 27:865–870.

Dingman, D. W. 1983. *Bacillus larvae:* Parameters involved with sporulation and characteristics of two bacteriophages. Ph.D. thesis, University of Iowa, U.S.A.

Dingman, D. W., N. Bakhiet, C. C. Field, and D. P. Stahly. 1984. Isolation of two bacteriophages from *Bacillus larvae*, PBL1 and PBL0.5, and partial characterization of PBL1. J. Gen. Virol. 65:1101–1105.

Dingman, D. W., and D. P. Stahly. 1983. Medium promoting sporulation of *Bacillus larvae* and metabolism of medium components. Appl. Environ. Microbiol. 46:860–869.

Dingman, D. W., and D. P. Stahly. 1984. Protection of *Bacillus larvae* from oxygen toxicity with emphasis on the role of catalase. Appl. Environ. Microbiol. 47:1228–1237.

Donovan, W. P., C. Dankocsik, and M. P. Gilbert. 1988a. Molecular characterization of a gene encoding a 72-kilodalton mosquito-toxic crystal protein from *Bacillus thuringiensis* subsp. *israelensis.* J. Bacteriol. 170:4732–4738.

Donovan, W. P., J. M. Gonzalez, Jr., M. P. Gilbert, and C. Dankocsik. 1988b. Isolation and characterization of EG2158, a new strain of *Bacillus thuringiensis* toxic to coleopteran larvae, and nucleotide sequence of the toxin gene. Mol. Gen. Genet. 214:365–372.

Drobniková, V., and J. Ludvik. 1982. Bacteriophage of *Bacillus larvae.* J. Apicultural Res. 21:53–56.

Dumbleton, L. J. 1945. Bacterial and nematode parasites of soil insects. New Zealand J. Science Technol. 27:76–81.

Dutky, S. R. 1940. Two new sporeforming bacteria causing milky diseases of Japanese beetle larvae. J. Agr. Res. 61:57–68.

Dutky, S. R. 1947. Preliminary observations on the growth requirements of *Bacillus popilliae* Dutky and *Bacillus lentimorbus* Dutky. J. Bacteriol. 54:267.

Dutky, S. R. 1963. The milky diseases, p. 75–115. In: E. A. Steinhaus (ed.), Insect pathology, an advanced treatise, vol. 2. Academic Press, New York.

Earp, D. J., and D. J. Ellar. 1987. *Bacillus thuringiensis* var. *morrisoni* strain PG14: Nucleotide sequence of a gene encoding a 27 kDa crystal protein. Nuc. Acids Res. 15:3619.

Ellis, B.-J., F. Obenchain, and R. Mehta. 1989. In vitro method for producing infective bacterial spores and spore-containing insecticidal compositions. United States Patent Number 4,824,671.

Fahmy, F., J. Flossdorf, and D. Claus. 1985. The DNA base composition of the type strains of the genus *Bacillus.* System. Appl. Microbiol. 6:60–65.

Fast, P. G. 1977. *Bacillus thuringiensis* delta-endotoxin: On the relative roles of spores and crystals in toxicity to spruce budworm (Lepidoptera: Tortricidae). Can. Entomol. 109:1515–1518.

Faust, R. M., J. Spizizen, V. Gage, and R. S. Travers. 1979. Extrachromosomal DNA in *Bacillus thuringiensis* var. *kurstaki*, var. *finitimus*, var. *sotto*, and in *Bacillus popilliae.* J. Invertebr. Pathol. 33:233–238.

Feng, X.-X., S.-C. Shing, and M.-H. Yang. 1982. Cultivation of *Bacillus popilliae* on artificial media. Acta Entomologica Sinica 25:156–159.

Ferro, D. N., and W. D. Gelernter. 1989. Toxicity of a new strain of *Bacillus thuringiensis* to Colorado potato beetle (*Coleoptera:* Chrysomelidae). J. Econ. Entomol. 82:750–755.

Fick, R. B., Jr., G. P. Naegel, S. U. Squier, R. E. Wood, J. B. L. Gee, and H. Y. Reynolds. 1984. Proteins of the cystic fibrosis respiratory tract fragmented immunoglobulin G opsonic antibody causing defective opsonophagocytosis. J. Clin. Invest. 74:236–248.

Fischer, H.-M., P. Lüthy, and S. Schweitzer. 1984. Introduction of plasmid pC194 into *Bacillus thuringiensis* by protoplast transformation and plasmid transfer. Arch. Microbiol. 139:213–217.

Fischhoff, D. A., K. S. Bowdish, F. J. Perlak, P. G. Marrone, S. M. McCormick, J. G. Niedermeyer, D. A. Dean, K. Kusano-Kretzmer, E. J. Mayer, D. E. Rochester, S. G. Rogers, and R. T. Fraley. 1987. Insect tolerant transgenic tomato plants. Bio-technology 5:807–813.

Fitz-James, P., and E. Young. 1969. Morphology of sporulation, p. 39–72. In: G. W. Gould and A. Hurst (ed.), The bacterial spore. Academic Press, New York.

Fleming, W. E. 1968. Biological control of the Japanese beetle. U.S. Department of Agriculture Technical Bulletin No. 1383. U.S. Department of Agriculture, Washington, D.C.

Foster, J. W., W. A. Hardwick, and B. Guirard. 1950. Antisporulation factors in complex organic media. I. Growth and sporulation studies on *Bacillus larvae*. J. Bacteriol. 59:463–470.

Fowler, M. 1972. A new milky disease organism from New Zealand. J. Invertebr. Pathol. 19:409–410.

Fowler, M. 1974. Milky disease (*Bacillus* spp.) occurrence and experimental infection in larvae of *Costelytra zealandica* and other Scarabaeidae. New Zealand J. Zool. 1:97–109.

Frank, M. E., and H. Hoffman. 1968. Origin of rigid bacterial "giant flagella" ("giant whips", riesenzöpfe). Can. J. Microbiol. 14:941–947.

Franke, A. E., and D. B. Clewell. 1981. Evidence for a chromosome-borne resistance transposon (Tn*916*) in *Streptococcus faecalis* that is capable of "conjugal" transfer in the absence of a conjugative plasmid. J. Bacteriol. 145:494–502.

Freese, E., and J. Heinze. 1983. Metabolic and genetic control of bacterial sporulation, p. 101–172. In: A. Hurst and G. W. Gould (ed.), The bacterial spore, vol. 2. Academic Press, New York.

Galjart, N. J., N. Sivasubramanian, and B. A. Federici. 1987. Plasmid location, cloning, and sequence analysis of the gene encoding a 27.3 kilodalton cytolytic protein from *Bacillus thuringiensis* subsp. *morrisoni* (PG-14). Curr. Microbiol. 16:171–177.

Gawron-Burke, C., and D. B. Clewell. 1982. A transposon in *Streptococcus faecalis* with fertility properties. Nature 300:281–284.

Geiser, M., S. Schweitzer, and C. Grimm. 1986. The hypervariable region in the genes coding for entomopathogenic crystal proteins of *Bacillus thuringiensis:* Nucleotide sequence of the *kurhd1* gene of subsp. *kurstaki* HD1. Gene 48:109–118.

Gherna, R. L. 1981. Preservation, p. 208–217. In: P. Gerhardt, R. G. E. Murray, R. N. Costilow, E. W. Nester, W. A. Wood, N. R. Krieg, and G. B. Phillips (ed.), Manual of methods for general bacteriology. American Society for Microbiology, Washington, D.C.

Giauffret, A., R. Sanchis, and Y. P. Taliercio. 1970. Comparison serologique de différents souches de *Bacillus* isolées de couvains d'abeilles atteints de loque. Bulletin Apicole de Documentation Scientifique et Technique et d'Information 13:25–34.

Gill, S. S., J. M. Hornung, J. E. Ibarra, G. J. P. Singh, and B. A. Federici. 1987. Cytolytic activity and immunological similarity of the *Bacillus thuringiensis* subsp. *israelensis* and *Bacillus thuringiensis* subsp. *morrisoni* isolate PG-14 toxins. Appl. Environ. Microbiol. 53:1251–1256.

Gochnauer, T. A. 1955. The isolation of a bacteriophage (bacterial virus) from *Bacillus larvae*. Bee World 36:101–103.

Gochnauer, T. A. 1969. Gel filtration of extracts from larvae with American foulbrood disease. J. Apicult. Res. 8:23–28.

Gochnauer, T. A. 1970. Some properties of a bacteriophage from *Bacillus larvae*. J. Invertebr. Pathol. 15:149–156.

Gochnauer, T. A. 1973. Growth, protease formation, and sporulation of *Bacillus larvae* in aerated broth culture. J. Invertebr. Pathol. 22:251–257.

Gochnauer, T. A. 1981. The distribution of *Bacillus larvae* spores in the environs of colonies infected with American foulbrood disease. Amer. Bee J. 121:332–335.

Gochnauer, T. A., and J. C. M. L'Arrivee. 1969a. Experimental infections with *Bacillus larvae*. I. A strain with a morphological marker. J. Invertebr. Pathol. 13:280–284.

Gochnauer, T. A., and J. C. M. L'Arrivee. 1969b. Experimental infections with *Bacillus larvae*. II. Bacteriophage production in the host. J. Invertebr. Pathol. 14:417–418.

Goldberg, L. J., and J. Margalit. 1977. A bacterial spore demonstrating rapid larvicidal activity against *Anopheles sergentii, Uranotaenia unguiculata, Culex univittattus, Aedes aegypti* and *Culex pipiens*. Mosq. News 37:355–358.

Gonzalez, J. M., B. J. Brown, and B. C. Carlton. 1982. Transfer of *Bacillus thuringiensis* plasmids coding for δ-endotoxin among strains of *B. thuringiensis* and *B. cereus*. Proc. Natl. Acad. Sci. USA 79:6951–6955.

Gonzalez, J. M., and B. C. Carlton. 1984. A large transmissible plasmid is required for crystal toxin production in *Bacillus thuringiensis* var. *israelensis*. Plasmid 11:28–38.

Gonzalez, J. M., H. T. Dulmage, and B. C. Carlton. 1981. Correlation between specific plasmids and δ-endotoxin production in *Bacillus thuringiensis*. Plasmid 5:351–365.

Gordon, R. E., W. C. Haynes, and C. H.-N. Pang. 1973. The genus *Bacillus*. U.S. Department of Agriculture Handbook No. 427. U.S. Department of Agriculture, Washington, D.C.

Granum, P. E., S. M. Pinnavaia, and D. J. Ellar. 1988. Comparison of the in vivo and in vitro activity of the delta-endotoxin of *Bacillus thuringiensis* var. *morrisoni* (HD-12) and two of its constituent proteins after cloning and expression in *Escherichia coli*. Eur. J. Biochem. 172:731–738.

Green, B. D., L. Battisti, and C. B. Thorne. 1989. Involvement of Tn*4430* in transfer of *Bacillus anthracis* plasmids mediated by *Bacillus thuringiensis* plasmid pX012. J. Bacteriol. 171:104–113.

Grossman, A. D., and R. Losick. 1988. Extracellular control of spore formation in *Bacillus subtilis*. Proc. Natl. Acac. Science USA 85:4369–4373.

Gupta, B. L., J. A. T. Dow, T. A. Hall, and W. R. Harvey. 1985. Electron probe X-ray microanalysis of the effects of *Bacillus thuringiensis* var. *kurstaki* crystal protein insecticide on ions in an electrogenic K⁺-transporting epithelium of the larval midgut in the lepidopteran *Manduca sexta*, in vitro. J. Cell Sci. 74:137–152.

Hachler, H., B. Berger-Bachi, and F. Kayser. 1987. Genetic characterization of a *Clostridium difficile* erythromycin-clindamycin determinant that is transferable to *Staphylococcus aureus*. Antimicrobial Agents and Chemotherapy 31:1039–1045.

Haider, M. Z., E. S. Ward, and D. J. Ellar. 1987. Cloning and heterologous expression of an insecticidal delta-endotoxin gene from *Bacillus thuringiensis* var. *aizawai* IC1 toxic to both lepidoptera and diptera. Gene 52:285–290.

Hannay, C. L. 1953. Crystalline inclusion in aerobic spore-forming bacteria. Nature (London) 172:1004.

Hannay, C. L., and P. Fitz-James. 1955. The protein crystals of *Bacillus thuringiensis* Berliner. Can. J. Microbiol. 1:694–710.

Hanson, R. S., J. Blicharska, and J. Szulmajster. 1964. Relationship between tricarboxylic acid cycle enzymes and sporulation of *B. subtilis*. Biochem. Biophys. Res. Commun. 17:1–7.

Hanson, R. S., V. R. Srinivasan, and H. O. Halvorson. 1963. Biochemistry of sporulation. I. Metabolism of acetate by vegetative and sporulating cells. J. Bacteriol. 85:451–460.

Harris, E. D., Jr. 1959. Observations on the occurrence of a milky disease among larvae of the northern masked chafer, *Cyclocephala borealis* Arrow. Fla. Entomol. 42:81–83.

Haseman, L. 1961. How long can spores of American foulbrood live? Amer. Bee J. 101:298–299.

Hawley, I. M., and G. F. White. 1935. Preliminary studies on the diseases of larvae of the Japanese beetle (*Popillia japonica* Newm.). New York Entomol. Soc. J. 1935. 43:405–412.

Haynes, W. C. 1972. The catalase test. An aid in the identification of *Bacillus larvae*. Amer. Bee J. 112:130–131.

Haynes, W. C., and C. D. Crowell. 1973. Sporogenicity of yeast autolyzates and casein hydrolyzates for *Bacillus popilliae* in liquid cultures. J. Invertebr. Pathol. 22:377–381.

Haynes, W. C., and L. Rhodes. 1963. A growth factor for *Bacillus lentimorbus* NRRL B-2522. Bacteriol. Proc., 10.

Haynes, W. C., and L. J. Rhodes. 1966. Spore formation by *Bacillus popilliae* in liquid medium containing activated carbon. J. Bacteriol. 91:2270–2274.

Haynes, W. C., and L. J. Rhodes. 1969. Course of sporulation of *Bacillus popilliae* in liquid medium containing activated carbon. J. Invertebr. Pathol. 13:161–166.

Haynes, W. C., G. St. Julian, Jr., M. C. Shekleton, H. H. Hall, and H. Tashiro. 1961. Preservation of infectious milky disease bacteria by lyophilization. J. Insect Pathol. 3:55–61.

Haynes, W. C., and L. J. Weih. 1972. Sporulation of *Bacillus popilliae* in liquid cultures. J. Invertebr. Pathol. 19:125–130.

Haynes, W. C., L. J. Weih, and C. Crowell. 1972. Sporulation of *Bacillus popilliae* in liquid medium as affected by kind of carbon and method of sterilization. Can. J. Microbiol. 18:515–518.

Heierson, A., R. Landén, A. Lövgren, G. Dalhammar, and H. G. Boman. 1987. Transformation of vegetative cells of *Bacillus thuringiensis* by plasmid DNA. J. Bacteriol. 169:1147–1152.

Held, G. A., L. A. Bulla, Jr., E. Ferrari, J. Hoch, A. I. Aronson, and S. A. Minnich. 1982. Cloning and localization of the lepidopteran protoxin gene of *Bacillus thuringiensis* subsp. *kurstaki*. Proc. Natl. Acad. Sci., USA 79:6065–6069.

Herrnstadt, C., T. E. Gilroy, D. A. Sobieski, B. D. Bennett, and F. H. Gaertner. 1987. Nucleotide sequence and deduced amino acid sequence of a coleopteran-active delta-endotoxin gene from *Bacillus thuringiensis* subsp. *san diego*. Gene 57:37–46.

Herrnstadt, C., G. G. Soares, E. R. Wilcox, and D. L. Edwards. 1986. A new strain of *Bacillus thuringiensis* with activity against coleopteran insects. Bio-technology 4:305–308.

Hess, A., R. Holländer, and W. Mannheim. 1979. Lipoquinones of some sporeforming rods, lactic-acid bacteria and actinomycetes. J. Gen. Microbiol. 115:247–252.

Hindley, J., and C. Berry. 1987. Identification, cloning, and sequence analysis of the *Bacillus sphaericus* 1593 41.9 kD larvicidal toxin gene. Molec. Microbiol. 1:187–194.

Hindley, J., and C. Berry, 1988. *Bacillus sphaericus* strain 2297: Nucleotide sequence of 41.9 kDa toxin gene. Nucleic Acids Res. 16:4168.

Hoage, T. R., and W. C. Rothenbuhler. 1966. Larval honey bee response to various doses of *Bacillus larvae* spores. J. Econ. Entomol. 59:42–45.

Hofmann, C., P. Lüthy, R. Hütter, and V. Pliska. 1988a. Binding of the delta endotoxin from *Bacillus thuringiensis* to brush-border membrane vesicles of the cabbage butterfly (*Pieris brassicae*). Eur. J. Biochem. 173:85–91.

Hofmann, C., H. Vanderbruggen, H. Höfte, J. Van Rie, S. Jansens, and H. Van Mellaert. 1988b. Specificity of *Bacillus thuringiensis* δ-endotoxins is correlated with the presence of high-affinity binding sites in the brush border membrane of target insect midguts. Proc. Natl. Acad. Sci. USA 85:7844–7848.

Höfte, H., H. deGreve, J. Seurinck, S. Jansens, J. Mahillon, C. Ampe, J. Vandekerckhove, H. Vanderbruggen, M. van Montagu, M. Zabeau, and M. Vaeck. 1986. Structural and functional analysis of a cloned delta endotoxin of *Bacillus thuringiensis* berliner 1715. Eur. J. Biochem. 161:273–280.

Höfte, J., J. Seurinck, A. Van Houtven, and M. Vaeck. 1987. Nucleotide sequence of a gene encoding an insecticidal protein of *Bacillus thuringiensis* var. *tenebrionis* toxic against Coleoptera. Nuc. Acids Res. 15:7183.

Höfte, H., and H. R. Whiteley. 1989. Insecticidal crystal proteins of *Bacillus thuringiensis*. Microbiol. Rev. 53:242–255.

Holder, I. A., and R. Wheeler. 1984. Experimental studies of the pathogenesis of infections owing to *Pseudomonas aeruginosa*: Elastase an immunoglobulin G protease. Can. J. Microbiol. 30:1118–1120.

Holst, E. C. 1945. An antibiotic from a bee pathogen. Science 102:593–594.

Holst, E. C. 1946. Newer knowledge of American foulbrood. Glean. Bee Cult. 74:138–139.

Holst, E. C., and A. P. Sturtevant. 1940. Relation of proteolytic enzymes to phase of life cycle of *Bacillus larvae*, and two new culture media for this organism. J. Bacteriol. 40:723–731.

Honigman, A., G. Nedjar-Pazerini, A. Yawetz, U. Oron, S. Schuster, M. Broza, and B. Sneh. 1986. Cloning and expression of the lepidopteran toxin produced by *Bacillus thuringiensis* var. *thuringiensis* in *Escherichia coli*. Gene 42:69–77.

Hrubant, G. R., and R. A. Rhodes. 1968. Agglutinability of sporeforming insect pathogens with antiglobulins to milky disease bacteria. J. Invertebr. Pathol. 11:371–376.

Hrubant, G. R., and R. A. Rhodes. 1969. Enzymatic lysis of vegetative cells of *Bacillus popilliae* and other sporeformers. Can. J. Microbiol. 15:827–833.

Huber, H. E., P. Lüthy, H. -R. Ebersold, and J. -L. Cordier. 1981. The subunits of the parasporal crystal of *Bacillus thuringiensis*: size, linkage and toxicity. Arch. Microbiol. 129:14–18.

Hultmark, D., A. Engström, K. Andersson, H. Steiner, H. Bennich, and H. G. Boman. 1983. Insect immunity. Attacins, a family of antibacterial proteins from *Hyalophora cecropia*. EMBO J. 2:571–576.

Hultmark, D., A. Engström, H. Bennich, R. Kapur, and H. G. Boman. 1982. Insect immunity: Isolation and structure of cecropin D and 4 minor antibacterial components from Cecropia pupae. Eur. J. Biochem. 127:207–217.

Hurlbert, R. E., J. E. Karlinsey, and K. D. Spence. 1985. Differential synthesis of bacteria-induced proteins of *Manduca sexta* larvae and pupae. J. Insect. Physiol. 31:205–215.

Hurley, J. M. 1989. Isolation of upstream control regions of genes expressed at different times during growth and sporulation in *Bacillus thuringiensis*. Ph.D. Dissertation, Iowa State University, Ames, Iowa.

Hurley, J. M., L. A. Bulla, Jr., and R. E. Andrews, Jr. 1987. Purification of the mosquitocidal and cytolytic proteins of *Bacillus thuringiensis* subsp. *israelensis*. Appl. Environ. Microbiol. 53:1316–1321.

Hurley, J. M., S. G. Lee, R. E. Andrews, Jr., M. J. Klowden, and L. A. Bulla, Jr. 1985. Separation of the cytolytic and mosquitocidal proteins of *Bacillus thuringiensis* subsp. *israelensis*. Biochem. Biophys. Res. Commun. 126:961–965.

Hurpin, B. 1966. Sur une "maladie laiteuse" des larves de *Melolontha melolontha* L. (*Coléopt.* Scarabeidae). Compte Rendus Soc. Biol. Paris 149:1966–1967.

Hurpin, B. 1967. Recherches épizootiologiques sur la maladie laiteuse à *Bacillus popilliae* "melolontha". Ann. Epiphyties. 18:127–173.

Iandolo, J. ., R. L. Powell, and L. A. Bulla, Jr. 1976. Dormant spores of *Bacillus thuringiensis* contain an inhibitor of RNA polymerase. Biochem. Biophys. Res. Commun. 69:237–244.

Ibarra, J. E., and B. A. Federici. 1986. Isolation of a relatively nontoxic 65-kilodalton protein inclusion from the parasporal body of *Bacillus thuringiensis* subsp. *israelensis*. J. Bacteriol. 165:527–533.

Insell, J. P., and P. C. Fitz-James. 1985. Composition and toxicity of the inclusion of *Bacillus thuringiensis* subsp. *israelensis*. Appl. Environ. Microbiol. 50:56–62.

Ishiwata, S. 1901. One of a kind of several flasherne (Sotto disease). Dainihan Sanbshi Kaiho 9:1–5 (In Japanese).

Ivins, B. E., S. L. Welkos, G. B. Knudson, and D. J. Lebland. 1987. Transposon Tn*916* mutagenesis in *Bacillus anthracis*. Infect. Immun. 56:176–181.

Jeliński, M. 1985. Some biochemical properties of *Bacillus larvae* White. Apidologie 16:69–76.

Johnson, D. E. 1976. Bacterial membrane transport of β-exotoxin, an antimetabolite of RNA synthesis. Nature 260:333–335.

Johnson, D. E. 1978. Inhibition of RNA polymerase from *Bacillus thuringiensis* and *Escherichia coli* by beta-exotoxin. Can. J. Microbiol. 24:537–543.

Johnson, D. E., L. A. Bulla, Jr., and K. W. Nickerson. 1975. Differential inhibition of β-exotoxin of vegetative- and sporulation-specific ribonucleic acid polymerases from *Bacillus thuringiensis* cells, p. 248–254. In: P. Gerhardt, R. N. Costilow, and H. L. Sadoff (ed.), Spores VI. American Society for Microbiology, Washington, D.C.

Jones, J. M., S. C. Yost, and P. A. Pattee. 1987. Transfer of the conjugal tetracylcine resistance transposon Tn*916* from *Streptococcus faecalis* to *Staphylococcus aureus* and identification of some insertion sites in the staphylococcal chromosome. J. Bacteriol. 169:2121–2131.

Kaaya, G. P., C. Flyg, and H. G. Boman. 1987. Insect immunity. Induction of cecropin and attacin-like antibacterial factors in the haemolymph of *Glossinia morsitans morsitans*. Insect Biochem. 17:309–315.

Kalfon, A., I. Larget-Thiery, J. -F. Charles, and H. de Barjac. 1983. Growth, sporulation, and larvicidal activity of *Bacillus sphaericus*. Eur. J. Appl. Microbiol. Biotechnol. 18:168–173.

Kalfon, A., M. Lugten, and J. Margalit. 1986. Development of selective media for *Bacillus sphaericus* and *Bacillus thuringiensis* var. *israelensis*. Appl. Microbiol. Biotechnol. 24:240–243.

Kaneda, T. 1969. Fatty acids in *Bacillus larvae*, *Bacillus lentimorbus*, and *Bacillus popilliae*. J. Bacteriol. 98:143–146.

Kaneda, T. 1977. Fatty acids of the genus *Bacillus*: An example of branched-chain preference. Bacteriol. Rev. 41:391–418.

Kaneko, T., R. Nozaki, and K. Aizawa. 1978. Deoxyribonucleic acid relatedness between *Bacillus anthracis*, *Bacillus cereus* and *Bacillus thuringiensis*. Microbiol. Immunol. 22:639–641.

Kathariou, S., P. Metz, H. Hof, and W. Goebel. 1987. Tn*916*-induced mutations in the hemolysin determinant affecting virulence of *Listeria monocytogenes*. J. Bacteriol. 169:1291–1297.

Katznelson, H., and A. G. Lochhead. 1948. Nutritional requirements of *Bacillus larvae*. J. Bacteriol. 55:763–764.

Kellen, W. R., T. Clark, J. Lindegren, B. Ho. M. Rogoff, and S. Singer. 1965. *Bacillus sphaericus* Neide as a pathogen of mosquitoes. J. Invertebr. Pathol. 7:442–448.

Kellen, W. R., and C. M. Meyer. 1964. *Bacillus sphaericus* Neide as a pathogen of mosquitoes. Proc. Annu. Conf. Calif. Mosq. Control Assn. 32:37.

Klein, M. G. 1981. Mass trapping for suppression of Japanese beetles, p. 183–190. In: E. R. Mitchell (ed.), Management of insect pests with semio-chemicals. Plenum Publishing Corp., New York.

Kim, Y. T., B. G. Gregory, and C. M. Ignoffo. 1972. The β-exotoxins of *Bacillus thuringiensis*. III. Effects on in vivo synthesis of macromolecules in an insect system. J. Invertebr. Pathol. 20:46–50.

Kleckner, N. 1981. Transposable elements in procaryotes. Ann. Rev. Genet. 15:341–404.

Klier, A., C. Bourgouin, and G. Rapoport. 1983. Mating between *Bacillus subtilis* and *Bacillus thuringiensis* and transfer of cloned crystal genes. Mol. Gen. Genet. 191:257–262.

Klier, A., F. Fargette, J. Ribier, and G. Rapoport. 1982. Cloning and expression of the crystal protein genes from *Bacillus thuringiensis* strain *berliner* 1715. EMBO J. 1:791–799.

Klier, A. F., and M.-M. Lecadet. 1976. Arguments based on hybridization-competition experiments in favor of the in vitro synthesis of sporulation-specific mRNAs by the RNA polymerase of *B. thuringiensis*. Biochem. Biophys. Res. Commun. 73:263–270.

Klier, A., D. Lereclus, J. Ribier, C. Bourgouin, G. Menou, M.-M. Lecadet, and G. Rapoport. 1985. Cloning and expression in *Escherichia coli* of the crystal protein gene from *Bacillus thuringiensis* strain *aizawa* 7–29 and comparison of the structural organization of genes from different serotypes, p. 217–224. In: J. A. Hoch, and P. Setlow (ed.), Molecular biology of microbial differentiation. American Society for Microbiology, Washington, D.C.

Klowden, M. J., and L. A. Bulla, Jr. 1984. Oral toxicity of *Bacillus thuringiensis* subsp. *israelensis* to adult mosquitoes. Appl. Environ. Microbiol. 48:665–667.

Klowden, M. J., L. A. Bulla, Jr., and R. L. Stoltz. 1985. Susceptibility of larval and adult *Simulium vittatum* (Diptera: Simuliidae) to the solubilized parasporal crystal of *Bacillus thuringiensis israelensis*. J. Med. Entomol. 22:466–467.

Knowles, B. H., and D. J. Ellar. 1987. Colloid-osmotic lysis is a general feature of the mechanism of action of *Bacillus thuringiensis* δ-endotoxins with different insect specificity. Biochim. Biophys. Acta 924:509–518.

Knowles, B. H., and D. J. Ellar. 1988. Differential specificity of two insecticidal toxins from *Bacillus thuringiensis*. var. *aizawai*. Mol. Microbiol. 2:153–157.

Knowles, B. H., and R. W. Farndale. 1988. Activation of insect cell adenylate cyclase by *Bacillus thuringiensis* δ-endotoxins and melittin. Toxicity is independent of cyclic AMP. Biochem. J. 253:235–241.

Knowles, B. H., W. E. Thomas, and D. J. Ellar. 1984. Lectin-like binding of *Bacillus thuringiensis* var. *kurstaki* lepidopteran-specific toxin is an initial step in insecticidal action. FEBS Lett. 168:197–202.

Krieg, A. 1961. Grundlagen der insektenpathologie. Steinkopff, Darmstadt.

Krieg, A. 1981. The genus *Bacillus*: Insect pathogens, p. 1743–1755. In: M. P. Starr, H. Stolp, H. G. Trüper, A. Balows, and H. G. Schlegel (ed.), The prokaryotes, vol. 2. Springer-Verlag, New York.

Krieg, A., A. M. Huger, and W. Schnetter. 1987a. *Bacillus thuringiensis* var. *san diego* strain M-7 is identical to the formerly in Germany isolated strain BI-256–82 *Bacillus thuringiensis* ssp. *tenebrionis* which is pathogenic to coleopteran insects. J. Appl. Entomol. 104:417–424.

Krieg, A., W. Schnetter, A. M. Huger, and G. A. Langenbruch. 1987b. *Bacillus thuringiensis* subsp. *tenebrionis*, strain BI256–82: a third pathotype within the H-serotype 8a8b. Syst. Appl. Microbiol. 9:138–141.

Kronstad, J. W., H. E. Schnepf, and H. R. Whiteley. 1983. Diversity of locations for the *Bacillus thuringiensis* crystal protein genes. J. Bacteriol. 154:419–428.

Kronstad, J. W., and H. R. Whiteley. 1984. Inverted repeat sequences flank a *Bacillus thuringiensis* crystal-protein gene. J. Bacteriol. 160:95–102.

Kronstad, J. W., and H. R. Whiteley. 1986. Three classes of homologous *Bacillus thuringiensis* crystal-protein genes. Gene 43:29–40.

Krych, V. K., J. L. Johnson, and A. A. Yousten. 1980. Deoxyribonucleic acid homologies among strains of *Bacillus sphaericus*. Int. J. Syst. Bacteriol. 30:476–484.

Krywienczyk, J. 1977. Antigenic comparison of δ-endotoxin as an aid in identification of *Bacillus thuringiensis* varieties. Publication IP-X-16. Insect Pathology Research Institute. Dept. of Fisheries and the Environment, Canadian Forestry Service, Sault Sainte Marie, Ontario.

Krywienczyk, J., H. T. Dulmage, and P. G. Fast. 1978. Occurrence of two serologically distinct groups within *Bacillus thuringiensis* serotype 3a, b var. *kurstaki*. J. Invertebr. Pathol. 31:372–375.

Krywienczyk, J., H. T. Dulmage, I. M. Hall, C. C. Beegle, K. Y. Arakawa, and P. G. Fast. 1981. Occurrence of *kurstaki* K-1 crystal activity in *Bacillus thuringiensis* subsp. *thuringiensis* serovar (H1). J. Invertebr. Pathol. 37:62–65.

Krywienczyk, J., and P. Lüthy. 1974. Serological relationship between three varieties of *Bacillus popilliae*. J. Invertebr. Pathol. 23:275–279.

Kuznetsov, L. E., and M. P. Khovrychev. 1984. Optimization of a synthetic nutrient medium for *Bacillus thuringiensis* cultivation. Mikrobiologiya 53:54–57.

Lacks, S., and B. Greenberg. 1977. Complementary specificity of restriction endonucleases of *Diplococcus pneumoniae* with respect to DNA methylation. J. Molec. Biol. 114:153–168.

Lecadet, M.-M., M.-O. Blondel, and J. Ribier. 1980. Generalized transduction in *Bacillus thuringiensis* var. *Berliner* 1715 using bacteriophage CP54 Ber. J. Gen. Microbiol. 121:203–212.

Lecadet, M.-M., V. Sanchis, G. Menou, P. Rabot, D. Lereclus, J. Chaufaux, and D. Martouret. 1988. Identification of a δ-endotoxin gene product specifically active against *Spodoptera littoralis* Bdv. among proteolysed fractions of the insecticidal crystals of *Bacillus thuringiensis* subsp. *aizawai* 7.29. Appl. Environ. Microbiol. 54:2689–2698.

Lereclus, D., M.-M. Lecadet, J. Ribier, and R. Dedonder. 1982. Molecular relationships among plasmids of *Bacillus thuringiensis*: Conserved sequences through 11 crystalliferous strains. Mol. Gen. Genet. 186:391–398.

Lereclus, D., J. Mahillon, G. Menou, and M.-M. Lecadet. 1986. Identification of Tn*4430*, a transposon of *Bacillus thuringiensis* functional in *Escherichia coli*. Mol. Gen. Genet. 204:52–57.

Lereclus, D., G. Menou, and M.-M. Lecadet. 1983. Isolation of a DNA sequence related to several plasmids from *Bacillus thuringiensis* after a mating involving the *Streptococcus faecalis* plasmid pAMβ-1. Mol. Gen. Genet. 191:307–313.

Lereclus, D., J. Ribier, A. Klier, G. Menou, and M.-M. Lecadet. 1984. A transposon-like structure related to the delta-endotoxin gene of *Bacillus thuringiensis*. EMBO J. 3:2561–2567.

Levin, M. D. 1984. Value of bee pollination to United States agriculture. Amer. Bee J. 124:184–186.

Lewis, L., and A. A. Yousten. 1987. Characterization of the surface protein layers of the mosquito pathogenic strains of *Bacillus sphaericus.* J. Bacteriol. 169:72–79.

Lewis, L., and A. A. Yousten. 1988. Bacteriophage attachment to the S-layer proteins of the mosquito-pathogenic strains of *Bacillus sphaericus.* Curr. Microbiol. 17:55–60.

Li, F., X. Pan, Y. Lu, R. Liu, and Q. Li. 1985. Studies on the extracellular polysaccharides of *Bacillus* sp. Acta Microbiologica Sinica 25:25–30.

Lilley, M., R. N. Ruffell, and H. J. Somerville. 1980. Purification of the insecticidal toxin in crystals of *Bacillus thuringiensis.* J. Gen. Microbiol. 118:1–11.

Lingg, A. J., K. J. McMahon, and C. Herzmann. 1967. Viability of *Bacillus popilliae* after lyophilization of liquid nitrogen frozen cells. Appl. Microbiol. 15:163–165.

Lochhead, A. G. 1928. Cultural studies of *Bacillus larvae* (White). Can. J. Agr. Sci. 9:80–89.

Lochhead, A. G. 1933. Semi-solid medium for the cultivation of *Bacillus larvae.* Bee World 14:114–115.

Lochhead, A. G. 1942. Growth factor requirements of *Bacillus larvae,* White. J. Bacteriol. 44:185–189.

Lodesani, M., C. Benassi, L. Grazia, and P. Ronchetti. 1985. A study on the sporulation of *Bacillus larvae.* J. Apicultural Res. 24:205–210.

Ludvik, J., V. Drobnikova, V. Vinter, O. Kofronova, and F. Smid. 1983. Ultrastructure of developmental stages of *Bacillus larvae.* 6th European cell cycle workshop—progress in cell cycle controls. [Published as an abstract in Spore Newsletter. 1984. 8:82–83.]

Lüthy, P. 1986. Insect pathogenic bacteria as pest control agents. Fortschr. Zool. 32:201–216.

Lüthy, P., P. Geiser, H. -R. Ebersold, and B. Trümpi. 1976. Use of insect cell cultures in the investigation of bacteria and rickettsiae pathogenic to insects. Proc. Int. Colloq. Invertebr. Pathol. 1:128–132.

Lüthy, P., and J. Krywienczyk. 1972. Serological comparison of three milky disease isolates. J. Invertebr. Pathol. 19:163–165.

Lüthy, P., C. Wyss, and L. Ettlinger. 1970. Behavior of milky disease organisms in a tissue culture system. J. Invertebr. Pathol. 16:325–330.

Lynch, M. J., and P. Baumann. 1985. Immunological comparisons of the crystal protein from strains of *Bacillus thuringiensis.* J. Invertebr. Pathol. 46:47–57.

Lynch, R. E., L. C. Lewis, and T. A. Brindley. 1976. Bacteria associated with eggs and 1st-instar larvae of the European corn borer; isolation techniques and pathogenicity. J. Invertebr. Pathol. 27:325–331.

Lysenko, O., E. Davidson, L. Lacey, and A. Yousten. 1985. Five new mosquito larvicidal strains of *Bacillus sphaericus* from non-mosquito origins. J. Am. Mosq. Control Assn. 1:369–371.

Maassen, A. 1906. Faulbrutseuche der bienen. Mitt. K. Biol. Anst. Land-u. Forstw. 2:28–29.

Maassen, A. 1908. Zur atiologie der sogenannten faulbrut der honigbienen. Arb. Biol. Reichsanstalt Anst. Land-u. Forstw. 6:53–70.

Mahillon, J., F. Hespel, A.-M. Pierssens, and J. Delcour. 1988. Cloning and partial characterization of three small cryptic plasmids from *Bacillus thuringiensis.* Plasmid 19:169–173.

Mahillon, J., and D. Lereclus. 1988. Structural and functional analysis of Tn*4430:* Identification of an integrase-like protein involved in the co-integrate-resolution process. EMBO J. 7:1515–1526.

Mahillon, J., and J. Seurinck. 1988. Complete nucleotide sequence of pGI-2, a *Bacillus thuringiensis* plasmid containing Tn*4430.* Nucl. Acid Res. 16:11827–11828.

Mahillon, J., J. Seurinck, J. Delcour, and M. Zabeau. 1987. Cloning and nucleotide sequence of different iso-IS*231* elements and their structural association with the Tn*4430* transposon in *Bacillus thuringiensis.* Gene 51:187–196.

Mahillon, J., J. Seurinck, L. Van Rompuy, J. Delcour, and M. Zabeau. 1985. Nucleotide sequence and structural organization of an insertion sequence element (IS*231*) from *Bacillus thuringiensis* strain *berliner* 1715. EMBO J. 4:3895–3899.

Manachini, P. L., A. Craveri, and A. Guicciardi. 1968. Compozione in basi dell' acido desossirobunleico di forme mesofile, termofacoltative e termofile del genere *Bacillus.* Ann. Microbiol. Enzimol. 18:1.

Maniatis, T., E. F. Fritsch, and J. Sambrook. 1982. Molecular cloning: A laboratory manual. Cold Spring Harbor Laboratory, Cold Spring Harbor, New York.

Margalit, J., and D. Dean. 1985. The story of *Bacillus thuringiensis* var. *israelensis.* J. Am. Mosq. Control Assoc. 1:1–7.

Margalit, J., A. Markus, and Z. Pelah. 1984. Effect of encapsulation on the persistence of *Bacillus thuringiensis* var. *israelensis* serotype H14. Appl. Microbiol. Biotechnol. 19:382–383.

Marinus, M. G. and R. Morris. 1973. Isolation of deoxyribonucleic acid methylase mutants of *Escherichia coli* K-12. J. Bacteriol. 114:1143–1150.

Martin, P. A. W., J. R. Lohr, and D. H. Dean. 1981. Transformation of *Bacillus thuringiensis* protoplasts by plasmid deoxyribonucleic acid. J. Bacteriol. 145:980–983.

Masson, L., P. Marcotte, G. Préfontaine, and R. Brousseau. 1989. Nucleotide sequence of a gene cloned from *Bacillus thuringiensis* subsp. *entomocidus* coding for an insecticidal protein toxic for *Bombyx mori.* Nuc. Acids Res. 17:446.

McDonald, K. O., and W. Burke. 1984. Plasmid transformation of *Bacillus sphaericus* 1593. J. Gen. Microbiol. 130:203–208.

McGaughey, W. H. 1985. Insect resistance to the biological insecticide *Bacillus thuringiensis.* Science 229:193–195.

McGaughey, W. H., and D. E. Johnson. 1987. Toxicity of different serotypes and toxins of *Bacillus thuringiensis* to resistant and susceptible Indian meal moths (*Lepidoptera: Pyralidae*). J. Econ. Entomol. 80:1122–1126.

McKay, L. L., A. Bhumiratana, and R. N. Costilow. 1971. Oxidation of acetate by various strains of *Bacillus popilliae.* Appl. Microbiol. 22:1070–1075.

McLinden, J. H., J. R. Sabourin, B. D. Clark, D. R. Gensler, W. E. Workman, and D. H. Dean. 1985. Cloning and expression of an insecticidal k-73 type crystal protein gene from *Bacillus thuringiensis* var. *kurstaki* into *Escherichia coli.* Appl. Environ. Microbiol. 50:623–628.

McPherson, S. A., F. J. Perlak, R. L. Fuchs, P. G. Marrone, P. B. Lavrik, and D. A. Fischhoff. 1988. Characterization of the coleopteran-specific protein gene of *Bacillus thuringiensis* var. *tenebrionis.* Biotechnol. 6:61–66.

Milner, R. J. 1974. A new variety of milky disease, *Bacillus popilliae* var. *rhopaea,* from *Rhopaea verreauxi.* Aust. J. Biol. Sci. 27:235–247.

Milner, R. J. 1977. A method for isolating milky disease, *Bacillus popilliae* var. *rhopaea* spores from the soil. J. Invertebr. Pathol. 30:283–287.

Milner, R. J. 1981a. Identification of the *Bacillus popilliae* group of insect pathogens, p. 44–59. In: H. D. Burges (ed.), Microbial control of pests and plant diseases 1970–1980. Acad. Press, New York.

Milner, R. J. 1981b. A novel milky disease organism from Australian Scarabaeids: Field occurrence, isolation and infectivity. J. Invertebr. Pathol. 37:304–309.

Milner, R. J., and C. D. Beaton. 1981. A novel milky disease organism from Australian scarabaeids: ultrastructure. J. Invertebr. Pathol. 37:310–318.

Miteva, V. I., N. I. Shivarova, and R. T. Grigorova. 1981. Transformation of *Bacillus thuringiensis* protoplasts by plasmid DNA. FEMS Microbiol. Lett. 12:253–256.

Mitruka, B. M., R. N. Costilow, S. H. Black, and R. E. Pepper. 1967. Comparisons of cells, refractile bodies, and spores of *Bacillus popilliae.* J. Bacteriol. 94:759–765.

Myers, P., and A. A. Yousten. 1980. Localization of a mosquito-larval toxin of *Bacillus sphaericus* 1593. Appl. Environ. Microbiol. 39:1205–1211.

Mylroie, R. L., and H. Katznelson. 1957. Carbohydrate metabolism of *Bacillus larvae.* J. Bacteriol. 74:217–222.

Naglich, J. G., and R. E. Andrews, Jr. 1988a. Introduction of the *Streptococcus faecalis* transposon Tn*916* into *Bacillus thuringiensis* subsp. *israelensis.* Plasmid 19:84–93.

Naglich, J. G., and R. E. Andrews, Jr. 1988b. Tn*916*-dependent conjugal transfer of pC194 and pUB110 from *Bacillus subtilis* into *Bacillus thuringiensis* subsp. *israelensis.* Plasmid 20:113–126.

Nakamura, L. K. 1984. *Bacillus pulvifaciens* sp. nov., nom. rev. Int. J. Syst. Bacteriol. 34:410–413.

Nakata, H. M., and H. O. Halvorson. 1960. Biochemical changes occurring during growth and sporulation of *Bacillus cereus.* J. Bacteriol. 80:801–810.

Nickerson, K. W., and L. A. Bulla, Jr. 1974. Physiology of sporeforming bacteria associated with insects: Minimal nutritional requirements for growth, sporulation, and parasporal crystal formation of *Bacillus thuringiensis.* Appl. Microbiol. 28:124–128.

Nida, K., and P. P. Cleary. 1983. Insertional inactivation of streptolysin S expression in *Streptococcus pyogenes.* J. Bacteriol. 155:1156–1161.

Norton, N. B., K. Orzech, and W. Burke. 1985. Construction and characterization of plasmid vectors for cloning in the entomocidal organism *Bacillus sphaericus* 1593. Plasmid 13:211–214.

Oeda, K., K. Oshie, M. Shimizu, K. Nakamura, H. Yamamoto, I. Nakayama, and H. Ohkawa. 1987. Nucleotide sequence of the insecticidal protein gene of *Bacillus thuringiensis* strain *aizawai* IPL7 and its high-level expression in *Escherichia coli.* Gene 53:113–119.

Ohba, M., and K. Aizawa. 1986. Distribution of *Bacillus thuringiensis* in soils of Japan. J. Invertebr. Pathol. 47:277–282.

Ohba, M., Y. M. Yu, and K. Aizawa. 1988. Occurrence of noninsectical *Bacillus thuringiensis* flagellar serotype 14 in the soil of Japan. Syst. Appl. Microbiol. 11:85–89.

Orzech, K. A., and W. Burke. 1984. Conjugal transfer of pAMB1 in *Bacillus sphaericus* 1593. FEMS Microbiol. Lett. 25:91–95.

Otte, E. 1973. Ein beitrag zur labordiagnose der bösartigen faulbrut der honigbiene unter besonderer berücksichtigung der immunofluoreszenzmethode. Apidologie 4:331–339.

Padua, L. E., M. Ohba and K. Aizawa. 1980. The isolates of *Bacillus thuringiensis* serotype 10 with a highly preferential toxicity to mosquito larvae. J. Invertebr. Pathol. 36:180–186.

Patel, N. G., and T. A. Gochnauer. 1959. The toxicity of extracts from foulbrood scale residues for honey-bee larvae maintained in vitro. J. Insect Pathol. 1:190–192.

Patel, N. G., and T. A. Gochnauer. 1972. The production and properties of *Bacillus larvae* proteases. Insect Biochem. 2:321–333.

Peng, Y.-S., and K.-Y. Peng. 1979. A study on the possible utilization of immunodiffusion and immunofluorescence techniques as the diagnostic methods for American foulbrood of honeybees (*Apis mellifera*). J. Invertebr. Pathol. 33:284–289.

Pepper, R. E., and R. N. Costilow. 1964. Glucose catabolism by *Bacillus popilliae* and *Bacillus lentimorbus.* J. Bacteriol. 87:303–310.

Pepper, R. E., and R. N. Costilow. 1965. Electron transport in *Bacillus popilliae.* J. Bacteriol. 89:271–276.

Pfannenstiel, M. A., G. A. Couche, E. J. Ross, and K. W. Nickerson. 1986. Immunological relationships among proteins making up the *Bacillus thuringiensis* subsp. *israelensis* crystalline toxin. Appl. Environ. Microbiol. 52:644–649.

Prasertphon, S., P. Areekul, and Y. Tanada. 1974. Sporulation of *Bacillus thuringiensis* in host cadavers. J. Invertebr. Pathol. 21:205–207.

Prefontaine, G., P. Fast, P. C. K. Lau, M. A. Hefford, Z. Hanna, and R. Brousseau. 1987. Use of oligonucleotide probes to study the relatedness of delta-endotoxin genes among *Bacillus thuringiensis* subspecies and strains. Appl. Environ. Microbiol. 53:2808–2814.

Priest, F. G., M. Goodfellow, and C. Todd. 1988. A numerical classification of the genus *Bacillus.* J. Gen. Microbiol. 134:1847–1882.

Rasnitsyn, S. P., A. A. Voitsik, and A. B. Zvantsov. 1988. Effect of the conditions of the development of mosquito larvae on their sensitivity to bacterial insecticides. Med. Parazitol. Parazit. Bolezni. 57:15–18.

Raun, E. S., G. R. Sutter, and M. A. Revelo. 1966. Ecological factors affecting the pathogenicity of *Bacillus thuringiensis* var. *thuringiensis* to the European corn borer and fall army worm. J. Invertebr. Pathol. 8:365–375.

Rhodes, R. A. 1965. Symposium on microbial insecticides. II. Milky disease of the Japanese beetle. Bacteriol. Rev. 29:373–381.

Rhodes, R. A., M. S. Roth, and G. R. Hrubant. 1965. Sporulation of *Bacillus popilliae* on solid media. Can. J. Microbiol. 11:779–783.

Rhodes, R. A., E. S. Sharpe, H. H. Hall, and R. W. Jackson. 1966. Characteristics of the vegetative growth of *Bacillus popilliae.* Appl. Microbiol. 14:189–195.

Roberts, M. C., and G. E. Kenny. 1987. Conjugal transfer of transposon Tn*916* from *Streptococcus faecalis* to *Mycoplasma hominis.* J. Bacteriol. 169:3836–3839.

Robinson, W. S., R. Nowogrodzki, and R. A. Morse. 1989a. The value of honey bees as pollinators of U.S. crops, Part 1. Amer. Bee J. 129:411–423.

Northern Idaho, USA. U.S. For. Serv. Div. State Prov. for Nor. Reg. Rep. 74:1–9.

Tyrell, D. J., L. A. Bulla, Jr., R. E. Andrews, Jr., K. J. Kramer, L. I. Davidson, and P. Nordin. 1981. Comparative biochemistry of entomocidal parasporal crystals of selected *Bacillus thuringiensis* strains. J. Bacteriol. 145:1052–1062.

Tyrell, D. J., L. I. Davidson, L. A. Bulla, Jr., and W. A. Ramoska. 1979. Toxicity of parasporal crystals of *Bacillus thuringiensis* subsp. *israelensis* to mosquitoes. Appl. Environ. Microbiol. 38:656–658.

Vankova, J., and K. Purrini. 1979. Natural epizootics caused by bacilli of the species *Bacillus thuringiensis* and *Bacillus cereus*. Z. Angew. Entomol. 88:216–221.

Visser, B., T. van der Salm, W. van den Brink, and G. Folkers. 1988. Genes from *Bacillus thuringiensis entomocidus* 60.5 coding for insect-specific crystal proteins. Mol. Gen. Genet. 212:219–224.

Vyas, H. G., D. P. Joshi, K. C. Patel, D. N. Yadav, and J. F. Dodia. 1986. Natural incidence of milky disease of white-grubs in Gujarat. Indian J. Agr. Science 56:213–214.

Waalwijk, C., A. M. Dullemans, M. E. S. van Workum, and B. Visser. 1985. Molecular cloning and the nucleotide sequence of the M_r 28,000 crystal protein gene of *Bacillus thuringiensis* subsp. *israelensis*. Nucl. Acids Res. 13:8207–8217.

Wabiko, H., G. A. Held, and L. A. Bulla, Jr. 1985. Only part of the protoxin gene of *Bacillus thuringiensis* subsp. *berliner* 1715 is necessary for insecticidal activity. Appl. Environ. Microbiol. 49:706–708.

Wabiko, H., K. C. Raymond, and L. A. Bulla, Jr. 1986. *Bacillus thuringiensis* entomocidal protoxin gene sequence and gene product analysis. DNA 5:305–314.

Wanger, A. R., and G. M. Dunny. 1985. Development of a system for genetic and molecular analysis of *Streptococcus agalactiae*. Res. Vet. Sci. 38:202–208.

Ward, E. S., and D. J. Ellar. 1987. Nucleotide sequence of a *Bacillus thuringiensis* var. *israelensis* gene encoding a 130 kDa delta-endotoxin. Nuc. Acids Res. 15:7195.

Ward, E. S., and D. J. Ellar. 1988. Cloning and expression of two homologous genes of *Bacillus thuringiensis* subsp. *israelensis* which encode 130-kilodalton mosquitocidal proteins. J. Bacteriol. 170:727–735.

Ward, E. S., D. J. Ellar, and J. A. Todd. 1984. Cloning and expression in *Escherichia coli* of the insecticidal delta-endotoxin gene of *Bacillus thuringiensis* var. *israelensis*. FEBS Lett. 175:377–382.

Weiner, B. A. 1978. Isolation and partial characterization of the parasporal body of *Bacillus popilliae*. Can. J. Microbiol. 24:1557–1561.

Weiner, B. A., W. F. Kwolek, G. St. Julian, H. H. Hall, and R. W. Jackson. 1966. Oxygen concentration in larval hemolymph of the Japanese beetle, *Popillia japonica*, infected with *Bacillus popilliae*. J. Invertebr. Pathol. 8:308–313.

Weiser, J. 1984. A mosquito-virulent *Bacillus sphaericus* in adult *Simulium damnosum* from northern Nigeria. Zbl. Mikrobiol. 139:57–60.

Weiser, J. N., and C. E. Rubens. 1987. Transposon mutagenesis of group B *Streptococcus* beta-hemolysin biosynthesis. Infect. Immun. 55:2314–2316.

West, A. W. 1984. Fate of the insecticidal proteinaceous parasporal crystal of *Bacillus thuringiensis* in soil. Soil Biol. Biochem. 16:357–360.

West, A. W., H. D. Burges, and C. H. Wyborn. 1984. Effect of incubation in natural and autoclaved soil upon potency and viability of *Bacillus thuringiensis*. J. Invertebr. Pathol. 44:121–127.

White, G. F. 1906. The bacteria of the apiary with special references to bee diseases. Tech. Ser. Bur. Entomol. U.S. Dept. Agr. No. 14.

White, G. F. 1907. The cause of American foulbrood. Bur. Ent. Circ. U.S. Dept. Agr. No. 94.

White, G. F. 1919. Unheated egg yolk media. Science 49:362.

White, P. J., and H. Lotay. 1980. Minimal nutritional requirements of *Bacillus sphaericus* NCTC 9602 and 26 other strains of this species: The majority grow and sporulate with acetate as sole major source of carbon. J. Gen. Microbiol. 118:13–19.

White, R. T. 1947. Milky disease infecting *Cyclocephala* larvae in the field. J. Econ. Entomol. 40:912–914.

Widner, W. R., and H. R. Whiteley. 1989. Two highly related insecticidal crystal proteins of *Bacillus thuringiensis* subsp. *kurstaki* possess different host range specificities. J. Bacteriol. 171:965–974.

Wie, S. I., R. E. Andrews, Jr., B. D. Hammock, R. M. Faust, and L. A. Bulla, Jr. 1982. Enzyme-linked immunosorbent assays for detection and quantitation of the entomocidal parasporal crystalline protein of *Bacillus thuringiensis* subspp. *kurstaki* and *israelensis*. Appl. Environ. Microbiol. 43:891–894.

Wie, S. L., B. D. Hammock, S. S. Gill, E. Grate, R. E. Andrews, Jr., R. M. Faust, L. A. Bulla, Jr., and C. H. Shaefer. 1984. An improved enzyme-linked immunoassay for detection and quantification of the entomocidal parasporal crystal proteins of *Bacillus thuringiensis* subspp. *kurstaki* and *israelensis*. J. Appl. Bacteriol. 57:447–454.

Wille, H. 1956. *Bacillus fribourgensis* n. sp., erreger einer "milky disease" im engerling von *Melolontha melolontha* L. Mitt. Schweiz. Entomol. Ges. 24:271–282.

Williston, B. K., and S. Singer. 1987. Initial studies of the use of amino peptidases for the differentiation of *Bacillus sphaericus* strains. J. Indust. Microbiol. 2:285–292.

Wu, D., and F. N. Chang. 1985. Synergism in the mosquitocidal activity of 26 and 65 kDa proteins from *Bacillus thuringiensis* subsp. *israelensis* crystal. FEBS Lett. 190:232–236.

Wyss, C. 1971. Sporulationsversuche mit drei varietäten von *Bacillus popilliae* Dutky. Zentralbl. Bacteriol., Parasitenk., Infektionskr. Hyg. II. 126:461–492.

Yelton, D. B., and C. B. Thorne. 1971. Comparison of *Bacillus cereus* bacteriophages CP51 and CP53. J. Virol. 8:242–253.

Yousten, A. A. 1984. Bacteriophage typing of mosquito pathogenic strains of *Bacillus sphaericus*. J. Invertebr. Pathol. 43:124.

Yousten, A. A., L. A. Bulla, Jr., and J. M. McCord. 1973. Superoxide dismutase in *Bacillus popilliae*, a catalase-less aerobe. J. Bacteriol. 113:524–525.

Yousten, A. A., and E. Davidson. 1982. Ultrastructural analysis of spores and parasporal crystals formed by *Bacillus sphaericus* 2297. Appl. Environ. Microbiol. 44:1449–1455.

Yousten, A. A., H. de Barjac, J. Hendrick, V. Cosmao-Dumanoir, and P. Myers. 1980. Comparison between bacteriophage typing and serotyping for the differentiation of *Bacillus sphaericus* strains. Ann. Microbiol. (Inst. Pasteur) 131B:297–308.

Yousten, A. A., S. Fretz, and S. Jelley. 1985. Selective medium for mosquito pathogenic strains of *Bacillus sphaericus*. Appl. Environ. Microbiol. 49:1532–1533.

Yousten, A. A., R. S. Hanson, L. A. Bulla, Jr., and G. St. Julian. 1974. Physiology of sporeforming bacteria associated with insects. V. Tricarboxylic acid cycle activ-

ity and adenosine triphosphate levels in *Bacillus popilliae* and *Bacillus thuringiensis*. Can. J. Microbiol. 20:1729–1734.

Yousten, A. A., and K. D. Nelson. 1976. A single form of superoxide dismutase found in *Bacillus popilliae* and in some other Gram-positive bacteria. J. Gen. Appl. Microbiol. 22:161–164.

CHAPTER 78

The Genus *Bacillus*—Medical

W. EDMUND FARRAR and ANNETTE C. REBOLI

Of the 34 species of the genus *Bacillus,* the two of greatest medical importance are *B. anthracis,* the causative agent of anthrax, and *B. cereus,* which causes food poisoning. Nonanthrax *Bacillus* species, including *B. cereus,* also cause a wide variety of other infections, and they are being recognized with increasing frequency as significant pathogens in humans. Species which have caused human disease include *B. cereus, B. alvei, B. megaterium, B. coagulans, B. laterosporus, B. subtilis, B. sphaericus, B. circulans, B. brevis, B. licheniformis, B. macerans, B. pumilus,* and *B. thuringiensis.* After *B. anthracis, B. cereus* is the most frequent human pathogen. *B. subtilis* has been used as a synonym for aerobic sporeformers other than *B. anthracis,* and many isolates described as *B. subtilis* in the early literature were probably *B. cereus* (Weinstein and Colburn, 1950).

Anthrax

Anthrax has a long and fascinating history. The disease in cattle is described in Egyptian and Mesopotamian writings of around 5000 B.C. (Dürst et al., 1986) and it may have been responsible for two of the plagues visited upon the Egyptians and their cattle in 1491 B.C. (the "murrain of beasts" and the "plague of boils and blains"), described in the ninth chapter of Exodus. In the *Georgics,* written around 37 B.C., Publius Vergilius Maro (Virgil) refers to anthrax infection in sheep, cattle, horses, and oxen, and in men handling wool from infected sheep (Dirckx, 1981). As a cause of "black bane" and "murrain," it caused heavy losses of cattle and sheep in England and Europe in Saxon and Medieval times, and may have contributed to the Black Death that ravaged Europe in the mid 14th century (Twigg, 1984). In 1613, there were approximately 60,000 human deaths from anthrax in southern Europe.

In a series of papers written in 1863 to 1864, Casimir Davaine presented evidence linking anthrax in animals and humans with certain bacteria found in the blood and tissues (Davaine, 1863a, 1863b, 1864; Davaine and Raimbert, 1864). In 1876, Robert Koch visualized anthrax bacilli in the blood and tissues of animals dying of anthrax, demonstrated that the bacillus could form spores that remained viable for long periods in hostile environments, and claimed credit for proving that the bacillus is the causative agent of anthrax (Koch, 1877). Louis Pasteur, in a series of papers beginning in 1877, isolated the anthrax bacillus in pure culture from animals, propagated the organism in serial cultures, and reproduced the disease in experimental animals (reprinted in Vallery-Radot, 1939). It was thus Pasteur who fulfilled "Koch's postulates" in establishing a causal relationship between the anthrax bacillus and anthrax (Carter, 1988).

Because of vaccination programs and improved practices of raising livestock, anthrax has become a rare disease in many industrialized countries; an average of only two or three cases per year have been reported in the United States during the past 20 years. Most cases in industrialized countries are associated with exposure to animal products, especially goat hair, imported from Iran, Turkey, Pakistan, and Sudan, where anthrax remains common among domestic livestock. Worldwide anthrax remains an important disease, with an estimated 20,000 to 100,000 human cases occurring each year (Glassman, 1958). Recent outbreaks of human infection have been reported from Zimbabwe (Davies, 1982, 1983, 1985; Turner, 1980); Turkey (Doganay et al., 1986); Thailand (Sirisanthana et al., 1984); Bangladesh (Samad and Hoque, 1986); Afghanistan (Arya et al., 1982); Tanzania (Webber, 1985); Ethiopia (Seboxa and Goldhagen, 1988); and Uganda (Ndyabahinduka et al., 1984). During the civil war in Zimbabwe in 1979 to 1980, when medical services and animal vaccination programs were disrupted, more than 6,000 human cases occurred, with approximately 100 deaths.

Because of the extraordinary stability of the spores of *B. anthracis* and their virulence following inhalation or ingestion, anthrax has received much consideration as a potential agent for use in biological warfare. In 1979, a large outbreak of anthrax, possibly involving more than 1,000 people, occurred in and around Sverdlovsk, a large industrial city in the Ural Mountains of the Soviet Union (Wade, 1980). It was suspected that this outbreak might have resulted from an accident of some kind, possibly an explosion at a secret military laboratory where a biological warfare agent was being prepared. This assertion was denied by the Soviet Union, and 10 years later a delegation of Soviet officials visited the National Academy of Sciences in Washington, D.C. to explain that the epidemic had consisted primarily of cases of intestinal anthrax, traced to a large lot of contaminated bone meal used as cattle feed. According to Soviet officials, the human victims became ill after eating infected meat purchased on the black market (Marshall, 1988).

Habitat and Ecology

The spores of *B. anthracis* are extremely resistant to heat, cold, desiccation, and chemical disinfection and may remain viable in the soil of pastures for many decades. Certain fields in Europe have been associated with repeated epizootics of anthrax in grazing animals for more than 100 years. Although the soil is the ultimate reservoir of *B. anthracis,* its ecology is complex (Van Ness, 1971). A vegetative phase may be required in which the organisms multiply to a density sufficient to infect grazing animals. In suitable limestone soils, with pH above 6.0 and ambient temperature above 15.5°C, a spore-vegetative cell-spore cycle may be maintained indefinitely and livestock grazing on such pastures may become infected. In unsuitable areas, perhaps because of acidity or intense bacterial competition, both organisms and spores may be eliminated rapidly. "Incubator areas" may develop in depressions where water has stood long enough to devitalize or kill grass. Flooding of these areas may result in dispersal of spores over a wide expanse. Cattle dying with anthrax commonly discharge large numbers of anthrax bacilli from the nose, mouth, and intestinal tract, thus returning organisms to the soil. Sporulation and dispersal of spores may also occur when the carcass of an animal dying of anthrax is opened in the field, either for necropsy or for butchering in countries where such animals provide meat for human consumption (McKendrick, 1980). Pasteur et al. (1880) showed that earthworms can return anthrax spores from the buried carcasses of animals dying of anthrax to the surface of the soil. Mechanical

dispersal of spores over short distances may occur via flies, and over long distances by vultures which have fed on the carcasses of infected animals (McKendrick, 1980; Davies, 1983).

Acquisition of anthrax by animals and humans occurs via spores; vegetative forms probably play no significant role in its transmission. Grazing animals often acquire anthrax during periods of drought, when the herds are forced to graze on spiny vegetation close to the ground and abrasions on the lips allow deposition of spores in the tissues; ingestion of spores along with vegetation is another route of infection. Fertilizers made from contaminated animal bones, hay, water, and animal feeds are additional sources of infection in animals. Biting flies may transmit anthrax among animals in some tropical countries but probably do not constitute an important source of infection worldwide.

Human cases of anthrax are acquired via contact with infected animals or animal products and may be classified as *agricultural* or *industrial.* Agricultural cases predominate in developing countries where there is an extensive agricultural economy and anthrax is enzootic in livestock. Outbreaks of human anthrax are almost always preceded by epizootics in cattle, as occurred in the large epidemic of human anthrax in Zimbabwe in 1979 to 1980. Nearly all cases of agricultural anthrax are of the cutaneous form and result from handling infected carcasses or animal products. In countries where animals dying of anthrax are consumed by humans, occasional cases of intestinal and oropharyngeal anthrax occur. An outbreak of 24 cases of oropharyngeal anthrax, and many cases of the cutaneous form, occurred in northern Thailand in 1982 as a result of individuals eating the meat of sick cattle and water buffaloes imported on the hoof from Burma (Sirisanthana et al., 1984). The role of biting insects in transmission of anthrax to humans remains controversial. Stable flies (*Stomoxys* spp.) and two species of *Aedes* mosquitoes have been shown to be capable of transmitting anthrax from infected guinea pigs to uninfected guinea pigs and mice (Turell and Knudson, 1987). During the 1979 to 1980 epidemic in Zimbabwe, several lines of evidence suggested that transmission occurred via the bites of stable flies and horse flies (Tabanidae): 1) the epidemic coincided with an explosive increase in the population of horse flies; 2) in children, who had little exposure to infected animal carcasses, there was an increased incidence of lesions on the head, neck, and face, where insect bites were common; and 3) a number of patients remembered an insect bite at the site where an anthrax lesion developed later (McKendrick, 1980; Davies, 1983). Person-to-person spread occurs very rarely if at all; there were no cases among the nursing staff and no evidence of cross-infection among patients during the Zimbabwe epidemic.

Industrial anthrax results from contact with spore-contaminated materials such as goat hair, animal hides, wool and bones used in manufacturing processes, or finished products such as shaving brushes, bongo drums, and saddle blankets. Most cases are of the cutaneous type, but occasional cases of inhalation anthrax (woolsorter's disease) are seen (Brachman, 1980). Nearly all such cases occur in workers exposed to aerosols of anthrax spores during early processing of imported goat hair or hides, or less commonly wool. Very rarely the disease occurs in laboratory workers or in artisans working with imported animal hair or skins. There have been only two cases of inhalation anthrax in the United States in the last 20 years (18 cases during this century).

Anthrax in Animals

Anthrax is primarily a disease of herbivorous animals, especially cattle, sheep, horses, goats, and wild herbivores, although nearly all mammals are susceptible to the disease to some degree. Omnivores such as man and swine, and carnivores such as dogs, possess greater natural resistance to anthrax than do herbivores (Knudson, 1986).

Three forms of anthrax, peracute, acute/subacute, and chronic, are recognized in animals (Hunter et al., 1989). The peracute form is seen mainly in ruminants, especially cattle, sheep, and goats. The animals may develop cerebral anoxia and pulmonary edema, and death may occur suddenly.

Acute and subacute anthrax are characterized by fever, depression, convulsions, dyspnea, and hemorrhage from the mouth, nose, and anus, with death occurring approximately 24 h after the initial signs of illness. This form of anthrax is seen primarily in cattle, sheep, and horses. Pathologic findings include hemorrhage and edema in many different organs. In cattle and sheep, the spleen is enlarged with a "blackberry jam" appearance.

In chronic anthrax, seen most often in swine but also in horses, dogs, and cattle, the typical clinical signs are edema of the tongue and pharyngeal tissues with dyspnea and serosanguinous discharge from the mouth. Death due to asphyxia may occur. In swine, an intestinal form of chronic anthrax has been reported. At necropsy, edema and hemorrhage of the involved tissues is found.

Anthrax in Humans

Cutaneous anthrax accounts for approximately 95% of cases. Spores of *B. anthracis* are deposited underneath the epidermis via a minor cut or abrasion or insect bite. These spores then germinate, and the vegetative cells produce toxin. Following an incubation period of 2 to 7 days, a small papule which is often pruritic appears at the site where spores have been inoculated. This enlarges and within 24 to 48 h develops into an ulcer surrounded by vesicles. Eventually the characteristic black eschar develops, surrounded by edema which is often striking, especially with lesions on the face and neck. (The name "anthrax" is derived from the Greek *anthrakos,* which means "coal," as does *charbon,* the French term for the disease.) The lesion is painless. Most patients are afebrile and the white blood cell count is usually normal or only slightly elevated. More than 90% of lesions occur on exposed areas such as the face, neck, hands, and arms. Lymphangitis and regional lymphadenopathy may be present.

Cutaneous anthrax is usually a self-limiting disease and lesions heal spontaneously in 80 to 90% of cases (Knudson, 1986). In untreated cases, death may occur in 10 to 20% due to airway obstruction (especially with lesions of the face or neck) (Davies, 1982), bacteremia with shock and renal failure, or meningitis. Early and appropriate antibiotic therapy reduces the case fatality rate to less than 1%.

Internal forms of anthrax infection are much more difficult to diagnose, and the case fatality rate is much higher than in the cutaneous form, even with prompt and appropriate antibiotic therapy. *Inhalation* anthrax accounts for up to 5% of cases (Brachman, 1980). Most cases result from the inhalation of spores of *B. anthracis* during the processing of contaminated goat hair or wool. The infectious dose is high, probably on the order of 50,000 spores. After inhalation, the spores are ingested by alveolar macrophages and carried to the hilar lymph nodes. Germination and multiplication of the vegetative forms, with production of toxin, occur in the hilar nodes, resulting in edema and hemorrhage in the nodes and surrounding mediastinal structures. Eventually the bacilli may gain access to the blood via the lymphatics, with sepsis and subsequent metastatic spread. The clinical picture is typically that of a biphasic disease. The initial stage, which lasts 2 or 3 days, has an insidious onset with mild fever, malaise, fatigue, myalgia, nonproductive cough, and sometimes a sensation of precordial pressure, and is often mistaken for a viral respiratory infection. The second stage begins suddenly with acute dyspnea, cyanosis, diaphoresis, and stridor with rapid progression to a moribund state and death within 24 h. The characteristic roentgenographic picture is one of mediastinal widening due to lymph node enlargement. At autopsy hemorrhagic mediastinitis and lymphadenitis are found, often with little involvement of the lung parenchyma. Of the 18 cases of inhalation anthrax reported in the American literature since 1900, 16 were fatal.

Intestinal anthrax is very rare in humans and has never been documented in the United States (Nalin et al., 1977; Jena, 1980; Ndyabahinduka et al., 1984; Bhat et al., 1985). The disease follows ingestion of heavily contaminated, undercooked meat. Multiplication of organisms and production of toxin in the bowel wall and in mesenteric lymph nodes leads to edematous and hemorrhagic enteritis and lymphadenitis. The clinical picture is variable. Some cases present with gaseous distention and multiple fluid levels suggesting small bowel obstruction, with development of peritonitis. At surgery or autopsy, severe edema and gangrene of the bowel with multiple perforations and black necrotic areas may be found. The mesentery may be grossly edematous and hemorrhagic mesenteric lymphadenitis may be present. Other patients present with rapid development of hypovolemic shock and massive hemorrhagic ascites. The case fatality rate is 25 to 50% regardless of treatment.

Anthrax meningitis, another rare form of the disease, may occur as a result of the bacteremia which can accompany cutaneous or inhalation anthrax (Koshi et al., 1981; Trivedy, 1981; Chandramukhi et al., 1983; Dürst et al., 1986). In approximately one-half the cases, meningitis results from hematogenous dissemination from a cutaneous lesion. Nearly one-half of patients with inhalation anthrax develop meningitis. The clinical picture is that of a rapidly progressive acute bacterial meningitis. The case fatality rate, even with antibiotic treatment, is nearly 100%. At autopsy an acute hemorrhagic leptomeningitis, sometimes with subarachnoid, cortical or deep hemorrhages, or diffuse encephalitis, is found.

Oropharyngeal anthrax has occurred following ingestion of infected beef or water buffalo meat (Sirisanthana et al., 1984; Doganay et al., 1986). The primary lesion on the tongue, tonsil, or pharyngeal wall exhibits central necrosis and ulceration with surrounding edema and congestion. The neck on the involved side is markedly swollen with enlargement of cervical lymph nodes. The case fatality rate in patients treated with penicillin has varied from 15 to 50%.

A few cases of anthrax have been reported in cases of attempted abortion with spore-infected date stems or twigs, or following delivery in a stable, leading to hemorrhagic metritis, parametritis, and oophoritis, most ending in septicemia and death (Dutz and Kohout-Dutz, 1981).

Treatment and Prevention

The prognosis in cutaneous anthrax is good; 80 to 90% of cases resolve without specific antibiotic therapy. However, because of the potential complications of malignant edema, septicemia, shock, renal failure, and meningitis, patients with anthrax should receive prompt treatment, on the basis of clinical suspicion or visualization of organisms in Gram stains of specimens taken from the lesions. Treatment should not be delayed until bacteriologic confirmation is obtained. Penicillin G is the drug of choice in the treatment of anthrax (Knudson, 1986). An appropriate dosage regimen for treatment of cutaneous anthrax is two million units intravenously every 6 h for 2 to 4 days, until substantial improvement has occurred. At this time therapy can be changed to oral penicillin V, and treatment continued for a total of 7 to 10 days. Cultures of vesicle fluid become negative within a few hours after initiation of therapy with penicillin, but the lesion continues to evolve to the eschar phase. Systemic manifestations and local edema usually resolve promptly. Appropriate alternative choices for patients allergic to penicillin are erythromycin, tetracycline, and trimethoprim/sulfamethoxazole. Corticosteroids are recommended by many investigators when massive edema is present, and tracheostomy should be performed in patients with airway obstruction due to massive edema. Local surgical manipulation is ineffective and is contraindicated.

In patients with inhalation or gastrointestinal anthrax, severe sepsis or anthrax meningitis, the prognosis is poor regardless of treatment. In guinea pigs with experimental infection, once the level of bacteremia reaches three million organisms per milliliter of blood, eradication of the infection with antibiotics does not prevent a fatal outcome due to toxemia. This observation probably holds true for humans with septicemic anthrax (Keppie et al., 1955). The best hope is to begin therapy with high doses of penicillin G (12–20 million units per day, given intravenously in divided doses every 4 h) as soon as this type of infection is suspected. There is some experimental evidence that the combination of penicillin plus streptomycin (1–2 g per day intramuscularly) may be more effective in septicemic anthrax than penicillin alone (Lincoln et al., 1964). Chloramphenicol is not effective in the treatment of experimental septicemic anthrax and should not be relied upon for the treatment of systemic forms of infection.

In most countries, livestock at risk are immunized annually with commercial vaccines consisting of viable spores of the Sterne strain of *B. anthracis,* a nonencapsulated toxigenic variant which lacks the pX02 plasmid (Hambleton et al., 1984). This vaccine is effective and safe for use in many domestic animals (cattle, sheep, pigs, camels, buffaloes, and elephants) but progressive disease due to the vaccine strain has been observed in goats and llamas (Cartwright et al., 1987; Sterne, 1988). The human anthrax vaccine licensed in the United States is produced by the Michigan Department of Public Health, and consists of aluminum hydroxide-adsorbed culture filtrate of the nonencapsulated toxigenic V770-NP1-R strain (composed primarily of protective antigen) (Ivins et al., 1986). Although guinea pigs immunized with the human vaccine produce high titers of antibody to protective antigen, only animals vaccinated with the Sterne strain are completely protected against challenge with highly virulent strains of *B. anthracis* (Little and Knudson, 1986). There appear to be at least two immunogenic vegetative cell-surface antigens present in the Sterne live vaccine which are not present in the human vaccine (Ezzell and Abshire, 1988). Modern studies have shown that the Pasteur vaccine strains (nontoxigenic, encapsulated) neither elicit antibody titers to any of the toxin components nor provide protection against challenge with virulent spores (Ivins et al., 1986). New approaches to development of more effective vaccines for human use include identification and incorporation of essential antigens and epitopes and transposon-induced mutagenesis to produce live vaccine candidate strains (Ivins and Welkos, 1988; Ivins et al., 1988).

When anthrax occurs in a herd, the source of infection must be determined and eliminated, and survivors should be treated prophylactically with antibiotics and the vaccination status brought up-to-date (Hunter, et al., 1989). The affected farm should be quarantined for a period of at least 2 weeks following the last death from anthrax. Milk from unvaccinated febrile dairy cattle in the herd should be discarded and sick animals should be isolated. Animals dying of anthrax should be cremated or buried deeply after being covered with lime. Necropsy of dead animals should not be carried out in the field because of the risk of contaminating the soil with spores. Cases of anthrax in animals and humans should be reported to the appropriate health authorities.

Virulence Factors of *B. anthracis*

Fully virulent strains of *B. anthracis* possess two unique virulence factors: a poly-D-glutamic acid capsule (Green et al., 1985), which inhibits phagocytosis, and a tripartite toxin composed of protective antigen (PA), edema factor (EF) and lethal factor (LF) (Stephen, 1981; Leppla, 1988). Capsules are produced by virulent strains of *B. anthracis* growing in vivo, and by cells grown on media containing serum or bicarbonate or both and incubated in a CO_2-enriched atmosphere. The colonies of organisms that produce capsules appear mucoid, whereas colonies of organisms grown in the absence of serum or bicarbonate fail to produce capsules and appear rough (Green et al., 1985).

The existence of an anthrax toxin was first clearly demonstrated by Smith et al. (1955), who produced local edema and death with injections of sterile plasma from infected guinea pigs. Studies by American and British investigators during the ensuing decade showed that the toxin contained three separate components (Stephen, 1981). The individual toxin components have no known biological effects when administered alone, but EF injected with PA into the skin of rabbits or guinea pigs causes local edema, and PA injected with LF into rats causes death in as little as 60 min. EF and LF are mutually inhibitory, suggesting that they compete for the same binding site on PA (Ezzell et al., 1984). Protective antigen, so-called because its injection into experimental animals results in protective immunity, binds to cell-surface receptors to produce an uptake system that can be used by both EF and LF to gain access to the cytoplasm. "Edema toxin" (EF + PA) and "lethal toxin" (LF + PA) thus resemble the A-B enzyme-binding structures characteristic of many well-studied bacterial toxins (Middlebrook and Dorland, 1984). After PA, analogous to the B chain, binds to a specific membrane receptor on the surface of a eukaryotic cell, it is cleaved at a single site, exposing a binding site for the other toxin component. The membrane-bound fragment of PA then binds to EF or LF and mediates the entry of the active moiety into the cytosol from an acidified endocytic vesicle. This fragment of PA has been shown to have channel-forming activity in planar phospholipid bilayers, and such channels may provide an aqueous pore in the endosomal membrane through which EF and LF enter the cell (Blaustein et al., 1989).

EF has been found to be a calmodulin-dependent adenylate cyclase that elevates cyclic

AMP levels approximately 200-fold above normal in Chinese hamster ovary cells (Leppla, 1982, 1984). Local edema, a typical sign of anthrax, may be directly related to adenylate cyclase activity associated with EF. The increase in intracellular cAMP caused by this toxin may lead to edema in a manner analogous to the loss of water in the intestine caused by cholera toxin, which also increases intracellular cAMP (Leppla, 1982). EF + PA also inhibits phagocytosis of anthrax bacilli by polymorphonuclear leukocytes, and blocks both particulate and phorbol myristate acetate-induced chemiluminescence in polymorphonuclear leukocytes (O'Brien et al., 1985). These effects are accompanied by an increase in intracellular cAMP levels. The findings suggest that EF + PA may increase host susceptibility to anthrax by suppressing polymorphonuclear leukocyte function. The dependence of EF activity on calmodulin, a substance found only in eukaryotic cells, suggests that EF did not evolve from a bacterial enzyme but from a eukaryotic adenylate cyclase, the gene for which was adventitiously transferred into *B. anthracis* and retained because it made the bacteria more virulent (Leppla, 1984). EF also exhibits substantial DNA homology and immunological relatedness with the *Bordetella pertussis* adenylate cyclase toxin (Escuyer et al., 1988). However, cytochalasin D, which interferes with endocytosis, and both ammonium chloride and chloroquine, which prevent acidification of endosomes, block intoxication of Chinese hamster ovary cells by EF, but not by pertussis adenylate cyclase toxin (Gordon et al., 1988).

The mechanism of action of LF is poorly understood. LF is lethal for many species of experimental animals and is assumed to be the major factor causing death in anthrax. No enzymatic activity has yet been associated with LF, and the nature of the intracellular target of LF is unknown. Certain macrophages and a mouse macrophage-like cell line are lysed at low concentrations of LF, but most cells are resistant (Singh et al., 1989). Resistant cells may either lack the intracellular target of LF or fail to process LF to an active form. There is evidence that PA may also be required for full activity of LF at a stage subsequent to endocytosis. LF is also calmodulin-dependent, and calcium is required at several steps for the expression of its effect (Bhatnagar et al., 1989).

Virulent strains of *B. anthracis* contain two large plasmids, pX01 and pX02 (Mikesell et al., 1983; Green et al., 1985; Kaspar and Robertson, 1987). Both plasmids are required for full pathogenicity, and strains which contain only one of these plasmids are avirulent. pX01 (174 kb pairs) encodes all three components of the anthrax toxin and pX02 (95 kb pairs) encodes the poly-D-glutamic acid capsule. The avirulent Sterne vaccine strain, which is pX01+/pX02-, produces toxin but no capsule and is used effectively as a live veterinary vaccine. The heat-attenuated Pasteur vaccine strains form capsules but are unable to produce toxin. Some rough variants of the Pasteur strains lack pX02, but others contain this plasmid. Reversion to mucoid colonies may be observed in strains that have retained pX02, but reversion of those cured of the plasmid is never observed. Pasteur probably cured his strains of plasmid pX01 by heat attenuation to produce his vaccine for immunization of cows and sheep.

The genes encoding all three toxin components have been cloned in *Escherichia coli* (Vodkin and Leppla, 1983; Robertson and Leppla, 1986; Mock et al., 1988; Tippetts and Robertson, 1988), and the base sequences for the PA and EF genes have been determined (Robertson et al., 1988; Welkos et al., 1988). The PA gene has been cloned in *B. subtilis,* and immunization with the live recombinant strains protected guinea pigs from lethal challenge with virulent *B. anthracis* spores (Ivins and Welkos, 1986).

Bacillus cereus Food Poisoning

B. cereus has been recognized as an agent of food poisoning for the past several decades (Hauge, 1955). Between 1972 and 1986, 52 outbreaks of food-borne disease associated with *B. cereus* (1.9% of the total) were reported to the Centers for Disease Control (CDC). *B. cereus* strains cause two types of food-poisoning syndromes. These are true intoxications rather than infections. One type is characterized by nausea and vomiting (100%) and abdominal cramps (100%) and has an incubation period of 1 to 6 h (Terranova and Blake, 1978). This type has been referred to as the "emetic syndrome" or the short-incubation form. It resembles *Staphylococcus aureus* food poisoning in symptomatology and the short incubation period.

The second type of *B. cereus* food poisoning is manifested primarily by abdominal cramps (75%) and diarrhea (96%) with an incubation period of 8 to 16 h (Terranova and Blake, 1978). Diarrhea may be small volume or profuse and watery. This type is the "diarrheal syndrome" or long-incubation form (Mortimer and McCann, 1974) and resembles food poisoning caused by *Clostridium perfringens*. The illness usually lasts less than 24 h. In a few patients symptoms may last longer (2–10 days) (Giannella and Brasile, 1979).

The short-incubation form is caused by a preformed heat-stable enterotoxin of molecular weight less than 5,000 which is produced by some *B. cereus* strains and which causes vomiting when fed to monkeys (Melling et al., 1976). The mechanism and site of action of this toxin are unknown (Turnbull et al., 1979b). The long-incubation form of illness is also enterotoxin-mediated. This toxin is produced in vivo (accounting for the longer incubation period), is heat-labile, and has a molecular weight of approximately 50,000 (Turnbull et al., 1979b). It activates intestinal adenylate cyclase and causes intestinal fluid secretion (Turnbull, 1976). It appears to have cytotoxic properties in rabbit small intestine and in guinea pig skin (Turnbull, 1976; Turnbull et al., 1979b).

B. cereus food poisoning occurs year round and is without any particular geographic distribution. The short-incubation form is most often associated with fried rice that has been cooked and then held at warm temperatures for several hours. The disease is often associated with Chinese restaurants, especially in Europe, where leftover rice is often held overnight at room temperature because cooked rice which has been refrigerated is difficult to toss into beaten eggs. Growth of *B. cereus* in rice under similar experimental conditions has been documented (Gilbert et al., 1974). In one reported outbreak, the vehicle was macaroni and cheese made from powdered milk that was the source of the organism (Holmes et al., 1981). The short-incubation type is most often caused by serotype 1 strains (Gilbert and Parry, 1977) probably because they are more heat resistant than strains of other serotypes (Parry and Gilbert, 1980).

Long-incubation *B. cereus* food poisoning is frequently associated with meat or vegetable-containing foods. *B. cereus* contaminates raw meats, vegetables, and milk products. It has been isolated from 50% of dried beans and cereals (Blakey and Priest, 1980) and from 25% of dried foods such as spices, seasoning mixes, and dried potatoes (Kim and Goepfert, 1971). An outbreak of the long-incubation form was traced to a "meals-on-wheels" program in which food was held at and above room temperature for a prolonged period (Jephcott et al., 1977).

The short-incubation or emetic form of *B. cereus* food poisoning is diagnosed by the isolation of *B. cereus* from the incriminated food. The long-incubation or diarrheal form is diagnosed by isolation of the organism from stool and food. Isolation from stools is not sufficient unless negative stool cultures are obtained from a control group (Terranova and Blake, 1978).

Fourteen percent of healthy adults have been reported to have transient gastrointestinal colonization with *B. cereus* (Ghosh, 1978). Serotyping may be of value in identifying a common source but is not readily available. Plasmid analysis recently proved useful in the epidemiological investigation of an outbreak of *B. cereus* gastroenteritis associated with the consumption of beef stew at a nursing home (DeBuono et al., 1988).

Because *B. cereus* gastroenteritis is generally a benign, self-limited illness, antimicrobial agents are of no value in management. Since bacteria grow best at temperatures ranging from 40 to 140°F, infection may be prevented if cold food is refrigerated and if hot food is held at greater than 140°F before serving.

Several methods have been described for production of the diarrheal toxin. Growth of *B. cereus* in brain heart infusion (BHI) broth with 0.1% glucose (pH=7.4) is simple (Spira and Goepfert, 1972). High speed centrifugation is used to remove organisms and the supernatant is examined serologically by a modified microslide gel double-diffusion method for the identification of *B. cereus* diarrheal antigen (Crowle, 1958). Conditions for production of the emetic toxin are not as well defined. Because of the association of rice-containing foods with the emetic form of *B. cereus* food poisoning, most media for demonstrating emetic activity have contained rice (Melling et al., 1976). Autoclaved rice is soaked in saline. An overnight culture of rice is liquefied by diastase then dialyzed for 24 h at 4°C against 10% polyethylene glycol (Melling et al., 1976). The addition of chicken, beef, or egg proteins or Casamino acids with B vitamins to rice media stimulate the growth of *B. cereus* (Morita and Woodburn, 1977). Serological identification of the emetic antigen has not been possible because of its low molecular weight, so biological assays are used. Rice mixtures containing *B. cereus* emetic toxin fed to monkeys cause emesis in 50% of the animals (Melling et al., 1976). Intravenous injection of kittens with concentrated culture fluids produced by culturing *B. cereus* in BHI broth supplemented with 0.1% glucose (pH=7.4) and heating at 100°C for 1.5 h has caused emesis (Bennett and Harmon, 1988).

Other *Bacillus* Infections

Other nonanthrax *Bacillus* infections have been classified as 1) local, usually involving the eye or an isolated organ that was previously damaged; 2) deep tissue or mixed, in which a *Bacillus* species is usually found in the company of other organisms; and 3) disseminated, in which the organism is consistently cultured from the blood or cerebrospinal fluid of a seriously ill person (Farrar, 1963).

Local Infections

The eye has been the organ most commonly infected by nonanthrax *Bacillus* species, especially *B. cereus*. They can cause conjunctivitis, keratitis, iridocyclitis, dacryocystitis, orbital abscess, and panophthalmitis. *Bacillus* species are most commonly isolated in the setting of penetrating nonsurgical trauma. In a review of nonsurgical post-traumatic endophthalmitis, *Bacillus* species ranked as the second most common pathogen in five of six series reviewed (Davey and Tauber, 1987). An intra-ocular foreign body such as a metal projectile is often present in this setting or the injury occurs in a rural or farm location where there is a greater risk of contamination with dust or soil. *B. cereus* is one of the most destructive organisms to infect the eye (Davey and Tauber, 1987). Panophthalmitis with *Bacillus* species also occurs in the setting of hematogenous dissemination, as in intravenous drug abusers (Davey and Tauber, 1987; Pearson, 1970; Young et al., 1980).

These patients present with a fulminant endophthalmitis or panophthalmitis that usually results in blindness and the need for enucleation. Although the exact pathogenesis is ill-defined, it is probably related to endogenous infection from contaminated drugs and paraphernalia (Shamsuddin et al., 1982). Approximately 50% of cultures of heroin and drug paraphernalia contain *Bacillus* species, making them the most common organism isolated (Tuazon et al., 1974). In one patient with panophthalmitis, *B. cereus* was isolated from a vitreous aspirate and also from the syringe used for injection (Young et al., 1980). Severe suppurative endogenous panophthalmitis caused by *B. cereus* has also resulted from injection of vitamin B (Bouza et al., 1979) and also after a blood transfusion (Kerkenezov, 1953).

Clinical features are fairly characteristic, with intense pain followed by chemosis, periorbital swelling, and severe proptosis. Within 48 h, a corneal ring abscess develops. This is a hallmark of an intraocular infection with *Bacillus* organisms. The end result is irreversible loss of vision and eventual loss of the eye (Davey and Tauber, 1987; O'Day et al., 1981). A characteristic feature of infection with *B. cereus* is its frequent association with fever and leukocytosis, which are not usually seen in panophthalmitis caused by other bacteria (Davey and Tauber, 1987). Early diagnosis is important. This organism

should be considered whenever ocular infection occurs in the setting of penetrating injury with probable soil contamination or in an intravenous drug user. Vitreous aspiration with Gram stain and culture is diagnostic.

Empiric therapy should begin before isolation of an organism and the availability of antibiotic susceptibility data. For treatment of severe ocular infections, early vitrectomy and intravitreal antibiotics, combined with systemic and periocular antibiotics, appear to offer the best chance for success (Davey and Tauber, 1987). Clindamycin or vancomycin should be combined with an aminoglycoside.

The first case of human infection with *B. thuringiensis* was reported by Samples and Buettner (1983) in a healthy farmer who splashed an insecticide containing the organism into his eye. The organism was recovered from a corneal ulcer. The clinical course was much less fulminant than with *B. cereus* infection.

Mixed Infections or Deep-Tissue Infections

This category includes cellulitis, traumatic wound infections, surgical wound infections, infected burns, infected necrotic tumors, necrotizing fasciitis, pyelonephritis, pericarditis, and pneumonia (Bias, 1927; Farrar, 1963; Ihde and Armstrong, 1973; Logan et al., 1985b; Pearson, 1970; Pennington et al., 1976; Tuazon et al., 1979). *Bacillus* species are rare pulmonary pathogens. *B. cereus* has caused pneumonia in patients with malignancy (Bekemeyer and Zimmerman, 1985; Pennington et al., 1976) and in those with no known immune compromise (Jonsson et al., 1983). Cavitating pneumonia has been reported (Leff et al., 1977). *B. sphaericus* caused a large pseudotumor of the lung in a patient with chronic asthma on corticosteroids (Isaacson et al., 1976). Symptoms are indistinguishable from other bacterial pneumonias. Complications include empyema, massive hemoptysis, acute respiratory failure, tension pneumothorax, and bronchopleural fistula (Bekemeyer and Zimmerman, 1985). Cases of necrotizing pneumonia and empyema may require resection of infected lung and decortication (Jonsson et al., 1983).

Bacillus species, especially *B. cereus,* can infect traumatic wounds in both the normal host and the immunocompromised host (Dryden and Kramer, 1987; Jaruratanasirikul et al., 1987). In the immunocompromised patient, the trauma can be as minor as scratch marks exposed to muddy water (Jaruratanasirikul et al., 1987). Severe toxicity may develop when *B. cereus* infects relatively minor burns (Attwood and Evans, 1983). In addition to traumatic wounds,

B. cereus has infected breast implants (Sliman et al., 1987) and pin sites following pin placement and plaster fixation of an open forearm fracture (Rutala et al., 1986). In the later situation, *B. cereus* was recovered from plaster-impregnated gauze rolls and tapwater samples. The pins may have facilitated infection by serving as a conduit for the plaster-associated bacteria to the pin insertion sites. Plaster samples were subsequently sterilized by steam or gas.

Primary cutaneous *B. cereus* infection was recently reported in neutropenic children (Henrickson et al., 1989). In these cases, there was a spring and summer seasonal predominance but no history or signs of injury to the skin. Vesicles or pustules with rapidly spreading cellulitis were seen only on the extremities. Although neutropenic and febrile, the children were not otherwise systemically ill. Cultures of vesicle fluid or wound drainage yielded pure growth of *B. cereus*. All cases responded to antibiotic therapy. Cutaneous *B. cereus* infection is clinically similar to cutaneous anthrax and has the same seasonality. This entity may be related to the secretion of the same or similar exotoxins.

Necrotizing fasciitis caused by *Bacillus* species has been reported in a leukemic patient and in a patient with sickle cell disease (Sliman et al., 1987; Tuazon et al., 1979). In both cases, *Bacillus* species grew in pure culture from deep tissue. *B. cereus* has been associated with myonecrosis postoperatively (Fitzpatrick et al., 1979) and following trauma (Johnson et al., 1984; Groschel et al., 1976). Treatment of these deep soft tissue infections requires surgical debridement and systemic antibiotics. Amputation may be necessary.

Streptococcal and other organisms are frequently isolated along with the *Bacillus* species. It is frequently unclear what role the *Bacillus* plays in these mixed infections. It may be a co-pathogen, a secondary invader, or a colonizer. Ihde and Armstrong (1973) found that whether the *Bacillus* was treated did not affect outcome in surgical wounds. However, they did note a change in the character of the drainage from sanguinous to serous after the *Bacillus* was no longer recovered in culture.

Disseminated Infections

The category of disseminated infections includes bacteremia, endocarditis, meningitis, and other such infections. Since *Bacillus* species are ubiquitous in the environment, it is not suprising that positive cultures for these organisms are fairly common. They are only rarely associated with actual infection, however. The prevalence of positive blood cultures for *Bacillus* species has ranged from 0.1 to 0.9% (Dalton and Allison, 1967; Kotin, 1952; Pearson, 1970). Isolation of *Bacillus* species from blood is clinically significant in 5 to 10% of cases (Weber et al., 1989).

Bacteremia is relatively common in *Bacillus* infections and the incidence is increasing. *B. cereus* is the predominant species. Bacteremia can be a complication of indwelling intravascular catheters (Banerjee et al., 1988; Cotton et al., 1987; Sliman et al., 1987). Scanning and transmission electron microscopy of a Hickman catheter after removal showed *Bacillus* organisms embedded in a biofilm composed of Gram-positive cocci and glycocalyx (Banerjee et al., 1988). In a study by Sliman et al. (1987), an intravascular device proved to be the source of the bacteremia in one-half of the cases, as evidenced by a positive culture of purulent drainage from a peripheral vein or the recovery of greater than 15 colonies from a semiquantitative culture of a catheter tip. Isolated bacteremias are usually eradicated easily. Bacteremia may clear after removal of an intravascular device in the absence of specific antibacterial therapy (Sliman et al., 1987). For *Bacillus* bacteremias in immunocompromised patients (especially those who are neutropenic) the catheter should be removed and systemic antibiotics administered (Banerjee et al., 1988; Cotton et al., 1987).

Bacteremia has also occurred in association with infections in a variety of body sites. Unlike those with primary bacteremia, patients with infections of solid organs frequently have high morbidity. Some of these cases of bacteremia may actually have been cases of endocarditis, and some were treated as such. A relatively common situation has been pneumonia with bacteremia in individuals with acute leukemia and neutropenia (Coonrod et al., 1971a; Ihde and Armstrong, 1973; Pennington et al., 1976; Sathmary, 1958). The neutropenic patient is at risk for life-threatening bacteremia following seemingly minor trauma (Guiot et al., 1986).

An outbreak of five cases of bacteremia occurred in a group of hemodialysis patients exposed to contaminated dialysis fluid and equipment (Curtis et al., 1967). Individual cases of bacteremia in association with volvulus and peritonitis (Sugar and McCloskey, 1977), infected ventriculojugular shunt cured only by removal of the shunt (Cox et al., 1959), and sepsis in a newborn infant, perhaps related to contaminated blood transfusion (Yow et al., 1949), have been reported.

There have been five outbreaks of pseudo-bacteremia associated with contaminated broth culture media (Crowley et al., 1983; Noble and Reeves, 1974), contaminated syringes (MacDonald, 1982) or alcohol swabs (Berger, 1983), and a contaminated needle in a radiometric blood culture analyzer (Gurevich et al., 1984).

Endocarditis due to *Bacillus* organisms is uncommon; only about 12 cases have been reported in the English-language literature (Agarwala et al., 1975; Block et al., 1978; Farrar, 1963; Oster and Kong, 1982; Yeh et al., 1967). More than one-half the cases have occurred in intravenous drug users (Craig et al., 1974; Reller, 1973; Tuazon et al., 1979; Weller et al., 1979), probably related to the frequent presence of *Bacillus* organisms on injection paraphernalia and in street heroin (Tuazon et al., 1974). *Bacillus* endocarditis cannot be distinguished from other bacterial causes of endocarditis on the basis of clinical features. The majority of patients have had low-grade fever and heart murmurs. There is a predominance of tricuspid valve involvement. Some have had hepatosplenomegaly, and a few have had peripheral stigmata. *B. cereus* has caused infection of a permanent ventricular pacemaker (Sliman et al., 1987). Echocardiogram revealed a vegetation on the pacemaker wire in the right ventricle. The organism was isolated from blood, the subcutaneous pacemaker pouch, and right ventricular tissue. There was persistence of the organism despite 10 days of appropriate antibiotic therapy. Cure was achieved only after removal of the pacemaker and 4 weeks of antibiotic therapy postoperatively.

In a study of 849 cerebrospinal fluid (CSF) cultures, nine were positive for nonanthrax *Bacillus* species. In seven patients the isolates represented contamination (Feder et al., 1988). There has been at least one outbreak of pseudomeningitis caused by contamination of broth culture media by *Bacillus* species (Lettau et al., 1988). True meningitis due to *Bacillus* organisms has been reported primarily in situations in which there is either direct access of exogenous organisms to the subarachnoid space (spinal anesthesia, head trauma, and neurosurgical procedures with and without foreign bodies), or in individuals whose immune function is depressed (newborn state, alcoholism, and hematologic malignancy), although at least one case has been reported in an otherwise healthy individual (Allen and Wilkinson, 1969; Boyette and Rights, 1952; Colpin et al., 1981; Farrar, 1963; Park et al., 1976; Patrick et al., 1989; Weinstein and Colburn, 1950; Weidermann, 1987). It has also occurred secondary to other infections such as otitis, mastoiditis, infected subdural hematoma, and hematogenous dissemination from a urinary tract infection.

Causative organisms have included *B. alvei, B. megaterium, B. subtilis, B. pumilus, B. circulans, B. sphaericus,* and *B. cereus.* Single cases of postoperative ventriculitis due to *B. licheniformis* and *B. cereus* have been reported (Young et al., 1982). The latter infection was in a patient with a ventriculoperitoneal shunt. Removal of any foreign body is essential for cure.

Brain abscess with *B. cereus* has been reported rarely (Ihde and Armstrong, 1973; Jenson et al., 1989). All of the patients were immunocompromised. In two of the patients, involvement of the brain parenchyma was probably via the hematogenous route from primary foci of *B. cereus* pneumonia (Ihde and Armstrong, 1973). Diagnosis was made postmortem. The other patient had multiple brain abscesses, concurrent meningitis and may have acquired the infection hematogenously from a potentially contaminated transfusion (Jenson et al., 1989). He was successfully treated with vancomycin and gentamicin, and vancomycin and rifampin for a total of 8 weeks. Mild memory deficit and mild residual paresis of the right upper extremity persisted as sequelae of the infection.

Acute and chronic osteomyelitis and septic arthritis caused by *Bacillus* species are rare. In a review of *Bacillus* infections, 10 cases of bone and joint infection were reported (Pearson, 1970). The organism was present in mixed culture in 9 of the 10 cases, and in 5 it was obtained from a wound or drainage. Most of the patients had previous trauma. Acute vertebral osteomyelitis caused by *B. cereus* has been reported in an intravenous drug user (Tuazon et al., 1979). An association between *Bacillus* species osteomyelitis and sickle cell disease has been noted (Reboli et al., 1989; Sliman et al., 1987; Solny et al., 1977). *B. alvei* caused infection of a prosthetic hip in a patient with sickle cell disease, probably via the hematogenous route (Reboli et al., 1989). In general, bacteremia in the setting of osteomyelitis is rare. *B. cereus* septic arthritis has occurred as a complication of arthrography (Robinson, 1979). *Bacillus* infections of bone and joints have been difficult to eradicate, require multiple surgical procedures, and cause substantial morbidity (Sliman et al., 1987).

Except for the characteristic signs and symptoms of panophthalmitis, most infections due to nonanthrax *Bacillus* species do not have distinctive clinical presentations, and although serious infections are being recognized with increasing frequency, endocarditis, meningitis, and sepsis are still sufficiently rare that the question of contaminant vs. pathogen is still valid. Diagnosis of true infection rests on the isolation of the organism from blood, normally sterile body fluids, or closed spaces (such as the pleural space) of a susceptible host in the appropriate clinical setting. A positive Gram stain and growth in pure culture are contributory evidence of the *Bacillus* being a true pathogen, whereas growth only in thioglycolate broth (indicating a small number of organisms) or the presence of *Bacillus* organisms in mixed culture with other species is suggestive of a contaminant. Although infection can occur in healthy persons, it is most often seen in immunocompromised patients such as those with hematologic malignancies, intravenous drug users, or patients with severe trauma (Ihde and Armstrong 1973; Pennington et al., 1976; Tuazon et al., 1979).

Hypersensitivity Reactions

Allergic reactions have been described in workers exposed to autolysates of *B. subtilis* used as enzymes in laundry products (Dubos, 1971; Flindt, 1969; Greenberg et al., 1970; Pepys et al., 1969). These reactions were mainly dermatitis and asthma. Those involved showed immediate and late reactions to inhalation and skin prick tests (Flindt, 1969). Hypersensitivity pneumonitis following exposure to wood dust contaminated with *B. subtilis* has been described (Johnson et al., 1980.)

Treatment and Prognosis

Many strains of *Bacillus* species are susceptible to the aminoglycosides, tetracycline, chloramphenicol, erythromycin, vancomycin, and clindamycin (Coonrod et al., 1971b; Tuazon et al., 1979). Susceptibility to penicillin, ampicillin, methicillin, and cephalosporins is usually species related and has been shown to be high for *B. subtilis* and low for *B. cereus*. *B. pumilus* is intermediate between *B. subtilis* and *B. cereus*. *B. cereus* is the most resistant species, showing resistance to the penicillins and cephalosporins that may be attributable to its ability to produce a β-lactamase that hydrolyzes these agents (Sabath and Abraham, 1965). In vitro susceptibility tests reveal all *B. cereus* strains to be susceptible to imipenem, vancomycin, chloramphenicol, gentamicin, and ciprofloxacin (Weber et al., 1988). Non-*B. cereus* strains were most susceptible to imipenem, vancomycin, daptomycin, and ciprofloxacin (Weber et al., 1988). Further studies are needed before imipenem and ciprofloxacin can be recommended for treatment of *Bacillus* infections. Clindamycin or vancomycin with or without an aminoglycoside may be the treatment of choice for endocarditis or sepsis with *B. cereus* (Sliman et al., 1987; Tuazon et al., 1979). Other species can be treated with penicillins or cephalosporins if they are found to be susceptible. Surgical drainage may be necessary in closed space and soft tissue infections. Infected prosthetic heart valves may need replacement but infected native valves usually can be treated with antibiotics alone.

Morbidity and mortality for disseminated infections were high prior to the antibiotic era. Except for panophthalmitis, which usually results in loss of vision and of the eye, other infections by *Bacillus* species have a good prognosis unless the underlying disease (such as malignancy) supervenes. *B. cereus* endocarditis in intravenous drug users has a relatively benign course and favorable prognosis. Of patients with meningitis, approximately 40% survived; however, most have had central nervous system sequelae.

Pathology and Pathogenesis

The pathologic changes in organs infected by nonanthrax *Bacillus* species are the inflammatory reactions caused by bacteria in general. There have not been any special or pathognomonic changes described in humans. Subcutaneous injection of *B. cereus* in mice, guinea pigs, and rabbits has caused deep, hemorrhagic, ulcerative skin lesions and death (Burdon et. al., 1967). Direct cytotoxicity of *B. cereus* culture filtrate for tissue culture cells has been noted (Bonventre, 1965). *B. cereus* culture filtrates were noted to kill mice and rats within minutes after intravenous injection (Bonventre and Eckert, 1963). Pathologically there were widespread thromboses. The filtrates were noted to be hemolytic and dermonecrotic. The pathologic changes included focal areas of coagulative necrosis and inflammatory cells in the pulmonary arterioles and capillaries. Thrombi were found in the pulmonary vessels. Because no organisms were seen in tissue sections, toxins were believed to be responsible for illness and death. This hypothesis was further supported by the finding that antitoxic serum had a protective effect.

In one case of neonatal meningitis, postmortem histopathologic examination revealed an invasive disease involving the brain parenchyma. The cellular composition of the CSF was significant for an increased number of mononuclear cells, implying a chronic infection (Patrick et al., 1989). The invasive and destructive nature of *B. cereus* has been demonstrated by microscopic examination of brain tissue in a patient with multiple brain abscesses (Jenson et al., 1989). There was extensive bacterial invasion of brain tissue and hemorrhagic necrosis.

Localized tissue infection with *B. cereus* is usually necrotizing due to the production of a potent exotoxin and usually results in rapid, fulminant tissue destruction.

Different *Bacillus* species produce a variety of enzymes including amylase, collagenase, hemolysin, lecithinase, phospholipase, protease and urease, antimicrobial substances (bacitracin, gramicidin, polymyxin, and tyrocidine), pigments, and toxins (Williams, 1981). *B. cereus* produces distinct extracellular products: β-lactamases, hemolysins, phospholipase C, and two lethal toxins (Coolbaugh and Williams, 1978; Gilbert and Kramer, 1984; Thompson et al., 1984; Turnbull et al., 1979a; Turnbull, 1981). The lethal toxins may contribute to the virulence of *B. cereus*. One has a molecular weight of approximately 50,000 and is a "loop fluid-inducing/skin test/necrotic toxin." The other toxin has a molecular weight of 55,000 to 56,000 and is cereolysin. It interacts with cholesterol as a receptor in the host cell membrane. It demonstrates vascular permeability activity which is difficult to distinguish from that of "loop fluid-inducing/skin test/necrotic toxin." Two hemolysins have been described. One has a molecular weight of 52,000 and causes immediate lysis of red blood cells (Coolbaugh and Williams, 1978). The other has a molecular weight of 31,000. *B. cereus* is not the only species that produces hemolysins; at least 18 *Bacillus* species have the capability (Bernheimer and Grushoff, 1967).

The lysins of some species resemble streptolysin O. Paper chromatography has demonstrated that hemolysins are not the same as phospholipases (Stein and Logan, 1965). Phospholipases and lecithinases may be identical, however. Phosphatidyl-choline hydrolase is also known as phospholipase C. It may have a role in ocular infections by disrupting cell membrane phospholipids after they have been exposed by the action of other toxins (Turnbull, 1981). Hemolysins may be responsible for changes in quality of drainage from infected wounds (Ihde and Armstrong, 1973).

Specimen Collection, Transport, Maintenance, and Microscopic Examination

Specimens and cultures which may contain *B. anthracis* require specific laboratory safety precautions. All work should be performed in a biological safety hood since aerosolization may carry a risk of infection by inhalation. Personnel should wear gloves while processing specimens and performing tests. Bench space should be disinfected with 1% sodium hypochlorite after work is completed and contaminated items should be autoclaved prior to disposal.

Vesicular fluid or exudate from cutaneous lesions suspected of being infected with *B. anthracis* should be sampled using dry sterile swabs (Doyle et al., 1985). Blood cultures should be taken and specimens of lymph nodes, spleen, peritoneal exudate, or other tissues obtained by biopsy or necropsy should be obtained for microscopic examination and culture. Specimens should be placed in secure sterile containers for transportation to the laboratory. The swabs can be used for inoculation of media for cultivation of the organism and for preparation of smears. Smears should be fixed for 3 to 5 min in Zenker's solution (2.5% aqueous potassium bichromate and 8% mercuric chloride) or 10% Formalin (37% formaldehyde) for 10 min to ensure inactivation of spores; even repeated passage of the slide through a flame may not kill all spores. Smears can be stained with either Gram or Giemsa stain. Microscopically, *B. anthracis* appears as a large Gram-positive rod with square ends, 1.0 to 1.3 μm wide and 3 to 10 μm long. Spores when present are oval, located centrally or subterminally, and produce little or no swelling of the cell. Capsules may often be seen in direct smears from infected tissues using capsule-staining methods, and specific fluorescent monoclonal antibody reagents may be obtained from the CDC, Atlanta, GA.

Clinical specimens such as blood, cerebrospinal fluid, and wound exudates for isolation of other *Bacillus* species do not require special handling. Prompt plating to prevent growth of contaminants is necessary. Tissue specimens should be macerated and inoculated into a peptone broth. As spores are viable for long periods, maintenance of cultures after initial isolation is usually not difficult. Preservation in sterile soil has been suggested, but nutrient agar with manganese sulfate is equally good for sporulation and maintenance and is more convenient than soil extract (Norris et al., 1981). *Bacillus* species may be stored on agar slants, frozen at $-70°$C, or freeze-dried for long-term maintenance. Repeated freezing and thawing should be avoided. Freeze-drying may alter the character of the organisms as can repeated subculturing at frequent intervals.

In cases of *B. cereus* food poisoning, food stuffs to be examined should be kept at or near 5°C until examined. They can be stored in a refrigerator for up to 48 h. If they are to be stored for longer periods of time they should be frozen. Twenty-five-gram portions of food are homogenized in Butterfield's buffer in a blender and diluted before testing (Bennett and Harmon, 1988).

Isolation and Identification

Cells of *Bacillus* organisms range from approximately 0.4 to 2 μm in width and from 3 to 12 μm in length. They are straight or slightly curved and have round or square ends. An initial Gram stain should be done on any suitable clinical material. It is the most rapid and easiest way to assess the quality of the sample and to have a preliminary categorization of the pathogen. If Gram-positive bacilli are seen, further testing will be needed to distinguish *Bacillus* species from other Gram-positive rods such as *Kurthia, Listeria, Rothia, Corynebacterium, Erysipelothrix,* and *Clostridium. Bacillus* species may at times resemble Gram-negative bacilli, especially cells from colonies that are older. In contrast to the nonfermenting Gram-negative rods, most *Bacillus* species do not usually grow on enteric agars. In addition, *Bacillus* species are susceptible to vancomycin, whereas Gram-negative nonfermenters are not, and a KOH test can be performed in which Gram-negative rods show a viscous thread (Gregersen, 1978). A colony is stirred into one or two drops of 3% KOH on a glass slide with a loop and the loop is slowly raised. Gram-negative bacteria make the KOH viscous, and a thread of this material follows the loop for 0.5 to 2 cm or more. Some *Bacillus* strains, particularly cells from older cultures that have lost their Gram positivity, may give a Gram-negative-type reaction; this test is of value for identification of *Bacillus* species only if no viscous thread is formed.

B. anthracis produces large, raised, opaque, grayish, granular colonies within 24 h when grown on ordinary laboratory media at 37°C. The borders of the colonies are irregular and may resemble a "Medusa head" or "comet tails." The cells adhere tenaciously to one another and if material is picked from the colony with an inoculating needle the remainder of the colony may assume a "stalagmite" shape.

Doyle et al. (1985) have provided a general approach for the identification of *B. anthracis.* In order of importance, they recommend seven procedures:

1. Culture the specimen on sheep blood agar. If the specimen is heavily contaminated use the selective PLET medium of Knisely, which contains polymyxin, lysozyme, EDTA, and thallous acetate. Each batch of this medium should be tested to demonstrate that it will allow growth of known strains of *B. anthracis,* while inhibiting growth of *B. cereus.*

2. Subculture the isolate on nutrient agar or BHI agar containing 0.5% sodium bicarbonate and supplemented with 0.7% bovine serum albumin, and incubated in 5% CO_2. Demonstrate the presence of a capsule by using either M'Fadyean stain or fluorescein-labelled anti-poly-D-glutamic acid. Fluorescent-antibody reagent may be used to detect organisms in tissue sections or from blood smears or culture. Fluorescein conjugates of heterologous sera and monoclonal antibody to capsule can be obtained from the CDC, Atlanta.

3. Determine the toxin-antitoxin reaction using a double diffusion assay with R-medium plus 1.5% agar (Ristroph and Ivins, 1983). A positive reaction confirms identification of *B. anthracis* but a negative reaction does not exclude it.

4. Test for susceptibility to penicillin; *B. anthracis* is the only *Bacillus* species which is consistently susceptible to penicillin G.

5. Determine the rates of hydrolysis of *p*-nitrophenyl-α-D-glucopyranoside and *p*-nitrophenyl-α-D-maltoside in the presence and absence of 1% Triton X-100. Alpha-glucosidase activity is increased in *B. anthracis* by incubation in the detergent; glucosidase activity in other members of the *B. cereus* group is diminished.

6. Test for agglutination of cells by soybean *(Glycine max)* lectin but not by the snail *Helix pomatia* lectin.

7. Determine susceptibility to gamma phage (Doyle et al., 1985; Rees et al. 1988).

The "string of pearls" test is based upon the susceptibility of *B. anthracis* to penicillin (Rees et al., 1988). A drop of young (12–18 h) broth culture is placed on two plates of nutrient agar that contains either 0.5 or 10 units of penicillin per milliliter. After incubating the inoculated media at 37° for 3 to 6 h, a cover slip is placed directly on the growth present on the medium. Microscopic examination of the underlying growth under low magnification reveals the presence of large rounded cells (or a "string of pearls"). Little or no growth should be seen on the medium containing 10 units of penicillin per milliliter. Approximately 95% of strains of *B. anthracis* give a positive test; only 8% of non-anthrax *Bacillus* strains are positive.

It is fairly easy to differentiate between *B. anthracis* and other *Bacillus* species with the exception of *B. cereus.* Table 1 shows important differential characteristics between *B. anthracis* and other pseudoanthrax bacilli. Unlike *B. anthracis, B. cereus* is usually resistant to penicillin, is resistant to gamma phage, does not en-

Table 1. Differences between *Bacillus anthracis* and the pseudoanthrax bacilli.

B. anthracis	Anthrax-like or pseudoanthrax bacilli
Nonmotile	Generally motile[a]
Capsulated	Noncapsulated[a]
Grows in long chains	Grow in short chains
No turbidity or pellicle in broth	Often turbidity and pellicle in broth
No growth on penicillin agar (10 μg/ml)	Usually good growth on penicillin agar
Inverted fir-tree growth in gelatin	Fir-tree growth absent or atypical[a]
Methylene blue reduced weakly	Methylene blue usually reduced strongly
Hemolysis of sheep cells weak	Hemolysis of sheep cells often strong
Liquefaction of gelatin slow	Liquefaction of gelatin usually rapid[a]
Lecithinase reaction weakly positive	Lecithinase reaction strongly positive[a]
Ferments salicin slowly	Often ferment salicin rapidly
Polysaccharide precipitin reaction strongly positive	Polysaccharide precipitin reaction weakly positive
Produces toxin, neutralized by *B. anthracis* antitoxin	Any toxic substances produced not neutralized by *B. anthracis* antitoxin
Pathogenic to laboratory animals	Mostly nonpathogenic. If pathogenic, produce disease unlike anthrax
Susceptible to phage	Insusceptible to phage
Culture filtrates nontoxic to tissue culture cells	Culture filtrates toxic to tissue culture cells[a]

[a]For *B. cereus*. From Reboli and Farrar (1988).

capsulate on bicarbonate agar, and does not show fluorescence of cell wall and capsule on fluorescent antibody-stained smears.

Bacillus cereus is a large-celled member of group 1. It has a cell width greater than 0.9 μm and spores which do not appreciably swell the sporangium. It is controversial whether *B. anthracis*, *B. mycoides*, and *B. thuringiensis* should have species status or be considered variant species of *B. cereus*. Within this species, strain heterogeneity is relatively common but typical characteristics are very stable. Four characteristics are useful in differentiating members of the *B. cereus* group (hemolytic activity, motility, rhizoid growth, and detection of toxin crystals) (Harmon, 1982).

Hemolytic Activity

Tryptic-soy-sheep blood agar is inoculated by lightly touching the agar surface with a loopful of culture. The plate is incubated at 30°C for 24 h and observed for hemolytic activity as indicated by a zone of complete hemolysis around the growth. *B. anthracis* is nonhemolytic; *B. cereus* is strongly hemolytic. *B. my-coides* and *B. thuringiensis* are weakly hemolytic or produce hemolysis only under the growth.

Motility Tests

Motility may be determined by direct microscopic examination or by using semisolid *B. cereus* motility medium (Harmon, 1982). The medium is inoculated by stabbing down the center and is incubated at 30°C for 18 to 24 h. Motile strains produce diffuse growth into the medium away from the stab; nonmotile strains except for *B. mycoides* grow only in and along the stab. Questionable results should be confirmed by the microscopic motility test which is performed by adding 0.2 ml of sterile distilled water to a nutrient agar slant, inoculating the slant with a loopful of culture, and incubating at 30°C for 6 to 8 h. A loopful of liquid culture from the base of the slant is suspended in a drop of water on a clean slide, covered with a cover slip, and examined immediately for motility (Bennett and Harmon, 1988). *B. cereus* and *B. thuringiensis* are motile; *B. anthracis* and the rhizoid strains of *B. mycoides* are nonmotile.

Rhizoid Growth

A pre-dried nutrient agar plate is inoculated centrally and incubated at 30°C for 1 to 2 days. If the culture is rhizoid, root-like structures will develop up to several centimeters from the point of inoculation. Rhizoid growth is characteristic only of *B. mycoides*.

Detection of Toxin Crystals

Endotoxin crystals of *B. thuringiensis* may be detected by phase contrast microscopy or by staining with 0.5% aqueous basic fuchsin or TB carbol fuchsin. The other members of the *B. cereus* group do not produce endotoxin crystals that can be detected by staining.

In certain clinical situations in which *Bacillus* infection is suspected, specimens can be suspended in distilled water or nonnutrient buffer and pasteurized (heated at 70–80°C for approximately 10 min) to kill contaminating vegetative cells. Spores will survive. Caution is needed, however, as organisms may be present only in the vegetative state and no growth will be achieved. In addition, heat treatment of spores may cause mutations. Air-drying and treatment with 50% ethanol for 1 h are also effective for selecting out sporulating organisms. Treatment with ethanol may be more desirable. Sporulation is often stimulated on esculin agar and may be facilitated on an acidified medium such as triple sugar iron agar (TSI).

Isolation is relatively easy, as most medically important *Bacillus* species grow well on blood agar or simple nutrient agar, which is composed of a meat extract in peptone with a pH of 6.8 to 7.0. Growth is sometimes improved by glucose but not by serum. Cultures should be incubated aerobically at 35 to 37°C for 24 h. Some strains require enrichment techniques using specific broth, incubation under specific conditions of pH, temperature, and salt concentration, or plating on specific agar media. Colony morphology and microscopic appearance form the basis of preliminary identification.

Nutrient Agar (pH = 6.8–7.0)

Agar	15.0 g
Peptone	5.0 g
Meat extract	3.0 g
Distilled water	To 1 liter

For sporulation, add 5 mg of hydrous manganese sulfate per 1 liter of medium.

Nutrient Broth (pH=6.8)

Peptone	5.0 g
Meat extract	3.0 g
Distilled water	To 1 liter

Colony morphology should be observed from young (18–24 h) cultures grown on nutrient agar under aerobic conditions. Single colonies are usually two to several mm in diameter and may have a finely granular appearance. Other colonies appear membranous and wrinkled (Wilson, 1983). *B. cereus* colonies vary from small, shiny, and compact to large, feathery, and spreading. On sheep blood agar, a lavender-colored colony with beta hemolysis is seen. Colonies of *B. subtilis* are usually large, dull, and flat, with a ground glass appearance. The microscopic appearance of cells is best observed on slides freshly coated with a thin layer of 2% water agar and examined with a phase-contrast microscope (Claus and Berkeley, 1986). Colonies suspected of being *B. cereus* should be subcultured and checked for motility, hemolysis, and susceptibility to penicillin. Serotyping with antibody directed against flagellar antigens should be employed for confirmation of *B. cereus*.

Four selective and differential plating media have been described for the isolation of *B. cereus* (Holbrook and Anderson, 1980; Kim and Goepfert, 1971; Mossell et al., 1967; Szabo et al., 1984). Three of them contain mannitol to enhance differentiation. The most common procedure for isolation and presumptive identification of *B. cereus* is based on the egg yolk reaction (Doyle et al., 1985). In this reaction, turbidity develops in egg yolk or in agar that contains egg yolk. It is caused by an extracellular substance (or substances) referred to as egg yolk turbidity factor, lecithinase, or phospholipase. The reaction may be due to a more com-

plex series of events than the action of a single enzyme (Kushner, 1957). Material to be cultured is placed onto blood agar (Harmon et al., 1984; Schiemann, 1978) containing polymyxin to suppress Gram-negative growth. Plates are incubated at 37°C overnight, and suspect colonies are then inoculated onto mannitol egg yolk agar by stabbing various sites on the plate and incubated at 35 to 37°C overnight. As *B. cereus* is lecithinase positive and does not ferment mannitol, plates are examined for a zone of precipitation or halos around sites of inoculation and for lack of evidence of fermentation. (If fermentation has occurred, the background color changes from violet to yellow.) Similar reactions may be seen with other members of the *B. cereus* group *(B. anthracis, B. thuringiensis,* and *B. mycoides).* Some strains of *B. cereus* give a weak or negative egg yolk reaction. On polymyxin pyruvate-egg-yolk-mannitol-bromothymol blue-agar (PEMBA) media, *B. cereus* has a distinct colonial appearance (Holbrook and Anderson, 1980). After incubation at 37°C for 24 to 48 h, colonies of *B. cereus* are rough, dry, and crenated or rhizoid. They have a distinctive bright blue color. (*B. thuringiensis* has the same appearance).

Egg-Yolk Reaction Medium

Tryptone	10 g
Disodium hydrogen phosphate	5 g
Potassium dihydrogen phosphate	1 g
Sodium chloride	2 g
Magnesium sulfate	0.1 g
Glucose	2 g
Distilled water	To 1 liter

Egg yolk (1.5 ml) is added to 100 ml of basal medium. The medium is held refrigerated overnight and the supernatant broth is aliquoted.

Measuring the Egg-Yolk Reaction

Cultures are inoculated into egg-yolk broth and into a control broth without egg yolk. After incubation for 1, 3, 5, and 7 days observe for the appearance of a heavy white precipitate in or on the surface of the egg-yolk medium.

PEMBA (Oxoid)(pH = 7.2)(Holbrook and Anderson, 1980)

Agar	18.0 g
Peptone	1.0 g
D-Mannitol	10.0 g
MgSO$_4$	0.1 g
NaCl	2.0 g
Na$_2$HPO$_4$	2.5 g
KH$_2$PO$_4$	0.25 g
Bromothymol blue	0.12 g
Distilled water	To 1 liter

Prior to use, the following are added to 90-ml aliquots of the basal medium:

| 20% (w/v) Sodium pyruvate | 5 ml |

Polymyxin	100 units/ml
Egg-yolk emulsion (Oxoid)	5 ml

Surface plating is usually employed to determine the presence of *B. cereus* in outbreak foods. The official method of the Association of Official Analytical Chemists (AOAC) for enumerating *B. cereus* in foods uses mannitol-egg yolk-polymyxin (MYP) media (Lancette and Harmon, 1980). Blood agar, PEMBA, and KG agar (Kim and Goepfert, 1971) are of similar efficacy and are used in some laboratories. If *B. cereus* counts are expected to be less than 1,000 per gram, the alternative AOAC most probable number (MPN) method using tryptic-soy-polymyxin broth is recommended (Lancette and Harmon, 1980). Presumptive colonies from the above plating media should be transferred to nutrient agar slants and subjected to biochemical tests.

The nature of members of this genus is such that strains may not exhibit the important characteristics of the genus, let alone the species. Because of the number of species and variant strains, it may not be possible to speciate quickly. Norris et al. (1981) suggest basing species identification on the results of two morphologic features, width of the rod, and parasporal body in the sporangium, and 11 tests: catalase production, Voges-Proskauer reaction, pH in Voges-Proskauer broth, growth in anaerobic agar, growth at 50°C, growth at 65°C, growth in 7% sodium chloride, production of acid and gas from glucose, reduction of nitrate to nitrite, hydrolysis of starch, and casein decomposition. Citrate utilization may not be useful because of disagreement in interpretation. Wilson (1983) advocates also testing for gelatin liquefaction, urease production, and growth in 4% sodium chloride. A scheme for identification of *Bacillus* species is provided in Table 2. Media recipes and performance of tests are based on the work of Gordon et al. (1973) and Smith et al. (1952).

Catalase Test

Cultures grown for 24 to 48 h on nutrient agar slants are flooded with 0.5 ml of 10% hydrogen peroxide. The reaction is positive if there is a rapid bubbling of gas. If no gas bubbles form, repeat, using growth on chocolate agar. The test is not for cultures grown on blood agar because of the catalase present in the red blood cells. It can be done on egg-yolk agar.

Voges-Proskauer Test Broth (pH = 6.5)

Proteose peptone	7 g
Glucose	5 g
Sodium chloride	5 g
Distilled water	To 1 liter

J broth for V-P Reaction (pH = 7.3–7.5)

Tryptone	5 g
Yeast extract	15 g
Distilled water	To 1 liter

Add aseptically glucose 5 g/liter sterilized separately.

V-P Reaction (Acetyl methyl carbinol production)

Tubes of V-P broth should be inoculated in triplicate and tested for acetyl-methyl-carbinol production after incubation for up to 7 days. Mix 3 ml of 40% (w/v) sodium hydroxide with the culture and add 0.5 to 1 mg of creatine. Observe for the production of a red color after 30 to 60 min at room temperature. Fastidious insect pathogens should be grown on J-broth medium. For pH testing, before cultures incubated for 7 days are tested for acetyl-methyl carbinol, the pH is measured.

Anaerobic Agar (pH = 7.2)

Trypticase	20 g
Glucose	10 g
Sodium chloride	5 g
Agar	15 g
Sodium thioglycolate	2 g
Sodium formaldehyde sulfoxylate	1 g
Distilled water	To 1 liter

Inoculate a tube of anaerobic agar with a small loopful of nutrient broth culture by stabbing to the bottom of the culture tube. At incubation temperatures below 45°C the growth should be recorded at 3 and 7 days.

Growth in Sodium Chloride

Tubes of nutrient broth containing 0, 4, 7, and 10% (w/v) sodium chloride are inoculated and incubated. Observe for growth during up to 14 days of incubation.

Nitrate Reduction Broth (pH = 7.0)

Peptone	5 g
Meat extract	3 g
Potassium nitrate	1 g
Distilled water	To 1 liter

Reduction of nitrate to nitrite

Grow cultures in the above broth. After 3 and 7 days incubation, test by moistening a strip of potassium iodide/starch paper with a few drops of 1 N HCl and then touching the paper with a loopful of the culture. Observe for the production of a purple color indicating the presence of nitrite and for the accumulation of nitrogen gas. For a more rapid test, nitrate broth is inoculated and incubated at 35°C for 24 h. Sulfanilic acid and alpha-naphthol (0.25 ml each) are added. The presence of nitrite is indicated by an orange color that develops within 10 min.

Starch Hydrolysis Agar

Potato starch	1 g/10 ml cold distilled water
Nutrient agar	100 ml

Starch Hydrolysis Test

Inoculate duplicate plates of starch agar and incubate. At 3 and 5 days, flood one of the plates with 95% ethanol. After 15 to 30 min the unchanged starch will become

Table 2. Characteristics useful to differentiate between some species of *Bacillus*.

Bacillus species	Rods 1.0 μm wide or wider	Parasporal bodies	Growth in anaerobic agar	Growth at 50°C	Growth at 65°C	Growth in 4% NaCl	Growth in 7% NaCl	Catalase	Voges-Proskauer reaction	Acid from glucose	Acid and gas in glucose	NO₃ reduced to NO₂	Starch hydrolyzed	pH in V-P medium <6.0	Hydrolysis of casein	Lecithinase
anthracis	+	−	+	−	−	+	+	+	+	+	−	+	−	+	+	+
megaterium	+	V	−	−	−	ND	+	+	−	+	−	V	−	V	+	−
cereus	+	+	+	−	−	ND	+	+	+	+	−	+	+	+	+	+
thuringiensis	+	+	+	+	−	+	+	+	+	+	−	+	+	+	+	+
lichenformis	−	−	+	+	−	+	+	+	+	+	−	+	+	V	+	−
subtilis	−	−	+	+	−	+	+	+	+	+	−	+	+	V	+	−
pumilus	−	−	−	+	−	+	+	+	−	+	−	−	−	+	+	−
firmus	−	−	+	+	−	−	−	+	+	+	−	+	+	−	V	−
coagulans	V	−	+	−	−	−	−	+	+	+	−	V	+	+	+	−
polymyxa	−	−	+	+	−	V	−	+	+	+	+	+	+	V	−	−
macerans	−	−	+	+	−	V	V	+	−	+	+	+	+	−	V	−
circulans	−	−	V	+	−	V	−	+	−	+	−	V	+	V	V	−
stearothermophilus	V	−	−	+	+	−	−	V	−	+	−	V	−	+	+	−
alvei	V	+	+	−	−	V	−	+	+	+	−	+	+	+	+	−
laterosporus	−	−	+	+	−	V	−	+	−	+	−	−	+	−	+	+
brevis	V	+	−	+	−	V	V	+	−	+	−	+	+	−	V	−
sphaericus	−	−	−	−	−	−	+	+	−	−	−	V	−	−	+	ND
larvae	−	−	+	−	−	V	+	−	+	+	−	−	−	−	−	ND
popilliae	−	+	+	−	−	+	−	−	−	+	−	V	−	−	−	ND
lentimorbus	−	−	+	−	−	+	−	−	−	+	−	−	−	−	−	−
pasteurii	−	−	+	V	−	−	+	ND	−	ND	−	+	−	ND	V	−
pantothenticus	−	−	+	−	−	+	+	+	−	+	−	V	−	+	V	−

ª+, greater than 85% of strains positive; −, greater than 85% of strains negative; V = variable character; ND, no data available. From Reboli and Farrar (1988).

white and opaque. Observe for a clear zone underneath and around the growth as an indicator of hydrolysis.

Casein Decomposition; Milk Agar

Agar   1 g in 50 ml of distilled water
Skim milk powder   5 g in 50 ml of distilled water

Casein Decomposition Test

Inoculate plates of milk agar with one streak of inoculum and examine after incubation at 7 and 14 days for clearing of the casein around and underneath the growth.

Citrate Utilization; Medium (pH = 6.8)

Trisodium citrate	1 g
Magnesium sulfate	1.2 g
Diammonium hydrogen phosphate	0.5 g
Potassium chloride	1 g
Agar	15 g
Distilled water	920 ml
Phenol red (0.04% w/v)	20 ml
Trace element solution*	40 ml

*$FeSO_4$, 200 mg; $ZnSO_4$, 10 mg; $MnCl_2$, 3 mg; H_3BO_3, 30 mg; $CoCl_2$, 20 mg; $CuCl_2$, 1 mg; $NiCl_2$, 2 mg; Na_2MoO_4, 3 mg; ethylene diamine tetraacetate, 500 mg; distilled water, 1 liter.

Citrate Utilization Reaction

Inoculate slants of citrate medium and incubate up to 14 days. Observe for production of a red (alkaline) color indicating utilization of organic acids.

Gelatin Liquefaction; Nutrient gelatin (pH = 7.0)

Use commercial nutrient gelatin medium or plain gelatin.

Gelatin	120 g
Distilled water	1 liter

As an alternative, nutrient agar supplemented with 0.4% gelatin may be used.

Gelatin Liquefaction Test

Inoculate tubes of nutrient gelatin and incubate at 28°C. Observe for liquefaction at 3 to 4 day intervals for 4 weeks. Before examination, hold the tubes at 20°C for about 4 h to allow unchanged gelatin to harden. If the gelatin does not harden at 20°C, liquefaction has occurred.

Lysozyme Resistance; Medium

Nutrient broth	99 ml
Lysozyme solution (10,000 enzyme units/ ml of distilled water)	1 ml

Lysozyme Test

Inoculate a loopful of a broth culture into a tube of the above medium and into a control tube of nutrient broth. After incubation for up to 7 to 14 days observe for growth or its absence. Members of the *B. cereus* group are resistant to lysozyme.

Tyrosine Decomposition; Agar

L-Tyrosine	0.5 g
Distilled water	10 ml
Sterile nutrient agar	100 ml

Tyrosine Test

Inoculate plates of the above agar with one streak of inoculum and incubate. Observe at up to 21 days for clearing of the tyrosine crystals around and below the growth. *B. cereus* and other members of the *B. cereus* group except for *B. anthracis* readily decompose tyrosine.

Phage typing using bacteriophage gamma has been used for typing *B. anthracis, B. thuringiensis, B. sphaericus,* and *B. stearothermophilus.*

Serologic tests (precipitation and complement fixation) have been used to separate *B. anthracis* from other *Bacillus* species, but antisera to *B. anthracis* react to some extent with other bacilli, limiting their usefulness. Agglutination tests have been found to be of little value, as these bacteria have the tendency to agglutinate spontaneously (Sievers and Zetterberg, 1940).

Bacillus anthracis and other *Bacillus* species can be differentiated by lectin agglutination assays (Cole et al., 1984). The agglutination can be carried out on a microscope slide. *B. anthracis* and *B. mycoides* give a positive agglutination with lectin from soybean *(Glycine max).* Agglutination with lectin from the snail *Helix pomatia* is positive only for *B. mycoides.* Spores have lectin receptors, and heating at 100°C decreases the agglutination.

A comparison of gas-liquid chromatography, DNA-DNA hybridization, API systems, and classic biochemical tests for speciation of some *Bacillus* species found API systems to be superior (O'Donnell et al., 1980). This impression was confirmed by Logan and Berkeley (1984) in their study of 600 *Bacillus* strains. API systems supplemented with spore examination, motility testing, and gas from carbohydrate fermentation were more rapid, accurate, and reproducible than classic biochemical tests. API and phage sensitivity testing are superior to other methods for distinguishing partially virulent and avirulent strains of *B. anthracis* from closely related species (Logan et al., 1985a).

Literature Cited

Agarwala, B., Baffes, T., and Grossman, A. 1975. Bacterial endocarditis following open heart surgery. Illinois Med. J. 148:38–39.

Allen, B. T., and Wilkinson, H. A. 1969. A case of meningitis and generalized Schwartzman reaction caused by *Bacillus sphaericus*. Johns Hopkins Med. J. 125:8–13.

Arya, L. S., Saidali, A., Qureshi, M. A., and Singh, M. 1982. Anthrax in infants and children in Afghanistan. Indian J. Pediatr. 49:529–534.

Attwood, A. I., and Evans, D. M. 1983. *Bacillus cereus* infection in burns. Burns Incl. Therm. Inj. 9:355–357.

Banerjee, C., Bustamante, C. I., Wharton, R., Talley, E., and Wade, J. C. 1988. *Bacillus* infections in patients with cancer. Arch Intern. Med. 148:1769–1774.

Bekemeyer, W. B., and Zimmerman, G. A. 1985. Life threatening complications associated with *Bacillus cereus* pneumonia. Am Rev. Respir. Dis. 131:466–469.

Bennett, R. W., and Harmon, S. M. 1988. *Bacillus cereus* food poisoning, p. 83–93. In: Balows, A., Hausler, W. J., and Lennette, E. H. (ed.), Laboratory diagnosis of infectious diseases: principles and practice. Springer-Verlag, New York.

Berger, S. A. 1983. Pseudobacteremia due to contaminated alcohol swabs. J. Clin. Microbiol. 18:974–975.

Bernheimer, A. W., and Grushoff, P. 1967. Extracellular hemolysins of aerobic sporogenic bacilli. J. Bacteriol. 93:1541–1543.

Bhat, P., Mohan, D. N., and Srinivasa, H. 1985. Intestinal anthrax with bacteriological investigations. J. Infect. Dis. 152:1357–1358.

Bhatnagar, R., Singh, Y., Leppla, S. H., and Friedlander, A. M. 1989. Calcium is required for the expression of anthrax lethal toxin activity in the macrophagelike cell line J774A1. Infect. Immun. 57:2107–2114.

Bias, W. J. 1927. A case of pathogenicity of *Bacillus subtilis*. J. Infect. Dis. 41:313–315.

Blakey, L. J., and Priest, F. G. 1980. The occurrence of *Bacillus cereus* in some dried foods including pulses and cereals. J. Appl. Bacteriol. 48:297.

Blaustein, R. O., Koehler, T. M., Collier, R. J., and Finkelstein, A. 1989. Anthrax toxin: Channel-forming activity of protective antigen in planar phospholipid bilayers. Proc. Natl. Acad. Sci. USA 86:2209–2213.

Block, C. S., Levy, M. L., and Fritz, V. U. 1978. *Bacillus cereus* endocarditis. A case report. S. Afr. Med J. 53:556–557.

Bonventre, P. F. 1965. Differential cytotoxicity of *Bacillus anthracis* and *Bacillus cereus* culture filtrates. J. Bacteriol. 90:284–285.

Bonventre, P. F., and Eckert, N. J. 1963. The biologic activities of *Bacillus anthracis* and *Bacillus cereus* culture filtrates. Am. J. Pathol. 43:201–212.

Bouza, E., Grant, S., Jordan, M. C., Yook, R. H., and Sulit, H. L. 1979. *Bacillus cereus* endogenous panophthalmitis. Arch. Ophthalmol. 97:498–499.

Boyette, D. P., and Rights, F. L. 1952. Heretofore undescribed aerobic sporeforming *Bacillus* in child with meningitis. J. Am. Med. Assoc. 148:1223–1224.

Brachman, P. S. 1980. Inhalation anthrax. Ann. N.Y. Acad. Sci. 353:83–93.

Burdon, K. L., Davis, J. S., and Wende, R. D. 1967. Experimental infection of mice with *Bacillus cereus*: studies of pathogenesis and pathologic changes. J. Infect. Dis. 117:307–316.

Carter, K. C. 1988. The Koch-Pasteur dispute on establishing the cause of anthrax. Bull. Hist. Med. 62:42–57.

Cartwright, M. E., McChesney, A. E., and Jones, R. L. 1987. Vaccination-related anthrax in three llamas. J. Am. Vet. Med. Assoc. 101:715–716.

Chandramukhi, A., Shankar, P., Rao, T. V., Sundararajan, S., and Swamy, H. S. 1983. Acute leptomeningitis due to *Bacillus anthracis*. A case report. Trop. Geogr. Med. 35:79–82.

Claus, D., and Berkeley, R. C. W. 1986. Genus *Bacillus*, p. 1105–1139. In: Sneath, P. H. A., Mair, N. S., Sharpe, M. E., and Holt, J. G. (ed.), Bergey's manual of systematic bacteriology, 1st ed., vol. 2. William & Wilkins Co., Baltimore.

Cole, H. B., Ezzell, J. W., Keller, K. F., and Doyle, R. J. 1984. Differentiation of *Bacillus anthracis* and other *Bacillus* species by lectins. J. Clin. Microbiol. 19:48–53.

Colpin, G. G. D., Guiot, H. F. L., Simonis, R. F. A., and Zwaan, F. E. 1981. *Bacillus cereus* meningitis in a patient under gnotobiotic care. Lancet 2:694–695.

Coolbaugh, J. C., and Williams, R. P. 1978. Production and characterization of two hemolysins of *Bacillus cereus*. Can. J. Microbiol. 24:1289–1295.

Coonrod, J. D., Leadley, P. J., and Eickhoff, T. C. 1971a. *Bacillus cereus* pneumonia and bacteremia. Am. Rev Respir. Dis. 103:711–714.

Coonrod, J. D., Leadley, P. J., and Eickhoff, T. C. 1971b. Antibiotic susceptibility of *Bacillus* species. J. Infect. Dis. 123:102–105.

Cotton, D. J., Gill, V. J., Marshall, D. J., Gress, J., Thaler, M., and Pizzo, P. A. 1987. Clinical features and therapeutic interventions in 17 cases of *Bacillus* bacteremia in an immunosuppressed patient population. J. Clin. Microbiol. 25:672–674.

Cox, R., Sockwell, G., and Landers, B. 1959. *Bacillus subtilis* septicemia. Report of a case and review of the literature. N. Engl. J. Med 261:894–896.

Craig, C. P., Lee, W. S., and Ho, M. 1974. *Bacillus cereus* endocarditis in an addict (letter). Ann. Intern. Med. 80:418.

Crowle, A. J. 1958. A simplified microdouble-diffusion agar precipitin technique. J. Lab. Clin. Med. 52:784–787.

Crowley, M. M., Shannon, R., and Spivak, M. 1983. Pseudobacteremia due to intrinsic contamination of blood culture media by *Bacillus* species (abstract). Am. J. Infect. Control. 11:150.

Curtis, J. R., Wing, A. J., and Coleman, J. C. 1967. *Bacillus cereus* bacteremia: a complication of intermittent hemodialysis. Lancet 1:136–138.

Dalton, H. P., and Allison, M. S. 1967. Etiology of bacteremia. Appl. Microbiol. 15:808–814.

Davaine, C. 1863a. Recherches sur les infusoires du sang dans la maladie connue sous le nom de *sang de rate*. Comptes Rendus de l'Academie des Sciences. 57:220–223.

Davaine, C. 1863b. Nouvelles recherches sur les infusoires du sang dans la maladie connue sous le nom de *sang de rate*. Comptes Rendus de l'Academie des Sciences. 57:351–53, 386–387.

Davaine, C. 1864. Nouvelles recherches sur la nature de la maladie charbonneuse connue sous le nom de *sang de rate*. Comptes Rendus de l'Academie des Sciences. 59:393–396.

Davaine, C., and Raimbert, L-A. 1864. Sur la presence des bacteridies dans la pustule maligne chez l'homme. Comptes Rendes de l'Academie des Sciences. 59:429–431.

Davey, R. T., and Tauber, W. B. 1987. Posttraumatic endophthalmitis: the emerging role of *Bacillus cereus* infection. Rev. Infect. Dis. 9:110–123.

Davies, J. C. A. 1982. A major epidemic of anthrax in Zimbabwe. Cent. Afr. J. Med. 28:291–298.

Davies, J. C. A. 1983. A major epidemic of anthrax in Zimbabwe. Cent. Afr. J. Med. 29:8–12.

Davies, J. C. A. 1985. A major epidemic of anthrax in Zimbabwe. The experience at the Beatrice Road Infectious Diseases Hospital, Harare. Cent. Afr. J. Med. 31:176–180.

DeBuono, B. A., Brondum, J., Kramer, J. M., Gilbert, R. J., and Opal, S. M. 1988. Plasmid, serotypic, and enterotoxin analysis of *Bacillus cereus* in an outbreak setting. J. Clin. Microbiol. 26:1571–1574.

Dirckx, J. H. 1981. Virgil on anthrax. Am. J. Dermatopathol. 3:191–195.

Doganay, M., Almac, A., and Hanagasi, R. 1986. Primary throat anthrax. A report of six cases. Scand. J. Infect. Dis. 18:415–419.

Doyle, R. J., Keller, K. F., and Ezzell, J. W. 1985. *Bacillus,* p. 211–215. In: Lennette, E. H., Balows, A., Hausler, W. J., and Shadomy, H. J. (ed.), Manual of clinical microbiology, 4th ed. American Society of Microbiology, Washington, D.C.

Dryden, M. S., and Kramer, J. M. 1987. Toxigenic *Bacillus cereus* as a cause of wound infections in the tropics. J. Infect. 15:207–212.

Dubos, R. 1971. Toxic factors in enzymes used in laundry products. Science 173:259–260.

Dürst, U. N., Bartenstein, J., Bühlmann, H., Wüst, J., and Spiegel, M. V. 1986. Anthraxmeningitis. Schweiz. Med. Wochenschr. 116:1222–1228.

Dutz, W., and Kohout-Dutz, E. 1981. Anthrax. Int. J. Dermatol. 20:203–206.

Escuyer, V., Duflot, E., Sezer, O., Danchin, A., and Mock, M. 1988. Structural homology between virulence-associated bacterial adenylate cyclases. Gene 71:293–298.

Ezzell, J. W., and Abshire, T. G. 1988. Immunological analysis of cell-associated antigens of *Bacillus anthracis.* Infect. Immun. 56:349–356.

Ezzell, J. W., Ivins, B. E., and Leppla, S. H. 1984. Immunoelectrophoretic analysis, toxicity and kinetics of in vitro production of the protective antigen and lethal factor components of *Bacillus anthracis* toxin. Infect. Immun. 45:761–767.

Farrar, W. E., Jr. 1963. Serious infections due to "non-pathogenic" organisms of the genus *Bacillus.* Review of their status as pathogens. Am. J. Med. 34:134–141.

Feder, H. M., Garibaldi, R. A., Nurse, B. A., and Kurker, R. 1988. *Bacillus* species isolates from cerebrospinal fluid in patients without shunts. Pediatrics 82:909–913.

Fitzpatrick, D. J., Turnbull, P. C., Keane, C. T., and English, L. F. 1979. Two gas-gangrene-like infections due to *Bacillus cereus.* Br. J. Surg. 66:577–579.

Flindt, M. L. H. 1969. Pulmonary disease due to inhalation of derivatives of *Bacillus subtilis* containing proteolytic enzyme. Lancet 1:1177–1181.

Ghosh, A. C. 1978. Prevalence of *Bacillus cereus* in the feces of healthy adults. J. Hyg. (London). 80:233–236.

Gianella, R. A., and Brasile, L. 1979. A hospital food-borne outbreak of diarrhea caused by *Bacillus cereus:* Clinical, epidemiologic and microbiologic studies. J. Infect. Dis. 139:366–370.

Gilbert, R. J., and Kramer, J. M. 1984. *Bacillus cereus* enterotoxins: Present status. Biochem. Soc Trans. 12:198–200.

Gilbert, R. J., and Parry, J. M. 1977. Serotypes of *Bacillus cereus* from outbreaks of food poisoning and from routine foods. J. Hyg. 78:69–74.

Gilbert, R. J., Stringer, M. F., and Peace, J. C. 1974. The survival and growth of *Bacillus cereus* in boiled and fried rice in relation to outbreaks of food poisoning. J. Hyg. Camb. 73:433–444.

Glassman, H. N. 1958. World incidence of anthrax in man. Public Health Rep. 73:22–24.

Gordon, R. E., Haynes, W. C., and Pang, C. H. N. 1973. The genus *Bacillus.* USDA Agriculture Handbook, No. 427, Washington DC.

Gordon, V. M., Leppla, S. H., and Hewlett, E. L. 1988. Inhibitors of receptor-mediated endocytosis block the entry of *Bacillus anthracis* adenylate cyclase toxin but not that of *Bordetella pertussis* adenylate cyclase toxin. Infect. Immun. 56:1066–1069.

Green, B. D., Battisti, L., Koehler, T. M., Thorne, C. B., and Ivins, B. E. 1985. Demonstration of a capsule plasmid in *Bacillus anthracis.* Infect. Immun. 49:291–297.

Greenberg, M., Milne, J. F., and Watt, A. 1970. Survey of workers exposed to dusts containing derivatives of *Bacillus subtilis.* Br. Med J. 2:629–633.

Gregersen, T. 1978. Rapid method for distinction of Gram-negative from Gram-positive bacteria. Eur. J. Appl. Microbiol. Biotechnol. 5:123–127.

Groschel, D., Burgess, M. A., and Bodey, G. P. 1976. Gas gangrene-like infection with *Bacillus cereus* in a lymphoma patient. Cancer 37:988–991.

Guiot, H. F. L., DePlanque, M. M. Richel, D. J., and van't Wout, J. W. 1986. *Bacillus cereus:* a snake in the grass for granulocytopenic patients. J. Infect. Dis. 153:1186.

Gurevich, I., Tafuro, P., Krystofiak, S. P., Kalter, R. D., and Cunha, B. A. 1984. Three clusters of *Bacillus* pseudobacteremia related to a radiometric blood culture analyzer. Infect. Control. 5:71–74.

Hambleton, P., Carman, J. A., and Melling, J. 1984. Anthrax: the disease in relation to vaccines. Vaccine 2:125–132.

Harmon, S. M. 1982. New methods for differentiating members of the *Bacillus cereus* group: collaborative study. J. Assoc. Off. Anal. Chem. 65:1133–1139.

Harmon, S. M., Kautter, D. A., and McClure, F. D. 1984. Comparison of selective plating media for enumeration of *Bacillus cereus* in foods. J. Food Prot. 47:65–67.

Hauge, S. 1955. Food poisoning caused by aerobic spore-forming bacilli. J. Appl. Bacteriol. 18:591–595.

Henrickson, K. J., Flynn, P. M., Shenep, J. L., and Pui, C. H. 1989. Primary cutaneous *Bacillus cereus* infection in neutropenic children. Lancet i: 601–603.

Holbrook, R., and Anderson, J. M. 1980. An improved selective and diagnostic medium for the isolation and enumeration of *Bacillus cereus* in foods. Can. J. Microbiol. 26:753–759.

Holmes, J. R., Plunkett, T., Pate, P., Roper, W. L., and Alexander, J. 1981. Emetic food poisoning caused by *Bacillus cereus.* Arch. Intern. Med. 141:766–767.

Hunter, L., Corbett, W., and Grindem, D. 1989. Anthrax. J. Am. Vet. Med. Assoc. 194:1028–1031.

Ihde, C., and Armstrong, D. 1973. Clinical spectrum of infection due to *Bacillus* species. Am J. Med 55:839–845.

Isaacson, P., Jacobs, P. H., Mackenzie, A. M. R., and Matthews, A. W. 1976. Pseudotumor of the lung caused by infection with *Bacillus sphaericus.* J. Clin. Pathol. 29:806–811.

Ivins, B. E., Ezzell, J. W., Jenski, J., Hedlund, K. W., Ristroph, J. D., and Leppla S. H. 1986. Immunization studies with attenuated strains of *Bacillus anthracis*. Infect. Immun. 52:454–458.

Ivins, B. E., and Welkos, S. L. 1986. Cloning and expression of the *Bacillus anthracis* protective antigen gene in *Bacillus subtilis*. Infect. Immun. 54:537–542.

Ivins, B. E., and Welkos, S. L. 1988. Recent advances in the development of an improved, human anthrax vaccine. Eur. J. Epidemiol. 4:12–19.

Ivins, B. E., Welkos, S. L., Knudson, G. B., and LeBlanc, D. J. 1988. Transposon Tn916 mutagenesis in *Bacillus anthracis*. Infect. Immun. 56:176–181.

Jaruratanasirikul, S., Kalnauwakul, S., and Lekhakula, A. 1987. Traumatic wound infection due to *Bacillus cereus* in an immunocompromised patient: a case report. Southeast Asian J. Trop. Med. Public Health. 18:112–114.

Jena, G. P. 1980. Intestinal anthrax in man. A case report. Centr. Afr. J. Med. 26:253–254.

Jenson, H. B., Levy, S. R., Duncan, C., and McIntosh, S. 1989. Treatment of multiple brain abscesses caused by *Bacillus cereus*. Pediatr. Infect. Dis. J. 8:795–798.

Jephcott, A. E., Barton, B. W., Gilbert, R. J., and Shearer, C. W. 1977. An unusual outbreak of food-poisoning associated with meals-on-wheels. Lancet ii: 129.

Johnson, C. L., Bernstein, I. L., Gallagher, J. S., Bonventre, P. F., and Brooks, S. M. 1980. Familial hypersensitivity pneumonitis induced by *Bacillus subtilis*. Am. Rev. Respir. Dis. 122:339–348.

Johnson, D. A., Aulicino, P. L., and Newby, J. G. 1984. *Bacillus cereus* induced myonecrosis. J. Trauma. 24:267–270.

Jonsson, S., Clarridge, J., and Young, E. J. 1983. Necrotizing pneumonia and empyema caused by *Bacillus cereus* and *Clostridium bifermentans*. Am. Rev. Respir. Dis. 127:357–359.

Kaspar, R. L., and Robertson, D. L. 1987. Purification and physical analysis of *Bacillus anthracis* plasmids pX01 and pX02. Biochem. Biophys. Res. Commun. 149:362–368.

Keppie, J., Smith, H., and Harris-Smith, P. W. 1955. The chemical basis of the virulence of *Bacillus anthracis*. III: The role of the terminal bacteremia in death of guinea-pigs from anthrax. Br. J. Exp. Pathol. 36:315–322.

Kerkenezov, N. 1953. Panophthalmitis after a blood transfusion. Br. J. Ophthalmol. 37:632–636.

Kim, H. U., and Goepfert, J. M. 1971. Enumeration and identification of *Bacillus cereus* in foods: 24-hour presumptive test medium. Appl. Microbiol. 22:581–587.

Knudson, G. B. 1986. Treatment of anthrax in man: History and current concepts. Milit. Med. 151:71–77.

Koch, R. 1877. Die Ätiologie der Milzbrandkrankheit, begründet auf die Entwicklungsgeschichte des *Bacillus anthracis*. Cohns Beiträge zur Biologie der Pflanzen. 2:277–310.

Koshi, G., Lalitha, M. K., Daniel, J., Chacko, A., and Pulimood, B. M. 1981. Anthrax meningitis, a clinical entity. J. Assoc. Physicians India 29:59–62.

Kotin, P. 1952. Techniques and interpretation of routine blood cultures: observations in 5000 consecutive patients. J. Am. Med. Assoc. 149:1273–1276.

Kushner, D. J. 1957. An evaluation of the egg yolk reaction as a test for lecithinase activity. J. Bacteriol. 73:297–302.

Lancette, G. A., and Harmon, S. M. 1980. Enumeration and confirmation of *Bacillus cereus* in foods: collaborative study. J. Assoc. Off. Anal. Chem. 63:581–586.

Leff, A., Jacobs, R., Gooding, V., Hauch, J., Conte, J., and Stulbarg, M. 1977. *Bacillus cereus* pneumonia. Survival in a patient with cavitary disease treated with gentamicin. Am. Rev. Respir. Dis. 115:151–154.

Leppla, S. H. 1982. Anthrax toxin edema factor: A bacterial adenylate cyclase that increases cyclic AMP concentrations in eukaryotic cells. Proc. Natl. Acad. Sci. USA. 79:3162–3166.

Leppla, S. H. 1984. *Bacillus anthracis* calmodulin-dependent adenylate cyclase: chemical and enzymatic properties and interactions with eucaryotic cells. Adv. Cyclic Nucleotide Protein Phosphorylation Res. 17:189–198.

Leppla, S. H. 1988. Production and purification of anthrax toxin. Methods Enzymol. 165:103–116.

Lettau, L. A., Benjamin, D., Cantrell, H. F., Potts, D. W., and Boggs, J. M. 1988. *Bacillus species* pseudomeningitis. Infect. Control Hosp. Epidemiol. 9:394–397.

Lincoln, R. E., Klein, F., Walker, J. S., Haines, B. W., Jones, W. I., Mahlandt, B. G., and Friedman, R. H. 1964. Successful treatment of Rhesus monkeys for septicemic anthrax. Antimicrob. Agents Chemother. pp. 759–763.

Little, S. F., and Knudson, G. B. 1986. Comparative efficacy of *Bacillus anthracis* live spore vaccine and protective antigen vaccine against anthrax in the guinea pig. Infect. Immun. 52:509–512.

Logan, N. A., and Berkeley, R. C. W. 1984. Identification of *Bacillus* strains using the API system. J. Gen. Microbiol. 130:1871–1882.

Logan, N. A., Carman, J. A., Melling, J., and Berkeley, R. C. W. 1985a. Identification of *Bacillus anthracis* by API tests. J. Med. Microbiol. 20:75–85.

Logan, N. A., Old, D. C., and Dick, H. M. 1985b. Isolation of *Bacillus circulans* from a wound infection. J. Clin. Pathol. 38:838–839.

MacDonald, N. 1982. Investigation of an outbreak of pseudobacteremia attributed to *Bacillus* species in a general hospital, p.83. In: Abstracts, 82nd Annual Meeting of the American Society for Microbiology, Atlanta, March 7–12, 1982.

Marshall, E. 1988. Sverdlovsk: anthrax capital. Science 240:383–385.

McKendrick, D. R. A. 1980. Anthrax and its transmission to humans. Cent. Afr. J. Med. 26:126–129.

Melling, J., Capel, B. J., Turnbull, P. C. B., and Gilbert, R. J. 1976. Identification of a novel enterotoxigenic activity associated with *Bacillus cereus*. J. Clin. Pathol. 29:938–940.

Middlebrook, J. L., and Dorland, R. B. 1984. Bacterial toxins: cellular mechanisms of action. Microbiol. Rev. 48:199–221.

Mikesell, P., Ivins, B. E., Ristroph, J. D. and Dreier, T. M. 1983. Evidence for plasmid-mediated toxin production in *Bacillus anthracis*. Infect. Immun. 39:371–376.

Mock, M., Labruyere, E., Glasser, P., Danchin, A., and Ullmann, A. 1988. Cloning and expression of the calmodulin-sensitive *Bacillus anthracis* adenylate cyclase in *Escherichia coli*. Gene 64:277–284.

Morita, T. N., and Woodburn, M. J. 1977. Stimulation of *Bacillus cereus* growth in cooked rice combinations. J. Food Sci. 42:1232–1235.

Mortimer, P. R., and McCann, G. 1974. Food-poisoning episodes associated with *Bacillus cereus* in fried rice. Lancet i:1043–1045.

Mossell, D. A. A., Koopman, M. J., and Jongerius, E. 1967. Enumeration of *Bacillus cereus* in foods. Appl. Microbiol. 15:650–653.

Nalin, D. R., Sultana, B., Sahunja, R., Islam, A. K., Rahim, M. A., Islam, M., Costa, B. S., Mawla, N., and Greenough, W. B. 1977. Survival of a patient with intestinal anthrax. Am. J. Med. 62:130–132.

Ndyabahinduka, D. G. K., Chu, I. H., Abdou, A. H., and Gaifuba, J. K. 1984. An outbreak of human gastrointestinal anthrax. Annali del Instituto Superiore di Sanita. 20:205–208.

Noble, R. C., and Reeves, S. A. 1974. *Bacillus* species pseudosepsis caused by contaminated commercial blood culture media. J. Am. Med. Assoc. 230:1002–1004.

Norris, J. R., Berkeley, R. C. W., Logan, N. A., and O'Donnell, A. G. 1981. The genera *Bacillus* and *Sporolactobacillus*, p. 1711–1742. In: Starr, M., Stolp, N., Truper, H., Balows, A., Schlegel, H., (ed.), The prokaryotes: a handbook on habitats, isolation, and identification, vol. 2. Springer-Verlag, Berlin.

O'Brien, J., Friedlander, A., Dreier, T., Ezzell, J., and Leppla, S. 1985. Effects of anthrax toxin components on human neutrophils. Infect. Immun. 47:306–310.

O'Day, D. M., Smith, R. S., Gregg, C. R., Turnbull, P. C. B., Head, W. S., Ives, J. A., and Ho, P. C. 1981. The problem of *Bacillus* species infection with special emphasis on the virulence of *Bacillus cereus*. Ophthalmology (Rochester) 88:833–838.

O'Donnell, A. G., Norris, J. R., Berkeley, R. C. W., Claus, D., Kaneko, T., Logan, N. A., and Nozaki, R. 1980. Characterization of *Bacillus subtilis, Bacillus pumilus, Bacillus licheniformis,* and *Bacillus amyloliquefaciens* by pyrolysis gas-liquid chromatography, deoxyribonucleic acid-deoxyribonucleic acid hybridization, biochemical tests and API systems. Int. J. Syst. Bacteriol. 30:448–459.

Oster, H. A., and Kong, T. Q. 1982. *Bacillus cereus* endocarditis involving a prosthetic valve. South. Med. J. 75:508–509.

Park, S. J., Chong, Y., and Lee, S. Y. 1976. Bacterial and fungal species isolated from cerebrospinal fluid in the past five years. Korean J. Pathol 10:137–142.

Parry, J. M., and Gilbert, R. J. 1980. Studies on the heat resistance of *Bacillus cereus* spores and growth of the organism in boiled rice. J. Hyg. (London). 84:77–82.

Pasteur, L., Chamberland, C. E., and Roux, P. T. E. 1880. Sur l'etiologie du charbon. Comptes Rendus de l'Academie des Sciences. 91:86–94.

Patrick, C. C., Langston, C., and Baker, C. J. 1989. *Bacillus* species infections in neonates. Rev. Infect. Dis. 11:612–615.

Pearson, H. E. 1970. Human infections caused by organisms of the *Bacillus* species. Am. J. Clin. Pathol. 53:506–515.

Pennington, J. E., Gibbons, N. D., Strobeck, J. E., Simpson, G. L., and Myerowitz, R. L. 1976. *Bacillus* species infection in patients with hematologic neoplasia. J. Am. Med. Assoc. 235:1473–1474.

Pepys, J., Hargreave, F. E., Longbottom, J. L., and Faux, J. 1969. Allergic reactions of the lungs to enzymes of *Bacillus subtilis*. Lancet 1:1181–1184.

Reboli, A. C., Bryan, C. S., and Farrar, W. E. 1989. Bacteremia and infection of a hip prosthesis caused by *Bacillus alvei*. J. Clin. Microbiol. 27:1395–1396.

Reboli, A., and Farrar, W. E. 1988. Infections due to nonanthrax *Bacillus* species, *Kurthia* and *Rothia*, p. 69–82. In: Balows, A., Hausler, W. J., Ohashi, M., and Turano, A., (ed.), Laboratory diagnosis of infectious diseases: principles and practice, vol. 1. Bacterial, mycotic, and parasitic diseases. Springer-Verlag, New York.

Rees, H. B., Martin, D. D., and Smith, M. A. 1988. Anthrax, p. 57–68. In: Balows, A., Hausler, W. J., Ohashi, M., and Turano, A., (ed.), Laboratory diagnosis of infectious diseases: principles and practices, vol. 1. Bacterial, mycotic and parasitic diseases. Springer-Verlag, New York.

Reller, L. B. 1973. Endocarditis caused by *Bacillus subtilis*. Am. J. Clin. Pathol. 60:714–718.

Ristroph, J. D., and Ivins, B. E. 1983. Elaboration of *Bacillus anthracis* antigens in a new, defined culture medium. Infect. Immun. 39:483–486.

Robertson, D. L., and Leppla, S. H. 1986. Molecular cloning and expression in *Escherichia coli* of the lethal factor gene of *Bacillus anthracis*. Gene 44:71–78.

Robertson, D. L., Tippetts, M. T., and Leppla, S. H. 1988. Nucleotide sequence of the *Bacillus anthracis* edema factor gene *(cya):* a calmodulin-dependent adenylate cyclase. Gene 73:363–371.

Robinson, S. C. 1979. *Bacillus cereus* septic arthritis following arthrography. Clin. Orthop. 145:237–238.

Rutala, W. A., Saviteer, S. M., Thomann, C. A., and Wilson, M. B. 1986. Plaster-associated *Bacillus cereus* wound infection. A case report. Orthopedics. 9:575–577.

Sabath, L. D., and Abraham, E. P. 1965. Cephalosporinase and penicillinase activity of *Bacillus cereus*. Antimicrob. Ag. Chemother. p. 392–397.

Samad, M. A., and Hoque, M. E. 1986. Anthrax in man and cattle in Bangladesh. J. Trop. Med. Hyg. 89:43–45.

Samples, J. R., and Buettner, H. 1983. Ocular infection caused by a biological insecticide. J. Infect. Dis. 148:614.

Sathmary, M. N. 1958. *Bacillus subtilis* septicemia and generalized aspergillosis in a patient with acute myeloblastic leukemia. N.Y. State J. Med 58:1870–1876.

Schiemann, D. A. 1978. Occurrence of *Bacillus cereus* and bacteriological quality of Chinese take-out food. J. Food Prot. 41:450–454.

Seboxa, T., and Goldhagen, J. 1988. Anthrax in Ethiopia. Trop. Geogr. Med. 41:108–112.

Shamsuddin D., Tuazon, C. U., Levy, C., and Curtin, J. 1982. *Bacillus cereus* panophthalmitis: Source of the organism. Rev. Infect. Dis. 4:97–103.

Sievers, O., and Zetterberg, B. 1940. A preliminary investigation into the antigenic character of spore forming, aerobic bacteria. J. Bacteriol. 40:45–56.

Singh, Y., Leppla, S. H., Bhatnagar, R., and Friedlander, A. M. 1989. Internalization and processing of *Bacillus anthracis* lethal toxin by toxin-sensitive and -resistant cells. J. Biol. Chem. 264:11099–11102.

Sirisanthana, T., Navacharoen, N., Tharavichitkul, P., Sirisanthana, V., and Brown, W. E. 1984. Outbreak of oraloropharyngeal anthrax: An unusual manifestation of human infection with *Bacillus anthracis*. Am. J. Trop. Med. Hyg. 33:144–150.

Sliman, R., Rehm, S., and Shlaes, D. M. 1987. Serious infections caused by *Bacillus* species. Medicine (Baltimore). 66:218–223.

Smith, H., Keppie, J., and Stanley, J. L. 1955. The chemical basis of the virulence of *Bacillus anthracis*. V. The specific toxin produced by *B. anthracis* in vivo. Br. J. Exp. Pathol. 36:460–472.

Smith, N. R., Gordon, R. E., and Clark, F. E. 1952. Aerobic spore forming bacteria. US Dept of Agric., Washington, DC. Agric. Monogr. 16.

Solny, M. N., Failing, G. R., and Borges, J. S. 1977. *Bacillus cereus* osteomyelitis (letter). Arch. Intern. Med. 137:401.

Spira, W. M., and Goepfert, J. M. 1972. *Bacillus cereus* induced fluid accumulation in rabbit ileal loops. Appl. Microbiol. 24:341–348.

Stein, M. W., and Logan, G. F. 1965. Characterization of the phospholipases of *Bacillus cereus* and their effects on erythrocytes, bone, and kidney cells. J. Bacteriol. 90:69–81.

Stephen, J. 1981. Anthrax toxin. Pharmacol. Ther. 12:501–513.

Sterne, M. 1988. Anthrax vaccines. J. Am. Vet. Med. Assoc. 192:141.

Sugar, A. M., and McCloskey, R. V. 1977. *Bacillus licheniformis* sepsis. J. Am. Med. Assoc. 238:1180–1181.

Szabo, R. A., Todd, E. C. D., and Rayman, M. K. 1984. Twenty-four hour isolation and confirmation of *Bacillus cereus* in foods. J. Food Prot. 47:856–860.

Terranova W., and Blake, P. A. 1978. *Bacillus cereus* food poisoning. N. Engl. J. Med. 298:143–144.

Thompson, N. E., Ketterhagen, M. J., Bergdoll, M. S., and Schantz, E. J. 1984. Isolation and some properties of an enterotoxin produced by *Bacillus cereus*. Infect. Immun. 43:887–894.

Tippetts, M. T., and Robertson, D. L. 1988. Molecular cloning and expression of the *Bacillus anthracis* edema factor toxin gene: a calmodulin-dependent adenylate cyclase. J. Bacteriol. 170:2263–2266.

Trivedy, J. R. 1981. Case of anthrax meningitis survival. Centr. Afr. J. Med. 27:166.

Tuazon, C. U., Hill, R., and Sheagren, J. N. 1974. Microbiologic study of street heroin and injection paraphernalia. J. Infect. Dis. 129:327–329.

Tuazon, C. U., Murray, H. W., Levy, C., Solny, M. N., Curtin, J. A., and Sheagren, J. N. 1979. Serious infections from *Bacillus* species. J. Am. Med. Assoc. 241:1137–1140.

Turell, M. J., and Knudson, G. B. 1987. Mechanical transmission of *Bacillus anthracis* by stable flies *(Stomoxys calcitrans)* and mosquitoes *(Aedes aegypti* and *Aedes taeniorhynchus)*. Infect. Immun. 55:1859–1861.

Turnbull, P. C. B. 1976. Studies on the production of enterotoxin by *Bacillus cereus*. J. Clin. Pathol. 29:941–948.

Turnbull, P. C. B. 1981. *Bacillus cereus* toxins. Pharmacol. Ther. 13:453–505.

Turnbull, P. C. B., Jorgensen, K., Kramer, J. M., Gilbert, R. J., and Parry, J. M. 1979a. Severe clinical conditions associated with *Bacillus cereus* and the apparent involvement of exotoxins. J. Clin. Pathol. 32:289–293.

Turnbull, P. C., Kramer, J. M., Jorgensen, K., Gilbert, R. J., and Melling, J. 1979b. Properties and production characteristics of vomiting, diarrheal, and necrotizing toxins of *Bacillus cereus*. Am. J. Clin. Nutr. 32:219–228.

Turner, M. 1980. Anthrax in humans in Zimbabwe. Centr. Afr. J. Med. 26:160–161.

Twigg, G. 1984. The black death: a reappraisal. Batsford, London. 254p.

Vallery-Radot, P. (ed.), *Oeuvres de Pasteur,* 7 vols. (Paris: Masson et Cie, 1939) 7:287–291.

Van Ness, G. B. 1971. Ecology of anthrax. Science 172:1303–1306.

Vodkin, M. H., and Leppla, S. H. 1983. Cloning of the protective antigen gene of *Bacillus anthracis*. Cell 34:693–697.

Wade, N. 1980. Death at Sverdlovsk: A critical diagnosis. Science. 209:1501–1502.

Webber, R. H. 1985. Anthrax in Lake Rukwa Valley, Tanzania: a persistent problem. J. Trop. Med. Hyg. 88:327–331.

Weber, D. J., Saviteer, S. M., Rutala, W. A., and Thomann, C. A. 1988. In vitro susceptibility of *Bacillus* spp. to selected antimicrobial agents. Antimicrob. Agents Chemother. 32:642–645.

Weber, D. J., Saviteer, S. M., Rutala, W. A., and Thomann, C. A. 1989. Clinical significance of *Bacillus* species isolated from blood cultures. South. Med. J. 82:705–709.

Weinstein, L., and Colburn, C. G. 1950. *Bacillus subtilis* meningitis and bacteremia. Arch. Intern. Med. 86:585–594.

Welkos, S. L., Lowe, J. R., Eden-McCutchan, F., Vodkin, M., Leppla, S. H., and Schmidt, J. J. 1988. Sequence and analysis of the DNA encoding protective antigen of *Bacillus anthracis*. Gene 69:287–300.

Weller, P. F., Nicholson, A., and Braslow, N. 1979. The spectrum of *Bacillus* bacteremias in heroin addicts. Arch. Intern. Med 139:293–294.

Wiedermann, B. L. 1987. Non-anthrax *Bacillus* infections in children. Pediatr. Infect. Dis. J. 6:218–220.

Williams, R. P. 1981. *Bacillus anthracis* and other aerobic spore forming bacilli. In: Braude A. (ed.), Medical microbiology and infectious diseases. WB Saunders, Philadelphia.

Wilson, G. 1983. *Bacillus:* the aerobic spore-bearing bacilli, p. 422–441. In: Wilson, G., Miles, A., Parker, M. T. (ed), Topley and Wilson's principles of bacteriology, virology and immunity. Williams & Wilkins, Baltimore.

Yeh, T. J., Anabtawi, I. N., Cornett, V. E., White, A., Stern, W. H., and Ellison, R. G. 1967. Bacterial endocarditis following open-heart surgery. Ann. Thorac. Surg. 3:29–36.

Young, E. J., Wallace, R. J., Ericsson, C. D., Harris, R. A., and Clarridge, J. 1980. Panophthalmitis due to *Bacillus cereus*. Arch. Intern. Med. 140:559–560.

Young, R. F., Yoshimori, R. N., Murray, D. L., and Chou, P. J. 1982. Postoperative neurosurgical infections due to *Bacillus* species. Surg. Neurol. 18:271–273.

Yow, M. D., Reinhart, J. B., and Butler, L. J. 1949. *Bacillus subtilis* septicemia treated with penicillin. J. Pediatr. 35:237–239.

Genera Related to the Genus *Bacillus—Sporolactobacillus, Sporosarcina, Planococcus, Filibacter,* and *Caryophanon*

DIETER CLAUS, DAGMAR FRITZE, and MILOSLAV KOCUR

Phylogenetic studies using comparisons of the oligonucleotide sequences of 16S rRNA have shown that taxa might group together which, according to classical taxonomy, would not have been found to be closely related. For instance, Fox et al. (1977) have shown that *Sporolactobacillus* may well be included in the genus *Bacillus,* and Pechman et al. (1976) and Fox et al. (1977) demonstrated that *Sporosarcina ureae* is more closely related to *Bacillus pasteurii* than are several other species of *Bacillus.* Similar studies suggest that a number of species or even genera, which, hitherto, have been placed in either nonsporeforming or Gram-negative taxa, should be placed close to the genus *Bacillus* and in particular to those species forming spherically shaped spores (Clausen et al., 1985; Stackebrandt et al., 1987). The dendrogram constructed according to these results shows a close neighboring relationship of *Filibacter limicola* to *Bacillus globisporus, Planococcus citreus* to *"B. aminovorans,"* and *Caryophanon latum* to *B. sphaericus.* The *Bacillus* species forming spherical spores showed a higher degree of relationship to these nonsporeforming organisms than to each other and, in particular, higher than to the *Bacillus* species forming oval to ellipsoidal spores.

Also, the genus *Marinococcus,* comprising three species, has been separated from the genus *Planococcus.* Ventosa et al. (1990) described *Salinicoccus,* a genus phenotypically similar to *Marinococcus* and *Planococcus.* Both genera will be discussed within this chapter although their phylogenetical relationships to the family Bacillaceae and in particular to *Planococcus* are not known.

The genus *Bacillus* seems to be phylogenetically incoherent, and the spherical-sporeforming *Bacillus* species may represent a separate evolutionary line within the genus (Bonde, 1981; Logan and Berkeley, 1981; Priest, 1981). This is supported by rRNA analyses (Stackebrandt et al., 1987).

The phenotypic differentiation of the genus *Bacillus* from other genera mainly utilizes the following features: aerobic, sporeforming, rod-shaped, and Gram-positive cell wall. Although these four characteristics seem to be unequivocal, it is not always easy to provide evidence for them.

The ability to form spores is sometimes difficult to demonstrate for a number of strains or even species of the various sporeforming bacteria. Some organisms originally described as nonsporeformers were reclassified after the detection of spores, e.g., *"Lineola longa"* as *"B. macroides"* (Bennet and Canale-Parola, 1965) and *"Lactobacillus cereale"* as *B. coagulans* (Gordon et al., 1973), and there may be more of these. Many strains of recognized species are known to easily lose their ability to form spores after only a few transfers on culture media. Special treatment is necessary to avoid this loss which is, in most cases, not a loss of genetic information but loss of the potency of expressing a property. A general procedure for the induction of spore formation in a culture, especially for regaining of the ability to form spores, cannot be given. Various methods have been described, each having its own advantages with certain strains or species. Methods vary from supplementation of media with trace elements, especially manganese ions or soil extract, to variations in the conditions of incubation and supply of nutrients. Most of these methods or references to them are given in reviews Claus and Berkeley (1986) and Kalakoutskii and Dobritsa (1984).

The requirement that the genus *Bacillus* be Gram-positive remains, although strains or species that are Gram-variable or Gram-negative are accepted. In many species, Gram-positive reactions are seen only in very young cultures. Wiegel (1981) proposed that a differentiation should be made between Gram reaction, a staining phenomenon and Gram type, which refers to the structure of the cell wall as seen in thin sections in the electron microscope. And, indeed, in many cases it has been shown that the cell wall of a Gram-negatively staining strain is actually Gram-positive in structure.

Minimal Descriptions of the Bacillaceae

In the following, a minimal description of the two sporeforming and the three nonsporeforming genera placed—on the basis of phylogenetic studies—into the family of the Bacillaceae is given in comparison with the genus *Bacillus* as currently defined.

Genus *Bacillus*

Staining mostly Gram-positive, rods, peritrichously flagellated, endospores ellipsoidal or spherical, swelling or not swelling the sporangium, aerobic to facultatively anaerobic, mostly catalase positive.

Genus *Sporolactobacillus*

Gram-positive, rods, sparse-peritrichously flagellated, ellipsoidal endospores, swelling the sporangium, aerobic to microaerobic, catalase negative.

Genus *Sporosarcina*

Gram-positive, spheres occurring in pairs, tetrads, or packages, often only one flagellum per cell, spherical endospores, aerobic, catalase positive.

Genus *Planococcus*

Gram-positive, spheres occurring in pairs or tetrads, one to three flagella per cell, asporogenous, aerobic, catalase positive.

Genus *Filibacter*

Stains Gram-negative but its Gram type is positive, multicellular filaments, motile through gliding, asporogenous, aerobic, catalase positive.

Genus *Caryophanon*

Gram-positive, multicellular large rods, peritrichously flagellated, asporogenous, aerobic, catalase positive.

The Genus *Sporolactobacillus*

During the course of studies on the distribution of microorganisms in assorted chicken feed, Kitahara and Suzuki (1963) isolated an unusual strain of a lactic acid bacterium possessing certain characters intermediate between those of the genera *Bacillus* and *Lactobacillus*. The rod-shaped strain was Gram-positive, motile, formed endospores, was microaerophilic and catalase negative, and showed a typical homofermentative metabolism, producing D(−)-lactic acid. The authors created the new subgenus *Sporolactobacillus* within the family Lactobacillaceae in order to accommodate this unusual bacterium. Later, *Sporolactobacillus* was transferred as an independent genus comprising only one species, *Sporolactobacillus inulinus*, into the family Bacillaceae (Kitahara and Toyota, 1972).

Similar strains were isolated by Nakayama and Yanoshi (1967a) and by Amemiya and Nakayama (1980). They were grouped with *S. inulinus* or were considered to belong to a new species for which the names "*S. laevas*," "*S. laevas* var. *intermedius*," and "*S. racemicus*" have been proposed (Yanagida et al., 1987a). The genus was reviewed by Norris (1981), Norris et al. (1981), and Kandler and Weiss (1986).

Habitats

The original isolate, classified as *Sporolactobacillus inulinus*, was from chicken feed (Kitahara and Suzuki, 1963). Nearly all other strains of sporolactobacilli have been isolated from the rhizosphere or from soil around root hairs of a variety of wild plants collected in Japan and Southeast Asia (Nakayama and Yanoshi, 1967b; Amemiya and Nakayama, 1980). From a total of about 700 samples collected in the United States (foods, feed, soil, environment), Doores and Westhoff (1983) were able to isolate only two strains of *Sporolactobacillus*, both from soil samples. These authors concluded, therefore, that the incidence of sporolactobacilli in the environment is extremely low. Nakayama and Yanoshi (1967a) have pointed out that the rhizosphere represents a habitat where nutrients necessary for the growth of lactic acid bacteria may be provided by excretion from root hairs. Since the rhizosphere has a tendency to undergo drying and heating by sunlight impinging on the soil surface, it would favor, besides other microorganisms, the growth of motile, sporeforming, lactic acid bacteria.

Isolation

Many *Sporolactobacillus* strains have been isolated by Nakayama and Yanoshi (1967b) and by Amemiya and Nakayama (1980) using the following method:

Isolation Method 1 (Nakayama and Yanoshi, 1967a)

The enrichment medium for *Sporolactobacillus* has the following composition (*Sporolactobacillus* broth):

Glucose	10.0 g
Polypeptone	10.0 g
Yeast Extract	10.0 g
Sodium citrate	0.027 g
KH_2PO_4	0.5 g
K_2HPO_4	0.5 g
$MgSO_4 \cdot 7H_2O$	0.3 g
NaCl	0.01 g
$MnSO_4 \cdot 5H_2O$	0.01 g
$CuSO_4 \cdot 5H_2O$	0.001 g
$CoCl_2 \cdot 6H_2O$	0.001 g
$FeSO_4 \cdot 7H_2O$	0.001 g
Soil extract (see below)	100 ml
Distilled water	900 ml

Soil extract is prepared by autoclaving 100 g of garden soil mixed with 200 ml distilled water for 20 min at 130°C. The mixture is centrifuged to obtain a clear supernatant. The glucose broth is adjusted to pH 6.4 and is sterilized for 15 min at 115°C. Root material with root hairs and adherent soil from plants like *Trifolium repens, Allium japonicum, Ranunculus scelerata* and others is collected. Small pieces of the specimen are placed in a test tube containing a few ml of the sterile enrichment medium. In order to kill asporogenous organisms the tubes are heated at 80°C for 20 min and are incubated anaerobically at 30°C. In most tubes, clostridia will develop. On further incubation, however, the pH of cultures may drop below 4.0. Material from such cultures is streaked onto *Sporolactobacillus* agar composed of the broth described above supplemented with 10 g of calcium carbonate and solidified by 15 g of agar per liter of medium. After streaking, the agar surface is covered by a polyvinylidene chloride film, sterilized by autoclaving between filter papers, in order to depress the growth of aerobic bacteria. The plates are incubated aerobically at 30 to 37°C. Pin-point colonies with a transparent halo are picked, purified, and tested for the absence of catalase and the presence of endospores.

A selective technique for the enrichment and isolation of strains of the genus *Sporolactobacillus* has been developed by Doores and Westhoff (1983). Since the incidence of sporolactobacilli in the environment seems to be low, enrichment for these organisms is more likely to be successful than direct plating.

Isolation Method 2 (Doores and Westhoff, 1983)

Soil samples of about 5 g are collected aseptically in sterile 100-ml screw-cap bottles. To the soil sample, 50 ml of the following modified MRS broth is added:

α-Methylglucoside	10.0 g
Proteose peptone	10.0 g
Beef extract	5.0 g
Yeast extract	5.0 g
Tween 80	1.0 g
Ammonium citrate	2.0 g
Sodium acetate	5.0 g
Magnesium sulfate	0.1 g
Manganese sulfate	0.05 g
Disodium phosphate	2.0 g
Potassium sorbate	1.0 g

Bromocresol green	0.0224 g
Distilled water	1 liter

The medium is adjusted to pH 5.5 with 1 N acetic acid. After vigorous shaking for 2 min, the soil suspension is allowed to settle for about 5 min. Thereafter, the supernatant is transferred to a sterile 100-ml bottle, and the pH readjusted to 5.5 with 1 N or 0.1 N acetic acid or sodium hydroxide. The medium is incubated for 7 days at 30°C in a carbon dioxide incubator with a carbon dioxide level of 5% and a relative humidity of 98%. Thereafter, a 2-ml sample is transferred to a sterile test tube, which is then heated for 5 min in a 80°C water bath to kill vegetative cells. Aliquots of 0.1 ml of the heat-treated sample are spread onto plates of modified MRS agar (modified MRS broth, pH 5.5 plus 15 g agar per liter of medium). The plates are incubated at 30°C. Colonies on the plates are screened for catalase production and benzidine reaction (both are negative for sporolactobacilli) as well as for the presence of spores and motility. The latter properties may be variable under the isolation conditions. Due to the specifity of the method, interference from environmental strains of *Bacillus* and *Lactobacillus* appears to be negligible.

Cultivation

Sporolactobacillus inulinus and other sporolactobacilli grow well on most media used for lactobacilli. In general, fermentable sugars are necessary for growth. Excellent growth has been observed on the GYP medium used by Kitahara and Suzuki (1963).

GYP Medium

Glucose	20.0 g
Yeast extract	5.0 g
Peptone	5.0 g
Agar	15.0 g
Distilled water	1 liter

Incubation should be at 30 to 37°C under an atmosphere containing 5% carbon dioxide (Doores and Westhoff, 1983). Under air, only feeble growth may occur.

A series of commercially available media also support good growth under the conditions described. These include ATP agar (BBL) and lactobacilli MRS agar (Difco). For growing a large number of strains of the sporolactobacilli, Yanagida et al. (1987a) successfully used the following medium:

Sporolactobacillus Growth Medium

Glucose	20.0 g
Yeast extract	10.0 g
Peptone	10.0 g
Sodium acetate	10.0 g
$MgSO_4 \cdot 7H_2O$	0.2 g
$MnSO_4 \cdot 4H_2O$	0.01 g
$FeSO_4 \cdot 7H_2O$	0.01 g
NaCl	0.01 g
Agar	20.0 g

Distilled water   1 liter
Adjust to pH 6.8. Incubate plates at 30°C.

Growth factor requirements have only been studied with a single strain which required biotin, pantothenic acid, leucine, and valine (Kitahara and Suzuki, 1963).

Maintenance and Preservation

Strains of *Sporolactobacillus* may be stored in liquid media for several months at −20°C if the acidity of cultures is neutralized after growth has ceased. Cultures can be freeze-dried by common procedures, for example, in 20% skim milk supplemented with 5% *myo*-inositol. Maximum viability of sporulated cultures is not known.

Cultural and Microscopic Appearance

On agar media, sporolactobacilli form pinpoint colonies which are circular with convex shape, grayish white, and glistening. Due to the formation of copious amounts of lactic acid, colonies on media containing calcium carbonate are surrounded by clear haloes. Growth is feeble on agar slants. In agar shake cultures, uniform growth can be observed except for a 5-mm zone near the surface.

Sporolactobacillus strains are Gram-positive or Gram-variable. They form slender rods which are sometimes slightly bent. Cells measure 0.7–0.8×3–5 μm and occur singly, in pairs, and rarely in short chains. They are motile by a small number of long, peritrichously inserted flagella. Endospores are formed in certain media (see below).

When cultured in tomato-meat (TM) medium (see below) devoid of ammonium sulfate and calcium carbonate, a large proportion of vegetative cells form tadpole-like structures.

Endospore Formation

In most media the formation of endospores is very rare. A sporulation rate of 10^{-4} to less than 10^{-6} has been observed. However, a medium in which sporulation takes place at a higher rate has been developed by Kitahara and Lai (1967). In this medium the sporulation frequency can be increased up to 10% by incubating cultures under 5% CO_2 at 37°C.

TM Medium for Sporulation of *S. inulinus*
(Kitahara and Lai, 1967)

Yeast extract	1.0 g
Meat extract	5.0 g
α-Methylglucoside	5.0 g
Ammonium sulfate	10.0 g
Tomato serum	200 ml

Distilled water   800 ml

Adjust the pH of the medium to 5.5 and add 20 g calcium carbonate (precipitated material). Doores and Westhoff (1981) found that the above medium without tomato serum was superior. Nakayama and Yanoshi (1967b) found that starch stimulated the formation of endospores.

Spores are ellipsoidal and 0.9 to 1.4×1.0–2.1 μm in size. They appear in the terminal to subterminal position. The sporangia are definitely swollen. As with other bacterial spore-formers, endospores contain dipicolinic acid.

Spores of only one *Sporolactobacillus* strain have been studied for heat resistance. They tolerate a treatment of 10 min at 70–80°C but not at 90°C. Decimal reduction times have been reported by Doores and Westhoff (1981).

Ultrastructure

Thin sections of vegetative cells show the usual cell wall type of Gram-positive bacteria (Kitahara and Toyota, 1972). The ultrastructure of spores is similar to those of *Bacillus* spores (Kitahara and Lai, 1967).

Chemotaxonomy

The cell wall of *Sporolactobacillus inulinus* contains *meso*-diaminopimelic acid, a large amount of polysaccharides, but no teichoic acid (Weiss et al., 1967; Okada et al., 1976). Uchida and Mogi (1973) examined the fatty acid composition of *Sporolactobacillus* strains and found odd-numbered, saturated anteiso- and iso-branched fatty acids. No unsaturated, even-numbered or cyclopropane fatty acids, which predominate in members of the genus *Lactobacillus*, have been detected.

Collins and Jones (1979) studied the isoprenoid quinone composition of strains of *S. inulinus*, "*S. laevas*," and "*S. racemicus*," as well as strains of some *Bacillus* species. In all strains, menaquinones with seven isoprene units (MK-7) predominate. In *Lactobacillus* strains, neither menaquinones nor ubiquinones were detected.

Genetic and Phylogenetic Relationships

Published values on the GC content of the DNA of *Sporolactobacillus inulinus* vary greatly. Using paper chromatography, Suzuki and Kitahara (1964) reported that the GC content of the type strain was 39.3 mol%. For the same strain, Kandler and Weiss (1986) found 38 mol% (T_m) and Miller et al. (1970) 47.3 mol%. This latter value was confirmed by Yanagida et al. (1987b) who found 46.8–47.0 mol% (T_m) for two strains of the species.

Miller et al. (1970) studied the DNA-DNA hybridization between strains of *S. inulinus, Lactobacillus plantarum,* and *Bacillus coagulans* and reported that there were no cross-reactions among these strains. Hybridization studies of Yanagida et al. (1987b), covering 33 *Sporolactobacillus* strains, have shown that the genus is genetically heterogenous, comprising five DNA homology groups (Table 1). Strain clusters found by a numerical taxonomic study of Yanagida et al. (1987a) did not correspond to the five DNA homology groups.

Results from cataloging the oligonucleotide pattern of ribosomal 16S RNA (Fox et al., 1977) have shown that *Sporolactobacillus* may well be included in the genus *Bacillus*. In later studies (Stackebrandt et al., 1987), it was found that *S. inulinus* was the only species possessing a phylogenetic position differing from that of the other aerobic or facultatively anaerobic Gram-positive sporeformers examined up to now.

The close relationship of *Sporolactobacillus* and *Bacillus* can also be seen from chemotaxonomical studies (see above).

Physiological and Biochemical Properties

Although the genus is described as microaerophilic, all strains studied can be grown under air, even on agar plates. Strains tested showed improved growth, however, in the presence of 5% carbon dioxide. Some strains have been described as facultatively anaerobic (Nakayama and Yanoshi, 1967b).

Sporolactobacillus strains grow between 15–20 and 40°C with an optimum temperature at around 30°C. Although growth-limiting pH values are not known, cultures grow at a pH as low as 5.0. Since the pH of the medium decreases during growth in glucose broth to 3.8 to 3.2 *(S. inulinus)* or to pH 4.4 (other sporolactobacilli), growth may also be possible at a starting pH of 4.5.

Sporolactobacillus strains do not form catalase. Most strains need fermentable carbohydrates for growth. Acid without gas is produced by nearly all strains of the different DNA homology groups from glucose, fructose, mannose, maltose, sucrose, and trehalose. Acid formation from other carbohydrates is variable. In contrast to other strains studied, strains belonging to *S. inulinus* do not form acid from galactose (Yanagida et al., 1987a).

The type strain of *S. inulinus* has been shown to ferment hexoses exclusively to D(−)-lactic acid (homolactic acid fermentation) producing less than 1% of volatile acids or ethanol (Kitahara and Suzuki, 1963). According to Yanagida et al. (1987a), however, most strains of the genus *Sporolactobacillus* however, form both D(−) and DL-lactic acid. These discrepancies have to be clarified.

Identification

It is not yet possible to group *Sporolactobacillus* isolates according to their phenotypic properties with the type species of the genus, *S. inulinus,* or with any of the other sporolactobacilli described. Also, isolates cannot be grouped by phenotypic properties with the different DNA homology groups described by Yanagida et al. (1987b). Additional studies on phenotypic properties of the large number of available sporolactobacilli are now necessary. Because of the close relationship of *Sporolactobacillus* to the genus *Bacillus,* it is important to study the group by the same methods used for the characterization of *Bacillus* strains (Claus and Berkeley, 1986).

It has to be emphasized that a clear distinction between the genera *Sporolactobacillus* and *Bacillus* is also not possible. *Sporolactobacillus* strains, which all are characterized by the formation of lactic acid, can be separated from lactic acid-forming *Bacillus* species *(B. coagulans, B. smithii, "B. laevolacticus," "B. racemilacticus")* only by their temperature relation-

Table 1. Properties of the DNA homology groups of the *Sporolactobacillus* strains.

Homology group	GC content (mol%)	Intragroup homology	Species (number of strains)
1	46.8–50.2	88–100%	*S. inulinus* (2) "*S. laevas* var. *intermedius*" (1)
2	42.5–47.4	70–100%	"*S. laevas*" (12) "*S. laevas* var. *intermedius*" (3)
3	42.9–46.2	55–100%	"*S. racemicus*" (7) "*S. laevas* var. *intermedius*" (2)
4	42.5–45.8	79–100%	"*S. racemicus*" (2) "*S. laevas* var. *intermedius*" (2)
5	43.0	100%	"*S. laevas*" (2)

Adapted from Yanagida et al. (1987b).

ships and/or by their negative catalase reaction. Other phenotypic properties which may be used in the differentiation from these *Bacillus* species have not been studied.

While the lack of catalase in *Sporolactobacillus* has been considered as the main criterion for its separation from the genus *Bacillus,* it should be noted that certain *Bacillus* species also lack this enzyme. These include *B. azotoformans* (Pichinoty et al., 1983), *B. larvae, B. lentimorbus, B. popilliae,* certain strains of *B. stearothermophilus* (Gordon et al., 1973), and *B. pulvifaciens* (Nakamura, 1984).

Applications

Polysaccharide production of strains of the genus *Sporolactobacillus* has been studied by Amemiya and Nakayama (1980). Water-soluble fructans are produced from sucrose which show antitumor activity.

The Genus *Sporosarcina*

From urea-containing enrichment cultures, Beijerinck (1901) repeatedly isolated packet-forming coccoid bacteria. In contrast to all other sarcinae known at that time, the isolates were found to be motile by flagella and formed endospores. Beijerinck described these isolates as the new species *"Planosarcina ureae."*

The species was later transferred by Löhnis (1911) to the genus *Sarcina,* which at that time comprised bacteria that were morphologically similar due to the formation of packets (sarcinae) of four or more cells. In physiological and biochemical properties, however, the various *Sarcina* species of Löhnis differed from each other substantially. Therefore, *"Sarcina ureae"* was later separated from the genus *Sarcina* and transferred to the genus *Sporosarcina* (Kocur and Martinec, 1963; MacDonald and MacDonald, 1962) as previously proposed by Orla-Jensen (1909) and by Kluyver and van Niel (1936).

Strains of two species of the genus *Sporosarcina* are available now in pure culture and have been extensively described: *S. ureae* (Beijerinck, 1901) Kluyver and van Niel, 1936[AL], and *S. halophila* Claus, Fahmy, Rolf and Tosunoglu, 1984[VP]. Beijerinck (1901) briefly mentioned another sporeforming sarcina which he named *"Urosarcina dimorpha,"* which developed only on media containing horse urine gelatin, but it is not certain whether pure cultures were isolated. Attempts to reisolate *"U. dimorpha"* have been unsuccessfull (D. Claus, unpublished observation). The only other described taxon of the genus is *"Sarcina pulmonum,"* a motile, sporeforming urea-degrading organism isolated from clinical material, but these strains were subsequently lost. According to Gibson (1935), these isolates were probably identical with *Sporosarcina ureae.*

S. ureae and *S. halophila* are considered to be nonpathogenic for humans, animals, and plants. The genus has been reviewed by Claus (1981), Norris (1981), and Claus and Fahmy (1986).

Habitats

Like other sporeforming bacteria, *Sporosarcina ureae* is widely distributed in soil. Fertile soils may contain up to 10^4 sarcinae/g (Gibson, 1935; Pregerson, 1973). The bacterium probably has also been isolated from liquid manure (Sames, 1898). A single isolate has also been reported from sea water (Wood, 1946). Pregerson (1973) has found *S. ureae* to be widely distributed in the United States and in various other parts of the world. According to her studies, the primary habitat of the organism appeared to be concentrated in certain urban soils closely associated with the activities of humans and, especially, dogs.

The prevalence of the bacterium in fertile soils and certain other urea-containing places, its resistance against the inhibitory effect of up to 5–10% urea, and its ability to produce urease suggest that *S. ureae* plays an active part in the decomposition of urea in natural habitats.

The sporeforming *"Sarcina pulmonum"* was isolated at the beginning of this century by several workers from sputum and from the respiratory tract in cases of phthisis. Although for some time it was considered to be responsible for severe infections of the lungs, its pathogenicity has been questioned (Lehmann and Neumann, 1927; Gibson, 1935). Recently, a bacterial isolate from a bronchial biopsy in a young child with cystic fibrosis has been identified as *Sporosarcina ureae.* It was considered, however, not to be directly related to the disease (Chomarat et al., 1990).

S. halophila has been isolated from salt marsh soils (Claus et al., 1983), saline soils, and ponds of solar salterns (Ventosa et al., 1983). Its moderately halophilic characteristics, together with its obligate requirements for sodium, magnesium, and chloride ions, as well as its optimum temperature for growth (about 30°C), suggest that salt marsh and saline soils are the natural habitat of this species.

Isolation

Selective methods for the enrichment of the two accepted *Sporosarcina* species in liquid media have not been reported. The selective enrichment method used by Beijerinck (1901) for the isolation of *S. ureae* does not seem to work well. Strains of both species, however, can be isolated by plating soil dilutions on appropriate agar media.

ISOLATION OF *SPOROSARCINA UREAE*. The presence of *S. ureae* in many soil samples has been demonstrated by a simple and effective method (Gibson, 1935). Since 10% urea, as used by Gibson, may inhibit the growth of *S. ureae* from many soils, urea should be used at a concentration of only 3–5%. At this concentration, the growth of a high percentage of the bacterial soil flora is strongly inhibited. This is specifically true for *Bacillus mycoides,* which will readily overgrow most soil dilution plates at urea concentrations below 3%.

Method 1 for the Isolation of *S. ureae* (Gibson, 1935; Modified by Claus, 1981)

> To 1 liter of nutrient agar, add 30, 50, or 100 g of urea. Sterilize the media at 121°C for 15 min and pour into petri dishes. Suspend about 5 g of a soil sample (fertile soil, air-dried) in 20 ml of sterile water and prepare soil dilutions (10^{-1} and 10^{-2}). Plate 0.1 ml of the soil suspension and of the two dilutions on the agar and incubate at about 25°C. The soil suspension may be heated, but there is little or no advantage to doing so because the majority of the organisms that develop on the media are sporeforming bacteria.
>
> After 3–5 days, examine only cream or pale yellow to bright orange colored colonies under low magnification and transmitted light. Colonies of *S. ureae* can be recognized as round and black at a magnification of about $10\times$, and by their coarsely granulated structure, especially at the edges of the colonies, at a magnification of about $50\times$. With some experience, colonies of *S. ureae* may be selected with rather high certainty. Similar types of colonies, however are often formed by strains of *Bacillus megaterium.* Therefore, it is advisable to compare the colony types on isolation plates with those of known strains of the two species.
>
> Prepare slides from selected colonies for observing the typical morphology of cells and confirm the provisional identification of isolated strains of *S. ureae* testing the motility and the production of spores (see below, under maintenance). Pure cultures are obtained by restreaking suspensions prepared from single colonies onto nutrient agar containing 2% urea.

Pregerson has isolated about 50 strains of *S. ureae* from 198 different soil samples and has proposed the following method for isolating the organism. A good source for isolation is soil from the base of trees where dogs have urinated.

Method 2 for the Isolation of *S. ureae* (Pregerson, 1973)

> Tryptic soy-yeast (TSY) agar containing (per liter of distilled water); 27.5 g Difco tryptic soy broth, 5.0 g Difco yeast extract, 5.0 g glucose, and 15.0 g Difco agar, is

adjusted with 1 N NaOH to pH 8.5 before autoclaving. Add aseptically filter-sterilized urea to give a final concentration of 1% (w/v) and pour the medium into petri dishes. Suspend 1 g of a soil sample in 15 ml of distilled water and mix the slurry with a Vortex mixer. Spread 0.1 ml of a series of dilutions on plates using a sterile bent glass rod. Prepare triplicate plates at 10^{-1} and 10^{-2} and duplicate plates at 10^{-3} and 10^{-4} dilutions.

> Incubate at 22°C and examine the plates on the third day and then daily with a dissecting microscope. Colonies of *S. ureae* show a uniform-surface granularity, smoothly opaque interiors, and an orange or cream color. Prepare slides from selected colonies for observing cell morphology and motility and the production of spores (see below under maintenance), and isolate pure cultures.

The isolation of *S. ureae* from a certain soil sample often is not reproducible due to the irregular background growth developing on dilutions plates. A more selective method suppressing such growth or a selective liquid enrichment method will allow the isolation of this species also from soils where it is only present in low numbers.

Isolation of *Sporosarcina halophila* (Claus et al., 1983)

> Two different agar media have been used for isolation. Bacto Marine Agar 2216 (Difco) is composed of:

> | | |
> |---|---|
> | Bacto peptone | 5.0 g |
> | Bacto yeast extract | 1.0 g |
> | Ferric citrate | 0.1 g |
> | Sodium chloride | 19.45 g |
> | Magnesium chloride | 8.8 g |
> | Sodium sulfate | 3.24 g |
> | Calcium chloride | 1.8 g |
> | Potassium chloride | 0.55 g |
> | Sodium bicarbonate | 0.16 g |
> | Potassium bromide | 0.08 g |
> | Strontium chloride | 0.034 g |
> | Boric acid | 0.022 g |
> | Sodium silicate | 0.004 g |
> | Sodium fluoride | 0.0024 g |
> | Ammonium nitrate | 0.0016 g |
> | Disodium phosphate | 0.008 g |
> | Agar | 15.0 g |
> | Distilled water | 1 liter |

> Adjust the pH to about 7.5. The medium is sterilized at 121°C for 15 min and poured into petri dishes.

Alternatively, the following medium may be used:

> | | |
> |---|---|
> | Peptone | 5.0 g |
> | Meat extract | 3.0 g |
> | Sodium chloride | 30.0 g |
> | Magnesium chloride | 5.0 g |
> | Agar | 15.0 g |
> | Distilled water | 1 liter |

> Adjust the pH to about 7.5

> Suspend about 5 g of a sample of salt marsh soil in 20 ml of sterile tap water. Heat the suspension at 70°C

for 10 min. Alternatively, the soil suspension may be treated with ethanol, final concentration 50% (v/v), for 1 h (to kill vegetative bacterial cells). Prepare soil dilutions in sterile water (10^{-1} to 10^{-4}). Plate 0.1 ml of the dilutions on agar medium.

After 3 days of incubation at 30°C, mainly pigmented (yellow, orange, or pink) colonies of various sizes of *Bacillus* species will have developed. Inspect the more crowded plates at a magnification of about 50× for round pin-point colonies which show, in transmitted light, a coarsely granulated structure at least at the edges of the colonies. The same colonies appear black under a magnification of only 10×. Check material of such colonies by phase contrast microscopy for the presence of sarcinae, which usually are motile or may have formed endospores.

For purification, suspend cell material taken from appropriate colonies in a drop of nutrient broth and streak onto plates of one of the agar preparations mentioned above. Colonies of *S. halophila* form an orange pigment which normally is not seen with the pin-point colonies developing on the isolation plates.

Fahmy et al. (1985) have observed that spores of *S. halophila* formed in pure culture will germinate only at a very low frequency in nutrient broth supplemented with 3% sodium chloride and 0.5% magnesium chloride, although growth of vegetative cells in such media is excellent. In contrast, a high percentage of spores readily germinate in nutrient broth supplemented with double-concentrated sea water (see below). This observation may be helpful in developing new media and methods for the selective enrichment and isolation of additional strains of *S. halophila* and to enumerate the species in natural samples.

Cultivation

Sporosarcina ureae grows well in normal nutrient broth. A more suitable medium contains, in addition, 1% urea or 0.5% ammonium chloride. In the latter case, the pH of the medium should be adjusted to 8.0 or 8.5. Since the organism is strictly aerobic, shaking is recommended to obtain high cell densities. The optimum temperature for growth is about 25°C.

A defined medium for *S. ureae* has been described by Goldman and Wilson (1977), which yields 5–6 g dry weight of bacteria per liter of culture.

S. ureae Growth Medium

L-Asparagine · H$_2$O, or L-Glutamine	30.0 g
KCl	3.4 g
K$_2$HPO$_4$	0.25 g
(NH$_4$)$_2$SO$_4$	0.2 g
MgSO$_4$·7H$_2$O	0.05 g
FeSO$_4$·7H$_2$O	0.0025 g
MnCl$_2$·4H$_2$O	0.00025 g
D-Biotin	0.001 g

L-Cysteine	0.005 g
Distilled water	1 liter

Prior to sterilization, the medium is adjusted to pH 8.7 with 1 M NaOH and supplemented with NaCl to a final concentration of 0.05 M Na$^+$. Sterilized solutions of biotin (0.4 mg/ml), L-cysteine (2 mg/ml), and (NH$_4$)$_2$SO$_4$ (80 mg/ml) were added to the separately sterilized medium.

Pregerson (1973) has pointed out that the growth factor requirements of strains of *S. ureae* may vary substantially. From 61 isolates studied, 13 grew on a mineral acetate medium without added growth factors, but 37 strains required biotin either singly or in combination with niacin and/or thiamine. Some of these strains also needed aspartate. Eleven strains showed more complex growth factor requirements. Most of the strains could utilize acetate, butyrate, or glutamate as a sole source of carbon and energy. Glutamate, however, generally could not be used by *S. ureae* as the sole nitrogen source.

According to the definition of Larsen (1962), *S. halophila* is a slightly halophilic bacterium. It is one of the few "marine" bacteria which shows an obligate growth requirement for sodium, magnesium, and chloride ions. Sodium cannot be replaced by potassium nor chloride by sulfate. The salt requirements of the strains apparently reflect the natural habitat of the species.

In the presence of 30 to 50 g NaCl (maximum 150 g) and 5 g MgCl$_2$ per liter, all strains of the species studied show profuse growth on a variety of complex media. The pH optimum for growth is between pH 7.0 and 9.0 with an optimum at about pH 7.8. Strains grow from 15 to 35 or 40°C. The optimum temperature of growth is around 30°C. The species is strictly aerobic. It grows well on the following medium:

S. halophila Growth Medium

Peptone from meat	5.0 g
Meat extract	3.0 g
NaCl	30.0 g
MgCl$_2$	5.0 g
Agar	15.0 g
Distilled water	1 liter
Adjust the pH to 7.8.	

Good growth is also observed on Bacto Marine Broth 2216 or Bacto Marine Agar 2216 (Difco). A chemically defined medium for *S. halophila* has not yet been developed.

Maintenance and Preservation

Vegetative cultures of *S. ureae* and of *S. halophila* may be kept for more than 8–12 weeks at about 4°C on nutrient agar slants if protected from drying. In tightly closed tubes, sporulated cultures (see below) may survive for several years if stored at 4°C. Sporulated cultures (see below) remain viable for more than one year.

For long-term maintenance, vegetative cells and spores of *S. ureae* and *S. halophila* can be preserved for years without significant loss in viability by lyophilization in skim milk (20% w/v) in the presence of 5% *myo*-inositol. Vegetative and sporulated cells can also be preserved for long periods in liquid nitrogen in the presence of a suitable cryoprotective agent like 5% (v/v) dimethylsulfoxide or 10% (v/v) glycerol (D. Claus, unpublished observations).

Cultural and Microscopic Appearance

Colonies of *Sporosarcina ureae* on agar are gray, opaque, circular, and slightly convex with an entire margin. On some media, a yellowish, brownish, or orange nondiffusible pigment may be produced. *S. halophila* forms round, smooth, and opaque colonies. Irrespective of the medium used, an orange nondiffusible pigment is formed.

The typical cell shape of both species is spherical. Often cells are modified by interfacial flattening in the cell aggregates. Especially in *S. halophila,* oval- to egg- or pear-shaped cells are also formed. The diameter of spherical cells is 1.0–2.5 μm. Oval cells of *S. halophila* measure 1–2 × 2–3 μm. Division walls usually are at right angles to one another.

In cultures of both species, various cell aggregate patterns can be found. Single cells, pairs, threes, tetrads, or packets are common or may predominate, depending on the medium used. In older cultures, aberrant forms often can be found: abnormally large cells appearing singly or within packets, short chains, or irregular packets. Apparently, this is due to an asynchronous cell division.

Cells and aggregates usually show a tumbling motility. With *S. ureae,* this is best observed using material from nutrient broth containing 1% urea (Kocur and Martinec, 1963). Most often, a single flagellum is formed per cell. It is difficult, however, to determine the precise number and location of flagella in cell packets (Sersen et al., 1983). *S. halophila* forms motile cells in young cultures grown on the agar media decribed above.

Endospore Formation

Both species form endospores under certain growth conditions. They are highly refractile, round, 0.5–1.5 μm in diameter, and located centrally or laterally. Like those of most *Bacillus* species, they resist heating for 10 min at up to 80°C or even more.

Spores of *S. ureae* can be obtained on one of the following sporulation media if the incubation temperature is lower than 22°C.

Sporulation Medium 1 for *S. ureae* (Gibson, 1935)

Peptone	5.0 g
Meat extract	5.0 g
Ammonium chloride	5.0 g
Agar	15.0 g
Distilled water	1 liter

Adjust the pH to 6.8–7.0.

Sporulation Medium 2 for *S. ureae* (Claus, 1981)

Peptone	5.0 g
Meat extract	3.0 g
Manganese chloride	50.0 mg
Agar	20.0 g
Distilled water	1 liter

The pH is not adjusted. After sterilization (20 min at 121°C) 20 ml of a filter-sterilized solution containing 10% (w/v) urea is added per liter of medium.

Sporulation Medium 3 for *S. ureae* (MacDonald and MacDonald, 1962)

Yeast extract	2.0 g
Peptone	3.0 g
Glucose	4.0 g
Malt extract	3.0 g
K_2HPO_4	1.0 g
$(NH_4)_2SO_4$	4.0 g
$MgSO_4$	0.8 g
$CaCl_2$	0.1 g
$MnSO_4 \cdot H_2O$	0.1 g
$FeSO_4 \cdot 7H_2O$	0.001 g
$ZnSO_4$	0.01 g
$CuSO_4 \cdot 5H_2O$	0.01 g
Agar	30.0 g
Water	1 liter

Adjust the pH before sterilization to 8.8–9.0.

At 22°C, all strains isolated by Pregerson (1973) showed production of spores on the latter medium. However, considerable variation in the onset and extent of sporulation was observed. Some strains produced only a few spores after 4 weeks of growth, while others produced an abundance of spores within 4 days. Some strains produced spores more readily when the peptone of the medium was replaced by casamino acids (Difco). Crowded conditions on plates apparently stimulated spore formation.

Spore formation of *S. halophila* initially was good on Difco Marine Agar 2216 (composition see above). This property was totally lost by all strains after purification and few transfers on this medium. However, with most strains, sporulation could be retained by growing cultures on the seawater agar of Lyman and Fleming

(1940) and by incubating the cultures at 22°C (Fahmy et al., 1985).

Seawater Agar

Bacto peptone (Difco)	5.0 g
Yeast extract	1.0 g
Ferrous phosphate·4H$_2$O	0.1 g
Manganese chloride	0.01 g
Agar	15.0 g
Synthetic seawater (see below)	1 liter

The pH of the medium is adjusted to 7.1 before sterilization.

Synthetic Seawater

Distilled water	1 liter
Sodium chloride	24.32 g
Magnesium chloride·6H$_2$O	10.99 g
Sodium sulfate	4.06 g
Calcium chloride·2H$_2$O	1.51 g
Potassium chloride	0.69 g
Sodium bicarbonate	0.20 g
Potassium bromide	0.10 g
Strontium chloride·6H$_2$O	0.042 g
Boric acid	0.027 g
Sodium silicate·9H$_2$O	0.005 g
Sodium fluoride	0.003 g
Ammonium nitrate	0.002 g
Ferrous phosphate·4H$_2$O	0.001 g

Like endospores of the genus *Bacillus,* the endospores of the two *Sporosarcina* species contain the spore-specific compound dipicolinic acid, which has never been found in vegetative bacterial cells (Fahmy et al., 1985; MacDonald and MacDonald, 1962; Thompson and Leadbetter, 1963).

Ultrastructure

Studies on the fine structure of vegetative cells, including regular surface layers, have been published for *S. ureae* (Beveridge, 1979; Beveridge, 1980; Engelhardt et al., 1986; Mazanec et al., 1965; Robinson and Spotts, 1983; Silva et al., 1973; Stewart and Beveridge, 1980). The ultrastructure of spores of both *Sporosarcina* species is similar to that of *Bacillus* spores (Silva et al., 1973; Robinson and Spotts, 1983; Fahmy et al., 1985).

Chemotaxonomy

Whereas the cortex of *Sporosarcina* spores, like all *Bacillus* species studied, contains *meso*-diaminopimelic acid, this compound has not been detected in the walls of vegetative cells of both *Sporosarcina* species (Claus et al., 1983; Ranftl, 1972; Schleifer and Kandler, 1972). The peptidoglycan types found in vegetative cells, however, differ with the two species. For *S. ureae,*

the lys-gly-D-glu type has been described (Linnett et al., 1974; Schleifer and Kandler, 1972), whereas in *S. halophila,* the orn-D-asp type is found (Kandler et al., 1983). Also in cell wall preparations of *S. halophila,* high amounts of a gamma-D-glutamyl polymer have been detected, but this compound is not present in *S. ureae* (Kandler et al., 1983).

The menaquinone system of both species is MK-7. This corresponds to the one found in most species of the genus *Bacillus* (Claus et al., 1983; Collins and Jones, 1981; Yamada et al., 1976).

Except for the observation that cell membranes of *S. ureae* contain phosphatidylethanolamine (Komura et al., 1975), other data of chemotaxonomic relevance have not been published.

Genetic and Phylogenetic Relationship

The GC content for 11 strains of *S. ureae* has been found to be in the range of 40.0–41.5 mol% (T$_m$) (Boháček et al., 1968b) and for three other strains, from 40.6–40.8 mol% (T$_m$) (Claus et al., 1983). DNA-DNA hybridization studies have been performed only with two strains of the species. The binding between these is 93% (Fahmy, unpublished observations).

The GC content of 22 strains of *S. halophila* was determined to be in about the same range (40.1–40.9 mol%; T$_m$). According to DNA-DNA hybridization studies, the species is homogenous; the homology of the DNA of the 22 strains studied is from 72–100%. The DNA homology between the type strains of the two species, however, is only 36% (Claus et al., 1983).

A close relationship of the sporeforming genera *Sporosarcina* and *Bacillus* was already suggested by Beijerinck (1901), who observed the occurrence of soil bacteria morphologically intermediate between *S. ureae* and *Bacillus megaterium.*

The genetic relationship of the two genera was first studied by Herndon and Bott (1969) using DNA-RNA hybridization methods. By comparative 16S rRNA cataloging, Pechman et al. (1976) and Fox et al. (1977) have shown that *S. ureae* is more closely related to *Bacillus pasteurii* than are several other species of *Bacillus.* These authors concluded that *S. ureae* should best be classified as a member of the genus *Bacillus.* In the eighth edition of *Bergey's Manual of Determinative Bacteriology,* however, the genus *Sporosarcina* was given generic rank in the family Bacillaceae (Gibson, 1974).

The close phylogenetic relationship of *S. ureae* to members of the genus *Bacillus* has been confirmed by Stackebrandt et al. (1987). *S.*

ureae forms spherical endospores and is found to cluster with other *Bacillus* species forming spherical spores (e.g., *B. pasteurii*, *B. sphaericus*). This group, which also includes some nonsporeforming bacteria, is separated from those *Bacillus* species forming ellipsoidal endospores (e.g., *B. subtilis*, *B. cereus*, *B. megaterium*, *B. pumilus*). These and other studies on the bacterial phylogeny suffer from the fact that very often the type strain of the species were not used.

The phylogenetic relationship of *S. halophila* to *S. ureae* or to members of the genus *Bacillus* has not been reported.

The nucleotide sequence of the 5S rRNA of *S. ureae* has been published by Park et al. (1988).

Physiological and Biochemical Properties

The general properties of *S. ureae* have been described by Kocur and Martinec (1963) and by MacDonald and MacDonald (1962). The results have been confirmed using the methods described by Gordon et al. (1973) for the characterization of *Bacillus* species (D. Claus, unpublished observations). Properties of *S. halophila* have been studied by Claus et al. (1983), using, as far as possible, the same methods. All media, however, were supplemented with 3% NaCl and 0.5% MgCl$_2$. The properties of the two species are listed in Tables 2 and 3.

Identification

The two *Sporosarcina* species can easily be differentiated by the series of phenotypic properties listed in Table 2. They can clearly be separated from morphologically similar cocci by the characters listed in Table 4.

Table 2. Differentiating physiological and biochemical properties of the two *Sporosarcina* species.

Character	S. ureae	S. halophila
Growth		
In nutrient broth	+	−
In nutrient broth plus 10% NaCl + 0.5% MgCl$_2$	−	+
Hydrolysis of:		
Starch	−	+
Pullulan	−	+
Casein	−	+
Gelatin	−	+
Nitrite from nitrate	+	−
Urease	+	−
Tyrosine decomposed	+	−

Symbols: +, positive; −, negative.
Adapted from Claus et al. (1983).

Table 3. Other properties of the two *Sporosarcina* species.

Character	S. ureae	S. halophila
Growth		
In nutrient broth	+	−
In nutrient broth plus 15% NaCl + 0.5% MgCl$_2$	−	+
At 30°C	+	+
37°C	V	+
40°C	−	V
45°C	−	−
At pH 5.7	−	−
Anaerobically	−	−
Catalase	+	+
Oxidase	+	+
Arginine dihydrolase	V	−
Phenylalanine deaminase	V	−
Phosphatase	−	V
Lecithinase	−	−
Indole	−	−
Voges-Proskauer test	−	−
Citrate utilized	−	−
Hydrolysis of		
Dextran	−	−
DNA	V (weak)	+
Chitin	−	−
Cellulose	−	−
Tween 80	V	−
Gas from nitrate	−	−

Symbols: +, positive; −, negative; V, variable.
Adapted from Claus et al. (1983).

Other Studies

Beijerinck (1901) and Gibson (1935) indicated that urea can be hydrolyzed by *Sporosarcina ureae* at concentrations of up to 5–10%. Urease was characterized as an exoenzyme by Pel'ttser (1969) and quantified by Kaltwasser et al. (1972). The presence of enzymes required for a functioning urea (ornithine) cycle in *S. ureae* has been demonstrated by Gruninger and Goldman (1988). A strain of *S. ureae* was included in studies on the esterase pattern of psychrotrophic *Bacillus* species (Higashi and Johnson, 1986).

A bacteriophage of *S. ureae* was first isolated by Kluckhohn and Spotts (1986). Jensen and Stenmark (1970) characterized the control mechanism of 3-deoxy-D-arabinoheptulosonate-7-phosphate synthetase and have shown that the control pattern resembled that of species in the genus *Bacillus*.

Applications

Intracellular L-phenylalanine dehydrogenase of *Sporosarcina ureae* can be used for the efficient enzymatic transformation of phenylpyruvate to L-phenylalanine (Asano and Nakazawa, 1985; Campagna and Bückmann, 1987). The enzyme

Table 4. Differentiating properties of the moderately halophilic cocci.

Character	Planococcus citreus[a]	Planococcus kocurii[b]	Marinococcus halophilus[c]	Marinococcus albus[c]	Marinococcus hispanicus[d]	Salinicoccus roseus[e]	Sporosarcina ureae[f]	Sporosarcina halophila[f]
Motility	+	+	+	+	−	−	+	+
Cell arrangement:								
Irregular clusters	−	−	−	−			−	−
Tetrads	+	+	+	+			+	+
Endospores	−	−	−	−	−	−	+	+
Pigment	YO	YO	YO	CW	RO	PR	D	YO
Growth in presence of:								
0% NaCl	+	+	−	−	−	−	+	−
15% NaCl	+	−	+	+	+	+	−	+
Oxidase	−	−	−	+	+	+	+	+
Urease	−	−	−	+	D	−	+	−
Hydrolysis of:								
Gelatin	+	+	+	−	+	+	−	+
DNA	+	+	−	+	D	+	+	D
Nitrate to nitrite	−	−	+	+	D	+	+	+
Acid from:								
Glucose	+	D	+	−	D	−	−	−
Maltose	−	−	+	−	D	−	−	−
Sucrose	−	−	−	−	−	−	−	−
GC content (mol%)	48–52	39–41	46.4	44.9	45.6–49.3	51.2	40.0–41.5	40.1–41.9
Menaquinone system	MK-7, MK-8	MK-7, MK-8	MK-7	MK-7	MK-7, MK-8	MK-6[g]	MK-7	MK-7
Cell wall type	L-Lys-D-glu	L-Lys-D-glu	meso-A$_2$pm	meso-A$_2$pm	meso-A$_2$pm	L-Lys-gly$_5$	L-Lys-gly-D-glu	Orn-D-asp

Symbols: +, positive; −, negative; D, differs among strains; pigment; YO, yellow-orange; CW, creamy white; RO, reddish orange; PR, pink-red.

[a]Kocur, 1986.
[b]Hao and Komagata, 1985.
[c]Hao et al., 1984.
[d]Marquez et al., 1990.
[e]Ventosa et al., 1990.
[f]Claus and Fahmy, 1986.
[g]B. Tindall, personal communication.

was highly purified and crystallized by Asano and Nakazawa (1985). It can also be used for the high yield synthesis of other L-amino acids from their keto analogs by combination with formate dehydrogenase (Asano and Nakazawa, 1987).

The Genus *Planococcus*

Motile cocci were first isolated by Ali-Cohen (1889) and were named *"Micrococcus agilis."* Migula (1894) established, for motile cocci, the new genus *"Planococcus,"* which he placed into the family Coccaceae (Migula, 1900). However, most bacterial taxonomists continued to include motile cocci in the genus *Micrococcus,* as they are very similar in cultural and biochemical characteristics to true micrococci. Bohácek et al. (1967, 1968a) determined the GC content of some micrococci and concluded that nonmotile cocci are not taxonomically related to motile strains. Kocur et al. (1970) studied the taxonomy of motile cocci and revived and amended the genus *Planococcus* Migula 1894 with the type species, *P. citreus.*

Only two species are now recognized in the genus, *P. citreus* Migula 1894, 236[AL] and *P. kocurii* Hao and Komagata 1986[VP]. As a result of chemotaxonomical studies, the species *P. halophilus* Novitski and Kushner 1976, 53[AL] was transferred to a new genus called *Marinococcus* (Hao et al., 1984).

Habitats

Planococci have been isolated from marine environments, sea water (ZoBell and Upham, 1944), marine clams (Leifson, 1964), fish-brining tanks (Georgala, 1957), and boiled and frozen shrimps and prawns (Alvarez, 1982; Hao and Komagata, 1985). They play a role in food hygiene because they may cause putrefaction of shrimps during storage (Alvarez, 1982).

Isolation

No selective medium for the isolation of planococci has been devised. For their isolation, seawater agar or nutrient agar supplemented with 10% NaCl may be used.

Seawater Agar

Beef extract	10 g
Peptone	10 g
Tap water	250 ml
Agar	20 g
Seawater	720 ml
Adjust to pH 7.20	

Cultivation

Planococci can be cultivated on seawater agar, on peptone-yeast agar, or on nutrient agar supplemented with up to 10% sodium chloride.

Maintenance and Preservation

Cultures of planococci may be maintained on slants of seawater agar at 4°C for about 6 months. Strains can be preserved in a freeze-dried state or in liquid nitrogen for long periods.

Cultural and Microscopic Appearance

Colonies of planococci are circular, slightly convex, smooth, and yellow-orange in color due to water-insoluble carotenoid pigments. Pigmentation as well as the type of carotenoids produced may be influenced by the concentration of sodium chloride in the medium and by the age of the culture (Thirkell and Summerfield, 1980). A hydrostatic pressure of 40 MPa has no influence on the pigment production of *P. citreus* (Courington and Goodwin, 1955).

The genus includes Gram-positive, nonsporforming, spherical cells of 1.0 to 1.2 μm in diameter, occurring singly, or in pairs, tetrads, or clumps. Motile cells usually have one to three flagella. Endospores are not formed.

Ultrastructure

The fine structure of the cells of *Planococcus citreus* is similar to that of other Gram-positive, catalase-positive cocci. The cell wall is double layered. Its thickness varies with the cell age from 25–35 nm (Kocur, 1986).

The concentration of salt in the medium affects the amount of membrane in the cell of *P. citreus.* Salt concentrations above and below the normal 3% of sea water apparently reduce the amount of membrane material present (Thirkell and Summerfield, 1977a).

Chemotaxonomy

The cell wall peptidoglycan of both *Planococcus* species is of the L-lys-D-glu type (Schleifer and Kandler, 1970; Hao and Komagata, 1985). No teichoic acid was found in the cell wall of planococci (Endresen and Oeding, 1973). *Marinococcus* species, in contrast, have a cell wall of the *meso*-diaminopimelic acid type.

In common with other Gram-positive bacteria, *P. citreus* contains *anteiso* $C_{15:0}$ as the major component of free fatty acids. Phospholipids detected in planococci are similar to that of *Sporosarcina* and include phosphatidylethanolamine, cardiolipin, and phosphatidylglycerol (Yamada et al., 1976; Thirkell and Summerfield, 1977b; Hao and Komagata, 1985).

Planococci contain normal menaquinones, with MK-8 and MK-7 in about equal amounts. A small amount of MK-6 is also present (Hao and Komagata, 1985).

Genetic and Phylogenetic Relations

The analysis of the GC content revealed two groups among *Planococcus* species. Strains of *P. citreus* show 47 to 52 mol% (T_m), while the DNA base composition of *P. kocurii* is in the range of 39–42 mol% (T_m) (Boháček et al., 1967, 1968; Hao and Komagata, 1985). Strains of *Marinococcus,* earlier grouped with *Planococcus,* have a GC content in the range of 43.9 to 46.6 mol% (T_m) (Hao et al., 1984). DNA-DNA hybridization studies have not been performed within the genus.

A comparative analysis of 16S rRNA sequences has shown that planococci bear no specific relationship to the genera *Micrococcus* and *Staphylococcus* (Stackebrandt and Woese, 1979). Planococci show a specific relationship to the genus *Bacillus,* particularly to *Bacillus pasteurii,* and to the genus *Sporosarcina* (Boháček et al., 1968b, Pechman et al., 1976; Stackebrandt and Woese, 1979). The data on the GC content and the chemical composition of the cell wall support these observations (Boháček et al., 1968b; Schleifer and Kandler, 1970). There is no antigenic relationship of planococci to staphylococci and micrococci (Oeding, 1971). In the light of the above facts, therefore, the genus *Planococcus* should be placed into the family Bacillaceae as a separate genus.

Physiological and Biochemical Properties

Strains of *Planococcus citreus* and *P. kocurii* can grow in the absence of sodium chloride but can also tolerate as much as 10% sodium chloride. They grow in the range of 5–30°C. Growth at 37°C is variable. Their biochemical activities are rather restricted (Table 4). They are catalase positive and hydrolyze DNA and gelatin (variably). Acid is formed from glycerol and glucose (variably), but not from other sugars or sugar alcohols.

Planococci are negative for the following reactions: oxidase, methyl red, Voges-Proskauer, production of indole, H_2S, nitrate reduction, phenylalanine deaminase, lysine decarboxylase, ornithine decarboxylase, arginine dihydrolase, urease, degradation of tyrosine, and hydrolysis of starch, esculin, and Tween 80. Assimilation of a series of organic acids is variable or negative (Hao and Komagata, 1985).

Planococci are susceptible to lysozyme, chloramphenicol, erythromycin, novobiocin, oleandomycin, penicillin, and tetracycline (Jeffries, 1969; Kocur et al., 1970).

P. citreus was found to be one of the few marine bacteria that reproduced at a hydrostatic pressure of 20 to 40 MPa (Oppenheimer and ZoBell, 1952).

Identification

Planococci may be clearly separated from morphologically similar cocci by the characters listed in Table 4. They differ from the species of the genus *Marinococcus* by their ability to grow on media without NaCl added and by their inability to grow on peptone-yeast agar supplemented with 15% NaCl. *Sporosarcina halophila* differs from planococci in endospore formation, inability to grow on media containing 15% NaCl, esculin hydrolysis, and several chemotaxonomic characters. The two *Planococcus* species can be differentiated phenotypically only by their ability to grow on media containing 15% NaCl, which is positive only for *P. citreus.*

However, differentiation is also possible through their DNA base composition, zymogram, and protein pattern. They can also be separated serologically, as shown by Oeding (1971).

The Genus *Marinococcus*

The genus *Marinococcus* includes Gram-positive, aerobic, and moderately halophilic cocci with *meso*-A_2pm acid in the cell wall. The genus was established by Hao et al. (1984) who transferred *Planococcus halophilus* Novitzki and Kushner (1976) into this genus and described the new species *Marinococcus albus.* Another new species, *M. hispanicus* has been isolated from a solar saltern (Marquez et al., 1990).

Habitat

Marinococci have only been isolated from marine environments, particularly from solar salterns (Ventosa et al., 1983; Marquez et al., 1990).

Isolation

No selective medium for the isolation of marinococci has been described. For their isolation, the salt complex medium of Ventosa et al. (1982) may be used.

Salt Complex Medium (Ventosa et al., 1982)

NaCl	8.1 g
MgCl₂	0.7 g
MgSO₄	0.96 g
CaCl₂	0.036 g
KCl	0.2 g
NaHCO₃	0.006 g
NaBr	0.026 g
Yeast extract	1.0 g
Proteose peptone No. 3	0.5 g
Glucose	0.1 g
Bacto-Agar (Difco)	20.0 g
Water	1 liter
Adjust to pH 7.5.	

Cultivation

Marinococci can be cultivated on salt complex medium (Ventosa et al. 1982).

Maintenance and Preservation

Cultures of *Marinococcus* may be kept for 6 months at about 4°C on salt complex medium (Ventosa et al., 1982) if protected from drying. Cultures can be preserved in the freeze-dried state or in liquid nitrogen for long periods.

Genetic and Phylogenetic Relationships

The GC content of marinococci is in the range of 44.9–49.3 mol% (Hao et al. 1984; Marquez et al., 1990). DNA-DNA hybridization studies have not been performed within the genus.

Physiological and Biochemical Properties

Strains of *Marinococcus* species grow well in media with 20% sodium chloride, but not on media without salt. They grow in the range of 15–37°C. Like the planococci, the biochemical activity of the marinococci is restricted. They are catalase positive but oxidase production varies with species. *M. halophilus* and *M. hispanicus* may produce acid from glycerol and hydrolyze gelatin and esculin.

Marinococci are negative for the following reactions: acid from lactose, arabinose, galactose, and fructose; production of acetoin and indole; hydrolysis of starch and Tween 80; lysine and ornithine decarboxylases, arginine dihydrolase, phenylalanine deaminase, phosphatase, and growth on nutrient agar without sodium chloride.

Identification

Marinococci may be clearly separated from morphologically similar genera by the characteristics listed in Table 4. The three *Marinococcus* species can be differentiated by several phenotypic properties which are also listed in Table 4.

The Genus *Salinicoccus*

The genus *Salinicoccus* has been described quite recently (Ventosa et al., 1990). Its characteristics are based on only one strain which was isolated from a solar saltern near Alicante (Spain). The properties of this moderately halophilic nonsporeforming coccus are significantly different from all other hitherto described cocci. This justifies its recognition as a new species in a new genus.

The characteristics which distinguish *Salinicoccus roseus* from the phenotypically similar Gram-positive cocci are given in Table 4. There are no data on the ultrastructure, or the genetic and phylogenetic relationships of *S. roseus*.

The Genus *Filibacter*

A study of filamentous bacteria in sediments of lakes (Maiden, 1983) resulted in the isolation of an organism that resembled members of the genus *Vitreoscilla* in that it was a multi-cellular, filamentous, gliding bacterium that was not pigmented. Major differences in cytochrome and DNA base composition, isoprenoid quinone content, and sensitivity to actinomycin D seemed to indicate a more close relationship to the Flexibacteriaceae. As the organism did not resemble any previously described taxon of Flexibacteriaceae and on the basis of differences from both, it was proposed that the organism be placed in a new genus, *Filibacter,* with the type species *Filibacter limicola* Maiden and Jones, 1985, 375[VP] (Maiden and Jones, 1984).

Habitat and Isolation

Filibacter limicola has only been isolated once, from sediments of an eutrophic freshwater lake in the English Lake District, Blelham Tarn (Maiden, 1983). For isolation, sediment cores are sampled and treated according to Maiden and Jones (1984) as follows:

Isolation Procedure for *F. limicola*

> The top 1 to 2-cm layer of a sediment core is taken and diluted in filtered lake water. 10 μl of the sample are used for inoculation of plates containing the following medium:

MYP Medium (Maiden and Jones, 1984)

Peptone (Oxoid, L37)	0.1 g
Yeast extract (Difco)	0.01 g
K₂HPO₄	0.028 g

MgCl$_2$·6H$_2$O	0.127 g
KNO$_3$	0.004 g
(NH$_4$)SO$_4$	0.06 g
MnSO$_4$·4H$_2$O	0.008 g
Ferric citrate	0.006 g
Agar	10.0 g
Distilled water	1 liter

The pH of the medium is adjusted with potassium hydrogen carbonate to give a final pH of 7.0. The plates are incubated at 20°C for two weeks. They are periodically examined for growth of filaments through a dissecting microscope at a magnification of about 30×. *Filibacter* filaments may occur singly, in small groups, or in circular formations at some distance from the inoculation point. Purification may be done by excising small agar blocks with a single or a few attached filaments and transferring them to fresh medium. This procedure is repeated until pure cultures are obtained.

Cultivation

Pure cultures may be grown at about 20°C on nutrient agar (Oxoid CM 3), tryptone soya agar (TSA, Oxoid CM 131), or on the defined medium used for the isolation. Good growth is also obtained on media containing certain amino acids and vitamins as the sole source of carbon and nitrogen source (Maiden and Jones, 1984).

Cultures, incubated at 20°C, can be maintained on slants of TSA at 10°C for some weeks.

Cultural and Microscopic Appearance

Growth on solid media is characterized by spreading whorls of growth and spiral colonies which are not pigmented. The organisms exhibit gliding motility and grow in filaments of 8–150 μm where the individual cells are 3–30 μm long and about 1 μm wide. The cells are long rods that are either straight, or curved with rounded ends. Junctions between individual cells are marked by constrictions.

Capsules and sheaths are not formed but slime is excreted. Endospores and flagella as well as branching are not observed. Granules staining with sudan black were demonstrated but the test of Oste and Holt (1982) for poly-β-hydroxybutyrate is negative. Volutin granules are not observed. The cells stain Gram-negatively, although the cell wall is about 40 μm thick, lacks an outer membrane, and therefore shows the characteristics typical of a Gram-positive organism. Electron micrographs of thin sections showed a five-layered cell wall (Clausen et al., 1985).

Chemotaxonomy

Membrane fatty acids of *F. limicola* are dominated by anteiso-C$_{15:0}$ and anteiso-C$_{17:0}$ components (Nichols et al., 1986). The cell wall of

Filibacter limicola contains no diaminopimelic acid and MK-7 is the major quinone isoprenolog. Cytochromes of the *c* type were determined to predominate, with lesser amounts of *b*-type cytochromes (Maiden and Jones, 1984).

Genetic and Phylogenetic Relationship

The GC content of the DNA of the type strain is 44 mol% (T$_m$) (Maiden and Jones, 1984). Comparative 16S rRNA cataloging studies show that *Filibacter* clusters with the Gram-positive species *Bacillus globisporus*, with which it forms a branch within a grouping that comprises also *Caryophanon latum, Sporosarcina ureae, Planococcus citreus*, and other spherical sporeforming *Bacillus* species. Several phenotypic properties support the allocation of this organism to the family Bacillaceae.

Physiological and Biochemical Properties

The organism is strictly aerobic in the temperature range around 20°C, with 4°C as a minimum and no growth at 30°C. Catalase and oxidase are present. Casein or starch are not hydrolyzed whereas gelatin is. Sulfur globules are not deposited intracellularly in the presence of sulfide. Urease is present. Carbohydrates are not degraded. The only compounds which are weakly used as carbon sources in the presence of amino acids are acetate, lactate, butyrate, and glycerol.

A similarly narrow range of carbon sources can be found with spherical sporeforming *Bacillus* strains such as *B. pasteurii* or *B. sphaericus*, with strains of the genus *Caryophanon*, or with *Planococcus* strains.

Identification

Filibacter limicola is morphologically similar to *Vitreoscilla stercoraria* but not to the two other *Vitreoscilla* species described. Differences to *V. stercoraria* are listed in Table 5. It should be emphasized, however, that only one strain of each of the two species has been compared.

Table 5. Differentiating properties of *Filibacter limicola* and *Vitreoscilla stercoraria*.

Character	F. limicola	V. stercoraria
Strict aerobe	+	+
Catalase	+	+
Oxidase	+	−
Growth at 4°C	+	−
Growth at 30°C	−	+
Hydrolysis of gelatin	+	−

Adapted from Maiden and Jones (1984).

The Genus *Caryophanon*

The genus name *Caryophanon* (karyon [Greek] = nucleus; phaneros [Greek] = bright, conspicuous, that which has a conspicuous nucleus) is a misnomer, arising from the original false conclusion that the cross-walls, nuclear material, and cytoplasm seen in stained trichomes was a nucleus. The true nature of the trichomes was shown by improved cytological techniques (Pringsheim and Robinow, 1947).

Historically, *C. latum* played a significant role in the development of bacterial cytology in the 1940s and 1950s: Because of the large size of its trichomes and its numerous stainable "nuclei," *Caryophanon* provided an ideal model in the "great mitosis" debate (Tuffery, 1955).

Today, two species are recognized, *Caryophanon latum* Peshkoff 1939, 244[AL] and *Caryophanon tenue* (Peshkoff) Trentini 1988, 220[VP] (effective publication: Trentini 1986, 1259). Affiliation of the genus to already known, specific taxa was always difficult. Trichome-forming bacteria are a diverse group of organisms that have as a common property an outer cell wall that holds together the individual cells of the trichome.

In the seventh edition of *Bergey's Manual of Determinative Bacteriology* (1957) the order Caryophanales was described to include trichome-forming organisms comprising three families: Caryophanaceae, Oscillospiraceae (sporeforming organisms, not in pure culture) and Arthromitaceae. Later, in the eighth edition of *Bergey's Manual* (1974), *Caryophanon* was listed under Part 16: Gram-Positive, Asporogenous Rod-shaped Bacteria: Genus of Uncertain Affiliation. In *Bergey's Manual of Systematic Bacteriology*, *Caryophanon* was placed into Section 14: Regular, Nonsporing, Gram-positive Rods. However, recent rRNA sequencing studies have revealed a close relationship of *Caryophanon* to the aerobic sporeforming bacteria (Stackebrandt et al., 1987).

Habitat

Caryophanon latum was first isolated in 1937 and *C. tenue* in 1938 from fresh cow manure (Peshkov, 1939). In most of the successive studies, cattle manure was used as the primary source of these organisms (Gershenfeld and Lam, 1953; Kele, 1970; Moran and Witter, 1976; Peshkov and Marek, 1973; Pringsheim and Robinow, 1947; Provost and Doetsch, 1962; Smith and Trentini, 1972; Trentini and Machen, 1973; Tufferey, 1955; Weeks and Kelley, 1958). Reports that *C. latum* has been found in or on other sources such as sewage (Tuffery, 1953), the oral cavity of dogs (Saphir and Carter, 1976), or decaying *Pleurotus* on the stump of a tree (R. E. Buchanan, unpublished observations) could not be verified. Other findings that *Caryophanon* occurs associated with manure of other animals (Kele, 1970; Trentini and Machen, 1973; Dean, 1973) may be doubted because of possible contamination of these sources with cattle manure.

Most samples taken from rumen fluid, cattle saliva, teeth scrapings, anal swabs, or rectal and intestinal samples from slaughtered cattle were negative for *Caryophanon*. Also, *Caryophanon* was not found in aseptically sampled cattle manure. It may, therefore, be concluded that these bacteria are not residents of the bovine digestive tract (Pringsheim and Robinow, 1947; Trentini and Machen, 1973; Kele, 1970; Dean, 1963).

Isolations of *Caryophanon* are easiest from 1- to 2-day-old droppings, whereas old field-dried cattle manure was usually negative for these organisms. Although the natural habitat of *Caryophanon* is not definitely known, it is assumed that these bacteria are secondary contaminants of cattle manure, dispersed to new droppings by air, by insects, or by the cattle itself.

Enrichment and Isolation

Caryophanon strains can be isolated only after an enrichment step. Good enrichment results are regularly obtained with fresh droppings from the barn gutters found with stanchioned cattle or with pasture manure after 1 to 2 days of field-aging as long as little or no rain has fallen during that time and the temperature has not fallen below 0°C.

Enrichment Procedure (Pringsheim and Robinow, 1947; Smith and Trentini, 1972)

> Cow dung about two days old is collected from pastures and brought to the laboratory as soon as possible. Samples of 600–800 ml are distributed into beakers and thoroughly mixed with distilled water so that the surface remains covered by a thin layer of water. The beakers are covered with aluminum foil. After 16–24 h of incubation at room temperature, top slurry material is viewed under the microscope for the typical, actively motile *Caryophanon* trichomes and trichome chains.
>
> Attempts at isolation should be made only on samples in which, at a magnification of 400×, *C. latum* occurs in numbers of at least three to four trichomes or trichome-chains per microscopic field. More concentrated samples with up to 20–30 trichomes per field may also be observed.

For isolation, plates may be inoculated directly from this surface material. More enhanced isolation may be achieved by removing as far as possible larger and smaller particles from the slurry such as plant material, protozoa, bacteria, and bacterial and fungal spores. It is necessary to combine several filtration and centrifugation steps:

Isolation Procedure

Skim off the top slurry material from samples with high numbers of trichomes and filter it through coarse filter paper (e.g.) coffee filter paper) to remove large particles (very small pores would soon be clogged by the very fine dung particles). Rinse the solid matter trapped by the filter paper with about 10 ml distilled water and filter again by squeezing it gently.

Centrifuge the combined filtrate using a rotor with approximately r= 15 cm at 1500 rpm. Under these conditions the relatively large trichomes will sediment in about 5 min. Discard the supernatant which includes a large proportion of the fine material. Resuspend the pellet in nutrient broth and repeat the filtration, centrifugation, and resuspension steps once or more.

By applying vacuum, filter the suspension through a membrane filter of about 8-μm pore diameter while adding repeatedly about 20 ml nutrient broth to a total amount of 400 ml. Take care to keep the filter moist throughout this procedure to prevent injury to the trichomes from drying. Remove the membrane filter and agitate it thoroughly along with 2 ml nutrient broth in a flask to wash off the adherent trichomes.

From the final suspension, prepare 10^{-2} to 10^{-5} dilutions and spread 0.1 ml of each dilution on cow dung agar plates containing 80 μg/ml streptomycin.

Cow Dung Agar (Smith and Trentini, 1972)

Cow dung	250 ml
Distilled water	750 ml
Agar	15 g

Mix fresh cow dung with distilled water for 2–3 min in a laboratory mixer to break up large dung particles and ensure thorough dispersion. Add additional water up to 1 liter and sterilize the medium for 20 min at 121°C. 10 ml of filter-sterilized streptomycin solution (0.8 g per 100 ml) is added after heat sterilization of the medium.

After incubation at room temperature for 48 h, typical colonies of Caryophanon colonies may develop. To obtain pure cultures, suspended material from single colonies is repeatedly restreaked onto cow dung agar with or without streptomycin.

Cultivation

Growth of Caryophanon strains is good on cow dung agar described above without the addition of streptomycin. Several semisynthetic media have been described for C. latum (Pringsheim and Robinow, 1947; Provost and Doetsch, 1962; Kele and McCoy, 1971; Smith and Trentini, 1973). However, they are all not as good as cow dung agar with respect to doubling time and maintenance of the original morphology.

Cultures usually grow best at room temperature or 25°C at a pH of 7.8–8.5. Moran and Whitter (1976), however, stated that 35°C is the optimal temperature for growth of C. latum, based upon the rate of change in colony diameter on a given medium.

Maintenance and Preservation

Active cultures of both Caryophanon species may be kept at 4–10°C on agar slants for more than 4–6 weeks if protected from drying. Cultures are readily preserved by standard techniques such as freezing in liquid nitrogen or lyophilization.

Cultural and Microscopic Appearance

Colonies of C. latum are pale yellow, about 1.5 mm in diameter, opaque, granular, lobate, and show a glistening surface after growth for 48 h on cow dung agar or cow dung agar containing 0.5% lactalbumin hydrolysate at room temperature. Colonies of C. tenue are similar to those of C. latum in overall appearance, except that they are only 0.5–1.0 mm in diameter and less irregular in shape under the same conditions.

Trichomes of C. latum are about 3 μm in width, and 10–20 μm in length. They are straight or slightly curved. Trichome chains may be formed especially in enrichment cultures. They contain two to six trichomes. Individual cells within a trichome are wider than they are long, and several growing septa at various stages of closure can be observed at the same time in each cell.

The size and shape of the organism vary with the culture conditions. As growth proceeds, transverse fissions may give rise to even-shorter, and even spherical, forms. Size frequently diminishes during laboratory cultivation, so that eventually most of the trichomes in a culture may not exceed 1 μm in diameter.

Trichomes of C. tenue are about 1.5 μm in width, 10–20 μm in length, straight or slightly curved, and are composed of fewer cells than the trichomes of C. latum. Individual cells within a trichome are longer than wide. Only one cross septum can be observed per single cell (Peshkov and Marek, 1973), which may be explained by the much slower growth of C. tenue in comparison to C. latum.

Caryophanon trichomes are Gram-positive, peritrichously flagellated, and show no branching. Their ends are round or slightly tapered. Under optimal growth conditions, individual cells exhibit a disk-like shape. Cross walls between cells show as dark lines, partly complete, partly developing by ingrowth from the external wall. Endospores, sheaths, and capsules are not formed.

In old cultures or under growth conditions which lead to extensive lysis, a small number of cell units is preserved as round bodies (spheroids). The significance of the spheroids is unknown but it has been suggested that they are part of a "life cycle" (Peshkov, 1939). Under good growth conditions, the spheroids are said to be able to grow into normal trichomes. It should be noted here that almost all investigations have been carried out with *Caryophanon latum*, the morphology of which is strikingly sensitive to growth conditions.

Morphology and motility of *Caryophanon* are difficult to maintain. Both characteristics are reported to be best preserved when 0.5 to 1% lactalbumin hydrolysate is added to cow dung medium. The most authentic morphology and motility of the trichomes, though, can still be observed in enrichment cultures. Optimal growth conditions that will maintain natural trichome morphology as well as mass or colony diameter increase in pure cultures have still to be developed.

Ultrastructure

In cells of *C. latum,* one or two superficial wall layers containing protein are present (Trentini and Gilleland, 1974). Mesosomes, nucleidosomes, and analogs of mitochondria were described by Shadrina et al. (1982).

Chemotaxonomy

Diaminopimelic acid is absent in *C. latum* (Becker et al., 1967). The peptidoglycan of the cell wall is composed of glutamic acid, alanine, lysine, and muramic acid in a molar ratio of 2:2:1:1. Wall material is sensitive to lysozyme (Trentini and Murray, 1975). Teichoic acids and *O*-acetyl groups are absent (W. Trentini, unpublished observations).

Genetic and Phylogenetic Relationships

Thirty-six strains of *C. latum* isolated from various geographic regions exhibit GC contents of 44.0–45.6 mol% DNA-DNA hybridization studies reveal—with a homology of 78–92%—a high homogeneity among the strains (Adcock et al., 1976). The genome size is 1,100–1,200 × 10^6 dalton.

Three strains of *C. tenue* examined for GC content exhibit values of 41.2–41.6 mol% (Adcock et al. 1976). The genome size is 900–1,000 × 10^6 dalton. DNA-DNA reassociation studies between *C. tenue* and *C. latum* revealed low homology values of 13–30%.

A close phylogenetic relationship of *Caryophanon* to members of the genus *Bacillus,* in particular to the spherical sporeforming *Bacillus* species, has been described by Stackebrandt et al. (1987). In a dendrogram based on rRNA sequencing studies, the closest relative of *Caryophanon* was *Bacillus sphaericus.* In the same cluster were found *Planococcus citreus, Filibacter limicola,* and *Sporosarcina ureae,* as well as other spherical sporeforming *Bacillus* species.

Physiological and Biochemical Properties

Studies on physiological and biochemical properties of the two *Caryophanon* species have been published mainly by Pringsheim and Robinow (1947), Provost and Doetsch (1962), and Adcock et al., (1976).

Both species are strict aerobes. Catalase is formed. The cytochrome-oxidase reaction is negative. Nutrition is chemoorganotrophic. Starch, gelatin, casein, and cellulose are not hydrolyzed, and indole is not produced. Acetate is used as a major carbon source. More recent studies (Rowenhagen, 1987) revealed that a number of other fatty acids (e.g.) butyrate, valerate, capronate, stearate, methylpropionate, and 2-methylbutyrate) are oxidized by most *Caryophanon* strains. Biotin has been found to be essential for growth, and thiamine seems to be stimulatory.

Strains of *C. latum* were shown not to degrade uric acid or urea and not to utilize glucose and other sugars. Poly-β-hydroxybutyrate is used as a carbon source. Tributyrin is only very weakly hydrolyzed. Nitrates are not reduced to nitrites. These organisms grow at 10 and 37°C but not at 45°C and at pH values between 6 and 8, with an optimum at pH 7.6–8 (Weeks and Kelley, 1958).

For both species, no information has been published on the major metabolic pathways, genetics, or antigenic structure. In almost all cultures with an unbalanced nutrition (e.g., older cultures) poly-β-hydroxybutyrate accumulates.

Table 6. Differentiating properties of *C. latum* and *C. tenue.*

Character	*C. latum*	*C. tenue*
Trichome width (μm)	2.8–3.2	1.4–2.0
Number of cross septa/cell unit	several	1
GC content (mol%)	44.0–45.6	41.2–41.6
Genome size (× 10^6 dalton)	1,100–1,200	900–1,000

Adapted from Trentini (1986).

Caryophanon strains are resistant to streptomycin, nalidixic acid, several sulfa drugs, and polymyxin B.

Drozd et al. (1987) described a case of concrete corrosion caused by a *Caryophanon* species.

Identification

The genus and the two recognized species can only be identified mainly through their morphological properties (Table 6).

A species-specific phage for *C. tenue* has been described (Peshkov et al., 1973; Peshkov et al., 1966).

Literature Cited

Adcock, K. A., R. J. Seidler, and W. C. Trentini. 1976. Deoxyribonucleic acid studies in the genus *Caryophanon*. Can. J. Microbiol. 22:1320–1327.

Ali-Cohen, C. H. 1889. Eigenbewegung bei Mikrokokken. Zentralbl. Bakteriol. Parasitenkd. Infektionskr. Hyg. Abt. 1 Orig. 6:33–36.

Alvarez, R. J. 1982. Role of *Planococcus citreus* in the spoilage of *Penaeus* shrimp. Zentralbl. Bakteriol. Parasitenkd. Infektionskr. Hyg., Abt. 1 Orig. Reihe C 3:503–512.

Amemiya, Y. and O. Nakayama. 1980. Polysaccharide formation by spore-bearing lactic acid bacteria. J. Gen. Appl. Microbiol. 26:159–166.

Asano, Y. and A. Nakazawa. 1985. Crystallization of phenylalanine dehydrogenase from *Sporosarcina ureae*. Agric. Biol. Chem. 49:3631–3632.

Asano, Y. and A. Nakazawa. 1987. High yield synthesis of L-amino acids by phenylalanine dehydrogenase from *Sporosarcina ureae*. Agric. Biol. Chem. 51:2035–2036.

Becker, B., E. M. Wortzel, and J. H. Nelson, III. 1967. Chemical composition of the cell wall of *Caryophanon latum*. Nature 213:300.

Beijerinck, M. W. 1901. Anhäufungsversuche mit Ureumbakterien, Ureumspaltung durch Urease und durch Katabolismus. Zentralbl. Bakteriol. Parasitenkd. Infektionskr. Hyg. Abt. 2 7:33–61.

Bennett, J. F. and E. Canale-Parola. 1965. The taxonomic status of *Lineola longa*. Arch. Microbiol. 52:197–205.

Beveridge, T. J. 1979. Surface arrays on the wall of *Sporosarcina ureae*. J. Bacteriol. 139:1039–1048.

Beveridge, T. J. 1980. Cell division in *Sporosarcina ureae*. Can. J. Microbiol. 26:235–242.

Boháček, J., M. Kocur, and T. Martinec. 1967. DNA base composition and taxonomy of some micrococci. J. Gen Microbiol. 46:369–376.

Boháček, J., M. Kocur, and T. Martinec. 1968a. Deoxyribonucleic acid base composition of some marine halophilic micrococci. J. Appl. Bacteriol. 31:215–219.

Boháček, J., M. Kocur, and T. Martinec. 1968b. Deoxyribonucleic acid base composition of *Sporosarcina ureae*. Arch. Microbiol. 64:23–28.

Bonde, G. J. 1981. *Bacillus* from marine habitats: allocation to phena established by numerical techniques, p. 181–215. In: R. C. W. Berkeley and M. Goodfellow (ed.),

The aerobic endosporeforming bacteria. Academic Press, London.

Campagna, R. and A. F. Bückmann. 1987. Comparison of the production of intracellular L-phenylalanine dehydrogenase by *Rhodococcus* M4 and *Sporosarcina ureae* at 50 liter scale. Appl. Microbiol. Biotechnol. 26:417–421.

Chomarat, M., M. de Montclos, J.-P. Flandrois, and F. Breysse. 1990. Isolation of *Sporosarcina ureae* from a bronchial biopsy from a child with cystic fibrosis. Europ. J. Clin. Microbiol. Infect. Dis. 9:302.

Claus, D. 1981. The genus *Sporosarcina*, p. 1804–1807. In: M. P. Starr, M. Stolp, H. G. Trüper, A. Balows, and H. G. Schlegel (ed.), The prokaryotes. Springer-Verlag, Berlin.

Claus, D. and R. C. W. Berkeley. 1986. Genus *Bacillus*, p. 1105–1139. In: P. H. A. Sneath, N. S. Mair, M. E. Sharpe, and J. G. Holt (ed.), Bergey's manual of systematic bacteriology, vol. 2. Williams and Wilkins, Baltimore.

Claus, D. and F. Fahmy. 1986. Genus *Sporosarcina* Kluyver and van Niel 1936, 401^{AL}, p. 1202–1206. In: P. H. A. Sneath, N. S. Mair, M. E. Sharpe, and J. G. Holt (ed.), Bergey's manual of systematic bacteriology, vol. 2. Williams and Wilkins, Baltimore.

Claus, D., F. Fahmy, H. J. Rolf, and N. Tosunoglu. 1983. *Sporosarcina halophila* sp. nov., an obligate, slightly halophilic bacterium from salt marsh soils. System. Appl. Microbiol. 4:496–506.

Clausen, V., J. G. Jones, and E. Stackebrandt. 1985. 16S ribosomal RNA analysis of *Filibacter limicola* indicates a close relationship to the genus *Bacillus*. J. Gen. Microbiol. 131:2659–2664.

Collins, M. D. and D. Jones. 1979. Isoprenoid quinone composition as a guide to the classification of *Sporolactobacillus* and possibly related bacteria. J. Appl. Bacteriol. 47:293–297.

Collins, M. D. and D. Jones. 1981. The distribution of isoprenoid quinone structural types in bacteria and their taxonomic implications. Microbiol. Rev. 45:316–354.

Courington, D. P. and T. W. Goodwin. 1955. A survey of the pigments of a number of chromogenic marine bacteria with special reference to the carotenoids. J. Bacteriol. 70:568–571.

Dean, D. S. 1963. Response of *Caryophanon latum* to oxygen. J. Bacteriol. 85:249–250.

Doores, S. and D. C. Westhoff. 1981. Heat resistance of *Sporolactobacillus inulinus*. J. Food Sci. 46:810–812.

Doores, S. and D. C. Westhoff. 1983. Selective method for the isolation of *Sporolactobacillus* from food and environmental sources. J. Appl. Bacteriol. 54:273–280.

Drozd, G. Ya., M. A. Sobol', and Yu. S. Varenko. 1987. Concrete corrosion caused by a *Caryophanon-* species. Mikrobiol. (Kiev) 49:61–64.

Endresen, C. and P. Oeding. 1973. Purification and characterization of serologically active cell wall substances from *Planococcus* strains. Acta Pathol. Microbiol. Scand. Sect. B 81:571–575.

Engelhardt, H., W. O. Saxton, and W. Baumeister. 1986. Three-dimensional structure of the tetragonal surface-layer of *Sporosarcina ureae*. J. Bacteriol. 168:309–317.

Fahmy, F., F. Mayer, and D. Claus. 1985. Endospores of *Sporosarcina halophila* characteristics and ultrastructure. Arch. Microbiol. 140:338–342.

Fox, G. E., K. R. Pechman, and C. R. Woese. 1977. Comparative cataloging of 16S ribosomal ribonucleic acid:

molecular approach to prokaryotic systematics. Int. J. System. Bacteriol. 27:44–57.

Fox, G. E., E. Stackebrandt, R. B. Hespell, J. Gibson, J. Maniloff, T. A. Dyer, R. S. Wolfe, W. E. Balch, R. S. Tanner, L. J. Magrum, L. B. Zablen, R. Blakemore, R. Gupta, L. Bonen, B. J. Lewis, D. A. Stahl, K. R. Luehrsen, K. Chen, and C. R. Woese. 1980. The phylogeny of prokaryotes. Science 209:457–463.

Georgala, D. L. 1957. Quantitative and qualitative aspects of the skin flora of North Sea cod and the effect thereon of handling on ship and on shore. Ph.D. Thesis. University of Aberdeen, Scotland

Gershenfeld, L. and G. T. Lam. 1953. The effect of certain antiseptics on *Caryophanon latum* Peshkoff. Am. J. Pharmacy 125:5–34.

Gibson, T. 1935. An investigation of *Sarcina ureae*, a sporeforming, motile coccus. Arch. Mikrobiol. 6:73–78.

Gibson, T. 1974. Genus *Sporosarcina*, p. 573–574. In: R. E. Buchanan and N. E. Gibbons (ed.), Bergey's manual of determinative bacteriology. Williams and Wilkins, Baltimore.

Goldman, M. and D. A. Wilson. 1977. Growth of *Sporosarcina ureae* in defined media. FEMS Microbiol. Lett. 2:113–115.

Gordon, R. E., W. C. Haynes, and C. H. Pang. 1973. The genus *Bacillus*. Agricultural Handbook No. 427. U.S. Department of Agriculture, Washington, DC.

Gruninger, S. E. and M. Goldman. 1988. Evidence for urea cycle activity in *Sporosarcina ureae*. Arch. Microbiol. 150:394–399.

Hao, M. V., M. Kocur, and K. Komagata. 1984. *Marinococcus* gen. nov., a new genus for motile cocci with *meso*-diaminopimelic acid in the cell wall; and *Marinococcus albus* sp. nov. and *Marinococcus halophilus* (Novitsky and Kushner) comb. nov. J. Gen. Appl. Microbiol. 30:449–459.

Hao, M. V. and K. Komagata. 1985. A new species of *Planococcus*, *P. kocurii*, isolated from fish, frozen foods, and fish curing brine. J. Gen. Appl. Microbiol. 31:441–455.

Hao, M. V. and K. Komagata. 1986. Validation of the publication of new names and new combinations previously effectively published outside the IJSB. List No. 22. Int. J. Syst. Bacteriol. 36:573.

Herndon, S. E., K. F. Bott. 1969. Genetic relationship between *Sarcina ureae* and members of the genus *Bacillus*. J. Bacteriol. 97:6–12.

Higashi, S. and R. M. Johnson. 1986. The effect of growth temperature on esterase patterns in psychrotrophic *Bacillus* species and other Gram-positive genera. Antonie van Leeuwenhoek J. Microbiol. 52:519–524.

Jeffries, L. 1969. Menaquinones in the classification of Micrococcaceae, with observations on the application of lysozyme and novobiocin sensitivity tests. Int. J. Syst. Bacteriol. 19:183–187.

Jensen, R. A. and S. L. Stenmark. 1970. Comparative allostery of 3-deoxy-D-arabino-heptulosonate-7-phosphate synthetase as a molecular basis for classification: two cases in point. J. Bacteriol. 101:763–769.

Kalakoutskii, L. V. and S. V. Dobritsa. 1984. Effect of nutrition on cellular differentiation in prokaryotic microorganisms and fungi, p. 17–121. In: A. I. Laskin and H. A. Lechevalier (ed.), CRC handbook of microbiology, vol. 6. CRC Press, Boca Raton, FL.

Kaltwasser, H., J. Krämer, and W. R. Conger. 1972. Control of urease formation in certain aerobic bacteria. Arch. Mikrobiol. 81:178–196.

Kandler, O., H. König, J. Wiegel, and D. Claus. 1983. Occurence of poly-gamma-D-glutamic acid and poly-alpha-L-glutamine in the genera *Xanthobacter*, *Flexithrix*, *Sporosarcina*, and *Planococcus*. System. Appl. Microbiol. 4:34–41.

Kandler, O. and N. Weiss. 1986b. Genus *Sporolactobacillus* Kitahara and Suzuki 1963, 69^AL, p. 1139–1141. In: P. H. A. Sneath, N. S. Mair, M. E. Sharpe, and J. G. Holt (ed.), Bergey's manual of systematic bacteriology, vol. 2. Williams and Wilkins, Baltimore.

Kele, R. A. 1970. Investigations on the nutrition, morphogenesis and habitat of *Caryophanon latum*. Ph.D. Thesis. University of Wisconsin, Madison, WI

Kele, R. A. and E. McCoy. 1971. Defined liquid minimal medium for *Caryophanon latum*. Appl. Microbiol. 22:728–729.

Kitahara, K. and C.-L. Lai, 1967. On the spore formation of *Sporolactobacillus inulinus*. J. Gen. Appl. Microbiol. 13:197–203.

Kitahara, K. and J. Suzuki. 1963. *Sporolactobacillus* nov. subgen. J. Gen. Appl. Microbiol. 9:59–71.

Kitahara, K. and T. Toyota. 1972. Auto-spheroplastization in *Sporolactobacillus inulinus*. J. Gen. Appl. Microbiol. 18:99–107.

Kluckhohn, L. W. and C. R. Spotts. 1986. A bacteriophage of *Sporosarcina ureae*. ASM Annual Meeting, Abstract I-5, p. 165.

Kluyver, A. J. and C. B. van Niel. 1936. Prospects for a natural system of classification of bacteria. Zentralbl. Bakteriol. Parasitenkd. Infektionskr. Hyg. Abt. 2 94:369–403.

Kocur, M. 1986. Genus *Planococcus*, p. 1011–1013. In: P. H. A. Sneath, N. S. Mair, M. E. Sharpe, and J. G. Holt (ed.), Bergey's manual of systematic bacteriology, vol. 2. Williams and Wilkins. Baltimore.

Kocur, M. and T. Martinec. 1963. The taxonomic status of *Sporosarcina ureae* (Beijerinck) Orla-Jensen. Int. Bull. Bacteriol. Nomencl. Taxon. 13:201–209.

Kocur, M., Z. Pácová, W. Hodgkiss, and T. Martinec. 1970. The taxonomic status of the genus *Planococcus* Migula 1894. Int. J. Syst. Bacteriol. 20:241–248.

Komura, I., K. Yamada, and K. Komagata. 1975. Taxonomic significance of phospholipid composition in aerobic Gram-positive cocci. J. Gen. Appl. Microbiol. 21:97–107.

Lehmann, K. B. and R. O. Neumann. 1927. Bakteriologie, insbesondere bakteriologische Diagnostik, vol. 2. Lehmanns, Munich.

Leifson, E. 1964. *Micrococcus eucinetus* n. sp. Int. J. Syst. Bacteriol. 14:41–44.

Linnett, P. E., R. J. Roberts, and J. L. Strominger. 1974. Biosynthesis and cross-linking of the gamma-glutamylglycine-containing peptidoglycan of vegetative cells of *Sporosarcina ureae*. J. Biol. Chem. 249:2497–2506.

Logan, N. A. and R. C. W. Berkeley. 1981. Classification and identification of members of the genus *Bacillus*, p. 105–140. In: R. C. W. Berkeley and M. Goodfellow (ed.), The aerobic endosporeforming bacteria. Academic Press, London.

Löhnis, F. 1911. Landwirtschaftlich-bakteriologisches Praktikum. Berlin: Verlag von Gebrüder Borntraeger.

Lyman, F. and R. H. Fleming. 1940. Composition of sea water. J. Mar. Res. 3:134–146.

MacDonald, R. E. and S. W. MacDonald. 1962. The physiology and natural relationships of the motile, sporeforming sarcinae. Can. J. Microbiol. 8:795–808.

Maiden, M. F. J. 1983. The biology of filamentous bacteria in freshwater sediments. Ph.D. dissertation, University of Wales.

Maiden, M. F. J. and J. G. Jones. 1984. A new filamentous, gliding bacterium, *Filibacter limicola* gen. nov. sp. nov., from lake sediment. J. Gen. Microbiol. 130:2943–2960.

Maiden, M. F. J. and J. G. Jones. 1985. In: Validation of the publication of new names and new combinations previously effectively published outside the IJSB. List No. 18. Int. J. Syst. Bacteriol. 35:375.

Marquez, M. C., A. Ventosa, and F. Ruiz-Berraquero. 1990. *Marinococcus hispanicus,* a new species of moderately halophilic Gram-positive cocci. Int. J. Syst. Bacteriol. 40:165–169.

Mazanec, K., M. Kocur, and T. Martinec. 1965. Electron microscopy of ultrathin sections of *Sporosarcina ureae.* J. Bacteriol. 90:808–816.

Migula, W. 1900. System der Bakterien, Band 2. Gustav Fischer, Jena.

Migula, W. 1894. Über ein neues System der Bakterien. Arb. Bakt. Inst. Karlsruhe 1:235–238.

Miller, A., III, W. E. Sandine and P. R. Elliker. 1970. Deoxyribonucleic acid base composition of lactobacilli determined by thermal denaturation. J. Bacteriol. 102:278–280.

Moran, J. W. and L. D. Witter. 1976. Effect of temperature and pH on the growth of *Caryophanon latum* colonies. Can. J. Microbiol. 22:1401–1403.

Nakamura, L. K. 1984. *Bacillus pulvifaciens* sp. nov., nom. rev. Int. J. Syst. Bacteriol. 34:410–413.

Nakayama, O. and M. Yanoshi. 1967a. Spore-bearing lactic acid bacteria isolated from rhizosphere. I. Taxonomic studies on *Bacillus laevolacticus* nov. sp. and *Bacillus racemilacticus* nov. sp. J. Gen. Appl. Microbiol. 13:139–153.

Nakayama, O. and M. Yanoshi. 1967b. Spore-bearing lactic acid bacteria isolated from rhizosphere. II. Taxonomic studies on the catalase-negative strains. J. Gen. Appl. Microbiol. 13:155–165.

Nichols, P., B. K. Stulp, J. G. Jones, and D. C. White. 1986. Comparison of fatty acid content and DNA homology of the filamentous gliding bacteria *Vitreoscilla, Flexibacter, Filibacter.* Arch. Microbiol. 146:1–6.

Norris, J. 1981. *Sporosarcina* and *Sporolactobacillus,* p. 337–357. In: R. C. W. Berkeley and M. Goodfellow (ed.), The aerobic, endosporeforming bacteria. Academic Press, London.

Norris, J. R., R. C. W. Berkeley, N. A. Logan, and A. G. O'Donell. 1981. The genera *Bacillus* and *Sporolactobacillus,* p. 1711–1742. In: M. P. Starr, H. Stolp, H. G. Trüper, A. Balows, and H. G. Schlegel (ed.), The prokaryotes. Springer-Verlag, Berlin.

Novitsky, T. J. and D. J. Kushner. 1976. *Planococcus halophilus* sp. nov. a facultatively halophilic coccus. Int. J. Syst. Bacteriol. 26:53–57.

Oeding, P. 1971. Serological investigations of *Planococcus* strains. Int. J. Syst. Bacteriol. 21:323–325.

Okada, S., T. Toyoda, M. Kozaki, and K. Kitahara. 1976. Studies on the cell wall of *Sporolactobacillus inulinus.* J. Agric. Chem. Soc. 50:259–263.

Oppenheimer, C. H. and C. E. ZoBell. 1952. The growth and viability of sixty-three species of marine bacteria as influenced by hydrostatic pressure. Sears Found. J. Marine Res. XI:10–18.

Orla-Jensen, O. 1909. Die Hauptlinien des natürlichen Bakteriensystems. Zentralbl. Bakteriol. Parasitenkd. Infektionskr. Hyg. Abt. 2 22:305–346.

Oste, A. G. and J. G. Holt. 1982. Nile Blue A as a fluorescent stain for poly-β-hydroxybutyrate. Appl. Environ. Microbiol. 44:238–241.

Park, J., H. Hori, and K. Komagata. 1988. Nucleotide sequences of 5S ribosomal RNAs of *Planococcus citreus, Planococcus kocurii* and *Sporosarcina ureae.* Nucleic Acids Res. 16:10358.

Pechmann, K. J., B. J. Lewis, and C. R. Woese. 1976. Phylogenetic status of *Sporosarcina ureae.* Int. J. Syst. Bacteriol. 26:305–310.

Pel'ttser, A. A. 1969. Isolation of the enzyme urease from the surrounding medium by the urobacteria *Sarcina ureae.* Isz. Timiryazev Se'skokhoz Akad. 3:230–231.

Peshkov, M. A. 1939. Cytology, karyology and cycle of development of new microbes—*Caryophanon latum* and *Caryophanon tenue.* Comptes Rendus (Doklady) de l'Académie des Sciences de l'URRS 25:244–247.

Peshkov, M. A. and B. I. Marek. 1973. Fine structure of *Caryophanon latum* and *Caryophanon tenue* Peshkoff. Microbiology 41:941–945.

Peshkov, M. A., B. I. Marek, and I. A. Shadrina. 1973. Intracellular development of phage in trichomes of *Caryophanon latum* and *Caryophanon tenue.* Microbiology 42:89–94.

Peshkov, M. A., A. S. Tikhonenko, and B. I. Marek. 1966. A bacteriophage against the multicellular microorganism *Caryophanon tenue* Peshkoff. Microbiology 35:577–581.

Pichinoty, F., H. de Barjac, M. Mandel, and J. Asselineau. 1983. Description of *Bacillus azotoformans* sp. nov. Int. J. Syst. Bacteriol. 33:660–662.

Pregerson, B. S. 1973. The distribution and physiology of *Sporosarcina ureae.* M.Sc. Thesis, California State University, Northridge, California.

Priest, F. G. 1981. DNA homology in the genus *Bacillus,* p. 33–57. In: R. C. W. Berkeley and M. Goodfellow (ed.), The aerobic endosporeforming bacteria. Academic Press, London.

Pringsheim, E. G. and C. F. Robinow. 1947. Observations on two very large bacteria, *Caryophanon latum* Peshkoff and *Lineola longa* (nomen provisorium). J. Gen. Microbiol. 1:267–278.

Provost, P. J. and R. N. Doetsch. 1962. An appraisal of *Caryophanon latum.* J. Gen. Microbiol. 28:547–557.

Ranftl, H. 1972. Zellwandzusammensetzung bei Bacillen und *Sporosarcina.* Ph.D. Thesis, Technical University, Munich.

Reanney, D. C. and H.-W. Ackermann. 1982. Comparative biology and evolution of bacteriophages, p. 205–280. In: M. A. Lauffes, F. B. Bang, K. Gavamorosch, and K. G. Smith (ed.), Advances in virus research, vol. 27. Academic Press. New York.

Robinson, R. W. and C. R. Spotts. 1983. The ultrastructure of sporulation in *Sporosarcina ureae.* Can. J. Microbiol. 29:807–814.

Rowenhagen, B. 1987. Fettsäureverwertung durch *Caryophanon latum* und *Caryophanon tenue* sowie morphologische und cytologische Unterschiede dieser Bakterien-Species. Ph.D. Thesis, University of Braunschweig.

Sames, T. 1898. Eine bewegliche Sarcine. Zentralbl. Bakteriol. Parasitenkd. Infektionskr. Hyg. Abt. 2, 4:664–669.

Saphir, D. A. and G. R. Carter. 1976. Gingival flora of the dog with special reference to bacteria associated with bites. J. Clin. Microbiol. 3:344–349.

Schleifer, K. H. and O. Kandler. 1970. Amino acid sequence of the murein of *Planococcus* and other Micrococcaceae. J. Bacteriol. 103:387–392.

Schleifer, K. H. and O. Kandler. 1972. Peptidoglycan types of bacterial cell walls and their taxonomic implications. Bacteriol. Rev. 36:407–477.

Sersen, C. D., R. N. Doetsch, and R. D. Sjoblad. 1983. Motility, behavior, and aggregation of *Sporosarcina ureae*. FEMS Microbiology Letters 17:201–204.

Shadrina, I. A., A. V. Mashkovtseva, N. A. Kostrikina, and V. I. Viryuzova. 1982. The heterogeneity of intracellular membranes in *Caryophanon latum*. Mikrobiologiya 51:809–814.

Silva, M. T., M. P. Lima, A. F. Fonseca, and J. F. C. Sousa. 1973. The fine structure of *Sporosarcina ureae* as related to its taxonomic position. J. Submicr. Cytol. 5:7–22.

Smith, D. L. and W. C. Trentini. 1972. Enrichment and selective isolation of *Caryophanon latum*. Can. J. Microbiol. 18:1197–1200.

Smith, D. L. and W. C. Trentini. 1973. On the Gram reaction of *Caryophanon latum*. Can. J. Microbiol. 19:757–760.

Sneath, P. H. A., N. S. Mair, M. E. Sharpe and J. G. Holt (ed.). 1986. Bergey's manual of systematic bacteriology, vol. 2. Williams and Wilkins, Baltimore.

Stackebrandt, E. and C. R. Woese. 1979. A phylogenetic dissection of the family Micrococcaceae. Curr. Microbiol. 2:317–322.

Stackebrandt, E., W. Ludwig, M. Weizenegger, S. Dorn, T. J. McGill, G. E. Fox, C. R. Woese, W. Schubert, and K. H. Schleifer. 1987. Comparative 16S RNA oligonucleotide analyses and murein types of round-spore-forming bacilli and non-spore forming relatives. J. Gen. Microbiol. 133:2523–2530.

Stewart, M. and T. J. Beveridge. 1980. Structure of the regular surface layer of *Sporosarcina ureae*. 142:302–309.

Suzuki, J. and K. Kitahara. 1964. Base composition of deoxyribonucleic acid in *Sporolactobacillus inulinus* and other lactic acid bacteria. J. Gen. Appl. Microbiol. 10:305–311.

Thirkell, D. and M. Summerfield. 1977a. The effect of varying sea salt concentration in the growth medium on the chemical composition of a purified membrane fraction from *Planococcus citreus* Migula. Antonie van Leeuwenhoek J. Microbiol. 43:37–42.

Thirkell, D. and M. Summerfield. 1977b. The membrane lipids of *Planococcus citreus* Migula from cells grown in the presence of three different concentrations of sea salt added to a basic medium. Antonie van Leeuwenhoek J. Microbiol. 43:43–54.

Thirkell, D. and M. Summerfield. 1980. Variation in pigment production by *Planococcus citreus* Migula with cultural age and with sea salt concentration in the medium. Antonie van Leeuwenhoek J. Microbiol. 46:51–57.

Thompson, R. S. and E. R. Leadbetter. 1963. On the isolation of dipicolinic acid from endospores of *Sarcina ureae*. Arch. Mikrobiol. 45:27–32.

Trentini, W. C. 1986. Genus *Caryophanon* Peshkoff 1939, 244^AL, p. 1259–1260. In: P. H. A. Sneath, N. S. Mair, M. E. Sharpe, and J. G. Holt (ed.), Bergey's manual of systematic bacteriology, vol. 2. Williams and Wilkins, Baltimore.

Trentini, W. C. 1988. Validation of the publication of new names and new combinations previously effectively published outside the IJSB. List No. 25. Int. J. Syst. Bacteriol. 38:220.

Trentini, W. C. and H. E. Gilleland Jr. 1974. Ultrastructure of the cell envelope and septation process in *Caryophanon latum* as revealed by thin section and freeze-etching techniques. Can. J. Microbiol. 20:1435–1442.

Trentini, W. C. and C. Machen. 1973. Natural habitat of *Caryophanon latum*. Can. J. Microbiol. 19:689–694.

Trentini, W. C. and R. G. E. Murray. 1975. Ultrastructural effects of lysozymes on the cell wall of *Caryophanon latum*. Can. J. Microbiol. 21:164–172.

Tuffery, A. A. 1953. The morphology and systematic position of *Caryophanon*. Atti del VI Congresso Internazionale di Microbiologia 1:104.

Tuffery, A. A. 1955. Nuclear changes in the growth cycle of *Caryophanon latum*. Exp. Cell Res. 9:182–185.

Uchida, K. and K. Mogi. 1973. Cellular fatty acid spectra of *Sporolactobacillus* and some other *Bacillus-Lactobacillus* intermediates as a guide to their taxonomy. J. Gen. Appl. Microbiol. 19:129–140.

Ventosa, A., M. C. Marquez, F. Ruiz-Berraquero, and M. Kocur. 1990. *Salinicoccus roseus* gen. nov., sp. nov., a new moderately halophilic Gram-positive coccus. Syst. Appl. Microbiol. 13:29–33.

Ventosa, A., E. Quesada, F. Rodriguez-Valera, F. Ruiz-Berraquero, and A. Ramos-Cormenzana. 1982. Numerical taxonomy of moderately halophilic Gram-negative rods. J. Gen. Microbiol. 128:1959–1968.

Ventosa, A., A. Ramos-Cormenzana, and M. Kocur. 1983. Moderately halophilic Gram-positive cocci from hypersaline environments. System. Appl. Microbiol. 4:564–570.

Weeks, O. B. and L. M. Kelley. 1958. Observations on the growth of the bacterium *Caryophanon latum*. J. Bacteriol. 75:326–330.

Weiss, N., R. Plapp, and O. Kandler. 1967. Die Aminosäuresequenz des DAP-haltigen Mureins von *Lactobacillus plantarum* und *Lactobacillus inulinus*. Arch. Mikrobiol. 58:313–323.

Wiegel, J. 1981. Distinction between the Gram reaction and the Gram type of bacteria. Int. J. Syst. Bacteriol. 31:88.

Wood, E. J. F. 1946. The isolation of *Sarcina ureae* (Beijerinck) Löhnis from sea water. J. Bacteriol. 51:287–289.

Yamada, Y., G. Inouye, Y. Tahara, and K. Kondo. 1976. The menaquinone system in the classification of aerobic Gram-positive cocci in the genera *Micrococcus, Staphylococcus, Planococcus,* and *Sporosarcina*. J. Gen. Appl. Microbiol. 22:227–236.

Yanagida, F., K.-I. Suzuki, T. Keneko, M. Kozaki, and K. Komagata. 1987a. Morphological, biochemical, and physiological characteristics of sporeforming lactic acid bacteria. J. Gen. Appl. Microbiol. 33:33–45.

Yanagida, F., K.-I. Suzuki, T. Keneko, M. Kozaki, and K. Komagata. 1987b. Deoxyribonucleic acid relatedness among some sporeforming lactic acid bacteria. J. Gen. Appl. Microbiol. 33:47–55.

ZoBell, C. E. and H. C. Upham. 1944. A list of marine bacteria including description of sixty new species. Bull. of the Scripps Inst. Oceanogr. Univ. Calif. Technical Series 5:239–292.

CHAPTER 80

The Genus *Desulfotomaculum*

FRIEDRICH WIDDEL

A summary of the diversity and biochemistry of bacteria that carry out a dissimilatory sulfate reduction has been given in Chapter 24. Sulfate-reducing bacteria that form heat-resistant endospores are classified in one genus, *Desulfotomaculum* (Campbell and Postgate, 1965). The first isolate of this genus, a moderate thermophile, was originally described as *Clostridium nigrificans* (Werkman and Weaver, 1927), later as *Sporovibrio desulfuricans* (Starkey, 1938), and finally was named *Desulfotomaculum nigrificans* (Campbell and Postgate, 1965). Subsequently, some other moderately thermophilic and some mesophilic *Desulfotomaculum* species were isolated (Table 1). The genus exhibits a great nutritional versatility, comparable to that of nonsporeforming sulfate reducers (see Chaper 24). H_2, alcohols, fatty acids, other aliphatic monocarboxylic or dicarboxylic acids, alanine, hexoses, or phenyl-substituted organic acids may be used as electron donors for dissimilatory sulfate reduction (Table 1). Utilization of polysaccharides or polypeptides has not been reported.

Figures 1 and 2 show morphological features of a few *Desulfotomaculum* species. They are true Gram-positive bacteria and thus phylogenetically separate from other sulfate-reducing eubacteria, as is evident from their cell wall structure (Nazina and Pivovarova, 1979; Sleytr et al., 1969; Fig. 2B) and from 16S rRNA analyses (Devereux et al., 1989; Fowler et al., 1986). *Desulfotomaculum* species cluster with the branch of Gram-positive bacteria with DNA of low GC content (Fig. 3). *Desulfotomaculum* species may be regarded as clostridia-like bacteria which have the additional capacity for dissimilatory sulfate reduction.

Habitats

Sporeforming sulfate reducers thrive essentially in the same habitats as nonsporeforming types. However, *Desulfotomaculum* species have been isolated less frequently than Gram-negative sulfate reducers from habitats that are permanently or usually anoxic. In these environments, sporeforming species are apparently less competitive than nonsporeforming ones. In contrast, if conditions are prevailing that are selective for bacteria with spores, *Desulfotomaculum* species appear to be the predominant sulfate reducers in the particular environment. In contrast to nonsporeforming sulfate reducers, because of its spores, *Desulfotomaculum* is able to survive dryness and oxic conditions for many months or even years. If the conditions turn again anoxic, *Desulfotomaculum* species are the first sulfate reducers to develop. Such a selection for *Desulfotomaculum* has been observed in rice paddies where oxic and anoxic conditions alternate due to seasonal flooding. Anoxic samples collected during the flooding period from the rhizosphere of rice contained *Desulfotomaculum* as the predominant sulfate reducer (V. A. Jacq, personal communication). Also, in some other cases, the source of isolation reflected the survival of *Desulfotomaculum* species as spores. *D. nigrificans* was found in canned food (Werkman and Weaver, 1927), and *D. orientis* (Adams and Postgate, 1959) and *D. sapomandens* (Cord-Ruwisch and Garcia, 1985) originate from soil. Actually, spores of lactate- or fatty acid-oxidizing *Desulfotomaculum* species are regularly detected in oxic humid or dry soil, even if it has never been anoxic (F. Widdel, unpublished observations). Spores are probably spread by wind or animals and remain viable for years.

In contrast to sulfate-reducing bacteria of the gamma subdivision, the genus *Desulfotomaculum* contains a number of moderately thermophilic species with temperature optima of 54 to 65°C (Table 1). A *D. nigrificans* strain was isolated from hot-oil-field water (Nazina and Rozanova, 1978). *D. geothermicum* and *D. kuznetsovii* were obtained from geothermal ground water, and *D. thermoacetoxidans* originated from a thermophilic biogas fermenter. Hence, sulfate reduction in habitats with temperatures between 50 and 65°C can mainly be due to *De-*

sulfotamaculum species. However, at temperatures above 60 to 65°C, *Thermodesulfobacterium* species may also play a role; however, they cannot oxidize acetate or fatty acids that are utilized by part of the thermophilic *Desulfotomaculum* species. If dissimilatory sulfate reduction occurs at temperatures of around 80°C or higher, this is probably due to archaebacterial sulfate reducers (see Chapter 31).

Most *Desulfotomaculum* species that have been described grow best at low salt concentrations and thus seem to be dwellers mainly of freshwater habitats or other aqueous environments with relatively low salt concentrations. However, *D. geothermicum* isolated from saline geothermal ground water, grew optimally with 20 to 30 g NaCl per liter (Daumas et al., 1988). A *D. nigrificans* strain from hot-oil-field water required 5 g NaCl per liter for optimum growth (Nazina and Pivovarova, 1979). Two other, unnamed types that were probably adapted to a saline environment were a benzoate-utilizing strain from an oil field (Cord-Ruwisch et al., 1986) and an acetate-utilizing strain from marine sediment (Keith et al., 1982).

D. ruminis has been isolated from sheep rumen (Coleman, 1960). However, with 10^2 cells/ml, this type was present in far lower numbers than *Desulfovibrio,* which exhibited an abundancy of 10^7 cells/ml (Howard and Hungate, 1976). *Desulfotomaculum acetoxidans* could be readily and repeatedly enriched from animal manure with butyrate as energy source (Widdel and Pfennig, 1977, 1981; F. Widdel, unpublished observations).

Cultivation Techniques and Media

For cultivation of *Desulfotomaculum* species, the same vessels, equipment, and anoxic techniques as described in detail for Gram-negative sulfate reducers (see Chapter 183) may be used. However, the choice of the reductant can be critical and needs special attention.

Use of Reductants

Desulfotomaculum nigrificans, D. orientis, D. ruminis, and some other incompletely oxidizing species that were originally isolated on the lactate medium with ascorbate and thioglycollate as reductants (see Chapter 183) show very poor growth in media reduced with commercial sodium sulfide (Klemps et al., 1985; F. Widdel, unpublished observations). The inhibition by Na_2S is probably not due to sulfide itself, since the added concentration (around 1.5 mM) is far below the concentration that is usually inhibitory to *Desulfotomaculum* species (7 to 10 mM; e.g. Coleman, 1960; Cord-Ruwisch and Garcia, 1985). It is more likely that certain impurities in the commercial sodium sulfide are inhibitory.

The indicated *Desulfotomaculum* species are therefore grown in the lactate medium for *Desulfovibrio,* which is prereduced with ascorbate and thioglycollate. Alternatively, the defined multipurpose medium (Chapter 183) may be used if ascorbate and thioglycollate are added instead of Na_2S. Bicarbonate/CO_2, various vitamins, and trace elements present in the multipurpose medium may be required by several species.

Dithionite has also been used as sole reductant for cultivation of *Desulfotomaculum* species (e.g., Cypionka and Pfennig, 1986). Even though some *Desulfotomaculum* species may be less sensitive toward this reducing agent than other anaerobes, dithionite is probably toxic if added at the same concentration as other reductants. It is therefore added immediately before inoculation and only in the smallest amount sufficient to reduce the medium. To assess the redox status, an indicator such as resazurine is added (from a 0.1% (wt/vol) solution of resazurine (sodium salt), 1 ml is added per liter of medium). Sodium dithionite is added from a fresh solution prepared under N_2 or as dry powder (Chapter 183) until resazurine is decolorized. Then, the culture vessel is inoculated and immediately sealed, either with a screw cap (in case of completely filled culture vessels) or with a stopper (in case of vessels with an anoxic head space).

Desulfotomaculum ruminis was isolated in sulfite agar and shown to tolerate sulfite up to a concentration of several mM. Even though the use of Na_2SO_3 as reductant is not very common, it should be considered for new media for *Desulfotomaculum.*

Desulfotomaculum guttoideum and species using fatty acids have been isolated in Na_2S-reduced media and can therefore be grown with this reductant. However, growth in the aforementioned media with other reductants may be also attempted. If Na_2S is used, dithionite (15 to 30 mg/l) as an additional reductant often significantly favors initial growth (Widdel and Pfennig, 1977, 1981).

Other Variations of Standard Media

A nonchelated trace element solution such as the solution 1 described in Chapter 183 has been used for cultivation of many *Desulfotomaculum* species (Cord-Ruwisch and Garcia,

Table 1. Properties of *Desulfotomaculum* species.

Species	Morphology	Width (μm)	Length (μm)	Motility[a]	GC content of DNA (mol%)	Major menaquinone[b]	Temperature optimum (°C)	Oxidation of organic substrates
Desulfotomaculum acetoxidans	Straight or curved rod	1–1.5	3.5–9	+ (sp)	38	MK-7	34–36	Complete
antarcticum	Rod	1–1.2	4–6	+ (pe)	ND	ND	20–30	Incomplete
geothermicum	Rod	0.5–0.8	2.3–2.5	+	50	ND	54	Complete
guttoideum	Rod, drop-shaped	1	2.3	+ (pe)	48	ND	31	Incomplete
kuznetsovii	Rod	1–1.4	3.5–5	+ (pe)	49	ND	60–65	Complete
nigrificans[c]	Rod	0.5–0.7	2–4	+ (pe)	45	MK-7	55	Incomplete
orientis	Straight or curved rod	0.7–1	3–5	+ (pe)	45	MK-7	37	Incomplete
ruminis	Rod	0.5–0.7	2–4	+ (pe)	49	MK-7	37	Incomplete
sapomandens	Rod	1.2–2	5–7	+	48	ND	38	Complete
thermoacetoxidans	Straight or curved rod	0.7	2–5	+	50	ND	55–60	Complete

[a]Flagellation pattern indicated in parentheses: pe, peritrichous; sp, single and polar.
[b]See Collins and Widdel (1986); ND, not determined or not reported.
[c]Type species.
[d]Symbols: +, utilized; +*, autotrophic growth; (+), poorly utilized; −, not utilized; ND, not determined.
[e]Not completely listed; for further substrates, see references.
[f]Symbols: bi, biotin; pa, *p*-aminobenzoate.
[g]Type strain does not require NaCl. Another strain requires 5 g NaCl per liter for optimum growth (Nazina and Rozanova, 1978).

1985; Daumas et al., 1988; Gogotova and Vainshtein, 1983; Widdel and Pfennig, 1977, 1981). Furthermore, trace elements chelated with EDTA (ethylenediaminetetraacetic acid, neutralized) or NTA (nitrilotriacetic acid, neutralized) have been used. The EDTA-containing solution (Nazina et al., 1988) had essentially the composition of solution 2. In case of NTA, the composition was as in solution 1, but with 12.8 g NTA (Cypionka and Pfennig, 1986), or as in solution 3, but with only 4.5 g NTA (Min and Zinder, 1990).

An excellent organic substrate for rapid growth of incompletely oxidizing, lactate-utilizing *Desulfotomaculum* species is pyruvate. Only half as much H_2S is formed per mol of pyruvate oxidized to acetate, as per mol of lactate oxidized. Therefore, high cell densities can be reached on pyruvate before H_2S becomes inhibitory. A stock solution of 2 M pyruvate is

H₂	Formate	Acetate	Fatty acids: C atoms	Ethanol	Lactate	Fumarate	Malate	Benzoate	Others[e]	Growth factor require- ment[f]	NaCl require- ment (g/l)	References
−	(+)	+	4–5	+	−	−	−	−	Butanol	bi	−	Widdel and Pfennig, 1977, 1981
ND	−	−	ND	ND	+	ND	ND	ND	Glucose	Unknown	−	Iizuka et al., 1969
+*	+	−	3–18	+	+	ND	ND	−	Fructose	Unknown	25	Daumas et al., 1988
+	−	−	ND	−	+	ND	−	ND	—	Unknown	−	Gogotova and Vainshtein, 1983
+*	+	+	3–16	+	+	+	+	−	Methanol, propanol, butanol	Unknown	−	Nazina et al., 1988
+	+	−	−	+	+	−	−	−	Fructose, alanine	Unknown	−	Campbell and Postgate, 1965; Klemps et al., 1985
+*	+	−	−	+	+	−	−	−	Methanol, 3,4,5-tri-methoxy-benzoate	Unknown	−g	Campbell and Postgate, 1965; Klemps et al., 1985
+	+	−	−	+	+	−	−	−	Alanine	bi, pa	−	Campbell and Postgate, 1965; Klemps et al., 1985
ND	+	(+)	4–18	+	−	+	+	+	Isobutyrate, 3-methylbu-tyrate, phenylace-tate	Unknown	−	Cord-Ruwisch and Garcia, 1985
+*	+	+	3–5	−	+	ND	+	−	Propanol, bu-tanol, alan-ine	Unknown	−	Min and Zin-der, 1990

Electron donors for sulfate reduction[d]

prepared from the sodium salt (22 g in 100 ml) or from the cheaper free acid. In the latter case, 14.4 ml (17.6 g) pyruvic acid is added to 40 ml distilled H_2O. Then, the solution is neutralized by slow addition of 4 M NaOH and finally of 1 M NaOH under stirring in an ice bath. H_2O is added to a final volume of 100 ml. The so-dium pyruvate solution is filter-sterilized and stored in the dark at 4°C. Per liter of medium, 15 to 20 ml pyruvate solution are added.

Desulfotomaculum nigrificans does not grow well in defined medium; this species is signifi-cantly stimulated by yeast extract that is present in the lactate medium (Chapter 183). If the mul-tipurpose medium is used, yeast extract is added at the same concentration (1 g/l). A somewhat weaker but distinct stimulation in the multipurpose medium was also achieved with a mixture of 60 mg alanine, 90 mg aspar-agine, 15 mg cysteine, and 85 g threonine in-

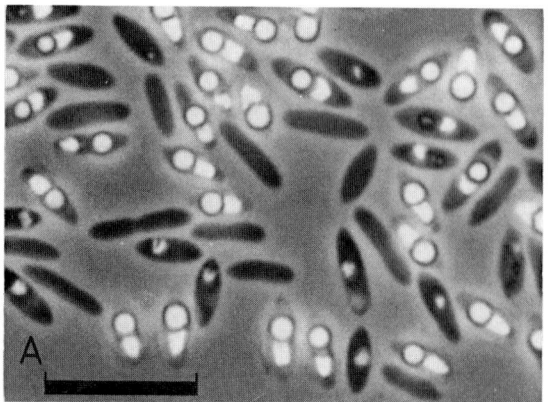

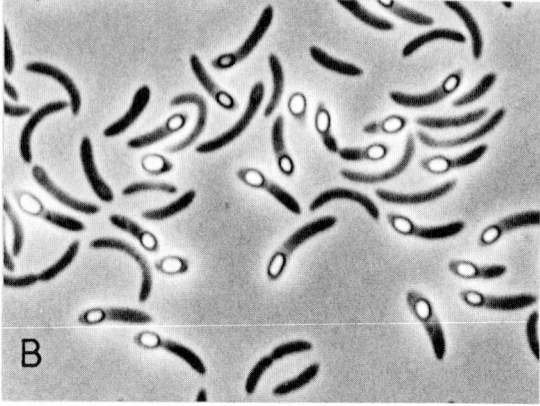

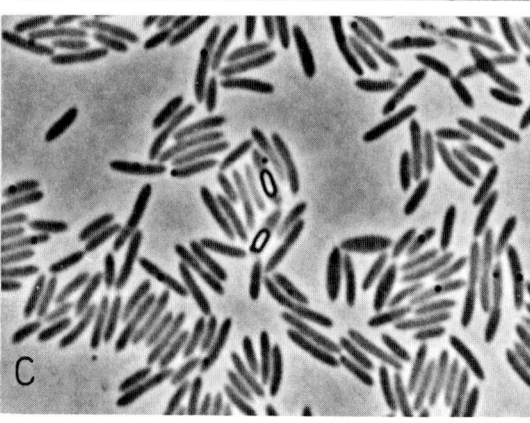

Fig. 1. Phase contrast photomicrographs of viable cells of sporeforming sulfate-reducing bacteria. Bar = 10 μm. (A) *D. acetoxidans* with spores, from a colony grown in agar with acetate. (B) *D. orientis* with spores, from a culture grown under sulfate limitation with H_2. (Courtesy of H. Cypionka.) (C) *D. ruminis* with two spores; *D. nigrificans* has the same size and morphology.

stead of yeast extract per liter of medium, and a threefold increased $FeSO_4$ concentration (F. Widdel, unpublished observations). Growth of other incompletely oxidizing *Desulfotomaculum* species is also promoted by yeast extract, but the effect is less pronounced that in the case of *D. nigrificans*. Yeast extract has no effect on the growth of *D. acetoxidans*.

Enrichment, Isolation, and Maintenance

Enrichment

The formation of heat- and drought-resistant spores in *Desulfotomaculum* species allows selective enrichment in the presence of other sulfate reducers. To eliminate nonsporeformers, samples from sediments or other anoxic habitats are pasteurized at 80°C for 10 to 20 min. To guarantee uniform heating, the sample volume is kept small (<10 ml). Pasteurization may be carried out in test tubes or bottles incubated in a thermostated water bath. Exclusion of air is not necessary.

Selective enrichment of *Desulfotomaculum* species is also possible from samples of oxic surface soil (see "Habitats," this chapter). Since nonsporeforming sulfate reducers are usually absent from such soil, *Desulfotomaculum* species are selectively enriched even from nonpasteurized soil samples.

In enrichments with nonpasteurized samples from organic-rich, anoxic habitats (sediments, sludge), *Desulfotomaculum* species are usually outcompeted by nonsporeforming sulfate reducers. An exception seem to be enrichments under moderately thermophilic conditions (50 to 60°C), which often yield *Desulfotomaculum* species even without preceding pasteurization. This is probably due to the lack of nonsporeforming sulfate reducers adapted to this temperature range in these kinds of habitats (see "Habitats," this chapter).

Besides temperature, selective factors for enrichment of various types of *Desulfotomaculum* include the electron donor and the salt concentration. *D. ruminis, D. orientis,* and nutritionally similar incomplete oxidizers may be enriched with lactate. Enrichment with an H_2/CO_2 mixture in the presence or absence of some acetate (1 mM) has not been tested but may be also successful. There is, however, no one method that guarantees selection of a particular species. Only *Desulfotomaculum acetoxidans* could be regularly enriched from cow or pig manure with acetate, butyrate, or isobutyrate as electron donors, even without preceding pasteurization (Widdel and Pfennig, 1977, 1981; F. Widdel, unpublished observations).

Various types of sporeforming sulfate reducers may be also obtained by serial dilution of soil or pasteurized mud samples in agar deeps (see below and Chapter 183).

Isolation

Desulfotomaculum species are isolated via serial dilutions in agar deeps or using one of the other procedures described in Chapter 183.

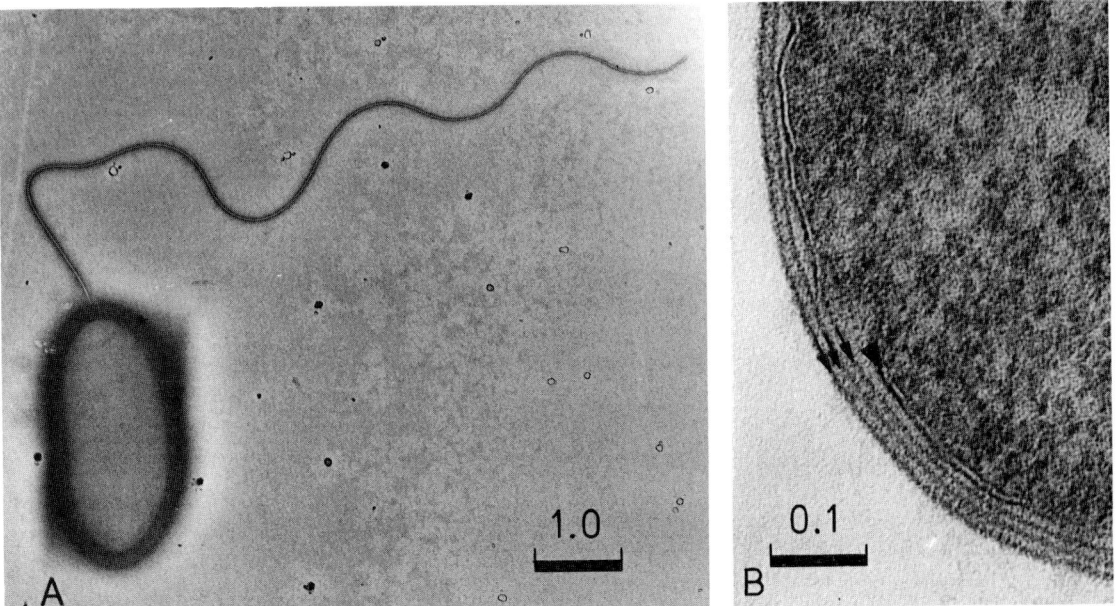

Fig. 2. Electron micrographs of sporeforming sulfate-reducing bacteria. (A) Negatively stained cell of *D. acetoxidans* showing the polar flagellum. (Courtesy of R. Lurz.) (B) Ultrathin section of a *D. nigrificans* strain from an oil field. The cytoplasmic membrane is indicated by the large arrow head, the murein wall by the small arrows. (Courtesy of T. N. Nazina and T. A. Pivovarova.)

Fig. 3. Relationship of *D. nigrificans* and *D. acetoxidans* to each other and to other bacteria, revealed by 16S rRNA oligonucleotide cataloging. Dendrogram based on the data of Fowler et al. (1986), Stackebrandt et al. (1987), and E. Stackebrandt (personal communication). Relationships of two other *Desulfotomaculum* species are shown in Fig. 6 of Chapter 183.

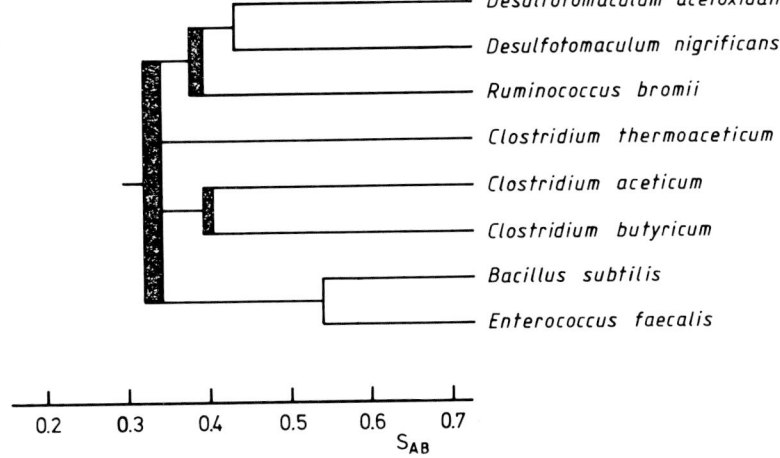

Maintenance

Like many sporeforming bacteria, vegetative cells of *Desulfotomaculum* species die off very soon after growth has ceased, especially if kept at optimal temperature. Many species that grow well lose their viability within 2 to 4 days. Since spores are usually not formed in cultures growing under optimal conditions, strains are easily lost upon prolonged incubation. The rate of cell death may be reduced by growth at suboptimal temperature (e.g., 20°C). However, the best method for preservation of *Desulfotomaculum*

cultures is in the sporulated state. Spore formation is often induced in colonies in agar deeps. Strains to be maintained are therefore diluted in agar as described for isolation of pure cultures (see Chapter 183). Agar tubes with fully grown colonies are incubated for several more weeks to allow completion of sporulation.

Formation of some spores in cultures of *Desulfomaculum acetoxidans* may also occur during slow growth on acetate at suboptimal temperature (20°C). This species never forms spores on butyrate, the preferred substrate for rapid germination of spores and for growth (Widdel and Pfennig, 1981).

Desulfotomaculum orientis forms spores if grown in sulfate-limited cultures with H$_2$ as electron donor (Cypionka and Pfennig, 1986; Fig. 1B).

Agar tubes with spore-containing colonies are refrigerated. For revival, colonies are picked and transferred to fresh liquid medium. Sporulated liquid cultures may be freeze-dried with skim milk or with the precipitated ferrous sulfate formed in the iron-rich lactate medium B (Chapter 183) and kept at −20°C.

Taxonomy and Identification

The genealogical affiliation of four *Desulfotomaculum* species with the branch of Gram-positive bacteria with low GC content has been demonstrated by oligonucleotide cataloging (Fowler et al., 1985; Fig. 3) and by the sequencing (Devereux et al., 1989) of 16S rRNA. Comparative genealogical studies on *Desulfotomaculum* species on a larger scale are lacking thus far. Hence, it is not known whether there is a coincidence of physiological and phylogenetic groupings within the genus. The physiological variations in the genus (Table 1) and the relatively low degree of relatedness even between the nutritionally similar *D. orientis* and *D. ruminis* (Devereux et al., 1989; Chapter 183, Fig. 6) indicates that the genus is less coherent than any other genus of Gram-negative sulfate reducers. However, as long as the genus *Clostridium*, with its multiple physiological types, is not subdivided, there is no reason to do so with the genus *Desulfotomaculum*. For the present, sporeforming sulfate reducers can be unequivocally identified as members of the genus *Desulfotomaculum*. A brief description of characteristics of the genus is given in the following. Properties of particular species are listed in Table 1. Figures 1 and 2 show morphological features of a few species.

Genus *Desulfotomaculum*

The common property of *Desulfotomaculum* species is dissimilatory sulfate reduction to sulfide and the formation of endospores. The location of spherical or oval spores may be central, subterminal, or terminal. In some species, gas vacuoles are formed simultaneously with spores (Daumas et al., 1988; Widdel and Pfennig, 1977, 1981; Fig. 1A). Spores are resistant to heat (80°C, 10 to 20 min), dryness, and oxic conditions. *Desulfotomaculum* species grow with H$_2$, alcohols, a few sugars, or simple organic acids including phenyl-substituted ones that serve as electron donors (Table 1). Some species oxidize organic electron donors incompletely to acetate. Species utilizing fatty acids or benzoate are capable of complete oxidation, even though these organic substrates may be partially converted to acetate (Daumas et al., 1988; Widdel and Pfennig, 1981). H$_2$-utilizing species may be capable of autotrophic growth (Brysch et al., 1987; Daumas et al., 1988; Klemps et al., 1985; Min and Zinder, 1990; Nazina et al., 1988). In the absence of sulfate, some species grow by acid fermentation of glucose, fructose, or pyruvate. Certain *Desulfotomaculum* species resemble homoacetogenic bacteria (see Chapter 21) in their ability to convert methanol, methoxyl groups of 3,4,5-trimethoxybenzoate, formate, and even H$_2$ and CO$_2$ to acetate (K. Hanselmann, personal communication; Klemps et al., 1985; Min and Zinder, 1990). This homoacetogenesis may allow weak growth in the absence of sulfate. Sulfite or thiosulfate can replace sulfate as electron acceptor in many species. *Desulfotomaculum* species are more sensitive toward the sulfide produced than most Gram-negative sulfate reducers. Whereas the latter often grow up to H$_2$S concentrations of around 20 mM, *Desulfotomaculum* species are inhibited by 7 to 10 mM H$_2$S. Desulfoviridin has never been found in the genus *Desulfotomaculum*; however, the sulfite reductase P582 is present, as shown by the characteristic carbon monoxide difference spectrum of reduced cell extracts (Trudinger, 1970; Widdel and Pfennig, 1977). It had originally been assumed that *Desulfotomaculum* species contain only cytochromes of the *b*-type, but *c*-type cytochromes have been detected as well (e.g., Daumas et al., 1988; Gogotova and Vainshtein, 1983; Jones, 1972).

Acknowledgments

This work was supported by a grant of the Deutsche Forschungsgemeinschaft.

Literature Cited

Adams, M. E., and J. R. Postgate. 1959. A new sulphate-reducing vibrio. J. Gen. Microbiol. 20:252–257.

Brysch, K., C. Schneider, G. Fuchs, and F. Widdel. 1987. Lithoautotrophic growth of sulfate-reducing bacteria, and description of *Desulfobacterium autotrophicum* gen. nov. sp. nov. Arch. Microbiol. 148:264–274.

Campbell, L. L., and J. R. Postgate. 1965. Classification of the spore-forming sulfate-reducing bacteria. Bacteriol. Rev. 29:359–363.

Coleman, G. S. 1960. A sulphate-reducing bacterium from the sheep rumen. J. Gen. Microbiol. 22:423–436.

Collins, M. D., and F. Widdel. 1986. Respiratory quinones of sulphate-reducing and sulphur-reducing bacteria: a systematic investigation. Syst. Appl. Microbiol. 8:8–18.

Cord-Ruwisch, R., and J. L. Garcia. 1985. Isolation and characterization of an anaerobic benzoate-degrading spore-forming sulfate-reducing bacterium, *Desulfotomaculum sapomandens* sp. nov. FEMS Microbiol. Lett. 29:325–330.

Cord-Ruwisch, R., W. Kleinitz, and F. Widdel. 1986. Sulfatreduzierende Bakterien in einem Erdölfeld—Arten und Wachstumsbedingungen. Erdöl, Erdgas, Kohle 102. Jahrg.:281–289.

Cypionka, H., and N. Pfennig. 1986. Growth yield of *Desulfotomaculum orientis* with hydrogen in chemostat culture. Arch. Microbiol. 143:396–399.

Daumas, S., R. Cord-Ruwisch, J. L. Garcia. 1988. *Desulfotomaculum geothermicum* sp. nov., a thermophilic, fatty acid-degrading, sulfate-reducing bacterium isolated with H_2 from geothermal ground water. Antonie van Leeuwenhoek 54:165–178.

Devereux, R., M. Delaney, F. Widdel, and D.A. Stahl. 1989. Natural relationships among sulfate-reducing eubacteria. J. Bacteriol. 171:6689–6695.

Fowler, V. J., F. Widdel, N. Pfennig, C. R. Woese, and E. Stackebrandt. 1986. Phylogenetic relationships of sulfate- and sulfur-reducing eubacteria. Syst. Appl. Microbiol. 8:32–41.

Gogotova, G. I., and M. B. Vainshtein. 1983. Spore-forming, sulfate-reducing bacterium *Desulfotomaculum guttoideum* sp. nov. Mikrobiologiya (USSR) 52:789–793.

Howard, B. H., and R. E. Hungate. 1976. *Desulfovibrio* of the sheep rumen. Appl. Environ. Microbiol. 32:598–602.

Iizuka, H., H. Okazaki, and N. Seto. 1969. A new sulfate-reducing bacterium isolated from Antarctica. J. Gen. Appl. Microbiol. (Tokyo) 15:11–18.

Jones, H. E. 1972. Cytochromes and other pigments of dissimilatory sulphate-reducing bacteria. Arch. Microbiol. 84:207–224.

Keith, S. M., R. A. Herbert, and C. G. Harfoot. 1982. Isolation of new types of sulphate-reducing bacteria from estuarine and marine sediments using chemostat enrichments. J. Appl. Bacteriol. 53:29–33.

Klemps, R., H. Cypionka, F. Widdel, and N. Pfennig. 1985. Growth with hydrogen, and further physiological characteristics of *Desulfotomaculum* species. Arch. Microbiol. 143:203–208.

Min, H., and S.H. Zinder. 1990. Isolation and characterization of a thermophilic sulfate-reducing bacterium *Desulfotomaculum thermoacetoxidans* sp. nov. Arch. Microbiol. 153:399–404.

Nazina, T. N., A. E. Ivanova, L. P. Kanchaveli, and E. P. Rozanova. 1988. *Desulfotomaculum kuznetsovii* sp. nov., a new spore-forming thermophilic methylotrophic sulfate-reducing bacterium. Mikrobiologiya (USSR) 57:823–827.

Nazina, T. N., and T. A. Pivovarova. 1979. Submicroscopic organization and spore formation in *Desulfotomaculum nigrificans*. Mikrobiologiya (USSR) 48:302–306.

Nazina, T. N., and E. P. Rozanova. 1978. Thermophilic sulfate-reducing bacterium from oil strata. Mikrobiologiya (USSR) 47:142–148.

Sleytr, U., H. Adam, and H. Klaushofer. 1969. Die Feinstruktur der Zellwand und Cytoplasmamembran von *Clostridium nigrificans*, dargestellt mit Hilfe der Gefrierätz- und Ultradünnschisstechnik. Arch. Microbiol. 66:40–58.

Stackebrandt, E., W. Ludwig, M. Weizenegger, S. Dorn, T. J. McGill, G. E. Fox, C. R. Woese, W. Schubert, and K. -H. Schleifer. 1987. Comparative 16S rRNA oligonucleotide analyses and murein types of round-spore-forming bacilli and non-spore-forming relatives. J. Gen Microbiol. 133:2523–2529.

Starkey, R. L. 1938. A study of spore formation and other morphological characteristics of *Vibrio desulfuricans*. Arch. Mikrobiol. 9:268–304.

Trudinger, P.A. 1970. Carbon monoxide reacting pigment from *Desulfotomaculum nigrificans* and its possible relevance to sulfite reduction. J. Bacteriol. 104:158–170.

Werkman, C. H., and H. J. Weaver. 1927. Studies in the bacteriology of sulphur stinker spoilage of canned sweet corn. Iowa State Coll. J. Sci. 2:57–67.

Widdel, F., and N. Pfennig. 1977. A new anaerobic, sporing, acetate-oxidizing, sulfate-reducing bacterium, *Desulfotomaculum* (emend.) *acetoxidans*. Arch. Microbiol. 112:119–122.

Widdel, F., and N. Pfennig. 1981. Sporulation and further nutritional characteristics of *Desulfotomaculum acetoxidans*. Arch. Microbiol. 129:401–402.

The Genus *Clostridium*—Nonmedical

HANS HIPPE, JAN R. ANDREESEN, and GERHARD GOTTSCHALK

The genus *Clostridium* was first proposed by A. Prazmowski in 1880. Since that time more than 100 bacterial species have been assigned to it, until today it represents one of the largest genera of the prokaryotes. This growth is not surprising because an organism has to meet only four criteria in order to be classified as a *Clostridium* species: 1) it must be able to form endospores; 2) it must obligatorily rely on an anaerobic energy metabolism; 3) it must be unable to carry out a dissimilatory reduction of sulfate; and 4) the cell wall must be of the Gram-positive type.

The first criterion differentiates the clostridia from all organisms unable to form endospores. The difficulty in applying this criterion is that some species may not sporulate under the culture conditions applied. Asporogenous mutants may also appear and proliferate, and it is, perhaps, not surprising that a high degree of genetic homology was found between the nonspore-forming "*Bacteroides trichoides*" and "*Eubacterium filamentosum*" on one hand and *Clostridium ramosum* on the other (see Cato et al., 1986). After spores were detected in strains of "*Bacteroides clostridiiformis* subsp. *clostridiiformis*," this species was classified as *Clostridium clostridiiforme* (Cato and Salmon, 1976; Kaneuchi et al., 1976).

The second criterion separates the clostridia from the *Bacillus* species that are aerobes and that contain electron-transport chains coupled to oxygen as the electron acceptor. Some *Bacillus* species, e.g., *B. polymyxa*, are facultative anaerobes. Members of both genera resemble one another with respect to cell shape and cell size. With the exception of *C. coccoides* (Kaneuchi et al., 1976), clostridial cells are straight or curved rods, 0.3–1.6 $\times$ 1–14 μm. A pronounced tendency to helical coiling of the cell chains is found in three species (Kaneuchi et al., 1979; Himelbloom and Canale-Parola, 1989). Unlike the appearance of cells of bacilli, clostridial cells that contain spores are often swollen. Those that form terminal spores look like drumsticks, and those that form subterminal or central spores, look like Chinese lanterns (Fig. 1). Although some clostridia tolerate oxygen and are able to grow on an agar surface under air, sporulation occurs only under anaerobic growth conditions. The reverse is true for the facultatively anaerobic species of *Bacillus*.

The third criterion (inability to carry out dissimilatory sulfate reduction) follows from the proposal by Campbell and Postgate (1965) to create the genus *Desulfotomaculum* which comprises the sporeforming sulfate-reducing bacteria.

The fourth criterion (Gram-positive cell wall type) is required because of the discovery of the genera *Sporomusa* (Möller et al., 1984) and *Sporohalobacter* (Oren et al., 1987). These anaerobic sporeformers possess a Gram-negative type of cell wall. Therefore, the definition of the genus *Clostridium* must include a Gram-positive wall type. The strictly anaerobic, saccharolytic sarcinas, *Sarcina ventriculi* and *S. maxima,* have been reported to form spores (Knöll and Horschak, 1973; Lowe et al., 1989). They can be differentiated from the clostridia by their unique appearance, spherical cells occuring in packets (see Chapter 83).

The genus *Clostridium* includes psychrophilic, mesophilic, and thermophilic species. Most of the clostridial species stain Gram-positive and are motile with peritrichous flagellation. The major role of these organisms in nature is in the degradation of organic material to acids, alcohols, CO_2, H_2, and minerals. Frequently, a butyric acid smell is associated with the genus *Clostridium*. This acid is produced by a number of species. However, several clostridia do not produce any butyrate but do form acetate, lactate, formate, or propionate as main fermentation products. The ability to form spores that resist dryness, heat, and aerobic conditions makes the clostridia ubiquitous. When the appropriate growth conditions are applied, they can be isolated from many kinds of material. They are frequently responsible for spoilage of food and dairy products and cause large economic losses.

A number of clostridia are pathogens. Because of their great importance, the pathogenic species are discussed in a separate chapter (Chapter 82). This separation is made primarily for the benefit of the reader and does not indicate that the pathogenicity or nonpathogenicity of clostridial species has important taxonomic significance. The information for toxin production may be encoded on extrachromosomal elements or prophage DNA, and, therefore, the differentiation of toxigenic from nontoxigenic species is not be warranted (Duncan et al., 1978).

Approved standards for the classification of the whole genus *Clostridium* do not exist. For many nonpathogenic species even the description is frequently incomplete and based on the study of only one strain, as in the cases of *C. aminovalericum, C. barkeri, C. cellobioparum, C. propionicum, C. sticklandii, C. thermoaceticum,* and about 20 other species. This descriptive poverty results from the fact that many clostridia were isolated in order to study the anaerobic degradation of certain compounds or the pathways leading to particular fermentation products, and not much emphasis has been put on their classification. Nevertheless, basic discoveries have been made using these organisms. Pasteur (1861) discovered "life without oxygen" while experimenting with a culture of a butyrate-forming microorganism—probably a strain of *Clostridium.* Winogradsky (1895, 1902) discovered nitrogen fixation in free-living bacteria using *C. pasteurianum.* The study of this organism also led to the discovery of ferredoxin (Mortenson et al., 1962) and of the first in vitro system that fixed molecular nitrogen (Mortenson, 1966). The first enzymatic reaction involving a vitamin B_{12} derivative was discovered in *C. tetanomorphum* (Barker et al., 1958), and investigations on *C. kluyveri* were important for an understanding of fatty acid synthesis and of the role of coenzyme A (Barker, 1956). The Stickland reaction, which is very important for many proteolytic clostridia, including many pathogens, was discovered in *C. sporogenes* (Stickland, 1934).

The genus *Clostridium* is very heterogeneous (Cato and Stackebrandt, 1989). It includes species that are moderately aerotolerant *(C. aerotolerans, C. carnis, C. durum, C. histolyticum,* and *C. tertium)* and others, such as *C. aminovalericum* (Rolfe et al., 1978), that are extremely fastidious. Some species contain cytochromes and quinones of unknown function (Gottwald et al., 1975; Das et al., 1989). The GC content of the DNA ranges from 21 to 54 mol%. Clostridial cell walls contain, in general, peptidoglycan of the *meso*-diaminopimelic acid direct-

linked type as well as teichoic acids, but exceptions have been observed (Cummins and Johnson, 1971; Schleifer and Kandler, 1972; Weiss et al., 1981).

Habitats

Restriction by Oxygen Sensitivity

Clostridia exhibit a more-or-less pronounced intolerance towards oxygen. This oxygen sensivity is either caused by a lack or a shortage of defense mechanisms against the toxic by-products of oxygen metabolism or by the interference of oxygen with some vital enzyme systems (see Chapter 12). The toxic by-products of oxygen metabolism are destroyed by the enzymes catalase (or a peroxidase) and superoxide dismutase (Cardenas, 1989). The levels of these enzymes were first reported to be zero or very low in most clostridia (McCord et al., 1971). However, appreciable activities of superoxide dismutase have been found in some strains of *C. perfringens* and *C. ramosum,* and its presence has been demonstrated for several other species (Gregory et al., 1978). The activity of superoxide dismutase in *C. sporogenes* and *C. bifermentans* can be increased up to 60-fold if these organisms are cultured under the stress of increased amounts of oxygen (Ashley and Shoesmith, 1977). Molecular oxygen may also interfere with NADH-oxidase activities; therefore, intermediary and biosynthetic metabolism may suffer from a general shortage of NADH in the presence of oxygen (Morris, 1975; O'Brien and Morris, 1971b; Uesugi and Yajima, 1978b).

The pioneering work on the effect of oxygen on clostridial growth was done with pathogens *(Clostridium botulinum, C. perfringens)* (see O'Brien and Morris, 1971b), but nonpathogenic species have also been studied. These studies indicate that the maximum oxygen concentration at which growth is still possible is not identical for all clostridia: *C. haemolyticum* requires a pO_2 lower than 0.5% for growth; *C. novyi* type A tolerates a pO_2 of up to 3% (Loesche, 1969). Many clostridia stop growing in the presence of molecular oxygen but resume growth when transferred back to anaerobic conditions. The vegetative cells of *C. acetobutylicum, C. butyricum, C. clostridiiforme,* and *C. ramosum* survive oxygen exposure for hours; *C. haemolyticum* and *C. novyi* type B survive only for minutes (Azova et al., 1970; O'Brien and Morris, 1971b; Stolp, 1955; Tally et al., 1977; Uesugi and Yajima, 1978a).

During growth, clostridia establish a characteristically low redox potential, between -400

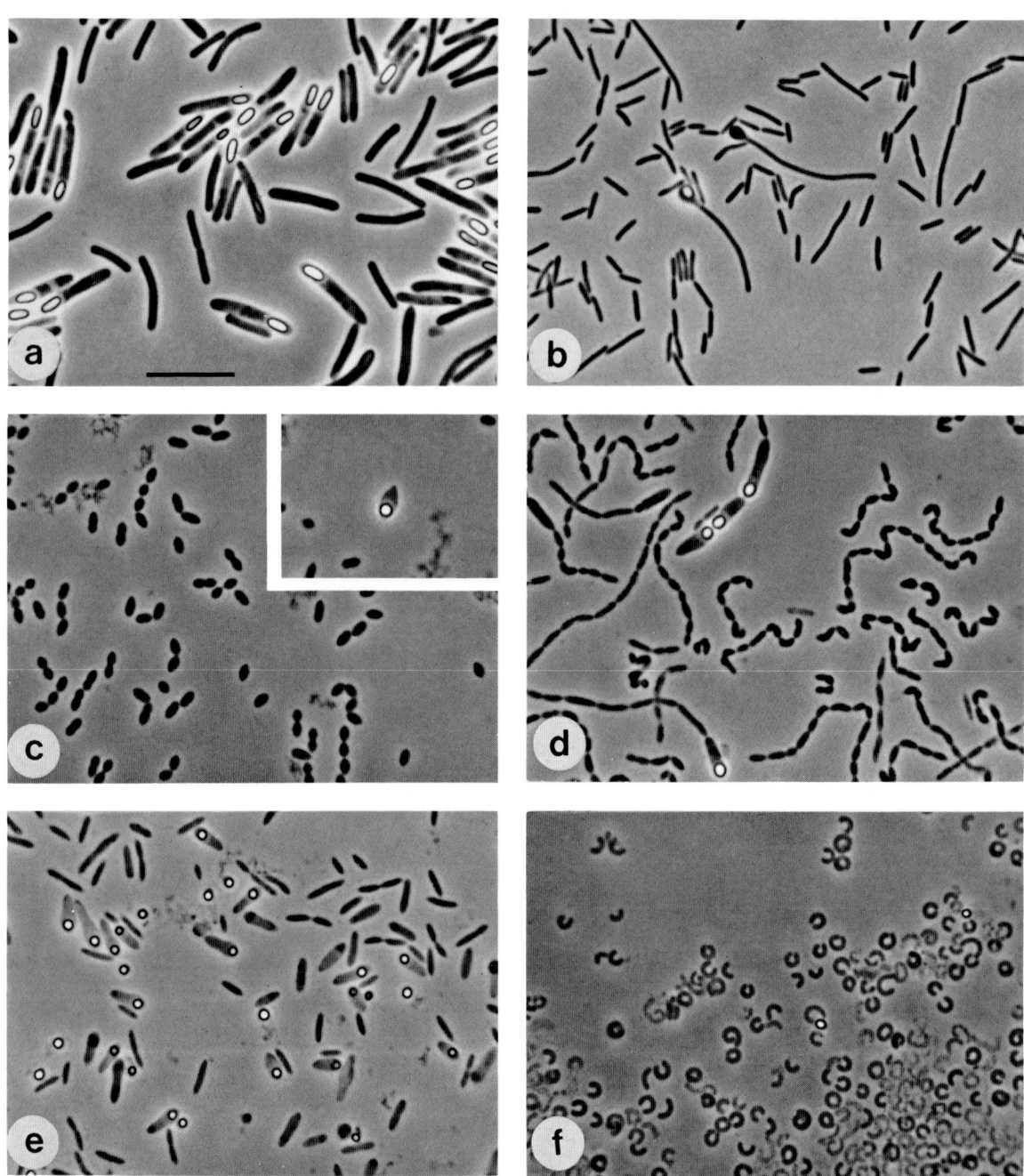

Fig. 1. Phase-contrast photomicrographs of some representatives of the genus *Clostridium*. Bar = 10 μm. Parts a to f (above) as follows: (a) *C. butyricum* DSM 552; yeast extract-peptone agar; large, straight or slightly curved rods with round ends; sporangial cells slightly distended; oval subterminal spores. (b) *C. thermocellum* DSM 1313; cellobiose medium; showing elongated sporulating cells with oval terminal spores. (c) *C. coccoides* DSM 935; CMC medium; coccobacilli to rod-shaped cells forming pairs and chains; round subterminal spores. (d) *C. oroticum* DSM 1287; CMC agar; rods occurring in long, tangled chains; enlarged sporangial cells with oval terminal-to-subterminal spores. (e) *C. indolis* DSM 755; CMC agar; with round terminal-to-subterminal spores. (f) *C. cocleatum* DSM 1551; CMC agar; showing preponderance of semicircular-to-circular cell forms; oval spores are subterminal to terminal.

Parts g to l (on facing page) as follows: (g) *C. bifermentans* DSM 631; CMC agar; central-to-subterminal oval spores not distending cells. (h) *C. sporogenes* DSM 767; peptone-yeast extract-glucose medium; oval subterminal spores swelling the cells slightly. (i) *C. acetobutylicum* DSM 792; milk agar; rods of varying length; sporeforming cells cigar-shaped; spores oval to cylindrical, subterminal. (j) *C. ramosum* DSM 1402; CMC agar; showing tendency to Y- and V-shaped cells. (k) *C. cadaveris* DSM 1284; CMC agar; rods with oval terminal spores. (l) *C. tetanomorphum* DSM 665; CMC agar; large, straight to slightly curved rods; nearly round terminal spores.

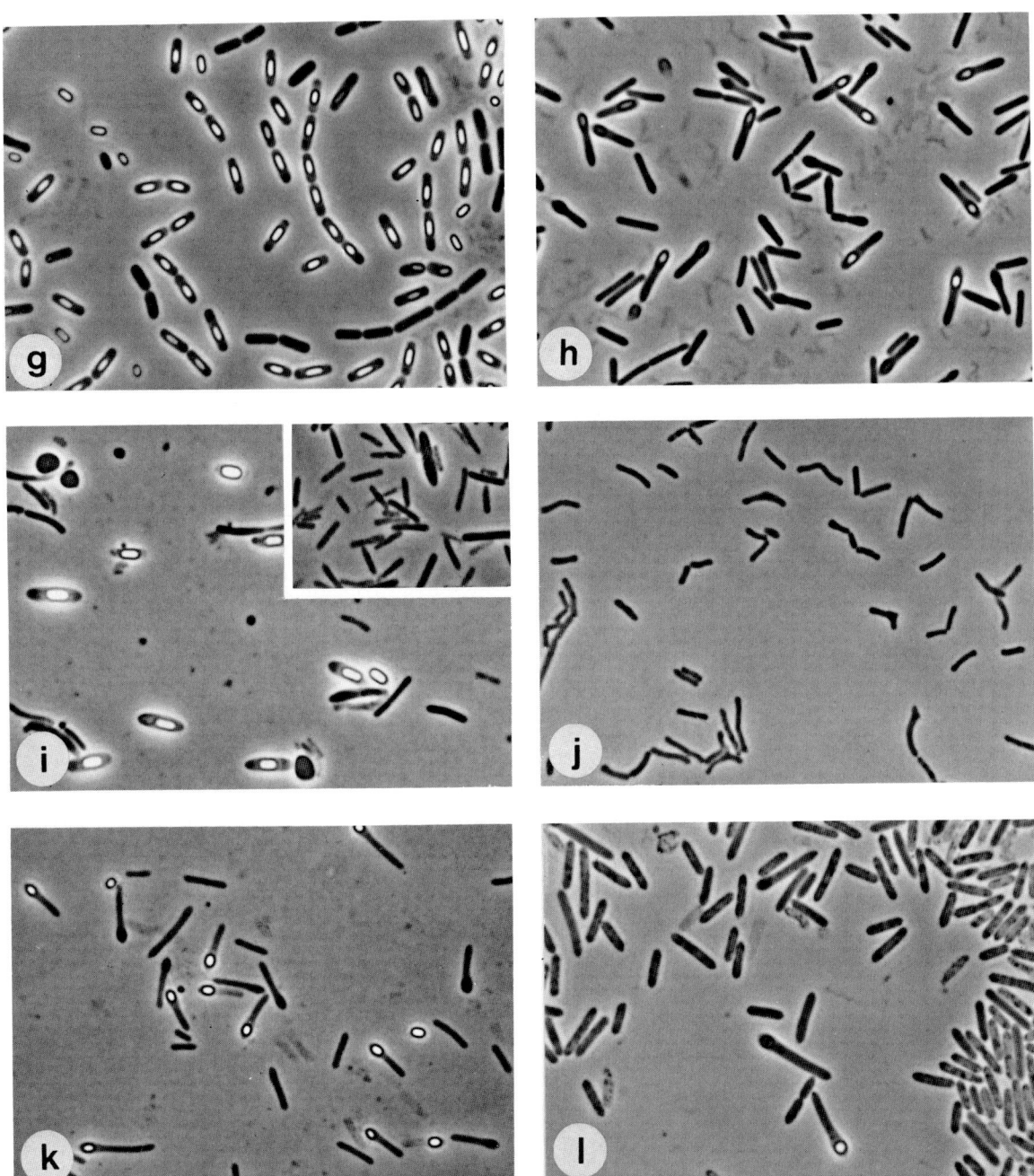

and −200 mV, in their environment (O'Brien and Morris, 1971b). Clostridia will not start growing when the E_h of their environment is above approximately +150 mV. Germination of spores is also affected by oxygen. Although spores are very resistant to oxygen and can be stored in distilled water or as dry material under air for years, their germination requires anaerobic conditions or conditions of low oxygen tension (Douglas et al., 1973; Sarathchandra et al., 1974). Some reports indicate that germination

at low oxygen tensions requires a larger supply of nutrients in the environment than germination under strictly anaerobic conditions (O'Brien and Morris, 1971b). In general, the process of germination is closely linked to the requirements of vegetative cells, perhaps because it would be useless for germination to occur under conditions unfavorable to growth.

Oxygen sensitivity restricts the habitat of the clostridia to anaerobic areas or areas with low oxygen tensions. Growing and dividing clostridial cells will, therefore, not be found in air-saturated surface layers of lakes and rivers or on the surface of organic material and soil. Clostridial spores, however, are present with high probability in these environments, and they will germinate when oxygen is exhausted and when appropriate nutrients are present. Habitats that contain sufficient amounts of organic material become anaerobic rather readily because the solubility of oxygen in aqueous solutions is low (ca. 9 mg per liter at 20°C under 1 bar air pressure) and because oxygen is removed rapidly by aerobes.

In permanently anaerobic habitats with a controlled supply of nutrients, such as in the rumen or in parts of the intestine, the clostridia may have an evolutionary disadvantage because of the energy cost of reduplicating the information for spore formation (Zamenhof and Eichhorn, 1967). This postulated disadvantage might explain the preponderance of nonspore-forming anaerobes in the above-mentioned habitats.

Nutrient Requirements

Due to the ubiquitous distribution of the clostridial spores and their resistance, the clostridia are potentially present anywhere. If the physical parameters (pO_2, E_h, pH, and temperature) are favorable and an organic compound is available as an energy source, then spores of some clostridial species will germinate and a clostridial population will be established. The substrate spectrum of the whole genus is extremely broad and covers a wide range of naturally occurring compounds. Many clostridia excrete exoenzymes which make various macromolecules accessible to them. On the basis of their preferred or characteristic substrates, four nutritional groups of clostridia can be distinguished: saccharolytic clostridia, proteolytic clostridia digesting meat, a combination of both, and specialists (Table 1).

SACCHAROLYTIC CLOSTRIDIA. These are usually nonpathogenic organisms able to grow on carbohydrates such as xylose, mannitol, glucose, fructose, lactose, and raffinose. This group includes species which utilize starch (e.g., *C. butyricum*), cellulose (e.g., *C. cellobioparum*), pectin (e.g., *C. felsineum*), and chitin (e.g., *C. sporogenes*) and which, therefore, are able to form the appropriate exoenzymes. (For references, see the list of clostridial species at the end of this chapter.)

PROTEOLYTIC CLOSTRIDIA. These organisms are able to excrete proteases and to digest proteins. Characteristic for this type of fermentation is

the formation of branched-chain fatty acids from the corresponding amino acids (Andreesen et al., 1989; Elsden and Hilton, 1978; Mead, 1971). Aromatic amino acids are either oxidatively decarboxylated or reduced, or the side chain is split off (Elsden et al., 1976). A number of proteolytic species are highly pathogenic (e.g., *C. botulinum, C. tetani*).

PROTEOLYTIC AND SACCHAROLYTIC CLOSTRIDIA. A typical representative of the non-toxin-producing species of this nutritional group is *C. oceanicum*. Most of the other species are toxin producers, such as *C. perfringens* and *C. sordellii*.

SPECIALISTS: NEITHER PROTEOLYTIC NOR SACCHAROLYTIC SPECIES. This group comprises organisms that have specialized on one or a few substrates. For example, *C. acidiurici* and *C. purinolyticum* grow on purines such as uric acid and adenine but not on sugars or amino acids. *C. kluyveri* has specialized on the fermentation of ethanol, acetate, and bicarbonate to butyrate, caproate, and molecular hydrogen. *C. propionicum* ferments only threonine and three-carbon compounds such as alanine, lactate, acrylate, serine, and cysteine. *C. cochlearium* degrades only glutamate, glutamine, and histidine.

Consequently, the genus *Clostridium* as a whole has a very high capacity for attacking a wide variety of organic compounds, and habitats that provide these compounds and fulfill the above-mentioned physical conditions will be very suitable for the germination of spores and the growth of vegetative cells. In the laboratory, most clostridial species are cultivated using rather complex media. Their actual nutritional requirements are not known in most cases and have been determined only for some species or certain strains of them (Table 2). However, growth of the organisms listed in Table 2 is generally stimulated if their media are supplemented with complex nutrients.

Isolation

Occurrence of Spores and Methods for the Elimination of Nonsporeformers

Clostridia differ widely in their readiness to produce spores. Although some media have been found to stimulate sporulation of many species (Gibbs and Hirsch, 1956; Perkins, 1965; Roberts, 1967), there is no one medium that is suitable as sporulation medium for the whole range of clostridia. Spores of several strongly saccharolytic species are preferentially produced in

Table 1. The saccharolytic and/or proteolytic clostridia and the specialists that grow on amino acids, purines, or special substrates.[a]

Saccharolytic species			Proteolytic species	Proteolytic and saccharolytic species	Specialists
C. aerotolerans	*C. fallax*	*C. rectum*	*C. argentinense*	*C. acetobutylicum*	*C. aciduirici*
C. absonum	*C. felsineum*	*C. roseum*	*C. collagenovorans*	*C. bifermentans*	*C. aminovalericum*
C. aceticum	*C. fervidum*	*C. saccharolyticum*	*C. ghoni*	*C. botulinum A, B, F (prot.)*	*C. bryantii*
C. arcticum	*C. formicoaceticum*	*C. sardiniensis*	*C. hastiforme*	*C. botulinum C, D*	*C. cochlearium*
C. aurantibutyricum	*C. glycolicum*	*C. sartagoforme*	*C. histolyticum*	*C. cadaveris*	*C. cylindrosporum*
C. baratii	*C. indolis*	*C. scatologenes*	*C. limosum*	*C. haemolyticum*	*C. irregularis*
C. barkeri	*C. innocuum*	*C. scindens*	*C. mangenotii*	*C. lituseburense*	*C. kluyveri*
C. beijerinckii	*C. intestinale*	*C. septicum*	*C. proteolyticum*	*C. novyi B, C*	*C. malenominatum*
C. botulinum B, E, F	*C. josui*	*C. sphenoides*	*C. putrefaciens*	*C. oceanicum*	*C. oxalicum*
C. butyricum	*C. lentocellum*	*C. spiroforme*	*C. subterminale*	*C. perfringens*	*C. pfennigii*
C. carnis	*C. leptum*	*C. stercorarium*	*C. tetani*	*C. putrefaciens*	*C. propionicum*
C. celatum	*C. magnum*	*C. symbiosum*		*C. putrificum*	*C. purinolyticum*
C. celerecrescens	*C. methylpentosum*	*C. tertium*		*C. sordellii*	*C. sporosphaeroides*
C. cellobioparum	*C. nexile*	*C. tetanomorphum*		*C. sporogenes*	*C. sticklandii*
C. cellulolyticum	*C. novyi A*	*C. thermoaceticum*			*C. villosum*
C. cellulovorans	*C. oroticum*	*C. thermoautotrophicum*			
C. chartatabidum	*C. papyrosolvens*	*C. thermobutyricum*			
C. chauvoei	*C. paraputrificum*	*C. thermocellum*			
C. clostridiiforme	*C. pasteurianum*	*C. thermocopriae*			
C. coccoides	*C. polysaccharolyticum*	*C. thermohydrosulfuricum*			
C. cocleatum	*C. populeti*	*C. thermolacticum*			
C. colinum	*C. puniceum*	*C. thermosaccharolyticum*			
C. difficile	*C. quercicolum*	*C. thermosulfurogenes*			
C. disporicum	*C. ramosum*	*C. tyrobutyricum*			
C. durum					

[a] A few species are listed in more than one column mainly because of strain differences.

Table 2. Nutritional requirements on synthetic media for some clostridia.

Organism	Vitamins	Amino acids	References
C. acetobutylicum	*p*-Aminobenzoate, biotin		Rubbo et al., 1941; Lampen and Peterson, 1943
C. acidiurici	Thiamine		Barker and Peterson, 1944; Schiefer-Ullrich et al., 1984
C. bifermentans	Biotin, nicotinamide, pantothenate, pyridoxine	Alanine, arginine, aspartate, cysteine, glutamate, glycine, histidine, isoleucine, leucine, methionine, phenylalanine, threonine, tryptophan, tyrosine	Holland and Cox, 1975
C. bryantii	—	—	Stieb and Schink, 1985
C. butyricum	Biotin	—	Cummins and Johnson, 1971
C. cellobioparum	Biotin	—	Hungate, 1944
C. collagenovorans	—	—	Jain and Zeikus, 1988
C. formicoaceticum	Pyridoxine	Lysine (or cadaverine or *m*-diaminopimelate), methionine	Leonhardt and Andreesen, 1977
C. glycolicum	Biotin, pantothenate	Arginine, glutamate, glycine, histidine, isoleucine, leucine, lysine, methionine, phenylalanine, proline, serine, threonine, tryptophan, tyrosine, valine	Gaston and Stadtman, 1963
C. kluyveri	*p*-Aminobenzoate, biotin	—	Bornstein and Barker, 1948a
C. magnum	—	—	Schink, 1984
C. methylpentosum	Biotin, vitamin B_{12}	—	Himelbloom and Canale-Parola, 1989
C. oxalicum	Vitamins (mixture)	—	Dehning and Schink, 1989
C. pasteurianum[a]	*p*-Aminobenzoate, biotin	—	Carnahan and Castle, 1958; Sergeant et al., 1968; Malette et al., 1974
C. proteolyticum	—	—	Jain and Zeikus, 1988
C. purinolyticum	Thiamine	—	Dürre et al., 1981
C. putrificum	*p*-Aminobenzoate, pantothenate	Arginine, glutamate, proline, serine, tyrosine, valine	Descrozailles et al., 1974
C. sporogenes	*p*-Aminobenzoate, biotin, nicotinate (folate, thiamine)[b]	Arginine, isoleucine, phenylalanine, tyrosine, valine (aspartate, glutamate, glycine, histidine, leucine, methionine, proline, serine, threonine, tryptophan)[b]	Shull et al., 1949; Campbell and Frank, 1956; Belokopytov et al., 1982; Lovitt et al., 1987
	p-Aminobenzoate, biotin, pantothenate, pyridoxine, thiamine	Arginine, glutamate, isoleucine, lysine, phenylalanine, tyrosine, valine	Chaigneau et al., 1974
C. tertium[c]	*p*-Aminobenzoate, biotin, nicotinamide, pantothenate, riboflavin, thiamine	Arginine, glutamate, histidine, isoleucine, leucine, lysine, methionine, serine, threonine, tryptophan, tyrosine, valine	Hasan and Hall, 1976
C. thermoaceticum	Nicotinate	—	Lundie and Drake, 1984
C. thermocellum	Biotin, pantothenate, pyridoxine, riboflavin, thiamine	—	McBee, 1950
	p-Aminobenzoate, biotin, folate, pantothenate, pyridoxine, riboflavin, thiamine	Cysteine, cystine, methionine, phenylalanine, tryptophan, tyrosine	Fleming and Quinn, 1971
	Biotin, pyridoxamine, B_{12}, *p*-aminobenzoic acid		Johnson et al., 1981

[a]Sulfate is also required.
[b]Requirement depends on strain and source of culture.
[c]Adenine is also required.

the presence of carbohydrates, whereas others sporulate better in their absence (Bèrgere and Hermier, 1970; Nasuno and Assai, 1960). Surface growth on agar media often gives higher sporulation rates than growth in liquid media. Incubation at a suboptimal temperature sometimes results in more spores than incubation at the optimal temperature. Special sporulation media have been developed for a few clostridia: e.g., *C. putrefaciens* (Roberts and Derrik, 1975), *C. thermosaccharolyticum* (Hsu and Ordal, 1969), *C. perfringens* (Duncan and Strong, 1968; Ellner, 1956; Nishida et al., 1969; Sacks and Thompson, 1978), *C. pasteurianum* (Emtsev, 1963; Mackey and Morris, 1971), *C. bifermentans* (Hitzman et al., 1957). For some species or strains, the initiation of spore formation is very difficult. This difficulty has long been known for *C. perfringens,* but it is also found in species such as *C. nexile* (Holdeman and Moore, 1974), *C. clostridiiforme* (Cato and Salmon, 1976), *C. leptum* (Moore et al., 1976), and *C. ramosum* (Holdeman et al., 1971).

In natural samples such as soil, anaerobic mud, sewage, feces, or material from other anaerobic habitats, the existence of spores of all the clostridial species present can be assumed. Therefore, enrichment cultures are usually started with spores, which has the advantage that nonsporeformers can be eliminated by pasteurization or alcohol treatment. Several clostridia produce extremely heat-resistant spores that survive when incubated at 100°C or more for several hours, but others show only a moderate or low heat resistance (Ingram, 1969; Roberts and Ingram, 1965). The use of comparatively high temperatures may also have selective properties (mutagenic effects), and strains with changed characteristics may be isolated as shown for *C. perfringens* (Hayase et al., 1974; Nishida et al., 1969). Therefore, the application of as low heat as possible during pasteurization is recommended. Incubation for 10 min at 70°C, 10 min at 80°C, or 10 min at 90–100°C is sufficient also for the elimination of most thermophilic nonsporeformers.

Pasteurization to Eliminate Nonsporeformers

Place 1 g of mud, soil, or other material to be used as inoculum in a sterile test tube and add 5 ml of sterile 0.9% NaCl solution. Flush tube with oxygen-free nitrogen gas, close it with a rubber stopper, and incubate it in a water bath at 80°C (or at the temperature desired). The temperature increase is monitored with a second tube that contains 6 ml of 0.9% NaCl solution and a thermometer. When the thermometer has reached 80°C, the tubes are held at this temperature for 10 min and then immediately cooled to room temperature. Samples pasteurized this way are used as inocula for enrichment cultures.

Treatment of the samples with ethanol has been suggested as an alternative procedure for killing the vegetative cells of the accompanying microflora (Johnston et al., 1964; Koransky et al., 1978). This method is an excellent means of facilitating the isolation of clostridia from samples in which they are far outnumbered by other facultative or obligate anaerobes. At the same time, any damage of the spores is minimized.

Alcohol Treatment for Eliminating Vegetative Cells (Johnston et al., 1964; Koransky et al., 1978)

Liquid samples or solid samples suspended in sterile water are mixed with an equal volume of absolute ethanol. After incubation for 60 min at room temperature, samples are used for dilution series. The absolute ethanol to be used should be filter-sterilized or autoclaved in closed tubes.

Provision of Anaerobic Conditions for Growth

Because the clostridial species exhibit varying degrees of oxygen intolerance, more or less rigid methods to exclude oxygen from the culture media will be recommended in the isolations to be described below. However, it should be pointed out that the Hungate technique—if the set-up is available—is the most convenient and effective method for cultivating clostridia.

The following methods can be used in the isolations:

ENRICHMENTS IN COMPLETELY FILLED GLASS-STOPPERED BOTTLES. This method is satisfactory for all enrichments because aerobes and facultative anaerobes present in the inoculum will remove all the oxygen and provide the conditions for growth of the obligate anaerobes. The addition of reducing compounds is only necessary when mineral media with a low content of complex organic nutrients are used in the primary enrichment steps. It is recommended that the medium be autoclaved. Autoclaving reduces the oxygen content of the medium, thereby preventing heavy growth of aerobes. Furthermore, it ensures that the anaerobes that grow originate from the inoculum and not from the medium.

GROWTH IN TUBES AND VOLUMETRIC FLASKS WITH A PYROGALLOL SEAL (WRIGHT, 1901; KÜRSTEINER, 1907; RITTER AND DORNER, 1932). The following points are important for culturing clostridial species: (1) the media should be boiled for a few minutes to remove dissolved oxygen; (2) the reducing agent (sodium thioglycolate, L-cysteine, sodium sulfide, etc.) must be added just before autoclaving; (3) the medium has to be cooled down rapidly after autoclaving; and (4) the tubes or flasks have to be sealed immediately after cooling. Sealing is done by pushing the cotton plug down the tube or the neck of the flask, adding some adsorbant wool, applying 20% (wt/vol) pyrogallol and potassium carbonate solutions (8 drops each per test tube), and sealing with a rubber stopper. If gas producers are cultivated, pyrogallol seals should be used with caution, and a vent is recommended.

GROWTH IN ANAEROBIC JARS (HOLDEMAN ET AL., 1977; WILLIS, 1977). Anaerobic jars are cylindrical vessels that are made of metal, glass, or plastic and that can be closed air-tight. They are convenient for the growth of clostridia on agar plates as well as in culture tubes. The early forms of anaerobic jars have now been replaced by cold-catalyst jars (BTL anaerobic jar, Baird and Tatlock; Whitley anaerobic jar, Don Whitley Scientific; BBL and Oxoid/Unipath anaerobic jars). Anaerobiosis inside the jars can be controlled by a commercially available indicator dye (methylene blue or resazurin). In the case of vented jars, the air can be quickly removed by evacuation and refilling with an appropriate mixture of oxygen-free cylinder gases, which include some hydrogen gas to remove the last traces of oxygen by reaction with the cold catalyst.

Operation of the jars with the GasKit (Don Whitley Scientific) or GasPak (BBL) disposable, hydrogen-carbon dioxide generator makes evacuation and refilling unnecessary and provides an oxygen-free nitrogen atmosphere that contains about 10% carbon dioxide. The carbon dioxide stimulates the germination of spores and favors the growth of clostridia (Holland et al., 1970; Roberts and Hobbs, 1968; Smith and Sullivan, 1989). The cold catalyst (palladium-coated alumina pellets) can be inactivated, especially by hydrogen sulfide, which may be produced during growth of many clostridia. Therefore, the catalyst in the jar has to be renewed before each use. The used catalyst can be reactivated by heating at 170°C for 2 h. Freshly prepared agar plates are predried by storing them in the anaerobic jar over silica gel desiccant for 2 days. No catalyst is required if the Anaerocult system of Merck, Darmstadt, FRG, is used for producing an oxygen-free environment in an anaerobic jar. The system contains components that, when reacted with water, chemically bind all the atmospheric oxygen present in the vessel and simultaneously produce carbon dioxide.

If the agar plates are to be inoculated soon after pouring, they can be dried by placing a sterile disk of filter paper and 2–3 drops of glyc-erol in the lid of the petri dish and incubating them in inverted position. Freshly inoculated plates should only be exposed to air for the shortest possible time.

GROWTH IN TUBES MADE ANAEROBIC ACCORDING TO HUNGATE (1969) AND MACY ET AL., (1972). These methods are extensively described in other sections of this Handbook (see Chapters 5 and 11). In this chapter, they will be described later for the direct isolation of C. sphenoides only.

General Media for Enrichment and Isolation

A number of complex media have been widely used for the enrichment and isolation of saccharolytic and proteolytic clostridia: reinforced clostridial medium (RCM) (Gibbs and Hirsch, 1956; Hirsch and Grinstedt, 1954); differential reinforced clostridial medium (DRCM) (Gibbs and Freame, 1965); cooked meat media (CM and CMC) (Holdeman et al., 1977; Robertson, 1915–1916); peptone-yeast extract medium (PY) (Holdeman et al., 1977); Viande-Levure medium (VL) (Beerens and Fievez, 1971; Willis, 1977). RCM and DRCM contain glucose and starch; the others can be supplemented with further ingredients, such as carbohydrates or certain amino acids, to make them more selective. DRCM is recommended especially for the detection and enumeration of clostridial spores in pasteurized samples of food (Freame and Fitzpatrick, 1971; Gibbs and Freame, 1965).

Any medium that is rich in carbohydrates and that contains some peptone and yeast extract as well as reducing agents is suitable for the enrichment of the common saccharolytic clostridia. A selective medium containing antimicrobial agents for the isolation of C. butyricum and a few other butyric acid-producing, saccharolytic clostridia from human feces has been described by Popoff (1984).

Potato mash medium, as described by Ruschmann (1928), has been used by several investigators in the past to study the clostridia in retting flax, in manure, in silage, and in milk (Dührsen, 1937; Glathe, 1934; Ritter, 1932; Ruschmann, 1928; Ruschmann and Bavendamm, 1925; Ruschmann and Harder, 1931). Its starch and pectin content favors the development of starch- and/or pectin-fermenting clostridia. Maize mash medium (Weizmann, 1919) and maize liver medium (McClung and McCoy, 1934) have been used for enrichment and isolation of the butanol-acetone-producing clostridia (Beesch, 1953; Weizmann, 1919; Weyer and Rettger, 1927), the pigment-producing strains (Hellinger, 1947; McClung, 1943), and other saccharolytic species (Gilliland and Vaughn, 1943; McClung and McCoy, 1934; Weizmann and Hellinger, 1940). Milk, without or with supplementations, has been found convenient for the enrichment of several clostridia from soil, wounds, and plant material or of the clostridia commonly present in the milk itself (Meyn, 1933; van Beynum and Pette, 1940; Weinzirl and Veldee, 1915; Weizmann and Hellinger, 1940; Winkler, 1961).

Potato tubers, stabbed and immersed in water, provide a simple method for the enrichment of common saccharolytic clostridia. The following procedure has been described by Veldkamp (1965).

Enrichment of Saccharolytic Clostridia with Potato Tubers (Veldkamp, 1965)

A potato is washed under the tap and subsequently stabbed once or twice with a knife. It is then placed in a beaker, and enough water is added just to cover the potato; the beaker is covered with a watch-glass and incubated at 37°C. The oxygen that might be introduced into the tissue is consumed by its cells. In the anaerobic environment *Clostridium* rapidly starts to decompose the pectin in between the plant cells. The tissue is thus macerated. When the tuber floats, due to profuse gas formation, the water is poured out of the glass; the potato is washed and dissected. Microscopic examination of the tuber contents invariably shows the typical pleomorphic clostridial cells; among these, spore-bearing spindle-shaped cells are often encountered.

Isolation can easily be achieved as follows. A sample of the macerated tissue is inoculated into yeast extract-glucose broth and after pasteurization (10 min at 80°C in a water bath) the culture is incubated at 37°C in N_2 atmosphere. A pure culture can be obtained by streaking a sample on yeast extract-glucose agar and incubating under N_2.

A double-layered polypectate agar medium has been used to isolate pectolytic *Clostridium* spp. from tubers and carrots and from suspensions of field and rhizosphere soils (Lund, 1972; Lund et al., 1981; Perry, 1982; Perry, 1985).

Media given by various authors for enrichment and growth of saccharolytic clostridia frequently show variations that are not essential for growth of these organisms. The following medium is recommended for the growth of many species.

Medium for Growth of Saccharolytic Clostridia

The medium contains:

1 M potassium phosphate, pH 7.5	30 ml
1 M MgSO₄	1 ml
Solution M (see below)	0.5 ml
0.2M FeSO₄ in 0.1 M H₂SO₄	0.2 ml
Trypticase	10 g
Yeast extract	6 g
Sodium thioglycolate	0.5 g
Energy source (sucrose or glucose)	10–20 g
Distilled water	1 liter

The pH is adjusted to 7.0–7.2. The medium is autoclaved for 20 min at 121°C.

Solution M contains:

MnCl₂	10 mM
CaCl₂	30 mM
CoCl₂	5 mM
Na₂MoO₄	5 mM

Since stock solutions of the salts can be stored in the laboratory, this medium can be prepared rather quickly. It is best to autoclave the sugar solutions separately.

The fixation of molecular nitrogen is a property shared by many representatives of the saccharolytic clostridia (Rosenblum and Wilson, 1949). Media that contain glucose or sucrose but lack a source of nitrogen have been used for the isolation of *C. pasteurianum* and other species. The comprehensive literature was summarized by Skinner (1971). Based on the method of Augier (1957), the following liquid medium for counting nitrogen-fixing clostridia in soil by the most probable number method (MPN) and for subsequent isolation has been recommended by Skinner (1971).

Isolation of Nitrogen-Fixing Saccharolytic Clostridia (Skinner, 1971)

The medium contains:

K₂HPO₄	0.8 g
KH₂PO₄	0.2 g
MgSO₄·7H₂O	0.2 g
NaCl	0.2 g
FeSO₄·7H₂O	10 mg
MnSO₄·4H₂O	10 mg
CaCl₂	10 mg
Na₂MoO₄·2H₂O	25 mg
Yeast extract	10 mg
Trace element solution (see below)	1 ml
Soil extract (see below)	10 ml
Glucose or sucrose	10.0 g
Sodium thioglycolate	1.0 g
Distilled water	1 liter

The pH is adjusted to 7.2.

The trace element solution contains:

Na₂B₄O₇·10H₂O	50 mg
CoNO₃·6H₂O	50 mg
CdSO₄·2H₂O	50 mg
CuSO₄·5H₂O	50 mg
ZnSO₄·7H₂O	50 mg
MnSO₄·H₂O	50 mg
Distilled water	1 liter

The stock solution should be saturated with CO_2.

The soil extract is prepared as follows (Augier, 1956): Equal weights of a neutral garden soil and water are combined and heated at 130°C for 1 h. After cooling, it is filtered through paper, bottled, autoclaved, and stored until use.

Sodium thioglycolate is added just before the medium is distributed in 10-ml portions into narrow test tubes fitted with Durham tubes, capped or plugged, and autoclaved at 121°C for 15 min. After inoculation with 1 ml of decimal dilutions of pasteurized soil, the tubes are incubated in anaerobic jars under a nitrogen atmosphere at 30°C for 2 weeks. Positive tubes show abundant gas formation, turbidity with or without a viscoid, whitish deposit or surface pellicle, and odor of butyric acid.

Isolations are made by streaking small inocula from positive tubes on plates prepared from the above medium plus 2% agar. The plates are incubated in anaerobic jars under an atmosphere of nitrogen. Colonies can be picked within 7 days at 30°C and purified by repeated streaking on nitrogen-free agar. Pasteurization can be applied at any stage, provided that spores are present.

The above enrichment procedure can be made rather specific for *C. pasteurianum* if the concentration of sucrose is increased to 15% (Spiegelberg, 1944; Witz et al., 1967). Inocula from such enrichments are again pasteurized and streaked on the above agar containing 2% glucose.

Clostridia, such as *C. butyricum* and *C. tyrobutyricum*, which ferment lactate plus acetate to butyrate, CO_2, and H_2, have been frequently found as the dominating anaerobic sporeformers in certain silages (Bryant and Burkey, 1956; Gibson, 1965; Gibson et al., 1958; Rosenberger, 1951, 1956) and milk products (Goudkov and Sharp, 1965). *C. tyrobutyricum* is responsible for spoilage of certain types of cheese by formation of gas ("late blowing" of hard cheese) or rancid odor (Goudkov and Sharp, 1966; Kutzner, 1963, 1966).

A medium originally developed by Bhat and Barker (1947) has been used for the detection and the isolation of lactate-utilizing sporeformers in silage (Rosenberger, 1951, 1956), in cheese (Kutzner, 1963, 1966), and in milk (Halligan and Fryer, 1976). Both *C. butyricum* and *C. tyrobutyricum* will grow in this medium; a distinction between these species is easily possible because of the striking differences in the utilization of carbohydrates: *C. butyricum* grows on maltose, raffinose, lactose, and starch, which are not utilized by *C. tyrobutyricum*.

Isolation of Clostridia That Ferment Lactate plus Acetate (Bhat and Barker, 1947)

The medium contains:

K_2HPO_4	0.5 g
$(NH_4)_2SO_4$	0.5 g
$MgSO_4 \cdot 7H_2O$	0.1 g
$FeSO_4 \cdot 7H_2O$	20 mg
Yeast extract	0.5 g
Biotin	0.1 μg
p-Aminobenzoate	100 μg

Sodium L-lactate	10.0 g
Sodium acetate	8.0 g
Sodium thioglycolate	0.5 g
Distilled water	1 liter

Sodium thioglycolate is added immediately before autoclaving. The pH is adjusted to 6.0–7.0, and the medium is autoclaved for 20 min at 121°C.

A slightly acidic pH value of the medium favors growth of *C. butyricum* and *C. tyrobutyricum*. The latter has been shown to start growing in media with a pH value of as low as 5.3 (Kutzner, 1963). When a sodium lactate solution is used for preparing the medium, the pH should be checked after autoclaving (it tends to become more acidic).

Enrichment cultures are set up in glass-stoppered bottles (filled to the neck). They are inoculated with pasteurized garden soil or other material and incubated at 37°C. Turbidity and gas formation are observed after a few days, accompanied by an increase of the pH. Two transfers into fresh medium (5–10% inoculum) are made before isolation is performed by streaking material on agar medium of the same composition as above plus 2% agar. The plates are incubated in anaerobic jars under a nitrogen atmosphere.

Using a slightly modified medium, *C. butyricum* was found to be the dominant lactate-fermenting sporeformer in the ensiling process of perennial ryegrass (Gibson et al., 1958). Clostridia other than *C. butyricum* and *C. tyrobutyricum* that are found occasionally in silage (*C. paraputrificum*, *C. tetanomorphum*, *C. perfringens*, and *C. sphenoides*) require more complex media or are not able to use lactate plus acetate as energy source.

Procedures for the detection of the very low numbers of spores of *C. tyrobutyricum* in milk used in making cheese have been described (Fryer and Halligan, 1976; Halligan and Fryer, 1976). First, an enrichment is done using a complex medium that contains calcium lactate. Then the presence of *C. tyrobutyricum* is confirmed by subculturing positive enrichments in a slightly modified medium as given above.

There have been many reports on the enrichment and isolation of cellulose-decomposing clostridia, but the cultures obtained were often not pure and subcultures are not available any more. In addition to the two classical cellulose-decomposing species *C. cellobioparum* (Hungate, 1944) and *C. thermocellum* (McBee, 1948, 1950; Viljoen et al., 1926), several new mesophilic and thermophilic species have been validly described (Van Gylswyk, 1980; Madden et al., 1982; Madden, 1983; Petitdemange et al., 1984; Sleat et al., 1984; LeRuyet et al., 1985; Sleat and Mah, 1985; Murray et al., 1986; Sukhumavasi et al., 1988; Palop et al., 1989).

Samples for the isolation of cellulolytic clostridia can be taken from soil, feces of herbivorous animals (horse manure has long been recognized as an excellent source for thermophilic strains also), compost, decayed plants, and from anaerobic sewage or mud. Except for the different incubation temperature, the procedures for enrichment and isolation of both mesophilic and thermophilic cellulolytic species are about the same.

Enrichment and Isolation of Cellulolytic Clostridia (Omelianski, 1902; Skinner, 1960, 1971)

For enrichment, the medium contains:

K₂HPO₄	1.0 g
(NH₄)₂SO₄	1.0 g
MgSO₄·7H₂O	0.5 g
CaCO₃	2.0 g
NaCl	0.5 g
Resazurin	1 mg
Cellulose (chopped filter paper)	20 g
Distilled water	1 liter

The final pH is adjusted to 7.1.

Enrichments are set up in completely filled, glass-stoppered bottles (50- or 100-ml), inoculated with 0.1 g of pasteurized sample. The sample material provides enough growth factors for the primary enrichment, and reducing conditions are established by the contaminating microflora. Incubations are made at 30–37°C. As soon as digestion of the filter paper becomes visible (1 or 2 weeks), isolation of the cellulose-degrading clostridia is performed by repeatedly streaking on cellulose agar medium.

The cellulose agar medium contains:

K₂HPO₄	7.5 g
KH₂PO₄	3.5 g
(NH₄)₂SO₄	0.5 g
NaCl	1.0 g
MgSO₄·7H₂O	50 mg
CaCl₂	50 mg
Resazurin	1 mg
Yeast extract	1.0 g
Cellulose	10.0 g
Agar	15.0 g
L-Cysteine hydrochloride	0.5 g
Distilled water	1 liter

The pH is adjusted to 7.1.

A suitable substrate for these enrichments is powdered cellulose (e.g., MN 300, Machery and Nagel; CF-11, Whatman) or a cellulose preparation made by wet-grinding of finely divided filter paper (e.g., Whatman No. 1 ashless cellulose paper) in a pebble mill (Hungate, 1950). To keep the cellulose particles suspended, Skinner (1960) recommended the addition of about 0.1% sodium carboxymethylcellulose (substitution range, 0.65–0.85) to the medium.

The cysteine hydrochloride is added to the medium shortly before autoclaving. The medium is autoclaved for 20 min at 121°C and distributed to petri dishes, which are kept in anaerobic jars under an atmosphere of 90% N₂ + 10% CO₂. Alternatively, screw-capped bottles (Skinner, 1971) or roll tubes (Hungate, 1950) may be used. Digestion of cellulose is visible after incubation at 30–37°C for 1 week or longer, depending on the ac-

tivity of the strains. Colonies are often very small, and material for transfers should be picked carefully under a dissecting microscope. Colonies that contain spores are suspended in cellulose-free basal medium and pasteurized (10 min at 75°C) before streaking again.

For purification, an alternate streaking on cellulose agar or on agar that contains 0.5% cellobiose instead of cellulose can be advantageous. Cellobiose is readily fermented by all the cellulolytic isolates described. However, spore production on cellobiose agar is not as pronounced as on cellulose agar. Therefore, colonies must be examined carefully before being pasteurized.

The "double-layer" method may be used at an early stage to detect cellulose-decomposing colonies by formation of clear halos. Petri dishes are filled with 20 ml of cellulose-free agar medium and subsequently overlayed with 5 ml of cellulose-containing agar medium.

The selective isolation of *C. sphenoides* from mud samples using citrate as the energy source for growth has been described (Walther et al., 1977). It has not been demonstrated whether this method is also applicable to the isolation of this organism from silage or human infections in which *C. sphenoides* has been reported to occur (Gibson, 1965; Smith, 1949). *C. sphenoides* is a representative of the relatively small group of saccharolytic bacteria that do not form butyrate but ferment carbohydrates to ethanol, acetate, CO₂, and H₂.

Direct Isolation of *Clostridium sphenoides* (Walther et al., 1977)

The medium contains:

K₂HPO₄	2.0 g
KH₂PO₄	3.4 g
MgSO₄·7H₂O	0.2 g
(NH₄)₂SO₄	0.3 g
NaCl	0.6 g
CaCl₂·2H₂O	60 mg
Yeast extract	4.0 g
Peptone	2.0 g
Trisodium citrate dihydrate	14.7 g
Resazurin	1 mg
Agar	15.0 g
Distilled water	1 liter

Adjust the pH to 6.7–7.0. 0.3 g of L-cysteine hydrochloride is added after the agar has been dissolved.

Applying the Hungate technique, 5-ml portions are added to 15-ml tubes. After autoclaving (20 min at 121°C), the tubes are kept in a water bath at 50°C. Then 0.3 ml of a pasteurized mud sample is used for the preparation of decimal serial dilutions. Roll tubes are prepared which are incubated at 37°C. After 48–72 h, single colonies that show an increased size as compared to the background growth of contaminating organisms are picked, inoculated into 10 ml of liquid citrate medium, and incubated at 37°C for 24 h. For further purification, serial dilutions in roll tubes containing citrate agar are repeated at least three times.

During isolation, the selected colonies are checked for the presence of wedge-shaped cells and of spherical, nearly terminally located spores, both of which are typical for *C. sphenoides.*

A number of single amino acids can be used by certain clostridial species as sources of energy, carbon, and nitrogen for growth (Andreesen et al., 1989). Media that contain a certain amino acid, small amounts of yeast extract, and minerals have been used for enrichment and isolation of some clostridial species. The use of L-alanine led to the isolation of *C. propionicum* (Cardon and Barker, 1946, 1947) and of γ-aminobutyrate and δ-aminovalerate to "*C. aminobutyricum*" and *C. aminovalericum,* respectively (Hardman and Stadtman, 1960a, 1960b). *C. tetanomorphum* has been isolated by Kornberg using L-histidine as the principal substrate (see Wachsman and Barker, 1955a), and the use of L-glutamate led to the isolation of *C. cochlearium* as well as of *C. tetanomorphum* (Barker, 1937, 1939). Lysine was used to enrich and isolate *Clostridium* SB4 (Costilow et al., 1966), which subsequently was identified as *C. subterminale* (H. A. Barker, personal communication). It appears that the possible advantage of using single amino acids as selective substrates for the isolation of certain clostridia has never been studied systematically. Investigations by Mead (1971), Elsden et al. (1976), and Elsden and Hilton (1978, 1979) have shown that a considerable number of clostridia are able to ferment single amino acids.

In the following procedure, the enrichment and isolation of *C. cochlearium* or *C. tetanomorphum* is described, using L-glutamate as the specific substrate.

Enrichment and Isolation of Glutamate-Fermenting Clostridia (Barker, 1937, 1939)

The medium contains:

K₂HPO₄	0.2 g
MgSO₄·7H₂O	0.1 g
Yeast extract	0.5 g
Sodium L-glutamate	10.0 g
Sodium thioglycolate	0.5 g
Tap water	1 liter

The final pH is ajusted to 7.6. Enrichments are set up in completely filled glass-stoppered bottles with pasteurized soil as inoculum (0.5 g soil per 100 ml). Within 1–2 days at 37°C, an abundant growth of sporeforming anaerobes is observed accompanied by a moderate production of gas. After transferring a volume of 2 ml once or twice to 100-ml bottles containing medium of the above composition, isolations are made by repeatedly streaking on L-glutamate agar plates or by diluting in agar roll tubes.

Laanbroek et al. (1979) described the enrichment of *C. cochlearium* from anaerobic sludge in an L-glutamate-limited anaerobic chemostat.

For mass culture of *C. cochlearium* or *C. tetanomorphum,* the "Medium for Growth of Saccharolytic Clostridia" (this chapter) may be used. Instead of sugars, sodium L-glutamate (17 g per 1 liter of medium) is added as energy source and Trypticase is omitted.

A selective isolation medium for *C. sporogenes* and related clostridia that are capable of growth by means of coupled oxidation-reduction reactions between appropriate pairs of amino acids has been developed by Fryer and Mead (1979). The semidefined medium contains alanine and proline as the principal energy sources and polymyxin. When samples from soil or mud were directly plated on agar medium, a high percentage of the isolates were strains of the *C. sporogenes/C. botulinum* group.

The anaerobic purine fermenters, *C. acidiurici* and *C. cylindrosporum,* were originally isolated by Barker and Beck (1942) from enrichment cultures with uric acid as the sole source of carbon, nitrogen, and energy. Several more strains of both species were obtained by Champion and Rabinowitz (1977) using the same method. Uric acid agar supplemented with tryptone, meat extract, liver extract, and chicken fecal extract was used by Barnes and Impey (1974) for the isolation of uric acid-decomposing anaerobes from avian cecum. The isolated strains of anaerobic sporeformers were designated *C. malenominatum.* Another purinolytic species, *C. purinolyticum,* was isolated using adenine as the selective substrate (Dürre et al., 1981).

Isolation of Purine-Fermenting Clostridia (Champion and Rabinowitz, 1977; Barker and Beck, 1942)

The medium contains:

Uric acid	2.0 g
10 N KOH	3.0 ml
K₂HPO₄·3H₂O (70% solution)	1.5 ml
MgSO₄·7H₂O	50 mg
CaCl₂·2H₂O	5 mg
FeSO₄·7H₂O	2 mg
Yeast extract	1.2 g
Resazurin	1 mg
Mercaptoacetic acid	1.5 ml
Distilled water (see below)	1 liter

The medium is prepared as follows: 500 ml distilled water plus KOH plus the K₂HPO₄ solution are brought to a boil. Then uric acid is slowly added. It goes into solution instantaneously. The solution is cooled down and the other ingredients of the medium are added. Mercaptoacetic acid is added shortly before autoclaving, and the pH is adjusted to 7.2 using a sterile, 60% K₂CO₃ solution. Other reducing agents may be used instead of mercaptoacetic acid: sodium thioglycolate (750 mg/liter) or cysteine hydrochloride (1 g/liter) plus di-thiothreitol (0.1 g/liter).

After autoclaving (20 min at 121°C), the medium is immediately used to completely fill 50-ml glass-stoppered bottles. After inoculation with 0.5 g of pasteurized soil or chicken feces, the bottles are incubated at 37°C. Growth occurs within 24–48 h and is accompanied by an increase of alkalinity. Utilization of uric acid can be monitored by the decrease in absorbancy at 290 nm of a 1 to 100 dilution of the growth medium. After one or two more transfers (0.5% inoculum) to 50-ml bottles containing the medium above, isolation is made under strict anaerobiosis by repeatedly streaking on uric acid agar plates incubated in anaerobic jars or on roll tubes.

The purinolytic clostridia require strictly anaerobic conditions for growth, and the application of stringent anaerobic culture techniques is necessary. The three accepted species, *C. acidiurici*, *C. cylindrosporum*, and *C. purinolyticum*, are clearly differentiated on the basis of DNA homology, whereas the form and position of their spores differ (Schiefer-Ullrich et al., 1984). *C. cylindrosporum* was named after its cylindrical spores that are variably situated and do not cause swelling of the mother cell (Barker and Beck, 1942; Andreesen et al., 1985).

The isolation of adenine-fermenting clostridia requires a medium of slightly different composition. Instead of uric acid, it contains 0.2% adenine, 0.2% sodium bicarbonate, and 10^{-7} M each of thiamine, sodium selenite, sodium molybdate, and sodium tungstate to replace yeast extract (Dürre et al., 1981).

C. kluyveri has specialized on the conversion of ethanol and acetate to butyrate, caproate, and H_2 (Barker and Taha, 1942; Bornstein and Barker, 1948a). It is unable to utilize carbohydrates or amino acids for growth and requires only biotin and *p*-aminobenzoate as growth factors. Therefore, the medium used for enrichment is very selective.

Isolation of *Clostridium kluyveri* (Bornstein and Barker, 1948a; Stadtman and Burton, 1955)

The medium contains:

K_2HPO_4	0.30 g
KH_2PO_4	0.20 g
NH_4Cl	0.25 g
$MgSO_4 \cdot 7H_2O$	0.20 g
$CaCl_2 \cdot 2H_2O$	10 mg
$FeSO_4 \cdot 7H_2O$	5 mg
$MnSO_4 \cdot 4H_2O$	2 mg
$Na_2MoO_4 \cdot 2H_2O$	2 mg
Biotin	10 µg
p-Aminobenzoate	200 µg
Ethanol	20 ml
Potassium acetate	5.0 g
Glacial acetic acid	2.5 ml
Resazurin	1 mg

Sodium thioglycolate	500 mg
Distilled water	1 liter

Thioglycolate is added shortly before autoclaving (20 min at 121°C). After autoclaving, the medium is rapidly cooled, and the pH is adjusted to 7.0 using a 60% K_2CO_3 solution autoclaved separately (about 8–10 ml are required).

The enrichment is made at 35°C in 100-ml glass-stoppered bottles completely filled with the anaerobically prepared sterile medium and inoculated with pasteurized mud from ensiled leaves of sugar beets (turnips), other decaying plant material, fresh water, or sewage digester. After 1 or 2 weeks, enrichment cultures that show gas production and that smell of butyric and caproic acids are used for inoculating fresh liquid medium (10% inoculum). After two to three transfers, serial dilutions are prepared in agar roll tubes using the above medium supplemented with 2% agar. The colonies of *C. kluyveri* that develop after several days are generally small (1–3 mm), fluffy, spherical or compact, and lens-shaped. The colonies should contain large cells, about 1×10 µm in size. Typical colonies are picked and transferred to a new series of roll tubes. If cells of selected colonies contain spores (oval, terminal), a pasteurization step can be applied after suspending material of a colony in a small volume of growth medium.

Identification

The identification of a bacterium as a member of *Clostridium* is generally relatively easy. A bacterium belongs to the genus *Clostridium* when it is restricted to an anaerobic energy metabolism, forms spores, is unable to carry out a dissimilatory sulfate reduction, and possesses a Gram-positive type cell wall structure. Some clostridial species do not sporulate readily (see "Occurrence of Spores and Methods for the Elimination of Nonsporeformers," this chapter). If spores are not detected, the strain is most probably a member of the genus *Eubacterium* (Chapter 86). To demonstrate spores, it may be necessary to inoculate slants that contain chopped meat agar. Appropriate procedures are described in the VPI *Anaerobe Laboratory Manual* (Holdeman et al., 1977).

To definitely identify an isolated strain as a certain clostridial species is more difficult. An appropriate identification key is given by Cato et al. (1986) in *Bergey's Manual of Systematic Bacteriology* (Sneath et al., (1986). It is based on numerous studies of many species and strains. Applying this key, the allocation of isolates to defined species will be possible in many cases. However, generally speaking, the taxonomy of the genus *Clostridium* is still in an unsatisfactory state. The descriptions of over 20 species are based on only one strain and that of several other species on very few strains. It is symptomatic for the situation that even experts in the field are often unable to assign a considerable percentage of their clostridial isolates to accepted species (Finne and Matches, 1974; Finegold et al., 1983; Matsuda et al., 1975; Salinatro et al., 1974; Timmis et al., 1974; Perry, 1985).

The genus *Clostridium* can physiologically be subdivided into three groups of saccharolytic and/or proteolytic species. Additional groups are formed by the specialist species, some of which do not grow on the media commonly employed for identification, and by the thermophiles. Nutritional specialists are *C. acidiurici, C. bryantii, C. cylindrosporum, C. kluyveri, C. oxalicum, C. pfennigii,* and *C. purinolyticum.* Eleven species are thermophiles whose optimum growth temperatures range between 55°C and 70°C. In *Bergey's Manual of Systematic Bacteriology* (Cato et al., 1986) utilization of glucose and gelatin hydrolysis are used to differentiate the clostridia into four groups. Because false-negative test results may occur in case of weakly gelatinolytic strains using the procedure recommended (Cato et al., 1986; Holdeman et al., 1977), gelatin hydrolysis should additionally be tested using a lower gelatin concentration or using the gelatin agar procedure (Willis and Williams, 1970; Nakamura et al., 1973; Lund et al., 1981; Whaley et al., 1982).

In order to assign an isolate to a certain species, a number of additional tests should be carried out:

1. Determination of aerotolerance. The following species grow on blood agar plates under air: *C. carnis, C. durum, C. histolyticum, C. intestinalis,* and *C. tertium.* Although *C. aerotolerans* has been claimed to tolerate oxygen, aerobic growth on agar plates was not reported (van Gylswyk and van der Toorn, 1987).
2. Determination of cell morphology and size as well as shape and position of mature endospores in the mother cell.
3. Determination of certain enzyme activities. This test includes assays for lecithinase, lipase, proteolysis, indole formation, reaction on milk, and nitrate reduction.
4. Detection of growth and acid production using media that contain carbohydrates as substrates. The main carbohydrates tested are glucose, fructose, mannose, mannitol, ribose, xylose, saccharose, maltose, lactose, melibiose, starch, and esculin.
5. Determination of the fermentation products by gas chromatography.

Holdeman et al. (1977) suggests using peptone-yeast extract-glucose medium, chopped meat-

carbohydrate medium, or any other medium suitable for a specific strain (Nunez-Montiel at al., 1983). The composition of the media and the kind of peptones used influence the pattern of fermentation products formed (Turton et al., 1983). Standardization of the growth conditions and the use of reference strains are both very important. The majority of the saccharolytic clostridia form butyrate as the predominant product. About 30 out of 70 purely saccharolytic species do not produce this acid but form some or all of the following products: acetate, formate, lactate, and ethanol. Only six *Clostridium* species form propionate as a major fermentation product. These species include: *C. arcticum, C. botulinum* type *D, C. haemolyticum, C. novyi, C. propionicum,* and *C. quercicolum.* Caproate is only formed by *C. kluyveri* and *C. scatologenes.* Proteolytic species are usually indicated by the appearance of branched-chain fatty acids among the fermentation products (Elsden and Hilton, 1978). There are some exceptions, however, such as proteolytic species that do not produce iso acids and nonproteolytic species fermenting amino acids that do produce iso acids.

Most clostridial species are motile. For 24 species, however, motility has never been reported. The observation of motility is helpful in classification; the observation of nonmotility is of little value because nonmotile strains of motile species are occasionally found.

In addition to the properties of the clostridial species that are used in identification keys, the GC content of the DNA gives an insight into their relationships. The GC content of the clostridia, as far as it is known, is summarized in Fig. 2. It is apparent from the graph (inset) that most species cluster around 28 mol%. However, about 21 species form a second group, which ranges in GC content from 40 to 54 mol%.

DNA-DNA and DNA-rRNA homology studied have been applied to solve a number of taxonomic problems, especially within the groups of butyric acid-producing and thermophilic clostridia (Johnson, 1970, 1973; Johnson and Francis, 1975; Kaneuchi et al., 1979; Matteuzzi et al., 1977, 1978; Nakamura et al., 1973, 1975, 1979).

Taxonomy and Phylogeny

The genus *Clostridium* has become one of the largest genera of the bacteria and at present contains just over 100 species. This is mainly due to the very simple definition of the genus, which basically has not changed since the first edition of *Bergey's Manual* was published. Cato et al.

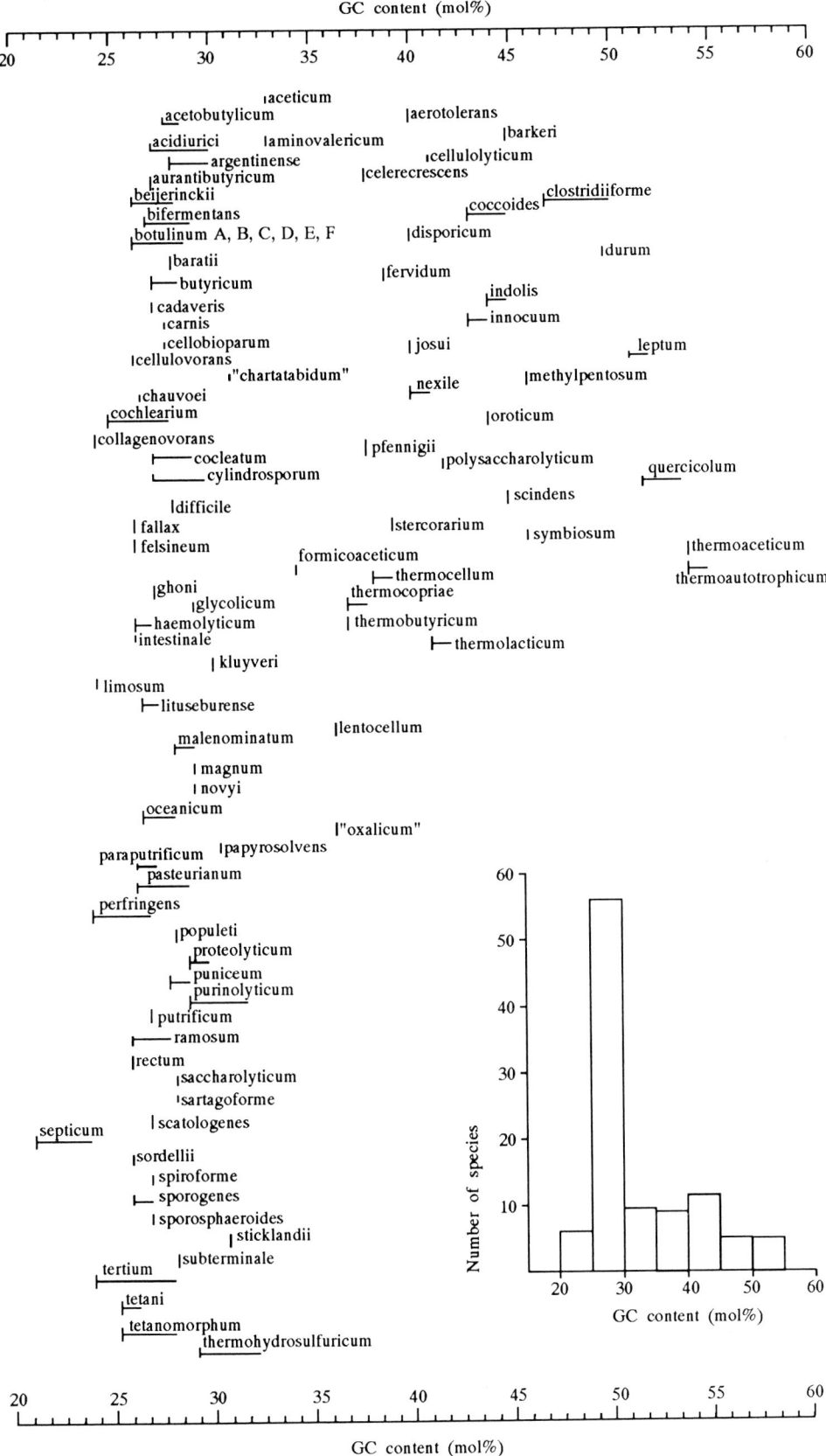

Fig. 2. Distribution of the GC content of the species of the genus *Clostridium*.

(1986) listed 83 species in *Bergey's Manual of Systematic Bacteriology* (Sneath et al., 1986), and since then about 25 additional species have been validly described.

New anaerobic sporeformers have also been described; they possess a Gram-negative cell wall structure and are phylogenetically distinct from *Clostridium.* They have been placed in the genera *Sporomusa* and *Sporohalobacter* (Möller et al., 1984; Oren, 1983; Stackebrandt et al., 1985; Oren et al., 1987).

Another new species, *Anaerobacter polyendosporus,* was found by comparative 5S rRNA analysis to be only distantly related to *C. butyricum,* the type species of the genus *Clostridium,* so that the authors proposed a separate genus (Duda et al., 1987). However, the organism is Gram-positive and, on the basis of the properties defined for the genus *Anaerobacter,* it is difficult to distinguish from the genus *Clostridium* (Table 3). Recently, *C. bryantii* was reclassified as the first species of the new genus *Syntrophospora* because of its distant relationship to the clostridia on the basis of 16S rRNA homology (Zhao et al., 1990).

Also, *Thermoanaerobacter finnii* has been described to produce heat-resistant endospores but evidence for true spore-forming ability based on ultrastructure and the presence of dipicolinic acid is still lacking (Schmid et al., 1986). Similarly, "atypical" heat-resistant spores were reported to occur in *Butyribacterium methylotrophicum* (Zeikus et al., 1980), an organism closely related to *Eubacterium limosum.*

The genus *Clostridium* exhibits a great diversity with respect to the morphological and metabolic properties, nutritional requirements, and GC content of its members (Andreesen et al., 1989). The classification to species level is based on data from traditional testing of morphological, physiological, and/or biochemical parameters. Groups of species that may be formed applying these data appear rather arbitrary and usually do not reflect closer genetic relationships. Several proposals made in the past to subdivide the genus have not been accepted (Chester, 1901; Weinberg, 1937; Hauduroy et al., 1937; Prévot, 1938, 1966).

Modern chemotaxonomic methods have hitherto gained only limited access to the classification of the clostridia. One reason for this is that all studies made in this field remained fragmentary, because they were restricted to selected species or groups and were not extended and continued actively to the other or to newly described species.

Gel Electrophoretic Cell Protein Patterns

Comparison of soluble cell protein pattern obtained by polyacrylamide gel electrophoresis (PAGE) is a valuable method for differentiation at the species level but is less suited to distinguish higher taxa. A PAGE analysis of native cell proteins applied to about 70 *Clostridium* species showed a good correlation of the protein patterns with available DNA homology data and phenotypic characteristics (Cato et al., 1982a). Strong indications were obtained that *C. perenne* and *C. paraperfringens* are synonyms of *C. baratii,* that *C. lentoputrescens* is synonymous with *C. cochlearium,* and that *C. pseudotetanicum* is identical with *C. beijerinckii* (Cato et al., 1982a, 1982b). On the other hand, several species which are phenotypically very similar could be easily differentiated by their very distinct protein patterns, e.g., *C. purinolyticum* from *C. cylindrosporum,* *C. sphenoides* from *C. indolis,* *C. beijerinckii* from *C. butyricum.* The usefulness of SDS-PAGE protein analysis for species differentiation has also been demonstrated by other investigators (Magot et al., 1983, Popoff et al., 1985; Bom et al., 1986; Jain and Zeikus, 1988; Wilde et al., 1989). An improvement of the resolution of the electrophoretic separation of soluble cell proteins was achieved by preventing sporulation of cultures and by the removal of the evolutionarily conservative ribosomal proteins from cell extracts, which are similar in all clostridia (Bom et al., 1986). Different toxigenic types of *C. botulinum* could then be easily distinguished.

Cell Walls Components

The peptidoglycan of the cell walls of most clostridia studied so far belongs to the *meso*-diaminopimelic acid (*m*-Dap) containing, directly linked type (Cummins and Johnson, 1971; Schleifer and Kandler, 1972; Weiss et al., 1981). There are about 20 species which possess LL-Dap, L-lysine, or L-ornithine instead of *m*-Dap, either glycine, L-serine, or D-asparate may form an interpeptide bridge (Schleifer and Kandler, 1972; Weiss et al., 1981). *C. barkeri* contains a rare peptidoglycan type of the cross-linking group B with lysine or ornithine forming the interpeptide bridge (Schleifer and Kandler, 1972; Tanner et al., 1981) and with L-serine and L-diaminobutyric acid at position one and four of the tetrapeptide subunit (Weiss et al., 1981). The peptidoglycan type of about 40 other *Clostridium* species has not yet been determined. Interestingly, the peptidoglycan both of *C. thermoaceticum* and *C. thermoautotrophicum* contains LL-Dap-glycine, which correlates well with the phenotypic and genetic similarity of both acetogenic thermophiles (Wiegel et al., 1981; Bateson et al., 1989). LL-Dap is also found, however, in some mesophilic species *(C. perfringens,*

Table 3. Differential characters of the genera of the obligately anaerobic sporeformers.

Character	Clostridium	Sarcina	Sporomusa	Desulfotomaculum	Syntrophospora	Sporohalobacter	"Anaerobacter"
Cell shape	Rod (curved, coiled, coccoid)[a]	Coccus in cubic packets	Curved rod (straight)[a]	Rod (curved)[a]	Curved rod	Rod	Rod (oval, spherical)
Number of spores per cell	1 (2)[a]	1	1	1	1	1	Up to 5
Motility	+/−	−	+	+	−	+	−
Gram staining reaction	+, (−)[a]	+	−	−, (+)[a]	V	−	ND
Gram type[b]	Positive	Positive	Negative	Positive	Positive	Negative	Positive
Lipopolysaccharide	−	−	+	−	−	ND	−
GC content (mol%)	21–54	28–31	41–49	37–52	38	29–32	29
Sulfate reduction to sulfide	−	−	−	+	−	−	−
Acetogenic growth with H$_2$ + CO$_2$	−, (+)[a]	−	+	−	−	−	−
Cytochromes	−, (type *b*)[a]	ND	Type *b*	Type *b*	ND	−	−
Menaquinone	−, (MK-7)[a]	ND	−	+ (MK-7)	ND	−	−
Peptidoglycan type	*m*-Dap (LL-Dap, L-Lys, L-Orn, L-Dab)[a]	LL-Dap	*m*-Dap	*m*-Dap	ND	ND	ND

Abbreviations: +, present; −, absent; V, variable; ND, no data.
[a]This is only found in a few species.
[b]As defined by cell wall ultrastructure.

C. carnis, C. fallax, C. putrefaciens) which are not only less related to the thermophilic clostridia but were even found to belong to different subgroups of the rRNA homology group I of Johnson and Francis (1975). Since the peptidoglycan type of many *Clostridium* species is not yet known, it remains unclear whether these cell wall data can contribute to the taxonomy of the clostridia as far as separation into different groups or genera is concerned.

Cytochromes and Quinones

Although only a few species have been investigated for the presence of quinones and cytochromes, it can be assumed that these compounds are not widely distributed among clostridia. Menaquinone and *b*-type cytochromes were found in the acetogenic species *C. formicoaceticum, C. thermoaceticum,* and *C. thermoautotrophicum* (Gottwald et al., 1975; Das et al., 1989), but not in *C. magnum* (Schink, 1984). *C. aceticum* possesses cytochrome *b* but obviously lacks menaquinone (Braun et al., 1981). The occurrence of these redox carriers is probably of little taxonomic significance.

Cellular Fatty Acids and Lipids

Mainly the proteolytic and amino acid-utilizing *Clostridium* species have been analyzed for their cellular fatty acid composition (for review, see O'Leary and Wilkinson, 1988). Earlier studies by Moss and Lewis (1967) showed that *C. perfringens, C. sporogenes,* and *C. bifermentans* each possess a different characteristic fatty acid profile which allowed a rapid differentiation of the three species from each other and from 10 other species. The fatty acids n-$C_{14:0}$ and n-$C_{16:0}$ predominated in the 21 species studied by Elsden et al. (1980). One group contained, in addition, iso- and anteiso-acids and consisted of species which metabolize amino acids to the corresponding volatile, branched-chain fatty acid products; a second group contained fatty acids only of the *n*-series and consisted of *C. histolyticum* and six glutamate-fermenting species which produce only acetic acid or mainly acetic acid and butyric acid as volatile metabolic products.

   C. thermoaceticum, C. thermocellum, C. thermohydrosulfuricum, and *C. thermolacticum* differ from each other in their fatty acid composition, while *C. thermosaccharolyticum* and *C. thermosulfurogenes* have similar fatty acid profiles (Mills et al., 1989). A preponderance of branched chain acids (from 50% up to 75% of the total fatty acids) was ascertained in *C. thermocellum* and *C. thermosaccharolyticum* (Herrero et al., 1982; Chan et al., 1971). In *C. ther-*

mohydrosulfuricum, a unique C_{30} dicarboxylic fatty acid was detected. This fatty acid is made up of two "head-to-head" condensed iso-C_{15} fatty acids analogous to the "head-to-head" condensed C_{40} biphytanes of the archaebacteria (Langworthy et al., 1989).

   Johnston and Goldfine (1988) examined 20 strains of butyric acid-producing, saccharolytic *Clostridium* species for phospholipid class composition, plasmalogens, and acyl and alk-1-enyl chains. Ethanolamine was found as the major moiety of the nitrogenous polar headgroups in the phospholipids of *C. butyrium* strains, while *C. beijerinckii* strains had *N*-methylethanolamine and ethanolamine in their phospholipids. *C. acetobutylicum* and *C. fallax* also contained ethanolamine as the major phospholipid base but differed from *C. butyricum* in their acyl and alk-1-enyl compositions. The lipid composition of *C. pasteurianum* was even more distinct. In this species, the acyl fatty acids are composed of over 90% n-$C_{16:0}$ and the alk-1-enyl acids of over 70% C_{17} cyclopropane acids (Johnston and Goldfine, 1983; Deo et al., 1979).

   Both the cellular fatty acids and the lipid composition can be regarded as very promising traits for use in the taxonomy of the genus *Clostridium*. However, first, standardization of the methods and extension of the studies to many more species are needed. In particular, data on lipid composition will be very helpful for sorting the clostridia into different groups.

   For the classification of clostridia to the species level, the MIDI Microbial Identification System (Microbial ID, Inc., Newark, DE, USA) in combination with Hewlett Packard gas chromatographs may be used; it is based on the automated analysis of cellular fatty acid composition and on the comparison of profiles with a computerized library of reference patterns. The system was originally developed using cell material grown on agar plates. Since many anaerobes, including several clostridia, do not give luxuriant growth on plates, a library has been developed at the Virgina Polytechnic Institute, Blacksburg, USA, that is based on cells grown in liquid media under standardized conditions. The latest library version titled Moores' 3 now includes almost 90 species plus several toxigenic types of species of medical importance, and the library will be expanded with time. It may provide an excellent system for the rapid identification of clostridia, and will also be helpful for better assignment of aberrant strains, such as nonmotile, nonsporeforming, or nontoxic variants, or for urease-, lecithinase-, or lipase-negative strains (Nakamura et al., 1976a; Popoff et al., 1985; Nakamura et al., 1976b; Skjelkvale et al., 1979).

Phylogeny

The phylogenetic position of the clostridia within the eubacteria and their relationship to each other has been studied by measurements of rRNA homologies (Johnson and Francis, 1975), by comparative 16S rRNA oligonucleotide analysis, and by comparison of 5S rRNA sequences (for reviews, see Woese, 1987; Cato and Stackebrandt, 1989). On the basis of rRNA cistron similarity, 51 species of the low GC-content DNA clostridia could be divided into three main groups, of which group I, including *C. butyricum,* was subdivided into 10 subgroups. Group II comprised 11 species and appeared relatively homogenous, whereas 6 species in group III had little or no rRNA nucleotide sequence similarity to the reference strains of group I and II or with *C. ramosum,* the reference strain of group III. Four species with a GC content above 40 mol% did not have any significant rRNA homology with *C. innocuum.* Some phenotypic characteristics correlated to some extent with the rRNA homology results, especially within group I, but most did not (Johnson and Francis, 1975).

Comparative 16S rRNA oligonucleotide studies have shown that clostridia, together with bacilli and some other nonsporeforming genera form one main group out of four subdivisions within the phylum of Gram-positive eubacteria (Woese, 1987). They are well separated from the second main group, embracing the actinomycetes and relatives, by a GC content below about 50 mol%. The Gram-positive low-GC bacteria form an ancient phylogenetic cluster; their deeply branched sublines, which are separated at S_{AB} values equal or below 0.35, are represented mainly by members of the present genus *Clostridium.* Several representatives of anaerobic, Gram-positive, nonsporeforming genera (e.g., *Eubacterium, Butyrivibrio, Acetobacterium, Acetogenium, Thermoanaerobium, Peptococcus, Peptostreptococcus,* anaerobic streptococci) were found to be related phylogenetically to certain clostridia and to belong to distinct sublines (Tanner et al., 1981, 1982; Woese, 1987; Ludwig et al., 1988). On the other hand, two species, *C. innocuum* and *C. ramosum,* have been found to be related to the wall-less mycoplasmas and its relatives within the *Bacillus* subline of descendent (Woese et al. 1980; Stackebrandt and Woese, 1981).

Although thermophilic clostridia such as *C. thermoautotrophicum* and *C. thermohydrosulfuricum* cluster tightly with *Thermoanaerobium brockii* and *Thermoanaerobacter ethanolicus,* as revealed by their 93–97% similarity of partial 16S rRNA sequences, others like *C. thermo-*

aceticum, C. thermosaccharolyticum, and *C. thermosulfurogenes* are remotely related (S_{AB} around 0.4) both to the other thermophiles and to each other (Bateson et al., 1989; Cato and Stackebrandt, 1989).

The nucleotide sequence of the 5S rRNA of only five clostridia, *C. butyricum, C. bifermentans, C. innocuum, C. pasteurianum* and *C. tyrobutyricum,* has been determined (Pribula et al., 1976, Rogers et al., 1985; Dams et al., 1987). For some clostridia, their locations in a 5S rRNA tree reconstructed from the sequences of 102 eubacteria, including 25 Gram-positive bacteria of low GC content, were rather different from those occupied in the 16S rRNA tree (Dams et al., 1987). The data did also not coincide with the relationships based on rRNA homologies of Johnson and Francis (1975).

A possible hierarchical structure for the Gram-positive, low-GC bacteria, combining the phylogenetic data based on available 16S rRNA and cellular rRNA homology data, has been proposed (Fox and Stackebrandt, 1987; Cato and Stackebrandt, 1989). In this tree clostridia along with their nonsporeforming relatives are distributed in six taxons of family rank. Some of these groups, defined by phylogenetic terms, appear homogeneous but others are still rather heterogeneous. The result of this is that the genus *Clostridium* would be retained only for about 30 clostridial species of the group containing the type species *C. butyricum* and corresponding to homology group I of Johnson and Francis (1975), whereas all other clostridial species would require a new taxonomic treatment. Any realization of this proposed split of the clostridia into different genera will dependent on the discovery of suitable characteristics which would allow clear phenotypic definition and differentiation of the new taxa. However, neither the chemotaxonomic or commonly realized features of clostridia indicate a fast solution at the present time.

Genetics

Several publications have appeared on the occurrence of plasmids in clostridia, on gene transfer, and on cloning of genes of clostridial origin (see Young et al., 1989a, 1989b).

Plasmids and Gene Transfer

A screening of 75 clostridial strains on plasmids was performed, and plasmids ranging in size from 3 to more than 100 kbp were detected in 26 strains representing 21 species (Lee et al., 1987). Most of these plasmids are cryptic, which is also true for those of *C. butyricum* (Minton and Morris, 1981; Luczak et al., 1985). Plasmids pIP401, pJIR25, and others of *C. perfringens* encode for tetracycline and chloramphenicol resistance and are conjugative (Brefort et al., 1977; Abraham et al., 1985). In *C. cochlearium,* a plasmid was detected which is involved in mercury resistance (Pan-Hou et al., 1980). Gene transfer to clostridia, especially to *C. acetobutylicum,* was reported, e.g. conjugative transfer of streptococcal plasmids (Oultram and Young, 1985; Reysset and Sebald, 1985; Yu and Pearce, 1986; Bertram and Dürre, 1989).

A natural transformation system seems not to be present in clostridia. Protoplasts can be generated from species such as *C. acetobutylicum, C. pasteurianum,* and *C. thermohydrosulfuricum* (Allcock et al., 1982; Reysset et al., 1987; Minton and Morris, 1983; Soutscheck-Bauer et al., 1985), and successful transformation of protoplasts has been reported (Lin and Blaschek, 1984b; Reid et al., 1983; Reysset et al., 1988). Electroporation has also been used to introduce DNA into *C. acetobutylicum* or *C. perfringens* (Oultram et al., 1988; Allen and Blaschek, 1988).

Cloning of Clostridial Genes

A number of genes involved in substrate utilization and in fermentative metabolism have been cloned from certain clostridia, especially from *C. acetobutylicum* and *C. thermocellum.* They include the genes for xylanase and α-amylase (Zappe et al., 1986, 1987; Verhasselt et al., 1989) from the former organism and five genes for endo-beta-1,4-glucanases from the latter (Joliff et al., 1986; Schwarz et al., 1988). For other cloned clostridial genes, see Young et al. (1989). Information on clostridial promotor sequences, nucleotide sequences of ribosome binding sites, and codon usage is also available (see Young et al., 1989).

Applications

A number of clostridial species are of interest for their applied aspects. This is because clostridia can produce interesting fermentation products, carry out bioconversions, and secrete useful enzymes and proteins.

Alcohols and Acids

Clostridial fermentation products of importance are *n*-butanol, acetone, propanediol-1,2, propanediol-1,3, ethanol, butyrate, and acetate. The acetone-butanol fermentation as carried out by *C. acetobutylicum* was used for approx-imately 30 years on a large scale for solvent production (Jones and Woods, 1986; Bahl and Gottschalk, 1988).

The industrial process for acetone and butanol production was operated as a batch process. Reviews covering the literature are cited under *C. acetobutylicum* (see later). More recent studies were devoted to the development of a continuous process (Bahl et al., 1982; Monot and Engasser, 1983; Groot et al., 1984; Bahl and Gottschalk, 1984), to a shift of the products towards *n*-butanol (Maddox, 1980; Bahl et al., 1986; Kim et al., 1984; Meyer et al., 1986), to increasing productivity by use of a cell recycle process (Schlote and Gottschalk, 1986; Afschar et al., 1985), to the recovery of the solvents by extraction (Roffler et al., 1987; Wayman and Parekh, 1987), and to the isolation of mutants resistant to higher solvent concentrations (Hermann et al., 1985; Lin and Blaschek, 1984a). Solvent production by *C. beijerinckii* has also been studied (Schoutens et al., 1985; George et al., 1983).

Propanediol-1,2 is produced from glucose by *C. sphenoides* and by *C. thermosaccharolyticum* when the cultures are subjected to phosphate limitation (Tran-Din and Gottschalk, 1985; Cameron and Cooney, 1986). Propanediol-1,3 is a major product of *C. acetobutylicum* when grown on glycerol (Forsberg, 1987). Acetogenic clostridia, including *C. thermoaceticum, C. thermoautotrophicum,* and *C. formicoaceticum,* have been studied as potential producers of calcium-magnesium acetate as deicer (Ljungdahl et al., 1985, 1989; Wiegel and Ljungdahl, 1986).

The potential of clostridia, especially of thermophilic species *(C. thermohydrosulfuricum* and *C. thermocellum)* for ethanol production has been investigated (Wiegel, 1980; Lovitt et al., 1984, 1988; Parkkinen, 1986; Kurose et al., 1988). *C. saccharolyticum* is also an ethanol producer of interest (Murray and Khan, 1983a; Asther and Khan, 1985). A general disadvantage of these organisms is that some acetate is also produced and that at high substrate concentrations, elevated fructose-1,6-bisphosphate levels lead to an activation of lactate dehydrogenase and subsequent lactate production (Germain et al., 1986).

Enzymes

Some clostridial species are a good source of stable enzymes, especially amylolytic and cellulolytic enzymes (Saha et al., 1989). α-Amylase and pullulanase of *C. thermosulfurogenes* and *C. thermohydrosulfuricum* have been isolated and characterized (Antranikian et al., 1987a, 1987b, 1987c; Melasniemi, 1987), and a β-amylase was detected in a strain of *C. thermosul-furogenes* (Hyun and Zeikus, 1985a). Cellulolytic enzymes in *C. thermocellum* form a large complex which is called the cellulosome (Lamed and Bayer, 1988). For a review on clostridial enzymes, see Saha et al. (1989).

Bioconversions

Some clostridial species may become important in the dechlorination of compounds such as herbicides. Lindane (γ-hexachlorocyclohexane) is converted to tetrachloro-1-cyclohexene by *C. butyricum, C. pasteurianum, C. sphenoides,* and *C. rectum* (Ohisa et al., 1980). Enoate reductases from *C. kluyveri* and *C. tyrobutyricum* allow the stereospecific reduction of acids such as cinnamate and 3-methyl-2-pentenoate. These reductions can be performed as electromicrobial reductions in which whole cells of the clostridia serve as catalysts, methylviologen as mediator, and electricity as a source of electrons (Simon et al., 1985).

Some conversion reactions of steroids have been observed, including 7-α-dehydroxylations by *C. perfringens* and *C. bifermentans* (Archer et al., 1981) and dehydrogenations in positions 4 and 5 (Owen, 1985). For a review, see Morris (1989).

Properties of the Species of Mostly Nonpathogenic Clostridia

A number of species that were formerly assigned to other species or genera are now assigned to established clostridia species or to two new genera; they are summarized in Table 4. In the following section the characteristics of the nonpathogenic species of clostridia are sum-marized. The species are listed in alphabetical order. A few pathogenic species, however, have been included because of the particular importance of their biochemistry.

CLOSTRIDIUM ABSONUM. Peptidoglycan cross-linkage via *m*-DAP; optimum growth temperature 30–45°C. Strains of this species were separated from *C. perfringens* and *C. baratii* ("*C. paraperfringens*") because of differences in lecithinase reaction, gelatin liquefaction, substrate spectrum, and DNA-DNA homology (Nakamura et al., 1973). The organism is saccharolytic and forms acetate, butyrate, and butanol as fermentation products. Although originally described as being nonmotile (Nakamura et al., 1973), weak motility and the presence of flagellae was reported by Cato et al. (1986).

CLOSTRIDIUM ACETICUM. GC content, 33 mol%; optimum growth temperature, 30°C. Converts $H_2 + CO_2$ to acetate, prefers to grow on carbohydrates at slightly alkaline pH (Karlsson et al., 1948; Wieringa, 1940; Adamse, 1980; Braun et al., 1981). Can fix nitrogen (Rosenblum and Wilson, 1949). The species was once thought to be lost but the original strain was revived (Braun et al., 1981). Ethanol seems to be a good substrate for *C. aceticum*-like organisms (Schink et al., 1985). The production of acetate from $H_2 + CO_2$ by unidentified spore-formers was reported by Ohwaki and Hungate (1977), Prins and Lankhorst (1977), Braun et al. (1979), and Conrad et al. (1989).

CLOSTRIDIUM ACETOBUTYLICUM. GC content, 28–29 mol%; peptidoglycan cross-linkage via *m*-DAP; optimum growth temperature, 37°C. Grows in a mineral medium that contains a carbohydrate as carbon and energy source, and biotin and *p*-aminobenzoate as growth factors (Lampen and Peterson, 1943; McCoy et al., 1926; Rubbo et al., 1941). Freshly isolated

Table 4. Assignment changes of some nonpathogenic clostridial species.

Previous name	New, valid name	Reference
"*C. amylolyticum*"	*C. beijerinckii*	Cummins and Johnson, 1971
C. botulinum type G	*C. argentinense*	Suen et al., 1988a
C. bryantii	*Syntrophospora bryantii*	Zhao et al., 1990
"*C. butylicum*"	*C. beijerinckii*	George et al., 1983
"*C. lactoacetophilum*"	*C. beijerinckii*	Cummins and Johnson, 1971
C. lentoputrescens	*C. cochlearium*	Nakamura et al., 1979
C. lortetii	*Sporohalobacter lortetii*	Oren et al., 1987
"*C. multifermentans*"	*C. beijerinckii*	Cummins and Johnson, 1971
"*C. oncolyticum*"	*C. sporogenes*	Cato et al., 1986
C. paraperfringens	*C. barati*	Cato et al., 1982b
C. perenne	*C. barati*	Cato et al., 1982b
"*C. plagarum*"	*C. perfringens*	Nakamura et al., 1976b
"*C. pseudotetanicum*"	*C. butyricum*	Johnson and Francis, 1975; Cato et al., 1982a
"*C. rubrum*"	*C. beijerinckii*	Cummins and Johnson, 1971
"*C. tartarivorum*"	*C. thermosaccharolyticum*	Matteuzzi et al., 1978

strains and cultures started from spores produce n-butanol, acetone, and CO_2, and reduced amounts of acetate, butyrate, and H_2 (McCoy and Fred, 1941). The formation of solvents starts in a later growth phase at slightly acidic pH (Davies and Stephenson, 1941). Conditions for solvent production in continuous culture were found, e.g., at pH 4.3 under phosphate limitation (Bahl et al., 1982). The effect of internal pH on solventogenesis was studied (Gottwald and Gottschalk, 1985). Under these conditions, asporogenous mutants are selected (Meinecke et al., 1984). The n-butanol:acetone ratio can be affected by medium composition (Maddox, 1983; Bahl et al., 1986) or by CO in the atmosphere (Datta and Zeikus, 1985). The process of acetone-butanol production has been reviewed (Abou-Zeid et al., 1978; Beesch, 1953; Spivey, 1978; Ennis et al., 1986; Bahl and Gottschalk, 1988). Glycerol is fermented to 1,3-propanediol and other products by some strains (Forsberg, 1987). In the presence of air, growth of C. acetobutylicum is stopped, but the cells are not killed and growth is resumed after transfer to anaerobic conditions (O'Brien and Morris, 1971b). Chloramphenicol is detoxified in a ferredoxin-dependent reaction (O'Brien and Morris, 1971a). The nitro group of metronidazole is likewise reduced (O'Brien and Morris, 1972). Cells of C. acetobutylicum have been used as a source for acetoacetate decarboxylase (Neece and Fridovich, 1967), crotonase (Waterson et al., 1972), thiolase (Wiesenborn et al., 1988), coenzyme A transferase (Wiesenborn et al., 1986b), phosphotransbutyrylase (Wiesenborn et al., 1989a), butyraldehyde dehydrogenase (Palosaari and Rogers, 1988), and NAD-dependent butanol dehydrogenase (Welch et al., 1989). Induction of enzymes involved in solvent formation was studied (Andersch et al., 1983; Hartmanis and Gatenbeck, 1984; Ballongue et al., 1985; Dürre et al., 1987). The formation of the following enzyme activities has been studied: NADH- and NADPH-ferredoxin oxidoreductases (Petitdemange et al., 1977), NADH-rubredoxin oxidoreductase (Marczak et al., 1984), L-lactate dehydrogenase (Freier and Gottschalk, 1987), α-amylase and glucoamylase (Ensley et al., 1975), proteolytic enzymes (Egorov et al., 1972), and stress- and growth phase-associated proteins (Terracciano et al., 1988). Bacteriocin production has been reported (Barber et al., 1979), and the ultrastructure of the cells has been studied (Cho and Doy, 1973). Biosynthesis of granulose was studied (Reysenbach et al., 1986) as well as sporulation (Long et al., 1983). Transformation and conjugation were described, and several genes were cloned (Bertram and Dürre, 1989; Janssen et al., 1988; Youngle-

son et al., 1989a, 1989b; Oultram et al., 1985, 1988). Regulation of solvent formation was studied on the genetic level by creation of mutants of the respective enzymes in the solventogenic and acetogenic branch, respectively (Bertram et al., 1990; Clark et al., 1989; Cary et al., 1988; Cueto and Mendez, 1990; Dürre et al., 1986).

CLOSTRIDIUM ACIDIURICI. GC content, 27–30 mol%; peptidoglycan cross-linkage via m-DAP; optimum growth temperature, 40–45°C. This organism is specialized on the fermentation of some purine derivatives (uric acid, xanthine, hypoxanthine, guanine) and will not ferment any carbohydrate or amino acid except glycine (Barker and Beck, 1941, 1942; Dürre et al., 1981; Dürre and Andreesen, 1983). It will grow in an entirely synthetic medium (Barker and Peterson, 1944). Fermentation products of C. acidiurici are acetate, CO_2, and ammonia (Barker, 1961; Schiefer-Ullrich et al., 1984). H_2 is not produced during growth. On the most reduced substrate (hypoxanthine), some acetate is formed by reduction of CO_2 (Schulman et al., 1972). The path of purine degradation to glycine has been elucidated by Rabinowitz (1963). Glycine is oxidized by glycine decarboxylase (Waber and Wood, 1979; Gariboldi and Drake, 1984) and reduced to acetate by glycine reductase (Dürre and Andreesen, 1983). The synthesis of formate dehydrogenase in C. acidiurici is stimulated by the presence of tungstate and selenite in the growth medium (Wagner and Andreesen, 1977). The isotope [185]W-tungsten is incorporated into formate dehydrogenase-containing fractions (Wagner and Andreesen, 1987). Selenite is also required for xanthine dehydrogenase synthesis (Wagner and Andreesen, 1979) and for growth in general (Schiefer-Ullrich et al., 1984). C. acidiurici has been used as a source for the purification of formyltetrahydrofolate synthetase (Rabinowitz and Pricer, 1962), pyruvate-ferredoxin oxidoreductase (Uyeda and Rabinowitz, 1971), ferredoxin (Champion and Rabinowitz, 1977), formate dehydrogenase (Kearny and Sagers, 1972), re-citrate synthase (Gottschalk and Dittbrenner, 1970), L-serine dehydratase (Carter and Sagers, 1972), and xanthine dehydrogenase (Wagner et al., 1984). The enzymes involved in the conversion of serine have also been studied (Sagers et al., 1961; Valentine et al., 1963). The folate coenzymes have been isolated (Curthoys et al., 1972), and the nucleotide sequence has been determined for the 10-formyltetrahydrofolate synthetase (Whitehead and Rabinowitz, 1988). The RNA polymerase is unique for eubacteria, being naturally rifampicin-resistant (Murray and Rabinowitz, 1981).

CLOSTRIDIUM AEROTOLERANS. GC content, 40 mol%; optimum growth temperature, 38°C. This xylan-digesting species was isolated from corn stover and from the rumen of sheep fed on corn stover. Its normal habitat is probably soil. The organism is remarkably aerotolerant; it is capable of growth in a shallow layer of liquid medium free of added reducing agents but exposed to air, due to its strong capability to create an anaerobic environment. A wide variety of carbohydrates is fermented, and ethanol, acetate, formate, lactate, carbon dioxide, and hydrogen are produced from xylan (van Gylswyk and van der Toorn, 1987).

"CLOSTRIDIUM AMINOBUTYRICUM." The organism exhibits a high degree of substrate specificity toward amino acids. It grows on γ-aminobutyrate, but not on closely related compounds, and forms acetate, butyrate, and ammonia. It ferments a variety of carbohydrates, but not below pH 7.0 (Hardman and Stadtman, 1960a). The enzymes involved in γ-aminobutyrate degradation (Hardman and Stadtman, 1963a) and the energetic balance of this fermentation (Hardman and Stadtman, 1963b) have been studied.

CLOSTRIDIUM AMINOVALERICUM. GC content, 33 mol%; peptidoglycan cross-linkage via *m*-DAP. The organism grows with δ-aminovalerate as substrate. This compound is fermented to acetate, propionate, valerate, and ammonia. Carbohydrates are also fermented, but not below pH 7.0 (Hardman and Stadtman, 1960b). Many of the enzymes involved in the degradation of δ-aminovalerate have been studied (Barker et al., 1987).

CLOSTRIDIUM ARCTICUM. Optimum growth temperature, 22–25°C. This N_2-fixing, saccharolytic species was found to be widely distributed in Arctic soils. Glucose is fermented mainly to propionate and acetate; only trace amounts of hydrogen are formed (Jordan and McNicol, 1979; Cato et al., 1986).

CLOSTRIDIUM AURANTIBUTYRICUM. GC content, 27–28 mol%; peptidoglycan cross-linkage via *m*-DAP. The orange-colored organism is saccharolytic and pectinolytic (Lund and Brocklehurst, 1978); it forms acetate, butyrate, lactate, ethanol, and butanol (Hellinger, 1947). Large amounts of butanol as well as acetone and isopropanol are formed (George et al., 1983). DNA homology studies indicate that it differs from all other butyrate-forming clostridia (Cummins and Johnson, 1971; Johnson and Francis, 1975; Matteuzzi et al., 1977).

CLOSTRIDIUM BARATII (*C. barati*, sic!). GC content, 28 mol%; peptidoglycan cross-linkage via *m*-DAP; optimum growth temperature, 35–40°C. The organism exhibits some relationship to *C. perfringens* and *C. absonum,* but clearly differs on a genetic basis and by gel electrophoretic cell protein patterns (Cato et al., 1982a). The former species *"C. paraperfringens"* (Nakamura et al., 1973) and *"C. perenne"* (McClung and McCoy, 1957) were found synonymous with *C. baratii* (Cato et al., 1982b). Casein and gelatin are not hydrolyzed; a toxin usually is not formed. A neurotoxin-producing strain causing type F infant botulism was described (Hall et al., 1985; Suen et al., 1988b). The organism is saccharolytic and forms mainly acetate, butyrate, and lactate, but no butanol.

CLOSTRIDIUM BARKERI. GC content, 45 mol%; unusual peptidoglycan cross-linkage involving the α-carboxyl group of glutamate and a diaminobutyric acid; optimum growth temperature, 30°C. Grows on carbohydrates, which are fermented to butyrate, lactate, CO_2, and some H_2. Kinetics of product formation has been studied in batch and continuous culture (Häggström, 1986). Characteristic is its ability to utilize nicotinic acid, which is degraded to propionate, acetate, CO_2, and ammonia (Stadtman et al., 1972). Breakdown of nicotinic acid is initiated by an $NADP^+$-dependent nicotinic acid dehydrogenase, a complex enzyme composed of selenium, molybdenum, iron/sulfur centers, and flavin to yield 6-hydroxynicotinate (Imhoff and Andreesen, 1979; Dilworth et al., 1982, 1983). Most of the enzymes involved in the further breakdown were also studied: the ferredoxin-dependent 6-hydroxynicotinate reductase (Holcenberg and Tsai, 1969), the adenosylcobalamin-dependent 2-methyleneglutarate mutase (Hartrampf and Buckel, 1986; Michel et al., 1989) and the 3-methylitaconate isomerase yielding dimethylmalate, which is hydrated by a specific hydratase and lyase (Eggerer, 1985). The biosynthesis of vitamin B_{12} has been studied on a comparative basis (Höllriegl et al., 1982). Spores of *C. barkeri* have been observed only during initial studies (Stadtman et al., 1972). *C. barkeri* forms a cluster with the acetogens *Eubacterium limosum* and *Acetobacterium woodii* according to 16S rRNA and murein structure (Tanner et al., 1982).

CLOSTRIDIUM BEIJERINCKII. GC content, 26–28 mol%; peptidoglycan cross-linkage via *m*-DAP; optimum growth temperature, 30°C. It differs from *C. butyricum* in its higher demand for growth factors such as amino acids. According to DNA-DNA homology studies, *C. beijerinckii* includes strains that originally were described as strains of *C. butylicum, C. butyricum, C. lactoacetophilum, C. multifermentans,* and *C. rubrum* (Cummins and Johnson, 1971; Johnson, 1973; Matteuzzi et al., 1977; Sjolander and McCoy, 1937). *C. beijerinckii* ferments glucose to butyrate, acetate, and butanol. Acetone or isopropanol is formed depending on the strain (George et al., 1983). Acidic conditions are not obligatory for onset of butanol formation (George and Chen, 1983), and its continuous production from whey using immobilized cells has been studied (Schoutens et al., 1985), as well as the expression of solvent-forming enzymes (Yan et al., 1988), especially of the two alcohol dehydrogenases involved (Hin et al., 1987). Lactate is a substrate in the presence of acetate (Bhat and Barker, 1947). The lactate racemase has been thoroughly studied (Pepple and Dennis, 1976). Glycerol is fermented mainly to 1,3-propanediol (Forsberg, 1987). The hydrogenase has been characterized as iron flavoprotein (Peck and Gest, 1957). The regulation of the tryptophan-synthesizing enzymes, especially of the anthranilate synthetase, has been reported (Baskerville and Twarog, 1972, 1974). The enzymes related to one carbon metabolism have been studied (Thauer et al., 1972), and a phosphotransbutyrylase was purified (Thompson and Chen, 1990). The importance of an intracellular polysaccharide for sporulation has been documented (Bergère et al., 1975). The organism contains a neuraminidase and an acylneuraminate lyase (Müller and Werner, 1974). The conditions for protoplast formation, L-colony growth, and regeneration have been reported (Birrer et al., 1989).

CLOSTRIDIUM BIFERMENTANS. GC content, 27 mol%; optimum growth temperature, 30–37°C; peptidoglycan cross-linkage via m-DAP. This proteolytic and lecithinase-positive species is widely distributed in nature and was also found in high numbers in marine sediments (Davis, 1969; Matches and Liston, 1974). Few sugars are fermented and several amino acids are metabolized (Mead, 1971; Elsden and Hilton, 1978; Elsden et al., 1976). Fermentation products include mainly acetate, formate, and hydrogen gas with smaller amounts of isobutyrate, isovalerate, isocaproate, propionate, and ethanol. Hydrocinnamic acid is produced in peptone-yeast extract-glucose media (Cato et al., 1986). Properties of purified leucine transaminase and α-keto-isocaproic acid decarboxylase, which are both involved in leucine oxidation to isovaleric acid, have been described (Britz and Wilkinson, 1983).

CLOSTRIDIUM BRYANTII. GC content, 37.6 mol%; optimum growth temperature, about 30°C. Strains of this species were isolated from marine and freshwater anaerobic mud (Stieb and Schink, 1985). The organism grows on both even- and odd-numbered fatty acids with 4 to 11 carbon atoms, and on 2-methylbutyrate in obligately synthrophic association

with hydrogen-scavenging anaerobes. No other organic substrates are utilized (with one exception, see below). Fatty acids are oxidized to acetate and hydrogen or acetate, propionate and hydrogen. The type strain originating from marine habitats does not require vitamins or any other growth factor. Growth is completely inhibited by phosphate concentrations higher than 10 mM (Stieb and Schink, 1985). Recently, *C. bryantii* could be adapted to grow in pure culture on crotonate with acetate and butyrate as major products. It was found to be phylogenetically closely related to *Syntrophomonas wolfei* but unrelated to any other *Clostridium*. It has been transferred, therefore, to a new genus, *Syntrophospora* (Zhao et al., 1990).

CLOSTRIDIUM BUTYRICUM. GC content, 27–28 mol%; peptidoglycan cross-linkage via m-DAP; optimum growth temperature, 25–37°C. This is the type species of the genus *Clostridium*. It was isolated by Prazmowski in 1880. Some strains of *C. butylicum, C. multifermentans,* and *C. fallax* belong to *C. butyricum* (Cummins and Johnson, 1971; Johnson, 1973). The ultrastructure and spore formation patterns have been studied (Rousseau et al., 1971a). *C. butyricum* is able to ferment a great variety of carbohydrates, including starch and pectin. Growth factors other than biotin are not required. The fermentation products are butyrate, acetate, CO_2, and H_2. Acetone is formed only in trace amounts; some strains produce isopropanol and butanol in substantial amounts (Kutzenock and Aschner, 1952; Sjolander and McCoy, 1937). The ratio in which acetate and butyrate are formed can be affected by the partial pressure of H_2 (Jungermann et al., 1973). H_2 production from glucose by immobilized cells has been studied (Karube et al., 1982). *C. butyricum* can be differentiated from *C. beijerinckii* by glycerol and ribose fermentation and by gas liquid chromatography. This differentiation is supported by studies on DNA homology and electrophoretic protein patterns (Magot et al., 1983). Butyrate is formed via butyryl phosphate; the enzymes involved, phosphotransbutyrylase and butyrate kinase, have been partially purified (Twarog and Wolfe, 1963; Valentine and Wolfe, 1960). In the presence of acetate, *C. butyricum* is able to ferment mannitol or L-lactate (Kutzner, 1963). *C. butyricum* has been used as a source for purification and study of ferredoxin (Benson et al., 1967), NAD(P)-ferredoxin oxidoreductase (Jungermann et al., 1971a), pyruvate cleavage (Hespell et al., 1969), pyruvate:formate lyase (Thauer et al., 1972; Wood and Jungermann, 1972), acetyl-CoA reductase (Day and Goldfine, 1978), pectin esterase, and polygalacturonate lyase (Sheiman et al., 1976). Product pattern and aggregate formation in continuous culture have been investigated, as well as H_2 evolution (van Andel et al., 1985; Zoutberg et al., 1989; Kayano et al., 1981). The phospholipids (Johnston and Goldfine, 1988), a butyricin (Clarke and Morris, 1976a), and the biosynthesis of nicotinic acid (Scott et al., 1969) have been studied. Oncolytic properties were described for a strain (M 55) that produces a bradykinin-degrading enzyme, protease, and nuclease (Brantner and Schwager, 1979; Möse and Möse, 1964); this strain has been incorrectly classified as *C. butyricum* or *"C. oncolyticum"* (Brantner et al., 1975) but was shown to belong to *C. sporogenes* (Cato et al., 1986). The β-lactamase produced by some strains has been characterized (Carlson et al., 1981; Magot, 1981), and two neurotoxigenic strains causing type E infant botulism were described (McCroskey et al., 1986; Suen et al., 1988b). Plasmids have been surveyed from clinical sources (Popoff and Truffaut, 1985).

CLOSTRIDIUM CADAVERIS. GC content, 27 mol%; optimum growth temperature, 30–37°C; peptidoglycan cross-linkage via *m*-DAP. The organism is proteolytic and digests meat. Glucose and fructose but no other sugars are used. Fermentation products in peptone-rich, glucose-containing media include butyrate, acetate, ethanol, butanol, hydrogen, and moderate amounts of isobutyrate and isovalerate (Cato et al., 1986). L-lysine decarboxylase was used in an enzyme electrode to assay L-lysine (Tran et al., 1983).

CLOSTRIDIUM CELATUM. Optimum growth temperature, 37°C. The organism was isolated from normal human feces and differentiated from phenotypically similar, saccharolytic and nitrite- and sulfide-producing organisms, such as *C. perfringens, C. absonum,* and *C. barati,* by the absence of lecithinase and hemolytic action and by the production of acetate, formate, ethanol, and a little butyrate from glucose. Toxins are not produced (Hauschild and Holdeman, 1974).

CLOSTRIDIUM CELERECRESCENS. GC content, 38 mol%; optimum growth temperature, 35°C. The organism was isolated from a methanogenic, cellulose-enriched culture started with a cow manure inoculum. Various carbohydrates, but no starch, are utilized, gelatin is hydrolyzed, and indole is formed. The fermentation products from cellulose or cellobiose in a yeast extract-containing medium are ethanol, acetate, formate, butyrate, isobutyrate, isovalerate, caproate, lactate, succinate, carbon dioxide and hydrogen. A high specific growth rate, 0.564 h^{-1}, attained on cellobiose distin-

guishes the organism from other cellulolytic clostridia (Palop et al., 1989).

CLOSTRIDIUM CELLOBIOPARUM. GC content, 28 mol%; peptidoglycan cross-linkage via *m*-DAP; optimum growth temperature, 30–37°C. This organism was isolated by Hungate (1944) in the course of studies on cellulose fermentation. In addition to cellulose, *C. cellobioparum* is able to ferment a great variety of carbohydrates. The products formed are ethanol, acetate, formate, CO_2, and H_2 (small amounts of lactate). In pure culture, growth and sugar breakdown are inhibited by molecular hydrogen (Chung, 1976). Only biotin is required as growth factor.

CLOSTRIDIUM CELLULOLYTICUM. GC content, 41 mol%; optimum growth temperature, 32–35°C. The organism was isolated from decayed grass compost. A limited number of carbohydrates is fermented. Ethanol, acetate, lactate, formate, carbon dioxide, and hydrogen are formed as major fermentation products from cellulose or cellobiose (Petitdemange et al., 1984).

CLOSTRIDIUM CELLULOVORANS. GC content, 26–27 mol%; optimum growth temperature, 37°C. The organism was isolated from a mesophilic, methanogenic digestor fermenting poplar wood. Cells stain Gram-negative, are nonmotile, and form oval to oblong spores of 2–4 μm in length. A few mono- and disaccharides, cellulose, xylan, and pectin are fermented, but starch is not utilized. The organism readily utilizes microcrystalline cellulose (Avicel). Acetate, butyrate, formate, lactate, carbon dioxide, and hydrogen are produced during fermentation of cellulose or cellobiose (Sleat et al., 1984).

"CLOSTRIDIUM CHARTATABIDUM." GC content, 31 mol%; optimum growth temperature, 39°C. The organism was isolated from the rumina of sheep and cattle where it occurs in numbers of 2×10^4 to 3×10^6 ml^{-1} rumen fluid. Rumen fluid is required for growth. Colonies on agar medium are strongly orange pigmented. Cells form oblong spores measuring 0.7×2–3 μm. Gelatin is hydrolyzed. A limited number of carbohydrates including cellulose is fermented but starch is not utilized. Growth on cellobiose, glucose and sucrose is rapid and characterized by equally rapid autolysis once the substrate is exhaused. Cellobiose is fermented to acetate, butyrate, ethanol and hydrogen (Kelly et al., 1987).

CLOSTRIDIUM CLOSTRIDIIFORME. GC content, 47–50 mol%; peptidoglycan cross-linkage via *m*-DAP; optimum growth temperature, 30–37°C. This species was created by Kaneuchi et al. (1976b) to replace *Bacteroides clostridiiformis* because most strains assigned to this species have been found to be sporeformers (see also Cato and Salmon, 1976). *C. clostridiiforme* ferments a number of carbohydrates and produces lactate, acetate, CO_2, and H_2 as the main products, but no butyrate; production of β-lactamase has been reported (Weinrich and Del Bene, 1976).

CLOSTRIDIUM COCCOIDES. GC content, 43–45 mol%,; peptidoglycan cross-linkage via *m*-DAP; optimum growth temperature, 37°C. The cells of this organism are nonmotile, coccobacillary to rodshaped, and occur singly, in pairs, and sometimes in short chains. A number of carbohydrates are fermented with succinate and acetate as the major product. Hydrogen and butyrate are not produced (Kaneuchi et al., 1976a). Found only in the feces of mice.

CLOSTRIDIUM COCHLEARIUM. GC content, 27–28 mol%; peptidoglycan cross-linkage via *m*-DAP; optimum growth temperature, 37–45°C. This organism does not ferment any carbohydrates. It grows only on glutamate, glutamine, and histidine (Barker, 1939), and can be isolated specifically in a chemostat using L-glutamate as substrate (Laanbroek et al., 1979). *C. cochlearium* uses the methyl-asparate pathway for glutamate breakdown (Buckel and Barker, 1974). The products formed are butyrate, acetate, CO_2, and H_2. DNA homology studies have shown that *C. cochlearium* is different from *C. tetanomorphum* (Nakamura et al., 1979; Wilde et al., 1989). In contrast to *C. tetanomorphum*, *C. cochlearium* does not acidify glucose-containing media and does not utilize ethanolamine (Möller et al, 1986; Wilde et al., 1989). The former type strain of *"C. lentoputrescens"* (Hartsell and Rettger, 1934) was found homologous with the type strain of *C. cochlearium* (Nakamura et al., 1979). The glycine reduction by NADH has been studied (Stadtman, 1962, 1978).

CLOSTRIDIUM COCLEATUM. GC content, 28–29 mol%; peptidoglycan cross-linkage via *m*-DAP. The organism is coiled and forms semicircular to circular forms and occasionally short, helical filaments. It ferments carbohydrates to acetate, but not to butyrate. It can be separated from *C. ramosum* and *C. spiroforme* on the basis of DNA-DNA homology studies (Kaneuchi et al., 1979).

CLOSTRIDIUM COLLAGENOVORANS. GC content, 24 mol%; optimum growth temperature, 30–37°C. The organism is proteolytic and was isolated from a sewage sludge digestor. It readily ferments gelatin, collagen, peptones, cooked meat, and slowly casein or serum albumin, but neither single amino acids, mixtures of them, or glycine-containing dipeptides, nor any carbohydrate. Major products formed from all the protein fermentations are acetate and carbon dioxide with only traces of isobutyrate, isovalerate, ethanol, hydrogen, and/or propionate. When the organism is grown on cooked meat, significant amounts of butyrate are also produced. Growth occurs in mineral medium on gelatin, and no vitamins or growth factors in yeast extract are required. An active collagenase is produced (Jain and Zeikus, 1988). Gelatin can be converted to methane via acetate by a stable coculture (Jain and Zeikus, 1989).

CLOSTRIDIUM CYLINDROSPORUM. GC content, 27–30 mol%; peptidoglycan cross-linkage via *m*-DAP; optimum growth temperature, 40–45°C. This species exhibits the same substrate specificity as *C. acidiurici*. Both species differ somewhat in spore morphology, in tungstate/molybdate requirement for formate dehydrogenase formation (Wagner and Andreesen, 1977; Schiefer-Ullrich et al., 1984), and in uptake of ^{185}W-tungsten (Wagner and Andreesen, 1987). In addition to acetate, *C. cylindrosporum* can produce larger amounts of formate and glycine than *C. acidiurici* under suboptimal growth conditions (Barker and Beck, 1941, 1942; Champion and Rabinowitz, 1977; Schiefer-Ullrich et al., 1984). However, using DNA-DNA hybridization, immunological methods, and 16S rRNA analysis an unequivocal differentiation of both species is possible (Andreesen et al., 1985). *C. cylindrosporum* has been used to elucidate sporulation and germination (Sacks and Smith, 1987; Smith and Sullivan, 1989), the pathway of purine degradation (Rabinowitz, 1963), and for purification and study of xanthine dehydrogenase (Bradshaw and Barker, 1960; Wagner and Andreesen, 1979), formyltetrahydrofolate synthetase (Rabinowitz and Pricer, 1962), ferredoxin (Champion and Rabinowitz, 1977), glycine formiminotranferase, formiminotetrahydrofolate cyclodeaminase, methylene-tetrahydrofolate dehydrogenase, serine hydroxymethyltransferase (Uyeda and Rabinowitz, 1965, 1967a, 1967b, 1968), and dihydrolipoamide dehydrogenase (Dietrichs and Andreesen, 1990).

CLOSTRIDIUM DISPORICUM. GC content, 40 mol%; two optimum growth temperatures were recorded: 25–26°C, 38.5–44°C. The unique characteristic of this nonmotile species is the formation of two spores in one and the same cell. Cells containing two spores are found at approximately 1 to 20% of the total population grown in SM10 broth (Barnes and Impey, 1974). Various carbohydrates including starch are fermented, mainly to acetate and lactate with trace amounts of butyrate and succinate. A single strain has been isolated from rat cecal content (Horn, 1987).

CLOSTRIDIUM DURUM. GC content, 50 mol%; optimum growth temperature, 30°C. The organism was found to be the most prominent species in the sediments of the Black Sea. It is one of the few aerotolerant clostridial species. Carbohydrates are fermented to acetate and formate; some lactate, ethanol, and propanol are also formed (Smith and Cato, 1974).

CLOSTRIDIUM FELSINEUM. GC content, 26 mol%; peptidoglycan cross-linkage via *m*-DAP; optimum growth temperature, 25–37°C. Ferments carbohydrates to butyrate, acetate, and some butanol. The organism is known to play an important role in the pectinolytic degradation of plant material during retting (Potter and McCoy, 1952). The formation of the pectin-degrading enzymes (endopectate lyase, pectate hydrolase, pectin esterase, endopolygalacturonase, and exopolygalacturonase) depends on the growth medium and the strain used (Lund and Brocklehurst, 1978; Vozniakovskaya et al., 1974).

CLOSTRIDIUM FERVIDUM (C. fervidus, sic!). GC content, 39 mol%; optimum growth temperature, 68°C. The organism was isolated from a geothermal spring. Cells stain Gram-negative. Growth occurs on Trypticase peptone or yeast extract but not on casamino acids. Presence of either of these components is required for growth on carbohydrates. Glucose, maltose, mannose, xylan, starch and pyruvate are fermented, but the amount of growth is only slightly enhanced by these carbohydrates. Serine is utilized as the sole carbon source. Glucose is fermented mainly to acetate. Minor quantities of valerate, butyrate, ethanol, and lactate are produced in Trypticase-containing media (Patel et al., 1987). Amino acid transport in membrane vesicles has been studied (Speelmans et al., 1989).

CLOSTRIDIUM FORMICOACETICUM. GC content, 34 mol%; pepidoglycan cross-linkage via *m*-DAP; optimum growth temperature, 30–37°C. Ferments some carbohydrates, but not glucose, to acetate. Almost three molecules of acetate are produced per molecule of fructose. Growth is largely stimulated by the presence of CO_2/bicarbonate or formate (Andreesen, et al., 1970; El Ghazzawi, 1967). Gluconate is metabolized by *C. formicoacetium* via 2-keto-3-deoxygluconate (Andreesen and Gottschalk, 1969). Fumarate is fermented to succinate, acetate, and CO_2 (Dorn et al., 1978a). Fumarate reductase of *C. formicoacetium* is a peripheral membrane protein (Dorn et al., 1978b). A cytochrome has been detected in this organism (Gottwald et al., 1975). Methoxylated aromatic compounds are utilized, and the aldehyde group of vanillin is oxidized to the carboxyl level, e.g.,

in the presence of CO as cometabolizable substrate (Lux et al., 1990). The carboxyl group is not decarboxylated under CO_2-limited conditions (Hsu et al., 1990). Various nonactivated carboxylates can be reduced to the corresponding alcohols by CO or formate in the presence of viologens (Fraisse and Simon, 1988). Acetic acid production by *C. formicoaceticum* from lactate and lactose has been studied (Tang et al., 1988, 1989). The organism can fix nitrogen (Bogdahn et al., 1983). The biochemistry of CO_2 fixation to acetate is quite similar to that of *C. thermoaceticum* and has recently been summarized (Fuchs, 1986; Ljungdahl, 1986; Wood et al., 1986; Spormann and Thauer, 1988). According to 16S rRNA analysis, *C. aceticum* is a closely related organism but *C. thermoaceticum* is not (Tanner et al., 1982).

CLOSTRIDIUM GLYCOLICUM. GC content, 29 mol%; peptidoglycan contains lysine and is cross-linked via an aspartate bridge; optimum growth temperature, 25–37°C. This organism grows on glucose, fructose, xylose, maltose, and sorbitol, and produces acetate, ethanol, H_2, and CO_2. It also decomposes cellulose slowly. Most characteristic is the ability to grow on ethylene glycol and 1,2-propylene glycol. Both compounds are stoichiometrically converted to equal amounts of the corresponding acid and alcohol. The diol dehydratase does not require a corrinoid (Hartmanis and Stadtman, 1986). *C. glycolicum* requires 15 amino acids, biotin, and *p*-aminobenzoate for growth (Gaston and Stadtman, 1963). Strains of uracil-degrading clostridia show great similarity to *C. glycolicum* (Mead et al., 1979), as does an acetogenic organism, a *Clostridium* sp. (Ohwaki and Hungate, 1977).

CLOSTRIDIUM INDOLIS. GC content, 44 mol%; optimum growth temperature, 37°C; peptidoglycan cross-linkage via *m*-DAP. The organism is saccharolytic and ferments a number of sugars with acetate, formate, ethanol, hydrogen, and carbon dioxide as fermentation products. Yields of ethanol amounting to 1.9 mol/mol glucose fermented have been reported (Corry, 1978). Pectin and pectate are fermented (Ng and Vaughn, 1963).

CLOSTRIDIUM INNOCUUM. GC content, 43–44 mol%; optimum growth temperature, 37°C; peptidoglycan cross-linkage via L-lysine-alanine-alanine (Weiss et al., 1981). A variety of carbohydrates including inulin but no starch is fermented, and butyrate, lactate, acetate, hydrogen, and carbon dioxide are produced (Cato et al., 1986).

CLOSTRIDIUM INTESTINALE (C. intestinalis, sic!). GC content, 26–28 mol%; optimum growth temperature, 37°C. The organism was isolated from the feces of cattle and pig. The species is saccharolytic and aerotolerant and grows on the surface of Eggerth-Gagnon agar incubated in air. It differs significantly from other aerotolerant *Clostridium* species, having a low GC content, by its phenotypic characteristics, and by lacking any DNA homology. The fermentation products formed during growth in peptone-yeast extract-glucose medium are major amounts of acetate and butyrate with minor amounts of lactate, formate, and succinate (Lee et al., 1989).

CLOSTRIDIUM IRREGULARE (*C. irregularis,* sic!). Optimum growth temperature, 30–37°C. This species is both nonsaccharolytic and nonproteolytic; gelatin, however, is hydrolyzed. Acetate and isovalerate are produced as the major fermentation products after growth in glucose containing yeast extract-peptone media (Cato et al., 1986). A 3-alpha-hydroxysteroid dehydrogenase was found in one strain (Mahoney et al., 1977).

CLOSTRIDIUM JOSUI. GC content, 40 mol%; optimum growth temperature, 45°C. The species is nonmotile and was isolated from Thai compost. Xylan and a few sugars are fermented; crystalline cellulose, rice straw, and other cellulosic material are hydrolyzed without any chemical pretreatment. Ethanol, acetate, carbon dioxide, and hydrogen are the main fermentation products formed during growth on cellulose or cellobiose (Sukhumavasi et al., 1988). Properties of purified endo-1,4-β-glucanase have been studied (Fujino et al., 1989).

CLOSTRIDIUM KLUYVERI. GC content, 30 mol%; peptidoglycan cross-linkage via *m*-DAP; optimum growth temperature, 37°C. This organism has specialized on the fermentation of ethanol and acetate to butyrate, caproate, and H_2. The ratio of products formed depends on the ratio of ethanol/acetate provided (Smith et al., 1985). Besides ethanol, 1-propanol can be utilized, and acetate can be replaced by succinate, crotonate, and a few other acids as electron acceptor (Kenealy and Waselefsky, 1985). Crotonate utilization requires an adaption period. Therefore, its utilization has been studied on a comparative basis (Bader et al., 1980). Carbohydrates and amino acids are not utilized for growth but some four carbon compounds (vinylacetate, 4-hydroxybutyrate [Bartsch and Barker, 1961]) can also serve as growth substrate. *C. kluyveri* requires CO_2/bicarbonate for growth and will develop in a mineral medium that contains, in addition to the substrates, only biotin and *p*-aminobenzoate (Bornstein and Barker, 1948a, 1948b); it can fix nitrogen (Kanamori et al., 1989). The ultrastructure of the organism has been revealed by electron mi-

croscopy (Cho and Doy, 1973). The reactions involved in butyrate formation by clostridia have been studied first in *C. kluyveri* (see Barker, 1956; Stadtman, 1976). Ethanol is oxidized to acetyl-CoA via acetaldehyde (Burton and Stadtman, 1953). The formation of ATP in *C. kluyveri* is coupled to the evolution of H_2. Per molecule of H_2 evolved, 0.5 molecules of ATP can be synthesized by substrate-level phosphorylation (Schoberth and Gottschalk, 1969; Thauer et al., 1968; for a model, see Gottschalk, 1986). H_2 is evolved from NADH via ferredoxin (Jungermann et al., 1971b). *C. kluyveri* synthesizes pyruvate by reductive carboxylation of acetyl-CoA (Andrew and Morris, 1965) and glutamate via citrate (Gottschalk and Barker, 1966; Tomlinson, 1954). Glyoxylate inhibits the pyruvate synthase system (Thauer et al., 1970). The synthesis of one carbon units (Jungermann et al., 1968), glycine formation from threonine and serine (Jungermann et al., 1970), and the synthesis of aromatic amino acids (Thauer et al., 1967) and of carbohydrates (Decker et al., 1966) have been studied. *C. kluyveri* has been used as source for the purification and study of acetaldehyde dehydrogenase (Burton and Stadtman, 1953; Hillmer and Gottschalk, 1974; Lurz et al., 1979; Smith and Kaplan, 1980), hydrogenase (Fredricks and Stadtman, 1965), diaphorase (Kaplan et al., 1969; Chenas et al., 1987), phosphotransacetylase (Bergmeyer et al., 1963; Henkin and Abeles, 1976; Kyrtopoulos and Satchell, 1972; Stadtman, 1955), coenzyme A transferase (Barker et al., 1955), NAD- and NADP-ferredoxin oxidoreductase (Jungermann et al., 1971b; Thauer et al., 1971), NADP-specific β-hydroxybutyryl-CoA dehydrogenase (Madan et al., 1973), and β-ketothiolase (Sliwkowski and Hartmanis, 1984). Other clostridial species contain a NAD-specific β-hydroxybutyryl-CoA dehydrogenase (von Hugo et al., 1972). The steady state concentrations of different nucleotides and of acetyl-CoA were determined in growing cells (Decker and Pfitzer, 1972; Rössle et al., 1981). Radioactive flavine nucleotides can be conveniently derived from the cells (Brühmüller and Decker, 1976). The cells have been used to carry out stereospecific biohydrogenation reactions of unsaturated compounds (Bader et al., 1978). The reactions of purified enoate reductase and of acryloyl-CoA reductase were described (Thanos et al., 1988; Sedlmeier and Simon, 1985).

CLOSTRIDIUM LENTOCELLUM. GC content, 36 mol%; optimum growth temperature, 40°C. The cells stain Gram-negative. Growth occurs optimally at pH 7.5–7.7 and is stimulated by hemin, vitamin K, and rumen fluid. A variety of carbohydrates including xylan, starch, glycogen, and cellulose are fermented, mainly to ethanol, acetate, carbon dioxide, and hydrogen. Digestion of cellulosic materials including microcrystalline cellulose is rather slow. The organism was isolated from an estuarine mud bank of a river (Murray et al., 1986).

CLOSTRIDIUM LEPTUM. GC content, 51–52 mol%; peptidoglycan cross-linkage via L-lysine and serine or glycine; optimum growth temperature, 37°C. This organism was isolated from feces. Some carbohydrates markedly stimulate growth. Acetate and ethanol are the fermentation products (Moore et al., 1976).

CLOSTRIDIUM LITUSEBURENSE. GC content, 27 mol%; optimum growth temperature, 30–37°C; peptidoglycan cross-linkage via L-lysine. The organism is proteolytic but also ferments a few sugars. Major fermentation products formed are butyrate, acetate, and isovalerate; minor products are formate, propionate, isobutyrate, ethanol, *n*-propanol, and isobutanol. Little or no hydrogen gas is formed (Cato et al., 1986).

CLOSTRIDIUM MAGNUM. GC content, 29 mol%; optimum growth temperature, 30–32°C. The organism was found in anoxic freshwater sediments and sewage sludge. Cells are long, fat, straight rods which often become spindle-shaped during growth in sugar-containing media. Cells are motile by polar and subpolar flagella and stain Gram-negative. No growth factors or vitamins are required. Growth is inhibited by 1% sodium chloride and 20 mM phosphate. The homoacetogenic organism ferments fructose, glucose, sucrose, xylose, malate, citrate, 2,3-butanediol, and acetoin completely to acetate with concomitant reduction of carbon dioxide (Schink, 1984).

CLOSTRIDIUM MALENOMINATUM. GC content, 28 mol%; optimum growth temperature, 37°C. Carbohydrates are not fermented, but peptone and yeast extract are. Acetate and butyrate are major fermentation products. According to rRNA homology studies, the organism forms a group with the acetate-producing peptolytic clostridia, *C. subterminale, C. histolyticum, C. argentinense* (formerly *C. botulinum* type G), and *C. limosum* (Johnson and Francis, 1975). A strain that decomposes uric acid was isolated from avian cecum. This strain utilized lactate as a carbon source with the formation of butyrate and propionate (Barnes and Impey, 1974).

CLOSTRIDIUM MANGENOTII. Optimum growth temperature, 30–37°C; peptidoglycan cross-linkage via *m*-DAP. The organism is non-saccharolytic but proteolytic and metabolizes a number of amino acids to the corresponding fatty acids. Fermentation products formed in glucose-yeast extract-peptone media mainly include acetate, formate, isocaproate, isovalerate, isobutyrate, and large amounts of hydrogen (Cato et al., 1986).

CLOSTRIDIUM METHYLPENTOSUM. GC content, 46 mol%; optimum growth temperature, 45°C. Single cells are ring-shaped, and tightly packed, left-handed, helical chains of cells are usually formed in cultures. In the latter respect the organism differs from *C. spiroforme,* which forms right-handed coils. Cells are surrounded by a large capsule. The organism was isolated from human feces. Among many compounds tested, only arabinose, rhamnose, fucose, and lyxose are utilized as substrates for growth in a mineral medium with biotin and vitamin B_{12}. The fermentation products include acetate, propionate, *n*-propanol, carbon dioxide, and hydrogen (Himelbloom and Canale-Parola, 1989).

CLOSTRIDIUM NEXILE. GC content, 40–41 mol%; optimum growth temperature, 37°C. The formation of heat-resistant spores has not been demonstrated, but the organism resists heating at 80°C for 10 min. Rods or nonmotile coccobacilli occurring in pairs or chains can be observed. Carbohydrates are fermented to acetate, formate, ethanol, and H_2; traces of lactate or succinate may be formed. The organism resembles *C. oroticum* morphologically, but does not hydrolyze orotic acid (Holdeman and Moore, 1974).

CLOSTRIDIUM OCEANICUM. GC content, 26–28 mol%; peptidoglycan cross-linkage via *m*-DAP; optimum growth temperature, 30–37°C. The organism was isolated from marine sediments and tolerates up to 4% NaCl. It is proteolytic, lecithinolytic, and slightly saccharolytic. From peptone-yeast extract, a variety of products are formed. Butyrate and lactate with smaller amounts of acetate are produced from glucose (Smith, 1970).

CLOSTRIDIUM OROTICUM. GC content, 44 mol%; peptidoglycan cross-linkage via *m*-DAP; optimum growth temperature, 30–37°C. This bacterium was originally described by Wachsman and Barker (1954) as *Zymobacterium oroticum.* Cato et al. (1968) demonstrated the formation of heat-resistant spores and transferred it to the genus *Clostridium. C. oroticum* is nonmotile and ferments a number of carbohydrates to ethanol, acetate, and CO_2 as the major products (small amounts of lactate and formate are also formed). Most characteristic for this organism is its growth on orotic acid, which is degraded to acetate, CO_2, ammonia, and a dicarboxylic acid. Enzyme preparations of *C. oroticum* have been used to study the breakdown of orotic acid (Liebermann and Kornberg, 1954). Growth on orotate increases the flavin content of the cells about three times (Kondo et al., 1960). The initial attack is carried out by an NAD-dependent dihydroorotate dehydrogenase, an iron-sulfur flavoprotein (Aleman and Handler, 1967). The ring-cleaving enzyme, dihydroorotase, contains zinc (Taylor et al., 1976), which can be reversibly removed (Pettigrew et al., 1985).

"CLOSTRIDIUM OXALICUM." GC content, 36.3 mol%; optimum growth temperature, 28–30°C. The organism was found in anoxic freshwater sediment and sewage sludge. It grows in vitamin-supplemented mineral medium on oxalate or oxamate as the sole energy substrate and with acetate as main carbon source. Oxalate is decarboxylated to formate. No other substrates are used. Growth rate and cell yields are enhanced by small amounts of yeast extract (Dehning and Schink, 1989).

CLOSTRIDIUM PAPYROSOLVENS. GC content, 30 mol%; optimum growth temperature, 20–30°C. A single strain has been isolated from an estuarine anaerobic sediment. The organism is saccharolytic and grows optimally in diluted-seawater medium at 1% salinity. Few sugars, glycerol, and esculin are fermented. Fermentation products from cellulose include acetate, lactate, ethanol, hydrogen, and carbon dioxide (Madden et al., 1982).

CLOSTRIDIUM PARAPUTRIFICUM. GC content, 26–27 mol%; optimum growth temperature, 30–37°C; peptidoglycan cross-linkage via L-lysine-serine-glycine. Various sugars are fermented mainly to acetate, butyrate, lactate, hydrogen and carbon dioxide. Lysis of Ehrlich ascites tumor tissue in mice was reported (Möse and Möse, 1964), but for another strain, a promotion of the formation of liver tumor in mice was observed (Mizutami and Mitsuoka, 1979). Degradation of creatine and creatinine to N-methylhydantoin has been studied (Szulmajster, 1958a, 1958b).

CLOSTRIDIUM PASTEURIANUM. GC content, 26–28 mol%; peptidoglycan cross-linkage via m-DAP; optimum growth temperature, 37°C. This organism was first described by Winogradsky in 1895. It grows in a defined medium (Mallette et al., 1974) and on a great variety of carbohydrates but not on starch. The products formed are butyrate, acetate, CO_2, and H_2. Part of the H_2 is derived from $NADH_2$ (Jungermann et al., 1973). High sugar content induces the formation of solvents (Harris et al., 1986). The fixation of nitrogen by free-living microorganisms was discovered through studying *C. pasteurianum* (Winogradsky, 1895, 1902). Ferredoxin was also discovered in this

organism (Mortenson et al., 1962). *C. pasteurianum* has been used for the purification and study of the following: ferredoxin (Graves et al., 1985; Schönheit et al., 1979; Moulis et al., 1984; Bonomi et al., 1985; Prince and Adams, 1987), flavodoxin (Knight and Hardy 1966; Mayhew, 1971; Przysiecki et al., 1985), rubredoxin (Lovenberg and Sobel, 1965; Vogel et al., 1977), a red, paramagnetic protein (Cardenas et al., 1976), two different hydrogenases (Adams et al., 1989; Chen and Blanchard, 1984; Chen and Mortenson, 1974; Kowal et al., 1989), NADH-ferredoxin reductase (Jungermann et al., 1971a; Petitdemange et al., 1972), pool sizes of pyridine nucleotides (Petitdemange et al., 1973b), ferredoxin-dependent reduction of thioredoxin (Hammel et al., 1983) or metronidazole (Loekerby et al., 1984), ferredoxin-linked sulfite reductase (Laishley et al., 1971), isotope fractionation during sulfite and sulfate reduction (Laishley and Krouse, 1978), discrimination of sulfur and selenium oxyanions (Bryant and Laishley, 1988) and a transport by proton symport (Bryant and Laishley, 1989), reduced ferredoxin:CO_2 oxidoreductase, a molybdenum-iron-sulfur protein (Scherer and Thauer, 1978; Liu and Mortenson, 1984), CO dehydrogenation (Fuchs et al., 1974), pyruvate dehydrogenase (Bush and Sauer, 1977; Westlake et al., 1961), the inhibition of pyruvate synthase reaction by glyoxylate (Thauer et al., 1970) and the utilization of CO_2 as active species (Thauer et al., 1975), glycolytic enzymes (Howell et al., 1969; Kotze, 1969), PEP phosphotransferase systems (von Hugo and Gottschalk, 1974a; Mitchell and Booth, 1984), 1-phosphofructokinase (du Toit et al., 1972; von Hugo and Gottschalk, 1974b), regulation of carbohydrate utilization (Mitchell et al., 1987; Roohi and Mitchell, 1987), pathway of gluconate degradation (Bender et al., 1971), gluconate dehydratase (Bender and Gottschalk, 1973), the uptake system for gluconate and galactose (Booth and Morris, 1975), the formation of an invertase (Laishley, 1975) and of the intracellular reserve polysaccharide (Brown et al., 1975; Darvill et al., 1977) by ADP-glucose pyrophosphorylase and granulose synthase (Robson et al., 1974) and its degradation by granulose phosphorylase (Robson and Morris, 1974), β-ketothiolase (Berndt and Schlegel, 1975), membrane-bound adenosine triphosphatase (Clarke et al., 1979; Clarke and Morris, 1976b, 1980; Riebeling et al., 1975), the exosporium (Mackey and Morris, 1972) and the occurrence of squalene in the spore membrane (Mercer et al., 1979), the differences in cellular fatty acids and lipid composition (Deo et al., 1979), the excretion of biosurfactants (Cooper et al., 1980), the enzymes involved in nitrogen metabolism (Kleiner and Fitzke, 1979), the biosynthesis of amino acids (Dainty and Peel, 1970), especially threonine aldolase (Dainty, 1970), serine hydroxymethyl-transferase (Stöcklein and Schmidt, 1985), glutamate synthase, and glutamine synthetase (Dainty, 1972), the effect of ammonium addition or sporulation on N_2-fixing cells (Kleiner, 1979; Vallespinos and Kleiner, 1980), the components of nitrogenase (Mortenson, 1978; Mortenson et al., 1967) such as an iron-molybdenum protein (Cramer et al., 1978; Rawlings et al., 1978; Zumft, 1978) and an iron protein (Tanaka et al., 1977), the kinetics of their formation (Seto and Mortenson, 1974; Hinton and Mortenson, 1985a) and the parallel uptake of molybdate ions (Elliot and Mortenson, 1976) and incorporation into the different molybdenum-containing proteins (Hinton and Mortenson, 1985b) involving molybdenum-binding/storage protein (Elliot and Mortenson, 1977; Hinton and Merritt, 1986), the immunological and genetic comparison of the nitrogenase components with corresponding proteins from other nitrogen-fixing organisms (Smith et al., 1976; Chen et al., 1986; Hinton et al., 1987), and NADH as a physiological electron donor (Jungermann et al., 1974). Crossing of the cytoplasmic membrane by undissociated acids such as butyric or acetic acid has also been studied (Kell et al., 1981).

CLOSTRIDIUM PERFRINGENS. GC content, 24–27 mol%; peptidoglycan cross-linkage via LL-DAP; optimum growth temperature, 45°C. The organism is highly pathogenic by the formation of phospholipase, hemolysins, and other toxins (Möllby et al., 1976; Hatheway, 1990). Lecithinase-negative strains, formerly described as *"C. plagarum,"* are now included in this species (Nakamura et al., 1976b; Cato et al., 1986). Both carbohydrates and proteins are fermented. The influence of carbohydrates on growth and sporulation in a defined medium has been studied (Sacks, 1983; Tortora, 1984; Goldner et al., 1985; Phillips, 1986; Ushijima et al., 1987), as well as spore-lytic enzymes (Gombas and Labbe, 1985). Conditions for production of protease and amines have been studied (Allison and Macfarlane, 1989a, 1989b). Some strains can utilize nitrate as electron sink, which results in a shift from butyrate to acetate fermentation (Ishimoto et al., 1974) and in an increase of the growth yield (Hasan and Hall, 1975). The ferredoxin-dependent nitrate reductase has been purified (Chiba and Ishimoto, 1973; Seki et al., 1987). The ferredoxin and rubredoxin of *C. perfringens* have been sequenced (Seki et al., 1989a, 1989b). The degradation of arginine (Schmidt et al., 1952), of glutamate by

glutamate decarboxylase (Cozzani et al., 1975), and the enzymes aspartate kinase (Kuramitsu and Watson, 1973), histidine decarboxylase (Recsei et al., 1983), α-L-fucosidase (Aminoff and Furakawa, 1970), N-acetylneuraminic acid aldolase (DeVries and Binkley, 1972), and phosphatidylserine synthase (Cousminer et al., 1982) have been studied. Some bacteriocins like sphingomyelinase and lecithinase are located on a plasmid, and the complete corresponding nucleotide sequence has been analyzed (Garnier and Cole, 1988; Saint-Joanis et al., 1989). Electroporation-mediated transformation has been shown (Scott and Rood, 1989).

CLOSTRIDIUM PFENNIGII. GC content, 38 mol%; optimum growth temperature, 36–38°C. This species was isolated from the rumen of a steer and requires 30% rumen fluid and 0.2% yeast extract for growth. Cells are motile by a lateral to subterminal flagellum. The organism is metabolically unique: it grows on pyruvate catabolizing it to acetate and carbon dioxide, on carbon monoxide producing acetate and butyrate, or by fermentation of methoxyl groups of monobenzenoids with the formation of the corresponding hydroxybenzoids and butyrate. No other substrates are utilized (Krumholz and Bryant, 1985).

CLOSTRIDIUM POLYSACCHAROLYTICUM. GC content, 41.6 mol%; optimum growth temperature, 30–38°C. Has been isolated from sheep rumen. Cellobiose, cellulose, starch, and xylan are fermented, and formate, butyrate, and hydrogen are the major metabolic products formed. Growth is stimulated by acetate and greatly stimulated by rumen fluid. CO_2 (bicarbonate) is required for growth. Stationary phase cultures lyse very readily (van Gylswyk, 1980). The organism was first classified as a member of the genus *Fusobacterium* but subsequently reclassified as a clostridium after spore formation was detected (van Gylswyk et al., 1980).

CLOSTRIDIUM POPULETI. GC content, 28 mol%; optimum growth temperature, 35°C. The organism was isolated from a batch-methanogenic fermentation of hybrid poplar wood. The cells stain Gram-negative. Gelatin is weakly hydrolyzed. Various carbohydrates including polymers like cellulose, pectin, xylan, and gum arabic are fermented. The fermentation products from cellulose or glucose are acetate, butyrate, lactate, carbon dioxide, and hydrogen (Sleat and Mah, 1985).

CLOSTRIDIUM PROPIONICUM. GC content, 35 mol% (B. Schink, personal communication); peptidoglycan cross-linkage via *m*-DAP; optimum growth temperature, 30–37°C. This organism does not utilize any carbohydrates. Growth substrates are: L-alanine, β-alanine, L- serine, L-lactate, acrylate, and L-threonine. Three carbon compounds are fermented to propionate, acetate, and CO_2 (and ammonia in the case of serine and alanine). Threonine is converted into propionate, butyrate, CO_2, and ammonia (Cardon and Barker, 1946, 1947). Propionate is formed from lactate via direct reduction of acrylyl-CoA and not via succinate as in propionibacteria (Leaver et al., 1955). Growth on β-alanine requires previous induction of α-alanine degradation and of acrylyl-CoA aminase (Goldfine and Stadtman, 1960; Vagelos et al., 1959). Acrylic acid can be produced by whole cells (Akedo et al., 1983). A dehydratase acting on lactyl-CoA has been purified and a radical mechanism is proposed (Kuchta et al., 1986; Schweiger and Buckel; 1985).

CLOSTRIDIUM PROTEOLYTICUM. GC content, 29.5 mol%; optimum growth temperature, 30–37°C. A single strain was isolated from a chicken manure digester. The organism is obligately proteolytic and utilizes gelatin, collagen, peptone, and meat; casein and serum albumin are slowly digested. Carbohydrates, single or mixtures of amino acids, and a number of glycine-containing dipeptides tested are not utilized. Fermentation products formed from gelatin are acetate and carbon dioxide with traces of hydrogen; the formation of other volatile products in trace amounts depends upon the type of proteinaceous material digested. Collagenase and a minor protease activity are produced. No vitamins or other growth factors are needed for growth. This species resembles *C. collagenovorans* in several respect but can be distinguished from the latter by the higher GC content of its DNA, absence of motility, and utilization of proline in gelatin. The gel electrophoretic cell protein pattern of both species are distinct (Jain and Zeikus, 1988).

CLOSTRIDIUM PUNICEUM. GC content, 28–29%; optimum growth temperature, 23–33°C. This pink-pigmented species isolated from spoiled potatoes and carrots is gelatinolytic, degrades casein, and ferments a number of carbohydrates including starch and pectate. Some strains utilize inulin and glycogen. Major fermentation products from glucose are butanol, butyrate, and acetate (Lund et al., 1981).

CLOSTRIDIUM PURINOLYTICUM. GC content, 29.0–29.3 mol%; peptidoglycan cross-linkage via *m*-DAP; optimum growth temperature, 36°C. The organism is specialized, like *C. acidurici* and *C. cylindrosporum,* on the fermentation of purines. However, a broad spectrum of purines and glycine-containing peptides are utilized (Dürre et al., 1981). Characteristic is its ability to use adenine and its lower opti-

mum growth temperature. Growth on glycine requires the presence of a purine besides selenium (Dürre and Andreesen, 1982c). The latter trace element is part of the glycine reductase (Dürre and Andreesen, 1983; Sliwkowski and Stadtman, 1988b) which forms acetate as its major end product, besides formate, CO_2, and NH_3. Xanthine is hydrolyzed to formiminoglycine via imidazole derivatives (Dürre and Andreesen, 1982d), whereas selenium-starved cells decompose uric acid by hydrolyzing the imidazole moiety (Dürre and Andreesen, 1982a). The organism has the capacity to reduce CO_2 to formate and further to acetate via glycine reductase (Dürre and Andreesen, 1982b).

CLOSTRIDIUM PUTREFACIENS. GC content, 22–25 mol%; peptidoglycan cross-linkage via LL-DAP-glycine; optimum growth temperature, 20–25°C; grows at 0–30°C, not at 37°C. This organism is the only named true psychrophilic *Clostridium* species (Roberts and Hobbs, 1968). The organism—first isolated by McBryde (1911) from spoiled ham and regarded by him as the causal agent of souring in brine-cured ham—and one strain isolated by E. Barnes were described as purely proteolytic (Ross, 1965). Other strains isolated from animal sources were found to ferment glucose and to form acetate, butyrate, and valerate as the fermentation products after growth in meat medium (Parsons and Sturges, 1927a, 1927b, 1927c; Sturges and Drake, 1927). The organism grows well in most media used routinely for culturing anaerobes but produces spores on lactose-egg yolk agar (Roberts and Derrik, 1975).

CLOSTRIDIUM QUERCICOLUM. GC content, 52–54 mol%; optimum growth temperature, 25–30°C. The organism was isolated from oak trees. It is motile, not proteolytic, does not attack cellulose or starch. Acetate, propionate, and H_2 are produced from glycerol, inositol, or fructose (Stankewich et al., 1971).

CLOSTRIDIUM RAMOSUM. GC content, 26–27 mol%; peptidoglycan cross-linkage via *m*-DAP. Spores are seen very seldom; therefore, the organism was described previously as *Nocardia, Actinomyces, Fusiformis, Bacteroides,* and *Ramibacterium ramosum.* Cells can occur in both "V" and "Y" arrangements. Carbohydrates are fermented to formate, acetate, lactate, and succinate (Holdeman et al., 1971); production of β-lactamase has been reported (Weinrich and Del Bene, 1976).

CLOSTRIDIUM ROSEUM. Optimum growth temperature, 37°C. The organism was isolated from German maize (McCoy and McClung, 1935). Colonies grown on beef peptone-yeast extract agar are pink to orange pigmented and become purplish black after exposure to air (McClung, 1943; McClung and McConnell, 1967; Cato et al., 1986). Gelatin is hydrolyzed and a number of carbohydrates including pectin are fermented. The major fermentation products are butyrate, acetate, and butanol. In addition, acetone and ethanol are formed in 5% corn mash medium (McCoy and McClung, 1935). Gluconate is fermented via 2-keto-3-deoxygluconate to pyruvate and glyceraldehyde-3-phosphate (Bender et al., 1971). This species has been taken for a strain of *C. felsineum* (Lund and Brocklehurst, 1978) but is now regarded as distinct from the latter (Cato et al., 1986).

CLOSTRIDIUM SACCHAROLYTICUM. GC content, 28 mol%; optimum growth temperature, 37°C. Cells stain Gram-negative and are nonmotile. The organism was isolated from a cellulose-enrichment culture, however, it does not ferment cellulose, but does ferment cellobiose and a broad variety of carbohydrates to ethanol, acetate, CO_2, H_2, and some lactate (Murray et al., 1982). The organism requires the addition of 1% yeast extract. Growth is not impaired by H_2 or 0.1 M acetate and 0.5 M ethanol. High substrate concentrations are tolerated (Murray and Khan, 1983a). The ratio of products formed is influenced by carbonate addition, headgas, incubation conditions, and temperature (Murray and Khan, 1983b). Ethanol tolerance and production were improved by a pyruvate-negative mutant to 80% of the theoretical value (Murray et al., 1983). Further improvements were accomplished by coculture with *Zymomonas anaerobia* (Asther and Khan, 1985).

CLOSTRIDIUM SARTAGOFORME (*C. sartagoformum,* sic!). GC content, 28 mol%; optimum growth temperature, 30–37°C. The organism is saccharolytic and produces acetate, butyrate, formate, hydrogen, and occasionally large amounts of lactate as the fermentation products (Cato et al., 1986).

CLOSTRIDIUM SCATOLOGENES. GC content, 27 mol%; optimum growth temperature, 30–37°C; peptidoglycan cross-linkage via m-DAP. Few sugars are fermented, and lactate, pyruvate, and various amino acids are utilized. In glucose-peptone media, acetate and butyrate are produced as the major fermentation products, and formate, propionate, isobutyrate, isovalerate, valerate, caproate, lactate, and succinate are formed as minor products (Cato et al., 1986). Skatole is produced which is responsible for the characteristic fecal odor of *C. scatologenes* (Fellers and Clough, 1925; Rosenberger, 1959). p-Cresol is formed from tyrosine and p-hydroxyphenylacetic acid (Elsden et al., 1976, Nunez-Montiel et al., 1983; Levett, 1987). Its

involvement in spoilage of pasteurized potato salads has been reported (Baumgart et al., 1984).

CLOSTRIDIUM SCINDENS. GC content, 45 mol%; optimum growth temperature, 45°C. Cells are nonmotile, but more than 40% of the cells of a population are fimbriated. The organism is saccharolytic and nonproteolytic and was found in human feces. The products of fermentation in peptone-yeast extract-glucose medium are acetate, ethanol, carbon dioxide, and hydrogen (Morris et al., 1985). At least four enzymes active on steroidal substrates are synthesized (Winter et al., 1984). A desmolase was described that cleaved the C_{17}-C_{20} carbon bond of a variety of 17-α-hydroxysteroids (Morris et al., 1985; Krafft et al., 1987). A novel form of 20-α-hydroxysteroid dehydrogenase was purified (Krafft and Hylemon, 1989).

CLOSTRIDIUM SPHENOIDES. GC content, 41–42 mol%; peptidoglycan cross-linkage via m-DAP; optimum growth temperature, 37°C. This species grows on a variety of carbohydrates and produces acetate, ethanol, CO_2, and H_2 as the major fermentation products. Characteristic is its ability to grow on citrate (Walther et al., 1977). Citrate lyase as key enzyme is inactivated by a configurational change, not by phosphorylation or deacetylation (Herzberg and Antranikian, 1986), although phosphorylation of proteins does occur in vivo (Antranikian et al., 1985). *C. sphenoides* has been shown to convert the insecticide hexachlorocyclohexane into tetrachlorocyclohexene (Heritage and MacRae, 1977). Formation of D(−)-1,2 propanediol from glucose under phosphate limitation has been studied (Tran-Din and Gottschalk, 1985).

CLOSTRIDIUM SPIROFORME. GC content, 27 mol%. The organism shares the ability with *C. ramosum* and *C. cocleatum* to form "Y"-branched cells and various degrees of coiling. After a heat shock, some strains become nearly straight and do not regain coiling. Carbohydrates are fermented to acetate; butyrate is not formed. The DNA homology to *C. ramosum* and *C. cocleatum* is rather high (about 50 mol%) (Kaneuchi et al., 1979). The helically coiled form results from semicircular cells joined end to end (Borriello et al., 1986). The species causes enterotoxemia in animals and was shown to produce a toxin similar to the *C. perfringens* type E iota toxin (Borriello and Carman, 1983).

CLOSTRIDIUM SPOROGENES. GC content, 26 mol%; peptidoglycan cross-linkage via m-DAP; optimum growth temperature, 30–40°C. This organism ferments glucose and fructose to butyrate, acetate, ethanol, CO_2, and H_2.

The pH value influences the pattern of organic acids produced (Montville et al., 1985). *C. sporogenes* grows in defined media (Belokopytov et al., 1982). Most of the amino acids required can be replaced by their analogous L-α-hydroxyacids. The range of carbohydrates used depends on the strain tested (Lovitt et al., 1987). CO_2 modulates growth energetics depending upon the growth medium (Dixon et al., 1987, 1988). The preferred substrates of *C. sporogenes* are amino acids, which are fermented by the "Stickland reaction." This reaction was discovered by Stickland in a study on the energy metabolism of *C. sporogenes*. In a series of publications (see Hoogerheide and Kocholaty, 1938; Stickland, 1935), it was established that, in the fermentation of pairs of amino acids, alanine, valine, leucine, and isoleucine serve as hydrogen donors and glycine, proline, hydroxyproline, ornithine, arginine, phenylalanine, and betaine serve as hydrogen acceptors (Marmelak and Quastel, 1953; Woods, 1936; Fryer and Mead, 1979; Naumann et al., 1983; Wildenauer and Winter, 1986; Winter et al., 1987). This reaction is of great importance to a number of clostridial species (see Barker, 1961; Nisman, 1954; Seto, 1980). The nature of the products formed depends on the amino acids fermented. In contrast to *C. sticklandii*, *C. sporogenes* reduces the aromatic amino acids to the corresponding arylpropionic acids (Elsden et al., 1976; Jellet et al., 1980). Of the branched chain amino acids, leucine can be oxidized to 3-methylbutyrate and be reduced to 4-methylvalerate. The latter compound is also formed by *C. botulinum*, *C. difficile*, *C. bifermentans*, *C. sordellii*, and *C. mangenotii* (Elsden and Hilton, 1978; Britz and Wilkinson, 1982). In this respect, these organisms are similar to *C. sporogenes* (Fryer and Mead, 1979). Common antigens with *C. botulinum* and *C. novyi* can be demonstrated (Poxton, 1984). The pattern of reduced products is due to the action of 2-oxocarboxylate reductase, 2-hydroxy acid dehydratase, and enoate reductase (Giesel and Simon, 1983b; Machacek-Pitsch et al., 1985). ATP formation is coupled to enoate reduction (Bader and Simon, 1983), whereas energy conservation can be achieved by vectorial proton ejection during proline reduction (Lovitt et al., 1986). Proline supresses glycine uptake and glycine reductase activity (Venugopalan, 1980). Pyrimidines can only be reduced to the dihydro-derivatives (Hilton et al., 1975). The enzymes of the arginine dihydrolase pathway (Venugopal and Nadkarni, 1977), the ornithine cyclase (deaminating) (Muth and Costilow, 1974), the proline dehydrogenase (Monticello and Costilow, 1981), the proline reductase and an MV-linked NAD(P)

reductase (James et al., 1988), the NADP-dependent dihydrolipoamide dehydrogenase (Dietrichs and Andreesen, 1990), the selenium-dependent glycine reduction (Costilow, 1977), and the B_{12}-coenzyme-dependent leucine 2,3-aminomutase (Poston, 1976) have been studied. By action of the last enzyme, a catabolism to acetate and isobutyrate via β-ketoisocaproate is possible (Poston, 1986). Production of thiaminase (Edwin et al., 1978; Kimura and Liao, 1964; Princewill, 1980) and the ability to degrade chitin (Timmis et al., 1974) have been demonstrated. Biosynthesis of valine (Monticello and Costilow, 1982) and isoleucine (Monticello et al., 1984) has been studied.

CLOSTRIDIUM SPOROSPHAEROIDES. GC content, 27 mol%; optimum growth temperature, 37–45°C; peptidoglycan cross-linkage via *m*-DAP. Gelatin is not hydrolyzed, and peptone media containing carbohydrates are not acidified. Fermentation products formed during growth in peptone media are mainly butyrate and acetate. Lactate, pyruvate, and citrate are utilized. Properties of a citrate lyase and a citrate lyase acetylating enzyme were studied (Antranikian et al., 1984; Quentmeier and Antranikian, 1985). Glutamate is fermented via the hydroxyglutarate pathway (Buckel, 1980a).

CLOSTRIDIUM STERCORARIUM. GC content, 39 mol%; optimum growth temperature, 65°C. The organism was isolated from compost material and ferments various sugars, starch, hemicelluloses, and cellulose. The fermentation products from cellulose include ethanol, acetate, lactate, carbon dioxide, and hydrogen (Madden, 1983). Endoglucanase, exoglucanase, and xylanase are produced during growth on cellulose (Creuzet and Frixon, 1983; Creuzet et al., 1983). Genes related to xylan hydrolysis have been cloned and expressed in *E. coli* (Sakka et al., 1990).

CLOSTRIDIUM STICKLANDII. GC content, 31 mol%; optimum growth temperature, 30–35°C. This organism was described by Stadtman and McClung in 1957. It shows only slight growth on some sugars (maltose, glucose, galactose, and ribose) and is specialized on the fermentation of pairs of amino acids, of which one is reduced (e.g., glycine or proline) and the other one is oxidatively decarboxylated and deaminated (e.g., the branched chain and the aromatic amino acids, arginine, ornithine, threonine, or serine) (Stickland reaction) (Elsden and Hilton, 1978, 1979). Formate increases the growth yield. Acetate can be formed from formate (Stadtman and White, 1954). Glycine is a preferred oxidant and is reduced to acetate by a membrane-bound glycine reductase, which consists of three proteins including a selenoprotein (see Stadtman, 1978, 1989; Sliwkowski and Stadtman, 1988a, 1988b; Arkowitz and Abeles, 1989; Tanaka and Stadtman, 1979). The membrane-bound proline reductase is composed differently (Seto, 1978, 1980). Glycine inhibits proline reduction (Schwartz et al., 1979). Alanine is not utilized but pyruvate. Threonine is catabolized via α-amino-β-ketobutyrate to glycine and acetyl-CoA (Golovchenko et al., 1983). Purines and pyrimidines are catabolized (Schäfer and Schwartz, 1976, 1980). *C. sticklandii* also grows with lysine as single amino acid, which is fermented to acetate, butyrate, and ammonia. The fermentation pathway involves a pyridoxal phosphate-dependent L-lysine-2,3-amino-mutase, a coenzyme B_{12}-dependent β-lysine-5,6-aminomutase, and a 3,5-diaminohexanoate dehydrogenase (see Baker and van der Drift, 1974; Stadtman, 1973). In contrast to *C. sporogenes,* ornithine is fermented by *C. sticklandii* to alanine, acetate, CO_2, and ammonia (Dyer and Costilow, 1968). The first steps in this fermentation are catalyzed by a coenzyme B_{12}-dependent aminomutase and a dehydrogenase, which yields 2-amino-4-ketopentanoate (Somack and Costilow, 1973a, 1973b; Tsuda and Friedmann, 1970). This compound is then cleaved to alanine and acetyl-CoA (Jeng et al., 1974). The organism contains ferredoxin, rubredoxin, and some flavoproteins (Stadtman, 1965) such as an electron-transferring flavoprotein involved in a reductase reaction (Dietrichs and Andreesen, 1990). A quinone-dependent *p*-nitrophenylphosphatase has been purified (Davis and Stadtman, 1985). Corrinoids are involved in the ribonucleotide reductase reaction (Abeles and Beck, 1967). In *C. sticklandii,* a selenium-containing tRNA[glu] is present (Ching, 1986). D-Selenocystine can be split into pyruvate, ammonia, and perselenide by a lyase present in *C. sticklandii,* but not in other species tested (Esaki et al., 1988).

CLOSTRIDIUM SUBTERMINALE. GC content, 28 mol%;, peptidoglycan cross-linkage via *m*-DAP; optimum growth temperature, 37°C. Like *C. sticklandii,* it will not ferment carbohydrates but branched-chain and aromatic amino acids are both oxidized (Elsden and Hilton, 1978, 1979). In contrast to *C. sticklandii,* this organism does not utilize threonine or proline, but proteins and other amino acids. It can be phenotypically differentiated from *C. hastiforme* by headspace gas chromatography (Niel et al., 1989). Lysine is fermented to acetate, butyrate, and ammonia. The pathway involves the mutase reactions mentioned for *C. sticklandii* (see Chirpich et al., 1970), a cleavage enzyme which converts 3-keto-5-aminohexanoate and acetyl-CoA to L-3-aminobutyryl-CoA

and acetoacetate (Yorifuji et al., 1977), an L-3-aminobutyryl-CoA deaminase (Jeng and Barker, 1974), and a butyryl-CoA:acetoacetate CoA transferase (Barker et al., 1978). An NAD-dependent glutamate dehydrogenase has been purified (Winnacker and Barker, 1970).

CLOSTRIDIUM SYMBIOSUM. GC content, 46 mol%; peptidoglycan cross-linkage via *m*-DAP. Because of difficulties in detecting spores and in Gram staining, the organism was originally described as *Fusiformis biacutus* (Beerens type II), *Fusobacterium symbiosum,* and *Bacteroides symbiosus.* The aminopeptidase assay is also not reliable for this organism (König et al., 1985). Carbohydrates are fermented to acetate, butyrate, and some lactate and alcohol (Kaneuchi et al., 1976b). The organism was first detected as a symbiont with *Entamoeba histolytica.* Both organisms share the ability to form pyrophosphate in the pyruvate-phosphate dikinase reaction (Reeves et al., 1968). Unlike most of the clostridia, glutamate is fermented via the α-hydroxyglutarate pathway (Buckel, 1980a). The dehydration reaction of (R)-2-hydroxyglutarate to (E)-glutaconate has been studied in detail (Buckel, 1980b).

CLOSTRIDIUM TETANOMORPHUM. GC content, 25–28 mol%; peptidoglycan cross-linkage via *m*-DAP; optimum growth temperature, 37°C. This organism has much in common with *C. cochlearium* (Barker, 1939; Nakamura et al., 1979). It differs in that it is able to ferment some carbohydrates (Anthony and Guest, 1968; Woods and Clifton, 1937; Nakamura et al., 1979; Wilde et al., 1989). Characteristic is its ability to grow on histidine (Wachsman and Barker, 1955a) and on L-glutamate, which is degraded to acetate, butyrate, CO_2, and H_2 (Wachsman and Barker, 1955b). The pathway of glutamate degradation involves a coenzyme B_{12}-dependent rearrangement of glutamate to β-methylaspartate (Barker, 1976; Barker et al., 1958). The ribonucleotide reductase is also dependent on B_{12} coenzyme (Abeles and Beck, 1967). The organism can form bactobilin, a blue pigment of the biliverdin type (Brumm et al., 1983). *C. tetanomorphum* has been used as a source for purification and study of glutamate mutase (Switzer and Barker, 1967), β-methylaspartase (Hsiang and Bright, 1969), L-citramalate hydrolyase (Wang and Barker, 1969), and citramalate lyase (Buckel and Bobi, 1976; Dimroth et al., 1977). The equilibrium constants of these reactions were determined (Buckel and Miller, 1987). The degradation of threonine involves an ADP-activated threonine deaminase (Vanquickenborne and Phillips, 1968; Whiteley and Tahara, 1966), a ferredoxin-dependent cleavage of α-oxobutyrate to propionyl-CoA, CO_2, and H_2, followed by a specific propionate kinase reaction (Tokushige and Hayaishi, 1972).

CLOSTRIDIUM THERMOACETICUM. GC content, 54 mol%; peptidoglycan cross-linkage via LL-DAP-glycine; optimum growth temperature, 60°C. No DNA-DNA or 16S rRNA homology is detectable to other thermophilic clostridia except *C. thermoautotrophicum* (Matteuzzi et al., 1978; Tanner et al., 1982; Bateson et al., 1989). This organism ferments glucose, fructose, xylose, and pyruvate to acetate (Fontaine et al., 1942). Growth on H_2 + CO_2 or CO is possible (Kerby and Zeikus, 1983). A minimally defined medium was developed (Lundie and Drake, 1984). 5-Aminolevulinic acid, a precursor of corrinoids and cytochromes in this organism, is formed via glutamate (Oh-hama et al., 1988). The presence of high concentrations of CO_2/bicarbonate is required (see also *C. formicoaceticum*). Per molecule of glucose, approximately 2.6 molecules of acetate are formed (Andreesen et al., 1973). Some of this acetate is synthesized by reduction of CO_2. The production of acetate has been evaluated and acid-tolerant, mutant strains were selected (Ljungdahl et al., 1985; Sugaya et al., 1986; Reed et al., 1987; Brumm, 1988). An electrogenic Na^+/H^+ antiporter is present (Terracciano et al., 1987). Utilization of certain aromatic compounds by demethylation or decarboxylation and the biotransformation of aromatic aldehydes has been described (Daniel et al., 1988; Wu et al., 1988; Hsu et al., 1990; Lux et al., 1990). *C. thermoaceticum* can form methanol from CO, and it can reduce enoates and alcanoates to the corresponding acids or alcohols (Simon et al., 1987; White et al., 1987). The carboxylic acid reductase is, besides formate dehydrogenase, the second genuine tungsten-containing enzyme (White et al., 1989). The biochemistry involved in acetate formation from CO_2 has been solved mostly by using *C. thermoaceticum* and is summarized by Fuchs (1986), Ljungdahl (1986), and Wood et al. (1986). All enzymes involved in this autotrophic pathway have been purified, some of them, e.g., formyltetrahydrofolate synthetase, methyl transferase, corrinoid/Fe-S protein, and CO dehydrogenase have been cloned (Lovell et al., 1988; Roberts et al., 1989). The last enzyme catalyzes the key reaction of the pathway. A controlled potential is necessary for activity of the proteins involved in these reactions (Lu et al., 1990). The functional groups involved in the catalysis of the different reactions taking place at the CO dehydrogenase have been studied (Shanmugasundaram et al., 1989). *C. thermoaceticum* contains a variety of electron carriers such as two rubredoxins, two ferredoxins, two types of cytochrome *b,* and a menaquinone (MK-7) (see Das et al., 1989). Energy conservation by electron transport has been postulated (Andreesen et al., 1973; Kellum and Drake, 1986), and ATPase has been characterized (Mayer et al., 1986).

CLOSTRIDIUM THERMOAUTOTROPHICUM. GC content, 53–55 mol%; optimum growth temperature, 56–60°C; peptidoglycan cross-linkage via LL-DAP and glycine. Found in mud and soils from various places. The homoacetogenic organism grows chemolithotrophically with hydrogen and carbon dioxide or CO, as well as chemoorganotrophically with a few carbohydrates or methanol, producing acetate exclusively (Wiegel et al., 1981; Berestovskaya et al., 1987). Strains of *C. thermoautotrophicum* form a very tight cluster with *C. thermoaceticum* (Bateson et al., 1989). A minimal medium has been developed (Savage and Drake, 1986). N_2 fixation and ammonium assimilation routes are known (Bogdahn and Kleiner, 1986a). The structure and function of a menaquinone (MK-7) involved in cytochrome *b*-mediated electron transport has been studied (Das et al., 1989). Membrane vesicles form an electrochemical proton gradient (Hugenholtz and Ljungdahl, 1989).

CLOSTRIDIUM THERMOBUTYRICUM. GC content, 37 mol%; optimum growth temperature, 55°C; peptidoglycan cross-linkage via *m*-DAP. The organism was isolated from horse manure. It differs from other thermophilic, cellulolytic species by the formation of oval, subterminal spores with little or no distension of the cells. Several mono- and disaccharides and cellulose are fermented but starch or other polysaccharides are not utilized. From glucose, the organism produces butyrate, carbon dioxide, and hydrogen, and minor amounts of acetate and lactate (Wiegel et al., 1989).

CLOSTRIDIUM THERMOCELLUM. GC content, 38–40 mol%; optimum growth temperature, 55–69°C. This organism ferments cellulose and cellobiose to ethanol, acetate, CO_2, and H_2 (some butyrate and lactate is also formed) (see McBee, 1950; Ng et al., 1977; Freier et al., 1988). In the presence of a methanogenic bacterium, the products are acetate, methane, and CO_2 (Weimer and Zeikus, 1977). The DNA-DNA homology to other thermophilic clostridia has been studied (Stainthorpe and Williams, 1988). The amino acid requirements are known (Vuillet et al., 1986). The fermentation of cellulosic substrates preferentially for ethanol production has been investigated (Saddler and Chan, 1982, 1983; Ljungdahl et al., 1983; Lynd et al., 1989; Tailliez et al., 1989), also the product inhibition by ethanol (Herrero and Gomez, 1980; Herrero et al., 1985). Differences in xylan degradation are observed compared to various noncellulolytic thermophiles (Wiegel et al., 1985). Ethanol-induced changes in the membrane lipid composition have been reported (Herrero et al., 1982). *C. thermocellum* has been used as source for purification and study of endo-β-1,4-glucanase (Ng and Zeikus, 1981; Petre et al., 1981; Ait et al., 1982), cellobiose phosphorylase (Alexander, 1968), and of malic enzyme (Lamed and Zeikus, 1981). Under certain conditions, glucose, fructose, and mannitol utilization seems to be possible. Enzymes of the Embden-Meyerhof pathway have been detected (Patni and Alexander, 1971a). Fructose is transported into the cells by the PEP phosphotransferase system (Patni and Alexander, 1971b). The cellulose gene can be expressed by *Bacillus subtilis* (Joliff et al., 1989). A mutant of *C. thermocellum* overproducing cellulase has been isolated (Mori, 1990).

CLOSTRIDIUM THERMOCOPRIAE. GC content, 36.7–37.8 mol%; optimum growth temperature, 60–65°C. Cells stain Gram-negative. Several strains have been isolated from animal feces, compost, soil, and a hot spring. Depending on the strain, cellulose, xylan, starch, glycogen, inulin and a variety of sugars are fermented. The major fermentation products are ethanol, acetate, butyrate, lactate, carbon dioxide, and hydrogen (Jin et al., 1988).

CLOSTRIDIUM THERMOHYDROSULFURICUM. GC content, 29.5–32.0 mol%; peptidoglycan cross-linkage via *m*-DAP; optimum growth temperature, 65–70°C, upper limit at 74–76°C. The organism was isolated from extraction juices of Austrian sugar beet factories (Hollaus and Klaushofer, 1973; Klaushofer and Parkkinen, 1965). Its occurrence in extraction juices causes the formation of H_2S, provided the supply water is acidified with sulfur dioxide. The organism is strongly saccharolytic and the pattern of fermentable carbohydrates is very similar to that of *C. thermosaccharolyticum*. It differs from the latter 1) by an about 10°C higher optimum and upper limit of growth temperature, 2) by a much more intensive formation of H_2S from sulfite and thiosulfate, and 3) by hexagonal cell wall surface structures in contrast to the rectangular pattern of *C. thermosaccharolyticum* (Hollaus and Sleytr, 1972; Sleytr and Glauert, 1976). Furthermore, both species can be separated by their low DNA relationship (Matteuzzi et al., 1978). Sporulation of several strains was found best in a soil extract medium that contained 0.1% D-xylose (Hollaus and Klaushofer, 1973). Spores are extremely heat resistant (Hyun et al., 1983). The glyco-

protein of the S-layer has been studied (Sára et al., 1989). Two different forms of lactate dehydrogenase are present in ethanol or lactate-producing cells (Turunen et al., 1987). Thermostable amylolytic enzymes are produced (Hyun and Zeikus, 1985b; Antranikian et al., 1987c; Melasniemi, 1987; Mathupala et al., 1990). Glucose is utilized prior to cellobiose (Ng and Zeikus, 1982). The biochemical basis for ethanol and hydrogen tolerance has been studied (Lovitt et al., 1988).

CLOSTRIDIUM THERMOLACTICUM. GC content, 41–42 mol%; optimum growth temperature, 60–65°C. Cells stain Gram-negative. The organism was isolated from methanogenic enrichment cultures on cellulose and xylan, respectively, that had been inoculated with composted cattle manure and duck weed, respectively (Le Ruyet et al., 1984). Xylan, starch, and various sugars are fermented with lactate as the main end product. One strain weakly ferments cellulose to produce a mixture of ethanol and acetate (Le Ruyet et al., 1985). Secretion of an endo-1,4-xylanase has been reported (Debeire et al., 1990).

CLOSTRIDIUM THERMOSACCHAROLYTICUM. GC content, 29–32 mol%; peptidoglycan cross-linkage via *m*-DAP; optimum growth temperature 55–60°C, upper limit at 67°C. It ferments a large number of carbohydrates, including glycogen, starch, and pectin (Hollaus and Klaushofer, 1973; Hollaus and Sleytr, 1972; McClung, 1935b; van Rijssel and Hansen, 1989) with acetate, butyrate, and lactate along with CO_2 and H_2 as the fermentation products (Sjolander, 1937). Resting cells form considerable amounts of ethanol during glucose fermentation (Lee and Ordal, 1967). The formation of *n*-butanol has also been reported (Freier-Schröder et al., 1989). Special sporulation media that contain L-arabinose or glucosides were developed by Pheil and Ordal (1967) and Hsu and Ordal (1969). Spores of this organism are extremely heat resistant, reaching D values of 4 min at 120°C (Ingram, 1969). *C. thermosaccharolyticum* has been subject of thorough study since it causes the thermophilic, "hard swell" type of spoilage of canned food. The thermophilic, tartrate-fermenting *"C. tartarivorum"* described by Mercer and Vaughn (1951) is phenotypically indistinguishable from *C. thermosaccharolyticum* (except the fermentation of tartrate) (Hollaus and Sleytr, 1972). By DNA-DNA homology, both organisms were found to belong to one genospecies; therefore, *"C. tartarivorum"* must be regarded as a tartrate-positive biotype of *C. thermosaccharolyticum* (Matteuzzi et al., 1978). The ferredoxins of *C. thermosaccharolyticum* and *"C. tartari-*

vorum" differ in two amino acids (Tanaka et al., 1973). The thermostability of glycolytic enzymes (Howell et al., 1969), especially of the fructose bisphosphate aldolase (Barnes et al., 1971), of pullulanase (Koch et al., 1987), and the fatty acid composition (Chan et al., 1971) have been compared.

CLOSTRIDIUM THERMOSULFURO-GENES. GC content, 33 mol%; optimum growth temperature, 60°C. Isolated from a thermal, volcanic, algalbacterial community and from sewage water of a fruit juice factory (Schink and Zeikus, 1983a; Madi et al., 1987). The organism is gelatinolytic and ferments numerous carbohydrates including starch and pectin. Fermentation products formed from glucose include ethanol, acetate, lactate, hydrogen, and carbon dioxide. A specific property of that organism is the capability to produce elemental sulfur from thiosulfate (Schink and Zeikus, 1983a). The production and excretion of thermostable α-amylase, β-amylase, pullulanase, and α-glucosidase during growth under starch limitation has been investigated (Antranikian et al., 1987a, 1987b; Nipkow et al., 1989) as well as pectinolytic enzymes (Schink and Zeikus, 1983b).

CLOSTRIDIUM TYROBUTYRICUM. GC content, 28 mol%; peptidoglycan cross-linkage via *m*-DAP; optimum growth temperature, 37°C. This organism ferments a few monosaccharides (glucose, fructose; some strains—galactose, arabinose, mannose, xylose, and mannitol) but no disaccharides or starch (Bryant and Burkey, 1956; Kutzner, 1963; Roux and Bergère, 1977; van Beynum and Pette, 1935). L-Lactate is fermented in the presence of acetate to butyrate, CO_2, and H_2 with little butanol formation. The limits of butyrate production in continuous culture have been determined (Michel-Savin et al., 1990). The organism plays an important role in causing cheese swells because it thrives at a lower pH than *C. butyricum* (Kutzner, 1963). Spore and coat formation have been studied (Rousseau et al., 1971b, 1971c), as have the regulation of NAD- and NADP-ferredoxin reductases (Petitdemange et al., 1973a) and the NADH-flavodoxin oxidoreductase (Petitdemange et al., 1979). The properties of acetate kinase and phosphotransacetylase, isolated from vegetative cells and spores, and their involvement in the initiation of spore germination have been investigated (Touraille and Bergère, 1974). *C. tyrobutyricum* strain La I (Giesel and Simon, 1983a) has been used to study a variety of stereospecific reduction reactions (Simon et al., 1985) which are mostly catalyzed by the enzyme enoate reductase (Kuno et al., 1985).

Comparative Studies on Clostridia

Various properties of the clostridia were discussed in a comprehensive handbook (Minton and Clarke, 1989). The clostridia were dealt with on a comparative basis in two symposia: *Annales de l'Institut Pasteur* 77:341–540, 1949; and the *Journal of Applied Bacteriology* 28:1–152, 1965. The *Bergey's Manual of Systematic Bacteriology* also summarizes relevant data (Cato et al., 1986). The genus *Clostridium* (McClung, 1956) and the enrichment and isolation procedures have been reviewed (Kutzner, 1965; Shapton and Board, 1971; Willis, 1977). The fermentation patterns of many clostridia (Holdeman et al., 1977; Moore et al., 1966), the cell wall composition (Cummins and Johnson, 1971; Schleifer and Kandler, 1972; Takumi and Kawate, 1976; Weiss et al., 1981), and the DNA base composition, DNA or RNA homologies, and 5S rRNA structure (Cato and Stackebrandt, 1989; Cummins and Johnson, 1971; Hill, 1966; Johnson, 1973; Johnson and Francis, 1975; Matteuzzi et al., 1977, 1978; Dams et al., 1987) have been summarized. The sensitivity to antibiotics and lysozyme and the production of DNAse, clostridiocines, and phages by different clostridia is well documented (Magot, 1984; Clarke et al., 1975; Döll, 1973; Johnson and Francis, 1975; Tagg et al., 1976; Nieves et al., 1981; Ogata and Hongo, 1979). The pectinolytic (Raynaud, 1949; Schink et al., 1981; Lund, 1982; Perry, 1985; Suzzi et al., 1987), chitinolytic (Timmis et al., 1974; Pel et al., 1989), pullulytic and amylolytic (Klingeberg et al., 1990; Antranikian, 1990), cellulolytic (Coughlan and Mayer, Chapter 20), psychrotrophic (Beerens et al., 1965), and thermophilic clostridia (Hollaus and Sleytr, 1972; McClung, 1935a) have been subject of special articles. The fatty acid and phospholipid compositions of selected groups of clostridia have been compared (Moss and Lewis, 1967; Chan et al., 1971; Elsden et al., 1980; Johnston and Goldfine, 1983). Involvement of clostridia in spoilage of meat, fish, vegetable foods, and dairy products as well their toxigenic properties has been reviewed (Roberts and Mead, 1986; Bergère and Accolas, 1986; Lund, 1982, 1986; Hatheway, 1990).

New species not validated so far have been described on the basis of the ability to form gas caps in connection with spore formation (Duda and Makareva, 1977; Krasilnikov et al., 1971). The energy-yielding processes of certain clostridia have been reviewed by Thauer et al. (1977), Andreesen et al. (1989), Bader and Simon (1983), and Morris (1986). The glycolytic enzymes and sugar transport of different clostridia have been compared (Kotze, 1969; Booth and Mitchell, 1987), especially the phospho-fructokinases (von Hugo and Gottschalk, 1974a), the ability to form 2-keto-3-deoxyglu-conate from gluconate (Bender et al., 1971), and the pentose metabolism (Cynkin and Gibbs, 1958; Ounine et al., 1983; Aduse-Opoku and Mitchell, 1988).

The presence and the properties of the hydrogenases (Mortenson and Chen, 1974; Adams et al., 1989), the ferredoxins and flavodoxins (Yoch and Valentine, 1972; Rabinowitz, 1972; Schönheit et al., 1978), the function of reduced pyridine nucleotide:ferredoxin oxidoreductases (Jungermann et al., 1973; Blusson et al., 1981) and their regulation (Petitdemange et al., 1976), the coenzyme specificity of different dehydrogenases (von Hugo et al., 1972), the function of pyruvate formate lyase (Thauer et al., 1972) and of phosphotransbutyrylase (Valentine and Wolfe, 1960), and the distribution of enoate reductases (Giesel and Simon, 1983b; Krause and Simon, 1989), NAD(H) degradation (Schuetz and Simon, 1986), and of acid phosphatase (Schallehn and Brandis, 1973; Ueno et al., 1970) have been studied. The pyrimidine and purine metabolism by clostridia has been investigated (Campbell, 1960; Champion and Rabinowitz, 1977; Hilton et al., 1975; Mead et al., 1979; Dürre and Andreesen, 1983) as well as the choline metabolism (Joblin et al., 1976). The amino acid-fermenting clostridia have been surveyed (Mead, 1971; Elsden and Hilton, 1979; Barker, 1981) with special attention to aromatic amino acids (Elsden et al., 1976) and the special products hydrocinnamic and phenylacetic acid (Moss et al., 1970; Mayraud and Bourgeau, 1982), to branched-chain amino acids (Elsden and Hilton, 1978; Britz and Wilkinson, 1982), and glutamate (Buckel and Barker, 1974; Buckel, 1980a, 1980b). A large number of clostridia has been screened for utilization of ethanolamine, betaine, choline, creatine, and creatinine (Möller et al., 1986). The ability for citrate utilization (Antranikian et al., 1984) and the presence and stereospecificity of citrate synthase has been studied on a comparative basis (Gottschalk and Barker, 1967), and aspects of leucine, porphyrin, and B_{12} biosynthesis have been studied (Wiegel and Schlegel, 1977; Brumm et al., 1982). Some species even excrete amino acids (Matteuzzi et al., 1978; Szech et al., 1989) or O-butylhomoserine (Uyeda et al., 1974). The physiological responses to stress factors were compared (Woods and Jones, 1986). The ability to fix molecular nitrogen is widely distributed among clostridia (Hammann and Ottow, 1976; Rosenblum and Wilson, 1949) and the enzymes involved in its assimilation were compared (Bogdahn and Kleiner, 1986a).

Clostridia were exposed to various artificial substrates, and their metabolic activities towards nitro- and chlorine-containing compounds (Edwards et al., 1973; Jagnow et al., 1977; Gälli and McCarty, 1989; Angermaier and Simon, 1983), and steroids (Mahony et al., 1977; Orpianesi et al., 1989) were compared. Hydroxysteroid dehydrogenases of different stereospecificity have been reported to occur in several clostridia, e.g., in *C. perfringens* (MacDonald et al., 1975, 1976), *C. innocuum* (Edenharder and Pfützner, 1989), *C. leptum* (Harris and Hylemon, 1977; Stellwag and Hylemon, 1977), *C. bifermentans* (Ferrari and Aragozzini, 1972; Ferrari et al., 1977), and in *C. tertium, C. indolis,* and *C. baratii ("paraperfringens")* (Goddard et al., 1975). The latter reduces the double bond in ring A of deoxycorticosterone (Capek et al., 1966). *C. scindens* cleaves off the side chain of steroids (Morris et al., 1985). Several intestinal clostridia can cleave flavanoids (Winter et al., 1989) or inactivate contraceptive steroid hormones (Bokkenheuser et al., 1983).

Acknowledgments

Work of the laboratories of the authors was supported by the Deutsche Forschungsgemeinschaft, the Akademie der Wissenschaften zu Göttingen, and the Fonds der Chemischen Industrie.

Literature Cited

Abeles, R. H., and Beck, W. S. 1967. The mechanism of action of cobamide coenzyme in the ribonucleotide reductase reaction. J. Biol.Chem. 242:3589–3593.

Abou-Zeid, A. A., Fouad, M., and Yassein, M. 1978. Microbial production of acetone-butanol by *Clostridium acetobutylicum.* Zentralbl. Bakteriol. Parasitenkd. Infektionskr. Hyg. Abt. 2 133:125–134.

Abraham, L. J., Wales, A. J., and Rood, J. I. 1985. Worldwide distribution of the conjugative *Clostridium perfringens* tetracycline resistance plasmid, pCW3. Plasmid 14:37–46.

Adamse, A. D. 1980. New isolation of *Clostridium aceticum* (Wieringa). Antonie van Leeuwenhoek J. Microbiol. Serol. 46:523–531.

Adams, M. W. W., Eccleston, E., and Howard, J. B. 1989. Iron-sulfur clusters of hydrogenase I and hydrogenase II of *Clostridium pasteurianum.* Proc. Natl. Acad. Sci. USA 86:4932–4936.

Aduse-Opoku, J., and Mitchell, W. J. 1988. Diauxic growth of *Clostridium thermosaccharolyticum* on glucose and xylose. FEMS Microbiol. Lett. 50:45–49.

Afschar, A. S., Biebl, H., Schaller, K., and Schügerl, K. 1985. Production of acetone and butanol by *Clostridium ace-*

tobutylicum in continuous culture with cell recycle. Appl. Microbiol. Biotechnol. 22:394–398.

Ahn, Byoung Kwon, and Shinsaku, H. 1990. Metabolic mechanism of ethanol-isolvaleric fermentation by a Clostridium saccharoperbutylacetonicum UV-mutant. Agric. Biol. Chem. 54:353–357.

Ait, N., Creuzet, N., and Cattaneo, J. 1982. Properties of β-glucosidase purified from Clostridium thermocellum. J. Gen. Microbiol. 128:569–577.

Akedo, M., Cooney, C. L., and Sinskey, A. J. 1983. Direct demonstration of lactate-acrylate interconversion in Clostridium propionicum. Biotechnol. 1:791–794.

Aleman, V., and Handler, P. 1967. Dihydroorotate dehydrogenase. I. General properties. J. Biol.Chem. 242:4087–4096.

Alexander, J. K. 1968:Purification and specificity of cellobiose phosphorylase from Clostridium thermocellum. J. Biol. Chem. 243:2899–2904.

Allcock, E. R., Reid, S. J., Jones, D. T., and Woods, D. R. 1982. Protoplast formation and regeneration in Clostridium acetobutylicum. Appl. Environ. Microbiol. 43:719–721.

Allen, S. P., and Blaschek, H. P. 1988. Electroporation-induced transformation of intact cells of Clostridium perfringens. Appl. Environ. Microbiol. 54:2322–2324.

Allison, C., and Macfarlane, G. T. 1989a. Influence of pH, mutant availability, and growth rate on amine production by Bacteroides fragilis and Clostridium perfringens. Appl. Environ. Microbiol. 55:2894–2898.

Allison, C., and Macfarlane, G. T. 1989b. Protease production by Clostridium perfringens in batch and continuous culture. Lett. Appl. Microbiol. 9:45–48.

Aminoff, D., and Furakawa, K. 1970. Enzymes that destroy blood group specificity. I. Purification and properties of α-L-fucosidase from Clostridium perfringens. J. Biol. Chem. 245:1659–1669.

Andersch, W., Bahl, H., and Gottschalk, G. 1983. Level of enzymes involved in acetate, butyrate, acetone and butanol formation by Clostridium acetobutylicum. Eur. J. Appl. Microbiol. Biotechnol. 18:327–332.

Andreesen, J. R., Bahl, H., and Gottschalk, G. 1989. Introduction to the physiology and biochemistry of the genus Clostridium, p. 27–62. In: N. P. Minton, D. J. Clarke (ed.), Clostridia, biotechnology handbooks 3. Plenum Press, New York.

Andreesen, J. R., and Gottschalk, G. 1969. The occurrence of a modified Entner-Doudoroff pathway in Clostridium aceticum. Arch. Mikrobiol. 69:160–170.

Andreesen, J. R., Gottschalk, G., and Schlegel, H. G. 1970. Clostridium formicoaceticum nov. spec. Isolation, description and distinction from C. aceticum and C. thermoaceticum. Arch. Mikrobiol. 72:154–174.

Andreesen, J. R., Schaupp, A., Neurauter, C., Brown, A., and Ljungdahl, L. G. 1973. Fermentation of glucose, fructose and xylose by Clostridium thermoaceticum. Effects of metals on growth yield, enzymes and the synthesis of acetate from CO₂. J. Bacteriol. 114:743–751.

Andreesen, J. R., Zindel, U., and Dürre, P. 1985. Clostridium cylindrosporum (ex Barker and Beck 1942) nom. rev. Int. J. Syst. Bacteriol. 35:206–207.

Andrew, J. G., and Morris, J. G. 1965. The biosynthesis of alanine by Clostridium kluyveri. Biochim. Biophys. Acta 97:176–179.

Angermaier, L., and Simon, H. 1983. On the reduction of aliphatic and aromatic nitro compounds by clostridia.

The role of ferredoxin and its stabilization. Hoppe Seyler's Z. Physiol. Chem. 364:961–975.

Anthony, C., and Guest, J. R. 1968. Deferred metabolism of glucose by Clostridium tetanomorphum. J. Gen. Microbiol. 54:277–286.

Antranikian, G. 1990. Physiology and enzymology of thermophilic anaerobic bacteria degrading starch. FEMS Microbiol. Rev. 75:201–218.

Antranikian, G., Friese, C., Quentmeier, A., Hippe, H., and Gottschalk, G. 1984. Distribution of the ability for citrate utilization amongst clostridia. Arch. Microbiol. 138:179–182.

Antranikian, G., Herzberg, C., and Gottschalk, G. 1985. In vivo phosphorylation of proteins in Clostridium sphenoides. FEMS Microbiol. Lett. 27:135–138.

Antranikian, G., Herzberg, C., and Gottschalk, G. 1987a. Production of thermostable α-amylase, pullulanase, and α-glucosidase in continuous culture by a new Clostridium isolate. Appl. Environ. Microbiol. 53:1668–1673.

Antranikian, G., Herzberg, C., Mayer, F., and Gottschalk, G. 1987b. Changes in the cell envelope stucture of Clostridium sp. strain EM1 during massive production of α-amylase and pullulanase. FEMS Microbiol. Lett. 41:193–197.

Antranikian, G., Zablowski, P., and Gottschalk, G. 1987c. Conditions for the overproduction and excretion of thermostable α-amylase and pullulanase from Clostridium thermohydrosulfuricum DSM 567. Appl. Microbiol. Biotechnol. 27:75–81.

Archer, R. H., Maddox, I. S., and Chong, R. 1981. 7α-Dehydroxylation of cholic acid by Clostridium bifermentans. Eur. J. Appl. Microbiol. Biotechnol. 12:46–52.

Arkowitz, R. A., and Abeles, R. H. 1989. Identification of acetyl phosphate as the product of clostridial glycine reductase:evidence for an acyl enzyme intermediate. Biochem. 28:4639–4644.

Ashley, N. V., and Shoesmith, J. G. 1977. Continuous culture of Clostridium sporogenes and Clostridium bifermentans in the presence of oxygen. Proc. Soc. Gen. Microbiol. 4:144.

Asther, M., and Khan, A. W. 1985. Improved fermentation of cellobiose, glucose, and xylose to ethanol by Zymomonas anaerobia and a high ethanol tolerant strain of Clostridium saccharolyticum. Appl. Microbiol. Biotechnol. 21:234–237.

Augier, J. 1956. A propos de la numeration des Azotobacter en milieu liquide. Annal. Inst. Pasteur 91:759.

Augier, J. 1957. A propos de la fixation biologique de l'azote atmosphérique et de la numération des Clostridium fixateurs dans les sols. Annal. Inst. Pasteur 92:817.

Azova, L. G., Gusev, M. V., and Ivoilov, V. S. 1970. Response to molecular oxygen in some members of Clostridium genus. Microbiology (English translation of Mikrobiologiya) 39:43–47.

Bader, J., Günther, H., Rambeck, B., and Simon, H. 1978. Properties of two clostridia strains acting as catalysts for the preparative stereospecific hydrogenation of 2-enoic acids and 2-alken-1-ols with hydrogen gas. Hoppe-Seyler's Z. Physiol. Chem. 359:19–27.

Bader, J., Günther, H., Schleicher, E., Simon, H., Pohl, S., and Mannheim, W. 1980. Utilization of (E)-2-butenoate (crotonate) by Clostridium kluyveri and some other Clostridium species. Arch. Microbiol. 125:159–165.

Bader, J., and Simon, H. 1983. ATP formation is coupled to the hydrogenation of 2-enoates in *Clostridium sporogenes.* FEMS Microbiol. Lett. 20:171–175.

Bahl, H., Andersch, W., and Gottschalk, G. 1982. Continuous production of acetone and butanol by *Clostridium acetobutylicum* in a two-stage phosphate limited chemostat. Eur. J. Appl. Microbiol. Biotechnol. 15:201–205.

Bahl, H., and Gottschalk, G. 1984. Parameters affecting solvent production by *Clostridium acetobutylicum* in continuous culture. Biotechnol. Bioengin. Symp. no. 14., 215–223.

Bahl, H., and Gottschalk, G. 1988. Microbial production of butanol/acetone. In: H. J. Rehm, and G. Reed (ed.), Biotechnology, vol. 6b. VCH Verlagsgesellschaft, Weinheim, Germany.

Bahl, H., Gottwald, M., Kuhn, A., Rale, V., Andersch, W., and Gottschalk, G. 1986. Nutritional factors affecting the ratio of solvents produced by *Clostridium acetobutylicum.* Appl. Environ. Microbiol. 52:169–172.

Baker, J. J., and van der Drift, C. 1974. Purification and properties of L-erythro-3, 5-diaminohexanoate dehydrogenase from *Clostridium sticklandii.* Biochem. 13:292–299.

Ballongue, J., Amine, J., Masion, E., Petitdemange, H., and Gay, R. 1985. Induction of acetoacetate decarboxylase in *Clostridium acetobutylicum.* FEMS Microbiol. Lett. 29:273–277.

Barber, J. M., Robb, F. T., Webster, J. R., and Woods, D. R. 1979. Bacteriocin production by *Clostridium acetobutylicum* in an industrial fermentation process. Appl. Environ. Microbiol. 37:433–437.

Barker, H. A. 1937. On the fermentation of glutamic acid. Enzymologia 2:175–182.

Barker, H. A. 1939. The use of glutamic acid for the isolation and identification of *Clostridium cochlearium* and *C. tetanomorphum.* Arch. Mikrobiol. 10:376–384.

Barker, H. A. 1956. Bacterial fermentations. John Wiley, New York.

Barker, H. A. 1961. Fermentations of nitrogenous organic compounds, p. 151–207. In: Gunsalus, I. C., Stanier, R. Y. (ed.), The bacteria, vol. 2. Academic Press, London.

Barker, H. A. 1976. Glutamate fermentation and the discovery of B₁₂ coenzymes, p. 95–104. In: Kornberg, A., Comudella, L., Horecker, B. L., and Oro, J. (ed.), Reflections on biochemistry. Pergamon Press, Oxford.

Barker, H. A. 1981. Amino acid degradation by anaerobic bacteria. Annu. Rev. Biochem. 50:23–40.

Barker, H. A., and Beck, J. V. 1941. The fermentative decomposition of purines by *Clostridium acidiurici* and *Clostridium cylindrosporum.* J. Biol.Chem. 141:3–27.

Barker, H. A., and Beck, J. V. 1942. *Clostridium acidiurici* and *Clostridium cylindrosporum,* organisms fermenting uric acid and some other purines. J. Bacteriol. 43:291–304.

Barker, H. A., D'Ari, L., and Kahn, J. 1987. Enzymatic reactions in the degradation of 5-aminovalerate by *Clostridium aminovalericum.* J. Biol. Chem. 262:8994–9003.

Barker, H. A., and Peterson, W. H. 1944. The nutritional requirements of *Clostridium kluyveri,* p. 599–602. In: Colowick, S. P., and Kaplan, N. O. (ed.), Methods in enzymology, vol. 1. Academic Press, London.

Barker, H. A., Stadtman, E. R., and Kornberg, A. 1955. Coenzyme A transphorase from *Clostridium kluyveri,* p. 599–602. In: Colowick, S. P., and Kaplan, N. O. (ed.), Methods in enzymology, vol. 1. Academic Press, London.

Barker, H. A., and Taha, S. M. 1942. *Clostridium kluyveri,* an organism concerned in the formation of caproic acid from ethyl alcohol. J. Bacteriol. 43:347–363.

Barker, H. A., Weissbach, H., and Smyth, R. D. 1958. A coenzyme containing pseudo-vitamin B₁₂. Proc. Natl. Acad. Sci. USA 44:1093–1097.

Barker, H. A., Jeng, I. M., Neff, N., Robertson, J. M., Tam, F. K., and Hosaka, S. 1978. Butyryl-CoA:acetoacetate CoA-transferase from a lysine-fermenting *Clostridium.* J. Biol. Chem. 253:1219–1225.

Barnes, E. M., Akagi, J. M., and Himes, R. H. 1971. Properties of fructose-1,6-diphosphate aldolase from two thermophilic and a mesophilic clostridia. Biochim. Biophys. Acta 227:199–203.

Barnes, E. M., and Impey, C. S. 1974. The occurrence and properties of uric acid decomposing anaerobic bacteria in the avium caecum. J. Appl. Bacteriol. 37:393–409.

Bartsch, R. G., and Barker, H. A. 1961. A vinylacetyl isomerase from *Clostridium kluyveri.* Arch. Biochem. Biophys. 92:122–132.

Baskerville, E. N., and Twarog, R. 1972. Regulation of the tryptophan synthetic enzymes in *Clostridium butyricum.* J. Bacteriol. 112:304–314.

Baskerville, E., and Twarog, R. 1974. Regulation of a ligand-mediated association-dissociation system of anthranilate synthesis in *Clostridium butyricum.* J. Bacteriol. 117:1184–1194.

Bateson, M., Wiegel, J., and Ward, D. M. 1989. Comparative analysis of 16S ribosomal RNA sequences of thermophilic fermentative bacteria isolated from hot spring cyanobacterial mats. Syst. Appl. Microbiol. 12:1–7.

Baumgart, H., Hippe, H., and Weber, B. 1984. Verderb pasteurisierter Feinkostsalate durch Clostridien. Chem. Mikrobiol. Technol. Lebensm. 8:109–114.

Beerens, H., and Fievez, L. 1971. Isolation of *Bacteroides fragilis* and *Sphaerophorus-Fusiformis* groups, p. 109–113. In: Shapton, D. A., and Board, R. G. (ed.), Isolation of anaerobes. Academic Press, London.

Beerens, H., Sugama, S., and Tahon-Castel, M. 1965. Psychrotrophic clostridia. J. Appl. Bacteriol. 28:36–48.

Beesch, S. C. 1953. A microbiological progress report. Aceton-butanol fermentation of starches. Appl. Microbiol. 1:85–95.

Belokopytov, B. F., Golovchenko, N. B., Krauzova, V. I., Chuvilskaya, N. A., Akimenio, V. K., and Lozinov, A. B. 1982. A chemical defined medium for cultivation of *Clostridium sporogenes.* Mikrobiologiya 51:354–361.

Bender, R., Andreesen, J. R., and Gottschalk, G. 1971. 2-Keto-3-deoxygluconate, an intermediate in the fermentation of gluconate by clostridia. J. Bacteriol. 107:570–573.

Bender, R., and Gottschalk, G. 1973. Purification and properties of D-gluconate dehydratase from *Clostridium pasteurianum.* Eur. J. Biochem. 40:309–321.

Benson, A. M., Mower, H. F., and Yasunobu, K. T. 1967. The amino acid sequence of *Clostridium butyricum* ferredoxin. Arch. Biochem. Biophys. 121:563–575.

Berestovskaya, Y. Y., Kryukov, V. R., Bodnar, I. V., and Pusheva, M. A. 1987. Growth of the homoacetic bacterium *Clostridium thermoautotrophicum*. Microbiologiya 56:642–647.

Bergère, J.-L., and Hermier, J. 1970. Spore properties of clostridia occurring in cheese. J. Appl. Bacteriol. 33:167–179.

Bergère, J.-L., Rousseau, M., and Mercier. C. 1975. Polyoside intracellulaire implique dans la sporulation de *Clostridium butyricum*. I. Cytologie, production et analyse enzymatique preliminaire. Ann. Microbiol. (Paris) 126A:295–314.

Bergère, J.-L., and Accolas, J.-P. 1986. Nonsporing and sporing anaerobes in diary products, p. 373–396. In: Barnes, E. M., and Mead, G. C. (ed.), Anaerobic bacteria in habitats other than man. The Soc. Appl. Bacteriol. Symp. Ser. no. 13, Blackwell Scientific Publ., Oxford.

Bergmeyer, H. U., Holz, G., Klotzsch, H., and Lang, G. 1963. Phosphotransacetylase aus *Clostridium kluyveri*. Zuechtung des Bakteriums, Isolierung, Kristallisation and Eigenschaften des Enzyms. Biochem. Z. 338:114–121.

Berndt, H., and Schlegel, H. G. 1975. Kinetics and properties of P-ketothiolase from *Clostridium pasteurianum*. Arch. Microbiol. 103:21–30.

Bertram, J., and Dürre, P. 1989. Conjugal transfer and expression of streptococcal transposons in *Clostridium acetobutylicum*. Arch. Microbiol. 151:551–557.

Bertram, J., Kuhn, A., and Dürre, P. 1990. Tn916-induced mutants of *Clostridium acetobutylicum* defective in regulation of solvent formation. Arch. Microbiol. 153:373–377.

Bhat, J. V., and Barker, H. A. 1947. *Clostridium lactoacetophilum* nov. spec. and the role of acetic acid in the butyric acid fermentation of lactate. J. Bacteriol. 54:381–391.

Birrer, G. A., Chesbro, W. R., and Zsigray, R. M. 1989. Protoplast formation, L-colony growth and regeneration of *Clostridium beijerinckii* NRRL B-592 and B-593 and *Clostridium acetobutylicum* ATCC 10132. J. Ind. Microbiol. 4:325–332.

Blusson, H., Petitdemange, H., and Gay, R. 1981. A new, fast, and sensitive assay for NADH-ferredoxin oxidoreductase detection in clostridia. Anal. Biochem. 110:176–181.

Bogdahn, M., Andreesen, J. R., and Kleiner, D. 1983. Pathways and regulation of N_2, ammonium, and glutamate assimilation by *Clostridium formicoaceticum*. Arch. Microbiol. 134:167–169.

Bogdahn, M., and Kleiner, D. 1986a. N_2 fixation and NH_4^+ assimilation in the thermophilic anaerobes *Clostridium thermosaccharolyticum* and *Clostridium thermoautotrophicum*. Arch. Microbiol. 144:102–104.

Bogdahn, M., and Kleiner, D. 1986b. Inorganic nitrogen metabolism in two cellulose-degrading clostridia. Arch. Microbiol. 145:159–161.

Bokkenheuser, V. D., Winter, J., Cohen, B. I., O'Rourke, S., and Mosbach, E. H. 1983. Inactivation of contraceptive steroid hormones by human intestinal clostridia. J. Clin. Microbiol. 18:500–504.

Bom, I. J., Smelt, J. P. P. M., Kersters, K., and Verrips, C. T. 1986. Identification and grouping of *Clostridium botulinum* strains by numerical analysis of their electrophoretic protein patterns. J. Appl. Bacteriol. 60:483–490.

Bonomi, F., Pagani, S., and Kurtz, D. M. 1985. Enzymatic synthesis of the 4Fe-4S clusters of *Clostridium pasteurianum* ferredoxin. Eur. J. Biochem. 148:67–73.

Booth, I. R., and Mitchell, W. J. 1987. Sugar transport and metabolism in the clostridia, p. 165. In: Reizer, J., and Peter Kofsky, A. (ed.), Sugar transport and metabolism in Gram-positive bacteria. Ellis Horwood, Chichester, U.K.

Booth, I. R., and Morris, J. G. 1975. Proton-motive force in the obligately anaerobic bacterium *Clostridium pasteurianum*. A role in the galactose and gluconate uptake. FEBS Lett. 59:153–157.

Bornstein, B. T., and Barker, H. A. 1948a. The nutrition of *Clostridium kluyveri*. J. Bacteriol. 55:223–230.

Bornstein, B. T., and Barker, H. A. 1948b. The energy metabolism of *Clostridium kluyveri* and the synthesis of fatty acids. J. Biol. Chem. 172:659–669.

Borriello, S. P., and Carman, R. J. 1983. Association of iota-like toxin and *Clostridium spiroforme* with both spontaneous and antibiotic-associated diarrhea and colitis in rabbits. J. Clin. Microbiol. 17:414–418.

Borriello, S. P., Davies, H. A., and Carman, R. J. 1986. Cellular morphology of *Clostridium spiroforme*. Vet. Microbiol. 11:191–195.

Bradshaw, W. H., and Barker, H. A. 1960. Purification and properties of xanthine dehydrogenase from *Clostridium cylindrosporum*. J. Biol. Chem. 235:3620–3629.

Brantner, H., Fischer, G., and Vivat, H. 1975. Studies on the cultivation of *Clostridium oncolyticum* M55, 5th communication: The influence of iron, zinc, and cobaltous ions on the growth and the kininase activity of *Clostridium oncolyticum* M55 ATCC 13.732. Zentralbl. Bakteriol. Parasitenkd. Infektionskr. Hyg. Abt. 1 Orig., A 233:253–260.

Brantner, H., and Schwager, J. 1979. Enzymatische Mechanismen der Onkolyse durch *Clostridium oncolyticum* M 55 ATCC 13732. Zentralbl. Bakteriol. Parasitenkd. Infektionskr. Hyg. I. Abt. Orig., Reihe A243:113–118.

Braun, M., Schoberth, S., and Gottschalk, G. 1979. Enumeration of bacteria forming acetate from H_2 and CO_2 in anaerobic habitats. Arch. Microbiol. 120:201–204.

Braun, M., Mayer, F., and Gottschalk, G. 1981. *Clostridium aceticum* (Wieringa), a microorganism producing acetic acid from molecular hydrogen and carbon dioxide. Arch. Microbiol. 128:288–293.

Breed, R. S., Murray, E. G. D., and Smith, N. R. 1957. Bergey's manual of determinative bacteriology, 7th ed. Williams & Wilkins, Baltimore.

Bréfort, G., Magot, M., Ionesco, H., and Sebald, M. 1977. Characterization and transferability of *Clostridium perfringens* plasmids. Plasmid 1:52–66.

Britz, M. L., and Wilkinson, R. G. 1982. Leucine dissimilation to isovaleric and isocaproic acids by cell suspensions of amino acid fermenting anaerobes: the Stickland reaction revisited. Can. J. Microbiol. 28:291–300.

Britz, M. L., and Wilkinson, R. G. 1983. Partial purification and characterization of two enzymes involved in isovaleric acid synthesis in *Clostridium bifermentans*. J. Gen. Microbiol. 129:3227–3237.

Brown, R. G., Lindberg, B., and Laishley, E. . 1975. Characterization of two reserve glucans from *Clostridium pasteurianum*. Can. J. Microbiol. 21:1136–1138.

Brühmüller, M., and Decker, K. 1976. A convenient biosynthetic method for the preparation of radioactive flavine nucleotides. Anal. Biochem. 71:550–554.

Brumm, P. J., Fried, J., and Friedmann, H. C. 1983. Bactobilin blue pigment isolated from *Clostridium tetanomorphum*. Proc. Natl. Acad. Sci. USA 80:3943–3947.

Brumm, P. J. 1988. Fermentation of single and mixed substrates by the parent and an acid- tolerant, mutant strain of *Clostridium thermoaceticum*. Biotechnol. Bioeng. 32:444–450.

Brumm, P. J., Thomas, G. A., and Friedmann, H. C. 1982. The role of 4,5-dioxovaleric acid in poryhyrin and vitamin B$_{12}$ formation by clostridia. Biochem. Biophys. Res. Commun. 104:814–822.

Bryant, M. P., and Burkey, L. A. 1956. The characteristics of lactate-fermenting sporeforming anaerobes from silage. J. Bacteriol. 71:43–46.

Bryant, M. P., and Laishley, E. J. 1988. Evidence for two transporters of sulfur and selenium oxyanions in *Clostridium pasteurianum*. Can. J. Microbiol. 34:700–703.

Bryant, M. P., and Laishley, E. J. 1989. Evidence for proton motive force dependent transport of selenite by *Clostridium pasteurianum*. Can. J. Microbiol. 35:481–486.

Buchanan, R. E., and Gibbons, N. E. 1974. Bergey's manual of determinative bacteriology, 8th ed. Williams & Wilkins, Baltimore.

Buckel, W., and Barker, H. A. 1974. Two pathways of glutamate fermentation by anaerobic bacteria. J. Bacteriol. 117:1248–1260.

Buckel, W., and Bobi, A. 1976. The enzyme complex citramalate lyase from *Clostridium tetanomorphum*. Eur. J. Biochem. 64:255–262.

Buckel, W. 1980a. Analysis of fermentation pathways of clostridia using double labelled glutamate. Arch. Microbiol. 127:167–169.

Buckel, W. 1980b. The reversible dehydration of (R)-2-hydroxyglutarate to (E)-glutaconate. Eur. J. Biochem. 106:439–447.

Buckel, W., and Miller, S. L. 1987. Equilibrium constants of several reactions involved in the fermentation of glutamate. Eur. J. Biochem. 164:565–569.

Burton, R. M., and Stadtman, E. R. 1953. The oxidation of acetaldehyde to acetyl-CoA by *Clostridium kluyveri*. J. Biol. Chem. 202:873–890.

Bush, R. S., and Sauer, F. D. 1977. Evidence for separate enzymes of pyruvate decarboxylation and pyruvate synthesis in soluble extracts of *Clostridium pasteurianum*. J. Biol. Chem. 252:2657–2661.

Cameron, D. C., and Cooney, C. L. 1986. A novel fermentation: The production of R(-)-1,2-propanediol and acetol by *Clostridium thermosaccharolyticum*. Biotechnology 4:651–654.

Campbell, L. L. 1960. Reductive degradation of pyrimidines. V. Enzymatic conversion of N-carbamyl-P-alanine to P-alanine, carbon dioxide, and ammonia. J. Biol. Chem. 235:2375–2378.

Campbell, L. L., and Frank, H. A. 1956. Nutritional requirements of some putrefactive anaerobic bacteria. J. Bacteriol. 71:267–269.

Campbell, L. L., and Postgate, J. R. 1965. Classification of the sporeforming sulfate-reducing bacteria. Bacteriol. Rev. 29:359–363.

Capek, A., Hanc. O., and Tadra, M. 1966. Microbial transformation of steroids, p. 115. Academia Publishing House of the Czechoslovak Academy of Sciences, Prague.

Cardenas, E. 1989. Biochemistry of oxygen toxicity. Annu. Rev. Biochem. 598:79–110.

Cardenas, J., Mortenson, L. E., and Yoch, D. C. 1976. Purification and properties of paramagnetic protein from *Clostridium pasteurianum* W 5. Biochim. Biophys. Acta 434:244–257.

Cardon, B. P., and Barker, H. A. 1946. Two new amino-acid-fermenting bacteria, *Clostridium propionicum* and *Diplococcus glycinophilus*. J. Bacteriol. 52:629–634.

Cardon, B. P., and Barker, H. A. 1947. Amino acid fermentations by *Clostridium propionicum* and *Diplococcus glycinophilus*. Arch. Biochem. 12:165–180.

Carlson, J. R., Sherrill, J. M., Rosenblatt, J. E., and McCarthy, L. R. 1981. Pencillinase activity in three strains of *Clostridium butyricum*. Curr. Microbiol. 5:251–254.

Carnahan, J. E., and Castle, J. E. 1958. Some requirements of biological nitrogen fixation. J. Bacteriol. 75:121–124.

Carter, J. E., and Sagers, R. D. 1972. Ferrous ion-dependent L-serine dehydratase from *Clostridium acidiurici*. J. Bacteriol. 109:757–763.

Cary, J. W., Petersen, D. J., Papoutsakis, E. T., and Bennett, G. N. 1988. Cloning and expression of *Clostridium acetobutylicum* phosphotransbutyrylase and butyrate kinase genes in *Escherichia coli*. J. Bacteriol. 170:4613–4618.

Cato, E. P., George, W. L., and Finegold, S. M. 1986. The genus *Clostridium*, pp. 1141–1200. In: H. A. Sneath, N. S. Mair, M. E. Sharpe, and J. G. Holt (ed.), Bergey's manual of systematic bacteriology, vol. 2. The Williams & Wilkins Co., Baltimore.

Cato, E. P., Hash, D. E., Holdeman, L. V., and Moore, W. E. C. 1982a. Electrophoretic study of *Clostridium* species. J. Clin. Microbiol. 15:688–702.

Cato, E. P., Holdeman, L. V., and Moore, W. E. C. 1982b. *Clostridium perenne* und *Clostridium paraperfringens*: Later subjective synonyms of *Clostridium barati*. Int. J. Syst. Bacteriol. 32:77–81

Cato, E., Moore, W. E. C., and Holdeman, L. V. 1968. *Clostridium oroticum* comb. nov. Amended description. Int. J. Syst. Bacteriol. 18:9–13.

Cato, E. P., and Salmon, C. W. 1976. Transfer of *Bacteroides clostridiiformis* subsp. *clostridiiformis* (Burri and Ankersmit) Holdeman and Moore and *Bacteroides clostridiiformis* subsp. *girans* (Prévot) Holdeman and Moore to the genus *Clostridium* as *Clostridium clostridiiforme* (Burri and Ankersmit) comb. nov. Emendation of description and designation of neotype strain. Int. J. Syst. Bacteriol. 26:205–211.

Cato, E. P., and Stackebrandt, E. 1989. Taxonomy and phylogeny, p. 1–26. In: Minton, N. P., and Clarke, D. J. (ed.), Clostridia, Biotechnology handbooks, vol. 3. Plenum Press, New York.

Chaigneau, M., Bory, J., Labarre, C., Cato, E., Moore, W. E. C., and Holdeman, L. V. 1968. *Clostridium oroticum* comb. nov. Amended description. Int. J. Syst. Bacteriol. 18:9–13.

Chaigneau, A. B., Bory, J., Labarre, C., and Descrozailles, J. 1974. Composition des gaz dégagés par *Welchia perfringens* et *Clostridium sporogenes* cultivés en différents milieux synthétiques. Ann. Pharmaceut. Francaises 32:619–622.

Champion, A. B., and Rabinowitz, J. C. 1977. Ferredoxin and formyltetrahydrofolate synthetase: Comparative studies with *Clostridium acidiurici*, *Clostridium cylindrosporum*, and newly isolated anaerobic uric acid-fermenting strains. J. Bacteriol. 132:1003–1020.

Chan, M., Himes, R. H., and Akagi, J. M. 1971. Fatty acid composition of thermophilic, mesophilic, and psychrophilic clostridia. J. Bacteriol. 106:876–881.

Chen, J. S., and Blanchard, D. K. 1984. Purification and properties of the H_2-oxidizing (uptake) hydrogenase of the N_2-fixing anaerobe *Clostridium pasteurianum* W5. Biochem. Biophys. Res. Commun. 122:9–16.

Chen, K. C.-K., Chen, J. S., and Johnson, J. L. 1986. Structural features of multiple nif H-like sequences and very biased codon usage in nitrogenase genes of *Clostridium pasteurianum*. J. Bacteriol. 166:162–172.

Chen, J. S., and Mortenson, L. E. 1974. Purification and properties of hydrogenase from *Clostridium pasteurianum* W 5. Biochim. Biophys. Acta 271:283–298.

Chenas, N. K., Vanozhinskis, Y. V., and Kulis, Y. Y. 1987. Kinetic principles of *Clostridium kluyveri* diaphorase. Biochem. (USSR), 52:57–63.

Chester, F. D. 1901. A manual of determinative bacteriology, p. 295–394. Macmillan, New York.

Chiba, S., and Ishimoto, M. 1973. Ferredoxin-linked nitrate reductase from *Clostridium perfringens*. J. Biochem. 73:1315–1318.

Ching, W. M. 1986. Characterization of selenium-containing tRNA$_{GLU}$ from *Clostridium sticklandii*. Arch. Biochem. Biophys. 244:137–146.

Chirpich, T. P., Zappia, V., Costilow, R. N., and Barker, H. A. 1970. Lysine 2,3-aminomutase. Purification and properties of a pyridoxal phosphate and S-adenosyl-methione-activated enzyme. J. Biol. Chem. 245:1778–1789.

Cho, K. Y., and Doy, C. H. 1973. Ultrastructure of the obligately anaerobic bacteria *Clostridium kluyveri* and *C. acetobutylicum*. Austral. J. Biol. Sci. 26:547–558.

Chung, K. T. 1976. Inhibitory effect of H_2 on growth of *Clostridium cellobioparum*. Appl. Environ. Microbiol. 31:342–348.

Clark, S. W., Bennet, G. N., and Rudolph, F. B. 1989. Isolation and characterization of mutants of *Clostridium acetobutylicum* ATCC 824 deficient in acetoacetyl-coenzyme A: acetate/butyrate: coenzyme A-transferase (CC 2.8.3.9.) and in other solvent pathway enzymes. Appl. Environ. Microbiol. 55:970–976.

Clarke, D. J., Fuller, F. M., and Morris, J. G. 1979. The membrane adenosine triphosphatase of *Clostridium pasteurianum*. Effects of key intermediates of glycolysis on its ATP phosphohydrolase activity. FEBS Lett. 100:52–56.

Clarke, D. J., and Morris, J. G. 1976a. Butyricin 7423: A bacteriocin produced by *Clostridium butyricum* NCIB 7423. J. Gen. Microbiol. 95:67–77.

Clarke, D. J., and Morris, J. G. 1976b. Partial purification of a dicyclohexylcarbodiimide-sensitive membrane adenosine triphosphatase complex from the obligately anaerobic bacterium *Clostridium pasteurianum*. Biochem. J. 154:725–729.

Clarke, D. J., and Morris, J. G. 1980. The mother-cell-membrane adenosine triphosphatase of sporulating *Clostridium pasteurianum*. Biochem. J. 186:191–199.

Clarke, D. J., Robson, R. M., and Morris, J. G. 1975. Purification of two *Clostridium* bacteriocins by procedures appropriate to hydrophobic proteins. Antimicrob. Agents Chemother. 7:256–264.

Conrad, R., Bak, F., Seitz, H. J., Thebrath, B., Mayer, H. P., and Schütz, H. 1989. Hydrogen turnover by psychrotrophic homoacetogenic and mesophilic methanogenic bacteria in anoxic paddy soil and lake sediment. FEMS Microbiol. Ecol. 62:285–294.

Cooper, D. G., Zajic, J. E., Gerson, D. F., and Manninen, K. I. 1980. Isolation and identification of biosurfactants produced during anaerobic growth of *Clostridium pasteurianum*. J. Ferment. Technol. 58:83–86.

Corry, J. E. L. 1978. Possible sources of ethanol ante- and postmortem: Its relationship to the biochemistry and microbiology of decomposition. J. Appl. Bacteriol. 44:1–56.

Costilow, R. N. 1977. Selenium requirement for the growth of *Clostridium sporogenes* with glycine as the oxidant in Stickland reaction. J. Bacteriol. 131:366–368.

Cousminer, J. J., Fischl, A. S., and Carman, G. M. 1982. Partial purification and properties of phosphatidylserine synthase from *Clostridium perfringens*. J. Bacteriol. 151:1372–1379.

Cozzani, I., Barsacchi, R., Dibenedetto, G., Saracchi, L., and Falcone, G. 1975. Regulation of breakdown and synthesis of L-glutamate decarboxylase in *Clostridium perfringens*. J. Bacteriol. 123:1115–1123.

Cramer, S. P., Hodgson, K. O., Gillum, W. O., and Mortenson, L. E. 1978. The Mo-site of nitrogenase. J. Am. Chem. Soc. 100:3398–3407.

Creuzet, N., Berenger, J.-F., and Frixon, C. 1983. Characterization of exoglucanase and synergistic hydrolysis of cellulose in *Clostridium stercorarium*. FEMS Microbiol. Lett. 20:347–350.

Creuzet, N., and Frixon, C. 1983. Purification and characterization of an endoglucanase from a newly isolated thermophilic anaerobic bacterium. Biochimie 65:149–156.

Cueto, P. H., and Méndez, B. S. 1990. Direct selection of *Clostridium acetobutylicum* fermentation mutants by a proton suicide method. Appl. Environ. Microbiol. 56:578–580.

Cummins, C. S., and Johnson, J. L. 1971. Taxonomy of the clostridia: Wall composition and DNA homologies in *Clostridium butyricum* and other butyric acid-producing clostridia. J. Gen. Microbiol. 67:33–46.

Curthoys, N. P., Scott, J. M., and Rabinowitz, J. C. 1972. Folate coenzymes of *Clostridium acidiurici*. The isolation of (1)-15, 10-methenyltetrahydropteroyltriglutamate, its conversion to tetrahydropteroyltriglutamate and (1)-10 [14C] formyltetrahydropteroyltriglutamate, and the synthesis of (1)-10-formyl-[6,7-^{3}H$_2$] tetrahydropteroyltriglutamate and (1)-[6, 7-^{3}H$_2$] tetrahydropteroyltriglutamate. J. Biol. Chem. 247:1959–1964.

Cynkin, M. A., Gibbs, M. 1958. Metabolism of pentoses by clostridia. II. The fermentation of C_{14}-labeled pentoses by *Clostridium perfringens*, *C. beijerinckii*, *C. butyricum*. J. Bacteriol. 75:335–338.

Dainty, R. H. 1970. Purification and properties of threonine aldolase from *Clostridium pasteurianum*. Biochem. J. 117:585–592.

Dainty, R. H. 1972. Glutamate biosynthesis in *Clostridium pasteurianum* and its significance in nitrogen metabolism. Biochem. J. 126:1055–1056.

Dainty, R. H., and Peel, J. L. 1970. Biosynthesis of amino acids in *Clostridium pasteurianum*. Biochem. J. 117:573–584.

Dams, E., Huysmans, E., Vandenberghe, A., and De Wachter, R. 1987. Structure of clostridial 5S ribosomal RNAs and bacterial evolution. Syst. Appl. Microbiol. 9:54–61.

Daniel, S. L., Wu, Z., and Drake, H. L. 1988. Growth of thermophilic acetogenic bacteria on methoxylated aromatic acids. FEMS Microbiol. Lett. 52:25–28.

Darvill, A. G., Hall, M. A., Fish, J. P., and Morris, J. G. 1977. The intracellular reserve polysaccharide of *Clostridium pasteurianum*. Can. J. Microbiol. 23:947–953.

Das, A., Hugenholtz, H., van Halbeek, H., and Ljungdahl, L. G. 1989. Structure and function of a menaquinone involved in electron transport in membranes of *Clostridium thermoaceticum* and *Clostridium thermoautotrophicum*. J. Bacteriol. 171:5823–5829.

Datta, R., and Zeikus, J. G. 1985. Modulation of acetone-butanol-ethanol fermentation by carbon monoxide and organic acids. Appl. Environ. Microbiol. 49:522–529.

Davies, R., and Stephenson, M. 1941. Studies on the acetone-butyl alcohol fermentation. I. Nutritional and other factors involved in the preparation of active suspensions of *Clostridium acetobutylicum* (Weizmann). Biochem. J. 35:1320–1331.

Davis, J. A. 1969. Isolation and identification of clostridia from North Sea sediments. J. Appl. Bacteriol. 32:164–169.

Davis, J. N., and Stadtman, T. C. 1985. Purification and properties of a quinone-dependent *p*-nitrophenylphosphatase from *Clostridium sticklandii*. Arch. Biochem. Biophys. 239:523–530.

Day, J. I. E., and Goldfine, H. 1978. Partial purification and properties of acyl-CoA reductase from *Clostridium butyricum*. Arch. Biochem. Biophys. 190:322–331.

Debeire, P., Priem, B., Strecker, G., and Vignon, M. 1990. Purification and properties of an endo-1,4-xylanase excreted by hydrolytic thermophilic anaerobe, *Clostridium thermolacticum*. Eur. J. Biochem. 187:573–580.

Decker, K., Barth, C., and Metz, H. 1966. Die Kohlenhydratsynthese in *Clostridium kluyveri*. II. Enzymatische Studien. Biochem. Z. 345:472–492.

Decker, K., and Pfitzer, S. 1972. Determination of steady-state concentrations of adenine nucleotides in growing *Clostridium kluyveri* cells by biosynthetic labeling. Anal. Biochem. 50:529–539.

Dehning, I., and Schink, B. 1989. Two new species of anaerobic oxalate-fermenting bacteria, *Oxalobacter vibrioformis* sp. nov. and *Clostridium oxalicum* sp. nov., from sediment samples. Arch. Microbiol. 153:79–84.

Deo, Y. M., Bryant, R. D., and Laishley, E. J. 1979. Differences in cellular fatty acids and lipid composition in *Clostridium pasteurianum* unter nitrogen- and non-nitrogen-fixing conditions. Curr. Microbiol. 3:55–58.

Descrozailles, J., Bory, J., Chaigneau, M., and Labarre, C. 1974. Composition des gaz dégagés par *Plectridium putrificum* cultivé en différents milieux synthétiques. Ann. Pharmaceut. Francaises 32:19–23.

DeVries, G. H., and Binkley, S. B. 1972. *N*-Acetylneuraminic acid aldolase of *Clostridium perfringens*. Purification, properties and mechanism of action. Arch. Biochem. Biophys. 151:234–242.

Dietrichs, D., and Andreesen, J. R. 1990. Purification and comparative studies on dihydrolipoamide dehydrogenase from anaerobic, glycine-utilizing bacteria *Peptostreptococcus glycinophilus*, *Clostridium cylindrosporum*, and *Clostridium sporogenes*. J. Bacteriol. 172:243–251.

Dilworth, G. L. 1982. Properties of the selenium-containing moiety of nicotinic acid hydroxylase from *Clostridium barkeri*. Arch. Biochem. Biophys. 219:30–38.

Dilworth, G. L. 1983. Occurrence of molybdenum in the nicotinic acid hydroxylase from *Clostridium barkeri*. Arch. Biochem. Biophys. 221:565–569.

Dimroth, P., Buckel, W., Loyal, R., and Eggerer, H. 1977. Isolation and function of the subunits of citramalate lyase and formation of hybrids with the subunits of citrate lyase. Eur. J. Biochem. 80:469–477.

Dixon, N. M., Lovitt, R. W., Kell, D. B., and Morris, J. G. 1987. Effects of p CO_2 on the growth and metabolism of *Clostridium sporogenes* NCIB 8053 in defined media. J. Appl. Bacteriol. 63:171–182.

Dixon, N. M., Lovitt, R. W., Morris, J. G., and Kell, D. B. 1988. Growth energetics of *Clostridium sporogenes* NCIB 8053: modulation by CO_2. J. Appl. Bacteriol. 65:119–133.

Döll, W. 1973. Untersuchungen ueber die DNase-Bildung von Clostridien. Zentralbl. Bakteriol. Parasitenkd. Infektionskr. Hyg. Abt. 1 Orig., Reihe A 224:115–119.

Dorn, M., Andreesen, J. R., and Gottschalk, G. 1978a. Fermentation of fumarate and L-malate by *Clostridium formicoaceticum*. J. Bacteriol. 133:26–32.

Dorn, M., Andreesen, J. R., and Gottschalk, G. 1978b. Fumarate reductase of *Clostridium formicoaceticum*. A peripheral membrane protein. Arch. Microbiol. 119:7–11.

Douglas, F., Hambleton, R., and Rigby, G. J. 1973. An investigation of the oxidation reduction potential and of the effect of oxygen on the germination and outgrowth of *Clostridium butyricum* spores, using platinum electrodes. J. Appl. Bacteriol. 36:625–633.

Duda, V. I., Lebedinsky, A. V., Mushegjan, M. S., and Mitjushina, L. L. 1987. A new anaerobic bacterium, forming up to five endospores per cell—*Anaerobacter polyendosporus* gen. et spec. nov. Arch. Microbiol. 148:121–127.

Duda, V. I., and Makareva, E. D. 1977. Morphogenesis and function of gas caps on spores of anaerobic bacteria of the genus *Clostridium*. Microbiology (English Translation of Mikrobiologiya) 46:563–569.

Dührsen, W. 1937. Untersuchungen über das Vorkommen von anaeroben Sporenbildnern in Milch unter Berücksichtigung ihrer sonstigen hygienischen Beschaffenheit. Zentralbl. Bakteriol. Parasitenkd. Infektionskr. Hyg. Abt. 2, 96:35–74.

Duncan, C. L., Rokos, E. A., Christenson, C. M., and Rood, J. I. 1978. Multiple plasmids in different toxigenic types of *Clostridium perfringens*—Possible control of beta-toxin production, p. 246–248. In: Schlessinger, D. (ed.), Microbiology, 1978. Am. Soc. Microbiology, Washington, DC.

Duncan, C. L., and Strong, D. H. 1968. Improved medium for sporulation of *Clostridium perfringens*. Appl. Microbiol. 16:82–89.

Dürre, P., Andersch, W., and Andreesen, J. R. 1981. Isolation and characterization of an adenine-utilizing, anaerobic sporeformer, *Clostridium purinolyticum* sp. nov. Int. J. Syst. Bacteriol. 31:184–194.

Dürre, P., and Andreesen, J. R. 1982a. Anaerobic degradation of uric acid via pyrimidine derivatives by selenium-starved cells of *Clostridium purinolyticum*. Arch. Microbiol. 131:255–260.

Dürre, P., and Andreesen, J. R. 1982b. Pathway of carbon dioxide reduction to acetate without a net energy requirement in *Clostridium purinolyticum*. FEMS Microbiol. Lett. 15:51–56.

Dürre, P., and Andreesen, J. R. 1982c. Selenium-dependent growth and glycine fermentation by *Clostridium purinolyticum*. J. Gen. Microbiol. 128:1457–1466.

Dürre, P., and Andreesen, J. R. 1982d. Separation and quantitation of purines and their anaerobic and aerobic degradation products by high-pressure liquid chromatography. Anal. Biochem. 123:32–40.

Dürre, P., and Andreesen, J. R. 1983. Purine and glycine metabolism by purinolytic clostridia. J. Bacteriol. 154:192–199.

Dürre, P., Kuhn, A., and Gottschalk, G. 1986. Treatment with allyl alcohol selects specifically for mutants of *Clostridium acetobutylicum* defective in butanol synthesis. FEMS Microbiol. Lett. 36:77–81.

Dürre, P., Kuhn, A., Gottwald, M., and Gottschalk, G. 1987. Enzymatic investigations on butanol dehydrogenase and butyraldehyde dehydrogenase in extracts of *Clostridium acetobutylicum*. Appl. Microbiol. Biotechnol. 26:268–272.

du Toit, P. J., Potgieter, D. J. J., and de Villiers, V. 1972. A study of the properties of 1-phosphofructokinase isolated from *Clostridium pasteurianum*. Enzymologia 43:285–300.

Dyer, J. K., and Costilow, R. N. 1968. Fermentation of ornithine by *Clostridium sticklandii*. J. Bacteriol. 96:1617–1622.

Edenharder, R., and Pfützner, M. 1989. Partial purification and characterization of an NAD-dependent 3β-hydroxysteroid dehydrogenase from *Clostridium innocuum*. Appl. Environ. Microbiol. 55:1656–1659.

Edwards, D. I., Dye, M., and Carne, H. 1973. The selective toxicity of antimicrobial nitroheterocyclic drugs. J. Gen.Microbiol. 76:135–145.

Edwin, E. E., Shreeve, J. E., and Jackman, R. 1978. A rapid colony test for thiaminase activity. J. Appl. Bacteriol. 44:305–312.

Eggerer, H. 1985. Completion of the degradation scheme for nicotinic acid by *Clostridium barkeri*. Curr. Top. Cell. Regulation 26:411–418.

Egorov, N. S., Loria, Z. K., and Vlasova, O. S. 1972. Effect of various carbon sources on synthesis of proteolytic enzymes by acetobutylic bacteria. Microbiology (English translation of Mikrobiologiya) 41:208–211.

El Ghazzawi, E. 1967. Neuisolierung von *Clostridium aceticum* Wieringa und stoffwechselphysiologische Untersuchungen. Arch. Mikrobiol. 57:1–19.

Elliott, B. B., and Mortenson, L. E. 1976. Regulation of molybdate transport by *Clostridium pasteurianum*. J. Bacteriol. 127:770–779.

Elliot, B. B., and Mortenson, L. E. 1977. Molybdenum storage component from *Clostridium pasteurianum*, p. 205–217. In: Newlon, W., Postgate, J. R., and Rodriquez-Barrueco, C. (ed.), Recent developments in nitrogen fixation. Academic Press, London.

Ellner, P. D. 1956. A medium promoting rapid quantitative sporulation in *Clostridium perfringens*. J. Bacteriol. 71:495–496.

Elsden, S. R., and Hilton, M. G. 1978. Volatile acid production from threonine, valine, leucine and isoleucine by clostridia. Arch. Microbiol. 117:165–172.

Elsden, S. R., and Hilton, M. G. 1979. Amino acid utilization patterns in clostridial taxonomy. Arch. Microbiol. 123 :137–141.

Elsden, S. R., Hilton, M. G., Parsley, K. R., and Self, R. 1980. The lipid fatty acids of proteolytic clostridia. J. Gen. Microbiol. 129:1075–1081.

Elsden, S. R., Hilton, M. G., Waller, J. M. 1976. The end products of the metabolism of aromatic amino acids by clostridia. Arch. Microbiol. 107:283–288.

Emtsev, V. T. 1963. Sporulation in *Clostridium pasteurianum*. Mikrobiologiya 32:434–438.

Ennis, B. M., Gutierrez, N. A. and Maddox, I. S. 1986. The acetone-butanol-ethanol fermentation: A current assessment. Process Biochem. October:131–147.

Ensley, B., McHugh, J. J., and Barton, L. L. 1975. Effect of carbon sources on formation of α-amylase and glucoamylase by *Clostridium acetobutylicum*. J. Gen. Appl. Microbiol. 21:51–59.

Esaki, N., Seraneeprakarn, V., Tanaka, H., and Soda, K. 1988. Purification and characterization of *Clostridium sticklandii* D-selenocystine α,β-lyase. J. Bacteriol. 170:751–756.

Fellers, C. R., and Clough, R. W. 1925. Indol and skatole determination of bacterial cultures. J. Bacteriol. 10:105–133.

Ferrari, A., and Aragozzini, F. 1972. Attivita di un ceppo di *Clostridium bifermentans* su alcuni acidi biliari. Ann.Microbiol. Enzimol. 22:131–136.

Ferrari. A., Scolastico, C., and Beretta, L. 1977. On the mechanism of cholic acid 7α-dehydroxylation by a *Clostridium bifermentans* cell-free extract. FEBS Letters 75:166–168.

Finegold, S. M., Sutter, V. L., and Mathisen, G. E. 1983. Normal indigenous intestinal flora, p. 3–31. In: Hentges, D. J. (ed.), Human Intestinal microflora in health and disease. Academic Press, New York.

Finne, G., and Matches, J. R. 1974. Low-temperature-growing clostridia from marine sediment. Can. J. Microbiol. 20:1639–1645.

Fleming, R. W., and Quinn, L. Y. 1971. Chemically defined medium for growth of *Clostridium thermocellum*, a cellulolytic thermophilic anaerobe. Appl. Microbiol. 21:967.

Fontaine, F., Peterson, W. H., McCoy, E., Johnson, M. J., and Ritter, G. J. 1942. A new type of glucose fermentation by *Clostridium thermoaceticum*, n. sp. J. Bacteriol. 43:701–715.

Forsberg, C. W. 1987. Production of 1,3-propanediol from glycerol by *Clostridium acetobutylicum* and other *Clostridium* species. Appl. Environ. Microbiol. 53:639–643.

Fox, G. E., and Stackebrandt, E. 1987. The application of 16S RNA cataloguing and 5S rRNA sequencing in bacterial systematics, p. 405–458. In: R. R. Colwell and R. Grigorova (ed.), Meth. Microbiol., vol. 19. Academic Press, Orlando, FL.

Fraisse, L., and Simon, H. 1988. Observations on the reduction of non-activated carboxylates by *Clostridium formicoaceticum* with carbon monoxide or formate and the influence of various viologens. Arch. Microbiol. 150:381–386.

Freame, B., and Fitzpatrick, B. W. 1971. The use of differential reinforced clostridial medium for the isolation and enumeration of clostridia from food, p. 49–55. In: Shapton, D. A., and Board, R. G. (ed.), Isolation of anaerobes. Academic Press, London.

Fredricks, W. W., and Stadtman, E. R. 1965. The role of ferredoxin in the hydrogenase system from *Clostridium kluyveri*. J. Biol. Chem. 240:4065–4071.

Freier, D., and Gottschalk, G. 1987. L(+)-Lactate dehydrogenase of *Clostridium acetobutylicum* is activated by fructose-1,6-bisphosphate. FEMS Microbiol. Lett. 43:229–233.

Freier, D., Mothershed, C. P., and Wiegel, J. 1988. Characterization of *Clostridium thermocellum* JW20. Appl. Environ. Microbiol. 54:204–211.

Freier-Schröder, D., Wiegel, J., and Gottschalk, G. 1989. Butanol formation by *Clostridium thermosaccharolyticum* at neutral pH. Biotechnol. Lett. 11:831–836.

Fryer, T. F., and Halligan, A. C. 1976. The detection of *Clostridium tyrobutyricum* in milk. New Zealand J. Dairy Sci. Technol. 11:132.

Fryer, T. F., and Mead, G. C. 1979. Development of a selective medium for the isolation of *Clostridium sporogenes* and related organisms. J. Appl. Bacteriol. 47:425–431.

Fuchs, G. 1986. CO_2 fixation in acetogenic bacteria: variations on a theme. FEMS Microbiol. Rev. 39:181–213.

Fuchs, G., Schnitker, U., and Thauer, R. K. 1974. Carbon monoxide oxidation by growing cultures of *Clostridium pasteurianum*. Eur. J. Biochem. 49:111–115.

Fujino, T., Sukhumvasi, J., Sasaki, T., Ohmiya, K., and Shimizu, S. 1989. Purification and properties of an endo-1,4-β-glucanase form *Clostridium josui*. J. Bacteriol. 171:4076–4079.

Gälli, R., and McCarty, P. L. 1989. Biotransformation of 1,1,1-trichloroethane, trichloromethane, and tetrachloromethane by a *Clostridium* sp.. Appl. Environ. Microbiol. 55:837–844.

Gariboldi, R. T., and Drake, H. L. 1984. Glycine synthase of the purinolytic bacterium *Clostridium acidiurici*. Purification of the glycine-CO_2 exchange system. J. Biol. Chem. 259:6085–6089.

Garnier, T., and Cole, S. T. 1988. Complete nucleotide sequence and genetic organization of the bacteriocinogenic plasmid, pIP 404, from *Clostridium perfringens*. Plasmid 19:134–150.

Gaston, L. W., and Stadtman, E. R. 1963. Fermentation of ethylene glycol by *Clostridium glycolicum*, sp. n. J. Bacteriol. 85:356–362.

George, H. A., and Chen, J. S. 1983. Acidic conditions are not obligatory for onset of butanol formation by *Clostridium beijerinckii* (synonym, *C. butylicum*). Appl. Environ. Microbiol. 46:321–327

George, H. A., Johnson, J. L., Moore, W. E. C., Holdeman, L. V., and Chen, J. S. 1983. Acetone, isopropanol, and butanol production by *Clostridium beijerinckii* (syn. *Clostridium butylicum*) and *Clostridium aurantibutyricum*. Appl. Environ. Microbiol. 45:1160–1163.

Germain, P., Toukourou, F., and Donaduzzi, L. 1986. Ethanol production by anaerobic thermophilic bacteria: regulation of lactate dehydrogenase activity in *Clostridium thermohydrosulfuricum*. Appl. Microbiol. Biotechnol. 24:300–305.

Gibbs, B. M., and Freame, B. 1965. Methods for the recovery of clostridia from foods. J. Appl. Bacteriol. 28:95–111.

Gibbs, B. M., and Hirsch, A. 1956. Spore formation by *Clostridium* species in an artificial medium. J. Appl. Bacteriol. 19:129–141.

Gibson, T. 1965. Clostridia in silage. J. Appl. Bacteriol. 28:56–62.

Gibson, T., Stirling, A. C., Keddi, R. M., and Rosenberger, R. F. 1958. Bacteriological changes in silage made at controlled temperatures. J. Gen. Microbiol. 19:112–129.

Giesel, H., Simon, H. 1983a. Immunological relationship of enoate reductases from different clostridia and the classification of *Clostridium* species La 1. FEMS Microbiol. Lett. 19:43–45.

Giesel, H., and Simon, H. 1983b. On the occurrence of enoate reductase and 2-oxo-carboxylate reductase in clostridia and some observations on the amino acid fermentation by *Peptostreptococcus anaerobius*. Arch. Microbiol. 135:51–57.

Gililland, J. R., and Vaughn, R. H. 1943. Characteristics of butyric acid bacteria from olives. J. Bacteriol. 46:315–322.

Glathe, H. 1934. Über die Rotte des Stalldüngers unter besonderer Berücksichtigung der Anaeroben-Flora. Zentralbl. Bakteriol. Parasitenkd. Infektionskr. Hyg. Abt. 2 91:65–101.

Goddard, P., Fernandez, F., West, B., Hill, M. J., and Barnes, P. J. 1975. The nuclear dehydrogenation of steroides by intestinal bacteria. J. Med. Microbiol. 8:429–435.

Goldfine, H., and Stadtman, E. R. 1960. Propionic acid metabolism. V. The conversion of β-alanine to propionic acid by cell-free extracts of *Clostridium propionicum*. J. Biol. Chem. 235:2238–2245.

Goldner, S. B., Solberg, M., and Post, L. S. 1985. Development of a minimal medium for *Clostridium perfringens* by using an anaerobic chemostat. Appl. Environ. Microbiol. 50:202–206.

Golovchenko, N. P., Belokopytov, B. F., and Akimenko, V. K. 1983. Threonine catabolism in the bacterium *Clostridium sticklandii*. Biochem. (USSR) 47:969–974.

Gombas, D. E., and Labbe, R. G. 1985. Purification and properties of spore-lytic enzymes from *Clostridium perfringens* type A spores. J. Gen. Microbiol. 131:1487–1496.

Gottschalk, G. 1986. Bacterial metabolism. New York, Heidelberg, Berlin: Springer Verlag.

Gottschalk, G., and Barker, H. A. 1966. Synthesis of glutamate and citrate by *Clostridium kluyveri*. A new type of citrate synthase. Biochem. 5:1125–1133.

Gottschalk, G., and Barker, H. A. 1967. Presence and stereospecificity of citrate synthase in anaerobic bacteria. Biochem. 6:1027–1034.

Gottschalk, G., and Dittbrenner, S. 1970. Properties of *(R)*-citrate synthase from *Clostridium acidiurici*. Hoppe-Seyler's Z. Physiol. Chem. 351:1183–1190.

Gottwald, M., Andreesen, J. R., Le Gall, J., and Ljungdahl, L. G. 1975. Presence of cytochrome and menaquinone in *Clostridium formicoaceticum* and *Clostridium thermoaceticum*. J. Bacteriol. 122:325–328.

Gottwald, M., and Gottschalk, G. 1985. The internal pH of *Clostridium acetobutylicum* and its effect on the shift from acid to solvent formation. Arch. Microbiol. 143:42–46.

Goudkov, A. V., and Sharp, M. E. 1965. Clostridia in dairying. J. Appl. Bacteriol. 28:63–73.

Goudkov, A. V., and Sharp, M. E. 1966. A preliminary investigation of the importance of clostridia in the production of rancid flavour in cheddar cheese. J. Dairy Res. 33:139–149.

Graves, M. C., Mullenbach, G. T., and Rabinowitz, J. C. 1985. Cloning and nucleotide sequence determination of the *Clostridium pasteurianum* ferredoxin gene. Proc. Natl. Acad. Sci. USA 82:1653–1657.

Gregory, E. M., Moore, W. E. C., and Holdeman, L. V. 1978. Superoxide dismutase in anaerobes: Survey. Appl. Environ. Microbiol. 35:988–991.

Groot, W. J., Schoutens, G. H., Van Beelen, P. N., Van den Oever, C. E., and Kossen, N. W. 1984. Increase of substrate conversion by pervaporation in the continuous butanol fermentation. Biotechnol. Lett. 6:789–792.

Häggström, L. 1986. Kinetics of product formation in batch and continuous culture of *Clostridium barkeri*. Appl. Microbiol. Biotechnol. 23:187–190.

Hall, J. D., McCroskey, L. M., Pincomb, B. J., and Hatheway. C. L. 1985. Isolation of an organism resembling *Clostridium barati* which produces type F botulinal toxin from an infant with botulism. J. Clin. Microbiol. 21:654–655.

Halligan, A. C., and Fryer, T. F. 1976. The development of a method for detecting spores of *Clostridium tyrobutyricum* in milk. New Zealand J. Dairy Sci. Technol. 11:100–106.

Hammann, R., and Ottow, J. C. G. 1976. Isolation and characterization of iron-reducing nitrogen-fixing saccharolytic clostridia from gley soils. Soil Biol. Biochem. 8:357–364.

Hammel, K. E., Cornwell, K. C., and Buchanan, B. B. 1983. Ferredoxin/flavoprotein-linked pathway for the reduction of thioredoxin. Proc. Natl. Acad. Sci. USA 80:3681–3685.

Hardman, J. K., and Stadtman, T. C. 1960a. Metabolism of ω-amino acids. I. Fermentation of γ-aminobutyric acid by *Clostridium aminobutyricum* n. sp.. J. Bacteriol. 79:544–548.

Hardman, J. K., and Stadtman, T. C. 1960b. Metabolism of ω-amino acids. II. Fermentation of δ-aminovaleric acid by *Clostridium aminovalericum* n. sp. J. Bacteriol. 79:549–552.

Hardman, J. K., and Stadtman, T. C. 1963a. Metabolism of ω-amino acids. IV. γ-Aminobutyrate fermentation by cell-free extracts of *Clostridium aminobutyricum*. J. Biol. Chem. 238:2088–2093.

Hardman, J. K., and Stadtman, T. C. 1963b. Metabolism of ω-amino acids. V. Energetics of the γ-aminobutyrate fermentation by *Clostridium aminobutyricum*. J. Bacteriol. 85:1326–1333.

Harris, J. N., and Hylemon, P. B. 1977. Purification and characterization of NADP-dependent 12-α-hydroxysteroid dehydrogenase from *Clostridium leptum* VPI 10900. Abstr. Ann. Meet. Am. Soc. Microbiol. 1977:197.

Harris, J., Mulden, R., Kell, D. B., Walter, R. P., and Morris, J. G. 1986. Solvent production by *Clostridium pasteurianum* in media of high sugar content. Biotechnol. Lett. 8:889–892.

Hartmanis, M. G. N., and Gatenbeck, S. 1984. Intermediary metabolism in *Clostridium acetobutylicum*: Levels of enzymes involved in the formation of acetate and butyrate. Appl. Environ. Microbiol. 46:1277–1283.

Hartmanis, M. G. N., and Stadtman, T. C. 1986. Diol metabolism and diol dehydratase in *Clostridium glycolicum*. Arch. Biochem. Biophys. 245:144–152.

Hartrampf, G., and Buckel, W. 1986. On the steric course of the adenosylcobalamin-dependent 2-methyleneglutarate mutase reaction in *Clostridium barkeri*. Eur. J. Biochem. 156:301–304.

Hartsell, S. E., and Rettger, L. F. 1934. A taxonomic study of *"Clostridium putrificum"* and its establishment as a definite entity *Clostridium lentoputrescens*, nov. spec. J. Bacteriol. 27:497–511.

Hasan, S. M., and Hall, J. B. 1975. The physiological function of nitrate reduction in *Clostridium perfringens*. J. Gen. Microbiol. 87:120–128.

Hasan, S. M., and Hall, J. B. 1976. Growth of *Clostridium tertium* and *Clostridium septicum* in chemically defined media. Appl. Environ. Microbiol. 31:442–443.

Hatheway, C. L. 1990. Toxigenic Clostridia. Clin. Microbiol. Rev. 3:66–98.

Hauduroy, P., Ehringer, G., Urbain, A., Guillot, G., and Magrou, J. 1937. Dictionnaire des bacteries pathogenes, p. 89–144. Masson et Cie, Paris.

Hauschild, A. H. W., and Holdeman, L. V. 1974. *Clostridium celatum* sp. nov., isolated from normal human feces. Int. J. Syst. Bacteriol. 24:478–481.

Hayase, M., Mitsui, N., Tamai, K., Nakamura, S., and Nishida, S. 1974. Isolation of *Clostridium absonum* and its cultural and biochemical properties. Infect. Immunity 9:15–19.

Hellinger, E. 1947. *Clostridium aurantibutyricum* (n. sp.): A pink butyric acid *Clostridium*. J. Gen. Microbiol. 1:203–210.

Henkin, J., and Abeles, R. H. 1976. Evidence against an acylenzyme intermediate in the reaction catalyzed by clostridial phosphotransacetylase. Biochem. 15:3472–3479.

Heritage, A. D., and MacRae, I. C. 1977. Degradation of lindane by cell-free preparations of *Clostridium sphenoides*. Appl. Environ. Microbiol. 34:222–224.

Hermann, M., Fayolle, F. Marchal, R., Podvin, L., Sebald, M., and Vandecasteele, J. P. 1985. Isolation and characterization of butanol-resistant mutants of *Clostridium acetobutylicum*. Appl. Environ. Microbiol. 50:1238–1243.

Herzberg, C., and Antranikian, G. 1986. Evidence for a unique pattern of citrate lyase inactivation in *Clostridium sphenoides*. Biochim. Biophys. Acta 871:107–120.

Herrero, A. A., and Gomez, R. F. 1980. Development of ethanol tolerance in *Clostridium thermocellum*: effect of growth temperature. Appl. Environ. Microbiol. 40:571–577.

Herrero, A. A., Gomez, R. F., and Roberts, M. F. 1982. Ethanol-induced changes in the membrane lipid composition of *Clostridium thermocellum*. Biochim. Biophys. Acta 693:195–204.

Herrero, A. A., Gomez, R. F., and Roberts, M. F. 1985. ^{31}P NMR studies of *Clostridium thermocellum*. J. Biol. Chem. 260:7442–7451.

Hespell, R. B., Joseph, R., and Mortlock, R. P. 1969. Requirement of coenzyme A in the phosphoroclastic reaction of anaerobic bacteria. J. Bacteriol. 100:1328–1334.

Hill, L. R. 1966. An index to deoxyribonucleic acid base compositions of bacterial species. J. Gen. Microbiol. 44:419–437.

Hillmer, P., and Gottschalk, G. 1974. Solubilization and partial characterization of particulate dehydrogenases from *Clostridium kluyveri*. Biochim. Biophys. Acta 334:12–23.

Hilton, M. G., Mead, G. C., and Elsden, S. R. 1975. The metabolism of pyrimidines by proteolytic clostridia. Arch. Microbiol. 102:145–149.

Himelbloom, B. H., and Canale-Parola, E. 1989. *Clostridium methylpentosum* sp. nov.: a ring-shaped intestinal bacterium that ferments only methylpentoses and pentoses. Arch. Microbiol. 151:287–293.

Hin, S. F., Zhu, C. X., Yan, R. T., and Chen, J. S. 1987. Butanol-ethanol dehydrogenase and butanol-ethanol-isopropanol dehydrogenase: different alcohol dehydrogenases in two strains of *Clostridium beijerinckii (Clostridium butylicum)*. Appl. Environ. Microbiol. 53:697–703.

Hinton, S. M., and Merritt, B. 1986. Purification and characterization of a molybdenum-pterin-binding protein (Mop) in *Clostridium pasteurianum* W5. J. Bacteriol. 168:688–693.

Hinton, S. M., and Mortenson, L. E. 1985a. Identification of molybdoproteins in *Clostridium pasteurianum*. J. Bacteriol. 162:477–484.

Hinton, S. M., and Mortenson, L. E. 1985b. Regulation and order of involvement of molybdoproteins during synthesis of molybdoenzymes in *Clostridium pasteurianum*. J. Bacteriol. 162:485–493.

Hinton, S. M., Slaughter, C., Eisner, W., and Fisher, W. 1987. The molybdenum-pterin binding protein is encoded by a multigene family in *Clostridium pasteurianum*. Gene 54:211–219.

Hirsch, A., and Grinsted, E. 1954. Methods for the growth and enumeration of anaerobic sporeformers from cheese, with observations on the effect of nisin. J. Dairy Res.21:101–110.

Hitzman, D. O., Halvorson, H. O., and Ukita, T. 1957. Requirements for production and germination of spores of anaerobic bacteria. J. Bacteriol. 74:1–7.

Holcenberg, J. S., and Tsai, L. 1969. Nicotinic acid metabolism. IV. Ferredoxin-dependent reduction of 6-hydroxynicotinic acid to 6-oxo-1,4,5,6-tetrahydronicotinic acid. J. Biol. Chem. 244:1204–1211.

Holdeman, L. V., Cato, E. P., and Moore, W. E. C. 1971. *Clostridium ramosum* (Vuillemin) comb. nov.: Emended description and proposed neotype strain. Int. J. Syst. Bacteriol. 21:35–39.

Holdeman, L. V., Cato, E. P., and Moore, W. E. C. 1977. Anaerobe laboratory manual, 4th ed. V.P.I. Anaerobe Laboratory, Virginia Polytechnic Institute and State University, Blacksburg, Virginia.

Holdeman, L. V., and Moore, W. E. C. 1974. New genus, *Coprococcus*, twelve new species, and emended descriptions of four previously described species of bacteria from human feces. Int. J. Syst. Bacteriol. 24:260–277.

Holland, D., Barker, A. N., and Wolf, J. 1970. The effect of carbon dioxide on spore germination in some clostridia. J. Appl. Bacteriol. 33:274–284.

Holland, K. T., and Cox, D. J. 1975. A synthetic medium for the growth of *Clostridium bifermentans*. J. Appl. Bacteriol. 38:193–198.

Hollaus, F., and Klaushofer, H. 1973. Identification of hyperthermophilic obligate anaerobic bacteria from extraction juices of beet sugar factories. Int. Sugar J. 75:237–241.

Hollaus, F., and Sleytr, U. 1972. On the taxonomy and fine structure of some hyperthermophilic saccharolytic clostridia. Arch. Mikrobiol. 86:129–146.

Höllriegl, V., Lamm, L., Rowold, J., Hörig, J., and Renz, P. 1982. Biosynthesis of vitamin B$_{12}$; different pathways in some aerobic and anaerobic microorganisms. Arch. Microbiol. 132:155–158.

Hoogerheide, J. C., and Kocholaty, W. 1938. Metabolism of the strict anaerobes (genus: *Clostridium*). II. Reduction of amino acids with gaseous hydrogen by suspensions of *Clostridium sporogenes*. Biochem. J. 32:949–957.

Horn, N. 1987. *Clostridium disporicum* sp. nov., a saccharolytic species able to form two spores per cell, isolated from rat cecum. Int. J. Syst. Bacteriol. 37:398–401.

Howell, N., Akagi, J. M., and Himes, R. H. 1969. Thermostability of glycolytic enzymes from thermophilic clostridia. Can. J. Microbiol. 15:461–464.

Hsiang, M. W., and Bright, H. J. 1969. β-Methylaspartase from *Clostridium tetanomorphum*, p. 347–353. In: Lowenstein, J. M. (ed.), Methods in enzymology, vol. 13. Academic Press, London.

Hsu, E. J., and Ordal, Z. J. 1969. Sporulation of *Clostridium thermosaccharolyticum*. Appl. Microbiol. 18:958–960.

Hsu, T., Daniels, S. L., Lux, M. F., and Drake, H. L. 1990. Biotransformations of carboxylated aromatic compounds by the acetogen *Clostridium thermoaceticum*: generation of growth-supportive CO equivalents under CO-limited conditions. J. Bacteriol. 172:212–217.

Hugenholtz, J., and Ljungdahl, L. G. 1989. Electron transport and electrochemical proton gradient in membrane vesicles of *Clostridium thermoautotrophicum*. J. Bacteriol. 171:2873–2875.

Hungate, R. E. 1944. Studies on cellulose fermentation. 1. The culture and physiology of an anaerobic cellulose digesting bacterium. J. Bacteriol. 48:499–513.

Hungate, R. E. 1950. The anaerobic mesophilic cellulolytic bacteria. Bacteriol. Rev. 14:1–49.

Hungate, R. E. 1969. A roll tube method for cultivation of strict anaerobes, p. 117–132. In: Norris, J. R., Ribbons, D. W. (ed.), Methods in microbiology, vol. 38. Academic Press, London.

Hyun, H. H., and Zeikus, J. G. 1985a. General biochemical characterization of thermostable extracellular β-amylase from *Clostridium thermosulfurogenes*. Appl. Environ. Microbiol. 49:1162–1167.

Hyun, H. H., and Zeikus, J. G. 1985b. General biochemical characterization of thermostable pullulanase and glucoamylase from *Clostridium thermohydrosulfuricum*. Appl. Environ. Microbiol. 49:1168–1173.

Hyun, H. H., Zeikus, J. G., Longin, R., Millet, J., and Ryter, A. 1983. Ultrastructure and extreme heat resistance of spores from thermophilic *Clostridium* species. J. Bacteriol. 156:1332–1337.

Imhoff, D., and Andreesen, J. R. 1979. Nicotinic acid hydroxylase from *Clostridium barkeri*. Selenium-dependent formation of active enzyme. FEMS Microbiol. Lett. 5:155–158.

Ingram, M. 1969. Sporeformers as food spoilage organisms, p. 549–610. In: Gould, G. W., and Hurst, A. (ed.), The bacterial spore. Academic Press, London.

Ishimoto, M., Umeyama, M., and Chiba, S. 1974. Alteration of fermentation products from butyrate to acetate by nitrate reduction in *Clostridium perfringens*. Z. Allg. Mikrobiol. 14:115–121.

Jagnow, G., Haider, K., and Ellwardt, P. C. 1977. Anaerobic dechlorination and degradation of hexachlorocyclohexane isomers by anaerobic and facultative anaerobic bacteria. Arch. Microbiol. 115:285–292.

Jain, M. K., and Zeikus, J. G. 1988. Taxonomic distinction of two new protein specific, hydrolytic anaerobes: Isolation and characterization of *Clostridium proteolyticum* sp. nov. and *Clostridium collagenovorans* sp. nov.. Syst. Appl. Microbiol. 10:134–141.

Jain, M. K., and Zeikus, J. G. 1989. Bioconversion of gelatin to methane by a coculture of *Clostridium collageno-*

vorans and *Methanosarcina barkeri.* Appl. Environ. Microbiol. 55:366–371.

James, E. W., Kell, D. B., Lovitt, R. W., and Morris, J. G. 1988. Electrosynthesis and electroanalysis using *Clostridium sporogenes.* Bioelectrochem. Bioenerg. 20:21–32.

Janssen, P. J., Jones, W. A., Jones, D. T., and Woods, D. R. 1988. Molecular analysis and regulation of the gln A gene of the Gram-positive anaerobe *Clostridium acetobutylicum.* J. Bacteriol. 170:400–408.

Jellet, J. J., Forrest, T. P., MacDonald, I. A., and Marrie, T. J. 1980. Production of indole-3-propanoic acid and 3-(*p*-hydroxyphenyl) propanoic acid by *Clostridium sporogenes:* a convenient thin-layer chromatography detection system. Can. J. Microbiol. 26:448–453.

Jeng, I. M., and Barker, H. A. 1974. Purification and properties of L-3-aminobutyryl coenzyme A deaminase from a lysine-fermenting *Clostridium.* J. Biol. Chem. 249:6578–6584.

Jeng, I. M., Somack, R., and Barker, H. A. 1974. Ornithine degradation in *Clostridium sticklandii,* pyridoxal phosphate and coenzyme A dependent thiolytic cleavage of 2-amino-4-ketopentanoate to alanine and acetyl coenzyme A. Biochemistry 13:2898–2903.

Jin, F., Yamasoto, K., and Toda, K. 1988. *Clostridium thermocopriae* sp. nov., a cellulolytic thermophile from animal feces, compost, soil, and a hot spring in Japan. Int. J. Syst. Bacteriol. 38:279–281.

Joblin, K. N., Johnson, A. W., Lappert, M. F., and Wallis, O. C. 1976. Coenzyme B_{12}-dependent reactions. Part IV. Observations on the purification of ethanolamine ammonia-lyase. Biochim. Biophys. Acta 452:262–270.

Johnson, E. A., Madia, A., and Demain, A. L. 1981. Chemically defined minimal medium for growth of the anaerobic cellulolytic thermophile *Clostridium thermocellum.* Appl. Environ. Microbiol. 41:1060–1062.

Johnson, J. L. 1970. Relationship of deoxyribonucleic acid homologies to cell wall structure. Int. J. Syst. Bacteriol. 20:421–424.

Johnson, J. L. 1973. Use of nucleic-acid homologies in the taxonomy of anaerobic bacteria. Int. J. Syst. Bacteriol. 23:308–315.

Johnson, J. L., and Francis, B. S. 1975. Taxonomy of the clostridia: Ribosomal ribonucleic acid homologies among the species. J. Gen. Microbiol. 88:229–244.

Johnston, N. C., and Goldfine, H. 1983. Lipid composition in the classification of the butyric acid-producing clostridia. J. Gen. Microbiol. 129:1075–1081.

Johnston, N. C., and Goldfine, H. 1988. Isolation and characterization of a novel four-chain ether lipid from *Clostridium butyricum:* the phosphatidylglycerol acetal of plasmenylethanolamine. Biochim. Biophys. Acta 961:1–12.

Johnston, R., Harmon, S., and Kautter, D. 1964. Method to facilitate the isolation of *Clostridium botulinum* type E. J. Bacteriol. 88:1521–1522.

Jones, D. T., and Woods, D. R. 1986. The acetone butanol fermentation revisited. Microbiol. Rev. 50:484–524.

Joliff, G., Edelman, A., Klier, A., and Rapoport, G. 1989. Inducible secretion of a cellulase from *Clostridium thermocellum* in *Bacillus subtilis.* Appl. Environ. Microbiol. 55:2739–2744.

Joliff, G., Béguin, P., and Aubert, J. P. 1986. Nucleotide sequence of the cellulase gene *cel*D encoding endo-

glucanase D of *Clostridium thermocellum.* Nucleic Acids Res. 14:8605–8613.

Jordan, D. C., and McNicol, P. J. 1979. A new nitrogen-fixing *Clostridium* species from a high Arctic ecosystem. Can. J. Microbiol. 25:947–948.

Jungermann, K., Kirchniawy, H., Katz, N., and Thauer, R. K. 1974. NADH, a physiological electron donor in clostridial nitrogen fixation. FEBS Lett. 43:203–206.

Jungermann, K., Leimenstoll, G., Rupprecht, E., and Thauer, R. K. 1971a. Demonstration of NADH-ferredoxin reductase in two saccharolytic clostridia. Arch. Mikrobiol. 80:370–372.

Jungerman, K., Rupprecht, E., Ohrloff, C., Thauer, R. K., and Decker, K. 1971b. Regulation of the reduced nicotinamide adenine dinucleotide-ferredoxin reductase system in *Clostridium kluyveri.* J. Biol. Chem. 246:960–963.

Jungerman, K., Thauer, R. K., and Decker, K. 1968. The synthesis of one-carbon units from CO_2 in *Clostridium kluyveri.* Eur. J. Biochem. 3:351–359.

Jungermann, K., Thauer, R. K., Leimenstoll, G., and Decker, K. 1973. Function of reduced pyridine nucleotide-ferredoxin oxidoreductases in saccharolytic clostridia. Biochim. Biophys. Acta 305:268–280.

Jungermann, K. A., Schmidt, W., Kirchniawy, F. H., Rupprecht, E. H., and Thauer, R. K. 1970. Glycine formation via threonine and serine aldolase. Its interrelation with the pyruvate formate lyase pathway of one-carbon unit synthesis in *Clostridium kluyveri.* Eur. J. Biochem. 16:424–429.

Kanamori, K., Weiss, R. L., and Roberts, J. D. 1989. Ammonia assimilation pathways in nitrogen-fixing *Clostridium kluyveri* and *Clostridium butyricum.* J. Bacteriol. 171:2148–2154.

Kaneuchi, C., Benno, Y., and Mitsuoka, T. 1976a. *Clostridium coccoides,* a new species from the feces of mice. Int. J. Syst.. Bacteriol. 26:482–486.

Kaneuchi, C., Miyazato, T., Shinjo, T., and Mitsuoka, T. 1979. Taxonomic study of helically coiled, sporeforming anaerobes isolated from the intestines of humans and other animals: *Clostridium cocleatum* sp. nov. and *Clostridium spiroforme* sp. nov. Int. J. Syst. Bacteriol. 29:1–12.

Kaneuchi, C., Watanabe, K., Terada, A., Benno, Y., and Mitsuoka, T. 1976b. Taxonomic study of *Bacteroides clostridiiformis* subsp. *clostridiiformis* (Burri and Ankersmit) Holdeman and Moore and of related organisms: Proposal of *Clostridium clostridiiformis* (Burri and Ankersmit) comb. nov. and *Clostridium symbiosum* (Stevens) comb. nov. Int. J. Syst. Bacteriol. 26:195–204.

Kaplan, F., Setlow, P., and Kaplan, N. O. 1969. Purification and properties of a DPHN-TPNH diaphorase from *Clostridium kluyveri.* Arch. Biochem. Biophys. 132:91–98.

Karlsson, J. L., Volcani, B. E., and Barker, H. A. 1948. The nutritional requirements of *Clostridium aceticum.* J. Bacteriol. 56:781–782.

Karube, I., Urano, N., Matsunaga, T., and Suzuki, S. 1982. Hydrogen production from glucose by immobilized growing cells of *Clostridium butyricum.* Eur. J. Appl. Microbiol. 16:5–9.

Kayano, H., Matsunaga, T., Karube, I., and Suzuki, S. 1981. Hydrogen evolution by co-immobilized *Chlorella vul-*

garis and *Clostridium butyricum* cells. Biochim. Biophys. Acta 638:80–85.

Kearny, J. J., and Sagers, R. D. 1972. Formate dehydrogenase from *Clostridium acidiurici*. J. Bacteriol. 109:152–161.

Kell, D. B., Peck, M. W., Rodger, G., and Morris, J. G. 1981. On the permeability to weak acids and bases of the cytoplasmic membrane of *Clostridium pasteurianum*. Biochem. Biophys. Res. Commun. 99:81–88.

Kellum, R., and Drake, H. L. 1986. Effects of carbon monoxide on one-carbon enzymes and energetics of *Clostridium thermoaceticum*. FEMS Microbiol. Lett. 34:41–45

Kelly, W. J., Asmundson, R. V., and Hopcroft, D. H. 1987. Isolation and characterization of a strictly anaerobic, cellulolytic sporeformer: *Clostridium chartatabidum* sp. nov. Arch. Microbiol. 147:169–173.

Kenaly, W. R., and Waselefsky, D. M. 1985. Studies on the substrate range of *Clostridium kluyveri;* the use of propanol and succinate. Arch. Microbiol. 141:187–194

Kerby, R., and Zeikus, J. G. 1983. Growth of *Clostridium thermoaceticum* on H_2/CO_2 or CO as energy source. Curr. Microbiol. 8:27–30.

Kim, B. H., Bellows, P., Datta, R., and Zeikus, J. G. 1984. Control of carbon and electron flow in *Clostridium acetobutylicum* fermentations: Utilization of carbon monoxide to inhibit hydrogen production and to enhance butanol yields. Appl. Environ. Microbiol. 48:764–770.

Kimura, R., and Liao, T. H. 1964. Taxonomic considerations on the *Clostridium thiaminolyticum* Kimura et Liao. Vitamins (Japan) 30:29–32.

Klaushofer, H., and Parkkinen, E. 1965. Zur Frage der Bedeutung aerober und anaerober thermophiler Sporenbildner als Infektionsursache in Rübenzuckerfabriken. 1. *Clostridium thermohydrosulfuricum*, eine neue Art eines saccharose-abbauenden, thermophilen, schwefelwasserstoffbildenden Clostridiums. Z. Zuckerind. 15:445–449.

Kleiner, D. 1979. Regulation of ammonium uptake and metabolism by nitrogen fixing bacteria. III. *Clostridium pasteurianum*. Arch. Microbiol. 120:263–270.

Kleiner, D., and Fitzke, E. 1979. Evidence for ammonia translocation by *Clostridium pasteurianum*. Biochem. Biophys. Res. Commun. 86:211–217

Klingeberg, M., Hippe, H., and Antranikian, G. 1990. Production of novel pullulanases at high concentrations of two newly isolated thermophilic clostridia. FEMS Microbiol. Lett. 69:145–152.

Knight, E., Jr., and Hardy, R. W. F. 1966. Isolation and characteristics of flavodoxin from nitrogen-fixing *Clostridium pasteurianum*. J. Biol. Chem. 241:2752–2756.

Knöll, H., and Horschak, R. 1973. Zur Ökologie der Gärungssarcinen *Sarcina ventriculi* und *Sarcina maxima*. Z. Allgem. Mikrobiol. 13:449–451.

Koch, R., Zablowski, P., and Antranikian, G. 1987. Highly active and thermostable amylases and pullulanases from various anaerobic thermophiles. Appl. Microbiol. Biotechnol. 27:192–198.

Kondo, H., Friedmann, H. C., and Vennesland, B. 1960. Flavin changes accompanying adaption of *Zymobacterium oroticum* to orotate. J. Biol. Chem. 235:1533–1535.

König, H., Buckel, W., and Langworthy, T. A. 1985. Ultrastructure of a cell envelope and amino acid composition of the murein of *Clostridium symbiosum*. FEMS Microbiol. Lett. 30:283–288.

Koransky, J. R., Allen, S. D., and Dowell, V. R., Jr. 1978. Use of ethanol for selective isolation of sporeforming microorganisms. Appl. Environ. Microbiol. 35:762–765.

Kotze, J. P. 1969. Glycolytic and related enzymes in clostridial classification. Appl. Microbiol. 18:744–747.

Kowal, A. T., Adams, M. W. W., and Johnson, M. K. 1989. Electron paramagnetic resonance studies of the low temperature photolytic behavior of oxidized hydrogenase I from *Clostridium pasteurianum*. J. Biol. Chem. 264:4342–4348.

Krafft, A. E., and Hylemon, P. B. 1989. Purification and characterization of a novel form of 20α-hydroxysteroid dehydrogenase from *Clostridium scindens*. J. Bacteriol. 171:2925–2932.

Krafft, A. E., Winter, J., Bokkenhauser, V. D., and Hylemon, P. B. 1987. Cofactor requirements of steroid-17–20-desmolase and 20α-hydroxysteroid dehydrogenase activities in cell extracts of *Clostridium scindens*. J. Steroid. Biochem. 28:49–54.

Krasilnikov, N. A., Pivovarov, G. E., and Duda, V. I. 1971 Physiological properties of anaerobic soil bacteria which form vesicular caps on their spores. Microbiology (English translation of Mikrobiologiya) 40:783–788.

Krause, G., and Simon, H. 1989. Design and applications of sensitive enzyme immuno assays specific for clostridial enoate reductases. Z. Naturforsch. 44c:345–352.

Krumholz, L. R., and Bryant, M. P. 1985. *Clostridium pfennigii* sp. nov., uses methoxyl groups of monobenzenoids and produces butyrate. Int. J. Syst. Bacteriol. 35:454–456.

Kuchta, R. D., Hanson, G. R., Holmquist, B., and Abeles, R. H. 1986. Fe-S centers in lactyl-CoA dehydratase. Biochem. 25:7301–7307.

Kuno, S., Bacher, A., and Simon, H. 1985. Structure of enoate reductase from a *Clostridium tyrobutyricum* (*C. spec.* La1). Hoppe-Seyler's Biol. Chem. 366:463–472.

Kuramitsu, H. K., and Watson, R. M. 1973. Regulation of aspartokinase activity in *Clostridium perfringens*. J. Bacteriol. 115:882–888.

Kurose, N., Kinoshita, S., Yagyu, J., Uchida, M., Hanai, S., and Obayashi, A. 1988. Improvement of ethanol production of thermophilic *Clostridium* sp. by mutation. J. Ferment. Technol. 66:467–472.

Kürsteiner, J. 1907. Beiträge zur Untersuchungstechnik obligat anaerober Bakterien, sowie zur Lehre der Anaerobiose überhaupt. Zentralbl. Bakteriol. Abt. 2 19:1–26, 97–115, 202–220, 385–394.

Kutzenock, A., and Aschner, M. 1952. Degenerative processes in a strain of *Clostridium butylicum*. J. Bacteriol. 64:829–836.

Kutzner, H. J. 1963. Untersuchungen an Clostridien mit besonderer Berücksichtigung der für die Milchwirtschaft wichtigen Arten. Zentralbl. Bakteriol. Parasitenkd. Infektionskr. Hyg. Abt. 1 191:441–450.

Kutzner, H. J. 1965. Prinzipien der Anreicherung und Isolierung von Clostridien. Zentralbl. Bakteriol. Parasitenkd. Infektionskr. Hyg. Abt. 1, Suppl. 1:363–394.

Kutzner, H. J. 1966. Untersuchungen über die Buttersäuregärung in Schnitt- und Hartkäse. 17. Int. Milchwirtschafts-Kongress München, Sektion D2, pp. 647–658.

Kyrtopoulos, S. A., and Satchell, D. P. N. 1972. The roles of univalent cations during catalysis by phosphate acetyl-transferase derived from *Clostridium kluyveri*. Biochem. J. 129:1163–1166.

Laanbroek, H. J., Smit, A. J., Klein Nulend, G., and Veldkamp, H. 1979. Competition for L-glutamate between specialised and versatile *Clostridium* species. Arch. Microbiol. 120:61–66.

Laishley, E. J. 1975. Regulation and properties of an invertase from *Clostridium pasteurianum*. Can. J. Microbiol. 21:1711–1718.

Laishley, E. J., and Krouse, H. R. 1978. Stable isotope fractionation by *Clostridium pasteurianum*. 2. Regulation of sulfite reductases by sulfite amino acids and their influence on sulfur isotope fractionation during SO_3^{2-} and SO_4^{2-} reduction. Can. J. Microbiol. 24:716–724.

Laishley, E. J., Lin, P. M., and Peck, H. D. 1971. A ferredoxin-linked sulfite reductase from *Clostridium pasteurianum*. Can. J. Microbiol. 17:889–895.

Lamed, R., and Bayer, E. A. 1988. The cellulosome of *Clostridium thermocellum*. Adv. Appl. Microbiol. 33:1–46.

Lamed, R., and Zeikus, J. G. 1981. Thermostable ammonium-activated malic enzyme of *Clostridium thermocellum*. Biochim. Biophys. Acta 660:251–255.

Lampen, J. H., and Peterson, E. H. 1943. Growth factor requirements of clostridia. Arch. Biochem. 2:443–449.

Langworthy, T. A., Holzer, G., Lovitt, R. W., and Zeikus, J. G. 1989. A unique C30 dicarboxylic fatty acid from *Clostridium thermohydrosulfuricum*. Ann. Meet. Am. Soc. Microbiol., Abstract K-20, p. 248.

Leaver, F. W., Wood, H. G., and Stjernholm, R. 1955. The fermentation of three carbon substrates by *Clostridium propionicum* and propionibacterium. J. Bacteriol. 70:521–530.

Lee, C. K., Dürre, P., Hippe, H., and Gottschalk, G. 1987. Screening for plasmids in the genus *Clostridium*. Arch. Microbiol. 148:107–114.

Lee, C. K., and Ordal, Z. J. 1967. Regulatory effect of pyruvate on the glucose metabolism of *Clostridium thermosaccharolyticum*. J. Bacteriol. 94:530–536.

Lee, W.-K., Fujisawa, T., Kawamura, S., Itoh, K., and Mitsuoka, T. 1989. *Clostridium intestinalis* sp. nov., an aerotolerant species isolated from the feces of cattle and pigs. Int. J. Syst. Bacteriol. 39:334–336.

Leonhardt, U., and Andreesen, J. R. 1977. Some properties of formate dehydrogenase, accumulation and incorporation of [185]W-tungsten into proteins of *Clostridium formicoaceticum*. Arch. Microbiol. 115:277–284.

Le Ruyet, P., Dubourguier, H. C., and Albagnac, G. 1984. Thermophilic fermentation of cellulose and xylan by methanogenic enrichment cultures: preliminary characterization of main species. Syst. Appl. Microbiol. 5:247–253.

Le Ruyet, P., Dubourguier, H. C., Albagnac, G., and Prensier, G. 1985. Characterization of *Clostridium thermolacticum* sp. nov., a hydrolytic thermophilic anaerobe producing high amounts of lactate. Syst. Appl. Microbiol. 6:196–202.

Levett, P. N. 1987. Production of *p*-cresol by *Clostridium difficile* on different basal media. Lett. Appl. Microbiol. 5:71–73.

Liebermann, I., and Kornberg, A. 1954. Enzymatic synthesis and breakdown of a pyrimidine, orotic acid. II. Dihydroorotic acid, ureidosuccinic acid and 5-carboxymethyl hydantoin. J. Biol. Chem. 207:911–924.

Lin, Y., and Blaschek, H. P. 1984a. Butanol production by a butanol-tolerant strain of *Clostridium acetobutylicum* in extruded corn broth. Appl. Environ. Microbiol. 45:966–973.

Lin, Y., and Blaschek, H. P. 1984b. Transformation of heat-treated *Clostridium acetobutylicum* with pUB110 plasmid DNA. Appl. Environ. Microbiol. 48:737–742.

Liu, C. L., and Mortenson, L. E. 1984. Formate dehydrogenase of *Clostridium pasteurianum*. J. Bacteriol. 159:375–380.

Ljungdahl, L. G. 1986. The autotrophic pathway of acetate synthesis in acetogenic bacteria. Annu. Rev. Microbiol. 40:415–450.

Ljungdahl, L. G., Carreira, L. H., Garrison, R. J., Rabek, N. E., and Wiegel, J. 1985. Comparison of three thermophilic acetogenic bacteria for production of calcium magnesium acetate. Biotechnol. Bioeng. Symp. 15:207–223.

Ljungdahl, L. G., Hugenholtz, J., and Wiegel, J. 1989. Acetogenic and acid-producing clostridia, p. 145–191. In: Minton, N. P., and Clarke, D. J. (ed.), Clostridia, biotechnology handbooks 3. Plenum Press, New York.

Ljungdahl, L. G., Pettersson, B., Eriksson, K. E., and Wiegel, J. 1983. A yellow affinity substance involved in the cellulolytic system of *Clostridium thermocellum*. Curr. Microbiol. 9:195–200.

Loekerby, D. L., Rabivc, H. R., Bryan, L. E., and Laishley, E. J. 1984. Ferredoxin-linked reduction of metronidazole in *Clostridium pasteurianum*. Antimicrob. Agents Chemother. 26:665–669.

Loesche, W. J. 1969. Oxygen sensitivity of various anaerobic bacteria. Appl. Microbiol. 18:723–727.

Long, S., Jones, D. T., and Woods, D. R. 1983. Sporulation of *Clostridium acetobutylicum* P262 in a defined medium. Appl. Environ. Microbiol. 45:1389–1393.

Lovell, C. R., Przybyla, A., and Ljungdahl, L. G. 1988. Cloning and expression in *Escherichia coli* of the *Clostridium thermoaceticum* gene encoding thermostable formyltetrahydrofolate synthetase. Arch. Microbiol. 149:280–285.

Lovenberg, W., and Sobel, B. E. 1965. Rubredoxin: A new electron transfer protein from *Clostridium pasteurianum*. Proc. Natl. Acad. Sci. USA 54:193–199.

Lovitt, R. W., Kell, D. B., and Morris, J. G. 1986. Proline reduction by *Clostridium sporogenes* is coupled to vectorial proton ejection. FEMS Microbiol. Lett. 36:269–273.

Lovitt, R. W., Longin, R., and Zeikus, J. G. 1984. Ethanol production by thermophilic bacteria: physiological comparison of solvent effects on parent and alcohol-tolerant strains of *Clostridium thermohydrosulfuricum*. Appl. Environ. Microbiol. 48:171–177.

Lovitt, R. W., Morris, J. G., and Kell, D. B. 1987. The growth and nutrition of *Clostridium sporogenes* NCIB 8053 in defined media. J. Appl. Bacteriol. 62:71–80.

Lovitt, R. W., Shen, G. J., and Zeikus, J. G. 1988. Ethanol production by thermophilic bacteria: biochemical basis for ethanol and hydrogen tolerance in *Clostridium thermohydrosulfuricum*. J. Bacteriol. 170:2809–2815.

Lowe, S. E., Pankratz, H. S., and Zeikus, J. G. 1989. Influence of pH extremes on sporulation and ultrastructure of *Sarcina ventriculi*. J. Bacteriol. 171:3775–3781.

Lu, W. P., Harder, S. C., and Ragsdale, S. W. 1990. Controlled potential enzymology of methyl transfer reactions involved in acetyl-CoA synthesis by CO dehydrogenase and the corrinoid/ironsulfur protein from *Clostridium thermoaceticum*. J. Biol. Chem. 265:3124–3133.

Luczak, H., Schwarzmoser, H., and Staudenbauer, W. L. 1985. Construction of *Clostridium butyricum* plasmids and transfer to *Bacillus subtilis*. Appl. Microbiol. Biotechnol. 23:114–122.

Ludwig, W., Weizenegger, M., Kilpper-Bälz, R., and Schleifer, K.-H. 1988. Phylogenetic relationship of anaerobic streptococci. Int. J. Syst. Bacteriol. 38:15–18.

Lund, B. M. 1972. Isolation of pectolytic clostridia from potatoes. J. Appl. Bacteriol. 35:609–614.

Lund, B. M., and Brocklehurst, T. F. 1978. Pectic enzymes of pigmented strains of *Clostridium*. J. Gen. Microbiol. 104:59–66.

Lund, B. M., and Brocklehurst, T. F., and Wyatt, G. M. 1981. Characterization of strains of *Clostridium puniceum* sp. nov., a pink-pigmented, pectolytic bacterium. J. Gen. Microbiol. 122:17–26.

Lund, B. 1982. Clostridia and plant disease: new pathogens?, p. 263–283. In: Mount, M. S., and Lacy, G. H. (ed.), Phytopathogenic prokaryotes, vol. 1. Academic Press, London.

Lund, B. M. 1986. Anaerobes in relation to food of plant origin, p. 351–372. In: Barnes, E. M., and Mead, G. C. (ed.), Anaerobic bacteria in habitats other than man. The Soc. Appl. Bacteriol. Symp. Ser. no. 13, Blackwell Scientific Publ., Oxford, U.K..

Lundie, L. L., and Drake, K. 1984. Development of a minimally defined medium for the acetogen *Clostridium thermoaceticum*. J. Bacteriol. 159:700–703.

Lurz, R., Mayer, F., and Gottschalk, G. 1979. Electron microscopic study on the quaternary structure of the isolated particulate alcohol-acetaldehyde dehydrogenase complex and its identity with polygonal bodies of *Clostridium kluyveri*. Arch. Microbiol. 120:255–262.

Lux, M. F., Keith, E., Hsu, T., and Drake, H. L. 1990. Biotransformations of aromatic aldehydes by acetogenic bacteria. FEMS Microbiol. Lett. 67:73–78.

Lynd, L. R., Grethlein, H. E., and Wolkin, R. H. 1989. Fermentation of cellulosic substrates in batch and continuous culture by *Clostridium thermocellum*. Appl. Environ. Microbiol. 55:3131–3139.

Machacek-Pitsch, C., Rauschenbach, P., and Simon, H. 1985. Observations on the elimination of water from 2-hydroxy acids in the metabolism of amino acids by *Clostridium sporogenes*. Hoppe-Seyler's Biol. Chem. 366:1057–1062.

MacDonald, I. A., Bishop, J. M., Mahony, D. E., and Williams, C. N. 1975. Convenient non-chromatographic assays for the microbial deconjugation and 7α-OH bioconversion of taurocholate. Appl. Microbiol. 30:530–535.

MacDonald, I. A., Meier, E. C., Mahony, D. E., and Constain, G. A. 1976. 3α-, 7α- and 12α-hydroxysteroid dehydrogenase activities from *Clostridium perfringens*. Biochim. Biophys. Acta 450:142–153.

Mackey, B. M., and Morris, J. G. 1971. Sporulation in *Clostridium pasteurianum*, p. 343. In: Barker, A. N., Gould, G. W., and Wolf, J. (ed.), Spore research 1971. Academic Press, London.

Mackey, B. M., and Morris, J. G. 1972. The exosporium of *Clostridium pasteurianum*. J. Gen. Microbiol. 73:325–338.

Macy, J. M., Snellen, J. E., and Hungate, R. E. 1972. Use of syringe methods for anaerobiosis. Am. J. Clin. Nutr. 25:1318–1323.

Madan, V. K., Hillmer, P., and Gottschalk, G. 1973. Purification and properties of NADP-dependent L(+)-3-hydroxybutyryl-CoA dehydrogenase from *Clostridium kluyveri*. Eur. J. Biochem. 32:51–56.

Madden, R. H. 1983. Isolation and characterization of *Clostridium stercorarium* sp. nov., cellulolytic thermophile. Int. J. Syst. Bacteriol. 33:837–840.

Madden, R. H., Bryder, M. J., and Poole, N. J. 1982. Isolation and characterization of an anaerobic, cellulolytic bacterium, *Clostridium papyrosolvens* sp. nov. Int. J. Syst. Bacteriol. 32:87–91.

Maddox, J. S. 1980. Production of *n*-butanol from whey filtrate using *Clostridium acetobutylicum* NCIB 2951. Biotechnol. Lett. 2:493–498.

Maddox, I. S. 1983. Use of silicalite for the adsorption of *n*-butanol from fermentation liquors. Biotechnol. Lett. 5:89–94.

Madi, E., Antranikian, G., Ohmiya, K., and Gottschalk, G. 1987. Thermostable amylolytic enzymes from a new *Clostridium* isolate. Appl. Environ. Microbiol. 53:1668–1673.

Magot, M. 1981. Some properties of the *Clostridium butyricum* group β-lactamase. J. Gen. Microbiol. 127:113–119.

Magot, M. 1984. L'antibiogramme des bactéries du genre *Clostridium*. Ann. Microbiol. (Paris) 135a:443–456.

Magot, M., Carlier, J.-P., and Popoff, M. R. 1983. Identification of *Clostridium butyricum* and *Clostridium beijerinckii* by gas liquid chromatography and sugar fermentation: correlation with DNA-homologies and electrophoretic patterns. J. Gen. Microbiol. 129:2837–2845.

Mahony, D. E., Meier, E. C., MacDonald, I. A., and Holdeman, L. V. 1977. Bile salt degradation by nonfermentative clostridia. Appl. Environ. Microbiol. 34:419–423.

Mallette, M. F., Reece, P., and Dawes, E. A. 1974. Culture of *Clostridium pasteurianum* in defined medium and growth as a function of sulfate concentration. Appl. Microbiol. 28:999–1003.

Marczak, R., Ballongue, J., Petitdemange, H., and Gay, R. 1984. Regulation of the biosynthesis of NADH-rubredoxin oxidoreductase in *Clostridium acetobutylicum*. Curr. Microbiol. 10:165–168.

Marmelak, R., and Quastel, J. H. 1953. Amino acid interactions in strict anaerobes *(Clostridium sporogenes)*. Biochim. Biophys. Acta 12:103–120.

Matches, J. R., and Liston, J. 1974. Mesophilic clostridia in Puget Sound. Can. J. Microbiol. 20:1–7

Mathupala, S., Saha, B. C., and Zeikus, J. G. 1990. Substrate competition and specifity at the active site of amylopullulanase from *Clostridium thermohydrosulfuricum*. Biochem. Biophys. Res. Commun. 166:126–132.

Matsuda, N., Matsumoto, N., Ushizawa, S., Kakegawa, Y., Kato, H., and Nishida, S. 1975. Specific distribution and heat resistance of mesophilic bacterial spores isolated from frozen raw meat used in canned meat manufacture. J. Food Hyg. Soc. Japan 16:253–257.

Matteuzzi, D., Crociani, F., and Emaldi, O. 1978. Amino acids produced by bifidobacteria and some clostridia. Ann. Microbiol. 129B:175–181.

Matteuzzi, D., Hollaus, F., and Biavati, B. 1978. Proposal of neotype for *Clostridium thermohydrosulfuricum* and the merging of *Clostridium tartarivorum* with *Clostridium thermosaccharolyticum*. Int. J. Syst. Bacteriol. 28:528–531.

Matteuzzi, D., Trovatelli, L. D., Biavati, B., and Zani, G. 1977. Clostridia from grana cheese. J. Appl. Bacteriol. 43:375–382.

Mayhew, S. G. 1971. Properties of two clostridial flavodoxins. Biochim. Biophys. Acta 235:276–288.

Mayer, F., Ivey, D. M., and Ljungdahl, L. G. 1986. Macromolecular organization of F$_1$-ATPase isolated from *Clostridium thermoaceticum* as revealed by electron microscopy. J. Bacteriol. 166:1128–1130.

Mayraud, D., and Bourgeau, G. 1982. Production of phenylacetic acid by anaerobes. J. Clin. Microbiol. 16:747–750.

McBee, R. H. 1948. The culture and physiology of a thermophilic cellulose-fermenting bacterium. J. Bacteriol. 56:653–663.

McBee, R. H. 1950. The anaerobic thermophilic cellulolytic bacteria. Bacteriol. Rev. 14:51–63.

McBryde, C. N. 1911. A bacteriological study of ham souring. U.S. Bureau of Animal Industry Bulletin No. 132.

McClung, L. S. 1935a. Studies on anaerobic bacteria. III. Historical review of certain thermophilic anaerobes. J. Bacteriol. 29:173–187.

McClung, L. S. 1935b. Studies on anaerobic bacteria. IV. Taxonomy of cultures of a thermophilic species causing "swells" of canned foods. J. Bacteriol. 29:189–202.

McClung, L. S. 1943. On the enrichment and purification of chromogenic sporeforming anaerobic bacteria. J. Bacteriol. 46:507–512.

McClung, L. S. 1956. The anaerobic bacteria with special reference to the genus *Clostridium*. Ann. Rev. Microbiol. 10:173–192.

McClung, L. S., and McCoy, E. 1934. Studies on anaerobic bacteria. I. A cornliver medium for the detection and dilution counts of various anaerobes. J. Bacteriol. 28:267–277.

McClung, L. S., and McCoy, E. 1957. Genus II. *Clostridium* Prazmowski, p. 634–693. In: R. S. Breed, E. G. D. Murray, and N. R. Smith (ed.), Bergey's manual of determinative bacteriology, 7th ed. The Williams and Wilkins Co., Baltimore.

McClung, L. S., and McConnell, J. D. 1967. Sporulation of *Clostridium roseum* and *Clostridium bifermentans*, p. 81–84. In: V. Fredette (ed.), The anaerobic bacteria. Proc. Int. Workshop, Montreal.

McCord, J. M., Keele, Jr., B. B., and Fridovich, I. 1971. An enzyme based theory of obligate anaerobiosis: The physiological function of superoxide dismutase. Proc. Natl. Acad. Sci. USA 68:1024–1027.

McCoy, E., and Fred, E. B. 1941. The stability of a culture for industrial fermentation. J. Bacteriol. 41:90–91.

McCoy, E., Fred, E. B., Peterson, W. H., and Hastings, E. G. 1926. A cultural study of the acetone butyl alcohol organism. J. Infect. Dis. 39:253–283.

McCoy, E., and McClung, L. S. 1935. Studies on anaerobic bacteria. VI. The nature and systematic position of a new chromogenic *Clostridium*. Arch. Mikrobiol. 6:230–238.

McCroskey, L. M., Hatheway, C. I., Fenicia, L., Pasolini, B., and Aureli, P. 1986. Characterization of an organism that produces type E botulinal toxin which resembles *Clostridium butyricum* from the feces of an infant with type E botulism. J. Clin. Microbiol. 23:201–202.

McNeil, B., and Kristiansen, B. 1986. The acetone-butanol fermentation. Adv. Appl. Microbiol. 31:61–92.

Mead, G. C. 1971. The amino acid-fermenting clostridia. J. Gen. Microbiol. 67:47–56.

Mead, G. C., Adams, B. W., Hilton, M. G., and Lord, P. G. 1979. Isolation and characterization of uracil-degrading clostridia from soil. J. Appl. Bacteriol. 46:465–472.

Meinecke, B., Bahl, H., and Gottschalk, G. 1984. Selection of an asporogenous strain of *Clostridium acetobutylicum* in continuous culture under phosphate limitation. Appl. Environ. Microbiol. 48:1064–1065.

Melasniemi, H. 1987. Characterization of α-amylase and pullulanase activities of *Clostridium thermohydrosulfuricum*. Biochem. J. 246:193–197.

Mercer, I., Modi, N., Clarke, D. J., and Morris, J. G. 1979. The occurrence and location of squalene in *Clostridium pasteurianum*. J. Gen. Microbiol. 111:437–440.

Mercer, W. A., and Vaughn, R. H. 1951. The characteristics of some thermophilic, tartrate-fermenting anaerobes. J. Bacteriol. 62:27–37.

Meyer, C. L., Roos, J. W., and Papoutsakis, E. T. 1986. Carbon monoxide gassing leads to alcohol production and butyrate uptake without acetone formation in continuous cultures of *Clostridium acetobutylicum*. Appl. Microbiol. Biotechnol. 24:159–167.

Meyn, A. 1933. Über das Vorkommen und den Nachweis anaerober Bazillen in der Milch. Milchwirtsch. Forschungen 15:426–432.

Michel, C., Hartrampf, G., and Buckel, W. 1989. Assay and purification of the adenosylcobalamin-dependent 2-methyleneglutarate mutase from *Clostridium barkeri*. Eur. J. Biochem. 184:103–107.

Michel-Savin, D., Marchal, R., and Vandecasteele, J. P. 1990. Butyrate production in continuous culture of *Clostridium tyrobutyricum*: effect of end-product inhibition. Appl. Microbiol. Biotechnol. 33:127–131.

Mills, C. K., Kefauver, M. B., and Gherna, R. L. 1989. Whole-cellular fatty acid analysis of thermophilic anaerobes and the aerobic thermophilic genera *Thermus* and *Rhodothermus*. Ann. Meet. Am. Soc. Microbiol., Abstract R-21, p. 283.

Minton, N. P., and Morris, J. G. 1981. Isolation and partial characterization of three cryptic plasmids from strains of *Clostridium butyricum*. J. Gen. Microbiol. 127:325–331.

Minton, N. P., and Morris, J. G. 1983. Regeneration of protoplasts of *Clostridium pasteurianum* ATCC 6013. J. Bacteriol. 155:432–434.

Minton, N. P., and Clark, D. J. 1989. Clostridia. Biotechnology handbooks, vol. 3. Plenum Press, London.

Mitchell, W. J., and Booth, I. R. 1984. Characterization of the *Clostridium pasteurianum* phosphotransferase system. J. Gen. Microbiol. 130:2193–2200.

Mitchell, W. J., Roohi, M. S., Mosley, M. J., and Booth, I. R. 1987. Regulation of carbohydrate utilization in *Clostridium pasteurianum*. J. Gen. Microbiol. 133:31–36.

Mizutami, T., and Mitsuoka, T. 1979. Effect of intestinal bacteria on incidence of liver tumors in gnotobiotic C3H/He male mice. J. Natl. Cancer Inst. 63:1365–1370.

Möllby, R., Holme, T., Nord, C. E., Smyth, C. J., and Wadström, T. 1976. Production of phospholipase C (alpha-toxin), haemolysins and lethal toxins by *Clostridium perfringens* types A to D. J. Gen. Microbiol. 96:137–144.

Möller, B., Hippe, H., and Gottschalk, G. 1986. Degradation of various amine compounds by mesophilic clostridia. Arch. Microbiol. 141:85–90.

Möller, B., Ossmer, R., Howard, B. H., Gottschalk, G., and Hippe, H. 1984. *Sporomusa,* a new genus of Gram-negative anaerobic bacteria including *Sporomusa sphaeroides* spec. nov. and *Sporomusa ovata* spec. nov. Arch. Microbiol. 139:388–396.

Monot, F., and Engasser, J. M. 1983. Continuous production of acetone butanol on an optimized synthetic medium. Eur. Appl. Microbiol. Biotechnol. 18:246–248.

Monticello, D. J., and Costilow, R. N. 1981. Purification and partial characterization of proline dehydrogenase from *Clostridium sporogenes.* Can. J. Microbiol. 27:942–948.

Monticello, D. J., and Costilow, R. N. 1982. Interconversion of valine and leucine by *Clostridium sporogenes.* J. Bacteriol. 151:946–949.

Monticello, D. J., Hadioetomo, R. S., and Costilow, R. N. 1984. Isoleucine synthesis by *Clostridium sporogenes* from propionate or α-methylbutyrate. J. Gen. Microbiol. 130:309–318.

Montville, T. J., Parris, N., and Conwaym, L. K. 1985. Influence of pH on organic acid production by *Clostridium sporogenes* in test tube and fermentor cultures. Appl. Environ. Microbiol. 49:733–736.

Moore, W. E. C., Cato, E. P., and Holdeman, L. V. 1966. Fermentation patterns of some *Clostridium* species. Int. J. Syst. Bacteriol. 16:383–415.

Moore, W. E. C., Johnson, J. L., and Holdeman, L. V. 1976. Emendation of *Bacteroidaceae* and *Butyrivibrio* and descriptions of *Desulfomonas* gen. nov. and ten new species in the genera *Desulfomonas, Butyrivibrio, Eubacterium, Clostridium,* and *Ruminococcus.* Int. J. Syst. Bacteriol. 26:238–252.

Mori, Y. 1990. Isolation of mutants of *Clostridium thermocellum* with enhanced cellulase production. Agric. Biol. Chem. 54:825–826.

Morris, G. N., Winter, J., Cato, E. P., Ritchie, A. E., and Bokkenheuser, V. D. 1985. *Clostridium scindens* sp. nov., a human intestinal bacterium with desmolytic activity on corticoids. Int. J. Syst. Bacteriol. 35:478–481.

Morris, J. G. 1975. The physiology of obligate anaerobiosis, p. 169–246. In: Rose, A. H., and Tempest, D. W. (ed.), Advances in microbial physiology, vol. 12. Academic Press, London.

Morris, J. G. 1986. Anaerobiosis and energy-yielding metabolism, p. 1–21. In: E. M. Barnes and G. C. Mead (ed.), Anaerobic bacteria in habitats other than man. The Soc. Appl. Bacteriol. Symp. Ser. no. 13. Blackwell Scientific Publ., Oxford, U.K.

Morris, J. G. 1989. Bioconversions, p. 193–225. In: Minton, N. P., Clarke, D. J., (ed.), Clostridia, biotechnology handbooks 3. Plenum Press, New York.

Mortenson, L. E. 1966. Components of cell-free extracts of *Clostridium pasteurianum* required for ATP-dependent H$_2$ evolution from dithionite and for N$_2$ fixation. Biochim. Biophys. Acta 127:18–25.

Mortenson, L. E. 1978. Regulation of nitrogen fixation, p. 179–232. In: Horecker, B. L., Stadtman, E. R. (ed.), Current topics in cellular regulation, vol. 13. Academic Press, New York.

Mortenson, L. E., and Chen, J. S. 1974. Hydrogenase, p. 231–282. In: Neilands, J B. (ed.), Microbial iron metabolism. A comprehensive treatise. Academic Press, New York.

Mortenson, L. E., Morris, J. A., and Jeng, D. Y. 1967. Purification, metal composition and properties of molybdoferredoxin and azoferredoxin, two of the components of the nitrogen-fixing system of *Clostridium pasteurianum.* Biochim. Biophys, Acta 141:516–522.

Mortenson, L. E., Valentine, R. C., and Carnahan, J. E. 1962. An electron transport factor from *Clostridium pasteurianum.* Biochem. Biophys. Res. Commun. 7:448–453.

Möse, J. R., and Möse, G. 1964. Oncolysis by clostridia. I. Activity of *Clostridium butyricum* (M-55) and other nonpathogenic clostridia against the Ehrlich carcinoma. Cancer Res. 24:212–216.

Moss, C. W., Lambert, M. A., and Goldsmith, D. H. 1970. Production of hydrocinnamic acid by clostridia. Appl. Microbiol. 19:375–378.

Moss, C. W., and Lewis, V. J. 1967. Characterization of clostridia by gas chromatography. I. Differentiation of species by cellular fatty acids. Appl. Microbiol. 15:390–397.

Moulis, J. M., Meyer, J., and Lutz, M. 1984. (4Fe-4x)$^{2+}$ (x = sulfur, selenium) clusters in *Clostridium pasteurianum* ferredoxin and in synthetic analogues: structural data from resonance raman spectroscopy. Biochem. 23:6605–6613.

Müller, H. E., and Werner, H. 1974. Occurrence of neuraminidase and acylneuraminate lyase in *Clostridium beijerinckii* and *Clostridium tertium.* Zentralbl. Bakteriol. Parasitenkd. Infektionskr. Hyg. Abt. 1 Orig. A 229:134–140.

Murray, C. L., and Rabinowitz, J. C. 1981. RNA polymerase from *Clostridium acidi-urici.* Characterization of a naturally occurring rifampicin-resistant bacterial enzyme. J. Biol. Chem. 256:5153–5161.

Murray, W. D., Hofmann, L., Campbell, N. L., and Madden, R. H. 1986. *Clostridium lentocellum* sp. nov., a cellulolytic species from river sediment containing paper-mill waste. Syst. Appl. Microbiol. 8:181–184.

Murray, W. D., Khan, A. W., and van den Bug, L. 1982. *Clostridium saccharolyticum* sp. nov., a saccharolytic species from sewage sludge. Int. J. Syst. Bacteriol. 32:132–135.

Murray, W. D., and Khan, A. W. 1983a. Ethanol production by a newly isolated anaerobe, *Clostridium saccharolyticum:* effects of culture medium and growth conditions. Can. J. Microbiol. 29:342–347.

Murray, W. D., and Khan, A. W. 1983b. Growth requirement of *Clostridium saccharolyticum,* an ethanologenic anaerobe. Can. J. Microbiol. 29:348–353.

Murray, W. D., Wemyss, K. B., and Khan, A. W. 1983. Increased ethanol production and tolerance by a pyruvate-negative mutant of *Clostridium saccharolyticum.* Eur. J. Appl. Microbiol. Biotechnol. 18:71–74.

Muth, W. L., and Costilow, R. N. 1974. Ornithine cyclase (deaminating). II. Properties of the homogeneous enzyme. J. Biol. Chem. 249:7457–7462.

Nakamura, S., Okado, I., Abe, T., and Nishida, S. 1979. Taxonomy of *Clostridium tetani* and related species. J. Gen. Microbiol. 113:29–35.

Nakamura, S., Sakurai, M., Nishida, S., Tatsuki, T., Yanagase, Y., Higashi, Y., and Amano, T. 1976b. Lecithinase-negative variants of *Clostridium perfringens;* the identity of *C. plagarum* with *C. perfringens.* Can. J. Microbiol. 22:1497–1501.

Nakamura, S., Shimamura, T., Hayase, M., and Nishida, S. 1973. Numerical taxonomy of saccharolytic clostridia, particularly *Clostridium perfringens* like strains: Description of *Clostridium absonum* sp. n. and *Clostridium paraperfringens.* Int. J. Syst. Bacteriol. 23:419–429.

Nakamura, S., Shimamura, T., Hayashi, H., and Nishida, S. 1975. Reinvestigation of the taxonomy of *Clostridium bifermentans* and *Clostridium sordellii.* J. Med. Microbiol. 8:299–309.

Nakamura, S., Shimamura, T., and Nishida, S. 1976a. Urease-negative strains of *Clostridium sordellii.* Can. J. Microbiol. 22:673–676.

Nasuno, S., and Assai, T. 1960. Some environmental factors affecting sporulation in butanol and butyric acid bacteria. J. Gen. Appl. Microbiol. 6:71–82.

Naumann, E., Hippe, H., and Gottschalk, G. 1983. Betaine: new oxidant in the Stickland reaction and methanogenesis from betaine and L-alanine by a *Clostridium sporogenes-Methanosarcina barkeri* coculture. Appl. Environ. Microbiol. 45:474–483.

Neece, M. S., and Fridovich, I. 1967. Acetoacetic decarboxylase, activation by heat. J. Biol.Chem. 242:2939–2944.

Ng, H., and Vaughn, R. H., 1963. *Clostridium rubrum* sp. n. and other pectinolytic clostridia from soil. J. Bacteriol. 85:1104–1113.

Ng, T. K., Weimer, P. J., and Zeikus, J. G. 1977. Cellulolytic and physiological properties of *Clostridium thermocellum.* Arch. Microbiol. 114:1–7.

Ng, T. K., and Zeikus, J. G. 1981. Purification and characterization of an endoglucanase (1,4-β-D-glucan glucano hydrolase) from *Clostridium thermocellum.* Biochem. J. 199:341–350.

Ng, T. K., and Zeikus, J. G. 1982. Differential metabolism of cellobiose and glucose by *Clostridium thermocellum* and *Clostridium thermohydrosulfuricum.* J. Bacteriol. 150:1391–1399.

Niel, P., Rimbault, A., Campion, G., and Leluan, G. 1989. Phenotypic differentiation between *Clostridium hastiforme* and *Clostridium subterminale* by headspace gas chromatography. Int. J. Syst. Bacteriol. 39:491–492.

Nieves, B. M., Gil, F., and Castillo, F. J. 1981. Growth inhibition activity and bacteriophage and bacteriocin like particles associated with different species of *Clostridium.* Can. J. Microbiol. 27:216–225.

Nipkow, A., Shen, G. J., and Zeikus, J. G. 1989. Continuous production of thermostable β-amylase with *Clostridium thermosulfurogenes:* effect of culture conditions and metabolite levels on enzyme synthesis and activity. Appl. Environ. Microbiol. 55:689–894.

Nishida, S., Seo, N., and Nakagawa, M. 1969. Sporulation, heat resistance, and biological properties of *Clostridium perfringens.* Appl. Microbiol. 17:303–309.

Nisman, B. 1954. The Stickland reaction. Bacteriol. Rev. 18:16–42.

Nunez-Montiel, O. L., Thompson, F. S., and Dowell, V. R. 1983. Norleucine-tyrosine broth for rapid identification for *Clostridium difficile* by gas-liquid chromatography. J. Clin. Microbiol. 17:382–385.

O'Brien, R. W., and Morris, J. G. 1971a. The ferredoxin-dependent reduction of chloramphenicol by *Clostridium acetobutylicum.* J. Gen. Microbiol. 67:265–271.

O'Brien, R. W., and Morris, J. G. 1971b. O₂ and the growth and metabolism of *Clostridium acetobutylicum.* J. Gen. Microbiol. 68:307–318.

O'Brien, R. W., and Morris, J. G. 1972. Effect of metronidazole on hydrogen production by *Clostridium acetobutylicum.* Arch. Mikrobiol. 84:225–233.

Ogata, S., and Hongo, M. 1979. Bacteriophages of the genus *Clostridium.* Adv. Appl. Microbiol. 25:241–273.

Oh-hama, T., Stolowich, N. J., and Scott, A. I. 1988. 5-Aminolevulinic acid formation from glutamate via the C₅ pathway in *Clostridium thermoaceticum.* FEBS Lett. 1:89–93.

Ohisa, N., Yamaguchi, M., and Kurihara, N. 1980. Lindane degradation by cell-free extracts of *Clostridium rectum.* Arch. Microbiol. 125:221–225.

Ohwaki, K., and Hungate, R. E. 1977. Hydrogen utilization by clostridia in sewage sludge. Appl. Environ. Microbiol. 33:1270–1274.

O'Leary, W. M., and Wilkinson, S. G. 1988. Part 2: Distribution of lipids. 5. Gram-positive bacteria, p. 117–201. In: C. Ratledge and S. G. Wilkinson (ed.), Microbial lipids, vol. 1. Academic Press, London.

Omelianski, W. 1902. Über die Gärung der Zellulose. Zentralbl. Bakteriol. Parasitenkd. Infektionskr. Hyg., Abt. 2 Orig. 8:225–231.

Oren, A. 1983. *Clostridium lortetii* sp. nov., a halophilic obligatory anaerobic bacterium producing endospores with attached gas vacuoles. Arch. Microbiol. 136:42–48.

Oren, A., Pohla, H., and Stackebrandt, E. 1987. Transfer of *Clostridium lortetii* to a new genus *Sporohalobacter* gen. nov. as *Sporohalobacter lortetii* comb. nov., and description of *Sporohalobacter marismortui* sp. nov. Syst. Appl. Microbiol. 9:239–246.

Orpianesi, C., Cresci, A., Trotta, F., La Rosa, F., and Mastrandrea, V. 1989. Molecular and chemotaxonomic studies on bile acid oxidizing clostridia. Syst. Appl. Microbiol. 12:134–140.

Oultram, J. D., Loughlin, M., Swinfield, T. J., Brehm, J. K., Thompson, D. E., and Minton, N. P. 1988. Introduction of plasmids into whole cells of *Clostridium acetobutylicum* by electroporation. FEMS Microbiol. Lett. 56:83–88.

Oultram, J. D., and Young, M. 1985. Conjugal transfer of plasmid pAMß1 from *Streptococcus lactis* and *Bacillus subtilis* to *Clostridium acetobutylicum.* FEMS Microbiol. Lett. 27:129–134.

Ounine, K., Petitdemange, H., Raval, G., and Gay, R. 1983. Acetone-butanol production from pentoses by *Clostridium acetobutylicum.* Biotechnol. Lett. 5:605–610.

Owen, R. W. 1985. Biotransformation of bile acids by clostridia. J. Med. Microbiol. 20:233–238.

Palopp, M. L., Valles, S., Pinaga, F., and Flors, A. 1989. Isolation and characterization of an anaerobic, cellulolytic bacterium, *Clostridium celerecrescens* sp. nov. Int. J. Syst. Bacteriol. 39:68–71.

Palosaari, N. R., and Rogers, P. 1988. Purification and properties of the inducible coenzyme A-linked butyraldehyde dehydrogenase from *Clostridium acetobutylicum.* J. Bacteriol. 170:2971–2976.

Pan-Hou, H. S. K., Hosono, M., and Imura, M. 1980. Plasmid-controlled mercury biotransformation by *Clostridium cochlearium* T2. Appl. Environ. Microbiol. 40:1007–1011.

Parkkinen, E. 1986. Conversion of starch into ethanol by *Clostridium thermohydrosulfuricum*. Appl. Microbiol. Biotechnol. 25:213–219.

Parsons, L. B., and Sturges, W. S. 1927a. Quantitative aspects of the metabolism of anaerobes. I. Proteolysis by *Clostridium putrefaciens* compared with that of other anaerobes. J. Bacteriol. 14:181–192.

Parsons, L. B., and Sturges, W. S. 1927b. Quantitative aspects of the metabolism of anaerobes. II. The relation between volatile acid and ammonia production during metabolism of *Clostridium putrefaciens*. J. Bacteriol.. 14:193–200.

Parsons, L. B., and Sturges, W. S. 1927c. Quantitative aspects of the metabolism of anaerobes. III.The volatile acids produced by *Clostridium putrefaciens* in cooked meat medium. J. Bacteriol. 14:201–215.

Pasteur, L. 1861. Animalcules infusoires vivant sans gaz oxygène libre et déterminant des fermentations. C.R. Acad. Sci. 52:344–347.

Patel, B. K. C., Monk, C., Littleworth, H., Morgan, H. W., and Daniel, R. M. 1987. *Clostridium fervidus* sp. nov., a new chemoorganotrophic acetogenic thermophile. Int. J. Syst. Bacteriol. 37:123–126.

Patni, N. J., and Alexander, J. K. 1971a. Utilization of glucose by *Clostridium thermocellum*. Presence of glucokinase and other glycolytic enzymes in cell extracts. J. Bacteriol. 105:220–225.

Patni, N. J., and Alexander, J. K. 1971b. Catabolism of fructose and mannitol in *Clostridium thermocellum*. Presence of phosphoenolpyruvate: fructose phosphotransferase, fructose 1-phosphate kinase, phosphoenolpyruvate: mannitol phosphotransferase, and mannitol 1-phosphate dehydrogenase in cell extracts. J. Bacteriol. 105:226–231.

Peck, H. D., and Gest, H. 1957. Hydrogenase of *Clostridium butylicum*. J. Bacteriol. 73:569–580.

Pel, R., Hessels, G., Aalfs, H., and Gottschal, J. C. 1989. Chitin degradation by *Clostridium* sp. strain 9.1 in mixed cultures with saccharolytic and sulfate-reducing bacteria. FEMS Microbiol. Ecol. 62:191–200.

Pepple, J. S., and Dennis, D. 1976. Lactate racemase. Hydroxylamine-dependent [18]O exchange of the α-hydroxyl of lactic acid. Biochim. Biophys. Acta 429:1036–1040.

Perkins, W. E. 1965. Production of clostridial spores. J. Appl. Bacteriol. 28:1–16.

Perry, D. A. 1982. Pectolytic *Clostridium* spp. in soils and rhizospheres of carrot and other arable crops in east Scotland. J. Appl. Bacteriol. 52:403–408.

Perry, D. A. 1985. Characteristics of pectolytic clostridia isolated from soil and plants in Scotland. J. Appl. Bacteriol. 58:293–302.

Petitdemange, H., Bengone, J. M., Bergère, J.-L., and Gay, R. 1973a. Regulation of the NAD and NADP-ferredoxin reductase activities in a Clostridium of the butyric acid group: *Clostridium tyrobutyricum*. Biochim. 55:1307–1310.

Petitdemange, E., Caillet, F., Giallo, J., and Gaudin, C. 1984. *Clostridium cellulolyticum* sp. nov., a cellulolytic, mesophilic species from decayed grass. Int. J. Syst. Bacteriol. 34:155–159.

Petitdemange, H., Cherrier, C., Bengone, J. M., and Gay, R. 1977. Etude des activités NADH et NADPH-ferrédoxine oxydoréductasiques chez *Clostridium acetobutylicum*. Can. J. Microbiol. 23:152–160.

Petitdemange, H., Cherrier, C., Raval, G., and Gay, R. 1976. Regulation of the NADH and NADPH-ferredoxin oxidoreductases in clostridia of the butyric group. Biochim. Biophys. Acta 421:334–347.

Petitdemange, H., Lambert, D., and Gay, R. 1972. Activités NAD et NADP ferrédoxine réductasique des extraits acellulaires de *Clostridium pastorianum*. C.R. Séances Soc. Biol. Fil. 166:1128–1132.

Petitdemange, H., Lambert, D., and Gay, R. 1973b. Determination of the reduced and oxidized pyridine nucleotides in *Clostridium pasteurianum*. C. R. des Séances Soc. Biol. Fil. 167:111–115.

Petitdemange, H., Marczak, R., and Gay, R. 1979. NADH-flavodoxin oxidoreductase activity in *Clostridium tyrobutyricum*. FEMS Microbiol. Lett. 5:291–294.

Petre, J., Longin, R., and Millet, J. 1981. Purification and properties of an endo-β-1,4 glucanase from *Clostridium thermocellum*. Biochim. 63:629–639.

Pettigrew, D. W., Bidigare, R. R., Mehta, B. J., Williams, M. I., and Sander, E. G. 1985. Dihydroorotase from *Clostridium oroticum*, purification and reversible removal of essential zinc. Biochem. J. 230:101–108.

Pheil, C. G., and Ordal, Z. J. 1967. Sporulation of the "thermophilic anaerobes." Appl. Microbiol. 15:893–898.

Phillips, K. D. 1986. A sporulation medium for *Clostridium perfringens*. Lett. Appl. Microbiol. 3:77–79.

Popoff, M. R. 1984. Selective medium for isolation of *Clostridium butyricum* from human feces. J. Clin. Microbiol. 20:417–420.

Popoff, M. R., Guillon, J.-P., and Carlier, J. P. 1985. Taxonomic position of lecithinase-negative strains of *Clostridium sordellii*. J. Gen. Microbiol. 131:1697–1703.

Popoff, M. R., and Truffaut, N. 1985. Survey of plasmids in *Clostridium butyricum* and *Clostridium beijerinckii* strains from different origins and different phenotypes. Curr. Microbiol. 12:151–156.

Poston, J. M. 1976. Leucine 2,3-aminomutase, an enzyme of leucine catabolism. J. Biol.Chem. 251:1859–1863.

Poston, J. M. 1986. βs-Leucine and the β-keto pathway of leucine metabolism. Adv. Enzymol. 58:173–189.

Potter, L. F., and McCoy, E. 1952. The fermentation of pectin and pectic acid by *Clostridium felsineum*. J. Bacteriol. 64:701–708.

Poxton, I. R. 1984. Demonstration of the common antigens of *Clostridium botulinum*, *C. sporogenes* and *C. novyi* by an enzyme-linked immunosorbent assay and electroblot transfer. J. Gen. Microbiol. 130:975–981.

Prazmowski, A. 1880. Untersuchungen über die Entwicklungsgeschichte und Fermentwirkung einer Bakterienart. Leipzig: Inaugural Dissertation. Hugo Voigt.

Prévot, A. R. 1938. Etudes de systematique bacterienne. IV. Critique de la conception actuelle du genre *Clostridium*. Ann. Inst. Pasteur (Paris) 61:72–91.

Prévot, A. R. 1957. Manual de classification et de détermination des bactéries anaérobies. Paris: Masson.

Prévot, A. R. 1966. Manual for the classification and determination of the anaerobic bacteria, 1st American ed. translated by V. Fredette. Philadelphia: Lea and Febiger.

Pribula, C. D., Fox, G. E., and Woese, C. R. 1976. Nucleotide sequences of *Clostridium pasteurianum* 5S rRNA. FEBS Lett. 64:350–352.

Prince, R. C., and Adams, M. W. W. 1987. Oxidation-reduction properties of the two Fe_4S_4 clusters in *Clostri-*

dium pasteurianum ferredoxin. J. Biol. Chem. 262:5125–5128.

Princewill, T. J. T. 1980. Thiaminase activity amongst strains of *Clostridium sporogenes*. J. Appl. Bacteriol. 48:249–252.

Prins, R. A., and Lankhorst, A. 1977. Synthesis of acetate from CO_2 in the cecum of some rodents. FEMS Microbiol. Lett. 1:255–258.

Przysiecki, C., Cheddar, G., Meyer, T. E., Tollin, G., and Cusanovich, M. A. 1985. Kinetics of reduction of high redox potential ferredoxins by the semiquinones of *Clostridium pasteurianum* flavodoxin and exogenous flavin mononucleotide. Electrostatic and redox potential effects. Biochemistry 24:5647–5652.

Quentmeier, A., and Antranikian, G. 1985. Characterization of citrate lyase from *Clostridium sporosphaeroides*. Arch. Microbiol. 141:85–90.

Rabinowitz, J. C. 1963. Intermediates in purine breakdown, p. 703–713. In: Colowick, S. P., and Kaplan, N. O. (ed.), Meth. Enzymol., vol. 6. Academic Press, New York.

Rabinowitz, J. C. 1972. Preparation and properties of clostridial ferredoxins. Methods Enzymol. 24:431–446.

Rabinowitz, J. C., and Pricer, W. E. 1962. Formyltetrahydrofolate synthetase. I. Isolation and crystallization of the enzyme. J. Biol.Chem. 237:2898–2902.

Rawlings, J., Shah, V. K., Chisnell, J. R., Brill, W. J., Zimmermann, R., Münck, E., and Orme-Johnson, W. H. 1978. Novel metal cluster in the iron-molybdenum cofactor of nitrogenase. J. Biol. Chem. 253:1001–1004.

Raynaud, M. 1949. Le bactéries anaérobies pectinolytiques. Ann. Inst. Pasteur (Paris) 77:434–470.

Recsei, P. A., Moore, W. M., and Snell, E. E. 1983. Pyruvoyl-dependent histidine decarboxylase from *Clostridium perfringens* and *Lactobacillus buchneri*. J. Biol. Chem. 258:439–444.

Reed, W. M., Keller, F. A., Kite, F. E., Bogdan, M. E., and Ganoung, J. S. 1987. Development of increased acetic acid tolerance in anaerobic homoacetogens through induced mutagenesis and continuous selection. Enz. Microbiol. Technol. 9:117–120.

Reeves, R. E., Manzies, R. A., and Hsu, D. S. 1968. The pyruvate-phosphate dikinase reaction. The fate of phosphate and the equilibrium. J. Biol.Chem. 243:5486–5491.

Reid, S. J., Allcock, E. R., Jones, D. T., and Woods, D. R. 1983. Transformation of *Clostridium acetobutylicum* protoplasts with bacteriophage DNA. Appl. Environ. Microbiol. 45:305–307.

Reysenbach, A. L., Ravenscroft, N., Long, S., Jones, D. T., and Woods, D. R. 1986. Characterization, biosynthesis, and regulation of granulose in *Clostridium acetobutylicum*. Appl. Environ. Microbiol. 52:185–190.

Reysset, G., and Sebald, M. 1985. Conjugal transfer of plasmid-mediated antibiotic resistance from streptococci to *Clostridium acetobutylicum*. Ann. Microbiol. Inst. Pasteur 136:275–282.

Reysset, G., Hubert, J., Podvin, L., and Sebald, M. 1987. Protoplast formation and regeneration of *Clostridium acetobutylicum* strain N1-4080. J. Gen. Microbiol. 133:2595–2600.

Reysset, G., Hubert, J., Podvin, L., and Sebald, M. 1988. Transfection and transformation of *Clostridium acetobutylicum* strain N1-4081 protoplasts. Biotechnol. Tech. 2:199–204.

Riebeling, V., Thauer, R. K., and Jungermann, K. 1975. The internal-alkaline pH gradient, sensitive to uncoupler and ATPase inhibitor, in gowing *Clostridium pasteurianum*. Eur. J. Biochem. 55:445–453.

Ritter, W. 1932. Eine Nachprüfung des Ruschmann'schen Kartoffelbreiverfahrens zum Nachweis von Buttersäurebazillen. Landwirtschaftliches Jahrbuch der Schweiz 46:601–608.

Ritter, W., and Dorner, W. 1932. Behebung eines wichtigen Nachteils des anaeroben Pyrogallolverschlusses. Zentralbl. Bakteriol. Parasitenkd. Infektionskr. Hyg. Abt. 1 Orig. 125:379–383.

Roberts, D. L., James-Hagstrom, J. E., Garvin, D. K., Gorst, C. M., Runquist, J. A., Baur, J. R., Haase, F. C., and Ragsdale, S. W. 1989. Cloning and expression of the gene cluster encoding key proteins involved in acetyl-CoA synthesis in *Clostridium thermoaceticum*: CO dehydrogenase, the corrinoid/Fe-S protein, and methyltransferase. Proc. Natl. Acad. Sci. 86:32–36.

Roberts, T. A. 1967. Sporulation of mesophilic clostridia. J. Appl. Bacteriol. 30:430–443.

Roberts, T. A., and Derrik, C. M. 1975. Sporulation of *Clostridium putrefaciens* and the resistance of the spores to heat, γ-radiation and curing salts. J. Appl. Bacteriol. 38:33–37.

Roberts, T. A., and Hobbs, G. 1968. Low temperature growth characteristics of clostridia. J. Appl. Bacteriol. 31:75–88.

Roberts, T. A., and Ingram, J. M. 1965. The resistance of spores of *Clostridium botulinum* type E to heat and radiation. J. Appl. Bacteriol. 28:125–138.

Roberts, T. A., and Mead, G. C. 1986. Involvement of intestinal anaerobes in the spoilage of red meats, poultry and fish, p. 333–349. In: E. M. Barnes and G. C. Mead (ed.), Anaerobic bacteria in habitats other than man. The Soc. Appl. Bacteriol. Symp. Ser. no. 13, Blackwell Scientific Publ., Oxford, U.K.

Robertson, M. 1915–1916. Notes upon certain anaerobes isolated from wounds. J. Path. Bacteriol. 20:327.

Robson, R. L., and Morris, J. G. 1974. Mobilization of granulose in *Clostridium pasteurianum*. Purification and properties of granulose phosphorylase. Biochem. J. 144:513–517.

Robson, R. L., Robson, R. M., and Morris, J. G. 1974. The biosynthesis of granulose by *Clostridium pasteurianum*. Biochem. J. 144:503–511.

Roffler, S. R., Blanch, H. W., and Wilke, C. R. 1987. Extractive fermentation of acetone and butanol: Process design and economic evaluation. Biotechnol. Prog. 3:131–140.

Rogers, M. J., Simmons, J., Walker, R. T., Weisburg, W. G., Woese, C. R., Tanner, R. S., Robinson, I. M., Stahl, D. A., Olsen, G. J., Leach, R. H., and Maniloff, J. 1985. Construction of the mycoplasma evolutionary tree from 5S rRNA sequence data. Proc. Natl. Acad. Sci. USA 82:1160–1164.

Rogers, P. 1986. Genetics and biochemistry of *Clostridium* relevant to development of fermentation processes. Adv. Appl. Microbiol. 31:1–60.

Rolfe, R. D., Hentges, D. J., Campbell, B. J., and Barrett, J. T. 1978. Factors related to the oxygen tolerance of anaerobic bacteria. Appl. Environ. Microbiol. 36:306–313.

Roohi, M. S., and Mitchell, W. J. 1987. Regulation of sorbitol metabolism by glucose in *Clostridium pasteur-*

ianum: a role for inducer exclusion. J. Gen. Microbiol. 133:2207–2215.

Rosenberger, R. F. 1951. The development of methods for the study of obligate anaerobes in silage. Proc. Soc. Appl. Bacteriol. 14:161–164.

Rosenberger, R. F. 1956. The isolation and cultivation of obligate anaerobes from silage. J. Appl. Bacteriol. 19:173–180.

Rosenberger, R. F. 1959. Obligate anaerobes which form skatole. J. Bacteriol. 77:517.

Rosenblum, E. D., and Wilson, P. W. 1949. Fixation of isotopic nitrogen by *Clostridium.* J. Bacteriol. 57:413–414.

Ross, H. E. 1965. *Clostridium putrefaciens.* A neglected anaerobe. J. Appl. Bacteriol. 28:49–51.

Rössle, M., Kreusch, J., and Decker, K. 1981. Acetyl coenzyme A and coenzyme A contents of growing *Clostridium kluyveri* as determined by isotope assays. Arch. Microbiol. 130:288–293.

Rousseau, M., Hermier, J., and Bergère, J.-L. 1971a. Structure de certains *Clostridium* du groupe butyrique. 1. Sporulation de *Clostridium butyricum* et *Clostridium saccharobutyricum.* Ann. Inst. Pasteur (Paris) 120:23–32.

Rousseau, M., Hermier, J., and Bergère, J.-L. 1971b. Structure de certains *Clostridium* du groupe butyrique. II. Sporulation de *Clostridium tyrobutyricum.* Ann. Inst. Pasteur (Paris) 120:33–41.

Rousseau, M., Hermier, J., and Bergère, J.-L. 1971c. Structure de certains *Clostridium* du groupe butyrique. III. Role de la membrane dans la formation de tuniques: Mise en évidence par l'analyse des formes anormales de sporulation. Ann. Inst. Pasteur (Paris) 121:3–12.

Roux, C., and Bergère, J.-L. 1977. Caractères taxonomiques de *Clostridium tyrobutyricum.* Ann. Microbiol. (Paris) 128a:267–276.

Rubbo, S. D., Maxwell, M., Fairbridge, R. A., and Gillespie, J. M. 1941. The bacteriology, growth factor requirements and fermentation reactions of *Clostridium acetobutylicum* (Weizmann). Australian J. Exp. Biol. Med. 19:185–198.

Ruschmann, G. 1928. Vergleichende biologische und chemische Untersuchungen an Stalldüngersorten. Mitteilung IV, Pferdemistsorten, II. Teil. Zentralbl. Bakteriol. Parasitenkd. Abt. 2 75:405–426.

Ruschmann, G., and Bavendamm, W. 1925. Zur Kenntnis der Rösterreger *Bacillus felsineus* Carbone und *Plectridium pectinovorum* (Bac. amylobacter A. M. et Bredemann). Zentralbl. Bakteriol. Parasitenkd., Abt. 2 64:340–394.

Ruschmann, G., and Harder, L. 1931. Die Buttersäuregärung im Silofutter und der Nachweis ihrer Erreger. Futterkonservierung 3:1–40.

Sacks, L. E. 1983. Influence of carbohydrates on growth and sporulation of *Clostridium perfringens* in a defined medium with or without guanosine. Appl. Environ. Microbiol. 46:1169–1175.

Sacks, L. E., and Smith, M. R. 1987. Sporulation of *Clostridium cylindrosporum* on a defined, low-manganese medium. Appl. Environ. Microbiol. 53:1696–1698.

Sacks, L. E., and Thompson, P. A. 1978. Clear, defined medium for the sporulation of *Clostridium perfringens.* Appl. Environ. Microbiol. 35:405–410.

Saddler, J. N., and Chan, M. K.-H. 1982. Optimization of *Clostridium thermocellum* growth on cellulose and pre-treated wood substrates. Eur. J. Appl. Microbiol. Biotechnol. 16:99–104.

Saddler, J. N., and Chan, M. K.-H. 1984. Conversion of pretreated lignocellulosic substrates to ethanol by *Clostridium thermocellum* in mono- and co-culture with *Clostridium thermosaccharolyticum* and *Clostridium thermohydrosulfuricum.* Can. J. Microbiol. 30:212–220.

Sagers, R. D., Benziman, M., and Gunsalus, I. C. 1961. Acetate formation in *Clostridium acidiurici:* acetokinase. J. Bacteriol. 82:233–238.

Saha, B. C., Lamed, R., and Zeikus, J. G. 1989. Clostridial enzymes, p. 227–263. In: Minton, N. P., and Clarke, D. J. (ed.), Clostridia, biotechnology handbooks 3. Plenum Press, New York.

Saint-Joanis, B., Garnier, T., and Cole, S. T. 1989. Gene cloning shows the alphatoxin of *Clostridium perfringens* to contain both sphingomyelinase and lecithinase activities. Mol. Gen. Genet. 219:453–460.

Sakka, K., Kojima, Y., Yoshikawa, K., and Shimada, K. 1990. Cloning and expression in *Escherichia coli* of *Clostridium stercorarium* strain F-9 genes related to xylan hydrolysis. Agric. Biol. Chem. 54:337–342.

Salanitro, J. P., Blake, I. G., Muirhead, P. A. 1974. Studies on the cecal microflora of commercial broiler chickens. Appl. Microbiol. 28:439–447.

Sára, M., Kupcü, S., and Sleytr, U. B. 1989. Localization of the carbohydrate residue of the S-layer glycoprotein from *Clostridium thermohydrosulfuricum* LIII-69. Arch. Microbiol. 151:416–420.

Sarathchandra, S. U., Barker, A. N., and Wolf, I. 1974. The effect of oxygen on the germination and outgrowth of three strains of *Clostridium,* p. 233–241. In: Barker, A. N., Gould, G. W., and Wolf, I. (ed.), Spore research, 1973. Academic Press, London.

Savage, M. D., and Drake, H. L. 1986. Adaptation of the acetogen *Clostridium thermoautotrophicum* to minimal medium. J. Bacteriol. 165:315–318.

Schäfer, R., and Schwartz, A. C. 1976. Catabolism of purines in *Clostridium sticklandii.* Zentralbl. Bakteriol. Parasitenkd. Infektionsk. Hyg. Abt. 1 Orig. 235:165–172.

Schäfer, R., and Schwartz, A. C. 1980. Degradation of pyrimidine bases in *Clostridium sticklandii.* Arch. Microbiol. 124:111–114.

Schallehn, G., and Brandis, H. 1973. Phosphatase-reagent for quick identification of *Clostridium perfringens.* Zentralbl. Bakteriol. Parasitenkd. Infektionskr. Hyg. Abt. 1 Orig., 225:343–345.

Scherer, P. A., and Thauer, R. K. 1978. Purification and properties of reduced ferredoxin: CO_2 oxidoreductase from *Clostridium pasteurianum,* a molybdenum iron-sulfur protein. Eur. J. Biochem. 85:125–135.

Schiefer-Ullrich, H., Wagner, R., Dürre, P., and Andreesen, J. R. 1984. Comparative studies on physiology and taxonomy of obligately purinolytic clostridia. Arch. Microbiol. 138:345–353.

Schildkraut, C. L., Marmur, J., and Doty, P. 1962. Base composition of deoxyribonucleic acid from its buoyant density in CsCl. J. Molec. Biol. 4:430–443.

Schink, B. 1984. *Clostridium magnum* sp. nov., a non-autotrophic homoacetogenic bacterium. Arch. Microbiol. 137:250–255.

Schink, B., Phelps, T. J., Eichler, B., and Zeikus, J. G. 1985. Comparison of ethanol degradation pathways in anoxic freshwater environments. J. Gen. Microbiol. 131:651–660.

Schink, B., Ward, J. C., and Zeikus, J. G. 1981. Microbiology of wetwood: importance of pectin degradation and *Clostridium* species in living trees. Appl. Environ. Microbiol. 42:526–532.

Schink, B., and Zeikus, J. G. 1983a. *Clostridium thermosulfurogenes* sp. nov., a new thermophile that produces elemental sulfur from thiosulfate. J. Gen. Microbiol. 129:1149–1158.

Schink, B., and Zeikus, J. G. 1983b. Characterization of pectinolytic enzymes of *Clostridium thermosulfurogenes*. FEMS Microbiol. Lett. 17:295–298.

Schleifer, K. H., and Kandler, O. 1972. Peptidoglycan types of bacterial cell walls and their taxonomic implications. Bacteriol. Rev. 36:407–477.

Schlote, D., and Gottschalk, G. 1986. Effect of cell recycle on continuous butanol/acetone fermentation with *Clostridium acetobutylicum* under phosphate limitation. Appl. Microbiol. Biotechnol. 24:1–5.

Schmidt, G. C., Logan, M. A., and Tytell, A. A. 1952. The degradation of arginine by *Clostridium perfringens* (BP 6K). J. Biol. Chem. 198:771–783.

Schmidt, U., Giesel, H., Schoberth, S. M., and Sahm, H. 1986. *Thermoanaerobacter finnii* spec. nov., a new ethanologenic sporogenous bacterium. Syst. Appl. Bacteriol. 8:80–85.

Schoberth, S., and Gottschalk, G. 1969. Considerations on the energy metabolism of *Clostridium kluyveri*. Arch. Mikrobiol. 65:318–328.

Schönheit, P., Brandis, A., and Thauer, R. K. 1979. Ferredoxin degradation in growing *Clostridium pasteurianum* during periods of iron deprivation. Arch. Microbiol. 120:73–76

Schönheit, P., Wäscher, C., and Thauer, R. K. 1978. A rapid procedure for the purification of ferredoxin from *Clostridium* using polyethyleneimine. FEBS Lett. 89:219–222.

Schoutens, G. H., Nieuwenhuizen, M. C. H., and Kossen, N. W. F. 1985. Continuous butanol production from whey permeate with immobilized *Clostridium beijerinckii* LMD 27.6. Appl. Microbiol. Biotechnol. 21:282–286.

Schuetz, H. J., and Simon, H. 1986. Degradation of NAD(H) by endogenous enzymes of yeasts and clostridia. Z. Naturforsch. 41c:172–178.

Schulman, M., Parker, D., Ljungdahl, L. G., and Wood, H. G. 1972. Total synthesis of acetate from CO_2. V. Determination by mass analysis of the different types of acetate formed from $^{13}CO_2$ by heterotrophic bacteria. J. Bacteriol. 109:633–644.

Schulman, M., Ghambeer, R. K., Ljungdahl, L. G., and Wood, H. G. 1973. Total synthesis of acetate from CO_2. VII. Evidence with *Clostridium thermoaceticum* that the carboxyl of acetate is derived from the carboxyl of pyruvate by transcarboxylation and not by fixation of CO_2. J. Biol. Chem. 248:6255–6261.

Schwartz, A. C., Quecke, W., and Brenschede, G. 1979. Inhibition by glycine of the catabolic reduction of proline in *Clostridium sticklandii;* evidence on the regulation of amino acid reduction. Z.Allgem. Mikrobiol. 19:211–220.

Schwarz, W. H., Schimming, S., Rücknagel, K. P., Burgschwaiger, S., Kreil, G., and Staudenbauer, W. L. 1988. Nucleotide sequence of the *celC* gene encoding endo-

glucanase C of *Clostridium thermocellum*. Gene 63:23–30.

Schweiger, G., and Buckel, W. 1985. Identification of acrylate, the product of the dehydration of (R)-lactate catalysed by cell-free-extracts from *Clostridium propionicum*. FEBS Lett. 185:253–256.

Scott, P. T., and Rood, J. I. 1989. Electroporation-mediated transformation of lysostaphin-treated *Clostridium perfringens*. Gene 82:327–333.

Scott, T. A., Bellion, E., and Martey, M. 1969. The conversion of *N*-formyl-L-aspartate into nicotinic acid by extracts of *Clostridium butylicum*. Eur. J. Biochem. 10:318–323.

Sedlmeier, H., and Simon, H. 1985. Purification and some properties of an acryloyl-CoA reductase of *Clostridium kluyveri*. Hoppe-Seyler's Biol. Chem. 366:953–961.

Seki, S., Hattori, Y., Hasegawa, T., Haraguchi, H., and Ishimoto, M. 1987. Studies on nitrate reductase of *Clostridium perfringens*. IV. Identification of metals, molybdenum cofactor, and iron-sulfur cluster. J. Biochem. 101:503–509.

Seki, Y., Seki, S., and Ishimoto, M. 1989a. The primary structure of *Clostridium perfringens* ferredoxin. J. Gen. Appl. Microbiol. 35:167–172.

Seki, Y., Seki, S., Satah, M., Ikeda, A., and Ishimoto, M. 1989b. Rubredoxin from *Clostridium perfringens:* complete amino acid sequence and participation in nitrate reduction. J. Biochem. 106:336–341.

Sergeant, K., Ford, J. W. S., and Longyear, V. M. C. 1968. Production of *Clostridium pasteurianum* in a defined medium. Appl. Microbiol. 16:296–300.

Seto, B. 1978. A pyruvate-containing peptide of proline reductase in *Clostridium sticklandii*. J. Biol. Chem. 253:4525–4529.

Seto, B. 1980. The stickland reaction, p. 49–64. In: C. J. Knowles (ed.), Diversity in bacterial respiratory systems, vol. 2. CRC Press, Boca Raton, FL.

Seto, B., and Mortenson, L. E. 1974. In vivo kinetics of nitrogenase formation in *Clostridium pasteurianum*. J. Bacteriol. 120:822–830.

Shanmugasundaram, T., Kumar, G. K., Shenoy, B. C., and Wood, H. G. 1989. Chemical modification of the functional arginine residues of carbon monoxide dehydrogenase from *Clostridium thermoaceticum*. Biochemistry 28:7112–7116.

Shapton, D. A., and Board, R. G. 1971. Isolation of anaerobes. Academic Press, London.

Sheiman, M. I., Macmillan, J. D., Miller, L., and Chase, T. 1976. Coordinated action of pectinesterase and polygalacturonate lyase complex of *Clostridium multifermentans*. Eur. J. Biochem. 64:565–572.

Shull, G. M., Thoma, R. W., and Peterson, W. H. 1949. Amino acid and unsaturated fatty acid requirements of *Clostridium sporogenes*. Arch. Biochem. 20:227–241.

Simon, H., Bader, J., Günther, H., Neumann, S., and Thanos, J. 1985. Chiral compounds synthesised by biocatalytic reductions. Angew. Chem. Int. Ed. Engl. 24:539–555.

Simon, H., White, H., Lebertz, H., and Thanos, I. 1987. Reduktion von 2-Enoaten und Alkanoaten mit Kohlenmonoxid oder Formiat, Viologenen und *Clostridium thermoaceticum* zu gesättigten Säuren und ungesättigten bzw. gesättigten Alkoholen. Angew. Chem. 99:785–787.

Sjolander, N. O. 1937. Studies on anaerobic bacteria. XII. The fermentation products of *Clostridium thermosaccharolyticum*. J. Bacteriol. 34:419–428.

Sjolander, N. O., and McCoy, E. 1937. Studies on anaerobic bacteria. A cultural study of some "butyric" anaerobes previously described in the literature. Zentralbl. Bakteriol. Parasitenkd. Infektionsk. Hyg. Abt. 2 97:314–324.

Skinner, F. A. 1960. The isolation of anaerobic cellulose-decomposing bacteria from soil. J. Gen. Microbiol. 22:539–554.

Skinner, F. A. 1971. The isolation of soil bacteria, p. 57–78. In: Shapton, D. A., and Board, R. G. (ed.), Isolation of anaerobes. Academic Press, London.

Skjelkvale, R., Stringer, M. F., and Smart, J. L. 1979. Enterotoxin production of lecithinase-positive and lecithinase-negative *Clostridium perfringens* isolated from food poisoning outbreaks and other sources. J. Appl. Bacteriol. 47:329–339.

Sleat, R., and Mah, R. A., 1985. *Clostridium populeti* sp. nov., a cellulolytic species from a woody-biomass digestor. Int. J. Syst. Bacteriol. 35:160–163.

Sleat, R., Mah, R. A., and Robinson, R. 1984. Isolation and characterization of an anaerobic, cellulolytic bacterium, *Clostridium cellulovorans* sp. nov. Appl. Environ. Microbiol. 48:88–93.

Sleytr, U. B., and Glauert, A. M. 1976. Ultrastructure of the cell walls of two closely related clostridia that possess different regular arrays of surface subunits. J. Bacteriol. 126:869–882.

Sliwkowski, M. X., and Hartmanis, M. G. N. 1984. Simultaneous single-step purification of thiolase and NAD-dependent 3-hydroxybutyryl-CoA dehydrogenase from *Clostridium kluyveri*. Anal. Biochem. 141:344–347.

Sliwkowski, M. X., and Stadtman, T. C. 1988a. Selenoprotein A of the clostridial glycine reductase complex: Purification and amino acid sequence of the selenocysteine-containing peptide. Proc. Natl. Acad. Sci. USA 85:368–371.

Sliwkowski, M. X., and Stadtman, T. C. 1988b. Selenium-dependent glycine reductase: differences in physicochemical properties and biological activities of selenoprotein A components isolated from *Clostridium sticklandii* and *Clostridium purinolyticum*. BioFactors 1:293–296.

Smith, B. E., Thorneley, R. N. F., Eady, R. R., and Mortenson, L. E. 1976. Nitrogenases from *Klebsiella pneumoniae* and *Clostridium pasteurianum*. Kinetic investigations of cross-reactions as a probe of the enzyme mechanism. Biochem. J. 157:439–447.

Smith, G. M., Kim, B. W., Franke, A. A., and Roberts, J. D. 1985. BC NMR studies of butyric fermentation in *Clostridium kluyveri*. J. Biol. Chem. 260:13509–13512.

Smith, L. DS. 1970. *Clostridium oceanicum*, sp. n., a spore-forming anaerobe isolated from marine sediments. J. Bacteriol. 103:811–813.

Smith, L. DS., and Cato, E. P. 1974. *Clostridium durum*, sp. nov., the predominant organism in a sediment core from the Black Sea. Can. J. Microbiol. 20:1393–1397.

Smith, L. DS., and Hobbs, G. 1974. Genus III. *Clostridium* Prazmowski 1880, p. 551–572. In: Buchanan, R. E., and Gibbons, N. E. (ed.), Bergey's manual of determinative bacteriology, 8th ed. Williams & Wilkins, Baltimore.

Smith, L. T., and Kaplan, N. O. 1980. Purification, properties, and kinetic mechanism of coenzyme A-linked aldehyde dehydrogenase from *Clostridium kluyveri*. Arch. Biochem. Biophys. 203:663–675.

Smith, M., and Sullivan, C. 1989. Germination of *Clostridium cylindrosporum* spores on medium containing uric acid. Appl. Environ. Microbiol. 55:1380–1385.

Sneath, P. H., Mair, N. S., and Sharpe, M. E. 1986. Bergey's manual of systematic bacteriology, vol. 2. Williams & Wilkins, Baltimore

Somack, R., and Costilow, R. N. 1973a. Purification and properties of a pyridoxal phosphate and coenzyme B_{12}-dependent D-α-ornithine 5,4-aminomutase. Biochemistry 12:2597–2604.

Somack, R., and Costilow, R. N. 1973b. 2,4-Diaminopentanoic acid C_4 dehydrogenase. J. Biol.Chem. 248:385–388.

Soutschek-Bauer, E., Hartl, L., and Staudenbauer, W. L. 1985. Transformation of *Clostridium thermohydrosulfuricum* DSM 568 with plasmid DNA. Biotechnol. Lett. 7:705–710.

Speelmans, G., De Vrij, W., and Konings, W. N. 1989. Characterization of amino acid transport in membrane vesicles from the thermophilic fermentative bacterium *Clostridium fervidus*. J. Bacteriol. 171:3788–3795.

Spiegelberg, C. H. 1944. Sugar and salt tolerance of *Clostridium pasteurianum* and some related anaerobes. J. Bacteriol. 48:13–30.

Spivey, M. J. 1978. The acetone/butanol/ethanol fermentation. Process Biochem. 13:2–4, 25.

Spormann, A. M., and Thauer, R. K. 1988. Anaerobic acetate oxidation to CO_2 by *Desulfotomaculum acetoxidans*. Arch. Microbiol. 150:374–380.

Stackebrandt, E., Pohla, H., Kroppenstedt, R., Hippe, H., and Woese, C. R. 1985. 16S rRNA analysis of *Sporomusa, Selenomonas,* and *Megasphaera:* On the phylogenetic origin of Gram-positive eubacteria. Arch. Microbiol. 143:270–276.

Stackebrandt, E., and Woese, C. R. 1981. The evolution of prokaryotes, p. 1–31. In: M. I. Carlile, I. F. Collins, and B. E. B. Moseley (ed.), Molecular and cellular aspects of microbial evolution. Cambridge University Press, Cambridge, U.K.

Stadtman, E. R. 1955. Phosphotransacetylase from *Clostridium kluyveri,* 596–599. In: Colowick, S. P., and Kaplan, N. O. (ed.), Methods in enzymology, vol. 1. Academic Press, New York.

Stadtman, E. R. 1976. The *Clostridium kluyveri*-acetyl-CoA epoch, p. 161–172. In: Kornberg, A., Horecker, B. L., Cornudella, L., and Oro, J. (ed.), Reflections on biochemistry. Pergamon Press, Oxford.

Stadtman, E. R., and Burton, R. M. 1955. Aldehyde dehydrogenase from *Clostridium kluyveri*, p. 581–583. In: Colowick, S. P., and Kaplan, N. O. (ed.), Methods in enzymology, vol. 1. Academic Press, New York.

Stadtman, E. R., Stadtman, T. C., Pastan, I., and Smith, L. 1972. *Clostridium barkeri* sp. n. J. Bacteriol. 110:758–760.

Stadtman, T. C. 1962. Studies on the enzymic reduction of amino acids. V. Coupling of a DPNH-generating system to glycine reduction. Arch. Biochem. Biophys. 99:36–44.

Stadtman, T. C. 1965. Electron transport proteins of *Clostridium sticklandii*, p. 439–445. In: San Pietro, A. (ed.), Nonheme-iron proteins: role in energy conversion. Antioch Press, Yellow Springs, Ohio.

Stadtman, T. C. 1973. Lysine metabolism by clostridia. Adv. Enzymol. 38:413–448.

Stadtman, T. C. 1978. Selenium-dependent clostridial glycine reductase, p. 373–382. In: Fleischer, S., and Packer, L. (ed.), Methods in enzymology, vol. 53. Academic Press, New York.

Stadtman, T. C., and McClung, L. S. 1957. *Clostridium sticklandii* nov. spec. J. Bacteriol. 73:218–219.

Stadtman, T. C., and White, F. H. 1954. Tracer studies on ornithine, lysine, and formate metabolism in an amino acid fermenting *Clostridium*. J. Bacteriol. 67:651–657.

Stadtman, T. C. 1989. Clostridial glycine reductase: Protein C, the acetyl group acceptor, catalyzes the arsenate-dependent decomposition of acetyl phosphate. Proc. Natl. Acad. Sci. USA, 86:7853–7856.

Stainthorpe, A. C., and Williams, R. A. D. 1988. Isolation and properties of *Clostridium thermocellum* from icelandic hot springs. Int. J. Syst. Bacteriol. 38:119–121.

Stankewich, J. P., Cosenza, B. J., and Shigo, A. L. 1971. *Clostridium quercicolum* sp. n., isolated from discolored tissues in living oak trees. Antonie van Leeuwenhoek J. Microbiol. Serol. 37:299–302.

Stellwag, E. J., and Hylemon, P. B. 1977. Characterization of 7-α-dehydroxylase in whole cells of *Clostridium leptum* VPI 10900. Abstr. Ann. Meet. Am. Soc. Microbiol. 1977:367.

Stickland, L. H. 1934. Studies in the metabolism of the strict anaerobes (genus *Clostridium*). I. The chemical reaction by which *Clostridium sporogenes* obtains energy. Biochem. J. 28:1746–1759.

Stickland, L. H. 1935. Studies in the metabolism of the strict anaerobes (genus *Clostridium*). III. The oxidation of alanine by *Clostridium sporogenes*. IV. The reduction of glycine by *Clostridium sporogenes*. Biochem. J. 29:889–898.

Stieb, M., and Schink, B. 1985. Anaerobic oxidation of fatty acids by *Clostridium bryantii* sp. nov., a sporeforming, obligately syntrophic bacterium. Arch. Microbiol. 140:387–390.

Stöcklein, W., and Schmidt, H. L. 1985. Evidence for L-threonine cleavage and allothreonine formation by different enzymes from *Clostridium pasteurianum:* threonine aldolase and serine hydroxymethyltransferase. Biochem. J. 232:621–622.

Stolp, H. 1955. Ernährungs- und entwicklungsphysiologische Untersuchungen an anaeroben Bakterien. II. Die Physiologie der Entwicklung von Clostridien unter besonderer Berücksichtigung des Reduktions-Oxydations-Potentials. Arch. Mikrobiol. 21:293–309.

Sturges, W. S., and Drake, E. T. 1927. A complete description of *Clostridium putrefaciens* (McBryde). J. Bacteriol. 14:175–179.

Suen, J. C., Hatheway, C. L., Steigerwaldt, A. G., and Brenner, D. J. 1988a. *Clostridium argentinense* sp. nov.: a genetically homogeneous group composed of all strains of *Clostridium botulinum* toxin type G and some nontoxigenic strains previously identified as *Clostridium subterminale* or *Clostridium hastiforme*. Int. J. Syst. Bacteriol. 38:375–381.

Suen, J. C., Steigerwaldt, A. G., and Brenner, D. J. 1988b. Genetic confirmation of identities of neurotoxigenic *Clostridium baratii* and *Clostridium butyricum* implicated as agents of infant botulism. J. Clin. Microbiol. 26:2191–2192.

Sugaya, D., Tuse, D., and Jones, J. L. 1986. Production of acetic acid by *Clostridium thermoaceticum* in batch and continuous fermentations. Biotechnol. Bioeng. 28:678–683.

Sukhumavasi, J., Ohmiya, K., Shimizu, S., and Ueno, K. 1988. *Clostridium josui* sp. nov., a cellulolytic, moderate thermophilic species from Thai compost. Int. J. Syst. Bacteriol. 38:179–182.

Suzzi, G., Papa, F., and Grazia, L. 1987. Pectolytic clostridia isolated from sugar pulp silages in Italy. J. Appl. Bacteriol. 63:481–485.

Switzer, R. L., and Barker, H. A. 1967. Purification and characterization of component S of glutamate mutase. J. Biol. Chem. 242:2658–2674.

Szech, U., Braun, M., and Kleiner, D. 1989. Uptake and excretion of amino acids by saccharolytic clostridia. FEMS Microbiol. Lett. 58:11–14.

Szulmajster, J. 1958a. Bacterial fermentation of creatinine. I. Isolation of N-methylhydantoin. J. Bacteriol. 75:633–639.

Szulmajster, J. 1958b. Bacterial degradation of creatinine. II. Creatinine desimidase. Biochim. Biophys. Acta 30:154–163.

Tagg, J. R., Dajani, A. S., and Wannamaker, L. W. 1976. Bacteriocins of Gram-positive bacteria. Bacteriol. Rev. 40:722–756.

Takumi, K., and Kawata, T. 1976. Quantitative chemical analyses and antigenic properties of peptidoglycans from *Clostridium botulinum* and other clostridia. Jap. J. Microbiol. 20:287–292.

Tally, F. P., Goldin, B. R., Jacobus, N. V., and Gorbach, S. L. 1977. Superoxide dismutase in anaerobic bacteria of clinical significance. Infect. Immunity 16:20–25.

Tanaka, H., and Stadtman, T. C. 1979. Selenium-dependent clostridial glycine reductase. Purification and characterization of the two membrane-associated protein components. J. Biol. Chem. 254:447–452.

Tanaka, M., Haniu, M., Yasunobu, K. T., Himes, R. H., and Akagi, J. M. 1973. The primary structure of *Clostridium thermosaccharolyticum* ferredoxin, a heat-stable ferredoxin. J. Biol. Chem. 248:5215–5217.

Tanaka, M., Haniu, M., Yasunobu, K. T., and Mortenson, L. E. 1977. The amino acid sequence of *Clostridium pasteurianum* iron protein, a component of nitrogenase. I. Tryptic peptides. II. Cyanogen peptides. III. The NH₂-terminal and COOH-terminal sequences, tryptic peptides of large cyanogen bromide peptides, and the complete sequences. J. Biol. Chem. 252:7081–7100.

Tang, I-C., Okos, M. R., and Yang, S.-T. 1989. Effects of pH and acetic acid on homoacetic fermentation of lactate by *Clostridium formicoaceticum*. Biotechnol. Bioeng. 34:1063–1074.

Tang, I-C., Yang, S.-T., and Okos, M. R. 1988. Acetic acid production from whey lactose by the co-culture of *Streptococcus lactis* and *Clostridium formicoaceticum*. Appl. Microbiol. Biotechnol. 28:138–143.

Tanner, R. S., Stackebrandt, E., Fox, G. E., and Woese, C. R. 1981. A phylogenetic analysis of *Acetobacterium woodii, Clostridium barkeri, Clostridium butyricum, Clostridium lituseburense, Eubacterium limosum,* and *Eubacterium tenue*. Curr. Microbiol. 5:35–38.

Tanner, R. S., Stackebrandt, E., Fox, G. E., Gupta, R., Magrum, L. J., and Woese, C. R. 1982. A phylogenetic analysis of anaerobic eubacteria capable of synthesizing acetate from carbon dioxide. Curr. Microbiol. 7:127–132.

Taylor, W. H., Taylor, M. L., Balch, W. E., and Gilchrist, P. S. 1976. Purification and properties of dihydroorotase, a zinc-containing metallo enzyme in *Clostridium oroticum*. J. Bacteriol. 127:863–873.

Teilliez, P., Girard, H., Millet, J., and Beguin, P. 1989. Enhanced cellulose fermentation by an asporogenous and ethanol-tolerant mutant of *Clostridium thermocellum*. Appl. Environ. Microbiol. 55:207–211.

Terracciano, J. S., Rapaport, E., and Kashket, E. R. 1988. Stress-and growth phase-associated proteins of *Clostridium acetobutylicum*. Appl. Environ. Microbiol. 54:1989–1995.

Terracciano, J. S., Schreurs, W. J. A., and Kashket, E. R. 1987. Membrane H⁺ conductance of *Clostridium thermoaceticum* and *Clostridium acetobutylicum:* evidence for electrogenic Na⁺/H⁺ antiporter in *Clostridium thermoaceticum*. Appl. Environ. Microbiol. 53:782–786.

Thanos, I., Deffner, A., and Simon, H. 1988. Reductions of 2-enals, dehydrogenation of saturated aldehydes and their racemisation. Hoppe-Seyler's Biol. Chem. 369:451–460.

Thauer, R. K., Jungermann, K., and Decker, K. 1967. A quantitative isotope method for regulation studies of aromatic amino acid synthesis under growth conditions. Eur. J. Biochem. 1:482–486.

Thauer, R. K., Jungermann, K., and Decker, K. 1977. Energy conservation in chemotrophic anaerobic bacteria. Bacteriol. Rev. 41:100–180.

Thauer, R. K., Jungermann, K., Henninger, H., Wenning, J., and Decker, K. 1968. The energy metabolism of *Clostridium kluyveri*. Eur. J. Biochem. 4:173–180.

Thauer, R. K., Käufer, B., and Scherer, P. 1975. The active species of "CO₂" utilized in ferredoxin-linked carboxylation reactions. Arch. Microbiol. 104:237–240.

Thauer, R. K., Kirchniawy, F. H., and Jungermann, K. A. 1972. Properties and function of the pyruvate-formatelyase reaction in clostridia. Eur. J. Biochem. 27:282–290.

Thauer, R. K., Rupprecht, E., and Jungermann, K. 1970. Glyoxylate inhibition of clostridial pyruvate synthase. FEBS Lett. 9:271–273.

Thauer, R. K., Rupprecht, E., Ohrloff, C., Jungermann, K., and Decker, K. 1971. Regulation of the reduced nicotinamide adenine dinucleotide phosphate-ferredoxin reductase system in *Clostridium kluyveri*. J. Biol.Chem. 246:954–959.

Thompson, D. K., and Chen, J.-S. 1990. Purification and properties of an acetoacetyl coenzyme A-reacting phosphotransbutyrylase from *Clostridium beijerinckii ("Clostridium butylicum")* NRRL B593. Appl. Environ. Microbiol. 56:607–613.

Timmis, K., Hobbs, G., and Berkeley, R. C. W. 1974. Chitinolytic clostridia isolated from marine mud. Can. J. Microbiol. 20:1284–1285.

Tokushige, M., and Hayaishi, O. 1972. Threonine metabolism and its regulation in *Clostridium tetanomorphum*. J. Biochem. 72:469–477.

Tomlinson, N. 1954. Carbon dioxide and acetate utilization by *Clostridium kluyveri*. III. A new part of glutamic acid synthesis. J. Biol.Chem. 209:605–609.

Tortora, J. C. O. 1984. Alternative medium for *Clostridium perfringens* sporulation. Appl. Environ. Microbiol. 47:1172–1174.

Touraille, C., and Bergère, J.-L. 1974. La germination de la spore de *Clostridium tyrobutyricum*. II. Démonstration de l'intervention de l'acetokinase et de la phosphotransacétylase par l'étude de leurs propriétés. Biochimie 56:404–422.

Tran, N. D., Romette, J. L., and Thomas, D. 1983. An enzyme electrode for specific determination of L-lysine: A real time control sensor. Biotechnol. Bioeng. 25:329–340.

Tran-Din, K., and Gottschalk, G. 1985. Formation of D(-)-1,2-propanediol and D(-)-lactate from glucose by *Clostridium sphenoides* under phosphate limitation. Arch. Microbiol. 142:87–92.

Tsuda, Y., and Friedmann, H. C. 1970. Ornithine metabolism by *Clostridium sticklandii*. Oxidation of ornithine to 2-amino-4-ketopentanoic acid via 2,4-diaminopentanoic acid; participation of B₁₂-coenzyme, pyridoxal phosphate, and pyridine nucleotide. J. Biol.Chem. 245:5914–5926.

Turton, L. J., Drucker, D. B., Hillier, V. F., and Ganguli, L. A. 1983. Effect of eight growth media upon fermentation profiles of ten anaerobic bacteria. J. Appl. Bacteriol. 54:295–304.

Turunen, M., Parkkinen, E., Londesborough, J., and Korhola, M. 1987. Distinct forms of lactate dehydrogenase purified from ethanol and lactate-producing cells of *Clostridium thermohydrosulfuricum*. J. Gen. Microbiol. 133:1865–1873.

Twarog, R., and Wolfe, R. S. 1963. Role of butyryl phosphate in the energy metabolism of *Clostridium tetanomorphum*. J. Bacteriol. 86:112–117.

Ueno, K., Fujii, H., Marni, F., Takahashi, J., Sugitani, T., Ushijima, T., and Suzuki, S. 1970. Acid phosphatase in *Clostridium perfringens*. A new rapid and simple identification method. Jap. J. Microbiol. 14:171–173.

Uesugi, I., and Yajima, M. 1978a. Oxygen and "strictly anaerobic" intestinal bacteria. I. The effect of dissolved oxygen on growth. Z. Allgem. Mikrobiol. 18:287–295.

Uesugi, I., and Yajima, M. 1978b. Oxygen and "strictly anaerobic" intestinal bacteria. II. Oxygen metabolism in strictly anaerobic bacteria. Z. Allgem. Mikrobiol. 18:593–601.

Ushijima, T., Sugitani, A., and Ozaki, Y. 1987. A pair of semisolid media facilitate detection of spore and enterotoxin of *Clostridium perfringens*. J. Microbiol. Methods 6:145–152.

Uyeda, K., and Rabinowitz, J. C. 1965. Metabolism of formiminoglycine. Glycine formiminotransferase. J. Biol. Chem. 240:1701–1710.

Uyeda, K., and Rabinowitz, J C. 1967a. Metabolism of formiminoglycine. Formiminotetrahydrofolate cyclodeaminase. J. Biol. Chem. 242:24–31.

Uyeda, K., and Rabinowitz, J. C. 1967b. Enzymes of clostridial purine fermentation. Methylenetetrahydrofolate dehydrogenase. J. Biol. Chem. 242:4378–4385.

Uyeda, K., and Rabinowitz, J. C. 1968. Enzymes of the clostridial purine fermentation: Serine hydroxymethyltransferase. Arch. Biochem. Biophys. 123:271–278.

Uyeda, K., and Rabinowitz, J. C. 1971. Pyruvate-ferredoxin oxido-reductase. III. Purification and properties of the enzyme. J. Biol.Chem. 246:3111–3119.

Uyeda, M., Gan, B. H., Takenobu, S., Ono, I., and Hongo, M. 1974. Extracellular formation of O-butylhomoserine by anaerobes. Agric. Biol. Chem. 38:1811–1818.

Vagelos, P. R., Earl, J. M., and Stadtman, E. R. 1959. Propionic acid metabolism. 1. The purification and properties of acrylyl coenzyme A aminase. J. Biol. Chem. 234:490–497.

Valentine, R. C., Brill, W. J., and Sagers, R. D. 1963. Ferredoxin-linked DPN reduction by pyruvate in extracts of Clostridium acidiurici. Biochem. Biophys. Res. Commun. 12:315–319.

Valentine, R. C., and Wolfe, R. S. 1960. Purification and role of phosphotransbutyrylase. J. Biol.Chem. 235:1948–1952.

Vallespinos, F., and Kleiner, D. 1980. Selective inactivation of nitrogenase and glutamate synthase during sporulation of Clostridium pasteurianum. J. Gen. Microbiol. 117:543–545.

van Andel, J. G., Zoutberg, G. R., Crabbendam, P. M., and Breure, A. M. 1985. Glucose fermentation by Clostridium butyricum grown under a self generated gas atmosphere in chemostat culture. Appl. Microbiol. Biotechnol. 23:21–26.

van Beynum, J., and Pette, J. W. 1935. Zuckervergärende und Laktatvergärende Buttersäurebacterien. Zentralbl. Bakteriol. Parasitenkd. Infektionsk. Hyg. Abt 2, 93:198–212.

van Beynum, J., and Pette, J. W. 1940. Een methode voor het aantoonen van boterzuurbacterien, speciaal geschikt voor het onderzoek van melk. Verslagen van Landbouwkundige Onderzoekingen 46C:379–396.

van Gylswyk, N. O. 1980. Fusobacterium polysaccharolyticum sp. nov., a Gram-negative rod from the rumen that produces butyrate and ferments cellulose and starch. J. Gen. Microbiol. 116:157–163.

van Gylswyk, N. O., Morris, E. J., and Els, H. J. 1980. Sporulation and cell wall structure of Clostridium polysaccharolyticum comb. nov. (formerly Fusobacterium polysaccharolyticum). J. Gen. Microbiol. 121:491–493.

van Gylswyk, N. O., and van der Toorn, J. J. T. K. 1987. Clostridium aerotolerans sp. nov., a xylanolytic bacterium from corn stover and from the rumina of sheep fed corn stover. Int. J. Syst. Bacteriol. 37:102–105.

Vanquickenborne, A., and Phillips, A. T. 1968. Purification and regulatory properties of the adenosine diphosphate-activated threonine dehydratase. J. Biol.Chem. 243:1312–1319.

van Rijssel, M., and Hansen, T. A. 1989. Fermentation of pectin by a newly isolated Clostridium thermosaccharolyticum strain. FEMS Microbiol. Lett. 61:41–46.

Veldkamp, H. 1965. Enrichment cultures of procaryotic organisms, p. 305–361. In: Norris, J. R., and Ribbons, D. W. (ed.), Methods in microbiology, vol. 3A. Academic Press, London.

Venugopal, V., and Nadkarni, G. B. 1977. Regulation of the arginine dihydrolase pathway in Clostridium sporogenes. J. Bacteriol. 131:693–695.

Venugopalan, V. 1980. Influence of growth conditions on glycine reductase of Clostridium sporogenes. J. Bacteriol. 141:386–388.

Verhasselt, P., Poncelet, F., Vits, K., van Gool, A., and Vanderleyden, J. 1989. Cloning and expression of a Clostridium acetobutylicum α-Amylase gene in Escherichia coli. FEMS Microbiol. Lett. 59:135–140.

Viljoen, J. A., Fred, E. B., and Peterson, W. H. 1926. The fermentation of cellulose by thermophilic bacteria. J. Agric. Sci. 16:1–17.

Vogel, H., Bruschi, M., and Le Gall, J. 1977. Phylogenetic studies of two rubredoxins from sulfate reducing bacteria. J. Mol. Evol. 9:111–119.

von Hugo, H., Schoberth, S., Madan, V. K., and Gottschalk, G. 1972. Coenzyme specificity of dehydrogenases and fermentation of pyruvate by clostridia. Arch. Microbiol. 87:189–202.

von Hugo, H., and Gottschalk, G. 1974a. Distribution of 1-phospho-fructokinase and PEP: fructose phosphotransferase activity in clostridia. FEBS Lett. 46:106–108.

von Hugo, H., and Gottschalk, G. 1974b. Purification and properties of 1-phosphofructokinase from Clostridium pasteurianum. Eur. J. Biochem. 48:455–463.

Vozniakovskaya, Y. M., Avrova, N. P., and Andronikashvili, E. D. 1974. Reproduction and synthesis of pectolytic enzymes by Clostridium felsineum on media with various carbon sources. Microbiology (English translation of Mikrobiologiya) 43:357–360.

Vuillet, S., Spinnler, H. E., and Blachere, H. 1986. Analysis of amino acid requirements of Clostridium thermocellum. Appl. Microbiol. Biotechnol. 23:496–498.

Waber, L. J., and Wood, H. G. 1979. Mechanism of acetate synthesis from CO_2 by Clostridium acidurici. J. Bacteriol. 140:468–478.

Wachsman, J. T., and Barker, H. A. 1954. Characterization of an orotic acid fermenting bacterium, Zymobacterium oroticum, nov. gen., nov. spec. J. Bacteriol. 68:400–404.

Wachsman, J. T., and Barker, H. A. 1955a. The accumulation of formamide during the fermentation of histidine by Clostridium tetanomorphum. J. Bacteriol. 69:83–88.

Wachsman, J. T., and Barker, H. A. 1955b. Tracer experiments on glutamate fermentation by Clostridium tetanomorphum. J. Biol.Chem. 217:695–702.

Wagner, R., and Andreesen, J. R. 1977. Differentiation between Clostridium acidiurici and Clostridium cylindrosporum on the basis of specific metal requirements for formate dehydrogenase formation. Arch. Microbiol. 114:219–224.

Wagner, R., and Andreesen, J. R. 1979. Selenium requirement for active xanthine dehydrogenase from Clostridium acidiurici and Clostridium cylindrosporum. Arch. Microbiol. 121:255–260.

Wagner, R., and Andreesen, J. R. 1987. Accumulation and incorporation of [185]W-tungsten into proteins of Clostridium acidurici and Clostridium cylindrosporum. Arch. Microbiol. 147:295–299.

Wagner, R., Commack, R., and Andreesen, J. R. 1984. Purification and characterization of xanthine dehydrogenase from Clostridium acidiurici grown in the presence of selenium. Biochim. Biophys.Acta 791:63–74.

Walther, R., Hippe, H., and Gottschalk, G. 1977. Citrate, a specific substrate for the isolation of Clostridium sphenoides. Appl. Environ. Microbiol. 33:955–962.

Wang, C. C., and Barker, H. A. 1969. Purification and properties of L-citramalate hydrolyase. J. Biol. Chem. 244:2516–2526.

Waterson, R. M., Castellino, F. J., Hass, G. M., and Hill, R. L. 1972. Purification and characterization of crotonase from Clostridium acetobutylicum. J. Biol. Chem. 247:5266–5271.

Wayman, M., and Parekh, R. 1987. Production of acetone-butanol by extractive fermentation using dibutylphthalate as extractant. J. Ferment. Technol. 65:295–300.

Weimer, P. J., and Zeikus, J. G. 1977. Fermentation of cellulose and cellobiose by *Clostridium thermocellum* in the absence and presence of *Methanobacterium thermoautotrophicum*. Appl. Environ. Microbiol. 33:289–297.

Weinberg, M., Nativelle, R., and Prevot, A. R. 1937. Les microbes anaerobies, p. 120–515. Masson et Cie, Paris.

Weinrich, A. E., and Del Bene, V. E. 1976. Beta-lactamase activity in anaerobic bacteria. Antimicrob. Agents Chemother. 10:106–111.

Weinzirl, J., and Veldee, M. V 1915. A bacteriological method for determining manurial pollution of milk. Am. J. Public Health 5:862–866.

Weiss, N., Schleifer, K.-H., and Kandler, O. 1981. The peptidoglycan types of Gram-positive anaerobic bacteria and their taxonomic implications. Rev. Inst. Pasteur (Lyon) 14:3–12.

Weizmann, C. 1919. Production of acetone and alcohol by a bacteriological process. U. S. Patent No. 1.3 15.585.

Weizmann, C., and Hellinger, E. 1940. Studies on some strains of butyric-acid-producing plectridia isolated from hemp, jute and flax. J. Bacteriol. 40:665–682.

Welch, R. W., Rudolph, F. B., and Rapoutsekis, E. T. 1989. Purification and characterization of the NADH-dependent butanol dehydrogenase from *Clostridium acetobutylicum* (ATCC 824). Arch. Biochem. Biophys. 273:309–318.

Westlake, D. W. S., Shug, A. L., and Wilson, P. W. 1961. The pyruvic dehydrogenase system of *Clostridium pasteurianum*. Can. J. Microbiol. 7:515–524.

Weyer, E. R., and Rettger, L. F. 1927. A comparative study of six different strains of the organism commonly concerned in large scale production of butyl alcohol and acetone by the biological process. J. Bacteriol. 14:399–424.

Whaley, D. N., Dowell, Jr. V. R., Wanderlinder, L. M., and Lombard, G. L. 1982. Gelatin agar medium for detecting gelatinase production by anaerobic bacteria. J. Clin. Microbiol. 16:224–229.

White, H., Lebertz, H., Thanos, I., and Simon, H. 1987. *Clostridium thermoaceticum* forms methanol from carbon monoxide in the presence of viologen dyes. FEMS Microbiol. Lett. 43:173–176.

White, H., Strobl, G., Feicht, R., and Simon, H. 1989. Carboxylic acid reductase: a new tungsten enzyme catalyses the reduction of nonactivated carboxylic acids to aldehydes. Eur. J. Biochem. 184:89–96.

Whitehead, T. R., and Rabinowitz, J. C. 1988. Nucleotide sequence of the *Clostridium acidiurici* ("*Clostridium acidi-urici*") gene from 10-formyltetrahydrofolate synthetase shows extensive amino acid homology with the trifunctional enzyme C₁-tetrahydrofolate synthase from *Saccharomyces cerevisiae*. J. Bacteriol. 170:3255–3261.

Whiteley, H. R., and Tahara, M. 1966. Threonine deaminase of *Clostridium tetanomorphum*. J. Biol. Chem. 241:4881–4889.

Wiegel, J. 1980. Formation of ethanol by bacteria. A pledge for the use of extreme thermophilic anaerobic bacteria in industrial ethanol fermentation processes. Experientia 36:1434–1446.

Wiegel, J., Braun, M., and Gottschalk, G. 1981. *Clostridium thermoautotrophicum* spec. nov., a thermophile producing acetate from molecular hydrogen and carbon dioxide. Curr. Microbiol. 5:255–260.

Wiegel, J., and Schlegel, H. G. 1977. Leucine biosynthesis: Effect of branched-chain amino acids and threonine on α-isopropyl-malate synthase activity from aerobic and anaerobic microorganisms. Biochem. Syst. Ecol. 5:169–176.

Wiegel, J., and Ljungdahl, L. G. 1986. The importance of thermophilic bacteria in biotechnology. Crit. Rev. Biotechnol. 3:39–108.

Wiegel, J., Kuk, S.-U., and Kohring, G. W. 1989. *Clostridium thermobutyricum* sp. nov., a moderate thermophile isolated from a cellulolytic culture, that produces butyrate as a major product. Int. J. Syst. Bacteriol. 39:199–204.

Wiegel, J., Mothershed, C. P., and Puls, J. 1985. Differences in xylan degradation by various noncellulolytic thermophilic anaerobes and *Clostridium thermocellum*. Appl. Environ. Microbiol. 49:656–659.

Wieringa, K. T. 1940. The formation of acetic acid from carbon dioxide and hydrogen by anaerobic sporeforming bacteria. Antonie van Leeuwenhoek J. Microbiol. Serol. 6:251–262.

Wiesenborn, D. P., Rudolph, F. B., and Papoutsakis, E. T. 1988. Thiolase from *Clostridium acetobutylicum* ATCC 824 and its role in the synthesis of acids and solvents. Appl. Environ. Microbiol. 54:2717–2722.

Wiesenborn, D. P., Rudolph, F. B., and Papoutsakis, E. T. 1989a. Phosphotransbutyrylase from *Clostridium acetobutylicum* ATCC 824 and its role in acidogenesis. Appl. Environ. Microbiol. 55:317–322.

Wiesenborn, D. P., Rudolph, F. B., and Papoutsakis, E. T. 1989b. Coenzyme A transferase from *Clostridium acetobutylicum* ATCC 824 and its role in the uptake of acids. Appl. Environ. Microbiol. 55:323–329.

Wilde, E., Hippe, H., Tosunoglu, N., Schallehn, G., Herwig, K., and Gottschalk, G. 1989. *Clostridium tetanomorphum* sp. nov., nom. rev. Int. J. Syst. Bacteriol. 39:127–134.

Wildenauer, F. X., and Winter, J. 1986. Fermentation of isoleucine and arginine by pure and syntrophic cultures of *Clostridium sporogenes*. FEMS Microbiol. Ecol. 38:373–379.

Willis, A. T. 1977. Anaerobic bacteriology: Clinical and laboratory practice, 3rd ed. London. Butterworths, Boston.

Willis, A. T., and Williams, K. 1970. Some culture reactions of *Clostridium tetani*. J. Med. Microbiol. 3:291–301.

Winkler, S. 1961. Vergleichende Untersuchungen an anaeroben Proben. Österreichische Milchwirtschaft 16:109–112.

Winnacker, E. L., and Barker, H. A. 1970. Purification and properties of a NAD-dependent glutamate dehydrogenase from *Clostridium* SB₄. Biochim. Biophys. Acta 212:225–242.

Winogradsky, S. 1895. Recherches sur l'assimilation de l'azote libre de l'atmosphère par les microbes. Archives des Sciences Biologiques (Leningrad) 3:297–352.

Winogradsky, S. 1902. *Clostridium pastorianum*, seine Morphologie und seine Eigenschaften als Buttersäureferment. Zentralbl. Bakteriol. Abt. 2, 9:43–54, 107–112.

Winter, J., Moore, L. H., Dowell, V. R., and Bokkenheuser, V. D. 1989. C-ring cleavage of flavonoids by human intestinal bacteria. Appl. Environ. Microbiol. 55:1203–1208.

Winter, J., Morris, G. N., O'Rourke-Locascio, S., Bokkenheuser, V. D., Mosbach, E. H., Cohen, B. I., and Hylemon, P. B. 1984. Mode of action of steroid desmolase and reductase sythesized by *Clostridium scindens* (formerly *Clostridium* strain 19). J. Lipid Res. 25:1124–1131.

Winter, J., Schindler, F., and Wildenauer, F. X. 1987. Fermentation of alanine and glycine by pure and syntrophic cultures of *Clostridium sporogenes*. FEMS Microbiol. Ecol. 45:153–161.

Witz, D. F., Detroy, R. W., and Wilson, P. W. 1967. Nitrogen fixation by growing cells and cell-free extracts of Bacillaceae. Arch. Mikrobiol. 55:369–381.

Woese, C. R., Maniloff, J., and Zablen, L. B. 1980. Phylogenetic analysis of the mycoplasmas. Proc. Natl. Acad. Sci. USA 77:494–498.

Woese, C. R. 1987. Bacterial evolution. Microbiol. Rev. 51:221–271.

Wood, H. G., Ragsdale, S. W., and Pezacka, E. 1986. The acetyl-CoA pathway of autotrophic growth. FEMS Microbiol. Rev. 39:345–362.

Wood, N. P., and Jungermann, K. 1972. Inactivation of the pyruvate formate lyase of *Clostridium butyricum*. FEBS Lett. 27:49–52.

Woods, D. D. 1936. Studies in the metabolism of the strict anaerobes (genus Clostridium). V. Further experiments on the coupled reactions between pairs of amino acids induced by *Clostridium sporogenes*. Biochem. J. 30:1934–1946

Woods, D. D., and Clifton, C. E. 1937. Studies in the metabolism of the strict anaerobes (genus *Clostridium*). VI. Hydrogen production and amino-acid utilization by *Clostridium tetanomorphum*. Biochem. J. 31:1774–1788.

Woods, D. R., and Jones, D. T. 1986. Physiological responses of *Bacteroides* and *Clostridium* strains to environmental stress factors. Adv. Microbiol. Physiol. 28:1–64.

Wright, J. H. 1901. A method for the cultivation of anaerobic bacteria. Zentralbl. Bakteriol. Abt. 1 Orig. 29:61.

Wu, Z., Daniel, S. L., and Drake, H. L. 1988. Characterization of a CO-dependent O-demethylating enzyme system from acetogen *Clostridium thermoaceticum*. J. Bacteriol. 170:5747–5750.

Yan, R. T., Zhu, C. X., Golemboski, C., and Chen, J. S. 1988. Expression of solvent-forming enzymes and onset of solvent production in batch cultures of *Clostridium beijerinckii* ("*Clostridium butylicum*"). Appl. Environ. Microbiol. 54:642–648.

Yoch, D. C., and Valentine, R. C. 1972. Ferredoxins and flavodoxins of bacteria. Annu. Rev. Microbiol. 26:139–162.

Yorifuji, T., Jeng, I. M., and Barker, H. A. 1977. Purification and properties of 3-keto-5-aminohexanoate cleaving enzyme from a lysine-fermenting *Clostridium*. J. Biol. Chem. 252:20–31.

Young, M., Minton, N. P., and Staudenbauer, W. L. 1989a. Recent advances in the genetics of the clostridia. FEMS Microbiol. Rev. 63:301–326.

Young, M., Staudenbauer, W. L., and Minton, N. P. 1989b. Genetics of *Clostridium*, p. 63–103. In: Minton, N. P., and Clarke, D. J. (ed.), Clostridia, biotechnology handbooks, vol. 3. Plenum Press, New York.

Youngleson, J. S., and Jones, D. T. 1989a. Homology between hydroxybutyryl and hydroxyacyl coenzyme A dehydrogenase enzymes from *Clostridium acetobutylicum* fermentation and vertebrate fatty acid β-oxidation pathways. J. Bacteriol. 171:6800–6807.

Youngleson, J. S., Jones, W. A., Jones, D. T., and Woods, D. R. 1989b. Molecular analysis and nucleotide sequence of the ADHI gene encoding and NADPH-dependent butanol dehydrogenase in the Gram-positive anaerobe *Clostridium acetobutylicum*. Gene 78:355–364.

Yu, P.-L., and Pearce, L. E. 1986. Conjugal transfer of streptococcal antibiotic resistance plasmids into *Clostridium acetobutylicum*. Biotechnol. Lett. 8:469–474.

Zamenhof, S., and Eichhorn, H. H. 1967. Study of microbial evolution through loss of biosynthetic functions: Establishment of "defective" mutants. Nature (London) 216:456–458.

Zappe, H., Jones, D. T., and Woods, D. R. 1986. Cloning and expression of *Clostridium acetobutylicum* endoglucanase, cellobiase and amino acid biosynthesis genes in *Escherichia coli*. J. Gen. Microbiol. 132:1367–1372.

Zappe, H., Jones, D. T., and Woods, D. R. 1987. Cloning and expression of a xylanase gene from *Clostridium acetobutylicum* P262 in *Escherichia coli*. Appl. Microbiol. Biotechnol. 27:57–63.

Zeikus, J. G., Lynd, L. H., Thompson, T. E., Krzycki, J. A., Weimer, P.-J., and Hegge, P. W. 1980. Isolation and characterization of a new methylotrophic acidogenic anaerobe, the Marburg strain. Curr. Microbiol. 3:381–386.

Zhao, H., Yang, D., Woese, C. R., and Bryant, M. P. 1990. Assignment of *Clostridium bryantii* to *Syntrophospora bryantii* gen. nov., comb. nov. on the basis of a 16S rRNA sequence analysis of its crotonate-grown pure culture. Int. J. Syst. Bacteriol. 40:40–44.

Zoutberg, G. R., Willemsberg, R., Smit, G., Teixeira de Mattos, M. J., and Neijssel, O. M. 1989. Aggregate-formation by *Clostridium butyricum*. Appl. Microbiol. Biotechnol. 32:17–21.

Zumft, W. G. 1978. Isolation of thiomolybdate compounds from the molybdenum-iron protein of clostridial nitrogenase. Eur. J. Biochem. 91:345–350.

CHAPTER 82

The Genus *Clostridium*—Medical

LOUIS D.S. SMITH

The pathogenic clostridia are Gram-positive, spore forming anaerobes that produce growth during toxins that necrotize body tissues or interfere with nerve transmission. They vary in their requirements for anaerobiosis, ranging from *Clostridium histolyticum,* which can grow in air, to *C. haemolyticum* which is exquisitively sensitive to traces of oxygen or to oxidized media. They all grow and produce their characteristic toxin on cooked meat medium. Further information concerning the growth of pathogenic clostridia is given by Smith and Williams (1984), Prevot et al, (1966) and, for the varieties of veterinary importance, by Sterne and Batty (1975).

The toxins of the pathogenic clostridia are traditionally referred to by Greek letters. The most poisonous toxin of each species is called the "alpha" toxin of that species, and the others (if present) are called the beta, gamma, etc. toxins. Consequently, the alpha toxin of one species probably has a quite different mode of action than the alpha toxin of another species. This scheme of nomenclature was devised when the toxins could be differentiated only by serological methods. Since then, however, several of the toxins have been found to be enzymes. Recently, a newly discovered toxin may be referred to by its properties, as the "enterotoxin" of *C. perfringens,* by an abbreviation of an outstanding property, such as the lethal toxin (LT) and the hemolytic toxin (HT) of *C. sordellii,* or simply by an alphabetical designation, as the A and B toxins of *C. difficile.*

In the following sections, where specific references are not given, the data are given in detail in Smith and Williams (1984).

Clostridium botulinum

This species consists of four groups, differing in their cultural characteristics, their optimal temperatures, and the kinds of toxin produced. The organisms in the four groups are similar in appearance, being Gram-positive rods with subterminal spores. They are basically soil organisms. Further information concerning them is given by Smith and Sugiyama (1988).

Cultural Groups

The organisms of *C. botulinum* in cultural group I have optimal temperatures of 37°C, are highly proteolytic, and produce A, B, or F toxins. They are straight to slightly curved rods with oval, subterminal spores distending the rods slightly. The spores have no exosporia or appendages. The rods are motile with peritrichous flagella. The GC content of the DNA is 26–28 mol%. *C. botulinum* tends to be found in soil that is neutral to alkaline in reaction and of low organic content. In the United States, type A is primarily found in the soil between the rise of the Rocky Mountains out of the Great Plains and the Pacific Coast, while type B is more common in the eastern part of the country, especially in the soil of the Allegheny mountains. These organisms seldom occur in the soil of Europe and Great Britain, but are more plentiful in the soil of Siberia and China.

The organisms in group I ferment glucose and fructose but are variable on maltose, salicin, and sorbitol. They do not ferment the other commonly used carbohydrates. Major fermentation products are acetic and butyric acids and minor products are isobutyric, isovaleric, isocaproic, propionic, and valeric acids, and several alcohols. Gelatin is liquefied, casein and proteins of cooked meat are digested, hydrogen sulfide is produced, lipase is produced, but not indol or urease. Growth factors required include biotin, thiamin, and *p*-aminobenzoic acid.

Species of group I share antigens. They are suscertible to erythromycin, metronidazole, penicillin, and rifampin but are resistant to gentamicin, naldixic acid, sulfamethoxazole, and trimethoprim. There is some strain variation of resistance to chloramphenicol, clindamycin, cefotoxin, and vancomycin (Swenson et al (1980).

The organisms in group II have an optimal temperature of 30°C, are not proteolytic, and produce B, E, or F toxins. Strains of type B of this group are found primarily in the soil of Europe and Great Britain. Type E strains seem to be world-wide in distribution, their major habitat being soil that is in contact with water.

They are straight rods with oval, subterminal spores. Spores of type E have characteristic appendages and exosporia; spores of type F have exosporia but no appendages. They are only moderately resistant to heat, not withstanding 10 min at 90°C. The rods are motile with peritrichous flagella, The GC content of the DNA is 26 to 28 mol%.

Organisms of group II ferment amygdalin, dextrin, fructose, galactose, glucose, glycogen, maltose, ribose, sorbitol, sucrose, and trehalose but not lactose, mannitol, melibiose, and salicin. Fermentation products are acetic and butyric acids with smaller amounts of formic, succinic, and lactic acids. Gelatin is liquefied, but casein and cooked meat are not digested. Indol and urease are not produced, nor is hydrogen sulfide formed. Nitrate is not reduced. Growth factors required are biotin, nicotinamide, thiamin, pyridoxin, and folic acid. Glycine and choline may be essential for toxin production and for normal morphology.

All organisms in this group share antigens. Strains are susceptible to chloramphenicol, clindamycin, erythromycin, penicillin, tetracycline, metronidazole, rifampin, cefoxithin, and vancomycin. They are resistant to nalidixic acid and gentamicin (Swenson et al., 1980).

The organisms in group III have an optimal temperature of 37 to 40°C, vary from being slightly proteolytic to not at all, and produce C_1, C_2, and D toxins. Organisms of this group are found primarily in soil that is moist or is in actual contact with water. Type C is apparently world-wide in distribution and outbreaks of botulism in wild ducks caused by this type have been reported from North America, South America, Europe, Africa, and Australia (Eklund and Dowell, 1987). They occur as straight rods with oval, subterminal spores. They are motile with peritrichous flagella. The GC content of the DNA is 26 to 28 mol%.

Strains of group III ferment glucose, glycerol, inositol, ribose, and xylose. They are variable on fructose, galactose, maltose, and melibiose (Oguma et al., 1986). The major fermentation products are propionic and butyric acids, with smaller amounts of acetic, valeric, isobutyric, and succinic acids. Gelatin is hydrolyzed; casein and cooked meat are digested if at all, very slowly. Hydrogen sulfide, indol, and urease are not produced. Lipase is produced on egg-yolk agar but lecithinase is not except for a very few strains.

All strains of group III are susceptible to cefotoxin, cephalothin, chloramphenicol, clindamycin, erythromycin, metronidazole, penicillin, rifampin, and tetracycline. They are resistant to nalidixic acid and gentamycin (Swenson et al., 1980)

Organisms in cultural group IV (sometimes called *C. argentinense*) have an optimal temperature of 37°C, although they will grow as high as 45°C and as low as 25°C. They are proteolytic, digesting casein and cooked meat, and producing type G toxin. They have been isolated twice from the soil of Argentina and five times from the soil of Switzerland (Sonnabend et al., 1987). They occur as straight rods with oval, subterminal spores. They are motile with peritrichous flagella. The GC conent of the DNA is 26 to 28 mol%.

Strains of group IV do not ferment any of the usual carbohydrates, although citrate may serve as an energy source, as do certain amino acids. The major fermentation products from the usual media are acetic acid, with smaller amounts of isobutyric, butyric, isovaleric, and lactic acids, and propyl and butyl alcohols. Group IV organisms produce hydrogen sulfide, but do not produce indol, urease, lipase on egg-yolk agar, or lecithinase.

Group IV strains are susceptible to cephalothin, cefoxitin, chloramphenicol, clindamycin, erythromycin, metronidazole, penicillin, rifampin, and tetracycline. They are resistant to vancomycin, nalidixic acid, and gentamicin (Swenson et al., 1980)

To the above should be added a few strains that do not fall into any of the *C. botulinum* cultural groups. One of these, an organism that produces type F toxin, culturally resembles *C. barati* (Hall et al., 1985). Another, isolated from two separate cases of infant botulism in Rome, produced type E toxin but resembled in its cultural characteristics *C. butyricum* (McCroskey et al., 1986).

Toxins

The botulinal neurotoxins are produced intracellularly with molecular weight M_r of 140,000 to 150,000. However, in foods and culture fluids, the toxins are M_r of 300,000 for the neurotoxins associate with a nontoxic protein of about the same size. They are then known as M (medium) toxins. Toxins of types A, B, and Ab associate with another hemagglutinating protein to form units of M_r 450,000, the L (large) toxins. Type A crystalline toxin is a dimer of two M_r 450,00 units and has M_r of 900,000. Full toxicity of the neurotoxins, all botulinal toxins except C_2, is not realized until a proteolytic enzyme cleaves (nicks) a peptide bond in the toxin molecule. This converts it into a dichain form whose two subunits are a heavy (H) subunit of about M_r 100,000 and light (L) subunit of about M_r 50,000.

The action of botulinal neurotoxin is rather complex. It prevents the release of acetyl choline at the neuromuscular junction. It does this as the end result of a three-step process (Simpson, 1987). The first step is the binding of the toxin to a tissue receptor of the nerve; the second is the translocation of the toxin to the interior of the nerve cell; the third is the poisoning action by which the release of the acetyl choline is prevented. The prevention of the release of the acetyl choline prevents stimuli from the nerves from reaching the muscle, which causes the paralysis of the muscle. As the paralysis of the muscles of the body progresses, the muscles involved in breathing fail and falter in their essential contractions, thus bringing about death.

Types C_1 and D toxins are induced by specific bacteriophages. If the cells are "cured" of their bacteriophage by ultraviolet treatment or by growing in a medium containing acridine orange, they become nontoxic. The bacteriophage governs the type of toxin produced, but does not affect the cultural properties of the organism.

The C_2 toxin, not governed by a bacteriophage, is a binary toxin; together, the two subunits are about 2,000 times as lethal as either of the subunits administered by itself. The H subunit binds to vulnerable cells, thus attracting the L subunit which has necrotizing action. The L subunit is an enzyme that mono-ribosylates certain amino acids (Simpson, 1984) according to the reaction:

L chain + substrate + NAD = L chain + ADP—ribose—substrate + nicotinamide.

The type G toxin was isolated as the L toxin, M_r 500,000 (Nukina et al., 1988)

The action of the botulinal neurotoxin can be prevented, but action cannot be reversed, by the action of specific antitoxin. Thus, type A anitoxin, for example, neutralizes only type A toxin, not type B toxin. Type B antitoxin, likewise neutralizes type B toxin only, not that of type A.

The Disease Botulism

Group I strains of *C. botulinum* primarily affect humans through their production of A, B, and F toxins. Some strains can grow in the body and are involved not only in food-borne botulism but also in infant and wound botulism. Food-borne botulism the best-known syndrome, is brought about by the ingestion of food that has been improperly processed and long stored. *C. botulinum* grows and produces toxin in such food and the ingestion of such preformed toxin is responsible for food-borne botulism.

In infant botulism, the organisms, following accidental ingestion, grow and produce toxin in the large intestine of infants usually less than 9 months of age. Older children and adults are seldom affected. For the years for which records have been maintained, there are about five times as many cases of infant botulism per year in the United States as there are cases of food-borne botulism.

In wound botulism, the organisms, along with other bacteria, inadvertently gain access into a penetrating wound and grow and produce botulinal toxin. Wound botulism is rare, and has been reported primarily in the western United States. Most cases are caused by type A.

Group II strains of *C. botulinum* are responsible for food-borne botulism by their production of types B, E, and F toxins. Because of their low-temperature optima they are unable to grow and produce toxin at body temperature; consequently, they do not cause infant or wound botulism. The mortality of botulism caused by group II strains is considerably lower than that induced by group I strains.

Group III strains of *C. botulinum* are responsible for the world-wide outbreaks of botulism in wild ducks, in localized outbreaks of botulism in other birds (Eklund and Dowell, 1987), and in botulism in a variety of animals, from cattle to lions. Ducks get botulism by ingesting maggots that have been living on the tissues of another duck that carries botulinal organisms and whose musculature has been invaded and became toxic after death. The ingestion of such toxic muscles does not injure the maggots themselves, but renders them highly toxic to any ducks that eat them. A similar situation holds for botulism in pheasants, for these birds are voracious eaters of maggots. In chickens, the situation seems to be more complex, involving as it does primarily broilers raised on deep litter (Dohms, 1987).

Clostridium carnis

The habitat of this organism, which rarely attacks humans, is the soil. It grows aerobically although slower and less profusely than it does anaerobically. It has an optimal temperature of 37°C, but does not grow appreciably at 25°C or at 45°C. It occurs as straight to slightly curved rods with oval, subterminal spores. It is motile with peritrichous flagella. The GC content of the DNA is 25 mol%.

It ferments glucose, fructose, amygdalin, cellobiose, maltose, mannose, salicin, and sucrose but not the other carbohydrates. The fermentation products are acetic and butyric acids with smaller amounts of formic and lactic acids. *C. carnis* slowly liquefies gelatin but does not digest casein or cooked meat. It does not produce hydrogen sulfide, lecithinase, lipase, or urease. Nitrate is not reduced to nitrite.

The toxin produced by *C. carnis* is hemolytic on the red cells of the species that are susceptible to infection by this organism. Besides an infrequent case in humans, it is pathogenic for muskrats, mink, cattle, guinea pigs, mice, rabbits, sheep, dogs, cats, sparrows, frogs, and axolotls; horses, pigeons, and chickens seem resistant to injection.

Clostridium chauvoei

This pleomorphic organism is microscopically unlike any of the other pathogenic clostridia. It seldom stains uniformly, some of the cells are definitely swollen, some appear lemon-shaped, and most show irregular staining. It apparently is a soil organism. It will grow at 37°C but not above 42°C. It occurs as straight to slightly curved rods with oval, subterminal spores. It is motile with peritrichous flagella. Most strains ferment glucose, fructose, maltose, lactose, and sucrose. Salicin may be slowly fermented by a few strains, but other carbohydrates are not attached. Fermentation products include considerable quantities of acetic and butyric acids and butanol. Biotin, nicotinic acid or amide, pantothenic acid, and pyridoxamine are required. Growth in most media seems to be related to the availability of cysteine. Large quantities of hydrogen are produced. Gelatin is liquefied, but casein and cooked meat are not digested. Nitrates are reduced to nitrites if glucose is present. Indol, lecithinase, lipase, and urease are not produced.

Toxins and Protective Antigens

The toxins and protective antigens of *C. chauvoei* are shown in Table 1. The alpha toxin apparently is not released as such, but rather as part of the soluble immunizing antigen. The M_r of the alpha toxin is about 27,000; that of the soluble immunizing antigen of which it is a part, is 53,000. The injection of this substance gives rise to protective immunity; that of the alpha toxin alone does not. The beta toxin of *C. chauvoei* is a heat-resistant deoxyribonuclease; the gamma toxin is a hyaluronidase that apparently is active in infection; the delta toxin is an ox-

Table 1. The toxins and protective antigens of *C. chauvoei*.

Toxin	Characteristics
Alpha	Lethal, necrotizing, hemolytic
Beta	Deoxyribonuclease
Gamma	Hyaluronidase
Delta	Oxygen-labile hemolysin
Soluble immunizing component	Heat-labile, protective

ygen-labile hemolysin that is related to those of other bacteria. Strains vary in their production of these substances. The alpha toxin is related to the lethal (alpha) toxin of *C. septicum*.

Diseases

C. chauvoei is primarily a pathogen of the even-toed ungulates. Various species of ungulates vary enormously in their susceptibility to infection by this organism. Cattle are by far most susceptible, followed by sheep, goats, swine, camels, and deer. Humans appear to be entirely resistant, as do horses, birds, dogs, cats, bears, coyotes, and foxes.

Most cases of *C. chauvoei* infection in cattle (conditions called Blackleg and black quarter) appear to be endogenous, arising in healthy well-fed animals which are rapidly gaining weight. *C. chauvoei* infection in sheep, however, tends to be a wound infection. The disease progresses swiftly, with death occurring in 12 to 36 h after the appearance of the first symptoms. In cattle, there are seldom any wounds or breaks in the skin. The principal lesion is always found in the voluntary muscle in some part of the body, being reddish brown to black in color and spongy because of the many small gas bubbles in it. The muscle itself may be surrounded by edema fluid. The number of *C. chauvoei* cells in the blood may rise to a million or more per milliliter.

Clostridium colinum

This organism, which causes ulcerative enteritis in upland game birds, sporulates poorly in culture and therefore has sometimes been identified as a *Eubacterium*. The spores, when they occur, are oval in shape and subterminal in position. *C. colinum* is a straight to slightly curved rod, motile with peritrichous flagella, and is difficult to isolate and carry in culture.

It ferments glucose, fructose, galactose, maltose, mannitol, mannose, raffinose, ribose, sucrose, and trehalose. Other carbohydrates are not fermented. The principal products of fermentation are acetic, formic, and propionic acids. *C. colinum* does not liquefy gelatin, nor digest casein or cooked meat. It does not reduce nitrate to nitrite, nor produce indol, hydrogen sulfide, urease, lecithinase, or lipase.

It is particularly pathogenic for young birds of the Virginia quail species (*Colinus virginianus*) which appear to be the most susceptible host under natural conditions of infection. It is also pathogenic for other species of quail, turkeys, chickens, partridges, and grouse but does not appear to cause disease in any mammalian species. Intravenous inoculation of 10^3 to 10^5 cells causes liver necrosis and ulcers in the intestinal tract of young quail with death in 18 h or less.

Clostridium difficile

C. difficile has its habitat in the human intestine. It is found in 90% of infants less than 1 year of age, but in only 30% of the infants of 2 years of age (Stark and Lee, 1982). Its occurrence in babies is dependent upon their food supply, for formula-fed infants are more apt to have *C. difficile* in their stools than breast-fed infants (Cooperstock et al., 1983). This organism is found in 7 to 14% of stools from adults.

It occurs as straight rods, with spores that are oval, subterminal in position in young cultures, becoming terminal in older ones. It is motile with peritrichous flagella. The GC content of the DNA is 28 mol%.

It ferments glucose, fructose, mannitol, and mannose. It is one of the few clostridia that is able to ferment mannitol but not maltose. Fermentation products include acetic, isobutyric, isovaleric, valeric, butyric and isocaproic acids, and ethanol and isobutanol. Gelatin is liquefied but there is no digestion of casein or cooked meat. Nitrate is not reduced to nitrite, nor is there production of indol, lecithinase, lipase, or urease.

C. difficile is susceptible to penicillin, bacitracin, metronidazole, vancomycin, sulfamethoxazole, trimethoprim, and micazole, It is resistant to cephalosporins, gentamicin, cefotoxin, streptomycin, cyclosterine, and spectinomycin, as well as to most other antimicrobial and antifungal agents. There is strain variation for sensitivity to chloramphenicol, clindamycin, erythromycin, rifamycin, and tetracycline (Wuest and Hardegger, 1988).

C. difficile produces two toxins, A and B (Bartlett et al, 1980). Toxin A, M_r 440,000 to 500,000, causes intestinal fluid accumulation (as measured in rabbit ileal loops); Toxin B, M_r 360,000 to 470,000 is primarily cytotoxic. Toxin

A cross-reacts with the hemorrhagic toxin of *C. sordellii* (Martinez and Wilkins, 1988) while toxin B of *C. difficile*, the cytotoxin, cross-reacts with the lethal (edema-producing) toxin of *C. sordellii* (Popoff, 1987)

C. difficile causes pseudomembranous colitis, chronic diarrhea, and necrotizing enterocolitis in humans. It is most common in persons after treatment with an antibiotic to which *C. difficile* is resistant but to which the other clostridia in the intestine are susceptible. Diagnosis is entirely clinical. Even the isolation of a toxin-producing strain of *C. difficile* is not evidence that it is responsible, as toxin has been found in the feces of normal infants (Donta and Myers, 1982).

Clostridium fallax

This pathogenic clostridium has occasionally been isolated from wounds, appendicitis, and chronic arthritis in humans and from sheep with a disease resembling blackleg. It is a relatively rare clostridium whose principal habitat appears to be the soil. It does not grow well in most media unless a fermentable carbohydrate is present. The optimal temperature is 37 to 45°C. It occurs as straight to curved rods with oval, eccentric to subterminal spores. It is motile with peritrichous flagella.

It ferments glucose, fructose, galactose, lactose, maltose, mannose, and ribose. It does not ferment any other carbohydrates. The principal fermentation products are acetic, butyric, and lactic acids. Gelatin is not liquefied and casein and cooked meat are not digested. Hydrogen sulfide is formed, but nitrate is not reduced to nitrite, nor is indol, lecithinase, lipase, or urease produced.

The wounds that it was isolated from were primarily edematous and described as having a fishy odor. In the laboratory, this organism rapidly loses virulence. Its toxin production was not studied.

Clostridium haemolyticum

Of all the pathogenic clostridia, members of this species are the most sensitive to oxygen. Its habitat is probably the soil and necrotic tissue in the livers of carrier animals. It is widely distributed in the world, probably as the result of the movement of carrier cattle from infected areas to uninfected areas.

C. haemolyticum is closely related to *C. novyi*. Indeed, some workers consider that it should belong to that species. True, it shares cultural characteristics and produces similar minor toxins. However, it is not pathogenic by the production of *C. novyi* alpha toxin; instead, its pathogenicity results from the production of a lecithinase, not produced by *C. novyi,* that affects the red blood cells and epthelial cells of the intestine of cattle. There is no cross-immunity between the diseases caused by *C. haemolyticum* and *C. novyi* and the vaccines for them are entirely different. For practical reasons, it is convenient to consider them as separate species.

C. haemolyticum has an optimal temperature of 37°C but does not grow at 15°C. It occurs as straight rods with oval, subterminal spores. It is motile with peritrichous flagella. The GC content of the DNA is 21 mol%. This organism ferments glucose, fructose, and mannose but not the other carbohydrates. However, strains tend to be variable on maltose. Fermentation products from glucose are acetic, butyric, and propionic acids. Lactate is converted to propionate. Gelatin is liquefied, but casein and cooked meat are not digested. Indole is formed in large amounts. Nitrate is not reduced, and hydrogen sulfide, lipase, and urease me not produced.

The lethal toxin of this organism is a calcium-requiring phospholipase C. Like the alpha toxin of *C. perfringens,* to which it shows a serological relationship, the *C. haemolyticum* toxin is inactivated by strong reducing agents and exhibits the same anomolous heat inactivation. Minor toxins produced by *C. haemolyticum* are a tropomyosinase (*C. novyi* eta toxin) and a lipase (*C. novyi* theta toxin) that is active on fats containing oleic acid.

Cattle are the only animal affected by *C. haemolyticum*. The majority of cases also involve the liver fluke (*Fasciola hepatica*). Apparently, the cattle ingest the spores which are taken up by the Kupffer cells of the hepatic sinusoids. The spores remain there unless they happen to lie in the path of an immature liver fluke blindly burrowing its way through the liver in search of a bile duct. Exposed to the necrotic tissue that results, the spores germinate and produce toxin. Although the resulting infarct occupies only a small part of the liver, enough toxin is produced to lyse a majority of the red cells and cause death.

Clostridium histolyticum

This organism, which grows aerobically as well as anaerobically, probably has its principal habitat in the soil, although it was isolated from the intestinal tract of humans. It has an optimal temperature of 37°C and occurs as a straight rod with oval, subterminal spores. There is no exosporium. It is sluggishly motile with pertrichous flagella. It is unable to ferment any of the usual carbohydrates, although it is active in attacking amino acids, especially glycine. The only fermentation product is acetic acid. Indol is not produced and nitrate is not reduced. Riboflavine is required; some strains also require thiamine. Amino acids required are histidine, arginine, methionine, proline, aspartic acid, isoleucine, leucine, tyrosine, phenylalanine, glutamic acid, valine, lysine, threonine, and cysteine. Some strains also require tryptophane and alanine.

Toxins

Alpha toxin is the major lethal factor produced by *C. histolyticum*. It is formed early in culture but is rapidly inactivated by the proteolytic enzymes. It is serologically related to the alpha toxin of *C. septicum*. The beta toxin consists of two or more collagenases, M_r 72,000 to 80,000. They are related serologically and require calcium for activity. They hydrolyze collagen as well as gelatin and are not inhibited by trypsin inhibitor. The gamma toxin consists of two proteases, M_r of 20,000 and 50,000, capable of hydrolyzing casein and other proteins. The delta toxin consists of one or more elastases. Epsilon toxin is an oxygen-sensitive hemolysin, similar to the oxygen-sensitive hemolysins of other bacteria.

Infections of wounds by *C. histolyticum* are rare and occur mostly in wounds of warfare. Recovery from a well-established wound infection is rare. The muscle is dark in color and may show by softening the effect of the collagen-digesting enzymes.

Clostridium limosum

This organism may have its principal habitat in the soil, although it was isolated only once from a sample of mud from the Ivory Coast. It was found in a few infections of a variety of species, including those of humans, cattle, chickens, pigs, alligators, cats, mules, and ducks. The optimal temperature for growth appears to be 37°C, although it grows at both 25 and 45°C. It occurs as a straight rod with oval, subterminal spores. It is motile with peritrichous flagella. The GC content of the DNA is 26 mol%. Like *C. histolyticum,* with which it was sometimes confused, *C. limosum* does not ferment any of the usual carbohydrates, getting its principal energy source from the fermentation of amino acids. The principal fermentation product is acetic acid. It liquefies gelatin, digests casein and cooked meat, and produces lecithinase as well as collagenase. It does not reduce nitrate to nitrite, nor produce indol, lipase, or urease.

The pathogenic action of this organism seems to be related to the action of the lecithinase or collagen. Nevertheless, most strains lose pathogenicity rapidly in the laboratory.

Clostridium novyi

C. novyi is divided into three types, A, B, and C, on the basis of the toxins produced (Table 2). *C. haemolyticum* is sometimes referred to as *C. novyi* type C.

Type A (*C. oedematiens*) strains of *C. novyi* are found in soil, marine sediments, and the intestinal contents of healthy cattle, sheep, and whales. The optimal temperature is high, 40 to 45°C. It occurs as straight rods with oval, subterminal spores that have no exposporia and no appendages. It is motile with peritrichous flagella. The GC content of the DNA is 23 mol%.

It ferments glucose, maltose, and ribose with strain variation on fructose, glycerol, and inositol. The products of fermentation are acetic, propionic, and butyric acids. Lactate is fermented to yield propionate. Gelatin is liquefied, but casein and cooked meat are not digested. Nitrate is not reduced to nitrite, nor is indol formed. However, lecithinase and lipase are produced.

Type A strains produce several biologically active substances including the lethal alpha toxin, and gamma, delta, and epsilon toxins. They are important in producing gas gangrene in humans. This has a long incubation period and a high mortality rate. It is marked by the quantity of edema fluid produced, often amounting to 2 liters per day.

Type B strains of *C. novyi* (sometimes called *C. gigas*) has been found in the soil only where it was contaminated by animals, particularly sheep. The optimal temperature is 37°C, but it will grow at 15 and 45°C. It occurs as large straight rods with oval, subterminal spores without appendages or exosporia. It is motile with peritrichous flagella. The GC content of the DNA is 26 mol%

Type B strains produce alpha, beta, zeta, and eta toxins. Although wound infections with type B occur, much more common is liver infection in sheep. Infectious necrotic hepatitis (black disease, bradsot) occurs when the path of the metacercariae of *Fasciola hepatica,* the liver fluke, happens to cross an area where spores of type B are lying. The spores then germinate and produce lethal amounts of alpha toxin. The name of the disease is derived from the black appearance of the inside of the hide, due to the loss of blood caused by the increased permeability of the arterioles and venules.

Type C strains (sometimes called *C. bubalorum*) were originally isolated from cases of osteomyelitis in water buffalo. This apparently is not a soil organism. The optimal temperature is high, about 45°C. It occurs as a straight rod with oval, subterminal spores, without exosporia or appendages. It is sluggishly motile with peritrichous flagella.

This organism weakly ferments glucose, inositol, and ribose. Fermentation products include large amounts of propionic and butyric acids with smaller amounts of acetic acid. Gelatin is liquefied by some strains, but casein and cooked meat are not digested. Nitrate is not reduced to nitrite. Indol is formed by a few strains, but hydrogen sulfide is not formed. This organism may produce small amounts of gamma toxin. Aside from chronic osteomyelitis, type C strains are not pathogenic.

Clostridium perfringens

This nonmotile organism is the most frequently isolated clostridium in clinical microbiology laboratories, although it seldom causes serious

Table 2. The toxins of *C. novyi*.

Toxin	Characteristics	Produced by type		
		A	B	C
Alpha	Necrotizing, lethal	++	++	−
Gamma	Necrotizing, lecithinolytic, hemolytic	+	tr	−
Delta	Oxygen-labile hemolysin	+	−	−
Epsilon	Lipolytic	+	−	−
Zeta	Hemolytic	−	+	−
Eta	Tropomyosinase	−	+	−
Theta	Lipase	−	tr	−

+, produced; ++, produced in lethal amounts; tr, trace amounts; −, not produced.

infections. It has been known under a number of names, but only one synonym, *C. welchii*, is still occasionally used. This species is divided into five types, A to E, on the basis of production of major lethal toxins (Table 3).

The habitat of *C. perfringens* depends upon the type. The principal habitats of type A are the soil and the intestines of humans, animals, and birds. The habitat of types B, C, D, and E appears to be the intestine of animals. The optimal temperature for the growth of strains of types A, D, and E is 44 to 45°C; at this temperature the organisms have a generation time of 9 to 12 min. Strains of types B and C have an optimal temperature of 37 to 40°C. *C. perfringens* ferments glucose, fructose, galactose, inositol, lactose, maltose, mannose, starch, and sucrose. There is strain variability on cellobiose, glycerol, inulin, raffinose, and salicin. The fermentation products include acetic and butyric acids, with or without butanol. Gelatin is liquefied, but casein and cooked meat are not digested. Nitrate is reduced, and hydrogen sulfide and indol are not produced. Lecithinase is produced but lipase and urease are not. However, a few strains of types B and E may produce urease. The GC content of the DNA is 25 to 27 mol%.

Toxins

The alpha toxin is a phospholipase C, splitting lecithin to phosphoryl choline and a diglyceride. This enzyme also attacks sphingomyelin and the phosphoglycerides of ethanolamine and serine. The presence of calcium is necessary for activity. Like many other lecithinases of bacterial origin, *C. perfringens* alpha toxin shows anomalous heat inactivation in culture fluids, being inactivated more rapidly at 55–75°C than at 100°C. Also, the toxin is inhibited by strong reducing agents. The M_r is about 30,000 and it has a major pI value of 5.49. The toxin as it is produced by the bacteria is inactive and becomes active after it unites with zinc. It is markedly hemolytic on the red cells of humans and most animals, but only slightly so on those of horses and goats.

Intravenous inoculation of the alpha toxin results in massive intravascular hemolysis. Because of the destruction of the platelets, there is at first a decrease in clotting time, followed by an abnormally great increase. Intramuscular injection of alpha toxin usually results in the fixation of the alpha toxin at the site of injection.

The beta toxin is produced by strains of types B and C. It has an M_r of about 30,000 and a major PI of 5.5. It is responsible, when it is produced in the intestine, for the wholesale loss of mucosa and the inhibition of intestinal movement. It is exquisitely sensitive to digestion by trypsin.

Delta toxin is produced by strains of types B and C. It is markedly hemolytic on the red cells of even-toed ungulates, but not on those of humans or other animals. It has M_r of 42,000 and is lethal for mice in a dose of 0.12 μ.

The epsilon toxin is produced by strains of types B and D as a slightly lethal prototoxin that has M_r of 33,000. It is activated by trypsin or some other proteolytic enzyme, becoming about 400 times more toxic. It is inactivated on longer contact with this enzyme.

The theta toxin (perfringolysin 0) is an oxygen-sensitive hemolysin, readily inactivated by certain steriods, especially cholesterol. Reactivation after mild exposure to oxygen is brought about by exposure to active reducing agents. It has an M_r of 74,000 and about 8,000 mouse lethal doses per mg of N.

Iota toxin is produced as a binary toxin by strains of type E. Individually, the two subunits have little activity, although when combined they are 64 times as toxic for mice. Iota alpha (Ia), one of the subunits, has an M_r of 47,500. Iota beta (Ib) has an M_r of 71,500 (Stiles and Wilkins, 1986). The light chain is an enzyme that mono-ADP-ribosylates certain amino acids. The heavy chain has little or no enzyme activity (Simpson et al., 1987).

The kappa toxin is a collagenase produced by strains of types A, D and E as well as by some strains of types B and C. It is an enzyme-hydrolyzing native collagen and gelatin. It has M_r of 80,000 and the intravenous lethal doses for mice is about 30 μg, causing hydrolysis of the reticulin framework of the lungs with massive hemorrhage.

Lambda, mu, and nu antigens produced by strains of *C. perfringens* are, respectively, a protease, a hyaluronidase, and a deoxyribonuclease. None of them are known to play an important part in infections caused by this organism. *C. perfringers* also produces a neuraminidase, M_r 64,000, and three sialidases with M_r of 310,000; 105,000; and 64,000.

Table 3. Major lethal toxins and types of *C. perfringens*.

Type	Toxins present			
	Alpha	Beta	Epsilon	Iota
A	+	−	−	−
B	+	+	+	−
C	+	+	−	−
D	+	−	+	−
E	+	−	−	+

+, present; −, not formed.

The enterotoxin is formed by most strains of type A when they sporulate. This is a protein of M_r 34,000 and a toxicity of 2,000 mouse lethal doses when injected intravenously. In humans, the mechanism of action is to reverse the transport of water and sodium chloride across the intestinal mucosa, with the result that these pass into the lumen of the intestine. Several media for the sporulation of *C. perfringens* have been described. One of the most recent is that of Phillips (1986), a blood agar base with a pH of 8.5, which contains 0.05% quinoline as well as bile and sodium bicarbonate.

Diseases

C. perfringens is the organism most commonly found in gas gangrene in humans. Although alpha toxin seems responsible for the initiation of the infection; it apparently does not get into the circulating blood. Death may be due to a lack of oxygen caused by the lowered blood volume and the ability of *C. perfringens* to remove oxygen from the circulating blood.

This organism is a common cause of food poisoning due to the formation of the enterotoxin in the intestine. The minimum dose to initiate food poisoning upon ingestion is 10^9 living vegetative cells that sporulate in the small intestine. *C. perfringens* food poisoning is seldom fatal, being marked by diarrhea and nausea, with no vomiting and no fever.

Necrotic enteritis (necrotizing jejunitis, pigbel, enteritis necroticans) is caused by strains of B and C in young animals and type C in humans. In newborn calves and lambs, the animals pick up the organisms from immune adults. In humans most cases are found in New Guinea, being associated with the eating of undercooked pork. In both newborn animals and in humans, the disease is associated with a lack of trypsin. Newborn ruminants seldom have sufficient trypsin in the small intestine for the first few weeks of life, and the underdone pork in New Guinea is eaten together with sweet potatoes which contain a heat-resistant trypsin inhibitor. Mortality in humans is 30 to 40%. This disease in animals is found in Europe, Africa, and the Middle East.

Type D enterotoxemia occurs frequently in young lambs, especially when they are single lambs on ewes that give copious quantities of milk. The disease is caused by the formation of epsilon toxin in the small intestine, growing on the excess food that the lamb takes in. The epsilon toxin increases capillary permeability; death results from edema of the brain.

Clostridium septicum

This organism (also called *Bacillus septicus, B. oedematis*) is found both in the soil and the intestinal contents of animals. It has an optimal temperature of 35 to 40°C. Although most often seen in culture as short, even-staining rods with oval, subterminal spores; it may occur as long cells in infected muscle. It often appears to swarm over the surface of blood agar instead of growing in individual colonies. It is motile with peritrichous flagella. The GC content of the DNA is 24 mol%.

C. septicum liquefies gelatin but does not digest casein or cooked meat. It does not form hydrogen sulfide or indol, lipase, lecithinase, or urease. It ferments glucose, fructose, cellobiose, maltose, lactose, and mannose. There is some strain variation on sucrose. Fermentation products are large amounts of acetic and butyric acids with smaller amounts of ethyl, isobutyl, and butyl alcohols. This organism has relatively complicated nutritional requirements. There is an absolute requirement for ammonium ion.

Toxins

The alpha toxin of *C. septicum* is the only lethal toxin produced by this organism. It lyses the red cells of man, cattle, sheep, pigs, and rabbits but not those of dogs, horses, guinea pigs, and chickens. In small doses it increases capillary permeability. Fundamentally, the alpha toxin of *C. septicum* is markedly similar in all its properties to the alpha toxin of *C. histolyticum* to which it is related serologically.

The beta toxin of *C. septicum* is a deoxyribonuclease related to the deoxyribonuclease of *C. chauvoei*. The gamma toxin is a hyaluronidase and the delta toxin is an oxygen-labile hemolysin. Other agents with biological activity are produced by this organism but are not important in any infection caused by it.

Diseases

C. septicum is sometimes the cause of gas gangrene in humans either alone or in company with other clostridia. The gas gangrene caused by this organism tends to have a short incubation period and the onset is abrupt. The muscles affected are bright red in color with purple mottling and copious edema. *C. septicum* infections are not uncommon in civilian life often being associated with malignancy, especially leukemia.

Cattle seem to be particularly susceptible to infections with this organism, especially following minor wounds, a condition involving muscle infection known as "malignant edema." However, this organism tends to invade all tissues *post-mortem* and may be recovered from lesions that it has not caused.

Other animals are susceptible to *C. septicum* infections including mink, seals, gnus, red lechee, eland, chickens, dogs, rabbits, goats, horses, and swine but to a lesser extent than cattle. In most of these animals, once an infection with *C. septicum* progresses, a condition similar to gas gangrene in humans occurs.

Clostridium sordellii

This organism is found primarily in the soil although it can take up residence in the large bowel of animals. It occurs as a straight rod with oval, subterminal spores with thick exosporia and some appendages. It is motile with peritrichous flagella. The GC content of the DNA is 24 mol%. *C. sordellii* is moderately proteolytic, liquefies gelatin and slowly digests casein and cooked meat but not collagen. Nitrate is not reduced, indol and lecithinase are produced but not lipase. Glucose, fructose, and maltose are fermented. There may be some strain variation on arabinose, glycerol, raffinose, ribose, and cellulose. The products of fermentation are largely acetic, isobutyric and isovaleric acid, and smaller amounts of propionic and isocaproic acids.

Several toxic substances are produced. The major lethal toxin (toxin LT) is the edema-producing toxin, M_r 250,000 to 280,000. This edema-producing toxin of *C. sordellii* is serologically similar to toxin B of *C. difficile* (Popoff, 1987). A hemorrhagic toxin (toxin HT) is also produced (Martinez and Wilkins, 1988) that has M_r 300,000. It is related to toxin A of *C. difficile*. Both the hemorrhagic toxin of *C. sordellii* and toxin A of *C. difficile* are enterotoxins causing the accumulation of fluid and mucus as well as hemorrhage in the rabbit ileal loop assay. These toxins are not identical, however. Also produced are a neuraminidase, a deoxyribonuclease, and an oxygen-labile hemolysin. None of these seems to be of any importance in the infections caused by *C. sordellii*.

This organism causes edematous wound infections in humans that may be confused with *C. novyi* infections. In cattle and sheep, liver infections may occur, particularly when induced by liver fluke infestation. Enterotoxemia in cattle, sheep, and foals may be a problem (Al Mashat and Taylor, 1983).

Clostridium spiroforme

This organism is unlike the other pathogenic clostridia morphologically for it occurs in the form of spirals and curved rods. It is one of the few pathogenic clostridia that is nonmotile. It has been isolated principally from cases of enterotoxemia in rabbits (Boriello and Carmun, 1983). The optimal temperature is 35 to 40°C. It is not proteolytic, does not liquefy gelatin or digest casein or cooked meat. It is almost inactive biochemically, does not form indol, reduce nitrate to nitrate, nor produce lecithinase or lipase. It ferments glucose, fructose, lactose, mannitol, mannose, sucrose, and trehalose. The products of fermentation are acetic and formic acids with smaller amounts of lactic and pyruvic acids. It is susceptible to chloramphenicol, erythromycin, and tetracycline. It is resistant to clindamycin and penicillin.

Like the C_2 toxin of *C. botulinum* (Ohishi and Tsuyama, 1986) and the iota toxin of *C. perfringens* (Stiles and Wilkins, 1986) the toxin of *C. spiroforme* is binary, consisting of two separate polypeptide chains (M_r 45,000 and 68,000). The lighter chain has enzymic activity, having mono ADP-ribosyl transferase activity (Simpson et al., 1989). The heavy chain does not have ADP-ribosyl transferase activity, nor does it enhance the activity of the light chain. The toxic activity in laboratory animals is neutralized by *C. perfringens* iota antitoxin.

It is the principal cause of juvenile eneritis in rabbits, although the etiology of the disease is complex. However, *C. spiroforme* is the principal pathogen isolated. It is carried by diarrheic rabbits (Peeters et al., 1986)

This pathogenic clostridia is not identical with the organism originally described as *C. spiroforme* that was isolated from human feces and that is nonpathogenic. Future study may suggest another name for the pathogenic species.

Clostridium tetani

This is primarily a soil organism, being found in about one-third of the soil samples examined for it in Brazil, Canada, Japan, and the United States. Optimal temperature for growth is around 37°C; there is no growth at 45°C. This organism occurs as straight rods with terminal spherical spores, without exosporia or appendages. Not all strains sporulate well. Most strains are motile with peritrichous flagella, but a few are not motile. Motile strains tend to swarm over the surface of solid media. The GC content of the DNA is 25 mol%.

There is considerable strain variation on the liquefaction of gelatin, on the production of indol, deoxyribonuclease, lipase, and fibrinolysin. There is also variation on the fermentation of glucose. The fermentation products from the usual carbohydrate media are acetic, butyric, and propionic acids and butanol. Fermentation of glutamic acid yields acetic and butyric acid, carbon dioxide, and ammonia. That of aspartic acid yields lactic acid and ethanol.

Growth factors required are biotin, folic acid, nicotinic acid, pantothenate, pyridoxamine, and uracil. Some strains also need adenine, oleic acid, riboflavine, and thiamin. The amino acids required include arginine, histidine, leucine, tryptophane, tyrosine, and valine.

Toxins

C. tetani produces two toxins, a spasmogenic neurotoxin and an oxygen-sensitive hemolysin (tetanolysin). The neurotoxin has M_r of 150,000 and 10^8 lethal dose for mice per mg of N. It is composed of two subchains which, after "nicking" and separation, consist of a heavy chain of M_r 100,000 and a light chain of M_r 50,000. The heavy chain is responsible for binding to the ganglioside in the membrane of the neuronal cells; the light chain is responsible for the toxicity. The action of the neurotoxin is to prevent the release of the neurotransmitter substance, glycine or gamma amino butyric acid, in the inhibitory nerve system of the spinal cord. It is this inhibitory system that prevents the contraction of a muscle when the muscle with opposite action contracts. Interference with the release of the inhibitory-transmitter substance allows the ungoverned propagation of impulses through all the ramifying connections of the motor neurons in the central nervous system, which leads to the spasmodic contraction of the muscles.

All animals are not equally sensitive to the toxin (Table 4). Descending tetanus is the form most often seen in humans and horses. The most susceptible nerve centers are those of the head and neck, followed by those of the upper part of the body. In humans the first symptom is often spasms of the jaw muscles, causing inability to open the mouth and giving rise to the name of "lockjaw."

The oxygen-sensitive hemolysin tetanolysin is serologically related to similar hemolysins produced by other bacteria including streptolysin O. It occurs as two hemolysins, M_r 48,000 and 53,000. Its action on red cells is markedly inhibited by cholesterol. Human red cells are the most sensitive to the oxygen-labile hemolysin, followed by those of pigeons, cattle, swine, rats, sheep, and chickens

Table 4. Amounts of tetanus toxin required to kill different species (equivalent body-weight basis).

Horse	1
Human	1*
Goat	6
Mouse	12
Rat	12
Rabbit	24
Monkey	48
Guinea Pig	72
Dog	600
Cat	7,200
Pigeon	12,000
Hen	300,000

*Estimated value.

Disease

Tetanus results in humans and the horse from small, trivial wounds, mostly. *C. tetani* is not a histotoxic organism and its growth is restricted to the tissue that was traumatized when the organism was introduced. The incubation period of tetanus varies from a few days to several weeks, with mortality being higher in those cases with shorter incubation periods. The symptoms are primarily those of spasmodic muscular contraction, including those of the breathing muscles. Diagnosis is made on clinical findings and careful history of patient, for *C. tetani* can be recovered from wounds in unimmunized patients in whom the disease does not occur and vice versa. Most cases in the United States occur following puncture wounds or lacerations of the leg. Crude obstetrical procedures involving the umbilical cord are sometimes responsible for tetanus in newborn infants in developing countries.

Literature Cited

Al Mashat, R. R., and D. J. Taylor. 1983. Production of diarrhea and enteric lesions in calves by the oral inoculation of pure cultures of *Clostridium sordellii*. Vet.Rec. 112:141–146.

Bartlett, J. G., N. S. Taylor, T-W Chang, and J. Dzink. 1980. Clinical and laboratory observations in *Clostridium difficile* colitis. Am. J. Clin. Nutr. 33:2521–2526.

Borriello, S. P., and R. J. Carman. 1983. Association of iota-like toxin and *Clostridium spiroforme* with spontaneous and antibiotic-associated diarrhea and colitis in rabbits. J. Clin. Microbiol. 17:414–418.

Cooperstock, M., L. Rigle, C. W. Woodruff, and A. Onderdonk. 1983. Influence of age, sex, and diet on asymptomatic colonization of infants with *Clostridium difficile* J. Clin. Microbiol. 17:830–833.

Dohms, J. E. 1987. Laboratory investigation of botulism in poultry, p. 295–314. In: M. E. Eklund and V. R., Jr., Dowell. (ed.), Avian botulism. C. C. Thomas, Springfield, IL.

Donta, S. T., and M. G. Myers. 1982. *Clostridium difficile* in asymptomatic neonates. J. Pediatr. 100:431–434.

Eklund, M. E. and V. R., Jr., Dowell. 1987. Avian botulism. C. C. Thomas, Springfield, IL.

Hall, J. D., L. M. McCroskey, B. . Pincomb, and C. L. Hatheway. 1985. Isolation of an organism resembling *Clostridium barati* which produces type F botulinal toxin from an infant with botulism. J.Clin. Microbiol. 21:654–655.

Martinez, R. D., and T. D. Wilkins. 1988. Purification and characterization of *Clostridium sordellii* hemorrhagic toxin and cross-reactivity with *Clostridium difficile* toxin A (enterotoxin). Infect. Immun. 56:1215–1221.

McCroskey, L. M., C. L. Hatheway, L. Fenicia, B. Pasolini, and P. Aureli. 1986. Characterization of an organism that produces type E botulinum toxin but which resembles *Clostridium butyricum* from the feces of an infant with type E botulism. J. Clin. Microbiol. 23:201–202.

Nukina, M., Y. Mochida, S. Sakaguchi, and G. Sakaguchi 1988. Purification of *Clostridium botulinum* type G progenitor. Zbl. Bakt. Hyg. A 268:220–227.

Oguma, K., T. Yamaguchi, K. Sudou, Y. Yokosawa, and Y. Fujikawa. 1986. Biochemical classification of *Clostridium botulinum* type C and D strains and their non-toxigenic derivatives. Appl. Environ. Microbiol. 51:256–260.

Ohishi, I., and S. Tsuyama. 1986. ADP-ribosylation of non-muscle actin with component I of C_2 toxin. Biochem. Biophys. Res. Commun. 136:802–806.

Peeters, J. E., R. Geeroms, R. J. Carman, and T. D. Wilkins. 1986. Significance of *Clostridium spiroforme* in the enteritis-complex of commercial rabbits. Vet. Microbiol. 13:25–31.

Phillips, K. D. 1986. A sporulation medium for *Clostridium perfringens*. Let. Appl. Microbiol. 3:77–79.

Popoff, M. R. 1987. Purification and characterization of *Clostridium sordellii* lethal toxin and cross-reactivity with *Clostridium difficile* cytotoxin. Infect. Immun. 55:35–43.

Prévot, A. R., A. Turpin, and P. Kaiser. 1966. Les bacteries anaerobies. Dunod, Paris.

Simpson, L. L. 1984. Molecular basis for the pharmacological actions of *Clostridium botulinum* type C_2 toxin. J. Pharmacol. Exp. Ther. 230:665–669.

Simpson, L. L., B. G. Stiles, H. H. Zepeda, and T. D. Wilkins. 1987. Molecular basis for the pathological actions of *Clostridium perfringens* iota toxin. Infect. Immun. 55:118–122.

Simpson, L. L., B. G. Stiles, H. Zepeda, and T. D. Wilkins. 1989. Production by *Clostridium spiroforme* of an iotalike toxin that possesses mono(ADP-ribosyl)-transferase activity: Identification of a novel class of ADP-ribosyltransferases. Infect. Immun. 57:255–261.

Smith, L. DS., and H. Sugiyama. 1988. Botulism. The organism, its toxins, the disease. C. C. Thomas, Springfield, IL.

Smith, L. DS., and B. L. Williams. 1984. The pathogenic anaerobic bacteria. C. C. Thomas, Springfield, IL.

Sonnabend, W. F., U. P. Sonnabend, and T. Krech. 1987. Isolation of *Clostridium botulinum* from Swiss soil specimens by using sequential steps in an identification scheme. Appl. Environ. Microbio. 53:1880–1884.

Stark, P. L., and A. Lee. 1982. The microbial ecology of the large bowel of breast-fed and formula-fed infants during the first year of life. J. Med. Microbiol.15:189–203.

Sterne, M., and I. Batty. 1975. Pathogenic clostridia. Butterworths, London.

Stiles, B. G., and T. D. Wilkins. 1986. Purification and characterization of *Clostridium perfringens* iota toxin: dependence on two linked proteins for biological activity. Infect. Immun. 54:683–688.

Swenson, J. M., C. Thornsberry, L. M. McCroskey, C. L. Hatheway, and V. R., Jr., Dowell. 1980. Susceptibility of *Clostridium botulinum* to 13 antimicrobial agents. Antimicrob. Agents Chemother. 18:13–19.

Wuest, J., and U. Hardegger, 1988. Studies on the resistance of *Clostridium difficile* to animicrobial agents. Zbl. Bakt. Hyg. A 267:383–394.

CHAPTER 83

The Anaerobic Gram-Positive Cocci

TAKAYUKI EZAKI, HIROSHI OYAIZU, and EIKO YABUUCHI

The anaerobic Gram-positive cocci discussed in this chapter are the peptococci, peptostreptococci, ruminococci, coprococci, and sarcinae. Most of these belong to the normal flora of human and animals and are isolated from various purulent infectious diseases. Sarcinae and some species of genus *Ruminococcus* are found in mud and soil.

The taxonomy of the anaerobic Gram-positive cocci has been revised during the last 10 years. Most members of the genus *Peptococcus* were transfered to the genera *Staphylococcus* (Kilpper-Bälz and Schleifer, 1981; Ludwig et al., 1981), *Streptococcus* (Cato, 1983), and in particular *Peptostreptococcus* (Ezaki and Yabuuchi, 1983), only one species, *Peptococcus niger*, remained in the genus *Peptococcus*. Members of the genus *Peptostreptococcus* are asaccharolytic except for one strongly saccharolytic species, *P. productus*. Asaccharolytic peptostreptococci have as a common characteristic the utilization of peptone as an energy source, but comparative genetic analyses suggest that they are a rather heterogeneous group. Six nonsaccharolytic species of the genus *Peptostreptococcus* studied by Huss et al. (1984) and Ludwig et al. (1981) were at least separated into five groups according to RNA-DNA hybridization and peptidoglycan analysis. Our recent preliminary 16S ribosomal sequence analysis among anaerobic Gram-positive cocci also suggests the heterogeneity of the peptostreptococci.

The genus *Ruminococcus* is also composed of a heterogenous group of organisms. Cellulose nondigesting ruminococci may be different from cellulolytic ruminococci (Bryant, 1986). Therefore, the current taxonomy of the Gram-positive anaerobic cocci is still unsatisfactory.

All genera discussed in this chapter were once placed in the family Peptococcaceae (Rogosa, 1971; Holdeman and Moore, 1974) and appeared in the *Approved Lists of Bacterial Names* (Skerman et al., 1980). In volume 2 of *Bergey's Manual of Systematic Bacteriology* (Sneath et al., 1986), however, the family name Peptococcaceae was no longer used and only the genera are listed (Holdeman-Moore et al., 1986a, 1986b).

16S rRNA and 5S rRNA sequence data are valuable tools for the phylogenic analysis of microorganisms (Balch et al., 1980; Fox et al., 1980; Tanner et al., 1982) and are having a strong impact on bacterial taxonomy. However, the sequences of anaerobic Gram-positive cocci have not yet been completed and thus their relationship at the genus and family level remains to be elucidated.

Habitats

Anaerobic cocci are part of the normal flora of the alimentary tract, skin, and vagina of humans and animals (Moore and Holdeman, 1974; Holdeman et al., 1976). The peptostreptococci are found in the mouth and upper respiratory tract and peptostreptococci, ruminococci, sarcinae, and coprococci are present in the lower small intestine and colon. *P. productus*, *Ruminococcus albus*, *R. flavefaciens*, and *R. bromii* are often isolated from the large intestine of humans.

Peptostreptococci and peptococci are constantly found in the vagina, in particular the species *Peptostreptococcus magnus*, *P. asaccharolyticus*, *P. prevotii*, "*P. hydrogenalis*," *P. anaerobius*, *P. tetradius*, *P. micros*, and *Peptococcus niger*. *Peptococcus niger* and peptostreptococci are often isolated from human skin (Wilkins and Jimenez-Ulate, 1975; Evans et al., 1978), whereas *Peptostreptococcus magnus*, *P. prevotii*, and *P. asaccharolyticus* are commonly found in the stool of humans and animals. "*Peptococcus glycinophilis*" (Cardon and Barker, 1946), later found to be identical with *Peptostreptococcus micros* (Cato et al., 1983) and *Ruminococcus pasteurii* are found in mud and creek sediments, respectively (Schink, 1984). Sarcinae are also found in soil (Canale-Parola, 1970).

Presence in Human Clinical Specimens

Members of the peptostreptococci are often found in various purulent discharges in humans (Thomas and Hare, 1954; Finegold, 1977). Foul smell from pus often suggests anaerobic infections because anaerobic bacteria often produce various volatile lower fatty acids. Many investigators have reported the isolation of anaebrobic cocci from the pus of ovarial, peritoneal, lung, liver, brain, kidney, and other soft-tissue abscesses (Anderson et al., 1972). Anaerobic cocci have been known to be associated with sepsis, septic thrombophlebitis, puerperal fever, sinusitis, and otitis media. In an early report of this century, anaerobic cocci or anaerobic streptococci were not referred specifically to as true "anaerobes," because capnophilic streptococci were once placed in the genus *Peptococcus* and were later transferred to the genus *Streptococcus* (Holdeman and Moore, 1974). Considering these changes of the taxonomy of the anaerobic cocci, a number of strains of Gram-positive anaerobic cocci isolated from various kinds of human clinical specimens were identified as members of the genus *Peptostreptococcus* in our laboratory. Among 278 strains of anaerobic Gram-positive cocci, only one strain was identified as *P. niger* and all others were assigned to the genus *Peptostreptococcus* (Table 1). Ruminococci, coprococci, and sarcinae were not found among these clinical isolates. Rosenblatt (1985) reported that among 1,340 anaerobic organisms isolated during 1983 at the Mayo Clinic, 26% were anaerobic cocci. Thirteen percent of these anaerobic cocci were identified as *P. magnus*. This predominance of peptostreptococci was also observed in our laboratory. Eighty four (27%) among 306 isolates were identified as *P. magnus*. One reason why this species is often isolated from human clinical specimens may be that it is the most aerotolerant organism among anaerobic cocci. Bourgault et al. (1980)

reported that *P. magnus* was a significant human pathogen but the pathogenic factors of these bacteria have not yet been studied.

Anaerobic bacteria are often isolated from mixed infections composed of different species of anaerobic bacteria or a mixture of aerobic and anaerobic organisms (Finegold, 1977; Sutter et al., 1975). However, anaerobic cocci were also isolated in pure culture: Finegold et al. (1968) reported on anaerobic cocci from 81 infections and one-third of the cases were isolated in pure culture. The anaerobic cocci were isolated as pure culture from brain, ovarial, peritoneal, and soft tissue abscesses. Among various human infections, gynecological infections are often associated with anaerobic cocci (Swenson et al., 1973; Sweet, 1975). In our laboratory, many isolates from gynecological specimens belonged to the species *Peptostreptococcus prevotii, P. asaccharolyticus, P. anaerobius, P. magnus,* and *P. tetradius.*

Anaerobic Gram-positive cocci are commonly involved in soft-tissue infections and many cases are mixed infections between aerobes and anaerobes. *Bacteroides melaninogenicus,* the *B. fragilis* group, and facultatively anaerobic Gram-negative bacteria such as *Escherichia coli* are often isolated together with anaerobic cocci. These infections are recognzied as bacterial synergistic gangrene (Anderson et al., 1972) and synergistic necrotizing cellulitis (Stone and Martin, 1972). Other human infections such as brain infections, pleuropulmonary infections, and abdominal abscesses associated with anaerobic cocci have been reported frequently (Finegold et al., 1968; Bourgault, et al., 1980; Rosenblatt, 1985). However, ruminococci and coprococci have not yet been associated with human infections although strains of *Sarcina ventriculi* have been found in vomitus of children.

Table 1. Anaerobic Gram-positive cocci isolated from human clinical specimens.

Organism	Rosenblatt (1985)	Marui (1981)
Peptostreptococcus magnus	176	126
P. asaccharolyticus	72	73
P. anaerobius	35	20
P. micros	31	10
P. prevotii	30	36
P. tetradoius		6
"*P. hydrogenalis*"		3
P. productus		2
P. indolicus		1
Peptococcus niger		1
Total	344 strains	278 strains

Isolation

Specimen Collection and Transport

Anaerobic cocci are inactivated by oxygen so that specimens must be protected from oxygen exposure and drying. Immediate plating of the specimen is most effective for the recovery of anaerobes. If immediate plating is not possible, anaerobic containers can be used for specimens during the transport process (Wilkins and Jimenz-Ulate, 1975) and various anaerobic transport containers are commercially available (Finegold, 1977; Hill, 1981).

Anaerobic Culture

Media for anaerobes should be kept under anaerobic conditions before use. Agar plates for the isolation of peptococci and peptostreptococci can be kept in anaerobic jars or anaerobic chambers. The plates for ruminococci, coprococci, and sarcinae should be kept under more reduced conditions than the plates for peptococci and peptostreptococci and these organisms only grow in well-maintained anaerobic chambers (such as anaerobic glove boxes) or in roll tubes. If an anaerobic chamber is used it should be filled with a gas mixture of 80% nitrogen, 10% hydrogen, and 10% CO_2. However, commercial gas mixtures are often contaminated with trace amounts of oxygen, so that a palladium catalyst should be placed in an anaerobic chamber. Since an anaerobic chamber provides a continuous oxygen-free environment during the processing of specimens and the culture of the bacteria, such a system is useful for the isolation of oxygen-sensitive bacteria as well as and for normal flora analysis. Several different models of anaerobic chambers are commercially available.

The roll tube method (Hungate, 1966) is a rather inexpensive alternative to the anaerobic chamber method. The method was originally developed by Hungate (1950) f/or the isolation of rumen bacteria and was later modified by Holdeman et al. (1977). The method is described in the VPI Anaerobe Laboratory Manual (Holdeman et al., 1977). Broth and solid media are placed in tubes with rubber stoppers to prevent oxygen exposure. Preparation of the media should be carried out under oxygen-free gas which is passed through a cold catalyst or heated copper catalyst. Media prepared by this technique are called prereduced anaerobically sterilized (PRAS) media; several kinds of PRAS agar and broth media are available commercially.

Another culture technique for small clinical laboratories is the anaerobic jar method. One jar can hold from 10 to 20 agar plates. The jar is commercially supplied with a gas-generation bag, an oxygen indicator, and a catalyst. (e.g., Oxoid, Merck, or BBL).

Primary Isolation Media for Peptococci and Peptostreptococci

Anaerobic cocci require complex nutrients. Growth factors may include vitamins, other cofactors, and amino acids. Tween 80 (final concentration in medium is 0.02%) enhances the growth of these organisms but is not required for their isolation. With the exception of *P. prod-*uctus, these cocci metabolize peptones and amino acids. Therefore, a complex medium such as blood agar or chopped meat medium is necessary for isolations. Basal media for blood agar plates include Brucella agar (BBL), Brucella HK agar (Kyokuto, Japan), Schaedler agar (BBL), Colombia agar (Difco), and Brain heart infusion agar supplemented with yeast extract. For the isolation of peptococci and peptostreptococci, media without blood such as chopped meat agar, modified GAM agar (Nissui, Japan), and ABCM agar (Eiken, Japan), are also useful. For the isolation of the anaerobic cocci from specimens contaminated with other organisms, a phenylethylalcohol blood agar (Difco, BBL) plate is recomended for their selective isolation.

Isolation Media for Ruminococci and Coprococci

Ruminococci require fermentable carbohydrates and ammonia (Bryant and Robinson, 1960, 1963) and the growth of coprococci is stimulated by carbohydrates. Thus, isolation media for these organisms should contain carbohydrates and ammonia. Selective media for the primary isolation of these organisms have not been developed, except for the isolation of cellulolytic organisms (e.g., *Ruminococcus flavefaciens,* Hungate, 1950). Unless special nutrients are added, commercial media do not support growth, although ruminococci and coprococci grow well on media containing 30–40% rumen fluid. The original Hungate anaerobic roll-tube method for the nonselective isolation of ruminococci has been modified by many workers. Holdeman et al. (1977) described the formula for PRAS roll tubes in detail.

Rumen Fluid-Glucose-Cellobiose Agar (Holdeman et al., 1977)

Glucose	0.25 g
Cellobiose	0.25 g
Soluble starch	0.5 g
$(NH_4)_2SO_4$	1.0 g
Rumen fluid	300.0 ml
Cysteine-HCl·H_2O	0.5 g
Resazurin solution (see below)	4.0 ml
Salts solution (see below)	500.0 ml
Hemin solution (see below)	10.0 ml
Agar	20.0
Distilled water	186.0 ml

Resazurin solution: 25 mg resazurin in 100 ml of distilled water.

Salts solution:

$CaCl_2$	0.2 g
$MgSO_4$	0.2 g
K_2HPO_4	1.0 g
KH_2PO_4	1.0 g
$NaHCO_3$	10.0 g

| NaCl | 2.0 g |
| Distilled water | 1000.0 ml |

Hemin solution: 50 mg hemin in 1 ml 1 N NaOH and make to 100 ml with distilled water.

Rumen fluid could be replaced with a volatile fatty acid (VFA) mixture which has been reported as medium 10 (Caldwell and Bryant 1966).

Medium 10 (Caldwell and Bryant, 1966)

Trypticase (BBL)	2.0 g
Yeast extract (Difco)	0.5 g
Hemin solution	2.0 ml
Glucose	0.5 g
Cellobiose	0.5 g
Soluble starch	0.5 g
Agar	18.0 g
Resazurin (0.1% solution)	1.0 ml
0.6% K$_2$HPO$_4$	37.5 ml
Mineral solution 1	37.5 ml
VFA mixture	3.1 ml
L-Cysteine (5% solution)	10.0 ml
Na$_2$CO$_3$ (8% solution)	50.0 ml
Ascorbic acid (25% solution)	2.0 ml
Distilled water	857.0 ml

Detailed procedures for the preparation of the original medium 10 roll tube is described in Holdeman et al. (1977).

Mineral solution 1

KH$_2$PO$_4$	0.6 g
(NH$_4$)$_2$SO$_4$	0.6 g
NaCl	1.2 g
MgSO$_4$·7H$_2$O	0.25 g
CaCl$_2$·2H$_2$O	0.16 g

VFA mixture

Acetic acid	17 ml
Propionic acid	6 ml
N-butyric acid	4 ml
N-valeric acid	1 ml
Isovaleric acid	1 ml
Isobutyric acid	1 ml
DL-α-Methylbutyric acid	1 ml

Selective isolation of cellulolytic ruminococci from rumen was discussed by Hungate (1950). Presumptive identification of cellulose digesters is possible on rumen fluid-amorphous cellulose-agar plates because they produce typical colonies surrounded with a clear zone due to cellulose digestion.

Primary Isolation of Sarcinae

The sarcinae are distinct in that their growth is stimulated by fermentable carbohydrates and then are able to grow at low pH (2.0–2.5), and selective media for their primary isolation is based on these two characteristics. Sarcinae are often found in feces of humans when isolation is done using Bifidobacterium medium, which

contains tomato juice and maltose (Eugon agar medium (BBL), 45.5 g; canned tomato juice, 400 ml; maltose, 10 g; hemin solution, 2 ml; and distilled water, 600 ml; pH 5–6). The selective isolation medium for sarcinae from soil was reported by Canale-Parola (1970). The medium contains maltose and malt extract broth (BBL) and the pH is adjusted to 2.2

Selective Enrichment Broth for the Isolation of Sarcinae from Soil (Canale-Parola, 1970)

Maltose	2.0 g
Malt extract broth (BBL)	5.0 g
Tap water	100.0 ml

Adjust pH 2.2 ± 0.1 with H$_2$SO$_4$.

Preservation of Cultures

Peptococci and peptostreptococci grow well in PRAS-chopped meat broth and can also be successfully subcultured in this broth.

PRAS-Chopped Meat Broth (Holdeman et al., 1977)

Ground beef (fat free)	500 g
Distilled water	1000 ml
1 N NaOH	25 ml

Mix the gredients, heat until boiling, then cool to room temperature. Skim off fat and filter the broth. Add distilled water to restore to 1–1 original volume and add 30 g trypticase, 5 g yeast extract, 5 g potassium phosphate, and 4 ml resazulin solution. Boil under oxygen-free nitrogen gas and add 0.5 g cysteine, 10 ml of hemin solution, and 1 μl vitamin K$_1$. Adjust pH 7.2 ± 0.2. Prepare test tubes containing meat particles (use one part meat particles to four to five parts fluid). Dispense 3–5 ml of the broth into the tubes filled with oxygen-free carbon dioxide gas, seal the tubes with rubber stoppers, then autoclave.

Since preparation of PRAS-chopped meat broth is time consuming, we often prepare semisolid GAM broth and modified semisolid Brucella medium (Kyokutou Seiyaku, Tokyo, Japan) in ordinary test tubes. Peptostreptococci grow well in this semisolid medium even under conditions that are not strictly anaerobic.

Modified Brucella Semisolid Medium for Maintenance of Peptococci and Peptostreptococci

Casein peptone	10.0 g
Beef peptone	10.0 g
Yeast extract	5.0 g
Soy peptone	4.0 g
Glucose	1.0 g
Sodium sulfite	0.1 g
Arginine	1.0 g
Vitamin K$_1$	1.0 μl
Hemin	0.01 g

Sodium pyruvate	1.0 g
Sodium fumarate	0.2 g
K$_2$HPO$_4$	0.4 g
KH$_2$PO$_4$	0.8 g
NaCl	2.5 g
Sodium thioglycolate	0.3 g
Soluble starch	5.0 g
Cysteine HCl	0.3 g
Agar	1.5 g
Distilled water	1000.0 ml

Oxygen contamination occurs only at the surface of the media and thus the tube can be used without filling with oxygen-free gas. These semisolid media can be used for at least 1 week after preparation. If an indicator such as resazurin is added to the semisolid media, an oxygenated area will be observed. For the maintenance of ruminococci, coprococci, and sarcinae, semisolid PRAS-chopped meat medium supplemented with various carbohydrates (Holdeman et al., 1977) is recommended. The rubber stopper of the test tube can be opened without flushing the inside of the tube with gas. After inoculation of organisms with a Pasteur pipette, the test tube should be sealed with a rubber stopper. Most ruminococci and coprococci may grow in this way, but they grow better if the tube is filled with a gas mixture (nitrogen, 80%; carbon dioxide, 10%; hydrogen, 10%). Scot II broth (Scot Lab) and Brucella HK semisolid (Kyokuto) are examples of commercially available prereduced media. We often apply ready-use semisolid media without filling the tubes with oxygen-free gas. Brucella HK and GAM semisolid media can support the growth of the anaerobic cocci other than *Peptococcus niger. P. niger* grows poorly in these commercial media but the addition of pyruvic acid will stimulate growth. Whenever growth is obtained, the anaerobic cocci die off so quickly in these semisolid maintenance media that they must be transferred into new media every week. Strains of *Peptostreptococcus tetradius* and *P. anaerobius* often die within 4 to 5 days after being transferred to new media.

Ruminococci and coprococci are maintained in PRAS-chopped meat broth supplemented with fermentable carbohydrate (glucose, cellobiose, or maltose) but they grow poorly in the above-mentioned commercial semisolid media. The concentration of the fermentable carbohydrate should be less than 0.2% in maintenance media because an excess amount of fermentable carbohydrate would cause quicker death of the organism. Sarcinae quickly die in acidic enrichment broth within 1 week.

These anaerobic cocci retain viability for at least several years at −80°C or in liquid nitrogen. Lyophilization of the anaerobic cocci is, however, recommended for long term-storage.

Basal Media for Identification

Basal prereduced peptone-yeast extract (PY) medium (Holdeman et al., 1977) is used for the characterization of anaerobic cocci. Most anaerobic cocci grow poorly in this medium. Thus a heavy inoculum of a young culture (5% of broth culture) is critical in this procedure. Commercial prereduced PY medium (Scot II) based on the description by Holdeman et al. (1977) is available.

Basal Prereduced Peptone-Yeast Extract (PY)
Medium (Holdeman et al., 1977)

Trypticase	5.0 g
Peptone	5.0 g
Yeast extract	10.0 g
Vitamin K$_1$	1.0 μl
Hemin	5.0 g
Cysteine HCl	0.5 g
Salts solution	40.0 ml
Distilled water	960.0 ml

Identification

Taxonomy

When the family Peptococcaceae was proposed by Rogosa (1971), *Peptococcus, Peptostreptococcus, Ruminococcus,* and *Sarcina* were assigned to this family. Later, in 1977, Holdeman and Moore (1974) placed the new genus *Coprococcus* in this family. The family name appeared in the *Approved Lists of Bacterial Names* (Skerman et al., 1980) but was omitted from the description in *Bergey's Manual of Systematic Bacteriology,* volume 2 (Sneath et al., 1986).

Most members of the genus *Peptostreptococcus* were once placed in the genus *Peptococcus* because they make pairs, or bunches of grape-like clusters. However, this morphological appearance is not a reliable character. Because the cell wall peptidoglycan type of "*Peptococcus saccharolyticus*" resembled the cell wall of staphylococci, "*P. saccharolyticus*" was transferred to the genus *Staphylococcus* (Kilpper-Bälz et al., 1980; Kilpper-Bälz and Schleifer, 1981). The GC content of the DNA of most anaerobic cocci is closer to the type species of the genus *Peptostreptococcus* than to that of the genus *Peptococcus* (Wilkins et al., 1975). Thus, most members of the genus *Peptococcus* were transferred to the genus *Peptostreptococcus.* "*Peptostreptococcus parvulus*" and *Peptostreptococcus productus,* two saccharolytic species of the genus *Peptostreptococcus,* have higher GC content (44–46%). *"P. parvulus"* was transferred to the genus *Streptococcus* because it produces lactate as its sole major metabolic end-product from glucose (Cato, 1983). However, recent studies showed that *"P. parvulus"* is not closely related to streptococci (Ludwig et al., 1988). The GC of content of the current species of the genus *Peptostreptococcus,* with the exception of *P. productus* (Ezaki, 1982; Ezaki and Yabuuchi, 1983), ranges from 28 to 37 mol% (see Table 2). The different peptidoglycan types (Schleifer and Nimmermann, 1973) and in particular DNA-rRNA hybridization (Huss et al., 1982, 1984) studies of members of the genus *Peptostreptococcus* suggest that it is not a well-defined genus and also contains unrelated organisms.

Partial sequence analysis of 16S rRNA of the *Peptostreptococcus* species supports this observation (Fig. 1). Among current members of the genus *Peptostreptococcus,* only two butyrate-producing groups of organisms were in the same rRNA homology group, *P. indolicus* and *P. asaccharolyticus,* which are indole positive. *P. prevotii* and *P. tetradius,* another highly related rRNA homology group, also form butyrate as a fermentation product but do not indole.

The name *"Gaffkya anaerobica"* was rejected (Judicial Commission, 1971) and the organism was transferred to the genus *Peptostreptococcus* as *P. tetradius* (Ezaki et al., 1983). Many strains identified as *P. asaccharolyticus* or *P. prevotii* are genetically quite different from their type strains (Ezaki et al 1983). *"P. hydrogenalis,"* previously reported as "unclassified group A-1," is an indole- and butyrate-positive species, and we thus expected that this species is related to the *P. asaccharolyticus* or *P. prevotii* 16S rRNA groups. (Fig. 1). *P. productus* was more closely related to ruminococci than to the peptostreptococci by 16S rRNA sequence analysis and this relationship is also supported by a similar GC content and other phenotypic characteristics.

The GC content of the DNA of *P. productus* is 44–45 mol%, a rather high value for a member of the genus *Peptostreptococcus* and actually close to the value of the ruminococci (39–46 mol%). *P. productus* is saccharolytic and differs in this respect from other members of the peptostreptococci, which are asaccharolytic and use peptones and amino acids as their energy sources. Ruminococci, on the other hand, are saccharolytic.

Strains of *P. productus* have been divided into two phenotypes, type I and type II (Holdeman et al., 1976; Moore and Holdeman, 1974). Type I strains produce formic acid as a fermentation product but type II strains do not. Type II strains were further subdivided into five subgroups. Genetic relatedness among these biotypes has not been studied.

P. heliotrinreducens (Lanigan, 1976), a relatively new member of the genus *Peptostreptococcus,* is more closely related to certain members of the genus *Streptococcus,* such as *S. parvulus* and *S. anginosus.* The organism was first proposed as a member of genus *Peptococcus* and was transfered to the genus *Peptostreptococcus* (Ezaki and Yabuuchi, 1986) because the GC content of the organism was in the range of that of the genus *Peptostreptococcus.*

According to our 16S rRNA analysis, *P. magnus* and *P. micros* form a cluster together with the more remotely related *P. anaerobius.* The data of Huss et al. (1984), on the other hand, suggest a relationship between *P. micros* and *S. parvulus.*

Four ovoid ruminococci (*R. torques, R. gnavus, R. lactaris,* and *R. callidus*) are in the same 16S rRNA homology group (Fig. 1) but are different from other ruminococci (unpublished observations). Human strains of *R. favefaciens* and *R. albus* do not hydrolyze cellobiose and may be different from cellulolytic strains (Bryant, 1986). However, the genetic relatedness between cellulolytic strains and non-cellullolytic strains has not yet been studied.

The morphology of sarcinae is a valuable character and helps to differentiate them from other anaerobic cocci (Canale-Parola, 1970, 1976). Strains that carry similar morphological characteristics are found in the genus *Sporosarcina* and withsome micrococci. However, the latter organisms are strictly aerobic and can be easily distinguished from the anaerobic sarcinae. Members of the genus *Sarcina* require a

Table 2. Differential features of five genera of anaerobic Gram-positive cocci.

	Peptococcus	*Peptostreptococcus*	*Ruminococcus*	*Coprococcus*	*Sarcina*
Peptone as energy source	+	+	−	−	−
Sugar fermented	−	d	+	+	+
Butyrate production	+	d	−	+	d
GC (mol%)	50–51	28–37, 44–45[a]	39–46	39–42	28–31

+, positive; −, negative; d, different among species.

[a]The GC content of the DNA of *Peptostreptococcus productus* ranges from 44–45 mol%, and others, from 28–37 mol%.

Fig. 1. 16S rRNA partial sequences of 21 anaerobic Gram-positive cocci. Dendrogram was drawn by unweighted average linkage using 157 bases (from 1220–1376) of each type strain. *R., Ruminococcus; P., Peptocostreptococcus; S., Streptococcus; C., Coprococcus.*

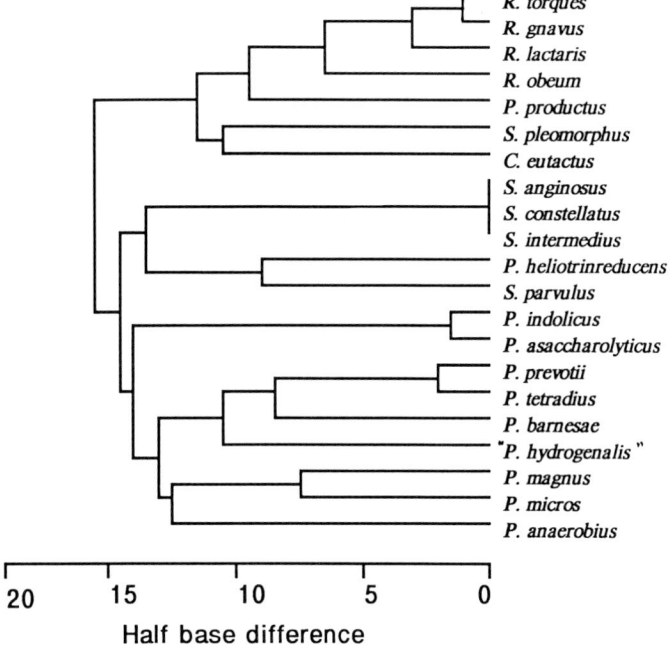

fermentable carbohydrate for a growth. The results of 16S rRNA sequence analysis (Fox et al., 1980) suggest that the strains of *Sarcina* species are related to clostridia, which are sporeforming anaerobic Gram-positive rods.

Identification

Differentiation of Anaerobic Cocci at the Genus Level

The first step for the identification of anaerobic cocci is to eliminate anaerobic strains of streptococci and staphylococci from "true anaerobic cocci". The genus *Streptococcus* contains three anaerobic species, *S. hansenii, S. pleomorphus*, and *S. parvulus* but none of these anaerobic streptococci are closely related to genuine streptococci, enterococci, or lactococci (Ludwig et al., 1988). Strains of *Gemella morbillorum* (Kilpper-Bälz and Schleifer, 1988), *S. anginosus, S. intermedius*, and *S. constellatus* often grow only under strictly anaerobic conditions but most of them will eventually grow in an oxygen-reduced environment or in a 5–10% CO_2 incubator after several transfers in the laboratory. *S. parvulus* was transferred from the genus *Peptostreptococcus* because it is saccharolytic and produces lactic acid as a sole major metabolic end product in PYG medium (Cato, 1983). It also grows on a blood agar if the inoculated plate is placed in a 10% CO_2 incubator. Anaerobic strains of streptococci and gemellae

solely produce lactic acid from PY medium containing 1% glucose. Members of the genera *Peptococcus, Peptostreptococcus, Ruminococcus, Coprococcus*, and *Sarcina* do not contain strains that produce lactic acid as the major or sole metabolic product from PYG medium.

Strains of *Staphylococcus saccharolyticus*, formerly assigned to the genus *Peptococcus*, are often found only on anaerobic plates (Evans et al., 1978). They often can grow under reduced oxygen conditions (e.g., in a 5–10% CO_2 incubator) or they grow slowly in aerobic primary culture. *Staphylococcus aureus* subsp. *anaerobius* (Fuente et al., 1985) also grows under reduced oxygen conditions but it is more aerotolerant than *S. saccharolyticus*. It is often difficult to differentiate *S. saccharolyticus* from other anearobes at the genus level. *S. saccharolyticus* produces acetic acid as a major metabolic product from PYG medium and thus must be differentiated from butyrate-negative peptostreptococci (e.g., *P. magnus* and *P. micros*). Simple biochemical tests, such as the nitrate reduction test and the urease test, are useful to differentiate *S. saccharolyticus* from the peptostreptococci.

Table 2 shows features that can be used for the differentiation of anaerobic Gram-positive cocci at the genus level. Peptococci and peptostreptococci use peptones and amino acids as their energy source but most strains do not use carbohydrates. These characteristics of peptococci and peptostreptococci are useful for differentiating them from ruminococci and copro-

Table 3. Differential characteristics of peptococci and peptostreptococci.[a]

	P. niger	Ps. anaerobius	Ps. indolicus	Ps. asaccharolyticus	"Ps. hydrogengenes"
Metabolic products:					
n-Capronate	+	−	−	−	−
Isocapronate	−	+	−	−	−
Butyrate	+	trace	+	+	+
Succinate from fumarate	−	NT	−	−	−
Propionate from lactate	−	−	+	−	−
Cell arrangements	Diplococci, clumps, clusters	Short chains	Clumps, diplococci, clusters	Clumps, diplococci, clusters	Clumps, diplococci, clusters
Gas from PYG	w	+	+	+	+
Indole	−	−	+	+	+
Nitrate	−	−	+	−	−
Coagulase	−	−	+	−	−
Urease	−	−	−	−	−
Catalase	−	−	−	−,w	−
Beta-gluconidase	−	−	−	−	−
Alpha-glucosidase	−	+	−	−	−
Alkaline phosphatase	−	−	+	−	+
Ammonium from:					−
Glutamate		−	+	+	−
Glycine		−	−	−	−
Acid from:					
Glucose	−	w,−	−	−,w	w
Lactose	−	−	−	−	−
Maltose	−	w,−	−	−	−
Mannose	−	−	−	−	−
Sucrose	−	−,w	−	−	−
Peptidoglycan[b]					
Position 1	Ala	Ala	Ala	Ala	NT
Position 3	Lys	Lys	Orn	Orn	NT
Interpeptide bridge	D-Asp	D-Asp	D-Glu	D-Glu	NT
GC (mol%)	50–51	33–34	32–34	30–34	30–31

w, weakly positive; (−), positive strains exist; NT, not tested; d, different reaction.

[a]P., Peptococcus; Ps., Peptostreptococcus.

[b]Adapted from Huss et al. (1982).

cocci. *Peptostreptococcus productus,* which has an exceptionally high GC content for a member of the genus *Peptostreptococcus,* is a saccharolytic species and does not use peptone, making it difficult to differentiate from ruminococci.

Peptococci and peptostreptococci cannot be distinguished on the basis of their morphological characteristics. Most strains of peptostreptococci form both short chains and grape-like clusters, as well as diplococci and tetracocci. To differentiate them, a determination of the GC content of the DNA can be done, since the GC of peptococci is 50–51 mol% and that of peptostreptococci ranges from 28–37 mol% (with the exception of *P. productus* (44–45 mol%).

Analysis of metabolic products from PYG medium is also helpful to differentiate peptococci and peptostreptococci. *Peptococcus niger,* currently the only species of this genus, produces *n*-caproic acid whereas no other anaerobic cocci produce this fatty acid (*Peptostrepto-*

coccus anaerobius produces iso-caproic acid). Ruminococci and coprococci are saccharolytic organisms and peptones are not used as energy sources, carbohydrates being required for growth. Coprococci produce butyric acid from PYG medium but ruminococci do not. Most ruminococci produce acetic acid, formic acid, and ethanol from PYG. Thus, the analysis of metabolic products from PYG offers the most reliable data to differentiate those two genera.

Ruminococcus pasteurii, a newly proposed tartrate-fermenting species (Schink, 1984) is the only exceptional species. It does not require fermentable carbohydrates but produces acetic acid, formic acid, and ethanol. The GC content of the DNA is 45.2 mol% and therefore the organism is tentatively assigned to the genus *Ruminococcus.*

Sarcinae are morphologically distinct from other anaerobic cocci. They occur in packets of eight or more and the division occurs in three

Ps. prevotii	Ps. tetradius	Ps. magnus	Ps. micros	Ps. productus	Ps. heliotrinreducens	Ps. barnesae
−	−	−	−	−	−	−
−	−	−	−	−	−	−
+	+	−	−	−	w	−
−	−	−	−	NT	+	NT
−	−	−	−	−	NT	NT
Diplococci, clumps, clusters	Tetrads, diplococci, short chain	Clumps, clusters, diplococci	Short chains, diplococci	Short chains, ovoid cells	Clumps, clusters	Clumps, clusters
+	+	−	−	+	−	−
−	−	−	−	−	−	−,(w)
−	−	−	−	−	+	−
(−)	+	−	−	d	−	−
−,w	−	d	−	−	−	−
(−)	+	−	−	−	−	NT
(−)	+	−	−	+	−	NT
+	−	−,w	+	−	−	NT
−	−	−	−	−	NT	−
−	−	+	+	−	NT	+
w,−	+	−	−	+	−	NT
−	−	−	−	+	−	NT
−,w	+	−	−	+	−	NT
−	+	−	−	+	−	NT
−	+	−	−	+	−	NT
Gly	NT	Gly	Gly	NT	NT	Gly
Lys	NT	Lys	Orn	NT	NT	Lys
D-Glu	NT	Gly	D-Asp	NT	NT	D-Asp
29–33	30–32	32–34	28–29	44–45	35–37	34–35

perpendicular planes. They require fermentable carbohydrates and thus resemble ruminococci and coprococci. The GC content of the DNA ranges from 28–30 mol% and is thus quite different from that of ruminococci and coprococci, in which the GC content is 39–46 mol%.

Identification of Species of the Genera *Peptococcus* and *Peptostreptococcus*

Biochemical characteristics used to differentiate each species are listed in Table 3. When obligately anaerobic Gram-positive cocci are isolated and their growth is not stimulated by carbohydrates, the first step for their identification is to analyze their metabolic products from PYG. Both *Peptococcus niger* and *Peptostreptococcus anaerobius* are easily identified by their products because they produce caproic acid, the former producing *n*-caproic acid and the latter iso-caproic acid (Wilkins et al., 1975). This in-

formation is often sufficient to identify both species because no other currently established species produces these fatty acids. Both species produce small amounts of H_2S in PYG medium. Unlike *Peptococcus niger*, *Peptostreptococcus anaerobius* grows confluently in commercial media and PYG. Isovaleric acid, isobutyric acid, and hydrogen gas are also produced by *P. anaerobius*. Other peptostreptococci are separated into two groups, the butyrate-positive and butyrate-negative groups.

Butyrate-producing peptostreptococci are differentiated by several simple biochemical tests: nitrate reduction, coagulase, urease, indole production, β-glucronidase, fermentation of sucrose, maltose, and lactose. This group of organisms can be divided into two subgroups by the indole test. *Peptostreptococcus asaccharolyticus* and *P. indolicus* belong to the indole-positive subgroup and *P. prevotii* and *P. tetradius* to the indole-negative subgroup.

P. asaccharolyticus is often isolated from vaginal discharges, stools, and various purulent specimens, while *P. indolicus* is usually not associated with human but with animal infections (Sørenson, 1973; Bourgault and Rosenblatt, 1979).

P. indolicus is differentiated from *P. asaccharolyticus* by the coagulase, and nitrate tests, and by propionate production from fumarate (Sørensen, 1973; Holdeman et al., 1977). Our previous study suggests that strains identified as *P. asaccharlyticus* are genetically heterogeneous but phenotypic tests to differentiate these genetically different organisms from established *P. asaccharolyticus* have not yet been found.

The indole-negative subgroup consists of two organisms, *P. prevotii* and *P. tetradius*. *P. prevotii* is more often associated with human infections than is *P. tetradius*. Strains identified as *P. prevotii* are also genetically heterogeneous (Ezaki, 1987). *P. tetradius* is more saccharolytic than *P. prevotii,* fermenting glucose, maltose, and sucrose whereas most strains of *P. prevotii* do not ferment these sugars. Urease and alkaline phosphatase activities of *P. tetradius* are also useful to differentiate it from *P. prevotii*. However, the type strain of *P. prevotii* is urease positive.

Four butyrate-negative strains of peptostreptococci are currently assigned to the genus *Peptostreptococcus: P. productus, P. magnus, P. micros,* and *P. barnesiae.* Their major metabolic product is acetic acid. Among them, *P. productus* is the only saccharolytic species. Type I strains of *P. productus* produce formic acid (Holdeman et al., 1976). Carbon monoxide stimulates growth (Lorowitz and Bryant, 1984). As mentioned before, this species is more related to ruminococci than to members of the genus *Peptostreptococcus.* Differentiation of this organisms from other members of the genus *Peptostreptococcus* is not difficult (Table 3). Other species of this group do not ferment carbohydrates and use peptones and amino acids as their energy source, commonly using glycine for growth. *P. magnus* and *P. micros* are often isolated from humans while *P. barnesae* is found in chicken feces. *P. magnus* and *P. micros* are morphologically quite different but they share many common characteristics. The cell size of *P. micros* is less than 0.6 μm (Holdeman et al., 1977) and they form small colonies on prereduced blood agar plate (<0.5 mm; usually described as pin-point colonies). However, biochemical tests to differentiate the two species are limited. *P. micros* produces alkaline phosphatase and has various peptidase activities (Ezaki and Yabuuchi, 1985). Determination of their cell wall peptidoglycan (Huss et al., 1984),

lipids profiles (Ezaki et al., 1985), and gel analysis by electrophoretic separation of soluble proteins (Cato et al., 1983) offer decisive information for the differentiation of the two species. The GC content of the DNA of *P. micros* ranges from 27–28 mol% and the value of *P. magnus* is from 32–34 mol%. Thus, the determination of GC content offers helpful information for the differentiation of the two species.

Peptostreptococcus barnesae was isolated from chicken feces and was proposed as a purine-utilizing new species of the genus *Peptostreptococcus* (Schiefer-Ullrich and Andreesen, 1985). The organism also utilizes glycine as a substrate, similar to *P. magnus* and *P. micros.* However, *P. barnesae* produces formic acid, acetic acid, ammonia, and CO_2. The cell wall peptidoglycan type is different from. *P. magnus* and *P. micros* and resembles that of the Hare group IX (Hare et al., 1952). However, this species cannot utilize arginine while Hare group IX strains do (Schiefer-Ullrich and Andreesen, 1985). Isolates from human specimens have not yet been reported. This organism grows poorly in prereduced chopped meat medium and produces small amounts of butyric acid. Pyrrolidone alkaloid were catabolized by this organism. It does not ferment carbohydrates and reduces fumarate to succinate. Nitrate is reduced and pyrrolidone arylamidase activity has been found (Ezaki and Yabuuchi, 1986).

Several kits for the rapid identification of peptostreptococci are commercially available (ABL, Anaerobic Identification System, Austin Biological Labs, TX; Rap-ID-ANA, Vitek System; API, Anident, France). These 2–4 h identification systems measure the presence of carbohydrate-digesting enzymes and oligopeptidases by using their chromogenic substrates. These enzyme activities obtained from organisms cultured in different media give variable results because different media contain various levels of amino acids, oligopeptides, and carbohydrates. These differences strongly influence the accuracy of the identification procedure. Thus, the media for the subculture of organisms should be carefully selected.

Identification of Species of the Genera *Ruminococcus, Coprococcus,* and *Sarcina*

Differentiating features of *Ruminococcus* species are listed in Table 4. Analysis of lower fatty acids and ethanol as their fermentation products is the first step for their identification. They do not produce butyric acid and require fermentable carbohydrates. Most species produce formic acid, whereas succinic acid is produced only by two species. These traits are important

Table 4. Differential characteristics of the nine species of the genus *Ruminococcus*.

	R. flavefaciens	R. albus	R. bromii	R. callidus	R. torques	R. gnavus	R. lactaris	R. obeum	R. pasteurii
Oval to pointed ends	−	−	−	+	+	+	+	+	−
Major fermentation product:									
Succinate	+	−	−	+	−	−	−	−	−
Lactate	d	−	−	+	+	−	+	−	−
Formate	d	d	d	d	d	+	+	−	+
Ethanol	−	+	+	−	+	+	−	+	+
Hydrolyze:									
Cellulose	d	d	−	NT	−	−	−	−	NT
Starch	−	−	+	−	−	d	−	−	NT
Ferment:									
Arabinose	d	d	−	−	−	+	−	+	NT
Xylose	d	d	−	−	−	+	−	+	NT
Cellobiose	+	+	−	+	−	−	−	−	NT
Fructose	−	d	+	−	+	+	+	+	+
Lactose	d	d	−	d	+	d	+	+	NT
Maltose	−	−	+	+	−	+	d	+	NT
Mannitol	−	−	−	−	−	−	−	d	NT
Raffinose	−	−	−	+	−	+	−	+	NT
Sucrose	d	d	−	+	−	d	−	+	NT
Glucose	d	d	d	d	+	+	+	+	NT
Production of H$_2$S	−	−	−	−	NT	d	−	−	NT
GC (mol%)	39–44	43–46	39–40	43	40–42	43	45	42	45

NT, not tested; d, different reaction.

to differentiate them from coprococci. *R. flavefaciens, R. albus* and *R. bromii* are morphologically different from other ovoid ruminococci. The type species of the genus *Ruminococcus* is a cellulose-hydrolysing species, *R. flavefaciens; R. albus* is another cellulolytic species. Human isolates, however, do not hydrolyze cellulose (Moore and Holdeman, 1974). Ovoid ruminococci are also isolated from human feces. These species are highly fermentative. A tartrate-fermenting species, *R. pasteurii,* proposed by Schink (1984), was isolated from fresh water creek sediment. L-Tartrate, citrate, pyruvate, and oxaloacetate were degraded and their fermentation products were formate, acetate, ethanol, and carbon dioxide. It grows in Brucella HK semisolid medium and thus the growth is not so fastidious.

Differential characteristics for three species of the genus *Coprococcus* are given in Table 5. All species produce butyric acid, which is a key factor for differentiating them from ruminococci. Carbohydrate fermentation of arabinose, maltose, xylose, and melezitose are useful for differentiating these three species.

The genus *Sarcina* consists of two species. Endospore formation by these organisms has been reported. *S. maxima* produces butyrate from PYG medium, whereas *S. ventriculi* does not (Canale-Parola, 1970). Characteristics for

Table 5. Differential characteristics of the three species of the genus *Coprococcus*.

	C. eutactus	C. comes	C. catus
Fermentation products:			
Formate	+	−	−
Butyrate	+	+	+
Ethanol	+	−	−
Ferment:			
Glucose	+	+	−,w
Arabinose	−	w	−
Cellobiose	+	−,w	−
Sucrose	+	+	−
Lactose	+	+,w	−
Melezitose	+	−	−
Xylose	−	+	−
GC (mol%)	41	40	39–42

w, weak reaction.

their differentiation are listed in Table 6. *S. ventriculi* may consist of different species because carbohydrate fermentation patterns of the strains isolated from human feces and soils are rather variable (unpublished observations).

Antibiotic Susceptibility

Anaerobic cocci are, in general, susceptible to antibiotics used for the treatment of anaerobic infections (Sutter and Finegold, 1976; Rosen-

Table 6. Differential characteristics of the two species of the genus *Sarcina*.

	S. ventriculi	*S. maxima*
Metabolic products:		
Butyrate	−	+
Acetate	+	+
Ethanol	+	−
Cellulose fermentation	+	−
Ferment:		
Glucose	+	+
Maltose	+	+
Sucrose	+	+
D-Arabinose	−	d
D-Xylose	−	+
Cellobiose	d	−
Lactose	+	d
Melibiose	+	−
Peptidoglycan, position 3	LL-A$_2$pm[a]	LL-A$_2$pm
Interpeptide bridge	Gly	Gly
GC (mol%)	28–31	28–29

[a]LL-A$_2$pm = LL-diaminopimelic acid.

blatt, 1985). Penicillin G and derivatives, cephem derivatives, lincomycin, and clindamicin are usually used to treat anaerobic infections. Peptostreptococci are often moderately resistant to tetracycline and metronidazole (Rosenblatt, 1985; Lorian, 1980). Techniques for the determination of minimum inhibitory concentration were worked out by Wilkins and Chalgren (1976). The medium they designed supported the growth of peptococci and peptostreptococci but most isolates of ruminococci, coprococci, and sarcina do not grow on this medium. New standard procedures to determine antibiotic susceptibility of anaerobes were published in the USA by the National Committee for Clinical Laboratory Standards (1985, 1986, and 1989).

Literature Cited

Anderson, C. B., Marr, J. J. and Jaffe, B. M. 1972. Anaerobic streptococcal infections simulating gas gangrene. Arch. Surgery 104:186–189.

Balch, W. E., Fox, G. E., Magrum, L. J., and Woese, C. W. 1979. Methanogens: reevaluation of a unique biological group, Microbiol. Rev. 43:260–296.

Bourgault, A. M., and Rosenblatt, J. E. 1979. First isolation of *Peptococcus indolicus* from a human clinical specimen. J. Clin. Microbiol. 9:549–550.

Bourgault, A. M., Rosenblatt, J. E. and Fittzgerald, R. H. 1980. *Peptococcus magnus*: a significant human pathogen. Ann. Intern. Med. 93:244–248.

Bryant, M. 1986. Genus *Ruminococcus* Sijpesteijn 1948, p. 1093–1099. In: P. H. A. Sneath, Mair, N. S., Sharpe, M. E. and Holt, J. E. (ed), Bergey's manual of systematic bacteriology, vol. 2. Williams & Wilkins, Baltimore.

Bryant, M. P., and Robinson, I. M. 1960. Some nutritional requirements of the genus *Ruminococcus* Appl. Microbiol. 9:91–95.

Bryant, M. P., and Robinson, I. M. 1963. Apparent incorporation of ammonia and amino acid carbon during during growth of selected species of ruminal bacteria. J. Diary Sci. 46:150–154.

Caldwell, D. R., and Bryant, M. P. 1966. Medium without rumen fluid for nonselective enumeration and isolation of rumen bacteria. Appl. Microbiol. 14:794–801.

Canale-Parola, E. 1970. Biology of the sugar-fermenting sarcinae. Bacteriol. Rev. 34:82–97.

Canale-Parola, E. 1986. Genus *Sarcina* Goodsir 1842, p. 1100–1103. In: P. H. A. Sneath, N. S. Mair, M. E. Sharpe, and J. G. Holt (ed), Bergey's mannual of systematic bacteriology, vol. 2. Wiliams & Wilkins, Baltimore.

Cardon, B. P., and Barker, H. A. 1946. Two new amino-acid-fermenting bacteria, *Clostridium propionicum* and *Diplococcus glycinophilus*. J. Bacteriol. 52:629–634.

Cato E. P. 1983. Transfer of *Peptostreptococcus parvulus* (Weinberg, Nativelle, and Prevot 1937) Smith 1957 to the genus *Streptococcus*: *Streptococcus parvulus* (Weinberg, Nativelle, and Prevot 1937) comb. nov., nom., emend. Int. J. Syst. Bacteriol. 33:82–84.

Cato, E. P., Johnson, J. L., Hash, D. E., and Holdeman, L. V. 1983. Synonomy of *Peptococcus glycinophilus* (Cardon and Barker 1946) Douglas 1957 with *Peptostreptococcus micros* (Prèvot 1933) Smith 1957 and electrophoretic differentiation of *Peptrostreptococcus micros* from *Peptococcus magnus* (Prèvot 1933) Holdeman and Moore 1972. Int. J. Syst. Bacteriol. 33:207–210.

Evans, C. A., Mattern, K. L., and Hallam, S. I. 1978. Isolation and identification of *Peptococcus saccharolyticus* from human skin. J. Clin. Microbiol. 7:261–264.

Ezaki, T. 1982. Mol % guanine plus cytosine of butyrate-producing anaerobic cocci and DNA/DNA hybridization relationships among them (manuscript in Japanese). Jpn. J. Bacteriol. 37:607–613.

Ezaki, T. 1987. Review on the classification of anaerobic Gram-positive cocci and their identification with special reference to peptococci and peptostreptococci (manuscript in Japanese). Jpn. J. Bacteriol. 42:471–478.

Ezaki, T., and Yabuuchi, E. 1983. Deoxyribonucleic acid base composition and DNA/DNA hybridization studies among the four species of *Peptostreptococcus* Kluyver and van Neil 1936. FEMS Microbiology. Letters 17:197–200.

Ezaki, T., and Yabuuchi, E. 1985. Oligopeptidase activity of Gram-positive anaerobic cocci used for rapid identification. J. Gen. Appl. Microbiol. 31:255–265.

Ezaki, T., and Yabuuchi, E. 1986. Transfer of *Peptococcus heliotrinreducens* to the genus *Peptostreptococcus heliotrinreducens* Lanigan 1983 comb. nov. Int. J. Syst. Bacteriol. 36:107–108.

Ezaki, T., Yamamoto, N., Ninomiya, K., Suzuki, S., and Yabuuchi, E. 1983. Transfer of *Peptococccus indolicus*, *Peptococcus asaccharolyticus*, *Peptococcus prevotii*, and *Peptococcus magnus* to the genus *Peptostreptococcus* and proposal of *Peptostreptococcus tetradius* sp. nov. Int. J. Syst. Bacteriol. 33:683–698.

Ezaki, T., Yabuuchi, E., Kaneda, K., and Yano, I. 1985. Extractable cellular lipids of Gram-positive anaerobic cocci. Proceedings of 15th annual meeting of the Japanese Association for Anaerobic Infection Research. 15:115–117.

Finegold., S. M. 1977. Anaerobic bacteria in human disease. Academic Press, New York.

Finegold, S. M., Miller, A. B., and Sutter, V. L. 1968. Anaerobic cocci in human infection. Bacteriological Proc. p. 94.

Fox, G. E., Stackebrandt, E., Hespell, R. B., Gibson, J., Maniloff, J., Dyer, T. A., Wolfe, R. S., Balch, W. E., Tanner, R. S., Magrum, L. J., Zablen, L. B., Blakemore, R., Gupta, R., Bonen, L., Lewis, B. J., Stahl, D. A., Luehrsen, K. R., Chen, K. R., Woese, C. R. 1980. The phylogeny of prokaryotes. Science. 209:457–463.

Fuente., R. D. L., Suarez, G., and Schleifer, K. H. 1985. Staphylococcus aureus subsp. anaerobius subsp. nov., the causal agent of abscess disease of sheep. Int. J. Syst. Bacteriol. 35:99–102.

Hare, R., Wildy, P., Billett, F. S., and Twort, D. N. 1952. The anaerobic cocci: gas formation, fermentation reactions, sensitivity to antibiotics and sulphonamides. Classification. J. Hyg. 50:295–319.

Hill, G. B. 1981. The anaerobic cocci, p. 1631–1658. In: M. P. Starr, H. Stolp, H. G. Truper, A. Balows, and H. G. Schegel (ed.), The prokaryotes. Springer-Verlag, Berlin.

Holdeman, L. V., and Moore, W. E. C. 1974. New genus, Coprococcus, Twelve new species, and emended descriptions of four previously described species of bacteria from human feces. Int. J. Syst. Bacteriol. 24:260–277.

Holdeman, L. V., Good, I. J., and Moore, W. E. C. 1976. Human fecal flora: variation in bacterial composition within individuals and a possible effect of emotional stress. Appl. Environ. Microb. 31:359–375.

Holdeman, L. V., Cato, E. P., and Moore, W. E. C. (ed). 1977. Anaerobe Laboratory Manual, 4th ed. Anaerobe Laboratory, Virginia Polytechnic Institute and State University, Blacksburg, Virginia.

Holdeman-Moore, L. V. H., Johnson, J. L., and Moore, W. E. C. 1986a. Genus Peptococcus Kluyver and van Niel 1936, p-1082–1083. In: Sneath, P. H. A., Mair, N. S., Sharpe, M. E., and Holt, J. E. (ed), Bergey's manual of systematic bacteriology., vol. 2. Williams & Wilkins, Baltimore.

Holdeman-Moore, W. E. C., Johnson, J. L., and Moore, W. E. C. 1986b. Genus Peptostreptococcus Kluyver and van Niel 1936. P 1083–1092. In: Sneath, P. H. A., Mair, M. S., Sharpe, M. E. and Holt, J. E.(ed), Bergey's manual of systematic bacteriology, vol. 2. Williams & Wilkins, Baltimore.

Hungate, R. E. 1950. The anaerobic mesophilic cellulolytic bacteria. Bacteriol. Rev. 14:1–49.

Hungate, R. E. 1966. The rumen and its microbes. Academic Press, New York.

Huss, V. A. R., Schleifer, K. H. Lindal, E., Schwan, O., and Smith, C. J. 1982. Peptidoglycan type, base composition of DNA, and DNA-DNA homology of Peptococcus indolicus and Peptococcus asaccharolyticus. FEMS Microbiol. Lett. 15:285–289.

Huss, V. A. R., Festl, H., and Schleifer, K. H. 1984. Nucleic acid hybridization studies and deoxyribonucleic acid base compositions of anaerobic, Gram-positive cocci. Int. J. Syst. Bacteriol. 34:95–101.

Judicial Commission of the International Committee on Systematic Bacteriology. 1971. Opinion 39. Rejection of the generic name Gaffkya Telvisan. Int. J. Syst. Bacteriol. 21:104–105.

Kilpper-Bälz, R., and Schleifer, K. H. 1981. Transfer of Peptococcus saccharolyticus Foubert and Douglas to the genus Staphylococcus: Staphylococcus saccharolyticus (Foubert and Douglas) comb. nov. Zentralblatt fur Bakteriologie. Parsitenkd. Infektionskr. Hygiene. Abt I. Originale C: 124–331.

Kilpper-Bälz, R., and Schleifer, K. H. 1988. Transfer of Streptococcus morbillorum to the genus Gemella as Gemella morbillorum comb. nov. Int. J. Syst. Bacteriol. 38:442–443.

Kilpper-Bälz, R., Buhl, U., and Schleifer, K. H. 1980. Nucleic acid homology studies between Peptococcus saccharolyticus and various anaerobic facultative anaerobic Gram-positive cocci. FEMS Microbiol. Lett. 8:205–210.

Lanigan, G. W. 1976. Peptococcus heliotrinreducens sp. nov., a cytochrome-reducing anaerobe which metabolizes pyrrolizidine alkaloids. J. Gen. Microbiol. 94:1–10.

Lorian V. (ed.). 1980. Antibiotics in laboratory medicine. Williams & Wilkins, Baltimore.

Lorowitz, W. H. and Bryant, M. P. 1984. Peptostreptococcus products strain that grows rapidly with CO as the energy source. Appl. Environ. Microb. 47:961–964.

Ludwig, W., Schleifer, K. H., Fox, G. E., Seewaldt, E., and Stackebrandt, E. 1981. A phylogenetic analysis of staphylococci. Peptococcus saccharolyticus, and Micrococcus mucilaginosus. J. Gen. Microbiol. 125:357–366.

Ludwig, W., Weizenegger, M., Kilpper-Bälz, R., and Schleifer, K. H. 1988. Phylogenetic relationships of anaerobic streptococci. Int. J. Syst. Bacteriol. 38:15–18.

Marui, T. 1981. Characterization of peptococci and peptostreptococci isolated from human clinical specimens and their taxonomical problems. Acta Scholae Medicin. Univ. GIFU 29:1071–1082.

Moore, W. E. C., and Holdeman, L. V. 1974. Human fecal flora: the normal flora of 20 Japanese-Hawaiians. Appl. Microbiol. 27:961–979.

National Committee for Clinical Laboratory Standards. 1985. Approved standard: Reference agar dilution procedure for antimicrobial susceptibility testing of anaerobic bacteria. NCCLS publication M11-A, Villanova, PA.

National Committee for Clinical Laboratory Standards. 1986. Proposed guideline: Alternative methods for antimicrobial susceptibility testing of anaerobic bacteria. NCCLS publication M17-P, Villanova, PA.

National Committee for Clinical Laboratory Standards. 1989. Methods for antimicrobial susceptibility testing of anaerobic bacteria-second edition; Tentative standard. NCCLS publication M11-T2, Villanova, PA.

Rogosa, M. 1971. Peptococcaceae, a new family to include the Gram-positive, anaerobic cocci of the genera Peptococcus, Peptostreptococcus, and Ruminococcus. Int. J. Syst. Bacteriol. 21:234–237.

Rosenblatt, J. E. 1985. Anaerobic cocci, p. 445–449. In: Lennette, E. H., Balows, A., Hausler, W. J. Jr., and Shadomy, H. J. (ed.), Manual of clinical microbiology, 4th ed. American Society for Microbiology.

Schink, B. 1984. Fermentation of tartrate enantiomers by anaerobic bacteria, and description of two new species of strict anaerobes, *Ruminococcus pasteurii* and *Illyobacter tartarcus*. Arch. Microbiol. 139:409–414.

Schiefer-Ullrich, H. and Andreesen, J. R. 1985. *Peptostreptococcus barnesae* sp. nov., a Gram-positive, anaerobic, obligately purine utilizing coccus from chicken feces. Arch. Microbiol. 143:26–31.

Schleifer, K. H. and Nimmermann, E. 1973. Peptidoglycan types of strains of the genus *Peptococcus*. Arch. Microbiol. 93:245–258.

Skerman, V. B. D., McGowan, V., and Sneath, P. H. A. (ed.). 1980. Approved lists of bacterial names. Int. J. Syst. Bacteriol. 30:225–420.

Sneath, P. H. A., Mair, N. S., Sharpe, M. E., and Holt, J. E. (ed.). 1986. Bergey's manual of systematic bacteriology, vol. 2. Williams & Wilkins, Baltimore.

Sørenson, G. H. 1973. *Micrococcus indolicus*: some biochemical properties, and the demonstration of six antigenically different types. Acta Vet. Scand. 14:301–326.

Stone, H. H., and Martin, J. D., Jr. 1972. Synergistic necrotizing cellulitis. Annl. Surg. 175:702–711.

Sutter, V. L. and Finegold, S. M. 1976. Susceptibility of anaerobic bacteria to 23 antimicrobial agents. Antibio. Agents Chem. 10:736–752.

Sutter, V. L, Vargo, V. L., Finegold, S. M. 1975. Wadsworth anaerobic bacteriology manual, 2nd ed. Los Angeles: Anaerobic Bacteriology Laboratory, Wadsworth Hospital Center and the School of Medicine at the University of California at Los Angeles.

Sweet, R. L. 1975. Anaerobic infections of the female genital tract. Am. J. Obstet. Gynecol. 122:891–901.

Swenson, R. M., Michaelson, T. C., Daly, M. ., Spaulding, E. H. 1973. Anaerobic bacterial infections of the female genital tract. Obstet. Gynecol. 42:538–541.

Tanner, R. S., Stackenbrandt, E., Fox, G. E., Gupta, R., Magrum, L. J. and Woese, C. R. 1982. A phylogenetic analysis of anaerobic eubacteria capable of synthesizing acetate from carbon dioxide. Curr. Microbiol. 7:127–132.

Thomas, C. G. A., and Hare, R. 1954. The classification of anaerobic cocci and their isolation in normal human beings and pathological processes. J. Clin. Pathol. 7:300–304.

Wilkins, T. D., and Chalgren, S. 1976. Medium for use in antibiotic susceptibility testing of anaerobic bacteria. Antimicrob. Agents Chemo. 10:926–928.

Wilkins, T. D., and Jimenez-Ulate, F. 1975. Anaerobic specimen transport device. J. Clin. Microbiol. 2:441–447.

Wilkins, T. D., Moore, W. E. C., West, S. E. H., and Holdeman, L. V. 1975. *Peptococcus niger* (Hall). Kluyver and van Niel 1936: emendation of description and designation of neotype strain. Int. J. Syst. Bacteriol. 25:47–49.

CHAPTER 84

The Genera *Haloanaerobium, Halobacteroides,* and *Sporohalobacter*

AHARON OREN

Botton sediments of hypersaline water bodies such as the Great Salt Lake, Utah; the Dead Sea; and marine salterns used for the production of solar salt can be expected to support a rich community of anaerobic halophilic bacteria, as the solubility of oxygen in hypersaline brines is low, and the amounts of organic matter available are often high. Therefore, it is surprising that still relatively little is known about anaerobic breakdown processes of organic matter in hypersaline sediments (Oren, 1988). This is especially true in the case of the terminal processes of breakdown which lead to sulfidogenesis and methanogenesis. No sulfate-reducing bacteria have yet been isolated that grow at salt concentrations exceeding 15–18%, and while a few halophilic methanogens have been characterized that grow on methylamines, no moderately or extremely halophilic methanogens are known that utilize hydrogen or acetate as energy sources.

It was only in 1983 that the first description appeared of the isolation of an obligately anaerobic fermentative bacterium growing at salt concentrations of 10–20% and higher: *Haloanaerobium praevalens,* obtained from the bottom sediments of the Great Salt Lake, Utah (Zeikus et al., 1983). Since that time, five moderately halophilic, obligately anaerobic chemoorganotrophic species have been described, classified in three genera:

1. *Haloanaerobium praevalens,* a nonmotile, rod-shaped bacterium growing at salt concentrations from 3–25% (optimally at 13%) (Zeikus, 1983; Zeikus et al., 1983). This isolate is similar to an organism called *"Bacteroides halosmophilus,"* isolated by Baumgartner (1937) from solar salt and from salted anchovies, but Baumgartner's isolate has been lost.
2. *Halobacteroides halobius,* a very long, slender, rod-shaped motile bacterium, growing at salt concentrations between 8–16% (Oren et al., 1984b), which has been shown to produce heat-resistant endospores (Oren, 1987).
3. *Halobacteroides acetoethylicus,* a short, motile, rod-shaped bacterium, growing at salt concentrations between 5–22% (optimally at 10%) (Rengpipat et al., 1988a).
4. A rod-shaped motile bacterium, producing terminal endospores with attached gas vacuoles. The isolate was originally named *Clostridium lortetii* (Oren, 1983) and later renamed *Sporohalobacter lortetii* (Oren et al., 1987). This strain requires NaCl concentrations between 6 and 12%.
5. *Sporohalobacter marismortui,* a rod-shaped motile bacterium, growing optimally at salt concentrations of 3–12%, sometimes observed to produce endospores, but without gas vacuoles (Oren et al., 1987).

All five isolates are similar in that they have a low GC content of their DNA (range 27–32 mol%), all stain Gram-negative and have a Gram-negative type of cell wall, and all are mesophilic, with optimal growth temperatures around 35–45°C.

The technique of 16S rRNA oligonucleotide cataloging has allowed the elucidation of the relationships between the isolates and the relationships of these anaerobic halophiles with the other bacterial groups. With the exception of *Halobacteroides acetoethylicus,* 16S rRNA catalogs of all strains have been analyzed, and the isolates were found to be eubacteria, related to each other and clustering in one group, for which a new family was created, the Haloanaerobiaceae (Oren, 1986a; Oren et al., 1984a). Within the eubacteria they do not show a clear relationship to any of the major subgroups of the kingdom, with the possible exception of the spirochetes, to which some relationship was suggested (Paster et al., 1984; Fox and Stackebrandt, 1987) (note that in the last-cited paper *Halobacteroides halobius* erroneously appears as *"Haloanaeromicrobium"*).

In addition to the species mentioned above, Shiba and Horikoshi (1988) and Shiba et al. (1989) have described a number of other anaerobic halophiles that probably belong to the Haloanaerobiaceae.

Habitats

Anaerobic halophilic bacteria of the family
Haloanaerobiaceae have been found in all anoxic hypersaline water bodies and sediments investigated. Habitats include bottom sediments
of the Great Salt Lake, Utah (Zeikus, 1983; Zeikus et al., 1983); the Dead Sea (Oren, 1987);
and sediments of a hypersaline sulfur spring
(Oren, 1987). *Halobacteroides acetoethylicus*
was recovered from such an exotic environment
as the brine waters associated with deep subsurface gas-bearing sandstones and injection
water filters on offshore oil rigs in the Gulf of
Mexico (Rengpipat et al., 1988a).

Environments not yet explored for the presence of anaerobic halophilic fermentative bacteria are hot hypersaline brines that are found
on several places on the sea bottom (Degens and
Ross, 1969), and sediments of natural or artificial solar ponds in which too a high salinity
is combined with high temperatures. Environments like these might well yield novel types of
bacteria.

Quantitative data on the occurrence of members of the Haloanaerobiaceae in hypersaline
anoxic environments are scarce. *Haloanaerobium praevalens* was reported to be present in
Great Salt Lake (Utah) surface sediment in
numbers of up to 10^8 /ml sediment (Zeikus,
1983; Zeikus et al., 1983), while 10^3–10^5 *Halobacteroides halobius* cells were counted per
milliliter of Dead Sea sediment (Oren et al.,
1984b).

A search for novel organisms by using a wider
variety of potential substrates and salt concentrations in enrichment cultures or in media for
direct isolation, or by extending the search to
previously unexplored habitats, will undoubtedly yield a much greater variety of anaerobic
fermentative bacteria than presently known,
and this may add important information to the
rather uncomplete picture we possess today of
anaerobic breakdown in hypersaline sediments
(Oren, 1988).

Isolation and Media
for Cultivation

Any anoxic reducing medium containing high
salt concentrations (5–25%) and containing a
suitable carbon source is a potential enrichment
and growth medium for anaerobic halophilic
bacteria of the family Haloanaerobiaceae. However, in enrichment cultures the development of
facultatively anaerobic moderately halophilic
bacteria should be avoided. Bacteria resembling

Vibrio costicola often develop in enrichment
cultures when salt concentrations in the lower
range (up to 10%) are used. Therefore, salt concentrations of 10–15% are recommended for the
enrichment of *Halobacteroides halobius* and
similar organisms.

As several species of the family Haloanaerobiaceae produce heat-resistant endospores, this
property can be used in a negative selection procedure (Oren, 1987): hypersaline sediment used
as inoculum is introduced into tubes containing
sterile growth medium, whereafter the tubes are
immediately submerged for 10–20 min in a
water bath at 80–100°C to kill vegetative cells,
followed by incubation of the tubes at the
growth temperature.

The members of the family Haloanaerobiaceae are obligate anaerobes. Though they are
not extremely sensitive to molecular oxygen, the
use of strict anaerobic techniques is recommended, which includes boiling the media under nitrogen, adding reducing agents such as
cysteine to the boiled media, and using anaerobic culture techniques such as described by
Balch et al. (1979) which employ anaerobic
pressure tubes with thick rubber stoppers
(Bellco Glass Co., Vineland, NJ, USA).

Two of the species described *(Sporohalobacter lortetii* and *S. marismortui)* have been isolated only once, and no time-tested enrichment
and isolation procedures were described for
these isolates. The conditions that originally led
to their successful isolation are described below.

All described species are mesophilic, growing
optimally at 35–45°C.

Isolation of *Haloanaerobium praevalens* from Great Salt Lake Sediment in Semi-defined Medium CS Medium (Zeikus et al., 1983)

Solution 1:

NaCl	130 g
MgSO$_4$· 7H$_2$O	5 g
KCl	1 g
Yeast extract	10 g
Trypticase (BBL)	10 g
Resazurin	0.01 g

Distilled water to a final volume of 920 ml

Solution 2:
Vitamin solution (Moench and Zeikus, 1983):

Biotin	2 mg
Folic acid	2 mg
Pyridoxine·HCl	10 mg
Thiamine·HCl	5 mg
Riboflavin	5 mg
Nicotinic acid	5 mg
Calcium pantothenate	5 mg
Vitamin B$_{12}$	0.1 mg
p- Aminobenzoic acid	5 mg

Lipoic acid   5 mg
Distilled water   1 liter

Solution 3:
Trace mineral solution (Moench and Zeikus, 1983):

Nitrilotriacetic acid	12.8 g
$FeCl_3 \cdot 4H_2O$	0.2 g
$MnCl_2 \cdot 4H_2O$	0.1 g
$CoCl_2 \cdot 6H_2O$	0.17 g
$CaCl_2 \cdot 2H_2O$	0.1 g
$ZnCl_2 \cdot$	0.1 g
$CuCl_2$	0.02 g
H_3BO_3	0.01 g
$NaMoO_4 \cdot 2H_2O$	0.01 g
$NiCl_2 \cdot 6H_2O$	0.026 g
NaCl	1 g
Na_2SeO_3	0.02 g
Distilled water	1 liter

Adjust pH to 6.5 with KOH.

Solution 4:

Sodium thioglycolate	0.25 g
Sodium ascorbate	0.25 g
Distilled water	25 ml

Sterilize by filtration.

Solution 5:

Glucose   100 g

Distilled water to a final volume of 1 liter
Remove oxygen by bubbling nitrogen and sterilize by autoclaving.

Solution 6:

NaOH   2 N

Remove oxygen by bubbling nitrogen and sterilize by autoclaving.

To 920 ml solution 1, 10 ml of each solutions 2 and 3 is added, and the mixture is boiled under nitrogen to remove molecular oxygen. Afterwards, 25 ml of solution 4 and 25 ml of solution 5 are added anaerobically, whereafter the pH is adjusted with solution 6 to 7.5. The medium is dispensed to culture tubes which are closed under nitrogen and autoclaved.

Zeikus et al. (1983) give an alternative complex salt-enrichment medium for *Haloanaerobium praevalens*.

Halo TYEG Medium

Solution 7:

Solar salts	30 g
NaCl	180 g
$MgSO_4 \cdot 7H_2O$	20 g
Yeast extract	1 g
Trypticase	5 g
Glucose	5 g
Sodium citrate	3 g
K_2SO_4	2 g
K_2HPO_4	1 g
$Fe(NH_4)_2SO_4 \cdot 6H_2O$	0.05 g
NH_4Cl	1 g
Vitamin solution (solution 2)	10 ml
Trace element solution (solution 3)	10 ml
Resazurin	0.01 g
Distilled water to a final volume of	975 ml

After adjustment of the pH to 7.6 with NaOH, the solution is boiled under N_2–CO_2 (95%:5%) and sterilized by autoclaving, whereafter 25 ml ascorbate-thioglycolate solution (solution 4) is added. Pure cultures are obtained by streaking on plates of the growth medium with 15 g agar/l, and incubating under anaerobic conditions.

Isolation of *Halobacteroides halobius* from the Dead Sea—and Other Sediments (Oren, 1987; Oren et al., 1984b)

Medium Used in the Original Isolation of *Halobacteroides halobius* from Dead Sea Sediments (Oren et al., 1984b)

Solution 8:

NH_4Cl	1 g
K_2SO_4	2 g
Sodium pyruvate	1 g
Yeast extract	1 g
$NaH_2PO_4 \cdot H_2O$	0.11 g
Ascorbic acid	0.5 g
Thioglycolic acid	0.1 g
Dead Sea water	800 ml
Distilled water	200 ml

Adjust pH to 7.0 with NaOH

Growth Medium for *Halobacteroides halobius* and *Sporohalobacter marismortui*

Solution 9:

NaCl	88, 100, or 140 g
$MgCl_2 \cdot 6H_2O$	20.3 g
$CaCl_2 \cdot 2H_2O$	7.35 g
KCl	3.7 g
Yeast extract	5 g
Resazurin	0.001 g

Distilled water to a final volume of 900 ml

Solution 10:

PIPES (Piperazine-*N,N'-bis*)-ethanesulfonic acid, sesquisodium salt	40 mM

pH 6.5–7.0

Remove oxygen by bubbling with nitrogen and sterilize by autoclaving.

Solution 9 is boiled under nitrogen, whereafter 0.5 g cysteine·HCl is added. The medium is dispensed in culture tubes under a gas phase of nitrogen and sterilized by autoclaving. The following components are then added aseptically: 50 ml glucose solution/1 (solution 5) and 50 ml solution 10.

The medium is suitable for direct isolation of *Halobacteroides halobius* when the NaCl concentration is increased to 140 g/l to inhibit the development of *Vibrio costicola*.

A more specific negative selection method was introduced (Oren, 1987), based on the heat resistance of the endospores: immediately after inoculation the tubes are submerged for 10–20 min in a water bath at 80–100°C, followed by incubation at the growth temperature.

Pure cultures can be isolated by streaking on plates of the growth medium enriched with $CaCO_3$ (5 g/l) and agar (20 g/l) in an anaerobic glove box, followed by incubation under nitrogen.

Medium for the Isolation of *Halobacteroides acetoethylicus* from Deep Subsurface Brine Waters Associated with Injection Water Filters on Offshore Oil Rigs (Rengpipat et al., 1988a)

Growth medium for *Halobacteroides acetoethylicus*

Solution 11:

NaCl	100 g
KH_2PO_4 + K_2HPO_4 buffer, pH 7	30 mM
NH_4Cl	1 g
Glucose	5 g
$MgCl_2 \cdot 6H_2O$	0.2 g
Trypticase	10 g
Yeast extract	0.5 g
Vitamin solution (solution 2)	10 ml
Trace element solution (solution 3)	10 ml
$Na_2S \cdot 9H_2O$	0.25 g
Resazurin	0.02 g

Distilled water to a final volume of 1 liter.

After adjustment of the pH to 7.3–7.4 with NaOH, the medium is boiled under N_2–CO_2 (95%:5%), dispensed into culture tubes, and sterilized by autoclaving. Glucose, Na_2S, yeast extract, phosphate buffer, and the vitamin solution are added aseptically to the autoclaved medium from concentrated sterile solutions. Pure cultures are isolated by streaking in anaerobic roll tubes with agar-supplemented growth medium.

Media for Isolation and Growth of *Sporohalobacter lortetii* from Dead Sea Sediments (Oren, 1983)

Medium Used in the Original Isolation of *Sporohalobacter lortetii*

Solution 12:

NH_4Cl	1 g
K_2SO_4	2 g
Sodium lactate	3.5 g
Yeast extract	1 g
KH_2PO_4	0.14 g
Ascorbic acid	0.5 g
Thioglycolic acid	0.1 g
Dead Sea water	800 ml
Distilled water	200 ml

The pH is adjusted to 6.5 with NaOH.

Glass-stoppered bottles are completely filled with the medium, inoculated with Dead Sea sediment and incubated at 30°C for 3 weeks. Pure cultures are then isolated by dilutions in tubes containing the above medium, but with 70% instead of 80% (v/v) Dead Sea water and enriched with 0.5 g $FeSO_4 \cdot 7H_2O/l$ and 15 g agar/l, molten at 45°C. After solidification, the agar is overlayered with paraffin wax. After incubation of 30°C, black colonies are investigated for the characteristic swollen cells with endospores and gas vacuoles.

Growth Medium for *Sporohalobacter lortetii* (Oren, 1983)

Solution 13:

Casamino acids	2 g
Nutrient broth	2 g

Yeast extract	2 g
L-Glutamic acid	4 g
NaCl	105 g
KCl	0.75 g
$FeSO_4 \cdot 7H_2O$	0.002 g
Vitamin solution (solution 2)	10 ml
Trace element solution (solution 3)	10 ml
Resazurin	0.001 g
L-Cysteine·HCl	0.5 g

Distilled water to a final volume of 900 ml.

Solution 14:

$MgCl_2 \cdot 6H_2O$	100 g/l
$CaCl_2 \cdot 2H_2O$	37 g/l

Remove oxygen by bubbling with nitrogen and sterilize by autoclaving.

The medium is boiled under nitrogen to remove molecular oxygen, whereafter the cysteine is added, the medium is dispensed to culture tubes under a gas phase of nitrogen, and sterilized by autoclaving. To the sterile medium 100 ml/l of $MgCl_2$-$CaCl_2$ solution (solution 14) is added, and the pH is adjusted to 6.5–7 with solution 6. Addition of 2 g glucose/l proved stimulatory.

Medium Used for the Original Isolation of *Sporohalobacter marismortui* from Dead Sea Sediments (Oren et al., 1987)

Medium for Isolation of *Sporohalobacter marismortui*

Solution 15:

NaCl	125 g
$MgCl_2 \cdot 6H_2O$	50 g
Sodium lactate	2 g
$MgSO_4 \cdot 7H_2O$	1 g
$CaSO_4 \cdot 2H_2O$	0.5 g
K_2HPO_4	1 g
NH_4Cl	1 g
Sodium thioglycolate	1 g
Yeast extract	0.5 g
$FeSO_4 \cdot 7H_2O$	0.1 g

The pH is adjusted to 7–7.2 with NaOH.

After incubation at 37°C, growth was obtained, and the strain was purified by streaking anaerobically on plates of the growth medium enriched with 5 g $CaCO_3/l$ and 20 g agar/l. Plates were incubated under nitrogen.

Preservation of Cultures

Several species of the family Haloanaerobiaceae, notably *Halobacteroides halobius* and *Sporohalobacter marismortui*, easily undergo autolysis, during which spherical degeneration forms are observed. Autolysis starts at the end of the exponential growth phase, especially at the highest temperatures enabling growth.

One possibility to avoid die-off of the culture is the use of media with reduced nutrient content and lower growth temperatures (20–25°C). In our experience, weekly transfers suffice in this case to maintain viable cultures.

The surest way to maintain cultures is by lyophilization. *Halobacteroides acetoethylicus* can be preserved by freezing anaerobic suspensions in 20% glycerol at $-80°C$ (Rengpipat et al., 1988a).

Identification

All described members of the family Haloanaerobiaceae are obligately anaerobic, chemoorganotrophic, moderate halophiles. They share a Gram-negative staining reaction, do not contain cytochromes, and are oxidase-and catalase-negative. Table 1 summarizes the main properties differentiating the five described species.

The property of endospore formation is often expressed only under special conditions, especially on solid growth media. Since *Halobacteroides halobius* was found to produce heat-sensitive endospores (Oren, 1987), the differences between the genera *Halobacteroides* and *Sporohalobacter* has become less clear, and a reclassification of the species into genera may be required in the future.

As all five described species are able to use carbohydrates, either as principal or as secondary carbon and energy source, the range of carbohydrates and other simple compounds utilized can be used to differentiate between the species (Table 2).

The technique of 16S rRNA nucleotide sequence analysis enabled the unification of an at-first-sight rather heterogeneous group of bacteria in a single family. Therefore, the surest way to ascertain the identity of new isolates with one of the described species of the Haloanaerobiaceae, or its identification as a new species within the family, is the characterization of its 16S rRNA, either by direct sequencing or by characterizing the oligonucleotides in a T1 RNAse digest.

Physiological Properties

While 16S rRNA analysis shows that they belong to the eubacterial kingdom, the representatives of the family Haloanaerobiaceae display several physiological and biochemical properties characteristic of the halophilic archaebacteria, rather than of the moderately halophilic aerobic eubacteria. While the halophilic eubacteria (aerobes, as well as anaerobic photosynthetic types) accumulate organic osmotic solutes such as glycine betaine and others to balance the cytoplasm osmotically with the surrounding medium, no organic osmotic solutes were found in the anaerobic halophilic eubacteria (Oren, 1986b; Rengpipat et al., 1988b). It was found that salt is not excluded from the cells, and high Na^+, K^+ and Cl^- concentrations were measured inside the cells of *Haloanaerobium praevalens, Halobacteroides halobius* (Oren, 1986b), and *Halobacteroides acetoethylicus* (Rengpipat et al., 1988b), high enough to be at least isotonic with the medium. Like that in the archaebacterial genus *Halobacterium,* the bulk cellular protein of the Haloanaerobiaceae is highly acidic (Oren, 1986b), and the enzymes retain their activity in the presence of high salt concentrations such as are present inside the cells (Rengpipat et al., 1988b).

Four out of the five anaerobic halophiles described are carbohydrate fermenters. *Sporohalobacter lortetii* is probably primarily an amino acid fermenter; glucose added to the growth medium is utilized only after other preferred medium components are exhausted (Oren, 1983).

Three out of the five species of the Haloanaerobiaceae *(Halobacteroides halobius, Sporohalobacter lortetii,* and *S. marismortui)* have been shown to be able to produce endospores; the existence of heat-resistant endospores has even been exploited in a selective enrichment procedure for *Halobacteroides halobius* (Oren, 1987). Whether all representatives of the family Haloanaerobiaceae have the ability to make endospores remains an open question. It should be noted here that spore formation in the two *Sporohalobacter* species was only observed on solid media, or in nutrient-poor liquid media. Evidence that *Haloanaerobium praevalens,* which has never been shown to make endospores, may share characteristics of the endospore-forming bacteria, came recently from the finding that all endospore-forming bacteria *(Bacillus* and *Clostridium)* hydrolyze the D-isomer of Nα-benzoyl-arginine-*p*-nitroanilide (BAPA), while nonsporeformers hydrolyze L-BAPA but not D-BAPA (Gofshtein-Gandman et al., 1988). Four representatives of the family Haloanaerobiaceae were tested for D-BAPA and L-BAPA hydrolysis, and it was found that all (with the exception of *Sporohalobacter lortetii*) hydrolyze D-BAPA, while L-BAPA was not hydrolyzed by any of them (Oren et al., 1989). If D-BAPA hydrolysis is indeed a chemotaxonomic signature for the ability of endospore formation or indicates relatedness to endospore-forming bacteria, *Haloanaerobium praevalens* also belongs to the group of endospore formers.

Applications

The use of anaerobic halophilic bacteria in the industrial fermentation of complex organic matter has been proposed, and preliminary tests

Table 1. Major characteristics of the described representatives of the family Haloanaerobiaceae.[a]

Genus and species	Cell size (μm)	Flagellation	GC content (mol%)	Fermentable substrates	Fermentation products	Endospores	Hydrolysis of D-BAPA[b]	NaCl concentration range (%)	NaCl concentration optimum (%)	Type strain[c]
Haloanaerobium praevalens	2–2.6 × 0.9–1.1	None	27	Carbohydrates, pectin, amino acids	Acetate, propionate, butyrate, H₂, CO₂	−	+	3–25	13	DSM 2228
Halobacteroides halobius	10–20 × 0.5	Peritrichous	30.7	Carbohydrates	Ethanol, acetate, H₂, CO₂	+(?)	+	8–17	9–15	ATCC 35273
Halobacteroides acetoethylicus	1–1.5 × 0.4–0.6	Peritrichous	32	Carbohydrates	Ethanol, acetate, H₂, CO₂	−	NR	6–20	10	ATCC 43120 DSM 3532
Sporohalobacter lortetii	2.5–10 × 0.5–0.6	Peritrichous	31.5	Amino acids (?)	Acetate, propionate, butyrate, H₂ (CO₂)	+	−	6–12	8–9	ATCC 35059 DSM 3070
Sporohalobacter marismortui	3–13 × 0.6	Peritrichous	29.6	Carbohydrates	Ethanol, acetate, formate, butyrate, H₂, CO₂	+	+	3–17	3–12	ATCC 35420

[a]+, present or detected; −, absent or not detected; NR, not reported.
[b]D-BAPA, D isomer of Nα-benzoyl-arginine-p-nitroanilide.
[c]DSM, Deutsche Sammlung für Mikroorganismen; ATCC, American Type Culture Collection.

Table 2. Carbohydrates and other components used as carbon and energy sources by members of the family Haloanaerobiaceae.[a]

Compound	Haloanaerobium praevalens	Halobacteroides halobius	Halobacteroides acetoethylicus	Sporohalobacter lortetii	Sporohalobacter marismortui
D-Glucose	+	+	+	+	+
D-Galactose	−	+	−	−	−
D-Mannose	+	+	+	NR	+
L-Rhamnose	NR	−	NR	NR	−
N-Acetylglucosamine	+	NR	+	NR	NR
Fructose	+	+	+	+	+
L-Fucose	NR	−	NR	NR	NR
D-Ribose	NR	−	NR	NR	±
D-Xylose	−	−	+	NR	−
L-Arabinose	NR	−	NR	NR	−
Sucrose	−	+	+	NR	+
Lactose	−	−	+	−	−
Cellobiose	−	−	+	−	−
D-Maltose	NR	+	NR	+	+
D-Raffinose	NR	+	NR	NR	−
Starch	−	+	−	+	+
Chitin	−	NR	−	NR	−
Pectin	+	NR	−	NR	NR
Pyruvate	−	+	+	NR	−
Lactate	−	NR	−	−	NR
Glycerol	−	−	−	NR	−

[a] +, compound utilized; −, compound not utilized; NR, not reported.

have been made (Wise, 1987), but possible applications of the organisms are still in an experimental stage.

Literature Cited

Balch, W. E., G. E. Fox, L. J. Magrum, C. R. Woese, and R. S. Wolfe. 1979. Methanogens: reevaluation of a unique biological group. Microbiol. Rev. 43:260–296.

Baumgartner, J. G. 1937. The salt limits and thermal stability of a new species of anaerobic halophile. Food Res. 2:321–329.

Degens, E. T., and D. A. Ross (ed.) 1969. Hot brines and recent heavy metal deposits in the Red Sea. Springer-Verlag, New York.

Fox, G. E., and E. Stackebrandt. 1987. The application of 16 S rRNA cataloguing and 5S rRNA sequencing in bacterial systematics, p. 405–458. In: R. R. Colwell, and R. Grigorova (ed.), Methods in microbiology, vol. 19. Academic Press, London.

Gofshtein-Gandman, L. V., A. Keynan, and Y. Milner. 1988. Bacteria of the genus Bacillus have a hydrolase stereospecific to the D isomer of benzoyl-arginine-p-nitroanilide. J. Bacteriol. 170:5895–5900.

Moench, T., and J. G. Zeikus. 1983. Nutritional growth requirements of Butyribacterium methylotrophicum on single carbon substrates and glucose. Curr. Microbiol. 9:151–154.

Oren, A. 1983. Clostridium lortetii sp. nov., a halophilic obligatory anaerobic bacterium producing endospores with attached gas vacuoles. Arch. Microbiol. 136:42–48.

Oren, A. 1986a. The ecology and taxonomy of anaerobic halophilic eubacteria. FEMS Microbiol. Rev. 39:23–29.

Oren, A. 1986b. Intracellular salt concentrations of the anaerobic halophilic eubacteria Haloanaerobium praevalens and Halobacteroides halobius. Can. J. Microbiol. 32:4–9.

Oren, A. 1987. A procedure for the selective enrichment of Halobacteroides halobius and related bacteria from hypersaline sediments. FEMS Microbiol. Lett. 42:201–204.

Oren, A. 1988. Anaerobic degradation of organic compounds at high salt concentrations. Antonie van Leeuwenhoek J. Microbiol. 54:267–277.

Oren, A., L. V. Gofshtein-Gandman, and A. Keynan. 1989. Hydrolysis of Nα-benzoyl-D-arginine-p-nitroanilide by members of the Haloanaerobiaceae: additional evidence that Haloanaerobium praevalens is related to endospore-forming bacteria. FEMS Microbiol. Lett. 58:5–10.

Oren, A., B. J. Paster, and C. R. Woese. 1984a. Haloanaerobiaceae: a new family of moderately halophilic, obligatory anaerobic bacteria. System. Appl. Microbiol. 5:71–80.

Oren, A., H. Pohla, and E. Stackebrandt. 1987. Transfer of Clostridium lortetii to a new genus Sporohalobacter gen. nov. as Sporohalobacter lortetii comb. nov., and description of Sporohalobacter marismortui sp. nov. System. Appl. Microbiol. 9:239–246.

Oren, A., W. G. Weisburg, M. Kessel, and C. R. Woese. 1984b. Halobacteroides halobius gen. nov., sp. nov., a moderately halophilic anaerobic bacterium from the bottom sediments of the Dead Sea. System. Appl. Microbiol. 5:58–70.

Paster, B. J., E. Stackebrandt, R. B. Hespell, C. M. Hahn, and C. R. Woese. 1984. The phylogeny of the spirochetes. System. Appl. Microbiol. 5:337–351.

Rengpipat, S., T. A. Langworthy, and J. G. Zeikus. 1988a. *Halobacteroides acetoethylicus* sp. nov., a new obligately anaerobic halophile isolated from deep subsurface hypersaline environments. System. Appl. Microbiol. 11:28–35.

Rengpipat, S., S. E. Lowe, and J. G. Zeikus. 1988b. Effect of extreme salt concentrations on the physiology and biochemistry of *Halobacteroides acetoethylicus*. J. Bacteriol. 170:3065–3071.

Shiba, H., and K. Horikoshi. 1988. Isolation and characterization of novel anaerobic, halophilic eubacteria from hypersaline environments of western America and Kenya. In: Proceedings of the FEMS symposium on the microbiology of extreme environments and its biotechnological potential, Portugal.

Shiba, H., H. Yamamoto, and K. Horikoshi. 1989. Isolation of strictly anaerobic halophiles from the aerobic surface sediment of hypersaline environments in California and Nevada. FEMS Microbiol. Lett. 57:191–196.

Wise, D. L. 1987. Meeting report—First international workshop on biogasification and biorefining of Texas lignite. Resources and Conservation 15:229–247.

Zeikus, J. G. 1983. Metabolic communication between biodegradative populations in nature, p. 423–462. In: J. H. Slater, R. Whittenbury, and J. W. T. Wimpenny (ed.), Microbes in their natural environments. Society for General Microbiology, Cambridge University Press, Cambridge.

Zeikus, J. G., T. E. Hegge, T. J. Thompson, T. J. Phelps, and T. A. Langworthy. 1983. Isolation and description of *Haloanaerobium praevalens* gen. nov. and sp. nov., an obligately anaerobic halophile common to Great Salt Lake sediments. Curr. Microbiol. 9:225–234.

CHAPTER 85

The Genera *Thermoanaerobacter, Thermoanaerobium,* and Other Thermoanaerobic Saccharolytic Bacteria of Uncertain Taxonomic Affiliation

MAHENDRA K. JAIN and J. GREGORY ZEIKUS

At present, the genus *Thermoanaerobacter* is comprised of two described species, *T. ethanolicus,* and *T. finnii* (Wiegel and Ljungdahl, 1981; Schmid et al., 1986). Two other strains assigned as *Thermoanaerobacter* await speciation and include strains B6A and LX11 (Weimer, 1985; Y. O. Lee, M. K. Jain, C. Lee, and J. G. Zeikus, unpublished observations). Strain JW200 has been designated the neotype strain of *T. ethanolicus.* The genus *Thermoanaerobium* is also comprised of only two described species, *T. brockii* and *T. lactoethylicum* (Zeikus et al., 1979; Kondratieva et al., 1989). *T. brockii* strain HTD4 is the neotype strain (Zeikus et al., 1979). These two genera share a common optimal temperature range (60–70°C), habitats, morphology, and energy-yielding metabolic pathways with each other and with certain thermophilic *Clostridium* species (i.e., *C. thermohydrosulfuricum* and *C. thermosulfurogenes*) of uncertain taxonomic affiliation (Cato and Stackebrandt, 1989). However, species of these genera differ in Gram-stain reactions, pH range for growth, fermentation substrates, end-product ratios, susceptibility to various antibiotics, motility, sporulation, sulfur metabolism, and cell size.

The generic names *Thermoanaerobacter* (Wiegel and Ljungdahl, 1981) and *Thermoanaerobium* (Zeikus et al., 1979) were used to describe the first thermophilic anaerobic rod-shaped, Gram-positive, nonsporeforming bacteria that produce ethanol and lactate as principal saccharide fermentation products. Numerous strains belonging to these two genera have been isolated in the last 10 years (Zeikus et al., 1979; Wiegel and Ljungdahl, 1981; Weimer et al., 1984; Weimer, 1985; Schmid et al., 1986; Kondratieva et al., 1989).

These two genera of thermophilic anaerobic bacteria, as well as spore forming species which grow near 70°C, have been isolated from hot springs and other diverse thermal environments. The discovery of these thermoanaerobic species has extended the known diversity of thermoanaerobes in nature. These bacteria can utilize a wide variety of substrates, including xylan, starch, glucose, cellobiose, maltose, sucrose, and lactose. Presumably, this broad spectrum arose from exposure of the organisms in their natural environment to carbohydrates derived from thermophilic algal- and bacterial-derived biomass (structural polysaccharides of photosynthetic microflora) or to structural polysaccharides from decaying higher plants which must occasionally enter the hot spring habitat via physical displacement. The wide variety of organic structures thus encountered in hot spring environments has probably selected for nutritionally versatile heterotrophic bacteria capable of degrading many carbohydrates.

The physiological and biochemical properties used previously to distinguish *Thermoanaerobium brockii* (Zeikus et al., 1979) from *Thermoanaerobacter ethanolicus* (Wiegel and Ljungdahl, 1981) were not adequate to assign these organisms to higher taxonomic groups or to distinguish each of these organisms from the other. New taxonomic assignments are reviewed here for these and similar sporeforming species based on molecular analysis data including DNA-DNA hybridization and 16S rRNA oligonucleotide cataloging.

Habitats

Thermophilic bacteria of a wide variety of types are common in self-heating soils and sediments and in volcanic habitats. *Thermoanaerobium brockii* strains as well as strains of *clostridium thermosulfurogenes* and *C. thermohydrosulfuricum* have been isolated from various hot spring environments (especially the thermophilic *Synechoccous/Chloroflexus* mat ecosystems such as that present in Octopus Spring in Yellowstone National Park, Wyoming, USA) (Zeikus et al., 1979, 1980; Schink and Zeikus, 1983). *Thermoanaerobium lactoethylicum* was isolated from waters of a hot spring in Kamchatka, USSR (Kondratieva et al., 1989). *Thermoanaerobium* sp. strain Tok6-B1 was isolated from fumaroles in New Zealand hot springs (Morgan et al., 1985; Patel et al., 1986). *Thermoanaerobacter ethanolicus* strain JW200 was isolated from water and mud samples (pH around 8.8, temperature 45–50°C) collected from an alkaline hot spring in White Creek opposite the Great Fountain Geyser located at Firehole Lake Drive in Yellowstone National Park, while strain JW 201 was isolated from water and mud samples (pH 5.5; temperature 55–60°C) from the outflow of the slightly acidic hot spring, Dragon's Mouth, of the Sulfur Caldron area in Yellowstone National Park (Wiegel and Ljungdahl, 1981). *Thermoanaerobacter finnii* was isolated from a methanogenic, continuous-culture-fermentor fluid which originated from the sediment sludge of Lake Kivu, an East African volcanic rift lake (Schmid et al., 1986; Schoberth and Tietze, 1978). *Thermoanaerobacter* strain B6A was isolated from an algal mat obtained from Big Spring, Thermopolis, Wyoming, USA (Weimer et al., 1984; Weimer, 1985).

Isolation

The enrichment procedures needed to obtain strains of *Thermoanaerobacter* and *Thermoanaerobium* include the use of anaerobic tube cultivation methods as developed by Hungate (1969), as modified for thermophiles by Wiegel (1986a) and Zeikus (Zeikus, 1977; Zeikus et al., 1980). Enrichments can be set up by inoculating appropriate anaerobic culture tubes with 5–10% of anoxic samples collected from thermal springs followed by incubation at 60–70°C. Xylose at 0.5–1% can be used as a substrate in a yeast extract-trypticase complex medium (Weimer and Zeikus, 1978) to enrich all the thermoanaerobic bacterial species mentioned except for *C. thermocellum*. The tubes and bottles should be checked within 24–36 hours for positive pressure (indicative of H_2 production) and visible turbidity and suitable tubes examined microscopically for the presence of rod-shaped cells. These enrichments should be transferred into homologous media and followed for growth. After 3–4 transfers, when sufficient enrichment has been achieved, a sample should be checked for fermentation products. With xylose as substrate, a fermentation product spectrum consisting of ethanol, acetate, lactate, H_2, and CO_2 is an indication of positive enrichment. The enrichment cultures can then be plated onto homologous media containing agar at 2% and with xylose as substrate in an anaerobic glove box. The plates can be incubated anaerobically at 60–65°C using a suitable container such as anaerobic gas jars or modified paint tanks (Belyaev et al., 1983). Samples from the enrichments can also be diluted into roll tubes containing appropriate media and 2% agar and incubated at 60–65°C. Alternatively, environmental samples can be plated by serial dilution onto an appropriate agar medium with xylose as substrate. After 2–4 days, the isolated colonies formed can be picked from plates or roll tubes and placed into liquid media tubes in the anaerobic glove box. The fully grown cultures are then checked for Gram stain, morphology, and fermentation product spectrum.

Several complex or semi-defined media have been used for isolation of the strains of *Thermoanaerobacter* and *Thermoanaerobium*. The recipes for these media are given below.

Enrichment and Isolation Medium for *Thermoanaerobacter ethanolicus* (Wiegel et al., 1979).
The following ingredients are added per liter:

KH_2PO_4	1.5 g
$Na_2HPO_4 \cdot 12H_2O$	4.2 g
NH_4Cl	0.5 g
$MgCl_2$	0.18 g
$CaCl_2$	0.05 g
$Co(NO_3)_2 \cdot 6H_2O$	2.9 mg
$Na_2MoO_4 \cdot 2H_2O$	2.4 mg
$Fe(NH_4)_2(SO_4)_2 \cdot 6H_2O$	39.2 mg
$MnCl_2 \cdot 4H_2O$	2.0 mg
Na_2SeO_3	0.17 mg
$ZnSO_4$	2.8 mg
Resazurin (redox indicator)	2.0 ng
Biotin	20 ng
Folic acid	20 ng
Pyridoxin hydrochloride	100 ng
Riboflavin	5 ng
Thiamine	50 ng
Nicotinic acid	50 ng
Pantothenic acid	50 ng
Vitamin B_{12}	1 ng
p-Aminobenzoic acid	50 ng
Thioctic acid	50 ng
Cysteine	0.5 g
$Na_2S \cdot 9H_2O$	0.5 g
Yeast extract (Difco)	2 g
Glucose	5 g

The medium is prepared under anaerobic conditions and stored under argon. When necessary, the pH (normally 7.3) can be adjusted to 7.3 with a sterile, anaerobic NaOH or HCl solution.

Medium M-1 for Enrichment and Isolation of *Thermoanaerobacter finnii* (Schmid et al., 1986)
The following ingredients are added per liter:

NH_4Cl	2.03 g
$MgCl_2 \cdot 6H_2O$	0.102 g
$CaCl_2 \cdot 2H_2O$	0.147 g
$FeCl_3 \cdot 6H_2O$	1.6 10^{-2} g
$Na_2HPO_4 \cdot 2H_2O$	1.02 g
$NaH_2PO_4 \cdot H_2O$	0.78 g

(NH$_4$)$_2$SO$_4$ 0.132 g
Minerals-1 (see below)   3 ml
Vitamins (Wolin et al., 1963)   10 ml
Resazurin   1 mg
Na$_2$S·9H$_2$O   0.08 g
Sodium thioglycolate   0.5 g
Sodium dithionite   0.04 g
Glacial acetic acid   0.566 ml
Lactic acid   0.17 ml
Glucose   5.95 g
Trypticase (BBL)   2.0 g
Yeast extract (Difco)   2.0 g
Fermenter fluid (see below)   10 ml
Minerals-2 (see below)   62.5 μl

The gas phase is 80% N$_2$ + 20% CO$_2$. The pH of the medium is adjusted to 6.8 with KHCO$_3$.

Minerals-1 (per liter of distilled water):

MnCl$_2$·4H$_2$O   0.475 g
FeSO$_4$·7H$_2$O   0.1 g
CoCl$_2$·6H$_2$O   0.17 g
CuSO$_4$·5H$_2$O   0.05 g
ZnSO$_4$·7H$_2$O   0.18 g
H$_3$BO$_3$ 0.01 g
NiCl$_2$·6H$_2$O   0.079 g
HCl (32%)   30 μl

Minerals-2 (per 5 ml of distilled water): Add 2 mg each of Na$_2$SeO$_3$·5H$_2$O, Na$_2$WO$_4$·2H$_2$O, and Na$_2$MoO$_4$·2H$_2$O.

Fermenter fluid: Liquid from the fermenter containing the thermophilic methanogenic consortium (Bochem et al., 1982) is centrifuged at 4000 × g for 20 min and the supernatant is stored frozen in polyethylene bottles. This processed liquid usually contains around 9 mmol of acetate per liter.

Basal Medium for Enrichment and Isolation of *Thermoanerobium brockii* (Zeikus et al., 1979)

The following ingredients are added per liter:

NH$_4$Cl   0.9 g
NaCl   0.9 g
MgCl$_2$ 0.2 g
KH$_2$PO$_4$ 0.75 g
K$_2$HPO$_4$ 1.5 g
Trace mineral solution (see below)   9 ml
10% FeSO$_4$ 0.03 ml
Resazurin (0.2%)   1 ml
Vitamin solution (Wolin et al., 1963)   5 ml
10 ml of 10% Na$_2$S·9H$_2$O is added after autoclaving.

Trace mineral solution (per liter of distilled water):

Nitrilotriacetic acid (neutralized to pH 6.5   12.8 g
   with KOH)
FeCl$_3$·4H$_2$O   0.2 g
MnCl$_2$·4H$_2$O   0.1 g
CoCl$_2$·6H$_2$O   0.17 g
CaCl$_2$·2H$_2$O   0.1 g
ZnCl$_2$ 0.1 g
CuCl$_2$ 0.02 g
H$_3$BO$_3$ 0.01 g
NaMoO$_4$·2H$_2$O   0.01 g
NaCl   1.0 g
Na$_2$SeO$_3$ 0.02 g

The pH of the medium is 7.2–7.4 and 95% N$_2$/5% CO$_2$ is used in the culture medium gas phase.

The basal medium is supplemented with yeast extract and various additions which are sterilized separately. TYEG medium consists of basal medium supplemented with 0.3% yeast extract, 1.0% tryptone, and 0.5% glucose (autoclaved separately).

Thermoanaerobium lactoethylicum Isolation Medium P2 (Kondratieva et al., 1989)

The following ingredients are added per liter:

KH$_2$PO$_4$ 2.0 g
K$_2$HPO$_4$ 3.0 g
NH$_4$Cl   2.0 g
MgCl$_2$·10H$_2$O   0.2 g
CaCl$_2$·6H$_2$O   0.05 g
Yeast extract (Difco)   2.0 g
Glucose   5.0 g

Na$_2$S·9H$_2$O is added after sterilization to final concentration of 0.05%. The initial pH varies between 3.0 and 9.0. The gas mixture is passed through a copper column (300°C) to remove traces of O$_2$. The growth on the solid media in petri dishes is obtained in anaerobic jars under 100% CO$_2$. Depending on the goal of the experiment, cultivation temperatures can vary between 20 and 80°C.

Cultivation

Thermoanaerobacter and *Thermoanaerobium* species can be cultivated very easily by growing them in anaerobic TYE medium (Zeikus et al., 1980) supplemented with either 1% xylose, starch, or glucose as substrate. The culture tubes can be inoculated with 2% inoculum and incubated at 60–70°C. Different species grow in the pH range of 4.0 to 8.0. These species also grow and form colonies anaerobically on plates containing the same medium but supplemented with 2% agar.

Maintenance and Preservation

Strains of *Thermoanaerobacter* and *Thermoanaerobium* can be maintained at room temperature for several days. An exponential phase culture can be removed from its growth temperature (40–75°C) and stored at low temperature (4°C) for 3–6 months. However, for long-term storage, the cultures should be stored in 20–30% glycerol (v/v) at −70°C or lyophilized. No systematic studies on survival of these strains under different preservation conditions are available. Wiegel (1986b) has reported that *T. ethanolicus* was recovered after 3 years when the culture was stored in 55–60% glycerol (v/v, final glycerol concentration) at −18°C under anaerobic conditions. Inoculation from the glyc-

erol-containing stock cultures into a pre-reduced medium yielded growing cells in less than
20 h. These cultures can be preserved as either
glycerol-containing stocks or in lyophilized
form.

Identification

Previously, strains both of *Thermoanaerobacter*
and of *Thermoanaerobium* were considered to
be strict anaerobes, motile or nonmotile rods,
Gram positive to Gram variable, and nonsporeforming (except for *T. finnii*), but forming
chains in which cells were of uneven length (or
even mini-cells). All species were saccharolytic
and degraded starch but not cellulose and
formed ethanol and/or lactate as major end
products with lower levels of H_2/CO_2 and acetate. Strains varied considerably in the amount
of ethanol or lactate formed from glucose. The
growth of both *Thermoanaerobium* and *Thermoanaerobacter* species was inhibited by oxygen, penicillin, cycloserine, streptomycin, tetracycline, and chloramphenicol. The GC
content of the DNA varied between 30–32
mol%. Consequently, it was not possible to adequately distinguish between *Thermoanaerobacter* and *Thermoanaerobium* strains based on
these morphological, cellular, and nutritional
properties alone.

Conclusive identification of *Thermoanaerobacter* and *Thermoanaerobium* species and
their distinction from *Clostridium* species is
complicated, very time-consuming, and requires DNA-DNA homology analysis for confirmation, but a preliminary identification can
be made by the diagnostic use of the key provided in Table 1. The procedure is to grow the
saccharolytic thermoanaerobic isolate at 60–
70°C, determine the simple metabolic response
to added thiosulfate, and perform chromatographic analysis on the crude lipid fraction ex-

tracted from cells. The practical utility of this
procedure for genus identification requires further documentation with other thermoanaerobic, saccharolytic, ethanologenic isolates.

A comparison of the cell structures of these
organisms is given in Figs. 1 to 4.

Physiological Properties

Examination of the enzymes and metabolic
pathways of *Thermoanaerobacter* and *Thermoanaerobium* indicates that the various species can utilize the same general pathways for
energy conservation during glucose fermentation and produce the same end products, i.e.,
ethanol, lactate, acetate, and H_2 plus CO_2. One
major difference between species is in the ratio
of ethanol to lactate. For example, in *T. ethanolicus*, ethanol is the major product of glucose
fermentation, whereas, lactate is the major
product in *T. brockii*. In addition, species vary
with respect to range of substrates fermented
and in certain catabolic and anabolic enzymes.
Tables 2 to 5 compare the fermentation substrate range and general physiology of some of
the well-characterized strains of *Thermoanaerobium* and *Thermoanaerobacter,* and of two
species of *Clostridium*.

Lamed and Zeikus (1980) showed that per
mole of glucose, 0.95 mol of ethanol, was produced by *Thermoanaerobium brockii* in media
containing 0.5% glucose and 0.05% yeast extract. However, the fermentation was shifted in
favor of lactate from ethanol when the yeast
extract concentration was raised to 0.1%.
Growth of *T. brockii* was inhibited at glucose
concentrations exceeding 1.5% (Sonnleitner et
al., 1984). However, the effect of high substrate
concentration was different in batch than in
continuous culture. In batch fermentations,
high glucose concentrations resulted in acetate
production. The yeast extract requirement was
shown to be about 60% of the glucose concentration, which is exceptionally high. The maximum specific ethanol formation rate by *T.
brockii* was only 0.7 g/g/hr (Sonnleitner et al.,
1984). *T. brockii* has both NAD- and NADP-
dependent alcohol dehydrogenase (ADH) activities (Lamed and Zeikus, 1980). The NAD-
linked enzyme is very sensitive to oxygen (Ben-
Bassat et al., 1981). *T. brockii* contains fructose-
1–6 bisphosphate-activated lactate dehydrogenase (Lamed and Zeikus, 1980).

The enzymes for pyruvate conversion to fermentation products in *T. brockii* differ significantly from those present in *C. thermocellum*
(Lamed and Zeikus, 1980). Especially, ethanol
is not produced by NADP-linked alcohol de-

Table 1. Diagnostic key for preliminary identification of
saccharolytic, ethanologenic thermoanaerobes.

   I. Lacks C_{30} dicarboxylic fatty acids, does not reduce
      thiosulfate
       Clostridium thermocellum
  II. Contains C_{30} dicarboxylic fatty acids, reduces
      thiosulfate to elemental sulfur
       Clostridium thermosulfurogenes
       Thermoanaerobacter strain LXII
       Thermoanaerobacter strain B6A
 III. Contains C_{30} dicarboxylic fatty acids, reduces
      thiosulfate to H_2S
       Thermoanaerobium brockii
       Thermoanaerobacter ethanolicus
       Clostridium thermohydrosulfuricum

Fig. 1. Phase contrast photomicrographs of *C. thermosulfurogenes.* (a) and (b) Sporeforming cells grown in LPBB, 0.2% xylose and 1.0% MOPS buffer medium. (c) and (d) Sulfur-depositing cultures grown on LPBB, 0.5% glucose, 0.1% yeast extract and 20 mM-$Na_2S_2O_3$. Note that phase-bright sulfur accumulates in the medium and on or within the cells. Bar = 5 μm. (From Schink and Zeikus, 1983.)

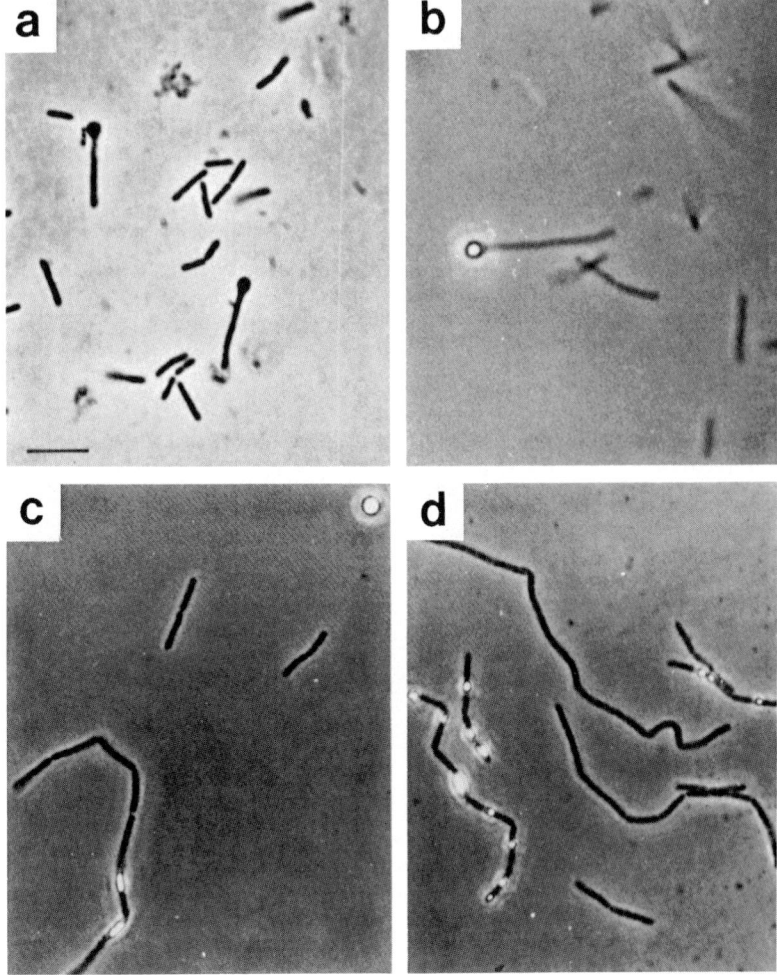

hydrogenase in *C. thermocellum.* This NADP-linked alcohol dehydrogenase is present in *Thermoanaerobium, Thermoanaerobacter, C. thermohydrosulfuricum,* and *C. thermosulfurogenes* (Lamed and Zeikus, 1981; Bryant and Ljungdahl, 1981; Hyun et al., 1985). The NADP-dependent alcohol dehydrogenase allows *T. brockii* to oxidize exogenous secondary alcohols to ketones or to reduce ketones such as acetone to their corresponding aldehydes or alcohols (Ben-Bassat et al., 1981). *T. brockii* can grow poorly in media devoid of added glucose or acetone and yeast extract.

An NADPH-dependent alcohol-aldehyde/ketone oxidoreductase was isolated from *T. brockii,* and this enzyme can reduce a broad spectrum of aldehydes and ketones or oxidize primary and secondary alcohols (Lamed and Zeikus, 1981; Lamed et al., 1981). When grown in continuous culture at 72°C, *T. brockii* reduced 3-ketoacid esters and 2-formyl-acid esters stereospecifically, in some cases yielding the opposite enantiomer compared with other

microbial systems (Sonnleitner et al., 1985). The ethyl esters of 3-keto-butanoic, 3-keto-pentanoic and 3-keto-hexanoic acid were reduced to give the resulting 3-S-(+)-hydroxyacid esters (Sonnleitner et al., 1985). The 2-formyl-substituted ethyl esters of propanoic, butanoic, 3-methyl-butanoic, and 2-phenyl-acetic acid were reduced to the corresponding nonracemic optically active 3-hydroxy esters.

T. brockii and *T. lactoethylicum* are similar in the sense that both utilize glucose via the Embden-Meyerhof pathway and that they are close with regard to their morphology and some additional physiological characters. However, *T. brockii* differs from *T. lactoethylicum* in lacking the pentose monophosphate-oxidizing pathway and the TCA cycle enzymes (Krasil'nikova et al., 1987).

T. brockii can be considered to be unsuitable for ethanol production mostly due to low ethanol and high lactate yields. However, studies of the metabolic pathways of this organism have shown that ethanol yields can be improved

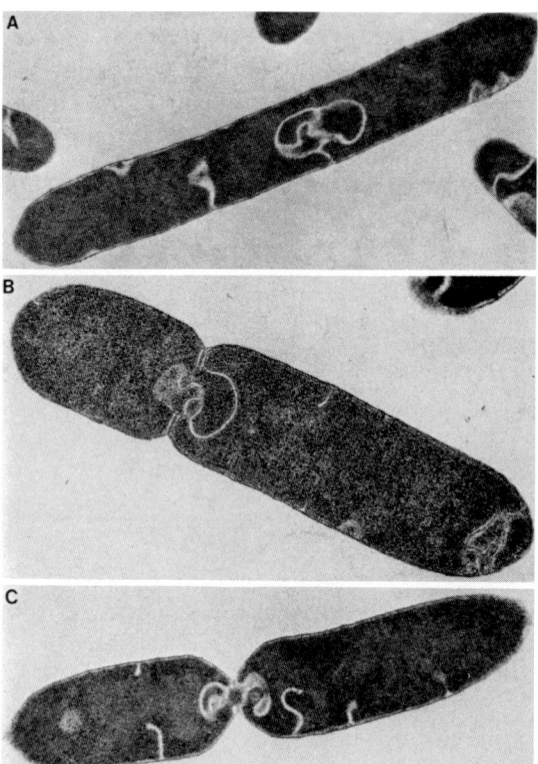

Fig. 2. Electron photomicrographs of thin sections of *Thermoanaerobium brockii* strain HTA1 grown in TYEG medium at 65°C. (A) Initiation of cell division. (B) During cell division. (C) Nearing end of cell division. (From Zeikus et al., 1979.)

by altering the normal electron flow during saccharide fermentation or by altering lactate dehydrogenase (LDH) activity (Ben-Bassat et al., 1981; Zeikus and Ng, 1982).

T. ethanolicus ferments 1% glucose or starch nearly quantitatively to ethanol, i.e., up to 1.95 mol of ethanol per mole of glucose, and grows optimally at pH 5.8–8.5 (Ljungdahl et al., 1981; Carreira et al., 1984). The fermentation balance over this range of pH or over a temperature range of 40–77°C does not vary significantly (Ljungdahl et al., 1981). It displays a saccharide fermentation nearly identical to *C. thermohydrosulfuricum* strain 39E (Ng et al., 1981; Lovitt et al., 1984). A higher specific growth rate and a high concentration of ethanol is produced when *T. ethanolicus* is grown at an initial pH of 8.0 as compared to 5–7.5 (Kannan and Muthrasan, 1985). When it is grown in media containing 10% glucose or less, ethanol is the main product, but ethanol production is reduced and lactate and acetate are increased when glucose concentrations exceed this amount (Ljungdahl et al., 1981). This organism can grow in media containing up to 20% sugars or starch, but fer-

mentation ceases when the ethanol level reaches approximately 0.6%. Growth is completely inhibited by 0.5 M acetate, 0.4 M lactate, or 1.3 M (6%) ethanol (Ljungdahl et al., 1981). However, using adaptation and selection techniques, mutants have been isolated that can tolerate up to 10% ethanol (Carriera and Ljungdahl, 1983). Although growth on xylose is as rapid as on glucose (about a 120-min doubling time), the pH range is more restricted; xylose does not support significant growth above pH 7.4 and below 6.0 (Wiegel and Puls, 1982). With 1% xylose, up to 1.4 mol ethanol per mole of xylose is formed, whereas it decreases even further with higher xylose concentrations (Wiegel and Puls, 1982). The highest ethanol formation rate measured in a growing culture on 2.3% xylose was below 10 mmoles/l/h (Wiegel et al., 1984).

In batch fermentation by *T. ethanolicus*, xylose concentration influenced the growth rate, product yield, and product distribution (Lacis and Lawford, 1988). In media containing 27 mM xylose, an ethanol yield of 1.3 mol ethanol/mol xylose (78% of maximum theoretical yield) was typically obtained.

The medium currently used to grow *T. ethanolicus* is complex and contains many vitamins, amino acids, and metals (Wiegel et al., 1979). Since zinc is a component of ADH, it must be supplied in the medium. The bacterium has not been cultured in the absence of yeast extract, which cannot be replaced by tryptone, casein hydrolysate, beef extract, or ashed yeast (Ljungdahl et al., 1981). Attempts to substitute yeast extract with corn steep liquor were unsuccessful (Kannan and Muthurasan, 1985).

The inhibition of *T. ethanolicus* fermentations appears to be related to controls at the enzyme levels, notably ADH and LDH. There are two ADHs in *T. ethanolicus* (Bryant and Ljungdahl, 1981; Bryant and Wiegel, 1983; Bryant et al., 1988), and these enzyme have been purified to homogeneity (Bryant et al., 1988). Both enzymes have high thermostability. The NADP-linked secondary alcohol dehydrogenase is postulated to be largely responsible for the formation of ethanol in fermentation of carbohydrates by *T. ethanolicus*. The sensitivity of the bacterium to high substrate and ethanol concentrations is apparently due to the regulation of several key enzymes involved in the fermentation, including alcohol dehydrogenases, lactate dehydrogenases, and acetate kinase (Carriera et al., 1982). An enzyme exhibiting both inorganic pyrophosphate (PPi) and ATP-acetate kinase activity has recently been purified from *T. ethanolicus* (Peck et al., 1983). Since *T. ethanolicus* has been found to utilize PPi as an energy source, it was suggested that this kinase may be involved, and that PPi could

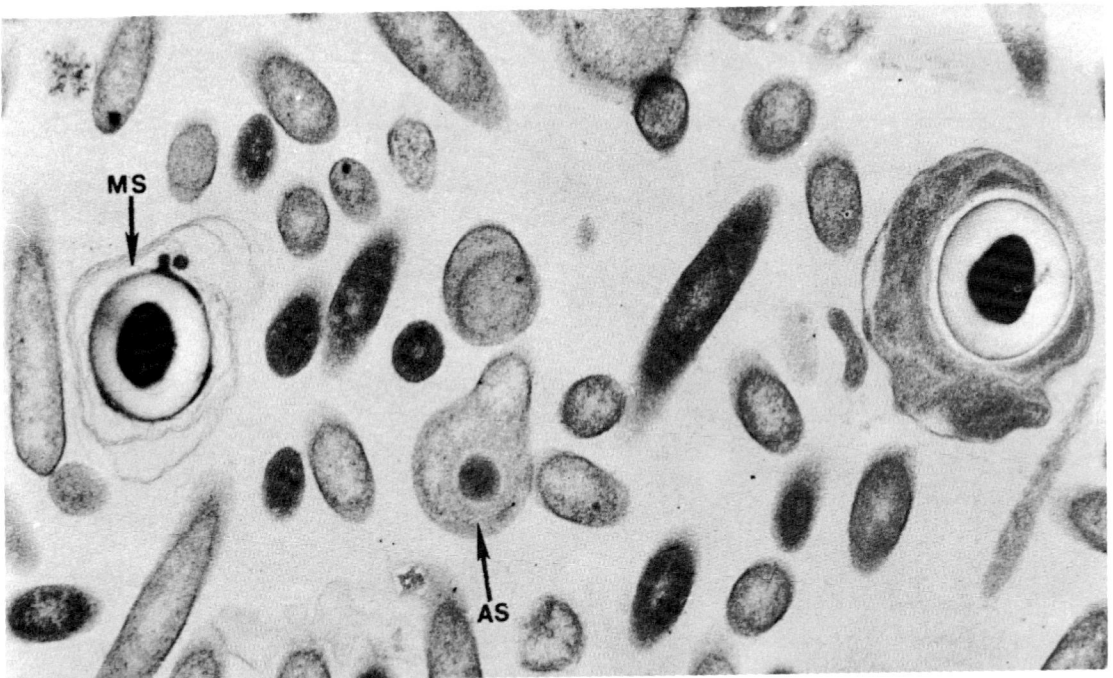

Fig. 3. Electron photomicrograph of thin sections of *C. thermohydrosulfuricum* prepared from a sporeforming culture. Both mature spores (MS) and abortive spores (AS) are shown by arrows. (From Hyun et al., 1983.)

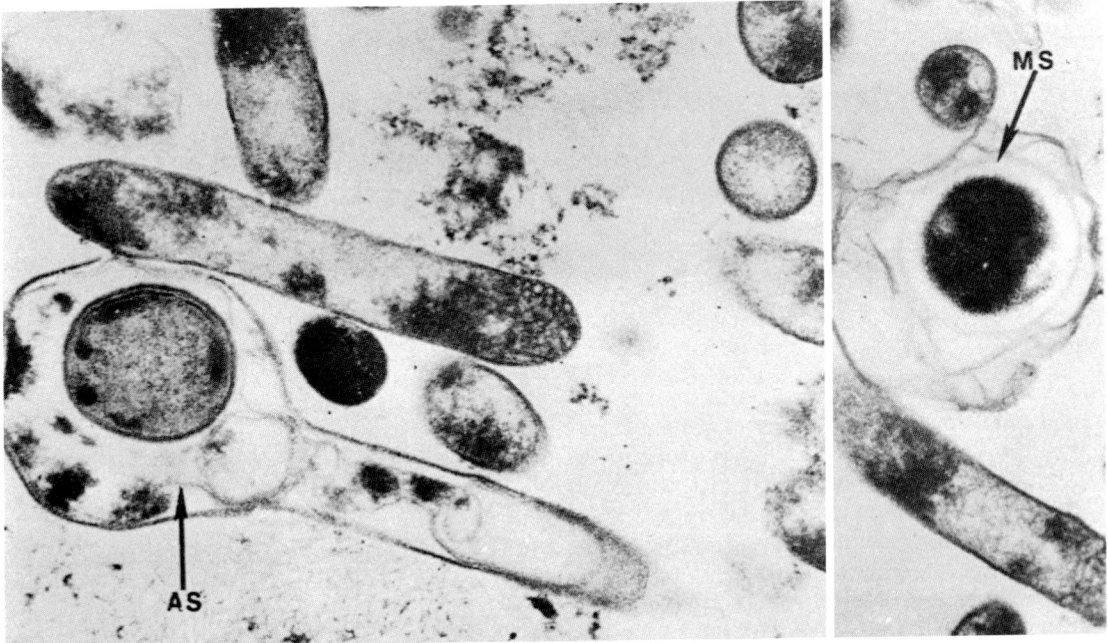

Fig. 4. Electron photomicrographs of thin sections of *C. thermosulfurogenes* prepared from a sporeforming culture. Both mature spores (MS) and abortive spores (AS) are shown by arrows. (From Hyun et al., 1983.)

possibly be used to modify the fermentation products.

Koch et al. (1987) were able to show production of high concentrations of amylases and pul-

lulanases by continuous cultivation of *T. finnii* and *T. ethanolicus* in chemostats under starch limitation. About 70% to 98% of these enzymes were transported and released into the culture

Table 2. Substrate utilization and product formation by three *Thermoanaerobium* species and by *Clostridium thermohydrosulfuricum*.

Substrate	*Thermoanaerobium brockii*	*Thermoanaerobium lactoethylicum*	*Thermoanaerobium* strain Tok6-B1	*Clostridium thermohydrosulfuricum*
Polymers				
Xylan	+	ND	ND	ND
Starch	+	+	+	+
Pectin	−	ND	+	−
Inulin	−	ND	ND	−
Dextrin	−	ND	ND	−
Cellulose	−	ND	ND	V
Sugars				
Glucose	+	+	ND	+
Cellobiose	+	+	+	+
Xylose	−	+	+	+
Maltose	+	+	+	+
Fructose	−	+	ND	V
Mannose	−	ND	ND	+
Sucrose	+	+	+	+
Ribose	−	ND	+	+
Lactose	+	+	+	d
Galactose	−	+	+	+
Raffinose	−	ND	+	d
Xylulose	−	ND	ND	ND
Arabinose	+	+	+	V
Melibiose	−	ND	ND	ND
Trehalose	−	ND	ND	ND
Rhamnose	ND	+	+	V
Acetate	−	ND	ND	ND
Pyruvate	+	ND	ND	ND
Mannitol	−	+	+	d
Sorbitol	ND	ND	+	d
Inositol	ND	ND	+	−
Products				
Ethanol	+	+	+	+
Acetic acid	+	+	+	+
Lactic acid	+	+	+	+
CO_2/H_2	+	+	+	+
Butyric acid	−	+	ND	ND
Propionic acid	ND	+	ND	ND
Isovaleric acid	ND	+	ND	ND

Symbols: +, present; −, absent; ND, not determined; d, detected; V, variable.

fluid. These extracellular enzymes were extremely thermostable under aerobic conditions and in the absence of substrate and metal ions. The enzymes had an optimal temperature of 90°C with pH optima ranging between 5 and 6. The addition of calcium ions in vitro significantly enhanced pullulanase activity from *T. finnii. Thermoanaerobium* strains also produce high levels of amylases and pullulanase (Coleman et al., 1987; Plant et al., 1987).

A novel amylopullulanase activity which breaks down starch has been characterized in *T. brockii* (Coleman et al., 1987), and this activity is also present in *C. thermohydrosulfuricum* (Hyun and Zeikus, 1985; Saha et al., 1988; Saha and Zeikus, 1989).

Thermoanaerobium and *Thermoanaerobacter* species degrade a wide variety of saccharides, including xylans and starch, but they do not ferment crystalline cellulose. This substrate range is shared by *C. thermosulfurogenes* and *C. thermohydrosulfuricum* but differs from that of *C. thermocellum* (see Table 2 and 3). Thermoanaerobic saccharolytic bacteria vary significantly with respect to utilization of thiosulfate as an electron acceptor (see Table 1). *T. brockii, T. ethanolicus,* and *C. thermohydrosulfuricum* contain a seroheme-coupled thiosulfate reductase and form H_2S. The addition of thiosulfate shifts, glucose fermentations of these organisms, resulting in acetate and H_2S as the major end products (Y. O Lee, M. K. Jain and J. G. Zeikus, unpublished observations). *C. thermosulfurogenes* and related species convert thiosulfate to elemental sulfur, which deposits on cells and in the medium (Schink and Zeikus,

Table 3. Fermentation substrate utilization and product formation by three *Thermoanaerobacter* species and *Clostridium thermosulfurogenes*.

Substrate	*Thermoanaerobacter ethanolicus*	*Thermoanaerobacter finnii*	*Thermoanaerobacter* strain B6A	*Clostridium thermosulfurogenes*
Polymers				
Xylan	+	ND	+	ND
Starch	+	ND	+	+
Pectin	+	ND	ND	+
Inulin	+	ND	ND	ND
Dextrin	+	ND	ND	ND
Cellulose	−	ND	ND	−
Sugars				
Glucose	+	+	+	+
Cellobiose	+	+	+	+
Xylose	+	+	+	+
Maltose	+	+	+	+
Fructose	+	+	+	ND
Mannose	+	+	+	+
Sucrose	+	+	+	+
Ribose	+	+	ND	−
Lactose	+	+	+	−
Galactose	+	+	+	+
Raffinose	−	ND	+	−
Xylulose	+	ND	ND	ND
Arabinose	−	−	ND	+
Melibiose	−	ND	−	+
Trehalose	−	ND	+	+
Rhamnose	−	−	+	−
Melezitose	ND	ND	−	−
Acetate	−	ND	ND	−
Pyruvate	+	+	ND	−
Mannitol	+	+	+	+
Xylitol	ND	−	ND	ND
Products				
Ethanol	+	+	+	+
Acetic acid	+	+	+	+
Lactic acid	+	+	+	+
CO_2/H_2	+	+	+	+

Symbols: +, present; −, absent; ND, not determined.

Table 4. Physiological properties of *Thermoanaerobium* species.

	Temperature Range (°C)					
	T_{min}	T_{opt}	T_{max}	pH range	Doubling time (min)	Ethanol ÷ glucose (mol/mol)
T. brockii	>35	65–70	<85	5.5–9.5	60	1
T. lactoethylicum	42	65	75	5.0–8.5	100	0.8
Thermoanaerobium strain Tok6-B1		70		5.0–9.0	60	

Table 5. Physiological properties of *Thermoanaerobacter* species.

	Temperature Range (°C)					Ethanol ÷ xylose (mol/mol)
	T_{min}	T_{opt}	T_{max}	pH range	Doubling time (min)	
T. ethanolicus	35	69	78	4.4–9.9	135	0.96
T. finnii	40	65	75	6.5–6.8		1.26
Thermoanaerobacter strain B6A		60		5.0–7.0		

1983). *C. thermocellum* does not use thiosulfate as an electron acceptor.

Applications

The potential utility of *Thermoanaerobacter, Thermoanaerobium,* and other saccharolytic thermoanaerobes has been proposed for industrial ethanol production of specialty alcohols, and for other genes that code for a variety of unique thermostable enzymes, such as alcohol dehydrogenase, amylases, and glucose isomerase, that could be cloned and over-produced in industrial hosts (Zeikus, 1979; Wiegel and Ljungdahl, 1981). To date only the alcohol dehydrogenase of *T. brockii* is produced commercially, and used as a biocatalyst for synthesis of Chiral alcohols.

The Present Taxonomic Situation

Saccharolytic thermoanaerobic bacteria that form ethanol and lactate, including species in the genera *Thermoanaerobium* and *Thermoanaerobacter,* as well as *Clostridium thermocellum, C. thermohydrosulfuricum,* and *C. thermosulfurogenes,* are currently not assigned to any family (Cato et al., 1986; Cato and Stackebrandt, 1989). *Thermoanaerobacter* is included in Section 15 of *Bergey's Manual of Systematic Bacteriology* under irregular, nonsporeforming, Gram-positive rods. The first thermoanaerobic nonsporeforming rods were isolated from Yellowstone National Park and identified as *Thermoanaerobium* strains HTD4 and JW200 (Zeikus et al., 1979; Wiegel and Ljundahl, 1979). Based primarily on differences in the ethanol/lactate ratio, Wiegel and Ljungdahl (1981) considered *Thermoanaerobium* strain JW200 to be different from *Thermoanaerobium brockii* (strain HTD4) and proposed the name *Thermoanaerobacter ethanolicus* (strain JW200) for it. Wiegel (1986b) first suggested that *T. brockii, T. ethanolicus,* and *Clostridium thermohydrosulfuricum* (strain 39E) could be close taxonomic relatives because they display a similar substrate range and temperature ranges for growth and a biphasic growth curve. Krasil'nikova et al. (1987) recognized the similarity of *Thermoanaerobium lactoethylicum* to both *T. brockii* and *T. ethanolicus* and suggested that the genus name *Thermoanaerobium* and not *Thermoanaerobacter* be used for these species.

Bateson et al. (1989) have reported that the 16S rRNA sequences of *T. brockii, T. ethanolicus,* and *C. thermohydrosulfuricum* were nearly identical but significantly different from *C. thermosulfurogenes.* Cato and Stackebrandt (1989) showed by 16S rRNA cataloging that sporeforming clostridia do not form one phylogenetically homogeneous family, but rather form six sublines which embrace both sporeforming and nonsporeforming species. Notably, *T. brockii* was in the same subline as the one that included *C. thermoaceticum* and *C. thermosaccharolyticum.*

Studies of Y. O. Lee, M. K. Jain, J. Wiegel, C. Lee and J. G. Zeikus (1990) (unpublished observations) on DNA-DNA hybridization analysis of saccharolytic ethanol lactate-producing thermoanaerobes indicate that the current taxonomic assignments and nomenclature of certain *Thermoanaerobacter, Thermoanaerobium,* and *Clostridium* species need to be changed. Three separate groups are recognizable (Table 6). Group and species assignments were based on greater than 50% and 90% homology, respectively. The three groups also differ in the mode of thiosulfate metabolism and the major lipid and fatty acid content, which can be used for a diagnostic key (see Table 1). *C. thermocellum* does not reduce thiosulfate and lacks C_{30} dicarboxylic acid-based lipids but contains typical clostridial fatty acids and hopanoids (Langworthy and Pond, 1986). Group II species reduce thiosulfate to elemental sulfur and contain C_{30} dicarboxylic acids (Langworthy and Pond, 1986). Group III species reduce thiosulfate to H_2S and contain C_{30} dicarboxylic acids (Langworthy and Pond, 1986).

By use of DNA-DNA homology, metabolic studies, and membrane lipid composition, taxonomic assignments can be made which support 16S rRNA cataloging studies that put sporeformers and nonsporeformers in the same family. In addition, the data indicate that sporeformers and nonsporeformers can be placed in the same genus and species. For example, *Thermoanaerobacter ethanolicus* is nearly identical to *C. thermohydrosulfuricum* strain 39E (the Yellowstone strain) and in the same group as *Thermoanaerobium brockii* and *C. thermohydrosulfuricum* E100–69 (the type strain). Thus, *T. ethanolicus* and *C. thermohydrosulfuricum* strain 39E probably belong to the same species (Table 6). Perhaps *T. ethanolicus* strain JW200 is an asporogenous mutant of *C. thermohydrosulfuricum* strain 39E.

Based on the data at hand, *Thermoanaerobium, Thermoanaerobacter,* and certain thermophilic *Clostridium* species should be reclassified taxonomically, and groups II and III (Table 6) be given proper genus assignments.

Table 6. Grouping of saccharolytic thermoanaerobic bacterial species of uncertain phylogenetic affiliation, based on DNA-DNA homologies.[a]

Group	Present name	Strain	Habitat
I	*Clostridium thermocellum*	LQRI	Farm soil
II	*Clostridium thermosulfurogenes*	AB (type)	Thermal spring
	Thermoanaerobacter sp.	LXII	Thermal spring
	Thermoanaerobacter sp.	B6A	Thermal spring
III	*Thermoanaerobium brockii*	HTD4 (type)	Thermal spring
	Clostridium thermohydrosulfuricum	E100-69 (type)	Farm soil
	Thermoanaerobacter ethanolicus[b]	JW200 (type)	Thermal spring
	Clostridium thermohydrosulfuricum[b]	39E	Thermal spring

[a]Strains belonging to the same group displayed greater than 50% homology and strains probably belonging to the same species displayed greater than 90% homology.
[b]Greater than 90% homology.

Literature Cited

Bateson, M. M., J. Wiegel, and D. M. Ward. 1989. Comparative analysis of 16S ribosomal RNA sequences of thermophilic fermentative bacteria isolated from hot spring cyanobacterial mats. Syst. Appl. Microbiol. 12:1–7.

Belyaev, S. S., R. Wolkin, W. R. Kenealy, M. J. DeNiro, S. Epstein, and J. G. Zeikus. 1983. Methanogenic bacteria from the Bondyuzhskoe oil field: General characterization and analysis of stable-carbon isotopic fractionation. Appl. Environ. Microbiol. 45:691–697.

Ben-Bassat, A., R. Lamed, and J. G. Zeikus. 1981. Ethanol production by thermophilic bacteria: metabolic control of end product formation in *Thermoanaerobium brockii*. J. Bacteriol. 146:192–199.

Bochem, H.-P., S. M. Schobert, B. Sprey, and P. Wengler. 1982. Thermophilic biomethanation of acetic acid: morphology and ultrastructure of a granular consortium. Can. J. Microbiol. 28:500–510.

Bryant, F. and L. G. Ljungdahl. 1981. Characterization of an alcohol dehydrogenase from *Thermoanaerobacter ethanolicus* active with ethanol and secondary alcohols. Biochem. Biophys. Res. Commun. 100:793–799.

Bryant, F. O., and J. Wiegel. 1983. Comparison of alcohol dehydrogenases from wild type JW-200 and mutant FE-4 strains of *Thermoanaerobacter ethanolicus*. 83rd Ann. Meet. Am. Soc. Microbiol. New Orleans, LA, March 6–11, 1983.

Bryant, F. O., J. Wiegel, and J. G. Ljungdahl. 1988. Purification and properties of primary and secondary alcohol dehydrogenases from *Thermoanaerobacter ethanolicus*. Appl. Environ. Microbiol. 54:460–465.

Carreira, L. H., and L. G. Ljungdahl. 1983. High ethanol producing derivatives of *Thermoanaerobacter ethanolicus*. 83rd Ann. Meet. Am. Soc. Microbiol. New Orleans, LA, March 6–11, 1983.

Carreira, L. H., L. G. Ljungdahl, F. Bryant, M. Szulczynski, and J. Wiegel. 1982. Control of products formation with *Thermoanaerobacter ethanolicus*, enzymology, and physiology, p. 351–356. In: Y. Ikeda and T. Beppu (ed.), Proceedings of the IVth International Symposium on Genetics of Industrial Microorganisms, Kyoto, Japan, Kodansha Ltd., Tokyo.

Carreira, L. H., N. E. Rabek, and L. G. Ljungdahl. 1984. Isolation and characterization of three new thermophilic clostridia from islandic hot springs. Ann. Meet. Am. Soc. Microbiol. Abstr. I35.

Cato, E. P., W. L. George, and S. M. Finegold. 1986. Genus *Clostridium* Prazmowski 1880, 23, p. 1141–1200. In: P. H. A. Sneath, N. S. Mair, M. E. Sharpe, and J. G. Holt (eds.), Bergey's manual of systematic bacteriology, vol. 2. Williams and Wilkins Co., Baltimore.

Cato, E. P., and E. Stackebrandt. 1989. Taxonomy and phylogeny, p. 1–26. In: N. P. Minton and D. J. Clarker (ed.), Clostridia. Plenum Press, New York.

Coleman, R. D., S.-S. Yang, and M. P. McAlister. 1987. Cloning of the debranching-enzyme gene from *Thermoanaerobium brockii* into *Escherichia coli* and *Bacillus subtilis*. J. Bacteriol. 169:4302–4307.

Hungate, R. E. 1969. A roll tube method for cultivation of strict anaerobes. Meth. Enzymol. 3B:117–132.

Hyun, H. H., G. J. Shen, and J. G. Zeikus. 1985. Differential amylosaccharide metabolism of *Clostridium thermosulfurogenes* and *Clostridium thermohydrosulfuricum*. J. Bacteriol. 164:1153–1161.

Hyun, H. H., and J. G. Zeikus. 1985. General biochemical characterization of thermostable pullulanase and glucoamylase from *Clostridium thermohydrosulfuricum*. Appl. Environ. Microbiol. 49:1168–1173.

Hyun, H. H., J. G. Zeikus, R. Longin, J. Millet, and A. Ryter. 1983. Ultrastructure and extreme heat resistance of spores from thermophilic *Clostridium* species. J. Bacteriol. 156:1332–1337.

Kannan, V., and R. Mutharasan. 1985. Ethanol fermentation characteristics of *Thermoanaerobacter ethanolicus*. Enzyme Microb. Technol. 7:87–89.

Koch, R., P. Zablowski, and G. Antranikian. 1987. Highly active and thermostable amylases and pullulanases from various anaerobic thermophiles. Appl. Microbiol. Biotechnol. 27:192–198.

Kondratieva, E. N., E. V. Zacharova, V. I. Duda, and V. V. Krivendo. 1989. *Thermoanaerobium lactoethylicum* spec. nov. a new anaerobic bacterium from a hot spring of Kamchatka. Arch. Microbiol. 151:117–122.

Krasil'nikova, E. N., N. A. Zorinn, and E. V. Zakharova. 1987. Glucose metabolism and hydrogenase activity in *Thermoanaerobium lactoethylicum* Mikrobiologiya 56:533–536.

Lacis, L. S., and H. G. Lawford. 1988. Ethanol production from xylsoe by *Thermoanaerobacter ethanolicus* in batch and continuous culture. Arch. Microbiol. 150:48–55.

Lamed, R. J., E. Keinan, and J. G. Zeikus. 1981. Potential applications of an alcohol-aldehyde/ketone oxidoreductase from thermophilic bacteria. Enzyme Microbiol. Technol. 3:144–148.

Lamed, R., and J. G. Zeikus. 1980. Ethanol production by thermophilic bacteria: Relationship between fermentation product yields of an catabolic enzyme activities in Clostridium thermocellum and Thermoanaerobium brockii. J. Bacteriol. 144:569–578.

Lamed, R. J., and J. G. Zeikus. 1981. Novel NADP-linked alcohol-aldehyde/ketone oxidoreductase in thermophilic ethanologenic bacteria. Biochem. J. 195:183–190.

Langworthy, T. A., and J. L. Pond. 1986. Membranes and lipids of thermophiles, p. 107–135. In T. D. Brock (ed.), Thermophiles. General, molecular and applied microbiology. John Wiley & Sons, New York.

Ljungdahl, L. G., F. Bryant, L. Carreira, T. Saiki, and J. Wiegel. 1980. Some aspects of thermophilic and extreme thermophilic anaerobic microorganisms, p. 397–419. In: A. Hollaender and R. Rabson (eds.), Trends in the biology of fermentations for fuels and chemicals. Plenum Publishing, New York.

Ljungdahl, L. G., L. Carreira, and J. Weigel. 1981. Production of ethanol from carbohydrates using anaerobic thermophilic bacteria. Ekman-Days Int. Symp. Wood Pulping Chem. 4:23–28.

Lovitt, R. W., R. Longin, and J. G. Zeikus. 1984. Ethanol production by thermophilic bacteria: physiological comparison of solvent effects on parent and alcohol-tolerant strains of Clostridium thermohydrosulfuricum. Appl. Environ. Microbiol. 48:171–177.

Morgan, H. W., B. K. C. Patel, and R. M. Daniel. 1985. Comparison of Thermoanaerobium sp. from a New Zealand hot spring with Thermoanaerobium brockii. FEMS (Fed. Eur. Microbiol. Soc.) Microbiol. Lett. 30:121–124.

Ng, T. K., A. Ben-Bassat, and J. G. Zeikus. 1981. Ethanol production by thermophilic bacteria: fermentation of cellulosic substrates by cocultures of Clostridium thermocellum and Clostridium thermohydrosulfuricum. Appl. Environ. Microbiol. 41:1337–1343.

Patel, B. K. C., H. W. Morgan, and R. M. Daniel. 1986. Studies on some thermophilic glycolytic anaerobic bacteria from New Zealand hot spring. Syst. Appl. Microbiol. 8:128–136.

Peck, H. D., C.-L. Liu, A. K. Varma, L. G. Ljungdahl, and X. Szulczynski. 1983. The utilization of inorganic pyrophosphate, tripolyphosphate and tetrapolyphosphate as energy sources for the growth of anaerobic bacteria, p. 317–348. In: A. Hollaender, A. I. Laskin, and P. Rogers (ed.), Basic biology of new developments in biotechnology. Plenum Press, New York.

Plant, A. R., R. M. Clemens, R. M. Daniels, and H. W. Morgan. 1987. Purification and preliminary characterization of an extracellular pullulanase from Thermoanaerobium Tok 6-B1. Appl. Microbiol. Biotech. 26:427–433.

Saha, B. C., S. P. Mathupala, and J. G. Zeikus. 1988. Purification and characterization of a highly thermostable pullulanase from Clostridium thermohydrosulfuricum. Biochem. J. 252:343–348.

Saha, B. C., and J. G. Zeikus. 1989. Novel highly thermostable pullulanase from thermophiles. Trends in Biotechnol. 7:234–239.

Schink, B., and J. G. Zeikus. 1983. Clostridium thermosulfurogenes sp. nov., a new thermophile that produces elemental sulphur from thiosulphate. J. Gen. Microbiol. 129:1149–1158.

Schmid, Y., H. Giesel, S. M. Schoberth, and H. Sahm. 1986. Thermoanaerobacter finnii new species a new ethanologenic sporogenous bacterium. Syst. Appl. Microbiol. 8:80–85.

Schoberth, S. and Tietze, K. 1978. Methanogenic bacteria from Lake Kivu sediments. Int. Congr. Microbiol. 12th, Munich, West Germany.

Sonnleitner, B., A. Fiechter, and F. Giovannini. 1984. Growth of Thermoanaerobium brockii in batch and continuous culture at supraoptimal temperatures. Appl. Microbiol. Biotechnol. 19:326–334.

Sonnleitner, B., F. Giovannini, and A. Fiechter. 1985. Stereospecific reductions of ketones and oxo acid esters using continuously growing cultures of Thermoanaerobium brockii. J. Biotechnol. 3:33–46.

Weimer, P. J. 1985. Thermophilic anaerobic fermentation of hemicellulose and hemicellulose-derived aldose sugars by Thermoanaerobacter strain B6A. Arch. Microbiol. 143:130–136.

Weimer, P. J., L. W. Wagner, S. Knowlton, and T. K. Ng. 1984. Thermophilic anaerobic bacteria which ferment hemicellulose characterization of organisms and identification of plasmids. Arch. Microbiol. 138:31–36.

Weimer, P. J., and J. G. Zeikus. 1978. One carbon metabolism in methanogenic bacteria: cellular characterization and growth of Methanosarcina barkeri. Arch. Microbiol. 119:49–57.

Wiegel, J. 1986a. Methods for isolation and study of thermophiles, p. 17–37. In: T. D. Brock (ed.), Thermophiles. General, molecular and applied microbiology. John Wiley & Sons, New York.

Wiegel, J. K. W. 1986b. Genus Thermoanaerobacter. Wiegel and Ljungdahl, 1982, 384, p. 1379–1383. In: P. H. A. Sneath, N. S. Mair, M. E. Sharpe, and J. G. Holt (ed.), Bergey's manual of systematic bacteriology, vol. 2. Williams and Wilkins Co., Baltimore.

Wiegel, J. and L. G. Ljungdahl, 1979. Isolation and characterization of a new extreme thermophilic anaerobic bacterium, I-63. Abstracts of the 79th Ann. Meet. Am. Soc. Microbiol., p. 105. Los Angeles, CA.

Wiegel, J., and L. G. Ljungdahl. 1981. Thermoanaerobacter ethanolicus gen. nov., spec. nov., a new, extreme thermophilic, anaerobic bacterium. Arch. Microbiol. 128:343–348.

Wiegel, J., and J. Puls. 1983. Production of ethanol from hemicelluloses of hardwoods and annual plants, p. 994. In: A. Strub, P. Chartier, and G. Schleser (ed.), Energy from biomass, 2nd European Community Conference. Applied Science Publishers, New York.

Wiegel, J., L. H. Carreira, C. P. Mothershed, and J. Puls. 1984. Production of ethanol from biopolymers by anaerobic, thermophilic, and extreme thermophilic bacteria. II. Thermoanaerobacter ethanolicus JW200 and its mutants in batch cultures and resting cell experiments. Biotechnol. Bioeng. Symp. 13:193–205.

Wiegel, J., L. G. Ljungdahl, and J. R. Rawson. 1979. Isolation from soil and properties of the extreme thermophile Clostridium thermohydrosulfuricum. J. Bacteriol. 139:800–810.

Wolin, E. A., M. J. Wolin, and R. S. Wolfe. 1963. Formation of methane by bacterial extracts. J. Biol. Chem. 238:2882–2886.

Zeikus, J. G. 1977. The biology of methanogenic bacteria. Bacteriol. Rev. 41:514–541.

Zeikus, J. G. 1979. Thermophilic bacteria: Ecology, physiology, and technology. Enzyme Microb. Technol. 1:243–252.

Zeikus, J. G., P. W. Hegge, and M. A. Anderson. 1979. *Thermoanaerobium brockii* gen. nov. and sp. nov., a new chemoorganotrophic, caldoactive anaerobic bacterium. Arch. Microbiol. 122:41–48.

Zeikus, J. G., A. Ben-Bassat, and P. W. Hegge. 1980. Microbiology of methanogenesis in thermal, volcanic environments. J. Bacteriol. 143:432–440.

Zeikus, J. G., and T. K. Ng. 1982. Thermophilic saccharide fermentations. Ann. Reports on Fermentation Processes 5:263–298.

CHAPTER 86

The Genus *Eubacterium*

JAN R. ANDREESEN

The genus name *Eubacterium* is somewhat confusing because of the use of the term "eubacterium" for the broad group of bacteria that are differentiated from the archaebacteria on the basis of the base sequence of their 16S rRNA (Woese, 1987; see also Chapter 1). However, the genus name *Eubacterium* was originally established by A. R. Prévot (1938) to describe a group of bacteria isolated from human feces that were considered to be beneficial. So far, no primary-disease-causing species of *Eubacterium* have been isolated, although some species might be considered to be opportunistic pathogens (Allen, 1985; Finegold and George, 1989; Hill et al., 1987). As defined, the genus *Eubacterium* includes those Gram-positive, nonspore-forming, obligately anaerobic rods that do not produce: 1) propionic acid as major acid (in contrast to *Propionibacterium*); 2) lactic acid alone (in contrast to *Lactobacillus*); 3) more acetic acid than lactic acid with and without formic acid (in contrast to *Bifidobacterium*); and 4) succinic acid (in the presence of carbon dioxide) and lactic acid with small amounts of acetic or formic acid (in contrast to *Actinomyces*) (Moore and Holdeman-Moore, 1986). Most of the saccharolytic species of *Eubacterium* described form butyric acid and hydrogen gas, as do the saccharolytic, sporeforming Gram-positive species of the genus *Clostridium*. In addition to ethanol, butanol is formed by *E. saburreum* and *E. tenue,* as also in some solvent-producing clostridia (Bahl and Gottschalk, 1988). Caproic acid is generally quite seldom encountered as end product, but can be a unique product of *E. alactolyticum, E. biforme,* and *E. limosum,* as well as of *Clostridium kluyveri, Peptococcus niger,* and *Megasphaera elsdenii* (Holdeman et al., 1977; Moore, and Holdeman-Moore 1986). *Eubacterium* spp. have the ability to degrade amino acids, as shown by the identification of isobutyric and isovaleric acids as products of *E. acidaminophilum, E. brachy, E. calanderi, E. combesii, E. limosum, E. tenue,* and *E. tucumanense* (Diaz et al., 1989; Moore and Holdeman-Moore,

1986; Mountfort et al., 1988; Zindel et al., 1988) and of phenylacetic acid as a product of *E. timidum* (Hill et al., 1987). As observed for clostridia (Andreesen et al., 1989), specialists able to degrade purines *(E. angustum),* trihydroxy-benzenoids *(E. oxidoreducens),* O-methoxylated aromatic acids *(E. callanderi),* or steroids *(E. desmolans)* have also been isolated (Beuscher and Andreesen, 1984; Krumholz and Bryant, 1986; Morris et al., 1986; Mountfort et al., 1988). Autotrophic growth on H_2 plus CO_2 or on one-carbon compounds such as CO, formate, and methanol is also possible in *E. limosum* and *E. tucumanense* (Diaz et al., 1989; Genthner and Bryant, 1982, 1987; Genthner et al., 1981). Therefore, clostridial strains that have lost the capacity to form spores or require very special conditions for induction of spore formation might erroneously be classified as Eubacterium species (see *E. combesii—C. subterminale*).

Some *Eubacterium* species (for instance *E. plautii*) are either Gram-negative (Hofstad and Aasjord, 1982) with very occasional weak Gram-positive areas in the cells or they decolorize rather easily (for instance, *E. cylindroides*). Therefore, these species can only be classified correctly after showing the absence of lipopolysaccharides, especially of 3-hydroxy fatty acids, and analysis of the outer membrane proteins (Baardsen et al., 1988). The chemical composition of the peptidoglycan has been proven to be a useful taxonomic marker. Although only a few species have been analyzed so far, they display a wide variety of types, such as the *meso*-diaminopimelic acid-direct type (Alγ according to the nomenclature of Schleifer and Kandler, 1972) in *E. acidaminophilum, E. angustum, E. alactolyticum, E. saburreum,* and *E. tenue;* the LL-diaminopimelic acid-glycine type (A3γ) in *E. combesii* and *E. lentum;* various unique structures in *E. nodatum* and *E. suis;* and the B2α type (found in *E. limosum* and *E. callanderi*) where peptide bridges are connected to the D-glutamic acid residue (Beuscher and Andreesen, 1984; Severin et al.,

1989a, 1989b; Tanner et al., 1981; Wegienek and Reddy, 1982; Weiss, 1981; Zindel et al., 1988). Holdeman et al. (1977) list some additional species that may contain *meso*-diaminopimelic acid, but no confirmation has been published.

The heterogeneity of the genus *Eubacterium* is further indicated by the wide range observed for the GC content of the DNA from 25 *(E. lentum)* to 55 mol% *(E. suis)*. Curiously, both of these organisms contain cytochromes and *E. lentum* even has an unsaturated menaquinone, although these electron carriers are not otherwise detected in species of this genus (Fernandez and Collins, 1987). For 17 of 34 species listed in *Bergey's Manual of Systematic Bacteriology* (Moore and Holdeman-Moore, 1986) the GC content is still unknown. In most cases this is due to the sparse growth of the respective species. 16S rRNA analysis has only been reported for *E. lentum* and *E. limosum* (Tanner et al., 1981), where a relationship is indicated to *Clostridium lituseburense* and *C. barkeri,* respectively.

Many synonyms were used for *Eubacterium* species in former times (Lewis and Sutter, 1981). Many of these species have not been studied in recent years or in detail, and the only information available is that they were recovered during studies of the rumen of fecal flora, subgingival peridontitis, or some other infectious material. Only the classical biochemical tests, the standard sugar utilization tests (acid production) and the fermentation product profile have been determined for most of the species. Therefore, much more detailed information has to be obtained before a reclassification of this genus is justified.

Habitats

All *Eubacterium* species described so far are mesophiles and grow optimally at 34 to 37°C. However, some of them (for instance, *E. biforme* and *E. brachy*) can not even grow at 30 or 45°C, which indicates an unusually narrow temperature spectrum. The latter condition may develop because these species live together with warm-blooded mammals or other animals. Thus, intestinal contents, the oral cavity, infections, and the rumen have been so far the main source for isolation of the species described. Correspondingly, the nutritional demands of these species can be rather complex or are often unresolved, in contrast to those of the few species isolated from sewage (Table 1).

In quantitative studies of the human intestinal flora, the *Eubacterium* species are the second most numerous with a median count of about 6×10^{10} organisms per g of dry weight of feces versus 10^{11} from *Bacteriodes* species (Finegold et al., 1983). *Eubacterium* species were recovered from 94% of humans regardless of diet, but only 58% of the isolates could be classified to a genus. *E. aerofaciens, E. contortum, E. cylindroides, E. lentum,* and *E. rectale* were identified in at least 20% of the cases, thus they are the most prominent species of *Eubacterium* in that habitat. *E. aerofaciens* was present in 40–60% of the cases, independent of the diet or state of health, whereas the presence of other species varied from three- to sixfold. Most of these studies were published between 1974 and 1977 and have been discussed in detail elsewhere (Lewis and Sutter, 1981). Detailed studies of the intestinal flora have resulted in the description of many new species, including *E. dolichum, E. eligens, E. formicigenerans, E. hadrum, E. hallii, E. ramulus,* and *E. siraeum* (Holdeman and Moore, 1974; Holdeman et al., 1976; Moore et al., 1976). Addition of Tween 80 to the growth medium proved to be mostly advantageous for those species recovered from intestine or feces. Although bile is an intestinal constituent, none of the species isolated was found to be stimulated in growth and acid production by the addition of bile. On the other hand, many *E. lentum*-like organisms are found to be quite active in transforming steroid compounds (Bokkenheuser and Winter, 1980), as will be discussed later in detail. *E. desmolans* even cleaves a carbon-carbon bond of 17-hydroxylated corticoids (Morris et al., 1986). *E. lentum* was the only organism isolated from the human colonic flora capable of inactivating digoxin, a widely used cardiac glycoside (Dobkin et al., 1983).

Most of the *Eubacterium* species isolated from (sub)gingival crevice, dental plaques, or calculus (Table 1) are rather asaccharolytic and hard to cultivate using the standard media. A pathogenic role might even be indicated by the high incidence (35 to 42% of samples cultured) and concentration (3 to 57% of cultivable flora) of *E. brachy, E. nodatum,* and *E. timidum* (Hill et al., 1987; Holdeman et al., 1980). All *E. yurii* subspecies form three-dimensional "test tube brush" arrangements from "*Actinomyces*-like" granulas, which are perpendicular to a central stalk (Margaret and Krywolop, 1986, 1988). *E. saburreum* is closely related to *E. yurii* and can be subdivided, like the latter species, into three serotypes (Kondo et al., 1979). *E. yurii* is especially sensitive to detergent residues on glassware and successful cultivation requires intensive rinsing of all glassware used. This might explain some of the problems experienced during cultivation of these organisms.

Table 1. Habitats of *Eubacterium* species.

Periodontal tissue	Intestine, feces	Rumen	Sewage	Soil	Infected materials, abscesses
E. alactolyticus	*E. aerofaciens*	*E. cellulosolvens*	*E. acidaminophilum*	*E. budayi*	*E. aerofaciens*
E. brachy	*E. biforme*	*E. limosum*	*E. angustum*	*E. combesii*	*E. alactolyticus*
E. nodatum	*E. contortum*	*E. oxidoreducens*	*E. callanderi*	*E. limosum*	*E. combesii*
E. saburreum	*E. cylindroides*	*E. ruminantium*	*E. limosum*	*E. moniliforme*	*E. contortum*
E. timidum	*E. desmolans*	*E. uniforme*	*E. tucumanense*	*E. multiforme*	*E. cylindroides*
E. yurii	*E. dolichum*			*E. nitritogenes*	*E. fossor*
	E. eligens			*E. tortosum*	*E. lentum*
	E. fissicatena				*E. limosum*
	E. formicigenerans				*E. moniliforme*
	E. hadrum				*E. multiforme*
	E. hallii				*E. nitritogenes*
	E. lentum				*E. nodatum*
	E. limosum				*E. plautii*
	E. plecicaudatum				*E. suis*
	E. ramulus				*E. tarantellus*
	E. rectale				*E. tenue*
	E. siraeum				*E. tortosum*
	E. tortosum				
	E. ventriosum				

Data were obtained from the original description of each species.

The seven recognized species reported to be present in soil or mud (Table 1) are also found in infectious material. This fact might be an indication for a secondary contamination of the host by soil, etc., for no *Eubacterium* species seems to be an agent causing a primary disease (Allen, 1985; Willis and Philipps, 1988). No systematic study has been performed to elucidate the presence and distribution of *Eubacterium* species in soil. They represent between 2 to 7% of the anaerobic isolates from clinical microbiology laboratories (Lewis and Sutter, 1981). Many species could be identified from different sources of infected tissue, but none of them seems to be pathogenic. As observed for the intestinal flora, most of the isolates (63%) could not be assigned to a known species. Again, *E. lentum* is the dominant species (Lewis and Sutter, 1981). So far, only a few species seem to be associated with a special host as *E. suis* is with pigs (Wegienek and Reddy, 1982), *E. tarantellus* with infected fish brain (Udey et al., 1977), *E. fossor* with the pharynx and tooth abscesses of horses (Bailey et al., 1986), and *E. plautii* as an endosymbiont of *Entamoeba histolytica* (Hofstad and Aasjord, 1982).

The rumen should be an ideal habitat for *Eubacterium* species due to its rich supply of organic materials (Stewart and Bryant, 1988). *E. ruminantium* seems to be well adapted to these conditions for it shows a requirement for sodium and rumen fluid and constitutes about 3–7% of the total isolates using a nonspecific medium (Bryant, 1959). *E. limosum* is one of the nutritionally most versatile species, degrading carbohydrates, amino acids, lactate, and methanol, and even growing autotrophically on $H_2 + CO_2$ or CO (Genthner et al., 1981; Genthner and Bryant, 1982). Strains of *E. limosum* have been isolated from the rumen fluid of sheep, but not of a steer. The fiber-degrading species *E. cellulosolvens, E. uniforme,* and *E. xylanophilum* can be specifically enriched using cellulose, cellobiose, or xylan as the selective substrate (van Gylswyk and van der Toorn, 1985, 1986). Xylan utilizers represent about 6% of the total viable count, with *E. uniforme* as the more versatile organism present at about 5% and *E. xylanophilum* as a specialist for cellobiose and xylan at about 1% of the xylan-degrading flora (van Gylswyk and van der Toorn, 1985). Lignins, tannins, and flavonoids are composed of trihydroxybenzenoids, so that the latter are common constituents of the rumen diet. *E. oxidoreducens* is specialized on gallate, pyrogallol, phloroglucinol, and quercetin degradation in the rumen. With all substrates except crotonate, *E. oxidoreducens* requires the presence of an electron donor such as H_2 or formate for decomposition of the substrate (Krumholz and Bryant, 1986). A pathway for gallate and phloroglucinol degradation to acetate and butyrate has been proposed (Krumholz et al., 1987), and the enzymes catalyzing unique reactions, like pyrogallol-phloroglucinol isomerase and phloroglucinol reductase, have been characterized (Haddock and Ferry, 1989; Krumholz and Bryant, 1988).

The anaerobic digesters of sewage plants represent the industrialized equivalent of the rumen, and *E. limosum,* as a versatile organism (Genthner et al., 1981), and the two quite similar species *E. callanderi* (Mountfort et al., 1988) and *E. tucumanense* (Diaz et al., 1989) have been recovered from that site. *E. callanderi* is somewhat specialized for the cleavage of phenylether bonds present in O-methoxylated aromatic substrates such as ferulate, sinapate, syringate, vanillate, or 3,4,5-trimethoxycinnamate (Mountfort et al., 1988). *E. tucumanese* grows readily on xylose and ethanol, consistent with the fact that it was isolated from a sludge bed reactor fed with stillage from distilleries (Diaz et al., 1989). *E. angustum* was isolated as the fastest growing anaerobe, utilizing only few purine compounds like uric acid, xanthine, guanine, and hypoxanthine (Beuscher and Andreesen, 1984). Purines form a major fraction of the cell constituents and are excreted as N-rich compound by birds, reptiles, insects, and other animals (Vogels and van der Drift, 1976). *E. acidaminophilum* is another nutritionally modest organism isolated from sludge, utilizing in pure culture only glycine and serine as single substrates. However, its substrate spectrum is extended to other amino acids or to some organic acids if a hydrogen-accepting organism or an acceptor like betaine, creatine, or sarcosine is also present. (Hormann and Andreesen, 1989; Zindel et al., 1988). Except for *E. oxidoreducens* (Krumholz and Bryant, 1986), substrate combinations have not been used for enrichment cultures. As in most other cases, no systematic study has been done on the anaerobic flora of this habitat. Most of the species were enriched using special biochemical abilities. Therefore, these species require no complex nutrients, but grow in mineral media supplemented with some vitamins, trace elements, and/or acetate.

Isolation

In most cases, the complex media used for isolation have been those recommended for all anaerobes by the Anaerobe Laboratory of the Virginia Polytechnic Institute (Holdeman et al., 1977), the Wadsworth Anaerobic Bacteriology Manual (Sutter et al., 1980), and the 9th edition of *Bergey's Manual of Systematic Bacteriology* (Moore and Holdeman-Moore, 1986). Due to the special, still-unresolved nutritional requirements of many species of *Eubacterium,* these media have to be supplemented with complex ingredients (vitamins, rumen fluid, volatile fatty acids, and/or special substrates such as cel-

lobiose or maltose) to obtain reasonable growth. Although these rich media appear to contain all necessary nutrients, growth of some species is still poor (Hill et al., 1987). To enhance growth of *E. acidaminophilum,* for example, selenium as a special micronutrient and a hydrogen acceptor or donor must be present (Zindel et al. 1988). If these are present in a balanced ratio, good growth will even occur in a simple defined medium. A salt solution, containing $CaCl_2$, $MgCl_2$, potassium phosphate, $NaHCO_3$, and NaCl (Holdeman et al., 1977), is a constituent of most media used to cultivate *Eubacterium* species. However, one should be aware of the fact that some species are inhibited by phosphate concentrations higher than 5 mM where the pH-buffering capacity is quite low (Zindel et al., 1988).

Most of the *Eubacterium* species are highly sensitive towards oxygen. Only *E. desmolans* can tolerate 4 h of exposure to atmospheric air (Morris et al., 1986). *E. suis* is reported to form barely discernible pinpoint colonies after incubation under air or 6% CO_2 (Wegienek and Reddy, 1982). This organism, however, was first described as a *Corynebacterium.* It contains cytochromes *b* and *c,* exhibits some similarities to anaerobic *Actinomyces* species, and has the highest GC content reported for a species of *Eubacterium.* Nitrate reduction is a property quite seldom encountered within the genus *Eubacterium,* especially as a marker valid for all strains (Bokkenheuser et al., 1979; Moore and Holdeman-Moore, 1986). This might indicate that most of the species within the genus are true anaerobes.

Identification

Chemotaxonomy is not yet applicable to the genus *Eubacterium* because too few species have been characterized in chemical terms. As emphasized before, the assignment of an organism to the genus *Eubacterium* depends on the nature of its metabolic products and their ratio on certain standard media and on the pH decrease observed during the fermentation of carbohydrates to the respective acids (Moore and Holdeman-Moore, 1986). The detection of acid production might be masked in species which are peptolytic and, therefore, alkalize the medium by ammonia formation. The nature and ratio of products formed can vary quite considerably even for a given organism. For *E. limosum,* they depend on the substrate offered and the presence or absence of additional electron acceptors or donors (such as bicarbonate, CO, fatty acids, or betaine). Acetate is the main product from H_2 plus CO_2 or from methanol in the presence of surplus bicarbonate, which favors a homoacetogenic fermentation (Genthner et al., 1981). However, butyrate will be formed at a low bicarbonate concentration, and even valerate, caproate, or higher homologs can be produced by *E. limosum* if propionate or butyrate are present in the growth medium above a threshold value (Lindley et al., 1987; Loubière et al., 1987). Caproate and H_2 are regularly formed from glucose, whereas 2-methylbutyrate and isobutyrate are the main products from isoleucine and valine, respectively (Genthner et al., 1981), and N,N'-dimethylglycine is a characteristic product of betaine fermentation (Müller et al., 1981). *E. limosum* even produces different products depending on the reducing agent used (Mariotto et al., 1989). Obviously, this great variability requires much more work to clarify *Eubacterium* chemotaxonomy. Only by using identical, well-specified conditions, can the species be differentiated by the identification key (Moore and Holdeman-Moore, 1986). Production (or absence) of butyric acid and acid formation leading to a sharp pH drop (or none) are the main lines of differentiation. The 10 newly described species can be included in the given scheme. However, especially the nonsaccharolytic species should be positively characterized by their individual abilities, as has been tried for *E. brachy, E. lentum, E. nodatum,* and *E. timidum* (Hill et al., 1987). Selected characteristics used for identification of Eubacterium species are given in Table 2.

Certain species of *Eubacterium* cannot be properly differentiated; for instance, *E. budayi* and *E. nitritogenes* are very closely related in their phenotypes. Further if they also hydrolyze gelatin, then they could be mistaken for *Clostridium perfringens,* a situation that might also obtain for *E. moniliforme* and *E. ventriosum* (Moore and Holdeman-Moore, 1986). Spore-like structures have been reported to be formed by *Butyribacterium methylotrophicum,* a close relative of *E. limosum* (Zeikus et al., 1980). For a strain of the latter organism, an increased heat resistance, but no spores have been observed (Genthner et al., 1981). These two species contain a special peptidoglycan of the B type, which is also present in *E. callanderi, Acetobacterium woodii,* and *Clostridium barkeri* (Mountfort et al., 1988; Tanner et al., 1981). *E. tucumanense* fits, at least physiologically, into this group of anaerobic one-carbon utilizers. Substrate concentration, pH value, and bicarbonate addition are important and critical variables to allow growth of these organisms on certain substrates (Buschhorn et al., 1989). Thus, differences observed so far among these species might dis-

appear in later studies. The newly described *E. yurii* seems to be closely related to *E. saburreum,* but it can be differentiated by DNA-DNA hybridization (Margaret and Krywolap, 1986). *E. tenue,* the *Eubacterium* species exhibiting the lowest GC content, belongs to a group that also includes *Clostridium lituseburense* and *Peptostreptococcus anaerobius,* according to 16S rRNA studies (Tanner et al., 1981). These species are related to the subgroup II-A of the genus *Clostridium* (e.g., to *C. bifermentans*) (Cato and Stackebrandt, 1989). *E. rectale* also shows relations to sporeformers (Moore and Holdeman-Moore, 1986). *E. ruminantium,* strains of *E. ventriosum* isolated from human intestine, and *Gemminger formicilis* have to be analyzed carefully to allow differentiation. The fast decolorization of *E. cylindroides* during the Gram-stain procedure might create problems and may lead to its identification as *Fusobacterium prausnitzii.* Variations in acid formation from maltose or sucrose and the amount of H_2 produced might lead to the identification of *Lachnospira multiparus* instead of *E. aerofaciens* or of *E. eligens* and *E. tarantellus* according to the current differentiation scheme (Moore and Holdeman-Moore, 1986). It should be evident from these examples that certain species can easily be misplaced. However, the identification scheme for the genus *Eubacterium* cannot do better for this group of anaerobes, because some of the named organisms actually do not even belong to this "catch-all" genus.

Applied Aspects

As emphasized above, *Eubacterium* species, especially *E. lentum,* represent a dominant group of organisms in the human gut. Although this species is normally inactive in most biochemical tests, organisms phenotypically related to it can catalyze a wide variety of transformation reactions on steroids (Akao et al., 1988; Bokkenheuser et al., 1979; Edenharder and Mielek, 1984; Hylemon, 1985; MacDonald et al., 1983) or are able to split off the side chain of cortisol, as is characteristic for *E. desmolans* (Bokkenheuser et al., 1986; Morris et al., 1986). So far, biotransformations of steroids carried out by anaerobes have not received any industrial attention (Sedlaczek, 1988), although the inability of these organisms to cleave the ring system of steroids might be regarded as an advantage.

Some *Eubacterium* species can reduce cholesterol to coprostanol. Most of these isolates require the presence of alkenyl ether lipids (plasmalogens) as a specific growth factor (Mott and Brinkley, 1979). Cortisol desmolase and 20-

Table 2. Biochemical reactions of species in the genus *Eubacterium*.[a]

Selected characteristics	E. acidaminophilum	E. aerofaciens	E. angustum	E. alactolyticum	E. biforme	E. brachy	E. budayi	E. calanderi	E. cellulosolvens	E. combesii	E. contortum	E. cylindroides	E. desmolans	E. dolichum	E. eligens	E. fissicatena	E. formicigenerans	E. fossor	E. hadrum	E. hallii	E. lentum	E. limosum
Problems experienced using ordinary complex media	+	−	+	−	−	+	−	−	−	−	−	−	(+)	+	(−)	−	−	−	−	−	+	−
Nutritional specialist	+	−	+	−	−	+	−	−	−	−	−	−	+	−	−	−	−	+	−	−	+	+
Utilization of sugars (restricted to few)	−	+	−	+	+	−	+	+	+	−	+	+	−	−	+	+	+	(+)	+	+	−	+
Utilization of amino acids or formation of branched chain fatty acids	+	−	−	−	−	+	−	+	−	+	−	−	−	−	−	−	−	−	−	−	+ (Arg)	+
Formation of:																						
Copious H_2	−	+	−	+	+	−	+	(+)	−	+	+	−	−	−	−	+	(+)	−	+	+	−	+
Butyrate	−	−	−	+	+	−	+	+	(+)	+	−	+	+	+	+	−	−	−	+	+	−	+
Caproate	−	(+)	−	+	(+)	−	−	−	−	−	−	−	−	−	−	−	−	−	−	−	−	+
Ethanol	(+)	−	−	−	−	−	−	−	−	(+)	+	−	−	−	+	+	+	−	−	−	−	−
Butanol	−	+	+	+	−	(+)	−	−	−	(+)	−	(+)	−	−	−	−	−	−	−	+	(+)	−
Formate	−	+	−	−	+	(+)	+	+	(+)	(+)	+	+	(+)	−	+	+	+	−	−	(+)	(+)	−
Lactate	−	+	−	−	−	−	+	(+)	+	+	−	(+)	−	(+)	−	(+)	+	+	+	(+)	(+)	−
Stereoisomer of lactate	(D/DL)	(D/DL)					(L)		(D)			(DL)										
Nitrate reduction	−	−	−	n.d.	−	n.d.	+	−	−	n.d.	−	−	−	n.d.	+	−	−	−	−	−	+	−
GC content (mol%)	44.0	n.d.	40.3	n.d.	32	n.d.	n.d.	47	50	n.d.	45	31	35	n.d.	36	45.5	40	43–46	32–33	36–38	n.d.	46–48
Cell wall type according to Schleifer and Kandler (1972)	m-A_2pm (A1γ)	−	m-A_2pm (A1γ)	m-A_2pm (A1γ)	−	−	−	B2α	−	LL-A_2pm-Gly (A3γ)	−	−	−	−	−	−	−	−	−	−	LL-A_2pm-Gly (A3γ)	B2α
Variable in Gram-stain	(+)	−	−	−	−	−	−	−	+	−	−	+	+	−	+	−	−	−	−	−	−	−
New species/facts, not included in scheme of Moore and Holdeman-Moore (1986)	+	−	+	−	−	+	−	+	+	−	−	−	+	−	−	−	−	+	−	−	+	+

(continued)

[a] Abbreviations: A_2pm, diaminopimelic acid; n.d., not determined; +, property present; −, property absent; (+), property variable.

Table 2. *Continued*

Selected characteristics	E. moniliforme	E. multiforme	E. nitritogenes	E. nodatum	E. oxidoreducens	E. plautii	E. plexicaudatum	E. ramulus	E. rectale	E. ruminantium	E. saburreum	E. siraeum	E. suis	E. tarantellus	E. tenue	E. timidum	E. tortuosum	E. tucumanense	E. uniforme	E. ventriosum	E. xylanophilum	E. yurii
Problems experienced using ordinary complex media	–	–	–	+	+	–	+	–	–	–	–	+	–	–	–	+	–	–	–	–	(+)	+
Nutritional specialist	–	–	–	+	+	–	–	–	–	(+)	–	–	–	–	–	+	–	–	–	–	+	+
Utilization of sugars (restricted to few)	+	+	+	–	–	+	+	+	+	+	+	+	+	+	+	–	+	+	+	+	(+)	(+)
Utilization of amino acids or formation of branched chain fatty acids	–	–	–	+	–	–	–	–	–	–	–	–	–	–	+	(+) Phe	–	+	–	–	–	–
Formation of: Copious H_2	+	+	+	–	–	–	+	+	+	–	+	+	–	+	+	–	(+)	+	–	–	–	–
Butyrate	+	+	+	+	+	+	+	+	+	+	+	–	–	–	+	–	+	+	–	+	+	+
Caproate	–	–	–	–	–	–	–	–	–	–	–	–	–	–	–	–	–	–	–	–	–	–
Ethanol	–	–	–	–	–	–	–	–	–	–	+	+	+	–	+	–	–	–	+	–	–	–
Butanol	+	–	–	–	–	–	+	–	–	–	+	–	+	–	+	–	–	–	–	–	–	–
Formate	(+)	(+)	(+)	(+)	–	–	–	+	–	+	(+)	–	+	+	+	(+)	+	–	+	+	+	–
Lactate	+	+	+	(+)	–	+	–	+	+	+	+	(+)	(+)	(+)	–	(+)	+	–	+	+	–	–
Stereoisomer of lactate	(L/DL)		(DL)							(DL)							(DL)		(L)	(D)		
Nitrate reduction	(+)	+	+	–	–	–	–	–	–	(+)	–	–	–	–	–	–	(+)	–	–	–	–	–
GC content (mol%)	n.d.	n.d.	n.d.	38–40	35.7	n.d.	44	39	30	n.d.	n.d.	45	55	n.d.	25.9	n.d.	n.d.	44	35	n.d.	39	32
Cell wall type according to Schleifer and Kandler (1972)	–	m-A_2pm (A1γ)	–	new Orn_2-Ala_2 type	–	no A_2pm	–	–	–	–	m-A_2pm (A1γ)	–	–	–	m-A_2pm (A1γ)	–	–	–	–	–	–	–
Variable in Gram-stain	–	–	+	–	–	+	+	–	–	+	+	–	–	–	–	+	–	–	–	–	–	–
New species/facts, not included in scheme of Moore and Holdeman-Moore (1986)	–	–	–	+	+	+	–	–	–	–	–	–	–	–	–	+	–	+	+	–	+	+

β-hydroxysteroid dehydrogenase are unique for *E. desmolans* (Bokkenheuser et al., 1986; Morris et al., 1986). A 21-dehydroxylating activity leading from 11-deoxycorticosterone (21-hydroxy-4-pregnene-3,20-dione) to progesterone is known for *E. lentum* (Bokkenheuser et al., 1977). This species can also produce a 16 α-dehydroxylase (Bokkenheuser and Winter, 1980), a 12 α- or 12 β-hydroxysteroid dehydrogenase (MacDonald et al., 1983), and bile acid 3 α-, 3 β-, 7 α-, and 12 α-hydroxysteroid dehydrogenase (Edenharder and Mielek, 1984), and forms ω-muricholic acid from the β-derivative (Eyssen et al., 1983). Epimerization reactions occur via the corresponding keto-compounds and might require the cometabolism of two species (Canzi et al., 1989; Edenharder and Schneider, 1985; MacDonald and Hutchinson, 1982).

The 7-dehydroxylase activity of *Eubacterium* sp. VPI 12 708 (closely related to *E. lentum*) seems to form a novel bile acid nucleotide in order to labilize the 7-hydroxy group to generate a 3-keto-$\Delta^{4,6}$-steroid nucleotide intermediate (Coleman et al., 1987). Proteins and even genes involved in the 7-dehydroxylation reaction have been characterized (Coleman et al., 1987; White et al., 1988a). A multigene family seems to be involved, and a 27-kDa protein exhibits extensive sequence homologies to alcohol/polyol dehydrogenases (White et al., 1988b). These seems to be the first reports applying molecular biology techniques to an organism of the genus *Eubacterium*.

The activity of the enzyme 16-hydroxyprogesterone dehydroxylase in combination with a 16-dehydroxyprogesterone reductase and progesterone reductase to form finally 17-isopregnanedione are 10-fold higher if H_2 or pyruvate are present (Glass and Burley, 1985). A study of many strains similar to *E. lentum* demonstrated that a few strains were also able to deconjugate both glycine and taurine conjugates of cholic acid and chenodeoxycholic acid. All strains studied could reduce one double bond of linoleic acid to *trans*-vaccenic acid (Eyssen and Verhulst, 1984). The reduction of the unsaturated lactone ring of digoxin, a cardiac glycoside, results in the formation of 99% pure 20R-dihydrodigoxin without hydrolysis of the glycosidic bond by *E. lentum* (Robertson et al., 1986). Also, the digitoxin derivatives are selectively reduced to the 20R-form during active growth (Chandrasekaran et al., 1987). Thus, many stereoselective biotransformation reactions can be performed by a variety of strains of *E. lentum*.

E. limosum is the other species that is of interest to biotechnologists due to its production of cobamides (Perlman and Semar, 1963; Vogt et al., 1988) and carboxylic acids (Bryant and Genthner, 1983; Lindley et al., 1987). The organism is a part of several highly efficient methanogenic consortia, due to its high metabolic versatility (Zellner and Winter, 1987).

Finally, it should be clear from this chapter that the genus *Eubacterium* is not well defined, and that much more attention should be given to most of its species in order to learn more about their physiological and biochemical characteristics and unique capabilities. To date, such studies have only been carried out with *E. lentum*. Now the genus *Eubacterium* is a collection of anaerobic organisms, some of which are totally unrelated. It is evident, that much more work is needed before a better understanding of the genus *Eubacterium* can be achieved.

Literature Cited

Akao, T., T. Akao, and K. Kobashi. 1988. Glycyrrhizin stimulates growth of *Eubacterium* sp. strain GLH, a human intestinal anaerobe. Appl. Environ. Microbiol. 54:2027–2030.

Allen, S. D. 1985. Gram-positive, nonsporeforming anaerobic bacilli, p. 461–472. In: E. H. Lennette, A. Balows, W. J. Hausler, and H. J. Shadomy (ed.), Manual of clinical microbiology, 4th ed. American Society for Microbiology, Washington.

Andreesen, J. R., H. Bahl, and G. Gottschalk. 1989. Introduction to the physiology and biochemistry of the genus *Clostridium*, p. 27–62. In: N. P. Minton, and D. J. Clarke (ed.), Clostridia. Plenum Press, New York.

Baardsen, R., V. Bakken, H. B. Jensen, and T. Hofstad. 1988. Outer membrane protein pattern of *Eubacterium plautii*. J. Gen. Microbiol. 134:1561–1564.

Bahl, H., and G. Gottschalk. 1988. Microbial production of butanol/acetone, p. 1–30. In: H. J. Rehm, and G. Reed (ed.), Biotechnology, vol. 6b. VCH Verlagsgesellschaft, Weinheim, Germany.

Bailey, G. D., and D. N. Love. 1986. *Eubacterium fossor* sp. nov., an agar-corroding organism from normal pharynx and oral and respiratory tract lesions of horses. Int. J. Syst. Bacteriol. 36:383–387.

Beuscher, H. U., and J. R. Andreesen. 1984. *Eubacterium angustum* sp. nov., a Gram-positive anaerobic, nonsporeforming, obligate purine fermenting organism. Arch. Microbiol. 140:2–8.

Bokkenheuser, V. D., and J. Winter. 1980. Biotransformation of steroid hormones by gut bacteria. Am. J. Clin. Nutr. 33:2502–2506.

Bokkenheuser, V. D., J. Winter, P. Dehazya, and W. G. Kelly. 1977. Isolation and characterization of human fecal bacteria capable of 21-dehydroxylating corticoids. Appl. Environ. Microbiol. 34:571–575.

Bokkenheuser, V. D., J. Winter, S. M. Finegold, V. L. Sutter, A. E. Ritchie, W. E. C. Moore, and L. V. Holdeman. 1979. New markers for *Eubacterium lentum*. Appl. Environ. Microbiol. 37:1001–1006.

Bokkenheuser, V. D., J. Winter, G. N. Morris, and S. Locascio. 1986. Steroid desmolase synthesis by *Eubacterium desmolans* and *Clostridium cadaveris.* Appl. Environ. Microbiol. 52:1153–1156.

Bryant, M. P. 1959. Bacterial species of the rumen. Bacteriol. Rev. 23:125–153.

Bryant, M. P., and B. R. S. Genthner. 1983. Microbial production of lower aliphatic carboxylic acids. U.S. Patent no. 4377638.

Buschhorn, H., P. Dürre, and G. Gottschalk. 1989. Production and utilization of ethanol by the homoacetogen *Acetobacterium woodii.* Appl. Environ. Microbiol. 55:1835–1840.

Canzi, E., E. Maconi, F. Aragozzini, and A. Ferrari. 1989. Cooperative 3-epimerization of chenodeoxycholic acid by *Clostridium innocuum* and *Eubacterium lentum.* Curr. Microbiol. 18:335–338.

Cato, E. P., and E. Stackebrandt. 1989. Taxonomy and phylogeny, p. 1–26. In: N. P. Minton, and D. J. Clarke (ed.), Clostridia. Plenum Press, New York.

Chandrasekaran, A., L. W. Robertson, and R. H. Reuning. 1987. Reductive inactivation of digitoxin by *Eubacterium lentum* cultures. Appl. Environ. Microbiol. 53:901–904.

Coleman, J. P., W. B. White, B. Egestad, J. Sjövall, and P. B. Hylemon. 1987. Biosynthesis of a novel bile acid nucleotide and mechanism of 7α-dehydroxylation by an intestinal *Eubacterium* species. J. Biol. Chem. 262:4701–4707.

Coleman, J. P., W. B. White, M. Lijewski, and P. B. Hylemon. 1988. Nucleotide sequence and regulation of a gene involved in bile acid 7-dehydroxylation by *Eubacterium* sp. strain VPI 12708. J. Bacteriol. 170:2070–2077.

Diaz, H. F., C. G. Nuñez, and F. Siñeriz. 1989. *Eubacterium tucumanense* sp. nov.: an anaerobic Gram-positive nonsporeformer isolated from an anaerobic digester. J. Gen. Microbiol. 135:2537–2541.

Dobkin, J. F., J. R. Saha, V. P. Butler, H. C. Neu, and J. Lindenbaum. 1983. Digoxin-inactivating bacteria: identification in human gut flora. Science 220:325–327.

Edenharder, R., and K. Mielek. 1984. Epimerization, oxidation and reduction of bile acids by *Eubacterium lentum.* Syst. Appl. Microbiol. 5:287–298.

Edenharder, R., and J. Schneider. 1985. 12 β-dehydrogenation of bile acids by *Clostridium paraputrificum, C. tertium,* and *C. difficile* and epimerization at carbon-12 of deoxycholic acid by cocultivation with 12 α-dehydrogenating *Eubacterium lentum.* Appl. Environ. Microbiol. 49:964–968.

Eyssen, H., G. de Pauw, J. Stragier, and A. Verhulst. 1983. Cooperative formation of ω-muricholic acid by intestinal microorganisms. Appl. Environ. Microbiol. 45:141–147.

Eyssen, H., and A. Verhulst. 1984. Biotransformation of linoleic acid and bile acids by *Eubacterium lentum.* Appl. Environ. Microbiol. 47:39–43.

Fernandez, F., and M. D. Collins. 1987. Vitamin K composition of anaerobic gut bacteria. FEMS Microbiol. Lett. 41:175–180.

Finegold, S. M., and W. L. George. 1989. Anaerobic infections in humans. Academic Press, San Diego.

Finegold, S. M., V. L. Sutter, and G. E. Mathisen. 1983. Normal indigenous intestinal flora, p. 3–31. In: D. J. Hentges (ed.), Human intestinal microflora in health and disease. Academic Press, New York.

Genthner, B. R. S., and M. P. Bryant. 1982. Growth of *Eubacterium limosum* with carbon monoxide as the energy source. Appl. Environ. Microbiol. 43:70–74.

Genthner, B. R. S., and M. P. Bryant. 1987. Additional characteristics of one-carbon-compound utilization by *Eubacterium limosum* and *Acetobacterium woodii.* Appl. Environ. Microbiol. 53:471–476.

Genthner, B. R. S., C. L. Davis, and M. P. Bryant. 1981. Features of rumen and sewage sludge strains of *Eubacterium limosum,* a methanol- and H_2-CO_2-utilizing species. Appl. Environ. Microbiol. 42:12–19.

Glass, T. L., and C. Z. Burley. 1985. Stimulation of 16-dehydroprogesterone and progesterone reductases of *Eubacterium* sp. strain 144 by hemin and hydrogen or pyruvate. Appl. Environ. Microbiol. 49:1146–1153.

Haddock, J. D., and J. G. Ferry. 1989. Purification and properties of phloroglucinol reductase from *Eubacterium oxidoreducens* G-41. J. Biol. Chem. 264:4423–4427.

Hill, G. B., O. M. Ayers, and A. P. Kohan. 1987. Characteristics and sites of infection of *Eubacterium nodatum, Eubacterium timidum, Eubacterium brachy,* and other asaccharolytic eubacteria. J. Clin. Microbiol. 25:1540–1545.

Hofstad, T., and P. Aasjord. 1982. *Eubacterium plautii* (Seguin 1928) comb. nov. Int. J. Syst. Bacteriol. 32:346–349.

Holdeman. L. V., E. P. Cato, J. A. Burmeister, and W. E. C. Moore. 1980. Descriptions of *Eubacterium timidum* sp. nov., *Eubacterium brachy* sp. nov., and *Eubacterium nodatum* sp. nov. isolated from human periodontitis. Int. J. Syst. Bacteriol. 30:163–169.

Holdeman, L. V., E. P. Cato, and W. E. C. Moore. 1977. Anaerobe Laboratory Manual, 4th ed., Anaerobe Laboratory, Virginia Polytechnic Institute and State University, Blacksburg, VA.

Holdeman, L. V., I. J. Good, and W. E. C. Moore. 1976. Human fecal flora: Variation in bacterial composition within individuals and a possible effect of emotional stress. Appl. Environ. Microbiol. 31:359–375.

Holdeman, L. V., and W. E. C. Moore. 1974. New genus, *Coprococcus,* twelve new species, and emended descriptions of four previously described species of bacteria from human feces. Int. J. Syst. Bacteriol. 24:260–277.

Hormann, K., and J. R. Andreesen. 1989. Reductive cleavage of sarcosine and betaine by *Eubacterium acidaminophilum* via enzyme systems different from glycine reductase. Arch. Microbiol. 153:50–59.

Hylemon, P. B. 1985. Metabolism of bile acids in intestinal microflora, p. 331–343. In: H. Danielsson, and J. Sjovall (ed.), Sterols and bile acids. Elsevier Sci. Publ. BV, Amsterdam.

Kondo, W., N. Sato, and T. Ito. 1979. Chemical structure of the polysaccharide antigen of *Eubacterium saburreum,* strain 02. Carbohydr. Res. 70:117–123.

Krumholz, L. R., and M. P. Bryant. 1986. *Eubacterium oxidoreducens* sp. nov. requiring H_2 or formate to degrade gallate, pyrogallol, phloroglucinol and quercetin. Arch. Microbiol. 144:8–14.

Krumholz, L. R., and M. P. Bryant. 1988. Characterization of the pyrogallol-phloroglucinol isomerase of *Eubacterium oxidoreducens.* J. Bacteriol. 170:2472–2479.

Krumholz, L. R., R. L. Crawford, M. E. Hemling, and M. P. Bryant. 1987. Metabolism of gallate and phloroglucinol in *Eubacterium oxidoreducens* via 3-hydroxy-5-oxohexanoate. J. Bacteriol. 169:1886–1890.

Lewis, R. P., and V. L. Sutter. 1981. The genus *Eubacterium,* p. 1903–1911. In: M. P. Starr, H. Stolp, H. G. Trüper, A. Balows, and H. G. Schlegel (ed.), The prokaryotes: a handbook on habitats, isolation, and identification of bacteria, vol. 2. Springer-Verlag, Berlin.

Lindley, N. D., P. Loubière, S. Pacaud, C. Mariotto, and G. Goma. 1987. Novel products of the acidogenic fermentation of methanol during growth of *Eubacterium limosum* in the presence of high concentrations of organic acids. J. Gen. Microbiol. 133:3557–3563.

Loubière, P., S. Pacaud, G. Goma, and N. D. Lindley. 1987. The effect of formate of the acidogenic fermentation of methanol by *Eubacterium limosum.* J. Gen. Appl. Microbiol. 33:463–470.

MacDonald, I. A., V. D. Bokkenheuser, J. Winter, A. M. McLernon, and E. H. Mosbach. 1983. Degradation of steroids in the human gut. J. Lipid. Res. 24:675–700.

MacDonald, I. A., and D. M. Hutchison. 1982. Epimerization versus dehydroxylation of the 7 α-hydroxylgroup of primary bile acids: competitive studies with *Clostridium absonum* and 7 α-dehydroxylating bacteria (*Eubacterium* sp.). J. Steroid Biochem. 17:295–303.

Margaret, B. S., and G. N. Krywolap. 1986. *Eubacterium yurii* subsp. *yurii* sp. nov. and *Eubacterium yurii* subsp. *margaretiae* subsp. nov.: test tube brush bacteria from subgingival dental plague. Int. J. Syst. Bacteriol. 36:145–149.

Margaret, B. S., and G. N. Krywolap. 1988. *Eubacterium yurii* subsp. *schtitka* subsp. nov.: test tube brush bacteria from subgingival dental plague. Int. J. Syst. Bacteriol. 38:207–208.

Mariotto, C., P. Loubière, G. Goma, and N. D. Lindley. 1989. Influence of various reducing agents on methylotrophic growth and organic acid production of *Eubacterium limosum.* Appl. Microbiol. Biotechnol. 32:193–198.

Moore, W. E. C., and L. V. Holdeman-Moore. 1986. Genus *Eubacterium,* p. 1353–1373. In: P. H. A. Sneath, N. S. Mair, M. E. Sharpe, and J. G. Holt (ed.), Bergey's manual of systematic bacteriology, vol. 2. Williams and Wilkins, Baltimore.

Moore, W. E. C., J. L. Johnson, and L. V. Holdeman. 1976. Emendation of *Bacteriodaceae* and *Butyrivibrio* and descriptions of *Desulfomonas* gen. nov. and ten new species in the genera *Desulfomonas, Butyrivibrio, Eubacterium, Clostridium* and *Ruminococcus.* Int. J. Syst. Bacteriol. 26:238–252.

Morris, G. N., J. Winter, E. P. Cato, A. E. Ritchie, and V. D. Bokkenheuser. 1986. *Eubacterium desmolans* sp. nov., a steroid desmolase producing species from cat fecal flora. Int. J. Syst. Bacteriol. 36:183–186.

Mott, G. E., and A. W. Brinkley. 1979. Plasmenylethanolamine: growth factor for cholesterol-reducing *Eubacterium.* J. Bacteriol. 139:755–760.

Mountfort, D. O., W. D. Grant, R. Clarke, and R. A. Asher. 1988. *Eubacterium callanderi* sp. nov. that demethoxylates O-methoxylated aromatic acids to volatile fatty acids. Int. J. Syst. Bacteriol. 38:254–258.

Müller, E., K. Fahlbusch, R. Walther, and G. Gottschalk. 1981. Formation of N,N-dimethylglycine, acetic acid, and butyric acid from betaine by *Eubacterium limosum.* Appl. Environ. Microbiol. 42:439–445.

Perlman, D., and J. B. Semar. 1963. Production of cobamides by *Butyribacterium rettgeri.* Biotechnol. Bioeng. 5:21–25.

Prévot, A. R. 1938. Etude de systématique bactérienne: III. Invalideté du genre *Bacteroides castellani* et Chalmers démembrement et reclassification. Ann. Inst. Pasteur 60:287–307.

Robertson, L. W., A. Chandrasekaran, R. H. Reuning, J. Hui, and B. D. Rawal. 1986. Reduction of digoxin to 20R-dihydrodigoxin by cultures of *Eubacterium lentum.* Appl. Environ. Microbiol. 51:1300–1303.

Schleifer, K. H., and O. Kandler. 1972. Peptidoglycan types of bacterial cell walls and their taxonomic implications. Bacteriol. Rev. 36:407–477.

Sedlaczek, L. 1988. Biotransformations of steroids. CRC Rev. Biotechnol. 7:186–236.

Severin, A. I., S. Kokeguchi, and K. Kato. 1989a. Chemical composition of *Eubacterium alactolyticum* cell wall peptidoglycan. Arch. Microbiol. 151:348–352.

Severin, A. I., S. Kokeguchi, and K. Kato. 1989b. Chemical composition of *Eubacterium nodatum* cell wall peptidoglycan. Arch. Microbiol. 151:353–358.

Stewart, C. S., and M. P. Bryant. 1988. The rumen bacteria, p. 21–&5. In: P. N. Hobson (ed.), The rumen microbial ecosystem. Elsevier Applied Science, London.

Sutter, V. L., D. M. Citron, and S. M. Finegold. 1980. Wadsworth anaerobic bacteriology manual. The C. V. Mosby Company, St. Louis, Michigan.

Tanner, R. S., E. Stackebrandt, G. E. Fox, and C. R. Woese. 1981. A phylogenetic analysis of *Acetobacterium woodii, Clostridium barkeri, Clostridium butyricum, Clostridium lituseburense, Eubacterium limosum,* and *Eubacterium tenue.* Curr. Microbiol. 5:35–38.

Udey, L. R., R. Young, and B. Sallman. 1977. Isolation and characterization of an anaerobic bacterium, *Eubacterium tarantellus* sp. nov., associated with striped mullet (*Mugil cephalus*) mortality in Biscayne Bay, Florida. J. Fish. Res. Board Can. 34:402–409.

van Gylswyk, N. O., and J. J. T. K. van der Toorn. 1985. *Eubacterium uniforme* sp. nov. and *Eubacterium xylanophilum* sp. nov., fiber-digesting bacteria from the rumina of sheep fed corn stover. Int. J. Syst. Bacteriol. 35:323–326.

van Gylswyk, N. O., and J. J. T. K. van der Toorn. 1986. Description and designation of a neotype strain of *Eubacterium cellulosolvens* (*Cillobacterium cellulosolvens* Bryant, Small, Bouma and Robinson) Holdeman and Moore. Int. J. Syst. Bacteriol. 36:275–277.

Vogels, G. D., and C. van der Drift. 1976. Degradation of purines and pyrimidines by microorganisms. Bacteriol. Rev. 40:403–468.

Vogt, J. R. A., L. Lamm-Kolonko, and P. Renz. 1988. Biosynthesis of vitamin B$_{12}$ in anaerobic bacteria. Experiments with *Eubacterium limosum* and D-erythrose ^{14}C-labeled in different positions. Eur. J. Biochem. 174:637–640.

Wegienek, J., and C. A. Reddy. 1982. Taxonomic study of "*Corynebacterium suis*" Soltyns and Spratling: proposal of *Eubacterium suis* (nov. rev.) comb. nov. Int. J. Syst. Bacteriol. 32:218–228.

White, W. B., J. P. Coleman, and P. B. Hylemon. 1988a. Molecular cloning of a gene encoding a 45.000 dalton polypeptide associated with bile acid 7-dehydroxylation in *Eubacterium* sp. strain VPI 12 708. J. Bacteriol. 170:611–616.

White, W. B., C. V. Franklund, J. P. Coleman, and P. B. Hylemon. 1988b. Evidence for a multigene family involved in bile acid 7-dehydroxylation in *Eubacterium* sp. strain VPI 12 708. J. Bacteriol. 170:4555–4561.

Weiss, N. 1981. Cell wall structure of anaerobic cocci. Rev. Inst. Pasteur Lyon 14:53–59.

Willis, A. T., and K. D. Phillips. 1988. Anaerobic infections. Clinical and laboratory practice, Public Health Laboratory Service, London.

Woese, C. R. 1987. Bacterial evolution. Microbiol. Rev. 51:221–271.

Zeikus, J. G., L. H. Lynd, T. E. Thompson, J. A. Krzycki, P. J. Weimer, and P. W. Hegge. 1980. Isolation and characterization of a new, methylotrophic, acidogenic anaerobe, the Marburg strain. Curr. Microbiol. 3:381–386.

Zellner, G., and J. Winter. 1987. Analysis of a highly efficient methanogenic consortium producing biogas from whey. Syst. Appl. Microbiol. 9:284–292.

Zindel, U., W. Freudenberg, M. Rieth, J. R. Andreesen, J. Schnell, and F. Widdel. 1988. *Eubacterium acidaminophilum* sp. nov., a versatile amino acid-degrading anaerobe producing or utilizing H_2 or formate. Description and enzymatic studies. Arch. Microbiol. 150:254–266.

CHAPTER 87

The Genera *Acetobacterium, Acetogenium, Acetoanaerobium,* and *Acetitomaculum*

BERNHARD SCHINK and MARTIN BOMAR

The genera *Acetobacterium, Acetoanaerobium,* and *Acetitomaculum* all comprise homacetogenic bacteria that are physiologically related to homoacetogenic representatives of other taxonomic groups, namely, *Clostridium, Eubacterium, Sporomusa, Peptococcus,* and *Syntrophococcus.*

The characteristic property of homoacetogenic bacteria is their ability to use carbon dioxide as an electron sink and to reduce it via the carbon monoxide dehydrogenase system, producing acetate as the typical fermentation product. The homoacetogenic metabolism is described in detail in Chapter 21. These bacteria should be clearly differentiated from bacteria of other metabolic types that are called "acetogenic" because they produce acetate as their main fermentation product, such as the obligately syntrophic proton-reducing fatty acid- or ethanol-oxidizing bacteria and some strains of anaerobes that can ferment aromatic compounds or glycine. To avoid confusion, therefore, the term "homoacetogenic" metabolism is preferred over "acetogenic" in this chapter.

The genus *Acetobacterium* was established in 1977 (Balch et al., 1977) to comprise Gram-positive homoacetogenic bacteria that did not form spores and therefore could not be grouped with the existing forms of homoacetogenic clostridia, namely, *Clostridium aceticum* (Wieringa, 1936, 1940), *C. thermoaceticum* (Fontaine et al., 1942), and *C. formicoaceticum* (Andreesen et al., 1970). The type species, *Acetobacterium woodii,* was first enriched and isolated in Woods Hole, Massachusetts, USA, from sediment of a marine estuary (Balch et al., 1977). Under conditions suitable for enriching hydrogen-utilizing methanogens at pH 6.7, acetate was formed in considerable amounts, and a new type of rod-shaped bacterium appeared. Similar bacteria were later also obtained in enrichments from freshwater sediment samples. An isolate obtained from sewage sludge was attributed to a new species, *A. wieringae* (Braun and Gottschalk, 1982). Studies on the degradation of primary alcohols and diols in freshwater sedi-

ments led to the isolation of *A. carbinolicum* (Eichler and Schink, 1984). Finally, a malate-utilizing strain was isolated from an enrichment with methoxyethanol as substrate, and placed into a new species, *A. malicum* (Tanaka and Pfennig, 1988).

After establishment of the genus *Acetobacterium,* other genera of homoacetogenic bacteria were defined. Their main properties are listed in Table 1, which illustrates why genera separate from *Acetobacterium* had to be established. The already-mentioned homoacetogenic clostridia differ from *Acetobacterium* by their ability to form spores. This is true as well for species that were isolated later, *C. thermoautotrophicum* (Wiegel et al., 1981), *C. magnum* (Schink, 1984), and *C. pfennigii* (Krumholtz and Bryant, 1985). Although *Sporomusa* species are also sporeformers, they stain definitively Gram-negative (Möller et al., 1984). *Eubacterium limosum* (Prévot, 1938), as well as *Butyribacterium rettgeri* (Barker and Haas, 1944) (which are taxonomically identical entities; Moore and Cato, 1965; Tanner et al., 1981), can clearly be differentiated from all other homoacetogens by their ability to produce butyrate and acetate. *Acetoanaerobium* differs from *Acetobacterium* by its Gram-stain (Gram-negative), although the cell wall architecture resembles that of a Gram-positive bacterium (Sleat et al., 1985). *Acetogenium kivui* is a thermophilic, non-sporeforming homacetogen (Leigh et al., 1981). *Peptostreptococcus productus* cells are coccoid (Lorowitz and Bryant, 1984), and *Acetitomaculum ruminis* cells are slightly curved and its DNA has a lower GC content (Greening and Leedle, 1989). *Syntrophococcus sucromutans* (Krumholtz and Bryant, 1986) carries out only an incomplete homoacetogenesis and depends either on a secondary substrate or on another organism as a syntrophic partner.

Comparison of 16S rRNA homologies has revealed that *Acetobacterium woodii* exhibits some phylogenetic relatedness to *Eubacterium limosum* and to some *Clostridium* species (Tanner et al., 1981). *Acetogenium kivui* is also re-

Table 1. Properties of described genera of homoacetogenic bacteria.

Genus	Cell shape	Spores	Gram type	GC content (mol%)	Optimum temperature (°C)	Reduced end product
Clostridium	Rods	+	+	29–54	30[a]	Acetate
Sporomusa	Rods	+	−	42–48	30	Acetate
Eubacterium (Butyrivibrio)	Rods	−	+	47–49	30–37	Acetate, butyrate
Acetobacterium	Rods	−	+	38–44	27–30	Acetate
Acetogenium	Rods	−	+[b]	38	66	Acetate
Acetoanaerobium	Rods	−	+[c]	36.8	37	Acetate
Acetitomaculum	Curved rods	−	+	32–36	38	Acetate
Peptostreptococcus	Cocci	−	+	44–46	37	Acetate
Syntrophococcus	Cocci	−	+	52	35–42	Acetate

[a]Most species grow best at 30° but two species are thermophilic (55–60°).
[b]Originally described as Gram-negative; in electron microscopic studies characterized as Gram-positive.
[c]Staining reaction, Gram-negative; in electron microscopic studies characterized as Gram-positive.

lated to some *Clostridium* species (Leigh et al., 1981), and the Gram-negative sporeformer *Sporomusa* was found to be related on the basis of 16S rRNA similarities to the clostridia as well (Stackebrandt et al., 1985). These findings indicate that the obvious physiological similarities between these main groups of homoacetogens may have a phylogenetic basis.

Habitats and Ecology

Detailed studies on the ecology of homoacetogenic bacteria are lacking so far and are not easy to perform, due to the metabolic versatility of these organisms (see below). Most known homoacetogenic bacteria were isolated from strictly anoxic environments, typically black sediments of estuaries, marine sources, freshwater ponds, and anaerobic sewage sludge. In the last, they were found to make up about 1% of the total hydrogen-oxidizing community (Braun et al., 1979). Homoacetogenic activities were also detected in the gastrointestinal tract of termites (Breznak and Switzer, 1986) and of higher animals and humans (Prins and Lankhorst, 1977; Lajoie et al., 1988). A first search for homoacetogens in the rumen was negative (Braun et al., 1979), but later considerable numbers (10^7–10^8 cells/ml) were detected in the rumen of steers (Greening and Leedle, 1989).

In all these environments, homoacetogens have to compete with other anaerobes for electron sources. Since utilization of carbon dioxide as an electron sink allows complete conversion of sugars to acetate, homoacetogens should have a considerable advantage over classical fermenters in carbohydrate fermentation. However, homoacetogens grow slower on sugars than *Escheria coli* and a classical *Clostridium* species do, and no homoacetogen has so far been de-

scribed that degrades polysaccharides, the main sources for sugars in most natural organic materials. In the oxidation of hydrogen and formate, homoacetogens compete with methanogens and sulfate reducers; due to their higher energy yields and substrate affinity, the latter usually win this competition (Schink, 1987; Zehnder and Stumm, 1988). Methanogens are similarly successful in utilization of methanol and methylamines (Winfrey and Ward, 1983; King et al., 1983; Lovley and Klug, 1983). Oxidation of primary alcohols and diols is more efficiently carried out by sulfate reducers and by fermenting bacteria cooperating syntrophically with methanogens (Eichler and Schink, 1985). In general, homoacetogens appear to be inferior to the respective specialists in every single case. Their success in a natural anoxic environment therefore, appears to be based mainly on their metabolic versatility, i.e., their ability to alternate between various substrates or to use them simultaneously. The only metabolic activity so far known in which they are specialists themselves is the demethylation of aromatic methylether compounds (Bache and Pfennig, 1981). This ability has therefore been used successfully for selective enrichment (see below).

During the oxidation of methyl compounds or other organic substrates, homoacetogenic bacteria produce hydrogen at low levels, and this hydrogen is either neutralized by the homoacetogens themselves (Seitz et al., 1988) or by other anaerobic hydrogen oxidizers. Interspecies hydrogen transfer to methanogenic or sulfate-reducing bacteria unable to use methanol was demonstrated in cocultures with *Sporomusa acidovorans* growing on methanol (Cord-Ruwisch and Ollivier, 1986). In other systems of interspecies hydrogen transfer, *Acetobacterium* species acted as hydrogen scavengers rather than hydrogen producers, e.g., in syntrophic oxidation of primary aliphatic alcohols (Schink and Stieb, 1983; Eichler and Schink, 1985). However, due to their broad metabolic versatility, homoacetogens probably do not depend to any major extent on cooperation with other trophic groups of anaerobes.

A very special case of syntrophy is represented by a recently described thermophilic homoacetogen that can both form acetate from CO_2 and H_2 and convert acetate to CO_2 and H_2, depending on the external hydrogen concentration (Zinder and Koch, 1984; Lee and Zinder, 1988). This bacterium has not been taxonomically characterized yet, but it may be related to the thermophilic *Acetogenium kivui* treated here.

Evidence has been provided that homoacetogenic bacteria may be able to outcompete methanogens for hydrogen and hydrogen equivalents under certain conditions, e.g., in mildly acidic lake sediments (pH 6.1; Phelps and Zeikus, 1984) or at low temperatures ($<20°C$; Conrad et al., 1989). The known species of methanogenic bacteria are not significantly active at pH<6.0 or at low temperatures; obviously, homoacetogens are less restricted in both respects. This may explain why they have a selective advantage in both cases, thus setting the stage for natural enrichments for homoacetogens in most freshwater sediments. Unfortunately, systematic enumeration studies on homoacetogens in such environments are lacking so far.

Of special interest is the relation of homoacetogens to salt content in its medium. The first strains of *A. woodii* were isolated from an estuarine sediment. All *Acetobacterium* strains isolated in our laboratory were rather independent of the prevailing salt concentration and grew equally well in freshwater, brackish water, or salt water medium. The recently discovered involvement of sodium ions in energy metabolism and the presence of sodium-proton antiporters in the cytoplasmic membrane (see Chapter 21) may allow these bacteria to convert proton motive force into sodium motive force and vice, versa and, with this, render them excellently adapted to environments of periodically changing salinity.

The role of homoacetogens in the digestive tracts of animals depends to a great extent on the particular animal studied. While reports on their occurrence in the cow rumen are contradictory (Braun et al., 1979; Greening and Leedle, 1989), homoacetogens are the predominant hydrogen utilizers in the hindgut of certain termites and cockroaches (Breznak and Switzer, 1986). It is not yet understood why homoace-

togens can outcompete the energetically more favored methanogens in the gastrointestinal system of these insects. Perhaps the insect itself has some influence on the nature of the anaerobic population in its gut by excretion of certain digestive enzymes or other kinds of substances, such as tensides, which may be inhibitory to methanogens.

In this section, we considered the ecology of homoacetogens in general rather than that of single genus or species. This is appropriate since most of them are common sediment bacteria that share the same ecological niches due to their similar physiological and cytological properties. *Acetitomaculum ruminis* was isolated from the rumen; it is too early to judge whether it is really typical for or restricted to this type of environment.

Acetogenium kivui may be restricted to thermal environments. It was originally isolated from sediments of Lake Kivu, Central Africa, with hydrogen/carbon dioxide as substrate, as were some strains of *Clostridium thermoaceticum,* with which it shares, among other properties, the ability to inhabit such a hot habitat.

Enrichment and Isolation

Homoacetogenic bacteria depend on the availability of sufficient amounts of carbon dioxide/bicarbonate in their growth medium. Lack of bicarbonate in phosphate-buffered media has prevented the isolation of homoacetogens for many years. For enrichment, isolation, and cultivation of the mesophilic homoacetogenic bacteria treated in this chapter, the following carbonate-buffered mineral medium is recommended (after Widdel and Pfennig, 1981; Schink and Pfennig, 1982):

Carbonate-Buffered Mineral Medium
Dissolve in 1 l of distilled water:

	Type of medium		
	Freshwater	Brackish water	Saltwater
$KH_2 \cdot PO_4$	0.2 g	0.2 g	0.2 g
NH_4Cl	0.5 g	0.5 g	0.5 g
NaCl	1.0 g	8.0 g	20.0 g
$MgCl_2 \cdot 6H_2O$	0.4 g	1.2 g	3.0 g
KCl	0.5 g	0.5 g	0.5 g
$CaCl_2 \cdot 2H_2O$	0.15 g	0.15 g	0.15 g

Autoclave the complete mineral medium in a vessel equipped with: 1) a filter inlet to allow flushing of the headspace with sterile oxygen-free gas; 2) screw cap inlet for addition of thermally unstable additives after autoclaving; 3) a silicon-tubing connection from the bottom of the vessel out to a dispensing tap (if possible with a protecting bell) for sterile dispensing of the medium (do not use latex tubing; it releases compounds which are highly toxic to many anaerobes); and 4) a stirring bar.

After autoclaving, connect the vessel with the still-hot medium to a line of oxygen-free nitrogen/carbon dioxide mixture (90% N_2/10% CO_2) at low pressure (<100 mbar), flush the headspace and cool it under this atmosphere to room temperature, perhaps with the help of a cooling water bath.

The mineral medium is amended with the following additions from stock solutions that have been sterilized separately (amounts per liter of medium); 1) 30 ml of 1 M $NaHCO_3$ solution (autoclaved in a *tightly closed* screw cap bottle with about 30% headspace; the bottle should be autoclaved in a further protecting vessel such as a polypropylene beaker, to avoid spills of carbonates if the bottle breaks in the autoclave); 2) 2 ml of 0.5 M $Na_2S \cdot 9H_2O$ solution (autoclaved separately under oxygen-free gas atmosphere as above); 3) 1 ml of trace element solution, such as SL 9 (Tschech and Pfennig, 1984) or SL 10 (Widdel et al., 1983); 4) 0.5 ml of 10-fold concentrated, filter-sterilized vitamin solution (Pfennig, 1978); and 5) adequate amounts of sterile 1 M HCl or 1 M Na_2CO_3 to adjust the pH to 7.1–7.3.

The complete medium is dispensed into either screw-cap bottles or screw-cap tubes that are filled completely to the top, leaving a lentil-sized air bubble for pressure equilibration. Enrichment cultures usually produce gas in the first enrichment stages and are better cultivated in half-filled serum bottles (50–100 ml volume) under a headspace of nitrogen/carbon dioxide mixture (90%/10%). Growth with hydrogen/carbon dioxide can be tested either in half-filled tubes sealed with butyl rubber stoppers (e.g., Hungate tubes) that are incubated horizontally or in serum bottles that are incubated horizontally on a slow shaker to assure sufficient gas supply in the culture fluid.

This mineral medium is amended with the respective organic substrates for enrichment and cultivation of pure cultures. The vitamin mixture is not really needed by all strains. For example, *Acetobacterium carbinolicum* does not require any vitamins, and others need only a few vitamins. *A. wieringae* has been reported to require yeast extract additions to the medium; with the medium described above, yeast extract is not required, but growth yields are higher in its presence. Special care should be taken with the trace element solutions used. We observed that traces of copper and borate in conventional trace element solutions can easily be toxic to *Acetobacterium* species; delays in growth initiation in freshwater medium could be overcome by enhancing the sodium chloride content to about 50 mM. *Acetoanaerobium noterae* uses yeast extract as an energy source (see below).

Enrichment Medium for *Acetogenium kivui*

For enrichment and cultivation of *A. Kivui,* the following medium is recommended (values in g/liter of medium):

K_2HPO_4	0.22 g
KH_2PO_4	0.22 g
NH_4Cl	0.31 g
$(NH_4)_2SO_4$	0.22 g
NaCl	0.45 g
$MgSO_4 \cdot 7H_2O$	0.09 g
$CaCl_2 \cdot 2H_2O$	0.006 g
$FeSO_4 \cdot 7H_2O$	0.002 g
$NaHCO_3$	4.5 g
L-cysteine$\cdot$HCl$\cdot$H_2O	0.5 g
$Na_2S \cdot 9H_2O$	0.5 g
$NaH_2PO_4 \cdot H_2O$	6.9 g
$Na_2HPO_4 \cdot 12H_2O$	17.9 g

It is advisable to autoclave cysteine, sodium sulfide, and the sodium phosphates separately as concentrated stock solutions in tightly closed vials. Cultures are grown with 2–4 bar of hydrogen/carbon dioxide mixture (70%/30%) or 1 bar of nitrogen/carbon dioxide mixture (70%/30%) at 60°C. Other media have not been tried with this bacterium; it is possible that the above-mentioned general medium could be used as well. In this case, the carbon dioxide partial pressure should be raised to 30% to maintain the pH.

For enrichment of the mesophilic homoacetogens treated here, a H_2/CO_2 mixture (80%/20%) in the culture headspace was usually a sufficiently selective enrichment substrate. Yeast extract or other complex medium additions should not be used in the enrichment stage to prevent growth of (often pathogenic!) peptolytic bacteria. To avoid growth of hydrogen-utilizing sulfate reducers, sulfate should be omitted from the medium. Growth of competing methanogens can be suppressed by lowering the pH to 6.0–6.5, or by addition of substances that are inhibitory for methanogens (e.g., 2 mM 2-bromoethanesulfonate in the medium, 0.5% acetylene or 2% ethylene in the headspace). Small amounts of dithionite have been used for the same purpose (Balch et al., 1977) but it did not select satisfactorily in our hands.

Also, ethylene glycol (10 mM) has proven to be a rather selective substrate for *Acetobacterium* species although it is also a good substrate for many *Pelobacter* species (see Chapter 186). *A. woodii* can be selectively enriched with methoxylated aromatic compounds such as vanillate, ferulate, trimethoxycinnamate, syringate, etc. (5–10 mM) (Bache and Pfennig, 1981). Although other *Acetobacterium* species can use these substrates as well, *A. woodii* appears to grow the fastest with them. *A. carbinolicum* can be easily enriched and isolated from freshwater and brackish water sediments with 10–20 mM ethanol, *n*-propanol, or *n*-butanol as sole substrates; methoxyethanol or malate select for *A.*

malicum. A selective enrichment procedure for *A. wieringae* is not known; the type strain was isolated with H_2/CO_2 from anoxic sewage sludge.

Acetoanaerobium noterae appears to be more fastidious than the *Acetobacterium* species treated here in its dependence on higher amounts of yeast extract (0.2% w/v). It was originally isolated from sediment of an oil-drilling site with pH 8.0, with H_2/CO_2 as substrate; perhaps the sampling site itself selected sufficiently for this organism, which resembles *Acetobacterium* in nearly all properties. The same may apply to *Acetitomaculum ruminis:* it is, besides *Eubacterium limosum,* the only nonsporeforming homoacetogen isolated from intestinal sources, and appears to require at least 1% rumen fluid in the medium.

It should be mentioned at this point that enrichments with many other substrates have yielded pure cultures of homoacetogenic bacteria resembling *Acetobacterium* strains both in their morphology and physiology. Among these substrates are triacetin (Emde and Schink, 1987), the formaldehyde-yielding compound hexamethylene tetramine (Schink, 1987), short-chain polyethylene glycols and their surface-active derivatives (Wagener and Schink, 1988), methoxyacetate (Schuppert and Schink, 1990), DL-mandelate, and furfuryl alcohol (B. Schink, unpublished observations). Obviously, the biochemical capacities of the members of this genus are considerably greater than originally assumed.

Isolation Procedure

After one to two transfers in the above-mentioned enrichment media, the presence or even the dominance of homoacetogens should be indicated by the accumulation of acetate, the typical fermentation product; hydrogen consumption without significant methane formation; and the appearance of rod-shaped, nonfluorescent cells of rather typical shape that display active, tumbling motility (see below). Purification can be carried out with organic substrates in an agar shake dilution series (Pfennig, 1978), or, especially with hydrogen as substrate, in roll tubes (Balch et al., 1979) either diluted in the agar or streaked on the agar surface. A convenient alternative to the roll tubes are flat bottle plates containing a thin agar film is kept under a defined gas mixture (Braun et al., 1979). At this stage, yeast extract or other undefined complex nutrients can be added to obtain as broad as possible a spectrum of homoacetogens. The use of calcium carbonate in the medium as an indicator of acid formation ("chalk agar") can

be helpful for differentiation of acetic acid formers from other anaerobes (Braun et al., 1979)

In agar dilution media, *Acetobacterium*-like bacteria form compact, yellowish, lens-shaped colonies of 0.5–1 mm diameter. Colonies of *A. woodii* on agar surfaces are circular and convex and are up to 1 mm in size. *Acetoanaerobium* colonies are fluffy, rhizoid, opaque, and granular. Colonies should be picked with either thin Pasteur pipettes or with a platinum inoculation loop, and purified again. Only after at least two subsequent dilution/streaking processes, can the resulting cultures be regarded as pure. Purity should be checked after growth in a complex medium such as "AC-medium" (Difco). However, in our experience, AC-medium prepared by the recommended recipe is so concentrated that it inhibits many sediment bacteria so that it should be used not only at its original concentration but also at a 1:5 or 1:10 dilution, provided that the agar content is corrected with additional agar. Instead of AC-medium, complex medium containing 0.1% (w/v) each of malt extract, yeast extract, and peptone can also be used.

Identification

As indicated in Table 2, cells of the various species of the genus *Acetobacterium* are basically all of the same size. In phase contrast photomicrographs, they all appear as straight, slightly pointed rods of varying length, very often in pairs or short chains. The single cells resemble an American football and are very easy to recognize; they do not differ significantly among the various species (Fig. 1). Identification of the various species has to be based mainly on the respective substrate ranges utilized, as outlined in Table 2. Cells of *Acetogenium kivui* are shown in Fig. 2. They are very similar to *Acetobacterium* cells, although the ends are less pointed. The shape and size of *Acetobacterium* cells can vary considerably with the growth conditions and the age of a culture: This is exemplified in Fig. 3 for *A. woodii* grown under various conditions. Cells of varying length are often encountered in cell chains; small cells in such chains can be coccoid (Fig. 3b and c).

The cells with swollen centers (Fig. 3b and c) that tend to appear after growth with sugars or in rich media have repeatedly been misinterpreted as stages of spore formation. The idea that *Acetobacterium* might be a degenerated *Clostridium* that has lost parts of the differentiation apparatus necessary for formation of real endospores is still intriguing; this idea is supported by the obvious relationship of these

Table 2. Cell size and substrates utilized for growth of seven homoacetogenic species.

Property	Acetobacterium				Acetogenium kivui	Acetoanaerobium noterae	Acetitomaculum ruminis
	woodii	wieringae	carbinolicum	malicum			
Cell size (µm)	1.0 × 2.0	1.0 × 2.0	0.8–1.0 × 1.5–2.5	1.0–1.3 × 1.8–4.0	0.7–0.8 × 2.0–7.5	0.8 × 1.0–5.0	0.8–1.0 × 2.0–4.0
Substrates used							
H₂/CO₂	+	+	+	+	+	+	+
Formate	+	+	+	(+)	+	−	+
Methanol	+	−	+	−	−	−	−
3,4,5-TMB[a]	+	−	+	+	ND	ND	ND
3,4,5-TMC[a]	+	−	+	+	ND	ND	ND
Ethanol	−[b]	−[c]	+	−	−	ND	ND
Propanol	−[c]	−[c]	+	−	−	ND	ND
Butanol	−[c]	−[c]	+	−	−	ND	ND
Pentanol	−	−	+	−	ND	ND	ND
Ethylene glycol	+	+	+	±	ND	ND	ND
1,2-Propanediol	+	+	+	+	ND	ND	ND
2,3-Butanediol	+	−	+	−	ND	ND	ND
Acetoin	+	+	+	+	ND	ND	ND
Glycerol	+	+	+	+	−	ND	−
Fumarate	−	−	−	+	−	−	ND
L-Malate	−	−	−	+	−	−	−
DL-Lactate	±	+	+	+	+	−	−
Pyruvate	+	−	+	+	+	−	−
Glucose	+	−	+	−	+	+	+
Fructose	+	+	+	±	+	−	±
Ethanolamine	−	+	−	−	ND	ND	ND
Choline	−	+	−	−	ND	ND	ND
Betain	+	+	+	+	ND	ND	−

+, good growth; (+), weak growth; ±, not all strains grow; −, no growth.

[a] TMB, trimethoxybenzoate; TMC, trimethoxycinnamate.

[b] Growth is possible only with at least 100 mM bicarbonate in the medium (Buschhorn et al., 1989).

[c] Substrate oxidation without growth occurs only in media with at least 100 mM bicarbonate (Buschhorn et al., 1989).

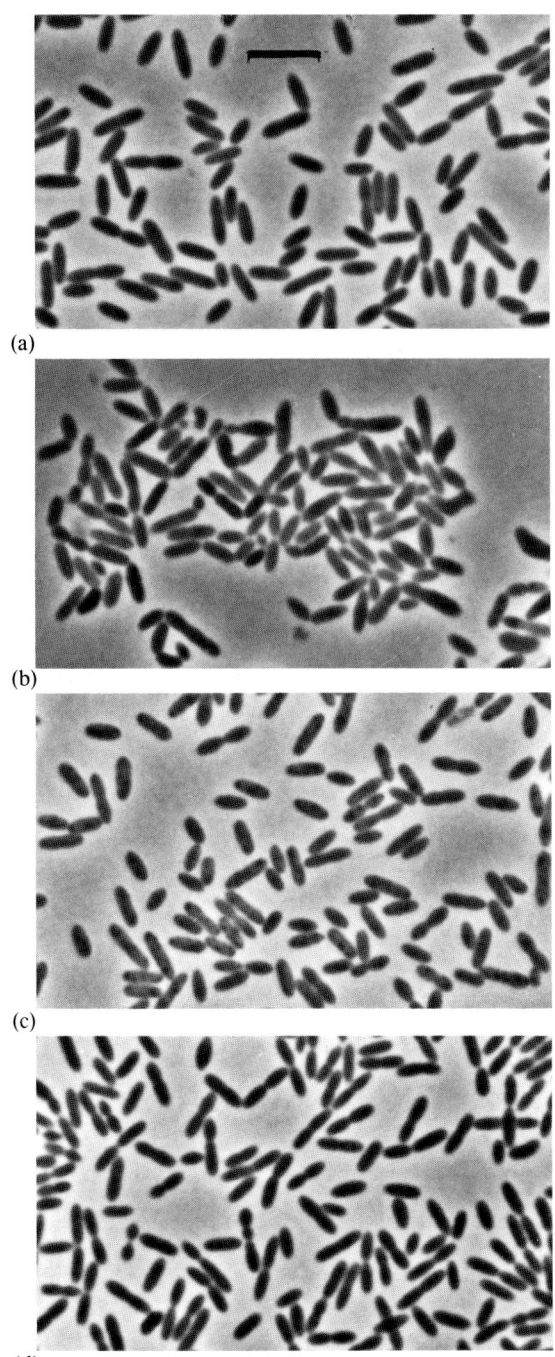

(a)

(b)

(c)

(d)

Fig. 1. Phase contrast photomicrographs of cells of *Acetobacterium* species grown in mineral medium with 10 mM glycerol as sole substrate. (a) *A. woodii;* (b) *A. wieringae;* (c) *A. carbinolicum;* (d) *A. malicum.* Bar = 5 μm.

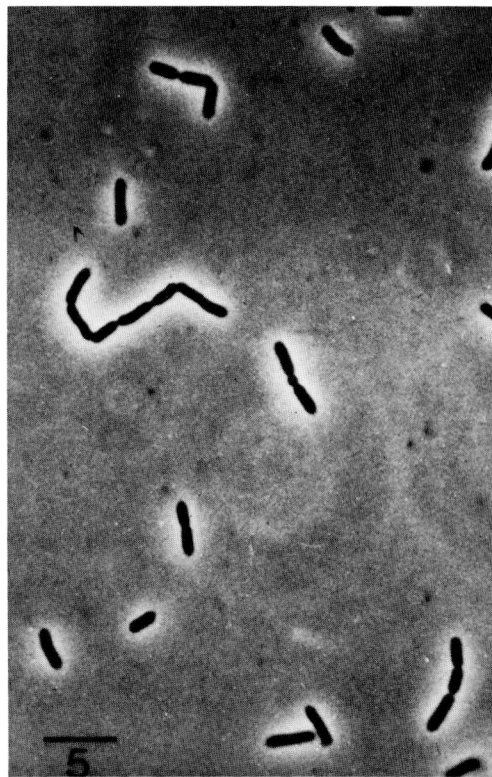

Fig. 2. Phase contrast photomicrograph of cells of *Acetogenium kivui.* Bar = 5 μm.

strains from pasteurized natural sediment samples have always failed.

The granular inclusions that are visible in phase contrast or electron microscopic pictures have been suggested several times to be storage material, e.g., poly-β-hydroxybutyrate, glycogen, or polyphosphates. However, a clear-cut chemical characterization of these materials is still missing. Accumulation of storage materials could help to explain the unusually low thresholds for hydrogen of *A. woodii,* in comparison to that of most methanogens or sulfate reducers (Seitz et al., 1988).

Young cells in fresh cultures of all *Acetobacterium* species are motile in a tumbling manner. Single, subpolar flagella were detected in electron micrographs of negatively stained cells (Balch et al., 1977; Mayer et al., 1977; Braun and Gottschalk, 1982). *Acetoanaerobium noterae* has 3–4 peritrichous flagella (Sleat et al., 1985), and motility has also been reported for some *Acetitomaculum ruminis* strains. Motility can be lost entirely in aging cultures. Only *Acetogenium kivui* is definitively nonmotile.

The type species *Acetobacterium woodii* was described originally as Gram-positive (Table 1). However, in conventional Gram-staining, *Acetobacterium* often behaves like a Gram-negative

bacteria and of *Eubacterium* species to certain *Clostridium* species (Tanner et al., 1981). However, no reliable proof for the presence of typical, thermoresistant endospores has ever been provided for *Acetobacterium* cells, and efforts in our laboratory to isolate *Acetobacterium*

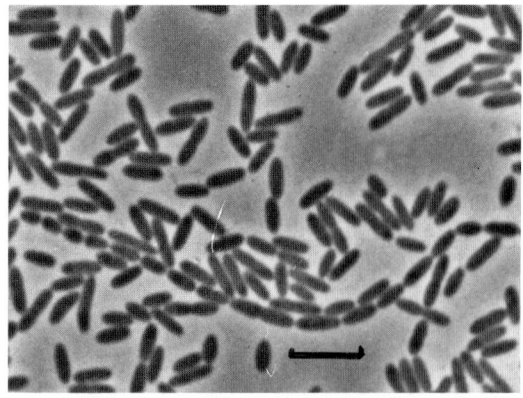

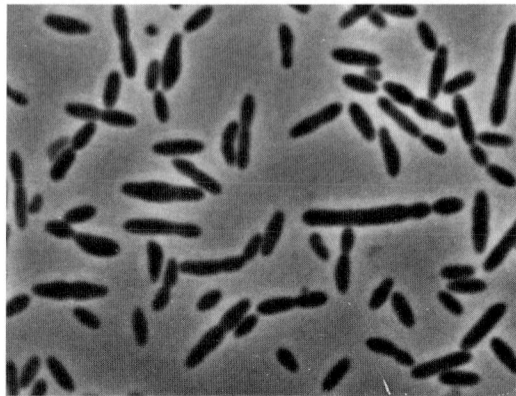

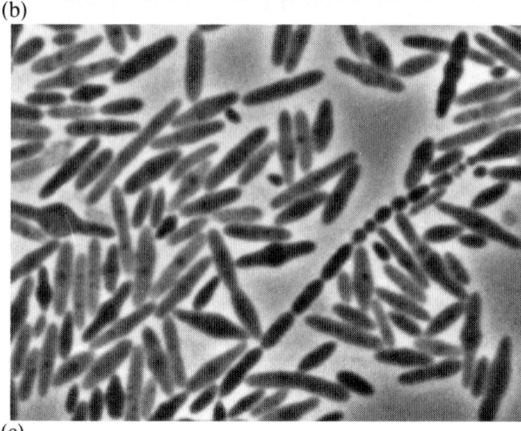

Fig. 3. Phase contrast photomicrograph of *A. woodii* cells grown in various media. Bar = 5 μm. (a) cells grown with 5 mM trimethoxybenzoate; (b) cells grown with 10 mM fructose; (c) old culture of cells grown with 10 mM fructose.

bacterium if the decolorization step is slightly extended and if a *Bacillus* or *Lactobacillus* strain with a very thick peptidoglycan layer is used as positive control. The peptidoglycan layer in the cell wall of *A. woodii* is comparatively thin and it is covered with an outer membrane of regularly arranged particles (Mayer et al., 1977). Similar cell wall structures have also been found in *Acetogenium kivui* (Rasch et al., 1984). These cell walls do not retain the crystal

violet-iodine complex efficiently and can easily be washed free of it. Therefore, *Acetobacterium* itself is an excellent Gram-positive control bacterium to ensure that the differentiation process is not extended too long if it still appears Gram-positive while an organism such as *E. coli* is Gram-negative, one can be sure that the decolorization step has been properly carried out. This is emphasized here in detail because both *Acetogenium* and *Acetoanaerobium,* as well as an unidentified, *Acetobacterium*-like homoacetogen (Samain et al., 1982), were originally described as Gram-negative, but careful electron microscopic reexamination revealed later that they actually have a Gram-positive cell wall (see Table 1). From this point of view, at least *Acetoanaerobium noterae* could be grouped with the genus *Acetobacterium.*

The peptidoglycan type of *A. woodii* is of the rather unusual crosslinkage type B, in which ornithyl residues function as interpeptide bridges, and the L-alanyl residue in position 1 of the peptide subunit is replaced by a L-seryl residue (Kandler and Schoberth, 1979). Peptidoglycan of the same type was also detected in *A. wieringae* (Braun and Gottschalk, 1982), *A. carbinolicum* (Eichler and Schink, 1984), and *Clostridium barkeri* and *Eubacterium limosum* (Schleifer and Kandler, 1972), again indicating a close resemblance between these groups of strict anaerobes.

Physiology

Homoacetogenic bacteria are experts in the utilization of one-carbon compounds. Moreover, they can carry out incomplete oxidations of reduced fermentation products released by other fermenting bacteria, and they can grow with monomeric sugars which are fermented completely to acetate with a comparably high ATP yield. The substrates utilized by the various species of mesophilic, nonsporeforming homoacetogens treated in this chapter are listed in Table 2. The data given are based on the respective original descriptions. Additional growth data for *A. woodii* and *A. wieringae* were provided in the descriptions of *A. carbinolicum* (Eichler and Schink, 1984) and *A. malicum* (Tanaka and Pfennig, 1988). The lists of growth characteristics of the other genera are relatively incomplete so far.

Most of the substrates listed are oxidized to acetate as sole product. This is true for all sugars tested; the list given in Table 2 does not include all sugars tested for growth with the various strains. Methoxylated aromatic compounds are demethylated to the corresponding phenols; the methyl residue is fermented to acetate, analogous to methanol. Some *A. woodii* strains reduce the acrylic acid side chain of caffeate derivates to the corresponding dihydrocaffeates; this process appears to be associated with electron transport phosphorylation (Tschech and Pfennig, 1984; Hansen et al., 1988). The published *Acetobacterium* strains demethylate only methoxylated aromatic compounds; cleavage of the methyl alkyl ether methoxyacetate was demonstrated only recently with a new *Acetobacterium* isolate (Schuppert and Schink 1990). Betain is demethylated to dimethylglycine, but no further. It still remains to be elucidated whether the demethylation reaction is carried out by the same enzyme system that also demethylates the phenylmethyl ethers.

Tests for utilization of methanol have mostly been carried out with excessively high substrate concentrations, yielding negative results. *A. woodii* was first reported to be unable to grow with this substrate (Balch et al., 1977); reevaluation of this finding revealed good growth with 5–10 mM methanol (Bache and Pfenning, 1981).

Ethanol, propanol, and butanol select for enrichment of *A. carbinolicum* from freshwater and brackish water sediments, and they are converted to the corresponding fatty acids with concomitant formation of acetate from carbon dioxide (Eichler and Schink, 1984). The other *Acetobacterium* species do not grow with these alcohols, in accordance with their original descriptions and later checks. It was recently found that *A. woodii* and *A. wieringae* can use these alcohols if the medium contains at least 100 mM bicarbonate buffer (Buschhorn et al., 1989). The ecological significance of this finding is questionable since sediments usually do not contain more than 60 mM bicarbonate. At limiting phosphate concentrations, *A. woodii, A. wieringae,* and *A. carbinolicum* fermented excessive glucose (50 mM) mainly to acetate and also to ethanol and alanine as side products that could later be reoxidized (Buschhorn et al., 1989).

Ethylene glycol and 1,2-propanediol are fermented to acetate or to acetate and propionate by *Acetobacterium* species. Fermentation of glycerol can lead to acetate as sole product (Eichler and Schink, 1984) or to acetate together with 1,3-propanediol, neither of which can be utilized any further (Emde and Schink, 1987). 1,3-Propanediol is probably formed via an unspecific diol dehydratase that also attacks the other diols mentioned.

Fatty acids with more than two carbon atoms cannot be oxidized by homoacetogenic bacteria. Therefore, the oxidation of propionate or bu-tyrate with carbon dioxide to acetate as sole fermentation product is an equilibrium reaction that cannot be coupled to ATP formation or growth.

Malate is utilized only by *A. malicum;* fumarate by none of the described species. *A. malicum* is also of interest for its ability to grow with methoxyethanol, an important industrial solvent. The substrate spectrum and the products formed indicate that this substrate is not degraded via demethylation to ethylene glycol but rather via a diol dehydratase-analogous reaction releasing methanol and acetaldehyde; the latter is further oxidized to acetate (Tanaka and Pfennig, 1988).

Catalase or oxidase activity has never been reported for any strain of homoacetogenic bacteria. The cytochromes that were found in *C. thermoaceticum* and *C. formicoaceticum* (Gottwald et al., 1975) have not been detected in any strain of the nonsporeforming homoacetogens treated in this chapter; however, definitive studies were carried out only with *A. woodii* (Tschech and Pfennig, 1984), *A. carbinolicum* (Eichler and Schink, 1984), and *A. malicum* (Tanaka and Pfennig, 1988).

All known *Acetobacterium* species and many other homoacetogenic bacteria have been found to be able to fix molecular nitrogen (M. Bomar and B. Schink, unpublished observations). The ecological importance of nitrogen fixation in the usually ammonia-rich environments typically inhabited by these bacteria is still a matter of speculation.

Applications

Homoacetogenic bacteria have been suggested as suitable catalysts for the formation of acetic acid from sugar and ethanol-containing waste materials (Zeikus, 1980). Since they form 3 mol acetate per mol hexose, instead of only 2 mol, as a classical yeast/*Acetobacter* combination would do, homoacetogenic bacteria should be more efficient for such applications. However, their free energy yield in the total conversion process is low, and most steps in acetate formation are reversible reactions that operate close to the thermodynamic equilibrium. This is exemplified clearly by the recently isolated thermophilic homoacetogen which can run the whole reaction chain to and from acetate in both directions with concomitant ATP formation (Lee and Zinder, 1988). Determination of pH values inside the cell as well as external pH values have shown that the internal pH follows the external one closely, both in the homoacetogenic *A. wieringae* and the aerobic *Acetobac-*

ter aceti (Menzel and Gottschalk, 1985). At pH 5.0, the acetic acid formed uncouples the membrane potential by free exchange across the cytoplasmic membrane (Baronofsky et al., 1984). Under these conditions, the energy metabolism of the homoacetogen is severely affected, whereas the aerobic acetic acid bacterium still has sufficient energy sources due to mainly irreversible ATP-yielding reactions (Menzel and Gottschalk, 1985). Homoacetogens are therefore less efficient in operation against high external acetate concentrations. A maximum concentration of 625 mM acetate ($=3.8\%$ v/v) was reached in a pH-controlled fermenter culture with the thermophilic organism *Acetogenium kivui* (Klemps et al., 1987). Only under these conditions *A. kivui* can grow at as high an acetate concentration as the aerobic *Acetobacter* species do. It appears from this comparison that the advantage of product efficiency of the anaerobe against the aerobe may be compensated by the thermodynamic and kinetic shortcomings of the less-exergonic anaerobic metabolism. Homoacetogens were recently found to also be able to reductively dechlorinate tetrachloromethane (Egli et al., 1988). It still remains to be determined whether they can be used efficiently in the degradation of such chemicals in industrial or municipal wastewater treatment systems.

Acknowledgments

The authors are thankful for the helpful contributions of John Leigh concerning *A. kivui,* as well as critical comments to this whole manuscript. They also gratefully acknowledge numerous fruitful discussions they had with N. Pfennig, R. Thauer, G. Diekert, and R. Conrad on the physiology, ecology, and biochemistry of homoacetogens.

Literature Cited

Andreesen, J. R., G. Gottschalk, and H. G. Schlegel. 1970. *Clostridium formicoaceticum* nov. spec. Isolation, description and distinction from *C. aceticum* and *C. thermoaceticum*. Arch. Mikrobiol. 72:154–174.

Bache, R., and N. Pfennig. 1981. Selective isolation of *Acetobacterium woodii* on methoxylated aromatic acids and determination of growth yields. Arch. Microbiol. 130:255–261.

Balch, W. E., S. Schoberth, R. S. Tanner, and R. S. Wolfe. 1977. *Acetobacterium*, a new genus of hydrogen-oxidizing, carbon dioxide-reducing, anaerobic bacteria. Int. J. Syst. Bacteriol. 27:355–361.

Balch, W. E., G. E. Fox, L. J. Magrum, C. R. Woese, and R. S. Wolfe. 1979. Methanogens: reevaluation of a unique biological group. Microbiol. Rev. 43:260–296.

Barker, H. A., and V. Haas. 1944. *Butyribacterium,* a new genus of Gram-positive, non-sporulating anaerobic bacteria of intestinal origin. J. Bacteriol. 47:301–305.

Baronofsky, J. J., W. J. A. Schreurs, and E. R. Kashket. 1984. Uncoupling by acetic acid limits growth of and acetogenesis by *Clostridium thermoaceticum*. Appl. Environ. Microbiol. 48:1134–1139.

Braun, M., and G. Gottschalk. 1982. *Acetobacterium wieringae* sp. nov., a new species producing acetic acid from molecular hydrogen and carbon dioxide. Zbl. Bakt. Hyg., I. Abt. Orig. C3:368–376.

Braun, M., S. Schoberth, and G. Gottschalk. 1979. Enumeration of bacteria forming acetate from H_2 and CO_2 in anaerobic habitats. Arch. Microbiol. 120:201–204.

Breznak, J. A., and J. M. Switzer. 1986. Acetate synthesis from H_2 plus CO_2 by termite gut microbes. Appl. Environ. Microbiol. 52:623–630.

Buschhorn, H., P. Dürre, and G. Gottschalk. 1989. Production and utilization of ethanol by the homoacetogen *Acetobacterium woodii*. Appl. Environ. Microbiol. 55:1835–1840.

Conrad, R., F. Bak, H. J. Seitz, B. Thebrath, H. P. Mayer, and H. Schütz. 1989. Hydrogen turnover by psychotrophic homoacetogenic and mesophilic methanogenic bacteria in anoxic paddy soil and lake sediment. FEMS Microbiol. Ecol. 62:285–294.

Cord-Ruwisch, R., and B. Ollivier. 1986. Interspecies hydrogen transfer during methanol degradation by *Sporomusa acidovorans* and hydrogenophilic anaerobes. Arch. Microbiol. 144:163–165.

Egli, C., T. Tschan, R. Scholtz, A. M. Cook, and Th. Leisinger. 1988. Transformation of tetrachloromethane to dichloromethane and carbon dioxide by *Acetobacterium woodii*. Appl. Environ. Microbiol. 54:2819–2824.

Eichler, B., and B. Schink. 1984. Oxidation of primary aliphatic alcohols by *Acetobacterium carbinolicum* sp. nov., a homoacetogenic anaerobe. Arch. Microbiol. 140:147–152.

Eichler, B., and B. Schink. 1985. Fermentation of primary alcohols and diols and pure culture of syntrophically alcohol-oxidizing anaerobes. Arch. Microbiol. 143:60–66.

Emde, R., and B. Schink. 1987. Fermentation of triacetin and glycerol by *Acetobacterium* sp. No energy is conserved by acetate excretion. Arch. Microbiol. 149:142–148.

Fontaine, F. E., W. H. Peterson, E. McCoy, M. J. Johnson, and G. J. Ritter. 1942. A new type of glucose fermentation by *Clostridium thermoaceticum* n. sp. J. Bacteriol. 43:701–715.

Gottwald, M., J. R. Andreesen, J. LeGall, and L. G. Ljungdahl. 1975. Presence of cytochrome and menaquinone in *Clostridium formicoaceticum* and *Clostridium thermoaceticum*. J. Bacteriol. 122:325–328.

Greening, R. C., and J. A. Z. Leedle. 1989. Enrichment and isolation of *Acetitomaculum ruminis* gen. nov., sp. nov.: acetogenic bacteria from the bovine rumen. Arch. Microbiol. 151:399–406.

Hansen, B., M. Bokranz, P. Schönheit, and A. Kröger. 1988. ATP formation coupled to caffeate reduction by H_2 in *Acetobacterium woodii* NZva16. Arch. Microbiol. 150:447–451.

Kandler, O., and S. Schoberth. 1979. Murein structure of *Acetobacterium woodii.* Arch. Microbiol. 120:181–183.

Klemps, R., S. M. Schoberth, and H. Sahm. 1987. Production of acetic acid by *Acetogenium kivui.* Appl. Microbiol. Biotechnol. 27:229–234.

King, G. M., M. J. Klug, and D. R. Loveley. 1983. Metabolism of acetate, methanol, and methylated amines in intertidal sediments of Lowes Cove, Maine. Appl. Environ. Microbiol. 45:1848–1853.

Krumholz, L. R., and M. P. Bryant. 1985. *Clostridium pfennigii* sp. nov. uses methoxyl groups of mono-benzenoids and produces butyrate. Int. J. Syst. Bacteriol. 35:454–456.

Krumholz, L. R., and M. P. Bryant. 1986. *Syntrophococcus sucromutans* sp. nov., gen nov. uses carbohydrates as electron donors and formate, monobezenoids or *Methanobrevibacter* as electron acceptor systems. Arch. Microbiol. 143:313–318.

Lajoie, S. F., S. Bank, T. L. Miller, and M. J. Wolin. 1988. Acetate production from hydrogen and [^{13}C] carbon dioxide by the microflora of human feces. Appl. Environ. Microbiol. 54:2723–2727.

Lee, M. J., and S. H. Zinder. 1988. Isolation and characterization of a thermophilic bacterium which oxidizes acetate in syntrophic association with a methanogen and which grows acetogenically on H_2-CO_2. Appl. Environ. Microbiol. 54:124–129.

Leigh, J. A., F. Mayer, and R. S. Wolfe. 1981. *Acetogenium kivui,* a new thermophilic hydrogen-oxidizing, acetogenic bacterium. Arch. Microbiol. 129:275–280.

Lorowitz, W. H., and M. P. Bryant. 1984. *Peptostreptococcus productus* strain that grows rapidly with CO as the energy source. Appl. Environ. Microbiol. 47:961–964.

Lovley, D. R., and M. J. Klug. 1983. Methanogenesis from methanol and methylamines and acetogenesis from hydrogen and carbon dioxide in the sediments of a eutrophic lake. Appl. Environ. Microbiol. 45:1310–1315.

Mayer, F., R. Lurz, and S. Schoberth. 1977. Electron microscopic investigation of the hydrogen-oxidizing acetate-forming anaerobic bacterium *Acetobacterium woodii.* Arch. Microbiol. 115:207–213.

Menzel, U., and G. Gottschalk. 1985. The internal pH of *Acetobacterium wieringae* and *Acetobacter aceti* during growth and production of acetic acid. Arch. Microbiol. 143:47–51.

Moller, B., R. Ossmer, B. H. Howard, G. Gottschalk, and H. Hippe. 1984. *Sporomusa,* a new genus of gram-negative anaerobic bacteria including *Sporomusa sphaeroides* spec. nov. and *Sporomusa ovata* spec. nov.. Arch. Microbiol. 139:388–396.

Moore, W. E. C., and E. P. Cato. 1965. Synonymy of *Eubacterium limosum* and *Butyribacterium rettgeri: Butyribacterium limosum* comb. nov. International Bulletin of Bacteriological Nomenclature and Taxonomy 15:69–80.

Pfennig, N. 1978. *Rhodocyclus purpureus* gen. nov. and sp. nov., a ring-shaped, vitamin B_{12} requiring member of the family Rhodospirillaceae. Int. J. Syst. Bacteriol. 28:283–288.

Phelps, T. J., and J. G. Zeikus. 1984. Influence of pH on terminal carbon metabolism in anoxic sediments from a mildly acidic lake. Appl. Environ. Microbiol. 48:1088–1095.

Prévot, A. R. 1938. Étude de systématique bactérienne. Ann. Inst. Pasteur (Paris) 60:285–307.

Prins, R. A., and A. Lankhorst. 1977. Synthesis of acetate from CO_2 in the cecum of some rodents. FEMS Microbiol. Lett. 1:255–258.

Rasch, M., W. O., Saxton, and W. Baumeister. 1984. The regular surface layer of *Acetogenium kivui:* some structural, developmental and evolutionary aspects. FEMS Microbiol. Lett. 24:285–290.

Samain, E., G. Albagnac, and H. C. Dubourguier. 1982. Characterization of a new propionic acid bacterium that ferments ethanol and displays a growth factor dependent association with a Gram-negative homoacetogen. FEMS Microbiol. Lett. 15:69–74.

Schink, B. 1984. *Clostridium magnum* sp. nov., a non-autotrophic homoacetogenic bacterium. Arch. Microbiol. 137:250–255.

Schink, B. 1987. Ecology of C_1-metabolizing anaerobes, p. 81–85. In: H. W. van Verseveld, J. A. Duine, (ed), Microbiol growth on C_1-compounds. Martinus Nijhoff Publ., Dordrecht, Netherlands.

Schink, B., and N. Pfening. 1982. Fermentation of trihydroxybenzenes by *Pelobacter acidigallici* gen. nov., sp. nov., a new strictly anaerobic, Gram-negative, nonsporeforming bacterium. Arch. Microbiol. 133:195–201.

Schink, B., and M. Stieb. 1983. Fermentative degradation of polyethylene glycol by a strictly anaerobic, nonsporeforming bacterium, *Pelobacter ventianus* sp. nov. Appl. Environ. Microbiol. 45:1905–1913.

Schleifer, K. H., and O. Kandler. 1972. Peptidoglycan types of bacterial cell walls and their taxonomic implications. Bacteriol. Rev. 36:407–477.

Schuppert, B., and B. Schink. 1990. Fermentation of methoxyacetate to glycolate and acetate by newly isolated strains of *Acetobacterium* sp. Arch. Microbiol. 153:200–204.

Seitz, H.- J., B. Schink, and R. Conrad. 1988. Thermodynamics of hydrogen metabolism in methanogenic cocultures degrading ethanol or lactate. FEMS Microbiol. Lett. 55:119–124.

Sleat, R., R. A. Mah, and R. Robinson. 1985. *Acetoanaerobium noterae* gen. nov., sp. nov.: an anaerobic bacterium that forms acetate from H_2 and CO_2. Int. J. System. Bacteriol. 35:10–15.

Stackebrandt, E., H. Pohla, R. Kroppenstedt, H. Hippe, and C. R. Woese. 1985. 16S rRNA analysis of *Sporomusa, Selenomonas,* and *Megasphaera:* on the phylogenetic origin of Gram-positive eubacteria. Arch. Microbiol. 143:270–276.

Tanaka, K., and Pfenning. 1988. Fermentation of 2-methoxyethanol by *Acetobacterium malicum* sp. nov. and *Pelobacter venetianus.* Arch. Microbiol. 149:181–187.

Tanner, R. S., E. Stackebrandt, G. E. Fox, and C. R. Woese. 1981. A phylogenetic analysis of *Acetobacterium woodii, Clostridium barkeri, Clostridium butyricum, Clostridium litusebureuse, Eubacterium limosum,* and *Eubacterium tenne.* Curr. Microbiol. 5:35–38.

Tschech, A., and N. Pfenning. 1984. Growth yield increase linked to caffeate reduction in *Acetobacterium woodii.* Arch. Microbiol. 137:163–167.

Wagener, S., and B. Schink. 1988. Fermentative degradation of nonionic surfactants by enrichment cultures and by pure cultures of homoacetogenic and propionate-forming bacteria. Appl. Environ. Microbiol. 54:561–565.

Widdel, F., and N. Pfennig. 1981. Studies on dissimilatory sulfate-reducing bacteria that decompose fatty acids. I.

Isolation of new sulfate-reducing bacteria enriched with acetate from saline environments. Description of *Desulfobacter postgatei* gen. nov., sp. nov. Arch. Microbiol. 129:395–400.

Widdel, F., G. W. Kohring, and F. Mayer. 1983 Studies on dissimilatory sulfate-reducing bacteria that decompose fatty acids. III. Characterization of the filamentous gliding *Desulfonema limicola* gen. nov., sp. nov., and *Desulfonema magnum* sp. nov. Arch. Microbiol. 134:286–294.

Wiegel, J., M. Braun, and G. Gottschalk. 1981. *Clostridium thermoautotrophicum* species novum (sic), a thermophile producing acetate from molecular hydrogen and carbon dioxide. Curr. Microbiol. 5:255–260.

Wieringa, K. T. 1936. Over net verdwijnen van waterstof en koolzuur onder anaerobe voorwarden. Antonie van Leeuwenhoek J. Microbiol. Serol. 3:263–273.

Wieringa, K. T. 1940. The formation of acetic acid from CO_2 and H_2 by anaerobic sporeforming bacteria. Antonie van Leeuwenhoek J. Microbiol. Serol. 6:251–262.

Winfrey, M. R., and D. M. Ward. 1983. Substrates for sulfate reduction and methane production in intertidal sediments. Appl. Environ. Microbiol. 45:193–199.

Zehnder, A. J. B., and W. Stumm. 1988. Geochemistry and biogeochemistry of anaerobic habitats, p. 1–38. In: A. J. B. Zehnder, (ed.), Biology of anaerobic microorganisms. John Wiley and Sons, New York.

Zeikus, J. G. 1980. Chemical and fuel production by anaerobic bacteria. Ann. Rev. Microbiol. 34:423–464.

Zinder, S. H., and M. Koch. 1984. Non-aceticlastic methanogenesis from acetate: acetate oxidation by a thermophilic syntrophic coculture. Arch. Microbiol. 138:263–272.

CHAPTER 88

The Genera *Mycoplasma, Ureaplasma, Acholeplasma, Anaeroplasma,* and *Asteroleplasma*

SHMUEL RAZIN

Introduction

The trivial term mycoplasmas is used for organisms comprising a very large group of prokaryotes distinguished phenotypically from other bacteria by their minute dimensions and total lack of cell walls (Razin and Freundt, 1984). Morphologically, mycoplasmas vary in shape from spherical or pear-shaped structures (0.3–0.8 μm in diameter) to branched or helical filaments. Genome replication precedes, but is not necessarily synchronized with, cell division. Thus, budding forms, filaments, and chains of beads may be observed (Fig. 1).

The total lack of a cell wall explains many of the unique properties of the mycoplasmas, such as sensitivity to osmotic shock and detergents, resistance to penicillin, and formation of peculiar fried-egg shape colonies (Razin and Oliver, 1961). Thin sections of mycoplasmas reveal that the cells are built essentially of three organelles: the cell membrane, ribosomes, and the characteristic prokaryotic genome. In this respect, the mycoplasmas are the closest to the concept of "minimal self-replicating cells" (Morowitz, 1984).

Taxonomically, the lack of cell walls is used to separate the mycoplasmas from other bacteria in a class named Mollicutes (*mollis,* soft; *cutis,* skin, in Latin). This class is the only one in the division Tenericutes, one of the four divisions of the kingdom Procaryotae, (Gibbons and Murray, 1978; Murray, 1984). The current classification of Mollicutes and the properties distinguishing the various taxa are presented in Table 1. This chapter deals with all the taxa of Mollicutes, except for the family Spiroplasmataceae, which is described in Chapter 89. While the trivial term mycoplasmas has been used to denote any species included in the Mollicutes, the trivial names acholeplasmas, ureaplasmas, anaeroplasmas, and spiroplasmas are commonly used for members of the corresponding genus. In order to keep the term mycoplasmas only for organisms included in the genus *Mycoplasma,* the trivial term mollicute(s) was introduced to describe all members of the class (Subcommittee, 1984). Nevertheless, the term mycoplasmas is still widely used in the broader sense and will be used as such in this chapter.

Since they are devoid of a cell wall, mycoplasmas are Gram-negative when stained by the Gram-stain. Most mycoplasmas are nonmotile and have no flagella. However, some *Mycoplasma* species, including the human and animal pathogens, *M. pneumoniae, M. genitalium, M. gallisepticum,* and *M. pulmonis,* exhibit gliding motility on liquid-covered surfaces (Bredt, 1979). In addition, *M. mobile,* isolated from fish gills (Kirchhoff et al., 1987a), is the fastest (2.0 to 4.5 μm/s) and most energetic glider, facilitating its use as a model for studies on factors influencing this peculiar type of motility. During gliding, *M. mobile* exhibits chemotactic and rheotactic behavior (Kirchhoff et al.,

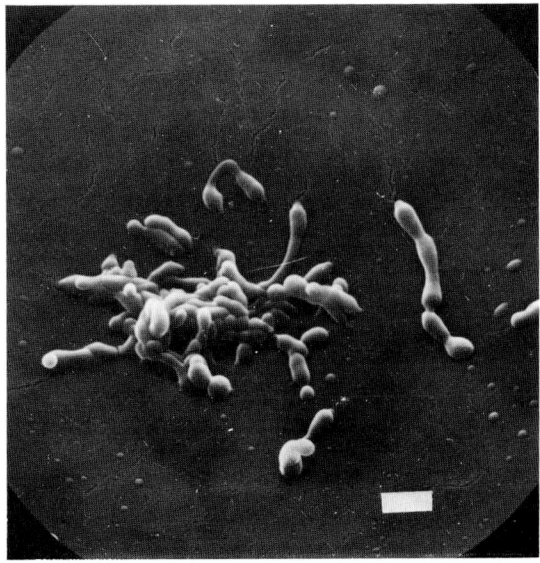

Fig. 1. Scanning electron micrograph of *Mycoplasma gallisepticum* cells showing the polymorphic shape of the organisms. Bar = 1 μm. (From Razin et al., 1980.)

Table 1. Taxonomy and properties of mycoplasmas (Class Mollicutes).

Classification	Current number of recognized species	Genome size (MDa)	GC content (mol%)	Cholesterol requirement	Distinctive properties	Habitat
Order I: Mycoplasmatales						
Family I: Mycoplasmataceae						
Genus I: *Mycoplasma*	87	400–800	23–41	+	−	Humans, animals, plants, insects
Genus II: *Ureaplasma*	5	500–700	27–30	+	Urease positive	Humans, animals
Family II: Spiroplasmataceae						
Genus I: *Spiroplasma*	11	1000	25–31	+	Helical filaments	Arthropods (including insects) and plants
Order II: Acholeplasmatales						
Family: Acholeplasmataceae						
Genus I: *Acholeplasma*	11	1000	27–36	−	−	Animals, plants, insects
Order III: Anaeroplasmatales						
Family: Anaeroplasmataceae						
Genus I: *Anaeroplasma*	4	1000	29–33	+	Obligate anaerobes	Bovine-ovine rumen
Genus II: *Asteroleplasma*	1	1000	40	−	(Oxygen-sensitive)	Bovine-ovine rumen

Adapted from Razin (1978).

1987b; Rosengarten et al., 1988b). Not only can the organisms glide on the surface of erythrocytes (Fischer et al., 1987), but they were actually observed to drag the erythrocytes while gliding on the glass surface (Rosengarten et al., 1988a). It is not surprising, therefore, that gliding motility requires energy (Rosengarten and Kirchhoff, 1987; 1988) and is apparently associated with cytoskeletal elements in the *M. mobile* cells (Piper et al., 1987).

The gliding *Mycoplasma* species usually possess a specialized tip structure (Fig. 2). These structures play an important role in adhesion of mycoplasmas to host cells (Razin, 1981, 1985b, 1986; Razin and Yogev, 1989; see "Disease Mechanisms", this chapter) and are also involved in the gliding motility of the mycoplasmas, since the leading direction of movement on the surface is always tip first (Bredt, 1979; Kirchhoff et al., 1987a). As will be described in Chapter 89, the helical filaments of spiroplasmas are usually motile with flexional and twitching movements, and often show an apparent rotatory motility.

Habitat

The mycoplasmas cultivated and identified so far are parasites of humans, animals, arthropods, and plants. The primary habitats of human and animal mycoplasmas are the mucous surfaces of the respiratory and urogenital tracts, and the eyes, alimentary canal, mammary glands, and joints in some animals (Razin and Barile, 1985). The obligatory anaerobic anaeroplasmas have so far been found only in the bovine and ovine rumen (Robinson, 1984). Spiroplasmas are widespread in the gut, hemocoel, and salivary glands of arthropods, and through sap-sucking insects the spiroplasmas may be introduced into the phloem tissues of plants and cause disease (see Chapter 89). Recent studies have indicated that sterol-nonrequiring *Acholeplasma* species as well as sterol-requiring *Mycoplasma* species can be found in the gut contents of a wide range of host insects, apparently forming the source for acholeplasmas isolated from plant surfaces (Tully et al., 1988; Williamson et al., 1990). It should be stressed that the *Acholeplasma* and *Mycoplasma* species isolated from insect guts are different from the species of these genera isolated from warm-blooded animals.

An apparently large group of mycoplasmas pathogenic for many plants and transmitted by insect vectors, is that of the poorly characterized Mycoplasma-like-organisms (MLOs). None of these organisms have been cultivated, but the application of monoclonal antibody and genetic probes (Lin and Chen, 1985; Kirkpatrick et al., 1987) indicates the genotypic diversity of the organisms included in this group.

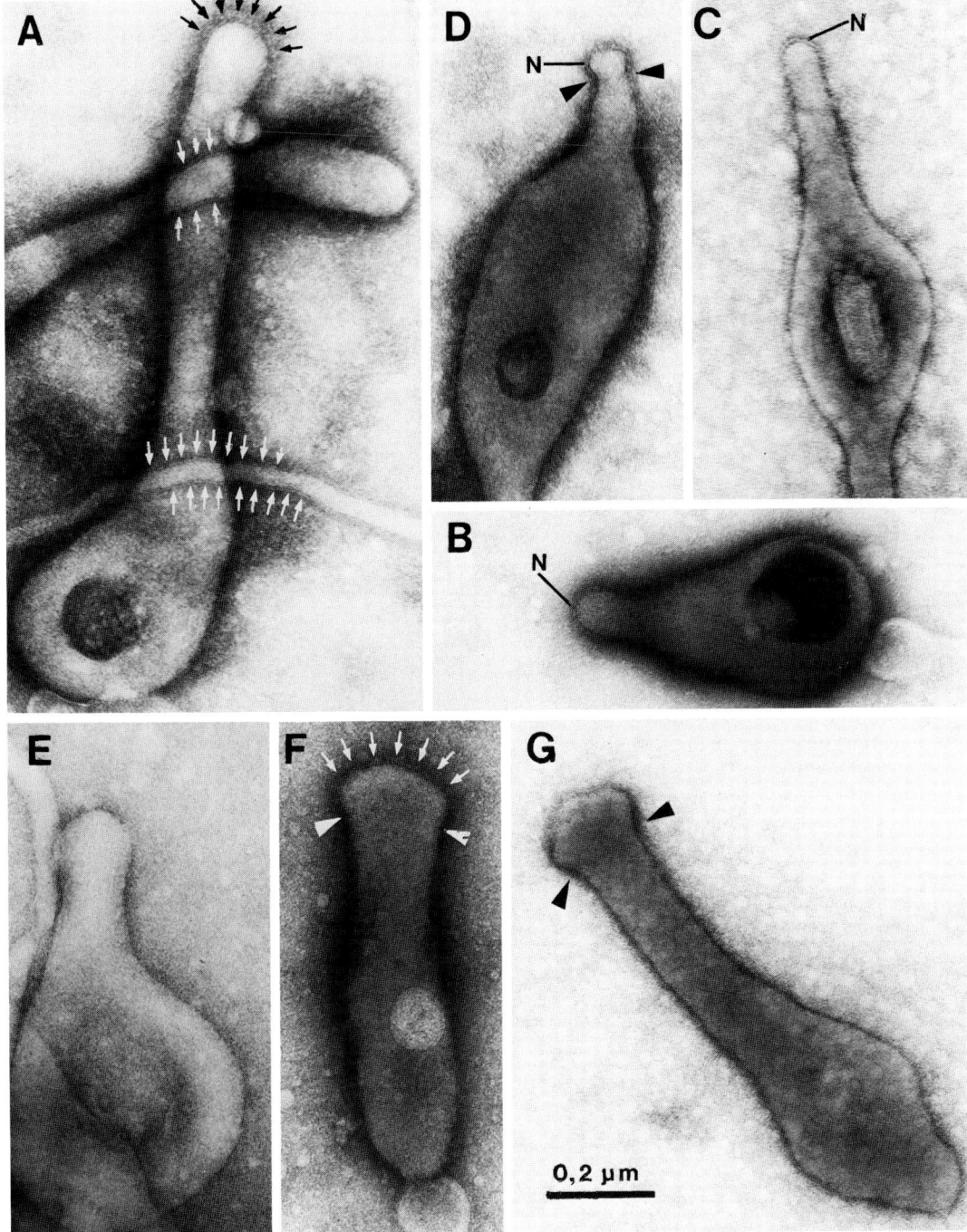

Fig. 2. Electron micrographs of negatively stained mycoplasmas with attachment tip structures. (A) *M. pulmonis* PG 34. Arrows show a thin surface layer on top of the protruding stalk. (B) *M. gallisepticum* PG 31, showing the distinct bleb structure. The particulate surface nap (N) is not restricted to the bleb structure. (C) *M. pneumoniae* FH. Note the tapered terminal structure covered by the nap (N). (D) *M. genitalium.* Arrowheads show the tip structure covered by a nap (N) consisting of small peplomer-like particles. (E) *M. alvi* Ilsley. The tip structure has no surface layer. (F) *M. sualvi* Mayfield B, with a broadened, blunt, terminal structure covered by a thin amorphous surface layer (arrows). (G) *M. mobile* strain 163K. Note the broadened headlike structure (arrowheads), with no surface layer. (All photos from Kirchhoff et al., 1984.)

Molecular Biology and Genetics

Major molecular properties distinguishing the Mollicutes from eubacteria are summarized in Table 2. The mycoplasmas have the smallest recorded genomes among self-replicating organisms. The reported genome sizes for members of the Mollicutes appeared to fall into two clusters: one composed of *Mycoplasma* and *Ureaplasma* species with a genome size of about 500 MDa, or approximately 750 kb, and the other of *Acholeplasma, Spiroplasma, Anaeroplasma,* and *Asteroleplasma* species, with a genome of about twice the size (Razin et al., 1983; Razin, 1985a). In most cases, determination of genome size was based on DNA renaturation kinetics (Carle and Bove, 1983). The few cases where measurements were done by direct electron microscopy of the surface-spread double-stranded genomic DNA, the values corresponded well with those obtained by renaturation kinetics. Some doubts about the validity of the above genome size data were recently raised by Pyle et al. (1988). They applied pulse-field electrophoresis to measure the size of fragments obtained by digestion of genomic DNA of several *Mycoplasma* and *Ureaplasma* species with restriction endonucleases. The sum of the sizes of the various fragments gave values that were considerably higher (900 to 1330 kb) than the expected 750 kb. These calculated high values of Pyle et al. (1988) were challenged by Maniloff (1989), who claimed that low GC mycoplasmal DNA may exhibit anomalous behavior in pulse-field electrophoresis, leading to slower than expected mobilities of the fragments and thus to incorrect calculated genome sizes. In addition, using a novel two-dimensional field-inversion gel electrophoresis has led Bautsch (1988) to conclude that the genome size of *M. mobile* is about 780 kb. Likewise, physical mapping of the *M. pneumoniae* genome by Wenzel and Herrmann (1988a) has also indicated that the genome size of this organism is about 800 kb. Yet, recent data (unpublished) confirm that some *Mycoplasma* and *Ureaplasma* species have genomes larger than 500 MDa (Table 1).

The minute size of the *Mycoplasma* and *Ureaplasma* genome implies a minimal number of genes. The estimated number of genes in species of these genera does not exceed 500, about one-fifth the number of genes in *Escherichia coli* (Muto, 1987). This is expressed by a small number of cell proteins (Kawauchi et al., 1982) and by the lack of many enzymatic activities and metabolic pathways, which is in accord with the parasitic mode of life and fastidious nature of mycoplasmas. Naturally, the small size of the mycoplasma genome makes it an attractive subject for complete deciphering of its genetic organization (Morowitz, 1984), and initial steps consisting of physical mapping of mycoplasmal genomes have already been taken (Pyle and Finch, 1988; Wenzel and Herrmann, 1988a).

The mycoplasma genome is not only very small, but its base composition is characterized

Table 2. Molecular properties distinguishing Mollicutes from other eubacteria.

Property	Mollicutes	Other eubacteria	References
Cell wall	Absent	Present	Razin (1969, 1981)
Peptidoglycan	Absent	Present	Razin (1969), Kandler and Konig (1985)
Genome size	500–1000 MDa	>1000 MDa	Razin et al. (1983)
GC content of genome	23–41 mol%	25–75% mol%	Razin and Tully (1983)
Number of detectable cell proteins	<400	>1000	Rodwell (1982), Kawauchi et al. (1982)
Number of rRNA operons	1–2	1–10	Razin (1985a)
5S rRNA length	104–113 nucleotides	>114 nucleotides	Rogers et al. (1985)
Number of elongation factor Tu *(tuf)* genes	1	1 or 2	Sela et al. (1989)
RNA polymerase	Rifampin resistant	Rifampin sensitive	Gadeau et al. (1986)
UGA codon usage	Tryptophan codon in *Mycoplasma* and *Spiroplasma* (not in *A. laidlawii*)	Stop codon	Yamao et al. (1985), Su et al. (1987), Renaudin et al. (1987), Tanaka et al. (1989)
DNA polymerase complex	One enzyme in *Mycoplasma* and *Ureaplasma*, three in *Acholeplasma* and *Spiroplasma*	Three enzymes	Mills et al. (1977), Maurel et al. (1989)

Adapted from Razin (1989).

by a low GC content (Razin, 1985a). The GC value of 24 mol% common to many mycoplasmas corresponds to the minimum theoretical value enabling coding for proteins with a normal amino acid composition (Elton, 1973). It is not surprising, therefore, that the adenine plus thymine (AT) content of mycoplasmal genes is high, and that of the intergenic spacer regions can reach values as high as 80 mol% AT (Muto et al., 1984; Muto and Osawa, 1987; Razin, 1985a). As a result, the mycoplasmas preferentially use codons rich in A and U in the third position (Muto et al., 1984; Ohkubo et al., 1987). Yet, computer analysis of the nucleotide sequence of the *M. pneumoniae* elongation factor *tuf* gene (Yogev et al., 1990), as well as the sequence of the P1 adhesin gene of *M. pneumoniae* (Su et al., 1987; Inamine et al., 1988a) indicates that the organism is utilizing nearly all codons of the genetic code. This finding appears to be in conflict with the low number of tRNA species found in mycoplasmas (Razin, 1985a). A possible answer to this apparent contradiction is provided by the finding that in *M. mycoides* at least some of the family codons are read by only one tRNA species, using an unconventional method that does not discriminate between the nucleotides in the third codon position, a finding similar to that reported for mitochondrial tRNA species (Samuelsson et al., 1987; Guindy et al., 1989).

A property of important phylogenetic and practical implications is the use by *Mycoplasma* and *Spiroplasma* species of UGA, the universal stop codon, as a tryptophan codon (Yamao et al., 1985; Schaper et al., 1987; Muto, 1987; Muto and Osawa, 1987; Su et al., 1987; Renaudin et al., 1987; Yamao et al., 1988; Bergemann et al., 1989). This outstanding property is shared by mycoplasmas and mitochondria and is thus of great phylogenetic interest. From the practical point of view, this property imposes a serious restriction on expression of cloned mycoplasmal genes in *E. coli*. Since *E. coli* regards UGA as a stop codon, translation of a mycoplasmal message in *E. coli* will stop where originally there should be tryptophan, so that mycoplasmal proteins expressed in *E. coli* may be truncated (Schaper et al., 1987; Trevino et al., 1986).

The question of whether the unconventional use of UGA as a tryptophan codon is shared by all the Mollicutes has recently been answered in the negative. *Acholeplasma laidlawii* was shown to have a single tryptophan tRNA species with an anticodon CCA that can translate only codon UGG, the normal tryptophan codon (Tanaka et al., 1989). This finding supports the notion that the change in the UGA codon as-

signment from stop to tryptophan occurred after separation of the *Spiroplasma* and *Mycoplasma* branches from the *Acholeplasma* branch (Rogers et al., 1985).

The regulatory signals of mycoplasmal gene expression appear to structurally resemble the corresponding eubacterial sequences. These include promoters, terminators (Razin, 1985a; Taschke and Herrmann, 1986, 1988; Christiansen, 1987a, 1987b; Gafny et al., 1988; Hyman et al., 1988; H. C. Hyman and S. Razin, unpublished observations), and Shine-Dalgarno sequences (Renaudin et al., 1987; Bové et al., 1989). Yet, the DNA-dependent RNA polymerases of mycoplasmas differ from those of eubacteria in being resistant to rifampin (Razin, 1985a), although the mycoplasmal enzyme has the same basic subunit structure as that of the eubacterial enzyme (Gadeau et al., 1986).

Study of the DNA polymerases of the Mollicutes also resulted in some interesting observations: while *Acholeplasma* and *Spiroplasma* species, with the 1000-MDa genomes, possess three DNA polymerases similar to eubacteria, *Mycoplasma* and *Ureaplasma* species have only a single enzyme (Mills et al., 1977; Maurel et al., 1989). This peculiar property of the mycoplasmas with a 500-MDa genome could, of course, be associated with the significant reduction in genome size that took place during the degenerative evolution of *Mollicutes* (see "Phylogeny," this chapter).

Studies on mycoplasma genetics were hindered at one time by the lack of adequate methods to genetically manipulate the organisms, including selectable markers (e.g., auxotrophic mutants), methods for DNA transfer (e.g., conjugation, transformation, and transduction), and vectors suitable for cloning (Razin, 1985a), but it has been demonstrated that an effective transposition into the *M. pulmonis* and *M. hyorhinis* chromosomes can be accomplished of the streptococcal transposon Tn916, carrying the tetracycline resistance determinant *tet*M (Dybvig and Cassell, 1987; Dybvig and Alderete, 1988). In fact, the *tet*M determinant was shown earlier to be responsible for tetracycline resistance in clinical isolates of *M. hominis* and *U. urealyticum* (Roberts et al., 1985; Roberts and Kenny, 1986). That *tet*M transfer could have occurred naturally by conjugation was demonstrated by Roberts and Kenny (1987). Spontaneous transfer of genes was also demonstrated by Barroso and Labarere (1988). Double resistance to vanadium oxide and arsenic acid could be transferred to *Spiroplasma citri* by facilitating contact of cells of two, single mutant strains with polyethylene glycol. This process seems to resemble gene transfer by fusion of protoplasts derived from Gram-positive bacteria.

The fact that Tn*916*, a transposon of Gram-positive bacteria, functions in mycoplasmas is perhaps not surprising in light of the widely accepted view that mycoplasmas are phylogenetically related to Gram-positive bacteria (see "Phylogeny," this chapter). Supporting this assumption is the finding that the nucleotide sequence of a small plasmid isolated from *M. mycoides* resembles in part the sequence of the staphylococcal plasmid pE194, encoding for erythromycin resistance (Bergemann et al., 1989). The above observations led Mahairas and Minion (1989a, 1989b) to attempt the insertion into the *M. pulmonis* genome of the staphylococcal transposon Tn*4001*, and a series of integrating plasmids constructed from cloned antibiotic resistance determinants, a plasmid-derived replicon and mycoplasmal chromosomal DNA. Successful transposition was mediated by polyethylene glycol. The finding that the plasmids integrated into the mycoplasma genome were stable in the absence of selection, and could be rescued in *E. coli* along with adjacent mycoplasmal DNA, led to the conclusion that they may be regarded as the first described *E. coli-M. pulmonis* shuttle vectors, and in this capacity may provide the tools for gene transfer from *E. coli* to mycoplasmas and among mycoplasmas themselves.

It could be expected that the great variety of plasmids and viruses infecting mycoplasmas (Razin et al., 1987; Maniloff, 1988) are responsible for chromosomal rearrangements and gene transfer. Integration of mycoplasma virus and plasmid DNA into the mycoplasma host chromosome is a rather common phenomenon (Dybvig and Maniloff, 1983; Mouches et al., 1984; Dickinson and Townsend, 1984; Nur et al., 1986, 1987; Ranhand et al., 1987). The existence of repetitive sequences in mycoplasma chromosomes (Ferrell et al., 1989; Taylor et al., 1988) can possibly be traced to integration of fragments of viral or plasmid DNA. Whether these repetitive sequences fulfil any biological function remains to be determined. Clearly, this mechanism constitutes a way by which genetic diversity in the Mollicutes species may occur. The recent finding of multiple copies of regions in the P1 cytadhesin gene of *M. pneumoniae* (see "Disease Mechanisms," this chapter) suggests additional mechanisms for gene conversion and chromosomal rearrangements in mycoplasmas (Su et al., 1988; Wenzel and Herrmann, 1988b).

Genes conserved during evolution serve as major phylogenetic markers (Woese, 1987) and as effective genetic probes for species and strain identification (see "Identification," this chapter). The rRNA genes have been the most studied conserved genes. The mycoplasma genome carries one or at most two copies of these genes, compared to 5–10 copies in most eubacteria (Amikam et al., 1984). The rRNA genes are usually linked in the mycoplasma in the classic order of eubacterial rRNA operons, i.e. 5′—16S—23S—5S—3′ (Sawada et al., 1981; Amikam et al., 1982, 1984), though in the case of *M. hyopneumoniae* the 5S rRNA gene is separated from the contiguous 16S and 23S genes (Taschke et al., 1986) and in *M. gallisepticum* the 16S rRNA gene is separated from the 23S and 5S genes in one of the two rRNA gene sets (Chen and Finch, 1989).

The *M. capricolum* rRNA genes (cloned in plasmid pMC5) have been widely used as probes for detecting mycoplasmas in contaminated cell cultures (Razin et al., 1984) and in Southern blots applied to distinguish species and strains (see "Identification," this chapter). The cloned rRNA genes themselves have been sequenced and the sequences used for the phylogenetic classification of the Mollicutes (see "Phylogeny," this chapter). Much of our knowledge about the regulatory signals of mycoplasmal genes also derives from studies on the rRNA operons (Taschke and Herrmann, 1988; Hyman et al., 1988).

The *tuf* gene is another conserved gene, coding for the elongation factor Tu, a protein that fulfills an essential role in protein synthesis in prokaryotes. Gram-negative bacteria carry two copies of this gene, while Gram-positive bacteria and mycoplasmas carry only one copy (Sela et al., 1989). The *tuf* gene of *M. pneumoniae* was recently cloned and its sequence revealed a high degree of homology with that of *tuf* genes of other prokaryotes, mitochondria, and chloroplasts (Yogev et al., 1990). As with the rRNA genes, the cloned *tuf* gene was shown to be an effective probe in Southern blot hybridization used to distinguish species and strains (see "Identification," this chapter).

Mention should be made of mycoplasmal tRNA genes, another class of highly conserved genes in prokaryotes for which relatively extensive information on gene structure, organization, and function, is available (Razin, 1985a; Rogers et al., 1987; Samuelsson et al., 1987). The genes of the proton-translocating ATPase complex of mycoplasmas also fall into the category of conserved genes (Rasmussen and Christiansen, 1987), as well as the *M. pneumoniae deo* C gene, coding for deoxyribose-phosphate aldolase, an enzyme acting in the pathway for catabolism of deoxyribonucleosides (Loechel et al., 1989).

Physiologic Properties

Based on their ability to metabolize carbohydrates, the mycoplasmas are divided into fermentative and nonfermentative organisms. Members of the fermentative group produce acid from carbohydrates, decreasing the pH of the growth medium. Those from the nonfermentative group may oxidize fatty acids and alcohols and do not significantly lower the pH of the culture medium. When arginine is present in large quantities, the nonfermentative mycoplasmas and some fermentative species use the arginine dihydrolase pathway and form ATP, CO_2, and ammonia, raising the pH of the culture medium (Razin, 1978). All the nonfermentative mycoplasmas lack hexokinase, phosphofructokinase, and aldolase, so that the Embden-Meyerhof pathway is not operating in these organisms (Desantis et al., 1989).

The mycoplasmas are usually facultative anaerobes, though the growth of some, such as *M. mycoides* and *M. pneumoniae*, appear to be improved by aeration. The *Anaeroplasma* and *Asteroleplasma* species are obligate anaerobes and are very sensitive to oxygen (Robinson, 1984; Robinson and Freundt, 1987). Extensive studies have definitely confirmed the lack of a complete tricarboxylic acid cycle from the many fermentative and nonfermentative species of mycoplasmas tested (Manolukas et al., 1988; Constantopoulos and McGarrity, 1987; Pollack et al., 1989).

All the mycoplasmas investigated so far have truncated respiratory systems and lack quinones and cytochromes, which rules out oxidative phosphorylation as an ATP-generating system (Razin, 1978). In fermentative mycoplasmas, ATP is formed during glycolysis, whereas in nonfermentative mycoplasmas, the arginine dihydrolase pathway has been proposed as a major source for cellular ATP. The presence of a potent carboxypeptidase in membranes of nonfermentative mycoplasmas suggests that this enzyme, by releasing arginine from peptides of the host, may supply the arginine substrate for the arginine dihydrolase pathway (Shibata and Watanabe, 1988).

Another possible mechanism for ATP generation in mycoplasmas is that proposed by Kahane et al. (1978), by which ATP is generated from acetylphosphate and adenosine-5'-diphosphate by acetate kinase, coupled with acetylphosphate formation from acetyl coenzyme A by phosphate acetyltransferase. Both enzymes are commonly found in fermentative and nonfermentative mycoplasmas.

The urea-splitting ureaplasmas constitute a special case. In these organisms neither glycolysis (Pollack, 1986; Robertson and Howard, 1987) nor arginine deiminase-ATP- or acetate kinase-ATP-generating pathways could be detected (Razin, 1978; Shepard and Masover, 1979). The unique dependence of ureaplasmas on urea for growth and the presence of a potent urease in these organisms (Shepard and Masover, 1979) has led to the hypothesis (Masover et al., 1977) that intracellular urea hydrolysis may be coupled to ATP synthesis through a chemiosmotic type of mechanism in which ammonium ions generated inside the cells by urease action diffuse across the membrane, producing an ion gradient and a membrane potential leading to ATP formation through the membrane-bound ATPase of this organism (Masover et al., 1977; Razin, 1978). Experimental support for this hypothesis was provided by Romano et al. (1980, 1986), who showed intracellular ATP formation in ureaplasma cells incubated with urea in buffer solution.

The extremely potent urease of *U. urealyticum* has been purified and characterized independently in several laboratories (Stemke et al., 1987; Precious et al., 1987; Blanchard et al., 1988; Saada and Kahane, 1988). The major subunit of this complex enzyme appears to be a polypeptide of about 70 kDa. The study of Thirkell et al. (1989) provided the most detailed information on the enzyme. It is composed of three subunits, 72, 14, and 11 kDa in mass. Since the active enzyme appears to have a molecular mass of about 190 kDa, the enzyme is probably built as a hexamer with equimolar ratios of the three subunits. The enzyme carries nickel, a property common to other ureases, both of eukaryotic and prokaryotic origin. Of great phylogenetic interest is the observation of Blanchard and Barile (1989), that the cloned *U. urealyticum* urease gene contains sequences homologous to those of urease genes from Gram-negative bacteria, adding this mycoplasmal gene to the list of genes conserved during evolution (see "Molecular Biology and Genetics," this chapter).

The membrane-associated adenosine triphosphatases (ATPases) are key enzymes in energy metabolism and solute transport in mycoplasmas. Most bacterial ATPases resemble the proton-translocating ATPases (F_1F_0) of mitochondria and chloroplasts, in which the F_1 moiety, responsible for hydrolytic activity, is located outside the lipid bilayer and can be easily detached from the membrane. In mycoplasmas, in contrast, the entire ATPase enzyme complex is associated with the membrane lipid bilayer and cannot be released unless detergents are applied to dissolve the membrane (Rottem and Razin, 1966; Ne'eman and Razin, 1975). Yet, the subunit of the mycoplasmal ATPase, which harbors the catalytic site, is highly conserved, as reflected by its antigenic cross-reactivity with antibody to the homologous β-subunit of the *E. coli* ATPase (Zilberstein et al., 1986) and by heterologous hybridization of the mycoplasmal ATPase genes with the cloned *E. coli atp* operon (Rasmussen and Christiansen, 1987).

Enzymes of purine and pyrimidine metabolism have been studied rather extensively in mycoplasmas with the aim of characterizing the metabolic pathways leading to nucleic acid precursor synthesis. In addition, the presence or absence of specific enzymes in the constructed pathways can be used as markers for taxonomic purposes (Cocks et al., 1988; McElwain et al., 1988).

Phylogeny

Being the smallest and simplest self-replicating prokaryotes, the mycoplasmas have attracted the attention of everyone interested in bacterial evolution and phylogeny. Are mycoplasmas the descendants of exceedingly primitive bacteria that existed before the development of a peptidoglycan based cell wall, or do they represent degenerate eubacterial forms that lost their cell walls? The balance of the molecular evidence, based largely on comparison of base sequences of the highly conserved 5S and 16S ribosomal RNA (rRNA) molecules, favors the hypothesis of degenerative evolution of mycoplasmas.

According to Woese and his colleagues (Woese et al., 1980; Rogers et al., 1985; Woese, 1987; reviewed by Razin, 1989) the mycoplasmas evolved as a branch of Gram-positive bacteria with genomes of low GC content, sharing the closest relatedness to two clostridia, *C. innocuum* and *C. ramosum* (Fig. 3). Yet, the great phenotypic and genotypic variability among mycoplasmas has led to the notion that they evolved from a variety of wall-covered bacteria; the mycoplasmas having accordingly a polyphyletic origin (Neimark, 1986). Those like Woese (1987) who support the evolution of mycoplasmas as a phylogenetic branch emerging from the progenitor(s) of the low GC Gram-positive bacteria explain the great variety of mycoplasmas as a process of rapid evolution.

Extensive sequencing of 16S rRNAs and their cloned genes obtained from about 50 species of mycoplasmas and their walled relatives (Weisburg et al., 1989) has led to the division of the Mollicutes into five phylogenetic units. These units are difficult to recognize on the basis of mutually exclusive phenotypic characters alone,

and do not correspond well with the conventional taxonomic units presented in Table 1.

Pathogenicity

Although some mycoplasmas are commensals and belong to the normal flora, many human and animal species are pathogenic, causing a variety of diseases that tend to run a chronic course. In humans, one species, *M. pneumoniae* has been established as the agent of primary atypical pneumonia, a common disease affecting mostly children and young adults. Post-infective complications, primarily in the nervous system, follow the pneumonia in some cases (Mansel et al., 1989). *Ureaplasma urealyticum* has been incriminated as a possible agent of nongonococcal urethritis in the male (Taylor-Robinson, 1985), urinary stone formation (Grenabo et al., 1988), and *in utero* infections of infants (Waites et al., 1988; Cassell et al., 1988). Another genital mycoplasma, *M. hominis,* a common inhabitant of the human genital tract, was shown to cause salpingitis in the female (Taylor-Robinson, 1985).

Many *Mycoplasma* species are established pathogens of farm animals, causing contagious pleuropneumonia, mastitis, and conjunctivitis in cattle, goats, and sheep, and chronic respiratory disease and arthritis in swine, chicken, and laboratory animals (Razin and Barile, 1985). Mycoplasmas are responsible for most-troublesome infections of cell cultures. These infections are persistent, difficult to cure, and frequently hard to detect and diagnose (McGarrity and Kotani, 1985). The damage caused by these infections to research laboratories and to biotechnological industries employing cell cultures cannot be overemphasized. The origin of contaminating mycoplasmas is either in components of the cell culture medium, particularly serum, or from the mycoplasma flora of the technician's mouth, spread by droplet infection. Ways to prevent, diagnose, and eliminate cell culture infections are described in detail in *Methods in Mycoplasmology* (Barile et al., 1983).

Disease Mechanisms

Most mycoplasmas infecting humans and animals adhere tenaciously to the epithelial linings of the respiratory or urogenital tracts and rarely invade the tissues and blood stream. Hence, they may be considered as typical surface parasites (Razin, 1985b, 1986; Razin and Yogev, 1989). Adherence is firm enough to prevent the

Fig. 3. Phylogenetic tree for the mycoplasmas and other members of the Gram-positive bacteria; based upon 16S rRNA sequence comparisons. (From Woese, 1987.)

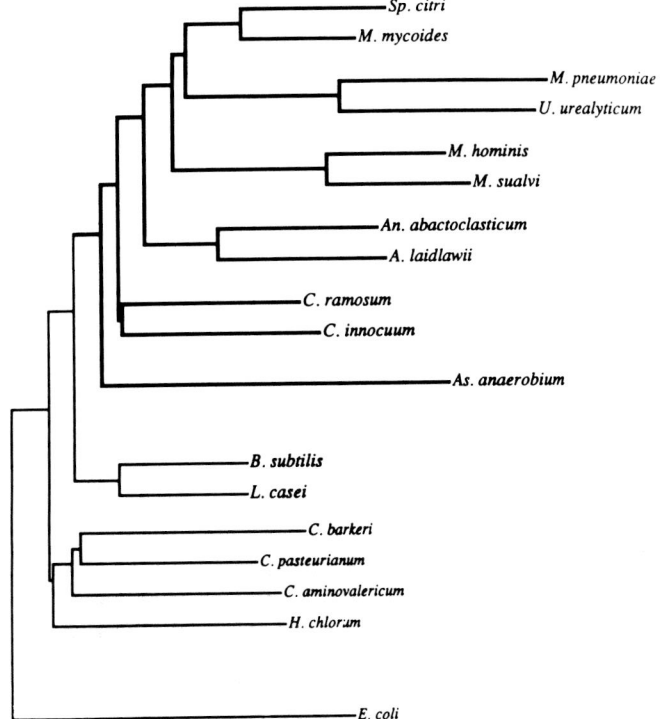

elimination of the parasites by the mucus secretions or by the urine stream. The lack of a cell wall from mycoplasmas and any of the appendages, like fimbriae, associated with adherence of other prokaryotes, indicate that mycoplasmal adhesins (*adhesin* is the parasitic cell component responsible for adhesion) must constitute part of the cell membrane. Moreover, the lack of a cell wall facilitates the contact of the mycoplasma membrane with that of the host. Although evidence supporting fusion of the two membranes is disputable, there are indications of antigenic exchange between the parasite and host membranes (Wise et al., 1978). Moreover, the intimate association between the adhering mycoplasmas and their host cells provides an environment in which local concentrations of mildly toxic by-products excreted by the parasite, such as H_2O_2 and ammonia, can build up and cause tissue damage (Razin, 1985b, 1986).

Extensive studies by Almagor et al. (1984, 1985, 1986) led to the formulation of the following pathological events of *M. pneumoniae* infection. The H_2O_2 and superoxide radicals (O_2^-) continuously generated by the adhering mycoplasmas penetrate into the host cell (Fig. 4). As a result of O_2^- accumulation, gradual irreversible inhibition of host cell catalase is induced, thereby causing intracellular H_2O_2 accumulation. This, in turn, may cause product inhibition of the host cell superoxide dismutase. The process is self-perpetuating and results in

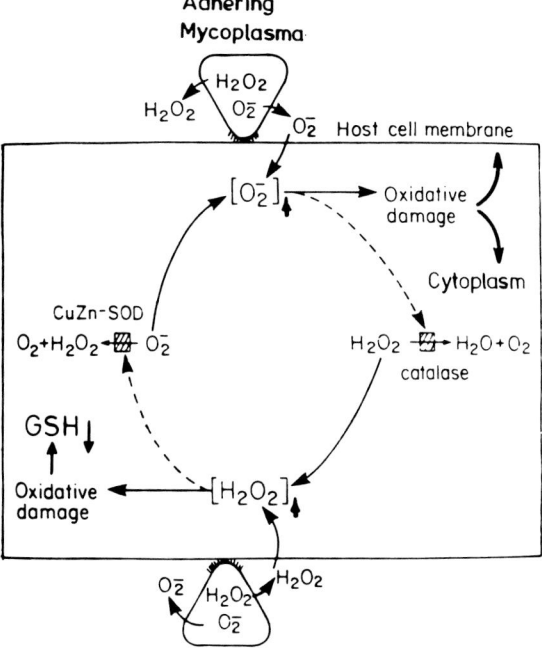

Fig. 4. Schematic presentation of the proposed mechanism for tissue damage by *M. pneumoniae*. GSH, glutathione; SOD, superoxide dismutase. (From Kahane, 1984.)

increased levels of O_2^- and H_2O_2, which induce progressive oxidative damage to vital cell constituents. Thus, malonyldialdehyde, a known oxidative product of membrane lipids, was shown to accumulate in cells infected by *M. pneumoniae* (Almagor et al., 1984).

Most significant strides have recently been made toward the definition of *M. pneumoniae* adhesins and the host cell receptors to which the bacteria adhere. A 170-kDa membrane protein, designated P1, was shown to cluster at the tip organelle of virulent mycoplasmas and function as the major adhesin of the pathogen (Baseman et al., 1982; Hu et al., 1982; Feldner et al., 1982). This protein is highly immunogenic and elicits a strong immune response in patients with primary atypical pneumonia and in experimental animals infected with *M. pneumoniae* (Hu et al., 1983; Leith et al., 1983; Vu et al., 1987; Jacobs et al., 1989). The P1 gene was recently cloned, sequenced, and the amino acid sequence of the corresponding protein was deduced (Su et al., 1987; Inamine et al., 1988a). The P1 gene is part of a three-gene operon (Inamine et al., 1988b). The large P1 protein apparently spans the mycoplasma membrane more than once (Su et al., 1987). As expected, monoclonal antibodies indicated a variety of epitopes on this large molecule (Dallo et al., 1988; Jacobs et al., 1989). It appears that the C-terminus of the P1 protein, but not the N-terminus, is involved in adherence, since antibodies to the C-terminus, but not to the N-terminus, inhibit adherence (Jacobs et al., 1987; Dallo et al., 1988).

The fact that P1 is a major immunogen and the findings that antibodies to it inhibit adherence have raised the hope that purified P1, or parts of it, may be used as a specific vaccinogen, protecting against *M. pneumoniae* infection. In order to pursue this direction of research, methods to obtain large quantities of pure P1 were needed. P1 had been isolated from *M. pneumoniae* cells by fractionation techniques (Jacobs et al., 1988b), but the yields were very low. Availability of the cloned P1 gene could obviously be used to produce the protein in transfected *E. coli*. However, the unusual usage by *M. pneumoniae* of the universal stop codon UGA as a tryptophan codon creates a problem (Schaper et al., 1987; Frydenberg et al., 1987; see also "Molecular Biology and Genetics," this chapter). In vitro conversion of the TGA triplets in the cloned P1 gene to the normal TGG tryptophan codon, or the use of suppressor *E. coli* strains that may overcome termination by UGA are possible ways to try and solve the problem. Nevertheless, findings by Jacobs et al. (1988a, 1988c), that guinea pigs preimmunized with purified P1 suffered from a more severe disease following challenge with the pathogen than nonimmunized animals, cast serious doubts about the effectiveness of a vaccine made of pure P1 protein.

This last disappointing finding points to the role of host reactions in pathogenesis of mycoplasma infections. The host may be responsible to a large extent for pneumonia symptoms through monitoring a local immunocyte and phagocytic response to the mycoplasmas (Fernald, 1979; Barile et al., 1988a, 1988b; Jacobs et al., 1988a, 1988c). An immunopathologic mechanism has also been suggested to explain the complications affecting organs distant from the respiratory tract in some *M. pneumoniae* patients. A variety of autoantibodies were detected in the serum of many of these patients, including cold agglutinins reacting on the erythrocyte I antigen, antibodies reacting with lymphocytes, smooth muscle, and antibodies to brain and lung antigens (Biberfeld, 1985). In addition, mycoplasmas are known to be potent activators of lymphoid cells and macrophages, a factor that should also be considered in any discussion of the mechanisms of disease by mycoplasmas (Cole et al., 1985).

Yet the protective effect of antibodies against mycoplasma infections should not be underestimated. Thus, resistance to bovine pleuropneumonia can be transferred by convalescent antiserum, and a very effective live vaccine against bovine pleuropneumonia consisting of *M. mycoides* organisms has been in use for many years (Barile, 1985; Barile et al., 1985). Secretory local antibodies such as IgA may be important in prevention of infection by interfering with the initial phase of mycoplasma colonization of the tracheal epithelium (Jacobs et al., 1988c). Moreover, though mycoplasmas can be phagocytized in the absence of specific antibody, the mycoplasmas remain viable within phagocytic cells and may be carried inside neutrophils to various parts of the body where they can initiate infection (Webster et al., 1988). Since specific antibodies by themselves inhibit mycoplasma growth in vitro, the role of antibodies is apparently to control growth of the mycoplasmas on mucosal surfaces, while neutrophils play no part in defense and may even aid dissemination of the infection. This helps to explain why patients with hypogammaglobulinemia are prone to systemic mycoplasma infections (Webster et al., 1988).

What is the nature of the receptors on the host cell surface responsible for the specific binding of the adhesins on the mycoplasma cell surface? While the nature of receptors has not been elucidated for most adhering mycoplasmas, involvement of host membrane glycoconjugates has long been implicated in the adhesion of *M. pneumoniae*, *M. gallisepticum*, *M. syn-*

oviae, and *M. genitalium* (Razin, 1985b, 1986). The rather intensive work carried out on receptors for *M. pneumoniae* has resulted in a relatively clear picture. The studies of Roberts et al. (1989) and Krivan et al. (1989) indicate that *M. pneumoniae* binds to two distinct types of receptors: one consists of sulfated glycolipids, and the other of 2–3-linked sialyl oligosaccharides on glycoproteins. Cultured cell lines and apparently the natural target tissue, the tracheal epithelium, carry the two receptor types on their surface. Hence, treatment of the tracheal epithelium with neuraminidase will affect only the sialylated type of receptors, but not the sulfated glycolipids, which explains the failure of neuraminidase treatment to completely block *M. pneumoniae* attachment (Razin, 1985b, 1986).

By using histochemical techniques with sequence-specific monoclonal antibodies and lectins directed to the specific carbohydrate moieties of receptors to *M. pneumoniae,* Loveless and Feizi (1989) have shown that the receptors are concentrated at the luminal aspects of the ciliated cells of human bronchial epithelium, but are absent on the secretory cells and the mucus they produce. These findings provide a biochemical basis for the way *M. pneumoniae* organisms evade the secreted mucus and adhere to the ciliated epithelium.

M. genitalium, first isolated from the urethral discharge of two men suffering from nongonococcal urethritis (Tully et al., 1981), has attracted much attention due to its possible pathogenicity to humans. Supporting this potential are the properties this mycoplasma shares with *M. pneumoniae,* including cell shape, cultural properties, motility, common antigens, cytadherence, cytopathogenicity, and pathogenicity to experimental animals (reviewed by Razin and Yogev, 1989). Yet, the suggestion that *M. genitalium* causes nongonococcal urethritis has not so far gained experimental support due to repeated failures to isolate the organism from additional urethral specimens. The recent isolation of *M. genitalium* from throat specimens (Baseman et al., 1988) points to another direction, suggesting that *M. genitalium* may colonize the respiratory tract and possible cause disease there.

The properties shared by *M. genitalium* and *M. pneumoniae* are apparently due to the existence of genomic sequences common to the two organisms (Yogev and Razin, 1986). Thus, the major adhesin of *M. genitalium,* named MgPa (a membrane protein of 140 kDa) shares regions of homology with the *M. pneumoniae* P1 adhesin, as detected at both the DNA and protein levels (Morrison-Plummer et al., 1987; Hu et al., 1987; Dallo et al., 1989a, 1989b; In-

amine et al., 1989). Whether or not *M. genitalium* is a human pathogen remains to be determined, but its sharing of genomic sequences and epitopes with *M. pneumoniae* and its apparent occurrence in the respiratory tract pose serious diagnostic problems (see "Identification," this chapter).

Culture

The limited biosynthetic abilities of mycoplasmas make them dependent on their hosts for the supply of many nutrients. This explains the great difficulties hampering the culture of mycoplasmas in the laboratory. There is a consensus that so far only a minor part of the mycoplasmas existing in nature has been cultivated. Thus, despite numerous efforts, none of the mycoplasma-like-organisms (MLOs) infecting plants has been cultivated (Lee and Davis, 1986; McCoy et al., 1989; see also Chapter 89). Some of the cultivable mycoplasmas, such as the human respiratory pathogen *M. pneumoniae,* grow very slowly (2–3 weeks incubation at 37°C), particularly on primary isolation. Some, like *Ureaplasma urealyticum,* grow very fast in vitro, but growth stops abruptly at titers of 10^6–10^7 colony-forming units per milliliter (cfu/ml) compared to 10^9–10^{10} cfu/ml in logarithmic cultures of wellgrowing mycoplasmas (Razin, 1978). *M. genitalium* grows so poorly in vitro that only a few successful isolations of it have been made so far (Tully et al., 1983; Baseman et al., 1988).

The complex media used for cultivation of mycoplasmas are usually based on beef heart infusion, peptone, yeast extract, and serum with various supplements (Table 3). The serum provides, among other nutrients, fatty acids, phospholipids, and cholesterol, required for membrane synthesis in an assimilable, nontoxic form (Rodwell and Mitchell, 1979; Razin, 1978). Serum may thus be substituted by liposomes made of phospholipids and cholesterol with bovine serum albumin to neutralize free fatty acid toxicity (Cluss and Somerson, 1986; Rodwell, 1983a).

Defined media were developed for only a few fermentative mycoplasmas. The media contain a complex assortment of amino acids, nucleic acid precursors, lipids, vitamins, inorganic ions, together with glucose as an energy source (Rodwell, 1983b; Hackett et al., 1987). Detailed composition of media for mycoplasmas can be found in *Methods in Mycoplasmology* (Razin and Tully, 1983). Table 3 provides the basic components of the most common culture media

Table 3. Composition of some common media for cultivation of mycoplasmas.

1. **Modified Hayflick medium (Freundt, 1983)**
   Supports good growth of the majority of *Mycoplasma* and *Acholeplasma* species

Heart infusion broth (Difco)	28.5 g
Deionized water	900 ml

   Sterilize by autoclaving (121°C for 20 min)
   Aseptically add the following sterile solutions:

Fresh yeast extract (25%, wt/vol)	100 ml
Horse serum	200 ml
Calf thymus DNA (0.2%, wt/vol)	12 ml
Thallium acetate (1%, wt/vol)	10 ml
Penicillin G (20,000 units/ml)	2.5 ml

   Adjust final pH to 7.8 with sterile NaOH solution.

2. **Medium SP-4 (Tully et al., 1977, 1979)**
   Supports growth of fastidious mycoplasmas, such as *M. pneumoniae* and *M. genitalium*

Mycoplasma broth base (BBL)	3.5 g
Tryptone (Difco)	10.0 g
Peptone (Difco)	5.3 g
Deionized water	615 ml

   Adjust pH to 7.8 and autoclave (121°C for 30 min).
   Aseptically add the following sterile solutions:

CMRL-1066 (10×, with glutamine, Gibco)	50 ml
Fresh yeast extract (25%)	35 ml
Yeastolate solution (2%, Difco)	100 ml
Fetal bovine serum (inactivated)	170 ml
Penicillin G (100,000 units/ml)	10 ml
Phenol red (0.1%)	20 ml

3. **U9C urea medium (Shepard and Lunceford, 1976)**
   Supports growth of ureaplasmas and indicates urea hydrolysis

Trypticase soy broth (BBL)	15.0 g
Magnesium chloride (MgCl$_2$·6H$_2$O)	0.2 g
Yeast extract (Difco)	1.0 g
Deionized water	900 ml

   Adjust pH to 5.5 and autoclave at 121°C for 15 min.
   Add aseptically the following sterile solutions:

Urea (10%, wt/vol)	3 ml
L-Cysteine-HCl (2%, wt/vol)	5 ml
GHL tripeptide (Gly-hist-lys, Calbiochem, 20 g/ml)	1 ml
Horse serum	100 ml
Penicillin G (100,000 units/ml)	10 ml
Phenol red (1% wt/vol)	1 ml

   The final pH of the medium should be 6.0±0.2.

4. **Serum-free medium (Tully, 1984)**
   For acholeplasmas and determination of sterol requirement

Mycoplasma broth base (BBL)	21 g
Deionized water	810 ml

   Adjust to pH 8.2 and autoclave (121°C for 30 min).
   Add aseptically the following filter sterilized solutions:

Fresh yeast extract (25%)	100 ml
Glucose (50%)	5 ml
Tween 80 (10%)	1 ml
Palmitic acid (10 mg/ml ethanol)	1 ml

   continued

Table 3. *Continued*

Bovine serum albumin (10%)	50 ml
Penicillin G (100,000 units/ml)	5 ml
Phenol red (0.1%)	20 ml

The final pH should be 7.8.

for *Mycoplasma, Ureaplasma,* and *Acholeplasma* species.

Although most mycoplasmas are facultative anaerobes, primary isolations frequently prefer anaerobic conditions provided by an atmosphere of 95% N$_2$ + 5% CO$_2$. For the obligate anaerobes, *Anaeroplasma* and *Asteroleplasma*, strict anaerobic conditions are required. The roll-tube technique is recommended with pre-reduced special culture medium maintained in a system excluding oxygen (Robinson, 1984).

The initial pH of the growth medium should be adjusted to about 8.0 for the fermentative mycoplasmas, and to 6.0–6.5 for the nonfermentative arginine-utilizing mycoplasmas and for the urea-hydrolysing ureaplasmas. Since the yield of mycoplasmas is usually very low, continuous culture of mycoplasmas in a pH-controlled chemostat has been applied to ureaplasmas (Masover et al., 1979) and to *M. mobile* (Krebs et al., 1989). The high cost of mycoplasma media and the danger of contamination are prohibitive factors, so that continuous culture systems are not commonly used.

The temperature range for mycoplasma growth varies according to species from about 20–40°C. The optimum temperature for most human and animal mycoplasmas is 36–37°C, while the optimum growth temperature for *M. mobile*, isolated from fish, is 25°C (Kirchhoff et al., 1987a). Likewise, the optimum temperature for growth of sterol-requiring mycoplasmas and sterol-nonrequiring acholeplasmas isolated from insects and plants lies between 25 and 30°C (Tully, 1989).

Preservation of Cultures

Cultures of mycoplasmas may retain viability at 4°C for weeks, particularly when glucose is omitted from the medium or when serum is replaced by bovine serum albumin and lipids (Taylor-Robinson and Behnke, 1987). However, prolonged storage of unfrozen cultures is unreliable, so that for long-term preservation, fast-freezing of young cultures to −70°C (Raccach et al., 1975) or freeze-drying are recommended. Detailed procedures have been described and evaluated by Leach (1983).

Identification

Mycoplasma identification has been based on the classical bacteriological tests, including morphology, cultural characteristics, physiological, and antigenic properties. A variety of tests based on the molecular analysis of genomic DNA, ribosomal RNAs, cell proteins, and lipids has been introduced (Razin, 1989). The first diagnostic kits based on specific DNA probes or monoclonal antibodies have already made their appearance, so that radical changes in identification procedures can be expected in the very near future.

The Committee on the Taxonomy of Mollicutes (a subcommittee of the International Committee on Systematic Bacteriology) issued recommended tests for mycoplasma identification as well as for description of new species of the class Mollicutes (Freundt et al., 1979). These recommendations have been followed rather strictly by mycoplasma taxonomists, improving considerably the quality of description of new species of the Mollicutes. The recommended tests include those required to define a new isolate at the higher taxonomic levels (class, order, and family) as well as tests for genus and species determination. Thus, a test for sterol requirement, based on growth promotion by cholesterol (Razin and Tully, 1970), is essential for separation of the sterol-nonrequiring *Acholeplasma* and *Asteroleplasma* species from all the other members of the Mollicutes that require sterol for growth. *Ureaplasma* identification is based on urea hydrolysis tests (Razin, 1983). Other key tests include those for sugar fermentation (Razin and Cirillo, 1983) and arginine hydrolysis (Barile, 1983). A great variety of serological tests have been employed, particularly in mycoplasma species and strain identification. The classical recommended tests include growth and metabolism inhibition by specific antisera (Clyde, 1983; Taylor-Robinson, 1983), as well as direct and indirect immunofluorescence tests applied to mycoplasma colonies (Gardella et al., 1983). More recently, more sensitive tests based on principles of enzyme-linked-immunosorbent assay (ELISA) (Kibe et al., 1985; Jacobs et al., 1986), immunoblotting (Bolske et al., 1987; Madson et al., 1987; Cimolai et al., 1988), immunobinding (Kotani and McGarrity, 1986), and immunoperoxidase tests (Imade et al., 1987) employing polyclonal or monoclonal antibodies, have been introduced. Due to the higher sensitivity and specificity of these tests, they are capable of identifying strains within a species.

The classical tests for mycoplasma identification and classification have been supplemented by a variety of tests based on genomic DNA analysis. While genome size determination (Carle and Bovè, 1983) has been critical for the classification of mycoplasmas at the higher taxonomic levels (Table 1), the procedures involved are cumbersome and suffer from a significant degree of variability (Bovè et al., 1989), so that genome size determination is not carried out routinely. On the other hand, determination of genomic GC content (for techniques see Carle et al., 1983; Kirchhoff and Flossdorf, 1987) has become an obligatory test required for definition of new Mollicutes species (Freundt et al., 1979). The easily obtained GC values are effective taxonomic measures. A difference in the GC content greater than 1.5–2.0 mol% between DNAs of two bacteria is considered sufficient to rule out their inclusion in the same species (Johnson, 1984). It should be emphasized, nevertheless, that identical or nearly identical GC values among microorganisms do not necessarily mean that the microorganisms are genetically related, because in many of these cases the base sequences, which determine genetic relatedness, differ significantly.

Since direct comparison of base sequences of entire prokaryotic genomes is still impractical, the proportion of common base sequences in DNAs of organisms tested for relatedness has been estimated indirectly by a variety of DNA-DNA or DNA-RNA hybridization techniques, the results being expressed as percent DNA homology. Considerable weight has been given to DNA homology values in determining genetic relatedness and establishing new species, as evidenced by classification of *Acholeplasma* (Stephens et al., 1983) *Anaeroplasma* (Stephens et al., 1985), and *Spiroplasma* (Bovè et al., 1989) species. Yet, data obtained by the various hybridization techniques may differ considerably (Razin, 1985a), stressing the need for standardization of procedures (Degorce-Dumas et al., 1983). Generally, the DNA homology data have supported the current speciation of mollicutes based on the classical tests. In addition, they revealed genetic heterogeneity within some established species, such as *M. hominis, A. laidlawii,* and *A. axanthum* (Razin, 1989).

DNA homology determinations are laborious and expensive. Consequently, simpler tests for assessing genetic relatedness have been investigated. Of these, electrophoresis patterns of mycoplasmal DNA digested by restriction endonucleases have gained wide use. The patterns produced by electrophoresis of the cleaved chromosomal fragments in agarose gels provide valuable information on the type and number of specific nucleotide sequences in the genome, and can be considered as genomic fingerprints, known also by the term restriction-fragment-length polymorphism (RFLP). The restriction patterns are particularly effective in determining genetic homogeneity or heterogeneity of strains within established species (Razin, 1989). Thus, restriction patterns were instrumental in the separation of the *U. urealyticum* serotypes into two genotypic clusters (Razin et al., 1983), indicated by DNA homology values (Christiansen et al., 1981) and by cell protein profiles (Mouches et al., 1981).

Further development of the restriction pattern approach was achieved by subjecting the DNA cleavage fragments, separated on agarose gels, to Southern blot hybridization using cloned conserved genes as probes. The genes first used as probes were the *M. capricolum* rRNA genes cloned in plasmid pMC5 (Amikam et al., 1982). The fact that the rRNA operons in the genomes of various mycoplasmas differ in restriction sites within the operon and in their flanking sequences (Amikam et al., 1984; Hyman et al., 1988) leads to the production of hybridization patterns peculiar to the different mycoplasma species, a finding utilized to identify mycoplasmas in infected cell cultures (Razin et al., 1984). As would be expected from the results of the DNA restriction patterns described above, the hybridization patterns with rRNA gene probes also revealed genotypic variation among strains of species such as *U. urealyticum* (Razin and Yogev, 1986), *M. hominis* (Christiansen et al., 1987b Christiansen and Andersen, 1988; Yogev et al., 1988a), *Acholeplasma laidlawii* (Yogev et al., 1988a), *M. gallisepticum,* and *M. synoviae* (Yogev et al., 1988b; Kleven et al., 1988). Of special interest is the ability of this technique to distinguish the live-vaccine strain F of *M. gallisepticum* from virulent field isolates (Yogev et al., 1988b; Kleven et al., 1988), indicating the value of hybridization patterns as epidemiological tools. In contrast to the genotypic heterogeneity of the above mentioned species, the hybridization patterns revealed a high degree of genotypic homogeneity among *M. pneumoniae* strains isolated from pneumonia patients during different epidemics (Yogev et al., 1988a).

Conserved genes other than rRNA genes can, of course, be used as probes in the hybridization tests. Thus, a cloned part of the ATPase *(atp)* operon from *E. coli* was used by Christiansen et al. (1987a) to reveal genotypic heterogeneity of *M. hominis* strains. The elongation factor *tuf* gene could also be used as an effective probe, yielding hybridization patterns that confirm the conclusions obtained with the rRNA gene probes (Yogev et al., 1988c).

To assess possible genetic relatedness between different species, a modification of the Southern blot technique was devised by Yogev and Razin (1986). In this modification, labeled total DNA of one organism is hybridized with the electrophoresed, digested DNA of the other. Appearance of hybridization bands additional to those detected by the specific rRNA gene probe pMC5, indicates the presence of genomic sequences shared by the two organisms (Fig. 5). In this way, the pairs *M. pneumoniae-M. genitalium* and *M. gallisepticum-M. synoviae* were shown to be genetically related (Yogev and Razin, 1986; Yogey et al., 1989) while *M. pirum* showed no relatedness to either *M. pneumoniae* or *M. genitalium* (Yogev et al., 1988d).

Development of DNA probes for identification of microorganisms appears to be the most promising approach for rapid diagnosis of infections caused by fastidious, hard-to-cultivate agents, and mycoplasmas fall well within this category (Siegler, 1989). Steps to develop a specific DNA probe usually include the cloning of restricted DNA of the target organism into a plasmid vector, which is easily maintained and produced in large quantities in an *E. coli* host. Clones carrying DNA fragments of the target organism are screened for size and tested by DNA hybridization (usually by the dot blot procedure) for cross-reactivity with DNAs of other mycoplasmas likely to be present in the natural microbial flora. In this way, species-specific DNA probes were selected for *M. pneumoniae* (Hyman et al., 1987), *M. genitalium* (Hyman et al., 1987; Risi et al., 1988), *M. gallisepticum* (Stipkovits et al., 1988; Geary et al, 1988; Hyman et al., 1989; Khan et al., 1989), and *M. synoviae* (Levisohn et al., 1989). The use of uncloned whole genomic DNA of *M. genitalium* as a probe to detect this mycoplasma in clinical specimens (Hooton et al., 1988) is open to criticism, since genomic DNA of *M. genitalium* shares common DNA sequences with *M. pneumoniae* (Yogev and Razin, 1986), so that unless very stringent hybridization conditions are employed, this probe may also react with *M. pneumoniae.*

Thus far most of the reported DNA probes were tested for sensitivity to and specificity on isolated homologous or heterologous DNAs or on cultured organisms. The reported data indicate that ^{32}P-labelled DNA probes are capable of detecting less than 1 ng DNA or 10^4–10^5 cfu (Razin et al., 1987). The effectiveness of the *M. gallisepticum* DNA probe in experimental infections and field outbreaks in chicken has also been tested. The probe enabled positive identification of *M. gallisepticum* at an early stage of infection, prior to development of a serolog-

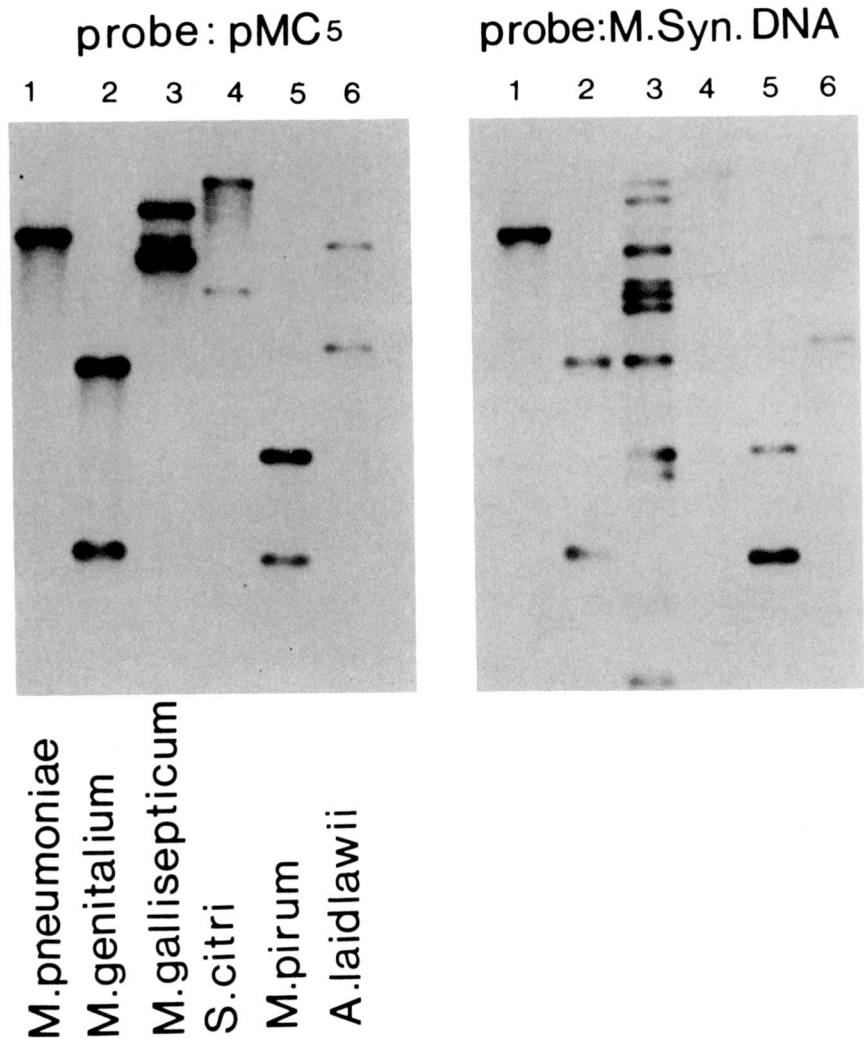

Fig. 5. Demonstration of genomic DNA sequences shared by *M. gallisepticum* and *M. synoviae*. The Southern blots shown are of *Eco*RI-digested DNAs of various mycoplasmas (numbered 1 to 6) hybridized with the rRNA gene probe pMC5 (left), or with total DNA of *M. synoviae* (right). The bands carrying the conserved rRNA gene sequences reacting with total genomic DNA of *M. synoviae* can be identified by comparison with the bands obtained with the specific rRNA gene probe pMC5. After hybridization with total *M. synoviae* DNA, only *M. gallisepticum* DNA reveals bands in addition to those revealed by pMC5, indicating the genetic relatedness of these two mycoplasmas. (From Yogev et al., 1989.)

ical response (Hyman et al., 1989; Levisohn et al., 1989). Clearly, successful application of the polymerase chain reaction (PCR) may increase sensitivity by increasing the amount of target DNA in the clinical specimen, but since the PCR methodology is new, few protocols for reproducible routine use exist at this time (Siegler, 1989).

An essential step in the development of diagnostic DNA probe kits involves the tagging of the probe by a nonradioactive label. The use of a biotin label appears to decrease sensitivity by a factor of about 10 as compared to the ^{32}P-labelled probe (Hyman et al., 1987; Geary et al., 1988; Stipkovits et al., 1988). Labeling by sul-

fonation of the *M. pneumoniae* DNA probe with the Chemiprobe kit (Orgenics, Inc., Israel) gave results almost equivalent in sensitivity and specificity to those obtained with the same probe labeled by ^{32}P (Fig. 6) . The first commercial diagnostic kit for *M. pneumoniae* detection produced by Gen-probe Corp. (San Diego, CA) has been subjected to testing in several diagnostic laboratories. The probe differs from those described above in consisting of ^{125}I-labelled cDNA specific to rRNA sequences of *M. pneumoniae*. Hence, this probe hybridizes with rRNA of the target organism rather than with its DNA. Moreover, hybridization is performed in liquid and not on a filter as in dot

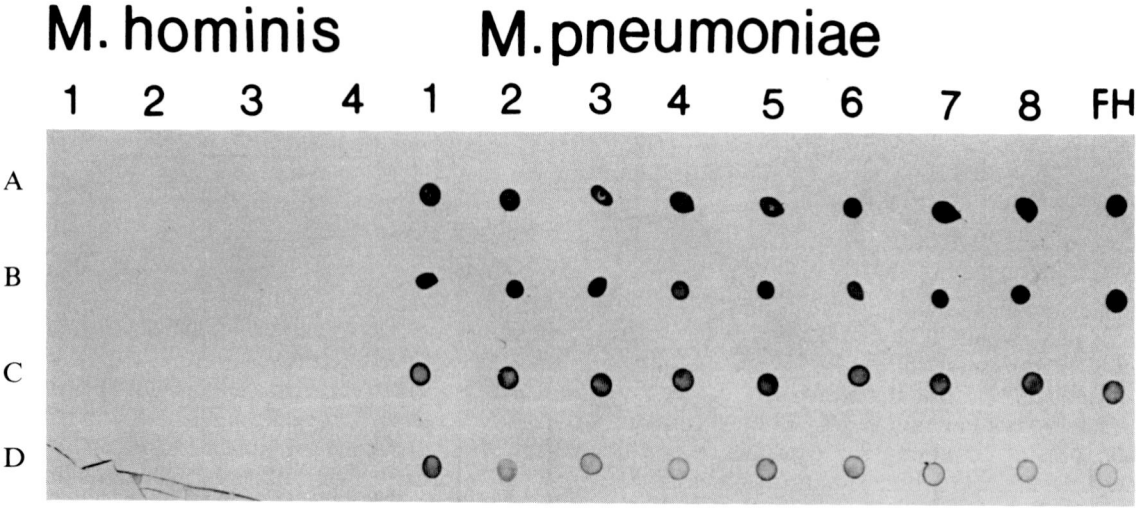

Fig. 6. Dot blot hybridization of the specific *M. pneumoniae* DNA probe pPN4, labeled by sulfonation (Hyman et al., 1987), with DNAs of four *M. hominis* strains, eight *M. pneumoniae* strains from different epidemics, and the FH laboratory strain of *M. pneumoniae*. The quantity of *M. pneumoniae* DNA in the dots was 50, 2, 5, and 1 ng in lines A, B, C, and D, respectively. The amounts of the *M. hominis* DNAs in the dots were 10 times higher than those of *M. pneumoniae* DNA on the corresponding lines. The blot shows both the specificity and sensitivity of the DNA probe. (From H. Hyman, D. Yogev, and S. Razin, unpublished observations.)

blots, shortening considerably the hybridization period. The results obtained on testing clinical specimens with this probe compare rather well with those obtained by the culture method (Dular et al., 1988; Tilton et al., 1988). Considering the fact that results by the probe procedure are available within a few hours, compared with 1–2 weeks or more required for culture, the diagnostic advantages of the probe technique are more than obvious. Yet, the fact that this probe is labelled by radioactive iodine can still be regarded as a disadvantage (Harris et al., 1988).

Applications

Most of the applications emanating from the molecular biology and genetic research of mycoplasmas have been in the area of development of diagnostic DNA probes and vaccines, described in preceding sections of this chapter. Extensive discussions concerning the development of mycoplasma vaccines can be found in Barile (1985) and Barile et al. (1985). While discussing biotechnological applications of mycoplasmas, mention should be made of a negative aspect concerning the troublesome infection of cell cultures by mycoplasma (see "Pathogenicity," this chapter). According to regulations of the USA Food and Drug Administration, every biological product produced in cell cultures should be tested for the presence of mycoplasmas, an ex-

pensive step that cannot and should not be avoided.

The high cost of mycoplasma cultivation and low yields of organisms do not encourage their use as a source for production of biochemicals, unless the material is unique. This seems to be the case for the peculiar DNA methylase produced by *Spiroplasma* sp. MQ-1. This enzyme was found in our laboratory (Nur et al., 1985) to methylate all cytosine residues located at the sequence cytosine-guanine, the only sequence methylated in eukaryotic DNA. To exploit this finding, the methylase gene of the spiroplasma was cloned, and a special system was developed to express it in *E. coli* and overcome the UGA problem (see "Molecular Biology and Genetics," this chapter). Expression in *E. coli* could be used to produce large quantities of the enzyme needed as a research reagent in laboratories working on eukaryotic DNA methylation.

Literature Cited

Almagor, M., I. Kahane, and S. Yatziv. 1984. Role of superoxide anion in host cell injury induced by *Mycoplasma pneumoniae* infection. A study in normal and trisomy 21 cells. J. Clin. Invest. 73:842–847.

Almagor, M., I. Kahane, J. M. Wiesel, and S. Yatziv. 1985. Human ciliated epithelial cells from nasal polyps as an experimental model for *Mycoplasma pneumoniae* infection. Infect. Immun. 48:552–555.

Almagor, M., I. Kahane, C. Gilon, and S. Yatziv. 1986. Protective effects of the glutathione redox cycle and vitamin E on cultured fibroblasts infected by *Mycoplasma pneumoniae*. Infect. Immun. 52:240–244.

Amikam, D., G. Glaser, and S. Razin. 1984. Mycoplasmas *(Mollicutes)* have a low number of ribosomal RNA genes. J. Bacteriol. 158:376–378.

Amikam, D., S. Razin, and G. Glaser. 1982. Ribosomal RNA genes in mycoplasma. Nucleic Acids Res. 10:4215–4222.

Barile, M. F. 1983. Arginine hydrolysis, p. 345–349. In: S. Razin and J. G. Tully (ed.), Methods in mycoplasmology, vol. 1. Academic Press, New York.

Barile, M. F. 1985. Immunization against mycoplasma infections, p. 451–492. In: S. Razin and M. F. Barile (ed.), The mycoplasmas, vol. 4: Mycoplasma pathogenicity. Academic Press, Orlando, FL.

Barile, M. F., J. M. Bove, J. M. Bradbury, G. H. Cassell, W. A. Clyde, Jr., G. S. Cottew, and P. Whittlestone. 1985. Control of mycoplasma diseases of man, animals, plants and insects. Bull. Inst. Pasteur 83:339–373.

Barile, M. F., D. K. F. Chandler, H. Yoshida, M. W. Grabowski, R. Harasawa, and S. Razin. 1988a. Parameters of *Mycoplasma pneumoniae* infection in Syrian hamsters. Infect. Immun. 56:2443–2449.

Barile, M. F., D. K. F. Chandler, H. Yoshida, M. W. Grabowski, and S. Razin. 1988b. Hamster challenge potency assay for evaluation of *Mycoplasma pneumoniae* vaccines. Infect. Immun. 56:2450–2457.

Barile, M. F., G. J. McGarrity, M. W. Grabowski, T. Steiner, V. Vanaman, E. M. Levine, S. N. Mueller, L. Gamon, and J. Sarama. 1983. Special techniques for isolation and identification of mycoplasmas from cell cultures, p. 155–208. In: J. G. Tully and S. Razin (ed.), Methods in mycoplasmology, vol. 2. Academic Press, New York.

Barroso G. and J. Labarere. 1988. Chromosomal gene transfer in *Spiroplasma citri*. Science 241:959–961.

Baseman, J. B., R. M. Cole, D. C. Krause, and D. K. Leith. 1982. Molecular basis for cytadsorption of *Mycoplasma pneumoniae*,. J. Bacteriol. 151:1514–1522.

Baseman J. B., S. F. Dallo, J. G. Tully, and D. L. Rose. 1988. Isolation and characterization of *Mycoplasma genitalium* strains from the human respiratory tract. J. Clin. Microbiol. 26:2266–2269.

Bautsch, W. 1988. Rapid physical mapping of the *Mycoplasma mobile* genome by two-dimensional field inversion gel electrophoresis techniques. Nucleic Acids Res. 16:11461–11467.

Bergemann, A. D., J. C. Whitley, and L. R. Finch. 1989. Homology of mycoplasma plasmid pADB201 and staphylococcal plasmid pE194. J. Bacteriol. 171:593–595.

Biberfeld, G. 1985. Infection sequelae and autoimmune reactions in *Mycoplasma pneumoniae* infection, p. 293–311. In: S. Razin and M. F. Barile (ed.), The mycoplasmas, vol. 4: Mycoplasma pathogenicity. Academic Press, Orlando, FL.

Blanchard, A., and M. F. Barile. 1989. Cloning of *Ureaplasma urealyticum* DNA sequences showing genetic homology with urease genes from gram-negative bacteria. Res. Microbiol. Inst. Pasteur 140:281–290.

Blanchard, A., S. Razin, G. E. Kenny, and M. F. Barile. 1988. Characteristics of the *Ureaplasma urealyticum* urease. J. Bacteriol. 170:2629–2697.

Bolske, G., M.-L. Strandberg, K. Bergstrom, and K.-E. Johansson. 1987. Species-specific antigens of *Mycoplasma hyopneumoniae* and cross-reactions with other porcine mycoplasmas. Curr. Microbiol. 15:233–239.

Bovè, J. M., P. Carle, M. Garnier, F. Laigret, J. Renaudin C. Saillard. 1989. Molecular and cellular biology of spiroplasmas, p. 243–364. In: R. F. Whitcomb and J. G. Tully (ed.), The mycoplasmas, vol. 5: Spiroplasmas, acholeplasmas, and mycoplasmas of plants and arthropods. Academic Press, San Diego, CA.

Bredt, W. 1979. Motility, p. 141–155. In: M. F. Barile and S. Razin (ed.), The mycoplasmas, vol. 1: Cell biology. Academic Press, New York.

Bredt, W., B. Kleinmann, and E. Jacobs. 1987. Antibodies in the sera of *Mycoplasma pneumoniae*–infected patients against proteins of *Mycoplasma genitalium* and other mycoplasmas of man. Zbl.Bakt.Hyg. A 266:32–42.

Carle, P., and J. M. Bove. 1983. Genome size determination, p. 309–311. In: S. Razin and J. G. Tully (ed.), Methods in mycoplasmology, vol. 1. Academic Press, New York.

Carle, P., C. Saillard, and J. M. Bove. 1983. Determination of guanine plus cytosine content of DNA, p. 301–308. In: S. Razin and J. G. Tully (ed.), Methods in mycoplasmology, vol. 1. Academic Press, New York.

Cassell, H. G., K. B. Waites, D. T. Crouse, P. T. Rudo, K. C. Canupp, S. Stagno, and G. R. Cutter. 1988. Association of *Ureaplasma urealyticum* infection of the lower respiratory tract with chronic lung disease and death in very-low-birth-weight infants. Lancet 2:240–244.

Chen, X. and L. R. Finch. 1989. Novel arrangements of rRNA genes in *Mycoplasma gallisepticum*: separation of the 16S gene of one set from the 23S and 5S genes. J. Bacteriol. 171:2876–2878.

Christiansen, C. 1987a. The mycoplasma genome part 1. Microbiol. Sci. 4:168–172.

Christiansen, C. 1987b. The mycoplasma genome part 2. Microbiol. Sci. 4:292–295.

Christiansen, C., F. T. Black, and E. A. Freundt. 1981. Hybridization experiments with deoxyribonucleic acid from *Ureaplasma urealyticum* serovars I to VIII. Int. J. Syst. Bacteriol. 31:259–262.

Christiansen, C., G. Christiansen, and O. F. Rasmussen. 1987a. Heterogeneity of *M. hominis* as detected by a probe for *atp* genes. Isr. J. Med. Sci. 23:591–594.

Christiansen, G., and H. Andersen. 1988. Heterogeneity among *Mycoplasma hominis* strains as detected by probes containing parts of ribosomal ribonucleic acid genes. Int. J. Syst. Bacteriol. 38:108–115.

Christiansen, G., H. Andersen, S. Birkelund, and E. A. Freundt. 1987b. Genomic and gene variation in *Mycoplasma hominis*. Isr. J. Med. Sci. 23:595–602.

Cimolai, N. A. Schryvers, L. E. Bryan, and D. E. Woods. 1988. Culture-amplified immunological detection of *Mycoplasma pneumoniae* in clinical specimens. Diagn. Microbiol. Infect. Dis. 9:207–212.

Cluss, R. G., and N. L. Somerson. 1986. Interaction of albumin and phospholipid: cholesterol liposomes in growth of *Mycoplasma* spp. App. Environ. Microbiol. 51:281–287.

Clyde, W. A., Jr. 1983. Growth inhibition tests. p, 405–410. In: S. Razin and J. G. Tully (ed.), Methods in mycoplasmology, vol. 1. Academic Press, New York.

Cocks, B. G., R. Youl, and L. R. Finch. 1988. Comparison of enzymes of nucleotide metabolism in two members of the *Mycoplasmataceae* family. Int. J. Syst. Bacteriol. 38:273–278.

Cole, B. C., Y. Naot, E. J. Stanbridge, and K. S. Wise. 1985. Interactions of mycoplasmas and their products with lymphoid cells in vitro, p.203–257. In: S. Razin and M. F. Barile (ed.), The mycoplasmas, vol. 4: Mycoplasma pathogenicity. Academic Press, Orlando, FL.

Constantopoulos, G., and G. J. McGarrity. 1987. Activities of oxidative enzymes in mycoplasmas. J. Bacteriol. 169:2012–2016.

Dallo, S. F., A. Chavoya, C. J. Su, and J. B. Baseman. 1989a. DNA and protein sequence homologies between the adhesins of *Mycoplasma genitalium* and *Mycoplasma pneumoniae*. Infect. Immun. 57:1059–1065.

Dallo, S. F., J. R. Horton, C. J. Su and J. B. Baseman. 1989b. Homologous regions shared by adhesin genes of *Mycoplasma pneumoniae* and *Mycoplasma genitalium*. Microbial Pathogenesis 6:69–73.

Dallo, S. F., C. J. Su, J. R. Horton, and J. B. Baseman. 1988. Identification of P1 gene domain containing epitope(s) mediating *Mycoplasma pneumoniae* cytadherence. J. Exp. Med. 167:718–723.

Degorce-Dumas, J. R., B. Richard, and J. M. Bove. 1983. Hybridization between mycoplasma DNAs, p. 319–325. In: S. Razin and J. G. Tully (ed.), Methods in mycoplasmology, vol. 1. Academic Press, New York.

Desantis, D., V. V. Tryon, and J. D. Pollack. 1989. Metabolism of Mollicutes: the Embden-Meyerhof-Parnas pathway and the hexose monophosphate shunt. J. Gen. Microbiol. 135:683–691.

Dickinson, M. J. and R. Townsend. 1984. The integration of a temperate phage infecting *Spiroplasma citri*. Isr. J. Med. Sci. 20:785–787.

Dular, R., R. Kajioka, and S. Kasatiya. 1988. Comparison of Gen-Probe commercial kit and culture technique for diagnosis of *Mycoplasma pneumoniae* infection. J. Clin. Microbiol. 26:1068–1069.

Dybvig, K., and J. Alderete. 1988. Transposition of *Mycoplasma pulmonis* and *Mycoplasma hyorhinis*: transposition of Tn916 and formation of cointegrate structures. Plasmid 20:33–41.

Dybvig, K., and G. H. Cassell. 1987. Transportation of Gram-positive Tn916 in *Acholeplasma laidlawii* and *Mycoplasma pulmonis*. Science 235:1392–1394.

Dybvig, K. and J. Maniloff. 1983. Integration and lysogeny by an enveloped mycoplasma virus. J. Gen. Virol. 64:1781–1785.

Elton, R. A. 1973. The relationship of DNA base composition and individual protein compositions in microorganisms. J. Mol. Evolution 2:263–276.

Feldner, J., U. Gobel, and W. Bredt. 1982. *Mycoplasma pneumoniae* adhesin localized to tip structure by monoclonal antibody. Nature 298:765–767.

Fernald, G. W. 1979. Humoral and cellular immune responses to mycoplasmas, p. 399–423. In: J. G. Tully and R. F. Whitcomb (ed.), The mycoplasmas, vol. 2: Human and animal mycoplasmas. Academic Press, New York.

Ferrell, R. V., M. B. Heidari, K. S. Wise, and M. A. McIntosh. 1989. A Mycoplasma genetic element resembling prokaryotic insertion sequences. Mol. Microbiol. 3:957–967.

Fischer, M., H. Kirchhoff, R. Rosengarten, G. Kerlen, and K.-H. Seack. 1987. Gliding movement of *Mycoplasma sp. nov.* strain 163K on erythrocytes. FEMS Microbiol. Lett. 40:321–324.

Freundt, E. A. 1983. Culture media for classic mycoplasmas, p.127–135. In: S. Razin and J. G. Tully (ed.), Methods in mycoplasmology, vol. 1. Academic Press, New York.

Freundt, E. A., R. F. Whitcomb, M. F. Barile, J. M. Bove, G. S. Cottew, H. Erno, R. Lemcke, J. Maniloff, H. Neimark, S. Razin, D. Taylor-Robinson, J. G. Tully, and P. Whittlestone. 1979. Proposal of minimal standards for descriptions of new species of the class Mollicutes. Int. J. Syst. Bacteriol. 29:172–180.

Frydenberg, J., K. Lind, and P. C. Hu. 1987. Cloning of *Mycoplasma pneumoniae* DNA and expression of P1-epitopes in *Escherichia coli*. Isr. J. Med. Sci. 23:759–762.

Gadeau, A.-P., C. Mouches, and J. M. Bove. 1986. Probable insensitivity of Mollicutes to rifampin and characterization of spiroplasmal DNA-dependent RNA polymerase. J. Bacteriol. 166:824–828.

Gafny, R., H. C. Hyman, S. Razin, and G. Glaser. 1988. Promoters of *Mycoplasma capricolum* ribosomal RNA operons: identical activities but different regulation in homologous and heterologous cells. Nucleic Acids Res. 16:61–76.

Gardella, R. S., R. A. DelGiudice, and J. G. Tully. 1983. Immunofluorescence, p. 431–439. In: S. Razin and J. G. Tully (ed.), Methods in mycoplasmology, vol. 1. Academic Press, New York.

Geary, S. J., R. Intres, and M. G. Gabridge. 1988. Species-specific biotinylated DNA probe for the detection of *Mycoplasma gallisepticum*. Mol. Cell. Probes 2:237–244.

Gibbons, N. E., and R. G. E. Murray. 1978. Proposals concerning the higher taxa of bacteria. Int. J. Syst. Bacteriol. 28:1–6.

Grenabo, L., H. Hedelin, and S. Pettersson. 1988. Urinary infection stones caused by *Ureaplasma urealyticum*: A review. Scan. J. Infect. Dis. Suppl. 53:46–49.

Guindy, Y. S., T. Samuelsson, and T.-I. Johansen. 1989. Unconventional codon reading by *Mycoplasma mycoides* tRNAs as revealed by partial sequence analysis. Biochem. J. 258:869–873.

Hackett, K. J., A. S. Ginsberg, S. Rottem, R. B. Henegar, and R. F. Whitcomb. 1987. A defined medium for a fastidious spiroplasma. Science 237:525–527.

Harris, R., B. P. Marmion, G. Varkanis, T. Kok, B. Lunn, and J. Martin. 1988. Laboratory diagnosis of *Mycoplasma pneumoniae* infection. 2. Comparison of methods for the direct detection of specific antigen or nucleic acid sequences in respiratory exudates. Epidem. Inf. 101:685–694.

Hooton, T. M., M. C. Roberts, P. L. Roberts, K. K. Holmes, W. E. Stamm, and G. E. Kenny. 1988. Prevalence of *Mycoplasma genitalium* determined by DNA probe in men with urethritis. Lancet 1:266–268.

Hu, P. C., R. M. Cole, Y. S. Huang, J. A. Graham, D. E. Gardner, A. M. Collier, and W. A. Clyde, Jr. 1982. *Mycoplasma pneumoniae* infection: role of a surface protein in the attachment organelle. Science 216:313–315.

Hu, P. C., C. H. Huang, A. M. Collier, and W. A. Clyde, Jr. 1983. Demonstration of antibodies to *Mycoplasma pneumoniae* attachment protein in human sera and respiratory secretions. Infect. Immun. 41:437–439.

Hu, P. C., U. Schaper, A. M. Collier, W. A. Clyde, Jr., M. Horikawa, Y. S. Huang, and M. F. Barile. 1987. A *Mycoplasma genitalium* protein resembling the *Mycoplasma pneumoniae* attachment protein. Infect. Immun. 55:1126–1131.

Hyman, H. C., R. Gafny, G. Glaser, and S. Razin. 1988. Promoter of the *Mycoplasma pneumoniae* rRNA operon. J. Bacteriol. 170:3262–3268.

Hyman, H. C., S. Levisohn, D. Yogev, and S. Razin. 1989. DNA probes for *Mycoplasma gallisepticum* and *Mycoplasma synoviae*: application in experimentally infected chickens. Vet. Microbiol. 20:323–337.

Hyman, H. C., D. Yogev, and S. Razin. 1987. DNA probes for detection and identification of *Mycoplasma pneumoniae* and *Mycoplasma genitalium*. J. Clin. Microbiol. 25:726–728.

Imada, Y., I. Uchida, and K. Hashimoto. 1987. Rapid identification of mycoplasmas by direct immunoperoxidase test using small square filter paper. J. Clin. Microbiol. 25:17–21.

Inamine, J. M., T. P. Denny, S. Loechel, U. Schaper, C.-H. Huang, K. F. Bott, and P. C. Hu, 1988a. Nucleotide sequence of the P1-attachment-protein gene of *Mycoplasma pneumoniae*. Gene 64:217–229.

Inamine, J. M., S. Loechel, and P. C. Hu, 1988b. Analysis of the nucleotide sequence of the P1-operon of *Mycoplasma pneumoniae*. Gene 73:175–183.

Inamine, J. M., S. Loechel, A. M. Collier, M. F. Barile, and P. C. Hu. 1989. Nucleotide sequence of *MgPa (mgp)* operon of *Mycoplasma genitalium* and comparison to the *P1 (mpp)* operon of *Mycoplasma pneumoniae*. Gene. 82:259–267.

Jacobs, E., M. Drews, A. Stuhlert, C. Buttner, P. J. Klein, M. Kist and W. Bredt. 1988a. Immunological reaction of guinea-pigs following intranasal *Mycoplasma pneumoniae* infection and immunization with the 168 kDa adherence protein. J. Gen. Microbiol. 134:473–479.

Jacobs, E., K. Fuchte, and W. Bredt. 1986. A 168-kilodalton protein of *Mycoplasma pneumoniae* used as antigen in a dot enzyme-linked immunosorbent assay. Eur. J. Clin. Microbiol. 5:435–440.

Jacobs, E., K. Fuchte, and W. Bredt. 1987. Amino acid sequence and antigenicity of the amino-terminus of the 168 kDa adherence protein of *Mycoplasma pneumoniae*, J. Gen. Microbiol. 133:2233–2236.

Jacobs, E., K. Fuchte, and W. Bredt. 1988b. Isolation of the adherence protein of *Mycoplasma pneumoniae* by fractionated solubilization and size exclusion chromatography. Biol. Chem. Hoppe-Seyler 369:1295–1299.

Jacobs, E., B. Gerstenecker, B. Mader, C.-H. Huang, P. C. Hu R. Halter, and W. Bredt. 1989. Binding sites of attachment-inhibiting monoclonal antibodies and antibodies from patient on peptide fragments of the *Mycoplasma pneumoniae* adhesin. Infect. Immun. 57:685–688.

Jacobs, E., A. Stuhlert, M. Drews, K. Pumpe, H. S. Shaefer, M. Kist, and W. Bredt. 1988c. Host reaction to *Mycoplasma pneumoniae* infections in guinea pigs preimmunized systemically with the adhesin of this pathogen. Microbial Pathogenesis 5:259–265.

Johnson, J. L. 1984. Nucleic acids in bacterial classification, p. 8–11. In: N. R. Krieg and J. G. Holt (ed.), Bergey's manual of systematic bacteriology, vol. 1. The Williams & Wilkins, Baltimore.

Kahane, I. 1984. *In vitro* studies on the mechanism of adherence and pathogenicity of mycoplasmas. Isr. J. Med. Sci. 20:874–877.

Kahane, I., S. Razin, and A. Muhlrad. 1978. Possible role of acetate kinase in ATP generation in *Mycoplasma hominis* and *Acholeplasma laidlawii*. FEMS Lett. 3:143–145

Kandler, O. and H. Konig. 1985. The cell envelopes of archaebacteria, p. 413–457 In: C. R. Woese and R. S. Wolfe (ed.), The bacteria, vol. 8. Academic Press, Orlando, FL.

Kawauchi, Y., A. Muto, and S. Osawa. 1982. The protein composition of *Mycoplasma capricolum*. Mol. Gen. Genet. 188:7–11.

Khan, M. I., B. C. Kirkpatrick, and R. Yamamoto. 1989. *Mycoplasma gallisepticum* species and strain-specific recombinant DNA probes. Avian Pathology 18:135–146.

Kibe, M. K., D. E. Bidwell, P. Turp, and G. R. Smith. 1985. Demonstration of cross-reactive antigens in F38 and related mycoplasmas by enzyme-linked immunosorbent assay (ELISA) and immunoblotting. J. Hyg. Camb. 95:95–106.

Kirchhoff, H., P. Beyene, M. Fischer, J. Flossdorf, J. Heitmann, B. Khattab, D. Lopatta, R. Rosengarten, G. Seidel, and C. Yousef. 1987a. *Mycoplasma mobile* sp. nov, a new species from fish. Int. J. Syst. Bacteriol. 37:192–197.

Kirchhoff, H., U. Boldt, R. Rosengarten, and A. Klein-Struckmeier. 1987b. Chemotactic response of a gliding mycoplasma. Curr. Microbiol. 15:57–60.

Kirchhoff, H. and J. Flossdorf. 1987. Determination of the guanine-plus-cytosine content of *Mollicutes* by isopycnic gradient centrifugation. Int. J. Syst. Bacteriol. 37:454–455.

Kirchhoff, H., R. Rosengarten, W. Lotz, M. Fischer, and D. Lopatta. 1984. Flask-shaped mycoplasmas: properties and pathogenicity for man and animals. Isr. J. Med. Sci. 20:848–853.

Kirkpatrick, B. C., D. C. Stenger, T. J. Morris, and A. H. Purcell. 1987. Cloning and detection of DNA from a noculturable plant pathogenic Mycoplasma-like Organism. Science 238:197–199.

Kleven, S. H., G. F. Browning, D. M. Bulach, E. Ghiocas, C. J. Morrow, and K. G. Whithear. 1988. Examination of *Mycoplasma gallisepticum* strains using restriction endonuclease DNA analysis and DNA-DNA hybridization. Avian Pathology 17:559–570.

Kotani, H. and G. J. McGarrity. 1986. Identification of mycoplasma colonies by immunobinding. J. Clin. Microbiol. 23:783–785.

Krebs, B., M. Schutz, M. Fischer, G. Sommer, and H. Kirchhoff. 1989. pH-controlled continuous cultivation of mycoplasma. Appl. Environ. Microbiol. 55:852–855.

Krivan, H. C., L. D. Olson, M. F. Barile, V. Ginsburg, and D. D. Roberts. 1989. Adhesion of *Mycoplasma pneumoniae* to sulfated glycolipids and inhibition by dextran sulfate. J. Biol. Chem. 264:9283–9288.

Leach, R. H. 1983. Preservation of mycoplasma cultures and culture collections, p. 197–204. In: S. Razin and J. G. Tully (ed.), Methods in mycoplasmology, vol. 1. Academic Press, New York.

Lee, I. M., and R. E. Davis. 1986. Prospects for in vitro culture of plant-pathogenic mycoplasmalike organisms. Annu. Rev. Phytopathol. 24:339–354.

Leith, D. K., L. B. Trevino, J. G. Tully, L. B. Senterfit, and J. B. Baseman. 1983. Host discrimination of *Mycoplasma pneumoniae* proteinaceous immunogens. J. Exp. Med. 157:502–514.

Levisohn, S., H. C. Hyman, D. Perlman, and S. Razin. 1989. The use of a specific DNA probe for detection of *Mycoplasma gallisepticum* in field outbreaks. Avian Pathol. 18:535–541.

Lin, C.-P., and T. A. Chen. 1985. Monoclonal antibodies against the aster yellow agent. Science 227:1233–1235.

Loechel, S., J. M. Inamine and P.-C. Hu. 1989. Nucleotide sequence of the *deo* C gene of *Mycoplasma pneumoniae*. Nucleic Acids Res. 17:801.

Loveless, R. W., and T. Feizi. 1989. Sialo-oligosaccharide receptor for *Mycoplasma pneumoniae* and related oligosaccharides of poly-N-acetyllactosamine series are polarized at the cilia and apical-microvillar domains of the ciliated cells in human bronchial epithelium. Infect. Immun. 57:1285–1289.

Madsen, R. D., F. A. Saeed, and S. R. Coats. 1987. The simultaneous direct detection of *Mycoplasma pneumoniae* and *Legionella pneumoniae* antigens in sputum specimens by a monoclonal antibody immunoblot assay. J. Immunol. Methods 103:205–210.

Mahairas, G. G., and F. C. Minion. 1989a. Transformation of *Mycoplasma pulmonis*: Demonstration of homologous recombination, introduction of cloned genes, and preliminary description of an integrating shuttle system. J. Bacteriol. 171:1775–1780.

Mahairas, G. G., and F. C. Minion. 1989b. Random insertion of the gentamicin resistance transposon Tn4001 in *Mycoplasma pulmonis*. Plasmid 21:43–47.

Maniloff, J. 1988. Mycoplasma viruses. Crit. Rev. Microbiol. 15:339–392.

Maniloff, J. 1989. Anomalous values of *Mycoplasma* genome sizes determined by pulse-field gel electrophoresis. Nucleic Acids Res. 17:1268.

Manolukas, J. T., M. F. Barile, D. K. F. Chandler, and J. D. Pollack. 1988. Presence of anaplerotic reactions and transamination, and the absence of the tricarboxylic acid cycle in Mollicutes. J. Gen. Microbiol. 134:791–800.

Mansel, J. K., E. C. Rosenow III, T. F. Smith, and J. W. Martin Jr. 1989. *Mycoplasma pneumoniae* pneumonia. Chest 95:639–646.

Masover, G. K., R. Perez, and A. Matin. 1979. Cultivation of *Ureaplasma urealyticum* in continuous culture. Infect. Immun. 23:175–177.

Masover, G. K., S. Razin, and L. Hayflick. 1977. Localization of enzymes in *Ureaplasma urealyticum* (T-strain mycoplasma). J. Bacteriol. 130:297–302.

Maurel, D., A. Charron, and C. Bebear. 1989. Mollicutes DNA polymerases: characterization of a single enzyme from *Mycoplasma mycoides* and *Ureaplasma urealyticum* and three enzymes from *Acholeplasma laidlawii*. Res. Microbiol. Inst. Pasteur 140:191–205.

McCoy, R. E. and IRPCM working team on mycoplasma-like-organisms of plants. 1989. plant diseases associated with mycoplasma-like-organisms, p. 545–640. In: R. F. Whitcomb and J. G. Tully (ed.), The mycoplasmas, vol. 5: spiroplasmas, acholeplasmas, and mycoplasmas of plants and arthropods. Academic Press, San Diego, CA.

McElwain, M. C., D. K. F. Chandler, M. F. Barile, T. F. Young, V. V. Tryon, J. W. Davis Jr., J. P. Petzel, C.-J.

Chang, M. V. Williams, and J. D. Pollack. 1988. Purine and pyrimidine metabolism in *Mollicutes* species. Int. J. Syst. Bacteriol. 38:417–423.

McGarrity, G. J., and H. Kotani. 1985. Cell culture mycoplasmas, p. 353–390. In: S. Razin and M. F. Barile (ed.), The mycoplasmas, vol. 4: Mycoplasma pathogenicity. Academic Press, Orlando, FL.

Mills, L. B., E. J. Stanbridge, W. D. Sedwick, and D. Korn. 1977. Purification and partial characterization of the principle deoxyribonucleic acid polymerase from *Mycoplasmatales*. J. Bacteriol. 132:641–649.

Morowitz, H. J. 1984. The completeness of molecular biology. Isr. J. Med. Sci. 20:750–753.

Morrison-Plummer, J., A. Lazzell, and J. B Baseman. 1987. Shared epitopes between *Mycoplasma pneumoniae* major adhesin protein P1 and a 140-kilodalton protein of *Mycoplasma genitalium*. Infect. Immun. 55:49–56.

Mouches, C., G. Barroso, A. Gadeau, and J. M. Bove. 1984. Characterization of two cryptic plasmids from *Spiroplasma citri* and occurrence of their DNA sequences among various spiroplasmas. Ann. Microbiol. (Inst. Pasteur) 135A:17–24.

Mouches, C., D. Taylor-Robinson, L. Stipkovits, and J. M. Bove. 1981. Comparison of human and animal ureaplasmas by one and two-dimensional protein analysis on polyacrylamide slab gel. Ann. Microbiol. (Inst. Pasteur) 132B:171–196.

Murray, R. G. E. 1984. The higher taxa, or, a place for everything . . . ? p. 31–34. In: N. R. Krieg and J. G. Holt (ed.), Bergey's manual for systematic bacteriology, vol. 1. Williams and Wilkins, Baltimore.

Muto, A., 1987. The genome structure of *Mycoplasma capricolum*. Isr. J. Med. Sci. 23:334–341.

Muto, A., Y. Kawauchi, F. Yamao, and S. Osawa. 1984. Preferential use of A- and U-rich codons for *Mycoplasma capricolum* ribosomal proteins S8 and L6. Nucleic Acids Res. 12:8209–8217.

Muto, A. and S. Osawa. 1987. The guanine and cytosine content of genomic DNA and bacterial evolution. Proc. Natl. Acad. Sci. USA 84:166–169.

Ne'eman, Z., and S. Razin. 1975. Characterization of the mycoplasma membrane proteins. V. Release and localization of membrane-bound enzymes in *Acholeplasma laidlawii*. Biochim. Biophys. Acta 375:54–68.

Neimark, H. 1986. Origins and evolution of wall-less prokaryotes, p. 21–42. In: S. Madoff, (ed.), The bacterial L-forms. Marcel Dekker Inc., New York.

Nur, I., G. Glaser, and S. Razin. 1986. Free and integrated plasmid DNA in spiroplasmas. Curr. Microbiol. 14:169–176.

Nur, I., D. J. LeBlanc, and J. G. Tully. 1987. Short, interspersed, and repetitive DNA sequences in spiroplasma species. Plasmid 17:110–116.

Nur, I., M. Szyf, A. Razin, G. Glaser, S. Rottem, and S. Razin. 1985. Procaryotic and eucaryotic traits of DNA methylation in spiroplasmas (mycoplasmas). J. Bacteriol. 164:19–24.

Ohkubo, S., A. Muto, Y. Kawauchi, F. Yamao, and S. Osawa. 1987. The ribosomal protein gene cluster of *Mycoplasma capricolum*. Mol. Gen. Genet. 210:314–322.

Piper, B., R. Rosengarten, and H. Kirchhoff. 1987. The influence of various substances on the gliding motility of *Mycoplasma mobile* 163K. J. Gen. Microbiol. 133:3193–3198.

Pollack, J. D. 1986. Metabolic distinctiveness of ureaplasmas. Pediatric Infect. Dis. 5: S305-S307.

Pollack, J. D., M. C. McElwain, D. DeSantis, J. T. Manolukas, J. G. Tully, C.-J. Chang, R. F. Whitcomb, K. J. Hackett, and M. V. Williams. 1989. Int. J. Syst. Bacteriol. 39:406-412.

Pyle, L. E., L. N. Corcoran, B. C. Cocks, A. D. Bergemann, J. C. Whitley, and L. R. Finch. 1988. Pulsed-field electrophoresis indicates larger-than-expected sizes for mycoplasma genomes. Nucleic Acids Res. 16:6015-6025.

Pyle, L. E., and L. R. Finch. 1988. The physical map of the genome of *Mycoplasma mycoides* subspecies *mycoides* Y with some functional loci. Nucleic Acids Res. 16:6027-6039.

Precious, B. L., D. Thirkell, and W. C. Russell. 1987. Preliminary characterization of the urease and a 96 kDa surface-expressed polypeptide of *Ureaplasma urealyticum*. J. Gen. Microbiol. 133:2659-2670.

Raccach, M., S. Rottem, and S. Razin. 1975. Survival of frozen mycoplasmas. Appl. Microbiol. 30:167-171.

Ranhand, J. M., I. Nur, D. L. Rose, and J. G. Tully. 1987. *Spiroplasma* species share common DNA sequences among their viruses, plasmids and genomes. Ann. Microbiol. (Inst. Pasteur) 138:509-522.

Rasmussen, O. F., and C. Christiansen. 1987. Identification of the proton ATPase operon in *Mycoplasma* strain FG50 by heterologous hybridization of *Mycoplasma* strain PG50. Isr. J. Med. Sci. 23:393-397.

Razin, S. 1969. Structure and function in mycoplasma. Annu. Rev. Microbiol. 23:317-356.

Razin, S. 1978. The mycoplasmas. Microbiol. Rev. 42:414-470.

Razin, S. 1981. The mycoplasma membrane, p. 165-250. In: B. K. Ghosh (ed.), Organization of prokaryotic cell membranes, vol. 1. CRC Press, Boca Raton, FL.

Razin, S. 1983. Urea hydrolysis, p. 351-353. In: S. Razin and J. G. Tully (ed.), Methods in mycoplasmology, vol. 1. Academic Press, New York.

Razin, S. 1985a. Molecular biology and genetics of mycoplasmas *(Mollicutes)*. Microbiol. Rev. 49:419-455.

Razin, S. 1985b. Mycoplasma adherence, p. 161-202. In: S. Razin and M. F. Barile (ed.), The mycoplasmas, vol. 4: Mycoplasma pathogenicity. Academic Press, Orlando, FL.

Razin, S. 1986. Mycoplasmal adhesins and lectins, p. 217-235. In: D. Mirelman (ed.), Microbial lectins and agglutinins. John Wiley and Sons, New York.

Razin, S. 1989. The molecular approach to mycoplasma phylogeny, p. 33-69. In: R. F. Whitcomb and J. G. Tully (ed.), The mycoplasmas, vol. 5: spiroplasmas, acholeplasmas, and mycoplasmas of plants and arthropods. Academic Press, San Diego, CA.

Razin, S., M. Banai, H. Gamliel, A. Polliack, W. Bredt, and I. Kahane. 1980. Scanning electron microscopy of mycoplasmas adhering to erythrocytes. Infect. Immun. 30:538-546.

Razin, S., M. F. Barile, R. Harasawa, D. Amikam, and G. Glaser. 1983. Characterization of the mycoplasma genome. Yale J. Biol. Med. 56:357-366.

Razin, S., and M. F. Barile (ed.). 1985. The mycoplasmas, vol. 4: Mycoplasma pathogenicity. Academic Press, Orlando, FL.

Razin, S., and V. P. Cirillo. 1983. Sugar fermentation, p. 337-343. In: S. Razin and J. G. Tully (ed.), Methods in mycoplasmology, vol. 1. Academic Press, New York.

Razin, S., and E. A. Freundt. 1984. The Mollicutes, Mycoplasmatales and Mycoplasmataceae, p. 740-742. In: N. R. Krieg and J. G. Holt (ed.), Bergey's manual of systematic bacteriology, vol. 1. Williams and Wilkins, Baltimore.

Razin, S., M. Gross, M. Wormser, Y. Pollack, and G. Glaser. 1984. Detection of mycoplasmas infecting cell cultures by DNA hybridization. In Vitro, 20:404-408.

Razin, S., R. Harasawa, and M. F. Barile. 1983. Cleavage patterns of the mycoplasma chromosomes, obtained by using restriction endonucleases as indicators of genetic relatedness among strains. Int. J. Syst. Bacteriol. 33:201-206.

Razin, S., H. C. Hyman, I. Nur, and D. Yogev. 1987. DNA probes for detection and identification of mycoplasmas (Mollicutes). Isr. J. Med. Sci. 23:735-741.

Razin, S., I. Nur, and G. Glaser. 1987. Spiroplasma plasmids. Isr. J. Med. Sci. 23:678-682.

Razin, S., and O. Oliver. 1961. Morphogenesis of mycoplasma and bacterial L-form colonies. J. Gen. Microbiol. 24:225-237.

Razin, S., and J. G. Tully. 1970. Cholesterol requirement of mycoplasmas. J. Bacteriol. 102:306-310.

Razin, S., and J. G. Tully (ed.). 1983. Methods in mycoplasmology, vol. 1. Academic Press, New York.

Razin, S., and D. Yogev. 1986. Genetic relatedness among *Ureaplasma urealyticum* serotypes (serovars). Pediatric Infect. Dis. 5:S300-S304.

Razin, S., and D. Yogev. 1989. Molecular approaches to characterization of mycoplasmal adhesins, p. 52-76. In:L. Switalski, M. Hook, and E. H. Beachy (ed.), Molecular mechanisms of microbial adhesion. Springer-Verlag, New York.

Renaudin, J., M.-C. Pascarel, and J. M. Bove. 1987. Spiroplasma virus 4: nucleotide sequence of the viral DNA, regulatory signals, and proposed genome organization. J. Bacteriol. 169:4950-4961.

Risi, G. F. Jr., D. H. Martin, J. A. Silberman, and J. C. Cohen. 1988. A DNA probe for detecting *Mycoplasma qenitalium* in clinical specimens. Mol. Cellular Probes. 2:327-335.

Roberts, D. D., L. D. Olson, M. F. Barile, V. Ginsburg, and H. C. Krivan. 1989. Sialic acid-dependent adhesion of *Mycoplasma pneumoniae* to purified glycoproteins. J. Biol. Chem. 264:9289-9293.

Roberts, M. C., and G. E. Kenny. 1986. Dissemination of the *tet*M tetracycline resistance determinant to *Ureaplasma urealyticum*. Antimicrob. Agents Chemother. 29:350-352.

Roberts, M. C. and G. E. Kenny. 1987. Conjugal transfer of transposon Tn916 from *Streptococcus faecalis* to *Mycoplasma hominis*. J. Bacteriol. 169:3836-3839.

Roberts, M. C., L. A. Koutsky, K. K. Holmes, D. J. LeBlanc, and G. E. Kenny. 1985. Tetracycline-resistant *Mycoplasma hominis*. strains contain streptococcal *tet*M sequences. Antimicrob. Agents Chemother. 28:141-143.

Robertson, J. A., and L. A. Howard. 1987. Effect of carbohydrates on growth of *Ureaplasma urealyticum* and *Mycoplasma hominis*. J. Clin. Microbiol. 25:160-161.

Robinson, I. M. 1984. Genus *Anaeroplasma*, p. 787-790. In: N. R. Krieg and J. G. Holt (ed.), Bergey's manual of systematic bacteriology, vol. 1. Williams and Wilkins, Baltimore.

Robinson, I. M., and E. A. Freundt. 1987. Proposal for an amended classification of anaerobic Mollicutes. Int. J. Syst. Bacteriol. 37:78–81.

Rodwell, A. W. 1982. The protein fingerprints of mycoplasmas. Rev. Infect. Dis. 4:S8-S17.

Rodwell, A. W. 1983a. Mycoplasma gallisepticum requires exogenous phospholipid for growth. FEMS Microbiol. Lett. 17:265–268.

Rodwell, A. W. 1983b. defined and partly defined media, p. 163–172. In: S. Razin and J. G. Tully (ed.), Methods in mycoplasmology, vol. 1. Academic Press, New York.

Rodwell, A. W., and A. Mitchell. 1979. Nutrition, Growth, and reproduction, p. 103–139. In: M. F. Barile and S. Razin (ed.), The mycoplasmas, vol. 1: Cell biology. Academic Press, New York.

Rogers, M. J., J. Simmons, R. T. Walker, W. G. Weisburg, C. R. Woese, R. S. Tanner, I. M. Robinson, D. A. Stahl, G. Olsen, R. H. Leach, and J. Maniloff. 1985. Construction of the mycoplasma evolutionary tree from 5S rRNA sequence data. Proc. Natl. Acad. Sci. USA 82:1160–1164.

Rogers, M. J., A. A Steinmetz, and R. T. Walker. 1987. Organization and structure of tRNA genes in Spiroplasma melliferum. Isr. J. Med. Sci. 23:357–360.

Romano, N., D. Russo Alesi, R. La Licata, and G. Tolone. 1986. Effects of urea phosphate, ammonium ions and pH on ureaplasma ATP synthesis. Microbiologica 9:405–413.

Romano, N., G. Tolone, F. Ajello, and R. La Licata 1980. Adenosine 5′-triphosphate synthesis induced by urea hydrolysis in Ureaplasma urealyticum. J. Bacteriol. 144:830–832.

Rosengarten, R., M. Fischer, H. Kirchhoff, G. Kerlen, and K.-H. Seack. 1988a. Transport of erythrocytes by gliding cells of Mycoplasma mobile 163K. Curr. Microbiol. 16:253–257.

Rosengarten, R., and H. Kirchhoff. 1987. Gliding motility of Mycoplasma sp. nov strain 163K. J. Bacteriol. 169:1891–1898.

Rosengarten, R., and H. Kirchhoff. 1988. Energetic aspects of the gliding motility of mycoplasmas. Curr. Microbiol. 16:247–252.

Rosengarten, R., A. Klein-Struckmeier, and H. Kirchhoff. 1988b. Rheotactic behavior of a gliding mycoplasma. J. Bacteriol. 170:989–990.

Rottem, S., and S. Razin. 1966. Adenosine triphosphatase activity of mycoplasma membranes. J. Bacteriol. 92:714–722.

Saada, A.-B., and I. Kahane. 1988. Purification and characterization of urease from Ureaplasma urealyticum. Zbl. Bakt. Hyg. A 269:160–167.

Samuelsson, T., Y. S. Guindy, F. Lustig, T. Boren, and U. Lagerkvist. 1987. Apparent lack of discrimination in the reading of certain codons in Mycoplasma mycoides. Proc. Natl. Acad. Sci. USA 84:3166–3170.

Sawada, M., S. Osawa, H. Kobayashi, H. Hori, and A. Muto. 1981. The number of ribosomal RNA genes in Mycoplasma capricolum. Mol. Gen. Genet. 182:502–504.

Schaper, U., J. S. Chapman, and P.-C. Hu. 1987. Preliminary indication of unusual codon usage in the DNA coding sequence of the attachment protein of Mycoplasma pneumoniae. Isr. J. Med. Sci. 23:361–367.

Sela, S., D. Yogev, S. Razin, and H. Bercovier. 1989. Duplication of the tuf gene: a new insight into the phylogeny of eubacteria. J. Bacteriol. 171:581–584.

Shepard, M. C., and C. D. Lunceford. 1976. Differential agar medium (A7) for identification of Ureaplasma urealyticum (human T mycoplasmas) in primary cultures of clinical material. J. Clin. Microbiol. 3:613–625.

Shepard, M. C., and G. K. Masover. 1979. Special features of ureaplasmas, p. 451–494. In: M. F. Barile and S. Razin (ed.), The mycoplasmas, vol. 1: Cell biology. Academic Press, New York.

Shibata, K.-I., and T. Watanabe. 1988. Purification and characterization of an arginine-specific carboxypeptidase from Mycoplasma salivarium. J. Bacteriol. 170:1795–1799.

Siegler, N. 1989. DNA-based testing: a progress report. ASM news 55:308–312.

Stemke, G. W., J. A. Robertson, and M. Nhan. 1987. Purification of urease from Ureaplasma urealyticum. Can. J. Microbiol. 33:857–862.

Stephens, E. B., G. S. Aulakh, D. L. Rose, J. G. Tully, and M. F. Barile. 1983. Intraspecies genetic relatedness among strains of Acholeplasma laidlawii and of Acholeplasma axanthum by nucleic acid hybridization. J. Gen. Microbiol. 129:1929–1934.

Stephens, E. B., I. M. Robinson, and M. F. Barile. 1985. Nucleic acid relationships among the anaerobic mycoplasmas. J. Gen. Microbiol. 131:1223–1227.

Stipkovits, L., S. Belak, B. S. McGwire, A. Ballagi-Pordany, and M. Santha. 1988. Rapid identification of Mycoplasma gallisepticum using a simple method of nucleic acid hybridization. Mol. Cell. Probes 2:339–344.

Su, C. J., A. Chavoya, and J. B. Baseman. 1988. Regions of Mycoplasma pneumoniae cytadhesin P1 structural gene exist as multiple copies. Infect. Immun. 56:3157–3161.

Su, C. J., V. V. Tryon, and J. B. Baseman. 1987. Cloning and sequence analysis of cytadhesin P1 gene from Mycoplasma pneumoniae. Infect. Immun. 55:3023–3029.

Subcommittee on the taxonomy of Mollicutes. 1984. Minutes of the 1980 meeting in Custer, S.D. Int. J. Syst. Bacteriol. 34:358–360.

Tanaka, R., A. Muto, and S. Osawa. 1989. Nucleotide sequence of tryptophan tRNA gene in Acholeplasma laidlawii. Nucleic Acids Res. 17:5842.

Taschke, C., and R. Herrmann. 1986. Analysis of transcription and processing signals of the 16S-23S rRNA operon of Mycoplasma hyopneumoniae. Mol. Gen. Genet. 205:434–441.

Taschke, C., and R. Herrmann. 1988. Analysis of transcription and processing signals in the 5′ regions of the two Mycoplasma capricolum rRNA operons. Mol. Gen. Genet. 212:522–530.

Taschke, C., M. Q. Klinkert, J. Wolters, and R. Herrmann. 1986. Organization of the ribosomal RNA genes in Mycoplasma hyopneumoniae: the 5S rRNA gene is separated from the 16S and 23S rRNA genes. Mol. Gen. Genet. 205:428–433.

Taylor, M. A., R. V. Ferrell, K. S. Wise, and M. A. McIntosh. 1988. Reiterated DNA sequences defining genomic diversity within the species Mycoplasma hyorhinis. Mol. Microbiol. 2:665–672.

Taylor-Robinson, D. 1983. Metabolism inhibition tests, p. 411–417. In: S. Razin and J. G. Tully (ed.), Methods in mycoplasmology, vol. 1. Academic Press, New York.

Taylor-Robinson, D. 1985. Mycoplasmal and mixed infections of the human male urogenital tract and their possible complications, p. 27–63. In: S. Razin and M. F.

Barile (ed.), The mycoplasmas, vol. 4: Mycoplasma pathogenicity. Academic Press, Orlando, FL.

Taylor-Robinson, D., and J. Behnke. 1987. The prolonged persistence of mycoplasmas in culture. J. Med. Microbiol. 23:89–92.

Thirkell, D., A. D. Myles, B. L. Precious, J. S. Frost, J. C. Woodall, M. G. Burdon, and W. C. Russell. 1989. The urease of *Ureaplasma urealyticum*. J. Gen. Microbiol. 135:315–323.

Tilton, R. C., F. Dias, H. Kidd, and R. W. Ryan. 1988. DNA probe versus culture for detection of *Mycoplasma pneumoniae* in clinical specimens. Diagn. Microbiol. Infect. Dis. 10:109–112.

Trevino, L. B., W. G. Haldenwang, and J. B. Baseman. 1986. Expression of *Mycoplasma pneumoniae* antigens in *Escherichia coli*. Infect. Immun. 53:129–134.

Tully, J. G. 1984. The family Acheloplasmataceae, genus *Acholeplasma*, p. 781–787. In: N. R. Krieg and J. G. Holt (ed.), Bergey's manual of systematic bacteriology, vol. 1. Williams and Wilkins, Baltimore.

Tully, J. G. 1989. Class Mollicutes: new perspectives from plant and arthropod studies, p. 1–31. In: R. F. Whitcomb and J. G. Tully (ed.), The mycoplasmas, vol. 5: spiroplasmas, acholeplasmas, and mycoplasmas of plants and arthropods. Academic Press, San Diego, CA.

Tully, J. G., D. L. Rose, R. F. Whitcomb, and R. P. Wenzel. 1979. Enhanced isolation of *Mycoplasma pneumoniae* from throat washings with a newly modified culture medium. J. Infect. Dis. 139:478–482.

Tully, J. G., D. L. Rose, P. Carle, J. M. Bove, K. J. Hackett, and R. F. Whitcomb 1988. *Acholeplasma entomophilum* sp. nov. from gut contents of a wide range of host insects. Int. J. Syst. Bacteriol. 38:164–167.

Tully, J. G., D. Taylor-Robinson, R. M. Cole, and D. L. Rose. 1981. A newly discovered mycoplasma in the human urogenital tract. Lancet 1, 1288–1291.

Tully, J. G., D. Taylor-Robinson, D. L. Rose, R. M. Cole, and J. M. Bove. 1983. *Mycoplasma genitalium*, a new species from the human urogenital tract. Int. J. Syst. Bacteriol. 33:387–396.

Tully, J. G., R. F. Whitcomb, H. F. Clark, and D. L. Williamson. 1977. Pathogenic mycoplasmas: cultivation and vertebrate pathogenicity of a new spiroplasma. Science 195:892–894.

Vu, A. C., H. M. Foy, F. D. Cartwright, and G. E. Kenny. 1987. The principal proteins of isolates of *Mycoplasma pneumoniae* measured by levels of immunoglobulin G in human serum are stable in strains collected over a 10-year period. Infect. Immun. 55:1830–1836.

Waites, K. B., P. T. Rudd, D. T. Crouse, K. C. Canupp, K. G. Nelson, C. Ramsey, and G. H. Cassell. 1988. Chronic *Ureaplasma urealyticum* and *Mycoplasma hominis* infections of central nervous systems in preterm infants. Lancet 1:17–23.

Webster, A. D. B., P. M. Furr, N. C. Hughes-Jones, B. D. Gorick, and D. Taylor-Robinson. 1988. Critical dependence on antibody for defence against mycoplasmas. Clin. Exp. Immunol. 71:383–387.

Weisburg, W. G., J. G. Tully, D. L. Rose, J. P. Petzel, H. Oyaizu, D. Yang, L. Mandelco, J. Sechrest, T. G. Lawrence, J. Van Etten, J. Maniloff, and C. R. Woese. 1989.

A phylogenetic analysis of the mycoplasmas: basis for their classification. J. Bacteriol. 171:6455–6467.

Wenzel, R., and R. Herrmann. 1988a. Physical mapping of the *Mycoplasma pneumoniae* genome. Nucleic Acids Res. 16:8323–8336.

Wenzel, R., and R. Herrmann. 1988b. Repetitive sequences in *Mycoplasma pneumoniae*. Nucleic Acids Res. 16:8337–8350.

Williamson, D. L., J. G. Tully, D. L. Rose, K. J. Hackett, R. Henegar, P. Carle, J. M. Bove, D. E. Colfesh, and R. F. Whitcomb. 1990. *Mycoplasma somnilux* sp. nov., *Mycoplasma luminosum* sp. nov. and *Mycoplasma lucivorax* sp. nov., new sterol-requiring mollicutes from firefly beetles *(Coleoptera: Lampyridae)*. Int. J. Syst. Bacteriol. 40:160–164.

Wise, K. S., G. H. Cassell, and P. T. Acton. 1978. Selective association of murine T lymphoblastoid cell surface alloantigens with *Mycoplasma hyorhinis*. Proc. Natl. Acad. Sci. USA 75:4479–4483.

Woese, C. R. 1987. Bacterial evolution. Microbiol. Rev. 51:221–271.

Woese, C. R., J. Maniloff, and L. B. Zablen. 1980. Phylogenetic analysis of the mycoplasmas. Proc. Natl. Acad. Sci. USA 77:494–498.

Yamao, F., S. Iwagami, Y. Azumi, A. Muto, S. Osawa, N. Fujita, and A. Ishihama. 1988. Evolutionary dynamics of tryptophan tRNAs in *Mycoplasma capricolum*. Mol. Gen. Genet. 212:364–369.

Yamao, F., A. Muto, Y. Kawauchi, M. Iwami, S. Iwagami, Y. Azumi, and S. Osawa. 1985. UGA is read as tryptophan in *Mycoplasma capricolum*. Proc. Natl. Acad. Sci. USA 82:2306–2309.

Yogev, D., D. Halachmi, G. E. Kenny, and S. Razin. 1988a. Distinction of species and strains of mycoplasma (Mollicutes) by genomic DNA fingerprints with an rRNA gene probe. J. Clin. Microbiol. 26:1198–1201.

Yogev, D., S. Levisohn, S. H. Kleven, D. Halachmi, and S. Razin. 1988b. Ribosomal RNA gene probes detect intraspecies heterogeneity in *Mycoplasma qallisepticum* and *M. synoviae*. Avian Dis. 32:220–231.

Yogev, D., S. Levisohn, and S. Razin. 1989. Genetic and antigenic relatedness between *Mycoplasma gallisepticum* and *Mycoplasma synoviae*. Veterinary Microbiol. 19:75–84.

Yogev, D. and S. Razin. 1986. Common deoxyribonucleic acid sequences in *Mycoplasma genitalium* and *Mycoplasma pneumoniae* genomes. Int. J. Syst. Bacteriol. 36:426–430.

Yogev, D., S. Sela, H. Bercovier, and S. Razin. 1988c. Elongation factor (EF-Tu) gene probe detects polymorphism in *Mycoplasma* strains. FEMS Microbiol. Lett. 50:145–149.

Yogev, D., S. Sela, H. Bercovier, and S. Razin. 1990. Nucleotide sequence and codon usage of the elongation factor Tu (EF-Tu) gene from *Mycoplasma pneumoniae*. Molecular Microbiol. 4:1303–1310.

Yogev, D., J. G. Tully, D. L. Rose, and S. Razin. 1988d. Genetic and antigenic distinction of *Mycoplasma pirum* from other mycoplasmas with specialized tip structures. Int. J. Syst. Bacteriol. 38:147–150.

Zilberstein, D., M. H. Shirvan, M. F. Barile, and S. Rottem. 1986. The -subunit of the F_1F_0-ATPase is conserved in mycoplasmas. J. Biol. Chem. 261:7109–7111.

The Genus *Spiroplasma*

JOSEPH G. TULLY and ROBERT F. WHITCOMB

Organisms now assigned to the genus *Spiroplasma* were initially confused, in their early history, with both spirochetes and viruses. This confusion seems understandable now when one considers that spiroplasmas—helical, filterable, wall-less prokaryotes—share a number of basic similarities with these other microbes.

The first spiroplasma to be cultivated on artificial media, obtained in pure culture, and characterized, was derived from citrus plants infected with "stubborn" disease, an infection initially thought to be of viral etiology (Calavan and Bové, 1989). The description of the morphologic and biologic features of the first named organism (*Spiroplasma citri*) in the genus (Saglio et al., 1973; Cole et al., 1973a) provided not only criteria for separation of spiroplasmas from other microbial forms but greatly stimulated further efforts to understand the place of these prokaryotes in the microbial world.

The first spiroplasma to be observed microscopically was the helical sex-ratio organism (SRO) that eliminates male progeny from certain neotropical species of *Drosophila* (Poulson and Sakaguchi, 1961). The pattern of vertical (transovarial) transmission of the trait suggested that an infectious agent was involved. Although conventional light microscopic techniques failed to reveal a microbe, phase-contrast and dark-field microscopy of fly hemolymph revealed numerous organisms that appeared to be spirochetes (Poulson and Sakaguchi, 1961). It was not until the ultrastructural features of the SRO were examined, revealing the absence of periplasmic fibrils (Williamson and Whitcomb, 1974), that one could clearly exclude the possibility that the organism was a spirochete. The SRO resisted vigorous efforts to cultivate it for years, but an approach involving a combination of insect tissue cells and cell-free medium was eventually successful (Hackett et al., 1986).

A second organism isolated in 1968 from rabbit ticks (*Haemaphysalis leporispalustris*) was also initially thought to be a spirochete. This microbe, the so-called 277F spirochete, was not only visualized by dark-field microscopy but

cultivated in a cell-free medium (Pickens et al., 1968). It was eventually identified as a spiroplasma on the basis of its ultrastructural features (Brinton and Burgdorfer, 1976).

A third organism discovered prior to *S. citri* was also under disguise as a virus. This agent, which was also isolated initially from rabbit ticks, was termed the "suckling mouse cataract agent" (SMCA) because of its ability to produce an experimental ocular disease in rodents (Clark, 1964). It was eventually identified as a wall-less prokaryote in 1974 (Bastardo et al., 1974; Zeigel and Clark, 1974), cultivated on artificial media and shown to be a spiroplasma in 1976 (Tully et al., 1976, 1977), and eventually named *S. mirum* (Tully et al., 1982).

While these observations revealed insects and other arthropods to be hosts for spiroplasmas, other developments indicated that plant hosts might also harbor a variety of wall-less prokaryotes. After the classic studies of Doi et al. (1967), which indicated that many plant diseases of the "yellows" type (see Chapter 88 and 228) might be induced by wall-less prokaryotes, the etiology of both corn stunt and citrus stubborn diseases was reexamined. In the case of corn stunt, wall-less prokaryotes were observed in tissues of infected plants and insects (Granados, 1969). In subsequent studies, motile, helical forms were observed by phase-contrast microscopy in sap from infected corn plants (Davis et al., 1972a, 1972b). The corn stunt agent also resisted cultivation for some time until successful media formulations were developed (Chen and Liao, 1975; Williamson and Whitcomb, 1975). The organism was recently designated *Spiroplasma kunkelii* (Whitcomb et al., 1986), and its biology and pathogenicity extensively reviewed (Whitcomb, 1989).

During the period of progress on corn stunt spiroplasma, several independent groups were investigating the etiological agent of citrus "stubborn" disease. Again, phloem tissues of affected plants were shown to contain wall-less prokaryotes (Laflèche and Bové, 1970; Igwegbe and Calavan, 1970). The causative organism was later cultivated in a cell-free medium (Fudl-Allah et al., 1972; Saglio et al., 1971, 1972) and shown to be capable of inducing disease in citrus (Markham et al., 1974). As a result of interchanges between mycoplasmologists and plant microbiologists at a meeting at the Ciba Foundation in London in early 1972, an international collaborative effort was initiated that eventually resulted in characterization and naming of the citrus stubborn disease agent (Saglio et al., 1973; Cole et al., 1973a). A more detailed history of the citrus disease and the biology of *S. citri* can be found in the review of Calavan and Bové (1989).

By 1975, spiroplasmas were regarded as arthropod-associated agents with pathological effects in some plant hosts and, rarely, in insect vectors (Whitcomb and Williamson, 1975). However, the discovery by Truman Clark (Clark, 1977) that honeybees frequently carried a spiroplasma that was lethal under some circumstances constituted a major breakthrough. Clark set forth to find the natural reservoir of the spiroplasmas and soon discovered that spiroplasmas occurred frequently on floral surfaces (Clark, 1978). Although these initial studies showed the spiroplasmal flora of flowers to be composed predominantly of organisms other than the bee agent, the flower niche eventually proved to be an important reservoir of spiroplasmas (Clark, 1978; Davis, 1978; Davis et al., 1979; McCoy et al., 1979; Vignault et al., 1980; Guo et al., 1990). The first honeybee spiroplasma was described as *S. melliferum* (Clark et al., 1985) and the first of the floral spiroplasmas to be named was designated *S. floricola* (Davis et al., 1981).

Discovery of the honeybee/flower habitat eventually led to exploration and discovery of spiroplasmas in a wide variety of other insect hosts (Clark, 1982). The seemingly universal involvement of insects and other arthropods as habitat for spiroplasmas has been amply confirmed (Hackett and Clark, 1989; Williamson et al., 1989a). Since many of the new isolates appear to represent putative species, the genus *Spiroplasma* has the potential to become the largest and most dominant group of wall-less prokaryotes.

Spiroplasma Taxonomy and Phylogeny

The type genus and species (*Spiroplasma citri*) was described and named in 1973 (Saglio et al., 1973; Cole et al., 1973a, 1973b), and a proposal made (Skripal, 1974, 1983) to elevate the genus

to family status (Spiroplasmataceae within the order Mycoplasmatales, class Mollicutes (see Whitcomb and Tully, 1984). These additions were endorsed by the International Committee on Systematic Bacteriology (ICSB) Subcommittee on the Taxonomy of Mollicutes (Subcommittee, 1977), especially since they followed closely the proposed minimum standards for description of new mollicutes (Subcommittee 1972, 1979). The suggested standards have been accorded wide recognition and future revisions are contemplated. Recommended standards have also greatly restrained proposals of premature binomials with inadequate supporting data. The description of *S. citri* in 1973 stimulated the search for new spiroplasmas in plants and arthropods, and it soon became obvious that a classification scheme was needed. Although a few of the new isolates were serologically distinct from *S. citri*, most other spiroplasmas exhibited significant levels of serologic or genetic relationship with this organism (Tully et al., 1973; Davis et al., 1974; Christiansen et al., 1979; Junca et al., 1980; Lee and Davis, 1980; Liao and Chen, 1981; Bové et al., 1982). A preliminary grouping scheme was proposed (Junca et al., 1980) in which serologic properties and genomic characteristics (DNA base composition and DNA-DNA relatedness values) were combined. Spiroplasma strains that were found to share serologic and other characteristics with *S. citri* were placed in four subgroups (I-1 to I-4), while unrelated spiroplasmas were assigned group numbers (II through V).

This classification scheme was expanded as new isolations were made and the spiroplasmas characterized (Whitcomb et al., 1982a, 1983; Bové et al., 1983a). Although the number of spiroplasma clusters assigned to group I subgroups increased to eight within 3 years, it became increasingly obvious that a large number of the new unclassified spiroplasma isolates from insects (Clark, 1982) represented putative new species, and that there would eventually be a rapid proliferation of new groups.

In the early 1980s, the Subcommittee (1984) recommended that, under certain specified conditions, subgroups might be elevated to species status. Especially important conditions were the existence of procedures for reliable laboratory identification of subgroup strain clusters and the importance of the cluster as pathogens or ecological models. Shortly thereafter, the Subcommittee (1985) also recognized the advantages of the grouping classification scheme for spiroplasmas, especially for putative new species for which no formal taxonomic description had been given. Although the Subcommittee encouraged workers describing new spiroplasma groups to complete a taxonomic description of the organism, they recommended continued use of the grouping scheme as an informal and interim mechanism for maintaining updated lists of currently recognized species and putative species. An ad hoc group appointed by the Subcommittee proposed a set of procedures for determination of morphological, serological, and genomic properties of these organisms (Whitcomb et al., 1987). These recommendations, therefore, essentially constitute minimal criteria for describing new *Spiroplasma* groups. Twenty four spiroplasma groups and eight subgroups have been recognized in the most recent version of the proposed classification scheme (Table 1) (Tully et al., 1987; Guo et al., 1990). Eleven of the groups or subgroups have been given species epithets (Table 1).

Significant information on the phylogeny and evolution of spiroplasmas and other mollicutes has appeared during the past decade. Earlier characterization of prokaryote 16S rRNA by oligonucleotide cataloging showed the mollicutes to be related to Gram-positive bacteria of low guanine + cytosine DNA composition, especially to a small subgroup of clostridia (*C. ramosum* and *C. innocuum*) (Woese et al., 1980). Later, sequence analysis of 5S rRNA from various mollicutes and *C. innocuum* confirmed this phylogenetic relationship and indicated that the mollicutes diverged from clostridial ancestors, first as an *Acholeplasma* branch (with a concomitant decrease in genome size from about 3000 to 1000 MDa), and then—with further division of this stem—to form the *Anaeroplasma* and *Spiroplasma* branches (Rogers et al., 1985). Information from this study also indicated that the *Spiroplasma* branch probably evolved further through independent genome reductions to about 500 MDa, to eventually yield the *Mycoplasma* and *Ureaplasma* lines. More recent comparisons of small sequence subunits of 16S ribosomal RNA from about 40 mollicutes and their walled ancestors have been reported (Weisburg et al., 1989). These studies support in general earlier conclusions on the evolution of mollicutes and provide further evidence for the pivotal ancestral role of spiroplasmas in the phylogeny of other sterol-requiring mollicutes. This work also demonstrated a close phylogenetic relationship between 10 selected *Spiroplasma* species with diverse phenotypic features and habitats.

Habitats of Spiroplasmas

As Flora of Insects

Spiroplasmas are found most frequently in the insect gut, less frequently in hemolymph, and occasionally in various organs (such as the salivary gland) (Clark, 1982; Hackett and Clark, 1989; Hackett et al., 1990). Spiroplasmas in the insect gut are generally acquired by natural feeding, either from plant tissue, nectar, or ingestion of other insects. When the organism is part of the normal gut flora, the same *Spiroplasma* species is often found in other members of the same insect species. Although numerous spiroplasma inhabitants of the insect gut have been cultivated on artificial media, many of the helical organisms observed in hemolymph are not yet cultivable (Clark, 1982).

The presence of spiroplasmas in the insect gut lumen can also lead to pathological consequences for the insect host and, in the case of insect-borne plant pathogens, for the plant host as well. Spiroplasmas involved in important vector-transmitted plant diseases (citrus stubborn and corn stunt) must pass through a complex biological cycle involving uptake from the sieve cells of the plant phloem with subsequent passage or multiplication in the insect alimentary tract, gut epithelium, basement membrane, hemocoel, and possibly some internal organs. Eventual passage of organisms from the hemocoel into the salivary cells and salivary duct, from which reinoculation of healthy plants takes place, is the final stage in the insect phase of the cycle. In the course of this cycle, induction of disease in the plant is the rule (Daniels, 1979). On the other hand, induction of disease in the insect may be related to unusual host-spiroplasma associations (Whitcomb et al., 1973; Whitcomb and Williamson, 1975, 1979; Madden and Nault, 1983; Whitcomb, 1989). Although spiroplasmas may occur intracellularly in their insect hosts (e.g., gut or salivary cells), their principal site of residence in leafhoppers is the hemolymph, where they may occur in considerable numbers.

In some insects, spiroplasmas that are ingested can prove highly invasive and lethal to the host. This occurs when the spiroplasma has the ability to penetrate the gut epithelium, thereby entering the hemolymph and inducing septicemia. These infections are observed in honeybees following exposure to either *S. melliferum* or *S. apis,* both of which are considered to be important insect pathogens (Clark, 1977; Clark et al., 1985; Mouches et al., 1982a). However, other spiroplasmas (such as the group II *Drosophila* organism) appear to be commensalistic residents of insect hemolymph. The organism is transovarially passed to ensuing generations and, although some decrease in the longevity of the infected host is observed, female flies frequently contain prodigious num-

Table 1. Group designations, species names, and characteristics of members of the genus *Spiroplasma*.

Binomal and/or common name	Group[a]	Strain designations[b]	GC content (mol%)	Glucose fermentation	Arginine hydrolysis	Principal host	Disease incited
Spiroplasma citri (Citrus stubborn spiroplasma)	I-1	Maroc-R8A2^T(27556); C189(27665); Israel	26	+	+	Dicots, leafhoppers	Citrus stubborn
S. melliferum (Honey bee spiroplasma)	I-2	BC-3^T(33219); AS 576(29416)	26	+	+	Bees	Honeybee spiroplasmosis
S. kunkelii (Corn stunt spiroplasma)	I-3	E275^T(29320); I-747(29051); B655(33289)	26	+	+	Maize, leafhoppers	Corn stunt
277F spiroplasma	I-4	277F(29761)	26	+	+	Rabbit ticks	None known
Green leaf bug spiroplasma	I-5	LB-12(33649)	26	+	+	Green leaf bugs	None known
Maryland flower spiroplasma	I-6	M55(33502); ET-1	28	+	+	Flowers, *Eristalis* flies	None known
Cocos spiroplasma	I-7	N525(33287); N628	26	+	+	Coconut palms	None known
S. phoeniceum (*Vinca* spiroplasma)	I-8	P40^T(43115)	26	+	+	*Catharanthus roseus*	Periwinkle disease
Sex-ratio spiroplasmas	II	DW-1(43153)	26	ND	ND	*Drosophila*	Sex ratio trait
S. floricola	III	23-6^T(29989); BNR1(33220); OBMG(33221)	26	+	−	Insects, flowers	None known
S. apis	IV	B31^T(33834); SR 3(33095); PPS1(33450)	30	+	+	Bees, flowers	"May disease"
S. mirum	V	SMCA^T(29335); GT-48(29334); TP-2(33503)	30	+	+	Rabbit ticks	Suckling mouse cataract disease
Ixodes spiroplasma	VI	Y32(33835)	25	+	−	*Ixodes pacificus* ticks	None known
Monobia spiroplasma	VII	MQ-1(33825)	28	+	−	*Monobia* wasps	None known
Syrphid spiroplasma	VIII	EA-1(33826)	30	+	+	*Eristalis arbustorum* flies	None known
Cotinus spiroplasma	IX	CN-5(33827)	29	+	+	*Cotinus* beetles	None known
S. culicicola (Mosquito spiroplasma)	X	AES-1^T(35112)	26	+	−	*Aedes* mosquitoes	None known

(continued)

Table 1. *Continued*

Binomal and/or common name	Group[a]	Strain designations[b]	GC content (mol%)	Glucose fermentation	Arginine hydrolysis	Principal host	Disease incited
Monobia spiroplasma	XI	MQ-4(35262)	26	+	+	*Monobia* wasps	None known
Cucumber beetle spiroplasma	XII	DU-1(43210)	25	+	−	*Diabrotica undecimpunctata* beetles	None known
S. sabaudiense (Mosquito spiroplasma)	XIII	Ar-1343[T](43303)	30	+	+	*Aedes* mosquitoes	None known
Ellychnia spiroplasma	XIV	EC-1(43212)	26	+	−	*Ellychnia corrusca* beetles	None known
Leafhopper spiroplasma	XV	I-25(43262)	26	+	−	*Cicadulina* leafhoppers	None known
Cantharis spiroplasma	XVI	CC-1(43207); Ar-1357; MQ-6	26	+	−	*Cantharis* beetles	None known
Deer fly spiroplasma	XVII	DF-1(43209)	29	+	+	*Chrysops* flies	None known
Tabanid spiroplasma	XVIII	TN-1(43211)	25	+	−	*Tabanus nigrovittatus*	None known
Firefly spiroplasma	XIX	PUP-1(43206)	26	+	−	*Photuris pennsylvanicus* beetles	None known
Colorado potato beetle spiroplasma	XX	LD-1(43213)	25	+	ND	*Leptinotarsa decemlineata*	None known
Flower spiroplasma	XXI	W115(43260)	24	+	ND	*Prunus* sp. flowers	None known
S. taiwanense (Mosquito spiroplasma)	XXII	CT-1[T](43302)	25	+	−	*Culex tritaeniorhynchus*	None known
Tabanid spiroplasma	XXIII	TG-1(43525)	26	+	−	*Tabanus gladiator*	None known
Chinese flower spiroplasma	XXIV	CCH(43960)	29	+	+	*Calystegia hederaceae*	None known

+, property present; −, property absent; ND, not done.
[a]Groups assigned on the basis of failure to cross-react in growth inhibition, metabolism inhibition, and deformation tests.
[b]Accession numbers from the American Type Culture Collection are given in parentheses.
Modified and reproduced with permission from International Journal of Systematic Bacteriology (Tully et al., 1987). (For detailed history of strains and references, see Williamson et al., 1989a; Guo et al., 1990).

bers (10^{10}–10^{11} per ml) of the organism in the hemolymph (Williamson and Poulson, 1979).

Spiroplasmas have been reported most often from species of the six evolutionarily advanced insect orders noted below. However, this information on host distribution is probably biased by current sampling techniques. One spiroplasma has been found in an insect representing an ancient order. The significance of this isolate (strain PALS-1) from a dragon fly (*Pachydiplax longipennis*: Odonata) is unclear (Hackett et al., 1990).

ORDER HYMENOPTERA. As noted above, the honeybee is the primary host for two important spiroplasma pathogens, *S. melliferum* (group I-2) and *S. apis* (group IV) (Clark et al., 1985; Mouches et al., 1983a). Bees other than the honeybee have been found to be carriers of spiroplasmas found predominately on floral surfaces, such as group I-6 (M55 strain) organisms and *S. floricola* (Hackett et al., 1984; Hackett and Clark, 1989). Several serologically distinct spiroplasmas have been isolated from the vespid wasp, including strain MQ-1 (group VII) from wasp hemolymph and strains MQ-4 (group XI) and MQ-6 (group XVI) from gut contents (Williamson et al., 1989a; Hackett and Clark, 1989).

ORDER COLEOPTERA. A number of spiroplasmas have been found in plant (or flower) feeding beetles, especially the green June beetle (*Cotinus nitida*) (strain CN-5: group IX), the soldier beetle (*Cantharis carolinus*) (strain CC-1: group XVI), the cucumber beetle (*Diabrotica undecimpunctata*) (strain DU-1: group XII), and the Colorado potato beetle (*Leptinotarsa decemlineata*) (strain LD-1: group XX) (Clark et al., 1982; Clark, 1984). Host specificity and current ecological information on beetle isolates have been reviewed recently (Hackett and Clark, 1989). Spiroplasma strain SLH, isolated from the scarabaeid beetle *Melolontha melolontha* and associated with so-called "lethargy disease" in beetles (Giannotti et al., 1981), was later shown to be closely related to *S. floricola* (Bové, 1981).

Firefly beetles and other members of the family Lampyridae have been found to harbor a number of distinct spiroplasmas (Clark, 1984), as well as other mollicutes (Tully et al., 1989; Williamson et al., 1990). The first group XIV representative (strain EC-1) was isolated from gut contents of *Ellychnia corrusca* (Clark, 1984). However, group XIV also occurs frequently in horseflies (Clark et al., 1984; Whitcomb et al., 1990; J. F. Anderson and J. G. Tully, unpublished observations), suggesting a common

feeding prey (including larval predation) by fireflies and horseflies (Hackett and Clark, 1989). A group XIX spiroplasma (strain PUP-1) was isolated initially from the gut contents of both larval and adult forms of *Photuris pennsylvanica* fireflies by Clark (1984). Isolates related but not identical to the PUP-1 strain have also been obtained from mosquitoes (Shaikh et al., 1987). Both larvae and adult *P. pennsylvanica* are predators; the PUP-1 strain could therefore be acquired from prey, including mosquitoes.

ORDER DIPTERA. Flower feeding flies are the habitat of two spiroplasma groups. Strain EA-1 was cultivated from the hemolymph of the syrphid fly *Eristalis arbustorum* and eventually characterized as the representative strain of group VIII (Clark, 1982; Whitcomb et al., 1983; Williamson et al., 1989a). Strain ET-1, from the gut of a second syrphid species (*Eristalis transversus*), was shown to be related to the Maryland flower spiroplasma group I-6 (Hackett et al., 1984).

Extensive searches for spiroplasmas in bloodsucking insects, such as mosquitoes and tabanids, have resulted in isolation of strains representing at least 11 groups. Mosquito isolates were first reported by Slaff and Chen (1982), and strain AES-1 (group X) from *Aedes sollicitans* was eventually characterized and described as the type strain of *S. culicicola* (Hung et al., 1987). Shortly thereafter, other new *Spiroplasma* species were found in mosquitos. These organisms included *S. sabaudiense* (group XIII) in *Aedes sticticus/vexans* (Chastel et al., 1985; Abalain-Colloc et al., 1987), and *S. taiwanense* (group XXII) in *Culex tritaeniorhynchus* (Clark et al., 1987; Abalain-Colloc et al., 1988). Group XVI spiroplasmas have also been found in a number of mosquito hosts (Chastel et al., 1989; Shaikh et al., 1987; D. L. Williamson, personal communication).

Occurrence of spiroplasmas in tabanid flies was established by Clark in 1984 when he isolated the DF-1 strain (group XVII) from gut contents of a female *Chrysops* deerfly and the TN-1 strain (group XVIII) from gut fluids of a female *Tabanus nigrovittatus* (Clark, 1984; Clark et al., 1984). Additional strains isolated from other tabanids were found to be distinct from group XVII and XVIII spiroplasmas. At least one of these strains (TG-1) has been given a group designation (XXIII), following appropriate serological comparisons (Clark et al., 1987; Tully et al., 1987). Ecological information indicates that many species of *Tabanus* in the USA harbor spiroplasmas, especially groups XIV and XXIII (Whitcomb et al., 1990; French et al., 1990). Spiroplasma assemblages in tabanids from various geographic regions differ (Whitcomb et al., 1990). Preliminary data indicates that horseflies may harbor additional spiroplasma groups (R. F. Whitcomb and F. E. French, personal observations).

As noted above, a complex of spiroplasmas (group II) occurs in a number of neotropical species of the genus *Drosophila* (order Diptera), in which they induce sex ratio abnormalities (Williamson and Poulson, 1979; Williamson et al., 1989a; McGarrity and Williamson, 1989).

ORDER LEPIDOPTERA. Butterflies frequently harbor group I-6 strains (Maryland flower spiroplasma) (Hackett et al., 1984). Less frequently, spiroplasmas isolated from butterflies appear to be related either to group X (*S. culicicola*), group XI (MQ-4), or group XVI (CC-1) spiroplasmas (Clark and Whitcomb, 1984). Since many of the insect hosts of these three spiroplasma groups feed on flower nectar, speculation has centered on a common feeding source for transfer of spiroplasmas to butterflies.

ORDER HOMOPTERA. *Spiroplasma citri* (group I-1) can be transmitted to susceptible citrus and other plants by at least six species in three genera of leafhoppers. These vectors include *Circulifer* (= *Neoaliturus*) *haematoceps, C. tenellus, Macrosteles fascifrons, Scaphytopius nitridus, S. acutus delongi,* and *S. californiensis* (Calavan and Bové, 1989). The corn stunt spiroplasma *S. kunkelii* (group I-2) is also transmitted under natural conditions to susceptible plants primarily by two leafhoppers, *Dalbulus maidis* and *D. elimatus*. Laboratory or greenhouse experiments have shown that 12 other leafhopper species are capable of transmitting *S. kunkelii* to maize (Nault and Bradfute, 1979; Whitcomb, 1989). *Macrosteles fascifrons* leafhoppers have been demonstrated (R. F. Whitcomb, unpublished observations) to be an experimental vector for the plant pathogenic *S. phoeniceum* (group I-8) organism, although the principal vector for natural transmission has not been established (Saillard et al., 1987).

Cicadulina bipunctella bipunctella leafhoppers have been identified as a host of group XV (strain I-25) spiroplasmas (C. Saillard, J. C. Vignault, and J. Bové, unpublished observations; see also Williamson et al., 1989a), and a strain (L89) of group IV honeybee spiroplasma (*S. apis*) was isolated from a froghopper (*Neophilaenus* sp.) in Corsica (Vignault et al., 1980; Tully et al., 1980).

ORDER HEMIPTERA. The group I-5 spiroplasma (strain LB-12) was isolated from the green leaf bug (*Trigonotylus ruficornis;* Hemiptera: Miri-

dae) in Taiwan (Lei et al., 1979). Few other attempts have been made to survey hemipteran insects, although unculturable helical organisms were observed in hemolymph of insects of at least four families (Clark, 1982; Hackett and Clark, 1989).

As Flora of Ticks

Organisms subsequently identified as spiroplasmas have been isolated from rabbit ticks (*Haemaphysalis leporispalustris*) collected in Georgia and Montana (Clark, 1964, 1974; Pickens et al., 1968). Although the Georgia strains (SMCA and GT-48) of *S. mirum* (group V) are serologically identical, they differ in their experimental pathogenicity for selected vertebrate hosts (Clark, 1974; Tully, 1982; McGarrity and Williamson, 1989). These differences were again emphasized when it was found recently that the SMCA but not the GT-48 strain could induce a malignant cell transformation in two vertebrate cell lines (Kotani et al., 1986). A third strain (TP-2) of this group, isolated from rabbit ticks collected in Maryland (Stiller et al., 1981), also differed from the Montana strains in experimental pathogenicity for vertebrates (Tully, 1982). A fourth isolation of *S. mirum* was recently made from rabbit ticks collected in New York state (J. F. Anderson, personal communication). The 277F strain, isolated from a pool of rabbit ticks collected in Montana, shared some serologic reactivity with several group I strains (Williamson et al., 1979). This implied relationship was confirmed by DNA hybridization tests (Junca et al., 1980), and the organism was eventually assigned a subgroup designation (group I-4) (Williamson et al., 1989a). No other representative of this group has been identified in insects, ticks, or other arthropds.

Eight serologically related isolates were cultured from pools of *Ixodes pacificus* ticks collected in Oregon in 1980 (Tully et al., 1981). These spiroplasmas proved to be serologically distinct from all established groups, and a representative strain (Y32) was designated as group VI in the current classification scheme (Whitcomb et al., 1982a; Tully et al., 1987). A limited number of other ticks, representing species in both Ixodidae (hard ticks) and Argasidae (soft ticks), have been examined for spiroplasmas, with negative results (Tully et al., 1983). Because surveys have been limited, the real extent of spiroplasma occurrence in ticks is unknown. We also do not know whether spiroplasma occurrence relates to developmental stages of the tick, or whether vertebrates play an essential role in maintaining spiroplasmas in ticks.

As Flora of Plants

Spiroplasmas are associated with plants in two ways. One role is played by organisms that invade the plant sieve tubes in the course of a biological cycle that also involves homopterous insects (Whitcomb and Williamson, 1979; Calavan and Bové, 1989; Whitcomb, 1989). The second role involves external contamination of floral parts, presumably through deposition of spiroplasmas by flower-visiting insects (Clark, 1978; McCoy et al., 1979; Vignault et al., 1980; Bové, 1984; Hackett and Clark, 1989).

Information on plant responses to spiroplasma infection has come primarily from studies on naturally infected plants (e.g., sweet orange trees and maize) or from experimental infections of plant hosts (Bové, 1984; Calavan and Bové, 1989; Whitcomb, 1989). The symptomatology of spiroplasmal plant diseases ranges widely but most frequently involves chlorosis, leaf mottling, proliferation of growing points, and general stunting of plants, usually with reduction in the size of leaves, flowers, and fruits. Spiroplasmas observed in sectioned sieve tubes of infected plants may appear sinusoidal or helical, depending on the thickness of the section. Plant symptoms appear to be related to the number of spiroplasmas in the sieve tubes (Davis et al., 1972a; Daniels, 1979). Efforts to elucidate the biochemical basis of spiroplasma pathogenicity for plants have been quite limited. Spiroplasma toxins or other metabolic products such as lactic acid have been thought to be elaborated by the organism (Daniels, 1979, 1983). The putative toxin, however, has not been purified or its chemical nature defined. Since spiroplasmas frequently carry extrachromosomal DNA in the form of plasmid or viral DNA (Bové et al., 1989), there has been speculation that these components may play some role in pathogenicity of the organisms for both plants and insects (McGarrity and Williamson, 1989; Bové et al., 1989; Calavan and Bové, 1989). However, it is not clear whether plant pathogenic effects of spiroplasmas are a consequence of direct action of spiroplasmal gene products or of spiroplasmal utilization of essential cellular or extracellular nutrients.

Natural *S. citri* infections have been found in more than 35 species of annual weeds and cultivated plants or flowers, including periwinkle, shortpod mustard, wild turnip, London rocket, a variety of brassicaceous crops (such as Chinese cabbage, broccoli, cabbage, Brussels sprouts, radish, etc.), China aster, lettuce, horseradish, cactus, cherry, peach and pear, and numerous varieties of citrus (Calavan and Oldfield, 1979; Calavan and Bové, 1989).

Experimental plant infections with *S. citri* have been reported in an additional 38 plant species (Calavan and Bové, 1989). *Spiroplasma kunkelii* appears to have a more restricted natural and experimental host range. In nature, *S. kunkelii* has been found only in maize and teosinte, and experimental infections have been demonstrated in perennial ryegrass (*Lolium perenne),* broad bean, periwinkle, mustard, pea, radish, and spinach (Whitcomb, 1989). The ability to study artificial infections of both *S. citri* and *S. kunkelii* has not only greatly enhanced our understanding of plant host responses but defined some important insect vector relationships with spiroplasmas (Golino and Oldfield, 1989).

Natural mixed infections of either *S. citri* or *S. kunkelii* with mycoplasma-like organisms (MLOs) (see McCoy et al., 1989 and Chapter 229) or viruses have also been observed in a number of plant species (Bové, 1984; Calavan and Bové, 1989; Whitcomb, 1989). In some *S. citri*-MLO coinfections, pathogenic effects of spiroplasmas on plants are limited and symptomatology may be either atypical or typical of the coinfecting MLO. Mixed infections with *S. citri* and some viruses (for example, in brittle-root disease of horseradish with mosaic virus or in citrus tristeza virus disease) have shown little influence on invasion by *S. citri* or in development of the viral plant disease (Calavan and Bové, 1989). Natural coinfections in corn with *S. kunkelii* and MLOs or viruses are very common (Nault and Bradfute, 1979; Whitcomb, 1989).

Much less is known about the sites of deposition or relationships of flower spiroplasmas to their plant hosts. Attempts to isolate spiroplasmas from internal tissues (sieve tubes) of surface-sterilized flowers and other plant parts have been unsuccessful (McCoy, 1979). This suggests that these organisms are part of the surface flora of plants. The ability of wall-less prokaryotes to exist on such surfaces in the absence of a close association with living cells may be surprising. However, some spiroplasmas (*S. citri* and *S. kunkelii*) remain viable for extended periods at room temperature when droplets of cultured organisms are dried on glass slides (R. F. Whitcomb, unpublished observations). Floral nectar is a more likely primary site for spiroplasma deposition by insects, since it offers a possible environment for multiplication of organisms and an infectious locus for transmission of spiroplasmas between insects (Clark, 1978, 1982; Whitcomb, 1981; Giannotti and Giannotti, 1986; Hackett et al., 1984; Hackett and Clark, 1989).

Isolation and Cultivation

Development of Cultivation Techniques

Since the citrus stubborn organism (*S. citri*) was considered initially to be a "mycoplasma-like organism", first attempts to cultivate the agent involved classical mycoplasma media containing a high-protein basal medium supplemented with fresh yeast extract and horse serum (Edward, 1947). This formulation, or variations of it, proved inadequate for primary isolation. Media were then developed that more closely resembled the chemical nature and osmolality of infected sieve tubes of plant phloem, utilizing high concentrations of sucrose and sorbitol. Medium SMC, and other minor modifications, provided conditions essential for the continuous cultivation (Saglio et al., 1971, 1972; Fudl-Allah et al., 1972) and characterization of the organism (*S. citri*) (Saglio et al., 1973; Cole et al., 1973a, 1973b). SMC medium was later modified to the BSR formulation, which permitted direct isolation of the organism from either plant material (citrus and periwinkle) or leafhopper vectors (Bové and Saillard, 1979; Bové et al., 1983b).

The SMC or BSR formulations, however, were inadequate for primary cultivation of *S. kunkelii*. After a large number of medium variations had been tried unsuccessfully, two groups of workers independently achieved the continuous cultivation of this organism on formulations designated C-3 (Chen and Liao, 1975) or M1 (Williamson and Whitcomb, 1975). The formulation of C-3, which was based upon the chemical composition of plant phloem, was later simplified to medium C-3G (Liao and Chen, 1977). The essential ingredients in M1 medium have been provided by the addition of insect or vertebrate cell culture supplements to the SMC formulation. M1 medium was later modified to M1A when the growth characteristics of both *S. citri* and the corn stunt agent were compared (Jones et al., 1977). These studies were, in turn, critical to formulation of the SP-4 culture medium that provided essential growth factors for cultivation of two fastidious tick-derived spiroplasmas: *S. mirum* strains from rabbit ticks (Tully et al., 1977) and strains isolated from *Ixodes pacificus* (Tully et al., 1981). Components of a vertebrate cell culture supplement (CMRL 1066, GIBCO, Grand Island, N.Y.) and fetal bovine serum appeared to be important factors for the success of SP-4 medium.

However, spiroplasmas observed microscopically in the hemolymph of *Drosophila* species (group II sex ratio organisms) and in gut fluids

of a beetle (group XX, Colorado potato beetle spiroplasma) resisted vigorous attempts at cultivation on a variety of artificial media (Williamson et al., 1983). Two developments eventually led to the successful cultivation of these agents. The first of these was the use of insect cell culture media (e.g., Schneider's *Drosophila* medium) as a supplement to media used for the corn stunt spiroplasma (Williamson and Whitcomb, 1975; Jones et al., 1977). Secondly, the development of both insect and vertebrate cell culture systems to study spiroplasma growth, attachment characteristics, and pathogenicity (Louis et al., 1978; Steiner et al., 1982, 1984; Garnier et al., 1984b; McGarrity and Kotani, 1984; Yunker et al., 1987) opened up new vistas for the use of insect cells as in vitro medium supplements. The first fastidious spiroplasma to be cultivated in these new formulations was the Colorado potato beetle spiroplasma (group XX), which was grown in co-cultivation with selected coleopteran and lepidopteran cell lines (Hackett and Lynn, 1985). Later, this spiroplasma was adapted to grow in broth in the absence of cells (Hackett et al., 1987). A similar approach also was successful in the cultivation of the *Drosophila* sex ratio spiroplasma (group II) (Hackett et al., 1986).

Most spiroplasmas isolated from plant surfaces, or from insect gut fluids, have been less fastidious in their growth needs. Many of these strains can be cultivated in simple formulations such as BSR, C-3, or C-3G (Liao and Chen, 1977), or in a chemically defined medium (CC-494) (Chang and Chen, 1982). Although these formulations might be successful in supporting spiroplasma growth, there is some evidence that omission of certain substrates such as arginine in media for primary isolation may lead to significant loss of some in vitro metabolic pathways (Clark et al., 1985).

Purification and Conservation of Cultures

There is substantial evidence that two or more serologically distinct spiroplasmas can occur in the same insect tissue (French et al., 1990) or on the same floral surface (Lei et al., 1979). Thus, any mollicute cultivated from plant or insect material should be purified by filter-cloning procedures prior to characterization (Tully, 1983b). In this procedure, a broth culture is passed through a membrane filter (220–300nm) and the diluted filtrate plated onto solid medium. Following incubation, a single colony is selected and transferred to fresh broth, and a logarithmic-phase culture again filter-cloned. This procedure should be completed at least three times. Some spiroplasma strains may not form distinct colonies on solid media (see below). In these instances, the purification technique can be performed with a broth culture and a microtiter plate, using a terminal dilution procedure that ensures a very low probability of more than one helical organism per well (Whitcomb and Hackett, 1986).

Spiroplasmas are readily preserved by lyophilization of logarithmic phase cultures dried directly in the medium used for their growth (Leach, 1983). This is the preferred method for long-term storage. Freezing broth cultures at $-70°C$ also provides adequate temporary storage conditions.

Factors Affecting Spiroplasma Isolation

Several host and cultural factors are important in primary isolation and maintenance of spiroplasmas during early passage levels (Tully, 1983c). These include the chemical and physical suitability of the culture medium environment in meeting the growth requirements of the organism, numbers of organisms in source materials, presence or absence of toxic or inhibitory substances in the inoculum, use of dilution and blind-passage techniques to reduce inhibitory materials, and dark-field monitoring of the cultures for growth.

Choice of an adequate medium for primary isolation and/or subsequent sustained cultivation of these organisms is still much more of an art than a science. As new information is obtained on the nutritional requirements and metabolism of the organisms (Chang, 1989; Pollack et al., 1989), improved culture medium formulations are expected. Preliminary trials for cultivation of new spiroplasmas should be performed on media such as SP-4 or M1D. These rich media have been successful in primary isolation of many spiroplasmas and their ability to support growth of most established strains is well documented (Whitcomb, 1983; Tully et al., 1987). Once an organism has been isolated, simpler medium formulations can often be used for maintenance and characterization. The use of antibiotics in culture media has made possible spiroplasma isolation in the presence of a variety of other competing organisms in plants, insects, and other arthropods. Penicillin (500–1000 units/ml) and polymyxin B (500–1000 units/ml) have been the drugs of choice for inhibition of either Gram-positive or Gram-negative prokaryotes, respectively (Whitcomb, 1983; Tully, 1983d).

Other cultural factors (e.g., atmospheric conditions, temperature of incubation, pH of medium) can also be important in primary isolation of mollicutes. For most spiroplasmas, aerobic incubation of broth cultures, anaerobic incubation of agar plate cultures, and an initial medium pH level of 7.5–7.8 are appropriate and successful. Although many spiroplasmas appear to grow best at or near 30°C, few strains or species have been examined in sufficient detail to establish either an optimum temperature or temperature range for growth. *Spiroplasma citri* and *S. phoeniceum* strains have been reported to grow best at 32°C (Saglio et al., 1973; Saillard et al., 1984, 1987). Strains of *S. mirum* are unique among most spiroplasmas, since they grow almost equally well at temperatures of 30°C to 37°C and are the only spiroplasmas currently known to grow at 37°C (Tully et al., 1982).

Isolation attempts for spiroplasmas from internal plant tissues (young leaves, stems, seeds, etc.) must necessarily involve some type of preparatory plant surface sterilization, since surface microorganisms pose a contamination threat. Frequently, a two minute immersion in 70% alcohol, a five minute immersion in 1% sodium hypochlorite, followed by two or three rinses in sterile distilled water are satisfactory (Bové et al., 1983b). Plant material is cut or chopped into small pieces (or homogenized) in 3–5 ml volumes of the selected broth medium. The fluid portion of the mixture can then be taken up with a syringe and needle, and the material filtered through a sterile 450 nm membrane filter. A small volume of this inoculum (0.2–0.3 ml) is added to about 1.8–2.7 ml of broth medium, and at least 5 to 10 serial 10-fold dilutions of the inoculum made in fresh medium. Some portion of the inoculum may also be used to inoculate solid media. The broth and agar cultures are incubated at both 30° and 32°C.

In attempts to isolate spiroplasmas from flower or plant surfaces, surface sterilization must be avoided. In these procedures, plant material should be washed for about 1 minute in 2–4 ml of broth medium, the fluid removed by pipette or syringe, and the wash fluid passed through a sterile membrane filter with an average pore diameter of 450 nm. Procedures similar to those outlined above for serial dilution of the inoculum in broth medium, inoculation of agar plates, and incubation should be performed (Bové et al., 1983b).

Techniques for isolation of spiroplasmas from insects or other arthropods have been described by Markham et al. (1983). Hemolymph from infected honeybees or leafhoppers frequently contains large numbers of spiroplasmas and few other microorganisms. Hemolymph removed from surface-sterilized insects can usually be added directly to broth media and serially diluted for enumeration of spiroplasmas. Jones et al. (1977) found that 0.1 μl of hemolymph collected from a single leafhopper infected with *S. kunkelii* and added to 1 ml of M1A medium gave nearly 100% isolation rates.

Whole insects or ticks are most often surface sterilized, as described above for plant material, with individual or pools of arthropods macerated in 1–2 ml volumes of the selected culture medium. Following filtration through 450 nm membrane filters, the inoculum is treated in a manner similar to that described for plant material.

Efforts to microscopically assess the spiroplasma content of insect or plant inocula prior to cultivation techniques, although very useful, are not entirely reliable. Dark-field microscopy of certain fluids (hemolymph or vascular plant sap) often reveals the presence of helical forms, while inocula such as homogenized insect fluids may contain particulate host material that interferes with optical detection of helical cells. In addition, not all viable spiroplasma cells are helical (Townsend et al., 1977, 1980a; Whitcomb and Williamson, 1975). Finally, it must be remembered that all helical forms observed are not necessarily spiroplasmas.

Spiroplasmas can often be isolated from diluted but not undiluted extracts of plant or insect tissues, suggesting that inhibitors are present in host tissues. For this reason, it is useful to perform serial dilutions of such extracts in fresh culture media in all primary isolation attempts (Bové et al., 1983b; Markham et al., 1983; Taylor-Robinson and Chen, 1983). These inhibitory substances have been shown to be either spiroplasmastatic or spiroplasmacidal (Liao and Chen, 1980).

The first suggestion of spiroplasma growth in primary cultures is often a slight to moderate acidification of the medium, and in most instances, slight turbidity. However, spiroplasmas differ greatly in their abilities to ferment carbohydrates, and some have active pathways for catabolism of arginine, so a decrease in medium pH cannot be relied upon as a strict indicator of growth. Although the arginine dihydrolase pathway is present in many spiroplasmas (Table 1), it is rarely dominant during primary isolation, when fermentative pathways are usually assertive.

Dark-field microscopy is an important technique for monitoring broth cultures containing inocula for primary isolation of spiroplasmas. An increase in the number of helical cells, especially those with 1–3 turns, suggests successful isolation. However, in some primary or early passage broth cultures cells may be deformed and few helical forms are seen. It is thought that

these morphologic changes might result from deficiencies in the culture medium. If these deformed cells persist in cultures passed two or three times in fresh medium, changes in the medium formulation are advisable. Some strains, such as the Y32 (group VI) spiroplasma, consist almost entirely of non-helical filaments or of very tightly coiled forms whose helicity is not apparent when viewed at the magnification employed in dark-field microscopy (Tully, 1983a).

Early broth passages of newly isolated spiroplasmas are likely to require prolonged incubation periods, in comparison to the shorter time required for strains well adapted to a particular medium. During this period, sudden "collapses" may occur, in which growth (as measured by turbidity and pH change) may be delayed for unusually long periods of time. Occasionally, broth cultures may have to be passaged in the absence of any visible changes in pH or turbidity (blind passage) to maintain the strain. Although these retrogressions have not been studied systematically, they often appear to be associated with production of high yields of virions of spiroplasmaviruses. It is thought that many spiroplasma strains carry viral genomic material that is not expressed by production of virions (Bové et al., 1989). Thus, preservation of early broth passages of an isolate by freezing at −70°C can provide a reserve of stock cultures for a line that may not survive an outbreak of virion production. As spiroplasma strains adapt to growth in fluid medium, overt virus infections tend to become less severe.

Some spiroplasma isolates may fail to produce colonies on solid media in early passages. Rapidly growing strains of higly motile spiroplasmas may produce diffuse growth over the entire surface of an agar plate; this often occurs when the agar content is less than 0.8%, a concentration commonly used for solid growth of mollicutes. Frequently, better colony growth, as exemplified by discrete, biphasic, fried egg-type colonies, occurs on media formulations that are less suitable for the organism (for example, on agar media containing horse rather than fetal bovine serum), or when the concentration of agar (specifically, washed Noble agar or agarose) is increased from about 0.8% to 1.6–2.25% (Whitcomb, 1983). Strains of *S. mirum* do not form fried egg colonies under any known cultural condition, although satellite colonies arise around a small central zone of growth when strains are grown on so-called "hard" agar (2.25%)(Tully et al., 1982). Likewise, spiroplasmas that have lost some helicity, such as the ASP-1 strain of *S. citri* (Townsend et al., 1980a) or *S. melliferum* motility mutants (Cohen et al., 1989) form classical fried egg colonies even on "soft" (0.8%) agar.

Maintenance of possible plant and/or insect pathogenicity of newly isolated spiroplasmas is also of immediate concern during adaptation of the organism to artificial media. Adapted strains of *S. citri* and *S. kunkelii* may lose pathogenicity after 10 to 20 in vitro passages. This change in virulence is related to the inability of the organism to complete its natural cycle involving plant and insect hosts (Whitcomb and Williamson, 1979; El-Bolok, 1981). For other spiroplasmas (e.g., *S. melliferum*, *S. apis*, *S. mirum*), pathogenicity for single hosts (in contrast to ability to complete an entire cycle) has been either retained for much longer periods of in vitro passage or drifts toward increased pathogenicity noted (Tully, 1982; McGarrity and Williamson, 1989; Williamson et al., 1989b). Whatever changes might occur in virulence characteristics, lyophilization of organisms at early passage levels is an important insurance against both strain loss and preservation of initial pathogenic features.

Isolation Media

A wide variety of medium formulations has been used to grow spiroplasmas, including media employed in cultivation of classical, nonhelical mollicutes. The use of these medium variations in primary isolation or maintenance has recently been summarized (Whitcomb, 1983). The specific details for three of the most widely used formulations are given here.

BSR medium is employed primarily in isolation of *Spiroplasma citri* from citrus and other plants (especially periwinkle), for cultivation of the organism from leafhopper vectors (Bové et al., 1978; Bové and Saillard, 1979), and for cultivation of a variety of other less fastidious spiroplasmas (Whitcomb, 1983).

Medium BSR (Bové et al., 1978)

Medium BSR contains, per liter:

Beef heart infusion (Difco)	25 g
Glucose	1 g
Fructose	1 g
Sucrose	10 g
Sorbitol	70 g
Phenol red	20 mg
Deionized water	900 ml

The basal medium is autoclaved at 121°C for 15 min. Horse serum (100 ml) is added as a sterile supplement to the cooled medium. The final pH should be about 7.6.

As noted above, *S. kunkelii* can be grown on a number of different medium formulations.

However, some isolates are more fastidious in growth requirements, particularly in primary isolation, and may not be easily grown on some simplified formulations. Anaerobic incubation has been reported to enhance primary isolation of *S. kunkelii* (Davis et al., 1984). M1 medium (Williamson and Whitcomb, 1975) was modified later to M1A (Jones et al., 1977) or M1B (fetal bovine serum replaced with bovine serum fraction)(Whitcomb, 1983). The current modification, medium M1D (Whitcomb, 1983), has been shown to support the primary isolation of *S. kunkelii* and the growth of a large number of other spiroplasmas (Whitcomb, 1983).

Medium M1D (Jones et al., 1977; Whitcomb, 1983)

The basal medium contains:

Mycoplasma Broth Base (BBL)	2.1 g
Tryptone (Difco)	1.0 g
Peptone (Difco)	1.8 g
Sucrose	1.0 g
Glucose	0.8 g
Fructose	0.8 g
Sorbitol	7.2 g
Phenol red (0.1% aqueous solution)	6.0 ml
Deionized water	80.0 ml

The basal medium is sterilized at 121°C for 15–20 min. The final pH should be about 7.8. The following sterile supplements are added to the cooled basal medium: 160 ml Schneider's *Drosophila* medium (GIBCO), 50 ml fetal bovine serum (Hyclone, Sterile Systems, Inc, Logan UT) heat inactivated at 56°C for 1 hr., and 2.5 ml of a stock penicillin solution (100,000 units/ml).

The tick spiroplasmas, as exemplified by *S. mirum* and the *Ixodes* group VI spiroplasma, can be adapted to grow on several spiroplasma or mollicute medium formulations. However, only the SP-4 medium has been shown to support the primary isolation of these organisms from chick embryo fluids (Tully et al., 1977) or tick suspensions (Stiller et al., 1981; J. F. Anderson, unpublished observations). This medium is also suitable for growth of most other known spiroplasmas, with the possible exception of *S. kunkelii* (group I-3), the *Drosophila* spiroplasma (group II), and the Colorado potato beetle spiroplasma (group XX).

Medium SP-4 (Tully et al., 1977; Whitcomb, 1983)

The basal medium contains:

Mycoplasma broth base (BBL)	1.4 g
Tryptone (Difco)	4.0 g
Peptone (Difco)	2.13 g
Phenol red (0.1% aqueous solution)	8.0 ml
Distilled water	262 ml

The basal medium is sterilized by autoclaving at 121°C for 15–20 min. The final pH should be 7.8. The following sterile supplements are added to the basal medium: 10x CMRL 1066 tissue culture supplement (with glutamine) (GIBCO), 20 ml; freshly prepared 25% yeast extract (GIBCO), 14 ml; 4% yeastolate solution (Difco), 20 ml; fetal bovine serum (Hyclone) (heat activated at 56°C for 1 h), 68 ml; glucose (50 % aqueous solution), 4 ml; penicillin (100,000 units/ml), 2 ml.

Identification

Morphological Characteristics

Spiroplasmas are motile, helical, wall-free prokaryotes (Whitcomb and Tully, 1984). These morphological features are readily demonstrated and are essential for preliminary identification. While helical filaments occur in broth cultures and in many plant and insect fluids, a number of conditions (osmolality of fluids, tissue fixatives, culture age, etc.) can cause spiroplasmas to assume a coccoid form. Dark-field microscopy at magnifications of 1,000–1,250× is the most useful means of monitoring spiroplasma broth cultures or for examination of insect hemolymph (Fig. 1). However, some spirochetes possess the morphology and movement observed in spiroplasmas, and it is presently not possible to absolutely distinguish these two types of microbes by direct light microscopy. Electron microscopic examination of both thick and thin sectioned preparations of spiroplasmas usually shows cells devoid of cell wall, periplasmic fibrils (axial filaments), and flagella—morphologic features that are seen in most spirochetes (see Chapter 191).

In young broth cultures of spiroplasmas, short helical forms (two-turn helix) predominate (Garnier et al., 1981). During the logarithmic phase of growth, the helical forms increase in length (most often to about 4 turns) and divide by constriction, releasing two-turn elementary helices (Garnier et al., 1981; Bové et al., 1989). In the stationary phase of growth, the helices may greatly elongate and then show progressive loss of helicity and motility, or they may aggregate into large "Medusa heads" that contain clumps of helical and non-helical organisms. Some *S. floricola* strains may become coccoid in the stationary phase (Whitcomb and Coan, 1980), which occurs rapidly with this fast-growing organism.

Newly formed elementary helices show morphological polarity, with a blunt and a tapered end (Garnier et al., 1984a, 1984b). *S. citri* strains demonstrated oxidoreduction sites and attachment to insect cell cultures only at the blunt end, whereas in other spiroplasmas there appears to be little relationship between mor-

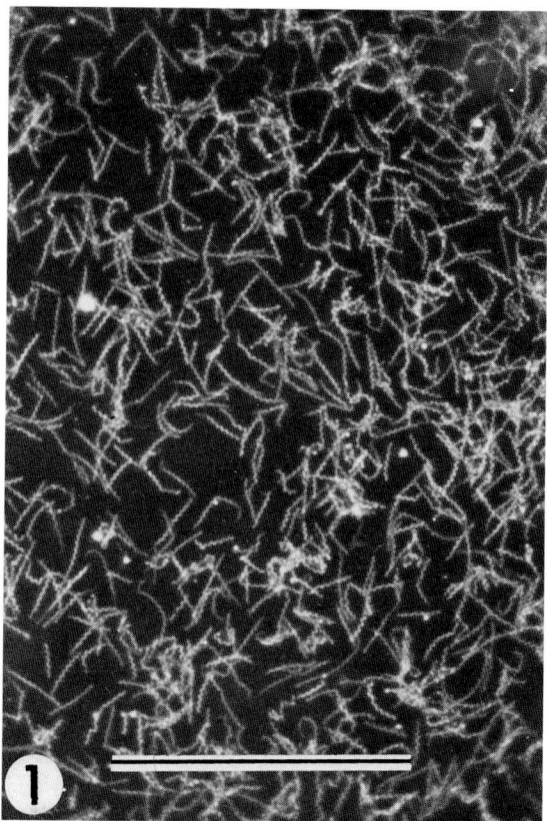

Fig. 1. Photomicrograph of a dark-field preparation of sex ratio spiroplasmas (group II) in the hemolymph of *Drosophila willistoni*. (Courtesy of D. L. Williamson.) Bar = 10 µm.

phologic polarity and cytadherence (Bové et al., 1989).

Motility in the logarithmic phase is characterized by a rapid spinning along the long axis of the cell and flexional or undulating movements. Both translational movement and chemotaxis have been demonstrated in spirolasmas (Davis et al., 1975; Daniels et al., 1980b; Daniels and Longland, 1984).

Negatively stained, logarithmic-phase spiroplasmas (if properly fixed) have a characteristic helical morphology. Cell pellets prepared from young broth cultures can be stained with a 3–6% solution of ammonium molybdate and examined by electron microscopy (Cole et al., 1973a, 1973b) (Fig. 2). Individual cells may vary in shape from pleomorphic spherical cells measuring 200–300 nm in diameter to complex branched or unbranched, helical or nonhelical filaments. The helical forms in logarithmic-phase cultures usually measure 100–200 nm in width and may be 3–12 µm in length. For more definitive information, fixed and sectioned cells should be examined by electron microscopy. Logarithmic phase broth cultures should be

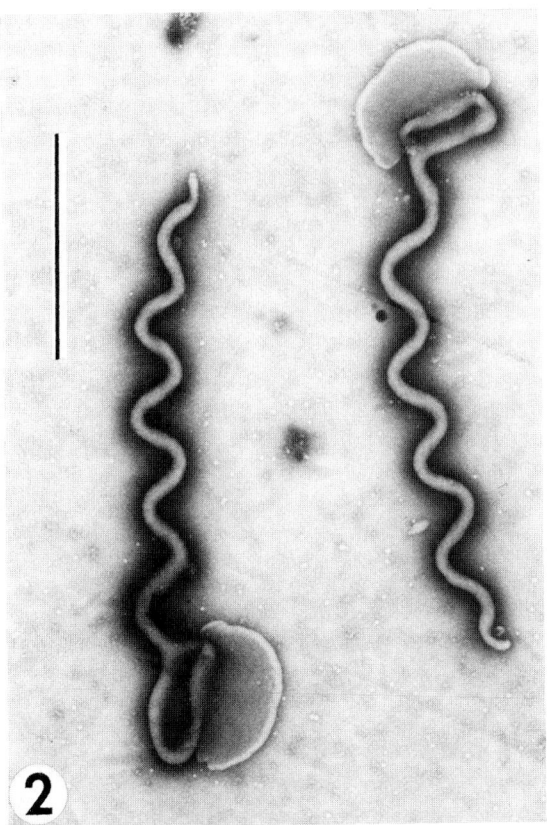

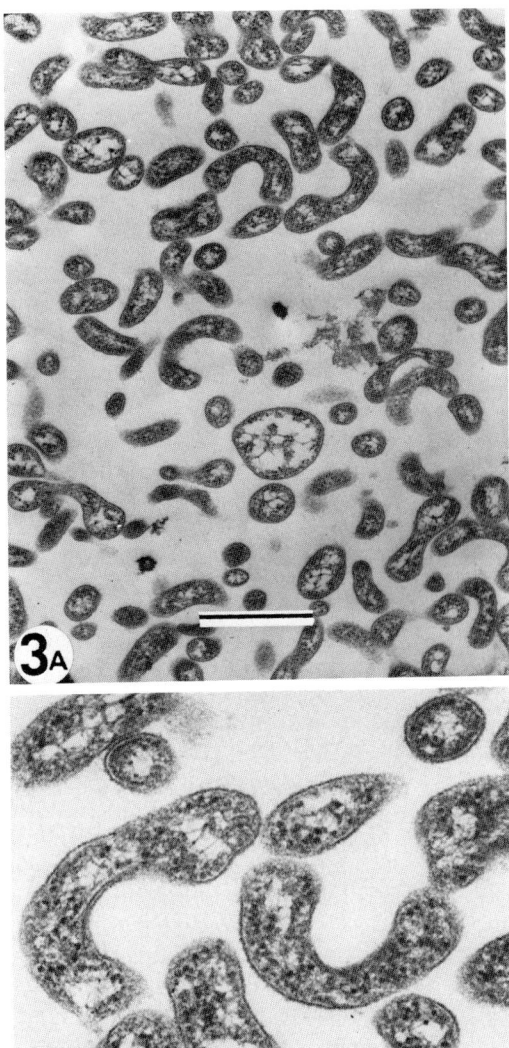

Fig. 2. Electron micrograph of negatively stained *Spiroplasma melliferum* (strain BC-3) from a honey bee. Spiroplasmas were grown in broth for 2 days at 25°C and then sedimented by centrifugation. The cell pellet was suspended in a small amount of 1% ammonium acetate, and the cells were then stained directly on the grid with 2% ammonium molybdate. (Courtesy of R. M. Cole.) Bar = 1 μm.

fixed in 3% glutaraldehyde, preferably by adding 6% fixative to an equal quantity of broth culture, for a period of 4–5 hrs. The fixed cells are sedimented by centrifugation, washed, and prepared for thick and thin sectioning and staining by conventional techniques (Cole et al., 1973a, 1973b). The slightly curved cells observed in sectioned pellets should show the presence of a typical trilaminar unit membrane with no outer cell wall or envelope, flagella, or axial filaments (Fig. 3).

Electron microscopy of negatively stained or thin-sectioned preparations may also show virions of spiroplasmaviruses. At least four morphologically distinct DNA viruses have been found in spiroplasmas (Cole, 1983; Bové et al., 1989). Infections with lytic viruses can apparently affect growth patterns of newly isolated spiroplasmas, and some viral DNA has been shown to be integrated into spiroplasma chromosomal DNA (Bové et al., 1989).

Fig. 3. Electron micrograph of thin section of corn stunt spiroplasma (*Spiroplasma kunkelii*). Cells were sedimented from broth and resuspended in M1 medium containing 1% glutaraldehyde, postfixed in 1% osmium tetroxide, dehydrated in a graded series of acetone, and then embedded in Epon. Sections were stained with 2% aqueous uranyl acetate and Reynold's lead citrate. (Courtesy of D. L. Williamson.) (A) Bar = 1.25 μm; (B) bar = 0.25 μm.

Biological and Molecular Characteristics

The major biological properties of spiroplasmas are summarized in Table 2. Since few of these features differ significantly within *Spiroplasma* species or group representatives, single characteristics have limited identification value. Despite impressive developments in the knowledge of the cell biology of spiroplasmas (Bové et al., 1989), there is no single genus-specific property that unequivocally and consistently identifies all members of the genus from other mollicutes.

Table 2. Biological characteristics of spiroplasmas.

Property	Characteristic
Helical morphology (logarithmic phase cultures)	+
Motility	+
Filterable (220–450 nm porosity membranes)	+
Glucose fermentation	+
Arginine hydrolysis	V
Urea hydrolysis	−
Temperature for growth of most strains	about 30°C
Cholesterol requirement	+
Colonies hemadsorb guinea pig red blood cells	V
Film and spot reaction	V
GC content of DNA (mol%)	24–30
Genome size (megadaltons)	about 1000
Association with arthropods	+

+, all type strains of species or group representatives show characteristic; −, none of the type strains of species or group representatives show characteristic; V, strains vary inter- or intra-specifically in their properties.

For references to specific techniques to assess biological features, see Subcommittee, 1979 or Tully et al., 1982.

Biological characteristics that are important in defining a spiroplasmas, in addition to morphologic and serologic properties, include determination of glucose and arginine metabolic pathways, filterability, growth requirement for cholesterol, and DNA base composition (Whitcomb et al., 1987; Williamson et al., 1989a). Detailed methodologies for measuring these properties and other biologic activities of spiroplasmas have been presented (Razin and Tully, 1983).

Electrophoretic analysis of spiroplasma proteins, using either one- or two-dimensional gels, has been used to study the nature and distribution of membrane proteins (Wróblewski et al., 1977; Mouches et al., 1979; Wróblewski et al., 1984) and to identify fibril proteins (Townsend et al., 1980b). When applied to spiroplasma identification, one-dimensional gels provided sufficient information for limited group identification of spiroplasmas (Mouches et al., 1979; Daniels et al., 1980a). Two-dimensional gels could distinguish group I subgroup members (Bové et al., 1983a) and various strains of group IV spiroplasmas (Mouches et al., 1982b). Whether these techniques could be used to identify individual members of the greatly expanded current list of group representatives (Table 1), has not been established.

The application of genetic probes to the identification of *Spiroplasma* species has been limited to date, although rapidly increasing information on the spiroplasma genome (Bové et al., 1989) suggests that genetic probes will play a

major role in future development of diagnostic techniques. A plasmid probe carrying DNA from a suspected plasmid derived from *S. citri* was shown to have high sensitivity in the detection of *S. citri* infections in plants and leafhoppers (Nur et al., 1986). Later, the so-called "plasmid" derived from *S. citri* was shown to be the replicative form of a type 1 spiroplasma virus (SpV1) present in many *S. citri* strains (Bové et al., 1989).

Serological Relationships

Serologic characterization of spiroplasmas is the most critical element in the identification of new isolates. A number of techniques have been developed to measure these serologic relationships (Williamson et al., 1989a; Chen et al., 1989), each of which has certain advantages or limitations. The deformation test (DF) (Williamson et al., 1978; Williamson, 1983) is very useful for rapid serologic screening of candidate isolates against antisera to group representatives (Williamson and Tully, 1982; Williamson, 1983; Williamson and Whitcomb, 1983; Tully et al., 1987). Isolates that exhibit a heterologous DF titer of 1:40 or greater with established spiroplasma groups are considered candidates for serovar (or subgroup) status. Resolution of the status of such strains requires that antiserum be prepared and reciprocal DF tests performed against all strains showing partial DF cross reactions. In such testing, usually one of two results will be observed: 1) reciprocal DF testing will show only one-way cross-reactions, in which case the new isolate is probably a new group; or 2) reciprocal DF testing shows significant two-way cross-reactions against one or more group antigens, in which case the candidate is designated a serovar or subgroup within the established group (Tully et al., 1987).

The spiroplasma growth inhibition (GI) test (Whitcomb et al., 1982b) is less sensitive than the DF test in establishing subgroup or serovar relationships, but the test is useful in serologic screening of isolates able to grow on solid media.

The metabolism inhibition test (MI) (Williamson et al., 1979; Williamson, 1983) has proved to be important in refined serologic analyses of spiroplasmas (Whitcomb et al., 1982a, 1983; Tully et al., 1987). Reciprocal MI tests are especially useful in showing subgroup or serovar relationships (Tully et al., 1987). When significant DF, MI, or GI one-way cross-reactions occur between new isolates and representative antisera, reciprocal MI tests will usually resolve the status of these isolates.

A spiroplasma enzyme-linked immunosorbent assay (ELISA) was first employed in the detection of the two principal plant pathogens (*S. citri* and *S. kunkelii*) in plant or insect material (Clark et al., 1978; Saillard et al., 1978; Raju and Nyland, 1981; Archer et al., 1982; Eden-Green, 1982). This technique has continued to be refined and standardized for epidemiologic studies (Saillard and Bové, 1983; Gordon et al., 1985). ELISA procedures have also been developed for assessing serologic relationships among cultivated spiroplasmas (Archer and Best, 1980; Tully et al., 1980; Hung et al., 1987). Sensitivity of the ELISA technique for detecting shared serologic responses appears to be comparable to the MI procedure (Tully et al., 1980).

A rapid modified ELISA (or immunobinding) test for serological analysis of spiroplasmas has been also described. In this technique, antigens and antisera are bound either to nitrocellulose paper (Kotani and McGarrity, 1985) or filter paper (Fletcher, 1987), and the usual subsequent enzyme-substrate color reactions measured. The modified test appears to be less sensitive than the usual ELISA procedure, but has the advantage of being rapid and applicable to the detection of *S. citri* in plant material or in single insect specimens (Fletcher, 1987).

Monoclonal antibodies to *S. citri* were first used to analyze specific spiroplasma membrane proteins (e.g., spiralin)(Mouches et al., 1983b). Later, monoclonal antibodies to both *S. citri* and *S. kunkelii* were employed in serologic identification of the respective plant pathogens or to study serological relationships among established spiroplasma groups (Lin and Chen, 1985a, 1985b; Chen et al., 1989; Jordan et al., 1989).

Literature Cited

Abalain-Colloc, M. L., C. Chastel, J. G. Tully, J. M. Bové, R. F. Whitcomb, B. Gilot, and D. L. Williamson. 1987. *Spiroplasma sabaudiense*, sp. nov., a new species from mosquitoes collected in France. Int. J. Syst. Bacteriol. 37:260–265.

Abalain-Colloc, M. L., L. Rosen, J. G. Tully, J. M. Bové, C. Chastel, and D. L. Williamson. 1988. *Spiroplasma taiwanense*, sp. nov. from *Culex tritaeniorhynchus* mosquitoes collected in Taiwan. Int. J. Syst. Bacteriol. 38:103–107.

Archer, D. B., and J. Best. 1980. Serological relatedness of spiroplasmas estimated by enzyme-linked immunosorbent assay and crossed immunoelectrophoresis. J. Gen. Microbiol. 119:413–422.

Archer, D. B., R. Townsend, and P. G. Markham. 1982. Detection of *Spiroplasma citri* in plants and insect hosts by ELISA. Plant Pathol. 31:299–306.

Bastardo, J. W., O. D. Ou, and R. H. Bussell. 1974. Biological and physical properties of the suckling mouse cataract agent grown in chick embryos. Infect. Immun. 9:444–451.

Bové, J. M. 1981. Mycoplasma infections of plants. Israel J. Med. Sci. 17:572–585.

Bové, J. M. 1984. Wall-less prokaryotes of plants. Annu. Rev. Phytopathology 22:361–396.

Bové, J. M., and C. Saillard. 1979. Cell biology of spiroplasmas, p. 83–153. In: R. F. Whitcomb and J. G. Tully (ed.), The mycoplasmas, vol. 3. Academic Press, New York.

Bové, J. M., J. C. Vignault, M. Garnier, C. Saillard, O. Garcia-Jurado, C. Bové, and A. Nhami. 1978. Mise en évidence de *Spiroplasma citri*, l'agent causal de la maladie du "stubborn" des agrumes, dans des pervenches (*Vinca rosea* L.) ornementales de la ville de Rabat. Compt. Rend. Acad. Sci. (Paris) 286:57–59.

Bové, J. M., C. Saillard, P. Junca, J. R. Degorce-Dumas, B. Ricard, A. Nhami, R. F. Whitcomb, D. Williamson, and J. G. Tully. 1982. Guanine-plus-cytosine content, hybridization percentages, and *Eco*R1 restriction enzyme profiles of spiroplasmal DNA. Rev. Infect. Dis. 4 (Suppl.):S129–S136.

Bové, J. M., C. Mouches, P. Carle-Junca, J. R. Degorce-Dumas, J. G. Tully, and R. F. Whitcomb. 1983a. Spiroplasmas of group I: the *Spiroplasma citri* cluster. Yale J. Biol. Med. 56:573–582.

Bové, J. M., R. F. Whitcomb, and R. E. McCoy. 1983b. Culture techniques for spiroplasmas from plants. Methods Mycoplasmol. 2:225–234.

Bové, J. M., P. Carle, M. Garnier, F. Laigret, J. Renaudin, and C. Saillard. 1989. Molecular and cellular biology of spiroplasmas, p. 243–364. In: R. F. Whitcomb and J. G. Tully (ed.), The mycoplasmas, vol. 5. Academic Press, San Diego, CA.

Brinton, L. P., and W. Burgdorfer. 1976. Cellular and subcellular organization of the 277F agent: a spiroplasma from the rabbit tick, *Haemaphysalis leporispalustris* (Acari:Ixodidae). Int. J. Syst. Bacteriol. 26:554–560

Calavan, E. C., and J. M. Bové. 1989. Ecology of *Spiroplasma citri*, p. 425–485. In: R. F. Whitcomb and J. G. Tully (ed.), The mycoplasmas, vol. 5. Academic Press, San Diego, CA.

Calavan, E. C., and G. N. Oldfield. 1979. Symptomatology of spiroplasmal plant diseases, p. 37–64. In: R. F. Whitcomb and J. G. Tully (ed.), The mycoplasmas, vol. 3. Academic Press, New York.

Chang, C-J. 1989. Nutrition and cultivation of spiroplasmas, p. 201–241. In: R. F. Whitcomb and J. G. Tully (ed.), The mycoplasmas, vol. 5. Academic Press, San Diego, CA.

Chang, C-J., and T. A. Chen. 1982. Spiroplasmas: cultivation in chemically defined medium. Science 215:1121–1122.

Chastel, C., B. Gilot, F. Le Goff, R. Gruffaz, and M. L. Abalain-Colloc. 1985. Isolement de spiroplasmes en France (Savoie, Alpes du Nord) à partir de Moustiques du genre *Aedes*. Compt. Rend. Acad. Sci. (Paris) 300:261–266.

Chastel, C., B. Gilot, F. Le Goff, B. Devau, G. Kerdraon, I. Humphery-Smith, R. Gruffaz, and A-M. Simitzis-Le Flohic. 1990. New developments in the ecology of mosquito spiroplasmas. Zbl. Bakt. Hyg. 20: (in press).

Chen, T. A., and C. H. Liao. 1975. Corn stunt spiroplasma: isolation, cultivation, and proof of pathogenicity. Science 188:1015–1017.

Chen, T. A., J. D. Lei, and C. P. Lin. 1989. Detection and identification of plant and insect mollicutes, p. 393–424. In: R. F. Whitcomb and J. G. Tully (ed.), The mycoplasmas, vol. 5. Academic Press, San Diego, CA.

Christiansen, C., G. Askaa, E. A. Freundt, and R. F. Whitcomb. 1979. Nucleic acid hybridization experiments with *Spiroplasma citri* and the corn stunt and suckling mouse cataract spiroplasmas. Curr. Microbiol. 2:323–326.

Clark, H F. 1964. Suckling mouse cataract agent. J. Infect. Dis. 114:476–487.

Clark, H F. 1974. The suckling mouse cataract agent (SMCA). Progr. Med. Virol. 18:307–322.

Clark, M. F., C. L. Flegg, M. Bar-Joseph, and S. Rottem. 1978. The detection of *Spiroplasma citri* by enzyme linked immunosorbent assay (ELISA). Phytopathology 92:332–337.

Clark, T. B. 1977. *Spiroplasma* sp., a new pathogen in honey bees. J. Invertebr. Pathol. 29:112–113.

Clark, T. B. 1978. Honey bee spiroplasmosis, a new problem for beekeepers. Amer. Bee J. 118:18–23.

Clark, T. B. 1982. Spiroplasmas: diversity of arthropod reservoirs and host-parasite relationships. Science 217:57–59.

Clark, T. B. 1984. Diversity of spiroplasma host-parasite relationships. Isr. J. Med. Sci. 20:995–997.

Clark, T. B., and R. F. Whitcomb. 1984. Pathogenicity of mollicutes for insects: possible use in biological control. Ann. Microbiol. 135A:141–150.

Clark, T. B., R. F. Whitcomb, and J. G. Tully. 1982. Spiroplasmas from coleopterous insects: new ecological dimensions. Microb. Ecol. 8:401–409.

Clark, T. B., B. V. Peterson, R. F. Whitcomb, R. B. Henegar, K. J. Hackett, and J. G. Tully. 1984. Spiroplasmas in the Tabanidae. Isr. J. Med. Sci. 20:1002–1005.

Clark, T. B., R. F. Whitcomb, J. G. Tully, C. Mouches, C. Saillard, J. M. Bové, H. Wróblewski, P. Carle, D. L. Rose, and D. L. Williamson. 1985. *Spiroplasma melliferum* sp. nov., a new species from the honeybee (*Apis mellifera*). Int. J. Syst. Bacteriol. 35:296–308.

Clark, T. B., R. B. Henegar, L. Rosen, K. J. Hackett, R. F. Whitcomb, C. Saillard, J. M. Bové, J. G. Tully, and D. L. Williamson. 1987. New spiroplasmas from insects and flowers: isolation, ecology, and host association. Isr. J. Med. Sci. 23:687–690.

Cohen, A. J., D. L. Williamson, and P. R. Brink. 1989. A motility mutant of *Spiroplasma melliferum* induced with nitrous acid. Curr. Microbiol. 18:219–222.

Cole, R. M. 1983. Isolation and characterization of spiroplasma viruses. Methods Mycoplasmol. 2:425–431.

Cole, R. M., J. G. Tully, T. J. Popkin, and J. M. Bové. 1973a. Morphology, ultrastructure, and bacteriophage infection of the helical mycoplasma-like organism (*Spiroplasma citri* gen. nov., sp. nov.) cultured from "stubborn" disease of citrus. J. Bacteriol. 115:367–386.

Cole, R. M., J. G. Tully, T. J. Popkin, and J. M. Bové. 1973b. Ultrastructure of the agent of citrus "stubborn" disease. Ann. N. Y. Acad. Sci. 225:471–493.

Daniels, M. J. 1979. Mechanisms of spiroplasma pathogenicity, p. 209–227. In: R. F. Whitcomb and J. G. Tully (ed.), The mycoplasmas, vol. 3. Academic Press, New York.

Daniels, M. J. 1983. Mechanisms of spiroplasma pathogenicity. Annu. Rev. Phytopathol. 21:29–43.

Daniels, M. J., and J. M. Longland. 1984. Chemotactic behavior of spiroplasmas. Curr. Microbiol. 10:191–193.

Daniels, M. J., D. B. Archer, M. A. Stephens, R. Townsend, J. M. Longland, and J. Best. 1980a. Comparison of spiroplasmas by polyacrylamide gel electrophoresis of cell proteins. Curr. Microbiol. 4:377–380.

Daniels, M. J., J. M. Longland, and J. Gilbart. 1980b. Aspects of motility and chemotaxis in spiroplasmas. J. Gen. Microbiol. 118:429–408.

Davis, R. E. 1978. Spiroplasma associated with flowers of the tulip tree (*Liriodendron tulipifera* L.). Canad. J. Microbiol. 24:954–959.

Davis, R. E., J. F. Worley, R. F. Whitcomb, T. Ishijima, and R. L. Steere. 1972a. Helical filaments produced by a mycoplasma-like organism associated with corn stunt disease. Science 176:521–523.

Davis, R. E., R. F. Whitcomb, T. A. Chen, and R. R. Granados. 1972b. Current status of the etiology of corn stunt disease, p. 205–214. In: K. Elliott and J. Birch (ed.), Ciba Foundation Symposium: Pathogenic Mycoplasmas. American Elsevier, New York.

Davis, R. E., G. Dupont, P. Saglio, B. Roy, J. C. Vignault, and J. M. Bové. 1974. Spiroplasmas: studies on the microorganism associated with corn stunt disease. Colloq. Inst. Natl. Sante Rech. Med. 33:187–194.

Davis, R. E., J. F. Worley, and M. Moseley. 1975. Spiroplasmas: primary isolation and cultivation in cystine-tryptone media and translational locomotion in semisolid versions. Proc. Am. Phytopathol. Soc. 2:56 (Abstract).

Davis, R. E., I. M. Lee, and L. K. Basciano. 1979. Spiroplasmas: serological grouping of strains associated with plants and insects. Canad. J. Microbiol. 25:861–866.

Davis, R. E., I. M. Lee, and J. F. Worley. 1981. *Spiroplasma floricola*, a new species isolated from surfaces of flowers of the tulip tree, *Liriodendron tulipifera* L. Int. J. Syst. Bacteriol. 31:456–464.

Davis, M. J., J. H. Tsai, and R. E. McCoy. 1984. Isolation of the corn stunt spiroplasma from maize in Florida. Plant Dis. 68:600–604.

Doi, Y., M. Teranaka, K. Yora, and H. Asuyama. 1967. Mycoplasma-or PLT group-like microrganisms found in the phloem elements of plants infected with mulberry dwarf, potato witches' broom, aster yellows, or *Paulownia* witches' broom. Ann. Phytopathol. Soc. (Japan) 33:279–266.

Eden-Green, S. J. 1982. Detection of corn stunt spiroplasma *in vivo* by ELISA using antisera to extracts from infected corn plants (*Zea mays*). Plant Pathol. 31:289–297.

Edward, D. G. ff 1947. A selective medium for pleuropneumonia-like organisms. J. Gen. Microbiol. 1:238–243.

El-Bolok, M. M. 1981. Specific and nonspecific transmission of spiroplasmas and mycoplasma-like organisms by leafhoppers (Cicadellidae:Homoptera) with implications for etiology of aster yellows disease. Ph.D. dissertion, University of Cairo, Giza.

Fletcher, J. 1987. Filter paper dot-immunobinding assay for detection of *Spiroplasma citri*. Appl. Environ. Microbiol. 53:183–184.

French, F. E., R. F. Whitcomb, J. G. Tully, K. J. Hackett, E. A. Clark, R. B. Henegar, A. G. Wagner, and D. L. Rose. 1990. Tabanid spiroplasmas of the southeast

USA: new groups and correlation with host life history strategy. Zbl. Bakt. Hyg. 20:(in press).

Fudl-Allah, A. A., E. C. Calavan, and E. C. K. Igwegbe. 1972. Culture of a mycoplasma-like organism associated with stubborn disease of citrus. Phytopathology 62:729–731.

Garnier, M., M. Clerc, and J. M. Bové. 1981. Growth and division of spiroplasmas: morphology of *Spiroplasma citri* during growth in liquid medium. J. Bacteriol. 147:642–652.

Garnier, M., M. Clerc, and J. M. Bové. 1984a. Growth and division of *Spiroplasma citri*: elongation of elementary helices. J. Bacteriol. 158:23–28.

Garnier, M., T. Steiner, G. Martin, and J. M. Bové. 1984b. Oxydo-reduction sites and relationships of spiroplasmas with insect cells in culture. Isr. J. Med. Sci. 20:840–842.

Giannotti, J., and D. Giannotti. 1986. Multiplication de mollicutes in situ et in vitro dans le nectar floral de différentes plantes. Compt. Rend. Acad. Sci. 302:669–674.

Giannotti, J., C. Vago, D. Giannotti, and C. Legoff. 1981. Etude comparée in vivo et in vitro des diverses formes de l'agent mollicute de la léthargie de Coléoptères. Compt. Rend. Acad. Sci. (Paris), Ser. D 292:1043–1049.

Golino, D. A., and G. N. Oldfield. 1989. Plant-pathogenic spiroplasmas and their leafhopper vectors, p. 267–299. In: K. F. Harris (ed.), Adv. Dis. Vector Res, Vol. 6. Springer-Verlag, New York.

Gordon, D. T., L. R. Nault, N. H. Gordon, and S. E. Heady. 1985. Serological detection of corn stunt spiroplasma and maize rayado fino virus in field-collected *Dalbulus* spp. from Mexico. Plant Dis. 69:108–111.

Granados, R. R. 1969. Electron microscopy of plants and insect vectors infected with corn stunt disease agent. Contrib. Boyce Thompson Inst. 24:173–188.

Guo, Y. H., T. A. Chen, R. F. Whitcomb, D. L. Rose, J. G. Tully, D. L. Williamson, X. D. Ye, and Y. X. Chen. 1990. *Spiroplasma chinensis* sp. nov. from flowers of *Calystegia hederaceae* in China. Int. J. Syst. Bacteriol. 40: (in press).

Hackett, K. J., and T. B. Clark. 1989. Ecology of spiroplasmas, p. 113–200. In: R. F. Whitcomb and J. G. Tully (ed.), The mycoplasmas, vol. 5. Academic Press, San Diego CA.

Hackett, K. J., and D. E. Lynn. 1985. Cell assisted growth of a fastidious spiroplasma. Science 230:825–827.

Hackett, K. J., T. B. Clark, A. Hicks, R. F. Whitcomb, E. Lowry, and S. W. T. Batra. 1984. Occurrence and frequency of subgroup I-6 spiroplasma in arthropods associated with old fields in Maryland and Virginia. Isr. J. Med. Sci. 20:1006–1008.

Hackett, K. J., D. E. Lynn, D. L. Williamson, A. S. Ginsberg, and R. F. Whitcomb. 1986. Cultivation of the *Drosophila* sex-ratio spiroplasma. Science 232:1253–1255.

Hackett, K. J., D. E. Lynn, A. S. Ginsberg, S. Rottem, R. B. Henegar, J. Adams, D. L. Williamson, and R. F. Whitcomb. 1987. Cell-assisted culture of fastidious spiroplasmas: initial analysis of growth factors. Isr. J. Med. Sci. 23:667–670.

Hackett, K. J., R. F. Whitcomb, R. B. Henegar, A. G. Wagner, E. A. Clark, J. G. Tully, F. Green, W. H. McKay, P. Santini, D. R. Rose, J. J. Anderson, and D. E. Lynn.

1990. Mollicute diversity in arthropod hosts. Zbl. Bakt. Hyg. (in press).

Hung, S. H. Y., T. A. Chen, R. F. Whitcomb, J. G. Tully, and Y. X. Chen. 1987. *Spiroplasma culicicola* sp. nov. from the salt marsh mosquito *Aedes sollicitans*. Int. J. Syst. Bacteriol. 37:365–370.

Igwegbe, E. C. K. and E. C. Calavan. 1970. Occurrence of mycoplasmalike bodies in phloem of stubborn-infected citrus seedlings. Phytopathology 60:1525–1526.

Jones, A. L., R. F. Whitcomb, D. L. Williamson, and M. E. Coan. 1977. Comparative growth and primary isolation of spiroplasmas in media based on insect tissue culture formulations. Phytopathology 47:738–746.

Jordan, R. L., M. Konai, I. M. Lee, and R. E. Davis. 1989. Species-specific and cross-reactive monoclonal antibodies to the plant-pathogenic spiroplasmas *Spiroplasma citri* and *S. kunkelii*. Phytopathology 79:880–887.

Junca, P., C. Saillard, J. G. Tully, O. Garcia-Jurado, J. R. Degorce-Dumas, C. Mouches, J. C. Vignault, R. Vogel, R. McCoy, R. F. Whitcomb, D. L. Williamson, J. Latrille, and J. M. Bové. 1980. Charactérisation de spiroplasmes isolés d'insectes et de fleurs de France continentale, de Corse et du Maroc. Proposition pour une classification des spiroplasmes. Compt. Rend. Acad. Sci. Ser. D. 290:1209–1212.

Kotani, H., and G. J. McGarrity. 1985. Rapid and simple identification of mycoplasmas by immunobinding. J. Immunol. Methods 85:257–267.

Kotani, H., D. Phillips, and G. J. McGarrity. 1986. Malignant transformation of NIH-3T3 and CV-1 cells by a helical mycoplasma, *Spiroplasma mirum*, strain SMCA. In Vitro Cell. Devel. Biol. 22:756–762.

Laflèche D., and J. M. Bové. 1970. Mycoplasmes dans les agrumes atteints de "Greening" et de "Stubborn" ou de maladies similaires. Fruits 25:455–465.

Leach, R. H. 1983. Preservation of mycoplasma cultures and culture collections. Methods Mycoplasmol. 1:197–204.

Lee, I. M., and R. E. Davis. 1980. DNA homology among diverse spiroplasma strains representing several serological groups. Canad. J. Microbiol. 26:1356–1363.

Lei, J. D., H. J. Su, and T. A. Chen. 1979. Spiroplasmas isolated from the green leafbug, *Trigonotylus ruficornis* Geoffroy, p. 89–97. In: Proceedings of the US-ROC Plant Mycoplasma Seminar. Taipei, Taiwan; National Science Council, Republic of China.

Liao, C. H., and T. A. Chen. 1977. Culture of corn stunt spiroplasmas in a simple medium. Phytopathology 67:802–807.

Liao, C. H., and T. A. Chen. 1980. Presence of spiroplasma-inhibitory substances in plant tissue extracts. Canad. J. Microbiol. 26:807–811.

Liao, C. H., and T. A. Chen. 1981. Deoxyribonucleic acid hybridization between *Spiroplasma citri* and the corn stunt spiroplasma. Curr. Microbiol. 5:83–86.

Lin, C. P., and T. A. Chen. 1985a. Production of monoclonal antibodies against *Spiroplasma citri*. Phytopathology 75:848–851.

Lin, C. P., and T. A. Chen. 1985b. Monoclonal antibodies against corn stunt spiroplasma. Can. J. Microbiol. 31:900–904.

Louis, C., J. M. Quiot, J. Giannotti, and C. Vago. 1978. Infection expérimentale d'une lignée cellulaire d'invertébré par le procaryote intravacuolaire de type mollicute, agent de la "léthargie des coléoptères". Ann. Microbiol. 129B:621–633.

Madden, L. V., and L. R. Nault. 1983. Differential pathogenicity of corn stunting mollicutes to leafhopper vectors in *Dalbulus* and *Baldulus* species. Phytopathology 73:1608–1614.

Markham, P. G., R. Townsend, M. Bar-Joseph, M. J. Daniels, A. Plaskitt, and B. M. Meddins. 1974. Spiroplasmas are the causal agents of citrus little leaf disease. Ann. Appl. Biol. 78:49–57.

Markham, P. G., T. B. Clark, and R. F. Whitcomb. 1983. Culture techniques for spiroplasmas from arthropods. Methods Mycoplasmol. 2:217–223.

McCoy, R. E. 1979. Mycoplasmas and yellows diseases, p. 229–264. In: R. F. Whitcomb and J. G. Tully (ed.), The mycoplasmas, vol. 3. Academic Press, New York.

McCoy, R. E., D. S. Williams, and D. L. Thomas. 1979. Isolation of mycoplasmas from flowers, p. 75–81. In: Proceedings of the US-ROC Plant Mycoplasma Seminar. Taipei, Taiwan; National Science Council, Republic of China.

McCoy, R. E., A. Caudwell, C. J. Chang, T. A. Chen, L. N. Chiykowski, M. T. Cousin, J. L. Dale, G. T. N. de Leeuw, D. A. Golino, K. J. Hackett, B. C. Kirkpatrick, R. Marwitz, H. Petzold, R. C. Sinha, M. Sugiura, R. F. Whitcomb, I. L. Yang, B. M. Zhu, and E. Seemüller. 1989. Plant diseases associated with mycoplasma-like organisms, p. 545–640. In: R. F. Whitcomb and J. G. Tully (ed.), The mycoplasmas, vol. 5. Academic Press, San Diego, CA.

McGarrity, G. J., and H. Kotani. 1984. Use of cell cultures to study spiroplasma infections. Isr. J. Med. Sci., 20:924–926.

McGarrity, G. J., and D. L. Williamson. 1989. Spiroplasma pathogenicity in vivo and in vitro, p. 365–392. In: R. F. Whitcomb and J. G. Tully (ed.), The mycoplasmas, vol.5. Academic Press, San Diego, CA.

Mouches, C., J. C. Vignault, J. G. Tully, R. F. Whitcomb, and J. M. Bové. 1979. Characterization of spiroplasmas by one- and two-dimensional protein analysis on polyacrylamide slab gels. Curr. Microbiol. 2:69–74.

Mouches, C., J. M. Bové, J. Albisetti, T. B. Clark, and J. G. Tully. 1982a. A spiroplasma of serogroup IV causes a May-disease-like disorder of honeybees in southwestern France. Microb. Ecol. 8:387–399.

Mouches, C., A. Menara, J. G. Tully, and J. M. Bové. 1982b. Polyacrylamide gel analysis of spiroplasmal proteins and its contribution to the taxonomy of spiroplasmas. Rev. Inf. Dis. 4 (Suppl.):S141–S147.

Mouches, C., J. M. Bové, J. G. Tully, D. L. Rose, R. E. McCoy, P. Carle-Junca, M. Garnier, and C. Saillard. 1983a. *Spiroplasma apis*, a new species from the honeybee (*Apis mellifera*). Ann. Inst. Microbiol. (Paris) 134A:383–397.

Mouches, C., T. Candresse, G. J. McGarrity, and J. M. Bové. 1983b. Analysis of spiroplasma proteins: contribution to the taxonomy of group IV spiroplasmas and the characterization of spiroplasma protein antigens. Yale J. Biol. Med. 56:431–437.

Nault, L. R., and O. E. Bradfute. 1979. Corn stunt: involvement of a complex of leafhopper-borne pathogens, p. 561–585. In: K. Maramorosch and K. F. Harris (ed.), Leafhopper vectors and plant disease agents. Academic Press, New York.

Nur, I., J. M. Bové, C. Saillard, S. Rottem, R. F. Whitcomb, and S. Razin. 1986. DNA probes in detection of spiroplasmas and mycoplasma-like organisms in plants and insects. FEMS Microbiol. Lett. 35:157–162.

Pickens, E. G., R. K. Gerloff, and W. Burgdorfer. 1968. Spirochete from rabbit tick, *Haemaphysalis leporispalustris* (Packard). J. Bacteriol. 95:291–299.

Pollack, J. D., M. C. McElwain, D. DeSantis, J. T. Manolukas, J. G. Tully, C. J. Chang, R. F. Whitcomb, K. J. Hackett, and M. V. Williams. 1989. Metabolism of the *Spiroplasmataceae*. Int. J. Syst. Bacteriol. 39:406–412.

Poulson, D. F., and B. Sakaguchi. 1961. Nature of the "sex ratio" agent in *Drosophila*. Science 133:1489–1490.

Raju, B. C., and G. Nyland. 1981. Enzyme-linked immunosorbent assay for the detection of corn stunt spiroplasma in plant and insect tissues. Curr. Microbiol. 5:101–104.

Razin, S., and J. G. Tully (ed.). 1983. Methods in Mycoplasmology, vol. 1, p. 504. Academic Press, New York.

Rogers, M. J., J. Simmons, R. T. Walker, W. G. Weisburg, C. R. Woese, R. J. Tanner, I. M. Robinson, D. A. Stahl, G. Olsen, R. H. Leach, and J. Maniloff. 1985. Construction of the mycoplasma evolutionary tree from 5S rRNA sequence data. Proc. Natl. Acad. Sci. (USA) 82:1160–1164.

Saglio, P., D. Laflèche, C. Bonissol, and J. M. Bové. 1971. Isolement, culture et observation au microscope électronique des structures de type mycoplasma associées à la maladie du Stubborn des agrumes et leur comparison avec les structures observées dans le cas de la maladie du Greening des agrumes. Physiol. Vegetale 9:569–582.

Saglio, P., D. Laflèche, M. Lhospital, G. Dupont, and J. M. Bové. 1972. Isolation and growth of citrus mycoplasmas, p. 187–203. In: K. Elliott and J. Birch (ed.), Ciba Foundation Symposium: Pathogenic mycoplasmas. American Elsevier, New York.

Saglio, P., M. Lhospital, D. Laflèche, G. Dupont, J. M. Bové, J. G. Tully, and E. A. Freundt. 1973. *Spiroplasma citri* gen. and sp. nov.: a mycoplasma-like organism associated with "stubborn" disease of citrus. Int. J. Syst. Bacteriol. 23:191–204.

Saillard, C., and J. M. Bové. 1983. Application of ELISA to spiroplasma detection and classification. Methods Mycoplasmol. 1:471–476.

Saillard, C., J. Dunez, O. Garcia-Jurado, A. Nhami, and J. M. Bové. 1978. Detection de *Spiroplasma citri* dans les agrumes et les pervenches par la technique immunoenzymatic "ELISA". Compt. Rend. Acad. Sci. Ser. D 286:1245–1248.

Saillard, C., J. C. Vignault, A. Gadeau, P. Carle, M. Garnier, A. Fos, J. M. Bové, J. G. Tully, and R. F. Whitcomb. 1984. Discovery of a new plant-pathogenic spiroplasma. Isr. J. Med. Sci. 20:1013–1015.

Saillard, C., J. C. Vignault, J. M. Bové, A. Raie, J. G. Tully, D. L. Williamson, A. Fos, M. Garnier, A. Gadeau, P. Carle, and R. F. Whitcomb. 1987. *Spiroplasma phoeniceum* sp. nov., a new plant-pathogenic species from Syria. Int. J. Syst. Bacteriol. 37:106–115.

Shaikh, A. A., W. E. Johnson, C. Stevens, and A. Y. Tang. 1987. The isolation of spiroplasmas from mosquitoes in Macon County, Alabama. J. Amer. Mosquito Control Assoc. 3:289–295.

Skripal, I. G. 1974. On improvement of taxonomy of the class Mollicutes and establishment in the order Mycoplasmatales of the new family Spiroplasmataceae, fam. nov. Mikrobiologii Zhurnal (Kiev) 36:462–467.

Skripal, I. G. 1983. Revival of the name *Spiroplasmataceae* fam. nov, nom. rev., omitted from the 1980 Approved Lists of Bacterial Names. Int. J. Syst. Bacteriol. 33:408.

Slaff, M., and T. A. Chen. 1982. The isolation of a spiroplasma from *Aedes sollicitans* (Walker) in New Jersey. J. Florida Anti-Mosquito Assoc. 53:19–21.

Steiner, T., G. J. McGarrity, and D. M. Phillips. 1982. Cultivation and partial characterization of spiroplasmas in cell cultures. Infect. Immun. 35:296–304.

Steiner, T., G. J. McGarrity, J. M. Bové, D. Phillips, and M. Garnier. 1984. Insect cell cultures in the study of attachment and pathogenicity of spiroplasmas and mycoplasmas. Ann. Microbiol. (Inst. Pasteur) 135A:47–53.

Stiller, D., R. F. Whitcomb, M. E. Coan, and J. G. Tully. 1981. Direct isolation in cell-free medium of a spiroplasma from *Haemaphysalis leporispalustris* (Acari: Ixodidae) in Maryland. Curr. Microbiol. 5:339–342.

Subcommittee on the Taxonomy of *Mollicutes*. 1972. Proposal for minimal standards for descriptions of new species of the order *Mycoplasmatales*. Int. J. Syst. Bacteriol. 22:184–188.

Subcommittee on the Taxonomy of *Mollicutes* 1977. Minutes of interim meeting, September 22, 1976. Int. J. Syst. Bacteriol. 27:393–394.

Subcommittee on the Taxonomy of *Mollicutes*. 1979. Proposals of minimum standards for descriptions of new species of the class *Mollicutes*. Int. J. Syst. Bacteriol. 29:172–180.

Subcommittee on the Taxonomy of *Mollicutes*. 1984. Minutes of the interim meeting, 30 August and 6 September 1982, Tokyo, Japan. Int. J. Syst. Bacteriol. 34:361–365.

Subcommittee on the Taxonomy of *Mollicutes*. 1985. Minutes of the interim meeting, 21 and 26 June 1984, Jerusalem, Israel. Int. J. Syst. Bacteriol. 35:378–381.

Taylor-Robinson, D., and T. A. Chen. 1983. Growth inhibitory factors in animal and plant tissues. Methods Mycoplasmol. 1:109–114.

Townsend, R., P. G. Markham, K. A. Plaskitt, and M. J. Daniels. 1977. Multiplication and morphology of *Spiroplasma citri* in the leafhopper *Euscelis plebejus*. Ann. Appl. Biol. 87:307–313.

Townsend, R., J. Burgess, and K. A. Plaskitt. 1980a. Morphology and ultrastructure of helical and nonhelical strains of *Spiroplasma citri*. J. Bacteriol. 142:973–981.

Townsend, R., D. B. Archer, and K. A. Plaskitt. 1980b. Purification and preliminary characterization of spiroplasma fibrils. J. Bacteriol. 142:694–700.

Tully, J. G. 1982. Interaction of spiroplasmas with plant, arthropod, and animal hosts. Rev. Inf. Dis. 4(Suppl.):S193–S198.

Tully, J. G. 1983a. Dark-field microscopy. Methods Mycoplasmol. 1:35–37.

Tully, J. G. 1983b. Cloning and filtration techniques for mycoplasmas. Methods Mycoplasmol. 1:173–177.

Tully, J. G. 1983c. General cultivation techniques for mycoplasmas and spiroplasmas. Methods Mycoplasmol. 1:99–101.

Tully, J. G. 1983d. Bacterial and fungal inhibitors in mycoplasma culture media. Methods Mycoplasmol. 1:205–209.

Tully, J. G., R. F. Whitcomb, J. M. Bové, and P. Saglio. 1973. Plant mycoplasmas: serological relation between

agents associated with citrus stubborn and corn stunt diseases. Science 182:827–829.

Tully, J. G., R. F. Whitcomb, D. L. Williamson, and H. F. Clark. 1976. Suckling mouse cataract agent is a helical wall-free prokaryote (spiroplasma) pathogenic for vertebrates. Nature 259:117–120.

Tully, J. G., R. F. Whitcomb, H. F. Clark, and D. L. Williamson. 1977. Pathogenic spiroplasmas: cultivation and vertebrate pathogenicity of a new spiroplasma. Science 195:892–894.

Tully, J. G., D. L. Rose, O. Garcia-Jurado, J. C. Vignault, C. Saillard, J. M. Bové, R. E. McCoy, and D. L. Williamson. 1980. Serological analysis of a new group of spiroplasmas. Curr. Microbiol. 3:369–372.

Tully, J. G., D. L. Rose, C. E. Yunker, J. Cory, R. F. Whitcomb, and D. L. Williamson. 1981. Helical mycoplasmas (spiroplasmas) in *Ixodes* ticks. Science 212:1043–1045.

Tully, J. G., R. F. Whitcomb, D. L. Rose, and J. M. Bové. 1982. *Spiroplasma mirum,* a new species from the rabbit tick (*Haemaphysalis leporispalustris*). Int. J. Syst. Bacteriol. 32:92–100.

Tully, J. G., R. F. Whitcomb, D. L. Rose, D. L. Williamson, and J. M. Bové. 1983. Characterization and taxonomic status of tick spiroplasmas: a review. Yale J. Biol. Med. 56:599–603.

Tully, J. G., D. L. Rose, E. Clark, P. Carle, J. M. Bovè, R. B. Henegar, R. F. Whitcomb, D. E. Colflesh, and D. L. Williamson. 1987. Revised group classification of the genus *Spiroplasma* (class *Mollicutes*), with proposed new groups XII to XXIII. Int. J. Syst. Bacteriol. 37:357–364.

Tully, J. G., D. L. Rose, K. J. Hackett, R. F. Whitcomb, P. Carle, J. M. Bové, D. E. Colflesh, and D. L. Williamson. 1989. *Mycoplasma ellychniae* sp. nov., a sterol-requiring mollicute from the firefly beetle *Ellychnia corrusca*. Int. J. Syst. Bacteriol. 39:284–289.

Vignault, J-C., J. M. Bové, C. Saillard, R. Vogel, A. Farro, L. Venegas, W. Stemmer, S. Aoki, R. McCoy, A. S. Al-Beldawi, J. Bonfils, G. Moutous, A. Fos, F. Poutiers, and G. Viennot-Bourgin. 1980. Mise en culture de spiroplasmes à partir de materiel végétal et d'insectes provenant de pays circum-mediterranées et du Proche-Orient. Compt. Rend. Acad. Sci. Ser. D. 290:775–778.

Weisburg, W. G., J. G. Tully, D. L. Rose, J. P. Petzel, H. Oyaizu, D. Yang, L. Mandelco, J. Sechrest, T. G. Lawrence, J. Van Etten, J. Maniloff, and C. R. Woese. 1989. A phylogenetic analysis of the mycoplasmas: basis for their classification. J. Bacteriol. 171:6455–6467.

Whitcomb, R. F. 1981. The biology of spiroplasmas. Annu. Rev. Entomol. 26:397–425.

Whitcomb, R. F. 1983. Culture media for spiroplasmas. Methods Mycoplasmol. 1:147–158.

Whitcomb, R. F. 1989. *Spiroplasma kunkelii:* biology and ecology, p. 487–544. In: R. F. Whitcomb and J. G. Tully (ed.), The mycoplasmas, vol. 5. Academic Press, San Diego, CA.

Whitcomb, R. F., and M. E. Coan. 1980. Comparative growth of flower, bee, and citrus spiroplasmas. p. 79 (Abstract). Proc. Amer. Soc. Microbiol.

Whitcomb, R. F., and K. J. Hackett. 1986. Cloning by limiting dilution in liquid media: an improved alternative for cloning mollicute species. p. 167 (Abstract). Proc. 6th Int. Org. Mycoplasmology Congress, Birmingham, AL.

Whitcomb, R. F., and J. G. Tully. 1984. Family *Spiroplasmataceae*, genus *Spiroplasma*, p. 781–787. In: N. R. Krieg and J. G. Holt (ed.), Bergey's manual of systematic bacteriology, vol. 1. Williams and Wilkins, Baltimore.

Whitcomb, R. F., and D. L. Williamson. 1975. Helical wall-free prokaryotes in insects: multiplication and pathogenicity. Ann. N. Y. Acad. Sci. 266:260–275.

Whitcomb, R. F., and D. L. Williamson. 1979. Pathogenicity of mycoplasmas for arthropods. Zentral. Bakteriol. Parasit. Infektion. Hyg., Abt. 1, Orig. Reihe A., 245:200–221.

Whitcomb, R. F., J. G. Tully, J. M. Bové, and P. Saglio. 1973. Spiroplasmas and acholeplasmas: multiplication in insects. Science 182:1251–1253.

Whitcomb, R. F., J. G. Tully, T. B. Clark, D. L. Williamson, and J. M. Bové. 1982a. Revised serological classification of spiroplasmas, new provisional groups and recommendations for serotyping of isolates. Curr. Microbiol. 7:291–296.

Whitcomb, R. F., J. G. Tully, P. McCawley, and D. L. Rose. 1982b. Application of the growth inhibition test to *Spiroplasma* taxonomy. Int. J. Syst. Bacteriol. 32:387–394.

Whitcomb, R. F., T. B. Clark, J. G. Tully, T. A. Chen, and J. M. Bové. 1983. Serological classification of spiroplasmas: current status. Yale J. Biol. Med. 56:453–459.

Whitcomb, R. F., T. A. Chen, D. L. Williamson, C. Liao, J. G. Tully, J. M. Bové, C. Mouches, D. L. Rose, M. E. Coan, and T. B. Clark. 1986. *Spiroplasma kunkelii* sp. nov., the etiological agent of corn stunt disease. Int. J. Syst. Bacteriol. 36:170–178.

Whitcomb, R. F., J. M. Bové, T. A. Chen, J. G. Tully, and D. L. Williamson. 1987. Proposed criteria for an interim serogroup classification for members of the genus *Spiroplasma* (class *Mollicutes*). Int. J. Syst. Bacteriol. 37:82–84.

Whitcomb, R. F., K. J. Hackett, J. G. Tully, E. A. Clark, F. E. French, R. B. Henegar, D. L. Rose, and A. G. Wagner. 1990. Tabanid spiroplasmas as a model for mollicute biogeography. Zbl. Bakt. Hyg. 20:(in press).

Williamson, D. L. 1983. The combined deformation-metabolism inhibition test. Methods Mycoplasmol. 1:477–483.

Williamson, D. L., and D. F. Poulson. 1979. Sex ratio organisms (spiroplasmas) of *Drosophila*, p. 175–208. In: R. F. Whitcomb and J. G. Tully (ed.), The mycoplasmas, vol. 3. Academic Press, New York.

Williamson, D. L., and J. G. Tully. 1982. Characterization of spiroplasmas by serology. Rev. Inf. Dis. 4(Suppl.):S137-S140.

Williamson, D. L., and R. F. Whitcomb. 1974. Helical, wall-free prokaryotes in *Drosophila*, leafhoppers and plants. Colloq. Inst. Nat. Santé Rech. Med. 33:283–290.

Williamson, D. L., and R. F. Whitcomb. 1975. Plant mycoplasmas: a cultivable spiroplasma causes corn stunt disease. Science 188:1018–1020.

Williamson, D. L., and R. F. Whitcomb. 1983. Special serological tests for spiroplasma identification. Methods Mycoplasmol. 2:249–259.

Williamson, D. L., R. F. Whitcomb, and J. G. Tully. 1978. The spiroplasma deformation test, a new serological method. Curr. Microbiol. 1:203–207.

Williamson, D. L., J. G. Tully, and R. F. Whitcomb. 1979. Serological relationships of spiroplasmas as shown by combined deformation and metabolism inhibition tests. Int. J. Syst. Bacteriol. 29:345–351.

Williamson, D. L., T. Steiner, and G. J. McGarrity. 1983. Spiroplasma taxonomy and identification of the sex ratio organisms: can they be cultivated? Yale J. Biol. Med., 56:583–592.

Williamson, D. L., J. G. Tully, and R. F. Whitcomb. 1989a. The genus *Spiroplasma*, p. 71–111. In: R. F. Whitcomb, J. G. Tully, (ed.), The mycoplasmas, vol. 5. Academic Press, San Diego, CA.

Williamson, D. L., K. J. Hackett, A. G. Wagner, and A. J. Cohen. 1989b. Pathogenicity of cultivated *Drosophila willistoni* spiroplasmas. Curr. Microbiol. 19:53–56.

Williamson, D. L., J. G. Tully, D. L. Rose, K. J. Hackett, R. Henegar, P. Carle, J. M. Bové, D. E. Colflesh, and R. F. Whitcomb 1990. *Mycoplasma somnilux* sp. nov., *Mycoplasma luminosum* sp. nov., and *Mycoplasma lucivorax* sp. nov., new sterol-requiring mollicutes from firefly beetles (Coleoptera: Lampyridae). Int. J. Syst. Bacteriol. 40:160–164.

Woese, C. R., J. Maniloff, and L. B. Zablen. 1980. Phylogenetic analysis of the mycoplasmas. Proc. Nat. Acad. Sci. (USA) 77:494–498.

Wróblewski, H., K. E. Johansson, and S. Hjérten. 1977. Purification and characterization of spiralin, the main protein of the *Spiroplasma citri* membrane. Biochim. Biophys. Acta 465:275–289.

Wróblewski, H., D. Robic, D. Thomas, and A. Blanchard. 1984. Comparison of the amino acid compositions and antigenic properties of spiralins purified from the plasma membranes of different spiroplasmas. Ann. Microbiol. 135A:73–82.

Yunker, C. E., J. G. Tully, and J. Cory. 1987. Arthropod cell lines in the isolation and propagation of tickborne spiroplasmas. Curr. Microbiol. 15:45–50.

Zeigel, R. F., and H. F. Clark. 1974. Electron microscopy of the suckling mouse cataract agent: a noncultivable animal pathogen possibly related to mycoplasma. Infect. Immun. 9:430–443.

CHAPTER 90

The Family Heliobacteriaceae

MICHAEL T. MADIGAN

General Properties of Heliobacteria

The family Heliobacteriaceae contains all the anoxygenic phototrophic bacteria that produce bacteriochlorophyll (Bchl) *g*. Two genera of heliobacteria are currently recognized, *Heliobacterium* and *Heliobacillus*. The genus *Heliobacterium* comprises the species *H. chlorum* (Gest and Favinger, 1983), *H. gestii* (Ormerod et al., 1990), and *H. fasciculum* (Ormerod et al., 1990), whereas *Heliobacillus* contains a single species, *H. mobilis* (Beer-Romero and Gest, 1987). The unique bacteriochlorophyll (Bchl *g*) of heliobacteria distinguishes them from the purple bacteria, which contain Bchl *a* or *b*, and from the green bacteria, which contain Bchl *c*, *c*$_s$, *d*, or *e* (and small amounts of Bchl *a*). Heliobacteria are also distinguished from all other anoxygenic phototrophs by lacking differentiated structures such as chlorosomes or intracytoplasmic membranes as sites of photosynthetic pigments; in heliobacteria, photopigment complexes reside within the cytoplasmic membrane (Gest and Favinger, 1983; Miller et al., 1986).

Bacteriochlorophyll *g* shows structural relationships to both chlorophyll *a* (Brockman and Lipinski, 1983; Michalski et al., 1987) and to the various bacteriochlorophylls of green and purple bacteria. Unlike all other bacteriochlorophylls, Bchl *g* contains a vinyl ($H_2C=CH_2$) group on ring I of the tetrapyrrole, as does green plant chlorophyll *a*. However, like bacteriochlorophylls *a* and *b* and unlike chlorophyll *a*, the pyrrole ring II of Bchl *g* is reduced. In addition, the esterifying alcohol of Bchl *g* is farnesol, the alcohol present in the bacteriochlorophylls of green sulfur bacteria (Gloe et al., 1975) instead of phytol or geranylgeraniol. Thus, Bchl *g* is clearly a molecular hybrid and because of this may have played an important role in the evolution of pigment diversity among photosynthetic microorganisms (Michalski et al., 1987).

The novel bacteriochlorophyll of heliobacteria is responsible for the unique absorption properties of these organisms. Heliobacteria absorb in the near infrared at 788 nm, distinctly away from regions of the spectrum absorbed by green bacteria and purple bacteria, which typically show infrared absorption maxima at 705–740 or 830–1100 nm, respectively (See Chapter 13). Absorption of radiation between about 760 and 800 nm is undoubtedly a survival strategy for heliobacteria. These wavelengths represent a "window" in the spectrum of virtually all phototrophic organisms (Pfennig, 1989). Hence, heliobacteria occupy a unique ecological position in the phototrophic world—absorbing strongly in a region of the electromagnetic spectrum that is not absorbed and is hence transmitted through the cells of other phototrophs. This, coupled with the fact that heliobacteria are primarily soil organisms (see below), suggests a very different ecological role for members of this group in nature.

All known heliobacteria are green in color, primarily due to Bchl *g*, but also because the major (if not sole) carotenoid present is the green pigment neurosporene (Gest and Favinger, 1983; van Dorssen et al., 1985). However, the green color of cells of heliobacteria may be more than just due to chance. The results of various biophysical studies have shown that several components of the photosynthetic electron transport chain of *H. chlorum* (and presumably other heliobacteria as well) are closely related to those of the green sulfur bacteria. For example, Bchl *c* or a related pigment is the primary acceptor of electrons from Bchl *g* (Amesz, 1989; Smit and Amesz, 1988). Once reduced by Bchl *g*, this pigment quickly proceeds to reduce a secondary electron acceptor that is apparently an iron-sulfur cluster similar to those of the green bacterium *Prosthechochloris aestuarii* (Smit and Amesz, 1988; Smit et al., 1989).

Redox studies of photochemical reactions in *H. chlorum* also reveal relationships between heliobacteria and green bacteria. Like *Chlorobium* and *Prosthecochloris*, the primary acceptor of electrons of *H. chlorum* reaction centers is at a much lower reduction potential than the primary acceptors of either purple bacteria or *Chloroflexus* (Fuller et al., 1985; Nuijs et al., 1985; Prince et al., 1985; Fischer, 1990; von Kan et al., 1990). The remainder of the electron transport chain supporting cyclic photophosphorylation in *H. chlorum,* in particular cytochrome c_{553} and the iron-sulfur centers, show strong molecular similarities to comparable components in green sulfur bacteria (Brok et al., 1986; Smit and Amesz, 1988; Smit et al., 1987a; 1987b; Vos et al., 1989).

By contrast, the reaction center of heliobacteria is very unlike that of green sulfur bacteria (Amesz, 1988, 1989). The reaction center of *H. chlorum* consists of a pair of Bchl *g* molecules that absorb at 798 nm (designated P798, Fuller et al., 1985; Prince et al., 1985; Smit and Amesz, 1988). At low temperature, at least three spectrally distinct forms of Bchl *g* can be resolved, showing absorption maxima at 778, 793, and 808 nm (van Dorssen et al., 1985). These forms of Bchl *g* serve as light-harvesting antennae bacteriochlorophyll and by so being funnel energy to the reaction center (van Dorssen et al., 1985; Smit et al., 1989). Reaction center Bchl *g* in *H. chlorum* is firmly embedded in a highly proteinaceous plasma membrane and contains tightly bound cytochrome *c* (Fuller et al., 1985). Several polypeptides associated with *H. chlorum* reaction centers stain for heme, indicating that cytochrome *b/c₁* complexes of the type found in other anoxygenic phototrophs are probably present in heliobacteria as well (Fuller et al., 1985). In *Heliobacillus mobilis,* a photoactive, reaction center-core antenna complex has been isolated that contains a 47 kDa polypeptide (Trost and Blankenship, 1989; van de Meent et al., 1990). This protein binds both the primary donor (a pigment called P800) and 20–25 antenna bacteriochlorophyll *g* molecules (Trost and Blankenship, 1989). Interestingly, the *H. mobilis* core complex resembles photosystem I of oxygenic phototrophs because both the primary donor species and antenna bacteriochlorophylls are bound to a single polypeptide. Thus, the structural similarities between bacteriochlorophyll *g* and chlorophyll *a* discussed above are reinforced by the finding that the pigment-protein complexes of heliobacteria closely resemble those of green plants.

Despite the fact that cells of heliobacteria stain Gram-negatively, the cell walls of these organisms are highly atypical. Electron micrographs of thin sections (Gest and Favinger, 1983) and freeze-etched (Miller *et al.,* 1986) cells of *H. chlorum* do not show the classical lipopolysaccharide layer of Gram-negative bacteria. Instead, the *H. chlorum* cell wall appears as a regular array of subunits approximately 11

nm in diameter (Fig. 1). Similar subunit structures have been observed in the walls of a few marine phototrophic bacteria (Remsen et al., 1970) and in the halophilic phototroph, *Rhodospirillum salexigens* (Evers et al., 1984). The chemical composition of the subunits in the *H. chlorum* cell wall are unknown but probably consist of protein. Thin sections of cells of *Heliobacterium gestii* and *Heliobacillus mobilis* reveal a cell wall structure similar to that of *H. chlorum* (Beer-Romero, 1986).

For some unknown reason, the cell walls of heliobacteria are extremely fragile. Because of this, cultures of heliobacteria tend to form spheroplasts and eventually lyse when approaching stationary phase, and the viability of more than one-week-old cultures is thus rather poor (Beer-Romero, 1986). The fact that lysis is probably due to cell wall peculiarities is suggested from studies of penicillin sensitivity. Heliobacteria are unusually sensitive to penicillin (Beer-Romero et al., 1988), an antibiotic that interferes with peptidoglycan crosslinking. Growth of *H. chlorum* is inhibited by as little as 2 ng/ml of penicillin G (Beer-Romero et al., 1988), approximately 10,000-fold less than that required to inhibit growth of a Gram-negative bacterium like *E. coli.* This suggests that either the amount of peptidoglycan synthesized by *H. chlorum* cells is very low (Beer-Romero *et al.,* 1988) or the nature of heliobacterial peptidoglycan is atypical of Gram-negative bacteria.

Preliminary chemical studies of heliobacterial cell walls suggest that the peptidoglycan of these organisms is indeed highly unusual. The peptidoglycan of *H. chlorum* contains L, L-diaminopimelic acid (instead of *meso*-diaminopimelic acid) as one of the crosslinking amino acids between muramic acid residues in peptidoglycan sheets, and also contains a glycine interbridge, typical of clostridia and several other Gram-*positive* bacteria (Beer-Romero et al., 1988). Whether the problem of lysis of heliobacteria is directly related to this unusual peptidoglycan is not known. However, it is likely that some component in the cell wall is inherently weak, and for unknown reasons becomes unstable in late stationary phase cells, triggering lytic events (Beer-Romero, 1986).

Several lines of evidence suggest similarities between heliobacteria and Gram-positive bacteria. Studies of the cell wall polysaccharides of *H. chlorum, H. gestii,* and *H. mobilis* showed that a polysaccharide-like material (in addition to peptidoglycan) was present in the cell walls of these species but that the material was clearly not lipopolysaccharide, a substance found in most Gram-negative bacteria (Beck et al., 1990). In addition, fatty acid analyses of heliobacterial

Fig. 1. Transmission electron micrograph of cross sections of cells of *Heliobacterium chlorum*. The studded subunit structure of the cell wall surface is visible. Bar = 0.25 µm. Micrograph courtesy of Rudi Turner and H. Gest.

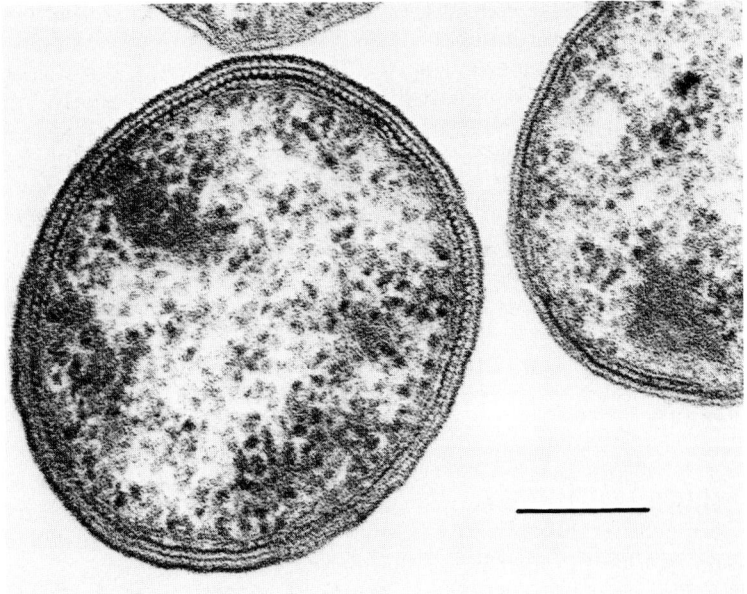

cells showed that they contained a high proportion of branched-chain fatty acids. As in Gram-positive bacteria, 40–60% of the total fatty acids in heliobacterial cells are branched-chained, with isopalmitoleic acid predominating (Beck et al., 1990).

Phylogenetic analyses of heliobacteria by 16S rRNA sequencing have shown that their "Gram-positive type" cell wall and fatty acid content is a reflection of their evolutionary lineage. Comparisons of their ribosomal RNA sequences show a specific relationship to Gram-positive bacteria, in particular to the "low GC" *(Bacillus/Clostridium)* subdivision (Woese et al., 1985). Hence, the apparently Gram-positive nature of heliobacterial cell walls is supported by molecular sequencing; heliobacteria likely arose from Gram-positive ancestry. Analyses of 16S rRNA from *Heliobacillus mobilis* (Beer-Romero and Gest, 1987) and *Heliobacterium gestii* (J. G. Ormerod, personal communication) show them to be very closely related to *Heliobacterium chlorum* and thus also to Gram-positive prokaryotes.

Physiologically, heliobacteria are *nonsulfur* phototrophic bacteria; their metabolism is photoheterotrophic, and high sulfide levels tend to inhibit growth (Beer-Romero, 1986; Beer-Romero and Gest, 1987; Beer-Romero et al., 1988). However, unlike purple nonsulfur bacteria, relatively few carbon sources are photometabolized by heliobacteria, pyruvate being the best carbon source for all species (Beer-Romero, 1986). *Heliobacillus mobilis,* the most nutritionally versatile species, uses lactate, acetate, and butyrate (the latter only in the presence of CO_2) in addition to pyruvate (Beer-Romero and Gest, 1987).

Photoheterotrophic growth of heliobacteria only occurs under strictly anaerobic conditions. As a consequence, media for growth of these organisms must be prereduced, and care must be taken when transferring cultures to avoid exposure to oxygen. Transfer of dense cultures can generally be made at the bench without elaborate anaerobic precautions. However, it is very important that the medium employed be highly reduced, either by addition of a reducing agent, or preferably by storage of sterile media for several days in equilibrium with the gas atmosphere of an anaerobic glove box (Beer-Romero, 1986).

Heliobacteria are apparently incapable of dark growth (Gest and Favinger, 1983; Beer-Romero, et al., 1988), although microaerobic dark growth has not been investigated. Autotrophic growth of heliobacteria, either phototrophically or lithotrophically, supported by H_2 or low levels of sulfide (Hansen and van Gemerden, 1972), has not been rigorously tested and also remains a potential growth option. Like most anoxygenic phototrophs, all heliobacteria tested have been shown to fix molecular nitrogen; assays have demonstrated acetylene reduction at significant rates by cultures of all three species (Beer-Romero, 1986). Ammonium salts as well as the amino acid glutamine are the only other nitrogen sources known to support growth of heliobacteria (Beer-Romero, 1986).

The only B vitamin required by heliobacteria is biotin (Beer-Romero, 1986). In addition, however, a reduced sulfur source (for biosynthetic purposes) is required by *H. gestii* and tends to stimulate growth of *H. mobilis* and *H. chlorum* (Beer-Romero, 1986). The latter two species apparently have assimilatory sulfate reductase pathways; both organisms will grow with sulfate as sole sulfur source (Beer-Romero, 1986). By contrast, *H. gestii* will not grow with sulfate as sole sulfur source and requires small amounts (1–2 mM) of cysteine, methionine, thioglycollate, or thiosulfate, or large amounts (> 0.1%) of yeast extract for good growth (Beer-Romero, 1986). Sulfide at 0.5–1 mM final concentration will also satisfy the biosynthetic sulfur requirements of *H. gestii* (Beer-Romero, 1986).

Habitats

The ecology and distribution of heliobacteria is poorly understood. However, based on isolations made to date, it appears that soil is the best, if not the only, source of these organisms. Heliobacteria have not as yet been isolated from aquatic habitats. This seems unusual considering that the ecology of all other anoxygenic phototrophs is closely linked to aquatic ecosystems (Madigan, 1988). If soil really is the major habitat of heliobacteria, it is likely that they are the dominant anoxygenic phototrophs in the soil environment and play a quite different ecological role in nature than do "typical" anoxygenic phototrophs.

Isolations of heliobacteria made thus far indicate that paddy soil is an excellent source of these organisms. J. G. Ormerod (personal communication) and Beer-Romero (1986) have isolated several heliobacteria from paddy soils in Thailand and Tanzania. For example, using a standard malate mineral salts medium for nonsulfur purple phototrophic bacteria (Ormerod et al., 1961) and no elaborate anaerobic precautions, heliobacteria could readily be isolated from paddy soil previously allowed to dry out thoroughly. By contrast, using the same enrichment protocol, a *moist* sample of the same paddy soil generally yielded purple bacteria only (J. G. Ormerod, personal communication). This suggests that drying decreases the viability of purple bacteria, but not heliobacteria, and that if both are present in the original inoculum, standard enrichment methods for nonsulfur phototrophic bacteria tend to favor development of purple bacteria instead of heliobacteria.

Table 1. Major properties of the Heliobacteriaceae.

	Heliobacterium chlorum	*Heliobacterium gestii*	*Heliobacillus mobilis*	*Heliobacterium fasciculum*
Cell shape	Rod	Spiral	Rod	Rod
Cell size (μm)	1×7–9	1×7–10	1×7–10	0.8–1 $\times$ 8–20, associate in parallel bundles
Pigments	Bacteriochlorophyll *g*, neurosporene[a]	Bacteriochlorophyll *g*, presumably neurosporene	Bacteriochlorophyll *g*, presumably neurosporene	Bacteriochlorophyll *g*, carotenoids unknown
Absorption peaks, in vivo (nm)	788, 718(S), 670, 575, 375	788, 718(S), 670, 575, 375	788, 718(S), 670, 575, 375	790
Motility type	Gliding	Single subpolar flagellum	Peritrichous flagella	Thick polar flagella
Carbon compounds photometabolized	Pyruvate, lactate	Pyruvate, lactate, butyrate	Pyruvate, lactate, butyrate, acetate	Unknown—after 2–3 transfers in pure culture, growth is sporadic but is stimulated by addition of a small amount of sterile enrichment medium
Required growth factors	Biotin	Biotin, reduced sulfur source	Biotin	Unknown
Optimum temperature (°C)	38–42	38–42	38–42	35–40
Optimum pH	6.2–7	6.2–7	6.2–7	7
GC content (mol%)	52	54.8	50.3	Unknown

[a]From van Dorssen et al. (1985). Although they are not apparent in spectra of intact cells (see Fig. 2), membrane preparations from *H. chlorum* examined at low temperature also contain peaks at 499, 463, and 440 nm due to absorption by neurosporene.

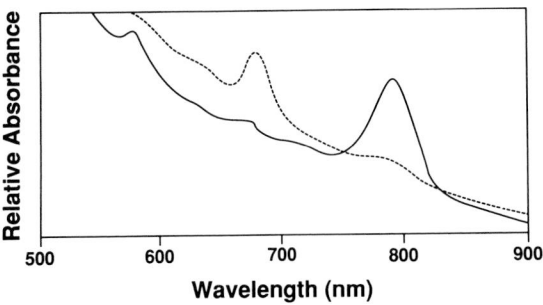

Fig. 2. Absorption spectra of intact cells of *Heliobacillus mobilis*. Solid line, spectrum of cells suspended in 60% sucrose containing 0.05% sodium ascorbate and kept under anaerobic conditions; dashed line, same cell suspension exposed to light and air for one hour.

The latter may explain why heliobacteria were overlooked for so long.

Nothing is known concerning the distribution of heliobacteria in soils of various physiochemical properties. However, it is perhaps noteworthy that *H. mobilis* (Beer-Romero and Gest, 1987), *H. gestii* and *H. fasciculum* (Ormerod et al., 1990) were all isolated from *dry* paddy soil. This suggests that these strictly anaerobic organisms have evolved strategies for surviving drying and any resultant exposure to oxygen. Of significance in this connection is the finding that even after heat pasteurization or boiling, soil samples still can be used as inocula for enrichment cultures of heliobacteria (J. G. Ormerod, personal communication).

The production of "resting" forms resembling bacterial endospores has been observed in certain heliobacteria and apparently accounts for the heat and desiccation resistance of these phototrophs. *H. gestii* and *H. fasciculum* both produce what appear to be true endospores which contain dipicolinic acid and elevated levels of calcium (Ormerod et al., 1990). Spores of *H. gestii* are subterminal, cylindrical, and form in small numbers in old cultures that sediment to the bottom in clumps. *H. fasciculum* sporulates abundantly, and the cylindrical endospores swell to distend the sporangium (Ormerod et al., 1990). Controlled experiments with suspensions of *H. gestii* endospores show that although spore viability decreases slightly with boiling, no loss of viability occurs in pasteurized (60°C for 30 min) suspensions of endospores or in suspensions treated at 85°C for 15 min (Ormerod et al., 1990).

Endospores represent a novel survival strategy for anoxygenic phototrophic bacteria and clearly impart new ecological significance to these organisms in nature. Heliobacterial endospores probably allow survival of heliobacteria during unfavorable growing conditions, for example during dry periods when soil zones become aerobic (heliobacterial endospores are presumably resistant to O_2 but this has not yet been demonstrated). Further work is necessary to determine what factors trigger sporulation in heliobacteria. Because of the phylogenetic link between heliobacteria and Gram-positive bacteria (Woese et al., 1985), the existence of endo-

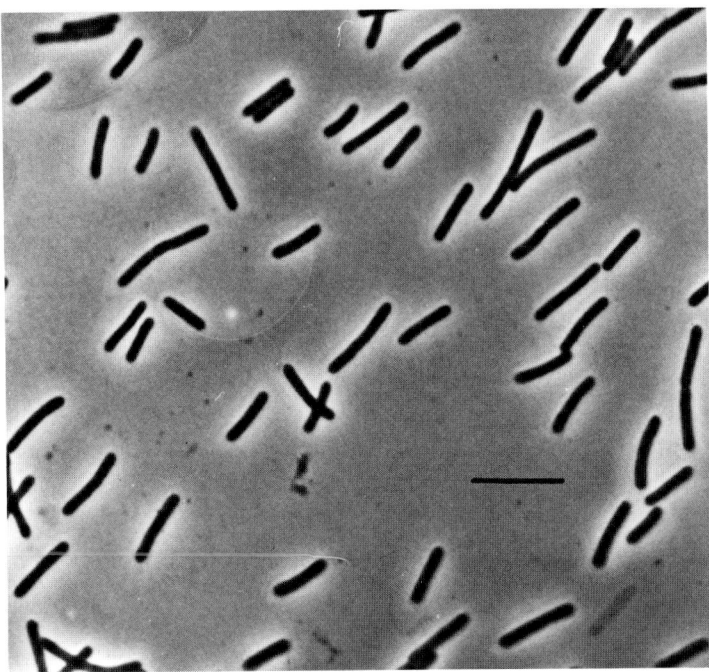

Fig. 3. Phase contrast micrograph of cells of *Heliobacterium chlorum*. Bar = 10 μm. Micrograph courtesy of J. Favinger and H. Gest.

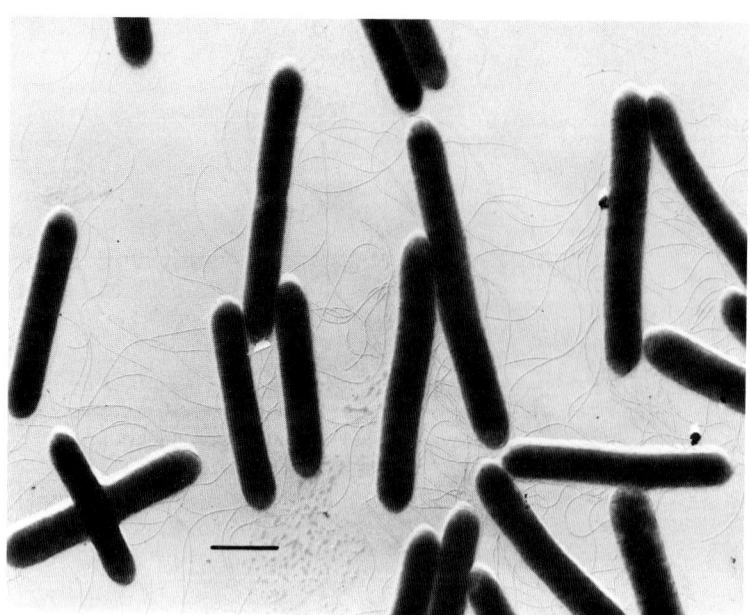

Fig. 4. Transmission electron micrograph of shadowed cells of *Heliobacillus mobilis*. Peritrichous flagella are visible. Bar = 2 μm. Micrograph courtesy of R. Turner and H. Gest.

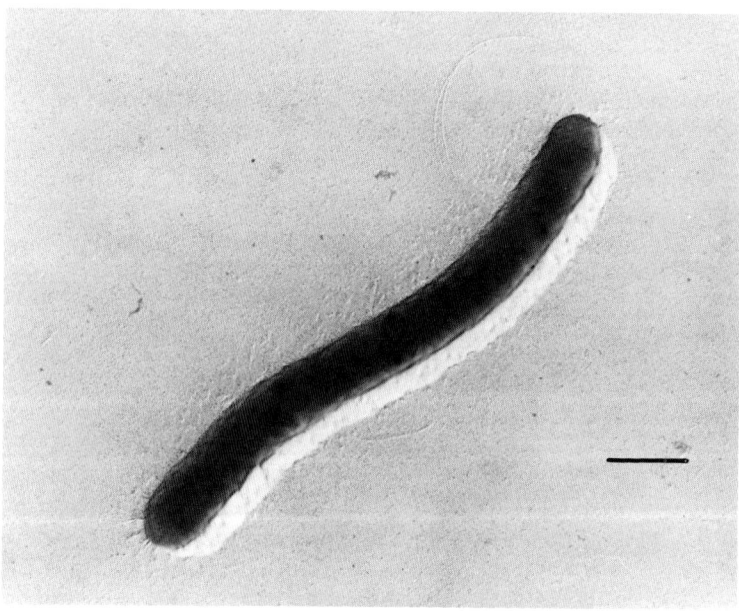

Fig. 5. Transmission electron micrograph of a shadowed cell of *Heliobacterium gestii*. Subpolar flagella are visible. Bar = 10 μm. Micrograph courtesy of R. Turner and H. Gest.

spore-forming heliobacteria is perhaps not surprising. However, heliobacterial endospores represent the first report of these structures outside of heterotrophic bacteria, and thus it would be premature to assume that the production of endospores in phototrophs occurs in response to the same environmental signals as in heterotrophs such as clostridia and bacilli.

It may also be significant to the ecology of heliobacteria that all species fix molecular nitrogen. In paddy soils, heliobacteria along with cyanobacteria may contribute fixed nitrogen to rice and other flooded soil cultivars. The connection between rice soils and anoxygenic pho-

totrophic bacteria has been extensively documented (Buresh et al., 1980; Habte and Alexander, 1980; Kobayashi, 1982; Kobayashi et al., 1967). However, the link between rice soils and specific, culturable phototrophs in most of these cases is lacking. The observation of nitrogen-fixing "*Thiospirillum*-like" phototrophs in lowland rice culture (Habte and Alexander, 1980) may really have been of N_2-fixing *Heliobacterium gestii*, because both organisms are large spirilla. One can imagine that a photoheterotrophic N_2-fixing lifestyle, such as that of heliobacteria, would be well suited to life in a rice-soil environment. Organic compounds

Fig. 6. Transmission electron micrograph of thin sections of low-light-grown cells of *Heliobacterium chlorum*. Note absence of intracytoplasmic membranes. Bar = 0.5 μm. Micrograph courtesy of R. Turner and H. Gest.

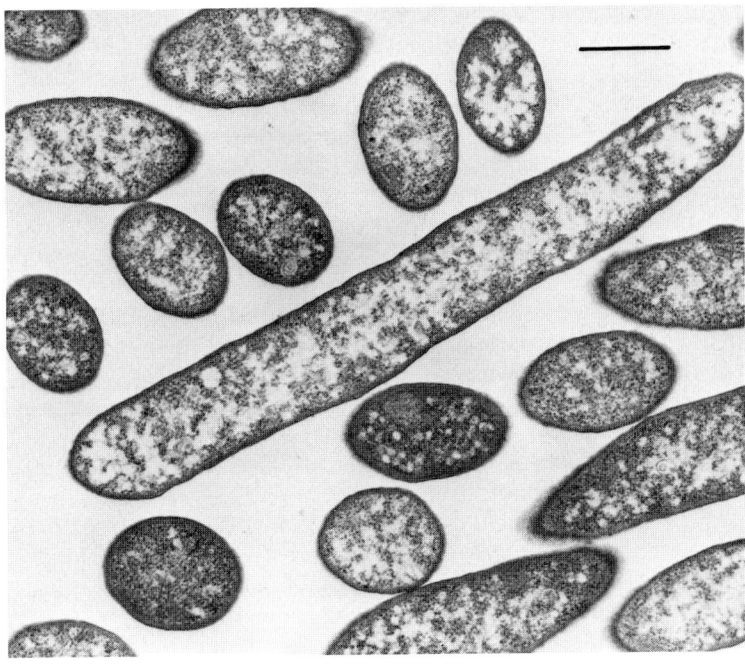

excreted by the roots of rice plants could serve as carbon sources and as reductants for N_2 fixation, the latter process driven in heliobacterial cells by the energy of sunlight. In return, heliobacteria could supply fixed N to the rice plant in a casual type of symbiosis. More work to document the association of heliobacteria with paddy soils and rice plants is obviously needed.

Enrichment and Isolation

Succesful enrichment of heliobacteria begins with the use of soil (rather than aquatic samples) as inocula. As noted above, dry soil may be advantageous. A second important consideration is adherence to strictly anaerobic conditions, at least when pure cultures are being isolated from crude enrichments. Light and temperature can also be selective enrichment factors. The studies of Beer-Romero (1986) showed that heliobacteria grow best at higher light intensities and at higher temperatures than are typically used for isolation of anoxygenic phototrophs; conditions of 10,000-lux incandescent illumination and 40°C are optimal for growth of heliobacteria. These conditions also discourage development of purple and green bacteria.

An enrichment protocol found useful for isolation of heliobacteria is one adapted from the general enrichment protocol of Gest et al. (1985) and is as follows: Twenty to thirty milliliters of mineral salts medium (for recipe, see below) is placed in a 50–70 ml serum vial, inoculated with soil, bubbled vigorously with N_2:CO_2 (95:5), and then sealed with a serum stopper under a stream of anaerobic gas. Alternatively, enrichments can be established in medium stored for several days in contact with the gas atmosphere of an anaerobic glove box (N_2:H_2:CO_2). The advantage of the latter system is that the medium can be dispensed to vials or bottles and inoculated and sealed with rubber stoppers (or other gas-tight closures) all within the confines of the anaerobic hood. This eliminates the need for alternative anaerobic techniques.

Positive enrichments for heliobacteria usually develop within 7–10 days and are brownish green to emerald green in color. Pure cultures of heliobacteria can be obtained by agar shake dilution or by streaking on agar plates. In either case, the extreme oxygen sensitivity of heliobacteria, especially freshly enriched isolates, mandates that agarbased isolation procedures be performed within an anaerobic glove box. After agar plates are streaked, they can be sealed immediately in anaerobic jars within the glove box and then the jar removed for incubation in a water bath or constant temperature room.

The use of light filters may be useful for specific enrichment of heliobacteria. Heliobacteria contain Bchl *g*, which absorbs (in vivo) near 790 nm (Gest and Favinger, 1983). This is well away from those regions of the spectrum where the pigments of green or purple bacteria absorb (Pfennig, 1989). Therefore, illuminating enrichments with light previously passed through a filter that transmits wavelengths only around 790 nm should in theory be highly selective for heliobacteria. In this connection, the Ealing Optics (S. Natick, MA) no. 35–4373 narrow-band, infrared-interference filter may be suitable. This filter has an optical "window" between 780 and 800 nm with maximal transmission at 790 nm, and it is probably the most specific filter available for enrichment studies.

Another successful protocol for isolation of heliobacteria is to pasteurize the inoculum. Experiments have shown that various heliobacteria survive heating to 80°C for 15 minutes; H. gestii was readily enriched from dry soil following this procedure (Ormerod et al., 1990). Other anoxygenic phototrophs would not be expected to survive such treatment, except perhaps for thermophilic forms like Chloroflexus and Chromatium tepidum (Pierson and Castenholz, 1974; Madigan, 1986). However, since the latter are not soil organisms, they would not be expected to interfere in this enrichment protocol.

Culture of Heliobacteria

Growth Media

For general enrichment of photoheterotrophic anoxygenic phototrophs, it is undesirable to use easily fermented organic compounds (like sugars) as carbon sources because they are rapidly catabolized by heterotrophic anaerobes (See Chapter 101; also Madigan, 1988). This is true for isolation of heliobacteria as well. However, pyruvate, a potentially fermentable organic acid, is used as carbon source by all known heliobacteria (Beer-Romero, 1986) and is a good enrichment substrate for these organisms. Enrichment experiments using pyruvate as the major carbon source anaerobically at 40°C and 10,000 lux have successfully yielded several strains of H. mobilis and H. gestii (Beer-Romero, 1986). Malate and succinate are also suitable for enrichment purposes (J. G. Ormerod, personal communication) despite the fact that they are not used by pure cultures of known heliobacteria (Beer-Romero et al., 1987). Presumably chemoheterotrophs in the enrichments metabolize these substrates to acetate or other compounds that can then support photoheterotrophic growth of heliobacteria.

Although it is usually not necessary, enrichments for heliobacteria can be made even more specific if N_2 rather than ammonia is employed as the sole source of nitrogen (Gest et al., 1985;

Beer-Romero, 1986). Enrichment cultures employing N_2 as the sole nitrogen source usually develop more slowly than enrichments containing ammonia, but the use of N_2 generally eliminates interference from the fermentative anaerobes that occasionally overgrow heliobacteria in pyruvate-based enrichments.

Pyruvate mineral salts (PMS) medium and mixed acid mineral salts (MAMS) medium have been successfully employed for enrichment culture of heliobacteria:

Pyruvate Mineral Salts (PMS) Medium

The following ingredients are added to 1 liter of distilled water:

Ethylenediaminetetraacetate sodium salt	10 mg
$MgSO_4 \cdot 7H_2O$	200 mg
$CaCl_2 \cdot 2H_2O$	75 mg
NH_4Cl	1 g
K_2HPO_4	0.9 g
KH_2PO_4	0.6 g
Sodium pyruvate	2.2 g
Trace element solution (see below)	1 ml
Yeast extract	0.1 g
Vitamin B_{12}	20 µg

Adjust pH to 6.8 with NaOH or HCl and sterilize by autoclaving for 20 min. After cooling briefly add either methionine, cysteine, or thiosulfate to a final concentration of 1 mM, and transfer to an anaerobic chamber until used.

Modifications of PMS Medium

LMS medium = PMS medium without pyruvate with sodium lactate added to a final concentration of 20 mM.

PMS-N medium = PMS medium without NH_4Cl. The gas in the headspace of the enrichment should be N_2: CO_2 (95:5).

AMS medium = PMS medium without pyruvate plus 20 mM sodium acetate and 0.1% $NaHCO_3$.

Mixed Acid Mineral Salts (MAMS) Medium

This enrichment medium avoids the use of pyruvate and employs a variety of organic acids as potential photoheterotrophic substrates. The following are added to 1 liter of distilled water:

Ethylenediaminetetraacetate sodium salt	10 mg
NaCl	0.4 g
$MgSO_4 \cdot 7H_2O$	200 mg
$CaCl_2 \cdot 2H_2O$	75 mg
NH_4Cl	0.8 g
K_2HPO_4	0.45 g
KH_2PO_4	0.3 g
$NaHCO_3$	1 g
Trace elements solution (see below)	1 ml
Yeast extract	0.2 g
Vitamin B_{12}	20 µg
Sodium acetate	1 g
Sodium malate	2 g
Sodium lactate	1 g

Adjust pH to 6.8 and sterilize by autoclaving for 20 min. When partially cooled, add NaHCO₃ from a filter-sterilized stock solution and add either methionine, cysteine, or thiosulfate to a final concentration of 1mM and transfer to an anaerobic chamber until used.

Modifications of MAMS Medium

MAMS-N medium = MAMS medium without NH₄Cl. Headspace of enrichment should be N₂:CO₂ (95:5).

Pyruvate-Yeast Extract (PYE) Medium

This medium is suitable for growth of pure cultures of heliobacteria but is unsuitable for enrichment purposes. The following are added to 1 liter of distilled water.

K₂HPO₄	1 g
MgSO₄·7H₂O	200 mg
CaCl₂·2H₂O	20 mg
Na₂S₂O₃·5H₂O	100 mg
Sodium pyruvate	2.2 g
Yeast extract	8 g

Adjust to pH 7, autoclave and store in anaerobic chamber until used.

Trace Element Solution for Above Media

The following are added to 1 liter of distilled water:

Ethylenediaminetetraacetate sodium salt	5.2 g
FeCl₂·4H₂O	1.5 g
ZnCl₂	70 mg
MnCl₂·4H₂O	100 mg
H₃BO₃	6 mg
CoCl₂·6H₂O	190 mg
CuCl₂·2H₂O	17 mg
NiCl₂·6H₂O	25 mg
Na₂MoO₄·2H₂O	188 mg
VoSO₄·2H₂O	30 mg
Na₂WO₄·2H₂O	2 mg

Add compounds in the above order; make sure that the EDTA is fully dissolved before adding remaining components. Store at 4°C.

Growth Vessels

Growth of pure cultures of heliobacteria can be accomplished in rubber-stoppered culture tubes (Bellco anaerobic culture tubes, 18 × 142 mm) or in screw-capped tubes or bottles. Sterile empty vessels can be filled aseptically with sterile prereduced media within the anaerobic chamber and then sealed with stoppers or screw caps before removal. Screw cap tubes or bottles should be filled completely and tightly capped. Because the seal of screw-capped vessels is rarely perfect, the addition of 0.05% (final concentration) of sodium ascorbate to media may be necessary for growth of some strains of heliobacteria in screw-capped tubes. Alternatively, nonsterile media in anaerobic culture tubes can be degassed with an anaerobic gas mixture (N₂:CO₂,95:5, previously passed through a hot, reduced copper furnace to re-move all traces of O₂), sealed under an atmosphere of anaerobic gas and autoclaved in an anaerobic tube press (Bellco). Properly sealed tubes can then be stored outside the anaerobic glove box and inoculated using standard Hungate-type techniques.

Storage of Pure Cultures

For short-term storage (weeks to months) of heliobacteria, "stabs" of medium PYE are inoculated and incubated in the light in anaerobic jars. Once fully grown, stab cultures can be stored for several months in anaerobic jars or in an anaerobic chamber exposed to low light. For unknown reasons stab cultures of heliobacteria retain viability much longer than liquid cultures stored under the same conditions. Freezing or lyophilizing cell suspensions can be used for long-term conservation of heliobacteria. For freezing, cultures grown to late exponential phase on medium PYE are mixed with an equal volume of fresh prereduced medium PYE containing 10% DMSO in sterile "snap-cap" disposable plastic tubes and kept anaerobically at 0° for 10 minutes. Tubes should then be stored at −80° (or lower) in a freezer or at −196° in liquid nitrogen. Lyophilization of dense suspensions of heliobacteria by the double vial method (Gherna, 1981) is also successful, and cells retain viability in this form for several years.

Identification

The major identifying feature of representatives of the family Heliobacteriaceae is the presence of Bchl g (Table 1). Organisms containing Bchl g show in vivo absorption maxima at 788–790 nm (Fig. 2; see also Table 1). However, one must be very careful in interpreting spectra of potential heliobacteria because of the ease with which Bchl g can be converted to a form of chlorophyll a upon exposure to oxygen and light, a phenomenon that greatly affects absorption spectra (Beer-Romero et al., 1988, see also Fig. 2). A small peak at 670 nm, always present in spectra of heliobacteria, can become the major peak following exposure of intact cells to light and air (Fig. 2). For spectral measurements, the best way to avoid conversion of Bchl g to the chlorophyll a-like pigment is to transfer cells to a prereduced (0.05% ascorbate) viscous medium (60% sucrose or 30% bovine serum albumin) within an anaerobic glove box, and then transfer the mixture to a curvette, seal, and remove the curvette, and immediately determine the absorption spectrum.

In shake tubes or on plates, colonies of heliobacteria are easily distinguished from purple or green sulfur bacteria. Colonies of heliobacteria are brown-green in color and later become bright emerald green. However, as they develop in shake tubes, colonies of heliobacteria are typically transparent, or nearly so, in contrast to the opaque appearance of colonies of purple or green bacteria. This feature helps in identifying even very young colonies of heliobacteria. In liquid media, cultures of heliobacteria are greenish brown in color, but dense cultures of *H. gestii* appear somewhat more yellow than do cultures of either *H. chlorum* or *H. mobilis.*

Morphologically, *H. chlorum* and *H. mobilis* are rods (Beer-Romero and Gest, 1987). Cells of *H. chlorum* are long rods, occasionally appearing slightly bent, approximately $1 \times 6–10$ μm (Fig. 3). Cells of *H. mobilis* can be slightly wider than those of *H. chlorum* but are generally similar in overall dimensions to cells of *H. chlorum* (Fig. 4). *H. mobilis* cells are motile by peritrichous flagella (Fig. 4) and motility is most reliably observed in young cultures. Cells of *H. gestii* are highly motile spirilla, approximately $1 \times 7–10$ μm with a single subpolar flagellum, and typically only a single wave per cell is seen (Fig. 5). Endospores are produced in old cultures of *H. gestii..* The single flagellum on each cell of *H. gestii* is inserted subpolarly (Fig. 5). Both *H. mobilis* and *H. gestii* respond to darkness by reversing the direction of swimming (photophobotaxis), and this phenomenon may be of help in deciding whether these organisms are present in initial enrichments. *Heliobacterium fasciculum* consists of motile rods that associate in parallel bundles, the entire bundle swimming with a rolling type of motion (Ormerod et al., 1990). Despite their affinity with Gram-positive bacteria, all heliobacteria stain Gram-negatively.

Electron micrographs of thin sections of heliobacteria (Fig. 6) show no highly developed intracytoplasmic membranes or chlorosomes, and this unusual situation is true of all representatives examined to date (Beer-Romero, 1986). The cell wall of heliobacteria has a "studded" appearance (Fig. 1) and consists of subunits of unknown composition. Cultures of all heliobacteria have a tendency to form spheroplasts in stationary phase, and this phenomenon is particularly acute in cultures of *H. chlorum.* Spheroplasted cultures become transparent, viscous, and emerald green in color, and show poor viability; cultures of *H. mobilis* are least susceptible to spontaneous lysis (Beer-Romero et al., 1988).

Type Strains and Genetic Properties of Heliobacteria

The type strains of heliobacteria are as follows: *Heliobacterium chlorum,* ATCC 35205ᵀ (GC content, 52 mol%); *Heliobacterium gestii,* strain "Ormerod" (GC content, 54.8 mol%); *Heliobacillus mobilis,* ATCC 43427ᵀ (GC content, 50.3 mol%); *Heliobacterium fasciculum,* strain "Ormerod" (GC content unknown) Cultures of *H. chlorum* and *H. mobilis* are available from the American Type Culture Collection, Rockville, MD.

The 16S ribosomal RNA sequence of *H. chlorum* has been published (Woese et al., 1985), and the results show a specific relationship between this organism and certain Gram-positive bacteria. The 16S rRNA sequence of *H. mobilis* is 98.3% homologous to the *H. chlorum* sequence, and the *H. gestii* sequence about 96% homologous to that of *H. chlorum.* The family Heliobacteriaceae is thus a tight phylogenetic group. A summary of the major properties of members of the family Heliobacteriaceae is given in Table 1.

Applications

The major application of heliobacteria thus far has been as experimental tools for biophysical studies of photosynthesis. However, agricultural applications of heliobacteria may arise from a better understanding of their nitrogen-fixing activities in soils. Heliobacteria are primarily soil organisms, and because of this they may be ecologically significant contributors of fixed nitrogen to soil ecosystems, paddy soils in particular. The ease of isolation of heliobacteria from paddy soils, even from thoroughly dried samples, suggests they are hardy organisms and probably abundant in tropical soils. It is possible that in paddy soils the photoheterotrophic metabolism of heliobacteria is driven by rice plant exudates and that in return the bacteria supply fixed nitrogen to the plant. If this can be experimentally confirmed, then heliobacteria would have great economic significance because rice is the major food staple for nearly one-half the world's population. Under any conditions, the fact that the ecology of heliobacteria is closely linked to soil (unlike that of purple and green bacteria) and the fact that all representatives fix molecular nitrogen, speaks well for the potential agricultural significance of this group.

Acknowledgments

I thank F. Rudi Turner and Howard Gest for supplying micrographs of various heliobacteria and John Ormerod for unpublished information on enrichment and isolation. Work of the author is supported by the United States Department of Agriculture.

Literature Cited

Amesz, J. 1988. Structural and functional properties of the reaction center of green bacteria and heliobacteria, p. 129–138. In: J. Breton and A. Vermeglio (ed.), The photosynthetic bacterial reaction center. Structure and dynamics. Plenum Press, New York.

Amesz, J. 1989. Energy transfer and electron transport in *Heliobacterium chlorum*. Photosynthetica. 23:403–410.

Beck, H., G. D. Hegeman, and D. White. 1990. Fatty acid and lipopolysaccharide analyses of three *Heliobacterium* spp. FEMS Microbiol. Letts. 69:229–232.

Beer-Romero, P. 1986. Comparative studies on *Heliobacterium chlorum, Heliospirillum gestii* and*Heliobacillus mobilis*. M.A. thesis, Indiana University, Bloomington.

Beer-Romero, P., and H. Gest. 1987. *Heliobacillus mobilis*, a peritrichously flagellated anoxyphototroph containing bacteriochlorophyll *g*. FEMS Microbiol. Letts. 41:109–114.

Beer-Romero, P. Favinger, J. L., and H. Gest. 1988. Distinctive properties of bacilliform photosynthetic heliobacteria. FEMS Microbiol. Letts. 49:451–454.

Brockmann, H., Jr., and A. Lipinski. 1983. Bacteriochlorophyll *g*. A new bacteriochlorophyll from *Heliobacterium chlorum*. Arch. Microbiol. 136:17–19.

Brok, M., H. Vasmel, J. T. G. Horikx, and A. J. Hoff. 1986. Electron transport components of *Heliobacterium chlorum* investigated by EPR spectroscopy at 9 and 35 GHz. FEBS Letts. 194:322–326.

Buresh, R. J., M. E. Casselman, and W. H. Patrick, Jr. 1980. Nitrogen fixation in flooded soil systems, a review. Adv. Agron. 33:149–192.

Evers, D. J., J. Weckesser, and G. Drews. 1984. Protein in the cell surface of the moderately halophilic phototrophic bacterium, *Rhodospirillum salexigens*. J. Bacteriol. 160:107–111.

Fischer, M. R. 1990. Photosynthetic electron transfer in *Heliobacterium chlorum* studied by EPR spectroscopy. Biochim. Biophys. Acta. 1015:471–481.

Fuller, R. C., S. G. Sprague, H. Gest, and R. E. Blankenship. 1985. A unique photosynthetic reaction center from *Heliobacterium chlorum*. FEBS Letts. 182:345–349.

Gest, H. and J. L. Favinger. 1983. *Heliobacterium chlorum*, an anoxygenic brownish-green photosynthetic bacterium containing a "new" form of bacteriochlorophyll. Arch. Microbiol. 136:11–16.

Gest, H. and J. L. Favinger. 1989. Genus *Heliobacterium*, p. 1707–1708. In: J. Staley (ed.), Bergey's manual of systematic bacteriology, vol. 3. Williams and Wilkins, Baltimore.

Gest, H., J. L. Favinger, and M. T. Madigan. 1985. Exploitation of N₂ fixation capacity for enrichment of anoxygenic photosynthetic bacteria in ecological studies. FEMS Microbiol. Ecol. 31:317–322.

Gherna, R. L. 1981. Preservation, p. 208–217. In: Gerhardt, P., R. G. E. Murry, R. N. Costilow, E. W. Nester, W. A. Wood, N. R. Krieg, and G. B. Phillips (ed.), Manual of methods for general bacteriology. American Society for Microbiology, Washington, DC.

Gloe, A., N. Pfennig, H. Brockman, Jr., and H. Trowitzsch. 1975. A new bacteriochlorophyll from brown-colored Chlorobiaceae. Arch. Microbiol. 102:103–109.

Habte, M., and M. Alexander. 1980. Nitrogen fixation by photosynthetic bacteria in lowland rice culture. Appl. Environ. Microbiol. 39:342–347.

Hansen, T. A., and H. van Gemerden. 1972. Sulfide utilization by purple non-sulfur bacteria. Arch. Mikrobiol. 86:49–56.

Kobayashi, M. 1982. The role of phototrophic bacteria in nature and their utilization, p. 643–661. In: N. S. Subbarao (ed.), Advances in agricultural microbiology. Butterworth Scientific, London.

Kobayashi, M., E. Takahashi, and E. Kawaguchi. 1967. Distribution of nitrogen-fixing microorganisms in paddy soils of Southeast Asia. Soil. Sci. 104:113–118.

Madigan, M. T. 1986. *Chromatium tepidum*, sp. nov., a thermophilic photosynthetic bacterium of the family *Chromatiaceae*. Intl. J. Syst. Bacteriol. 36:222–227.

Madigan, M. T. 1988. Microbiology, physiology, and ecology of phototrophic bacteria, p. 39–111. In: A. J. B. Zehnder (ed.), Biology of anaerobic microorganisms. John Wiley & Sons, New York.

Michalski, T. J., J. E. Hunt, M. K. Bowman, K. Bardeen, H. Gest, J. R. Norris, and J. J. Katz. 1987. Bacteriopheophytin *g*: Properties and some speculations on a possible primary role for bacteriochlorophylls *b* and *g* in the biosynthesis of chlorophylls. Proc. Natl. Acad. Sci. (U.S.A.) 84:2570–2574.

Miller, K. R., J. S. Jacob, U. Smith, S. Kolaczkowski, and M. K. Bowman. 1986. *Heliobacterium chlorum*: cell organization and structure. Arch. Microbiol. 146:111–114.

Nuijs, A. M., R. J. van Dorssen, L. N. M. Duysens, and J. Amesz. 1985. Excited states and primary photochemical reactions in the photosynthetic bacterium *Heliobacterium chlorum*. Proc. Natl. Acad. Sci. (U.S.A.) 82:6865–6868.

Ormerod, J., T. Nesbakken, and Y. Torgersen. 1990. Phototrophic bacteria that form heat-resistant endospores. In: M. Baltscheffsky (ed.), Proceedings of the VIII International Congress on Photosynthesis. Kluwer, Academic Publishers, Dordrecht, The Netherlands. (in press)

Ormerod, J. G., K. S. Ormerod, and H. Gest. 1961. Light-dependent utilization of organic compounds and photoproduction of molecular hydrogen by photosynthetic bacteria; relationships with nitrogen metabolism. Arch. Biochem. Biophys. 94:449–463.

Pfennig, N. 1989. Ecology of phototrophic purple and green sulfur bacteria, p. 97–116. In: H. G. Schelegel and B. Bowien (ed.), Autotrophic bacteria. Science Tech Publishers, Madison, WI and Springer-Verlag, New York.

Pierson, B. K. and R. W. Castenholz. 1974. A phototrophic, gliding filamentous bacterium of hot springs, *Chloroflexus aurantiacus*, gen. and sp. nov. Arch. Microbiol. 100:5–24.

Prince, R. C., H. Gest, and R. E. Blankenship. 1985. Thermodynamic properties of the photochemical reaction center of *Heliobacterium chlorum.* Biochim. Biophys. Acta 810:377–384.

Remsen, C. C., S. W. Watson, and H. G. Trüper. 1970. Macromolecular subunits in the walls of marine photosynthetic bacteria. J. Bacteriol. 103:255–258.

Smit, H. W. J., and J. Amesz. 1988. Electron transfer in the reaction center of green sulfur bacteria and *Heliobacterium chlorum,* p. 97–108. In: J. M. Olson, J. G. Ormerod, J. Amesz, E. Stackebrandt, and H. G. Trüper (ed.), Green photosynthetic bacteria. Plenum Press, New York.

Smit, H. W. J., J. Amesz, and M. F. R. van der Hoeven. 1987a. Electron transport and triplet formation in membranes of the photosynthetic bacterium *Heliobacterium chlorum.* Biochim. Biophys. Acta. 893:232–240.

Smit, H. W. J., J. Amesz, M. F. R. van der Hoeven, and L. N. M. Duysens. 1987b. Electron transport in *Heliobacterium chlorum,* p. 189–192. In: J. Biggins (ed.), Progress in photosynthesis research. Martinus Nijhoff Publishers, Dordrecht, The Netherlands.

Smit, H. W. J., R. J. van Dorssen, and J. Amesz. 1989. Charge separation and trapping efficiency in membranes of *Heliobacterium chlorum* at low temperature. Biochim. Biophys. Acta. 973:212–219.

Trost, J. T., and R. E. Blankenship. 1989. Isolation of a photoactive photosynthetic reaction center-core antenna complex from *Heliobacillus mobilis.* Biochemistry. 28:9898–9904.

van Dorssen, R. J., H. Vasmel, and J. Amesz. 1985. Antenna organization and energy transfer in membranes of *Heliobacterium chlorum.* Biochim. Biophys. Acta. 809:199–203.

van Kan, P. J. M., T. J. Aartsma, and J. Amesz. 1989. Primary photosynthetic processes in *Heliobacterium chlorum* at 15K. Photosyn. Res. 22:61–68.

van de Meent, E. J., F. A. M. Kleinherenbrink, and J. Amesz. 1990. Properties of a solubilized and purified antenna-reaction center complex from heliobacteria, p. 145–148. In: M. Baltscheffsky (ed.), Current research in photosynthesis. Vol II. Kluwer Academic Publishers, Dordrecht, The Netherlands.

Vos, M. H., H. E. Klaassen, and H. J. van Gorkom. 1989. Electron transport in *Heliobacterium chlorum* whole cells studied by electroluminescence and absorbance difference spectroscopy. Biochim. Biophys. Acta. 973:163–169.

Woese, C. R., B. A. Debrunner-Vossbrinck, H. Oyaizu, E. Stackebrandt, and W. Ludwig. 1985. Gram-positive bacteria: Possible photosynthetic ancestry. Science 229:762–765.

CHAPTER 91

The Genera *Pectinatus* and *Megasphaera*

AULI HAIKARA

Introduction

Pectinatus and *Megasphaera* are both anaerobic contaminants of packaged beer which were first isolated only in the 1970s. They produce unpleasant off-flavors (hydrogen sulfide, fatty acids). *Pectinatus* was first described by Lee et al. (1978). They concluded that the isolate did not fit into any known genus and proposed a new genus and species, *Pectinatus cerevisiiphilus*. The first isolate of a *Megasphaera* sp. was made from German beer by Weiss et al. (1979). A new species, *Megasphaera cerevisiae* sp. nov., was proposed by Engelmann and Weiss in 1985. Soon after the discovery of *Pectinatus* in the USA, this beer-spoilage bacterium was also observed in Germany, Sweden, Norway, Finland, and Japan (Back et al., 1979; Haikara et al., 1981b; Haukeli, 1980; Kirchner et al., 1980; Takahashi, 1983). The appearance of *Megasphaera* in breweries outside Germany has been described by Haikara (1985b) and Haikara and Lounatmaa (1987).

The simultaneous observation of *Pectinatus* and *Megasphaera* species in beer in many countries is mainly explained by their anaerobic nature. The dissolved oxygen content of beer as well as the volume of air in the head space has decreased considerably in recent years. This development is due to advances in brewing technology leading to minimal access of oxygen to the product in order to improve the chemical stability of beer. Therefore, contaminations caused by these strictly anaerobic bacteria are mainly encountered in large, modern breweries equipped with effective filling technology (Chelack and Ingledew, 1987; Seidel, 1979).

Habitats

Pectinatus and *Megasphaera* are contaminants of unpasteurized packaged beer. *Pectinatus* contaminations in beer have been reported to occur in the United States, Germany, Sweden, Nor-

way, Japan, and Finland. The *Megasphaera* sp. is apparently less widespread and has only been reported in Germany and Finland. All the strains described and characterized to date have been isolated from spoiled beer; however, Back (1981) reported the occurrence of *Pectinatus* and *Megasphaera* in beer before filtration. He also detected *Pectinatus* in a pitching yeast sample and in the rinse water of filling lines. Moreover, Seidel (1985) referred to the detection of *Pectinatus* and *Megasphaera* strains in yeast cultures used to initiate the brewing process.

The natural environments of anaerobic beer spoilage organisms and the source of contaminations are not known. In addition to the habitats described above, *Pectinatus* has sporadically been found in lubrication oil mixed with beer, in drainage and water pipe systems, in air of the filling hall and in the filling machine, on the floor of the filling hall, in condensed water on the ceiling, in chain lubricants, and in steeping water of malt before milling (Back et al., 1988; Dürr, 1983; Haukeli, 1980; Lee et al., 1980; Soberka and Warzecha, 1986). With the exception of some cases of isolation from pitching yeast and secondary fermentation, the *Pectinatus* findings in Finland have also been concentrated in the filling hall, including samples from unfilled bottles after the washing machine, from the floor, from rinse water in the CO_2 channel of the filling machine, from the CO_2 membrane, and from the CO_2 line. On the basis of these detection results, water seems to be one of the most likely sources of contamination. Moreover, it is obvious that *Pectinatus,* in spite of its anaerobic nature, can survive in aerosols and be transferred via the air into beer. On the other hand, the detection of *Pectinatus* sp. in air near unclean bottles in filling halls indicates the possibility that air in the filling area may be contaminated by dirty bottles (Dürr, 1983; A. Haikara, unpublished observations).

The frequency of contaminations caused by anaerobic beer-spoilage organisms has increased significantly during recent years. According to the data collected by Back (1981) and

Back et al. (1988). *Pectinatus* and *Megasphaera* contaminations accounted for 1–2% of the total number of brewery contaminations in 1981. In 1986, corresponding figures of 3% for *Pectinatus* and 2% for *Megasphaera* were recorded, and in 1987 frequencies of 7% were calculated for both types of contaminant. It is possible that this increasing tendency will continue in the future as a result of the reduced oxygen level in beer caused by modern, effective filling techniques.

Growth of *Pectinatus* and *Megasphaera* in Beer

It has been observed in practice and in laboratory tests that beer with a low alcohol content ($< 2.25\%$ w/v) is more prone to *Pectinatus* and especially *Megasphaera* contaminations than beer with higher alcohol content (Haikara, 1984; Haikara and Lounatmaa 1987). The rate of spoilage is inversely dependent on the alcohol content of beer. The growth of *Megasphaera* is already restricted in commercial beer with an alcohol content of 3.5% w/v. *Pectinatus* bacteria are more alcohol tolerant, growing rather well in beer with higher alcohol content (3.7–4.4% w/v) although slower than in low-alcohol beer. As reported earlier, *Pectinatus* does not grow in strong beer with an alcohol content exceeding 5.2% w/v (Haikara et al., 1981b; Haukeli, 1980; Kirchner et al., 1980; Seidel et al., 1979).

The low pH of beer is one of most important characteristics affecting the growth of microorganisms. The higher the pH of beer, the more susceptible it is to *Megasphaera* contamination. *Pectinatus* is very acid-tolerant. A pH value of 4.1 is needed for retardation of its growth, whereas the normal beer pH of 4.5 already restricts the growth of *Megasphaera* (Haikara, 1984; Haikara and Lounatmaa, 1987). In German Alt beer (pH 4.0) the growth of *Megasphaera* is prevented but *Pectinatus* can still grow slowly (Kirchner et al., 1980; Seidel et al., 1979).

Hop bitter substances generally restrict the growth of microorganisms in beer. However, both *Pectinatus* and *Megasphaera* have proved to be rather tolerant to hops, since growth has been observed in beer with bitterness in the range of 33–38 European Brewery Convention units (Back, 1981; Kirchner et al., 1980; Seidel et al., 1979). The growth of *Pectinatus* and *Megasphaera* bacteria in beer is dependent on the combined effects of several factors, the oxygen content of beer being the most decisive. The dissolved oxygen content as well as the volume of air in the head space have decreased considerably in recent years due to advances in filling technology. At present an air volume of 1 ml

or less in the head space can easily be achieved, as can an oxygen content as low as 0.3 mg/l (Seidel et al., 1979).

Isolation

Cultivation Procedures for *Pectinatus*

Many agar media and broths can be used for cultivation and isolation of *Pectinatus* strains. Thioglycolate medium (Difco) with 1% glucose and 1.5% agar (Lee, 1984) and de Man-Rogosa-Sharpe (MRS) lactobacilli broth or agar (Oxoid, Merck) have been most commonly used. LL-agar (lactate-lead acetate) is the only medium designed specifically for selective cultivation and isolation of *Pectinatus* strains (Lee et al., 1981). Lactate is the sole source of carbon in this medium for differentiation of *Pectinatus* strains and lactic acid bacteria from other brewery microorganisms. Lead acetate is used to differentiate lactic acid bacteria from the H_2S-producing *Pectinatus* species on the basis of black coloration of *Pectinatus* colonies. Phenethyl alcohol is included as a component of the medium because it is inhibitory to aerobic Gram-negative bacteria and yeast (Lilley and Brewer, 1953). The use of phenethyl alcohol is somewhat controversial, because in our experiments it seems to restrict the growth of *Pectinatus* strains (unpublished observations).

Lactate-Lead Acetate Agar (Lee et al., 1981)
   LL-agar (Lee et al., 1981) contains, per liter:

Yeast extract	5 g
Beef extract	3 g
Sodium lactate (60% syrup)	17.0 ml
Ascorbic acid	1 g
$Na_2S_2O_3 \times 5H_2O$	0.1 g
Lead acetate	0.2 g
Methylene blue	0.002 g
Phenylethyl alcohol	2.0 ml
Agar 15 g	

The incubation temperature is 30–32°C for 5–10 days.

More important than the media used for cultivation of *Pectinatus* species is the maintenance of strictly anaerobic cultivation and incubation conditions. The anaerobic glove box, the Hungate roll tube method, and the Gas Pak system have been used for cultivation and isolation of *Pectinatus* species (Back et al., 1979; Haikara et al., 1981a,b; Hungate, 1969; Kirchner et al., 1980; Lee et al., 1978). Moreover, Ogg et al. (1979) devised a special Lee tube for the cultivation and enumeration of *Pectinatus* sp. and other anaerobic organisms. The use of prereduced media and special reducing diluents (Holdeman et al., 1977) has been reported to enhance the growth of *Pectinatus* in GasPak anaerobic jars (Back et al., 1979; Haikara et al., 1981b; Kirchner et al., 1980). Cultivation without an anaerobic atmosphere in broth with a minimum headspace is possible provided that the inoculum is heavy (2%) and that oxygen is removed by boiling before use (Haikara et al., 1981a, 1981b; Lee, 1984). The anaerobic glove box provides optimum growth conditions for *Pectinatus* bacteria. Because this equipment is not generally available in brewery quality-control laboratories, anaerobic jars can be regarded as the most convenient and reliable for system cultivation of *Pectinatus* species.

Cultivation Procedures for *Megasphaera*

Due to the inability of *Megasphaera* sp. to break down glucose, replacement of glucose with fructose or lactate in peptone-yeast extract medium is needed for maximal growth of *Megasphaera* species (Engelmann and Weiss, 1985).

Peptone-Yeast Extract-Fructose (PYF) Medium (Engelmann and Weiss, 1985)

PYF medium contains, per liter:

Peptone	5 g
Tryptone	5 g
Yeast extract	10 g
Na₂HPO₄	2 g
Tween 80	1 ml
Cysteine-HCl	0.5 g
Fructose	5 g
Adjust to pH 7.0.	

A simple PYF broth (1% peptone, 1% yeast extract, 2% fructose) has also been successfully used for cultivation of *Megasphaera* species (Haikara and Lounatmaa, 1987).

Due to the strictly anaerobic nature of *Megasphaera cerevisiae,* the anaerobic cultivation conditions recommended in the previous paragraph for *Pectinatus* are also valid for *Megasphaera.*

Preservation of *Pectinatus* and *Megasphaera* Cultures

Working cultures can be maintained by subculturing (30°C, 1–2 days) at least every 2 weeks in PYG broth (1% peptone, 1% yeast extract, 2% glucose supplemented with 0.5 g/l cysteine hydrochloride as a reducing agent). For cultivation of *Megasphaera* fructose is used instead of glucose. The *Pectinatus* strains isolated from yeast cultures lose their viability rapidly and subculturing every week is necessary. After in-

cubation the cultures are stored at 4°C in anaerobic jars. The working culture can also be frozen at −75°C, for example, in commercial anaerobic Portagerm vials containing a solid, reduced, buffered medium. The freshly grown cell culture (1 ml) is supplemented with 10 or 15% glycerol and injected into the vials.

For long-term preservation of strains isolated from beer, conventional freeze-drying methods using 20% skim milk as protective agent as well as freezing in liquid nitrogen at −196°C using 5% glycerol as protective agent has been used. However, the *Pectinatus* strains isolated from yeast cultures are very sensitive to both these methods. The recovery of freeze-dried cultures after storage for 2 years in very poor. The only successful method for these strains has been freezing in liquid nitrogen using 5% dimethyl sulfoxide as protective agent. The polypropylene straws are packed after filling (0.1 ml) into 2 ml plastic screw cap ampoules (e.g., Nunc) and placed directly into liquid nitrogen (Suihko and Haikara, 1990).

Identification

Pectinatus

MORPHOLOGY The most distinctive feature of *P. cerevisiiphilus* is the comb-like flagellation on only one side of the cell (Fig. 1). The number of flagella per cell depends on the cell size and its condition, but generally ranges from one to 23 (Lee 1984). The cells are slightly curved with rounded ends, 0.4 to 0.8 μm by 2 to 32 μm or more in length (Back et al., 1979; Haikara et al., 1981b; Lee et al., 1978). They normally occur singly or in pairs, only rarely in chains. In old cultures very elongated cells with a helical shape may be found (Fig. 2), and round cell forms are also observed. The active, young cells form an "X" shape during movement whereas old cells are characterized by a snakelike slow movement (Lee et al., 1978). One strain forms ring-shaped cells not found in cultures of the other strains (Haikara et al., 1981a). In beer, the cells are thin and have a heterogeneous appearance (Seidel et al., 1979).

Colonies of *P. cerevisiiphilus* are circular, entire, beige to white, glistening, and opaque, although one strain has been reported to produce a rough colony (Back et al., 1979). The growth of this strain also differs in liquid culture, since the turbidity decreases at the end of the exponential growth phase due to the deposition of a heavy precipitate on the bottom of the bottle (Seidel et al., 1979).

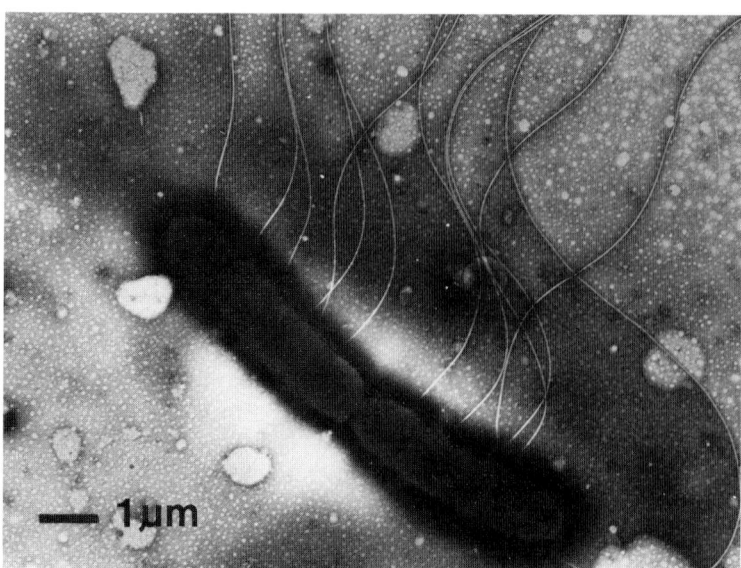

Fig. 1. Electron micrograph of *Pectinatus cerevisiiphilus* (ATCC 3332). (From Haikara et al., 1981b, with permission.)

Fig. 2. Dark-field micrograph of *Pectinatus cerevisiiphilus* (ATCC 3332).

ELECTRON MICROSCOPY. The unique comb-like flagellar arrangement of *P. cerevisiiphilus* observed by Lee et al. (1978) using scanning electron microscopy was confirmed by other workers in negative stained preparations (Back et al., 1979; Haikara et al., 1981b; Kirchner et al., 1980). The cell envelope structure seen in such sections revealed a multilayered cell wall typical of Gram-negative bacteria (Lee et al., 1978). The peptidoglycan layer of *P. cerevisiiphilus* is very thick (30 nm), almost filling the periplasmic space of the cell envelope (Haikara et al., 1981a, 1981b). The freeze-fracture technique has also been used in electron microscopical investigations of *Pectinatus* strains (Haikara et al., 1981a, 1981b). These studies have demonstrated the similarity of the structures of the cell envelopes in different *Pectinatus* strains (Haikara et al., 1981a). The thick peptidoglycan layer and the invaginations and the mesosomes of the cytoplasmic membrane are typical features of Gram-positive bacteria. On the other hand, the outer membrane typical of Gram-negative bacteria makes *P. cerevisiiphilus* an interesting intermediate form between these two groups.

TAXONOMY AND DIFFERENTIATION. Anaerobic, nonsporeforming Gram-negative rods were assigned in *Bergey's Manual of Determinative Bacteriology* to the family Bacteroidaceae (Holdeman and Moore, 1974). When the location of flagella, the metabolic products, and the GC content of the DNA of the genus *Pectinatus* were compared with that of isolates from the nine genera belonging to the family Bacteroidaceae, it was observed that most of these genera possessed several major characteristics distinguishing them from *Pectinatus* (Lee et al., 1978). Species producing propionic acid were found only among the genera *Bacteroides* and *Selenomonas*. On the basis of the lateral flagellation, the *Selenomonas* species most closely resemble *Pectinatus* species. However, selemonads have a tuft attachment of flagella at the center of the concave side of the cell, compared with a comb-like flagellar arrangement along the long axis of the cell of *Pectinatus*. In addition the GC content of the DNA of selenomonads (53–61 mol%) is much higher than the reported value of 39.8 mol% for *Pectinatus cerevisiiphilus*. The most prominent differences between the genera *Pectinatus* and *Bacteroides* are in the arrangement of the flagella and especially in the mechanism of propionic acid formation (Haikara et al., 1981a). *Pectinatus cerevisiiphilus* was approved (Skerman et al., 1980) and described in *Bergey's Manual of Systematic Bacteriology* by Lee (1984).

CHROMATOGRAPHIC IDENTIFICATION. *P. cerevisiiphilus* is the only bacterium known to produce large quantities of propionic acid in beer, hence gas chromatography has proved to be a useful tool for identification purposes (Back et al., 1979; Haikara et al., 1981b; Lee et al., 1978; Schisler et al., 1979). The production of high levels of hydrogen sulfide by *Pectinatus* can also be used for identification purposes. In lager breweries no other contaminants are able to produce H₂S. In British ale, the genus *Zymomonas* is known as an H₂S-producing spoilage organism. However, the simultaneous production of acetaldehyde by this bacterium produces an odor of rotten apples in the ale. This is easily distinguishable, even without any instrument, from the odor of rotten eggs caused by the growth of *Pectinatus* in beer.

SEROLOGICAL CHARACTERISTICS. The immunological properties of *Pectinatus* strains can be used for identification of this genus. The fluorescent antibody staining technique has revealed that sera prepared against different *Pectinatus* strains have no cross-reactions with other brewery contaminants (Haikara et al., 1981a). According to gel immunodiffusion and immuno-electrophoresis tests, *Pectinatus* strains were serologically different and could be assigned to three different groups (I, II, and III) (Haikara et al., 1981b; Haikara, 1983). Groups I and III were apparently more closely related to each other than to the more distant group II. Moreover, the results obtained with the purified lipopolysaccharide (LPS) preparations supported this immunological grouping of *Pectinatus* strains. The LPSs of different strains reacted only with the corresponding antisera or with antisera produced against strains belonging to the same immunological group (Haikara, 1984).

CELL WALL COMPOSITION. The division of *P. cerevisiiphilus* strains on the basis of cell-surface-protein pattern has revealed two main

groups (A, B), which correspond to the serological groups I and II (Hakalehto et al., 1984).

The chemical composition of LPS, especially that of the lipid part, can be used as a taxonomic criterion because it is known to differ substantially among the distinct bacterial genera, whereas little variation occurs among the species of one genus (Rietschel et al., 1982). Helander et al. (1983) reported that the chemical composition of LPSs isolated from five *P. cerevisiiphilus* strains had identical fatty acid patterns, which differed from those of most Gram-negative bacteria. The closest resemblance was with the fatty acid pattern of the genus *Veillonella*, (Hewett et al., 1971) but the pattern was fundamentally different from that of *Bacteroides*, again supporting the separate taxonomic grouping of these two genera.

The biological activities of LPSs from *P. cerevisiiphilus* have been shown to be of the same order of magnitude as those of LPSs from *Escherichia coli* (Helander et al., 1984).

Megasphaera

MORPHOLOGY. *Megasphaera cerevisiae* sp. nov. is a Gram-negative coccus (Figs. 3 and 4) . The spherical or slightly oval cells, 1.3 to 1.6 μm in diameter, occur singly or in short chains (Engelmann and Weiss, 1985). In fixed or stained preparations the cell diameter is 1.0–1.2 μm (Haikara and Lounatmaa, 1987). Cell size is one differential characteristic between the brewery species and *M. elsdenii*. In wet mounts, the diameter of *M. elsdenii* is 2.4–2.6 μm and in fixed preparations 1.2–1.9 μm, in accordance with its Latin name—big sphere (Rogosa, 1984). The ultrastructure of the cell surface of the *Megasphaera* strains isolated from beer and of *M. elsdenii* has been reported to be very uniform (Haikara and Lounatmaa, 1987). *M. cerevisiae* forms whitish, smooth, opaque, flat and shiny colonies, 2–5 mm in diameter (Engelmann and Weiss, 1985; Weiss et al., 1979).

TAXONOMIC DIFFERENTATION. Anaerobic Gram-negative cocci were placed in *Bergey's Manual* in the family Veillonellaceae (Rogosa, 1974). On the basis of differential characteristics of the genera *Veillonella*, *Acidaminococcus* and *Megasphaera* of this family, the beer isolates could be assigned to the genus *Megasphaera* (Weiss et al., 1979). The ability to ferment carbohydrates and produce volatile fatty acids containing 5–6 carbon atoms are the most distinctive characteristics of the genus *Megasphaera*. The only previously known species in this genus was a rumen microbe, *M. elsdenii* (Rogosa, 1984). The GC content of the DNA of

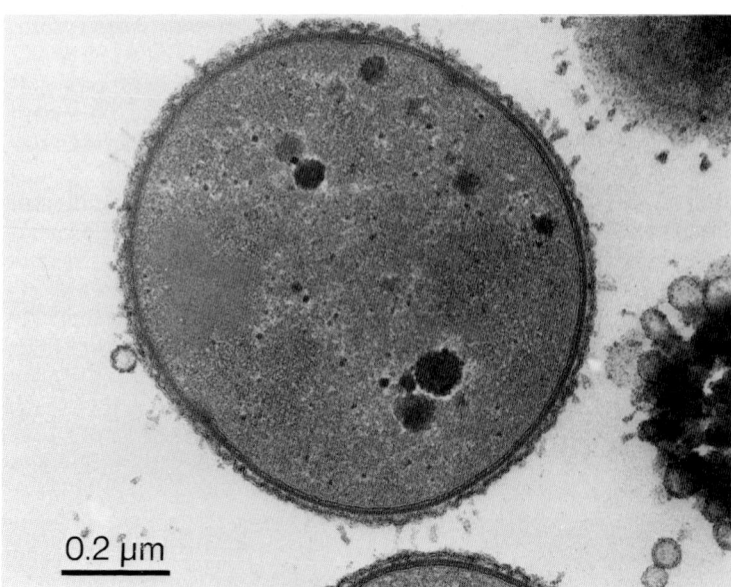

Fig. 3. Electron micrograph of *Megasphaera cerevisiae* (VTT-E-84195). (Courtesy of K. Lounatmaa.)

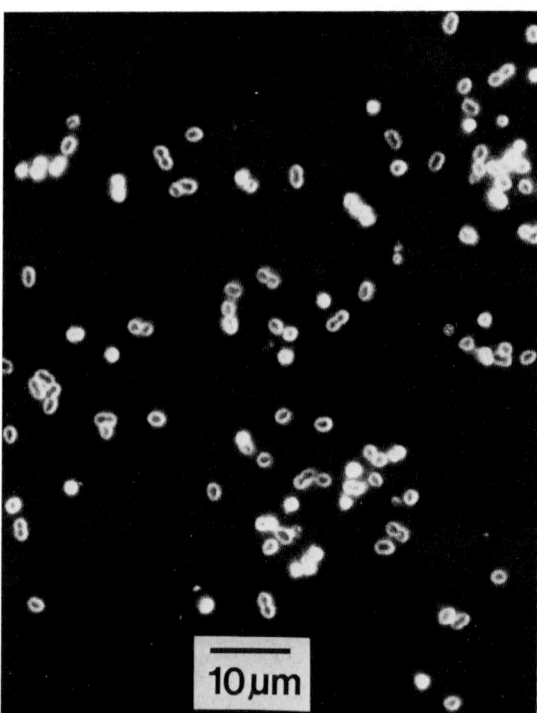

Fig. 4. Dark-field micrograph of *Megasphaera cerevisiae* (VTT-E-84195).

genospecies (Engelmann and Weiss, 1985), the homology within this group being over 72%. Moreover, no genomic relatedness was detectable between *Veillonella parvula* and the beer isolates, and hence a new species named *M. cerevisiae* was proposed by Engelmann and Weiss (1985).

CELL WALL COMPOSITION. The cell surface protein pattern and the chemical composition of *M. cerevisiae* LPS is not known. However, Engelmann and Weiss (1985) analyzed the amino acid composition of the cell wall peptidoglycans of *Megasphaera* spp. The peptidoglycan of *M. cerevisiae* as well as that of *M. elsdenii* was of the meso-diaminopimelic acid direct type and contained putrescine residues.

CHROMATOGRAPHIC IDENTIFICATION. The distinctive metabolic profile of *Megasphaera* species (Haikara, 1985b; Weiss et al., 1979) provides a useful tool for their identification. Moreover, the detection of butyric, valeric, and caproic acids in brewery samples can be regarded as proof of the presence of *Megasphaera*. The identification can be complemented with the detection of hydrogen sulfide produced by *Megasphaera*.

Physiological And Biochemical Properties of *Pectinatus*

Only very small physiological and biochemical differences have been shown to exist between different *Pectinatus* isolates (Back et al., 1979; Haikara et al., 1981a, 1981b; Kirchner et al., 1980; Lee et al., 1978). The GC content of DNA

this species is 53.6 mol%. The corresponding values of 12 brewery isolates ranged from 42.4 to 44.8 mol% GC (Engelmann and Weiss, 1985), indicating that there is no genomic relationship between the *Megasphaera* species isolated from beer and *M. elsdenii*. DNA/DNA hybridization studies also supported this finding and showed that all isolates from beer belong to a single of the type strain is 39.8 mol%. Differences of only 2–5 mol% in the DNA base composition of other *Pectinatus* strains have been reported, indicating that the strains described are closely releated (Back et al., 1979; Haikara et al., 1981b; Kirchner et al., 1980).

P. cerevisiiphilus is catalase- and cytochrome oxidase-negative, and does not produce indole, liquify gelatin, reduce nitrate, or hydrolyze arginine. The Voges-Proskauer test is positive for all strains (Back et al., 1979; Haikara et al., 1981b; Kirchner et al., 1980; Lee et al., 1978). Only the urease test has given variable results for different strains (Back et al., 1979; Haikara et al., 1981b).

P. cerevisiiphilus grows at temperatures between 15 and 40°C, the optimum being 30 to 32°C (Back et al., 1979; Haikara et al., 1981b; Kirchner et al., 1980; Lee et al., 1978).

Pectinatus strains grow well in the pH range 4.5–8.5 (Back et al., 1979; Haikara et al., 1981b), the optimum pH being between 6.0 and 7.0 (Kirchner et al., 1980; Takahashi, 1983). However, rather good growth occurs even in the pH range 4.0 to 4.5 (Kirchner et al., 1980). The acid tolerance of *P. cerevisiiphilus* was also verified in growth tests in beer (Haikara, 1984). The normal pH of Finnish beer (average 4.4) did not restrict the growth of *Pectinatus* sp. Some retardation of growth was detected at pH 4.1. In German Alt beer (pH 4.0), the growth of *Pectinatus* was considerably reduced but still not totally prevented (Kirchner et al., 1980; Seidel et al., 1979). In the German beer type called Berliner Weisse, with a pH of 3.1, no growth occurred (Kirchner et al., 1980).

Alcohol tolerance tests have indicated that *Pectinatus* is capable of growing in nutrient broth containing up to 10% v/v ethanol (Haukeli, 1980). However, no growth has been observed in beers with an alcohol content exceeding 5.2% w/v (Haikara et al., 1981b; Haukeli, 1980; Kirchner et al., 1980; Seidel et al., 1979).

Metabolism

Some differences exist in the utilization of carbon sources by *Pectinatus* strains (Back et al., 1979; Haikara et al., 1981b; Kirchner et al., 1980; Lee et al., 1979; Takahashi, 1983). In contrast to other brewery contaminants, *Pectinatus* strains are incapable of utilizing maltose, the main carbohydrate of wort. On the other hand they can utilize lactate. Thus, the presence of lactic acid bacteria can effectively facilitate the growth of *Pectinatus*.

Schleifer et al. (1990) have found two phenotypically distinguishable groups among *Pectinatus* strains. These two groups differ clearly on the basis of utilization of xylose, cellobiose, melibiose, N-acetyl-glucosamine, and inositol (Table 1). Schleifer et al. (1990) therefore propose division of the genus *Pectinatus* into two species, *P. cerevisiiphilus* and *P. frisingensis*.

The main metabolic products of *P. cerevisiiphilus* from glucose (Table 2) are propionic, acetic, succinic, and lactic acids, and acetoin (Back et al., 1979; Haikara et al., 1981b; Kirchner et al., 1980; Lee et al., 1978). Back et al. (1979) and Kirchner et al. (1980) did not observe production of lactate. The relative amounts of metabolites are dependent on the substrate utilized by *Pectinatus* (Back et al., 1979; Lee et al., 1978). However, no differences in this respect were found between different

Table 1. Utilization of various carbon sources by *Pectinatus* species.

Substrate	*P. cerevisiiphilus*	*P. frisingensis* sp. nov.
Arabinose, fructose, mannose, glucose, erythritol, galactose, glycerol, rhamnose, lactate, ribose, sorbitol, mannitol	+	+
Glycogen, lactose, raffinose, starch, sucrose, trehalose, dextrin, inulin, melezitose, ethanol, salicin, maltose	−	−
Adonitol	+	v
Esculin	−	v
Xylitol	−	v
Dulcitol	v	+
Xylose	+	−
Melibiose	+	−
Cellobiose	−	+
Inositol	−	+
N-Acetyl-glucosamine	−	+

+, 75% or more of the strains were positive; −, 75% or more were negative; v, 26–74% positive.

Table 2. Metabolic products formed by *Pectinatus* strains.

Strain	Product concentration (mmol/dm³)[a]				
	Acetic acid	Propionic acid	Lactic acid	Succinic acid	Acetoin
VTT-E-79100	12	116	1.0	0.4	17
ATCC 29359	12	57	0.7	0.4	15
DSM 20465	17	78	0.8	1.3	15
DSM 20466	24	74	0.8	0.9	12

[a]Experiments were carried out with resting cells in a 2% glucose solution in phosphate buffer, pH 6.0. Incubation time was 1 day.
Adapted from Haikara et al. (1981b).

strains (Haikara et al., 1981b). The most abundant acid produced by *P. cerevisiiphilus* is propionic acid. The amount determined in contaminated beer is quite high, often more than 1,000 mg/l (Haikara et al., 1981a; Takahashi, 1983).

The mechanism of propionic acid synthesis by *P. cerevisiiphilus* has been studied by Haikara et al. (1981a, 1981b). *Pectinatus* bacteria use the same pathway for production of propionic acid as do propionibacteria. In this pathway, succinate oxidoreductase catalyzes the reduction of fumarate to succinate (Hettinga and Reinhold, 1972). Malonate is a specific competitive inhibitor of this enzyme and can thus inhibit the production of propionic acid, leading to a corresponding increase in acetic acid production. This was shown to occur in different *Pectinatus* strains and also in *Propionibacterium* species (Haikara et al., 1981a, 1981b). Taxonomically this was a significant observation, since *Bacteroides* species are known to use the acrylate pathway for propionate synthesis (Wallnöfer et al., 1967).

A very specific feature of the metabolism of *Pectinatus* sp. is the production of organic sulfur compounds (Haikara et al., 1981a). In addition to hydrogen sulfide, detected by Lee et al. (1980), production of methylmercaptan and dimethyl-trisulfide has been observed. Gas chromatographic analysis did not reveal the formation of dimethyldisulfide, dimethylsulfide, or thiolacetate.

Physiological and Biochemical Properties of *Megasphaera*

Megasphaera elsdenii, has been used as a reference species in the physiological and biochemical tests of the beer isolate, (Weiss et al., 1979; Haikara and Lounatmaa, 1987). The optimum temperature for *Megasphaera* beer isolates is 30–37°C and for *M. elsdenii* 37–42°C. *M. elsdenii* does not grow at room temperature but the brewery isolates can even grow at 15°C. A slightly lower optimum temperature of 28°C

has been reported for the type strain *M. cerevisiae* (Engelmann and Weiss, 1985).

Megasphaera sp. is more sensitive to low pH than *Pectinatus* sp., especially in beer. Increase in the pH of beer from 4.1 to 4.7 has been found to accelerate the growth of *Megasphaera* sp. (Haikara and Lounatmaa, 1987). No growth occurs at pH 4.1. Seidel et al. (1979) also reported the growth of *Megasphaera* sp. in beer of pH from 4.38 to 4.57, but not at pH 4.0.

In addition to low pH, alcohol is the most important factor inhibiting the growth of *Megasphaera* in beer. Very small amounts of ethanol retard the growth of *Megasphaera* sp. (Haikara and Lounatmaa, 1987). The dependence of the rate of spoilage on the alcohol content was also demonstrated in experiments in which growth of *Megasphaera* sp. was followed in commercial beers with different alcohol contents (Haikara, 1986). Growth did not occur in beer with an alcohol content of 4.3% w/v. Correspondingly no growth has been observed in German beers with alcohol contents over 5.5 and 6.5% v/v (Back 1981; Seidel et al., 1979).

Utilization of different carbon sources by *Megasphaera* species is presented in Table 3. The carbohydrate spectrum of *Megasphaera cerevisiae* is very narrow compared with that of *M. elsdenii*. The brewery isolates do not utilize glucose or maltose, but like *M. elsdenii* they can grow in sugar-free medium although growth is very poor (Engelmann and Weiss, 1985; Weiss et al., 1979). With respect to the fermentation of carbohydrates, the brewery isolates of *Megasphaera* form a very uniform group. Like *P. cerevisiiphilus*, they have the ability to utilize lactate and, hence, lactic acid bacteria could in practice facilitate their growth.

Metabolic end products of *Megasphaera* spp. are iso- and *n*-butyric acid, iso- and *n*-valeric acid, acetic, propionic, and caproic acids (Engelmann and Weiss, 1985; Weiss et al., 1979). The fatty acids produced are dependent on the carbon source available in the medium (Engelmann and Weiss, 1985; Haikara, 1985b; Weiss et al., 1979). The predominant end product

Table 3. Utilization of different carbon sources by *Megasphaera* species.

	Utilization of carbon sources[a]								
	Glucose	Fructose	Mannitol	Maltose	Saccharose	Arabinose	Lactate	Pyruvate	None
M. cerevisiae									
DMS 20461	−	+	−	−	−	−	+	+	(+)
DMS 20462[b]	−	+	−	−	−	+	−	+	(+)
Megasphaera species									
VTT-E-84195	−	+	−	−	−	+	+	+	(+)
VTT-E-85230	−	+	−	−	−	+	+	+	(+)
M. elsdenii									
ATCC 25940	+	+	+	+	+	−	+	+	(+)

+, growth; (+), slight growth; −, no additional growth caused by sugar.
[a]None of strains used xylose, glycerol, cellobiose, mannose, melezitose, raffinose, sorbitol, rhamnose, lactose, or trehalose.
[b]The proposed type strain for *M. cerevisiae*.
Modified from Haikara and Lounatmaa (1987).

Table 4. Amount of metabolic products formed by resting cells of *Megasphaera* strains.

		Product concentration (mmol/mmol substrate)						
Strain	Substrate	Acetic acid	Propionic acid	Butyric acid	Valeric acid	Caproic acid	H_2	CO_2
20462	Fructose	0.10	0.07	0.27	0.17	0.24	0.38	0.65
	Lactate	0.15	0.11	0.17	0.42	0.03	0.04	0.22
20461	Fructose	0.10	0.02	0.23	0.09	0.29	0.39	0.63
	Lactate	0.19	0.15	0.11	0.37	<0.01	0.04	0.19

Modified from Engelmann and Weiss (1985).

from lactate is *n*-valeric acid (Table 4). In beer, the main metabolic product is *n*-butyric acid (Haikara and Lounatmaa, 1987). The production of caproic acid is a typical feature of the genus *Megasphaera*. All *Megasphaera* species produce hydrogen sulfide (Engelmann and Weiss, 1985; Weiss et al., 1979). Due to the mixture of different fatty acids and H_2S, the flavor of contaminated beer is particularly unpleasant.

All *Megasphaera* strains isolated from beer are catalase-, benzidine-, and Voges-Proskauer-negative, do not hydrolyze urea, arginine, gelatin, or esculin, do not reduce nitrate, and do not form indole (Weiss et al., 1979).

Detection and Prevention of *Pectinatus* and *Megasphaera* Contamination

The presence of bacteria and yeast in filtered or final beer is normally detected using membrane filtration and anaerobic incubation on an appropriate medium. Haikara (1985a) demonstrated the unsuitability of conventional membrane filtration for the detection of *Pectinatus* bacteria in beer. This was anticipated due to the sensitivity of these bacteria to oxygen. The average recovery of *Pectinatus* cells was 5–19% depending on the physiological state of the cells. The use of prereduced media or a carbon dioxide atmosphere during filtration did not im-

prove the recovery significantly (Haikara, 1985a). The use of an anaerobic cabinet could solve this problem. However, this equipment is not generally available in brewery laboratories. Ogg et al. (1979) devised a special Lee tube for cultivation and enumeration of *Pectinatus* sp. and other anaerobic organisms. This tube does not, however, solve the detection problems in breweries because of its small sample volume of only one milliliter. Hence, forcing tests or enrichment are the only practical methods available in quality control of beer. In forcing tests, the development of turbidity is followed in filled, closed beer bottles incubated at 30°C for 6 weeks. In the enrichment method, 15–20 ml of sterilized, concentrated MRS-broth (72 g/l) supplemented with fructose (10 g/l) is added to the head space of the bottle immediately after filling and capping. The bottle is reclosed with a sterile cap and incubated at 30°C for four weeks. After development of turbidity, the presence of *Pectinatus* and *Megasphaera* is examined microscopically and they are identified morphologically and by smell. In practice an incubation time of 2–3 weeks is usually needed for *Pectinatus* growth and 3–4 weeks for *Megasphaera*.

Shortening of the forcing time has been achieved by direct staining of cells on the membrane after visualization with the immunofluorescence technique or the direct epifluorescent filter technique (DEFT) with acridine orange (Haikara, 1985a,b). Direct staining has revealed contaminations one to four days earlier than the development of visual turbidity.

Due to the distinctive metabolic profiles of *Pectinatus* and *Megasphaera,* gas chromatography has also been tested as a detection method (Haikara et al., 1981a; Haikara, 1985b). Detectable amounts of propionic acid were formed without visible growth of *Pectinatus,* one day before the turbidity appeared. Correspondingly, n-butyric or n-valeric acid produced by *Megasphaera* could be detected one day before the beer became visually turbid.

Winnewisser and Donhauser (1987) have developed an indirect enzymatic immunotest for detection of *Pectinatus* in beer after membrane filtration. The detection limit of this assay was 50 cells/ml beer. Sample enrichment is necessary to reach the required cell density, since normally only a few cells are present in one bottle of beer.

Viable *Pectinatus* and *Megasphaera* cells do not occur in beer pasteurized after bottling. Treatment at 58–60°C for one minute has been shown to be sufficient to kill all *Pectinatus* cells (Haukeli, 1980; Lee et al., 1981).

The elimination of contaminations caused by anaerobic beer spoilers requires adequate cleaning, disinfection, and sterilization of equipment known to be prone to contamination. Lee et al. (1981) reported that iodine and chlorine were effective disinfectants for the control of *Pectinatus* sp. In our experiments, in which different disinfectants were tested, peracetic acid (0.75%, 1 min) and formaldehyde (0.3%, 1 min) were the most efficient (Haikara, 1984). Disinfectants with iodine as the active compound proved to be ineffective. *Megasphaera* was killed most easily with oxidizing agents, iodophor having no effect (Haikara, 1986). Quaternary ammonium compound (0.2%) killed *Megasphaera* cells in one minute.

The detection of *Pectinatus* in the air of the handling area of dirty bottles has suggested one simple way to eliminate at least this source of contamination; the treatment of dirty bottles and cases with disinfectant before their entry to the filling hall could diminish the risk of airborne contamination of beer (Anonymous, 1988).

Future Taxonomic Rearrangement of the Anaerobic Bacteria Isolated from Brewing Processes

As was stated above, only slight differences have been found between the physiological and bio-chemical characteristics of different *Pectinatus* strains described in the literature. However, clear serological differences between strains were found, on the basis of which they could be divided into three immunological groups, two of these being very closely related (Haikara, 1983). Moreover, on the basis of the cell-surface-protein patterns two main groups were distinguished among *Pectinatus* strains (Hakalehto et al., 1984). Also DNA/DNA hybridization studies have revealed two distinct homology groups among *Pectinatus* strains isolated from beer (Haikara, 1989). The homology percentage between these two groups was very low (<10%), but a high degree of homology (>70%) was observed between strains belonging to the same group.

The extensive studies carried out by Schleifer et al. (1990) have revealed a new species, *P. frisingensis,* among beer isolates. In addition to a low degree of genomic homology between this new species and *P. cerevisiiphilus,* many characteristic differences were observed in the utilization of carbon sources by these species (see Table 1). The establishment of this new species seems to be well motivated.

As was mentioned earlier in this chapter all *Pectinatus* strains hitherto described have been isolated from spoiled beer. In 1981 Back reported the apparent occurrence of *Pectinatus* in a pitching yeast sample. In Finland two new "*Pectinatus*" strains were isolated in 1984 from pitching yeast. In the above-mentioned work of Schleifer et al., a new species of the genus *Selenomonas, S. lacticifex,* and a new genus *Zymophilus*—comprising two species, *Z. raffinosivorans* and *Z. paucivorans*—have been proposed for the strains isolated from pitching yeast. *Pectinatus, Selenomonas,* and *Zymophilus* species can be differentiated genetically and phenotypically. The GC content of *S. lacticifex* is 51–52 mol% versus 38–42 mol% for *Pectinatus* and *Zymophilus* species. DNA/DNA hybridization studies have revealed a low degree of genomic homology between *Pectinatus, Selenomonas,* and *Zymophilus* species. *S. lacticifex* could also be differentiated on the basis of lactic acid production. Moreover, all species differed clearly in the utilization of several carbohydrates.

The two new strains isolated in Finland from pitching yeast, preliminarily identified as *Pectinatus* bacteria, proved to possess no genomic relatedness to *Pectinatus* strains isolated from beer (Haikara, 1989). Moreover, immunological and several physiological differences between these two groups were found. On the basis of the proposed taxonomic rearrangement of *Pectinatus, Selenomonas,* and *Zymophilus* species, the yeast isolates could be identified as *S. lacticifex.*

It is obvious that the taxonomic rearrangement of anaerobic bacteria isolated from brewing process will be unavoidable in the future.

Literature Cited

Anonymous. 1988. Wachstum von *Megasphaera* und *Pectinatus.* Brauwelt 128:1217–1218.

Back, W. 1981. Nachweis und Identifizierung gramnegativer bierschädlicher Bakterien. Brauwissenschaft 34:197–204.

Back, W., S. Breu, and C. Weigand. 1988. Infektionsursachen im Jahre 1987. Brauwelt 128:1358–1362.

Back, W., N. Weiss, and H. Seidel. 1979. Isolierung und systematische Zuordnung bierschädlicher gramnegativer Bakterien. II: Gramnegative anaerobe Stäbchen/Anhang: Aus Bier isolierte gramnegative fakultativ Stäbchen. Brauwissenschaft 32:233–238.

Chelack, B. J., and W. M. Ingledew. 1987. Anaerobic Gramnegative bacteria in brewing—A review. J. Am. Soc. Brew. Chem. 45:123–127.

Dürr, P. 1983. Luftkeimindikation bierschädlicher Bakterien, Neue Methode mittels Luftkeimsammelgerät und Luftkeimindikator. Brauwelt 123:1652–1655.

Engelmann, U., and N. Weiss. 1985. *Megasphaera cerevisiae* sp. nov.: A new Gram-negative obligately anaerobic coccus isolated from spoiled beer. System. Appl. Microbiol. 6:287–290.

Haikara, A. 1983. Immunological characterization of *Pectinatus cerevisiophilus* strains. Appl. Environ. Microbiol. 46:1054–1058.

Haikara, A. 1984. Beer spoilage organisms. Occurrence and detection with particular reference to a new genus *Pectinatus.* Ph.D. thesis, Technical Research Centre of Finland, Publication 14.

Haikara, A. 1985a. Detection of *Pectinatus* contaminants in beer. J. Am. Soc. Brew. Chem. 43:43–46.

Haikara, A. 1985b. Detection of anaerobic, Gram-negative bacteria in beer. Monatsschr. für Brauwissenschaft 38:239–243.

Haikara, A. 1986. Uudet panimokontaminantit. Mallas ja Olut 69–77.

Haikara, A. 1989. Invasion of anaerobic bacteria into pitching yeast. Proc. 22nd Congr. Eur. Brew. Conv. 537–544.

Haikara, A., T.-M. Enari, and K. Lounatmaa 1981a. The genus *Pectinatus,* a new group of anaerobic beer spoilage bacteria. Proc. 18th Congr. Eur. Brew. Conv. 229–240.

Haikara, A., and K. Lounatmaa. 1987. Characterization of *Megasphaera* sp., a new anaerobic beer spoilage coccus. Proc. 21st Congr. Eur. Brew. Conv. 473–480.

Haikara, A., L. Penttilä, T.-M. Enari, and K. Lounatmaa. 1981b. Microbiological, biochemical, and electron microscopic characterization of a *Pectinatus* strain. Appl. Environ. Microbiol. 41:511–517.

Hakalehto, E., A. Haikara, T.-M. Enari, and K. Lounatmaa. 1984. Hydrochloric acid extractable protein patterns of *Pectinatus cerevisiophilus* strains. Food Microbiol. 1:209–216.

Haukeli, A. D. 1980. En ny ølskadelig bakterie i tappet øl. Referat från det 18. Skandinaviska Bryggeritekniska Mötet, Stockholm. 112–122.

Helander, I., E. Hakalehto, J. Ahvenainen, and A. Haikara. 1983. Characterization of lipopolysaccharides of *Pectinatus cerevisiophilus.* FEMS Microbiol. Lett. 18:223–226.

Helander, I., K. Saukkonen, E. Hakalehto and M. Vaara. 1984. Biological activities of lipopolysaccharides from *Pectinatus cerevisiophilus.* FEMS Microbiol. Lett. 24:39–42.

Hettinga, D. H. and G. W. Reinhold. 1972. The propionic acid bacteria—a review, II Metabolism. J. Milk Food Technol. 35:358–372.

Hewett, M. J., K. W. Knox and D. G. Bishop. 1971. Biochemical studies on lipopolysaccharides of *Veillonella.* Eur. J. Biochem. 19:169–175.

Holdeman, L. V., and W. E. C. Moore. 1974. Gram-negative anaerobic rods: Family I. *Bacteroidaceae* p. 384–426. In: R. E. Buchanan and N. E. Gibbons (ed.), Bergey's Manual of Determinative Bacteriology, 8th ed. Williams & Wilkins Co., Baltimore.

Holdeman, L. V., E. P. Cato, and W. E. C. Moore. 1977. Anaerobe laboratory manual, 4th ed. p. 145. Virginia Polytechnic Institute and State University, Blacksburg.

Hungate, R. E. 1969. A roll tube method for the cultivation of strict anaerobes, p. 117–132. In: J. R. Norris and D. W. Ribbons (ed.), Methods in microbiology, vol. 3B. Academic Press, Inc., New York.

Kirchner, G., R. Lurz, and K. Matsuzawa. 1980. Biertrübungen durch Bakterien der Gattung *Bacteroides.* Monatsschr. Brauwissenschaft 33:461–467.

Lee, S. Y. 1984. Genus XI *Pectinatus,* p. 655–658. In: N. R. Krieg, and J. G. Holt (ed.), Bergey's manual of systematic bacteriology, vol. 1. Williams & Wilkins, Baltimore.

Lee, S. Y., M. S. Mabee, and N. O. Jangaard. 1978. *Pectinatus,* a new genus of the family Bacteroidaceae. Int. J. Syst. Bacteriol. 28:582–594.

Lee, S. Y., M. S. Mabee, N. O. Jangaard, and E. K. Horiuchi. 1980. *Pectinatus,* a new genus of bacteria capable of growth in hopped beer. J. Inst. Brew. 86:28–30.

Lee, S. Y., S. E. Moore, and M. S. Mabee. 1981. Selective-differential medium for isolation and differentiation of *Pectinatus* from other brewery microorganisms. Appl. Environ. Microbiol. 41:386–387.

Lilley, B. D., and J. H. Brewer. 1953. The selective antibacterial action of phenylethyl alcohol. J. Am. Pharm. Assoc. 42:6–8.

Ogg, J. E., S. Y. Lee, and B. J. Ogg. 1979. A modified tube method for the cultivation and enumeration of anaerobic bacteria. Can. J. Microbiol. 25:987–990.

Rietschel, E. Th., C. Galanos, O. Lüderitz, and O. Westphal. 1982. In: D. R. Webb, (ed.), Immunopharmacology. Marcel Dekker, New York.

Rogosa, M. 1974. Gram-negative anaerobic cocci. Genus III *Megasphaera,* p. 448–449. In: R. E. Buchanan, and N. E. Gibbons (ed.), Bergey's manual of determinative bacteriology, 8th ed. Williams & Wilkins, Baltimore.

Rogosa, M. 1984. Anaerobic Gram-negative cocci. Genus III *Megasphaera,* p. 685. In: N. R. Krieg, and J. G. Holt (ed.), Bergey's manual of systematic bacteriology, vol. 1. Williams & Wilkins, Baltimore.

Schisler, D. O., M. S. Mabee, and C. W. Hahn. 1979. Rapid identification of important beer microorganisms using gas chromatography. J. Am. Soc. Brew. Chem. 37:69–76.

Schleifer, K. H., M. Leuteritz, N. Weiss, W. Ludwig, G. Kirchhof, and H. Seidel-Rüfer. 1990. Taxonomic study of anaerobic, Gram-negative, rod-shaped bacteria from breweries: Emended description of *Pectinatus cerevisiiphilus* and description of *Pectinatus frisingensis* sp. nov., *Selenomonas lacticifex* sp. nov., *Zymophilus raffinosivorans* gen. nov., sp. nov., and *Zymophilus paucivorans* gen. nov., sp. nov. Int. J. Syst. Bacteriol. 40:19–27.

Seidel, H. 1985. 100 Jahre biologische Brauerei-Betriebskontrolle, Alte und neue Probleme. Brauwelt 125:1954–1958.

Seidel, H., W. Back, and N. Weiss. 1979. Isolierung und systematische Zuordnung bierschädlicher gramnegativer Bakterien III: Welche Gefahr stellen die in den beiden vorausgegangenen Mitteilungen vorgestellten gramnegativen Kokken und Stäbchen für das Bier dar. Brauwissenschaft 32:262–270.

Skerman, V. R. D., V. McGowan, and P. H. A. Sneath. 1980. Approved lists of bacterial names. Int. J. Syst. Bacteriol. 30:341.

Soberka, R., and A. Warzecha. 1986. Influence de certains facteurs sur le taux d'oxygène dissous au cours de la fabrication de la bière. Bios 17:31–40.

Suihko, M.-L., and A. Haikara. 1990. Maintenance of the anaerobic beer spoilage bacteria *Pectinatus* and *Megasphaera*. Food Microbiology, in press.

Takahashi, N. 1983. Pressumed *Pectinatus* strain isolated from Japanese beer. Bull. Brew. Sci. 28:11–14.

Wallnöfer, P., and R. L. Baldwin. 1967. Pathway of propionate formation in *Bacteroides ruminicole*. J. Bacteriol. 93:504–505.

Weiss, N., H. Seidel, and W. Back. 1979. Isolierung und systematische Zuordnung bierschädlicher gramnegativer Bakterien. I: Gramnegative strikt anaérobe Kokken. Brauwissenschaft 32:189–194.

Winnewisser, W., and S. Donhauser. 1987. Enzymimmuntest zum Nachweis von *Pectinatus* in Bier. Proc. 21st Congr. Eur. Brew. Conv. 481–488.

CHAPTER 92

The Genus *Selenomonas*

ROBERT B. HESPELL, BRUCE J. PASTER, and FLOYD E. DEWHIRST

Species of *Selenomonas* are defined as anaerobic, Gram-negative, curved or crescent-shaped rods that are motile by means of a tuft of flagella originating from the inner curvature of the cell. Selenomonads have been isolated from the rumen and ceca of mammals, and the human oral cavity. Depending upon the health or diet of the host, these bacteria can constitute a significant proportion of the total microbial population. In general, selenomonads are obligately saccharolytic, although some strains ferment lactate or amino acids. It has been suggested that the role of ruminal and intestinal selenomonads involves the fermentation of soluble sugars and lactate in their natural environments. Oral selenomonads may play a role in the pathogenesis of periodontal disease in humans. The first classification of selenomonads was by Miller (1887), who designated strains from the human mouth as *Spirillum sputigenum*, but the present classification system places these organisms in the genus *Selenomonas* as *S. sputigena* (Bryant, 1984; Johnson et al., 1985). As a result of recent findings, however, at least six species of oral selenomonad have been recognized (Moore et al., 1987). Phylogenetic studies based on 16S ribosomal RNA sequence analysis have shown that species of *Selenomonas* are closely related to the bacteria *Centipeda periodontii, Pectinatus cerevisiiphilus,* and *Sporomusa paucivorans.* The selenomonads are more distantly related to anaerobic Gram-negative cocci of the genera *Veillonella* and *Megasphaera.* Selenomonads, *Veillonella* and related bacteria comprise a phylogenetic grouping which is more closely related to Gram-positive bacteria than to typical Gram-negative bacteria.

Habitats

Selenomonads have been observed in and isolated primarily from the rumen, the human mouth, and the cecum of mammals (Table 1). Gram-negative organisms having the cell morphology and flagellar arrangement typical for selenomonads have also been observed in river water (Leifson, 1960), and recently selenomonads have been isolated from anaerobic sewage sludge (Nanninga et al., 1987) and bog water

Table 1. Strains, sources, and DNA base composition of *Selenomonas* species.

Species	Natural source	Strain[a]	GC content (mol%)
S. acidaminophilia	Sewage sludge	DSM 3853	48.0
S. ruminantium	Rumen	ATCC 12561	49.0
S. ruminantium	Rumen	HD1	50.5
S. ruminantium	Rumen	HD4	53.5
S. ruminantium	Bog water	—	51.6
S. palpitans	Guinea pig cecum	—	—
S. artemidis	Gingival crevice	ATCC 43528	58.0
S. noxia	Gingival crevice	ATCC 43542	57.0
S. flueggei	Gingival crevice	ATCC 43531	56.0
S. infelix	Gingival crevice	ATCC 43532	58.0
S. sputigena	Gingival crevice	ATCC 35185	57.0
S. dianae	Gingival crevice	ATCC 43527	53.0

[a]ATTC = American Type Culture Collection, Rockville, MD, U.S.A.; DSM = Deutsche Sammlung von Mikroorganismen, Göttingen, FRG; Strains HD1 and HD4 from M. P. Bryant.
Data from Nanninga et al. (1987); Kingsley and Hoeniger (1973); and Moore et al. (1987).

(Harborth and Hanert, 1982). Selenomonads were probably first observed by Antonie van Leeuwenhoek in gingival scrapings from the human mouth (Dobell, 1960). Traditionally, speciation within *Selenomonas* has been based upon the habitat from which the strain was isolated (Buchanan and Gibbons, 1974; Lessel and Breed, 1954). Although helpful, this criterion has been largely supplanted by more adequate cytological, biochemical, and molecular criteria.

Selenomonads appear to be part of the normal indigenous microflora of human gingival crevices and are often more abundant in those persons having clinically detectable gingivitis or peridontal disease. Oral selenomonads may play a role in peridontal disease, inasmuch as lipopolysaccharide purified from these bacterial species has been shown to possess several endotoxic properties in mice (Kurimoto et al., 1986). Some occurrences of *S. sputigena* and other selenomonads in human septicemia have been reported (MacCarthy and Carlson, 1981; Pomeroy et al., 1987).

Selenomonads isolated from the rumen are usually shown to be strains of *S. ruminantium*. These organisms are routinely observed and isolated from rumen contents of cows and sheep. (Bryant, 1956; Hobson and Mann, 1961; Prins, 1971). In general selenomonads are more numerous in animals fed rations such as grains, which contain rapidly fermentable carbohydrates, than they are in animals fed silage or straw (Caldwell and Bryant, 1966). In addition to the fermentation of soluble carbohydrates, rumen selenomonads have other important roles in the rumen. These organisms are among the most important members of the glycerol-fermenting species of ruminal bacteria in sheep (Hobson and Mann, 1961) and in cattle (Bryant, 1956). While lactate is fermented by only a few species of ruminal bacteria, many ruminal selenomonad strains ferment lactate. Often, these strains are designated as *S. ruminantium* subspecies *lactilytica* and are almost phylogenetically identical to *S. ruminantium* (Fig. 1). Lower ruminal pH and increased lactate formation resulting from bloat or high-grain feeding, can lead to a substantial increase of ruminal selenomonads. Urea present in saliva and feedstuffs is degraded in the rumen by urease to form ammonia, a major nitrogen source for growth of many rumen microorganisms. *S. ruminantium* probably contributes to the hydrolysis of urea because urease-producing strains are commonly isolated (Wozny et al., 1977).

Selenomonad strains that have been phenotypically classified as *S. ruminantium* constitute a significant part of the intestinal microbial populations of swine. Approximately 21% of the total bacterial isolates from cecal contents of healthy swine were strains of *S. ruminantium* (Robinson et al., 1981). *S. ruminantium* strains were found to represent about 5% and 15% of

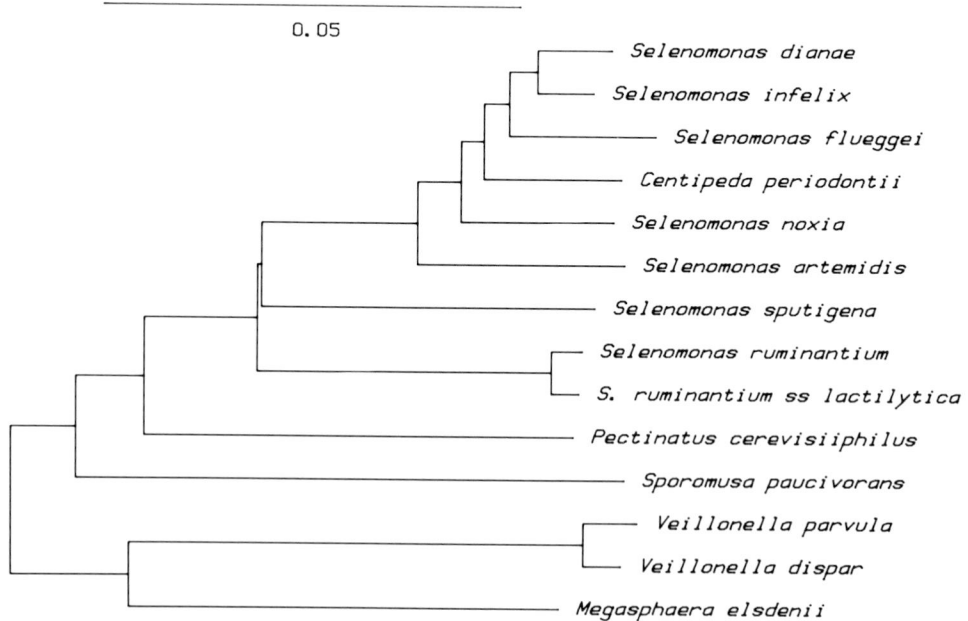

Fig. 1. Phylogenetic tree for the selenomonads and related bacteria. The scale indicates a 5% difference in nucleotide sequence, as determined by taking the sum of all branch lengths connecting two species. (Sequences for *S. paucivorans* and *M. elsdenii* courtesy of C. R. Woese.)

the total bacteria associated with the epithelial tissue of the colon in healthy and dysentery-infected swine, respectively (Robinson et al., 1984). Although this last study did not find selenomonad strains in luminal contents of the colon, these organisms have been detected in swine feces (Salanitro et al., 1977). Presumably, a functional role for selenomonads in the swine intestine involves fermentation of soluble sugars as occurs with these organisms in the rumen. The dominance of *S. ruminantium* strains in the swine intestine may be related to the availability of lactate. The swine of the small intestine can contain high levels of lactate (15 to 20 mM) and lactate levels decrease almost 10-fold upon reaching the cecum (Imoto and Namioka, 1978). In this regard, all of the selenomonad strains isolated from cecal contents have been lactate-fermenting strains, namely *S. ruminantium* subsp. *lactilytica* (Robinson et al., 1981). Lactate may also be available to *Selenomonas* strains that adhere to the colon wall since other adherent bacteria include lactate-producing species such as *Lactobacillus and Enterococcus* (Robinson et al., 1984).

Selenomonas species have frequently been observed and isolated from the cecum of a number of small rodents. The study by Ogimoto (1972) indicated that about 5% of the total bacterial isolates from rat cecal contents were selenomonads. All their rat strains could ferment any one of a number of sugars, including cellobiose; propionate was the major fermentation acid. With freeze-fracture electron microscopic techniques, selenomonads could be seen in the rat cecum submucosa. Gram-negative organisms that have the cell morphology and flagellar arrangement typical for selenomonads have been observed in cecal contents of squirrels. These organisms were enumerated and isolated from cecal contents of the 13-lined ground squirrel (*Citellus tridecemlineatus*) and shown to be as high as 1×10^9 to 1×10^{10} cells per gram of cecal contents (Barnes and Burton, 1970). In addition, it was shown that these selenomonads constituted 18% or more of the total viable cells of the cecal contents of both active and hibernating squirrels. All selenomonad strains from squirrels could ferment glucose and some strains could ferment starch, but further biochemical characteristics of these strains have not been determined. Many investigators have observed selenomonads in the cecal contents of guinea pigs (Kingsley and Hoeniger, 1973; Robinow, 1954). These organisms have been designated as *S. palpitans* (Simons, 1922), but since the organisms have not been isolated or grown in pure culture, this classification is questionable. However, electron micrographs of

"*S. palpitans*" indicate it is a selenomonad which differs in some respects in cell morphology from *S. ruminantium* (Kingsley and Hoeniger, 1973).

The isolation of selenomonads from nonmammalian environments has been reported only on one occasion; using lactate and sulfate enrichment cultures, a selenomonad strain was isolated from ditch water from a bog habitat (Harborth and Hanert, 1982). This strain was phenotypically similar to *S. ruminantium* except it has a lower optimal growth temperature (25°C) and produces catalase. The authors suggested it be considered *S. ruminantium* subspecies *psychrocataligenes*. Selenomonad strain DSM 3853 (strain DKglu16) was isolated from glutamate- plus aspartate-limited chemostat cultures inoculated with anaerobic sewage sludge (Nanninga et al., 1987). Because this strain fermented only amino acids and differed in many other ways from other *Selenomonas* species, it was named *S. acidaminophila*. These studies suggest that selenomonads may exist in a number of natural, anaerobic habitats.

Isolation

Selective Enrichment and Isolation

Enrichment and selection procedures have been effectively developed only for the isolation of selenomonad strains from ruminal contents. Ruminal organisms can be selectively isolated using SS medium that has mannitol as the only added carbohydrate (Tiwari et al., 1969). The selective factors of SS medium are: 1) the use of mannitol as the main energy source, (since few species of ruminal bacteria can ferment this sugar); 2) the pH of the medium is 6.0, which does not affect selenomonad growth, but inhibits growth of many other species; and 3) the medium contains no branched-chain volatile fatty acids or heme, and either or both of these compounds are required for the growth of many other ruminal bacteria. Substitution of glycerol or lactate for mannitol in SS medium might allow for preferential isolation of selenomonad strains commonly designated as *S. ruminantium* subspecies *lactilytica*.

Selective Medium (SS) for Isolation of
Selenomonas ruminantium (Modified from Tiwari et al., 1969)

Mannitol	0.2 g
Trypticase	0.5 g
Yeast extract	0.1 g
Sodium acetate	0.1 g
FeSO$_4$·7H$_2$O	0.1 g

Mineral S solution (see below)	4.1 ml
n-Valeric acid	0.05 ml
Distilled water	92.5 ml
L-Cysteine·HCl solution (2.5%) (see below)	1.0 ml
Sodium carbonate solution (8%) (see below)	2.5 ml

The L-cysteine solution is prepared with oxygen-free distilled water under a nitrogen-gas phase. Both the sodium carbonate solution and final SS medium are prepared and equilibrated under a carbon dioxide gas phase. The L-cysteine and sodium carbonate solutions are autoclaved separately and added to the cooled medium. The composition of the mineral S solution is: monobasic potassium phosphate, 12.0 g; ammonium sulfate, 6 g; sodium chloride, 12.0 g; magnesium sulfate heptahydrate, 2.5 g; calcium chloride dihydrate, 1.6 g; and distilled water to a final volume of one liter. This solution is stable when stored at 5°C.

In regard to other species of *Selenomonas,* some enrichments for these organisms appear possible from samples of anaerobic sludge, bog water, or highly eutrophic waters. With such samples, the use of glutamate or aspartate as the main carbon source may lead to enrichment of *S. acidaminophila* strains (Nanninga et al., 1982), whereas use of lactate may result in enrichment of *S. ruminantium* strains (Harborth and Hanert, 1982).

MacDonald and Madlener (1957) examined several methods for isolating oral selenomonads. These authors found that complex media containing sodium lauryl sulfate (0.01%) or sodium oleate (0.15%) in addition to sheep serum (10%) would inhibit the growth of many microorganisms present in gingival scrapings, but would not affect selenomonad growth. Nonselective media, such as commercially available blood agar, can sometimes be used for the isolation of oral selenomonads. In certain subgingival sites that possess clinical signs of periodontitis, oral selenomonads can represent 10–30% of the total bacterial population (S. S. Socransky, personal communication). Appropriate dilutions of clinical samples are plated onto agar media and incubated under an atmosphere of 80% N_2/10% CO_2/10% H_2. Identification of these motile bacteria is initially based upon colony morphology. Since cells are able to migrate through solid medium with lowered concentrations of agar, colonies with a spreading, fuzzy morphology form. Some species of *Selenomonas* or *Centipeda* form a hazy zone of growth on the surface of the agar medium (Lai et al., 1983).

Cultivation Media

The nutrient requirements of *S. ruminantium* strains are rather simple, and most, if not all, strains can be grown anaerobically in chemically defined media containing glucose, minerals, B vitamins, ammonia, sulfide, and a volatile fatty acid (usually n-valerate), under a carbon dioxide atmosphere (Bryant and Robinson, 1962; Kanegasaki and Takahashi, 1967; John et al., 1974; Tiwari et al., 1969). More complex media containing yeast extract and trypticase, such as those used for *Succinivibrio dextrinosolvens* (see Chapter 222), will support growth of *S. ruminantium* strains. These complex media probably can support growth of *S. acidaminophila,* but glutamate, aspartate, pyruvate, or lactate would be needed as the energy source (Nanninga et al., 1987). For *S. ruminantium,* sulfide or cysteine serve as sole sulfur sources. Substances that can serve as sole nitrogen sources include ammonia, urea, certain single amino acids (cysteine, serine, threonine, aspartate, histidine, glutamate, and valine), or the purines adenine and uric acid, Some strains utilize urea or ammonia as a sole nitrogen source, but many strains do not. When lactate is the energy source, biotin and p-aminobenzoic acid satisfy the vitamin requirements; aspartate, malate, or fumarate are required for growth and aromatic amino acids may be stimulatory to growth (Linehan et al., 1978).

The nutritional features of most human oral species of *Selenomonas* have not been studied in detail, as has been done with *S. ruminantium.* Most of these species have been isolated on very complex media, such as brain-heart infusion agar or on this medium supplemented with 5% rabbit blood or serum. Rich media such as peptone-yeast extract-glucose often support growth of many of these species (Moore et al., 1987).

Preservation of Cultures

Most species can be maintained for long periods by storing cultures in liquid nitrogen or ultracold freezers (Hespell and Canale-Parola, 1970). Depending upon the strain, preservation by lyophilization under anaerobic conditions may be possible. Alternatively, short-term storage (6 to 15 months) may be possible by placing glycerol-containing cultures in normal ($-20°C$) freezers (Teather, 1982).

Identification

Phenotypic Properties

Quite often, newly isolated bacterial strains can be identified as *Selenomonas* based on the characteristic cell shape and tumbling motility of these species. All species are obligately anaerobic, motile, nonsporeforming, Gram-negative rods. The cells are usually curved or cresent-shaped and have a tuft of flagella that originates from the concave side of the cell, as revealed by Leifson flagella staining or by electron microscopy. Strains of the related genus *Pectinatus* have a similar curved-cell morphology, but the flagella are arranged linearly along the entire length of the concave side of the cell (Lee et al., 1978). *Centipeda* strains are Gram-negative, curved or helical rods. However, the flagella *Centibeda* are inserted in a line or stop that spirals around the cell, resulting in bundles that arise from both cell sides, giving a centipede-like appearance to the cell (Lai et al., 1983). A major characteristic of all known *Selenomonas* species is that they produce both acetate and propionate as major fermentation acids. Many strains also form small amounts of lactate and/or succinate (Table 2).

Species of *Selenomonas* have varying types of colony morphologies and cell sizes. Most strains of *S. ruminantium* produce large colonies (3- to 6-mm diameter) that are smooth, entire, slightly convex, and light tan to white in color. Often, these colonies have a gray to black appearance (due to hydrogen sulfide production) starting in the center of the colony. The cells are usually 1 μm by 2.0 to 4.0 μm long, but *S. ruminantium* subspecies *lactilytica* strains can be 2.0–3.0 μm

by 5.0–10.0 μm long. Many *S. ruminantium* strains have carbohydrate granules in the cytoplasm, and cells may be strongly iodophilic (Prins, 1971) but no capsular material is present.

With newly isolated strains, *S. sputigena* colonies on blood-agar media are generally small (0.5 to 1.2 mm in diameter), smooth, convex, and gray to gray-yellow in color. Larger colonies tend to have an irregular edge and translucent appearance. Most of the other oral selenomonad species form minute colonies (0.5–1.0 mm in diameter) that are shiny, smooth, and colorless to white. A spreading growth over the entire plate is not uncommon for several species. The cell sizes of most oral selenomonad species range from about 1.0–1.4 μm wide by 3.0–5.5 μm long.

Species of *Selenomonas*, *Pectinatus*, and *Centipeda* can be differentiated from one another on the basis of a number of phenotypic traits (see Table 2). *C. periodontii* is the only species listed that has a bilateral flagella arrangement and that does not form acetate as a major fermentation acid, although trace amounts of acetate can be made by some strains. *P. cerivisiiphilus* and *S. ruminantium* are the only listed species to produce hydrogen sulfide. *P. cerivisiiphilus* differs from *S. ruminantium* by its lin-

Table 2. Characteristics that differentiate species of *Selenomonas, Pectinatus,* and *Centipeda.*[a]

Characteristic	*S. acidiaminophila*[b]	*S. ruminantium*[c]	*S. artemidis*[d]	*S. noxia*[d]	*S. flueggei*[d]	*S. infelix*[d]	*S. sputigena*[d,f]	*S. dianae*[d]	*P. cerevisiiphilus*[e]	*C. periodontii*[f]
Acid from:										
Lactate	+	V	−	−	−	−	NK	−	+	+
Pyruvate	+	−	+	−	+	+	NK	−	NK	NK
Glutamate	+	−	NK	NK	NK	NK	NK	NK	NK	NK
Sucrose	−	+	+	−	+	+	+	+	−	+
Cellobiose	−	+	−	−	−	−	−	−	+	V
Mannitol	−	+	+	−	+	+	−	+	+	+
Lactose	−	V	−	−	+	+	+	+	−	+
Trehalose	−	V	−	−	−	−	−	+	−	V
Esculin hydrolysis	−	+	−	−	−	+	−	+	−	−
Hydrogen sulfide production	−	+	−	−	−	−	−	−	+	−
Nitrate reduction	−	V	+	−	+	+	v	+	−	+
Gelatin hydrolysis	+	−	−	−	+	−	−	−	−	−
Fermentation acids	A,P,s	A,P	A,P,L,s	A,P	A,P,l	A,P,l	A,P,l,s	A,P,L	A,P,l,s	a,P,l,s

[a]Positive = +; negative = −; variable = V; not known = NK. Major acid = upper case; minor acid = lower case: acetate = A; propionate = P; lactate = L; succinate = S.
[b]Data from Nanninga et al. (1987).
[c]Data from Bryant (1956).
[d]Data from Moore et al. (1987).
[e]Data from Lee et al. (1978).
[f]Data from Lai et al. (1983).

ear array of flagella and its inabilities to ferment sucrose and hydrolyze esculin (Lee et al., 1978). *S. acidaminophila* differs from all other listed species by its inability to ferment sugars, by using only lactate, pyruvate, glutamate, and aspartate, and by being able to hydrolyze gelatin. *S. ruminantium* is the only *Selenomonas* species capable of fermenting cellobiose. *S. infelix* and *S. dianae* are the only oral *Selenomonas* species capable of esculin hydrolysis, and these two species can be separated on the basis of acid production on trehalose. With respect to the esculin-negative oral *Selenomonas* species, *S. flueggei* and *S. sputigena* are both positive for acid from lactose and can be separated on the basis of acid from mannitol, whereas *S. artemidis* is negative for acid from lactose. *S. noxia* is negative for all of the previously mentioned traits of *Selenomonas* species, but growth is abundant in a peptone-yeast extract medium containing glucose, mannose, sorbitol, or sorbose.

Phylogeny

The GC content of the DNA from *Selenomonas* species ranges from 48 to 58 mol% (see Table 1). *S. acidaminophila* has the lowest value, whereas *S. ruminantium* strains vary from 49 to almost 54 mol%. The oral species of *Selenomonas* have only a narrow range of 53 to 58 mol%, but it is clear from the results of DNA-DNA hybridization (Table 3) that these strains are, in fact, separate species.

The phylogeny of species of *Selenomonas* and related bacteria has been determined by using 16S rRNA sequence analysis (Dewhirst et al., 1989). Complete 16S rRNA sequences of these bacteria were compared with the rRNA sequences of over 250 other bacterial species (Paster and Dewhirst, unpublished observations). From these data, a phylogenetic tree was constructed (Fig. 1). The microorganisms tested fall into two major groups—*Selenomonas* and related bacteria occupy one branch, and *Veillonella* and related bacteria are on the other. The *Selenomonas* group is phylogenetically coherent with interspecies homology levels of 90 to 99%. *Selenomonas dianae, S. infelix, S. flueggei, S. noxia, S. artemidis,* and *Centipeda periodontii* form a very tight cluster with a homology range of 96 to 99%. *Selenomonas sputigena* and *S. ruminantium* have an average sequence homology of 94% with members of this cluster. Aside from its unusual flagellation, *C. periodontii* is phenotypically similar to other members of the genus *Selenomonas*.

Pectinatus cerevisiiphilus, a Gram-negative, anaerobic, motile rod originally isolated from spoiled beer, is related to the *Selenomonas* group, but with an average homology value of only 91% (for further information see Chapter 91). *Sporomusa paucivorans* is an anaerobic, Gram-negative, sporeforming rod. This organism is related to the selenomonads at an average homology level of 90%.

In the other major branch of this tree, *Veillonella dispar* and *V. parvula*—anaerobic, Gram-negative cocci isolated from humans—are very closely related to each other with 99% sequence homology. *Megasphaera elsdenii* (see Chapter 91), an anaerobic, large, Gram-negative coccus isolated from humans and the ovine rumen, is related to *Veillonella* at a level of 92% homology. Species within these two bacterial genera are related to members of the *Selenomonas* branch with an average homology of 88%. The close relationship between motile, curved rods and nonmotile cocci may seem unusual, but there are phenotypic traits that unify this diverse group. One of the more convincing characteristics is that species of *Selenomonas, Sporomusa, Veillonella,* and *Megasphaera* all possess the one of the diamines, either cadaverine or putrescine, which are covalently bound to their peptidoglycan (Stackebrandt et al., 1985).

It has been previously shown that the genera *Selenomonas* and *Veillonella* share a branch with the Gram-positive bacteria (Stackebrandt et al., 1985). Members of these two groups both have an average 16S rRNA sequence homology of 85% with the Gram-positive bacteria, as represented by species of *Clostridium, Bacillus,* and *Enterococcus,* but only 79% sequence homology with Gram-negative bacteria such as *E. coli* and related bacteria (Fig. 2). Other bacterial groups, such as those containing the spirochetes, bacteroides, and radiation-resistant bacteria, branch deeper in the tree (unpublished observations) with even lower sequence homologies (e.g., <75%) with the selenomanads. From these data, it is evident that the selenomonads should be removed from the *Bacteroidaceae,* their current taxonomic placement in *Bergey's Manual of Systematic Bacteriology.* In general, a sequence homology of less than 80% indicates that the bacteria compared are in different major taxonomic divisions.

Now that 16S rRNA sequences are known (see below), it will be possible to develop short DNA probes targeted for signature regions of 16S rRNA. These probes can be used to identify new strains or species isolated from environmental samples. Ultimately, these probes will allow for the rapid identification of selenomonads directly from samples without in vitro cultivation. Family-specific and species-specific DNA probes have been already used to identify

Table 3. Percentage of homology from DNA-DNA hybridizations of oral *Selenomonas* species.

Species	S. artemidis	S. noxia	S. flueggei	S. infelix	S. sputigena	S. dianae
S. artemidis	100					
S. noxia	24	100				
S. flueggei	13	22	100			
S. infelix	15	20	18	100		
S. sputigena	4–8	13	5	6	100	
S. dianae	14	28	15	30	4–8	100

Data from Moore et al. (1987).

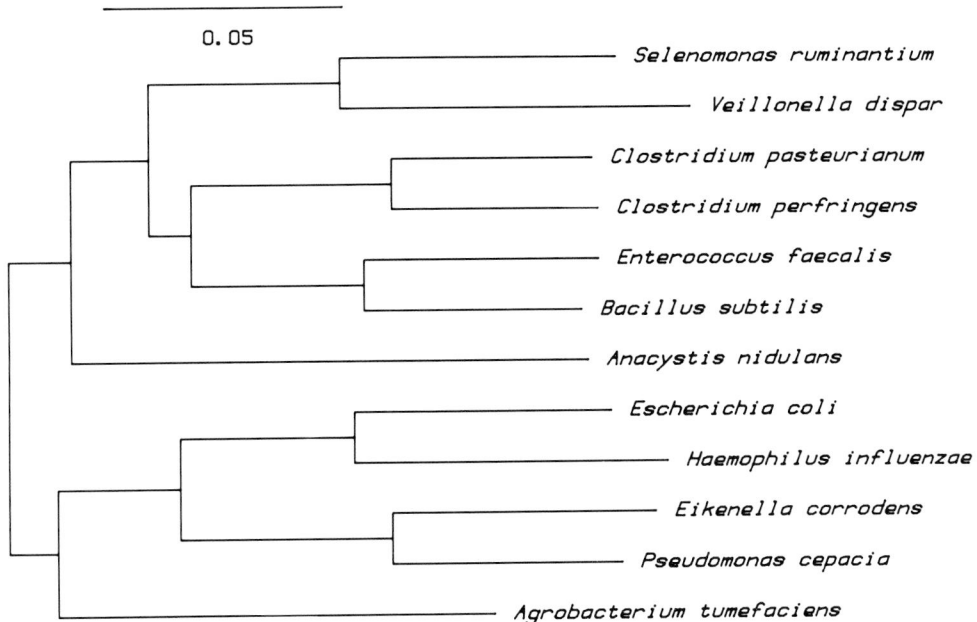

Fig. 2. Phylogenetic tree for the selenomonads, related bacteria, and distantly related bacteria. (Sequences for *C. perfringens* and *E. faecalis* courtesy of C. R. Woese.)

RNA isolated from selenomonads (Dewhirst et al., 1989).

Physiological Traits

Many biochemical and physiological studies have been done with *S. ruminantium,* but not with the oral species of *Selenomonas.* Early studies on isolation of ruminal bacteria did not indicate that *S. ruminantium* produced hydrogen gas as a fermentation product. However, when this organism was grown in the presence of a hydrogen-utilizing bacterium, namely a methanogen, (Scheifinger et al., 1975b). As a consequence, considerably less lactate and propionate were produced from glucose fermentation. *S. ruminantium* also changes the array of fermentation acids made as function of growth rate, with acetate and propionate dominating at low growth rates and lactate at high growth rates (Hobson, 1965). These shifts in fermentation products are correlated with less

ATP formation and can be largely explained by the effects of pyruvate on the activity of the intracellular lactic acid dehydrogenase (Scheifinger et al., 1975a; Wallace, 1978). At higher growth rates, intracellular pyruvate pools probably increase and this compound causes a homotropic activation of lactate acid dehydrogenase, resulting in higher rates of lactate formation. The other major fermentation acid, propionate, has been shown to be formed via the succinate pathway (Paytner and Elsden, 1970). Propionate formation most likely involves generation of ATP via formation of a proton-motive force, since *S. ruminantium* was one of the first anaerobes to be shown to possess cytochromes, mainly of the cytochrome-*b* type (DeVries et al., 1974).

Ammonia and urea play central roles in ruminal nitrogen metabolism. *S. ruminantium* is one of the major ureolytic bacteria in the rumen (Wozny et al., 1977). Studies on ammonia assimilation and glutamate formation in *S. ru-*

minantium indicate this organism possesses both the glutamate dehydrogenase and glutamine synthetase pathways for ammonia assimilation (Smith et al., 1980). In fact, this study was the first to show presence of a nonadenylylation control mechanism for glutamine synthetase in a Gram-negative organism. However, it was shown later that this enzyme and urease were coordinately controlled (Smith et al., 1981), as in other bacteria.

One of the first detailed studies on survival of anaerobic bacteria under nongrowing conditions was with *S. ruminantium*. Washed cell suspensions obtained from glucose-limited chemostats were found to lose viability very rapidly, with about a 50% loss in 2.5 hours (Mink and Hespell, 1981). The viability losses could not be attributed to cell lysis but correlated with rapid declines in cellular DNA, RNA, and protein. Similar effects were found with cells grown under nitrogen-limited conditions but better survival times were observed that correlated with growth rates (Mink et al., 1982). During starvation, the cells produced trace amounts of acetate as the only fermentation product. In addition, it was shown that the cellular levels of urease, glutamine synthetase, and glutamate dehydrogenase remained relatively stable despite loss of cell viability.

Applications

At present, no biotechnological applications have been made with *Selenomonas* species. However, a recent paper on the kinetics of glucose fermentation by *S. ruminantium* suggests that this organism might have commercial application for the production of lactic acid (Shimizu et al., 1989). A new species of the genus *Selenomonas* has been described by Schleifer et al. (1990). *Selenomonas lacticifex* was isolated from pitching yeast and can ferment glucose to lactic acid as major product.

Literature Cited

Barnes, E. M., and G. C. Burton. 1970. The effect of hibernation on the caecal flora of the thirteen-lined ground squirrel (*Citellus tridecemlineatus*). J. Appl. Bacteriol. 33: 505–514.

Bryant, M. P. 1956. The characteristics of strains of *Selenomonas* isolated from bovine rumen contents. J. Bacteriol. 72:162–167.

Bryant, M. P. 1984. Genus IX. *Selenomonas* pp. 650–653. In: N. R. Krieg and J. G. Holt (ed.), Bergey's manual of systematic bacteriology, vol. 1. Williams and Wilkins, Batimore.

Bryant, M. P., and I. M. Robinson. 1962. Some nutritional characteristics of predominant culturable ruminal bacteria. J. Bacteriol. 84:605–614.

Buchanan, R. E., and N. E. Gibbons (ed.). 1974. Bergey's manual of determinative bacteriology, 8th ed. Williams and Wilkins, Baltimore.

Caldwell, D. R., and M. P. Bryant. 1966. Medium without rumen fluid for nonselective enumeration and isolation of rumen bacteria. Appl. Microbiol. 14:794–801.

DeVries, W., W. M. C. Wijck-Kapteyn, and S. K. H. Oosterhuis. 1974. The presence and function of cytochromes in *Selenomonas ruminantium, Anaerovibrio lipolytica*, and *Veillonella alcalescens*. J. Gen. Microbiol. 81:69–78.

Dewhirst, F. E., B. J. Paster, P. L. Bright, D. A. Pelletier, J. B. Edwards, and J. M. Ago. 1989. *Selenomonas, Centipeda* and *Veillonella*: 16S rRNA sequences, probes and phylogeny. J. Dent. Res. 68:255–261.

Dobell, C. 1960. Antonie van Leeuwenhoek and his little animals. Dover Publications, New York.

Harborth, P. B., and H. H. Hanert. 1982. Isolation of *Selenomonas ruminantium* from an aquatic ecosystem. Arch. Microbiol. 132:135–140.

Hespell, R. B., and E. Canale-Parola. 1970. *Spirochaeta litoralis* sp. n., a strictly anaerobic marine spirochete. Archiv. Mikrobiol. 74:1–18.

Hobson, P. N. 1965. Continuous culture of some anaerobic and facultatively anaerobic rumen bacteria. J. Gen. Microbiol. 38:167–180.

Hobson, P. N., and S. O. Mann. 1961. The isolation of glycerol-fermenting and lipolytic bacteria from the rumen of the sheep. J. Gen. Microbiol. 25:227–240.

Hobson, P. N., S. O. Mann, and W. Smith. 1962. Serological tests of a relationship between rumen selenomonads in vitro and in vivo. J. Gen. Microbiol. 29:265–270.

Imoto, S. and S. Namioka. 1978. VFA production in the pig's large intestine. J. Anim. Sci. 47:467–471.

John, A., H. R. Isaacson, and M. P. Bryant. 1974. Isolation and characteristics of a ureolytic strain of *Selenomonas ruminantium*. J. Dairy Sci. 57:1003–1014.

Johnson, J. L., L. V. Holdeman, and W. E. C. Moore. 1985. Replacement of the Type strain of *Selenomonas sputigena* under rule 18g. Int. J. System. Bacteriol. 35:371–374.

Kanegasaki, S., and H. Takahashi. 1967. Function of growth factors for rumen microorganisms. I. Nutritional characteristics of *Selenomonas ruminantium*. J. Bacteriol. 93:456–463.

Kingsley, V. V., and J. F. M. Hoeniger. 1973. Growth, structure, and classification of *Selenomonas*. Bacteriol. Rev. 37:479–521.

Kurimoto, T. C., S. M. Tachibana, and T. Watanabe. 1986. Biological and chemical characterization of lipopolysaccharide from *Selenomonas* spp. in human periodontal pockets. Infect. Immun. 51:969–971.

Lai, C. H., B. M. Males, P. A. Dougherty, P. Berthold, and M. A. Listgarten. 1983. *Centipeda periodontii* gen. nov., sp. nov. from human periodontal lestions. Intern. J. Syst. Bacteriol. 33:628–635.

Lee, S. Y., M. S. Mabee, and N. O. Jangaard. 1978. *Pectinatus*, a new genus of the family Bacteroidaceae. Int. J. Syst. Bacteriol. 28:582–594.

Leifson, E. 1960. Atlas of bacterial flagellation. Academic Press, New York.

Lessel, E. F., and R. S. Breed. 1954. *Selenomonas* Boskamp, 1922—a genus that includes species showing an unusual type of flagellation. Bacteriol. Rev. 18:165–168.

Linehan, B., C. V. Scheifinger, and M. J. Wolin. 1978. Nutritional requirements of *Selenomonas ruminantium* for growth on lactate, glycerol, or glucose. Appl. Environ. Microbiol. 35:317–322.

MacCarthy, L. R., and J. R. Carlson. 1981. *Selenomonas sputigena* septicemia. J. Clin. Microbiol. 14:684–685.

MacDonald, J. B., and E. M. Madlener. 1957. Studies on the isolation of *Spirillum sputigenum*. Can. J. Microbiol. 3:679–686.

Miller, W. D. 1887. Über pathogene Mundpilze. Inaugural dissertation, Berlin.

Mink, R. W., and R. B. Hespell. 1981. Long-term nutrient starvation of continuously cultured (glucose-limited) *Selenomonas ruminantium*. J. Bacteriol. 148:541–550.

Mink, R. W., J. A. Patterson, and R. B. Hespell. 1982. Changes in viability, cell composition, and enzyme levels during starvation of continuously cultured (ammonia-limited) *Selenomonas ruminantium*. Appl. Environ. Microbiol. 44:913–922.

Moore, L. V., J. L. Johnson, and W. E. C. Moore. 1987. *Selenomonas noxia* sp. nov., *Selenomonas flueggei* sp. nov., *Selenomonas infelix* sp. nov., *Selenomonas dianae* sp. nov., and *Selenomonas artemidis* sp. nov. from the human gingival crevice. Intern J. Syst. Bacteriol. 36:271–280.

Nanninga, H. J., W. J. Drent, and J. C. Gottschal. 1987. Fermentation of glutamate by *Selenomonas acidaminophila* sp. nov. Arch. Microbiol. 147:152–157.

Ogimoto, K. 1972. Über *Selenomonas* aus dem Caecum von Ratten. Zentralbl. Bakteriol. Parasitenk. Infektions. Hyg. Abt. 1 Orig. Reihe A 221:467–473.

M. J. B., and S. R. Elsden. 1970. Mechanism of propionate formation by *Selenomonas ruminantium*, a rumen microorganism. J. Gen. Microbiol. 61:1–7.

Pomeroy, C., C. J. Shanholtzer, and L. R. Peterson. 1987. *Selenomonas* bacteraemia—case report and review of the literature. J. Infect. 15:237–242.

Prins, R. A. 1971. Isolation, culture, and fermentation characteristics of *Selenomonas ruminantium* var. *bryanti* var. n. from the rumen of sheep. J. Bacteriol. 105:802–825.

Robinow, C. F. 1954. Addendum to: *Selenomonas* Boskamp 1922—a genus that includes species showing an unusual type of flagellation, by E. F. Lessel, Jr. and R. S. Breed. Bacteriol. Rev. 18:168.

Robinson, I. M., M. J. Allison, and J. A. Bucklin. 1981. Characterization of the cecal bacteria of normal pigs. Appl. Environ. Microbiol. 41:950–955.

Robinson, I. M., S. C. Whipp, J. A. Bucklin, and M. J. Allison. 1984. Characterization of predominant bacteria from the colons of normal and dysenteric pigs. Appl. Environ. Microbiol. 48:964–969.

Salanitro, J. P., I. G. Blake, and P. A. Muirhead. 1977. Isolation and identification of fecal bacteria from adult swine. Appl. Environ. Microbiol. 33:79–84.

Scheifinger, C. C., M. J. Latham, and M. J. Wolin. 1975a. Relationship of lactate dehydrogenase specifity and growth rate to lactate metabolism by *Selenomonas ruminantium*. Appl. Microbiol. 30:916–921.

Scheifinger, C. C., B. Linehan, and M. J. Wolin. 1975b. H_2 production by *Selenomonas ruminantium* in the absence and presence of methanogenic bacteria. Appl. Microbiol. 29:480–483.

Schleifer, K. H., M. Leutwitz, N. Weiss, W. Ludwig, G. Kirchhof, and H. Seidel-Rüfer. 1990. Taxonomic study of anaerobic, Gram-negative, rod-shaped bacteria from breweries: emended description of *Pectinatus cerevisiiphilus* and description of *Pectinatus frisingensis* sp. nov., *Selenomonas lacticifex* sp. nov., *Zymophilus raffinosivorans* gen. nov., sp. nov. and *Zymophilus paucivorans* sp. nov. Intern J. Syst. Bacteriol. 40:19–27.

Shimizu, G., M. A. Cotta, and R. J. Bothast. 1989. The kinetics of glucose fermentations by *Selenomonas ruminantium* HD4 grown in continous culture. Biotechnol. Lett. 11:67–72.

Simons, H. 1922. Ueber *Selenomonas palpitans* n. sp. Zentralbl. Bakteriol. Parasitenkd Infektionskr. Hyg. Abt. 1 Orig. 87:50.

Smith, C. J., R. B. Hespell, and M. P. Bryant. 1980. Ammonia assimilation and glutamate formation in the anaerobe *Selenomonas ruminantium*. 141:593–602.

Smith, C. J., R. B. Hespell, and M. P. Bryant. 1981. Regulation of urease and ammonia assimilatory enzymes in *Selenomonas ruminantium*. Appl. Environ. Microbiol. 42:89–96.

Stackebrandt, E., H. Pohla, R. Kroppenstedt, H. Hippe, and C. R. Woese. 1985. 16S rRNA analysis of *Sporomusa*, *Selenomonas*, and *Megasphaera* on the phylogenetic origin of Gram-positive eubacteria. Arch. Microbiol. 143:270–276.

Teather, R. M. 1982. Maintenance of laboratory strains of obligately anaerobic rumen bacteria. Appl. Environ. Microbiol. 44:499–501.

Tiwari, A. D., M. P. Bryant, and R. S. Wolfe. 1969. Simple method for isolation of *Selenomonas ruminantium* and some nutritional characteristics of the species. J. Dairy Sci. 52:2054–2056.

Wallace, R. J. 1978. Control of lactate production by *Selenomonas ruminantium*: Homotropic activation of lactate dehydrogenase by pyruvate. J. Gen. Microbiol. 107:45–52.

Wozny, M. A., M. P. Bryant, L. V. Holdeman, and W. E. C. Moore. 1977. Urease assay and urease-producing species of anaerobes in the bovine rumen and human feces. Appl. Environ. Microbiol. 33:1097–1104.

CHAPTER 93

The Genus *Sporomusa*

JOHN A. BREZNAK

The genus *Sporomusa* has only recently been established (Möller et al., 1984). It consists of anaerobic, fermentative, straight to "banana-shaped" Gram-negative rods (0.4–0.9×2–8 μm), most species of which have the ability to form endospores (Breznak et al., 1988; Dehning et al., 1989; Hermann et al., 1987; Ollivier et al., 1985). Acetate is the sole or major product formed from a variety of substrates, including one-carbon substrates (e.g., $H_2 + CO_2$, methanol, CO, formate), various organic acids and alcohols, certain sugars, N-methyl compounds (e.g., betaine), and methoxylated aromatics. Not only do cells give a negative reaction with the Gram-stain procedure, they also display a true Gram-negative cell wall morphology, which includes the presence of a lipopolysaccharide-containing outer membrane (Möller et al., 1984). To this writer's knowledge, *Sporomusa* is one of only two named genera of endosporeforming bacteria currently represented in culture whose cells are truly Gram-negative (the other is *Sporohalobacter*, see "Identification" section in this chapter). This property distinguishes them from members of the genus *Clostridium*. Moreover, none has been shown to respire anaerobically with sulfate. This distinguishes them from the dissimilatory sulfate-reducing *Desulfotomaculum*, whose cells also stain Gram-negative (Campbell and Singleton, Jr., 1986), but whose cell wall ultrastructure is of the Gram-positive type (Nazina and Pivovarova, 1979; Sleytr et al., 1969).

Studies of the natural (i.e., phylogenetic) relationship of *Sporomusa* to other bacteria have been limited, although analysis of 16S ribosomal RNA oligonucleotide catalogs from *S. sphaeroides* and *S. ovata* indicate a distinct, but remote, relationship to the Gram-negative *Selenomonas* and *Megasphaera* and to Gram-positive bacteria of the "*Clostridium* subdivision" (Stackebrandt et al., 1985). However, assignment of *Sporomusa* to a new or currently existing family has not yet been proposed.

Habitats

Sporomusa has been isolated from anoxic sediments of freshwater rivers, lakes, creeks, and ditches (Dehning et al., 1989; Hermann et al., 1987; Möller et al., 1984); from soil and silage (Möller et al., 1984); from sugar beet factory and distillery wastewater (Möller et al., 1984; Ollivier et al., 1985); from horse and cattle dung (Möller et al., 1984); and from the gut of wood-eating termites wherein their abilty to form acetate from $H_2 + CO_2$ is important to termite nutrition (Breznak and Switzer, 1986; Breznak et al., 1988). Inocula for the isolations cited above were obtained from many parts of the world. No reports exist of *Sporomusa* isolated from brackish or marine habitats, although there is no reason to think such strains should not exist. For example, *S. malonica* was found to grow in a "brackish water" medium containing 10 g NaCl and 1.5 g $MgCl_2 \cdot 6 H_2O$ per liter (Dehning et al., 1989).

Isolation

Selective Enrichment

At present there does not exist an enrichment/isolation procedure that has proven to be entirely specific for *Sporomusa*. However, selectivity can be enhanced by exploiting the resistance to heat or ethanol that is typical of bacterial endospores, followed by inoculation of pretreated source material into an anoxic medium that is low in sulfate (to discourage growth of sulfate-reducing bacteria) and that contains a substrate that is reasonably selective for *Sporomusa*. Heat treatment is done by pasteurizing aqueous suspensions of anoxic mud or other material at 80°C for 10–15 minutes. Treatment with ethanol is done at a final concentration of 50% (v/v) sterile ethanol for 1 h (Koransky et al., 1978), followed by centrifugation and two washings with distilled water prior to inoculation. A variety of organic substrates can be used as energy sources by *Sporomusa,* including various sugars, organic acids, and alcohols. However, in liquid enrichment cultures *Sporomusa* might have difficulty competing with clostridia for many of these compounds. Accordingly, substrates that are fairly selective for *Sporomusa* are H_2/CO_2 (80:20 or 66:33, v/v) or methanol/CO_2 (molar ratio $\leqslant 2$) which are used by all species of the genus. When methanol is used, the concentration in the medium should be $\leqslant$ 10 mM to avoid toxicity. When CO_2 is included in the gas phase, it should be balanced by an appropriate amount of HCO_3^- (or other buffer system) in the liquid phase to maintain the pH near 7, which is at or near the pH optimum of all strains tested. In fact, CO_2/HCO_3^- is an excellent "natural" buffer system for cultivation of many anaerobes (Hungate, 1969). An incubation temperature of 30–35°C will permit growth of all known species of *Sporomusa.*

Formate is also used by *Sporomusa* but most species grow poorly with this substrate. Consequently, growth and/or acetate production above that of control (no substrate) enrichments may be difficult to detect. Specific enrichment and isolation of *S. malonica, S. termitida,* and *S. acidovorans* is, in principle, possible by using sodium succinate as an energy source. These species can grow by decarboxylation of succinate to propionate $+$ CO_2. *S. malonica* and *S. termitida* will also grow by decarboxylation of malonate to acetate $+$ CO_2 (see "Physiology and Ecology" in this chapter). Use of N-methyl compounds as substrates may also provide some selectivity. For example, betaine was used by pure cultures of all species tested, i.e., *S. sphaeroides* and *S. ovata* (Möller et al., 1984), *S. paucivorans* (Hermann et al., 1987), and *S. termitida* (Breznak et al., 1988).

S. paucivorans is the only species of the genus that has not been shown to form endospores (Hermann et al., 1987). Consequently, pasteurization or ethanol treatment of inocula must be avoided for enrichment or direct isolation of this species. However, the fact is that with the exception of *S. sphaeroides* and *S. ovata,* most other species were isolated from enrichment cultures established with untreated inocula and H_2/CO_2 or methanol/CO_2 as substrates. *S. malonica* was obtained from a glutarate-degrading consortium established with an untreated inoculum and then isolated in agar medium with crotonate as a substrate (Dehning et al., 1989).

Following inoculation of liquid enrichments, cultures are monitored for growth of putative *Sporomusa* by noting the development of turbidity, by microscopic examination for cells with a morphology typical of *Sporomusa* (Figure 1), and by assay for acetate production (or propionate production if succinate is the substrate). The latter can be done by gas chromatography or by high-performance liquid chromatography of acidified samples of culture supernatant fluid (Breznak and Switzer, 1986). If H_2/CO_2 is used as substrate, consumption of this gas mixture is also presumptive evidence of *Sporomusa* growth and results in the development of negative pressure in the headspace of culture vessels. This can be tested by inserting an H_2/CO_2-filled syringe (5 cc or larger; equipped with a 25 gauge needle) through the rubber stopper of the vessel and noting the extent to which the plunger moves downward (this simultaneously replenishes the gas used). A potential confounding factor in the enrichment of *Sporomusa* with H_2/CO_2, methanol/CO_2, or N-methyl compounds is the outcompetition of *Sporomusa* by methanogens which also use these substrates, particularly if inocula are not pretreated with heat or ethanol as described above. To circumvent this possibility, one might try including a specific inhibitor of the growth of methanogens in the enrichment medium, eg. 2-bromoethanesulfonate (BES) at 1 to 50 mM final concentration (Oremland, 1988). However, as yet no systematic studies have been made of the efficacy of BES in media for enrichment of *Sporomusa.*

Isolation

Isolation of *Sporomusa* is done by streaking the surface of anoxic agar media or by diluting samples through a series of "roll tubes" or "shake tubes" (see "Media and Cultivation Techniques," this chapter). Isolation media are then examined for colonies of a size (0.5–4 mm) and morphology (round with an entire margin; white to dark brown) similar to that exhibited by known species of *Sporomusa.* Suspected *Sporomusa* colonies are picked, suspended in a small amount of sterile, anoxic medium, and examined by phase-contrast microscopy for typical *Sporomusa* cells (Fig. 1a). Smears can also be heat-fixed and Gram-stained for brightfield microscopic examination. Presumptive *Sporomusa* colonies are then used as inocula to repeat the isolation procedure until the purity of the cultures is confirmed and the isolates can be completely characterized. If agar media containing H_2/CO_2 as substrate are held in rubber-stoppered tubes or bottles, efforts may be focused on those colonies present in vessels that develop negative pressure in the headspace. Inclusion of a pH indicator in agar media (e.g., 0.01% bromocresol green) may aid in recognizing acid-producing colonies of *Sporomusa* (Braun et al., 1979).

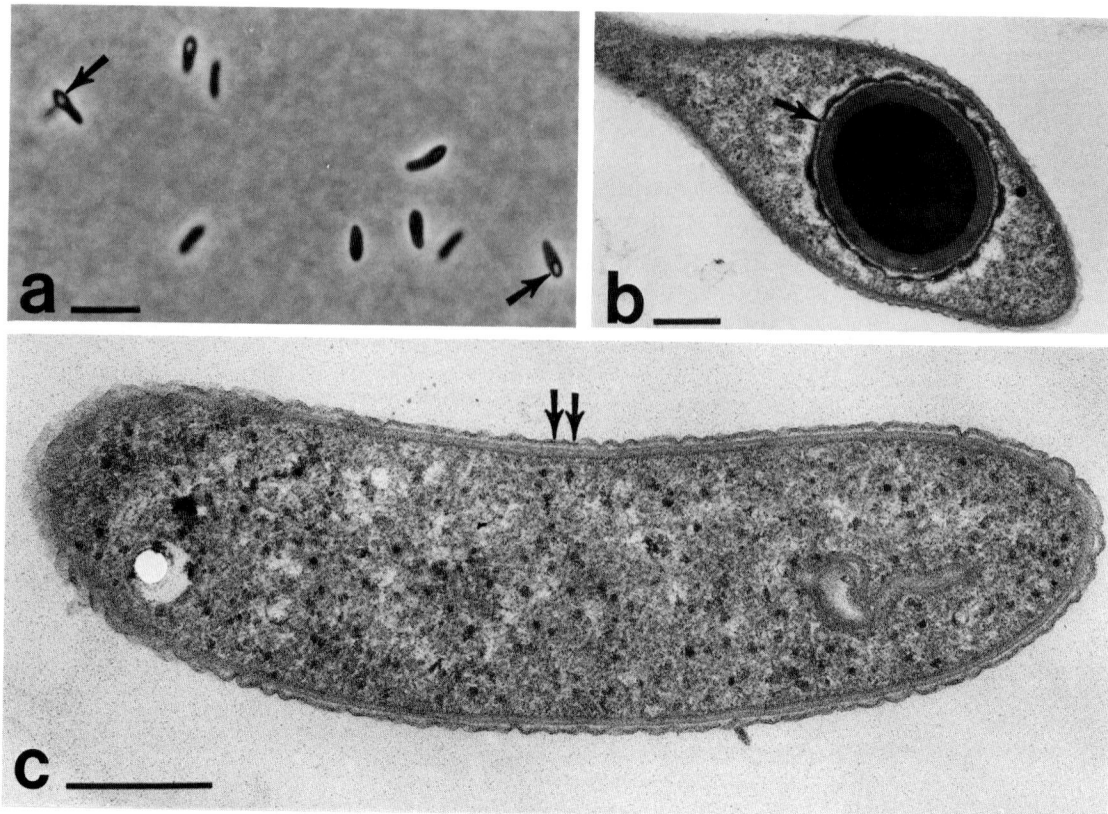

Fig. 1. Morphology of *Sporomusa termitida* strain JSN-2. (a) Phase-contrast micrograph. Bar = 10 μm. (b), (c) Transmission-electron micrographs of thin sections. Bars = 0.25 μm. Note endospores (a, b, single arrows) and the outer membrane of the cell wall (c, double arrows). (From Breznak et al., 1988.)

Media and Cultivation Techniques

Various culture media have been used for *Sporomusa* and all are more or less similar in composition. As yet no detailed nutritional studies have been done to define the vitamin and/or trace element requirement of individual species. The composition of a basal medium used to enrich for, isolate, and cultivate *S. sphaeroides* and *S. ovata* is presented below as an example. It would probably, with little or no modification, support the growth of all known species.

Basal Medium for *S. sphaeroides* and *S. ovata* (Möller et al., 1984)

K$_2$HPO$_4$	0.348 g
KH$_2$PO$_4$	0.227 g
NH$_4$Cl	0.500 g
MgSO$_4$·7H$_2$O	0.500 g
CaCl$_2$·2H$_2$O	0.025 g
NaCl	2.250 g
FeSO$_4$·7H$_2$O	2 mg
NaHSeO$_3$	15 μg
SL-10 trace element solution (see below)	3 ml

Resazurin	1 mg
Distilled water	905 ml

The mixture is boiled for 5 min and cooled in an ice bath under a stream of N$_2$/CO$_2$ (80:20, v/v) until room temperature is reached. At this point are added 80 ml NaHCO$_3$ solution (4 g NaHCO$_3$ in 80 ml distilled water; gassed with N$_2$/CO$_2$ for 20 min) and 2 ml of a 10X stock vitamin solution (see below). The medium is then dispensed in tubes or bottles in 9.9 ml amounts (or multiples thereof), sealed under N$_2$/CO$_2$ with rubber (preferably butyl rubber) stoppers, and sterilized by autoclaving at 121°C for 20 min. Before use, the medium is reduced by injecting through the stopper 0.1 ml of Na$_2$S solution (30 g Na$_2$S · 9H$_2$O per liter distilled water; sterilized by autoclaving; final concentration in the medium is 1.25 mM) or cysteine solution (30 g L-cysteine hydrochloride monohydrate per liter distilled water; sterilized by autoclaving; final concentration in the medium is 1.71 mM) per 9.9 ml of basal medium. This should result in decolorization of resorufin (formed from resazurin upon heating) indicating that the redox potential of the medium is below −51 mV.

SL-10 trace mineral solution (Widdel et al., 1983) contains:

HCl (25%)	10 ml
FeCl$_2$·4H$_2$O	1.5 g

CoCl$_2$·6H$_2$O	190 mg
MnCl$_2$·4H$_2$O	100 mg
ZnCl$_2$	70 mg
H$_3$BO$_3$	6 mg
Na$_2$MoO$_4$·2H$_2$O	36 mg
NiCl$_2$·6H$_2$O	24 mg
CuCl$_2$·2H$_2$O	2 mg
Distilled water	990 ml

The FeCl$_2$ is dissolved in the HCl first, then distilled water is added and the other salts are sequentially dissolved.

10X stock vitamin solution (Wolin et al., 1964) contains:

Biotin	20 mg
Folic acid	20 mg
Pyridoxine hydrochloride	100 mg
Thiamine hydrochloride	50 mg
Riboflavin	50 mg
Nicotinic acid	50 mg
Calcium pantothenate	50 mg
Para-aminobenzoic acid	50 mg
Thioctic acid	10 mg
Vitamin B$_{12}$	0.1 mg
Distilled water	1000 ml

When H$_2$/CO$_2$ is used as substrate, the medium can be dispensed into culture tubes or bottles under H$_2$/CO$_2$ instead of N$_2$/CO$_2$ before sterilization. When other substrates are used as energy sources, they are prepared separately as concentrated stock solutions, sterilized by autoclaving or filtration, and added to the medium at a final concentration of 5–50 mM. In this case, a slight adjustment may be made in the initial volume of distilled water used to prepare the basal medium so as to accommodate the volume of stock substrate added later. For solid medium, agar is included at a final concentration of 15–20 g per liter.

Culture tubes and bottles are those used for cultivation of methanogens and other anaerobic bacteria and are available commercially (Balch and Wolfe, 1976; Miller and Wolin, 1974). Agar media for colony isolation can be dispensed in rubber-stoppered "bottle plates," specially designed for streaking anaerobes (Hermann et al., 1986). Alternatively, colony isolation can be done by using "roll tubes" or "shake tubes," which are, in effect, pour plates in a tube. For the former, a tube containing inoculated and mixed molten agar is held horizontally under cold water and rolled as the agar solidifies, resulting in a shell of agar medium on the inside surface of the tube. For the latter, the agar is allowed to solidify while the tube is held upright in a vertical position. A detailed description of the roll tube technique is given by Hungate (1969), who also discusses the theoretical and practical aspects of preparation of anoxic media, as well as methods for cultivation and manipulation of strict anaerobes.

Additional Comments on Isolation and Cultivation

Most species of *Sporomusa* require, or are greatly stimulated by, yeast extract and a source of amino acids. Consequently, inclusion in the medium of yeast extract and trypticase, casitone, or casamino acids, at a concentration of 0.5–2 g per liter, may well increase the chances for successful enrichment and isolation of pure cultures.

Although Na$_2$S or cysteine (at 1–2 mM final concentation) are the reducing agents most often used for cultivation of *Sporomusa*, *S. termitida* could not be enriched in media containing these compounds or various alternatives (e.g., titanium citrate, dithiothreitol, amorphous ferrous sulfide, thioglycollate, or ascorbic acid). Successful enrichment and isolation of this species was only achieved by using PdCl$_2$ powder (330 mg per liter) as a catalyst for reduction of the medium by H$_2$, which was a component of the substrate mixture of H$_2$ + CO$_2$. After pure cultures were obtained, they could be adapted to grow with dithiothreitol (1 mM final concentration) or other reducing agents (Breznak et al., 1988). Thus, unsuccessful enrichment or isolation of *Sporomusa* might be the result of the reducing agent being used. It is noteworthy that growth of *S. ovata* is inhibited at sulfide concentrations of 2 mM or greater (Heijthuijsen and Hansen, 1986).

Preservation of Cultures

The ability of most species of *Sporomusa* to form endospores suggests that sporulated cultures may be stored without special precaution for extended periods. This writer has obtained viable subcultures of *S. termitida* from broth cultures kept at room temperature for over a year, cultures in which only endospores and ghostlike remnants of vegetative cells were seen. Hermann et al. (1987) maintained stock cultures of the nonsporing *S. paucivorans* by freezing cells at −80°C in growth medium containing 20% glycerol. This writer has also retrieved viable *S. termitida* from 10-fold concentrated cell suspensions frozen at −60°C in fresh medium containing 4% (v/v) dimethyl sulfoxide.

Identification

Identification of *Sporomusa* is based on: 1) the size (0.4–0.9 × 2–8 μm) and usually curved shape of cells; 2) the formation of endospores by most species; 3) the true Gram-negative character of the cell wall layer; and 4) their production of acetic acid as sole or major product of the fermentation of most substrates, including one-carbon compounds (e.g., H$_2$ + CO$_2$; methanol [+CO$_2$]). None of the species respires anaerobically with NO$_3^-$ or SO$_4^{-2}$. *S. paucivorans* can reduce S^0 and cystine, but does not carry out energy-yielding dissimilatory reduction of these compounds (Hermann et al., 1987).

The morphology of *S. termitida* is shown in Fig. 1 as an example. Endospores of *Sporomusa* are either spherical or oval, terminal to subterminal in location, and may distend the sporangium (Figs. 1a and c). They are heat-resistant (80°C for 10 min) and contain dipicolinic acid (Breznak et al., 1988). However, it is not yet known whether they contain lipopolysaccharide. Vegetative cells are motile by means of up to 15 flagella inserted laterally. This arrangement imparts a characteristic tumbling type of motility to cells. Membrane-bound cytochromes of the *b*-type (Breznak et al., 1988; Dehning et al., 1989; Möller et al., 1984) and *c*-type (Möller et al., 1984) have been observed in some species. In addition, *S. ovata* has been found to possess a high content of a unique corrinoid, *p*-cresolylcobamide, which may function as a methyl group carrier in the catabolic conversion of methanol to acetate (Stupperich et al., 1988; Stupperich and Eisinger, 1989). Preliminary analysis of *S. malonica* suggests that it too may contain this type of corrinoid (Dehning et al., 1989).

Some properties useful in distinguishing the various species of *Sporomusa* are presented in Table 1. It is obvious that some species are currently distinguished from others on the basis of seemingly minor differences, such as the utilization of particular substrates. Thus, if a comprehensive reevaluation of the genus is undertaken, consolidation of some of these species may eventually occur.

Kane and Breznak (1989) have isolated an H_2/CO_2-utilizing acetogen from gut contents of the dry-wood termite *Pterotermes occidentis*. The bacterium, strain APO-1, is also a truly Gram-negative endosporeformer, but its cells consist of long, thin rods (0.3–0.4×6–60 μm) that are morphologically quite distinct from those of *Sporomusa* species. Moreover, the organism forms abundant amounts of butyrate during glucose fermentation (approximately 1 mol butyrate per mol glucose fermented). These data, as well as recent analysis of 16S rRNA nucleotide sequences, suggest that it represents a new genus (manuscript in preparation).

Another genus of Gram-negative endospore-forming bacteria is *Sporohalobacter* (Oren et al., 1987). These bacteria are halophilic anaerobes that produce a variety of products from glucose fermentation and have a relatively low GC content in their DNA (29.6–31.5 mol%). Analysis of 16S rRNA oligonucleotide catalogs indicates that they are not closely related to *Sporomusa*.

Physiology and Ecology

Sporomusa species have been available in pure culture since 1984, yet our current understanding of their physiology and ecology is still meager. Their pathway for acetate formation from one-carbon substrates (e.g, $H_2 + CO_2$) remains a matter of speculation, although two likely possibilities are the "acetyl CoA pathway" Fuchs, 1986; Wood et al., 1986; also referred to as the "Wood pathway," Ljungdahl, 1986) and a path-

Table 1. Properties useful in differentiating species of the genus *Sporomusa*.

Property	S. sphaeroides	S. ovata	S. acidovorans	S. paucivorans	S. termitida	S. malonica
Cell size (μm)	0.5–0.8 × 2–4	0.7–0.9 × 3–5	0.7 × 5	0.4–0.7 × 2–3	0.5–0.8 × 2–8	0.7 × 2.6–4.8
Endospores	+	+	+	−	+	+
Motility	+	+	+	+	+	+
Catalase	+	(−)	ND	−	+	−
Utilization of:						
$H_2 + CO_2$	+	+	+	+	+	+
Methanol	+	+	+	+	+	+
Formate	+	+	+	+	+	+
Betaine	+	+	ND	+	+	+
Fructose	−	+	+	−	−	+
Mannitol	−	−	ND	−	+	ND
Glycerol	+	−	+	+	−	−
Propanol	+	+	ND	+	−	+
Citrate	−	−	−	−	+	+
Fumarate	−	−	+	−	−	+
Malate	ND	ND	+	−	−	+
Succinate	−	−	+	−	+	+
GC content (mol%)	46.7–47.4	41.3–43.3	42	47.1	48.6	44.1

Symbols: +, positive; −, negative; (−), negative or very weak reaction; ND, not determined.

way involving the synthesis of glycine and its reduction by the glycine reductase system (Gottschalk, 1986). Both of these are functionally equivalent and result in a total synthesis of acetate from $H_2 + CO_2$ as depicted in equation 1,

$$4H_2 + 2\ HCO_3^- + H^+ \rightarrow acetate^- \qquad (1)$$
$$+ 4H_2O\ (\Delta\ G^{0\prime} = -104.6\ kJ)$$

although the latter involves the intermediate synthesis of glycine and the cycling of NH_4^+. In this regard, it is of interest thaDthree N-methyl glycine derivatives (i.e., betaine, N,N-dimethylglycine, and sarcosine) as acetogenic substrates, suggesting that cells may reductively cleave the N–CH_2 bond of the glycine moiety (Möller et al., 1984). However, of the three species tested neither *S. sphaeroides, S. ovata,* nor *S. paucivorans* could grow on glycine itself as an energy source (Hermann et al., 1987; Möller et al., 1984).

All species of *Sporomusa* ferment methanol to acetate, but exogenous CO_2/HCO_3^- is required as an electron acceptor for good growth and fermentation of this substrate, as depicted in equation 2.

$$4CH_3OH + 2HCO_3^- \rightarrow 3\ acetate^- \qquad (2)$$
$$+ H^+ + 4H_2O\ (\Delta\ G^{0\prime} = -219.8\ kJ)$$

In theory, fermentation of methanol should be possible without exogenous CO_2/HCO_3^- (as shown in equation 3)

$$4CH_3OH \rightarrow 2\ acetate^- + 4H_2 \qquad (3)$$
$$+ 2H^+\ (\Delta\ G^{0\prime} = -115.2\ kJ)$$

if excess reducing equivalents were evolved as H_2. However, assuming that the carboxyl group of acetate arises from oxidation of part of the methanol via tetrahydrofolate (THF) intermediates (Heijthuijsen and Hansen, 1986), the reduction of protons to H_2 ($E_0^\prime = -414\ mV$) with electrons derived from the oxidation of methyl-THF to methylene-THF ($E_0^\prime = -120\ mV$) may require too much ATP to make this option energetically feasible for growth. This concept is discussed in greater detail by Heijthuijsen and Hansen (1986). By contrast, growth of *Sporomusa* on methanol in the absence or presence of CO_2/HCO_3^- is possible by co-culture with an H_2-consuming methanogen or sulfidogen (Cord-Ruwisch and Ollivier, 1986; Heijthuijsen and Hansen, 1986; Hermann et al., 1987). Under such conditions, a greater fraction of the methanol is completely oxidized to CO_2 by the *Sporomusa* and proportionally less (or no) acetate is formed by the co-culture. Such interspecies transfer of H_2 to methanogens has also been observed for *S. paucivorans* during glycerol fer-

mentation (Hermann et al., 1987). Syntrophic oxidation of methanol and other organic substrates by *Sporomusa* may be important in natural habitats.

A fascinating aspect of the metabolism of some species of *Sporomusa* is their ability to obtain energy for growth by decarboxylation of organic acids. *S. malonica* and *S. termitida* can grow by decarboxylation of malonate to acetate + CO_2 (equation 4; Breznak et al., 1988; Dehning et al., 1989).

$$Malonate^{2-} + H_2O \rightarrow acetate^- \qquad (4)$$
$$+ HCO_3^-\ (\Delta\ G^{0\prime} = -17.4\ kJ)$$

These two species, as well as *S. acidovorans,* can also grow by decarboxylation of succinate to propionate + CO_2 as depicted in equation 5 (Breznak et al., 1988; Dehning et al., 1989).

$$Succinate^{2-} + H_2O \rightarrow propionate^- \qquad (5)$$
$$+ HCO_3^-\ (\Delta\ G^{0\prime} = -20\ kJ)$$

In these respects, they resemble *Malonomonas rubra* (Dehning and Schink, 1989) and *Propionigenium modestum* (Schink and Pfennig, 1982), which also grow by the above overall reactions even though the associated free energy changes are very small.

Although acetate is the sole or major product from most substrates, propionate is the major product formed from succinate by some species (see equation 5). In addition, *S. malonica* forms both propionate and acetate from malate and fumarate (Dehning et al., 1989). *S. malonica* and *S. paucivorans* form acetate + propionate during fermentation of propanol or 1,2-propanediol, and butyrate + acetate during the fermentation of butanol (Dehning et al., 1989; Hermann et al., 1987), and so resemble *Acetobacterium carbinolicum* (Eichler and Schink, 1984). In addition, *S. paucivorans* forms isobutyrate + acetate from isobutanol (Hermann et al., 1987).

In the gut of termites, *S. termitida* is one of a number of bacteria whose ability to form acetate from $H_2 + CO_2$ is of major importance to termite nutrition. In fact, up to one-third of the insect's respiratory requirement can be met by oxidation of acetate produced by gut microbes from $H_2 + CO_2$ (Breznak and Switzer, 1986; Breznak et al., 1988). What remains unclear is the basis by which *S. termitida* and other acetogens outcompete resident methanogens for the bulk of the $H_2 + CO_2$ produced as a result of microbial fermentation in the termite hindgut. Thermodynamic considerations alone suggest that H_2/CO_2 methanogenesis is more exergonic than is H_2/CO_2 acetogenesis, three times more so at concentrations of substrates and products that are typically found in anaerobic habitats (Dolfing, 1988). One factor favoring the acetogens was thought to be their intrinsic affinity for H_2, but the K_m (6 μM) and V_{max} (380 nmol $\cdot$ min^{-1} $\cdot$ mg protein^{-1}) for H_2 uptake by cells of *S. termitida* was not unusually lower or higher, respectively, than that of most methanogens examined (Breznak et al., 1988). Moreover, the H_2 threshold of *S. acidovorans* and *S. termitida* (430 and 830 ppm H_2, respectively; Cord-Ruwisch et al., 1988) is about 10-fold greater than that of hydrogenotrophic methanogens (Cord-Ruwisch et al., 1988; Lovley, 1985; Lovley et al., 1984). Obviously, the competetive success of *S. termitida* in the termite gut must depend on factors other than its affinity for H_2 alone. The ability of *S. termitida* to grow mixotrophically (i.e., by simultaneous use of H_2 + CO_2 + organic compounds for energy) may have a bearing on its competetiveness for H_2 in the termite gut (Breznak and Switzer, 1989). However, more research is needed to clarify this puzzling issue.

Literature Cited

Balch, W. E. and R. S. Wolfe. 1976. New approach to the cultivation of methanogenic bacteria: 2-mercaptoethanesulfonic acid (HS-CoM)-dependent growth of *Methanobacterium ruminantium* in a pressurized atmosphere. Appl. Environ. Microbiol. 32:781–791.

Braun, M., S. Schoberth, and G. Gottschalk. 1979. Enumeration of bacteria forming acetate from H_2 and CO_2 in anaerobic habitats. Arch. Microbiol. 120:201–204.

Breznak, J. A., and J. M. Switzer. 1986. Acetate synthesis from H_2 plus CO_2 by termite gut microbes. Appl. Environ. Microbiol. 52:623–630.

Breznak, J. A., and J. M. Switzer. 1989. Mixotrophy in the termite gut acetogen, *Sporomusa termitida*. Abstr. Ann. Mtg. Amer. Soc. Microbiol., p. 239.

Breznak, J. A., J. M. Switzer, and H.-J. Seitz. 1988. *Sporomusa termitida* sp. nov., an H_2/CO_2-utilizing acetogen isolated from termites. Arch. Microbiol. 150:282–288.

Campbell, L. L., and R. Singleton, Jr. 1986. Genus IV. *Desulfotomaculum* Campbell and Postgate 1965, 361[AL], p. 1200–1202. In: P. H. A. Sneath, N. S. Mair, M. E. Sharpe, and J. G. Holt (ed.), Bergey's manual of systematic bacteriology, vol. 2. Williams and Wilkins, Baltimore.

Cord-Ruwisch, R. and B. Ollivier. 1986. Interspecific hydrogen transfer during methanol degradation by *Sporomusa acidovorans* and hydrogenophilic anaerobes. Arch. Microbiol. 144:163–165.

Cord-Ruwisch, R., H.-J. Seitz, and R. Conrad. 1988. The capacity of hydrogenotrophic anaerobic bacteria to compete for traces of hydrogen depends on the redox potential of the terminal electron acceptor. Arch. Microbiol. 149:350–357.

Dehning, I., and B. Schink. 1989. *Malonomonas rubra* gen. nov., sp. nov., a microaerotolerant anaerobic bacterium

growing by decarboxylation of malonate. Arch. Microbiol. 151:427–433.

Dehning, I., M. Stieb, and B. Schink. 1989. *Sporomusa malonica* sp. nov., a homoacetogenic bacterium growing by decarboxylation of malonate or succinate. Arch. Microbiol. 151:421–426.

Dolfing, J. 1988. Acetogenesis, p. 417–468. In: A. J. B. Zehnder (ed.), Biology of anaerobic microorganisms. John Wiley and Sons, New York.

Eichler, B., and B. Schink. 1984. Oxidation of primary aliphatic alcohols by *Acetobacterium carbinolicum* sp. nov., a homoacetogenic anaerobe. Arch. Microbiol. 140:147–152.

Fuchs, G. 1986. CO_2 fixation in acetogenic bacteria: variations on a theme. FEMS Microbiol. Rev. 39:181–213.

Gottschalk, G. 1986. Bacterial metabolism, 2nd ed. Springer-Verlag, New York.

Heijthuijsen, J. H. F. G., and T. A. Hansen. 1986. Interspecies hydrogen transfer in co-cultures of methanol-utilizing acidogens and sulfate-reducing or methanogenic bacteria. FEMS Microbiol. Ecol. 38:57–64.

Hermann, M., K. M. Noll, and R. S. Wolfe. 1986. Improved agar bottle plate for isolation of methanogens or other anaerobes in a defined gas atmosphere. Appl. Environ. Microbiol. 51:1124–1126.

Hermann, M., M.-R. Popoff, and M. Sebald. 1987. *Sporomusa paucivorans* sp. nov., a methylotrophic bacterium that forms acetic acid from hydrogen and carbon dioxide. Int. J. Syst. Bacteriol. 37:93–101.

Hungate, R. E. 1969. A roll tube method for cultivation of strict anaerobes, p. 117–132. In: J. R. Norris and D. W. Ribbons (ed.), Methods in microbiology, vol. 3B. Academic Press, New York.

Kane, M. D., and J. A. Breznak. 1989. H_2/CO_2 acetogenic bacteria: new isolates from termite guts. Abstr. Ann. Mtg. Amer. Soc. Microbiol., p. 234.

Koransky, J. R., S. D. Allen, and V. R. Dowell, Jr. 1978. Use of ethanol for selective isolation of sporeforming microorganisms. Appl. Environ. Microbiol. 35:762–765.

Ljungdahl, L. G. 1986. The autotrophic pathway of acetate synthesis in acetogenic bacteria. Ann. Rev. Microbiol. 40:415–450.

Lovley, D. R. 1985. Minimum threshold for hydrogen metabolism in methanogenic bacteria. Appl. Environ. Microbiol. 49:1530–1531.

Lovley, D. R., R. C. Greening, and J. G. Ferry. 1984. Rapidly growing rumen methanogenic organism that synthesizes coenzyme M and has a high affinity for formate. Appl. Environ. Microbiol. 48:81–87.

Miller, T. L., and M. J. Wolin. 1974. A serum bottle modification of the Hungate technique for cultivating obligate anaerobes. Appl. Microbiol. 27:985–987.

Möller, B., R. Ossmer, B. H. Howard, G. Gottschalk, and H. Hippe. 1984. *Sporomusa*, a new genus of Gram-negative anaerobic bacteria including *Sporomusa sphaeroides* spec. nov. and *Sporomusa ovata* spec. nov. Arch. Microbiol. 139:388–396.

Nazina, T. N., and T. A. Pivovarova. 1979. Submicroscopic organization and sporulation in *Desulfotomaculum nigrificans*. Microbiology (Engl. Transl. Mikrobiologiya) 48:241–245.

Ollivier, B., R. Cordruwisch, A. Lombardo, and J.-L. Garcia. 1985. Isolation and characterization of *Sporomusa acidovorans* sp. nov., a methylotrophic homoacetogenic bacterium. Arch. Microbiol. 142:307–310.

Oremland, R. S. 1988. Biogeochemistry of methanogenic bacteria, p. 641–705. In: A. J. B. Zehnder (ed.), Biology of anaerobic microorganisms. John Wiley and Sons, New York.

Oren, A., H. Pohla, and E. Stackebrandt. 1987. Transfer of *Clostridium lortetii* to a new genus *Sporohalobacter* gen. nov. as *Sporohalobacter lortetii* comb. nov., and description of *Sporohalobacter marismortui* sp. nov. System. Appl. Microbiol. 9:239–246.

Schink, B., and N. Pfennig. 1982. *Propionigenium modestum* gen. nov. sp. nov., a new strictly anaerobic, nonsporing bacterium growing on succinate. Arch. Microbiol. 133:209–216.

Sleytr, U., H. Adam, and H. Klaushofer. 1969. Die Feinstruktur der Zellwand und Cytoplasmamembran von *Clostridium nigrificans,* dargestellt mit Hilfe der Gefrierätz- und Ultradünnschnittechnik. Arch. Mikrobiol. 66:40–58.

Stackebrandt, E., H. Pohla, R. Kroppenstedt, H. Hippe, and C. R. Woese. 1985. 16S rRNA analysis of *Sporomusa, Selenomonas,* and *Megasphaera:* on the phylogenetic origin of Gram-positive eubacteria. Arch. Microbiol. 143:270–276.

Stupperich, E., and H. J. Eisinger. 1989. Biosynthesis of *para*-cresolyl cobamide in *Sporomusa ovata.* Arch. Microbiol. 151:372–377.

Stupperich, E., H. J. Eisinger, and B. Kräutler. 1988. Diversity of corrinoids in acetogenic bacteria: *p*-cresolylcobamide from *Sporomusa ovata,* 5-methoxy-6-methylbenzimidazolylcobamide from *Clostridium formicoaceticum* and vitamin B_{12} from *Acetobacterium woodii.* Eur. J. Biochem. 172:459–464.

Widdel, F., G.-W. Kohring, and F. Mayer. 1983. Studies on dissimilatory sulfate-reducing bacteria that decompose fatty acids. III. Characterization of the filamentous gliding *Desulfonema limicola* gen. nov., sp. nov., and *Desulfonema magnum* sp. nov. Arch. Microbiol. 134:286–294.

Wolin, E. A., R. S. Wolfe, and M. J. Wolin. 1964. Viologen dye inhibition of methane formation by *Methanobacillus omelianskii.* J. Bacteriol. 87:993–998.

Wood, H. G., S. W. Ragsdale, and E. Pezacka. 1986. The acetyl-CoA pathway of autotrophic growth. FEMS Microbiol. Rev. 39:345–362.

CHAPTER 94

The Genera *Butyrivibrio, Lachnospira,* and *Roseburia*

ROBERT B. HESPELL

Members of the genera *Butyrivibrio, Lachnospira,* and *Roseburia* are anaerobic bacteria that are usually isolated from the gastrointestinal tract of mammals. These organisms are motile, curved rods, and usually stain Gram negative or weakly Gram positive. However, this staining may not be reflective of the true nature of the cell wall structure. Strains of *Butyrivibrio* may have cell wall structures similar to either Grampositive (Cheng and Costerton, 1977) or Gramnegative types (Fig. 1D; Dibbayawan et al., 1985). Since ruminal butyrivibrios characteristically produce butyric acid and degrade plant fibers such as xylans, the name *B. fibrisolvens* is quite descriptive. *Roseburia* also produces butyric acid and is named after Theodor Rosebury, an American microbiologist who described and studied microorganisms indigenous to humans. Because this organism was isolated from the mouse cecum, the species name *cecicola* has been proposed (Stanton and Savage, 1983a). Strains of *Lachnospira* from a variety of fermentation products, but do not produce butyric acid. These organisms form colonies that take on the appearance of woolly balls (Bryant and Small, 1965b), hence the name *L. multiparus,* meaning woolly colony-forming spirals that make many products.

Although the members of *Butyrivibrio, Lachnospira,* and *Roseburia* appear to be easily differentiated based on a variety of phenotypic traits, other traits such as cell morphology or kinds of fermentation products suggest that these bacteria may have some common relationships. Based on comparisons of partial 16S rRNA sequences, these bacteria form a large group of species and strains that are part of the Gram-positive cluster (Fig. 2), indicating that they are phylogenetically related to Gram-positive bacteria. In addition, DNA-DNA hybridization data indicate that B. fibrisolvens strains apparently are a genetically heterologous group of organisms comprised of several species and numerous unrelated strains (Table 1; Mannarelli, 1988). However, these strains obviously have enough similar phenotypic traits and these specific traits have been used to presumptively identify new isolates such as *B. fibrisolvens.* The variability of *B. fibrisolvens* isolates are not entirely surprising since it has been known for sometime that *B. fibrisolvens* strains differ greatly in serological properties (Margherita and Hungate, 1963; Margherita et al., 1964; Hazelwood et al., 1986) and in nutritional properties (Bryant and Small, 1956a; Bryant and Robinson, 1962).

Habitats

Butyrivibrio

Butyrivibrios are common inhabitants of the bovine rumen and are present in the gastrointestinal tract of variety of animals. They have been isolated from animals in numerous geographical areas throughout the world. Butyrivibrios have been shown to represent 22 to 30% of the bacterial population in the rumen of high Arctic Svalbard reindeer (Orpin et al., 1985) and are one of the more numerous kinds of bacteria found in the cecum of sheep (Lewis and Dehority, 1985). Butyrivibrios have also been shown to be present in high numbers (10^6 to 10^7/g) in fecal material obtained from rabbits, horses, and humans (Brown and Moore, 1960). *Butyrivibrio* strains isolated from human fecal material have been designated as *B. crossatus* because these strains differ in substrate utilization and flagellar arrangement from *B. fibrisolvens* (Moore et al., 1976). Finally, it should be noted that butyrivibrios can occur in environments outside of animals, and recently strains have been isolated from napiergrass-fed anaerobic digestors (Sewell et al., 1988).

Almost all strains of *B. fibrisolvens* are quite versatile in that a wide range of sugars are fermented, extracellular proteases are made (Cotta and Hespell, 1986), and both cellular and extracellular esterase activities are produced (Hespell and O'Bryan-Shah, 1988). Although all characterized strains of *B. fibrisolvens* are highly

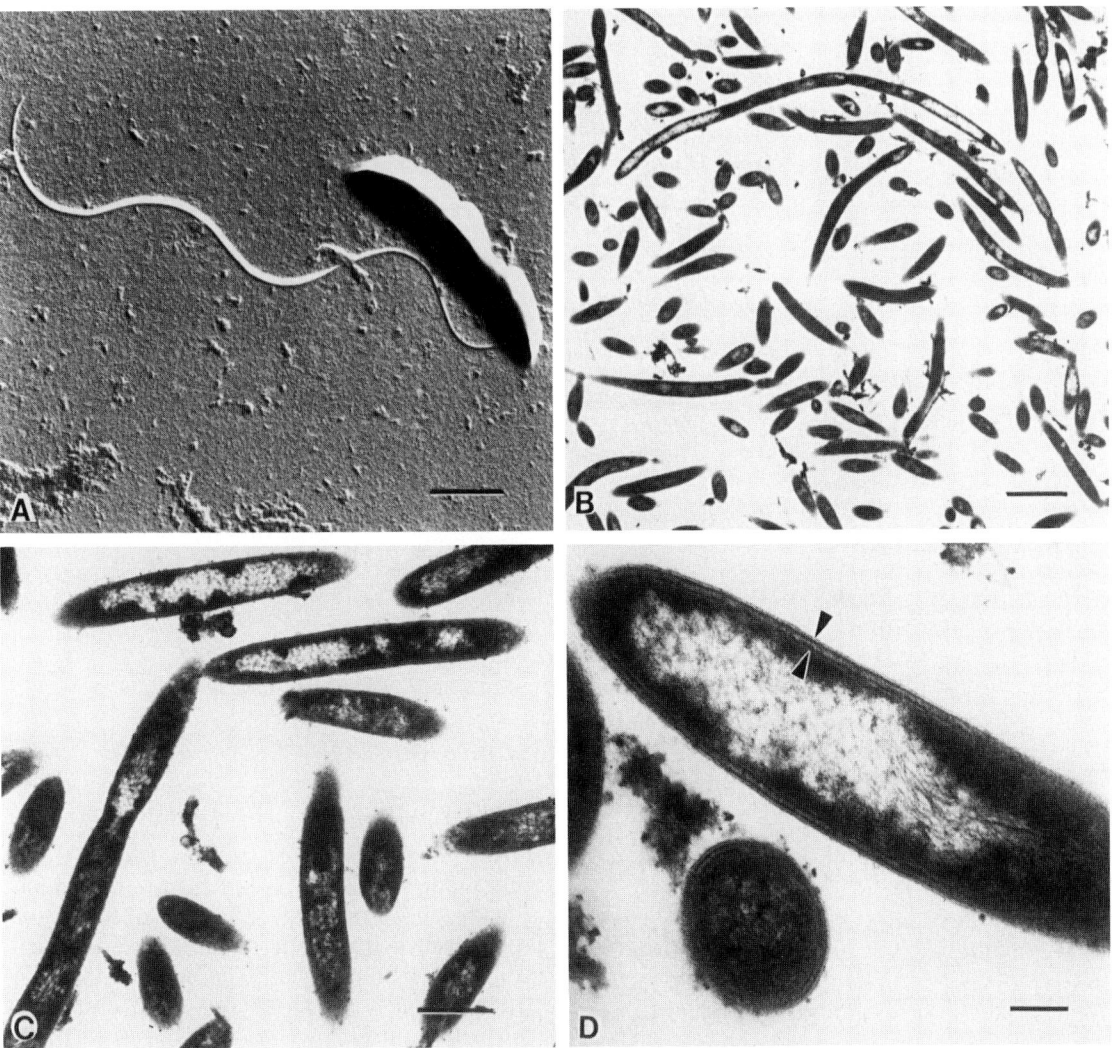

Fig. 1. Electron micrographs of *B. fibrisolvens* 113. (A) Whole cell, platinum-shadowed. Bar = 0.5 μm. (B) Thin section of cells, showing filamentous aspects of attached cells after cross-wall formation. Bar = 1.0 μm. (C) Thin section of cells, showing intracellular, electron-lucent inclusions. Bar = 0.5 μm. (D) Thin section of cells, high-magnification micrograph. The cell envelope contains two thin membranes separated by a thick layer (arrows). Bar = 0.1 μm. (From Sewell et al., 1988.)

xylanolytic, only about 10% of these strains are capable of good growth on cellulose. Many strains are also pectinolytic and amylolytic. A few strains are capable of anaerobically degrading complex heterocyclic compounds of the bioflavanoid type (Cheng et al., 1969, 1970). Some strains can deacetylate trichothecenes such as T-2 toxin (Westlake et al., 1987). Given this wide biochemical diversity of *Butyrivibrio* strains, these bacteria are presumably involved in many important ecological roles in degradation of fibrous plant materials, starches, pectins, and proteins in the rumen and probably also in the cecum and large intestine of many mammals.

Lachnospira

Lachnospira multiparus is considered to be an ecologically important species of ruminal bacteria and can be a major constituent of the bacterial population under certain dietary conditions. *Lachnospira* strains are the primary pectin fermenters isolated from ruminal contents of animals fed diets high in pectin (e.g., lush legumes, citrus pulps). With animals fed rich ladino clover, 16 to 31% of the total bacterial isolates were *Lachnospira* strains (Bryant et al., 1960). Strains have also been isolated from ruminal contents of animals fed alfalfa hay when media containing pectin as the sole carbohydrate source are used. Compared to bu-

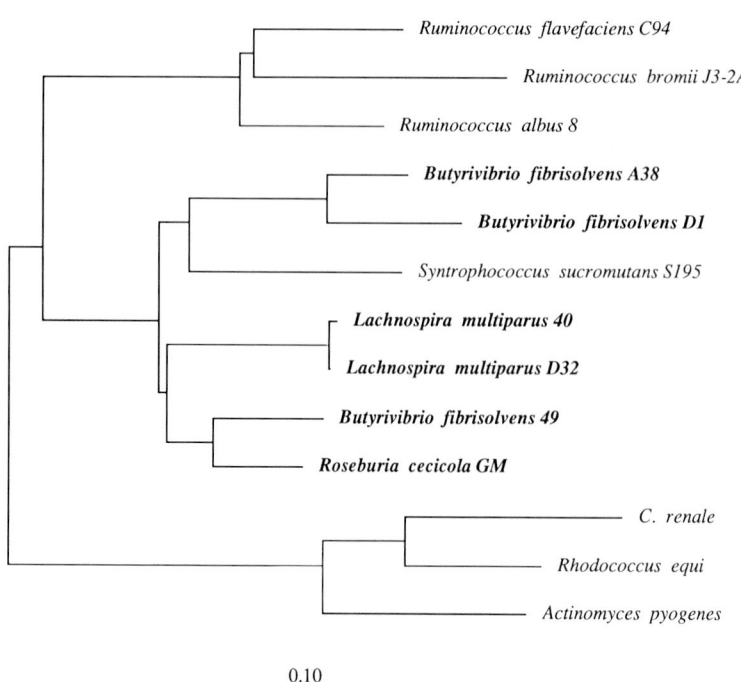

Fig. 2. Phylogenetic tree for *Butyrivibrio, Lachnospira, Roseburia,* and related bacteria. The major division within the tree corresponds to the division between the high and low GC groups of Gram-positive bacteria. The tree was inferred by the distance-matrix method and the scale bar represents 0.1 nucleotide changes per sequence position. Approximately 600 positions were compared, corresponding to positions 316 to 489, 633 to 879, and 1080 to 1376 of the *E. coli* 16S rRNA molecule.

tyrivibrios, lachnospiras have a limited range of carbohydrates that are fermented. Besides pectin, most strains ferment sucrose, fructose, and glucose, sugars which are usually present at substantial levels in fresh, young forage. *L. multiparus* is capable of invading and causing maceration of these types of plant materials (Cheng et. al., 1979).

Roseburia

Roseburia cecicola has been isolated from mucosal scrapings taken from the ceca of conventional laboratory mice (Stanton and Savage, 1983a), but the occurrence of this species in other animals is not known at this time. Like *B. fibrisolvens, R. cecicola* also produces large amounts of butyric acid, but it ferments only a limited range of sugars and is not xylanolytic. *R. cecicola* is motile by means of flagella that are present in a fascicle inserted subapically. Motility and chemotaxis often have been suggested as important characteristics for microorganisms that colonize mucosa and other animal tissues. In mucosal scrapings, motile bacteria outnumber nonmotile bacteria by twofold (Stanton and Savage, 1983b). In this regard, it has been shown that motility may be essential for *R. cecicola* to colonize the ceca of conventional mice (Stanton and Savage, 1983b). Antibodies made to the flagellin protein of *R. cecicola* were used to show that this organism is associated with the mucuosal epithelium (Martin and Savage, 1984).

Isolation

Selective Enrichment and Isolation

Butyrivibrio fibrisolvens strains have been isolated using a wide variety of culture media and growth conditions, but definitive selective isolation or enrichment procedures have not been published. Isolation using media containing xylan as the only added carbohydrate source has shown that, strains of *B. fibrisolvens* and *Bacteroides ruminicola* are the predominant organisms in ruminal contents (Dehority, 1966). Using media with ball-milled or finely ground cellulose as the sole carbohydrate source, butyrivibrios are frequently isolated from ruminal (Shane et al., 1969) or cecal (Lewis and Dehority, 1985) contents. With these cellulose-containing media, colonies of butyrivibrios are often surrounded by weak zones of cellulose degradation compared to the extensive clearing zones of cellulolytic cocci (usually *Ruminococcus* species) or of normotile, pleomorphic rods (usually *Fibrobacter succinogenes;* formerly called *Bacteroides succinogenes*). *Bacteroides* and *Butyrivibrio* strains constitute the majority of bacterial isolates obtained from ruminal contents when media containing plant saponins are used (Gutierrez et al., 1959). A selective medium for *Butyrivibrio fibrisolvens* strains might be achieved with xylan as the sole carbohydrate source and nalidixic acid as an inhibitor, since well-characterized *B. fibrisolvens* strains (see Mannarelli, 1988 for listing) are uniformly xy-

Table 1. Grouping of *Butyrivibrio fibrisolvens* strains by DNA relatedness and extracellular polysaccharide composition.[a]

Group	Type[b]	Strains	Polysaccharide sugars								
			Rhamnose	Fucose	Altrose	Mannose	Galactose	Glucose	Lactyl galactose	Lactyl glucose	Lactyl rhamnose
A-1	G	D1; AcTF2; 114	+[d]	+	−	+	+	+	−	−	−
A-2	P	A38	+	+	−	+	+	+	−	−	−
B-1	G	113; 112; 110 E21c; 1L631 NOR37; 835; E9a	+	−	−	+	+	+	−	−	+
B-2	P	D16f; LM8/1b S2; C3	+	−	−	+	+	+	−	−	−
C-1	G	CF3; CF1B; CF2d CF3a; CF3c; CF4c	tr.[c]	−	+	+	tr.	+	+	+	−
D-1	G	49; H17c	tr.	−	−	−	+	+	+	−	+
		12; CE51; CE52	tr.	−	−	−	+	+	+	−	−
D-2	P	H10b; D30g	tr.	−	−	−	+	+	+	−	−
E-1	G	R28; P17	+	−	+	−	+	+	+	−	−
E-2	P	H4a; E46a	+	+	+	+	+	+	+	−	−
		ARD22a; C14; PI26	+	+	+	−	+	+	−	−	−

[a] DNA relatedness data compiled from Mannarelli (1988) and polysaccharide data compiled from Stack (1988).
[b] Type based primarily on DNA relatedness (G) or polysaccharide composition (P).
[c] Trace.
[d] +, sugar present; −, sugar absent.

lanolytic and are resistant to high levels (30 to 500 µg/ml) of nalidixic acid (Hespell, unpublished observations). However, in agreement with early studies (Fulghum et al., 1968), most strains are quite sensitive to many antibiotics, especially those that affect cell wall synthesis (e.g., penicillin, bacitracin).

All known *Lachnospira multiparus* strains are pectinolytic and can be readily isolated from ruminal contents with media containing pectin as the only added energy source (Dehority, 1969). These media are only partially selective, because colonies of other pectin-fermenting bacteria such as *Butyrivibrio fibrisolvens* and *Bacteroides ruminicola* are also obtained. More effective selective isolation media or enrichment procedures for *L. multiparus* are not known.

Roseburia cecicola was isolated using procedures that selected for motile microorganisms present in mucosal scrapings (Stanton and Savage, 1983a). One approach involved inoculating a capillary tube filled with growth media. After incubation, the end of the tube opposite the inoculation end was broken off and some of the contents were removed and used to inoculate agar media to obtain individual colonies. This selection procedure is based on the concept that motile bacteria would migrate through the tube whereas nonmotile bacteria would remain near the tube inoculation end. *R. cecicola*, like *B. fibrisolvens*, is resistant to nalidixic acid and inclusion of this antibiotic in isolation media might make this previous procedure even more selective for *R. cecicola*, although this has not yet been examined.

Cultivation Media

Members of *Butyrivibrio, Lachnospira*, and *Roseburia* are all strictly anaerobic bacteria. The complex medium (RGM, Table 2) can be used to grow strains of these bacteria as well as many other species of ruminal bacteria (Hespell et al., 1987). These bacteria can be grown at 37°C, but usually little or no growth takes place at 30°C or 45°C. Many *B. fibrisolvens* strains are not nutritionally fastidous and can be grown on a chemically defined medium (Table 2), which has been used in studies of proteolytic activities with these strains (Cotta and Hespell, 1986). Most *B. fibrisolvens* strains can use ammonia as a sole nitrogen source and many strains can also use urea. Peptides in the form of trypticase or other peptones are not required, but often stimulate growth when added to ammonia-containing media. Similar stimulatory effects have been observed with *L. multiparus* (Bryant and Robinson, 1962). The precise vitamin and other

Table 2. Media for *Butyrivibrio, Lachnospira,* and *Roseburia.*[a]

Ingredient	Amount per 100 ml medium	
	Complex— RGM	Defined— *B. fibrisolvens*
Carbohydrate (5%)[b,c]	5.0 ml	5.0 ml
Trypticase	0.3 g	—
Yeast extract	0.2 g	—
Ammonium chloride (5.3%)	1.0 ml	1.0 ml
Vitamin solution[b,d]	—	1.0 ml
IVI VFA solution[e]	3.0 ml	3.0 ml
Mineral A[f]	4.0 ml	4.0 ml
Mineral B[g]	4.0 ml	4.0 ml
Trace minerals[h]	0.1 ml	0.1 ml
Hemin/naphthoquinone solution[i]	1.0 ml	1.0 ml
Sodium carbonate solution (8.0%)[b,j]	5.0 ml	5.0 ml
L-cysteine·HCl (5.0%)[b,c]	1.0 ml	1.0 ml
Resazurin (0.1%)	0.1 ml	0.1 ml
Distilled water	76.0 ml	75.0 ml

[a]Prepared anaerobically under carbon dioxide gas phase; final pH 6.8.

[b]Prepared separately and added to cooled, autoclaved medium.

[c]Prepared, autoclaved, and stored under a nitrogen gas phase.

[d]Containing per 100 ml *N*-2-hydroxyethylpiperazine-*N'*-2-ethanesulfonic acid buffer (5 mM; pH 7.5): biotin, folic acid, *p*-aminobenzoic acid, and cyanocobalamin (2.5 mg each); calcium pantothenate, nicotinamide, riboflavin, thiamine hydrochloride, and pyridoxamine (20 mg each). The resulting solution is filter sterilized (pore size, 0.22 µm) and stored under nitrogen in a dark container.

[e]Prepared by adding 7-ml acetate, 3-ml propionate, 2-ml butyrate, and 0.6-ml each of isobutyrate, 2-methyl-butyrate, isovalerate, and *n*-valerate to 700 ml of 0.2 M NaOH. Adjust to pH 7.0 with NaOH and to final volume of 1 l.

[f]Mineral A = 0.5% K_2HPO_4.

[g]Mineral B = 1.0% KH_2PO_4, 1.2% NaCl, 0.58% $NaSO_4$, 0.16% $Cacl_2·2H_2O$, and 0.25% $MgSO_4·7H_2O$.

[h]Prepared as described by Hespell and Canale-Parola (1970) by adding 25 ml of modified Pfennig's metals to 3.75 g $CaCl_2·2H_2O$, 12.5 g $MgCl_2·6H_2O$, 0.5 g $FeSO_4·7H_2O$, and 1.0 g Na_4EDTA dissolved in 75 ml distilled water. Modified Pfennig's metals are prepared by separately dissolving (heat if needed) each of the following compounds in distilled water and adding (in order) to 1 l distilled water: 0.5 g $AlCl_3$, 0.25 g KI, 0.25 g KBr, 0.25 g LiCl, 3.5 g $MnCl_2·4H_2O$, 5.5 g H_3BO_3, 0.5 g $ZnCl_2$, 0.5 g $CuCl_2·2H_2O$, 0.5 g $NiCl_2·6H_2O$, 0.5 g $CoCl_2·6H_2O$, 0.15 g $SnCl_2·2H_2O$, 0.15 g $BaCl_2·2H_2O$, 0.25 g $NaMoO_4·2H_2O$, and 0.05 g $NaVO_3$. The resultant solution is adjusted to pH 3.0 to 4.0 with HCl and brought to a final volume of 1.8 l with distilled water.

[i]Made by dissolving 25 mg 1,4-naphthoquinone in 2 ml 95% ethanol, adding 48 ml distilled water and 10 mg hemin dissolved in 50 ml of 0.01 M NaOH.

[j]Autoclaved separately; prepared, equilibrated, and stored under a carbon dioxide gas phase.

growth requirements for most strains of *Butyrivibrio, Lachnospira,* and *Roseburia* have not been determined.

Preservation of Cultures

Most strains can be maintained for long times by storage of cultures in liquid nitrogen or in ultracold freezers (Hespell and Canale-Parola, 1970). Preservation by lyophilization under anaerobic conditions may be possible, but this has not been examined and success will likely vary with strain. Short-term storage (6 to 15 months) is possible by placing glycerol-containing cultures in normal ($-20°C$) freezers (Teather, 1982a).

Identification

Phenotypic Properties

BUTYRIVIBRIO. On complex, carbohydrate agar-containing media, many *B. fibrisolvens* strains produce surface colonies that are 2 to 4 mm in diameter, entire, slightly convex, translucent, and light tan to white in color. Some strains produce rough colonies that have filamentous edges and might represent strains that produce little extracellular polysaccharide material. In contrast, strains such as the CF types (Group C-1, Table 1) produce large amounts of extracellular polysaccharide, the colonies are quite mucoid and gas bubbles emanate from the colonies. Subsurface colonies of most strains are lenticular to "Y" shaped, and double lens-shaped colony formations are not uncommon. In cellulose-containing media, the cellulose-digesting strains are surrounded by zones of cellulose degradation, variable both in zone size and extent of digestion (usually weak). Similar observations also occur when media containing natural xylans are used. However, if a dyed xylan (e.g., Remazol brilliant blue xylan) is used, the colonies of all strains are surrounded by large clearing zones. Most strains grow rapidly in broth culture and produce a flocculent sediment, whereas some strains produce a granular sediment that adheres to glass surfaces. Cultures of these latter strains and those that produce large amounts of extracellular polysaccharides can become quite thick and viscous when the media contain large amounts (0.5% or greater) of the energy source. Since growth of most strains is rapid (doubling times of 2 h or less) and much acid production occurs, cell characteristics and motility should be monitored with young cultures and in media with a low amount of an added energy source. Finally, it should be noted that variations in colony formations and in broth culture appearances have not yet been clearly correlated with the known genetic and polysaccharide groups (Table 2).

B. fibrisolvens appears classically as a small, motile, slightly curved rod (0.4 to 0.6 μm $\times$ 2 to 5 μm long) with tapered ends (Figs. 1A and B). However, there is considerable variation among strains. The cells of some strains are almost spindle-shaped and have bluntly pointed ends, while cells of other strains are quite curved and can form helices composed of two to four cells. The cells show translational motility characterized by rapid or intense vibrating movement. Motility is by means of montrichous flagellation, with the flagellum attached subterminally. Although often only a few cells in a wet mount preparation may show motility, truly nonmotile strains have not been well documented. The lack of motility in an isolated strain is often due to cultural conditions employed. Bryant and Small (1956a) reported that 15 strains were nonmotile when grown in a ruminal fluid-glucose-cellobiose medium, but all were motile if cellobiose was deleted and the glucose concentration decreased. These findings suggest that a low pH due to acid production may inhibit motility, particularly since low pH also inhibits growth (Gill and King, 1958).

The initial study of *Butyrivibrio* (Bryant and Small, 1956a) and many subsequent studies (Brown and Moore, 1960; Cheng et al., 1969; Shane et al., 1969; Sewell et al., 1988) have consistently reported that *Butyrivibrio* strains appear as Gram-negative cells when stained by conventional procedures and viewed by light microscopy. However, some *B. fibrisolvens* strains have significant Gram-positive characteristics. Lipoteichoic acids are present only in Gram-positive bacteria, and *B. fibrisolvens* strains 49, NOR37, and 1L631 were also found to possess lipoteichoic acids (Hewett et al., 1976). Later, Cheng and Costerton (1977) clearly established by electron microscopic observations that *B. fibrisolvens* strain D1 had a Gram-positive cell wall structure. However, the peptidoglycan layer was very thin (12 to 15 nm) compared to normal Gram-positive bacteria (30 to 50 nm). Thus, it was thought that the thinness of the peptidoglycan might be the basis for the negative Gram stain. In contrast to *B. fibrisolvens* D1, strains NOR37, 49, 835, and IM8/1B appear to have an outer envelope layering that is similar to, but not identical to, the outer membrane typical of the Gram-negative cell wall structure (Sharpe et. al., 1975; Dibbayawan et al., 1985). Strain 113 also has this type of cell wall structure (Figs. 1C and D; Sewell et al., 1988). A preliminary examination of the peptidoglycans from strains representative of the various *B. fibrisolvens* groups listed in Table 1 indicates that this peptidoglycan layer contains *meso*-diaminopimelic acid units are directly cross-linked as that found in many Gram-negative bacteria (R. Hespell, N. Weiss, and K. H. Schleifer, unpublished observations). Overall, it would appear that *B. fibrisolvens* strains possess a multilayered cell wall, which is atypical of the classical Gram-positive or Gram-negative wall structures. Regardless of the exact nature of the cell wall, all *B. fibrisolvens* strains are quite sensitive to penicillin or bacitracin and are resistant to nalidixic acid, properties more typical of Gram-positive bacteria.

The production of large amounts of *n*-butyric acid from the fermentation of carbohydrates is a major characteristic, along with cell morphology and Gram-staining which can be used to place a newly isolated strain in the genus *Butyrivibrio*. Smaller amounts of formate and hydrogen gas are made by all strains and many strains produce trace amounts of ethanol, but no strain produces succinate. Although some strains require carbon dioxide to initiate growth, all studied strains produce carbon dioxide. *B. fibrisolvens* strains generally exhibit two major fermentation acid patterns. One type of *B. fibrisolvens* produces large amounts of lactate and removes acetate added to the medium while the other type produces little or no lactate, but substantial amounts of acetate are made. These fermentation patterns have been used to separate butyrivibrio-like isolates (Shane et al., 1969). Hungate (1966) proposed another species, *B. alactacidigens,* for strains that do not produce lactate. At present, however, correlations between fermentation patterns and genetic groupings (Table 1) cannot be shown.

In contrast to many other species of ruminal bacteria, any given *B. fibrisolvens* strain is capable of fermenting a wide variety of carbohydrates. In addition all strains are xylanolytic and the majority of strains ferment starch and pectin. Some strains degrade cellulose weakly. While it was thought that degradation of rutin was a common trait in *B. fibrisolvens* strains (Hespell and Bryant 1981), this property is actually found in only a few strains. Almost all strains can ferment 15 to 20 different soluble carbohyrates and the most commonly used ones include glucose, maltose, sucrose, fructose, cellobiose, xylose, and arabinose. While it seems reasonable that *B. fibrisolvens* strains could be separated on the basis of specific carbohyrate usage into groups that would be consistent with genetic groups (Table 1), sufficient information is not yet available to create this determinative classification. At present, the best phenotypic trait that allows for placement of newly isolated *Butyrivibrio*-like strains into one of these genetic groups is neutral sugar composition of the extracellular polysaccharides (Table 1). These polysaccharides are easily isolated from culture fluids and can be analyzed by well-established techniques (Stack, 1988).

With respect to nitrogen requirements, ammonia can serve as the sole nitrogen source for most *B. fibrisolvens* strains. Many strains can also use mixtures of amino acids or more complex nitrogen sources (trypticase, casein hydrolysate, or peptone). However, ammonia supports considerably more growth than an equivalent amount of peptide or amino acid nitrogen when these compounds are provided at growth-limiting concentrations (Bryant, 1973). When measured by gelatin liquefaction, many strains appear to be nonproteolytic. This is probably a misleading conclusion, particularly if high levels (e.g., 10 to 12%) of gelatin are used in the medium for determination for activity. A survey of a large number of strains indicated they were proteolytic when this activity was measured by azocasein hydrolysis (Cotta and Hespell, 1986). In fact, it appears that constitutive production of protease as well as esterase/lipase (Hespell and O'Bryan-Shah, 1988) activities are very common traits in *B. fibrisolvens* strains.

In addition to *B. fibrisolvens,* the only other species presently in this genus is *Butyrivibrio crossatus.* This species can be isolated from human rectal or fecal material and differs considerably from *B. fibrisolvens* (Moore et al., 1976). Strains of this bacterium are motile by means of polar to subpolar lophotrichous flagella. Surface colonies on brain heart infusion agar are quite small (0.2- to 1.0-mm diameter), circular, convex, and translucent. Slightly larger subsurface colonies are produced and are lenticular in shape. The cell morphology is similar to *B. fibrisolvens* and cells stain Gram negative. *B. crossatus* has a narrow range of fermentable substrates that include maltose, starch, glycogen, and dextrin. Large amounts of butyrate are made and some hydrogen gas is formed by most strains. Tests for lecithinase, lipase, and ammonia production from peptone were negative. As pointed out by Moore and Holdeman (1974), *Eubacterium rectale* is part of the normal human fecal flora and many strains are very similar to butyrivibrios. The genetic relatedness of *B. crossatus* to *B. fibrisolvens* strains is not known at this time.

LACHNOSPIRA. All isolated strains are considered to constitute a single species: *Lachnospira multiparus* (Bryant, 1974). Some biochemically atypical strains have been described (Akkado and Blackburn, 1963;), but the exact relationship(s) of these strains to *Lachnospira* has not been firmly established. With nutrient-rich rumen fluid-carbohydrate media, surface colonies of *L. multiparus* are large (3 to 5 mm in diameter), flat, white, and filamentous. Colonies within agar media are quite distinctive, appearing as white woolly balls. The typical appearance of *Lachnospira* is a curved rod (0.4- to 0.6-μm $\times$ 2- to 4-μm long) with bluntly pointed ends. Cells generally stain weakly Gram-positive, but can be Gram-negative in older cultures. However, the ultrastructure of the cell wall appears to be that of a Gram-positive bacterium (Cheng et al., 1979). In wet mount preparations, the organisms appear singly or in pairs of motile, curved rods. Most cells show some translational motility characterized by frequent tumbling and movement in circular patterns. Although originally described as having polar, monotrichous flagellation (Bryant and Small, 1956a), it was later shown that the flagellum is subterminally or laterally attached (Bryant, 1974).

Many *L. multiparus* strains are capable of very rapid and extensive growth in carbohydrate-peptone-yeast extract media. Under these conditions, the cells often form very long chains or filaments. The cells in these structures are only slightly curved and have rounded ends. Quite often these structures become entwined with one another, causing growth in liquid media to appear as a flocculent material that readily settles. This floc may be quite difficult to disperse and may adhere to glass surfaces. Growth of some strains can result in viscous cultures, suggesting the production of a polysaccharide material.

While all strains do not grow on cellulose or xylans and some strains may show weak growth on starch, pectin and polygalacturonate are rapidly fermented by all *L. multiparus* strains. The range of soluble carbohyrates used is limited and includes glucose, fructose, sucrose, cellobiose, esculin, salicin, and usually glucuronate. Xylose is weakly or variably fermented. The main fermentation products made from glucose are lactate, formate, acetate, ethanol, carbon dioxide, and some hydrogen gas. Small amounts of acetoin or propionate may be formed, but no butyrate or succinate is made. Biochemical tests for nitrate reduction, catalase, indole, or hydrogen sulfide are negative.

ROSEBURIA. The information available on characteristics of *R. cecicola* is based primarily on the description by Stanton and Savage (1983a). Surface colonies of *R. cecicola* on agar media are circular with a smooth edge, white, and 1.5 to 3.0 mm in diameter. The colonies have a granular appearance and are mucoid in texture. Subsurface colonies appear lenticular in shape, brownish white, and are about 1.0 mm in diameter. *R. cecicola* cells are curved rods (0.5 μm $\times$ 2.5- to 5.0-μm long) with rounded ends. Cells are actively motile and in wet mount preparations a single, large fascicle of flagella can be seen to arise from the concave cell side. The fascicle is subterminally located and consists of 20 to 35 flagella. The cells stain Gram-negative and electron microscopic observations of thin sections indicate this bacterium has a multilayered cell wall structure.

R. cecicola does not ferment cellulose, xylan, or pectin, but does grow with starch as the energy source. Dextrin, amylopectin, and glycogen also support growth. Cellobiose, maltose, sucrose, raffinose, glucose, galactose, glucuronate, xylose, glycerol, and sorbitol are fermented. During fermentation of glucose, equimolar amounts of acetate are utilized and large amounts of butyrate and carbon dioxide are formed. Small amounts of ethanol are also made and hydrogen gas has been detected as a product. These fermentation products as well as the flagella arrangement clearly distinguish *R. cecicola* from other bacterial species in the genera *Selenomonas* and *Pectinatus.*

Phylogenetic Traits

The phylogenetic status of *Butyrivibrio, Lachnospira,* and *Roseburia* species is currently being developed. Partial 16S rRNA sequences have been determined for some strains and the results indicate these bacteria are of Gram-positive origin (Fig. 2; data and figure provided by D. Stahl). These bacteria belong to a large group designated as the Gram-positive cluster. This cluster includes the major ruminal cellulolytic species, *Ruminococcus flavefaciens* and *R. albus.* Interestingly, the ruminal acetogen, *Syntrophococcus sucromutans* (Krumholtz and Bryant, 1986) also appears to be in this cluster and is partially related to the group of *Butyrivibrio fibrisolvens* represented by strains D1 and A38. Although the 16S rRNA data indicate these two strains represent one grouping of *B. fibrisolvens* strains, the findings from DNA-DNA hybridizations indicate that these strains have very low homology (Mannarelli, 1988). However, the placement of *B. fibrisolvens* 49 in a group separate from strain D1 (Fig. 2) is consistent with the DNA-DNA hybridization data. The phenotypic traits of *B. fibrisolvens* and *R. cecicola,* as discussed previously, suggest these species might be related, which is borne out by the 16S rRNA data (Fig. 2). The possible relationship of *Lachnospira multiparus* to *B. fibrisolvens* is not expected given the major phenotypic properties of these species. With continued research on the biochemistry and genetics of these bacteria, a clear understanding of the phenotypic and genetic relationships between the various strains of these three genera should soon emerge. These data allow for development of a systematic classification of these rather interesting bacteria.

Biochemical and Physiological Traits

BUTYRIVIBRIO. B. fibrisolvens strains are invariably found in the rumen of animals fed a wide range of diets, but these bacteria usually are found in considerable numbers with animals fed diets high in forage materials that are generally poorly digested (Thorley et al., 1968; Shane et al., 1969). Although not as effective at fiber digestion as species of *Ruminococcus* or *Fibrobacter,* many *Butyrivibrio* strains are capable of solubilizing about 20 to 45% of the entire plant cell wall (Morris and Gylswyk, 1980). With respect to the plant cell fibers such as hemicelluloses or xylans, solubilization occurs at a faster rate than utilization of the breakdown products (Dehority, 1967). Initial studies with several *Butyrivibrio* strains indicated that the xylanase activity, which is of major importance in this solubilization, is located extracellularly, whereas xylosidase activity, which degrades soluble xylooligosaccharides, is cell bound (Hespell et al., 1987). This situation appears to be the case with all strains examined thus far. Pectin, starch, and proteins can be major components of plant cells and most *B. fibrisolvens* strains can degrade these materials, indicating that these bacteria can degrade most plant cell materials.

B. fibrisolvens strains characteristically produce butyric acid as the main product from fermentation of hexoses (Bryant and Small, 1956a) or pentoses (Hespell et al., 1987). However, the formation of the other fermentation acids varies considerably, depending upon strain. Shane et al. (1969) subdivided strains of *Butyrivibrio* into two groups. Group 1 strains utilize acetate and make small amounts of formate and large amounts of lactate. Group 2 strains produce acetate and make large amounts of formate and small amounts of lactate. Unfortunately, this grouping of strains based on fermentation product patterns does not correlate with genetic grouping Table 1 and may be more related to cultural conditions. Utilization of acetate is stimulated by a low availability of carbon dioxide, the presence of lactate, a and growth-lim-

itation of energy source (Latham and Legakis, 1976; Jarvis et al., 1978). In addition, acetate utilization is affected by acetate levels added to the medium as well as by the growth rate of the culture. The acetate utilized is incorporated primarily into butyrate on an almost equal-molar basis with acetate formed from the fermented carbohydrate source.

Biochemical and enzymatic studies with *B. fibrisolvens* have not been extensive. The key enzymes of the Emden-Meyerhof pathway plus lactate dehydrogenase, phosphotransacetylase, and acetate kinase have been detected at high levels (Joyner and Baldwin, 1966). Based on studies employing radioactively labelled substrates, pyruvate synthase and pyruvate-formate lyase activities are present in *B. fibrisolvens* (van Gylswyk, 1976). The formation of butyryl-CoA from acetyl-CoA involves the sequential actions of thiolase, β-hydroxybutyryl-CoA dehydrogenase, crotonase, and crotonyl-CoA reductase (Miller and Jensel, 1979). Butyryl-CoA is converted to butyrate by phosphotransbutyrylase and butyrate kinase. In regard to biosynthetic metabolism, oxaloacetate is formed by the action of phosphoenolpyruvate carboxykinase (van Gylswyk, 1979). Glutamate, succinate, malate, and isocitrate dehydrogenases appear to be present (Joyner and Baldwin, 1966).

The production of a number of extracellular enzymes is a common trait among *B. fibrisolvens* strains. Culture fluids have been found to contain xylanase and xylosidase (Hespell et al., 1987), protease (Cotta and Hespell, 1986), arabinofuranosidase (Hespell, unpublished observations), exopectate lyase and pectinesterase (Wojciechowicz et al., 1982), and lipase/esterase and acetyl xylan esterase activities (Hespell and O'Bryan-Shah, 1988). Cell-bound esterase activities are also present in many strains. With strain S2, these activities have been characterized as phospholipase and galactolipase (Hazlewood et al., 1983). Protease has been the most studied extracellular activity (Cotta and Hespell, 1986). Most *B. fibrisolvens* strains produce proteolytic activity equal to or greater than that found with other proteolytic ruminal bacteria such as *Ruminobacter amylophilus* or *Bacteroides ruminicola*. With *Butyrivibrio fibrisolvens* 49, the activity is constitutively made, but the level of activity is modulated by growth parameters. Energy substrates do not affect activity levels. The highest level of activity is produced with casamino acids but is reduced by 70 to 85% when trypticase or ammonium chloride is used as a sole nitrogen source. Based on the effects of inhibitors, the protease was determined to be of a serine protease type. The data from initial purification attempts indicate that the activity might be catalyzed by a low molecular weight protein (about 30 kDa or less) that associates with extracellular polysaccharides. *B. fibrisolvens* H17c is very closely related to strain 49 (Mannarelli, 1988) and was reported to possess 10 extracellular serine proteases (Strydom et al., 1986). It is unlikely that this organism has this many proteases. The occurrence of multiple protease bands on polyacrylamide gels might be due to different associations of protease and extracellular polysaccharide materials.

Genetic studies with *B. fibrisolvens* have been limited. The development of genetic exchange systems and cloning of genes from these strains has just begun. The first evidence for the presence of plasmids in *B. fibrisolvens* was reported by Teather (1982b), who found large plasmid bands by gel electrophoresis. Mann et al. (1986) were the first to characterize a butyrivibrio plasmid. This cryptic plasmid, pOM1, was isolated from strain 49 and was 2.8-kb pairs in size. The first gene to be cloned from a *B. fibrisolvens* strain was a xylosidase from strain 113 (Sewell et al., 1989). The gene was localized to a 3.8kb DNA fragment in pUC18 and in recombinant *Escherichia coli,* a new 60,000-kDa protein was observed. The recombinant and native xylosidases appeared similar and both enzymes hydrolyzed xylooligosaccharides, but not xylan.

LACHNOSPIRA. Biochemical and physiological studies with *L. multiparus* have mostly dealt with pectin degradation. This organism is capable of extensive maceration of clover and grass leaves, which can contain 2 to 8% pectin (Cheng et al., 1979). During pectin fermentation, methanol is formed as a product of pectin methyl esterase activity. In the rumen, this compound can be cross-fed to methanol-utilizing bacteria such as *Eubacterium limosum.* Evidence for this type of cross-feeding has been demonostrated with cocultures of these species (Rode et al., 1981). An extracellular polygalacturonate lyase has been partially purified from strain 685 (Wojciechowicz, 1980). The enzyme required calcium for activity and was most active toward polygalacturonate, producing mainly the unsaturated digalacturonate from random attack on this substrate. Culture fluids of this strain also were found to a contain an exopolygalacturonase. In contrast, more recent studies with *L. multiparus* D15d and 2389 found no evidence for this last type of activity (Silley, 1985). However, both pectin lyase and pectinesterase activities were detected with these two strains.

ROSEBURIA. Few studies have yet been done with *R. cecicola.* The flagella, isolated from this organism using mechanical shearing (Martin and Savage, 1985), are composed of a single protein, flagellin, with an estimated molecular weight of 42,000. The gene coding for this flagellin protein has been cloned and sequenced (Martin and Savage, 1988a). Based on the sequence data, the molecular weight was estimated to be 31,370 and comparisons with sequences of other bacterial flagellins indicated conserved regions. An interesting observation with *R. cecicola* is that high-molecular-weight DNA cannot be prepared from this organism if *R. roseburia* is exposed to oxygen prior to cell lysis (Martin and Savage, 1988b). Similar problems might be encountered in isolating DNA from other strictly anaerobic bacteria.

Applications

At present, no significant biotechnological applications have been made with strains of *Butyrivibrio, Lachnospira,* or *Roseburia.* However, the presence of polysaccharides with unusual sugar compositions (Stack, 1988) suggests these organisms may be of industrial importance. Similarly, the large array of extracellular polysaccharide-degrading enzymes made by *B. fibrisolvens* strains represents a rich source of these enzymes and of genes for cloning. Applications of *B. fibrisolvens* to biomass fermentors may also be possible, since this organism constituted a major portion of the xylanolytic isolates obtained from napiergrass-fed anaerobic digesters (Sewell et al., 1988).

Acknowledgments

I thank Donna Williams for preparation of the electron micrographs used in Fig. 1 and Michael A. Cotta for his helpful discussions and comments.

Literature Cited

Akkada, A. R. A., and Blackburn, T. H. 1963. Some observations on the nitrogen metabolism of rumen proteolytic bacteria. J. Gen. Microbiol. 31:461–469.

Brown, D. W., and Moore, W. E. C. 1960. Distribution of *Butyrivibrio fibrisolvens* in nature. J. Dairy Sci. 43:1570–1574.

Bryant, M. P. 1973. Nutritional requirements of the predominant rumen cellulolytic bacteria. Fed. Proceed. 32:1809–1813.

Bryant, M. P. 1974. Genus XIII. *Lachnospira* Bryant and Small 1956, 24, p.661–662. In: N. R. Krieg (ed.), Bergey's manual of systematic bacteriology, vol. 1. Williams and Wilkins, Baltimore.

Bryant, M. P., Barrentine, B. F., Sykes, J. F., Robinson, I. M., Shawver, C. V., and Williams, L. W. 1960. Predominant bacteria in the rumen of cattle on bloat-provoking Ladino clover pasture. J. Dairy Sci. 43:1435–1444.

Bryant, M. P., and Robinson, I. M. 1962. Some nutritional characteristics of predominant culturable ruminal bacteria. J. Bacteriol. 84:605–614.

Bryant, M. P., and Small, N. 1956a. The anaerobic montrichous butyric acid-producing curved rod-shaped bacteria of the rumen. J. Bacteriol. 72:16–21.

Bryant, M. P., and Small, N. 1956b. Characteristics of two new genera of anaerobic curved rods isolated from the rumen of cattle. J. Bacteriol. 72:22–26.

Cheng, K.-J., and Costerton, J. W. 1977. Ultrastructure of Butyrivibrio fibrisolvens: A Gram-positive bacterium? J. Bacteriol. 129:1506–1512.

Cheng, K.-J., Dinsdale, D., and Stewart, C. S. 1979. Maceration of clover and grass leaves by Lachnospira multiparus. Appl. Environ. Microbiol. 38:723–729.

Cheng, K.-J., Jones, G. A., Simpson, F. J., and Bryant, M. P. 1969. Isolation and identification of rumen bacteria capable of anaerobic rutin degradation. Can. J. Microbiol. 15:1365–1371.

Cheng, K.-J., Krishnamurty, H. G., Jones, G. A., and Simpson, F. J. 1970. Identification of products produced by the anaerobic degradation of naringin by Butyrivibrio sp. C₃. Can. J. Microbiol. 16:129–131.

Cotta, M. A., and Hespell, R. B. 1986. Proteolytic activity of the ruminal bacterium Butyrivibrio fibrisolvens. Appl. Environ. Microbiol. 52:51–58.

Dehority, B. A. 1966. Characterization of several bovine rumen bacteria isolated with a xylan medium. J. Bacteriol. 91:1724–1729.

Dehority, B. A. 1967. Rate of hemicellulose degradation and utilization by pure cultures of rumen bacteria. Appl. Microbiol. 15:987–993.

Dehority, B. A. 1969. Pectin-fermenting bacteria isolated from the bovine rumen. J. Bacteriol. 99:189–196.

Dibbayawan, T., Cox, G., Cho, K. Y., and Dwarte, D. M. 1985. Cell wall and plasma membrane architecture of Butyrivibrio spp. J. Ultrastruct. Res. 90:286–293.

Fulghum, R. S., Baldwin, B. B., and Williams, P. P. 1968. Antibiotic susceptibility of anaerobic ruminal bacteria. Appl. Microbiol. 16:301–307.

Gill, J. W., and King, K. W. 1958. Nutritional characteristics of a Butyrivibrio. J. Bacteriol. 75:666–673.

Gutierrez, J., Davis, R. E., and Lindahl, I. L. 1959. Characteristics of saponin-utilizing bacteria from the rumen of cattle. Appl. Microbiol. 7:304–308.

Hazlewood, G. P., Theodorou, M. K., Hutchings, A., Jordon, D. J., and Galfre, G. 1986. Preparation and characterization of monoclonal antibodies to a Butyrivibrio sp. and their potential use in the identification of rumen Butyrivibrios using an enzyme-linked immunosorbent assay. J. Gen. Microbiol. 132:43–52.

Hazlewood, G. P., Cho, K. Y., Dawson, R. M. C., and Munn, E. A. 1983. Subcellular fractionation of the gram negative rumen bacterium, Butyrivibrio S2, by protoplast formation, and localisation of lipolytic enzymes in the plasma membrane. J. Appl. Bacteriol. 55:337–347.

Hespell, R. B., and Bryant, M. P. 1981. The Genera Butyrivibrio, Succinivibrio, Succinimonas, Lachnospira, and Selenomonas, p.1479–1494. In: M. P. Stan, H. Stolp, H. G. Truper, A. Balows, and H. G. Schlegel (ed.), The prokaryotes. Springer-Verlag, Berlin.

Hespell, R. B., Wolf, R., and Bothast, R. J. 1987. Fermentation of xylans by Butyrivibrio fibrisolvens and other ruminal bacteria. Appl. Environ. Microbiol. 53:2849–2853.

Hespell, R. B., and O'Bryan-Shah, P. J. 1988. Esterase activities in Butyrivibrio fibrisolvens strains. Appl. Environ. Microbiol. 54:1917–1922.

Hespell, R. B., and Canale-Parola, E. 1970. Spirochaeta litoralis sp. n., a strictly anaerobic marine spirochete. Arch. Mikrobiol. 74:1–18.

Hewett, M. J., Wicken, A. J., Knox, K. W., and Sharpe, M. E. 1976. Isolation of lipoteichoic acids from Butyrivibrio fibrisolvens. J. Gen. Microbiol. 94:126–130.

Hungate, R. E. 1966. The rumen and its microbes. New York: Academic Press.

Jarvis, B. D. W., Henderson, C., and Asmundson, R. V. 1978. The role of carbonate in the metabolism of glucose by Butyrivibrio fibrisolvens. J. Gen. Microbiol. 105:287–295.

Joyner, A. E., and Baldwin, R. L. 1966. Enzymatic studies of pure cultures of rumen microorganisms. J. Bacteriol. 92:1321–1330.

Krumholz, L. R., and Bryant, M. P. 1986. Syntrophococcus sucromutans sp. nov., gen. nov., uses carbohydrates as electron donors and formate, methoxybenzenoids or Methanobrevibacter as the electron acceptor system. Arch. Microbiol. 143:313–318.

Latham, M. J., and Legakis, N. J. 1976. Cultural factors influencing the utilization or production of acetate by Butyrivibrio fibrisolvens. J. Gen. Microbiol. 94:380–388.

Lewis, S. M., and Dehority, B. A. 1985. Microbiology and ration digestibility in the hindgut of the ovine. Appl. Environ. Microbiol. 50:356–363.

Mann, S. P., Hazlewood, G. P., and Orpin, C. G. 1986. Characterization of a cryptic plasmid (pOM1) in Butyrivibrio fibrisolvens by restriction endonuclease analysis and its cloning in Escherichia coli. Curr. Microbiol. 13:17–22.

Mannarelli, B. M. 1988. Deoxyribonucleic acid relatedness among strains of the species Butyrivibrio fibrisolvens. Intern. J. Syst. Bacteriol. 38:340–347.

Margherita, S. S., and Hungate, R. E. 1963. Serological analysis of Butyrivibrio from the bovine rumen. J. Bacteriol. 86:855–860.

Margherita, S. S., Hungate, R. E., and Storz, H. 1964. Variation in rumen Butyrivibrio strains. J. Bacteriol. 87:1304–1308.

Martin, J., and Savage, D. C. 1984. Habitat of a motile, obligatorily anaerobic bacterium indigenous to the murine gastrointestinal tract. Microecology and Therapy 14:293–294.

Martin, J. H., and Savage, D. C. 1985. Purification and characterization of flagella from Roseburia cecicola, an obligately anaerobic bacterium. J. Gen. Microbiol. 131:2075–2078.

Martin, J. H., and Savage, D. C. 1988a. Cloning, nucleotide sequence, and taxonomic implications of the flagellin gene of Roseburia cecicola. J. Bacteriol. 170:2612–2617.

Martin, J. H., and Savage, D. C. 1988b. Degradation of DNA in cells and extracts of the obligately anaerobic bacterium *Roseburia cecicola* upon exposure to air. Appl. Environ. Microbiol. 54:1619–1621.

Miller, T. L., and Jenesel, S. E. 1979. Enzymology of butyrate formation by *Butyrivibrio fibrisolvens.* J. Bacteriol. 138:99–104.

Moore, W. E. C., and Holdeman, L. V. 1974. Human fecal flora: The normal flora of 20 Japanese-Hawaiians. Appl. Microbiol. 27:961–979.

Moore, W. E. C., Johnson, J. L. and Holdeman, L. V. 1976. Emendation of *Bacteroidaceae* and *Butyrivibrio* and descriptions of *Desulfomonas* gen. nov. and ten new species in the genera *Desulfomonas, Butyrivibrio, Eubacterium, Clostridium,* and *Ruminococcus.* Intern. J. System. Bacteriol. 26:238–252.

Morris, E. J., and van Gylswyk, N. O. 1980. Comparison of the action of rumen bacteria on cell walls from *Eragrostis tef.* J. Agric. Sci., Camb. 95:313–323.

Orpin, C. G., Mathiesen, S. D., Greenwood, Y., and Blix, A. S. 1985. Seasonal changes in the ruminal microflora of the high-Arctic Svalbard reindeer (*Rangifer tarandus platyrhynchus*). Appl. Environ. Microbiol. 50:144–151.

Rode, L. M., Genthner, B. R. S., and Bryant, M. P. 1981. Syntrophic association by cultures of the methanol- and CO₂-H₂-utilizing species *Eubacterium limosum* and pectin-fermenting *Lachnospira multiparus* during growth in a pectin medium. Appl. Environ. Microbiol. 42:20–22.

Sewell, G. W., Aldrich, H. C., Williams, D., Mannarelli, B., Wilkie, A., Hespell, R. B., Smith, P. H., and Ingram, L. O. 1988. Isolation and characterization of xylan-degrading strains of *Butyrivibrio fibrisolvens* from a Napier grass-fed anaerobic digestor. Appl. Environ. Microbiol. 54:1085–1090.

Sewell, G. W., Utt, E. A., Hespell, R. B., Mackenzie, K. F., and Ingram, L. O. 1989. Identification of the *Butyrivibrio fibrisolvens* xylosidase gene (xylB) coding region and its expression in *Escherichia coli.* Appl. Environ. Microbiol. 55:306–311.

Shane, B. S., Gouws, L., and Kistner, A. 1969. Cellulolytic bacteria occurring in the rumen of sheep conditioned to low-protein teff hay. J. Gen. Microbiol. 55:445–457.

Sharpe, M. E., Brock, J. H., and Phillips, B. A. 1975. Glycerol teichoic acid as an antigenic determinant in a gram-negative bacterium *Butyrivibrio fibrisolvens.* J. Gen. Microbiol. 88:355–363.

Silley, P. 1985. A note on the pectinolytic enzymes of *Lachnospira multiparus.* J. Appl. Bacteriol. 58:145–149.

Stack, R. J. 1988. Neutral sugar composition of extracellular polysaccharides produced by strains of *Butyrivibrio fibrisolvens.* Appl. Environ. Microbiol. 54:878–883.

Stack, R. J., Stein, T. M., and Plattner, R. D. 1988. 4-0-(1-carboxyethyl)-D-galactose: A new acidic sugar from the extracellular polysaccharide produced by *Butyrivibrio fibrisolvens* strain 49. Biochem. J. 256:769–773.

Stack, R. J., and Weisleder, D. 1989. 4-0-(1-carboxyethyl)-L-rhamnose: A second unique acidic sugar found in an extracellular polysaccharide from *Butyrivibrio fibrisolvens* strain 49. (Submitted to Biochemical Journal).

Stanton, T. B., and Savage, D. C. 1983a. *Roseburia cecicola* gen. nov., sp. nov., a motile, obligately anaerobic bacterium from a mouse cecum. Intern. J. Syst. Bacteriol. 33:618–627.

Stanton, T. B., and Savage, D. C. 1983b. Colonization of gnotobiotic mice by *Roseburia cecicola,* a motile, obligately anaerobic bacterium from murine ceca. Appl. Environ. Microbiol. 45:1677–1684.

Strydom, E., Mackie, R. I, and Woods, D. R. 1986. Detection and characterization of extracellular proteases in *Butyrivibrio fibrisolvens* H17c. Appl. Microbiol. Biotechnol. 24:214–217.

Teather, R. M. 1982a. Maintenance of laboratory strains of obligately anaerobic rumen bacteria. Appl. Environ. Microbiol. 44:499–501.

Teather, R. M. 1982b. Isolation of plasmid DNA from *Butyrivibrio fibrisolvens.* Appl. Environ. Microbiol. 43:298–302.

Thorley, C. M., Sharpe, M. E., and Bryant, M. P. 1968. Modification of the rumen bacterial flora by feeding cattle ground and pelleted roughage as determined with culture media with and without rumen fluid. J. Dairy Science 51:1811–1816.

van Gylswyk, N. O. 1976. Some aspects of the metabolism of *Butyrivibrio fibrisolvens.* J. Gen. Microbiol. 97:105–111.

van Gylswyk, N. O. 1979. Oxaloacetate synthesis in *Butyrivibrio fibrisolvens.* Appl. Environ. Microbiol. 37:1245–1247.

Westlake, K., Mackie, R. I., and Dutton, M. F. 1987. T-2 toxin metabolism by ruminal bacteria and its effect on their growth. Appl. Environ. Microbiol. 53:587–592.

Wojciechowicz, M., Heinrichova, K., and Ziolecki, A. 1980. A polygalacturonate lyase produced by *Lachnospira multiparus* isolated from the bovine rumen. J. Gen. Microbiol. 117:193–199.

Wojciechowicz, M., Heinrichova, K., and Ziolecki, A. 1982. An exopectate lyase of *Butyrivibrio fibrisolvens* from the bovine rumen. J. Gen. Microbiol. 128:2661–2665.

CHAPTER 95

The Genus *Veillonella*

PAUL E. KOLENBRANDER and LILLIAN V. H. MOORE

Introduction

Bacteria of the genus *Veillonella* Prévot are anaerobic, Gram-negative cocci (Rogosa, 1984). The type species was originally described by Veillon and Zuber (1898) as *Staphylococcus parvulus* and renamed by Prévot (1933) as *Veillonella parvula*. Although all strains of *Veillonella* are phenotypically very similar, seven species are recognized by DNA homology analysis (Mays et al., 1982). They are found in the alimentary canal of humans, rodents, and ruminants and can constitute a major fraction of the total bacterial population on the epithelial surfaces of the human oral cavity.

Veillonellae are characterized by their unusual metabolism. They generally are unable to ferment carbohydrates including glucose but grow well anaerobically on lactate, pyruvate, malate, or fumarate. They posses methylmalonyl-CoA decarboxylase, which belongs to a unique class of vectorial catalysts that convert the free energy of decarboxylation reactions into an electrochemical gradient of sodium ions (Dimroth, 1985). The enzyme catalyzes an essential reaction in the fermentation of lactate and provides an energy conservation mechanism, which contributes to the high cell yields per mole of lactate fermented.

Human oral veillonellae live in a flowing or lotic environment and have developed mechanisms to colonize the exposed surfaces. Veillonellae themselves are only weakly adherent to hard and soft tissue surfaces, but they can adhere to other genera of oral bacteria (Gibbons and Nygaard, 1970; Hughes et al., 1988) and thus become part of interbacterial networks (intergeneric coaggregations; Kolenbrander, 1988; Kolenbrander, 1989). Many of these coaggregation partners are early colonizers that are capable of primary attachment to both soft and hard tissue surfaces. Intergeneric coaggregation is very specific in that veillonellae from one econiche adhere to other bacteria found in that econiche, e.g., tongue dorsum, but usually do not bind to bacteria from other econiches, e.g.,

subgingival plaque (Hughes et al., 1988). Thus, interactions among oral bacteria including veillonellae appear to be important determinants in bacterial colonization of different surfaces.

Because of their unusual metabolic capabilities, the veillonellae form an essential link in a natural food chain, since they use the lactic acid produced by other oral bacteria from carbohydrate fermentation. Their rise in numbers in the oral ecosystem immediately follows or parallels the proliferation of lactic acid-producing streptococci and actinomyces, both primary colonizers of the tooth surface and mucosal epithelium and both coaggregation partners of veillonellae. These two properties, intergeneric coaggregation and metabolic communication, may contribute significantly to the high numbers of veillonellae found at all times in dental plaque, whether obtained from healty sites or diseased sites, which otherwise have very different bacterial populations (Moore et al., 1985; Moore et al., 1987; Dzink et al., 1988).

Habitats

Normal Flora

Veillonellae are the most numerous anaerobes in human saliva (Sutter, 1984), where they are found at concentrations of 1.7 to 6.9×10^7/ml (Richardson and Jones, 1958; Rogosa, et al., 1958), and constitute from 5% (Liljemark and Gibbons, 1971) to about 16% (Hardie and Bowden, 1974) of the cultivable anaerobic flora. Almost 10% of the total cultivable anaerobic flora from the tongue dorsum consists of veillonellae (Liljemark and Gibbons, 1971). In supragingival plaque, estimates of the percent veillonellae of the total bacteria have ranged from about 1% (Liljemark and Gibbons, 1971) to 5% (Moore et al., 1987; Moore et al., 1985). Direct comparison of the subgingival and supragingival plaque bacteria isolated from healthy sites indicates that about 4% and 5%, respectively, are veillonellae (Moore et al., 1987; Moore et al.,

1985). While the veillonellae found on the tongue dorsum are either strains of *V. atypica* or *V. dispar* (Hughes et al., 1988), the ones found in subgingival plaque samples are primarily *V. parvula* (Moore et al., 1987; Moore et al., 1985). The veillonellae isolated from saliva and buccal mucosa are mostly *V. atypica* and *V. dispar* (Hughes et al., 1988). Of the seven species recognized (Rogosa, 1984), the above three are indigenous to the respiratory and intestinal tracts of humans, and the other species are distributed among ruminants, rodents, and pigs.

Human Infection

Veillonellae are much less frequently isolated from infection sites than the anaerobic streptococci (*Streptococcus, Peptococcus, Peptostreptococcus*). They have been isolated as the only bacterial type from patients with endocarditis (Loewe et al., 1946; Greaves and Kaiser, 1984), hepatic abscesses (Lambe et al., 1974), and pleuropulmonary infections (Bartlett and Finegold, 1972). They were present in 14 of 62 specimens of chronic sinusitis (Heineman and Braude, 1963) and were the only genus isolated in some cases (Frederick and Braude, 1974). They have also been found in bitewound infections (Goldstein et al., 1984), gynecological infections (Chow et al., 1975), intraabdominal abscesses (Moore et al., 1969), pelvic abscesses (Williams, 1977), septicemia and osteomyelitis (Borchardt et al., 1977; Barnhart et al., 1983), pleuropulmonary infections (Bartlett et al., 1974; Martin, 1974), and periodontal disease (Moore, 1987). When isolated from blood cultures, they usually are mixed with other species and probably occur mixed with other species in most isolations from most sources.

Thirteen strains of *Veillonella* were among the 601 clinical isolates of anaerobic bacteria tested for minimal inhibitory concentrations of 10 antimicrobial agents (Martin et al., 1972). They were like the other anaerobes in their susceptibility to the various antimicrobials, and many were resistant to tetracycline at concentrations greater than 10 µg/ml. In periodontal patients undergoing tetracycline therapy, *Veillonella* species, along with *Streptococcus* and *Neisseria,* were found consistently resistant to tetracycline, with minimal inhibitory concentrations as high as 128 ug/ml (Williams et al., 1979). The DNA from these species, as well as from tetracycline-resistant *Fusobacterium nucleatum* and *Peptostreptococcus anaerobius*, hybridized with a TetM probe, prepared from one of the streptococcal tetracycline resistance determinants (Roberts and Moncla, 1988). These results suggest that this determinant can exist and confer tetracycline resistance in either aerobic or anaerobic Gram-negative and Gram-positive bacteria.

As compared to the normal flora of the oral cavity, veillonellae in patients with severe xerostomia due to Sjögren's syndrome are greatly reduced in number, whereas the numbers of *Candida* spp. and *Staphylococcus aureus* are increased (MacFarlane, 1984). Veillonellae have been isolated from root canal specimens and may be important in endodontic infections (Burnett and Schuster, 1978). *V. dispar* was found in much higher numbers and greater frequency than *V. parvula* in subgingival plaque obtained from patients with non-insulin-dependent diabetes mellitus (Zambon et al., 1988). In a study of dental decay of children's molars shortly after eruption, *S. mutans* and lactobacilli were associated, whereas veillonellae, although consistently present, could not be linked to the development of decay (Loesche et al., 1984). Veillonellae were found in nearly all samples from older adults with a high root-surface-caries risk, but no correlation with the disease was noted (Ellen et al., 1985). Thus, the occurrence of veillonellae in human infection may be a consequence of ecological changes in the infection site rather than a direct pathogenic property of these bacteria.

Experimental Animal Models

Veillonellae are among the predominant microbes of the saliva, tongue dorsum, buccal mucosa, and gingival crevice of both the plaque-susceptible and plaque-resistant rats that were derived from the Wistar Kyoto strain (Isogai et al., 1985). On the basis of total anaerobic cultivable flora, veillonellae were the most numerous bacteria from subgingival plaque of both STR/N periodontitis-susceptible mice and Swiss-Webster non-periodontitis-susceptible mice (Wolff et al., 1985). Their numbers remained constant from 3 to 13 months, the time interval during which other microbial changes occurred along with the advancing periodontitis in the STR/N mice. Thus, the number and distribution of veillonellae in these experimental animals are very similar to those observed in the human oral cavity.

Ecological Succession in the Human Oral Cavity

Before and After Tooth Eruption

Shortly after birth, veillonellae appear in the human mouth. They are infrequently detectable in newborns one to eight days old, but are found in 75% of infants that were 101 days old and in 100% of one-year-old children (McCarthy et al., 1965). The increase in numbers of veillonellae occurs following the colonization of the predominant oral bacterial species, *Streptococcus salivarius*. Veillonellae are not known to adhere well to buccal epithelial cells (Liljemark and Gibbons, 1971), whereas *S. salivarius* adheres very well (Gibbons and van Houte, 1971). Some veillonellae coaggregate with *S. salivarius*, which may provide a mechanism for veillonellae to adhere in the oral cavity (Hughes et al., 1988).

From Healthy Gingiva to Periodontal Disease

V. parvula is the predominant veillonella in human dental plaque and constitutes from 93 to 98% of the cultivable veillonellae in healthy subgingival sites (Moore, et al., 1985), and its numbers increase in gingivitis as compared with health (Moore et al., 1987b; Slots et al., 1978). Although *V. atypica* and *V. dispar* are present consistently in plaque, they normally occur in low numbers and are not associated with any oral diseases.

Two early studies were conducted to examine shifts in the bacterial population of cleaned hard surfaces in the adult human oral cavity. In the first investigation (Slack and Bowden, 1965), an artificial device was positioned to simulate an interproximal site and, in the other study (Ritz, 1967), a professionally cleaned tooth surface was examined. Both studies indicated that veillonellae were detectable along with streptococci after only 24 hours, and that after 3 days, fusobacteria, actinomyces, and a few other bacterial types also were present. Even when the plaque composition on a cleaned tooth surface was monitored at minute or hourly intervals for the first day, veillonellae were rarely found in samples during the first 24 hours (Socransky et al., 1977; Theilade et al. 1982) but constituted about 20% of the strains isolated from 1- and 3-day-old dental plaque (Theilade et al. 1982).

Other surveys indicate that veillonellae remain a significant part of the population in both healthy and diseased sites (Syed and Loesche, 1978; Loesche and Syed, 1978; Dzink et al., 1985) and are among the 10 most numerous species of subgingival plaque bacteria under all conditions (Moore et al., 1985, 1987b; Dzink et al. 1988), including experimental gingivitis (Moore et al., 1982a), localized juvenile periodontitis (Moore et al., 1985; Williams et al., 1985), adult chronic moderate periodontitis

(Moore et al., 1983), and rapidly progressing generalized periodontitis (Moore et al., 1982b). In all of these extensive investigations of the microbial composition of dental plaque, veillonellae were always found as significant members of a consortium of bacterial species. Cleaned tooth surfaces are repopulated by streptococci and actinomyces, and both of these are coaggregation partners of *V. parvula*, the predominant veillonella in plaque (Hughes et al., 1988). Populations in periodontally diseased sites are dominated by fusobacteria, which are coaggregation partners of all three human veillonella species (Kolenbrander et al., 1989).

Metabolic Communication

Metabolism

The unusual physiology and special energy-conservation mechanism of veillonellae have been reviewed (Delwiche et al., 1985; Dimroth, 1985). The veillonellae are unable to use glucose to support growth, and they do not have a functional hexokinase (Rogosa et al., 1965). Radiolabeled glucose is not incorporated into bacterial cell compounds (Winter and Delwiche, 1975), suggesting that the glucose phosphotransferase system is also absent. Ribose and fructose can be incorporated into nucleic acid (Kafkewitz and Delwiche, 1972) or lipopolysaccharide (Winter and Delwiche, 1975; Tortorello and Delwiche, 1983) but are not fermented. Veillonellae can utilize lactate as carbon and energy source for growth. Pyruvate, fumarate, malate, α-ketoglutarate, and some purines also are substrates for growth. Propionic acid is the major end product along with acetic acid, carbon dioxide, and hydrogen. Propionic acid but not acetic acid is toxic to cultured human gingival fibroblasts and may play a role in host tissue damage (Singer and Buckner, 1981).

Fermentation of lactate proceeds through four-carbon dicarboxylic acid intermediates and involves a novel energy conversion mechanism of biotin-dependent, sodium-transport, methylmalonyl-CoA decarboxylase (Hilpert and Dimroth, 1983). The decarboxylase is membrane bound, is specifically activated by sodium ions, and converts part of the energy of the highly exergonic decarboxylation reaction into a sodium ion gradient. The sodium gradient may be used to drive the active transport of the growth substrate. This methylmalonyl-CoA decarboxylase, which is found in both veillonellae and *Selenomonas ruminantium* (Melville et al. 1988), the oxaloacetate decarboxylase from *Klebsiella aerogenes* (Dimroth, 1981), and the glutaconyl-CoA decarboxylase from *Acidaminococcus fermentans* (Buckel and Semmler, 1982) have been found only in anaerobic bacteria, where they are induced by their respective fermentation pathway substrates (e.g., lactate with the veillonellae). The physiological significance of these enzymes is twofold: 1) they are essential for growth on appropriate substrates, and 2) they offer a mechanism for energy conservation and thus provide additional energy to that obtained from the fermentation itself.

Gluconeogenesis occurs by a reversed glycolytic sequence to glucose-6-phosphate from lactate (Rogosa et al., 1965), or by an alternate pathway through malate and glyoxylate to 3-phosphoglycerate (Pestka and Delwiche, 1983), rather than going directly from lactate through 2-phosphoglycerate to 3-phosphoglycerate. Lactate carbon is incorporated into six-carbon sugars (Ng and Hamilton, 1974).

The peptidoglycan of some veillonellae contains putrescine or cadaverine, which are required for normal growth. These organic bases are linked to the α-carboxyl group of the D-glutamic acid residue, and, during normal growth, they occupy over 40% of the residues (Kamio and Nakamura, 1987). Other normal constituents of the peptidoglycan are N-acetylglucosamine, N-acetylmuramic acid, L-alanine, D-glutamic acid, *meso*-diaminopimelic acid, and D-alanine. The lipopolysaccharide composition includes glucosamine, tridecanoic acid, and 3-hydroxytridecanoic acid in the lipid moiety and contains 2-keto-3-deoxyoctonic acid, glyceromannoheptose, and variable components including galactose, rhamnose, glucosamine, and glucose in the polysaccharide moiety (Hofstad and Kristoffersen, 1970; Hewett et al., 1971). The lipopolysaccharide has been shown to be an endotoxin (Mergenhagen et al., 1961).

Purines may be fermented with the production of acetate, propionate, ammonia, urea, CO_2, and H_2 (Whiteley and Douglas, 1951). Large amounts of gas may be formed. Nitrate is reduced through nitrite and hydroxylamine to ammonia, which is assimilated (Inderlied and Delwiche, 1973; Ruoff and Delwiche, 1977; Yordy and Delwiche, 1979).

Communication

Digestion of proteins by protease-secreting bacterial inhabitants of anaerobic consortia would provide amino acids such as L-serine, which is known to enhance lactate metabolism of veillonellae (Hoshino, 1987). Menaquinones, a potential source of vitamin K for other bacteria, are present in some veillonellae (Ramotar et al., 1984). Mixtures of *V. parvula* and *Haemophilus parainfluenzae*, both isolated from the gastric juice of achlorhydric patients, accumulated nitrite during nitrate reduction, and the nitrites could be decreased by including the nitrite-reducing species *Neisseria subflava* or *Streptococcus sanguis* (Forsythe and Cole, 1987). Co-culturing amylolytic rumen bacteria and lactate-utilizing bacteria (veillonellae) on starch usually produced higher growth yields (Marounek and Bartos, 1987). Growth of *Veillonella* with the lactate-producing strains of *S. mutans* (Distler and Kroncke, 1980), with lactate and succinate-producing *Actinomyces* species (Distler and Kroncke, 1981), or with *Eubacterium saburreum* (Mashimo et al., 1981) has been demonstrated.

A consortium of nine oral bacteria in a glucose-limited chemostat was monitored with respect to changes in the composition of the bacterial population accompanying pH changes (McDermid et al., 1986). At pH 7.0, veillonellae were most numerous along with streptococci, and, at pH 4.1, they were numerically second only to *L. casei*. They regained their same predominance with the streptococci when the pH was returned to 7.0. These experiments show that veillonellae are able to effectively utilize lactic acid produced by a variety of oral bacteria including streptococci and lactobacilli (Mikx and van der Hoeven, 1975; Delwiche et al., 1985) in either acidic or neutral environments.

That similar metabolic communications may occur in vivo is suggested by the observation that incipient carious lesions in the teeth of children have an increased number of *S. mutans* and *Lactobacillus* spp. along with increased numbers of *Veillonella* spp. (Boyar and Bowden, 1985). Also, veillonellae are associated with the dense infections of *S. mutans* on susceptible surfaces related to developing lesions of nursing caries in young children (Milnes and Bowden, 1985). While the veillonellae may protect the tooth surface by removal of the stronger lactic acid and replacement with its own metabolic end products, the weaker acetic and propionic acids, veillonellae may also function in a protective role if their numbers can increase quickly enough in response to small amounts of lactic acid. In accord with this hypothesis, the predominant supragingival flora of caries-free Tanzanian children consisted of unusually high proportions of *Veillonella* sp. (30% to 40% of the total cultivable microflora) along with the oral streptococci (Kilian et al., 1979). Plaque from caries-free adults also contain higher levels of *Veillonella* species than does plaque from caries-susceptible adults (Minah et al., 1981).

With gnotobiotic animals, it was shown that *S. sanguis* or *S. mutans* was more cariogenic when an animal was mono-infected with either species than when the streptococci were associated with veillonellae (Mikx et al., 1972). This caries reduction was interpreted as the result of the conversion of streptococcal-produced lactic acid to the weaker acids, acetic and propionic acids, by the veillonellae. A reduction in lactic acid and an increase in acetic acid was shown to occur in dental plaque of gnotobiotic rats hosting an experimental symbiosis of veillonellae and cariogenic streptococci (van der Hoeven et al., 1978). Compared to the high numbers of *S. mutans* found in nursing caries of children, the number of *S. mutans* required to cause caries in gnotobiotic animals is quite low. The reduction in caries by association with veillonellae may occur through the rapid growth response by veillonellae to small amounts of lactic acid, while the dense streptococcal population in nursing caries maintains a high lactic acid concentration, and little significant reduction in caries occurs even in the presence of veillonellae.

Adherence

Nichrome Steel Wire Model

The first observation that veillonellae could participate in a special adherence arrangement with other oral bacteria involved the relationship between *V. parvula* V5 (previously called *V. alcalescens*) and *Actinomyces viscosus* T6 (*"Odontomyces viscosus"*) on nichrome steel wires that were suspended in broth cultures (Bladen et al., 1970). The veillonellae were unable to adhere to the solid supports, whereas the actinomyces growing in sucrose-containing medium formed primary plaque on the wires. Transfer of the primary plaque-containing wires to lactate-based medium inoculated with *V. parvula* resulted in a large increase in the amount of deposited plaque due to the veillonellae, which could grow in the lactate medium and which were apparently able to attach to the initial colonizers.

The requirement for sucrose during primary plaque formation suggested that glucosyltransferase, an extracellular enzyme that catalyzes the incorporation of the glucose moiety from sucrose into a water-insoluble polysaccharide polymer (Robrish et al., 1972), may be involved and that direct cell-to-cell contact may not be necessary. The enzyme is elaborated by many oral streptococci including *S. salivarius,* whose extracellular glucosyltransferase can bind di-

rectly to *V. parvula* cells (Wittenberger et al., 1977). Such enzyme-bound veillonellae can adhere in large numbers to the smooth wire surface in the presence of sucrose and in the absence of other organisms (McCabe and Donkersloot, 1977). Thus, adherence of one type of cell (veillonella), with no innate ability to adhere to a smooth surface, can be the result of the effect of an extracellular enzyme that is produced by a cell of a different type (streptococcus) and that catalyzes the formation of an insoluble, adherence-mediating, polysaccharide.

Intergeneric Coaggregation

LIST OF PARTNERS Direct adherence between veillonellae and oral bacteria of other genera was first demonstrated by Gibbons and Nygaard (1970), who noted coaggregation between *V. parvula* V5 (*V. alcalescens*) and *A. viscosus* T6, coccobacillus 26, and *Neisseria* 17. *V. parvula* V1 (*V. alcalescens*) and *V. parvula* V4 were coaggregation partners of many of the 46 human oral *Streptococcus salivarius* isolates examined for other adherence properties (Weerkamp and McBride, 1980). Coaggregation between *V. parvula* V1 and fibrillar strains of *S. salivarius* was much stronger than with fimbriate strains of *S. salivarius* after one hour of incubation, but after 24 hours both interactions were equally strong (Handley et al., 1987). A bald mutant, *S. salivarius* HB-B, devoid of surface structures, was unable to coaggregate with *V. parvula* V1 (Harty and Handley, 1988). Coculturing of *V. parvula* and *Eubacterium saburreum* resulted in coccus-filament associations that remained adherent even after vigorous vortex mixing (Mashimo et al., 1981).

All of the nearly 200 isolates of *Veillonella* spp. tested in a recent survey coaggregated with *Fusobacterium nucleatum;* some also coaggregated with *Actinomyces israelii, A. naeslundii, A. odontolyticus, A. viscosus, Gemella morbillorum, Streptococcus salivarius, S. sanguis, Rothia dentocariosa,* and *Propionibacterium acnes* (Hughes et al., 1988). The patterns of coaggregation between the veillonellae and their partners delineated four coaggregation groups. None of the veillonellae coaggregated with other veillonellae, which illustrates the predilection for intergeneric coaggregation rather than intrageneric interactions.

MECHANISMS OF COAGGREGATION. The mechanisms by which veillonellae participate in intergeneric coaggregations are unknown. Many of the coaggregations are inhibited by lactose

(60 mM), and all of the coaggregations with streptococci and actinomyces are prevented by heat (85°C/30 min) or protease treatment of the veillonella (Hughes et al., 1988). The same treatment of the partner has no effect on the ability of the partner to coaggregate. On the basis of the properties of numerous other coaggregations among oral bacteria (Kolenbrander, 1988), it is likely that many coaggregations involving veillonellae are mediated by lectin-carbohydrate interactions. The lectin (protease-sensitive surface component) is expressed on the surface of one cell type (e.g., veillonella) and a complementary carbohydrate is present on the other cell type (partner).

RELATIONSHIP TO COLONIZATION SITE. Almost 100% (58 of 59 strains) of the *Veillonella* isolates obtained from the tongue dorsum coaggregated with *Streptococcus salivarius,* a predominant inhabitant of the tongue (Hughes et al., 1988). But the 58 tongue isolates which were all *V. atypica* or *V. dispar* did not coaggregate with *Actinomyces viscosus, A. naeslundii, A. israelii,* and *Streptococcus sanguis* isolated from subgingival plaque. In contrast, 24 subgingival veillonellae strains, of which 20 were *V. parvula,* coaggregated with the actinomyces, streptococci, and other normal inhabitants of subgingival plaque, but they exhibited no coaggregation with strains of *S. salivarius.* Five other subgingival veillonellae failed to coaggregate with subgingival bacteria but did coaggregate with *S. salivarius.* Greater than 80% (87 of 105 strains) of the veillonellae isolated from buccal mucosa and saliva coaggregated with *S. salivarius,* the predominant streptococcus found in such samples. These results indicate the potential for a functional role for interbacterial adherence in mediating bacterial colonization of different habitats within the oral cavity.

An early observation of interbacterial interactions was the presence of "corn cob" and "test tube brush" morphologies of bacteria in dental plaque (Jones, 1972; Listgarten, 1976). A central rod- or filamentous-shaped cell was surrounded by clusters of spherical-shaped cells in the former and by shorter rod-shaped cells in the latter. These cellular arrangements were commonly observed at the edges of developing plaque rather than in the center or near the tooth surface. Veillonellae easily form corn cob structures with fusobacteria when the numbers of veillonellae are in a large excess (Fig. 1A). When the numbers of the two cell types are equal, large coaggregates are formed not only with fusobacteria (Fig. 1B), but also with streptococci (Fig. 1C), and actinomyces (Fig. 1D). In each pairing, the large coaggregates are composed of an interacting network of both cell types. Smaller coaggregates are also visible (large arrows, Figs. 1C and 1D), and even when the partners have the same shape (Fig. 1C), the interaction between the two cell types is clearly seen.

The oral cavity is a lotic environment, which requires mechanisms of adherence for the inhabitants to colonize the surfaces. Saliva lubricates and cleanses the mouth and contains lysozyme and bicarbonate ions that can lyse veillonellae (Tortosa, et al., 1981). Adherence to other bacteria that are already attached to either the tooth surface or to mucosal surfaces may protect veillonellae from these lytic activities as well as be their primary means of accretion.

Gnotobiotic Animal Model

Coaggregation in vivo between *V. parvula* V1 (*V. alcalescens*) and *S. mutans* was demonstrated with a gnotobiotic rat model system (McBride and van der Hoeven, 1981). The veillonella strain could not colonize the teeth but the two strains of *S. mutans* attached and colonized the smooth surface. The veillonella coaggregated with only one of the strains of *S. mutans.* If the coaggregation-positive strain of *S. mutans* was allowed to colonize the teeth before infection with the veillonellae, then the veillonellae attached and colonized the tooth surface. In contrast, no colonization of veillonellae above control values was seen when the coaggregation-negative *S. mutans* was used. The validity of the animal model was enhanced by the results of administering another coaggregating pair, whose adherence could be inhibited by lactose (van der Hoeven et al., 1985). One of the pair was allowed to colonize. Its partner was then added to the lactose-containing drinking water. This reduced the initial adherence of the coaggregating partner to control levels found when the partner was given alone to the animal. These results indicate that intergeneric coaggregation among oral bacteria is sensitive in vivo to similar perturbations (e. g., lactose inhibition) in the gnotobiotic animal as is found in vitro by mixing suspensions of the two partner strains (Kolenbrander and Andersen, 1986; Kolenbrander, 1988).

Earlier studies had demonstrated that human strains of *Veillonella parvula (V. alcalescens)* could not establish in monoinfected germ-free mice, whereas *Streptococcus mitis* did colonize by itself (Gibbons et al., 1964). However, when inoculated with several other human strains including *Streptococcus mitis, Veillonella* readily became part of the oral flora, suggesting that

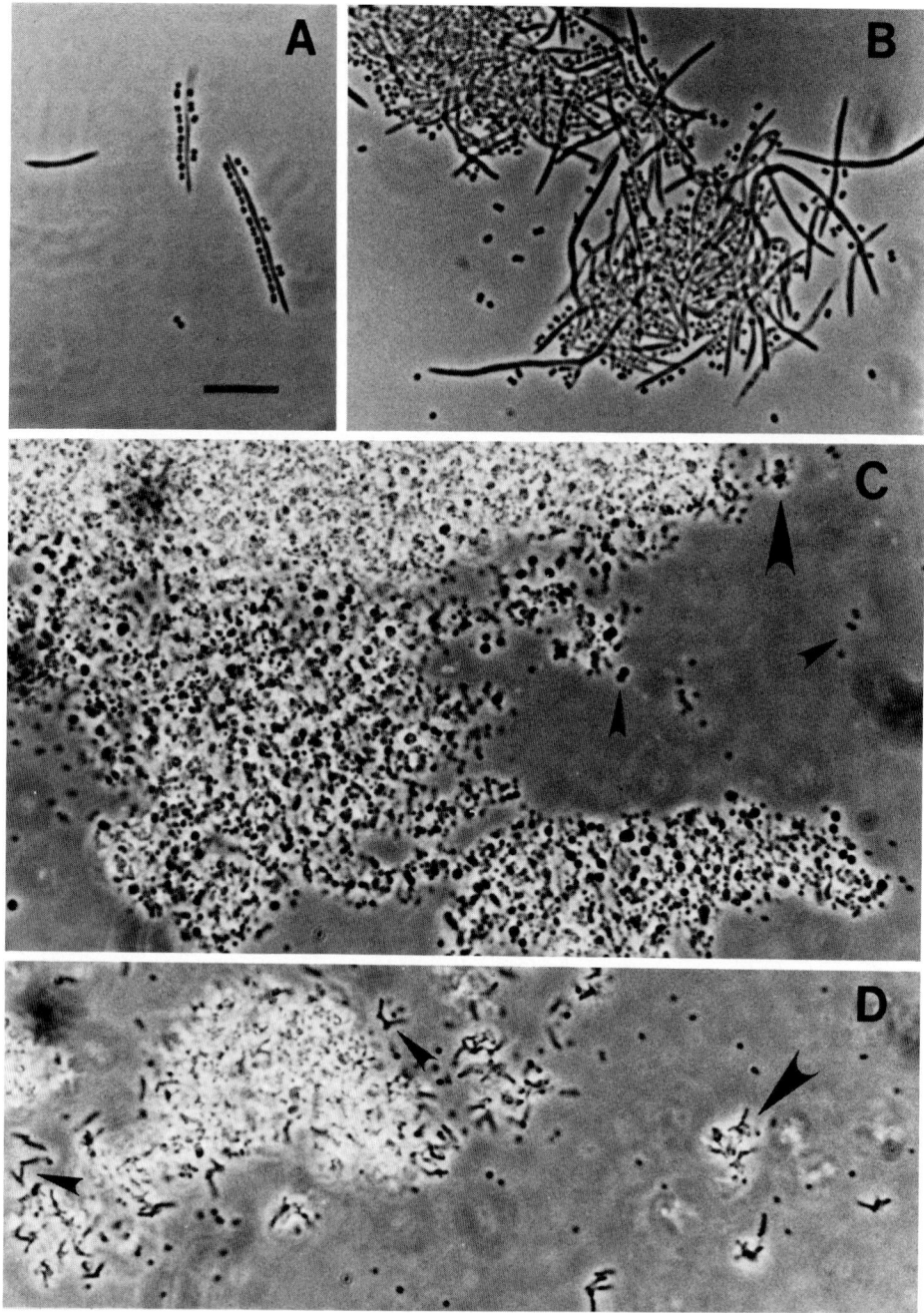

Fig. 1. Intergeneric coaggregation between *Veillonella atypica* PK1910 and other oral bacteria. When the ratio of *Veillonella atypica* PK1910 to *Fusobacterium nucleatum* PK1594 is 10 to 1, (A) coaggregates appear as "corn cobs" where the coccal-shaped veillonellae line up along the length of the slender rod-shaped fusobacteria. Coaggregates appear as clumps of mixed cell-types when equal numbers of: (B) fusobacteria, (C) *Streptococcus sanguis* 34, or (D) *Actinomyces viscosus* T14V are combined with the veillonellae. The large arrows in Figs. 1C and 1D indicate small coaggregates composed of a few cells of each cell type; the small arrows in Fig. 1C indicate the large phase-dark streptococcal spheres and the smaller and lighter gray veillonellae; the small arrows in Fig. 1D indicate the characteristic V-shaped or irregularly shaped actinomyces. Bar = 10 μm. All figures are at the same magnification.

the veillonellae adhere to primary colonizers but not directly to the host tissues. It is also possible that lactic acid is in limiting supply in the gnotobiotic animal and that, thus, veillonellae colonize poorly in the absence of an adequate nutrient supply. In either situation, veillonellae require other bacteria to establish themselves in the oral cavity.

Human Oral Cavity

Veillonellae are considered to be among the earliest colonizers of freshly cleaned enamel although they adhere poorly to a cleaned, human tooth surface (Liljemark and Gibbons, 1971) and to spheroidal hydroxyapatite (McBride and van der Hoeven, 1981), an extensively used model surface for adherence of oral bacteria to in vivo enamel surfaces (Clark, et al., 1978). Their numbers increase in parallel with the numbers of actinomyces and only after the streptococci are established (Ritz, 1967). Their inability to adhere well to smooth surfaces is in sharp contrast to their ability to bind to the surface of initial adherent cells such as *S. sanguis, A. viscosus, A. naeslundii,* and *A. israelii,* which are known to bind to spheroidal hydroxyapatite (Clark et al., 1978; Clark et al., 1981; Kolenbrander and Celesk, 1983).

Veillonellae are found in proportionally very high numbers on the tongue dorsum as compared to their numbers in buccal mucosa samples or dental plaque (Liljemark and Gibbons, 1971). In this study, streptomycin-resistant strains of veillonella were introduced into the mouths of human volunteers for 5 minutes, samples were taken from the tongue dorsum and tooth surface after 45 minutes and plated on streptomycin-containing agar (Rogosa et al., 1958), and the resulting colonies (number of veillonellae) were counted. The distribution of veillonellae on the two surfaces was according to their proportions found naturally.

The ecological relationship of coaggregation to colonization site was examined by testing veillonellae from subgingival plaque or the dorsum of the tongue for their ability to coaggregate with other bacteria isolated from the tongue or from subgingival plaque (Hughes et al., 1988 also "Relationship to Colonization Site," this chapter). Results from the study of Hughes et al. indicate that bacteria occupy the same site as their coaggregation partners and suggest a direct link between coaggregation and colonization in the oral cavity.

Isolation and Preservation

Identification

Thin sections of veillonellae viewed by electron microscopy exhibit typical Gram-negative surface layers consisting of an outer membrane, peptidoglycan layer, and cytoplasmic membrane (Bladen and Mergenhagen, 1964; Bladen et al., 1967; Tortosa et al., 1981). Phase contrast photomicrographs of the three human species of *Veillonella* are shown in Figure 2. The cells are spherical and are either single or in short chains of two to four cells. Most of the cells are 0.3 to 0.5 μm in diameter.

Identifying fresh isolates to the genus level of *Veillonella* is relatively straightforward. Gram-negative, small, spherical cells that reduce nitrate and grow on lactate or pyruvate but not on glucose under anaerobic atmosphere are the primary characteristics. Gas chromatographic analysis of the fermentation end products indicates acetic and propionic acids are the major products, and H_2 and CO_2 are also produced (Holdeman et al., 1977). Pyrolysis gas chromatography of lipopolysaccharides has been reported to be useful in differentiating *Veillonella, Fusobacterium,* and *Bacteroides* (Dahlén and Ericsson, 1983).

Classification to the species level is labor intensive and relies on numerous negative reactions in standard biochemical tests. Recently,

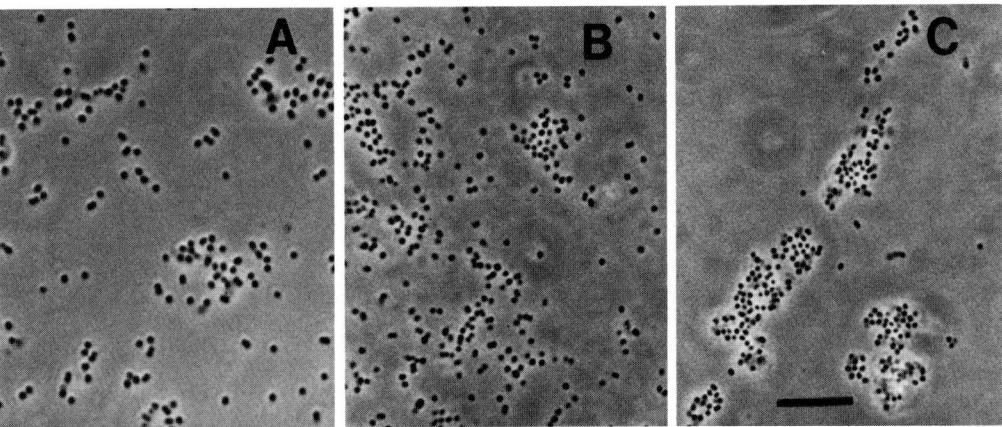

Fig. 2. Phase contrast photomicrographs of (A) *Veillonella atypica* PK1910, (B) *Veillonella dispar* PK1913, and (C) *Veillonella parvula* PK1914 isolates from the human oral cavity. Bar = 10 μm. All figures are at the same magnification.

methods involving analyses of cellular fatty acids have been developed (Moore et al., 1987a) and found useful in distinguishing the three human species, *V. atypica, V. dispar,* and *V. parvula* (W. E. C. Moore and L. V. H. Moore, unpublished observations; Hughes et al., 1988).

Selective Enrichment

Selective Medium for *Veillonella* Species

The following selective medium was designed by Rogosa (1956) and modified by Rogosa et al. (1958) and has been used successfully in our laboratories to isolate veillonellae from the human oral cavity (Hughes et al., 1988):

Tryptone	5.0 g
Yeast extract	3.0 g
Sodium thioglycollate	0.75 g
Basic fuchsin	0.002 g
Tween 80	1.0 g
Sodium lactate (60%)	21.0 ml
Agar	15.0 g

Distilled water is added to make 1 liter and the pH is adjusted to 7.5 before autoclaving. Vancomycin is added at 7.5μg per ml before pouring plates. The plates are incubated anaerobically under an atmosphere of 80% N_2, 10% CO_2, and 10% H_2 at 37°C for 48 hours. A commercial medium based on this recipe is also available (Bacto Veillonella Agar, Difco, Detroit, Michigan).

Cultivation

Some nonselective media that have been successfully used to determine the bacterial flora (including veillonellae) of healthy subgingival sites or of periodontally diseased sites are: Brucella agar supplemented with sheep blood and menadione (Williams et al., 1976), MM10 medium (Slots, 1977), trypticase soy blood agar (Newman et al., 1978), and supplemented brain heart infusion blood agar (Moore et al., 1982a).

Preservation

Strains survive freezing at −80°C, especially if they are suspended in blood; they also survive lyophilization.

Taxonomy

The following are among the few positive phenotypic traits that are recognized for strains of veillonellae: small (0.3 to 0.5 μm), Gram-negative cocci; usually found in pairs; anaerobic growth on lactate or pyruvate, with accompanying production of propionic and acetic acids, CO_2 and H_2; and nitrate reduction (Rogosa, 1964). Known variations of these properties include cells arranged in masses, short chains or single cells, and resistance to Gram-stain de-

colorization. Aerobic respiration of lactate or oxaloacetate occurs (Rogosa, 1964), and some strains of *V. parvula* can grow aerobically in static culture on lactate (C. Hughes and P. Kolenbrander, unpublished observations) indicating that not all veillonellae are strict anaerobes. Some strains require putrescine or cadaverine (Rogosa and Bishop, 1964a) and some produce H_2S (Rogosa and Bishop, 1964b). Carbohydrates are not fermented (except for fructose) by most strains of *V. criceti,* which are found in the oral cavity of hamsters (Mays et al., 1982).

Serological groupings were useful characteristics to aid in distinguishing among strains of animal and human origin (Rogosa, 1965). Many of these strains were included in the 116 veillonellae examined for DNA homology (Mays et al., 1982). Seven DNA homology groups distinct at the species level were found (Table 1) and these correspond to the seven species recognized in *Bergey's Manual of Systematic Bacteriology* (Rogosa, 1984). Ribosomal RNA homology (16S and 23S) studies revealed three clusters having average intracluster homology values of more than 90% and intercluster homology values of 60 to 70% (Johnson and Harich, 1983). *V. atypica, V. caviae, V. dispar, V. rodentium,* and *V. parvula* belonged to one cluster. *V. criceti* and *V. ratti* were in a second cluster, and the third cluster consisted of two strains of the "3312A" homology group, which comprise an unnamed species.

Veillonella are nonmotile and nonsporeforming. They do not use amino acids as a primary energy source and produce only sparse growth in a complex medium without added lactate, pyruvate, or other metabolizable intermediates. Together with *Acidaminococcus and Megasphaera,* they comprise the family Veillonellaceae (Rogosa, 1984). The GC content of the DNA of *Veillonella* species is 40.3 to 44.4 mol %, for *Acidaminococcus* it is 56.6 mol %, and for *Megasphaera* it is 53.1 to 54.1 mol %. Unlike veillonellae, acidaminococci degrade amino acids, forming acetate and butyrate, while megasphaerae ferment both carbohydrates and or-

Table 1. Characteristics of the type strains of the seven recognized species of the genus *Veillonella.*

Species	ATCC no.	Serogroup	Origin
V. atypica	17744	V	Human
V. caviae	33540	None	Guinea pig
V. criceti	17747	I	Hamster
V. dispar	17748	VII	Human
V. parvula	10790	VI	Human
V. ratti	17746	III	Rat
V. rodentium	17743	II	Hamster

ganic acids but are inactive on amino acids (Rogosa, 1971). The end products of *Megasphaera* fermentation in a complex medium are acetate, butyrate, caproate, valerate, propionate, isobutyrate, and isovalerate. Colonies of *Veillonella* but not *Acidaminococcus* or *Megasphaera* exhibit a pink to red fluorescence on brain heart infusion agar when exposed to long-wave (366 nm) UV light (Chow et al., 1975b). All species of *Veillonella* exhibit the fluorescence on brain heart infusion agar containing either sheep or horse blood (Brazier and Riley, 1988). The common human species, *V. atypica, V. dispar,* and *V. parvula,* do not fluoresce on blood agar base containing either type of blood unless the agar contains δ-aminolevulinic acid. The fluorescence fades rapidly (5 to 10 min.) on exposure to air, and the pigment is typical of metal-free porphyrins.

Differentiation among the species of veillonellae on the basis of commonly used phenotypic tests is difficult, if not impossible. Decomposition of hydrogen peroxide, once used to differentiate between *V. parvula* and *V. alcalescens,* has been reported to vary among subcultures of the same strain (Mays et al., 1982). However, polyacrylamide gel electrophoretic patterns of soluble proteins are distinct among the species (Mays et al., 1982).

Bacteriophages

Both temperate (Shimizu, 1968) and virulent (Totsuka, 1976) phages for human oral veillonellae have been isolated. The phage receptor for one of the veillonellophages is a lipopolysaccharide (Totsuka, 1988).

Plasmids

Plasmids are found in about 50% of the human oral strains, and most have several plasmids whose molecular sizes range from 1.1 to 28 megadaltons (Arai et al., 1984). It has been suggested that the fructose fermentation by some strains of *V. criceti* may be associated with a plasmid (Mays et al., 1982).

Acknowledgments

This chapter is dedicated to our friend and colleague, Dr. Morrison Rogosa, who died March 28, 1989. His pioneering work in developing selective media and his later contributions towards the clarification of the taxonomy of the genus *Veillonella* are widely recognized.

Literature Cited

Arai, T., A. Kusakabe, and S. Komatsu. 1984. A survey of plasmids in *Veillonella* strains isolated from human oral cavity. Kitasato Arch. Exp. Med. 57:233–237.

Barnhart, R. A., M. R. Weitekamp, and R. C. Aber. 1983. Osteomyelitis caused by veillonella. Amer. J. Med. 74:902–904.

Bartlett, J. G., and S. M. Finegold. 1972. Anaerobic pleuropulmonary infections. Medicine 51:413–450.

Bartlett, J. G., S. L. Gorbach, and S. M. Finegold. 1974. The bacteriology of aspiration pneumonia. Amer. J. Med. 56:202–207.

Bladen, H. A., H. Gewurz, and S. E. Mergenhagen. 1967. Interactions of the complement system with the surface and endotoxic lipopolysaccharide of *Veillonella alcalescens.* J. Exp. Med. 125:767–786.

Bladen, H., G. Hageage, F. Pollock, and R. Harr. 1970. Plaque formation in vitro on wires by Gram-negative oral microorganisms (*Veillonella*). Arch. Oral Biol. 15:127–133.

Bladen, H. A., and S. E. Mergenhagen. 1964. Ultrastructure of *Veillonella* and morphological correlation of an outer membrane with particles associated with endotoxic activity. J. Bacteriol. 88:1482–1492.

Borchardt, K. A., M. Baker, and R. Gelber. 1977. *Veillonella parvula* septicemia and osteomyelitis. Ann. Intern. Med. 86:63–64.

Boyar, R. M., and G. H. Bowden. 1985. The microflora associated with the progression of incipient carious lesions in teeth of children living in a water-fluoridated area. Caries Res. 19:298–306.

Brazier, J. S., and T. V. Riley. 1988. UV red fluorescence of *Veillonella* spp. J. Clin. Microbiol. 26:383–384.

Buckel, W., and R. Semmler. 1982. A biotin dependent sodium pump glutaconyl coenzyme A decarboxylase from *Acidaminococcus fermentans.* FEBS Lett. 148:35–38.

Burnett, G. W., and G. S. Schuster. 1978. Infections of the tooth pulp and periapical tissues, p. 244–253. In: Oral microbiology and infectious diseases. Student edition. Williams and Wilkins, Baltimore.

Chow, A. W., J. R. Marshall, and L. B. Guze. 1975a. Anaerobic infections of the female genital tract: Prospects and perspectives. Obstetrical Gynecological Survey 30:477–494.

Chow, A. W., V. Patten, and L. B. Guze. 1975b. Rapid screening of *Veillonella* by ultraviolet fluorescence. J. Clin. Microbiol. 2:546–548.

Clark, W. B., L. L. Baumann, and R. J. Gibbons. 1978. Comparative estimates of bacterial affinities and adsorption sites on hydroxyapatite surfaces. Infect. Immun. 19:846–853.

Clark, W. B., E. L. Webb, T. T. Wheeler, W. Fischlschweiger, D. C. Birdsell, and B. J. Mansheim. 1981. Role of surface fimbriae (fibrils) in the adsorption of *Actinomyces* species to saliva-treated hydroxyapatite surfaces. Infect. Immun. 33:908–917.

Dahlén, G., and I. Ericsson. 1983. Differentiation between Gram-negative anaerobic bacteria by pyrolysis gas chromatography of lipopolysaccharides. J. Gen. Microbiol. 129:557–563.

Delwiche, E. A., J. J. Pestka, and M. L. Tortorello. 1985. The veillonellae: Gram-negative cocci with a unique physiology. Ann. Rev. Microbiol. 39:175–193.

Dimroth, P. 1981. Characterization of a membrane-bound biotin-containing enzyme: oxaloacetate decarboxylase from *Klebsiella aerogenes*. Eur. J. Biochem. 115:353–358.

Dimroth, P. 1985. Biotin-dependent decarboxylases as energy transducing systems. Ann. N.Y. Acad. Sci. 447:72–85.

Distler, W., and A. Kröncke. 1980. Acid formation by mixed cultures of cariogenic strains of *Streptococcus mutans* and *Veillonella alcalescens*. Arch. Oral Biol. 25:655–658.

Distler, W., and A. Kröncke. 1981. Acid formation by mixed cultures of dental plaque bacteria *Actinomyces* and *Veillonella*. Arch. Oral Biol. 26:123–126.

Dzink, J. L., S. S. Socransky, and A. D. Haffajee. 1988. The predominant cultivable microbiota of active and inactive lesions of destructive periodontal diseases. J. Clin. Periodontol. 15:316–323.

Dzink, J. L., A. C. R. Tanner, A. D. Haffajee, and S. S. Socransky. 1985. Gram-negative species associated with active destructive periodontal lesions. J. Clin. Periodontol. 12:648–659.

Ellen, R. P., D. W. Banting, and E. D. Fillery. 1985. Longitudinal microbiological investigation of a hospitalized population of older adults with a high root surface caries risk. J. Dent. Res 64:1377–1381.

Forsythe, S. J., and J. A. Cole. 1987. Nitrite accumulation during anaerobic nitrate reduction by binary suspensions of bacteria isolated from the achlorhydric stomach. J. Gen. Microbiol. 133:1845–1849.

Frederick, J., and A. I. Braude. 1974. Anaerobic infection of the paranasal sinuses. N. Engl. J. Med. 290:135–137.

Gibbons, R. J., and M. Nygaard. 1970. Interbacterial aggregation of plaque bacteria. Arch. Oral Biol. 15:1397–1400.

Gibbons, R. J., S. S. Socransky, and B. Kapsimalis. 1964. Establishment of human indigenous bacteria in germ-free mice. J. Bacteriol. 88:1316–1323.

Gibbons, R. J., and J. Van Houte. 1971. Selective bacterial adherence to oral epithelial surfaces and its role as an ecological determinant. Infect. Immun. 3:567–573.

Goldstein, E. J. C., D. M. Citron, and S. M. Finegold. 1984. Role of anaerobic bacteria in bite-wound infections. Rev. Infect. Dis. 6:S177–S183.

Greaves, W. L., and A. B. Kaiser. 1984. Endocarditis due to *Veillonella alcalescens*. South. Med J. 77:1211–1212.

Handley, P. S., D. W. S. Harty, J. E. Wyatt, C. R. Brown, J. P. Doran, and A. C. C. Gibbs. 1987. A comparison of the adhesion, coaggregation and cell-surface hydrophobicity properties of fibrillar and fimbriate strains of *Streptococcus salivarius*. J. Gen. Microbiol. 133:3207–3217.

Hardie, J. M., and G. H. Bowden. 1974. The normal microbial flora of the mouth, p. 47–83. In: F. A. Skinner and J. G. Carr (ed.), The normal microbial flora of man. Academic Press, London.

Harty, D. W. S., and P. S. Handley. 1988. Fermentation products, amino acid utilization, maintenance energies and growth yields for the fibrillar *Streptococcus salivarius* HB and a non-fibrillar mutant HB-B grown in continuous culture under glucose limitation. J. Appl. Bacteriol. 65:143–152.

Heineman, H. S., and A. I. Braude. 1963. Anaerobic infection of the brain. Amer. J. Med. 35:682–697.

Hewett, M. J., K. W. Knox, and D. G. Bishop. 1971. Biochemical studies on lipopolysaccharides of *Veillonella*. Eur. J. Biochem. 19:169–175.

Hilpert, W., and P. Dimroth. 1983. Purification and characterization of a new sodium-transport decarboxylase. Methylmalonyl-CoA decarboxylase from *Veillonella alcalescens*. Eur. J. Biochem. 132:579–587.

Hofstad, T., and T. Kristoffersen. 1970. Chemical composition of endotoxin from oral *Veillonella*. Acta Pathol. Microbiol, Scand Sec. B. 78:760–764.

Holdeman, L. V., E. P. Cato, and W. E. C. Moore (ed.). 1977. Anaerobe laboratory manual, 4th ed. Virginia Polytechnic Institute and State University, Blacksburg.

Hoshino, E. 1987. L-serine enhances the anaerobic lactate metabolism of *Veillonella dispar* ATCC 17745. J. Dent. Res. 66:1162–1165.

Hughes, C. V., P. E. Kolenbrander, R. N. Andersen, and L. V. H. Moore. 1988. Coaggregation properties of human oral *Veillonella* spp.: relationship to colonization site and oral ecology. Appl. Environ. Microbiol. 54:1957–1963.

Inderlied, C. B., and E. A. Delwiche. 1973. Nitrate reduction and the growth of *Veillonella alcalescens*. J. Bacteriol. 114:1206–1212.

Isogai, E., H. Isogai, H. Sawada, H. Kaneko, and N. Ito. 1985. Microbial ecology of plaque in rats with naturally occuring gingivitis. Infect. Immun. 48:520–527.

Johnson, J. L., and B. Harich. 1983. Ribosomal ribonucleic acid homology among species of the genus *Veillonella* Prévot. Int. J. Syst. Bacteriol. 33:760–764.

Jones, S. J. 1972. A special relationship between spherical and filamentous microorganisms in mature human dental plaque. Arch. Oral Biol. 17:613–616.

Kafkewitz, D., and E. A. Delwiche. 1972. Ribose utilization by *Veillonella alcalescens*. J. Bacteriol. 109:1144–1148.

Kamio, Y., and K. Nakamura. 1987. Putrescine and cadaverine are constituents of peptidoglycan in *Veillonella alcalescens* and *Veillonella parvula*. J. Bacteriol. 169:2881–2884.

Kilian, M., A. Thylstrup, and O. Fejerskov. 1979. Predominant plaque flora of Tanzanian children exposed to high and low water fluoride concentrations. Caries Res. 13:330–343.

Kolenbrander, P. E. 1988. Intergeneric coaggregation among human oral bacteria and ecology of dental plaque. Annu. Rev. Microbiol. 42:627–656.

Kolenbrander, P. E. 1989. Surface recognition among oral bacteria: multigeneric coaggregations and their mediators. Crit. Rev. Microbiol. 17:137–159.

Kolenbrander, P. E., and R. N. Andersen. 1986. Multigeneric aggregations among oral bacteria: a network of independent cell-to-cell interactions. J. Bacteriol. 168:851–859.

Kolenbrander, P. E., R. N. Andersen, and L. V. H. Moore. 1989. Coaggregation of *Fusobacterium nucleatum*, *Selenomonas flueggei*, *Selenomonas infelix*, *Selenomonas noxia*, and *Selenomonas sputigena* with strains from eleven genera of oral bacteria. Infect. Immun. 57:3194–3203.

Kolenbrander, P. E., and R. A. Celesk. 1983. Coaggregation of human oral *Cytophaga* species and *Actinomyces israelii*. Infect. Immun. 40:1178–1185.

Lambe, D. W., Jr., D. H. Vroon, and C. W. Rietz. 1974. Infections due to anaerobic cocci, p. 585–599. In: A. Balows, R. M. DeHaan, V. R. Dowell, Jr., and L. B.

Guze, (ed), Anaerobic bacteria: role in disease. Charles C. Thomas, Springfield, IL.

Liljemark, W. F., and R. J. Gibbons. 1971. Ability of *Veillonella* and *Neisseria* species to attach to oral surfaces and their proportions present indigenously. Infect. Immun. 4:264–268.

Lisgarten, M. A. 1976. Structure of the microbial flora associated with peridontal health and disease in man. J. Periodontol. 47:1–18.

Loesche, W. J., S. Eklund, R. Earnest, and B. Burt. 1984. Longitudinal investigation of bacteriology of human fissure decay: epidemiological studies in molars shortly after eruption. Infect. Immun. 46:765–772.

Loesche, W. J., and S. A. Syed. 1978. Bacteriology of human experimental gingivitis: effect of plaque and gingivitis score. Infect. Immun. 21:830–839.

Loewe, L., P. Rosenblatt, and E. Alture-Werber. 1946. A refractory case of subacute bacterial endocarditis due to *Veillonella gazogenes* clinically arrested by a combination of pencillin, sodium para-aminohippurate, and heparin. Amer. Heart J. 32:327–338.

MacFarlane, T. W. 1984. The oral ecology of patients with severe Sjogren's syndrome. Microbios 41:99–106.

McBride, B. C., and J. S. van der Hoeven. 1981. Role of interbacterial adherence in colonization of the oral cavities of gnotobiotic rats infected with *Streptococcus mutans* and *Veillonella alcalescens*. Infect. Immun. 33:467–472.

McCabe, R. M., and J. A. Donkersloot. 1977. Adherence of *Veillonella* species mediated by extracellular glucosyltransferase from *Streptococcus salivarius*. Infect. Immun. 18:726–734.

McCarthy, C., M. L. Snyder, and R. B. Parker. 1965. The indigenous oral flora of man. I. The newborn to the 1-year-old infant. Arch. Oral Biol. 10:61–70.

McDermid, A. S., A. S. McKee, D. C. Ellwood, and P. D. Marsh. 1986. The effect of lowering the pH on the composition and metabolism of a community of nine oral bacteria grown in a chemostat. J. Gen. Microbiol. 132:1205–1214.

Marounek, M., and S. Bartos. 1987. Interactions between rumen amylolytic and lactate-utilizing bacteria in growth on starch. J. Appl. Bacteriol. 63:233–238.

Martin, W. J. 1974. Isolation and identification of anaerobic bacteria in the clinical laboratory: a 2-year experience. Mayo Clin. Proc. 49:300–308.

Martin, W. J., M. Gardner, and J. A. Washington II. 1972. In vitro antimicrobial susceptibility of anaerobic bacteria isolated from clinical specimens. Antimicrob. Ag. Chemother. 1:148–158.

Mashimo, P. A., Y. Murayama, H. Reynolds, C. Mouton, S. A. Ellison, and R. J. Genco. 1981. *Eubacterium saburreum* and *Veillonella parvula*: a symbiotic association of oral strains. J. Periodontol. 52:374–379.

Mays, T. D., L. V. Holdeman, W. E. C. Moore, M. Rogosa, and J. L. Johnson. 1982. Taxonomy of the genus *Veillonella* Prévot. Int. J. Syst. Bacteriol. 32:28–36.

Melville, S. B., T. A. Michel, and J. M. Macy. 1988. Pathway and sites for energy conservation in the metabolism of glucose by *Selenomonas ruminantium*. J. Bacteriol. 170:5298–5304.

Mergenhagen, S. E., E. G. Hampp, and H. W. Scherp. 1961. Preparation and biological activities of endotoxins from oral bacteria. J. Infect. Dis. 108:304–310.

Mikx, F. H. M., and J. S. van der Hoeven. 1975. Symbiosis of *Streptococcus mutans* and *Veillonella alcalescens* in mixed continuous cultures. Arch. Oral Biol. 20:407–410.

Mikx, F. H. M., J. S. van der Hoeven, K. G. König, A. J. M. Plasschaert, and B. Guggenheim. 1972. Establishment of defined microbial ecosystems in germ-free rats. I. The effect of the interaction of *Streptococcus mutans* or *Streptococcus sanguis* with *Veillonella alcalescens* on plaque formation and caries activity. Caries Res. 6:211–223.

Milnes, A. R., and G. H. W. Bowden. 1985. The microflora associated with developing lesions of nursing caries. Caries Res. 19:289–297.

Minah, G. E., G. B. Lovekin, and J. P. Finney. 1981. Sucrose-induced ecological response of experimental dental plaques from caries-free and caries-susceptible human volunteers. Infect. Immun. 34:662–675.

Moore, L. V. H., J. L. Johnson, and W. E. C. Moore. 1987. *Selenomonas noxia* sp. nov., *Selenomonas flueggei* sp. nov., *Selenomonas infelix* sp. nov., *Selenomonas dianae* sp. nov., and *Selenomonas artemidis* sp. nov., from the human gingival crevice. Int. J. Syst. Bacteriol. 37:271–280.

Moore, L. V. H., W. E. C. Moore, E. P. Cato, R. M. Smibert, J. A. Burmeister, A. M. Best, and R. R. Ranney. 1987. Bacteriology of human gingivitis. J. Dent. Res. 66:989–995.

Moore, W. E. C. 1987. Microbiology of periodontal disease. J. Periodont. Res. 22:335–341.

Moore, W. E. C., E. P. Cato, and L. V. Holdeman. 1969. Anaerobic bacteria of the gastrointestinal flora and their occurrence in clinical infections. J. Infect. Dis. 119:641–649.

Moore, W. E. C., L. V. Holdeman, E. P. Cato, R. M. Smibert, J. A. Burmeister, and R. R. Ranney. 1983. Bacteriology of moderate (chronic) periodontitis in mature adult humans. Infect. Immun. 42:510–515.

Moore, W. E. C., L. V. Holdeman, E. P. Cato, R. M. Smibert, J. A. Burmeister, K. G. Palcanis, and R. R. Ranney. 1985. Comparative bacteriology of juvenile periodontitis. Infect. Immun. 48:507–519.

Moore, W. E. C., L. V. Holdeman, R. M. Smibert, I. J. Good, J. A. Burmeister, K. G. Palcanis, and R. R. Ranney. 1982a. Bacteriology of experimental gingivitis in young adult humans. Infect. Immun. 38:651–667.

Moore, W. E. C., L. V. Holdeman, R. M. Smibert, D. E. Hash, J. A. Burmeister, and R. R. Ranney. 1982b. Bacteriology of severe periodontitis in young adult humans. Infect. Immun. 38:1137–1148.

Newman, M. G., V. Grinenco, M. Weiner, I. Angel, H. Karge, and R. Nisengard. 1978. Predominant microbiota associated with periodontal health in the aged. J. Periodontol. 49:553–559.

Ng, S. K. C., and I. R. Hamilton. 1974. Gluconeogenesis of *V. parvula* strain M_4 evidence for the indirect conversion of pyruvate to P-enolpyruvate. Can. J. Microbiol. 20:19–28.

Pestka, J. J., and E. A. Delwiche. 1983. An alternative pathway for 3-phosphoglycerate generation in *Veillonella*. Can J. Microbiol. 29:218–224.

Prévot, A. R. 1933. Études de systématique bactérienne. I. Lois generales. II. Cocci anaérobies. Ann. Sci. Nat. Bot. 15:23–260.

Ramotar, K., J. M. Conly, H. Chubb, and T. J. Louie. 1984. Production of menaquinones by intestinal anaerobes. J. Infect. Dis. 150:213–218.

Richardson, R. L., and M. Jones. 1958. A bacteriologic census of human saliva. J. Dent. Res. 37:697–709.

Ritz, H. L. 1967. Microbial population shifts in developing human dental plaque. Arch. Oral Biol. 12:1561–1568.

Roberts, M. C., and B. J. Moncla. 1988. Tetracycline resistance and TetM in oral anaerobic bacteria and *Neisseria perflava-N. sicca.* Antimicrob. Agents and Chemother. 32:1271–1273.

Robrish, S. A., W. Reid, and M. I. Krichevsky. 1972. Distribution of enzymes forming polysaccharides from sucrose and the composition of extracellular polysaccharide synthesized by *Streptococcus mutans.* Appl. Microbiol. 24:184–190.

Rogosa, M. 1956. A selective medium for the isolation and enumeration of the veillonella from the oral cavity. J. Bacteriol. 72:533–536.

Rogosa, M. 1964. The genus *Veillonella.* I. General cultural, ecological, and biochemical considerations. J. Bacteriol. 87:162–170.

Rogosa, M. 1965. The genus *Veillonella.* IV. Serological groupings and genus and species emendations. J. Bacteriol. 90:704–709.

Rogosa, M. 1971. Transfer of *Veillonella* Prévot and *Acidaminococcus* Rogosa from Neisseriaceae to Veillonellaceae fam. nov., and the inclusion of *Megasphaera* Rogosa in Veillonellaceae. Int. J. Syst. Bacteriol. 21:231–233.

Rogosa, M. 1984. Anaerobic Gram-negative cocci: Family 1. Veillonellaceae Rogosa 1971. p. 680–685. *In* N. R. Krieg and J. G. Holt, (ed.), Bergey's manual of systematic bacteriology, vol. 1, Williams and Wilkins, Baltimore.

Rogosa, M., and F. S. Bishop. 1964a. The genus *Veillonella.* II. Nutritional studies. J. Bacteriol. 87:574–580.

Rogosa, M., and F. S. Bishop. 1964b. The genus *Veillonella.* III. Hydrogen sulfide production by growing cultures. J. Bacteriol. 88:37–41.

Rogosa, M., R. J. Fitzgerald, M. E. MacKintosh, and A. J. Beaman. 1958. Improved medium for selective isolation of *Veillonella.* J. Bacteriol. 76:455–456.

Rogosa, M., M. I. Krichevsky, and F. S. Bishop. 1965. Truncated glycolytic system in *Veillonella.* J. Bacteriol. 90:164–171.

Ruoff, K. L., and E. A. Delwiche. 1977. Nitrate-reductase electrontransport cofactors in *Veillonella alcalescens.* Can. J. Microbiol. 23:1562–1567.

Shimizu, Y. 1968. Experimental studies on the bacteriophages of the *Veillonella* strains isolated from the oral cavity. Odontology (Tokyo) 55:533–541.

Singer, R. E., and B. A. Buckner. 1981. Butyrate and propionate: important components of toxic dental plaque extracts. Infect. Immun. 32:458–463.

Slack, G. L., and G. H. Bowden. 1965. Preliminary studies of experimental dental plaque in vivo. Adv. Fluorine Res. Dent. Caries Prev. 3:193–215.

Slots. J. 1977. Microflora in the healthy gingival sulcus in man. Scand. J. Dent. Res. 85:247–254.

Slots, J., D. Möenbo, J. Langebaek, and A. Frandsen. 1978. Microbiota of gingivitis in man. Scand. J. Dent. Res. 86:174–181.

Socransky, S. S., A. D. Manganiello, D. Propas, V. Oram, and J. van Houte. 1977. Bacteriological studies of developing supragingival dental plaque. J. Periodont. Res. 12:90–106.

Sutter, V. L. 1984. Anaerobes as normal oral flora. Rev. Infect. Dis. 6:S62–S66.

Syed, S. A., and W. J. Loesche. 1978. Bacteriology of human experimental gingivitis: effect of plaque age. Infect. Immun. 21:821–829.

Theilade, E., J. Theilade, and L. Mikkelsen. 1982. Microbiological studies on early dento-gingival plaque on teeth and mylar strips in humans. J. Periodont. Res. 17:12–25.

Tortorello, M. L., and E. A. Delwiche. 1983. Utilization of fructose and ribose in lipopolysaccharide synthesis by *Veillonella parvula.* Infect. Immun. 41:423–425.

Tortosa, M., M.-I. Cho, T. J. Wilkens, V. J. Iacono, and J. J. Pollock. 1981. Bacteriolysis of *Veillonella alcalescens* by lysozyme and inorganic anions present in saliva. Infect. Immun. 32:1261–1273.

Totsuka, M. 1976. Studies on veillonellophages isolated from washings of human oral cavity. Bull. Tokyo Med. Dent. Univ. 23:261–273.

Totsuka, M. 1988. Phage-receptor on the cell wall of *Veillonella rodentium.* Antonie van Leeuwenhoek 54:229–233.

Van der Hoeven, J. S., M. H. de Jong, and P. E. Kolenbrander. 1985. In vivo studies of microbial adherence in dental plaque, p. 220–227. In: S. E. Mergenhagen and B. Rosan (ed.), Molecular basis of oral microbial adhesion. American Society for Microbiology, Washington, D.C.

Van der Hoeven, J. S., A. I. Toorop, and F. H. M. Mikx. 1978. Symbiotic relationship of *Veillonella alcalescens* and *Streptococcus mutans* in dental plaque in gnotobiotic rats. Caries Res. 12:142–147.

Veillon, A., and M. M. Zuber. 1898. II. Recherches sur quelques microbes strictement anaérobies et leur role en pathologie. Arch. Med. Exp. 10:517–545.

Weerkamp, A. H., and B. C. McBride. 1980. Characterization of the adherence properties of *Streptococcus salivarius.* Infect. Immun. 29:459–468.

Whiteley, H. R., and H. C. Douglas. 1951. The fermentation of purines by *Micrococcus lactilyticus.* J. Bacteriol. 61:605–616.

Williams, B. L., J. L. Ebersole, M. D. Spektor, R. C. Page. 1985. Assessment of serum antibody patterns and analysis of subgingival microflora of members of a family with a high prevalence of early-onset periodontitis. Infect. Immun. 49:742–750.

Williams, B. L., S. K.-Å. Osterberg, and J. Jorgensen. 1979. Subgingival microflora of periodontal patients on tetracycline therapy. J. Clin. Periodontol. 6:210–221.

Williams, B. L., R. M. Pantalone, and J. C. Sherris. 1976. Subgingival microflora and periodontitis. J. Periodont. Res. 11:1–18.

Williams. B. T. 1977. Pelvic abscess associated with repeated recovery of veillonella. Am. J. Obstet. Gynecol. 129:342–343.

Winter, P. F., and E. A. Delwiche. 1975. Cell wall composition and incorporation of radio-labelled compounds by *Veillonella alcalescens.* Can. J. Microbiol. 21:2039–2047.

Wittenberger, C. L., A. J. Beaman, L. N. Lee, R. M. McCabe, and J. A. Donkersloot. 1977. Possible role of *Streptococcus salivarius* glucosyltransferase in adherence of *Veillonella* to smooth surfaces, p. 417–421. In:

D. Schlessinger (ed.), Microbiology 1977. American Society for Microbiology, Washington, D.C.

Wolff, L. F., M. J. Krupp, and W. F. Liljemark. 1985. Microbial changes associated with advancing periodontitis in STR/N mice. J. Periodont. Res. 20:378–385.

Yordy, D. M., and E. A. Delwiche. 1979. Nitrite reduction in *Veillonella alcalescens*. J. Bacteriol. 137:905–911.

Zambon, J. J., H. Reynolds, J. G. Fisher, M. Shlossman, R. Dunford, and R. J. Genco. 1988. Microbiological and immunological studies of adult periodontitis in patients with noninsulin-dependent diabetes mellitus. J. Periodontol. 59:23–31.

CHAPTER 96

The Genus *Syntrophomonas* and Other Syntrophic Anaerobes

MICHAEL J. MCINERNEY

In methanogenic ecosystems, syntrophic bacteria (see Chapter 11) degrade propionate, longer chain fatty acids, and some aromatic acids, with the production of acetate, CO_2, H_2, and possibly formate (McInerney, 1986). The degradation of these compounds is thermodynamically unfavorable unless the reactions are coupled to the use of H_2 and/or formate by bacteria such as methanogens. Table 1 shows the free energy yield of some of these reactions. Four genera of syntrophic bacteria have been described, *Syntrophobacter* (Boone and Bryant, 1980), *Syntrophus* (Mountfort et al., 1984), *Syntrophospora* (formerly *Clostridium*) (Stieb and Schink, 1985), and *Syntrophomonas* (McInerney et al., 1981a). *Syntrophomonas* includes several species and subspecies. Several other unnamed syntrophic organisms, probably representing either new genera or new species, have been obtained in coculture, but their taxonomic status has not been determined.

The syntrophic bacteria are very specialized organisms with respect both to the compounds that serve as their energy sources and to the reactions that are used to reoxidize their reduced cofactors. Some syntrophic bacteria are known to use only one compound as an energy source, while others use only a certain class of compounds, such as fatty acids of 4 to 18 carbons in length. In general, they do not use carbohydrates, proteinaceous materials, or other common bacterial energy sources. Because of the reduced state of carbon in the substrates used by these organisms, it is not possible to generate an oxidized intermediate which will serve as an electron acceptor, such as occurs for the degradation of carbohydrates and amino acids by fermentative bacteria. Syntrophic bacteria also lack the enzymatic ability to reduce CO_2 to acetate and to use alternate electron acceptors, such as oxyanions of sulfur and nitrogen, or organic compounds, such as fumarate. Thus, the reduction of protons to H_2 or of bicarbonate to formate is the only mechanism available to reoxidize their reduced cofactors. Because these reactions are thermodynamically

favorable only when the concentrations of H_2 and formate are maintained at low levels, the metabolism of syntrophic bacteria in their natural habitat is obligately dependent on the activity of methanogens (see also Chapter 11). Syntrophic bacteria cannot be isolated in pure culture with their natural substrates as the energy source, but must be grown in monoaxenic (two-membered) cultures in association with H_2/formate users. A few syntrophic species can be selected to use more oxidized carbon compounds (for example, crotonate), which generate intermediates that can serve as electrons acceptors. This has allowed the isolation of these species in pure culture (Beaty and McInerney, 1987; Zhao et al., 1990).

Some species of other physiological groups of anaerobic bacteria require syntrophic interactions to degrade certain compounds. Some strains of the genera *Desulfovibrio*, *Pelobacter*, *Thermoanaerobium*, and *Bacteroides* degrade primary alcohols only in syntrophic association with H_2/formate-using bacteria (Ben-Bassat et al., 1981; Bryant et al., 1977; Dwyer and Tiedje, 1986; Eichler and Schink, 1985; Schink, 1984; Schink and Stieb, 1983). The S organism, which was the first syntrophic bacterium discovered (Bryant et al., 1967; Reddy et al., 1972), is morphologically similar to *Pelobacter* species. *Syntrophococcus sucromutans* requires an exogenous electron acceptor, either formate, cinnamate derivatives, methoxybenzenoids, or a H_2/formate-using bacterium, to oxidize carbohydrates (Krumholz and Bryant, 1986).

Habitats

The syntrophic bacteria are found in environments where organic matter is completely degraded to CH_4 and CO_2. In these environments, oxygen, oxyanions of sulfur and nitrogen, and ferrous iron are not readily available and the reduction of CO_2 to CH_4 is the dominant electron-accepting reaction (McInerney, 1986). Examples of such environments include meso-

Table 1. The major reactions involved in the anaerobic degradation of organic compounds by obligately syntrophic bacteria

Reaction	Free energy yield $\Delta G^{0\prime}$ (kJ/mol)
I. Some reactions of methanogens:	
1. $4H_2 + HCO_3^- + H^+ \leftrightarrow CH_4 + 3H_2O$	-135.6
2. $4HCO_2^- + H_2O + H^+ \leftrightarrow CH_4 + 3HCO_3^-$	-130.4
II. Some reactions of obligately syntrophic bacteria:	
A. Without H_2/formate users	
3a. $CH_3CH_2OH + H_2O \leftrightarrow CH_3COO^- + H^+ + 2H_2$	$+ 9.6$
b. $CH_3CH_2OH + 2HCO_3^- \leftrightarrow CH_3COO^- + H^+ + 2HCO_2^-$	$+ 7.0$
4. $CH_3CH_2CH_2COO^- + 2H_2O \leftrightarrow 2CH_3COO^- + H^+ + 2H_2$	$+ 48.1$
5. $CH_3CH_2CH_2CH_2COO^- + H_2O \leftrightarrow CH_3COO^- + CH_3CH_2COO^- + H^+ + 2H_2$	$+ 48.1$
6. $CH_3CH_2COO^- + 3H_2O \leftrightarrow CH_3COO^- + HCO_3^- + H^+ + 3H_2$	$+ 76.1$
7. $C_7H_5O_2^- + 7H_2O \leftrightarrow 3CH_3COO^- + HCO_3^- + 3H^+ + 3H_2$	$+ 70.6$
B. With H_2/formate users	
8. $2CH_3CH_2OH + HCO_3^- \leftrightarrow 2CH_3COO^- + CH_4 + H_2O + H^+$	-116.4
9. $2CH_3CH_2CH_2COO^- + HCO_3^- + H_2O \leftrightarrow 4CH_3COO^- + CH_4 + H^+$	$- 39.4$
10. $2CH_3CH_2CH_2CH_2COO^- + HCO_3^- + H_2O \leftrightarrow 2CH_3CH_2COO^- + 2CH_3COO^- + CH_4 + H^+$	$- 39.4$
11. $4CH_3CH_2COO^- + 3H_2O \leftrightarrow 4CH_3COO^- + 3CH_4 + HCO_3^- + H^+$	-102.4
12. $4C_7H_5O_2^- + 19H_2O \leftrightarrow 12CH_3COO^- + 3CH_4 + HCO_3^- + 9H^+$	-124.4

Data are from Kaiser and Hanselman (1982) and Thauer et al. (1977) or were calculated from data therein.

philic and thermophilic sludge digestors, freshwater aquatic sediments, and flooded soils. In these methanogenic environments, the syntrophic bacteria constitute an essential trophic level that converts the metabolites produced by fermentative bacteria from polysaccharides, proteins, and lipids to the methanogenic substrates—acetate, H_2, and formate. In marine and estuarine sediments, sulfate reduction rather than methanogenesis is the dominant electron-accepting process and fatty acids and aromatic acids are degraded by sulfate reducers without the involvement of interspecies electron transfer reactions (Banat and Nedwell, 1983). Fatty acid- and aromatic-degrading syntrophic bacteria are present (Stieb and Schink, 1985; Tschech and Schink, 1985, 1986), but represent only a small portion of the microbial population in sulfate-reducing environments.

Enumeration studies and adaptation experiments support this conclusion. Most probable numbers of anaerobic, syntrophic, butyrate-degrading bacteria in mesophilic and thermophilic sludge digestors and in freshwater sediments were 4.5×10^6 per g (wet weight) (McInerney et al., 1979; Henson and Smith, 1985) while the numbers of butyrate-degrading, sulfate-reducing bacteria were 1,000-fold lower (M. J. McInerney, unpublished observations). The syntrophic, stearate- or oleate-degrading bacteria were found in similar levels in sludge digestors (Lorowitz et al., 1989; Roy et al., 1986). The number of sporeforming, syntrophic, fatty acid-degrading bacteria was about 200 times higher in freshwater than marine sed-

iments (Stieb and Schink, 1985). Thermophilic degradation of butyrate to CH_4 was detected without a lag with inocula obtained from anaerobic digestors and freshwater sediments (Henson and Smith, 1985). However, methanogenesis from butyrate occurred with a two-week lag when inocula from marine sediments were used, and not at all with inocula from hypersaline environments. Exceptions were the isovalerate-degrading, syntrophic bacteria, which are about 200 times more numerous in marine versus freshwater sediments (Stieb and Schink, 1986).

Eichler and Schink (1985) found that the numbers of alcohol degraders was 10 to 100 times higher in anaerobic creek sediments than in sewage sludge digestors. In these creek sediments, homoacetogenic bacteria or bacteria that form propionate as the reduced product predominated over syntrophic alcohol degraders. In sewage sludge, syntrophic alcohol degraders similar to *Pelobacter carbinolicus* were more numerous (Eichler and Schink, 1985; Dubourguier et al., 1986).

In the rumen and other parts of the gastrointestinal tract of warm-blooded animals, including humans, syntrophic bacteria grow too slowly to be maintained in the system. The fatty acids are not degraded to any significant extent. However, bacteria similar to *Syntrophomonas wolfei* subsp. *wolfei* and *S. wolfei* subsp. *saponavida* are present in low numbers in bovine rumen fluid (McInerney et al., 1981b; Lorowitz et al., 1989).

Isolation

The syntrophic bacteria are strict anaerobes (see Chapter 11). Thus, the growth of these bacteria requires the use of a method for removing and excluding oxygen from all environments in which the bacteria are exposed and the maintenance in liquid and solid media of low oxidation-reduction potentials, usually by the inclusion of strong chemical reducing agent(s), such as combinations of L-cysteine and sodium sulfide. Because of the slow growth rates of syntrophic cocultures, the medium must be very carefully prepared to insure that it remains reduced for long periods of time. We use the Bryant (1972) modification of the Hungate technique to prepare anaerobic media and solutions. Once dispensed and sterilized, additions, inoculations, and sampling of media are done with syringes and needles, using the procedures outlined by Miller and Wolin (1974) and Balch and Wolfe (1976). This reduces the probability of introducing oxygen into reduced media or accidentally contaminating the medium.

Selective Enrichment

Syntrophic associations that degrade fatty acids, primary alcohols, and some aromatic compounds to CH_4 and CO_2 can be selectively enriched using an anaerobic medium that lacks electron acceptors—such as oxyanions of nitrogen and sulfur, elemental sulfur, or ferrous iron—and incubating the cultures in the dark. Either a mineral salts medium with B-vitamins and rumen fluid or minimal media without organic nutrients other than the substrate is used (McInerney et al., 1979; Stieb and Schink, 1985). Because of the marked substrate specificity exhibited by the syntrophic bacteria (Table 2), the growth of certain species can be selected by addition of the appropriate substrate. For instance, S. wolinii uses only propionate and not other fatty acids (Boone and Bryant, 1980); Syntrophus buswellii uses only benzoate (Mountfort and Bryant, 1982). The addition of long-chain fatty acids to the medium selects for Syntrophomonas sapovorans and S. wolfei subsp. saponivida (Lorowitz et al., 1989; Roy et al., 1986). The use of caproate rather than butyrate will select for the sporeforming, fatty acid degrader Syntrophospora bryantii, rather than Syntrophomonas species (Steib and Schink, 1985).

Inocula are obtained from methanogenic environments where organic matter is completely fermented to the gaseous products CH_4 and CO_2. Marine and brackish water sediments that are high in organic matter have also been used.

The enrichments are maintained by transferring 20 to 50% of the volume every 1 to 2 weeks. Substrate degradation is followed by comparing methane production in substrate-emended tubes to controls without substrate.

Growth Media and Solutions

The following basal medium can be used to grow many different syntrophic cocultures by changing the substrate added to the medium, i.e., an alcohol, a fatty acid, etc. (McInerney et al., 1979):

Basal Medium

The total volume is adjusted to 1 liter with distilled water.

Substrate	0.1 to 2 g
Mineral solution (see below)	50 ml
Trace metal solution (see below)	10 ml
Vitamin solution (see below)	5 ml
Clarified rumen fluid	50 ml
Resazurin	0.001 g
$NaHCO_3$	3.5 g
Cysteine-sulfide reducing solution (see below)	20 ml

All the medium ingredients, except the reducing solution and $NaHCO_3$, are dissolved in distilled water, and the mixture is adjusted to a pH of 7.2 to 7.4 with NaOH. The medium is brought to volume (minus the total volume of the components to be added later), heated to a boil under a stream of an O_2-free, 80% N_2:20% CO_2 gas mixture. Solid $NaHCO_3$ is added slowly to prevent excessive boiling of the medium. The medium is dispensed under a continuous stream of the above anaerobic gas mixture into serum tubes (18 by 150 mm). The tubes are fitted with black rubber septum-type stoppers, which are held in place during autoclaving by aluminum crimp seals. The cysteine-sulfide reducing solution is added to individual tubes no more than several hours to a day before use. This ensures that each tube will remain reduced for the long periods of time that are required to grow syntrophic bacteria. Solid medium for roll tubes contains 2% agar for use at mesophilic temperatures and at least 3% agar for use at higher temperatures. The medium is dispensed into 18-by-150-mm culture tubes fitted with no. 1 black rubber stoppers. Slants and shake tubes are prepared by addition of 1% agar to the basal medium.

Mineral Solution

KH_2PO_4	10.0 g
$MgCl_2 \cdot 6H_2O$	6.6 g
NaCl	8.0 g
NH_4Cl	8.0 g
$CaCl_2 \cdot 2H_2O$	1.0 g

Trace Metal Solution

MnSO$_4$·H$_2$O	1.0 g
Fe(NH$_4$)$_2$(SO$_4$)$_2$·6H$_2$O	0.8 g
CoCl$_2$·6H$_2$O	0.2 g
ZnSO$_4$·7H$_2$O	0.2 g
CuCl$_2$·2H$_2$O	0.02 g
NiCl$_2$·6H$_2$O	0.02 g
Na$_2$MoO$_4$·2H$_2$O	0.02 g
Na$_2$SeO$_4$	0.02 g
Na$_2$WO$_4$	0.02 g
Nitrilotriacetic acid	2.0 g
pH adjusted to 6.0 with KOH	

Vitamin solution

Nicotinic acid	20 mg
Cyanocobalamin	20 mg
Thiamin·HCl	10 mg
p-Aminobenzoic acid	10 mg
Pyridoxine·HCl	50 mg
Calcium pantothenate	5 mg

The salinity of the basal medium can be adjusted by increasing the concentrations of NaCl and MgCl$_2$·6H$_2$O (Widdel and Pfennig, 1981). For sulfate-reducing cocultures, 3 g/liter of Na$_2$SO$_4$ is added. Clarified rumen fluid is prepared as described previously (Bryant and Robinson, 1961). Clarified anaerobic digestor fluid has also been used as an alternative to rumen fluid (Henson and Smith, 1985).

The cysteine-sulfide reducing solution contains 2.5% each of L-cysteine·HCl and NaS$_2$·9H$_2$O (Bryant and Robinson, 1961). For enrichment medium, a solution containing only sodium sulfide is used. The solution is prepared inside the anaerobic chamber and dispensed into serum tubes (18 by 150 mm) (Bellco Glass) fitted with black rubber septum-like stoppers and sealed with aluminum closures. The tubes are sterilized by autoclaving. The solution can be stored inside the anaerobic chamber for many months, but should be disgarded when a white precipitate (L-cystine) forms.

The substrate is added at a concentration which will not inhibit growth. For short-chain fatty acids, a concentration of 20 mM is used while lower concentrations (less than 5 mM) are used for long-chain fatty acids and aromatic compounds. Higher concentrations of long-chain fatty acids require the addition of calcium in equimolar amounts for unsaturated fatty acids and in at least one-half of the molar amount for saturated fatty acids (Roy et al., 1985). Phenolic compounds should be added at much lower concentrations (0.5 mM).

Growth of the H$_2$/Formate User

The H$_2$/formate-using bacteria are grown in the basal medium without the addition of a substrate. The tubes are pressurized to 300 kPa with 80% H$_2$:20% CO$_2$ gas phase, using the gas-sing apparatus described by Balch and Wolfe (1976). The NaHCO$_3$ concentration of the basal medium is increased to 0.5 g per liter to maintain a neutral pH. For the growth of H$_2$-using sulfate-reducing bacteria, 3 g per liter of NaSO$_4$ is added to the basal medium. The cultures are repressurized two to three times daily with the gas mixture. The reader should consult the chapters on the methanogenic bacteria and the sulfate-reducing bacteria for additional information on the cultivation and properties of these organisms.

Mesophilic syntrophic bacteria have been obtained in coculture with hydrogen-using strains of *Desulfovibrio* and *Methanospirillum hungatei* (Boone and Bryant, 1980; McInerney et al., 1979; Mountfort and Bryant, 1982; Stieb and Shink, 1985). Thermophilic species have been obtained in coculture with strains of *Methanobacterium thermoautotrophicum* or related strains (Ahring and Westermann, 1987; Henson and Smith, 1985; Zinder and Koch, 1984). The choice of the appropriate H$_2$/formate user is critical in order to obtain cocultures of syntrophic bacteria. Sulfate-reducing, syntrophic cocultures grow faster and to higher yields than the respective methanogenic coculture (McInerney et al., 1979), probably due to the more favorable thermodynamics and kinetics of sulfate reduction compared to methanogenesis (Robinson and Tiedje, 1984; Thauer et al., 1977). Also, the production of sulfide maintains favorable anaerobic conditions during the extended incubations that are required for the growth of syntrophic bacteria. *M. hungatei* is used because it is almost always the most numerous H$_2$-using bacterium in the enrichments (McInerney et al., 1979). *Wolinella succinogenes* was used to obtain cocultures of two syntrophic bacteria that catabolize benzenoids (Barik et al., 1985). However, *W. succinogenes* can grow by oxidizing the sulfide present in the medium using fumarate as the electron acceptor. Thus, the growth of *W. succinogenes* is not dependent on the metabolism of the syntrophic bacterium, and coupling between the two organisms may not occur (M. J. McInerney, unpublished observations).

Isolation of Cocultures

The enrichment is serially diluted 10-fold in an anaerobic dilution solution that has the same composition as the enrichment medium with the organic components deleted. The highest dilutions are inoculated into tubes of molten agar medium. The H$_2$/formate user is added to each roll tube or to the last few dilution tubes prior to their inoculation by the syntrophic enrich-

Table 2. Characteristics of the obligately syntrophic bacteria.

Species	Cell shape	Cell size (μm)	Cell wall type	Flagellation pattern	Spore formation	Substrate used Coculture	Substrate used Alone
Syntrophobacter wolinii	Rod	0.6–1.0 by 1–4.5	–[a]	–	–	C_3[d]	
Syntrophomonas wolfei subsp. wolfei	Curved rod	0.7 by 3.7	–	2–8, laterally concave side	–	C_4–C_8,iC_7	C_4–C_6,UFA
subsp. saponivida	Curved rod	0.4–0.6 by 2	–	2–4, laterally concave side	–		
saponivorans	Curved rod	0.5 by 2.5	–	2–4, laterally concave side		C_4–C_{18},O,E,L	
Strain NSF-2	Curved rod	0.2–0.3 by 2–4	–		–	C_4–C_6,2MB	
Thermophiles	Curved rod	0.5–0.8 by 2–3	–		–	C_4	
Thermophile strain AOR	Rod	0.4–0.6 by 2–3	+	–	–	C_2Et,EG	EG,Py,B,H_2,F
Syntrophospora bryantii	Rod	0.4 by 3–6	+		+	C_4–C_{11},2MB	Crot
Strain SF-1	Rod	0.4–0.6 by 3–5	+		+	C_4–C_6,2MB,iC_4	
Strain GraIval	Rod	1 by 1.25	+	Single polar	–	iC_5	
Strain GröIval	Rod	1.75 by 2.5	+	–	–	iC_5	
Syntrophus buswellii	Rod	0.8 by 1–2	–	Single polar	–	Bz	
Strain HQ Göl	Rod	0.8–1.3 by 1.6	–		–	Bz	diOHB,diOHBz
Strain KNO32	Rod	0.8 by 1.5–3	–		–	Bz,OHBz	
Strain WoO21	Coccoid rod	0.8 by 0.8–1.7	–		–	Bz,OHBz	
Clostridium strain MA266	Rod	0.8–1 by 2–3.5	+	+[c]	+	diOHB,diOHBz	
PA-1	Rod	0.8 by 1–2	–[b]	+[c]	–	Many	PA
P-2	Rod	0.6 by 1	–[b]	+[c]	–	Bz,PA,PHC	

Abbreviations: C, carbon number of the fatty acid; UFA, unsaturated fatty acid, Crot, crotonate; iC, iso-branched fatty acid; 2MB, 2-methylbutyrate; O, oleate; E, elaidate; L, linoleate; Bz, benzoate; OHBz, hydroxybenzoate; OHB, hydroxybenzene; diOHBz, dihydroxybenzoate; diOHB, dihydroxybenzene; PA, phenylacetate; P, phenol; HC, hydrocinnamate; EG, ethylene glycol; Py, pyruvate; B, betaine; F, formate.

[a]For cell wall type, + and – refer to Gram-positive and Gram-negative cell wall types, respectively. Otherwise, these signs indicate the presence or absence of a trait.

[b]Not confirmed by electron microscopy.

[c]Motile, but type of flagellation not determined.

Data compiled from Ahring and Westermann (1987); D.A. Amos and M.J. McInerney, unpublished observations; Barik et al. (1985); Beaty and McInerney (1987); Boone and Bryant (1980); Henson and Smith (1985); Lee and Zinder (1988a); Lorowitz et al. (1989); McInerney et al. (1979, 1981a); Mountfort and Bryant (1982); Mountfort et al. (1984); Roy et al. (1985, 1986); Shelton and Tiedje (1984); Stieb and Schink (1985, 1986); Szewzyk and Schink (1989); Tschech and Schink (1985, 1986); Zhao et al. (1990); and Zinder and Koch (1984).

ment. After inoculation, the roll tubes are rolled in ice water and then placed at the incubation temperature in an upright position. The shake tube method of Pfenning (1978) has also been successfully used to obtain syntrophic cocultures. This method is preferred when lower agar concentrations are needed to allow the diffusion of less-soluble substrates such as long-chain fatty acids.

Colonies of a syntrophic bacterium with its partner will take weeks to several months to appear. Thus, those colonies which appear within the first two weeks of incubation are probably not syntrophic cocultures and are marked so that late-forming colonies can be easily detected. Colonies are picked to slants containing the same medium without exposure to air, by using bent Pasteur pipets and a mouth tube. When a sulfate reducer is used as the syntrophic partner, black or darkbrown colonies will appear within about 3 weeks, for cocultures of butyrate-degrading syntrophic bacteria, to as long as 6 to 8 weeks for cocultures of propionate or benzoate degraders. Colonies of syntrophic bacteria with methanogens will appear several weeks later than that observed under sulfate-reducing conditions. Monoaxenic cultures (cocultures) containing only the syntrophic bacterium and its partner can be obtained by repetition of the above steps.

Several of the syntrophic bacteria can be isolated in pure culture by selecting for growth with crotonate (Beaty and McInerney, 1987). The basal medium containing original substrate and 20 mM crotonate is used. Growth occurs first using the original substrate and, after several weeks to months, further growth may be observed as a result of crotonate degradation. The syntrophic bacterium is isolated by serial dilution and inoculation of roll tubes as described above.

Routine Cultivation

It is possible to grow syntrophic bacteria in any size of tube or bottle as long as adequate precautions are taken to avoid contamination by oxygen or other organisms. Serum tubes containing 5 to 10 ml or serum bottles containing 20 to 100 ml of the basal medium with the appropriate substrate provide the most convenient method of cultivation. Larger volumes are grown in thick-walled glass bottles or carboys (Wofford et al., 1986). A large inoculum, about 1 to 20%, is used. Cultures are incubated in a static position at temperatures of 25 to 35°C for mesophilic strains and 55 to 60°C for thermophilic strains.

Maintenance of Cultures

Cocultures are routinely transferred every 2 weeks to 1 month. After growth, slants of syntrophic bacteria remain viable for several months at room temperature. Long-term viability is maintained by storage in liquid nitrogen in sterile, anaerobic medium containing either 20% (wt/vol) glycerol or 5% (vol/vol) dimethylsulfoxide (filter-sterilized) as a cryoprotectant.

Selective Lysis and Separation of Syntrophic Cocultures

Cell-free extracts of *Syntrophomonas wolfei* subsp. *wolfei* grown in coculture with *M. hungatei* in butyrate basal medium can be prepared by selectively lysing cells of the syntrophic bacterium with lysozyme and removing unlysed cells, including those of the methanogen, by centrifugation (Wofford et al., 1986). Cells of *S. wolfei* subsp. *wolfei* (Beaty et al., 1987) and *Syntrophus buswelli* (P. S. Beaty and M. J. McInerney, unpublished observations) have been separated from those of the H_2/formate user by centrifugation in Percoll density gradients.

Identification

Morphological and physiological properties are important in the identification of the syntrophic bacteria. It is important that appropriate controls are performed to determine whether a characteristic of the coculture is due to the presence of the syntrophic bacterium or of the H_2/formate-using bacterium. In general, the syntrophic bacteria are characterized by the limited number of compounds that support growth alone or in coculture with H_2/formate using-bacteria (Table 2). With a few exceptions, the presence of a H_2/formate removal system is required if the degradation of the substrate is to be thermodynamically favorable. *Syntrophospora bryantii* and the subspecies of *S. wolfei* can be selected to grow with crotonate without the need for a H_2/formate removal system (D. A. Amos and M. J. McInerney, unpublished observations; Beaty and McInerney, 1987; Lorowitz et al., 1989; Zhao et al., 1990).

Syntrophobacter wolinii is a nonmotile, Gram-negative rod that degrades propionate only in coculture (Boone and Bryant, 1980). Members of the genus *Syntrophomonas* are Gram-negative, medium-sized, curved to helical rods (Fig. 1A) with two to eight flagella laterally inserted on the concave side of the cell (Lorowitz et al., 1989; McInerney et al., 1979, 1981a; Roy et al., 1986). The species and sub-

Fig. 1. Photomicrographs of (A) *Syntrophomonas wolfei* subsp. *wolfei* grown in pure culture with crotonate and (B) *Syntrophus buswellii* (arrows) grown in coculture with *Desulfovibrio* strain G11 with benzoate. Bar = 10 μm. The latter is a composite picture of several fields of view.

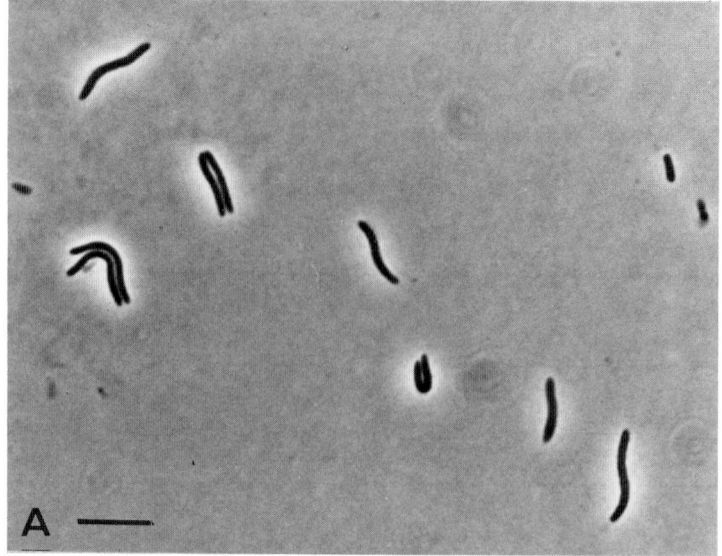

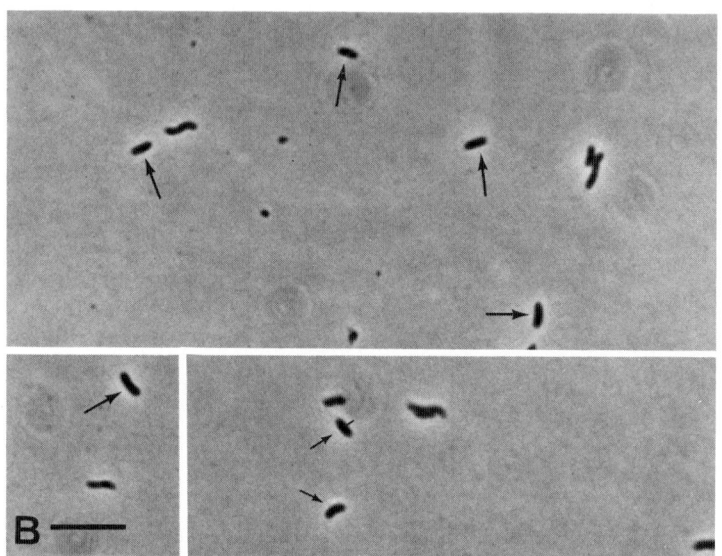

species of *Syntrophomonas* are differentiated from each other based on their substrate utilization patterns (Table 2). *S. bryantii* is a Gram-positive, sporeforming rod which uses C_4 to C_{11} saturated fatty acids and 2-methylbutyrate only in coculture (Stieb and Schink, 1985; Zhao et al., 1990). *Syntrophus buswellii* (Fig. 1B) is a short, motile, Gram-negative rod, which only uses benzoate in coculture and not any other aromatic compounds or fatty acids (Mountfort and Bryant, 1982; Mountfort et al., 1984). *Clostridium* strain Ma 366 is a short, sporeforming, motile rod which uses one type of dihydroxybenzene and two types of dihydroxybenzoates in coculture (Tschech and Schink, 1986). Several other strains have been obtained in coculture with H_2/formate users; they are unnamed, but probably represent new species or genera

(Table 2). Also, a bacterium that grows by the reductive removal of hydroxyl moieties from certain aromatic compounds in pure culture can degrade benzoate in coculture (Szewzyk and Schink, 1989). Zinder and Koch (1984) obtained a thermophilic bacterium that oxidizes acetate to CO_2 only when grown in association with a methanogen. This organism was subsequently shown to be an acetogenic bacterium (Lee and Zinder, 1988a).

16S rRNA sequence analysis of pure cultures of the *S. wolfei* subspecies and of *S. bryantii* show that these organisms are Gram-positive, and are closely related to each other but not to other species of the genus *Clostridium* (Zhao et al., 1990). Based on the sequence analysis, the syntrophic, Gram-positive, sporeforming, fatty acid degraders are now placed in a new genus,

Syntrophospora. Although the subspecies of *S. wolfei* are phylogenetically closely related to *S. bryantii,* they differ from *S. bryantii* in their cell wall ultrastructure and in spore formation, and thus they remain in a separate genus.

Physiological Properties

The syntrophic bacteria are usually grown in association with bacteria that can use either H_2 or formate. It is not possible to determine which compound is the intermediate since the stoichiometry of the overall reaction is the same whether H_2 or formate is the intermediate. The ability of the syntrophic bacteria to produce H_2 has been shown by coculturing these bacteria with methanogens that do not use formate (McInerney et al., 1981a; Lee and Zinder, 1988b). Also, butyrate oxidation by *S. wolfei* subsp. *wolfei* grown in coculture with *M. hungatei* occurs in the absence of methanogenesis when a hydrogenation catalyst (Pd-BaSO$_4$) is present to couple the reaction to the reduction of ethylene to ethane (Kasper et al., 1987). Kinetic analysis of H_2 turnover and diffusion (Conrad et al., 1985; Boone et al., 1989) indicate the H_2 may not be the major intermediate transferred in natural syntrophic associations. Theile and Zeikus (1988) provide evidence that reducing equivalents may be transferred as formate rather than H_2. It is likely that syntrophic bacteria produce either or both compounds as shown for the amino acid-degrading anaerobe *Eubacterium acidaminophilum,* which grows in coculture with either a H_2-using bacterium or *Desulfovibrio baarsii,* which uses formate, but not hydrogen (Zindel et al., 1988).

The production of acetate from even-numbered fatty acids and acetate and propionate from odd-numbered fatty acids (Table 1) by *Syntrophomonas* species and *S. bryantii* indicates that these compounds are β-oxidized. Wofford et al. (1986) showed that *S. wolfei* subsp. *wolfei* has high activities of the enzymes needed to β-oxidize fatty acids and to generate ATP by substrate-level phosphorylation, and that the activation of the fatty acid occurs by a coenzyme A (CoA) transferase reaction. This pathway predicts that one ATP is produced per butyrate degraded. However, the change in free energy of syntrophic butyrate degradation (-20kJ/mol) is too low to allow the net formation of one ATP per butyrate (Thauer et al., 1977). Thauer and Morris (1984) proposed that about two-thirds of the ATP is used to produce H_2 (or formate) from the electrons generated in the oxidation of the acyl-Co A intermediate to the enoyl-Co A intermediate. Thus, only about one-third of an

ATP is made per butyrate degraded, which is consistent with the small change in free energy of the reaction. Enzymatic studies of the isovalerate-degrading bacterium, strain Gralval support a pathway that does not provide for the net synthesis of ATP by substrate-level phosphorylation, indicating that the bacterium must use ion gradients for energy (Steib and Schink, 1986).

The syntrophic degradation of propionate has been studied in enrichments and in fresh samples from lake sediments and digestor sludge using propionate labeled in specific positions (Koch et al., 1983; Schink, 1985). Randomization of the label occurred in the product, acetate, and in propionate, suggesting that propionate is degraded by a pathway generating a symmetrical intermediate. The use of high-resolution ^{13}C-nuclear magnetic resonance spectroscopy demonstrated the production of succinate (Houwen et al., 1987).

Many, if not most, aromatic compounds that are capable of being mineralized by anaerobic consortia are degraded by pathways that converge on phenol and benzoate. Prior to ring cleavage, degradation of substituents occurs in a stepwise fashion, with each step catalyzed by a different specialized organism (Young, 1984). Evidence also suggests that phenol is carboxylated prior to ring cleavage (Knoll and Winter, 1987). The anaerobic metabolism of benzoate proceeds first by reducing the delocalization energy of the pi-electrons by hydrogenation reactions and then by catalyzing nonoxidative ring cleavage (Evans, 1977; Keith et al., 1978).

Biotechnological applications of syntrophic bacteria include their use in anaerobic reactors to treat industrial wastes and the stimulation of their activity to clean up contaminated environments.

Literature Cited

Ahring, B. K., and P. Westermann. 1987. Thermophilic anaerobic degradation of butyrate by a butyrate-utilizing bacterium in coculture and triculture with methanogenic bacteria. Appl. Environ. Microbiol. 53:429–433.

Balch, W. E., and R. S. Wolfe. 1976. New approach to the cultivation of methanogenic bacteria: 2-mercapchethanesulfonic acid (HS-CoM)-dependent growth of *Methanobacterium ruminantium* in a pressurized atmosphere. Appl. Environ. Microbiol. 32:781–791.

Banat, I. M., and D. B. Nedwell. 1983. Mechanism of turnover of C_2-C_4 fatty acids in high-sulfate and low-sulfate anaerobic sediments. FEMS Microbiol. Lett. 17:107–110.

Barik, S., W. J. Brulla, and M. P. Bryant. 1985. PA-1, a versatile anaerobe obtained in pure culture, catabolizes benzenoids and other compounds in syntrophy with hydrogenotrophs, and P-2 plus *Wolinella* sp. degrades benzenoids. Appl. Environ. Microbiol. 50:304–310.

Beaty, P. S., and M. J. McInerney. 1987. Growth of *Syntrophomonas wolfei* in pure culture. Arch. Microbiol. 147:389–393.

Beaty, P. S., N. Q. Wofford, and M. J. McInerney. 1987. Separation of *Syntrophomonas wolfei* from *Methanospirillum hungatei* in syntrophic cocultures by using Percoll gradients. Appl. Environ. Microbiol. 53:1183–1185.

Ben-Bassat, A., R. Lamed, and J. G. Zeikus. 1981. Ethanol production by thermophilic bacteria: Metabolic control of end product formation in *Thermoanaerobium brockii*. J. Bacteriol. 146:192–199.

Boone, D. R., and M. P. Bryant. 1980. Propionate-degrading bacterium *Syntrophobacter wolinii* sp. nov., gen. nov., from methanogenic ecosystems. Appl. Environ. Microbiol. 40:626–632.

Boone, D. R., R. L. Johnson, and Y. Liu. 1989. Diffusion of the interspecies electron carriers H_2 and formate in methanogenic ecosystems and its implications in the measurement of K_m for H_2 and formate uptake. Appl. Environ. Microbiol. 55:1735–1741.

Bryant, M. P. 1972. Commentary on the Hungate technique for the culture of anaerobic bacteria. Am. J. Clin. Nutri. 25:1324–1328.

Bryant, M. P., L. L. Campbell, C. A. Reddy, and M. R. Crabill. 1977. Growth of *Desulfovibrio* in lactate or ethanol media low in sulfate in association with H_2-utilizing methanogenic bacteria. Appl. Environ. Microbiol. 33:1162–1169.

Bryant, M. P., and I. M. Robinson. 1961. An improved nonselective culture medium for ruminal bacteria and its use in determining diurnal variation in numbers of bacteria in the rumen. J. Dairy Sci. 44:1446–1456.

Bryant, M. P., E. A. Wolin, M. J. Wolin, and R. S. Wolfe. 1967. *Methanobacillus omelianskii*, a symbiotic association of two species of bacteria. Arch. Mikrobiol. 59:20–31.

Conrad, R., F. S. Lupton, and J. G. Zeikus. 1985. Gas metabolism in support of juxtapositioning of hydrogen-producing and methanogenic bacteria in sewage sludge and lake sediments. Appl. Environ. Microbiol. 50:595–601.

Dubourguier, H.-C., E. Samain, G. Prendier, and G. Albagnac. 1986. Characterization of two strains of *Pelobacter carbinolicus* isolated from anaerobic digestors. Arch. Microbiol. 145:248–253.

Dwyer, D. F., and J. M. Tiedje. 1986. Metabolism of polyethylene glycol by two anaerobic bacteria, *Desulfovibrio desulfuricans* and a *Bacteroides* sp. Appl. Environ. Microbiol. 52:852–856.

Eichler, B., and B. Schink. 1985. Fermentation of primary alcohols and diols and pure culture of syntrophically alcohol-oxidizing anaerobes. Arch. Microbiol. 143:60–66.

Evans, W. C. 1977. Biochemistry of the bacterial catabolism of aromatic compounds in anaerobic environments. Nature 270:17–22.

Henson, J. M., and P. H. Smith. 1985. Isolation of a butyrate-utilizing bacterium in coculture with *Methanobacterium thermoautotrophicum* from a thermophilic digestor. Appl. Environ. Micrbiol. 49:1461–1466.

Houwen, F. P., C. Dijkema, C. H. H. Schoenmakers, A. J. M. Stams, and A. J. B. Zehnder. 1987. ^{13}C-NMR

study of propionate degradation by a methanogenic coculture. FEMS Micobiol. Lett. 41:269–274.

Kaiser, J.-L., and K. W. Hanselmann. 1982. Fermentative metabolism of substituted monoaromatic compounds by a bacterial community from anaerobic sediments. Arch. Microbiol. 133:185–194.

Kasper, H. F., A. J. Holland, and D. O. Mountfort. 1987. Simultaneous butyrate oxidation by *Syntrophomonas wolfei* and catalytic olefin reduction in absence of interspecies hydrogen transfer. Arch. Microbiol. 147:334–339.

Keith C. L., R. L. Bridges, L. R. F. Fina, K. L. Iverson, and J. A. Cloran. 1978. The anaerobic decomposition of benzoic acid during methane formation. IV. Dearomatization of the ring and volatile fatty acids formed on ring rupture. Arch. Microbiol. 118:173–177.

Knoll, G., and J. Winter. 1987. Anaerobic degradation of phenol in sewage sludge. Benzoate formation from phenol and CO_2 in the presence of hydrogen. Appl. Microbiol. Biotechnol. 25:384–391.

Koch, M., J. Dolfing, K. Wurhmann, and A. J. B. Zehnder. 1983. Pathways of propionate degradation by enriched methanoganic cultures. Appl. Environ. Microbiol. 45:1411–1414.

Krumholz, L. R., and M. P. Bryant. 1986. *Syntrophococcus sucromutans* sp. nov., gen. nov., uses carbohydrates as electron donors and formate, methoxymonobenzenoids, or *Methanobrevibacter* as electron acceptor systems. Arch. Microbiol. 143:313–318.

Lee, M. J., and S. Zinder. 1988a. Isolation and characterization of a thermophilic bacterium which oxidizes acetate in syntrophic association with a methanogen and which grows acetogenically on H_2/CO_2. Appl. Environ. Microbiol. 54:124–129.

Lee, M. J., and S. Zinder. 1988b. Hydrogen partial pressures in a thermophilic acetate-oxidizing methanogenic coculture. Appl. Environ. Microbiol. 54:1457–1461.

Lorowitz, W. H., H. Zhao, and M. P. Bryant. 1989. *Syntrophomonas wolfei* subsp. *saponavida* subsp. nov., a long-chain fatty acid-degrading, anaerobic, syntrophic bacterium; *Syntrophomonas wolfei* subsp. *wolfei* subsp. nov.; and emended descriptions of the genus and species. Int. J. Syst. Bacteriol. 39:122–126.

McInerney, M. J. 1986. Transient and persistent association among prokaryotes, p. 293–338. In: J. S. Poindexter and E. R. Leadbetter (ed.), Bacteria in nature, vol. 2. Plenum Press, New York.

McInerney, M. J., M. P. Bryant, R. B. Hespell, and J. W. Costerton. 1981a. *Syntrophomonas wolfei* gen. nov., sp. nov., an anaerobic, syntrophic, fatty acid-oxidizing bacterium. Appl. Environ. Microbiol. 41:1029–1039.

McInerney, M. J., M. P. Bryant, and N. Pfennig. 1979. Anaerobic bacterium that degrades fatty acids in syntrophic association with methanogens. Arch. Microbiol. 122:129–135.

McInerney, M. J., R. I. Mackie, and M. P. Bryant. 1981b. Syntrophic association of a butyrate-degrading bacterium and *Methanosarcina* enriched from bovine rumen fluid. Appl. Environ. Microbiol. 41:826–828.

Miller, T. L., and M. J. Wolin. 1974. A serum bottle modification of the Hungate technique for culturing obligate anaerobes. Appl. Microbiol. 27:985–987.

Mountfort, D. O., W. J. Brulla, L. R. Krumholz, and M. P. Bryant. 1984. *Syntrophus buswellii* gen. nov., sp. nov.:

a benzoate catabolizer from methanogenic ecosystems. Int. J. Syst. Bacteriol. 34:216–217.

Mountfort, D. O., and M. P. Bryant. 1982. Isolation and characterization of an anaerobic syntrophic benzoate-degrading bacterium from sewage sludge. Arch. Microbiol. 133:249–256.

Pfennig, N. 1978. *Rhodocyclus purpureus* gen. nov. and sp. nov., a ring shaped, vitamin B_{12} requiring member of the family Rhodospirillaceae. Int. J. Syst. Bacteriol. 28:283–288.

Reddy, C. A., M. P. Bryant, and M. J. Wolin. 1972. Characteristics of S organism isolated from *Methanobacillus omelianskii*. J. Bacteriol. 109:539–545.

Robinson, J. A., and J. M. Tiedje. 1984. Competition between sulfate-reducing and methanogenic bacteria for H_2 under resting and growing conditions. Arch. Microbiol. 137:26–32.

Roy, F., G. Albagnac, and E. Samain. 1985. Influence of calcium addition on growth of highly purified syntrophic cultures degrading long-chain fatty acids. Appl. Environ. Microbiol. 49:702–702.

Roy, F., E. Samain, H. C. Dubrouguier, and G. Albagnac. 1986. *Syntrophomonas sapovorans* sp. nov., a new obligately proton reducing anaerobe oxidizing saturated and unsaturated long chain fatty acids. Arch. Microbiol. 145:142–147.

Schink, B. 1984. Fermentation of 2,3-butanediol by new strains of Gram-negative, nonsporeforming anaerobes, *Pelobacter carbinolicus* sp. nov. and *Pelobacter propionicus* sp. nov. and evidence for formation of propionate from C_2 compounds. Arch. Microbiol. 137:33–41.

Schink, B. 1985. Mechanisms and kinetics of succinate and propionate degradation in anoxic freshwater sediments and sewage sludge. J. Gen. Microbiol. 131:643–650.

Schink, B., and M. Stieb. 1983. Fermentative degradation of polyethylene glycol by a strictly anaerobic, Gram-negative, nonsporeforming bacterium, *Pelobacter venetianus,* sp. nov. Appl. Environ. Microbiol. 45:1905–1913.

Shelton, D. R., and J. M. Tiedje. 1984. Isolation and partial characterization of bacteria in an anaerobic consortium that mineralizes 3-chlorobenzoic acid. Appl. Environ. Microbiol. 48:840–848.

Stieb, M., and B. Schink. 1985. Anaerobic oxidation of fatty acids by *Clostridium bryantii* sp. nov., a sporeforming, obligately syntrophic bacterium. Arch. Microbiol. 140:387–390.

Stieb, M., and B. Schink. 1986. Anaerobic degradation of isovalerate by a defined methanogenic coculture. Arch. Microbiol. 144:291–295.

Szewzyk, U., and B. Schink. 1989. Degradation of hydroquinone, gentisate, and benzoate, by a fermenting bacterium in pure culture or defined mixed culture. Arch. Microbiol. 151:541–545.

Thauer, R. K., K. Jungermann, and K. Decker. 1977. Energy conservation in chemotrophic bacteria. Bacteriol. Rev. 41:100–180.

Thauer, R. K., and J. G. Morris. 1984. Metabolism of chemotrophic anaerobes: old views and new aspects. Symp. Soc. Gen. Microbiol. 36:123–374.

Thiele, J. H., and J. G. Zeikus. 1988. Control of interspecies electron flow during anaerobic digestion: significance of formate transfer versus hydrogen transfer during syntrophic methanogenesis in flocs. Appl. Environ. Microbiol. 54:20–29.

Tschech, A., and B. Schink. 1985. Fermentative degradation of resorcinol and resorcylic acids. Arch. Microbiol. 143:52–59.

Tschech, A., and B. Schink. 1986. Fermentative metabolism of monohydroxybenzoates by defined syntrophic cocultures. Arch. Microbiol. 145:396–402.

Widdel, F., and N. Pfennig. 1981. Studies on dissimilatory sulfate-reducing bacteria that decompose fatty acids. I. Isolation of new sulfate-reducing bacteria enriched with acetate from saline environments. Description of *Desulfobacter postgatei* gen. nov., sp. nov. Arch. Microbiol. 134:286–294.

Wofford, N. Q., P. S. Beaty, and M. J. McInerney. 1986. Preparation of cell extracts and the enzymes involved in fatty acid metabolism in *Syntrophomonas wolfei*. J. Bacteriol. 167:179–185.

Young, L. Y. 1984. Anaerobic degradation of aromatic compounds, p. 487–523. In: D. T. Gibson (ed.), Microbial degradation of organic compounds. Marcel Dekker, New York.

Zhao, H., D. Yeng, C. R. Woese, and M. P. Bryant. 1990. Assignment of *Clostridium bryantii* to *Syntrophospora bryantii* gen. nov. comb., based on 16S rRNA sequence analysis of its crotonate-grown pure culture. Int. J. Syst. Bacteriol. 40:40–44.

Zindel, U., W. Freudenberg, M. Rieth, J. R. Andressen, J. Schnell, and F. Widdel. 1988. *Eubacterium acidaminophilum* sp. nov., a versatile amino acid-degrading anaerobe producing or utilizing H_2 or formate. Arch. Microbiol. 150:254–266.

Zinder, S. H., and M. Koch. 1984. Non-acetoclastic methanogenesis from acetate: acetate oxidation by a thermophilic syntrophic coculture. Arch. Microbiol. 138:263–272.

CHAPTER 97

The Cyanobacteria—Isolation, Purification, and Identification

JOHN B. WATERBURY

The oxygenic photosynthetic bacteria comprise two groups, the cyanobacteria and the prochlorophytes, that are distinguished by their photosynthetic pigment composition. The cyanobacteria contain chlorophyll a and phycobiliproteins as their primary photosynthetic pigments (see Chapter 98) while the prochlorophytes contain chlorophylls a and b and lack phycobiliproteins (see Chapter 99).

The cyanobacteria are morphologically and developmentally one of the most diverse groups of prokaryotes. They range from simple unicellular forms that reproduce by binary fission to complex filamentous forms that possess a variety of highly differentiated cell types. Some of the filamentous forms are capable of true branching, and some are even truly multicellular, as the result of cellular differentiation and functional specialization between vegetative cells and heterocysts, the sites of oxygenic photosynthesis and dinitrogen fixation, respectively.

Traditionally this group of phototrophic prokaryotes has been classified as a group of algae under the aegis of the Botanical Code (Stafleu et al., 1972). Phycologists have developed a system of classification for these organisms based on their morphological, developmental, and ecological characters, as determined not on pure cultures but on natural samples. This system, which contains about 150 genera and 1000 species, has proven to be successful for the classification of cyanophytes in natural material but is inadequate in many instances for the classification of cyanobacteria maintained in axenic culture.

During the last 25 years, following the unequivocal demonstration of the prokaryotic nature of these organisms, a number of bacteriologists have become interested in and have applied traditional microbiological techniques to the study of cyanobacteria. As a result of these studies there are currently several hundred isolates of cyanobacteria maintained in pure culture. These isolates include representatives of many, but certainly not all, of the major groups of cyanobacteria described in the botanical literature.

The classification of cultured cyanobacteria has been problematic. It was hoped that phenotypic and genotypic characters, made accessible by the availability of pure cultures, would facilitate the classification of the group. However, until recently, this, in large part, had not proven to be the case. As in the traditional botanical taxonomic treatments, morphological and developmental features form the bases for the description of taxa at the level of genera and above (Castenholz, 1989a, 1989b, 1989c; Rippka, 1988b; Waterbury, 1989). Exceptions occur among some unicellular cyanobacteria that lack adequate morphological and developmental complexity to permit genera to be defined by these criteria alone (Waterbury and Rippka, 1989). However, because of the importance of structural and developmental characters for the classification of both field and cultured material, it will be possible for the system of classification currently being developed for the cyanobacteria, under the aegis of the Bacteriological Code (Lapage, 1975) to represent, to a large degree, a logical extension and refinement of the classical botanical system.

Major Groups of Cyanobacteria

The cyanobacteria studied in pure culture are currently placed in five orders (Castenholz, 1989a, 1989b, 1989c; Waterbury, 1989; Waterbury and Rippka, 1989) described below that correspond closely to the five groups (Sections) used by Stanier and his collaborators (Rippka, 1988b; Rippka et al., 1979, 1981b).

Chroococcales

Members of the order Chroococcales are unicellular cyanobacteria that reproduce by binary fission or budding. Division occurs in one, two, or three planes at right angles to one another or in irregular planes. Cells can range in size from

0.5–30 μm and can occur as single cocci and rods or as cell aggregates. The form and size of cell aggregates depend on the planes of division and on the presence of extracellular slime or structured sheaths that hold the cells together.

Pleurocapsales

All members of the order Pleurocapsales reproduce by multiple fission, a feature that distinguishes them from other cyanobacteria. Pleurocapsalean cyanobacteria range from unicellular forms that divide exclusively by multiple fission to forms that produce cell aggregates by vegetative binary fission. Such cell aggregates range in complexity from groups of a few cells to complex pseudofilamentous cell assemblages. Following aggregate formation, some cells in the assemblage undergo multiple fission and release unicellular structures called baeocytes. Baeocytes immediately initiate growth leading to the next vegetative cell cycle.

Oscillatoriales

The order Oscillatoriales includes all the undifferentiated filamentous cyanobacteria. Cell division occurs by binary fission in one plane at right angles to the long axis of the trichomes. Reproduction occurs by trichome fragmentation or by the production of undifferentiated hormogonia released from the ends of trichomes. Although cell size and shape and trichome length vary widely among members of the order, within individual organisms, cell diameter and cell shape are quite constant. Cell diameters vary from 0.5–100 μm, and cell dimensions range from being much longer than wide (rod-shaped) to being much wider than long (disk-shaped).

Nostocales

The order Nostocales includes filamentous cyanobacteria that are capable of cell differentiation and that divide by binary fission in one plane at right angles to the long axis of the trichomes. Differentiation may result in the production of several types of specialized cells: 1) heterocysts; 2) akinetes; 3) specialized reproductive trichomes (hormogonia) whose cells are morphologically distinguishable from vegetative cells; and 4) tapered trichomes, formed usually in response to a nitrogen gradient caused by terminally located heterocysts.

Stigonematales

Members of the order Stigonematales are filamentous cyanobacteria capable of the same degree of cellular differentiation as members of the Nostocales, but in addition they are able to divide by binary fission in multiple planes. The resulting thalli may display true branching and possess both uniseriate and multiseriate trichomes.

Phylogeny

Until the advent of molecular sequencing it was not possible to quantitatively determine phylogenetic relationships among prokaryotes. Woese and his colleagues have revolutionized the study of prokaryotic phylogeny using ribosomal RNA sequence analyses. Using both partial and complete 16S rRNA sequences, Woese has divided the prokaryotes into two major subgroups, the archaebacteria and the eubacteria (Woese, 1987). The cyanobacteria, as well as the chloroplasts of higher plants and the prochlorophytes (Turner et al., 1988) all fall within a common lineage that comprises one of 10 major eubacterial taxa (Woese, 1987). At the time of Woese's review, the cyanobacteria were circumscribed by eight rRNA sequences from isolates representing only a minor portion of cyanobacterial diversity (Bonen et al., 1979).

A more detailed study by Giovannoni et al. (1988) analyzed the 16S rRNAs of 29 strains of cyanobacteria, including key reference strains from each of the five orders. This study permitted its authors to delineate the major phylogenetic patterns of the cyanobacteria currently available in pure culture. They found that the rRNA sequence diversity within the cyanobacteria is considerably less than the diversity that separates other major eubacterial taxa, indicating that relatively close phylogenetic relationships underlie the extensive morphological diversity that occurs within the cyanobacteria. In addition, many of the cyanobacterial lineages have similar branching depths (Fig. 1), indicating that the modern groups arose from an expansive radiation. The combination of relatively small sequence diversity and the fanlike radiation of the lineages makes it impossible at present to determine the precise branching orders of many of the cyanobacterial groups. Members of the Chroococcales and the Oscillatoriales are dispersed throughout the phylogenetic tree, indicating that these two orders, as presently constituted, do not represent coherent evolutionary lineages (Fig. 1). Members of the Pleurocapsales fall within a single lineage indicating that reproduction by multiple fission is of monophyletic origin (Fig. 1). Members of the Nostocales and Stigonematales fall within a single coherent lineage, suggesting that the ordinal

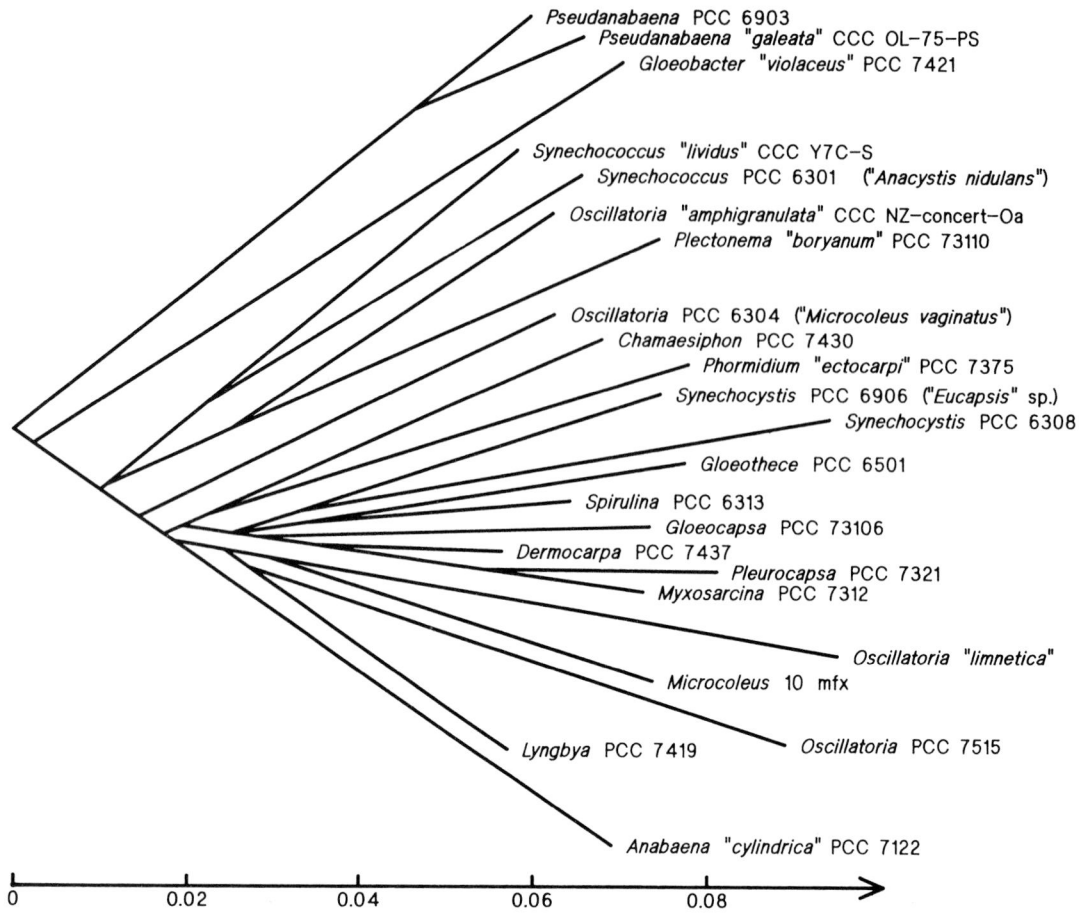

Fig. 1. Rooted-tree topology illustrating evolutionary relationships among 16S rRNAs from cyanobacteria. Evolutionary distances are proportional to the horizontal component of segment length in this representation. *A. tumefaciens, B. subtilis,* and *P. testosteroni* 16S rRNA sequences were used to locate the root. Only one heterocystous cyanobacterium, *Anabaena* sp. strain PCC 7122, is included in this tree. The scale is in units of fixed point mutations per sequence position. (From Giovannoni et al., 1988.)

separation of these two groups may be unwarranted (Fig. 2).

Physiological Properties Relating to Habitat Diversity

Cyanobacteria possess properties that allow them to successfully inhabit and often dominate a wide variety of illuminated habitats. Their dominant mode of nutrition is photoautotrophy, using a photosynthetic apparatus that carries out oxygenic plantlike photosynthesis with chlorophyll *a* as the primary photosynthetic pigment and phycobiliproteins as auxilliary light-harvesting pigments (Bryant, 1986; van Liere and Walsby, 1982; see also Chapter 98). Some cyanobacteria are capable of switching from oxygenic photosynthesis to anoxygenic bacterial-type photosynthesis when they occur

in environments where hydrogen sulfide is present at relatively high concentrations (3 mM) (Cohen et al., 1986; Padan and Cohen, 1982).

Cyanobacteria can exploit niches varying from very dimly lit caves to salt-marsh algal mats exposed to full sunlight (Wyman and Fay, 1987). Adaptations to light include the ability to alter their photosynthetic apparatus in response to light quantity and quality by changing the surface area of their photosynthetic thylakoids, and by changing the size and phycobiliprotein composition of their phycobilisomes (Cohen-Bazire and Bryant, 1982; Glazer, 1981; Kana and Glibert, 1987a, 1987b). In addition, some of the phycoerythrin-containing cyanobacteria can adapt to changes in light quality by complementary chromatic adaptation, a process in which phycobiliprotein synthesis is under light-wavelength control (Tandeau de Marsac, 1977, 1983). Cyanobacteria also protect themselves from growth-inhibiting light intens-

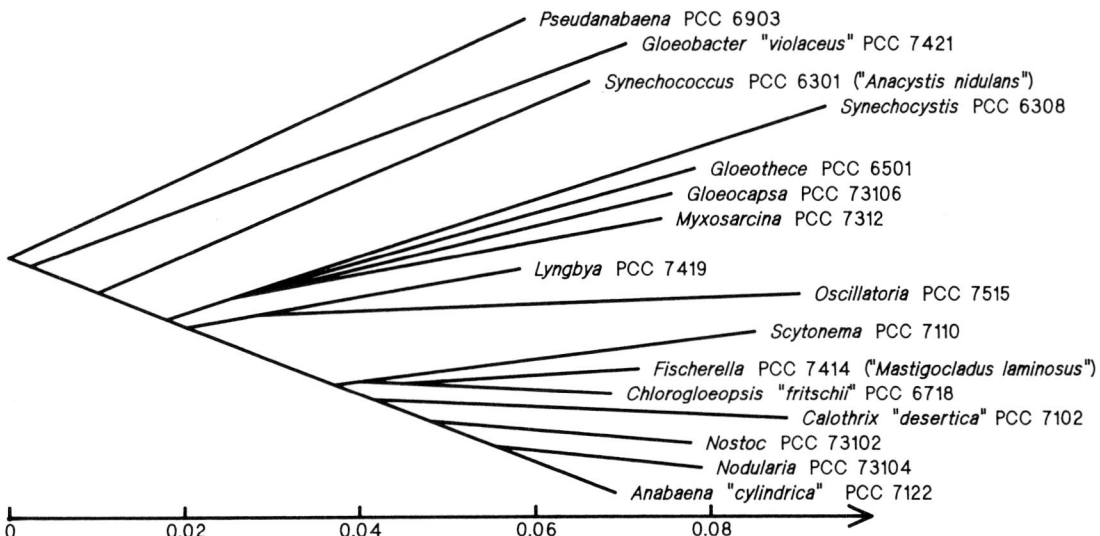

Fig. 2. Rooted-tree topology illustrating evolutionary relationships among 16S rRNAs from heterocystous cyanobacteria. See the caption of Fig. 1 for explanation. (From Giovannoni et al., 1988.)

ities with a variety of pigments that adsorb harmful radiation (e.g., carotenoids and xanthophylls) and possibly through structures that scatter light (e.g., gas vacuoles and calcified sheaths) (van Liere and Walsby, 1982). Finally, some cyanobacteria have the ability to position themselves in a light field through buoyancy regulation or active motility. Cyanobacteria derive buoyancy from the possession of gas vesicles and are thought to regulate their position in a water column through the use of one or more of three mechanisms: 1) regulation of gas vesicle formation; 2) gas vesicle collapse resulting from high cell turgor pressure; and 3) changes in cell density caused by the temporal cycling of the synthesis and degradation of carbohydrate reserves (van Liere and Walsby, 1982; Walsby, 1978, 1987). Gliding motility coupled with phototaxis permits many filamentous cyanobacteria to move to areas of optimal light conditions (Castenholz, 1982). In microbial mats, for example, filamentous cyanobacteria can migrate to different levels within the mat in response to light intensity.

Cyanobacteria are predominantly photoautotrophic although some are capable of heterotrophy when grown under laboratory conditions. The ability to grow photoheterotrophically using simple organic compounds as sole carbon sources and light as a source of energy is relatively common among cultured cyanobacteria, whereas the ability to grow chemoheterotrophically in the dark using an organic compound as a sole source of carbon and energy is more restricted (Rippka et al., 1979). The significance of photoheterotrophy

and chemoheterotrophy to cyanobacteria in natural environments is still unresolved (Smith, 1982). However, it is unlikely that either process will be found to be important to the metabolism of free-living cyanobacteria, since the range of substrates is limited to a restricted number of simple sugars that can be metabolized via the oxidative pentose-phosphate cycle, and the concentrations of these compounds are usually too low in nature to contribute significantly to cyanobacterial carbon metabolism

Cyanobacteria are also capable of mixotrophy, a process in which a variety of organic compounds, such as amino acids, that cannot serve as sole carbon sources, are assimilated as a supplement to autotrophic CO_2 fixation. This use of organic compounds is potentially more important than either photo- or chemoheterotrophy to the metabolism of free-living cyanobacteria in nature, but it is poorly documented at present.

The ability of many cyanobacteria to fix dinitrogen permits them to exploit habitats low in combined nitrogen. One of the key physiological properties of nitrogenase, the enzyme system responsible for dinitrogen fixation, is its sensitivity to oxygen inactivation and repression. This presents an acute problem for nitrogen-fixing cyanobacteria that not only live in oxygenated environments but also produce oxygen during photosynthesis. Some filamentous cyanobacteria have solved this problem by spatially separating the processes of oxygenic photosynthesis and dinitrogen fixation through the production of highly differentiated cells, known as heterocysts, that are specialized sites of dinitrogen fixation (Bothe, 1982). Heterocysts lack the ability to carry out oxygenic photosynthesis. There are also a growing number of non-heterocystous, filamentous (Pearson et al., 1979; Stahl and Krumbein, 1985) and unicellular (Gallon et al., 1974; Huang and Chow, 1986, 1988; Mitsui et al., 1987) cyanobacteria that are capable of fixing dinitrogen by temporally separating these processes. In these organisms, photosynthesis occurs during the light period, while dinitrogen fixation occurs during the dark at the expense of carbohydrate reserves (glycogen) built up during the photosynthetic period.

Cyanobacteria are found in environments with quite different temperature ranges. Most cyanobacteria are mesophilic and live in environments where temperature may range from freezing to 40°C. They typically have growth optima between 20 and 35°C and maximum temperatures permitting growth below 45°C. Cyanobacteria isolated from the open oceans, where temperature ranges are more moderate, often have temperature maxima near 30°C (Waterbury et al., 1986). One tropical, marine, unicellular, nitrogen-fixing cyanobacterium has a temperature range permitting growth only between 26°C and 32°C, which is one of the narrowest temperature ranges known for a free-living mesophilic prokaryote.

Cyanobacteria are also commonly found in more extreme environments. In Antarctica, they are present as cryptoendoliths in rocks in the cold dry deserts and in the plankton and microbial mats of lakes. With the exception of one strain of *Chroococcidiopsis* sp. that is a psychrophile, the cyanobacterial isolates from the Antarctic desert rocks are mesophiles, with temperature optima near 35°C (I. Friedmann, personal communication). On the other hand, many of the cyanobacteria isolated from Antarctic lakes are psychrophiles, having maximum temperatures permitting growth of 20°C (Seaburg, 1981).

Cyanobacteria, including representatives of each of the five orders, are conspicuous inhabitants of hot springs where they occur at temperatures up to 74°C. Thermophilic strains of cyanobacteria have growth optima above 45°C and often fail to grow, but can survive, at room temperature (Castenholz, 1981).

Cyanobacteria occur in habitats of widely differing salinity. Freshwater habitats contain diverse and often prominent populations of cyanobacteria. In marine habitats, cyanobacterial isolates can be divided into two categories based on their major ionic requirements for growth. Some are halotolerant and grow equally well on a medium with either a seawater or

freshwater base. Others have obligate requirements for concentrations of sodium, magnesium, calcium, and chloride that reflect the chemistry of seawater. These requirements are not met by supplementing freshwater media with sodium chloride alone (Waterbury et al., 1986). Many cyanobacteria isolated from soils are tolerant of salt concentrations in excess of 1 M NaCl, whereas marine isolates usually fail to grow at this concentration, probably because the salinity of seawater has been relatively constant for long periods.

The occurrence of cyanobacteria in hypersaline environments is well documented in the classical descriptive literature (Hof and Frémy, 1933) but only a very restricted group is truly halophilic. Individual strains of *Aphanothece halophytica* isolated from a salt evaporation pond (Yopp et al., 1978), from Great Salt Lake, Utah (Brock, 1976), and the Solar Lake, Sinai, Israel (Cohen, 1975) each have major ionic requirements for growth that reflect the chemistry of their individual habitats. The isolates from the salt evaporation pond and from Great Salt Lake are truly halophilic. They grow optimally in approximately 2 M NaCl and have minimum NaCl requirements for growth of 0.7 M and 1.0 M NaCl, respectively.

Cyanobacteria are most prominent in habitats with neutral to alkaline pH. They are excluded from highly acidic environments but are characteristic inhabitants of acidic hot springs and peat bogs where the pH is above 5 (Castenholz, 1981, 1988b). To date, isolates from acidic hot springs and peat bogs grow optimally at neutral pH, indicating that they are mildly acid-tolerant rather than acidophilic (Rippka et al., 1981a). A restricted number of cyanobacteria are characteristic of highly alkaline habitats; for example, *Spirulina platensis,* a dominant cyanobacterium in highly alkaline lakes, has a pH optima for growth between 8 and 11 (Ciferri, 1982).

Cyanobacteria are often found under conditions of extreme desiccation in habitats ranging from soils to both tropical and Antarctic deserts (Fogg et al., 1973). Viability can be maintained for extended periods under desiccated conditions but active growth is limited to wet seasons (de Winder et al., 1989).

Although cyanobacteria are present, and in many instances, conspicuous, under extreme conditions of temperature, salinity, pH, and desiccation, cultures isolated from these habitats often possess optima for growth that are more moderate. Thus, many cyanobacteria seem to be able to tolerate extreme environments but relatively few are obligate extremophiles.

In addition to the diverse habitats occupied by free-living forms, cyanobacteria also occur in symbiosis with a wide variety of eukaryotes (see Chapter 212). The ability of cyanobacteria to fix dinitrogen usually plays an important role in these associations.

Isolation and Purification

There have been numerous reviews detailing procedures for the isolation and purification of cyanobacteria. General treatments include Castenholz (1988a), Castenholz and Waterbury (1989), and Rippka (1988a). Articles concerning specific groups of cyanobacteria include Castenholz (1988b) for thermophiles, Wolk (1988) for filamentous nitrogen fixers, and Mitsui and Cao (1988), Waterbury et al. (1986), and Waterbury and Willey (1988) for marine cyanobacteria. In addition, the chapters on cyanobacteria included in the first edition of *The Prokaryotes* (Starr et al., 1981) contain methods and media that are still timely. The discussion here is a synopsis of the literature cited above.

Collection and Treatment of Samples

Many cyanobacteria are conspicuous in nature, a feature that greatly facilitates sample collection. They are conspicuous as major components of microbial mats in both freshwater and marine marshes, in desert microbial crusts, on moist rocks in terrestrial habitats, and in the marine intertidal zone and as epiphytic colonies on terrestrial and freshwater plants and marine macroalgae. They are often less conspicuous in aquatic environments but characteristic species of planktonic cyanobacteria form dense blooms in both freshwater and marine environments (Walsby, 1981). Even in aquatic environments where blooms are not evident, small unicellular cyanobacteria are often abundant and important components of the microbial food webs in freshwater lakes (Caron et al., 1985; Pick and Caron, 1987) and in the open ocean (Waterbury et al., 1986).

The first step in isolating cyanobacteria should be a careful description of the sampling site. In addition to the site location and description, parameters such as light intensity, temperature, pH, and salinity, should be recorded. This information is not only useful for describing the natural habitat of the cyanobacteria to be isolated but can be critically important when designing culture media and conditions of incubation.

Samples for enrichment of cyanobacteria that are not growing submerged in water can be sampled by removing rock chips, microbial tufts, etc., and placing them, in small containers. The collected samples should not be immersed in water but should be kept moist by the addition of wetted paper to avoid growth of contaminating microorganisms. Precaution should be taken to avoid contamination in collecting and storing water samples from which cyanobacteria are to be cultured. In extreme cases, for example, to isolate cyanobacteria from very oligotrophic waters, it is necessary to rigorously clean water samplers, storage containers, and culture vessels with acid followed by repeated washings with ultrapure deionized water. Samples of natural populations should be transported to the laboratory for processing as quickly as possible, taking care to avoid exposing them to extremes of temperature or light.

Prior to culturing, samples should be examined carefully and if possible photographed using light microscopy to determine and document the identity and relative abundances of the cyanobacteria present and to assess the extent of contamination by undesirable microorganisms. The identification of cyanobacteria in field samples is best achieved using the botanical descriptive literature. The treatise of Geitler (1932) or later treatises that follow the "Geitlerian School" have proven to be the most useful (Anagnostidis and Komárek, 1985, 1988; Bourrelly, 1985; Desikachary, 1959; Komárek and Anagnostidis, 1986).

Portions of nonaqueous natural samples in which the cyanobacteria are conspicuous should be preserved (usually by drying) and deposited in a recognized herbarium to satisfy the botanical requirement for type material. In practice, good photomicrographs of the natural material will provide excellent supplementary documentation of the type material and will aid in correlating the identity of the cyanobacteria in the natural material with the identity of cultures isolated from the sample.

Except in rare instances where the natural sample appears to be monospecific, it will be necessary to physically manipulate the sample both to break up compact clumps of organisms and to separate the cyanobacterium of particular interest from other cyanobacteria and contaminating microorganisms. A good method of dispersal in samples that contain unicellular forms is to gently grind the natural material, wetted with sterile medium, between two sterile microscope slides. Samples containing filamentous forms can be dispersed using sterile dissecting needles to tease apart clumps or tufts of natural material that have been placed in drops of sterile medium in the bottom of petri dishes. This process is facilitated by monitoring and making the manipulations using a dissecting microscope.

Once the samples have been dispersed it is often possible to remove some of the contaminating bacteria by washing with sterile medium. This can be accomplished for samples containing unicellular cyanobacteria by repeatedly washing the cells in a small sterile filter apparatus set up with a membrane filter with a pore size slightly smaller than the cyanobacterial cell. With the aid of a dissecting microscope, samples containing filamentous cyanobacteria can often be washed by picking individual trichomes with a platinum needle or a drawn-out Pasteur pipette. The individual trichomes are then passed through successive large drops of sterile liquid medium in the bottom of a petri dish or they can be dragged through solidified agar medium to physically remove attached contaminating organisms.

Media and Growth Conditions

Many media have been described for cyanobacteria. Included here are nine different media, some of which have proven successful for the isolation and maintenance of a wide variety of cyanobacteria and some designed for the growth of more specialized groups (Tables 1 and 2).

Since they are photoautotrophs, cyanobacteria can be grown in simple mineral media. Vitamin B_{12} is the only growth factor known and is only required by a few marine isolates (Waterbury and Stanier, 1978). Media should be designed to mimic the chemical composition of the environment from which cyanobacteria are to be isolated. In addition, they must be supplemented with essential nutrients needed to support cell growth, including sources of nitrogen, phosphorus, trace elements, etc. When cyanobacteria are being isolated from extreme environments, it is prudent to use two different media: one having the characteristics of the extreme environment and one in which the extreme variable(s) is moderated. Nutrient levels in growth and maintenance media are usually high to support large cell yields. It is often advantageous to dramatically lower the nutrient levels in media that are used for enrichment and isolation. For example, compare the high nutrient levels in medium BG-11 with those in Aquil (Table 1).

To date, several hundred cyanobacteria, including representatives from most of the major groups, have been isolated and purified using standard microbiological techniques. Despite this success there still are many cyanobacteria that have not yet been cultured. Of these, cyanobacteria characteristically found in the oligotrophic waters of the oceans and freshwater lakes have been particularly problematic.

It is likely that our inability to culture certain cyanobacteria is not because we fail to add something to the medium that the organisms require, but rather because we inadvertently add constituents that are toxic. Toxicity may derive from contaminants left on the surface of inadequately cleaned samplers and culture vessels, from contaminants present in either the chemicals or water used to make culture media, or from the intentional addition of necessary compounds to culture media in quantities high enough to become toxic.

The addition of copper to cyanobacterial media is an interesting example. Copper is a nearly universal component of trace element solutions used to supplement culture media for a wide variety of microorganisms, including cyanobacteria (Table 2). However, copper has also been shown to be extremely toxic. Rueter et al. (1979) examined the effects of copper on *Trichodesmium thiebautii*, a prominent and, at that time, unculturable oceanic cyanobacterium. They found that the addition of 10^{-8} M copper caused a marked decrease in the rate of CO_2 fixation by natural populations of *T. thiebautii*. Waterbury and Stanier (1978) used a marine medium (MN) to isolate and culture a wide variety of intertidal marine cyanobacteria. Medium MN had a natural seawater base and used nutrient additions and trace metals from medium BG-11 (Tables 1 and 2). It proved quite successful for the isolation of coastal marine cyanobacteria but would not support the growth of open-ocean strains of marine *Synechococcus* spp. (Waterbury et al., 1979, 1986). The fact that medium MN contained 3.2×10^{-7} M copper, more than was necessary to inhibit photosynthesis in *Trichodesmium* sp., led us to design medium SN, whose principal difference from medium MN and was the omission of copper from its trace metal solution (Table 2). Medium SN and its more dilute form, SNAX, have been used successfully for the maintenance and isolation of many oceanic and coastal marine cyanobacteria.

Ohki et al. (1986) succeeded in culturing *Trichodesmium thiebautii* and *T. erythraeum* using the artificial seawater medium Aquil (Morel et al., 1979; see also Table 1) in which the phosphate was reduced to 0.5 μg/l and copper was deleted from the trace element colution (PIV) (Table 2). Aquil was designed to minimize heavy metal contamination by passing its major nutrients through Chelex 100 columns (Bio-Rad Laboratories, Richmond, CA). Two factors seem to have contributed to the successful cul-

Table 1. Nine recipes for cyanobacterial media.

Ingredient (amount per liter)	Medium designation								
	BG-11[a]	Z8[b]	D[c]	AO[d]	Aquil[e]	SN[f]	SNAX[g]	RC[h]	YOPP[i]
Deionized water (ml)	1000	1000	1000	1000	1000	250	250	1000	1000
Seawater (ml)						750	750		
NaCl (g)			0.008	1.0	24.36			125	116.9
$MgSO_4 \cdot 7H_2O$ (g)	0.075	0.25	0.1	0.2				3.5	10.0
$MgCl_2 \cdot 6H_2O$ (g)					11.03			10.0	10.68
KCl (g)					0.7			2.5	2.0
Na_2SO_4 (g)					4.09				
K_2SO_4 (mg)				1000					
$CaSO_4 \cdot 2H_2O$ (mg)			60						
$CaCl_2 \cdot 2H_2O$ (mg)	36			40	1000–1350				500
$NaNO_3$ (mg)	1500	467	700	2500	8.5	750	75	750	
KNO_3 (mg)			100						
$Ca(NO_3)_2 \cdot 4H_2O$ (mg)		59							1000
NH_4Cl (mg)		31					5.3		
K_2HPO_4 (mg)	30			500[j]		15	1.5	15	
KH_2PO_4 (mg)									50
Na_2HPO_4 (mg)			110						
$NaH_2PO_4 \cdot H_2O$ (mg)					0.5				
Na_2CO_3 (g)	0.02	0.02		4.03[j]		0.01	0.001	0.02	
$NaHCO_3$ (g)				13.61[j]	0.2				
KBr (mg)					10.0				
NaF (mg)					3.0				
Na_2 EDTA·$2H_2O$ (mg)	1.0		80		5	0.5	0.5		
NTA (mg)			100						
Ferric ammonium citrate (brown crystals) (mg)	6.0							3.0	
Citric acid (mg)	6.0								
$FeCl_3$ (mg)		0.3							
$FeSO_4 \cdot 7H_2O$ (mg)				10					
Ferric EDTA (acid 10% Fe) (mg)									5.0
Fe EDTA[l] (ml)		10							
Glycylglycine buffer (mg)								3.0	
Micronutrients (ml)	1.0	1.0	0.5	1.0[k]	0.5	1.0	0.1	1.0	1.0
Micronutrient mix used (see Table 2)	A5 + Co	Gaffron	D micro	Gaffron or A5 + CO	PIV	Cyano	Cyano	A5 + Co	Sheridan and Castenholz
Final pH after autoclaving	7.1	—	7.5[m]	9.4–9.8		7.8–8.2	7.8–8.2	—	7.8[n]

[a]From Allen (1968).
[b]From Kotai (1972).
[c]From Castenholz (1981).
[d]From Aiba and Ogawa (1977).
[e]From Morel et al. (1979).
[f]From Waterbury and Willey (1988).
[g]From Waterbury et al (1986).
[h]From van Rijn and Cohen (1983).
[i]From Yopp et al. (1978).
[j]These three ingredients are autoclaved as a separate solution and added to the medium after cooling.
[k]Original medium used trace metal mixes A5 + B6 (see Rippka, 1988a). Gaffron's trace metal mix is quite similar and has been substituted for A5 + B6.
[l]Solution A, 2.8 g $FeCl_3$ in 100 ml 0.1 N HCl; solution B, 3.9 gm EDTA-disodium in 100 ml 0.1 N NaOH. Add 10.0 ml solution A and 9.5 ml solution B plus water to 1 liter.
[m]pH adjusted to 8.2 with NaOH before autoclaving.
[n]pH adjusted to 7.8 with NaOH before autoclaving.

Table 2. Six recipes for micronutrient solutions.[a]

Ingredient (amount per liter)	A5 + Co[b]	D micro[c]	Sheridan and Castenholz[d]	Cyano[e]	Gaffron[f]	PIV[g]
Deionized H_2O (ml)	1000	1000	1000	1000	1000	1000
H_2SO_4 (conc.) (ml)		0.5				
HCl (conc.) (ml)			3.0			
H_3BO_3 (g)	2.86	0.5	0.5		3.1	
$MnCl_2 \cdot 4H_2O$ (g)	1.81		2.0	1.4		0.041
$MnSO_4 \cdot 4H_2O$ (g)		3.01			2.23	
$ZnSO_4 \cdot 7H_2O$ (g)	0.22	0.5		0.22	0.287	
$ZnNO_3 \cdot 6H_2O$ (g)			0.5			
ZnCl (g)						0.005
$NaMoO_4 \cdot 2H_2O$ (g)	0.39	0.025	0.025	0.39		0.004
$(NH_4)_6Mo_7O_{24} \cdot 4H_2O$ (g)					0.088	
$CuSO_4 \cdot 5H_2O$ (g)	0.079	0.025				
$CuCl_2 \cdot 2H_2O$ (g)			0.025			
$Co(NO_3)_2 \cdot 6H_2O$ (g)	0.049		0.025	0.025	0.146	
$CoCl_2 \cdot 6H_2O$ (g)		0.045				0.002
$VOSO_4 \cdot 6H_2O$ (g)			0.025		0.054	
$Al_2(SO_4)_3 K_2SO_4 \cdot 2H_2O$ (g)					0.474	
$NiSO_4(NH_4)_2SO_4 \cdot 6H_2O$ (g)					0.198	
$Cd(NO_3)_2 \cdot 4H_2O$ (g)					0.154	
$Cr(NO_3)_3 \cdot 7H_2O$ (g)					0.037	
$Na_2WO_4 \cdot 2H_2O$ (g)					0.033	
KBr (g)					0.119	
KI (g)					0.083	
$FeCl_3 \cdot 6H_2O$ (g)						0.097
Disodium EDTA (g)						0.75[h]
Citric acid·H_2O (g)				6.25		
Ferric ammonium citrate (brown crystals) (g)				6.00		
Amount added per liter of medium	1.0 ml	1.0 ml	1.0 ml	0.1–1.0 ml	0.1 ml	0.5 ml

[a]It is usually best to dissolve each compound separately and then add dissolved components together and bring to 1 liter.
[b]From Rippka et al. (1979).
[c]From Castenholz (1981).
[d]From Waterbury and Stanier (1981).
[e]From Waterbury et al. (1986).
[f]From Hughes et al. (1958).
[g]From Starr (1978).
[h]The disodium EDTA should be added first.

turing of *Trichodesmium* spp.: the removal of heavy metals from the nutrients added to Aquil and the omission of copper as a trace element.

Copper may also have played a role in the culturing of freshwater planktonic cyanobacteria. Medium Z8, originally described by Kotai (1972), has been used by Skulberg and his colleagues (Skullberg, 1983) to isolate and maintain a large collection of freshwater planktonic cyanobacteria. Skulberg used Gaffron's trace element solution as described by Hughes et al. (1958). In the paper by Hughes et al. (1958), Gaffron's trace element solution did not contain copper, but by 1960 (Zehnder and Gorham, 1960) copper had been added to Gaffron's trace metal solution.

These examples suggest that it may be beneficial to omit copper from all trace element solutions used to grow cyanobacteria. If cyanobacteria have a copper requirement for growth, it can probably be met by traces present as contaminants in the water and the chemicals used to make the medium.

The quality of water and the purity of chemicals used to make media are critically important for the successful cultivation of cyanobacteria. Freshwater media and nutrient stock solutions should be made with either double-distilled water or, preferably, with single-distilled water which has been subject to ion exchange and to activated charcoal columns such as the Millipore-Q water purifier. Marine cyanobacteria usually do best in media made with natural seawater. Seawater should be filtered through glass fiber filters at the time of collection and may be stored for extended periods prior to use in carefully cleaned plastic carboys. Chemicals used to prepare both basal media and nutrient solutions should be of the highest quality. In situations where heavy metal contamination is critical, for example, in culturing oligotrophic cyanobacteria such as *Trichodesmium* spp., both the basal medium and nutrient solutions can be treated with Chelex 100 (Morel et al., 1979).

Care should also be taken to assure that glass and plastic ware that are used to culture cyanobacteria are carefully cleaned. In situations where cleanliness is critical, for example, to avoid trace metal contamination, the following protocol that is a modification of one proposed by Fitzwater et al. (1982) can be used. Glass or plastic ware are soaked in Micro detergent (International Products Corporation, Trenton, NJ 08602-0118) at a final concentration of the detergent of 20 ml per 1 of Super-Q water (Millipore) for one week with occasional agitation. The vessels are then rinsed eight times in Super-Q water and then soaked in high-quality 0.5 N HCl for an additional week, followed by eight washes in Super-Q water.

Liquid Media

Medium BG-11 (Table 1), originally described by Hughes et al. (1958) and modified by Allen (1968), has been widely used for the isolation and maintenance of many cyanobacteria. When it is used as an enrichment medium it is beneficial to reduce the nitrate concentration by as much as a factor of 10 (Rippka, 1988a). A variant of this medium called BG-11$_0$ lacks sodium nitrate and is used for cyanobacteria capable of dinitrogen fixation. Sodium lost by the removal of sodium nitrate should be replaced by the addition of 1.0 g/liter NaCl to medium BG-11$_0$.

Medium Z8, originally described by Kotai (1972), has been used by Skulberg (1983) to isolate and maintain a large collection of freshwater planktonic cyanobacteria. The omission of copper from its trace element solution may be the key to its ability to support the growth of freshwater oligotrophic cyanobacteria.

Medium D was developed by Castenholz (1981) to isolate and maintain thermophilic cyanobacteria. Variants of this medium (Castenholz, 1988a) have also been used to culture a wide variety of nonthermophilic cyanobacteria.

Medium AO is a very alkaline medium used by Aiba and Ogawa (1977) for the isolation and maintenance of *Arthrospira platensis,* a filamentous cyanobacterium characteristic of alkaline lakes.

Aquil is a defined marine medium designed by Morel et al. (1979) for trace metal studies

with marine phytoplankton. Ohki et al. (1986) used Aquil modified by a threefold reduction in the phosphate concentration to successfully culture *Trichodesmium* spp., an important oceanic dinitrogen-fixing cyanobacterium that had hitherto been unculturable. Detailed instructions for the preparation of Aquil are given in Morel et al. (1979).

Medium SN and its more dilute counterpart SNAX are marine media that have been used to maintain and isolate a wide variety of open-ocean unicellular cyanobacteria (Waterbury et al., 1986; Waterbury and Willey, 1988). They have replaced medium MN (Waterbury and Stanier, 1978, 1981), which had been used to culture coastal isolates but failed to support growth of the open-ocean cyanobacteria. The principal difference between them is that medium SN contains no added copper in the Cyano trace element solution (Table 2).

Medium RC is a medium used by van Rijn and Cohen (1983) to support the growth of a halotolerant strain of *Aphanothece halophytica* isolated from Solar Lake, Sinai, Israel. This strain grew optimally at salinities between 0.5 and 1 M NaCl but could tolerate salinities approaching 3 M NaCl.

The YOPP medium was designed to support the growth of an obligately halophilic strain of *Aphanothece halophytica* that was isolated from a salt evaporation pond in the southern San Franciso Bay (Yopp et al., 1978). Their strain grew optimally in approximately 2 M NaCl and had a minimum NaCl requirement for growth of 0.7 M.

Solid Media

Solid media should be prepared by separate autoclaving of the agar and mineral solutions, and subsequent mixing after they have cooled to 50°C (Allen, 1968). The agar concentration should be kept as low as possible, maintaining just enough firmness to permit streaking. Growth of cyanobacteria is often slow, requiring long incubation times. To prevent the agar plates from drying they should contain approximately 40 ml of medium per petri dish and should be incubated in clear plastic boxes (available in stores as vegetable crispers or sweater boxes).

The general protocol described has been modified as follows for more fastidious marine unicellular cyanobacteria (Waterbury et al., 1986; Waterbury and Willey, 1988): Solid media are prepared using Difco Bacto-agar that has been further purified as follows: 100 g of agar is washed by stirring with 3 liters of double-distilled water in a 4-liter beaker. After 30 min of stirring, the agar is allowed to settle, the wash water is siphoned off, and the agar is filtered onto Whatman F4 filter paper in a Büchner funnel. This procedure is repeated once more or until the filtrate is clear. The agar is then washed with 3 liters of 95% ethanol followed by a final 3-liter wash with analytical grade acetone. The agar is then dried at 50°C in glass baking dishes for 2-3 days and stored in a tightly covered container. Solid media prepared with the purified agar at a final concentration of 0.7% are sufficiently stable for streaking.

To prepare 40 agar plates of medium SN from 1 liter of medium, the following three solutions are prepared and autoclaved separately: 1) 750 ml of filtered seawater in a Teflon bottle; 2) 7.0 g super-clean agar in 200 ml of double-distilled water in a 2-liter glass flask; and 3) the mineral salts for 1 liter of medium in 50 ml of double-distilled water in a 125-ml glass flask. After autoclaving, the seawater and minerals are added to the flask of agar. Sterile sodium sulfite (2 mM final concentration) is added aseptically to the hot agar solution, which is then cooled to 50° before the plates are poured. It is critical that the surface of agar plates be dry prior to streaking.

Conditions of Incubation

LIGHT. Although cyanobacteria occur naturally under a wide range of light conditions, they usually grow best when cultured at light intensities varying from $10-75$ μEin·m^{-2}·sec^{-1} supplied by warm- or cool-white fluorescent lamps. Traditionally, cyanobacteria have been cultured under conditions of constant illumination; however, some cyanobacteria, for example, nonheterocystous nitrogen fixers that temporally separate dinitrogen fixation and photosynthesis, require a light-dark cycle. We recommend that during isolation and purification, cultures be incubated in a light-dark cycle with a 14-h light period and a 10-h dark period.

TEMPERATURE. The temperature of incubation during isolation and purification should be chosen on the basis of the temperature range of the natural environment where the sample was taken. During the initial stages of isolation it is prudent to incubate samples at several temperatures until the range that will support good growth is determined.

Techniques of Isolation and Purification

Isolation and purification are facilitated by the fact that cyanobacteria are prominent and visible components of many habitats. Natural samples can be collected, dispersed, and the individual cyanobacteria isolated directly. Samples in which unicellular forms predominant can be streaked on agar plates. Samples rich in gliding filamentous forms can be teased apart and small portions of inoculum placed in the center of agar plates. Whole trichomes or hormogonia will glide away from the central inoculum, and these migrating trichomes can be removed with a platinum spade by cutting out a small block of agar and transferring it to a new agar plate. In addition to the direct isolations on agar plates, secondary cultures for each sample should be established in the appropriate liquid medium. Aqueous samples that are too dilute to isolate colonies directly on agar plates should be enriched in a liquid medium. Serial dilution of aqueous samples during the enrichment process often aids in isolating cyanobacteria.

Although successive streaking or transferring from agar plate to agar plate is the standard method for purifying cyanobacteria, there are enumerable variations that can be tried, many of which are described in the reviews cited at the beginning of the section on isolation and purification.

Purification of unicellular cyanobacteria can be achieved by streaking on solid media. It is usually advisable to transfer isolated colonies picked from agar plates into liquid medium and to allow them to grow up before restreaking. Motile filamentous cyanobacteria can be purified by transferring individual trichomes or hormogonia to new agar plates and allowing them to glide away from the contaminants before transferring them to a successive plate.

Small heterotrophic bacteria are frequently nearly invisible contaminants of cyanobacterial cultures, and success at purifying cyanobacteria requires diligence. Agar plates should be examined daily with a dissecting microscope. The time-window within which it is possible to detect isolated cyanobacterial colonies and transfer them before they become overrun by heterotrophic bacterial contaminants can be quite short. Some of the rapidly gliding filamentous forms can clean themselves of contaminants within hours of inoculation.

Careful microscopic examination of stationary phase cultures, using a combination of phase contrast and epifluorescence illumination, is the most rigorous test of purity.

Maintenance of Pure Cultures

Stock Cultures

The majority of cyanobacteria currently in pure culture are best maintained on agar slants. However, some of the planktonic cyanobacteria, especially the heavily gas-vacuolated forms, and rapidly gliding filamentous forms should be maintained in liquid cultures. Nitrogen fixers should be maintained on media devoid of combined nitrogen.

The intervals between successive transfers of stock cultures are variable. Some of the more sensitive strains must be transferred every two weeks, while others on slants may last months. Stock cultures should be incubated in low light at a temperature appropriate for the individual strains. Stock cultures of cyanobacteria should never be stored in a dark refrigerator!

Long-Term Preservation

Storage in liquid nitrogen is the best method for the long-term preservation of cyanobacteria. Dimethylsulfoxide (5-10% v/v) is added as a cryoprotectant to stationary phase cultures that have been concentrated by centrifugation. The cell concentrate is dispensed into plastic cryogenic ampules and frozen slowly at 1°C/min until they reach −80°C. The slow freezing is satisfactorily accomplished by placing the ampules in a −80°C freezer for one hour. The ampules are then placed in either the vapor or liquid phase of the liquid nitrogen storage container. Samples are reactivated by placing them in a 37°C water bath and transferring them into fresh liquid medium immediately upon thawing. Heavily gas-vacuolated cyanobacteria can be successfully frozen in liquid nitrogen if their gas vacuoles are collapsed by centrifugation prior to freezing (Rippka et al., 1981a).

The Identification of Cyanobacteria in Pure Culture

The keys and descriptions of cyanobacteria in pure culture used here are based on the treatises of Rippka et al. (1979, 1981b) and Rippka (1988b), and on the first treatment of the cyanobacteria in *Bergey's Manual of Systematic Bacteriology* (Castenholz, 1989a, 1989b, 1989c; Waterbury, 1989; Waterbury and Rippka, 1989). The descriptions of many of the genera and "groups" are provisional. Even though several hundred different cyanobacteria have been obtained in pure culture and partially characterized, there remains an enormous amount of work before a satisfactory system of classification for this important group of photosynthetic prokaryotes is achieved. The following key may be useful, however.

Key to the Orders of Cyanobacteria

A. Unicellular or nonfilamentous aggregates of cells held together by extracellular slime or sheath layers .... B

A. Filamentous: Range of morphology from simple trichomes to highly differentiated branching thalli C

B. Reproduction by binary fission in one, two, or three planes, or by budding .... Order Chroococcales

B. Cell division by multiple fission or by a combination of multiple fission and binary fission .... Order Pleurocapsales

C. Cell division by binary fission in one plane at right angles to filament axis. Cell differentiation absent (terminal cells may be differentiated). Reproduction by trichome fragmentation or undifferentiated hormogonia .... Order Oscillatoriales

C. Cell division by binary fission in one or more planes. Differentiation occurs resulting in specialized cells and structures .... D

D. Cell division in one plane at right angles to trichome axis .... Order Nostocales

D. Cell division in one or more planes resulting in complex thalli .... Order Stigonematales

Comments on Genera and Groups of the Order Chroococcales

The order Chroococcales contains all the unicellular cyanobacteria, including those that divide by binary fission (family Chroococcaceae) and by budding (family Chamaesiphonaceae) (Fig. 3). Several of the genera shown in Fig. 3 are provisional and are more properly called "groups" (see below). The presence or absence and characteristics of cell aggregates and the number and regularity of the planes of division are the major characteristics used to delineate

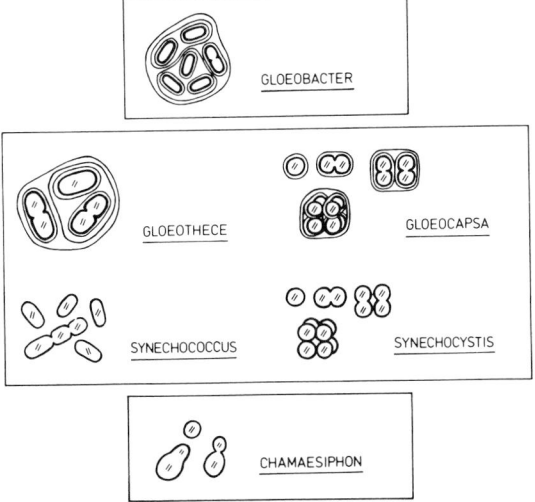

Fig. 3. Schematic presentation of the genera assigned to the order Chroococcales. The thin lines surrounding cells designate sheath material; the presence of thylakoids is indicated by the symbol //.

genera and groups of unicellular cyanobacteria in culture and to describe genera in the traditional botanical literature, as shown in the key below.

Key to the Genera and Groups of the Order Chroococcales*

A. Reproduction by binary fission .... B
A. Reproduction by repeated budding from the apical pole of the cell; cells ovoid; thylakoids present ....
.... Genus *Chamaesiphon*
B. Thylakoids present C
B. Thylakoids absent; division in one plane; cells rod-shaped; sheath present .... Genus *Gloeobacter*
C. Division in one plane .... D
C. Division in two or three planes .... F
D. Cell diameter of >3 μm; capable of aerobic N₂ fixation or nitrogenase produced anaerobically .... E
D. Cell diameter of <3 μm .... *Synechococcus* group
E. Cells rod-shaped; sheath present .... Genus *Gloeothece*
E. Cells rod-shaped; sheath absent .... *Cyanothece* group
F. Cells coccoid to hemispherical, held together in aggregates by multilaminated sheath material ...... *Gloeocapsa* group
F. Cells coccoid, occurring singly, in pairs, or in aggregates held together by amorphous capsular material ....
.... *Synechocystis* group

Cell aggregates in the Chroococcales are held together by multilaminated sheath material or by amorphous slime or capsular material (Fig. 3). The possession of extracellular sheath layers has proven to be a stable feature of many cyanobacterial groups in culture and is a primary character used in the description of several unicellular taxa including: *Gloeobacter, Gloeothece,* and the *Gloecapsa* group. Other unicellular cyanobacteria, primarily some members of the *Synechocystis* group, occur in cell aggregates in nature (and, more rarely, in culture) that are held together by amorphous slime or capsular material. Slime production has proven to be an unreliable taxonomic character because its production in culture is affected by the growth phase of the cyanobacteria and the conditions under which they were grown.

In the Chroococcales, transverse binary fission occurs in one, two, or three successive planes at right angles to one another or in irregular planes. The number and regularity of the successive planes of division are stable features of individual cyanobacteria that should in principle be readily determinable from cultured material. However, in practice, it is often difficult to determine the number of successive planes of division, with the distinction between

*Adapted from Waterbury and Rippka (1989).

division in two or three planes being especially problematical. As a consequence, the planes of division have not been determined for many of the strains currently in pure culture, with the result that some of the loosely defined groups, particularly the *Synechocystis* group, contain some strains that divide in two successive planes and some that divide in three successive planes.

The generic description of *Gloeothece* conforms to traditional botanical usage. Members of this genus are rod-shaped cyanobacteria that divide by transverse binary fission in a single plane. The cells occur in aggregates held together by well-defined sheath layers. The cells contain intracellular photosynthetic thylakoids, a feature that distinguishes them from *Gloeobacter*. All the strains currently placed in this genus are capable of dinitrogen fixation. *Chamaesiphon* contains unicellular cyanobacteria that divide by budding (Fig. 3). This definition conforms to botanical usage except for a difference in terminology. In the botanical literature, the smaller daughter cell (bud) resulting from unequal binary fission is termed an "exospore." *Gloeobacter* is a new genus that is not described in the botanical literature and is distinguished from other cyanobacteria by the lack of intracellular photosynthetic thylakoids (Rippka et al., 1974).

The four groups included in the Chroococcales by Waterbury and Rippka (1989) reflect the provisional nature of the taxonomy of cyanobacteria based on pure cultures. The groups are divided into "clusters," each equivalent to but not given the formal status of a genus.

The *Synechococcus* group contains coccoid to rod-shaped cyanobacteria with cells smaller than 3 μm in diameter that divide by binary fission in a single plane and that lack structured sheaths. Morphologically, all members of this group appear deceptively similar. The true extent of heterogeneity within the group is apparent from the span of the GC contents of their DNAs, which range from 39–71 mol%. Waterbury and Rippka (1989) divided the *Synechococcus* group into six clusters, three marine and three freshwater. Further subdivision and arrangement of these clusters will occur before genus designations are proposed and formalized. For example, in a phylogenetic study of the cyanobacteria, Giovannoni et al. (1988) showed that two strains included in the *Synechococcus* cluster, PCC 6301 (*"Anacystis nidulans"*), a freshwater clone, and Y7C-S (*"Synechococcus lividus"*), a thermophilic clone, are on independent deep branches in the cyanobacterial phylogenetic tree (Fig. 1). The phylogenetic distance between these two strains indicates that they should be assigned to two genera.

The *Cyanothece* group contains coccoid to rod-shaped cyanobacteria with cells larger than 3 μm in diameter that divide by binary fission in one plane and that lack sheaths. This group is superficially distinguished from the *Synechococcus* group by cell size and from *Gloeothece* and *Gloeobacter* by the absence of well-defined sheath layers. Many of the strains assigned to this group by Waterbury and Rippka (1989) are either capable of dinitrogen fixation or synthesize nitrogenase under anaerobic conditions. The heterogeneity of the group is evidenced by the fact that it currently contains freshwater, marine, and halophilic isolates.

The *Gloeocapsa* group contains cyanobacteria that divide by binary fission in two or three planes at right angles to one another, resulting in cell aggregates held together by multilaminated sheath material. In the traditional botanical literature, members of this group would be placed in either *Gloeocapsa* or *Chroococcus*.

The *Synechocystis* group contains unicellular cyanobacteria that divide by binary fission in two or three planes at right angles to one another. The cells typically occur singly or in pairs in culture; but in nature and, rarely, in culture, some can also occur in aggregates held together by amorphous capsular material. Waterbury and Rippka (1989) provisionally subdivided the group into four clusters, one containing marine isolates and three containing freshwater isolates.

Some representatives of the Chroococcales are illustrated in Chapter 13, Figs. 8 and 10.

Comments on Genera and Groups of the Order Pleurocapsales

The order Pleurocapsales contains cyanobacteria that reproduce by the formation of small spherical cells (called baeocytes) produced through multiple fission. The unicellular genera of the order divide exclusively by multiple fission. In other genera, cell aggregates are produced by binary fission, after which some or all of the cells in the aggregates undergo multiple fission and release baeocytes (Fig. 4). The small vegetative cells that are the products of multiple fission, termed "endospores" in the botanical literature, were renamed baeocytes (Greek for small cells) by Waterbury and Stanier (1978) to avoid confusion with the bacterial endospore. The ordinal and generic definitions used here and in *Bergey's Manual of Systematic Bacteriology* (Waterbury, 1989) are those proposed by Waterbury and Stanier (1978). Some of the botanical generic definitions were modified to in-

corporate new properties and the reinterpretation of features from the botanical descriptive literature that were documented during the study of developmental cycles in culture.

Key to the Genera and Groups of the Order Pleurocapsales*

A. Cell division solely by multiple fission .... B
A. Cell division by a combination of binary and multiple fission C
B. Baeocytes motile .... Genus *Dermocarpa*
B. Baeocytes nonmotile .... Genus *Xenococcus*
C. Baeocyte development leads to the formation of a vegetative cell that undergoes one to three binary fissions, producing a single apical cell which divides by multiple fission and releases baeocytes; the basal cell subsequently enlarges and repeats the cycle .... Genus *Dermocarpella*
C. Baeocyte development, followed by repeated binary fission, to produce cell aggregates of varying size and complexity .... D
D. Binary fission occurs in three planes at right angles to one another, producing a cubical aggregate of cells, all of which normally undergo multiple fission .... E
D. Binary fission occurs in many different planes, to produce irregular, sometimes pseudo-filamentous aggregates of cells. Some or all of the cells in the aggregate undergo multiple fission .... *Pleurocapsa* group
E. Baeocytes motile .... Genus *Myxosarcina*
E. Baeocytes nonmotile .... Genus *Chroococcidiopsis*

The genera *Dermocarpa* and *Xenococcus* are unicellular cyanobacteria that divide solely by multiple fission (Fig. 4). Their developmental patterns are similar, differing only with respect to baeocyte motility. The baeocytes of *Dermocarpa* are capable of gliding motility for a short period following their release from the parental cell, whereas in *Xenococcus*, the baeocytes are nonmotile. *Dermocarpella* represents the simplest pleurocapsalean cyanobacterium, whose developmental cycle incorporates division by both binary and multiple fissions. The baeocyte enlarges asymmetrically into an ovoid vegetative cell, which then undergoes binary fission to form a large apical cell and from one to three smaller basal cells. The large apical cell then undergoes multiple fission and releases motile baeocytes. Subsequently the basal cells enlarge and undergo binary fission and repeat the cycle. The developmental patterns of *Myxosarcina* and *Chroococcidiopsis* are very similar, differing only with respect to baeocyte motility (Fig. 4). The baeocyte enlarges symmetrically into a vegetative cell of fixed strain-specific size. The vegetative cell then begins to divide by bi-

---

*Adapted from Waterbury (1989).

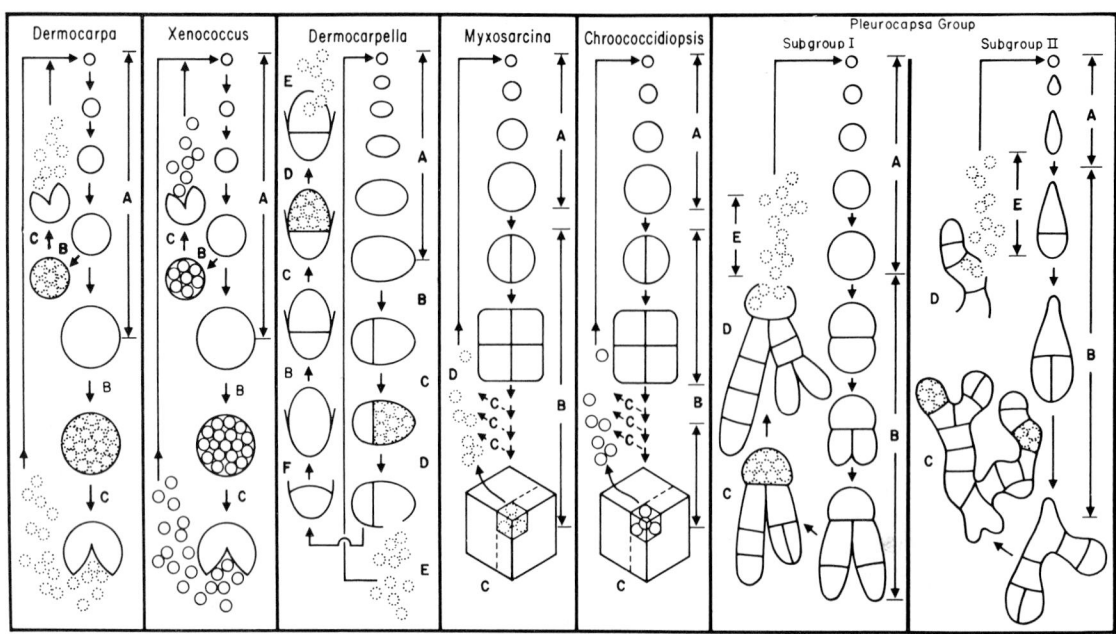

Fig. 4. Schematic presentation of genera assigned to the order Pleurocapsales. Baeocytes that are not surrounded by a fibrous (F) layer at the time of release and are, consequently, motile are symbolized by dotted circles to distinguish them from baeocytes that are surrounded by an F layer (solid circles). *Dermocarpa:* A, symmetric baeocyte enlargement; B, multiple fission, leading to baeocyte formation; C, baeocyte release followed by a brief period of baeocyte motility. *Xenococcus:* A, symmetric baeocyte enlargement; B, multiple fission leading to baeocyte formation; C, release of immotile baeocytes. *Dermocarpella:* A, asymmetric baeocyte enlargement; B, binary fission, giving rise to a small basal cell and a larger apical cell; C, multiple fission of the apical cell, leading to baeocyte formation; D, baeocyte release; E, period of baeocyte motility; F, enlargement of basal cell. *Myxosarcina:* A, symmetric baeocyte enlargement to a predetermined size; B, repeated binary fission in three regular planes; C, multiple fission of almost all the cells in the aggregate, followed by baeocyte release; D, period of baeocyte motility. *Chroococcidiopsis:* A, symmetric baeocyte enlargement to a predetermined size; B, repeated binary fissions in three regular planes; C, multiple fission of almost all the cells in the aggregate, followed by the release of immotile baeocytes. *Pleurocapsa* subgroup I: A, symmetric baeocyte enlargement; B, binary fissions in many irregular planes; C, multiple fission of some vegetative cells; D, baeocyte release; E, period of baeocyte motility. *Pleurocapsa* subgroup II: A, asymmetric baeocyte enlargement; B, binary fission in many irregular planes; C, multiple fission of some vegetative cells; D, baeocyte release; E, period of baeocyte motility. (From Waterbury and Stanier, 1978.)

nary fission in three alternating planes at right angles to one another to produce a large, approximately cubical, cell aggregate. Multiple fission, when it occurs, is massive; almost all the cells in the aggregate undergo cleavage and release baeocytes.

The *Pleurocapsa* group includes a large number of internally diverse strains. Their development involves baeocyte enlargement, followed by binary fission in irregular planes to produce cell aggregates that differ widely in size and complexity in different strains (Fig. 4). They range from small, compact masses of cells to large structures, consisting of a central mass of cells from which radiate more or less extensive pseudofilamentous outgrowths. Eventually, some cells in the aggregate undergo multiple fission and produce motile baeocytes. The baeocyte enlarges symmetrically, into a spherical vegetative cell (strains of subgroup I), or asymmetrically, into an elongated vegetative cell

(strains of subgroup II), before the onset of binary fission.

A representative of the genus *Dermocarpa* and a member of the Pleurocapsa group are illustrated in Chapter 13, Figs. 11 and 12.

Comments on Genera and Groups of the Order Oscillatoriales

The order Oscillatoriales contains filamentous cyanobacteria that divide by binary fission in a single plane (Fig. 5). Trichomes are made up solely of vegetative cells; differentiation, if it occurs, is limited to morphological changes in the terminal cells of filaments.

Phycological treatments to delineate genera of this order rely heavily on the presence or absence and the characteristics of extracellular sheath material. Although sheaths are still used as a character to define oscillatorian genera in pure culture, primary emphasis is placed on the

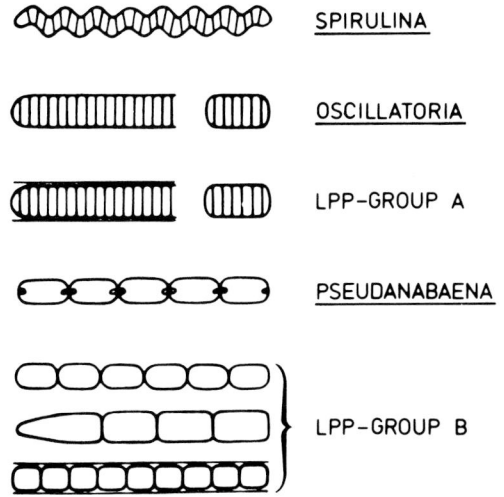

SPIRULINA

OSCILLATORIA

LPP-GROUP A

PSEUDANABAENA

LPP-GROUP B

Fig. 5. Schematic presentation of the genera assigned to the order Oscillatoriales. The thin lines surrounding trichomes designate sheath material. LPP-group A consists of the genus *Lyngbya*. The polar bodies in *Pseudanabaena* are gas vacuoles.

size, shape, and arrangement of the cells within filaments (Castenholz, 1989a; Rippka, 1988b).

Key to the Genera and Groups of the Order Oscillatoriales

A. Trichomes straight or loosely coiled for a portion of their length C
A. Trichomes helically coiled .... B
B. Cross-walls between vegetative cells invisible by light microscopy .... Genus *Spirulina*
B. Cross-walls visible between vegetative cells ....
.... Genus *Arthrospira*
C. Cells disk-shaped (wider then long) .... D
C. Cells isodiametric or cylindrical .... E
D. Trichomes with distinct sheath .... Genus *Lyngbya*
D. Trichomes without or with very thin sheath ....
.... Genus *Oscillatoria*
E. Cells cylindrical, several trichomes harbored within a common sheath .... Genus *Microcoleus*
E. Trichomes not within a common sheath .... F
F. Cells cylindrical, constriction occurs at cross-walls and gas vacuoles occur at cell poles .... Genus *Pseudanabaena*
F. Cells isodiametric or cylindrical, with or without sheaths, motile or nonmotile .... LPP group

The genera *Spirulina* and *Arthrospira* are filamentous cyanobacteria with helically coiled trichomes (Fig. 5). In *Spirulina* the trichomes are tightly coiled, and the cross-walls between adjacent vegetative cells are not visible by light microscopy. The trichomes in *Arthrospira* are loosely coiled and the cross-walls are evident by light microscopy. Ultrastructural studies by

Guglielmi and Cohen-Bazire (1982) demonstrated that the arrangement of pores in the peptidoglycan layer of the cell wall was different in *Spirulina* and *Arthrospira* but the arrangement in *Arthrospira* and members of the genus *Oscillatoria* was similar. In some botanical treatments (Bourrelly, 1985), *Arthrospira* is included within the genus *Oscillatoria*. Further study is needed to circumscribe these genera.

The genera *Oscillatoria* and *Lyngbya* (both sensu Rippka, 1988b) both include filamentous cyanobacteria with straight trichomes composed of disk-shaped cells (cells much wider than they are long) (Fig. 5). In *Oscillatoria,* the trichomes lack sheaths or are very lightly sheathed and exhibit gliding motility involving trichome rotation during translocation. The trichomes of *Lyngbya* are heavily sheathed and nonmotile. Undifferentiated hormogonia are released from the ends of the sheath and are motile for a period until new sheath material is synthesized. These definitions are more restrictive than those used by Castenholz (1989a), which also include forms with cells longer than they are wide.

The genus *Microcoleus* contains filamentous cyanobacteria with straight trichomes composed of cylindrical cells. Although this genus is not currently recognized by Rippka (1988b), because the common sheath is often not produced in culture, the common sheath is very characteristic in field material.

The genus *Pseudanabaena* contains filamentous cyanobacteria with unsheathed motile trichomes composed of cylindrical to barrel-shaped cells (Fig. 5). The cells usually contain polar gas vacuoles, and the cell walls are normally constricted at the junction between adjacent vegetative cells. They are capable of gliding motility but translocation is not accompanied by trichome rotation.

The LPP group (sensu Rippka, 1988b) contains a heterogeneous assemblage of filamentous cyanobacteria including forms that would be assigned to genera such as *Phormidium* and *Plectonema* in the botanical literature. LPP-group A is actually *Lyngbya* (Fig. 5). A satisfactory resolution of this group awaits more detailed study.

Some representatives of the Oscillatoriales are illustrated in Chapter 13, Figs. 9 and 13.

Comments on the Genera of the Orders Nostocales and Stigonematales

Members of the orders Nostocales and Stigonematales are distinguished from the filamentous cyanobacteria in the order Oscillatoriales by their capacity for cellular differentiation

(Figs. 6 and 7). In the absence of combined nitrogen, some vegetative cells in or at the ends of trichomes develop into heterocysts. Mature heterocysts are distinguishable from vegetative cells by their thick cell walls, relatively weak pigmentation, and refractile polar granules at points of attachment to adjacent vegetative cells. When mature, a heterocyst can neither divide nor revert to a vegetative cell; it is the specific cellular site of nitrogen fixation under aerobic conditions (Wolk, 1982). Members of the Nostocales and the Stigonematales are truly multicellular as a result of cellular differentiation and functional specialization between vegetative cells and heterocysts.

Many members of these two orders also produce thick-walled resting cells known as akinetes. Mature akinetes are usually larger than vegetative cells, have a lower phycobiliprotein content, and may produce pigments that give them a brownish appearance. They also contain large amounts of reserve material (Herdman, 1987; Nichols and Adams, 1982). They are formed as cultures approach the stationary phase of growth and are usually not dependent on the nature of the nitrogen source, except in those cyanobacteria that only form akinetes adjacent to heterocysts (e.g., *Cylindrospermum*). In such cases, repression of heterocyst formation by combined nitrogen is coupled with the repression of akinete formation. Akinetes are resistant to desiccation and to cold, but not to heat. They germinate under favorable growth conditions and give rise to new filaments.

The number of planes of division is the sole character that separates the orders Nostocales and Stigonematales. In members of the Nostocales, cell division always occurs in a single plane at right angles to the long axis of the trichomes, which are consequently uniseriate and unbranched (Fig. 6). Members of the Stigonematales are capable of division in more than one plane, giving rise to true branches and both multiseriate and uniseriate trichomes (Fig. 7).

Key to the Genera of the Order Nostocales

A. Reproduction by trichome breakage (hormogonia not produced) or by germination of akinetes (if they are produced) .... B

A. Reproduction by hormogonia that are morphologically distinguishable from vegetative trichomes; reproduction may also occur by trichome breakage and akinete germination .... D

B. Heterocysts are terminal, may occur at both ends of trichomes. Akinetes (when present) are directly adjacent to heterocysts .... Genus *Cylindrospermum*

B. Heterocysts intercalary or terminal; location of akinetes (when present) variable C

C. Vegetative cells spherical, ovoid, or cylindrical .... Genus *Anabaena*

C. Vegetative cells shorter than wide to disk-shaped .... Genus *Nodularia*

D. Heterocysts are terminal; mature trichomes taper from base near heterocyst to tip .... Genus *Calothrix*

D. Trichomes do not taper .... E

E. Trichomes heavily sheathed, false branching may occur usually at site of intercalary heterocyst. Hormogonia produce heterocysts at one end of filaments only .... Genus *Scytonema*

E. Trichomes not sheathed, false branching absent. Hormogonia produce heterocysts at both ends of young filaments .... Genus *Nostoc*

Nostocalean genera currently available in pure culture can be divided into two groups by their capacity to produce hormogonia: One group, the genera *Cylindrospermum, Nodularia,* and *Anabaena,* do not form hormogonia (Fig. 6). Reproduction in these genera is by trichome breakage or by the germination of akinetes if they are produced. The definitions of the genera *Cylindrospermum* and *Nodularia* correspond to traditional botanical definitions: Members of the genus *Cylindrospermum* produce untapered trichomes with a terminal heterocyst. Akinetes are always produced adjacent to heterocysts. Members of the genus *Nodularia* produce untapered trichomes composed of cells that are wider than long (disk-shaped). Heterocysts in this genus are differentiated from both terminal and intercalary cells. The distinction between the genera *Anabaena* and *Nostoc* is problematic. In the botanical literature, the distinction between the two genera is made on the basis of slime production; in the genus *Nostoc,* this results in characteristic macroscopic colonies when observed in natural material. In culture, the morphology of filaments in the two genera is similar; there, their distinction rests on the production of hormogonia by *Nostoc* and their absence in *Anabaena,* a distinction that may be difficult to make because some strains of *Nostoc* only produce hormogonia erratically (Lachance, 1981; Rippka, 1988b).

Members of the genera *Scytonema* and *Calothrix* share with *Nostoc* the ability to form differentiated hormogonia (Fig. 6). However, the pattern of development of hormogonia into mature filaments differs in each genus. The hormogonia in all three genera are initially devoid of heterocysts. In *Nostoc,* the hormogonia form two heterocysts, one at each end of the filament. Subsequent growth gives rise to mature trichomes in which intercalary heterocysts are also produced. In *Scytonema* and *Calothrix,* heter-

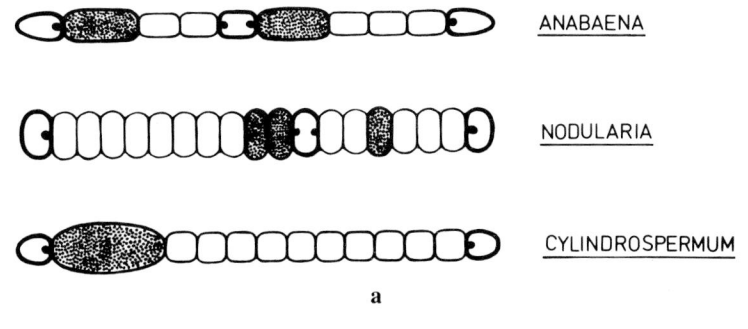

ANABAENA
NODULARIA
CYLINDROSPERMUM
a
hormogonium
young trichome
trichome in the mature state
Genus
NOSTOC
SCYTONEMA
CALOTHRIX
b

Fig. 6. Schematic presentation of the genera assigned to the order Nostocales. (a) Without developmental cycle; (b) with developmental cycle. Heavy-walled cells with polar granules represent heterocysts; heavy-walled cells that are stippled represent akinetes; thin lines surrounding trichomes designate sheath material.

hormogonium
young trichome
culture in the mature state
CHLOROGLOEOPSIS
FISCHERELLA

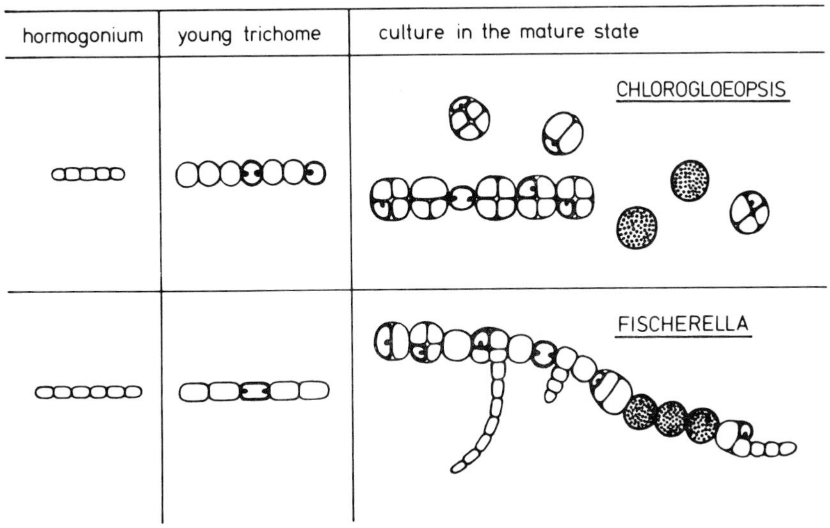

Fig. 7. Schematic presentation of two of the genera assigned to the order Stigonematales. Heavy-walled cells with polar granules represent heterocysts; heavy-walled cells that are stippled represent akinetes; thin lines surrounding groups of cells designate sheath material.

ocyst formation in the developing hormogonia is restricted to the terminal position at only one end of the cellular chain.

In *Scytonema,* growth and elongation of hormogonia give rise to heavily sheathed trichomes of constant width, in which heterocysts are predominantly intercalary. In *Calothrix,* hormogonia that are composed of cells of constant width give rise to tapered trichomes. The direction of tapering that is characteristic of *Calothrix* is determined by the location of the first heterocyst produced. Members of the genus *Calothrix* (sensu Rippka, 1988b) are often included in a number of traditional genera such as *Gloeotrichia, Rivularia, Isactis,* and *Dichothrix.*

Key to the Genera of the Order Stigonematales

A. Cells of mature trichome are spherical and divide in multiple planes. The resulting cell aggregates are irregular in shape and appear more unicellular than filamentous. Some cells undergo a series of divisions in one plane to produce short motile hormogonia .... Genus *Chlorogloeopsis*

A. True branching results from lateral division along main axis of the filamcent .... B

B. Cells of lateral branches are morphologically different from the cells in the main axis .... Genus *Fischerella*

B. Cells of lateral branches are not morphologically different from the cells in the main axis C

C. Trichomes uniseriate throughout .... Genus *Hapalosiphon*

C. Trichomes multiseriate, at least in part ....
.... Genus *Stigonema*

The genus *Chlorogloeopsis* is a member of the Stigonematales even though it never displays a branched filamentous habit characteristic of the other genera in this order (Fig. 7). It does however produce typical, short-chained, motile hormogonia that differentiate into multiseriate-cell aggregates that appear more unicellular than filamentous.

The genera *Fischerella* and *Hapalosiphon* are both capable of true branching (Fig. 7). In *Fischerella* the cells of the lateral branches differ in size from the cells in the main axis and are uniseriate, whereas the main axis may become multiseriate. In *Hapalosiphon* both the main axis and lateral branches are uniseriate and composed of similar sized cells.

Members of the genus *Stigonema* are also capable of true branching. *Stigonema* is distinguished from *Fischerella* and *Hapalosiphon* by the complexity of its thallus structure resulting from an ability to form both multiseriate and uniseriate branches. Only one strain of *Stigonema minutum* has been cultured (Zehnder,

1985). There are also a number of genera in this order that have not yet been cultured which will make fascinating subjects for future study (Geitler, 1932).

Representatives of the genera *Anabaena, Cylindrospermum, Calothrix,* and *Fischerella* are illustrated in Chapter 13, Figs. 14–17.

Literature Cited

Aiba, S., and T. Ogawa. 1977. Assessment of growth yield of a blue-green alga, *Spirulina platensis,* in axenic and continuous culture. J. Gen. Microbiol. 102:179–182.

Allen, M. M. 1968. Simple conditions for the growth of unicellular blue-green algae on plates. J. Phycol. 4:1–4.

Anagnostidis, K., and J. Komárek. 1985. Modern approach to the classification system of cyanophytes. 1. Introduction. Arch. Hydrobiol. Suppl. 71:291–302.

Anagnostidis, K., and J. Komárek. 1988. Modern approach to the classification system of cyanophytes. III. Oscillatoriales. Arch. Hydrobiol. Suppl. 80:327–472.

Bonen, L., W. F. Doolittle, and G. E. Fox. 1979. Cyanobacterial evolution: Results of 16S ribosomal ribonucleic acid sequence analyses. Can. J. Biochem. 57:879–888.

Bothe, H. 1982. Nitrogen fixation, p. 87–104. In: N. G. Carr and B. A. Whitton (ed.), The biology of cyanobacteria. University of California Press, Berkeley.

Bourrelly, P. 1985. Les algues d'eau douce. III. Les algues bleues et rouges, les Eugleniens, Peridiniens et Cryptomonadines. H. Boubée, Paris.

Brock, T. D. 1976. Halophilic blue-green algae. Arch. Microbiol. 107:109–111.

Bryant, D. A. 1986. The cyanobacterial photosynthetic apparatus: comparisons to those of higher plants and photosynthetic bacteria. In: T. Platt and W. K. W. Li, (ed.), Photosynthetic picoplankton. Can. Bull. Fish. Aquat. Sci. 214:71–120.

Caron, D. A., F. R. Pick, and D. R. S. Lean. 1985. Chroococcoid cyanobacteria in Lake Ontario: vertical and seasonal distributions during 1982. J. Phycol. 21:171–175.

Castenholz, R. W. 1981. Isolation and cultivation of thermophilic cyanobacteria, p. 236–246. In: M. P. Starr, H. Stolp, H. G. Trüper, A. Balows, and H. G. Schlegel (ed.), The prokaryotes, vol. 1. Springer-Verlag, Berlin.

Castenholz, R. 1982. Motility and taxis, p. 413–439. In: N. G. Carr and B. A. Whitton (ed.), The biology of cyanobacteria. University of California Press, Berkeley.

Castenholz, R. W. 1988a. Culturing methods for cyanobacteria, p. 68–93. In: L. Packer and A. N. Glazer (ed.), Methods in enzymology, vol. 167. Academic Press, Inc., New York.

Castenholz, R. W. 1988b. Thermophilic cyanobacteria: special problems, p. 96–100. In: L. Packer and A. N. Glazer (ed.), Methods in enzymology, vol. 167. Academic Press, Inc., New York.

Castenholz, R. W. 1989a. Order Oscillatoriales, p. 1771–1780. In: J. T. Staley, M. P. Bryant, N. Pfennig, and J. G. Holt (ed.), Bergey's manual of systematic bacteriology, vol. 3. Williams and Wilkins, Baltimore.

Castenholz, R. W. 1989b. Order Nostocales, p. 1780–1798. In: J. T. Staley, M. P. Bryant, N. Pfennig, and J. G. Holt (ed.), Bergey's manual of systematic bacteriology, vol. 3. Williams and Wilkins, Baltimore.

Castenholz, R. W. 1989c. Order Stigonematales, p. 1794–1799. In: J. T. Staley, M. P. Bryant, N. Pfennig, and

J. G. Holt (ed.), Bergey's manual of systematic bacteriology, vol. 3. Williams and Wilkins, Baltimore.

Castenholz, R. W., and J. B. Waterbury. 1989. Oxygenic photosynthetic bacteria: Group I. Cyanobacteria, p. 1710–1727. In: J. T. Staley, M. P. Bryant, N. Pfennig, and J. G. Holt (ed.), Bergey's manual of systematic bacteriology, vol. 3. Williams and Wilkins, Baltimore.

Ciferri, O. 1983. *Spirulina*, the edible microorganism. Microbiol. Rev. 47:551–578.

Cohen, Y. 1975. Dynamics of prokaryotic photosynthetic communities of the Solar Lake. Ph.D. thesis. Hebrew University, Jerusalem, Israel.

Cohen, Y., B. B. Jørgensen, M. P. Revsbech, and R. Poplawski. 1986. Adaptation to hydrogen sulfide of oxygenic and anoxygenic photosynthesis among cyanobacteria. Appl. Environ. Microbiol. 51:398–407.

Cohen-Bazire, G., and D. A. Bryant. 1982. Phycobilisomes: composition and structure, p. 143–190. In: N. G. Carr and B. A. Whitton (ed.), The biology of cyanobacteria. University of California Press, Berkeley.

Desikachary, T. V. 1959. Cyanophyta. Indian Council of Agricultural Research, New Delhi, India.

de Winder, B., H. C. P. Matthijs, and L. R. Mur. 1989. The role of water retaining substrata in the photosynthetic response of three drought tolerant phototrophic microorganisms isolated from a terrestrial habitat. Arch. Microbiol. 152:485–462.

Fitzwater, S. E., G. A. Knauer, and J. H. Martin. 1982. Metal contamination and its effect on primary production measurements. Limnol. Oceanogr. 27:544–551.

Fogg, G. E., W. D. P. Stewart, P. Fay, and A. E. Walsby. 1973. The blue-green algae. Academic Press, London.

Gallon, J. R., T. A. LaRue, and W. G. W. Kurz. 1974. Photosynthesis and nitrogenase activity in the blue-green alga *Gloeocapsa*. Can. J. Microbiol. 20:1633–1637.

Geitler, L. 1932. Cyanophyceae. In: Rabenhorst (ed.), Kryptogamenflora von Deutschland, Österreich und der Schweiz, vol. XIV. Akademische Verlags, Leipzig. Reprinted (1971): Johnson Reprint Co., New York.

Giovannoni, S. J., S. Turner, G. J. Olsen, S. Barns, D. J. Lane, and N. R. Pace. 1988. Evolutionary relationships among cyanobacteria and green chloroplasts. J. Bacteriol. 170:3584–3592.

Glazer, A. N. 1981. Photosynthetic accessory proteins with bilin prosthetic groups, p. 51–96. In: M. D. Hatch and N. K. Boardman (ed.), The biochemistry of plants, vol. 8. Academic Press, New York.

Guglielmi, G., and G. Cohen-Bazire. 1982. Structure et distribution des pores et des perforations de l'enveloppe de peptidoglycane chez quelques cyanobactéries. Protistologica 18:151–165.

Herdmann, M. 1987. Akinetes: structure and function, p. 227–250. In: P. Fay and C. Van Baalen (ed.), The cyanobacteria. Elsevier, Amsterdam.

Hof, T., and P. Frémy. 1933. On Myxophyceae living in strong brines. Rec. Trav. Bot. Neerl. 30:140–162.

Huang, T. C., and T. J. Chow. 1986. New type of N_2-fixing unicellular cyanobacterium (blue-green alga). FEMS Microbiol. Letts. 36:109–110.

Huang, T. C., and T. J. Chow. 1988. Comparative studies of some nitrogen-fixing unicellular cyanobacteria isolated from rice fields. J. Gen. Microbiol. 134:3089–3097.

Hughes, E. O., P. R. Gorham, and A. Zehnder. 1958. Toxicity of a unialgal culture of *Microcystis aeruginosa*. Can. J. Microbiol. 4:225–236.

Kana, T. M., and P. M. Glibert. 1987a. Effect of irradiances up to 2000 uE $m^{-2}s^{-1}$ on marine *Synechococcus* WH7803. I. Growth, pigmentation, and cell composition. Deep-Sea Res. 34:479–495.

Kana, T. M., and P. M. Glibert. 1987b. Effect of irradiances up to 2000 uE $m^{-2}s^{-1}$ on marine *Synechococcus* WH7803. II. Photosynthetic responses and mechanisms. Deep-Sea Res. 34:497–516.

Komárek, J., and K. Anagnostidis. 1986. Modern approach to the classification system of cyanophytes. 2. Chroococcales. Arch. Hydrobiol. Suppl. 73:157–226.

Kotai, J. 1972. Instructions for preparation of modified nutrient solution Z8 for algae. Publication B-11/69, Norwegian Institute for Water Research, Blindern, Oslo 3, Norway.

Lachance, M.-A. 1981. Genetic relatedness of heterocystous cyanobacteria by deoxyribonucleic acid-deoxyribonucleic acid reassociation. Internat. J. Syst. Bacteriol. 31:139–147.

LaPage, S. P., P. H. Sneath, E. F. Lessel, V. B. D. Skerman, H. P. R. Seeliger, and W. A. Clark (ed.). 1975. International Code of Nomenclature of Bacteria. American Society for Microbiology, Washington, DC.

Mitsui, A., and S. Cao. 1988. Isolation and culture of marine nitrogen-fixing unicellular cyanobacteria *Synechococcus*, p. 105–113. In: L. Packer and A. N. Glazer (ed.), Methods in enzymology, vol. 167. Academic Press, Inc., New York.

Mitsui, A., S. Cao, A. Takahashi, and T. Arai. 1987. Growth synchrony and cellular parameters of the unicellular nitrogen-fixing marine cyanobacterium, *Synechococcus* sp. strain Miami BG 043511 under continuous illumination. Physiol. Plantarum 69:1–8.

Morel, F. M N., J. G. Rueter, D. M. Anderson, and R. R. L. Guillard. 1979. Aquil: A chemically defined phytoplankton culture medium for trace metal studies. J. Phycol. 15:135–141.

Nichols, J. M., and D. G. Adams. 1982. Akinetes, p. 387–412. In: N. G. Carr and B. A. Whitton (ed.), The biology of cyanobacteria. University of California Press, Berkeley.

Ohki, K., J. G. Rueter, and Y. Fujita. 1986. Cultures of the pelagic cyanophytes *Trichodesmium erythraeum* and *T. thiebautii* in synthetic medium. Mar. Biol. 91:9–13.

Padan, E., and Y. Cohen. 1982. Anoxygenic photosynthesis, p. 215–235. In: N. G. Carr and B. A. Whitton (ed.), The biology of cyanobacteria. University of California Press, Berkeley.

Pearson, H. W., R. Howsley, C. K. Kjeldsen, and A. E. Walsby. 1979. Aerobic nitrogenase activity associated with a non-heterocystous filamentous cyanobacterium. FEMS Microbiol. Lett. 5:163–167.

Pick, F. R., and D. A. Caron. 1987. Picoplankton and nanoplankton biomass in Lake Ontario: relative contribution of phototrophic and heterotrophic communities. Can. J. Fish. Aqua. Sci. 44:2164–2172.

Rippka, R. 1988a. Isolation and purification of cyanobacteria, p. 3–27. In: L. Packer and A. N. Glazer (ed.), Methods in enzymology, vol. 167. Academic Press, Inc., New York.

Rippka, R. 1988b. Recognition and identification of cyanobacteria, p. 28–67. In: L. Packer and A. N. Glazer (ed.), Methods in enzymology, vol. 167. Academic Press, Inc., New York.

Rippka, R., J. Deruelles, J. B. Waterbury, M. Herdman, and R. Y. Stanier. 1979. Generic assignments, strain histories and properties of pure cultures of cyanobacteria. J. Gen. Microbiol. 111:1–61.

Rippka, R., J. B. Waterbury, and G. Cohen-Bazire. 1974. A cyanobacterium which lacks thylakoids. Arch. Microbiol. 100:419–436.

Rippka, R., J. B. Waterbury, and R. Y. Stanier. 1981a. Isolation and purification of cyanobacteria: Some general principles, p. 212–220. In: M. P. Starr, H. Stolp, H. G. Trüper, A. Balows, and H. G. Schlegel (ed.), The prokaryotes, vol. 1. Springer-Verlag, Berlin.

Rippka, R., J. B. Waterbury, and R. Y. Stanier. 1981b. Provisional generic assignment for cyanobacteria in pure culture, p. 247–256. In: M. Starr, P. H. Stolp, H. G. Trüper, A. Balows, and H. G. Schlegel (ed.), The prokaryotes, vol. 1. Springer-Verlag, Berlin.

Rueter, J. G., J. J. McCarthy, and E. J. Carpenter. 1979. The toxic effect of copper on *Oscillatoria (Trichodesmium) thiebautii*. Limnol. Oceanogr. 24:558–562.

Seaburg, K. G., B. C. Parker, R. A. Wharton, Jr., and G. M. Simmons, Jr. 1981. Temperature-growth responses of algal isolates from antarctic oases. J. Phycol. 17:353–360.

Skulberg, O. M. 1983. Culture collection of algae at Norwegian Institute for water research. Norwegian Institute for Water Research, Blindern, Oslo 3, Norway.

Smith, A. J. 1982. Mode of cyanobacterial carbon metabolism, p. 47–85. In: N. G. Carr and B. A. Whitton (ed.), The biology of cyanobacteria. University of California Press, Berkeley.

Stafleu, F. A., C. E. B. Bonner, R. McVaugh, R. D. Meikle, R. C. Rollins, R. Ross, and E. G. Voss (ed.). 1972. International Code of Botanical Nomenclature. A. Oosthoek, Utrecht, The Netherlands.

Stal, L. J., and W. E. Krumbein. 1985. Nitrogenase activity in the non-heterocystous cyanobacterium *Oscillatoria* sp. grown under alternating light-dark cycles. Arch. Microbiol. 143:72–76.

Starr, M. P., H. Stolp, H. G. Trüper, A. Balows, and H. G. Schlegel (ed.). 1981. Section B: The phototrophic prokaryotes, p. 197–256. The prokaryotes. vol. 1. Springer-Verlag, Berlin.

Starr, R. C. 1978. The culture collection of algae at the University of Texas at Austin. J. Phycol. 14 (suppl.):47–100.

Tandeau de Marsac, N. 1977. Occurrence and nature of chromatic adaptation in cyanobacteria. J. Bacteriol. 130:82–91.

Tandeau de Marsac, N. 1983. Phycobilisomes and complementary chromatic adaptation in cyanobacteria. Bull. Inst. Pasteur 81:201–254.

Turner, S., T. Burger-Wiersma, S. J. Giovannoni, L. R. Mur, and N. R. Pace. 1988. The relationship of a procholophyte *Prochlorothrix hollandica* to green chloroplasts. Nature 337:380–382.

van Liere, L., and A. E. Walsby. 1982. Interactions of cyanobacteria with light, p. 9–45. In: N. G. Carr and B. A. Whitton (ed.), The biology of the cyanobacteria. University of California Press, Berkeley.

van Rijn, J., and Y. Cohen. 1983. Ecophysiology of the cyanobacterium *Dactylococcopsis salina:* Effect of light intensity, sulphide and temperature. J. Gen. Microbiol. 129:1849–1856.

Walsby, A. E. 1978. The gas vesicles of aquatic prokaryotes. Symp. Soc. Gen. Microbiol. 28:327–357.

Walsby, A. E. 1981. Cyanobacteria: Planktonic gas-vacuolate forms, p. 224–235. In: M. P. Starr, H. Stolp, H. G. Trüper, A. Balows, and H. G. Schlegel (ed.), The prokaryotes, vol. 1. Springer-Verlag, Berlin.

Walsby, A. E. 1987. Mechanisms of buoyancy regulation by planktonic cyanobacteria with gas vesicles, p. 377–392. In: P. Fay and C. Van Baalen (ed.), The cyanobacteria. Elsevier, Amsterdam.

Waterbury, J. B. 1989. Order Pleurocapsales Geitler 1925, emend. Waterbury and Stanier 1978, p. 1746–1770. In: J. T. Staley, M. P. Bryant, N. Pfennig, and J. G. Holts (ed.), Bergey's manual of systematic bacteriology, vol. 3. Williams and Wilkins, Baltimore.

Waterbury, J. B., and R. Rippka. 1989. The order Chroococcales Wettstein 1924, emend. Rippka et al., 1979, p. 1728–1746. In: J. T. Staley, M. P. Bryant, N. Pfennig, and J. G. Holts (ed.), Bergey's manual of systematic bacteriology, vol. 3. Williams and Wilkins, Baltimore.

Waterbury, J. B., and R. Y. Stanier. 1978. Patterns of growth and development in pleurocapsalean cyanobacteria. Microbiol. Rev. 42:2–44.

Waterbury, J. B., and R. Y. Stanier. 1981. Isolation and growth of cyanobacteria from marine and hypersaline environments, p. 221–223. In: M. P. Starr, H. Stolp, H. G. Trüper, A. Balows, and H. G. Schlegel (ed.), The prokaryotes, vol. 1. Springer-Verlag, Berlin.

Waterbury, J. B., S. W. Watson, R. R. Guillard, and L. E. Brand. 1979. Widespread occurrence of a unicellular, marine, planktonic cyanobacterium. Nature 277:293–294.

Waterbury, J. B., S. W. Watson, F. W. Valois, and D. G. Franks. 1986. Biological and ecological characterization of the marine unicellular cyanobacterium *Synechococcus*. In: T. Platt and W. K. W. Li (ed.), Photosynthetic picoplankton. Can. Bull. Fish. Aquat. Sci. 214:71–120.

Waterbury, J. B., and J. M. Willey. 1988. Isolation and growth of marine planktonic cyanobacteria, p. 101–105. In: L. Packer and A. N. Glazer (ed.), Methods in enzymology, vol. 167. Academic Press, Inc., New York.

Woese, C. R. 1987. Bacterial evolution. Microbiol. Rev. 51:221–271.

Wolk, C. P. 1982. Heterocysts, p. 359–386. In: N. G. Carr and B. A. Whitton (ed.), The biology of cyanobacteria. University of California Press, Berkeley.

Wolk, C. P. 1988. Purification and storage of nitrogen-fixing filamentous cyanobacteria, p. 93–95. In: L. Packer and A. N. Glazer (ed.), Methods in enzymology, vol. 167. Academic Press, Inc., New York.

Wyman, M., and P. Fay. 1987. Acclimation to the natural light climate, p. 347–376. In: P. Fay and C. Van Baalen (ed.), The cyanobacteria. Elsevier, Amsterdam.

Yopp, J. H., D. R. Tindall, D. R. Miller, and D. M. Schmid. 1978. Isolation, purification and evidence of the obligate halophilic nature of the blue-green alga *Aphanothece halophytica* Frémy (Chroococcales). Phycol. 17:172–177.

Zehnder, A. 1985. Isolation and cultivation of large cyanophytes for taxonomic purposes. Arch. Hydrobiol. Suppl. 71:281–289.

Zehnder, A., and P. R. Gorham. 1960. Factors influencing the growth of *Microcystis aeruginosa* Kütz. emend. Elenkin. Can. J. Microbiol. 6:645–660.

CHAPTER 98

The Cyanobacteria—Ecology, Physiology, and Molecular Genetics

YEHUDA COHEN and MICHAEL GUREVITZ

Introduction

The cyanobacteria and prochloroaceae are the only prokaryotic groups that share the use of photosystems I and II and hence the ability to carry out oxygenic photosynthesis with all photosynthetic eukaryotic organisms (Stanier, 1977). The structure of the reaction center complexes seems to be evolutionarily conserved in all these organisms, but there is a large diversity in their antenna chlorophyll complexes (Glazer, 1983). Cyanobacteria are unique since they are also capable of anaerobic metabolism, and several groups among the cyanobacteria share with the prokaryotic anoxygenic phototrophic bacteria the ability to perform anoxygenic photosynthesis using reduced electron donors (Cohen et al., 1975a; Padan and Cohen, 1982; Cohen et al., 1986; de Wit and van Gemerden, 1989; Garcia-Pichel and Castenholz, 1990). They also share with many archaebacteria the ability to use elemental sulfur for anaerobic dark respiration (Oren and Shilo, 1979).

Cyanobacteria are a diverse group of oxygenic photosynthetic prokaryotes, exhibiting versatile physiology and wide ecological tolerance that contribute to their competitive success over broad spectrum of environments both planktonic and benthic (Shilo, 1989). They are dominant in a broad spectrum of terrestrial habitats, including deserts, where they are found as endoliths in the antarctic dry valleys and cold deserts (Friedman, 1982); hot deserts in the Sahel (Potts and Friedmann, 1981); tropical rain forests; and mangrove swamps. Planktonic cyanobacterial blooms are common in freshwater bodies, especially where eutrophic conditions exist (hyperscums of *Microcystis aeruginosa* over 1 m thick have been reported, Zohary and Cohen, 1991), as well as in oligotrophic oceans (where *Synechococcus* spp. may account for 50% of the chlorophyll *a* biomass, Waterbury et al., 1979; Olson et al., 1990). Among this widespread group of photosynthetic prokaryotes are species that are able to grow as aerobic photoautotrophs; anaerobic photoautotrophs using either H_2S, thiosulfate, or H_2 as alternative electron donors; photoheterotrophs; chemoheterotrophs; and as anaerobic or aerobic dinitrogen fixers (Allen, 1985; Paerl et al., 1989; Stal et al., 1989b).

Historically, cyanobacteria were long classified as blue-green algae according to the Botanical Code, and it was only in the eighth addition of *Bergey's Manual of Determinative Bacteriology* (Buchanan and Gibbons, 1974) that they were first assigned to a separate division of the prokaryotes.

Cyanobacteria are unique in that they are Gram-negative prokaryotes and yet perform oxygenic photosynthesis very similar to that of higher plants. Thus, they may serve as a model system to resolve biological questions difficult to approach in higher plants, and they can be target organisms for research not directly related to photosynthesis. Hence, not only questions associated with the photosynthetic apparatus and function, carbon fixation, light-regulated gene expression, but also cell differentiation and resistance to environmental factors or stress may be easier to address with the power of molecular genetics in cyanobacteria, rather than in higher plants.

The key to the genetic study and manipulation of cyanobacteria is the development of a transformation system (either natural or artificial) for DNA uptake. Such a system of cellular competence could be used to internalize exogenous DNA with or without a special treatment to the recipient cells prior to their incubation with the donor DNA. The majority of cyanobacterial strains capable of DNA-mediated transformation possess physiological, or natural, competence for DNA uptake. The mechanism involved in such a process is still unknown but the phenomenon has been extensively used to transform cyanobacterial cells either chromosomally or with plasmids.

Chromosomal transformation is achieved by recombination and internalization of donor chromosomal DNA with homologous DNA in the recipient cell. By using donor DNA that has been subjected prior to the transformation to either in vivo mutagenesis or to recombinant

DNA manipulations in vitro, it is possible to create and easily select for mutants according to their phenotype or a genetic marker. It is thus possible to produce insertion or deletion mutations in cyanobacterial genomes, permitting the analysis of gene structure, function, and organization.

Plasmid transformation involves the introduction of a plasmid DNA molecule able to replicate in the recipient cell. This allows development of plasmid vectors for cyanobacteria that can be used to produce cyanobacterial partial diploids or meroploids, as well as to introduce foreign genes whose function is analyzed in the recipient cell.

Cyanobacteria can also be transformed with nonreplicating plasmids containing chromosomal-homologous sequences. In such a case, except for homologous recombination between the plasmid and the chromosome, another outcome may be the incorporation of the entire plasmid into the cyanobacterial chromosome if a single recombination event occurs between the homologous sequences. By constructing such mutants, it is possible to create heterogenotes in which a foreign mutant gene is tandemly placed next to the wild-type gene in the chromosome.

So far, only unicellular strains have been found to possess a natural competence for DNA uptake. However, the development of conjugal DNA-mediated transfer for filamentous cyanobacteria has circumvented this problem and opened a way to introduce genetic material into cells that were previously restricted for genetic transformation. The shuttle vectors developed for such transformation are capable of replicating in both *Escherichia coli,* in which DNA manipulations can be performed, and in the recipient cyanobacterium. Another way to introduce DNA into cyanobacteria could exploit cyanophages as candidate vectors. Despite the fact that several cyanophages have been described, no gene transduction has been reported in cyanobacteria.

Our present understanding of the ecology, physiology, and molecular genetics of cyanobacteria allows the exploration of the use of cyanobacteria for several biotechnological uses, such as the production of specific photosynthetic pigments and herbicides and the use of cyanobacteria for agricultural dinitrogen fixation.

Cyanobacterial Mats

Cyanobacterial mats are stratified benthic microbial communities in which cyanobacteria are dominant, developing at sediment-water interfaces in a wide spectrum of shallow aquatic systems. They form laminated multilayers (biofilms) embedded in copious amounts of polysaccharides excreted by the benthic microbial community (Krumbein et al., 1977). The laminated microbial community develops in response to chemical microgradients at the mat surface. Such gradients can be measured by the use of specific microelectrodes for pO_2, pH, and pS^{2-} (Jørgensen et al., 1979b, 1983; Revsbech et al., 1983; Revsbech and Jørgensen, 1986). Steep microgradients of oxygen from zero to 1000 μM, pH from 7.0 to over 11.0, and H_2S from zero to 500 μM can be found in the upper 2 mm of the cyanobacterial mat. These chemical microgradients are largely the result of intensive oxygenic photosynthesis, which takes place at the very surface of the cyanobacterial mat. Light is rapidly absorbed by the benthic cyanobacterial community, leading to a steep gradient of photosynthetically available light (Jørgensen et al., 1987; Jørgensen, 1989). Sulfide production and oxidation is tightly coupled to the excretion of organic photosynthates by the cyanobacteria, resulting in maximal rates of sulfate reduction within the photic zone at midday (Cohen, 1984b). The concomitant activities of oxygenic photosynthesis, together with anaerobic metabolism of anoxygenic photosynthesis by associated anoxygenic phototrophic bacteria, such as *Chloroflexus* spp., *Ectothiorhodosira*-like spp., and *Thiocapsa* spp. (Stolz, 1984; D'Amelio et al., 1987), as well as sulfate reduction (Jørgensen and Cohen, 1977), and methane production (Cohen, 1986), may be possible by H_2 production through photoreduction of water by the cyanobacteria at high light intensity and under CO_2 limitation. It is probably CO_2 limitation at the photosynthetically active zone of the cyanobacterial mat that causes isotope fractionation for $^{13}C/^{12}C$ in the organic matter in shallow-water hypersaline microbial mats to be relatively heavy ($-4‰$ to $-11‰$) (Des Marais et al., 1989). Addition of CO_2 results in a drastic enhancement of the rate of CO_2 photoassimilation by the cyanobacterial mat community (Rothschild and Mancinelli, 1990).

Because the chemical microgradients are largely light-dependent (Cohen, 1984b), the cyanobacterial mat becomes completely anoxic at night and thus exposed to elevated H_2S concentration at about neutral pH. Cyanobacteria may cope with these diurnal fluctuations of oxygen and sulfide by either migrating vertically in the mat and thus avoiding the exposure to toxic levels of sulfide at lower light intensity (Castenholz, 1973; Richardson and Castenholz, 1989) or by shifting their photosynthetic mode to adapt to life under sulfide (Cohen, 1984a).

Highly developed cyanobacterial mats are common in hot springs at temperatures of less than 74°C and at pH levels above 5.0 (Castenholz, 1984; Ward et al., 1989), in alkaline lakes, and in marine (e.g., Bauld, 1984) and hypersaline ecosystems at salinities of up to 20% salt (Des Marais et al., 1989). Cyanobacterial mats at salinities lower than 4.5% rarely accumulate because they are heavily grazed by a variety of higher organisms. Only under extreme environmental conditions of elevated temperatures and/or high salinity do cyanobacterial mats accumulate (to over 1 m thick in in environments such as the hypersaline Solar Lake on the coast of the Gulf of Aqaba in Sinai; Krumbein et al., 1977).

Cyanobacterial mats are probably the oldest form of known life on earth, as shown by the oldest known microfossils which were found in lithified microbial mats (stromatolites) in Central Australia and have been dated at 3.56 billion years (Knoll, 1989). Stromatolites are the most dominant sedimentary structures in rocks throughout the Precambrian era, together with vast deposits of Banded Iron Formation, both of which are associated with widespread development of cyanobacterial mats. While most researchers agree that these sedimentary rocks are associated with the development of microbial mats, many dispute the interpretation of cyanobacterial mat communities, mainly because 16S rRNA analyses of cyanobacterial cultures suggest these organisms are more recently evolved (Woese, 1987; Turner et al., 1989).

Marine and hypersaline cyanobacterial communities are often covered with thin layers of eukaryotic diatoms, *Navicula*, *Nitzchia*, and *Amphora* spp. The cyanobacterial mats in marine and hypersaline environments consist primarily of the filamentous cyanobacteria *Microcoleus chthonoplastes*, *Oscillatoria limnetica*, *Phormidium* spp., *Oscillatoria* spp., *Lyngbia* spp., *Calotrix* spp., *Spirulina* spp., and *Schytonema* spp. They are often associated with a variety of unicellular cyanobacteria, of which the dominant types are *Synechococcus* spp., *Synechocystis* spp., *Pleurocapsa* spp., and *Dermocarpa* spp. (Jørgensen et al., 1983; Stolz, 1984; D'Amelio et al., 1989). The unicellular forms are more abundant at salinities higher than 10%. These cyanobacterial communities occupy the photic zone, which may range from less than 2 mm to over 6 cm. These cyanobacteria are closely associated with a variety of anoxygenic phototrophs, many of which are not yet available in pure culture, including several morphotypes of the green gliding phototrophic bacteria (*Chloroflexus* spp.) and filamentous purple bacteria resembling *Ectothiorhodospira*

(D'Amelio et al., 1987, 1989). Other anoxygenic phototrophic bacteria found in hypersaline mats include *Chromatium* spp., *Thiocapsa pfennigii* (Stolz, 1984), and *Ectothiorhodospira halochloris* (Imhoff et al., 1978). The assemblage of the various anoxygenic phototrophic bacteria are found within the cyanobacterial layer or right below at the deeper part of the photic zone, where a distinct olive-green and/or purple layer can often be found. Right below this layer, a distinct microzone of chemotrophic *Beggiatoa* spp. and *Thioploca* spp. is characteristic for the chemocline, where both oxygen and sulfide coexist (D'Amelio et al., 1989).

In hot spring cyanobacterial mats, where chemical composition and temperature vary widely (Brock, 1978), cyanobacteria are excluded at temperatures above 74°C and at acidic pH (Brock, 1973). *Synechococcus lividus* forms a thick layer over a distinct accretion of *Chloroflexus auranticus* at temperatures of 74 to 60°C. Only at lower temperature can a deep brownish red, c-phycoerythrin-containing cyanobacterium, *Oscillatoria terebriformis*, be found. At temperatures below 45°C, the moderately thermophilic cyanobacteria *Phormidium*, *Pseudoanabaena*, *Synechococcus*, *Calotrix*, and *Mastigocladus laminosus* are found (Castenholz, 1984).

Nitrogen fixation was demonstrated in the mat-forming cyanobacteria *Oscillatoria limosa* (Stal and Krumbein, 1985; Stal et al., 1989) and *Microcoleus chthonoplastes* (Pearson et al., 1981).

Cyanobacterial mats developing in extreme conditions of elevated temperature or at high salinity are clearly laminated microbial communities that can serve as good models for the study of metabolic and genetic interactions in a multilayered microbial biofilm. The various aspects of the ecology, physiology, and molecular biology of both natural communities of cyanobacterial mats as well as isolated pure cultures from microbial mats are discussed in Cohen et al. (1984b) and Cohen and Rosenberg (1989).

Anoxygenic Photosynthesis in Cyanobacteria

Cyanobacteria are often found to develop in environments rich in sulfide. These environment are also rich in available nitrogen and phosphate and CO_2 sources as the result of the anaerobic breakdown of organic matter (Cohen, 1984b). Sulfidic environments may also provide a low-oxygen environment suitable for an optimal operation of ribulose bisphosphate carboxylase (rubisco), the key enzyme of CO_2 photoassimilation (Avron, 1989). Yet H_2S is highly toxic and brief exposure to low sulfide concentrations may irreversibly inhibit most eukaryotic phototrophs. Even sulfide-producing prokaryotes, such as sulfate-reducing bacteria, are inhibited at elevated H_2S concentrations.

Nakamura (1938) was the first to observe growth of *Oscillatoria* in 1 mM H_2S without the detection of oxygen production and the deposition of sulfur in the cells, suggesting that H_2S was being used as an electron donor for CO_2 photoassimilation. Cohen et al. (1975a, 1975b) confirmed this observation in cultures of *Oscillatoria limnetica* isolated from the hypolimnion of a stratified hypersaline lake in Sinai, where this cyanobacterium was the dominant phototroph at up to 3 mM H_2S (Cohen et al., 1977). This cyanobacterium was found to carry out anoxygenic CO_2 photoassimilation with sulfide as an alternative electron donor to water in a photosystem I (PS-I)-driven reaction similar to that carried out by anoxygenic photosynthetic bacteria. Assimilation of CO_2 in the presence of sulfide was found to involve a stoichiometric oxidation of H_2S to elemental sulfur, which was deposited outside the cells (Cohen et al., 1975). This process was independent of photosystem II (PS-II), which was found to be fully inhibited at low sulfide concentrations of 0.1-0.2 mM. Anoxygenic photosynthesis in this organism required an induction period of 2 h in the light and in the presence of sulfide. Photosynthetic efficiency of H_2S-driven anoxygenic photosynthesis was found to be even higher than oxygenic photosynthesis, especially when exposed to red light, and could sustain growth at a doubling rate of about 6 h (Oren et al., 1977; Oren and Padan, 1978). A specific sulfide-quinone reductase was needed for the induction of anoxygenic photosynthesis (Arieli et al., 1991). These workers demonstrated that photosynthetic thylakoids isolated from *Oscillatoria limnetica* grown under sulfide have the capacity in the dark to catalyze electron transfer from sulfide to externally added quinones. They therefore proposed that a membrane-bound sulfide-quinone reductase is the sulfide-induced factor which enabled the use of sulfide for anoxygenic photosynthesis in *Oscillatoria limnetica*. The detailed molecular mechanism of electron transport coupled to proton translocation in cytochrome b_6f/bc_1 complexes is not yet understood, but cytochrome b_6f may be involved in anoxygenic photosynthesis. Arieli et al. (1991) hypothesize the existence of several electron carriers which mediate electron transfer from different donors to the plastoquinone-b_6f complexes. These carriers are thought to have a common site, which interacts with the plastoquinone or cytochrome b_6, as well as a different site, which interacts with the various different electron donors.

Sulfide was found to be oxidized to thiosulfate in *Microcoleus chthonoplastes* (de Wit and van Gemerden, 1987, 1989). Thiosulfate was found to be an electron donor (Castenholz and Utkilen, 1984). Sulfide could be replaced with molecular hydrogen as another alternative electron donor for H_2-dependent photoreduction of CO_2 in another PS-I-dependent anoxygenic photosynthesis in *Oscillatoria limnetica* and *Aphanotheca halophytica* (Belkin et al., 1988).

Depending on the reaction conditions, the electrons from sulfide in *O. limnetica* may be channelled to CO_2 photoassimilation, dinitrogen fixation, or hydrogen evolution under CO_2 limitation (Belkin, Shahak, and Padan, 1988). Sulfide stimulation of hydrogen production was also found in *Nostoc muscorum* by Weisshaar and Boeger (1983).

Cyanobacteria may evolve different mechanisms to grow in the presence of sulfide. *Oscillatoria* spp. grown in sulfide-rich hot springs in Utah were found to be able to carry out oxygenic photosynthesis even in the presence of 2 mM sulfide (Cohen et al., 1986), and similar behavior was found in cyanobacteria grown in sulfide-rich hot springs in New Zealand. In these organisms, PS-II is considerably less sensitive to sulfide toxicity compared to *Oscillatoria limnetica*, but the mechanism of sulfide detoxification is not as yet known. These organisms do not have the capacity for anoxygenic photosynthesis.

Microcoleus chthonoplastes, a cosmopolitan mat-building cyanobacterium found in marine and hypersaline environments, has been demonstrated to have both the capacity to carry out oxygenic photosynthesis at 1 mM sulfide concentration and to be able to carry out anoxygenic photosynthesis using sulfide as an electron donor (Cohen et al., 1986; Jorgensen et al., 1986; de Wit and van Gemerden, 1989).

Other Anaerobic Metabolism Pathways in Cyanobacteria

Nitrogen Fixation

Anaerobic conditions enhance dinitrogen fixation in cyanobacteria because of the toxicity of oxygen to the nitrogenase enzyme. Some cyanobacteria have developed specialized cells called heterocysts where PS-II is absent, but many cyanobacteria can fix dinitrogen aerobically without the need for heterocysts. Two main mechanisms for keeping oxygen away from the nitrogenase sites have been described: 1) Formation of oxygen-poor microniches in aggregates of cyanobacteria, where PS-II and oxygen evolution take place in the periphery of the aggregate while dinitrogen fixation occurs in the center of the cyanobacterial clump. Such a mechanism was described for aggregates of *Trichodesmium* (taxonomically, *Oscillatoria*), a widespread marine filamentous cyanobacterium common in nitrogen-poor tropical oceans (Paerl et al., 1989b; Currin et al., 1990). Using immunoassays, Paerl and his colleagues have shown the localization of nitrogenase in these organisms in the center of the *Trichodesmium* aggregate (Paerl et al., 1989a). 2) Temporal separation of oxygen evolution and nitrogenase activity was demonstrated in several unicellular cyanobacteria, including the widespread picoplanktonic *Synechococcus* in tropical oceans (Mitsui et al., 1986) and the mat-forming cyanobacterium *Oscillatoria limosa* (Stal et al., 1989).

Anaerobic Dark Respiration and Fermentation

Several mat-forming cyanobacteria have been found to survive better under anaerobic conditions in the dark. This phenomenon was described for mats of *Microcoleus chthonoplastes* (Jørgensen et al., 1988), *Oscillatoria terebriformis* (Richardson and Castenholz, 1987), and *O. limosa* (Stal et al., 1989). Glycogen was found to be fermented in a heterofermentative mode producing equal amounts of lactate, ethanol, and CO_2 in *O. limosa* (Stal et al., 1989), while only lactate was found in anaerobic fermentation in *O. limnetica* (Oren and Shilo, 1979). Acetate production in anerobically grown *O. limosa* in the dark was postulated to be the result of anerobic degradation of trehalose (Stal et al., 1989), which serves as an osmolite in several cyanobacteria (Hershkovitz et al., 1991).

Addition of elemental sulfur to anaerobically grown *O. limnetica* and *O. limosa* resulted in anaerobic respiration and reduction of the elemental sulfur to sulfide in the dark (Oren and Shilo, 1979; Stal et al., 1989). The reduction of elemental sulfur in the dark to sulfide was demonstrated in the chemocline of Solar Lake, Sinai (Jøorgensen et al., 1979), where a bloom of the unicellular cyanobacterium *Dactylococcopsis salina* (Walsby et al., 1983; van Rijn and Cohen, 1983) and *Aphanotheca halophytica* were dominant.

Given the widespread distribution of cyanobacteria in a spectrum of sulfide-rich environments, such as hot springs and cyanobacterial mats in marine and hypersaline environments, a variety of other low-redox, sulfur-dependent metabolic pathways are expected to be discovered in the future.

Cyanobacterial DNA

Like other prokaryotes, cyanobacteria contain two types of DNA, chromosomal DNA and smaller, autonomous, extrachromosomal molecules. Both types of DNA can be used for gene transfer and manipulation, each having both advantages and disadvantages. In both instances, transformability of the cell is a prerequisite for gene manipulations. The introduction of foreign DNA can be monitored either by following a genetic marker or by observing some phenotypic alteration indicating a successful transformation. Targeting of the foreign DNA into specific sites in the acceptor DNA and the copy number of the introduced gene may both be of great importance for its expression. Mobilizing DNA sequences into or out of a cyanobacterial cell is not always easier when using extrachromosomal self-replicating molecules. So far, no specific function has been assigned to cyanobacterial plasmids, and in some instances, curing the cells of their plasmids did not have any deleterious effect on the cells, nor was their viability harmed. Nevertheless, various procedures for transforming cyanobacteria have been developed using both types of cyanobacterial DNA.

Genome Size

The genome size of 128 strains representing all major taxonomic groups of cyanobacteria has been measured from the kinetics of renaturation of DNA. The range of size is between 1.6×10^9 Da in the unicellular strains, which is comparable in size to those of other bacterial genomes (Gillis et al., 1970; Wallace and Morovitz, 1973), and 8.6×10^9 Da, in the larger filamentous strains, which greatly exceeds the largest genome previously described in prokaryotes. Even though a larger genome might be anticipated in organisms capable of fixing nitrogen and exhibiting morphological differentiation, the great excess of DNA in these strains raises the possibility that some DNA sequences do not possess a coding capacity (Doolittle, 1979). Genetic mapping of the chromosome of cyanobacteria may shed light on such enigmas as well as establish a physical map of the chromosome. Wolk and coworkers have initiated such a study with *Anabaena variabilis* by producing a set of serially overlapping cosmid clones (Herrero and Wolk, 1986).

The genome sizes of cyanobacteria are discontinuously distributed into four distinct groups with means of 2.2×10^9, 3.6×10^9, 5.0×10^9, and 7.4×10^9 Da. This suggests that genome evolution in cyanobacteria occurred by a series of duplications of a small ancestral genome (Herdman et al., 1979).

Cyanobacteria contain several identical chromosomes in each cell (Williams, 1988), which suggests the possibility of interactions between them. Even though genome interaction seems very likely, no direct molecular evidence for recombination between chromosomes in cyanobacteria has been documented. An intrachromosomal gene conversion mechanism involving the psbA gene family in Synechococcus PCC 7942 was elucidated by Brusslan and Haselkorn (personal communication). Interchromosomal recombination was observed in Synechocystis PCC 6803 by Gurevitz and co-workers (unpublished observations). Thus, the general assumption favors the existence of such mechanisms in cyanobacteria. The recA gene was identified in some cyanobacterial species (Murphy et al., 1987), and in many instances, recombination between chromosomal DNA and plasmid DNA has been proposed. To minimize fortuitous results, this phenomenon should be taken into account whenever introduction of foreign or engineered genes into cyanobacteria is considered.

Plasmids

The existence of nonchromosomal DNA molecules in cyanobacteria was first observed in Anacystis nidulans (Synechococcus sp.) (Asato and Ginoza, 1973). Ever since, it has been evident that nearly all cyanobacteria possess endogenous plasmids (Rebiere et al., 1986), suggesting that it might be feasible to use these molecules as tools in molecular genetic studies. However, no function encoded by cyanobacterial plasmids has been identified so far, and the regulatory mechanism of their replication is still unknown. Several workers have tried to attribute various functions to plasmids, such as gas vacuolation (Walsby, 1977), toxin production (Hauman, 1981; Schwabe et al., 1988), resistance to high salt concentrations or heavy metals (Van den Hondel et al., 1979), resistance to antibiotics (Flores and Wolk, 1985), and synthesis of restriction/modification enzymes (Whitehead and Brown, 1985). It was observed that spontaneous loss of plasmids from different unicellular strains, such as Synechococcus sp. PCC 6301 and 73109 (Whitehead and Brown, 1985) and Synechocystis sp. PCC 6803 (Tandeau de

Marsac and Houmard, 1987), could occur without causing any obvious phenotypic change.

Still, several research groups have produced various cloning vectors or shuttle vectors (E. coli-cyanobacterium) by utilizing these self-replicating molecules. Nevertheless, this approach is limited since the parameters related to mechanisms involved in plasmid replication and control of plasmid copy number in cyanobacteria remain obscure.

The number of autonomous plasmids per cyanobacterial cell may vary from one up to eight, with sizes ranging from 1.3 kb to about 130 kb (Tandeau de Marsac and Houmard, 1987).

In a few instances, different strains of cyanobacteria possess plasmids of identical size and endonucleolytic digestion pattern (Walsby, 1977; Hauman, 1981; Van den Hondel et al., 1979) or sequence homologies (Schwabe et al., 1988). Since some of these strains are of different geographical origins, an implication can be made that interspecific or even intergeneric plasmid transfer may occur in nature.

In other cases, multimeric forms of a single plasmid species were elucidated by digestion with restriction enzymes. Reaston et al. (1980) found in Nostoc sp. PCC 7524, three plasmids, pDU1, pDU2, and pDU3 (6.1, 11.8, and 37.3 kb respectively), where pDU2 is a dimeric form of pDU1. This raises questions in regard to the function and preservation of such forms in the cell, particularly if an active recombination mechanism followed by segregation is considered. Even though the plasmids obtained from different cyanobacterial species are generally similar when purified by different researchers, there are exceptions that in few instances were attributed in part to the extraction protocol used in the various laboratories. However, the very extensively studied cyanobacterium Calothrix sp. PCC 7601 (Fremyella diplosiphon) is a real exception since different restriction patterns and different plasmid numbers were obtained in various laboratories (Simon, 1978; Bogorad et al., 1983; Tandeau de Marsac and Houmard, 1987). These variations could not be explained or related to phenotypic changes of the cells. However, these results, together with previous observations (Tandeau de Marsac, 1983), that spontaneous pigment mutants arise with high frequency (10^{-3} to 10^{-4}) in this organism led the de Marsac group (Tandeau de Marsac and Houmard, 1987) to suggest the existence of mobile genetic elements in the Calothrix strain. So far, whether these elements might be transposons, insertion elements, or Mu-type phages has not been determined. If mobile DNA elements do exist, one would expect high frequencies of genome rearrangements in cyanobacteria. Nevertheless, the only rearrangement documented so far has been the well-analyzed rearrangement of the *nif* region in the filamentous dinitrogen-fixing strain *Anabaena* sp. PCC 7120, which accompanies the differentiation of vegetative cells into heterocysts (Golden et al., 1985a, 1985b).

Restriction/Modification of Cyanobacterial DNA

Chromosomal DNA from various unicellular and filamentous cyanobacteria is considered highly resistant to cleavage by a number of restriction endonucleases, although only a modest repertoire of restriction enzymes was found in these organisms (Van den Honel et al., 1983; Herrero et al., 1984; Lambert and Carr, 1984). Still, more than 50 strains from nine different genera contain from one to five restriction endonucleases (Tandeau de Marsac and Houmard, 1987). Some of the endonucleases, such as the AvaI or AvaII type, are isoschizomers that are widely distributed among different genera. Presumably, this may be related to a common evolutionary origin. Together with their modification methylase associates, they could be acquired by the transfer of plasmid- or cyanophage-encoded genes. However, no correlation among the plasmid content of cyanobacterial strains, their phage sensitivity, and their restriction endonuclease content has been observed.

A similar pattern of DNA modification was found by Lambert and Carr (1984) in diverse filamentous strains that contain (*Gleotrichia* and *Plectonema*) or do not contain (*Nostoc*) different types of restriction endonucleases. Concomitantly, they observed a great variation in the susceptibility of genomes from five unicellular strains, to cleavage by a group of restriction enzymes. By using isoschizomers, Geier and Modrich (1979) were able to demonstrate that DNA resistance to cleavage could originate either from a host-controlled restriction/modification system, as found in other groups of bacteria, or due to a *dam* (DNA adenine methylase) enzyme. Examples of possible restriction/modification systems operating in vivo were depicted for *Anabaena* sp. PCC 7937 (Currier and Wolk, 1979) and for *Synechococcus* sp. PCC 6301 (Szekers, 1981; Szekers et al., 1983). Cyanophage N-1 derived from *Anabaena* sp. PCC 7120 was able to form plaques on *Anabaena* sp. PCC 7937 at a very low efficiency, but the surviving progeny of the phage increased its plaque formation efficiency when reinfecting the same strain (*Anabaena* sp. PCC 7937). The conclusion from this experiment was that the cyanophage N-1 DNA that escaped restriction in *An-*

abaena sp. PCC 7937 could have been modified by the host cell. On the other hand, in experiments performed in vitro by Szekers (1981), it was shown that DNA from *Synechococcus* sp. PCC 6301 was cleaved by an endonuclease produced by AS-1 cyanophage-infected cells. Moreover, the cyanophage DNA was unprotected after cloning in a plasmid vector propagated in an *E. coli* strain. These results indicated that a restriction/modification system was induced in *Synechococcus* sp. PCC 6301 by AS-1 cyanophage; the modification mechanism involved remains unknown.

Another example is the modification enzyme M.NspMACI, which was purified from *Nostoc* sp. PCC 8009. This strain normally contains an isoschizomer of BglII (Reaston and Carr, 1985) called NspMACI, which is unable to cleave DNA previously modified by M.NspMACI while BglII does. This is an indication that the modification sites affected by M.BglII and M.NspMACI are different.

Several explanations for the resistance to cleavage of cyanobacterial DNA were proposed (Lambert and Carr, 1984; Jager and Potts, 1988), including the presence of specific inhibitors like DNA-binding proteins, unusual bases (other than MeC and MeA), and the absence of specific recognition sequences in the genomic DNA of certain cyanobacteria. Whether such modification processes play a role in cyanobacterial gene expression and cell differentiation requires further examination.

Genetic Recombination

Cyanobacteria are susceptible to ultraviolet (UV) irradiation, and therefore they are expected to possess an efficient mechanism to permit repair of UV-induced damage (by genetic exchange and gap-filling). The occurrence of homologous recombination in cyanobacteria explains the high efficiency of foreign DNA incorporation into the genome of unicellular strains like *Synechococcus* PCC 7942 and *Synechocystis* PCC 6803. Recombination activity in cyanobacteria is advantageous for genetic engineering when utilized for the internalization of exogenous DNA into the chromosome. It may be considered a disadvantage when such activity may destroy engineered constructs mobilized into cyanobacteria via plasmid vectors. On one hand, homologous recombination has become a useful way to achieve insertions or deletions at various genes. However, one of the major problems associated with the introduction of cloning vehicles into cyanobacteria is recombination between vector and resident plasmids. For the stable maintenance of hybrid plasmids in *Synechocystis* PCC 6803, some Rec⁻ mutants deficient in homologous recombination were employed. In these mutants, no transformation with either chromosomal markers or integrative vectors occurred, while the autonomously replicative vector pSE76 transformed such recipients at high frequency (Shestakov et al., 1985). Although all transformable cyanobacteria are recombination-proficient, no documentation in respect to the molecular mechanism involved has been reported. However, the *recA* gene from *Synechococcus* PCC 7002 was identified, and its sequence determined (Murphy et al., 1987). This problem is significant and should be addressed when planning experiments of complementation, of DNA transfer (when recombination activity should be avoided), and of studies related to dominance of genes (Haselkorn, 1985). Cyanobacteria possess several identical chromosomes per cell. When transformed with exogenous DNA, in most cases, all the chromosomes will contain the foreign DNA after a few cell cycles, due to segregation. Whether this is followed by interchromosomal gene conversion is still unknown. A concomitant unsolved question is related to the possible occurrence of intrachromosomal gene conversion in cyanobacteria. In addition, it is still unknown whether there exists a mechanism for gene duplication or amplification in cyanobacteria. It is suspected that such a mechanism does exist, particularly as a response to environmental stimuli like an excess of light. Such processes rely on recombination activity, and their better understanding is obviously necessary.

Gene Transfer in Cyanobacteria

To mobilize genetic material into cyanobacteria, the target cyanobacterium should: 1) be transformable by exogenous DNA; 2) have no host restriction; 3) be Rec⁻ when stabilization of autonomous vector is required; 4) allow the generation of meroploids; 5) express genetic markers; and 6) recognize regulatory elements required for transcriptional and translational activities.

A few unicellular strains were found to be competent for genetic transformation and can be readily transformed with either circular or linear DNA. The donor DNA may contain a cyanobacterial origin of replication (*ori*), thus becoming a replicon within the transformed cells, or be lacking an *ori*. In either case, the fate of this DNA inside the transformed cell (recipient) may be different. A more sophisticated approach was required for the development of gene transfer mechanisms in filamentous strains.

A recombinant plasmid, pCH1 (Apᴿ), derived from the endogenous pUH24 plasmid of *Synechococcus* PCC 7942 and possessing the Tn*901* transposon, was the first used by Van den Hondel and co-workers (1980) to transform *Synechocystis* PCC 7942. Transformed cyanobacteria grew in presence of ampicillin but the vectors used were unable to replicate in *E. coli*, which limited their construction and utilization. Since recombinant DNA technology has been largely developed using *E. coli*, it became a great advantage to have shuttle vectors capable of replicating both in *E. coli* and in cyanobacteria. A few cyanobacterial plasmids were used to construct a variety of shuttle vectors (Pouwels et al., 1985; Tandeau de Marsac and Houmard, 1987).

The pCH1 plasmid was further used to create a deletion derivative (pUC1) lacking one of the inverted repeats of Tn*901*. This construction stabilized the marker and prevented any further transpositions. From pUC1 (Apᴿ), a family of shuttle vectors was prepared by Van Arkel and colleagues (Kuhlemeier et al., 1981) by combining it with the *E. coli* plasmid pACYC184 (Cmᴿ). The resulting shuttle vector pUC104 (Apᴿ, Cmᴿ), was stably maintained in *Synechococcus* PCC 7942. Shuttle vector pUC104 purified from *E. coli* was less efficient in transforming the cyanobacterium as compared to pUC104 purified from the cyanobacterium itself. It was hypothesized that there was a difference in their post-replicational DNA modifications (Kuhlmeier et al., 1981). This idea was supported by Gallagher and Burke (1985), who found a sequence-specific endonuclease, AniI, that could restrict foreign DNA. Plasmid pUC104 and lambda DNA were combined to construct a cosmid vector (pPUC29) capable of being packaged in vitro into lambda phage particles and of being introduced into *E. coli* by infection. A further step to optimize the host-vector system for gene cloning in *Synechococcus* PCC 7942 was carried out by Kuhlemeier et al. (1983), who isolated a host strain (R2-SPc) cured of the endogenous plasmid pUH24. A new construct, pUC303, has allowed cloning of several genes, including genes involved with nitrate reductase activity.

A limitation to gene cloning in autonomously replicating vectors may arise from recombination between two homologous genes located in the plasmid and in the chromosome. This limitation can be avoided by dual selection for both the vector and for the chromosomal insert (Kuhlemeier et al., 1985).

Other shuttle vectors for *Synechococcus* PCC 7942 were constructed with pUH24 plasmid and *E. coli* plasmid of the pBR family. The hybrid shuttle vector pLS103 (Cm^R) was constructed from pUH24 and pBR322 by Sherman and Van de Putte (1982). The plasmid pECAN1 (Ap^R, Cm^R), based on pBR325, was constructed by Gendel et al. (1983).

Most of these shuttle vectors contain just a few unique cloning sites. Therefore, a series of hybrid plasmids with multiple cloning sites were produced by Gendel et al. (1983). These vectors, the pPLANB series, provided five or seven unique sites and were able to transform both *E. coli* and *Synechococcus* PCC 7942 with high frequency (10^{-5} transformants/cell). Similarly, Lau and Straus (1985) constructed several small, versatile, shuttle vectors: Plasmid pXB7 (Ap^R) with 10 unique cloning sites; pECAN8 (Ap^R), with four sites within the *lacZ* gene, which permits easy detection of recombinants in the presence of X-gal in some *E. coli* strains and pKBX (Km^R) with 10 cloning sites.

Since the large plasmid of *Synechococcus* PCC 7942, pUH25, contains an origin of replication compatible with that of pUH24, Laudenbach et al. (1985) constructed the pANLO1a and pANLO1b shuttle vectors. Cotransformation with both types of replicons allows simultaneous cloning of foreign genes into the cyanobacterium.

Although great attention was directed towards the cyanobacterium *Synechococcus* PCC 7942, hybrid plasmids were constructed with the small, cryptic plasmid pAQ1 from *Agmenellum quadruplicatum* PR-6 as well. Buzby et al. (1983) produced the pAQE2 and pAQE10 from pAQ1 and pBR322 and pBR325 respectively. The PR-6 strain contains the restriction enzyme AquI (AvaI) and, as demonstrated with the R-2 strain, shuttle vectors purified from PR-6 had higher transformation efficiencies than vectors propagated in *E. coli*. It is likely that the cyanobacterial DNA was modified to resist AquI. A significant difference between PR-6 and R-2 strains is the fact that dimeric and trimeric forms of biphasic plasmids yielded higher transformation frequencies than the analogous monomeric forms (Buzby et al., 1983). By contrast, in R2 strain, monomers are more efficient for transformation. The shuttle vector pAQE17 constructed by Buzby et al. (1985) was used to express allophycocyanin genes from the cyanelle *Cyanophora paradoxa* in *Synechococcus* PCC 7002.

The similarity of the photosynthetic apparatus between cyanobacteria and higher plants is of great advantage for a molecular genetic approach to various constituents involved in this system. Thus, an heterotrophic organism could be beneficial when photosynthetic gene manipulation is required without deleterious effects on the investigated organism. Since the cyanobacterium *Synechocystis* PCC 6803 is a facultative photoheterotroph and amenable for genetic transformation, measures to develop a host-vector system with this organism were taken. The small endogenous plasmid, pSS2, derived from this strain was used to create shuttle cloning vectors (Shestakov et al., 1985; Chauvat et al., 1986). The plasmid pSE176 (Cm^R, Km^R), was constructed from pSS2, pACYC184, pUC4K and contained several restriction sites at the Km^R region plus the EcoRI site in the Cm^R gene. Chauvat and co-workers (1986) have also used pSS2 but with pACYC177 and constructed the hybrid plasmid pFCLV7 (Cm^R, Km^R).

Attempts to transfer plasmids of *E. coli* into cyanobacteria were unsuccessful, probably due to the fact that the origin of replication of these plasmids was not recognized by cyanobacterial replication factors. However, McFadden and coworkers reported the incorporation of pBR322 plasmid into permeaplasts of *Synechococcus* PCC 6301 (Daniell et al., 1986). Permeaplasts are potentially viable cells with high permeability and capacity for cell-wall regeneration. Therefore, they may take up DNA at elevated ratios and subsequently repair and divide. In another experiment, this group demonstrated the uptake and expression of foreign DNA within whole cells of *Synechococcus* PCC 6301 (McFadden and Daniell, 1988). The mechanism involved in the incorporation and expression of the foreign DNA in these cyanobacteria has to be clarified.

A simple method to introduce foreign DNA into the chromosome of unicellular cyanobacteria was developed by Williams and Szalay (1983). They constructed chimeric DNA in an *E. coli* vector consisting of a DNA fragment derived from *Synechococcus* PCC 7942 that had been interrupted by the foreign (donor) DNA fragment aimed for introduction into the cyanobacterium. This construct could propagate in *E. coli* and transform cyanobacteria. The transformation was achieved via double recombination between homologous chromosomal and plasmid DNA sequences. This allows site-directed insertion of foreign DNA into the cyanobacterial chromosome. Addition of a selectable marker to the foreign DNA sequence enabled an easy selection of recombinant cells. Furthermore, in a case where a single recombination event took place, the entire vector could integrate into the chromosome. Distinction between both cases could be achieved by growing the transformants in the presence of one antibiotic (double recombination) or two antibiotics (single recombination). This method was adapted for *Synechocystis* PCC 6803 (Williams, 1988; Shestakov et al., 1985) and is currently widely utilized for studying a variety of cyanobacterial genes via their insertional inactivation, deletion, or modification, or the expression of foreign genes in cyanobacteria.

DNA from *E. coli* can be transferred into unicellular cyanobacteria in various ways that depend on the natural competence of the recipient strains. Some of the filamentous strains, such as *Anabaena* and *Nostoc*, possess unique systems related to differentiation and nitrogen fixation, which are very attractive for molecular studies. Unfortunately, the amenability of these systems to genetic manipulations is limited. For such strains, a gene-transfer system is of great advantage. Wolk and co-workers developed a very elegant conjugal system for transfer of exogenous DNA into *Anabaena* PCC 7120 (Elhai and Wolk, 1988). Three elements were required for the conjugation process: 1) a shuttle vector capable of replicating in *E. coli* cells and cyanobacteria which carries a genetic marker expressible in cyanobacteria and without too many AvaI and AvaII restriction sites; 2) a colicin K or colicin D plasmid capable of mobilizing the shuttle vector in *trans*; and 3) an IncP plasmid, such as RP4, R702, R751, and R7K, capable of mediating the transfer of DNA into a wide range of Gram-negative bacteria including cyanobacteria. This process necessitates conjugal contact since the DNA is transferred via the pili of the conjugants. Practicably, a colicin-containing strain of *E. coli* is transformed with the shuttle plasmid, then mated with the RP4-containing *E. coli* strain. The mating mixture is put on top of the recipient cyanobacteria and transferred after one day to selective medium. Green recombinant colonies can be detected within 10 days.

The DNA transfer by conjugation method has been applied to other strains as well. Bullerjahn and Sherman (1985) constructed a plasmid containing the colE1 *ori*, RP4 *tra* functions, and transposon Tn*501* (conferring resistance to mercuric ions). In *E. coli*, low concentrations of Hg^{2+} promote transposition and the same phenomenon is seen in *Synechocystis* PCC 6714; there, a low level of Hg^{2+} present during conjugation resulted in a fivefold increase in the frequency of obtaining Hg^{2+}-resistant exconjugants. The integration of Tn*501* into the cyanobacterial chromosome was verified by Southern blotting. This demonstrated the feasibility of using the conjugal DNA-transfer method with unicellular cyanobacteria.

Isolation of Cyanobacterial Genes

To analyze gene structure and function, it is crucial to isolate the gene of interest. The following methods have been employed for the isolation of cyanobacterial genes in recent years.

Direct Cloning by Phenotypic Complementation

This method utilizes the phenotypic complementation of *Escherichia coli* or cyanobacterial mutants. The approach involves selection with mutants maintained under permissive growth conditions, e.g., either permissive temperature for temperature-sensitive mutations or photoheterotrophic or heterotrophic conditions for strains capable of these metabolic modes. Shotgun-cloned genomic DNA fragments from the wild-type strain can be introduced into the mutant by using a suitable vector. After a certain period under nonselective conditions to allow expression of marker genes associated with the introduced DNA, the cells are shifted to nonpermissive, selective conditions. Complementary DNA fragments may then be identified due to their ability to allow growth of mutants under otherwise nonpermissive conditions. Complementation is then verified by isolation of the recombinant plasmid and its introduction into the mutant, which should result in the phenotypic difference previously observed.

This general strategy was employed in cloning DNA fragments conferring herbicide resistance in *Synechococcus* sp. PCC 7002 (Buzby et al., 1987) or DNA fragments that complemented photosynthetically impaired mutants of *Synechocystis* sp. PCC 6803 (Dzelzkalns and Bogorad, 1987). A distinct advantage of this method, that is not available in eukaryotic systems, is that it can be used to isolate DNA fragments complementing mutations in components very difficult to identify and isolate. Also, it can be used to identify genes whose products are not structural cell components and therefore cannot be detected in typical purified preparations of cellular fractions or complexes. In rare instances, genes encoding peripheral polypeptides to the photosynthetic process may be cloned by complementation of characterized mutations in *Escherichia coli*. As an example, this strategy was employed in cloning the phosphoenolpyruvate carboxylase (*pcc*) gene of *Synechococcus* sp. PCC 6301 (Kodaki et al., 1985), and the *glnA* gene of *Anabaena* sp. PCC 7120 (Fisher et al., 1981).

Heterologous Hybridization

This method for isolation of cyanobacterial genes uses heterologous hybridization with cloned DNA fragments from prokaryotes or eukaryotes. It is probably the most widely applied method simply because of the sequence conservation within many genes of various phylogenetic origins whose product is functionally similar, thus preserving common structural characteristics and DNA sequence.

In higher plants, many components of the photosynthetic apparatus are encoded by chloroplast genes. The complete nucleotide sequence of two chloroplast genomes was determined, and the open reading frames (ORFs) and unidentified reading frames (URFs) were extensively analyzed (Ohyama et al., 1986; Shinozaki et al., 1986). Also, many nuclear-encoded polypeptides which are part of the photosynthetic apparatus were identified, and their genes or cDNA clones isolated (Coruzzi et al., 1983; Smeekens et al., 1985a, 1985b; Tittgen et al., 1986). The availability of characterized genes or cDNA clones from higher plants that play a role in photosynthesis provides a ready source of materials for isolation and subsequent manipulation of their closely homologous cyanobacterial counterparts. A large number of genes encoding components of the photosynthetic apparatus in cyanobacteria were cloned using heterologous hybridizing probes. In most cases, these probes were derived from cloned genes related to the chloroplast genome. For example: psaA and psaB (Cantrell and Bryant, 1987b; Lambert et al., 1985) related to the PS-I complex; psbA (Curtis and Haselkorn, 1984), psbB (Vermaas et al., 1987), psbC and psbD (Williams and Chisholm, 1987), psbE and psbF (Pakrasi et al., 1988) related to the PS-II complex; petA, petB, petD (Kallas et al., 1987), and petF (Van der Plas et al., 1986) related to the electron transfer chain between both photosystems; atpA, atpB, atpE, atpH, and atpI (Cozens and Walker, 1987; Curtis, 1987; Lambert et al., 1985) related to the ATP synthase complex in the photosynthetic membrane; and rbcL and rbcS (Curtis and Haselkorn, 1983; Shinozaki et al., 1983; Nierzwicky-Bauer et al., 1984; Shinozaki and Sugiura, 1983) for rubisco.

Low-stringency hybridization with heterologous probes was also successfully employed to isolate members of the phycobiliprotein or linker-polypeptide multigene families (Conley et al., 1985, 1986; Houmard et al., 1986; Mazel et al., 1986; Dubbs and Bryant, 1987; Belknap and Haselkorn, 1987; Lemaux and Grossman, 1985). With this type of hybridization, sometimes it is possible to obtain positive results with sequences whose polypeptide product is not immunologically cross-reactive (Dubbs and Bryant, 1987; Belknap and Haselkron, 1987)

Hybridization between psbA and psbD genes also occurs at very low stringency, although the gene products are antigenically distinct (Nixon et al., 1986). This hybridizable feature is usually dependent on several easy controlled factors: the probe should be a purified gene from an internal DNA fragment that is conserved in a functional and structural sense (Bryant and Tandeau de Marsac, 1988); stringent washes of blots should be avoided; low-ionic-strength washes at the hybridization temperature should not be performed. Under such conditions, even 45–55% nucleotide sequence similarities were sufficient for isolating genes (Dubbs and Bryant, 1987; Lemaux and Grossman, 1985; Murphy et al., 1987).

Synthetic Oligonucleotides

Cloning genes by utilization of synthetic oligonucleotide probes has become an increasingly popular approach in molecular biology. The advanced technology of DNA synthesis of fragments up to 100-200 base pairs in length allows screening of DNA libraries for a specific sequence even when only a small fragment of amino acid sequence of a protein product was determined.

However, due to the degeneracy of the genetic code, only in extremely rare instances will such experiments yield the desired DNA sequence. Three strategies can be employed to overcome this problem: In the first, if only a few potential sequences exist, all corresponding sequences can be synthesized and examined individually as hybridization probes. This strategy enabled the cloning of the cpcA gene of Synechococcus sp. PCC 7002 (De Lorimier et al., 1984) and the cpcB gene of Cyanophora paradoxa (Bryant et al., 1985). In the second, all potential oligonucleotide sequences are synthesized as a mixture. This mixture is radiolabeled and used for probing. The conditions for such hybridization should be determined empirically to allow only the putative perfect match to hybridize and to reduce strong competition by closely related sequences that might interfere or prevent detection of the desired signal or generate false positive signals. This strategy was used to clone genes such as the cpcB gene of Synechococcus sp. PCC 7002 (Pilot and Fox, 1984), the petF gene encoding ferredoxin in Anabaena sp. PCC 7120 (Alam et al., 1986) and Synechococcus sp. PCC 7942 (Reith et al., 1986), and the cpeA gene encoding the α subunit of phycoerythrin in Calothrix sp. PCC 7601 (Mazel et al., 1986). By using two independent oligonucleotide mixtures, it is possible to minimize false-positive signals normally obtained in such experiments.

Success in such experiment requires careful planning of the cloning procedure, fine tuning of the hybridization conditions, empirical determination of stringency to allow hybridization of only a perfect match sequence, and use of independent probe mixtures for a particular sequence. The third strategy utilizes unique-sequence oligonucleotide probes 30-50 base pairs in length. Codon usage data are used to predict the most likely base to be at a degenerate position, and deoxyinosine may be inserted at ambiguous codon positions (Takahashi et al., 1985). By careful selection of target sequences, the probes can be synthesized with approximately 75–80% resemblance, and hybridization may result in positive signals under restrictive conditions, as with heterologous probes. This strategy was successfully used to clone the gvpA gene of Calothrix sp. PCC 7601 (Tandeau de Marsac et al., 1985) and the psbE and psbF genes of Synechocystis sp. PCC 6803 (Pakrasi et al., 1988) and Cyanophora paradoxa (Cantrell and Bryant, 1987a).

This third method has now gained a strong momentum due to the development of the polymerase chain reaction technique (PCR), which simplifies a technically tedious protocol and makes it possible to isolate even rare sequences present at very low frequency within the largest genomes of organisms. This technique diminished concern for the region of a known polypeptide to be used to minimize the number of different oligonucleotide sequences in the synthesized mixture (due to the ambiguity of codon usage). The 5'-regions of the oligonucleotides can be designed to contain restriction sites for an easy cloning of the PCR product into a vector.

Immunological Screening

This method was developed for immunological screening of expression libraries containing the genes of interest from various sources. Antiserum is applied to the expressed gene product derived from the library and may signal the desired clone by immunological reaction followed by radiolabeled or colorimetric visualization. Since many cyanobacterial genes may be weakly expressed or not expressed at all from their own promoters in Escherichia coli, the screening may be facilitated by using expression vectors such as pUC plasmids (Vieira and Messing, 1982) or lambda phages (Young and Davies, 1983). This immunological approach was employed successfully in the isolation of clones encoding the cpcB gene of Cyanophora paradoxa (Lemaux and Grossman, 1983), the linker phycobiliprotein of Nostoc sp. PCC 8009 (Zilinskas

and Howell, 1987), and the 35-kDa extrinsic protein of the oxygen evolution complex (OEC) of PS-II from Synechocystis sp. PCC 6803 (Philbrick and Zilinskas, 1988). In the latter case, antiserum raised against the spinach 33-kDa extrinsic protein of OEC cross-reacted with cyanobacterial the polypeptide.

This technique demonstrates the structural similarity between components of the photosynthetic apparatus of cyanobacteria and higher plants. The alpha, beta, and gamma subunits of the ATPase synthase are closely related to their plant homologs, as indicated by immunodecoration (Hicks et al., 1986). This similarity was confirmed by nucleotide sequence analysis of the corresponding genes from Synechococcus sp. PCC 6301 (Cozens and Walker, 1987) and Anabaena sp. PCC 7120 (Curtis, 1987). Similar results were obtained for components of both photosystems, the plastoquinol-plastocyanin reductase, and other enzymatic and soluble electron-transfer proteins (Vermaas et al., 1986; Nechushtai et al., 1983; Van der Vies et al., 1986).

RNA-DNA Hybridizations

This approach can use as probes only conserved genes and stable RNA molecules. Ribosomal RNAs (rRNA) are stable molecules and are widely used for determination of evolutional development or relationships between various organisms. Chloroplast rRNA probes were successfully used to clone the rRNA genes (rnn) of Synechococcus sp. PCC 6301 (Anacystis nidulans) (Tomioka et al., 1981; Tomioka and Sugiura, 1983, 1984; Kumano et al., 1983; Douglas and Doolittle, 1984a, 1984b). In this study it was discovered that the tRNA and rRNA primary and secondary structures of Synechococcus sp. PCC 6301 bear closer resemblance to those of chloroplast RNAs than to their counterparts in Escherichia coli. These similarities strengthen the endosymbiotic hypothesis, implying that eukaryotic chloroplasts are derived from an ancestral photosynthetic prokaryote (Gray and Doolittle, 1982). In addition, 14 out of 15 nucleotides at the 3'-end of the 16S rRNA are identical in Escherichia coli, Synechococcus sp. PCC 6301, and tobacco chloroplasts. Hence, it is very likely that transcripts in these three phylogenetic variable organisms have the same ribosome binding sites (Tandeau de Marsac and Houmard, 1987).

Regulation of Gene Expression

Promoter Sequences, Codon Usage, and Ribosome Binding Sites

From the cyanobacterial promoter regions elucidated so far, some homology between Pribnow boxes in *E. coli* and cyanobacteria was observed. This suggests common characteristics of RNA polymerase binding sites in both organisms. This assumption is supported by the fact that several prokaryotic genes such as *cat, lacZ, npt,* and *lux* are well expressed in cyanobacteria. The major variation in the Pribnow consensus is at the −35 region (Reith et al., 1986). This may be partly explained by the fact that regulation of the expression of some cyanobacterial genes involves complex processes and mechanisms like chromatic adaptation and heterocyst differentiation that may be under developmental and/or environmental regulation.

Codon usage in cyanobacteria can be evaluated by comparing the nucleotide sequence to that of the amino acid product (Shinozaki et al., 1983). Such a survey showed that codon usage is variable between unicellular and filamentous strains; correlation between the GC content in cyanobacterial DNA and the utilization of GC-enriched codons is maintained, particularly in *Synechococcus* PCC 6301; on a quantitative basis, the codon usage in cyanobacteria and *E. coli* is compatible, with the exceptions of the codons for leucine and proline.

From comparisons between the ribosome binding region in *E. coli* and the sequence of the 16S RNA of *Synechococcus* PCC 6301, some complementary sequences were identified. Also, the binding and activity of cyanobacterial ribosomes are affected by specific inhibitors of prokaryotic ribosomal functioning. Still, in other instances, the eukaryotic-like initiation via scanning of cyanobacterial ribosomal functioning is also encountered. Thus, the mechanism used for the initiation of translation in cyanobacteria still remains an open question.

Regulation of Cyanobacterial Gene Expression

Promoter sequences and DNA binding factors have been looked at in order to understand gene regulation in cyanobacteria. Most of the genes under study belong to the major complexes related to light harvesting, photosynthesis, and nitrogen fixation, but genes involved with ATP formation and cell differentiation have also been investigated. Haselkorn and co-workers cloned the genes for a putative sigma factor and for the gamma subunit of RNA polymerase from vegetative cells of *Anabaena* PCC 7120. Both genes possess homology to their *E. coli* counterparts. Study of their inactivation may provide insights into the regulation of transcription in this cyanobacterium. The RNA polymerase of *Anabaena* PCC 7120 contains five

different subunits: an α subunit that is related to an *E. coli* counterpart; a β subunit that contains the nucleotide binding site and is strongly related to the *E. coli* subunit; a β′ subunit that is weakly related to the β′ subunit of *E. coli*; a 66-kDa polypeptide that is related to the β′ subunit of *E. coli*; and a 52-kDa polypeptide that is related to another *E. coli* subunit. More than 10 cyanobacterial species were tested and found to contain a unique subunit of RNA polymerase that is not found in all other eubacteria; this places them, as a group, closer to archaebacteria (Haselkorn et al., 1983).

As mentioned in "Chromatic Adaptation," the synthesis of phycobiliproteins is regulated by light quality. Similarly, the synthesis of some of the linker polypeptides that are noncolored proteins required for assembly of phycobilisomes, depends on the light spectrum. Color mutants induced by UV mutagenesis in *Synechocystis* PCC 6701 were analyzed by Anderson and co-workers (1984). Characterization of rod and core assembly intermediates accumulating in such mutants enabled specific models to be proposed for both the assembly pathway and the final structure of the phycobilisomes in this cyanobacterium. From the various analyses performed so far with *Fremyella, Cyanophora, Agmenellum,* and *Synechocystis,* it is clear that sets of genes encoding phycobiliproteins that are polycistronically transcribed exist in cyanobacteria. The response of the various genes to alterations in light quality represents another regulatory mechanism. However, the DNA sequences involved in turning on/off the transcription of a certain gene or the binding of secondary affectors like DNA-binding proteins that inhibit/enhance expression, are still unknown.

Nitrogen fixation has been studied extensively in *Anabaena* PCC 7120 (Haselkorn, 1986). The arrangement of the genes encoding the protein components of the nitrogenase complex in *Anabaena* is different from that found in *Klebsiella, Azotobacter,* and *Rhodopseudomonas.* In the latter bacteria, *nif* H, *nif* D, and *nif* K encoding dinitrogenase reductase and dinitrogenase α and β subunits, respectively, are contiguous and cotranscribed from a promoter next to *nif* H. The same gene organization is observed in *Anabaena* vegetative cell DNA, except that there is just over 11 kb of DNA between *nif* D and *nif* K. Most of this DNA is not transcribed either in cells growing on ammonia or induced for nitrogenase formation. Golden et al. (1985a, 1985b) have elucidated a unique process of DNA excision during heterocyst differentiation in *Anabaena.* An 11-kb DNA fragment between *nif* K and *nif* D that contains the *xis*A gene is excised, followed by recombination between directly repeated, identical 11-bp sequences located within the coding regions of both genes. A consequence of the excision is that the ORF of *nif* D is fused to the 5'-flanking sequence of *nif* K, thus replacing 26 amino acids by 43 different amino acids. The polycistronic transcript produced by the new operon is 5 kb long and contains sequences for all three *nif* genes.

Another rearrangement occurring during *Anabaena* heterocyst formation involves the *nif* S gene, which is required for nitrogenase maturation. A DNA region adjacent to *nif* S gene (located 3' to *nif* H in *Anabaena* vegetative-cell DNA) is rearranged via a site-specific recombination. Interestingly, the sequences at the junctions involved in the recombination differ from the sequences at the recombination junctions of *nif* D. It is suggested that two different recombination enzymes are involved in the two rearrangements.

A number of cloned genes that are turned on or off during heterocyst differentiation in *Anabaena* were used to demonstrate that the corresponding protein levels are regulated at the transcriptional level (Haselkorn et al., 1983; Wealand et al., 1989).

Machray et al. (1988) discovered an entire transposable element, IS2, near the *rbc* genes of the heterocystous cyanobacterium *Chlorogloeopsis fritschii* CCAP1411/1b. This finding provides evidence of genetic transfer between the Gram-negative *E. coli* and cyanobacteria and may have significance in nucleotide sequence rearrangements known to occur adjacent to the *rbc* and *nif* genes in some nitrogen-fixing cyanobacteria.

Gene Families

Gene inactivation is a useful tool for assessing the function of individual genes of a multigene family. This method was used to engineer a variety of cyanobacterial strains in which the functionality of members of gene families was analyzed. *Synechococcus* sp. PCC 7942 was examined for its three *psb*A genes by constructing mutants in which one or two out of the three genes were inactivated by an antibiotic-resistance cassette (Golden et al., 1986). This demonstrated that all three *psb*A genes are functional and that each is capable of producing sufficient Q_B protein to support the function of PS-II and photoautotrophic growth.

Golden et al. (1986) reported that out of the three members of this gene family, two functional forms of the Q_B protein are detected. Form I, produced by *psb*AI, differs by 25 residues from form II originating from *psb*AII and *psb*AIII, mostly in the amino terminus of the protein. By constructing translational gene fusions between the individual *psb*A genes and a *lacZ* gene from *Escherichia coli,* the expression of each of the *psb*A members could be followed in vivo under various illumination conditions. This experiment indicated a differential expression of the *psb*A gene family dependent on the light availability. Expression of *psb*AI is 500-fold and 50-fold greater than expression of *psb*AII and *psb*AIII, respectively, under similar illumination. If light intensity was decreased there was increased expression of the *psb*AI reporter and decreased expression of the other two genes. These results are supported by earlier reports on the relative abundance of the *psb*A transcripts in this organism when probed with antisense RNA from upstream untranslated regions of the three genes (Brusslan and Haselkorn, 1989).

Analysis of the putative promoter regions of the three *psb*A genes indicated an *E. coli*-like consensus sequence preceding the *psb*AII and *psb*AIII coding regions, whereas the *psb*AI putative promoter lacks a conserved −10 sequence (Golden et al., 1986). This difference is strengthened by the finding that the *psb*AI-*lacZ* fusion that is highly expressed in the cyanobacterium failed to produce the blue colony phenotype in *E. coli* plated in presence of the β-galactosidase indicator 5-bromo-4-chloro-3-indolyl β-D-galactopyranoside (X-Gal).

These observations indicate that some cyanobacterial promoters are regulated differently than *E. coli*-like promoters and that they can be enhanced by environmental factors like high light intensity. The mechanism involved in such regulation is still unknown.

The function of two *psb*D genes coding for the D2 protein of PS-II was similarly analyzed. Since *psb*DI and *psb*C genes overlap, an additional *psb*C gene was inserted into the *psb*DII locus, thus creating a strain in which *psb*DI alone could be inactivated. This experiment demonstrated that, although not essential for viability, expression from the *psb*DI locus is required for optimal growth (Golden et al., 1989) while *psb*DII is dispensable under laboratory growth conditions.

Expression of Cyanobacterial Genes in *Escherichia coli*

A number of cyanobacterial genes were cloned, and in some instances expressed, in *E. coli* (Gurevitz et al., 1985; Gatenby et al., 1985). In most of these experiments, the genes were placed under an *E. coli* promoter to ensure binding of RNA polymerase. However, the phosphatase gene from the nitrogen-fixing cyanobacterium *Nostoc commune* UTEX 584 was cloned and was most probably expressed from its own promoter in *E. coli* (Xie et al., 1989).

Chromatic Adaptation

The response of photosynthetic organisms to light has been a major issue in photosynthesis research. The correlation between the ability to sense and capture light to many outcoming processes, including the onset of gene expression, biochemical and metabolic pathways, and the synthesis and assembly of building blocks for specific subcellular structures, was studied extensively. Cyanobacteria have developed a unique mechanism for optimizing their light-harvesting potential when exposed to changes in the spectral quality of light. This is achieved by adjusting their photosynthetic pigment system via an increase or decrease of the constituents of the photosynthetic apparatus and an adaptive rearrangement of various components involved in light capture and its transformation into biochemical energy (Post et al., 1989). Variations in the ambient light regime are reflected in various adaptation features existing in many eukaryotic algae and prokaryotic photosynthesizers. Yet, cyanobacteria are unique in that they possess phycobilisomes with which they are able to adjust their light-harvesting potential by changing the molar ratio of the phycobilins, the pigmented proteins within the phycobilisomes. This process was termed "complementary chromatic adaptation" by Bennet and Bogorad (1973) and Tandeau de Marsac (1977). It involves a complete turnover of the phycobilisome structure, and its genetic control is currently under extensive study by cyanobacterial molecular biologists. An understanding of phycobilin synthesis will allow the recognition of genes involved in a regulatory mechanism representative of those involved in adaptation to environmental variations.

The cyanobacterium *Calothrix* sp. PCC 7601 (*Fremyella diplosiphon*) was the focus for the studies related to the chromatic adaptation. Conley et al. (1986) elucidated the genes for the synthesis of the phycobiliproteins. Three sets of the alpha and beta allophycocyanin (APC) genes, plus two sets of the alpha and beta phycocyanin (PC) genes, are all transcribed off the same strand of a 13-kb cluster on the cyanobacterial genome. The order of their arrangement seems to be conserved, as was previously found in the cyanelle DNA of the eukaryotic alga *Cyanophora paradoxa* (Lemaux et al.,

1983, 1985). Pilot and Fox (1984) isolated the phycobiliprotein genes from *Agmenellum quadruplicatum* by using oligodeoxynucleotide probes based on the amino acid sequence of PC of *Agmenellum*. They succeeded in isolating the gene cluster of these proteins, determined their order to be PCβ, PCα, and found that these genes are cotranscribed in white light from an *E. coli*-like promoter. Similar results were obtained by De Lorimier et al. (1984), who used a replicating plasmid to introduce the cloned genes APCα and β from *Cyanophora* into *Agmenellum*. The phycobilisomes were purified from the recombinant organism (De Lorimier et al., 1987), and isoelectric focusing of the phycobiliproteins permitted discrimination between *Agmenellum* and *Cyanophora* APC. Some *Cyanophora* APC could be detected in assembled bilisomes, indicating that bilin attachment to apoprotein and correct assembly into bilisome cores is possible in a heterologous host. Such an experiment paves the way for an in vitro mutagenesis program for the study of assembly and structure/function relationships of these proteins.

Differentiation

Formation of heterocysts in some filamentous cyanobacterial strains provides an excellent system for studying cell differentiation. The heterocysts are physiologically, biochemically, and structurally distinct from the vegetative cells. They can express genes that are inactive in the vegetative cells, and synthesize constituents for new metabolic pathways functioning in the assimilation of molecular nitrogen. Concomitantly, these specialized cells lose their photosynthetic capacity (structures and function) and become the nitrogen suppliers of the filament. It should be extremely interesting to probe with a molecular marker the early detection of genes which turn on before any obvious changes are observed.

For genetic analysis of filamentous cyanobacteria, Schmetterer et al. (1986) constructed mobilizable plasmid vectors able to express the luciferase gene in *Anabaena* species. These *lux* genes may be utilized as promoter probes for studies of cell differentiation in the filaments. Indeed, this approach was recently rewarded when Holland et al. (1989) followed the induction of the *het*A gene in the premature heterocysts of *Anabaena* PCC 7120. Wolk and Elhai (personal communication) developed a remarkable technology in which transcriptional fusions put the expression of luciferase under the control of *het*A transcriptional signals. Light emitted by individual cells within filaments could be monitored by an image processing system attached to a conventional microscope. In this experiment, 9 h after nitrogen deprivation, cells emitting light were observed at intervals along the cyanobacterial filament. Only later, these cells became structurally distinguishable from the rest of the vegetative cells. It was concluded that hetA is a gene that is expressed early in the differentiating filament of *Anabaena*. This approach provides a breakthrough in the understanding of the biochemical and genetic control of heterocysts differentiation in nitrogen-fixing cyanobacteria and may be used in the characterization of developmentally impaired mutants (Wolk et al., 1988).

Metabolism

Photosynthetic Genes

Cyanobacteria appear to be ideal organisms for the genetic study of photosynthesis since they perform photosynthesis similar to that of higher plants (Ho and Krogmann, 1982), and some species are available for genetic manipulations (Williams, 1988). However, most of the cyanobacterial strains studied are obligate autotrophs, and therefore most mutations in genes involved in photosynthesis are lethal. This problem may be overcome in two ways: 1) by the isolation of conditional lethal mutations; or 2) by the use of a facultatively heterotrophic strains which are capable of growing on an exogenous source of organic carbon (Rippka, 1972). Pakrasi et al. (1988) developed a genetic technique for molecular analysis of electron transport in PS-II in *Synechocystis* PCC 6803. Their methodology involves deletion of specific genes from the cyanobacterial genome and replacement with mutant genes prepared by site-directed mutagenesis. In this way it is possible to investigate the role of specific amino acid residues of a given polypeptide involved in binding pigments or the role of various cofactors in the overall activity of this complex. Two important characteristics of *Synechocystis* strain PCC 6803 permit such studies: 1) their naturally occurring genetic transformation system (Grigorieva and Shestakov, 1982); and 2) their ability to grow photoheterotrophically on glucose, which is necessary for the propagation of PS-II mutants that are incapable of photosynthesis (Rippka, 1972; Jansson et al., 1987). This strain was recently used in an elegant molecular experiment aimed at the elucidation of the identity of Z, the primary electron donor to the PS-II reaction center, and D, an oxidizable PS-II component

structurally resembling Z. For many years, Z and D were assumed to be plastoquinols based on a variety of physical measurements performed on the oxidized donors. However, since tyrosine radicals may yield similar physical signals to those seen in PS-II, the strongly conserved tyrosine 160 in the D2 polypeptide adjacent to the reactive histidine residue associated with P680-binding was changed to a phenylalanine (Debus et al., 1988; Vermaas et al., 1988) by site-directed mutagenesis. This change resulted in the disappearance of the electron paramagnetic response signal and led to the conclusion that tyrosine 160 in the D2 protein is the electron donor D and, by analogy, tyrosine 161 in the D1 polypeptide is electron donor Z. This discovery was achieved by site-directed modification in a cloned psbD gene, followed by its introduction back into the cyanobacterial genome of *Synechocystis* PCC 6803. In a different study, Pakrasi et al. (1988) examined the role of the psbE, psbF, psbI, and psbJ gene products in PS-II of *Synechocystis* PCC 6803. They used site-directed mutagenesis of cloned ORFs of the psbEFIJ operon and demonstrated a complete loss of PS-II assembly and activity when the entire operon was deleted.

Sugar Metabolism

The primary pathway for the catabolism of endogenous glucose in cyanobacteria is the oxidative pentose phosphate cycle. This pathway is important for maintaining ATP and reducing power levels during darkness. In addition, in nitrogen-fixing cyanobacteria, this cycle may contribute reducing power for nitrogenase activity. The levels of glucose-6-phosphate dehydrogenase and 6-phosphogluconate dehydrogenase are sevenfold higher in heterocysts than in vegetative cells, suggesting that there is a genetic control of these enzymes in cyanobacteria which is linked to environmental effects on sugar metabolism. During the transition of a *Synechococcus* PCC 7942 culture from exponential growth to stationary phase, an increase in the specific activity of the enzyme 6-phosphogluconate dehydrogenase (6PGD) is observed. By fusing the gnd gene with a promoterless lacZ gene and transforming a gnd⁻ *Synechococcus* mutant, the growth-phase-dependent induction of 6PGD synthesis was found to be regulated at the transcriptional level. Although cyanobacteria are known to assimilate inorganic carbon via the C-3 reductive photosynthetic pathway, they also fix large amounts of carbon in the light as C-4 acids. Malate and aspartate are the major products of phosphoenolpyruvate carboxylase (PEPCase) activity. It is possible that PEPCase activity replenishes the TCA cycle with intermediates that are used for biosynthetic purposes. A genetic approach to determine the role of the enzyme in cyanobacteria may confirm this assumption. Such an experiment should involve the isolation of the *ppc* gene, followed by insertional inactivation that will indicate whether this gene is dispensable.

RUBISCO. The study of rubisco structure and function was for many years an attractive challenge due to its central role in photosynthesis and CO_2 fixation. Some of the progress gained recently has been achieved by a molecular genetic approach that involves *rbc* genes derived from cyanobacteria. Since the large subunit of the enzyme is coded in higher plants by the chloroplast genome, and chloroplasts are still untransformable, it is impossible to genetically modify the large subunit of the enzyme (*rbc*L) in higher plants. Attempts to express the holoenzyme in vitro by introducing the *rbc*L and the *rbc*S genes in various ways into *E. coli* were unsuccessful. The *rbc* gene from the photosynthetic anaerobe *Rhodospirillum rubrum* has been expressed in *E. coli* and various mutations have been introduced, but the *R. rubrum* enzyme is quite different from the enzyme of higher plants. On the other hand, cyanobacteria possess a rubisco enzyme very homologous to that of higher plants, which has the substantial advantage of carrying both *rbc* genes on a single operon. This feature permitted the successful expression of rubisco from a variety of cyanobacteria in *E. coli* (Gatenby et al., 1985; Gurevitz et al., 1985; Tabita, 1988). Although many technical difficulties still limit the development of a reliable selection system to study structure/function relationships of rubisco, it may perhaps be easier to approach in transformable cyanobacteria. Meanwhile, polypeptidic chaperonins (factors assumed to play a role in the assembly of rubisco) were proposed and their involvement demonstrated by using expression vectors containing *rbc* genes from *Synechococcus* PCC 6301 (Goloubinoff et al., 1989).

Ci PUMP. In cyanobacteria, the dependence of photosynthesis on external inorganic carbon is regulated by the presence of an inducible active inorganic carbon (Ci) transport system. This system operates to elevate the intracellular concentration of CO_2 so that the relatively low-affinity cyanobacterial rubisco may function at an efficiency close to that occurring at CO_2 saturation levels. Four key operational elements have been defined for this concentrating mechanism: 1) a transport system; 2) a means to en-

ergize the transport by photosynthesis; 3) a leak barrier to reduce the back flux of CO_2; and 4) a mechanism to rapidly interconvert inorganic carbon species within the cell (Reinhold et al., 1987). Interconversion of inorganic carbon species is necessary so that CO_2, the substrate required by rubisco, is formed from HCO_3^-, the carbon species that is delivered to the cell. Recently, a molecular approach to resolve the mechanism of this interconversion was initiated by several groups trying to resolve the various constituents of the concentrating mechanism. The strategy employed is the creation of mutants requiring high levels of CO_2 and using these mutants in genetic complementation experiments aimed at identifying components and genes of the system. Price and Badger (1989) succeeded in expressing a carbonic anhydrase derived from human tissues in the cytosol of *Synechococcus* PCC 7942 and created a high-CO_2 requirer. This was probably due to the conversion of cytosolic HCO_3^- into CO_2, which leaked out from the cell and prevented the accumulation of HCO_3^- within the carboxysomes. This loss of the ability to accumulate internal Ci confirmed the previous assumption that the carbonic anhydrase enzyme resided only within the carboxysome. Ogawa et al. (1987) isolated a high-CO_2 requirer (RK1) from *Synechococcus* PCC 7942, pinpointing a 42-kDa protein not synthesized in the mutant and obtained such a mutant (RKb) with *Synechocystis* PCC 6803 (J. Pierce, personal communication). Kaplan and co-workers isolated the E1 and 0221 mutants of *Synechococcus* PCC 7942, which require elevated CO_2 levels for growth. Recently, they oriented DNA sequences relative to the *rbc* locus, which seem to play a role in carboxysome formation, thus indirectly creating a requirement for higher CO_2 concentration (Friedberg et al., 1989).

Stress

Adaptation of cyanobacteria to stress conditions has been documented in several instances. *Microcystis firma*, like most other cyanobacteria, is unable to adapt to higher salt concentrations in the dark but in response to salt stress in the light, an osmoregulant, glucosylglycerol, found thus far only in cyanobacteria, is synthesized from glycogen. When the glycogen pool is depleted in the dark, glucosylglycerol is not produced in sufficient amounts. In the light, the salt-dependent accumulation of glucosylglycerol is characterized by negligible turnover in salt-adapted cells and by small but continuous leakage of this substance into the medium (Hagemann et al., 1987).

Iron is an essential component of photosynthetic cytochromes and of nonheme iron-sulfur proteins. In photosynthetic organisms, ferredoxin functions primarily as the terminal electron acceptor of the photosynthetic electron transport chain. Under conditions of moderate iron limitation, the Fe-S protein ferredoxin is replaced by the flavoprotein flavodoxin. The genes encoding ferredoxin and flavodoxin proteins were cloned from *Synechococcus* PCC 7942. Whereas the gene encoding ferredoxin is constitutively transcribed, flavodoxin was found to be transcriptionally regulated by the availability of iron. The flavodoxin mRNA was observed only in a medium low in iron, and it disappeared upon addition of iron (Laudenbach et al., 1988).

Phycocyanins represent approximately 35% of the total cell protein of *Calothrix* sp. PCC 7601. Three phycocyanin operons were characterized in this cyanobacterium by Tandeau de Marsac et al. (1988). Regulation of the third operon, *cpc*3, represents a novel response to environmental stress. The expression of the *cpc*3 operon is turned on under sulfur limitation, while *cpc*1 and *cpc*2 are switched off. The protein product of this operon lacks sulfur-containing amino acids except for those at chromophore-binding sites. This adaptation allows survival of the cell under extreme growth conditions represented by sulfur limitation (Tandeau de Marsac et al., 1988).

The availability and turnover of phosphorus may play a key role in determining the development of water blooms or of economically important nitrogen-fixing communities such as those found in rice fields. The gene *iph* (for indole phosphate hydrolase) may be useful for the study of phosphate regulation in cyanobacteria and may prove useful in the study of cyanobacterial promoter function, a subject on which there are limited data (Schneider et al., 1987).

A new UV-A/B-absorbing pigment bound to a polysaccharide core which had maxima at 312 and 330 nm was found in the cosmopolitan terrestrial cyanobacterium *Nostoc commune*. The pigment is found in high amounts (up to 10% of dry weight) in colonies grown under solar UV radiation but only in low concentrations in laboratory cultures illuminated by artificial light without UV. Synthesis of the pigment is induced by UV light, and the pigment protects *Nostoc* from UV radiation. *Nostoc* is also capable of withstanding extreme water stress (drought). Apparently, the UV pigment can participate in water storage, since it is a polysaccharide and is located outside the cell in the polysaccharide matrix (Scherer et al., 1988).

Biotechnology

There is increasing interest in the use of cyanobacteria for biotechnology for two main reasons: 1) Because of the great similarity of their photosynthetic apparatus to that of higher plants, cyanobacteria are excellent model systems for studying oxygenic photosynthesis. 2) Mass cultivation of cyanobacteria has become a promising route for the production of large quantities of natural products of biotechnological and agricultural values.

Model Systems

Resistance to herbicides of the triazine and urea types in higher plants is a chloroplastic trait. Such resistance is a single-gene trait and may be utilized in the future for genetic engineering of crop plants. However, no reliable procedure for stable transformation of chloroplasts is available. Assuming that such a procedure will become feasible, it is necessary to better understand the phenotypic expression of the triazine-resistance gene. Thus, identical and diverse mutations have been introduced into the *psb*A gene encoding the D1 protein in the cyanobacterium *Synechococcus* PCC 7942. This reaction center II polypeptide was found to be the primary site for mutations conferring triazine resistance. The power of cyanobacterial molecular genetics was employed in the introduction of a variety of mutations to enable the elucidation of the D1 architecture in the thylakoid membranes, and the construction of a *Synechococcus* PCC 7942 strain that is partially diploid for *psb*A and heterozygous for triazine resistance (Pecker et al., 1987). From this experiment it was concluded that triazine resistance is a recessive trait in cyanobacteria. On the other hand, diuron resistance in *Synechococcus* PCC 7942 was reported to be a dominant trait by Brusslan and Haselkorn (1989).

Another important facet of future uses for cyanobacterial biotechnology may be related to rubisco. This key enzyme of CO_2 fixation controls the rate of net photosynthesis. Its complex structure and assembly, and the involvement of various factors in holoenzyme synthesis make it an extremely difficult target for genetic modification in higher plants. The simpler pathway of expression of a very similar enzyme in cyanobacteria makes it accessible to genetic manipulations which may lead in the future to better understanding of the relations between its structure and its kinetic properties.

Applications

FOOD AND NATURAL PRODUCTS. The possibility that mass cultures of specific cyanobacterial strains may be suitable for human or animal feeding is not new. The filamentous nonheterocystous cyanobacterium *Spirulina* has been part of some human diets in Africa for centuries, providing a rich source of protein, vitamins (particularly B_{12}), and the essential fatty acid gamma-linoleic acid (Cohen et al., 1987). This cyanobacterium can be grown on marginal land using saline water unsuitable for conventional agriculture. The filamentous nature of the cyanobacterium permits simple harvesting by filtration through screens. To make *Spirulina* a major food or animal feed product, its cost of production should be reduced while it is grown efficiently on a large scale. However, molecular genetics of *Spirulina* has not yet developed due to the lack of systems for transformation, transfection, or conjugation.

Cyanobacteria may be utilized for water treatment functions not practical by current mechanical or chemical means, such as purification via the incorporation of heavy metal ions.

Cyanobacteria grown on a large scale could be utilized for the production of various natural products such as polysaccharides, carotenoids, antioxidants, antibiotics, and precursors of pharmaceuticals.

Insect Control

The larvicidal gene of *Bacillus sphaericus* has been cloned and introduced into *Synechococcus* PCC 7942 by an autonomously replicating vector (Tandeau de Marsac et al., 1987). The recombinant cells produced active toxin whose level of activity against *Culex* mosquito larvae was found to be the same either in *E. coli* or in the cyanobacterium. Very often spores of *B. sphaericus* settle rapidly onto bottom mud and are removed from the larval feeding area. Introduction of such a cloned toxin into cyanobacteria, which grow on the upper layers of aquatic habitats, may provide an elegant solution for mosquito control, since cyanobacteria will persist longer in such an environment and reach target insect larvae more effectively than the spores do.

Literature Cited

Alam, J., Whitaker, D. W., Krogmann, D. W., and Curtis, S. E. 1986. Isolation and sequence of the gene for ferredoxin I from the cyanobacterium *Anabaena* sp. strain PCC 7120. Journal of Bacteriology 168:1265–1271.

Allen, M. M. 1985. Oxygenic photosynthesis in cyanobacteria, p. 133–153. In: Leadbetter, E. R., and Poindexter, J. S. (ed.), Bacteria in nature. Plenum Press, New York.

Anderson, L. K., Rayner, M. C., and Eiserling, F. A. 1984. Ultraviolet mutagenesis of *Synechocystis* sp. 6701; mutations in chromatic adaptation and phycobilisome assembly. Archives of Microbiology 138:237–243.

Arieli, B., Padan, E., and Shahak, Y. 1991. Sulfide-induced sulfide-quinone reductase activity in thylakoids of *Oscillatoria limnetica*. Journal of Biological Chemistry 266:104–111.

Asato, Y., and Ginoza, H. S. 1973. Separation of small circular DNA molecules from the blue-green alga *Anacystis nidulans*. Nature 244:132–133.

Avron, M. 1989. Efficiency of biosolar energy conversion by aquatic photosynthetic organisms, p. 387–389. In: Cohen, Y., and Rosenberg, E. (ed.), Microbial mat: Physiological ecology of benthic microbial communities. American Society for Microbiology, Washington, DC.

Bauld, J. 1984. Microbial mats in marginal marine environments: Shark Bay, Western Australia and Spencer Gulf, South Australia, p. 39–58. In: Cohen, Y., Castenholz, R. W., and Halvorson, H. O. (ed.), Microbial mats: Stromatolites. Alan R. Liss, New York.

Belkin, S., Shahak, Y., and Padan, E. 1988. Anoxygenic photosynthetic electron transport. Methods of Enzymology 167:380–386.

Belknap, W. R., and Haselkorn, R. 1987. Cloning and light regulation of expression of the phycocyanin operon of the cyanobacterium *Anabaena*. EMBO Journal 6:871–884.

Bennet, A., and Bogorad, L. 1973. Complementary chromatic adaptation in a filamentous blue green alga. Journal of Cell Biology 58:419–435.

Bogorad, L., Gendel, S. M., Haury, J. H., and Koller, K. P. 1983. Photomorphogenesis and complementary chromatic adaptation in *Fremyella diplosiphon*, p. 119–126. In: Papageorgiou, G. C. and Packer, L. (ed.), Photosynthetic prokaryotes: Cell differentiation and function. Elsevier, New York.

Brock, T. D. 1973. Lower pH limit for the existence of blue-green algae. Science 179:480–482.

Brock, T. D. 1978. Thermophilic microorganisms and life at high temperatures. Springer-Verlag, New York.

Brusslan, J. A., and Haselkorn, R. 1989. Resistance to the photosystem II herbicide diuron is dominant to sensitivity in the cyanobacterium *Synechococcus* sp. PCC 7942. EMBO Journal 8:1237–1245.

Bryant, D. A., de Lorimier, R., Lambert, D. H., Dubbs, J. M., Stirewalt, V. L., Stevens, S. E., Jr., Porter, R. D., Tam, J., and Jay, E. 1985. Molecular cloning and nucleotide sequence of the α and β subunits of allophycocyanin from the organelle genome of *Cyanophora paradoxa*. Proceedings National Academy Science USA 81:3242–3246.

Bryant, D. A., and Tandeau de Marsac, N. 1988. Isolation of genes encoding components of photosynthetic apparatus. Methods in Enzymology 167:755–765.

Buchanan, R. E., and Gibbons, N. E. 1974. Bergey's manual of determinative bacteriology, 8th ed. Williams and Wilkins, Baltimore.

Buzby, J. S., Mumma, R. O., Bryant, D. A., Gingrick, J., Hamilton, R. H., Porter, R. D., Mullin, C. A., and Stevens, S. E., Jr. 1987. Genes with mutations causing herbicide resistance from the cyanobacterium *Synechococcus* sp. PCC 7002, p. 757–760. In: J. Biggins (ed.), Progress in photosynthesis research, vol. 4. Martinus Nijhoff Publishers, Dordrecht, The Netherlands.

Buzby, J. S., Porter, R. D., and Stevens, S. E. 1983. Plasmid transformation and characterization of the *Escherichia coli* DNA adenine methylase (*dam*) gene. Nucleic Acids Research 11:837–851.

Buzby, J. S., Porter, R. D., and Stevens, S. E., Jr. 1985. Expression of the *E. coli lacZ* gene on a plasmid vector in a cyanobacterium. Science 230:805–807.

Cantrell, A., and Bryant, D. A. 1987a. Molecular cloning and nucleotide sequence of the genes encoding cytochrome b_{559} from the cyanelle genome of *Cyanophora paradoxa*, p. 659–662. In: J. Biggins (ed.), Progress in photosynthesis research, vol. 4. Martinus Nijhoff Publishers, Dordrecht, The Netherlands.

Cantrell, A., and Bryant, D. A. 1987b. Molecular cloning and nucleotide sequence of the *psaA* and *psaB* genes of the cyanobacterium *Synechococcus* sp. PCC 7002. Plant Molecular Biology 9:453–468.

Castenholz, R. W. 1973. Movements, p. 111–128. In: Fogg, G. E., Stewart, W. D. P., Fay, P., and Walsbey, A. E. (ed.), The blue-green algae. Academic Press, London.

Castenholz, R. W. 1984. Composition of hot spring microbial mats: a summary, p. 101–119. In: Cohen, Y., Castenholz, R. W., and Halvorson, H. O. (ed.), Microbial mats: Stromatolites. Alan R. Liss, New York.

Castenholz, R. W., and Utkilen, H. C. 1984. Physiology of sulfide tolerance in a thermophilic Oscillitoria. Archives of Microbiology 138:299–305.

Chauvat, F., De Vries, L., Van den Ende, A., and Van Arkel, G. A. 1986. A host vector system for gene cloning in the cyanobacterium *Synechocystis* PCC6803. Molecular General Genetics 204:185–191.

Cohen, Y. 1984a. The Solar Lake cyanobacterial mats: strategies of photosynthetic life under sulfide, p. 133–148. In: Cohen, Y., Castenholz, R. W., and Halvorson, H. O. (ed.), Microbial mats: Stromatolites. Alan R. Liss, New York.

Cohen, Y. 1984b. Oxygenic photosynthesis, anoxygenic photosynthesis, and sulfate reduction in cyanobacterial mats, p. 435–441. In: Klug, M. J., and Reddy, C. A. (ed.), Current perspectives in microbial ecology. American Society for Microbiology, Washington, DC.

Cohen, Y. 1986. Interaction of cycles of C, O, S, and Fe in hypersaline cyanobacterial mats, p. 213–217. In: Megusar, F., and Gantar, M. (ed.), Perspectives in microbial ecology. Slovene Society for Microbiology, Ljubljana.

Cohen, Y., Castenholz, R. W., and Halvorson, H. O. (ed.). 1984. Microbial mats: stromatolites. Alan R. Liss, New York.

Cohen, Y., and Ideses, R. 1989. Hypersaline cyanobacterial mats: environmental manipulations for the production of specific carotenoids and specific polysaccharides, p. 135–138. In: Miyachi S., Karube I., and Ishida Y. (ed.), Current topics in marine biotechnology. Japanese Society for Marine Biotechnology, Tokyo.

Cohen, Y., Jorgensen, B. B., Revsbech, N. P., and Poplawski, R. 1986. Adaptation to hydrogen sulfide of oxygenic and anoxygenic photosynthesis among cyanobacteria. Applied and Environmental Microbiology 51:398–407.

Cohen, Y., Jorgensen, B. B., Padan, E., and Shilo, M. 1975a. Sulfide-dependent anoxygenic photosynthesis in the cyanobacterium *Oscillatoria limnetica*. Nature 257:489–491.

Cohen, Y., Krumbein, W.E., and Shilo, M. 1977. Solar Lake (Sinai) 2. Distribution of photosynthetic microorganisms and primary production. Limnology and Oceanography 22:609–620.

Cohen, Y., Padan, E., and Shilo, M. 1975b. Facultative anoxygenic photosynthesis in the cyanobacterium *Oscillatoria limnetica*. Journal of Bacteriology 122:855–861.

Cohen, Y., and Rosenberg, E. (ed.). 1989. Microbial mat: Physiological ecology of benthic microbial communities. American Society for Microbiology, Washington, DC.

Cohen, Z., Vonshak, A., and Richmond, A. 1987. Fatty acid composition of *Spirulina* strains grown under various environmental conditions. Phytochemistry 26:2255–2258.

Conley, P. B., Lemaux, P. G., and Grossman, A. R. 1985. Cyanobacterial light-harvesting complex subunits are encoded in two red-light induced transcripts. Science 230:550–553.

Conley, P. B., Lemaux, P. G., Lomax, T. L., and Grossman, A. R. 1986. Genes encoding major light-harvesting polypeptides are clustered on the genome of the cyanobacterium *Fremyella diplosiphon*. Proceedings National Academy Science USA 83:3924–3928.

Coruzzi, G., Broglie, R., Cashmore, A., and Chua, N.-H. 1983. Nucleotide sequence of two pea cDNA clones encoding the small subunit of ribulose 1,5-bisphosphate carboxylase and the major chlorophyll a/b-binding thylakoid polypeptide. Journal of Biological Chemistry 258:1399–1402.

Cozens, A., and Walker, J. E. 1987. The organization and sequence of the genes for ATP synthase subunits in the cyanobacterium *Synechococcus* 6301: support for an endosymbiotic origin of chloroplasts. Journal of Molecular Biology 194:359–383.

Currier, T. C., and Wolk, C. P. 1979. Characteristics of *Anabaena variabilis* influencing plaque formation by cyanophage N-1. Journal of Bacteriology 139:88–92.

Currin, C. A., Paerl, H. W., Suba, G. K., and Alberte, R. S. 1990. Immunofluorescence detection and characterization of N_2-fixing microorganisms from aquatic environments. Limnology and Oceanography 35:59–71.

Curtis, S. E. 1987. Genes encoding the beta and epsilon subunits of the proton-translocating ATPase from *Anabaena* sp. strain PCC 7120. Journal of Bacteriology 169:80–86.

Curtis, S. E., and Haselkorn, R. 1983. Isolation and sequence of the gene for the large subunit of ribulose 1,5-biphosphate from the cyanobacterium *Anabaena* 7120. Proceedings National Academy Science USA 80:1835–1839.

Curtis, S. E., and Haselkorn, R. 1984. Isolation, sequence and expression of two members of the 32kd thylakoid membrane protein gene family from the cyanobacterium *Anabaena* 7120. Plant Molecular Biology 3:249–258.

D'Amelio, E. D., Cohen, Y., and Des Marais, D. J. 1987. Association of a new type of gliding, filamentous purple phototrophic bacterium inside bundles of *Microcoleus chthonoplastes* in hypersaline cyanobacterial mats. Archives for Microbiology 147:213–220.

D'Amelio, E. D., Cohen, Y., and Des Marais, D. J. 1989. Comparative functional ultrastructure of two hypersaline submerged cyanobacterial mats: Guerrero Negro, Baja California Sur, Mexico, and Solar Lake, Sinai,

Egypt, p. 97–113. In: Cohen, Y., and Rosenberg, E. (ed.), Microbial mat: physiological ecology of benthic microbial communities. American Society for Microbiology, Washington, DC.

Daniell, H., Sarojini, G., and McFadden, B. A. 1986. Transformation of the cyanobacterium *Anacystis nidulans* 6301 with the *Escherichia coli* plasmid pBR322. Proceedings National Academy Science USA 83:2546–2550.

Debus, R. J., Barry, B. A., Babcock, G. T., and McIntosh, L. 1988. Site directed mutagenesis identifies a tyrosine radical involved in the photosynthetic oxygen evolving system. Proceedings National Academy Science USA 85:427–430.

De Lorimier, R., Bryant, D. A., Porter, R. D., Liu, W. Y., Jay, E., and Stevens, S. E., Jr. 1984. Genes for the α and β subunits of phycocyanin.. Proceedings National Academy Science USA 81:7946–7950.

De Lorimier, R., Guglielmi, G., Bryant, D. A., and Stevens, S. E. 1987. Functional expression of plastid allophycocyanin genes in a cyanobacterium. Journal of Bacteriology 169:1830–1835.

Des Marais, D. J., Cohen, Y., Nguyen, H., Cheatham, M., Cheatham, T., and Manuez, E. 1989. Carbon isotopic trends in hypersaline ponds and microbial mats at Guerrero Negro, Baja California Sur, Mexico: Implications for Precambrian stromatolites, p. 191–206. In: Cohen, Y., and Rosenberg, E. (ed.), Microbial mat: physiological ecology of benthic microbial communities. American Society for Microbiology, Washington, DC.

De Wit, R., and van Gemerden, H. 1987. Oxidation of sulfide and thiosulfide by *Microcoleus chthonoplastes*. FEMS Microbial Ecology 45:7–13.

De Wit, R., and van Gemerden, H. 1989. Growth responses of the cyanobacterium *Microcoleus chthonoplastes* with sulfide as an electron donor, p. 320–325. In: Cohen, Y., and Rosenberg, E. (ed.), Microbial mat: Physiological ecology of benthic microbial communities. American Society for Microbiology, Washington, DC.

Doolittle, W. F. 1979. The cyanobacterial genome, its expression, and the control of that expression. Advances in Microbial Physiology 20:1–102.

Douglas, S. E., and Doolittle, W. F. 1984a. Complete nucleotide sequence of the 23S rRNA gene of the cyanobacterium *Anacystis nidulans*. Nucleic Acids Research 12:3373–3386.

Douglas, S. E., and Doolittle, W. F. 1984b. Nucleotide sequence of the 5S rRNA gene and flanking regions in the cyanobacterium *Anacystis nidulans*. FEBS Letters 166:307–310.

Dubbs, J. M., and Bryant, D. A. 1987. Organization of the genes encoding phycoerythrin and the two differentially expressed phycocyanins in the cyanobacterium *Pseudanabaena* PCC7409, p. 765–768. In: J. Biggins (ed.), Progress in photosynthesis research, vol. 4. Martinus Nijhoff Publishers, Dordrecht, The Netherlands.

Dzelzkalns, V. A., and V. A., and Bogorad, L. 1987. Genetic and biochemical analysis of cyanobacteria defective in photosynthetic oxygen evolution, p. 841–844. In: J. Biggins (ed.), Progress in photosynthesis research, vol. 4. Martinus Nijhoff Publishers, Dordrecht, Netherlands.

Elhai, J., and Wolk, C. P. 1988. Conjugal transfer of DNA to cyanobacteria. Methods in Enzymology 167:747–754.

Fisher, R., Tuli, R., and Haselkorn, R. 1981. A cloned cyanobacterial gene for glutamine synthetase functions in *Escherichia coli,* but the enzyme is not adenylated. Proceedings National Academy Science USA 78:3393–3397.

Flores, E., and Wolk, C. P. 1985. Production of bacteriocins by filamentous nitrogen-fixing cyanobacteria, p. 210. In: Fifth International Symposium on Photosynthetic Prokaryotes, Grindelwald, Switzerland.

Friedberg, D., Kaplan, A., Ariel, R., Kessel, M., and Seijffers, J. 1989. The 5′-flanking region of the gene encoding the large subunit of ribulose 1,5-bisphosphate carboxylase/oxygenase is crucial for growth of the cyanobacterium *Synechococcus* sp. strain PCC 7942 at the level of CO_2 in air. Journal of Bacteriology 171:6069–6076.

Friedmann, E. I. 1982. Endolithic microorganisms in the antarctic cold desert. Science 215:1045–1053.

Gallagher, M. L., and Burke, W. F. 1985. Sequence-specific endonuclease from the cyanobacterium *Anacystis nidulans* R2. FEMS Microbial Letters 26:317–321.

Garcia-Pichel, F., and Castenholz, R. W. 1990. Comparative anoxygenic photosynthetic capacity of seven strains of a thermophilic cyanobacterium. Archives of Microbiology 153:344–351.

Gatenby, A. A., Van der Vies, S. M., and Bradley, D. 1985. Assembly in *E. coli* of a functional multi-subunit ribulose bisphosphate carboxylase from a blue green alga. Nature 314:617–620.

Geier, G. E., and Modrich, P. 1979. Recognition sequence of the *dam* methylase of *Escherichia coli* k12 and mode of cleavage of DpnI endonuclease. Journal of Biological Chemistry 254:1408–1413.

Gendel, S., Straus, N., Pulleyblank, D., and Williams, J. 1983. A novel shuttle cloning vector for the cyanobacterium *Anacystis nidulans*. FEMS Microbial Letters 19:291–294.

Gillis, M., De Ley, J., and De Cleene, M. 1970. The determination of molecular weight of bacterial genome DNA from renaturation rates. European Journal of Biochemistry 12:143–153.

Glazer, A. N. 1983. Comparative biochemistry of photosynthetic light-harvesting systems. Annual Reviews of Biochemistry 52:125–157.

Golden, J. W., Lammers, P. J., Mulligan, M. E., and Haselkorn, R. 1985a. Rearrangement of *Anabaena* nitrogen fixation genes, p. 121. In: Fifth International Symposium on Photosynthetic Prokaryotes, Grindenwald, Switzerland.

Golden, J. W., Robinson, S. J. and Haselkorn, R. 1985b. Rearrangement of nitrogen fixation genes during heterocystis differentiation in the cyanobacterium *Anabaena*. Nature 314:419–423.

Golden, S. S., Brusslan, J., and Haselkorn, R. 1986. Expression of a family of *psbA* genes encoding a photosystem II polypeptide in the cyanobacterium *Anacystis nidulans* R2. EMBO Journal 5:2789–2798.

Golden, S. S., Cho, D.-S. C., and Natly, M. S. 1989. Two functional *psbD* genes in the cyanobacterium *Synechococcus* sp. strain PCC 7942. Journal of Bacteriology 171:4707–4713.

Goloubinoff, P., Gatenby, A. A., and Lorimer, G. H. 1989. GroE heat-shock proteins promote assembly of foreign prokaryotic ribulose bisphosphate carboxylase oligomers in *Escherichia coli*. Nature 337:44–47.

Gray, M. W., and Doolittle, W. F. 1982. Has the endosymbiont hypothesis been proven? Microbiological Reviews 46:1–42.

Grigorieva, G., and Shestakov, S. 1982. Transformation in the cyanobacterium *Synechocystis* sp. 6803. FEMS Microbiological Letters 13:367–370.

Gurevitz, M., Somerville, C. R., and McIntosh, L. 1985. Pathway of assembly of ribulose bisphosphate carboxylase/oxygenase from *Anabaena* 7120 expressed in *Escherichia coli*. Proceedings Natural Academy Science USA 82:6546–6550.

Hagemann, M., Erdman, N., and Wittenburg, E. 1987. Synthesis of glucosylglycerol in salt-stressed cells of the cyanobacterium *Microcystis firma*. Archives of Microbiology 148:275–279.

Haselkorn, R. 1985. Genes and gene transfer in cyanobacteria. Plant Molecular Biology Reporter 3:24–32.

Haselkorn, R. 1986. Organization of the genes for nitrogen fixation in photosynthetic bacteria and cyanobacteria. Annual Review of Microbiology 40:525–547.

Haselkorn, R., Rice, D., Curtis, S. E., and Robinson, J. 1983. Organization and transcription of genes important in *Anabaena* heterocyst differentiation. Annual Microbiology Institute Pasteur 1384:181–193.

Hauman, J. H. 1981. Is a plasmid(s) involved in the toxicity of *Microcystis aeruginosa*? p. 97–102. In: W. W. Carmichael (ed.), The water environment: Algal toxins and health, vol. 20. Plenum, New York.

Herdman, M., Janvier, M., Rippka, R., and Stanier, R. Y. 1979. Genome size of cyanobacteria. Journal of General Microbiology 111:73–85.

Herrero, A., Elhai, J., Hohn, B., and Wolk, C. P. 1984. Infrequent cleavage of cloned *Anabaena variabilis* DNA by restriction endonucleases from *A. variabilis*. Journal of Bacteriology 160:871–874.

Herrero, A., and Wolk, C. P. 1986. Genetic mapping of the chromosome of the cyanobacterium *Anabaena variabilis*. Journal of Biological Chemistry 261:7748–7754.

Hershkovitz, N., Oren, A., and Cohen, Y. 1991. Accumulation of trehalose and sucrose in cyanobacteria exposed to matric water stress. Applied and Environmental Microbiology. 57:645–648.

Hicks, D. B., Nelson, N., and Yocum, C. F. 1986. Cyanobacterial and chloroplast F_1-ATP-ases: cross-reconstitution of photophosphorylation and subunit immunological relationships. Biochimica and Biophysica Acta 851:217–222.

Ho, K. K., and Krogmann, D. W. 1982. Photosynthesis, p. 191–214. In: N. G. Carr and B. A. Whitton (ed.), The biology of cyanobacteria. Blackwell Scientific Publications, Oxford.

Holland, D., Elhai, J., and Wolk, C. P. 1989. Characterization of *het*A, a gene involved in heterocyst differentiation: its sequence, time of transcription, and localization of expression along filaments. In: Molecular biology of cyanobacteria workshop. Toronto, Canada.

Houmard, J., Mazel, D., Moguet, C., Bryant, D. A., and Tandeau de Marsac, N. 1986. Organization and nucleotide sequence of genes encoding core components of the phycobilisomes from *Synechococcus* 6301. Molecular General Genetics 205:404–410.

Imhoff, J. F., Hashwa, F., and Trueper, H. G. 1978. Isolation of extremely halophilic phototrophic bacteria from alkaline Wadi Natrun. Archives of Hydrobiology 84:381–388.

Jager, K., and Potts, M. 1988. Distinct fractions of genomic DNA from cyanobacterium *Nostoc commune* that differ in the degree of methylation. Gene 74:197–201.

Jansson, C., Debus, R. J., Osiewacz, H. D., Gurevitz, M., and McIntosh, L. 1987. Construction of an obligate photoheterotrophic mutant of the cyanobacterium *Synechocystis* 6803. Plant Physiology 85:1021–1025.

Jørgensen, B. B. 1989. Light penetration, absorption and action spectra in cyanobacterial mats, p. 123–137. In: Cohen, Y., and Rosenberg, E. (ed.), Microbial mat: Physiological ecology of benthic microbial communities. American Society for Microbiology, Washington.

Jørgensen, B. B., and Cohen, Y. 1977. Solar Lake (Sinai) 5. The sulfur cycle of the benthic cyanobacterial mats. Limnology and Oceanography 22:657–666.

Jørgensen, B. B., Cohen, Y., and Des Marais, P. J. 1987. Photosynthetic action spectra adaptation to special light distribution in a benthic cyanobacterial mat. Applied and Environmental Microbiology 53:879–886.

Jørgensen, B. B., Cohen, Y., and Revsbech, N. P. 1986. Transition from anoxygenic to oxygenic photosynthesis in a *Microcoleus chthonoplastes* cyanobacterial mat. Applied and Environmental Microbiology 51:408–417.

Jørgensen, B. B., Cohen, Y., and Revsbech, N. P. 1988. Photosynthetic potential and light dependent oxygen consumption in a cyanobacterial mat. Applied and Environmental Microbiology 54:176–182.

Jørgensen, B. B., Kuenen, J. G., and Cohen, Y. 1979a. Microbial transformation of sulfur compounds in a stratified lake (Solar Lake, Sinai). Limnology and Oceanography 24:189–194.

Jørgensen, B. B., Revsbech, N. P., Blackburn, T. H., and Cohen, Y. 1979b. Diurnal cycle of oxygen and sulfide microgradients and microbial photosynthesis in a cyanobacterial mat sediment. Applied and Environmental Microbiology 38:46–58.

Jørgensen, B. B., Revsbech, N. P., and Cohen, Y. 1983. Photosynthesis and structure of benthic microbial mats: Microelectrode and SEM studies of four cyanobacterial communities. Limnology and Oceanography 28:1075–1093.

Kallas, T., Spiller, S., and Malkin, R. 1987. Cyanobacterial genes for the cytochrome b_6f complex: sequence homology with plastid and bacterial genes but divergence of operon structure, p. 12.801–803. In: J. Biggins (ed.), Progress in photosynthesis research, vol. 4. Martinus Nijhoff Publishers, Dordrecht, The Netherlands.

Knoll, A. H. 1989. The paleomicrobiological information in Proterozoic rocks. p. 469–484. In Cohen, Y. and Rosenberg, E. (ed.). 1989. Microbial mat: Physiological ecology of benthic microbial communities. American Society for Microbiology, Washington, DC.

Kodaki, T., Katagiri, F., Asano, M., Izui, K., and Katsuki, H. 1985. Cloning of phosphoenolpyruvate carboxylase gene from a cyanobacterium *Anacystis nidulans* in *Escherichia coli*. Journal of Biochemistry 97:533–539.

Krumbein, W. E., Cohen, Y., and Shilo, M. 1977. Solar Lake (Sinai) 4. Stromatolitic cyanobacterial mats. Limnology and Oceanography 22:635–656.

Kuhlemeier, C. J., Borrias, W. E., Van den Hondel, C. A. M. J. J., and Van Arkel, G. A. 1981. Construction and characterization of two hybrid plasmids capable of transformation to *Anacystis nidulans* R2 and *Esche-*

richia coli K12. Molecular General Genetics 184:249–254.

Kuhlemeier, C. J., Hardon, E. M., Van Arkel, G. A., and Van de Vate, C. 1985. Self cloning in the cyanobacterium *Anacystis nidulans* R2: Fate of cloned gene after reintroduction. Plasmid 14:200–208.

Kuhlemeier, C. J., Thomas, A. A. M., Van der Ende, A., Van Leen, R. W., Borrias, W. E., Van den Hondel, C. A. M. J. J. and Van Arkel, G. A. 1983. A host-vector system for gene cloning in the cyanobacterium *Anacystis nidulans* R2. Plasmid 10:156–163.

Kumano, M., Tomioka, N., and Sugiura, M. 1983. The complete nucleotide sequence of 23S rRNA gene from a blue-green alga, *Anacystis nidulans*. Gene 24:219–225.

Lambert, D. H., Bryant, D. A., Stirewalt, V. L., Dubbs, J. M., Stevens, S. E., Jr., and Porter, R. D. 1985. Gene map for the *Cyanophora paradoxa* cyanelle. Journal of Bacteriology 164:659–664.

Lambert, G. R., and Carr, N. G. 1984. Resistance of DNA from filamentous and unicellular cyanobacteria to restriction endonuclease cleavage. Biochimica and Biophysica Acta 781:45–55.

Lau, R. H., and Straus, H. A. 1985. Versatile shuttle vectors for the unicellular cyanobacterium *Anacystis nidulans* R2. FEMS Microbiological Letters 27:253–256.

Laudenbach, D. E., Reith, M. E., and Straus, N. A. 1988. Isolation, sequence analysis, and transcriptional studies of the flavodoxin gene from *Anacystis nidulans* R2. Journal of Bacteriology 170:258–265.

Laudenbach, D. E., Straus, N. A., and Williams, J. P. 1985. Evidence for two distinct origins of replication in the large endogenous plasmid of *Anacystis nidulans* R2. Molecular General Genetics 199:300–305.

Lemaux, P. G., and Grossman, A. R. 1983. Isolation and characterization of a gene for a major light-harvesting polypeptide from *Cyanophora paradoxa*. Proceedings National Academy Science USA 81:4100–4104.

Lemaux, P. G., and Grossman, A. R. 1985. Major light-harvesting polypeptides encoded in polycistronic transcripts in a eukaryotic alga. EMBO Journal 4:1911–1919.

Machray, G. C., Vakeria, D., Codd, G. A., and Stewart, W. D. P. 1988. Insertion sequence IS2 in the cyanobacterium *Chlorogloeopsis fritschii*. Gene 67:301–305.

Mazel, D., Guglielmi, G., Houmard, J., Sidler, W., Bryant, D. A., and Tandeau de Marsac, N. 1986. Green light induces transcription of the phycoerythrin operon in the cyanobacterium *Calothrix* 7601. Nucleic Acids Research 14:8279–8290.

McFadden, B. A., and Daniell, H. 1988. Binding, uptake and expression of foreign DNA by cyanobacteria and isolated etioplasts. Photosynthesis Research 19:23–37.

Meiners, S., and Schindler, M. 1987. Immunological evidence for gap junction polypeptide in plant cells. Journal of Biological Chemistry 262:951–953.

Mitsui, A., Kumazawa, S., Takahashi, A., Ikemoto, S., and Arai, T. 1986. Strategy by which nitrogen-fixing unicellular cyanobacteria grow autotrophically. Nature 323:720–722.

Murphy, R. C., Bryant, D. A., Porter, R. D., and Tandeau de Marsac, N. 1987. Molecular cloning and characterization of the *rec*A gene from the cyanobacterium *Synechococcus* PCC 7002. Journal of Bacteriology 169:2739–2747.

Nakamura, H. 1938. Uber die Kohlenaesureassimilation bei niederen algen in Anwesenheit des Schwefelwasserstaffes. Acta Phytochemica (Tokyo) 10:271–281.

Nechushtai, R., Muster, P., Binder, A., Liveanu, V., and Nelson, N. 1983. Photosystem I reaction center from the thermophilic cyanobacterium *Mastigocladus laminosus*. Proceedings National Academy Science USA 80:1179–1183.

Nierzwicki-Bauer, S. A., Curtis, S. E., and Haselkorn, R. 1984. Cotranscription of genes encoding the small and large subunits of ribulose 1,5-bisphosphate carboxylase in the cyanobacterium *Anabaena* 7120. Proceedings National Academy Science USA 81:5961–5965.

Nixon, P. J., Dyer, T. A., Barber, J., and Hunter, C. N. 1986. Immunological evidence for the presence of the D1 and D2 proteins in PSII cores of higher plants. FEBS Letters 209:83–86.

Ogawa, T., Kaneda, T., and Omata, T. 1987. A mutant of *Synechococcus* PCC7942 incapable of adapting to low CO_2 concentration. Plant Physiology 84:711–715.

Ohyama, K., Fukuzawa, H., Kohchi, T., shirai, H., Sano, T., Sano, S., Umesono, K., Shiki, Y., Tekeuchi, M., Chang, Z., Aota, S., Inokuchi, H., and Ozeki, H. 1986. Chloroplast gene organization deduced from the complete sequence of liverwort *Marchantia polymorpha* chloroplast DNA. Nature 322:572–574.

Olson, R. J., Chisholm, S. W., Zettler, E. R., and Armbrust, E. V. 1990. Pigments, size, and distribution of *Synechococcus* in the North Atlantic and Pacific oceans. Limnology and Oceanography 35:45–58.

Oren, A., and Padan, E. 1978. Induction of anaerobic photoautotrophic growth in the cyanobacterium *Oscillatoria limnetica*. Journal of Bacteriology 133:558–563.

Oren, A., Padan, E., and Avron, M. 1977. Quantum yields for oxygenic and anoxygenic photosynthesis in the cyanobacterium *Oscillatoria limnetica*. Proceedings of the National Academy of Sciences USA 74:2152–2156.

Oren, A., and Shilo, M. 1979. Anaerobic heterotrophic dark metabolism in the cyanobacterium *Oscillatoria limnetica*: sulfur respiration and lactate fermentation. Archives of Microbiology 122:77–84.

Padan, E., and Cohen, Y. 1982. Anoxygenic photosynthesis, p. 215–235. In: Carr, N. C., and Whitton, B. A. (ed.), The biology of cyanobacteria. Blackwell Scientific, Oxford.

Paerl, H. W., Prisco, J. C., and Brawner, D. L. 1989. Immunochemical localization of nitrogenase in marine Trichodesmium aggregates: Relationship to N₂ fixation potential. Applied and Environmental Microbiology 55:2965–2975.

Paerl, H. W., Bebout, B. M., and Prufert, L. E. 1989. Naturally occurring patterns of oxygenic photosynthesis and N₂ fixation in a marine microbial mat, p. 326–341. In: Cohen, Y., and Rosenberg, E. (ed.), Microbial mat: Physiological ecology of benthic microbial communities. American Society for Microbiology, Washington, DC.

Pakrasi, H. B., Williams, J. G. K., and Arntzen, C. J. 1988. Targeted mutagenesis of the *psb*E and *psb*F genes blocks photosynthetic electron transport: evidence for a functional role of cytochrome b_{559} in photosystem II. EMBO Journal 7:325–332.

Pearson, H. W., Malin, G., and Hawsley, R. 1981. Physiological studies on in situ nitrogenase activity by axenic cultures of the blue-green alga *Microcoleus chthono-plastes.* British Phycological Journal 16:139.

Pecker, I., Ohad, N., and Hischberg, J. 1987. The chloro-plast encoded type of herbicide resistance is a recessive trait in cyanobacteria, p. 811–814. In: J. Biggins, (ed.), Progress in photosynthesis research, vol. 3. Martinus Nijhoff Publishers, Dordrecht, The Netherlands.

Philbrick, J.B., and Zilinskas, B. A. 1988. Cloning, nucleo-tide sequence and mutational analysis of the gene en-coding the Photosystem II manganese-stabilizing poly-peptide of *Synechocystis* 6803. Molecular General Genetics 212:418–425.

Pilot, T. J., and Fox, J. L. 1984. Cloning and sequencing of the genes encoding for the α and β subunits of C-phy-cocyanin from the cyanobacterium *Agmenellum quad-ruplicatum.* Proceedings National Academy Science USA 81:6983–6987.

Post, A. E., Zwart, G., Sweers, J. P., Veen, A., Rensman, D., Van der Heuvel, A., and Mur, L. R. 1989. Chromatic regulation of photosynthesis in cyanobacteria, p. 305–312. In: Cohen, Y., and Rosenberg, E. (ed.), Microbial mats: Physiological ecology of benthic microbial com-munities. American Society for Microbiology, Wash-ington, DC.

Potts, M., and Friedmann, E. I. 1981. Effects of water stress on cryptoendolithic cyanobacteria from hot desert rocks. Archives of Microbiology 130:267–271.

Pouwels, P. H., Enger-Valk, B. A., and Brammer, W. J. 1985. Vectors for cyanobacteria, sec. IX.1–IX.5. In: Cloning vectors, a laboratory manual. Elsevier, Amsterdam.

Price, G. D., and Badger, M. R. 1989. Expression of human carbonic anhydrase in the cyanobacterium *Synecho-coccus* PCC7942 creates a high CO_2-requiring pheno-type. Plant Physiology 91:505–513.

Reaston, J., and Carr, N. G. 1985. Purification of the mod-ification enzyme M. *Nsp* MACI from the filamentous cyanobacterium *Nostoc* PCC 8009, p. 332. In: Fifth International Symposium on Photosynthetic Prokar-yotes, Grindenwald, Switzerland.

Reaston, J. R., Van den Hondel, C. A. M. J. J., Vanden Ende, A., Van Arkel, G. A., Stewart, W. D. P., and Herdman, M. 1980. Comparison of plasmids from the cyanobacterium *Nostoc* PCC7524 with two mutant strains unable to form heterocysts. FEMS Microbio-logical Letters 9:185–188.

Rebiere, M. C., Castets, A. M., Houmard, J., and Tandeau de Marsac, N. 1986. Plasmid distribution among uni-cellular and filamentous cyanobacteria: occurrence of large and megaplasmids. FEMS Microbiological Let-ters 37:269–275.

Reinhold, L., Zviman, M., and Kaplan, A. 1987. Inorganic carbon fluxes and photosynthesis in cyanobacteria—a quantitative model, p. 6289–6296. In: J. Biggins (ed.), Progress in photosynthesis research, vol. 4. Martinus Nijhoff Publishers, Dordrecht, The Netherlands.

Reith, M. E., Laudenbach, D. E., and Straus, N. A. 1986. Isolation and nucleotide sequence analysis of the fer-redoxin I gene from the cyanobacterium *Anacystis ni-dulans* R2. Journal of Bacteriology 168:1319–1324.

Revsbech, N. P., and Jorgensen, B. B. 1986. Microelec-trodes: their use in microbial ecology. Advances in Mi-crobial Ecology 9:293–352.

Revsbech, N. P., Jorgensen, B. B., Blackburn, T. H., and Cohen, Y. 1983. Microelectrode studies of the photo-

synthetic and O_2, H_2S and pH profiles of a microbial mat. Limnology and Oceanography 28:1062–1074.

Richardson, L. L., and Castenholz, R. W. 1987. Enhanced survival of the cyanobacterium *Oscillatoria terebrifor-mis* in darkness under anaerobic conditions. Applied and Environmental Microbiology 53:2151–2158.

Richardson, L. L., and Castenholz, R. W. 1989. Chemoki-netic motility responses of the cyanobacterium *Oscil-latoria terebriformis.* Applied and Environmental Mi-crobiology 55:261–263.

Rippka, R. 1972. Photoheterotrophy and chemoheterotro-phy among unicellular blue-green algae. Archiv für Mikrobiologie 87:93–98.

Rothchild, L. J., and Mancinelli, R. L. 1990. Model of car-bon fixation in microbial mats from 3,500 Myr ago to the present. Nature 345:710–712.

Scherer, S., Almon, H., and Boeger, P. 1988. Interaction of photosynthesis, respiration and nitrogen fixation in cy-anobacteria. Photosynthesis Research 15:95–114.

Scherer, S., Chen, T. W., and Boger, P. 1988. A new UV-A/B protecting pigment in the terrestrial cyanobacter-ium *Nostoc commune.* Plant Physiology 88:1055–1057.

Schmetterer, G., Wolk, C. P., and Elhai, J. 1986. Expression of luciferases from *Vibrio harveyi* and *Vibrio fischeri* in filamentous cyanobacteria. Journal of Bacteriology 167:411–414.

Schneider, G. J., Tumer, N. E., Richaud, C., Borbely, G., and Haselkorn, R. 1987. Purification and characteri-zation of RNA polymerase from the cyanobacterium *Anabaena* 7120. Journal of Biological Chemistry 262:14633–14679.

Schwabe, W., Weihe, A., Borner, T., Henning, M., and Kohl, J. G. 1988. Plasmids in toxic and non-toxic strains of the cyanobacterium *Microcystis aeruginosa.* Current Microbiology 17:133–137.

Sherman, L. A., and Van de Putte, P. 1982. Construction of a hybrid plasmid capable of replication in the bacte-rium *Escherichia coli* and the cyanobacterium *Anacys-tis nidulans.* Journal of Bacteriology 150:410–413.

Shestakov, S., Elanskaya, I., and Bibikova, M. 1985. Vectors for gene cloning in *Synechocystis* sp. 6803, p. 109. In: Abstracts, Fifth International Symposium Photosyn-thetic Prokaryotes, Grindenwald, Switzerland.

Shilo, M. 1989. The unique characteristics of benthic cy-anobacteria, p. 207–213. In: Cohen, Y., and Rosenberg, E. (ed.), Microbial mat: Physiological ecology of benthic microbial communities. American Society for Microbiology, Washington, DC.

Shinozaki, K., Ohme, M., Tanaka, M., Wakasugi, T., Hay-ashida, N., Matsubayashi, T., Zaita, N., Chunwongse, J., Obokata, J., Yamagychi-Shinozaki, K., Ohto, C., Torazawa, K., Meng, B. Y., Sugita, M., Deno, H., Ka-mogashira, T., Yamada, K., Kusuda, J., Takaiwa, F., Kato, A., Tohdon, N., Shimada, H., and Sugiura, M. 1986. The complete nucleotide sequence of the tobacco chloroplast genome: its gene organization and expres-sion. EMBO Journal 5:2043–2049.

Shinozaki, K., and Sugiura, M. 1983. The gene for the small subunit of ribulose 1,5-bisphosphate carboxylase/oxy-genase is located close to the gene of the large subunit in the cyanobacterium *Anacystis nidulans* 6301. Nu-cleic Acids Research 11:6957–6964.

Shinozaki, K., Yamada, C., Takakata, N., and Sugiura, M. 1983. Molecular cloning and sequence analysis of the cyanobacterial gene for the large subunit of ribulose

1,5-bisphosphate carboxylase/oxygenase. Proceedings National Academy Science USA 80:4050–4054.

Simon, R. D. 1978. Survey of extrachromosomal DNA found in the filamentous cyanobacteria. Journal of Bacteriology 136:414–418.

Smeekens, S., De Groot, M., Van Binsbergen, J., and Weisbeek, P. 1985b. Sequence of the precursor of the chloroplast thylakoid lumen protein plastocyanin. Nature 317:456–458.

Smeekens, S., Van Binsbergen, J., and Weisbeek, P. 1985a. The plant ferredoxin precursor nucleotide sequence of a full length cDNA clone. Nucleic Acids Research 13:3179–3194.

Stal, L., Heyer, H., Bekker, S., Villbrandt, M., and Krumbein, W. E. 1989. Aerobic-anaerobic metabolism in the cyanobacterium Oscillatoria limosa. In: Cohen, Y., and Rosenberg, E. (ed.), Microbial mat: Physiological ecology of benthic microbial communities. American Society for Microbiology, Washington, DC.

Stal, L. J., and Krumbein, W. E. 1985. Oxygen protection of nitrogenase in aerobically nitrogen-fixing, non-heterocystous cyanobacterium Oscillatoria sp. Archives for Microbiology 143:72–76.

Stanier, R. Y., and Cohen-Bazire, G. 1977. Phototrophic prokaryotes. The cyanobacteria. Annual Reviews of Microbiology 31:225–274.

Stolz, J. F. 1984. Fine structure of the stratified microbial community at Laguna Figuera, Baja California, Mexico: II: Transmission electron microscopy as a diagnostic tool in studying microbial communities in situ, p. 23–38. In: Cohen, Y., Castenholz, R. W., and Harvorson, H. O. (ed.), Microbial mats: Stromatolites. Alan R. Liss, New York.

Szekers, M. 1981. Phage-induced development of a site-specific endonuclease in Anacystis nidulans, a cyanobacterium. Virology 111:1–10.

Szekers, M., Szmidt, A. E., and Torok, I. 1983. Evidence for a restriction/modification-like system in Anacystis nidulans infected by cyanophage AS-I. European Journal of Biochemistry 131:137–141.

Tabita, F. R. 1988. Molecular and cellular regulation of autotrophic carbon dioxide fixation in microorganisms. Microbiological Reviews 52:155–189.

Takahashi, Y., Kato, K., Hayashizaki, Y., Wakabayashi, T., Ohtsuka, E., Matsuki, S., Ikehara, M., and Matsubara, K. 1985. Molecular cloning of the human cholecystokinin gene by use of a synthetic probe containing deoxyinosine. Proceedings National Academy Science USA 82:1931–1935.

Tandeau de Marsac, N. 1977. Occurrence and nature of chromatic adaptation in cyanobacteria. Journal of Bacteriology 130:82–91.

Tandeau, de Marsac, N. 1983. Phycobilisomes and complementary chromatic adaptation. Bulletin Institute Pasteur 81:201–254.

Tandeau de Marsac, N., De la Torre, F., and Szulmajster, J. 1987. Expression of the larvicidal gene of Bacillus sphaericus 1593M in the cyanobacterium Anacystis nidulans R2. Molecular General Genetics 209:396–398.

Tandeau de Marsac, N., and Houmard, J. 1987. Advances in cyanobacterial molecular genetics, p. 251–302. In: Fay P., and Van Baalen, C. (ed.), The cyanobacteria. Elsevier, Amsterdam.

Tandeau de Marsac, N., Mazel, D., Bryant, D. A., and Houmard, J. 1985. Molecular cloning and nucleotide sequence of a developmentally regulated gene from the cyanobacterium Calothrix PCC7601: a gas vesicle protein gene. Nucleic Acids Research 13:7223–7236.

Tandeau de Marsac, N., Mazel, D., Damerval, T., Guglielmi, G., Capuano, V., and Houmard, J. 1988. Photoregulation of gene expression in the filamentous cyanobacterium Calothrix sp. PCC 7601: light-harvesting complexes and cell differentiation. Photosynthesis Research 18:99–132.

Tittgen, J., Hermans, J., Steppuhn, J., Jansen, T., Jansson, C., Andersson, B., Nechushtai, R., Nelson, N., and Hermann, R. G. 1986. Isolation of cDNA clones for fourteen nuclear-encoded thylakoid membrane proteins. Molecular General Genetics 204:258–265.

Tomioka, N., Shinozaki, K., and Sugiura, M. 1981. Molecular cloning and characterization of ribisomal RNA genes from a blue-green alga, Anacystis nidulans. Molecular General Genetics 184:359–363.

Tomioka, N., and Sugiura, M. 1983. The complete nucleotide sequence of a 16S ribosomal RNA gene from a blue-green alga, Anacystis nidulans. Molecular General Genetics 191:46–50.

Tomioka, N., and Sugiura, M. 1984. Nucleotide sequence of the 16S-23S spacer region in the rrnA operon from a blue-green alga, Anacystis nidulans. Molecular General Genetics 193:427–430.

Turner, S., DeLong, E. F., Giovannoni, S. J., Olsen, G. J., and Pace, N. R. 1989. Phylogenetic analysis of microorganisms and natural populations using rRNA sequences, p. 390–401. In: Cohen, Y., and Rosenberg, E. (ed.), Microbial mat: Physiological ecology of benthic microbial communities. American Society for Microbiology, Washington, DC.

Van den Hondel, C. A. M. J. J., Keegstra, W., Borrias, W. E., and Van Arkel, G. A. 1979. Homology of plasmids in strains of unicellular cyanobacteria. Plasmid 2:323–333.

Van den Hondel, C. A. M. J. J., Van Leen, R. W., Van Arkel, G. A., Duyvesteyn, M., and De Waard, A. 1983. Sequence-specific nucleases from the cyanobacterium Fremyella diplosiphon, and a peculiar resistance of its chromosomal DNA towards cleavage by other restriction enzymes. FEMS Microbiological Letters 16:7–12.

Van den Hondel, C. A. M. J. J., Verbeek, S., Van der Ende, A., Weisbeek, P. J., Borrias, W. E., and Van Arkel, G. A. 1980. Introduction of transposon Tn901 into a plasmid of Anacystis nidulans: preparation for cloning in Tn901 into a plasmid of Anacystis nidulans: preparation for cloning in cyanobacteria. Proceedings National Academy Science USA 77:1570–1574.

Van der Plas, J., de Groot, R. P., Woortman, M. R., Weisbeek, P. J., and Van Arkel, G. A. 1986. Coding sequence of a ferredoxin gene from Anacystis nidulans R2 (Synechococcus PCC7942). Nucleic Acids Research 14:7804.

Van der Vies, S. M., Bradley, D., and Gatenby, A. A. 1986. Assembly of cyanobacterial and higher plant ribulose bisphosphate carboxylase subunits into functional homologous and heterologous enzyme molecules in Escherichia coli. EMBO Journal 5:2439–2444.

Van Rijn, J., and Cohen, Y. 1983. Ecophysiology of the cyanobacterium Dactylococcopsis salina: Effects of light intensity, sulfide and temperature. Journal of General Microbiology 129:1849–1856.

Vermass, W. F. J., Rutherford, A. W., and Hansson, O. 1988. Site directed mutagenesis in photosystem II of cyanobacterium *Synechocystis* sp. PCC6803: Donor D is a tyrosine residue in the D2 protein. Proceedings National Academy Science USA 85:8477–8481.

Vermass, W. F. J., Williams, J. G. K., and Arntzen, C. J. 1987. Sequencing and modification of *psb*B, the gene encoding the CP-47 protein of photosystem II, in the cyanobacterium *Synechocystis* 6803. Plant Molecular Biology 8:317–326.

Vermaas, W. F. J., Williams, J. G. K., Rutherford, A. W., Mathis, P., and Arntzen, C. J. 1986. Genetically engineered mutant of the cyanobacterium *Synechocystis* 6803 lacks the photosystem II chlorophyll-binding protein CP-47. Proceedings National Academy Science USA 83:9474–9477.

Vieira, J., and Messing, J. 1982. The pUC plasmids, an M13mp7-derived system for insertion mutagenesis and sequencing with synthetic universal primers. Gene 19:259–268.

Wallace, D. C., and Morovitz, H. J. 1973. Genome size and evolution. Chromosoma 40:121–126.

Walsby, A. E. 1977. Absence of gas vesicle protein in a mutant of *Anabaena flos-aquae*. Archives of Microbiology 114:167–170.

Walsby, A. E., Van Rijn, J., and Cohen, Y. 1983. The biology of a new gas-vaculated cyanobacterium, *Dactylococcopsis salina* sp. nov., in the Solar Lake. Proceedings of the Royal Society London B 217:417–447.

Ward, D. M., Weller, R., Shiea, J., Castenholz, R., and Cohen, Y. 1989. Hot spring microbial mats: Anoxygenic and oxygenic mats of possible evolutionary importance, p. 3–15. In: Cohen, Y., and Rosenberg, E. (ed.), Microbial mat: Physiological ecology of benthic microbial communities. American Society for Microbiology, Washington, DC.

Waterbury, J. B., Watson, S. W., Guillard, R. R., and Brand, L. E. 1979. Widespread occurrence of unicellular marine, planktonic cyanobacterium. Nature 277:293–294.

Wealand, J. L., Myers, J. A., and Hirschberg, R. 1989. Changes in gene expression during nitrogen starvation in *Anabaena variabilis* ATCC 29413. Journal of Bacteriology 171:1309–1313.

Weisshaar, H., and Boeger, P. 1983. Sulfide stimulation of light-induced hydrogen production by the cyanobacterium *Nostoc muscorum*. Naturforschung, Section c, Bioscience 38:237–242.

Whitehead, P. R., and Brown, N. L. 1985. Three restriction endonucleases from *Anabaena flos-aquae*. Journal General Microbiology 131:951–958.

Williams, J. G. K. 1988. Construction of specific mutations in photosystem II photosynthetic reaction center by genetic engineering methods in *Synechocystis* 6803. Methods in Enzymology 167:766–778.

Williams, J. G. K., and Chisholm, D. A. 1987. Nucleotide sequences of both *psb*D genes from the cyanobacterium *Synechocystis* 6803, p. 12.809–812. In: J. Biggins (ed.), Progress in photosynthesis research, vol. 4. Martinus Nijhoff Publishers, Dordrecht, The Netherlands.

Williams, J. G. K., and Szalay, A. A. 1983. Stable integration of foreign DNA into the chromosome of the cyanobacterium *Synechococcus* R2. Gene 24:37–51.

Woese, C. R. 1987. Bacterial evolution. Microbiological Reviews 51:221–171.

Wolk, C. P., Cai, Y., Cardemil, L., Flores, E., Hohn, B., Murry, M., Schmetterer, G., Schrautemeier, B., and Wilson, R. 1988. Isolation and complementation of mutants of *Anabaena* sp. strain PCC 7120 unable to grow aerobically on dinitrogen. Journal of Bacteriology 170:1239–1244.

Xie, W. Q., Whitton, B. A., Simon, J. W., Jager, K., Reed, D., and Potts, M. 1989. *Nostoc commune* UTEX584 gene indole phosphate hydroxylase activity in *Escherichia coli*. Journal of Bacteriology 171:708–713.

Young, R. A., and Davis, R. W. 1983. Efficient isolation of genes using antibody probes, Proceedings National Academy Science USA 80:1194–1198.

Zilinskas, B. A., and Howell, D. A. 1987. Comparative immunology of the phycobilisome linker polypeptides, p. 1161–1164. In: J. Biggins (ed.), Progress in photosynthesis research, vol. 2. Martinus Nijhoff Publishers, Dordrecht, The Netherlands.

Zohary, T., and Cohen, Y. 1991. Photosynthesis, photoinhibition, photooxidation, and maturation of a *Microcystis aeruginosa* hyperscum. Applied and Environmental Microbiology. In press.

CHAPTER 99

The Order Prochlorales

LUUC R. MUR and TINEKE BURGER-WIERSMA

The order Prochlorales accommodates unicellular and filamentous species that perform oxygenic photosynthesis and contain chlorophyll a and chlorophyll b as the primary photosynthetic pigments. Phycobiliproteins are absent. The cell wall possesses both Gram-positive and Gram-negative structural features. All the known species are either free living or are associated with invertebrate hosts.

The order Prochlorales was proposed by Florenzano et al. (1986) and emended by Burger-Wiersma et al. (1989). Thus far, two families are distinguished, the unicellular, coccoid organisms of the family Prochloraceae, with one symbiotic species, and the filamentous, free-living organisms of the family Prochlorotrichaceae. Burger-Wiersma et al. (1989) referred to the Prochlorales as oxychlorobacteria rather than prochlorophytes, as proposed by Lewin (1976).

Lewin (1975) and Lewin and Withers (1975) were the first to describe prokaryotic, oxygenic, phototrophic organisms with thylakoids containing chlorophyll a and b, but lacking phycobiliproteins. The first unicellular spherical organism discovered was associated with ascidians. Because this organism could neither be assigned to the cyanobacteria nor to the Chlorophyta, Lewin proposed a new division, the Prochlorophyta, with one species, *Prochloron didemni* (Lewin, 1976, 1977).

Antia (1977) argued that the differences in pigment composition were insufficient to demarcate this organism from the cyanobacteria. He suggested that the criteria defining the cyanobacteria be made less rigid in order to avoid unnecessary proliferation of the prokaryotic divisions.

Because of their prokaryotic nature, Florenzano et al. (1986) proposed to assign the Prochlorales according to the International Code of Nomenclature of Bacteria, by placing the genus Prochloron in the family Prochloraceae, order Prochlorales, in the class Photobacteria.

Burger-Wiersma et al. (1986) isolated a second representative of the Prochlorales. This organism, *Prochlorothrix hollandica*, is filamentous and free-living. It was isolated from a eutrophic, shallow, freshwater lake in the Netherlands and grows well in liquid or solidified mineral medium. *Prochlorothrix* and *Prochloron* contain identical pigments: chlorophyll a and b, zeaxanthin, and β-carotene. To accommodate this new organism in the order of Prochlorales a new family, Prochlorotrichaceae, and a new genus, *Prochlorothrix*, were proposed (Burger-Wiersma et al., 1989).

Chisholm et al. (1988) discovered another chlorophyll a- and b-containing, spherical, free-living prokaryote in the Pacific and the North Atlantic Oceans that bears a number of similarities with *Prochloron* and *Prochlorothrix*. Interestingly, the bulk of both chlorophyll a and b of this organism are divinyl chlorophylls, and the most abundant carotenoids are zeaxanthin and α-carotene (Chisholm et al., 1988; S. W. Chisholm, personal communication). At present, further details are lacking and the systematic allocation of this organism is not yet clear.

The relationship between the oxychlorobacteria and the cyanobacteria has been discussed in a number of publications. Stam et al. (1985) found only a poor DNA-DNA reassociation between a number of strains of *Prochloron* and the cyanobacterium *Synechocystis*. In contrast, Seewaldt and Stackebrandt (1982) found a high degree of similarity in the 16S ribosomal RNA catalogs of *Prochloron* and of cyanobacteria. From this, they concluded that *Prochloron* is closely related to the cyanobacteria. This conclusion was supported by Herdman (1981) who compared the genome sizes of cyanobacteria and *Prochloron*. After a phylogenetic analysis based on 16S rRNA, *Prochlorothrix hollandica* was found to be strongly related to the unicellular cyanobacteria (Turner et al., 1989). Thus, the principal differences between the oxychlorobacteria and the cyanobacteria are the presence in the former of chlorophyll b and the absence of phycobiliproteins.

Habitat

Prochloron cells have been found in extracellular symbiosis with 20 species of ascidians, chiefly didemnids and one holothurian (Cox, 1986). The associations were mainly found throughout the tropical Pacific and Indian Oceans, mostly at cryptic, shaded habitats, but sometimes on the open reef flat (Kott, 1980). The *Prochloron* cells inhabit three regions of the colonies of the ascidians: 1) on the external surface of the test (Figs. 1 and 2); 2) embedded in the test; or 3) in the common cloacal cavity of the colony. It has been suggested that these host-symbiont relationships closely correlate with three different morphological types of *Prochloron*. The three morphological types were distinguished from each other by the shape and arrangement of the thylakoids, vacuolation, and the presence and location of polyhedral bodies.

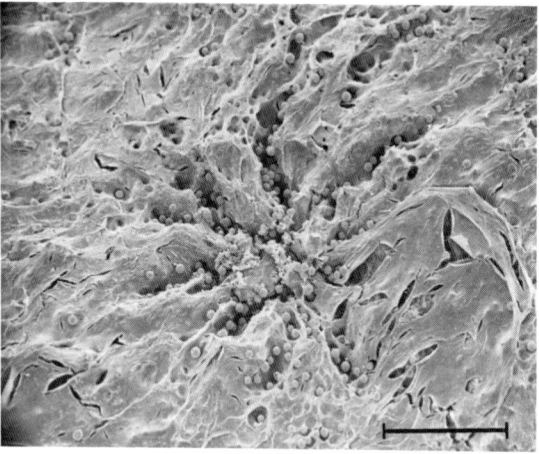

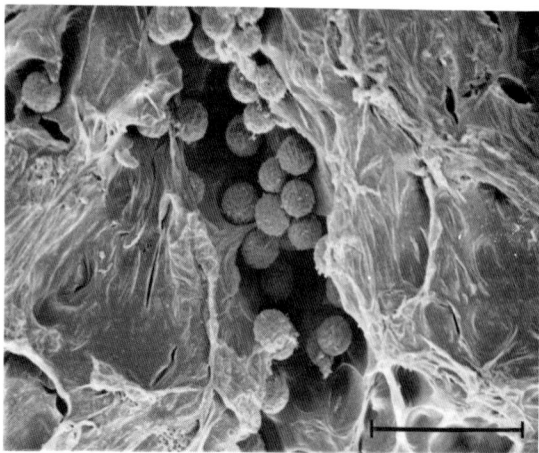

Figs. 1 and 2. Scanning electron micrographs of *Prochloron* cells in oral grooves on the surface of the ascidian *Didemnum carneolentum.* Bar = 250 μm in Fig. 1 (top) and 50 μm in Fig. 2 (bottom). (Courtesy of L. Cheng.)

However, it is impossible to judge whether these various types are phenotypically or genotypically determined until *Prochloron* can be grown in culture. Therefore, it seems reasonable to accept the conclusions of Stam et al. (1985) and Stackebrandt et al. (1982), based on DNA-DNA reassociation and 16S rRNA analyses of a number of samples of *Prochloron,* that all strains examined belonged to one species.

Olsen (1986) found a good relationship between the behavior and growth rate of the ascidian colonies and the amount of incident light. At 475–1200 $\mu E \cdot m^{-2} \cdot sec^{-1}$, the ascidian colonies grew 40% faster than in the dark. Colonies kept in the dark and at low-light intensities showed a tendency to ascend the walls of the growth chambers in which they were kept and flattened out to a more encrusting morphology. These data show that, although ascidians can grow without *Prochloron,* their growth is enhanced by the photosynthetic activity of *Prochloron.*

Up to now, *Prochlorothrix hollandica* has only been isolated from the shallow, highly eutrophic, freshwater Loosdrecht Lake in the Netherlands. However, there is evidence of significant numbers of this species in several comparable lakes in the Netherlands (Burger-Wiersma et al., 1989). The Loosdrecht Lake originated from peat excavation and consists of several interconnected subsystems (Van Liere et al., 1984). The average depth of the lake is 2 m and the average Secchi disk depth (a measure of water clarity) ranges from 0.2 m in summer to 0.8 m in winter. There is no stratification, and filaments of *Prochlorothrix* are homogenously mixed in the water column. The highest population densities of *Prochlorothix hollandica* are found in summer when the temperature of the entire water column is 15 to 25°C. The pH ranges between 8 and 10. Although the organism lives under low-light conditions, it is very likely that during extended parts of the growth season, phosphorus is the limiting growth factor.

Enrichment

Attempts to culture *Prochloron* have been generally unsuccessful. Patterson and Withers (1982) described the growth of *Prochloron* in culture for several generations at a pH of 5.5 after addition of L-tryptophan or a combination of indole and serine. However, these data have not been confirmed by other publications. Alberte (1989) made some suggestions for culturing *Prochloron* and recommended a high CO_2 level in combination with a low O_2 concentration.

Prochlorothrix hollandica can be isolated from lake water by micromanipulation or plating. It grows well in a number of growth media used for cyanobacteria. The composition of one mineral medium used to isolate *Prochlorothrix* is the following:

FPG medium (Burger-Wiersma et al., 1989).

$NaNO_3$	500 mg
K_2HPO_4	25 mg
$MgSO_4 \cdot 2H_2O$	50 mg
$CaCl_2 \cdot 2H_2O$	13 mg
Water	1 liter

Adjust to pH 8.2 with NaOH, and autoclave. After autoclaving add 2 ml each of filter-sterilized solutions of trace elements, $NaHCO_3$, and $EDTA/FeCl_2$ (see below).

Solution of trace elements.

H_3BO_3	850 mg
$MnCl_2 \cdot 4H_2O$	530 mg
$ZnSO_4 \cdot 7H_2O$	66 mg
$(NH_4)_6Mo_7O_{24} \cdot 24H_2O$	0.6 mg
$CuSO_4 \cdot 5H_2O$	24 mg
$Co(NO_3)_2 \cdot 6H_2O$	24 mg
NH_4VO_3	3 mg
Water	1 liter

Solution of $NaHCO_3$

$NaHCO_3$	60 g
Water	1 liter

Solution of $EDTA/FeCl_2$

$FeCl_2 \cdot 2H_2O$	800 mg
EDTA-Na	660 mg
Water	1 liter

The organism also grows well in medium BG11 (Rippka and Waterbury, 1977). Judged empirically, the purity of the water and the chemicals used is of decisive importance, and glass distilled water and chemicals of analytical grade are required. For solidified media, the agar was washed three times with distilled water, rinsed with 96% ethanol and 100% acetone, and dried in air. Optimal growth was obtained using medium BG11 with 0.5% agar. On agar plates, the organism forms a monolayer of trichomes but it does not move actively.

The organism grows in continuous light or in light-dark cycles at photon flux densities of 40–100 $\mu E \cdot m^{-2} \cdot sec^{-1}$. Optimum growth was found at a pH of 8.4 and a temperature of 25°C. The growth rate drops dramatically at temperatures $\mu E \cdot m^{-2} \cdot sec^{-1}$ below 20°C (Burger-Wiersma et al., 1989).

Identification

Prochloron forms green patches on the surfaces and around the common cloaca of the calcified colonial ascidians (Lewin, 1975). The cells have a diameter of 10 to 30 μm, are bright green, generally spherical, and have no visible mucilaginous sheath (Lewin, 1976; Lewin and Cheng, 1989). Neither gas vacuoles nor cyanophycin have been found (Fig. 3). The cells show a typical prokaryotic ultrastructure but have a chlorophyll composition like green algae, as mentioned earlier. Depending on the morphological type, the cells have numerous small vacuoles or a very large central vacuole which fills the middle of the cell. The *Prochloron* cells found in the cloacal cavity of the genus *Didemnum* have only very little vacuolation. Burger-Wiersma and Matthijs (1989) argued that the presence of these numerous small vacuoles might be due to the formation of excessive amounts of phenolic compounds during growth under unfavorable conditions.

The thylakoid membranes, which are lacking phycobilisomes, are often paired or stacked (Fig. 3), a phenomenon known from the thylakoids of chloroplasts but absent in cyanobacteria. Polyhedral bodies, probably carboxysomes, are found in the central parts of the cells, in a zone of thylakoid-free cytoplasm which underlies the peripheral cytoplasm, or in the peripheral thylakoid band.

Prochlorothrix hollandica is a filamentous organism with cylindrical cells which are 0.5–1.5 μm wide and 3–10 μm long (Fig. 4). Unfavorable growth conditions may increase length and di-

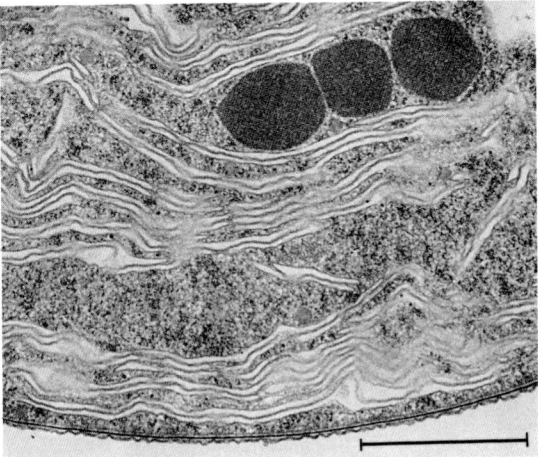

Fig. 3. Transmission electron micrograph of a portion of a *Prochloron* cell in the ciliated cloacal duct of *Diplosoma virens*. Visible are the multilayered wall, paired thylakoids, and polyhedral bodies. Bar = 1 μm. (Courtesy of K. W. Lee.)

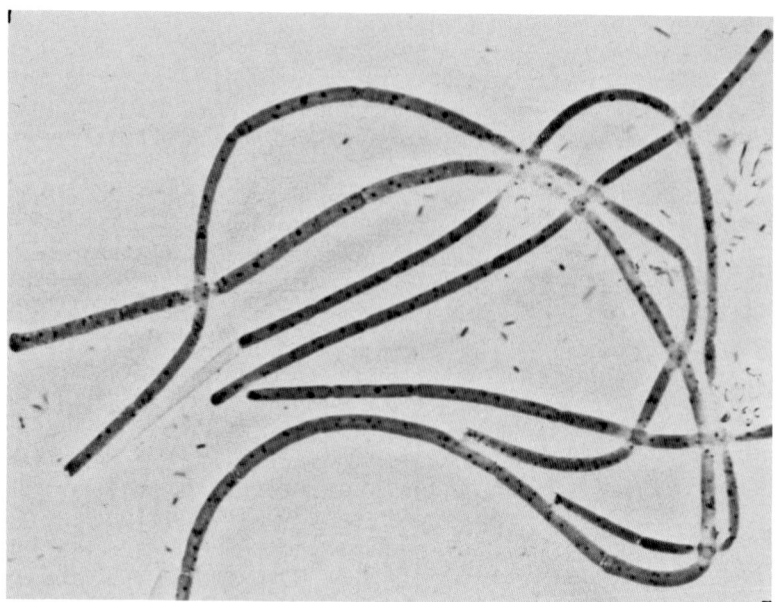

Fig. 4. Natural plankton sample from Loosdrecht Lake containing *P. hollandica*. At the end of some trichomes, the cell walls of empty cells can be seen.

ameter. The straight undifferentiated trichomes consist of from two to more than 100 cells and are nonmotile. Periodic constrictions along the trichomes mark the location of cross-walls between adjacent cells. Under the microscope, this species is very difficult to distinguish from some *Oscillatoria*-like cyanobacteria with the same diameter. Because *Prochlorothrix* is lacking phycocyanin, it can be recognized under a fluorescence microscope by the absence of the bright-orange to red fluorescence exhibited by cyanobacteria after excitation by light with a wavelength of 540 nm (Burger-Wiersma et al., 1989).

In the cells of *Prochlorothrix*, carboxysomes are found in the cytoplasm in the vicinity of the thylakoids (Fig. 5). The thylakoid membranes are arranged in more or less stacked and unstacked regions (Miller et al., 1988). No vacuoles have been found but in contrast to *Prochloron*, *Prochlorothrix* contains gas vesicles (Fig. 6) (Golecki and Jürgens, 1989).

The cells of the free-living species found in the tropical oceans are spherical to rod-shaped with a diameter of 0.6 to 0.8 μm. More details have not been published (Chisholm et al., 1988).

Physiology

Despite the lack of cultures physiological data on *Prochloron* have been obtained, especially concerning photosynthesis, respiration, and nitrogen assimilation.

As shown below, the chlorophyll content of the cells seems to be strongly related to the light conditions. However, the ratio of chlorophyll *a* to *b* seems also to be dependent on the *Prochloron* host (Paerl et al., 1984). The chlorophyll *a:b* ratio ranges from 3 to 20 (Alberte et al., 1986; Paerl et al., 1984; Thorne et al., 1977). The data on the influence of light conditions on photosynthesis come partly from freshly collected *Prochloron* (Alberte et al., 1986) and partly from experiments with *Prochloron*-ascidian symbionts as a whole (Alberte et al., 1987; Fisher and Trench, 1980; Pardy, 1984). From both types of experiments, it can be concluded that *Prochloron* has a good adaptation to the light climate. Cells from low-light conditions possess two times more chlorophyll than cells from high-light conditions, 5.5 and 2.7 $\times$ 10^{-9} mg per cell, respectively, and a slightly lower chlorophyll *a:b* ratio (2.6 and 3.4 from low and high light, respectively). Photosynthesis in cells from low-light conditions involved a relatively small number of large photosynthetic units (PSUs), resulting in a high efficiency for light expressed per unit of chlorophyll. Cells from high-light conditions had a relatively large number of small PSUs, resulting in a high maximal photosynthetic rate. Light adaptation did not influence the quantitative ratio between the two photosystems. Optimum photosynthesis was found at temperatures between 28 and 30°C. Between 15 and 30°C, the Q_{10} value for photosynthesis was 3.47 (Alberte et al., 1986). From the photosynthetic rates of *Prochloron* and the respiration rates of the ascidian, it could be concluded that the carbon contribution from the symbiont to the host was up to 56% (Alberte, et al., 1987). However, the nature of the translocated compounds is not yet clear.

Fig. 5. Electron micrograph of cells of *P. hollandica* showing carboxysomes and both stacked and unstacked thylakoids.

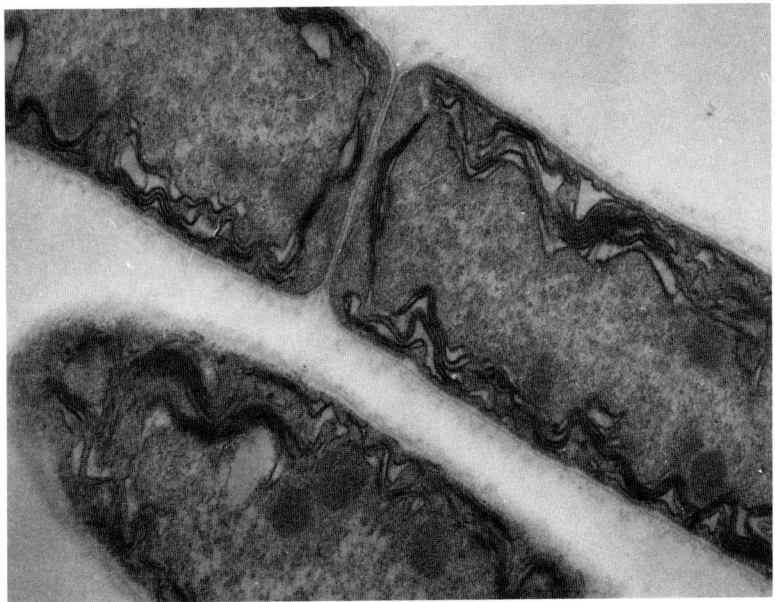

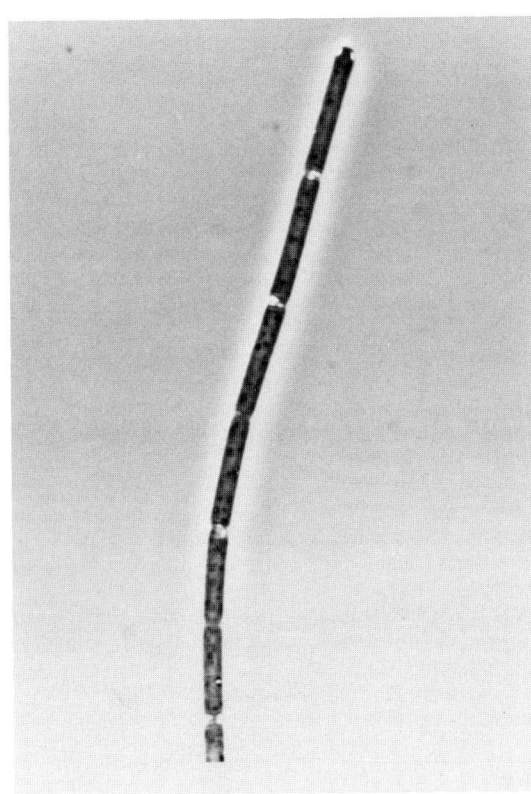

Fig. 6. A trichome of *P. hollandica.* At the poles of the cells, small gas vesicles are visible.

Ammonium seems to be the most important nitrogen source for *Prochloron*. For ammonium assimilation, the glutamine synthetase/glutamate synthase (GOGAT) pathway seems to be most important (Parry, 1985). Because no ni-

trate reductase was found in freshly isolated cells, it is unlikely that in natural situations, nitrate is an important nitrogen source, especially, because after a two- to four-h exposure with nitrate, nitrate reductase was induced. Paerl (1984) found that only *Prochloron* in combination with the ascidian *Lissoclinum* showed nitrogen fixation and that nitrogen fixation was light-mediated in this association. Neither the isolated cells nor the ascidians free from *Prochloron* or other bacteria revealed nitrogenase activity.

Prochlorothrix hollandica seems to have the same light adaptation pattern as *Prochloron.* Here, we also find an increase in pigment content when the cells are grown at low-light conditions. However, the changes are much more pronounced. The chlorophyll *a* content increases by a factor of six, and the chlorophyll *b* content by a factor of 2.5 when the cells are brought from 200 to 8 μE·m^{-2}·sec^{-1}. The chlorophyll *a:b* ratio increases from 7 to 18. The efficiency for light expressed per unit of chlorophyll is found to be lower under low-light conditions. According to Burger-Wiersma and Post (1989), this might be caused by the self-shading effect of the chlorophyll. The photosynthetic capacity per unit of chlorophyll is found to be two times higher at high-light intensities. In contrast to *Prochloron,* however, the ratio of the two photosynthetic reaction centers, RC I:RC II, increased at low-light conditions (Burger-Wiersma and Post, 1989).

P. hollandica can assimilate either ammonium or nitrate. Attempts to induce nitrogenase activity were unsuccessful (Burger-Wiersma et al., 1989). NaCl concentrations higher than 25 mM inhibited the growth of *P. hollandica*.

Literature Cited

Alberte, R. S., L. Cheng, and R. A. Lewin. 1986. Photosynthetic characteristics of *Prochloron*-ascidian symbiosis. I. Light and temperature responses of the algal symbiont of *Lissoclinum patella*. Mar. Biol. 90:575–587.

Alberte, R. S., L. Cheng, and R. A. Lewin. 1987. Characteristics of *Prochloron*-ascidian symbiosis. II. Photosynthesis-irradiance relationships and carbon balance of associations from Palau, Micronesia. Symbiosis 4:147–170.

Alberte, R. S. 1989. Physiological and cellular features of *Prochloron*, p. 31–52. In: R. A. Lewin and L. Cheng (ed.), *Prochloron*, a microbial enigma. Chapman and Hall, New York.

Antia, N. J. 1977. A critical appraisal of Lewin's Prochlorophyta. British Phycol. J. 12:271–276.

Burger-Wiersma, T. and H. C. P. Matthijs. 1989. The biology of the Prochlorales, p. 1–24. In: G. A. Codd, F. R. Tabita, and L. Dijkhuizen (ed.), Advances in autotrophic microbiology and one-carbon metabolism, vol. 1. Kluwer Acad. Publ., Dordrecht, The Netherlands.

Burger-Wiersma, T. and A. F. Post. 1989. Functional analysis of the photosynthetic apparatus of *Prochlorothrix hollandica* (Prochlorales), a chlorophyll *b* containing procaryote. Plant Physiol. 91:770–774.

Burger-Wiersma, T., L. J. Stal, and L. R. Mur. 1989. *Prochlorothrix hollandica* gen. nov.: a filamentous oxygenic photoautotrophic prokaryote containing chlorophyll *a* and *b*: assignment to Prochlorotrichaceae fam. nov. and order Prochlorales Florenzano, Balloni and Materassi 1986, with emendation of the ordial description. Int. J. Syst. Bacteriol. 39:250–257.

Burger-Wiersma, T., M. Veenhuis, H. J. Korthals, C. C. M. Van de Wiel, and L. R. Mur. 1986. A new prokaryote containing chlorophyll *a* and *b*. Nature 320:262–264.

Chisholm, S. W., R. J. Olsen, E. R. Zettler, R. Goerlicke, J. B. Waterbury, and N. A. Welschmeyer. 1988. A novel free-living prochlorophyte abundant in the oceanic euphotic zone. Nature 334:340–343.

Cox, G. 1986. Comparison of *Prochloron* from different hosts I. Structural and ultrastructural characteristics. New Phytol. 104:429–445.

Fisher, C. R., and R. K. Trench. 1980. In vitro carbon fixation by *Prochloron* sp. isolated from *Diplosoma virens* Biol. Bull. 159:636–648.

Florenzano, G., W. Balloni, and R. Materassi. 1986. Nomenclature of *Prochloron didemni* (Lewin 1977) sp. nov., nom. rev., *Prochloron* (Lewin 1976) gen. nov., nom. rev., Prochloraceae fam. nov., Prochlorales ord. nov., nom. rev. in the class Photobacteria Gibbons and Murray 1978. Int. J. Syst. Bacteriol. 36:351–353.

Golecki, J. R., and U. J. Jürgens. 1989. Ultrastructural studies on the membrane systems and cell inclusions of the filamentous prochlorophyte *Prochlorothrix hollandica*. Arch. Microbiol. 152:77–82.

Herdman, M. 1981. Desoxyribonucleic acid base composition and genome size of *Prochloron*. Arch. Microbiol. 129:314–316.

Kott, P. 1980. Algal-bearing didemnid ascidians in the Indo-west Pacific. Mem. Qd. Mus. 20:1–47.

Lewin, R. A. 1975. A marine *Synechocystis* (Cyanophyta, Chroococcales) epizoic on ascidians. Phycologia 14:153–160.

Lewin, R. A. 1976. Prochlorophyta as a proposed new division of algae. Nature 261:697–698.

Lewin, R. A. 1977. *Prochloron*, type genus of the Prochlorophyta. Phycologia 16:217.

Lewin, R. A. 1981. The prochlorophytes, p. 257–266. In: M. P. Starr, H. Stolp, H. G. Trüper, A. Balows, and H. G. Schlegel (ed.), The prokaryotes. Springer-Verlag, New York.

Lewin, R. A., and L. Cheng (ed.). 1989. *Prochloron*, a micobial enigma. Chapman and Hall, New York.

Lewin, R. A. and N. W. Withers. 1975. Extraordinary pigment composition of a prokaryotic alga. Nature 256:735–737.

Miller, K. R., J. S. Jacob, T. Burger-Wiersma, and H. C. P. Matthijs. 1988. Photosynthetic membrane structure in *Prochlorothrix hollandica*: a chlorophyll *b*-containing procaryote. J. Cell Sci. 91:577–586.

Olsen, R. R. 1986. Light enhanced growth of the ascidian *Didemnum molle/Prochloron* sp. symbiosis. Mar. Biol. 93:437–442.

Paerl, H. W. 1984. N₂ fixation (nitrogenase activity) attributable to a specific *Prochloron* (Prochlorophyta)-ascidian association in Palau, Micronesia. Mar. Biol. 81:251–254.

Paerl, H. W., R. A. Lewin, and L. Cheng. 1984. Variations in chlorophyll and carotenoid pigmentation among *Prochloron* (Prochlorophyta) symbionts in diverse marine ascidians. Bot. Mar. 27:257–264.

Pardy, R. L. 1984. Oxygen consumption and production by tropical ascidians symbiotic with *Prochloron*. Comp. Biochem. Physiol. 79A:345–348.

Parry, D. L. 1985. Nitrogen assimilation in the symbiotic marine algae *Prochloron* spp. Mar. Biol. 87:219–222.

Patterson, G. M. and N. W. Withers. 1982. Laboratory cultivation of *Prochloron* a tryptophan auxotroph. Science 217:1934–1935.

Rippka, R. and J. B. Waterbury. 1977. The synthesis nitrogenase by non-heterocystous cyanobacteria. FEMS Microbiol. Letters 2:83–86.

Seewaldt, E. and E. Stackebrandt. 1982. Partial sequence of 16S ribosomal RNA and the phylogeny of *Prochloron*. Nature 295:618–620.

Stackebrandt, E., E. Seewaldt, V. J. Fowler, and K. Scheifer. 1982. The relateness of *Prochloron* sp. isolated from different didemnid ascidian hosts. Arch. Microbiol. 132:216–217.

Stam, W. T., S. A. Boele-Bos, and B. K. Stulp. 1985. Genetic relationships between *Prochloron* samples from different locations and hosts as determined from DNA-DNA reassociation. Arch. Microbiol. 142:340–341.

Thorne, S. W., E. H. Newcomb, and C. B. Osmond. 1977. Identification of chlorophyll *b* in extracts of prokaryotic algae by fluorescence spectroscopy. Proc. Natl. Acad. Sci. USA 74:575–578.

Turner, S., T. Burger-Wiersma, S. J. Giovanni, L. R. Mur, and N. R. Pace. 1989. The relationship of a prochlorophyte *Prochlorothrix hollandica* to green chloroplasts. Nature 337:380–382.

Van Liere, L., S. Parma, L. R. Mur, P. Leentvaar, and G. B. Engelen. 1984. Loosdrecht Lakes restoration project, an introduction. Verh. Internat. Verein. Limnol. 22:829–834.

CHAPTER 100

The Proteobacteria: Ribosomal RNA Cistron Similarities and Bacterial Taxonomy

JOZEF DE LEY

Unravelling (Supra)generic Relationships Within the Proteobacteria

To fully understand the relationships between bacteria, the use of one method is not enough and a polyphasic approach is required. Fig. 1 shows the seat and the nature of most of the methods used. Comparisons of the total genome DNA between two or more microorganisms involve studies of, for example, the chemical composition (mean mole GC content), base sequence (DNA-DNA hybridizations, restriction fragment patterns), total genome length, and heterologous transformation.

Some cistrons, in particular those concerned with the mechanism of protein biosynthesis (rRNA, tRNA) and with terminal electron transfer (cytochromes), have been conserved throughout evolutionary history. Comparisons of amino acid sequences of bacterial cytochromes (cyt c), for example, are a useful check on phylogenetic or taxonomic conclusions from other methods, and are valid within a certain taxonomic span (Fig. 2). Unfortunately, there are several classes of cyt c, which limits their usefulness; the methods for isolation and complete sequencing of cytochromes are time-consuming; and the cytochromes are relatively small molecules with about 100 amino acids.

Comparisons of rRNA are carried out by one of the following methods: DNA:-(16 or 23S) rRNA hybridization, 16S rRNA cataloging, or complete sequencing of either 16S or 5S rRNA (Fig. 1).

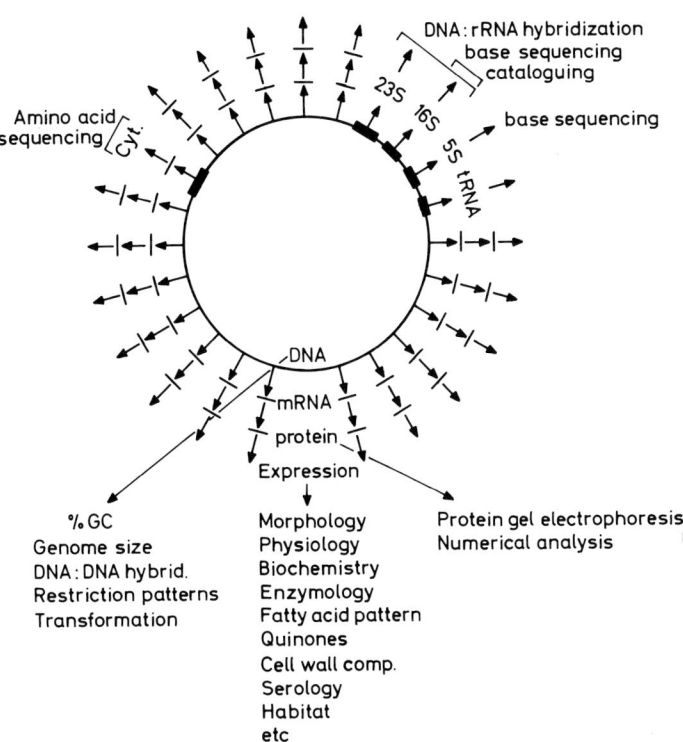

Fig. 1. The seat and the nature of most of the genotypic and phenotypic methods used for monophasic studies or polyphasic comparisons of bacteria.

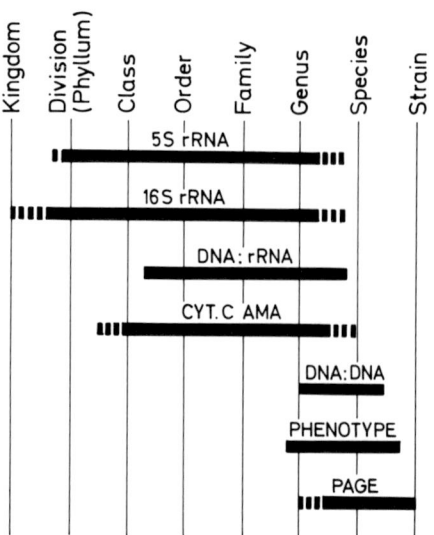

Fig. 2. Taxonomic validity range of several methods: 5S rRNA sequencing; 16S rRNA cataloging is valid up to about a class or a division, complete sequencing extends the validity range to about a kingdom; DNA-rRNA hybridization; cytochrome *c* amino acid sequencing; DNA-DNA hybridization; numerical analysis of phenotypic features; numerical analysis of protein gel electrophoresis (PAGE).

16S rRNA cataloging has been extensively applied (see Woese, 1987). Similarities between pairs of strains have been expressed either by the coefficient S_{AB} (Fox et al., 1977) obtained by comparing sets of oligonucleotide sequences from 16S rRNAs, by "signatures" (bases or sets of short base sequences occurring within a group of organisms but preferably lacking in others, e.g., Woese et al., 1984a, 1985b), or by unrooted phylogenetic trees of evolutionary distances. A number of data have been published on total 16S rRNA base sequences (see Chapter 1; see also Oyaizu and Woese, 1985). The rRNA cataloging method has the advantage that its validity span is wide: from about a subgenus to about a subclass or a class; complete sequencing extends the validity range to higher taxonomic levels (Fig. 2). There are, however, also some disadvantages to the method. Several reasons why S_{AB} analyses are inadequate in some conditions have been extensively discussed (e.g., Fischer et al., 1985; Fowler et al., 1986; Hespell et al., 1984; Ludwig et al., 1983a; Woese, 1987). A better interpretation is possible through the distribution comparison of signature sequences, and, where possible, through comparison of complete sequences (Oyaizu and Woese, 1985). Another weakness of the 16S rRNA cataloging and sequencing methods is that frequently only one strain of a species has been included (in early papers sometimes without strain numbers or acronyms, and no type strain. It is obviously

unsafe to draw taxonomic conclusions from one single strain, even from a type strain—which rather frequently may be atypical for its taxon. It is recommended, for partial or complete rRNA sequencing that several representative strains be used, covering the heterogeneity of each taxon, strains which are only selected after DNA homology grouping and numerical analysis (Sokal and Sneath, 1963) of a large number of phenotypic features and protein gel electrophoregrams of a considerable number of strains (Fig. 1). In addition, Sneath (1989) showed that the statistical sampling error may cause considerable uncertainties concerning the level at which two or more groups, derived from rRNA cataloging or sequencing, are linked together.

A rather large number of complete 5S rRNA sequences from various bacteria has been clustered using a dissimilarity coefficient D_{AB} (Huysmans and De Wachter, 1986).

Hybridizations between labeled 16S or 23S rRNA from a great number of carefully selected type or reference strains were carried out with DNA from a large number of strains (e.g., De Ley et al., 1986, 1987; De Vos et al., 1989). During the last 18 years, our group has carried out many thousands of DNA-rRNA hybridizations, involving a very large number of Gram-negative bacteria from many hundreds of species belonging to many genera, so that each taxon can be represented by many strains. The most significant and useful parameter of a DNA-rRNA hybrid is $T_{m(e)}$, the temperature at which half of the DNA-rRNA hybrid is denatured in a solution with a defined salt composition and concentration. This parameter is a measure of the base sequence similarities between rRNA cistrons. The percentage of rRNA bound to filter-fixed DNA (percent rRNA binding) is useful to differentiate between taxa with the same $T_{m(e)}$ value. The results have been presented mainly in $T_{m(e)}$ dendrograms (see later, Figs. 5, 6, and 7). In these dendrograms, an "rRNA branch" consists of strains which form DNA-rRNA hybrids with a $\Delta T_{m(e)}$ of at most 12°C versus the $T_{m(e)}$ of the homologous duplex of the reference organism. Each rRNA branch is named after its reference taxon. Frequently, a number of rRNA branches link at a $\Delta T_{m(e)}$ of about 5 to 8°C from the homologous reference duplexes: we call this an "rRNA complex" or an "rRNA cluster." In some cases, smaller rRNA subclusters are distinguishable. Some rRNA branches or rRNA complexes correspond to a higher taxonomic category e.g., the families Enterobacteriaceae (Fig. 5), Alcaligenaceae (Fig. 6), and Acetobacteraceae (Fig. 7). A number of rRNA branches and rRNA complexes link together in larger units at a $\Delta T_{m(e)}$ of about 12 to 14°C which we call "rRNA superfamilies." Fig. 4 shows the $T_{m(e)}$ levels at which the superfamilies link together. The term "rRNA superfamily" is not intended to have a taxonomic meaning, but it very likely corresponds to a taxonomic order. The disadvantage of the DNA-rRNA hybridization method is that its range of validity is limited from a taxonomic order to about a subgenus (Fig. 2). The advantages of the hybridization method are that: 1) it is easy, reliable, and fast, allowing the inclusion of many strains of a taxon; 2) it readily identifies strains with a suitable set of reference rRNAs; and 3) it can be used to detect many taxonomically misidentified and misnamed organisms. It is our experience that at least 20% of named strains are generically or specifically misnamed; in some groups, such as the pseudomonads and the various spirilla, up to about 60% of the strains are generically misnamed. In the text below straight brackets [ ] enclose a misnamed taxon. We made no difference between valid and nonvalid names.

Fig. 3 shows the correlation between the parameters of four genotypic methods: %D of DNA homology, $T_{m(e)}$ of DNA-rRNA hybridizations, S_{AB} of 16S rRNA cataloging, and D_{AB} of 5S rRNA sequencing. DNA homology values from DNA-DNA hybridization methods are valid only in about the top 5°C Tm(e), thus in the range of genus—species. The correlation $T_{m(e)}$–D_{AB} is about linear; the correlation Tm(e)–S_{AB} is slightly S-shaped.

In the early seventies it became clear that the methods used, however valuable, such as numerical analysis of numerous phenotypic features, comparative metabolism, DNA base composition and DNA-DNA hybridization, were useful only in the upper ramifications of the relatedness tree of bacteria. In order to look deeper into their evolutionary funnel, the presence of a common orthologous agent was required with a very basic function in the cell, e.g. in protein biosynthesis, anabolic metabolism, or energy production or transport. In addition this agent had to have a sufficient information content and a relatively slow rate of evolution (conservation), versus the bulk of the genome, so that a reliable minimum of structural similarities remained. One possible choice was rRNA, as Moore and McCarthy (1967) had shown that it is conserved. Two groups independently and at about the same time started unraveling the deeper phylogenetic relationships between bacteria on the basis of rRNA comparisons. Our research group explored the possibilities of DNA-(16 or 23S) rRNA hybridizations and polyphasic comparisons. Woese et al. used rRNA cataloging and later sequencing. In general the concordance between the two methods is excellent. One of the main conclusions was that a phenotypically and genotypically diverse group of Gram-negative bacteria, inappropriately named "purple bacteria and their relatives," nevertheless showed a coherent tree-like, deep evolutionary relationship. This group was given class status as the Proteobacteria (Stackebrandt et al., 1988c). It consists of several major subgroups named either rRNA superfamilies I, II, III and IV (De Ley, 1978) or alpha, beta, gamma and delta (Woese et al., 1984a, 1984b, 1985b; Woese 1987). The alpha and beta groups correspond to superfamilies IV and III, respectively, I + II constitute the major part of the gamma group.

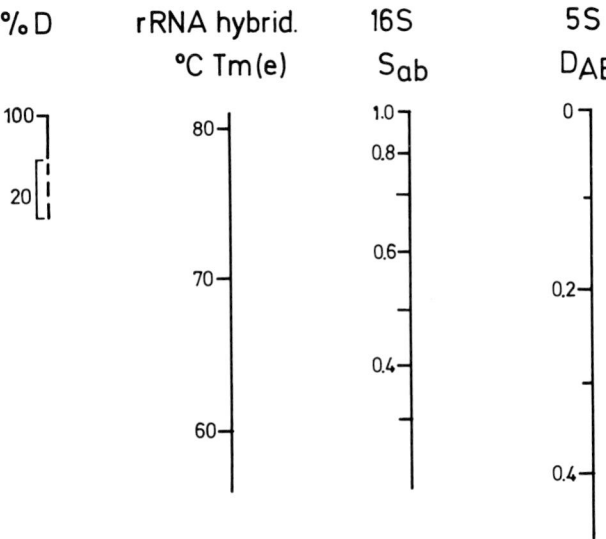

Fig. 3. The correlation between the parameters of four genotypic methods: %D of DNA homology, $T_{m(e)}$ of DNA-rRNA hybridization, S_{AB} of 16S rRNA cataloging, and D_{AB} of 5S rRNA sequencing.

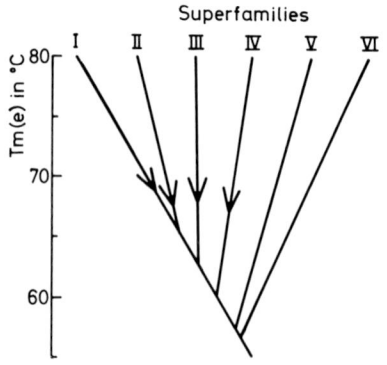

Fig. 4. The $T_{m(e)}$ levels at which the rRNA superfamilies I to VI link together. The superfamilies I to IV belong in the Proteobacteria. The arrow heads show the approximate $T_{m(e)}$ levels where the different rRNA branches and rRNA clusters link within the superfamilies.

A number of general dendrograms, representing the relationships within the Proteobacteria, and deduced from either DNA-rRNA hybridization, 16S rRNA cataloging, and 16S or 5S rRNA sequencing, have been shown in the literature (De Ley, 1978, 1981, 1984; De Vos et al., 1989; De Wachter et al., 1985; Fox et al., 1980; Gibson et al., 1979; Huysmans and De Wachter, 1986; Stackebrandt et al., 1988a; Woese, 1987; Woese et al., 1984a, 1984b, 1985b). Many separate reports which present detailed fragments of the relatedness tree have been published so far. It is the purpose of the present contribution to attempt—at the generic and suprageneric levels—to unite the results of the main methods of rRNA comparisons with Proteobacteria into one picture, to point out similarities and disagreements, and to correlate, if possible, this tree based on the comparisons of a few cistrons, with the multi-cistron expression of the phenotype.

The Gamma Group

The name "gamma group" was proposed by Woese et al. (1985b).

The rRNA Superfamily I

The name "rRNA superfamily I" was proposed by De Ley (1978). Our rRNA superfamily I corresponds to part of the gamma-3 subgroup of the Proteobacteria (Woese et al., 1985b). In working out the dendrogram of this superfamily (Fig. 5), we included about 300 strains belonging to numerous species. The rRNA superfamily I consists of four taxonomic families: the Enterobacteriaceae, the Aeromonadaceae, the Vibrionaceae (J. De Ley and R. Tytgat, unpub-

lished observations), and the Pasteurellaceae (De Ley et al., 1990), and some additional genera. These different clusters of strains link together at an average Tm(e) value of 68.6°C. Extensive genotypic and phenotypic data from the literature and from our research groups strongly support the reality of this relatedness tree.

Woese et al. (1985b) included the following genera (total of 12 strains) in the gamma group: *Escherichia, Enterobacter, Serratia, Yersinia, Proteus, Pasteurella, Aeromonas, Vibrio,* and *"Photobacter" (Vibrio* or *Photobacterium?*); *Alteromonas putrefaciens* is generically misnamed and belongs in *Shewanella.* Evidence for considerable similarities between the rRNA cistrons of the Enterobacteriaceae, the Vibrionaceae, and the Aeromonadaceae was deduced from DNA-rRNA competition hybridizations (Baumann and Baumann, 1976). There is a considerable similarity between the 5S rRNAs from genera in the Enterobacteriaceae, the Vibrionaceae, and the Aeromonadaceae (Colwell et al., 1986; Luehrsen and Fox, 1981; Woese et al., 1975). Baumann and Schubert (1984) listed various genotypic and phenotypic features suggesting a common evolutionary origin for the Vibrionaceae and the Enterobacteriaceae. Alignment and grouping by the weighted-pairgroup method of the sequences of a variety of 5S rRNAs from organisms of rRNA superfamily I showed that they group together in nearly the same fashion as in our rRNA hybridization results, except for some small differences in linkage levels (Huysmans and De Wachter, 1986): the Enterobacteriaceae and Aeromonadaceae link together, and they are closely followed by the Vibrionaceae and by *Colwellia.*

THE FAMILY ENTEROBACTERIACEAE. This family now contains about 25 genera, defined mainly on the basis of phenotyping and DNA-DNA hybridizations (Brenner, 1984). J. De Ley, and R. Tytgat (unpublished observations), using DNA-rRNA hybridizations versus labeled rRNA from a reference strain of *Escherichia coli,* examined quite a number of strains from one or more species of the following genera: *Buttiauxella (agrestis), Cedecea (davisae), Edwardsiella (hoshinae, ictaluri, tarda), Enterobacter (aerogenes, cloacae, sakazakii), Erwinia (amylovora, aroideae, carotovora, chrysanthemi, herbicola, milletiae, nigrifluens, rhapontici, stewartii, uredovora), Escherichia (coli, hermannii), Ewingella (americana), Klebsiella (aerogenes, pneumoniae, rubiacearum), Kluyvera (ascorbata), Koserella (trabulsii), Leclercia (adecarboxylata), Morganella (morganii), Obesumbacterium (proteus), Plesiomonas (shigelloides), Proteus (mirabilis, myxofaciens, vulgaris), Prov-*

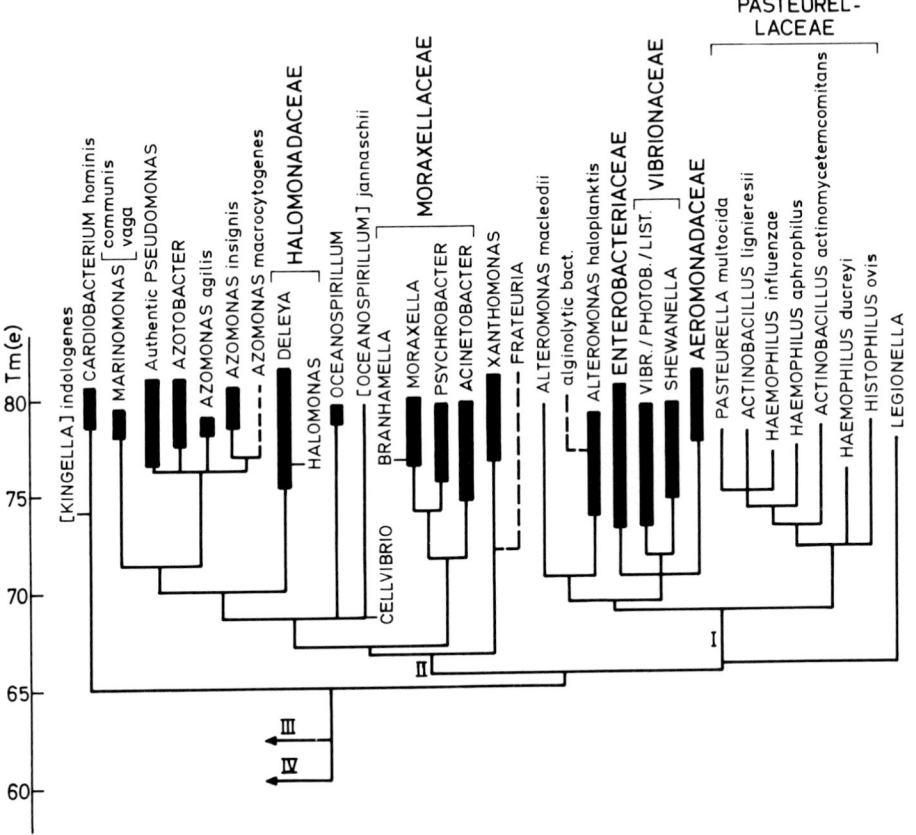

Fig. 5. Condensed rRNA cistron similarity dendrogram showing the major bacterial taxa from rRNA superfamilies I and II. All data are from DNA-rRNA hybridizations done in our laboratory and are expressed as °C of $T_{m(e)}$. Most of the species included have not been shown here. Names of rRNA branches are given. The roman numerals indicate the roots of the rRNA superfamilies. Solid bar: $T_{m(e)}$ range encompassing all strains and taxa of the rRNA branch. Abbreviations: List., *Listonella*; Photob., *Photobacterium*; Vibr., *Vibrio*; broken line: known rRNA branch but labeled rRNA not yet available. For discussion and references, see text.

idencia (alcalifaciens, friedericiana, rettgeri, rustigiana, stuartii), Rahnella (aquatilis), Salmonella (choleraesuis, typhimurium), Serratia (ficaria, fonticola, marcescens, odorifera, rubidaea), Tatumella (ptyseos), Xenorhabdus (luminescens, nematophilus), and *Yersinia (pseudotuberculosis, ruckeri)*.

Some intergeneric rRNA cistron similarities have been examined in addition to the ones reported above. There is a 98% similarity between the rRNA catalogs of *Escherichia coli* and *[Aerobacter] aerogenes* (Woese et al., 1974). According to the data of Ehlers et al. (1988), *Xenorhabdus* is either very close to or a border case of the Enterobacteriaceae, but in our hands this genus is quite clearly a member of the Enterobacteriaceae at 74.0 to 75.4°C $T_{m(e)}$. The close evolutionary distance between 5S rRNAs from *Erwinea aroideae, Escherichia coli, Plesiomonas shigelloides, Proteus mirabilis, Proteus vulgaris, [Aerobacter] aerogenes, Salmonella typhimurium*, and *Serratia marcescens* has

been established (Huysmans and De Wachter, 1986).

We are not aware that representative strains of the following genera have been tested for their rRNA cistron similarities with other members of the Enterobacteriaceae: *Arizona, Budvicia, Citrobacter, Hafnia, Leminorella, Levinia, Pragia*, and *Shigella*. Although, on the basis of DNA-DNA homology for some (Brenner, 1984) and phenotypic similarities for others, one may be confident that these genera are indeed members in good standing of the family discussed here, it would nevertheless be reassuring to have data on their rRNA cistron similarities.

The numerous representatives of the Enterobacteriaceae, which we included in our DNA-rRNA hybridizations, cluster together on the *E. coli* rRNA branch in a $T_{m(e)}$ range from 72.1°C to 80.5°C. We found a similar $\Delta T_{m(e)}$ range of about 8°C from the top down in other taxonomic families (see below). As all are rather closely together on the *E. coli* rRNA branch, and as the internal taxonomic structure of this family is already well known, we did not attempt to add other rRNA branches, except for *Yersinia pseudotuberculosis*. Hybridizations with labeled rRNA from a reliable reference strain of *E. coli* is an excellent and fast way to determine with certainty whether an unknown organism belongs in this family or not: absolutely all other organisms have $T_{m(e)}$ values below 70.5°C. In this fashion we were able to identify or to confirm a number of misnamed or unnamed strains in the Enterobacteriaceae. *[Pseudomonas trifolii]* (Komagata et al., 1968), *[Agrobacterium gypsophilae]*, Ruiter's strains (De Smedt and De Ley, 1979), and the leaf nodule organism (De Vos et al., 1980) are all members of *Erwinia herbicola*. *[Flavobacterium] proteus* is a member of *Obesumbacterium* (see above). *Plesiomonas* is still classified in the Vibrionaceae, but removal from this family has been recommended (MacDonell and Colwell, 1985a); our data show that *Plesiomonas* should be moved into the Enterobacteriaceae; this agrees with the fact that *Plesiomonas* produces the typical enterobacterial common antigen (Ramia et al., 1982). The misnamed *[Pasteurella]* strains NCTC 10547, 10549, and 10553, and the generically misnamed *[Flavobacterium] acidificum* also belong in the Enterobacteriaceae.

The correct location of these well-known members of the Enterobacteriaceae in one separate rRNA cluster inspires confidence in the value of the DNA-rRNA hybridization method and its applications and conclusions in other more obscure areas of bacterial taxonomy. The internal structure of this family at the inter- or intra-generic levels is better unravelled by methods (Fig. 1 and 2) such as DNA-DNA hybridization (Brenner, 1984), computer-assisted numerical analysis of numerous phenotypic features (Mergaert et al., 1984; Sokal and Sneath, 1963; Verdonck et al., 1987) and protein gel electrophoregrams (Izard et al., 1981; Kersters and De Ley, 1975), determinations of GC content, similarities between ribosomal proteins (Hori and Osawa, 1978), and serology.

THE FAMILY AEROMONADACEAE. This family, proposed by Colwell et al. (1986), contains only one genus, *Aeromonas*. DNA-rRNA hybridizations (J. De Ley and R. Tytgat, unpublished observations) indicate that there are two main clusters in this genus. One is in the top 4°C of the *Aeromonas* rRNA branch and contains the majority of the named *Aeromonas* strains and the misnamed *[Haemophilus] piscium* type strain ATCC 10801^T; the second cluster is located at about 8°C from the top and contains

a number (but not all) of *[Aeromonas] formicans* strains. The latter group might be a second genus.

The genus *Colwellia* contains obligately barophilic bacteria (Deming et al., 1988). The taxonomic position of this genus is not yet clear; it varies with the clustering method applied to the 5S rRNA base sequences: it is either a member of the Aeromonadaceae (Colwell et al., 1986), or equidistantly removed from the Aeromonadaceae and the Enterobacteriaceae (MacDonell and Colwell, 1985a), or outside the three large families Enterobacteriaceae, Vibrionaceae, and Aeromonadaceae (Deming et al., 1988; MacDonell and Colwell, 1985a). 5S rRNA sequence grouping (Huysmans and De Wachter, 1986) supports the latter conclusion.

THE FAMILY VIBRIONACEAE. This family now appears to contain the four genera *Vibrio*, *Photobacterium*, *Listonella*, and *Shewanella*, with a total of about 30 species (Baumann and Baumann, 1984; Baumann et al., 1984a; MacDonell and Colwell, 1985a). Baumann et al. (1983a) established and compared levels of phenotypic and genotypic relationships within and between *Vibrio* and *Photobacterium*. In our DNA-rRNA hybridizations (J. De Ley and R. Tytgat, unpublished observations) we included nine species of *Vibrio* (*albensis*, *alginolyticus*, *fischeri*, *harveyi*, *marinus*, *natriegens*, *nigripulchritudo*, *parahaemolyticus* and *proteolyticus*), four of *Photobacterium* (*leognathi*, *logei*, *mandapamensis*, and *phosphoreum*), two of *Listonella* (*anguillara* and *pelagia*), and three of *Shewanella* (*putrefaciens*, *rubescens*, and *hanedai*). The Tm(e) values range over 8°C versus the reference *Vibrio* rRNA. Our method detected several generically misnamed strains and taxa which belong in the Vibrionaceae, such as *[Alteromonas] putrefaciens* and *[Achromobacter] iophagus*. Clustering of 5S rRNA base sequences allows a more refined insight in the structure of this family (Colwell et al., 1986; MacDonell and Colwell, 1985a, 1985b) although each type of clustering requires careful interpretation.

Allomonas, a new genus of polarly flagellated rods from human feces and contaminated river water or sewage, was described by Kalina et al. (1984). It was proposed as a member of the Vibrionaceae. rRNA similarities are required for an exact taxonomic location.

THE FAMILY PASTEURELLACEAE. This family stands at 68.5°C $T_{m(e)}$ from all the above taxa. Officially (Krieg and Holt, 1984), it contains three genera—*Pasteurella*, *Haemophilus*, and *Actinobacillus*—and about 30 species. From DNA-rRNA hybridizations (De Ley et al.,

1990), it appears that this family is much more complex. Seven reference rRNAs, each one from a different and representative DNA homology group in this family, reveal only part of the heterogeneity of the Pasteurellaceae. The results show that this family is as heterogeneous as the Enterobacteriaceae and the Vibrionaceae—$\Delta T_{m(e)}$ of 7.5°C—and that, in addition to the three accepted genera, there is room for at least four other new genera around each of the following rRNA branches: *Histophilus, [Haemophilus] ducreyi, [Haemophilus] aphrophilus,* and *[Actinobacillus] actinomycetemcomitans.*

THE GENUS *ALTEROMONAS.* The members of *Alteromonas* are straight or curved rods, with a single polar flagellum. In contrast to the other members of rRNA superfamily I, *Alteromonas* contains strictly aerobic chemoorganotrophs; they require sea water for growth. DNA-rRNA hybridizations (Van Landschoot and De Ley, 1983) revealed that *Alteromonas* is heterogeneous and that it contains some misnamed organisms. We found that *Alteromonas* consists of two rRNA branches that are 9°C $\Delta T_{m(e)}$ removed from each other. The *Alt. haloplanktis* rRNA branch contains most of the named *Alteromonas* species: *Alt. espejiana, Alt. undina, Alt. aurantia, Alt. rubra,* and *Alt. luteoviolacea.* In addition, this rRNA branch contains a number of generically misnamed *[Pseudomonas]* species such as *[Ps.] marinoglutinosa, [Ps.] nigrifaciens, [Ps.] carrageenovora, [Ps.] piscicida,* and *[Ps.] atlantica* ATCC 19262, and some unnamed alginolytic bacteria. The second *Alteromonas* cluster contains only the species *Alt. macleodii.* A limited number of phenotypic differences between both branches have been found so far; therefore, both clusters should temporarily be retained together in one genus. Two other species, *[Alt.] vaga* and *[Alt.] communis,* belong in the genus *Marinomonas* (see rRNA superfamily II below). *[Alt.] putrefaciens, [Ps.] rubescens,* and *[Alt.] hanedai* are generically misnamed and belong in the border genus *Shewanella* of the Vibrionaceae (see above). Recently proposed *Alteromonas* species such as *Alt. denitrificans* (Enger et al., 1987) and *Alt. colwelliana* (Weine et al., 1988) have not yet been examined for rRNA cistron similarities.

THE GENUS *RUMINOBACTER.* *[Bacteroides] amylophilus* was located by 16S rRNA cataloging somewhere at the base of rRNA superfamily I (Woese et al., 1985b). However, this species is generically misidentified and was later renamed *Ruminobacter* (Stackebrandt and Hippe, 1986). Its exact position is uncertain. It consists of obligately anaerobic small rods with a CO_2-requiring succinate-type fermentation.

SOME COMMON FEATURES IN rRNA SUPERFAMILY I. The DNAs of all members of this superfamily form hybrids with reference-labeled rRNAs with a Tm(e) of at least 69.5°C. Phenotypically the five main clusters (Enterobacteriaceae, Vibrionaceae, Aeromonadaceae, Pasteurellaceae, and Alteromonas) share a limited number of features as expected: they are all either rather small, straight or curved rods (0.2–1.5 × 0.3–3.5, up to 6 μm) or coccoids (Pasteurellaceae). They do not form endospores or microcysts. Many strains require extraneous organic growth factors (vitamins, amino acids, NAD); several Pasteurellaceae have lost the genomic information for biosynthesis of these growth factors during adaptation to parasitic life. They are chemoorganotrophic. They all grow and metabolize aerobically with O_2 as electron acceptor. The four families have all a fermentative metabolism: a variety of sugars and sugar derivatives are catabolized to pyruvate, probably mainly through the Embden-Meyerhof pathway (glycolysis) and partly through the shunt pathway. A variety of metabolic routes fan out from pyruvate according to the enzymes formed in the course of evolution: production and accumulation of lactate, acetate, formate, α-ketoglutarate, ethanol, 2,3-butanediol, 1,3-propanediol, acetoin, H_2, and CO_2 gas. In a limited number of Enterobacteriaceae, Vibrionaceae, and Aeromonadaceae, details of the metabolic pathways are well known; insufficient data are available for the Pasteurellaceae. *Alteromonas* is a notable exception as it is unable to carry out a fermentation. There are clear indications of a sporadic use of the Entner-Doudoroff (ED) mechanism in carbohydrate metabolism: an induced ED system for the catabolism of gluconate by members of the Vibrionaceae, of glucose and fructose in *Alteromonas* and of galacturonate by *Erwinia.* In view of the close phylogenetic relationships between *Alteromonas* and the other rRNA branches, one may expect more metabolic similarities between them than are known today; a search for common and similar metabolic pathways looks promising. Many representatives of the four families specialized as pathogens for man and a great variety of animals; they probably contain the heaviest concentration of pathogens in the bacterial world: the agents for salmonellosis, shigellosis, plague, cholera, various forms of diarrhoea and gastroenteritis, fish furonculosis, Egyptian conjunctivitis, fowl cholera, "wooden tongue" in cattle, and many more diseases.

The rRNA Superfamily II

The name rRNA superfamily II was proposed by De Ley (1978). This superfamily II considerably extends part of the gamma-3 subgroup (Woese et al., 1985b). It links at a Tm(e) of 65.6°C with its closest neighbor, the rRNA superfamily I. The rRNA superfamily II consists now of at least 14 rRNA branches; in working it out we included several hundred strains from over a hundred species. All these organisms link together at or above 66.7°C. A brief summary of our data is seen in Fig. 5.

Woese et al. (1985b) included representatives of the following genera (total of 18 strains): *Pseudomonas* (section I), *Oceanospirillum*, *Serpens*, *Xanthomonas*, *Lysobacter*, and *Acinetobacter*. Alignment of 5S rRNA base sequences (Huysmans and De Wachter, 1986) grouped strains of rRNA superfamily II together and showed them to be closest to our rRNA superfamily I.

THE AUTHENTIC GENUS *PSEUDOMONAS*. Since the original creation of the genus *Pseudomonas* Migula 1894, numerus incompletely described strains of aerobic, polarly flagellated, Gram-negative, rodlike bacteria have been placed in this genus. The main result was an uncontrolled proliferation of species and an unjustified enhancement of the genus. The Berkeley research group (Stanier et al., 1966; Palleroni, 1984) improved and clarified the situation but did not bring a final taxonomic solution. In *Bergey's Manual of Systematic Bacteriology*, the genus *Pseudomonas* is divided in three sections I, II, and III, with a total of about 27 species (plus 41 pathovars in *Pseudomonas syringae*), and two other sections, IV and V, with 65 species of uncertain affiliation (Palleroni, 1984). These sections reflect the rRNA clusters detected by competitive rRNA hybridizations, using a limited group of 35 *Pseudomonas* and 3 *Xanthomonas* strains (Palleroni et al., 1973). Because almost no representatives of other bacterial genera were included, it was impossible to establish the exact position of each of these rRNA clusters within the general framework of Gram-negative taxa. Our research group, using DNA-rRNA hybridization, reexamined the genotypic relationships among about 240 representative strains of over 130 *Pseudomonas* taxa (including all available type and reference strains) (De Vos, 1981; De Vos and De Ley, 1983; De Vos et al., 1985a, 1989). These results prove that the named *Pseudomonas* species are widely dispersed over the total rRNA dendrogram (Figs. 5, 6, and 7). Even the 27 formal *Pseudomonas*

species from sections I, II, and III are distributed over three different rRNA superfamilies and certainly have to be redivided over at least three different genera. About two-thirds of the named *Pseudomonas* species are generically misnamed and are distributed over at least seven genera; they all belong in rRNA superfamilies other than II (see Fig. 1 in De Vos et al., 1989). These organisms need to be reclassified and generically renamed. We shall return to a number of them below. The largest group of species clusters is the *Pseudomonas fluorescens* rRNA branch in rRNA superfamily II; here belongs also the type species *Ps aeruginosa*. This cluster thus represents the authentic genus *Pseudomonas*. According to our findings, it contains 45 named *Pseudomonas* species (including the 12 species from Section I in *Bergey's Manual of Systematic Bacteriology*) and the pathovars from *Ps. syringae;* it also contains several misnamed [*Achromobacter*] species such as [*Achr.*] *albus*, [*Achr.*] *butyri*, [*Achr.*] *cycloclastes* LMG 2841, [*Achr.*] *formosus*, [*Achr.*] *liquefaciens*, and [*Achr.*] *venosus*, the generically misnamed [*Agrobacterium*] *agile*, [*Flavobacterium*] *lutescens*, and [*Fl.*] *denitrificans*, and a small number of unnamed strains. A decisive criterion in the redefinition of the authentic genus *Pseudomonas* is that all members have a $\Delta T_{m(e)}$ of $\leq 6°C$ versus the type strain of *Pseudomonas fluorescens*. Ubiquinone Q-9 is to be expected in bacteria of the authentic genus *Pseudomonas* (Auling et al., 1988). More criteria, such as DNA-DNA hybridization and computer-assisted numerical analysis of phenotypic features, of protein gel electrophoregrams, and of fatty acid distribution, will be required to clarify the internal taxonomic structure of the newly defined authentic genus *Pseudomonas*.

The genus *Serpens*. Oligonucleotide cataloging has shown that *Serpens flexibilis* is another taxon which is probably phylogenetically highly related to the authentic genus *Pseudomonas*: it links with a S_{ab} value of 0.89 to a strain of *Ps. pseudoalcaligenes* (Woese et al., 1982). *Serpens* shows a number of interesting differences with the authentic pseudomonads: it is up to 25 μm long and is very flexible; it does not use carbohydrates; lactate is the main effective carbon source and energy supplier through a very active TCA cycle.

THE GENERA *AZOTOBACTER* AND *AZOMONAS*. De Smedt et al. (1980) showed that genera of heterotrophic, free-living, aerobic, N$_2$-fixing bacteria do not belong in one family but are widely dispersed over different rRNA superfamilies. All the species of the genera *Azotobacter* and *Azomonas* cluster in the top 5.5°C

$T_{m(e)}$ when compared with the reference rRNAs of *Azotobacter*, and *Azomonas*. All the cyst-forming organisms, which are phenotypically very similar, have very similar rRNA cistrons. For these reasons, Tchan (1984) limited the family Azotobacteraceae to both genera. The rRNA cistrons of both genera are most closely related to those of the authentic *Pseudomonas*. This seems rather surprising as the organisms are quite different (rather large, blunt rods to ovals with a diameter of at least 2 μm, fixing N_2 abundantly; *Azotobacter* forms cysts). Nevertheless, this relatedness is supported by a comparison of the structure of other gene products: Ambler (1973) found close similarities between the amino acid sequences of cytochrome *c*-551 of five species from the authentic genus *Pseudomonas* and *c*-551 of *Azotobacter vinelandii*. These results suggest that considerably more similarities between both taxa may exist. Biological N_2-fixation can be performed by a great variety of prokaryotic microorganisms; in rRNA superfamily I, for example, *Klebsiella* is a well-studied N_2-fixer. The dinitrogenase genes (Ruvkun and Ausubel, 1980) and the amino acid sequences of the proteins are very similar (Triplett et al., 1989).

THE GENUS *MARINOMONAS*. *[Alteromonas] communis* and *[Alt.] vaga* are quite far removed from the bulk of the members of the genus *Alteromonas* which is in rRNA superfamily I (Van Landschoot and De Ley, 1983); therefore, a new genus *Marinomonas* was created. Bacteria of this genus metabolize glucose and fructose through an inducible Entner-Doudoroff pathway and they metabolize aromatic compounds such as *m*- and *p*-hydroxybenzoate after *meta*-cleavage (Baumann et al., 1984b). Both species have been isolated from seawater off the Hawaiian Archipelago. *Marinomonas* links at 71.1°C $T_{m(e)}$ with the authentic *Pseudomonas—Azotobacter—Azomonas* cluster.

THE FAMILY HALOMONADACEAE. This family was proposed to unite a number of halotolerant, rodlike aerobic chemoorganotrophs requiring at least 75 mM sodium for growth (Franzmann et al., 1988). The rRNA catalogs of four representative strains are at about 0.65 S_{ab} from each other. This family consists so far of two genera, *Halomonas* and *Deleya*. This agrees very well with our DNA-rRNA hybridization data, where these taxa group together in the top 7°C $\Delta T_{m(e)}$ on the *Deleya* rRNA branch. The position of this family is not quite clear, however: in the $T_{m(e)}$ dendrogram, it is closest to the authentic *Pseudomonas* group at 70°C (K. Kersters and P. Segers, unpublished observations) whereas

rRNA cataloging of four strains (Franzmann et al., 1988) locates them equally far removed from rRNA superfamilies I and II. More research is needed.

The genus *Deleya* was created to unite a number of generically misnamed, peritrichously flagellated, marine *[Alcaligenes]* strains (Baumann et al., 1983b). In addition to the species originally included (*D. aesta, D. cupida, D. halophila, D. marina, D. pacifica,* and *D. venusta*), there is evidence that still other generically misnamed taxa such as *[Pseudomonas] bathycetes* and *[Ps.] beijerinckii* (De Vos et al., 1989), *[P.] halosaccharolytica,* several misnamed *[Alcaligenes],* and some other organisms also belong in this genus.

Halomonas is a genus of polarly and peritrichously flagellated, halotolerant rods and frequently elongated flexible filaments (Vreeland et al., 1980). The species so far are *H. elongata, H.* (formerly *Flavobacterium) halmephila,* and *H. subglaciescola.* They are on the *Deleya* rRNA branch and 5–6°C $\Delta T_{m(e)}$ removed from the *Deleya* reference strain.

Whether *Halovibrio* (Fendrich, 1988) fits into the Halomonadaceae or not remains to be established.

THE GENUS *OCEANOSPIRILLUM*. The named spirilla are very heterogeneous in general and are dispersed over the rRNA superfamilies II, III, and IV. A comprehensive study of the taxonomic positions and the relationships between named *Oceanospirillum* species was carried out by a polyphasic approach (Pot et al., 1989; see also Chapter 172). Five species, including the type species (*O. beijerinckii, O. japonicum, O. linum, O. maris, O. multiglobuliferum,* and two new subspecies, *O. maris* subsp. *hiroshimense* and *O. beijerinckii* subsp. *pelagicum*), constitute a separate authentic *Oceanospirillum* rRNA branch. Our findings are in agreement with cataloging results (Woese et al., 1982, 1985b) that showed that four of the above species constitute a separate branch in the gamma group. The *Oceanospirillum* rRNA branch is the nearest neighbor at a $T_{m(e)}$ of 68.8°C of the large cluster containing the authentic *Pseudomonas, Azotobacter, Azomonas, Marinomonas,* and *Deleya.* All other oceanospirilla are generically misnamed. Each of the following three taxa—*[O.] jannaschii, [O.] kriegii,* and *[O.] minutulum*—constitutes a separate rRNA branch in rRNA superfamily II, and separates at the same level of 68.8°C $T_{m(e)}$. It seems very likely that each one represents a new genus of spirilla, still to be named. At this same level of 68.8°C, we located four other taxa: the cellulolytic *Cellvibrio mixtus* subsp. *mixtus,* and three generically misnamed pseudomonads *[Ps.] nautica, [Ps.] elongata,* and *[Ps.] atlantica* LMG 2139 (De Vos et al., 1989). Whether they are on the same rRNA branch or on a different one from *[Oceanospirillum] minutulum* is not yet known. *[O.] pusillum* is also generically misnamed and belongs in rRNA superfamily IV (Pot et al., 1989).

THE *ACINETOBACTER, MORAXELLA, BRANHAMELLA,* AND *PSYCHROBACTER* RRNA CLUSTER. These taxa cluster together at or above a $T_{m(e)}$ of 71.3°C. They link to all the other members of rRNA superfamily II at 66.7 to 67.0°C $T_{m(e)}$. The same level of relationship between *Acinetobacter* and members of rRNA superfamily II (including the *Xanthomonas* rRNA branch) was also detected by oligonucleotide cataloging (Woese et al., 1985b) and by sequence comparison of 16S rRNAs (Woese, 1987). The three taxa *Acinetobacter, Moraxella,* and *Branhamella,* together with *Neisseria* and *Kingella,* have been united in the family Neisseriaceae (Bøvre, 1984). DNA-rRNA hybridizations show, however, that reality is different since both *Neisseria* and *Kingella* belong in rRNA superfamily III. Therefore a considerably emended family Neisseriaceae has been proposed (Rossau et al., 1989a) (see below). *Acinetobacter, Moraxella, Branhamella,* and *Psychrobacter* belong to a new family called the Moraxellaceae (R. Rossau, M. Gillis, and J. De Ley, unpublished observations). The oxidase-negative genus *Acinetobacter* separates from the three other genera at a Tm(e) of about 71.0°C. The exact taxonomic position of *Acinetobacter* in the Tm(e) dendrogram was established by Van Landschoot et al. (1986) using and locating numerous strains. The generically misnamed *[Pseudomonas] cruciviae* and *[Ps.] pavonaceae* are also on the *Acinetobacter* rRNA branch. The rest of the cluster under consideration consists of two main branches, the *Moraxella* branch and the *Psychrobacter* branch. The *Moraxella* rRNA branch contains the classical moraxellae, generically misnamed *Alysiella* strains, and *Branhamella* with the false neisseriae, *[Br.] catarrhalis, [Br.] caviae, [Br.] ovis,* and *[Br.] cuniculi.* This branch contains many parasites of humans and animals. The *Psychrobacter* branch contains not only *Psy. immobilis* (Juni and Heym, 1986), but also the generically misnamed *[Moraxella] phenylpyruvica* and related organisms. Strains of this branch are widespread on biological materials, the marine environment, and clinical material.

The genus *Xanthomonas.* The *Xanthomonas* rRNA branch is quite separate; at a $T_{m(e)}$ of 66.7°C, it is farthest removed from all the other clusters in rRNA superfamily II. DNA-rRNA

hybridization (De Vos, 1981; De Vos and De Ley, 1983) has revealed that nearly all formally accepted *Xanthomonas* species are located on the *Xanthomonas* rRNA branch in a tight cluster, including *X. campestris, X. fragariae, X. albilineans,* and *X. axonopodis.* Very few *Xanthomonas* strains have been misnamed generically such as the species *[X.] ampelina* which was removed to the new genus *Xylophilus* (Willems et al., 1987) within rRNA superfamily III (see below). Quite a number of generically misnamed species from other genera really belong in *Xanthomonas* (De Vos et al., 1985a). One such case is *[Aplanobacter] populi,* agent of the poplar canker, which is now firmly a new species of *Xanthomonas* (De Vos and De Ley, 1983). Other cases of generically misnamed species, most of them phytopathogens, are *[Pseudomonas] betle, [Ps.] boreopolis, [Ps.] gardneri, [Ps.] hibiscicola, [Ps.] mangiferaeindicae, [Ps.] pictorum, [Ps.] viticola, [Ps.] vitiswoodrowii,* and *[Beneckea] hyperoptica* ATCC 15803. Some of these taxa had already been allocated to *Xanthomonas* on phenotypic grounds. Their exact taxonomic position will be reported separately (M. Van De Mooter et al., unpublished observations). Many genotypic and phenotypic criteria have proved convincingly that strains from nosocomial infections, soil, plants, etc. that are named *Pseudomonas maltophilia* had to be renamed *Xanthomonas maltophilia* (Swings et al., 1983). The main reason why quite a number of *"Pseudomonas"* species ultimately proved to be members of *Xanthomonas* and not vice versa is that the differentiation between both genera is rather difficult on phenotypic grounds, particularly for *Xanthomonas* strains not isolated from diseased plants. The inclusion in the latter genus is largely decided on phytopathogenic grounds, the presence of xanthomonadins, and the requirement for growth factors. Unequivocal identification in *Xanthomonas* is now easily possible through DNA-rRNA hybridizations: to the generic definition of *Xanthomonas* should be added that all strains group together in the $T_{m(e)}$ range from 75.5 to 81°C versus the type strain of *Xanthomonas campestris* and that all other taxa have a $T_{m(e)}$ versus the same type strain of less than 72°C.

The genus *Frateuria.* This small genus separates from the *Xanthomonas* rRNA branch at 71.8°C $T_{m(e)}$. These organisms were isolated from *Lilium auratum* and fruits of *Rubus parvifolius* in Japan. Phenotypically they resemble the acetic acid bacteria from rRNA superfamily IV but genotypically they are far removed from it (Swings et al., 1980).

The genus *Lysobacter.* This is a genus of pigmented, thin, gliding, flexing cells, strongly proteolytic, often degrading chitin, alginate, pectate, and lysing a variety of microorganisms. These features seem to suggest eventual relationships with the Myxobacteriales (Christensen and Cook, 1978); however, *Lysobacter* strain ATCC 27796 is at 0.75 S_{AB} from *Xanthomonas* (Woese et al., 1985b).

The genus *Xylella*. *Xylella* is a genus (one species, *Xyl. fastidiosa*) of rods to long filaments, which are nutritionally very fastidious. Their habitat is the xylem of plant tissue. They are associated with a great variety of diseases on a wide variety of plants. Partial sequencing of 16S rRNA and signature analysis relates them to *Lysobacter* and most closely to *Xanthomonas* (Wells et al., 1987).

SOME COMMON FEATURES IN rRNA SUPERFAMILY II. These organisms share a limited number of common features. Quite a number of them occur in water and soil, some in marine environments. A number of them are opportunistic parasites or even pathogens for either plants, animals, or man. On the whole, however, they are not as fatal for man or animals as are numerous strains from rRNA superfamily I. They are nearly all strictly aerobic. Many of them prefer to break down carbohydrates, frequently with the production of end products other than CO_2 and water (various sugar acids and keto-sugar acids, acetic acid, etc.). In the taxa which have been examined for it, the carbohydrate catabolism proceeds frequently by way of the Entner-Doudoroff and the shunt pathways. In a number of cases *(Xanthomonas* and authentic *Pseudomonas)* only part of the glycolytic system is present. It would be useful to examine more taxa of this superfamily for these features and to include also other basic features of metabolism, such as amino acid sequences of enzymes engaged in electron transfer, key enzymes in amino acid biosynthesis, and carbohydrate breakdown. Busse and Auling (1988) examined the distribution of polyamines in the Proteobacteria: for the gamma group, it was reported that members of the authentic genus *Pseudomonas* (Fig. 5) contain putrescine, spermidine, and ubiquinone Q-9. *Acinetobacter* specifically contains diaminopropanol.

OTHER TAXA IN THE GAMMA GROUP. A number of other taxa have been located in this very large and heterogeneous group, below rRNA superfamilies I and II.

THE GENUS *LEGIONELLA*. Members of the genus *Legionella* are implicated in human pneumonia. They are characterized as follows: amino acids but not carbohydrates used as C source; branched fatty acids in cell wall; require cysteine and iron. They stand apart as a separate branch, both by DNA-rRNA hybridizations, having a $T_{m(e)}$ of about 66°C (J. De Ley, W. Mannheim et al., unpublished observations) and by 16S rRNA cataloging (Ludwig and Stackebrandt, 1983b). According to the latter authors, the S_{AB} values of four *Legionella* strains locate them as an independent branch with a specific remote relationship (S_{AB} 0.35 to 0.42) to the alpha group; this is, however, not supported by oligonucleotide signatures. Partial sequence analysis of 16S rRNA supported a generic separation between *Legionella* and *Tatlockia* (Fox and Brown, 1989).

THE GENUS *FRANCISELLA*. These organisms, the causative agent of tularemia in humans and animals, constitute another isolated branch (J. De Ley, W. Mannheim et al., unpublished observations).

THE GENUS *CARDIOBACTERIUM*. These are facultatively anaerobic rods, fermenting glucose mainly to lactic acid; a normal inhabitant of the human oral, nasal, and pharyngeal mucosa; and the causative agent of endocarditis. It constitutes still another separate rRNA branch. This branch also carries the generically misnamed *[Kingella] indologenes* (J. De Ley, W. Mannheim et al., unpublished observations).

PHOTOTROPHIC BACTERIA. The gamma group contains two important families of phototrophic bacteria: the Chromatiaceae and the Ectothiorhodospiraceae (Woese et al., 1985b). Both are anaerobic, photolithotropic, purple sulfur bacteria, incorporating and reducing CO_2 completely or partly through the Calvin cycle; the ATP required is ultimately derived from light quanta, and protons and electrons are derived from inorganic reduced sulfur compounds (H_2S, etc.). The former family deposits elemental sulfur inside the cell, the latter family deposits it outside the cell and is also very halophilic and alkalophilic. 16S rRNA cataloging established (Fowler et al., 1984; Stackebrandt et al., 1984) that the purple sulfur bacteria of the above two families linked together at 0.47 S_{AB}. Strains of the genera *Chromatium, Amoebobacter, Lamprocystis, Thiocapsa, Thiocystis, Thiodictyon,* and *Thiospirillum* form a phylogenetically coherent cluster in the Chromatiaceae at $S_{AB} \approx 0.66$. The genus *Ectothiorhodospira* is somewhat more heterogeneous at $S_{AB} \approx 0.53$. There is a deep phylogenetic separation at $S_{AB} \approx 0.33$ between the purple sulfur and the purple nonsulfur phototrophic bacteria. This is also apparent from 5S rRNA sequence comparisons (R. De Wachter personal communication).

SOME COLORLESS SULFUR-OXIDIZING AND RELATED ORGANISMS. 5S rRNA sequencing (Lane et al., 1985; Stahl et al., 1987) showed that some aerobic sulfur bacteria are phylogenetically and taxonomically very heterogeneous. The following taxa (usually only one strain per species examined) are members of the gamma group and represent various peripheral rRNA branches near its basis: *Beggiatoa leptomitiformis*, *Beggiatoa alba* and its neighbor *Vitreoscilla beggiatoides*, *Leucothrix mucor*, and the generically misnamed obligate chemolithotrophic species *[Thiobacillus] neapolitanus* and *[Th.] ferrooxidans* m1 (for other *Thiobacillus* species see also rRNA superfamilies III and IV below). Some, such as the *Calyptogena magnifica* symbiont and *"Thiomicrospira"* L-12 might be at the base of rRNA superfamily I; *Thiothrix nivea* and its neighbor, the sulfide-oxidizing procaryotic symbiont of the marine bivalve *Solemya velum*, and *Thiomicrospira pelophila* might be at the base of rRNA superfamily II. Different clustering methods locate the taxa sometimes in different positions (Stahl et al., 1987; R. De Wachter, personal communication). The above symbionts constitute an interesting group of chemoautotrophic bacteria which live in a marine invertebrate host; the symbionts oxidize reduced sulfur compounds from the environment and fix CO_2, while the host gets part of the organic carbon. Partial sequencing of 16S rRNA revealed that the gill symbionts from the lucinid clams *Lucinoma* and *Codakia orbicularis*, and the trophosome symbiont from *Riftia pachyptila*, a tube worm from deep-sea hydrothermal vents, constitute one cluster in the gamma group. The gill symbionts from the bivalves *Calyptogena* and *Bathymodiolus*, also from deep-sea hydrothermal vents, constitute another cluster (Distel et al., 1988).

METHYLOMONAS ALBA. One strain of methane-utilizing bacteria containing the ribulose monophosphate (RuMP) pathway was examined through 5S rRNA sequencing. It is phylogenetically very far removed from the methanolotrophs (see rRNA superfamily III) (Wolfrum and Stolp, 1987) and the facultative methylotrophic genus *Methylobacterium* (see rRNA superfamily IV). The rRNA base sequence most closely resembles these of *Thiomicrospira*.

The rRNA superfamily III. The Beta Group.

The name rRNA superfamily III was proposed by De Ley (1978) for the large group to be discussed now. The name beta group was proposed

by Woese et al. (1984b). Both DNA-rRNA hybridizations and 16S rRNA cataloging yield clusters containing approximately the same organisms. Occasionally, there are some differences in the position of a taxon, probably due to the different technical approaches followed, and to the nature and the number of the strains included. rRNA superfamily III (Fig. 6) links at 62°C $T_{m(e)}$ with rRNA superfamilies I and II and at 60°C $T_{m(e)}$ with rRNA superfamily IV. The beta group links at about 0.34 S_{AB} with the alpha and gamma groups. The rRNA superfamily III contains at least 22 rRNA branches grouped in a number of rRNA clusters. The numerous genera and species in the superfamily group together at or above 68°C $T_{m(e)}$.

The Emended Family Neisseriaceae

DNA-rRNA hybridization (Rossau et al., 1986, 1989a; Van Landschoot et al., 1986) has shown that the bacterial family Neisseriaceae as described in *Bergey's Manual* (Bøvre, 1984) and in the first edition of *The Prokaryotes* (Bøvre and Hagen, 1981) cannot be maintained as such and should be redefined. Our data, involving over a hundred strains of numerous taxa and numerous labeled reference rRNAs, showed beyond doubt that the emended Neisseriaceae consists of the following taxa distributed over at least four rRNA branches: 1) the true *Neisseria* (*gonorrhoeae, meningitidis*, etc., but not the "false" neisserias *ovis, caviae*, and *cuniculi*); 2) the agar-pitting *Eikenella* from human oral and intestinal origin; 3) two genera of unique, flat, multicellular gliding bacteria called *Simonsiella* from the oral cavities of humans and animals and *Alysiella* from sheep; 4) *Kingella kingae* and 5) *K. denitrificans* (not the generically misnamed *[Kingella] indologenes*) from the human upper respiratory tract; and 6) the Centers for Disease Control (CDC) groups EF-4 and M-5, common in the oral flora of cats and/or dogs. All these organisms are closely related with a $T_{m(e)}$ range from 73.2 to 80.8°C. The positioning of *Alysiella* with *Acinetobacter* (Stakebrandt, 1986) is mistaken.

DNA-rRNA hybridization in the Neisseriaceae and the complete sequence of *Neisseria gonorrhoeae* 16S rRNA led to the development of rRNA-derived DNA probes which can be used to differentiate between organisms at the species or subspecies level (Rossau et al., 1988, 1989b).

Clustering of 16S rRNA sequences from one or a few strains of *Eikenella corrodens*, *Simonsiella muelleri*, and *Neisseria gonorrhoeae* and some reference strains from rRNA superfamilies I to III confirmed our findings that these

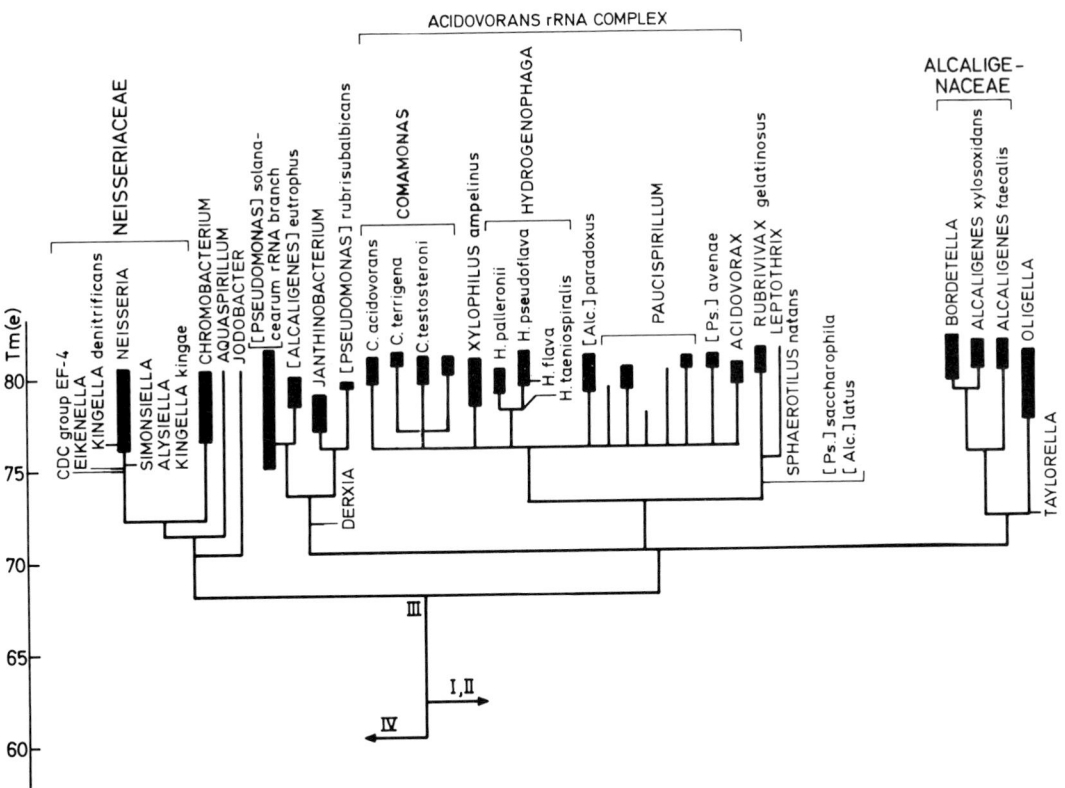

Fig. 6. Condensed rRNA cistron similarity dendrogram showing the major bacterial taxa from rRNA superfamily III. Abbreviations: Alc., *Alcaligenes;* C., *Comamonas;* H., *Hydrogenophaga;* and Ps., *Pseudomonas.* Other explanations as for Fig. 5.

organisms constitute one group (Dewhirst et al., 1989).

The Genus *Chromobacterium.*

The violet bacteria, originally limited to one genus, *Chromobacterium,* were studied by DNA-rRNA hybridization and split in two distinctly different genera (De Ley et al., 1978). Only the emended genus of the violet, facultatively aerobic, mesophilic, occasionally septicemic or pyogenic *Chromobacterium violaceum* concerns us here (for *Janthinobacterium* and *Iodobacter* see below). *Chromobacterium* is the closest neighbor of the Neisseriaceae at a $T_{m(e)}$ of 72.0°C (Rossau et al., 1989a). Whether it should eventually be included in the latter family can not yet be decided.

The Authentic Genus *Aquaspirillum*

The spirilla in general are extremely heterogeneous and are distributed over at least 18 rRNA branches in the $T_{m(e)}$ dendrogram (B. Pot, M. Gillis, J. De Ley, unpublished observations). The presently accepted genus *Aquaspirillum* in *Bergey's Manual* (1984) contains 17 species (Krieg, 1984). However, DNA-rRNA hybridi-

zation data (Chapter 131) show that the aquaspirilla are distributed over at least 10 different rRNA branches. According to our data, the emended authentic genus *Aquaspirillum* contains only the type species *Aq. serpens, Aq. bengal,* and *Aq. fasciculus.* This genus links at a $T_{m(e)}$ of 71.1°C to the above mentioned Neisseriaceae and *Chromobacterium* taxa. rRNA cataloging (Woese et al., 1984b) correctly put *Aq. serpens* and *Aq. bengal* together, in agreement with our data and with the data from slot blot hybridizations which showed that *Aq. bengal* is but a subjective synonym of *Aq. serpens* (Boivin et al., 1985). However, the rRNA cataloging method added *Aq. dispar* to the same branch, whereas we found it to be on a separate branch. An S_{AB} value of 0.56 separates *Chromobacterium* from *Aquaspirillum.*

Iodobacter fluviatilis

These organisms have originally been isolated from river water and named *Chromobacterium fluviatile* (Moss and Bryant, 1982). They separate at 71.0°C $T_{m(e)}$ from the authentic genus *Chromobacterium,* at about the same level (70.0°C) from the authentic genus *Aquaspiril-*

lum and at about 67°C $T_{m(e)}$ from *Janthinobacterium* (B. Pot, M. Gillis, J. De Ley, unpublished observations). Phenotypically it is a separate group. For all these reasons it was renamed *Iodobacter* (Logan, 1989). 16S rRNA sequences of the type strains of *C. violaceum* and *Iodobacter fluviatilis* and of the Neisseriaceae (Dewhirst et al., 1989) have the same relative positions when clustered as they have in the Tm(e) dendrogram.

The Solanacearum rRNA Cluster

DNA-rRNA hybridization showed this to be a large cluster, consisting mostly of some 13 generically misnamed *[Pseudomonas]* species (e.g., *alliicola*, *andropogonis*, *glumae*, *solanacearum*, etc.) and several generically misnamed *[Alcaligenes]* species (e.g., *eutrophus* and *hydrogenophilus*) (De Vos and De Ley, 1983; De Vos et al., 1985a, 1989). They are all located at or above 73.0°C $T_{m(e)}$ versus the reference rRNAs. Woese et al., (1984b) examined only two strains from this cluster. The correct internal structure within this rRNA cluster remains to be worked out. We estimate that at least two new genera will have to be created to accommodate all these species. Jenni et al. (1988) established the close relationships between the H_2-lithotrophic species *[Alcaligenes] eutrophus*, which fixes CO_2 through ribulose-bisphosphate carboxylase and the Calvin cycle, and related oxalate-oxidizing non-H_2-lithotrophic strains.

The *Janthinobacterium—[Pseudomonas] rubrisubalbicans* rRNA Cluster

The third genus of the violet bacteria mentioned above is the monospecific, strictly aerobic, "psychrophilic," nonpathogenic *Janthinobacterium* which grows in temperate regions. It is in a different rRNA cluster than *Chromobacterium* (De Ley et al., 1978) and *Iodobacter*. Linked to it at 76.0°C $T_{m(e)}$ is *[Ps.] rubrisubalbicans*, the generically misnamed causative organisms of "mottled stripe" disease on sugarcane (Goor et al., 1986). Also linked to it are some bacteria of clinical origin. The numerous strains of the solanacearum rRNA cluster and the *Janthinobacterium* rRNA cluster link at a $T_{m(e)}$ of 73.0°C. In rRNA cataloging, three strains from both clusters link at 0.63 S_{AB} (Woese et al., 1984b).

The Genus *Derxia*

In the same cluster, but a little lower, at about 72°.0C $T_{m(e)}$ we place the genus *Derxia* (De Smedt et al., 1980). It is phenotypically quite different from the two above taxa; it consists of

pleomorphic rods and is mainly found in tropical regions. It is able to fix N_2 and to grow as a facultative H_2-autotroph; the nitrogenase can apparently function even aerobically because it is protected against O_2 by the large gummy capsule. It is genotypically quite different from other free-living nitrogen-fixers such as *Azotobacter*, *Azomonas*, and *Beijerinckia*.

The Acidovorans rRNA Cluster.

This is a very large rRNA cluster consisting of at least 15 rRNA branches which link at about 76°C Tm(e). Taxonomically, this cluster contains at least five genera and a number of generically misnamed species (A. Willems, M. Gillis, J. De Ley, unpublished observations). Genotypic and phenotypic data have led to a revival of the genus *Comamonas* (De Vos et al., 1985b); it is a genus of Gram-negative bacteria with bipolar tufts of flagella; these bacteria rarely attack carbohydrates, but grow well on organic or amino acids; they have been isolated from soil and human blood. Later, the generically misnamed *[Pseudomonas] acidovorans* and *[Ps.] testosteroni* were included in this genus as *Comamonas acidovorans* and *C. testosteroni* (Tamaoka et al., 1987). A second genus in this rRNA cluster is the very homogeneous *Xylophilus* (Willems et al., 1987): these organisms are slow-growing and rod-shaped, they cause bacterial necrosis of grape vines mainly in the Mediterranean area. This genus contains only one species, *X. ampelinus*. A third genus is *Hydrogenophaga* (Willems et al., 1989); it encompasses yellow-pigmented, hydrogen-oxidizing rods, growing well on organic and amino acids. A fourth genus has provisionally been named *Paucispirillum* (A. Willems, B. Pot, M. Goor, M. Gillis, J. De Ley, unpublished observations). It consists of six rRNA branches. These branches correspond to the following species *Paucispirillum anulus*, *Pa. delicatum*, *Pa. giesbergeri*, *Pa. gracile*, *Pa. metamorphum*, *Pa. psychrophilum* and *Pa. sinuosum*. These species are presently still attributed to *Aquaspirillum*. They grow poorly on most media, preferring organic acids. They have been isolated from water. A fifth genus, *Acidovorax* (A. Willems et al., 1990), forms a separate rRNA branch. Most of these organisms are of clinical origin, a minority are from soil. They constitute a group of aerobic, chemoorganotrophic organisms growing well on a variety of organic acids and amino acids. Several strains can live autotrophically using H_2 as energy source. This genus consists of three species *Acidovorax facilis* (the renamed *[Ps.] facilis*), *Av. delafieldii* (the renamed *[Ps.] delafieldii*) and *Av. temperans*. Other taxa which belong in the rRNA cluster are several generically mis-named *Pseudomonas* species and subspecies and two generically misnamed *Alcaligenes* spe-cies. One strain each of *[Pseudomonas] acido-vorans, [Ps.] testosteroni, Comamonas terri-gena, [Aq.] aquaticum,* and *[Aq.] gracile* link at 0.75–0.89 S_{AB} value in rRNA cataloging (Woese et al., 1984b).

The *Rubrivivax gelatinosus—Leptothrix* rRNA Branches. The taxon *Rhodocyclus gela-tinosus* was created provisionally (Imhoff et al., 1984). More genotypic data for the evaluation of its taxonomic position is now available. It is genotypically and phenotypically quite different from the other two species of this genus, the type species *Rhc. purpureus* and *Rhc. tenuis*. A new genus name *Rubrivivax* (Willems et al., 1991) has been proposed for the genetically sep-arate *[Rhodocyclus] gelatinosus*. Several other taxa branch off at about 75°C $T_{m(e)}$ from the gelatinosus rRNA branch, deep enough to prove generic differences. A separate branch is con-stituted by the genus *Leptothrix* with the species *discophora* and *cholodnii*. *Sphaerotilus natans,* and the facultatively autotrophic species *[A.] la-tus* and *[Ps.] saccharophila* link near the base of the *Leptothrix* rRNA branch; their relation-ship is still unknown. 16S rRNA cataloging like-wise indicated a relationship between *Rubrivi-vax gelatinosus* and *Sphaerotilus natans* at S_{AB} 0.78 (Woese et al., 1984b). All the above taxa are closest to the acidovorans rRNA cluster at about 73°C $T_{m(e)}$. The sheathed heterotrophic bacteria *Leptothrix* and *Sphaerotilus* deposit Fe- and sometimes Mn-oxides in and on the sheaths. *Vitreoscilla* appears also to be closely related to *Rubrivivax* (Stahl et al., 1987). Both *Rhodocyclus purpureus* and *Rhc. tenuis,* al-though still members of RNA superfamily III, are at its base with a $T_{m(e)}$ of about 68°C (A. Willems, M. Gillis, J. De Ley, unpublished ob-servations). 16S rRNA cataloging locates *Rhc. tenuis* somewhere in the beta-2 subgroup (Woese et al., 1984b), at the base of our solan-acearum-*Janthinobacterium* rRNA cluster.

The Genus *Vitreoscilla*

5S rRNA base sequencing showed (Stahl et al., 1987) showed that the genus *Vitreoscilla* is het-erogeneous, as *Vi. filiformis* is somewhat closer to *Rubrivivax gelatinosus* than *Vi. stercoraria,* whereas *Vi. beggiatoides* is in a group together with *Beggiatoa alba* somewhere at the base at or already outside the gamma group. Prelimi-nary DNA-rRNA hybridization data confirm the heterogeneity of *Vitreoscilla* (A. Willems, M. Gillis, and J. De Ley, unpublished obser-vations). The relationships within and between

these taxa are certainly more complex than is presently known. 16S rRNA sequencing locates *Vi. stercoraria* below the base of the emended Neisseriaceae (Dewhirst et al., 1989).

The *Alcaligenaceae—Oligella—Taylorella* rRNA Cluster

The delineation of the genus *Alcaligenes* has been extremely blurred for about five decades. At one time *Alcaligenes* was a "genus" with a GC range of about 30 to about 70 mol%, and even contained Gram-positive strains or coc-coid organisms. Extensive numerical analysis of phenotypic features and genotypic studies of a very large number of strains identified the real genus (Kersters and De Ley, 1984a). The au-thentic, now well-described, genus *Alcaligenes* is limited to a group of organisms with a $T_{m(e)}$ of at least 75.5°C versus reference strains of either *Alc. faecalis* or *Alc. xylosoxidans* (De Ley et al., 1986). It had been suspected on some phenotypic grounds that *Bordetella* was also re-lated to *Alcaligenes*. *B. pertussis* is the causal agent of whooping cough in humans; *B. bron-chiseptica,* the causal agent of severe respiratory infections in dogs and swine; and *B. avium,* the causal agent of rhinotracheitis in turkey poults (Kersters et al., 1984). Our DNA-rRNA hy-bridization showed *Bordetella* indeed to be a close neighbor at a $T_{m(e)}$ of 75.5°C versus *Alc. faecalis* and 79°C versus *Alc. xylosoxidans,* so close that a new and separate family of Alca-ligenaceae is justified (De Ley et al., 1986). On extensive phenotypic and genotypic grounds, two other genera, *Oligella* and *Taylorella,* are neighbors of this family (Rossau et al., 1987). Both genera might eventually later be assimi-lated in the Alcaligenaceae, or might constitute a new family, but more data are required. The genus *Oligella* consists of small, fastidious rods which use only a few organic and amino acids and are otherwise biochemically rather inert. So far, it consists of two species, *Ol. urethralis* and *Ol. ureolytica,* with probably low pathogenicity for the urinary tract. *Oligella* links at a $T_{m(e)}$ of 71.7°C to the Alcaligenaceae. *Taylorella equi-genitalis* is the causal agent of endometritis and cervitis in mares. It links at about the same level of 71.7°C $T_{m(e)}$ with the Alcaligenaceae and *Oli-gella.*

The Genus *Thiobacillus*

A number of these chemoautotrophic bacteria derive their carbon from CO_2, and their ATP and reducing equivalents from the aerobic ox-idation of reduced inorganic sulfur compounds. 5S rRNA sequences show the thiobacilli to be genotypically very heterogeneous; most of them are members of the beta group, a few others are located in the alpha and gamma groups. So far, one or a few strains each of the following six species have been found to be members of the beta group, mainly by 5S rRNA sequencing (Lane et al., 1985; Stahl et al., 1987), a few by 16S rRNA cataloging (Woese et al., 1984b): the mixotrophic *Th. perometabolis, Th. intermedius,* the obligate chemolithotrophs *Th. thioparus, Th. thiooxidans, Th. denitrificans,* and *Th. ferrooxidans* ATCC 19859. As *Th. thioparus* is the type species, it is the nucleus of the authentic genus *Thiobacillus;* all the *Thiobacillus* species in the alpha and gamma groups are thus generically misnamed. In the beta group, *Th. intermedius* and *Th. perometabolis* from a separate rRNA branch, they also share a number of similar growth features. *Th. thioparus* forms a second rRNA branch, and the acidophilic *Th. thiooxidans* and *Th. ferrooxidans* form a third rRNA branch (Lane et al., 1985). The species *Th. ferrooxidans* is heterogeneous since strain ml belongs in the gamma group. *Th. neapolitanus* and *Thiomicrospira* also belong in the gamma group, while three facultatively chemolithotrophic *Thiobacillus* species belong in the alpha group (Lane et al., 1985; Stahl et al., 1987). The inclusion of more strains and correlations with other features are needed for a taxonomic reevaluation of the present genus *Thiobacillus.*

Some Obligate *Methanolotrophic RuMP-type* Bacteria

The taxonomic position of these organisms is all but unknown. 5S rRNA sequence determinations of one strain each of the methanol-utilizers *Methylomonas clara, Methanolomonas glucoseoxidans,* and *Methanomonas methylovora* showed that they are members of the beta group (rRNA superfamily III). D_{AB} values indicate that they are closest to each other (D_{AB} 0.0087 to 0.1419), followed by *Thiobacillus* (D_{AB} down to 0.21−0.27) at the base of this group. A methanotrophic strain of *Methylomonas alba* is located in the gamma group (see above) (Wolfrum and Stolp, 1987), and the facultatively methylotrophic genus *Methylobacterium* is rRNA superfamily IV.

The Ammonia to Nitrite Oxidizers

These are strictly aerobic, obligate chemolithotrophic bacteria. They obtain ATP and reducing equivalents (required for the fixation of CO_2 through the Calvin cycle) from the oxidation of ammonia to nitrite. There are several morphological cell types. 16S rRNA cataloging was carried out on one strain each of the following taxa:

Nitrosococcus mobilis, Nitrosomonas europeae, Nitrosovibrio tenuis, Nitrosolobus multiformis, and *Nitrosospira* species (Woese et al., 1984b). All these strains link together with a S_{AB} of about 0.55; they are located at the base of the beta group with an S_{AB} of about 0.45. The latter three taxa have similar rRNA cistrons; the other two form deeply different branches.

Spirillum volutans

At about the same average level of S_{AB} equal to 0.50, *Spirillum volutans* constitutes a separate rRNA branch with the above "nitroso"-bacteria as neighbors (Woese et al., 1982, 1984b). It is one of the very large (up to 60 μm long) freshwater spirilla with visible bipolar flagella. They are unable to metabolize carbohydrates but live preferentially at the expense of succinate in microaerophilic conditions.

Some Common Features in rRNA Superfamily III

Except for their similarities in rRNA cistrons, few other common features are known so far in this extremely heterogeneous superfamily. Busse and Auling (1988) showed that the unusual polyamine 2-hydroxyputrescine is possibly ubiquitous in this superfamily and that putrescine is the dominant amine, but more strains need to be examined. There are a few photoorganotrophs and some chemoautotrophs. The majority of taxa are chemoorganotrophic; they are aerobic and do not ferment; many of them are soil and water bacteria; they prefer on the whole to use organic acids and amino acids as carbon and energy source; their ability to use carbohydrates are much more limited than in rRNA superfamilies I and II. In a number of cases the presence of an active Entner-Doudoroff pathway and hexose monophosphate shunt has been established, as well as a weak or incomplete glycolytic system. Much more research on the comparative biochemistry and enzymology of these organisms is called for. There are also a number of organisms able to oxidize H_2 or to grow chemoautotrophically on it. As in all other rRNA superfamilies, there are some N_2-fixers.

The rRNA Superfamily IV. The Alpha Group

The name "rRNA superfamily IV" was proposed by De Ley (1978) and De Smedt et al. (1980) for this large group. The synonymous name "alpha group" was proposed by Woese et al. (1984a). This very large group of bacteria is phenotypically again very heterogeneous; most photoautotrophs of the Rhodospirillaceae family, some chemolithotrophs, and many chemoorganotrophs are placed here. DNA-rRNA hybridizations (Fig. 7) and 16S rRNA cataloging yield essentially the same results. The former method reveals at least six large rRNA, clusters with a total of about 20 rRNA branches in rRNA superfamily IV. All these clusters link at an average $T_{m(e)}$ of 70°C. This superfamily links at about 60°C to the three above mentioned ones. rRNA cataloging reveals three clusters in the alpha group (alpha-1, -2, and -3) and a single strain of *Erythrobacter longus;* they link together at an S_{AB} value of about 0.44; this group links at an average S_{AB} of 0.32 with the beta and gamma groups (Woese et al., 1984a).

The *Agrobacterium—Rhizobium—Phyllobacterium—Brucella—Ochrobactrum* rRNA Cluster

In rRNA cataloging, this large rRNA cluster is represented by one strain each of *Agrobacterium tumefaciens* and *Rhizobium leguminosarum,* and by the rickettsia-like *Rochalimaea* (Weisburg et al., 1985; Woese, 1987). Extensive phenotypic and genotypic studies of hundreds of strains from this cluster have been performed (De Ley, 1972, 1974; De Ley et al., 1987; De Smedt and De Ley, 1977; Holmes et al., 1988; Jarvis et al., 1986; Kersters and De Ley, 1975, 1984b; Kersters et al. 1973), and Fig. 7 presents a simplified outline of the relationships within this complex cluster. The dense nucleus of this cluster is constituted by the fast-growing, peritrichously flagellated, root-nodulating rhizobia and the phytopathogenic agrobacteria. The different rRNA branches link at 74.4°C $T_{m(e)}$. Phenotypic analysis and DNA-DNA hybridization revealed that *Agrobacterium* consists largely of two main clusters: cluster 1 encompassing 3-ketolactose-positive, tumorigenic, and avirulent strains; and cluster 2 encompassing 3-ketolactose negative, tumorigenic, rhizogenic, and avirulent strains. The same two clusters are revealed by DNA-rRNA hybridization. In the $T_{m(e)}$ dendrogram (Fig. 7), both *Agrobacterium* rRNA branches appear in the midst of the fast-growing *Rhizobium.* The latter genus consists of several rRNA branches. One of them is the *Rhizobium meliloti—Rh. leguminosarum—Rh. fredii* rRNA branch (Jarvis et al., 1986). It was recently proposed (Chen et al., 1988) that a new genus, *Sinorhizobium* be created for the species *Rhizobium fredii.* A second, quite distinct rRNA branch at 74.4°C Tm(e) contains the *Rhizobium loti* strains; this group of fast-grow-

ing organisms nodulates trefoils (*Lotus* sp.); if it has sufficiently differentiating phenotypic features, it may become a new genus. The meliloti—leguminosarum rRNA group is about as far removed from *Agrobacterium* cluster 2 as from the as-yet unnamed *Galega* rhizobia rRNA branch (not shown in Fig. 7). The latter group might represent another new taxon. Rothe et al. (1987) located, by rRNA cataloging, the budding water-bacteria *Blastobacter aggregatus* close to *Agrobacterium tumefaciens* at 0.84 S_{AB} (see also *Blastobacter* below). Also in the middle of the *Rhizobium—Agrobacterium* cluster at 77°C $T_{m(e)}$, we found the genus *Mycoplana*, which are motile, branching soil bacteria able to attack phenol. The slow-growing genus *Bradyrhizobium* will be discussed below in its correct genotypic surroundings. The genus *Phyllobacterium* links at 73.1°C $T_{m(e)}$ with the *Agrobacterium—Rhizobium* cluster (De Smedt and De Ley, 1977). It was created for aerobic, chemoorganotrophic bacteria isolated from leaf nodules of Rubiaceae, Myrsinaceae, etc. (Knösel, 1984). These three genera of bacteria fit genotypically and phenotypically in the family Rhizobiaceae. Their genetic similarity is emphasized by the acceptance and expression of the *Agrobacterium* Ti plasmid and the *Rhizobium* Sym plasmid in *Phyllobacterium* (van Veen et al., 1988). In view of their considerable genotypic and phenotypic similarities, it is almost certain that the taxa of the Rhizobiaceae have the same phylogenetic origin. It was, however, very surprising to find *Brucella* (De Ley et al., 1987) and *Ochrobactrum* (Holmes et al., 1988), together with *Phyllobacterium*, on one rRNA branch. *Brucella* is the cause of brucellosis, a widespread and dangerous zoonosis which also produces an incapacitating disease in humans. *Ochrobactrum,* a genus of peritrichous, strictly aerobic rods, was created to harbor generically misnamed *Achromobacter* strains and bacteria of the CDC group Vd, predominantly of human clinical origin. rRNA cataloging has shown (Weisburg et al., 1985) considerable rRNA similarities between *Agrobacterium tumefaciens* and *Rochalimaea quintana,* a member of the Rickettsiaceae and the etiological agent of trench fever in humans. The closeness of the human and animal pathogen *Brucella* and its clinical neighbor *Ochrobactrum* strongly suggests to us that the possibility of a close relationship between *Rochalimaea* and the latter two genera should be examined. Finally, the CO-oxidizing and carboxydotrophic bacteria are very heterogeneous: they are distributed over several rRNA clusters. The generically misnamed *[Alcaligenes] carboxydus* is located in the *Agrobacterium—Rhizobium* clus-

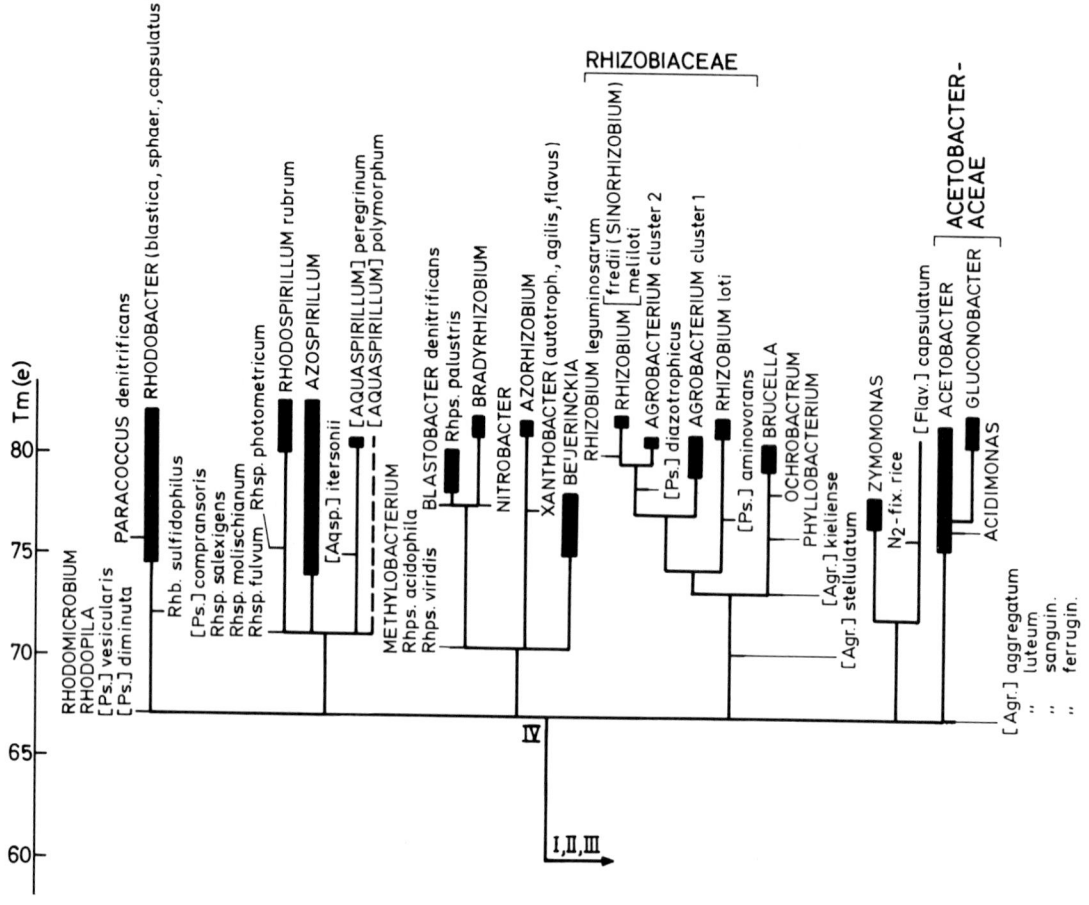

Fig. 7. Condensed rRNA cistron similarity dendrogram showing the major bacterial taxa from rRNA superfamily IV. Abbreviations: Agr., *Agrobacterium;* Aqsp., *Aquaspirillum;* Flav., *Flavobacterium;* Ps., *Pseudomonas;* Rhb., *Rhodobacter;* Rhps., *Rhodopseudomonas;* Rhsp., *Rhodospirillum.* Other explanations as for Fig. 5.

ter 0.58 S_{AB} from *Agrobacterium* and *Rhizobium;* it is at the same distance from the nucleus of the cluster as *Rochalimaea; [Alc.] carboxydus* is thus only distantly related to the above three genera (Auling et al., 1988).

Within or close to the same cluster at various $T_{m(e)}$ levels between 70°C and 78°C, there are several generically misnamed organisms (De Ley et al., 1987): the N_2-fixing *[Pseudomonas] diazotrophicus* from the root of wetland rice (Watanabe, 1987) has been located at 78°C just below the *Rhizobium–Sinorhizobium–Agrobacterium* cluster 1 branching; *[Achromobacter] cycloclastes* LMG 1127 is located at 76°C; *[Azotomonas] fluorescens* at 74°C; the budding, amine-utilizing, facultatively methylotrophic *[Ps.] aminovorans* is on the *Rhizobium loti* rRNA branch at 77°C (Green and Gillis, 1989); *[Agr.] kieliense* is just below the cluster at 73°C, and *[Agrobacterium] stellatum* is at about 71°C $T_{m(e)}$. The activities and functions of a number of these organisms are not well understood; nomenclatural changes will be required.

The Family Acetobacteraceae

These strictly aerobic chemoorganotrophic organisms prefer to live in or on plant material. They are biochemically characterized by a great variety of incomplete oxidations (aliphatic alcohols to acids, aldoses to aldonic and keto-aldonic acids, Bertrand-Hudson oxidations to keto-functions, etc.) and by a double system of enzymes, one in the cytoplasm and a very active one in the cytoplasmic membrane. This family consists of two genera, *Acetobacter* and *Gluconobacter,* and each genus constitutes a separate rRNA branch; both branches link together at 77°C $T_{m(e)}$ and their root is at the basis of rRNA superfamily IV. It is most likely that *Acidomonas,* a genus of acidophilic, facultatively methylotrophic bacteria (Urakami et al., 1989) also belongs in this rRNA cluster, as the DNA-rRNA hybrids of the type strain are indistinguishable from those of *Acetobacter* (M. Gillis and J. De Ley, unpublished observations):

The *Zymomonas–[Flavobacterium] capsulatum* rRNA Cluster

Zymomonas is genotypically a very dense and homogeneous cluster. It originates in the root of rRNA superfamily IV. It is a genus of preferably anaerobic rods which ferment a few sugars mainly to ethanol and CO_2. They carry out this fermentation through the Entner-Doudoroff pathway. This and their genotypic position in the midst of strictly aerobic bacteria of rRNA superfamily IV (Gillis and De Ley, 1980) strongly suggest that these bacteria are of a strictly aerobic phylogenetic origin and that the loss of some electron-transferring enzyme(s) obliged them to live on and ferment carbohydrates anaerobically. The *Zymomonas* rRNA branch links at 71.8°C with the capsulatum rRNA branch (Bauwens and De Ley, 1981; De Smedt et al., 1980). The latter branch contains several species which are generally yellow-pigmented; they are all generically misnamed. The reference rRNA is from *[Flavobacterium] capsulatum,* a yellow, capsulated, rapidly growing organism, originally isolated from distilled water. The following species are located in the range 74.5–76°C Tm(e). *[Pseudomonas] azotocolligans* (De Vos et al., 1989) is not a N_2-fixer, contrary to the original description. The citrate-using and N_2-fixing *[Chromobacterium] folium* has been detected in germinated seeds and in leaf nodules of *Psychotria* and *Ardisia* (see above, *Phyllobacterium*). *[Pseudomonas] paucimobilis* is mainly of nosocomial and clinical origin, and *[Flavobacterium] devorans* has been isolated from water. Both are close neighbors on our $T_{m(e)}$ dendrogram (not shown in Fig. 7); this is in excellent agreement with their phenotypic near-identity, and the conclusion that *[Flavobacterium] devorans* as a name should disappear since it is a later synonym of *[Pseudomonas] paucimobilis* (Yabuuchi et al., 1979). N_2-fixing bacteria from the rhizosphere of rice have also been located at 76°C $T_{m(e)}$ close to *Ps. paucimobilis* (Bally R. et al., 1985). All these organisms require a suitable genus name.

The *Beijerinckia–Azorhizobium–Bradyrhizobium–Rhodopseudomonas palustris* rRNA Cluster

This cluster consists of several main rRNA branches which link at 70.4°C $T_{m(e)}$. Using the hybridization method, this cluster originates in the root of rRNA superfamily IV. It is represented by three strains in rRNA cataloging in group alpha-2 at about 0.48 S_{AB}, close to the root of the alpha group (Woese et al., 1984a).

THE *BEIJERINCKIA* rRNA BRANCH. The genus *Beijerinckia* consists of acid-tolerant, free-living, strictly aerobic, N_2-fixing rods, producing copious and cohesive slime, and living mainly in tropical, lateritic soils. We located strains of all *Beijerinckia* species in the top 3°C $T_{m(e)}$ of this rRNA branch. There are no other genera in the immediate vicinity of *Beijerinckia,* suggesting its phylogenetically rather isolated position (De Smedt et al., 1980).

THE *AZORHIZOBIUM* rRNA BRANCH. Genotypically and phenotypically, the genus *Azorhizobium* is quite different from other nodulating taxa such as *Rhizobium* and *Bradyrhizobium.* (Dreyfus et al., 1988) *Azorhizobium* contains free-living, obligately aerobic, small rods that fix N2 microaerobically. Lactate and succinate are their favorite carbon sources. Their main feature is that they effectively nodulate stems and roots of the African tropical legume *Sesbania rostrata.* The *Azorhizobium* rRNA branch links with both other taxa in this rRNA cluster at 70.4°C $T_{m(e)}$.

THE GENUS *XANTHOBACTER*. The closest genotypic neighbor of *Azorhizobium* is *Xanthobacter* which is on the same rRNA branch at 77.2–78.0°C $T_{m(e)}$ (Dreyfus et al., 1988). We included several species, *Xb. autotrophicus, Xb. agilis,* and *Xb. flavus* (the former *[Mycobacterium] flavum*). These organisms are also obligately aerobic, N_2-fixing rods, isolated from water and soil. Typical also is their chemolithotrophic nature, their ability to oxidize H_2, and the presence of the yellow zeaxanthine dirhamnoside.

THE *RHODOPSEUDOMONAS PALUSTRIS–BRADYRHIZOBIUM–NITROBACTER* rRNA SUBCLUSTER. The photoautotrophic / chemoorganotrophic genus *Rhodopseudomonas,* the chemoautotrophic *Nitrobacter,* the carboxydotrophic *[Ps.] carboxydovorans,* and the chemoorganotrophic *Bradyrhizobium,* and a *Blastobacter* species are located close together in this subcluster. It is a fascinating meeting point for three biochemically and bioenergetically different ways of life which appear to have a common phylogenetic origin. One should also expect considerable similarities in the structure and function of their proteins.

The *Bradyrhizobium* rRNA branch links with the *Rhps. palustris* rRNA branch at 77.5°C $T_{m(e)}$ (Jarvis et al., 1986) by DNA-rRNA hybridization, or at 0.55 S_{ab} by rRNA cataloging (Hennecke et al., 1985). *Bradyrhizobium* is phenotypically quite distinct from the other legume-nodulating rhizobia (Jordan, 1984). This genus consists of slow-growing, aerobic rods, inducing N_2-fixing nodule production on roots of tropical and some temperate zone leguminous plants.

The *Rhodopseudomonas palustris* rRNA branch. DNA-rRNA hybridization shows that the genus *Rhodopseudomonas* is very heterogeneous, with *Rhps. viridis* and *Rhps. acidophila* at the bottom of the entire cluster, about $10°C \Delta T_{m(e)}$ below the reference *Rhps. palustris* rRNA and even further away than the genera *Bradyrhizobium* and *Nitrobacter*. Phenotypic expressions of heterogeneity in *Rhodopseudomonas* have been summarized by Imhoff et al. (1984): there are differences in the size of cytochrome c_2, the nature of the cytochromes, lipopolysaccharides, and quinones, and the pathway of sulfate assimilation.

The genus *Nitrobacter* contains heterotrophic and chemolithotrophic bacteria; the latter use the aerobic oxidation of nitrite to nitrate as a source of energy and reducing equivalents to incorporate CO_2 into cell material. Three strains of *Nitrobacter* link with *Rhps. palustris* at $0.77 S_{AB}$ in rRNA cataloguing (Seewaldt et al., 1982), and two strains link at $75°C T_{m(e)}$ with *Rhps. palustris* reference rRNA in DNA-rRNA hybridization (M. Gillis and J. De Ley, unpublished observations). The close phylogenetic relationship between both genera, detected by rRNA cistron similarities, is also supported by the similarity in a number of phenotypic features (Seewaldt et al., 1982).

The genus *Blastobacter* consists of budding, aerobic, chemoorganotrophic rods isolated from surface water from brooks, ponds, lakes, etc. This genus appears to be heterogeneous. *Blastobacter denitrificans* was localized at $77.5°C T_{m(e)}$ in the same cluster as the above three genera (Green and Gillis, 1989). rRNA cataloging localized *Blastobacter aggregatus* close to *Agrobacterium tumefaciens* (Rothe et al., 1987). More *Blastobacter* strains will have to be included.

The carboxydotrophic *[Pseudomonas] carboxydovorans* is highly related to the above four genera at $0.80–0.87 S_{AB}$ (Auling et al., 1988). Two other carboxydotrophic organisms are briefly mentioned in rRNA superfamily IV.

THE GENUS *METHYLOBACTERIUM*. This group contains pink-pigmented, facultatively methylotrophic, Gram-negative, pleomorphic rods (Green and Bousfield, 1983). So far we included two strains only in our DNA-rRNA hybridization: *Mb. radiotolerans* (the former *[Ps.] radiora*) and an unspecified strain. They link at the bottom of the *Beijerinckia–Azorhizobium–Bradyrhizobium–Rhps. palustris* rRNA

cluster at about $70.5°C T_{m(e)}$. Other methylotrophic taxa have been briefly mentioned in the gamma group and in rRNA superfamily III (see above).

THE GENUS *PHENYLOBACTERIUM*. This taxon (Lingens et al., 1985) constitutes an isolated branch at $S_{AB} 0.39–0.51$ in the alpha-2 subcluster (Ludwig et al., 1984). Metabolically, they are rather unusual aerobic organisms which grow preferentially by opening and breaking down the benzene ring of chloridazon and related pyridazone derivatives, and of the pyrazone derivatives pyramidone and antipyrine.

The *Rhodospirillum–Azospirillum–[Aquaspirillum]* rRNA Cluster

This rRNA cluster contains at least four rRNA branches which link together at $71°C T_{m(e)}$. It corresponds to the alpha-1 subgroup of Woese et al. (1984a).

THE GENUS *RHODOSPIRILLUM*. The photoorganotrophic genus *Rhodospirillum* is heterogeneous. In rRNA hybridizations or rRNA cataloging, *R. rubrum* and *R. photometricum* link at $75.5°C T_{m(e)}$ or $0.74 S_{AB}$, respectively. Both species have a large type of cytochrome c_2. Other *Rhodospirillum* species such as *molischianum, fulvum*, and *salexigens* are at the bottom of the entire cluster. The exact position of *Rhodopila globiformis* is undecided: S_{AB} values locate it in the alpha-1 group, whereas our hybridization data locate it at the bottom of the rRNA superfamily IV.

THE *AZOSPIRILLUM* rRNA BRANCH. Organisms of this genus are widely distributed in the rhizosphere of tropical and subtropical grasses; it is generally accepted that these bacteria enhance the growth of plants. We examined this genus extensively by rRNA hybridization (De Ley, 1978; De Smedt et al., 1980; Reinhold et al., 1987). The species *Azsp. brasilense* and *lipoferum* are close together in the top $2.5°C T_{m(e)}$ of the rRNA branch and the species *Azsp. amazonense* and *halopraeferens* are around $74°C T_{m(e)}$. Our findings on the three former species agree very well with the conclusions from the DNA and rRNA homology studies of Falk et al. (1986). The latter authors also showed that *Conglomeromonas largomobilis* subsp. *largomobilis* (microaerophilic, N_2 fixing organisms isolated from freshwater sources) is a subjective synonym of *Azospirillum lipoferum* but that the subsp. *parooensis* is quite different; its taxonomic position has not yet been established.

THE GENERICALLY MISNAMED [*AQUASPIRILLUM*] rRNA BRANCHES. *[Aquaspirillum] peregrinum* and *[Aqsp.] itersonii* (both subspecies *itersonii* and *vulgatum*) constitute one rRNA branch. Another rRNA branch contains *[Aqsp.] polymorphum* (see Chapter 131).

[*PSEUDOMONAS*] *COMPRANSORIS*. This is one of the aerobic, CO-oxidizing bacteria (see above) which is generically misnamed. It is located, both by rRNA hybridization (De Vos et al., 1985; M. Gillis and J. De Ley, unpublished observations) and by rRNA cataloging (Auling et al., 1988) at the bottom of the *Rhodospirillum–Azospirillum–[Aquaspirillum]* rRNA cluster, at 71°C $T_{m(e)}$ and 0.44 S_{AB} respectively.

Prosthecate and Budding Bacteria

Many taxa of budding and/or prosthecate bacteria have been located in rRNA superfamily IV (Stackebrandt et al., 1988b). They are: *Ancalomicrobium, Blastobacter, Caulobacter, Dichotomicrobium, Filomicrobium, Gemmobacter, Hyphomicrobium, Hyphomonas, Nitrobacter winogradskyi, Pedomicrobium, Prosthecomicrobium, Rhodomicrobium, Rhodopseudomonas palustris, Rhps. viridis, Rhps. acidophila,* and *Stella.*

THE *HYPHOMICROBIUM–FILOMICROBIUM–PEDOMICROBIUM–DICHOTOMICROBIUM* GROUP. The phylogenetic position of this group of bacteria was determined by Stackebrandt et al. (1988b). *Hyphomicrobium* contains hyphal, prosthecate, and budding aerobic bacteria preferring one-carbon compounds (e.g., methanol) as both carbon and energy source. Likewise, *Pedomicrobium* encompasses aerobic, prosthecate, and budding bacteria. They produce lateral hyphae from the main cell body. Both *Pedomicrobium* and *Filomicrobium* prefer C-1 compounds as carbon sources but use complex organic acids as well. *Dichotomicrobium* is morphologically unique, with tetrahedral cells and dichotomously branched hyphae; it is moderately thermophilic and halotolerant; it uses some organic acids from the Krebs cycle as carbon source and some amino acids as nitrogen source (Hirsch and Hoffmann, 1989). Moore (1977) showed that the C-1-utilizing, prosthecate and budding bacteria constitute a group of related genera. This group has, as relatively close neighbors, other prosthecate and budding, amino acid-utilizing or photosynthetic organisms such as *Rhodomicrobium, Prosthecomicrobium,* and *Caulobacter.* Roggentin and Hirsch (1989) showed by DNA-rRNA hybridization that *Hyphomicrobium* is very hetero-

geneous and consists of at least five deeply different groups which may eventually be elevated to the generic level. As the hybridization method was different from ours, we cannot definitely localize these taxa in our $T_{m(e)}$ dendrogram. 16S rRNA cataloging indicates that six strains of *Hyphomicrobium* (including *Hm. vulgare*), and one strain each of *Filomicrobium, Pedomicrobium,* and *Dichotomicrobium* constitute a separate rRNA cluster in rRNA superfamily IV. In sharp contrast, 5S rRNA sequencing locates *Hm. vulgare* closest to the *Agrobacterium* rRNA cluster (Stackebrandt et al., 1988b). *Hm. vulgare* links at 0.45 S_{AB} with the *Rhps. palustris* cluster (Rothe et al., 1987).

THE GENUS *HYPHOMONAS*. This group encompasses hyphal, prosthecate, and budding aerobic bacteria. They form relatively more intercalary buds. Typical features are that they require amino acids as source of carbon and energy; C-1 compounds and carbohydrates are not used. *Hyphomonas polymorpha* is a separate branch in the alpha group (Stackebrandt et al., 1988b).

THE GENUS *CAULOBACTER*. This is a well-known genus of chemoorganotrophic, strictly aerobic, prosthecate, nonbudding bacteria, mainly found in fresh water, marine environments or soil; they divide by asymmetric binary fission. They can grow on complex organic compounds; carbohydrates are assimilated mainly by the Entner-Doudoroff and the TCA cycle. *C. crescentus* forms a separate branch in rRNA superfamily IV (Stackebrandt et al., 1988b).

THE GENERA *PROSTHECOMICROBIUM* AND *ANCALOMICROBIUM*. These genera contain unicellular, chemoorganotrophic bacteria with two to many tapering prosthecae. They grow well on some sugars, sugar alcohols, and organic acids as carbon source. The former genus is nonfermentative; the latter is a facultative anaerobe. They constitute separate branches in this rRNA superfamily (for preliminary information, see Stackebrandt et al., 1988b).

THE GENUS *RHODOMICROBIUM*. This genus contains hyphal, prosthecate, and budding bacteria. They are photoorganotrophs. Some strains grow anaerobically in the light, utilizing simple organic compounds as electron and hydrogen sources for CO_2 reduction; in the dark, they have a microaerophilic to aerobic oxidative metabolism. 16S rRNA cataloging locates this genus at the lower border of the alpha-2 subgroup (Woese et al., 1984a). Our rRNA hybridization method locates it near the bottom of rRNA superfamily IV (M. Gillis, unpublished observations).

THE GENUS *STELLA*. The cells of *Stella* are flat stars with six triangular prosthecae; they multiply by fission, not by budding (see Chapter 104). They occur in soils and aquatic habitats. They do not use sugars and sugar alcohols, but amino acids and TCA intermediates as carbon and energy source. The position of *Stella* in the phylogenetic dendrogram differs with use of different techniques. The values of S_{AB} and signature oligonucleotides indicate that there are some uncertainties concerning the position of this genus. Fischer et al. (1985) concluded that *Stella* is a member of the alpha group and probably a member of the *Rhodospirillum—Azospirillum—[Aquaspirillum]* rRNA cluster. Auling et al. (1988) located this genus within the latter cluster at 0.51 S_{AB}. Another 16S rRNA cataloging dendrogram represents *Stella* as a separate rRNA branch, different from the spirilla cluster and the *Agrobacterium—Rhizobium* cluster, whereas an unrooted 5S rRNA tree does locate it within or close to the latter cluster (Stackebrandt et al., 1988b).

The *Rhodobacter—Paracoccus* rRNA Cluster

This cluster is rooted at 67°C Tm(e) in rRNA superfamily IV. It corresponds to the alpha-3 subgroup in rRNA cataloging (Woese et al., 1984a).

SOME PURPLE BACTERIA. In the top 7.5°C $T_{m(e)}$ are located the anaerobic photoorganotrophic species *Rhodobacter capsulatus*, *Rhb. sphaeroides*, and *Rhodopseudomonas blastica*. We presume that the latter species has been generically misnamed and should be relocated in *Rhodobacter*. Somewhat lower, at 10°C $\Delta T_{m(e)}$, we located the marine anaerobic photoorganotrophic *Rhodobacter sulphidophilus*.

PARACOCCUS DENITRIFICANS. This is metabolically a quite interesting organism as it can grow chemolithotrophically with H_2 and CO_2, or chemoorganotrophically with various organic compounds as sole carbon source. On the basis of previous data, Dickerson et al. (1976) suggested that *Paracoccus denitrificans*, certain aerobic bacteria, and the ancestors of mitochondria arose from the Rhodospirillaceae by loss of photophosphorylation. DNA-rRNA hybridization locates *Paracoccus denitrificans* with 75.5°C $T_{m(e)}$ at the lower border of the genus *Rhodobacter* (M. Gillis and J. De Ley, unpublished observations); by 16S rRNA cataloging, *Rho-*

dobacter capsulatus and *Rhb. sphaeroides* are the closest neighbors (Fox et al., 1980). When the 5S rRNA nucleotide sequences of *Paracoccus denitrificans* and *Rhodobacter sphaeroides* were compared, only 9 bases out of 117 were different (Villanueva et al., 1985). A number of phenotypic similarities between both genera are known already. 5S rRNA sequencing located another facultative chemolithotrophic organism, *Thiobacillus versutus*, in the vicinity of *Paracoccus* (Lane et al., 1985). A search for more similarities at the biochemical and molecular-enzymatic levels may be rewarding.

THE GENUS *GEMMOBACTER*. This genus of aerobic to facultatively anaerobic, heterotrophic, budding rods is related to *Rhodobacter capsulatus* and *Rhb. sphaeroides* at 0.72 S_{AB} (Rothe et al., 1987).

Erythrobacter longus

These orange- and pink-pigmented bacteria were isolated from the marine environment. They are aerobic chemoorganotrophs; no autotrophic growth occurs (Shiba and Simidu, 1982). They display a distinct phenotypical similarity to the Rhodospirillaceae as they contain bacteriochlorophyll *a*, apparently an ancestral remnant. Nevertheless, this taxon represents a quite separate rRNA branch in the alpha group (M. Gillis and J. De Ley, unpublished observations; Stackebrandt et al., 1988b; Woese et al., 1984a).

The Genus *Acidiphilium*

This is a genus of aerobic, acidophilic (pH 2.5 to 5.9), chemoorganotrophic rods that are common in acid mineral environments. Their 5S rRNA is very similar to that of *Thiobacillus acidophilus* (Lane et al., 1985), a species containing facultative chemolithotrophic organisms which are also acidophilic (pH ≤3), and which grow autotrophically on elemental sulfur, but also heterotrophically on some sugars or organic acids. The next neighbor in 5S RNA similarity is *Thiobacillus novellus*, a species of aerobic, neutrophilic organisms which can grow efficiently either chemolithotrophically on a sulfur source such as thiosulfate or chemoorganotrophically. For other authentic thiobacilli, see above.

Nonpositioned Members of rRNA Superfamily IV

The following taxa are unrelated to any of the above-mentioned rRNA clusters or rRNA branches. They remain temporarily at about

67°C Tm(e), the root of rRNA superfamily IV. They are mainly generically misnamed species: the yellow-pigmented *[Agrobacterium] luteum*, the brown-pigmented *[Agrobacterium] ferrugineum*, the red-pigmented *[Agrobacterium] sanguineum*, and *[Agrobacterium] aggregatum*. They are nearly all star-forming, salt-requiring rods from the western Baltic Sea (Ahrens, 1968). We failed to find close neighbors of the generically misnamed *[Pseudomonas] diminuta* and *[Pseudomonas] vesicularis*. Both species have many phenotypic features which set them apart from authentic *Pseudomonas* species. 5S rRNA sequencing also shows *[Ps.] diminuta* to be a separate branch (Lane et al., 1985) which, according to Pace et al., (1985), is near the origin of the *Thiobacillus novellus* branch. On the basis of 16S rRNA cataloging, *[Ps.] diminuta* is a separate branch in the alpha-2 subgroup, close to its root (Woese et al., 1984a). For *Rhodopila globiformis* and *Rhodomicrobium vannielli*, see above.

Some Common Features of rRNA Superfamily IV

This is a very large and heterogeneous superfamily. There are some features which are common to several taxa. Many bacterial taxa live in the phytosphere; of these many play a role in plant-bacterium relations, either as saprophyte, as symbiont, or as pathogen: e.g., *Acetobacter*, *Agrobacterium*, *Azorhizobium*, *Azospirillum*, *Bradyrhizobium*, *[Chromobacterium] folium*, *Gluconobacter*, *Phyllobacterium*, *Nitrobacter*, *Rhizobium*, *Sinorhizobium*, and *Zymomonas*.

Many of these fix N₂: *Acetobacter diazotrophicus* (Gillis et al., 1989), *[Aquaspirillum] peregrinum*, *Azorhizobium*, *Azospirillum*, *Beijerinckia*, *Bradyrhizobium*, *[Chromobacterium] folium*, *Rhizobium*, the *Rhodospirillaceae*, *Sinorhizobium*, *Xanthobacter*, and the N₂-fixers from the rhizophere of rice. Quite a number of taxa, but still less than half, are budding and/or prosthecate.

There are distinct amino acid sequence similarities in the cytochrome *c* from *Agrobacterium* (Tempst, 1983), *Paracoccus*, *Rhodobacter*, *Rhodomicrobium*, *Rhodopseudomonas*, and *Rhodospirillum* (Gibson et al., 1979).

The presence of ubiquinone Q-10 as the major quinone (all strains from superfamily IV examined so far have Q-10), *sym*-homospermidine as a major polyamine, and *cis*-vaccenic acid as the major fatty acid are possible criteria for membership in rRNA superfamily IV (Auling et al., 1988). Busse and Auling (1988) examined polyamine patterns as possible chemotaxonomic markers in the Proteobacteria.

The alpha-1 group (the *Rhodospirillum—Azospirillum* rRNA cluster, see above) displays a heterogeneous polyamine pattern. The *sym*-homospermidine is a good marker for the alpha-2 group (the large *Rhizobium*, *Agrobacterium*, *Phyllobacterium*, *Brucella*, *Rhodopseudomonas*, *Azorhizobium*, etc. rRNA clusters), except for *Agrobacterium*. The alpha-3 group (the *Rhodobacter—Paracoccus* rRNA cluster) produces spermidine.

In these organisms where it was tested (the acetic acid bacteria, *Agrobacterium*, *Rhizobium*, *Rhodobacter*, *Zymomonas*) the Entner-Doudoroff and shunt pathways are used.

The Delta Group

This group was defined by Woese (1987) and Stackebrandt et al. (1988c). It consists mainly of three subgroups: 1) the dissimilatory sulfate- and sulfur-reducing bacteria, 2) *Bdellovibrio*, and 3) six representatives of the order Myxococcales. These subgroups are about as far removed from each other as the large groups of the Proteobacteria are from each other. Due to the inadequacies of the cataloging method, specific relationships between these organisms and the other groups of Proteobacteria cannot always be detected with certainty through S_{AB} clustering; one has to resort to signature comparisons (Hespell et al., 1984) or even to complete sequencing of 16S rRNA (Oyaizu and Woese, 1985). These three subgroups of organisms differ so much from each other phenotypically that it is not possible, except for some basic universal metabolic pathways, to see significant similar features.

The Obligate Anaerobic Dissimilatory Sulfate- and Sulfur-Reducing Bacteria

These organisms carry out the complete or partial oxidation of small organic molecules coupled to the reduction of sulfur or sulfate as electron and hydrogen acceptors. 16S rRNA cataloging was carried out on a total of 12 species from eight genera (Fowler et al., 1986). The sporeforming genus of *Desulfotomaculum* can be excluded at once as it is more related to the clostridia. The genera *Desulfovibrio* (effecting an incomplete oxidation of lactate to CO_2), *Desulfobacter* (complete oxidation of acetate), *Desulfobulbus* (incomplete oxidation of propionate), *Desulfococcus* (complete oxidation of fatty acids, lactate, alcohols, and aromatic carboxylic acids), and *Desulfuromonas* (the only obligate sulfur-reducer linked to complete oxidation of acetate) link low, at S_{AB} of about 0.35, in a loose and heterogeneous cluster. *Desulfosarcina* and the gliding *Desulfonema* (both completely oxidizing small organic molecules) link together somewhat higher at about 0.53 S_{AB}, which is in the same range as the species linkage within the genera *Desulfovibrio* and *Desulfuromonas* (S_{AB} of 0.47 to 0.64).

The Genus *Pelobacter*

This genus harbors strictly anaerobic rods which are unable to use carbohydrates; most of the species ferment a variety of aliphatic or aromatic polyhydroxy compounds (Schink and Pfennig, 1982). 16S rRNA cataloging located three species as very close neighbours to *Desulfuromonas* (Stackebrandt et al., 1989). *Pelobacter acidigallici*, an organism fermenting trihydroxybenzenes (gallic acid, pyrogallol, etc.) to acetate and CO_2, links at S_{AB} 0.66 most closely to *Desulfuromonas succinoxidans*, an organism oxidizing acetate and succinate completely to CO_2. Two other species, *Pelobacter venetianus* (fermenting polyethylene glycol to acetate and ethanol) and *Pelobacter carbinolicus* (fermenting some C-2 and C-3 diols to acetate and ethanol), are closest to the acetate-oxidizing *Desulfuromonas acetexigens* and *Dm. acetoxidans*. In both taxa, organic compounds are the source of carbon, energy, protons, and electrons, while only the electron and proton acceptor systems differ.

The Genus *Bdellovibrio*

The organisms of this genus are quite different from the previous ones: they are small predacious rods that carry out most of their unique biphasic life cycle multiplying within and at the expense of their prey host bacterium. Five strains have been used for 16S rRNA cataloging (Hespell et al., 1984). The nucleotide catalogs of *Bd. bacteriovorus* and of *Bd. starrii* are virtually identical; one strain of *Bd. stolpii* and the unspecified strain BM4 are remotely related with S_{ab} of about 0.47. However, a specific relationship between both groups could not be detected, because this is one of the cases where the S_{ab} values and signature comparisons are inadequate.

The Myxobacteria

This is a large group of chemoorganotrophic, strictly aerobic, gliding, generally pigmented, unicellular rods, which aggregate and form fruiting bodies, containing myxospores or microcysts. They frequently lyse other procaryotes. Using DNA-rRNA hybridization, Moore and McCarthy (1967) showed that the rRNA

cistrons of *Myxococcus xanthus* and *Myxococcus fulvus* were almost identical. 16S rRNA cataloging (Ludwig et al., 1983a) showed that one strain each of *Myxococcus fulvus*, *Stigmatella aurantiaca*, and *Cystobacter fuscus* all belonging to the same suborder Cystobacterineae (Reichenbach and Dworkin, 1981), are close to each other at S_{ab} of 0.78. *Sorangium cellulosum* and *Nannocystis exedens* are separated from each other and from the representatives of the latter suborder at S_{ab} of about 0.41 to 0.44. The myxobacteria are not related to other gliding organisms such as *Cytophaga*. Moore and McCarthy (1967) have shown that a strain of *Sorangium* differed more in Tm(e) from *Myxococcus* than *E. coli* does from *Pseudomonas aeruginosa*.

Some General Considerations

In the "Gram-negative" bacteria or Proteobacteria, a number of basic biochemical mechanisms are heterogeneously distributed over the different groups or rRNA superfamilies. They are: 1) the photosynthetic mechanisms needed to produce ATP, electrons, and protons from photons; 2) the chemoautotrophic mechanisms needed to produce ATP, electrons, and protons from various inorganic reactions (*Nitrobacter*, the nitrosobacteria, *Paracoccus*, etc); 3) the reductive pentose phosphate cycle in the purple bacteria, the reductive tricarboxylic acid cycle in the green bacteria, and other carboxylation reactions for autotrophic CO_2 incorporation; 4) the carbohydrate catabolic and anabolic pathways, such as glycolysis, the shunt (pentose phosphate oxidative cycle, the Entner-Doudoroff pathway, and the TCA cycle: 5) the nitrogenases; 6) the cytochromes and ferredoxins; and 7) (perhaps) the hydrogenases. The above features are probably ancient ones and have persisted in some and disappeared in other taxa.

The current phylogenetic relatedness tree with the various groups of the Proteobacteria confronts us with a bewildering range of phenotypic features apparently indicating independent and uncoordinated evolutionary modifications. The frequent grouping together of phenotypically disparate taxa presents a confusing problem for general and applied microbiology (e.g., identification keys; clinical microbiology; phytobacteriology, etc.). A wide variety of physiological and biochemical reactions (sugar oxidations, nitrate reduction, gelatine liquefaction, etc.) appear to be present at random in many genera and species in the dendrograms. This may be because we look only at the active centers of the enzymes, whereas the real difference will be hidden in other amino acid sequences of homologous proteins. Simple, fast, and reliable tests for diagnostic purposes will be needed. Another problem may be that similar phenotypic features arose independently and more than once, such as complete heterotrophy, morphological similarities, etc. It will be a great challenge for the further development of microbiology to understand, unravel, and reproduce the mechanisms of these complex genotypic changes and their phenotypic expressions.

Some quantitative aspects of the relatedness tree can probably improve the present taxonomy of bacteria. For example, it is striking that genera which have been grouped together in families on phenotypic grounds, such as the Enterobacteriaceae, the Vibrionaceae, the Aeromonadaceae, and the Pasteurellaceae, also group together in individual rRNA branches or rRNA clusters. Other cases of phenotypically well-justified families forming an rRNA cluster are the Alcaligenaceae, the Acetobacteraceae, and the Neisseriaceae. Their lower $T_{m(e)}$ limits are in the range 72 to 76°C. The data shown in Figs. 5 to 7 might reveal other future bacterial families if confirmed by extensive phenotypic evidence, e.g., *Xanthomonas—Frateuria*, the authentic *Pseudomonas—Azotobacter—Azomonas, Rhodospirillum—Azospirillum,* and some misnamed aquaspirilla. On the other hand, a taxonomic order might be delineated at 67 to 68°C $T_{m(e)}$, if sufficient phenotypic evidence were available.

Acknowledgments

The author is indebted to the Belgian Nationaal Fonds voor Wetenschappelijk Onderzoek (NFWO) and the Belgian Fonds voor Geneeskundig Wetenschappelijk Onderzoek (FGWO) for personnel and research grants over a period of a few decades; he is also indebted to the Instituut tot Aanmoediging van het Wetenschappelijk Onderzoek in Nijverheid en Landbouw (IWONL; Belgium) for grants for the extensive genotypic and phenotypic study of the phytopathogenic bacteria. The author is particularly grateful to the numerous permanent and temporary collaborators in his department over the last three decades for their research, creativity, hard work, and enthusiasm in elucidating fundamental areas of the genotypic and phenotypic relationships within the Proteobacteria.

Literature Cited

Ahrens, R. 1968. Taxonomische Untersuchungen an sternbildenden Agrobacterium-Arten aus der westlichen Ostsee. Kieler Meeresforschungen. 24:147–173.

Ambler, R. P. 1973. Bacterial cytochromes *c* and molecular evolution. Syst. Zool. 22:554–565.

Auling, G., J. Busse, M. Hahn, H. Hennecke, R.-M. Kroppenstedt, A. Probst, and E. Stackebrandt. 1988. Phylogentic heterogeneity and chemotaxonomic properties of certain Gram-negative aerobic carboxydobacteria. Syst. Appl. Microbiol. 10:264–272.

Bally R., M. Rahman, R. Tytgat, D. Thomas Bauzon, and J. Balandreau. 1985. Taxonomie moléculaire des bactéries fixatrices d'azote isolées de la rhizosphère du riz. Soc. Franç. Microbiol. Paris, Abstr.

Baumann, L., R. D. Bowditch, and P. Baumann. 1983b. Description of *Deleya* gen. nov. created to accomodate the marine species *Alcaligenes aestus, A. pacificus, A. cupidus, A. venustus,* and *Pseudomonas marina*. Int. J. Syst. Bacteriol. 33:793–802.

Baumann, L., and P. Baumann. 1976. Study of the relationship among marine and terrestrial enterobacteria by means of in vitro DNA-ribosomal RNA hybridization. Microbios Letters 3:11–20.

Baumann, P., and L. Baumann. 1984. Genus II. *Photobacterium* Beijerinck 1889, 401, p. 539–545. In: N. R. Krieg and J. G. Holt (ed.), Bergey's manual of systematic bacteriology, vol. 1. Williams & Wilkins, Baltimore.

Baumann, P., L. Baumann, M. J. Woolkalis, and S. S. Bang. 1983a. Evolutionary relationships in *Vibrio* and *Photobacterium*: a basis for a natural classification. Ann. Rev. Microbiol. 37:369–398.

Baumann, P., A. L. Furniss, and J. V. Lee. 1984a. Genus I. *Vibrio* Pacini 1984, 411, p. 518–538. In: N. R. Krieg and J. G. Holt (ed.), Bergey's manual of systematic bacteriology. vol. 1. Williams & Wilkins, Baltimore.

Baumann, P., M. J. Gauthier, and L. Baumann. 1984b. Genus *Alteromonas* Baumann, Baumann, Mandel and Allen 1972, 418, p. 343–352. In: N. R. Krieg and J. G. Holt (ed.), Bergey's manual of systematic bacteriology. vol. 1. Williams & Wilkins, Baltimore.

Baumann, P., and R. H. W. Schubert. 1984. Family II. Vibrionaceae Véron 1965, 5245, p. 516–517. In: N. R. Krieg and J. G. Holt (ed.), Bergey's manual of systematic bacteriology, vol. 1. Williams & Wilkins, Baltimore.

Bauwens, M., and J. De Ley. 1981. Improvements in the taxonomy of *Flavobacterium* by DNA-rRNA hybridizations, p. 27–31. In: H. Reichenbach and O. B. Weeks (ed.), The *Flavobacterium-Cytophaga* group. Gesellschaft für Biotechnologische Forschung mbH, Braunschweig-Stöckheim.

Boivin, M. F., V. L. Morris, E. C. M. Lee-Chan, and R. G. E. Murray. 1985. Deoxyribonucleic acid relatedness between selected members of the genus *Aquaspirillum* by slot blot hybridization: *Aquaspirillum serpens* (Mueller 1786) Hylemon, Wells, Krieg, and Jannasch 1973 emended to include *Aquaspirillum bengal* as a subjective synonym. Int. J. Syst. Bacteriol. 35:512–517.

Bøvre, K. 1984. Family VIII. Neisseriaceae Prévot 1933, 119, p. 288–290. In: N. R. Krieg and J. G. Holt (ed.), Bergey's manual of systematic bacteriology, vol. 1. Williams & Wilkins, Baltimore, MD.

Bøvre, K., and N. Hagen. 1981. The family Neisseriaceae: rod-shaped species of the genera *Moraxella, Acinetobacter, Kingella,* and *Neisseria,* and the *Branhamella* group of cocci, p. 1506–1529. In: M. P. Starr, H. Stolp,

H. G. Trüper, A. Balows, and H. G. Schlegel (ed.), The prokaryotes, vol. 1. Springer-Verlag, Berlin.

Brenner, D. J. 1984. Family I. Enterobacteriaceae Rahn 1937, nom. fam. cons. Opin. 15, Jud. Comm. 1958, 73; Ewing, Farmer, and Brenner 1980, 674; Judicial Commission 1981, 104, p. 408–420. In: N. R. Krieg and J. G. Holt (ed.), Bergey's manual of systematic bacteriology, vol. 1. Williams & Wilkins, Baltimore.

Busse, J., and G. Auling. 1988. Polyamine pattern as a chemotaxonomic marker within the Proteobacteria. Syst. Appl. Microbiol. 11:1–8.

Chen, W. X., G. H. Yan, and J. L. Li. 1988. Numerical taxonomic study of fast-growing soybean rhizobia and a proposal that Rhizobium fredii be assigned to Sinorhizobium gen. nov. Int. J. Syst. Bacteriol. 38:392–397.

Christensen, P., and F. D. Cook. 1978. Lysobacter, a new genus of nonfruiting, gliding bacteria with a high base ratio. Int. J. Syst. Bacteriol. 28:367–393.

Colwell, R. R., M. T. MacDonell, and J. De Ley. 1986. Proposal to recognize the family Aeromonadaceae fam. nov. Int. J. Syst. Bacteriol. 36:473–477.

De Ley, J. 1972. Agrobacterium: intrageneric relationships and evolution. Proc. Third Int. Conf. Plant Pathogenic Bacteria, Wageningen. 251–259.

De Ley, J. 1974. Phylogeny of procaryotes. Taxon. 23:291–300.

De Ley, J. 1978. Modern molecular methods in bacterial taxonomy: evaluation, application, prospects, p. 347–357. In: Proc. 4th Int. Conf. Plant Pathogenic bacteria, vol. 1. Gilbert-Clarey, Tours, France.

De Ley, J. 1981. Évolution des cistrons codant pour l'ARN-r bactérien. Symbioses. 13:109–121.

De Ley, J. 1984. DNA-rRNA hybridizations in bacterial taxonomy, p. 3–9. In: A. Sanna and G. Morace (ed.), New horizons in microbiology. Elsevier, Amsterdam.

De Ley, J., W. Mannheim, R. Mutters, K. Piechulla, R. Tytgat, P. Segers, M. Bisgaard, W. Frederiksen, K.-H. Hinz, and M. Vanhoucke. 1990. Inter-and intrafamilial similarities of rRNA cistrons of the Pasteurellaceae. Int. J. Syst. Bacteriol. 40:126–137.

De Ley, J., W. Mannheim, P. Segers, A. Lievens, M. Denijn, M. Vanhoucke, and M. Gillis. 1987. Ribosomal ribonucleic acid cistron similarities and taxonomic neighborhood of Brucella and CDC group Vd. Int. J. Syst. Bacteriol. 37:35–42.

De Ley, J., P. Segers, and M. Gillis. 1978. Intra- and intergeneric similarities of Chromobacterium and Janthinobacterium ribosomal ribonucleic acid cistrons. Int. J. Syst. Bacteriol. 28:154–168.

De Ley, J., P. Segers, K. Kersters, W. Mannheim, and A. Lievens. 1986. Intra- and intergeneric similarities of the Bordetella ribosomal ribonucleic acid cistrons: proposal for a new family Alcaligenaceae. Int. J. Syst. Bacteriol. 36:405–414.

Deming, J. W., L. K. Somers, W. L. Straube, D. G. Swartz, and M. T. MacDonell. 1988. Isolation of an obligately barophilic bacterium and description of a new genus, Colwellia gen. nov. Syst. Appl. Microbiol. 10:152–160.

De Smedt, J., M. Bauwens, R. Tytgat, and J. De Ley. 1980. Intra- and intergeneric similarities of ribosomal ribonucleic acid cistrons of free-living, nitrogen-fixing bacteria. Int. J. Syst. Bacteriol. 30:106–122.

De Smedt, J., and J. De Ley. 1977. Intra- and intergeneric similarities of Agrobacterium ribosomal ribonucleic acid cistrons. Int. J. Syst. Bacteriol. 27:222–240.

De Smedt, J., and J. De Ley. 1979. Identification of Ruiter's strains, isolated from browned marinated herring, as members of Erwinia herbicola. Int. J. Syst. Bacteriol. 29:183–187.

De Vos, P. 1981. Intra- and intergeneric similarities of ribosomal ribonucleic acid cistrons in and with the genus Pseudomonas. Academiae Analecta. 43:24–60.

De Vos, P., and J. De Ley. 1983. Intra- and intergeneric similarities of Pseudomonas and Xanthomonas ribosomal ribonucleic acid cistrons. Int. J. Syst. Bacteriol. 33:487–509.

De Vos, P., M. Goor, M. Gillis, and J. De Ley. 1985a. Ribosomal ribonucleic acid cistron similarities of phytopathogenic Pseudomonas species. Int. J. Syst. Bacteriol. 35:169–184.

De Vos, P., K. Kersters, and J. De Ley. 1980. Identification of the leaf nodule bacterial strain PeH₂20 as Erwinia herbicola subsp. herbicola. Zbl. Bakt. Hyg., I Abt. Orig. C 1:237–242.

De Vos, P., K. Kersters, E. Falsen, B. Pot, M. Gillis, P. Segers, and J. De Ley. 1985b. Comamonas Davis and Park 1962 gen. nov., nom. rev. emend., and Comamonas terrigena Hugh 1962 sp. nov., nom. rev. Int. J. Syst. Bacteriol. 35:443–453.

De Vos, P., A. Van Landschoot, P. Segers, R. Tytgat, M. Gillis, M. Bauwens, R. Rossau, M. Goor, B. Pot, K. Kersters, P. Lizzaraga, and J. De Ley. 1989. Genotypic relationships and taxonomic localization of unclassified Pseudomonas and Pseudomonas-like strains by deoxyribonucleic acid-ribosomal ribonucleic acid hybridizations. Int. J. Syst. Bacteriol. 39:35–49.

De Wachter, R., E. Huysmans, and A. Vandenberghe. 1985. 5 S ribosomal RNA as a tool for studying evolution, p. 115–141. In: K. H. Schleifer and E. Stackebrandt (ed.), Evolution of prokaryotes. Academic Press, London.

Dewhirst, F. E., B. J. Paster, and P. L. Bright. 1989. Chromobacterium, Eikenella, Kingella, Neisseria, Simonsiella, and Vitreoscilla species comprise a major branch of the beta group Proteobacteria by 16S ribosomal ribonucleic acid sequence comparison: transfer of Eikenella and Simonsiella to the family Neisseriaceae (emend.). Int. J. Syst. Bacteriol. 39:258–266.

Dickerson, R. E., R. Timkovich, and R. J. Almassy. 1976. The cytochrome fold and the evolution of bacterial energy metabolism. J. Mol. Biol. 100:473–491.

Distel, D. L., D. J. Lane, G. J. Olsen, S. J. Giovannoni, B. Pace, N. R. Pace, D. A. Stahl, and H. Felbeck. 1988. Sulfur-oxidizing bacterial endosymbionts: analysis of phylogeny and specificity by 16S rRNA sequences. J. Bacteriol. 170:2506–2510.

Dreyfus, B., J. L. Garcia, and M. Gillis. 1988. Characterization of Azorhizobium caulinodans gen. nov., sp. nov., a stem-nodulating nitrogen-fixing bacterium isolated from Sesbania rostrata. Int. J. Syst. Bacteriol. 38:89–98.

Ehlers, R.-U., U. Wyss, and E. Stackebrandt. 1988. 16S rRNA cataloguing and the phylogenetic position of the genus Xenorhabdus. Syst. Appl. Microbiol. 10:121–125.

Enger, O, H. Nygaard, M. Solberg, G. Schei, J. Nielsen, and I. Dundas. 1987. Characterization of Alteromonas denitrificans sp. nov. Int. J. Syst. Bacteriol. 37:416–421.

Falk, E. C., J. L. Johnson, V. L. D. Baldani, J. Döbereiner, and N. R. Krieg. 1986. Deoxyribonucleic and ribonucleic acid homology studies of the genera Azospirillum and *Conglomeromonas*. Int. J. Syst. Bacteriol. 36:80–85.

Fendrich, C. 1988. *Halovibrio variabilis* gen. nov. sp. nov., *Pseudomonas halophila* sp. nov. and a new halophilic aerobic coccoid eubacterium from Great Salt Lake, Utah, USA. Syst. Appl. Microbiol. 11:36–43.

Fischer, A., T. Roggentin, H. Schlesner, and E. Stackebrandt. 1985. 16S ribosomal RNA oligonucleotide cataloging and the phylogenetic position of *Stella humosa*. Syst. Appl. Microbiol. 6:43–47.

Fowler, V. J., N. Pfennig, W. Schubert, and E. Stackebrandt. 1984. Towards a phylogeny of phototrophic purple sulfur bacteria—16S rRNA oligonucleotide cataloging of 11 species of Chromatiaceae. Arch. Microbiol. 139:382–387.

Fowler, V. J., F. Widdel, N. Pfennig, C. F. Woese, and E. Stackebrandt. 1986. Phylogenetic relationships of sulfate- and sulfur-reducing eubacteria. Syst. Appl. Microbiol. 8:32–41.

Fox, G. E., K. J. Pechman, and C. R. Woese 1977. Comparative cataloging of 16S ribosomal ribonucleic acid: molecular approach to prokaryotic systematics. Int. J. Syst. Bacteriol. 27:44–57.

Fox, G. E., E. Stackebrandt, R. B. Hespell, J. Gibson, J. Maniloff, T. A. Dyer, R. S. Wolfe, W. E. Balch, R. S. Tanner, L. J. Magrum, L. B. Zablen, R. Blakemore, R. Gupta, L. Bonen, B. J. Lewis, D. A. Stahl, K. R. Luehrsen, K. N. Chen, and C. R. Woese. 1980. The phylogeny of prokaryotes. Science. 209:457–463.

Fox, K. F., and A. Brown. 1989. Partial sequence analysis of the 16S rRNA of legionellae: taxonomic implications. Syst. Appl. Microbiol. 11:135–139.

Franzmann, P. D., U. Wehmeyer, and E. Stackebrandt. 1988. Halomonadaceae fam. nov., a new family of the class Proteobacteria to accomodate the genera *Halomonas* and *Deleya*. Syst. Appl. Microbiol. 11:16–19.

Gibson, J., E. Stackebrandt, L. B. Zablen, R. Gupta, and C. R. Woese. 1979. A phylogenetic analysis of the purple photosynthetic bacteria. Curr. Microbiol. 3:59–64.

Gillis, M., and J. De Ley. 1980. Intra- and intergeneric similarities of the ribosomal ribonucleic acid cistrons of *Acetobacter* and *Gluconobacter*. Int. J. Syst. Bacteriol. 30:7–27.

Gillis, M., K. Kersters, B. Hoste, D. Janssens, R. M. Kroppenstedt, M. P. Stephan, K. R. S. Teixeira, J. Döbereiner, and J. De Ley. 1989. *Acetobacter diazotrophicus* sp. nov., a nitrogen-fixing acetic acid bacterium associated with sugercane. Int. J. Syst. Bacteriol. 39:361–364.

Goor, M., E. Falsen, B. Pot, M. Gillis, K. Kersters, and J. De Ley. 1986. Taxonomic position of the phytopathogen *Pseudomonas rubrisubalbicans* and related clinical isolates. XIV Intern. Congress Microbiol. Manchester, P.B1–17.

Green, P. N., and I. J. Bousfield. 1983. Emendation of *Methylobacterium* Patt, Cole, and Hanson 1976; *Methylobacterium rhodinum* (Heumann 1962) comb. nov. corrig.; *Methylobacterium radiotolerans* (Ito and Iizuka 1971) comb. nov. corrig.; and *Methylobacterium mesophilicum* (Austin and Goodfellow 1979) comb. nov. Int. J. Syst. Bacteriol. 33:875–877.

Green, P. N., and M. Gillis. 1989. Classification of *Pseudomonas aminovorans* and some related methylated amine utilizing bacteria. J. Gen. Microbiol. 135:2071–2076.

Hennecke, H., K. Kaluza, B. Thöny, M. Fuhrman, W. Ludwig, and E. Stackebrandt. 1985. Concurrent evolution of nitrogenase genes and 16S rRNA in *Rhizobium* species and other nitrogen fixing bacteria. Arch. Microbiol. 142:342–348.

Hespell, R. B., B. J. Paster, T. J. Macke, and C. R. Woese. 1984. The origin and phylogeny of the bdellovibrios. Syst. Appl. Microbiol. 5:196–203.

Hirsch, P., and B. Hoffmann. 1989. *Dichotomicrobium thermohalophilum*, gen. nov., sp. nov., budding prosthecate bacteria from the Solar Lake (Sinai) and some related strains. Syst. Appl. Microbiol. 11:291–301.

Holmes, B., M. Popoff, M. Kiredjian, and K. Kersters. 1988. *Ochrobactrum antropi* gen. nov., sp. nov. from human clinical specimens and previously known as group Vd. Int. J. Syst. Bacteriol. 38:406–416.

Hori, H., and S. Osawa. 1978. Evolution of ribosomal proteins in Enterobacteriaceae. J. Bacteriol. 133:1089–1095.

Huysmans, E., and R. De Wachter. 1986. The distribution of 5S ribosomal RNA sequences in phenetic hyperspace. Implications for eubacterial, eukaryotic, archaebacterial and early biotic evolution. Endocyt. Cell Res. 3:133–155.

Imhoff, J. F., H. G. Trüper, and N. Pfennig. 1984. Rearrangement of the species and genera of the phototrophic "purple nonsulfur bacteria." Int. J. Syst. Bacteriol. 34:340–343.

Izard, D., C. Ferragut, F. Gavini, K. Kersters, J. De Ley, and H. Leclerc. 1981. *Klebsiella terrigena*, a new species from soil and water. Int. J. Syst. Bacteriol. 31:116–127.

Jarvis, B. D. W., M. Gillis, and J. De Ley. 1986. Intra- and intergeneric similarities between the ribosomal ribonucleic acid cistrons of *Rhizobium* and *Bradyrhizobium* species and some related bacteria. Int. J. Syst. Bacteriol. 36:129–138.

Jenni, B., L. Realini, M. Aragno, and A. Ü. Tamer. 1988. Taxonomy of non H₂-lithotrophic, oxalate-oxidizing bacteria related to *Alcaligenes eutrophus*. Syst. Appl. Microbiol. 10:126–133.

Jordan, D. C. 1984. Genus II. *Bradyrhizobium* Jordan 1982, 137, p. 242–244. In: N. R. Krieg and J. G. Holt (ed.), Bergey's manual of systematic bacteriology, vol. 1. Williams & Wilkins, Baltimore, MD.

Juni, E., and G. A. Heym. 1986. *Psychrobacter immobilis* gen. nov., sp. nov.: genospecies composed of Gram-negative aerobic, oxidase-positive coccobacilli. Int. J. Syst. Bacteriol. 36:388–391.

Kalina, G. P., A. S. Antonov, T. P. Turova, and T. I. Grafova. 1984. *Allomonas enterica* gen. nov., sp. nov.: deoxyribonucleic acid homology between *Allomonas* and some other members of the *Vibrionaceae*. Int. J. Syst. Bacteriol. 34:150–154.

Kersters, K., and J. De Ley. 1975. Identification and grouping of bacteria by numerical analysis of their electrophoretic protein patterns. J. gen. Microbiol. 87:333–342.

Kersters, K., and J. De Ley. 1984a. Genus *Alcaligenes* Castellani and Chalmers 1919, 936, p. 361–373. In: N. R. Krieg and J. G. Holt (ed.), Bergey's manual of systematic bacteriology, vol. 1. Williams & Wilkins, Baltimore, MD

Kersters, K., and J. De Ley. 1984b. Genus III. *Agrobacterium* Conn 1942, 359, p. 244–254. In: N. R. Krieg and

J. G. Holt (ed.), Bergey's manual of systematic bacteriology, vol. 1. Williams & Wilkins, Baltimore, MD.

Kersters, K., J. De Ley, P. H. A. Sneath, and M. Sackin. 1973. Numerical taxonomic analysis of *Agrobacterium*. J. gen. Microbiol. 78:227–239.

Kersters, K., K.-H. Hinz, A. Hertle, P. Segers, A. Lievens, O. Siegmann, and J. De Ley. 1984. *Bordetella avium* sp. nov., isolated from the respiratory tracts of turkeys and other birds. Int. J. Syst. Bacteriol. 34:56–70.

Knösel, D. H. 1984. Genus IV *Phyllobacterium* (ex Knösel) nom. rev. p. 254–256. In: N. R. Krieg and J. G. Holt (ed.), Bergey's manual of systematic bacteriology, vol. 1. Williams & Wilkins, Baltimore, MD.

Komagata, K., Y. Tamagawa, and H. Iizuka. 1968. Characteristics of *Erwinia herbicola*. J. Gen. Appl. Microbiol. 14:19–37.

Krieg, N. R. 1984. Genus *Aquaspirillum* Hylemon, Wells, Krieg and Jannasch 1973, 361, p. 72–90. In: N. R. Krieg and J. G. Holt (ed.), Bergey's manual of systematic bacteriology, vol. 1. Williams & Wilkins, Baltimore, MD.

Krieg, N. R., and J. G. Holt (ed.). 1984. Bergey's manual of systematic bacteriology, vol. 1. Williams & Wilkins, Baltimore, MD.

Lane, D. J., D. A. Stahl, G. J. Olsen, D. J. Heller, and N. R. Pace. 1985. Phylogenetic analysis of the genera *Thiobacillus* and *Thiomicrospira* by 5S ribosomal RNA sequences. J. Bacteriol. 163:75–81.

Lingens, F., R. Blecher, H. Blecher, F. Blobel, J. Eberspächer, C. Fröhner, H. Görisch, H. Görisch, and G. Layh. 1985. *Phenylobacterium immobile* gen. nov., sp. nov., a Gram-negative bacterium that degrades the herbicide Chloridazon. Int. J. Syst. Bacteriol. 35:26–39.

Logan, N. A. 1989. Numerical taxonomy of violet-pigmented, Gram-negative bacteria and description of *Iodobacter fluviatile* gen. nov., comb. nov. Int. J. Syst. Bacteriol. 39:450–456.

Ludwig, W., J. Eberspächer, F. Lingens, and E. Stackebrandt. 1984. 16S ribosomal RNA studies on the relationship of a chloridazon-degrading Gram-negative eubacterium. Syst. Appl. Microbiol. 5:241–246.

Ludwig, W., K.-H. Schleifer, H. Reichenbach, and E. Stackebrandt. 1983a. A phylogenetic analysis of the myxobacteria *Myxococcus fulvus*, *Stigmatella aurantiaca*, *Cystobacter fuscus*, *Sorangium cellulosum* and *Nannocystis exedens*. Arch. Microbiol. 135:58–62.

Ludwig, W., and E. Stackebrandt. 1983b. A phylogenetic analysis of *Legionella*. Arch. Microbiol. 135:45–50.

Luehrsen, K. R., and G. E. Fox. 1981. The nucleotide sequence of *Beneckea harveyi* 5S rRNA. J. Mol. Evol. 17:52–55.

MacDonell, M. T., and R. R. Colwell. 1985a. The phylogeny of the Vibrionaceae and recommendation for two new genera, *Listonella* and *Shewanella*. Syst. Appl. Microbiol. 6:171–182.

MacDonell, M. T., and R. R. Colwell. 1985b. Nuclease S1 analysis of eubacterial 5S rRNA secondary structure. J. Mol. Evol. 22:237–242.

Mergaert, J., L. Verdonck, K. Kersters, J. Swings, J.-M. Boeufgras, and J. De Ley. 1984. Numerical taxonomy of *Erwinia* species using API systems. J. gen. Microbiol. 130:1893–1910.

Moore, R. L. 1977. Ribosomal ribonucleic acid cistron homologies among *Hyphomicrobium* and various other bacteria. Can. J. Microbiol. 23:478–481.

Moore, R. L., and B. J. McCarthy. 1967. Comparative study of ribosomal ribonucleic acid cistrons in Enterobacteria and Myxobacteria. J. Bacteriol. 94:1066–1074.

Moss, M. O., and T. N. Bryant. 1982. DNA:rRNA hybridization studies of *Chromobacterium fluviatile*. J. gen. Microbiol. 128:829–834.

Oyaizu, H., and C. R. Woese. 1985. Phylogenetic relationships among the sulfate respiring bacteria, myxobacteria and purple bacteria, Syst. Appl. Microbiol. 6:257–263.

Pace, N. R., D. A. Stahl, D. J. Lane, and G. J. Olsen. 1985. Analyzing natural microbial populations by rRNA sequences. ASM News 51:4–12.

Palleroni, N. J. 1984. Genus I. *Pseudomonas* Migula 1894, 237, p. 141–199. In: N. R. Krieg and J. G. Holt (ed.), Bergey's manual of systematic bacteriology, vol. 1. Williams & Wilkins, Baltimore, MD.

Palleroni, N. J., R. Kunisawa, R. Contopoulou, and M. Doudoroff. 1973. Nucleic acid homologies in the genus *Pseudomonas*. Int. J. Syst. Bacteriol. 23:333–339.

Pot, B., M. Gillis, C. Aerts, and J. De Ley. 1984. Genetic heterogeneity within *Aquaspirillum* and *Oceanospirillum*. FEMS Symposium on Evolution of Prokaryotes, Abstr. München.

Pot, B., M. Gillis, B. Hoste, A. Van de Velde, F. Bekaert, K. Kersters, and J. De Ley. 1989. Intra- and intergeneric relationships of the genus *Oceanospirillum*. Int. J. Syst. Bacteriol. 39:23–34.

Ramia, S., E. Neter, and D. J. Brenner. 1982. Production of enterobacterial common antigen as an aid to classification of newly identified species of the families Enterobacteriaceae and Vibrionaceae. Int. J. Syst. Bacteriol. 32:395–398.

Reichenbach, H., and M. Dworkin. 1981. The order *Myxobacterales*, p. 328–355. In: M. P. Starr, H. Stolp, H. G. Trüper, A. Balows and H. G. Schlegel (ed.), The prokaryotes: a handbook on habitats, isolation and identification of bacteria. Springer-Verlag, Berlin.

Reinhold, B., T. Hurek, I. Fendrik, B. Pot, M. Gillis, K. Kersters, S. Thielemans, and J. De Ley. 1987. *Azospirillum halopraeferens* sp. nov., a nitrogen-fixing organism associated with roots of kalar grass (*Leptochloa fusca* (L.) Kunth). Int. J. Sys. Bacteriol. 37:43–51.

Roggentin, T., and P. Hirsch. 1989. Ribosomal RNA cistron similarities among *Hyphomicrobium* species and several other hyphal, budding bacteria. Syst. Appl. Microbiol. 11:140–147.

Rossau, R., K. Kersters, E. Falsen, E. Jantzen, P. Segers, A. Union, L. Nehls, and J. De Ley. 1987. *Oligella*, a new genus including *Oligella urethralis* comb. nov. (formerly *Moraxella urethralis*) and *Oligella ureolytica* sp. nov. (formerly CDC group IVe): relationship to *Taylorella equigenitalis* and related taxa. Int. J. Syst. Bacteriol. 37:198–210.

Rossau, R., L. Heyndrickx, and H. Van Heuverswijn. 1988. Nucleotide sequence of a 16S ribosomal RNA gene from *Neisseria gonorrhoeae*. Nucleic Acid Research. 16:6227.

Rossau, R., G. Vandenbussche, S. Thielemans, P. Segers, H. Grosch, E. Göthe, W. Mannheim, and J. De Ley. 1989a. Ribosomal ribonucleic acid cistron similarities and deoxyribonucleic acid homologies of *Neisseria, Kingella, Eikenella, Simonsiella, Alysiella,* and Centers for Disease Control groups EF-4 and M-5 in the emended family Neisseriaceae. Int. J. Syst. Bacteriol. 39:185–198.

Rossau, R., A. Van Landschoot, W. Mannheim, and J. De Ley. 1986. Inter- and intrageneric similarities of ribosomal ribonucleic acid cistrons of the Neisseriaceae. Int. J. Syst. Bacteriol. 36:323–332.

Rossau, R., E. Vanmechelen, J. De Ley, and H. Van Heuverswijn. 1989b. Specific Neisseria gonorrhoeae DNA-probes derived from ribosomal RNA. J. gen. Microbiol. 135:1735–1745.

Rothe, B., A. Fischer, P. Hirsch, M. Sittig, and E. Stackebrandt. 1987. The phylogenetic position of the budding bacteria Blastobacter aggregatus and Gemmobacter aquatilis gen. nov., sp. nov. Arch. Microbiol. 147:92–99.

Ruvkun, G. B., and F. M. Ausubel. 1980. Interspecies homology of nitrogenase genes. Proc. Natl. Acad. Sci. USA 77:191–195.

Schink, B., and N. Pfennig. 1982. Fermentation of trihydroxybenzenes by Pelobacter acidigallici gen. nov., sp. nov., a strictly anaerobic non-sporeforming bacterium. Arch. Microbiol. 133:195–201.

Seewaldt, E., K.-H. Schleifer, E. Bock, and E. Stackebrandt. 1982. The close phylogenetic relationship of Nitrobacter and Rhodopseudomonas palustris. Arch. Microbiol. 131:287–290.

Shiba, T., and U. Simidu. 1982. Erythrobacter longus gen. nov., sp. nov., an aerobic bacterium which contains bacteriochlorophyll a. Int. J. Syst. Bacteriol. 32:211–217.

Sneath, P. H. A. 1989. Analysis and interpretation of sequence data for bacterial systematics: the view of a numerical taxonomist. Syst. Appl. Microbiol. 12:15–31.

Sokal, R. R., and P. H. A. Sneath. 1963. Principles of numerical taxonomy. W. H. Freeman and Co., San Francisco.

Stackebrandt, E. 1986. Das hierarchische System der Eubakterien: Problem und Lösungsansätze. Forum Mikrobiologie 5:255–260.

Stackebrandt, E., M. Embley, and J. Weckesser. 1988a. Phylogenetic, evolutionary, and taxonomic aspects of phototrophic eubacteria. p. 201–215. In: J. M. Olson, J. G. Ormerod, J. Amesz, E. Stackebrandt, and H. G. Trüper (ed.), Green photosynthetic bacteria. Plenum Publishing Corp.

Stackebrandt, E., A. Fischer, T. Roggentin, U. Wehmeyer, D. Bomar, and J. Smida. 1988b. A phylogenetic survey of budding and/or prosthecate, non-phototrophic eubacteria: membership of Hyphomicrobium, Hyphomonas, Pedomicrobium, Filomicrobium, Caulobacter and "Dichotomicrobium" to the alpha subdivision of purple non-sulfur bacteria. Arch. Microbiol. 149:547–556.

Stackebrandt, E., V. J. Fowler, W. Schubert, and J. F. Imhoff. 1984. Towards a phylogeny of phototrophic purple sulfur bacteria—the genus Ectothiorhodospira. Arch. Microbiol. 137:366–370.

Stackebrandt, E., and H. Hippe. 1986. Transfer of Bacteroides amylophilus to a new genus Ruminobacter gen. nov., nom. rev. as Ruminobacter amylophilus comb. nov. Syst. Appl. Microbiol. 8:204–207.

Stackebrandt, E., R. G. E. Murray, and H. G. Trüper. 1988c. Proteobacteria classis nov., a name for the phylogenetic taxon that includes the "purple bacteria and their relatives." Int. J. Syst. Bacteriol. 38:321–325.

Stackebrandt, E., U. Wehmeyer, and B. Schink. 1989. The phylogenetic status of Pelobacter acidigallici, Pelobacter venetianus, and Pelobacter carbinolicus. Syst. Appl. Microbiol. 11:257–260.

Stahl, D. A., D. J. Lane, G. J. Olsen, D. J. Heller, T. M. Schmidt, and N. R. Pace. 1987. Phylogenetic analysis of certain sulfide-oxidizing and related morphologically conspicuous bacteria by 5S ribosomal ribonucleic acid sequences. Int. J. Syst. Bacteriol. 37:116–122.

Stanier, R. Y., N. J. Palleroni, and M. Doudoroff. 1966. The aerobic pseudomonads: a taxonomic study. J. Gen. Microbiol. 43:159–271.

Swings, J., P. De Vos, M. Van den Mooter, and J. De Ley. 1983. Transfer of Pseudomonas maltophilia Hugh 1981 to the genus Xanthomonas as Xanthomonas maltophilia (Hugh 1981) comb. nov. Int. J. Syst. Bacteriol. 33:409–413.

Swings, J., M. Gillis, K. Kersters, P. De Vos, F. Gosselé, and J. De Ley. 1980. Frateuria, a new genus for "Acetobacter aurantius." Int. J. Syst. Bacteriol. 30:547–556.

Tamaoka, J., D.-M. Ha, and K. Komagata. 1987. Reclassification of Pseudomonas acidovorans den Dooren de Jong 1926 and Pseudomonas testosteroni Marcus and Talalay 1956 as Comamonas acidovorans comb. nov. and Comamonas testosteroni comb. nov., with an emended description of the genus Comamonas. Int. J. Syst. Bacteriol. 37:52–59.

Tchan, Y.-T. 1984. Family II. Azotobacteraceae Pribram 1933, 5, p. 219–220. In: N. R. Krieg and J. G. Holt (ed.), Bergey's manual of systematic bacteriology, vol. 1. Williams & Wilkins, Baltimore, MD.

Tempst, P. 1983. Amino acid sequences and molecular evolution of three Agrobacterium cytochromes c-556. Academiae Analecta. 45:109–135.

Triplett, E. W., G. P. Roberts, P. W. Ludden, and J. Handelsman. 1989. What's new in nitrogen fixation. ASM News 55:15–21.

Urakami, T., J. Tamaoka, K.-I Suzuki, and K. Komagata. 1989. Acidomonas gen. nov., incorporating Acetobacter methanolicus as Acidomonas methanolica comb. nov. Int. J. Syst. Bacteriol. 39:50–55.

Van Landschoot, A., and J. De Ley. 1983. Intra- and intergeneric similarities of the rRNA cistrons of Alteromonas, Marinomonas (gen. nov.) and some other Gram-negative bacteria. J. Gen. Microbiol. 129:3057–3074.

Van Landschoot, A., R. Rossau, and J. De Ley. 1986. Intra- and intergeneric similarities of the ribosomal ribonucleic acid cistrons of Acinetobacter. Int. J. Syst. Bacteriol. 36:150–160.

van Veen, R. J. M., H. den Dulk-Ras, T. Bisseling, R. A. Schilperoort, and P. J. J. Hooykaas. 1988. Crown gall tumor and root nodule formation of the bacterium Phyllobacterium myrsinacearum after the introduction of an Agrobacterium Ti plasmid or a Rhizobium Sym plasmid. Molecular Plant-Microbe Interactions. 1:231–234.

Verdonck, L., J. Mergaert, C. Rijckaert, J. Swings, K. Kersters, and J. De Ley. 1987. Genus Erwinia: numerical analysis of phenotypic features. Int. J. Syst. Bacteriol. 37:4–18.

Villanueva, E., K. R. Luehrsen, J. Gibson, N. Delihas, and G. E. Fox. 1985. Phylogenetic origins of the plant mitochondrion based on a comparative analysis of 5S ribosomal RNA sequences. J. Mol. Evol. 22:46–52.

Vreeland, R. H., C. D. Litchfield, E. L. Martin, and E. Elliot. 1980. *Halomonas elongata*, a new genus and species of extremely salt-tolerant bacteria. Int. J. Syst. Bacteriol. 30:485–495.

Watanabe, I., R. So, J. K. Ladha, Y. Katayama-Fujimura,
and H. Kuraishi. 1987. A new nitrogen-fixing species
of pseudomonas: *Pseudomonas diazotrophicus* sp. nov.
isolated from the root of wetland rice. Can J. Microbiol.
33:670–678.

Weine, R. M., V. E. Coyne, P. Brayton, P. West, and S. F.
Raiken. 1988. *Alteromonas collwelliana* sp. nov., an isolate from oyster habitats. Int. J. Syst. Bacteriol. 38:240–
244.

Weisburg, W. G., C. R. Woese, M. E. Dobson, and E. Weiss.
1985. A common origin of Rickettsiae and certain plant
pathogens. Science 230:556–558.

Wells, J. M., B. C. Raju, H.-Y. Hung, W. G. Weisburg, L.
Mandelco-Paul, and D. J. Brenner. 1987. *Xylella fastidiosa* gen. nov., sp. nov.: Gram-negative, xylem-lim-
ited, fastidious plant bacteria related to *Xanthomonas*
spp. Int. J. Syst. Bacteriol. 37:136–143.

Willems, A., J. Busse, M. Goor, B. Pot, E. Falsen, E. Jantzen,
B. Hoste, M. Gillis, K. Kersters, G. Auling, and J. De
Ley. 1989. *Hydrogenophaga*, a new genus of hydrogen-
oxidizing bacteria that includes *Hydrogenophaga flava*
comb. nov. (formerly *Pseudomonas flava*), *Hydrogenophaga palleronii* (formerly *Pseudomonas palleronii*),
Hydrogenophaga pseudoflava (formerly *Pseudomonas
pseudoflava* and "*Pseudomonas carboxydoflava*"),
and *Hydrogenophaga taeniospiralis* (formerly *Pseudomonas taeniospiralis*). Int. J. Syst. Bacteriol. 39:319–
333.

Willems, A., E. Falsen, B. Pot, E. Jantzen, B. Hoste, P.
Vandamme, M. Gillis, K. Kersters, and J. De Ley. 1990.
Acidovorax, a new genus for *Pseudomonas facilis*, *Pseudomonas delafieldii*, E. Falsen (EF) group 13, EF group
16, and several clinical isolates, with the species *Acidovorax facilis* comb. nov., *Acidovorax delafieldii*
comb. nov., and *Acidovorax temperans* sp. nov. Int. J.
Syst. Bacteriol. 40:384–398.

Willems, A., M. Gillis, and J. De Ley. 1991. Transfer of
Rhodocyclus gelatinosus to *Rubrivivax gelatinosus* gen.
nov., comb. nov. phylogenetic relationships with *Leptothrix*, *Sphaerotilus natans*, *Pseudomonas saccharophila* and *Alcaligenes latus*. Int. J. Syst. Bacteriol. 41:
(in press).

Willems, A., M. Gillis, K. Kersters, L. Van Den Broecke,
and J. De Ley. 1987. Transfer of *Xanthomonas ampelina* Panagopoulos 1969 to a new genus, *Xylophilus*
gen. nov. as *Xylophilus ampelinus* (Panagopoulos 1969)
comb. nov. Int. J. Syst. Bacteriol. 37:422–430.

Woese, C. R. 1987. Bacterial evolution. Microbiol. Revs.
51:221–271.

Woese, C. R., P. Blanz, R. B. Hespell, and C. M. Hahn.
1982. Phylogenetic relationships among various helical
bacteria. Curr. Microbiol. 7:119–124.

Woese, C. R., C. D. Pribula, G. E. Fox, and L. B. Zablen.
1975. The nucleotide sequence of the 5S ribosomal
RNA from a *Photobacterium*. J. Mol. Evol. 5:35–46.

Woese, C. R., M. L. Sogin, and L. A. Sutton. 1974. Prokaryote phylogeny. I. Concerning the relatedness of *Aero-
bacter aerogenes* to *Escherichia coli*. J. Mol. Evol.
3:293–299.

Woese, C. R., E. Stackebrandt, T. J. Macke, and G. E. Fox.
1985a. A phylogenetic definition of the major eubacterial taxa. Syst. Appl. Microbiol. 6:143–151.

Woese, C. R., E. Stackebrandt, W. G. Weisburg, B. J. Paster,
M. T. Madigan, V. J. Fowler, C. M. Hahn, P. Blanz, R.
Gupta, K. H. Nealson, and G. E. Fox. 1984a. The phy-
logeny of purple bacteria: the alpha subdivision. Syst.
Appl. Microbiol. 5:315–326.

Woese, C. R., W. G. Weisburg, C. M. Hahn, B. J. Paster,
L. B. Zablen, B. J. Lewis, T. J. Macke, W. Ludwig, and
E. Stackebrandt. 1985b. The phylogeny of purple bacteria: the gamma subdivision. Syst. Appl. Microbiol.
6:25–33.

Woese, C. R., W. G. Weisburg, B. J. Paster, C. M. Hahn,
R. S. Tanner, N. R. Krieg, H.-P. Koops, H. Harms, and
E. Stackebrandt. 1984b. The phylogeny of purple bacteria: the beta subdivision. Syst. Appl. Microbiol.
5:327–336.

Wolfrum, T., and H. Stolp. 1987. Comparative studies on
5S RNA sequences of RuMP-type methylotrophic bacteria. Syst. Appl. Microbiol. 9:273–276.

Yabuuchi, E., E. Tanimura, A. Ohyama, I. Yano, and A.
Yamamoto. 1979. *Flavobacterium devorans* ATCC
10829: a strain of *Pseudomonas paucimobilis*. J. Gen.
Appl. Microbiol. 25:95–107.

Index Volumes I–IV

C₁ toxin, *Clostridium botulinum,* 1868–1869

C₂ toxin, *Clostridium botulinum,* 1868

C-3 medium, *Spiroplasma,* 1968

CA2 agar
  Cytophagales, 3643, 3646
  Lysobacter, 3258
  myxobacteria, 3433

CA13 agar, Cytophagales, 3643, 3646

Cabbage
  Azotobacteraceae, 3164
  Listeria, 1598
  Spiroplasma, 1967

Cabbage butterfly, *Bacillus,* 1717–1718

Cabbage chloranty mycoplasma-like organism, 4057

CAC medium, myxobacteria, 3437

Cactus
  myxobacteria, 3422
  Spiroplasma, 1967

Cadaverine production, *Lactobacillus,* 1550

Cadmium accumulation
  Citrobacter, 2750–2751
  Zoogloea, 3961–3962

Caduceia symbionts, 3856

Caedibacter, 3871–3872, 3874–3875

Caedibacter caryophila, 3870–3871, 3874–3875, 3877–3878

Caedibacter chlorellopellens, see Pseudocaedibacter

Caedibacter paraconjugatus, 3874, 3878

Caedibacter pseudomutans, 3874, 3877

Caedibacter taeniospiralis, 3865, 3868–3869, 3871–3872, 3874, 3877

Caedibacter varicaedens, 3867–3869, 3871, 3874, 3878

Caenomorpha medusula, 3886

Caenomorpha universalis, 3886

Caffeate, 665

Caiman, *Erysipelothrix,* 1630

CAL agar, *Yersinia,* 2874–2875

Calacarea symbiont, 3822

Calathrix, 2093

Calcium binding, *Zoogloea,* 3961

Calcium carbonate inclusion
  Achromatium, 3937–3938
  Macromonas, 3939

Calcium malate agar
  actinoplanetes, 1042, 1049
  Streptosporangiaceae, 1125

Calcium requirement
  cellulolytic bacteria, 472–473, 475
  methanogens, 742

Caldariella, see Sulfolobus

Caldariellaquinone, 689, 694

Calderobacterium hydrogenophilum, 344, 3924, 3926–3929

Caldocellum saccharolyticum, 479, 481, 484, 491

Caldolysin, 3751

Calduplex, 31

Calf scours, 2678, 2855

Calothrix, 26, 538, 549, 2074, 2076, 2081, 2084, 2089–2090, 2096, 3822, 3825–3826

Calothrix crustacea, 3825

Calothrix pulvinata, 3825

Caloxanthin, *Erythrobacter,* 2487

Calvin cycle, 58–59, 332
  ammonia-oxidizing bacteria, 2634
  colorless sulfur bacteria, 387, 389

Calyptogena symbiont, 29, 403, 2122, 3896–3898, 3902–3903

Camel, *Coxiella,* 2473

Camelina sativa pathogen, 660

CAMP test
  Actinobacillus, 3346–3347
  Corynebacterium, 1178, 1180
  Haemophilus, 3320–3321
  Listeria, 1602–1603
  Streptococcus, 1456–1457

Campy BAP medium, *Campylobacter,* 3501

Campylobacter, 29, 33, 35–36, 168, 399, 564, 910, 2565–2566, 2715, 3386–3387, 3488–3505, 3512

Campylobacter cinaedi, 3488, 3491–3492, 3494, 3497–3498, 3501

Campylobacter coli, 3488–3490, 3492–3495, 3498–3499, 3501, 3503–3504, 3513

Campylobacter concisus, 3488, 3490–3491, 3493–3496, 3498, 3503, 3513, 3517–3520

Campylobacter cryaerophila, 558, 2565, 3488, 3491, 3494–3495, 3502–3503

Campylobacter fennelliae, 3488, 3491–3492, 3494, 3497–3498, 3501

Campylobacter fetus, 3488–3492, 3494, 3501, 3504–3505, 3513

Campylobacter hyointestinalis, 3488, 3490, 3493–3494, 3501–3502, 3513

Campylobacter jejuni, 2566, 3488–3490, 3492–3495, 3498, 3501, 3503–3504, 3513

Campylobacter laridis, 3488, 3490, 3492–3493, 3495, 3498, 3504, 3513

Campylobacter mucosalis, 558, 3488, 3490, 3493–3494, 3501–3503

Campylobacter mustelae, see Helicobacter mustelae

Campylobacter nitrofigilis, 536, 539, 2565, 3488, 3491, 3494–3496, 3501–3502

Campylobacter pylori, see Helicobacter pylori

Campylobacter sputorum, 558, 3488, 3491, 3493–3495, 3501, 3513, 3519–3520

"*Campylobacter upsaliensis,*" 3488, 3490, 3492–3494, 3498, 3501

Canadian aster yellows mycoplasma-like organism, 4056–4057, 4060

Canadian clover phyllody mycoplasma-like organism, 4056–4057, 4060

Caniculitis, lacrimal, *see* Lacrimal canaliculitis

Candicidin, 977

Candicin, 978

Candida, 2035

Candida mycoderma, 2274

Candida pulcherrima, 956

Candiplanecin, 1054

Candle jar, *Campylobacter,* 3503

Cane gall, 2220, 2225

Canine, *Rickettsia,* 2408

Canker
  Erwinia, 668, 2907, 2910
  grapevine, 3133–3135
  phytopathogenic pseudomonads, 3104
  poinsettia, 667
  poplar, 667, 2680
  tomato, 667, 1359
  tulip, 667

Canned food, 84
  Clostridium, 1837

CAPD peritonitis, *Corynebacterium,* 1182

Capillary microscopy, 266

Capillary-tube agglutination test, *Coxiella,* 2476

Capnocytophaga, 919, 3340, 3594, 3631, 3645–3654, 3657–3665, 4120

Capnocytophaga canimorsus, 3665

Capnocytophaga cynodegmi, 3665

Capnocytophaga gingivalis, 3664–3665

Capnocytophaga ochracea, 3658, 3664

Capnocytophaga sputigena, 3664–3665

Capnoid
  Cytophagales, 3654
  Lysobacter, 3264

Capreomycin, 1054

Capric acid utilization
  Alcaligenes, 2549
  Janthinobacterium, 2597
  Psychrobacter, 3243
  rhizobia, 2200
  Serratia, 2836

Caprifig, *Serratia,* 2825

Caproic acid production, 82
  anaerobic Gram-positive cocci, 1886–1887
  Clostridium, 1814, 1825, 1828
  Eubacterium, 1914, 1918–1920
  Megasphaera, 2000–2001, 2043

Caproic acid utilization
  Azotobacteraceae, 3149–3150, 3153
  Psychrobacter, 3243
  purple nonsulfur bacteria, 2149
  rhizobia, 2200
  Rhodocyclus, 2558
  sulfate-reducing bacteria, 3363

Capronate utilization, Azotobacteraceae, 3150

Caprylate utilization
  Azotobacteraceae, 3149–3150, 3153
  Psychrobacter, 3243
  purple nonsulfur bacteria, 2149
  Rhodocyclus, 2558
  Serratia, 2836
  sulfate-reducing bacteria, 3363

Caprylate-thallous (CT) agar medium, *Serratia,* 2825, 2833–2834, 2836

G